Integration Formulas

1. $\displaystyle\int u^n\,du = \frac{u^{n+1}}{n+1} + C,\ n \neq -1$

2. $\displaystyle\int \frac{1}{u}\,du = \ln|u| + C$

3. $\displaystyle\int e^u\,du = e^u + C$

4. $\displaystyle\int b^u\,du = \frac{1}{\ln b}\,b^u + C$

5. $\displaystyle\int \sin u\,du = -\cos u + C$

6. $\displaystyle\int \cos u\,du = \sin u + C$

7. $\displaystyle\int \sec^2 u\,du = \tan u + C$

8. $\displaystyle\int \csc^2 u\,du = -\cot u + C$

9. $\displaystyle\int \sec u \tan u\,du = \sec u + C$

10. $\displaystyle\int \csc u \cot u\,du = -\csc u + C$

11. $\displaystyle\int \tan u\,du = -\ln|\cos u| + C$

12. $\displaystyle\int \cot u\,du = \ln|\sin u| + C$

13. $\displaystyle\int \sec u\,du = \ln|\sec u + \tan u| + C$

14. $\displaystyle\int \csc u\,du = \ln|\csc u - \cot u| + C$

15. $\displaystyle\int u \sin u\,du = \sin u - u\cos u + C$

16. $\displaystyle\int u \cos u\,du = \cos u + u\sin u + C$

17. $\displaystyle\int \sin^2 u\,du = \tfrac{1}{2}u - \tfrac{1}{4}\sin 2u + C$

18. $\displaystyle\int \cos^2 u\,du = \tfrac{1}{2}u + \tfrac{1}{4}\sin 2u + C$

19. $\displaystyle\int \sin au \sin bu\,du = \frac{\sin(a-b)u}{2(a-b)} - \frac{\sin(a+b)u}{2(a+b)} + C$

20. $\displaystyle\int \cos au \cos bu\,du = \frac{\sin(a-b)u}{2(a-b)} + \frac{\sin(a+b)u}{2(a+b)} + C$

21. $\displaystyle\int e^{au} \sin bu\,du = \frac{e^{au}}{a^2+b^2}(a\sin bu - b\cos bu) + C$

22. $\displaystyle\int e^{au} \cos bu\,du = \frac{e^{au}}{a^2+b^2}(a\cos bu + b\sin bu) + C$

23. $\displaystyle\int \sinh u\,du = \cosh u + C$

24. $\displaystyle\int \cosh u\,du = \sinh u + C$

25. $\displaystyle\int \operatorname{sech}^2 u\,du = \tanh u + C$

26. $\displaystyle\int \operatorname{csch}^2 u\,du = -\coth u + C$

27. $\displaystyle\int \tanh u\,du = \ln(\cosh u) + C$

28. $\displaystyle\int \coth u\,du = \ln|\sinh u| + C$

29. $\displaystyle\int \ln u\,du = u\ln u - u + C$

30. $\displaystyle\int u \ln u\,du = \tfrac{1}{2}u^2 \ln u - \tfrac{1}{4}u^2 + C$

31. $\displaystyle\int \frac{1}{\sqrt{a^2-u^2}}\,du = \sin^{-1}\frac{u}{a} + C$

32. $\displaystyle\int \frac{1}{\sqrt{a^2+u^2}}\,du = \ln\left|u + \sqrt{a^2+u^2}\right| + C$

33. $\displaystyle\int \sqrt{a^2-u^2}\,du = \frac{u}{2}\sqrt{a^2-u^2} + \frac{a^2}{2}\sin^{-1}\frac{u}{a} + C$

34. $\displaystyle\int \sqrt{a^2+u^2}\,du = \frac{u}{2}\sqrt{a^2+u^2} + \frac{a^2}{2}\ln\left|u + \sqrt{a^2+u^2}\right| + C$

35. $\displaystyle\int \frac{1}{a^2-u^2}\,du = \frac{1}{2a}\ln\left|\frac{a+u}{a-u}\right| + C$

36. $\displaystyle\int \frac{1}{a^2+u^2}\,du = \frac{1}{a}\tan^{-1}\frac{u}{a} + C$

37. $\displaystyle\int \frac{1}{\sqrt{u^2-a^2}}\,du = \ln\left|u + \sqrt{u^2-a^2}\right| + C$

38. $\displaystyle\int \sqrt{u^2-a^2}\,du = \frac{u}{2}\sqrt{u^2-a^2} - \frac{a^2}{2}\ln\left|u + \sqrt{u^2-a^2}\right| + C$

ADVANCED
Engineering
Mathematics

SEVENTH EDITION

Dennis G. Zill
Loyola Marymount University

JONES & BARTLETT
LEARNING

World Headquarters
Jones & Bartlett Learning
5 Wall Street
Burlington, MA 01803
978-443-5000
info@jblearning.com
www.jblearning.com

Jones & Bartlett Learning books and products are available through most bookstores and online booksellers. To contact Jones & Bartlett Learning directly, call 800-832-0034, fax 978-443-8000, or visit our website, www.jblearning.com.

20797-2

Production Credits
VP, Product Management: Christine Emerton
Director of Product Management: Laura Pagluica
Product Manager: Edward Hinman
Content Strategist: Melissa Duffy
Content Coordinator: Paula-Yuan Gregory
Project Manager: Lori Mortimer
Project Specialist: Jennifer Risden
Senior Digital Project Specialist: Angela Dooley
Director of Marketing: Andrea DeFronzo
Production Services Manager: Colleen Lamy
Product Fulfillment Manager: Wendy Kilborn
Composition: Exela Technologies
Text and Cover Design: Scott Moden
Media Development Editor: Faith Brosnan
Rights Specialist: James Fortney
Cover Image: Domestic: © Pol.mch/Shutterstock; International: © Nuno Valente Fotografia/Shutterstock
Printing and Binding: LSC Communications

Library of Congress Cataloging-in-Publication Data
Names: Zill, Dennis G., author.
Title: Advanced engineering mathematics / Dennis G. Zill.
Description: Seventh edition. | Burlington, MA : Jones & Bartlett Learning,
 [2022] | Includes index. |
Identifiers: LCCN 2020022145 | ISBN 9781284206241 (hardcover)
Subjects: LCSH: Engineering mathematics.
Classification: LCC TA330 .Z55 2022 | DDC 620.001/51–dc23
LC record available at https://lccn.loc.gov/2020022145

6048

Printed in the United States of America
24 23 22 21 20 10 9 8 7 6 5 4 3 2 1

Contents

4 The Laplace Transform — 216

5 Series Solutions of Linear Equations — 270

Preface

In courses such as *calculus* or *differential equations*, the content is fairly standardized, but the content of a course entitled *engineering mathematics* often varies considerably between two different academic institutions. Therefore a text entitled *Advanced Engineering Mathematics* is a compendium of many mathematical topics, all of which are loosely related by the expedient of either being needed or useful in courses in science and engineering or in subsequent careers in these areas. There is literally no upper bound to the number of topics that could be included in a text such as this. Consequently, this book represents the author's opinion of what constitutes *engineering mathematics*.

Content of the Text

For flexibility in topic selection this text is divided into five major parts. As can be seen from the titles of these various parts, it should be obvious that it is my belief that the backbone of science/engineering-related mathematics is the theory and applications of ordinary and partial differential equations.

PART 1: Ordinary Differential Equations (Chapters 1–6)

The six chapters in Part 1 constitute a complete short course in ordinary differential equations. These chapters, with some modifications, correspond to Chapters 1, 2, 3, 4, 5, 6, 7, and 9 in the text *A First Course in Differential Equations with Modeling Applications*, *Eleventh Edition*, by Dennis G. Zill (Cengage Learning). In Chapter 2 the focus is on methods for solving first-order differential equations and their applications. Chapter 3 deals mainly with linear second-order differential equations and their applications. Chapter 4 is devoted to the solution of differential equations and systems of differential equations by the important Laplace transform.

PART 2: Vectors, Matrices, and Vector Calculus (Chapters 7–9)

Chapter 7, *Vectors*, and Chapter 9, *Vector Calculus*, include the standard topics that are usually covered in the third semester of a calculus sequence: vectors in 2- and 3-space, vector functions, directional derivatives, line integrals, double and triple integrals, surface integrals, Green's theorem, Stokes' theorem, and the divergence theorem. In Section 7.6 the vector concept is generalized; by defining vectors analytically we lose their geometric interpretation but keep many of their properties in *n*-dimensional and infinite-dimensional vector spaces. Chapter 8, *Matrices*, is an introduction to systems of algebraic equations, determinants, and matrix algebra, with special emphasis on those types of matrices that are useful in solving systems of linear differential equations. Optional sections on cryptography, error correcting codes, the method of least squares, and discrete compartmental models are presented as applications of matrix algebra.

PART 3: Systems of Differential Equations (Chapters 10 and 11)

There are two chapters in Part 3. Chapter 10, *Systems of Linear Differential Equations*, and Chapter 11, *Systems of Nonlinear Differential Equations*, draw heavily on the matrix material presented in Chapter 8 of Part 2. In Chapter 10, systems of linear first-order equations are solved utilizing the concepts of eigenvalues and eigenvectors, diagonalization, and by means of a matrix exponential function. In Chapter 11, qualitative aspects of autonomous linear and nonlinear systems are considered in depth.

PART 4: Partial Differential Equations (Chapters 12–16)

The core material on Fourier series and boundary-value problems involving second-order partial differential equations was drawn from the text *Differential Equations with Boundary-Value Problems*, *Ninth Edition*, by Dennis G. Zill (Cengage Learning). In Chapter 12, *Fourier Series*, the fundamental topics of sets of orthogonal functions and expansions of functions in terms of an infinite series of orthogonal functions are presented. These topics are then utilized in Chapters 13 and 14 where boundary-value problems in rectangular, polar, cylindrical, and spherical coordinates are solved using the method of separation of variables. In Chapter 15, *Integral Transforms*, boundary-value problems are solved by means of the Laplace and Fourier integral transforms.

PART 5: Complex Analysis (Chapters 17–20)

The final four chapters of the hardbound text cover topics ranging from the basic complex number system through applications of conformal mappings in the solution of Dirichlet's problem. This material by itself could easily serve as a one quarter introductory course in complex variables. This material was taken from *Complex Analysis: A First Course with Applications*, *Third Edition*, by Dennis G. Zill and Patrick D. Shanahan (Jones & Bartlett Learning).

Design of the Text

For the benefit of those instructors and students who have not used the preceding edition, a word about the design of the text is in order. Each chapter opens with its own table of contents and a brief introduction to the material covered in that chapter. Because of the great number of figures, definitions, and theorems throughout this text, I use a double-decimal numeration system. For example, the interpretation of "Figure 1.2.3" is

<div align="center">

Chapter Section of Chapter 1

↓ ↓

1.2.3 ← Third figure in Section 1.2

</div>

I think that this kind of numeration makes it easier to find, say, a theorem or figure when it is referred to in a later section or chapter. In addition, to better link a figure with the text, the *first* textual reference to each figure is done in the same font style and color as the figure number. For example, the first reference to the second figure in Section 5.7 is given as **FIGURE 5.7.2** and all subsequent references to that figure are written in the traditional style Figure 5.7.2.

Features of the Seventh Edition

- One of the goals of this revision was to emphasize applications throughout the text. So, the application problems contributed to previous editions

 Air Exchange, **Exercises 2.7**

 Potassium-40 Decay, **Exercises 2.9**

 Potassium–Argon Dating, **Exercises 2.9**

 Invasion of the Marine Toads, **Chapter 2 in Review**

 Temperature of a Fluid, **Exercises 3.6**

 Blowing in the Wind, **Exercises 3.9**

 The Paris Guns, **Chapter 3 in Review**

 have been retained.

- Section 15.5, *Finite Fourier Transforms,* is new to the text.

- New examples and many new problems (conceptual and applied) have been added throughout the text.

- New figures and photos have been added to highlight some of the older problems and discussions.

- New **REMARKS** have been added to some sections, and some of the older **REMARKS** have been expanded.

- A bit of history has been added following the proper name of a person associated with a certain differential equation or problem.

- Several application problems in chapter review exercises have been moved to the appropriate section exercises.

- Parts of several sections have been rewritten in an attempt to improve clarity.

- Appendix A, *Integral-Defined Functions,* is new to the text.

- The table of Laplace transforms in Appendix C has been expanded.

Supplements

For Instructors

- *Complete Solutions Manual* (*CSM*) by Warren S. Wright and Roberto Martinez
- Test Bank
- Slides in PowerPoint format
- Image Bank
- WebAssign: WebAssign is a flexible and fully customizable online instructional system that puts powerful tools in the hands of teachers, enabling them to deploy assignments, instantly assess individual student performance, and realize their teaching goals. Much more than just a homework grading system, WebAssign delivers secure online testing, customizable precoded questions directly from exercises in this textbook, and unparalleled customer service. Instructors who adopt this program for their classroom use will have access to a digital version of this textbook. Students who purchase an access code for WebAssign will also have access to the digital version of the printed text.

 With WebAssign instructors can:

 · Create and distribute algorithmic assignments using questions specific to this textbook

 · Grade, record, and analyze student responses and performance instantly

 · Offer more practice exercises, quizzes, and homework

 · Upload resources to share and communicate with students seamlessly

 For more detailed information and to sign up for free faculty access, please visit webassign.com. For information on how students can purchase access to WebAssign bundled with this textbook, please contact your Jones & Bartlett account representative at go.jblearning.com/findmyrep.

Designated instructor materials are for qualified instructors only. Jones & Bartlett Learning reserves the right to evaluate all requests. For detailed information and to request access to instructor resources, please visit:

go.jblearning.com/ZillAEM7e.

For Students

■ A WebAssign Student Access Code can be bundled with a copy of this text at a discount when requested by the adopting instructor. It may also be purchased separately online when WebAssign is required by the student's instructor or institution. The student access code provides the student with access to his or her specific classroom assignments in WebAssign and access to a digital version of this text.

■ A *Student Solutions Manual (SSM)* prepared by Warren S. Wright and Roberto Martinez provides solutions to selected problems from the text.

■ Access to the Student Companion Website and Projects Center, available at go.jblearning.com/ZillAEM7e, is included with each new copy of the text. This site includes the following resources to enhance student learning:

 · Chapter 21 Probability

 · Chapter 22 Statistics

 · Additional projects and essays that appeared in earlier editions of this text, including:

 Two Properties of the Sphere

 Vibration Control: Vibration Isolation

 Vibration Control: Vibration Absorbers

 Minimal Surfaces

 Road Mirages

 Two Ports in Electrical Circuits

 The Hydrogen Atom

 Instabilities of Numerical Methods

 A Matrix Model for Environmental Life Cycle Assessment

 Steady Transonic Flow Past Thin Airfoils

 Making Waves: Convection, Diffusion, and Traffic Flow

 When Differential Equations Invaded Geometry: Inverse Tangent Problem of the 17th Century

 Tricky Time: The Isochrones of Huygens and Leibniz

 The Uncertainty Inequality in Signal Processing

 Traffic Flow

 Temperature Dependence of Resistivity

 Fraunhofer Diffraction by a Circular Aperture

 The Collapse of the Tacoma Narrow Bridge: A Modern Viewpoint

 Atmospheric Drag and the Decay of Satellite Orbits

 Forebody Drag of Bluff Bodies

Acknowledgments

The task of compiling a text this size is, to say the least, difficult and many people have put much time and energy into this revision. So I would like to take this opportunity to express my sincerest appreciation to everyone—most of them unknown to me—at Jones & Bartlett Learning who were involved in the publication of this edition. A special word of thanks goes to my editor Laura Pagluica and production editor Jennifer Risden for their guidance in putting all the pieces of a large puzzle together.

Over the years I have been very fortunate to receive valuable input, solicited and unsolicited, from students and my academic colleagues. An occasional word of support is always appreciated, but it is the criticisms and suggestions for improvement that have enhanced each edition. So it is fitting that I once again recognize and thank the following reviewers for sharing their expertise and insights:

Raul M. Aguilar
Massachusetts Maritime Academy

A. Alton
Augustana University

Yuri Antipov
Louisiana State University

Victor Argueta
Alma College

Mahboub Baccouch
University of Nebraska at Omaha

Ken Bosworth
Idaho State University

Kristen Campbell
Elgin Community College

Han-Taw Chen
National Cheng Kung University

Larry Chien
Santa Clara University

John T. Van Cleve
Jacksonville State University

William Criminale
University of Washington

Juan F. Diaz, Jr.
Mount Aloysius College

Vlad Dobrushkin
University of Rhode Island

Jeff Dodd
Jacksonville State University

Victor Elias
University of Western Ontario

Robert E. Fennell
Clemson University

Seferino Fierroz
Oxbridge Academy

Franklin Fondjo Fotou
Langston University

Stan Friedlander
Bronx Community College

Michelle Ghrist
Gonzaga University

David Gilliam
Texas Tech University

Paul Goethals
United States Military Academy

Seckin Gokaltun
The College of New Jersey

Stewart Goldenberg
California Polytechnic State University

Herman Gollwitzer
Drexel University

Reinaldo J. Gonzalez
University of Houston

Ronald B. Gunther
Oregon State University

Daniel Hallinan, Jr.
Florida A&M University–Florida State University College of Engineering

Noel Harbetson
California State University

Angela Hare
Messiah College

Parameswar Harikumar
University of Tulsa

Donald Hartig
California Polytechnic State University

Sonia Henckel
Lawrence Technological University

Robert W. Hunt
Humbolt State University

David Keyes
Columbia University

Mario Klaric
Midlands Technical College

Vuryl Klassen
California State University, Fullerton

Cecilia Knoll
Florida Institute of Technology

Daniel Kraus
SUNY Oswego

Myren Krom
California State University, Sacramento

Jeffrey J. Leader
Rose-Human Institute of Technology

David O. Lomen
University of Arizona

Maria Ludu
Embry-Riddle Aeronautical University

Lewis D. Ludwig
Denison University

Tony Mastroberardino
Penn State Erie, The Behrend College

Osvaldo Mendez
University of Texas, El Paso

Kelley B. Mohrmann
U.S. Military Academy

James L. Moseley
West Virginia University

Dr. Lloyd Moyo
Henderson State University

Gregory E. Muleski
University of Missouri, Kansas City

Charles P. Neumann
Carnegie Mellon University

Evgeni Nikolaev
Rutgers University

Zulkifli Mohd Nopiah
University of Kebangsaan Malaysia

Justin Oelgoetz
Austin Peay State University

Bruce O'Neill
Milwaukee School of Engineering

Sang June Oh
California State University, Fullerton

Dale Peterson
US Air Force Academy

Christopher S. Raymond
University of Delaware

Geoffrey Recktenwald
Michigan State University

Thomas N. Roe
South Dakota State University

Gary Stout
Indiana University of Pennsylvania

Ahlam Tannouri
Morgan State University

Jeremy L. Thompson
US Air Force Academy

David Titley-Peloquin
McGill University

Benjamin Varela
Rochester Institute of Technology

Jun Xu
Tarleton State University

Tian-Shiang Yang
National Cheng Kung University

Bashkim Zendeli
Lawrence Technological University

Although many eyes have scanned the thousands of symbols and hundreds of equations in the text, it is a surety that some errors persist. I apologize for this in advance and I would certainly appreciate hearing about any errors that you may find, either in the text proper or in the supplemental manuals. In order to expedite their correction, contact my editor at: **EHinman@jblearning.com**

Dennis G. Zill

PART 1

Ordinary Differential Equations

Introduction to Differential Equations

The purpose of this short chapter is twofold: to introduce the basic terminology of **differential equations** and to briefly examine how differential equations arise in an attempt to describe or **model** physical phenomena in mathematical terms.

1.1 Definitions and Terminology

INTRODUCTION The words *differential* and *equation* certainly suggest solving some kind of equation that contains derivatives. But before you start solving anything, you must learn some of the basic definitions and terminology of the subject.

■ **A Definition** The derivative dy/dx of a function $y = \phi(x)$ is itself another function $\phi'(x)$ found by an appropriate rule. For example, the function $y = e^{0.1x^2}$ is differentiable on the interval $(-\infty, \infty)$, and its derivative is $dy/dx = 0.2xe^{0.1x^2}$. If we replace $e^{0.1x^2}$ in the last equation by the symbol y, we obtain

$$\frac{dy}{dx} = 0.2xy. \tag{1}$$

Now imagine that a friend of yours simply hands you the **differential equation** in (1), and that you have no idea how it was constructed. Your friend asks: "What is the function represented by the symbol y?" You are now face-to-face with *one* of the basic problems in a course in differential equations:

How do you solve such an equation for the unknown function $y = \phi(x)$?

The problem is loosely equivalent to the familiar reverse problem of differential calculus: Given a derivative, find an antiderivative.

Before proceeding any further, let us give a more precise definition of the concept of a differential equation.

DEFINITION 1.1.1 Differential Equation

An equation containing the derivatives of one or more dependent variables, with respect to one or more independent variables, is said to be a **differential equation (DE)**.

In order to talk about them, we will classify a differential equation by **type**, **order**, and **linearity**.

■ **Classification by Type** If a differential equation contains only ordinary derivatives of one or more functions with respect to a *single* independent variable it is said to be an **ordinary differential equation (ODE)**. An equation involving only partial derivatives of one or more functions of two or more independent variables is called a **partial differential equation (PDE)**. Our first example illustrates several of each type of differential equation.

EXAMPLE 1 Types of Differential Equations

(a) The equations

an ODE can contain more
than one dependent variable
↓ ↓

$$\frac{dy}{dx} + 6y = e^{-x}, \quad \frac{d^2y}{dx^2} + \frac{dy}{dx} - 12y = 0, \quad \text{and} \quad \frac{dx}{dt} + \frac{dy}{dt} = 3x + 2y \tag{2}$$

are examples of ordinary differential equations.

(b) The equations

$$\frac{\partial^2 u}{\partial x^2} + \frac{\partial^2 u}{\partial y^2} = 0, \quad \frac{\partial^2 u}{\partial x^2} = \frac{\partial^2 u}{\partial t^2} - \frac{\partial u}{\partial t}, \quad \frac{\partial u}{\partial y} = -\frac{\partial v}{\partial x} \tag{3}$$

are examples of partial differential equations. Notice in the third equation that there are two dependent variables and two independent variables in the PDE. This indicates that u and v must be functions of *two or more* independent variables. ≡

Notation Throughout this text, ordinary derivatives will be written using either the **Leibniz notation** dy/dx, d^2y/dx^2, $d^3y/dx^3, \ldots$, or the **prime notation** y', y'', $y''', \ldots$. Using the latter notation, the first two differential equations in (2) can be written a little more compactly as $y' + 6y = e^{-x}$ and $y'' + y' - 12y = 0$, respectively. Actually, the prime notation is used to denote only the first three derivatives; the fourth derivative is written $y^{(4)}$ instead of y''''. In general, the nth derivative is $d^n y/dx^n$ or $y^{(n)}$. Although less convenient to write and to typeset, the Leibniz notation has an advantage over the prime notation in that it clearly displays both the dependent and independent variables. For example, in the differential equation $d^2x/dt^2 + 16x = 0$, it is immediately seen that the symbol x now represents a dependent variable, whereas the independent variable is t. You should also be aware that in physical sciences and engineering, **Newton's dot notation** (derogatively referred to by some as the "flyspeck" notation) is sometimes used to denote derivatives with respect to time t. Thus the differential equation $d^2s/dt^2 = -32$ becomes $\ddot{s} = -32$. Partial derivatives are often denoted by a **subscript notation** indicating the independent variables. For example, the first and second equations in (3) can be written, in turn, as $u_{xx} + u_{yy} = 0$ and $u_{xx} = u_{tt} - u_t$.

■ Classification by Order The **order of a differential equation** (ODE or PDE) is the order of the highest derivative in the equation.

EXAMPLE 2 Order of a Differential Equation

The differential equations

$$\overset{\text{highest order}}{\underset{\downarrow}{\frac{d^2y}{dx^2}}} + 5\left(\frac{dy}{dx}\right)^3 - 4y = e^x, \qquad 2\overset{\text{highest order}}{\underset{\downarrow}{\frac{\partial^4 u}{\partial x^4}}} + \frac{\partial^2 u}{\partial t^2} = 0$$

are examples of a second-order ordinary differential equation and a fourth-order partial differential equation, respectively. ≡

A first-order ordinary differential equation is sometimes written in the **differential form**

$$M(x, y)\,dx + N(x, y)\,dy = 0.$$

EXAMPLE 3 Differential Form of a First-Order ODE

If we assume that y is the dependent variable in a first-order ODE, then recall from calculus that the differential dy is defined to be $dy = y'\,dx$.

(a) By dividing by the differential dx an alternative form of the equation

$$(y - x)\,dx + 4x\,dy = 0$$

is given by

$$y - x + 4x\,\frac{dy}{dx} = 0 \quad \text{or equivalently} \quad 4x\,\frac{dy}{dx} + y = x.$$

(b) By multiplying the differential equation

$$6xy\,\frac{dy}{dx} + x^2 + y^2 = 0$$

by dx we see that the equation has the alternative differential form

$$(x^2 + y^2)\,dx + 6xy\,dy = 0.$$ ≡

In symbols, we can express an nth-order ordinary differential equation in one dependent variable by the **general form**

$$F(x, y, y', \ldots, y^{(n)}) = 0, \tag{4}$$

where F is a real-valued function of $n + 2$ variables: $x, y, y', \ldots, y^{(n)}$. For both practical and theoretical reasons, we shall also make the assumption hereafter that it is possible to solve an ordinary differential equation in the form (4) uniquely for the highest derivative $y^{(n)}$ in terms of the remaining $n + 1$ variables. The differential equation

$$\frac{d^n y}{dx^n} = f(x, y, y', \ldots, y^{(n-1)}), \tag{5}$$

where f is a real-valued continuous function, is referred to as the **normal form** of (4). Thus, when it suits our purposes, we shall use the normal forms

$$\frac{dy}{dx} = f(x, y) \quad \text{and} \quad \frac{d^2 y}{dx^2} = f(x, y, y')$$

to represent general first- and second-order ordinary differential equations.

EXAMPLE 4 Normal Form of an ODE

(a) By solving for the derivative dy/dx the normal form of the first-order differential equation

$$4x \frac{dy}{dx} + y = x \quad \text{is} \quad \frac{dy}{dx} = \frac{x - y}{4x}.$$

(b) By solving for the derivative y'' the normal form of the second-order differential equation

$$y'' - y' + 6y = 0 \quad \text{is} \quad y'' = y' - 6y. \qquad \equiv$$

■ Classification by Linearity An nth-order ordinary differential equation (4) is said to be **linear** in the variable y if F is linear in $y, y', \ldots, y^{(n)}$. This means that an nth-order ODE is linear when (4) is $a_n(x)y^{(n)} + a_{n-1}(x)y^{(n-1)} + \cdots + a_1(x)y' + a_0(x)y - g(x) = 0$ or

$$a_n(x) \frac{d^n y}{dx^n} + a_{n-1}(x) \frac{d^{n-1} y}{dx^{n-1}} + \cdots + a_1(x) \frac{dy}{dx} + a_0(x)y = g(x). \tag{6}$$

Two important special cases of (6) are **linear first-order** ($n = 1$) and **linear second-order** ($n = 2$) ODEs.

$$a_1(x) \frac{dy}{dx} + a_0(x)y = g(x) \quad \text{and} \quad a_2(x) \frac{d^2 y}{dx^2} + a_1(x) \frac{dy}{dx} + a_0(x)y = g(x). \tag{7}$$

In the additive combination on the left-hand side of (6) we see that the characteristic two properties of a linear ODE are

Remember these two characteristics of a linear ODE.

- The dependent variable y and all its derivatives $y', y'', \ldots, y^{(n)}$ are of the first degree; that is, the power of each term involving y is 1.
- The coefficients $a_0, a_1, \ldots, a_n$ of $y, y', \ldots, y^{(n)}$ depend at most on the independent variable x.

A **nonlinear** ordinary differential equation is simply one that is not linear. If the coefficients of $y, y', \ldots, y^{(n)}$ contain the dependent variable y or its derivatives or if powers of $y, y', \ldots, y^{(n)}$, such as $(y')^2$, appear in the equation, then the DE is nonlinear. Also, nonlinear functions of the dependent variable or its derivatives, such as $\sin y$ or $e^{y'}$, cannot appear in a linear equation.

EXAMPLE 5 Linear and Nonlinear Differential Equations

(a) The equations

$$(y - x)\, dx + 4x\, dy = 0, \quad y'' - 2y' + y = 0, \quad x^3 \frac{d^3 y}{dx^3} + 3x \frac{dy}{dx} - 5y = e^x$$

are, in turn, examples of *linear* first-, second-, and third-order ordinary differential equations. We have just demonstrated in part (a) of Example 3 that the first equation is linear in y by writing it in the alternative form $4xy' + y = x$.

1.1 Definitions and Terminology | **5**

(b) The equations

$$\underset{\underset{\text{coefficient depends on } y}{\uparrow}}{\overset{\text{nonlinear term:}}{}}$$

$$(1 - y)y' + 2y = e^x, \qquad \underset{\underset{\text{nonlinear function of } y}{\uparrow}}{\overset{\text{nonlinear term:}}{}} \frac{d^2y}{dx^2} + \sin y = 0, \qquad \underset{\underset{\text{power not 1}}{\uparrow}}{\overset{\text{nonlinear term:}}{}} \frac{d^4y}{dx^4} + y^2 = 0,$$

are examples of *nonlinear* first-, second-, and fourth-order ordinary differential equations, respectively.

(c) By using the quadratic formula the nonlinear first-order differential equation $(y')^2 + 2xy' - y = 0$ can be written as two nonlinear first-order equations in normal form

$$y' = -x + \sqrt{x^2 + y} \quad \text{and} \quad y' = -x - \sqrt{x^2 + y}. \qquad \equiv$$

■ **Solution** As stated before, one of our goals in this course is to solve—or find solutions of—differential equations. The concept of a solution of an ordinary differential equation is defined next.

DEFINITION **1.1.2** Solution of an ODE

Any function ϕ, defined on an interval I and possessing at least n derivatives that are continuous on I, which when substituted into an nth-order ordinary differential equation reduces the equation to an identity, is said to be a **solution** of the equation on the interval.

In other words, a solution of an nth-order ordinary differential equation (4) is a function ϕ that possesses at least n derivatives and

$$F(x, \phi(x), \phi'(x), \ldots, \phi^{(n)}(x)) = 0 \text{ for all } x \text{ in } I.$$

We say that ϕ *satisfies* the differential equation on I. For our purposes, we shall also assume that a solution ϕ is a real-valued function. In our initial discussion we have already seen that $y = e^{0.1x^2}$ is a solution of $dy/dx = 0.2xy$ on the interval $(-\infty, \infty)$.

Occasionally it will be convenient to denote a solution by the alternative symbol $y(x)$.

■ **Interval of Definition** You can't think *solution* of an ordinary differential equation without simultaneously thinking *interval*. The interval I in Definition 1.1.2 is variously called the **interval of definition**, the **interval of validity**, or the **domain of the solution** and can be an open interval (a, b), a closed interval $[a, b]$, an infinite interval (a, ∞), and so on.

EXAMPLE 6 Verification of a Solution

Verify that the indicated function is a solution of the given differential equation on the interval $(-\infty, \infty)$.

(a) $\dfrac{dy}{dx} = xy^{1/2}; \quad y = \frac{1}{16}x^4$ **(b)** $y'' - 2y' + y = 0; \quad y = xe^x$

SOLUTION One way of verifying that the given function is a solution is to see, after substituting, whether each side of the equation is the same for every x in the interval $(-\infty, \infty)$.

(a) From *left-hand side:* $\dfrac{dy}{dx} = 4 \cdot \dfrac{x^3}{16} = \dfrac{x^3}{4}$

right-hand side: $xy^{1/2} = x \cdot \left(\dfrac{x^4}{16}\right)^{1/2} = x \cdot \dfrac{x^2}{4} = \dfrac{x^3}{4},$

we see that each side of the equation is the same for every real number x. Note that $y^{1/2} = \frac{1}{4}x^2$ is, by definition, the nonnegative square root of $\frac{1}{16}x^4$.

(b) From the derivatives $y' = xe^x + e^x$ and $y'' = xe^x + 2e^x$ we have for every real number x,

$$\text{left-hand side: } y'' - 2y' + y = (xe^x + 2e^x) - 2(xe^x + e^x) + xe^x = 0$$

$$\text{right-hand side: } 0.$$

≡

Note, too, that in Example 6 each differential equation possesses the constant solution $y = 0$, defined on $(-\infty, \infty)$. A solution of a differential equation that is identically zero on an interval I is said to be a **trivial solution**.

■ Solution Curve
The graph of a solution ϕ of an ODE is called a **solution curve**. Since ϕ is a differentiable function, it is continuous on its interval I of definition. Thus there may be a difference between the graph of the *function* ϕ and the graph of the *solution* ϕ. Put another way, the domain of the function ϕ does not need to be the same as the interval I of definition (or domain) of the solution ϕ.

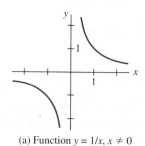

(a) Function $y = 1/x, x \neq 0$

(b) Solution $y = 1/x, (0, \infty)$

FIGURE 1.1.1 Example 7 illustrates the difference between the function $y = 1/x$ and the solution $y = 1/x$

EXAMPLE 7 **Function vs. Solution**

(a) Considered simply as a *function*, the domain of $y = 1/x$ is the set of all real numbers x except 0. When we graph $y = 1/x$, we plot points in the xy-plane corresponding to a judicious sampling of numbers taken from its domain. The rational function $y = 1/x$ is discontinuous at 0, and its graph, in a neighborhood of the origin, is given in **FIGURE 1.1.1(a)**. The function $y = 1/x$ is not differentiable at $x = 0$ since the y-axis (whose equation is $x = 0$) is a vertical asymptote of the graph.

(b) Now $y = 1/x$ is also a solution of the linear first-order differential equation $xy' + y = 0$ (verify). But when we say $y = 1/x$ is a *solution* of this DE we mean it is a function defined on an interval I on which it is differentiable and satisfies the equation. In other words, $y = 1/x$ is a solution of the DE on *any* interval not containing 0, such as $(-3, -1)$, $(\frac{1}{2}, 10)$, $(-\infty, 0)$, or $(0, \infty)$. Because the solution curves defined by $y = 1/x$ on the intervals $(-3, -1)$ and on $(\frac{1}{2}, 10)$ are simply segments or pieces of the solution curves defined by $y = 1/x$ on $(-\infty, 0)$ and $(0, \infty)$, respectively, it makes sense to take the interval I to be as large as possible. Thus we would take I to be either $(-\infty, 0)$ or $(0, \infty)$. The solution curve on the interval $(0, \infty)$ is shown in Figure 1.1.1(b).

≡

■ Explicit and Implicit Solutions
You should be familiar with the terms *explicit* and *implicit functions* from your study of calculus. A solution in which the dependent variable is expressed solely in terms of the independent variable and constants is said to be an **explicit solution**. For our purposes, let us think of an explicit solution as an explicit formula $y = \phi(x)$ that we can manipulate, evaluate, and differentiate using the standard rules. We have just seen in the last two examples that $y = \frac{1}{16} x^4$, $y = xe^x$, and $y = 1/x$ are, in turn, explicit solutions of $dy/dx = xy^{1/2}$, $y'' - 2y' + y = 0$, and $xy' + y = 0$. Moreover, the trivial solution $y = 0$ is an explicit solution of all three equations. We shall see when we get down to the business of actually solving some ordinary differential equations that methods of solution do not always lead directly to an explicit solution $y = \phi(x)$. This is particularly true when attempting to solve nonlinear first-order differential equations. Often we have to be content with a relation or expression $G(x, y) = 0$ that defines a solution ϕ implicitly.

DEFINITION 1.1.3 **Implicit Solution of an ODE**

A relation $G(x, y) = 0$ is said to be an **implicit solution** of an ordinary differential equation (4) on an interval I provided there exists at least one function ϕ that satisfies the relation as well as the differential equation on I.

It is beyond the scope of this course to investigate the conditions under which a relation $G(x, y) = 0$ defines a differentiable function ϕ. So we shall assume that if the formal implementation of a method of solution leads to a relation $G(x, y) = 0$, then there exists at least one function ϕ that satisfies both the relation (that is, $G(x, \phi(x)) = 0$) and the differential equation on an

interval I. If the implicit solution $G(x, y) = 0$ is fairly simple, we may be able to solve for y in terms of x and obtain one or more explicit solutions. See (*iv*) in the *Remarks*.

EXAMPLE 8 Verification of an Implicit Solution

The relation $x^2 + y^2 = 25$ is an implicit solution of the nonlinear differential equation

$$\frac{dy}{dx} = -\frac{x}{y} \tag{8}$$

on the interval defined by $-5 < x < 5$. By implicit differentiation we obtain

$$\frac{d}{dx}x^2 + \frac{d}{dx}y^2 = \frac{d}{dx}25 \quad \text{or} \quad 2x + 2y\frac{dy}{dx} = 0. \tag{9}$$

Solving the last equation in (9) for the symbol dy/dx gives (8). Moreover, solving $x^2 + y^2 = 25$ for y in terms of x yields $y = \pm\sqrt{25 - x^2}$. The two functions $y = \phi_1(x) = \sqrt{25 - x^2}$ and $y = \phi_2(x) = -\sqrt{25 - x^2}$ satisfy the relation (that is, $x^2 + \phi_1^2 = 25$ and $x^2 + \phi_2^2 = 25$) and are explicit solutions defined on the interval $(-5, 5)$. The solution curves given in **FIGURE 1.1.2(b)** and 1.1.2(c) are segments of the graph of the implicit solution in Figure 1.1.2(a).

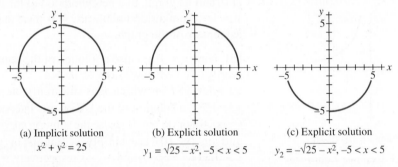

(a) Implicit solution
$x^2 + y^2 = 25$

(b) Explicit solution
$y_1 = \sqrt{25 - x^2},\ -5 < x < 5$

(c) Explicit solution
$y_2 = -\sqrt{25 - x^2},\ -5 < x < 5$

FIGURE 1.1.2 An implicit solution and two explicit solutions in Example 8

Any relation of the form $x^2 + y^2 - c = 0$ *formally* satisfies (8) for any constant c. However, it is understood that the relation should always make sense in the real number system; thus, for example, we cannot say that $x^2 + y^2 + 25 = 0$ is an implicit solution of the equation. Why not?

Because the distinction between an explicit solution and an implicit solution should be intuitively clear, we will not belabor the issue by always saying, "Here is an explicit (implicit) solution."

■ Families of Solutions The study of differential equations is similar to that of integral calculus. When evaluating an antiderivative or indefinite integral in calculus, we use a single constant c of integration. Analogously, when solving a first-order differential equation $F(x, y, y') = 0$, we *usually* obtain a solution containing a single arbitrary constant or parameter c. A solution containing an arbitrary constant represents a set $G(x, y, c) = 0$ of solutions called a **one-parameter family of solutions**. When solving an nth-order differential equation $F(x, y, y', \dots, y^{(n)}) = 0$, we seek an **$n$-parameter family of solutions** $G(x, y, c_1, c_2, \dots, c_n) = 0$. This means that a single differential equation can possess an infinite number of solutions corresponding to the unlimited number of choices for the parameter(s). A solution of a differential equation that is free of arbitrary parameters is called a **particular solution**.

EXAMPLE 9 Particular Solution

(a) For all values of c, the one-parameter family $y = cx - x\cos x$ is an explicit solution of the linear first-order differential equation

$$xy' - y = x^2 \sin x$$

on the interval $(-\infty, \infty)$. **FIGURE 1.1.3** shows the graphs of some particular solutions in this family for various choices of c. The solution $y = -x\cos x$, the red curve in the figure, is a particular solution corresponding to $c = 0$.

FIGURE 1.1.3 Some solutions of DE in part (a) of Example 9

FIGURE 1.1.4 Some solutions of DE in part (b) of Example 9

(b) The two-parameter family $y = c_1e^x + c_2xe^x$ is an explicit solution of the linear second-order differential equation

$$y'' - 2y' + y = 0$$

in part (b) of Example 6. **FIGURE 1.1.4** shows seven of the "double infinity" of solutions in this family. The solution curves in red, green, and blue are the graphs of the particular solutions $y = 5xe^x$ ($c_1 = 0$, $c_2 = 5$), $y = 3e^x$ ($c_1 = 3$, $c_2 = 0$), and $y = 5e^x - 2xe^x$ ($c_1 = 5$, $c_2 = -2$) respectively. ≡

In all the preceding examples, we have used x and y to denote the independent and dependent variables, respectively. But you should become accustomed to seeing and working with other symbols to denote these variables. For example, we could denote the independent variable by t and the dependent variable by x.

EXAMPLE 10 Using Different Symbols

The functions $x = c_1 \cos 4t$ and $x = c_2 \sin 4t$, where c_1 and c_2 are arbitrary constants or parameters, are both solutions of the linear differential equation

$$x'' + 16x = 0.$$

For $x = c_1 \cos 4t$, the first two derivatives with respect to t are $x' = -4c_1 \sin 4t$ and $x'' = -16c_1 \cos 4t$. Substituting x'' and x then gives

$$x'' + 16x = -16c_1 \cos 4t + 16(c_1 \cos 4t) = 0.$$

In like manner, for $x = c_2 \sin 4t$ we have $x'' = -16c_2 \sin 4t$, and so

$$x'' + 16x = -16c_2 \sin 4t + 16(c_2 \sin 4t) = 0.$$

Finally, it is straightforward to verify that the linear combination of solutions for the two-parameter family $x = c_1 \cos 4t + c_2 \sin 4t$ is also a solution of the differential equation. ≡

The next example shows that a solution of a differential equation can be a piecewise-defined function.

EXAMPLE 11 A Piecewise-Defined Solution

You should verify that the one-parameter family $y = cx^4$ is a one-parameter family of solutions of the linear differential equation $xy' - 4y = 0$ on the interval $(-\infty, \infty)$. See **FIGURE 1.1.5(a)**. The piecewise-defined differentiable function

$$y = \begin{cases} -x^4, & x < 0 \\ x^4, & x \geq 0 \end{cases}$$

is a particular solution of the equation but cannot be obtained from the family $y = cx^4$ by a single choice of c; the solution is constructed from the family by choosing $c = -1$ for $x < 0$ and $c = 1$ for $x \geq 0$. See **FIGURE 1.1.5(b)**. ≡

(a)

(b)

FIGURE 1.1.5 Some solutions of $xy' - 4y = 0$ in Example 11

■ **Singular Solution** Sometimes an nth-order differential equation possesses a solution that is not a member of an n-parameter family of solutions of the equation—that is, a solution that cannot be obtained by specializing *any* of the parameters in the family of solutions. Such a solution is called a **singular solution**.*

EXAMPLE 12 Singular Solution

We saw on pages 6 and 7 that the functions $y = \frac{1}{16}x^4$ and $y = 0$ are solutions of the differential equation $dy/dx = xy^{\frac{1}{2}}$ on $(-\infty, \infty)$. In Section 2.2 we shall demonstrate, by actually solving it, that the differential equation $dy/dx = xy^{\frac{1}{2}}$ possesses the one-parameter family of solutions

*There is a bit more to the definition of a singular solution, but it is beyond the intended level of this text.

$y = (\frac{1}{4}x^2 + c)^2$, $c \geq 0$. When $c = 0$ the resulting particular solution is $y = \frac{1}{16}x^4$. But the trivial solution $y = 0$ is a singular solution since it is not a member of the family $y = (\frac{1}{4}x^2 + c)^2$; there is no way of assigning a value to the constant c to obtain $y = 0$. ≡

■ **Systems of Differential Equations** Up to this point we have been discussing single differential equations containing one unknown function. But often in theory, as well as in many applications, we must deal with systems of differential equations. A **system of ordinary differential equations** is two or more equations involving the derivatives of two or more unknown functions of a single independent variable. For example, if x and y denote dependent variables and t the independent variable, then a system of two first-order differential equations is given by

$$\frac{dx}{dt} = f(t, x, y)$$

$$\frac{dy}{dt} = g(t, x, y). \tag{10}$$

A **solution** of a system such as (10) is a pair of differentiable functions $x = \phi_1(t)$, $y = \phi_2(t)$ defined on a common interval I that satisfy each equation of the system on this interval. See Problems 49 and 50 in Exercises 1.1.

REMARKS

(*i*) It might not be apparent whether a first-order ODE written in differential form $M(x, y) \, dx + N(x, y) \, dy = 0$ is linear or nonlinear because there is nothing in this form that tells us which symbol denotes the dependent variable. See Problems 11 and 12 in Exercises 1.1.

(*ii*) We will see in the chapters that follow that a solution of a differential equation may involve an **integral-defined function**. One way of defining a function F of a single variable x by means of a definite integral is

$$F(x) = \int_a^x g(t) \, dt. \tag{11}$$

If the integrand g in (11) is continuous on an interval $[a, b]$ and $a \leq x \leq b$, then the derivative form of the Fundamental Theorem of Calculus states that F is differentiable on (a, b) and

$$F'(x) = \frac{d}{dx} \int_a^x g(t) \, dt = g(x). \tag{12}$$

The integral in (11) is often **nonelementary**, that is, an integral of a function g that does not have an elementary-function antiderivative. Elementary functions include the familiar functions studied in a typical precalculus course:

constant, polynomial, rational, exponential, logarithmic, trigonometric, and inverse trigonometric functions,

as well as rational powers of these functions, finite combinations of these functions using addition, subtraction, multiplication, division, and function compositions. For example, even though e^{-t^2}, $\sqrt{1 + t^3}$, and $\cos t^2$ are elementary functions, the integrals $\int e^{-t^2} dt$, $\int \sqrt{1 + t^3} \, dt$, and $\int \cos t^2 \, dt$ are nonelementary. See Problems 27–30 in Exercises 1.1.

(*iii*) Although the concept of a solution of a differential equation has been emphasized in this section, you should be aware that a DE does not necessarily have to possess a solution. See Problem 51 in Exercises 1.1. The question of whether a solution exists will be touched on in the next section.

(*iv*) A few last words about implicit solutions of differential equations are in order. In Example 8 we were able to solve the relation $x^2 + y^2 = 25$ for y in terms of x to get two explicit solutions, $\phi_1(x) = \sqrt{25 - x^2}$ and $\phi_2(x) = -\sqrt{25 - x^2}$, of the differential equation (8). But don't read too much into this one example. Unless it is easy, obvious, or important, or you are

instructed to, there is usually no need to try to solve an implicit solution $G(x, y) = 0$ for y explicitly in terms of x. Also do not misinterpret the second sentence following Definition 1.1.3. An implicit solution $G(x, y) = 0$ can define a perfectly good differentiable function ϕ that is a solution of a DE, but yet we may not be able to solve $G(x, y) = 0$ using analytical methods such as algebra. The solution curve of ϕ may be a segment or piece of the graph of $G(x, y) = 0$. See Problems 57 and 58 in Exercises 1.1.

(*v*) If *every* solution of an *n*th-order ODE $F(x, y, y', \ldots, y^{(n)}) = 0$ on an interval I can be obtained from an *n*-parameter family $G(x, y, c_1, c_2, \ldots, c_n) = 0$ by appropriate choices of the parameters $c_i, i = 1, 2, \ldots, n$, we then say that the family is the **general solution** of the DE. In solving linear ODEs, we shall impose relatively simple restrictions on the coefficients of the equation; with these restrictions one can be assured that not only does a solution exist on an interval but also that a family of solutions yields all possible solutions. Nonlinear equations, with the exception of some first-order DEs, are usually difficult or even impossible to solve in terms of familiar elementary functions. Furthermore, if we happen to obtain a family of solutions for a nonlinear equation, it is not evident whether this family contains all solutions. On a practical level, then, the designation "general solution" is applied only to linear DEs. Don't be concerned about this concept at this point but store the words *general solution* in the back of your mind—we will come back to this notion in Section 2.3 and again in Chapter 3.

1.1 Exercises Answers to selected odd-numbered problems begin on page ANS-1.

In Problems 1–10, state the order of the given ordinary differential equation. Determine whether the equation is linear or nonlinear by matching it with (6).

1. $(1 - x)y'' - 4xy' + 5y = \cos x$

2. $x\dfrac{d^3y}{dx^3} - \left(\dfrac{dy}{dx}\right)^4 + y = 0$

3. $t^5 y^{(4)} - t^3 y'' + 6y = 0$

4. $\dfrac{d^2u}{dr^2} + \dfrac{du}{dr} + u = \cos(r + u)$

5. $\dfrac{d^2y}{dx^2} = \sqrt{1 + \left(\dfrac{dy}{dx}\right)^2}$

6. $\dfrac{d^2R}{dt^2} = -\dfrac{k}{R^2}$

7. $(\sin \theta)y''' - (\cos \theta)y' = 2$

8. $\ddot{x} - \left(1 - \tfrac{1}{3}\dot{x}^2\right)\dot{x} + x = 0$

9. $\sin\left(\dfrac{dy}{dx}\right) = y + x$

10. $\dfrac{dx}{dy} + y^3 x = \sin y$

In Problems 11 and 12, determine whether the given first-order differential equation is linear in the indicated dependent variable by matching it with the first differential equation given in (7).

11. $(y^2 - 1)dx + x\,dy = 0$; in y; in x

12. $u\,dv + (v + uv - ue^u)\,du = 0$; in v; in u

In Problems 13–16, verify that the indicated function is an explicit solution of the given differential equation. Assume an appropriate interval I of definition for each solution.

13. $2y' + y = 0$; $\quad y = e^{-x/2}$

14. $\dfrac{dy}{dt} + 20y = 24$; $\quad y = \tfrac{6}{5} - \tfrac{6}{5}e^{-20t}$

15. $y'' - 6y' + 13y = 0$; $\quad y = e^{3x}\cos 2x$

16. $y'' + y = \tan x$; $\quad y = -(\cos x)\ln(\sec x + \tan x)$

In Problems 17–20, verify that the indicated function $y = \phi(x)$ is an explicit solution of the given first-order differential equation. Proceed as in Example 7, by considering ϕ simply as a *function*, give its domain. Then by considering ϕ as a *solution* of the differential equation, give at least one interval I of definition.

17. $(y - x)y' = y - x + 8$; $\quad y = x + 4\sqrt{x + 2}$

18. $y' = 25 + y^2$; $\quad y = 5\tan 5x$

19. $y' = 2xy^2$; $\quad y = 1/(4 - x^2)$

20. $2y' = y^3 \cos x$; $\quad y = (1 - \sin x)^{-1/2}$

In Problems 21 and 22, verify that the indicated expression is an implicit solution of the given first-order differential equation. Find at least one explicit solution $y = \phi(x)$ in each case. Use a graphing utility to obtain the graph of an explicit solution. Give an interval I of definition of each solution ϕ.

21. $\dfrac{dX}{dt} = (X - 1)(1 - 2X)$; $\quad \ln\left(\dfrac{2X - 1}{X - 1}\right) = t$

22. $2xy\,dx + (x^2 - y)\,dy = 0$; $\quad -2x^2y + y^2 = 1$

In Problems 23–26, verify that the indicated family of functions is a solution of the given differential equation. Assume an appropriate interval I of definition for each solution.

23. $\dfrac{dP}{dt} = P(1 - P); \quad P = \dfrac{c_1 e^t}{1 + c_1 e^t}$

24. $\dfrac{dy}{dx} + 4xy = 8x^3; \quad y = 2x^2 - 1 + c_1 e^{-2x^2}$

25. $\dfrac{d^2 y}{dx^2} - 4\dfrac{dy}{dx} + 4y = 0; \quad y = c_1 e^{2x} + c_2 x e^{2x}$

26. $x^3 \dfrac{d^3 y}{dx^3} + 2x^2 \dfrac{d^2 y}{dx^2} - x\dfrac{dy}{dx} + y = 12x^2;$

$y = c_1 x^{-1} + c_2 x + c_3 x \ln x + 4x^2$

In Problems 27–30, use (12) to verify that the indicated function is a solution of the given differential equation. Assume an appropriate interval I of definition of each solution.

27. $x\dfrac{dy}{dx} - 3xy = 1; \quad y = e^{3x} \displaystyle\int_1^x \dfrac{e^{-3t}}{t}\, dt$

28. $2x\dfrac{dy}{dx} - y = 2x \cos x; \quad y = \sqrt{x} \displaystyle\int_4^x \dfrac{\cos t}{\sqrt{t}}\, dt$

29. $x^2 \dfrac{dy}{dx} + xy = 10 \sin x; \quad y = \dfrac{5}{x} + \dfrac{10}{x} \displaystyle\int_1^x \dfrac{\sin t}{t}\, dt$

30. $\dfrac{dy}{dx} + 2xy = 1; \quad y = e^{-x^2} + e^{-x^2} \displaystyle\int_0^x e^{t^2}\, dt$

31. Verify that the piecewise-defined function

$$y = \begin{cases} -x^2, & x < 0 \\ x^2, & x \geq 0 \end{cases}$$

is a solution of the differential equation $xy' - 2y = 0$ on the interval $(-\infty, \infty)$.

32. In Example 8 we saw that $y = \phi_1(x) = \sqrt{25 - x^2}$ and $y = \phi_2(x) = -\sqrt{25 - x^2}$ are solutions of $dy/dx = -x/y$ on the interval $(-5, 5)$. Explain why the piecewise-defined function

$$y = \begin{cases} \sqrt{25 - x^2}, & -5 < x < 0 \\ -\sqrt{25 - x^2}, & 0 \leq x < 5 \end{cases}$$

is not a solution of the differential equation on the interval $(-5, 5)$.

In Problems 33–36, find values of m so that the function $y = e^{mx}$ is a solution of the given differential equation.

33. $y' + 2y = 0$ **34.** $3y' = 4y$

35. $y'' - 5y' + 6y = 0$ **36.** $2y'' + 9y' - 5y = 0$

In Problems 37–40, find values of m so that the function $y = x^m$ is a solution of the given differential equation.

37. $xy'' + 2y' = 0$ **38.** $4x^2 y'' + y = 0$

39. $x^2 y'' - 7xy' + 15y = 0$ **40.** $x^2 y''' - 3xy'' + 3y' = 0$

In Problems 41–44, use the concept that $y = c$, $-\infty < x < \infty$, is a constant function if and only if $y' = 0$ to determine whether the given differential equation possesses constant solutions.

41. $3xy' + 5y = 10$ **42.** $y' = y^2 + 2y - 3$

43. $(y - 1)y' = 1$ **44.** $y'' + 4y' + 6y = 10$

In Problems 45–48, verify that the one-parameter family is a solution of the given differential equation. Find at least one singular solution of the DE.

45. $y = (x + c_1)^2; \quad \left(\dfrac{dy}{dx}\right)^2 = 4y$

46. $y = 3 \sin(x + c_1); \quad \left(\dfrac{dy}{dx}\right)^2 = 9 - y^2$

47. $x - \sqrt{16 - y^2} = c_1; \quad y\dfrac{dy}{dx} + \sqrt{16 - y^2} = 0$

48. $y = x - (x - c_1)^2; \quad \left(\dfrac{dy}{dx}\right)^2 - 2\dfrac{dy}{dx} + 4y = 4x - 1$

In Problems 49 and 50, verify that the indicated pair of functions is a solution of the given system of differential equations on the interval $(-\infty, \infty)$.

49. $\dfrac{dx}{dt} = x + 3y$

$\dfrac{dy}{dt} = 5x + 3y;$

$x = e^{-2t} + 3e^{6t},$

$y = -e^{-2t} + 5e^{6t}$

50. $\dfrac{d^2 x}{dt^2} = 4y + e^t$

$\dfrac{d^2 y}{dt^2} = 4x - e^t;$

$x = \cos 2t + \sin 2t + \tfrac{1}{5}e^t,$

$y = -\cos 2t - \sin 2t - \tfrac{1}{5}e^t$

Discussion Problems

51. Make up a differential equation that does not possess any real solutions.

52. Make up a differential equation that you feel confident possesses only the trivial solution $y = 0$. Explain your reasoning.

53. What function do you know from calculus is such that its first derivative is itself? Its first derivative is a constant multiple k of itself? Write each answer in the form of a first-order differential equation with a solution.

54. What function (or functions) do you know from calculus is such that its second derivative is itself? Its second derivative is the negative of itself? Write each answer in the form of a second-order differential equation with a solution.

55. Given that $y = \sin x$ is an explicit solution of the first-order differential equation $dy/dx = \sqrt{1 - y^2}$. Find an interval I of definition. [*Hint: I* is *not* the interval $(-\infty, \infty)$.]

56. Discuss why it makes intuitive sense to presume that the linear differential equation $y'' + 2y' + 4y = 5 \sin t$ has a solution of the form $y = A \sin t + B \cos t$, where A and B are constants. Then find specific constants A and B so that $y = A \sin t + B \cos t$ is a particular solution of the DE.

In Problems 57 and 58, the given figure represents the graph of an implicit solution $G(x, y) = 0$ of a differential equation $dy/dx = f(x, y)$. In each case the relation $G(x, y) = 0$ implicitly defines several solutions of the DE. Carefully reproduce each figure on a piece of paper. Use different colored pencils to mark off segments, or pieces, on each graph that correspond to graphs of solutions. Keep in mind that a solution ϕ must be a function and differentiable. Use the solution curve to estimate the interval I of definition of each solution ϕ.

57.

58.

FIGURE 1.1.6 Graph for Problem 57

FIGURE 1.1.7 Graph for Problem 58

59. The graphs of the members of the one-parameter family $x^3 + y^3 = 3cxy$ are called **folia of Descartes** after the French mathematician and inventor of analytic geometry, **René Descartes** (1596–1650). Verify that this family is an implicit solution of the first-order differential equation.

$$\frac{dy}{dx} = \frac{y(y^3 - 2x^3)}{x(2y^3 - x^3)}.$$

60. The graph in **FIGURE 1.1.7** is the member of the family of folia in Problem 59 corresponding to $c = 1$. Discuss: How can the DE in Problem 59 help in finding points on the graph of $x^3 + y^3 = 3xy$ where the tangent line is vertical? How does knowing where a tangent line is vertical help in determining an interval I of definition of a solution ϕ of the DE? Carry out your ideas and compare with your estimates of the intervals in Problem 58.

61. In Example 8, the largest interval I over which the explicit solutions $y = \phi_1(x)$ and $y = \phi_2(x)$ are defined is the open interval $(-5, 5)$. Why can't the interval I of definition be the closed interval $[-5, 5]$?

62. In Problem 23, a one-parameter family of solutions of the DE $P' = P(1 - P)$ is given. Does any solution curve pass through the point $(0, 3)$? Through the point $(0, 1)$?

63. Discuss, and illustrate with examples, how to solve differential equations of the forms $dy/dx = f(x)$ and $d^2y/dx^2 = f(x)$.

64. The differential equation $x(y')^2 - 4y' - 12x^3 = 0$ has the form given in (4). Determine whether the equation can be put into the normal form $dy/dx = f(x, y)$.

65. The normal form (5) of an nth-order differential equation is equivalent to (4) whenever both forms have exactly the same solutions. Make up a first-order differential equation for which $F(x, y, y') = 0$ is not equivalent to the normal form $dy/dx = f(x, y)$.

66. Find a linear second-order differential equation $F(x, y, y', y'') = 0$ for which $y = c_1x + c_2x^2$ is a two-parameter family of solutions. Make sure that your equation is free of the arbitrary parameters c_1 and c_2.

Qualitative information about a solution $y = \phi(x)$ of a differential equation can often be obtained from the equation itself. Before working Problems 67–70, recall the geometric significance of the derivatives dy/dx and d^2y/dx^2.

67. Consider the differential equation $dy/dx = e^{-x^2}$.
 (a) Explain why a solution of the DE must be an increasing function on any interval of the x-axis.
 (b) What are $\lim_{x \to -\infty} dy/dx$ and $\lim_{x \to \infty} dy/dx$? What does this suggest about a solution curve as $x \to \pm\infty$?
 (c) Determine an interval over which a solution curve is concave down and an interval over which the curve is concave up.
 (d) Sketch the graph of a solution $y = \phi(x)$ of the differential equation whose shape is suggested by parts (a)–(c).

68. Consider the differential equation $dy/dx = 5 - y$.
 (a) Either by inspection, or by the method suggested in Problems 41–44, find a constant solution of the DE.
 (b) Using only the differential equation, find intervals on the y-axis on which a nonconstant solution $y = \phi(x)$ is increasing. Find intervals on the y-axis on which $y = \phi(x)$ is decreasing.

69. Consider the differential equation $dy/dx = y(a - by)$, where a and b are positive constants.
 (a) Either by inspection, or by the method suggested in Problems 41–44, find two constant solutions of the DE.
 (b) Using only the differential equation, find intervals on the y-axis on which a nonconstant solution $y = \phi(x)$ is increasing. On which $y = \phi(x)$ is decreasing.
 (c) Using only the differential equation, explain why $y = a/2b$ is the y-coordinate of a point of inflection of the graph of a nonconstant solution $y = \phi(x)$.
 (d) On the same coordinate axes, sketch the graphs of the two constant solutions found in part (a). These constant solutions partition the xy-plane into three regions. In each region, sketch the graph of a nonconstant solution $y = \phi(x)$ whose shape is suggested by the results in parts (b) and (c).

70. Consider the differential equation $y' = y^2 + 4$.
 (a) Explain why there exist no constant solutions of the DE.
 (b) Describe the graph of a solution $y = \phi(x)$. For example, can a solution curve have any relative extrema?
 (c) Explain why $y = 0$ is the y-coordinate of a point of inflection of a solution curve.
 (d) Sketch the graph of a solution $y = \phi(x)$ of the differential equation whose shape is suggested by parts (a)–(c).

Computer Lab Assignments

In Problems 71 and 72, use a CAS to compute all derivatives and to carry out the simplifications needed to verify that the indicated function is a particular solution of the given differential equation.

71. $y^{(4)} - 20y''' + 158y'' - 580y' + 841y = 0;$

$$y = xe^{5x}\cos 2x$$

72. $x^3y''' + 2x^2y'' + 20xy' - 78y = 0;$

$$y = 20\,\frac{\cos(5\ln x)}{x} - 3\,\frac{\sin(5\ln x)}{x}$$

1.2 Initial-Value Problems

INTRODUCTION We are often interested in problems in which we seek a solution $y(x)$ of a differential equation so that $y(x)$ satisfies prescribed side conditions—that is, conditions that are imposed on the unknown $y(x)$ or on its derivatives. In this section we examine one such problem called an *initial-value problem*.

■ **Initial-Value Problem** On some interval I containing x_0, the problem

$$\textit{Solve:}\qquad \frac{d^n y}{dx^n} = f(x, y, y', \dots, y^{(n-1)})$$

$$\textit{Subject to:}\quad y(x_0) = y_0, y'(x_0) = y_1, \dots, y^{(n-1)}(x_0) = y_{n-1},\tag{1}$$

where $y_0, y_1, \dots, y_{n-1}$ are arbitrarily specified real constants, is called an **initial-value problem (IVP)**. The values of $y(x)$ and its first $n-1$ derivatives at a single point x_0: $y(x_0) = y_0$, $y'(x_0) = y_1, \dots,$ $y^{(n-1)}(x_0) = y_{n-1}$, are called **initial conditions (IC)**.

■ **First- and Second-Order IVPs** The problem given in (1) is also called an ***n*th-order initial-value problem**. For example,

$$\textit{Solve:}\qquad \frac{dy}{dx} = f(x, y)$$

$$\textit{Subject to:}\quad y(x_0) = y_0\tag{2}$$

and

$$\textit{Solve:}\qquad \frac{d^2 y}{dx^2} = f(x, y, y')$$

$$\textit{Subject to:}\quad y(x_0) = y_0, y'(x_0) = y_1\tag{3}$$

FIGURE 1.2.1 First-order IVP

FIGURE 1.2.2 Second-order IVP

are **first-** and **second-order initial-value problems**, respectively. These two problems are easy to interpret in geometric terms. For (2) we are seeking a solution of the differential equation on an interval I containing x_0 so that a solution curve passes through the prescribed point (x_0, y_0). See **FIGURE 1.2.1**. For (3) we want to find a solution of the differential equation whose graph not only passes through (x_0, y_0) but passes through so that the slope of the curve at this point is y_1. See **FIGURE 1.2.2**. The term *initial condition* derives from physical systems where the independent variable is time t and where $y(t_0) = y_0$ and $y'(t_0) = y_1$ represent, respectively, the position and velocity of an object at some beginning, or initial, time t_0.

Solving an *n*th-order initial-value problem frequently entails using an *n*-parameter family of solutions of the given differential equation to find *n* specialized constants so that the resulting particular solution of the equation also "fits"—that is, satisfies—the *n* initial conditions.

EXAMPLE 1 First-Order IVPs

(a) It is readily verified that $y = ce^x$ is a one-parameter family of solutions of the simple first-order equation $y' = y$ on the interval $(-\infty, \infty)$. If we specify an initial condition, say, $y(0) = 3$, then substituting $x = 0$, $y = 3$ in the family determines the constant $3 = ce^0 = c$. Thus the function $y = 3e^x$ is a solution of the initial-value problem

$$y' = y, \quad y(0) = 3.$$

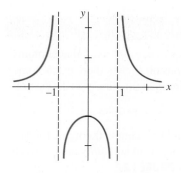

(a) Function defined for all x
except $x = \pm 1$

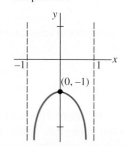

(b) Solution defined on interval
containing $x = 0$

FIGURE 1.2.4 Graphs of function
and solution of IVP in Example 2

FIGURE 1.2.3 Solutions of IVPs in
Example 1

(b) Now if we demand that a solution of the differential equation pass through the point $(1, -2)$ rather than $(0, 3)$, then $y(1) = -2$ will yield $-2 = ce$ or $c = -2e^{-1}$. The function $y = -2e^{x-1}$ is a solution of the initial-value problem

$$y' = y, \quad y(1) = -2.$$

The graphs of these two solutions are shown in blue in **FIGURE 1.2.3**.

The next example illustrates another first-order initial-value problem. In this example, notice how the interval I of definition of the solution $y(x)$ depends on the initial condition $y(x_0) = y_0$.

EXAMPLE 2 **Interval I of Definition of a Solution**

In Problem 6 of Exercises 2.2 you will be asked to show that a one-parameter family of solutions of the first-order differential equation $y' + 2xy^2 = 0$ is $y = 1/(x^2 + c)$. If we impose the initial condition $y(0) = -1$, then substituting $x = 0$ and $y = -1$ into the family of solutions gives $-1 = 1/c$ or $c = -1$. Thus, $y = 1/(x^2 - 1)$. We now emphasize the following three distinctions.

- Considered as a *function*, the domain of $y = 1/(x^2 - 1)$ is the set of real numbers x for which $y(x)$ is defined; this is the set of all real numbers except $x = -1$ and $x = 1$. See **FIGURE 1.2.4(a)**.
- Considered as a *solution of the differential equation* $y' + 2xy^2 = 0$, the interval I of definition of $y = 1/(x^2 - 1)$ could be taken to be any interval over which $y(x)$ is defined and differentiable. As can be seen in Figure 1.2.4(a), the largest intervals on which $y = 1/(x^2 - 1)$ is a solution are $(-\infty, -1), (-1, 1),$ and $(1, \infty)$.
- Considered as a *solution of the initial-value problem* $y' + 2xy^2 = 0, y(0) = -1$, the interval I of definition of $y = 1/(x^2 - 1)$ could be taken to be any interval over which $y(x)$ is defined, differentiable, and contains the initial point $x = 0$; the largest interval for which this is true is $(-1, 1)$. See Figure 1.2.4(b).

See Problems 3–6 in Exercises 1.2 for a continuation of Example 2.

EXAMPLE 3 **Second-Order IVP**

In Example 10 of Section 1.1 we saw that $x = c_1 \cos 4t + c_2 \sin 4t$ is a two-parameter family of solutions of $x'' + 16x = 0$. Find a solution of the initial-value problem

$$x'' + 16x = 0, \quad x(\pi/2) = -2, \quad x'(\pi/2) = 1. \tag{4}$$

SOLUTION We first apply $x(\pi/2) = -2$ to the given family of solutions: $c_1 \cos 2\pi + c_2 \sin 2\pi = -2$. Since $\cos 2\pi = 1$ and $\sin 2\pi = 0$, we find that $c_1 = -2$. We next apply $x'(\pi/2) = 1$ to the one-parameter family $x(t) = -2 \cos 4t + c_2 \sin 4t$. Differentiating and then setting $t = \pi/2$ and $x' = 1$ gives $8 \sin 2\pi + 4c_2 \cos 2\pi = 1$, from which we see that $c_2 = \frac{1}{4}$. Hence $x = -2 \cos 4t + \frac{1}{4} \sin 4t$ is a solution of (4). ≡

Existence and Uniqueness Two fundamental questions arise in considering an initial-value problem:

Does a solution of the problem exist? If a solution exists, is it unique?

For a first-order initial-value problem such as (2), we ask:

Existence $\begin{cases} \textit{Does the differential equation } dy/dx = f(x, y) \textit{ possess solutions?} \\ \textit{Do any of the solution curves pass through the point } (x_0, y_0)? \end{cases}$

Uniqueness $\begin{cases} \textit{When can we be certain that there is precisely one solution curve passing} \\ \textit{through the point } (x_0, y_0)? \end{cases}$

Note that in Examples 1 and 3, the phrase "*a* solution" is used rather than "*the* solution" of the problem. The indefinite article "a" is used deliberately to suggest the possibility that other solutions may exist. At this point it has not been demonstrated that there is a single solution of each problem. The next example illustrates an initial-value problem with two solutions.

FIGURE 1.2.5 Two solutions of the same IVP in Example 4

EXAMPLE 4 **An IVP Can Have Several Solutions**

Each of the functions $y = 0$ and $y = \frac{1}{16}x^4$ satisfies the differential equation $dy/dx = xy^{1/2}$ and the initial condition $y(0) = 0$, and so the initial-value problem $dy/dx = xy^{1/2}$, $y(0) = 0$, has at least two solutions. As illustrated in **FIGURE 1.2.5**, the graphs of both functions pass through the same point $(0, 0)$. ≡

Within the safe confines of a formal course in differential equations one can be fairly confident that *most* differential equations will have solutions and that solutions of initial-value problems will *probably* be unique. Real life, however, is not so idyllic. Thus it is desirable to know in advance of trying to solve an initial-value problem whether a solution exists and, when it does, whether it is the only solution of the problem. Since we are going to consider first-order differential equations in the next two chapters, we state here without proof a straightforward theorem that gives conditions that are sufficient to guarantee the existence and uniqueness of a solution of a first-order initial-value problem of the form given in (2). We shall wait until Chapter 3 to address the question of existence and uniqueness of a second-order initial-value problem.

THEOREM **1.2.1** **Existence of a Unique Solution**

Let R be a rectangular region in the xy-plane defined by $a \leq x \leq b$, $c \leq y \leq d$, that contains the point (x_0, y_0) in its interior. If $f(x, y)$ and $\partial f/\partial y$ are continuous on R, then there exists some interval I_0: $(x_0 - h, x_0 + h)$, $h > 0$, contained in $[a, b]$, and a unique function $y(x)$ defined on I_0 that is a solution of the initial-value problem (2).

FIGURE 1.2.6 Rectangular region R

The foregoing result is one of the most popular existence and uniqueness theorems for first-order differential equations, because the criteria of continuity of $f(x, y)$ and $\partial f/\partial y$ are relatively easy to check. The geometry of Theorem 1.2.1 is illustrated in **FIGURE 1.2.6**.

EXAMPLE 5 **Example 4 Revisited**

We saw in Example 4 that the differential equation $dy/dx = xy^{1/2}$ possesses at least two solutions whose graphs pass through $(0, 0)$. Inspection of the functions

$$f(x, y) = xy^{1/2} \quad \text{and} \quad \frac{\partial f}{\partial y} = \frac{x}{2y^{1/2}}$$

shows that they are continuous in the upper half-plane defined by $y > 0$. Hence Theorem 1.2.1 enables us to conclude that through any point (x_0, y_0), $y_0 > 0$, in the upper half-plane there is some interval centered at x_0 on which the given differential equation has a unique solution. Thus, for example, even without solving it we know that there exists some interval centered at 2 on which the initial-value problem $dy/dx = xy^{1/2}$, $y(2) = 1$, has a unique solution. ≡

In Example 1, Theorem 1.2.1 guarantees that there are no other solutions of the initial-value problems $y' = y$, $y(0) = 3$, and $y' = y$, $y(1) = -2$, other than $y = 3e^x$ and $y = -2e^{x-1}$, respectively. This follows from the fact that $f(x, y) = y$ and $\partial f/\partial y = 1$ are continuous throughout the entire xy-plane. It can be further shown that the interval I on which each solution is defined is $(-\infty, \infty)$.

■ **Interval of Existence/Uniqueness** Suppose $y(x)$ represents a solution of the initial-value problem (2). The following three sets on the real x-axis may not be the same: the domain of the function $y(x)$, the interval I over which the solution $y(x)$ is defined or exists, and the interval I_0 of existence *and* uniqueness. In Example 7 of Section 1.1 we illustrated the difference between the domain of a function and the interval I of definition. Now suppose (x_0, y_0) is a point in the interior of the rectangular region R in Theorem 1.2.1. It turns out that the continuity of the function $f(x, y)$ on R by itself is sufficient to guarantee the existence of at least one solution of $dy/dx = f(x, y)$, $y(x_0) = y_0$, defined on some interval I. The interval I of definition for this initial-value problem is usually taken to be the largest interval containing x_0 over which

the solution $y(x)$ is defined and differentiable. The interval I depends on both $f(x, y)$ and the initial condition $y(x_0) = y_0$. See Problems 31–34 in Exercises 1.2. The extra condition of continuity of the first partial derivative $\partial f/\partial y$ on R enables us to say that not only does a solution exist on some interval I_0 containing x_0, but it also is the *only* solution satisfying $y(x_0) = y_0$. However, Theorem 1.2.1 does not give any indication of the sizes of the intervals I and I_0; *the interval I of definition need not be as wide as the region R, and the interval I_0 of existence and uniqueness may not be as large as I.* The number $h > 0$ that defines the interval I_0: $(x_0 - h, x_0 + h)$, could be very small, and so it is best to think that the solution $y(x)$ is *unique in a local sense*, that is, a solution defined near the point (x_0, y_0). See Problem 50 in Exercises 1.2.

REMARKS

(*i*) The conditions in Theorem 1.2.1 are sufficient but not necessary. When $f(x, y)$ and $\partial f/\partial y$ are continuous on a rectangular region R, it must always follow that a solution of (2) exists and is unique whenever (x_0, y_0) is a point interior to R. However, if the conditions stated in the hypotheses of Theorem 1.2.1 do not hold, then anything could happen: Problem (2) *may* still have a solution and this solution *may* be unique, or (2) *may* have several solutions, or it may have no solution at all. A rereading of Example 4 reveals that the hypotheses of Theorem 1.2.1 do not hold on the line $y = 0$ for the differential equation $dy/dx = xy^{1/2}$, and so it is not surprising, as we saw in Example 4 of this section, that there are two solutions defined on a common interval $(-h, h)$ satisfying $y(0) = 0$. On the other hand, the hypotheses of Theorem 1.2.1 do not hold on the line $y = 1$ for the differential equation $dy/dx = |y - 1|$. Nevertheless, it can be proved that the solution of the initial-value problem $dy/dx = |y - 1|$, $y(0) = 1$, is unique. Can you guess this solution?

(*ii*) You are encouraged to read, think about, work, and then keep in mind Problem 49 in Exercises 1.2.

1.2 Exercises Answers to selected odd-numbered problems begin on page ANS-1.

In Problems 1 and 2, $y = 1/(1 + c_1 e^{-x})$ is a one-parameter family of solutions of the first-order DE $y' = y - y^2$. Find a solution of the first-order IVP consisting of this differential equation and the given initial condition.

1. $y(0) = -\frac{1}{3}$ **2.** $y(-1) = 2$

In Problems 3–6, $y = 1/(x^2 + c)$ is a one-parameter family of solutions of the first-order DE $y' + 2xy^2 = 0$. Find a solution of the first-order IVP consisting of this differential equation and the given initial condition. Give the largest interval I over which the solution is defined.

3. $y(2) = \frac{1}{3}$ **4.** $y(-2) = \frac{1}{2}$

5. $y(0) = 1$ **6.** $y\left(\frac{1}{2}\right) = -4$

In Problems 7–10, $x = c_1 \cos t + c_2 \sin t$ is a two-parameter family of solutions of the second-order DE $x'' + x = 0$. Find a solution of the second-order IVP consisting of this differential equation and the given initial conditions.

7. $x(0) = -1$, $x'(0) = 8$

8. $x(\pi/2) = 0$, $x'(\pi/2) = 1$

9. $x(\pi/6) = \frac{1}{2}$, $x'(\pi/6) = 0$

10. $x(\pi/4) = \sqrt{2}$, $x'(\pi/4) = 2\sqrt{2}$

In Problems 11–14, $y = c_1 e^x + c_2 e^{-x}$ is a two-parameter family of solutions of the second-order DE $y'' - y = 0$.

Find a solution of the second-order IVP consisting of this differential equation and the given initial conditions.

11. $y(0) = 1$, $y'(0) = 2$ **12.** $y(1) = 0$, $y'(1) = e$

13. $y(-1) = 5$, $y'(-1) = -5$ **14.** $y(0) = 0$, $y'(0) = 0$

In Problems 15 and 16, determine by inspection at least two solutions of the given first-order IVP.

15. $y' = 3y^{2/3}$, $y(0) = 0$ **16.** $xy' = 2y$, $y(0) = 0$

In Problems 17–24, determine a region of the xy-plane for which the given differential equation would have a unique solution whose graph passes through a point (x_0, y_0) in the region.

17. $\dfrac{dy}{dx} = y^{2/3}$ **18.** $\dfrac{dy}{dx} = \sqrt{xy}$

19. $x\dfrac{dy}{dx} = y$ **20.** $\dfrac{dy}{dx} - y = x$

21. $(4 - y^2)y' = x^2$ **22.** $(1 + y^3)y' = x^2$

23. $(x^2 + y^2)y' = y^2$ **24.** $(y - x)y' = y + x$

In Problems 25–28, determine whether Theorem 1.2.1 guarantees that the differential equation $y' = \sqrt{y^2 - 9}$ possesses a unique solution through the given point.

25. $(1, 4)$ **26.** $(5, 3)$

27. $(2, -3)$ **28.** $(-1, 1)$

29. (a) By inspection, find a one-parameter family of solutions of the differential equation $xy' = y$. Verify that each member of the family is a solution of the initial-value problem $xy' = y$, $y(0) = 0$.

(b) Explain part (a) by determining a region R in the xy-plane for which the differential equation $xy' = y$ would have a unique solution through a point (x_0, y_0) in R.

(c) Verify that the piecewise-defined function

$$y = \begin{cases} 0, & x < 0 \\ x, & x \geq 0 \end{cases}$$

satisfies the condition $y(0) = 0$. Determine whether this function is also a solution of the initial-value problem in part (a).

30. (a) Verify that $y = \tan(x + c)$ is a one-parameter family of solutions of the differential equation $y' = 1 + y^2$.

(b) Since $f(x, y) = 1 + y^2$ and $\partial f/\partial y = 2y$ are continuous everywhere, the region R in Theorem 1.2.1 can be taken to be the entire xy-plane. Use the family of solutions in part (a) to find an explicit solution of the first-order initial-value problem $y' = 1 + y^2$, $y(0) = 0$. Even though $x_0 = 0$ is in the interval $(-2, 2)$, explain why the solution is not defined on this interval.

(c) Determine the largest interval I of definition for the solution of the initial-value problem in part (b).

31. (a) Verify that $y = -1/(x + c)$ is a one-parameter family of solutions of the differential equation $y' = y^2$.

(b) Since $f(x, y) = y^2$ and $\partial f/\partial y = 2y$ are continuous everywhere, the region R in Theorem 1.2.1 can be taken to be the entire xy-plane. Find a solution from the family in part (a) that satisfies $y(0) = 1$. Find a solution from the family in part (a) that satisfies $y(0) = -1$. Determine the largest interval I of definition for the solution of each initial-value problem.

32. (a) Find a solution from the family in part (a) of Problem 31 that satisfies $y' = y^2$, $y(0) = y_0$, where $y_0 \neq 0$. Explain why the largest interval I of definition for this solution is either $(-\infty, 1/y_0)$ or $(1/y_0, \infty)$.

(b) Determine the largest interval I of definition for the solution of the first-order initial-value problem $y' = y^2$, $y(0) = 0$.

33. (a) Verify that $3x^2 - y^2 = c$ is a one-parameter family of solutions of the differential equation $y\,dy/dx = 3x$.

(b) By hand, sketch the graph of the implicit solution $3x^2 - y^2 = 3$. Find all explicit solutions $y = \phi(x)$ of the DE in part (a) defined by this relation. Give the interval I of definition of each explicit solution.

(c) The point $(-2, 3)$ is on the graph of $3x^2 - y^2 = 3$, but which of the explicit solutions in part (b) satisfies $y(-2) = 3$?

34. (a) Use the family of solutions in part (a) of Problem 33 to find an implicit solution of the initial-value problem $y\,dy/dx = 3x$, $y(2) = -4$. Then, by hand, sketch the graph of the explicit solution of this problem and give its interval I of definition.

(b) Are there any explicit solutions of $y\,dy/dx = 3x$ that pass through the origin?

In Problems 35–38, the graph of a member of a family of solutions of a second-order differential equation $d^2y/dx^2 = f(x, y, y')$ is given. Match the solution curve with at least one pair of the following initial conditions.

(a) $y(1) = 1, y'(1) = -2$ **(b)** $y(-1) = 0, y'(-1) = -4$
(c) $y(1) = 1, y'(1) = 2$ **(d)** $y(0) = -1, y'(0) = 2$
(e) $y(0) = -1, y'(0) = 0$ **(f)** $y(0) = -4, y'(0) = -2$

35.

FIGURE 1.2.7 Graph for Problem 35

36.

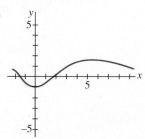

FIGURE 1.2.8 Graph for Problem 36

37.

FIGURE 1.2.9 Graph for Problem 37

38.

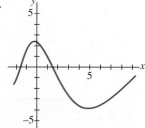

FIGURE 1.2.10 Graph for Problem 38

In Problems 39–44, $y = c_1 \cos 3x + c_2 \sin 3x$ is a two-parameter family of solutions of the second-order DE $y'' + 9y = 0$. If possible, find a solution of the differential equation that satisfies the given side conditions. The conditions specified at two different points are called boundary conditions.

39. $y(0) = 0, y(\pi/6) = -1$ **40.** $y(0) = 0, y(\pi) = 0$
41. $y'(0) = 0, y'(\pi/4) = 0$ **42.** $y(0) = 1, y'(\pi) = 5$
43. $y(0) = 0, y(\pi) = 4$ **44.** $y'(\pi/3) = 1, y'(\pi) = 0$

Discussion Problems

In Problems 45 and 46, use Problem 63 in Exercises 1.1 and (2) and (3) of this section.

45. Find a function $y = f(x)$ whose graph at each point (x, y) has the slope given by $8e^{2x} + 6x$ and has the y-intercept $(0, 9)$.

46. Find a function $y = f(x)$ whose second derivative is $y'' = 12x - 2$ at each point (x, y) on its graph and $y = -x + 5$ is tangent to the graph at the point corresponding to $x = 1$.

47. Consider the initial-value problem $y' = x - 2y$, $y(0) = \frac{1}{2}$. Determine which of the two curves shown in **FIGURE 1.2.11** is the only plausible solution curve. Explain your reasoning.

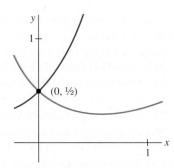

FIGURE 1.2.11 Graph for Problem 47

48. Without attempting to solve the initial-value problem $y' = x^2 + y^2$, $y(0) = 1$, find the values of $y'(0)$ and $y''(0)$.

49. Suppose that the first-order differential equation $dy/dx = f(x, y)$ possesses a one-parameter family of solutions and that $f(x, y)$ satisfies the hypotheses of Theorem 1.2.1 in some rectangular region R of the xy-plane. Explain why two different solution curves cannot intersect or be tangent to each other at a point (x_0, y_0) in R.

50. The functions

$$y(x) = \tfrac{1}{16}x^4, \quad -\infty < x < \infty$$

and

$$y(x) = \begin{cases} 0, & x < 0 \\ \tfrac{1}{16}x^4, & x \geq 0 \end{cases}$$

have the same domain but are clearly different. See **FIGURES 1.2.12(a)** and 1.2.12(b), respectively. Show that both functions are solutions of the initial-value problem $dy/dx = xy^{1/2}$, $y(2) = 1$ on the interval $(-\infty, \infty)$. Resolve the apparent contradiction between this fact and the last sentence in Example 5.

FIGURE 1.2.12 Two solutions of the IVP in Problem 50

51. Show that

$$x = \int_0^y \frac{1}{\sqrt{t^3 + 1}}\, dt$$

is an implicit solution of the initial-value problem

$$2\frac{d^2y}{dx^2} - 3y^2 = 0, \quad y(0) = 0, y'(0) = 1.$$

Assume that $y \geq 0$. [*Hint*: The integral is nonelementary. See (*ii*) in the *Remarks* at the end of Section 1.1.]

1.3 Differential Equations as Mathematical Models

INTRODUCTION In this section we introduce the notion of a *mathematical model*. Roughly speaking, a mathematical model is a mathematical description of something. This description could be as simple as a function. For example, **Leonardo da Vinci** (1452–1519) was able to deduce the speed v of a falling body by examining a sequence. Leonardo allowed water drops to fall, at equally spaced intervals of time, between two boards covered with blotting paper. When a spring mechanism was disengaged, the boards were clapped together. See **FIGURE 1.3.1**. By carefully examining the sequence of water blots, Leonardo discovered that the distances between consecutive drops increased in "a continuous arithmetic proportion." In this manner he discovered the formula $v = gt$.

Although there are many kinds of mathematical models, in this section we focus only on differential equations and discuss some specific differential-equation models in biology, physics, and chemistry. Once we have studied some methods for solving DEs, in Chapters 2 and 3 we return to, and solve, some of these models.

FIGURE 1.3.1 Da Vinci's apparatus for determining the speed of falling body

Mathematical Models It is often desirable to describe the behavior of some real-life system or phenomenon, whether physical, sociological, or even economic, in mathematical terms. The mathematical description of a system or a phenomenon is called a **mathematical model** and is constructed with certain goals in mind. For example, we may wish to understand the mechanisms of a certain ecosystem by studying the growth of animal populations in that system, or we may wish to date fossils by means of analyzing the decay of a radioactive substance either in the fossil or in the stratum in which it was discovered.

Construction of a mathematical model of a system starts with *identification of the variables* that are responsible for changing the system. We may choose not to incorporate all these variables into the model at first. In this first step we are specifying the **level of resolution** of the model. Next, *we make a set of reasonable assumptions* or hypotheses about the system we are trying to describe. These assumptions will also include any empirical laws that may be applicable to the system.

For some purposes it may be perfectly within reason to be content with low-resolution models. For example, you may already be aware that in modeling the motion of a body falling near the surface of the Earth, the retarding force of air friction is sometimes ignored in beginning physics courses; but if you are a scientist whose job it is to accurately predict the flight path of a long-range projectile, air resistance and other factors such as the curvature of the Earth have to be taken into account.

Since the assumptions made about a system frequently involve *a rate of change* of one or more of the variables, the mathematical depiction of all these assumptions may be one or more equations involving *derivatives*. In other words, the mathematical model may be a differential equation or a system of differential equations.

Once we have formulated a mathematical model that is either a differential equation or a system of differential equations, we are faced with the not insignificant problem of trying to solve it. *If* we can solve it, then we deem the model to be reasonable if its solution is consistent with either experimental data or known facts about the behavior of the system. But if the predictions produced by the solution are poor, we can either increase the level of resolution of the model or make alternative assumptions about the mechanisms for change in the system. The steps of the modeling process are then repeated as shown in **FIGURE 1.3.2**.

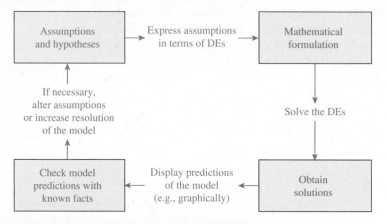

FIGURE 1.3.2 Steps in the modeling process

Of course, by increasing the resolution we add to the complexity of the mathematical model and increase the likelihood that we cannot obtain an explicit solution.

A mathematical model of a physical system will often involve the variable time t. A solution of the model then gives the **state of the system**; in other words, for appropriate values of t, the values of the dependent variable (or variables) describe the system in the past, present, and future.

■ Population Dynamics

One of the earliest attempts to model human **population growth** by means of mathematics was by the English economist **Thomas Robert Malthus** (1776–1834) in 1798. Basically, the idea of the Malthusian model is the assumption that the rate at which a population of a country grows at a certain time is proportional* to the total population of the country at that time. In other words, the more people there are at time t, the more there are going to be in the future. In mathematical terms, if $P(t)$ denotes the total population at time t, then this assumption can be expressed as

$$\frac{dP}{dt} \propto P \quad \text{or} \quad \frac{dP}{dt} = kP, \tag{1}$$

where k is a constant of proportionality. This simple model, which fails to take into account many factors (immigration and emigration, for example) that can influence human populations to either grow or decline, nevertheless turned out to be fairly accurate in predicting the population of the United States during the years 1790–1860. Populations that grow at a rate described by (1) are rare; nevertheless, (1) is still used to model *growth of small populations over short intervals of time*, for example, bacteria growing in a petri dish.

*If two quantities u and v are proportional, we write $u \propto v$. This means one quantity is a constant multiple of the other: $u = kv$.

Radioactive Decay

The nucleus of an atom consists of combinations of protons and neutrons. Many of these combinations of protons and neutrons are unstable; that is, the atoms decay or transmute into the atoms of another substance. Such nuclei are said to be radioactive. For example, over time, the highly radioactive radium, Ra-226, transmutes into the radioactive gas radon, Rn-222. In modeling the phenomenon of **radioactive decay**, it is assumed that the rate dA/dt at which the nuclei of a substance decay is proportional to the amount (more precisely, the number of nuclei) $A(t)$ of the substance remaining at time t:

$$\frac{dA}{dt} \propto A \quad \text{or} \quad \frac{dA}{dt} = kA. \tag{2}$$

Of course equations (1) and (2) are exactly the same; the difference is only in the interpretation of the symbols and the constants of proportionality. For growth, as we expect in (1), $k > 0$, and in the case of (2) and decay, $k < 0$.

The model (1) for growth can be seen as the equation $dS/dt = rS$, which describes the growth of capital S when an annual rate of interest r is compounded continuously. The model (2) for decay also occurs in a biological setting, such as determining the half-life of a drug—the time that it takes for 50% of a drug to be eliminated from a body by excretion or metabolism. In chemistry, the decay model (2) appears as the mathematical description of a first-order chemical reaction. The point is this:

A single differential equation can serve as a mathematical model for many different phenomena.

Mathematical models are often accompanied by certain side conditions. For example, in (1) and (2) we would expect to know, in turn, an initial population P_0 and an initial amount of radioactive substance A_0 that is on hand. If this initial point in time is taken to be $t = 0$, then we know that $P(0) = P_0$ and $A(0) = A_0$. In other words, a mathematical model can consist of either an initial-value problem or, as we shall see later in Section 3.9, a boundary-value problem.

Newton's Law of Cooling/Warming

According to **Newton's empirical law of cooling**—or warming—the rate at which the temperature of a body changes is proportional to the difference between the temperature of the body and the temperature of the surrounding medium, the so-called ambient temperature. If $T(t)$ represents the temperature of a body at time t, T_m the temperature of the surrounding medium, and dT/dt the rate at which the temperature of the body changes, then Newton's law of cooling/warming translates into the mathematical statement

$$\frac{dT}{dt} \propto T - T_m \quad \text{or} \quad \frac{dT}{dt} = k(T - T_m), \tag{3}$$

where k is a constant of proportionality. In either case, cooling or warming, if T_m is a constant, it stands to reason that $k < 0$.

Spread of a Disease

A contagious disease—for example, a flu virus—is spread throughout a community by people coming into contact with other people. Let $x(t)$ denote the number of people who have contracted the disease and $y(t)$ the number of people who have not yet been exposed. It seems reasonable to assume that the rate dx/dt at which the disease spreads is proportional to the number of encounters or *interactions* between these two groups of people. If we assume that the number of interactions is jointly proportional to $x(t)$ and $y(t)$, that is, proportional to the product xy, then

$$\frac{dx}{dt} = kxy, \tag{4}$$

where k is the usual constant of proportionality. Suppose a small community has a fixed population of n people. If one infected person is introduced into this community, then it could be argued that $x(t)$ and $y(t)$ are related by $x + y = n + 1$. Using this last equation to eliminate y in (4) gives us the model

$$\frac{dx}{dt} = kx(n + 1 - x). \tag{5}$$

An obvious initial condition accompanying equation (5) is $x(0) = 1$.

Chemical Reactions

The disintegration of a radioactive substance, governed by the differential equation (2), is said to be a **first-order reaction**. In chemistry, a few reactions follow this same empirical law: If the molecules of substance A decompose into smaller molecules, it

is a natural assumption that the rate at which this decomposition takes place is proportional to the amount of the first substance that has not undergone conversion; that is, if $X(t)$ is the amount of substance A remaining at any time, then $dX/dt = kX$, where k is a negative constant since X is decreasing. An example of a first-order chemical reaction is the conversion of t-butyl chloride into t-butyl alcohol:

$$(CH_3)_3CCl + NaOH \rightarrow (CH_3)_3COH + NaCl.$$

Only the concentration of the t-butyl chloride controls the rate of reaction. But in the reaction

$$CH_3Cl + NaOH \rightarrow CH_3OH + NaCl,$$

for every molecule of methyl chloride, one molecule of sodium hydroxide is consumed, thus forming one molecule of methyl alcohol and one molecule of sodium chloride. In this case the rate at which the reaction proceeds is proportional to the product of the remaining concentrations of CH_3Cl and of $NaOH$. If X denotes the amount of CH_3OH formed and α and β are the given amounts of the first two chemicals A and B, then the instantaneous amounts not converted to chemical C are $\alpha - X$ and $\beta - X$, respectively. Hence the rate of formation of C is given by

$$\frac{dX}{dt} = k(\alpha - X)(\beta - X), \tag{6}$$

where k is a constant of proportionality. A reaction whose model is equation (6) is said to be **second order**.

▐ Mixtures

The mixing of two salt solutions of differing concentrations gives rise to a first-order differential equation for the amount of salt contained in the mixture. Let us suppose that a large mixing tank initially holds 300 gallons of brine (that is, water in which a certain number of pounds of salt has been dissolved). Another brine solution is pumped into the large tank at a rate of 3 gallons per minute; the concentration of the salt in this inflow is 2 pounds of salt per gallon. When the solution in the tank is well stirred, it is pumped out at the same rate as the entering solution. See **FIGURE 1.3.3**. If $A(t)$ denotes the amount of salt (measured in pounds) in the tank at time t, then the rate at which $A(t)$ changes is a net rate:

$$\frac{dA}{dt} = \left(\begin{array}{c}\text{input rate}\\ \text{of salt}\end{array}\right) - \left(\begin{array}{c}\text{output rate}\\ \text{of salt}\end{array}\right) = R_{in} - R_{out}. \tag{7}$$

The input rate R_{in} at which the salt enters the tank is the product of the inflow concentration of salt and the inflow rate of fluid. Note that R_{in} is measured in pounds per minute:

$$\underset{\substack{\downarrow\\ \text{concentration}\\ \text{of salt}\\ \text{in inflow}}}{} \quad \underset{\substack{\downarrow\\ \text{input rate}\\ \text{of brine}}}{} \quad \underset{\substack{\downarrow\\ \text{input rate}\\ \text{of salt}}}{}$$

$$R_{in} = (2 \text{ lb/gal}) \cdot (3 \text{ gal/min}) = (6 \text{ lb/min}).$$

Now, since the solution is being pumped out of the tank at the same rate that it is pumped in, the number of gallons of brine in the tank at time t is a constant 300 gallons. Hence the concentration of the salt in the tank, as well as in the outflow, is $c(t) = A(t)/300$ lb/gal, and so the output rate R_{out} of salt is

$$\underset{\substack{\downarrow\\ \text{concentration}\\ \text{of salt}\\ \text{in outflow}}}{} \quad \underset{\substack{\downarrow\\ \text{output rate}\\ \text{of brine}}}{} \quad \underset{\substack{\downarrow\\ \text{output rate}\\ \text{of salt}}}{}$$

$$R_{out} = \left(\frac{A(t)}{300} \text{ lb/gal}\right) \cdot (3 \text{ gal/min}) = \frac{A(t)}{100} \text{ lb/min}.$$

The net rate (7) then becomes

$$\frac{dA}{dt} = 6 - \frac{A}{100} \quad \text{or} \quad \frac{dA}{dt} + \frac{1}{100}A = 6. \tag{8}$$

If r_{in} and r_{out} denote general input and output rates of the brine solutions,* respectively, then there are three possibilities: $r_{in} = r_{out}$, $r_{in} > r_{out}$, and $r_{in} < r_{out}$. In the analysis leading to (8) we have assumed that $r_{in} = r_{out}$. In the latter two cases, the number of gallons of brine in the tank is

input rate of brine
3 gal/min

constant
300 gal

output rate of brine
3 gal/min

FIGURE 1.3.3 Mixing tank

*Don't confuse these symbols with R_{in} and R_{out}, which are input and output rates of *salt*.

either increasing ($r_{in} > r_{out}$) or decreasing ($r_{in} < r_{out}$) at the net rate $r_{in} - r_{out}$. See Problems 10–12 in Exercises 1.3.

Draining a Tank

Evangelista Torricelli (1608–1647) was an Italian physicist who invented the barometer and was a student of Galileo Galilei. In hydrodynamics, **Torricelli's law** states that the speed v of efflux of water through a sharp-edged hole at the bottom of a tank filled to a depth h is the same as the speed that a body (in this case a drop of water) would acquire in falling freely from a height h; that is, $v = \sqrt{2gh}$, where g is the acceleration due to gravity. This last expression comes from equating the kinetic energy $\frac{1}{2}mv^2$ with the potential energy mgh and solving for v. Suppose a tank filled with water is allowed to drain through a hole under the influence of gravity. We would like to find the depth h of water remaining in the tank at time t. Consider the tank shown in **FIGURE 1.3.4**. If the area of the hole is A_h (in ft^2) and the speed of the water leaving the tank is $v = \sqrt{2gh}$ (in ft/s), then the volume of water leaving the tank per second is $A_h\sqrt{2gh}$ (in ft^3/s). Thus if $V(t)$ denotes the volume of water in the tank at time t,

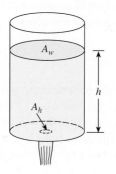

FIGURE 1.3.4 Water draining from a tank

$$\frac{dV}{dt} = -A_h\sqrt{2gh}, \tag{9}$$

where the minus sign indicates that V is decreasing. Note here that we are ignoring the possibility of friction at the hole that might cause a reduction of the rate of flow there. Now if the tank is such that the volume of water in it at time t can be written $V(t) = A_w h$, where A_w (in ft^2) is the *constant* area of the upper surface of the water (see Figure 1.3.4), then $dV/dt = A_w dh/dt$. Substituting this last expression into (9) gives us the desired differential equation for the height of the water at time t:

$$\frac{dh}{dt} = -\frac{A_h}{A_w}\sqrt{2gh}. \tag{10}$$

It is interesting to note that (10) remains valid even when A_w is not constant. In this case we must express the upper surface area of the water as a function of h; that is, $A_w = A(h)$. See Problem 14 in Exercises 1.3.

Series Circuits

The mathematical analysis of electrical circuits and networks is relatively straightforward, using two laws formulated by the German physicist **Gustav Robert Kirchhoff** (1824–1887) in 1845 while he was still a student. Consider the single-loop *LRC*-series circuit containing an inductor, resistor, and capacitor shown in **FIGURE 1.3.5(a)**. The current in a circuit after a switch is closed is denoted by $i(t)$; the charge on a capacitor at time t is denoted by $q(t)$. The letters L, R, and C are known as inductance, resistance, and capacitance, respectively, and are generally constants. Now according to **Kirchhoff's second law**, the impressed voltage $E(t)$ on a closed loop must equal the sum of the voltage drops in the loop. Figure 1.3.5(b) also shows the symbols and the formulas for the respective voltage drops across an inductor, a resistor, and a capacitor. Since current $i(t)$ is related to charge $q(t)$ on the capacitor by $i = dq/dt$, by adding the three voltage drops

(a) *LRC*-series circuit

(b) Symbols and voltage drops

FIGURE 1.3.5 Current $i(t)$ and charge $q(t)$ are measured in amperes (A) and coulombs (C), respectively

Inductor Resistor Capacitor
$$L\frac{di}{dt} = L\frac{d^2q}{dt^2}, \qquad iR = R\frac{dq}{dt}, \qquad \frac{1}{C}q$$

and equating the sum to the impressed voltage, we obtain a second-order differential equation

$$L\frac{d^2q}{dt^2} + R\frac{dq}{dt} + \frac{1}{C}q = E(t). \tag{11}$$

We will examine a differential equation analogous to (11) in great detail in Section 3.8.

Falling Bodies

In constructing a mathematical model of the motion of a body moving in a force field, one often starts with Newton's second law of motion. Recall from elementary physics that **Newton's first law of motion** states that a body will either remain at rest or will continue to move with a constant velocity unless acted upon by an external force. In each case this is equivalent to saying that when the sum of the forces $F = \Sigma F_k$—that is, the *net* or resultant force—acting on the body is zero, then the acceleration a of the body is zero. **Newton's second law of motion** indicates that when the net force acting on a body is not zero, then the net force is proportional to its acceleration a, or more precisely, $F = ma$, where m is the mass of the body.

FIGURE 1.3.6 Position of rock measured from ground level

FIGURE 1.3.7 Falling body of mass m

(a) Telephone wires

(b) Suspension bridge

FIGURE 1.3.8 Cables suspended between vertical supports

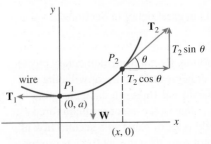

FIGURE 1.3.9 Element of cable

Now suppose a rock is tossed upward from a roof of a building as illustrated in **FIGURE 1.3.6**. What is the position $s(t)$ of the rock relative to the ground at time t? The acceleration of the rock is the second derivative d^2s/dt^2. If we assume that the upward direction is positive and that no force acts on the rock other than the force of gravity, then Newton's second law gives

$$m\frac{d^2s}{dt^2} = -mg \quad \text{or} \quad \frac{d^2s}{dt^2} = -g. \quad (12)$$

In other words, the net force is simply the weight $F = F_1 = -W$ of the rock near the surface of the Earth. Recall that the magnitude of the weight is $W = mg$, where m is the mass of the body and g is the acceleration due to gravity. The minus sign in (12) is used because the weight of the rock is a force directed downward, which is opposite to the positive direction. If the height of the building is s_0 and the initial velocity of the rock is v_0, then s is determined from the second-order initial-value problem

$$\frac{d^2s}{dt^2} = -g, \quad s(0) = s_0, \quad s'(0) = v_0. \quad (13)$$

Although we have not stressed solutions of the equations we have constructed, we note that (13) can be solved by integrating the constant $-g$ twice with respect to t. The initial conditions determine the two constants of integration. You might recognize the solution of (13) from elementary physics as the formula $s(t) = -\frac{1}{2}gt^2 + v_0t + s_0$.

Falling Bodies and Air Resistance Prior to the famous experiment by Italian mathematician and physicist **Galileo Galilei** (1564–1642) from the Leaning Tower of Pisa, it was generally believed that heavier objects in free fall, such as a cannonball, fell with a greater acceleration than lighter objects, such as a feather. Obviously a cannonball and a feather, when dropped simultaneously from the same height, *do* fall at different rates, but it is not because a cannonball is heavier. The difference in rates is due to air resistance. The resistive force of air was ignored in the model given in (13). Under some circumstances a falling body of mass m—such as a feather with low density and irregular shape—encounters air resistance proportional to its instantaneous velocity v. If we take, in this circumstance, the positive direction to be oriented downward, then the net force acting on the mass is given by $F = F_1 + F_2 = mg - kv$, where the weight $F_1 = mg$ of the body is a force acting in the positive direction and air resistance $F_2 = -kv$ is a force, called **viscous damping**, or **drag**, acting in the opposite or upward direction. See **FIGURE 1.3.7**. Now since v is related to acceleration a by $a = dv/dt$, Newton's second law becomes $F = ma = m\,dv/dt$. By equating the net force to this form of Newton's second law, we obtain a first-order differential equation for the velocity $v(t)$ of the body at time t,

$$m\frac{dv}{dt} = mg - kv. \quad (14)$$

Here k is a positive constant of proportionality called the **drag coefficient**. If $s(t)$ is the distance the body falls in time t from its initial point of release, then $v = ds/dt$ and $a = dv/dt = d^2s/dt^2$. In terms of s, (14) is a second-order differential equation

$$m\frac{d^2s}{dt^2} = mg - k\frac{ds}{dt} \quad \text{or} \quad m\frac{d^2s}{dt^2} + k\frac{ds}{dt} = mg. \quad (15)$$

Suspended Cables Suppose a flexible cable, wire, or heavy rope is suspended between two vertical supports. Physical examples of this could be a long telephone wire strung between two posts as shown in red in **FIGURE 1.3.8(a)**, or one of the two cables supporting the roadbed of a suspension bridge shown in red in Figure 1.3.8(b). Our goal is to construct a mathematical model that describes the shape that such a cable assumes.

To begin, let's agree to examine only a portion or element of the cable between its lowest point P_1 and any arbitrary point P_2. As drawn in blue in **FIGURE 1.3.9**, this element of the cable is the curve in a rectangular coordinate system with the y-axis chosen to pass through the lowest point P_1 on the curve and the x-axis chosen a units below P_1. Three forces are acting on the cable: the tensions $\mathbf{T}_1$ and $\mathbf{T}_2$ in the cable that are tangent to the cable at P_1 and P_2, respectively, and the portion $\mathbf{W}$ of the total vertical load between the points P_1 and P_2. Let $T_1 = |\mathbf{T}_1|$, $T_2 = |\mathbf{T}_2|$, and $W = |\mathbf{W}|$ denote the magnitudes of these vectors. Now the tension $\mathbf{T}_2$ resolves

into horizontal and vertical components $T_2 \cos \theta$ and $T_2 \sin \theta$. Because of static equilibrium, we can write

$$T_1 = T_2 \cos \theta \quad \text{and} \quad W = T_2 \sin \theta.$$

By dividing the last equation by the first, we eliminate T_2 and get $\tan \theta = W/T_1$. But since $dy/dx = \tan \theta$, we arrive at

$$\frac{dy}{dx} = \frac{W}{T_1}. \tag{16}$$

This simple first-order differential equation serves as a model for both the shape of a flexible wire, such as a telephone wire hanging under its own weight, as well as the shape of the cables that support the roadbed. We will come back to equation (16) in Exercises 2.2 and in Section 3.11.

REMARKS

Except for equation (16), the differential equations derived in this section have described a *dynamical system*—a system that changes or evolves over time. Since the study of dynamical systems is a branch of mathematics currently in vogue, we shall occasionally relate the terminology of that field to the discussion at hand.

In more precise terms, a **dynamical system** consists of a set of time-dependent variables, called **state variables**, together with a rule that enables us to determine (without ambiguity) the state of the system (this may be past, present, or future states) in terms of a state prescribed at some time t_0. Dynamical systems are classified as either discrete-time systems or continuous-time systems. In this course we shall be concerned only with continuous-time dynamical systems—systems in which *all* variables are defined over a continuous range of time. The rule or the mathematical model in a continuous-time dynamical system is a differential equation or a system of differential equations. The **state of the system** at a time t is the value of the state variables at that time; the specified state of the system at a time t_0 is simply the initial conditions that accompany the mathematical model. The solution of the initial-value problem is referred to as the **response of the system**. For example, in the preceding case of radioactive decay, the rule is $dA/dt = kA$. Now if the quantity of a radioactive substance at some time t_0 is known, say $A(t_0) = A_0$, then by solving the rule, the response of the system for $t \geq t_0$ is found to be $A(t) = A_0 e^{t - t_0}$ (see Section 2.7). The response $A(t)$ is the single-state variable for this system. In the case of the rock tossed from the roof of the building, the response of the system, the solution of the differential equation $d^2 s/dt^2 = -g$ subject to the initial state $s(0) = s_0$, $s'(0) = v_0$, is the function $s(t) = -\frac{1}{2} g t^2 + v_0 t + s_0$, $0 \leq t \leq T$, where the symbol T represents the time when the rock hits the ground. The state variables are $s(t)$ and $s'(t)$, which are, respectively, the vertical position of the rock above ground and its velocity at time t. The acceleration $s''(t)$ is *not* a state variable since we only have to know any initial position and initial velocity at a time t_0 to uniquely determine the rock's position $s(t)$ and velocity $s'(t) = v(t)$ for any time in the interval $[t_0, T]$. The acceleration $s''(t) = a(t)$ is, of course, given by the differential equation $s''(t) = -g$, $0 < t < T$.

One last point: Not every system studied in this text is a dynamical system. We shall also examine some static systems in which the model is a differential equation.

1.3 Exercises Answers to selected odd-numbered problems begin on page ANS-1.

Population Dynamics

1. Under the same assumptions underlying the model in (1), determine a differential equation governing the growing population $P(t)$ of a country when individuals are allowed to immigrate into the country at a constant rate $r > 0$. What is the differential equation for the population $P(t)$ of the country when individuals are allowed to emigrate at a constant rate $r > 0$?

2. The population model given in (1) fails to take death into consideration; the growth rate equals the birth rate. In another model of a changing population of a community, it is assumed that the rate at which the population changes is a *net* rate—that is, the difference between the rate of births and the rate of deaths in the community. Determine a model for the population $P(t)$ if both the birth rate and the death rate are proportional to the population present at time t.

3. Using the concept of a net rate introduced in Problem 2, determine a differential equation governing a population $P(t)$ if the birth rate is proportional to the population present at time t but the death rate is proportional to the square of the population present at time t.

4. Modify the model in Problem 3 for the net rate at which the population $P(t)$ of a certain kind of fish changes by also assuming that the fish are harvested at a constant rate $h > 0$.

Newton's Law of Cooling/Warming

5. A cup of coffee cools according to Newton's law of cooling (3). Use data from the graph of the temperature $T(t)$ in **FIGURE 1.3.10** to estimate the constants T_m, T_0, and k in a model of the form of the first-order initial-value problem

$$\frac{dT}{dt} = k(T - T_m), \quad T(0) = T_0.$$

FIGURE 1.3.10 Cooling curve in Problem 5

6. The ambient temperature T_m in (3) could be a function of time t. Suppose that in an artificially controlled environment, $T_m(t)$ is periodic with a 24-hour period, as illustrated in **FIGURE 1.3.11**. Devise a mathematical model for the temperature $T(t)$ of a body within this environment.

FIGURE 1.3.11 Ambient temperature in Problem 6

Spread of a Disease/Technology

7. Suppose a student carrying a flu virus returns to an isolated college campus of 1000 students. Determine a differential equation governing the number of students $x(t)$ who have contracted the flu if the rate at which the disease spreads is proportional to the number of interactions between the number of students with the flu and the number of students who have not yet been exposed to it.

8. At a time $t = 0$, a technological innovation is introduced into a community with a fixed population of n people. Determine a differential equation governing the number of people $x(t)$ who have adopted the innovation at time t if it is assumed that the rate at which the innovation spreads through the community is jointly proportional to the number of people who have adopted it and the number of people who have not adopted it.

Mixtures

9. Suppose that a large mixing tank initially holds 300 gallons of water in which 50 pounds of salt has been dissolved. Pure water is pumped into the tank at a rate of 3 gal/min, and when the solution is well stirred, it is pumped out at the same rate. Determine a differential equation for the amount $A(t)$ of salt in the tank at time t. What is $A(0)$?

10. Suppose that a large mixing tank initially holds 300 gallons of water in which 50 pounds of salt has been dissolved. Another brine solution is pumped into the tank at a rate of 3 gal/min, and when the solution is well stirred, it is pumped out at a *slower* rate of 2 gal/min. If the concentration of the solution entering is 2 lb/gal, determine a differential equation for the amount $A(t)$ of salt in the tank at time t.

11. What is the differential equation in Problem 10 if the well-stirred solution is pumped out at a *faster* rate of 3.5 gal/min?

12. Generalize the model given in (8) of this section by assuming that the large tank initially contains N_0 number of gallons of brine, r_{in} and r_{out} are the input and output rates of the brine, respectively (measured in gallons per minute), c_{in} is the concentration of the salt in the inflow, $c(t)$ is the concentration of the salt in the tank as well as in the outflow at time t (measured in pounds of salt per gallon), and $A(t)$ is the amount of salt in the tank at time t.

Draining a Tank

13. Suppose water is leaking from a tank through a circular hole of area A_h at its bottom. When water leaks through a hole, friction and contraction of the stream near the hole reduce the volume of the water leaving the tank per second to $cA_h\sqrt{2gh}$ where $c\,(0 < c < 1)$ is an empirical constant. Determine a differential equation for the height h of water at time t for the cubical tank in **FIGURE 1.3.12**. The radius of the hole is 2 in. and $g = 32$ ft/s^2.

FIGURE 1.3.12 Cubical tank in Problem 13

14. The right-circular conical tank shown in **FIGURE 1.3.13** loses water out of a circular hole at its bottom. Determine a differential equation for the height of the water h at time t. The radius of the hole is 2 in., $g = 32$ ft/s^2, and the friction/contraction factor introduced in Problem 13 is $c = 0.6$.

FIGURE 1.3.13 Conical tank in Problem 14

Series Circuits

15. A series circuit contains a resistor and an inductor as shown in **FIGURE 1.3.14**. Determine a differential equation for the current $i(t)$ if the resistance is R, the inductance is L, and the impressed voltage is $E(t)$.

FIGURE 1.3.14 *LR*-series circuit in Problem 15

16. A series circuit contains a resistor and a capacitor as shown in **FIGURE 1.3.15**. Determine a differential equation for the charge $q(t)$ on the capacitor if the resistance is R, the capacitance is C, and the impressed voltage is $E(t)$.

FIGURE 1.3.15 *RC*-series circuit in Problem 16

Falling Bodies and Air Resistance

17. For high-speed motion through the air—such as the skydiver shown in **FIGURE 1.3.16** falling before the parachute is opened—air resistance is closer to a power of the instantaneous velocity $v(t)$. Determine a differential equation for the velocity $v(t)$ of a falling body of mass m if air resistance is proportional to the square of the instantaneous velocity.

FIGURE 1.3.16 Air resistance proportional to square of velocity in Problem 17

Newton's Second Law and Archimedes' Principle

18. A cylindrical barrel s ft in diameter of weight w lb is floating in water as shown in **FIGURE 1.3.17(a)**. After an initial depression, the barrel exhibits an up-and-down bobbing motion along a vertical line. Using Figure 1.3.17(b), determine a differential equation for the vertical displacement $y(t)$ if the origin is taken to be on the vertical axis at the surface of the water when the barrel is at rest. Assume the downward direction is positive, that the weight density of the water is 62.4 lb/ft^3, and that there is no resistance between the barrel and the water. Use **Archimedes' principle**: Buoyancy, or

the upward force of the water on the barrel, is equal to the weight of the water displaced. **Archimedes of Syracuse** (287 BCE–212 BCE) was arguably one of the greatest scientists/mathematicians of antiquity. Using his approximation of the number π, he found the area of a circle as well as the surface area and volume of a sphere.

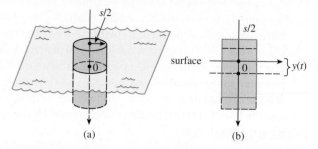

FIGURE 1.3.17 Bobbing motion of floating barrel in Problem 18

Newton's Second Law and Hooke's Law

19. After a mass m is attached to a spring, it stretches s units and then hangs at rest in the equilibrium position as shown in **FIGURE 1.3.18(b)**. After the spring/mass system has been set in motion, let $x(t)$ denote the directed distance of the mass beyond the equilibrium position. As indicated in Figure 1.3.18(c), assume that the downward direction is positive, that the motion takes place in a vertical straight line through the center of gravity of the mass, and that the only forces acting on the system are the weight of the mass and the restoring force of the stretched spring. Use **Hooke's law**: The restoring force of a spring is proportional to its total elongation. Determine a differential equation for the displacement $x(t)$ at time t.

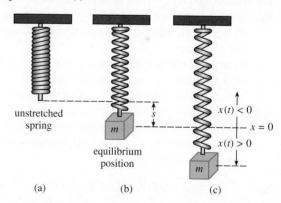

FIGURE 1.3.18 Spring/mass system in Problem 19

20. In Problem 19, what is a differential equation for the displacement $x(t)$ if the motion takes place in a medium that imparts a damping force on the spring/mass system that is proportional to the instantaneous velocity of the mass and acts in a direction opposite to that of motion?

Newton's Second Law and Variable Mass

When the mass m of a body moving through a force field is variable, **Newton's second law of motion** takes on the following form: If the net force acting on a body is not zero, then the net force F is equal to the time rate of change of momentum of the body. That is,

$$F = \frac{d}{dt}(mv), ^* \tag{17}$$

where mv is momentum. Use this formulation of Newton's second law in Problems 21 and 22.

21. Consider a single-stage rocket that is launched vertically upward as shown in the accompanying photo. Let $m(t)$ denote the total mass of the rocket at time t (which is the sum of three masses: the constant mass of the payload, the constant mass of the vehicle, and the variable amount of fuel). Assume that the positive direction is upward, air resistance is proportional to the instantaneous velocity v of the rocket, and R is the upward thrust or force generated by the propulsion system. Use (17) to find a mathematical model for the velocity $v(t)$ of the rocket.

© Sebastian Kaulitzki/Shutterstock
Rocket in Problem 21

22. In Problem 21, suppose $m(t) = m_p + m_v + m_f(t)$ where m_p is constant mass of the payload, m_v is the constant mass of the vehicle, and $m_f(t)$ is the variable amount of fuel.
 (a) Show that the rate at which the total mass of the rocket changes is the same as the rate at which the mass of the fuel changes.
 (b) If the rocket consumes its fuel at a constant rate λ, find $m(t)$. Then rewrite the differential equation in Problem 21 in terms of λ and the initial total mass $m(0) = m_0$.
 (c) Under the assumption in part (b), show that the burnout time $t_b > 0$ of the rocket, or the time at which all the fuel is consumed, is $t_b = m_f(0)/\lambda$, where $m_f(0)$ is the initial mass of the fuel.

Newton's Second Law and the Law of Universal Gravitation

23. By **Newton's law of universal gravitation**, the free-fall acceleration a of a body, such as the satellite shown in **FIGURE 1.3.19**, falling a great distance to the surface is *not* the constant g. Rather, the acceleration a is inversely proportional to the square of the distance from the center of the Earth, $a = k/r^2$, where k is the constant of proportionality. Use the fact that at the surface of the Earth $r = R$ and $a = g$ to determine k. If the positive direction is upward, use Newton's second law and his universal law of gravitation to find a differential equation for the distance r.

satellite of mass m

surface

r

R

Earth of mass M

FIGURE 1.3.19 Satellite in Problem 23

24. Suppose a hole is drilled through the center of the Earth and a bowling ball of mass m is dropped into the hole, as shown in **FIGURE 1.3.20**. Construct a mathematical model that describes the motion of the ball. At time t let r denote the distance from the center of the Earth to the mass m, M denote the mass of the Earth, M_r denote the mass of that portion of the Earth within a sphere of radius r, and δ denote the constant density of the Earth.

surface

m

r

R

FIGURE 1.3.20 Hole through Earth in Problem 24

Additional Mathematical Models

25. **Learning Theory** In the theory of learning, the rate at which a subject is memorized is assumed to be proportional to the amount that is left to be memorized. Suppose M denotes the total amount of a subject to be memorized and $A(t)$ is the amount memorized in time t. Determine a differential equation for the amount $A(t)$.

26. **Forgetfulness** In Problem 25, assume that the rate at which material is *forgotten* is proportional to the amount memorized in time t. Determine a differential equation for $A(t)$ when forgetfulness is taken into account.

27. **Infusion of a Drug** A drug is infused into a patient's bloodstream at a constant rate of r grams per second. Simultaneously, the drug is removed at a rate proportional to the amount $x(t)$ of the drug present at time t. Determine a differential equation governing the amount $x(t)$.

28. **Tractrix** A motorboat starts at the origin and moves in the direction of the positive x-axis, pulling a waterskier along a curve C called a **tractrix**. See **FIGURE 1.3.21**. The waterskier, initially located on the y-axis at the point $(0, s)$, is pulled by keeping a rope of constant length s, which is kept taut throughout the motion. At time $t > 0$ the waterskier is at the point $P(x, y)$. Find the differential equation of the path of motion C.

*Note that when m is constant, this is the same as $F = ma$.

FIGURE 1.3.21 Tractrix curve in Problem 28

29. **Reflecting Surface** Assume that when the plane curve C shown in **FIGURE 1.3.22** is revolved about the x-axis it generates a surface of revolution with the property that all light rays L parallel to the x-axis striking the surface are reflected to a single point O (the origin). Use the fact that the angle of incidence is equal to the angle of reflection to determine a differential equation that describes the shape of the curve C. Such a curve C is important in applications ranging from construction of telescopes to satellite antennas, automobile headlights, and solar collectors. [*Hint*: Inspection of the figure shows that we can write $\phi = 2\theta$. Why? Now use an appropriate trigonometric identity.]

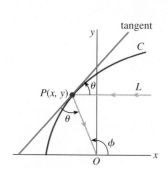

FIGURE 1.3.22 Reflecting surface in Problem 29

© George Dukin/Shutterstock
Satellite dish antenna

Discussion Problems

30. Reread Problem 53 in Exercises 1.1 and then give an explicit solution $P(t)$ for equation (1). Find a one-parameter family of solutions of (1).

31. Reread the sentence following equation (3) and assume that T_m is a positive constant. Discuss why we would expect $k < 0$ in (3) in both cases of cooling and warming. You might start by interpreting, say, $T(t) > T_m$ in a graphical manner.

32. Reread the discussion leading up to equation (8). If we assume that initially the tank holds, say, 50 lb of salt, it stands to reason that since salt is being added to the tank continuously for $t > 0$, that $A(t)$ should be an increasing function. Discuss how you might determine from the DE, without actually solving it, the number of pounds of salt in the tank after a long period of time.

33. **Population Model** The differential equation $dP/dt = (k \cos t)P$, where k is a positive constant, is a model of human population $P(t)$ of a certain community. Discuss an interpretation for the solution of this equation; in other words, what kind of population do you think the differential equation describes?

34. **Rotating Fluid** As shown in **FIGURE 1.3.23(a)**, a right-circular cylinder partially filled with fluid is rotated with a constant angular velocity ω about a vertical y-axis through its center. The rotating fluid is a surface of revolution S. To identify S, we first establish a coordinate system consisting of a vertical plane determined by the y-axis and an x-axis drawn perpendicular to the y-axis such that the point of intersection of the axes (the origin) is located at the lowest point on the surface S. We then seek a function $y = f(x)$, which represents the curve C of intersection of the surface S and the vertical coordinate plane. Let the point $P(x, y)$ denote the position of a particle of the rotating fluid of mass m in the coordinate plane. See Figure 1.3.23(b).

(a) At P, there is a reaction force of magnitude F due to the other particles of the fluid, which is normal to the surface S. By Newton's second law the magnitude of the net force acting on the particle is $m\omega^2 x$. What is this force? Use Figure 1.3.23(b) to discuss the nature and origin of the equations

$$F \cos \theta = mg, \qquad F \sin \theta = m\omega^2 x.$$

(b) Use part (a) to find a first-order differential equation that defines the function $y = f(x)$.

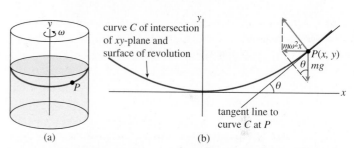

FIGURE 1.3.23 Rotating fluid in Problem 34

35. **Falling Body** In Problem 23, suppose $r = R + s$, where s is the distance from the surface of the Earth to the falling body. What does the differential equation obtained in Problem 23 become when s is very small compared to R?

36. **Raindrops Keep Falling** In meteorology, the term *virga* refers to falling raindrops or ice particles that evaporate before they reach the ground. Assume that a typical raindrop is spherical in shape. Starting at some time, which we can designate as $t = 0$, the raindrop of radius r_0 falls from rest from a cloud and begins to evaporate.

(a) If it is assumed that a raindrop evaporates in such a manner that its shape remains spherical, then it also makes sense to assume that the rate at which the raindrop evaporates—that is, the rate at which it loses mass—is proportional to its surface area. Show that this latter assumption implies that the rate at which the radius r of the raindrop decreases is a constant. Find $r(t)$. [*Hint*: See Problem 63 in Exercises 1.1.]

(b) If the positive direction is downward, construct a mathematical model for the velocity v of the falling raindrop at time t. Ignore air resistance. [*Hint*: Use the form of Newton's second law as given in (17).]

37. Let It Snow The "snowplow problem" is a classic and appears in many differential equations texts but was probably made famous by Ralph Palmer Agnew:

One day it started snowing at a heavy and steady rate. A snowplow started out at noon, going 2 miles the first hour and 1 mile the second hour. What time did it start snowing?

If possible, find the text *Differential Equations*, Ralph Palmer Agnew, McGraw-Hill, and then discuss the construction and solution of the mathematical model.

© aetb/iStock/Thinkstock

Snowplow in Problem 37

38. Reread this section and classify each mathematical model as linear or nonlinear.

39. Population Dynamics Suppose that $P'(t) = 0.15\,P(t)$ represents a mathematical model for the growth of a certain cell culture, where $P(t)$ is the size of the culture (measured in millions of cells) at time t (measured in hours). How fast is the culture growing at the time t when the size of the culture reaches 2 million cells?

40. Radioactive Decay Suppose that

$$A'(t) = -0.0004332\,A(t)$$

represents a mathematical model for the decay of radium-226, where $A(t)$ is the amount of radium (measured in grams) remaining at time t (measured in years). How much of the radium sample remains at time t when the sample is decaying at a rate of 0.002 grams per year?

1 Chapter in Review Answers to selected odd-numbered problems begin on page ANS-1.

In Problems 1 and 2, fill in the blank and then write this result as a linear first-order differential equation that is free of the symbol c_1 and has the form $dy/dx = f(x, y)$. The symbols c_1 and k represent constants.

1. $\dfrac{d}{dx} c_1 e^{kx} = $ _____

2. $\dfrac{d}{dx}(5 + c_1 e^{-2x}) = $ _____

In Problems 3 and 4, fill in the blank and then write this result as a linear second-order differential equation that is free of the symbols c_1 and c_2 and has the form $F(y, y'') = 0$. The symbols c_1, c_2, and k represent constants.

3. $\dfrac{d^2}{dx^2}(c_1 \cos kx + c_2 \sin kx) = $ _____

4. $\dfrac{d^2}{dx^2}(c_1 \cosh kx + c_2 \sinh kx) = $ _____

In Problems 5 and 6, compute y' and y'' and then combine these derivatives with y as a linear second-order differential equation that is free of the symbols c_1 and c_2 and has the form $F(y, y', y'') = 0$. The symbols c_1 and c_2 represent constants.

5. $y = c_1 e^x + c_2 x e^x$ **6.** $y = c_1 e^x \cos x + c_2 e^x \sin x$

In Problems 7–12, match each of the given differential equations with one or more of these solutions:
(a) $y = 0$, **(b)** $y = 2$, **(c)** $y = 2x$, **(d)** $y = 2x^2$.

7. $xy' = 2y$ **8.** $y' = 2$

9. $y' = 2y - 4$ **10.** $xy' = y$

11. $y'' + 9y = 18$ **12.** $xy'' - y' = 0$

In Problems 13 and 14, determine by inspection at least one solution of the given differential equation.

13. $y'' = y'$ **14.** $y' = y(y - 3)$

In Problems 15 and 16, interpret each statement as a differential equation.

15. On the graph of $y = \phi(x)$, the slope of the tangent line at a point $P(x, y)$ is the square of the distance from $P(x, y)$ to the origin.

16. On the graph of $y = \phi(x)$, the rate at which the slope changes with respect to x at a point $P(x, y)$ is the negative of the slope of the tangent line at $P(x, y)$.

17. (a) Give the domain of the function $y = x^{2/3}$.
 (b) Give the largest interval I of definition over which $y = x^{2/3}$ is a solution of the differential equation $3xy' - 2y = 0$.

18. (a) Verify that the one-parameter family $y^2 - 2y = x^2 - x + c$ is an implicit solution of the differential equation $(2y - 2)y' = 2x - 1$.
 (b) Find a member of the one-parameter family in part (a) that satisfies the initial condition $y(0) = 1$.
 (c) Use your result in part (b) to find an explicit *function* $y = \phi(x)$ that satisfies $y(0) = 1$. Give the domain of ϕ. Is $y = \phi(x)$ a *solution* of the initial-value problem? If so, give its interval I of definition; if not, explain.

19. Given that $y = -\dfrac{2}{x} + x$ is a solution of the DE $xy' + y = 2x$. Find x_0 and the largest interval I for which $y(x)$ is a solution of the IVP

$$xy' + y = 2x, \qquad y(x_0) = 1.$$

20. Suppose that $y(x)$ denotes a solution of the initial-value problem $y' = x^2 + y^2$, $y(1) = -1$ and that $y(x)$ possesses at least a second derivative at $x = 1$. In some neighborhood of $x = 1$, use the DE to determine whether $y(x)$ is increasing or decreasing, and whether the graph $y(x)$ is concave up or concave down.

21. A differential equation may possess more than one family of solutions.
 (a) Plot different members of the families $y = \phi_1(x) = x^2 + c_1$ and $y = \phi_2(x) = -x^2 + c_2$.
 (b) Verify that $y = \phi_1(x)$ and $y = \phi_2(x)$ are two solutions of the nonlinear first-order differential equation $(y')^2 = 4x^2$.
 (c) Construct a piecewise-defined function that is a solution of the nonlinear DE in part (b) but is not a member of either family of solutions in part (a).

22. What is the slope of the tangent line to the graph of the solution of $y' = 6\sqrt{y} + 5x^3$ that passes through $(-1, 4)$?

In Problems 23–26, verify that the indicated function is an explicit solution of the given differential equation. Give an interval of definition I for each solution.

23. $y'' + y = 2\cos x - 2\sin x$; $\quad y = x\sin x + x\cos x$

24. $y'' + y = \sec x$; $\quad y = x\sin x + (\cos x)\ln(\cos x)$

25. $x^2y'' + xy' + y = 0$; $\quad y = \sin(\ln x)$

26. $x^2y'' + xy' + y = \sec(\ln x)$;
 $y = \cos(\ln x)\ln(\cos(\ln x)) + (\ln x)\sin(\ln x)$

In Problems 27–30, use (12) of Section 1.1 to verify that the indicated function is a solution of the given differential equation. Assume an appropriate interval I of definition of each solution.

27. $\dfrac{dy}{dx} + (\sin x)y = x$; $\quad y = e^{\cos x}\displaystyle\int_0^x te^{-\cos t}\,dt$

28. $\dfrac{dy}{dx} - 2xy = e^x$; $\quad y = e^{x^2}\displaystyle\int_0^x e^{t - t^2}\,dt$

29. $x^2y'' + (x^2 - x)y' + (1 - x)y = 0$; $\quad y = x\displaystyle\int_1^x \dfrac{e^{-t}}{t}\,dt$

30. $y'' + y = e^{x^2}$; $\quad y = \sin x\displaystyle\int_0^x e^{t^2}\cos t\,dt - \cos x\displaystyle\int_0^x e^{t^2}\sin t\,dt$

In Problems 31–34, verify that the indicated expression is an implicit solution of the given differential equation.

31. $x\dfrac{dy}{dx} + y = \dfrac{1}{y^2}$; $\quad x^3y^3 = x^3 + 5$

32. $\left(\dfrac{dy}{dx}\right)^2 + 1 = \dfrac{1}{y^2}$; $\quad (x - 7)^2 + y^2 = 1$

33. $y'' = 2y(y')^3$; $\quad y^3 + 3y = 2 - 3x$

34. $(1 + xy)y' + y^2 = 0$; $\quad y = e^{-xy}$

35. Find a constant c_1 such that $y = c_1 + \cos 3x$ is a solution of the differential equation $y'' + 9y = 5$.

36. Find constants c_1 and c_2 such that $y = c_1 + c_2x$ is a solution of the differential equation $y' + 2y = 3x$.

37. If c is an arbitrary constant, find a first-order differential equation for which $y = ce^{-x} + 4x - 6$ is a solution. [*Hint:* Differentiate and eliminate c between the two equations.]

38. Find a function $y = f(x)$ whose graph passes through $(0, 0)$ and whose slope at any point (x, y) in the xy-plane is $6 - 2x$.

In Problems 39–42, $y = c_1e^{-3x} + c_2e^x + 4x$ is a two-parameter family of the second-order differential equation $y'' + 2y' - 3y = -12x + 8$. Find a solution of the second-order initial-value problem consisting of this differential equation and the given initial conditions.

39. $y(0) = 0$, $y'(0) = 0$ $\qquad$ **40.** $y(0) = 5$, $y'(0) = -11$

41. $y(1) = -2$, $y'(1) = 4$ $\quad$ **42.** $y(-1) = 1$, $y'(-1) = 1$

In Problems 43 and 44, verify that the function defined by the definite integral is a particular solution of the given differential equation. In both problems, use **Leibniz's rule** for the derivative of an integral:

$$\dfrac{d}{dx}\int_{u(x)}^{v(x)} F(x, t)\,dt = F(x, v(x))\dfrac{dv}{dx} - F(x, u(x))\dfrac{du}{dx} + \int_{u(x)}^{v(x)} \dfrac{\partial}{\partial x}F(x, t)\,dt.$$

43. $y'' + 9y = f(x)$; $\quad y(x) = \dfrac{1}{3}\displaystyle\int_0^x f(t)\sin 3(x - t)\,dt$

44. $xy'' + y' - xy = 0;$ $y = \displaystyle\int_0^\pi e^{x\cos t}\, dt$ [*Hint*: After computing y' use integration by parts with respect to t.]

45. The graph of a solution of a second-order initial-value problem $d^2y/dx^2 = f(x, y, y')$, $y(2) = y_0$, $y'(2) = y_1$, is given in **FIGURE 1.R.1**. Use the graph to estimate the values of y_0 and y_1.

FIGURE 1.R.1 Graph for Problem 45

46. A tank in the form of a right-circular cylinder of radius 2 ft and height 10 ft is standing on end. If the tank is initially full of water, and water leaks from a circular hole of radius $\frac{1}{2}$ in. at its bottom, determine a differential equation for the height h of the water at time t. Ignore friction and contraction of water at the hole.

47. A uniform 10-foot-long heavy rope is coiled loosely on the ground. As shown in **FIGURE 1.R.2** one end of the rope is pulled vertically upward by means of a constant force of 5 lb. The rope weighs 1 lb/ft. Use Newton's second law in the form given in (17) in Exercises 1.3 to determine a differential equation for the height $x(t)$ of the end above ground level at time t. Assume that the positive direction is upward.

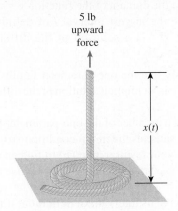

FIGURE 1.R.2 Rope pulled upward in Problem 47

First-Order Differential Equations

We begin our study of differential equations (DEs) with first-order equations. In this chapter we illustrate the three different ways DEs can be studied: qualitatively (Section 2.1), analytically (Sections 2.2–2.5), and numerically (Section 2.6). The chapter ends with an introduction to mathematical modeling with DEs (Sections 2.7–2.9).

2.1 Solution Curves Without a Solution

INTRODUCTION Some differential equations do not possess any solutions. For example, there is no real function that satisfies $(y')^2 + 1 = 0$. Some differential equations possess solutions that can be found **analytically**, that is, solutions in explicit or implicit form found by implementing an equation-specific method of solution. These solution methods may involve certain manipulations, such as a substitution, and procedures, such as integration. Some differential equations possess solutions, but the differential equation cannot be solved analytically. In other words, when we say that a solution of a DE exists, we do not mean that there also exists a method of solution that will produce explicit or implicit solutions. Over a time span of centuries, mathematicians have devised ingenious procedures for solving some very specialized equations, so there are, not surprisingly, a large number of differential equations that can be solved analytically. Although we shall study some of these methods of solution for first-order equations in the subsequent sections of this chapter, let us imagine for the moment that we have in front of us a first-order differential equation in normal form $dy/dx = f(x, y)$, and let us further imagine that we can neither find nor invent a method for solving it analytically. This is not as bad a predicament as one might think, since the differential equation itself can sometimes "tell" us specifics about how its solutions "behave." We have seen in Section 1.2 that whenever $f(x, y)$ and $\partial f/\partial y$ satisfy certain continuity conditions, **qualitative** questions about existence and uniqueness of solutions can be answered. In this section we shall see that other qualitative questions about properties of solutions—such as, How does a solution behave near a certain point? or How does a solution behave as $x \to \infty$?—can often be answered when the function f depends solely on the variable y.

We begin our study of first-order differential equations with two ways of analyzing a DE qualitatively. Both these ways enable us to determine, in an approximate sense, what a solution curve must look like without actually solving the equation.

2.1.1 Direction Fields

Slope We begin with a simple concept from calculus: A derivative dy/dx of a differentiable function $y = y(x)$ gives slopes of tangent lines at points on its graph. Because a solution $y = y(x)$ of a first-order differential equation $dy/dx = f(x, y)$ is necessarily a differentiable function on its interval I of definition, it must also be continuous on I. Thus the corresponding solution curve on I must have no breaks and must possess a tangent line at each point $(x, y(x))$. The slope of the tangent line at $(x, y(x))$ on a solution curve is the value of the first derivative dy/dx at this point, and this we know from the differential equation $f(x, y(x))$. Now suppose that (x, y) represents any point in a region of the xy-plane over which the function f is defined. The value $f(x, y)$ that the function f assigns to the point represents the slope of a line, or as we shall envision it, a line segment called a **lineal element**. For example, consider the equation $dy/dx = 0.2xy$, where $f(x, y) = 0.2xy$. At, say, the point $(2, 3)$, the slope of a lineal element is $f(2, 3) = 0.2(2)(3) = 1.2$. **FIGURE 2.1.1(a)** shows a line segment with slope 1.2 passing through $(2, 3)$. As shown in Figure 2.1.1(b), *if* a solution curve also passes through the point $(2, 3)$, it does so tangent to this line segment; in other words, the lineal element is a miniature tangent line at that point.

Direction Field If we systematically evaluate f over a rectangular grid of points in the xy-plane and draw a lineal element at each point (x, y) of the grid with slope $f(x, y)$, then the collection of all these lineal elements is called a **direction field** or a **slope field** of the differential equation $dy/dx = f(x, y)$. Visually, the direction field suggests the appearance or shape of a family of solution curves of the differential equation, and consequently it may be possible to see at a glance certain qualitative aspects of the solutions—regions in the plane, for example, in which a solution exhibits an unusual behavior. A single solution curve that passes through a direction field must follow the flow pattern of the field; it is tangent to a lineal element when it intersects a point in the grid.

(a) $f(2, 3) = 1.2$ is slope of lineal element at (2, 3)

(b) A solution curve passing through (2, 3)

FIGURE 2.1.1 Solution curve is tangent to lineal element at (2, 3)

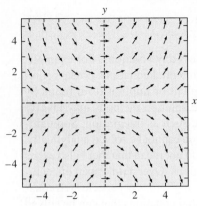

(a) Direction field for $dy/dx = 0.2xy$

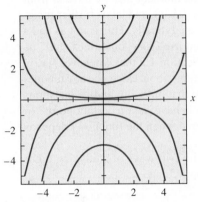

(b) Some solution curves in the family $y = ce^{0.1x^2}$

FIGURE 2.1.2 Direction field and solution curves in Example 1

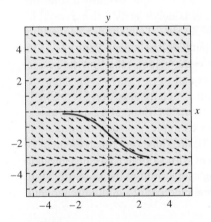

FIGURE 2.1.3 Direction field for $dy/dx = \sin y$ in Example 2

EXAMPLE 1 **Direction Field**

The direction field for the differential equation $dy/dx = 0.2xy$ shown in **FIGURE 2.1.2(a)** was obtained using computer software in which a 5×5 grid of points (mh, nh), m and n integers, was defined by letting $-5 \leq m \leq 5$, $-5 \leq n \leq 5$ and $h = 1$. Notice in Figure 2.1.2(a) that at any point along the x-axis $(y = 0)$ and the y-axis $(x = 0)$ the slopes are $f(x, 0) = 0$ and $f(0, y) = 0$, respectively, so the lineal elements are horizontal. Moreover, observe in the first quadrant that for a fixed value of x, the values of $f(x, y) = 0.2xy$ increase as y increases; similarly, for a fixed y, the values of $f(x, y) = 0.2xy$ increase as x increases. This means that as both x and y increase, the lineal elements become almost vertical and have positive slope ($f(x, y) = 0.2xy > 0$ for $x > 0$, $y > 0$). In the second quadrant, $|f(x, y)|$ increases as $|x|$ and y increase, and so the lineal elements again become almost vertical but this time have negative slope ($f(x, y) = 0.2xy < 0$ for $x < 0$, $y > 0$). Reading left to right, imagine a solution curve starts at a point in the second quadrant, moves steeply downward, becomes flat as it passes through the y-axis, and then as it enters the first quadrant moves steeply upward—in other words, its shape would be concave upward and similar to a horseshoe. From this it could be surmised that $y \rightarrow \infty$ as $x \rightarrow \pm\infty$. Now in the third and fourth quadrants, since $f(x, y) = 0.2xy > 0$ and $f(x, y) = 0.2xy < 0$, respectively, the situation is reversed; a solution curve increases and then decreases as we move from left to right.

We saw in (1) of Section 1.1 that $y = e^{0.1x^2}$ is an explicit solution of the differential equation $dy/dx = 0.2xy$; you should verify that a one-parameter family of solutions of the same equation is given by $y = ce^{0.1x^2}$. For purposes of comparison with Figure 2.1.2(a) some representative graphs of members of this family are shown in Figure 2.1.2(b). ≡

EXAMPLE 2 **Direction Field**

Use a direction field to sketch an approximate solution curve for the initial-value problem $dy/dx = \sin y$, $y(0) = -\frac{3}{2}$.

SOLUTION Before proceeding, recall that from the continuity of $f(x, y) = \sin y$ and $\partial f/\partial y = \cos y$, Theorem 1.2.1 guarantees the existence of a unique solution curve passing through any specified point (x_0, y_0) in the plane. Now we set our computer software again for a 5×5 rectangular region and specify (because of the initial condition) points in that region with vertical and horizontal separation of $\frac{1}{2}$ unit—that is, at points (mh, nh), $h = \frac{1}{2}$, m and n integers such that $-10 \leq m \leq 10$, $-10 \leq n \leq 10$. The result is shown in **FIGURE 2.1.3**. Since the right-hand side of $dy/dx = \sin y$ is 0 at $y = 0$ and at $y = -\pi$, the lineal elements are horizontal at all points whose second coordinates are $y = 0$ or $y = -\pi$. It makes sense then that a solution curve passing through the initial point $(0, -\frac{3}{2})$ has the shape shown in color in the figure. ≡

■ **Increasing/Decreasing** Interpretation of the derivative dy/dx as a function that gives slope plays the key role in the construction of a direction field. Another telling property of the first derivative will be used next, namely, if $dy/dx > 0$ (or $dy/dx < 0$) for all x in an interval I, then a differentiable function $y = y(x)$ is increasing (or decreasing) on I.

REMARKS

Sketching a direction field by hand is straightforward but time consuming; it is probably one of those tasks about which an argument can be made for doing it once or twice in a lifetime, but it is overall most efficiently carried out by means of computer software. Prior to calculators, PCs, and software, the **method of isoclines** was used to facilitate sketching a direction field by hand. For the DE $dy/dx = f(x, y)$, any member of the family of curves $f(x, y) = c$, c a constant, is called an **isocline**. Lineal elements drawn through points on a specific isocline, say, $f(x, y) = c_1$, all have the same slope c_1. In Problem 15 in Exercises 2.1, you have your two opportunities to sketch a direction field by hand.

2.1.2 Autonomous First-Order DEs

■ DEs Free of the Independent Variable In Section 1.1 we divided the class of ordinary differential equations into two types: linear and nonlinear. We now consider briefly another kind of classification of ordinary differential equations, a classification that is of particular importance in the qualitative investigation of differential equations. An ordinary differential equation in which the independent variable does not appear explicitly is said to be **autonomous**. If the symbol x denotes the independent variable, then an autonomous first-order differential equation can be written in general form as $F(y, y') = 0$ or in normal form as

$$\frac{dy}{dx} = f(y). \tag{1}$$

We shall assume throughout the discussion that follows that f in (1) and its derivative f' are continuous functions of y on some interval I. The first-order equations

$$\overset{f(y)}{\underset{\downarrow}{}} \qquad \overset{f(x, y)}{\underset{\downarrow}{}}$$
$$\frac{dy}{dx} = 1 + y^2 \quad \text{and} \quad \frac{dy}{dx} = 0.2xy$$

are autonomous and nonautonomous, respectively.

Many differential equations encountered in applications, or equations that are models of physical laws that do not change over time, are autonomous. As we have already seen in Section 1.3, in an applied context, symbols other than y and x are routinely used to represent the dependent and independent variables. For example, if t represents time, then inspection of

$$\frac{dA}{dt} = kA, \quad \frac{dx}{dt} = kx(n + 1 - x), \quad \frac{dT}{dt} = k(T - T_m), \quad \frac{dA}{dt} = 6 - \frac{1}{100}A,$$

where k, n, and T_m are constants, shows that each equation is time-independent. Indeed, *all* of the first-order differential equations introduced in Section 1.3 are time-independent and so are autonomous.

■ Critical Points The zeros of the function f in (1) are of special importance. We say that a real number c is a **critical point** of the autonomous differential equation (1) if it is a zero of f, that is, $f(c) = 0$. A critical point is also called an **equilibrium point** or **stationary point**. Now observe that if we substitute the constant function $y(x) = c$ into (1), then both sides of the equation equal zero. This means

> *If c is a critical point of (1), then $y(x) = c$ is a constant solution of the autonomous differential equation.*

A constant solution $y(x) = c$ of (1) is called an **equilibrium solution**; equilibria are the *only* constant solutions of (1).

As already mentioned, we can tell when a nonconstant solution $y = y(x)$ of (1) is increasing or decreasing by determining the algebraic sign of the derivative dy/dx; in the case of (1) we do this by identifying the intervals on the y-axis over which the function $f(y)$ is positive or negative.

EXAMPLE 3 **An Autonomous DE**

The differential equation

$$\frac{dP}{dt} = P(a - bP),$$

where a and b are positive constants, has the normal form $dP/dt = f(P)$, which is (1) with t and P playing the parts of x and y, respectively, and hence is autonomous. From $f(P) = P(a - bP) = 0$, we see that 0 and a/b are critical points of the equation and so the equilibrium solutions are $P(t) = 0$ and $P(t) = a/b$. By putting the critical points on a vertical line, we divide

P-axis

FIGURE 2.1.4 Phase portrait for Example 3

the line into three intervals defined by $-\infty < P < 0, 0 < P < a/b, a/b < P < \infty$. The arrows on the line shown in **FIGURE 2.1.4** indicate the algebraic sign of $f(P) = P(a - bP)$ on these intervals and whether a nonconstant solution $P(t)$ is increasing or decreasing on an interval. The following table explains the figure.

Interval	Sign of $f(P)$	$P(t)$	Arrow
$(-\infty, 0)$	minus	decreasing	points down
$(0, a/b)$	plus	increasing	points up
$(a/b, \infty)$	minus	decreasing	points down

Figure 2.1.4 is called a **one-dimensional phase portrait**, or simply **phase portrait**, of the differential equation $dP/dt = P(a - bP)$. The vertical line is called a **phase line**. ≡

Solution Curves Without solving an autonomous differential equation, we can usually say a great deal about its solution curves. Since the function f in (1) is independent of the variable x, we can consider f defined for $-\infty < x < \infty$ or for $0 \le x < \infty$. Also, since f and its derivative f' are continuous functions of y on some interval I of the y-axis, the fundamental results of Theorem 1.2.1 hold in some horizontal strip or region R in the xy-plane corresponding to I, and so through any point (x_0, y_0) in R there passes only one solution curve of (1). See **FIGURE 2.1.5(a)**. For the sake of discussion, let us suppose that (1) possesses exactly two critical points, c_1 and c_2, and that $c_1 < c_2$. The graphs of the equilibrium solutions $y(x) = c_1$ and $y(x) = c_2$ are horizontal lines, and these lines partition the region R into three subregions R_1, R_2, and R_3 as illustrated in Figure 2.1.5(b). Without proof, here are some conclusions that we can draw about a nonconstant solution $y(x)$ of (1):

(a) Region R

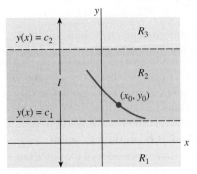

(b) Subregions R_1, R_2, and R_3

FIGURE 2.1.5 Lines $y(x) = c_1$ and $y(x) = c_2$ partition R into three horizontal subregions

- If (x_0, y_0) is in a subregion R_i, $i = 1, 2, 3$, and $y(x)$ is a solution whose graph passes through this point, then $y(x)$ remains in the subregion R_i for all x. As illustrated in Figure 2.1.5(b), the solution $y(x)$ in R_2 is bounded below by c_1 and above by c_2; that is, $c_1 < y(x) < c_2$ for all x. The solution curve stays within R_2 for all x because the graph of a nonconstant solution of (1) cannot cross the graph of either equilibrium solution $y(x) = c_1$ or $y(x) = c_2$. See Problem 33 in Exercises 2.1.
- By continuity of f we must then have either $f(y) > 0$ or $f(y) < 0$ for all x in a subregion R_i, $i = 1, 2, 3$. In other words, $f(y)$ cannot change signs in a subregion. See Problem 33 in Exercises 2.1.
- Since $dy/dx = f(y(x))$ is either positive or negative in a subregion R_i, $i = 1, 2, 3$, a solution $y(x)$ is strictly monotonic—that is, $y(x)$ is either increasing or decreasing in a subregion R_i. Therefore $y(x)$ cannot be oscillatory, nor can it have a relative extremum (maximum or minimum). See Problem 33 in Exercises 2.1.
- If $y(x)$ is *bounded above* by a critical point c_1 (as in subregion R_1 where $y(x) < c_1$ for all x), then the graph of $y(x)$ must approach the graph of the equilibrium solution $y(x) = c_1$ either as $x \to \infty$ or as $x \to -\infty$. If $y(x)$ is *bounded*, that is, bounded above and below by two consecutive critical points (as in subregion R_2 where $c_1 < y(x) < c_2$ for all x), then the graph of $y(x)$ must approach the graphs of the equilibrium solutions $y(x) = c_1$ and $y(x) = c_2$, one as $x \to \infty$ and the other as $x \to -\infty$. If $y(x)$ is *bounded below* by a critical point (as in subregion R_3 where $c_2 < y(x)$ for all x), then the graph of $y(x)$ must approach the graph of the equilibrium solution $y(x) = c_2$ either as $x \to \infty$ or as $x \to -\infty$. See Problem 34 in Exercises 2.1.

With the foregoing facts in mind, let us reexamine the differential equation in Example 3.

EXAMPLE 4 **Example 3 Revisited**

The three intervals determined on the P-axis or phase line by the critical points $P = 0$ and $P = a/b$ now correspond in the tP-plane to three subregions:

$$R_1: -\infty < P < 0, \qquad R_2: 0 < P < a/b, \qquad R_3: a/b < P < \infty,$$

where $-\infty < t < \infty$. The phase portrait in Figure 2.1.4 tells us that $P(t)$ is decreasing in R_1, increasing in R_2, and decreasing in R_3. If $P(0) = P_0$ is an initial value, then in R_1, R_2, and R_3, we have, respectively, the following:

(i) For $P_0 < 0$, $P(t)$ is bounded above. Since $P(t)$ is decreasing, $P(t)$ decreases without bound for increasing t and so $P(t) \to 0$ as $t \to -\infty$. This means the negative t-axis, the graph of the equilibrium solution $P(t) = 0$, is a horizontal asymptote for a solution curve.

(ii) For $0 < P_0 < a/b$, $P(t)$ is bounded. Since $P(t)$ is increasing, $P(t) \to a/b$ as $t \to \infty$ and $P(t) \to 0$ as $t \to -\infty$. The graphs of the two equilibrium solutions, $P(t) = 0$ and $P(t) = a/b$, are horizontal lines that are horizontal asymptotes for any solution curve starting in this subregion.

(iii) For $P_0 > a/b$, $P(t)$ is bounded below. Since $P(t)$ is decreasing, $P(t) \to a/b$ as $t \to \infty$. The graph of the equilibrium solution $P(t) = a/b$ is a horizontal asymptote for a solution curve.

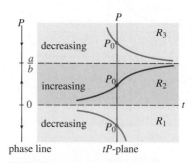

FIGURE 2.1.6 Phase portrait and solution curves in each of the three subregions in Example 4

In **FIGURE 2.1.6**, the phase line is the P-axis in the tP-plane. For clarity, the original phase line from Figure 2.1.4 is reproduced to the left of the plane in which the subregions R_1, R_2, and R_3 are shaded. The graphs of the equilibrium solutions $P(t) = a/b$ and $P(t) = 0$ (the t-axis) are shown in the figure as blue dashed lines; the solid graphs represent typical graphs of $P(t)$ illustrating the three cases just discussed. ≡

In a subregion such as R_1 in Example 4, where $P(t)$ is decreasing and unbounded below, we must necessarily have $P(t) \to -\infty$. Do *not* interpret this last statement to mean $P(t) \to -\infty$ as $t \to \infty$; we could have $P(t) \to -\infty$ as $t \to T$, where $T > 0$ is a finite number that depends on the initial condition $P(t_0) = P_0$. Thinking in dynamic terms, $P(t)$ could "blow up" in finite time; thinking graphically, $P(t)$ could have a vertical asymptote at $t = T > 0$. A similar remark holds for the subregion R_3.

The differential equation $dy/dx = \sin y$ in Example 2 is autonomous and has an infinite number of critical points since $\sin y = 0$ at $y = n\pi$, n an integer. Moreover, we now know that because the solution $y(x)$ that passes through $(0, -\frac{3}{2})$ is bounded above and below by two consecutive critical points $(-\pi < y(x) < 0)$ and is decreasing ($\sin y < 0$ for $-\pi < y < 0$), the graph of $y(x)$ must approach the graphs of the equilibrium solutions as horizontal asymptotes: $y(x) \to -\pi$ as $x \to \infty$ and $y(x) \to 0$ as $x \to -\infty$.

EXAMPLE 5 **Solution Curves of an Autonomous DE**

The autonomous equation $dy/dx = (y - 1)^2$ possesses the single critical point 1. From the phase portrait in **FIGURE 2.1.7(a)**, we conclude that a solution $y(x)$ is an increasing function in the subregions defined by $-\infty < y < 1$ and $1 < y < \infty$, where $-\infty < x < \infty$. For an initial condition $y(0) = y_0 < 1$, a solution $y(x)$ is increasing and bounded above by 1, and so $y(x) \to 1$ as $x \to \infty$; for $y(0) = y_0 > 1$, a solution $y(x)$ is increasing and unbounded.

Now $y(x) = 1 - 1/(x + c)$ is a one-parameter family of solutions of the differential equation. (See Problem 4 in Exercises 2.2.) A given initial condition determines a value for c. For the initial conditions, say, $y(0) = -1 < 1$ and $y(0) = 2 > 1$, we find, in turn, that $y(x) = 1 - 1/(x + \frac{1}{2})$ and so $y(x) = 1 - 1/(x - 1)$. As shown in Figure 2.1.7(b) and 2.1.7(c), the graph of each of these rational functions possesses a vertical asymptote. But bear in mind that the solutions of the IVPs

$$\frac{dy}{dx} = (y - 1)^2, \ y(0) = -1 \quad \text{and} \quad \frac{dy}{dx} = (y - 1)^2, \ y(0) = 2$$

are defined on special intervals. The two solutions are, respectively,

$$y(x) = 1 - \frac{1}{x + \frac{1}{2}}, \ -\frac{1}{2} < x < \infty \quad \text{and} \quad y(x) = 1 - \frac{1}{x - 1}, \ -\infty < x < 1.$$

The solution curves are the portions of the graphs in Figures 2.1.7(b) and 2.1.7(c) shown in blue. As predicted by the phase portrait, for the solution curve in Figure 2.1.7(b), $y(x) \to 1$ as $x \to \infty$; for the solution curve in Figure 2.1.7(c), $y(x) \to \infty$ as $x \to 1$ from the left.

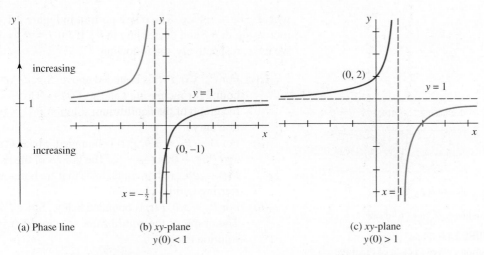

increasing

increasing

$y = 1$

$(0, -1)$

$x = -\frac{1}{2}$

(a) Phase line

(b) *xy*-plane
$y(0) < 1$

$(0, 2)$

$y = 1$

$x = 1$

(c) *xy*-plane
$y(0) > 1$

FIGURE 2.1.7 Behavior of solutions near $y = 1$ in Example 5

≡

Attractors and Repellers Suppose $y(x)$ is a nonconstant solution of the autonomous differential equation given in (1) and that c is a critical point of the DE. There are basically three types of behavior $y(x)$ can exhibit near c. In **FIGURE 2.1.8** we have placed c on four vertical phase lines. When both arrowheads on either side of the dot labeled c point *toward* c, as in Figure 2.1.8(a), all solutions $y(x)$ of (1) that start from an initial point (x_0, y_0) sufficiently near c exhibit the asymptotic behavior $\lim_{x \to \infty} y(x) = c$. For this reason the critical point c is said to be **asymptotically stable**. Using a physical analogy, a solution that starts near c is like a charged particle that, over time, is drawn to a particle of opposite charge, and so c is also referred to as an **attractor**. When both arrowheads on either side of the dot labeled c point *away* from c, as in Figure 2.1.8(b), all solutions $y(x)$ of (1) that start from an initial point (x_0, y_0) move away from c as x increases. In this case the critical point c is said to be **unstable**. An unstable critical point is also called a **repeller**, for obvious reasons. The critical point c illustrated in Figures 2.1.8(c) and 2.1.8(d) is neither an attractor nor a repeller. But since c exhibits characteristics of both an attractor and a repeller—that is, a solution starting from an initial point (x_0, y_0) sufficiently near c is attracted to c from one side and repelled from the other side—we say that the critical point c is **semi-stable**. In Example 3, the critical point a/b is asymptotically stable (an attractor) and the critical point 0 is unstable (a repeller). The critical point 1 in Example 5 is semi-stable.

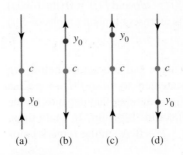

FIGURE 2.1.8 Critical point c is an attractor in (a), a repeller in (b), and semi-stable in (c) and (d)

| EXAMPLE 6 | **Classifying Critical Points** |

Locate and classify all critical points of $\dfrac{dy}{dx} = 4y - y^3$.

SOLUTION Rewriting the differential equation as

$$\frac{dy}{dx} = y(4 - y^2) = y(2 - y)(2 + y)$$

we see from $y(2 - y)(2 + y) = 0$ that $y = 0$, $y = 2$, and $y = -2$ are critical points of the DE. Now by examining, in turn, the algebraic signs of dy/dx on intervals of the *y*-axis determined by the critical points, we see from the phase portrait in **FIGURE 2.1.9** that:

$dy/dx > 0 \quad dy/dx < 0$

$(-\infty, -2), (-2, 0),$ implies $y = -2$ is asymptotically stable (attractor),

$dy/dx < 0 \quad dy/dx > 0$

$(-2, 0), (0, 2),$ implies $y = 0$ is unstable (repeller),

$dy/dx > 0 \quad dy/dx < 0$

$(0, 2), (2, \infty),$ implies $y = 2$ is asymptotically stable (attractor).

See Problems 21–28 in Exercises 2.1.

≡

FIGURE 2.1.9 Phase portrait of DE in Example 6

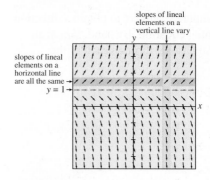

slopes of lineal elements on a vertical line vary

slopes of lineal elements on a horizontal line are all the same →

$y = 1$

FIGURE 2.1.10 Direction field for an autonomous DE

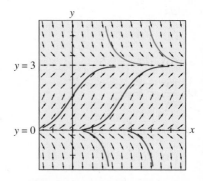

$y = 3$

$y = 0$

FIGURE 2.1.11 Translated solution curves of an autonomous DE

■ **Autonomous DEs and Direction Fields** If a first-order differential equation is autonomous, then we see from the right-hand side of its normal form $dy/dx = f(y)$ that slopes of lineal elements through points in the rectangular grid used to construct a direction field for the DE depend solely on the y-coordinate of the points. Put another way, lineal elements passing through points on any *horizontal* line must all have the same slope and therefore are parallel; slopes of lineal elements along any *vertical* line will, of course, vary. These facts are apparent from inspection of the horizontal gray strip and vertical blue strip in **FIGURE 2.1.10**. The figure exhibits a direction field for the autonomous equation $dy/dx = 2(y - 1)$. The red lineal elements in Figure 2.1.10 have zero slope because they lie along the graph of the equilibrium solution $y = 1$.

■ **Translation Property** Recall from precalculus mathematics that the graph of a function $y = f(x - k)$, where k is a constant, is the graph of $y = f(x)$ rigidly translated or shifted horizontally along the x-axis by an amount $|k|$; the translation is to the right if $k > 0$ and to the left if $k < 0$.

It turns out that under the assumptions stated after equation (1), solution curves of an autonomous first-order DE are related by the concept of translation. To see this, let's consider the differential equation $dy/dx = y(3 - y)$, which is a special case of the autonomous equation considered in Examples 3 and 4. Since $y = 0$ and $y = 3$ are equilibrium solutions of the DE, their graphs divide the xy-plane into subregions R_1, R_2, and R_3, defined by the three inequalities:

$$R_1: -\infty < y < 0, \qquad R_2: 0 < y < 3, \qquad R_3: 3 < y < \infty.$$

In **FIGURE 2.1.11** we have superimposed on a direction field of the DE six solution curves. The figure illustrates that all solution curves of the same color, that is, solution curves lying within a particular subregion R_i, all look alike. This is no coincidence but is a natural consequence of the fact that lineal elements passing through points on any horizontal line are parallel. That said, the following **translation property** of an autonomous DE should make sense:

> If $y(x)$ is a solution of an autonomous differential equation $dy/dx = f(y)$, then $y_1(x) = y(x - k)$, k a constant, is also a solution.

Hence, if $y(x)$ is a solution of the initial-value problem $dy/dx = f(y)$, $y(0) = y_0$, then $y_1(x) = y(x - x_0)$ is a solution of the IVP $dy/dx = f(y)$, $y(x_0) = y_0$. For example, it is easy to verify that $y(x) = e^x$, $-\infty < x < \infty$, is a solution of the IVP $dy/dx = y$, $y(0) = 1$ and so a solution $y_1(x)$ of, say, $dy/dx = y$, $y(4) = 1$ is $y(x) = e^x$ translated 4 units to the right:

$$y_1(x) = y(x - 4) = e^{x-4}, \quad -\infty < x < \infty.$$

2.1 Exercises Answers to selected odd-numbered problems begin on page ANS-2.

2.1.1 Direction Fields

In Problems 1–4, reproduce the given computer-generated direction field. Then sketch, by hand, an approximate solution curve that passes through each of the indicated points. Use different colored pencils for each solution curve.

1. $\dfrac{dy}{dx} = x^2 - y^2$

(a) $y(-2) = 1$ (b) $y(3) = 0$

(c) $y(0) = 2$ (d) $y(0) = 0$

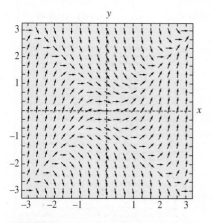

FIGURE 2.1.12 Direction field for Problem 1

2. $\dfrac{dy}{dx} = e^{-0.01xy^2}$

(a) $y(-6) = 0$ (b) $y(0) = 1$
(c) $y(0) = -4$ (d) $y(8) = -4$

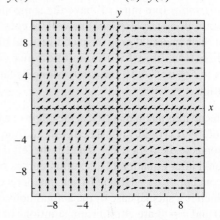

FIGURE 2.1.13 Direction field for Problem 2

3. $\dfrac{dy}{dx} = 1 - xy$

(a) $y(0) = 0$ (b) $y(-1) = 0$
(c) $y(2) = 2$ (d) $y(0) = -4$

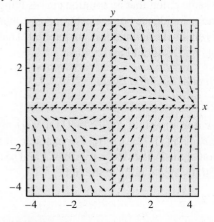

FIGURE 2.1.14 Direction field for Problem 3

4. $\dfrac{dy}{dx} = (\sin x)\cos y$

(a) $y(0) = 1$ (b) $y(1) = 0$
(c) $y(3) = 3$ (d) $y(0) = -\frac{5}{2}$

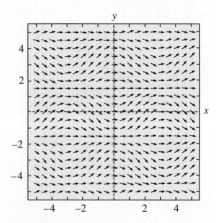

FIGURE 2.1.15 Direction field for Problem 4

In Problems 5–12, use computer software to obtain a direction field for the given differential equation. By hand, sketch an approximate solution curve passing through each of the given points.

5. $y' = x$

(a) $y(0) = 0$
(b) $y(0) = -3$

6. $y' = x + y$

(a) $y(-2) = 2$
(b) $y(1) = -3$

7. $y\dfrac{dy}{dx} = -x$

(a) $y(1) = 1$
(b) $y(0) = 4$

8. $\dfrac{dy}{dx} = \dfrac{1}{y}$

(a) $y(0) = 1$
(b) $y(-2) = -1$

9. $\dfrac{dy}{dx} = 0.2x^2 + y$

(a) $y(0) = \dfrac{1}{2}$
(b) $y(2) = -1$

10. $\dfrac{dy}{dx} = xe^y$

(a) $y(0) = -2$
(b) $y(1) = 2.5$

11. $y' = y - \cos\dfrac{\pi}{2}x$

(a) $y(2) = 2$
(b) $y(-1) = 0$

12. $\dfrac{dy}{dx} = 1 - \dfrac{y}{x}$

(a) $y(-\frac{1}{2}) = 2$
(b) $y(\frac{3}{2}) = 0$

In Problems 13 and 14, the given figures represent the graph of $f(y)$ and $f(x)$, respectively. By hand, sketch a direction field over an appropriate grid for $dy/dx = f(y)$ (Problem 13) and then for $dy/dx = f(x)$ (Problem 14).

13.

FIGURE 2.1.16 Graph for Problem 13

14.

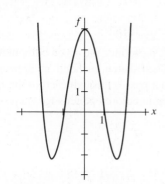

FIGURE 2.1.17 Graph for Problem 14

15. In parts (a) and (b) sketch **isoclines** $f(x, y) = c$ (see the *Remarks* on page 36) for the given differential equation using the indicated values of c. Construct a direction field over a grid by carefully drawing lineal elements with the appropriate

slope at chosen points on each isocline. In each case, use this rough direction field to sketch an approximate solution curve for the IVP consisting of the DE and the initial condition $y(0) = 1$.

(a) $dy/dx = x + y$; c an integer satisfying $-5 \le c \le 5$

(b) $dy/dx = x^2 + y^2$; $c = \frac{1}{4}, c = 1, c = \frac{9}{4}, c = 4$

Discussion Problems

16. (a) Consider the direction field of the differential equation $dy/dx = x(y - 4)^2 - 2$, but do not use technology to obtain it. Describe the slopes of the lineal elements on the lines $x = 0$, $y = 3$, $y = 4$, and $y = 5$.

 (b) Consider the IVP $dy/dx = x(y - 4)^2 - 2$, $y(0) = y_0$, where $y_0 < 4$. Can a solution $y(x) \to \infty$ as $x \to \infty$? Based on the information in part (a), discuss.

17. For a first-order DE $dy/dx = f(x, y)$, a curve in the plane defined by $f(x, y) = 0$ is called a **nullcline** of the equation, since a lineal element at a point on the curve has zero slope. Use computer software to obtain a direction field over a rectangular grid of points for $dy/dx = x^2 - 2y$, and then superimpose the graph of the nullcline $y = \frac{1}{2}x^2$ over the direction field. Discuss the behavior of solution curves in regions of the plane defined by $y < \frac{1}{2}x^2$ and by $y > \frac{1}{2}x^2$. Sketch some approximate solution curves. Try to generalize your observations.

18. (a) Identify the nullclines (see Problem 17) in Problems 1, 3, and 4. With a colored pencil, circle any lineal elements in **FIGURES 2.1.12, 2.1.14**, and **2.1.15** that you think may be a lineal element at a point on a nullcline.

 (b) What are the nullclines of an autonomous first-order DE?

2.1.2 Autonomous First-Order DEs

19. Consider the autonomous first-order differential equation $dy/dx = y - y^3$ and the initial condition $y(0) = y_0$. By hand, sketch the graph of a typical solution $y(x)$ when y_0 has the given values.

 (a) $y_0 > 1$ (b) $0 < y_0 < 1$
 (c) $-1 < y_0 < 0$ (d) $y_0 < -1$

20. Consider the autonomous first-order differential equation $dy/dx = y^2 - y^4$ and the initial condition $y(0) = y_0$. By hand, sketch the graph of a typical solution $y(x)$ when y_0 has the given values.

 (a) $y_0 > 1$ (b) $0 < y_0 < 1$
 (c) $-1 < y_0 < 0$ (d) $y_0 < -1$

In Problems 21–28, find the critical points and phase portrait of the given autonomous first-order differential equation. Classify each critical point as asymptotically stable, unstable, or semi-stable. By hand, sketch typical solution curves in the regions in the xy-plane determined by the graphs of the equilibrium solutions.

21. $\dfrac{dy}{dx} = y^2 - 3y$

22. $\dfrac{dy}{dx} = y^2 - y^3$

23. $\dfrac{dy}{dx} = (y - 2)^4$

24. $\dfrac{dy}{dx} = 10 + 3y - y^2$

25. $\dfrac{dy}{dx} = y^2(4 - y^2)$

26. $\dfrac{dy}{dx} = y(2 - y)(4 - y)$

27. $\dfrac{dy}{dx} = y \ln(y + 2)$

28. $\dfrac{dy}{dx} = \dfrac{ye^y - 9y}{e^y}$

In Problems 29 and 30, consider the autonomous differential equation $dy/dx = f(y)$, where the graph of f is given. Use the graph to locate the critical points of each differential equation. Sketch a phase portrait of each differential equation. By hand, sketch typical solution curves in the subregions in the xy-plane determined by the graphs of the equilibrium solutions.

29.

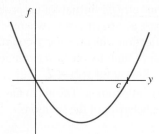

FIGURE 2.1.18 Graph for Problem 29

30.

FIGURE 2.1.19 Graph for Problem 30

Discussion Problems

31. Consider the autonomous DE $dy/dx = (2/\pi)y - \sin y$. Determine the critical points of the equation. Discuss a way of obtaining a phase portrait of the equation. Classify the critical points as asymptotically stable, unstable, or semi-stable.

32. A critical point c of an autonomous first-order DE is said to be **isolated** if there exists some open interval that contains c but no other critical point. Discuss: Can there exist an autonomous DE of the form given in (1) for which *every* critical point is nonisolated? Do not think profound thoughts.

33. Suppose that $y(x)$ is a nonconstant solution of the autonomous equation $dy/dx = f(y)$ and that c is a critical point of the DE. Discuss: Why can't the graph of $y(x)$ cross the graph of the equilibrium solution $y = c$? Why can't $f(y)$ change signs in one of the subregions discussed on page 38? Why can't $y(x)$ be oscillatory or have a relative extremum (maximum or minimum)?

34. Suppose that $y(x)$ is a solution of the autonomous equation $dy/dx = f(y)$ and is bounded above and below by two consecutive critical points $c_1 < c_2$, as in subregion R_2 of

Figure 2.1.5(b). If $f(y) > 0$ in the region, then $\lim_{x \to \infty} y(x) = c_2$. Discuss why there cannot exist a number $L < c_2$ such that $\lim_{x \to \infty} y(x) = L$. As part of your discussion, consider what happens to $y'(x)$ as $x \to \infty$.

35. Using the autonomous equation (1), discuss how it is possible to obtain information about the location of points of inflection of a solution curve.

36. Consider the autonomous DE $dy/dx = y^2 - y - 6$. Use your ideas from Problem 35 to find intervals on the y-axis for which solution curves are concave up and intervals for which solution curves are concave down. Discuss why *each* solution curve of an initial-value problem of the form $dy/dx = y^2 - y - 6$, $y(0) = y_0$, where $-2 < y_0 < 3$, has a point of inflection with the same y-coordinate. What is that y-coordinate? Carefully sketch the solution curve for which $y(0) = -1$. Repeat for $y(2) = 2$.

37. Suppose the autonomous DE in (1) has no critical points. Discuss the behavior of the solutions.

Mathematical Models

38. **Population Model** The differential equation in Example 3 is a well-known population model. Suppose the DE is changed to

$$\frac{dP}{dt} = P(aP - b),$$

where a and b are positive constants. Discuss what happens to the population P as time t increases.

39. **Population Model** Another population model is given by

$$\frac{dP}{dt} = kP - h,$$

where $h > 0$ and $k > 0$ are constants. For what initial values $P(0) = P_0$ does this model predict that the population will go extinct?

40. **Terminal Velocity** The autonomous differential equation

$$m\frac{dv}{dt} = mg - kv,$$

where k is a positive constant of proportionality called the drag coefficient and g is the acceleration due to gravity, is a model for the instantaneous velocity v of a body of mass m that is falling under the influence of gravity. Because the term $-kv$ represents air resistance or drag, the velocity of a body falling from a great height does not increase without bound as time t increases. Use a phase portrait of the differential equation to find the limiting, or terminal, velocity of the body. Explain your reasoning. See page 24.

41. **Terminal Velocity** In Problem 17 of Exercises 1.3, we indicated that for high-speed motion of a body, air resistance is taken to be proportional to a power of its instantaneous velocity v. If we take air resistance to be proportional to v^2, then the mathematical model for the instantaneous velocity of a falling body of mass m in Problem 40 becomes

$$m\frac{dv}{dt} = mg - kv^2,$$

where $k > 0$. Use a phase portrait to find the terminal velocity of the body. Explain your reasoning. See page 27.

42. **Chemical Reactions** When certain kinds of chemicals are combined, the rate at which a new compound is formed is governed by the differential equation

$$\frac{dX}{dt} = k(\alpha - X)(\beta - X),$$

where $k > 0$ is a constant of proportionality and $\beta > \alpha > 0$. Here $X(t)$ denotes the number of grams of the new compound formed in time t. See pages 21 and 22.

(a) Use a phase portrait of the differential equation to predict the behavior of X as $t \to \infty$.

(b) Consider the case when $\alpha = \beta$. Use a phase portrait of the differential equation to predict the behavior of X as $t \to \infty$ when $X(0) < \alpha$. When $X(0) > \alpha$.

(c) Verify that an explicit solution of the DE in the case when $k = 1$ and $\alpha = \beta$ is $X(t) = \alpha - 1/(t + c)$. Find a solution satisfying $X(0) = \alpha/2$. Find a solution satisfying $X(0) = 2\alpha$. Graph these two solutions. Does the behavior of the solutions as $t \to \infty$ agree with your answers to part (b)?

2.2 Separable Equations

INTRODUCTION Consider the first-order equations $dy/dx = f(x, y)$. When f does not depend on the variable y, that is, $f(x, y) = g(x)$, the differential equation

$$\frac{dy}{dx} = g(x) \tag{1}$$

can be solved by integration. If $g(x)$ is a continuous function, then integrating both sides of (1) gives the solution $y = \int g(x)\,dx = G(x) + c$, where $G(x)$ is an antiderivative (indefinite integral) of $g(x)$. For example, if $dy/dx = 1 + e^{2x}$, then $y = \int (1 + e^{2x})\,dx$ or $y = x + \frac{1}{2}e^{2x} + c$ defined on $(-\infty, \infty)$.

A Definition Equation (1), as well as its method of solution, is just a special case when f in $dy/dx = f(x, y)$ is a product of a function of x and a function of y.

DEFINITION 2.2.1 Separable Equation

A first-order differential equation of the form

$$\frac{dy}{dx} = g(x)h(y)$$

is said to be **separable** or to have **separable variables**.

For example, the differential equations

$$\frac{dy}{dx} = x^2 y^4 e^{5x-3y} \quad \text{and} \quad \frac{dy}{dx} = y + \cos x$$

are separable and nonseparable, respectively. To see this, note that in the first equation we can factor $f(x, y) = x^2 y^4 e^{5x-3y}$ as

$$f(x, y) = x^2 y^4 e^{5x-3y} = \overbrace{(x^2 e^{5x})}^{g(x)}\overbrace{(y^4 e^{-3y})}^{h(y)},$$

but in the second equation there is no way of writing $y + \cos x$ as a product of a function of x times a function of y.

■ **Method of Solution** Observe that by dividing by the function $h(y)$, a separable equation can be written as

$$p(y)\frac{dy}{dx} = g(x), \tag{2}$$

where, for convenience, we have denoted $1/h(y)$ by $p(y)$. From this last form we can see immediately that (2) reduces to (1) when $h(y) = 1$.

Now if $y = \phi(x)$ represents a solution of (2), we must have $p(\phi(x))\phi'(x) = g(x)$, and therefore,

$$\int p(\phi(x))\phi'(x)\, dx = \int g(x)\, dx. \tag{3}$$

But $dy = \phi'(x)\, dx$, and so (3) is the same as

$$\int p(y)\, dy = \int g(x)\, dx \quad \text{or} \quad H(y) = G(x) + c, \tag{4}$$

where $H(y)$ and $G(x)$ are antiderivatives of $p(y) = 1/h(y)$ and $g(x)$, respectively.

Equation (4) indicates the procedure for solving separable equations. A one-parameter family of solutions, usually given implicitly, is obtained by integrating both sides of the differential form $p(y)\, dy = g(x)\, dx$.

In solving first-order DEs, use only one constant. ▶ There is no need to use two constants in the integration of a separable equation, because if we write $H(y) + c_1 = G(x) + c_2$, then the difference $c_2 - c_1$ can be replaced by a single constant c, as in (4). In many instances throughout the chapters that follow, we will relabel constants in a manner convenient to a given equation. For example, multiples of constants or combinations of constants can sometimes be replaced by a single constant.

EXAMPLE 1 Solving a Separable DE

Solve $(1 + x)\, dy - y\, dx = 0.$

SOLUTION Dividing by $(1 + x)y$, we can write $dy/y = dx/(1 + x)$, from which it follows that

$$\int \frac{dy}{y} = \int \frac{dx}{1 + x}$$

$$\ln|y| = \ln|1 + x| + c_1$$

$$|y| = e^{\ln|1+x|+c_1} = e^{\ln|1+x|} \cdot e^{c_1} \qquad \leftarrow \text{laws of exponents}$$

$$= |1 + x|e^{c_1} \qquad \leftarrow \begin{cases} |1 + x| = 1 + x, & x \geq -1 \\ |1 + x| = -(1 + x), & x < -1 \end{cases}$$

and so $\qquad\qquad y = \pm e^{c_1}(1 + x).$

Relabeling $\pm e^{c_1}$ by c then gives $y = c(1 + x)$ defined on $(-\infty, \infty)$. $\qquad \equiv$

In the solution of Example 1, because each integral results in a logarithm, a judicious choice for the constant of integration is $\ln|c|$ rather than c. Rewriting the second line of the solution as $\ln|y| = \ln|1 + x| + \ln|c|$ enables us to combine the terms on the right-hand side by the properties of logarithms. From $\ln|y| = \ln|c(1 + x)|$, we immediately get $y = c(1 + x)$. Even if the indefinite integrals are not *all* logarithms, it may still be advantageous to use $\ln|c|$. However, no firm rule can be given.

In Section 1.1 we have already seen that a solution curve may be only a segment or an arc of the graph of an implicit solution $G(x, y) = 0$.

FIGURE 2.2.1 Solution curve for IVP in Example 2

EXAMPLE 2 Solution Curve

Solve the initial-value problem $\dfrac{dy}{dx} = -\dfrac{x}{y},\quad y(4) = -3.$

SOLUTION By rewriting the equation as $y\,dy = -x\,dx$ we get

$$\int y\,dy = -\int x\,dx \quad \text{and} \quad \frac{y^2}{2} = -\frac{x^2}{2} + c_1.$$

We can write the result of the integration as $x^2 + y^2 = c^2$ by replacing the constant $2c_1$ by c^2. This solution of the differential equation represents a one-parameter family of concentric circles centered at the origin.

Now when $x = 4$, $y = -3$, so that $16 + 9 = 25 = c^2$. Thus the initial-value problem determines the circle $x^2 + y^2 = 25$ with radius 5. Because of its simplicity, we can solve this implicit solution for an explicit solution that satisfies the initial condition. We have seen this solution as $y = \phi_2(x)$ or $y = -\sqrt{25 - x^2}$, $-5 < x < 5$ in Example 8 of Section 1.1. A solution curve is the graph of a differentiable *function*. In this case the solution curve is the lower semicircle, shown in blue in **FIGURE 2.2.1**, that contains the point $(4, -3)$. $\qquad \equiv$

EXAMPLE 3 Initial-Value Problem

Solve the initial-value problem $\dfrac{dy}{dx} = 1 + y^2,\quad y(0) = 0.$

SOLUTION Rewriting the equation as $\dfrac{dy}{1 + y^2} = dx$ we see that

$$\int \frac{dy}{1 + y^2} = \int dx \quad \text{gives} \quad \tan^{-1}y = x + c.$$

Now $y(0) = 0$ in the last equation gives $\tan^{-1}0 = 0 + c$ and so $c = 0$. The solution is $\tan^{-1}y = x$ or $y = \tan x$ with interval of definition $(-\pi/2, \pi/2)$. $\qquad \equiv$

Most of us would write the solution $\tan^{-1}y = x + c$ in Example 3 as $y = \tan(x + c)$. But we have to be a bit careful here. If we simply specify a value of c, then $y = \tan(x + c)$ actually defines an infinite number of particular solutions corresponding to an infinite number of intervals of definition. For example, for $c = 0$

$$\dots y = \tan x, (-3\pi/2, -\pi/2), \ y = \tan x, (-\pi/2, \pi/2), \ y = \tan x, (\pi/2, 3\pi/2), \dots$$

are solutions of the differential equation. See **FIGURE 2.2.2**. Only the red graph is the solution of the IVP in Example 3.

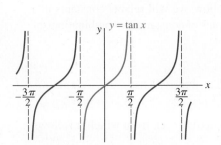

FIGURE 2.2.2 Solutions of DE in Example 3

■ **Losing a Solution** Some care should be exercised when separating variables, since the variable divisors could be zero at a point. Specifically, if r is a zero of the function $h(y)$, then

substituting $y = r$ into $dy/dx = g(x) h(y)$ makes both sides zero; in other words, $y = r$ is a constant solution of the differential equation. But after separating variables, observe that the left side of $dy/h(y) = g(x) dx$ is undefined at r. As a consequence, $y = r$ may not show up in the family of solutions obtained after integration and simplification. Recall, such a solution is called a singular solution.

EXAMPLE 4 **Losing a Solution**

Solve $\dfrac{dy}{dx} = y^2 - 4$.

SOLUTION We put the equation in the form

$$\frac{dy}{y^2 - 4} = dx \quad \text{or} \quad \left[\frac{\frac{1}{4}}{y - 2} - \frac{\frac{1}{4}}{y + 2} \right] dy = dx. \tag{5}$$

The second equation in (5) is the result of using partial fractions on the left side of the first equation. Integrating and using the laws of logarithms gives

$$\frac{1}{4} \ln|y - 2| - \frac{1}{4} \ln|y + 2| = x + c_1 \quad \text{or} \quad \ln\left|\frac{y - 2}{y + 2}\right| = 4x + c_2 \quad \text{or} \quad \frac{y - 2}{y + 2} = e^{4x + c_2}.$$

Here we have replaced $4c_1$ by c_2. Finally, after replacing e^{c_2} by c and solving the last equation for y, we get the one-parameter family of solutions

$$y = 2 \frac{1 + ce^{4x}}{1 - ce^{4x}}. \tag{6}$$

Now if we factor the right side of the differential equation as $dy/dx = (y - 2)(y + 2)$, we know from the discussion in Section 2.1 that $y = 2$ and $y = -2$ are two constant (equilibrium) solutions. The solution $y = 2$ is a member of the family of solutions defined by (6) corresponding to the value $c = 0$. However, $y = -2$ is a singular solution; it cannot be obtained from (6) for any choice of the parameter c. This latter solution was lost early on in the solution process. Inspection of (5) clearly indicates that we must preclude $y = \pm 2$ in these steps. ≡

EXAMPLE 5 **An Initial-Value Problem**

Solve the initial-value problem

$$\cos x(e^{2y} - y) \frac{dy}{dx} = e^y \sin 2x, \quad y(0) = 0.$$

SOLUTION Dividing the equation by $e^y \cos x$ gives

$$\frac{e^{2y} - y}{e^y} dy = \frac{\sin 2x}{\cos x} dx.$$

Before integrating, we use termwise division on the left side and the trigonometric identity $\sin 2x = 2 \sin x \cos x$ on the right side. Then

$$\text{integration by parts} \rightarrow \quad \int (e^y - ye^{-y}) \, dy = 2 \int \sin x \, dx$$

yields

$$e^y + ye^{-y} + e^{-y} = -2\cos x + c. \tag{7}$$

The initial condition $y = 0$ when $x = 0$ implies $c = 4$. Thus a solution of the initial-value problem is

$$e^y + ye^{-y} + e^{-y} = 4 - 2\cos x. \tag{8} \equiv$$

Use of Computers In the *Remarks* at the end of Section 1.1 we mentioned that it may be difficult to use an implicit solution $G(x, y) = 0$ to find an explicit solution $y = \phi(x)$. Equation (8) shows that the task of solving for y in terms of x may present more problems than just the drudgery of symbol pushing—it simply can't be done! Implicit solutions such as (8) are

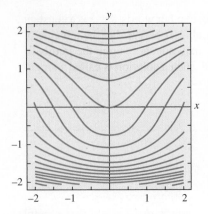

FIGURE 2.2.3 Level curves $G(x, y) = c$, where $G(x,y) = e^y + ye^{-y} + e^{-y} + 2\cos x$

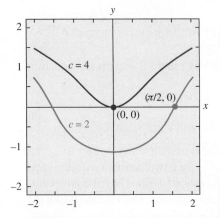

FIGURE 2.2.4 Level curves $c = 2$ and $c = 4$

FIGURE 2.2.5 Piecewise-defined solutions of (9)

somewhat frustrating; neither the graph of the equation nor an interval over which a solution satisfying $y(0) = 0$ is defined is apparent. The problem of "seeing" what an implicit solution looks like can be overcome in some cases by means of technology. One way* of proceeding is to use the contour plot application of a CAS. Recall from multivariate calculus that for a function of two variables $z = G(x, y)$ the *two-dimensional* curves defined by $G(x, y) = c$, where c is constant, are called the *level curves* of the function. With the aid of a CAS we have illustrated in **FIGURE 2.2.3** some of the level curves of the function $G(x, y) = e^y + ye^{-y} + e^{-y} + 2 \cos x$. The family of solutions defined by (7) are the level curves $G(x, y) = c$. **FIGURE 2.2.4** illustrates, in blue, the level curve $G(x, y) = 4$, which is the particular solution (8). The red curve in Figure 2.2.4 is the level curve $G(x, y) = 2$, which is the member of the family $G(x, y) = c$ that satisfies $y(\pi/2) = 0$.

If an initial condition leads to a particular solution by finding a specific value of the parameter c in a family of solutions for a first-order differential equation, it is a natural inclination for most students (and instructors) to relax and be content. However, a solution of an initial-value problem may not be unique. We saw in Example 4 of Section 1.2 that the initial-value problem

$$\frac{dy}{dx} = xy^{1/2}, \quad y(0) = 0, \tag{9}$$

has at least two solutions, $y = 0$ and $y = \frac{1}{16}x^4$. We are now in a position to solve the equation.

Separating variables and integrating $y^{-\frac{1}{2}} dy = x\, dx$ gives $2y^{\frac{1}{2}} = \frac{1}{2}x^2 + c_1$. Solving for y and replacing $\frac{1}{2}c_1$ by the symbol c yields

$$y = (\tfrac{1}{4}x^2 + c)^2. \tag{10}$$

Each of the functions in the family given in (10) is a solution of equation (9) on the interval $(-\infty, \infty)$ provided we take $c \geq 0$. See Problem 56 in Exercises 2.2. Now when we substitute $x = 0$, $y = 0$ in (10) we see that $c = 0$. Therefore $y = \frac{1}{16}x^4$ is a solution of the IVP. But note that the solution $y = 0$ is not a member of the family of solutions (10). This singular solution was lost in the solution process by dividing by $y^{\frac{1}{2}}$. The initial-value problem (9) actually possesses many more solutions, since for any choice of the parameter $a \geq 0$ the piecewise-defined function

$$y = \begin{cases} 0, & x < a \\ \frac{1}{16}(x^2 - a^2)^2, & x \geq a \end{cases}$$

satisfies both the differential equation and the initial condition. See **FIGURE 2.2.5**.

■ An Integral-Defined Function

In (*ii*) of the *Remarks* at the end of Section 1.1 it was pointed out that a solution method for a certain kind of differential equation may lead to an integral-defined function. This is especially true for separable differential equations because integration *is* the method of solution. For example, if g is continuous on some interval I containing x_0 and x, then a solution of the simple initial-value problem $dy/dx = g(x)$, $y(x_0) = y_0$ defined on I is given by

$$y(x) = y_0 + \int_{x_0}^{x} g(t)\, dt.$$

To see this, we have immediately from (12) of Section 1.1 that $dy/dx = g(x)$ and $y(x_0) = y_0$ because $\int_{x_0}^{x_0} g(t)\, dt = 0$. When $\int g(t)\, dt$ is nonelementary, that is, cannot be expressed in terms of elementary functions, the form $y(x) = y_0 + \int_{x_0}^{x} g(t)\, dt$ may be the best we can do in obtaining an explicit solution of an IVP. The next example illustrates this idea.

> **EXAMPLE 6** **An Initial-Value Problem**

Solve $\dfrac{dy}{dx} = e^{-x^2}, \quad y(2) = 6.$

*In Section 2.6 we discuss several other ways of proceeding that are based on the concept of a numerical solver.

SOLUTION The function $g(x) = e^{-x^2}$ is continuous on the interval $(-\infty, \infty)$ but its antiderivative is not an elementary function. Using t as a dummy variable of integration, we integrate both sides of the given differential equation:

$$\int_2^x \frac{dy}{dt}\, dt = \int_2^x e^{-t^2}\, dt$$

$$y(t)\Big]_2^x = \int_2^x e^{-t^2}\, dt$$

$$y(x) - y(2) = \int_2^x e^{-t^2}\, dt$$

$$y(x) = y(2) + \int_2^x e^{-t^2}\, dt.$$

Using the initial condition $y(2) = 6$ we obtain the solution $y(x) = 6 + \int_2^x e^{-t^2}\, dt$. ≡

The procedure illustrated in Example 6 works equally well on separable equations $dy/dx = g(x)f(y)$ where, say, $f(y)$ possesses an elementary antiderivative but $g(x)$ does not possess an elementary antiderivative. See Problems 31–34 in Exercises 2.2.

REMARKS

In some of the preceding examples we saw that the constant in the one-parameter family of solutions for a first-order differential equation can be relabeled when convenient. Also, it can easily happen that two individuals solving the same equation correctly arrive at dissimilar expressions for their answers. For example, by separation of variables, we can show that one-parameter families of solutions for the DE $(1 + y^2)\, dx + (1 + x^2)\, dy = 0$ are

$$\arctan x + \arctan y = c \quad \text{or} \quad \frac{x + y}{1 - xy} = c.$$

As you work your way through the next several sections, keep in mind that families of solutions may be equivalent in the sense that one family may be obtained from another by either relabeling the constant or applying algebra and trigonometry. See Problems 27 and 28 in Exercises 2.2.

2.2 Exercises Answers to selected odd-numbered problems begin on page ANS-2.

In Problems 1–22, solve the given differential equation by separation of variables.

1. $\dfrac{dy}{dx} = \sin 5x$

2. $\dfrac{dy}{dx} = (x + 1)^2$

3. $dx + e^{3x}dy = 0$

4. $dy - (y - 1)^2 dx = 0$

5. $x\dfrac{dy}{dx} = 4y$

6. $\dfrac{dy}{dx} + 2xy^2 = 0$

7. $\dfrac{dy}{dx} = e^{3x+2y}$

8. $e^x y\dfrac{dy}{dx} = e^{-y} + e^{-2x-y}$

9. $y \ln x \dfrac{dx}{dy} = \left(\dfrac{y + 1}{x}\right)^2$

10. $\dfrac{dy}{dx} = \left(\dfrac{2y + 3}{4x + 5}\right)^2$

11. $\csc y\, dx + \sec^2 x\, dy = 0$

12. $\sin 3x\, dx + 2y \cos^3 3x\, dy = 0$

13. $(e^y + 1)^2 e^{-y} dx + (e^x + 1)^3 e^{-x} dy = 0$

14. $x(1 + y^2)^{1/2} dx = y(1 + x^2)^{1/2} dy$

15. $\dfrac{dS}{dr} = kS$

16. $\dfrac{dQ}{dt} = k(Q - 70)$

17. $\dfrac{dP}{dt} = P - P^2$

18. $\dfrac{dN}{dt} + N = Nte^{t+2}$

19. $\dfrac{dy}{dx} = \dfrac{xy + 3x - y - 3}{xy - 2x + 4y - 8}$

20. $\dfrac{dy}{dx} = \dfrac{xy + 2y - x - 2}{xy - 3y + x - 3}$

21. $\dfrac{dy}{dx} = x\sqrt{1 - y^2}$

22. $(e^x + e^{-x})\dfrac{dy}{dx} = y^2$

In Problems 23–30, find an explicit solution of the given initial-value problem.

23. $\dfrac{dx}{dt} = 4(x^2 + 1), \quad x(\pi/4) = 1$

24. $\dfrac{dy}{dx} = \dfrac{y^2 - 1}{x^2 - 1}, \quad y(2) = 2$

25. $x^2 \dfrac{dy}{dx} = y - xy, \quad y(-1) = -1$

26. $\dfrac{dy}{dt} + 2y = 1, \quad y(0) = \frac{5}{2}$

27. $\sqrt{1 - y^2}\,dx - \sqrt{1 - x^2}\,dy = 0, \quad y(0) = \sqrt{3}/2$

28. $(1 + x^4)\,dy + x(1 + 4y^2)\,dx = 0, \quad y(1) = 0$

29. $\dfrac{dy}{dx} = -y \ln y, \quad y(0) = e$

30. $x \sinh y \dfrac{dy}{dx} = \cosh y, \quad y(1) = 0$

In Problems 31–34, proceed as in Example 6 and find an explicit solution of the given initial-value problem.

31. $\dfrac{dy}{dx} = ye^{-x^2}, \quad y(4) = 1$

32. $\dfrac{dy}{dx} = y^2 \sin x^2, \quad y(-2) = \frac{1}{3}$

33. $\dfrac{dy}{dx} = (1 + y^2)\sqrt{1 + \cos x^3}, \quad y(1) = 1$

34. $\dfrac{dy}{dx} = \dfrac{e^{-2y}\sin x}{1 + x^2}, \quad y(0) = 0$

In Problems 35–38, find an explicit solution of the given initial-value problem. Determine the exact interval I of definition of each solution by analytical methods. Use a graphing utility to plot the graph of each solution.

35. $\dfrac{dy}{dx} = \dfrac{2x + 1}{2y}, \quad y(-2) = -1$

36. $(2y - 2)\dfrac{dy}{dx} = 3x^2 + 4x + 2, \quad y(1) = -2$

37. $e^y dx - e^{-x} dy = 0, \quad y(0) = 0$

38. $\sin x\,dx + y\,dy = 0, \quad y(0) = 1$

39. (a) Find a solution of the initial-value problem consisting of the differential equation in Example 4 and the initial conditions $y(0) = 2$, $y(0) = -2$, $y(\frac{1}{4}) = 1$.

 (b) Find the solution of the differential equation in Example 4 when $\ln c_1$ is used as the constant of integration on the *left-hand* side in the solution and $4 \ln c_1$ is replaced by $\ln c$. Then solve the same initial-value problems in part (a).

40. Find a solution of $x \dfrac{dy}{dx} = y^2 - y$ that passes through the indicated points.

 (a) $(0, 1)$ **(b)** $(0, 0)$ **(c)** $(\frac{1}{2}, \frac{1}{2})$ **(d)** $(2, \frac{1}{4})$

41. Find a singular solution of Problem 21. Of Problem 22.

42. Show that an implicit solution of

$$2x \sin^2 y\,dx - (x^2 + 10)\cos y\,dy = 0$$

is given by $\ln(x^2 + 10)\csc y = c$. Find the constant solutions, if any, that were lost in the solution of the differential equation.

Often a radical change in the form of the solution of a differential equation corresponds to a very small change in either the initial condition or the equation itself. In Problems 43–46, find an explicit solution of the given initial-value problem. Use a graphing utility to plot the graph of each solution. Compare each solution curve in a neighborhood of $(0, 1)$.

43. $\dfrac{dy}{dx} = (y - 1)^2, \quad y(0) = 1$

44. $\dfrac{dy}{dx} = (y - 1)^2, \quad y(0) = 1.01$

45. $\dfrac{dy}{dx} = (y - 1)^2 + 0.01, \quad y(0) = 1$

46. $\dfrac{dy}{dx} = (y - 1)^2 - 0.01, \quad y(0) = 1$

47. Every autonomous first-order equation $dy/dx = f(y)$ is separable. Find explicit solutions $y_1(x), y_2(x), y_3(x)$, and $y_4(x)$ of the differential equation $dy/dx = y - y^3$ that satisfy, in turn, the initial conditions $y_1(0) = 2$, $y_2(0) = \frac{1}{2}$, $y_3(0) = -\frac{1}{2}$, and $y_4(0) = -2$. Use a graphing utility to plot the graphs of each solution. Compare these graphs with those predicted in Problem 19 of Exercises 2.1. Give the exact interval of definition for each solution.

48. (a) The autonomous first-order differential equation $dy/dx = 1/(y - 3)$ has no critical points. Nevertheless, place 3 on a phase line and obtain a phase portrait of the equation. Compute d^2y/dx^2 to determine where solution curves are concave up and where they are concave down (see Problems 35 and 36 in Exercises 2.1). Use the phase portrait and concavity to sketch, by hand, some typical solution curves.

 (b) Find explicit solutions $y_1(x), y_2(x), y_3(x)$, and $y_4(x)$ of the differential equation in part (a) that satisfy, in turn, the initial conditions $y_1(0) = 4$, $y_2(0) = 2$, $y_3(1) = 2$, and $y_4(-1) = 4$. Graph each solution and compare with your sketches in part (a). Give the exact interval of definition for each solution.

In Problems 49–54, use a technique of integration or a substitution to find an explicit solution of the given differential equation or initial-value problem.

49. $\dfrac{dy}{dx} = \dfrac{1}{1 + \sin x}$

50. $\dfrac{dy}{dx} = \dfrac{\sin \sqrt{x}}{\sqrt{y}}$

51. $(\sqrt{x} + x)\dfrac{dy}{dx} = \sqrt{y} + y$

52. $\dfrac{dy}{dx} = y^{2/3} - y$

53. $\dfrac{dy}{dx} = \dfrac{e^{\sqrt{x}}}{y}, \quad y(1) = 4$

54. $\dfrac{dy}{dx} = \dfrac{x \tan^{-1} x}{y}, \quad y(0) = 3$

Discussion Problems

55. (a) Explain why the interval of definition of the explicit solution $y = \phi_2(x)$ of the initial-value problem in Example 2 is the *open* interval $(-5, 5)$.
 (b) Can any solution of the differential equation cross the x-axis? Do you think that $x^2 + y^2 = 1$ is an implicit solution of the initial-value problem $dy/dx = -x/y$, $y(1) = 0$?

56. On page 48 we showed that a one-parameter family of solutions of the first-order differential equation $dy/dx = xy^{\frac{1}{2}}$ is $y = (\frac{1}{4}x^4 + c)^2$ for $c \geq 0$. Each solution in this family is defined on the interval $(-\infty, \infty)$. The last statement is not true if we choose c to be negative. For $c = -1$, explain why $y = (\frac{1}{4}x^4 - 1)^2$ *is not* a solution of the DE on $(-\infty, \infty)$. Find an interval of definition I on which $y = (\frac{1}{4}x^4 - 1)^2$ *is* a solution of the DE.

57. In Problems 47 and 48 we saw that every autonomous first-order differential equation $dy/dx = f(y)$ is separable. Does this fact help in the solution of the initial-value problem
$$\frac{dy}{dx} = \sqrt{1 + y^2}\sin^2 y, \quad y(0) = \tfrac{1}{2}?$$
Discuss. Sketch, by hand, a plausible solution curve of the problem.

58. (a) Solve the two initial-value problems:
$$\frac{dy}{dx} = y, \; y(0) = 1 \quad \text{and} \quad \frac{dy}{dx} = y + \frac{y}{x \ln x}, \; y(e) = 1.$$
 (b) Show that there are more than 1.65 million digits in the y-coordinate of the point of intersection of the two solution curves in part (a).

59. Find a function whose square plus the square of its derivative is 1.

60. (a) The differential equation in Problem 27 is equivalent to the normal form
$$\frac{dy}{dx} = \sqrt{\frac{1 - y^2}{1 - x^2}}$$
 in the square region in the xy-plane defined by $|x| < 1$, $|y| < 1$. But the quantity under the radical is nonnegative also in the regions defined by $|x| > 1$, $|y| > 1$. Sketch all regions in the xy-plane for which this differential equation possesses real solutions.
 (b) Solve the DE in part (a) in the regions defined by $|x| > 1$, $|y| > 1$. Then find an implicit and an explicit solution of the differential equation subject to $y(2) = 2$.

Mathematical Models

61. **Hawking Radiation** According to the English theoretical physicist **Stephen Hawking** (1942–2018), the mass $m(t)$ of a black hole decreases over time due to radiation. A model for the mass is given by
$$\frac{dm}{dt} = -\frac{k}{m^2}.$$
Solve for $m(t)$ when $m(0) = m_0$.

62. **Hanging Chain** Suppose a flexible steel chain is hanging under its own weight and its ends are attached at the same height. See the photo below.
 (a) When determining the shape of the chain, one has to first solve the differential equation
$$\frac{du}{dx} = a\sqrt{1 + u^2},$$
 where a is a constant. Solve for $u(x)$ when $u(0) = 0$.
 (b) The function $u(x)$ found in part (a) is related to the shape of the chain $y(x)$ by the differential equation
$$\frac{dy}{dx} = u.$$
 Solve for $y(x)$ when $y(0) = 1/a$. Identify the shape of the hanging chain.

© Stoyan G. Ivanov/Shutterstock
Hanging chain in Problem 62

63. **Tautochrone Problem** The tautochrone problem consists of finding a curve C such that the time of descent of a bead sliding without friction on C to its lowest point is independent of its starting point. If the curve C in **FIGURE 2.2.6** is a tautochrone, then the three beads starting from different positions corresponding to x-coordinates x_1, x_2, and x_3 will reach the origin $(0, 0)$ in the same time. The Dutch mathematician and astronomer **Christiaan Huygens** (1629–1695) was the first to publish the solution of the tautochrone problem in 1673.

In determining such a curve, the following differential equation must be solved:
$$1 + \left(\frac{dx}{dy}\right)^2 = \frac{a}{y},$$
where a is a constant. Find parametric equations for C by solving the DE in the form
$$dx = \sqrt{\frac{a - y}{y}}\, dy.$$
Use the substitution $y = a \sin^2 \phi$.

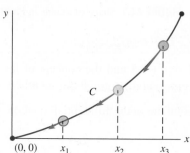

FIGURE 2.2.6 Tautochrone in Problem 63

64. Chemical Reaction In a certain second-order chemical reaction involving a single reactant, the rate of reaction is given by

$$\frac{dA}{dt} = k(A - x)^2,$$

where k and A are constants and $x(t)$ is the number of grams of the new chemical formed in time t. Solve for $x(t)$ when $x(0) = 0$.

65. Suspension Bridge In (16) of Section 1.3 we saw that a mathematical model for the shape of a flexible cable strung between two vertical supports is

$$\frac{dy}{dx} = \frac{W}{T_1}, \qquad (11)$$

where W denotes the portion of the total vertical load between the points P_1 and P_2 shown in Figure 1.3.9. The DE (11) is separable under the following conditions that describe a suspension bridge.

Let us assume that the x- and y-axes are as shown in **FIGURE 2.2.7**—that is, the x-axis runs along the horizontal roadbed, and the y-axis passes through $(0, a)$, which is the lowest point on one cable over the span of the bridge, coinciding with the interval $[-L/2, L/2]$. In the case of a suspension bridge, the usual assumption is that the vertical load in (11) is only a uniform roadbed distributed along the horizontal axis. In other words, it is assumed that the weight of all cables is negligible in comparison to the weight of the roadbed and that the weight per unit length of the roadbed (say, pounds per horizontal foot) is a constant ρ. Use this information to set up and solve an appropriate initial-value problem from which the shape (a curve with equation $y = \phi(x)$) of each of the two cables in a suspension bridge is determined. Express your solution of the IVP in terms of the sag h and span L shown in Figure 2.2.7.

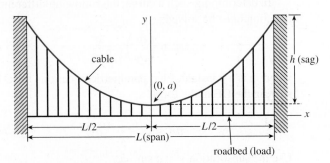

FIGURE 2.2.7 Shape of a cable in Problem 65

Computer Lab Assignments

66. (a) Use a CAS and the concept of level curves to plot representative graphs of members of the family of solutions of the differential equation $\dfrac{dy}{dx} = -\dfrac{8x + 5}{3y^2 + 1}$.

Experiment with different numbers of level curves as well as various rectangular regions defined by $a \le x \le b, c \le y \le d$.

(b) On separate coordinate axes plot the graphs of the particular solutions corresponding to the initial conditions: $y(0) = -1; y(0) = 2; y(-1) = 4; y(-1) = -3$.

67. (a) Find an implicit solution of the IVP

$$(2y + 2)dy - (4x^3 + 6x)dx = 0, \quad y(0) = -3.$$

(b) Use part (a) to find an explicit solution $y = \phi(x)$ of the IVP.

(c) Consider your answer to part (b) as a *function* only. Use a graphing utility or a CAS to graph this function, and then use the graph to estimate its domain.

(d) With the aid of a root-finding application of a CAS, determine the approximate largest interval I of definition of the *solution* $y = \phi(x)$ in part (b). Use a graphing utility or a CAS to graph the solution curve for the IVP on this interval.

68. (a) Use a CAS and the concept of level curves to plot representative graphs of members of the family of solutions of the differential equation $\dfrac{dy}{dx} = \dfrac{x(1 - x)}{y(-2 + y)}$.

Experiment with different numbers of level curves as well as various rectangular regions in the xy-plane until your result resembles **FIGURE 2.2.8**.

(b) On separate coordinate axes, plot the graph of the implicit solution corresponding to the initial condition $y(0) = \frac{3}{2}$. Use a colored pencil to mark off that segment of the graph that corresponds to the solution curve of a solution ϕ that satisfies the initial condition. With the aid of a root-finding application of a CAS, determine the approximate largest interval I of definition of the solution ϕ. [*Hint*: First find the points on the curve in part (a) where the tangent is vertical.]

(c) Repeat part (b) for the initial condition $y(0) = -2$.

FIGURE 2.2.8 Level curves in Problem 68

2.3 Linear Equations

INTRODUCTION We continue our search for solutions of first-order DEs by next examining linear equations. Linear differential equations are an especially "friendly" family of differential equations in that, given a linear equation, whether first-order or a higher-order kin, there is always a good possibility that we can find some sort of solution of the equation that we can look at.

■ **A Definition** The form of a linear first-order DE was given in (7) of Section 1.1. This form, the case when $n = 1$ in (6) of that section, is reproduced here for convenience.

DEFINITION 2.3.1 Linear Equation

A first-order differential equation of the form

$$a_1(x)\frac{dy}{dx} + a_0(x)y = g(x) \tag{1}$$

is said to be a **linear equation** in the dependent variable y.

When $g(x) = 0$, the linear equation (1) is said to be **homogeneous**; otherwise, it is **nonhomogeneous**.

■ **Standard Form** By dividing both sides of (1) by the lead coefficient $a_1(x)$ we obtain a more useful form, the **standard form**, of a linear first-order equation

$$\frac{dy}{dx} + P(x)y = f(x). \tag{2}$$

We seek a solution of (2) on an interval I for which both functions P and f are continuous.

In the discussion that follows, we illustrate a property and a procedure and end up with a formula representing the form that every solution of (2) must have. But more than the formula, the property and the procedure are important, because these two concepts carry over to linear equations of higher order.

■ **The Property** The differential equation (2) has the property that its solution is the sum of the two solutions, $y = y_c + y_p$, where y_c is a solution of the associated homogeneous equation

$$\frac{dy}{dx} + P(x)y = 0 \tag{3}$$

and y_p is a particular solution of the nonhomogeneous equation (2). To see this, observe

$$\frac{d}{dx}[y_c + y_p] + P(x)[y_c + y_p] = \underbrace{\left[\frac{dy_c}{dx} + P(x)y_c\right]}_{0} + \underbrace{\left[\frac{dy_p}{dx} + P(x)y_p\right]}_{f(x)} = f(x).$$

■ **The Homogeneous DE** The homogeneous equation (3) is also separable. This fact enables us to find y_c by writing (3) as

$$\frac{dy}{y} + P(x)\,dx = 0$$

and integrating. Solving for y gives $y_c = ce^{-\int P(x)\,dx}$. For convenience let us write $y_c = cy_1(x)$, where $y_1 = e^{-\int P(x)\,dx}$. The fact that $dy_1/dx + P(x)y_1 = 0$ will be used next to determine y_p.

■ **The Nonhomogeneous DE** We can now find a particular solution of equation (2) by a procedure known as **variation of parameters**. The basic idea here is to find a function u so that $y_p = u(x)y_1(x) = u(x)e^{-\int P(x)\,dx}$ is a solution of (2). In other words, our assumption for y_p is the same as $y_c = cy_1(x)$ except that c is replaced by the "variable parameter" u. Substituting $y_p = uy_1$ into (2) gives

$$\underset{\text{Product Rule}}{\underbrace{\downarrow}}$$

$$u\frac{dy_1}{dx} + y_1\frac{du}{dx} + P(x)uy_1 = f(x) \quad \text{or} \quad u\left[\frac{dy_1}{dx} + P(x)y_1\right] + y_1\frac{du}{dx} = f(x)$$

so that

$$y_1\frac{du}{dx} = f(x).$$

Separating variables and integrating then gives

$$du = \frac{f(x)}{y_1(x)}\,dx \quad \text{and} \quad u = \int \frac{f(x)}{y_1(x)}\,dx.$$

From the definition of $y_1(x)$, we see $1/y_1(x) = e^{\int P(x)\,dx}$. Therefore

$$y_p = uy_1 = \left(\int \frac{f(x)}{y_1(x)}\,dx\right)e^{-\int P(x)\,dx} = e^{-\int P(x)\,dx}\int e^{\int P(x)\,dx}f(x)\,dx,$$

and

$$y = y_c + y_p = ce^{-\int P(x)\,dx} + e^{-\int P(x)\,dx}\int e^{\int P(x)\,dx}f(x)\,dx. \tag{4}$$

Hence if (2) has a solution, it must be of form (4). Conversely, it is a straightforward exercise in differentiation to verify that (4) constitutes a one-parameter family of solutions of equation (2).

You should **not memorize** the formula given in (4). There is an equivalent but easier way of solving (2). If (4) is multiplied by

$$e^{\int P(x)\,dx} \tag{5}$$

and then

$$e^{\int P(x)\,dx}\,y = c + \int e^{\int P(x)\,dx}f(x)\,dx \tag{6}$$

is differentiated,

$$\frac{d}{dx}[e^{\int P(x)\,dx}y] = e^{\int P(x)\,dx}f(x), \tag{7}$$

we get

$$e^{\int P(x)\,dx}\frac{dy}{dx} + P(x)e^{\int P(x)\,dx}y = e^{\int P(x)\,dx}f(x). \tag{8}$$

Dividing the last result by $e^{\int P(x)\,dx}$ gives (2).

■ **Method of Solution** The recommended method of solving (2) actually consists of (6)–(8) worked in reverse order. In other words, if (2) is multiplied by (5), we get (8). The left side of (8) is recognized as the derivative of the product of $e^{\int P(x)\,dx}$ and y. This gets us to (7). We then integrate both sides of (7) to get the solution (6). Because we can solve (2) by integration after multiplication by $e^{\int P(x)\,dx}$, we call this function an **integrating factor** for the differential equation. For convenience we summarize these results. We again emphasize that you should not memorize formula (4) but work through the following two-step procedure each time.

Guidelines for Solving a Linear First-Order Equation

(*i*) Put a linear first-order equation of form (1) into standard form (2) and then determine $P(x)$ and the integrating factor $e^{\int P(x)\,dx}$.

(*ii*) Multiply (2) by the integrating factor. The left side of the resulting equation is automatically the derivative of the product of the integrating factor and y. Write

$$\frac{d}{dx}[e^{\int P(x)\,dx}y] = e^{\int P(x)\,dx}f(x)$$

and then integrate both sides of this equation.

EXAMPLE 1 Solving a Linear DE

Solve $\dfrac{dy}{dx} - 3y = 6$.

SOLUTION This linear equation can be solved by separation of variables. Alternatively, since the equation is already in the standard form (2), we see that the integrating factor is $e^{\int (-3)\,dx} = e^{-3x}$. We multiply the equation by this factor and recognize that

$$e^{-3x}\frac{dy}{dx} - 3e^{-3x}y = 6e^{-3x} \quad \text{is the same as} \quad \frac{d}{dx}\left[e^{-3x}y\right] = 6e^{-3x}.$$

Integrating both sides of the last equation,

$$\int \frac{d}{dx}\left[e^{-3x}y\right] = \int 6e^{-3x}\,dx$$

gives $e^{-3x}y = -2e^{-3x} + c$. Thus a solution of the differential equation is $y = -2 + ce^{3x}$, $-\infty < x < \infty$. ≡

When a_1, a_0, and g in (1) are constants, the differential equation is autonomous. In Example 1, you can verify from the normal form $dy/dx = 3(y + 2)$ that -2 is a critical point and that it is unstable and a repeller. Thus a solution curve with an initial point either above or below the graph of the equilibrium solution $y = -2$ pushes away from this horizontal line as x increases.

■ Constant of Integration

Notice in the general discussion and in Example 1 we disregarded a constant of integration in the evaluation of the indefinite integral in the exponent of $e^{\int P(x)\,dx}$. If you think about the laws of exponents and the fact that the integrating factor multiplies both sides of the differential equation, you should be able to answer why writing $\int P(x)\,dx + c$ is unnecessary. See Problem 55 in Exercises 2.3.

■ General Solution

Suppose again that the functions P and f in (2) are continuous on a common interval I. In the steps leading to (4) we showed that if (2) has a solution on I, then it must be of the form given in (4). Conversely, it is a straightforward exercise in differentiation to verify that any function of the form given in (4) is a solution of the differential equation (2) on I. In other words, (4) is a one-parameter family of solutions of equation (2), and *every* solution of (2) defined on I is a member of this family. Consequently, we are justified in calling (4) the **general solution** of the differential equation on the interval I. Now by writing (2) in the normal form $y' = F(x, y)$ we can identify $F(x, y) = -P(x)y + f(x)$ and $\partial F/\partial y = -P(x)$. From the continuity of P and f on the interval I, we see that F and $\partial F/\partial y$ are also continuous on I. With Theorem 1.2.1 as our justification, we conclude that there exists one and only one solution of the first-order initial-value problem

$$\frac{dy}{dx} + P(x)y = f(x), \quad y(x_0) = y_0 \tag{9}$$

defined on *some* interval I_0 containing x_0. But when x_0 is in I, finding a solution of (9) is just a matter of finding an appropriate value of c in (4); that is, for each x_0 in I there corresponds a distinct c. In other words, the interval I_0 of existence and uniqueness in Theorem 1.2.1 for the initial-value problem (9) is the entire interval I.

EXAMPLE 2 General Solution

Solve $x\dfrac{dy}{dx} - 4y = x^6 e^x$.

SOLUTION By dividing by x we get the standard form

$$\frac{dy}{dx} - \frac{4}{x}y = x^5 e^x. \tag{10}$$

From this form we identify $P(x) = -4/x$ and $f(x) = x^5 e^x$ and observe that P and f are continuous on the interval $(0, \infty)$. Hence the integrating factor is

we can use $\ln x$ instead of $\ln |x|$ since $x > 0$
$$\downarrow$$
$$e^{-4\int dx/x} = e^{-4\ln x} = e^{\ln x^{-4}} = x^{-4}.$$

Here we have used the basic identity $b^{\log_b N} = N$, $N > 0$. Now we multiply (10) by x^{-4},

$$x^{-4}\frac{dy}{dx} - 4x^{-5}y = xe^x, \quad \text{and obtain} \quad \frac{d}{dx}\left[x^{-4}y\right] = xe^x.$$

It follows from integration by parts that $\int xe^x dx = xe^x - e^x$ and so

$$\int \frac{d}{dx}[x^{-4}y] = \int xe^x dx$$

gives $x^{-4}y = xe^x - e^x + c$. Thus the general solution defined on $(0, \infty)$ is $y = x^5 e^x - x^4 e^x + cx^4$. $\equiv$

■ Singular Points Except in the case when the lead coefficient is 1, the recasting of equation (1) into the standard form (2) requires division by $a_1(x)$. Values of x for which $a_1(x) = 0$ are called **singular points** of the equation. Singular points are potentially troublesome. Specifically in (2), if $P(x)$ (formed by dividing $a_0(x)$ by $a_1(x)$) is discontinuous at a point, the discontinuity may carry over to functions in the general solution of the differential equation.

EXAMPLE 3 **General Solution**

Find the general solution of $(x^2 - 9)\dfrac{dy}{dx} + xy = 0$.

SOLUTION We write the differential equation in standard form

$$\frac{dy}{dx} + \frac{x}{x^2 - 9}y = 0 \tag{11}$$

and identify $P(x) = x/(x^2 - 9)$. Although P is continuous on $(-\infty, -3)$, on $(-3, 3)$, and on $(3, \infty)$, we shall solve the equation on the first and third intervals. On these intervals the integrating factor is

$$e^{\int x\,dx/(x^2-9)} = e^{\frac{1}{2}\int 2x\,dx/(x^2-9)} = e^{\frac{1}{2}\ln|x^2-9|} = \sqrt{x^2 - 9}.$$

After multiplying the standard form (11) by this factor, we get

$$\frac{d}{dx}\left[\sqrt{x^2 - 9}\,y\right] = 0 \quad \text{and integrating gives} \quad \sqrt{x^2 - 9}\,y = c.$$

Thus on either $(-\infty, -3)$ or $(3, \infty)$, the general solution of the equation is $y = c/\sqrt{x^2 - 9}$. $\equiv$

Notice in the preceding example that $x = 3$ and $x = -3$ are singular points of the equation and that every function in the general solution $y = c/\sqrt{x^2 - 9}$ is discontinuous at these points. On the other hand, $x = 0$ is a singular point of the differential equation in Example 2, but the general solution $y = x^5 e^x - x^4 e^x + cx^4$ is noteworthy in that every function in this one-parameter family is continuous at $x = 0$ and is defined on the interval $(-\infty, \infty)$ and not just on $(0, \infty)$ as stated in the solution. However, the family $y = x^5 e^x - x^4 e^x + cx^4$ defined on $(-\infty, \infty)$ cannot be considered the general solution of the DE, since the singular point $x = 0$ still causes a problem. See Problems 50 and 51 in Exercises 2.3. We will study singular points for linear differential equations in greater depth in Section 5.2.

EXAMPLE 4 **An Initial-Value Problem**

Solve the initial-value problem $\dfrac{dy}{dx} + y = x, \quad y(0) = 4.$

SOLUTION The equation is in standard form, and $P(x) = 1$ and $f(x) = x$ are continuous on the interval $(-\infty, \infty)$. The integrating factor is $e^{\int dx} = e^x$, and so integrating

$$\frac{d}{dx}[e^x y] = xe^x$$

gives $e^x y = xe^x - e^x + c$. Solving this last equation for y yields the general solution $y = x - 1 + ce^{-x}$. But from the initial condition we know that $y = 4$ when $x = 0$. Substituting these values in the general solution implies $c = 5$. Hence the solution of the problem on the interval $(-\infty, \infty)$ is

$$y = x - 1 + 5e^{-x}. \qquad (12) \equiv$$

FIGURE 2.3.1 Some solutions of the DE in Example 4

Recall that the general solution of every linear first-order differential equation is a sum of two special solutions: y_c, the general solution of the associated homogeneous equation (3), and y_p, a particular solution of the nonhomogeneous equation (2). In Example 4 we identify $y_c = ce^{-x}$ and $y_p = x - 1$. **FIGURE 2.3.1**, obtained with the aid of a graphing utility, shows (12) in blue along with other representative solutions in the family $y = x - 1 + ce^{-x}$. It is interesting to observe that as x gets large, the graphs of *all* members of the family are close to the graph of $y_p = x - 1$, which is shown in green in Figure 2.3.1. This is because the contribution of $y_c = ce^{-x}$ to the values of a solution becomes negligible for increasing values of x. We say that $y_c = ce^{-x}$ is a **transient term** since $y_c \to 0$ as $x \to \infty$. While this behavior is not a characteristic of all general solutions of linear equations (see Example 2), the notion of a transient is often important in applied problems.

▪ Piecewise-Linear Differential Equation

In the construction of mathematical models (especially in the biological sciences and engineering) it can happen that one or more coefficients in a differential equation is a piecewise-defined function. In particular, when either $P(x)$ or $f(x)$ in (2) is a piecewise-defined function the equation is then referred to as a **piecewise-linear differential equation**. In the next example, $f(x)$ is piecewise continuous on the interval $[0, \infty)$ with a single jump discontinuity at $x = 1$. The basic idea is to solve the initial-value problem in two parts corresponding to the two intervals over which $f(x)$ is defined; each part consists of a linear equation solvable by the method of this section. As we will see, it is then possible to piece the two solutions together at $x = 1$ so that $y(x)$ is continuous on $[0, \infty)$. See Problems 33–38 in Exercises 2.3.

EXAMPLE 5 **An Initial-Value Problem**

Solve $\dfrac{dy}{dx} + y = f(x)$, $y(0) = 0$ where $f(x) = \begin{cases} 1, & 0 \le x \le 1 \\ 0, & x > 1. \end{cases}$

SOLUTION The graph of the discontinuous function f is shown in **FIGURE 2.3.2**. We solve the DE for $y(x)$ first on the interval $[0, 1]$ and then on the interval $(1, \infty)$. For $0 \le x \le 1$ we have

$$\frac{dy}{dx} + y = 1 \quad \text{or, equivalently,} \quad \frac{d}{dx}[e^x y] = e^x.$$

Integrating this last equation and solving for y gives $y = 1 + c_1 e^{-x}$. Since $y(0) = 0$, we must have $c_1 = -1$, and therefore $y = 1 - e^{-x}$, $0 \le x \le 1$. Then for $x > 1$, the equation

$$\frac{dy}{dx} + y = 0$$

FIGURE 2.3.2 Discontinuous $f(x)$ in Example 5

leads to $y = c_2 e^{-x}$. Hence we can write

$$y = \begin{cases} 1 - e^{-x}, & 0 \le x \le 1 \\ c_2 e^{-x}, & x > 1. \end{cases}$$

By appealing to the definition of continuity at a point it is possible to determine c_2 so that the foregoing function is continuous at $x = 1$. The requirement that $\lim_{x \to 1^+} y(x) = y(1)$ implies that $c_2 e^{-1} = 1 - e^{-1}$ or $c_2 = e - 1$. As seen in **FIGURE 2.3.3**, the piecewise defined function

$$y = \begin{cases} 1 - e^{-x}, & 0 \le x \le 1 \\ (e - 1)e^{-x}, & x > 1 \end{cases} \qquad (13)$$

is continuous on the interval $[0, \infty)$. $\equiv$

FIGURE 2.3.3 Graph of function in (13) of Example 5

It is worthwhile to think about (13) and Figure 2.3.3 a little bit; you are urged to read and answer Problem 53 in Exercises 2.3.

■ Error Function

In mathematics, science, and engineering, some important functions are defined in terms of nonelementary integrals. Two such special functions are the **error function** and **complementary error function**:

See Appendix A. ▶

$$\operatorname{erf}(x) = \frac{2}{\sqrt{\pi}} \int_0^x e^{-t^2} dt \quad \text{and} \quad \operatorname{erfc}(x) = \frac{2}{\sqrt{\pi}} \int_x^\infty e^{-t^2} dt. \tag{14}$$

From the known result $\int_0^\infty e^{-t^2} dt = \sqrt{\pi}/2$, we can write $(2/\sqrt{\pi}) \int_0^\infty e^{-t^2} dt = 1$. Using the additive interval property of definite integrals $\int_0^\infty = \int_0^x + \int_x^\infty$ we can rewrite the last result in the alternative form

$$\frac{2}{\sqrt{\pi}} \int_0^\infty e^{-t^2} dt = \overbrace{\frac{2}{\sqrt{\pi}} \int_0^x e^{-t^2} dt}^{\operatorname{erf}(x)} + \overbrace{\frac{2}{\sqrt{\pi}} \int_x^\infty e^{-t^2} dt}^{\operatorname{erfc}(x)} = 1. \tag{15}$$

It is seen from (15) that the error function $\operatorname{erf}(x)$ and complementary error function $\operatorname{erfc}(x)$ are related by the identity

$$\operatorname{erf}(x) + \operatorname{erfc}(x) = 1. \tag{16}$$

Because of its importance in probability, statistics, and applied partial differential equations, the error function has been extensively tabulated. Note that $\operatorname{erf}(0) = 0$ is one obvious function value. Numerical values of $\operatorname{erf}(x)$ can also be found using a CAS such as *Mathematica*.

If we are solving an initial-value problem (9) and recognize that indefinite integration of the right-hand side of (7) would lead to a nonelementary integral, then as we saw in Example 6 of Section 2.2 it is convenient to use instead *definite* integration over the interval $[x_0, x]$. The last example illustrates that this procedure automatically incorporates the initial condition at x_0 into the solution of the DE, in other words, we do not have to solve for the constant c in its general solution.

EXAMPLE 6 The Error Function

Solve the initial-value problem $\dfrac{dy}{dx} - 2xy = 2, \quad y(0) = 1.$

SOLUTION The differential equation is already in standard form, and so we see that the integrating factor is $e^{\int(-2x\,dx)} = e^{-x^2}$. Multiplying both sides of the equation by this factor then gives $e^{-x^2}\dfrac{dy}{dx} - 2xe^{-x^2}y = 2e^{-x^2}$, which is the same as

$$\frac{d}{dx}[e^{-x^2}y] = 2e^{-x^2}. \tag{17}$$

Because indefinite integration of both sides of equation (17) leads to the nonelementary integral $\int e^{-x^2} dx$, we identify $x_0 = 0$ and use definite integration over the interval $[0, x]$:

$$\int_0^x \frac{d}{dt}[e^{-t^2}y(t)]\,dt = 2\int_0^x e^{-t^2}\,dt \quad \text{or} \quad e^{-x^2}y(x) - y(0) = 2\int_0^x e^{-t^2}\,dt.$$

Using the initial condition $y(0) = 1$ the last expression yields the solution

$$y = e^{x^2} + 2e^{x^2}\int_0^x e^{-t^2}\,dt. \tag{18}$$

Then by inserting the factor $\sqrt{\pi}/\sqrt{\pi}$ into this solution in the following manner:

$$y = e^{x^2} + 2e^{x^2}\int_0^x e^{-t^2}\,dt = e^{x^2}\left[1 + \sqrt{\pi}\left(\overbrace{\frac{2}{\sqrt{\pi}}\int_0^x e^{-t^2}\,dt}^{\operatorname{erf}(x)}\right)\right]$$

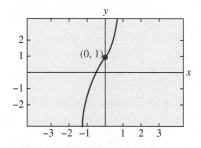

FIGURE 2.3.4 Graph of (19) in Example 6

we see from (14) that (18) can be rewritten in terms of the error function as

$$y = e^{x^2}[1 + \sqrt{\pi}\,\text{erf}(x)]. \tag{19}$$

The graph of solution (19), shown in **FIGURE 2.3.4**, was obtained with the aid of a CAS. ☰

See Problems 41–48 in Exercises 2.3.

Use of Computers Some computer algebra systems are capable of producing explicit solutions for some kinds of differential equations. For example, to solve the equation $y' + 2y = x$, we use the input commands

$$\textbf{DSolve[y'[x] + 2 y[x] == x, y[x], x]} \qquad \text{(in } \textit{Mathematica)}$$

and $\qquad$ **dsolve(diff(y(x), x) + 2*y(x) = x, y(x));** $\qquad$ (in *Maple*)

Translated into standard symbols, the output of each program is $y = -\frac{1}{4} + \frac{1}{2}x + ce^{-2x}$.

REMARKS

(*i*) Occasionally a first-order differential equation is not linear in one variable but is linear in the other variable. For example, the differential equation

$$\frac{dy}{dx} = \frac{1}{x + y^2}$$

is not linear in the variable y. But its reciprocal

$$\frac{dx}{dy} = x + y^2 \quad \text{or} \quad \frac{dx}{dy} - x = y^2$$

is recognized as linear in the variable x. You should verify that the integrating factor $e^{\int(-1)dy} = e^{-y}$ and integration by parts yield an implicit solution of the first equation: $x = -y^2 - 2y - 2 + ce^y$.

(*ii*) Because mathematicians thought they were appropriately descriptive, certain words were "adopted" from engineering and made their own. The word *transient*, used earlier, is one of these terms. In future discussions the words *input* and *output* will occasionally pop up. The function f in (2) is called the **input** or **driving function**; a solution of the differential equation for a given input is called the **output** or **response**.

(*iii*) The term *special functions* mentioned in conjunction with the error function on page 58 also applies to the sine integral function and the Fresnel sine integral function introduced in Problems 47 and 48 in Exercises 2.3. "Special Functions" is actually a well-defined field of study in mathematics. The Fresnel sine integral function is one of two such special integral functions that the French engineer/physicist **Augustin-Jean Fresnel** (1788–1827) encountered in his study of optics. More special functions are studied in Section 5.3.

2.3 Exercises Answers to selected odd-numbered problems begin on page ANS-2.

In Problems 1–24, find the general solution of the given differential equation. Give the largest interval over which the general solution is defined. Determine whether there are any transient terms in the general solution.

1. $\dfrac{dy}{dx} = 5y$

2. $\dfrac{dy}{dx} + 2y = 0$

3. $\dfrac{dy}{dx} + y = e^{3x}$

4. $3\dfrac{dy}{dx} + 12y = 4$

5. $y' + 3x^2y = x^2$

6. $y' + 2xy = x^3$

7. $x^2y' + xy = 1$

8. $y' = 2y + x^2 + 5$

9. $x\dfrac{dy}{dx} - y = x^2\sin x$

10. $x\dfrac{dy}{dx} + 2y = 3$

11. $x\dfrac{dy}{dx} + 4y = x^3 - x$

12. $(1 + x)\dfrac{dy}{dx} - xy = x + x^2$

13. $x^2y' + x(x + 2)y = e^x$

14. $xy' + (1 + x)y = e^{-x}\sin 2x$

15. $y\,dx - 4(x + y^6)\,dy = 0$

16. $y\,dx = (ye^y - 2x)\,dy$

17. $\cos x\dfrac{dy}{dx} + (\sin x)y = 1$

18. $\cos^2 x \sin x \dfrac{dy}{dx} + (\cos^3 x)y = 1$

19. $(x + 1)\dfrac{dy}{dx} + (x + 2)y = 2xe^{-x}$

20. $(x + 2)^2 \dfrac{dy}{dx} = 5 - 8y - 4xy$

21. $\dfrac{dr}{d\theta} + r\sec\theta = \cos\theta$

22. $\dfrac{dP}{dt} + 2tP = P + 4t - 2$

23. $x\dfrac{dy}{dx} + (3x + 1)y = e^{-3x}$

24. $(x^2 - 1)\dfrac{dy}{dx} + 2y = (x + 1)^2$

In Problems 25–32, solve the given initial-value problem. Give the largest interval I over which the solution is defined.

25. $xy' + y = e^x, \quad y(1) = 2$

26. $y\dfrac{dx}{dy} - x = 2y^2, \quad y(1) = 5$

27. $L\dfrac{di}{dt} + Ri = E; \quad i(0) = i_0, L, R, E,$ and i_0 constants

28. $\dfrac{dT}{dt} = k(T - T_m); \quad T(0) = T_0, K, T_m,$ and T_0 constants

29. $(x + 1)\dfrac{dy}{dx} + y = \ln x, \quad y(1) = 10$

30. $y' + (\tan x)y = \cos^2 x, \quad y(0) = -1$

31. $\left(\dfrac{e^{-2\sqrt{x}} - y}{\sqrt{x}}\right)\dfrac{dx}{dy} = 1, \quad y(1) = 1$

32. $(1 + t^2)\dfrac{dx}{dt} + x = \tan^{-1}t, \quad x(0) = 4$

[*Hint*: In your solution let $u = \tan^{-1}t$.]

In Problems 33–36, proceed as in Example 5 to solve the given initial-value problem. Use a graphing utility to graph the continuous function $y(x)$.

33. $\dfrac{dy}{dx} + 2y = f(x), \quad y(0) = 0,$ where

$$f(x) = \begin{cases} 1, & 0 \le x \le 3 \\ 0, & x > 3 \end{cases}$$

34. $\dfrac{dy}{dx} + y = f(x), \quad y(0) = 1,$ where

$$f(x) = \begin{cases} 1, & 0 \le x \le 1 \\ -1, & x > 1 \end{cases}$$

35. $\dfrac{dy}{dx} + 2xy = f(x), \quad y(0) = 2,$ where

$$f(x) = \begin{cases} x, & 0 \le x < 1 \\ 0, & x \ge 1 \end{cases}$$

36. $(1 + x^2)\dfrac{dy}{dx} + 2xy = f(x), \quad y(0) = 0,$ where

$$f(x) = \begin{cases} x, & 0 \le x < 1 \\ -x, & x \ge 1 \end{cases}$$

In Problems 37 and 38, proceed as in Example 5 to solve the given initial-value problem. Use a graphing utility to graph the continuous function $y(x)$.

37. $\dfrac{dy}{dx} + P(x)y = 4x, \quad y(0) = 3,$ where

$$P(x) = \begin{cases} 2, & 0 \le x \le 1 \\ -\dfrac{2}{x}, & x > 1 \end{cases}$$

38. $\dfrac{dy}{dx} + P(x)y = 0, \quad y(0) = 4,$ where

$$P(x) = \begin{cases} 1, & 0 \le x \le 2 \\ 5, & x > 2 \end{cases}$$

In Problems 39 and 40, the given differential equation is not linear in y. Nevertheless, find a general solution of the equation.

39. $dx = (x + y^2)\,dy$

40. $y\,dx + (2x + xy - 3)\,dy = 0$

In Problems 41 and 42, proceed as in Example 6 and express the solution of the given initial-value problem in terms of erf(x) (Problem 41) and erfc(x) (Problem 42).

41. $\dfrac{dy}{dx} - 2xy = 1, \quad y(1) = 1$

42. $\dfrac{dy}{dx} - 2xy = 1, \quad y(0) = \sqrt{\pi}/2$

In Problems 43–46, proceed as in Example 6 and express the solution of the given initial-value problem in terms of an integral-defined function.

43. $\dfrac{dy}{dx} + e^x y = 1, \quad y(0) = 1$

44. $x^2 \dfrac{dy}{dx} - y = x^3, \quad y(1) = 0$

45. $\dfrac{dy}{dx} + xy = \cos x, \quad y(0) = 2$

46. $\dfrac{dy}{dx} + (\sin x)y = 10x, \quad y(0) = 0$

47. The **sine integral function** is defined as

$$\text{Si}(x) = \int_0^x \dfrac{\sin t}{t}\,dt,$$

where the integrand is defined to be 1 at $x = 0$. Express the solution of the initial-value problem

$$x^3 \dfrac{dy}{dx} + 2x^2 y = 10\sin x, \quad y(1) = 0$$

in terms of Si(x).

48. The **Fresnel sine integral function** is defined as

$$S(x) = \int_0^x \sin\left(\frac{\pi}{2}t^2\right) dt.$$

Express the solution of the initial-value problem

$$\frac{dy}{dx} - (\sin x^2)y = 0, \quad y(0) = 5$$

in terms of $S(x)$.

Discussion Problems

49. Reread the discussion following Example 1. Construct a linear first-order differential equation for which all nonconstant solutions approach the horizontal asymptote $y = 4$ as $x \to \infty$.

50. Reread Example 2 and then discuss, with reference to Theorem 1.2.1, the existence and uniqueness of a solution of the initial-value problem consisting of $xy' - 4y = x^6 e^x$ and the given initial condition.
 (a) $y(0) = 0$
 (b) $y(0) = y_0, \quad y_0 > 0$
 (c) $y(x_0) = y_0, \quad x_0 > 0, \quad y_0 > 0$

51. Reread Example 3 and then find the general solution of the differential equation on the interval $(-3, 3)$.

52. Reread the discussion following Example 4. Construct a linear first-order differential equation for which all solutions are asymptotic to the line $y = 3x - 5$ as $x \to \infty$.

53. Reread Example 5 and then discuss why it is technically incorrect to say that the function in (13) is a *solution* of the IVP on the interval $[0, \infty)$.

54. (a) Construct a linear first-order differential equation of the form $xy' + a_0(x)y = g(x)$ for which $y_c = c/x^3$ and $y_p = x^3$. Give an interval on which $y = x^3 + c/x^3$ is the general solution of the DE.
 (b) Give an initial condition $y(x_0) = y_0$ for the DE found in part (a) so that the solution of the IVP is $y = x^3 - 1/x^3$. Repeat if the solution is $y = x^3 + 2/x^3$. Give an interval I of definition of each of these solutions. Graph the solution curves. Is there an initial-value problem whose solution is defined on the interval $(-\infty, \infty)$?
 (c) Is each IVP found in part (b) unique? That is, can there be more than one IVP for which, say, $y = x^3 - 1/x^3$, x in some interval I is the solution?

55. In determining the integrating factor (5), there is no need to use a constant of integration in the evaluation of $\int P(x)\,dx$. Explain why using $\int P(x)\,dx + c$ has no effect on the solution of (2).

Mathematical Models

56. Radioactive Decay Series The following system of differential equations is encountered in the study of the decay of a special type of radioactive series of elements:

$$\frac{dx}{dt} = -\lambda_1 x,$$

$$\frac{dy}{dt} = \lambda_1 x - \lambda_2 y,$$

where λ_1 and λ_2 are constants. Discuss how to solve this system subject to $x(0) = x_0$, $y(0) = y_0$. Carry out your ideas.

57. Heart Pacemaker A heart pacemaker consists of a switch, a battery of constant voltage E_0, a capacitor with constant capacitance C, and the heart as a resistor with constant resistance R. When the switch is closed, the capacitor charges; when the switch is open, the capacitor discharges, sending an electrical stimulus to the heart. During the time the heart is being stimulated, the voltage E across the heart satisfies the linear differential equation

$$\frac{dE}{dt} = -\frac{1}{RC}E.$$

Solve the DE subject to $E(4) = E_0$.

Computer Lab Assignments

58. (a) Use a CAS to graph the solution curve of the initial-value problem in Problem 42 on the interval $(-\infty, \infty)$.
 (b) Use tables or a CAS to find the value $y(2)$.

59. (a) Use a CAS to graph the solution curve of the initial-value problem in Problem 47 on the interval $[0, \infty)$.
 (b) Use a CAS to find the value of the absolute maximum of the solution $y(x)$ on the interval.

60. (a) Use a CAS to graph the solution curve of the initial-value problem in Problem 48 on the interval $(-\infty, \infty)$.
 (b) It is known that Fresnel sine integral $S(x) \to \frac{1}{2}$ as $x \to \infty$ and $S(x) \to -\frac{1}{2}$ as $x \to -\infty$. What does the solution $y(x)$ approach as $x \to \infty$? As $x \to -\infty$?
 (c) Use a CAS to find the values of the absolute maximum and the absolute minimum of the solution $y(x)$ on the interval.

2.4 Exact Equations

INTRODUCTION Although the simple differential equation $y\,dx + x\,dy = 0$ is separable, we can solve it in an alternative manner by recognizing that the left-hand side is equivalent to the differential of the product of x and y; that is, $y\,dx + x\,dy = d(xy)$. By integrating both sides of the equation we immediately obtain the implicit solution $xy = c$.

Differential of a Function of Two Variables If $z = f(x, y)$ is a function of two variables with continuous first partial derivatives in a region R of the xy-plane, then its **differential** (also called the total differential) is

$$dz = \frac{\partial f}{\partial x} dx + \frac{\partial f}{\partial y} dy.$$ (1)

Now if $f(x, y) = c$, it follows from (1) that

$$\frac{\partial f}{\partial x} dx + \frac{\partial f}{\partial y} dy = 0.$$ (2)

In other words, given a one-parameter family of curves $f(x, y) = c$, we can generate a first-order differential equation by computing the differential. For example, if $x^2 - 5xy + y^3 = c$, then (2) gives

$$(2x - 5y)\, dx + (-5x + 3y^2)\, dy = 0.$$ (3)

For our purposes it is more important to turn the problem around; namely, given a first-order DE such as (3), can we recognize that it is equivalent to the differential $d(x^2 - 5xy + y^3) = 0$?

DEFINITION 2.4.1 Exact Equation

A differential expression $M(x, y)\, dx + N(x, y)\, dy$ is an **exact differential** in a region R of the xy-plane if it corresponds to the differential of some function $f(x, y)$. A first-order differential equation of the form

$$M(x, y)\, dx + N(x, y)\, dy = 0$$

is said to be an **exact equation** if the expression on the left side is an exact differential.

For example, the equation $x^2y^3\, dx + x^3y^2\, dy = 0$ is exact, because the left side is $d(\frac{1}{3}x^3y^3) = x^2y^3\, dx + x^3y^2\, dy$. Notice that if $M(x, y) = x^2y^3$ and $N(x, y) = x^3y^2$, then $\partial M/\partial y = 3x^2y^2 = \partial N/\partial x$. Theorem 2.4.1 shows that the equality of these partial derivatives is no coincidence.

THEOREM 2.4.1 Criterion for an Exact Differential

Let $M(x, y)$ and $N(x, y)$ be continuous and have continuous first partial derivatives in a rectangular region R defined by $a < x < b, c < y < d$. Then a necessary and sufficient condition that $M(x, y)\, dx + N(x, y)\, dy$ be an exact differential is

$$\frac{\partial M}{\partial y} = \frac{\partial N}{\partial x}.$$ (4)

PROOF: (Proof of the Necessity) For simplicity let us assume that $M(x, y)$ and $N(x, y)$ have continuous first partial derivatives for all (x, y). Now if the expression $M(x, y)\, dx + N(x, y)\, dy$ is exact, there exists some function f such that for all x in R,

$$M(x, y)\, dx + N(x, y)\, dy = \frac{\partial f}{\partial x} dx + \frac{\partial f}{\partial y} dy.$$

Therefore,

$$M(x, y) = \frac{\partial f}{\partial x}, \quad N(x, y) = \frac{\partial f}{\partial y},$$

and

$$\frac{\partial M}{\partial y} = \frac{\partial}{\partial y}\left(\frac{\partial f}{\partial x}\right) = \frac{\partial^2 f}{\partial y\, \partial x} = \frac{\partial}{\partial x}\left(\frac{\partial f}{\partial y}\right) = \frac{\partial N}{\partial x}.$$

The equality of the mixed partials is a consequence of the continuity of the first partial derivatives of $M(x, y)$ and $N(x, y)$. ≡

The sufficiency part of Theorem 2.4.1 consists of showing that there exists a function f for which $\partial f/\partial x = M(x, y)$ and $\partial f/\partial y = N(x, y)$ whenever (4) holds. The construction of the function f actually reflects a basic procedure for solving exact equations.

■ **Method of Solution** Given an equation of the form $M(x, y)\,dx + N(x, y)\,dy = 0$, determine whether the equality in (4) holds. If it does, then there exists a function f for which

$$\frac{\partial f}{\partial x} = M(x, y).$$

We can find f by integrating $M(x, y)$ with respect to x, while holding y constant:

$$f(x, y) = \int M(x, y)\,dx + g(y), \tag{5}$$

where the arbitrary function $g(y)$ is the "constant" of integration. Now differentiate (5) with respect to y and assume $\partial f/\partial y = N(x, y)$:

$$\frac{\partial f}{\partial y} = \frac{\partial}{\partial y}\int M(x, y)\,dx + g'(y) = N(x, y).$$

This gives

$$g'(y) = N(x, y) - \frac{\partial}{\partial y}\int M(x, y)\,dx. \tag{6}$$

Finally, integrate (6) with respect to y and substitute the result in (5). The implicit solution of the equation is $f(x, y) = c$.

Some observations are in order. First, it is important to realize that the expression $N(x, y) - (\partial/\partial y)\int M(x, y)\,dx$ in (6) is independent of x, because

$$\frac{\partial}{\partial x}\left[N(x, y) - \frac{\partial}{\partial y}\int M(x, y)\,dx\right] = \frac{\partial N}{\partial x} - \frac{\partial}{\partial y}\left(\frac{\partial}{\partial x}\int M(x, y)\,dx\right) = \frac{\partial N}{\partial x} - \frac{\partial M}{\partial y} = 0.$$

Second, we could just as well start the foregoing procedure with the assumption that $\partial f/\partial y = N(x, y)$. After integrating N with respect to y and then differentiating that result, we would find the analogues of (5) and (6) to be, respectively,

$$f(x, y) = \int N(x, y)\,dy + h(x) \quad \text{and} \quad h'(x) = M(x, y) - \frac{\partial}{\partial x}\int N(x, y)\,dy.$$

If you find that integration of $\partial f/\partial x = M(x, y)$ with respect to x is difficult, then try integrating $\partial f/\partial y = N(x, y)$ with respect to y. In either case *none of these formulas should be memorized.*

■ **EXAMPLE 1** **Solving an Exact DE**

Solve $2xy\,dx + (x^2 - 1)\,dy = 0$.

SOLUTION With $M(x, y) = 2xy$ and $N(x, y) = x^2 - 1$ we have

$$\frac{\partial M}{\partial y} = 2x = \frac{\partial N}{\partial x}.$$

Thus the equation is exact, and so, by Theorem 2.4.1, there exists a function $f(x, y)$ such that

$$\frac{\partial f}{\partial x} = 2xy \quad \text{and} \quad \frac{\partial f}{\partial y} = x^2 - 1.$$

From the first of these equations we obtain, after integrating,

$$f(x, y) = x^2 y + g(y).$$

Taking the partial derivative of the last expression with respect to y and setting the result equal to $N(x, y)$ gives

$$\frac{\partial f}{\partial y} = x^2 + g'(y) = x^2 - 1. \leftarrow N(x, y)$$

It follows that $\qquad g'(y) = -1 \quad$ and $\quad g(y) = -y.$

Hence, $f(x, y) = x^2 y - y$, and so the solution of the equation in implicit form is $x^2 y - y = c$. The explicit form of the solution is easily seen to be $y = c/(x^2 - 1)$ and is defined on any interval not containing either $x = 1$ or $x = -1$. ≡

The solution of the DE in Example 1 is *not* $f(x, y) = x^2 y - y$. Rather it is $f(x, y) = c$, or if a constant is used in the integration of $g'(y)$, we can then write the solution as $f(x, y) = 0$. Note, too, that the equation could be solved by separation of variables.

> Note the form of the solution. It is $f(x, y) = c$.

EXAMPLE 2 Solving an Exact DE

Solve $\quad (e^{2y} - y \cos xy)\, dx + (2xe^{2y} - x \cos xy + 2y)\, dy = 0.$

SOLUTION The equation is exact because

$$\frac{\partial M}{\partial y} = 2e^{2y} + xy\sin xy - \cos xy = \frac{\partial N}{\partial x}.$$

Hence a function $f(x, y)$ exists for which

$$M(x, y) = \frac{\partial f}{\partial x} \quad \text{and} \quad N(x, y) = \frac{\partial f}{\partial y}.$$

Now for variety we shall start with the assumption that $\partial f/\partial y = N(x, y)$;

that is, $\qquad\qquad \dfrac{\partial f}{\partial y} = 2xe^{2y} - x\cos xy + 2y$

$$f(x, y) = 2x \int e^{2y}\, dy - x \int \cos xy\, dy + 2 \int y\, dy + h(x)$$

Remember, the reason x can come out in front of the symbol $\int$ is that in the integration with respect to y, x is treated as an ordinary constant. It follows that

$$f(x, y) = xe^{2y} - \sin xy + y^2 + h(x)$$

$$\frac{\partial f}{\partial x} = e^{2y} - y\cos xy + h'(x) = e^{2y} - y\cos xy \leftarrow M(x, y)$$

and so $h'(x) = 0$ or $h(x) = c$. A family of solutions is $xe^{2y} - \sin xy + y^2 + c = 0$. ≡

EXAMPLE 3 An Initial-Value Problem

Solve the initial-value problem $\quad \dfrac{dy}{dx} = \dfrac{xy^2 - \cos x \sin x}{y(1 - x^2)}, \quad y(0) = 2.$

SOLUTION By writing the differential equation in the form

$$(\cos x \sin x - xy^2)\, dx + y(1 - x^2)\, dy = 0$$

we recognize that the equation is exact because

$$\frac{\partial M}{\partial y} = -2xy = \frac{\partial N}{\partial x}.$$

Now $\qquad\qquad \dfrac{\partial f}{\partial y} = y(1 - x^2)$

$$f(x, y) = \frac{y^2}{2}(1 - x^2) + h(x)$$

$$\frac{\partial f}{\partial x} = -xy^2 + h'(x) = \cos x \sin x - xy^2.$$

The last equation implies that $h'(x) = \cos x \sin x$. Integrating gives

$$h(x) = -\int (\cos x)(-\sin x\, dx) = -\frac{1}{2}\cos^2 x.$$

Thus $\qquad \dfrac{y^2}{2}(1 - x^2) - \dfrac{1}{2}\cos^2 x = c_1 \quad$ or $\quad y^2(1 - x^2) - \cos^2 x = c,$ (7)

where $2c_1$ has been replaced by c. The initial condition $y = 2$ when $x = 0$ demands that $4(1) - \cos^2(0) = c$ and so $c = 3$. An implicit solution of the problem is then $y^2(1 - x^2) - \cos^2 x = 3.$ ≡

FIGURE 2.4.1 Some solution curves in the family (7)

The solution curve of the IVP in Example 3 is part of an interesting family of curves and is the curve drawn in blue in **FIGURE 2.4.1**. The graphs of the members of the one-parameter family of solutions given in (7) can be obtained in several ways, two of which are using software to graph level curves as discussed in the last section, or using a graphing utility and carefully graphing the explicit functions obtained for various values of c by solving $y^2 = (c + \cos^2 x)/(1 - x^2)$ for y.

■ **Integrating Factors** Recall from the last section that the left-hand side of the linear equation $y' + P(x)y = f(x)$ can be transformed into a derivative when we multiply the equation by an integrating factor. The same basic idea sometimes works for a nonexact differential equation $M(x, y)\, dx + N(x, y)\, dy = 0$. That is, it is sometimes possible to find an **integrating factor** $\mu(x, y)$ so that after multiplying, the left-hand side of

$$\mu(x, y)\, M(x, y)\, dx + \mu(x, y)\, N(x, y)\, dy = 0 \tag{8}$$

is an exact differential. In an attempt to find μ we turn to the criterion (4) for exactness. Equation (8) is exact if and only if $(\mu M)_y = (\mu N)_x$, where the subscripts denote partial derivatives. By the Product Rule of differentiation the last equation is the same as $\mu M_y + \mu_y M = \mu N_x + \mu_x N$ or

$$\mu_x N - \mu_y M = (M_y - N_x)\mu. \tag{9}$$

Although M, N, M_y, N_x are known functions of x and y, the difficulty here in determining the unknown $\mu(x, y)$ from (9) is that we must solve a partial differential equation. Since we are not prepared to do that we make a simplifying assumption. Suppose μ is a function of one variable; say that μ depends only upon x. In this case $\mu_x = du/dx$ and (9) can be written as

$$\frac{d\mu}{dx} = \frac{M_y - N_x}{N}\mu. \tag{10}$$

We are still at an impasse if the quotient $(M_y - N_x)/N$ depends upon both x and y. However, if after all obvious algebraic simplifications are made, the quotient $(M_y - N_x)/N$ turns out to depend solely on the variable x, then (10) is a first-order ordinary differential equation. We can finally determine μ because (10) is *separable* as well as *linear*. It follows from either Section 2.2 or Section 2.3 that $\mu(x) = e^{\int ((M_y - N_x)/N)\, dx}$. In like manner it follows from (9) that if μ depends only on the variable y, then

$$\frac{d\mu}{dy} = \frac{N_x - M_y}{M}\mu. \tag{11}$$

In this case, if $(N_x - M_y)/M$ is a function of y, only then we can solve (11) for μ.

We summarize the results for the differential equation

$$M(x, y)\, dx + N(x, y)\, dy = 0. \tag{12}$$

• If $(M_y - N_x)/N$ is a function of x alone, then an integrating factor for equation (12) is

$$\mu(x) = e^{\int \frac{M_y - N_x}{N}\, dx}. \tag{13}$$

- If $(N_x - M_y)/M$ is a function of y alone, then an integrating factor for equation (12) is

$$\mu(y) = e^{\int \frac{N_x - M_y}{M} dy}. \tag{14}$$

A Nonexact DE Made Exact

The nonlinear first-order differential equation $xy\, dx + (2x^2 + 3y^2 - 20)\, dy = 0$ is not exact. With the identifications $M = xy$, $N = 2x^2 + 3y^2 - 20$ we find the partial derivatives $M_y = x$ and $N_x = 4x$. The first quotient from (13) gets us nowhere since

$$\frac{M_y - N_x}{N} = \frac{x - 4x}{2x^2 + 3y^2 - 20} = \frac{-3x}{2x^2 + 3y^2 - 20}$$

depends on x and y. However (14) yields a quotient that depends only on y:

$$\frac{N_x - M_y}{M} = \frac{4x - x}{xy} = \frac{3x}{xy} = \frac{3}{y}.$$

The integrating factor is then $e^{\int 3\, dy/y} = e^{3\ln y} = e^{\ln y^3} = y^3$. After multiplying the given DE by $\mu(y) = y^3$ the resulting equation is

$$xy^4\, dx + (2x^2y^3 + 3y^5 - 20y^3)\, dy = 0.$$

You should verify that the last equation is now exact as well as show, using the method of this section, that a family of solutions is $\frac{1}{2}x^2y^4 + \frac{1}{2}y^6 - 5y^4 = c$. ≡

REMARKS

(*i*) When testing an equation for exactness, make sure it is of the precise form $M(x, y)\, dx + N(x, y)\, dy = 0$. Sometimes a differential equation is written $G(x, y)\, dx = H(x, y)\, dy$. In this case, first rewrite it as $G(x, y)\, dx - H(x, y)\, dy = 0$, and then identify $M(x, y) = G(x, y)$ and $N(x, y) = -H(x, y)$ before using (4).

(*ii*) In some texts on differential equations the study of exact equations precedes that of linear DEs. If this were so, the method for finding integrating factors just discussed can be used to derive an integrating factor for $y' + P(x)y = f(x)$. By rewriting the last equation in the differential form $(P(x)y - f(x))\, dx + dy = 0$ we see that

$$\frac{M_y - N_x}{N} = P(x).$$

From (13) we arrive at the already familiar integrating factor $e^{\int P(x)\, dx}$ used in Section 2.3.

2.4 Exercises Answers to selected odd-numbered problems begin on page ANS-2.

In Problems 1–20, determine whether the given differential equation is exact. If it is exact, solve it.

1. $(2x - 1)\, dx + (3y + 7)\, dy = 0$

2. $(2x + y)\, dx - (x + 6y)\, dy = 0$

3. $(5x + 4y)\, dx + (4x - 8y^3)\, dy = 0$

4. $(\sin y - y \sin x)\, dx + (\cos x + x \cos y - y)\, dy = 0$

5. $(2xy^2 - 3)\, dx + (2x^2y + 4)\, dy = 0$

6. $\left(2y - \dfrac{1}{x} + \cos 3x\right)\dfrac{dy}{dx} + \dfrac{y}{x^2} - 4x^3 + 3y \sin 3x = 0$

7. $(x^2 - y^2)\, dx + (x^2 - 2xy)\, dy = 0$

8. $\left(1 + \ln x + \dfrac{y}{x}\right) dx = (1 - \ln x)\, dy$

9. $(x - y^3 + y^2 \sin x)\, dx = (3xy^2 + 2y \cos x)\, dy$

10. $(x^3 + y^3)\, dx + 3xy^2\, dy = 0$

11. $(y \ln y - e^{-xy})\, dx + \left(\dfrac{1}{y} + x \ln y\right) dy = 0$

12. $(3x^2y + e^y)\, dx + (x^3 + xe^y - 2y)\, dy = 0$

13. $x\dfrac{dy}{dx} = 2xe^x - y + 6x^2$

14. $\left(1 - \dfrac{3}{y} + x\right)\dfrac{dy}{dx} + y = \dfrac{3}{x} - 1$

15. $\left(x^2 y^3 - \dfrac{1}{1 + 9x^2}\right)\dfrac{dx}{dy} + x^3 y^2 = 0$

16. $(5y - 2x)y' - 2y = 0$

17. $(\tan x - \sin x \sin y)\, dx + \cos x \cos y\, dy = 0$

18. $(2y \sin x \cos x - y + 2y^2 e^{xy^2})\, dx = (x - \sin^2 x - 4xy e^{xy^2})\, dy$

19. $(4t^3 y - 15t^2 - y)\, dt + (t^4 + 3y^2 - t)\, dy = 0$

20. $\left(\dfrac{1}{t} + \dfrac{1}{t^2} - \dfrac{y}{t^2 + y^2}\right)dt + \left(ye^y + \dfrac{1}{t^2 + y^2}\right)dy = 0$

In Problems 21–26, solve the given initial-value problem.

21. $(x + y)^2\, dx + (2xy + x^2 - 1)\, dy = 0, \quad y(1) = 1$

22. $(e^x + y)\, dx + (2 + x + ye^y)\, dy = 0, \quad y(0) = 1$

23. $(4y + 2t - 5)\, dt + (6y + 4t - 1)\, dy = 0, \quad y(-1) = 2$

24. $\left(\dfrac{3y^2 - t^2}{y^5}\right)\dfrac{dy}{dt} + \dfrac{t}{2y^4} = 0, \quad y(1) = 1$

25. $(y^2 \cos x - 3x^2 y - 2x)\, dx + (2y \sin x - x^3 + \ln y)\, dy = 0,$
$y(0) = e$

26. $\left(\dfrac{1}{1 + y^2} + \cos x - 2xy\right)\dfrac{dy}{dx} = y(y + \sin x), \quad y(0) = 1$

In Problems 27 and 28, find the value of k so that the given differential equation is exact.

27. $(y^3 + kxy^4 - 2x)\, dx + (3xy^2 + 20x^2 y^3)\, dy = 0$

28. $(6xy^3 + \cos y)\, dx + (2kx^2 y^2 - x \sin y)\, dy = 0$

In Problems 29–32, verify that the given differential equation is not exact. Multiply the given differential equation by the indicated integrating factor $\mu(x, y)$ and verify that the new equation is exact. Solve.

29. $(-xy \sin x + 2y \cos x)\, dx + 2x \cos x\, dy = 0; \quad \mu(x, y) = xy$

30. $(x^2 + 2xy - y^2)\, dx + (y^2 + 2xy - x^2)\, dy = 0;$
$\mu(x, y) = (x + y)^{-2}$

31. $(x^3 y^2 + y)\, dx + x\, dy = 0; \quad \mu(x, y) = 1/x^2 y^2$

32. $x\, dx + (y - x^2 - y^2)\, dy = 0; \quad \mu(x, y) = 1/(x^2 + y^2)$

In Problems 33–38, solve the given differential equation by finding, as in Example 4, an appropriate integrating factor.

33. $(2y^2 + 3x)\, dx + 2xy\, dy = 0$

34. $y(x + y + 1)\, dx + (x + 2y)\, dy = 0$

35. $6xy\, dx + (4y + 9x^2)\, dy = 0$

36. $\cos x\, dx + \left(1 + \dfrac{2}{y}\right)\sin x\, dy = 0$

37. $(10 - 6y + e^{-3x})\, dx - 2\, dy = 0$

38. $(y^2 + xy^3)\, dx + (5y^2 - xy + y^3 \sin y)\, dy = 0$

In Problems 39 and 40, solve the given initial-value problem by finding, as in Example 4, an appropriate integrating factor.

39. $x\, dx + (x^2 y + 4y)\, dy = 0, \quad y(4) = 0$

40. $(x^2 + y^2 - 5)\, dx = (y + xy)\, dy, \quad y(0) = 1$

41. (a) Show that a one-parameter family of solutions of the equation

$$(4xy + 3x^2)\, dx + (2y + 2x^2)\, dy = 0$$

is $x^3 + 2x^2 y + y^2 = c$.

(b) Show that the initial conditions $y(0) = -2$ and $y(1) = 1$ determine the same implicit solution.

(c) Find explicit solutions $y_1(x)$ and $y_2(x)$ of the differential equation in part (a) such that $y_1(0) = -2$ and $y_2(1) = 1$. Use a graphing utility to graph $y_1(x)$ and $y_2(x)$.

Discussion Problems

42. Consider the concept of an integrating factor used in Problems 29–40. Are the two equations

$$M\, dx + N\, dy = 0 \quad \text{and} \quad \mu M\, dx + \mu N\, dy = 0$$

necessarily equivalent in the sense that a solution of one is also a solution of the other? Discuss.

43. Reread Example 3 and then discuss why we can conclude that the interval of definition of the explicit solution of the IVP (the blue curve in Figure 2.4.1) is $(-1, 1)$.

44. Discuss how the functions $M(x, y)$ and $N(x, y)$ can be found so that each differential equation is exact. Carry out your ideas.

(a) $M(x, y)\, dx + \left(xe^{xy} + 2xy + \dfrac{1}{x}\right)dy = 0$

(b) $\left(x^{-1/2} y^{1/2} + \dfrac{x}{x^2 + y}\right)dx + N(x, y)\, dy = 0$

45. Differential equations are sometimes solved by having a clever idea. Here is a little exercise in cleverness: Although the differential equation

$$(x - \sqrt{x^2 + y^2})\, dx + y\, dy = 0$$

is not exact, show how the rearrangement

$$\dfrac{x\, dx + y\, dy}{\sqrt{x^2 + y^2}} = dx$$

and the observation $\frac{1}{2}d(x^2 + y^2) = x\, dx + y\, dy$ can lead to a solution.

46. True or False: Every separable first-order equation $dy/dx = g(x)h(y)$ is exact.

Mathematical Model

47. Falling Chain A portion of a uniform chain of length 8 ft is loosely coiled around a peg at the edge of a high horizontal platform and the remaining portion of the chain hangs at rest over the edge of the platform. See **FIGURE 2.4.2**. Suppose the length of the overhang is 3 ft, that the chain weighs 2 lb/ft, and that the positive direction is downward. Starting at $t = 0$ seconds, the weight of the overhanging portion causes the chain on the table to uncoil smoothly and fall to the floor. If $x(t)$ denotes the length of the chain overhanging the table at time $t > 0$, then $v = dx/dt$ is its

velocity. When all resistive forces are ignored, it can be shown that a mathematical model relating v and x is

$$xv\frac{dv}{dx} + v^2 = 32x.$$

(a) Rewrite the model in differential form. Proceed as in Problems 33–38 and solve the DE by finding an appropriate integrating factor. Find an explicit solution $v(x)$.

(b) Determine the velocity with which the chain leaves the platform.

peg

platform edge

$x(t)$

FIGURE 2.4.2 Uncoiling chain in Problem 47

Computer Lab Assignment

48. Streamlines **(a)** The solution of the differential equation

$$\frac{2xy}{(x^2 + y^2)^2}\,dx + \left[1 + \frac{y^2 - x^2}{(x^2 + y^2)^2}\right]dy = 0$$

is a family of curves that can be interpreted as streamlines of a fluid flow around a circular object whose boundary is described by the equation $x^2 + y^2 = 1$. Solve this DE and note the solution $f(x, y) = c$ for $c = 0$.

(b) Use a CAS to plot the streamlines for $c = 0, \pm 0.2, \pm 0.4, \pm 0.6$, and ± 0.8 in three different ways. First, use the **contourplot** of a CAS. Second, solve for x in terms of the variable y. Plot the resulting two functions of y for the given values of c, and then combine the graphs. Third, use the CAS to solve a cubic equation for y in terms of x.

2.5 Solutions by Substitutions

INTRODUCTION We usually solve a differential equation by recognizing it as a certain kind of equation (say, separable) and then carrying out a procedure, consisting of equation-specific mathematical steps, that yields a function that satisfies the equation. Often the first step in solving a given differential equation consists of transforming it into another differential equation by means of a **substitution**. For example, suppose we wish to transform the first-order equation $dy/dx = f(x, y)$ by the substitution $y = g(x, u)$, where u is regarded as a function of the variable x.

If g possesses first-partial derivatives, then the Chain Rule gives

$$\frac{dy}{dx} = g_x(x, u) + g_u(x, u)\frac{du}{dx}.$$

By replacing dy/dx by $f(x, y)$ and y by $g(x, u)$ in the foregoing derivative, we get the new first-order differential equation

$$f(x, g(x, u)) = g_x(x, u) + g_u(x, u)\frac{du}{dx},$$

which, after solving for du/dx, has the form $du/dx = F(x, u)$. If we can determine a solution $u = \phi(x)$ of this second equation, then a solution of the original differential equation is $y = g(x, \phi(x))$.

Homogeneous Equations If a function f possesses the property $f(tx, ty) = t^\alpha f(x, y)$ for some real number α, then f is said to be a **homogeneous function** of degree α. For example, $f(x, y) = x^3 + y^3$ is a homogeneous function of degree 3 since

$$f(tx, ty) = (tx)^3 + (ty)^3 = t^3(x^3 + y^3) = t^3 f(x, y),$$

whereas $f(x, y) = x^3 + y^3 + 1$ is seen not to be homogeneous. The functions

$$f(x, y) = \frac{x - 2y}{3x + y} \quad \text{and} \quad g(x, y) = \frac{1}{\sqrt{x^2 + y^2}}$$

are homogeneous of degrees 0 and -1, respectively, because

$$f(tx, ty) = \frac{tx - 2ty}{3tx + ty} = \frac{t(x - 2y)}{t(3x + y)} = \frac{x - 2y}{3x + y} = t^0 f(x, y)$$

$$g(tx, ty) = \frac{1}{\sqrt{t^2 x^2 + t^2 y^2}} = \frac{1}{t\sqrt{x^2 + y^2}} = t^{-1} \frac{1}{\sqrt{x^2 + y^2}} = t^{-1} g(x, y).$$

A first-order DE written in differential form

$$M(x, y)\, dx + N(x, y)\, dy = 0 \tag{1}$$

is said to be a **homogeneous equation** if the coefficient functions $M(x, y)$ and $N(x, y)$ are homogeneous functions of the *same* degree. In other words, (1) is a homogeneous equation if

$$M(tx, ty) = t^\alpha M(x, y) \quad \text{and} \quad N(tx, ty) = t^\alpha N(x, y).$$

A linear first-order DE $a_1 y' + a_0 y = g(x)$ is homogeneous when $g(x) = 0$. ▶

The word *homogeneous* as used here does not mean the same as it does when applied to linear differential equations. See Sections 2.3 and 3.1.

If M and N are homogeneous functions of degree α, we can also write

$$M(x, y) = x^\alpha M(1, u) \quad \text{and} \quad N(x, y) = x^\alpha N(1, u) \quad \text{where } u = y/x, \tag{2}$$

and $\qquad\quad M(x, y) = y^\alpha M(v, 1) \quad \text{and} \quad N(x, y) = y^\alpha N(v, 1) \quad \text{where } v = x/y. \tag{3}$

See Problem 33 in Exercises 2.5. Properties (2) and (3) suggest the substitutions that can be used to solve a homogeneous differential equation. Specifically, *either* of the substitutions $y = ux$ or $x = vy$, where u and v are new dependent variables, will reduce a homogeneous equation to a *separable* first-order differential equation. To show this, observe that as a consequence of (2) a homogeneous equation $M(x, y)\, dx + N(x, y)\, dy = 0$ can be rewritten as

$$x^\alpha M(1, u)\, dx + x^\alpha N(1, u)\, dy = 0 \quad \text{or} \quad M(1, u)\, dx + N(1, u)\, dy = 0,$$

where $u = y/x$ or $y = ux$. By substituting the differential $dy = u\, dx + x\, du$ into the last equation and gathering terms, we obtain a separable DE in the variables u and x:

$$M(1, u)\, dx + N(1, u)[u\, dx + x\, du] = 0$$

$$[M(1, u) + uN(1, u)]\, dx + xN(1, u)\, du = 0$$

or $\qquad\qquad \dfrac{dx}{x} + \dfrac{N(1, u)\, du}{M(1, u) + uN(1, u)} = 0.$

We hasten to point out that the preceding formula should not be memorized; rather, *the procedure should be worked through each time*. The proof that the substitutions $x = vy$ and $dx = v\, dy + y\, dv$ also lead to a separable equation follows in an analogous manner from (3).

EXAMPLE 1 Solving a Homogeneous DE

Solve $\quad (x^2 + y^2)\, dx + (x^2 - xy)\, dy = 0.$

SOLUTION Inspection of $M(x, y) = x^2 + y^2$ and $N(x, y) = x^2 - xy$ shows that these coefficients are homogeneous functions of degree 2. If we let $y = ux$, then $dy = u\, dx + x\, du$ so that, after substituting, the given equation becomes

$$(x^2 + u^2 x^2)\, dx + (x^2 - ux^2)[u\, dx + x\, du] = 0$$

$$x^2(1 + u)\, dx + x^3(1 - u)\, du = 0$$

$$\frac{1 - u}{1 + u}\, du + \frac{dx}{x} = 0$$

$$\left[-1 + \frac{2}{1 + u} \right] du + \frac{dx}{x} = 0. \qquad \leftarrow \text{long division}$$

After integration the last line gives

$$-u + 2 \ln|1 + u| + \ln|x| = \ln|c|$$

$$-\frac{y}{x} + 2 \ln\left|1 + \frac{y}{x}\right| + \ln|x| = \ln|c|. \qquad \leftarrow \text{resubstituting } u = y/x$$

Using the properties of logarithms, we can write the preceding solution as

$$\ln\left|\frac{(x + y)^2}{cx}\right| = \frac{y}{x} \quad \text{or} \quad (x + y)^2 = cxe^{y/x}. \qquad \equiv$$

Although either of the indicated substitutions can be used for every homogeneous differential equation, in practice we try $x = vy$ whenever the function $M(x, y)$ is simpler than $N(x, y)$. Also it could happen that after using one substitution, we may encounter integrals that are difficult or impossible to evaluate in closed form; switching substitutions may result in an easier problem.

▮ Bernoulli's Equation The differential equation

$$\frac{dy}{dx} + P(x)y = f(x)y^n, \qquad (4)$$

where n is any real number, is called **Bernoulli's equation** after the Swiss mathematician **Jacob Bernoulli** (1654–1705), who studied the equation around 1695. Note that for $n = 0$ equation (4) is linear, and for $n = 1$ equation (4) is separable and linear. In 1696 the German polymath **Gottfried Wilhelm Leibniz** (1646–1716) showed that for $n \neq 0$ and $n \neq 1$, any differential equation of the form given in (4) can be reduced to a linear equation by means of the substitution $u = y^{1-n}$. The next example illustrates this substitution method.

EXAMPLE 2 Solving a Bernoulli DE

Solve $x\dfrac{dy}{dx} + y = x^2y^2.$

SOLUTION We begin by rewriting the differential equation in the form given in (4) by dividing by x:

$$\frac{dy}{dx} + \frac{1}{x}y = xy^2.$$

With $n = 2$ and $u = y^{-1}$, we next substitute $y = u^{-1}$ and

$$\frac{dy}{dx} = -u^{-2}\frac{du}{dx} \qquad \leftarrow \text{Chain Rule}$$

into the given equation and simplify. The result is

$$\frac{du}{dx} - \frac{1}{x}u = -x.$$

The integrating factor for this linear equation on, say, $(0, \infty)$ is

$$e^{-\int dx/x} = e^{-\ln x} = e^{\ln x^{-1}} = x^{-1}.$$

Integrating $\dfrac{d}{dx}[x^{-1}u] = -1$

gives $x^{-1}u = -x + c$ or $u = -x^2 + cx$. Since $u = y^{-1}$ we have $y = 1/u$, and so a solution of the given equation is $y = 1/(-x^2 + cx)$. $\equiv$

Note that we have not obtained the general solution of the original nonlinear differential equation in Example 2, since $y = 0$ is a singular solution of the equation.

Reduction to Separation of Variables A differential equation of the form

$$\frac{dy}{dx} = f(Ax + By + C) \qquad (5)$$

can always be reduced to an equation with separable variables by means of the substitution $u = Ax + By + C$, $B \neq 0$. Example 3 illustrates the technique.

EXAMPLE 3 **An Initial-Value Problem**

Solve the initial-value problem $\dfrac{dy}{dx} = (-2x + y)^2 + 7$, $y(0) = 0$.

SOLUTION If we let $u = -2x + y$, then $du/dx = -2 + dy/dx$, and so the differential equation is transformed into

$$\frac{du}{dx} + 2 = u^2 - 7 \qquad \text{or} \qquad \frac{du}{dx} = u^2 - 9.$$

The last equation is separable. Using partial fractions,

$$\frac{du}{(u - 3)(u + 3)} = dx \qquad \text{or} \qquad \frac{1}{6}\left[\frac{1}{u - 3} - \frac{1}{u + 3}\right] du = dx$$

and integrating, then yields

$$\frac{1}{6}\ln\left|\frac{u - 3}{u + 3}\right| = x + c_1 \qquad \text{or} \qquad \frac{u - 3}{u + 3} = e^{6x + 6c_1} = ce^{6x}. \quad \leftarrow \text{replace } e^{6c_1} \text{ by } c$$

Solving the last equation for u and then resubstituting gives the solution

$$u = \frac{3(1 + ce^{6x})}{1 - ce^{6x}} \qquad \text{or} \qquad y = 2x + \frac{3(1 + ce^{6x})}{1 - ce^{6x}}. \qquad (6)$$

Finally, applying the initial condition $y(0) = 0$ to the last equation in (6) gives $c = -1$. With the aid of a graphing utility we have shown in **FIGURE 2.5.1** the graph of the particular solution

$$y = 2x + \frac{3(1 - e^{6x})}{1 + e^{6x}}$$

in blue along with the graphs of some other members of the family solutions (6). ≡

FIGURE 2.5.1 Some solutions of the DE in Example 3

2.5 Exercises Answers to selected odd-numbered problems begin on page ANS-3.

Each DE in Problems 1–14 is homogeneous.

In Problems 1–10, solve the given differential equation by using an appropriate substitution.

1. $(x - y)\, dx + x\, dy = 0$

2. $(x + y)\, dx + x\, dy = 0$

3. $x\, dx + (y - 2x)\, dy = 0$

4. $y\, dx = 2(x + y)\, dy$

5. $(y^2 + yx)\, dx - x^2\, dy = 0$

6. $(y^2 + yx)\, dx + x^2\, dy = 0$

7. $\dfrac{dy}{dx} = \dfrac{y - x}{y + x}$

8. $\dfrac{dy}{dx} = \dfrac{x + 3y}{3x + y}$

9. $-y\, dx + (x + \sqrt{xy})\, dy = 0$

10. $x\dfrac{dy}{dx} = y + \sqrt{x^2 - y^2}, x > 0$

In Problems 11–14, solve the given initial-value problem.

11. $xy^2\dfrac{dy}{dx} = y^3 - x^3$, $y(1) = 2$

12. $(x^2 + 2y^2)\dfrac{dx}{dy} = xy$, $y(-1) = 1$

13. $(x + ye^{y/x})\, dx - xe^{y/x}\, dy = 0$, $y(1) = 0$

14. $y\, dx + x(\ln x - \ln y - 1)\, dy = 0$, $y(1) = e$

Each DE in Problems 15–22 is a Bernoulli equation.

In Problems 15–20, solve the given differential equation by using an appropriate substitution.

15. $x\dfrac{dy}{dx} + y = \dfrac{1}{y^2}$

16. $\dfrac{dy}{dx} - y = e^x y^2$

17. $\dfrac{dy}{dx} = y(xy^3 - 1)$ **18.** $x\dfrac{dy}{dx} - (1 + x)y = xy^2$

19. $t^2\dfrac{dy}{dt} + y^2 = ty$ **20.** $3(1 + t^2)\dfrac{dy}{dt} = 2ty(y^3 - 1)$

In Problems 21 and 22, solve the given initial-value problem.

21. $x^2\dfrac{dy}{dx} - 2xy = 3y^4, \quad y(1) = \frac{1}{2}$

22. $y^{1/2}\dfrac{dy}{dx} + y^{3/2} = 1, \quad y(0) = 4$

Each DE in Problems 23–30 *is of the form given in* (5).

In Problems 23–28, solve the given differential equation by using an appropriate substitution.

23. $\dfrac{dy}{dx} = (x + y + 1)^2$ **24.** $\dfrac{dy}{dx} = \dfrac{1 - x - y}{x + y}$

25. $\dfrac{dy}{dx} = \tan^2(x + y)$ **26.** $\dfrac{dy}{dx} = \sin(x + y)$

27. $\dfrac{dy}{dx} = 2 + \sqrt{y - 2x + 3}$

28. $\dfrac{dy}{dx} = 1 + e^{y - x + 5}$

In Problems 29 and 30, solve the given initial-value problem.

29. $\dfrac{dy}{dx} = \cos(x + y), \quad y(0) = \pi/4$

30. $\dfrac{dy}{dx} = \dfrac{3x + 2y}{3x + 2y + 2}, \quad y(-1) = -1$

Discussion Problems

Some second-order differential equations can be reduced to a first-order equation by means of a substitution. In Problems 31 and 32, use the substitution $u = y'$ to reduce the given differential equation to a first-order equation, and then solve that equation by a method discussed in this section. Then use $u = y'$ again to find the solution y of the given equation.

31. $y'' + (y')^2 = 0$ **32.** $y'' - 2y' = 4x$

33. **(a)** Show that it is always possible to express any homogeneous differential equation $M(x, y)\,dx + N(x, y)\,dy = 0$ in the form

$$\dfrac{dy}{dx} = F\left(\dfrac{y}{x}\right).$$

(b) Show that the substitution $u = y/x$ in the last differential equation in part (a) yields a separable equation.

34. Put the homogeneous differential equation

$$(5x^2 + 2y^2)\,dx - xy\,dy = 0$$

in the form $dy/dx = F(y/x)$ given in part (a) of Problem 33. Then solve the equation using the substitution $u = y/x$.

35. **(a)** Determine two singular solutions of the DE in Problem 10.

(b) If the initial condition $y(5) = 0$ is as prescribed in Problem 10, then what is the largest interval I over which the solution is defined? Use a graphing utility to plot the solution curve for the IVP.

36. In Example 3, the solution $y(x)$ becomes unbounded as $x \to \pm\infty$. Nevertheless $y(x)$ is asymptotic to a curve as $x \to -\infty$ and to a different curve as $x \to \infty$. Find the equations of these curves.

37. The differential equation

$$\dfrac{dy}{dx} = P(x) + Q(x)y + R(x)y^2$$

is called a **Riccati equation** after the Venetian jurist and mathematician **Jacopo Francesco Riccati** (1676–1754) who was able to solve several DEs of this type.

(a) A Riccati equation can be solved by a succession of two substitutions *provided* we know a particular solution y_1 of the equation. Show that the substitution $y = y_1 + u$ reduces Riccati's equation to a Bernoulli equation (4) with $n = 2$. The Bernoulli equation can then be reduced to a linear equation by the substitution $w = u^{-1}$.

(b) Find a one-parameter family of solutions for the differential equation

$$\dfrac{dy}{dx} = -\dfrac{4}{x^2} - \dfrac{1}{x}y + y^2,$$

where $y_1 = 2/x$ is a known solution of the equation.

38. Devise an appropriate substitution to solve

$$xy' = y\ln(xy).$$

Mathematical Models

39. **Population Growth** In the study of population dynamics one of the most famous models for a growing but bounded population is the **logistic equation**

$$\dfrac{dP}{dt} = P(a - bP),$$

where a and b are positive constants. Although we will come back to this equation and solve it by an alternative method in Section 2.8, solve the DE this first time using the fact that it is a Bernoulli equation.

40. **Falling Chain** In Problem 47 in Exercises 2.4 we saw that a mathematical model for the velocity v of a chain slipping off the edge of a high horizontal platform is

$$xv\dfrac{dv}{dx} + v^2 = 32x.$$

In that problem you were asked to solve the DE by converting it into an exact equation using an integrating factor. This time solve the DE using the fact that it is a Bernoulli equation.

2.6 A Numerical Method

INTRODUCTION In Section 2.1 we saw that we could glean *qualitative* information from a first-order DE about its solutions even before we attempted to solve the equation. In Sections 2.2–2.5 we examined first-order DEs *analytically*; that is, we developed procedures for actually obtaining explicit and implicit solutions. But many differential equations possess solutions and yet these solutions cannot be obtained analytically. In this case we "solve" the differential equation *numerically*; this means that the DE is used as the cornerstone of an algorithm for *approximating* the unknown solution. It is common practice to refer to the algorithm as a *numerical method*, the approximate solution as a *numerical solution*, and the graph of a numerical solution as a *numerical solution curve*.

In this section we are going to consider only the simplest of numerical methods. A more extensive treatment of this subject is found in Chapter 6.

Using the Tangent Line Let us assume that the first-order initial-value problem

$$y' = f(x, y), \quad y(x_0) = y_0 \tag{1}$$

possesses a solution. One of the simplest techniques for approximating this solution is to use tangent lines. For example, let $y(x)$ denote the unknown solution of the first-order initial-value problem $y' = 0.1\sqrt{y} + 0.4x^2$, $y(2) = 4$. The nonlinear differential equation cannot be solved directly by the methods considered in Sections 2.2, 2.4, and 2.5; nevertheless, we can still find approximate numerical values of the unknown $y(x)$. Specifically, suppose we wish to know the value of $y(2.5)$. The IVP has a solution, and, as the flow of the direction field in **FIGURE 2.6.1(a)** suggests, a solution curve must have a shape similar to the curve shown in blue.

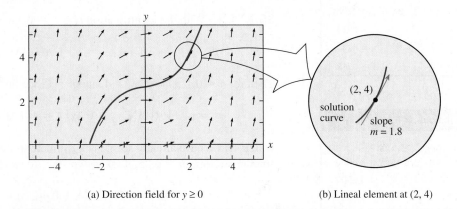

(a) Direction field for $y \geq 0$ (b) Lineal element at (2, 4)

FIGURE 2.6.1 Magnification of a neighborhood about the point (2, 4)

The direction field in Figure 2.6.1(a) was generated so that the lineal elements pass through points in a grid with integer coordinates. As the solution curve passes through the initial point (2, 4), the lineal element at this point is a tangent line with slope given by $f(2, 4) = 0.1\sqrt{4} + 0.4(2)^2 = 1.8$. As is apparent in Figure 2.6.1(a) and the "zoom in" in Figure 2.6.1(b), when x is close to 2 the points on the solution curve are close to the points on the tangent line (the lineal element). Using the point (2, 4), the slope $f(2, 4) = 1.8$, and the point-slope form of a line, we find that an equation of the tangent line is $y = L(x)$, where $L(x) = 1.8x + 0.4$. This last equation, called a **linearization** of $y(x)$ at $x = 2$, can be used to approximate values $y(x)$ within a small neighborhood of $x = 2$. If $y_1 = L(x_1)$ denotes the value of the y-coordinate on the tangent line and $y(x_1)$ is the y-coordinate on the solution curve corresponding to an x-coordinate x_1 that is close to $x = 2$, then $y(x_1) \approx y_1$. If we choose, say, $x_1 = 2.1$, then $y_1 = L(2.1) = 1.8(2.1) + 0.4 = 4.18$, and so $y(2.1) \approx 4.18$.

Euler's Method To generalize the procedure just illustrated, we use the linearization of the unknown solution $y(x)$ of (1) at $x = x_0$:

$$L(x) = f(x_0, y_0)(x - x_0) + y_0. \tag{2}$$

solution curve

$(x_1, y(x_1))$

error

(x_1, y_1)

slope $= f(x_0, y_0)$

(x_0, y_0)

$L(x)$

x_0 $x_1 = x_0 + h$

h

FIGURE 2.6.2 Approximating $y(x_1)$ using a tangent line

The graph of this linearization is a straight line tangent to the graph of $y = y(x)$ at the point (x_0, y_0). We now let h be a positive increment of the x-axis, as shown in **FIGURE 2.6.2**. Then by replacing x by $x_1 = x_0 + h$ in (2) we get

$$L(x_1) = f(x_0, y_0)(x_0 + h - x_0) + y_0 \quad \text{or} \quad y_1 = y_0 + hf(x_0, y_0),$$

where $y_1 = L(x_1)$. The point (x_1, y_1) on the tangent line is an approximation to the point $(x_1, y(x_1))$ on the solution curve. Of course the accuracy of the approximation $y_1 \approx y(x_1)$ depends heavily on the size of the increment h. Usually we must choose this **step size** to be "reasonably small." We now repeat the process using a second "tangent line" at (x_1, y_1).* By replacing (x_0, y_0) in the above discussion with the new starting point (x_1, y_1), we obtain an approximation $y_2 \approx y(x_2)$ corresponding to two steps of length h from x_0, that is, $x_2 = x_1 + h = x_0 + 2h$ and

$$y(x_2) = y(x_0 + 2h) = y(x_1 + h) \approx y_2 = y_1 + hf(x_1, y_1).$$

Continuing in this manner, we see that $y_1, y_2, y_3, \ldots$, can be defined recursively by the general formula

$$y_{n+1} = y_n + hf(x_n, y_n), \tag{3}$$

where $x_n = x_0 + nh$, $n = 0, 1, 2, \ldots$. This procedure of using successive "tangent lines" is called **Euler's method**.

EXAMPLE 1 Euler's Method

Consider the initial-value problem $y' = 0.1\sqrt{y} + 0.4x^2$, $y(2) = 4$. Use Euler's method to obtain an approximation to $y(2.5)$ using first $h = 0.1$ and then $h = 0.05$.

SOLUTION With the identification $f(x, y) = 0.1\sqrt{y} + 0.4x^2$, (3) becomes

$$y_{n+1} = y_n + h(0.1\sqrt{y_n} + 0.4x_n^2).$$

Then for $h = 0.1$, $x_0 = 2$, $y_0 = 4$, and $n = 0$, we find

$$y_1 = y_0 + h(0.1\sqrt{y_0} + 0.4x_0^2) = 4 + 0.1(0.1\sqrt{4} + 0.4(2)^2) = 4.18,$$

which, as we have already seen, is an estimate to the value of $y(2.1)$. However, if we use the smaller step size $h = 0.05$, it takes two steps to reach $x = 2.1$. From

$$y_1 = 4 + 0.05(0.1\sqrt{4} + 0.4(2)^2) = 4.09$$

$$y_2 = 4.09 + 0.05(0.1\sqrt{4.09} + 0.4(2.05)^2) = 4.18416187$$

we have $y_1 \approx y(2.05)$ and $y_2 \approx y(2.1)$. The remainder of the calculations were carried out using software; the results are summarized in Tables 2.6.1 and 2.6.2. We see in Tables 2.6.1 and 2.6.2 that it takes five steps with $h = 0.1$ and ten steps with $h = 0.05$, respectively, to get to $x = 2.5$. Also, each entry has been rounded to four decimal places. ≡

In Example 2 we apply Euler's method to a differential equation for which we have already found a solution. We do this to compare the values of the approximations y_n at each step with the true values of the solution $y(x_n)$ of the initial-value problem.

EXAMPLE 2 Comparison of Approximate and Exact Values

Consider the initial-value problem $y' = 0.2xy$, $y(1) = 1$. Use Euler's method to obtain an approximation to $y(1.5)$ using first $h = 0.1$ and then $h = 0.05$.

TABLE 2.6.1 $h = 0.1$

x_n	y_n
2.00	4.0000
2.10	4.1800
2.20	4.3768
2.30	4.5914
2.40	4.8244
2.50	5.0768

TABLE 2.6.2 $h = 0.05$

x_n	y_n
2.00	4.0000
2.05	4.0900
2.10	4.1842
2.15	4.2826
2.20	4.3854
2.25	4.4927
2.30	4.6045
2.35	4.7210
2.40	4.8423
2.45	4.9686
2.50	5.0997

*This is not an actual tangent line since (x_1, y_1) lies on the first tangent and not on the solution curve.

SOLUTION With the identification $f(x, y) = 0.2xy$, (3) becomes

$$y_{n+1} = y_n + h(0.2x_n y_n),$$

where $x_0 = 1$ and $y_0 = 1$. Again with the aid of computer software we obtain the values in Tables 2.6.3 and 2.6.4.

TABLE 2.6.3 $h = 0.1$

x_n	y_n	Actual Value	Absolute Error	% Rel. Error
1.00	1.0000	1.0000	0.0000	0.00
1.10	1.0200	1.0212	0.0012	0.12
1.20	1.0424	1.0450	0.0025	0.24
1.30	1.0675	1.0714	0.0040	0.37
1.40	1.0952	1.1008	0.0055	0.50
1.50	1.1259	1.1331	0.0073	0.64

TABLE 2.6.4 $h = 0.05$

x_n	y_n	Actual Value	Absolute Error	% Rel. Error
1.00	1.0000	1.0000	0.0000	0.00
1.05	1.0100	1.0103	0.0003	0.03
1.10	1.0206	1.0212	0.0006	0.06
1.15	1.0318	1.0328	0.0009	0.09
1.20	1.0437	1.0450	0.0013	0.12
1.25	1.0562	1.0579	0.0016	0.16
1.30	1.0694	1.0714	0.0020	0.19
1.35	1.0833	1.0857	0.0024	0.22
1.40	1.0980	1.1008	0.0028	0.25
1.45	1.1133	1.1166	0.0032	0.29
1.50	1.1295	1.1331	0.0037	0.32

In Example 1, the true values were calculated from the known solution $y = e^{0.1(x^2 - 1)}$ (verify). Also, the **absolute error** is defined to be

$$|true\ value - approximation|.$$

The **relative error** and **percentage relative error** are, in turn,

$$\frac{absolute\ error}{|true\ value|} \quad \text{and} \quad \frac{absolute\ error}{|true\ value|} \times 100.$$

By comparing the last two columns in Tables 2.6.3 and 2.6.4, it is clear that the accuracy of the approximations improve as the step size h decreases. Also, we see that even though the percentage relative error is growing with each step, it does not appear to be that bad. But you should not be deceived by one example. If we simply change the coefficient of the right side of the DE in Example 2 from 0.2 to 2, then at $x_n = 1.5$ the percentage relative errors increase dramatically. See Problem 4 in Exercises 2.6.

■ **A Caveat** Euler's method is just one of many different ways a solution of a differential equation can be approximated. Although attractive for its simplicity, *Euler's method is seldom used in serious calculations.* We have introduced this topic simply to give you a first taste of numerical methods. We will go into greater detail and discuss methods that give significantly greater accuracy, notably the **fourth-order Runge–Kutta method**, in Chapter 6. We shall refer to this important numerical method as the **RK4 method**.

■ **Numerical Solvers** Regardless of whether we can actually find an explicit or implicit solution, if a solution of a differential equation exists, it represents a smooth curve in the Cartesian plane. The basic idea behind *any* numerical method for ordinary differential equations is to somehow approximate the y-values of a solution for preselected values of x. We start at a specified initial point (x_0, y_0) on a solution curve and proceed to calculate in a step-by-step fashion a sequence of points $(x_1, y_1), (x_2, y_2), \ldots, (x_n, y_n)$ whose y-coordinates y_i approximate the y-coordinates $y(x_i)$ of points $(x_1, y(x_1)), (x_2, y(x_2)), \ldots, (x_n, y(x_n))$ that lie on the graph of the usually unknown solution $y(x)$. By taking the x-coordinates close together (that is, for small values of h) and by joining the points $(x_1, y_1), (x_2, y_2), \ldots, (x_n, y_n)$ with

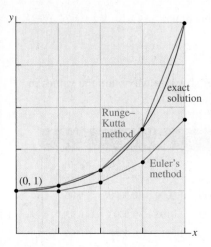

FIGURE 2.6.3 Comparison of numerical methods

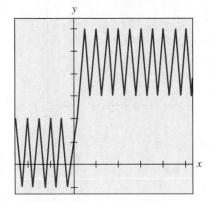

FIGURE 2.6.4 A not very helpful numerical solution curve

short line segments, we obtain a polygonal curve that appears smooth and whose qualitative characteristics we hope are close to those of an actual solution curve. Drawing curves is something well suited to a computer. A computer program written to either implement a numerical method or to render a visual representation of an approximate solution curve fitting the numerical data produced by this method is referred to as a **numerical solver**. There are many different numerical solvers commercially available, either embedded in a larger software package such as a computer algebra system or as a stand-alone package. Some software packages simply plot the generated numerical approximations, whereas others generate both hard numerical data as well as the corresponding approximate or **numerical solution curves**. As an illustration of the connect-the-dots nature of the graphs produced by a numerical solver, the two red polygonal graphs in **FIGURE 2.6.3** are numerical solution curves for the initial-value problem $y' = 0.2xy$, $y(0) = 1$, on the interval $[0, 4]$ obtained from Euler's method and the RK4 method using the step size $h = 1$. The blue smooth curve is the graph of the exact solution $y = e^{0.1x^2}$ of the IVP. Notice in Figure 2.6.3 that even with the ridiculously large step size of $h = 1$, the RK4 method produces the more believable "solution curve." The numerical solution curve obtained from the RK4 method is indistinguishable from the actual solution curve on the interval $[0, 4]$ when a more typical step size of $h = 0.1$ is used.

■ **Using a Numerical Solver** Knowledge of the various numerical methods is not necessary in order to use a numerical solver. A solver usually requires that the differential equation be expressed in normal form $dy/dx = f(x, y)$. Numerical solvers that generate only curves usually require that you supply $f(x, y)$ and the initial data x_0 and y_0 and specify the desired numerical method. If the idea is to approximate the numerical value of $y(a)$, then a solver may additionally require that you state a value for h, or, equivalently, require the number of steps that you want to take to get from $x = x_0$ to $x = a$. For example, if we want to approximate $y(4)$ for the IVP illustrated in Figure 2.6.3, then, starting at $x = 0$, it takes four steps to reach $x = 4$ with a step size of $h = 1$; 40 steps is equivalent to a step size of $h = 0.1$. Although it is not our intention here to delve into the many problems that one can encounter when attempting to approximate mathematical quantities, you should be at least aware of the fact that a numerical solver may break down near certain points or give an incomplete or misleading picture when applied to some first-order differential equations in the normal form. **FIGURE 2.6.4** illustrates the numerical solution curve obtained by applying Euler's method to a certain first-order initial value problem $dy/dx = f(x, y)$, $y(0) = 1$. Equivalent results were obtained using three different commercial numerical solvers, yet the graph is hardly a plausible solution curve. (Why?) There are several avenues of recourse when a numerical solver has difficulties; three of the more obvious are decrease the step size, use another numerical method, or try a different numerical solver.

2.6 Exercises Answers to selected odd-numbered problems begin on page ANS-3.

In Problems 1 and 2, use Euler's method to obtain a four-decimal approximation of the indicated value. Carry out the recursion of (3) by hand, first using $h = 0.1$ and then using $h = 0.05$.

1. $y' = 2x - 3y + 1$, $y(1) = 5$; $y(1.2)$
2. $y' = x + y^2$, $y(0) = 0$; $y(0.2)$

In Problems 3 and 4, use Euler's method to obtain a four-decimal approximation of the indicated value. First use $h = 0.1$ and then use $h = 0.05$. Find an explicit solution for each initial-value problem and then construct tables similar to Tables 2.6.3 and 2.6.4.

3. $y' = y$, $y(0) = 1$; $y(1.0)$
4. $y' = 2xy$, $y(1) = 1$; $y(1.5)$

In Problems 5–10, use a numerical solver and Euler's method to obtain a four-decimal approximation of the indicated value. First use $h = 0.1$ and then use $h = 0.05$.

5. $y' = e^{-y}$, $y(0) = 0$; $y(0.5)$
6. $y' = x^2 + y^2$, $y(0) = 1$; $y(0.5)$
7. $y' = (x - y)^2$, $y(0) = 0.5$; $y(0.5)$
8. $y' = xy + \sqrt{y}$, $y(0) = 1$; $y(0.5)$
9. $y' = xy^2 - \dfrac{y}{x}$, $y(1) = 1$; $y(1.5)$
10. $y' = y - y^2$, $y(0) = 0.5$; $y(0.5)$

In Problems 11 and 12, use a numerical solver to obtain a numerical solution curve for the given initial-value problem. First use Euler's method and then the RK4 method. Use

$h = 0.25$ in each case. Superimpose both solution curves on the same coordinate axes. If possible, use a different color for each curve. Repeat, using $h = 0.1$ and $h = 0.05$.

11. $y' = 2(\cos x)y, \quad y(0) = 1$

12. $y' = y(10 - 2y), \quad y(0) = 1$

Discussion Problem

13. Use a numerical solver and Euler's method to approximate $y(1.0)$, where $y(x)$ is the solution to $y' = 2xy^2$, $y(0) = 1$. First use $h = 0.1$ and then $h = 0.05$. Repeat using the RK4 method. Discuss what might cause the approximations of $y(1.0)$ to differ so greatly.

2.7 Linear Models

INTRODUCTION In this section we solve some of the linear first-order models that were introduced in Section 1.3.

Growth and Decay The initial-value problem

$$\frac{dx}{dt} = kx, \quad x(t_0) = x_0, \tag{1}$$

where k is the constant of proportionality, serves as a model for diverse phenomena involving either **growth** or **decay**. We have seen in Section 1.3 that in biology, over short periods of time, the rate of growth of certain populations (bacteria, small animals) is observed to be proportional to the population present at time t. If a population at some initial time t_0 is known, then the solution of (1) can be used to predict the population in the future—that is, at times $t > t_0$. The constant of proportionality k in (1) can be determined from the solution of the initial-value problem using a subsequent measurement of x at some time $t_1 > t_0$. In physics and chemistry, (1) is seen in the form of a *first-order reaction*, that is, a reaction whose rate or velocity dx/dt is directly proportional to the first power of the reactant concentration x at time t. The decomposition or decay of U-238 (uranium) by radioactivity into Th-234 (thorium) is a first-order reaction.

EXAMPLE 1 Bacterial Growth

A culture initially has P_0 number of bacteria. At $t = 1$ h the number of bacteria is measured to be $\frac{3}{2}P_0$. If the rate of growth is proportional to the number of bacteria $P(t)$ present at time t, determine the time necessary for the number of bacteria to triple.

SOLUTION We first solve the differential equation in (1) with the symbol x replaced by P. With $t_0 = 0$ the initial condition is $P(0) = P_0$. We then use the empirical observation that $P(1) = \frac{3}{2}P_0$ to determine the constant of proportionality k.

Notice that the differential equation $dP/dt = kP$ is both separable and linear. When it is put in the standard form of a linear first-order DE,

$$\frac{dP}{dt} - kP = 0,$$

we can see by inspection that the integrating factor is e^{-kt}. Multiplying both sides of the equation by this term immediately gives

$$\frac{d}{dt}[e^{-kt}P] = 0.$$

Integrating both sides of the last equation yields $e^{-kt}P = c$ or $P(t) = ce^{kt}$. At $t = 0$ it follows that $P_0 = ce^0 = c$, and so $P(t) = P_0e^{kt}$. At $t = 1$ we have $\frac{3}{2}P_0 = P_0e^k$ or $e^k = \frac{3}{2}$. From the last equation we get $k = \ln\frac{3}{2} = 0.4055$. Thus $P(t) = P_0e^{0.4055t}$. To find the time at which the number of bacteria has tripled, we solve $3P_0 = P_0e^{0.4055t}$ for t. It follows that $0.4055t = \ln 3$, and so

$$t = \frac{\ln 3}{0.4055} \approx 2.71\,\text{h}.$$

FIGURE 2.7.1 Time in which initial population triples in Example 1

See **FIGURE 2.7.1**.

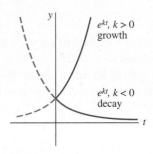

y

$e^{kt}, k > 0$
growth

$e^{kt}, k < 0$
decay

t

FIGURE 2.7.2 Growth ($k > 0$) and decay ($k < 0$)

Notice in Example 1 that the actual number P_0 of bacteria present at time $t = 0$ played no part in determining the time required for the number in the culture to triple. The time necessary for an initial population of, say, 100 or 1,000,000 bacteria to triple is still approximately 2.71 hours.

As shown in **FIGURE 2.7.2**, the exponential function e^{kt} increases as t increases for $k > 0$ and decreases as t increases for $k < 0$. Thus problems describing growth (whether of populations, bacteria, or even capital) are characterized by a positive value of k, whereas problems involving decay (as in radioactive disintegration) yield a negative k value. Accordingly, we say that k is either a **growth constant** ($k > 0$) or a **decay constant** ($k < 0$).

Half-Life
In physics the **half-life** is a measure of the stability of a radioactive substance. The half-life is simply the time it takes for one-half of the atoms in an initial amount A_0 of a substance to transmute into the atoms of another element. The longer the half-life of a substance, the more stable it is. For example, the most commonly occurring uranium isotope, U-238, has a half-life of approximately 4,500,000,000 years. In about 4.5 billion years, one-half of a quantity of U-238 is transmuted into lead, Pb-206. In contrast, the isotope Ra-223 of radium has a half-life of 11.4 days. Of the 33 isotopes of radium, Ra-226 is the most stable, with a half-life of approximately 1600 years. The nucleus of an atom of Ra-226 is transmuted into the nucleus of a radon atom Rn-222. Radon is a heavy, odorless, colorless, and highly radioactive gas that usually originates in the ground. Because radon can penetrate a sealed concrete floor, it frequently accumulates in the basements of new and highly insulated homes. Some medical organizations claim that after cigarette smoking, exposure to radon gas is the second leading cause of lung cancer. Because of its short half-life, the isotope Ra-223 is frequently used in nuclear medicine to treat forms of cancer, especially metastatic bone cancer and advanced prostate cancer.

EXAMPLE 2 — Half-Life of Plutonium

A breeder reactor converts relatively stable uranium-238 into the isotope plutonium-239. After 15 years it is determined that 0.043% of the initial amount A_0 of the plutonium has disintegrated. Find the half-life of this isotope if the rate of disintegration is proportional to the amount remaining.

SOLUTION Let $A(t)$ denote the amount of plutonium remaining at any time. As in Example 1, the solution of the initial-value problem

$$\frac{dA}{dt} = kA, \quad A(0) = A_0 \tag{2}$$

is $A(t) = A_0 e^{kt}$. If 0.043% of the atoms of A_0 have disintegrated, then 99.957% of the substance remains. To find the decay constant k, we use $0.99957A_0 = A(15)$; that is, $0.99957A_0 = A_0 e^{15k}$. Solving for k then gives $k = \frac{1}{15}\ln 0.99957 = -0.00002867$. Hence $A(t) = A_0 e^{-0.00002867t}$. Now the half-life is the corresponding value of time at which $A(t) = \frac{1}{2}A_0$. Solving for t gives $\frac{1}{2}A_0 = A_0 e^{-0.00002867t}$ or $\frac{1}{2} = e^{-0.00002867t}$. The last equation yields

$$t = \frac{\ln 2}{0.00002867} \approx 24{,}180 \text{ years.}$$

≡

Carbon Dating
About 1950, a team of scientists at the University of Chicago led by the American physical chemist **Willard Frank Libby** (1908–1980) devised a method using a radioactive isotope of carbon as a means of determining the approximate ages of carbonaceous fossilized matter. The theory of carbon dating is based on the fact that the radioisotope carbon-14 is produced in the atmosphere by the action of cosmic radiation on nitrogen-14. The ratio of the amount of C-14 to the stable C-12 in the atmosphere appears to be a constant, and as a consequence the proportionate amount of the isotope present in all living organisms is the same as that in the atmosphere. When a living organism dies, the absorption of C-14, by breathing, eating, or photosynthesis, ceases. Thus by comparing the proportionate amount of C-14, say, in a fossil with the constant amount ratio found in the atmosphere, it is possible to obtain a reasonable estimation of its age. The method is based on the knowledge of the half-life of C-14. Libby's calculated value for the half-life of C-14 was approximately 5600 years and is called the **Libby half-life**. Today the commonly accepted value for the half-life of C-14 is the **Cambridge half-life** that is close to 5730 years. For his work,

© Jack Fields/Science Source/Getty Images
Willard Frank Libby

Libby was awarded the Nobel Prize for chemistry in 1960. Libby's method has been used to date wooden furniture in Egyptian tombs, the woven flax wrappings of the Dead Sea Scrolls, and the cloth of the enigmatic Shroud of Turin.

EXAMPLE 3 **Age of a Fossil**

A fossilized bone is found to contain 0.1% of its original amount of C-14. Determine the age of the fossil.

SOLUTION As in Example 2 the starting point is $A(t) = A_0 e^{kt}$. To determine the value of the decay constant k we use the fact that $\frac{1}{2}A_0 = A(5730)$ or $\frac{1}{2}A_0 = A_0 e^{5730k}$. The last equation implies $5730k = \ln \frac{1}{2} = -\ln 2$ and so we get $k = -(\ln 2)/5730 = -0.00012097$. Therefore $A(t) = A_0 e^{-0.00012097t}$. With $A(t) = 0.001A_0$ we have $0.001A_0 = A_0 e^{-0.00012097t}$ and $-0.00012097t = \ln(0.001) = -\ln 1000$. Thus

$$t = \frac{\ln 1000}{0.00012097} \approx 57{,}103 \text{ years.} \qquad \equiv$$

The date found in Example 3 is really at the border of accuracy of this method. The usual carbon-14 technique is limited to about 10 half-lives of the isotope, or roughly 60,000 years. One fundamental reason for this limitation is the relatively short half-life of C-14. There are other problems as well; the chemical analysis needed to obtain an accurate measurement of the remaining C-14 becomes somewhat formidable around the point $0.001A_0$. Moreover, this analysis requires the destruction of a rather large sample of the specimen. If this measurement is accomplished indirectly, based on the actual radioactivity of the specimen, then it is very difficult to distinguish between the radiation from the specimen and the normal background radiation. But recently the use of a particle accelerator has enabled scientists to separate the C-14 from the stable C-12 directly. When the precise value of the ratio of C-14 to C-12 is computed, the accuracy can be extended to 70,000–100,000 years. For these reasons and the fact that the C-14 dating is restricted to organic materials this method is used mainly by archaeologists. On the other hand, geologists who are interested in questions about the age of rocks or the age of the Earth use **radiometric dating** techniques. Radiometric dating, invented by the New Zealand–born physicist and Nobel laureate **Ernest Rutherford** (1871–1937) around 1905, is based on the radioactive decay of a naturally occurring radioactive isotope with a very long half-life and a comparison between a measured quantity of this decaying isotope and one of its decay products using known decay rates. Radiometric methods using potassium-argon, rubidium-strontium, or uranium-lead can give ages of certain kinds of rocks of several billion years. The potassium-argon method of dating is also used to date terrestrial materials, such as minerals and lava, and extra-terrestrial materials, such as meteorites and lunar rocks. The age of a fossil, such as a dinosaur bone, cannot be dated by the C-14 method since it has very little organic material left in it. Its age can be estimated by determining the age of the rock strata in which it was found. See Problems 5 and 6 in Exercises 2.9 for a discussion of the **potassium-argon method**.

The size and location of the sample caused major difficulties when a team of scientists were invited to use carbon-14 dating on the Shroud of Turin in 1988. See Problem 14 in Exercises 2.7.

Ernest Rutherford

Newton's Law of Cooling/Warming

In equation (3) of Section 1.3 we saw that the mathematical formulation of Newton's empirical law of cooling of an object is given by the linear first-order differential equation

$$\frac{dT}{dt} = k(T - T_m), \qquad (3)$$

where k is a constant of proportionality, $T(t)$ is the temperature of the object for $t > 0$, and T_m is the ambient temperature—that is, the temperature of the medium around the object. In Example 4 we assume that T_m is constant.

EXAMPLE 4 **Cooling of a Cake**

When a cake is removed from an oven, its temperature is measured at 300°F. Three minutes later its temperature is 200°F. How long will it take for the cake to cool off to a room temperature of 70°F?

(a)

$T(t)$	t (in min.)
75°	20.1
74°	21.3
73°	22.8
72°	24.9
71°	28.6
70.5°	32.3

(b)

FIGURE 2.7.3 Temperature of cooling cake in Example 4

SOLUTION In (3) we make the identification $T_m = 70$. We must then solve the initial-value problem

$$\frac{dT}{dt} = k(T - 70), \quad T(0) = 300 \tag{4}$$

and determine the value of k so that $T(3) = 200$.

Equation (4) is both linear and separable. Separating variables,

$$\frac{dT}{T - 70} = k\,dt,$$

yields $\ln|T - 70| = kt + c_1$, and so $T = 70 + c_2 e^{kt}$. When $t = 0$, $T = 300$, so that $300 = 70 + c_2$ gives $c_2 = 230$, and, therefore, $T = 70 + 230e^{kt}$. Finally, the measurement $T(3) = 200$ leads to $e^{3k} = \frac{13}{23}$ or $k = \frac{1}{3}\ln\frac{13}{23} = -0.19018$. Thus

$$T(t) = 70 + 230e^{-0.19018t}. \tag{5}$$

We note that (5) furnishes no solution to $T(t) = 70$ since $\lim_{t\to\infty} T(t) = 70$. Yet intuitively we expect the cake to reach the room temperature after a reasonably long period of time. How long is "long"? Of course, we should not be disturbed by the fact that the model (4) does not quite live up to our physical intuition. Parts (a) and (b) of **FIGURE 2.7.3** clearly show that the cake will be approximately at room temperature in about one-half hour. ≡

▬ **Mixtures** The mixing of two fluids sometimes gives rise to a linear first-order differential equation. When we discussed the mixing of two brine solutions in Section 1.3, we assumed that the rate $x'(t)$ at which the amount of salt in the mixing tank changes was a net rate:

$$\frac{dx}{dt} = \left(\begin{array}{c}\text{input rate}\\\text{of salt}\end{array}\right) - \left(\begin{array}{c}\text{output rate}\\\text{of salt}\end{array}\right) = R_{in} - R_{out}. \tag{6}$$

In Example 5 we solve equation (8) of Section 1.3.

EXAMPLE 5	**Mixture of Two Salt Solutions**

Recall that the large tank considered in Section 1.3 held 300 gallons of a brine solution. Salt was entering and leaving the tank; a brine solution was being pumped into the tank at the rate of 3 gal/min, mixed with the solution there, and then the mixture was pumped out at the rate of 3 gal/min. The concentration of the salt in the inflow, or solution entering, was 2 lb/gal, and so salt was entering the tank at the rate $R_{in} = (2\text{ lb/gal}) \cdot (3\text{ gal/min}) = 6$ lb/min and leaving the tank at the rate $R_{out} = (x/300\text{ lb/gal}) \cdot (3\text{ gal/min}) = x/100$ lb/min. From this data and (6) we get equation (8) of Section 1.3. Let us pose the question: If there were 50 lb of salt dissolved initially in the 300 gallons, how much salt is in the tank after a long time?

SOLUTION To find the amount of salt $x(t)$ in the tank at time t, we solve the initial-value problem

$$\frac{dx}{dt} + \frac{1}{100}x = 6, \quad x(0) = 50.$$

Note here that the side condition is the initial amount of salt, $x(0) = 50$ in the tank, and not the initial amount of liquid in the tank. Now since the integrating factor of the linear differential equation is $e^{t/100}$, we can write the equation as

$$\frac{d}{dt}[e^{t/100}x] = 6e^{t/100}.$$

Integrating the last equation and solving for x gives the general solution $x(t) = 600 + ce^{-t/100}$. When $t = 0$, $x = 50$, so we find that $c = -550$. Thus the amount of salt in the tank at any time t is given by

$$x(t) = 600 - 550e^{-t/100}. \tag{7}$$

t (min)	x (lb)
50	266.41
100	397.67
150	477.27
200	525.57
300	572.62
400	589.93

(b)

FIGURE 2.7.4 Pounds of salt in tank as a function of time in Example 5

The solution (7) was used to construct the table in **FIGURE 2.7.4(b)**. Also, it can be seen from (7) and Figure 2.7.4(a) that $x(t) \rightarrow 600$ as $t \rightarrow \infty$. Of course, this is what we would expect in this case; over a long time the number of pounds of salt in the solution must be (300 gal)(2 lb/gal) = 600 lb. ≡

In Example 5 we assumed that the rate at which the solution was pumped in was the same as the rate at which the solution was pumped out. However, this need not be the situation; the mixed brine solution could be pumped out at a rate r_{out} faster or slower than the rate r_{in} at which the other brine solution was pumped in.

Example 5 Revisited

If the well-stirred solution in Example 5 is pumped out at the slower rate of $r_{out} = 2$ gallons per minute, then liquid accumulates in the tank at a rate of

$$r_{in} - r_{out} = (3 - 2) \text{ gal/min} = 1 \text{ gal/min}.$$

After t minutes there are $300 + t$ gallons of brine in the tank and so the concentration of the outflow is $c(t) = x/(300 + t)$. The output rate of salt is then $R_{out} = c(t) \cdot r_{out}$ or

$$R_{out} = \left(\frac{x}{300 + t} \text{ lb/gal} \right) \cdot (2 \text{ gal/min}) = \frac{2x}{300 + t} \text{ lb/min}.$$

Hence equation (6) becomes

$$\frac{dx}{dt} = 6 - \frac{2x}{300 + t} \quad \text{or} \quad \frac{dx}{dt} + \frac{2}{300 + t}x = 6.$$

Multiplying the last equation by the integrating factor

$$e^{\int \frac{2}{300+t}dt} = e^{\ln(300 + t)^2} = (300 + t)^2$$

yields

$$\frac{d}{dt}[(300 + t)^2 x] = 6(300 + t)^2.$$

By integrating and applying the initial condition $x(0) = 50$ we obtain the solution $x(t) = 600 + 2t - (4.95 \times 10^7)(300 + t)^{-2}$. See the discussion following (8) of Section 1.3, Problem 12 in Exercises 1.3, and Problems 27–30 in Exercises 2.7. ≡

■ **Series Circuits** For a series circuit containing only a resistor and an inductor, Kirchhoff's second law states that the sum of the voltage drop across the inductor ($L(di/dt)$) and the voltage drop across the resistor (iR) is the same as the impressed voltage ($E(t)$) on the circuit. See **FIGURE 2.7.5**. Thus we obtain the linear differential equation for the current $i(t)$,

$$L\frac{di}{dt} + Ri = E(t), \tag{8}$$

FIGURE 2.7.5 LR-series circuit

where L and R are known as the inductance and the resistance, respectively. The current $i(t)$ is also called the **response** of the system.

The voltage drop across a capacitor with capacitance C is given by $q(t)/C$, where q is the charge on the capacitor. Hence, for the series circuit shown in **FIGURE 2.7.6**, Kirchhoff's second law gives

$$Ri + \frac{1}{C}q = E(t). \tag{9}$$

FIGURE 2.7.6 RC-series circuit

But current i and charge q are related by $i = dq/dt$, so (9) becomes the linear differential equation

$$R\frac{dq}{dt} + \frac{1}{C}q = E(t). \tag{10}$$

EXAMPLE 7 *LR*-Series Circuit

A 12-volt battery is connected to an *LR*-series circuit in which the inductance is $\frac{1}{2}$ henry and the resistance is 10 ohms. Determine the current *i* if the initial current is zero.

SOLUTION From (8) we see that we must solve the equation

$$\frac{1}{2}\frac{di}{dt} + 10i = 12$$

subject to $i(0) = 0$. First, we multiply the differential equation by 2 and read off the integrating factor e^{20t}. We then obtain

$$\frac{d}{dt}[e^{20t}i] = 24e^{20t}.$$

Integrating each side of the last equation and solving for *i* gives $i(t) = \frac{6}{5} + ce^{-20t}$. Now $i(0) = 0$ implies $0 = \frac{6}{5} + c$ or $c = -\frac{6}{5}$. Therefore the response is $i(t) = \frac{6}{5} - \frac{6}{5}e^{-20t}$. ≡

From (4) of Section 2.3 we can write a general solution of (8):

$$i(t) = \frac{e^{-(R/L)t}}{L}\int e^{(R/L)t}E(t)\,dt + ce^{-(R/L)t}. \tag{11}$$

In particular, when $E(t) = E_0$ is a constant, (11) becomes

$$i(t) = \frac{E_0}{R} + ce^{-(R/L)t}. \tag{12}$$

Note that as $t \to \infty$, the second term in (12) approaches zero. Such a term is usually called a **transient term**; any remaining terms are called the **steady-state** part of the solution. In this case E_0/R is also called the **steady-state current**; for large values of time it then appears that the current in the circuit is simply governed by Ohm's law ($E = iR$). **Georg Simon Ohm** (1789–1854) was a physicist and mathematician born in Erlangen, Principality of Bayreuth, located then in the Holy Roman Empire.

REMARKS

The solution $P(t) = P_0e^{0.4055t}$ of the initial-value problem in Example 1 described the population of a colony of bacteria at any time $t \geq 0$. Of course, $P(t)$ is a continuous function that takes on *all* real numbers in the interval defined by $P_0 \leq P < \infty$. But since we are talking about a population, common sense dictates that P can take on only positive integer values. Moreover, we would not expect the population to grow continuously—that is, every second, every microsecond, and so on—as predicted by our solution; there may be intervals of time $[t_1, t_2]$ over which there is no growth at all. Perhaps, then, the graph shown in **FIGURE 2.7.7(a)** is a more realistic description of P than is the graph of an exponential function. Using a continuous function to describe a discrete phenomenon is often more a matter of convenience than of accuracy. However, for some purposes we may be satisfied if our model describes the system fairly closely when viewed macroscopically in time, as in Figures 2.7.7(b) and 2.7.7(c), rather than microscopically, as in Figure 2.7.7(a). Keep firmly in mind, a mathematical model is not reality.

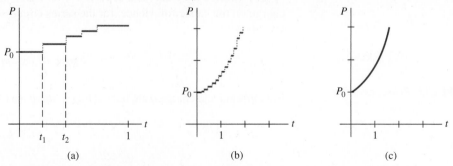

FIGURE 2.7.7 Population growth is a discrete process

Growth and Decay

1. The population of a community is known to increase at a rate proportional to the number of people present at time t. If an initial population P_0 has doubled in 5 years, how long will it take to triple? To quadruple?

2. Suppose it is known that the population of the community in Problem 1 is 10,000 after 3 years. What was the initial population P_0? What will the population be in 10 years? How fast is the population growing at $t = 10$?

3. The population of a town grows at a rate proportional to the population present at time t. The initial population of 500 increases by 15% in 10 years. What will the population be in 30 years? How fast is the population growing at $t = 30$?

4. The population of bacteria in a culture grows at a rate proportional to the number of bacteria present at time t. After 3 hours it is observed that 400 bacteria are present. After 10 hours 2000 bacteria are present. What was the initial number of bacteria?

5. The radioactive isotope of lead, Pb-209, decays at a rate proportional to the amount present at time t and has a half-life of 3.3 hours. If 1 gram of this isotope is present initially, how long will it take for 90% of the lead to decay?

6. Initially, 100 milligrams of a radioactive substance was present. After 6 hours the mass had decreased by 3%. If the rate of decay is proportional to the amount of the substance present at time t, find the amount remaining after 24 hours.

7. Determine the half-life of the radioactive substance described in Problem 6.

8. **(a)** Consider the initial-value problem $dA/dt = kA$, $A(0) = A_0$, as the model for the decay of a radioactive substance. Show that, in general, the half-life T of the substance is $T = -(\ln 2)/k$.
 (b) Show that the solution of the initial-value problem in part (a) can be written $A(t) = A_0 2^{-t/T}$.
 (c) If a radioactive substance has a half-life T given in part (a), how long will it take an initial amount A_0 of the substance to decay to $\frac{1}{8}A_0$?

9. The amount $A(t)$ of a radioactive substance remaining at any time can be determined directly from the initial amount A_0 and the amount A_1 of the substance present at a time $t_1 > 0$. Show that

$$A(t) = A_0\left(\frac{A_1}{A_0}\right)^{t/t_1}.$$

10. Using the result in Problem 9, show that the half-life T of a radioactive substance is

$$T = \frac{t_1 \ln 2}{\ln A_0 - \ln A_1}.$$

Use this result to verify the half-life in Problem 7.

11. When a vertical beam of light passes through a transparent medium, the rate at which its intensity I decreases is proportional to $I(t)$, where t represents the thickness of the medium (in feet). In clear seawater, the intensity 3 feet below the surface is 25% of the initial intensity I_0 of the incident beam. What is the intensity of the beam 15 feet below the surface?

12. When interest is compounded continuously, the amount of money increases at a rate proportional to the amount S present at time t, that is, $dS/dt = rS$, where r is the annual rate of interest.
 (a) Find the amount of money accrued at the end of 5 years when $5000 is deposited in a savings account drawing $5\frac{3}{4}\%$ annual interest compounded continuously.
 (b) In how many years will the initial sum deposited have doubled?
 (c) Use a calculator to compare the amount obtained in part (a) with the amount $S = 5000(1 + \frac{1}{4}(0.0575))^{5(4)}$ that is accrued when interest is compounded quarterly.

Carbon Dating

13. Archaeologists used pieces of burned wood, or charcoal, found at the site to date prehistoric paintings and drawings on walls and ceilings in a cave in Lascaux, France. See the photo below. Use the information on pages 78 and 79 to determine the approximate age of a piece of burned wood, if it was found that 85.5% of the C-14 found in living trees of the same type had decayed.

© thipjang/Shutterstock
Cave wall painting in Problem 13

14. The **Shroud of Turin**, which shows the negative image of the body of a man who appears to have been crucified, is believed by many to be the burial shroud of Jesus of Nazareth. See the photo on the next page. In 1988 the Vatican granted permission to have the shroud carbon dated. Three independent scientific laboratories analyzed the cloth and concluded that the shroud was approximately 660 years old,* an age consistent with its historical appearance. Using this age,

*Some scholars have disagreed with the finding. For more information on this fascinating mystery, see the Shroud of Turin website home page at www.shroud.com.

determine what percentage of the original amount of C-14 remained in the cloth as of 1988.

© imageBROKER/Shutterstock
Shroud image in Problem 14

Newton's Law of Cooling/Warming

15. A thermometer is removed from a room where the temperature is 70°F and is taken outside, where the air temperature is 10°F. After one-half minute the thermometer reads 50°F. What is the reading of the thermometer at $t = 1$ min? How long will it take for the thermometer to reach 15°F?

16. A thermometer is taken from an inside room to the outside, where the air temperature is 5°F. After 1 minute the thermometer reads 55°F, and after 5 minutes it reads 30°F. What is the initial temperature of the inside room?

17. A small metal bar, whose initial temperature was 20°C, is dropped into a large container of boiling water. How long will it take the bar to reach 90°C if it is known that its temperature increased 2° in 1 second? How long will it take the bar to reach 98°C?

18. Two large containers A and B of the same size are filled with different fluids. The fluids in containers A and B are maintained at 0°C and 100°C, respectively. A small metal bar, whose initial temperature is 100°C, is lowered into container A. After 1 minute the temperature of the bar is 90°C. After 2 minutes the bar is removed and instantly transferred to the other container. After 1 minute in container B the temperature of the bar rises 10°. How long, measured from the start of the entire process, will it take the bar to reach 99.9°C?

19. A thermometer reading 70°F is placed in an oven preheated to a constant temperature. Through a glass window in the oven door, an observer records that the thermometer read 110°F after $\frac{1}{2}$ minute and 145°F after 1 minute. How hot is the oven?

20. At $t = 0$ a sealed test tube containing a chemical is immersed in a liquid bath. The initial temperature of the chemical in the test tube is 80°F. The liquid bath has a controlled temperature (measured in degrees Fahrenheit) given by $T_m(t) = 100 - 40e^{-0.1t}$, $t \geq 0$, where t is measured in minutes.

(a) Assume that $k = -0.1$ in (2). Before solving the IVP, describe in words what you expect the temperature $T(t)$ of the chemical to be like in the short term. In the long term.

(b) Solve the initial-value problem. Use a graphing utility to plot the graph of $T(t)$ on time intervals of various lengths. Do the graphs agree with your predictions in part (a)?

21. A dead body was found within a closed room of a house where the temperature was a constant 70°F. At the time of discovery, the core temperature of the body was determined to be 85°F. One hour later a second measurement showed that the core temperature of the body was 80°F. Assume that the time of death corresponds to $t = 0$ and that the core temperature at that time was 98.6°F. Determine how many hours elapsed before the body was found.

22. Repeat Problem 21 if evidence indicated that the dead person was running a fever of 102°F at the time of death.

Mixtures

23. A tank contains 200 liters of fluid in which 30 grams of salt is dissolved. Brine containing 1 gram of salt per liter is then pumped into the tank at a rate of 4 L/min; the well-mixed solution is pumped out at the same rate. Find the number $A(t)$ of grams of salt in the tank at time t.

24. Solve Problem 23 assuming that pure water is pumped into the tank.

25. A large tank is filled to capacity with 500 gallons of pure water. Brine containing 2 pounds of salt per gallon is pumped into the tank at a rate of 5 gal/min. The well-mixed solution is pumped out at the same rate. Find the number $A(t)$ of pounds of salt in the tank at time t.

26. In Problem 25, what is the concentration $c(t)$ of the salt in the tank at time t? At $t = 5$ min? What is the concentration of the salt in the tank after a long time; that is, as $t \rightarrow \infty$? At what time is the concentration of the salt in the tank equal to one-half this limiting value?

27. Solve Problem 25 under the assumption that the solution is pumped out at a faster rate of 10 gal/min. When is the tank empty?

28. Determine the amount of salt in the tank at time t in Example 5 if the concentration of salt in the inflow is variable and given by $c_{in}(t) = 2 + \sin(t/4)$ lb/gal. Without actually graphing, conjecture what the solution curve of the IVP should look like. Then use a graphing utility to plot the graph of the solution on the interval [0, 300]. Repeat for the interval [0, 600] and compare your graph with that in Figure 2.7.4(a).

29. A large tank is partially filled with 100 gallons of fluid in which 10 pounds of salt is dissolved. Brine containing $\frac{1}{2}$ pound of salt per gallon is pumped into the tank at a rate of 6 gal/min. The well-mixed solution is then pumped out at a slower rate of 4 gal/min. Find the number of pounds of salt in the tank after 30 minutes.

30. In Example 5 the size of the tank containing the salt mixture was not given. Suppose, as in the discussion following Example 5, that the rate at which brine is pumped into the tank is 3 gal/min but that the well-stirred solution is pumped out at a rate of 2 gal/min. It stands to reason that since brine is accumulating in the tank at the rate of 1 gal/min, any finite tank must eventually overflow. Now suppose that the tank has an open top and has a total capacity of 400 gallons.

(a) When will the tank overflow?

(b) What will be the number of pounds of salt in the tank at the instant it overflows?

(c) Assume that, although the tank is overflowing, the brine solution continues to be pumped in at a rate of 3 gal/min and the well-stirred solution continues to be pumped out at a rate of 2 gal/min. Devise a method for determining the number of pounds of salt in the tank at $t = 150$ min.

(d) Determine the number of pounds of salt in the tank as $t \rightarrow \infty$. Does your answer agree with your intuition?

(e) Use a graphing utility to plot the graph $A(t)$ on the interval [0, 500).

Series Circuits

31. A 30-volt electromotive force is applied to an *LR*-series circuit in which the inductance is 0.1 henry and the resistance is 50 ohms. Find the current $i(t)$ if $i(0) = 0$. Determine the current as $t \rightarrow \infty$.

32. Solve equation (8) under the assumption that $E(t) = E_0 \sin \omega t$ and $i(0) = i_0$.

33. A 100-volt electromotive force is applied to an *RC*-series circuit in which the resistance is 200 ohms and the capacitance is 10^{-4} farad. Find the charge $q(t)$ on the capacitor if $q(0) = 0$. Find the current $i(t)$.

34. A 200-volt electromotive force is applied to an *RC*-series circuit in which the resistance is 1000 ohms and the capacitance is 5×10^{-6} farad. Find the charge $q(t)$ on the capacitor if $i(0) = 0.4$. Determine the charge and current at $t = 0.005$ s. Determine the charge as $t \rightarrow \infty$.

35. An electromotive force

$$E(t) = \begin{cases} 120, & 0 \leq t \leq 20 \\ 0, & t > 20 \end{cases}$$

is applied to an *LR*-series circuit in which the inductance is 20 henries and the resistance is 2 ohms. Find the current $i(t)$ if $i(0) = 0$.

36. An *LR*-series circuit has a variable inductor with the inductance defined by

$$L(t) = \begin{cases} 1 - \frac{1}{10}t, & 0 \leq t < 10 \\ 0, & t > 10. \end{cases}$$

Find the current $i(t)$ if the resistance is 0.2 ohm, the impressed voltage is $E(t) = 4$, and $i(0) = 0$. Graph $i(t)$.

Additional Linear Models

37. Air Resistance In (14) of Section 1.3 we saw that a differential equation describing the velocity v of a falling mass subject to air resistance proportional to the instantaneous velocity is

$$m\frac{dv}{dt} = mg - kv,$$

where $k > 0$ is a constant of proportionality called the drag coefficient. The positive direction is downward.

(a) Solve the equation subject to the initial condition $v(0) = v_0$.

(b) Use the solution in part (a) to determine the limiting, or terminal, velocity of the mass. We saw how to determine the terminal velocity without solving the DE in Problem 40 in Exercises 2.1.

(c) If the distance s, measured from the point where the mass was released above ground, is related to velocity v by $ds/dt = v$, find an explicit expression for $s(t)$ if $s(0) = 0$.

38. How High?—No Air Resistance Suppose a small cannonball weighing 16 pounds is shot vertically upward with an initial velocity $v_0 = 300$ ft/s. The answer to the question, "How high does the cannonball go?" depends on whether we take air resistance into account.

(a) Suppose air resistance is ignored. If the positive direction is upward, then a model for the state of the cannonball is given by $d^2s/dt^2 = -g$ (equation (12) of Section 1.3). Since $ds/dt = v(t)$ the last differential equation is the same as $dv/dt = -g$, where we take $g = 32$ ft/s². Find the velocity $v(t)$ of the cannonball at time t.

(b) Use the result obtained in part (a) to determine the height $s(t)$ of the cannonball measured from ground level. Find the maximum height attained by the cannonball.

39. How High?—Linear Air Resistance Repeat Problem 38, but this time assume that air resistance is proportional to instantaneous velocity. It stands to reason that the maximum height attained by the cannonball must be *less* than that in part (b) of Problem 38. Show this by supposing that the drag coefficient is $k = 0.0025$. [*Hint*: Slightly modify the DE in Problem 37.]

40. Skydiving A skydiver weighs 125 pounds, and her parachute and equipment combined weigh another 35 pounds. After exiting from a plane at an altitude of 15,000 feet, she waits 15 seconds and opens her parachute. See **FIGURE 2.7.8**. Assume the constant of proportionality in the model in Problem 37 has the value $k = 0.5$ during free fall and $k = 10$ after the parachute is opened. Assume that her initial velocity on leaving the plane is zero. What is her velocity and how far has she traveled 20 seconds after leaving the plane? How does her velocity at 20 seconds compare with

her terminal velocity? How long does it take her to reach the ground? [*Hint:* Think in terms of two distinct IVPs.]

FIGURE 2.7.8 Skydiver in Problem 40

41. **Evaporating Raindrop** As a raindrop falls, it evaporates while retaining its spherical shape. If we make the further assumptions that the rate at which the raindrop evaporates is proportional to its surface area and that air resistance is negligible, then a model for the velocity $v(t)$ of the raindrop is

$$\frac{dv}{dt} + \frac{3(k/\rho)}{(k/\rho)t + r_0} v = g.$$

Here ρ is the density of water, r_0 is the radius of the raindrop at $t = 0$, $k < 0$ is the constant of proportionality, and the downward direction is taken to be the positive direction.
(a) Solve for $v(t)$ if the raindrop falls from rest.
(b) Reread Problem 36 of Exercises 1.3 and then show that the radius of the raindrop at time t is $r(t) = (k/\rho)t + r_0$.
(c) If $r_0 = 0.01$ ft and $r = 0.007$ ft 10 seconds after the raindrop falls from a cloud, determine the time at which the raindrop has evaporated completely.

42. **Fluctuating Population** The differential equation $dP/dt = (k \cos t)P$, where k is a positive constant, is a mathematical model for a population $P(t)$ that undergoes yearly seasonal fluctuations. Solve the equation subject to $P(0) = P_0$. Use a graphing utility to obtain the graph of the solution for different choices of P_0.

43. **Population Model** In one model of the changing population $P(t)$ of a community, it is assumed that

$$\frac{dP}{dt} = \frac{dB}{dt} - \frac{dD}{dt},$$

where dB/dt and dD/dt are the birth and death rates, respectively.
(a) Solve for $P(t)$ if $dB/dt = k_1P$ and $dD/dt = k_2P$.
(b) Analyze the cases $k_1 > k_2$, $k_1 = k_2$, and $k_1 < k_2$.

44. **Memorization** When forgetfulness is taken into account, the rate of memorization of a subject is given by

$$\frac{dA}{dt} = k_1(M - A) - k_2 A,$$

where $k_1 > 0$, $k_2 > 0$, $A(t)$ is the amount to be memorized in time t, M is the total amount to be memorized, and

$M - A$ is the amount remaining to be memorized. See Problems 25 and 26 in Exercises 1.3.
(a) Since the DE is autonomous, use the phase portrait concept of Section 2.1 to find the limiting value of $A(t)$ as $t \to \infty$. Interpret the result.
(b) Solve for $A(t)$ subject to $A(0) = 0$. Sketch the graph of $A(t)$ and verify your prediction in part (a).

45. **Drug Dissemination** A mathematical model for the rate at which a drug disseminates into the bloodstream is given by $dx/dt = r - kx$, where r and k are positive constants. The function $x(t)$ describes the concentration of the drug in the bloodstream at time t.
(a) Since the DE is autonomous, use the phase portrait concept of Section 2.1 to find the limiting value of $x(t)$ as $t \to \infty$.
(b) Solve the DE subject to $x(0) = 0$. Sketch the graph of $x(t)$ and verify your prediction in part (a). At what time is the concentration one-half this limiting value?

46. **Rocket Motion** Suppose a small single-stage rocket of total mass $m(t)$ is launched vertically and that the rocket consumes its fuel at a constant rate. If the positive direction is upward and if we take air resistance to be linear, then a differential equation for its velocity $v(t)$ is given by

$$\frac{dv}{dt} + \frac{k - \lambda}{m_0 - \lambda t} v = -g + \frac{R}{m_0 - \lambda t},$$

where k is the drag coefficient, λ is the rate at which fuel is consumed, R is the thrust of the rocket, m_0 is the total mass of the rocket at $t = 0$, and g is the acceleration due to gravity. See Problem 21 in Exercises 1.3.
(a) Find the velocity $v(t)$ of the rocket if $m_0 = 200$ kg, $R = 2000$ N, $\lambda = 1$ kg/s, $g = 9.8$ m/s^2, $k = 3$ kg/s, and $v(0) = 0$.
(b) Use $ds/dt = v$ and the result in part (a) to find the height $s(t)$ of the rocket at time t.

47. **Rocket Motion—Continued** In Problem 46, suppose that of the rocket's initial mass m_0, 50 kg is the mass of the fuel.
(a) What is the burnout time t_b, or the time at which all the fuel is consumed? See Problem 22 in Exercises 1.3.
(b) What is the velocity of the rocket at burnout?
(c) What is the height of the rocket at burnout?
(d) Why would you expect the rocket to attain an altitude higher than the number in part (b)?
(e) After burnout what is a mathematical model for the velocity of the rocket?

Discussion Problems

48. **Cooling and Warming** A small metal bar is removed from an oven whose temperature is a constant 300°F into a room whose temperature is a constant 70°F. Simultaneously, an identical metal bar is removed from the room and placed into the oven. Assume that time t is measured in minutes. Discuss: Why is there a future value of time, call it $t^* > 0$, at which the temperature of each bar is the same?

49. Heart Pacemaker A heart pacemaker, shown in **FIGURE 2.7.9**, consists of a switch, a battery, a capacitor, and the heart as a resistor. When the switch S is at P, the capacitor charges; when S is at Q, the capacitor discharges, sending an electrical stimulus to the heart. In Problem 57 in Exercises 2.3, we saw that during the time the electrical stimulus is being applied to the heart, the voltage E across the heart satisfies the linear DE

$$\frac{dE}{dt} = -\frac{1}{RC}E.$$

(a) Let us assume that over the time interval of length t_1, $(0, t_1)$, the switch S is at position P shown in Figure 2.7.9 and the capacitor is being charged. When the switch is moved to position Q at time t_1 the capacitor discharges, sending an impulse to the heart over the time interval of length t_2: $[t_1, t_1 + t_2)$. Thus, over the initial charging/discharging interval $(0, t_1 + t_2)$ the voltage to the heart is actually modeled by the piecewise-linear differential equation

$$\frac{dE}{dt} = \begin{cases} 0, & 0 \le t < t_1 \\ -\dfrac{1}{RC}E, & t_1 \le t < t_1 + t_2. \end{cases}$$

By moving S between P and Q, the charging and discharging over time intervals of lengths t_1 and t_2 is repeated indefinitely. Suppose $t_1 = 4$ s, $t_2 = 2$ s, $E_0 = 12$ V, and $E(0) = 0$, $E(4) = 12$, $E(6) = 0$, $E(10) = 12$, $E(12) = 0$, and so on. Solve for $E(t)$ for $0 \le t \le 24$.

(b) Suppose for the sake of illustration that $R = C = 1$. Use a graphing utility to graph the solution for the IVP in part (a) for $0 \le t \le 24$.

FIGURE 2.7.9 Model of a pacemaker in Problem 49

50. Sliding Box (a) A box of mass m slides down an inclined plane that makes an angle θ with the horizontal as shown in **FIGURE 2.7.10**. Find a differential equation for the velocity $v(t)$ of the box at time t in each of the following three cases:

(*i*) No sliding friction and no air resistance
(*ii*) With sliding friction and no air resistance
(*iii*) With sliding friction and air resistance

In cases (*ii*) and (*iii*), use the fact that the force of friction opposing the motion of the box is μN, where μ is the coefficient of sliding friction and N is the normal

component of the weight of the box. In case (*iii*) assume that air resistance is proportional to the instantaneous velocity.

(b) In part (a), suppose that the box weighs 96 lb, that the angle of inclination of the plane is $\theta = 30°$, that the coefficient of sliding friction is $\mu = \sqrt{3}/4$, and that the additional retarding force due to air resistance is numerically equal to $\frac{1}{4}v$. Solve the differential equation in each of the three cases, assuming that the box starts from rest from the highest point 50 ft above ground.

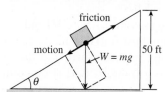

FIGURE 2.7.10 Box sliding down inclined plane in Problem 50

Contributed Problem

Pierre Gharghouri, Professor Emeritus
Jean-Paul Pascal, Associate Professor
Department of Mathematics, Ryerson University, Toronto, Canada

51. Air Exchange A large room has a volume of 2000 m³. The air in this room contains 0.25% by volume of carbon dioxide (CO_2). Starting at 9:00 A.M. fresh air containing 0.04% by volume of CO_2 is circulated into the room at the rate of 400 m³/min. Assume that the stale air leaves the room at the same rate as the incoming fresh air and that the stale air and fresh air mix immediately in the room. See **FIGURE 2.7.11**.

(a) If $v(t)$ denotes the volume of CO_2 in the room at time t, what is $v(0)$? Find $v(t)$ for $t > 0$. What is the percentage of CO_2 in the air of the room at 9:05 A.M?

(b) When does the air in the room contain 0.06% by volume of CO_2?

(c) What should be the flow rate of the incoming fresh air if it is required to reduce the level of CO_2 in the room to 0.08% in 4 minutes?

FIGURE 2.7.11 Air exchange in Problem 51

Computer Lab Assignments

52. Sliding Box—Continued (a) In Problem 50, let $s(t)$ be the distance measured down the inclined plane from the highest point. Use $ds/dt = v(t)$ and the solution for each of the three cases in part (b) of Problem 50 to find the time that it takes the box to slide completely down the

inclined plane. A root-finding application of a CAS may be useful here.

(b) In the case in which there is friction ($\mu \neq 0$) but no air resistance, explain why the box will not slide down the plane starting from *rest* from the highest point above ground when the inclination angle θ satisfies $\tan \theta \leq \mu$.

(c) The box *will* slide downward on the plane when $\tan \theta \leq \mu$ if it is given an initial velocity $v(0) = v_0 > 0$. Suppose that $\mu = \sqrt{3}/4$ and $\theta = 23°$. Verify that $\tan \theta \leq \mu$. How far will the box slide down the plane if $v_0 = 1$ ft/s?

(d) Using the values $\mu = \sqrt{3}/4$ and $\theta = 23°$, approximate the smallest initial velocity v_0 that can be given to the box so that, starting at the highest point 50 ft above ground, it will slide completely down the

inclined plane. Then find the corresponding time it takes to slide down the plane.

53. What Goes Up **(a)** It is well known that the model in which air resistance is ignored, part (a) of Problem 38, predicts that the time t_a it takes the cannonball to attain its maximum height is the same as the time t_d it takes the cannonball to fall from the maximum height to the ground. Moreover, the magnitude of the impact velocity v_i will be the same as the initial velocity v_0 of the cannonball. Verify both of these results.

(b) Then, using the model in Problem 39 that takes linear air resistance into account, compare the value of t_a with t_d and the value of the magnitude of v_i with v_0. A root-finding application of a CAS (or graphic calculator) may be useful here.

2.8 Nonlinear Models

INTRODUCTION We finish our discussion of single first-order differential equations by examining some nonlinear mathematical models.

■ Population Dynamics If $P(t)$ denotes the size of a population at time t, the model for exponential growth begins with the assumption that $dP/dt = kP$ for some $k > 0$. In this model the **relative**, or **specific**, **growth rate** defined by

$$\frac{dP/dt}{P} \tag{1}$$

is assumed to be a constant k. True cases of exponential growth over long periods of time are hard to find, because the limited resources of the environment will at some time exert restrictions on the growth of a population. Thus (1) can be expected to decrease as P increases in size.

The assumption that the rate at which a population grows (or declines) is dependent only on the number present and not on any time-dependent mechanisms such as seasonal phenomena (see Problem 33 in Exercises 1.3) can be stated as

$$\frac{dP/dt}{P} = f(P) \quad \text{or} \quad \frac{dP}{dt} = Pf(P). \tag{2}$$

The differential equation in (2), which is widely assumed in models of animal populations, is called the **density-dependent hypothesis**.

■ Logistic Equation Suppose an environment is capable of sustaining no more than a fixed number of K individuals in its population. The quantity K is called the **carrying capacity** of the environment. Hence, for the function f in (2) we have $f(K) = 0$, and we simply let $f(0) = r$. **FIGURE 2.8.1** shows three functions f that satisfy these two conditions. The simplest assumption that we can make is that $f(P)$ is linear—that is, $f(P) = c_1 P + c_2$. If we use the conditions $f(0) = r$ and $f(K) = 0$, we find, in turn, $c_2 = r$, $c_1 = -r/K$, and so f takes on the form $f(P) = r - (r/K)P$. Equation (2) becomes

$$\frac{dP}{dt} = P\left(r - \frac{r}{K}P\right). \tag{3}$$

Relabeling constants $a = r$ and $b = r/K$, the nonlinear equation (3) is the same as

$$\frac{dP}{dt} = P(a - bP). \tag{4}$$

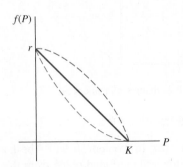

FIGURE 2.8.1 Simplest assumption for $f(P)$ is a straight line

Around 1840 the Belgian mathematician/biologist **Pierre François Verhulst** (1804–1849) was concerned with mathematical models for predicting the human population of various countries. One of the equations he studied was (4), where $a > 0$, $b > 0$. Equation (4) came to be known as the **logistic equation**, and its solution is called the **logistic function**. The graph of a logistic function is called a **logistic curve**.

The linear differential equation $dP/dt = kP$ does not provide a very accurate model for population when the population itself is very large. Overcrowded conditions, with the resulting detrimental effects on the environment, such as pollution and excessive and competitive demands for food and fuel, can have an inhibiting effect on population growth. As we shall now see, a solution of (4) that satisfies an initial condition $P(0) = P_0$, where $0 < P_0 < a/b$, is bounded as $t \to \infty$. If we rewrite (4) as $dP/dt = aP - bP^2$, the nonlinear term $-bP^2$, $b > 0$, can be interpreted as an "inhibition" or "competition" term. Also, in most applications, the positive constant a is much larger than the constant b.

Logistic curves have proved to be quite accurate in predicting the growth patterns, in a limited space, of certain types of bacteria, protozoa, water fleas (*Daphnia*), and fruit flies (*Drosophila*).

■ Solution of the Logistic Equation

One method of solving (4) is by separation of variables. Decomposing the left side of $dP/P(a - bP) = dt$ into partial fractions and integrating gives

$$\left(\frac{1/a}{P} + \frac{b/a}{a - bP} \right) dP = dt$$

$$\frac{1}{a} \ln |P| - \frac{1}{a} \ln |a - bP| = t + c$$

$$\ln \left| \frac{P}{a - bP} \right| = at + ac$$

$$\frac{P}{a - bP} = c_1 e^{at}.$$

It follows from the last equation that

$$P(t) = \frac{ac_1 e^{at}}{1 + bc_1 e^{at}} = \frac{ac_1}{bc_1 + e^{-at}}.$$

If $P(0) = P_0$, $P_0 \neq a/b$, we find $c_1 = P_0/(a - bP_0)$, and so, after substituting and simplifying, the solution becomes

$$P(t) = \frac{aP_0}{bP_0 + (a - bP_0)e^{-at}}. \tag{5}$$

■ Graphs of $P(t)$

The basic shape of the graph of the logistic function $P(t)$ can be obtained without too much effort. Although the variable t usually represents time and we are seldom concerned with applications in which $t < 0$, it is nonetheless of some interest to include this interval when displaying the various graphs of P. From (5) we see that

$$P(t) \to \frac{aP_0}{bP_0} = \frac{a}{b} \text{ as } t \to \infty \quad \text{and} \quad P(t) \to 0 \text{ as } t \to -\infty.$$

The dashed line $P = a/2b$ shown in **FIGURE 2.8.2** corresponds to the y-coordinate of a point of inflection of the logistic curve. To show this, we differentiate (4) by the Product Rule:

$$\frac{d^2P}{dt^2} = P\left(-b\frac{dP}{dt} \right) + (a - bP)\frac{dP}{dt} = \frac{dP}{dt}(a - 2bP)$$

$$= P(a - bP)(a - 2bP)$$

$$= 2b^2 P\left(P - \frac{a}{b} \right)\left(P - \frac{a}{2b} \right).$$

(a)

(b)

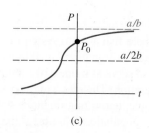

(c)

FIGURE 2.8.2 Logistic curves for different initial conditions

From calculus, recall that the points where $d^2P/dt^2 = 0$ are possible points of inflection, but $P = 0$ and $P = a/b$ can obviously be ruled out. Hence $P = a/2b$ is the only possible ordinate value at which the concavity of the graph can change. For $0 < P < a/2b$ it follows that $P'' > 0$, and $a/2b < P < a/b$ implies $P'' < 0$. Thus, as we read from left to right, the graph changes from concave up to concave down at the point corresponding to $P = a/2b$. When the initial value satisfies $0 < P_0 < a/2b$, the graph of $P(t)$ assumes the shape of an S, as we see in Figure 2.8.2(b). For $a/2b < P_0 < a/b$ the graph is still S-shaped, but the point of inflection occurs at a negative value of t, as shown in Figure 2.8.2(c).

We have already seen equation (4) above in (5) of Section 1.3 in the form $dx/dt = kx(n + 1 - x)$, $k > 0$. This differential equation provides a reasonable model for describing the spread of an epidemic brought about initially by introducing an infected individual into a static population. The solution $x(t)$ represents the number of individuals infected with the disease at time t.

EXAMPLE 1 **Logistic Growth**

Suppose a student carrying a flu virus returns to an isolated college campus of 1000 students. If it is assumed that the rate at which the virus spreads is proportional not only to the number x of infected students but also to the number of students not infected, determine the number of infected students after 6 days if it is further observed that after 4 days $x(4) = 50$.

SOLUTION Assuming that no one leaves the campus throughout the duration of the disease, we must solve the initial-value problem

$$\frac{dx}{dt} = kx(1000 - x), \quad x(0) = 1.$$

By making the identifications $a = 1000k$ and $b = k$, we have immediately from (5) that

$$x(t) = \frac{1000k}{k + 999ke^{-1000kt}} = \frac{1000}{1 + 999e^{-1000kt}}.$$

Now, using the information $x(4) = 50$, we determine k from

$$50 = \frac{1000}{1 + 999e^{-4000k}}.$$

We find $-1000k = \frac{1}{4}\ln\frac{19}{999} = -0.9906$. Thus

$$x(t) = \frac{1000}{1 + 999e^{-0.9906t}}.$$

Finally, $\qquad x(6) = \frac{1000}{1 + 999e^{-5.9436}} = 276$ students.

Additional calculated values of $x(t)$ are given in the table in **FIGURE 2.8.3(b)**.

(a)

t (days)	x (number infected)
4	50 (observed)
5	124
6	276
7	507
8	735
9	882
10	953

(b)

FIGURE 2.8.3 Number of infected students in Example 1

Modifications of the Logistic Equation There are many variations of the logistic equation. For example, the differential equations

$$\frac{dP}{dt} = P(a - bP) - h \quad \text{and} \quad \frac{dP}{dt} = P(a - bP) + h \tag{6}$$

could serve, in turn, as models for the population in a fishery where fish are **harvested** or are **restocked** at rate h. When $h > 0$ is a constant, the DEs in (6) can be readily analyzed qualitatively or solved by separation of variables. The equations in (6) could also serve as models of a human population either increased by immigration or decreased by emigration. The rate h in (6) could be a function of time t or may be population dependent; for example, harvesting might be done periodically over time or may be done at a rate proportional to the population P at time t. In the latter instance, the model would look like $P' = P(a - bP) - cP$, $c > 0$. A human population of a community might change due to immigration in such a manner that the contribution due to immigration is large when the population P of the community is itself small, but then the immigration

contribution might be small when P is large; a reasonable model for the population of the community is then $P' = P(a - bP) + ce^{-kP}$, $c > 0$, $k > 0$. Another equation of the form given in (2),

$$\frac{dP}{dt} = P(a - b \ln P), \tag{7}$$

is a modification of the logistic equation known as the **Gompertz differential equation** named after the English self-educated mathematician **Benjamin Gompertz** (1779–1865). This DE is sometimes used as a model in the study of the growth or decline of population, in the growth of solid tumors, and in certain kinds of actuarial predictions. See Problems 5–10 in Exercises 2.8.

Chemical Reactions

Suppose that a grams of chemical A are combined with b grams of chemical B. If there are M parts of A and N parts of B formed in the compound and $X(t)$ is the number of grams of chemical C formed, then the numbers of grams of chemicals A and B remaining at any time are, respectively,

$$a - \frac{M}{M + N} X \quad \text{and} \quad b - \frac{N}{M + N} X.$$

By the law of mass action, the rate of the reaction satisfies

$$\frac{dX}{dt} \propto \left(a - \frac{M}{M + N} X \right)\left(b - \frac{N}{M + N} X \right). \tag{8}$$

If we factor out $M/(M + N)$ from the first factor and $N/(M + N)$ from the second and introduce a constant $k > 0$ of proportionality, (8) has the form

$$\frac{dX}{dt} = k(\alpha - X)(\beta - X), \tag{9}$$

where $\alpha = a(M + N)/M$ and $\beta = b(M + N)/N$. Recall from (6) of Section 1.3 that a chemical reaction governed by the nonlinear differential equation (9) is said to be a **second-order reaction**.

EXAMPLE 2 Second-Order Chemical Reaction

A compound C is formed when two chemicals A and B are combined. The resulting reaction between the two chemicals is such that for each gram of A, 4 grams of B are used. It is observed that 30 grams of the compound C are formed in 10 minutes. Determine the amount of C at time t if the rate of the reaction is proportional to the amounts of A and B remaining and if initially there are 50 grams of A and 32 grams of B. How much of the compound C is present at 15 minutes? Interpret the solution as $t \to \infty$.

SOLUTION Let $X(t)$ denote the number of grams of the compound C present at time t. Clearly $X(0) = 0$ g and $X(10) = 30$ g.

If, for example, 2 grams of compound C are present, we must have used, say, a grams of A and b grams of B so that $a + b = 2$ and $b = 4a$. Thus we must use $a = \frac{2}{5} = 2(\frac{1}{5})$ g of chemical A and $b = \frac{8}{5} = 2(\frac{4}{5})$ g of B. In general, for X grams of C we must use

$$\frac{1}{5} X \text{ grams of } A \quad \text{and} \quad \frac{4}{5} X \text{ grams of } B.$$

The amounts of A and B remaining at any time are then

$$50 - \frac{1}{5} X \quad \text{and} \quad 32 - \frac{4}{5} X,$$

respectively.

Now we know that the rate at which compound C is formed satisfies

$$\frac{dX}{dt} \propto \left(50 - \frac{1}{5} X \right)\left(32 - \frac{4}{5} X \right).$$

To simplify the subsequent algebra, we factor $\frac{1}{5}$ from the first term and $\frac{4}{5}$ from the second and then introduce the constant of proportionality:

$$\frac{dX}{dt} = k(250 - X)(40 - X).$$

By separation of variables and partial fractions we can write

$$-\frac{\frac{1}{210}}{250 - X}\,dX + \frac{\frac{1}{210}}{40 - X}\,dX = k\,dt.$$

Integrating gives

$$\ln\left|\frac{250 - X}{40 - X}\right| = 210kt + c_1 \quad \text{or} \quad \frac{250 - X}{40 - X} = c_2 e^{210kt}. \tag{10}$$

When $t = 0$, $X = 0$, so it follows at this point that $c_2 = \frac{25}{4}$. Using $X = 30$ g at $t = 10$, we find $210k = \frac{1}{10}\ln\frac{88}{25} = 0.1258$. With this information we solve the last equation in (10) for X:

$$X(t) = 1000\frac{1 - e^{-0.1258t}}{25 - 4e^{-0.1258t}}. \tag{11}$$

The behavior of X as a function of time is displayed in **FIGURE 2.8.4**. It is clear from the accompanying table and (11) that $X \to 40$ as $t \to \infty$. This means that 40 grams of compound C are formed, leaving

$$50 - \frac{1}{5}(40) = 42 \text{ g of } A \quad \text{and} \quad 32 - \frac{4}{5}(40) = 0 \text{ g of } B. \qquad \equiv$$

(a)

t (min)	X (g)
10	30 (measured)
15	34.78
20	37.25
25	38.54
30	39.22
35	39.59

(b)

FIGURE 2.8.4 Amount of compound C in Example 2

REMARKS

The indefinite integral $\displaystyle\int \frac{du}{a^2 - u^2}$ can be evaluated in terms of logarithms, the inverse hyperbolic tangent, or the inverse hyperbolic cotangent. For example, of the two results,

$$\int \frac{du}{a^2 - u^2} = \frac{1}{a}\tanh^{-1}\frac{u}{a} + c, \qquad |u| < a \tag{12}$$

$$\int \frac{du}{a^2 - u^2} = \frac{1}{2a}\ln\left|\frac{a + u}{a - u}\right| + c, \qquad |u| \neq a \tag{13}$$

(12) may be convenient for Problems 19 and 31 in Exercises 2.8, whereas (13) may be preferable in Problem 32.

2.8 Exercises
Answers to selected odd-numbered problems begin on page ANS-3.

Logistic Equation

1. The number $N(t)$ of supermarkets throughout the country that are using a computerized checkout system is described by the initial-value problem

$$\frac{dN}{dt} = N(1 - 0.0005N), \quad N(0) = 1.$$

(a) Use the phase portrait concept of Section 2.1 to predict how many supermarkets are expected to adopt the new

procedure over a long period of time. By hand, sketch a solution curve of the given initial-value problem.

(b) Solve the initial-value problem and then use a graphing utility to verify the solution curve in part (a). How many companies are expected to adopt the new technology when $t = 10$?

2. The number $N(t)$ of people in a community who are exposed to a particular advertisement is governed by the logistic equation. Initially $N(0) = 500$, and it is observed that $N(1) = 1000$. Solve for $N(t)$ if it is predicted that the

limiting number of people in the community who will see the advertisement is 50,000.

3. A model for the population $P(t)$ in a suburb of a large city is given by the initial-value problem

$$\frac{dP}{dt} = P(10^{-1} - 10^{-7}P), \quad P(0) = 5000,$$

where t is measured in months. What is the limiting value of the population? At what time will the population be equal to one-half of this limiting value?

4. (a) Census data for the United States between 1790 and 1950 are given in the following table. Construct a logistic population model using the data from 1790, 1850, and 1910.

Year	Population (in millions)
1790	3.929
1800	5.308
1810	7.240
1820	9.638
1830	12.866
1840	17.069
1850	23.192
1860	31.433
1870	38.558
1880	50.156
1890	62.948
1900	75.996
1910	91.972
1920	105.711
1930	122.775
1940	131.669
1950	150.697

 (b) Construct a table comparing actual census population with the population predicted by the model in part (a). Compute the error and the percentage error for each entry pair.

Modifications of the Logistic Equation

5. If a constant number h of fish is harvested from a fishery per unit time, then a model for the population $P(t)$ of the fishery at time t is given by

$$\frac{dP}{dt} = P(a - bP) - h, \quad P(0) = P_0,$$

where $a, b, h,$ and P_0 are positive constants. Suppose $a = 5$, $b = 1$, and $h = 4$. Show that the differential equation becomes

$$\frac{dP}{dt} = -(P - 4)(P - 1).$$

Since the DE is autonomous, use the phase portrait concept of Section 2.1 to sketch representative solution curves corresponding to the cases $P_0 > 4, 1 < P_0 < 4,$ and $0 < P_0 < 1$. Determine the long-term behavior of the population in each case.

6. Solve the IVP in Problem 5. Verify your results of your phase portrait in Problem 5 by using a graphing utility to plot the graph of $P(t)$ with an initial condition taken from each of the three intervals given.

7. Use the information in Problems 5 and 6 to determine whether the fishery population $P(t)$ becomes extinct in finite time. If so, find that time.

8. Investigate the harvesting model in Problems 5–7 both qualitatively and analytically in the case $a = 5$, $b = 1$, $h = \frac{25}{4}$. Determine whether the population becomes extinct in finite time. If so, find that time.

9. Repeat Problem 8 in the case $a = 5, b = 1, h = 7$.

10. (a) Suppose $a = b = 1$ in the Gompertz differential equation (7). Since the DE is autonomous, use the phase portrait concept of Section 2.1 to sketch representative solution curves corresponding to the cases $P_0 > e$ and $0 < P_0 < e$.

 (b) Suppose $a = 1, b = -1$ in (7). Use a new phase portrait to sketch representative solution curves corresponding to the cases $P_0 > e^{-1}$ and $0 < P_0 < e^{-1}$.

11. Find an explicit solution of equation (7) subject to $P(0) = P_0$.

12. **The Allee Effect** For an initial population P_0, where $P_0 > K$ the logistic population model (3) predicts that population cannot sustain itself over time so it decreases but yet never falls below the carrying capacity K of the ecosystem. Moreover, for $0 < P_0 < K$, the same model predicts that, regardless of how small P_0 is, the population increases over time and does not surpass the carrying capacity K. See Figure 2.8.2, where $a/b = K$. But the American ecologist **Warder Clyde Allee** (1885–1955) showed that by depleting certain fisheries beyond a certain level, the fishery population never recovers. How would you modify the differential equation (3) to describe a population P that has these same two characteristics of (3) but additionally has a **threshold level** A, $0 < A < K$, below which the population cannot sustain itself and becomes extinct. [*Hint*: Construct a phase portrait of what you want and then form a DE.]

Chemical Reactions

13. Two chemicals A and B are combined to form a chemical C. The rate, or velocity, of the reaction is proportional to the product of the instantaneous amounts of A and B not converted to chemical C. Initially there are 40 grams of A and 50 grams of B, and for each gram of B, 2 grams of A are used. It is observed that 10 grams of C are formed in 5 minutes. How much is formed in 20 minutes? What is the limiting amount of C after a long time? How much of chemicals A and B remains after a long time?

14. Solve Problem 13 if 100 grams of chemical A are present initially. At what time is chemical C half-formed?

Additional Nonlinear Models

15. Leaking Cylindrical Tank A tank in the form of a right-circular cylinder standing on end is leaking water through a circular hole in its bottom. As we saw in (10) of Section 1.3, when friction and contraction of water at the hole are ignored, the height h of water in the tank is described by

$$\frac{dh}{dt} = -\frac{A_h}{A_w}\sqrt{2gh},$$

where A_w and A_h are the cross-sectional areas of the water and the hole, respectively.

(a) Solve for $h(t)$ if the initial height of the water is H. By hand, sketch the graph of $h(t)$ and give its interval I of definition in terms of the symbols A_w, A_h, and H. Use $g = 32$ ft/s^2.

(b) Suppose the tank is 10 ft high and has radius 2 ft and the circular hole has radius $\frac{1}{2}$ in. If the tank is initially full, how long will it take to empty?

16. Leaking Cylindrical Tank—Continued When friction and contraction of the water at the hole are taken into account, the model in Problem 15 becomes

$$\frac{dh}{dt} = -c\frac{A_h}{A_w}\sqrt{2gh},$$

where $0 < c < 1$. How long will it take the tank in Problem 15(b) to empty if $c = 0.6$? See Problem 13 in Exercises 1.3.

17. Leaking Conical Tank A tank in the form of a right-circular cone standing on end, vertex down, is leaking water through a circular hole in its bottom.

(a) Suppose the tank is 20 ft high and has radius 8 ft and the circular hole has radius 2 in. In Problem 14 in Exercises 1.3 you were asked to show that the differential equation governing the height h of water leaking from a tank is

$$\frac{dh}{dt} = -\frac{5}{6h^{3/2}}.$$

In this model, friction and contraction of the water at the hole were taken into account with $c = 0.6$, and g was taken to be 32 ft/s^2. See Figure 1.3.13. If the tank is initially full, how long will it take the tank to empty?

(b) Suppose the tank has a vertex angle of $60°$, and the circular hole has radius 2 in. Determine the differential equation governing the height h of water. Use $c = 0.6$ and $g = 32$ ft/s^2. If the height of the water is initially 9 ft, how long will it take the tank to empty?

18. Inverted Conical Tank Suppose that the conical tank in Problem 17(a) is inverted, as shown in **FIGURE 2.8.5**, and that water leaks out a circular hole of radius 2 in. in the center of the circular base. Is the time it takes to empty a full tank the same as for the tank with vertex down in Problem 17? Take the friction/contraction coefficient to be $c = 0.6$ and $g = 32$ ft/s^2.

FIGURE 2.8.5 Inverted conical tank in Problem 18

19. Air Resistance A differential equation governing the velocity v of a falling mass m subjected to air resistance proportional to the square of the instantaneous velocity is

$$m\frac{dv}{dt} = mg - kv^2,$$

where $k > 0$ is the drag coefficient. The positive direction is downward.

(a) Solve this equation subject to the initial condition $v(0) = v_0$.

(b) Use the solution in part (a) to determine the limiting, or terminal, velocity of the mass. We saw how to determine the terminal velocity without solving the DE in Problem 41 in Exercises 2.1.

(c) If distance s, measured from the point where the mass was released above ground, is related to velocity v by $ds/dt = v(t)$, find an explicit expression for $s(t)$ if $s(0) = 0$.

20. How High?—Nonlinear Air Resistance Consider the 16-pound cannonball shot vertically upward in Problems 38 and 39 in Exercises 2.7 with an initial velocity $v_0 = 300$ ft/s. Determine the maximum height attained by the cannonball if air resistance is assumed to be proportional to the square of the instantaneous velocity. Assume the positive direction is upward and take the drag coefficient to be $k = 0.0003$. [*Hint*: Slightly modify the DE in Problem 19.]

21. That Sinking Feeling (a) Determine a differential equation for the velocity $v(t)$ of a mass m sinking in water that imparts a resistance proportional to the square of the instantaneous velocity and also exerts an upward buoyant force whose magnitude is given by Archimedes' principle. See Problem 18 in Exercises 1.3. Assume that the positive direction is downward.

(b) Solve the differential equation in part (a).

(c) Determine the limiting, or terminal, velocity of the sinking mass.

22. Solar Collector The differential equation

$$\frac{dy}{dx} = \frac{-x + \sqrt{x^2 + y^2}}{y}$$

describes the shape of a plane curve C that will reflect all incoming light beams to the same point and could be a model for the mirror of a reflecting telescope, a satellite antenna,

or a solar collector. See Problem 29 in Exercises 1.3. There are several ways of solving this DE.

(a) Verify that the differential equation is homogeneous (see Section 2.5). Show that the substitution $y = ux$ yields

$$\frac{u\,du}{\sqrt{1 + u^2}(1 - \sqrt{1 + u^2})} = \frac{dx}{x}.$$

Use a CAS (or another judicious substitution) to integrate the left-hand side of the equation. Show that the curve C must be a parabola with focus at the origin and is symmetric with respect to the x-axis.

(b) Show that the first differential equation can also be solved by means of the substitution $u = x^2 + y^2$.

23. **Tsunami** (a) A simple model for the shape of a tsunami is given by

$$\frac{dW}{dx} = W\sqrt{4 - 2W},$$

where $W(x) > 0$ is the height of the wave expressed as a function of its position relative to a point offshore. By inspection, find all constant solutions of the DE.

(b) Solve the differential equation in part (a). A CAS may be useful for integration.

(c) Use a graphing utility to obtain the graphs of all solutions that satisfy the initial condition $W(0) = 2$.

24. **Evaporation** An outdoor decorative pond in the shape of a hemispherical tank is to be filled with water pumped into the tank through an inlet in its bottom. Suppose that the radius of the tank is $R = 10$ ft, that water is pumped in at a rate of π ft^3/min, and that the tank is initially empty. See **FIGURE 2.8.6**. As the tank fills, it loses water through evaporation. Assume that the rate of evaporation is proportional to the area A of the surface of the water and that the constant of proportionality is $k = 0.01$.

(a) The rate of change dV/dt of the volume of the water at time t is a net rate. Use this net rate to determine a differential equation for the height h of the water at time t. The volume of the water shown in the figure is $V = \pi R h^2 - \frac{1}{3}\pi h^3$, where $R = 10$. Express the area of the surface of the water $A = \pi r^2$ in terms of h.

(b) Solve the differential equation in part (a). Graph the solution.

Output: water evaporates at rate proportional to area A of surface

Input: water pumped in at rate of π ft^3/min

(a) Hemispherical tank (b) Cross section of tank

FIGURE 2.8.6 Pond in Problem 24

(c) If there were no evaporation, how long would it take the tank to fill?

(d) With evaporation, what is the depth of the water at the time found in part (c)? Will the tank ever be filled? Prove your assertion.

25. **Sawing Wood** A long *uniform* piece of wood (cross sections are the same) is cut through perpendicular to its length by a vertical saw blade. See **FIGURE 2.8.7**. If the friction between the sides of the saw blade and the wood through which the blade passes is ignored, then it can be assumed that the rate at which the saw blade moves through the piece of wood is inversely proportional to the width of the wood in contact with its cutting edge. As the blade advances through the wood (moving left to right), the width of a cross section changes as a nonnegative continuous function w. If a cross section of the wood is described as a region in the xy-plane defined over an interval $[a, b]$, then as shown in Figure 2.8.7(c) the position x of the saw blade is a function of time t and the vertical cut made by the blade can be represented by a vertical line segment. The length of this vertical line is the width $w(x)$ of the wood at that point. Thus the position $x(t)$ of the saw blade and the rate dx/dt at which it moves to the right are related to $w(x)$ by

$$w(x)\frac{dx}{dt} = k, \quad x(0) = a. \tag{14}$$

Here k represents the number of square units of the material removed by the saw blade per unit time. In the problems that follow, we assume that the saw can be programmed so that $k = 1$. Find an *explicit* solution $x(t)$ of the initial-value problem (14) when a cross section of the piece of wood is triangular and is bounded by the graphs of $y = x$, $x = 1$, and $y = 0$. How long does it take the saw to cut through this piece of wood?

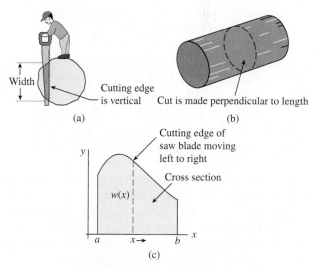

Width
Cutting edge is vertical

Cut is made perpendicular to length

(a) (b)

Cutting edge of saw blade moving left to right

Cross section

$w(x)$

(c)

FIGURE 2.8.7 Sawing a log in Problem 25

26. (a) Find an *implicit* solution of the initial-value problem (14) in Problem 25 when the piece of wood is a circular log. Assume a cross section is a circle of radius 2

centered at $(0,0)$. [*Hint*: To save time, see formula 33 in the table of integrals given on the right page inside the front cover.]

(b) Solve the implicit solution obtained in part (b) for time t as a function of x. Graph the function $t(x)$. With the aid of the graph, approximate the time that it takes the saw to cut through this piece of wood. Then find the exact value of this time.

27. Solve the initial-value problem (14) in Problem 25 when a cross section of a uniform piece of wood is the triangular region given in **FIGURE 2.8.8**. Assume again that $k = 1$. How long does it take to cut through this piece of wood?

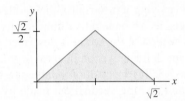

FIGURE 2.8.8 Triangular cross section in Problem 27

Computer Lab Assignments

28. **Regression Line** Read the documentation for your CAS on *scatter plots* (or *scatter diagrams*) and *least-squares linear fit*. The straight line that best fits a set of data points is called a **regression line** or a **least squares line**. Your task is to construct a logistic model for the population of the United States, defining $f(P)$ in (2) as an equation of a regression line based on the population data in the table in Problem 4. One way of doing this is to approximate the left-hand side $\dfrac{1}{P}\dfrac{dP}{dt}$ of the first equation in (2) using the forward difference quotient in place of dP/dt:

$$Q(t) = \frac{1}{P(t)}\frac{P(t+h) - P(t)}{h}.$$

(a) Make a table of the values t, $P(t)$, and $Q(t)$ using $t = 0$, $10, 20, \ldots, 160$, and $h = 10$. For example, the first line of the table should contain $t = 0$, $P(0)$, and $Q(0)$. With $P(0) = 3.929$ and $P(10) = 5.308$,

$$Q(0) = \frac{1}{P(0)}\frac{P(10) - P(0)}{10} = 0.035.$$

Note that $Q(160)$ depends on the 1960 census population $P(170)$. Look up this value.

(b) Use a CAS to obtain a scatter plot of the data $(P(t), Q(t))$ computed in part (a). Also use a CAS to find an equation of the regression line and to superimpose its graph on the scatter plot.

(c) Construct a logistic model $dP/dt = Pf(P)$, where $f(P)$ is the equation of the regression line found in part (b).

(d) Solve the model in part (c) using the initial condition $P(0) = 3.929$.

(e) Use a CAS to obtain another scatter plot, this time of the ordered pairs $(t, P(t))$ from your table in part (a).

Use your CAS to superimpose the graph of the solution in part (d) on the scatter plot.

(f) Look up the U.S. census data for 1970, 1980, and 1990. What population does the logistic model in part (c) predict for these years? What does the model predict for the U.S. population $P(t)$ as $t \to \infty$?

29. **Immigration Model** **(a)** In Examples 3 and 4 of Section 2.1, we saw that any solution $P(t)$ of (4) possesses the asymptotic behavior $P(t) \to a/b$ as $t \to \infty$ for $P_0 > a/b$ and for $0 < P_0 < a/b$; as a consequence, the equilibrium solution $P = a/b$ is called an attractor. Use a root-finding application of a CAS (or a graphic calculator) to approximate the equilibrium solution of the immigration model

$$\frac{dP}{dt} = P(1 - P) + 0.3e^{-P}.$$

(b) Use a graphing utility to graph the function $F(P) = P(1 - P) + 0.3e^{-P}$. Explain how this graph can be used to determine whether the number found in part (a) is an attractor.

(c) Use a numerical solver to compare the solution curves for the IVPs

$$\frac{dP}{dt} = P(1 - P), \quad P(0) = P_0$$

for $P_0 = 0.2$ and $P_0 = 1.2$ with the solution curves for the IVPs

$$\frac{dP}{dt} = P(1 - P) + 0.3e^{-P}, \quad P(0) = P_0$$

for $P_0 = 0.2$ and $P_0 = 1.2$. Superimpose all curves on the same coordinate axes but, if possible, use a different color for the curves of the second initial-value problem. Over a long period of time, what percentage increase does the immigration model predict in the population compared to the logistic model?

30. **What Goes Up . . .** In Problem 20 let t_a be the time it takes the cannonball to attain its maximum height and let t_d be the time it takes the cannonball to fall from the maximum height to the ground. Compare the value of t_a with the value of t_d and compare the magnitude of the impact velocity v_i with the initial velocity v_0. See Problem 53 in Exercises 2.7. A root-finding application of a CAS may be useful here. [*Hint*: Use the model in Problem 19 when the cannonball is falling.]

31. **Skydiving** A skydiver is equipped with a stopwatch and an altimeter. She opens her parachute 25 seconds after exiting a plane flying at an altitude of 20,000 ft and observes that her altitude is 14,800 ft. Assume that air resistance is proportional to the square of the instantaneous velocity, her initial velocity upon leaving the plane is zero, and $g = 32$ ft/s^2.

(a) Find the distance $s(t)$, measured from the plane, that the skydiver has traveled during free fall in time t. [*Hint*: The constant of proportionality k in the model

given in Problem 19 is not specified. Use the expression for terminal velocity v_t obtained in part (b) of Problem 19 to eliminate k from the IVP. Then eventually solve for v_t.]

(b) How far does the skydiver fall and what is her velocity at $t = 15$ s?

32. Hitting Bottom A helicopter hovers 500 feet above a large open tank full of liquid (not water). A dense compact object weighing 160 pounds is dropped (released from rest) from the helicopter into the liquid. Assume that air resistance is proportional to instantaneous velocity v while the object is in the air and that viscous damping is proportional to v^2 after the object has entered the liquid. For air, take $k = \frac{1}{4}$, and for the liquid, $k = 0.1$. Assume that the positive direction is downward. If the tank is 75 feet high, determine the time and the impact velocity when the object hits the bottom of the tank. [*Hint*: Think in terms of two distinct IVPs. If you use (13), be careful in removing the absolute value sign. You might compare the velocity when the object hits the liquid—the initial velocity for the second problem—with the terminal velocity v_t of the object falling through the liquid.]

2.9 Modeling with Systems of First-Order DEs

INTRODUCTION In this section we are going to discuss mathematical models based on some of the topics that we have already discussed in the preceding two sections. This section will be similar to Section 1.3 in that we are only going to discuss systems of first-order differential equations as mathematical models and we are not going to solve any of these models. There are two good reasons for not solving systems at this point: First, we do not as yet possess the necessary mathematical tools for solving systems, and second, some of the systems that we discuss cannot be solved analytically. We shall examine solution methods for systems of linear first-order DEs in Chapter 10 and for systems of linear higher-order DEs in Chapters 3 and 4.

■ **Systems** Up to now all the mathematical models that we have considered were single differential equations. A single differential equation could describe a single population in an environment, but if there are, say, two interacting and perhaps competing species living in the same environment (for example, rabbits and foxes), then a model for their populations $x(t)$ and $y(t)$ *might* be a system of two first-order differential equations such as

$$\begin{aligned} \frac{dx}{dt} &= g_1(t, x, y) \\ \frac{dy}{dt} &= g_2(t, x, y). \end{aligned} \tag{1}$$

When g_1 and g_2 are linear in the variables x and y, that is,

$$g_1(t, x, y) = c_1 x + c_2 y + f_1(t) \quad \text{and} \quad g_2(t, x, y) = c_3 x + c_4 y + f_2(t),$$

then (1) is said to be a **linear system**. A system of differential equations that is not linear is said to be **nonlinear**.

■ **Radioactive Series** In the discussion of radioactive decay in Sections 1.3 and 2.7, we assumed that the rate of decay was proportional to the number $A(t)$ of nuclei of the substance present at time t. When a substance decays by radioactivity, it usually doesn't just transmute into one stable substance and then the process stops. Rather, the first substance decays into another radioactive substance, this substance in turn decays into yet a third substance, and so on. This process, called a **radioactive decay series**, continues until a stable element is reached. For example, the uranium decay series is U-238 $\rightarrow$ Th-234 $\rightarrow \cdots \rightarrow$ Pb-206, where Pb-206 is a stable isotope of lead. The half-lives of the various elements in a radioactive series can range from billions of years (4.5×10^9 years for U-238) to a fraction of a second. Suppose a radioactive series is described schematically by $X \xrightarrow{-\lambda_1} Y \xrightarrow{-\lambda_2} Z$, where $k_1 = -\lambda_1 < 0$ and $k_2 = -\lambda_2 < 0$ are the decay constants for substances X and Y, respectively, and Z is a stable element. Suppose too, that $x(t)$, $y(t)$, and $z(t)$ denote amounts of substances X, Y, and Z, respectively, remaining at time t. The decay of element X is described by

$$\frac{dx}{dt} = -\lambda_1 x,$$

whereas the rate at which the second element Y decays is the net rate,

$$\frac{dy}{dt} = \lambda_1 x - \lambda_2 y,$$

since it is *gaining* atoms from the decay of X and at the same time *losing* atoms due to its own decay. Since Z is a stable element, it is simply gaining atoms from the decay of element Y:

$$\frac{dz}{dt} = \lambda_2 y.$$

In other words, a model of the radioactive decay series for three elements is the linear system of three first-order differential equations

$$
\begin{aligned}
\frac{dx}{dt} &= -\lambda_1 x \\
\frac{dy}{dt} &= \lambda_1 x - \lambda_2 y \\
\frac{dz}{dt} &= \lambda_2 y.
\end{aligned}
\tag{2}
$$

pure water
3 gal/min

mixture
1 gal/min

mixture
4 gal/min

mixture
3 gal/min

FIGURE 2.9.1 Connected mixing tanks

■ **Mixtures** Consider the two tanks shown in **FIGURE 2.9.1**. Let us suppose for the sake of discussion that tank A contains 50 gallons of water in which 25 pounds of salt is dissolved. Suppose tank B contains 50 gallons of pure water. Liquid is pumped in and out of the tanks as indicated in the figure; the mixture exchanged between the two tanks and the liquid pumped out of tank B is assumed to be well stirred. We wish to construct a mathematical model that describes the number of pounds $x_1(t)$ and $x_2(t)$ of salt in tanks A and B, respectively, at time t.

By an analysis similar to that on page 22 in Section 1.3 and Example 5 of Section 2.7, we see for tank A that the net rate of change of $x_1(t)$ is

$$
\frac{dx_1}{dt} = \overbrace{(3 \text{ gal/min}) \cdot (0 \text{ lb/gal}) + (1 \text{ gal/min}) \cdot \left(\frac{x_2}{50} \text{ lb/gal}\right)}^{\text{input rate of salt}} - \overbrace{(4 \text{ gal/min}) \cdot \left(\frac{x_1}{50} \text{ lb/gal}\right)}^{\text{output rate of salt}}
$$

$$
= -\frac{2}{25} x_1 + \frac{1}{50} x_2.
$$

Similarly, for tank B, the net rate of change of $x_2(t)$ is

$$
\frac{dx_2}{dt} = 4 \cdot \frac{x_1}{50} - 3 \cdot \frac{x_2}{50} - 1 \cdot \frac{x_2}{50} = \frac{2}{25} x_1 - \frac{2}{25} x_2.
$$

Thus we obtain the linear system

$$
\begin{aligned}
\frac{dx_1}{dt} &= -\frac{2}{25} x_1 + \frac{1}{50} x_2 \\
\frac{dx_2}{dt} &= \frac{2}{25} x_1 - \frac{2}{25} x_2.
\end{aligned}
\tag{3}
$$

Observe that the foregoing system is accompanied by the initial conditions $x_1(0) = 25$, $x_2(0) = 0$.

■ **A Predator–Prey Model** Suppose that two different species of animals interact within the same environment or ecosystem, and suppose further that the first species eats only vegetation and the second eats only the first species. In other words, one species is a predator and the other is a prey. For example, wolves hunt grass-eating caribou, sharks devour little fish, and the snowy owl pursues an arctic rodent called the lemming. For the sake of discussion, let us imagine that the predators are foxes and the prey are rabbits.

Let $x(t)$ and $y(t)$ denote, respectively, the fox and rabbit populations at time t. If there were no rabbits, then one might expect that the foxes, lacking an adequate food supply, would decline in number according to

$$\frac{dx}{dt} = -ax, \quad a > 0. \tag{4}$$

When rabbits are present in the environment, however, it seems reasonable that the number of encounters or interactions between these two species per unit time is jointly proportional to their populations x and y; that is, proportional to the product xy. Thus when rabbits are present there is a supply of food, and so foxes are added to the system at a rate bxy, $b > 0$. Adding this last rate to (4) gives a model for the fox population:

$$\frac{dx}{dt} = -ax + bxy. \tag{5}$$

On the other hand, were there no foxes, then the rabbits would, with an added assumption of unlimited food supply, grow at a rate that is proportional to the number of rabbits present at time t:

$$\frac{dy}{dt} = dy, \quad d > 0. \tag{6}$$

But when foxes are present, a model for the rabbit population is (6) decreased by cxy, $c > 0$; that is, decreased by the rate at which the rabbits are eaten during their encounters with the foxes:

$$\frac{dy}{dt} = dy - cxy. \tag{7}$$

Equations (5) and (7) constitute a system of nonlinear differential equations

$$\frac{dx}{dt} = -ax + bxy = x(-a + by)$$
$$\frac{dy}{dt} = dy - cxy = y(d - cx), \tag{8}$$

where a, b, c, and d are positive constants. This famous system of equations is known as the **Lotka–Volterra predator–prey model**.

Except for two constant solutions, $x(t) = 0$, $y(t) = 0$ and $x(t) = d/c$, $y(t) = a/b$, the nonlinear system (8) cannot be solved in terms of elementary functions. However, we can analyze such systems quantitatively and qualitatively. See Chapters 6 and 11.

<div style="background:#888;color:#fff;display:inline-block;padding:2px 8px;">EXAMPLE **1**</div> **Predator–Prey Model**

Suppose

$$\frac{dx}{dt} = -0.16x + 0.08xy$$
$$\frac{dy}{dt} = 4.5y - 0.9xy$$

represents a predator–prey model. Since we are dealing with populations, we have $x(t) \geq 0$, $y(t) \geq 0$. **FIGURE 2.9.2**, obtained with the aid of a numerical solver, shows typical population curves of the predators and prey for this model superimposed on the same coordinate axes. The initial conditions used were $x(0) = 4$, $y(0) = 4$. The curve in red represents the population $x(t)$ of the predator (foxes), and the blue curve is the population $y(t)$ of the prey (rabbits). Observe that the model seems to predict that both populations $x(t)$ and $y(t)$ are periodic in time. This makes intuitive sense since, as the number of prey decreases, the predator population eventually decreases because of a diminished food supply; but attendant to a decrease in the number of predators is an increase in the number of prey; this in turn gives rise to an increased number of predators, which ultimately brings about another decrease in the number of prey. ≡

FIGURE 2.9.2 Population of predators (red) and prey (blue) appear to be periodic in Example 1

■ **Competition Models** Now suppose two different species of animals occupy the same ecosystem, not as predator and prey but rather as competitors for the same resources (such as food and living space) in the system. In the absence of the other, let us assume that the rate at which each population grows is given by

$$\frac{dx}{dt} = ax \quad \text{and} \quad \frac{dy}{dt} = cy, \tag{9}$$

respectively.

Since the two species compete, another assumption might be that each of these rates is diminished simply by the influence, or existence, of the other population. Thus a model for the two populations is given by the linear system

$$\frac{dx}{dt} = ax - by$$
$$\frac{dy}{dt} = cy - dx, \tag{10}$$

where a, b, c, and d are positive constants.

On the other hand, we might assume, as we did in (5), that each growth rate in (9) should be reduced by a rate proportional to the number of interactions between the two species:

$$\frac{dx}{dt} = ax - bxy$$
$$\frac{dy}{dt} = cy - dxy. \tag{11}$$

Inspection shows that this nonlinear system is similar to the Lotka–Volterra predator–prey model. Last, it might be more realistic to replace the rates in (9), which indicate that the population of each species in isolation grows exponentially, with rates indicating that each population grows logistically (that is, over a long time the population is bounded):

$$\frac{dx}{dt} = a_1 x - b_1 x^2 \quad \text{and} \quad \frac{dy}{dt} = a_2 y - b_2 y^2. \tag{12}$$

When these new rates are decreased by rates proportional to the number of interactions, we obtain another nonlinear model

$$\frac{dx}{dt} = a_1 x - b_1 x^2 - c_1 xy = x(a_1 - b_1 x - c_1 y)$$
$$\frac{dy}{dt} = a_2 y - b_2 y^2 - c_2 xy = y(a_2 - b_2 y - c_2 x), \tag{13}$$

where all coefficients are positive. The linear system (10) and the nonlinear systems (11) and (13) are, of course, called **competition models**.

Networks

An electrical network having more than one loop also gives rise to simultaneous differential equations. As shown in **FIGURE 2.9.3**, the current $i_1(t)$ splits in the directions shown at point B_1, called a *branch point* of the network. By the first of the two circuit laws formulated by Gustav Robert Kirchhoff, we can write

$$i_1(t) = i_2(t) + i_3(t). \tag{14}$$

In addition, we can also apply **Kirchhoff's second law** to each loop. For loop $A_1 B_1 B_2 A_2 A_1$, summing the voltage drops across each part of the loop gives

$$E(t) = i_1 R_1 + L_1 \frac{di_2}{dt} + i_2 R_2. \tag{15}$$

Similarly, for loop $A_1 B_1 C_1 C_2 B_2 A_2 A_1$ we find

$$E(t) = i_1 R_1 + L_2 \frac{di_3}{dt}. \tag{16}$$

Using (14) to eliminate i_1 in (15) and (16) yields a system of two linear first-order equations for the currents $i_2(t)$ and $i_3(t)$:

FIGURE 2.9.3 Network whose model is given in (17)

$$L_1 \frac{di_2}{dt} + (R_1 + R_2)i_2 + R_1 i_3 = E(t)$$

$$L_2 \frac{di_3}{dt} + R_1 i_2 + R_1 i_3 = E(t). \qquad (17)$$

FIGURE 2.9.4 Network whose model is given in (18)

We leave it as an exercise (see Problem 16 in Exercises 2.9) to show that the system of differential equations describing the currents $i_1(t)$ and $i_2(t)$ in the network containing a resistor, an inductor, and a capacitor shown in **FIGURE 2.9.4** is

$$L \frac{di_1}{dt} + Ri_2 = E(t)$$

$$RC \frac{di_2}{dt} + i_2 - i_1 = 0. \qquad (18)$$

2.9 Exercises Answers to selected odd-numbered problems begin on page ANS-4.

Radioactive Series

1. We have not discussed methods by which systems of first-order differential equations can be solved. Nevertheless, systems such as (2) can be solved with no knowledge other than how to solve a single linear first-order equation. Find a solution of (2) subject to the initial conditions $x(0) = x_0$, $y(0) = 0$, $z(0) = 0$.

2. In Problem 1, suppose that time is measured in days, that the decay constants are $k_1 = -0.138629$ and $k_2 = -0.004951$, and that $x_0 = 20$. Use a graphing utility to obtain the graphs of the solutions $x(t)$, $y(t)$, and $z(t)$ on the same set of coordinate axes. Use the graphs to approximate the half-lives of substances X and Y.

3. Use the graphs in Problem 2 to approximate the times when the amounts $x(t)$ and $y(t)$ are the same, the times when the amounts $x(t)$ and $z(t)$ are the same, and the times when the amounts $y(t)$ and $z(t)$ are the same. Why does the time determined when the amounts $y(t)$ and $z(t)$ are the same make intuitive sense?

4. Construct a mathematical model for a radioactive series of four elements W, X, Y, and Z, where Z is a stable element.

Contributed Problem

Jeff Dodd, Professor
Department of Mathematical Sciences, Jacksonville State University

5. **Potassium-40 Decay** The mineral potassium, whose chemical symbol is K, is the eighth most abundant element in the Earth's crust, making up about 2% of it by weight, and one of its naturally occurring isotopes, K-40, is radioactive. The radioactive decay of K-40 is more complex than that of carbon-14 because each of its atoms decays through one of two different nuclear decay reactions into one of two different substances: the mineral calcium-40 (Ca-40) or the gas argon-40 (Ar-40). Dating methods have been developed using both of these decay products. In each case, the age of a sample is calculated using the ratio of two numbers: the amount of the *parent* isotope K-40 in the sample and the amount of the *daughter* isotope (Ca-40 or Ar-40) in the sample that is **radiogenic**, in other words, the substance

that originates from the decay of the parent isotope after the formation of the rock.

An igneous rock is solidified magma

The amount of K-40 in a sample is easy to calculate. K-40 comprises 1.17% of naturally occurring potassium, and this small percentage is distributed quite uniformly, so that the mass of K-40 in the sample is just 1.17% of the total mass of potassium in the sample, which can be measured. But for several reasons it is complicated, and sometimes problematic, to determine how much of the Ca-40 in a sample is radiogenic. In contrast, when an igneous rock is formed by volcanic activity, all of the argon (and other) gas previously trapped in the rock is driven away by the intense heat. At the moment when the rock cools and solidifies, the gas trapped inside the rock has the same composition as the atmosphere. There are three stable isotopes of argon, and in the atmosphere they occur in the following relative abundances: 0.063% Ar-38, 0.337% Ar-36, and 99.60% Ar-40. Of these, just one, Ar-36, is not created radiogenically by the decay of any element, so any Ar-40 in excess of $99.60/(0.337) = 295.5$ times the amount of Ar-36 must be radiogenic. So the amount of radiogenic Ar-40 in the sample can be determined from the amounts of Ar-38 and Ar-36 in the sample, which can be measured.

Assuming that we have a sample of rock for which the amount of K-40 and the amount of radiogenic Ar-40 have been determined, how can we calculate the age of the rock? Let $P(t)$ be the amount of K-40, $A(t)$ the amount of radiogenic

Ar-40, and $C(t)$ the amount of radiogenic Ca-40 in the sample as functions of time t in years since the formation of the rock. Then a mathematical model for the decay of K-40 is the system of linear first-order differential equations

$$\frac{dA}{dt} = \lambda_A P$$

$$\frac{dC}{dt} = \lambda_C P$$

$$\frac{dP}{dt} = -(\lambda_A + \lambda_C)P,$$

where $\lambda_A = 0.581 \times 10^{-10}$ and $\lambda_C = 4.962 \times 10^{-10}$.

(a) From the system of differential equations find $P(t)$ if $P(0) = P_0$.

(b) Determine the half-life of K-40.

(c) Use $P(t)$ from part (a) to find $A(t)$ and $C(t)$ if $A(0) = 0$ and $C(0) = 0$.

(d) Use your solution for $A(t)$ in part (c) to determine the percentage of an initial amount P_0 of K-40 that decays into Ar-40 over a very long period of time (that is, $t \to \infty$). What percentage of P_0 decays into Ca-40?

6. **Potassium–Argon Dating** The knowledge of how K-40 decays can be used to determine the age of very old igneous rocks.

(a) Use the solutions obtained in parts (a) and (c) of Problem 5 to show that

$$\frac{A(t)}{P(t)} = \frac{\lambda_A}{\lambda_A + \lambda_C}[e^{(\lambda_A + \lambda_C)t} - 1].$$

(b) Solve the expression in part (a) for t in terms $A(t)$, $P(t)$, λ_A, and λ_C.

(c) Suppose it is found that each gram of a rock sample contains 8.6×10^{-7} grams of radiogenic Ar-40 and 5.3×10^{-6} grams of K-40. Use the equation obtained in part (b) to determine the approximate age of the rock.

Mixtures

7. Consider two tanks A and B, with liquid being pumped in and out at the same rates, as described by the system of equations (3). What is the system of differential equations if, instead of pure water, a brine solution containing 2 pounds of salt per gallon is pumped into tank A?

8. Use the information given in **FIGURE 2.9.5** to construct a mathematical model for the number of pounds of salt $x_1(t)$, $x_2(t)$, and $x_3(t)$ at time t in tanks A, B, and C, respectively.

FIGURE 2.9.5 Mixing tanks in Problem 8

9. Two very large tanks A and B are each partially filled with 100 gallons of brine. Initially, 100 pounds of salt is dissolved in the solution in tank A and 50 pounds of salt is dissolved in the solution in tank B. The system is closed in that the well-stirred liquid is pumped only between the tanks, as shown in **FIGURE 2.9.6**.

(a) Use the information given in the figure to construct a mathematical model for the number of pounds of salt $x_1(t)$ and $x_2(t)$ at time t in tanks A and B, respectively.

(b) Find a relationship between the variables $x_1(t)$ and $x_2(t)$ that holds at time t. Explain why this relationship makes intuitive sense. Use this relationship to help find the amount of salt in tank B at $t = 30$ min.

FIGURE 2.9.6 Mixing tanks in Problem 9

10. Three large tanks contain brine, as shown in **FIGURE 2.9.7**. Use the information in the figure to construct a mathematical model for the number of pounds of salt $x_1(t)$, $x_2(t)$, and $x_3(t)$ at time t in tanks A, B, and C, respectively. Without solving the system, predict limiting values of $x_1(t)$, $x_2(t)$, and $x_3(t)$ as $t \to \infty$.

FIGURE 2.9.7 Mixing tanks in Problem 10

Predator–Prey Models

11. Consider the Lotka–Volterra predator–prey model defined by

$$\frac{dx}{dt} = -0.1x + 0.02xy$$

$$\frac{dy}{dt} = 0.2y - 0.025xy,$$

where the populations $x(t)$ (predators) and $y(t)$ (prey) are measured in the thousands. Suppose $x(0) = 6$ and $y(0) = 6$. Use a numerical solver to graph $x(t)$ and $y(t)$. Use the graphs to approximate the time $t > 0$ when the two populations are first equal. Use the graphs to approximate the period of each population.

Competition Models

12. Consider the competition model defined by

$$\frac{dx}{dt} = x(2 - 0.4x - 0.3y)$$

$$\frac{dy}{dt} = y(1 - 0.1y - 0.3x),$$

where the populations $x(t)$ and $y(t)$ are measured in the thousands and t in years. Use a numerical solver to analyze the populations over a long period of time for each of the cases:

(a) $x(0) = 1.5$, $y(0) = 3.5$ (b) $x(0) = 1$, $y(0) = 1$
(c) $x(0) = 2$, $y(0) = 7$ (d) $x(0) = 4.5$, $y(0) = 0.5$

13. Consider the competition model defined by

$$\frac{dx}{dt} = x(1 - 0.1x - 0.05y)$$

$$\frac{dy}{dt} = y(1.7 - 0.1y - 0.15x),$$

where the populations $x(t)$ and $y(t)$ are measured in the thousands and t in years. Use a numerical solver to analyze the populations over a long period of time for each of the cases:

(a) $x(0) = 1$, $y(0) = 1$ (b) $x(0) = 4$, $y(0) = 10$
(c) $x(0) = 9$, $y(0) = 4$ (d) $x(0) = 5.5$, $y(0) = 3.5$

Networks

14. Show that a system of differential equations that describes the currents $i_2(t)$ and $i_3(t)$ in the electrical network shown in **FIGURE 2.9.8** is

$$L\frac{di_2}{dt} + L\frac{di_3}{dt} + R_1 i_2 = E(t)$$

$$-R_1\frac{di_2}{dt} + R_2\frac{di_3}{dt} + \frac{1}{C}i_3 = 0.$$

FIGURE 2.9.8 Network in Problem 14

15. Determine a system of first-order differential equations that describe the currents $i_2(t)$ and $i_3(t)$ in the electrical network shown in **FIGURE 2.9.9**.

FIGURE 2.9.9 Network in Problem 15

16. Show that the linear system given in (18) describes the currents $i_1(t)$ and $i_2(t)$ in the network shown in Figure 2.9.4. [*Hint*: $dq/dt = i_3$.]

Additional Mathematical Models

17. SIR Model A communicable disease is spread throughout a small community, with a fixed population of n people, by contact between infected individuals and people who are susceptible to the disease. Suppose initially that everyone is susceptible to the disease and that no one leaves the community while the epidemic is spreading. At time t, let $s(t)$, $i(t)$, and $r(t)$ denote, in turn, the number of people in the community (measured in hundreds) who are *susceptible* to the disease but not yet infected with it, the number of people who are *infected* with the disease, and the number of people who have *recovered* from the disease. Explain why the system of differential equations

$$\frac{ds}{dt} = -k_1 si$$

$$\frac{di}{dt} = -k_2 i + k_1 si$$

$$\frac{dr}{dt} = k_2 i,$$

where k_1 (called the *infection rate*) and k_2 (called the *removal rate*) are positive constants, is a reasonable mathematical model, commonly called a **SIR model**, for the spread of the epidemic throughout the community. Give plausible initial conditions associated with this system of equations. Show that the system implies that

$$\frac{d}{dt}(s + i + r) = 0.$$

Why is this consistent with the assumptions?

18. (a) In Problem 17 explain why it is sufficient to analyze only

$$\frac{ds}{dt} = -k_1 si$$

$$\frac{di}{dt} = -k_2 i + k_1 si.$$

(b) Suppose $k_1 = 0.2$, $k_2 = 0.7$, and $n = 10$. Choose various values of $i(0) = i_0$, $0 < i_0 < 10$. Use a numerical solver to determine what the model predicts about the epidemic in the two cases $s_0 > k_2/k_1$ and $s_0 \le k_2/k_1$. In the case of an epidemic, estimate the number of people who are eventually infected.

In Problems 1–10, fill in the blanks or answer true/false.

1. The DE $y' - ky = A$, where k and A are constants, is autonomous. The critical point _____ of the equation is a(n) _____ (attractor or repeller) for $k > 0$ and a(n) _____ (attractor or repeller) for $k < 0$.

2. The initial-value problem $x \dfrac{dy}{dx} - 4y = 0$, $y(0) = k$, has an infinite number of solutions for $k =$ _____ and no solution for $k =$ _____.

3. By inspection, two solutions of the differential equation $y' + |y| = 5$ are _____.

4. Every autonomous DE $y' = f(y)$ is separable. _____

5. The linear DE $y' + k_1 y = k_2$, where k_1 and k_2 are nonzero constants, possesses a constant solution. _____

6. The first-order DE $\dfrac{d\theta}{dr} = r\theta + r + \theta + 1$ is not separable. _____

7. If $y' = e^x y$ then $y =$ _____.

8. The linear DE $a_1(x)y' + a_0(x)y = 0$ is also separable. _____

9. The DE $y' + y^2 = 1$ is a second-order equation. _____

10. $y = e^{\cos x} \int_0^x te^{-\cos t}\, dt$ is a solution of the linear first-order differential equation _____.

In Problems 11 and 12, construct an autonomous first-order differential equation $dy/dx = f(y)$ whose phase portrait is consistent with the given figure.

11.

12.

FIGURE 2.R.1 Phase portrait in Problem 11 **FIGURE 2.R.2** Phase portrait in Problem 12

13. The number 0 is a critical point of the autonomous differential equation $dx/dt = x^n$, where n is a positive integer. For what values of n is 0 asymptotically stable? Semi-stable? Unstable? Repeat for the equation $dx/dt = -x^n$.

14. Consider the differential equation

$$\frac{dP}{dt} = f(P), \quad \text{where} \quad f(P) = -0.5P^3 - 1.7P + 3.4.$$

The function $f(P)$ has one real zero, as shown in **FIGURE 2.R.3**. Without attempting to solve the differential equation, estimate the value of $\lim_{t \to \infty} P(t)$.

FIGURE 2.R.3 Graph for Problem 14

15. **FIGURE 2.R.4** is a portion of the direction field of a differential equation $dy/dx = f(x, y)$. By hand, sketch two different solution curves, one that is tangent to the lineal element shown in black and the other that is tangent to the lineal element shown in red.

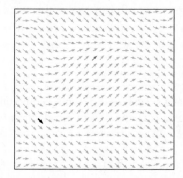

FIGURE 2.R.4 Direction field for Problem 15

16. Classify each differential equation as separable, exact, linear, homogeneous, or Bernoulli. Some equations may be more than one kind. Do not solve.

(a) $\dfrac{dy}{dx} = \dfrac{x - y}{x}$ (b) $\dfrac{dy}{dx} = \dfrac{1}{y - x}$

(c) $(x + 1)\dfrac{dy}{dx} = -y + 10$ (d) $\dfrac{dy}{dx} = \dfrac{1}{x(x - y)}$

(e) $\dfrac{dy}{dx} = \dfrac{y^2 + y}{x^2 + x}$ (f) $\dfrac{dy}{dx} = 5y + y^2$

(g) $y\, dx = (y - xy^2)\, dy$ (h) $x\dfrac{dy}{dx} = ye^{xy} - x$

(i) $xyy' + y^2 = 2x$ (j) $2xyy' + y^2 = 2x^2$

(k) $y\, dx + x\, dy = 0$

(l) $\left(x^2 + \dfrac{2y}{x}\right) dx = (3 - \ln x^2)\, dy$

(m) $\dfrac{dy}{dx} = \dfrac{x}{y} + \dfrac{y}{x} + 1$ (n) $\dfrac{y}{x^2}\dfrac{dy}{dx} + e^{2x^3 + y^2} = 0$

In Problems 17–24, solve the given differential equation.

17. $(y^2 + 1)\, dx = y\sec^2 x\, dy$

18. $y(\ln x - \ln y)\,dx = (x\ln x - x\ln y - y)\,dy$

19. $(6x + 1)y^2\dfrac{dy}{dx} + 3x^2 + 2y^3 = 0$

20. $\dfrac{dx}{dy} = -\dfrac{4y^2 + 6xy}{3y^2 + 2x}$

21. $t\dfrac{dQ}{dt} + Q = t^4\ln t$

22. $(2x + y + 1)y' = 1$

23. $(x^2 + 4)\,dy = (2x - 8xy)\,dx$

24. $(2r^2\cos\theta\sin\theta + r\cos\theta)\,d\theta + (4r + \sin\theta - 2r\cos^2\theta)\,dr = 0$

In Problems 25–28, express the solution of the given initial-value problem in terms of an integral-defined function.

25. $2\dfrac{dy}{dx} + (4\cos x)y = x, \quad y(0) = 1$

26. $\dfrac{dy}{dx} - 4xy = \sin x^2, \quad y(0) = 7$

27. $x\dfrac{dy}{dx} + 2y = xe^{x^2}, \quad y(1) = 3$

28. $x\dfrac{dy}{dx} + (\sin x)y = 0, \quad y(0) = 10$

In Problems 29 and 30, solve the given initial-value problem.

29. $\dfrac{dy}{dx} + y = f(x), \quad y(0) = 5,$ where

$$f(x) = \begin{cases} e^{-x}, & 0 \le x < 1 \\ 0, & x \ge 1 \end{cases}$$

30. $\dfrac{dy}{dx} + P(x)y = e^x, \quad y(0) = -1,$ where

$$P(x) = \begin{cases} 1, & 0 \le x < 1 \\ -1, & x \ge 1 \end{cases}$$

In Problems 31 and 32, solve the given initial-value problem and give the largest interval I on which the solution is defined.

31. $\sin x\dfrac{dy}{dx} + (\cos x)y = 0, \quad y(7\pi/6) = -2$

32. $\dfrac{dy}{dt} + 2(t + 1)y^2 = 0, \quad y(0) = -\tfrac{1}{8}$

33. (a) Without solving, explain why the initial-value problem

$$\dfrac{dy}{dx} = \sqrt{y}, \quad y(x_0) = y_0,$$

has no solution for $y_0 < 0$.

(b) Solve the initial-value problem in part (a) for $y_0 > 0$ and find the largest interval I on which the solution is defined.

34. (a) Find an implicit solution of the initial-value problem

$$\dfrac{dy}{dx} = \dfrac{y^2 - x^2}{xy}, \quad y(1) = -\sqrt{2}.$$

(b) Find an explicit solution of the problem in part (a) and give the largest interval I over which the solution is defined. A graphing utility may be helpful here.

35. Graphs of some members of a family of solutions for a first-order differential equation $dy/dx = f(x, y)$ are shown in **FIGURE 2.R.5**. The graph of an implicit solution $G(x, y) = 0$ that passes through the points $(1, -1)$ and $(-1, 3)$ is shown in red. With colored pencils, trace out the solution curves of the solutions $y = y_1(x)$ and $y = y_2(x)$ defined by the implicit solution such that $y_1(1) = -1$ and $y_2(-1) = 3$. Estimate the interval I on which each solution is defined.

FIGURE 2.R.5 Graph for Problem 35

36. Use Euler's method with step size $h = 0.1$ to approximate $y(1.2)$ where $y(x)$ is a solution of the initial-value problem $y' = 1 + x\sqrt{y}, y(1) = 9$.

37. In March 1976, the world population reached 4 billion. A popular news magazine predicted that with an average yearly growth rate of 1.8%, the world population would be 8 billion in 45 years. How does this value compare with that predicted by the model that says the rate of increase is proportional to the population at any time t?

38. Iodine-131 is a radioactive liquid used in the treatment of cancer of the thyroid. After one day in storage, analysis shows that initial amount of iodine-131 in a sample has decreased by 8.3%.

(a) Find the amount of iodine-131 remaining in the sample after 8 days.

(b) What is the significance of your answer in part (a)?

39. Ötzi the Iceman In 1991 hikers found a well-preserved body of a man partially frozen in a glacier in the Ötztal Alps on the border between Austria and Italy. From the cuts and bruises found on his body, the trauma caused by a blow to his head, and an arrowhead found in his left shoulder, it is likely that Ötzi the Iceman—as he came to be called—was murdered. Through carbon-dating techniques, it was found that the body of Ötzi the Iceman contained 53% as much C-14 as found in a living person. Assume that the Iceman was carbon dated in 1991.

(a) Using the Cambridge half-life of C-14, give an educated guess as to the date of his death.

(b) Then use the technique illustrated in Example 3 of Section 2.7 to calculate the approximate date of his death.

© dpa/Corbis
The Iceman in Problem 39

40. Air containing 0.06% carbon dioxide is pumped into a room whose volume is 8000 ft^3. The air is pumped in at a rate of 2000 ft^3/min, and the circulated air is then pumped out at the same rate. If there is an initial concentration of 0.2% carbon dioxide, determine the subsequent amount in the room at any time. What is the concentration at 10 minutes? What is the steady-state, or equilibrium, concentration of carbon dioxide?

41. Solve the differential equation

$$\frac{dy}{dx} = -\frac{y}{\sqrt{s^2 - y^2}}$$

of the tractrix. See Problem 28 in Exercises 1.3. Assume that the initial point on the y-axis is (0, 10) and that the length of the rope is $x = 10$ ft.

42. Suppose a cell is suspended in a solution containing a solute of constant concentration C_s. Suppose further that the cell has constant volume V and that the area of its permeable membrane is the constant A. By **Fick's law of diffusion**, introduced in 1855 by the German physician/physiologist **Adolf Eugen Fick** (1829–1901), the rate of change of its mass m is directly proportional to the area A and the difference $C_s - C(t)$, where $C(t)$ is the concentration of the solute inside the cell at any time t. Find $C(t)$ if $m = V \cdot C(t)$ and $C(0) = C_0$. See **FIGURE 2.R.6**.

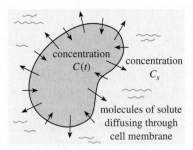

FIGURE 2.R.6 Cell in Problem 42

43. Suppose that as a body cools, the temperature of the surrounding medium increases because it completely absorbs the heat being lost by the body. Let $T(t)$ and $T_m(t)$ be the temperatures of the body and the medium at time t, respectively. If the initial temperature of the body is T_1 and the initial temperature of the medium is T_2, then it can

be shown in this case that Newton's law of cooling is $dT/dt = k(T - T_m)$, $k < 0$, where $T_m = T_2 + B(T_1 - T)$, $B > 0$ is a constant.
(a) The foregoing DE is autonomous. Use the phase portrait concept of Section 2.1 to determine the limiting value of the temperature $T(t)$ as $t \to \infty$. What is the limiting value of $T_m(t)$ as $t \to \infty$?
(b) Verify your answers in part (a) by actually solving the differential equation.
(c) Discuss a physical interpretation of your answers in part (a).

44. According to **Stefan's law of radiation**, the absolute temperature T of a body cooling in a medium at constant temperature T_m is given by

$$\frac{dT}{dt} = k(T^4 - T_m^4),$$

where k is a constant. **Josef Stefan** (1835–1893), a mathematical physicist born in the Austrian Empire, published his law of radiation in 1879. Stefan's law can be used over a greater temperature range than Newton's law of cooling.
(a) Solve the differential equation.
(b) Show that when $T - T_m$ is small compared to T_m then Newton's law of cooling approximates Stefan's law. [*Hint*: Think binomial series of the right-hand side of the DE.]

45. Suppose an RC-series circuit has a variable resistor. If the resistance at time t is defined by $R(t) = k_1 + k_2 t$, where k_1 and k_2 are known positive constants, then the differential equation in (10) of Section 2.7 becomes

$$(k_1 + k_2 t)\frac{dq}{dt} + \frac{1}{C}q = E(t),$$

where C is a constant. If $E(t) = E_0$ and $q(0) = q_0$, where E_0 and q_0 are constants, then show that

$$q(t) = E_0 C + (q_0 - E_0 C)\left(\frac{k_1}{k_1 + k_2 t}\right)^{1/Ck_2}.$$

46. A classical problem in the calculus of variations is to find the shape of a curve $\mathscr{C}$ such that a bead, under the influence of gravity, will slide from point $A(0, 0)$ to point $B(x_1, y_1)$ in the least time. See **FIGURE 2.R.7**. It can be shown that a nonlinear differential equation for the shape $y(x)$ of the path is $y[1 + (y')^2] = k$, where k is a constant. First solve for dx in terms of y and dy, and then use the substitution $y = k \sin^2\theta$ to obtain a parametric form of the solution. The curve $\mathscr{C}$ turns out to be a **cycloid**.

FIGURE 2.R.7 Sliding bead in Problem 46

The **clepsydra**, or water clock, was a device used by the ancient Egyptians, Greeks, Romans, and Chinese to measure the passage of time by observing the change in the height of water that was permitted to flow out of a small hole in the bottom of a container or tank. In Problems 47–50, use the differential equation (see Problems 15–18 in Exercises 2.8)

$$\frac{dh}{dt} = -c\frac{A_h}{A_w}\sqrt{2gh}$$

as a model for the height h of water in a tank at time t. Assume in each of these problems that $h(0) = 2$ ft corresponds to water filled to the top of the tank, the hole in the bottom is circular with radius $\frac{1}{32}$ in., $g = 32$ ft/s^2, and $c = 0.6$.

47. Suppose that a tank is made of glass and has the shape of a right-circular cylinder of radius 1 ft. Find the height $h(t)$ of the water.

48. For the tank in Problem 47, how far up from its bottom should a mark be made on its side, as shown in **FIGURE 2.R.8**, that corresponds to the passage of 1 hour? Continue and determine where to place the marks corresponding to the passage of 2 h, 3 h, ..., 12 h. Explain why these marks are not evenly spaced.

FIGURE 2.R.8 Clepsydra in Problem 48

49. Suppose that the glass tank has the shape of a cone with circular cross sections as shown in **FIGURE 2.R.9**. Can this water clock measure 12 time intervals of length equal to 1 hour? Explain using sound mathematics.

FIGURE 2.R.9 Clepsydra in Problem 49

50. Suppose that $r = f(h)$ defines the shape of a water clock for which the time marks are equally spaced. Use the above differential equation to find $f(h)$ and sketch a typical graph

of h as a function of r. Assume that the cross-sectional area A_h of the hole is constant. [*Hint*: In this situation, $dh/dt = -a$, where $a > 0$ is a constant.]

51. A model for the populations of two interacting species of animals is

$$\frac{dx}{dt} = k_1 x(\alpha - x)$$

$$\frac{dy}{dt} = k_2 xy.$$

Solve for x and y in terms of t.

52. Initially, two large tanks A and B each hold 100 gallons of brine. The well-stirred liquid is pumped between the tanks as shown in **FIGURE 2.R.10**. Use the information given in the figure to construct a mathematical model for the number of pounds of salt $x_1(t)$ and $x_2(t)$ at time t in tanks A and B, respectively.

FIGURE 2.R.10 Mixing tanks in Problem 52

53. It is estimated that the ecosystem of Yellowstone National Park can sustain a grey wolf population of 450. An initial population in 1997 was 40 grey wolves, and it was subsequently determined that the population grew to 95 wolves after 15 years. How many wolves does the mathematical model

$$\frac{dP}{dt} = kP\ln\frac{450}{P}$$

predict there will be in the park 30 years after their introduction?

54. (a) Use a graphing utility to graph the wolf population $P(t)$ found in Problem 53.
 (b) Use the solution $P(t)$ in Problem 53 to find $\lim_{t\to\infty} P(t)$.
 (c) Show that the differential equation in Problem 53 is a special case of Gompertz's equation ((7) in Section 2.8).

When all the curves in a family $G(x, y, c_1) = 0$ intersect orthogonally all the curves in another family $H(x, y, c_2) = 0$, the families are said to be **orthogonal trajectories** of each other. See **FIGURE 2.R.11**. If $dy/dx = f(x, y)$ is the differential equation of one family, then the differential equation for the orthogonal trajectories of this family is $dy/dx = -1/f(x, y)$. In Problems 55–58, find the differential equation of the given family. Find the orthogonal trajectories of this family. Use a graphing utility to graph both families on the same set of coordinate axes.

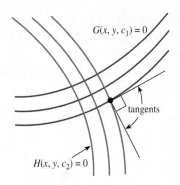

$G(x, y, c_1) = 0$

tangents

$H(x, y, c_2) = 0$

FIGURE 2.R.11 Orthogonal trajectories

55. $y = c_1 x$

56. $x^2 - 2y^2 = c_1$

57. $y = -x - 1 + c_1 e^x$

58. $y = \dfrac{1}{x + c_1}$

When the orthogonal trajectories of a one-parameter family of curves $G(x, y, c_1) = 0$ are itself, then this family is said to be **self-orthogonal**. In Problems 59 and 60, use the given method to show that the family of confocal parabolas $y^2 = 4c(x + c)$ is self-orthogonal.

59. Suppose the family $y^2 = 4c(x + c)$ is defined in the piecewise manner:

$y^2 = 4p(x + p),\ p > 0$ (parabolas opening to the right),

$y^2 = -4q(x - q),\ q > 0$ (parabolas opening to the left).

Find the points of intersection of the curves defined by these two equations. Then show that the tangent lines are perpendicular at the points of intersection.

60. Show that the differential equation that defines the family $y^2 = 4c(x + c)$ and the differential equation that defines the orthogonal trajectories of the given family are the same.

Contributed Problem

Rick Wicklin, PhD
Senior Researcher in Computational Statistics, SAS Institute Inc.

61. Invasion of the Marine Toads* In 1935, the poisonous American marine toad (*Bufo marinus*) was introduced, against the advice of ecologists, into some of the coastal sugar cane districts in Queensland, Australia, as a means of controlling sugar cane beetles. Due to lack of natural predators and the existence of an abundant food supply, the toad population grew and spread into regions far from the original districts. The survey data given in the accompanying table indicate how the toads expanded their territorial bounds within a 40-year period. Our goal in this problem is to find a population model of the form $P(t_i)$ but we want to construct the model that *best* fits the given data. Note that the data are not given as *number of toads* at 5-year intervals since this kind of information would be virtually impossible to obtain.

*This problem is based on the article "Teaching Differential Equations with a Dynamical Systems Viewpoint" by Paul Blanchard, *The College Mathematics Journal* 25 (1994), 385–395.

© Ryan M. Bolton/ShutterStock, Inc.
Marine toad (*Bufo marinus*)

Year	Area Occupied
1939	32,800
1944	55,800
1949	73,600
1954	138,000
1959	202,000
1964	257,000
1969	301,000
1974	584,000

(a) For ease of computation, let's assume that, on the average, there is one toad per square kilometer. We will also count the toads in units of thousands and measure time in years with $t = 0$ corresponding to 1939. One way to model the data in the table is to use the initial condition $P_0 = 32.8$ and to search for a value of k so that the graph of $P_0 e^{kt}$ appears to fit the data points. Experiment, using a graphic calculator or a CAS, by varying the **birth rate** k until the graph of $P_0 e^{kt}$ appears to fit the data well over the time period $0 \le t \le 35$.

Alternatively, it is also possible to solve analytically for a value of k that will guarantee that the curve passes through exactly two of the data points. Find a value of k so that $P(5) = 55.8$. Find a different value of k so that $P(35) = 584$.

(b) In practice, a mathematical model rarely passes through every experimentally obtained data point, and so statistical methods must be used to find values of the model's parameters that best fit experimental data. Specifically, we will use **linear regression** to find a value of k that describes the given data points:

- Use the table to obtain a new data set of the form $(t_i, \ln P(t_i))$, where $P(t_i)$ is the given population at the times $t_1 = 0, t_2 = 5, \ldots$.
- Most graphic calculators have a built-in routine to find the line of least squares that fits this data. The routine gives an equation of the form $\ln P(t) = mt + b$, where m and b are, respectively, the slope and intercept corresponding to the line of best fit. (Most calculators also give the value of the correlation coefficient that indicates how well the data are approximated by a line; a correlation coefficient of 1 or -1 means perfect correlation. A correlation coefficient near 0 may mean that the data do not appear to be fit by an exponential model.)
- Solving $\ln P(t) = mt + b$ gives $P(t) = e^{mt+b}$ or $P(t) = e^b e^{mt}$. Matching the last form with $P_0 e^{kt}$, we see that e^b is an approximate initial population, and m is the value of the birth rate that best fits the given data.

(c) So far you have produced four different values of the birth rate k. Do your four values of k agree closely with each other? Should they? Which of the four values do you think is the best model for the growth of the toad population during the years for which we have data? Use this birth rate to predict the toad's range in the year 2039.

Given that the area of Australia is 7,619,000 km^2, how confident are you of this prediction? Explain your reasoning.

62. **Invasion of the Marine Toads—Continued** In part (a) of Problem 61, we made the assumption that there was an average of one toad per square kilometer. But suppose we are wrong and there were actually an average of two toads per square kilometer. As before, solve analytically for a value of k that will guarantee that the curve passes through exactly two of the data points. In particular, if we now assume that $P(0) = 65.6$, find a value of k so that $P(5) = 111.6$, and a different value of k so that $P(35) = 1168$. How do these values of k compare with the values you found previously? What does this tell us? Discuss the importance of knowing the exact average density of the toad population.

Higher-Order Differential Equations

We turn now to differential equations of order two or higher. In the first six sections of this chapter we examine the underlying theory and solution methods for certain kinds of linear DEs. The difficulties that surround higher-order nonlinear DEs and the few methods for solving such equations are examined next (Section 3.7). The chapter concludes with higher-order linear and nonlinear mathematical models (Sections 3.8, 3.9, and 3.11) and the first of several methods on how to solve systems of linear DEs (Section 3.12).

3.1 Theory of Linear Equations

INTRODUCTION We turn now to differential equations of order two or higher. In this section we will examine some of the underlying theory of linear DEs. Then in the five sections that follow we learn how to solve linear higher-order differential equations.

3.1.1 Initial-Value and Boundary-Value Problems

Initial-Value Problem In Section 1.2 we defined an initial-value problem for a general nth-order differential equation. For a linear differential equation, an **nth-order initial-value problem (IVP)** is

$$\textbf{Solve:} \quad a_n(x)\frac{d^n y}{dx^n} + a_{n-1}(x)\frac{d^{n-1}y}{dx^{n-1}} + \cdots + a_1(x)\frac{dy}{dx} + a_0(x)y = g(x)$$

(1)

$$\textbf{Subject to:} \quad y(x_0) = y_0, \; y'(x_0) = y_1, \ldots, \; y^{(n-1)}(x_0) = y_{n-1}.$$

Recall that for a problem such as this, we seek a function defined on some interval I containing x_0 that satisfies the differential equation and the n initial conditions specified at x_0: $y(x_0) = y_0$, $y'(x_0) = y_1, \ldots, y^{(n-1)}(x_0) = y_{n-1}$. We have already seen that in the case of a second-order initial-value problem, a solution curve must pass through the point (x_0, y_0) and have slope y_1 at this point.

Existence and Uniqueness In Section 1.2 we stated a theorem that gave conditions under which the existence and uniqueness of a solution of a first-order initial-value problem were guaranteed. The theorem that follows gives sufficient conditions for the existence of a unique solution of the problem in (1).

THEOREM 3.1.1 Existence of a Unique Solution

Let $a_n(x)$, $a_{n-1}(x)$, $\ldots$, $a_1(x)$, $a_0(x)$, and $g(x)$ be continuous on an interval I, and let $a_n(x) \neq 0$ for every x in this interval. If $x = x_0$ is any point in this interval, then a solution $y(x)$ of the initial-value problem (1) exists on the interval and is unique.

EXAMPLE 1 Unique Solution of an IVP

The initial-value problem

$$3y''' + 5y'' - y' + 7y = 0, \quad y(1) = 0, \quad y'(1) = 0, \quad y''(1) = 0$$

possesses the trivial solution $y = 0$. Since the third-order equation is linear with constant coefficients, it follows that all the conditions of Theorem 3.1.1 are fulfilled. Hence $y = 0$ is the *only* solution on any interval containing $x = 1$. ≡

EXAMPLE 2 Unique Solution of an IVP

You should verify that the function $y = 3e^{2x} + e^{-2x} - 3x$ is a solution of the initial-value problem

$$y'' - 4y = 12x, \quad y(0) = 4, \quad y'(0) = 1.$$

Now the differential equation is linear, the coefficients as well as $g(x) = 12x$ are continuous, and $a_2(x) = 1 \neq 0$ on any interval I containing $x = 0$. We conclude from Theorem 3.1.1 that the given function is the unique solution on I. ≡

The requirements in Theorem 3.1.1 that $a_i(x)$, $i = 0, 1, 2, \ldots, n$ be continuous and $a_n(x) \neq 0$ for every x in I are both important. Specifically, if $a_n(x) = 0$ for some x in the interval, then the solution of a linear initial-value problem may not be unique or even exist. For example, you should verify that the function $y = cx^2 + x + 3$ is a solution of the initial-value problem

$$x^2 y'' - 2xy' + 2y = 6, \quad y(0) = 3, \quad y'(0) = 1$$

on the interval $(-\infty, \infty)$ for any choice of the parameter c. In other words, there is no unique solution of the problem. Although most of the conditions of Theorem 3.1.1 are satisfied, the obvious difficulties are that $a_2(x) = x^2$ is zero at $x = 0$ and that the initial conditions are also imposed at $x = 0$.

■ Boundary-Value Problem
Another type of problem consists of solving a linear differential equation of order two or greater in which the dependent variable y or its derivatives are specified at *different points*. A problem such as

Solve: $a_2(x)\dfrac{d^2y}{dx^2} + a_1(x)\dfrac{dy}{dx} + a_0(x)y = g(x)$

Subject to: $y(a) = y_0, \; y(b) = y_1$

is called a **two-point boundary-value problem**, or simply a **boundary-value problem (BVP)**. The prescribed values $y(a) = y_0$ and $y(b) = y_1$ are called **boundary conditions (BC)**. A solution of the foregoing problem is a function satisfying the differential equation on some interval I, containing a and b, whose graph passes through the two points (a, y_0) and (b, y_1). See **FIGURE 3.1.1**.

For a second-order differential equation, other pairs of boundary conditions could be

$$y'(a) = y_0, \quad y(b) = y_1$$
$$y(a) = y_0, \quad y'(b) = y_1$$
$$y'(a) = y_0, \quad y'(b) = y_1,$$

where y_0 and y_1 denote arbitrary constants. These three pairs of conditions are just special cases of the general boundary conditions

$$A_1y(a) + B_1y'(a) = C_1$$
$$A_2y(b) + B_2y'(b) = C_2.$$

The next example shows that even when the conditions of Theorem 3.1.1 are fulfilled, a boundary-value problem may have several solutions (as suggested in Figure 3.1.1), a unique solution, or no solution at all.

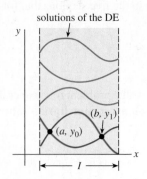

FIGURE 3.1.1 Colored curves are solutions of a BVP

EXAMPLE 3 A BVP Can Have Many, One, or No Solutions

In Example 10 of Section 1.1 we saw that the two-parameter family of solutions of the differential equation $x'' + 16x = 0$ is

$$x = c_1 \cos 4t + c_2 \sin 4t. \tag{2}$$

(a) Suppose we now wish to determine that solution of the equation that further satisfies the boundary conditions $x(0) = 0, x(\pi/2) = 0$. Observe that the first condition $0 = c_1 \cos 0 + c_2 \sin 0$ implies $c_1 = 0$, so that $x = c_2 \sin 4t$. But when $t = \pi/2, 0 = c_2 \sin 2\pi$ is satisfied for any choice of c_2 since $\sin 2\pi = 0$. Hence the boundary-value problem

$$x'' + 16x = 0, \quad x(0) = 0, \quad x(\pi/2) = 0 \tag{3}$$

has infinitely many solutions. **FIGURE 3.1.2** shows the graphs of some of the members of the one-parameter family $x = c_2 \sin 4t$ that pass through the two points $(0, 0)$ and $(\pi/2, 0)$.

(b) If the boundary-value problem in (3) is changed to

$$x'' + 16x = 0, \quad x(0) = 0, \quad x(\pi/8) = 0, \tag{4}$$

then $x(0) = 0$ still requires $c_1 = 0$ in the solution (2). But applying $x(\pi/8) = 0$ to $x = c_2 \sin 4t$ demands that $0 = c_2 \sin(\pi/2) = c_2 \cdot 1$. Hence $x = 0$ is a solution of this new boundary-value problem. Indeed, it can be proved that $x = 0$ is the *only* solution of (4).

(c) Finally, if we change the problem to

$$x'' + 16x = 0, \quad x(0) = 0, \quad x(\pi/2) = 1, \tag{5}$$

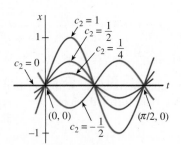

FIGURE 3.1.2 The BVP in (3) of Example 3 has many solutions

we find again that $c_1 = 0$ from $x(0) = 0$, but that applying $x(\pi/2) = 1$ to $x = c_2 \sin 4t$ leads to the contradiction $1 = c_2 \sin 2\pi = c_2 \cdot 0 = 0$. Hence the boundary-value problem (5) has no solution.

≡

3.1.2 Homogeneous Equations

A linear nth-order differential equation of the form

Note $y = 0$ is always a solution of a homogeneous linear equation.

$$a_n(x)\frac{d^n y}{dx^n} + a_{n-1}(x)\frac{d^{n-1}y}{dx^{n-1}} + \cdots + a_1(x)\frac{dy}{dx} + a_0(x)y = 0 \tag{6}$$

is said to be **homogeneous**, whereas an equation

$$a_n(x)\frac{d^n y}{dx^n} + a_{n-1}(x)\frac{d^{n-1}y}{dx^{n-1}} + \cdots + a_1(x)\frac{dy}{dx} + a_0(x)y = g(x) \tag{7}$$

with $g(x)$ not identically zero is said to be **nonhomogeneous**. For example, $2y'' + 3y' - 5y = 0$ is a homogeneous linear second-order differential equation, whereas $x^2 y''' + 6y' + 10y = e^x$ is a nonhomogeneous linear third-order differential equation. The word *homogeneous* in this context does not refer to coefficients that are homogeneous functions as in Section 2.5; rather, the word has exactly the same meaning as in Section 2.3.

We shall see that in order to solve a nonhomogeneous linear equation (7), we must first be able to solve the **associated homogeneous equation** (6).

To avoid needless repetition throughout the remainder of this section, we shall, as a matter of course, make the following important assumptions when stating definitions and theorems about the linear equations (6) and (7). On some common interval I,

Remember these assumptions in the definitions and theorems of this chapter.

- the coefficients $a_i(x)$, $i = 0, 1, 2, \ldots, n$, are continuous;
- the right-hand member $g(x)$ is continuous; and
- $a_n(x) \neq 0$ for every x in the interval.

�썸 Differential Operators

In calculus, differentiation is often denoted by the capital letter D; that is, $dy/dx = Dy$. The symbol D is called a **differential operator** because it transforms a differentiable function into another function. For example, $D(\cos 4x) = -4\sin 4x$, and $D(5x^3 - 6x^2) = 15x^2 - 12x$. Higher-order derivatives can be expressed in terms of D in a natural manner:

$$\frac{d}{dx}\left(\frac{dy}{dx}\right) = \frac{d^2 y}{dx^2} = D(Dy) = D^2 y \quad \text{and in general} \quad \frac{d^n y}{dx^n} = D^n y,$$

where y represents a sufficiently differentiable function. Polynomial expressions involving D, such as $D + 3$, $D^2 + 3D - 4$, and $5x^3 D^3 - 6x^2 D^2 + 4xD + 9$, are also differential operators. In general, we define an **nth-order differential operator** to be

$$L = a_n(x)D^n + a_{n-1}(x)D^{n-1} + \cdots + a_1(x)D + a_0(x). \tag{8}$$

As a consequence of two basic properties of differentiation, $D(cf(x)) = c\, Df(x)$, c a constant, and $D\{f(x) + g(x)\} = Df(x) + Dg(x)$, the differential operator L possesses a **linearity property**; that is, L operating on a linear combination of two differentiable functions is the same as the linear combination of L operating on the individual functions. In symbols, this means

$$L\{\alpha f(x) + \beta g(x)\} = \alpha L(f(x)) + \beta L(g(x)), \tag{9}$$

where α and β are constants. Because of (9) we say that the nth-order differential operator L is a **linear operator**.

▝ Differential Equations

Any linear differential equation can be expressed in terms of the D notation. For example, the differential equation $y'' + 5y' + 6y = 5x - 3$ can be written as $D^2 y + 5Dy + 6y = 5x - 3$ or $(D^2 + 5D + 6)y = 5x - 3$. Using (8), the linear nth-order differential equations (6) and (7) can be written compactly as

$$L(y) = 0 \quad \text{and} \quad L(y) = g(x),$$

respectively.

Superposition Principle

In the next theorem we see that the sum, or **superposition**, of two or more solutions of a homogeneous linear differential equation is also a solution.

THEOREM 3.1.2 Superposition Principle—Homogeneous Equations

Let $y_1, y_2, \ldots, y_k$ be solutions of the homogeneous nth-order differential equation (6) on an interval I. Then the linear combination

$$y = c_1 y_1(x) + c_2 y_2(x) + \cdots + c_k y_k(x),$$

where the c_i, $i = 1, 2, \ldots, k$ are arbitrary constants, is also a solution on the interval.

PROOF: We prove the case $k = 2$. Let L be the differential operator defined in (8), and let $y_1(x)$ and $y_2(x)$ be solutions of the homogeneous equation $L(y) = 0$. If we define $y = c_1 y_1(x) + c_2 y_2(x)$, then by linearity of L we have

$$L(y) = L\{c_1 y_1(x) + c_2 y_2(x)\} = c_1 L(y_1) + c_2 L(y_2) = c_1 \cdot 0 + c_2 \cdot 0 = 0. \qquad \equiv$$

Corollaries to Theorem 3.1.2

(a) A constant multiple $y = c_1 y_1(x)$ of a solution $y_1(x)$ of a homogeneous linear differential equation is also a solution.

(b) A homogeneous linear differential equation always possesses the trivial solution $y = 0$.

EXAMPLE 4 Superposition—Homogeneous DE

The functions $y_1 = x^2$ and $y_2 = x^2 \ln x$ are both solutions of the homogeneous linear equation $x^3 y''' - 2xy' + 4y = 0$ on the interval $(0, \infty)$. By the superposition principle, the linear combination

$$y = c_1 x^2 + c_2 x^2 \ln x$$

is also a solution of the equation on the interval. $\qquad \equiv$

The function $y = e^{7x}$ is a solution of $y'' - 9y' + 14y = 0$. Since the differential equation is linear and homogeneous, the constant multiple $y = ce^{7x}$ is also a solution. For various values of c we see that $y = 9e^{7x}$, $y = 0$, $y = -\sqrt{5}e^{7x}$, ..., are all solutions of the equation.

Linear Dependence and Linear Independence

The next two concepts are basic to the study of linear differential equations.

DEFINITION 3.1.1 Linear Dependence/Independence

A set of functions $f_1(x), f_2(x), \ldots, f_n(x)$ is said to be **linearly dependent** on an interval I if there exist constants $c_1, c_2, \ldots, c_n$, not all zero, such that

$$c_1 f_1(x) + c_2 f_2(x) + \cdots + c_n f_n(x) = 0$$

for every x in the interval. If the set of functions is not linearly dependent on the interval, it is said to be **linearly independent**.

In other words, a set of functions is linearly independent on an interval if the only constants for which

$$c_1 f_1(x) + c_2 f_2(x) + \cdots + c_n f_n(x) = 0$$

for every x in the interval are $c_1 = c_2 = \cdots = c_n = 0$.

It is easy to understand these definitions in the case of two functions $f_1(x)$ and $f_2(x)$. If the functions are linearly dependent on an interval, then there exist constants c_1 and c_2 that are not both zero such that for every x in the interval $c_1 f_1(x) + c_2 f_2(x) = 0$. Therefore, if we assume that $c_1 \neq 0$, it follows that $f_1(x) = (-c_2/c_1)f_2(x)$; that is,

If two functions are linearly dependent, then one is simply a constant multiple of the other.

(a)

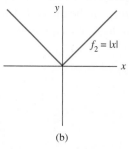

(b)

FIGURE 3.1.3 The set consisting of f_1 and f_2 is linearly independent on $(-\infty, \infty)$

Conversely, if $f_1(x) = c_2 f_2(x)$ for some constant c_2, then $(-1) \cdot f_1(x) + c_2 f_2(x) = 0$ for every x on some interval. Hence the functions are linearly dependent, since at least one of the constants (namely, $c_1 = -1$) is not zero. We conclude that:

Two functions are linearly independent when neither is a constant multiple of the other on an interval.

For example, the functions $f_1(x) = \sin 2x$ and $f_2(x) = \sin x \cos x$ are linearly dependent on $(-\infty, \infty)$ because $f_1(x)$ is a constant multiple of $f_2(x)$. Recall from the double angle formula for the sine that $\sin 2x = 2 \sin x \cos x$. On the other hand, the functions $f_1(x) = x$ and $f_2(x) = |x|$ are linearly independent on $(-\infty, \infty)$. Inspection of **FIGURE 3.1.3** should convince you that neither function is a constant multiple of the other on the interval.

It follows from the preceding discussion that the ratio $f_2(x)/f_1(x)$ is not a constant on an interval on which $f_1(x)$ and $f_2(x)$ are linearly independent. This little fact will be used in the next section.

EXAMPLE 5 Linearly Dependent Functions

The functions $f_1(x) = \cos^2 x$, $f_2(x) = \sin^2 x$, $f_3(x) = \sec^2 x$, $f_4(x) = \tan^2 x$ are linearly dependent on the interval $(-\pi/2, \pi/2)$ since

$$c_1 \cos^2 x + c_2 \sin^2 x + c_3 \sec^2 x + c_4 \tan^2 x = 0,$$

when $c_1 = c_2 = 1$, $c_3 = -1$, $c_4 = 1$. We used here $\cos^2 x + \sin^2 x = 1$ and $1 + \tan^2 x = \sec^2 x$ for every number x in the interval. $\equiv$

A set of n functions $f_1(x), f_2(x), \ldots, f_n(x)$ is linearly dependent on an interval I if at least one of the functions can be expressed as a linear combination of the remaining functions. For example, three functions $f_1(x), f_2(x)$, and $f_3(x)$ are linearly dependent on I if at least one of these functions is a linear combination of the other two, say,

$$f_3(x) = c_1 f_1(x) + c_2 f_2(x)$$

for all x in I. A set of n functions is linearly independent on I if no one function is a linear combination of the other functions.

EXAMPLE 6 Linearly Dependent Functions

The functions $f_1(x) = \sqrt{x} + 5$, $f_2(x) = \sqrt{x} + 5x$, $f_3(x) = x - 1$, $f_4(x) = x^2$ are linearly dependent on the interval $(0, \infty)$ since f_2 can be written as a linear combination of f_1, f_3, and f_4. Observe that

$$f_2(x) = 1 \cdot f_1(x) + 5 \cdot f_3(x) + 0 \cdot f_4(x)$$

for every x in the interval $(0, \infty)$. $\equiv$

■ Solutions of Differential Equations

We are primarily interested in linearly independent functions or, more to the point, linearly independent solutions of a linear differential equation. Although we could always appeal directly to Definition 3.1.1, it turns out that the question of whether n solutions $y_1, y_2, \ldots, y_n$ of a homogeneous linear nth-order differential equation (6) are linearly independent can be settled somewhat mechanically using a determinant.

DEFINITION 3.1.2 Wronskian

Suppose each of the functions $f_1(x), f_2(x), \ldots, f_n(x)$ possesses at least $n - 1$ derivatives. The determinant

$$W(f_1, f_2, \ldots, f_n) = \begin{vmatrix} f_1 & f_2 & \cdots & f_n \\ f_1' & f_2' & \cdots & f_n' \\ \vdots & \vdots & & \vdots \\ f_1^{(n-1)} & f_2^{(n-1)} & \cdots & f_n^{(n-1)} \end{vmatrix},$$

where the primes denote derivatives, is called the **Wronskian** of the functions.

The Wronskian determinant is named after the Polish philosopher, inventor, lawyer, physicist, and mathematician **Józef Maria Hoëné-Wronski** (1776–1853).

THEOREM 3.1.3 **Criterion for Linearly Independent Solutions**

Let $y_1, y_2, \ldots, y_n$ be n solutions of the homogeneous linear nth-order differential equation (6) on an interval I. Then the set of solutions is **linearly independent** on I if and only if $W(y_1, y_2, \ldots, y_n) \neq 0$ for every x in the interval.

It follows from Theorem 3.1.3 that when $y_1, y_2, \ldots, y_n$ are n solutions of (6) on an interval I, the Wronskian $W(y_1, y_2, \ldots, y_n)$ is either identically zero or never zero on the interval. Thus, if we can show that $W(y_1, y_2, \ldots, y_n) \neq 0$ for *some* x_0 in I, then the solutions $y_1, y_2, \ldots, y_n$ are linearly independent on I. For example, the functions

$$y_1(x) = \frac{\cos(2 \ln x)}{x^3}, \quad y_2(x) = \frac{\sin(2 \ln x)}{x^3}$$

are solutions of the differential equation

$$x^2 y'' + 7xy' + 13y = 0$$

on the interval $(0, \infty)$. Note that the coefficient functions $a_2(x) = x^2$, $a_1(x) = 7x$, and $a_0(x) = 13$ are continuous on $(0, \infty)$ and that $a_2(x) \neq 0$ for every value of x in the interval. The Wronskian is

$$W(y_1(x), y_2(x)) = \begin{vmatrix} \dfrac{\cos(2 \ln x)}{x^3} & \dfrac{\sin(2 \ln x)}{x^3} \\[2ex] \dfrac{-2x^2 \sin(2 \ln x) - 3x^2 \cos(2 \ln x)}{x^6} & \dfrac{2x^2 \cos(2 \ln x) - 3x^2 \sin(2 \ln x)}{x^6} \end{vmatrix}.$$

Rather than expanding this unwieldy determinant, we choose $x = 1$ in the interval $(0, \infty)$ and find

$$W(y_1(1), y_2(1)) = \begin{vmatrix} 1 & 0 \\ -3 & 2 \end{vmatrix} = 2.$$

The fact that $W(y_1(1), y_2(1)) = 2 \neq 0$ is sufficient to conclude that $y_1(x)$ and $y_2(x)$ are linearly independent on $(0, \infty)$.

A set of n linearly independent solutions of a homogeneous linear nth-order differential equation is given a special name.

DEFINITION 3.1.3 **Fundamental Set of Solutions**

Any set $y_1, y_2, \ldots, y_n$ of n linearly independent solutions of the homogeneous linear nth-order differential equation (6) on an interval I is said to be a **fundamental set of solutions** on the interval.

The basic question of whether a fundamental set of solutions exists for a linear equation is answered in the next theorem.

THEOREM 3.1.4 **Existence of a Fundamental Set**

There exists a fundamental set of solutions for the homogeneous linear nth-order differential equation (6) on an interval I.

Analogous to the fact that any vector in three dimensions can be expressed uniquely as a linear combination of the *linearly independent* vectors **i, j, k**, any solution of an nth-order homogeneous linear differential equation on an interval I can be expressed uniquely as a linear combination of n linearly independent solutions on I. In other words, n linearly independent solutions $y_1, y_2, \ldots, y_n$ are the basic building blocks for the general solution of the equation.

THEOREM 3.1.5 **General Solution—Homogeneous Equations**

Let $y_1, y_2, \ldots, y_n$ be a fundamental set of solutions of the homogeneous linear nth-order differential equation (6) on an interval I. Then the **general solution** of the equation on the interval is

$$y = c_1 y_1(x) + c_2 y_2(x) + \cdots + c_n y_n(x),$$

where c_i, $i = 1, 2, \ldots, n$ are arbitrary constants.

Theorem 3.1.5 states that if $Y(x)$ is any solution of (6) on the interval, then constants $C_1, C_2, \ldots,$ C_n can always be found so that

$$Y(x) = C_1 y_1(x) + C_2 y_2(x) + \cdots + C_n y_n(x).$$

We will prove the case when $n = 2$.

PROOF: Let Y be a solution and y_1 and y_2 be linearly independent solutions of $a_2 y'' + a_1 y' + a_0 y = 0$ on an interval I. Suppose $x = t$ is a point in I for which $W(y_1(t), y_2(t)) \neq 0$. Suppose also that $Y(t) = k_1$ and $Y'(t) = k_2$. If we now examine the equations

$$C_1 y_1(t) + C_2 y_2(t) = k_1$$
$$C_1 y_1'(t) + C_2 y_2'(t) = k_2,$$

it follows that we can determine C_1 and C_2 uniquely, provided that the determinant of the coefficients satisfies

$$\begin{vmatrix} y_1(t) & y_2(t) \\ y_1'(t) & y_2'(t) \end{vmatrix} \neq 0.$$

But this determinant is simply the Wronskian evaluated at $x = t$, and, by assumption, $W \neq 0$. If we define $G(x) = C_1 y_1(x) + C_2 y_2(x)$, we observe that $G(x)$ satisfies the differential equation, since it is a superposition of two known solutions; $G(x)$ satisfies the initial conditions

$$G(t) = C_1 y_1(t) + C_2 y_2(t) = k_1 \quad \text{and} \quad G'(t) = C_1 y_1'(t) + C_2 y_2'(t) = k_2;$$

$Y(x)$ satisfies the *same* linear equation and the *same* initial conditions. Since the solution of this linear initial-value problem is unique (Theorem 3.1.1), we have $Y(x) = G(x)$ or $Y(x) = C_1 y_1(x) + C_2 y_2(x)$. $\equiv$

EXAMPLE 7 **General Solution of a Homogeneous DE**

The functions $y_1 = e^{3x}$ and $y_2 = e^{-3x}$ are both solutions of the homogeneous linear equation $y'' - 9y = 0$ on the interval $(-\infty, \infty)$. By inspection, the solutions are linearly independent on the x-axis. This fact can be corroborated by observing that the Wronskian

$$W(e^{3x}, e^{-3x}) = \begin{vmatrix} e^{3x} & e^{-3x} \\ 3e^{3x} & -3e^{-3x} \end{vmatrix} = -6 \neq 0$$

for every x. We conclude that y_1 and y_2 form a fundamental set of solutions, and consequently $y = c_1 e^{3x} + c_2 e^{-3x}$ is the general solution of the equation on the interval $(-\infty, \infty)$. $\equiv$

EXAMPLE 8 **A Solution Obtained from a General Solution**

The function $y = 4 \sinh 3x - 5e^{3x}$ is a solution of the differential equation $y'' - 9y = 0$ in Example 7. (Verify this.) In view of Theorem 3.1.5, we must be able to obtain this solution from the general solution $y = c_1 e^{3x} + c_2 e^{-3x}$. Observe that if we choose $c_1 = 2$ and $c_2 = -7$, then $y = 2e^{3x} - 7e^{-3x}$ can be rewritten as

$$y = 2e^{3x} - 2e^{-3x} - 5e^{-3x} = 4\left(\frac{e^{3x} - e^{-3x}}{2}\right) - 5e^{-3x}.$$

The last expression is recognized as $y = 4 \sinh 3x - 5e^{-3x}$. $\equiv$

EXAMPLE 9 **General Solution of a Homogeneous DE**

The functions $y_1 = e^x$, $y_2 = e^{2x}$, and $y_3 = e^{3x}$ satisfy the third-order equation

$$y''' - 6y'' + 11y' - 6y = 0.$$

Since

$$W(e^x, e^{2x}, e^{3x}) = \begin{vmatrix} e^x & e^{2x} & e^{3x} \\ e^x & 2e^{2x} & 3e^{3x} \\ e^x & 4e^{2x} & 9e^{3x} \end{vmatrix} = 2e^{6x} \neq 0$$

for every real value of x, the functions y_1, y_2, and y_3 form a fundamental set of solutions on $(-\infty, \infty)$. We conclude that $y = c_1 e^x + c_2 e^{2x} + c_3 e^{3x}$ is the general solution of the differential equation on the interval $(-\infty, \infty)$. ≡

3.1.3 Nonhomogeneous Equations

Any function y_p free of arbitrary parameters that satisfies (7) is said to be a **particular solution** of the equation. For example, it is a straightforward task to show that the constant function $y_p = 3$ is a particular solution of the nonhomogeneous equation $y'' + 9y = 27$.

Now if $y_1, y_2, \ldots, y_k$ are solutions of (6) on an interval I and y_p is any particular solution of (7) on I, then the linear combination

$$y = c_1 y_1(x) + c_2 y_2(x) + \cdots + c_k y_k(x) + y_p(x) \tag{10}$$

is also a solution of the nonhomogeneous equation (7). If you think about it, this makes sense, because the linear combination $c_1 y_1(x) + c_2 y_2(x) + \cdots + c_k y_k(x)$ is mapped into 0 by the operator $L = a_n D^n + a_{n-1} D^{n-1} + \cdots + a_1 D + a_0$, whereas y_p is mapped into $g(x)$. If we use $k = n$ linearly independent solutions of the nth-order equation (6), then the expression in (10) becomes the general solution of (7).

THEOREM 3.1.6 **General Solution—Nonhomogeneous Equations**

Let y_p be any particular solution of the nonhomogeneous linear nth-order differential equation (7) on an interval I, and let $y_1, y_2, \ldots, y_n$ be a fundamental set of solutions of the associated homogeneous differential equation (6) on I. Then the **general solution** of the equation on the interval is

$$y = c_1 y_1(x) + c_2 y_2(x) + \cdots + c_n y_n(x) + y_p(x),$$

where the c_i, $i = 1, 2, \ldots, n$ are arbitrary constants.

PROOF: Let L be the differential operator defined in (8), and let $Y(x)$ and $y_p(x)$ be particular solutions of the nonhomogeneous equation $L(y) = g(x)$. If we define $u(x) = Y(x) - y_p(x)$, then by linearity of L we have

$$L(u) = L\{Y(x) - y_p(x)\} = L(Y(x)) - L(y_p(x)) = g(x) - g(x) = 0.$$

This shows that $u(x)$ is a solution of the homogeneous equation $L(y) = 0$. Hence, by Theorem 3.1.5, $u(x) = c_1 y_1(x) + c_2 y_2(x) + \cdots + c_n y_n(x)$, and so

$$Y(x) - y_p(x) = c_1 y_1(x) + c_2 y_2(x) + \cdots + c_n y_n(x)$$

or
$$Y(x) = c_1 y_1(x) + c_2 y_2(x) + \cdots + c_n y_n(x) + y_p(x). \qquad ≡$$

■ **Complementary Function** We see in Theorem 3.1.6 that the general solution of a nonhomogeneous linear equation consists of the sum of two functions:

$$y = c_1 y_1(x) + c_2 y_2(x) + \cdots + c_n y_n(x) + y_p(x) = y_c(x) + y_p(x).$$

The linear combination $y_c(x) = c_1 y_1(x) + c_2 y_2(x) + \cdots + c_n y_n(x)$, which is the general solution of (6), is called the **complementary function** for equation (7). In other words, to solve a nonhomogeneous linear differential equation we first solve the associated homogeneous equation and

then find any particular solution of the nonhomogeneous equation. The general solution of the nonhomogeneous equation is then

$$y = complementary\ function + any\ particular\ solution$$
$$= y_c + y_p.$$

EXAMPLE 10 General Solution of a Nonhomogeneous DE

By substitution, the function $y_p = -\frac{11}{12} - \frac{1}{2}x$ is readily shown to be a particular solution of the nonhomogeneous equation

$$y''' - 6y'' + 11y' - 6y = 3x. \tag{11}$$

In order to write the general solution of (11), we must also be able to solve the associated homogeneous equation

$$y''' - 6y'' + 11y' - 6y = 0.$$

But in Example 9 we saw that the general solution of this latter equation on the interval $(-\infty, \infty)$ was $y_c = c_1e^x + c_2e^{2x} + c_3e^{3x}$. Hence the general solution of (11) on the interval is

$$y = y_c + y_p = c_1e^x + c_2e^{2x} + c_3e^{3x} - \frac{11}{12} - \frac{1}{2}x. \qquad\equiv$$

■ Another Superposition Principle

The last theorem of this discussion will be useful in Section 3.4, when we consider a method for finding particular solutions of nonhomogeneous equations.

THEOREM 3.1.7 Superposition Principle—Nonhomogeneous Equations

Let $y_{p_1}, y_{p_2}, \ldots, y_{p_k}$ be k particular solutions of the nonhomogeneous linear nth-order differential equation (7) on an interval I corresponding, in turn, to k distinct functions $g_1, g_2, \ldots, g_k$. That is, suppose y_{p_i} denotes a particular solution of the corresponding differential equation

$$a_n(x)y^{(n)} + a_{n-1}(x)y^{(n-1)} + \cdots + a_1(x)y' + a_0(x)y = g_i(x), \tag{12}$$

where $i = 1, 2, \ldots, k$. Then

$$y_p(x) = y_{p_1}(x) + y_{p_2}(x) + \cdots + y_{p_k}(x) \tag{13}$$

is a particular solution of

$$a_n(x)y^{(n)} + a_{n-1}(x)y^{(n-1)} + \cdots + a_1(x)y' + a_0(x)y$$
$$= g_1(x) + g_2(x) + \cdots + g_k(x). \tag{14}$$

PROOF: We prove the case $k = 2$. Let L be the differential operator defined in (8), and let $y_{p_1}(x)$ and $y_{p_2}(x)$ be particular solutions of the nonhomogeneous equations $L(y) = g_1(x)$ and $L(y) = g_2(x)$, respectively. If we define $y_p(x) = y_{p_1}(x) + y_{p_2}(x)$, we want to show that y_p is a particular solution of $L(y) = g_1(x) + g_2(x)$. The result follows again by the linearity of the operator L:

$$L(y_p) = L\{y_{p_1}(x) + y_{p_2}(x)\} = L(y_{p_1}(x)) + L(y_{p_2}(x)) = g_1(x) + g_2(x). \qquad\equiv$$

EXAMPLE 11 Superposition—Nonhomogeneous DE

You should verify that

$$\begin{array}{lll} y_{p_1} = -4x^2 & \text{is a particular solution of} & y'' - 3y' + 4y = -16x^2 + 24x - 8, \\ y_{p_2} = e^{2x} & \text{is a particular solution of} & y'' - 3y' + 4y = 2e^{2x}, \\ y_{p_3} = xe^x & \text{is a particular solution of} & y'' - 3y' + 4y = 2xe^x - e^x. \end{array}$$

It follows from Theorem 3.1.7 that the superposition of y_{p_1}, y_{p_2}, and y_{p_3},

$$y = y_{p_1} + y_{p_2} + y_{p_3} = -4x^2 + e^{2x} + xe^x,$$

is a solution of

$$y'' - 3y' + 4y = \underbrace{-16x^2 + 24x - 8}_{g_1(x)} + \underbrace{2e^{2x}}_{g_2(x)} + \underbrace{2xe^x - e^x}_{g_3(x)}. \qquad\equiv$$

This sentence is a generalization of Theorem 3.1.7. ▶

If the y_{p_i} are particular solutions of (12) for $i = 1, 2, \ldots, k$, then the linear combination

$$y_p = c_1 y_{p_1} + c_2 y_{p_2} + \cdots + c_k y_{p_k},$$

where the c_i are constants, is also a particular solution of (14) when the right-hand member of the equation is the linear combination

$$c_1 g_1(x) + c_2 g_2(x) + \cdots + c_k g_k(x).$$

Before we actually start solving homogeneous and nonhomogeneous linear differential equations, we need one additional bit of theory presented in the next section.

REMARKS

This remark is a continuation of the brief discussion of dynamical systems given at the end of Section 1.3.

A dynamical system whose rule or mathematical model is a linear nth-order differential equation

$$a_n(t)y^{(n)} + a_{n-1}(t)y^{(n-1)} + \cdots + a_1(t)y' + a_0(t)y = g(t)$$

is said to be a **linear system**. The set of n time-dependent functions $y(t), y'(t), \ldots, y^{(n-1)}(t)$ are the **state variables** of the system. Recall, their values at some time t give the **state** of the system. The function g is variously called the **input function** or **forcing function**. A solution $y(t)$ of the differential equation is said to be the **output** or **response** of the system. Under the conditions stated in Theorem 3.1.1, the output or response $y(t)$ is uniquely determined by the input and the state of the system prescribed at a time t_0; that is, by the initial conditions $y(t_0), y'(t_0), \ldots, y^{(n-1)}(t_0)$.

In order that a dynamical system be a linear system, it is necessary that the superposition principle (Theorem 3.1.7) hold in the system; that is, the response of the system to a superposition of inputs is a superposition of outputs. We have already examined some simple linear systems in Section 2.7 (linear first-order equations); in Section 3.8 we examine linear systems in which the mathematical models are second-order differential equations.

3.1 Exercises Answers to selected odd-numbered problems begin on page ANS-5.

3.1.1 Initial-Value and Boundary-Value Problems

In Problems 1–4, the given family of functions is the general solution of the differential equation on the indicated interval. Find a member of the family that is a solution of the initial-value problem.

1. $y = c_1 e^x + c_2 e^{-x}$, $(-\infty, \infty)$; $y'' - y = 0$, $y(0) = 0$, $y'(0) = 1$

2. $y = c_1 e^{4x} + c_2 e^{-x}$, $(-\infty, \infty)$; $y'' - 3y' - 4y = 0$, $y(0) = 1$, $y'(0) = 2$

3. $y = c_1 x + c_2 x \ln x$, $(0, \infty)$; $x^2 y'' - xy' + y = 0$, $y(1) = 3$, $y'(1) = -1$

4. $y = c_1 + c_2 \cos x + c_3 \sin x$, $(-\infty, \infty)$; $y''' + y' = 0$, $y(\pi) = 0$, $y'(\pi) = 2$, $y''(\pi) = -1$

5. Given that $y = c_1 + c_2 x^2$ is a two-parameter family of solutions of $xy'' - y' = 0$ on the interval $(-\infty, \infty)$, show that constants c_1 and c_2 cannot be found so that a member of the family satisfies the initial conditions $y(0) = 0$, $y'(0) = 1$. Explain why this does not violate Theorem 3.1.1.

6. Find two members of the family of solutions in Problem 5 that satisfy the initial conditions $y(0) = 0$, $y'(0) = 0$.

7. Given that $x(t) = c_1 \cos \omega t + c_2 \sin \omega t$ is the general solution of $x'' + \omega^2 x = 0$ on the interval $(-\infty, \infty)$, show that a solution satisfying the initial conditions $x(0) = x_0$, $x'(0) = x_1$, is given by

$$x(t) = x_0 \cos \omega t + \frac{x_1}{\omega} \sin \omega t.$$

8. Use the general solution of $x'' + \omega^2 x = 0$ given in Problem 7 to show that a solution satisfying the initial conditions $x(t_0) = x_0$, $x'(t_0) = x_1$, is the solution given in Problem 7 shifted by an amount t_0:

$$x(t) = x_0 \cos \omega(t - t_0) + \frac{x_1}{\omega} \sin \omega(t - t_0).$$

In Problems 9 and 10, find an interval centered about $x = 0$ for which the given initial-value problem has a unique solution.

9. $(x - 2)y'' + 3y = x$, $y(0) = 0$, $y'(0) = 1$

10. $y'' + (\tan x)y = e^x$, $y(0) = 1$, $y'(0) = 0$

11. (a) Use the family in Problem 1 to find a solution of $y'' - y = 0$ that satisfies the boundary conditions $y(0) = 0$, $y(1) = 1$.

(b) The DE in part (a) has the alternative general solution $y = c_3 \cosh x + c_4 \sinh x$ on $(-\infty, \infty)$. Use this family to find a solution that satisfies the boundary conditions in part (a).

(c) Show that the solutions in parts (a) and (b) are equivalent.

12. Use the family in Problem 5 to find a solution of $xy'' - y' = 0$ that satisfies the boundary conditions $y(0) = 1$, $y'(1) = 6$.

In Problems 13 and 14, the given two-parameter family is a solution of the indicated differential equation on the interval $(-\infty, \infty)$. Determine whether a member of the family can be found that satisfies the boundary conditions.

13. $y = c_1 e^x \cos x + c_2 e^x \sin x$; $y'' - 2y' + 2y = 0$

(a) $y(0) = 1$, $y'(\pi) = 0$ **(b)** $y(0) = 1$, $y(\pi) = -1$
(c) $y(0) = 1$, $y(\pi/2) = 1$ **(d)** $y(0) = 0$, $y(\pi) = 0$

14. $y = c_1 x^2 + c_2 x^4 + 3$; $x^2 y'' - 5xy' + 8y = 24$

(a) $y(-1) = 0$, $y(1) = 4$ **(b)** $y(0) = 1$, $y(1) = 2$
(c) $y(0) = 3$, $y(1) = 0$ **(d)** $y(1) = 3$, $y(2) = 15$

3.1.2 Homogeneous Equations

In Problems 15–22, determine whether the given set of functions is linearly dependent or linearly independent on the interval $(-\infty, \infty)$.

15. $f_1(x) = x$, $f_2(x) = x^2$, $f_3(x) = 4x - 3x^2$

16. $f_1(x) = 0$, $f_2(x) = x$, $f_3(x) = e^x$

17. $f_1(x) = 5$, $f_2(x) = \cos^2 x$, $f_3(x) = \sin^2 x$

18. $f_1(x) = \cos 2x$, $f_2(x) = 1$, $f_3(x) = \cos^2 x$

19. $f_1(x) = x$, $f_2(x) = x - 1$, $f_3(x) = x + 3$

20. $f_1(x) = 2 + x$, $f_2(x) = 2 + |x|$

21. $f_1(x) = 1 + x$, $f_2(x) = x$, $f_3(x) = x^2$

22. $f_1(x) = e^x$, $f_2(x) = e^{-x}$, $f_3(x) = \sinh x$

In Problems 23–30, verify that the given functions form a fundamental set of solutions of the differential equation on the indicated interval. Form the general solution of the equation.

23. $y'' - y' - 12y = 0$; $e^{-3x}, e^{4x}, (-\infty, \infty)$

24. $y'' - 4y = 0$; $\cosh 2x, \sinh 2x, (-\infty, \infty)$

25. $y'' - 2y' + 5y = 0$; $e^x \cos 2x, e^x \sin 2x, (-\infty, \infty)$

26. $4y'' - 4y' + y = 0$; $e^{x/2}, xe^{x/2}, (-\infty, \infty)$

27. $x^2 y'' - 6xy' + 12y = 0$; $x^3, x^4, (0, \infty)$

28. $x^2 y'' + xy' + y = 0$; $\cos(\ln x), \sin(\ln x), (0, \infty)$

29. $x^3 y''' + 6x^2 y'' + 4xy' - 4y = 0$; $x, x^{-2}, x^{-2} \ln x, (0, \infty)$

30. $y^{(4)} + y'' = 0$; $1, x, \cos x, \sin x, (-\infty, \infty)$

3.1.3 Nonhomogeneous Equations

In Problems 31–34, verify that the given two-parameter family of functions is the general solution of the nonhomogeneous differential equation on the indicated interval.

31. $y'' - 7y' + 10y = 24e^x$;
$y = c_1 e^{2x} + c_2 e^{5x} + 6e^x, (-\infty, \infty)$

32. $y'' + y = \sec x$;
$y = c_1 \cos x + c_2 \sin x + x \sin x + (\cos x) \ln(\cos x)$,
$(-\pi/2, \pi/2)$

33. $y'' - 4y' + 4y = 2e^{2x} + 4x - 12$;
$y = c_1 e^{2x} + c_2 x e^{2x} + x^2 e^{2x} + x - 2, (-\infty, \infty)$

34. $2x^2 y'' + 5xy' + y = x^2 - x$;
$y = c_1 x^{-1/2} + c_2 x^{-1} + \frac{1}{15}x^2 - \frac{1}{6}x, (0, \infty)$

35. (a) Verify that $y_{p_1} = 3e^{2x}$ and $y_{p_2} = x^2 + 3x$ are, respectively, particular solutions of

$$y'' - 6y' + 5y = -9e^{2x}$$

and $y'' - 6y' + 5y = 5x^2 + 3x - 16.$

(b) Use part (a) to find particular solutions of

$$y'' - 6y' + 5y = 5x^2 + 3x - 16 - 9e^{2x}$$

and $y'' - 6y' + 5y = -10x^2 - 6x + 32 + e^{2x}.$

36. (a) By inspection, find a particular solution of

$$y'' + 2y = 10.$$

(b) By inspection, find a particular solution of

$$y'' + 2y = -4x.$$

(c) Find a particular solution of $y'' + 2y = -4x + 10$.

(d) Find a particular solution of $y'' + 2y = 8x + 5$.

Discussion Problems

37. Let $n = 1, 2, 3, \ldots$. Discuss how the observations $D^n x^{n-1} = 0$ and $D^n x^n = n!$ can be used to find the general solutions of the given differential equations.

(a) $y'' = 0$ **(b)** $y''' = 0$
(c) $y^{(4)} = 0$ **(d)** $y'' = 2$
(e) $y''' = 6$ **(f)** $y^{(4)} = 24$

38. Suppose that $y_1 = e^x$ and $y_2 = e^{-x}$ are two solutions of a homogeneous linear differential equation. Explain why $y_3 = \cosh x$ and $y_4 = \sinh x$ are also solutions of the equation.

39. (a) Verify that $y_1 = x^3$ and $y_2 = |x|^3$ are linearly independent solutions of the differential equation $x^2 y'' - 4xy' + 6y = 0$ on the interval $(-\infty, \infty)$.

(b) For the functions y_1 and y_2 in part (a), show that $W(y_1, y_2) = 0$ for every real number x. Does this result violate Theorem 3.1.3? Explain.

(c) Verify that $Y_1 = x^3$ and $Y_2 = x^2$ are also linearly independent solutions of the differential equation in part (a) on the interval $(-\infty, \infty)$.

(d) Besides the functions y_1, y_2, Y_1, and Y_2 in parts (a) and (c), find a solution of the differential equation that satisfies $y(0) = 0$, $y'(0) = 0$.

(e) By the superposition principle, Theorem 3.1.2, both linear combinations $y = c_1 y_1 + c_2 y_2$ and $Y = c_1 Y_1 + c_2 Y_2$ are solutions of the differential equation. Discuss whether one, both, or neither of the linear combinations is a general solution of the differential equation on the interval $(-\infty, \infty)$.

40. Is the set of functions $f_1(x) = e^{x+2}, f_2(x) = e^{x-3}$ linearly dependent or linearly independent on the interval $(-\infty, \infty)$? Discuss.

41. By substituting $y = e^{mx}$ into the differential equation

$$\frac{d^4y}{dx^4} - 10\frac{d^2y}{dx^2} + 9y = 0$$

find four linearly independent solutions of the equation. Give the general solution of the equation on $(-\infty, \infty)$.

42. Suppose $y_1, y_2, \ldots, y_k$ are k linearly independent solutions on $(-\infty, \infty)$ of a homogeneous linear nth-order differential equation with constant coefficients. By Theorem 3.1.2 it

follows that $y_{k+1} = 0$ is also a solution of the differential equation. Is the set of solutions $y_1, y_2, \ldots, y_k, y_{k+1}$ linearly dependent or linearly independent on $(-\infty, \infty)$? Discuss.

43. Suppose that $y_1, y_2, \ldots, y_k$ are k nontrivial solutions of a homogeneous linear nth-order differential equation with constant coefficients and that $k = n + 1$. Is the set of solutions $y_1, y_2, \ldots, y_k$ linearly dependent or linearly independent on $(-\infty, \infty)$? Discuss.

44. If y_{p_1} and y_{p_2} are particular solutions of the nonhomogeneous linear differential equation (7), then show that the difference of these particular solutions $y = y_{p_1} - y_{p_2}$ is a solution of the associated homogeneous differential equation (6).

3.2 Reduction of Order

INTRODUCTION In Section 3.1 we saw that the general solution of a homogeneous linear second-order differential equation

$$a_2(x)y'' + a_1(x)y' + a_0(x)y = 0 \tag{1}$$

was a linear combination $y = c_1y_1 + c_2y_2$, where y_1 and y_2 are solutions that constitute a linearly independent set on some interval I. Beginning in the next section we examine a method for determining these solutions when the coefficients of the DE in (1) are constants. This method, which is a straightforward exercise in algebra, breaks down in a few cases and yields only a single solution y_1 of the DE. It turns out that we can construct a second solution y_2 of a homogeneous equation (1) (even when the coefficients in (1) are variable) provided that we know one nontrivial solution y_1 of the DE. The basic idea described in this section is that the linear *second-order* equation (1) can be reduced to a linear *first-order* DE by means of a substitution involving the known solution y_1. A second solution, y_2 of (1), is apparent after this first-order DE is solved.

■ Reduction of Order Suppose $y(x)$ denotes a known solution of equation (1). We seek a second solution $y_2(x)$ of (1) so that y_1 and y_2 are linearly independent on some interval I. Recall that if y_1 and y_2 are linearly independent, then their ratio y_2/y_1 is nonconstant on I; that is, $y_2/y_1 = u(x)$ or $y_2(x) = u(x)y_1(x)$. The idea is to find $u(x)$ by substituting $y_2(x) = u(x)y_1(x)$ into the given differential equation. This method is called **reduction of order** since we must solve a first-order equation to find u.

The first example illustrates the basic technique.

EXAMPLE 1 **Finding a Second Solution**

Given that $y_1 = e^x$ is a solution of $y'' - y = 0$ on the interval $(-\infty, \infty)$, use reduction of order to find a second solution y_2.

SOLUTION If $y = u(x)y_1(x) = u(x)e^x$, then the first two derivatives of y are obtained from the product rule:

$$y' = ue^x + e^xu', \quad y'' = ue^x + 2e^xu' + e^xu''.$$

By substituting y and y'' into the original DE, it simplifies to

$$y'' - y = e^x(u'' + 2u') = 0.$$

Since $e^x \neq 0$, the last equation requires $u'' + 2u' = 0$. If we make the substitution $w = u'$, this linear second-order equation in u becomes $w' + 2w = 0$, which is a linear first-order

equation in w. Using the integrating factor e^{2x}, we can write $d/dx\,[e^{2x}w] = 0$. After integrating we get $w = c_1 e^{-2x}$ or $u' = c_1 e^{-2x}$. Integrating again then yields $u = -\frac{1}{2}c_1 e^{-2x} + c_2$. Thus

$$y = u(x)e^x = -\frac{c_1}{2}e^{-x} + c_2 e^x. \tag{2}$$

By choosing $c_2 = 0$ and $c_1 = -2$ we obtain the desired second solution, $y_2 = e^{-x}$. Because $W(e^x, e^{-x}) \neq 0$ for every x, the solutions are linearly independent on $(-\infty, \infty)$. $\equiv$

Since we have shown that $y_1 = e^x$ and $y_2 = e^{-x}$ are linearly independent solutions of a linear second-order equation, the expression in (2) is actually the general solution of $y'' - y = 0$ on the interval $(-\infty, \infty)$.

■ **General Case** Suppose we divide by $a_2(x)$ in order to put equation (1) in the **standard form**

$$y'' + P(x)y' + Q(x)y = 0, \tag{3}$$

where $P(x)$ and $Q(x)$ are continuous on some interval I. Let us suppose further that $y_1(x)$ is a known solution of (3) on I and that $y_1(x) \neq 0$ for every x in the interval. If we define $y = u(x)y_1(x)$, it follows that

$$y' = uy_1' + y_1 u', \qquad y'' = uy_1'' + 2y_1'u' + y_1 u''$$

$$y'' + Py' + Qy = u[\underbrace{y_1'' + Py_1' + Qy_1}_{\text{zero}}] + y_1 u'' + (2y_1' + Py_1)u' = 0.$$

This implies that we must have

$$y_1 u'' + (2y_1' + Py_1)u' = 0 \quad \text{or} \quad y_1 w' + (2y_1' + Py_1)w = 0, \tag{4}$$

where we have let $w = u'$. Observe that the last equation in (4) is both linear and separable. Separating variables and integrating, we obtain

$$\frac{dw}{w} + 2\frac{y_1'}{y_1}dx + P\,dx = 0$$

$$\ln|wy_1^2| = -\int P\,dx + c \quad \text{or} \quad wy_1^2 = c_1 e^{-\int P\,dx}.$$

We solve the last equation for w, use $w = u'$, and integrate again:

$$u = c_1 \int \frac{e^{-\int P\,dx}}{y_1^2}\,dx + c_2.$$

By choosing $c_1 = 1$ and $c_2 = 0$, we find that $y = u(x)y_1(x)$ is a second solution of equation (3). We summarize the result.

THEOREM 3.2.1 **Reduction of Order**

Let $y_1(x)$ be a solution of the homogeneous differential equation (3) on an interval I and that $y_1(x) \neq 0$ for all x in I. Then

$$y_2(x) = y_1(x)\int \frac{e^{-\int P(x)\,dx}}{y_1^2(x)}\,dx \tag{5}$$

is a second solution of (3) on the interval I.

It makes a good review of differentiation to verify that the function $y_2(x)$ defined in (5) satisfies equation (3). Furthermore, if $y_2(x)$ is the solution in (5), then the functions $y_1(x)$ and $y_2(x)$ are linearly independent on any interval I on which $y_1(x)$ is not zero. To prove linear independence, we use Theorem 3.1.3. The Wronskian of the functions is

$$W(y_1, y_2) = \begin{vmatrix} y_1(x) & y_2(x) \\ y_1'(x) & y_2'(x) \end{vmatrix} = \begin{vmatrix} y_1(x) & y_1(x)\displaystyle\int \frac{e^{-\int P(x)\,dx}}{y_1^2(x)}\,dx \\ y_1'(x) & \dfrac{e^{-\int P(x)\,dx}}{y_1(x)} + y_1'(x)\displaystyle\int \frac{e^{-\int P(x)\,dx}}{y_1^2(x)}\,dx \end{vmatrix} = e^{-\int P(x)\,dx} \neq 0.$$

The general solution of (3) on I is the linear combination $y(x) = c_1 y_1(x) + c_2 y_2(x)$.

EXAMPLE 2 A Second Solution by Formula (5)

The function $y_1 = x^2$ is a solution of $x^2 y'' - 3xy' + 4y = 0$. Find the general solution on the interval $(0, \infty)$.

SOLUTION From the standard form of the equation

$$y'' - \frac{3}{x} y' + \frac{4}{x^2} y = 0,$$

we find from (5)
$$y_2 = x^2 \int \frac{e^{3\int dx/x}}{x^4}\,dx \quad \leftarrow e^{3\int dx/x} = e^{\ln x^3} = x^3$$

$$= x^2 \int \frac{dx}{x} = x^2 \ln x.$$

The general solution on the interval $(0, \infty)$ is given by $y = c_1 y_1 + c_2 y_2$; that is, $y = c_1 x^2 + c_2 x^2 \ln x$. ≡

REMARKS

(*i*) We have derived and illustrated how to use (5) because this formula appears again in the next section and in Section 5.2. We use (5) simply to save time in obtaining a desired result. Your instructor will tell you whether you should memorize (5) or whether you should know the first principles of reduction of order.

(*ii*) The integral in (5) may be nonelementary. In this case we simply write the second solution in terms of an integral-defined function:

$$y_2(x) = y_1(x) \int_{x_0}^x \frac{e^{-\int P(t)\,dt}}{y_1^2(t)}\,dt,$$

where we assume that the integrand is continuous on the interval $[x_0, x]$. See Problems 23 and 24 in Exercises 3.2.

3.2 Exercises Answers to selected odd-numbered problems begin on page ANS-5.

In Problems 1–18, the indicated function $y_1(x)$ is a solution of the given equation. Use reduction of order or formula (5), as instructed, to find a second solution $y_2(x)$.

1. $y'' - 4y' + 4y = 0$; $y_1 = e^{2x}$
2. $y'' + 2y' + y = 0$; $y_1 = xe^{-x}$
3. $y'' + 16y = 0$; $y_1 = \cos 4x$
4. $y'' + 9y = 0$; $y_1 = \sin 3x$
5. $y'' - y = 0$; $y_1 = \cosh x$

6. $y'' - 25y = 0$; $y_1 = e^{5x}$
7. $9y'' - 12y' + 4y = 0$; $y_1 = e^{2x/3}$
8. $6y'' + y' - y = 0$; $y_1 = e^{x/3}$
9. $x^2 y'' - 7xy' + 16y = 0$; $y_1 = x^4$
10. $x^2 y'' + 2xy' - 6y = 0$; $y_1 = x^2$
11. $xy'' + y' = 0$; $y_1 = \ln x$
12. $4x^2 y'' + y = 0$; $y_1 = x^{1/2} \ln x$
13. $x^2 y'' - xy' + 2y = 0$; $y_1 = x \sin(\ln x)$

14. $x^2y'' - 3xy' + 5y = 0;$ $\qquad y_1 = x^2\cos(\ln x)$

15. $(1 - 2x - x^2)y'' + 2(1 + x)y' - 2y = 0;$ $\quad y_1 = x + 1$

16. $(1 - x^2)y'' + 2xy' = 0;$ $\qquad y_1 = 1$

17. $xy'' - (2x + 1)y' + (x + 1)y = 0;$ $\qquad y_1 = e^x$

18. $y'' + 2(\tan x)y' - y = 0;$ $\qquad y_1 = \sin x$

In Problems 19–22, the indicated function $y_1(x)$ is a solution of the associated homogeneous equation. Use the method of reduction of order to find a second solution $y_2(x)$ of the homogeneous equation and a particular solution $y_p(x)$ of the given nonhomogeneous equation.

19. $y'' - 4y = 2;$ $\qquad y_1 = e^{-2x}$

20. $y'' + y' = 1;$ $\qquad y_1 = 1$

21. $y'' - 3y' + 2y = 5e^{3x};$ $\qquad y_1 = e^x$

22. $y'' - 4y' + 3y = x;$ $\qquad y_1 = e^x$

In Problems 23 and 24, the indicated function $y_1(x)$ is a solution of the given differential equation. Use formula (5) to find a second solution $y_2(x)$ expressed in terms of an integral-defined function. See (*ii*) in the *Remarks* at the end of this section.

23. $x^2y'' + (x^2 - x)y' + (1 - x)y = 0;$ $\quad y_1 = x$

24. $2xy'' - (2x + 1)y' + y = 0;$ $\quad y_1 = e^x$

Discussion Problems

25. (a) Give a convincing demonstration that the second-order equation $ay'' + by' + cy = 0$, a, b, and c constants, always possesses at least one solution of the form $y_1 = e^{m_1x}$, m_1 a constant.

(b) Explain why the differential equation in part (a) must then have a second solution, either of the form $y_2 = e^{m_2x}$ or of the form $y_2 = xe^{m_1x}$, m_1 and m_2 constants.

(c) Reexamine Problems 1–8. Can you explain why the statements in parts (a) and (b) above are not contradicted by the answers to Problems 3–5?

26. Verify that $y_1(x) = x$ is a solution of $xy'' - xy' + y = 0$. Use reduction of order to find a second solution $y_2(x)$ in the form of an infinite series. Conjecture an interval of definition for $y_2(x)$.

Computer Lab Assignment

27. (a) Verify that $y_1(x) = e^x$ is a solution of

$$xy'' - (x + 10)y' + 10y = 0.$$

(b) Use (5) to find a second solution $y_2(x)$. Use a CAS to carry out the required integration.

(c) Explain, using Corollary (a) of Theorem 3.1.2, why the second solution can be written compactly as

$$y_2(x) = \sum_{n=0}^{10} \frac{1}{n!}x^n.$$

3.3 Linear Equations with Constant Coefficients

INTRODUCTION We have seen that the linear first-order DE $y' + ay = 0$, where a is a constant, possesses the exponential solution $y = c_1e^{-ax}$ on the interval $(-\infty, \infty)$. Therefore, it is natural to ask whether exponential solutions exist for homogeneous linear higher-order DEs

$$a_ny^{(n)} + a_{n-1}y^{(n-1)} + \cdots + a_1y' + a_0y = 0, \tag{1}$$

where the coefficients a_i, $i = 0, 1, \ldots, n$ are real constants and $a_n \neq 0$. The surprising fact is that *all* solutions of these higher-order equations are either exponential functions or are constructed out of exponential functions.

■ **Auxiliary Equation** We begin by considering the special case of a homogeneous linear second-order equation

$$ay'' + by' + cy = 0. \tag{2}$$

If we try a solution of the form $y = e^{mx}$, then after substituting $y' = me^{mx}$ and $y'' = m^2e^{mx}$ equation (2) becomes

$$am^2e^{mx} + bme^{mx} + ce^{mx} = 0 \quad \text{or} \quad e^{mx}(am^2 + bm + c) = 0.$$

Since e^{mx} is never zero for real values of x, it is apparent that the only way that this exponential function can satisfy the differential equation (2) is to choose m as a root of the quadratic equation

$$am^2 + bm + c = 0. \tag{3}$$

This last equation is called the **auxiliary equation** of the differential equation (2). Since the two roots of (3) are $m_1 = (-b + \sqrt{b^2 - 4ac})/2a$ and $m_2 = (-b - \sqrt{b^2 - 4ac})/2a$, there will be three forms of the general solution of (1) corresponding to the three cases:

- m_1 and m_2 are real and distinct ($b^2 - 4ac > 0$),
- m_1 and m_2 are real and equal ($b^2 - 4ac = 0$), and
- m_1 and m_2 are conjugate complex numbers ($b^2 - 4ac < 0$).

We discuss each of these cases in turn.

Case I: Distinct Real Roots Under the assumption that the auxiliary equation (3) has two unequal real roots m_1 and m_2, we find two solutions, $y_1 = e^{m_1 x}$ and $y_2 = e^{m_2 x}$, respectively. We see that these functions are linearly independent on $(-\infty, \infty)$ and hence form a fundamental set. It follows that the general solution of (2) on this interval is

$$y = c_1 e^{m_1 x} + c_2 e^{m_2 x}. \tag{4}$$

Case II: Repeated Real Roots When $m_1 = m_2$ we necessarily obtain only one exponential solution, $y_1 = e^{m_1 x}$. From the quadratic formula we find that $m_1 = -b/2a$ since the only way to have $m_1 = m_2$ is to have $b^2 - 4ac = 0$. It follows from the discussion in Section 3.2 that a second solution of the equation is

$$y_2 = e^{m_1 x} \int \frac{e^{2m_1 x}}{e^{2m_1 x}} \, dx = e^{m_1 x} \int dx = x e^{m_1 x}. \tag{5}$$

In (5) we have used the fact that $-b/a = 2m_1$. The general solution is then

$$y = c_1 e^{m_1 x} + c_2 x e^{m_1 x}. \tag{6}$$

Case III: Conjugate Complex Roots If m_1 and m_2 are complex, then we can write $m_1 = \alpha + i\beta$ and $m_2 = \alpha - i\beta$, where α and $\beta > 0$ are real and $i^2 = -1$. Formally, there is no difference between this case and Case I, hence

$$y = C_1 e^{(\alpha + i\beta)x} + C_2 e^{(\alpha - i\beta)x}.$$

However, in practice we prefer to work with real functions instead of complex exponentials. To this end we use **Euler's formula**:

$$e^{i\theta} = \cos \theta + i \sin \theta,$$

where θ is any real number.* It follows from this formula that

$$e^{i\beta x} = \cos \beta x + i \sin \beta x \quad \text{and} \quad e^{-i\beta x} = \cos \beta x - i \sin \beta x, \tag{7}$$

where we have used $\cos(-\beta x) = \cos \beta x$ and $\sin(-\beta x) = -\sin \beta x$. Note that by first adding and then subtracting the two equations in (7), we obtain, respectively,

$$e^{i\beta x} + e^{-i\beta x} = 2 \cos \beta x \quad \text{and} \quad e^{i\beta x} - e^{-i\beta x} = 2i \sin \beta x.$$

Since $y = C_1 e^{(\alpha + i\beta)x} + C_2 e^{(\alpha - i\beta)x}$ is a solution of (2) for any choice of the constants C_1 and C_2, the choices $C_1 = C_2 = 1$ and $C_1 = 1, C_2 = -1$ give, in turn, two solutions:

$$y_1 = e^{(\alpha + i\beta)x} + e^{(\alpha - i\beta)x} \quad \text{and} \quad y_2 = e^{(\alpha + i\beta)x} - e^{(\alpha - i\beta)x}.$$

But

$$y_1 = e^{\alpha x}(e^{i\beta x} + e^{-i\beta x}) = 2e^{\alpha x} \cos \beta x$$

and

$$y_2 = e^{\alpha x}(e^{i\beta x} - e^{-i\beta x}) = 2i e^{\alpha x} \sin \beta x.$$

Hence from Corollary (a) of Theorem 3.1.2 the last two results show that $e^{\alpha x} \cos \beta x$ and $e^{\alpha x} \sin \beta x$ are *real* solutions of (2). Moreover, these solutions form a fundamental set on $(-\infty, \infty)$. Consequently, the general solution of (2) on $(-\infty, \infty)$ is

$$y = c_1 e^{\alpha x} \cos \beta x + c_2 e^{\alpha x} \sin \beta x = e^{\alpha x}(c_1 \cos \beta x + c_2 \sin \beta x). \tag{8}$$

*A formal derivation of Euler's formula can be obtained from the Maclaurin series $e^x = \sum_{n=0}^{\infty} x^n/n!$ by substituting $x = i\theta$, using $i^2 = -1, i^3 = -i, \ldots$, and then separating the series into real and imaginary parts. The plausibility thus established, we can adopt $\cos \theta + i \sin \theta$ as the *definition* of $e^{i\theta}$.

EXAMPLE 1 Second-Order DEs

Solve the following differential equations.

(a) $2y'' - 5y' - 3y = 0$ **(b)** $y'' - 10y' + 25y = 0$ **(c)** $y'' + 4y' + 7y = 0$

SOLUTION We give the auxiliary equations, the roots, and the corresponding general solutions.

(a) $2m^2 - 5m - 3 = (2m + 1)(m - 3) = 0$, $m_1 = -\frac{1}{2}$, $m_2 = 3$. From (4),

$$y = c_1 e^{-x/2} + c_2 e^{3x}.$$

(b) $m^2 - 10m + 25 = (m - 5)^2 = 0$, $m_1 = m_2 = 5$. From (6),

$$y = c_1 e^{5x} + c_2 x e^{5x}.$$

(c) $m^2 + 4m + 7 = 0$, $m_1 = -2 + \sqrt{3}i$, $m_2 = -2 - \sqrt{3}i$. From (8) with $\alpha = -2$, $\beta = \sqrt{3}$, we have

$$y = e^{-2x}(c_1 \cos \sqrt{3}x + c_2 \sin \sqrt{3}x). \qquad \equiv$$

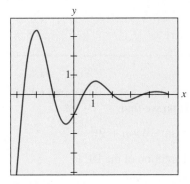

FIGURE 3.3.1 Graph of solution of IVP in Example 2

EXAMPLE 2 **An Initial-Value Problem**

Solve the initial-value problem $4y'' + 4y' + 17y = 0$, $y(0) = -1$, $y'(0) = 2$.

SOLUTION By the quadratic formula we find that the roots of the auxiliary equation $4m^2 + 4m + 17 = 0$ are $m_1 = -\frac{1}{2} + 2i$ and $m_2 = -\frac{1}{2} - 2i$. Thus from (8) we have $y = e^{-x/2}(c_1 \cos 2x + c_2 \sin 2x)$. Applying the condition $y(0) = -1$, we see from $e^0(c_1 \cos 0 + c_2 \sin 0) = -1$ that $c_1 = -1$. Differentiating $y = e^{-x/2}(-\cos 2x + c_2 \sin 2x)$ and then using $y'(0) = 2$ gives $2c_2 + \frac{1}{2} = 2$ or $c_2 = \frac{3}{4}$. Hence the solution of the IVP is $y = e^{-x/2}(-\cos 2x + \frac{3}{4} \sin 2x)$. In **FIGURE 3.3.1** we see that the solution is oscillatory but $y \to 0$ as $x \to \infty$. $\qquad \equiv$

■ **Two Equations Worth Knowing** The two linear differential equations

$$y'' + k^2 y = 0 \quad \text{and} \quad y'' - k^2 y = 0,$$

k real, are important in applied mathematics. For $y'' + k^2 y = 0$, the auxiliary equation $m^2 + k^2 = 0$ has imaginary roots $m_1 = ki$ and $m_2 = -ki$. With $\alpha = 0$ and $\beta = k$ in (8), the general solution of the DE is seen to be

$$y = c_1 \cos kx + c_2 \sin kx. \tag{9}$$

On the other hand, the auxiliary equation $m^2 - k^2 = 0$ for $y'' - k^2 y = 0$ has distinct real roots $m_1 = k$ and $m_2 = -k$ and so by (4) the general solution of the DE is

$$y = c_1 e^{kx} + c_2 e^{-kx}. \tag{10}$$

Notice that if we choose $c_1 = c_2 = \frac{1}{2}$ and $c_1 = \frac{1}{2}$, $c_2 = -\frac{1}{2}$ in (10), we get the particular solutions $y = \frac{1}{2}(e^{kx} + e^{-kx}) = \cosh kx$ and $y = \frac{1}{2}(e^{kx} - e^{-kx}) = \sinh kx$. Since $\cosh kx$ and $\sinh kx$ are linearly independent on any interval of the x-axis, an alternative form for the general solution of $y'' - k^2 y = 0$ is

$$y = c_1 \cosh kx + c_2 \sinh kx. \tag{11}$$

See Problems 43, 44, and 66 in Exercises 3.3.

■ **Higher-Order Equations** In general, to solve an nth-order differential equation

$$a_n y^{(n)} + a_{n-1} y^{(n-1)} + \cdots + a_2 y'' + a_1 y' + a_0 y = 0, \tag{12}$$

where the a_i, $i = 0, 1, \ldots, n$ are real constants, we must solve an nth-degree polynomial equation

$$a_n m^n + a_{n-1} m^{n-1} + \cdots + a_2 m^2 + a_1 m + a_0 = 0. \tag{13}$$

If all the roots of (13) are real and distinct, then the general solution of (12) is

$$y = c_1 e^{m_1 x} + c_2 e^{m_2 x} + \cdots + c_n e^{m_n x}.$$

It is somewhat harder to summarize the analogues of Cases II and III because the roots of an auxiliary equation of degree greater than two can occur in many combinations. For example,

a fifth-degree equation could have five distinct real roots, or three distinct real and two complex roots, or one real and four complex roots, or five real but equal roots, or five real roots but with two of them equal, and so on. When m_1 is a root of multiplicity k of an nth-degree auxiliary equation (that is, k roots are equal to m_1), it can be shown that the linearly independent solutions are

$$e^{m_1x}, \quad xe^{m_1x}, \quad x^2e^{m_1x}, \ldots, \quad x^{k-1}e^{m_1x}$$

and the general solution must contain the linear combination

$$c_1e^{m_1x} + c_2xe^{m_1x} + c_3x^2e^{m_1x} + \cdots + c_kx^{k-1}e^{m_1x}.$$

Lastly, it should be remembered that when the coefficients are real, complex roots of an auxiliary equation always appear in conjugate pairs. Thus, for example, a cubic polynomial equation can have at most two complex roots.

EXAMPLE 3 **Third-Order DE**

Solve $y''' + 3y'' - 4y = 0$.

SOLUTION It should be apparent from inspection of $m^3 + 3m^2 - 4 = 0$ that one root is $m_1 = 1$ and so $m - 1$ is a factor of $m^3 + 3m^2 - 4$. By division we find

$$m^3 + 3m^2 - 4 = (m - 1)(m^2 + 4m + 4) = (m - 1)(m + 2)^2,$$

and so the other roots are $m_2 = m_3 = -2$. Thus the general solution of the DE is

$$y = c_1e^x + c_2e^{-2x} + c_3xe^{-2x}. \qquad\qquad \equiv$$

EXAMPLE 4 **Fourth-Order DE**

Solve $\dfrac{d^4y}{dx^4} + 2\dfrac{d^2y}{dx^2} + y = 0$.

SOLUTION The auxiliary equation $m^4 + 2m^2 + 1 = (m^2 + 1)^2 = 0$ has roots $m_1 = m_3 = i$ and $m_2 = m_4 = -i$. Thus from Case II the solution is

$$y = C_1e^{ix} + C_2e^{-ix} + C_3xe^{ix} + C_4xe^{-ix}.$$

By Euler's formula the grouping $C_1e^{ix} + C_2e^{-ix}$ can be rewritten as $c_1\cos x + c_2\sin x$ after a relabeling of constants. Similarly, $x(C_3e^{ix} + C_4e^{-ix})$ can be expressed as $x(c_3\cos x + c_4\sin x)$. Hence the general solution is $y = c_1\cos x + c_2\sin x + c_3x\cos x + c_4x\sin x$. $\quad\equiv$

■ **Repeated Complex Roots** Example 4 illustrates a special case when the auxiliary equation has repeated complex roots. In general, if $m_1 = \alpha + i\beta$, $\beta > 0$, is a complex root of multiplicity k of an auxiliary equation with real coefficients, then its conjugate $m_2 = \alpha - i\beta$ is also a root of multiplicity k. From the $2k$ complex-valued solutions

$$e^{(\alpha+i\beta)x}, \quad xe^{(\alpha+i\beta)x}, \quad x^2e^{(\alpha+i\beta)x}, \ldots, \quad x^{k-1}e^{(\alpha+i\beta)x}$$

$$e^{(\alpha-i\beta)x}, \quad xe^{(\alpha-i\beta)x}, \quad x^2e^{(\alpha-i\beta)x}, \ldots, \quad x^{k-1}e^{(\alpha-i\beta)x}$$

we conclude, with the aid of Euler's formula, that the general solution of the corresponding differential equation must then contain a linear combination of the $2k$ real linearly independent solutions

$$e^{\alpha x}\cos\beta x, \quad xe^{\alpha x}\cos\beta x, \quad x^2e^{\alpha x}\cos\beta x, \ldots, \quad x^{k-1}e^{\alpha x}\cos\beta x$$

$$e^{\alpha x}\sin\beta x, \quad xe^{\alpha x}\sin\beta x, \quad x^2e^{\alpha x}\sin\beta x, \ldots, \quad x^{k-1}e^{\alpha x}\sin\beta x.$$

In Example 4 we identify $k = 2$, $\alpha = 0$, and $\beta = 1$.

■ **Rational Roots** Of course the most difficult aspect of solving constant-coefficient differential equations is finding roots of auxiliary equations of degree greater than two. Something we can try is to test the auxiliary equation for rational roots. Recall from precalculus mathematics, if $m_1 = p/q$ is a rational root (expressed in lowest terms) of a polynomial

equation $a_n m^n + \cdots + a_1 m + a_0 = 0$ with integer coefficients, then the integer p is a factor of the constant term a_0 and the integer q is a factor of the leading coefficient a_n.

<hr>

EXAMPLE 5 **Finding Rational Roots**

Solve $3y''' + 5y'' + 10y' - 4y = 0$.

SOLUTION To solve the equation we must solve the cubic polynomial auxiliary equation $3m^3 + 5m^2 + 10m - 4 = 0$. With the identifications $a_0 = -4$ and $a_3 = 3$ then the integer factors of a_0 and a_3 are, respectively, $p: \pm 1, \pm 2, \pm 4$ and $q: \pm 1, \pm 3$. So the possible rational roots of the cubic equation are

$$\frac{p}{q}: \pm 1, \pm 2, \pm 4, \pm \frac{1}{3}, \pm \frac{2}{3}, \pm \frac{4}{3}.$$

Each of these numbers can then be tested—say, by synthetic division. In this way we discover both the root $m_1 = \frac{1}{3}$ and the factorization

$$3m^3 + 5m^2 + 10m - 4 = (m - \tfrac{1}{3})(3m^2 + 6m + 12).$$

The quadratic formula applied to $3m^2 + 6m + 12 = 0$ then yields the remaining two roots $m_2 = -1 - \sqrt{3}i$ and $m_3 = -1 + \sqrt{3}i$. Therefore the general solution of the given differential equation is $y = c_1 e^{x/3} + e^{-x}(c_2 \cos\sqrt{3}x + c_3 \sin\sqrt{3}x)$. ≡

■ Use of Computers Finding roots or approximations of roots of polynomial equations is a routine problem with an appropriate calculator or computer software. The computer algebra systems *Mathematica* and *Maple* can solve polynomial equations (in one variable) of degree less than five in terms of algebraic formulas. For the auxiliary equation in the preceding paragraph, the commands

Solve[3 m∧3 + 5 m∧2 + 10 m − 4 == 0, m] (in *Mathematica*)

solve(3*m∧3 + 5*m∧2 + 10*m − 4, m); (in *Maple*)

yield immediately their representations of the roots $\frac{1}{3}, -1 + \sqrt{3}i, -1 - \sqrt{3}i$. For auxiliary equations of higher degree it may be necessary to resort to numerical commands such as **NSolve** and **FindRoot** in *Mathematica*. Because of their capability of solving polynomial equations, it is not surprising that some computer algebra systems are also able to give explicit solutions of homogeneous linear constant-coefficient differential equations. For example, the inputs

DSolve [y''[x] + 2 y'[x] + 2 y[x] == 0, y[x], x] (in *Mathematica*)

dsolve(diff(y(x), x\$2) + 2*diff(y(x), x) + 2*y(x) = 0, y(x)); (in *Maple*)

give, respectively,

$$y[x] \; -> \; \frac{\textbf{C[2] Cos [x]} \; - \; \textbf{C[1] Sin [x]}}{\textbf{E}^x} \tag{14}$$

and **y(x) = _C1 exp(–x) sin(x) – _C2 exp(–x) cos(x)**

Translated, this means $y = c_2 e^{-x} \cos x + c_1 e^{-x} \sin x$ is a solution of $y'' + 2y' + 2y = 0$.

In the classic text *Differential Equations* by Ralph Palmer Agnew* (used by the author as a student), the following statement is made:

> *It is not reasonable to expect students in this course to have computing skills and equipment necessary for efficient solving of equations such as*

$$4.317 \frac{d^4y}{dx^4} + 2.179 \frac{d^3y}{dx^3} + 1.416 \frac{d^2y}{dx^2} + 1.295 \frac{dy}{dx} + 3.169y = 0. \tag{15}$$

Although it is debatable whether computing skills have improved in the intervening years, it is a certainty that technology has. If one has access to a computer algebra system, equation (15) could

<hr>

*McGraw-Hill, New York, 1960.

be considered reasonable. After simplification and some relabeling of the output, *Mathematica* yields the (approximate) general solution

$$y = c_1 e^{-0.728852x} \cos(0.618605x) + c_2 e^{-0.728852x} \sin(0.618605x)$$

$$+ \, c_3 e^{-0.476478x} \cos(0.759081x) + c_4 e^{-0.476478x} \sin(0.759081x).$$

We note in passing that the **DSolve** and **dsolve** commands in *Mathematica* and *Maple*, like most aspects of any CAS, have their limitations.

Finally, if we are faced with an initial-value problem consisting of, say, a fourth-order differential equation, then to fit the general solution of the DE to the four initial conditions we must solve a system of four linear equations in four unknowns (the c_1, c_2, c_3, c_4 in the general solution). Using a CAS to solve the system can save lots of time. See Problems 37, 38, 75, and 76 in Exercises 3.3.

REMARKS

In case you are wondering, the method of this section also works for homogeneous linear *first-order* differential equations $ay' + by = 0$ with constant coefficients. For example, to solve, say, $2y' + 7y = 0$, we substitute $y = e^{mx}$ into the DE to obtain the auxiliary equation $2m + 7 = 0$. Using $m = -\frac{7}{2}$, the general solution of the DE is then $y = c_1 e^{-7x/2}$.

3.3 Exercises Answers to selected odd-numbered problems begin on page ANS-5.

In Problems 1–14, find the general solution of the given second-order differential equation.

1. $4y'' + y' = 0$

2. $y'' - 36y = 0$

3. $y'' - y' - 6y = 0$

4. $y'' - 3y' + 2y = 0$

5. $y'' + 8y' + 16y = 0$

6. $y'' - 10y' + 25y = 0$

7. $12y'' - 5y' - 2y = 0$

8. $y'' + 4y' - y = 0$

9. $y'' + 9y = 0$

10. $3y'' + y = 0$

11. $y'' - 4y' + 5y = 0$

12. $2y'' + 2y' + y = 0$

13. $3y'' + 2y' + y = 0$

14. $2y'' - 3y' + 4y = 0$

In Problems 15–30, find the general solution of the given higher-order differential equation.

15. $y''' - 4y'' - 5y' = 0$

16. $y''' - y = 0$

17. $y''' - 5y'' + 3y' + 9y = 0$

18. $y''' + 3y'' - 4y' - 12y = 0$

19. $\dfrac{d^3u}{dt^3} + \dfrac{d^2u}{dt^2} - 2u = 0$

20. $\dfrac{d^3x}{dt^3} - \dfrac{d^2x}{dt^2} - 4x = 0$

21. $y''' + 3y'' + 3y' + y = 0$

22. $y''' - 6y'' + 12y' - 8y = 0$

23. $y^{(4)} + y''' + y'' = 0$

24. $y^{(4)} - 2y'' + y = 0$

25. $16\dfrac{d^4y}{dx^4} + 24\dfrac{d^2y}{dx^2} + 9y = 0$

26. $\dfrac{d^4y}{dx^4} - 7\dfrac{d^2y}{dx^2} - 18y = 0$

27. $\dfrac{d^5u}{dr^5} + 5\dfrac{d^4u}{dr^4} - 2\dfrac{d^3u}{dr^3} - 10\dfrac{d^2u}{dr^2} + \dfrac{du}{dr} + 5u = 0$

28. $2\dfrac{d^5x}{ds^5} - 7\dfrac{d^4x}{ds^4} + 12\dfrac{d^3x}{ds^3} + 8\dfrac{d^2x}{ds^2} = 0$

29. $\dfrac{d^6y}{dx^6} - 2\dfrac{d^5y}{dx^5} + 5\dfrac{d^4y}{dx^4} = 0$

30. $\dfrac{d^6y}{dx^6} - 3\dfrac{d^4y}{dx^4} + 3\dfrac{d^2y}{dx^2} - y = 0$

In Problems 31–38, solve the given initial-value problem.

31. $y'' + 16y = 0$, $y(0) = 2$, $y'(0) = -2$

32. $\dfrac{d^2y}{d\theta^2} + y = 0$, $y(\pi/3) = 0$, $y'(\pi/3) = 2$

33. $\dfrac{d^2y}{dt^2} - 4\dfrac{dy}{dt} - 5y = 0$, $y(1) = 0$, $y'(1) = 2$

34. $4y'' - 4y' - 3y = 0$, $y(0) = 1$, $y'(0) = 5$

35. $y'' + y' + 2y = 0$, $y(0) = y'(0) = 0$

36. $y'' - 2y' + y = 0$, $y(0) = 5$, $y'(0) = 10$

37. $y''' + 12y'' + 36y' = 0$, $y(0) = 0$, $y'(0) = 1$, $y''(0) = -7$

38. $y''' + 2y'' - 5y' - 6y = 0$, $y(0) = y'(0) = 0$, $y''(0) = 1$

In Problems 39–42, solve the given boundary-value problem.

39. $y'' - 10y' + 25y = 0$, $y(0) = 1$, $y(1) = 0$

40. $y'' + 4y = 0$, $y(0) = 0$, $y(\pi) = 0$

41. $y'' + y = 0$, $y'(0) = 0$, $y'(\pi/2) = 0$

42. $y'' - 2y' + 2y = 0$, $y(0) = 1$, $y(\pi) = 1$

In Problems 43 and 44, solve the given problem first using the form of the general solution given in (10). Solve again, this time using the form given in (11).

43. $y'' - 3y = 0$, $y(0) = 1$, $y'(0) = 5$

44. $y'' - y = 0$, $y(0) = 1$, $y'(1) = 0$

In Problems 45 and 46, find all numbers $\alpha > 0$ so that the given boundary-value problem possesses nontrivial (that is, nonzero) solutions.

45. $y'' + \alpha^2 y = 0$, $y(0) = 0$, $y(2) = 0$

46. $y'' + \alpha^2 y = 0$, $y'(0) = 0$, $y(4) = 0$

In Problems 47–52, each figure represents the graph of a particular solution of one of the following differential equations:

(a) $y'' - 3y' - 4y = 0$ (b) $y'' + 4y = 0$
(c) $y'' + 2y' + y = 0$ (d) $y'' + y = 0$
(e) $y'' + 2y' + 2y = 0$ (f) $y'' - 3y' + 2y = 0$

Match a solution curve with one of the differential equations. Explain your reasoning.

47.

FIGURE 3.3.2 Graph for Problem 47

48.

FIGURE 3.3.3 Graph for Problem 48

49.

FIGURE 3.3.4 Graph for Problem 49

50.

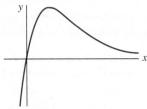

FIGURE 3.3.5 Graph for Problem 50

51.

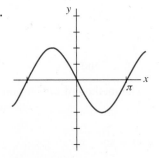

FIGURE 3.3.6 Graph for Problem 51

52.

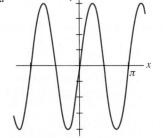

FIGURE 3.3.7 Graph for Problem 52

In Problems 53–62, find a homogeneous linear differential equation with constant coefficients whose general solution is given.

53. $y = c_1 e^x + c_2 e^{6x}$

54. $y = c_1 e^{-5x} + c_2 e^{-4x}$

55. $y = c_1 + c_2 e^{3x}$

56. $y = c_1 e^{-10x} + c_2 x e^{-10x}$

57. $y = c_1 \cos 8x + c_2 \sin 8x$

58. $y = c_1 \cosh \frac{1}{2}x + c_2 \sinh \frac{1}{2}x$

59. $y = c_1 e^x \cos x + c_2 e^x \sin x$

60. $y = c_1 + c_2 e^{-2x} \cos 5x + c_3 e^{-2x} \sin 5x$

61. $y = c_1 + c_2 x + c_3 e^{7x}$

62. $y = c_1 \cos x + c_2 \sin x + c_3 \cos 3x + c_4 \sin 3x$

Discussion Problems

63. Two roots of a cubic auxiliary equation with real coefficients are $m_1 = -\frac{1}{2}$ and $m_2 = 3 + i$. What is the corresponding homogeneous linear differential equation?

64. Find the general solution of $y''' + 6y'' + y' - 34y = 0$ if it is known that $y_1 = e^{-4x} \cos x$ is one solution.

65. To solve $y^{(4)} + y = 0$ we must find the roots of $m^4 + 1 = 0$. This is a trivial problem using a CAS, but it can also be done by hand working with complex numbers. Observe that $m^4 + 1 = (m^2 + 1)^2 - 2m^2$. How does this help? Solve the differential equation.

66. Verify that $y = \sinh x - 2 \cos (x + \pi/6)$ is a particular solution of $y^{(4)} - y = 0$. Reconcile this particular solution with the general solution of the DE.

67. The complex number $3 + 2i$ is a root of multiplicity 3 of the auxiliary equation of a homogeneous linear sixth-order differential equation with real constant coefficients. Find the general solution of the differential equation.

68. Find the general solution of the differential equation

$$y^{(4)} - 8y''' + 18y'' - 8y' + y = 0$$

if the auxiliary equation is

$$m^4 - 8m^3 + 18m^2 - 8m + 1 = (m^2 - 4m + 1)^2 = 0.$$

69. Consider the boundary-value problem $y'' + \lambda y = 0$, $y(0) = 0$, $y(\pi/2) = 0$. Discuss: Is it possible to determine real values of λ so that the problem possesses (a) trivial solutions? (b) nontrivial solutions?

70. In the study of techniques of integration in calculus, certain indefinite integrals of the form $\int e^{ax} f(x)\, dx$ could be evaluated by applying integration by parts twice, recovering the original integral on the right-hand side, solving for the original integral, and obtaining a constant multiple $k \int e^{ax} f(x)\, dx$ on the left-hand side. Then the value of the integral is found by dividing by k. Discuss: For what kinds of functions f does the described procedure work? Your solution should lead to a differential equation. Carefully analyze this equation and solve for f.

Computer Lab Assignments

In Problems 71–74, use a computer either as an aid in solving the auxiliary equation or as a means of directly obtaining the general solution of the given differential equation. If you use a CAS to obtain the general solution, simplify the output and, if necessary, write the solution in terms of real functions.

71. $y''' - 6y'' + 2y' + y = 0$

72. $6.11y''' + 8.59y'' + 7.93y' + 0.778y = 0$

73. $3.15y^{(4)} - 5.34y'' + 6.33y' - 2.03y = 0$

74. $y^{(4)} + 2y'' - y' + 2y = 0$

In Problems 75 and 76, use a CAS as an aid in solving the auxiliary equation. Form the general solution of the differential equation. Then use a CAS as an aid in solving the system of equations for the coefficients c_i, $i = 1, 2, 3, 4$ that result when the initial conditions are applied to the general solution.

75. $2y^{(4)} + 3y''' - 16y'' + 15y' - 4y = 0$,
$y(0) = -2, y'(0) = 6, y''(0) = 3, y'''(0) = \frac{1}{2}$

76. $y^{(4)} - 3y''' + 3y'' - y' = 0$,
$y(0) = y'(0) = 0, y''(0) = y'''(0) = 1$

3.4 Undetermined Coefficients

INTRODUCTION To solve a nonhomogeneous linear differential equation

$$a_n y^{(n)} + a_{n-1} y^{(n-1)} + \cdots + a_1 y' + a_0 y = g(x) \tag{1}$$

we must do two things: (*i*) find the complementary function y_c; and (*ii*) find *any* particular solution y_p of the nonhomogeneous equation. Then, as discussed in Section 3.1, the general solution of (1) on an interval I is $y = y_c + y_p$.

The complementary function y_c is the general solution of the associated homogeneous DE of (1), that is,

$$a_n y^{(n)} + a_{n-1} y^{(n-1)} + \cdots + a_1 y' + a_0 y = 0.$$

In the last section we saw how to solve these kinds of equations when the coefficients were constants. Our goal then in the present section is to examine a method for obtaining particular solutions.

■ **Method of Undetermined Coefficients** The first of two ways we shall consider for obtaining a particular solution y_p is called the **method of undetermined coefficients**. The underlying idea in this method is a conjecture, an educated guess really, about the form of y_p motivated by the kinds of functions that make up the input function $g(x)$. The general method is limited to nonhomogeneous linear DEs such as (1) where

- the coefficients, a_i, $i = 0, 1, \ldots, n$ are constants, and
- where $g(x)$ is a constant, a polynomial function, exponential function $e^{\alpha x}$, sine or cosine functions $\sin \beta x$ or $\cos \beta x$, or finite sums and products of these functions.

A constant k is a polynomial function of degree 0. ▶

Strictly speaking, $g(x) = k$ (a constant) is a polynomial function. Since a constant function is probably not the first thing that comes to mind when you think of polynomial functions, for emphasis we shall continue to use the redundancy "constant functions, polynomial functions,"

The following functions are some examples of the types of inputs $g(x)$ that are appropriate for this discussion:

$$g(x) = 10, \quad g(x) = x^2 - 5x, \quad g(x) = 15x - 6 + 8e^{-x}$$

$$g(x) = \sin 3x - 5x \cos 2x, \quad g(x) = xe^x \sin x + (3x^2 - 1)e^{-4x}.$$

That is, $g(x)$ is a linear combination of functions of the type

$$P(x) = a_n x^n + a_{n-1} x^{n-1} + \cdots + a_1 x + a_0, \quad P(x)e^{\alpha x}, \quad P(x)e^{\alpha x} \sin \beta x, \quad \text{and} \quad P(x)e^{\alpha x} \cos \beta x,$$

where n is a nonnegative integer and α and β are real numbers. The method of undetermined coefficients is not applicable to equations of form (1) when

$$g(x) = \ln x, \quad g(x) = \frac{1}{x}, \quad g(x) = \tan x, \quad g(x) = \sin^{-1} x,$$

and so on. Differential equations in which the input $g(x)$ is a function of this last kind will be considered in Section 3.5.

The set of functions that consists of constants, polynomials, exponentials $e^{\alpha x}$, sines, and cosines has the remarkable property that derivatives of their sums and products are again sums and products of constants, polynomials, exponentials $e^{\alpha x}$, sines, and cosines. Since the linear combination of derivatives $a_n y_p^{(n)} + a_{n-1} y_p^{(n-1)} + \cdots + a_1 y_p' + a_0 y_p$ must be identical to $g(x)$, it seems reasonable to assume that y_p *has the same form as $g(x)$*.

The next two examples illustrate the basic method.

EXAMPLE **1** **General Solution Using Undetermined Coefficients**

Solve $y'' + 4y' - 2y = 2x^2 - 3x + 6$. (2)

SOLUTION Step 1 We first solve the associated homogeneous equation $y'' + 4y' - 2y = 0$. From the quadratic formula we find that the roots of the auxiliary equation $m^2 + 4m - 2 = 0$ are $m_1 = -2 - \sqrt{6}$ and $m_2 = -2 + \sqrt{6}$. Hence the complementary function is

$$y_c = c_1 e^{-(2+\sqrt{6})x} + c_2 e^{(-2+\sqrt{6})x}.$$

Step 2 Now, since the function $g(x)$ is a quadratic polynomial, let us assume a particular solution that is also in the form of a quadratic polynomial:

$$y_p = Ax^2 + Bx + C.$$

We seek to determine *specific* coefficients A, B, and C for which y_p is a solution of (2). Substituting y_p and the derivatives $y_p' = 2Ax + B$ and $y_p'' = 2A$ into the given differential equation (2), we get

$$y_p'' + 4y_p' - 2y_p = 2A + 8Ax + 4B - 2Ax^2 - 2Bx - 2C$$

$$= 2x^2 - 3x + 6.$$

Since the last equation is supposed to be an identity, the coefficients of like powers of x must be equal:

That is,

$$-2A = 2, \quad 8A - 2B = -3, \quad 2A + 4B - 2C = 6.$$

Solving this system of equations leads to the values $A = -1$, $B = -\frac{5}{2}$, and $C = -9$. Thus a particular solution is

$$y_p = -x^2 - \frac{5}{2}x - 9.$$

Step 3 The general solution of the given equation is

$$y = y_c + y_p = c_1 e^{-(2+\sqrt{6})x} + c_2 e^{(-2+\sqrt{6})x} - x^2 - \frac{5}{2}x - 9.$$ ≡

EXAMPLE **2** **Particular Solution Using Undetermined Coefficients**

Find a particular solution of $y'' - y' + y = 2 \sin 3x$.

SOLUTION A natural first guess for a particular solution would be $A \sin 3x$. But since successive differentiations of $\sin 3x$ produce $\sin 3x$ *and* $\cos 3x$, we are prompted instead to assume a particular solution that includes both of these terms:

$$y_p = A \cos 3x + B \sin 3x.$$

Differentiating y_p and substituting the results into the differential equation gives, after regrouping,

$$y_p'' - y_p' + y_p = (-8A - 3B) \cos 3x + (3A - 8B) \sin 3x = 2 \sin 3x$$

or

From the resulting system of equations,

$$-8A - 3B = 0, \qquad 3A - 8B = 2,$$

we get $A = \frac{6}{73}$ and $B = -\frac{16}{73}$. A particular solution of the equation is

$$y_p = \frac{6}{73}\cos 3x - \frac{16}{73}\sin 3x.$$

$\equiv$

As we mentioned, the form that we assume for the particular solution y_p is an educated guess; it is not a blind guess. This educated guess must take into consideration not only the types of functions that make up $g(x)$ but also, as we shall see in Example 4, the functions that make up the complementary function y_c.

EXAMPLE 3 **Forming y_p by Superposition**

Solve $y'' - 2y' - 3y = 4x - 5 + 6xe^{2x}$. (3)

SOLUTION **Step 1** First, the solution of the associated homogeneous equation $y'' - 2y' - 3y = 0$ is found to be $y_c = c_1 e^{-x} + c_2 e^{3x}$.

Step 2 Next, the presence of $4x - 5$ in $g(x)$ suggests that the particular solution includes a linear polynomial. Furthermore, since the derivative of the product xe^{2x} produces $2xe^{2x}$ and e^{2x}, we also assume that the particular solution includes both xe^{2x} and e^{2x}. In other words, g is the sum of two basic kinds of functions:

$$g(x) = g_1(x) + g_2(x) = polynomial + exponentials.$$

How to use Theorem 3.1.7 in the solution of Example 3. ▶

Correspondingly, the superposition principle for nonhomogeneous equations (Theorem 3.1.7) suggests that we seek a particular solution

$$y_p = y_{p_1} + y_{p_2},$$

where $y_{p_1} = Ax + B$ and $y_{p_2} = Cxe^{2x} + Ee^{2x}$. Substituting

$$y_p = Ax + B + Cxe^{2x} + Ee^{2x}$$

into the given equation (3) and grouping like terms gives

$$y_p'' - 2y_p' - 3y_p = -3Ax - 2A - 3B - 3Cxe^{2x} + (2C - 3E)e^{2x} = 4x - 5 + 6xe^{2x}.$$ (4)

From this identity we obtain the four equations

$$-3A = 4, \qquad -2A - 3B = -5, \qquad -3C = 6, \qquad 2C - 3E = 0.$$

The last equation in this system results from the interpretation that the coefficient of e^{2x} in the right member of (4) is zero. Solving, we find $A = -\frac{4}{3}$, $B = \frac{23}{9}$, $C = -2$, and $E = -\frac{4}{3}$. Consequently,

$$y_p = -\frac{4}{3}x + \frac{23}{9} - 2xe^{2x} - \frac{4}{3}e^{2x}.$$

Step 3 The general solution of the equation is

$$y = c_1 e^{-x} + c_2 e^{3x} - \frac{4}{3}x + \frac{23}{9} - \left(2x + \frac{4}{3}\right)e^{2x}.$$

$\equiv$

In light of the superposition principle (Theorem 3.1.7), we can also approach Example 3 from the viewpoint of solving two simpler problems. You should verify that substituting

$$y_{p_1} = Ax + B \qquad \text{into} \qquad y'' - 2y' - 3y = 4x - 5$$

and $$y_{p_2} = Cxe^{2x} + Ee^{2x} \qquad \text{into} \qquad y'' - 2y' - 3y = 6xe^{2x}$$

yields, in turn, $y_{p_1} = -\frac{4}{3}x + \frac{23}{9}$ and $y_{p_2} = -(2x + \frac{4}{3})e^{2x}$. A particular solution of (3) is then $y_p = y_{p_1} + y_{p_2}$.

The next example illustrates that sometimes the "obvious" assumption for the form of y_p is not a correct assumption.

EXAMPLE 4 **A Glitch in the Method**

Find a particular solution of $\quad y'' - 5y' + 4y = 8e^x$.

SOLUTION Differentiation of e^x produces no new functions. Thus, proceeding as we did in the earlier examples, we can reasonably assume a particular solution of the form $y_p = Ae^x$. But substitution of this expression into the differential equation yields the contradictory statement $0 = 8e^x$, and so we have clearly made the wrong guess for y_p.

The difficulty here is apparent upon examining the complementary function $y_c = c_1 e^x + c_2 e^{4x}$. Observe that our assumption Ae^x is already present in y_c. This means that e^x is a solution of the associated homogeneous differential equation, and a constant multiple Ae^x when substituted into the differential equation necessarily produces zero.

What then should be the form of y_p? Inspired by Case II of Section 3.3, let's see whether we can find a particular solution of the form

$$y_p = Axe^x.$$

Substituting $y_p' = Axe^x + Ae^x$ and $y_p'' = Axe^x + 2Ae^x$ into the differential equation and simplifying gives

$$y_p'' - 5y_p' + 4y_p = -3Ae^x = 8e^x.$$

From the last equality we see that the value of A is now determined as $A = -\frac{8}{3}$. Therefore a particular solution of the given equation is $y_p = -\frac{8}{3}xe^x$. ≡

The difference in the procedures used in Examples 1–3 and in Example 4 suggests that we consider two cases. The first case reflects the situation in Examples 1–3.

***Case I*:** No function in the assumed particular solution is a solution of the associated homogeneous differential equation.

In Table 3.4.1 we illustrate some specific examples of $g(x)$ in (1) along with the corresponding form of the particular solution. We are, of course, taking for granted that no function in the assumed particular solution y_p is duplicated by a function in the complementary function y_c.

TABLE 3.4.1 Trial Particular Solutions

$g(x)$	Form of y_p
1. 1 (any constant)	A
2. $5x + 7$	$Ax + B$
3. $3x^2 - 2$	$Ax^2 + Bx + C$
4. $x^3 - x + 1$	$Ax^3 + Bx^2 + Cx + E$
5. $\sin 4x$	$A \cos 4x + B \sin 4x$
6. $\cos 4x$	$A \cos 4x + B \sin 4x$
7. e^{5x}	Ae^{5x}
8. $(9x - 2)e^{5x}$	$(Ax + B)e^{5x}$
9. $x^2 e^{5x}$	$(Ax^2 + Bx + C)e^{5x}$
10. $e^{3x} \sin 4x$	$Ae^{3x} \cos 4x + Be^{3x} \sin 4x$
11. $5x^2 \sin 4x$	$(Ax^2 + Bx + C) \cos 4x + (Ex^2 + Fx + G) \sin 4x$
12. $xe^{3x} \cos 4x$	$(Ax + B)e^{3x} \cos 4x + (Cx + E)e^{3x} \sin 4x$

EXAMPLE 5 **Forms of Particular Solutions—Case I**

Determine the form of a particular solution of

(a) $y'' - 8y' + 25y = 5x^3 e^{-x} - 7e^{-x}$ **(b)** $y'' + 4y = x \cos x$.

SOLUTION **(a)** We can write $g(x) = (5x^3 - 7)e^{-x}$. Using entry 9 in Table 3.4.1 as a model, we assume a particular solution of the form

$$y_p = (Ax^3 + Bx^2 + Cx + E)e^{-x}.$$

Note that there is no duplication between any of the terms in y_p and the terms in the complementary function $y_c = e^{4x}(c_1 \cos 3x + c_2 \sin 3x)$.

(b) The function $g(x) = x \cos x$ is similar to entry 11 in Table 3.4.1 except, of course, that we use a linear rather than a quadratic polynomial and $\cos x$ and $\sin x$ instead of $\cos 4x$ and $\sin 4x$ in the form of y_p:

$$y_p = (Ax + B) \cos x + (Cx + E) \sin x.$$

Again observe that there is no duplication of terms between y_p and $y_c = c_1 \cos 2x + c_2 \sin 2x$. $\equiv$

If $g(x)$ consists of a sum of, say, m terms of the kind listed in the table, then (as in Example 3) the assumption for a particular solution y_p consists of the sum of the trial forms $y_{p_1}, y_{p_2}, \ldots, y_{p_m}$ corresponding to these terms:

$$y_p = y_{p_1} + y_{p_2} + \cdots + y_{p_m}.$$

The foregoing sentence can be put another way.

Form Rule for Case I: *The form of y_p is a linear combination of all linearly independent functions that are generated by repeated differentiations of $g(x)$.*

> EXAMPLE 6 **Finding y_p by Superposition—Case I**

Determine the form of a particular solution of

$$y'' - 9y' + 14y = 3x^2 - 5 \sin 2x + 8xe^{6x}.$$

SOLUTION The right-hand side of the equation $g(x) = 3x^2 - 5 \sin 2x + 8xe^{6x}$ consists of three different types of functions x^2, $\sin 2x$, and xe^{6x}. The derivatives of these functions yield, in turn, the additional functions x, 1; $\cos 2x$; and e^{6x}. Therefore:

corresponding to x^2 we assume $y_{p_1} = Ax^2 + Bx + C$,

corresponding to $\sin 2x$ we assume $y_{p_2} = E \cos 2x + F \sin 2x$,

corresponding to xe^{6x} we assume $y_{p_3} = Gxe^{6x} + He^{6x}$.

The assumption for a particular solution of the given nonhomogeneous differential equation is then

$$y_p = y_{p_1} + y_{p_2} + y_{p_3} = Ax^2 + Bx + C + E \cos 2x + F \sin 2x + (Gx + H)e^{6x}.$$

Note that none of the seven terms in this assumption for y_p duplicates a term in the complementary function $y_c = c_1 e^{2x} + c_2 e^{7x}$. $\equiv$

> ***Case II:*** A function in the assumed particular solution is also a solution of the associated homogeneous differential equation.

The next example is similar to Example 4.

> EXAMPLE 7 **Particular Solution—Case II**

Find a particular solution of $y'' - 2y' + y = e^x$.

SOLUTION The complementary function is $y_c = c_1 e^x + c_2 x e^x$. As in Example 4, the assumption $y_p = Ae^x$ will fail since it is apparent from y_c that e^x is a solution of the associated homogeneous equation $y'' - 2y' + y = 0$. Moreover, we will not be able to find a particular solution of the form $y_p = Axe^x$ since the term xe^x is also duplicated in y_c. We next try

$$y_p = Ax^2 e^x.$$

Substituting into the given differential equation yields $2Ae^x = e^x$ and so $A = \frac{1}{2}$. Thus a particular solution is $y_p = \frac{1}{2}x^2 e^x$. $\equiv$

Suppose again that $g(x)$ consists of m terms of the kind given in Table 3.4.1, and suppose further that the usual assumption for a particular solution is

$$y_p = y_{p_1} + y_{p_2} + \cdots + y_{p_m},$$

where the y_{p_i}, $i = 1, 2, \ldots, m$ are the trial particular solution forms corresponding to these terms. Under the circumstances described in Case II, we can make up the following general rule.

Multiplication Rule for Case II: *If any y_{p_i} contains terms that duplicate terms in y_c, then that y_{p_i} must be multiplied by x^n, where n is the smallest positive integer that eliminates that duplication.*

EXAMPLE 8 An Initial-Value Problem

Solve the initial-value problem $y'' + y = 4x + 10 \sin x$, $y(\pi) = 0$, $y'(\pi) = 2$.

SOLUTION The solution of the associated homogeneous equation $y'' + y = 0$ is $y_c = c_1 \cos x + c_2 \sin x$. Since $g(x) = 4x + 10 \sin x$ is the sum of a linear polynomial and a sine function, our normal assumption for y_p, from entries 2 and 5 of Table 3.4.1, would be the sum of $y_{p_1} = Ax + B$ and $y_{p_2} = C \cos x + E \sin x$:

$$y_p = Ax + B + C \cos x + E \sin x. \tag{5}$$

But there is an obvious duplication of the terms $\cos x$ and $\sin x$ in this assumed form and two terms in the complementary function. This duplication can be eliminated by simply multiplying y_{p_2} by x. Instead of (5) we now use

$$y_p = Ax + B + Cx \cos x + Ex \sin x.$$

Differentiating this expression and substituting the results into the differential equation gives

$$y_p'' + y_p = Ax + B - 2C \sin x + 2E \cos x = 4x + 10 \sin x, \tag{6}$$

and so $A = 4$, $B = 0$, $-2C = 10$, $2E = 0$. The solutions of the system are immediate: $A = 4$, $B = 0$, $C = -5$, and $E = 0$. Therefore from (6) we obtain $y_p = 4x - 5x \cos x$. The general solution of the given equation is

$$y = y_c + y_p = c_1 \cos x + c_2 \sin x + 4x - 5x \cos x.$$

We now apply the prescribed initial conditions to the general solution of the equation. First, $y(\pi) = c_1 \cos \pi + c_2 \sin \pi + 4\pi - 5\pi \cos \pi = 0$ yields $c_1 = 9\pi$ since $\cos \pi = -1$ and $\sin \pi = 0$. Next, from the derivative

$$y' = -9\pi \sin x + c_2 \cos x + 4 + 5x \sin x - 5 \cos x$$

and $y'(\pi) = -9\pi \sin \pi + c_2 \cos \pi + 4 + 5\pi \sin \pi - 5 \cos \pi = 2$

we find $c_2 = 7$. The solution of the initial value is then

$$y = 9\pi \cos x + 7 \sin x + 4x - 5x \cos x. \qquad \equiv$$

EXAMPLE 9 Using the Multiplication Rule

Solve $y'' - 6y' + 9y = 6x^2 + 2 - 12e^{3x}$.

SOLUTION The complementary function is $y_c = c_1 e^{3x} + c_2 x e^{3x}$. And so, based on entries 3 and 7 of Table 3.4.1, the usual assumption for a particular solution would be

$$y_p = \underbrace{Ax^2 + Bx + C}_{y_{p_1}} + \underbrace{Ee^{3x}}_{y_{p_2}}.$$

Inspection of these functions shows that the one term in y_{p_2} is duplicated in y_c. If we multiply y_{p_2} by x, we note that the term xe^{3x} is still part of y_c. But multiplying y_{p_2} by x^2 eliminates all duplications. Thus the operative form of a particular solution is

$$y_p = Ax^2 + Bx + C + Ex^2 e^{3x}.$$

Differentiating this last form, substituting into the differential equation, and collecting like terms gives

$$y_p'' - 6y_p' + 9y_p = 9Ax^2 + (-12A + 9B)x + 2A - 6B + 9C + 2Ee^{3x} = 6x^2 + 2 - 12e^{3x}.$$

It follows from this identity that $A = \frac{2}{3}$, $B = \frac{8}{9}$, $C = \frac{2}{3}$, and $E = -6$. Hence the general solution $y = y_c + y_p$ is

$$y = c_1 e^{3x} + c_2 x e^{3x} + \frac{2}{3}x^2 + \frac{8}{9}x + \frac{2}{3} - 6x^2 e^{3x}. \qquad \equiv$$

EXAMPLE 10 **Third-Order DE—Case I**

Solve $y''' + y'' = e^x \cos x$.

SOLUTION From the characteristic equation $m^3 + m^2 = 0$ we find $m_1 = m_2 = 0$ and $m_3 = -1$. Hence the complementary function of the equation is $y_c = c_1 + c_2 x + c_3 e^{-x}$. With $g(x) = e^x \cos x$, we see from entry 10 of Table 3.4.1 that we should assume

$$y_p = Ae^x \cos x + Be^x \sin x.$$

Since there are no functions in y_p that duplicate functions in the complementary solution, we proceed in the usual manner. From

$$y_p''' + y_p'' = (-2A + 4B)e^x \cos x + (-4A - 2B)e^x \sin x = e^x \cos x$$

we get $-2A + 4B = 1$, $-4A - 2B = 0$. This system gives $A = -\frac{1}{10}$ and $B = \frac{1}{5}$, so that a particular solution is $y_p = -\frac{1}{10}e^x \cos x + \frac{1}{5}e^x \sin x$. The general solution of the equation is

$$y = y_c + y_p = c_1 + c_2 x + c_3 e^{-x} - \frac{1}{10}e^x \cos x + \frac{1}{5}e^x \sin x. \qquad \equiv$$

EXAMPLE 11 **Fourth-Order DE—Case II**

Determine the form of a particular solution of $y^{(4)} + y''' = 1 - x^2 e^{-x}$.

SOLUTION Comparing $y_c = c_1 + c_2 x + c_3 x^2 + c_4 e^{-x}$ with our normal assumption for a particular solution

$$y_p = \underbrace{A}_{y_{p_1}} + \underbrace{Bx^2 e^{-x} + Cx e^{-x} + Ee^{-x}}_{y_{p_2}},$$

we see that the duplications between y_c and y_p are eliminated when y_{p_1} is multiplied by x^3 and y_{p_2} is multiplied by x. Thus the correct assumption for a particular solution is

$$y_p = Ax^3 + Bx^3 e^{-x} + Cx^2 e^{-x} + Ex e^{-x}. \qquad \equiv$$

REMARKS

(*i*) In Problems 27–36 of Exercises 3.4, you are asked to solve initial-value problems, and in Problems 37–40, boundary-value problems. As illustrated in Example 8, be sure to apply the initial conditions or the boundary conditions to the general solution $y = y_c + y_p$. Students often make the mistake of applying these conditions only to the complementary function y_c since it is that part of the solution that contains the constants.

(*ii*) From the "Form Rule for Case I" on page 136 of this section you see why the method of undetermined coefficients is not well suited to nonhomogeneous linear DEs when the input function $g(x)$ is something other than the four basic types listed in red on page 132. If $P(x)$ is a polynomial, continued differentiation of $P(x)e^{\alpha x} \sin \beta x$ will generate an independent set containing only a *finite* number of functions—all of the same type, namely, polynomials times $e^{\alpha x} \sin \beta x$ or $e^{\alpha x} \cos \beta x$. On the other hand, repeated differentiations of input functions such as $g(x) = \ln x$ or $g(x) = \tan^{-1} x$ generate an independent set containing an *infinite* number of functions:

$$\text{derivatives of } \ln x: \qquad \frac{1}{x}, \frac{-1}{x^2}, \frac{2}{x^3}, \ldots,$$

$$\text{derivatives of } \tan^{-1} x: \qquad \frac{1}{1 + x^2}, \frac{-2x}{(1 + x^2)^2}, \frac{-2 + 6x^2}{(1 + x^2)^3}, \ldots.$$

In Problems 1–26, solve the given differential equation by undetermined coefficients.

1. $y'' + 3y' + 2y = 6$

2. $4y'' + 9y = 15$

3. $y'' - 10y' + 25y = 30x + 3$

4. $y'' + y' - 6y = 2x$

5. $\frac{1}{4}y'' + y' + y = x^2 - 2x$

6. $y'' - 8y' + 20y = 100x^2 - 26xe^x$

7. $y'' + 3y = -48x^2e^{3x}$

8. $4y'' - 4y' - 3y = \cos 2x$

9. $y'' - y' = -3$

10. $y'' + 2y' = 2x + 5 - e^{-2x}$

11. $y'' - y' + \frac{1}{4}y = 3 + e^{x/2}$

12. $y'' - 16y = 2e^{4x}$

13. $y'' + 4y = 3\sin 2x$

14. $y'' - 4y = (x^2 - 3)\sin 2x$

15. $y'' + y = 2x \sin x$

16. $y'' - 5y' = 2x^3 - 4x^2 - x + 6$

17. $y'' - 2y' + 5y = e^x \cos 2x$

18. $y'' - 2y' + 2y = e^{2x}(\cos x - 3\sin x)$

19. $y'' + 2y' + y = \sin x + 3\cos 2x$

20. $y'' + 2y' - 24y = 16 - (x + 2)e^{4x}$

21. $y''' - 6y'' = 3 - \cos x$

22. $y''' - 2y'' - 4y' + 8y = 6xe^{2x}$

23. $y''' - 3y'' + 3y' - y = x - 4e^x$

24. $y''' - y'' - 4y' + 4y = 5 - e^x + e^{2x}$

25. $y^{(4)} + 2y'' + y = (x - 1)^2$

26. $y^{(4)} - y'' = 4x + 2xe^{-x}$

In Problems 27–36, solve the given initial-value problem.

27. $y'' + 4y = -2$, $y(\pi/8) = \frac{1}{2}$, $y'(\pi/8) = 2$

28. $2y'' + 3y' - 2y = 14x^2 - 4x - 11$,
$y(0) = 0$, $y'(0) = 0$

29. $5y'' + y' = -6x$, $y(0) = 0$, $y'(0) = -10$

30. $y'' + 4y' + 4y = (3 + x)e^{-2x}$, $y(0) = 2$, $y'(0) = 5$

31. $y'' + 4y' + 5y = 35e^{-4x}$, $y(0) = -3$, $y'(0) = 1$

32. $y'' - y = \cosh x$, $y(0) = 2$, $y'(0) = 12$

33. $\dfrac{d^2x}{dt^2} + \omega^2 x = F_0 \sin \omega t$, $x(0) = 0$, $x'(0) = 0$

34. $\dfrac{d^2x}{dt^2} + \omega^2 x = F_0 \cos \gamma t$, $x(0) = 0$, $x'(0) = 0$

35. $y''' - 2y'' + y' = 2 - 24e^x + 40e^{5x}$, $y(0) = \frac{1}{2}$, $y'(0) = \frac{5}{2}$, $y''(0) = -\frac{9}{2}$

36. $y''' + 8y = 2x - 5 + 8e^{-2x}$, $y(0) = -5$, $y'(0) = 3$, $y''(0) = -4$

In Problems 37–40, solve the given boundary-value problem.

37. $y'' + y = x^2 + 1$, $y(0) = 5$, $y(1) = 0$

38. $y'' - 2y' + 2y = 2x - 2$, $y(0) = 0$, $y(\pi) = \pi$

39. $y'' + 3y = 6x$, $y(0) = 0$, $y(1) + y'(1) = 0$

40. $y'' + 3y = 6x$, $y(0) + y'(0) = 0$, $y(1) = 0$

In Problems 41 and 42, solve the given initial-value problem in which the input function $g(x)$ is discontinuous. [*Hint*: Solve each problem on two intervals, and then find a solution so that y and y' are continuous at $x = \pi/2$ (Problem 41) and at $x = \pi$ (Problem 42).]

41. $y'' + 4y = g(x)$, $y(0) = 1$, $y'(0) = 2$, where
$$g(x) = \begin{cases} \sin x, & 0 \le x \le \pi/2 \\ 0, & x > \pi/2 \end{cases}$$

42. $y'' - 2y' + 10y = g(x)$, $y(0) = 0$, $y'(0) = 0$, where
$$g(x) = \begin{cases} 20, & 0 \le x \le \pi \\ 0, & x > \pi \end{cases}$$

Discussion Problems

43. Consider the differential equation $ay'' + by' + cy = e^{kx}$, where a, b, c, and k are constants. The auxiliary equation of the associated homogeneous equation is
$$am^2 + bm + c = 0.$$

(a) If k is not a root of the auxiliary equation, show that we can find a particular solution of the form $y_p = Ae^{kx}$, where $A = 1/(ak^2 + bk + c)$.

(b) If k is a root of the auxiliary equation of multiplicity one, show that we can find a particular solution of the form $y_p = Axe^{kx}$, where $A = 1/(2ak + b)$. Explain how we know that $k \ne -b/(2a)$.

(c) If k is a root of the auxiliary equation of multiplicity two, show that we can find a particular solution of the form $y = Ax^2e^{kx}$, where $A = 1/(2a)$.

44. Discuss how the method of this section can be used to find a particular solution of $y'' + y = \sin x \cos 2x$. Carry out your idea.

In Problems 45–48, without solving match a solution curve of $y'' + y = f(x)$ shown in each figure with one of the following functions:

(i) $f(x) = 1$, *(ii)* $f(x) = e^{-x}$,
(iii) $f(x) = e^x$, *(iv)* $f(x) = \sin 2x$,
(v) $f(x) = e^x \sin x$, *(vi)* $f(x) = \sin x$.

Briefly discuss your reasoning.

45.

FIGURE 3.4.1 Graph for Problem 45

46.

FIGURE 3.4.2 Graph for Problem 46

47.

FIGURE 3.4.3 Graph for Problem 47

48.

FIGURE 3.4.4 Graph for Problem 48

Computer Lab Assignments

In Problems 49 and 50, find a particular solution of the given differential equation. Use a CAS as an aid in carrying out differentiations, simplifications, and algebra.

49. $y'' - 4y' + 8y = (2x^2 - 3x)e^{2x}\cos 2x$
$$+ (10x^2 - x - 1)e^{2x}\sin 2x$$

50. $y^{(4)} + 2y'' + y = 2\cos x - 3x\sin x$

3.5 Variation of Parameters

INTRODUCTION It was pointed out in the *Remarks* at the end of Section 3.4 that the method of undetermined coefficients has an inherent weakness that limits its wider application to linear equations: The DE must have constant coefficients, and the input function $g(x)$ must be of the type listed in Table 3.4.1. In this section we examine a method for determining a particular solution y_p of a nonhomogeneous linear DE that has, in theory, no such restrictions on it. This method, due to the eminent Italian astronomer and mathematician **Joseph-Louis Lagrange** (1736–1813), is known as **variation of parameters**.

■ **Some Assumptions** To adapt the method of variation of parameters to a linear second-order differential equation

$$a_2(x)y'' + a_1(x)y' + a_0(x)y = g(x), \tag{1}$$

we begin as we did in Section 3.2—we put (1) in the **standard form**

$$y'' + P(x)y' + Q(x)y = f(x) \tag{2}$$

by dividing through by the lead coefficient $a_2(x)$. Equation (2) is the second-order analogue of the linear first-order equation $dy/dx + P(x)y = f(x)$. In (2) we shall assume $P(x)$, $Q(x)$, and $f(x)$ are continuous on some common interval I. As we have already seen in Section 3.3, there is no difficulty in obtaining the complementary function y_c of (2) when the coefficients are constants.

■ **Method of Variation of Parameters** Corresponding to the substitution $y_p = u_1(x)y_1(x)$ that we used in Section 2.3 to find a particular solution y_p of $dy/dx + P(x)y = f(x)$, for the linear second-order DE (2) we seek a solution of the form

$$y_p(x) = u_1(x)y_1(x) + u_2(x)y_2(x), \tag{3}$$

where y_1 and y_2 form a fundamental set of solutions on I of the associated homogeneous form of (1). Using the product rule to differentiate y_p twice, we get

$$y_p' = u_1 y_1' + y_1 u_1' + u_2 y_2' + y_2 u_2'$$
$$y_p'' = u_1 y_1'' + y_1' u_1' + y_1 u_1'' + u_1' y_1' + u_2 y_2'' + y_2' u_2' + y_2 u_2'' + u_2' y_2'.$$

Substituting (3) and the foregoing derivatives into (2) and grouping terms yields

$$
\begin{aligned}
y_p'' + P(x)y_p' + Q(x)y_p &= u_1 \overbrace{[y_1'' + Py_1' + Qy_1]}^{\text{zero}} + u_2 \overbrace{[y_2'' + Py_2' + Qy_2]}^{\text{zero}} \\
&\quad + y_1 u_1'' + u_1' y_1' + y_2 u_2'' + u_2' y_2' + P[y_1 u_1' + y_2 u_2'] + y_1' u_1' + y_2' u_2' \\
&= \frac{d}{dx}[y_1 u_1'] + \frac{d}{dx}[y_2 u_2'] + P[y_1 u_1' + y_2 u_2'] + y_1' u_1' + y_2' u_2' \\
&= \frac{d}{dx}[y_1 u_1' + y_2 u_2'] + P[y_1 u_1' + y_2 u_2'] + y_1' u_1' + y_2' u_2' = f(x). \tag{4}
\end{aligned}
$$

Because we seek to determine two unknown functions u_1 and u_2, reason dictates that we need two equations. We can obtain these equations by making the further assumption that the functions u_1 and u_2 satisfy $y_1 u_1' + y_2 u_2' = 0$. This assumption does not come out of the blue but is prompted by the first two terms in (4), since, if we demand that $y_1 u_1' + y_2 u_2' = 0$, then (4) reduces to $y_1' u_1' + y_2' u_2' = f(x)$. We now have our desired two equations, albeit two equations for determining the derivatives u_1' and u_2'. By Cramer's rule, the solution of the system

If you are unfamiliar with Cramer's rule, see Section 8.7.

$$y_1 u_1' + y_2 u_2' = 0$$
$$y_1' u_1' + y_2' u_2' = f(x)$$

can be expressed in terms of determinants:

$$u_1' = \frac{W_1}{W} = -\frac{y_2 f(x)}{W} \quad \text{and} \quad u_2' = \frac{W_2}{W} = \frac{y_1 f(x)}{W} \tag{5}$$

where

$$W = \begin{vmatrix} y_1 & y_2 \\ y_1' & y_2' \end{vmatrix}, \quad W_1 = \begin{vmatrix} 0 & y_2 \\ f(x) & y_2' \end{vmatrix}, \quad W_2 = \begin{vmatrix} y_1 & 0 \\ y_1' & f(x) \end{vmatrix}. \tag{6}$$

The functions u_1 and u_2 are found by integrating the results in (5). The determinant W is recognized as the Wronskian of y_1 and y_2. By linear independence of y_1 and y_2 on I, we know that $W(y_1(x), y_2(x)) \neq 0$ for every x in the interval.

■ **Summary of the Method** Usually it is not a good idea to memorize formulas in lieu of understanding a procedure. However, the foregoing procedure is too long and complicated to use each time we wish to solve a differential equation. In this case it is more efficient simply to use the formulas in (5). Thus to solve $a_2 y'' + a_1 y' + a_0 y = g(x)$, first find the complementary function $y_c = c_1 y_1 + c_2 y_2$ and then compute the Wronskian $W(y_1(x), y_2(x))$. By dividing by a_2, we put the equation into the standard form $y'' + Py' + Qy = f(x)$ to determine $f(x)$. We find u_1 and u_2 by integrating $u_1' = W_1/W$ and $u_2' = W_2/W$, where W_1 and W_2 are defined as in (6). A particular solution is $y_p = u_1 y_1 + u_2 y_2$. The general solution of equation (1) is then $y = y_c + y_p$.

> **EXAMPLE 1** **General Solution Using Variation of Parameters**

Solve $y'' - 4y' + 4y = (x + 1)e^{2x}$.

SOLUTION From auxiliary equation $m^2 - 4m + 4 = (m - 2)^2 = 0$ we have $y_c = c_1 e^{2x} + c_2 x e^{2x}$. With the identifications $y_1 = e^{2x}$ and $y_2 = x e^{2x}$, we next compute the Wronskian:

$$W(e^{2x}, xe^{2x}) = \begin{vmatrix} e^{2x} & xe^{2x} \\ 2e^{2x} & 2xe^{2x} + e^{2x} \end{vmatrix} = e^{4x}.$$

Since the given differential equation is already in form (2) (that is, the coefficient of y'' is 1), we identify $f(x) = (x + 1)e^{2x}$. From (6) we obtain

$$W_1 = \begin{vmatrix} 0 & xe^{2x} \\ (x + 1)e^{2x} & 2xe^{2x} + e^{2x} \end{vmatrix} = -(x + 1)xe^{4x}, \quad W_2 = \begin{vmatrix} e^{2x} & 0 \\ 2e^{2x} & (x + 1)e^{2x} \end{vmatrix} = (x + 1)e^{4x},$$

and so from (5)

$$u_1' = -\frac{(x + 1)xe^{4x}}{e^{4x}} = -x^2 - x, \quad u_2' = \frac{(x + 1)e^{4x}}{e^{4x}} = x + 1.$$

Integrating the foregoing derivatives gives

$$u_1 = -\frac{1}{3}x^3 - \frac{1}{2}x^2 \quad \text{and} \quad u_2 = \frac{1}{2}x^2 + x.$$

Hence from (3) we have

$$y_p = \left(-\frac{1}{3}x^3 - \frac{1}{2}x^2\right)e^{2x} + \left(\frac{1}{2}x^2 + x\right)xe^{2x} = \frac{1}{6}x^3e^{2x} + \frac{1}{2}x^2e^{2x}$$

and $\qquad y = y_c + y_p = c_1e^{2x} + c_2xe^{2x} + \frac{1}{6}x^3e^{2x} + \frac{1}{2}x^2e^{2x}.$ ≡

EXAMPLE 2 **General Solution Using Variation of Parameters**

Solve $4y'' + 36y = \csc 3x$.

SOLUTION We first put the equation in the standard form (2) by dividing by 4:

$$y'' + 9y = \frac{1}{4}\csc 3x.$$

Since the roots of the auxiliary equation $m^2 + 9 = 0$ are $m_1 = 3i$ and $m_2 = -3i$, the complementary function is $y_c = c_1 \cos 3x + c_2 \sin 3x$. Using $y_1 = \cos 3x$, $y_2 = \sin 3x$, and $f(x) = \frac{1}{4}\csc 3x$, we obtain

$$W(\cos 3x, \sin 3x) = \begin{vmatrix} \cos 3x & \sin 3x \\ -3\sin 3x & 3\cos 3x \end{vmatrix} = 3,$$

$$W_1 = \begin{vmatrix} 0 & \sin 3x \\ \frac{1}{4}\csc 3x & 3\cos 3x \end{vmatrix} = -\frac{1}{4}, \quad W_2 = \begin{vmatrix} \cos 3x & 0 \\ -3\sin 3x & \frac{1}{4}\csc 3x \end{vmatrix} = \frac{1}{4}\frac{\cos 3x}{\sin 3x}.$$

Integrating

$$u_1' = \frac{W_1}{W} = -\frac{1}{12} \quad \text{and} \quad u_2' = \frac{W_2}{W} = \frac{1}{12}\frac{\cos 3x}{\sin 3x}$$

gives

$$u_1 = -\frac{1}{12}x \quad \text{and} \quad u_2 = \frac{1}{36}\ln|\sin 3x|.$$

Thus from (3) a particular solution is

$$y_p = -\frac{1}{12}x \cos 3x + \frac{1}{36}(\sin 3x)\ln|\sin 3x|.$$

The general solution of the equation is

$$y = y_c + y_p = c_1 \cos 3x + c_2 \sin 3x - \frac{1}{12}x \cos 3x + \frac{1}{36}(\sin 3x)\ln|\sin 3x|. \qquad (7) \equiv$$

Equation (7) represents the general solution of the differential equation on, say, the interval $(0, \pi/3)$.

Constants of Integration

When computing the indefinite integrals of u_1' and u_2', we need not introduce any constants. This is because

$$y = y_c + y_p = c_1 y_1 + c_2 y_2 + (u_1 + a_1) y_1 + (u_2 + b_1) y_2$$
$$= (c_1 + a_1) y_1 + (c_2 + b_1) y_2 + u_1 y_1 + u_2 y_2$$
$$= C_1 y_1 + C_2 y_2 + u_1 y_1 + u_2 y_2.$$

Integral-Defined Functions

We have seen several times in the preceding sections and chapters that when a solution method involves integration we may encounter nonelementary integrals. As the next example shows, sometimes the best we can do in constructing a particular solution (3) of a linear second-order differential equation is to use the integral-defined functions.

$$u_1(x) = -\int_{x_0}^{x} \frac{y_2(t)f(t)}{W(t)}\,dt \quad \text{and} \quad u_2(x) = \int_{x_0}^{x} \frac{y_1(t)f(t)}{W(t)}\,dt.$$

Here we assume that the integrand is continuous on the interval $[x_0, x]$. See Problems 27–30 in Exercises 3.5.

EXAMPLE 3 **General Solution Using Variation of Parameters**

Solve $y'' - y = 1/x$.

SOLUTION The auxiliary equation $m^2 - 1 = 0$ yields $m_1 = -1$ and $m_2 = 1$. Therefore $y_c = c_1 e^x + c_2 e^{-x}$. Now $W(e^x, e^{-x}) = -2$ and

$$u_1' = -\frac{e^{-x}(1/x)}{-2}, \quad u_2' = \frac{e^{x}(1/x)}{-2}$$

$$u_1 = \frac{1}{2}\int_{x_0}^{x} \frac{e^{-t}}{t}\,dt, \quad u_2 = -\frac{1}{2}\int_{x_0}^{x} \frac{e^{t}}{t}\,dt.$$

Since the foregoing integrals are nonelementary, we are forced to write

$$y_p = \frac{1}{2}e^x \int_{x_0}^{x} \frac{e^{-t}}{t}\,dt - \frac{1}{2}e^{-x}\int_{x_0}^{x} \frac{e^{t}}{t}\,dt,$$

and so

$$y = y_c + y_p = c_1 e^x + c_2 e^{-x} + \frac{1}{2}e^x \int_{x_0}^{x} \frac{e^{-t}}{t}\,dt - \frac{1}{2}e^{-x}\int_{x_0}^{x} \frac{e^{t}}{t}\,dt. \qquad \equiv$$

In Example 3 we can integrate on any interval $[x_0, x]$ not containing the origin. Also see Examples 2 and 3 in Section 3.10.

Higher-Order Equations

The method we have just examined for nonhomogeneous second-order differential equations can be generalized to linear nth-order equations that have been put into the standard form

$$y^{(n)} + P_{n-1}(x)y^{(n-1)} + \cdots + P_1(x)y' + P_0(x)y = f(x). \tag{8}$$

If $y_c = c_1 y_1 + c_2 y_2 + \cdots + c_n y_n$ is the complementary function for (8), then a particular solution is

$$y_p(x) = u_1(x)y_1(x) + u_2(x)y_2(x) + \cdots + u_n(x)y_n(x),$$

where the u_k', $k = 1, 2, \ldots, n$, are determined by the n equations

$$\begin{aligned}
y_1 u_1' + \quad y_2 u_2' + \cdots + \quad y_n u_n' &= 0 \\
y_1' u_1' + \quad y_2' u_2' + \cdots + \quad y_n' u_n' &= 0 \\
\vdots \qquad\qquad\qquad \vdots & \\
y_1^{(n-1)} u_1' + y_2^{(n-1)} u_2' + \cdots + y_n^{(n-1)} u_n' &= f(x).
\end{aligned} \tag{9}$$

The first $n - 1$ equations in this system, like $y_1 u_1' + y_2 u_2' = 0$ in (4), are assumptions made to simplify the resulting equation after $y_p = u_1(x)y_1(x) + \cdots + u_n(x)y_n(x)$ is substituted in (8). In this case, Cramer's rule gives

$$u_k' = \frac{W_k}{W}, \quad k = 1, 2, \ldots, n,$$

where W is the Wronskian of $y_1, y_2, \ldots, y_n$ and W_k is the determinant obtained by replacing the kth column of the Wronskian by the column consisting of the right-hand side of (9), that is, the column $(0, 0, \ldots, f(x))$. When $n = 2$ we get (5). When $n = 3$, the particular solution is $y_p = u_1 y_1 + u_2 y_2 + u_3 y_3$, where $y_1, y_2,$ and y_3 constitute a linearly independent set of solutions of the associated homogeneous DE, and u_1, u_2, u_3 are determined from

$$u_1' = \frac{W_1}{W}, \quad u_2' = \frac{W_2}{W}, \quad u_3' = \frac{W_3}{W}, \tag{10}$$

$$W = \begin{vmatrix} y_1 & y_2 & y_3 \\ y_1' & y_2' & y_3' \\ y_1'' & y_2'' & y_3'' \end{vmatrix}, \quad W_1 = \begin{vmatrix} 0 & y_2 & y_3 \\ 0 & y_2' & y_3' \\ f(x) & y_2'' & y_3'' \end{vmatrix}, \quad W_2 = \begin{vmatrix} y_1 & 0 & y_3 \\ y_1' & 0 & y_3' \\ y_1'' & f(x) & y_3'' \end{vmatrix}, \text{ and } W_3 = \begin{vmatrix} y_1 & y_2 & 0 \\ y_1' & y_2' & 0 \\ y_1'' & y_2'' & f(x) \end{vmatrix}.$$

See Problems 33 and 34 in Exercises 3.5.

■ **Particular Solution May Not Be Unique** In the problems in Exercises 3.5, do not hesitate to simplify, if possible, the form of the particular solution y_p found by variation of parameters. Depending on how the antiderivatives of u_1' and u_2' are found, you might not obtain the same y_p given in the answer section. For example, in Problem 3 in Exercises 3.5, both $y_p = \frac{1}{2}\sin x - \frac{1}{2}x\cos x$ and $y_p = -\frac{1}{2}x\cos x$ are particular solutions of the differential equation $y'' + y = \sin x$. If we use the first particular solution, then a general solution of the differential equation is

$$y = c_1 \cos x + c_2 \sin x + \tfrac{1}{2}\sin x - \tfrac{1}{2}x\cos x.$$

But note that by writing $y = c_1 \cos x + (c_2 + \frac{1}{2})\sin x - \frac{1}{2}x\cos x$ and defining $c_3 = c_2 + \frac{1}{2}$ another perfectly valid general solution is

$$y = c_1 \cos x + c_3 \sin x - \tfrac{1}{2}x\cos x.$$

3.5 Exercises Answers to selected odd-numbered problems begin on page ANS-6.

In Problems 1–20, solve each differential equation by variation of parameters.

1. $y'' + y = \sec x$

2. $y'' + y = \tan x$

3. $y'' + y = \sin x$

4. $y'' + y = \sec \theta \tan \theta$

5. $y'' + y = \cos^2 x$

6. $y'' + y = \sec^2 x$

7. $y'' - y = \cosh x$

8. $y'' - y = \sinh 2x$

9. $y'' - 9y = \dfrac{9x}{e^{3x}}$

10. $4y'' - y = e^{x/2} + 3$

11. $y'' + y' = 3$

12. $y'' - 4y' + 3y = e^x$

13. $y'' + 3y' + 2y = \dfrac{1}{1 + e^x}$

14. $y'' - 2y' + y = \dfrac{e^x}{1 + x^2}$

15. $y'' + 3y' + 2y = \sin e^x$ **16.** $y'' - 2y' + y = e^t \arctan t$

17. $y'' + 2y' + y = e^{-t} \ln t$ **18.** $2y'' + y' = 6x$

19. $3y'' - 6y' + 6y = e^x \sec x$

20. $4y'' - 4y' + y = e^{x/2}\sqrt{1 - x^2}$

In Problems 21–26, solve each differential equation by variation of parameters subject to the initial conditions $y(0) = 1$, $y'(0) = 0$.

21. $4y'' - y = xe^{x/2}$ **22.** $y'' + y' = e^{-x}$

23. $y'' - 2y' + y = e^x \sec^2 x$

24. $2y'' + y' - y = x + 1$

25. $y'' + 2y' - 8y = 2e^{-2x} - e^{-x}$

26. $y'' - 4y' + 4y = (12x^2 - 6x)e^{2x}$

In Problems 27–30, proceed as in Example 3 and solve each differential equation by variation of parameters.

27. $y'' + y = e^{x^2}$ **28.** $y'' - 4y = \dfrac{e^{2x}}{x}$

29. $y'' + y' - 2y = \ln x$ **30.** $2y'' + 2y' + y = 4\sqrt{x}$

In Problems 31 and 32, the indicated functions are known linearly independent solutions of the associated homogeneous differential equation on the interval $(0, \infty)$. Find the general solution of the given nonhomogeneous equation.

31. $x^2 y'' + xy' + (x^2 - \frac{1}{4})y = x^{3/2}$; $y_1 = x^{-1/2}\cos x$, $y_2 = x^{-1/2}\sin x$

32. $x^2 y'' + xy' + y = \sec(\ln x)$; $y_1 = \cos(\ln x)$, $y_2 = \sin(\ln x)$

In Problems 33–36, solve the given third-order differential equation by variation of parameters.

33. $y''' + y' = \tan x$ **34.** $y''' + 4y' = \sec 2x$

35. $y''' - 2y'' - y' + 2y = e^{4x}$

36. $y''' - 3y'' + 2y' = \dfrac{e^{2x}}{1 + e^x}$

Discussion Problems

In Problems 37 and 38, discuss how the methods of undetermined coefficients and variation of parameters can be combined to solve the given differential equation. Carry out your ideas.

37. $3y'' - 6y' + 30y = 15 \sin x + e^x \tan 3x$

38. $y'' - 2y' + y = 4x^2 - 3 + x^{-1}e^x$

39. What are the intervals of definition of the general solutions in Problems 1, 7, 17, and 20? Discuss why the interval of definition of the general solution in Problem 32 is *not* $(0, \infty)$.

40. Find the general solution of $x^4 y'' + x^3 y' - 4x^2 y = 1$ given that $y_1 = x^2$ is a solution of the associated homogeneous equation.

Computer Lab Assignments

In Problems 41 and 42, the indefinite integrals of the equations in (5) are nonelementary. Use a CAS to find the first four nonzero terms of a Maclaurin series of each integrand and then integrate the result. Find a particular solution of the given differential equation.

41. $y'' + y = \sqrt{1 + x^2}$ **42.** $4y'' - y = e^{x^2}$

3.6 Cauchy–Euler Equations

INTRODUCTION The relative ease with which we were able to find explicit solutions of linear higher-order differential equations with *constant coefficients* in the preceding sections does not, in general, carry over to linear equations with *variable coefficients*. We shall see in Chapter 5 that when a linear differential equation has variable coefficients, the best that we can usually expect is to find a solution in the form of an infinite series. However, the type of differential equation considered in this section is an exception to this rule; it is an equation with variable coefficients whose general solution can always be expressed in terms of powers of x, sines, cosines, logarithmic, and exponential functions. Moreover, its method of solution is quite similar to that for constant equations.

© Colport/Alamy Stock Photo
Augustin-Louis Cauchy

© Science History Images/Alamy Stock Photo
Leonhard Euler

■ **Cauchy–Euler Equation** A linear differential equation of the form

$$a_n x^n \frac{d^n y}{dx^n} + a_{n-1} x^{n-1} \frac{d^{n-1}y}{dx^{n-1}} + \cdots + a_1 x \frac{dy}{dx} + a_0 y = g(x),$$

where the coefficients $a_n, a_{n-1}, \ldots, a_0$ are constants, is known as a **Cauchy–Euler equation**. The differential equation is named in honor of two of the most esteemed mathematicians of all time, the French baron **Augustin-Louis Cauchy** (1789–1857) and the prolific Swiss **Leonhard Euler** (1707–1783). Cauchy's name is one of 72 French mathematicians, scientists, and engineers engraved on the Eiffel Tower in Paris in recognition of their achievements. The observable characteristic of this type of differential equation is that the degree $k = n, n - 1, \ldots, 1, 0$ of the monomial coefficients x^k matches the order k of differentiation $d^k y/dx^k$:

<div align="center">

same same

↓ ↓ ↓ ↓

</div>

$$a_n x^n \frac{d^n y}{dx^n} + a_{n-1} x^{n-1} \frac{d^{n-1}y}{dx^{n-1}} + \cdots.$$

As in Section 3.3, we start the discussion with a detailed examination of the forms of the general solutions of the homogeneous second-order equation

$$ax^2 \frac{d^2 y}{dx^2} + bx \frac{dy}{dx} + cy = 0. \tag{1}$$

The solution of higher-order equations follows analogously. Also, we can solve the nonhomogeneous equation $ax^2 y'' + bxy' + cy = g(x)$ by variation of parameters, once we have determined the complementary function $y_c(x)$ of (1).

The coefficient of d^2y/dx^2 is zero at $x = 0$. Hence, in order to guarantee that the fundamental results of Theorem 3.1.1 are applicable to the Cauchy–Euler equation, we confine our attention to finding the general solution on the interval $(0, \infty)$. Solutions on the interval $(-\infty, 0)$ can be obtained by substituting $t = -x$ into the differential equation. See Problems 49 and 50 in Exercises 3.6.

■ Method of Solution

We try a solution of the form $y = x^m$, where m is to be determined. Analogous to what happened when we substituted e^{mx} into a linear equation with constant coefficients, after substituting x^m each term of a Cauchy–Euler equation becomes a polynomial in m times x^m since

$$a_k x^k \frac{d^k y}{dx^k} = a_k x^k m(m - 1)(m - 2) \cdots (m - k + 1)x^{m-k}$$
$$= a_k m(m - 1)(m - 2) \cdots (m - k + 1)x^m.$$

For example, by substituting $y = x^m$ the second-order equation (1) becomes

$$ax^2 \frac{d^2y}{dx^2} + bx \frac{dy}{dx} + cy = am(m - 1)x^m + bmx^m + cx^m = (am(m - 1) + bm + c)x^m = 0.$$

Thus $y = x^m$ is a solution of the differential equation whenever m is a solution of the **auxiliary equation**

$$am(m - 1) + bm + c = 0 \quad \text{or} \quad am^2 + (b - a)m + c = 0. \tag{2}$$

There are three different cases to be considered, depending on whether the roots of this quadratic equation are real and distinct, real and equal, or complex. In the last case the roots appear as a conjugate pair.

> **Case I: Distinct Real Roots** Let m_1 and m_2 denote the real roots of (2) such that $m_1 \neq m_2$. Then $y_1 = x^{m_1}$ and $y_2 = x^{m_2}$ form a fundamental set of solutions. Hence the general solution of (1) is
>
> $$y = c_1 x^{m_1} + c_2 x^{m_2}. \tag{3}$$

EXAMPLE 1 **Distinct Roots**

Solve $\quad x^2 \dfrac{d^2y}{dx^2} - 2x \dfrac{dy}{dx} - 4y = 0.$

SOLUTION Rather than just memorizing equation (2), it is preferable to assume $y = x^m$ as the solution a few times in order to understand the origin and the difference between this new form of the auxiliary equation and that obtained in Section 3.3. Differentiate twice,

$$\frac{dy}{dx} = mx^{m-1}, \quad \frac{d^2y}{dx^2} = m(m - 1)x^{m-2},$$

and substitute back into the differential equation:

$$x^2 \frac{d^2y}{dx^2} - 2x \frac{dy}{dx} - 4y = x^2 \cdot m(m - 1)x^{m-2} - 2x \cdot mx^{m-1} - 4x^m$$
$$= x^m(m(m - 1) - 2m - 4) = x^m(m^2 - 3m - 4) = 0$$

if $m^2 - 3m - 4 = 0$. Now $(m + 1)(m - 4) = 0$ implies $m_1 = -1$, $m_2 = 4$ and so (3) yields the general solution $y = c_1 x^{-1} + c_2 x^4$. ≡

> **Case II: Repeated Real Roots** If the roots of (2) are repeated (that is, $m_1 = m_2$), then we obtain only one solution, namely, $y = x^{m_1}$. When the roots of the quadratic equation $am^2 + (b - a)m + c = 0$ are equal, the discriminant of the coefficients is necessarily zero. It follows from the quadratic formula that the root must be $m_1 = -(b - a)/2a$.
> Now we can construct a second solution y_2, using (5) of Section 3.2. We first write the Cauchy–Euler equation in the standard form

$$\frac{d^2y}{dx^2} + \frac{b}{ax}\frac{dy}{dx} + \frac{c}{ax^2}y = 0$$

and make the identifications $P(x) = b/ax$ and $\int (b/ax)\, dx = (b/a) \ln x$. Thus

$$y_2 = x^{m_1} \int \frac{e^{-(b/a)\ln x}}{x^{2m_1}}\, dx$$

$$= x^{m_1} \int x^{-b/a} \cdot x^{-2m_1}\, dx \qquad \leftarrow e^{-(b/a)\ln x} = e^{\ln x^{-(b/a)}} = x^{-b/a}$$

$$= x^{m_1} \int x^{-b/a} \cdot x^{(b-a)/a}\, dx \qquad \leftarrow -2m_1 = (b-a)/a$$

$$= x^{m_1} \int \frac{dx}{x} = x^{m_1} \ln x.$$

The general solution of (1) is then

$$y = c_1 x^{m_1} + c_2 x^{m_1} \ln x. \tag{4}$$

EXAMPLE 2 Repeated Roots

Solve $\quad 4x^2 \dfrac{d^2y}{dx^2} + 8x \dfrac{dy}{dx} + y = 0.$

SOLUTION The substitution $y = x^m$ yields

$$4x^2 \frac{d^2y}{dx^2} + 8x \frac{dy}{dx} + y = x^m(4m(m-1) + 8m + 1) = x^m(4m^2 + 4m + 1) = 0$$

when $4m^2 + 4m + 1 = 0$ or $(2m + 1)^2 = 0$. Since $m_1 = -\frac{1}{2}$ is a repeated root, (4) gives the general solution $y = c_1 x^{-1/2} + c_2 x^{-1/2} \ln x$. $\equiv$

For higher-order equations, if m_1 is a root of multiplicity k, then it can be shown that

$$x^{m_1}, \quad x^{m_1} \ln x, \quad x^{m_1}(\ln x)^2, \quad \ldots, \quad x^{m_1}(\ln x)^{k-1}$$

are k linearly independent solutions. Correspondingly, the general solution of the differential equation must then contain a linear combination of these k solutions.

Case III : Conjugate Complex Roots If the roots of (2) are the conjugate pair $m_1 = \alpha + i\beta$, $m_2 = \alpha - i\beta$, where α and $\beta > 0$ are real, then a solution is $y = C_1 x^{\alpha+i\beta} + C_2 x^{\alpha-i\beta}$. But when the roots of the auxiliary equation are complex, as in the case of equations with constant coefficients, we wish to write the solution in terms of real functions only. We note the identity

$$x^{i\beta} = (e^{\ln x})^{i\beta} = e^{i\beta \ln x},$$

which, by Euler's formula, is the same as

$$x^{i\beta} = \cos(\beta \ln x) + i \sin(\beta \ln x).$$

Similarly, $\qquad\qquad\qquad x^{-i\beta} = \cos(\beta \ln x) - i \sin(\beta \ln x).$

Adding and subtracting the last two results yields

$$x^{i\beta} + x^{-i\beta} = 2\cos(\beta \ln x) \qquad \text{and} \qquad x^{i\beta} - x^{-i\beta} = 2i\sin(\beta \ln x),$$

respectively. From the fact that $y = C_1 x^{\alpha+i\beta} + C_2 x^{\alpha-i\beta}$ is a solution for any values of the constants, we see, in turn, for $C_1 = C_2 = 1$ and $C_1 = 1, C_2 = -1$ that

$$y_1 = x^\alpha(x^{i\beta} + x^{-i\beta}) \qquad \text{and} \qquad y_2 = x^\alpha(x^{i\beta} - x^{-i\beta})$$

or $\qquad\qquad\qquad y_1 = 2x^\alpha \cos(\beta \ln x) \qquad \text{and} \qquad y_2 = 2ix^\alpha \sin(\beta \ln x)$

are also solutions. Since $W(x^\alpha \cos(\beta \ln x), x^\alpha \sin(\beta \ln x)) = \beta x^{2\alpha - 1} \neq 0$, $\beta > 0$, on the interval $(0, \infty)$, we conclude that

$$y_1 = x^\alpha \cos(\beta \ln x) \quad \text{and} \quad y_2 = x^\alpha \sin(\beta \ln x)$$

constitute a fundamental set of real solutions of the differential equation. Hence the general solution of (1) is

$$y = x^\alpha [c_1 \cos(\beta \ln x) + c_2 \sin(\beta \ln x)]. \tag{5}$$

EXAMPLE 3 An Initial-Value Problem

Solve the initial-value problem $\quad 4x^2 y'' + 17y = 0, \quad y(1) = -1, \quad y'(1) = -\frac{1}{2}$.

SOLUTION The y' term is missing in the given Cauchy–Euler equation; nevertheless, the substitution $y = x^m$ yields

$$4x^2 y'' + 17y = x^m(4m(m - 1) + 17) = x^m(4m^2 - 4m + 17) = 0$$

when $4m^2 - 4m + 17 = 0$. From the quadratic formula we find that the roots are $m_1 = \frac{1}{2} + 2i$ and $m_2 = \frac{1}{2} - 2i$. With the identifications $\alpha = \frac{1}{2}$ and $\beta = 2$, we see from (5) that the general solution of the differential equation on the interval $(0, \infty)$ is

$$y = x^{1/2}[c_1 \cos(2 \ln x) + c_2 \sin(2 \ln x)].$$

By applying the initial conditions $y(1) = -1$, $y'(1) = 0$ to the foregoing solution and using $\ln 1 = 0$ we then find, in turn, that $c_1 = -1$ and $c_2 = 0$. Hence the solution of the initial-value problem is $y = -x^{1/2}\cos(2 \ln x)$. The graph of this function, obtained with the aid of computer software, is given in **FIGURE 3.6.1**. The particular solution is seen to be oscillatory and unbounded as $x \to \infty$. $\equiv$

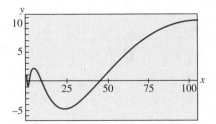

FIGURE 3.6.1 Graph of solution of IVP in Example 3

The next example illustrates the solution of a third-order Cauchy–Euler equation.

EXAMPLE 4 Third-Order Equation

Solve $\quad x^3 \dfrac{d^3 y}{dx^3} + 5x^2 \dfrac{d^2 y}{dx^2} + 7x \dfrac{dy}{dx} + 8y = 0$.

SOLUTION The first three derivatives of $y = x^m$ are

$$\frac{dy}{dx} = mx^{m-1}, \quad \frac{d^2 y}{dx^2} = m(m - 1)x^{m-2}, \quad \frac{d^3 y}{dx^3} = m(m - 1)(m - 2)x^{m-3}$$

so that the given differential equation becomes

$$x^3 \frac{d^3 y}{dx^3} + 5x^2 \frac{d^2 y}{dx^2} + 7x \frac{dy}{dx} + 8y = x^3 m(m-1)(m-2)x^{m-3} + 5x^2 m(m-1)x^{m-2} + 7xmx^{m-1} + 8x^m$$

$$= x^m(m(m - 1)(m - 2) + 5m(m - 1) + 7m + 8)$$

$$= x^m(m^3 + 2m^2 + 4m + 8) = x^m(m + 2)(m^2 + 4) = 0.$$

In this case we see that $y = x^m$ will be a solution of the differential equation for $m_1 = -2$, $m_2 = 2i$, and $m_3 = -2i$. Hence the general solution on the interval $(0, \infty)$ is

$$y = c_1 x^{-2} + c_2 \cos(2 \ln x) + c_3 \sin(2 \ln x). \qquad \equiv$$

▪ Nonhomogeneous Equations

The method of undetermined coefficients as described in Section 3.4 does not carry over, *in general*, to linear differential equations with variable coefficients. Consequently, in the following example the method of variation of parameters is employed.

EXAMPLE 5 **Variation of Parameters**

Solve $x^2 y'' - 3xy' + 3y = 2x^4 e^x$.

SOLUTION Since the equation is nonhomogeneous, we first solve the associated homogeneous equation. From the auxiliary equation $(m - 1)(m - 3) = 0$ we find $y_c = c_1 x + c_2 x^3$. Now before using variation of parameters to find a particular solution $y_p = u_1 y_1 + u_2 y_2$, recall that the formulas $u_1' = W_1/W$ and $u_2' = W_2/W$, where W_1, W_2, and W are the determinants defined on page 141, and were derived under the assumption that the differential equation has been put into the standard form $y'' + P(x)y' + Q(x)y = f(x)$. Therefore we divide the given equation by x^2, and from

$$y'' - \frac{3}{x}y' + \frac{3}{x^2}y = 2x^2 e^x$$

we make the identification $f(x) = 2x^2 e^x$. Now with $y_1 = x$, $y_2 = x^3$ and

$$W = \begin{vmatrix} x & x^3 \\ 1 & 3x^2 \end{vmatrix} = 2x^3, \quad W_1 = \begin{vmatrix} 0 & x^3 \\ 2x^2 e^x & 3x^2 \end{vmatrix} = -2x^5 e^x, \quad W_2 = \begin{vmatrix} x & 0 \\ 1 & 2x^2 e^x \end{vmatrix} = 2x^3 e^x$$

we find $u_1' = -\dfrac{2x^5 e^x}{2x^3} = -x^2 e^x$ and $u_2' = \dfrac{2x^3 e^x}{2x^3} = e^x.$

The integral of the latter function is immediate, but in the case of u_1' we integrate by parts twice. The results are $u_1 = -x^2 e^x + 2xe^x - 2e^x$ and $u_2 = e^x$. Hence

$$y_p = u_1 y_1 + u_2 y_2 = (-x^2 e^x + 2xe^x - 2e^x)x + e^x x^3 = 2x^2 e^x - 2xe^x.$$

Finally the general solution is $y = y_c + y_p = c_1 x + c_2 x^3 + 2x^2 e^x - 2xe^x.$ ≡

■ **A Generalization** The second-order differential equation

$$a(x - x_0)^2 \frac{d^2 y}{dx^2} + b(x - x_0)\frac{dy}{dx} + cy = 0 \tag{6}$$

is a generalization of equation (1). Note that (6) reduces to (1) when $x_0 = 0$. We can solve Cauchy–Euler equations of the form given in (6) exactly as we did in (1), namely, by seeking solutions $y = (x - x_0)^m$ and using

$$\frac{dy}{dx} = m(x - x_0)^{m-1} \quad \text{and} \quad \frac{d^2 y}{dx^2} = m(m - 1)(x - x_0)^{m-2}.$$

See Problems 39–42 in Exercises 3.6.

■ **Reduction to Constant Coefficients** The similarities between the forms of solutions of Cauchy–Euler equations and solutions of linear equations with constant coefficients is not just a coincidence. For example, when the roots of the auxiliary equations for $ay'' + by' + cy = 0$ and $ax^2 y'' + bxy' + cy = 0$ are distinct and real, the respective general solutions are

$$y = c_1 e^{m_1 x} + c_2 e^{m_2 x} \quad \text{and} \quad y = c_1 x^{m_1} + c_2 x^{m_2}, \; x > 0. \tag{7}$$

In view of the identity $e^{\ln x} = x, x > 0$, the second solution given in (7) can be expressed in the same form as the first solution:

$$y = c_1 e^{m_1 \ln x} + c_2 e^{m_2 \ln x} = c_1 e^{m_1 t} + c_2 e^{m_2 t},$$

where $t = \ln x$. This last result illustrates the fact that any Cauchy–Euler equation can *always* be rewritten as a linear differential equation with constant coefficients by means of the substitution $x = e^t$ and its equivalent $t = \ln x$. The idea is to solve the new differential equation in terms of the variable t, using the methods of the previous sections, and once the general solution is obtained, resubstitute $x = e^t$ and $t = \ln x$.

This method illustrated in the last example requires the use of the Chain Rule of differentiation.

EXAMPLE 6 **Changing the DE to Constant Coefficients**

Solve $\quad x^2\dfrac{d^2y}{dx^2} - x\dfrac{dy}{dx} + y = \ln x.$

SOLUTION With the substitution $t = \ln x$, it follows from the Chain Rule and Product Rule that

$$\frac{dy}{dx} = \frac{dy}{dt}\frac{dt}{dx} = \frac{1}{x}\frac{dy}{dt}$$

$$\frac{d^2y}{dx^2} = \frac{1}{x}\frac{d}{dx}\left(\frac{dy}{dt}\right) + \frac{dy}{dt}\left(-\frac{1}{x^2}\right)$$

$$= \frac{1}{x}\left(\frac{d^2y}{dt^2}\frac{1}{x}\right) + \frac{dy}{dt}\left(-\frac{1}{x^2}\right)$$

$$= \frac{1}{x^2}\left(\frac{d^2y}{dt^2} - \frac{dy}{dt}\right).$$

Substituting the foregoing derivatives in the original DE and simplifying yields

$$x^2\frac{1}{x^2}\left(\frac{d^2y}{dt^2} - \frac{dy}{dt}\right) - x\left(\frac{1}{x}\frac{dy}{dt}\right) + y = t$$

or $\qquad\qquad\qquad\qquad \dfrac{d^2y}{dt^2} - 2\dfrac{dy}{dt} + y = t.$

Since the last equation has constant coefficients, its auxiliary equation is $m^2 - 2m + 1 = 0$ or $(m - 1)^2 = 0$. Thus $y_c = c_1e^t + c_1te^t$.

Then by undetermined coefficients we try a particular solution of the form $y_p = A + Bt$. This assumption leads to $-2B + A + Bt = t$, so $A = 2$ and $B = 1$. Using $y = y_c + y_p$, we get $y = c_1e^t + c_2te^t + 2 + t$. By resubstituting $e^t = x$ and $t = \ln x$ we see that the general solution of the original differential equation on the interval $(0, \infty)$ is $y = c_1x + c_2x\ln x + 2 + \ln x.$ $\equiv$

See Problems 43–48 in Exercises 3.6.

Solutions for $x < 0$ In the preceding discussions, we have solved Cauchy–Euler equations for $x > 0$. One way of solving a Cauchy–Euler equation for $x < 0$ is to change the independent variable by means of the substitution $t = -x$ (which implies $t > 0$) and using the Chain Rule:

$$\frac{dy}{dx} = \frac{dy}{dt}\frac{dt}{dx} = -\frac{dy}{dt} \quad \text{and} \quad \frac{d^2y}{dx^2} = \frac{d}{dt}\left(-\frac{dy}{dt}\right)\frac{dt}{dx} = \frac{d^2y}{dt^2}.$$

See Problems 49 and 50 in Exercises 3.6.

3.6 Exercises Answers to selected odd-numbered problems begin on page ANS-6.

In Problems 1–18, solve the given differential equation.

1. $x^2y'' - 2y = 0$
2. $4x^2y'' + y = 0$
3. $xy'' + y' = 0$
4. $xy'' - 3y' = 0$
5. $x^2y'' + xy' + 4y = 0$
6. $x^2y'' + 5xy' + 3y = 0$
7. $x^2y'' - 3xy' - 2y = 0$
8. $x^2y'' + 3xy' - 4y = 0$
9. $25x^2y'' + 25xy' + y = 0$
10. $4x^2y'' + 4xy' - y = 0$
11. $x^2y'' + 5xy' + 4y = 0$
12. $x^2y'' + 8xy' + 6y = 0$
13. $3x^2y'' + 6xy' + y = 0$
14. $x^2y'' - 7xy' + 41y = 0$
15. $x^3y''' - 6y = 0$
16. $x^3y''' + xy' - y = 0$
17. $xy^{(4)} + 6y''' = 0$
18. $x^4y^{(4)} + 6x^3y''' + 9x^2y'' + 3xy' + y = 0$

In Problems 19–24, solve the given differential equation by variation of parameters.

19. $xy'' - 4y' = x^4$
20. $2x^2y'' + 5xy' + y = x^2 - x$
21. $x^2y'' - xy' + y = 2x$
22. $x^2y'' - 2xy' + 2y = x^4e^x$
23. $x^2y'' + xy' - y = \ln x$
24. $x^2y'' + xy' - y = \dfrac{1}{x + 1}$

In Problems 25–30, solve the given initial-value problem. Use a graphing utility to graph the solution curve.

25. $x^2y'' + 3xy' = 0$, $y(1) = 0$, $y'(1) = 4$

26. $x^2y'' - 5xy' + 8y = 0$, $y(2) = 32$, $y'(2) = 0$

27. $x^2y'' + xy' + y = 0$, $y(1) = 1$, $y'(1) = 2$

28. $x^2y'' - 3xy' + 4y = 0$, $y(1) = 5$, $y'(1) = 3$

29. $xy'' + y' = x$, $y(1) = 1$, $y'(1) = -\frac{1}{2}$

30. $x^2y'' - 5xy' + 8y = 8x^6$, $y(\frac{1}{2}) = 0$, $y'(\frac{1}{2}) = 0$

In Problems 31 and 32, solve the given boundary-value problem.

31. $xy'' - 7xy' + 12y = 0$, $y(0) = 0$, $y(1) = 0$

32. $x^2y'' - 3xy' + 5y = 0$, $y(1) = 0$, $y(e) = 1$

In Problems 33–38, find a homogeneous Cauchy–Euler differential equation whose general solution is given.

33. $y = c_1x^4 + c_2x^{-2}$

34. $y = c_1 + c_2x^5$

35. $y = c_1x^{-3} + c_2x^{-3}\ln x$

36. $y = c_1 + c_2x + c_3x\ln x$

37. $y = c_1\cos(\ln x) + c_2\sin(\ln x)$

38. $y = c_1x^{1/2}\cos(\frac{1}{2}\ln x) + c_2x^{1/2}\sin(\frac{1}{2}\ln x)$

In Problems 39–42, use the substitution $y = (x - x_0)^m$ to solve the given equation.

39. $(x + 3)^2y'' - 8(x + 3)y' + 14y = 0$

40. $(x - 1)^2y'' - (x - 1)y' + 5y = 0$

41. $(x + 2)^2y'' + (x + 2)y' + y = 0$

42. $(x - 4)^2y'' - 5(x - 4)y' + 9y = 0$

In Problems 43–48, use the substitution $t = \ln x$ to transform the given Cauchy–Euler equation to a differential equation with constant coefficients. Solve the original equation by solving the new equation using the procedures in Sections 3.3–3.5.

43. $x^2y'' + 9xy' - 20y = 0$ **44.** $x^2y'' - 9xy' + 25y = 0$

45. $x^2y'' + 10xy' + 8y = x^2$ **46.** $x^2y'' - 4xy' + 6y = \ln x^2$

47. $x^2y'' - 3xy' + 13y = 4 + 3x$

48. $x^3y''' - 3x^2y'' + 6xy' - 6y = 3 + \ln x^3$

In Problems 49 and 50, use the substitution $t = -x$ to solve the given initial-value problem on the interval $(-\infty, 0)$.

49. $4x^2y'' + y = 0$, $y(-1) = 2$, $y'(-1) = 4$

50. $x^2y'' - 4xy' + 6y = 0$, $y(-2) = 8$, $y'(-2) = 0$

Mathematical Models

Contributed Problem

Pierre Gharghouri, Professor Emeritus, Jean-Paul Pascal, Associate Professor, Department of Mathematics Ryerson University, Toronto, Canada

51. Temperature of a Fluid A very long cylindrical shell is formed by two concentric circular cylinders of different radii. A chemically reactive fluid fills the space between the concentric cylinders as shown in green in **FIGURE 3.6.2**.

The inner cylinder has a radius of 1 and is thermally insulated, while the outer cylinder has a radius of 2 and is maintained at a constant temperature T_0. The rate of heat generation in the fluid due to the chemical reactions is proportional to T/r^2, where $T(r)$ is the temperature of the fluid within the space bounded between the cylinders defined by $1 < r < 2$. Under these conditions the temperature of the fluid is defined by the following boundary-value problem:

$$\frac{d^2T}{dr^2} + \frac{1}{r}\frac{dT}{dr} = \frac{1}{r^2}T, \quad 1 < r < 2,$$

$$\left.\frac{dT}{dr}\right|_{r=1} = 0, \quad T(2) = T_0.$$

Find the temperature distribution $T(r)$ within the fluid.

FIGURE 3.6.2 Cylindrical shell in Problem 51

52. In Problem 51, find the minimum and maximum values of $T(r)$ on the interval defined by $1 \le r \le 2$. Why do these values make intuitive sense?

53. Bending of a Circular Plate In the analysis of the bending of a uniformly loaded circular plate, the equation $w(r)$ of the deflection curve of the plate can be shown to satisfy the third-order differential equation

$$\frac{d^3w}{dr^3} + \frac{1}{r}\frac{d^2w}{dr^2} - \frac{1}{r^2}\frac{dw}{dr} = \frac{q}{2D}r, \qquad (8)$$

where q and D are constants. Here r is the radial distance from a point on the circular plate to its center.

(a) Use the method of this section along with variation of parameters as given in (10) of Section 3.5 to find the general solution of equation (8).

(b) Find a solution of (8) that satisfies the boundary conditions

$$w'(0) = 0, \quad w(a) = 0, \quad w'(a) = 0,$$

where $a > 0$ is the radius of the plate. [*Hint*: The condition $w'(0) = 0$ is correct. Use this condition to determine one of the constants in the general solution of (8) found in part (a).]

54. In the engineering textbook where the differential equation in Problem 53 was found, the author states that the differential equation is readily solved by integration. True, but you have to realize that equation (8) can be written in the form given next.

(a) Verify that equation (8) can be written in the alternative form

$$\frac{d}{dr}\left[\frac{1}{r}\frac{d}{dr}\left(r\frac{dw}{dr}\right)\right] = \frac{q}{2D}r. \qquad (9)$$

(b) Solve equation (9) using only integration with respect to r. Show your result is equivalent to the solution obtained in part (a) of Problem 53.

Discussion Problems

55. Give the largest interval over which the general solution of Problem 42 is defined.

56. Find a Cauchy–Euler differential equation of lowest order with real coefficients if it is known that 2 and $1 - i$ are two roots of its auxiliary equation.

57. The initial conditions $y(0) = y_0$, $y'(0) = y_1$, apply to each of the following differential equations:

$$x^2y'' = 0,$$
$$x^2y'' - 2xy' + 2y = 0,$$
$$x^2y'' - 4xy' + 6y = 0.$$

For what values of y_0 and y_1 does each initial-value problem have a solution?

58. What are the x-intercepts of the solution curve shown in Figure 3.6.1? How many x-intercepts are there in the interval defined by $0 < x < \frac{1}{2}$?

Computer Lab Assignments

In Problems 59–62, solve the given differential equation by using a CAS to find the (approximate) roots of the auxiliary equation.

59. $2x^3y''' - 10.98x^2y'' + 8.5xy' + 1.3y = 0$

60. $x^3y''' + 4x^2y'' + 5xy' - 9y = 0$

61. $x^4y^{(4)} + 6x^3y''' + 3x^2y'' - 3xy' + 4y = 0$

62. $x^4y^{(4)} - 6x^3y''' + 33x^2y'' - 105xy' + 169y = 0$

In Problems 63 and 64, use a CAS as an aid in computing roots of the auxiliary equation, the determinants given in (10) of Section 3.5, and integrations.

63. $x^3y''' - x^2y'' - 2xy' + 6y = x^2$

64. $x^3y''' + 2x^2y'' - 8xy' + 12y = x^{-4}$

3.7 Nonlinear Equations

INTRODUCTION The difficulties that surround higher-order *nonlinear* DEs and the few methods that yield analytic solutions are examined next.

■ **Some Differences** There are several significant differences between linear and nonlinear differential equations. We saw in Section 3.1 that homogeneous linear equations of order two or higher have the property that a linear combination of solutions is also a solution (Theorem 3.1.2). Nonlinear equations do not possess this property of superposability. For example, on the interval $(-\infty, \infty)$, $y_1 = e^x$, $y_2 = e^{-x}$, $y_3 = \cos x$, and $y_4 = \sin x$ are four linearly independent solutions of the nonlinear second-order differential equation $(y'')^2 - y^2 = 0$. But linear combinations such as $y = c_1e^x + c_3 \cos x$, $y = c_2e^{-x} + c_4 \sin x$, $y = c_1e^x + c_2e^{-x} + c_3 \cos x + c_4 \sin x$ are not solutions of the equation for arbitrary nonzero constants c_i. See Problem 1 in Exercises 3.7.

In Chapter 2 we saw that we could solve a few nonlinear first-order differential equations by recognizing them as separable, exact, homogeneous, or perhaps Bernoulli equations. Even though the solutions of these equations were in the form of a one-parameter family, this family did not, as a rule, represent the general solution of the differential equation. On the other hand, by paying attention to certain continuity conditions, we obtained general solutions of linear first-order equations. Stated another way, nonlinear first-order differential equations can possess singular solutions whereas linear equations cannot. But the major difference between linear and nonlinear equations of order two or higher lies in the realm of solvability. Given a linear equation there is a chance that we can find some form of a solution that we can look at, an explicit solution or perhaps a solution in the form of an infinite series. On the other hand, nonlinear higher-order differential equations virtually defy solution. This does not mean that a nonlinear higher-order differential equation has no solution but rather that there are no analytical methods whereby either an explicit or implicit solution can be found.

Although this sounds disheartening, there are still things that can be done; we can always analyze a nonlinear DE qualitatively and numerically.

Let us make it clear at the outset that nonlinear higher-order differential equations are important—dare we say even more important than linear equations?—because as we fine-tune the mathematical model of, say, a physical system, we also increase the likelihood that this higher-resolution model will be nonlinear.

We begin by illustrating an analytical method that occasionally enables us to find explicit/implicit solutions of special kinds of nonlinear second-order differential equations.

Reduction of Order

Nonlinear second-order differential equations $F(x, y', y'') = 0$, where the dependent variable y is missing, and $F(y, y', y'') = 0$, where the independent variable x is missing, can sometimes be solved using first-order methods. Each equation can be reduced to a first-order equation by means of the substitution $u = y'$.

Dependent Variable Missing

The next example illustrates the substitution technique for an equation of the form $F(x, y', y'') = 0$. If $u = y'$, then the differential equation becomes $F(x, u, u') = 0$. If we can solve this last equation for u, we can find y by integration. Note that since we are solving a second-order equation, its solution will contain two arbitrary constants.

EXAMPLE 1 **Dependent Variable y Is Missing**

Solve $y'' = 2x(y')^2$.

SOLUTION If we let $u = y'$, then $du/dx = y''$. After substituting, the second-order equation reduces to a first-order equation with separable variables; the independent variable is x and the dependent variable is u:

$$\frac{du}{dx} = 2xu^2 \quad \text{or} \quad \frac{du}{u^2} = 2x\,dx$$

$$\int u^{-2}du = \int 2x\,dx$$

$$-u^{-1} = x^2 + c_1^2.$$

The constant of integration is written as c_1^2 for convenience. The reason should be obvious in the next few steps. Since $u^{-1} = 1/y'$, it follows that

$$\frac{dy}{dx} = -\frac{1}{x^2 + c_1^2}$$

and so $$y = -\int \frac{dx}{x^2 + c_1^2} \quad \text{or} \quad y = -\frac{1}{c_1}\tan^{-1}\frac{x}{c_1} + c_2.$$ ≡

Independent Variable Missing

Next we show how to solve an equation that has the form $F(y, y', y'') = 0$. Once more we let $u = y'$, but since the independent variable x is missing, we use this substitution to transform the differential equation into one in which the independent variable is y and the dependent variable is u. To this end we use the Chain Rule to compute the second derivative of y:

$$y'' = \frac{du}{dx} = \frac{du}{dy}\frac{dy}{dx} = u\frac{du}{dy}.$$

In this case the first-order equation that we must now solve is $F(y, u, u\,du/dy) = 0$.

EXAMPLE 2 **Independent Variable x Is Missing**

Solve $yy'' = (y')^2$.

SOLUTION With the aid of $u = y'$, the Chain Rule shown above, and separation of variables, the given differential equation becomes

$$y\left(u\frac{du}{dy}\right) = u^2 \quad \text{or} \quad \frac{du}{u} = \frac{dy}{y}.$$

Integrating the last equation then yields $\ln |u| = \ln |y| + c_1$, which, in turn, gives $u = c_2 y$, where the constant $\pm e^{c_1}$ has been relabeled as c_2. We now resubstitute $u = dy/dx$, separate variables once again, integrate, and relabel constants a second time:

$$\int \frac{dy}{y} = c_2 \int dx \quad \text{or} \quad \ln |y| = c_2 x + c_3 \quad \text{or} \quad y = c_4 e^{c_2 x}.$$ ≡

Use of Taylor Series In some instances a solution of a nonlinear initial-value problem, in which the initial conditions are specified at x_0, can be approximated by a Taylor series centered at x_0.

EXAMPLE 3 **Taylor Series Solution of an IVP**

Let us assume that a solution of the initial-value problem

$$y'' = x + y - y^2, \quad y(0) = -1, \quad y'(0) = 1 \qquad (1)$$

exists. If we further assume that the solution $y(x)$ of the problem is analytic at 0, then $y(x)$ possesses a Taylor series expansion centered at 0:

$$y(x) = y(0) + \frac{y'(0)}{1!}x + \frac{y''(0)}{2!}x^2 + \frac{y'''(0)}{3!}x^3 + \frac{y^{(4)}(0)}{4!}x^4 + \frac{y^{(5)}(0)}{5!}x^5 + \cdots. \qquad (2)$$

Note that the value of the first and second terms in the series (2) are known since those values are the specified initial conditions $y(0) = -1$, $y'(0) = 1$. Moreover, the differential equation itself defines the value of the second derivative at 0: $y''(0) = 0 + y(0) - y(0)^2 = 0 + (-1) - (-1)^2 = -2$. We can then find expressions for the higher derivatives y''', $y^{(4)}$, ..., by calculating the successive derivatives of the differential equation:

$$y'''(x) = \frac{d}{dx}(x + y - y^2) = 1 + y' - 2yy' \qquad (3)$$

$$y^{(4)}(x) = \frac{d}{dx}(1 + y' - 2yy') = y'' - 2yy'' - 2(y')^2 \qquad (4)$$

$$y^{(5)}(x) = \frac{d}{dx}(y'' - 2yy'' - 2(y')^2) = y''' - 2yy''' - 6y'y'' \qquad (5)$$

and so on. Now using $y(0) = -1$ and $y'(0) = 1$ we find from (3) that $y'''(0) = 4$. From the values $y(0) = -1$, $y'(0) = 1$, and $y''(0) = -2$, we find $y^{(4)}(0) = -8$ from (4). With the additional information that $y'''(0) = 4$, we then see from (5) that $y^{(5)}(0) = 24$. Hence from (2), the first six terms of a series solution of the initial-value problem (1) are

$$y(x) = -1 + x - x^2 + \frac{2}{3}x^3 - \frac{1}{3}x^4 + \frac{1}{5}x^5 + \cdots. \qquad \equiv$$

Use of a Numerical Solver Numerical methods, such as Euler's method or a Runge–Kutta method, are developed solely for first-order differential equations and then are extended to systems of first-order equations. In order to analyze an nth-order initial-value problem numerically, we express the nth-order ODE as a system of n first-order equations. In brief, here is how it is done for a second-order initial-value problem: First, solve for y''; that is, put the DE into normal form $y'' = f(x, y, y')$, and then let $y' = u$. For example, if we substitute $y' = u$ in

$$\frac{d^2y}{dx^2} = f(x, y, y'), \quad y(x_0) = y_0, \quad y'(x_0) = u_0, \qquad (6)$$

then $y'' = u'$ and $y'(x_0) = u(x_0)$ so that the initial-value problem (6) becomes

Solve: $\begin{cases} y' = u \\ u' = f(x, y, u) \end{cases}$

Subject to: $y(x_0) = y_0, \quad u(x_0) = u_0.$

However, it should be noted that a commercial numerical solver *may not* require* that you supply the system.

*Some numerical solvers require only that a second-order differential equation be expressed in normal form $y'' = f(x, y, y')$. The translation of the single equation into a system of two equations is then built into the computer program, since the first equation of the system is always $y' = u$ and the second equation is $u' = f(x, y, u)$.

y

Taylor polynomial

solution curve
generated by a
numerical solver

x

FIGURE 3.7.1 Comparison of two
approximate solutions in Example 4

y

10

x

FIGURE 3.7.2 Numerical solution
curve of IVP in (1) of Example 3

EXAMPLE 4 **Graphical Analysis of Example 3**

Following the foregoing procedure, the second-order initial-value problem in Example 3 is equivalent to

$$\frac{dy}{dx} = u$$

$$\frac{du}{dx} = x + y - y^2$$

with initial conditions $y(0) = -1$, $u(0) = 1$. With the aid of a numerical solver we get the solution curve shown in blue in **FIGURE 3.7.1**. For comparison, the curve shown in red is the graph of the fifth-degree Taylor polynomial $T_5(x) = -1 + x - x^2 + \frac{2}{3}x^3 - \frac{1}{3}x^4 + \frac{1}{5}x^5$. Although we do not know the interval of convergence of the Taylor series obtained in Example 3, the closeness of the two curves in the neighborhood of the origin suggests that the power series may converge on the interval $(-1, 1)$. $\equiv$

Qualitative Questions The blue graph in Figure 3.7.1 raises some questions of a qualitative nature: Is the solution of the original initial-value problem oscillatory as $x \to \infty$? The graph generated by a numerical solver on the larger interval shown in **FIGURE 3.7.2** would seem to *suggest* that the answer is yes. But this single example, or even an assortment of examples, does not answer the basic question of whether *all* solutions of the differential equation $y'' = x + y - y^2$ are oscillatory in nature. Also, what is happening to the solution curves in Figure 3.7.2 when x is near -1? What is the behavior of solutions of the differential equation as $x \to -\infty$? Are solutions bounded as $x \to \infty$? Questions such as these are not easily answered, in general, for nonlinear second-order differential equations. But certain kinds of second-order equations lend themselves to a systematic qualitative analysis, and these, like their first-order relatives encountered in Section 2.1, are the kind that have no explicit dependence on the independent variable. Second-order ODEs of the form

$$F(y, y', y'') = 0 \quad \text{or} \quad \frac{d^2y}{dx^2} = f(y, y'),$$

that is, equations free of the independent variable x, are called **autonomous**. The differential equation in Example 2 is autonomous, and because of the presence of the x term on its right side, the equation in Example 3 is nonautonomous. For an in-depth treatment of the topic of stability of autonomous second-order differential equations and autonomous systems of differential equations, the reader is referred to Chapter 11.

3.7 Exercises Answers to selected odd-numbered problems begin on page ANS-6.

In Problems 1 and 2, verify that y_1 and y_2 are solutions of the given differential equation but that $y = c_1y_1 + c_2y_2$ is, in general, not a solution.

1. $(y'')^2 = y^2$; $y_1 = e^x$, $y_2 = \cos x$

2. $yy'' = \frac{1}{2}(y')^2$; $y_1 = 1$, $y_2 = x^2$

In Problems 3–6, the dependent variable y is missing in the given differential equation. Proceed as in Example 1 and solve the equation by using the substitution $u = y'$.

3. $y'' + (y')^2 + 1 = 0$ **4.** $y'' = 1 + (y')^2$

5. $x^2y'' + (y')^2 = 0$ **6.** $e^{-x}y'' = (y')^2$

In Problems 7–10, the independent variable x is missing in the given differential equation. Proceed as in Example 2 and solve the equation by using the substitution $u = y'$.

7. $yy'' + (y')^2 + 1 = 0$ **8.** $(y + 1)y'' = (y')^2$

9. $y'' + 2y(y')^3 = 0$ **10.** $y^2y'' = y'$

In Problems 11 and 12, solve the given initial-value problem.

11. $2y'y'' = 1$, $y(0) = 2$, $y'(0) = 1$

12. $y'' + x(y')^2 = 0$, $y(1) = 4$, $y'(1) = 2$

13. Consider the initial-value problem

$$y'' + yy' = 0, \quad y(0) = 1, \quad y'(0) = -1.$$

(a) Use the DE and a numerical solver to graph the solution curve.

(b) Find an explicit solution of the IVP. Use a graphing utility to graph this solution.

(c) Find an interval of definition for the solution in part (b).

14. Find two solutions of the initial-value problem

$$(y'')^2 + (y')^2 = 1, \quad y(\pi/2) = \frac{1}{2}, \quad y'(\pi/2) = \sqrt{3}/2.$$

Use a numerical solver to graph the solution curves.

In Problems 15 and 16, show that the substitution $u = y'$ leads to a Bernoulli equation. Solve this equation (see Section 2.5).

15. $xy'' = y' + (y')^3$ **16.** $xy'' = y' + x(y')^2$

In Problems 17–20, proceed as in Example 3 and obtain the first six nonzero terms of a Taylor series solution, centered at 0, of the given initial-value problem. Use a numerical solver and a graphing utility to compare the solution curve with the graph of the Taylor polynomial.

17. $y'' = x + y^2$, $y(0) = 1$, $y'(0) = 1$

18. $y'' + y^2 = 1$, $y(0) = 2$, $y'(0) = 3$

19. $y'' = x^2 + y^2 - 2y'$, $y(0) = 1$, $y'(0) = 1$

20. $y'' = e^y$, $y(0) = 0$, $y'(0) = -1$

21. In calculus, the curvature of a curve that is defined by a function $y = f(x)$ is defined as

$$\kappa = \frac{y''}{[1 + (y')^2]^{3/2}}.$$

Find $y = f(x)$ for which $\kappa = 1$. [*Hint*: For simplicity, ignore constants of integration.]

Discussion Problems

22. In Problem 1 we saw that $\cos x$ and e^x were solutions of the nonlinear equation $(y'')^2 - y^2 = 0$. Verify that $\sin x$ and e^{-x} are also solutions. Without attempting to solve the differential equation, discuss how these explicit solutions can be found by using knowledge about linear equations. Without attempting to verify, discuss why the linear combinations $y = c_1 e^x + c_2 e^{-x} + c_3 \cos x + c_4 \sin x$ and $y = c_2 e^{-x} + c_4 \sin x$ are not, in general, solutions, but the two special linear combinations $y = c_1 e^x + c_2 e^{-x}$ and $y = c_3 \cos x + c_4 \sin x$ *must* satisfy the differential equation.

23. Discuss how the method of reduction of order considered in this section can be applied to the third-order differential equation $y''' = \sqrt{1 + (y'')^2}$. Carry out your ideas and solve the equation.

24. Discuss how to find an alternative two-parameter family of solutions for the nonlinear differential equation $y'' = 2x(y')^2$ in Example 1. [*Hint*: Suppose that $-c_1^2$ is used as the constant of integration instead of $+c_1^2$.]

Mathematical Models

25. **Motion in a Force Field** A mathematical model for the position $x(t)$ of a body moving rectilinearly on the x-axis in an inverse-square force field is given by

$$\frac{d^2x}{dt^2} = -\frac{k^2}{x^2}.$$

Suppose that at $t = 0$ the body starts from rest from the position $x = x_0$, $x_0 > 0$. Show that the velocity of the body at time t is given by $v^2 = 2k^2(1/x - 1/x_0)$. Use the last expression and a CAS to carry out the integration to express time t in terms of x.

26. A mathematical model for the position $x(t)$ of a moving object is

$$\frac{d^2x}{dt^2} + \sin x = 0.$$

Use a numerical solver to investigate graphically the solutions of the equation subject to $x(0) = 0$, $x'(0) = x_1$, $x_1 \geq 0$. Discuss the motion of the object for $t \geq 0$ and for various choices of x_1. Investigate the equation

$$\frac{d^2x}{dt^2} + \frac{dx}{dt} + \sin x = 0$$

in the same manner. Give a possible physical interpretation of the dx/dt term.

3.8 Linear Models: Initial-Value Problems

INTRODUCTION In this section we are going to consider several linear dynamical systems in which each mathematical model is a linear second-order differential equation with constant coefficients along with initial conditions specified at time t_0:

$$a_2 \frac{d^2y}{dt^2} + a_1 \frac{dy}{dt} + a_0 y = g(t), \quad y(t_0) = y_0, \quad y'(t_0) = y_1.$$

The function g is variously called the **input**, **driving**, or **forcing function** of the system. The **output** or **response** of the system is a function $y(t)$ defined on an I interval containing t_0 that satisfies both the differential equation and the initial conditions on the interval I.

3.8.1 Spring/Mass Systems: Free Undamped Motion

We begin this first section by examining the motion of a mass m attached to a spring. The word *spring* in this discussion will refer to a flexible coil, or helical, spring. See the *Remarks* at the end of this section.

■ **Hooke's Law** Suppose a spring/mass system consists of a flexible spring suspended vertically from a rigid support with a mass m attached to its free end. The amount of stretch s, or

elongation, of the spring will of course depend on the mass; masses with different weights stretch the spring by differing amounts. **Hooke's law**, devised by the English natural philosopher and polymath **Robert Hooke** (1635–1703) around 1660, states that the spring itself exerts a restoring force F opposite to the direction of elongation and proportional to the amount of elongation s. Simply stated, $F = -ks$, where $k > 0$ is a constant of proportionality called the **spring constant** or **stiffness** of the spring. Thus a spring is essentially characterized by the number k. Using $|F| = k|s|$, we see that if a mass weighing 10 lb stretches a spring $\frac{1}{2}$ ft, then $10 = k(\frac{1}{2})$ implies $k = 20$ lb/ft. Necessarily then, a mass weighing, say, 16 lb stretches the same spring $\frac{4}{5}$ ft.

FIGURE 3.8.1 Spring/mass system

unstretched

equilibrium position
$mg - ks = 0$

motion

(a) (b) (c)

FIGURE 3.8.2 Positive direction is below equilibrium position

$x < 0$

$x = 0$

$x > 0$

Newton's Second Law
After a mass m is attached to a spring, it stretches the spring by an amount s and attains a position of equilibrium at which its weight W is balanced by the restoring force ks. Recall that weight is defined by $W = mg$, where mass is measured in slugs, kilograms, or grams and $g = 32$ ft/s², 9.8 m/s², or 980 cm/s², respectively. As indicated in **FIGURE 3.8.1(b)**, the condition of equilibrium is $mg = ks$ or $mg - ks = 0$. If the mass is displaced by an amount x from its equilibrium position, the restoring force of the spring is then $k(x + s)$. Assuming that there are no retarding forces acting on the system and assuming that the mass vibrates free of other external forces—**free motion**—we can equate Newton's second law with the net, or resultant, force of the restoring force and the weight:

$$m \frac{d^2 x}{dt^2} = -k(s + x) + mg = -kx + \underbrace{mg - ks}_{\text{zero}} = -kx. \tag{1}$$

The negative sign in (1) indicates that the restoring force of the spring acts opposite to the direction of motion. Furthermore, we can adopt the convention that displacements measured *below* the equilibrium position are positive. See **FIGURE 3.8.2**.

DE of Free Undamped Motion
By dividing (1) by the mass m we obtain the second-order differential equation $d^2 x/dt^2 + (k/m)x = 0$ or

$$\frac{d^2 x}{dt^2} + \omega^2 x = 0, \tag{2}$$

where $\omega^2 = k/m$. Equation (2) is said to describe **simple harmonic motion** or **free undamped motion**. Two obvious initial conditions associated with (2) are $x(0) = x_0$, the amount of initial displacement, and $x'(0) = x_1$, the initial velocity of the mass. For example, if $x_0 > 0$, $x_1 < 0$, the mass starts from a point *below* the equilibrium position with an imparted *upward* velocity. When $x_1 = 0$ the mass is said to be released from *rest*. For example, if $x_0 < 0$, $x_1 = 0$, the mass is released from rest from a point $|x_0|$ units *above* the equilibrium position.

Solution and Equation of Motion
To solve equation (2) we note that the solutions of the auxiliary equation $m^2 + \omega^2 = 0$ are the complex numbers $m_1 = \omega i$, $m_2 = -\omega i$. Thus from (8) of Section 3.3 we find the general solution of (2) to be

$$x(t) = c_1 \cos \omega t + c_2 \sin \omega t. \tag{3}$$

The **period of free vibrations** described by (3) is $T = 2\pi/\omega$, and the **frequency** is $f = 1/T = \omega/2\pi$. For example, for $x(t) = 2 \cos 3t - 4 \sin 3t$ the period is $2\pi/3$ and the frequency is $3/2\pi$. The former number means that the graph of $x(t)$ repeats every $2\pi/3$ units; the latter number means that there are three cycles of the graph every 2π units or, equivalently, that the mass undergoes $3/2\pi$ complete vibrations per unit time. In addition, it can be shown that the period $2\pi/\omega$ is the time interval between two successive maxima of $x(t)$. Keep in mind that a maximum of $x(t)$ is a positive displacement corresponding to the mass's attaining a maximum distance *below* the equilibrium position, whereas a minimum of $x(t)$ is a negative displacement corresponding to the mass's attaining a maximum height *above* the equilibrium position. We refer to either case as an **extreme displacement** of the mass. Finally, when the initial conditions are used to determine the constants c_1 and c_2 in (3), we say that the resulting particular solution or response is the **equation of motion**.

EXAMPLE 1 **Free Undamped Motion**

A mass weighing 2 pounds stretches a spring 6 inches. At $t = 0$ the mass is released from a point 8 inches below the equilibrium position with an upward velocity of $\frac{4}{3}$ ft/s. Determine the equation of free motion.

SOLUTION Because we are using the engineering system of units, the measurements given in terms of inches must be converted into feet: 6 in. $= \frac{1}{2}$ ft; 8 in. $= \frac{2}{3}$ ft. In addition, we must convert the units of weight given in pounds into units of mass. From $m = W/g$ we have $m = \frac{2}{32} = \frac{1}{16}$ slug. Also, from Hooke's law, $2 = k(\frac{1}{2})$ implies that the spring constant is $k = 4$ lb/ft. Hence (1) gives

$$\frac{1}{16}\frac{d^2x}{dt^2} = -4x \quad \text{or} \quad \frac{d^2x}{dt^2} + 64x = 0.$$

The initial displacement and initial velocity are $x(0) = \frac{2}{3}$, $x'(0) = -\frac{4}{3}$, where the negative sign in the last condition is a consequence of the fact that the mass is given an initial velocity in the negative, or upward, direction.

Now $\omega^2 = 64$ or $\omega = 8$, so that the general solution of the differential equation is

$$x(t) = c_1 \cos 8t + c_2 \sin 8t. \tag{4}$$

Applying the initial conditions to $x(t)$ and $x'(t)$ gives $c_1 = \frac{2}{3}$ and $c_2 = -\frac{1}{6}$. Thus the equation of motion is

$$x(t) = \frac{2}{3}\cos 8t - \frac{1}{6}\sin 8t. \tag{5} \equiv$$

■ **Alternative Form of $x(t)$** When $c_1 \neq 0$ and $c_2 \neq 0$, the actual **amplitude** A of free vibrations is not obvious from inspection of equation (3). For example, although the mass in Example 1 is initially displaced $\frac{2}{3}$ foot beyond the equilibrium position, the amplitude of vibrations is a number larger than $\frac{2}{3}$. Hence it is often convenient to convert an equation of simple harmonic motion of the form given in (3) into the form of a shifted sine function (6) or a shifted cosine function (6′). In both (6) and (6′) the number $A = \sqrt{c_1^2 + c_2^2}$ is the **amplitude of free vibrations**, and ϕ is a **phase angle**. Note carefully that ϕ is defined in a slightly different manner in (7) and (7′).

$$y = A\sin(\omega t + \phi) \tag{6} \qquad\qquad y = A\cos(\omega t - \phi) \tag{6′}$$

where where

$$\left.\begin{array}{l} \sin\phi = \dfrac{c_1}{A} \\[2mm] \cos\phi = \dfrac{c_2}{A} \end{array}\right\} \tan\phi = \dfrac{c_1}{c_2} \quad (7) \qquad \left.\begin{array}{l} \sin\phi = \dfrac{c_2}{A} \\[2mm] \cos\phi = \dfrac{c_1}{A} \end{array}\right\} \tan\phi = \dfrac{c_2}{c_1} \quad (7′)$$

To verify (6), we expand $\sin(\omega t + \phi)$ by the addition formula for the sine function:

$$A\sin(\omega t + \phi) = A\sin\omega t\cos\phi + A\cos\omega t\sin\phi = (A\sin\phi)\cos\omega t + (A\cos\phi)\sin\omega t. \tag{8}$$

FIGURE 3.8.3 A relationship between $c_1 > 0$, $c_2 > 0$, and phase angle ϕ

It follows from **FIGURE 3.8.3** that if ϕ is defined by

$$\sin\phi = \frac{c_1}{\sqrt{c_1^2 + c_2^2}} = \frac{c_1}{A}, \quad \cos\phi = \frac{c_2}{\sqrt{c_1^2 + c_2^2}} = \frac{c_2}{A},$$

then (8) becomes

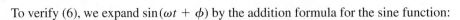

$$A\frac{c_1}{A}\cos\omega t + A\frac{c_2}{A}\sin\omega t = c_1\cos\omega t + c_2\sin\omega t = x(t).$$

EXAMPLE 2 **Alternative Form of Solution (5)**

In view of the foregoing discussion, we can write the solution (5) in the alternative forms (6) and (6′). In both cases the amplitude is $A = \sqrt{(\frac{2}{3})^2 + (-\frac{1}{6})^2} = \sqrt{\frac{17}{36}} \approx 0.69$ ft. But some care should be exercised when computing a phase angle ϕ.

Be careful in the computation of the phase angle ϕ. ▶

(a) With $c_1 = \frac{2}{3}$ and $c_2 = -\frac{1}{6}$ we find from (7) that $\tan \phi = -4$. A calculator then gives $\tan^{-1}(-4) = -1.326$ rad. However, this is *not* the phase angle since $\tan^{-1}(-4)$ is located in the *fourth quadrant* and therefore contradicts the fact that $\sin \phi > 0$ and $\cos \phi < 0$ because $c_1 > 0$ and $c_2 < 0$. Hence we must take ϕ to be the *second-quadrant* angle $\phi = \pi + (1.326) = 1.816$ rad. Thus, (6) gives

$$x(t) = \frac{\sqrt{17}}{6} \sin(8t + 1.816). \qquad (9)$$

The period of this function is $T = 2\pi/8 = \pi/4$.

(b) Now with $c_1 = \frac{2}{3}$ and $c_2 = -\frac{1}{6}$, we see that (7′) indicates that $\sin \phi < 0$ and $\cos \phi > 0$ and so the angle ϕ lies in the fourth quadrant. Hence from $\tan \phi = -\frac{1}{4}$ we can take $\phi = \tan^{-1}(-\frac{1}{4}) = -0.245$ rad. A second alternative form of solution (5) is then

$$x(t) = \frac{\sqrt{17}}{6} \cos(8t - (-0.245)) \quad \text{or} \quad x(t) = \frac{\sqrt{17}}{6} \cos(8t + 0.245). \qquad (9') \quad \equiv$$

▪ **Graphical Interpretation** FIGURE 3.8.4(a) illustrates the mass in Example 2 going through approximately two complete cycles of motion. Reading left to right, the first five positions marked with black dots in the figure correspond, respectively,

- to the initial position $(x = \frac{2}{3})$ of the mass below the equilibrium position,
- to the mass passing through the equilibrium position $(x = 0)$ for the first time heading upward,
- to the mass at its extreme displacement $(x = -\frac{\sqrt{17}}{6})$ above the equilibrium position,
- to the mass passing through the equilibrium position $(x = 0)$ for the second time heading downward, and
- to the mass at its extreme displacement $(x = \frac{\sqrt{17}}{6})$ below the equilibrium position.

The dots on the graph of (9) given in Figure 3.8.4(b) also agree with the five positions just given. Note, however, that in Figure 3.8.4(b) the positive direction in the tx-plane is the usual upward

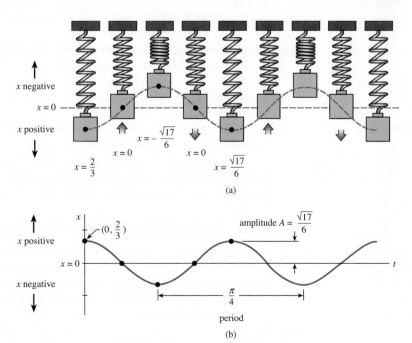

(a)

(b)

FIGURE 3.8.4 Simple harmonic motion

direction and so is opposite to the positive direction indicated in Figure 3.8.4(a). Hence the blue graph representing the motion of the mass in Figure 3.8.4(b) is the mirror image through the t-axis of the red dashed curve in Figure 3.8.4(a).

Form (6) is very useful since it is easy to find values of time for which the graph of $x(t)$ crosses the positive t-axis (the line $x = 0$). We observe that $\sin(\omega t + \phi) = 0$ when $\omega t + \phi = n\pi$, where n is a nonnegative integer.

FIGURE 3.8.5 Parallel springs

rigid support

k_1 k_2

m

FIGURE 3.8.5 Parallel springs

▪ Double Spring Systems

Suppose two *parallel* springs, with constants k_1 and k_2, are attached to a common rigid support and then to a metal plate of negligible mass. A single mass m is attached to the center of the plate in the double-spring arrangement as shown in **FIGURE 3.8.5**. If the mass is displaced from its equilibrium position, the displacement x is the same for both springs and so the net restoring force of the spring in (1) is simply $-k_1 x - k_2 x = -(k_1 + k_2)x$. We say that

$$k_{\text{eff}} = k_1 + k_2$$

is the **effective spring constant** of the system.

On the other hand, suppose two springs supporting a single *mass m* are in *series*, that is, the springs are attached end-to-end as shown in **FIGURE 3.8.6**. In this case, a displacement x of the mass from its equilibrium consists of the sum $x = x_1 + x_2$, where x_1 and x_2 are the displacements of the respective springs. But the restoring force is the same for both springs so if k_{eff} is the **effective spring constant** of the system we have

$$-k_{\text{eff}}(x_1 + x_2) = -k_1 x_1 = -k_2 x_2.$$

rigid support

k_1

k_2

m

FIGURE 3.8.6 Springs in series

From $k_1 x_1 = k_2 x_2$ we see $x_1 = (k_2/k_1)x_2$ and so $-k_{\text{eff}}(x_1 + x_2) = -k_2 x_2$ is the same as

$$k_{\text{eff}}\left(\frac{k_2}{k_1}x_2 + x_2\right) = k_2 x_2.$$

Solving the last equation for k_{eff} yields

$$k_{\text{eff}} = \frac{k_1 k_2}{k_1 + k_2}.$$

So in either of the above cases, the differential equation of motion is (1) with k replaced by k_{eff}. See Problems 13–18 in Exercises 3.8.

▪ Systems with Variable Spring Constants

In the model discussed above, we assumed an ideal world, a world in which the physical characteristics of the spring do not change over time. In the nonideal world, however, it seems reasonable to expect that when a spring/mass system is in motion for a long period the spring would weaken; in other words, the "spring constant" would vary, or, more specifically, decay with time. In one model for an **aging spring**, the spring constant k in (1) is replaced by the decreasing function $K(t) = ke^{-\alpha t}$, $k > 0$, $\alpha > 0$. The linear differential equation $mx'' + ke^{-\alpha t}x = 0$ cannot be solved by the methods considered in this chapter. Nevertheless, we can obtain two linearly independent solutions using the methods in Chapter 5. See Problem 19 in Exercises 3.8.

When a spring/mass system is subjected to an environment in which the temperature is rapidly decreasing, it might make sense to replace the constant k with $K(t) = kt$, $k > 0$, a function that increases with time. The resulting model, $mx'' + ktx = 0$, is a form of **Airy's differential equation**. Like the equation for an aging spring, Airy's equation can be solved by the methods of Chapter 5. See Problem 20 in Exercises 3.8.

m

(a)

m

(b)

FIGURE 3.8.7 Damping devices

3.8.2 Spring/Mass Systems: Free Damped Motion

The concept of free harmonic motion is somewhat unrealistic since the motion described by equation (1) assumes that there are no retarding forces acting on the moving mass. Unless the mass is suspended in a perfect vacuum, there will be at least a resisting force due to the surrounding medium. As **FIGURE 3.8.7** shows, the mass could be suspended in a viscous medium or connected to a dashpot damping device.

■ DE of Free Damped Motion

In the study of mechanics, damping forces acting on a body are considered to be proportional to a power of the instantaneous velocity. In particular, we shall assume throughout the subsequent discussion that this force is given by a constant multiple of dx/dt. When no other external forces are impressed on the system, it follows from Newton's second law that

$$m\frac{d^2x}{dt^2} = -kx - \beta\frac{dx}{dt}, \tag{10}$$

where β is a positive *damping constant* and the negative sign is a consequence of the fact that the damping force acts in a direction opposite to the motion.

Dividing (10) by the mass m, we find the differential equation of **free damped motion** is $d^2x/dt^2 + (\beta/m)\,dx/dt + (k/m)x = 0$ or

$$\frac{d^2x}{dt^2} + 2\lambda\frac{dx}{dt} + \omega^2 x = 0, \tag{11}$$

where

$$2\lambda = \frac{\beta}{m}, \quad \omega^2 = \frac{k}{m}. \tag{12}$$

The symbol 2λ is used only for algebraic convenience, since the auxiliary equation is $m^2 + 2\lambda m + \omega^2 = 0$ and the corresponding roots are then

$$m_1 = -\lambda + \sqrt{\lambda^2 - \omega^2}, \quad m_2 = -\lambda - \sqrt{\lambda^2 - \omega^2}.$$

We can now distinguish three possible cases depending on the algebraic sign of $\lambda^2 - \omega^2$. Since each solution contains the *damping factor* $e^{-\lambda t}$, $\lambda > 0$, the displacements of the mass become negligible over a long period of time.

Case I: $\lambda^2 - \omega^2 > 0$ In this situation the system is said to be **overdamped** because the damping coefficient β is large when compared to the spring constant k. The corresponding solution of (11) is $x(t) = c_1 e^{m_1 t} + c_2 e^{m_2 t}$ or

$$x(t) = e^{-\lambda t}(c_1 e^{\sqrt{\lambda^2 - \omega^2}\,t} + c_2 e^{-\sqrt{\lambda^2 - \omega^2}\,t}). \tag{13}$$

FIGURE 3.8.8 Motion of an overdamped system

Equation 13 represents a smooth and nonoscillatory motion. **FIGURE 3.8.8** shows two possible graphs of $x(t)$.

Case II: $\lambda^2 - \omega^2 = 0$ The system is said to be **critically damped** because any slight decrease in the damping force would result in oscillatory motion. The general solution of (11) is $x(t) = c_1 e^{m_1 t} + c_2 t e^{m_1 t}$ or

$$x(t) = e^{-\lambda t}(c_1 + c_2 t). \tag{14}$$

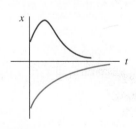

FIGURE 3.8.9 Motion of a critically damped system

Some graphs of typical motion are given in **FIGURE 3.8.9**. Notice that the motion is quite similar to that of an overdamped system. It is also apparent from (14) that the mass can pass through the equilibrium position at most one time.

Case III: $\lambda^2 - \omega^2 < 0$ In this case the system is said to be **underdamped** because the damping coefficient is small compared to the spring constant. The roots m_1 and m_2 are now complex:

$$m_1 = -\lambda + \sqrt{\omega^2 - \lambda^2}\,i, \quad m_2 = -\lambda - \sqrt{\omega^2 - \lambda^2}\,i.$$

Thus the general solution of equation (11) is

$$x(t) = e^{-\lambda t}(c_1 \cos\sqrt{\omega^2 - \lambda^2}\,t + c_2 \sin\sqrt{\omega^2 - \lambda^2}\,t). \tag{15}$$

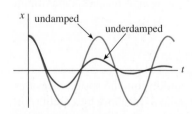

FIGURE 3.8.10 Motion of an underdamped system

As indicated in **FIGURE 3.8.10**, the motion described by (15) is oscillatory, but because of the coefficient $e^{-\lambda t}$, the amplitudes of vibration $\to 0$ as $t \to \infty$.

EXAMPLE 3 Overdamped Motion

It is readily verified that the solution of the initial-value problem

$$\frac{d^2x}{dt^2} + 5\frac{dx}{dt} + 4x = 0, \quad x(0) = 1, \quad x'(0) = 1$$

is

$$x(t) = \frac{5}{3}e^{-t} - \frac{2}{3}e^{-4t}. \tag{16}$$

The problem can be interpreted as representing the overdamped motion of a mass on a spring. The mass starts from a position 1 unit *below* the equilibrium position with a *downward* velocity of 1 ft/s.

To graph $x(t)$ we find the value of t for which the function has an extremum—that is, the value of time for which the first derivative (velocity) is zero. Differentiating (16) gives $x'(t) = -\frac{5}{3}e^{-t} + \frac{8}{3}e^{-4t}$ so that $x'(t) = 0$ implies $e^{3t} = \frac{8}{5}$ or $t = \frac{1}{3}\ln\frac{8}{5} = 0.157$. It follows from the first derivative test, as well as our intuition, that $x(0.157) = 1.069$ ft is actually a maximum. In other words, the mass attains an extreme displacement of 1.069 feet below the equilibrium position.

We should also check to see whether the graph crosses the t-axis, that is, whether the mass passes through the equilibrium position. This cannot happen in this instance since the equation $x(t) = 0$, or $e^{3t} = \frac{2}{5}$, has the physically irrelevant solution $t = \frac{1}{3}\ln\frac{2}{5} = -0.305$.

The graph of $x(t)$, along with some other pertinent data, is given in **FIGURE 3.8.11**. ≡

FIGURE 3.8.11 Overdamped system in Example 3

t	$x(t)$
1	0.601
1.5	0.370
2	0.225
2.5	0.137
3	0.083

(b)

EXAMPLE 4 Critically Damped Motion

An 8-pound weight stretches a spring 2 feet. Assuming that a damping force numerically equal to two times the instantaneous velocity acts on the system, determine the equation of motion if the weight is released from the equilibrium position with an upward velocity of 3 ft/s.

SOLUTION From Hooke's law we see that $8 = k(2)$ gives $k = 4$ lb/ft and that $W = mg$ gives $m = \frac{8}{32} = \frac{1}{4}$ slug. The differential equation of motion is then

$$\frac{1}{4}\frac{d^2x}{dt^2} = -4x - 2\frac{dx}{dt} \quad \text{or} \quad \frac{d^2x}{dt^2} + 8\frac{dx}{dt} + 16x = 0. \tag{17}$$

The auxiliary equation for (17) is $m^2 + 8m + 16 = (m + 4)^2 = 0$ so that $m_1 = m_2 = -4$. Hence the system is critically damped and

$$x(t) = c_1e^{-4t} + c_2te^{-4t}. \tag{18}$$

Applying the initial conditions $x(0) = 0$ and $x'(0) = -3$, we find, in turn, that $c_1 = 0$ and $c_2 = -3$. Thus the equation of motion is

$$x(t) = -3te^{-4t}. \tag{19}$$

To graph $x(t)$ we proceed as in Example 3. From $x'(t) = -3e^{-4t}(1 - 4t)$ we see that $x'(t) = 0$ when $t = \frac{1}{4}$. The corresponding extreme displacement is $x(\frac{1}{4}) = -3(\frac{1}{4})e^{-1} = -0.276$ ft. As shown in **FIGURE 3.8.12**, we interpret this value to mean that the weight reaches a maximum height of 0.276 foot above the equilibrium position. ≡

FIGURE 3.8.12 Critically damped system in Example 4

EXAMPLE 5 Underdamped Motion

A 16-pound weight is attached to a 5-foot-long spring. At equilibrium the spring measures 8.2 feet. If the weight is pushed up and released from rest at a point 2 feet above the equilibrium position, find the displacements $x(t)$ if it is further known that the surrounding medium offers a resistance numerically equal to the instantaneous velocity.

SOLUTION The elongation of the spring after the weight is attached is $8.2 - 5 = 3.2$ ft, so it follows from Hooke's law that $16 = k(3.2)$ or $k = 5$ lb/ft. In addition, $m = \frac{16}{32} = \frac{1}{2}$ slug so that the differential equation is given by

$$\frac{1}{2}\frac{d^2x}{dt^2} = -5x - \frac{dx}{dt} \quad \text{or} \quad \frac{d^2x}{dt^2} + 2\frac{dx}{dt} + 10x = 0. \tag{20}$$

Proceeding, we find that the roots of $m^2 + 2m + 10 = 0$ are $m_1 = -1 + 3i$ and $m_2 = -1 - 3i$, which then implies the system is underdamped and

$$x(t) = e^{-t}(c_1 \cos 3t + c_2 \sin 3t). \tag{21}$$

Finally, the initial conditions $x(0) = -2$ and $x'(0) = 0$ yield $c_1 = -2$ and $c_2 = -\frac{2}{3}$, so the equation of motion is

$$x(t) = e^{-t}\left(-2\cos 3t - \frac{2}{3}\sin 3t\right). \tag{22} \quad \equiv$$

■ **Alternative Form of x(t)** In a manner identical to the procedure used on page 158, we can write any solution

$$x(t) = e^{-\lambda t}(c_1 \cos\sqrt{\omega^2 - \lambda^2}\,t + c_2 \sin\sqrt{\omega^2 - \lambda^2}\,t)$$

in the alternative form

$$x(t) = Ae^{-\lambda t}\sin(\sqrt{\omega^2 - \lambda^2}\,t + \phi), \tag{23}$$

where $A = \sqrt{c_1^2 + c_2^2}$ and the phase angle ϕ is determined from the equations

$$\sin\phi = \frac{c_1}{A}, \quad \cos\phi = \frac{c_2}{A}, \quad \tan\phi = \frac{c_1}{c_2}.$$

The coefficient $Ae^{-\lambda t}$ is sometimes called the **damped amplitude** of vibrations. Because (23) is not a periodic function, the number $2\pi/\sqrt{\omega^2 - \lambda^2}$ is called the **quasi period** and $\sqrt{\omega^2 - \lambda^2}/2\pi$ is the **quasi frequency**. The quasi period is the time interval between two successive maxima of $x(t)$. You should verify, for the equation of motion in Example 5, that $A = 2\sqrt{10}/3$ and $\phi = 4.391$. Therefore an equivalent form of (22) is

$$x(t) = \frac{2\sqrt{10}}{3}e^{-t}\sin(3t + 4.391).$$

3.8.3 Spring/Mass Systems: Driven Motion

■ **DE of Driven Motion with Damping** Suppose we now take into consideration an external force $f(t)$ acting on a vibrating mass on a spring. For example, $f(t)$ could represent a driving force causing an oscillatory vertical motion of the support of the spring. See **FIGURE 3.8.13**. The inclusion of $f(t)$ in the formulation of Newton's second law gives the differential equation of **driven** or **forced motion**:

$$m\frac{d^2x}{dt^2} = -kx - \beta\frac{dx}{dt} + f(t). \tag{24}$$

Dividing (24) by m gives

$$\frac{d^2x}{dt^2} + 2\lambda\frac{dx}{dt} + \omega^2 x = F(t), \tag{25}$$

where $F(t) = f(t)/m$ and, as in the preceding section, $2\lambda = \beta/m$, $\omega^2 = k/m$. To solve the latter nonhomogeneous equation we can use either the method of undetermined coefficients or variation of parameters.

FIGURE 3.8.13 Oscillatory vertical motion of the support

EXAMPLE 6 **Interpretation of an Initial-Value Problem**

Interpret and solve the initial-value problem

$$\frac{1}{5}\frac{d^2x}{dt^2} + 1.2\frac{dx}{dt} + 2x = 5\cos 4t, \quad x(0) = \frac{1}{2}, \quad x'(0) = 0. \tag{26}$$

SOLUTION We can interpret the problem to represent a vibrational system consisting of a mass ($m = \frac{1}{5}$ slug or kilogram) attached to a spring ($k = 2$ lb/ft or N/m). The mass is released from rest $\frac{1}{2}$ unit (foot or meter) below the equilibrium position. The motion is damped ($\beta = 1.2$) and is being driven by an external periodic ($T = \pi/2$ s) force beginning at $t = 0$. Intuitively we would expect that even with damping, the system would remain in motion until such time as the forcing function was "turned off," in which case the amplitudes would diminish. However, as the problem is given, $f(t) = 5\cos 4t$ will remain "on" forever.

We first multiply the differential equation in (26) by 5 and solve

$$\frac{dx^2}{dt^2} + 6\frac{dx}{dt} + 10x = 0$$

by the usual methods. Since $m_1 = -3 + i, m_2 = -3 - i$, it follows that

$$x_c(t) = e^{-3t}(c_1\cos t + c_2\sin t).$$

Using the method of undetermined coefficients, we assume a particular solution of the form $x_p(t) = A\cos 4t + B\sin 4t$. Differentiating $x_p(t)$ and substituting into the DE gives

$$x_p'' + 6x_p' + 10x_p = (-6A + 24B)\cos 4t + (-24A - 6B)\sin 4t = 25\cos 4t.$$

The resulting system of equations

$$-6A + 24B = 25, \quad -24A - 6B = 0$$

yields $A = -\frac{25}{102}$ and $B = \frac{50}{51}$. It follows that

$$x(t) = e^{-3t}(c_1\cos t + c_2\sin t) - \frac{25}{102}\cos 4t + \frac{50}{51}\sin 4t. \tag{27}$$

When we set $t = 0$ in the above equation, we obtain $c_1 = \frac{38}{51}$. By differentiating the expression and then setting $t = 0$, we also find that $c_2 = -\frac{86}{51}$. Therefore the equation of motion is

$$x(t) = e^{-3t}\left(\frac{38}{51}\cos t - \frac{86}{51}\sin t\right) - \frac{25}{102}\cos 4t + \frac{50}{51}\sin 4t. \tag{28} \equiv$$

■ Transient and Steady-State Terms When F is a periodic function, such as $F(t) = F_0\sin\gamma t$ or $F(t) = F_0\cos\gamma t$, the general solution of (25) for $\lambda > 0$ is the sum of a nonperiodic function $x_c(t)$ and a periodic function $x_p(t)$. Moreover, $x_c(t)$ dies off as time increases; that is, $\lim_{t\to\infty} x_c(t) = 0$. Thus for a long period of time, the displacements of the mass are closely approximated by the particular solution $x_p(t)$. The complementary function $x_c(t)$ is said to be a **transient term** or **transient solution**, and the function $x_p(t)$, the part of the solution that remains after an interval of time, is called a **steady-state term** or **steady-state solution**. Note therefore that the effect of the initial conditions on a spring/mass system driven by F is transient. In the particular solution (28), $e^{-3t}(\frac{38}{51}\cos t - \frac{86}{51}\sin t)$ is a transient term and $x_p(t) = -\frac{25}{102}\cos 4t + \frac{50}{51}\sin 4t$ is a steady-state term. The graphs of these two terms and the solution (28) are given in **FIGURES 3.8.14(a)** and 3.8.14(b), respectively.

(a)

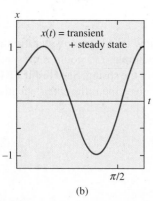

(b)

FIGURE 3.8.14 Graph of solution (28) in Example 6

EXAMPLE 7 **Transient/Steady-State Solutions**

The solution of the initial-value problem

$$\frac{d^2x}{dt^2} + 2\frac{dx}{dt} + 2x = 4\cos t + 2\sin t, \quad x(0) = 0, \quad x'(0) = x_1,$$

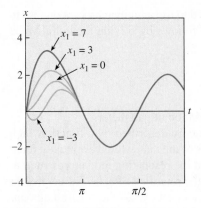

FIGURE 3.8.15 Graphs of solution in Example 7 for various values of x_1

where x_1 is constant, is given by

$$x(t) = \underbrace{(x_1 - 2)e^{-t} \sin t}_{\text{transient}} + \underbrace{2 \sin t}_{\text{steady state}}.$$

Solution curves for selected values of the initial velocity x_1 are shown in **FIGURE 3.8.15**. The graphs show that the influence of the transient term is negligible for about $t > 3\pi/2$. ≡

DE of Driven Motion Without Damping

With a periodic impressed force and no damping force, there is no transient term in the solution of a problem. Also, we shall see that a periodic impressed force with a frequency near or the same as the frequency of free undamped vibrations can cause a severe problem in any oscillatory mechanical system.

EXAMPLE 8 **Undamped Forced Motion**

Solve the initial-value problem

$$\frac{d^2x}{dt^2} + \omega^2 x = F_0 \sin \gamma t, \quad x(0) = 0, \quad x'(0) = 0, \tag{29}$$

where F_0 is a constant and $\gamma \neq \omega$.

SOLUTION The complementary function is $x_c(t) = c_1 \cos \omega t + c_2 \sin \omega t$. To obtain a particular solution we assume $x_p(t) = A \cos \gamma t + B \sin \gamma t$ so that

$$x_p'' + \omega^2 x_p = A(\omega^2 - \gamma^2) \cos \gamma t + B(\omega^2 - \gamma^2) \sin \gamma t = F_0 \sin \gamma t.$$

Equating coefficients immediately gives $A = 0$ and $B = F_0/(\omega^2 - \gamma^2)$. Therefore

$$x_p(t) = \frac{F_0}{\omega^2 - \gamma^2} \sin \gamma t.$$

Applying the given initial conditions to the general solution

$$x(t) = c_1 \cos \omega t + c_2 \sin \omega t + \frac{F_0}{\omega^2 - \gamma^2} \sin \gamma t$$

yields $c_1 = 0$ and $c_2 = -\gamma F_0/\omega(\omega^2 - \gamma^2)$. Thus the solution is

$$x(t) = \frac{F_0}{\omega(\omega^2 - \gamma^2)} (-\gamma \sin \omega t + \omega \sin \gamma t), \quad \gamma \neq \omega. \tag{30} \equiv$$

Pure Resonance

Although equation (30) is not defined for $\gamma = \omega$, it is interesting to observe that its limiting value as $\gamma \rightarrow \omega$ can be obtained by applying L'Hôpital's rule. This limiting process is analogous to "tuning in" the frequency of the driving force ($\gamma/2\pi$) to the frequency of free vibrations ($\omega/2\pi$). Intuitively we expect that over a length of time we should be able to increase the amplitudes of vibration substantially. For $\gamma = \omega$ we define the solution to be

$$x(t) = \lim_{\gamma \to \omega} F_0 \frac{-\gamma \sin \omega t + \omega \sin \gamma t}{\omega(\omega^2 - \gamma^2)} = F_0 \lim_{\gamma \to \omega} \frac{\frac{d}{d\gamma}(-\gamma \sin \omega t + \omega \sin \gamma t)}{\frac{d}{d\gamma}(\omega^3 - \omega\gamma^2)}$$

$$= F_0 \lim_{\gamma \to \omega} \frac{-\sin \omega t + \omega t \cos \gamma t}{-2\omega\gamma}$$

$$= F_0 \frac{-\sin \omega t + \omega t \cos \omega t}{-2\omega^2}$$

$$= \frac{F_0}{2\omega^2} \sin \omega t - \frac{F_0}{2\omega} t \cos \omega t. \tag{31}$$

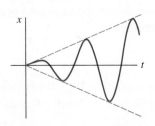

FIGURE 3.8.16 Graph of solution in (31) illustrating pure resonance

As suspected, when $t \rightarrow \infty$ the displacements become large; in fact, $|x(t_n)| \rightarrow \infty$ when $t_n = n\pi/\omega$, $n = 1, 2, \ldots$. The phenomenon we have just described is known as **pure resonance**. The graph given in **FIGURE 3.8.16** shows typical motion in this case.

In conclusion, it should be noted that there is no actual need to use a limiting process on (30) to obtain the solution for $\gamma = \omega$. Alternatively, equation (31) follows by solving the initial-value problem

$$\frac{d^2x}{dt^2} + \omega^2 x = F_0 \sin \omega t, \quad x(0) = 0, \quad x'(0) = 0$$

directly by the methods of undetermined coefficients or variation of parameters.

If the displacements of a spring/mass system were actually described by a function such as (31), the system would necessarily fail. Large oscillations of the mass would eventually force the spring beyond its elastic limit. One might argue too that the resonating model presented in Figure 3.8.16 is completely unrealistic because it ignores the retarding effects of ever-present damping forces. Although it is true that pure resonance cannot occur when the smallest amount of damping is taken into consideration, large and equally destructive amplitudes of vibration (although bounded as $t \to \infty$) can occur. See Problem 49 in Exercises 3.8.

3.8.4 Series Circuit Analogue

■ **LRC-Series Circuits** As mentioned in the introduction to this chapter, many different physical systems can be described by a linear second-order differential equation similar to the differential equation of forced motion with damping:

$$m\frac{d^2x}{dt^2} + \beta\frac{dx}{dt} + kx = f(t). \tag{32}$$

FIGURE 3.8.17 *LRC*-series circuit

If $i(t)$ denotes current in the **LRC-series electrical circuit** shown in **FIGURE 3.8.17**, then the voltage drops across the inductor, resistor, and capacitor are as shown in Figure 1.3.5. By Kirchhoff's second law, the sum of these voltages equals the voltage $E(t)$ impressed on the circuit; that is,

$$L\frac{di}{dt} + Ri + \frac{1}{C}q = E(t). \tag{33}$$

But the charge $q(t)$ on the capacitor is related to the current $i(t)$ by $i = dq/dt$, and so (33) becomes the linear second-order differential equation

$$L\frac{d^2q}{dt^2} + R\frac{dq}{dt} + \frac{1}{C}q = E(t). \tag{34}$$

The nomenclature used in the analysis of circuits is similar to that used to describe spring/mass systems.

If $E(t) = 0$, the **electrical vibrations** of the circuit are said to be **free**. Since the auxiliary equation for (34) is $Lm^2 + Rm + 1/C = 0$, there will be three forms of the solution with $R \neq 0$, depending on the value of the discriminant $R^2 - 4L/C$. We say that the circuit is

	overdamped if	$R^2 - 4L/C > 0$,
	critically damped if	$R^2 - 4L/C = 0$,
and	**underdamped** if	$R^2 - 4L/C < 0$.

In each of these three cases the general solution of (34) contains the factor $e^{-Rt/2L}$, and so $q(t) \to 0$ as $t \to \infty$. In the underdamped case when $q(0) = q_0$, the charge on the capacitor oscillates as it decays; in other words, the capacitor is charging and discharging as $t \to \infty$. When $E(t) = 0$ and $R = 0$, the circuit is said to be undamped and the electrical vibrations do not approach zero as t increases without bound; the response of the circuit is **simple harmonic**.

 Underdamped Series Circuit

Find the charge $q(t)$ on the capacitor in an *LRC*-series circuit when $L = 0.25$ henry (h), $R = 10$ ohms (Ω), $C = 0.001$ farad (F), $E(t) = 0$ volts (V), $q(0) = q_0$ coulombs (C), and $i(0) = 0$ amperes (A).

SOLUTION Since $1/C = 1000$, equation (34) becomes

$$\frac{1}{4} q'' + 10q' + 1000q = 0 \quad \text{or} \quad q'' + 40q' + 4000q = 0.$$

Solving this homogeneous equation in the usual manner, we find that the circuit is underdamped and $q(t) = e^{-20t}(c_1 \cos 60t + c_2 \sin 60t)$. Applying the initial conditions, we find $c_1 = q_0$ and $c_2 = q_0/3$. Thus $q(t) = q_0 e^{-20t}(\cos 60t + \frac{1}{3} \sin 60t)$. Using (23), we can write the foregoing solution as

$$q(t) = \frac{q_0 \sqrt{10}}{3} e^{-20t} \sin(60t + 1.249). \qquad \equiv$$

When there is an impressed voltage $E(t)$ on the circuit, the electrical vibrations are said to be **forced**. In the case when $R \neq 0$, the complementary function $q_c(t)$ of (34) is called a **transient solution**. If $E(t)$ is periodic or a constant, then the particular solution $q_p(t)$ of (34) is a **steady-state solution**.

EXAMPLE 10 **Steady-State Current**

Find the steady-state solution $q_p(t)$ and the **steady-state current** in an *LRC*-series circuit when the impressed voltage is $E(t) = E_0 \sin \gamma t$.

SOLUTION The steady-state solution $q_p(t)$ is a particular solution of the differential equation

$$L\frac{d^2q}{dt^2} + R\frac{dq}{dt} + \frac{1}{C}q = E_0 \sin \gamma t.$$

Using the method of undetermined coefficients, we assume a particular solution of the form $q_p(t) = A \sin \gamma t + B \cos \gamma t$. Substituting this expression into the differential equation, simplifying, and equating coefficients gives

$$A = \frac{E_0 \left(L\gamma - \dfrac{1}{C\gamma} \right)}{-\gamma \left(L^2\gamma^2 - \dfrac{2L}{C} + \dfrac{1}{C^2\gamma^2} + R^2 \right)}, \quad B = \frac{E_0 R}{-\gamma \left(L^2\gamma^2 - \dfrac{2L}{C} + \dfrac{1}{C^2\gamma^2} + R^2 \right)}.$$

It is convenient to express A and B in terms of some new symbols.

$$\text{If } X = L\gamma - \frac{1}{C\gamma}, \quad \text{then} \quad X^2 = L^2\gamma^2 - \frac{2L}{C} + \frac{1}{C^2\gamma^2}.$$

$$\text{If } Z = \sqrt{X^2 + R^2}, \quad \text{then} \quad Z^2 = L^2\gamma^2 - \frac{2L}{C} + \frac{1}{C^2\gamma^2} + R^2.$$

Therefore $A = E_0 X/(-\gamma Z^2)$ and $B = E_0 R/(-\gamma Z^2)$, so the steady-state charge is

$$q_p(t) = -\frac{E_0 X}{\gamma Z^2} \sin \gamma t - \frac{E_0 R}{\gamma Z^2} \cos \gamma t.$$

Now the steady-state current is given by $i_p(t) = q_p'(t)$:

$$i_p(t) = \frac{E_0}{Z}\left(\frac{R}{Z} \sin \gamma t - \frac{X}{Z} \cos \gamma t \right). \qquad (35) \quad \equiv$$

The quantities $X = L\gamma - 1/(C\gamma)$ and $Z = \sqrt{X^2 + R^2}$ defined in Example 10 are called, respectively, the **reactance** and **impedance** of the circuit. Both the reactance and the impedance are measured in ohms.

© Image Source/Getty Images
Mousetrap

REMARKS

(*i*) A coil spring that is twisted rather than pulled is referred to as a **torsion spring**. The mathematics of a torsion spring is basically the same as that of a stretched spring. The differential equation

$$I\frac{d^2\theta}{dt^2} + C\frac{d\theta}{dt} + k\theta = \tau(t) \tag{36}$$

is analogous to (24). Here $\theta(t)$ represents the angle of twist, $C\,d\theta/dt$ is angular damping, $k\theta$ is the torque exerted by the spring (the angular form of Hooke's law), and $\tau(t)$ is the driving torque. A simple example of a torsion spring is the spring in a mousetrap shown in the first photo on the left. Garage door openers employ one or two torsion springs depending on the weight of the door. When the door is closed the springs are twisted, so when the door is opened the springs unwind to help lift the door. The springs are above the two door windows in the second photo on the left.

© Konstantin L/Shutterstock
Garage door openers

(*ii*) Acoustic vibrations can be equally as destructive as large mechanical vibrations. Jazz singers have inflicted destruction on the lowly wine glass. See the third photo on the left. Sounds from organs and piccolos have been known to crack windows.

> *As the horns blew, the people began to shout. When they heard the signal horn they raised a tremendous shout. The wall collapsed....* (Joshua 6:20)

Did the power of acoustic resonance cause the walls of Jericho to tumble down? This is the conjecture of some contemporary scholars.

© ImagineGolf/E+/Getty Images
Shattered wine glass

(*iii*) If you have ever looked out of a window while in flight, you have probably observed that the wings of an airplane are not perfectly rigid. A reasonable amount of flex or flutter is not only tolerated but necessary to prevent the wing from snapping like a piece of peppermint candy. In late 1959 and early 1960, two commercial plane crashes occurred involving the relatively new turboprop Lockheed L-188 Electra, illustrating the destructive effects of large mechanical oscillations.

The unusual aspect of these crashes was that they occurred in mid-flight. Barring midair collisions, the safest period during any flight is when the plane has attained its cruising altitude. It is well known that a plane is most vulnerable to an accident when it is least maneuverable—namely, during either takeoff or landing. So, having two planes simply fall out of the sky was not only a tragedy but an embarrassment to the aircraft industry and a thoroughly puzzling problem to aerodynamic engineers. In crashes of this sort, a structural failure of some kind is immediately suspected. After a massive technical investigation, the problem was eventually traced in each case to an outboard engine and its engine housing. Roughly, it was determined that when each plane surpassed a critical speed of approximately 400 mph, a propeller and engine began to wobble, causing a gyroscopic force that could not be quelled or damped by the engine housing. This external vibrational force was then transferred to the already oscillating wing. This, in itself, need not have been destructively dangerous, since aircraft wings are designed to withstand the stress of unusual and excessive forces. (In fact, the particular wing in question was so incredibly strong that test engineers and pilots who were deliberately trying to snap a wing under every conceivable flight condition failed to do so.) But, unfortunately, after a short period of time during which the engine wobbled rapidly, the frequency of the impressed force actually slowed to a point at which it approached and finally coincided with the maximum frequency of wing flutter (around 3 cycles per second). The resulting resonance situation finally accomplished what test engineers could not do: namely, the amplitudes of wing flutter became large enough to snap the wing. See **FIGURE 3.8.18**.

© Antony Nettle/Alamy Stock Photo
Lockheed L-188 Electra

The problem was solved in two steps. All models of this particular plane were required to fly at speeds substantially below 400 mph until each plane could be modified by considerably strengthening (or stiffening) the engine housings. A strengthened engine housing was shown to be able to impart a damping effect capable of preventing the critical resonance phenomenon, even in the unlikely event of a subsequent engine wobble.

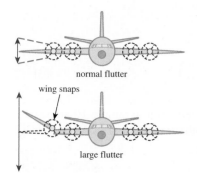

normal flutter

wing snaps

large flutter

FIGURE 3.8.18 Wing snaps in two

3.8.1 Spring/Mass Systems: Free Undamped Motion

1. A mass weighing 4 pounds is attached to a spring whose spring constant is 16 lb/ft. What is the period of simple harmonic motion?

2. A 20-kilogram mass is attached to a spring. If the frequency of simple harmonic motion is $2/\pi$ cycles/s, what is the spring constant k? What is the frequency of simple harmonic motion if the original mass is replaced with an 80-kilogram mass?

3. A mass weighing 24 pounds, attached to the end of a spring, stretches it 4 inches. Initially, the mass is released from rest from a point 3 inches above the equilibrium position. Find the equation of motion.

4. Determine the equation of motion if the mass in Problem 3 is initially released from the equilibrium position with an initial downward velocity of 2 ft/s.

5. A mass weighing 20 pounds stretches a spring 6 inches. The mass is initially released from rest from a point 6 inches below the equilibrium position.
 (a) Find the position of the mass at the times $t = \pi/12$, $\pi/8$, $\pi/6$, $\pi/4$, and $9\pi/32$ s.
 (b) What is the velocity of the mass when $t = 3\pi/16$ s? In which direction is the mass heading at this instant?
 (c) At what times does the mass pass through the equilibrium position?

6. A force of 400 newtons stretches a spring 2 meters. A mass of 50 kilograms is attached to the end of the spring and is initially released from the equilibrium position with an upward velocity of 10 m/s. Find the equation of motion.

7. Another spring whose constant is 20 N/m is suspended from the same rigid support but parallel to the spring/mass system in Problem 6. A mass of 20 kilograms is attached to the second spring, and both masses are initially released from the equilibrium position with an upward velocity of 10 m/s.
 (a) Which mass exhibits the greater amplitude of motion?
 (b) Which mass is moving faster at $t = \pi/4$ s? At $\pi/2$ s?
 (c) At what times are the two masses in the same position? Where are the masses at these times? In which directions are they moving?

8. A mass weighing 32 pounds stretches a spring 2 feet.
 (a) Determine the amplitude and period of motion if the mass is initially released from a point 1 foot above the equilibrium position with an upward velocity of 2 ft/s.
 (b) How many complete cycles will the mass have made at the end of 4π seconds?

9. A mass weighing 8 pounds is attached to a spring. When set in motion, the spring/mass system exhibits simple harmonic motion.
 (a) Determine the equation of motion if the spring constant is 1 lb/ft and the mass is initially released from a point

6 inches below the equilibrium position with a downward velocity of $\frac{3}{2}$ ft/s.
 (b) Express the equation of motion in the form of a shifted sine function given in (6).
 (c) Express the equation of motion in the form of a shifted cosine function given in (6′).

10. A mass weighing 10 pounds stretches a spring $\frac{1}{4}$ foot. This mass is removed and replaced with a mass of 1.6 slugs, which is initially released from a point $\frac{1}{3}$ foot above the equilibrium position with a downward velocity of $\frac{5}{4}$ ft/s.
 (a) Express the equation of motion in the form of a shifted sine function given in (6).
 (b) Express the equation of motion in the form of a shifted cosine function given in (6′).
 (c) Use one of the solutions obtained in parts (a) and (b) to determine the times the mass attains a displacement below the equilibrium position numerically equal to $\frac{1}{2}$ the amplitude of motion.

11. A mass weighing 64 pounds stretches a spring 0.32 foot. The mass is initially released from a point 8 inches above the equilibrium position with a downward velocity of 5 ft/s.
 (a) Find the equation of motion.
 (b) What are the amplitude and period of motion?
 (c) How many complete cycles will the mass have completed at the end of 3π seconds?
 (d) At what time does the mass pass through the equilibrium position heading downward for the second time?
 (e) At what time does the mass attain its extreme displacement on either side of the equilibrium position?
 (f) What is the position of the mass at $t = 3$ s?
 (g) What is the instantaneous velocity at $t = 3$ s?
 (h) What is the acceleration at $t = 3$ s?
 (i) What is the instantaneous velocity at the times when the mass passes through the equilibrium position?
 (j) At what times is the mass 5 inches below the equilibrium position?
 (k) At what times is the mass 5 inches below the equilibrium position heading in the upward direction?

12. A mass of 1 slug is suspended from a spring whose spring constant is 9 lb/ft. The mass is initially released from a point 1 foot above the equilibrium position with an upward velocity of $\sqrt{3}$ ft/s. Find the times for which the mass is heading downward at a velocity of 3 ft/s.

13. A mass weighing 20 pounds stretches one spring 6 inches and another spring 2 inches. The springs are attached to a common rigid support and then to a metal plate of negligible mass as shown in Figure 3.8.5; the mass is attached to the center of the plate in the double-spring arrangement. Determine the effective spring constant of this system. Find the equation of motion if the mass is initially released from the equilibrium position with a downward velocity of 2 ft/s.

14. A certain weight stretches one spring $\frac{1}{3}$ foot and another spring $\frac{1}{2}$ foot. The two springs are attached in parallel to a common rigid support in a manner indicated in Problem 13 and Figure 3.8.5. The first weight is set aside, and a mass weighing 8 pounds is attached to the double-spring arrangement and the system is set in motion. If the period of motion is $\pi/15$ second, determine how much the first mass weighs.

15. Solve Problem 13 again, but this time assume that the springs are in series as shown in Figure 3.8.6.

16. Solve Problem 14 again, but this time assume that the springs are in series as shown in Figure 3.8.6.

17. Find the effective spring constant of the parallel-spring system shown in Figure 3.8.5 when both springs have the spring constant k. Give a physical interpretation of this result.

18. Find the effective spring constant of the series-spring system shown in Figure 3.8.6 when both springs have the spring constant k. Give a physical interpretation of this result.

19. A model of a spring/mass system is $4x'' + e^{-0.1t}x = 0$. By inspection of the differential equation only, discuss the behavior of the system over a long period of time.

20. A model of a spring/mass system is $4x'' + tx = 0$. By inspection of the differential equation only, discuss the behavior of the system over a long period of time.

3.8.2 Spring/Mass Systems: Free Damped Motion

In Problems 21–24, the given figure represents the graph of an equation of motion for a damped spring/mass system. Use the graph to determine

(a) whether the initial displacement is above or below the equilibrium position, and

(b) whether the mass is initially released from rest, heading downward, or heading upward.

21.

FIGURE 3.8.19 Graph for Problem 21

22.

FIGURE 3.8.20 Graph for Problem 22

23.

FIGURE 3.8.21 Graph for Problem 23

24.

FIGURE 3.8.22 Graph for Problem 24

25. A mass weighing 4 pounds is attached to a spring whose constant is 2 lb/ft. The medium offers a damping force that

is numerically equal to the instantaneous velocity. The mass is initially released from a point 1 foot above the equilibrium position with a downward velocity of 8 ft/s. Determine the time at which the mass passes through the equilibrium position. Find the time at which the mass attains its extreme displacement from the equilibrium position. What is the position of the mass at this instant?

26. A 4-foot spring measures 8 feet long after a mass weighing 8 pounds is attached to it. The medium through which the mass moves offers a damping force numerically equal to $\sqrt{2}$ times the instantaneous velocity. Find the equation of motion if the mass is initially released from the equilibrium position with a downward velocity of 5 ft/s. Find the time at which the mass attains its extreme displacement from the equilibrium position. What is the position of the mass at this instant?

27. A 1-kilogram mass is attached to a spring whose constant is 16 N/m, and the entire system is then submerged in a liquid that imparts a damping force numerically equal to 10 times the instantaneous velocity. Determine the equations of motion if

(a) the mass is initially released from rest from a point 1 meter below the equilibrium position, and then

(b) the mass is initially released from a point 1 meter below the equilibrium position with an upward velocity of 12 m/s.

28. In parts (a) and (b) of Problem 27, determine whether the mass passes through the equilibrium position. In each case find the time at which the mass attains its extreme displacement from the equilibrium position. What is the position of the mass at this instant?

29. A force of 2 pounds stretches a spring 1 foot. A mass weighing 3.2 pounds is attached to the spring, and the system is then immersed in a medium that offers a damping force numerically equal to 0.4 times the instantaneous velocity.

(a) Find the equation of motion if the mass is initially released from rest from a point 1 foot above the equilibrium position.

(b) Express the equation of motion in the form given in (23).

(c) Find the first time at which the mass passes through the equilibrium position heading upward.

30. After a mass weighing 10 pounds is attached to a 5-foot spring, the spring measures 7 feet. This mass is removed and replaced with another mass that weighs 8 pounds. The entire system is placed in a medium that offers a damping force numerically equal to the instantaneous velocity.

(a) Find the equation of motion if the mass is initially released from a point $\frac{1}{2}$ foot below the equilibrium position with a downward velocity of 1 ft/s.

(b) Express the equation of motion in the form given in (23).

(c) Find the times at which the mass passes through the equilibrium position heading downward.

(d) Graph the equation of motion.

31. A mass weighing 10 pounds stretches a spring 2 feet. The mass is attached to a dashpot damping device that offers a damping force numerically equal to β ($\beta > 0$) times the instantaneous velocity. Determine the values of the damping constant β so that the subsequent motion is (a) overdamped, (b) critically damped, and (c) underdamped.

32. A mass weighing 24 pounds stretches a spring 4 feet. The subsequent motion takes place in a medium that offers a damping force numerically equal to β ($\beta > 0$) times the instantaneous velocity. If the mass is initially released from the equilibrium position with an upward velocity of 2 ft/s, show that when $\beta > 3\sqrt{2}$ the equation of motion is

$$ x(t) = \frac{-3}{\sqrt{\beta^2 - 18}} e^{-2\beta t/3} \sinh \frac{2}{3} \sqrt{\beta^2 - 18}\, t. $$

3.8.3 Spring/Mass Systems: Driven Motion

33. A mass weighing 16 pounds stretches a spring $\frac{8}{3}$ feet. The mass is initially released from rest from a point 2 feet below the equilibrium position, and the subsequent motion takes place in a medium that offers a damping force numerically equal to one-half the instantaneous velocity. Find the equation of motion if the mass is driven by an external force equal to $f(t) = 10 \cos 3t$.

34. A mass of 1 slug is attached to a spring whose constant is 5 lb/ft. Initially the mass is released 1 foot below the equilibrium position with a downward velocity of 5 ft/s, and the subsequent motion takes place in a medium that offers a damping force numerically equal to two times the instantaneous velocity.
 (a) Find the equation of motion if the mass is driven by an external force equal to $f(t) = 12 \cos 2t + 3 \sin 2t$.
 (b) Graph the transient and steady-state solutions on the same coordinate axes.
 (c) Graph the equation of motion.

35. A mass of 1 slug, when attached to a spring, stretches it 2 feet and then comes to rest in the equilibrium position. Starting at $t = 0$, an external force equal to $f(t) = 8 \sin 4t$ is applied to the system. Find the equation of motion if the surrounding medium offers a damping force numerically equal to eight times the instantaneous velocity.

36. In Problem 35 determine the equation of motion if the external force is $f(t) = e^{-t} \sin 4t$. Analyze the displacements for $t \to \infty$.

37. When a mass of 2 kilograms is attached to a spring whose constant is 32 N/m, it comes to rest in the equilibrium position. Starting at $t = 0$, a force equal to $f(t) = 68e^{-2t} \cos 4t$ is applied to the system. Find the equation of motion in the absence of damping.

38. In Problem 37 write the equation of motion in the form $x(t) = A \sin(\omega t + \phi) + Be^{-2t} \sin(4t + \theta)$. What is the amplitude of vibrations after a very long time?

39. A mass m is attached to the end of a spring whose constant is k. After the mass reaches equilibrium, its support begins to oscillate vertically about a horizontal line L according to a formula $h(t)$. The value of h represents the distance in feet measured from L. See FIGURE 3.8.23.
 (a) Determine the differential equation of motion if the entire system moves through a medium offering a damping force numerically equal to $\beta(dx/dt)$.
 (b) Solve the differential equation in part (a) if the spring is stretched 4 feet by a weight of 16 pounds and $\beta = 2$, $h(t) = 5 \cos t$, $x(0) = x'(0) = 0$.

FIGURE 3.8.23 Oscillating support in Problem 39

40. A mass of 100 grams is attached to a spring whose constant is 1600 dynes/cm. After the mass reaches equilibrium, its support oscillates according to the formula $h(t) = \sin 8t$, where h represents displacement from its original position. See Problem 39 and Figure 3.8.23.
 (a) In the absence of damping, determine the equation of motion if the mass starts from rest from the equilibrium position.
 (b) At what times does the mass pass through the equilibrium position?
 (c) At what times does the mass attain its extreme displacements?
 (d) What are the maximum and minimum displacements?
 (e) Graph the equation of motion.

In Problems 41 and 42, describe in words a spring/mass system whose model is the given differential equation. Assume that all constants are positive.

41. $m \dfrac{d^2 x}{dt^2} + \beta \dfrac{dx}{dt} + ke^{-\alpha t} x = f(t)$

42. $m \dfrac{d^2 x}{dt^2} + \beta \dfrac{dx}{dt} + (k_1 + (k - k_1)e^{-\alpha t}) x = A \cos(2\pi t)$, $0 < k_1 < k$

In Problems 43 and 44, solve the given initial-value problem.

43. $\dfrac{d^2 x}{dt^2} + 4x = -5 \sin 2t + 3 \cos 2t$, $x(0) = -1$, $x'(0) = 1$

44. $\dfrac{d^2 x}{dt^2} + 9x = 5 \sin 3t$, $x(0) = 2$, $x'(0) = 0$

45. (a) Show that the solution of the initial-value problem
 $$ \frac{d^2 x}{dt^2} + \omega^2 x = F_0 \cos \gamma t, \quad x(0) = 0, \quad x'(0) = 0 $$
 is
 $$ x(t) = \frac{F_0}{\omega^2 - \gamma^2} (\cos \gamma t - \cos \omega t). $$
 (b) Evaluate $\displaystyle \lim_{\gamma \to \omega} \frac{F_0}{\omega^2 - \gamma^2} (\cos \gamma t - \cos \omega t)$.

46. Compare the result obtained in part (b) of Problem 45 with the solution obtained using variation of parameters when the external force is $F_0 \cos \omega t$.

47. (a) Show that $x(t)$ given in part (a) of Problem 45 can be written in the form

$$x(t) = \frac{-2F_0}{\omega^2 - \gamma^2} \sin \frac{1}{2}(\gamma - \omega)t \sin \frac{1}{2}(\gamma + \omega)t.$$

(b) If we define $\epsilon = \frac{1}{2}(\gamma - \omega)$, show that when ϵ is small, an approximate solution is

$$x(t) = \frac{F_0}{2\epsilon\gamma} \sin \epsilon t \sin \gamma t.$$

When ϵ is small the frequency $\gamma/2\pi$ of the impressed force is close to the frequency $\omega/2\pi$ of free vibrations. When this occurs, the motion is as indicated in **FIGURE 3.8.24**. Oscillations of this kind are called **beats** and are due to the fact that the frequency of $\sin \epsilon t$ is quite small in comparison to the frequency of $\sin \gamma t$. The red dashed curves, or envelope of the graph of $x(t)$, are obtained from the graphs of $\pm(F_0/2\epsilon\gamma) \sin \epsilon t$. Use a graphing utility with various values of F_0, ϵ, and γ to verify the graph in Figure 3.8.24.

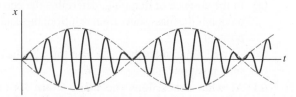

FIGURE 3.8.24 Beats phenomenon in Problem 47

Computer Lab Assignments

48. Can there be beats when a damping force is added to the model in part (a) of Problem 45? Defend your position with graphs obtained either from the explicit solution of the problem

$$\frac{d^2x}{dt^2} + 2\lambda \frac{dx}{dt} + \omega^2 x = F_0 \cos \gamma t, \quad x(0) = 0, \quad x'(0) = 0$$

or from solution curves obtained using a numerical solver.

49. (a) Show that the general solution of

$$\frac{d^2x}{dt^2} + 2\lambda \frac{dx}{dt} + \omega^2 x = F_0 \sin \gamma t$$

is

$$x(t) = Ae^{-\lambda t} \sin\left(\sqrt{\omega^2 - \lambda^2}t + \phi\right)$$

$$+ \frac{F_0}{\sqrt{(\omega^2 - \gamma^2)^2 + 4\lambda^2\gamma^2}} \sin(\gamma t + \theta),$$

where $A = \sqrt{c_1^2 + c_2^2}$ and the phase angles ϕ and θ are, respectively, defined by $\sin \phi = c_1/A$, $\cos \phi = c_2/A$ and

$$\sin \theta = \frac{-2\lambda\gamma}{\sqrt{(\omega^2 - \gamma^2)^2 + 4\lambda^2\gamma^2}},$$

$$\cos \theta = \frac{\omega^2 - \gamma^2}{\sqrt{(\omega^2 - \gamma^2)^2 + 4\lambda^2\gamma^2}}.$$

(b) The solution in part (a) has the form $x(t) = x_c(t) + x_p(t)$. Inspection shows that $x_c(t)$ is transient, and hence for large values of time, the solution is approximated by $x_p(t) = g(\gamma) \sin(\gamma t + \theta)$, where

$$g(\gamma) = \frac{F_0}{\sqrt{(\omega^2 - \gamma^2)^2 + 4\lambda^2\gamma^2}}.$$

Although the amplitude $g(\gamma)$ of $x_p(t)$ is bounded as $t \to \infty$, show that the maximum oscillations will occur at the value $\gamma_1 = \sqrt{\omega^2 - 2\lambda^2}$. What is the maximum value of g? The number $\sqrt{\omega^2 - 2\lambda^2}/2\pi$ is said to be the **resonance frequency** of the system.

(c) When $F_0 = 2$, $m = 1$, and $k = 4$, g becomes

$$g(\gamma) = \frac{2}{\sqrt{(4 - \gamma^2)^2 + \beta^2\gamma^2}}.$$

Construct a table of the values of γ_1 and $g(\gamma_1)$ corresponding to the damping coefficients $\beta = 2$, $\beta = 1$, $\beta = \frac{3}{4}$, $\beta = \frac{1}{2}$, and $\beta = \frac{1}{4}$. Use a graphing utility to obtain the graphs of g corresponding to these damping coefficients. Use the same coordinate axes. This family of graphs is called the **resonance curve** or **frequency response curve** of the system. What is γ_1 approaching as $\beta \to 0$? What is happening to the resonance curve as $\beta \to 0$?

50. Consider a driven undamped spring/mass system described by the initial-value problem

$$\frac{d^2x}{dt^2} + \omega^2 x = F_0 \sin^n \gamma t, \quad x(0) = 0, \quad x'(0) = 0.$$

(a) For $n = 2$, discuss why there is a single frequency $\gamma_1/2\pi$ at which the system is in pure resonance.

(b) For $n = 3$, discuss why there are two frequencies $\gamma_1/2\pi$ and $\gamma_2/2\pi$ at which the system is in pure resonance.

(c) Suppose $\omega = 1$ and $F_0 = 1$. Use a numerical solver to obtain the graph of the solution of the initial-value problem for $n = 2$ and $\gamma = \gamma_1$ in part (a). Obtain the graph of the solution of the initial-value problem for $n = 3$ corresponding, in turn, to $\gamma = \gamma_1$ and $\gamma = \gamma_2$ in part (b).

3.8.4 Series Circuit Analogue

51. Find the charge on the capacitor in an LRC-series circuit at $t = 0.01$ s when $L = 0.05$ h, $R = 2\,\Omega$, $C = 0.01$ F, $E(t) = 0$ V, $q(0) = 5$ C, and $i(0) = 0$ A. Determine the first time at which the charge on the capacitor is equal to zero.

52. Find the charge on the capacitor in an LRC-series circuit when $L = \frac{1}{4}$ h, $R = 20\,\Omega$, $C = \frac{1}{300}$ F, $E(t) = 0$ V, $q(0) = 4$ C, and $i(0) = 0$ A. Is the charge on the capacitor ever equal to zero?

In Problems 53 and 54, find the charge on the capacitor and the current in the given *LRC*-series circuit. Find the maximum charge on the capacitor.

53. $L = \frac{5}{3}$ h, $R = 10\ \Omega$, $C = \frac{1}{30}$ F, $E(t) = 300$ V, $q(0) = 0$ C, $i(0) = 0$ A

54. $L = 1$ h, $R = 100\ \Omega$, $C = 0.0004$ F, $E(t) = 30$ V, $q(0) = 0$ C, $i(0) = 2$ A

55. Find the steady-state charge and the steady-state current in an *LRC*-series circuit when $L = 1$ h, $R = 2\ \Omega$, $C = 0.25$ F, and $E(t) = 50 \cos t$ V.

56. Show that the amplitude of the steady-state current in the *LRC*-series circuit in Example 10 is given by E_0/Z, where Z is the impedance of the circuit.

57. Use Problem 56 to show that the steady-state current in an *LRC*-series circuit when $L = \frac{1}{2}$ h, $R = 20\ \Omega$, $C = 0.001$ F, and $E(t) = 100 \sin 60t$ V is given by $i_p(t) = 4.160 \sin(60t - 0.588)$.

58. Find the steady-state current in an *LRC*-series circuit when $L = \frac{1}{2}$ h, $R = 20\ \Omega$, $C = 0.001$ F, and $E(t) = 100 \sin 60t + 200 \cos 40t$ V.

59. Find the charge on the capacitor in an *LRC*-series circuit when $L = \frac{1}{2}$ h, $R = 10\ \Omega$, $C = 0.01$ F, $E(t) = 150$ V, $q(0) = 1$ C, and $i(0) = 0$ A. What is the charge on the capacitor after a long time?

60. Show that if L, R, C, and E_0 are constant, then the amplitude of the steady-state current in Example 10 is a maximum when $\gamma = 1/\sqrt{LC}$. What is the maximum amplitude?

61. Show that if L, R, E_0, and γ are constant, then the amplitude of the steady-state current in Example 10 is a maximum when the capacitance is $C = 1/L\gamma^2$.

62. Find the charge on the capacitor and the current in an *LC*-circuit when $L = 0.1$ h, $C = 0.1$ F, $E(t) = 100 \sin \gamma t$ V, $q(0) = 0$ C, and $i(0) = 0$ A.

63. Find the charge on the capacitor and the current in an *LC*-circuit when $E(t) = E_0 \cos \gamma t$ V, $q(0) = q_0$ C, and $i(0) = i_0$ A.

64. In Problem 63, find the current when the circuit is in resonance.

3.9 Linear Models: Boundary-Value Problems

INTRODUCTION The preceding section was devoted to dynamic physical systems each described by a mathematical model consisting of a linear second-order differential equation accompanied by prescribed *initial conditions*—that is, side conditions that are specified on the unknown function and its first derivative at a single point. But often the mathematical description of a steady-state phenomenon or a static physical system demands that we solve a linear differential equation subject to *boundary conditions*—that is, conditions specified on the unknown function, or on one of its derivatives, or even on a linear combination of the unknown function and one of its derivatives, at two different points. By and large, the number of specified boundary conditions matches the order of the linear DE. We begin this section with an application of a relatively simple linear fourth-order differential equation associated with four boundary conditions.

■ **Deflection of a Beam** Many structures are constructed using girders, or beams, and these beams deflect or distort under their own weight or under the influence of some external force. As we shall now see, this deflection $y(x)$ is governed by a relatively simple linear fourth-order differential equation.

To begin, let us assume that a beam of length L is homogeneous and has uniform cross sections along its length. In the absence of any load on the beam (including its weight), a curve joining the centroids of all its cross sections is a straight line called the **axis of symmetry**. See **FIGURE 3.9.1(a)**. If a load is applied to the beam in a vertical plane containing the axis of symmetry, the beam, as shown in Figure 3.9.1(b), undergoes a distortion, and the curve connecting the centroids of all cross sections is called the **deflection curve** or **elastic curve**. The deflection curve approximates the shape of the beam. Now suppose that the x-axis coincides with the axis of symmetry and that the deflection $y(x)$, measured from this axis, is positive if downward. In the theory of elasticity it is shown that the bending moment $M(x)$ at a point x along the beam is related to the load per unit length $w(x)$ by the equation

$$\frac{d^2M}{dx^2} = w(x). \tag{1}$$

In addition, the bending moment $M(x)$ is proportional to the curvature κ of the elastic curve

$$M(x) = EI\kappa, \tag{2}$$

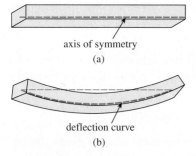

axis of symmetry
(a)

deflection curve
(b)

FIGURE 3.9.1 Deflection of a homogeneous beam

where E and I are constants; E is Young's modulus of elasticity of the material of the beam, and I is the moment of inertia of a cross section of the beam (about an axis known as the neutral axis). The product EI is called the **flexural rigidity** of the beam.

Now, from calculus, curvature is given by $\kappa = y''/[1 + (y')^2]^{3/2}$. When the deflection $y(x)$ is small, the slope $y' \approx 0$ and so $[1 + (y')^2]^{3/2} \approx 1$. If we let $\kappa \approx y''$, equation (2) becomes $M = EI\,y''$. The second derivative of this last expression is

$$\frac{d^2 M}{dx^2} = EI\frac{d^2}{dx^2}y'' = EI\frac{d^4 y}{dx^4}. \tag{3}$$

Using the given result in (1) to replace $d^2 M/dx^2$ in (3), we see that the deflection $y(x)$ satisfies the fourth-order differential equation

$$EI\frac{d^4 y}{dx^4} = w(x). \tag{4}$$

Boundary conditions associated with equation (4) depend on how the ends of the beam are supported. A cantilever beam is **embedded** or **clamped** at one end and **free** at the other. A diving board, an outstretched arm, an airplane wing, and a balcony are common examples of such beams, but even trees, flagpoles, skyscrapers, and the George Washington monument can act as cantilever beams, because they are embedded at one end and are subject to the bending force of the wind. For a cantilever beam, the deflection $y(x)$ must satisfy the following two conditions at the embedded end $x = 0$:

- $y(0) = 0$ since there is no deflection, and
- $y'(0) = 0$ since the deflection curve is tangent to the x-axis (in other words, the slope of the deflection curve is zero at this point).

At $x = L$ the free-end conditions are

- $y''(L) = 0$ since the bending moment is zero, and
- $y'''(L) = 0$ since the shear force is zero.

The function $F(x) = dM/dx = EI\,d^3 y/dx^3$ is called the shear force. If an end of a beam is **simply supported** or **hinged** (also called **pin supported**, and **fulcrum supported**), then we must have $y = 0$ and $y'' = 0$ at that end. Table 3.9.1 summarizes the boundary conditions that are associated with (4). See **FIGURE 3.9.2**.

$x = 0$ $x = L$

(a) Embedded at both ends

$x = 0$ $x = L$

(b) Cantilever beam: embedded at the left end, free at the right end

$x = 0$ $x = L$

(c) Simply supported at both ends

FIGURE 3.9.2 Beams with various end conditions

TABLE 3.9.1

Ends of the Beam	Boundary Conditions
Embedded	$y = 0, y' = 0$
Free	$y'' = 0, y''' = 0$
Simply supported or hinged	$y = 0, y'' = 0$

EXAMPLE 1 **An Embedded Beam**

A beam of length L is embedded at both ends. Find the deflection of the beam if a constant load w_0 is uniformly distributed along its length—that is, $w(x) = w_0, 0 < x < L$.

SOLUTION From (4), we see that the deflection $y(x)$ satisfies

$$EI\frac{d^4 y}{dx^4} = w_0.$$

Because the beam is embedded at both its left end ($x = 0$) and right end ($x = L$), there is no vertical deflection and the line of deflection is horizontal at these points. Thus the boundary conditions are

$$y(0) = 0, \quad y'(0) = 0, \quad y(L) = 0, \quad y'(L) = 0.$$

We can solve the nonhomogeneous differential equation in the usual manner (find y_c by observing that $m = 0$ is a root of multiplicity four of the auxiliary equation $m^4 = 0$, and then find a particular solution y_p by undetermined coefficients), or we can simply integrate the equation $d^4 y/dx^4 = w_0/EI$ four times in succession. Either way, we find the general solution of the equation $y = y_c + y_p$ to be

$$y(x) = c_1 + c_2 x + c_3 x^2 + c_4 x^3 + \frac{w_0}{24EI}x^4.$$

Now the conditions $y(0) = 0$ and $y'(0) = 0$ give, in turn, $c_1 = 0$ and $c_2 = 0$, whereas the remaining conditions $y(L) = 0$ and $y'(L) = 0$ applied to $y(x) = c_3 x^2 + c_4 x^3 + \dfrac{w_0}{24EI} x^4$ yield the simultaneous equations

$$c_3 L^2 + c_4 L^3 + \frac{w_0}{24EI} L^4 = 0$$

$$2c_3 L + 3c_4 L^2 + \frac{w_0}{6EI} L^3 = 0.$$

Solving this system gives $c_3 = w_0 L^2 / 24EI$ and $c_4 = -w_0 L / 12EI$. Thus the deflection is

$$y(x) = \frac{w_0 L^2}{24EI} x^2 - \frac{w_0 L}{12EI} x^3 + \frac{w_0}{24EI} x^4$$

or $y(x) = \dfrac{w_0}{24EI} x^2 (x - L)^2$. By choosing $w_0 = 24EI$, and $L = 1$, we obtain the graph of the deflection curve in **FIGURE 3.9.3**. ≡

FIGURE 3.9.3 Deflection curve for Example 1

The discussion of the beam notwithstanding, a physical system that is described by a two-point boundary-value problem usually involves a second-order differential equation. Hence, for the remainder of the discussion in this section we are concerned with boundary-value problems of the type

> **Solve:** $\quad a_2(x)\dfrac{d^2 y}{dx^2} + a_1(x)\dfrac{dy}{dx} + a_0(x)y = g(x), \quad a < x < b \qquad$ (5)

> **Subject to:** $\quad \begin{aligned} A_1 y(a) + B_1 y'(a) &= C_1 \\ A_2 y(b) + B_2 y'(b) &= C_2. \end{aligned} \qquad$ (6)

In (5) we assume that the coefficients $a_0(x)$, $a_1(x)$, $a_2(x)$, and $g(x)$ are continuous on the interval $[a, b]$ and that $a_2(x) \neq 0$ for all x in the interval. In (6) we assume that A_1 and B_1 are not both zero and A_2 and B_2 are not both zero. When $g(x) = 0$ for all x in $[a, b]$ and C_1 *and* C_2 are 0, we say that the boundary-value problem is **homogeneous**. Otherwise, we say that the boundary-value problem is **nonhomogeneous**. For example, the BVP $y'' + y = 0$, $y(0) = 0$, $y(\pi) = 0$ is homogeneous, whereas the BVP $y'' + y = 1$, $y(0) = 0$, $y(2\pi) = 0$ is nonhomogeneous.

■ **Eigenvalues and Eigenfunctions** In applications involving homogeneous boundary-value problems, one or more of the coefficients in the differential equation (5) may depend on a constant parameter λ. As a consequence the solutions $y_1(x)$ and $y_2(x)$ of the homogeneous DE (5) also depend on λ. We often wish to determine those values of the parameter for which the boundary-value problem has **nontrivial solutions**. The next example illustrates this idea.

> **EXAMPLE 2** **Nontrivial Solutions of a BVP**

Solve the homogeneous boundary-value problem

$$y'' + \lambda y = 0, \quad y(0) = 0, \quad y(L) = 0.$$

SOLUTION We consider three cases: $\lambda = 0$, $\lambda < 0$, and $\lambda > 0$.

Case I: For $\lambda = 0$, the solution of the DE $y'' = 0$ is $y = c_1 x + c_2$. Applying the boundary conditions $y(0) = 0$ and $y(L) = 0$ to this solution yields, in turn, $c_2 = 0$ and $c_1 = 0$. Hence for $\lambda = 0$, the only solution of the boundary-value problem is the trivial solution $y = 0$.

Case II: For $\lambda < 0$, it is convenient to write $\lambda = -\alpha^2$, where $\alpha > 0$. With this new notation the auxiliary equation is $m^2 - \alpha^2 = 0$ and has roots $m_1 = \alpha$ and $m_2 = -\alpha$. Because the interval on which we are working is finite, we choose to write the general solution of $y'' - \alpha^2 y = 0$ in the hyperbolic form $y = c_1 \cosh \alpha x + c_2 \sinh \alpha x$. From $y(0) = 0$ we see

$$y(0) = c_1 \cosh 0 + c_2 \sinh 0 = c_1 \cdot 1 + c_2 \cdot 0 = c_1$$

implies $c_1 = 0$. Hence $y = c_2 \sinh \alpha x$. The second boundary condition $y(L) = 0$ then requires $c_2 \sinh \alpha L = 0$. When $\alpha \neq 0$, $\sinh \alpha L \neq 0$, and so we are forced to choose $c_2 = 0$. Once again the only solution of the BVP is the trivial solution $y = 0$.

Case III: For $\lambda > 0$ we write $\lambda = \alpha^2$, where $\alpha > 0$. The auxiliary equation $m^2 + \alpha^2 = 0$ now has complex roots $m_1 = i\alpha$ and $m_2 = -i\alpha$, and so the general solution of the DE $y'' + \alpha^2 y = 0$ is $y = c_1 \cos \alpha x + c_2 \sin \alpha x$. As before, $y(0) = 0$ yields $c_1 = 0$ and so $y = c_2 \sin \alpha x$. Then $y(L) = 0$ implies

$$c_2 \sin \alpha L = 0.$$

If $c_2 = 0$, then necessarily $y = 0$. But this time we can require $c_2 \neq 0$ since $\sin \alpha L = 0$ is satisfied whenever αL is an integer multiple of π:

$$\alpha L = n\pi \quad \text{or} \quad \alpha = \frac{n\pi}{L} \quad \text{or} \quad \lambda_n = \alpha_n^2 = \left(\frac{n\pi}{L}\right)^2, \quad n = 1, 2, 3, \ldots.$$

Therefore for any real nonzero c_2, $y_n = c_2 \sin(n\pi x/L)$ is a solution of the problem for each positive integer n. Since the differential equation is homogeneous, any constant multiple of a solution is also a solution. Thus we may, if desired, simply take $c_2 = 1$. In other words, for each number in the sequence

$$\lambda_1 = \frac{\pi^2}{L^2}, \quad \lambda_2 = \frac{4\pi^2}{L^2}, \quad \lambda_3 = \frac{9\pi^2}{L^2}, \ldots,$$

the *corresponding* function in the sequence

$$y_1 = \sin\frac{\pi}{L}x, \quad y_2 = \sin\frac{2\pi}{L}x, \quad y_3 = \sin\frac{3\pi}{L}x, \ldots,$$

is a nontrivial solution of the original problem. ≡

FIGURE 3.9.4 Graphs of eigenfunctions $y_n = \sin(n\pi x/L)$, for $n = 1, 2, 3, 4, 5$

The numbers $\lambda_n = n^2\pi^2/L^2$, $n = 1, 2, 3, \ldots$ for which the boundary-value problem in Example 2 possesses nontrivial solutions are known as **eigenvalues**. The nontrivial solutions that depend on these values of λ_n, $y_n = c_2 \sin(n\pi x/L)$ or simply $y_n = \sin(n\pi x/L)$ are called **eigenfunctions**. The graphs of the eigenfunctions for $n = 1, 2, 3, 4, 5$ are shown in **FIGURE 3.9.4**. Note that each of the five graphs passes through the two points $(0, 0)$ and $(0, L)$.

EXAMPLE 3　　**Example 2 Revisited**

If we choose $L = \pi$ in Example 2, then the eigenvalues of the problem

$$y'' + \lambda y = 0, \quad y(0) = 0, \quad y(\pi) = 0$$

are $\lambda_n = n^2$, $n = 1, 2, 3, \ldots$. It follows then that the boundary-value problem

$$y'' + 10y = 0, \quad y(0) = 0, \quad y(\pi) = 0$$

possesses only the trivial solution $y = 0$ because 10 is *not* an eigenvalue; that is, 10 is not the square of a positive integer. ≡

■ Buckling of a Thin Vertical Column In the eighteenth century Leonhard Euler was one of the first mathematicians to study an eigenvalue problem in analyzing how a thin elastic column buckles under a compressive axial force.

Consider a long slender vertical column of uniform cross section and length L. Let $y(x)$ denote the deflection of the column when a constant vertical compressive force, or load, P is applied to its top, as shown in **FIGURE 3.9.5**. By comparing bending moments at any point along the column we obtain

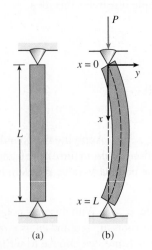

FIGURE 3.9.5 Elastic column buckling under a compressive force

$$EI\frac{d^2y}{dx^2} = -Py \quad \text{or} \quad EI\frac{d^2y}{dx^2} + Py = 0, \tag{7}$$

where E is Young's modulus of elasticity and I is the moment of inertia of a cross section about a vertical line through its centroid.

EXAMPLE 4 **The Euler Load**

Find the deflection of a thin vertical homogeneous column of length L subjected to a constant axial load P if the column is simply supported or hinged at both ends.

SOLUTION The boundary-value problem to be solved is

$$EI\frac{d^2y}{dx^2} + Py = 0, \quad y(0) = 0, \quad y(L) = 0.$$

First note that $y = 0$ is a perfectly good solution of this problem. This solution has a simple intuitive interpretation: If the load P is not great enough, there is no deflection. The question then is this: For what values of P will the column bend? In mathematical terms: For what values of P does the given boundary-value problem possess nontrivial solutions?

By writing $\lambda = P/EI$ we see that

$$y'' + \lambda y = 0, \quad y(0) = 0, \quad y(L) = 0$$

is identical to the problem in Example 2. From Case III of that discussion we see that the deflection curves are $y_n(x) = c_2 \sin(n\pi x/L)$, corresponding to the eigenvalues $\lambda_n = P_n/EI = n^2\pi^2/L^2$, $n = 1, 2, 3, \ldots$. Physically this means that the column will buckle or deflect only when the compressive force is one of the values $P_n = n^2\pi^2 EI/L^2$, $n = 1, 2, 3, \ldots$. These different forces are called **critical loads**. The deflection curve corresponding to the smallest critical load $P_1 = \pi^2 EI/L^2$, called the **Euler load**, is $y_1(x) = c_2 \sin(\pi x/L)$ and is known as the **first buckling mode**.

≡

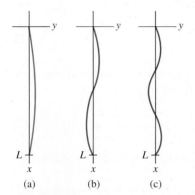

The deflection curves in Example 4 corresponding to $n = 1$, $n = 2$, and $n = 3$ are shown in **FIGURE 3.9.6**. Note that if the original column has some sort of physical restraint put on it at $x = L/2$, then the smallest critical load will be $P_2 = 4\pi^2 EI/L^2$ and the deflection curve will be as shown in Figure 3.9.6(b). If restraints are put on the column at $x = L/3$ and at $x = 2L/3$, then the column will not buckle until the critical load $P_3 = 9\pi^2 EI/L^2$ is applied and the deflection curve will be as shown in Figure 3.9.6(c). See Problem 25 in Exercises 3.9.

FIGURE 3.9.6 Deflection curves for compressive forces P_1, P_2, P_3

■ **Rotating String** The simple linear second-order differential equation

$$y'' + \lambda y = 0 \tag{8}$$

occurs again and again as a mathematical model. In Section 3.8 we saw (8) in the forms $d^2x/dt^2 + (k/m)x = 0$ and $d^2q/dt^2 + (1/LC)q = 0$ as models for, respectively, the simple harmonic motion of a spring/mass system and the simple harmonic response of a series circuit. It is apparent when the model for the deflection of a thin column in (7) is written as $d^2y/dx^2 + (P/EI)y = 0$ that it is the same as (8). We encounter the basic equation (8) one more time in this section: as a model that defines the deflection curve or the shape $y(x)$ assumed by a rotating string. The physical situation is analogous to when two persons hold a jump rope and twirl it in a synchronous manner. See **FIGURE 3.9.7** parts (a) and (b).

Suppose a string of length L with constant linear density ρ (mass per unit length) is stretched along the x-axis and fixed at $x = 0$ and $x = L$. Suppose the string is then rotated about that axis at a constant angular speed ω. Consider a portion of the string on the interval $[x, x + \Delta x]$, where Δx is small. If the magnitude T of the tension $\mathbf{T}$, acting tangential to the string, is constant along the string, then the desired differential equation can be obtained by equating two different formulations of the net force acting on the string on the interval $[x, x + \Delta x]$. First, we see from Figure 3.9.7(c) that the net vertical force is

$$F = T \sin\theta_2 - T \sin\theta_1. \tag{9}$$

(a)

(b)

(c)

FIGURE 3.9.7 Rotating rope and forces acting on it

When angles θ_1 and θ_2 (measured in radians) are small, we have $\sin\theta_2 \approx \tan\theta_2$ and $\sin\theta_1 \approx \tan\theta_1$. Moreover, since $\tan\theta_2$ and $\tan\theta_1$ are, in turn, slopes of the lines containing the vectors $\mathbf{T}_2$ and $\mathbf{T}_1$, we can also write

$$\tan\theta_2 = y'(x + \Delta x) \quad \text{and} \quad \tan\theta_1 = y'(x).$$

Thus (9) becomes

$$F \approx T[\,y'(x + \Delta x) - y'(x)].$$ (10)

Second, we can obtain a different form of this same net force using Newton's second law, $F = ma$. Here the mass of string on the interval is $m = \rho \Delta x$; the centripetal acceleration of a body rotating with angular speed ω in a circle of radius r is $a = r\omega^2$. With Δx small we take $r = y$. Thus the net vertical force is also approximated by

$$F \approx -(\rho \Delta x)y\omega^2,$$ (11)

where the minus sign comes from the fact that the acceleration points in the direction opposite to the positive y-direction. Now by equating (10) and (11) we have

difference quotient
$\downarrow$

$$T[y'(x + \Delta x) - y'(x)] = -(\rho \Delta x)y\omega^2 \quad \text{or} \quad T\,\frac{y'(x + \Delta x) - y'(x)}{\Delta x} + \rho\omega^2 y = 0.$$ (12)

For Δx close to zero the difference quotient in (12) is approximately the second derivative d^2y/dx^2. Finally we arrive at the model

$$T\,\frac{d^2y}{dx^2} + \rho\omega^2 y = 0.$$ (13)

Since the string is anchored at its ends $x = 0$ and $x = L$, we expect that the solution $y(x)$ of equation (13) should also satisfy the boundary conditions $y(0) = 0$ and $y(L) = 0$.

REMARKS

(*i*) We will pursue the subject of eigenvalues and eigenfunctions for linear second-order differential equations in greater detail in Section 12.5.

(*ii*) Eigenvalues are not always easily found as they were in Example 2; you may have to approximate roots of equations such as $\tan x = -x$ or $\cos x \cosh x = 1$. See Problems 32 and 38–42 in Exercises 3.9.

(*iii*) Boundary conditions can lead to a homogeneous algebraic system of linear equations where the unknowns are the coefficients c_i in the general solution of the DE. Such a system is always consistent, but in order to possess a nontrivial solution (in the case when the number of equations equals the number of unknowns) we must have the determinant of the coefficients equal to zero. See Problems 21 and 22 in Exercises 3.9.

3.9 Exercises Answers to selected odd-numbered problems begin on page ANS-7.

Deflection of a Beam

In Problems 1–5, solve equation (4) subject to the appropriate boundary conditions. The beam is of length L, and w_0 is a constant.

1. (a) The beam is embedded at its left end and free at its right end, and $w(x) = w_0$, $0 < x < L$.

 (b) Use a graphing utility to graph the deflection curve when $w_0 = 24EI$ and $L = 1$.

2. (a) The beam is simply supported at both ends, and $w(x) = w_0$, $0 < x < L$.

 (b) Use a graphing utility to graph the deflection curve when $w_0 = 24EI$ and $L = 1$.

3. (a) The beam is embedded at its left end and simply supported at its right end, and $w(x) = w_0$, $0 < x < L$.

 (b) Use a graphing utility to graph the deflection curve when $w_0 = 48EI$ and $L = 1$.

4. (a) The beam is embedded at its left end and simply supported at its right end, and $w(x) = w_0 \sin(\pi x/L)$, $0 < x < L$.

 (b) Use a graphing utility to graph the deflection curve when $w_0 = 2\pi^3 EI$ and $L = 1$.

 (c) Use a root-finding application of a CAS (or a graphic calculator) to approximate the point in the graph in part (b) at which the maximum deflection occurs. What is the maximum deflection?

5. (a) The beam is simply supported at both ends, and $w(x) = w_0 x$, $0 < x < L$.

 (b) Use a graphing utility to graph the deflection curve when $w_0 = 36EI$ and $L = 1$.

(c) Use a root-finding application of a CAS (or a graphic calculator) to approximate the point in the graph in part (b) at which the maximum deflection occurs. What is the maximum deflection?

6. (a) Find the maximum deflection of the cantilever beam in Problem 1.

(b) How does the maximum deflection of a beam that is half as long compare with the value in part (a)?

(c) Find the maximum deflection of the simply supported beam in Problem 2.

(d) How does the maximum deflection of the simply supported beam in part (c) compare with the value of maximum deflection of the embedded beam in Example 1?

7. A cantilever beam of length L is embedded at its right end, and a horizontal tensile force of P pounds is applied to its free left end. When the origin is taken at its free end, as shown in **FIGURE 3.9.8**, the deflection $y(x)$ of the beam can be shown to satisfy the differential equation

$$EIy'' = Py - w(x)\frac{x}{2}.$$

Find the deflection of the cantilever beam if $w(x) = w_0x$, $0 < x < L$, and $y(0) = 0$, $y'(L) = 0$.

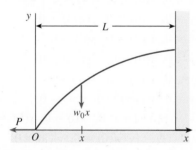

FIGURE 3.9.8 Deflection of cantilever beam in Problem 7

8. When a compressive instead of a tensile force is applied at the free end of the beam in Problem 7, the differential equation of the deflection is

$$EIy'' = -Py - w(x)\frac{x}{2}.$$

Solve this equation if $w(x) = w_0x$, $0 < x < L$, and $y(0) = 0$, $y'(L) = 0$.

Contributed Problem

Rick Wicklin, PhD
Senior Researcher in Computational Statistics, SAS Institute Inc.

9. Blowing in the Wind In September 1989, Hurricane Hugo hammered the coast of South Carolina with winds estimated at times to be as high as 60.4 m/s (135 mi/h). Of the billions of dollars in damage, approximately $420 million of this was due to the market value of loblolly pine (*Pinus taeda*) lumber in the Francis Marion National Forest. One image from that storm remains hauntingly

bizarre: All through the forest and surrounding region, thousands upon thousands of pine trees lay pointing exactly in the same direction, and all the trees were broken 5–8 meters from their base. In September 1996, Hurricane Fran destroyed over 8.2 million acres of timber forest in eastern North Carolina. As happened seven years earlier, the planted loblolly trees all broke at approximately the same height. This seems to be a reproducible phenomenon, brought on by the fact that the trees in these planted forests are approximately the same age and size.

Courtesy of Texas Forest Service
Wind damage to loblolly pines

In this problem, we are going to examine a mathematical model for the bending of loblolly pines in strong winds, and then use the model to predict the height at which a tree will break in hurricane-force winds.*

Wind hitting the branches of a tree transmits a force to the trunk of the tree. The trunk is approximately a big cylindrical beam of length L, and so we will model the deflection $y(x)$ of the tree with the static beam equation $EIy^{(4)} = w(x)$ (equation (4) in this section), where x is distance measured in meters from ground level. Since the tree is rooted into the ground, the accompanying boundary conditions are those of a cantilevered beam: $y(0) = 0$, $y'(0) = 0$ at the rooted end, and $y''(L) = 0$, $y'''(L) = 0$ at the free end, which is the top of the tree.

(a) Loblolly pines in the forest have the majority of their crown (that is, branches and needles) in the upper 50% of their length, so let's ignore the force of the wind on the lower portion of the tree. Furthermore, let's assume that the wind hitting the tree's crown results in a uniform load per unit length w_0. In other words, the load on the tree is modeled by

$$w(x) = \begin{cases} 0, & 0 \le x < L/2 \\ w_0, & L/2 \le x \le L \end{cases}$$

We can determine $y(x)$ by integrating both sides of $EIy^{(4)} = w(x)$. Integrate $w(x)$ on $[0, L/2]$ and then on $[L/2, L]$ to find an expression for $EIy'''(x)$ on each of these intervals. Let c_1 be the constant of integration on $[0, L/2]$ and c_2 be the constant of integration on $[L/2, L]$. Apply the boundary condition $y'''(L) = 0$ and solve for c_2. Then find the value of c_1 that ensures continuity of the third derivative y''' at the point $x = L/2$.

*For further information about the bending of trees in high winds, see the articles by W. Kubinec (*Phys. Teacher,* March, 1990), or F. Mergen (*J. Forest.*) 52(2), 1954.

(b) Following the same procedure as in part (a) show that

$$EIy'(x) = \begin{cases} \dfrac{w_0}{8}(-2Lx^2 + 3L^2x), & 0 \le x \le L/2 \\[2mm] \dfrac{w_0}{48}(8x^3 - 24Lx^2 + 24L^2x - L^3), & L/2 \le x \le L. \end{cases}$$

Integrate EIy' to obtain the deflection $y(x)$.

(c) Note that in our model $y(L)$ describes the maximum amount by which the loblolly will bend. Compute this quantity in terms of the problem's parameters.

10. Blowing in the Wind—Continued By making some assumptions about the density of crown's foliage, the total force F on the tree can be calculated using a formula from physics: $F = \rho A v^2/6$ where $\rho \approx 1.225$ kg/m^3 is the density of air, v is the wind speed in m/s, and A is the cross-sectional area of the tree's crown. If we assume that the crown is roughly cylindrical, then its cross section is a rectangle of area $A = (2R)(L/2) = RL$, where R is the average radius of the cylinder. Then the total force *per unit length* is then $w_0 = F/(L/2) = 0.408Rv^2$.

The cross-sectional moment of inertia for a uniform cylindrical beam is $I = \frac{1}{4}\pi r^4$, where r is the radius of the cylinder (tree trunk).

By your answer to part (c) in Problem 9 and the explanations above, the amount that a loblolly will bend depends on each of the parameters in the following table.

Symbol	Description	Typical values
r	radius of trunk	0.15–0.25 m
R	radius of crown	3–4 m
L	height of pine	15–20 m
E	modulus of elasticity	11–14 N/m^2
v	Hugo wind speeds	40–60 m/s (90–135 mi/h)

(a) Mathematically show how each parameter affects the bending, and explain in physical terms why this makes sense. (For example, a large value of E results in less bending since E appears in the denominator of $y(L)$. This means that hard wood like oak bends less than a soft wood such as palm.)

(b) Graph $y(x)$ for 40 m/s winds for an "average" tree by choosing average values of each parameter (for example, $r = 0.2$, $R = 3.5$, and so on). Graph $y(x)$ for 60 m/s winds for a "tall" (but otherwise average) pine.

(c) Recall that in the derivation of the beam equation on pages 173–174 it was assumed that the deflection of the beam was small. What is the *largest* possible value of $y(L)$ that is predicted by the model if all parameters are chosen from the given table? Is this prediction realistic, or is the mathematical model no longer valid for parameters in this range?

(d) The beam equation always predicts that a beam will bend, even if the load and flexural rigidity reflect an elephant standing on a toothpick! Different methods are used by engineers to predict when and where a beam will break. In particular, a beam subject to a load will

break at the location where the stress function $y''(x)/I(x)$ reaches a maximum. Differentiate the function in part (b) of Problem 9 to obtain EIy'', and use this to obtain the stress $y''(x)/I(x)$.

(e) Real pine tree trunks are not uniformly wide; they taper as they approach the top of the tree. Substitute $r(x) = 0.2 - x/(15L)$ into the equation for I and then use a graphing utility to graph the resulting stress as a function of height for an average loblolly. Where does the maximum stress occur? Does this location depend on the speed of the wind? On the radius of the crown? On the height of the pine tree? Compare the model to observed data from Hurricane Hugo.

(f) A mathematical model is *sensitive* to an assumption if small changes in the assumption lead to widely different predictions for the model. Repeat part (e) using $r(x) = 0.2 - x/(20L)$ and $r(x) = 0.2 - x/(10L)$ as formulas that describe the radius of a pine tree trunk that tapers. Is our model sensitive to our choice for these formulas?

Eigenvalues and Eigenfunctions

In Problems 11–20, find the eigenvalues and eigenfunctions for the given boundary-value problem.

11. $y'' + \lambda y = 0$, $y(0) = 0$, $y(\pi) = 0$

12. $y'' + \lambda y = 0$, $y'(0) = 0$, $y(\pi/4) = 0$

13. $y'' + \lambda y = 0$, $y'(0) = 0$, $y(L) = 0$

14. $y'' + \lambda y = 0$, $y(0) = 0$, $y'(\pi/2) = 0$

15. $y'' + \lambda y = 0$, $y'(0) = 0$, $y(\pi) = 0$

16. $y'' + \lambda y = 0$, $y(-\pi) = 0$, $y(\pi) = 0$

17. $y'' + 2y' + (\lambda + 1)y = 0$, $y(0) = 0$, $y(5) = 0$

18. $y'' + (\lambda + 1)y = 0$, $y'(0) = 0$, $y'(1) = 0$

19. $x^2y'' + xy' + \lambda y = 0$, $y(1) = 0$, $y(e^\pi) = 0$

20. $x^2y'' + xy' + \lambda y = 0$, $y'(e^{-1}) = 0$, $y(1) = 0$

In Problems 21 and 22, find the eigenvalues and eigenfunctions for the given boundary-value problem. Consider only the case $\lambda = \alpha^4$, $\alpha > 0$. [*Hint*: Read (*iii*) in the *Remarks*.]

21. $y^{(4)} - \lambda y = 0$, $y(0) = 0$, $y''(0) = 0$, $y(1) = 0$, $y''(1) = 0$

22. $y^{(4)} - \lambda y = 0$, $y'(0) = 0$, $y'''(0) = 0$, $y(\pi) = 0$, $y''(\pi) = 0$

Buckling of a Thin Column

23. Consider Figure 3.9.6. Where should physical restraints be placed on the column if we want the critical load to be P_4? Sketch the deflection curve corresponding to this load.

24. The critical loads of thin columns depend on the end conditions of the column. The value of the Euler load P_1 in Example 4 was derived under the assumption that the column was hinged at both ends. Suppose that a thin vertical homogeneous column is embedded at its base ($x = 0$) and free at its top ($x = L$) and that a constant axial load P is applied to its free

end. This load either causes a small deflection δ as shown in **FIGURE 3.9.9** or does not cause such a deflection. In either case the differential equation for the deflection $y(x)$ is

$$EI\frac{d^2y}{dx^2} + Py = P\delta.$$

(a) What is the predicted deflection when $\delta = 0$?

(b) When $\delta \neq 0$, show that the Euler load for this column is one-fourth of the Euler load for the hinged column in Example 4.

FIGURE 3.9.9 Deflection of vertical column in Problem 24

25. As was mentioned in Problem 24, the differential equation (7) that governs the deflection $y(x)$ of a thin elastic column subject to a constant compressive axial force P is valid only when the ends of the column are hinged. In general, the differential equation governing the deflection of the column is given by

$$\frac{d^2}{dx^2}\left(EI\frac{d^2y}{dx^2}\right) + P\frac{d^2y}{dx^2} = 0.$$

Assume that the column is uniform (EI is a constant) and that the ends of the column are hinged. Show that the solution of this fourth-order differential equation subject to the boundary conditions $y(0) = 0$, $y''(0) = 0$, $y(L) = 0$, $y''(L) = 0$ is equivalent to the analysis in Example 4.

26. Suppose that a uniform thin elastic column is hinged at the end $x = 0$ and embedded at the end $x = L$.

(a) Use the fourth-order differential equation given in Problem 25 to find the eigenvalues λ_n, the critical loads P_n, the Euler load P_1, and the deflections $y_n(x)$.

(b) Use a graphing utility to graph the first buckling mode.

Rotating String

27. Consider the boundary-value problem introduced in the construction of the mathematical model for the shape of a rotating string:

$$T\frac{d^2y}{dx^2} + \rho\omega^2 y = 0, \quad y(0) = 0, \quad y(L) = 0.$$

For constant T and ρ, define the **critical speeds** of angular rotation ω_n as the values of ω for which the boundary-value problem has nontrivial solutions. Find the critical speeds ω_n and the corresponding deflections $y_n(x)$.

28. When the magnitude of tension T is not constant, then a model for the deflection curve or shape $y(x)$ assumed by a rotating string is given by

$$\frac{d}{dx}\left[T(x)\frac{dy}{dx}\right] + \rho\omega^2 y = 0.$$

Suppose that $1 < x < e$ and that $T(x) = x^2$.

(a) If $y(1) = 0$, $y(e) = 0$, and $\rho\omega^2 > 0.25$, show that the critical speeds of angular rotation are

$$\omega_n = \tfrac{1}{2}\sqrt{(4n^2\pi^2 + 1)/\rho}$$

and the corresponding deflections are

$$y_n(x) = c_2 x^{-1/2}\sin(n\pi\ln x), \quad n = 1, 2, 3, \ldots.$$

(b) Use a graphing utility to graph the deflection curves on the interval $[1, e]$ for $n = 1, 2, 3$. Choose $c_2 = 1$.

Additional Boundary-Value Problems

29. Temperature in a Sphere Consider two concentric spheres of radius $r = a$ and $r = b$, $a < b$. See **FIGURE 3.9.10**. The temperature $u(r)$ in the region between the spheres is determined from the boundary-value problem

$$r\frac{d^2u}{dr^2} + 2\frac{du}{dr} = 0, \quad u(a) = u_0, \quad u(b) = u_1,$$

where u_0 and u_1 are constants. Solve for $u(r)$.

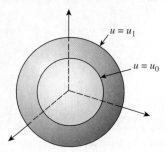

FIGURE 3.9.10 Concentric spheres in Problem 29

30. Temperature in a Ring The temperature $u(r)$ in the circular ring or annulus shown in **FIGURE 3.9.11** is determined from the boundary-value problem

$$r\frac{d^2u}{dr^2} + \frac{du}{dr} = 0, \quad u(a) = u_0, \quad u(b) = u_1,$$

where u_0 and u_1 are constants. Show that

$$u(r) = \frac{u_0\ln(r/b) - u_1\ln(r/a)}{\ln(a/b)}.$$

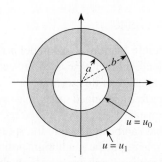

FIGURE 3.9.11 Circular ring in Problem 30

31. Rotation of a Shaft Suppose the x-axis on the interval $[0, L]$ is the geometric center of a long straight shaft, such as the propeller shaft of a ship. When the shaft is rotating at a constant angular speed ω about this axis the deflection $y(x)$ of the shaft satisfies the differential equation

$$EI\frac{d^4y}{dx^4} - \rho\omega^2 y = 0,$$

where ρ is density per unit length. If the shaft is simply supported, or hinged, at both ends the boundary conditions are then

$$y(0) = 0, \quad y''(0) = 0, \quad y(L) = 0, \quad y''(L) = 0.$$

(a) If $\lambda = \alpha^4 = \rho\omega^2/EI$, then find the eigenvalues and eigenfunctions for this boundary-value problem.

(b) Use the eigenvalues λ_n in part (a) to find corresponding angular speeds ω_n. The values ω_n are called **critical speeds**. The value ω_1 is called the **fundamental critical speed** and, analogous to Example 4, at this speed the shaft changes shape from $y = 0$ to a deflection given by $y_1(x)$.

Courtesy of the National Archives and Records Administration
Propeller shaft of the battleship USS *Missouri*

32. In Problem 31, suppose $L = 1$. If the shaft is fixed at both ends then the boundary conditions are

$$y(0) = 0, \quad y'(0) = 0, \quad y(1) = 0, \quad y'(1) = 0.$$

(a) Show that the eigenvalues $\lambda_n = \alpha_n^4$ are defined by the positive roots of $\cos\alpha \cosh\alpha = 1$. [*Hint*: See the instructions to Problems 21 and 22.]

(b) Show that the eigenfunctions are

$$y_n(x) = (-\sin\alpha_n + \sinh\alpha_n)(\cos\alpha_n x - \cosh\alpha_n x)$$
$$+ (\cos\alpha_n - \cosh\alpha_n)(\sin\alpha_n x - \sinh\alpha_n x).$$

Discussion Problems

33. Simple Harmonic Motion The model $mx'' + kx = 0$ for simple harmonic motion, discussed in Section 3.8, can be related to Example 2 of this section.

Consider a free undamped spring/mass system for which the spring constant is, say, $k = 10$ lb/ft. Determine those masses m_n that can be attached to the spring so that when each mass is released at the equilibrium position at $t = 0$ with a nonzero velocity v_0, it will then pass through the equilibrium position at $t = 1$ second. How many times will each mass m_n pass through the equilibrium position in the time interval $0 < t < 1$?

34. Damped Motion Assume that the model for the spring/mass system in Problem 33 is replaced by $mx'' + 2x' + kx = 0$. In other words, the system is free but is subjected to damping numerically equal to two times the instantaneous velocity. With the same initial conditions and spring constant as in Problem 33, investigate whether a mass m can be found that will pass through the equilibrium position at $t = 1$ second.

In Problems 35 and 36, determine whether it is possible to find values y_0 and y_1 (Problem 35) and values of $L > 0$ (Problem 36) so that the given boundary-value problem has **(a)** precisely one nontrivial solution, **(b)** more than one solution, **(c)** no solution, and **(d)** the trivial solution.

35. $y'' + 16y = 0, \quad y(0) = y_0, \quad y(\pi/2) = y_1$

36. $y'' + 16y = 0, \quad y(0) = 1, \quad y(L) = 1$

37. Consider the boundary-value problem

$$y'' + \lambda y = 0, \quad y(-\pi) = y(\pi), \quad y'(-\pi) = y'(\pi).$$

(a) The type of boundary conditions specified are called **periodic boundary conditions**. Give a geometric interpretation of these conditions.

(b) Find the eigenvalues and eigenfunctions of the problem.

(c) Use a graphing utility to graph some of the eigenfunctions. Verify your geometric interpretation of the boundary conditions given in part (a).

38. Show that the eigenvalues and eigenfunctions of the boundary-value problem

$$y'' + \lambda y = 0, \quad y(0) = 0, \quad y(1) + y'(1) = 0$$

are $\lambda_n = \alpha_n^2$ and $y_n = \sin\alpha_n x$, respectively, where α_n, $n = 1$, 2, 3, . . . are the consecutive positive roots of the equation $\tan\alpha = -\alpha$.

Computer Lab Assignments

39. Use a CAS to plot graphs to convince yourself that the equation $\tan\alpha = -\alpha$ in Problem 38 has an infinite number of roots. Explain why the negative roots of the equation can be ignored. Explain why $\lambda = 0$ is not an eigenvalue even though $\alpha = 0$ is an obvious solution of the equation $\tan\alpha = -\alpha$.

40. Use a root-finding application of a CAS to approximate the first four eigenvalues λ_1, λ_2, λ_3, and λ_4 for the BVP in Problem 38.

41. Use a CAS to approximate the eigenvalues λ_1, λ_2, λ_3, and λ_4 of the boundary-value problem:

$$y'' + \lambda y = 0, \quad y(0) = 0, \quad y(1) - \tfrac{1}{2}y'(1) = 0.$$

Give the corresponding approximate eigenfunctions $y_1(x)$, $y_2(x)$, $y_3(x)$, and $y_4(x)$.

42. Use a CAS to approximate the eigenvalues λ_1, λ_2, λ_3, and λ_4 defined by the equation in part (a) of Problem 32.

3.10 Green's Functions

INTRODUCTION We have seen in Section 3.8 that the linear second-order differential equation

$$a_2(x)\frac{d^2y}{dx^2} + a_1(x)\frac{dy}{dx} + a_0(x)y = g(x) \tag{1}$$

plays an important role in applications. In the mathematical analysis of physical systems it is often desirable to express the response or output $y(x)$ of (1) subject to either initial conditions or boundary conditions directly in terms of the forcing function or input $g(x)$. In this manner the response of the system can quickly be analyzed for different forcing functions.

To see how this is done we start by examining solutions of initial-value problems in which the DE (1) has been put into the standard form

$$y'' + P(x)y' + Q(x)y = f(x) \tag{2}$$

by dividing the equation by the lead coefficient $a_2(x)$. We also assume throughout this section that the coefficient functions $P(x)$, $Q(x)$, and $f(x)$ are continuous on some common interval I.

3.10.1 Initial-Value Problems

Three Initial-Value Problems We will see as the discussion unfolds that the solution of the second-order initial-value problem

$$y'' + P(x)y' + Q(x)y = f(x), \quad y(x_0) = y_0, \quad y'(x_0) = y_1 \tag{3}$$

can be expressed as the superposition of two solutions: the solution y_h of the associated homogeneous DE with nonhomogeneous initial conditions

$$y'' + P(x)y' + Q(x)y = 0, \quad y(x_0) = y_0, \quad y'(x_0) = y_1, \tag{4}$$

and the solution y_p of the nonhomogeneous DE with homogeneous (that is, zero) initial conditions

$$y'' + P(x)y' + Q(x)y = f(x), \quad y(x_0) = 0, \quad y'(x_0) = 0. \tag{5}$$

> Here at least one of the numbers y_0 or y_1 is assumed to be nonzero. If both y_0 and y_1 are 0, then the solution of the IVP is $y = 0$.

As we have seen in the preceding sections of this chapter, in the case where P and Q are constants the solution of the IVP (4) presents no difficulties: We use the methods of Section 3.3 to find the general solution of the homogeneous DE and then use the given initial conditions to determine the two constants in that solution. So we will focus on the solution of the IVP (5). Because of the zero initial conditions, the solution of (5) could describe a physical system that is initially at rest and so is sometimes called a **rest solution**.

Green's Function If $y_1(x)$ and $y_2(x)$ form a fundamental set of solutions on the interval I of the associated homogeneous form of (2), then a particular solution of the nonhomogeneous equation (2) on the interval I can be found by variation of parameters. Recall from (3) of Section 3.5, the form of this solution is

$$y_p(x) = u_1(x)y_1(x) + u_2(x)y_2(x). \tag{6}$$

The variable coefficients $u_1(x)$ and $u_2(x)$ in (6) are defined by (5) of Section 3.5:

$$u_1'(x) = -\frac{y_2(x)f(x)}{W}, \quad u_2'(x) = \frac{y_1(x)f(x)}{W}. \tag{7}$$

The linear independence of $y_1(x)$ and $y_2(x)$ on the interval I guarantees that the Wronskian $W = W(y_1(x), y_2(x)) \neq 0$ for all x in I. If x and x_0 are numbers in I, then integrating the derivatives in (7) on the interval $[x_0, x]$ and substituting the results in (6) give

> Because $y_1(x)$ and $y_2(x)$ are constant with respect to the integration on t, we can move these functions inside the definite integrals.

$$y_p(x) = y_1(x)\int_{x_0}^x \frac{-y_2(t)f(t)}{W(t)}\,dt + y_2(x)\int_{x_0}^x \frac{y_1(t)f(t)}{W(t)}\,dt$$

$$= \int_{x_0}^x \frac{-y_1(x)y_2(t)}{W(t)}f(t)\,dt + \int_{x_0}^x \frac{y_1(t)y_2(x)}{W(t)}f(t)\,dt, \tag{8}$$

where
$$W(t) = W(y_1(t), y_2(t)) = \begin{vmatrix} y_1(t) & y_2(t) \\ y_1'(t) & y_2'(t) \end{vmatrix}.$$

From the properties of the definite integral, the two integrals in the second line of (8) can be rewritten as a single integral

$$y_p(x) = \int_{x_0}^{x} G(x, t) f(t) \, dt. \tag{9}$$

The function $G(x, t)$ in (9),

$$G(x, t) = \frac{y_1(t)y_2(x) - y_1(x)y_2(t)}{W(t)}, \tag{10}$$

is called the **Green's function** for the differential equation (2). Green's function is named after the self-educated English mathematician and physicist **George Green** (1793–1841), who developed this concept around 1830.

Important. Read this paragraph a second time. ▶

Observe that a Green's function (10) depends only on the fundamental solutions, $y_1(x)$ and $y_2(x)$ of the associated homogeneous differential equation for (2), and *not* on the forcing function $f(x)$. Therefore all linear second-order differential equations (2) with the same left-hand side but different forcing functions have the same Green's function. So an alternative title for (10) is the **Green's function for the second-order differential operator** $L = D^2 + P(x)D + Q(x)$.

EXAMPLE 1 Particular Solution

Use (9) and (10) to find a particular solution of $y'' - y = f(x)$.

SOLUTION The solutions of the associated homogeneous equation $y'' - y = 0$ are $y_1 = e^x$, $y_2 = e^{-x}$, and $W(y_1(t), y_2(t)) = -2$. It follows from (10) that the Green's function is

$$G(x, t) = \frac{e^t e^{-x} - e^x e^{-t}}{-2} = \frac{e^{x-t} - e^{-(x-t)}}{2} = \sinh(x - t). \tag{11}$$

Thus from (9), a particular solution of the DE is

$$y_p(x) = \int_{x_0}^{x} \sinh(x - t) f(t) \, dt. \tag{12} \equiv$$

EXAMPLE 2 General Solutions

Find the general solution of the following nonhomogeneous differential equations.

(a) $y'' - y = 1/x$ **(b)** $y'' - y = e^{2x}$

SOLUTION From Example 1, both DEs possess the same complementary function $y_c = c_1 e^{-x} + c_2 e^x$. Moreover, as pointed out in the paragraph preceding Example 1, the Green's function for both differential equations is (11).

(a) With the identifications $f(x) = 1/x$ and $f(t) = 1/t$ we see from (12) that a particular solution of $y'' - y = 1/x$ is $y_p(x) = \int_{x_0}^{x} \frac{\sinh(x - t)}{t} \, dt$. Thus the general solution $y = y_c + y_p$ of the given DE on any interval $[x_0, x]$ not containing the origin is

$$y = c_1 e^x + c_2 e^{-x} + \int_{x_0}^{x} \frac{\sinh(x - t)}{t} \, dt. \tag{13}$$

You should compare this solution with that found in Example 3 of Section 3.5.

(b) With $f(x) = e^{2x}$ in (12), a particular solution of $y'' - y = e^{2x}$ is

$$y_p(x) = \int_{x_0}^{x} \sinh(x - t) e^{2t} \, dt.$$

The general solution $y = y_c + y_p$ is then

$$y = c_1 e^x + c_2 e^{-x} + \int_{x_0}^{x} \sinh(x - t) e^{2t} dt. \tag{14} \equiv$$

Now consider the special initial-value problem (5) with homogeneous initial conditions. One way of solving the problem when $f(x) \neq 0$ has already been illustrated in Sections 3.4 and 3.5; that is, apply the initial conditions $y(x_0) = 0$, $y'(x_0) = 0$ to the general solution of the nonhomogeneous DE. But there is no actual need to do this because we already have solution of the IVP at hand; it is the function defined in (9).

THEOREM 3.10.1 Solution of the IVP (5)

The function $y_p(x)$ defined in (9) in the solution of the initial-value problem (5).

PROOF: By construction we know that $y_p(x)$ in (9) satisfies the nonhomogeneous DE. Next, because a definite integral has the property $\int_a^a = 0$ we have

$$y_p(x_0) = \int_{x_0}^{x_0} G(x_0, t) f(t)\, dt = 0.$$

Finally, to show that $y'_p(x_0) = 0$ we utilize the Leibniz formula* for the derivative of an integral:

$$y'_p(x) = \overbrace{G(x, x)}^{\text{0 from (10)}} f(x) + \int_{x_0}^{x} \frac{y_1(t)y'_2(x) - y'_1(x)y_2(t)}{W(t)} f(t)\, dt.$$

Hence,

$$y'_p(x_0) = \int_{x_0}^{x_0} \frac{y_1(t)y'_2(x_0) - y'_1(x_0)y_2(t)}{W(t)} f(t)\, dt = 0. \qquad\qquad \equiv$$

EXAMPLE 3 **Example 2 Revisited**

Solve the initial-value problems
(a) $y'' - y = 1/x$, $\quad y(1) = 0$, $\quad y'(1) = 0$ **(b)** $y'' - y = e^{2x}$, $\quad y(0) = 0$, $\quad y'(0) = 0$.

SOLUTION **(a)** With $x_0 = 1$ and $f(t) = 1/t$, it follows from (13) of Example 2 and Theorem 3.10.1 that the solution of the initial-value problem is

$$y_p(x) = \int_1^x \frac{\sinh(x - t)}{t}\, dt,$$

where $[1, x]$, $x > 0$.

(b) Identifying $x_0 = 0$ and $f(t) = e^{2t}$, we see from (14) that the solution of the IVP is

$$y_p(x) = \int_0^x \sinh(x - t) e^{2t} dt. \tag{15} \equiv$$

In Part (b) of Example 3, we can carry out the integration in (15), but bear in mind that x is held constant throughout the integration with respect to t:

$$y_p(x) = \int_0^x \sinh(x - t) e^{2t} dt = \int_0^x \frac{e^{x-t} - e^{-(x-t)}}{2} e^{2t} dt$$

$$= \frac{1}{2} e^x \int_0^x e^t dt - \frac{1}{2} e^{-x} \int_0^x e^{3t} dt$$

$$= \frac{1}{3} e^{2x} - \frac{1}{2} e^x + \frac{1}{6} e^{-x}.$$

*See Problems 43 and 44 on pages 31 and 32.

EXAMPLE 4 **Another IVP**

Solve the initial-value problem

$$y'' + 4y = x, \quad y(0) = 0, \quad y'(0) = 0.$$

SOLUTION We begin by constructing the Green's function for the given differential equation.

The two linearly independent solutions of $y'' + 4y = 0$ are $y_1(x) = \cos 2x$ and $y_2(x) = \sin 2x$. From (10), with $W(\cos 2t, \sin 2t) = 2$, we find

$$G(x, t) = \frac{\cos 2t \sin 2x - \cos 2x \sin 2t}{2} = \frac{1}{2} \sin 2(x - t).$$

With the identification $x_0 = 0$, a solution of the given initial-value problem is

$$y_p(x) = \frac{1}{2} \int_0^x t \sin 2(x - t) \, dt.$$

If we wish to evaluate the integral, we first write

> Here we have used the trigonometric identity $\sin(2x - 2t) = \sin 2x \cos 2t - \cos 2x \sin 2t$.

$$y_p(x) = \frac{1}{2} \sin 2x \int_0^x t \cos 2t \, dt - \frac{1}{2} \cos 2x \int_0^x t \sin 2t \, dt$$

and then use integration by parts:

$$y_p(x) = \frac{1}{2} \sin 2x \left[\frac{1}{2} t \sin 2t + \frac{1}{4} \cos 2t \right]_0^x - \frac{1}{2} \cos 2x \left[-\frac{1}{2} t \cos 2t + \frac{1}{4} \sin 2t \right]_0^x$$

or

$$y_p(x) = \frac{1}{4} x - \frac{1}{8} \sin 2x. \qquad \equiv$$

Initial-Value Problems—Continued

We can now make use of Theorem 3.10.1 to find the solution of the initial-value problem posed in (3).

THEOREM 3.10.2 Solution of the IVP (3)

If y_h is the solution of the initial-value problem (4) and y_p is the solution (9) of the initial-value problem (5) on the interval I, then

$$y = y_h + y_p \qquad (16)$$

is the solution of the initial-value problem (3).

PROOF: Because y_h is a linear combination of the fundamental solutions, it follows from (10) of Section 3.1 that $y = y_h + y_p$ is a solution of the nonhomogeneous DE. Moreover, since y_h satisfies the initial conditions in (4) and y_p satisfies the initial conditions in (5), we have

$$y(x_0) = y_h(x_0) + y_p(x_0) = y_0 + 0 = y_0$$

$$y'(x_0) = y'_h(x_0) + y'_p(x_0) = y_1 + 0 = y_1. \qquad \equiv$$

Keeping in mind the absence of a forcing function in (4) and the presence of such a term in (5), we see from (16) that the response $y(x)$ of a physical system described by the initial-value problem (3) can be separated into two different responses:

$$y(x) = \underbrace{y_h(x)}_{\substack{\text{response of system} \\ \text{due to initial conditions} \\ y(x_0) = y_0, \quad y'(x_0) = y_1}} + \underbrace{y_p(x)}_{\substack{\text{response of system} \\ \text{due to the forcing} \\ \text{function } f}} \qquad (17)$$

In different symbols, the following initial-value problem represents the pure resonance situation for a vibrating spring/mass system. See pages 165 and 166.

EXAMPLE 5 **Using Theorem 3.10.2**

Solve the initial-value problem

$$y'' + 4y = \sin 2x, \quad y(0) = 1, \quad y'(0) = -2.$$

SOLUTION We solve two initial-value problems.

First, we solve $y'' + 4y = 0$, $y(0) = 1$, $y'(0) = -2$. By applying the initial conditions to the general solution $y(x) = c_1 \cos 2x + c_2 \sin 2x$ of the homogeneous DE, we find that $c_1 = 1$ and $c_2 = -1$. Therefore, $y_h(x) = \cos 2x - \sin 2x$.

Next we solve $y'' + 4y = \sin 2x$, $y(0) = 0$, $y'(0) = 0$. Since the left-hand side of the differential equation is the same as the DE in Example 4, the Green's function is the same; namely, $G(x, t) = \frac{1}{2} \sin 2(x - t)$. With $f(t) = \sin 2t$ we see from (9) that the solution of this second problem is $y_p(x) = \frac{1}{2} \int_0^x \sin 2(x - t) \sin 2t \, dt$.

Finally, in view of (16) in Theorem 3.10.2, the solution of the original IVP is

$$y(x) = y_h(x) + y_p(x) = \cos 2x - \sin 2x + \frac{1}{2} \int_0^x \sin 2(x - t) \sin 2t \, dt. \qquad \textbf{(18)} \equiv$$

If desired, we can integrate the definite integral in (18) by using the trigonometric identity

$$\sin A \sin B = \frac{1}{2}\left[\cos(A - B) - \cos(A + B)\right]$$

with $A = 2(x - t)$ and $B = 2t$:

$$
\begin{aligned}
y_p(x) &= \frac{1}{2} \int_0^x \sin 2(x - t) \sin 2t \, dt \\
&= \frac{1}{4} \int_0^x [\cos(2x - 4t) - \cos 2x] \, dt \\
&= \frac{1}{4}\left[-\frac{1}{4} \sin(2x - 4t) - t \cos 2x\right]_0^x \\
&= \frac{1}{8} \sin 2x - \frac{1}{4} x \cos 2x.
\end{aligned}
\qquad \textbf{(19)}
$$

Hence, the solution (18) can be rewritten as

$$y(x) = y_h(x) + y_p(x) = \cos 2x - \sin 2x + \left(\frac{1}{8} \sin 2x - \frac{1}{4} x \cos 2x\right),$$

or

$$y(x) = \cos 2x - \frac{7}{8} \sin 2x - \frac{1}{4} x \cos 2x. \qquad \textbf{(20)}$$

Note that the physical significance indicated in (17) is lost in (20) after combining like terms in the two parts of the solution $y(x) = y_h(x) + y_p(x)$.

The beauty of the solution given in (18) is that we can immediately write down the response of a system if the initial conditions remain the same but the forcing function is changed. For example, if the problem in Example 5 is changed to

$$y'' + 4y = x, \quad y(0) = 1, \quad y'(0) = -2,$$

we simply replace $\sin 2t$ in the integral in (18) by t and the solution is then

$$
\begin{aligned}
y(x) &= y_h(x) + y_p(x) \\
&= \cos 2x - \sin 2x + \frac{1}{2} \int_0^x t \sin 2(x - t) \, dt \quad \leftarrow \text{see Example 4} \\
&= \frac{1}{4} x + \cos 2x - \frac{9}{8} \sin 2x.
\end{aligned}
$$

Because the forcing function f is isolated in the particular solution $y_p(x) = \int_{x_0}^{x} G(x, t)f(t)\,dt$, the solution in (16) is useful when f is piecewise defined. The next example illustrates this idea.

EXAMPLE 6 **An Initial-Value Problem**

Solve the initial-value problem

$$y'' + 4y = f(x), \quad y(0) = 1, \quad y'(0) = -2,$$

when the forcing function f is piecewise defined:

$$f(x) = \begin{cases} 0, & x < 0 \\ \sin 2x, & 0 \le x \le 2\pi \\ 0, & x > 2\pi. \end{cases}$$

SOLUTION From (18), with $\sin 2t$ replaced by $f(t)$, we can write

$$y(x) = \cos 2x - \sin 2x + \frac{1}{2}\int_{0}^{x} \sin 2(x - t)f(t)\,dt.$$

Because f is defined in three pieces, we consider three cases in the evaluation of the definite integral. For $x < 0$,

$$y_p(x) = \frac{1}{2}\int_{0}^{x} \sin 2(x - t)0\,dt = 0,$$

for $0 \le x \le 2\pi$,

$$y_p(x) = \frac{1}{2}\int_{0}^{x} \sin 2(x - t)\sin 2t\,dt \quad \leftarrow \text{using the integration in (19)}$$

$$= \frac{1}{8}\sin 2x - \frac{1}{4}x\cos 2x,$$

and finally for $x > 2\pi$, we can use the integration following Example 5:

$$y_p(x) = \frac{1}{2}\int_{0}^{2\pi} \sin 2(x - t)\sin 2t\,dt + \frac{1}{2}\int_{2\pi}^{x} \sin 2(x - t)0\,dt$$

$$= \frac{1}{2}\int_{0}^{2\pi} \sin 2(x - t)\sin 2t\,dt$$

$$= \frac{1}{4}\left[-\frac{1}{4}\sin(2x - 4t) - t\cos 2x \right]_{0}^{2\pi} \quad \leftarrow \text{using the integration in (19)}$$

$$= -\frac{1}{16}\sin(2x - 8\pi) - \frac{1}{2}\pi\cos 2x + \frac{1}{16}\sin 2x$$

$$= -\frac{1}{2}\pi\cos 2x.$$

Hence $y_p(x)$ is

$$y_p(x) = \begin{cases} 0, & x < 0 \\ \frac{1}{8}\sin 2x - \frac{1}{4}x\cos 2x, & 0 \le x \le 2\pi \\ -\frac{1}{2}\pi\cos 2x, & x > 2\pi \end{cases}$$

and so

$$y(x) = y_h(x) + y_p(x) = \cos 2x - \sin 2x + y_p(x).$$

Putting all the pieces together we get

$$y(x) = \begin{cases} \cos 2x - \sin 2x, & x < 0 \\ (1 - \frac{1}{4}x)\cos 2x - \frac{7}{8}\sin 2x, & 0 \le x \le 2\pi \\ (1 - \frac{1}{2}\pi)\cos 2x - \sin 2x, & x > 2\pi. \end{cases}$$

The graph $y(x)$ is given in **FIGURE 3.10.1**.

FIGURE 3.10.1 Graph of $y(x)$ in Example 6

We next examine how a boundary-value problem (BVP) can be solved using a different kind of Green's function.

3.10.2 Boundary-Value Problems

In contrast to a second-order IVP in which $y(x)$ and $y'(x)$ are specified at the same point, a BVP for a second-order DE involves conditions on $y(x)$ and $y'(x)$ that are specified at two different points $x = a$ and $x = b$. Conditions such as

$$y(a) = 0, y(b) = 0; \quad y(a) = 0, y'(b) = 0; \quad y'(a) = 0, \ y'(b) = 0$$

are just special cases of the more general homogeneous boundary conditions

$$A_1 y(a) + B_1 y'(a) = 0 \tag{21}$$

and

$$A_2 y(b) + B_2 y'(b) = 0, \tag{22}$$

where $A_1, A_2, B_1,$ and B_2 are constants. Specifically, our goal is to find an integral solution $y_p(x)$ that is analogous to (9) for nonhomogeneous boundary-value problems of the form

$$\begin{aligned} y'' + P(x)y' + Q(x)y &= f(x), \\ A_1 y(a) + B_1 y'(a) &= 0 \\ A_2 y(b) + B_2 y'(b) &= 0. \end{aligned} \tag{23}$$

In addition to the usual assumptions that $P(x)$, $Q(x)$, and $f(x)$ are continuous on $[a, b]$, we assume that the homogeneous problem

$$\begin{aligned} y'' + P(x)y' + Q(x)y &= 0, \\ A_1 y(a) + B_1 y'(a) &= 0 \\ A_2 y(b) + B_2 y'(b) &= 0, \end{aligned}$$

possesses only the trivial solution $y = 0$. This latter assumption is sufficient to guarantee that a unique solution of (23) exists and is given by an integral $y_p(x) = \int_a^b G(x, t)f(t)\,dt$, where $G(x, t)$ is a Green's function.

The starting point in the construction of $G(x, t)$ is again the variation of parameters formulas (6) and (7).

■ **Another Green's Function** Suppose $y_1(x)$ and $y_2(x)$ are linearly independent solutions on $[a, b]$ of the associated homogeneous form of the DE in (23) and that x is a number in the interval $[a, b]$. Unlike the construction of (8) where we started by integrating the derivatives in (7) over the same interval, we now integrate the first equation in (7) on $[b, x]$ and the second equation in (7) on $[a, x]$:

$$u_1(x) = -\int_b^x \frac{y_2(t)f(t)}{W(t)}\,dt \quad \text{and} \quad u_2(x) = \int_a^x \frac{y_1(t)f(t)}{W(t)}\,dt. \tag{24}$$

The reason for integrating $u_1'(x)$ and $u_2'(x)$ over different intervals will become clear shortly. From (24), a particular solution $y_p(x) = u_1(x)y_1(x) + u_2(x)y_2(x)$ of the DE is

$$y_p(x) = y_1(x) \overbrace{\int_x^b \frac{y_2(t)f(t)}{W(t)}\,dt}^{\substack{\text{here we used the minus} \\ \text{sign in (24) to reverse} \\ \text{the limits of integration}}} + y_2(x) \int_a^x \frac{y_1(t)f(t)}{W(t)}\,dt$$

or

$$y_p(x) = \int_a^x \frac{y_2(x)y_1(t)}{W(t)}f(t)\,dt + \int_x^b \frac{y_1(x)y_2(t)}{W(t)}f(t)\,dt. \tag{25}$$

The right-hand side of (25) can be written compactly as a single integral

$$y_P(x) = \int_a^b G(x, t) f(t) \, dt,\tag{26}$$

where the function $G(x, t)$ is

$$G(x, t) = \begin{cases} \dfrac{y_1(t)y_2(x)}{W(t)}, & a \le t \le x \\[2ex] \dfrac{y_1(x)y_2(t)}{W(t)}, & x \le t \le b. \end{cases}\tag{27}$$

The piecewise-defined function (27) is called a **Green's function** for the boundary-value problem (23). It can be proved that $G(x, t)$ is a continuous function of x on the interval $[a, b]$.

Now if the solutions $y_1(x)$ and $y_2(x)$ used in the construction of (27) are chosen in such a manner that at $x = a$, $y_1(x)$ satisfies $A_1 y_1(a) + B_1 y_1'(a) = 0$, and at $x = b$, $y_2(x)$ satisfies $A_2 y_2(b) + B_2 y_2'(b) = 0$, then, wondrously, $y_p(x)$ defined in (26) satisfies both homogeneous boundary conditions in (23).

To see this we will need

$$y_p(x) = u_1(x)y_1(x) + u_2(x)y_2(x)\tag{28}$$

and

$$\begin{aligned} y_p'(x) &= u_1(x)y_1'(x) + y_1(x)u_1'(x) + u_2(x)y_2'(x) + y_2(x)u_2'(x) \\ &= u_1(x)y_1'(x) + u_2(x)y_2'(x). \end{aligned}\tag{29}$$

The last line in (29) results from the fact that
$y_1(x)u_1'(x) + y_2(x)u_2'(x) = 0$.
See the discussion in Section 3.5 following (4).

Before proceeding, observe in (24) that $u_1(b) = 0$ and $u_2(a) = 0$. In view of the second of these two properties we can show that $y_p(x)$ satisfies (21) whenever $y_1(x)$ satisfies the same boundary condition. From (28) and (29) we have

$$A_1 y_p(a) + B_1 y_p'(a) = A_1[u_1(a)y_1(a) + \overbrace{u_2(a)}^{0}y_2(a)] + B_1[u_1(a)y_1'(a) + \overbrace{u_2(a)}^{0}y_2'(a)]$$

$$= u_1(a)\underbrace{[A_1 y_1(a) + B_1 y_1'(a)]}_{0 \text{ from (21)}} = 0.$$

Likewise, $u_1(b) = 0$ implies that whenever $y_2(x)$ satisfies (22) so does $y_p(x)$:

$$A_2 y_p(b) + B_2 y_p'(b) = A_2[\overbrace{u_1(b)}^{0}y_1(b) + u_2(b)y_2(b)] + B_2[\overbrace{u_1(b)}^{0}y_1'(b) + u_2(b)y_2'(b)]$$

$$= u_2(b)\underbrace{[A_2 y_2(b) + B_2 y_2'(b)]}_{0 \text{ from (22)}} = 0.$$

The next theorem summarizes these results.

THEOREM 3.10.3 Solution of the BVP (23)

Let $y_1(x)$ and $y_2(x)$ be linearly independent solutions of

$$y'' + P(x)y' + Q(x)y = 0$$

on $[a, b]$, and suppose $y_1(x)$ and $y_2(x)$ satisfy boundary conditions (21) and (22), respectively. Then the function $y_p(x)$ defined in (26) is a solution of the boundary-value problem (23).

The boundary condition $y'(0) = 0$ is a special case of (21) with $a = 0$, $A_1 = 0$, and $B_1 = 1$. The boundary condition $y(\pi/2) = 0$ is a special case of (22) with $b = \pi/2$, $A_2 = 1$, and $B_2 = 0$.

▶

EXAMPLE 7 **Using Theorem 3.10.3**

Solve the boundary-value problem

$$y'' + 4y = 3, \quad y'(0) = 0, \quad y(\pi/2) = 0.$$

SOLUTION The solutions of the associated homogeneous equation $y'' + 4y = 0$ are $y_1(x) = \cos 2x$ and $y_2(x) = \sin 2x$ and $y_1(x)$ satisfies $y'(0) = 0$, whereas $y_2(x)$ satisfies $y(\pi/2) = 0$. The Wronskian is $W(y_1, y_2) = 2$, and so from (27) we see that the Green's function for the boundary-value problem is

$$G(x, t) = \begin{cases} \frac{1}{2}\cos 2t \sin 2x, & 0 \le t \le x \\ \frac{1}{2}\cos 2x \sin 2t, & x \le t \le \pi/2. \end{cases}$$

It follows from Theorem 3.10.3 that a solution of the BVP is (26) with the identifications $a = 0$, $b = \pi/2$, and $f(t) = 3$:

$$y_p(x) = 3 \int_0^{\pi/2} G(x, t)\, dt$$

$$= 3 \cdot \frac{1}{2}\sin 2x \int_0^x \cos 2t\, dt + 3 \cdot \frac{1}{2}\cos 2x \int_x^{\pi/2} \sin 2t\, dt,$$

or after evaluating the definite integrals, $y_p(x) = \frac{3}{4} + \frac{3}{4}\cos 2x$. ≡

Don't infer from the preceding example that the demand that $y_1(x)$ satisfy (21) and $y_2(x)$ satisfy (22) uniquely determines these functions. As we see in the last example, there is a certain arbitrariness in the selection of these functions.

EXAMPLE 8 **A Boundary-Value Problem**

Solve the boundary-value problem

$$x^2 y'' - 3xy' + 3y = 24x^5, \quad y(1) = 0, \quad y(2) = 0.$$

SOLUTION The differential equation is recognized as a Cauchy–Euler DE.

From the auxiliary equation $m(m - 1) - 3m + 3 = (m - 1)(m - 3) = 0$ the general solution of the associated homogeneous equation is $y = c_1 x + c_2 x^3$. Applying $y(1) = 0$ to this solution implies $c_1 + c_2 = 0$ or $c_1 = -c_2$. By choosing $c_2 = -1$ we get $c_1 = 1$ and $y_1 = x - x^3$. On the other hand, $y(2) = 0$ applied to the general solution shows $2c_1 + 8c_2 = 0$ or $c_1 = -4c_2$. The choice $c_2 = -1$ now gives $c_1 = 4$ and so $y_2(x) = 4x - x^3$. The Wronskian of these two functions is

$$W(y_1(x), y_2(x)) = \begin{vmatrix} x - x^3 & 4x - x^3 \\ 1 - 3x^2 & 4 - 3x^2 \end{vmatrix} = 6x^3.$$

Hence, the Green's function for the boundary-value problem is

$$G(x, t) = \begin{cases} \dfrac{(t - t^3)(4x - x^3)}{6t^3}, & 1 \le t \le x \\ \dfrac{(x - x^3)(4t - t^3)}{6t^3}, & x \le t \le 2. \end{cases}$$

In order to identify the correct forcing function f we must write the DE in standard form:

$$y'' - \frac{3}{x}y' + \frac{3}{x^2}y = 24x^3.$$

From this equation we see that $f(t) = 24t^3$ and so (26) becomes

$$y_p(x) = 24 \int_1^2 G(x, t) t^3 dt$$

$$= 4(4x - x^3) \int_1^x (t - t^3) \, dt + 4(x - x^3) \int_x^2 (4t - t^3) \, dt.$$

Straightforward definite integration and algebraic simplification yields the solution
$y_p(x) = 12x - 15x^3 + 3x^5.$ ≡

3.10 Exercises Answers to selected odd-numbered problems begin on page ANS-8.

3.10.1 Initial-Value Problems

In Problems 1–6, proceed as in Example 1 to find a particular solution $y_p(x)$ of the given differential equation in the integral form (9).

1. $y'' - 16y = f(x)$
2. $y'' + 3y' - 10y = f(x)$
3. $y'' + 2y' + y = f(x)$
4. $4y'' - 4y' + y = f(x)$
5. $y'' + 9y = f(x)$
6. $y'' - 2y' + 2y = f(x)$

In Problems 7–12, proceed as in Example 2 to find the general solution of the given differential equation. Use the results obtained in Problems 1–6. Do not evaluate the integral that defines $y_p(x)$.

7. $y'' - 16y = xe^{-2x}$
8. $y'' + 3y' - 10y = x^2$
9. $y'' + 2y' + y = e^{-x}$
10. $4y'' - 4y' + y = \arctan x$
11. $y'' + 9y = x + \sin x$
12. $y'' - 2y' + 2y = \cos^2 x$

In Problems 13–18, proceed as in Example 3 to find the solution of the given initial-value problem. Evaluate the integral that defines $y_p(x)$.

13. $y'' - 4y = e^{2x}$, $y(0) = 0$, $y'(0) = 0$
14. $y'' - y' = 1$, $y(0) = 0$, $y'(0) = 0$
15. $y'' - 10y' + 25y = e^{5x}$, $y(0) = 0$, $y'(0) = 0$
16. $y'' + 6y' + 9y = x$, $y(0) = 0$, $y'(0) = 0$
17. $y'' + y = \csc x \cot x$, $y(\pi/2) = 0$, $y'(\pi/2) = 0$
18. $y'' + y = \sec^2 x$, $y(\pi) = 0$, $y'(\pi) = 0$

In Problems 19–30, proceed as in Example 5 to find a solution of the given initial-value problem.

19. $y'' - 4y = e^{2x}$, $y(0) = 1$, $y'(0) = -4$
20. $y'' - y' = 1$, $y(0) = 10$, $y'(0) = 1$
21. $y'' - 10y' + 25y = e^{5x}$, $y(0) = -1$, $y'(0) = 1$
22. $y'' + 6y' + 9y = x$, $y(0) = 1$, $y'(0) = -3$

23. $y'' + y = \csc x \cot x$, $y(\pi/2) = -\pi/2$, $y'(\pi/2) = -1$
24. $y'' + y = \sec^2 x$, $y(\pi) = \frac{1}{2}$, $y'(\pi) = -1$
25. $y'' + 3y' + 2y = \sin e^x$, $y(0) = -1$, $y'(0) = 0$
26. $y'' + 3y' + 2y = \dfrac{1}{1 + e^x}$, $y(0) = 0$, $y'(0) = 1$
27. $x^2 y'' - 2xy' + 2y = x$, $y(1) = 2$, $y'(1) = -1$
28. $x^2 y'' - 2xy' + 2y = x \ln x$, $y(1) = 1$, $y'(1) = 0$
29. $x^2 y'' - 6y = \ln x$, $y(1) = 1$, $y'(1) = 3$
30. $x^2 y'' - xy' + y = x^2$, $y(1) = 4$, $y'(1) = 3$

In Problems 31–34, proceed as in Example 6 to find a solution of the initial-value problem with the given piecewise-defined forcing function.

31. $y'' - y = f(x)$, $y(0) = 8$, $y'(0) = 2$,

where $f(x) = \begin{cases} -1, & x < 0 \\ 1, & x \ge 0 \end{cases}$

32. $y'' - y = f(x)$, $y(0) = 3$, $y'(0) = 2$,

where $f(x) = \begin{cases} 0, & x < 0 \\ x, & x \ge 0 \end{cases}$

33. $y'' + y = f(x)$, $y(0) = 1$, $y'(0) = -1$,

where $f(x) = \begin{cases} 0, & x < 0 \\ 10, & 0 \le x \le 3\pi \\ 0, & x > 3\pi \end{cases}$

34. $y'' + y = f(x)$, $y(0) = 0$, $y'(0) = 1$,

where $f(x) = \begin{cases} 0, & x < 0 \\ \cos x, & 0 \le x \le 4\pi \\ 0, & x > 4\pi \end{cases}$

3.10.2 Boundary-Value Problems

In Problems 35 and 36, (a) use (25) and (26) to find a solution of the boundary-value problem. (b) Verify that function $y_p(x)$ satisfies the differential equations and both boundary conditions.

35. $y'' = f(x)$, $y(0) = 0$, $y(1) = 0$
36. $y'' = f(x)$, $y(0) = 0$, $y(1) + y'(1) = 0$

37. In Problem 35 find a solution of the BVP when $f(x) = 1$.

38. In Problem 36 find a solution of the BVP when $f(x) = x$.

In Problems 39–44, proceed as in Examples 7 and 8 to find a solution of the given boundary-value problem.

39. $y'' + y = 1$, $y(0) = 0$, $y(1) = 0$

40. $y'' + 9y = 1$, $y(0) = 0$, $y'(\pi) = 0$

41. $y'' - 2y' + 2y = e^x$, $y(0) = 0$, $y(\pi/2) = 0$

42. $y'' - y' = e^{2x}$, $y(0) = 0$, $y(1) = 0$

43. $x^2 y'' + xy' = 1$, $y(e^{-1}) = 0$, $y(1) = 0$

44. $x^2 y'' - 4xy' + 6y = x^4$, $y(1) - y'(1) = 0$, $y(3) = 0$

Discussion Problems

45. Suppose the solution of the boundary-value problem

$$y'' + Py' + Qy = f(x), \quad y(a) = 0, y(b) = 0,$$

$a < b$, is given by $y_p(x) = \int_a^b G(x, t) f(t)\, dt$ where $y_1(x)$ and $y_2(x)$ are solutions of the associated homogeneous differential equation chosen in the construction of $G(x, t)$ so that $y_1(a) = 0$ and $y_2(b) = 0$. Prove that the solution of the boundary-value problem with nonhomogeneous DE and boundary conditions,

$$y'' + Py' + Qy = f(x), \quad y(a) = A, y(b) = B$$

is given by

$$y(x) = y_p(x) + \frac{B}{y_1(b)} y_1(x) + \frac{A}{y_2(a)} y_2(x).$$

[*Hint*: In your proof, you will have to show that $y_1(b) \neq 0$ and $y_2(a) \neq 0$. Reread the assumptions following (23).]

46. Use the result in Problem 45 to solve

$$y'' + y = 1, \quad y(0) = 5, y(1) = -10.$$

3.11 Nonlinear Models

INTRODUCTION In this section we examine some nonlinear higher-order mathematical models. We are able to solve some of these models using the substitution method introduced on page 153. In some cases where the model cannot be solved, we show how a nonlinear DE can be replaced by a linear DE through a process called *linearization*.

■ **Nonlinear Springs** The mathematical model in (1) of Section 3.8 has the form

$$m\frac{d^2x}{dt^2} + F(x) = 0, \tag{1}$$

where $F(x) = kx$. Since x denotes the displacement of the mass from its equilibrium position, $F(x) = kx$ is Hooke's law, that is, the force exerted by the spring that tends to restore the mass to the equilibrium position. A spring acting under a linear restoring force $F(x) = kx$ is naturally referred to as a **linear spring**. But springs are seldom perfectly linear. Depending on how it is constructed and the material used, a spring can range from "mushy" or soft to "stiff" or hard, so that its restorative force may vary from something below to something above that given by the linear law. In the case of free motion, if we assume that a nonaging spring possesses some nonlinear characteristics, then it might be reasonable to assume that the restorative force $F(x)$ of a spring is proportional to, say, the cube of the displacement x of the mass beyond its equilibrium position or that $F(x)$ is a linear combination of powers of the displacement such as that given by the nonlinear function $F(x) = kx + k_1 x^3$. A spring whose mathematical model incorporates a nonlinear restorative force, such as

$$m\frac{d^2x}{dt^2} + kx^3 = 0 \quad \text{or} \quad m\frac{d^2x}{dt^2} + kx + k_1 x^3 = 0, \tag{2}$$

is called a **nonlinear spring**. In addition, we examined mathematical models in which damping imparted to the motion was proportional to the instantaneous velocity dx/dt, and the restoring force of a spring was given by the linear function $F(x) = kx$. But these were simply assumptions; in more realistic situations damping could be proportional to some power of the instantaneous velocity dx/dt. The nonlinear differential equation

$$m\frac{d^2x}{dt^2} + \beta \left|\frac{dx}{dt}\right| \frac{dx}{dt} + kx = 0 \tag{3}$$

is one model of a free spring/mass system with damping proportional to the square of the velocity. One can then envision other kinds of models: linear damping and nonlinear restoring force,

nonlinear damping and nonlinear restoring force, and so on. The point is, nonlinear characteristics of a physical system lead to a mathematical model that is nonlinear.

Notice in (2) that both $F(x) = kx^3$ and $F(x) = kx + k_1x^3$ are odd functions of x. To see why a polynomial function containing only odd powers of x provides a reasonable model for the restoring force, let us express F as a power series centered at the equilibrium position $x = 0$:

$$F(x) = c_0 + c_1x + c_2x^2 + c_3x^3 + \cdots.$$

When the displacements x are small, the values of x^n are negligible for n sufficiently large. If we truncate the power series with, say, the fourth term, then

$$F(x) = c_0 + c_1x + c_2x^2 + c_3x^3.$$

In order for the force at $x > 0$ ($F(x) = c_0 + c_1x + c_2x^2 + c_3x^3$) and the force at $-x < 0$ ($F(-x) = c_0 - c_1x + c_2x^2 - c_3x^3$) to have the same magnitude but act in the opposite directions, we must have $F(-x) = -F(x)$. Since this means F is an odd function, we must have $c_0 = 0$ and $c_2 = 0$, and so $F(x) = c_1x + c_3x^3$. Had we used only the first two terms in the series, the same argument yields the linear function $F(x) = c_1x$. For discussion purposes we shall write $c_1 = k$ and $c_2 = k_1$. A restoring force with mixed powers such as $F(x) = kx + k_1x^2$, and the corresponding vibrations, are said to be **unsymmetrical**.

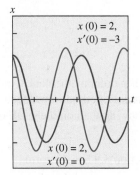

FIGURE 3.11.1 Hard and soft springs

■ **Hard and Soft Springs** Let us take a closer look at the equation in (1) in the case where the restoring force is given by $F(x) = kx + k_1x^3$, $k > 0$. The spring is said to be **hard** if $k_1 > 0$ and **soft** if $k_1 < 0$. Graphs of three types of restoring forces are illustrated in **FIGURE 3.11.1**. The next example illustrates these two special cases of the differential equation $m\,d^2x/dt^2 + kx + k_1x^3 = 0$, $m > 0$, $k > 0$.

EXAMPLE 1 **Comparison of Hard and Soft Springs**

The differential equations

$$\frac{d^2x}{dt^2} + x + x^3 = 0 \qquad (4)$$

and

$$\frac{d^2x}{dt^2} + x - x^3 = 0 \qquad (5)$$

are special cases of (2) and are models of a hard spring and soft spring, respectively. **FIGURE 3.11.2(a)** shows two solutions of (4) and Figure 3.11.2(b) shows two solutions of (5) obtained from a numerical solver. The curves shown in red are solutions satisfying the initial conditions $x(0) = 2$, $x'(0) = -3$; the two curves in blue are solutions satisfying $x(0) = 2$, $x'(0) = 0$. These solution curves certainly suggest that the motion of a mass on the hard spring is oscillatory, whereas motion of a mass on the soft spring is not oscillatory. But we must be careful about drawing conclusions based on a couple of solution curves. A more complete picture of the nature of the solutions of both equations can be obtained from the qualitative analysis discussed in Chapter 11.

≡

■ **Nonlinear Pendulum** Any object that swings back and forth is called a **physical pendulum**. The **simple pendulum** is a special case of the physical pendulum and consists of a rod of length l to which a mass m is attached at one end. In describing the motion of a simple pendulum in a vertical plane, we make the simplifying assumptions that the mass of the rod is negligible and that no external damping or driving forces act on the system. The displacement angle θ of the pendulum, measured from the vertical as shown in **FIGURE 3.11.3**, is considered positive when measured to the right of OP and negative to the left of OP. Now recall that the arc s of a circle of radius l is related to the central angle θ by the formula $s = l\theta$. Hence angular acceleration is

$$a = \frac{d^2s}{dt^2} = l\frac{d^2\theta}{dt^2}.$$

From Newton's second law we then have

$$F = ma = ml\frac{d^2\theta}{dt^2}.$$

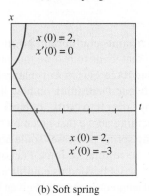

FIGURE 3.11.2 Numerical solution curves

(a) Hard spring

(b) Soft spring

FIGURE 3.11.3 Simple pendulum

From Figure 3.11.3 we see that the magnitude of the tangential component of the force due to the weight W is $mg \sin \theta$. In direction this force is $-mg \sin \theta$, since it points to the left for $\theta > 0$ and to the right for $\theta < 0$. We equate the two different versions of the tangential force to obtain $ml \, d^2\theta/dt^2 = -mg \sin \theta$ or

$$\frac{d^2\theta}{dt^2} + \frac{g}{l} \sin \theta = 0. \tag{6}$$

■ **Linearization** Because of the presence of $\sin \theta$, the model in (6) is nonlinear. In an attempt to understand the behavior of the solutions of nonlinear higher-order differential equations, one sometimes tries to simplify the problem by replacing nonlinear terms by certain approximations. For example, the Maclaurin series for $\sin \theta$ is given by

$$\sin \theta = \theta - \frac{\theta^3}{3!} + \frac{\theta^5}{5!} - \cdots,$$

and so if we use the approximation $\sin \theta \approx \theta - \theta^3/6$, equation (6) becomes $d^2\theta/dt^2 + (g/l)\theta - (g/6l)\theta^3 = 0$. Observe that this last equation is the same as the second nonlinear equation in (2) with $m = 1$, $k = g/l$, and $k_1 = -g/6l$. However, if we assume that the displacements θ are small enough to justify using the replacement $\sin \theta \approx \theta$, then (6) becomes

$$\frac{d^2\theta}{dt^2} + \frac{g}{l} \theta = 0. \tag{7}$$

See Problem 24 in Exercises 3.11. If we set $\omega^2 = g/l$, we recognize (7) as the differential equation (2) of Section 3.8 that is a model for the free undamped vibrations of a linear spring/mass system. In other words, (7) is again the basic linear equation $y'' + \lambda y = 0$ discussed on pages 175–176 of Section 3.9. As a consequence, we say that equation (7) is a **linearization** of equation (6). Since the general solution of (7) is $\theta(t) = c_1 \cos \omega t + c_2 \sin \omega t$, this linearization suggests that for initial conditions amenable to small oscillations the motion of the pendulum described by (6) will be periodic.

EXAMPLE 2 **Two Initial-Value Problems**

The graphs in **FIGURE 3.11.4(a)** were obtained with the aid of a numerical solver and represent solution curves of equation (6) when $\omega^2 = 1$. The blue curve depicts the solution of (6) that satisfies the initial conditions $\theta(0) = \frac{1}{2}$, $\theta'(0) = \frac{1}{2}$ whereas the red curve is the solution of (6) that satisfies $\theta(0) = \frac{1}{2}$, $\theta'(0) = 2$. The blue curve represents a periodic solution—the pendulum oscillating back and forth as shown in Figure 3.11.4(b) with an apparent amplitude $A \leq 1$. The red curve shows that θ increases without bound as time increases—the pendulum, starting from the same initial displacement, is given an initial velocity of magnitude great enough to send it over the top; in other words, the pendulum is whirling about its pivot as shown in Figure 3.11.4(c). In the absence of damping the motion in each case is continued indefinitely.

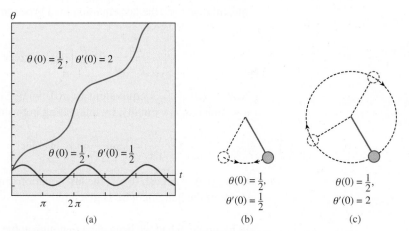

FIGURE 3.11.4 Numerical solution curves in (a); oscillating pendulum in (b); whirling pendulum in (c) in Example 2

Telephone Wires The first-order differential equation

$$\frac{dy}{dx} = \frac{W}{T_1}$$

is equation (16) of Section 1.3. This differential equation, established with the aid of Figure 1.3.9 on page 24, serves as a mathematical model for the shape of a flexible cable suspended between two vertical supports when the cable is carrying a vertical load. In Exercises 2.2, you may have solved this simple DE under the assumption that the vertical load carried by the cables of a suspension bridge was the weight of a horizontal roadbed distributed evenly along the *x*-axis. With $W = \rho w$, ρ the weight per unit length of the roadbed, the shape of each cable between the vertical supports turned out to be parabolic. We are now in a position to determine the shape of a uniform flexible cable hanging under its own weight, such as a wire strung between two telephone posts. The vertical load is now the wire itself, and so if ρ is the linear density of the wire (measured, say, in lb/ft) and s is the length of the segment P_1P_2 in Figure 1.3.9, then $W = \rho s$. Hence,

$$\frac{dy}{dx} = \frac{\rho s}{T_1}. \tag{8}$$

Since the arc length between points P_1 and P_2 is given by

$$s = \int_0^x \sqrt{1 + \left(\frac{dy}{dx}\right)^2}\, dx, \tag{9}$$

it follows from the Fundamental Theorem of Calculus that the derivative of (9) is

$$\frac{ds}{dx} = \sqrt{1 + \left(\frac{dy}{dx}\right)^2}. \tag{10}$$

Differentiating (8) with respect to x and using (10) leads to the nonlinear second-order equation

$$\frac{d^2y}{dx^2} = \frac{\rho}{T_1}\frac{ds}{dx} \quad \text{or} \quad \frac{d^2y}{dx^2} = \frac{\rho}{T_1}\sqrt{1 + \left(\frac{dy}{dx}\right)^2}. \tag{11}$$

In the example that follows, we solve (11) and show that the curve assumed by the suspended cable is a **catenary**. Before proceeding, observe that the nonlinear second-order differential equation (11) is one of those equations having the form $F(x, y', y'') = 0$ discussed in Section 3.7. Recall, we have a chance of solving an equation of this type by reducing the order of the equation by means of the substitution $u = y'$.

© Tad Denson/Shutterstock

Shape of hanging telephone wires is a catenary

EXAMPLE 3 **An Initial-Value Problem**

From the position of the *y*-axis in Figure 1.3.9 on page 24 it is apparent that initial conditions associated with the second differential equation in (11) are $y(0) = a$ and $y'(0) = 0$. If we substitute $u = y'$, the last equation in (11) becomes $\dfrac{du}{dx} = \dfrac{\rho}{T_1}\sqrt{1 + u^2}$. Separating variables,

$$\int \frac{du}{\sqrt{1 + u^2}} = \frac{\rho}{T_1}\int dx \quad \text{gives} \quad \sinh^{-1}u = \frac{\rho}{T_1}x + c_1.$$

Now, $y'(0) = 0$ is equivalent to $u(0) = 0$. Since $\sinh^{-1} 0 = 0$, we find $c_1 = 0$ and so $u = \sinh(\rho x/T_1)$. Finally, by integrating both sides of

$$\frac{dy}{dx} = \sinh\frac{\rho}{T_1}x \quad \text{we get} \quad y = \frac{T_1}{\rho}\cosh\frac{\rho}{T_1}x + c_2.$$

Using $y(0) = a$, $\cosh 0 = 1$, the last equation implies that $c_2 = a - T_1/\rho$. Thus we see that the shape of the hanging wire is given by $y = (T_1/\rho)\cosh(\rho x/T_1) + a - T_1/\rho$. ≡

In Example 3, had we been clever enough at the start to choose $a = T_1/\rho$, then the solution of the problem would have been simply the hyperbolic cosine $y = (T_1/\rho)\cosh(\rho x/T_1)$.

FIGURE 3.11.5 Distance to rocket is large compared to R

Rocket Motion

In Section 1.3 we saw that the differential equation of a free-falling body of mass m near the surface of the Earth is given by

$$m \frac{d^2s}{dt^2} = -mg \qquad \text{or simply} \qquad \frac{d^2s}{dt^2} = -g,$$

where s represents the distance from the surface of the Earth to the object and the positive direction is considered to be upward. In other words, the underlying assumption here is that the distance s to the object is small when compared with the radius R of the Earth; put yet another way, the distance y from the center of the Earth to the object is approximately the same as R. If, on the other hand, the distance y to an object, such as a rocket or a space probe, is large compared to R, then we combine Newton's second law of motion and his universal law of gravitation to derive a differential equation in the variable y.

Suppose a rocket is launched vertically upward from the ground as shown in **FIGURE 3.11.5**. If the positive direction is upward and air resistance is ignored, then the differential equation of motion after fuel burnout is

$$m \frac{d^2y}{dt^2} = -k \frac{Mm}{y^2} \qquad \text{or} \qquad \frac{d^2y}{dt^2} = -k \frac{M}{y^2}, \tag{12}$$

where k is a constant of proportionality, y is the distance from the center of the Earth to the rocket, M is the mass of the Earth, and m is the mass of the rocket. To determine the constant k, we use the fact that when $y = R$, $kMm/R^2 = mg$ or $k = gR^2/M$. Thus the last equation in (12) becomes

$$\frac{d^2y}{dt^2} = -g \frac{R^2}{y^2}. \tag{13}$$

See Problem 14 in Exercises 3.11.

Variable Mass

Notice in the preceding discussion that we described the motion of the rocket after it has burned all its fuel, when presumably its mass m is constant. Of course during its powered ascent, the total mass of the rocket varies as its fuel is being expended. The second law of motion, as originally advanced by Newton, states that when a body of mass m moves through a force field with velocity v, the time rate of change of the momentum mv of the body is equal to applied or net force F acting on the body:

$$F = \frac{d}{dt}(mv). \tag{14}$$

See Problems 21 and 22 in Exercises 1.3. If m is constant, then (14) yields the more familiar form $F = m\,dv/dt = ma$, where a is acceleration. We use the form of Newton's second law given in (14) in the next example, in which the mass m of the body is variable.

| EXAMPLE 4 | Rope Pulled Upward by a Constant Force |

A uniform 10-foot-long heavy rope is coiled loosely on the ground. One end of the rope is pulled vertically upward by means of a constant force of 5 lb. The rope weighs 1 lb per foot. Determine the height $x(t)$ of the end above ground level at time t. See **FIGURE 1.R.2** and Problem 47 in Chapter 1 in Review.

SOLUTION Let us suppose that $x = x(t)$ denotes the height of the end of the rope in the air at time t, $v = dx/dt$, and that the positive direction is upward. For that portion of the rope in the air at time t we have the following variable quantities:

$$\begin{aligned} \textit{weight:} \quad & W = (x\text{ ft}) \cdot (1\text{ lb/ft}) = x, \\ \textit{mass:} \quad & m = W/g = x/32, \\ \textit{net force:} \quad & F = 5 - W = 5 - x. \end{aligned}$$

Thus from (14) we have

Product Rule
↓

$$\frac{d}{dt}\left(\frac{x}{32}v\right) = 5 - x \qquad \text{or} \qquad x\frac{dv}{dt} + v\frac{dx}{dt} = 160 - 32x. \qquad (15)$$

Since $v = dx/dt$ the last equation becomes

$$x\frac{d^2x}{dt^2} + \left(\frac{dx}{dt}\right)^2 + 32x = 160. \qquad (16)$$

The nonlinear second-order differential equation (16) has the form $F(x, x', x'') = 0$, which is the second of the two forms considered in Section 3.7 that can possibly be solved by reduction of order. In order to solve (16), we revert back to (15) and use $v = x'$ along with the Chain Rule. From $\dfrac{dv}{dt} = \dfrac{dv}{dx}\dfrac{dx}{dt} = v\dfrac{dv}{dx}$ the second equation in (15) can be rewritten as

$$xv\frac{dv}{dx} + v^2 = 160 - 32x. \qquad (17)$$

On inspection (17) might appear intractable, since it cannot be characterized as any of the first-order equations that were solved in Chapter 2. However, by rewriting (17) in differential form $M(x, v)\, dx + N(x, v)\, dv = 0$, we observe that the nonexact equation

$$(v^2 + 32x - 160)\, dx + xv\, dv = 0 \qquad (18)$$

can be transformed into an exact equation by multiplying it by an *integrating factor.** When (18) is multiplied by $\mu(x) = x$, the resulting equation is exact (verify). If we identify $\partial f/\partial x = xv^2 + 32x^2 - 160x$, $\partial f/\partial v = x^2v$, and then proceed as in Section 2.4, we arrive at

$$\frac{1}{2}x^2v^2 + \frac{32}{3}x^3 - 80x^2 = c_1. \qquad (19)$$

From the initial condition $x(0) = 0$ it follows that $c_1 = 0$. Now by solving $\frac{1}{2}x^2v^2 + \frac{32}{3}x^3 - 80x^2 = 0$ for $v = dx/dt > 0$ we get another differential equation,

$$\frac{dx}{dt} = \sqrt{160 - \frac{64}{3}x},$$

which can be solved by separation of variables. You should verify that

$$-\frac{3}{32}\left(160 - \frac{64}{3}x\right)^{1/2} = t + c_2. \qquad (20)$$

This time the initial condition $x(0) = 0$ implies $c_2 = -3\sqrt{10}/8$. Finally, by squaring both sides of (20) and solving for x we arrive at the desired result,

$$x(t) = \frac{15}{2} - \frac{15}{2}\left(1 - \frac{4\sqrt{10}}{15}t\right)^2. \qquad (21) \equiv$$

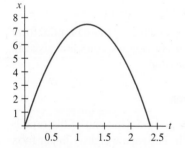

FIGURE 3.11.6 Graph of (21) in Example 4

The graph of (21) given in **FIGURE 3.11.6** should not, on physical grounds, be taken at face value. See Problems 15 and 16 in Exercises 3.11.

*See pages 65–66 in Section 2.4.

To the Instructor

In addition to Problems 24 and 25, all or portions of Problems 1–6, 8–13, and 23 could serve as *Computer Lab Assignments.*

Nonlinear Springs

In Problems 1–4, the given differential equation is a model of an undamped spring/mass system in which the restoring force $F(x)$ in (1) is nonlinear. For each equation use a numerical solver to plot the solution curves satisfying the given initial conditions. If the solutions appear to be periodic, use the solution curve to estimate the period T of oscillations.

1. $\dfrac{d^2x}{dt^2} + x^3 = 0,$

$x(0) = 1,\ x'(0) = 1;\quad x(0) = \tfrac{1}{2},\ x'(0) = -1$

2. $\dfrac{d^2x}{dt^2} + 4x - 16x^3 = 0,$

$x(0) = 1,\ x'(0) = 1;\quad x(0) = -2,\ x'(0) = 2$

3. $\dfrac{d^2x}{dt^2} + 2x - x^2 = 0,$

$x(0) = 1,\ x'(0) = 1;\quad x(0) = \tfrac{3}{2},\ x'(0) = -1$

4. $\dfrac{d^2x}{dt^2} + xe^{0.01x} = 0,$

$x(0) = 1,\ x'(0) = 1;\quad x(0) = 3,\ x'(0) = -1$

5. In Problem 3, suppose the mass is released from the initial position $x(0) = 1$ with an initial velocity $x'(0) = x_1$. Use a numerical solver to estimate the smallest value of $|x_1|$ at which the motion of the mass is nonperiodic.

6. In Problem 3, suppose the mass is released from an initial position $x(0) = x_0$ with the initial velocity $x'(0) = 1$. Use a numerical solver to estimate an interval $a \leq x_0 \leq b$ for which the motion is oscillatory.

7. Find a linearization of the differential equation in Problem 4.

8. Consider the model of an undamped nonlinear spring/mass system given by $x'' + 8x - 6x^3 + x^5 = 0$. Use a numerical solver to discuss the nature of the oscillations of the system corresponding to the initial conditions:

$x(0) = 1,\quad x'(0) = 1;\qquad x(0) = -2,\quad x'(0) = \tfrac{1}{2};$

$x(0) = \sqrt{2},\quad x'(0) = 1;\qquad x(0) = 2,\quad x'(0) = \tfrac{1}{2};$

$x(0) = 2,\quad x'(0) = 0;\qquad x(0) = -\sqrt{2},\quad x'(0) = -1.$

In Problems 9 and 10, the given differential equation is a model of a damped nonlinear spring/mass system. Predict the behavior of each system as $t \to \infty$. For each equation use a numerical solver to obtain the solution curves satisfying the given initial conditions.

9. $\dfrac{d^2x}{dt^2} + \dfrac{dx}{dt} + x + x^3 = 0,$

$x(0) = -3,\quad x'(0) = 4;\quad x(0) = 0,\quad x'(0) = -8$

10. $\dfrac{d^2x}{dt^2} + \dfrac{dx}{dt} + x - x^3 = 0,$

$x(0) = 0,\quad x'(0) = \tfrac{3}{2};\quad x(0) = -1,\quad x'(0) = 1$

11. The model $mx'' + kx + k_1x^3 = F_0 \cos \omega t$ of an undamped periodically driven spring/mass system is called **Duffing's differential equation** after the German engineer **Georg Duffing** (1861–1944). Consider the initial-value problem $x'' + x + k_1x^3 = 5 \cos t, x(0) = 1, x'(0) = 0$. Use a numerical solver to investigate the behavior of the system for values of $k_1 > 0$ ranging from $k_1 = 0.01$ to $k_1 = 100$. State your conclusions.

12. **(a)** Find values of $k_1 < 0$ for which the system in Problem 11 is oscillatory.

(b) Consider the initial-value problem

$$x'' + x + k_1x^3 = \cos \tfrac{3}{2}t, \quad x(0) = 0, \quad x'(0) = 0.$$

Find values for $k_1 < 0$ for which the system is oscillatory.

Nonlinear Pendulum

13. Consider the model of the free damped nonlinear pendulum given by

$$\frac{d^2\theta}{dt^2} + 2\lambda\frac{d\theta}{dt} + \omega^2\sin\theta = 0.$$

Use a numerical solver to investigate whether the motion in the two cases $\lambda^2 - \omega^2 > 0$ and $\lambda^2 - \omega^2 < 0$ corresponds, respectively, to the overdamped and underdamped cases discussed in Section 3.8 for spring/mass systems. Choose appropriate initial conditions and values of λ and ω.

Rocket Motion

14. **(a)** Use the substitution $v = dy/dt$ to solve (13) for v in terms of y. Assume that the velocity of the rocket at burnout is $v = v_0$ and that $y \approx R$ at that instant; show that the approximate value of the constant c of integration is $c = -gR + \tfrac{1}{2}v_0^2$.

(b) Use the solution for v in part (a) to show that the escape velocity of the rocket is given by $v_0 = \sqrt{2gR}$. [*Hint:* Take $y \to \infty$ and assume $v > 0$ for all time t.]

(c) The result in part (b) holds for any body in the solar system. Use the values $g = 32$ ft/s^2 and $R = 4000$ mi to show that the escape velocity from the Earth is (approximately) $v_0 = 25{,}000$ mi/h.

(d) Find the escape velocity from the Moon if the acceleration of gravity is $0.165g$ and $R = 1080$ mi.

Variable Mass

15. **(a)** In Example 4, show that equation (16) possesses a constant solution $x(t) = k > 0$.

(b) Why does the value of k in part (a) make intuitive sense?

(c) What is the initial velocity of the rope?

16. **(a)** In Example 4, what does (21) predict to be the maximum amount of rope lifted by the constant force?

(b) Explain any difference between your answer to part (a) in this problem and your answer to part (a) of Problem 15.

(c) Why would you expect the solution $x(t)$ of the problem in Example 4 to be oscillatory?

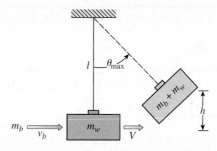

FIGURE 3.11.7 Ballistic pendulum in Problem 17

17. **Ballistic Pendulum** Historically, in order to maintain quality control over munitions (bullets) produced by an assembly line, the manufacturer would use a ballistic pendulum to determine the muzzle velocity of a gun, that is, the speed of a bullet as it leaves the barrel. The ballistic pendulum, invented in 1742 by the British mathematician and military engineer **Benjamin Robins** (1707–1751), is simply a plane pendulum consisting of a rod of negligible mass to which a block of wood of mass m_w is attached. The system is set in motion by the impact of a bullet that is moving horizontally at the unknown muzzle velocity v_b; at the time of the impact, $t = 0$, the combined mass is $m_w + m_b$, where m_b is the mass of the bullet embedded in the wood. We have seen in (7) of this section, that in the case of small oscillations, the angular displacement $\theta(t)$ of a plane pendulum shown in Figure 3.11.3 is given by the linear DE $\theta'' + (g/l)\theta = 0$, where $\theta > 0$ corresponds to motion to the right of vertical. The velocity v_b can be found by measuring the height h of the mass $m_w + m_b$ at the maximum displacement angle θ_{max} shown in **FIGURE 3.11.7**.

Intuitively, the horizontal velocity V of the combined mass $m_w + m_b$ after impact is only a fraction of the velocity v_b of the bullet, that is,

$$V = \left(\frac{m_b}{m_w + m_b}\right)v_b.$$

Now recall, a distance s traveled by a particle moving along a circular path is related to the radius l and central angle θ by the formula $s = l\theta$. By differentiating the last formula with respect to time t, it follows that the angular velocity ω of the mass and its linear velocity v are related by $v = l\omega$. Thus the initial angular velocity ω_0 at the time t at which the bullet impacts the wood block is related to V by $V = l\omega_0$ or

$$\omega_0 = \left(\frac{m_b}{m_w + m_b}\right)\frac{v_b}{l}.$$

(a) Solve the initial-value problem

$$\frac{d^2\theta}{dt^2} + \frac{g}{l}\theta = 0, \quad \theta(0) = 0, \quad \theta'(0) = \omega_0.$$

(b) Use the result from part (a) to show that

$$v_b = \left(\frac{m_w + m_b}{m_b}\right)\sqrt{lg}\,\theta_{max}.$$

(c) Use Figure 3.11.7 to express $\cos\theta_{max}$ in terms of l and h. Then use the first two terms of the Maclaurin series for $\cos\theta$ to express θ_{max} in terms of l and h. Finally, show that v_b is given (approximately) by

$$v_b = \left(\frac{m_w + m_b}{m_b}\right)\sqrt{2gh}.$$

18. **Ballistic Pendulum—Continued** Use the result in Problem 17 to find the muzzle velocity v_b when $m_b = 5\,g$, $m_w = 1\,kg$, and $h = 6\,cm$.

19. **Pursuit Curve** In a naval exercise, a ship S_1 is pursued by a submarine S_2 as shown in **FIGURE 3.11.8**. Ship S_1 departs point $(0, 0)$ at $t = 0$ and proceeds along a straight-line course (the y-axis) at a constant speed v_1. The submarine S_2 keeps ship S_1 in visual contact, indicated by the straight dashed line L in the figure, while traveling at a constant speed v_2 along a curve C. Assume that S_2 starts at the point $(a, 0)$, $a > 0$, at $t = 0$ and that L is tangent to C.

(a) Determine a mathematical model that describes the curve C. [*Hint:* $\dfrac{dt}{dx} = \dfrac{dt}{ds}\dfrac{ds}{dx}$ where s is arc length measured along C.].

(b) Find an explicit solution of the differential equation. For convenience define $r = v_1/v_2$.

(c) Determine whether the paths of S_1 and S_2 will ever intersect by considering the cases $r > 1$, $r < 1$, and $r = 1$.

FIGURE 3.11.8 Pursuit curve in Problem 19

20. **Pursuit Curve** In another naval exercise, a destroyer S_1 pursues a submerged submarine S_2. Suppose that S_1 at $(9, 0)$ on the x-axis detects S_2 at $(0, 0)$ and that S_2 simultaneously detects S_1. The captain of the destroyer S_1 assumes that the submarine will take immediate evasive action and conjectures that its likely new course is the straight line indicated in **FIGURE 3.11.9**. When S_1 is at $(3, 0)$ it changes from its straight-line course toward the origin to a pursuit curve C. Assume that the speed of the destroyer is, at all times, a constant 30 mi/h and the submarine's speed is a constant 15 mi/h.

(a) Explain why the captain waits until S_1 reaches $(3, 0)$ before ordering a course change to C.

(b) Using polar coordinates, find an equation $r = f(\theta)$ for the curve C.

(c) Let T denote the time, measured from the initial detection, at which the destroyer intercepts the submarine. Find an upper bound for T.

FIGURE 3.11.9 Pursuit curve in Problem 20

Discussion Problems

21. Discuss why the damping term in equation (3) is written as

$$\beta \left| \frac{dx}{dt} \right| \frac{dx}{dt} \quad \text{instead of} \quad \beta \left(\frac{dx}{dt} \right)^2.$$

22. (a) Experiment with a calculator to find an interval $0 \le \theta < \theta_1$, where θ is measured in radians, for which you think $\sin \theta \approx \theta$ is a fairly good estimate. Then use a graphing utility to plot the graphs of $y = x$ and $y = \sin x$ on the same coordinate axes for $0 \le x \le \pi/2$. Do the graphs confirm your observations with the calculator?

(b) Use a numerical solver to plot the solution curves of the initial-value problems

$$\frac{d^2\theta}{dt^2} + \sin \theta = 0, \quad \theta(0) = \theta_0, \quad \theta'(0) = 0$$

and $\quad \dfrac{d^2\theta}{dt^2} + \theta = 0, \quad \theta(0) = \theta_0, \quad \theta'(0) = 0$

for several values of θ_0 in the interval $0 \le \theta < \theta_1$ found in part (a). Then plot solution curves of the initial-value problems for several values of θ_0 for which $\theta_0 > \theta_1$.

23. (a) Consider the nonlinear pendulum whose oscillations are defined by (6). Use a numerical solver as an aid to determine whether a pendulum of length l will oscillate faster on the Earth or on the Moon. Use the same initial conditions, but choose these initial conditions so that the pendulum oscillates back and forth.

(b) For which location in part (a) does the pendulum have greater amplitude?

(c) Are the conclusions in parts (a) and (b) the same when the linear model (7) is used?

Computer Lab Assignments

24. Consider the initial-value problem

$$\frac{d^2\theta}{dt^2} + \sin \theta = 0, \quad \theta(0) = \frac{\pi}{12}, \quad \theta'(0) = -\frac{1}{3}$$

for the nonlinear pendulum. Since we cannot solve the differential equation, we can find no explicit solution of this problem. But suppose we wish to determine the first time $t_1 > 0$ for which the pendulum in Figure 3.11.3, starting from its initial position to the right, reaches the position OP—that is, find the first positive root of $\theta(t) = 0$. In this problem and the next we examine several ways to proceed.

(a) Approximate t_1 by solving the linear problem

$$\frac{d^2\theta}{dt^2} + \theta = 0, \quad \theta(0) = \frac{\pi}{12}, \quad \theta'(0) = -\frac{1}{3}.$$

(b) Use the method illustrated in Example 3 of Section 3.7 to find the first four nonzero terms of a Taylor series solution $\theta(t)$ centered at 0 for the nonlinear initial-value problem. Give the exact values of all coefficients.

(c) Use the first two terms of the Taylor series in part (b) to approximate t_1.

(d) Use the first three terms of the Taylor series in part (b) to approximate t_1.

(e) Use a root-finding application of a CAS (or a graphing calculator) and the first four terms of the Taylor series in part (b) to approximate t_1.

(f) In this part of the problem you are led through the commands in *Mathematica* that enable you to approximate the root t_1. The procedure is easily modified so that any root of $\theta(t) = 0$ can be approximated. (*If you do not have* Mathematica, *adapt the given procedure by finding the corresponding syntax for the CAS you have on hand.*) Reproduce and then, in turn, execute each line in the given sequence of commands.

```
sol = NDSolve[{y''[t] + Sin[y[t]} == 0,
      y[0] == Pi/12, y'[0] == −1/3},
      y, { t, 0, 5} ]//Flatten
solution = y[t]/.sol
Clear[y]
y[t_]: = Evaluate[solution]
y[t]
gr1 = Plot[y[t], { t, 0, 5} ]
root = FindRoot[y[t] == 0, { t, 1} ]
```

(g) Appropriately modify the syntax in part (f) and find the next two positive roots of $\theta(t) = 0$.

25. Consider a pendulum that is released from rest from an initial displacement of θ_0 radians. Solving the linear model (7) subject to the initial conditions $\theta(0) = \theta_0, \theta'(0) = 0$ gives $\theta(t) = \theta_0 \cos \sqrt{g/l}\, t$. The period of oscillations predicted by this model is given by the familiar formula $T = 2\pi/\sqrt{g/l} = 2\pi\sqrt{l/g}$. The interesting thing about this formula for T is that it does not depend on the magnitude of the initial displacement θ_0. In other words, the linear model predicts that the time that it would take the pendulum to swing from an initial displacement of, say, $\theta_0 = \pi/2$ (= 90°) to $-\pi/2$ and back again would be exactly the same time to

cycle from, say, $\theta_0 = \pi/360$ (= 0.5°) to $-\pi/360$. This is intuitively unreasonable; the actual period must depend on θ_0.

If we assume that $g = 32$ ft/s^2 and $l = 32$ ft, then the period of oscillation of the linear model is $T = 2\pi$ s. Let us compare this last number with the period predicted by the nonlinear model when $\theta_0 = \pi/4$. Using a numerical solver that is capable of generating hard data, approximate the solution of

$$\frac{d^2\theta}{dt^2} + \sin\theta = 0, \quad \theta(0) = \frac{\pi}{4}, \quad \theta'(0) = 0$$

for $0 \le t \le 2$. As in Problem 24, if t_1 denotes the first time the pendulum reaches the position OP in Figure 3.11.3, then the period of the nonlinear pendulum is $4t_1$. Here is another way of solving the equation $\theta(t) = 0$. Experiment with small step sizes and advance the time starting at $t = 0$ and ending at $t = 2$. From your hard data, observe the time t_1 when $\theta(t)$ changes, for the first time, from positive to negative. Use the value t_1 to determine the true value of the period of the nonlinear pendulum. Compute the percentage relative error in the period estimated by $T = 2\pi$.

3.12 Solving Systems of Linear DEs

INTRODUCTION We conclude this chapter as we did in Chapter 2 with systems of differential equations. But unlike Section 2.9, we will actually solve systems in the discussion that follows.

Coupled Systems/Coupled DEs
In Section 2.9 we briefly examined some mathematical models that were systems of linear and nonlinear first-order ODEs. In Section 3.8 we saw that the mathematical model describing the displacement of a mass on a single spring, current in a series circuit, and charge on a capacitor in a series circuit consisted of a *single* differential equation. When physical systems are coupled—for example, when two or more mixing tanks are connected, when two or more spring/mass systems are attached, or when circuits are joined to form a network—the mathematical model of the system usually consists of a set of coupled differential equations, in other words, a system of differential equations.

We did not attempt to solve any of the systems considered in Section 2.9. The same remarks made in Sections 3.7 and 3.11 pertain as well to systems of nonlinear ODEs; that is, it is nearly impossible to solve such systems analytically. However, *linear* systems with constant coefficients can be solved. The method that we shall examine in this section for solving linear systems with constant coefficients simply uncouples the system into distinct linear ODEs in each dependent variable. Thus, this section gives you an opportunity to practice what you learned earlier in the chapter.

Before proceeding, let us continue in the same vein as Section 3.8 by considering a spring/mass system, but this time we derive a mathematical model that describes the vertical displacements of two masses in a coupled spring/mass system.

(a) Equilibrium (b) Motion (c) Forces

FIGURE 3.12.1 Coupled spring/mass system

Coupled Spring/Mass System
Suppose two masses m_1 and m_2 are connected to two springs A and B of negligible mass having spring constants k_1 and k_2, respectively. As shown in **FIGURE 3.12.1(a)**, spring A is attached to a rigid support and spring B is attached to the bottom of mass m_1. Let $x_1(t)$ and $x_2(t)$ denote the vertical displacements of the masses from their equilibrium positions. When the system is in motion, Figure 3.12.1(b), spring B is subject to both an elongation and a compression; hence its net elongation is $x_2 - x_1$. Therefore it follows from Hooke's law that springs A and B exert forces $-k_1x_1$ and $k_2(x_2 - x_1)$, respectively, on m_1. If no damping is present and no external force is impressed on the system, then the net force on m_1 is $-k_1x_1 + k_2(x_2 - x_1)$. By Newton's second law we can write

$$m_1 \frac{d^2x_1}{dt^2} = -k_1x_1 + k_2(x_2 - x_1).$$

Similarly, the net force exerted on mass m_2 is due solely to the net elongation of spring B, that is, $-k_2(x_2 - x_1)$. Hence we have

$$m_2 \frac{d^2x_2}{dt^2} = -k_2(x_2 - x_1).$$

In other words, the motion of the coupled system is represented by the system of linear second-order equations

$$m_1 x_1'' = -k_1 x_1 + k_2(x_2 - x_1)$$
$$m_2 x_2'' = -k_2(x_2 - x_1). \tag{1}$$

After we have illustrated the main idea of this section, we will return to system (1).

■ Systematic Elimination

The method of **systematic elimination** for solving systems of linear equations with constant coefficients is based on the algebraic principle of elimination of variables. The analogue of *multiplying* an algebraic equation by a constant is *operating* on an ODE with some combination of derivatives. The elimination process is expedited by rewriting each equation in a system using differential operator notation. Recall from Section 3.1 that a single linear equation

$$a_n y^{(n)} + a_{n-1} y^{(n-1)} + \cdots + a_1 y' + a_0 y = g(t),$$

where the coefficients a_i, $i = 0, 1, \ldots, n$ are constants, can be written as

$$(a_n D^n + a_{n-1} D^{n-1} + \cdots + a_1 D + a_0) y = g(t).$$

If an nth-order differential operator $a_n D^n + a_{n-1} D^{n-1} + \cdots + a_1 D + a_0$ factors into differential operators of lower order, then the factors commute. Now, for example, to rewrite the system

$$x'' + 2x' + y'' = x + 3y + \sin t$$
$$x' + y' = -4x + 2y + e^{-t}$$

in terms of the operator D, we first bring all terms involving the dependent variables to one side and group the same variables:

$$x'' + 2x' - x + y'' - 3y = \sin t \qquad \qquad (D^2 + 2D - 1)x + (D^2 - 3)y = \sin t$$
$$x' + 4x + y' - 2y = e^{-t} \qquad \text{so that} \qquad (D + 4)x + (D - 2)y = e^{-t}.$$

■ Solution of a System

A **solution** of a system of differential equations is a set of sufficiently differentiable functions $x = \phi_1(t)$, $y = \phi_2(t)$, $z = \phi_3(t)$, and so on, that satisfies each equation in the system on some common interval I.

■ Method of Solution

Consider the simple system of linear first-order equations

$$\frac{dx}{dt} = 3y \qquad \qquad \qquad Dx - 3y = 0$$
$$\qquad \qquad \text{or, equivalently,} \qquad \qquad \qquad \qquad \tag{2}$$
$$\frac{dy}{dt} = 2x \qquad \qquad \qquad 2x - Dy = 0.$$

Operating on the first equation in (2) by D while multiplying the second by -3 and then adding eliminates y from the system and gives $D^2 x - 6x = 0$. Since the roots of the auxiliary equation of the last DE are $m_1 = \sqrt{6}$ and $m_2 = -\sqrt{6}$, we obtain

$$x(t) = c_1 e^{-\sqrt{6}t} + c_2 e^{\sqrt{6}t}. \tag{3}$$

Multiplying the first equation in (2) by 2 while operating on the second by D and then subtracting gives the differential equation for y, $D^2 y - 6y = 0$. It follows immediately that

$$y(t) = c_3 e^{-\sqrt{6}t} + c_4 e^{\sqrt{6}t}. \tag{4}$$

This is important. ▶ Now, (3) and (4) do not satisfy the system (2) for every choice of c_1, c_2, c_3, and c_4 because the system itself puts a constraint on the number of parameters in a solution that can be chosen arbitrarily. To see this, observe that after substituting $x(t)$ and $y(t)$ into the first equation of the original system, (2) gives, after simplification,

$$(-\sqrt{6}c_1 - 3c_3)e^{-\sqrt{6}t} + (\sqrt{6}c_2 - 3c_4)e^{\sqrt{6}t} = 0.$$

Since the latter expression is to be zero for all values of t, we must have $-\sqrt{6}c_1 - 3c_3 = 0$ and $\sqrt{6}c_2 - 3c_4 = 0$. Thus we can write c_3 as a multiple of c_1 and c_4 as a multiple of c_2:

$$c_3 = -\frac{\sqrt{6}}{3}c_1 \quad \text{and} \quad c_4 = \frac{\sqrt{6}}{3}c_2. \tag{5}$$

Hence we conclude that a solution of the system must be

$$x(t) = c_1 e^{-\sqrt{6}t} + c_2 e^{\sqrt{6}t}, \quad y(t) = -\frac{\sqrt{6}}{3}c_1 e^{-\sqrt{6}t} + \frac{\sqrt{6}}{3}c_2 e^{\sqrt{6}t}.$$

You are urged to substitute (3) and (4) into the second equation of (2) and verify that the same relationship (5) holds between the constants.

EXAMPLE 1 Solution by Elimination

Solve

$$\begin{aligned} Dx + (D+2)y &= 0 \\ (D-3)x - \quad\quad 2y &= 0. \end{aligned} \tag{6}$$

SOLUTION Operating on the first equation by $D - 3$ and on the second by D and then subtracting eliminates x from the system. It follows that the differential equation for y is

$$[(D-3)(D+2) + 2D]y = 0 \quad \text{or} \quad (D^2 + D - 6)y = 0.$$

Since the characteristic equation of this last differential equation is $m^2 + m - 6 = (m-2)(m+3) = 0$, we obtain the solution

$$y(t) = c_1 e^{2t} + c_2 e^{-3t}. \tag{7}$$

Eliminating y in a similar manner yields $(D^2 + D - 6)x = 0$, from which we find

$$x(t) = c_3 e^{2t} + c_4 e^{-3t}. \tag{8}$$

As we noted in the foregoing discussion, a solution of (6) does not contain four independent constants. Substituting (7) and (8) into the first equation of (6) gives

$$(4c_1 + 2c_3)e^{2t} + (-c_2 - 3c_4)e^{-3t} = 0.$$

From $4c_1 + 2c_3 = 0$ and $-c_2 - 3c_4 = 0$ we get $c_3 = -2c_1$ and $c_4 = -\frac{1}{3}c_2$. Accordingly, a solution of the system is

$$x(t) = -2c_1 e^{2t} - \frac{1}{3}c_2 e^{-3t}, \quad y(t) = c_1 e^{2t} + c_2 e^{-3t}. \qquad \equiv$$

Since we could just as easily solve for c_3 and c_4 in terms of c_1 and c_2, the solution in Example 1 can be written in the alternative form

$$x(t) = c_3 e^{2t} + c_4 e^{-3t}, \quad y(t) = -\frac{1}{2}c_3 e^{2t} - 3c_4 e^{-3t}.$$

Watch for a shortcut. ▶

It sometimes pays to keep one's eyes open when solving systems. Had we solved for x first, then y could be found, along with the relationship between the constants, by using the last equation in (6). You should verify that substituting $x(t)$ into $y = \frac{1}{2}(Dx - 3x)$ yields $y = -\frac{1}{2}c_3 e^{2t} - 3c_4 e^{-3t}$.

EXAMPLE 2 Solution by Elimination

Solve

$$\begin{aligned} x' - 4x + y'' &= t^2 \\ x' + x + y' &= 0. \end{aligned} \tag{9}$$

SOLUTION First we write the system in differential operator notation:

$$\begin{aligned} (D-4)x + D^2 y &= t^2 \\ (D+1)x + Dy &= 0. \end{aligned} \tag{10}$$

Then, by eliminating x, we obtain

$$[(D + 1)D^2 - (D - 4)D]y = (D + 1)t^2 - (D - 4)0$$

or

$$(D^3 + 4D)y = t^2 + 2t.$$

Since the roots of the auxiliary equation $m(m^2 + 4) = 0$ are $m_1 = 0$, $m_2 = 2i$, and $m_3 = -2i$, the complementary function is

$$y_c = c_1 + c_2 \cos 2t + c_3 \sin 2t.$$

To determine the particular solution y_p we use undetermined coefficients by assuming $y_p = At^3 + Bt^2 + Ct$. Therefore

$$y_p' = 3At^2 + 2Bt + C, \quad y_p'' = 6At + 2B, \quad y_p''' = 6A,$$

$$y_p''' + 4y_p' = 12At^2 + 8Bt + 6A + 4C = t^2 + 2t.$$

The last equality implies $12A = 1$, $8B = 2$, $6A + 4C = 0$, and hence $A = \frac{1}{12}$, $B = \frac{1}{4}$, $C = -\frac{1}{8}$. Thus

$$y = y_c + y_p = c_1 + c_2 \cos 2t + c_3 \sin 2t + \frac{1}{12}t^3 + \frac{1}{4}t^2 - \frac{1}{8}t. \tag{11}$$

Eliminating y from the system (9) leads to

$$[(D - 4) - D(D + 1)]x = t^2 \quad \text{or} \quad (D^2 + 4)x = -t^2.$$

It should be obvious that

$$x_c = c_4 \cos 2t + c_5 \sin 2t$$

and that undetermined coefficients can be applied to obtain a particular solution of the form $x_p = At^2 + Bt + C$. In this case the usual differentiations and algebra yield $x_p = -\frac{1}{4}t^2 + \frac{1}{8}$, and so

$$x = x_c + x_p = c_4 \cos 2t + c_5 \sin 2t - \frac{1}{4}t^2 + \frac{1}{8}. \tag{12}$$

Now c_4 and c_5 can be expressed in terms of c_2 and c_3 by substituting (11) and (12) into either equation of (9). By using the second equation, we find, after combining terms,

$$(c_5 - 2c_4 - 2c_2) \sin 2t + (2c_5 + c_4 + 2c_3) \cos 2t = 0$$

so that $c_5 - 2c_4 - 2c_2 = 0$ and $2c_5 + c_4 + 2c_3 = 0$. Solving for c_4 and c_5 in terms of c_2 and c_3 gives $c_4 = -\frac{1}{5}(4c_2 + 2c_3)$ and $c_5 = \frac{1}{5}(2c_2 - 4c_3)$. Finally, a solution of (9) is found to be

$$x(t) = -\frac{1}{5}(4c_2 + 2c_3) \cos 2t + \frac{1}{5}(2c_2 - 4c_3) \sin 2t - \frac{1}{4}t^2 + \frac{1}{8},$$

$$y(t) = c_1 + c_2 \cos 2t + c_3 \sin 2t + \frac{1}{12}t^3 + \frac{1}{4}t^2 - \frac{1}{8}t. \qquad \equiv$$

EXAMPLE 3 **A Mathematical Model Revisited**

In (3) of Section 2.9 we saw that a system of linear first-order differential equations described the number of pounds of salt $x_1(t)$ and $x_2(t)$ of a brine mixture that flows between two tanks. See Figure 2.9.1. At that time we were not able to solve the system. But now, in terms of differential operators, the system is

$$\left(D + \frac{2}{25}\right)x_1 - \frac{1}{50}x_2 = 0$$

$$-\frac{2}{25}x_1 + \left(D + \frac{2}{25}\right)x_2 = 0.$$

Operating on the first equation by $D + \frac{2}{25}$, multiplying the second equation by $\frac{1}{50}$, adding, and then simplifying, gives

$$(625D^2 + 100D + 3)x_1 = 0.$$

From the auxiliary equation $625m^2 + 100m + 3 = (25m + 1)(25m + 3) = 0$ we see immediately that

$$x_1(t) = c_1 e^{-t/25} + c_2 e^{-3t/25}.$$

In like manner we find

$$(625D^2 + 100D + 3)x_2 = 0$$

and so

$$x_2(t) = c_3 e^{-t/25} + c_4 e^{-3t/25}.$$

Substituting $x_1(t)$ and $x_2(t)$ into, say, the first equation of the system then gives

$$(2c_1 - c_3)e^{-t/25} + (-2c_2 - c_4)e^{-3t/25} = 0.$$

From this last equation we find $c_3 = 2c_1$ and $c_4 = -2c_2$. Thus a solution of the system is

$$x_1(t) = c_1 e^{-t/25} + c_2 e^{-3t/25}, \qquad x_2(t) = 2c_1 e^{-t/25} - 2c_2 e^{-3t/25}.$$

In the original discussion we assumed that initial conditions were $x_1(0) = 25$ and $x_2(0) = 0$. Applying these conditions to the solution yields $c_1 + c_2 = 25$ and $2c_1 - 2c_2 = 0$. Solving these equations simultaneously gives $c_1 = c_2 = \frac{25}{2}$. Finally, a solution of the initial-value problem is

$$x_1(t) = \frac{25}{2} e^{-t/25} + \frac{25}{2} e^{-3t/25}, \qquad x_2(t) = 25e^{-t/25} - 25e^{-3t/25}.$$

FIGURE 3.12.2 Pounds of salt in tanks in Example 3

The graphs of $x_1(t)$ and $x_2(t)$ are given in **FIGURE 3.12.2**. Notice that even though the number of pounds $x_2(t)$ of salt in tank B starts initially at 0 lb it quickly increases and surpasses the number of pounds $x_1(t)$ of salt in tank A. $\equiv$

In our next example we solve system (1) under the assumption that $k_1 = 6$, $k_2 = 4$, $m_1 = 1$, and $m_2 = 1$.

EXAMPLE 4 **A Special Case of System (1)**

Solve

$$\begin{aligned} x_1'' + 10x_1 \quad\; - 4x_2 &= 0 \\ -4x_1 + x_2'' + 4x_2 &= 0 \end{aligned} \tag{13}$$

subject to $x_1(0) = 0$, $x_1'(0) = 1$, $x_2(0) = 0$, $x_2'(0) = -1$.

SOLUTION Using elimination on the equivalent form of the system

$$\begin{aligned} (D^2 + 10)x_1 \quad\; - 4x_2 &= 0 \\ -4x_1 + (D^2 + 4)x_2 &= 0 \end{aligned}$$

we find that x_1 and x_2 satisfy, respectively,

$$(D^2 + 2)(D^2 + 12)x_1 = 0 \quad \text{and} \quad (D^2 + 2)(D^2 + 12)x_2 = 0.$$

Thus we find

$$x_1(t) = c_1 \cos \sqrt{2}t + c_2 \sin \sqrt{2}t + c_3 \cos 2\sqrt{3}t + c_4 \sin 2\sqrt{3}t$$
$$x_2(t) = c_5 \cos \sqrt{2}t + c_6 \sin \sqrt{2}t + c_7 \cos 2\sqrt{3}t + c_8 \sin 2\sqrt{3}t.$$

Substituting both expressions into the first equation of (13) and simplifying eventually yields $c_5 = 2c_1$, $c_6 = 2c_2$, $c_7 = -\frac{1}{2}c_3$, $c_8 = -\frac{1}{2}c_4$. Thus, a solution of (13) is

(a) $x_1(t)$

(b) $x_2(t)$

FIGURE 3.12.3 Displacements of the two masses in Example 4

$$x_1(t) = c_1 \cos \sqrt{2}t + c_2 \sin \sqrt{2}t + c_3 \cos 2\sqrt{3}t + c_4 \sin 2\sqrt{3}t$$
$$x_2(t) = 2c_1 \cos \sqrt{2}t + 2c_2 \sin \sqrt{2}t - \tfrac{1}{2}c_3 \cos 2\sqrt{3}t - \tfrac{1}{2}c_4 \sin 2\sqrt{3}t.$$

The stipulated initial conditions then imply $c_1 = 0$, $c_2 = -\sqrt{2}/10$, $c_3 = 0$, $c_4 = \sqrt{3}/5$. And so the solution of the initial-value problem is

$$x_1(t) = -\frac{\sqrt{2}}{10}\sin\sqrt{2}t + \frac{\sqrt{3}}{5}\sin 2\sqrt{3}t$$

$$x_2(t) = -\frac{\sqrt{2}}{5}\sin\sqrt{2}t - \frac{\sqrt{3}}{10}\sin 2\sqrt{3}t. \tag{14}$$

The graphs of x_1 and x_2 in **FIGURE 3.12.3** reveal the complicated oscillatory motion of each mass. ≡

We will revisit Example 4 in Section 4.6, where we will solve the system in (13) by means of the Laplace transform.

3.12 Exercises Answers to selected odd-numbered problems begin on page ANS-8.

In Problems 1–20, solve the given system of differential equations by systematic elimination.

1. $\dfrac{dx}{dt} = 2x - y$

$\dfrac{dy}{dt} = x$

2. $\dfrac{dx}{dt} = 4x + 7y$

$\dfrac{dy}{dt} = x - 2y$

3. $\dfrac{dx}{dt} = -y + t$

$\dfrac{dy}{dt} = x - t$

4. $\dfrac{dx}{dt} - 4y = 1$

$\dfrac{dy}{dt} + x = 2$

5. $(D^2 + 5)x - \quad 2y = 0$

$-2x + (D^2 + 2)y = 0$

6. $(D + 1)x + (D - 1)y = 2$

$3x + (D + 2)y = -1$

7. $\dfrac{d^2x}{dt^2} = 4y + e^t$

$\dfrac{d^2y}{dt^2} = 4x - e^t$

8. $\dfrac{d^2x}{dt^2} + \dfrac{dy}{dt} = -5x$

$\dfrac{dx}{dt} + \dfrac{dy}{dt} = -x + 4y$

9. $Dx + D^2y = e^{3t}$

$(D + 1)x + (D - 1)y = 4e^{3t}$

10. $D^2x - Dy = t$

$(D + 3)x + (D + 3)y = 2$

11. $(D^2 - 1)x - y = 0$

$(D - 1)x + Dy = 0$

12. $(2D^2 - D - 1)x - (2D + 1)y = 1$

$(D - 1)x + \quad\quad Dy = -1$

13. $2\dfrac{dx}{dt} - 5x + \dfrac{dy}{dt} = e^t$

$\dfrac{dx}{dt} - x + \dfrac{dy}{dt} = 5e^t$

14. $\dfrac{dx}{dt} + \dfrac{dy}{dt} = e^t$

$-\dfrac{d^2x}{dt^2} + \dfrac{dx}{dt} + x + y = 0$

15. $(D - 1)x + (D^2 + 1)y = 1$

$(D^2 - 1)x + (D + 1)y = 2$

16. $D^2x - 2(D^2 + D)y = \sin t$

$x + \quad\quad Dy = 0$

17. $Dx = y$

$Dy = z$

$Dz = x$

18. $Dx + \quad\quad z = e^t$

$(D - 1)x + Dy + Dz = 0$

$x + 2y + Dz = e^t$

19. $\dfrac{dx}{dt} = 6y$

$\dfrac{dy}{dt} = x + z$

$\dfrac{dz}{dt} = x + y$

20. $\dfrac{dx}{dt} = -x + z$

$\dfrac{dy}{dt} = -y + z$

$\dfrac{dz}{dt} = -x + y$

In Problems 21 and 22, solve the given initial-value problem.

21. $\dfrac{dx}{dt} = -5x - y$

$\dfrac{dy}{dt} = 4x - y$

$x(1) = 0,\ y(1) = 1$

22. $\dfrac{dx}{dt} = y - 1$

$\dfrac{dy}{dt} = -3x + 2y$

$x(0) = 0,\ y(0) = 0$

Mathematical Models

23. Projectile Motion A projectile shot from a gun has weight $w = mg$ and velocity $\mathbf{v}$ tangent to its path of motion or trajectory. Ignoring air resistance and all other forces acting on the projectile except its weight, determine a system of differential equations that describes its path of motion. See **FIGURE 3.12.4**. Solve the system. [*Hint*: Use Newton's second law of motion in the x and y directions.]

FIGURE 3.12.4 Path of projectile in Problem 23

24. Projectile Motion with Air Resistance Determine a system of differential equations that describes the path of motion in Problem 23 if linear air resistance is a retarding force $\mathbf{k}$ (of magnitude k) acting tangent to the path of the projectile but opposite to its motion. See **FIGURE 3.12.5**. Solve the system. [*Hint*: $\mathbf{k}$ is a multiple of velocity, say, $\beta\mathbf{v}$.]

FIGURE 3.12.5 Forces in Problem 24

Computer Lab Assignments

25. Consider the solution $x_1(t)$ and $x_2(t)$ of the initial-value problem given at the end of Example 3. Use a CAS to graph $x_1(t)$ and $x_2(t)$ in the same coordinate plane on the interval [0, 100]. In Example 3, $x_1(t)$ denotes the number of pounds of salt in tank A at time t, and $x_2(t)$ the number of pounds of salt in tank B at time t. See Figure 2.9.1. Use a root-finding application to determine when tank B contains more salt than tank A.

26. (a) Reread Problem 10 of Exercises 2.9. In that problem you were asked to show that the system of differential equations

$$\frac{dx_1}{dt} = -\frac{1}{50}x_1$$

$$\frac{dx_2}{dt} = \frac{1}{50}x_1 - \frac{2}{75}x_2$$

$$\frac{dx_3}{dt} = \frac{2}{75}x_2 - \frac{1}{25}x_3$$

is a model for the amounts of salt in the connected mixing tanks A, B, and C shown in Figure 2.9.7. Solve the system subject to $x_1(0) = 15$, $x_2(0) = 10$, $x_3(0) = 5$.

(b) Use a CAS to graph $x_1(t)$, $x_2(t)$, and $x_3(t)$ in the same coordinate plane on the interval [0, 200].

(c) Since only pure water is pumped into tank A, it stands to reason that the salt will eventually be flushed out of all three tanks. Use a root-finding application of a CAS to determine the time when the amount of salt in each tank is less than or equal to 0.5 pounds. When will the amounts of salt $x_1(t)$, $x_2(t)$, and $x_3(t)$ be simultaneously less than or equal to 0.5 pounds?

27. (a) Use systematic elimination to solve the system (1) for the coupled spring/mass system when $k_1 = 4$, $k_2 = 2$, $m_1 = 2$, and $m_2 = 1$ and with initial conditions $x_1(0) = 2$, $x_1'(0) = 1$, $x_2(0) = -1$, $x_2'(0) = 1$.

(b) Use a CAS to plot the graphs of $x_1(t)$ and $x_2(t)$ in the tx-plane. What is the fundamental difference in the motions of the masses m_1 and m_2 in this problem and that of the masses illustrated in Figure 3.12.3?

(c) As parametric equations, plot $x_1(t)$ and $x_2(t)$ in the x_1x_2-plane. The curve defined by these parametric equations is called a **Lissajous curve**.

3 Chapter in Review Answers to selected odd-numbered problems begin on page ANS-8.

Answer Problems 1−8 without referring back to the text. Fill in the blank or answer true/false.

1. The only solution of the initial-value problem $y'' + x^2y = 0$, $y(0) = 0$, $y'(0) = 0$ is _____.

2. For the method of undetermined coefficients, the assumed form of the particular solution y_p for $y'' - y = 1 + e^x$ is _____.

3. A constant multiple of a solution of a linear differential equation is also a solution. _____

4. If f_1 and f_2 are linearly independent functions on an interval I, then their Wronskian $W(f_1, f_2) \neq 0$ for all x in I. _____

5. If a 10-pound weight stretches a spring 2.5 feet, a 32-pound weight will stretch it _____ feet.

6. The period of simple harmonic motion of an 8-pound weight attached to a spring whose constant is 6.25 lb/ft is _____ seconds.

7. The differential equation describing the motion of a mass attached to a spring is $x'' + 16x = 0$. If the mass is released at $t = 0$ from 1 meter above the equilibrium position with a downward velocity of 3 m/s, the amplitude of vibrations is _____ meters.

8. If simple harmonic motion is described by $x(t) = (\sqrt{2}/2)\sin(2t + \phi)$, the phase angle ϕ is _____ when $x(0) = -\frac{1}{2}$ and $x'(0) = 1$.

9. Give an interval over which $f_1(x) = x^2$ and $f_2(x) = x\,|x|$ are linearly independent. Then give an interval on which f_1 and f_2 are linearly dependent.

10. Without the aid of the Wronskian determine whether the given set of functions is linearly independent or linearly dependent on the indicated interval.

 (a) $f_1(x) = \ln x$, $f_2(x) = \ln x^2$, $(0, \infty)$
 (b) $f_1(x) = x^n$, $f_2(x) = x^{n+1}$, $n = 1, 2, \ldots, (-\infty, \infty)$
 (c) $f_1(x) = x$, $f_2(x) = x + 1$, $(-\infty, \infty)$
 (d) $f_1(x) = \cos(x + \pi/2)$, $f_2(x) = \sin x$, $(-\infty, \infty)$
 (e) $f_1(x) = 0$, $f_2(x) = x$, $(-5, 5)$
 (f) $f_1(x) = 2$, $f_2(x) = 2x$, $(-\infty, \infty)$
 (g) $f_1(x) = x^2$, $f_2(x) = 1 - x^2$, $f_3(x) = 2 + x^2$, $(-\infty, \infty)$
 (h) $f_1(x) = xe^{x+1}$, $f_2(x) = (4x - 5)e^x$, $f_3(x) = xe^x$, $(-\infty, \infty)$

11. Suppose $m_1 = 3$, $m_2 = -5$, and $m_3 = 1$ are roots of multiplicity one, two, and three, respectively, of an auxiliary equation. Write down the general solution of the corresponding homogeneous linear DE if it is
 (a) an equation with constant coefficients,
 (b) a Cauchy–Euler equation.

12. Find a Cauchy–Euler differential equation $ax^2y'' + bxy' + cy = 0$, where a, b, and c are real constants, if it is known that
 (a) $m_1 = 3$ and $m_2 = -1$ are roots of its auxiliary equation,
 (b) $m_1 = i$ is a complex root of its auxiliary equation.

In Problems 13 and 14, verify that the indicated function $y_1(x)$ is a solution of the given differential equation. Use formula (5) of Section 3.2 to find the general solution of the equation on the interval $(0, \infty)$.

13. $y'' - 2xy' + 2y = 0$; $y_1(x) = x$

14. $x^2y'' + 3xy' + y = 0$; $y_1(x) = x^{-1}$

In Problems 15–30, use the procedures developed in this chapter to find the general solution of each differential equation.

15. $y'' - 2y' - 2y = 0$

16. $2y'' + 2y' + 3y = 0$

17. $y''' + 10y'' + 25y' = 0$

18. $2y''' + 9y'' + 12y' + 5y = 0$

19. $3y''' + 10y'' + 15y' + 4y = 0$

20. $2y^{(4)} + 3y''' + 2y'' + 6y' - 4y = 0$

21. $y'' - 3y' + 5y = 4x^3 - 2x$

22. $y'' - 2y' + y = x^2e^x$

23. $y''' - 5y'' + 6y' = 8 + 2\sin x$

24. $y''' - y'' = 6$

25. $y'' - 2y' + 2y = e^x \tan x$

26. $y'' - y = \dfrac{2e^x}{e^x + e^{-x}}$

27. $6x^2y'' + 5xy' - y = 0$

28. $2x^3y''' + 19x^2y'' + 39xy' + 9y = 0$

29. $x^2y'' - 4xy' + 6y = 2x^4 + x^2$

30. $x^2y'' - xy' + y = x^3$

31. Write down the form of the general solution $y = y_c + y_p$ of the given differential equation in the two cases $\omega \neq \alpha$ and $\omega = \alpha$. Do not determine the coefficients in y_p.
 (a) $y'' + \omega^2 y = \sin \alpha x$ (b) $y'' - \omega^2 y = e^{\alpha x}$

32. (a) Given that $y = \sin x$ is a solution of $y^{(4)} + 2y''' + 11y'' + 2y' + 10y = 0$, find the general solution of the DE *without the aid of a calculator or a computer.*
 (b) Find a linear second-order differential equation with constant coefficients for which $y_1 = 1$ and $y_2 = e^{-x}$ are solutions of the associated homogeneous equation and $y_p = \frac{1}{2}x^2 - x$ is a particular solution of the nonhomogeneous equation.

33. (a) Write the general solution of the fourth-order DE $y^{(4)} - 2y'' + y = 0$ entirely in terms of hyperbolic functions.
 (b) Write down the form of a particular solution of $y^{(4)} - 2y'' + y = \sinh x$.

34. Consider the differential equation $x^2y'' - (x^2 + 2x)y' + (x + 2)y = x^3$. Verify that $y_1 = x$ is one solution of the associated homogeneous equation. Then show that the method of reduction of order discussed in Section 3.2 leads both to a second solution y_2 of the homogeneous equation and to a particular solution y_p of the nonhomogeneous equation. Form the general solution of the DE on the interval $(0, \infty)$.

In Problems 35–40, solve the given differential equation subject to the indicated conditions.

35. $y'' - 2y' + 2y = 0$, $y(\pi/2) = 0$, $y(\pi) = -1$

36. $y'' + 2y' + y = 0$, $y(-1) = 0$, $y'(0) = 0$

37. $y'' - y = x + \sin x$, $y(0) = 2$, $y'(0) = 3$

38. $y'' + y = \sec^3 x$, $y(0) = 1$, $y'(0) = \frac{1}{2}$

39. $y'y'' = 4x$, $y(1) = 5$, $y'(1) = 2$

40. $2y'' = 3y^2$, $y(0) = 1$, $y'(0) = 1$

41. (a) Use a CAS as an aid in finding the roots of the auxiliary equation for $12y^{(4)} + 64y''' + 59y'' - 23y' - 12y = 0$. Give the general solution of the equation.
 (b) Solve the DE in part (a) subject to the initial conditions $y(0) = -1, y'(0) = 2, y''(0) = 5, y'''(0) = 0$. Use a CAS as an aid in solving the resulting systems of four equations in four unknowns.

42. Find a member of the family of solutions of

$$xy'' + y' + \sqrt{x} = 0$$

whose graph is tangent to the x-axis at $x = 1$. Use a graphing utility to obtain the solution curve.

In Problems 43–46, use systematic elimination to solve the given system of differential equations.

43. $\dfrac{dx}{dt} + \dfrac{dy}{dt} = 2x + 2y + 1$ 44. $\dfrac{dx}{dt} = 2x + y + t - 2$

 $\dfrac{dx}{dt} + 2\dfrac{dy}{dt} = y + 3$ $\dfrac{dy}{dt} = 3x + 4y - 4t$

45. $(D - 2)x \qquad\quad -y = -e^t$

 $-3x + (D - 4)y = -7e^t$

46. $(D + 2)x + (D + 1)y = \sin 2t$

 $5x + (D + 3)y = \cos 2t$

47. A free undamped spring/mass system oscillates with a period of 3 s. When 8 lb is removed from the spring, the system then has a period of 2 s. What was the weight of the original mass on the spring?

48. A 12-pound weight stretches a spring 2 feet. The weight is released from a point 1 foot below the equilibrium position with an upward velocity of 4 ft/s.
 (a) Find the equation describing the resulting simple harmonic motion.
 (b) What are the amplitude, period, and frequency of motion?
 (c) At what times does the weight return to the point 1 foot below the equilibrium position?
 (d) At what times does the weight pass through the equilibrium position moving upward? Moving downward?
 (e) What is the velocity of the weight at $t = 3\pi/16$ s?
 (f) At what times is the velocity zero?

49. A spring with constant $k = 2$ is suspended in a liquid that offers a damping force numerically equal to four times the instantaneous velocity. If a mass m is suspended from the spring, determine the values of m for which the subsequent free motion is nonoscillatory.

50. A 32-pound weight stretches a spring 6 inches. The weight moves through a medium offering a damping force numerically equal to β times the instantaneous velocity. Determine the values of β for which the system will exhibit oscillatory motion.

51. A series circuit contains an inductance of $L = 1$ h, a capacitance of $C = 10^{-4}$ F, and an electromotive force of $E(t) = 100 \sin 50t$ V. Initially the charge q and current i are zero.
 (a) Find the equation for the charge at time t.
 (b) Find the equation for the current at time t.
 (c) Find the times for which the charge on the capacitor is zero.

52. Show that the current $i(t)$ in an LRC-series circuit satisfies the differential equation

$$L\frac{d^2i}{dt^2} + R\frac{di}{dt} + \frac{1}{C}i = E'(t),$$

where $E'(t)$ denotes the derivative of $E(t)$.

53. Consider the boundary-value problem

$$y'' + \lambda y = 0, \quad y(0) = y(2\pi), \quad y'(0) = y'(2\pi).$$

Show that except for the case $\lambda = 0$, there are two independent eigenfunctions corresponding to each eigenvalue.

54. Sliding Bead A bead is constrained to slide along a frictionless rod of length L. The rod is rotating in a vertical plane with a constant angular velocity ω about a pivot P fixed at the midpoint of the rod, but the design of the pivot allows the bead to move along the entire length of the rod. Let $r(t)$ denote the position of the bead relative to this rotating coordinate system, as shown in **FIGURE 3.R.1**. In order to apply Newton's second law of motion to this rotating frame of reference it is necessary to use the fact that the net force acting on the bead is the sum of the real forces (in this case, the force due to gravity) and the inertial forces (coriolis, transverse, and centrifugal). The mathematics is a little complicated, so we give just the resulting differential equation for r,

$$m\frac{d^2r}{dt^2} = m\omega^2 r - mg\sin(\omega t).$$

 (a) Solve the foregoing DE subject to the initial conditions $r(0) = r_0$, $r'(0) = v_0$.
 (b) Determine initial conditions for which the bead exhibits simple harmonic motion. What is the minimum length L of the rod for which it can accommodate simple harmonic motion of the bead?
 (c) For initial conditions other than those obtained in part (b), the bead must eventually fly off the rod. Explain using the solution $r(t)$ in part (a).
 (d) Suppose $\omega = 1$ rad/s. Use a graphing utility to plot the graph of the solution $r(t)$ for the initial conditions $r(0) = 0$, $r'(0) = v_0$, where v_0 is 0, 10, 15, 16, 16.1, and 17.
 (e) Suppose the length of the rod is $L = 40$ ft. For each pair of initial conditions in part (d), use a root-finding application to find the total time that the bead stays on the rod.

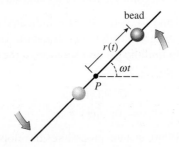

FIGURE 3.R.1 Rotating rod in Problem 54

55. Suppose a mass m lying on a flat, dry, frictionless surface is attached to the free end of a spring whose constant is k. In **FIGURE 3.R.2(a)** the mass is shown at the equilibrium position $x = 0$; that is, the spring is neither stretched nor compressed. As shown in Figure 3.R.2(b), the displacement $x(t)$ of the mass to the right of the equilibrium position is positive and negative to the left. Derive a differential equation for the free horizontal (sliding) motion of the mass. Discuss the difference between the derivation of this DE and the analysis leading to (1) of Section 3.8.

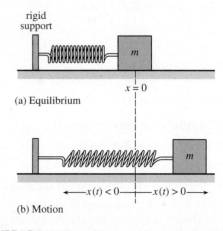

FIGURE 3.R.2 Sliding spring/mass system in Problem 55

56. What is the differential equation of motion in Problem 55 if kinetic friction (but no other damping forces) acts on the sliding mass? [*Hint*: Assume that the magnitude of the force of kinetic friction is $f_k = \mu mg$, where mg is the weight of the mass and the constant $\mu > 0$ is the coefficient of kinetic friction. Then consider two cases: $x' > 0$ and $x' < 0$. Interpret these cases physically.]

In Problems 57 and 58, use a Green's function to solve the given initial-value problem.

57. $y'' + y = \tan x$, $y(0) = 2$, $y'(0) = -5$

58. $x^2 y'' - 3xy + 4y = \ln x$, $y(1) = 0$, $y'(1) = 0$

59. The Caught Pendulum Suppose the massless rod in the discussion of the nonlinear pendulum in Section 3.11 is actually a string of length l. A mass m is attached to the end of the string and the pendulum is released from rest at a small displacement angle $\theta_0 > 0$. When the pendulum reaches the equilibrium position, the dashed red line in **FIGURE 3.R.3**, the string hits a nail and gets caught at this point $l/4$ above the mass. The mass oscillates from this new pivot point as shown in the figure.

(a) Construct and solve a linear initial-value problem that gives the displacement angle, denote it $\theta_1(t)$, for $0 \le t < T$, where T represents the time when the string first hits the nail.

(b) Find the time T in part (a).

(c) Construct and solve a linear initial-value problem that gives the displacement angle, denote it $\theta_2(t)$, for $t \ge T$, where T is the time in part (a). Compare the amplitude and period of oscillations in this case with that predicted by the initial-value problem in part (a).

FIGURE 3.R.3 Pendulum in Problem 59

60. The Caught Pendulum—Continued (a) Use a graphing utility to obtain the graphs of the displacement angles $\theta_1(t)$ and $\theta_2(t)$ in parts (a) and (c) of Problem 59. Use the same coordinate axes. Take $\theta_0 = 0.2$ radians and $l = 2$ ft.

(b) Then use a numerical solver or a CAS to obtain the graphs of the solutions $\theta_1(t)$ and $\theta_2(t)$ of the corresponding *nonlinear* initial-value problems and compare with part (a). Take $\theta_0 = 0.2$ radians and $l = 2$ ft.

(c) Experiment with values of θ_0 until you discern a noticeable difference between the solutions of the linear and nonlinear initial-value problems.

61. Use a Maclaurin series to show that a power series solution of the initial-value problem

$$\frac{d^2\theta}{dt^2} + \frac{g}{l}\sin\theta = 0, \quad \theta(0) = \frac{\pi}{6}, \quad \theta'(0) = 0$$

is given by

$$\theta(t) = \frac{\pi}{6} - \frac{g}{4l}t^2 + \frac{\sqrt{3}g^2}{96l^2}t^4 + \cdots.$$

[*Hint*: See Example 3 in Section 3.7.]

62. Spring Pendulum The rotational form of Newton's second law of motion is:

The time rate of change of angular momentum about a point is equal to the moment of the resultant force (torque).

In the absence of damping or other external forces, an analogue of (14) in Section 3.11 for the pendulum shown in Figure 3.11.3 is then

$$\frac{d}{dt}\left(ml^2\frac{d\theta}{dt}\right) = -mgl\sin\theta. \tag{1}$$

(a) When m and l are constant show that (1) reduces to (6) of Section 3.11.

(b) Now suppose the rod in Figure 3.11.3 is replaced with a spring of negligible mass. When a mass m is attached to its free end the spring hangs in the vertical equilibrium position shown in **FIGURE 3.R.4** and has length l_0. When the spring pendulum is set in motion we assume that the motion takes place in a vertical plane and the spring is stiff enough not to bend. For $t > 0$ the length of the spring is then $l(t) = l_0 + x(t)$, where x is the displacement from the equilibrium position. Find the differential equation for the displacement angle $\theta(t)$ defined by (1).

FIGURE 3.R.4 Spring pendulum in Problem 62

Contributed Problem

Jeff Dodd, Professor
Department of Mathematical Sciences, Jacksonville State University

63. The Paris Guns The first mathematically correct theory of projectile motion was originally formulated by the Italian astronomer/mathematician **Galileo Galilei** (1564–1642). Galileo's younger Italian collaborators **Bonaventura Cavalieri** (1598–1647) and **Evangelista Torricelli** (1608–1647) then clarified and extended his theory. Galileo's theory was based on two simple hypotheses suggested

by experimental observations: that a projectile moves with constant horizontal velocity and with constant downward vertical acceleration. Galileo, Cavalieri, and Torricelli did not have calculus at their disposal, so their arguments were largely geometric, but we can reproduce their results using a system of differential equations. Suppose that a projectile is launched from ground level at an angle θ with respect to the horizontal and with an initial velocity of magnitude $\|\mathbf{v}_0\| = v_0$ m/s. Let the projectile's height above the ground be y meters and its horizontal distance from the launch site be x meters, and for convenience take the launch site to be the origin in the xy-plane. Then Galileo's hypotheses can be represented by the following initial-value problem:

$$\frac{d^2x}{dt^2} = 0$$
$$\frac{d^2y}{dt^2} = -g, \tag{2}$$

where $g = 9.8$ m/s^2, $x(0) = 0$, $y(0) = 0$, $x'(0) = v_0\cos\theta$ (the x-component of the initial velocity), and $y'(0) = v_0\sin\theta$ (the y-component of the initial velocity). See **FIGURE 3.R.5** and Problem 23 in Exercises 3.12.

(a) Note that the system of equations in (2) is decoupled; that is, it consists of separate differential equations for $x(t)$ and $y(t)$. Moreover, each of these differential equations can be solved simply by antidifferentiating twice. Solve (2) to obtain explicit formulas for $x(t)$ and $y(t)$ in terms of v_0 and θ. Then algebraically eliminate t to show that the trajectory of the projectile in the xy-plane is a parabola.

(b) A central question throughout the history of ballistics has been this: Given a gun that fires a projectile with a certain initial speed v_0, at what angle with respect to the horizontal should the gun be fired to maximize its range? The range is the horizontal distance traversed by a projectile before it hits the ground. Show that according to (2), the range of the projectile is $(v_0^2/g)\sin 2\theta$, so that a maximum range v_0^2/g is achieved for the launch angle $\theta = \pi/4 = 45°$.

(c) Show that the maximum height attained by the projectile if launched with $\theta = 45°$ for maximum range is $v_0^2/(4g)$.

FIGURE 3.R.5 Ballistic projectile in Problem 63

64. **The Paris Guns—Continued** Mathematically, Galileo's model in Problem 63 is perfect. But in practice it is only as accurate as the hypotheses upon which it is based. The motion of a real projectile is resisted to some extent by the air, and the stronger this effect, the less realistic are the

hypotheses of constant horizontal velocity and constant vertical acceleration, as well as the resulting independence of the projectile's motion in the x and y directions.

The first successful model of air resistance was formulated by the Dutch scientist **Christiaan Huygens** (1629–1695) and by the English polymath **Sir Isaac Newton** (1643–1727). It was based not so much on a detailed mathematical formulation of the underlying physics involved, which was beyond what anyone could manage at the time, but on physical intuition and groundbreaking experimental work. Newton's version, which is known to this day as **Newton's law of air resistance**, states that the resisting force or **drag force** on an object moving through a resisting medium acts in the direction opposite to the direction of the object's motion with a magnitude proportional to the density ρ of the medium, the cross sectional area A of the object taken perpendicular to the direction of motion, and the square of the speed $\|\mathbf{v}\|$ of the object. In modern vector notation, the drag force is a vector $\mathbf{f}_D$ given in terms of the velocity $\mathbf{v}$ of the object:

$$\mathbf{f}_D = -\frac{1}{2}C\rho A\|\mathbf{v}\|^2\frac{\mathbf{v}}{\|\mathbf{v}\|} = -\frac{1}{2}C\rho A\|\mathbf{v}\|\,\mathbf{v},$$

where in the coordinate system of Problem 63 the velocity is the vector $\mathbf{v} = \langle dx/dt, dy/dt\rangle$. When the force of gravity and this drag force are combined according to Newton's second law of motion we get:

$$m\left\langle \frac{d^2x}{dt^2}, \frac{d^2y}{dt^2}\right\rangle = \langle 0, -mg\rangle + \mathbf{f}_D.$$

The initial-value problem (2) is then modified to read:

$$m\frac{d^2x}{dt^2} = -\frac{1}{2}C\rho A\left[\sqrt{\left(\frac{dx}{dt}\right)^2 + \left(\frac{dx}{dt}\right)^2}\right]\frac{dx}{dt}$$
$$m\frac{d^2y}{dt^2} = -mg - \frac{1}{2}C\rho A\left[\sqrt{\left(\frac{dx}{dt}\right)^2 + \left(\frac{dx}{dt}\right)^2}\right]\frac{dy}{dt}, \tag{3}$$

where $x(0) = 0$, $y(0) = 0$, $x'(0) = v_0\cos\theta$, $y'(0) = v_0\sin\theta$.

Huygens seemed to believe that the proportionality of drag force to the square of the speed was universal, but Newton suspected that multiple different physical effects contributed to drag force, not all of which behaved that way. He turned out to be correct. For example, when the speed of an object is low enough compared to the **viscosity** (internal resistance to flow) of the medium, drag force ends up being approximately proportional to its speed (not the square of its speed), a relationship known as **Stokes' law of air resistance**. Even today, there is no quick recipe for predicting the drag force for all objects under all conditions. The modeling of air resistance is complicated and is done in practice by a combination of theoretical and empirical methods. The coefficient C in (3) is called the **drag coefficient**. It is dimensionless (that is, it is the same no matter what units are used to measure mass, distance, and time) and it can usually be regarded as depending only on the

shape of a projectile and not on its size. The drag coefficient is such a convenient index for measuring how much air resistance is felt by a projectile of a given shape that it is now defined in terms of the drag force to be $2\|\mathbf{f}_D\|/(\rho A\|\mathbf{v}\|^2)$ even when this ratio cannot be regarded as constant. For example, under Stokes' law of air resistance, C would be proportional to the reciprocal of the speed. Of greater concern to us is the fact that the drag coefficient of a projectile in air increases sharply as its speed approaches the speed of sound (approximately 340 m/s in air), then decreases gradually for even higher speeds, becoming nearly constant again for speeds several times the speed of sound. This was first discovered by the military engineer Benjamin Robins, whose book *Principles of Gunnery* is generally regarded as inaugurating the modern age of artillery and of the science of ballistics in general. As guns were used to shoot projectiles further and further with greater and greater initial speeds throughout the eighteenth and nineteenth centuries, the dependence of the drag coefficient on speed took on greater and greater practical importance. Moreover, as these projectiles went higher and higher, the fact that the density of the air decreases with increasing altitude also became important. By World War I, the density of the air as a function of altitude y above sea level in meters was commonly modeled by the function:

$$\rho(y) = 1.225e^{-0.00010361y}\,\text{kg/m}^3$$

and military engineers were accustomed to incorporating into (3) the dependence of C on velocity and of ρ on y. But one last major surprise was stumbled upon by German engineers during World War I. Our version of this story is based on the book *Paris Kanonen—the Paris Guns (Wilhelmgeschütze) and Project HARP* by Gerald V. Bull (Verlag E. S. Mittler & Sohn GmbH, Herford, 1988).

In the fall of 1914, the German Navy charged the Friedrich Krupp engineering firm with designing a system (gun and shells) capable of bombing the English port of Dover from the French coast. This would require firing a shell approximately 37 kilometers, a range some 16 kilometers greater than had ever been achieved before. Krupp was ready for this challenge because it had already succeeded in designing and building shells with innovative shapes that had lower drag coefficients than any prewar shells. The drag coefficient for one of these shells can be well approximated by the following piecewise linear function, where the speed v is in m/s:

$$C(v) = \begin{cases} 0.2, & 0 < v < 306 \\ 0.2 + (v - 306)/340, & 306 \le v < 374, \\ 0.4 - (v - 374)/3230, & 374 \le v < 1020 \\ 0.2, & v > 1020. \end{cases}$$

In addition, Krupp's engineers already had built an experimental gun having a 35.5 cm diameter barrel that could fire 535 kg shells with an initial speed of 940 m/s. They calculated that if they built one of their new low-drag shells with that diameter and mass and used this gun

to launch it at a 43° angle to the horizontal, the shell should have a range of about 39 km. The shell was built, and a test firing was conducted on October 21, 1914. For the results, we quote a first-hand account by Professor Fritz Rausenberger, managing director of the Krupp firm at the time (from pages 24–25 in Bull's book):

After the firing of the first shot, with a top zone propelling charge and at 43° elevation, we all waited with anxiety for the spotting report to be telephoned back to us giving the location of the inert shell impact. The anxiety was that normally associated when trying to reach a range never before achieved. But the spotter's report on impact never came. None of the observers located along the full length of the range had observed impact

Since no observation posts had been established beyond the 40 km mark, any impact outside of the area would have to be located and reported by local inhabitants using normal telephone communication between the neighbouring farms and villages. Thus it took several hours before the range staff received notification that the shell had impacted in a garden (without causing damage) some 49 km down-range from the battery. This was an unexpectedly favorable result but raised the question of how the range increase of 25% over that predicted using standard exterior ballistic techniques occurred After careful study of the method of calculating range, it was clear that in the computations an average, constant air density was used which was larger than the average along the trajectory. The method of calculating trajectory was therefore changed to allow for variation of density along the trajectory. This was done by dividing the atmosphere into 3 km bands from the Earth's surface upwards. For each band an average air density value was determined and applied over that portion of the trajectory falling in this band. This step-by-step calculation technique was carried out from the muzzle until impact. The resultant calculated trajectory, using the drag coefficient as determined from small calibre firings, matched closely the experimental results from the 21st of October Meppen firing.

Note that Rausenberger does not tell us what "average, constant" air density was used in the faulty calculation, or how it was determined. Actually, there is a logical problem here in that it is not possible to know how low the air density will become along the path of a trajectory without already knowing how high the trajectory will go. Nevertheless the engineers were confident of their calculations, so it seems likely that they did not regard the air density as a critical parameter when their concern was only to find an approximate range. (After all, they had never shot anything so high before.)

(a) Use a computer algebra system to write a routine that can numerically solve (3) with the piecewise defined

C and exponentially decaying ρ given above and graph the resulting trajectory in the xy-plane. (You may need to rewrite (3) as a first-order system. See Section 6.4.) The area A is that of a circle with the diameter of the shell. Test your routine on the case $\rho = 0$, which was solved analytically in Problem 63.

(b) Suppose that, as Krupp engineers, we use the results of part (c) in Problem 63 to calculate the maximum height M that would have been attained by the test shell had it been launched in a vacuum ($\rho = 0$), then figure that the real test shell might reach about half that height, and finally settle on a "constant, average" value for ρ of $(\rho(M/2) + \rho(0))/2$. Plot the resulting trajectory, and show that the resulting range is uncannily close to that predicted by the Krupp engineers for the October 14, 1914, test.

(c) Note that the launch angle for this test was not the 45° angle that yields maximum range in a vacuum, but a smaller 43° angle. Does this smaller launch angle lead to greater range under the constant air density assumption that you used in part (b)? To the nearest degree, what launch angle yields maximum range according to this model?

(d) Now plot the trajectory of the test shell using the proper exponentially decaying ρ. What happens? (In evaluating the results, it should make you feel better that we are not pretending to take everything into account in this model. For example, a missile moving at an angle relative to its axis of symmetry can experience a substantial lift force of the same sort that makes airplanes fly. Our model does not account for the possibility of lift, the curvature and rotation of the Earth, or numerous other effects.)

Rausenberger notes that once his engineers realized how important the exponentially decaying ρ was in calculating range, they did a series of calculations and found that the maximum range for their test would actually have been achieved with a launch angle of 50° to 55°. In hindsight, they realized that this was because a larger launch angle would result in the shell traveling higher and therefore in less dense air.

(e) Check this by plotting the trajectory of the test shell with exponentially decaying ρ every two degrees from 43° to 55°. What do you find?

After these surprising results, engineers at Krupp became interested in the challenge of attaining even larger ranges. The only way they could think of to talk the German High Command into committing to the trouble and expense of pursuing this goal was to sell them on the possibility of bombing Paris from behind the German front line, which would require a range of some 120 km. The German High Command quickly approved this idea, and after several years of work Krupp produced what are now known as the **Paris Guns**. These guns were designed to launch a 106 kg shell having a diameter of 216 mm with an initial velocity of 1646 m/s. At a 50° launch angle, such a shell was predicted to travel over 120 km.

© Chronicle/Alamy Stock Photo

A Paris Gun and shells

(f) Simulate the trajectory of a shell from a Paris Gun using (2) with exponentially decaying ρ. Evaluate the results keeping in mind the caveats about our modeling noted in part (d). How high does the shell go? Now change the launch angle from 50° to 45°. What happens?

Seven Paris Guns were built, but only three were used. They fired a total of 351 shells towards Paris between March 23 and August 9 of 1918. The damage and casualties that they caused were not tactically significant. They were never intended to be so; there was no control over where the shells would fall in Paris, and the amount of explosive carried by each shell was quite small. Instead, they were intended as a form of intimidation, a "scare tactic." However, military historians agree that they were not effective in that sense either. Their significance turned out to be more scientific than military. The shells that they launched were the first man-made objects to reach the stratosphere, initiating the space age in the science of ballistics.

CHAPTER 4

The Laplace Transform

In Chapter 3 we solved linear mathematical models such as $mx'' + \beta x' + \kappa x = f(t)$ and $Lq'' + Rq' + q/C = E(t)$ where the driving functions $f(t)$ and $E(t)$ were continuous. However, in practice, discontinuous driving functions are not uncommon. Although we solved piecewise linear DEs in Chapter 2, the Laplace transform discussed in this chapter is an especially valuable tool that simplifies the solution of such equations.

4.1 Definition of the Laplace Transform

INTRODUCTION In elementary calculus you learned that differentiation and integration are *transforms*—this means, roughly speaking, that these operations transform a function into another function. For example, the function $f(x) = x^2$ is transformed, in turn, into a linear function, a family of cubic polynomial functions, and a constant by the operations of differentiation, indefinite integration, and definite integration:

$$\frac{d}{dx}x^2 = 2x, \quad \int x^2\, dx = \frac{1}{3}x^3 + c, \quad \int_0^3 x^2\, dx = 9.$$

Moreover, these two transforms possess the **linearity property**; this means the transform of a linear combination of functions is a linear combination of the transforms. For α and β constants,

$$\frac{d}{dx}[\alpha f(x) + \beta g(x)] = \alpha f'(x) + \beta g'(x)$$

$$\int [\alpha f(x) + \beta g(x)]\, dx = \alpha \int f(x)\, dx + \beta \int g(x)\, dx$$

and

$$\int_a^b [\alpha f(x) + \beta g(x)]\, dx = \alpha \int_a^b f(x)\, dx + \beta \int_a^b g(x)\, dx$$

provided each derivative and integral exists. In this section we will examine a special type of integral transform called the **Laplace transform**. In addition to possessing the linearity property, the Laplace transform has many other interesting properties that make it very useful in solving linear initial-value problems.

■ Integral Transform If $f(x, y)$ is a function of two variables, then a definite integral of f with respect to one of the variables leads to a function of the other variable. For example, by holding y constant we see that $\int_1^2 2xy^2\, dx = 3y^2$. Similarly, a definite integral such as $\int_a^b K(s, t)f(t)\, dt$ transforms a function f of the variable t into a function F of the variable s. We are particularly interested in an **integral transform**, where the interval of integration is the unbounded interval $[0, \infty)$. If $f(t)$ is defined for $t \geq 0$, then the improper integral $\int_0^\infty K(s, t)f(t)\, dt$ is defined as a limit

$$\int_0^\infty K(s, t) f(t)\, dt = \lim_{b \to \infty} \int_0^b K(s, t) f(t)\, dt. \tag{1}$$

If the limit in (1) exists, we say that the integral exists or is **convergent**; if the limit does not exist, the integral does not exist and is **divergent**. The limit in (1) will, in general, exist for only certain values of the variable s.

■ A Definition The function $K(s, t)$ in (1) is said to be the **kernel** of the transform. The choice $K(s, t) = e^{-st}$ as the kernel gives us an especially important integral transform.

DEFINITION 4.1.1 Laplace Transform

Let f be a function defined on $[0, \infty)$. Then the function F defined by

$$F(s) = \int_0^\infty e^{-st} f(t)\, dt \tag{2}$$

is said to be the **Laplace transform** of f. The domain of $F(s)$ is the set of values of s for which the improper integral (2) converges.

The Laplace transform of a function f is also denoted by $\mathscr{L}\{f(t)\}$. In general discussion and applications, we shall use a lowercase letter to denote the function being transformed and the corresponding capital letter to denote its Laplace transform. For example,

$$\mathscr{L}\{f(t)\} = F(s), \quad \mathscr{L}\{g(t)\} = G(s), \quad \mathscr{L}\{i(t)\} = I(s).$$

© ART Collection/Alamy Stock Photo
Pierre-Simon Marquis de Laplace

The Laplace transform is named in honor of the French mathematician, astronomer, and physicist **Pierre-Simon Marquis de Laplace** (1749–1827), who used the transform in his investigations of probability theory. Most likely the Laplace transform was invented by Leonard Euler. For his services to the Bourbons in restoring the monarchy following the fall of Napoleon Bonaparte in 1814, King Louis XVIII awarded Laplace the title of Marquis.

In a general setting, the variable s could be a complex number. But throughout this chapter we shall assume that s is real. As the next four examples show, the domain of $F(s)$ depends on the function $f(t)$.

EXAMPLE 1 **Using Definition 4.1.1**

Find $\mathscr{L}\{1\}$.

SOLUTION From (2),

$$\mathscr{L}\{1\} = \int_0^\infty e^{-st}(1)\, dt = \lim_{b\to\infty} \int_0^b e^{-st}\, dt$$

$$= \lim_{b\to\infty} \frac{-e^{-st}}{s}\bigg|_0^b = \lim_{b\to\infty} \frac{-e^{-sb}+1}{s} = \frac{1}{s}$$

provided $s > 0$. In other words, when $s > 0$, the exponent $-sb$ is negative and $e^{-sb} \to 0$ as $b \to \infty$. The integral $\int_0^\infty e^{-st}\, dt$ diverges for $s < 0$, and so the domain of $F(s) = 1/s$ is defined by $s > 0$. ≡

The use of the limit sign becomes somewhat tedious, so we shall adopt the notation $\big|_0^\infty$ as a shorthand to writing $\lim_{b\to\infty}(\)\big|_0^b$. For example,

$$\mathscr{L}\{1\} = \int_0^\infty e^{-st}(1)\, dt = \frac{-e^{-st}}{s}\bigg|_0^\infty = \frac{1}{s}, \quad s > 0.$$

At the upper limit, it is understood we mean $e^{-st} \to 0$ as $t \to \infty$ for $s > 0$.

EXAMPLE 2 **Using Definition 4.1.1**

Find $\mathscr{L}\{t\}$.

SOLUTION From Definition 4.1.1, we have $\mathscr{L}\{t\} = \int_0^\infty e^{-st}t\, dt$. Integrating by parts and using $\lim_{t\to\infty} te^{-st} = 0$, $s > 0$, along with the result from Example 1, we obtain

$$\mathscr{L}\{t\} = \frac{-te^{-st}}{s}\bigg|_0^\infty + \frac{1}{s}\int_0^\infty e^{-st}\, dt = \frac{1}{s}\mathscr{L}\{1\} = \frac{1}{s}\left(\frac{1}{s}\right) = \frac{1}{s^2}. \qquad \equiv$$

EXAMPLE 3 **Using Definition 4.1.1**

Find **(a)** $\mathscr{L}\{e^{-3t}\}$ **(b)** $\mathscr{L}\{e^{6t}\}$.

SOLUTION In each case we use Definition 4.1.1.

(a) $\mathscr{L}\{e^{-3t}\} = \int_0^\infty e^{-3t}e^{-st}\, dt = \int_0^\infty e^{-(s+3)t}\, dt$

$$= \frac{-e^{-(s+3)t}}{s+3}\bigg|_0^\infty$$

$$= \frac{1}{s+3}.$$

The last result is valid for $s > -3$ because in order to have $\lim_{t\to\infty} e^{-(s+3)t} = 0$ we must require that $s + 3 > 0$ or $s > -3$.

(b) $\mathscr{L}\{e^{6t}\} = \displaystyle\int_0^\infty e^{6t} e^{-st} dt = \int_0^\infty e^{-(s-6)t} dt$

$$= \frac{-e^{-(s-6)t}}{s-6} \bigg|_0^\infty$$

$$= \frac{1}{s-6}.$$

In contrast to part (a), this result is valid for $s > 6$ because $\lim_{t\to\infty} e^{-(s-6)t} = 0$ demands $s - 6 > 0$ or $s > 6$. ≡

EXAMPLE 4　　**Using Definition 4.1.1**

Find　$\mathscr{L}\{\sin 2t\}$.

SOLUTION　From Definition 4.1.1 and integration by parts we have

$$\mathscr{L}\{\sin 2t\} = \int_0^\infty e^{-st} \sin 2t \, dt = \frac{-e^{-st} \sin 2t}{s} \bigg|_0^\infty + \frac{2}{s} \int_0^\infty e^{-st} \cos 2t \, dt \qquad (3)$$

$$= \frac{2}{s} \int_0^\infty e^{-st} \cos 2t \, dt, \quad s > 0 \qquad (4)$$

$$\underset{\substack{\downarrow \\ \lim_{t\to\infty} e^{-st}\cos 2t = 0,\, s > 0}}{} \qquad \underset{\substack{\downarrow \\ \text{Laplace transform of } \sin 2t}}{}$$

$$= \frac{2}{s}\left[\frac{-e^{-st}\cos 2t}{s} \bigg|_0^\infty - \frac{2}{s} \int_0^\infty e^{-st} \sin 2t \, dt \right]$$

$$= \frac{2}{s^2} - \frac{4}{s^2} \mathscr{L}\{\sin 2t\}.$$

At this point we have an equation with $\mathscr{L}\{\sin 2t\}$ on both sides of the equality. Solving for that quantity yields the result

$$\mathscr{L}\{\sin 2t\} = \frac{2}{s^2 + 4}, \quad s > 0.$$
≡

Note that the integral in (4) of Example 4 is the Laplace transform of $\cos 2t$. So (3) and (4) show that Laplace transform of $\sin 2t$ is a multiple of the Laplace transform of $\cos 2t$:

$$\mathscr{L}\{\sin 2t\} = \frac{2}{s} \mathscr{L}\{\cos 2t\}.$$

Thus, with no further work we obtain the additional result

$$\mathscr{L}\{\cos 2t\} = \frac{s}{2} \mathscr{L}\{\sin 2t\} = \frac{s}{2} \cdot \frac{2}{s^2 + 4} = \frac{s}{s^2 + 4}, \quad s > 0.$$

■ Linearity of the Laplace Transform

Suppose the functions f and g possess Laplace transforms for $s > c_1$ and $s > c_2$, respectively. If c denotes the maximum of the two numbers c_1 and c_2, then for $s > c$ and constants α and β we can write

$$\int_0^\infty e^{-st}[\alpha f(t) + \beta g(t)] \, dt = \alpha \int_0^\infty e^{-st} f(t) \, dt + \beta \int_0^\infty e^{-st} g(t) \, dt.$$

In other words,

$$\mathscr{L}\{\alpha f(t) + \beta g(t)\} = \alpha \mathscr{L}\{f(t)\} + \beta \mathscr{L}\{g(t)\}. \qquad (5)$$

Because of the property given in (5), $\mathscr{L}$ is said to be a **linear transform**. Furthermore, by the properties of the definite integral, the transform of any finite linear combination of functions $f_1(t), f_2(t), \ldots, f_n(t)$ is the sum of the transforms $\mathscr{L}\{f_1(t)\}, \mathscr{L}\{f_2(t)\}, \ldots, \mathscr{L}\{f_n(t)\}$ provided each transform exists on some common interval of the s-axis.

EXAMPLE 5 **Linearity of the Laplace Transform**

In this example we use the results of the preceding examples to illustrate the linearity of the Laplace transform.

(a) From Examples 1 and 2 we know that both $\mathscr{L}\{1\}$ and $\mathscr{L}\{t\}$ exist on the interval defined by $s > 0$. Hence, for $s > 0$ we can write

$$\mathscr{L}\{1 + 5t\} = \mathscr{L}\{1\} + 5\mathscr{L}\{t\} = \frac{1}{s} + \frac{5}{s^2}.$$

(b) From Example 3 we saw that $\mathscr{L}\{e^{6t}\}$ exists on the interval defined by $s > 6$, and in Example 4 we saw that $\mathscr{L}\{\sin 2t\}$ exists on the interval defined by $s > 0$. Thus both transforms exist for the common values of s defined by $s > 6$, and we can write

$$\mathscr{L}\{2e^{6t} - 15\sin 2t\} = 2\mathscr{L}\{e^{6t}\} - 15\mathscr{L}\{\sin 2t\} = \frac{2}{s - 6} - \frac{15}{s^2 + 4}.$$

(c) From Examples 1, 2, and 3 we have for $s > 0$,

$$\mathscr{L}\{10e^{-3t} - 5t + 8\} = 10\mathscr{L}\{e^{-3t}\} - 5\mathscr{L}\{t\} + 8\mathscr{L}\{1\} = \frac{10}{s + 3} - \frac{5}{s^2} + \frac{8}{s}. \qquad \equiv$$

EXAMPLE 6 **Linearity of the Laplace Transform**

Find $\mathscr{L}\{\sinh t\}$.

SOLUTION Recall, the definition of the hyperbolic sine is

$$\sinh t = \frac{e^t - e^{-t}}{2} = \frac{1}{2}e^t - \frac{1}{2}e^{-t}.$$

Now, you should be able to modify the work in parts (b) and (a) of Example 3 to show that

$$\mathscr{L}\{e^t\} = \frac{1}{s - 1}, \quad \text{for } s > 1 \quad \text{and} \quad \mathscr{L}\{e^{-t}\} = \frac{1}{s + 1}, \quad \text{for } s > -1.$$

Hence by the linearity property (5), we have for $s > 1$,

$$\mathscr{L}\{\sinh t\} = \frac{1}{2}\mathscr{L}\{e^t\} - \frac{1}{2}\mathscr{L}\{e^{-t}\} = \frac{1}{2}\left(\frac{1}{s - 1} - \frac{1}{s + 1}\right) = \frac{1}{s^2 - 1}. \qquad \equiv$$

We state the generalization of some of the preceding examples by means of the next theorem. From this point on we shall also refrain from stating any restrictions on s; it is understood that s is sufficiently restricted to guarantee the convergence of the appropriate Laplace transform.

THEOREM 4.1.1 **Transforms of Some Basic Functions**

$$\textbf{(a)} \quad \mathscr{L}\{1\} = \frac{1}{s}$$

$$\textbf{(b)} \quad \mathscr{L}\{t^n\} = \frac{n!}{s^{n+1}}, n = 1, 2, 3, \dots \qquad\qquad \textbf{(c)} \quad \mathscr{L}\{e^{at}\} = \frac{1}{s - a}$$

$$\textbf{(d)} \quad \mathscr{L}\{\sin kt\} = \frac{k}{s^2 + k^2} \qquad\qquad \textbf{(e)} \quad \mathscr{L}\{\cos kt\} = \frac{s}{s^2 + k^2}$$

$$\textbf{(f)} \quad \mathscr{L}\{\sinh kt\} = \frac{k}{s^2 - k^2} \qquad\qquad \textbf{(g)} \quad \mathscr{L}\{\cosh kt\} = \frac{s}{s^2 - k^2}$$

The result in (b) of Theorem 4.1.1 can be formally justified for n a positive integer using integration by parts to first show that

$$\mathscr{L}\{t^n\} = \frac{n}{s}\mathscr{L}\{t^{n-1}\}.$$

Then for $n = 1$, 2, and 3, we have, respectively,

$$\mathscr{L}\{t\} = \frac{1}{s} \cdot \mathscr{L}\{1\} = \frac{1}{s} \cdot \frac{1}{s} = \frac{1}{s^2}$$

$$\mathscr{L}\{t^2\} = \frac{2}{s} \cdot \mathscr{L}\{t\} = \frac{2}{s} \cdot \frac{1}{s^2} = \frac{2 \cdot 1}{s^3}$$

$$\mathscr{L}\{t^3\} = \frac{3}{s} \cdot \mathscr{L}\{t^2\} = \frac{3}{s} \cdot \frac{2 \cdot 1}{s^3} = \frac{3 \cdot 2 \cdot 1}{s^4}.$$

If we carry on in this manner, you should be convinced that

$$\mathscr{L}\{t^n\} = \frac{n \cdots 3 \cdot 2 \cdot 1}{s^{n+1}} = \frac{n!}{s^{n+1}}.$$

A more extensive list of functions and their Laplace transforms is given in Appendix C.

The Laplace transforms of trigonometric functions can often be obtained with the help of trigonometric identities and (d) and (e) of Theorem 4.1.1.

FIGURE 4.1.1 Piecewise-continuous function

| EXAMPLE 7 | Using a Trigonometric Identity |

Find $\mathscr{L}\{\cos^2 t\}$.

SOLUTION Using the half-angle formula $\cos^2 t = \frac{1}{2}(1 + \cos 2t)$, the linearity property (5), and (a) and (e) of Theorem 4.1.1, we obtain for $s > 0$,

$$\mathscr{L}\{\cos^2 t\} = \mathscr{L}\left\{\frac{1}{2}(1 + \cos 2t)\right\} = \frac{1}{2}\mathscr{L}\{1\} + \frac{1}{2}\mathscr{L}\{\cos 2t\}$$

$$= \frac{1}{2}\left(\frac{1}{s}\right) + \frac{1}{2}\left(\frac{s}{s^2 + 4}\right) = \frac{s^2 + 2}{s(s^2 + 4)}. \qquad \equiv$$

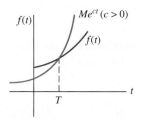

FIGURE 4.1.2 Function f is of exponential order

▇ Sufficient Conditions for Existence of $\mathscr{L}\{f(t)\}$

The integral that defines the Laplace transform does not have to converge. For example, neither $\mathscr{L}\{1/t\}$ nor $\mathscr{L}\{e^{t^2}\}$ exists. Sufficient conditions guaranteeing the existence of $\mathscr{L}\{f(t)\}$ are that f be piecewise continuous on $[0, \infty)$ and that f be of exponential order for $t > T$. Recall that a function f is piecewise continuous on $[0, \infty)$ if, in any interval defined by $0 \le a \le t \le b$, there are at most a finite number of points t_k, $k = 1, 2, \ldots, n$ $(t_{k-1} < t_k)$, at which f has finite discontinuities and is continuous on each open interval defined by $t_{k-1} < t < t_k$. See **FIGURE 4.1.1**. The concept of exponential order is defined in the following manner.

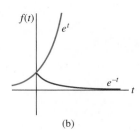

DEFINITION 4.1.2 Exponential Order

A function f is said to be of **exponential order** if there exist constants c, $M > 0$, and $T > 0$ such that $|f(t)| \le Me^{ct}$ for all $t > T$.

If f is an *increasing* function, then the condition $|f(t)| \le Me^{ct}$, $t > T$, simply states that the graph of f on the interval (T, ∞) does not grow faster than the graph of the exponential function Me^{ct}, where c is a positive constant. See **FIGURE 4.1.2**. The functions $f(t) = t$, $f(t) = e^{-t}$, and $f(t) = 2\cos t$ are all of exponential order $c = 1$ for $t > 0$ since we have, respectively,

$$|t| \le e^t, \quad |e^{-t}| \le e^t, \quad |2\cos t| \le 2e^t.$$

A comparison of the graphs on the interval $[0, \infty)$ is given in **FIGURE 4.1.3**.

A positive integral power of t is always of exponential order since, for $c > 0$,

$$|t^n| \le Me^{ct} \quad \text{or} \quad \left|\frac{t^n}{e^{ct}}\right| \le M \quad \text{for} \quad t > T$$

FIGURE 4.1.3 Functions with blue graphs are of exponential order

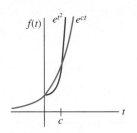

FIGURE 4.1.4 $f(t) = e^{t^2}$ is not of exponential order

is equivalent to showing that $\lim_{t\to\infty} t^n/e^{ct}$ is finite for $n = 1, 2, 3, \ldots$. The result follows by n applications of L'Hôpital's rule. A function such as $f(t) = e^{t^2}$ is not of exponential order since, as shown in **FIGURE 4.1.4**, e^{t^2} grows faster than any positive linear power of e for $t > c > 0$. This can also be seen from

$$\left| \frac{e^{t^2}}{e^{ct}} \right| = e^{t^2 - ct} = e^{t(t-c)} \to \infty \text{ as } t \to \infty$$

for *any* value of c. By the same reasoning, $e^{-st}e^{t^2} \to \infty$ as $t \to \infty$ for any s and so the improper integral $\int_0^\infty e^{-st}e^{t^2}\,dt$ diverges. In other words, $\mathscr{L}\{e^{t^2}\}$ does not exist.

THEOREM 4.1.2 Sufficient Conditions for Existence

If $f(t)$ is piecewise continuous on the interval $[0, \infty)$ and of exponential order, then $\mathscr{L}\{f(t)\}$ exists for $s > c$.

PROOF: By the additive interval property of definite integrals,

$$\mathscr{L}\{f(t)\} = \int_0^T e^{-st}f(t)\,dt + \int_T^\infty e^{-st}f(t)\,dt = I_1 + I_2.$$

The integral I_1 exists because it can be written as a sum of integrals over intervals on which $e^{-st}f(t)$ is continuous. Now f is of exponential order, so there exists constants c, $M > 0$, $T > 0$ so that $|f(t)| \le Me^{ct}$ for $t > T$. We can then write

$$|I_2| \le \int_T^\infty |e^{-st}f(t)|\,dt \le M\int_T^\infty e^{-st}e^{ct}\,dt = M\int_T^\infty e^{-(s-c)t}\,dt = M\frac{e^{-(s-c)T}}{s-c}$$

for $s > c$. Since $\int_T^\infty Me^{-(s-c)t}\,dt$ converges, the integral $\int_T^\infty |e^{-st}f(t)|\,dt$ converges by the comparison test for improper integrals. This, in turn, implies that I_2 exists for $s > c$. The existence of I_1 and I_2 implies that $\mathscr{L}\{f(t)\} = \int_0^\infty e^{-st}f(t)\,dt$ exists for $s > c$. ≡

EXAMPLE 8 **Transform of a Piecewise-Continuous Function**

Find $\mathscr{L}\{f(t)\}$ for $f(t) = \begin{cases} 0, & 0 \le t < 3 \\ 2, & t \ge 3. \end{cases}$

FIGURE 4.1.5 Piecewise-continuous function in Example 8

SOLUTION This piecewise-continuous function appears in **FIGURE 4.1.5**. Since f is defined in two pieces, $\mathscr{L}\{f(t)\}$ is expressed as the sum of two integrals:

$$\mathscr{L}\{f(t)\} = \int_0^\infty e^{-st}f(t)\,dt = \int_0^3 e^{-st}(0)\,dt + \int_3^\infty e^{-st}(2)\,dt$$

$$= -\frac{2e^{-st}}{s}\bigg|_3^\infty$$

$$= \frac{2e^{-3s}}{s}, \quad s > 0. \qquad ≡$$

In the remaining sections of this chapter we shall refrain from stating any restrictions on s in the examples; it is understood that s is sufficiently restricted to guarantee the convergence of the appropriate Laplace transform.

REMARKS

Throughout this chapter we shall be concerned primarily with functions that are both piecewise continuous and of exponential order. We note, however, that these two conditions are sufficient but not necessary for the existence of a Laplace transform. The function $f(t) = t^{-1/2}$ is not piecewise continuous on the interval $[0, \infty)$; nevertheless, its Laplace transform exists. See Problem 47 in Exercises 4.1.

In Problems 1–18, use Definition 4.1.1 to find $\mathscr{L}\{f(t)\}$.

1. $f(t) = \begin{cases} -1, & 0 \le t < 1 \\ 1, & t \ge 1 \end{cases}$

2. $f(t) = \begin{cases} 4, & 0 \le t < 2 \\ 0, & t \ge 2 \end{cases}$

3. $f(t) = \begin{cases} t, & 0 \le t < 1 \\ 1, & t \ge 1 \end{cases}$

4. $f(t) = \begin{cases} 2t + 1, & 0 \le t < 1 \\ 0, & t \ge 1 \end{cases}$

5. $f(t) = \begin{cases} \sin t, & 0 \le t < \pi \\ 0, & t \ge \pi \end{cases}$

6. $f(t) = \begin{cases} \sin t, & 0 \le t < \pi/2 \\ 0, & t \ge \pi/2 \end{cases}$

7.

FIGURE 4.1.6 Graph for Problem 7

8.

FIGURE 4.1.7 Graph for Problem 8

9.

FIGURE 4.1.8 Graph for Problem 9

10.

FIGURE 4.1.9 Graph for Problem 10

11. $f(t) = e^{t+7}$

12. $f(t) = e^{-2t-5}$

13. $f(t) = te^{4t}$

14. $f(t) = t^2 e^{-2t}$

15. $f(t) = e^{-t} \sin t$

16. $f(t) = e^t \cos t$

17. $f(t) = t \cos t$

18. $f(t) = t \sin t$

In Problems 19–38, use Theorem 4.1.1 to find $\mathscr{L}\{f(t)\}$.

19. $f(t) = 2t^4$

20. $f(t) = t^5$

21. $f(t) = 4t - 10$

22. $f(t) = 7t + 3$

23. $f(t) = t^2 + 6t - 3$

24. $f(t) = -4t^2 + 16t + 9$

25. $f(t) = (t + 1)^3$

26. $f(t) = (2t - 1)^3$

27. $f(t) = 1 + e^{4t}$

28. $f(t) = t^2 - e^{-9t} + 5$

29. $f(t) = (1 + e^{2t})^2$

30. $f(t) = (e^t - e^{-t})^2$

31. $f(t) = 4t^2 - 5 \sin 3t$

32. $f(t) = \cos 5t + \sin 2t$

33. $f(t) = \sinh kt$

34. $f(t) = \cosh kt$

35. $f(t) = e^t \sinh t$

36. $f(t) = e^{-t} \cosh t$

37. $f(t) = \cosh^2 kt$

38. $f(t) = \sinh^2 kt$

In Problems 39–44, find $\mathscr{L}\{f(t)\}$ by first using a trigonometric identity.

39. $f(t) = \cos^2 t - \sin^2 t$

40. $f(t) = 2 \sin^2 \dfrac{t}{2}$

41. $f(t) = \sin 3t \cos 3t$

42. $f(t) = \sin^4 t$

43. $f(t) = \sqrt{2} \sin\left(2t + \dfrac{\pi}{4}\right)$

44. $f(t) = 10 \cos\left(t - \dfrac{\pi}{6}\right)$

45. One definition of the **gamma function** $\Gamma(\alpha)$ is given by the improper integral

$$\Gamma(\alpha) = \int_0^\infty t^{\alpha-1} e^{-t}\, dt, \quad \alpha > 0.$$

Use this definition to show that $\Gamma(\alpha + 1) = \alpha\Gamma(\alpha)$.

46. Use Problem 45 to show that

$$\mathscr{L}\{t^\alpha\} = \frac{\Gamma(\alpha + 1)}{s^{\alpha+1}}, \quad \alpha > -1.$$

This result is a generalization of Theorem 4.1.1(b).

In Problems 47–50, use the results in Problems 45 and 46 and the fact that $\Gamma(\tfrac{1}{2}) = \sqrt{\pi}$ to find the Laplace transform of the given function.

47. $f(t) = t^{-1/2}$

48. $f(t) = t^{1/2}$

49. $f(t) = t^{3/2}$

50. $f(t) = 6t^{1/2} - 24t^{5/2}$

Discussion Problems

51. For what values of s is the following true

$$\mathscr{L}\{2e^{3t} - e^{-t} + 5\} = 2\mathscr{L}\{e^{3t}\} - \mathscr{L}\{e^{-t}\} + 5\mathscr{L}\{1\}?$$

52. Figure 4.1.4 suggests, but does not prove, that the function $f(t) = e^{t^2}$ is not of exponential order. How does the observation that $t^2 > \ln M + ct$, for $M > 0$ and t sufficiently large, show that $e^{t^2} > Me^{ct}$ for any c?

53. Use part (c) of Theorem 4.1.1 to show that

$$\mathscr{L}\{e^{(a+ib)t}\} = \frac{s - a + ib}{(s - a)^2 + b^2},$$

where a and b are real and $i^2 = -1$. Show how Euler's formula (page 126) can then be used to deduce the results

$$\mathscr{L}\{e^{at} \cos bt\} = \frac{s - a}{(s - a)^2 + b^2}$$

and

$$\mathscr{L}\{e^{at} \sin bt\} = \frac{b}{(s - a)^2 + b^2}.$$

54. Under what conditions is a linear function

$$f(x) = mx + b,$$

$m \ne 0$, a linear transform?

55. The function $f(t) = 2te^{t^2}\cos e^{t^2}$ is not of exponential order. Nevertheless, show that the Laplace transform $\mathscr{L}\{2te^{t^2}\cos e^{t^2}\}$ exists. [*Hint*: Use integration by parts.]

56. Explain why the function

$$f(t) = \begin{cases} t, & 0 \le t < 2 \\ 4, & 2 < t < 4 \\ 1/(t - 4), & t > 4 \end{cases}$$

is not piecewise continuous on $[0, \infty)$.

57. Show that the function $f(t) = 1/t^2$ does not possess a Laplace transform. [*Hint:* Write $\mathscr{L}\{1/t^2\}$ as two improper integrals:

$$\mathscr{L}\{1/t^2\} = \int_0^1 \frac{e^{-st}}{t^2}\, dt + \int_1^\infty \frac{e^{-st}}{t^2}\, dt = I_1 + I_2.$$

Show that I_1 diverges.]

58. If $\mathscr{L}\{f(t)\} = F(s)$ and $a > 0$ is a constant, show that

$$\mathscr{L}\{f(at)\} = \frac{1}{a} F\left(\frac{s}{a}\right).$$

This result is known as the **change of scale theorem**.

In Problems 59–62, use the given Laplace transform and the result in Problem 58 to find the indicated Laplace transform. Assume that a and k are positive constants.

59. $\mathscr{L}\{e^t\} = \dfrac{1}{s-1}; \ \mathscr{L}\{e^{at}\}$

60. $\mathscr{L}\{\cos t\} = \dfrac{s}{s^2+1}; \ \mathscr{L}\{\cos kt\}$

61. $\mathscr{L}\{t - \sin t\} = \dfrac{1}{s^2(s^2+1)}; \ \mathscr{L}\{kt - \sin kt\}$

62. $\mathscr{L}\{\cos t \, \sinh t\} = \dfrac{s^2-2}{s^4+4}; \ \mathscr{L}\{\cos kt \, \sinh kt\}$

4.2 Inverse Transforms and Transforms of Derivatives

INTRODUCTION In this section we take a few small steps into an investigation of how the Laplace transform can be used to solve certain types of equations. After we discuss the concept of the *inverse* Laplace transform and examine the transforms of derivatives we then use the Laplace transform to solve some simple ordinary differential equations.

4.2.1 Inverse Transforms

■ **The Inverse Problem** If $F(s)$ represents the Laplace transform of a function $f(t)$, that is, $\mathscr{L}\{f(t)\} = F(s)$, we then say $f(t)$ is the **inverse Laplace transform** of $F(s)$ and write $f(t) = \mathscr{L}^{-1}\{F(s)\}$. For example, from Examples 1, 2, and 3 in Section 4.1 we have, respectively,

$$1 = \mathscr{L}^{-1}\left\{\frac{1}{s}\right\}, \quad t = \mathscr{L}^{-1}\left\{\frac{1}{s^2}\right\}, \quad \text{and} \quad e^{-3t} = \mathscr{L}^{-1}\left\{\frac{1}{s+3}\right\}.$$

The analogue of Theorem 4.1.1 for the inverse transform is presented next.

THEOREM 4.2.1 Some Inverse Transforms

$$\textbf{(a)} \ \ 1 = \mathscr{L}^{-1}\left\{\frac{1}{s}\right\}$$

(b) $t^n = \mathscr{L}^{-1}\left\{\dfrac{n!}{s^{n+1}}\right\}, n = 1, 2, 3,\ldots$ **(c)** $e^{at} = \mathscr{L}^{-1}\left\{\dfrac{1}{s-a}\right\}$

(d) $\sin kt = \mathscr{L}^{-1}\left\{\dfrac{k}{s^2+k^2}\right\}$ **(e)** $\cos kt = \mathscr{L}^{-1}\left\{\dfrac{s}{s^2+k^2}\right\}$

(f) $\sinh kt = \mathscr{L}^{-1}\left\{\dfrac{k}{s^2-k^2}\right\}$ **(g)** $\cosh kt = \mathscr{L}^{-1}\left\{\dfrac{s}{s^2-k^2}\right\}$

When evaluating inverse transforms, it often happens that a function of s under consideration does not match *exactly* the form of a Laplace transform $F(s)$ given in a table. It may be necessary to "fix up" the function of s by multiplying and dividing by an appropriate constant.

EXAMPLE 1 **Applying Theorem 4.2.1**

Find **(a)** $\mathscr{L}^{-1}\left\{\dfrac{1}{s^5}\right\}$ **(b)** $\mathscr{L}^{-1}\left\{\dfrac{1}{s^2+7}\right\}.$

SOLUTION **(a)** To match the form given in part (b) of Theorem 4.2.1, we identify $n + 1 = 5$ or $n = 4$ and then multiply and divide by 4!:

$$\mathscr{L}^{-1}\left\{\frac{1}{s^5}\right\} = \frac{1}{4!}\mathscr{L}^{-1}\left\{\frac{4!}{s^5}\right\} = \frac{1}{24}t^4.$$

(b) To match the form given in part (d) of Theorem 4.2.1, we identify $k^2 = 7$ and so $k = \sqrt{7}$. We fix up the expression by multiplying and dividing by $\sqrt{7}$:

$$\mathscr{L}^{-1}\left\{\frac{1}{s^2 + 7}\right\} = \frac{1}{\sqrt{7}}\mathscr{L}^{-1}\left\{\frac{\sqrt{7}}{s^2 + 7}\right\} = \frac{1}{\sqrt{7}}\sin\sqrt{7}t. \qquad \equiv$$

■ Linearity of the Inverse Laplace Transformation

Analogous to (5) of Section 4.1, the inverse Laplace transform is also a linear transform. Suppose the functions $f(t) = \mathscr{L}^{-1}\{F(s)\}$ and $g(t) = \mathscr{L}^{-1}\{G(s)\}$ are piecewise continuous on $[0, \infty)$ and of exponential order. Then for constants α and β we can write

$$\mathscr{L}^{-1}\{\alpha F(s) + \beta G(s)\} = \alpha\mathscr{L}^{-1}\{F(s)\} + \beta\mathscr{L}^{-1}\{G(s)\}. \qquad (1)$$

Like (5) of Section 4.1, (1) extends to any finite linear combination of Laplace transforms.

EXAMPLE 2 **Termwise Division and Linearity**

Find $\mathscr{L}^{-1}\left\{\dfrac{-2s + 6}{s^2 + 4}\right\}$.

SOLUTION We first rewrite the given function of s as two expressions by means of termwise division and then use (1):

$$\underset{\downarrow}{\text{termwise division}} \qquad \underset{\downarrow}{\text{linearity and fixing up constants}}$$

$$\mathscr{L}^{-1}\left\{\frac{-2s + 6}{s^2 + 4}\right\} = \mathscr{L}^{-1}\left\{\frac{-2s}{s^2 + 4} + \frac{6}{s^2 + 4}\right\} = -2\mathscr{L}^{-1}\left\{\frac{s}{s^2 + 4}\right\} + \frac{6}{2}\mathscr{L}^{-1}\left\{\frac{2}{s^2 + 4}\right\} \quad (2)$$

$$= -2\cos 2t + 3\sin 2t. \quad \leftarrow \text{parts (e) and (d) of Theorem 4.2.1 with } k = 2 \qquad \equiv$$

■ Partial Fractions

Partial fractions play an important role in finding inverse Laplace transforms. The decomposition of a rational expression into component fractions can be done quickly by means of a single command on most computer algebra systems. Indeed, some CASs have packages that implement Laplace transform and inverse Laplace transform commands. But for those of you without access to such software, we will review in this and subsequent sections of this chapter some of the basic algebra in the important cases in which the denominator of a Laplace transform $F(s)$ contains (*i*) distinct linear factors, (*ii*) repeated linear factors, and (*iii*) quadratic polynomials with no real factors.

Although we shall examine each of these cases as this chapter develops, it still might be a good idea for you to consult either a calculus text or a current precalculus text for a more comprehensive review of this theory.

The following example illustrates partial fraction decomposition in the case when the denominator of $F(s)$ is factorable into distinct linear factors.

EXAMPLE 3 **Partial Fractions and Linearity**

Partial fractions: distinct linear factors in denominator ▶

Find $\mathscr{L}^{-1}\left\{\dfrac{s^2 + 6s + 9}{(s - 1)(s - 2)(s + 4)}\right\}$.

SOLUTION There exist unique constants A, B, and C such that

$$\frac{s^2 + 6s + 9}{(s - 1)(s - 2)(s + 4)} = \frac{A}{s - 1} + \frac{B}{s - 2} + \frac{C}{s + 4}$$

$$= \frac{A(s - 2)(s + 4) + B(s - 1)(s + 4) + C(s - 1)(s - 2)}{(s - 1)(s - 2)(s + 4)}.$$

Since the denominators are identical, the numerators are identical:

$$s^2 + 6s + 9 = A(s - 2)(s + 4) + B(s - 1)(s + 4) + C(s - 1)(s - 2). \qquad (3)$$

By comparing coefficients of powers of s on both sides of the equality, we know that (3) is equivalent to a system of three equations in the three unknowns A, B, and C. However, recall that there is a shortcut for determining these unknowns. If we set $s = 1$, $s = 2$, and $s = -4$ in (3) we obtain, respectively,*

$$16 = A(-1)(5), \quad 25 = B(1)(6), \quad 1 = C(-5)(-6),$$

and so $A = -\frac{16}{5}$, $B = \frac{25}{6}$, $C = \frac{1}{30}$. Hence the partial fraction decomposition is

$$\frac{s^2 + 6s + 9}{(s - 1)(s - 2)(s + 4)} = -\frac{16/5}{s - 1} + \frac{25/6}{s - 2} + \frac{1/30}{s + 4}, \qquad (4)$$

and thus, from the linearity of $\mathscr{L}^{-1}$ and part (c) of Theorem 4.2.1,

$$\mathscr{L}^{-1}\left\{\frac{s^2 + 6s + 9}{(s - 1)(s - 2)(s + 4)}\right\} = -\frac{16}{5}\mathscr{L}^{-1}\left\{\frac{1}{s - 1}\right\} + \frac{25}{6}\mathscr{L}^{-1}\left\{\frac{1}{s - 2}\right\} + \frac{1}{30}\mathscr{L}^{-1}\left\{\frac{1}{s + 4}\right\}$$

$$= -\frac{16}{5}e^t + \frac{25}{6}e^{2t} + \frac{1}{30}e^{-4t}. \qquad (5) \equiv$$

4.2.2 Transforms of Derivatives

Transform of a Derivative As pointed out in the introduction to this chapter, our immediate goal is to use the Laplace transform to solve differential equations. To that end we need to evaluate quantities such as $\mathscr{L}\{dy/dt\}$ and $\mathscr{L}\{d^2y/dt^2\}$.

To begin, we consider the case when the function f is continuous, f' is piecewise continuous on $[0, \infty)$, and both functions are of exponential order with c as specified in Definition 4.1.2. For simplicity, we also assume that f' has a single point of discontinuity t_0 on the interval $[0, \infty)$. Then integration by parts gives

$$\mathscr{L}\{f'(t)\} = \int_0^{t_0} e^{-st}f'(t)\, dt + \int_{t_0}^{\infty} e^{-st}f'(t)\, dt$$

$$= \left[e^{-st}f(t) \Big|_0^{t_0} + s\int_0^{t_0} e^{-st}f(t)\, dt \right] + \left[e^{-st}f(t) \Big|_{t_0}^{\infty} + s\int_{t_0}^{\infty} e^{-st}f(t)\, dt \right]$$

$$= -f(0) + \lim_{t\to\infty} e^{-st}f(t) + s\left[\int_0^{t_0} e^{-st}f(t)\, dt + \int_{t_0}^{\infty} e^{-st}f(t)\, dt \right]$$

$$= -f(0) + \lim_{t\to\infty} e^{-st}f(t) + s\int_0^{\infty} e^{-st}f(t)\, dt.$$

Because the function f is of exponential order, the term $e^{-st}f(t) \to 0$ as $t \to \infty$ for $s > c$, and the last term is recognized as $\mathscr{L}\{f(t)\}$. Using the notation $\mathscr{L}\{f(t)\} = F(s)$ we have shown that

$$\mathscr{L}\{f'(t)\} = sF(s) - f(0). \qquad (6)$$

The result in (6) can be used recursively to find Laplace transforms of higher derivatives. For example, if $f'(t)$ and $f(t)$ in $\mathscr{L}\{f'(t)\} = s\mathscr{L}\{f(t)\} - f(0)$ are replaced by $f''(t)$ and $f'(t)$ we get

$$\mathscr{L}\{f''(t)\} = s\mathscr{L}\{f'(t)\} - f'(0) = s[s\mathscr{L}\{f(t)\} - f(0)] - f'(0)$$

*The numbers 1, 2, and -4 are the zeros of the common denominator $(s - 1)(s - 2)(s + 4)$.

or
$$\mathcal{L}\{f''(t)\} = s^2 F(s) - sf(0) - f'(0). \tag{7}$$

Similarly, $\mathcal{L}\{f'''(t)\} = s\mathcal{L}\{f''(t)\} - f''(0)$ yields

$$\mathcal{L}\{f'''(t)\} = s^3 F(s) - s^2 f(0) - sf'(0) - f''(0). \tag{8}$$

The results in (6), (7), and (8) are special cases of the following theorem.

THEOREM 4.2.2 Transform of a Derivative

Suppose $f, f', \ldots, f^{(n-1)}$ are continuous on $[0, \infty)$, $f^{(n)}$ is piecewise continuous on $[0, \infty)$, and $f, f', \ldots, f^{(n-1)}, f^{(n)}$ are of exponential order with c as specified in Definition 4.1.2. Then for $s > c$,

$$\mathcal{L}\{f^{(n)}(t)\} = s^n F(s) - s^{n-1}f(0) - s^{n-2}f'(0) - \cdots - f^{(n-1)}(0),$$

where $F(s) = \mathcal{L}\{f(t)\}$.

■ **Solving Linear ODEs** It is apparent from the general result given in Theorem 4.2.2 that $\mathcal{L}\{d^n y/dt^n\}$ depends on $Y(s) = \mathcal{L}\{y(t)\}$ and the $n - 1$ derivatives of $y(t)$ evaluated at $t = 0$. This property makes the Laplace transform ideally suited for solving linear initial-value problems in which the differential equation has *constant coefficients*. Such a differential equation is simply a linear combination of terms $y, y', y'', \ldots, y^{(n)}$:

$$a_n \frac{d^n y}{dt^n} + a_{n-1} \frac{d^{n-1}y}{dt^{n-1}} + \cdots + a_0 y = g(t),$$

$$y(0) = y_0, y'(0) = y_1, \ldots, y^{(n-1)}(0) = y_{n-1},$$

where the coefficients a_i, $i = 0, 1, \ldots, n$ and $y_0, y_1, \ldots, y_{n-1}$ are constants. By the linearity property, the Laplace transform of this linear combination is a linear combination of Laplace transforms:

$$a_n \mathcal{L}\left\{\frac{d^n y}{dt^n}\right\} + a_{n-1} \mathcal{L}\left\{\frac{d^{n-1}y}{dt^{n-1}}\right\} + \cdots + a_0 \mathcal{L}\{y\} = \mathcal{L}\{g(t)\}. \tag{9}$$

From Theorem 4.2.2, (9) becomes

$$a_n[s^n Y(s) - s^{n-1}y(0) - \cdots - y^{(n-1)}(0)]$$
$$+ a_{n-1}[s^{n-1}Y(s) - s^{n-2}y(0) - \cdots - y^{(n-2)}(0)] + \cdots + a_0 Y(s) = G(s), \tag{10}$$

where $\mathcal{L}\{y(t)\} = Y(s)$ and $\mathcal{L}\{g(t)\} = G(s)$. In other words:

The Laplace transform of a linear differential equation with constant coefficients becomes an algebraic equation in $Y(s)$.

If we solve the general transformed equation (10) for the symbol $Y(s)$, we first obtain $P(s)Y(s) = Q(s) + G(s)$, and then write

$$Y(s) = \frac{Q(s)}{P(s)} + \frac{G(s)}{P(s)}, \tag{11}$$

where $P(s) = a_n s^n + a_{n-1}s^{n-1} + \cdots + a_0$, $Q(s)$ is a polynomial in s of degree less than or equal to $n - 1$ consisting of the various products of the coefficients a_i, $i = 1, \ldots, n$, and the prescribed initial conditions $y_0, y_1, \ldots, y_{n-1}$, and $G(s)$ is the Laplace transform of $g(t)$.* Typically we put the two terms in (11) over the least common denominator and then decompose the expression into two or more partial fractions. Finally, the solution $y(t)$ of the original initial-value problem is $y(t) = \mathcal{L}^{-1}\{Y(s)\}$, where the inverse transform is done term by term.

*The polynomial $P(s)$ is the same as the nth degree auxiliary polynomial in (13) in Section 3.3, with the usual symbol m replaced by s.

The procedure is summarized in **FIGURE 4.2.1**.

FIGURE 4.2.1 Steps in solving an IVP by the Laplace transform

The next example illustrates the foregoing method of solving DEs.

EXAMPLE 4 **Solving a First-Order IVP**

Use the Laplace transform to solve the initial-value problem

$$\frac{dy}{dt} + 3y = 13 \sin 2t, \quad y(0) = 6.$$

SOLUTION We first take the transform of each member of the differential equation:

$$\mathscr{L}\left\{\frac{dy}{dt}\right\} + 3\mathscr{L}\{y\} = 13\,\mathscr{L}\{\sin 2t\}. \tag{12}$$

But from (6), $\mathscr{L}\{dy/dt\} = sY(s) - y(0) = sY(s) - 6$, and from part (d) of Theorem 4.1.1, $\mathscr{L}\{\sin 2t\} = 2/(s^2 + 4)$, and so (12) is the same as

$$sY(s) - 6 + 3Y(s) = \frac{26}{s^2 + 4} \quad \text{or} \quad (s + 3)Y(s) = 6 + \frac{26}{s^2 + 4}.$$

Solving the last equation for $Y(s)$, we get

$$Y(s) = \frac{6}{s + 3} + \frac{26}{(s + 3)(s^2 + 4)} = \frac{6s^2 + 50}{(s + 3)(s^2 + 4)}. \tag{13}$$

▶ Partial fractions: quadratic polynomial with no real factors

Since the quadratic polynomial $s^2 + 4$ does not factor using real numbers, its assumed numerator in the partial fraction decomposition is a linear polynomial in s:

$$\frac{6s^2 + 50}{(s + 3)(s^2 + 4)} = \frac{A}{s + 3} + \frac{Bs + C}{s^2 + 4}.$$

Putting the right side of the equality over a common denominator and equating numerators gives $6s^2 + 50 = A(s^2 + 4) + (Bs + C)(s + 3)$. Setting $s = -3$ then yields immediately $A = 8$. Since the denominator has no more real zeros, we equate the coefficients of s^2 and s: $6 = A + B$ and $0 = 3B + C$. Using the value of A in the first equation gives $B = -2$, and then using this last value in the second equation gives $C = 6$. Thus

$$Y(s) = \frac{6s^2 + 50}{(s + 3)(s^2 + 4)} = \frac{8}{s + 3} + \frac{-2s + 6}{s^2 + 4}.$$

We are not quite finished because the last rational expression still has to be written as two fractions. But this was done by termwise division in Example 2. From (2) of that example,

$$y(t) = 8\mathscr{L}^{-1}\left\{\frac{1}{s + 3}\right\} - 2\mathscr{L}^{-1}\left\{\frac{s}{s^2 + 4}\right\} + 3\mathscr{L}^{-1}\left\{\frac{2}{s^2 + 4}\right\}.$$

It follows from parts (c), (d), and (e) of Theorem 4.2.1 that the solution of the initial-value problem is $y(t) = 8e^{-3t} - 2\cos 2t + 3\sin 2t$. ≡

EXAMPLE 5 Solving a Second-Order IVP

Solve $y'' - 3y' + 2y = e^{-4t}$, $y(0) = 1$, $y'(0) = 5$.

SOLUTION Proceeding as in Example 4, we transform the DE by taking the sum of the transforms of each term, use (6) and (7), use the given initial conditions, use part (c) of Theorem 4.1.1, and then solve for $Y(s)$:

$$\mathcal{L}\left\{\frac{d^2y}{dt^2}\right\} - 3\mathcal{L}\left\{\frac{dy}{dt}\right\} + 2\mathcal{L}\{y\} = \mathcal{L}\{e^{-4t}\}$$

$$s^2Y(s) - sy(0) - y'(0) - 3[sY(s) - y(0)] + 2Y(s) = \frac{1}{s+4}$$

$$(s^2 - 3s + 2)Y(s) = s + 2 + \frac{1}{s+4}$$

$$Y(s) = \frac{s+2}{s^2 - 3s + 2} + \frac{1}{(s^2 - 3s + 2)(s+4)} = \frac{s^2 + 6s + 9}{(s-1)(s-2)(s+4)} \qquad (14)$$

and so $y(t) = \mathcal{L}^{-1}\{Y(s)\}$. The details of the decomposition of $Y(s)$ in (14) into partial fractions have already been carried out in Example 3. In view of (4) and (5) the solution of the initial-value problem is $y(t) = -\frac{16}{5}e^t + \frac{25}{6}e^{2t} + \frac{1}{30}e^{-4t}$. ≡

Examples 4 and 5 illustrate the basic procedure for using the Laplace transform to solve a linear initial-value problem, but these examples may appear to demonstrate a method that is not much better than the approach to such problems outlined in Sections 2.3 and 3.3–3.6. Don't draw any negative conclusions from the two examples. Yes, there is a lot of algebra inherent in the use of the Laplace transform, *but* observe that we do not have to use variation of parameters or worry about the cases and algebra in the method of undetermined coefficients. Moreover, since the method incorporates the prescribed initial conditions directly into the solution, there is no need for the separate operations of applying the initial conditions to the general solution $y = c_1y_1 + c_2y_2 + \cdots + c_ny_n + y_p$ of the DE to find specific constants in a particular solution of the IVP.

The Laplace transform has many operational properties. We will examine some of these properties and illustrate how they enable us to solve problems of greater complexity in the sections that follow.

We conclude this section with a little bit of additional theory related to the types of functions of s that we will generally be working with. The next theorem indicates that not every arbitrary function of s is a Laplace transform of a piecewise-continuous function of exponential order.

THEOREM 4.2.3 Behavior of $F(s)$ as $s \to \infty$

If a function f is piecewise continuous on $[0, \infty)$ and of exponential order with c as specified in Definition 4.1.2 and $\mathcal{L}\{f(t)\} = F(s)$, then $\lim\limits_{s \to \infty} F(s) = 0$.

PROOF: Since $f(t)$ is piecewise continuous on the closed interval $[0, T]$, it is necessarily bounded on the interval. That is, $|f(t)| \le M_1 = M_1e^{0t}$. Also, because f is assumed to be of exponential order, there exist constants γ, $M_2 > 0$, and $T > 0$, such that $|f(t)| \le M_2e^{\gamma t}$ for $t > T$. If M denotes the maximum of $\{M_1, M_2\}$ and c denotes the maximum of $\{0, \gamma\}$, then

$$|\mathcal{L}\{f(t)\}| \le \int_0^\infty e^{-st}|f(t)|\,dt \le M\int_0^\infty e^{-st} \cdot e^{ct}\,dt = -M\frac{e^{-(s-c)t}}{s-c}\Big|_0^\infty = \frac{M}{s-c}$$

for $s > c$. As $s \to \infty$, we have $|\mathcal{L}\{f(t)\}| \to 0$, and so $\mathcal{L}\{f(t)\} \to 0$. ≡

As a consequence of Theorem 4.2.3 we can say that functions of s such as $F_1(s) = 1$ and $F_2(s) = s/(s+1)$ are not the Laplace transforms of piecewise-continuous functions of exponential order since $F_1(s) \not\to 0$ and $F_2(s) \not\to 0$ as $s \to \infty$. But you should not conclude from this that $F_1(s)$ and $F_2(s)$ are not Laplace transforms. There are other kinds of functions.

REMARKS

(*i*) The inverse Laplace transform of a function $F(s)$ may not be unique; in other words, it is possible that $\mathcal{L}\{f_1(t)\} = \mathcal{L}\{f_2(t)\}$ and yet $f_1 \neq f_2$. For our purposes this is not anything to be concerned about. If f_1 and f_2 are piecewise continuous on $[0, \infty)$ and of exponential order, then f_1 and f_2 are *essentially* the same. See Problem 56 in Exercises 4.2. However, if f_1 and f_2 are continuous on $[0, \infty)$ and $\mathcal{L}\{f_1(t)\} = \mathcal{L}\{f_2(t)\}$, then $f_1 = f_2$ on the interval.

(*ii*) This remark is for those of you who will be required to do partial fraction decompositions by hand. There is another way of determining the coefficients in a partial fraction decomposition in the special case when $\mathcal{L}\{f(t)\} = F(s)$ is a rational function of s and the denominator of F is a product of *distinct* linear factors. Let us illustrate by reexamining Example 3. Suppose we multiply both sides of the assumed decomposition

$$\frac{s^2 + 6s + 9}{(s-1)(s-2)(s+4)} = \frac{A}{s-1} + \frac{B}{s-2} + \frac{C}{s+4} \tag{15}$$

by, say, $s - 1$, simplify, and then set $s = 1$. Since the coefficients of B and C on the right side of the equality are zero, we get

$$\left.\frac{s^2 + 6s + 9}{(s-2)(s+4)}\right|_{s=1} = A \quad \text{or} \quad A = -\frac{16}{5}.$$

Written another way,

$$\left.\frac{s^2 + 6s + 9}{\boxed{(s-1)}(s-2)(s+4)}\right|_{s=1} = -\frac{16}{5} = A,$$

where we have colored or *covered up* the factor that canceled when the left side was multiplied by $s - 1$. Now to obtain B and C we simply evaluate the left-hand side of (15) while covering up, in turn, $s - 2$ and $s + 4$:

$$\left.\frac{s^2 + 6s + 9}{(s-1)\boxed{(s-2)}(s+4)}\right|_{s=2} = \frac{25}{6} = B \quad \text{and} \quad \left.\frac{s^2 + 6s + 9}{(s-1)(s-2)\boxed{(s+4)}}\right|_{s=-4} = \frac{1}{30} = C.$$

The desired decomposition (15) is given in (4). This special technique for determining coefficients is naturally known as the **cover-up method**.

(*iii*) In this remark we continue our introduction to the terminology of dynamical systems. Because of (9) and (10) the Laplace transform is well adapted to *linear* dynamical systems. In (11) the polynomial $P(s) = a_n s^n + a_{n-1} s^{n-1} + \cdots + a_0$ is the total coefficient of $Y(s)$ in (10) and is simply the left-hand side of the DE with the derivatives $d^k y/dt^k$ replaced by powers s^k, $k = 0, 1, \ldots, n$. It is usual practice to call the reciprocal of $P(s)$, namely, $W(s) = 1/P(s)$, the **transfer function** of the system and write (11) as

$$Y(s) = W(s)Q(s) + W(s)G(s). \tag{16}$$

In this manner we have separated, in an additive sense, the effects on the response that are due to the initial conditions (that is, $W(s)Q(s)$) and to the input function g (that is, $W(s)G(s)$). See (13) and (14). Hence the response $y(t)$ of the system is a superposition of two responses

$$y(t) = \mathcal{L}^{-1}\{W(s)Q(s)\} + \mathcal{L}^{-1}\{W(s)G(s)\} = y_0(t) + y_1(t).$$

If the input is $g(t) = 0$, then the solution of the problem is $y_0(t) = \mathcal{L}^{-1}\{W(s)Q(s)\}$. This solution is called the **zero-input response** of the system. On the other hand, the function $y_1(t) = \mathcal{L}^{-1}\{W(s)G(s)\}$ is the output due to the input $g(t)$. Now if the initial state of the system is the zero state (all the initial conditions are zero), then $Q(s) = 0$, and so the only solution of the initial-value problem is $y_1(t)$. The latter solution is called the **zero-state response** of the system. Both $y_0(t)$ and $y_1(t)$ are particular solutions: $y_0(t)$ is a solution of the IVP consisting of the associated homogeneous equation with the given initial conditions, and $y_1(t)$ is a solution of the IVP consisting of the nonhomogeneous equation with zero initial

conditions. In Example 5, we see from (14) that the transfer function is $W(s) = 1/(s^2 - 3s + 2)$, the zero-input response is

$$y_0(t) = \mathscr{L}^{-1}\left\{\frac{s+2}{(s-1)(s-2)}\right\} = -3e^t + 4e^{2t},$$

and the zero-state response is

$$y_1(t) = \mathscr{L}^{-1}\left\{\frac{1}{(s-1)(s-2)(s+4)}\right\} = -\frac{1}{5}e^t + \frac{1}{6}e^{2t} + \frac{1}{30}e^{-4t}.$$

Verify that the sum of $y_0(t)$ and $y_1(t)$ is the solution $y(t)$ in that example and that $y_0(0) = 1$, $y_0'(0) = 5$, whereas $y_1(0) = 0$, $y_1'(0) = 0$.

4.2 Exercises Answers to selected odd-numbered problems begin on page ANS-9.

4.2.1 Inverse Transforms

In Problems 1–30, use Theorem 4.2.1 to find the given inverse transform.

1. $\mathscr{L}^{-1}\left\{\dfrac{1}{s^3}\right\}$

2. $\mathscr{L}^{-1}\left\{\dfrac{1}{s^4}\right\}$

3. $\mathscr{L}^{-1}\left\{\dfrac{1}{s^2} - \dfrac{48}{s^5}\right\}$

4. $\mathscr{L}^{-1}\left\{\left(\dfrac{2}{s} - \dfrac{1}{s^3}\right)^2\right\}$

5. $\mathscr{L}^{-1}\left\{\dfrac{(s+1)^3}{s^4}\right\}$

6. $\mathscr{L}^{-1}\left\{\dfrac{(s+2)^2}{s^3}\right\}$

7. $\mathscr{L}^{-1}\left\{\dfrac{1}{s^2} - \dfrac{1}{s} + \dfrac{1}{s-2}\right\}$

8. $\mathscr{L}^{-1}\left\{\dfrac{4}{s} + \dfrac{6}{s^5} - \dfrac{1}{s+8}\right\}$

9. $\mathscr{L}^{-1}\left\{\dfrac{1}{4s+1}\right\}$

10. $\mathscr{L}^{-1}\left\{\dfrac{1}{5s-2}\right\}$

11. $\mathscr{L}^{-1}\left\{\dfrac{5}{s^2+49}\right\}$

12. $\mathscr{L}^{-1}\left\{\dfrac{10s}{s^2+16}\right\}$

13. $\mathscr{L}^{-1}\left\{\dfrac{4s}{4s^2+1}\right\}$

14. $\mathscr{L}^{-1}\left\{\dfrac{1}{4s^2+1}\right\}$

15. $\mathscr{L}^{-1}\left\{\dfrac{2s-6}{s^2+9}\right\}$

16. $\mathscr{L}^{-1}\left\{\dfrac{s+1}{s^2+2}\right\}$

17. $\mathscr{L}^{-1}\left\{\dfrac{1}{s^2+3s}\right\}$

18. $\mathscr{L}^{-1}\left\{\dfrac{s+1}{s^2-4s}\right\}$

19. $\mathscr{L}^{-1}\left\{\dfrac{s}{s^2+2s-3}\right\}$

20. $\mathscr{L}^{-1}\left\{\dfrac{1}{s^2+s-20}\right\}$

21. $\mathscr{L}^{-1}\left\{\dfrac{0.9s}{(s-0.1)(s+0.2)}\right\}$

22. $\mathscr{L}^{-1}\left\{\dfrac{s-3}{(s-\sqrt{3})(s+\sqrt{3})}\right\}$

23. $\mathscr{L}^{-1}\left\{\dfrac{s}{(s-2)(s-3)(s-6)}\right\}$

24. $\mathscr{L}^{-1}\left\{\dfrac{s^2+1}{s(s-1)(s+1)(s-2)}\right\}$

25. $\mathscr{L}^{-1}\left\{\dfrac{1}{s^3+5s}\right\}$

26. $\mathscr{L}^{-1}\left\{\dfrac{s}{(s+2)(s^2+4)}\right\}$

27. $\mathscr{L}^{-1}\left\{\dfrac{2s-4}{(s^2+s)(s^2+1)}\right\}$

28. $\mathscr{L}^{-1}\left\{\dfrac{1}{s^4-9}\right\}$

29. $\mathscr{L}^{-1}\left\{\dfrac{1}{(s^2+1)(s^2+4)}\right\}$

30. $\mathscr{L}^{-1}\left\{\dfrac{6s+3}{s^4+5s^2+4}\right\}$

In Problems 31 and 32, find the given inverse Laplace transform by finding the Laplace transform of the indicated function f.

31. $\mathscr{L}^{-1}\left\{\dfrac{1}{(s^2+a^2)(s^2+b^2)}\right\}$; $f(t) = a\sin bt - b\sin at$

32. $\mathscr{L}^{-1}\left\{\dfrac{s}{(s^2+a^2)(s^2+b^2)}\right\}$; $f(t) = \cos bt - \cos at$

4.2.2 Transforms of Derivatives

In Problems 33–46, use the Laplace transform to solve the given initial-value problem.

33. $\dfrac{dy}{dt} - y = 1$, $y(0) = 0$

34. $2\dfrac{dy}{dt} + y = 0$, $y(0) = -3$

35. $y' + 6y = e^{4t}$, $y(0) = 2$

36. $y' - y = 2\cos 5t$, $y(0) = 0$

37. $y'' + 5y' + 4y = 0$, $y(0) = 1$, $y'(0) = 0$

38. $y'' - 4y' = 6e^{3t} - 3e^{-t}$, $y(0) = 1$, $y'(0) = -1$

39. $y'' + y = \sqrt{2}\sin\sqrt{2}t$, $y(0) = 10$, $y'(0) = 0$

40. $y'' + 9y = e^t$, $y(0) = 0$, $y'(0) = 0$

41. $2y''' + 3y'' - 3y' - 2y = e^{-t}$, $y(0) = 0$, $y'(0) = 0$, $y''(0) = 1$

42. $y''' + 2y'' - y' - 2y = \sin 3t$, $y(0) = 0$, $y'(0) = 0$, $y''(0) = 1$

43. $y^{(4)} - y = 0$, $y(0) = 0$, $y'(0) = 0$, $y''(0) = 0$, $y'''(0) = 2$

44. $y^{(4)} - 9y = 0$, $y(0) = 1$, $y'(0) = 0$, $y''(0) = 3$, $y'''(0) = 0$

45. $y^{(4)} - y = 0$, $y(0) = 1$, $y'(0) = 0$, $y''(0) = 0$, $y'''(0) = 0$

46. $y^{(4)} + y'' = 1$, $y(0) = 0$, $y'(0) = 0$, $y''(0) = 0$, $y'''(0) = 0$

In Problems 47 and 48, use one of the inverse Laplace transforms found in Problems 31 and 32 to solve the given initial-value problem.

47. $y'' + 4y = 10 \cos 5t$, $\quad y(0) = 0$, $\quad y'(0) = 0$

48. $y'' + 2y = \sin t$, $\quad y(0) = 0$, $\quad y'(0) = 0$

The inverse forms of the results in Problem 53 in Exercises 4.1 are

$$\mathscr{L}^{-1}\left\{\frac{s-a}{(s-a)^2 + b^2}\right\} = e^{at}\cos bt$$

and $\quad \mathscr{L}^{-1}\left\{\frac{b}{(s-a)^2 + b^2}\right\} = e^{at}\sin bt.$

In Problems 49 and 50, use the Laplace transform and these inverses to solve the given initial-value problem.

49. $y' + y = e^{-3t}\cos 2t$, $\quad y(0) = 0$

50. $y'' - 2y' + 5y = 0$, $\quad y(0) = 1$, $\quad y'(0) = 3$

In Problems 51 and 52, use the table of Laplace transforms in Appendix C to solve the given initial-value problem.

51. $y^{(4)} + 16y = 0$, $y(0) = 0$, $y'(0) = 0$, $y''(0) = 1$, $y'''(0) = 0$

52. $y^{(4)} + 2y'' + y = 0$, $y(0) = 0$, $y'(0) = 1$, $y''(0) = 0$, $y'''(0) = 0$

Discussion Problems

53. Using $F(s) = \mathscr{L}\{f(t)\}$ the transform in (6) is the same as

$$\mathscr{L}\{f'(t)\} = s\mathscr{L}\{f(t)\} - f(0).$$

With $f(t) = te^{at}$, discuss how this result in conjunction with (c) of Theorem 4.1.1 can be used to find $\mathscr{L}\{te^{at}\}$.

54. Proceed as in Problem 53, but this time discuss how to use (7) with $f(t) = t \sin kt$ in conjunction with (d) and (e) of Theorem 4.1.1 to find $\mathscr{L}\{t \sin kt\}$.

55. Suppose $f(t)$ is continuous for $t \geq 0$ and of exponential order. If $g(t) = \int_0^t f(\tau)\,d\tau$, then use (6) to show that $\mathscr{L}\{\int_0^t f(\tau)\,d\tau\} = F(s)/s$, where $F(s) = \mathscr{L}\{f(t)\}$.

56. Make up two functions f_1 and f_2 that have the same Laplace transform. Do not think profound thoughts.

4.3 Translation Theorems

INTRODUCTION It is not convenient to use Definition 4.1.1 each time we wish to find the Laplace transform of a given function $f(t)$. For example, the integration by parts required to determine the transform of, say, $f(t) = e^t t^2 \sin 3t$ is formidable to say the least when done by hand. In this section and the next we present several labor-saving theorems that enable us to build up a more extensive list of transforms (see the table in Appendix C) without the necessity of using the definition of the Laplace transform.

4.3.1 Translation on the *s*-axis

Evaluating transforms such as $\mathscr{L}\{e^{5t}t^3\}$ and $\mathscr{L}\{e^{-2t}\cos 4t\}$ is straightforward provided we know $\mathscr{L}\{t^3\}$ and $\mathscr{L}\{\cos 4t\}$, which we do. In general, if we know $\mathscr{L}\{f(t)\} = F(s)$ it is possible to compute the Laplace transform of an exponential multiple of the function f, that is, $\mathscr{L}\{e^{at}f(t)\}$, with no additional effort other than *translating*, or *shifting*, $F(s)$ to $F(s - a)$. This result is known as the **first translation theorem** or **first shifting theorem**.

> **THEOREM 4.3.1 First Translation Theorem**
>
> If $\mathscr{L}\{f(t)\} = F(s)$ exists for $s > c$ and a is any constant, then
>
> $$\mathscr{L}\{e^{at}f(t)\} = F(s - a)$$
>
> for $s > a + c$.

PROOF: The proof is immediate, since by Definition 4.1.1

$$\mathscr{L}\{e^{at}f(t)\} = \int_0^\infty e^{-st}e^{at}f(t)\,dt = \int_0^\infty e^{-(s-a)t}f(t)\,dt = F(s - a).$$

Because $F(s)$ exists for $s > c$, $F(s - a)$ exists for $s - a > c$ or $s > a + c$. $\quad\equiv$

If we consider s a real variable, then the graph of $F(s - a)$ is the graph of $F(s)$ shifted on the s-axis by the amount $|a|$. If $a > 0$, the graph of $F(s)$ is shifted a units to the right, whereas if $a < 0$, the graph is shifted $|a|$ units to the left. See **FIGURE 4.3.1**.

FIGURE 4.3.1 Shift on *s*-axis

For emphasis it is sometimes useful to use the symbolism

$$\mathscr{L}\{e^{at}f(t)\} = \mathscr{L}\{f(t)\}_{s \to s-a},$$

where $s \to s - a$ means that in the Laplace transform $F(s)$ of $f(t)$ we replace the symbol s where it appears by $s - a$.

EXAMPLE 1 Using the First Translation Theorem

Find **(a)** $\mathscr{L}\{e^{5t}t^3\}$ **(b)** $\mathscr{L}\{e^{-2t}\cos 4t\}$.

SOLUTION The results follow from Theorems 4.1.1 and 4.3.1.

(a) $\mathscr{L}\{e^{5t}t^3\} = \mathscr{L}\{t^3\}_{s \to s-5} = \dfrac{3!}{s^4}\bigg|_{s \to s-5} = \dfrac{6}{(s-5)^4}$

(b) $\mathscr{L}\{e^{-2t}\cos 4t\} = \mathscr{L}\{\cos 4t\}_{s \to s-(-2)} = \dfrac{s}{s^2+16}\bigg|_{s \to s+2} = \dfrac{s+2}{(s+2)^2+16}$ ≡

▓ Inverse Form of Theorem 4.3.1

To compute the inverse of $F(s - a)$ we must recognize $F(s)$, find $f(t)$ by taking the inverse Laplace transform of $F(s)$, and then multiply $f(t)$ by the exponential function e^{at}. This procedure can be summarized symbolically in the following manner:

$$\mathscr{L}^{-1}\{F(s-a)\} = \mathscr{L}^{-1}\{F(s)|_{s \to s-a}\} = e^{at}f(t), \tag{1}$$

where $f(t) = \mathscr{L}^{-1}\{F(s)\}$.

EXAMPLE 2 Partial Fractions and Completing the Square

▶ *Partial fractions: repeated linear factors in (a).*

Find **(a)** $\mathscr{L}^{-1}\left\{\dfrac{2s+5}{(s-3)^2}\right\}$ **(b)** $\mathscr{L}^{-1}\left\{\dfrac{s/2+5/3}{s^2+4s+6}\right\}$.

SOLUTION **(a)** A repeated linear factor is a term $(s - a)^n$, where a is a real number and n is a positive integer ≥ 2. Recall that if $(s - a)^n$ appears in the denominator of a rational expression, then the assumed decomposition contains n partial fractions with constant numerators and denominators $s - a, (s - a)^2, \dots, (s - a)^n$. Hence with $a = 3$ and $n = 2$ we write

$$\frac{2s+5}{(s-3)^2} = \frac{A}{s-3} + \frac{B}{(s-3)^2}.$$

The numerator obtained by putting the two terms on the right over a common denominator is $2s + 5 = A(s - 3) + B$, and this identity yields $A = 2$ and $B = 11$. Therefore,

$$\frac{2s+5}{(s-3)^2} = \frac{2}{s-3} + \frac{11}{(s-3)^2} \tag{2}$$

and

$$\mathscr{L}^{-1}\left\{\frac{2s+5}{(s-3)^2}\right\} = 2\mathscr{L}^{-1}\left\{\frac{1}{s-3}\right\} + 11\mathscr{L}^{-1}\left\{\frac{1}{(s-3)^2}\right\}. \tag{3}$$

Now $1/(s - 3)^2$ is $F(s) = 1/s^2$ shifted 3 units to the right. Since $\mathscr{L}^{-1}\{1/s^2\} = t$, it follows from (1) that

$$\mathscr{L}^{-1}\left\{\frac{1}{(s-3)^2}\right\} = \mathscr{L}^{-1}\left\{\frac{1}{s^2}\bigg|_{s \to s-3}\right\} = e^{3t}t.$$

Finally, (3) is

$$\mathscr{L}^{-1}\left\{\frac{2s+5}{(s-3)^2}\right\} = 2e^{3t} + 11e^{3t}t. \tag{4}$$

(b) To start, observe that the quadratic polynomial $s^2 + 4s + 6$ does not have real zeros and so has no real linear factors. In this situation we complete the square:

$$\frac{s/2+5/3}{s^2+4s+6} = \frac{s/2+5/3}{(s+2)^2+2}. \tag{5}$$

Our goal here is to recognize the expression on the right as some Laplace transform $F(s)$ in which s has been replaced throughout by $s + 2$. What we are trying to do here is analogous to working part (b) of Example 1 backward. The denominator in (5) is already in the correct form; that is, $s^2 + 2$ with s replaced by $s + 2$. However, we must fix up the numerator by manipulating the constants: $\frac{1}{2}s + \frac{5}{3} = \frac{1}{2}(s + 2) + \frac{5}{3} - \frac{2}{2} = \frac{1}{2}(s + 2) + \frac{2}{3}$.

Now by termwise division, the linearity of $\mathscr{L}^{-1}$, parts (d) and (e) of Theorem 4.2.1, and finally from (1),

$$\frac{s/2 + 5/3}{(s + 2)^2 + 2} = \frac{(1/2)(s + 2) + 2/3}{(s + 2)^2 + 2} = \frac{1}{2}\frac{s + 2}{(s + 2)^2 + 2} + \frac{2}{3}\frac{1}{(s + 2)^2 + 2}$$

$$\mathscr{L}^{-1}\left\{\frac{s/2 + 5/3}{s^2 + 4s + 6}\right\} = \frac{1}{2}\mathscr{L}^{-1}\left\{\frac{s + 2}{(s + 2)^2 + 2}\right\} + \frac{2}{3}\mathscr{L}^{-1}\left\{\frac{1}{(s + 2)^2 + 2}\right\}$$

$$= \frac{1}{2}\mathscr{L}^{-1}\left\{\frac{s}{s^2 + 2}\bigg|_{s \to s+2}\right\} + \frac{2}{3\sqrt{2}}\mathscr{L}^{-1}\left\{\frac{\sqrt{2}}{s^2 + 2}\bigg|_{s \to s+2}\right\} \quad \text{(6)}$$

$$= \frac{1}{2}e^{-2t}\cos\sqrt{2}t + \frac{\sqrt{2}}{3}e^{-2t}\sin\sqrt{2}t. \quad \text{(7)} \equiv$$

EXAMPLE 3 Initial-Value Problem

Solve $y'' - 6y' + 9y = t^2e^{3t}$, $y(0) = 2$, $y'(0) = 17$.

SOLUTION Before transforming the DE note that its right-hand side is similar to the function in part (a) of Example 1. We use Theorem 4.3.1, the initial conditions, simplify, and then solve for $Y(s) = \mathscr{L}\{f(t)\}$:

$$\mathscr{L}\{y''\} - 6\mathscr{L}\{y'\} + 9\mathscr{L}\{y\} = \mathscr{L}\{t^2e^{3t}\}$$

$$s^2Y(s) - sy(0) - y'(0) - 6[sY(s) - y(0)] + 9Y(s) = \frac{2}{(s - 3)^3}$$

$$(s^2 - 6s + 9)Y(s) = 2s + 5 + \frac{2}{(s - 3)^3}$$

$$(s - 3)^2Y(s) = 2s + 5 + \frac{2}{(s - 3)^3}$$

$$Y(s) = \frac{2s + 5}{(s - 3)^2} + \frac{2}{(s - 3)^5}.$$

The first term on the right has already been decomposed into individual partial fractions in (2) in part (a) of Example 2:

$$Y(s) = \frac{2}{s - 3} + \frac{11}{(s - 3)^2} + \frac{2}{(s - 3)^5}.$$

Thus $$y(t) = 2\mathscr{L}^{-1}\left\{\frac{1}{s - 3}\right\} + 11\mathscr{L}^{-1}\left\{\frac{1}{(s - 3)^2}\right\} + \frac{2}{4!}\mathscr{L}^{-1}\left\{\frac{4!}{(s - 3)^5}\right\}. \quad \text{(8)}$$

From the inverse form (1) of Theorem 4.3.1, the last two terms are

$$\mathscr{L}^{-1}\left\{\frac{1}{s^2}\bigg|_{s \to s-3}\right\} = te^{3t} \quad \text{and} \quad \mathscr{L}^{-1}\left\{\frac{4!}{s^5}\bigg|_{s \to s-3}\right\} = t^4e^{3t},$$

and so (8) is $y(t) = 2e^{3t} + 11te^{3t} + \frac{1}{12}t^4e^{3t}$. $\equiv$

EXAMPLE 4 An Initial-Value Problem

Solve $y'' + 4y' + 6y = 1 + e^{-t}$, $y(0) = 0$, $y'(0) = 0$.

SOLUTION

$$\mathscr{L}\{y''\} + 4\mathscr{L}\{y'\} + 6\mathscr{L}\{y\} = \mathscr{L}\{1\} + \mathscr{L}\{e^{-t}\}$$

$$s^2 Y(s) - sy(0) - y'(0) + 4[sY(s) - y(0)] + 6Y(s) = \frac{1}{s} + \frac{1}{s+1}$$

$$(s^2 + 4s + 6)Y(s) = \frac{2s+1}{s(s+1)}$$

$$Y(s) = \frac{2s+1}{s(s+1)(s^2+4s+6)}.$$

Since the quadratic term in the denominator does not factor into real linear factors, the partial fraction decomposition for $Y(s)$ is found to be

$$Y(s) = \frac{1/6}{s} + \frac{1/3}{s+1} - \frac{s/2 + 5/3}{s^2 + 4s + 6}.$$

Moreover, in preparation for taking the inverse transform, we have already manipulated the last term into the necessary form in part (b) of Example 2. So in view of the results in (6) and (7) we have the solution

$$y(t) = \frac{1}{6}\mathscr{L}^{-1}\left\{\frac{1}{s}\right\} + \frac{1}{3}\mathscr{L}^{-1}\left\{\frac{1}{s+1}\right\} - \frac{1}{2}\mathscr{L}^{-1}\left\{\frac{s+2}{(s+2)^2 + 2}\right\} - \frac{2}{3\sqrt{2}}\mathscr{L}^{-1}\left\{\frac{\sqrt{2}}{(s+2)^2 + 2}\right\}$$

$$= \frac{1}{6} + \frac{1}{3}e^{-t} - \frac{1}{2}e^{-2t}\cos\sqrt{2}t - \frac{\sqrt{2}}{3}e^{-2t}\sin\sqrt{2}t. \qquad \equiv$$

4.3.2 Translation on the *t*-axis

■ Unit Step Function In engineering, one frequently encounters functions that are either "off" or "on." For example, an external force acting on a mechanical system or a voltage impressed on a circuit can be turned off after a period of time. It is convenient, then, to define a special function that is the number 0 (off) up to a certain time $t = a$ and then the number 1 (on) after that time. This function is called the **unit step function** or the **Heaviside function**. See the *Remarks* at the end of this section.

FIGURE 4.3.2 Graphs of unit step functions

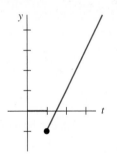

FIGURE 4.3.3 Function can be written $f(t) = (2t - 3)\mathscr{U}(t - 1)$

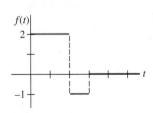

FIGURE 4.3.4 Function can be written $f(t) = 2 - 3\mathscr{U}(t - 2) + \mathscr{U}(t - 3)$

DEFINITION 4.3.1 Unit Step Function

The **unit step function** $\mathscr{U}(t - a)$ is defined to be

$$\mathscr{U}(t - a) = \begin{cases} 0, & 0 \le t < a \\ 1, & t \ge a. \end{cases}$$

Notice that we define $\mathscr{U}(t - a)$ only on the nonnegative *t*-axis since this is all that we are concerned with in the study of the Laplace transform. In a broader sense $\mathscr{U}(t - a) = 0$ for $t < a$. The graphs of $\mathscr{U}(t - a)$ and $\mathscr{U}(t)$ are given in (a) and (b), respectively, of **FIGURE 4.3.2**.

When a function f defined for $t \ge 0$ is multiplied by $\mathscr{U}(t - a)$, the unit step function "turns off" a portion of the graph of that function. For example, consider the function $f(t) = 2t - 3$. To "turn off" the portion of the graph of f on, say, the interval $0 \le t < 1$, we simply form the product $(2t - 3)\mathscr{U}(t - 1)$. See **FIGURE 4.3.3**. In general, the graph of $f(t)\mathscr{U}(t - a)$ is 0 (off) for $0 \le t < a$ and is the portion of the graph of f (on) for $t \ge a$.

The unit step function can also be used to write piecewise-defined functions in a compact form. For example, by considering $0 \le t < 2$, $2 \le t < 3$, $t \ge 3$, and the corresponding values of $\mathscr{U}(t - 2)$ and $\mathscr{U}(t - 3)$, it should be apparent that the piecewise-defined function shown in **FIGURE 4.3.4** is the same as $f(t) = 2 - 3\mathscr{U}(t - 2) + \mathscr{U}(t - 3)$. Also, a general piecewise-defined function of the type

$$f(t) = \begin{cases} g(t), & 0 \le t < a \\ h(t), & t \ge a \end{cases} \qquad (9)$$

is the same as

$$f(t) = g(t) - g(t)\,\mathcal{U}(t - a) + h(t)\,\mathcal{U}(t - a). \tag{10}$$

Similarly, a function of the type

$$f(t) = \begin{cases} 0, & 0 \le t < a \\ g(t), & a \le t < b \\ 0, & t \ge b \end{cases} \tag{11}$$

can be written
$$f(t) = g(t)[\mathcal{U}(t - a) - \mathcal{U}(t - b)]. \tag{12}$$

FIGURE 4.3.5 Function in Example 5

EXAMPLE 5 A Piecewise-Defined Function

Express $f(t) = \begin{cases} 20t, & 0 \le t < 5 \\ 0, & t \ge 5 \end{cases}$ in terms of unit step functions. Graph.

SOLUTION The graph of f is given in **FIGURE 4.3.5**. Now from (9) and (10) with $a = 5$, $g(t) = 20t$, and $h(t) = 0$, we get $f(t) = 20t - 20t\,\mathcal{U}(t - 5)$. ≡

Consider a general function $y = f(t)$ defined for $t \ge 0$. The piecewise-defined function

$$f(t - a)\,\mathcal{U}(t - a) = \begin{cases} 0, & 0 \le t < a \\ f(t - a), & t \ge a \end{cases} \tag{13}$$

plays a significant role in the discussion that follows. As shown in **FIGURE 4.3.6**, for $a > 0$ the graph of the function $y = f(t - a)\,\mathcal{U}(t - a)$ coincides with the graph of $y = f(t - a)$ for $t \ge a$ (which is the *entire* graph of $y = f(t)$, $t \ge 0$, shifted a units to the right on the t-axis) but is identically zero for $0 \le t < a$.

We saw in Theorem 4.3.1 that an exponential multiple of $f(t)$ results in a translation of the transform $F(s)$ on the s-axis. As a consequence of the next theorem we see that whenever $F(s)$ is multiplied by an exponential function e^{-as}, $a > 0$, the inverse transform of the product $e^{-as}F(s)$ is the function f shifted along the t-axis in the manner illustrated in Figure 4.3.6(b). This result, presented next in its direct transform version, is called the **second translation theorem** or **second shifting theorem**.

THEOREM 4.3.2 Second Translation Theorem

If $\mathscr{L}\{f(t)\} = F(s)$ exists for $s > c \ge 0$, and a is a positive constant, then

$$\mathscr{L}\{f(t - a)\,\mathcal{U}(t - a)\} = e^{-as}F(s)$$

for $s > c$.

PROOF: By the additive interval property of integrals, $\int_0^\infty e^{-st}f(t - a)\,\mathcal{U}(t - a)\,dt$ can be written as two integrals:

$$\mathscr{L}\{f(t - a)\,\mathcal{U}(t - a)\} = \underbrace{\int_0^a e^{-st}f(t - a)\,\mathcal{U}(t - a)\,dt}_{\text{zero for } 0 \le t < a} + \underbrace{\int_a^\infty e^{-st}f(t - a)\,\mathcal{U}(t - a)\,dt}_{\text{one for } t \ge a}$$

$$= \int_a^\infty e^{-st}f(t - a)\,dt.$$

Now if we let $v = t - a$, $dv = dt$ in the last integral, then

$$\mathscr{L}\{f(t - a)\,\mathcal{U}(t - a)\} = \int_0^\infty e^{-s(v + a)}f(v)\,dv$$

$$= e^{-as}\int_0^\infty e^{-sv}f(v)\,dv = e^{-as}\mathscr{L}\{f(t)\} = e^{-as}F(s). \quad ≡$$

FIGURE 4.3.6 Shift on t-axis

(a) $f(t)$, $t \ge 0$

(b) $f(t - a)\,\mathcal{U}(t - a)$

EXAMPLE 6 **Using the Second Translation Theorem**

Find $\mathscr{L}\{\sin 4t\, \mathscr{U}(t - \pi/2)\}$.

SOLUTION Because the function $f(t) = \sin 4t$ has period $\pi/2$ we can write

$$f(t) = \sin 4t = f(t - \pi/2) = \sin 4(t - \pi/2).$$

Then with the identification $a = \pi/2$, it follows from Theorem 4.3.2 that

$$\mathscr{L}\{\sin 4t\, \mathscr{U}(t - \pi/2)\} = \mathscr{L}\{\sin 4(t - \pi/2)\, \mathscr{U}(t - \pi/2)\}$$

$$= e^{-\pi s/2}\mathscr{L}\{\sin 4t\} = e^{-\pi s/2}\,\frac{4}{s^2 + 16}. \qquad \equiv$$

We often wish to find the Laplace transform of just a unit step function. This can be found from either Definition 4.1.1 or Theorem 4.3.2. If we identify $f(t) = 1$ in Theorem 4.3.2, then $f(t - a) = 1$, $F(s) = \mathscr{L}\{1\} = 1/s$, and so

$$\mathscr{L}\{\mathscr{U}(t - a)\} = \frac{e^{-as}}{s}. \qquad (14)$$

EXAMPLE 7 **Figure 4.3.4 Revisited**

Find the Laplace transform of the function f whose graph is given in Figure 4.3.4.

SOLUTION We use f expressed in terms of the unit step function

$$f(t) = 2 - 3\mathscr{U}(t - 2) + \mathscr{U}(t - 3)$$

and the result given in (14):

$$\mathscr{L}\{f(t)\} = 2\mathscr{L}\{1\} - 3\mathscr{L}\{\mathscr{U}(t - 2)\} + \mathscr{L}\{\mathscr{U}(t - 3)\}$$

$$= \frac{2}{s} - \frac{3e^{-2s}}{s} + \frac{e^{-3s}}{s}. \qquad \equiv$$

■ **Inverse Form of Theorem 4.3.2** If $f(t) = \mathscr{L}^{-1}\{F(s)\}$, the inverse form of Theorem 4.3.2, with $a > 0$, is

$$\mathscr{L}^{-1}\{e^{-as} F(s)\} = f(t - a)\mathscr{U}(t - a). \qquad (15)$$

EXAMPLE 8 **Using Formula (15)**

Find **(a)** $\mathscr{L}^{-1}\left\{\dfrac{1}{s - 4}e^{-2s}\right\}$ **(b)** $\mathscr{L}^{-1}\left\{\dfrac{s}{s^2 + 9}e^{-\pi s/2}\right\}$.

SOLUTION **(a)** With the identifications $a = 2$, $F(s) = 1/(s - 4)$, $\mathscr{L}^{-1}\{F(s)\} = e^{4t}$, we have from (15)

$$\mathscr{L}^{-1}\left\{\frac{1}{s - 4}e^{-2s}\right\} = e^{4(t - 2)}\mathscr{U}(t - 2).$$

(b) With $a = \pi/2$, $F(s) = s/(s^2 + 9)$, $\mathscr{L}^{-1}\{F(s)\} = \cos 3t$, (15) yields

$$\mathscr{L}^{-1}\left\{\frac{s}{s^2 + 9}e^{-\pi s/2}\right\} = \cos 3(t - \pi/2)\,\mathscr{U}(t - \pi/2).$$

The last expression can be simplified somewhat using the addition formula for the cosine. Verify that the result is the same as $-\sin 3t\,\mathscr{U}(t - \pi/2)$. $\qquad \equiv$

■ **Alternative Form of Theorem 4.3.2** We are frequently confronted with the problem of finding the Laplace transform of a product of a function g and a unit step function $\mathscr{U}(t - a)$ where the function g lacks the precise shifted form $f(t - a)$ in Theorem 4.3.2. To find the Laplace transform of $g(t)\mathscr{U}(t - a)$, it is possible to fix up $g(t)$ into the required form $f(t - a)$ by algebraic

manipulations. For example, if we want to use Theorem 4.3.2 to find the Laplace transform of $t^2\mathcal{U}(t-2)$, we would have to force $g(t) = t^2$ into the form $f(t-2)$. You should work through the details and verify that

$$t^2 = (t-2)^2 + 4(t-2) + 4$$

is an identity. Therefore

$$\mathcal{L}\{t^2\mathcal{U}(t-2)\} = \mathcal{L}\{(t-2)^2\mathcal{U}(t-2) + 4(t-2)\mathcal{U}(t-2) + 4\mathcal{U}(t-2)\}$$

$$= \mathcal{L}\{(t-2)^2\mathcal{U}(t-2)\} + \mathcal{L}\{4(t-2)\mathcal{U}(t-2)\} + \mathcal{L}\{4\mathcal{U}(t-2)\}$$

$$= e^{-2s}\mathcal{L}\{t^2\} + 4e^{-2s}\mathcal{L}\{t\} + 4e^{-2s}\mathcal{L}\{1\}$$

$$= \frac{2e^{-2s}}{s^3} + \frac{4e^{-2s}}{s^2} + \frac{4e^{-2s}}{s}$$

by Theorem 4.3.2. But since these manipulations are time-consuming and often not obvious, it is simpler to devise an alternative version of Theorem 4.3.2. Using Definition 4.1.1, the definition of $\mathcal{U}(t-a)$, and the substitution $u = t - a$, we obtain

$$\mathcal{L}\{g(t)\mathcal{U}(t-a)\} = \int_a^\infty e^{-st}g(t)\,dt = \int_0^\infty e^{-s(u+a)}g(u+a)\,du.$$

That is,
$$\mathcal{L}\{g(t)\mathcal{U}(t-a)\} = e^{-as}\mathcal{L}\{g(t+a)\}. \tag{16}$$

With $a = 2$ and $g(t) = t^2$ we have by (16),

$$\mathcal{L}\{t^2\mathcal{U}(t-2)\} = e^{-2s}\mathcal{L}\{(t+2)^2\}$$

$$= e^{-2s}\mathcal{L}\{t^2 + 4t + 4\}$$

$$= e^{-2s}\left(\frac{2}{s^3} + \frac{4}{s^2} + \frac{4}{s}\right),$$

which agrees with our previous calculation.

EXAMPLE 9 Second Translation Theorem—Alternative Form

Find $\mathcal{U}\{\cos t\,\mathcal{U}(t-\pi)\}$.

SOLUTION With $g(t) = \cos t$, $a = \pi$, then $g(t+\pi) = \cos(t+\pi) = -\cos t$ by the addition formula for the cosine function. Hence by (16),

$$\mathcal{L}\{\cos t\,\mathcal{U}(t-\pi)\} = -e^{-\pi s}\mathcal{L}\{\cos t\} = -\frac{s}{s^2+1}e^{-\pi s}.$$

In the next two examples, we solve in turn an initial-value problem and a boundary-value problem. Both problems involve a piecewise-linear differential equation.

EXAMPLE 10 An Initial-Value Problem

Solve $y' + y = f(t)$, $y(0) = 5$, where $f(t) = \begin{cases} 0, & 0 \le t < \pi \\ 3\cos t, & t \ge \pi. \end{cases}$

SOLUTION The function f can be written as $f(t) = 3\cos t\,\mathcal{U}(t-\pi)$ and so by linearity, the results of Example 9, and the usual partial fractions, we have

$$\mathcal{L}\{y'\} + \mathcal{L}\{y\} = 3\mathcal{L}\{\cos t\,\mathcal{U}(t-\pi)\}$$

$$sY(s) - y(0) + Y(s) = -3\frac{s}{s^2+1}e^{-\pi s}$$

$$(s+1)Y(s) = 5 - \frac{3s}{s^2+1}e^{-\pi s}$$

$$Y(s) = \frac{5}{s+1} - \frac{3}{2}\left[-\frac{1}{s+1}e^{-\pi s} + \frac{1}{s^2+1}e^{-\pi s} + \frac{s}{s^2+1}e^{-\pi s}\right]. \tag{17}$$

Now proceeding as we did in Example 8, it follows from (15) with $a = \pi$ that the inverses of the terms in the bracket are

$$\mathscr{L}^{-1}\left\{\frac{1}{s+1}e^{-\pi s}\right\} = e^{-(t-\pi)}\,\mathscr{U}(t-\pi), \quad \mathscr{L}^{-1}\left\{\frac{1}{s^2+1}e^{-\pi s}\right\} = \sin(t-\pi)\,\mathscr{U}(t-\pi),$$

and

$$\mathscr{L}^{-1}\left\{\frac{s}{s^2+1}e^{-\pi s}\right\} = \cos(t-\pi)\,\mathscr{U}(t-\pi).$$

Thus the inverse of (17) is

$$y(t) = 5e^{-t} + \frac{3}{2}e^{-(t-\pi)}\,\mathscr{U}(t-\pi) - \frac{3}{2}\sin(t-\pi)\,\mathscr{U}(t-\pi) - \frac{3}{2}\cos(t-\pi)\,\mathscr{U}(t-\pi)$$

$$= 5e^{-t} + \frac{3}{2}\left[e^{-(t-\pi)} + \sin t + \cos t\right]\mathscr{U}(t-\pi) \leftarrow \text{trigonometric identities}$$

$$= \begin{cases} 5e^{-t}, & 0 \le t < \pi \\ 5e^{-t} + \frac{3}{2}e^{-(t-\pi)} + \frac{3}{2}\sin t + \frac{3}{2}\cos t, & t \ge \pi. \end{cases} \tag{18}$$

With the aid of a graphing utility we get the graph of (18), shown in **FIGURE 4.3.7**. ≡

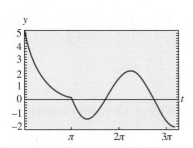

FIGURE 4.3.7 Graph of function (18) in Example 10

■ Beams In Section 3.9 we saw that the static deflection $y(x)$ of a uniform beam of length L carrying load $w(x)$ per unit length is found from the linear fourth-order differential equation

$$EI\frac{d^4 y}{dx^4} = w(x), \tag{19}$$

where E is Young's modulus of elasticity and I is a moment of inertia of a cross section of the beam. The Laplace transform is particularly useful when $w(x)$ is piecewise defined, but in order to use the transform, we must tacitly assume that $y(x)$ and $w(x)$ are defined on $(0, \infty)$ rather than on $(0, L)$. Note, too, that the next example is a boundary-value problem rather than an initial-value problem.

EXAMPLE 11 **A Boundary-Value Problem**

A beam of length L is embedded at both ends as shown in **FIGURE 4.3.8**. Find the deflection of the beam when the load is given by

$$w(x) = \begin{cases} w_0\left(1 - \dfrac{2}{L}x\right), & 0 < x \le L/2 \\[2mm] 0, & L/2 < x < L, \end{cases}$$

where w_0 is a constant.

FIGURE 4.3.8 Embedded beam with a variable load in Example 11

SOLUTION Recall that, since the beam is embedded at both ends, the boundary conditions are $y(0) = 0$, $y'(0) = 0$, $y(L) = 0$, $y'(L) = 0$. Now by (10) we can express $w(x)$ in terms of the unit step function:

$$w(x) = w_0\left(1 - \frac{2}{L}x\right) - w_0\left(1 - \frac{2}{L}x\right)\mathscr{U}\left(x - \frac{L}{2}\right)$$

$$= \frac{2w_0}{L}\left[\frac{L}{2} - x + \left(x - \frac{L}{2}\right)\mathscr{U}\left(x - \frac{L}{2}\right)\right].$$

Transforming (19) with respect to the variable x gives

$$EI(s^4 Y(s) - s^3 y(0) - s^2 y'(0) - s y''(0) - y'''(0)) = \frac{2w_0}{L}\left[\frac{L/2}{s} - \frac{1}{s^2} + \frac{1}{s^2}e^{-Ls/2}\right]$$

or

$$s^4 Y(s) - s y''(0) - y'''(0) = \frac{2w_0}{EIL}\left[\frac{L/2}{s} - \frac{1}{s^2} + \frac{1}{s^2}e^{-Ls/2}\right].$$

If we let $c_1 = y''(0)$ and $c_2 = y'''(0)$, then

$$Y(s) = \frac{c_1}{s^3} + \frac{c_2}{s^4} + \frac{2w_0}{EIL}\left[\frac{L/2}{s^5} - \frac{1}{s^6} + \frac{1}{s^6}e^{-Ls/2}\right],$$

and consequently

$$y(x) = \frac{c_1}{2!}\mathscr{L}^{-1}\left\{\frac{2!}{s^3}\right\} + \frac{c_2}{3!}\mathscr{L}^{-1}\left\{\frac{3!}{s^4}\right\}$$

$$+ \frac{2w_0}{EIL}\left[\frac{L/2}{4!}\mathscr{L}^{-1}\left\{\frac{4!}{s^5}\right\} - \frac{1}{5!}\mathscr{L}^{-1}\left\{\frac{5!}{s^6}\right\} + \frac{1}{5!}\mathscr{L}^{-1}\left\{\frac{5!}{s^6}e^{-Ls/2}\right\}\right]$$

$$= \frac{c_1}{2}x^2 + \frac{c_2}{6}x^3 + \frac{w_0}{60EIL}\left[\frac{5L}{2}x^4 - x^5 + \left(x - \frac{L}{2}\right)^5\mathscr{U}\left(x - \frac{L}{2}\right)\right].$$

Applying the conditions $y(L) = 0$ and $y'(L) = 0$ to the last result yields a system of equations for c_1 and c_2:

$$c_1\frac{L^2}{2} + c_2\frac{L^3}{6} + \frac{49w_0L^4}{1920EI} = 0$$

$$c_1L + c_2\frac{L^2}{2} + \frac{85w_0L^3}{960EI} = 0.$$

Solving, we find $c_1 = 23w_0L^2/960EI$ and $c_2 = -9w_0L/40EI$. Thus the deflection is

$$y(x) = \frac{23w_0L^2}{1920EI}x^2 - \frac{3w_0L}{80EI}x^3 + \frac{w_0}{60EIL}\left[\frac{5L}{2}x^4 - x^5 + \left(x - \frac{L}{2}\right)^5\mathscr{U}\left(x - \frac{L}{2}\right)\right]. \qquad \equiv$$

REMARKS

Many of the results presented in this chapter were devised by the English electrical engineer **Oliver Heaviside** (1850–1925) and are set forth in his 1899 treatise *Electromagnetic Theory*. The Heaviside operational calculus, as his procedures came to be called, were a means of solving linear differential equations with constant coefficients that arose in his investigation of transmission lines. Since many of his results lacked formal proof, Heaviside's procedures initially met with scorn from mathematicians. A curmudgeon, Heaviside, in turn, called these "establishment" mathematicians "woodenheaded." When Heaviside, using his symbolic methods, was able to obtain answers to problems that mathematicians could not solve, their scorn turned to censure and his papers were no longer published in mathematics journals. In 1902 Heaviside predicted the existence of a layer of ionized gas in the upper atmosphere that would reflect radio waves back to Earth. After the discovery of this layer in 1924 it came to be called the *Heaviside layer*. Heaviside lived the last years of his life as a recluse in poverty, forgotten by the scientific community. He died in an unheated house in 1925.

In an attempt to put Heaviside's operational calculus on a sound foundation, mathematicians discovered that the rules of his calculus matched many properties of the Laplace transform. Over time, Heaviside's operational calculus disappeared, to be replaced by the theory and applications of the Laplace transform.

Oliver Heaviside

4.3 Exercises Answers to selected odd-numbered problems begin on page ANS-10.

4.3.1 Translation on the s-axis

In Problems 1–20, find either $F(s)$ or $f(t)$, as indicated.

1. $\mathscr{L}\{te^{10t}\}$

2. $\mathscr{L}\{te^{-6t}\}$

3. $\mathscr{L}\{t^3e^{-2t}\}$

4. $\mathscr{L}\{t^{10}e^{-7t}\}$

5. $\mathscr{L}\{t(e^t + e^{2t})^2\}$

6. $\mathscr{L}\{e^{2t}(t-1)^2\}$

7. $\mathscr{L}\{e^t \sin 3t\}$

8. $\mathscr{L}\{e^{-2t}\cos 4t\}$

9. $\mathscr{L}\{(1 - e^t + 3e^{-4t})\cos 5t\}$

10. $\mathscr{L}\left\{e^{3t}\left(9 - 4t + 10\sin\frac{t}{2}\right)\right\}$

11. $\mathscr{L}^{-1}\left\{\dfrac{1}{(s+2)^3}\right\}$ **12.** $\mathscr{L}^{-1}\left\{\dfrac{1}{(s-1)^4}\right\}$

13. $\mathscr{L}^{-1}\left\{\dfrac{1}{s^2-6s+10}\right\}$ **14.** $\mathscr{L}^{-1}\left\{\dfrac{1}{s^2+2s+5}\right\}$

15. $\mathscr{L}^{-1}\left\{\dfrac{s}{s^2+4s+5}\right\}$ **16.** $\mathscr{L}^{-1}\left\{\dfrac{2s+5}{s^2+6s+34}\right\}$

17. $\mathscr{L}^{-1}\left\{\dfrac{s}{(s+1)^2}\right\}$ **18.** $\mathscr{L}^{-1}\left\{\dfrac{5s}{(s-2)^2}\right\}$

19. $\mathscr{L}^{-1}\left\{\dfrac{2s-1}{s^2(s+1)^3}\right\}$ **20.** $\mathscr{L}^{-1}\left\{\dfrac{(s+1)^2}{(s+2)^4}\right\}$

In Problems 21–24, write the hyperbolic functions in terms of exponential functions, and then find $F(s)$ using Theorem 4.3.1.

21. $\mathscr{L}\{\sin kt \sinh kt\}$ **22.** $\mathscr{L}\{\sin kt \cosh kt\}$

23. $\mathscr{L}\{\cos kt \sinh kt\}$ **24.** $\mathscr{L}\{\cos kt \cosh kt\}$

In Problems 25–34, use the Laplace transform to solve the given initial-value problem.

25. $y' + 4y = e^{-4t}$, $y(0) = 2$

26. $y' - y = 1 + te^t$, $y(0) = 0$

27. $y'' + 2y' + y = 0$, $y(0) = 1$, $y'(0) = 1$

28. $y'' - 4y' + 4y = t^3 e^{2t}$, $y(0) = 0$, $y'(0) = 0$

29. $y'' - 6y' + 9y = t$, $y(0) = 0$, $y'(0) = 1$

30. $y'' - 4y' + 4y = t^3$, $y(0) = 1$, $y'(0) = 0$

31. $y'' - 6y' + 13y = 0$, $y(0) = 0$, $y'(0) = -3$

32. $2y'' + 20y' + 51y = 0$, $y(0) = 2$, $y'(0) = 0$

33. $y'' - y' = e^t \cos t$, $y(0) = 0$, $y'(0) = 0$

34. $y'' - 2y' + 5y = 1 + t$, $y(0) = 0$, $y'(0) = 4$

In Problems 35 and 36, use the Laplace transform and the procedure outlined in Example 11 to solve the given boundary-value problem.

35. $y'' + 2y' + y = 0$, $y'(0) = 2$, $y(1) = 2$

36. $y'' + 8y' + 20y = 0$, $y(0) = 0$, $y'(\pi) = 0$

37. A 4-lb weight stretches a spring 2 ft. The weight is released from rest 18 in. above the equilibrium position, and the resulting motion takes place in a medium offering a damping force numerically equal to $\frac{7}{8}$ times the instantaneous velocity. Use the Laplace transform to find the equation of motion $x(t)$.

38. Recall that the differential equation for the instantaneous charge $q(t)$ on the capacitor in an *LRC*-series circuit is

$$L\frac{d^2 q}{dt^2} + R\frac{dq}{dt} + \frac{1}{C}q = E(t). \qquad (20)$$

See Section 3.8. Use the Laplace transform to find $q(t)$ when $L = 1$ h, $R = 20\ \Omega$, $C = 0.005$ F, $E(t) = 150$ V, $t > 0$, $q(0) = 0$, and $i(0) = 0$. What is the current $i(t)$?

39. Consider the battery of constant voltage E_0 that charges the capacitor shown in **FIGURE 4.3.9**. Divide equation (20) by L

and define $2\lambda = R/L$ and $\omega^2 = 1/LC$. Use the Laplace transform to show that the solution $q(t)$ of $q'' + 2\lambda q' + \omega^2 q = E_0/L$, subject to $q(0) = 0$, $i(0) = 0$, is

$$q(t) = \begin{cases} E_0 C\left[1 - e^{-\lambda t}\left(\cosh\sqrt{\lambda^2 - \omega^2}\,t \right.\right. \\ \quad \left.\left. + \dfrac{\lambda}{\sqrt{\lambda^2 - \omega^2}}\sinh\sqrt{\lambda^2 - \omega^2}\,t\right)\right], \ \lambda > \omega \\[2mm] E_0 C\left[1 - e^{-\lambda t}(1 + \lambda t)\right], \quad\quad\quad \lambda = \omega \\[2mm] E_0 C\left[1 - e^{-\lambda t}\left(\cos\sqrt{\omega^2 - \lambda^2}\,t \right.\right. \\ \quad \left.\left. + \dfrac{\lambda}{\sqrt{\omega^2 - \lambda^2}}\sin\sqrt{\omega^2 - \lambda^2}\,t\right)\right], \ \lambda < \omega. \end{cases}$$

FIGURE 4.3.9 Circuit in Problem 39

40. Use the Laplace transform to find the charge $q(t)$ in an *RC*-series when $q(0) = 0$ and $E(t) = E_0 e^{-kt}$, $k > 0$. Consider two cases: $k \neq 1/RC$ and $k = 1/RC$.

4.3.2 Translation on the *t*-axis

In Problems 41–52, find either $F(s)$ or $f(t)$, as indicated.

41. $\mathscr{L}\{(t-1)\,\mathscr{U}(t-1)\}$ **42.** $\mathscr{L}\{e^{2-t}\,\mathscr{U}(t-2)\}$

43. $\mathscr{L}\{t\,\mathscr{U}(t-2)\}$ **44.** $\mathscr{L}\{(3t+1)\,\mathscr{U}(t-1)\}$

45. $\mathscr{L}\{\cos 2t\,\mathscr{U}(t-\pi)\}$ **46.** $\mathscr{L}\{\sin t\,\mathscr{U}(t-\pi/2)\}$

47. $\mathscr{L}^{-1}\left\{\dfrac{e^{-2s}}{s^3}\right\}$ **48.** $\mathscr{L}^{-1}\left\{\dfrac{(1+e^{-2s})^2}{s+2}\right\}$

49. $\mathscr{L}^{-1}\left\{\dfrac{e^{-\pi s}}{s^2+1}\right\}$ **50.** $\mathscr{L}^{-1}\left\{\dfrac{se^{-\pi s/2}}{s^2+4}\right\}$

51. $\mathscr{L}^{-1}\left\{\dfrac{e^{-s}}{s(s+1)}\right\}$ **52.** $\mathscr{L}^{-1}\left\{\dfrac{e^{-2s}}{s^2(s-1)}\right\}$

In Problems 53–58, match the given graph with one of the given functions in (a)–(f). The graph of $f(t)$ is given in **FIGURE 4.3.10**.

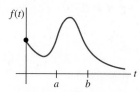

FIGURE 4.3.10 Graph for Problems 53–58

(a) $f(t) - f(t)\,\mathscr{U}(t-a)$
(b) $f(t-b)\,\mathscr{U}(t-b)$
(c) $f(t)\,\mathscr{U}(t-a)$
(d) $f(t) - f(t)\,\mathscr{U}(t-b)$
(e) $f(t)\,\mathscr{U}(t-a) - f(t)\,\mathscr{U}(t-b)$
(f) $f(t-a)\,\mathscr{U}(t-a) - f(t-a)\,\mathscr{U}(t-b)$

53.

FIGURE 4.3.11 Graph for Problem 53

54.

FIGURE 4.3.12 Graph for Problem 54

55.

FIGURE 4.3.13 Graph for Problem 55

56.

FIGURE 4.3.14 Graph for Problem 56

57.

FIGURE 4.3.15 Graph for Problem 57

58.

FIGURE 4.3.16 Graph for Problem 58

In Problems 59–70, write each function in terms of unit step functions. Find the Laplace transform of the given function.

59. $f(t) = \begin{cases} 2, & 0 \le t < 3 \\ -2, & t \ge 3 \end{cases}$

60. $f(t) = \begin{cases} 1, & 0 \le t < 4 \\ 0, & 4 \le t < 5 \\ 1, & t \ge 5 \end{cases}$

61. $f(t) = \begin{cases} 0, & 0 \le t < 1 \\ t^2, & t \ge 1 \end{cases}$

62. $f(t) = \begin{cases} 0, & 0 \le t < 3\pi/2 \\ \sin t, & t \ge 3\pi/2 \end{cases}$

63. $f(t) = \begin{cases} t, & 0 \le t < 2 \\ 0, & t \ge 2 \end{cases}$

64. $f(t) = \begin{cases} \sin t, & 0 \le t < 2\pi \\ 0, & t \ge 2\pi \end{cases}$

65. $f(t) = \begin{cases} t, & 0 \le t < 1 \\ 0, & 1 \le t < 2 \\ \sin \pi t, & t \ge 2 \end{cases}$

66. $f(t) = \begin{cases} \sin t, & 0 \le t < 2\pi \\ 2 \sin t, & 2\pi \le t < 4\pi \\ 3 \sin t, & t \ge 4\pi \end{cases}$

67.

FIGURE 4.3.17 Graph for Problem 67

rectangular pulse

68.

staircase function

FIGURE 4.3.18 Graph for Problem 68

69.

FIGURE 4.3.19 Graph for Problem 69

70.

FIGURE 4.3.20 Graph for Problem 70

In Problems 71–78, use the Laplace transform to solve the given initial-value problem.

71. $y' + y = f(t)$, $y(0) = 0$, where $f(t) = \begin{cases} 0, & 0 \le t < 1 \\ 5, & t \ge 1 \end{cases}$

72. $y' + y = f(t)$, $y(0) = 0$, where $f(t) = \begin{cases} 1, & 0 \le t < 1 \\ -1, & t \ge 1 \end{cases}$

73. $y' + 2y = f(t)$, $y(0) = 0$, where $f(t) = \begin{cases} t, & 0 \le t < 1 \\ 0, & t \ge 1 \end{cases}$

74. $y'' + 4y = f(t)$, $y(0) = 0$, $y'(0) = -1$, where

$$f(t) = \begin{cases} 1, & 0 \le t < 1 \\ 0, & t \ge 1 \end{cases}$$

75. $y'' + 4y = \sin t\, \mathcal{U}(t - 2\pi)$, $y(0) = 1$, $y'(0) = 0$

76. $y'' - 5y' + 6y = \mathcal{U}(t - 1)$, $y(0) = 0$, $y'(0) = 1$

77. $y'' + y = f(t)$, $y(0) = 0$, $y'(0) = 1$, where

$$f(t) = \begin{cases} 0, & 0 \le t < \pi \\ 1, & \pi \le t < 2\pi \\ 0, & t \ge 2\pi \end{cases}$$

78. $y'' + 4y' + 3y = 1 - \mathcal{U}(t - 2) - \mathcal{U}(t - 4) + \mathcal{U}(t - 6)$, $y(0) = 0$, $y'(0) = 0$

79. Suppose a mass weighing 32 lb stretches a spring 2 ft. If the weight is released from rest at the equilibrium position, find the equation of motion $x(t)$ if an impressed force $f(t) = 20t$ acts on the system for $0 \le t < 5$ and is then removed (see Example 5). Ignore any damping forces. Use a graphing utility to obtain the graph $x(t)$ on the interval [0, 10].

80. Solve Problem 79 if the impressed force $f(t) = \sin t$ acts on the system for $0 \le t < 2\pi$ and is then removed.

In Problems 81 and 82, use the Laplace transform to find the charge $q(t)$ on the capacitor in an RC-series circuit subject to the given conditions.

81. $q(0) = 0$, $R = 2.5\ \Omega$,
$C = 0.08$ F, $E(t)$ given
in **FIGURE 4.3.21**

82. $q(0) = q_0$, $R = 10\ \Omega$, $C = 0.1$ F,
$E(t)$ given in **FIGURE 4.3.22**

FIGURE 4.3.21 $E(t)$ in Problem 81

FIGURE 4.3.22 $E(t)$ in Problem 82

83. (a) Use the Laplace transform to find the current $i(t)$ in a single-loop LR-series circuit when $i(0) = 0$, $L = 1$ h, $R = 10\ \Omega$, and $E(t)$ is as given in **FIGURE 4.3.23**.
(b) Use a computer graphing program to graph $i(t)$ for $0 \le t \le 6$. Use the graph to estimate $i_{\max}$ and $i_{\min}$, the maximum and minimum values of the current, respectively.

FIGURE 4.3.23 $E(t)$ in Problem 83

84. (a) Use the Laplace transform to find the charge $q(t)$ on the capacitor in an RC-series circuit when $q(0) = 0$, $R = 50\ \Omega$, $C = 0.01$ F, and $E(t)$ is as given in **FIGURE 4.3.24**.
(b) Assume $E_0 = 100$ V. Use a computer graphing program to graph $q(t)$ for $0 \le t \le 6$. Use the graph to estimate $q_{\max}$, the maximum value of the charge.

FIGURE 4.3.24 $E(t)$ in Problem 84

85. A cantilever beam is embedded at its left end and free at its right end. Use the Laplace transform to find the deflection $y(x)$ when the load is given by

$$w(x) = \begin{cases} w_0, & 0 < x < L/2 \\ 0, & L/2 \le x < L. \end{cases}$$

86. Solve Problem 85 when the load is given by

$$w(x) = \begin{cases} 0, & 0 < x < L/3 \\ w_0, & L/3 \le x < 2L/3 \\ 0, & 2L/3 \le x < L. \end{cases}$$

87. Find the deflection $y(x)$ of a cantilever beam embedded at its left end and free at its right end when the load is as given in Example 11.

88. A beam is embedded at its left end and simply supported at its right end. Find the deflection $y(x)$ when the load is as given in Problem 85.

89. Cake Inside an Oven Reread Example 4 in Section 2.7 on the cooling of a cake that is taken out of an oven.
(a) Devise a mathematical model for the temperature of a cake while it is *inside* the oven based on the following assumptions: At $t = 0$ the cake mixture is at the room temperature of 70°; the oven is not preheated so that at $t = 0$, when the cake mixture is placed into the oven, the temperature inside the oven is also 70°; the temperature of the oven increases linearly until $t = 4$ minutes, when the desired temperature of 300° is attained; the oven temperature is a constant 300° for $t \ge 4$.
(b) Use the Laplace transform to solve the initial-value problem in part (a).

Discussion Problems

90. Discuss how you would fix up each of the following functions so that Theorem 4.3.2 could be used directly to find the given Laplace transform. Check your answers using (16) of this section.
(a) $\mathcal{L}\{(2t + 1)\,\mathcal{U}(t - 1)\}$
(b) $\mathcal{L}\{e^t\,\mathcal{U}(t - 5)\}$
(c) $\mathcal{L}\{\cos t\,\mathcal{U}(t - \pi)\}$
(d) $\mathcal{L}\{(t^2 - 3t)\,\mathcal{U}(t - 2)\}$

91. Assume that Theorem 4.3.1 holds when the symbol a is replaced by ki, where k is a real number and $i^2 = -1$. Show that $\mathcal{L}\{te^{kti}\}$ can be used to deduce

$$\mathcal{L}\{t \cos kt\} = \frac{s^2 - k^2}{(s^2 + k^2)^2}$$

and

$$\mathcal{L}\{t \sin kt\} = \frac{2ks}{(s^2 + k^2)^2}.$$

92. Use the Laplace transform to solve the initial-value problem

$$x'' + \omega^2 x = \cos \omega t, \quad x(0) = 0, \ x'(0) = 0.$$

4.4 Additional Operational Properties

INTRODUCTION In this section we develop several more operational properties of the Laplace transform. Specifically, we shall see how to find the transform of a function $f(t)$ that is multiplied by a monomial t^n, the transform of a special type of integral, and the transform of a periodic function. The last two transform properties allow us to solve some equations that we have not encountered up to this point: Volterra integral equations, integrodifferential equations, and ordinary differential equations in which the input function is a periodic piecewise-defined function.

4.4.1 Derivatives of Transforms

■ **Multiplying a Function by t^n** The Laplace transform of the product of a function $f(t)$ with t can be found by differentiating the Laplace transform of $f(t)$. If $F(s) = \mathscr{L}\{f(t)\}$ and if we assume that interchanging of differentiation and integration is possible, then

$$\frac{d}{ds} F(s) = \frac{d}{ds} \int_0^\infty e^{-st} f(t)\, dt = \int_0^\infty \frac{\partial}{\partial s} [e^{-st} f(t)]\, dt = -\int_0^\infty e^{-st} t f(t)\, dt = -\mathscr{L}\{tf(t)\};$$

that is,

$$\mathscr{L}\{tf(t)\} = -\frac{d}{ds}\mathscr{L}\{f(t)\}.$$

Similarly,

$$\mathscr{L}\{t^2 f(t)\} = \mathscr{L}\{t \cdot tf(t)\} = -\frac{d}{ds}\mathscr{L}\{tf(t)\}$$

$$= -\frac{d}{ds}\left(-\frac{d}{ds}\mathscr{L}\{f(t)\}\right) = \frac{d^2}{ds^2}\mathscr{L}\{f(t)\}.$$

The preceding two cases suggest the following general result for $\mathscr{L}\{t^n f(t)\}$.

THEOREM 4.4.1 Derivatives of Transforms

If the function f is piecewise continuous on $[0, \infty)$ and of exponential order with c as specified in Definition 4.1.2 and $\mathscr{L}\{f(t)\} = F(s)$, then for $n = 1, 2, 3, \ldots$ and $s > c$,

$$\mathscr{L}\{t^n f(t)\} = (-1)^n \frac{d^n}{ds^n} F(s).$$

EXAMPLE 1 Using Theorem 4.4.1

Find $\mathscr{L}\{t \sin kt\}$.

SOLUTION With $f(t) = \sin kt$, $F(s) = k/(s^2 + k^2)$, and $n = 1$, Theorem 4.4.1 gives

$$\mathscr{L}\{t \sin kt\} = -\frac{d}{ds}\mathscr{L}\{\sin kt\} = -\frac{d}{ds}\left(\frac{k}{s^2 + k^2}\right) = \frac{2ks}{(s^2 + k^2)^2}.$$ ≣

If we wish to find the Laplace transforms $\mathscr{L}\{t^2 \sin kt\}$ and $\mathscr{L}\{t^3 \sin kt\}$, all we need do is take the negative of the derivative with respect to s of $\mathscr{L}\{t \sin kt\}$, found in Example 1, followed by the negative of the derivative with respect to s of $\mathscr{L}\{t^2 \sin kt\}$:

$$\mathscr{L}\{t^2 \sin kt\} = -\frac{d}{ds}\mathscr{L}\{t \sin kt\} = -\frac{d}{ds}\left(\frac{2ks}{(s^2 + k^2)^2}\right) = \frac{6ks^2 - 2k^3}{(s^2 + k^2)^3}$$

$$\mathscr{L}\{t^3 \sin kt\} = -\frac{d}{ds}\mathscr{L}\{t^2 \sin kt\} = -\frac{d}{ds}\left(\frac{6ks^2 - 2k^3}{(s^2 + k^2)^3}\right) = \frac{24ks^3 - 24k^3 s}{(s^2 + k^2)^4}.$$

To find transforms of functions $t^n e^{at}$ we can use either the first translation theorem or Theorem 4.4.1. For example,

$$\text{Theorem 4.3.1: } \mathcal{L}\{te^{3t}\} = \mathcal{L}\{t\}_{s \to s-3} = \frac{1}{s^2}\bigg|_{s \to s-3} = \frac{1}{(s-3)^2}$$

Note that either theorem could be used. ▶

$$\text{Theorem 4.4.1: } \mathcal{L}\{te^{3t}\} = -\frac{d}{ds}\mathcal{L}\{e^{3t}\} = -\frac{d}{ds}\frac{1}{s-3} = (s-3)^{-2} = \frac{1}{(s-3)^2}.$$

EXAMPLE 2 An Initial-Value Problem

Solve $x'' + 16x = \cos 4t$, $x(0) = 0$, $x'(0) = 1$.

SOLUTION The initial-value problem could describe the forced, undamped, and resonant motion of a mass on a spring. The mass starts with an initial velocity of 1 foot per second in the downward direction from the equilibrium position.

Transforming the differential equation gives

$$(s^2 + 16)X(s) = 1 + \frac{s}{s^2 + 16} \quad \text{or} \quad X(s) = \frac{1}{s^2 + 16} + \frac{s}{(s^2 + 16)^2}.$$

Now we have just learned in Example 1 that

$$\mathcal{L}^{-1}\left\{\frac{2ks}{(s^2 + k^2)^2}\right\} = t\sin kt, \tag{1}$$

and so with the identification $k = 4$ in (1) and in part (d) of Theorem 4.2.1, we obtain

$$x(t) = \frac{1}{4}\mathcal{L}^{-1}\left\{\frac{4}{s^2 + 16}\right\} + \frac{1}{8}\mathcal{L}^{-1}\left\{\frac{8s}{(s^2 + 16)^2}\right\}$$

$$= \frac{1}{4}\sin 4t + \frac{1}{8}t\sin 4t. \qquad\qquad\equiv$$

Theorem 4.4.1 is sometimes useful in solving initial-value problems involving differential equations with variable coefficients that are monomials at^m, $m = 0, 1, 2, \ldots$. For example, if a linear second-order differential equation contains the term $ty''(t)$, then from Theorem 4.4.1 with $n = 1$ its Laplace transform is

$$\mathcal{L}\{ty''(t)\} = -\frac{d}{ds}\mathcal{L}\{y''(t)\}$$

$$= -\frac{d}{ds}(s^2 Y(s) - sy(0) - y'(0)) \tag{2}$$

$$= -s^2 Y'(s) - 2sY(s) + y(0).$$

The effect of the last term in (2) is to change a second-order DE in t to a first-order DE in s. The next example illustrates the procedure.

EXAMPLE 3 An Initial-Value Problem

Solve $ty'' + y' = t^3 + 1$, $y(0) = 0$, $y'(0) = 1$.

SOLUTION Using (2) above and (7) of Section 4.2 the Laplace transform of the differential equation is

$$\mathcal{L}\{ty''\} + \mathcal{L}\{y'\} = \mathcal{L}\{t^3\} + \mathcal{L}\{1\}$$

$$-\frac{d}{ds}(s^2 Y(s) - sy(0) - y'(0)) + (sY(s) - y(0)) = \frac{3!}{s^4} + \frac{1}{s}.$$

$$-s^2 Y'(s) - 2sY(s) + y(0) + sY(s) - y(0) = \frac{3!}{s^4} + \frac{1}{s}.$$

Simplifying, we obtain a linear first-order differential equation

$$sY'(s) + Y(s) = -\frac{6}{s^5} - \frac{1}{s^2} \quad \text{or} \quad \frac{d}{ds}[sY(s)] = -\frac{6}{s^5} - \frac{1}{s^2}.$$

Integrating with respect to s and solving for $Y(s)$ then gives

$$Y(s) = \frac{3}{2}\frac{1}{s^5} + \frac{1}{s^2} + \frac{c}{s}.$$

The inverse Laplace transform is then

$$y(t) = \frac{3}{2}\frac{1}{4!}\mathscr{L}^{-1}\left\{\frac{4!}{s^5}\right\} + \mathscr{L}^{-1}\left\{\frac{1}{s^2}\right\} + c\mathscr{L}^{-1}\left\{\frac{1}{s}\right\}$$

$$= \frac{1}{16}t^4 + t + c.$$

Applying the given initial condition $y(0) = 0$ to the last expression gives $c = 0$.
The solution of the IVP is then $y = \frac{1}{16}t^4 + t$. ≡

4.4.2 Transforms of Integrals

■ **Convolution** If functions f and g are piecewise continuous on the interval $[0, \infty)$, then the **convolution** of f and g, denoted by the symbol $f * g$, is a function defined by the integral

$$f * g = \int_0^t f(\tau)\, g(t - \tau)\, d\tau. \tag{3}$$

Because we are integrating in (3) with respect to the variable τ (the lowercase Greek letter *tau*), the convolution $f * g$ is a function of t. To emphasize this fact, (3) is also written $(f * g)(t)$.

EXAMPLE 4 **Convolution of Two Functions**

Find **(a)** $e^t * \sin t$ **(b)** $\mathscr{L}\{e^t * \sin t\}$.

SOLUTION **(a)** With the identifications

$$f(t) = e^t,\ g(t) = \sin t \quad \text{and} \quad f(\tau) = e^\tau,\ g(t - \tau) = \sin(t - \tau)$$

it follows from (3) and integration by parts that

$$e^t * \sin t = \int_0^t e^\tau \sin(t - \tau)\, d\tau$$

$$= \frac{1}{2}[e^\tau \sin(t - \tau) + e^\tau \cos(t - \tau)]_0^t$$

$$= \frac{1}{2}(-\sin t - \cos t + e^t). \tag{4}$$

(b) Then from (4) and parts (c), (d), and (e) of Theorem 4.1.1 we find

$$\mathscr{L}\{e^t * \sin t\} = -\frac{1}{2}\mathscr{L}\{\sin t\} - \frac{1}{2}\mathscr{L}\{\cos t\} + \frac{1}{2}\mathscr{L}\{e^t\}$$

$$= -\frac{1}{2}\frac{1}{s^2 + 1} - \frac{1}{2}\frac{s}{s^2 + 1} + \frac{1}{2}\frac{1}{s - 1}$$

$$= \frac{1}{(s - 1)(s^2 + 1)}.$$ ≡

■ **Properties of the Convolution** As the notation $f * g$ suggests, the convolution (3) is often interpreted as a *generalized product* of two functions f and g. Indeed, convolution has some

properties that are similar to ordinary multiplication. If f, g, and h are piecewise continuous functions on $[0, \infty)$, then

$$
\begin{aligned}
f * g &= g * f && \text{(commutative law)}\\
f * (g * h) &= (f * g) * h && \text{(associative law)}\\
f * (g + h) &= (f * g) + (f * h) && \text{(distributive law)}\\
(cf) * g &= f * (cg) = c(f * g) && (c = \text{constant})
\end{aligned}
$$

The commutative law $f * g = g * f$ simply states that two integrals are equal. To prove this, we use the change of variables $u = t - \tau$, $du = -d\tau$ in (3),

$$
f * g = -\int_t^0 f(t - u)g(u)\, du = \int_0^t g(u)f(t - u)\, du = g * f.
$$

But one has to be careful. For example, if the dot $\cdot$ denotes multiplication and $*$ convolution, then $f \cdot 1 = f$ but $f * 1 \neq f$.

■ **Convolution Theorem**　We have seen that if f and g are both piecewise continuous for $t \geq 0$, then the Laplace transform of a sum $f + g$ is the sum of the individual Laplace transforms. While it is *not* true that the Laplace transform of the product fg is the product of the Laplace transforms, we see in the next theorem—called the **convolution theorem**—that the Laplace transform of the generalized product $f * g$ is the product of the Laplace transforms of f and g.

THEOREM 4.4.2　Convolution Theorem

If the functions f and g are piecewise continuous on $[0, \infty)$ and of exponential order with c as specified in Definition 4.1.2 and $\mathscr{L}\{f(t)\} = F(s)$, $\mathscr{L}\{g(t)\} = G(s)$, then for $s > c$,

$$
\mathscr{L}\{f * g\} = \mathscr{L}\{f(t)\}\, \mathscr{L}\{g(t)\} = F(s)G(s).
$$

PROOF: Let

$$
F(s) = \mathscr{L}\{f(t)\} = \int_0^\infty e^{-s\tau} f(\tau)\, d\tau \quad \text{and} \quad G(s) = \mathscr{L}\{g(t)\} = \int_0^\infty e^{-s\beta} g(\beta)\, d\beta.
$$

Proceeding formally, we have

$$
F(s)G(s) = \left(\int_0^\infty e^{-s\tau} f(\tau)\, d\tau\right)\left(\int_0^\infty e^{-s\beta} g(\beta)\, d\beta\right)
$$

$$
= \int_0^\infty \int_0^\infty e^{-s(\tau + \beta)} f(\tau)g(\beta)\, d\tau\, d\beta
$$

$$
= \int_0^\infty f(\tau)\, d\tau \int_0^\infty e^{-s(\tau + \beta)} g(\beta)\, d\beta.
$$

Holding τ fixed, we let $t = \tau + \beta$, $dt = d\beta$, so that

$$
F(s)G(s) = \int_0^\infty f(\tau)\, d\tau \int_\tau^\infty e^{-st} g(t - \tau)\, dt.
$$

In the $t\tau$-plane we are integrating over the shaded region in **FIGURE 4.4.1**. Since f and g are piecewise continuous on $[0, \infty)$ and of exponential order, it is possible to interchange the order of integration:

$$
F(s)G(s) = \int_0^\infty e^{-st}\, dt \int_0^t f(\tau)g(t - \tau)\, d\tau = \int_0^\infty e^{-st}\left\{\int_0^t f(\tau)g(t - \tau)\, d\tau\right\} dt = \mathscr{L}\{f * g\}. \quad \equiv
$$

Theorem 4.4.2 shows that we can find the Laplace transform of the convolution $f * g$ of two functions without actually evaluating the definite integral $\int_0^t f(\tau)g(t - \tau)\, d\tau$ as we did in (4). The next example illustrates the idea.

FIGURE 4.4.1 Changing order of integration from t first to τ first

EXAMPLE 5 **Using Theorem 4.4.2**

Find $\mathscr{L}\left\{\displaystyle\int_0^t e^\tau \sin(t-\tau)\,d\tau\right\}$.

SOLUTION This is the same as the transform $\mathscr{L}\{e^t * \sin t\}$ that we found in part (b) of Example 4. This time we use Theorem 4.4.2 that the Laplace transform of the convolution of f and g is the product of their Laplace transforms:

$$\mathscr{L}\left\{\int_0^t e^\tau \sin(t-\tau)\,d\tau\right\} = \mathscr{L}\{e^t * \sin t\}$$

$$= \mathscr{L}\{e^t\}\cdot\mathscr{L}\{\sin t\}$$

$$= \frac{1}{s-1}\cdot\frac{1}{s^2+1}$$

$$= \frac{1}{(s-1)(s^2+1)}. \qquad \equiv$$

■ **Inverse Form of Theorem 4.4.2** The convolution theorem is sometimes useful in finding the inverse Laplace transforms of the product of two Laplace transforms. From Theorem 4.4.2 we have

$$\mathscr{L}^{-1}\{F(s)G(s)\} = f * g. \tag{5}$$

Many of the results in the table of Laplace transforms in Appendix C can be derived using (5). For instance, in the next example we obtain entry 25 of the table:

$$\mathscr{L}\{\sin kt - kt\cos kt\} = \frac{2k^3}{(s^2+k^2)^2}. \tag{6}$$

EXAMPLE 6 **Inverse Transform as a Convolution**

Find $\mathscr{L}^{-1}\left\{\dfrac{1}{(s^2+k^2)^2}\right\}$.

SOLUTION Let $F(s) = G(s) = \dfrac{1}{s^2+k^2}$

so that $f(t) = g(t) = \dfrac{1}{k}\mathscr{L}^{-1}\left\{\dfrac{k}{s^2+k^2}\right\} = \dfrac{1}{k}\sin kt.$

In this case (5) gives

$$\mathscr{L}^{-1}\left\{\frac{1}{(s^2+k^2)^2}\right\} = \frac{1}{k^2}\int_0^t \sin k\tau \sin k(t-\tau)\,dt. \tag{7}$$

Now recall from trigonometry that

$$\sin A \sin B = \frac{1}{2}[\cos(A-B) - \cos(A+B)].$$

If we set $A = k\tau$ and $B = k(t-\tau)$, we can carry out the integration in (7):

$$\mathscr{L}^{-1}\left\{\frac{1}{(s^2+k^2)^2}\right\} = \frac{1}{2k^2}\int_0^t [\cos k(2\tau - t) - \cos kt]\,d\tau$$

$$= \frac{1}{2k^2}\left[\frac{1}{2k}\sin k(2\tau - t) - \tau\cos kt\right]_0^t$$

$$= \frac{\sin kt - kt\cos kt}{2k^3}.$$

Multiplying both sides by $2k^3$ gives the inverse form of (6). $\equiv$

Transform of an Integral When $g(t) = 1$ and $\mathscr{L}\{g(t)\} = G(s) = 1/s$, the convolution theorem implies that the Laplace transform of the integral of f is

$$\mathscr{L}\left\{\int_0^t f(\tau)\,d\tau\right\} = \frac{F(s)}{s}. \tag{8}$$

The inverse form of (8),

$$\int_0^t f(\tau)\,d\tau = \mathscr{L}^{-1}\left\{\frac{F(s)}{s}\right\}, \tag{9}$$

can be used in lieu of partial fractions when s^n is a factor of the denominator and $f(t) = \mathscr{L}^{-1}\{F(s)\}$ is easy to integrate. For example, we know for $f(t) = \sin t$ that $F(s) = 1/(s^2 + 1)$, and so by (9)

$$\mathscr{L}^{-1}\left\{\frac{1}{s(s^2 + 1)}\right\} = \mathscr{L}^{-1}\left\{\frac{1/(s^2 + 1)}{s}\right\} = \int_0^t \sin\tau\,d\tau = 1 - \cos t$$

$$\mathscr{L}^{-1}\left\{\frac{1}{s^2(s^2 + 1)}\right\} = \mathscr{L}^{-1}\left\{\frac{1/s(s^2 + 1)}{s}\right\} = \int_0^t (1 - \cos\tau)\,d\tau = t - \sin t$$

$$\mathscr{L}^{-1}\left\{\frac{1}{s^3(s^2 + 1)}\right\} = \mathscr{L}^{-1}\left\{\frac{1/s^2(s^2 + 1)}{s}\right\} = \int_0^t (\tau - \sin\tau)\,d\tau = \frac{1}{2}t^2 - 1 + \cos t$$

and so on.

Volterra Integral Equation The convolution theorem and the result in (8) are useful in solving other types of equations in which an unknown function appears under an integral sign. In the next example, we solve a **Volterra integral equation** for $f(t)$,

$$f(t) = g(t) + \int_0^t f(\tau)\,h(t - \tau)\,d\tau. \tag{10}$$

The functions $g(t)$ and $h(t)$ are known. Notice that the integral in (10) has the convolution form (3) with the symbol h playing the part of g.

EXAMPLE 7 **An Integral Equation**

Solve $f(t) = 3t^2 - e^{-t} - \displaystyle\int_0^t f(\tau)\,e^{t-\tau}\,d\tau$ for $f(t)$.

SOLUTION In the integral we identify $h(t - \tau) = e^{t-\tau}$ so that $h(t) = e^t$. We take the Laplace transform of each term; in particular, by Theorem 4.4.2 the transform of the integral is the product of $\mathscr{L}\{f(t)\} = F(s)$ and $\mathscr{L}\{e^t\} = 1/(s - 1)$:

$$F(s) = 3 \cdot \frac{2}{s^3} - \frac{1}{s + 1} - F(s) \cdot \frac{1}{s - 1}.$$

After solving the last equation for $F(s)$ and carrying out the partial fraction decomposition, we find

$$F(s) = \frac{6}{s^3} - \frac{6}{s^4} + \frac{1}{s} - \frac{2}{s + 1}.$$

The inverse transform then gives

$$f(t) = 3\mathscr{L}^{-1}\left\{\frac{2!}{s^3}\right\} - \mathscr{L}^{-1}\left\{\frac{3!}{s^4}\right\} + \mathscr{L}^{-1}\left\{\frac{1}{s}\right\} - 2\mathscr{L}^{-1}\left\{\frac{1}{s + 1}\right\}$$

$$= 3t^2 - t^3 + 1 - 2e^{-t}.$$

≡

Series Circuits In a single-loop or series circuit, Kirchhoff's second law states that the sum of the voltage drops across an inductor, resistor, and capacitor is equal to the voltage $E(t)$ impressed on the circuit. Now the voltage drops across an inductor, resistor, and capacitor are, respectively,

$$L\frac{di}{dt}, \quad Ri(t), \quad \text{and} \quad \frac{1}{C}q(t),$$

where $i(t)$ is the current and $q(t)$ is the charge on the capacitor. The constants L, R, and C are called, in turn, the inductance, resistance, and capacitance. Hence, in an LR-series circuit, such as that shown in **FIGURE 4.4.2(a)**, the sum of the voltage drops across the inductance and the resistor is the differential equation

$$L\frac{di}{dt} + Ri = E(t) \tag{11}$$

for the current $i(t)$. The sum of the voltage drops across the resistor and the capacitor for the RC-series circuit in Figure 4.4.2(b) is

$$Ri + \frac{1}{C}q = E(t). \tag{12}$$

Because $i(t)$ and $q(t)$ are related by $dq/dt = i$, we can write $q = \int_0^t i(\tau)\,d\tau$. Substituting the last expression in (12) yields an **integral equation** for the current $i(t)$:

$$Ri + \frac{1}{C}\int_0^t i(\tau)\,d\tau = E(t). \tag{13}$$

For the LC-series circuit shown in Figure 4.4.2(c), the sum of the voltage drops gives an equation that involves both an integral and the derivative of $i(t)$:

$$L\frac{di}{dt} + \frac{1}{C}\int_0^t i(\tau)\,d\tau = E(t). \tag{14}$$

Finally, for the LRC-series circuit in Figure 4.4.2(d), we have

$$L\frac{di}{dt} + Ri + \frac{1}{C}\int_0^t i(\tau)\,d\tau = E(t). \tag{15}$$

Equations (14) and (15) are called **integrodifferential equations**. In view of the transform (8), we can use the Laplace transform to solve equations such as (11), (13), (14), and (15).

(a) LR-series circuit (b) RC-series circuit (c) LC-series circuit (d) LRC-series circuit

FIGURE 4.4.2 Series circuits

EXAMPLE 8 **An Integrodifferential Equation**

Determine the current $i(t)$ in a single-loop LRC-circuit when $L = 0.1$ h, $R = 2\ \Omega$, $C = 0.1$ F, $i(0) = 0$, and the impressed voltage is

$$E(t) = 120t - 120t\,\mathcal{U}(t - 1).$$

SOLUTION Using the given data, equation (15) becomes

$$0.1\frac{di}{dt} + 2i + 10\int_0^t i(\tau)\,d\tau = 120t - 120t\,\mathcal{U}(t - 1).$$

Now by (8), $\mathscr{L}\{\int_0^t i(\tau)\,d\tau\} = I(s)/s$, where $I(s) = \mathscr{L}\{i(t)\}$. Thus the Laplace transform of the integrodifferential equation is

$$0.1sI(s) + 2I(s) + 10\frac{I(s)}{s} = 120\left[\frac{1}{s^2} - \frac{1}{s^2}e^{-s} - \frac{1}{s}e^{-s}\right]. \leftarrow \text{by (16) of Section 4.3}$$

Multiplying this equation by $10s$, using $s^2 + 20s + 100 = (s + 10)^2$, and then solving for $I(s)$ gives

$$I(s) = 1200\left[\frac{1}{s(s+10)^2} - \frac{1}{s(s+10)^2}e^{-s} - \frac{1}{(s+10)^2}e^{-s}\right].$$

By partial fractions,

$$I(s) = 1200\left[\frac{1/100}{s} - \frac{1/100}{s+10} - \frac{1/10}{(s+10)^2} - \frac{1/100}{s}e^{-s}\right.$$

$$\left. + \frac{1/100}{s+10}e^{-s} + \frac{1/10}{(s+10)^2}e^{-s} - \frac{1}{(s+10)^2}e^{-s}\right].$$

From the inverse form of the second translation theorem, (15) of Section 4.3, we finally obtain

$$i(t) = 12[1 - \mathscr{U}(t-1)] - 12[e^{-10t} - e^{-10(t-1)}\mathscr{U}(t-1)]$$

$$- 120te^{-10t} - 1080(t-1)e^{-10(t-1)}\mathscr{U}(t-1).$$

Written as a piecewise-defined function, the current is

$$i(t) = \begin{cases} 12 - 12e^{-10t} - 120te^{-10t}, & 0 \le t < 1 \\ -12e^{-10t} + 12e^{-10(t-1)} - 120te^{-10t} - 1080(t-1)e^{-10(t-1)}, & t \ge 1. \end{cases} \quad (16)$$

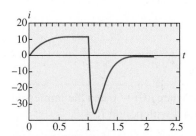

FIGURE 4.4.3 Graph of current $i(t)$ in (16) of Example 8

Using the last form of the solution and a CAS, we graph $i(t)$ on each of the two intervals and then combine the graphs. Note in **FIGURE 4.4.3** that even though the input $E(t)$ is discontinuous, the output or response $i(t)$ is a continuous function. $\equiv$

4.4.3 Transform of a Periodic Function

■ **Periodic Function** If a periodic function f has period T, $T > 0$, then $f(t + T) = f(t)$. The Laplace transform of a periodic function can be obtained by integration over one period.

THEOREM 4.4.3 **Transform of a Periodic Function**

If $f(t)$ is piecewise continuous on $[0, \infty)$, of exponential order, and periodic with period T, then

$$\mathscr{L}\{f(t)\} = \frac{1}{1 - e^{-sT}}\int_0^T e^{-st}f(t)\,dt.$$

PROOF: Write the Laplace transform of f as two integrals:

$$\mathscr{L}\{f(t)\} = \int_0^T e^{-st}f(t)\,dt + \int_T^\infty e^{-st}f(t)\,dt.$$

When we let $t = u + T$, the last integral becomes

$$\int_T^\infty e^{-st}f(t)\,dt = \int_0^\infty e^{-s(u+T)}f(u+T)\,du = e^{-sT}\int_0^\infty e^{-su}f(u)\,du = e^{-sT}\mathscr{L}\{f(t)\}.$$

Therefore

$$\mathscr{L}\{f(t)\} = \int_0^T e^{-st}f(t)\,dt + e^{-sT}\mathscr{L}\{f(t)\}.$$

Solving the equation in the last line for $\mathscr{L}\{f(t)\}$ proves the theorem. $\equiv$

$f(t)$

1

a　$2a$　$3a$　$4a$　t

FIGURE 4.4.4 Square wave in
Example 9

EXAMPLE 9 **Transform of a Periodic Function**

Find the Laplace transform of the periodic function shown in **FIGURE 4.4.4**.

SOLUTION The function $f(t)$ in the figure is called a **square wave** and has period $T = 2a$. For $0 \leq t < 2a$, $f(t)$ can be defined by

$$f(t) = \begin{cases} 1, & 0 \leq t < a \\ 0, & a \leq t < 2a \end{cases}$$

and outside the interval by $f(t + 2a) = f(t)$. Now from Theorem 4.4.3,

$$\mathscr{L}\{f(t)\} = \frac{1}{1 - e^{-2sa}} \int_0^{2a} e^{-st} f(t)\, dt$$

$$= \frac{1}{1 - e^{-2sa}} \int_0^a e^{-st} \cdot 1\, dt + \frac{1}{1 - e^{-2sa}} \int_a^{2a} e^{-st} \cdot 0\, dt$$

$$= \frac{1}{1 - e^{-2sa}} \frac{1 - e^{-sa}}{s} \quad \leftarrow 1 - e^{-2sa} = (1 + e^{-sa})(1 - e^{-sa})$$

$$= \frac{1}{s(1 + e^{-sa})}. \tag{17} \equiv$$

EXAMPLE 10 **A Periodic Impressed Voltage**

Determine the current $i(t)$ in a single loop LR-series circuit when $i(0) = 0$ and the impressed voltage $E(t)$ is the square-wave function given in Figure 4.4.4 with period $T = 2$.

SOLUTION The differential equation for the current $i(t)$ is given in (11). If we use the result in (17) of the preceding example with $a = 1$, then the Laplace transform of the DE is

$$Ls\,I(s) + R\,I(s) = \frac{1}{s(1 + e^{-s})} \quad \text{or} \quad I(s) = \frac{1/L}{s(s + R/L)} \cdot \frac{1}{1 + e^{-s}}. \tag{18}$$

To find the inverse Laplace transform of the last function, we first make use of geometric series. With the identification $x = e^{-s}$, $s > 0$, the geometric series

$$\frac{1}{1 + x} = 1 - x + x^2 - x^3 + \cdots \quad \text{becomes} \quad \frac{1}{1 + e^{-s}} = 1 - e^{-s} + e^{-2s} - e^{-3s} + \cdots.$$

From

$$\frac{1}{s(s + R/L)} = \frac{L/R}{s} - \frac{L/R}{s + R/L},$$

we can then rewrite (18) as

$$I(s) = \frac{1}{R}\left(\frac{1}{s} - \frac{1}{s + R/L}\right)(1 - e^{-s} + e^{-2s} - e^{-3s} + \cdots)$$

$$= \frac{1}{R}\left(\frac{1}{s} - \frac{e^{-s}}{s} + \frac{e^{-2s}}{s} - \frac{e^{-3s}}{s} + \cdots\right) - \frac{1}{R}\left(\frac{1}{s + R/L} - \frac{e^{-s}}{s + R/L} + \frac{e^{-2s}}{s + R/L} - \frac{e^{-3s}}{s + R/L} + \cdots\right).$$

By applying the form of the second translation theorem to each term of both series we obtain

$$i(t) = \frac{1}{R}\Big(1 - \mathscr{U}(t - 1) + \mathscr{U}(t - 2) - \mathscr{U}(t - 3) + \cdots\Big)$$

$$- \frac{1}{R}\Big(e^{-Rt/L} - e^{-R(t-1)/L}\mathscr{U}(t - 1) + e^{-R(t-2)/L}\mathscr{U}(t - 2) - e^{-R(t-3)/L}\mathscr{U}(t - 3) + \cdots\Big)$$

or, equivalently,

$$i(t) = \frac{1}{R}(1 - e^{-Rt/L}) + \frac{1}{R}\sum_{n=1}^{\infty}(-1)^n(1 - e^{-R(t-n)/L})\mathcal{U}(t - n).$$

To interpret the solution, let us assume for the sake of illustration that $R = 1$, $L = 1$, and $0 \le t < 4$. In this case

$$i(t) = 1 - e^{-t} - (1 - e^{t-1})\mathcal{U}(t - 1) + (1 - e^{-(t-2)})\mathcal{U}(t - 2) - (1 - e^{-(t-3)})\mathcal{U}(t - 3);$$

in other words,

$$i(t) = \begin{cases} 1 - e^{-t}, & 0 \le t < 1 \\ -e^{-t} + e^{-(t-1)}, & 1 \le t < 2 \\ 1 - e^{-t} + e^{-(t-1)} - e^{-(t-2)}, & 2 \le t < 3 \\ -e^{-t} + e^{-(t-1)} - e^{-(t-2)} + e^{-(t-3)}, & 3 \le t < 4. \end{cases} \quad (19)$$

FIGURE 4.4.5 Graph of current $i(t)$ in Example 10

The graph of $i(t)$ for $0 \le t < 4$, given in **FIGURE 4.4.5**, was obtained with the help of a CAS. ☰

REMARKS

You should verify either by substitution in the differential equation or by the methods of Section 2.3 that $y(t) = e^{-t}\int_0^t e^{u^2+u}\,du$ is a perfectly good solution of the linear initial-value problem $y' + y = e^{t^2}$, $y(0) = 0$. We now solve the same equation with a formal application of the Laplace transform. If we denote $\mathcal{L}\{y(t)\} = Y(s)$ and $\mathcal{L}\{e^{t^2}\} = F(s)$, then the transform of the equation is

$$sY(s) - y(0) + Y(s) = F(s) \quad \text{or} \quad Y(s) = \frac{F(s)}{s + 1}.$$

Using $\mathcal{L}^{-1}\{F(s)\} = e^{t^2}$ and $\mathcal{L}^{-1}\left\{\dfrac{1}{s+1}\right\} = e^{-t}$ it follows from the inverse form (5) of the convolution theorem that the solution of the initial-value problem is

$$y(t) = \mathcal{L}^{-1}\left\{F(s) \cdot \frac{1}{s+1}\right\} = \int_0^t e^{\tau^2} \cdot e^{-(t-\tau)}\,d\tau = e^{-t}\int_0^t e^{\tau^2+\tau}\,d\tau.$$

With τ playing the part of u, this is the solution as first given.

Unfortunately, the Laplace transform "solution method" of the foregoing initial-value problem is pure nonsense. Read the discussion preceding Theorem 4.1.2. Also, see Problem 63 in Exercises 4.4.

4.4 Exercises Answers to selected odd-numbered problems begin on page ANS-10.

4.4.1 Derivatives of Transforms

In Problems 1–8, use Theorem 4.4.1 to evaluate the given Laplace transform.

1. $\mathcal{L}\{te^{-10t}\}$
2. $\mathcal{L}\{t^3e^t\}$
3. $\mathcal{L}\{t\cos 2t\}$
4. $\mathcal{L}\{t\sinh 3t\}$
5. $\mathcal{L}\{t\sinh t\}$
6. $\mathcal{L}\{t^2\cos t\}$
7. $\mathcal{L}\{te^{2t}\sin 6t\}$
8. $\mathcal{L}\{te^{-3t}\cos 3t\}$

In Problems 9–14, use the Laplace transform to solve the given initial-value problem. Use the table of Laplace transforms in Appendix C as needed.

9. $y' + y = t\sin t$, $y(0) = 0$

10. $y' - y = te^t\sin t$, $y(0) = 0$
11. $y'' + 9y = \cos 3t$, $y(0) = 2$, $y'(0) = 5$
12. $y'' + y = \sin t$, $y(0) = 1$, $y'(0) = -1$
13. $y'' + 16y = f(t)$, $y(0) = 0$, $y'(0) = 1$, where

$$f(t) = \begin{cases} \cos 4t, & 0 \le t < \pi \\ 0, & t \ge \pi \end{cases}$$

14. $y'' + y = f(t)$, $y(0) = 1$, $y'(0) = 0$, where

$$f(t) = \begin{cases} 1, & 0 \le t < \pi/2 \\ \sin t, & t \ge \pi/2 \end{cases}$$

In Problems 15 and 16, use a graphing utility to graph the indicated solution.

15. $y(t)$ of Problem 13 for $0 \le t < 2\pi$

16. $y(t)$ of Problem 14 for $0 \le t < 3\pi$

In Problems 17–20, use Theorem 4.4.1 to reduce the given differential to a linear first-order DE in the transformed function $Y(s) = \mathcal{L}\{y(t)\}$. Solve the first-order DE and then find $y(t) = \mathcal{L}^{-1}\{Y(s)\}$. Use the table of Laplace transforms in Appendix C as needed. *Note*: Your solution may contain an arbitrary constant.

17. $ty'' - y' = 2t^2$, $y(0) = 0, y'(0) = 0$

18. $ty'' - 4ty' - 4y = 0$, $y(0) = 0, y'(0) = 5$

19. $ty'' - 2y' + ty = 0$, $y(0) = 0, y'(0) = 0$

20. $ty'' + (t - 3)y' + y = 0$, $y(0) = 0, y'(0) = 0$

In Problems 21 and 22, proceed as in Problems 17–20 to solve the given differential equation. But in each of these problems, you will have to use Theorem 4.2.3 on $Y(s)$ to eliminate a term containing a constant c of integration.

21. $2y'' + ty' - 2y = 10$, $y(0) = 0, y'(0) = 0$

22. $y'' + ty' - 2y = 8$, $y(0) = 0, y'(0) = 0$

4.4.2 Transforms of Integrals

In Problems 23–26, proceed as in Example 4 and find the convolution $f * g$ of the given functions. After integrating find the Laplace transform of $f * g$.

23. $f(t) = 4t, g(t) = 3t^2$ **24.** $f(t) = t, g(t) = e^{-t}$

25. $f(t) = e^{-t}, g(t) = e^t$ **26.** $f(t) = \cos 2t, g(t) = e^t$

In Problems 27–38, proceed as in Example 5 and find the Laplace transform of $f * g$ using Theorem 4.4.2. Do not evaluate the convolution integral before transforming.

27. $\mathcal{L}\{1 * t^3\}$ **28.** $\mathcal{L}\{t^2 * te^t\}$

29. $\mathcal{L}\{e^{-t} * e^t \cos t\}$ **30.** $\mathcal{L}\{e^{2t} * \sin t\}$

31. $\mathcal{L}\left\{\int_0^t e^\tau \, d\tau\right\}$ **32.** $\mathcal{L}\left\{\int_0^t \cos \tau \, d\tau\right\}$

33. $\mathcal{L}\left\{\int_0^t e^{-\tau} \cos \tau \, d\tau\right\}$ **34.** $\mathcal{L}\left\{\int_0^t \tau \sin \tau \, d\tau\right\}$

35. $\mathcal{L}\left\{\int_0^t \tau e^{t-\tau} \, d\tau\right\}$ **36.** $\mathcal{L}\left\{\int_0^t \sin \tau \cos(t - \tau) \, d\tau\right\}$

37. $\mathcal{L}\left\{t\int_0^t \sin \tau \, d\tau\right\}$ **38.** $\mathcal{L}\left\{t\int_0^t \tau e^{-\tau} \, d\tau\right\}$

In Problems 39–42, use (9) to evaluate the given inverse transform.

39. $\mathcal{L}^{-1}\left\{\dfrac{1}{s(s - 1)}\right\}$ **40.** $\mathcal{L}^{-1}\left\{\dfrac{1}{s^2(s - 1)}\right\}$

41. $\mathcal{L}^{-1}\left\{\dfrac{1}{s^3(s - 1)}\right\}$ **42.** $\mathcal{L}^{-1}\left\{\dfrac{1}{s(s - a)^2}\right\}$

In Problems 43 and 44, use (3) and (5) to solve the given initial-value problem. Let $\mathcal{L}\{f(t)\} = F(s)$.

43. $y'' + y = f(t)$, $y(0) = 1, y'(0) = 1$

44. $y'' - 4y' + 4y = f(t)$, $y(0) = 0, y'(0) = 5$

45. The table in Appendix C does not contain an entry for

$$\mathcal{L}^{-1}\left\{\frac{8k^3 s}{(s^2 + k^2)^3}\right\}.$$

 (a) Use (5) along with the result in (6) to evaluate this inverse transform. Use a CAS as an aid in evaluating the convolution integral.

 (b) Reexamine your answer to part (a). Could you have obtained the result in a different manner?

46. Use the Laplace transform and the result of Problem 45 to solve the initial-value problem

$$y'' + y = \sin t + t \sin t, \quad y(0) = 0, \ y'(0) = 0.$$

Use a graphing utility to graph the solution.

In Problems 47–58, use the Laplace transform to solve the given integral equation or integrodifferential equation.

47. $f(t) + \displaystyle\int_0^t (t - \tau)f(\tau) \, d\tau = t$

48. $f(t) = 2t - 4\displaystyle\int_0^t \sin \tau f(t - \tau) \, d\tau$

49. $f(t) = te^t + \displaystyle\int_0^t \tau f(t - \tau) \, d\tau$

50. $f(t) + 2\displaystyle\int_0^t f(\tau) \cos(t - \tau) \, d\tau = 4e^{-t} + \sin t$

51. $f(t) + \displaystyle\int_0^t f(\tau) \, d\tau = 1$

52. $f(t) = \cos t + \displaystyle\int_0^t e^{-\tau} f(t - \tau) \, d\tau$

53. $f(t) = 1 + t - \dfrac{8}{3}\displaystyle\int_0^t (\tau - t)^3 f(\tau) \, d\tau$

54. $t - 2f(t) = \displaystyle\int_0^t (e^\tau - e^{-\tau}) f(t - \tau) \, d\tau$

55. $y'(t) = 1 - \sin t - \displaystyle\int_0^t y(\tau) \, d\tau$, $y(0) = 0$

56. $\dfrac{dy}{dt} + 6y(t) + 9\displaystyle\int_0^t y(\tau) \, d\tau = 1$, $y(0) = 0$

57. $\dfrac{dy}{dt} = 10 - \displaystyle\int_0^t e^{-4t} y(t - \tau) \, d\tau$, $y(0) = 5$

58. $\dfrac{dy}{dt} + 2y + \displaystyle\int_0^t y(\tau) \, d\tau = t^3 e^{-t}$, $y(0) = 0$

59. Solve equation (13) subject to $i(0) = 0$ with
$$R = 50 \, \Omega, \, C = 0.01 \, \text{F},$$
$$E(t) = 100[\mathcal{U}(t - 1) - \mathcal{U}(t - 3)].$$
Use a graphing utility to graph the solution for $0 \le t \le 5$.

60. Solve equation (14) subject to $i(0) = 0$ with
 $L = 1$ h, $C = 0.04$ F,
 $$E(t) = 100t - 100(t - 1)\mathcal{U}(t - 1) - 100\mathcal{U}(t - 4).$$
 Use a graphing utility to graph the solution for $0 \le t \le 5$.

In Problems 61 and 62, solve equation (15) subject to $i(0) = 0$ with L, R, C, and $E(t)$ as given. Use a graphing utility to graph the solution for $0 \le t \le 3$.

61. $L = 0.1$ h, $R = 3\ \Omega$, $C = 0.05$ F,
 $$E(t) = 100[\mathcal{U}(t - 1) - \mathcal{U}(t - 2)]$$

62. $L = 0.005$ h, $R = 1\ \Omega$, $C = 0.02$ F,
 $$E(t) = 100[t - (t - 1)\mathcal{U}(t - 1)]$$

63. The Laplace transform $\mathscr{L}\{e^{-t^2}\}$ exists, but without finding it solve the initial-value problem

$$y'' + 9y = 3e^{-t^2}, \quad y(0) = 0, \quad y'(0) = 0.$$

64. Solve the integral equation

$$f(t) = e^t + e^t \int_0^t e^{-\tau} f(\tau)\, d\tau.$$

4.4.3 Transform of a Periodic Function

In Problems 65–70, use Theorem 4.4.3 to find the Laplace transform of the given periodic function.

65.

Meander function

FIGURE 4.4.6 Graph for Problem 65

66.

Square wave

FIGURE 4.4.7 Graph for Problem 66

67.

Sawtooth function

FIGURE 4.4.8 Graph for Problem 67

68.

Triangular wave

FIGURE 4.4.9 Graph for Problem 68

69.

Full-wave rectification of $\sin t$

FIGURE 4.4.10 Graph for Problem 69

70.

Half-wave rectification of $\sin t$

FIGURE 4.4.11 Graph for Problem 70

In Problems 71 and 72, solve equation (11) subject to $i(0) = 0$ with $E(t)$ as given. Use a graphing utility to graph the solution for $0 \le t \le 4$ in the case when $L = 1$ and $R = 1$.

71. $E(t)$ is the meander function in Problem 65 with amplitude 1 and $a = 1$.

72. $E(t)$ is the sawtooth function in Problem 67 with amplitude 1 and $b = 1$.

In Problems 73 and 74, solve the model for a driven spring/mass system with damping

$$m\frac{d^2x}{dt^2} + \beta\frac{dx}{dt} + kx = f(t), \quad x(0) = 0, \quad x'(0) = 0,$$

where the driving function f is as specified. Use a graphing utility to graph $x(t)$ for the indicated values of t.

73. $m = \frac{1}{2}$, $\beta = 1$, $k = 5$, f is the meander function in Problem 65 with amplitude 10, and $a = \pi$, $0 \le t \le 2\pi$.

74. $m = 1$, $\beta = 2$, $k = 1$, f is the square wave in Figure 4.4.4 with amplitude 5, and $a = \pi$, $0 \le t \le 4\pi$.

Discussion Problems

75. Show how to use the Laplace transform to find the numerical value of the improper integral $\int_0^\infty te^{-2t}\sin 4t\, dt$.

76. In Problem 63 we were able to solve an initial-value problem without knowing the Laplace transform $\mathscr{L}\{e^{-t^2}\}$. In this problem you are asked to find the actual transformed function $Y(s) = \mathscr{L}\{e^{-t^2}\}$ by solving another initial-value problem.

(a) If $y = e^{-t^2}$, then show that y is a solution of the initial-value problem

$$\frac{dy}{dt} + 2ty = 0, \quad y(0) = 1.$$

(b) Find $Y(s) = \mathscr{L}\{e^{-t^2}\}$ by using the Laplace transform to solve the problem in part (a). [*Hint*: First find $Y(0)$ by rereading the material on the error function in Section 2.3. Then in the solution of the resulting linear first-order DE in $Y(s)$ integrate on the interval $[0, s]$. It also helps to use a dummy variable of integration.]

In Problems 77 and 78, use Theorem 4.4.1 with $n = 1$,

$$\mathscr{L}\{tf(t)\} = -F'(s),$$

where $\mathscr{L}\{f(t)\} = F(s)$, to find the inverse Laplace transform of the given function.

77. $F(s) = \ln\left(\dfrac{s-3}{s+1}\right)$ **78.** $F(s) = \arctan\left(\dfrac{2}{s}\right)$

In Problems 79 and 80, use the known result

$$\mathscr{L}\left\{\frac{f(t)}{t}\right\} = \int_s^\infty F(u)\,du,$$

where $\mathscr{L}\{f(t)\} = F(s)$, to find the Laplace transform of the given function. The symbols a and k are positive constants.

79. $f(t) = \dfrac{\sin at}{t}$ **80.** $f(t) = \dfrac{2(1 - \cos kt)}{t}$

Computer Lab Assignments

81. In this problem you are led through the commands in *Mathematica* that enable you to obtain the symbolic Laplace transform of a differential equation and the solution of the initial-value problem by finding the inverse transform. In *Mathematica* the Laplace transform of a function $y(t)$ is obtained using **LaplaceTransform [y[t], t, s]**. In line two of the syntax, we replace **LaplaceTransform [y[t], t, s]** by the symbol **Y**. (*If you do not have* Mathematica, *then adapt the given procedure by finding the corresponding syntax for the CAS you have on hand.*)

Consider the initial-value problem

$$y'' + 6y' + 9y = t\sin t, \quad y(0) = 2, \ y'(0) = -1.$$

Precisely reproduce and then, in turn, execute each line in the given sequence of commands. Either copy the output by hand or print out the results.

> **diffequat = y″[t] + 6y′[t] + 9y[t] == t Sin[t]**
> **transformdeq = LaplaceTransform [diffequat, t, s]/.**
> **{y[0] − > 2, y′[0] − > −1,**
> **LaplaceTransform [y[t], t, s] − > Y}**
> **soln = Solve[transformdeq, Y] // Flatten**
> **Y = Y/. soln**
> **InverseLaplaceTransform[Y, s, t]**

82. Appropriately modify the procedure of Problem 81 to find a solution of

$$y''' + 3y' - 4y = 0, \quad y(0) = 0, \ y'(0) = 0, \ y''(0) = 1.$$

83. The charge $q(t)$ on a capacitor in an *LC*-series circuit is given by

$$\frac{d^2q}{dt^2} + q = 1 - 4\,\mathscr{U}(t - \pi) + 6\,\mathscr{U}(t - 3\pi), q(0) = 0, q'(0) = 0.$$

Appropriately modify the procedure of Problem 81 to find $q(t)$. Graph your solution.

4.5 Dirac Delta Function

INTRODUCTION Just before the *Remarks* on page 229, we indicated that as an immediate consequence of Theorem 4.2.3, $F(s) = 1$ cannot be the Laplace transform of a function f that is piecewise continuous on $[0, \infty)$ and of exponential order. In the discussion that follows we are going to introduce a function that is very different from the kinds that you have studied in previous courses. We shall see that there does indeed exist a function, or more precisely a *generalized* function, whose Laplace transform is $F(s) = 1$.

Unit Impulse Mechanical systems are often acted on by an external force (or emf in an electrical circuit) of large magnitude that acts only for a very short period of time. For example, a vibrating airplane wing could be struck by lightning, a mass on a spring could be given a sharp blow by a ball peen hammer, a ball (baseball, golf ball, tennis ball) could be sent soaring when struck violently by some kind of club (baseball bat, golf club, tennis racket). See **FIGURE 4.5.1**. The graph of the piecewise-defined function

FIGURE 4.5.1 A baseball bat applies a force of large magnitude on the ball for a very short period of time

$$\delta_a(t - t_0) = \begin{cases} 0, & 0 \le t < t_0 - a \\ \dfrac{1}{2a}, & t_0 - a \le t < t_0 + a \\ 0, & t \ge t_0 + a, \end{cases} \tag{1}$$

(b) Behavior of δ_a as $a \to 0$

FIGURE 4.5.2 Unit impulse

$a > 0$, $t_0 > 0$, shown in **FIGURE 4.5.2(a)**, could serve as a model for such a force. For a small value of a, $\delta_a(t - t_0)$ is essentially a constant function of large magnitude that is "on" for just a very short period of time, around t_0. The behavior of $\delta_a(t - t_0)$ as $a \to 0$ is illustrated in Figure 4.5.2(b). The function $\delta_a(t - t_0)$ is called a **unit impulse** since it possesses the integration property $\int_0^\infty \delta_a(t - t_0)\, dt = 1$.

The Dirac Delta Function

In practice it is convenient to work with another type of unit impulse, a "function" that approximates $\delta_a(t - t_0)$ and is defined by the limit

$$\delta(t - t_0) = \lim_{a \to 0} \delta_a(t - t_0). \tag{2}$$

The latter expression, which is not a function at all, can be characterized by the two properties

$$(i) \quad \delta(t - t_0) = \begin{cases} \infty, & t = t_0 \\ 0, & t \neq t_0 \end{cases} \quad \text{and} \quad (ii) \int_0^\infty \delta(t - t_0)\, dt = 1.$$

The unit impulse $\delta(t - t_0)$ is called the **Dirac delta function** after its inventor, the English theoretical physicist **Paul Adrien Maurice Dirac** (1902–1984). Dirac shared the Nobel Prize in Physics in 1933.

It is possible to obtain the Laplace transform of the Dirac delta function by the formal assumption that $\mathscr{L}\{\delta(t - t_0)\} = \lim_{a \to 0} \mathscr{L}\{\delta_a(t - t_0)\}$.

THEOREM 4.5.1 Transform of the Dirac Delta Function

For $t_0 > 0$, $\qquad\qquad \mathscr{L}\{\delta(t - t_0)\} = e^{-st_0}. \tag{3}$

PROOF: To begin, we can write $\delta_a(t - t_0)$ in terms of the unit step function by virtue of (11) and (12) of Section 4.3:

$$\delta_a(t - t_0) = \frac{1}{2a}\left[\mathscr{U}(t - (t_0 - a)) - \mathscr{U}(t - (t_0 + a))\right].$$

By linearity and (14) of Section 4.3, the Laplace transform of this last expression is

$$\mathscr{L}\{\delta_a(t - t_0)\} = \frac{1}{2a}\left[\frac{e^{-s(t_0 - a)}}{s} - \frac{e^{-s(t_0 + a)}}{s}\right] = e^{-st_0}\left(\frac{e^{sa} - e^{-sa}}{2sa}\right). \tag{4}$$

Since (4) has the indeterminate form 0/0 as $a \to 0$, we apply L'Hôpital's rule:

$$\mathscr{L}\{\delta(t - t_0)\} = \lim_{a \to 0} \mathscr{L}\{\delta_a(t - t_0)\} = e^{-st_0} \lim_{a \to 0}\left(\frac{e^{sa} - e^{-sa}}{2sa}\right) = e^{-st_0}. \qquad \equiv$$

Now when $t_0 = 0$, it seems plausible to conclude from (3) that

$$\mathscr{L}\{\delta(t)\} = 1.$$

The last result emphasizes the fact that $\delta(t)$ is not the usual type of function that we have been considering since we expect from Theorem 4.2.3 that $\mathscr{L}\{f(t)\} \to 0$ as $s \to \infty$.

EXAMPLE 1 Two Initial-Value Problems

Solve $y'' + y = 4\delta(t - 2\pi)$ subject to

(a) $y(0) = 1$, $y'(0) = 0$ **(b)** $y(0) = 0$, $y'(0) = 0$.

The two initial-value problems could serve as models for describing the motion of a mass on a spring moving in a medium in which damping is negligible. At $t = 2\pi$ the mass is given a sharp blow. In part (a) the mass is released from rest 1 unit below the equilibrium position. In part (b) the mass is at rest in the equilibrium position.

SOLUTION **(a)** From (3) the Laplace transform of the differential equation is

$$s^2 Y(s) - s + Y(s) = 4e^{-2\pi s} \quad \text{or} \quad Y(s) = \frac{s}{s^2 + 1} + \frac{4e^{-2\pi s}}{s^2 + 1}.$$

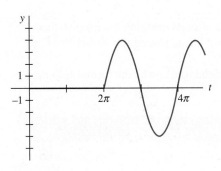

FIGURE 4.5.3 In Example 1(a), moving mass is struck at $t = 2\pi$

FIGURE 4.5.4 In Example 1(b), mass is at rest until struck at $t = 2\pi$

Using the inverse form of the second translation theorem, (15) of Section 4.3, we find

$$y(t) = \cos t + 4\sin(t - 2\pi)\,\mathcal{U}(t - 2\pi).$$

Since $\sin(t - 2\pi) = \sin t$, the foregoing solution can be written as

$$y(t) = \begin{cases} \cos t, & 0 \le t < 2\pi \\ \cos t + 4\sin t, & t \ge 2\pi. \end{cases} \tag{5}$$

In **FIGURE 4.5.3** we see from the graph of (5) that the mass is exhibiting simple harmonic motion until it is struck at $t = 2\pi$. The influence of the unit impulse is to increase the amplitude of vibration to $\sqrt{17}$ for $t > 2\pi$.

(b) In this case the transform of the equation is simply

$$Y(s) = \frac{4e^{-2\pi s}}{s^2 + 1},$$

and so

$$y(t) = 4\sin(t - 2\pi)\,\mathcal{U}(t - 2\pi)$$

$$= \begin{cases} 0, & 0 \le t < 2\pi \\ 4\sin t, & t \ge 2\pi. \end{cases} \tag{6}$$

The graph of (6) in **FIGURE 4.5.4** shows, as we would expect from the initial conditions, that the mass exhibits no motion until it is struck at $t = 2\pi$. ≡

Alternative Definition If $\delta(t - t_0)$ were a function in the traditional sense, then property (*i*) on page 257 would imply $\int_0^\infty \delta(t - t_0)\, dt = 0$ rather than $\int_0^\infty \delta(t - t_0)\, dt = 1$.

Because the Dirac delta function did not "behave" like an ordinary function, even though its users produced correct results, it was initially met with great scorn by mathematicians. However, in the 1940s Dirac's controversial function was put on a rigorous footing by the French mathematician **Laurent Schwartz** (1915–2002) in his book *Théorie des distributions*, and this, in turn, led to an entirely new branch of mathematics known as the **theory of distributions** or **generalized functions**. In this theory (2) is not the accepted definition of $\delta(t - t_0)$, nor does one speak of a function whose values are either ∞ or 0. Although we shall not pursue this topic any further, suffice it to say that the Dirac delta function is best characterized by its effect on other functions. If f is a continuous function, then

$$\int_0^\infty f(t)\,\delta(t - t_0)\, dt = f(t_0) \tag{7}$$

can be taken as the *definition* of $\delta(t - t_0)$. This result is known as the **sifting property** since $\delta(t - t_0)$ has the effect of sifting the value $f(t_0)$ out of the set of values of f on $[0, \infty)$. Note that with $f(t) = 1$ then (7) gives

$$\int_0^\infty \delta(t - t_0)\, dt = 1. \tag{8}$$

Similarly, if $f(t) = e^{-st}$ then (7) gives immediately

$$\int_0^\infty e^{-st}\delta(t - t_0)\, dt = e^{-st_0}. \tag{9}$$

The results in (8) and (9) are consistent with property (*ii*) and (3) on page 257. See Problems 19 and 20 in Exercises 4.5.

REMARKS

In the *Remarks* in Section 4.2 we indicated that the transfer function of a general linear *n*th-order differential equation with constant coefficients is $W(s) = 1/P(s)$, where $P(s) = a_n s^n + a_{n-1} s^{n-1} + \cdots + a_0$. The transfer function is the Laplace transform of function $w(t)$, called the **weight function** of a linear system. But $w(t)$ can be characterized in terms of the discussion at hand. For simplicity let us consider a second-order linear system in which the input is a unit impulse at $t = 0$:

$$a_2 y'' + a_1 y' + a_0 y = \delta(t), \quad y(0) = 0, \quad y'(0) = 0.$$

Applying the Laplace transform and using $\mathcal{L}\{\delta(t)\} = 1$ shows that the transform of the response y in this case is the transfer function

$$Y(s) = \frac{1}{a_2 s^2 + a_1 s + a_0} = \frac{1}{P(s)} = W(s) \quad \text{and so} \quad y = \mathcal{L}^{-1}\left\{\frac{1}{P(s)}\right\} = w(t).$$

From this we can see, in general, that the weight function $y = w(t)$ of an nth-order linear system is the zero-state response of the system to a unit impulse. For this reason $w(t)$ is called as well the **impulse response** of the system.

4.5 Exercises Answers to selected odd-numbered problems begin on page ANS-11.

In Problems 1–14, use the Laplace transform to solve the given differential equation subject to the indicated initial conditions.

1. $y' - 3y = \delta(t - 2), \quad y(0) = 0$

2. $y' + y = \delta(t - 1), \quad y(0) = 2$

3. $y'' + y = \delta(t - 2\pi), \quad y(0) = 0, y'(0) = 1$

4. $y'' + 16y = \delta(t - 2\pi), \quad y(0) = 0, y'(0) = 0$

5. $y'' + y = \delta(t - \pi/2) + \delta(t - 3\pi/2), \quad y(0) = 0, y'(0) = 0$

6. $y'' + y = \delta(t - 2\pi) + \delta(t - 4\pi), \quad y(0) = 1, y'(0) = 0$

7. $y'' + 2y' = \delta(t - 1), \quad y(0) = 0, y'(0) = 1$

8. $y'' - 2y' = 1 + \delta(t - 2), \quad y(0) = 0, y'(0) = 1$

9. $y'' + 4y' + 5y = \delta(t - 2\pi), \quad y(0) = 0, y'(0) = 0$

10. $y'' + 2y' + y = \delta(t - 1), \quad y(0) = 0, y'(0) = 0$

11. $y'' + 4y' + 13y = \delta(t - \pi) + \delta(t - 3\pi), \quad y(0) = 1, y'(0) = 0$

12. $y'' - 7y' + 6y = e^t + \delta(t - 2) + \delta(t - 4), \quad y(0) = 0, y'(0) = 0$

13. $y'' + y = \mathcal{U}(t - 1) + \delta(t - 2) - \mathcal{U}(t - 3), \quad y(0) = 0, \\ y'(0) = 0$

14. $y'' + 2y' + y = (t - 1)\mathcal{U}(t - 1) + \delta(t - 5), \quad y(0) = 0, \\ y'(0) = 0$

In Problems 15 and 16, use the Laplace transform to solve the given initial-value problem. Graph your solution on the interval $[0, 8\pi]$.

15. $y'' + y = \sum_{k=1}^{\infty} \delta(t - k\pi), \quad y(0) = 0, y'(0) = 1$

16. $y'' + y = \sum_{k=1}^{\infty} \delta(t - 2k\pi), \quad y(0) = 0, y'(0) = 1$

In Problems 17 and 18, a uniform beam of length L carries a concentrated load w_0 at $x = \frac{1}{2}L$. Solve the differential equation

$$EI \frac{d^4 y}{dx^4} = w_0 \delta(x - \tfrac{1}{2}L), \quad 0 < x < L,$$

subject to the given boundary conditions.

17. $y(0) = 0, y'(0) = 0, y''(L) = 0, y'''(L) = 0$

FIGURE 4.5.5 Beam embedded at its left end and free at its right end

18. $y(0) = 0, y'(0) = 0, y(L) = 0, y'(L) = 0$

FIGURE 4.5.6 Beam embedded at both ends

In Problems 19 and 20, use the Laplace transform and (7) to solve the given initial-value problem. Use a graphing utility to graph the solution.

19. $y'' + y = 2\cos t\, \delta(t - \pi) + 6\cos t\, \delta(t - 2\pi), \quad y(0) = 1, \\ y'(0) = 0$

20. $y'' + 4y' + 3y = e^t \delta(t - 1), \quad y(0) = 0, y'(0) = 2$

4.6 Systems of Linear Differential Equations

INTRODUCTION When initial conditions are specified, the Laplace transform reduces a system of linear differential equations with constant coefficients to a set of simultaneous algebraic equations in the transformed functions.

Coupled Springs In our first example we solve the model

$$m_1 x_1'' = -k_1 x_1 + k_2(x_2 - x_1)$$
$$m_2 x_2'' = -k_2(x_2 - x_1)$$

(1)

that describes the motion of two masses m_1 and m_2 in the coupled spring/mass system shown in Figure 3.12.1 of Section 3.12.

EXAMPLE 1 **Example 4 of Section 3.12 Revisited**

Use the Laplace transform to solve

$$x_1'' + 10x_1 - 4x_2 = 0$$
$$-4x_1 + x_2'' + 4x_2 = 0$$

(2)

subject to $x_1(0) = 0$, $x_1'(0) = 1$, $x_2(0) = 0$, $x_2'(0) = -1$. This is system (1) with $k_1 = 6$, $k_2 = 4$, $m_1 = 1$, and $m_2 = 1$.

SOLUTION The transform of each equation is

$$s^2 X_1(s) - s x_1(0) - x_1'(0) + 10 X_1(s) - 4 X_2(s) = 0$$

$$-4 X_1(s) + s^2 X_2(s) - s x_2(0) - x_2'(0) + 4 X_2(s) = 0,$$

where $X_1(s) = \mathscr{L}\{x_1(t)\}$ and $X_2(s) = \mathscr{L}\{x_2(t)\}$. The preceding system is the same as

$$(s^2 + 10)X_1(s) - \qquad\qquad 4 X_2(s) = 1$$
$$-4 X_1(s) + (s^2 + 4)X_2(s) = -1.$$

(3)

Solving (3) for $X_1(s)$ and using partial fractions on the result yields

$$X_1(s) = \frac{s^2}{(s^2 + 2)(s^2 + 12)} = -\frac{1/5}{s^2 + 2} + \frac{6/5}{s^2 + 12},$$

and therefore

$$x_1(t) = -\frac{1}{5\sqrt{2}} \mathscr{L}^{-1}\left\{\frac{\sqrt{2}}{s^2 + 2}\right\} + \frac{6}{5\sqrt{12}} \mathscr{L}^{-1}\left\{\frac{\sqrt{12}}{s^2 + 12}\right\}$$

$$= -\frac{\sqrt{2}}{10} \sin \sqrt{2}t + \frac{\sqrt{3}}{5} \sin 2\sqrt{3}t.$$

Substituting the expression for $X_1(s)$ into the first equation of (3) gives us

$$X_2(s) = -\frac{s^2 + 6}{(s^2 + 2)(s^2 + 12)} = -\frac{2/5}{s^2 + 2} - \frac{3/5}{s^2 + 12}$$

and

$$x_2(t) = -\frac{2}{5\sqrt{2}} \mathscr{L}^{-1}\left\{\frac{\sqrt{2}}{s^2 + 2}\right\} - \frac{3}{5\sqrt{12}} \mathscr{L}^{-1}\left\{\frac{\sqrt{12}}{s^2 + 12}\right\}$$

$$= -\frac{\sqrt{2}}{5} \sin \sqrt{2}t - \frac{\sqrt{3}}{10} \sin 2\sqrt{3}t.$$

Finally, the solution to the given system (2) is

$$x_1(t) = -\frac{\sqrt{2}}{10} \sin \sqrt{2}t + \frac{\sqrt{3}}{5} \sin 2\sqrt{3}t$$

(4)

$$x_2(t) = -\frac{\sqrt{2}}{5} \sin \sqrt{2}t - \frac{\sqrt{3}}{10} \sin 2\sqrt{3}t.$$

The solution (4) is the same as (14) of Section 3.12.

≡

Of course, if n masses m_i, $i = 1, 2, \ldots, n$ are connected to n springs, then the system of simultaneous second-order differential equations (1) generalizes to

$$
\begin{aligned}
m_1 x_1'' &= -k_1 x_1 + k_2(x_2 - x_1) \\
m_2 x_2'' &= -k_2(x_2 - x_1) + k_3(x_3 - x_2) \\
m_3 x_3'' &= -k_3(x_3 - x_2) + k_4(x_4 - x_3) \\
&\;\;\vdots \qquad\qquad\qquad \vdots \\
m_n x_n'' &= -k_n(x_n - x_{n-1}).
\end{aligned}
\tag{5}
$$

This system of equations serves as a model for another vibrating physical system. See the *Remarks* at the end of this section.

■ **Networks** In (18) of Section 2.9 we saw that currents $i_1(t)$ and $i_2(t)$ in the network containing an inductor, a resistor, and a capacitor shown in **FIGURE 4.6.1** were governed by the system of first-order differential equations

$$
L\frac{di_1}{dt} + Ri_2 = E(t)
$$

$$
RC\frac{di_2}{dt} + i_2 - i_1 = 0.
\tag{6}
$$

FIGURE 4.6.1 Electrical network

We solve this system by the Laplace transform in the next example.

EXAMPLE 2 **An Electrical Network**

Solve the system in (6) under the conditions $E(t) = 60$ V, $L = 1$ h, $R = 50\ \Omega$, $C = 10^{-4}$ F, and the currents i_1 and i_2 are initially zero.

SOLUTION We must solve

$$
\frac{di_1}{dt} + 50i_2 = 60
$$

$$
50(10^{-4})\frac{di_2}{dt} + i_2 - i_1 = 0
$$

subject to $i_1(0) = 0$, $i_2(0) = 0$.

Applying the Laplace transform to each equation of the system and simplifying gives

$$
sI_1(s) + \qquad 50I_2(s) = \frac{60}{s}
$$

$$
-200I_1(s) + (s + 200)I_2(s) = 0,
$$

where $I_1(s) = \mathscr{L}\{i_1(t)\}$ and $I_2(s) = \mathscr{L}\{i_2(t)\}$. Solving the system for I_1 and I_2 and decomposing the results into partial fractions gives

$$
I_1(s) = \frac{60s + 12{,}000}{s(s + 100)^2} = \frac{6/5}{s} - \frac{6/5}{s + 100} - \frac{60}{(s + 100)^2}
$$

$$
I_2(s) = \frac{12{,}000}{s(s + 100)^2} = \frac{6/5}{s} - \frac{6/5}{s + 100} - \frac{120}{(s + 100)^2}.
$$

Taking the inverse Laplace transform, we find the currents to be

$$
i_1(t) = \frac{6}{5} - \frac{6}{5}e^{-100t} - 60te^{-100t}
$$

$$
i_2(t) = \frac{6}{5} - \frac{6}{5}e^{-100t} - 120te^{-100t}.
$$

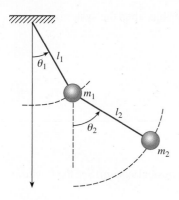

FIGURE 4.6.2 Double pendulum

Note that both $i_1(t)$ and $i_2(t)$ in Example 2 tend toward the value $E/R = \frac{6}{5}$ as $t \to \infty$. Furthermore, since the current through the capacitor is $i_3(t) = i_1(t) - i_2(t) = 60te^{-100t}$, we observe that $i_3(t) \to 0$ as $t \to \infty$.

■ **Double Pendulum** As shown in **FIGURE 4.6.2**, a double pendulum oscillates in a vertical plane under the influence of gravity. For small displacements $\theta_1(t)$ and $\theta_2(t)$, it can be shown that the system of differential equations describing the motion is

$$(m_1 + m_2)l_1^2 \theta_1'' + m_2 l_1 l_2 \theta_2'' + (m_1 + m_2)l_1 g \theta_1 = 0$$

$$m_2 l_2^2 \theta_2'' + m_2 l_1 l_2 \theta_1'' + m_2 l_2 g \theta_2 = 0. \tag{7}$$

As indicated in Figure 4.6.2, θ_1 is measured (in radians) from a vertical line extending downward from the pivot of the system, and θ_2 is measured from a vertical line extending downward from the center of mass m_1. The positive direction is to the right, and the negative direction is to the left.

EXAMPLE 3 **Double Pendulum**

It is left as an exercise to fill in the details of using the Laplace transform to solve system (7) when $m_1 = 3$, $m_2 = 1$, $l_1 = l_2 = 16$, $g = 32$, $\theta_1(0) = 1$, $\theta_2(0) = -1$, $\theta_1'(0) = 0$, and $\theta_2'(0) = 0$. You should find that

$$\theta_1(t) = \frac{1}{4}\cos\frac{2}{\sqrt{3}}t + \frac{3}{4}\cos 2t$$

$$\theta_2(t) = \frac{1}{2}\cos\frac{2}{\sqrt{3}}t - \frac{3}{2}\cos 2t. \tag{8}$$

With the aid of a CAS, the positions of the two masses at $t = 0$ and at subsequent times are shown in **FIGURE 4.6.3**. See Problem 24 in Exercises 4.6.

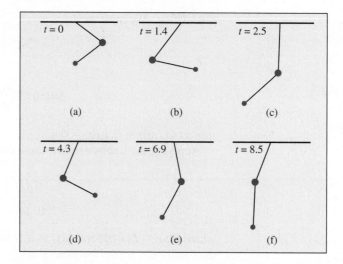

FIGURE 4.6.3 Positions of masses at various times in Example 3

REMARKS

System (5) can also be interpreted as a mathematical model for the horizontal displacements of successive floors of a multistory building due to an earthquake. If the horizontal position of the center of the ith floor and its mass are denoted by x_i and m_i, respectively, then $x_{i+1} - x_i$ is the horizontal displacement of the $(i + 1)$st floor relative to the ith floor. Assuming Hooke's law holds, the restoring force between those two floors is $k_i(x_{i+1} - x_i)$, where k_i is a constant. In the case of a three-story building, the displacements are given by a system of three second-order differential equations

$$m_1 x_1'' = -k_0 x_1 + k_1(x_2 - x_1)$$
$$m_2 x_2'' = -k_1(x_2 - x_1) + k_2(x_3 - x_2)$$
$$m_3 x_3'' = -k_2(x_3 - x_2). \tag{9}$$

See **FIGURE 4.6.4**.

© Nigel Spiers/Shutterstock
Building destroyed by an earthquake

(a) Floors of a building (b) Horizontal forces acting on a floor

FIGURE 4.6.4 Three-story building

4.6 Exercises Answers to selected odd-numbered problems begin on page ANS-11.

In Problems 1–12, use the Laplace transform to solve the given system of differential equations.

1. $\dfrac{dx}{dt} = -x + y$

$\dfrac{dy}{dt} = 2x$

$x(0) = 0, \ y(0) = 1$

2. $\dfrac{dx}{dt} = 2y + e^t$

$\dfrac{dy}{dt} = 8x - t$

$x(0) = 1, \ y(0) = 1$

3. $\dfrac{dx}{dt} = x - 2y$

$\dfrac{dy}{dt} = 5x - y$

$x(0) = -1, \ y(0) = 2$

4. $\dfrac{dx}{dt} + 3x + \dfrac{dy}{dt} = 1$

$\dfrac{dx}{dt} - x + \dfrac{dy}{dt} - y = e^t$

$x(0) = 0, \ y(0) = 0$

5. $2\dfrac{dx}{dt} + \dfrac{dy}{dt} - 2x = 1$

$\dfrac{dx}{dt} + \dfrac{dy}{dt} - 3x - 3y = 2$

$x(0) = 0, \ y(0) = 0$

6. $\dfrac{dx}{dt} + x - \dfrac{dy}{dt} + y = 0$

$\dfrac{dx}{dt} + \dfrac{dy}{dt} + 2y = 0$

$x(0) = 0, \ y(0) = 1$

7. $\dfrac{d^2x}{dt^2} + x - y = 0$

$\dfrac{d^2x}{dt^2} + y - x = 0$

$x(0) = 0, \ x'(0) = -2,$
$y(0) = 0, \ y'(0) = 1$

8. $\dfrac{d^2x}{dt^2} + \dfrac{dx}{dt} + \dfrac{dy}{dt} = 0$

$\dfrac{d^2y}{dt^2} + \dfrac{dy}{dt} - 4\dfrac{dx}{dt} = 0$

$x(0) = 1, \ x'(0) = 0,$
$y(0) = -1, \ y'(0) = 5$

9. $\dfrac{d^2x}{dt^2} + \dfrac{d^2y}{dt^2} = t^2$

$\dfrac{d^2x}{dt^2} - \dfrac{d^2y}{dt^2} = 4t$

$x(0) = 8, \ x'(0) = 0,$
$y(0) = 0, \ y'(0) = 0$

10. $\dfrac{dx}{dt} - 4x + \dfrac{d^3y}{dt^3} = 6\sin t$

$\dfrac{dx}{dt} + 2x - 2\dfrac{d^3y}{dt^3} = 0$

$x(0) = 0, \ y(0) = 0,$
$y'(0) = 0, \ y''(0) = 0$

11. $\dfrac{d^2x}{dt^2} + 3\dfrac{dy}{dt} + 3y = 0$

$\dfrac{d^2x}{dt^2} + 3y = te^{-t}$

$x(0) = 0, \ x'(0) = 2,$
$y(0) = 0$

12. $\dfrac{dx}{dt} = 4x - 2y + 2\,\mathcal{U}(t - 1)$

$\dfrac{dy}{dt} = 3x - y + \mathcal{U}(t - 1)$

$x(0) = 0, \ y(0) = \frac{1}{2}$

13. Solve system (1) when $k_1 = 3, \ k_2 = 2, \ m_1 = 1, \ m_2 = 1$ and $x_1(0) = 0, \ x_1'(0) = 1, \ x_2(0) = 1, \ x_2'(0) = 0$.

14. Derive the system of differential equations describing the straight-line vertical motion of the coupled springs shown in equilibrium in **FIGURE 4.6.5**. Use the Laplace transform to solve the system when $k_1 = 1, k_2 = 1, k_3 = 1, m_1 = 1, m_2 = 1$ and $x_1(0) = 0, \ x_1'(0) = -1, \ x_2(0) = 0, \ x_2'(0) = 1$.

FIGURE 4.6.5 Coupled springs in Problem 14

15. (a) Show that the system of differential equations for the currents $i_2(t)$ and $i_3(t)$ in the electrical network shown in **FIGURE 4.6.6** is

$$L_1 \frac{di_2}{dt} + Ri_2 + Ri_3 = E(t)$$

$$L_2 \frac{di_3}{dt} + Ri_2 + Ri_3 = E(t).$$

(b) Solve the system in part (a) if $R = 5\ \Omega, L_1 = 0.01$ h, $L_2 = 0.0125$ h, $E = 100$ V, $i_2(0) = 0$, and $i_3(0) = 0$.

(c) Determine the current $i_1(t)$.

FIGURE 4.6.6 Network in Problem 15

FIGURE 4.6.8 Network in Problem 20

16. (a) In Problem 14 in Exercises 2.9 you were asked to show that the currents $i_2(t)$ and $i_3(t)$ in the electrical network shown in **FIGURE 4.6.7** satisfy

$$L\frac{di_2}{dt} + L\frac{di_3}{dt} + R_1 i_2 = E(t)$$

$$-R_1\frac{di_2}{dt} + R_2\frac{di_3}{dt} + \frac{1}{C}i_3 = 0.$$

Solve the system if $R_1 = 10\ \Omega$, $R_2 = 5\ \Omega$, $L = 1$ h, $C = 0.2$ F,

$$E(t) = \begin{cases} 120, & 0 \le t < 2 \\ 0, & t \ge 2, \end{cases}$$

$i_2(0) = 0$, and $i_3(0) = 0$.

(b) Determine the current $i_1(t)$.

FIGURE 4.6.7 Network in Problem 16

17. Solve the system given in (17) of Section 2.9 when $R_1 = 6\ \Omega$, $R_2 = 5\ \Omega$, $L_1 = 1$ h, $L_2 = 1$ h, $E(t) = 50 \sin t$ V, $i_2(0) = 0$, and $i_3(0) = 0$.

18. Solve (6) when $E = 60$ V, $L = \frac{1}{2}$ h, $R = 50\ \Omega$, $C = 10^{-4}$ F, $i_1(0) = 0$, and $i_2(0) = 0$.

19. Solve (6) when $E = 60$ V, $L = 2$ h, $R = 50\ \Omega$, $C = 10^{-4}$ F, $i_1(0) = 0$, and $i_2(0) = 0$.

20. (a) Show that the system of differential equations for the charge on the capacitor $q(t)$ and the current $i_3(t)$ in the electrical network shown in **FIGURE 4.6.8** is

$$R_1\frac{dq}{dt} + \frac{1}{C}q + R_1 i_3 = E(t)$$

$$L\frac{di_3}{dt} + R_2 i_3 - \frac{1}{C}q = 0.$$

(b) Find the charge on the capacitor when $L = 1$ h, $R_1 = 1\ \Omega$, $R_2 = 1\ \Omega$, $C = 1$ F,

$$E(t) = \begin{cases} 0, & 0 < t < 1 \\ 50e^{-t}, & t \ge 1, \end{cases}$$

$i_3(0) = 0$, and $q(0) = 0$.

Mathematical Models

21. Range of a Projectile—No Air Resistance If you worked Problem 23 in Exercises 3.12, you saw that when air resistance and all other forces except its weight $w = mg$ are ignored, the path of motion of a ballistic projectile, such as a cannon shell, is described by the system of linear differential equations

$$m\frac{d^2x}{dt^2} = 0$$

$$m\frac{d^2y}{dt^2} = -mg. \tag{10}$$

(a) If the projectile is launched from level ground with an initial velocity $\mathbf{v}_0$ assumed to be tangent to its path of motion or trajectory, then the initial conditions accompanying the system are $x(0) = 0$, $x'(0) = v_0\cos\theta$, $y(0) = 0$, $y'(0) = v_0\sin\theta$ where $v_0 = \|\mathbf{v}_0\|$ the initial speed is constant and θ is the constant angle of elevation of the cannon. See Figure 3.R.5 in *The Paris Guns* problem on page 212. Use the Laplace transform to solve system (10).

(b) The solutions $x(t)$ and $y(t)$ of the system in part (a) are parametric equations of the trajectory of the projectile. By using $x(t)$ to eliminate the parameter t in $y(t)$ show that the trajectory is parabolic.

(c) Use the results of part (b) to show that the horizontal range R of the projectile is given by

$$R = \frac{v_0^2}{g}\sin 2\theta. \tag{11}$$

From (11) we not only see that R is a maximum when $\theta = \pi/4$ but also that a projectile launched at distinct complementary angles θ and $\pi/2 - \theta$ has the same submaximum range. See **FIGURE 4.6.9**. Use a trigonometric identity to prove this last result.

(d) Show that the maximum height H of the projectile is given by

$$H = \frac{v_0^2}{2g}\sin^2\theta. \tag{12}$$

(e) Suppose $g = 32$ ft/s^2, $\theta = 38°$, and $v_0 = 300$ ft/s. Use (11) and (12) to find the horizontal range and maximum height of the projectile. Repeat with $\theta = 52°$ and $v_0 = 300$ ft/s.

(f) Because formulas (11) and (12) are not valid in all cases (see Problem 22), it is advantageous to you to remember that the range and maximum height of a ballistic

projectile can be obtained by working directly with $x(t)$ and $y(t)$, that is, by solving $y(t) = 0$ and $y'(t) = 0$. The first equation gives the time when the projectile hits the ground and the second gives the time when $y(t)$ is a maximum. Find these times and verify the range and maximum height obtained in part (e) for the trajectory with $\theta = 38°$ and $v_0 = 300$ ft/s. Repeat with $\theta = 52°$.

(g) With $g = 32$ ft/s^2, $\theta = 38°$ and $v_0 = 300$ ft/s use a graphing utility or CAS to plot the trajectory of the projectile defined by the parametric equations $x(t)$ and $y(t)$ in part (a). Repeat with $\theta = 52°$. Using different colors superimpose both curves on the same coordinate system.

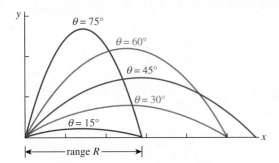

FIGURE 4.6.9 Projectiles in Problem 21

22. **Range of a Projectile—With Linear Air Resistance** In *The Paris Guns* problem on pages 209–214, the effect that nonlinear air resistance has on the trajectory of a cannon shell was examined numerically. In this problem we consider linear air resistance on a projectile.

(a) Suppose that air resistance is a retarding force tangent to the path of the projectile but acts opposite to the motion. If we take air resistance to be proportional to the velocity of the projectile, then from Problem 24 of Exercises 3.12 the motion of the projectile is described by the system of linear differential equations

$$m\frac{d^2x}{dt^2} = -\beta\frac{dx}{dt}$$

$$m\frac{d^2y}{dt^2} = -mg - \beta\frac{dy}{dt},$$

(13)

where $\beta > 0$ is a constant. Use the Laplace transform to solve system (13) subject to the initial conditions in part (a) of Problem 21.

(b) Suppose $m = \frac{1}{4}$ slug, $g = 32$ ft/s^2, $\beta = 0.02$, $\theta = 38°$, and $v_0 = 300$ ft/s. Use a calculator or CAS to approximate the time when the projectile hits the ground and then compute $x(t)$ to find its corresponding horizontal range.

(c) The complementary-angle property in part (c) of Problem 21 does not hold when air resistance is taken into consideration. To show this, repeat part (b) using the complementary angle $\theta = 52°$ and compare the horizontal range with that found in part (b).

(d) With $m = \frac{1}{4}$ slug, $g = 32$ ft/s^2, $\beta = 0.02$, $\theta = 38°$, and $v_0 = 300$ ft/s, use a graphing utility or CAS to plot the trajectory of the projectile defined by the parametric equations $x(t)$ and $y(t)$. Repeat with $\theta = 52°$. Using different colors, superimpose both of these curves along with the two curves in part (g) of Problem 21 on the same coordinate system.

23. **Spring-Coupled Pendulums** (a) Suppose two identical pendulums are coupled by means of a spring with constant k. See **FIGURE 4.6.10**. Under the same assumptions made in the discussion preceding Example 3 in Section 4.6, it can be shown that when the displacement angles $\theta_1(t)$ and $\theta_2(t)$ are small, the system of differential equations describing the motion is

$$\theta_1'' + \frac{g}{l}\theta_1 = -\frac{k}{m}(\theta_1 - \theta_2)$$

$$\theta_2'' + \frac{g}{l}\theta_2 = \frac{k}{m}(\theta_1 - \theta_2).$$

Use the Laplace transform to solve the system when $\theta_1(0) = \theta_0$, $\theta_1'(0) = 0$, $\theta_2(0) = \psi_0$, $\theta_2'(0) = 0$, where θ_0 and ψ_0 are constants. For convenience, let $\omega^2 = g/l$ and $K = k/m$.

(b) Use the solution in part (a) to discuss the motion of the coupled pendulums in the special case when the initial conditions are $\theta_1(0) = \theta_0$, $\theta_1'(0) = 0$, $\theta_2(0) = \theta_0$, $\theta_2'(0) = 0$.

(c) Repeat part (b) when the initial conditions are $\theta_1(0) = \theta_0$, $\theta_1'(0) = 0$, $\theta_2(0) = -\theta_0$, $\theta_2'(0) = 0$.

FIGURE 4.6.10 Coupled pendulums in Problem 23

Computer Lab Assignment

24. (a) Use the Laplace transform and the information given in Example 3 to obtain the solution (8) of the system given in (7).

(b) Use a graphing utility to plot the graphs of $\theta_1(t)$ and $\theta_2(t)$ in the $t\theta$-plane. Which mass has extreme displacements of greater magnitude? Use the graphs to estimate the first time that each mass passes through its equilibrium position. Discuss whether the motion of the pendulums is periodic.

(c) As parametric equations, graph $\theta_1(t)$ and $\theta_2(t)$ in the $\theta_1\theta_2$-plane. The curve defined by these parametric equations is called a **Lissajous curve**.

(d) The position of the masses at $t = 0$ is given in Figure 4.6.3(a). Note that we have used 1 radian $\approx 57.3°$. Use a calculator or a table application in a CAS to construct a table of values of the angles θ_1 and θ_2 for $t = 1, 2, \ldots, 10$ seconds. Then plot the positions of the two masses at these times.

(e) Use a CAS to find the first time that $\theta_1(t) = \theta_2(t)$ and compute the corresponding angular value. Plot the positions of the two masses at these times.

(f) Utilize a CAS to also draw appropriate lines to simulate the pendulum rods as in Figure 4.6.3. Use the animation capability of your CAS to make a "movie" of the motion of the double pendulum from $t = 0$ to $t = 10$ using a time increment of 0.1. [*Hint*: Express the coordinates $(x_1(t), y_1(t))$ and $(x_2(t), y_2(t))$ of the masses m_1 and m_2, respectively, in terms of $\theta_1(t)$ and $\theta_2(t)$.]

4 Chapter in Review Answers to selected odd-numbered problems begin on page ANS-11.

In Problems 1 and 2, use the definition of the Laplace transform to find $\mathcal{L}\{f(t)\}$.

1. $f(t) = \begin{cases} t, & 0 \le t < 1 \\ 2 - t, & t \ge 1 \end{cases}$ **2.** $f(t) = \begin{cases} 0, & 0 \le t < 2 \\ 1, & 2 \le t < 4 \\ 0, & t \ge 4 \end{cases}$

In Problems 3–36, fill in the blanks or answer true/false.

3. If f is not piecewise continuous on $[0, \infty)$, then $\mathcal{L}\{f(t)\}$ will not exist. _____

4. The function $f(t) = (e^t)^{10}$ is not of exponential order. _____

5. $F(s) = s^2/(s^2 + 4)$ is not the Laplace transform of a function that is piecewise continuous and of exponential order. _____

6. If $\mathcal{L}\{f(t)\} = F(s)$ and $\mathcal{L}\{g(t)\} = G(s)$, then $\mathcal{L}^{-1}\{F(s)G(s)\} = f(t)g(t)$. _____

7. $\mathcal{L}\{e^{-7t}\} =$ _____ **8.** $\mathcal{L}\{te^{-7t}\} =$ _____

9. $\mathcal{L}\{\sin 2t\} =$ _____ **10.** $\mathcal{L}\{e^{-3t}\sin 2t\} =$ _____

11. $\mathcal{L}\{t\sin 2t\} =$ _____ **12.** $\mathcal{L}\{\sin 2t\,\mathcal{U}(t - \pi)\} =$ _____

13. $\mathcal{L}\{\sin^2 t\} =$ _____ **14.** $\mathcal{L}\{\cosh 6t\} =$ _____

15. $\mathcal{L}\{\int_0^t e^\tau \sin(t - \tau)\, d\tau\} =$ _____

16. $\mathcal{L}\{\int_0^t (t - \tau)^2 \cos \tau\, d\tau\} =$ _____

17. $\mathcal{L}\{e^{3(t-4)}\mathcal{U}(t - 4)\} =$ _____

18. $\mathcal{L}\{1 - \delta(t - 5)\} =$ _____

19. $\mathcal{L}^{-1}\left\{\dfrac{20}{s^6}\right\} =$ _____

20. $\mathcal{L}^{-1}\left\{\dfrac{1}{3s - 1}\right\} =$ _____

21. $\mathcal{L}^{-1}\left\{\dfrac{1}{(s - 5)^3}\right\} =$ _____

22. $\mathcal{L}^{-1}\left\{\dfrac{1}{s^2 - 5}\right\} =$ _____

23. $\mathcal{L}^{-1}\left\{\dfrac{1}{s^2 + s}\right\} =$ _____

24. $\mathcal{L}^{-1}\left\{\dfrac{1}{s^2 + s - 2}\right\} =$ _____

25. $\mathcal{L}^{-1}\left\{\dfrac{s}{s^2 - 10s + 29}\right\} =$ _____

26. $\mathcal{L}^{-1}\left\{\dfrac{1}{s^2 - 12\pi s + 37\pi^2}\right\} =$ _____

27. $\mathcal{L}^{-1}\left\{\dfrac{4s - 86}{s^2 + 2s + 101}\right\} =$ _____

28. $\mathcal{L}^{-1}\left\{\dfrac{e^{-5s}}{s^2}\right\} =$ _____

29. $\mathcal{L}^{-1}\left\{\dfrac{s + \pi}{s^2 + \pi^2}e^{-s}\right\} =$ _____

30. $\mathcal{L}^{-1}\left\{\dfrac{1}{L^2 s^2 + n^2\pi^2}\right\} =$ _____

31. $\mathcal{L}\{e^{-5t}\}$ exists for $s >$ _____.

32. If $\mathcal{L}\{f(t)\} = F(s)$, then $\mathcal{L}\{te^{8t}f(t)\} =$ _____.

33. If $\mathcal{L}\{f(t)\} = F(s)$ and $k > 0$, then $\mathcal{L}\{e^{at}f(t - k)\,\mathcal{U}(t - k)\} =$ _____.

34. $\mathcal{L}\left\{\int_0^t e^{a\tau}f(\tau)\, d\tau\right\} =$ _____ whereas $\mathcal{L}\left\{e^{at}\int_0^t f(\tau)\, d\tau\right\} =$ _____.

35. $1 * 1 =$ _____

36. $t^2 * t =$ _____

37. Given that $\mathcal{L}^{-1}\left\{\dfrac{1}{s^2 + 3s + \frac{5}{2}}\right\} = 2e^{-3t/2}\sin\dfrac{t}{2}$ use (6) of Section 4.2 to find $\mathcal{L}^{-1}\left\{\dfrac{s}{s^2 + 3s + \frac{5}{2}}\right\}$.

38. Given that $\mathcal{L}^{-1}\left\{\dfrac{1}{s^2 - 7s + 12}\right\} = e^{4t} - e^{3t}$ use (9) of Section 4.4 to find $\mathcal{L}^{-1}\left\{\dfrac{1}{s(s^2 - 7s + 12)}\right\}$.

In Problems 39–42, use the unit step function to write down an equation for each graph in terms of the function $y = f(t)$ whose graph is given in **FIGURE 4.R.1**.

FIGURE 4.R.1 Graph for Problems 39–42

39.

FIGURE 4.R.2 Graph for Problem 39

40.

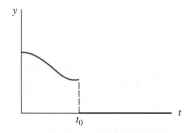

FIGURE 4.R.3 Graph for Problem 40

41.

FIGURE 4.R.4 Graph for Problem 41

42.

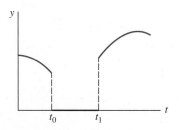

FIGURE 4.R.5 Graph for Problem 42

In Problems 43–46, express f in terms of unit step functions. Find $\mathscr{L}\{f(t)\}$ and $\mathscr{L}\{e^t f(t)\}$.

43.

FIGURE 4.R.6 Graph for Problem 43

44.

FIGURE 4.R.7 Graph for Problem 44

45.

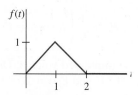

FIGURE 4.R.8 Graph for Problem 45

46.

FIGURE 4.R.9 Graph for Problem 46

In Problems 47 and 48, sketch the graph of the given function. Find $\mathscr{L}\{f(t)\}$.

47. $f(t) = -1 + 2\sum_{k=1}^{\infty}(-1)^{k+1}\mathscr{U}(t - k)$

48. $f(t) = \sum_{k=0}^{\infty}(2k + 1 - t)[\mathscr{U}(t - 2k) - \mathscr{U}(t - 2k - 1)]$

In Problems 49–58, use the Laplace transform to solve the given equation.

49. $y'' - 2y' + y = e^t, \quad y(0) = 0, \quad y'(0) = 5$

50. $y'' - 8y' + 20y = te^t, \quad y(0) = 0, \quad y'(0) = 0$

51. $y'' + 6y' + 5y = t - t\mathscr{U}(t - 2), \quad y(0) = 1, \quad y'(0) = 0$

52. $y' - 5y = f(t)$, where $f(t) = \begin{cases} t^2, & 0 \le t < 1 \\ 0, & t \ge 1 \end{cases}$, $y(0) = 1$

53. $y' + 2y = f(t), \quad y(0) = 1$, where the graph of $f(t)$ is given in **FIGURE 4.R.10**

FIGURE 4.R.10 Graph for Problem 53

54. $y'' + 5y' + 4y = f(t), \quad y(0) = 0, \quad y'(0) = 3$, where
$$f(t) = 12\sum_{k=0}^{\infty}(-1)^k \mathscr{U}(t - k)$$

55. $y'(t) = \cos t + \int_0^t y(\tau)\cos(t - \tau)\, d\tau, \quad y(0) = 1$

56. $y' + y = \int_0^t e^{-\tau}y(t - \tau)\, d\tau, \quad y(0) = 1$

57. $y(t) = e^{2t} + \int_0^t y(\tau) \sin(t - \tau) \, d\tau$

58. $\int_0^t y(\tau) y(t - \tau) \, d\tau = 6t^3$

In Problems 59 and 60, use the Laplace transform to solve each system.

59. $x' + y = t$
$4x + y' = 0$
$x(0) = 1, \ y(0) = 2$

60. $x'' + y'' = e^{2t}$
$2x' + y'' = -e^{2t}$
$x(0) = 0, \ y(0) = 0$
$x'(0) = 0, \ y'(0) = 0$

61. The current $i(t)$ in an RC-series circuit can be determined from the integral equation

$$Ri + \frac{1}{C} \int_0^t i(\tau) \, d\tau = E(t),$$

where $E(t)$ is the impressed voltage. Determine $i(t)$ when $R = 10 \ \Omega$, $C = 0.5$ F, and $E(t) = 2(t^2 + t)$.

62. A series circuit contains an inductor, a resistor, and a capacitor for which $L = \frac{1}{2}$ h, $R = 10 \ \Omega$, and $C = 0.01$ F, respectively. The voltage

$$E(t) = \begin{cases} 10, & 0 \le t < 5 \\ 0, & t \ge 5 \end{cases}$$

is applied to the circuit. Determine the instantaneous charge $q(t)$ on the capacitor for $t > 0$ if $q(0) = 0$ and $q'(0) = 0$.

63. A uniform cantilever beam of length L is embedded at its left end ($x = 0$) and is free at its right end. Find the deflection $y(x)$ if the load per unit length is given by

$$w(x) = \frac{2w_0}{L} \left[\frac{L}{2} - x + \left(x - \frac{L}{2} \right) \mathcal{U} \left(x - \frac{L}{2} \right) \right].$$

64. When a uniform beam of length L is supported on an elastic foundation the differential equation for its deflection $y(x)$ is given by

$$EI \frac{d^4 y}{dx^4} + ky = w(x),$$

where k is the modulus of the foundation and $-ky$ is the restoring force of the foundation that acts in the direction opposite to that of the load $w(x)$. For algebraic convenience, suppose the differential equation is written as

$$\frac{d^4 y}{dx^4} + 4a^4 y = \frac{w(x)}{EI},$$

where $a = (k/4EI)^{1/4}$. Assume $L = \pi$ and $a = 1$. Use the Laplace transform and the table of Laplace transforms in Appendix C to find the deflection $y(x)$ of a beam that is supported on an elastic foundation when the beam is simply supported at both ends and the load $w(x)$ is a constant w_0 uniformly distributed along its length. See **FIGURE 4.R.11**. [*Hint*: You will also need the factorization

$$s^4 + 4 = (s^2 - 2s + 2)(s^2 + 2s + 2).]$$

$w(x) = w_0$

elastic foundation

FIGURE 4.R.11 Beam in Problem 64

65. Use the Laplace transform and the table of Laplace transforms in Appendix C to find the deflection $y(x)$ of a beam that is supported on an elastic foundation as in Problem 64 when the beam is embedded at both ends and the load $w(x)$ is a constant w_0 concentrated at $x = \pi/2$.

Series Solutions of Linear Equations

Up to this point we primarily have solved linear differential equations of order two (or higher) with constant coefficients. The Cauchy–Euler equations in Section 3.6 were the only exception. In applications, linear high-order equations with variable coefficients are just as important as DEs with constant coefficients. In this chapter we will see that the same ease with which we solved Cauchy–Euler equations does not carry over to a simple second-order equation with variable coefficients such as $y'' - xy = 0$. We will see that solutions of this equation are functions defined by **infinite series**.

5.1 Solutions about Ordinary Points

INTRODUCTION In Section 3.3 we saw that solving a homogeneous linear DE with *constant coefficients* was essentially a problem in algebra. By finding the roots of the auxiliary equation we could write a general solution of the DE as a linear combination of the elementary functions x^k, $x^k e^{\alpha x}$, $x^k e^{\alpha x} \cos \beta x$, and $x^k e^{\alpha x} \sin \beta x$, k a nonnegative integer. But as pointed out in the introduction to Section 3.6, *most* linear higher-order DEs with *variable coefficients* cannot be solved in terms of elementary functions. A usual course of action for equations of this sort is to assume a solution in the form of infinite series and proceed in a manner similar to the method of undetermined coefficients (Section 3.4). In this section we consider linear second-order DEs with variable coefficients that possess solutions in the form of *power series*.

5.1.1 Review of Power Series

Recall from calculus that a power series in $x - a$ is an infinite series of the form

$$\sum_{n=0}^{\infty} c_n(x - a)^n = c_0 + c_1(x - a) + c_2(x - a)^2 + \cdots.$$

Such a series is also said to be a **power series centered at a**. For example, the power series $\sum_{n=0}^{\infty}(x + 1)^n$ is centered at $a = -1$. In this section we are concerned mainly with power series in x; in other words, power series such as $\sum_{n=1}^{\infty} 2^{n-1}x^n = x + 2x^2 + 4x^3 + \cdots$ that are centered at $a = 0$. The following list summarizes some important facts about power series.

(a) $(a - R, a + R)$, series diverges at both endpoints

(b) $[a - R, a + R]$, series converges at both endpoints

(c) $(a - R, a + R]$, series converges at right endpoint and diverges at left endpoint

(d) $[a - R, a + R)$, series converges at left endpoint and diverges at right endpoint

FIGURE 5.1.1 Intervals of convergence for $R > 0$

- **Convergence** A power series $\sum_{n=0}^{\infty} c_n(x - a)^n$ is convergent at a specified value of x if its sequence of partial sums $\{S_N(x)\}$ converges; that is, if $\lim_{N \to \infty} S_N(x) = \lim_{N \to \infty} \sum_{n=0}^{N} c_n(x - a)^n$ exists. If the limit does not exist at x, the series is said to be divergent.
- **Interval of Convergence** Every power series has an interval of convergence. The interval of convergence is the set of all real numbers x for which the series converges.
- **Radius of Convergence** Every power series has a radius of convergence R. If $R > 0$, then a power series $\sum_{n=0}^{\infty} c_n(x - a)^n$ converges for $|x - a| < R$ and diverges for $|x - a| > R$. If the series converges only at its center a, then $R = 0$. If the series converges for all x, then we write $R = \infty$. Recall that the absolute-value inequality $|x - a| < R$ is equivalent to the simultaneous inequality $a - R < x < a + R$. A power series may or may not converge at the endpoints $a - R$ and $a + R$ of this interval. **FIGURE 5.1.1** shows four possible intervals of convergence for $R > 0$.
- **Absolute Convergence** Within its interval of convergence a power series converges absolutely. In other words, if x is a number in the interval of convergence and is not an endpoint of the interval, then the series of absolute values $\sum_{n=0}^{\infty} |c_n(x - a)^n|$ converges.
- **Ratio Test** Convergence of power series can often be determined by the ratio test. Suppose that $c_n \neq 0$ for all n, and that

$$\lim_{n \to \infty} \left| \frac{c_{n+1}(x - a)^{n+1}}{c_n(x - a)^n} \right| = |x - a| \lim_{n \to \infty} \left| \frac{c_{n+1}}{c_n} \right| = L.$$

If $L < 1$ the series converges absolutely, if $L > 1$ the series diverges, and if $L = 1$ the test is inconclusive. For example, for the power series $\sum_{n=1}^{\infty} (x - 3)^n/2^n n$ the ratio test gives

$$\lim_{n \to \infty} \left| \frac{(x - 3)^{n+1}/2^{n+1}(n + 1)}{(x - 3)^n/2^n n} \right| = |x - 3| \lim_{n \to \infty} \frac{n}{2(n + 1)} = \frac{1}{2}|x - 3|.$$

The series converges absolutely for $\frac{1}{2}|x - 3| < 1$ or $|x - 3| < 2$ or $1 < x < 5$. This last interval is referred to as the *open* interval of convergence. The series diverges for $|x - 3| > 2$, that is, for $x > 5$ or $x < 1$. At the left endpoint $x = 1$ of the open interval of convergence, the series of constants $\sum_{n=1}^{\infty} ((-1)^n/n)$ is convergent by the alternating series test. At the right endpoint $x = 5$, the series $\sum_{n=1}^{\infty} (1/n)$ is the divergent harmonic series. The interval of convergence of the series is $[1, 5)$ and the radius of convergence is $R = 2$.

- **A Power Series Defines a Function** A power series defines a function $f(x) = \sum_{n=0}^{\infty} c_n(x - a)^n$ whose domain is the interval of convergence of the series. If the radius of convergence is $R > 0$, then f is continuous, differentiable, and integrable on the interval $(a - R, a + R)$. Moreover, $f'(x)$ and $\int f(x)\, dx$ can be found by term-by-term differentiation and integration. Convergence at an endpoint may be either lost by differentiation or gained through integration. If $y = \sum_{n=0}^{\infty} c_n x^n$ is a power series in x, then the first two derivatives are $y' = \sum_{n=0}^{\infty} nx^{n-1}$ and $y'' = \sum_{n=0}^{\infty} n(n - 1)x^{n-2}$. Notice that the first term in the first derivative and the first two terms in the second derivative are zero. We omit these zero terms and write

$$y' = \sum_{n=1}^{\infty} c_n nx^{n-1} = c_1 + 2c_2 x + 3c_3 x^2 + 4c_4 x^3 + \cdots$$

$$y'' = \sum_{n=2}^{\infty} c_n n(n - 1)x^{n-2} = 2c_2 + 6c_3 x + 12c_4 x^2 + \cdots .$$

(1)

These results are important and will be used shortly.
- **Identity Property** If $\sum_{n=0}^{\infty} c_n(x - a)^n = 0$, $R > 0$, for all numbers x in the interval of convergence, then $c_n = 0$ for all n.
- **Analytic at a Point** A function f is said to be analytic at a point a if it can be represented by a power series in $x - a$ with either a positive or an infinite radius of convergence. In calculus it is seen that infinitely differentiable functions, such as e^x, $\cos x$, $\sin x$, $\ln(1 + x)$, and so on, can be represented by Taylor series

$$\sum_{n=0}^{\infty} \frac{f^{(n)}(a)}{n!}(x - a)^n = f(a) + \frac{f'(a)}{1!}(x - a) + \frac{f''(a)}{2!}(x - a)^2 + \cdots .$$

You might remember some of the following power series representations.

Power Series	Interval of Convergence
$e^x = 1 + \dfrac{x}{1!} + \dfrac{x^2}{2!} + \dfrac{x^3}{3!} + \cdots = \sum_{n=0}^{\infty} \dfrac{1}{n!}x^n$	$(-\infty, \infty)$
$\cos x = 1 - \dfrac{x^2}{2!} + \dfrac{x^4}{4!} - \dfrac{x^6}{6!} + \cdots = \sum_{n=0}^{\infty} \dfrac{(-1)^n}{(2n)!}x^{2n}$	$(-\infty, \infty)$
$\sin x = x - \dfrac{x^3}{3!} + \dfrac{x^5}{5!} - \dfrac{x^7}{7!} + \cdots = \sum_{n=0}^{\infty} \dfrac{(-1)^n}{(2n + 1)!}x^{2n+1}$	$(-\infty, \infty)$
$\ln(1 + x) = x - \dfrac{x^2}{2} + \dfrac{x^3}{3} - \dfrac{x^4}{4} + \cdots = \sum_{n=1}^{\infty} \dfrac{(-1)^{n+1}}{n}x^n$	$(-1, 1]$
$\dfrac{1}{1 - x} = 1 + x + x^2 + x^3 + \cdots = \sum_{n=1}^{\infty} x^n$	$(-1, 1)$

(2)

These Taylor series centered at $a = 0$, also called Maclaurin series, show that the five functions in (2) are analytic at $a = 0$.

The results in (2) can also be used to obtain power series representations of other functions. For example, if we wish to find the Maclaurin series representation of, say, e^{x^2}, we need only replace x in the Maclaurin series for e^x with x^2:

$$e^{x^2} = 1 + \frac{x^2}{1!} + \frac{x^4}{2!} + \frac{x^6}{3!} + \cdots = \sum_{n=0}^{\infty} \frac{1}{n!}x^{2n}.$$

Similarly, to obtain a Taylor series representation of $\ln x$ centered at $a = 1$ we replace x by $x - 1$ in the Maclaurin series for $\ln(1 + x)$:

$$\ln x = \ln(1 + (x - 1)) = (x - 1) - \frac{(x - 1)^2}{2} + \frac{(x - 1)^3}{3} - \frac{(x - 1)^4}{4} + \cdots = \sum_{n=1}^{\infty} \frac{(-1)^{n+1}}{n}(x - 1)^n.$$

The interval of convergence for the power series representation of e^{x^2} is the same as e^x in (2), that is, $(-\infty, \infty)$. But the interval of convergence of the Taylor series of $\ln x$ is now $(0, 2]$; this interval is the interval $(-1, 1]$ given in (2) shifted 1 unit to the right.

- **Arithmetic of Power Series** Power series can be combined through the operations of addition, multiplication, and division. The procedures for power series are similar to the way in which two polynomials are added, multiplied, and divided—that is, we add coefficients of like powers of x, use the distributive law and collect like terms, and perform long division. For example, using the series in (2),

$$
e^x \sin x = \left(1 + x + \frac{x^2}{2} + \frac{x^3}{6} + \frac{x^4}{24} + \cdots\right)\left(x - \frac{x^3}{6} + \frac{x^5}{120} - \frac{x^7}{5040} + \cdots\right)
$$

$$
= (1)x + (1)x^2 + \left(-\frac{1}{6} + \frac{1}{2}\right)x^3 + \left(-\frac{1}{6} + \frac{1}{6}\right)x^4 + \left(\frac{1}{120} - \frac{1}{12} + \frac{1}{24}\right)x^5 + \cdots
$$

$$
= x + x^2 + \frac{x^3}{3} - \frac{x^5}{30} - \cdots.
$$

Since the power series for e^x and $\sin x$ converge for $|x| < \infty$, the product series converges on the same interval. Problems involving multiplication or division of power series can be done with minimal fuss using a computer algebra system.

■ **Shifting the Summation Index** For the remainder of this section, as well as this chapter, it is important that you become adept at simplifying the sum of two or more power series, each series expressed in summation (sigma) notation, to an expression with a single Σ. As the next example illustrates, combining two or more summations as a single summation often requires a reindexing, that is, a shift in the index of summation.

EXAMPLE 1 Adding Two Power Series

Write

$$
\sum_{n=2}^{\infty} n(n-1)c_n x^{n-2} - \sum_{n=0}^{\infty} c_n x^{n+1}
$$

as one power series.

Important. ▶ **SOLUTION** In order to add the two series, it is necessary that *both summation indices start with the same number and that the powers of x in each series be "in phase"*; that is, if one series starts with a multiple of, say, x to the first power, then we want the other series to start with the same power. Note that in the given problem, the first series starts with x^0, whereas the second series starts with x^1. By writing the first term of the first series outside of the summation notation,

$$
\underset{n=2}{\overset{\infty}{\sum}} n(n-1)c_n x^{n-2} - \underset{n=0}{\overset{\infty}{\sum}} c_n x^{n+1} = 2\cdot 1 c_2 x^0 + \underset{\substack{n=3 \\ \uparrow \\ \text{series starts with } x \text{ for } n=3}}{\overset{\infty}{\sum}} n(n-1)c_n x^{n-2} - \underset{\substack{n=0 \\ \uparrow \\ \text{series starts with } x \text{ for } n=0}}{\overset{\infty}{\sum}} c_n x^{n+1},
$$

we see that both series on the right side start with the same power of x, namely, x^1. Now to get the same summation index we are inspired by the exponents of x; we let $k = n - 2$ in the first series and at the same time let $k = n + 1$ in the second series. The right side becomes

$$
2c_2 + \underset{k=1}{\overset{\infty}{\sum}} (k+2)(k+1)c_{k+2} x^k - \underset{k=1}{\overset{\infty}{\sum}} c_{k-1} x^k. \tag{3}
$$

Remember, the summation index is a "dummy" variable; the fact that $k = n - 1$ in one case and $k = n + 1$ in the other should cause no confusion if you keep in mind that it is the *value* of the summation index that is important. In both cases k takes on the same successive

values $k = 1, 2, 3, \ldots$ when n takes on the values $n = 2, 3, 4, \ldots$ for $k = n - 1$ and $n = 0, 1, 2, \ldots$ for $k = n + 1$. We are now in a position to add the series in (3) term by term:

$$\sum_{n=2}^{\infty} n(n-1)c_n x^{n-2} - \sum_{n=0}^{\infty} c_n x^{n+1} = 2c_2 + \sum_{k=1}^{\infty} [(k+2)(k+1)c_{k+2} - c_{k-1}]x^k. \quad (4) \ \equiv$$

If you are not convinced of the result in (4), then write out a few terms on both sides of the equality.

5.1.2 Power Series Solutions

■ **A Definition** Suppose the linear second-order differential equation

$$a_2(x)y'' + a_1(x)y' + a_0(x)y = 0 \quad (5)$$

is put into standard form

$$y'' + P(x)y' + Q(x)y = 0 \quad (6)$$

by dividing by the leading coefficient $a_2(x)$. We make the following definition.

DEFINITION 5.1.1 Ordinary and Singular Points

A point x_0 is said to be an **ordinary point** of the differential equation (5) if both $P(x)$ and $Q(x)$ in the standard form (6) are analytic at x_0. A point that is not an ordinary point is said to be a **singular point** of the equation.

Every finite value of x is an ordinary point of $y'' + (e^x)y' + (\sin x)y = 0$. In particular, $x = 0$ is an ordinary point, since, as we have already seen in (2), both e^x and $\sin x$ are analytic at this point. The negation in the second sentence of Definition 5.1.1 stipulates that if at least one of the functions $P(x)$ and $Q(x)$ in (6) fails to be analytic at x_0, then x_0 is a singular point. Note that $x = 0$ is a singular point of the differential equation $y'' + (e^x)y' + (\ln x)y = 0$, since $Q(x) = \ln x$ is discontinuous at $x = 0$ and so cannot be represented by a power series in x.

■ **Polynomial Coefficients** We shall be interested primarily in the case in which (5) has polynomial coefficients. A polynomial is analytic at any value x, and a rational function is analytic *except* at points where its denominator is zero. Thus, if $a_2(x)$, $a_1(x)$, and $a_0(x)$ are polynomials with no common factors, then both rational functions $P(x) = a_1(x)/a_2(x)$ and $Q(x) = a_0(x)/a_2(x)$ are analytic except where $a_2(x) = 0$.

It follows, then, that $x = x_0$ is an ordinary point of (5) if $a_2(x_0) \neq 0$, whereas $x = x_0$ is a singular point of (5) if $a_2(x_0) = 0$.

For example, the only singular points of the equation $(x^2 - 1)y'' + 2xy' + 6y = 0$ are solutions of $x^2 - 1 = 0$ or $x = \pm 1$. All other finite values* of x are ordinary points. Inspection of the Cauchy–Euler equation $ax^2y'' + bxy' + cy = 0$ shows that it has a singular point at $x = 0$. Singular points need not be real numbers. The equation $(x^2 + 1)y'' + xy' - y = 0$ has singular points at the solutions of $x^2 + 1 = 0$; namely, $x = \pm i$. All other values of x, real or complex, are ordinary points.

We state the following theorem about the existence of power series solutions without proof.

THEOREM 5.1.1 Existence of Power Series Solutions

If $x = x_0$ is an ordinary point of the differential equation (5), we can always find two linearly independent solutions in the form of a power series centered at x_0; that is, $y = \sum_{n=0}^{\infty} c_n(x - x_0)^n$. A series solution converges at least on some interval defined by $|x - x_0| < R$, where R is the distance from x_0 to the closest singular point.

*For our purposes, ordinary points and singular points will always be finite points. It is possible for an ODE to have, say, a singular point at infinity.

FIGURE 5.1.2 Distance from ordinary point 0 to singular points

A solution of the form $y = \sum_{n=0}^{\infty} c_n(x - x_0)^n$ is said to be a **solution about the ordinary point** x_0. The distance R in Theorem 5.1.1 is the *minimum* value or **lower bound** for the radius of convergence. For example, the complex numbers $1 \pm 2i$ are singular points of $(x^2 - 2x + 5)y'' + xy' - y = 0$, but since $x = 0$ is an ordinary point of the equation, Theorem 5.1.1 guarantees that we can find two power series solutions centered at 0. That is, the solutions look like $y = \sum_{n=0}^{\infty} c_n x^n$, and, moreover, we know without actually finding these solutions that each series must converge *at least* for $|x| < \sqrt{5}$, where $R = \sqrt{5}$ is the distance in the complex plane from 0 to either of the numbers $1 + 2i$ or $1 - 2i$. See **FIGURE 5.1.2**. However, the differential equation has a solution that is valid for much larger values of x; indeed, this solution is valid on the interval $(-\infty, \infty)$ because it can be shown that one of the two solutions is a polynomial.

In the examples that follow, as well as in Exercises 5.1, we shall, for the sake of simplicity, only find power series solutions about the ordinary point $x = 0$. If it is necessary to find a power series solution of an ODE about an ordinary point $x_0 \neq 0$, we can simply make the change of variable $t = x - x_0$ in the equation (this translates $x = x_0$ to $t = 0$), find solutions of the new equation of the form $y = \sum_{n=0}^{\infty} c_n t^n$, and then resubstitute $t = x - x_0$.

> All the power series solutions will be centered at 0.

Finding a power series solution of a homogeneous linear second-order ODE has been accurately described as "the method of undetermined *series* coefficients," since the procedure is quite analogous to what we did in Section 3.4. In brief, here is the idea: We substitute $y = \sum_{n=0}^{\infty} c_n x^n$ into the differential equation, combine series as we did in Example 1, and then equate all coefficients to the right side of the equation to determine the coefficients c_n. But since the right side is zero, the last step requires, by the identity property in the preceding bulleted list, that all coefficients of x must be equated to zero. No, this does *not* mean that all coefficients *are* zero; this would not make sense, because after all, Theorem 5.1.1 guarantees that we can find two linearly independent solutions. Example 2 illustrates how the single assumption that $y = \sum_{n=0}^{\infty} c_n x^n = c_0 + c_1 x + c_2 x^2 + \cdots$ leads to two sets of coefficients so that we have two distinct power series $y_1(x)$ and $y_2(x)$, both expanded about the ordinary point $x = 0$. The general solution of the differential equation is $y = C_1 y_1(x) + C_2 y_2(x)$; indeed, if $y_1(0) = 1, y_1'(0) = 0$, and $y_2(0) = 0, y_2'(0) = 1$, then $C_1 = c_0$ and $C_2 = c_1$.

EXAMPLE 2 Power Series Solutions

Solve $y'' - xy = 0$.

SOLUTION Since there are no finite singular points, Theorem 5.1.1 guarantees two power series solutions, centered at 0, convergent for $|x| < \infty$. Substituting $y = \sum_{n=0}^{\infty} c_n x^n$ and the second derivative $y'' = \sum_{n=2}^{\infty} n(n - 1)c_n x^{n-2}$ (see (1)) into the differential equation gives

$$y'' - xy = \sum_{n=2}^{\infty} c_n n(n - 1)x^{n-2} - x\sum_{n=0}^{\infty} c_n x^n = \sum_{n=2}^{\infty} c_n n(n - 1)x^{n-2} - \sum_{n=0}^{\infty} c_n x^{n+1}. \quad (7)$$

Now we have already added the last two series on the right side of the equality in (7) by shifting the summation index in Example 1. From the result given in (4),

$$y'' - xy = 2c_2 + \sum_{k=1}^{\infty} [(k + 1)(k + 2)c_{k+2} - c_{k-1}]x^k = 0. \quad (8)$$

At this point we invoke the identity property. Since (8) is identically zero, it is necessary that the coefficient of each power of x be set equal to zero; that is, $2c_2 = 0$ (it is the coefficient of x^0), and

$$(k + 1)(k + 2)c_{k+2} - c_{k-1} = 0, \quad k = 1, 2, 3, \dots. \quad (9)$$

Now $2c_2 = 0$ obviously dictates that $c_2 = 0$. But the expression in (9), called a **recurrence relation**, determines the c_k in such a manner that we can choose a certain subset of the set of coefficients to be *nonzero*. Since $(k + 1)(k + 2) \neq 0$ for all values of k, we can solve (9) for c_{k+2} in terms of c_{k-1}:

> This formula is called a *two-term recurrence relation*.

$$c_{k+2} = \frac{c_{k-1}}{(k + 1)(k + 2)}, \quad k = 1, 2, 3, \dots. \quad (10)$$

This relation generates consecutive coefficients of the assumed solution one at a time as we let k take on the successive integers indicated in (10):

$$k = 1, \qquad c_3 = \frac{c_0}{2 \cdot 3}$$

$$k = 2, \qquad c_4 = \frac{c_1}{3 \cdot 4}$$

$$k = 3, \qquad c_5 = \frac{c_2}{4 \cdot 5} = 0 \quad \leftarrow c_2 \text{ is zero}$$

$$k = 4, \qquad c_6 = \frac{c_3}{5 \cdot 6} = \frac{1}{2 \cdot 3 \cdot 5 \cdot 6} c_0$$

$$k = 5, \qquad c_7 = \frac{c_4}{6 \cdot 7} = \frac{1}{3 \cdot 4 \cdot 6 \cdot 7} c_1$$

$$k = 6, \qquad c_8 = \frac{c_5}{7 \cdot 8} = 0 \quad \leftarrow c_5 \text{ is zero}$$

$$k = 7, \qquad c_9 = \frac{c_6}{8 \cdot 9} = \frac{1}{2 \cdot 3 \cdot 5 \cdot 6 \cdot 8 \cdot 9} c_0$$

$$k = 8, \qquad c_{10} = \frac{c_7}{9 \cdot 10} = \frac{1}{3 \cdot 4 \cdot 6 \cdot 7 \cdot 9 \cdot 10} c_1$$

$$k = 9, \qquad c_{11} = \frac{c_8}{10 \cdot 11} = 0 \quad \leftarrow c_8 \text{ is zero}$$

and so on. Now substituting the coefficients just obtained into the original assumption

$$y = c_0 + c_1 x + c_2 x^2 + c_3 x^3 + c_4 x^4 + c_5 x^5 + c_6 x^6 + c_7 x^7 + c_8 x^8 + c_9 x^9 + c_{10} x^{10} + c_{11} x^{11} + \cdots$$

we get

$$y = c_0 + c_1 x + 0 + \frac{c_0}{2 \cdot 3} x^3 + \frac{c_1}{3 \cdot 4} x^4 + 0 + \frac{c_0}{2 \cdot 3 \cdot 5 \cdot 6} x^6$$

$$+ \frac{c_1}{3 \cdot 4 \cdot 6 \cdot 7} x^7 + 0 + \frac{c_0}{2 \cdot 3 \cdot 5 \cdot 6 \cdot 8 \cdot 9} x^9 + \frac{c_1}{3 \cdot 4 \cdot 6 \cdot 7 \cdot 9 \cdot 10} x^{10} + 0 + \cdots.$$

After grouping the terms containing c_0 and the terms containing c_1, we obtain $y = c_0 y_1(x) + c_1 y_2(x)$, where

$$y_1(x) = 1 + \frac{1}{2 \cdot 3} x^3 + \frac{1}{2 \cdot 3 \cdot 5 \cdot 6} x^6 + \frac{1}{2 \cdot 3 \cdot 5 \cdot 6 \cdot 8 \cdot 9} x^9 + \cdots = 1 + \sum_{k=1}^{\infty} \frac{1}{2 \cdot 3 \cdots (3k-1)(3k)} x^{3k}$$

$$y_2(x) = x + \frac{1}{3 \cdot 4} x^4 + \frac{1}{3 \cdot 4 \cdot 6 \cdot 7} x^7 + \frac{1}{3 \cdot 4 \cdot 6 \cdot 7 \cdot 9 \cdot 10} x^{10} + \cdots = x + \sum_{k=1}^{\infty} \frac{1}{3 \cdot 4 \cdots (3k)(3k+1)} x^{3k+1}.$$

Since the recursive use of (10) leaves c_0 and c_1 completely undetermined, they can be chosen arbitrarily. As mentioned prior to this example, the linear combination $y = c_0 y_1(x) + c_1 y_2(x)$ actually represents the general solution of the differential equation. Although we know from Theorem 5.1.1 that each series solution converges for $|x| < \infty$, this fact can also be verified by the ratio test. $\equiv$

The differential equation in Example 2 is called **Airy's equation** and is named after the English astronomer and mathematician **Sir George Biddell Airy** (1801–1892). Airy's equation is encountered in the study of diffraction of light, diffraction of radio waves around the surface of the Earth, aerodynamics, and the deflection of a uniform thin vertical column that bends under its own weight. Other common forms of Airy's equation are $y'' + xy = 0$ and $y'' \pm \alpha^2 xy = 0$. See Problem 47 in Exercises 5.3 for an application of the last equation. It should also be

apparent by making a sign change in (10) of Example 2 that the general solution of $y'' + xy = 0$ is $y = c_0 y_1(x) + c_1 y_2(x)$, where the series solutions are in this case

$$y_1(x) = 1 - \frac{1}{2 \cdot 3} x^3 + \frac{1}{2 \cdot 3 \cdot 5 \cdot 6} x^6 - \frac{1}{2 \cdot 3 \cdot 5 \cdot 6 \cdot 8 \cdot 9} x^9 + \cdots = 1 + \sum_{k=1}^{\infty} \frac{(-1)^k}{2 \cdot 3 \cdots (3k-1)(3k)} x^{3k}$$

$$y_2(x) = x - \frac{1}{3 \cdot 4} x^4 + \frac{1}{3 \cdot 4 \cdot 6 \cdot 7} x^7 - \frac{1}{3 \cdot 4 \cdot 6 \cdot 7 \cdot 9 \cdot 10} x^{10} + \cdots = x + \sum_{k=1}^{\infty} \frac{(-1)^k}{3 \cdot 4 \cdots (3k)(3k+1)} x^{3k+1}.$$

EXAMPLE 3 **Power Series Solution**

Solve $(x^2 + 1)y'' + xy' - y = 0$.

SOLUTION As we have already seen on page 274, the given differential equation has singular points at $x = \pm i$, and so a power series solution centered at 0 will converge at least for $|x| < 1$, where 1 is the distance in the complex plane from 0 to either i or $-i$. The assumption $y = \sum_{n=0}^{\infty} c_n x^n$ and its first two derivatives (see (1)) lead to

$$(x^2 + 1) \sum_{n=2}^{\infty} n(n-1)c_n x^{n-2} + x \sum_{n=1}^{\infty} n c_n x^{n-1} - \sum_{n=0}^{\infty} c_n x^n$$

$$= \sum_{n=2}^{\infty} n(n-1)c_n x^n + \sum_{n=2}^{\infty} n(n-1)c_n x^{n-2} + \sum_{n=1}^{\infty} n c_n x^n - \sum_{n=0}^{\infty} c_n x^n$$

$$= 2c_2 x^0 - c_0 x^0 + 6c_3 x + c_1 x - c_1 x + \underbrace{\sum_{n=2}^{\infty} n(n-1)c_n x^n}_{k=n}$$

$$+ \underbrace{\sum_{n=4}^{\infty} n(n-1)c_n x^{n-2}}_{k=n-2} + \underbrace{\sum_{n=2}^{\infty} n c_n x^n}_{k=n} - \underbrace{\sum_{n=2}^{\infty} c_n x^n}_{k=n}$$

$$= 2c_2 - c_0 + 6c_3 x + \sum_{k=2}^{\infty} [k(k-1)c_k + (k+2)(k+1)c_{k+2} + k c_k - c_k] x^k$$

$$= 2c_2 - c_0 + 6c_3 x + \sum_{k=2}^{\infty} [(k+1)(k-1)c_k + (k+2)(k+1)c_{k+2}] x^k = 0.$$

From this last identity we conclude that $2c_2 - c_0 = 0$, $6c_3 = 0$, and

$$(k+1)(k-1)c_k + (k+2)(k+1)c_{k+2} = 0.$$

Thus,
$$c_2 = \frac{1}{2} c_0, \qquad c_3 = 0,$$

$$c_{k+2} = \frac{1-k}{k+2} c_k, \qquad k = 2, 3, 4, \ldots.$$

Substituting $k = 2, 3, 4, \ldots$ into the last formula gives

$$c_4 = -\frac{1}{4} c_2 = -\frac{1}{2 \cdot 4} c_0 = -\frac{1}{2^2 2!} c_0$$

$$c_5 = -\frac{2}{5} c_3 = 0 \quad \leftarrow c_3 \text{ is zero}$$

$$c_6 = -\frac{3}{6} c_4 = \frac{3}{2 \cdot 4 \cdot 6} c_0 = \frac{1 \cdot 3}{2^3 3!} c_0$$

$$c_7 = -\frac{4}{7} c_5 = 0 \quad \leftarrow c_5 \text{ is zero}$$

$$c_8 = -\frac{5}{8} c_6 = -\frac{3 \cdot 5}{2 \cdot 4 \cdot 6 \cdot 8} c_0 = -\frac{1 \cdot 3 \cdot 5}{2^4 4!} c_0$$

$$c_9 = -\frac{6}{9} c_7 = 0 \quad \leftarrow c_7 \text{ is zero}$$

$$c_{10} = -\frac{7}{10} c_8 = \frac{3 \cdot 5 \cdot 7}{2 \cdot 4 \cdot 6 \cdot 8 \cdot 10} c_0 = -\frac{1 \cdot 3 \cdot 5 \cdot 7}{2^5 5!} c_0$$

and so on. Therefore,

$$y = c_0 + c_1 x + c_2 x^2 + c_3 x^3 + c_4 x^4 + c_5 x^5 + c_6 x^6 + c_7 x^7 + c_8 x^8 + c_9 x^9 + c_{10} x^{10} + \cdots$$

$$= c_0 \left[1 + \frac{1}{2} x^2 - \frac{1}{2^2 2!} x^4 + \frac{1 \cdot 3}{2^3 3!} x^6 - \frac{1 \cdot 3 \cdot 5}{2^4 4!} x^8 + \frac{1 \cdot 3 \cdot 5 \cdot 7}{2^5 5!} x^{10} - \cdots \right] + c_1 x$$

$$= c_0 y_1(x) + c_1 y_2(x).$$

The solutions are the polynomial $y_2(x) = x$ and the power series

$$y_1(x) = 1 + \frac{1}{2} x^2 + \sum_{n=2}^{\infty} (-1)^{n-1} \frac{1 \cdot 3 \cdot 5 \cdots (2n-3)}{2^n n!} x^{2n}, \quad |x| < 1. \qquad \equiv$$

EXAMPLE 4 Three-Term Recurrence Relation

If we seek a power series solution $y = \sum_{n=0}^{\infty} c_n x^n$ for the differential equation

$$y'' - (1 + x)y = 0,$$

we obtain $c_2 = c_0/2$ and the recurrence relation

This formula is called a ▶
*three-term recurrence
relation.*

$$c_{k+2} = \frac{c_k + c_{k-1}}{(k+1)(k+2)}, \quad k = 1, 2, 3, \ldots.$$

Examination of the formula shows that the coefficients $c_3, c_4, c_5, \ldots$ are expressed in terms of *both* c_1 and c_0, and moreover, the algebra required to do this becomes a little messy. To simplify life we can first choose $c_0 \neq 0$, $c_1 = 0$; this yields consecutive coefficients for one solution that are expressed entirely in terms of c_0. If we next choose $c_0 = 0$, $c_1 \neq 0$, then the coefficients for the other solution are expressed in terms of c_1. Using $c_2 = \frac{1}{2} c_0$ in both cases, the recurrence relation for $k = 1, 2, 3, \ldots$ gives

$c_0 \neq 0, c_1 = 0$

$$c_2 = \frac{1}{2} c_0$$

$$c_3 = \frac{c_1 + c_0}{2 \cdot 3} = \frac{c_0}{2 \cdot 3} = \frac{1}{6} c_0$$

$$c_4 = \frac{c_2 + c_1}{3 \cdot 4} = \frac{c_0}{2 \cdot 3 \cdot 4} = \frac{1}{24} c_0$$

$$c_5 = \frac{c_3 + c_2}{4 \cdot 5} = \frac{c_0}{4 \cdot 5} \left[\frac{1}{6} + \frac{1}{2} \right] = \frac{1}{30} c_0$$

$c_0 = 0, c_1 \neq 0$

$$c_2 = \frac{1}{2} c_0 = 0$$

$$c_3 = \frac{c_1 + c_0}{2 \cdot 3} = \frac{c_1}{2 \cdot 3} = \frac{1}{6} c_1$$

$$c_4 = \frac{c_2 + c_1}{3 \cdot 4} = \frac{c_1}{3 \cdot 4} = \frac{1}{12} c_1$$

$$c_5 = \frac{c_3 + c_2}{4 \cdot 5} = \frac{c_1}{4 \cdot 5 \cdot 6} = \frac{1}{120} c_1$$

and so on. Finally, we see that the general solution of the equation is $y = c_0 y_1(x) + c_1 y_2(x)$, where

$$y_1(x) = 1 + \frac{1}{2}x^2 + \frac{1}{6}x^3 + \frac{1}{24}x^4 + \frac{1}{30}x^5 + \cdots$$

and

$$y_2(x) = x + \frac{1}{6}x^3 + \frac{1}{12}x^4 + \frac{1}{120}x^5 + \cdots.$$

Each series converges for all finite values of x.

Nonpolynomial Coefficients

The next example illustrates how to find a power series solution about the ordinary point $x_0 = 0$ of a differential equation when its coefficients are not polynomials. In this example we see an application of multiplication of two power series.

EXAMPLE 5 ODE with Nonpolynomial Coefficients

Solve $y'' + (\cos x)y = 0$.

SOLUTION We see $x = 0$ is an ordinary point of the equation because, as we have already seen, $\cos x$ is analytic at that point. Using the Maclaurin series for $\cos x$ given in (2), along with the usual assumption $y = \sum_{n=0}^{\infty} c_n x^n$ and the results in (1), we find

$$y'' + (\cos x)y = \sum_{n=2}^{\infty} n(n-1)c_n x^{n-2} + \left(1 - \frac{x^2}{2!} + \frac{x^4}{4!} - \frac{x^6}{6!} + \cdots\right)\sum_{n=0}^{\infty} c_n x^n$$

$$= 2c_2 + 6c_3 x + 12c_4 x^2 + 20c_5 x^3 + \cdots + \left(1 - \frac{x^2}{2!} + \frac{x^4}{4!} + \cdots\right)(c_0 + c_1 x + c_2 x^2 + c_3 x^3 + \cdots)$$

$$= 2c_2 + c_0 + (6c_3 + c_1)x + \left(12c_4 + c_2 - \frac{1}{2}c_0\right)x^2 + \left(20c_5 + c_3 - \frac{1}{2}c_1\right)x^3 + \cdots = 0.$$

It follows that

$$2c_2 + c_0 = 0, \quad 6c_3 + c_1 = 0, \quad 12c_4 + c_2 - \frac{1}{2}c_0 = 0, \quad 20c_5 + c_3 - \frac{1}{2}c_1 = 0,$$

and so on. This gives $c_2 = -\frac{1}{2}c_0$, $c_3 = -\frac{1}{6}c_1$, $c_4 = \frac{1}{12}c_0$, $c_5 = \frac{1}{30}c_1, \ldots$. By grouping terms we arrive at the general solution $y = c_0 y_1(x) + c_1 y_2(x)$, where

$$y_1(x) = 1 - \frac{1}{2}x^2 + \frac{1}{12}x^4 - \cdots \quad \text{and} \quad y_2(x) = x - \frac{1}{6}x^3 + \frac{1}{30}x^5 - \cdots.$$

Since the differential equation has no finite singular points, both power series converge for $|x| < \infty$.

Solution Curves

The approximate graph of a power series solution $y(x) = \sum_{n=0}^{\infty} c_n x^n$ can be obtained in several ways. We can always resort to graphing the terms in the sequence of partial sums of the series, in other words, the graphs of the polynomials $S_N(x) = \sum_{n=0}^{N} c_n x^n$. For large values of N, $S_N(x)$ should give us an indication of the behavior of $y(x)$ near the ordinary point $x = 0$. We can also obtain an approximate solution curve by using a numerical solver as we did in Section 3.7. For example, if you carefully scrutinize the series solutions of Airy's equation in Example 2, you should see that $y_1(x)$ and $y_2(x)$ are, in turn, the solutions of the initial-value problems

$$y'' - xy = 0, \quad y(0) = 1, \quad y'(0) = 0,$$
$$y'' - xy = 0, \quad y(0) = 0, \quad y'(0) = 1. \tag{11}$$

The specified initial conditions "pick out" the solutions $y_1(x)$ and $y_2(x)$ from $y = c_0 y_1(x) + c_1 y_2(x)$, since it should be apparent from our basic series assumption $y = \sum_{n=0}^{\infty} c_n x^n$ that $y(0) = c_0$ and $y'(0) = c_1$. Now if your numerical solver requires a system of equations, the substitution $y' = u$ in $y'' - xy = 0$ gives $y'' = u' = xy$, and so a system of two first-order equations equivalent to Airy's equation is

$$y' = u$$
$$u' = xy. \tag{12}$$

(a) Plot of $y_1(x)$

(b) Plot of $y_2(x)$

FIGURE 5.1.3 Solutions of Airy's equation

Initial conditions for the system in (12) are the two sets of initial conditions in (11) but rewritten as $y(0) = 1$, $u(0) = 0$, and $y(0) = 0$, $u(0) = 1$. The graphs of $y_1(x)$ and $y_2(x)$ shown in **FIGURE 5.1.3** were obtained with the aid of a numerical solver using the fourth-order Runge–Kutta method with a step size of $h = 0.1$.

REMARKS

(*i*) In the problems that follow, do not expect to be able to write a solution in terms of summation notation in each case. Even though we can generate as many terms as desired in a series solution $y = \sum_{n=0}^{\infty} c_n x^n$ either through the use of a recurrence relation or, as in Example 5, by multiplication, it may not be possible to deduce any general term for the coefficients c_n. We may have to settle, as we did in Examples 4 and 5, for just writing out the first few terms of the series.

(*ii*) A point x_0 is an ordinary point of a *nonhomogeneous* linear second-order DE $y'' + P(x)y' + Q(x)y = f(x)$ if $P(x)$, $Q(x)$, and $f(x)$ are analytic at x_0. Moreover, Theorem 5.1.1 extends to such DEs—in other words, we can find power series solutions $y = \sum_{n=0}^{\infty} c_n(x - x_0)^n$ of nonhomogeneous linear DEs in the same manner as in Examples 2–5. See Problem 38 in Exercises 5.1.

5.1 Exercises Answers to selected odd-numbered problems begin on page ANS-12.

5.1.1 Review of Power Series

In Problems 1–4, find the radius of convergence and interval of convergence for the given power series.

1. $\displaystyle\sum_{n=1}^{\infty} \frac{2^n}{n} x^n$

2. $\displaystyle\sum_{n=0}^{\infty} \frac{(100)^n}{n!} (x + 7)^n$

3. $\displaystyle\sum_{k=1}^{\infty} \frac{(-1)^k}{10^k} (x - 5)^k$

4. $\displaystyle\sum_{k=0}^{\infty} k!(x - 1)^k$

In Problems 5 and 6, the given function is analytic at $x = 0$. Find the first four terms of a power series in x. Perform the multiplication by hand or use a CAS, as instructed.

5. $\sin x \cos x$

6. $e^{-x} \cos x$

In Problems 7 and 8, the given function is analytic at $x = 0$. Find the first four terms of a power series in x. Perform the long division by hand or use a CAS, as instructed. Give the open interval of convergence.

7. $\dfrac{1}{\cos x}$

8. $\dfrac{1 - x}{2 + x}$

In Problems 9 and 10, rewrite the given power series so that its general term involves x^k.

9. $\displaystyle\sum_{n=1}^{\infty} nc_n x^{n+2}$

10. $\displaystyle\sum_{n=3}^{\infty} (2n - 1)c_n x^{n-3}$

In Problems 11 and 12, rewrite the given expression as a single power series whose general term involves x^k.

11. $\displaystyle\sum_{n=1}^{\infty} 2nc_n x^{n-1} + \sum_{n=0}^{\infty} 6c_n x^{n+1}$

12. $\displaystyle\sum_{n=2}^{\infty} n(n - 1)c_n x^n + 2\sum_{n=2}^{\infty} n(n - 1)c_n x^{n-2} + 3\sum_{n=1}^{\infty} nc_n x^n$

In Problems 13–16, verify by direct substitution that the given power series is a particular solution of the indicated differential equation.

13. $y = \displaystyle\sum_{n=0}^{\infty} \frac{(-1)^n}{n!} x^{2n}$, $\quad y' + 2xy = 0$

14. $y = \displaystyle\sum_{n=0}^{\infty} (-1)^n x^{2n}$, $\quad (1 + x^2)y' + 2xy = 0$

15. $y = \displaystyle\sum_{n=1}^{\infty} \frac{(-1)^{n+1}}{n} x^n$, $\quad (x + 1)y'' + y' = 0$

16. $y = \displaystyle\sum_{n=0}^{\infty} \frac{(-1)^n}{2^{2n}(n!)^2} x^{2n}$, $\quad xy'' + y' + xy = 0$

5.1.2 Power Series Solutions

In Problems 17 and 18, without actually solving the given differential equation, find a lower bound for the radius of convergence of power series solutions about the ordinary point $x = 0$. About the ordinary point $x = 1$.

17. $(x^2 - 25)y'' + 2xy' + y = 0$

18. $(x^2 - 2x + 10)y'' + xy' - 4y = 0$

In Problems 19–30, find two power series solutions of the given differential equation about the ordinary point $x = 0$.

19. $y'' - 3xy = 0$

20. $y'' + x^2 y = 0$

21. $y'' - 2xy' + y = 0$

22. $y'' - xy' + 2y = 0$

23. $y'' + x^2 y' + xy = 0$

24. $y'' + 2xy' + 2y = 0$

25. $(x - 1)y'' + y' = 0$

26. $(x + 2)y'' + xy' - y = 0$

27. $y'' - (x + 1)y' - y = 0$

28. $(x^2 + 1)y'' - 6y = 0$

29. $(x^2 + 2)y'' + 3xy' - y = 0$

30. $(x^2 - 1)y'' + xy' - y = 0$

In Problems 31–34, use the power series method to solve the given initial-value problem.

31. $(x - 1)y'' - xy' + y = 0$, $y(0) = -2$, $y'(0) = 6$

32. $(x + 1)y'' - (2 - x)y' + y = 0$, $y(0) = 2$, $y'(0) = -1$

33. $y'' - 2xy' + 8y = 0$, $y(0) = 3$, $y'(0) = 0$

34. $(x^2 + 1)y'' + 2xy' = 0$, $y(0) = 0$, $y'(0) = 1$

In Problems 35 and 36, use the procedure in Example 5 to find two power series solutions of the given differential equation about the ordinary point $x = 0$.

35. $y'' + (\sin x)y = 0$ **36.** $y'' + e^x y' - y = 0$

Discussion Problems

37. Without actually solving the differential equation $(\cos x)y'' + y' + 5y = 0$, find a lower bound for the radius of convergence of power series solutions about $x = 0$. About $x = 1$.

38. How can the method described in this section be used to find a power series solution of the *nonhomogeneous* equation $y'' - xy = 1$ about the ordinary point $x = 0$? Of $y'' - 4xy' - 4y = e^x$? Carry out your ideas by solving both DEs.

39. Is $x = 0$ an ordinary or a singular point of the differential equation $xy'' + (\sin x)y = 0$? Defend your answer with sound mathematics.

40. For purposes of this problem, ignore the graphs given in Figure 5.1.3. If Airy's differential equation is written as $y'' = xy$, what can we say about the shape of a solution curve if $x > 0$ and $y > 0$? If $x < 0$ and $y > 0$? If $x < 0$ and $y < 0$?

Computer Lab Assignments

41. (a) Find two power series solutions for $y'' + xy' + y = 0$ and express the solutions $y_1(x)$ and $y_2(x)$ in terms of summation notation.

 (b) Use a CAS to graph the partial sums $S_N(x)$ for $y_1(x)$. Use $N = 2, 3, 5, 6, 8, 10$. Repeat using the partial sums $S_N(x)$ for $y_2(x)$.

 (c) Compare the graphs obtained in part (b) with the curve obtained using a numerical solver. Use the initial conditions $y_1(0) = 1$, $y_1'(0) = 0$, and $y_2(0) = 0$, $y_2'(0) = 1$.

 (d) Reexamine the solution $y_1(x)$ in part (a). Express this series as an elementary function. Then use (5) of Section 3.2 to find a second solution of the equation. Verify that this second solution is the same as the power series solution $y_2(x)$.

42. (a) Find one more nonzero term for each of the solutions $y_1(x)$ and $y_2(x)$ in Example 5.

 (b) Find a series solution $y(x)$ of the initial-value problem $y'' + (\cos x)y = 0$, $y(0) = 1$, $y'(0) = 1$.

 (c) Use a CAS to graph the partial sums $S_N(x)$ for the solution $y(x)$ in part (b). Use $N = 2, 3, 4, 5, 6, 7$.

 (d) Compare the graphs obtained in part (c) with the curve obtained using a numerical solver for the initial-value problem in part (b).

5.2 Solutions about Singular Points

INTRODUCTION The two differential equations $y'' - xy = 0$ and $xy'' + y = 0$ are similar only in that they are both examples of simple linear second-order DEs with variable coefficients. That is all they have in common. Since $x = 0$ is an *ordinary point* of the first equation, we saw in the preceding section that there was no problem in finding two distinct power series solutions centered at that point. In contrast, because $x = 0$ is a *singular point* of the second DE, finding two infinite series solutions—notice we did not say "power series solutions"—of the equation about that point becomes a more difficult task.

▌ **A Definition** A singular point $x = x_0$ of a linear differential equation

$$a_2(x)y'' + a_1(x)y' + a_0(x)y = 0 \tag{1}$$

is further classified as either regular or irregular. The classification again depends on the functions P and Q in the standard form

$$y'' + P(x)y' + Q(x)y = 0. \tag{2}$$

DEFINITION 5.2.1 Regular/Irregular Singular Points

A singular point x_0 is said to be a **regular singular point** of the differential equation (1) if the functions $p(x) = (x - x_0)P(x)$ and $q(x) = (x - x_0)^2 Q(x)$ are both analytic at x_0. A singular point that is not regular is said to be an **irregular singular point** of the equation.

The second sentence in Definition 5.2.1 indicates that if one or both of the functions $p(x) = (x - x_0)P(x)$ and $q(x) = (x - x_0)^2 Q(x)$ fails to be analytic at x_0, then x_0 is an irregular singular point.

Polynomial Coefficients As in Section 5.1, we are mainly interested in linear equations (1) where the coefficients $a_2(x)$, $a_1(x)$, and $a_0(x)$ are polynomials with no common factors. We have already seen that if $a_2(x_0) = 0$, then $x = x_0$ is a singular point of (1) since at least one of the rational functions $P(x) = a_1(x)/a_2(x)$ and $Q(x) = a_0(x)/a_2(x)$ in the standard form (2) fails to be analytic at that point. But since $a_2(x)$ is a polynomial and x_0 is one of its zeros, it follows from the Factor Theorem of algebra that $x - x_0$ is a factor of $a_2(x)$. This means that after $a_1(x)/a_2(x)$ and $a_0(x)/a_2(x)$ are reduced to lowest terms, the factor $x - x_0$ must remain, to some positive integer power, in one or both denominators. Now suppose that $x = x_0$ is a singular point of (1) but that both the functions defined by the products $p(x) = (x - x_0)P(x)$ and $q(x) = (x - x_0)^2 Q(x)$ are analytic at x_0. We are led to the conclusion that multiplying $P(x)$ by $x - x_0$ and $Q(x)$ by $(x - x_0)^2$ has the effect (through cancellation) that $x - x_0$ no longer appears in either denominator. We can now determine whether x_0 is regular by a quick visual check of denominators:

If $x - x_0$ appears at most to the first power in the denominator of $P(x)$ and at most to the second power in the denominator of $Q(x)$, then $x = x_0$ is a regular singular point.

Moreover, observe that if $x = x_0$ is a regular singular point and we multiply (2) by $(x - x_0)^2$, then the original DE can be put into the form

$$(x - x_0)^2 y'' + (x - x_0)p(x)y' + q(x)y = 0, \tag{3}$$

where p and q are analytic at $x = x_0$.

EXAMPLE 1 **Classification of Singular Points**

It should be clear that $x = 2$ and $x = -2$ are singular points of

$$(x^2 - 4)^2 y'' + 3(x - 2)y' + 5y = 0.$$

After dividing the equation by $(x^2 - 4)^2 = (x - 2)^2 (x + 2)^2$ and reducing the coefficients to lowest terms, we find that

$$P(x) = \frac{3}{(x - 2)(x + 2)^2} \quad \text{and} \quad Q(x) = \frac{5}{(x - 2)^2 (x + 2)^2}.$$

We now test $P(x)$ and $Q(x)$ at each singular point.

In order for $x = 2$ to be a regular singular point, the factor $x - 2$ can appear at most to the first power in the denominator of $P(x)$ and at most to the second power in the denominator of $Q(x)$. A check of the denominators of $P(x)$ and $Q(x)$ shows that both these conditions are satisfied, so $x = 2$ is a regular singular point. Alternatively, we are led to the same conclusion by noting that both rational functions

$$p(x) = (x - 2)P(x) = \frac{3}{(x + 2)^2} \quad \text{and} \quad q(x) = (x - 2)^2 Q(x) = \frac{5}{(x + 2)^2}$$

are analytic at $x = 2$.

Now since the factor $x - (-2) = x + 2$ appears to the second power in the denominator of $P(x)$, we can conclude immediately that $x = -2$ is an irregular singular point of the equation. This also follows from the fact that $p(x) = (x + 2)P(x) = 3/[(x - 2)(x + 2)]$ is not analytic at $x = -2$. ≡

In Example 1, notice that since $x = 2$ is a regular singular point, the original equation can be written as

$$p(x) \text{ analytic at } x = 2 \qquad q(x) \text{ analytic at } x = 2$$
$$\downarrow \qquad\qquad \downarrow$$
$$(x - 2)^2 y'' + (x - 2)\frac{3}{(x + 2)^2}y' + \frac{5}{(x + 2)^2}y = 0.$$

As another example, we can see that $x = 0$ is an irregular singular point of $x^3 y'' - 2xy' + 8y = 0$ by inspection of the denominators of $P(x) = -2/x^2$ and $Q(x) = 8/x^3$. On the other hand, $x = 0$ is a regular singular point of $xy'' - 2xy' + 8y = 0$ since $x - 0$ and $(x - 0)^2$ do not even

appear in the respective denominators of $P(x) = -2$ and $Q(x) = 8/x$. For a singular point $x = x_0$, any nonnegative power of $x - x_0$ less than one (namely, zero) and any nonnegative power less than two (namely, zero and one) in the denominators of $P(x)$ and $Q(x)$, respectively, imply that x_0 is a regular singular point. A singular point can be a complex number. You should verify that $x = 3i$ and $x = -3i$ are two regular singular points of $(x^2 + 9)y'' - 3xy' + (1 - x)y = 0$.

Any second-order Cauchy–Euler equation $ax^2y'' + bxy' + cy = 0$, with a, b, and c real constants, has a regular singular point at $x = 0$. You should verify that two solutions of the Cauchy–Euler equation $x^2y'' - 3xy' + 4y = 0$ on the interval $(0, \infty)$ are $y_1 = x^2$ and $y_2 = x^2 \ln x$. If we attempt to find a power series solution about the regular singular point $x = 0$, namely, $y = \sum_{n=0}^{\infty} c_n x^n$, we would succeed in obtaining only the polynomial solution $y_1 = x^2$. The fact that we would not obtain the second solution is not surprising since $\ln x$, and consequently, $y_2 = x^2 \ln x$, is not analytic at $x = 0$; that is, y_2 does not possess a Taylor series expansion centered at $x = 0$.

■ **Method of Frobenius** To solve a differential equation (1) about a regular singular point, we employ the following theorem due to the eminent German mathematician **Ferdinand Georg Frobenius** (1849–1917).

THEOREM 5.2.1 **Frobenius' Theorem**

If $x = x_0$ is a regular singular point of the differential equation (1), then there exists at least one nonzero solution of the form

$$y = (x - x_0)^r \sum_{n=0}^{\infty} c_n(x - x_0)^n = \sum_{n=0}^{\infty} c_n(x - x_0)^{n+r}, \qquad (4)$$

where the number r is a constant to be determined. The series will converge at least on some interval defined by $0 < x - x_0 < R$.

Notice the words *at least* in the first sentence of Theorem 5.2.1. This means that, in contrast to Theorem 5.1.1, we have no assurance that two series solutions of the type indicated in (4) can be found. The **method of Frobenius**, finding series solutions about a regular singular point x_0, is similar to the "method of undetermined series coefficients" of the preceding section in that we substitute $y = \sum_{n=0}^{\infty} c_n(x - x_0)^{n+r}$ into the given differential equation and determine the unknown coefficients c_n by a recursion relation. However, we have an additional task in this procedure; before determining the coefficients, we must first find the unknown exponent r. If r is found to be a number that is not a nonnegative integer, then the corresponding solution $y = \sum_{n=0}^{\infty} c_n(x - x_0)^{n+r}$ is not a power series.

▶ As we did in the discussion of solutions about ordinary points, we shall always assume, for the sake of simplicity in solving differential equations, that the regular singular point is $x = 0$.

All solutions will be about the regular singular point 0.

EXAMPLE 2 **Two Series Solutions**

Because $x = 0$ is a regular singular point of the differential equation

$$3xy'' + y' - y = 0, \qquad (5)$$

we try to find a solution of the form $y = \sum_{n=0}^{\infty} c_n x^{n+r}$. Now

$$y' = \sum_{n=0}^{\infty} (n + r)c_n x^{n+r-1} \quad \text{and} \quad y'' = \sum_{n=0}^{\infty} (n + r)(n + r - 1)c_n x^{n+r-2}$$

so that

$$3xy'' + y' - y = 3 \sum_{n=0}^{\infty} (n + r)(n + r - 1)c_n x^{n+r-1} + \sum_{n=0}^{\infty} (n + r)c_n x^{n+r-1} - \sum_{n=0}^{\infty} c_n x^{n+r}$$

$$= \sum_{n=0}^{\infty} (n + r)(3n + 3r - 2)c_n x^{n+r-1} - \sum_{n=0}^{\infty} c_n x^{n+r}$$

$$= x^r \left[r(3r - 2)c_0 x^{-1} + \sum_{n=1}^{\infty} (n + r)(3n + 3r - 2)c_n x^{n-1} - \sum_{n=0}^{\infty} c_n x^n \right]$$

$$\underbrace{}_{k = n - 1} \underbrace{}_{k = n}$$

$$= x^r \left[r(3r - 2)c_0 x^{-1} + \sum_{k=0}^{\infty} [(k + r + 1)(3k + 3r + 1)c_{k+1} - c_k]x^k \right] = 0,$$

which implies $r(3r - 2)c_0 = 0$

$$(k + r + 1)(3k + 3r + 1)c_{k+1} - c_k = 0, \quad k = 0, 1, 2, \dots.$$

Since nothing is gained by taking $c_0 = 0$, we must then have

$$r(3r - 2) = 0, \tag{6}$$

and

$$c_{k+1} = \frac{c_k}{(k + r + 1)(3k + 3r + 1)}, \quad k = 0, 1, 2,\dots. \tag{7}$$

The two values of r that satisfy the quadratic equation (6), $r_1 = \frac{2}{3}$ and $r_2 = 0$, when substituted in (7), give two different recurrence relations:

$$r_1 = \tfrac{2}{3}, \quad c_{k+1} = \frac{c_k}{(3k + 5)(k + 1)}, \quad k = 0, 1, 2, \dots \tag{8}$$

$$r_2 = 0, \quad c_{k+1} = \frac{c_k}{(k + 1)(3k + 1)}, \quad k = 0, 1, 2, \dots. \tag{9}$$

From (8) we find:

$$c_1 = \frac{c_0}{5 \cdot 1}$$

$$c_2 = \frac{c_1}{8 \cdot 2} = \frac{c_0}{2! \, 5 \cdot 8}$$

$$c_3 = \frac{c_2}{11 \cdot 3} = \frac{c_0}{3! \, 5 \cdot 8 \cdot 11}$$

$$c_4 = \frac{c_3}{14 \cdot 4} = \frac{c_0}{4! \, 5 \cdot 8 \cdot 11 \cdot 14}$$

$$\vdots$$

$$c_n = \frac{c_0}{n! \, 5 \cdot 8 \cdot 11 \cdots (3n + 2)}.$$

From (9) we find:

$$c_1 = \frac{c_0}{1 \cdot 1}$$

$$c_2 = \frac{c_1}{2 \cdot 4} = \frac{c_0}{2! \, 1 \cdot 4}$$

$$c_3 = \frac{c_2}{3 \cdot 7} = \frac{c_0}{3! \, 1 \cdot 4 \cdot 7}$$

$$c_4 = \frac{c_3}{4 \cdot 10} = \frac{c_0}{4! \, 1 \cdot 4 \cdot 7 \cdot 10}$$

$$\vdots$$

$$c_n = \frac{(-1)^n c_0}{n! \, 1 \cdot 4 \cdot 7 \cdots (3n - 2)}.$$

Here we encounter something that did not happen when we obtained solutions about an ordinary point; we have what looks to be two different sets of coefficients, but each set contains the *same* multiple c_0. Omitting this term, the series solutions are

$$y_1(x) = x^{2/3} \left[1 + \sum_{n=1}^{\infty} \frac{1}{n! \, 5 \cdot 8 \cdot 11 \cdots (3n + 2)} x^n \right] \tag{10}$$

$$y_2(x) = x^0 \left[1 + \sum_{n=1}^{\infty} \frac{1}{n! \, 1 \cdot 4 \cdot 7 \cdots (3n - 2)} x^n \right]. \tag{11}$$

By the ratio test it can be demonstrated that both (10) and (11) converge for all finite values of x; that is, $|x| < \infty$. Also, it should be apparent from the form of these solutions that neither series is a constant multiple of the other and, therefore, $y_1(x)$ and $y_2(x)$ are linearly independent on the entire x-axis. Hence by the superposition principle $y = C_1 y_1(x) + C_2 y_2(x)$ is another solution of (5). On any interval not containing the origin, such as $(0, \infty)$, this linear combination represents the general solution of the differential equation. $\equiv$

Indicial Equation Equation (6) is called the **indicial equation** of the problem, and the values $r_1 = \frac{2}{3}$ and $r_2 = 0$ are called the **indicial roots**, or **exponents**, of the singularity $x = 0$. In general, after substituting $y = \sum_{n=0}^{\infty} c_n x^{n+r}$ into the given differential equation and simplifying, the indicial equation is a quadratic equation in r that results from equating the *total coefficient of the lowest power of x to zero*. We solve for the two values of r and substitute these values into a recurrence relation such as (7). Theorem 5.2.1 guarantees that at least one nonzero solution of the assumed series form can be found.

It is possible to obtain the indicial equation in advance of substituting $y = \sum_{n=0}^{\infty} c_n x^{n+r}$ into the differential equation. If $x = 0$ is a regular singular point of (1), then by Definition 5.2.1 both functions $p(x) = xP(x)$ and $q(x) = x^2 Q(x)$, where P and Q are defined by the standard form (2), are analytic at $x = 0$; that is, the power series expansions

$$p(x) = xP(x) = a_0 + a_1 x + a_2 x^2 + \cdots \quad \text{and} \quad q(x) = x^2 Q(x) = b_0 + b_1 x + b_2 x^2 + \cdots \quad (12)$$

are valid on intervals that have a positive radius of convergence. By multiplying (2) by x^2, we get the form given in (3):

$$x^2 y'' + x[xP(x)]\, y' + [x^2 Q(x)]\, y = 0. \quad (13)$$

After substituting $y = \sum_{n=0}^{\infty} c_n x^{n+r}$ and the two series in (12) into (13) and carrying out the multiplication of series, we find the general indicial equation to be

$$r(r-1) + a_0 r + b_0 = 0, \quad (14)$$

where a_0 and b_0 are defined in (12). See Problems 13 and 14 in Exercises 5.2.

EXAMPLE 3 Two Series Solutions

Solve $2xy'' + (1 + x)y' + y = 0$.

SOLUTION Substituting $y = \sum_{n=0}^{\infty} c_n x^{n+r}$ gives

$$2xy'' + (1+x)y' + y = 2\sum_{n=0}^{\infty}(n+r)(n+r-1)c_n x^{n+r-1} + \sum_{n=0}^{\infty}(n+r)c_n x^{n+r-1}$$

$$+ \sum_{n=0}^{\infty}(n+r)c_n x^{n+r} + \sum_{n=0}^{\infty} c_n x^{n+r}$$

$$= \sum_{n=0}^{\infty}(n+r)(2n+2r-1)c_n x^{n+r-1} + \sum_{n=0}^{\infty}(n+r+1)c_n x^{n+r}$$

$$= x^r \left[r(2r-1)c_0 x^{-1} + \underbrace{\sum_{n=1}^{\infty}(n+r)(2n+2r-1)c_n x^{n-1}}_{k=n-1} + \underbrace{\sum_{n=0}^{\infty}(n+r+1)c_n x^n}_{k=n} \right]$$

$$= x^r \left[r(2r-1)c_0 x^{-1} + \sum_{k=0}^{\infty}[(k+r+1)(2k+2r+1)c_{k+1} + (k+r+1)c_k]x^k \right],$$

which implies

$$r(2r-1) = 0 \quad (15)$$

$$(k+r+1)(2k+2r+1)c_{k+1} + (k+r+1)c_k = 0, \quad k = 0, 1, 2, \dots. \quad (16)$$

From (15) we see that the indicial roots are $r_1 = \frac{1}{2}$ and $r_2 = 0$.

For $r_1 = \frac{1}{2}$, we can divide by $k + \frac{3}{2}$ in (16) to obtain

$$c_{k+1} = \frac{-c_k}{2(k+1)}, \quad k = 0, 1, 2, \dots, \quad (17)$$

whereas for $r_2 = 0$, (16) becomes

$$c_{k+1} = \frac{-c_k}{2k+1}, \quad k = 0, 1, 2, \dots. \quad (18)$$

From (17):	From (18):
$c_1 = \dfrac{-c_0}{2 \cdot 1}$	$c_1 = \dfrac{-c_0}{1}$
$c_2 = \dfrac{-c_1}{2 \cdot 2} = \dfrac{c_0}{2^2 \cdot 2!}$	$c_2 = \dfrac{-c_1}{3} = \dfrac{c_0}{1 \cdot 3}$
$c_3 = \dfrac{-c_2}{2 \cdot 3} = \dfrac{-c_0}{2^3 \cdot 3!}$	$c_3 = \dfrac{-c_2}{5} = \dfrac{-c_0}{1 \cdot 3 \cdot 5}$
$c_4 = \dfrac{-c_3}{2 \cdot 4} = \dfrac{c_0}{2^4 \cdot 4!}$	$c_4 = \dfrac{-c_3}{7} = \dfrac{c_0}{1 \cdot 3 \cdot 5 \cdot 7}$
$\vdots$	$\vdots$
$c_n = \dfrac{(-1)^n c_0}{2^n n!}.$	$c_n = \dfrac{(-1)^n c_0}{1 \cdot 3 \cdot 5 \cdot 7 \cdots (2n-1)}.$

Thus for the indicial root $r_1 = \frac{1}{2}$ we obtain the solution

$$y_1(x) = x^{1/2}\left[1 + \sum_{n=1}^{\infty} \frac{(-1)^n}{2^n n!} x^n\right] = \sum_{n=0}^{\infty} \frac{(-1)^n}{2^n n!} x^{n+1/2},$$

where we have again omitted c_0. The series converges for $x \geq 0$; as given, the series is not defined for negative values of x because of the presence of $x^{1/2}$. For $r_2 = 0$, a second solution is

$$y_2(x) = 1 + \sum_{n=1}^{\infty} \frac{(-1)^n}{1 \cdot 3 \cdot 5 \cdot 7 \cdots (2n-1)} x^n, \quad |x| < \infty.$$

On the interval $(0, \infty)$, the general solution is $y = C_1 y_1(x) + C_2 y_2(x)$. ≡

EXAMPLE 4　　**Only One Series Solution**

Solve　$xy'' + y = 0$.

SOLUTION　From $xP(x) = 0$ and $x^2 Q(x) = x$ and the fact that 0 and x are their own power series centered at 0, we conclude $a_0 = 0$ and $b_0 = 0$ and so from (14) the indicial equation is $r(r-1) = 0$. You should verify that the two recurrence relations corresponding to the indicial roots $r_1 = 1$ and $r_2 = 0$ yield exactly the same set of coefficients. In other words, in this case the method of Frobenius produces only a single series solution

$$y_1(x) = \sum_{n=0}^{\infty} \frac{(-1)^n}{n!(n+1)!} x^{n+1} = x - \frac{1}{2} x^2 + \frac{1}{12} x^3 - \frac{1}{144} x^4 + \cdots. \qquad ≡$$

■ **Three Cases**　For the sake of discussion, let us again suppose that $x = 0$ is a regular singular point of equation (1) and that indicial roots r_1 and r_2 of the singularity are real, with r_1 denoting the largest root. When using the method of Frobenius, we distinguish three cases corresponding to the nature of the indicial roots r_1 and r_2.

Case I:　If r_1 and r_2 are distinct and do not differ by an integer, there exist two linearly independent solutions of the form

$$y_1(x) = \sum_{n=0}^{\infty} c_n x^{n+r_1} \quad \text{and} \quad y_2(x) = \sum_{n=0}^{\infty} b_n x^{n+r_2}.$$

This is the case illustrated in Examples 2 and 3.

In the next case, we see that when the difference of indicial roots $r_1 - r_2$ is a positive integer, the second solution *may* contain a logarithm.

Case II: If $r_1 - r_2 = N$, where N is a positive integer, then there exist two linearly independent solutions of equation (1) of the form

$$y_1(x) = \sum_{n=0}^{\infty} c_n x^{n+r_1}, \quad c_0 \neq 0, \tag{19}$$

$$y_2(x) = C y_1(x) \ln x + \sum_{n=0}^{\infty} b_n x^{n+r_2}, \quad b_0 \neq 0. \tag{20}$$

where C is a constant that could be zero.

Finally, in the last case, the case when the indicial roots r_1 and r_2 are equal, a second solution will *always* contain a logarithm. The situation is analogous to the solution of a Cauchy–Euler equation when the roots of the auxiliary equation are equal.

Case III: If $r_1 = r_2$, then there always exist two linearly independent solutions of equation (1) of the form

$$y_1(x) = \sum_{n=0}^{\infty} c_n x^{n+r_1}, \quad c_0 \neq 0, \tag{21}$$

$$y_2(x) = y_1(x) \ln x + \sum_{n=0}^{\infty} b_n x^{n+r_2}. \tag{22}$$

▇ Finding a Second Solution

When the difference $r_1 - r_2$ is a positive integer (Case II), we *may* or *may not* be able to find two solutions having the form $y = \sum_{n=0}^{\infty} c_n x^{n+r}$. This is something we do not know in advance but is determined after we have found the indicial roots and have carefully examined the recurrence relation that defines the coefficients c_n. We just may be lucky enough to find two solutions that involve only powers of x; that is, $y_1(x) = \sum_{n=0}^{\infty} c_n x^{n+r_1}$ (equation (19)) and $y_2 = \sum_{n=0}^{\infty} b_n x^{n+r_2}$ (equation (20) with $C = 0$). See Problem 31 in Exercises 5.2. On the other hand, in Example 4 we see that the difference of indicial roots is a positive integer $(r_1 - r_2 = 1)$ and the method of Frobenius failed to give a second series solution. In this situation, equation (20), with $C \neq 0$, indicates what the second solution looks like. Finally, when the difference $r_1 - r_2$ is a zero (Case III), the method of Frobenius fails to give a second series solution; the second solution (22) always contains a logarithm and is actually (20) with $C = 1$. One way to obtain this second solution with the logarithmic term is to use the fact that

This is (5) in Section 3.2. ▶

$$y_2(x) = y_1(x) \int \frac{e^{-\int P(x)\,dx}}{y_1^2(x)} \, dx \tag{23}$$

is also a solution of $y'' + P(x)y' + Q(x)y = 0$ whenever $y_1(x)$ is the known solution. We will illustrate how to use (23) in the next example.

▇ **EXAMPLE 5** **Example 4 Revisited—Using a CAS**

Find the general solution of $xy'' + y = 0$.

SOLUTION From the known solution given in Example 4,

$$y_1(x) = x - \frac{1}{2}x^2 + \frac{1}{12}x^3 - \frac{1}{144}x^4 + \cdots,$$

we can construct a second solution $y_2(x)$ using formula (23). For those with the time, energy, and patience, the drudgery of squaring a series, long division, and integration of the quotient can be carried out by hand. But all these operations can be done with relative ease with the help of a CAS. We give the results:

Here is a good place to
use a computer algebra
system. ▶

$$y_2(x) = y_1(x) \int \frac{e^{-\int 0\,dx}}{[y_1(x)]^2}\,dx = y_1(x) \int \frac{dx}{\left[x - \frac{1}{2}x^2 + \frac{1}{12}x^3 - \frac{1}{144}x^4 + \cdots\right]^2}$$

$$= y_1(x) \int \frac{dx}{\left[x^2 - x^3 + \frac{5}{12}x^4 - \frac{7}{72}x^5 + \cdots\right]} \qquad \leftarrow \text{after squaring}$$

$$= y_1(x) \int \left[\frac{1}{x^2} + \frac{1}{x} + \frac{7}{12} + \frac{19}{72}x + \cdots\right]dx \qquad \leftarrow \text{after long division}$$

$$= y_1(x)\left[-\frac{1}{x} + \ln x + \frac{7}{12}x + \frac{19}{144}x^2 + \cdots\right] \qquad \leftarrow \text{after integrating}$$

$$= y_1(x)\ln x + y_1(x)\left[-\frac{1}{x} + \frac{7}{12}x + \frac{19}{144}x^2 + \cdots\right] \qquad \leftarrow \text{multiply out}$$

or

$$y_2(x) = y_1(x)\ln x + \left[-1 - \frac{1}{2}x + \frac{1}{2}x^2 + \cdots\right].$$

On the interval $(0, \infty)$, the general solution is $y = C_1 y_1(x) + C_2 y_2(x)$.

REMARKS

(*i*) The three different forms of a linear second-order differential equation in (1), (2), and (3) were used to discuss various theoretical concepts. But on a practical level, when it comes to actually solving a differential equation using the method of Frobenius, it is advisable to work with the form of the DE given in (1).

(*ii*) When the difference of indicial roots $r_1 - r_2$ is a positive integer ($r_1 > r_2$), it sometimes pays to iterate the recurrence relation using the smaller root r_2 first. See Problems 31 and 32 in Exercises 5.2.

(*iii*) Since an indicial root r is a root of a quadratic equation, it could be complex. We shall not, however, investigate this case.

(*iv*) If $x = 0$ is an irregular singular point, we may not be able to find *any* solution of the DE of form $y = \sum_{n=0}^{\infty} c_n x^{n+r}$.

5.2 Exercises Answers to selected odd-numbered problems begin on page ANS-13.

In Problems 1–10, determine the singular points of the given differential equation. Classify each singular point as regular or irregular.

1. $x^3 y'' + 4x^2 y' + 3y = 0$
2. $x(x + 3)^2 y'' - y = 0$
3. $(x^2 - 9)^2 y'' + (x + 3)y' + 2y = 0$
4. $y'' - \frac{1}{x}y' + \frac{1}{(x-1)^3}y = 0$
5. $(x^3 + 4x)y'' - 2xy' + 6y = 0$
6. $x^2(x - 5)^2 y'' + 4xy' + (x^2 - 25)y = 0$
7. $(x^2 + x - 6)y'' + (x + 3)y' + (x - 2)y = 0$
8. $x(x^2 + 1)^2 y'' + y = 0$
9. $x^3(x^2 - 25)(x - 2)^2 y'' + 3x(x - 2)y' + 7(x + 5)y = 0$
10. $(x^3 - 2x^2 - 3x)^2 y'' + x(x - 3)^2 y' - (x + 1)y = 0$

In Problems 11 and 12, put the given differential equation into the form (3) for each regular singular point of the equation. Identify the functions $p(x)$ and $q(x)$.

11. $(x^2 - 1)y'' + 5(x + 1)y' + (x^2 - x)y = 0$
12. $xy'' + (x + 3)y' + 7x^2 y = 0$

In Problems 13 and 14, $x = 0$ is a regular singular point of the given differential equation. Use the general form of the indicial equation in (14) to find the indicial roots of the singularity. Without solving, discuss the number of series solutions you would expect to find using the method of Frobenius.

13. $x^2 y'' + (\frac{5}{3}x + x^2)y' - \frac{1}{3}y = 0$
14. $xy'' + y' + 10y = 0$

In Problems 15–24, $x = 0$ is a regular singular point of the given differential equation. Show that the indicial roots of the singularity do not differ by an integer. Use the method of Frobenius to obtain two linearly independent series solutions about $x = 0$. Form the general solution on the interval $(0, \infty)$.

15. $2xy'' - y' + 2y = 0$
16. $2xy'' + 5y' + xy = 0$
17. $4xy'' + \frac{1}{2}y' + y = 0$

18. $2x^2y'' - xy' + (x^2 + 1)y = 0$

19. $3xy'' + (2 - x)y' - y = 0$

20. $x^2y'' - (x - \frac{2}{9})y = 0$

21. $2xy'' - (3 + 2x)y' + y = 0$

22. $x^2y'' + xy' + (x^2 - \frac{4}{9})y = 0$

23. $9x^2y'' + 9x^2y' + 2y = 0$

24. $2x^2y'' + 3xy' + (2x - 1)y = 0$

In Problems 25–30, $x = 0$ is a regular singular point of the given differential equation. Show that the indicial roots of the singularity differ by an integer. Use the method of Frobenius to obtain at least one series solution about $x = 0$. Use (21) where necessary and a CAS, if instructed, to find a second solution. Form the general solution on the interval $(0, \infty)$.

25. $xy'' + 2y' - xy = 0$

26. $x^2y'' + xy' + (x^2 - \frac{1}{4})y = 0$

27. $xy'' - xy' + y = 0$

28. $y'' + \dfrac{3}{x}y' - 2y = 0$

29. $xy'' + (1 - x)y' - y = 0$

30. $xy'' + y' + y = 0$

In Problems 31 and 32, $x = 0$ is a regular singular point of the given differential equation. Show that the indicial roots of the singularity differ by an integer. Use the recurrence relation found by the method of Frobenius first with the largest root r_1. How many solutions did you find? Next use the recurrence relation with the smaller root r_2. How many solutions did you find?

31. $xy'' + (x - 6)y' - 3y = 0$

32. $x(x - 1)y'' + 3y' - 2y = 0$

33. **(a)** The differential equation $x^4y'' + \lambda y = 0$ has an irregular singular point at $x = 0$. Show that the substitution $t = 1/x$ yields the differential equation

$$\frac{d^2y}{dt^2} + \frac{2}{t}\frac{dy}{dt} + \lambda y = 0,$$

which now has a regular singular point at $t = 0$.

(b) Use the method of this section to find two series solutions of the second equation in part (a) about the singular point $t = 0$.

(c) Express each series solution of the original equation in terms of elementary functions.

Mathematical Model

34. **Buckling of a Tapered Column** In Example 4 of Section 3.9, we saw that when a constant vertical compressive force or load P was applied to a thin column of uniform cross section, the deflection $y(x)$ satisfied the boundary-value problem

$$EI\frac{d^2y}{dx^2} + Py = 0, \quad y(0) = 0, \quad y(L) = 0.$$

The assumption here is that the column is hinged at both ends. The column will buckle or deflect only when the compressive force is a critical load P_n.

(a) In this problem let us assume that the column is of length L, is hinged at both ends, has circular cross sections, and is tapered as shown in **FIGURE 5.2.1(a)**. If the column, a truncated cone, has a linear taper $y = cx$ as shown in cross section in Figure 5.2.1(b), the moment of inertia of a cross section with respect to an axis perpendicular to the xy-plane is $I = \frac{1}{4}\pi r^4$, where $r = y$ and $y = cx$. Hence we can write $I(x) = I_0(x/b)^4$, where $I_0 = I(b) = \frac{1}{4}\pi(cb)^4$. Substituting $I(x)$ into the differential equation in (24), we see that the deflection in this case is determined from the BVP

$$x^4\frac{d^2y}{dx^2} + \lambda y = 0, \quad y(a) = 0, \quad y(b) = 0,$$

where $\lambda = Pb^4/EI_0$. Use the results of Problem 33 to find the critical loads P_n for the tapered column. Use an appropriate identity to express the buckling modes $y_n(x)$ as a single function.

(b) Use a CAS to plot the graph of the first buckling mode $y_1(x)$ corresponding to the Euler load P_1 when $b = 11$ and $a = 1$.

FIGURE 5.2.1 Tapered column in Problem 34

Discussion Problems

35. Discuss how you would define a regular singular point for the linear third-order differential equation

$$a_3(x)y''' + a_2(x)y'' + a_1(x)y' + a_0(x)y = 0.$$

36. Each of the differential equations

$$x^3y'' + y = 0 \quad \text{and} \quad x^2y'' + (3x - 1)y' + y = 0$$

has an irregular singular point at $x = 0$. Determine whether the method of Frobenius yields a series solution of each differential equation about $x = 0$. Discuss and explain your findings.

37. We have seen that $x = 0$ is a regular singular point of any Cauchy–Euler equation $ax^2y'' + bxy' + cy = 0$. Are the indicial equation (14) for a Cauchy–Euler equation and its auxiliary equation related? Discuss.

38. Use the method of Frobenius to show that the general solution of the differential equation $4xy'' + 2y' + y = 0$ on $(0, \infty)$ is

$$y = C_1 \sin\sqrt{x} + C_2 \cos\sqrt{x}.$$

5.3 Special Functions

INTRODUCTION The two differential equations

$$x^2 y'' + xy' + (x^2 - \nu^2)y = 0 \tag{1}$$

$$(1 - x^2)y'' - 2xy' + n(n + 1)y = 0 \tag{2}$$

FRIEDRICH WILHELM BESSEL
1784 Astronom und Mathematiker 1846
1984

© PjrStamps/Alamy Stock Photo
Friedrich Wilhelm Bessel

often arise in the advanced studies of applied mathematics, physics, and engineering. Equation (1) is called **Bessel's equation of order** ν and is named after the noted German astronomer and mathematician **Friedrich Wilhelm Bessel** (1784–1846), who was the first person to determine the accurate distance from the Sun to another star. Bessel first encountered a special form of equation (1) in his study of elliptical planetary motion and eventually carried out a systematic study of the properties of the solutions of the general equation. Differential equation (2) is known as **Legendre's equation of order** n and is named after the French mathematician **Adrien-Marie Legendre** (1752–1833). When we solve (1), we will assume that $\nu \geq 0$; whereas in (2), we will consider only the case where n is a nonnegative integer.

5.3.1 Bessel Functions

■ **The Solution** Because $x = 0$ is a regular singular point of Bessel's equation, we know that there exists at least one solution of the form $y = \sum_{n=0}^{\infty} c_n x^{n+r}$. Substituting the last expression into (1) then gives

$$x^2 y'' + xy' + (x^2 - \nu^2)y = \sum_{n=0}^{\infty} c_n (n + r)(n + r - 1)x^{n+r} + \sum_{n=0}^{\infty} c_n (n + r)x^{n+r} + \sum_{n=0}^{\infty} c_n x^{n+r+2} - \nu^2 \sum_{n=0}^{\infty} c_n x^{n+r}$$

$$= c_0(r^2 - r + r - \nu^2)x^r$$

$$+ x^r \sum_{n=1}^{\infty} c_n [(n + r)(n + r - 1) + (n + r) - \nu^2]x^n + x^r \sum_{n=0}^{\infty} c_n x^{n+2}$$

$$= c_0(r^2 - \nu^2)x^r + x^r \sum_{n=1}^{\infty} c_n [(n + r)^2 - \nu^2]x^n + x^r \sum_{n=0}^{\infty} c_n x^{n+2}. \tag{3}$$

From (3) we see that the indicial equation is $r^2 - \nu^2 = 0$ so that the indicial roots are $r_1 = \nu$ and $r_2 = -\nu$. When $r_1 = \nu$, (3) becomes

$$x^\nu \sum_{n=1}^{\infty} c_n n(n + 2\nu)x^n + x^\nu \sum_{n=0}^{\infty} c_n x^{n+2}$$

$$= x^\nu \left[(1 + 2\nu)c_1 x + \underbrace{\sum_{n=2}^{\infty} c_n n(n + 2\nu)x^n}_{k = n - 2} + \underbrace{\sum_{n=0}^{\infty} c_n x^{n+2}}_{k = n} \right]$$

$$= x^\nu \left[(1 + 2\nu)c_1 x + \sum_{k=0}^{\infty} [(k + 2)(k + 2 + 2\nu)c_{k+2} + c_k]x^{k+2} \right] = 0.$$

Therefore, by the usual argument we can write $(1 + 2\nu)c_1 = 0$ and

$$(k + 2)(k + 2 + 2\nu)c_{k+2} + c_k = 0$$

or

$$c_{k+2} = \frac{-c_k}{(k + 2)(k + 2 + 2\nu)}, \quad k = 0, 1, 2, \dots. \tag{4}$$

The choice $c_1 = 0$ in (4) implies $c_3 = c_5 = c_7 = \cdots = 0$, so for $k = 0, 2, 4, \dots$ we find, after letting $k + 2 = 2n$, $n = 1, 2, 3, \dots$, that

$$c_{2n} = -\frac{c_{2n-2}}{2^2 n(n + \nu)}. \tag{5}$$

Thus

$$c_2 = -\frac{c_0}{2^2 \cdot 1 \cdot (1 + \nu)}$$

$$c_4 = -\frac{c_2}{2^2 \cdot 2(2 + \nu)} = \frac{c_0}{2^4 \cdot 1 \cdot 2(1 + \nu)(2 + \nu)}$$

$$c_6 = -\frac{c_4}{2^2 \cdot 3(3 + \nu)} = -\frac{c_0}{2^6 \cdot 1 \cdot 2 \cdot 3(1 + \nu)(2 + \nu)(3 + \nu)}$$

$$\vdots$$

$$c_{2n} = \frac{(-1)^n c_0}{2^{2n} n!(1 + \nu)(2 + \nu) \cdots (n + \nu)}, \quad n = 1, 2, 3, \ldots. \tag{6}$$

It is standard practice to choose c_0 to be a specific value—namely,

$$c_0 = \frac{1}{2^\nu \Gamma(1 + \nu)},$$

where $\Gamma(1 + \nu)$ is the gamma function. See Appendix A. Since this latter function possesses the convenient property $\Gamma(1 + \alpha) = \alpha\Gamma(\alpha)$, we can reduce the indicated product in the denominator of (6) to one term. For example,

$$\Gamma(1 + \nu + 1) = (1 + \nu)\Gamma(1 + \nu)$$

$$\Gamma(1 + \nu + 2) = (2 + \nu)\Gamma(2 + \nu) = (2 + \nu)(1 + \nu)\Gamma(1 + \nu).$$

Hence we can write (6) as

$$c_{2n} = \frac{(-1)^n}{2^{2n+\nu} n!(1 + \nu)(2 + \nu) \cdots (n + \nu)\Gamma(1 + \nu)} = \frac{(-1)^n}{2^{2n+\nu} n!\Gamma(1 + \nu + n)}$$

for $n = 0, 1, 2, \ldots.$

▨ Bessel Functions of the First Kind

The series solution $y = \sum_{n=0}^{\infty} c_{2n} x^{2n+\nu}$ is usually denoted by $J_\nu(x)$:

$$J_\nu(x) = \sum_{n=0}^{\infty} \frac{(-1)^n}{n!\Gamma(1 + \nu + n)} \left(\frac{x}{2}\right)^{2n+\nu}. \tag{7}$$

If $\nu \geq 0$, the series converges at least on the interval $[0, \infty)$. Also, for the second exponent $r_2 = -\nu$ we obtain, in exactly the same manner,

$$J_{-\nu}(x) = \sum_{n=0}^{\infty} \frac{(-1)^n}{n!\Gamma(1 - \nu + n)} \left(\frac{x}{2}\right)^{2n-\nu}. \tag{8}$$

The functions $J_\nu(x)$ and $J_{-\nu}(x)$ are called **Bessel functions of the first kind** of order ν and $-\nu$, respectively. Depending on the value of ν, (8) may contain negative powers of x and hence converge on the interval $(0, \infty)$.*

Now some care must be taken in writing the general solution of (1). When $\nu = 0$, it is apparent that (7) and (8) are the same. If $\nu > 0$ and $r_1 - r_2 = \nu - (-\nu) = 2\nu$ is not a positive integer, it follows from Case I of Section 5.2 that $J_\nu(x)$ and $J_{-\nu}(x)$ are linearly independent solutions of (1) on $(0, \infty)$, and so the general solution on the interval is $y = c_1 J_\nu(x) + c_2 J_{-\nu}(x)$. But we also know from Case II of Section 5.2 that when $r_1 - r_2 = 2\nu$ is a positive integer, a second series solution of (1) *may* exist. In this second case we distinguish two possibilities. When $\nu = m = $ positive integer, $J_{-m}(x)$ defined by (8) and $J_m(x)$ are not linearly independent solutions. It can be shown that J_{-m} is a constant multiple of J_m (see Property (*i*) on page 295). In addition, $r_1 - r_2 = 2\nu$ can be a positive integer when ν is half an odd positive integer. It can be shown in this latter event that $J_\nu(x)$ and $J_{-\nu}(x)$ are linearly independent. In other words, the general solution of (1) on $(0, \infty)$ is

$$y = c_1 J_\nu(x) + c_2 J_{-\nu}(x), \quad \nu \neq \text{integer}. \tag{9}$$

The graphs of $y = J_0(x)$ (blue) and $y = J_1(x)$ (red) are given in **FIGURE 5.3.1**.

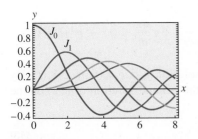

FIGURE 5.3.1 Bessel functions of the first kind for $n = 0, 1, 2, 3, 4$

*When we replace x by $|x|$, the series given in (7) and (8) converge for $0 < |x| < \infty$.

EXAMPLE 1 General Solution: ν Not an Integer

By identifying $\nu^2 = \frac{1}{4}$ and $\nu = \frac{1}{2}$ we can see from (9) that the general solution of the equation $x^2 y'' + xy' + (x^2 - \frac{1}{4})y = 0$ on $(0, \infty)$ is $y = c_1 J_{1/2}(x) + c_2 J_{-1/2}(x)$. ≡

Bessel Functions of the Second Kind

If $\nu \neq$ integer, the function defined by the linear combination

$$Y_\nu(x) = \frac{\cos \nu\pi J_\nu(x) - J_{-\nu}(x)}{\sin \nu\pi} \tag{10}$$

and the function $J_\nu(x)$ are linearly independent solutions of (1). Thus another form of the general solution of (1) is $y = c_1 J_\nu(x) + c_2 Y_\nu(x)$, provided $\nu \neq$ integer. As $\nu \to m$, m an integer, (10) has the indeterminate form 0/0. However, it can be shown by L'Hôpital's rule that $\lim_{\nu \to m} Y_\nu(x)$ exists. Moreover, the function

$$Y_m(x) = \lim_{\nu \to m} Y_\nu(x)$$

and $J_m(x)$ are linearly independent solutions of $x^2 y'' + xy' + (x^2 - m^2)y = 0$. Hence for *any* value of ν the general solution of (1) on the interval $(0, \infty)$ can be written as

$$y = c_1 J_\nu(x) + c_2 Y_\nu(x). \tag{11}$$

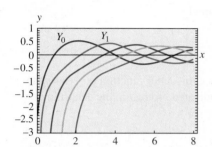

FIGURE 5.3.2 Bessel functions of the second kind for $n = 0, 1, 2, 3, 4$

$Y_\nu(x)$ is called the **Bessel function of the second kind** of order ν. **FIGURE 5.3.2** shows the graphs of $Y_0(x)$ (blue) and $Y_1(x)$ (red). Because the Bessel function of the second kind of order ν was first introduced in 1867 by the German mathematician **Carl Gottfried Neumann** (1832–1925), it is also referred to as the **Neumann function** of order ν. The Neumann function is often denoted by N_ν.

EXAMPLE 2 General Solution: ν an Integer

By identifying $\nu^2 = 9$ and $\nu = 3$ we see from (11) that the general solution of the equation $x^2 y'' + xy' + (x^2 - 9)y = 0$ on $(0, \infty)$ is $y = c_1 J_3(x) + c_2 Y_3(x)$. ≡

DEs Solvable in Terms of Bessel Functions

Sometimes it is possible to transform a differential equation into equation (1) by means of a change of variable. We can then express the solution of the original equation in terms of Bessel functions. For example, if we let $t = \alpha x$, $\alpha > 0$, in

$$x^2 y'' + xy' + (\alpha^2 x^2 - \nu^2)y = 0, \tag{12}$$

then by the Chain Rule,

$$\frac{dy}{dx} = \frac{dy}{dt}\frac{dt}{dx} = \alpha \frac{dy}{dt} \quad \text{and} \quad \frac{d^2 y}{dx^2} = \frac{d}{dt}\left(\frac{dy}{dx}\right)\frac{dt}{dx} = \alpha^2 \frac{d^2 y}{dt^2}.$$

Accordingly (12) becomes

$$\left(\frac{t}{\alpha}\right)^2 \alpha^2 \frac{d^2 y}{dt^2} + \left(\frac{t}{\alpha}\right)\alpha \frac{dy}{dt} + (t^2 - \nu^2)y = 0 \quad \text{or} \quad t^2 \frac{d^2 y}{dt^2} + t \frac{dy}{dt} + (t^2 - \nu^2)y = 0.$$

The last equation is Bessel's equation of order ν with solution $y = c_1 J_\nu(t) + c_2 Y_\nu(t)$. By resubstituting $t = \alpha x$ in the last expression we find that the general solution of (12) on the interval $(0, \infty)$ is

$$y = c_1 J_\nu(\alpha x) + c_2 Y_\nu(\alpha x). \tag{13}$$

Equation (12), called the **parametric Bessel equation of order** ν, and its general solution (13) are very important in the study of certain boundary-value problems involving partial differential equations that are expressed in cylindrical coordinates.

■ Modified Bessel Functions

Another equation that bears a resemblance to (1) is the **modified Bessel equation of order** ν,

$$x^2 y'' + xy' - (x^2 + \nu^2)y = 0. \tag{14}$$

This DE can be solved in the manner just illustrated for (12). This time if we let $t = ix$, where $i^2 = -1$, then (14) becomes

$$t^2 \frac{d^2 y}{dt^2} + t \frac{dy}{dt} + (t^2 - \nu^2)y = 0.$$

Since solutions of the last DE are $J_\nu(t)$ and $Y_\nu(t)$, *complex-valued* solutions of equation (14) are $J_\nu(ix)$ and $Y_\nu(ix)$. A real-valued solution, called the **modified Bessel function of the first kind** of order ν, is defined in terms of $J_\nu(ix)$:

$$I_\nu(x) = i^{-\nu} J_\nu(ix). \tag{15}$$

See Problem 21 in Exercises 5.3. The general solution of (14) is

$$y = c_1 I_\nu(x) + c_2 I_{-\nu}(x), \quad \nu \neq \text{integer.} \tag{15}$$

When ν is an integer n the functions $I_n(x)$ and $I_{-n}(x)$ are not linearly independent on the interval $(0, \infty)$. So, analogous to (10), we define the **modified Bessel function of the second kind** of order $\nu \neq$ integer to be

$$K_\nu(x) = \frac{\pi}{2} \frac{I_{-\nu}(x) - I_\nu(x)}{\sin \nu \pi}, \tag{16}$$

and for integral $\nu = n$,

$$K_n(x) = \lim_{\nu \to n} K_\nu(x).$$

Because I_ν and K_ν are linearly independent on the interval $(0, \infty)$ for any value of ν, the general solution of (14) is

$$y = c_1 I_\nu(x) + c_2 K_\nu(x). \tag{17}$$

FIGURE 5.3.3 Modified Bessel function of the first kind for $n = 0, 1, 2, 3, 4$

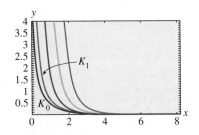

FIGURE 5.3.4 Modified Bessel function of the second kind for $n = 0, 1, 2, 3, 4$

The graphs of $I_0(x)$ (blue) and $I_1(x)$ (red) are given in **FIGURE 5.3.3** and the graphs $K_0(x)$ (blue) and $K_1(x)$ (red) are shown in **FIGURE 5.3.4**. Unlike the Bessel functions of the first and second kinds, the graphs of the modified Bessel functions of the first kind and second kind are not oscillatory. Moreover, the graphs in Figures 5.3.3 and 5.3.4 illustrate the fact that the modified Bessel functions $I_n(x)$ and $K_n(x)$, $n = 0, 1, 2, \ldots$ have no real zeros in the interval $(0, \infty)$. Also, note that $K_n(x) \to \infty$ as $x \to 0^+$.

Proceeding as we did in (12) and (13), we see that the general solution of the **parametric form of the modified Bessel equation of order** ν

$$x^2 y'' + xy' - (\alpha^2 x^2 + \nu^2)y = 0 \tag{18}$$

on the interval $(0, \infty)$ is

$$y = c_1 I_\nu(\alpha x) + c_2 K_\nu(\alpha x). \tag{19}$$

EXAMPLE 3 Parametric Modified Bessel Equation

By identifying $\alpha^2 = 25$, $\nu^2 = 4$, $\alpha = 5$, and $\nu = 2$ it follows from (19) that the general solution of the equation $x^2 y'' + xy' - (25x^2 + 4)y = 0$ on $(0, \infty)$ is $y = c_1 I_2(5x) + c_2 K_2(5x)$. ☰

Yet another equation, important because many differential equations fit into its form by appropriate choices of the parameters, is

$$y'' + \frac{1 - 2a}{x} y' + \left(b^2 c^2 x^{2c-2} + \frac{a^2 - p^2 c^2}{x^2} \right) y = 0, \quad p \geq 0. \tag{20}$$

Although we shall not supply the details, the general solution of (20),

$$y = x^a \left[c_1 J_p(bx^c) + c_2 Y_p(bx^c) \right], \tag{21}$$

can be found by means of a change in both the independent and the dependent variables: $z = bx^c$, $y(x) = \left(\dfrac{z}{b}\right)^{a/c} w(z)$. If p is not an integer, then Y_p in (21) can be replaced by J_{-p}.

EXAMPLE 4 **Using (20)**

Find the general solution of $xy'' + 3y' + 9y = 0$ on $(0, \infty)$.

SOLUTION By writing the given DE as

$$y'' + \frac{3}{x}y' + \frac{9}{x}y = 0$$

we can make the following identifications with (20):

$$1 - 2a = 3, \quad b^2c^2 = 9, \quad 2c - 2 = -1, \quad \text{and} \quad a^2 - p^2c^2 = 0.$$

The first and third equations imply $a = -1$ and $c = \frac{1}{2}$. With these values the second and fourth equations are satisfied by taking $b = 6$ and $p = 2$. From (21) we find that the general solution of the given DE on the interval $(0, \infty)$ is $y = x^{-1}\left[c_1 J_2(6x^{1/2}) + c_2 Y_2(6x^{1/2})\right]$. $\equiv$

■ **Aging Spring and Bessel Functions** In Section 3.8 we saw that a mathematical model for the free undamped motion of a mass m on an aging spring is given by $mx'' + ke^{-\alpha t}x = 0$, $\alpha > 0$. We are now in a position to find the general solution of this differential equation.

EXAMPLE 5 **Aging Spring Revisited**

Use the substitution $s = \dfrac{2}{\alpha}\sqrt{\dfrac{k}{m}}e^{-\alpha t/2}$ to solve the differential equation $mx'' + ke^{-\alpha t}x = 0$.

SOLUTION By the Chain Rule,

$$\frac{dx}{dt} = \frac{dx}{ds}\frac{ds}{dt} = \frac{dx}{ds}\left[\frac{2}{\alpha}\sqrt{\frac{k}{m}}\left(-\frac{\alpha}{2}\right)e^{-\alpha t/2}\right] = \frac{dx}{ds}\left(-\sqrt{\frac{k}{m}}e^{-\alpha t/2}\right)$$

$$\frac{d^2x}{dt^2} = \frac{d}{dt}\left(\frac{dx}{dt}\right) = \frac{dx}{ds}\left(\frac{\alpha}{2}\sqrt{\frac{k}{m}}e^{-\alpha t/2}\right) + \frac{d}{dt}\left(\frac{dx}{ds}\right)\left(-\sqrt{\frac{k}{m}}e^{-\alpha t/2}\right)$$

$$= \frac{dx}{ds}\left(\frac{\alpha}{2}\sqrt{\frac{k}{m}}e^{-\alpha t/2}\right) + \frac{d^2x}{ds^2}\frac{ds}{dt}\left(-\sqrt{\frac{k}{m}}e^{-\alpha t/2}\right)$$

$$= \left(\frac{k}{m}e^{-\alpha t}\right)\frac{d^2x}{ds^2} + \left(\frac{\alpha}{2}\sqrt{\frac{k}{m}}e^{-\alpha t/2}\right)\frac{dx}{ds}.$$

Thus $\quad m\dfrac{d^2x}{dt^2} + ke^{-\alpha t}x = ke^{-\alpha t}\dfrac{d^2x}{ds^2} + \dfrac{\alpha}{2}\sqrt{mk}\,e^{-\alpha t/2}\dfrac{dx}{ds} + ke^{-\alpha t}x = 0.$

Multiplying the last equation by $2^2/\alpha^2 m$ then gives

$$\left(\frac{2^2}{\alpha^2}\frac{k}{m}e^{-\alpha t}\right)\frac{d^2x}{ds^2} + \left(\frac{2}{\alpha}\sqrt{\frac{k}{m}}e^{-\alpha t/2}\right)\frac{dx}{ds} + \left(\frac{2^2}{\alpha^2}\frac{k}{m}e^{-\alpha t}\right)x = 0$$

or $\qquad\qquad\qquad s^2\dfrac{d^2x}{ds^2} + s\dfrac{dx}{ds} + s^2x = 0.$

The last equation is recognized as (1) with $\nu = 0$ and where the symbols x and s play the roles of y and x, respectively. Hence from (11), the general solution of the last differential equation is $x = c_1 J_0(s) + c_2 Y_0(s)$. Finally, if we substitute s, then the general solution of $mx'' + ke^{-\alpha t}x = 0$ is seen to be

$$x(t) = c_1 J_0\left(\frac{2}{\alpha}\sqrt{\frac{k}{m}}e^{-\alpha t/2}\right) + c_2 Y_0\left(\frac{2}{\alpha}\sqrt{\frac{k}{m}}e^{-\alpha t/2}\right).$$

$\equiv$

The other model discussed in Section 5.1 of a spring whose characteristics change with time was $mx'' + ktx = 0$. By dividing through by m we see that the equation $x'' + (k/m)tx = 0$ is Airy's equation, $y'' + \alpha^2 xy = 0$. See the discussion following Example 2 in Section 5.1. The general solution of Airy's differential equation can also be written in terms of Bessel functions. See Problems 34, 35, and 46 in Exercises 5.3.

■ **Properties** We list below a few of the more useful properties of Bessel functions of the first and second kinds of order m, $m = 0, 1, 2, \ldots$:

$$(i) \quad J_{-m}(x) = (-1)^m J_m(x) \qquad\qquad (ii) \quad J_m(-x) = (-1)^m J_m(x)$$

$$(iii) \quad J_m(0) = \begin{cases} 0, & m > 0 \\ 1, & m = 0 \end{cases} \qquad (iv) \quad \lim_{x \to 0^+} Y_m(x) = -\infty.$$

Note that Property (ii) indicates that $J_m(x)$ is an even function if m is an even integer and an odd function if m is an odd integer. The graphs of $Y_0(x)$ and $Y_1(x)$ in Figure 5.3.2 illustrate Property (iv): $Y_m(x)$ is unbounded at the origin. This last fact is not obvious from (10). The solutions of the Bessel equation of order 0 can be obtained using the solutions $y_1(x)$ in (21) and $y_2(x)$ in (22) of Section 5.2. It can be shown that (21) of Section 5.2 is $y_1(x) = J_0(x)$, whereas (22) of that section is

$$y_2(x) = J_0(x) \ln x - \sum_{k=1}^{\infty} \frac{(-1)^k}{(k!)^2}\left(1 + \frac{1}{2} + \cdots + \frac{1}{k}\right)\left(\frac{x}{2}\right)^{2k}.$$

The Bessel function of the second kind of order 0, $Y_0(x)$, is then defined to be the linear combination $Y_0(x) = \frac{2}{\pi}(\gamma - \ln 2)\, y_1(x) + \frac{2}{\pi}\, y_2(x)$ for $x > 0$. That is,

$$Y_0(x) = \frac{2}{\pi} J_0(x)\left[\gamma + \ln\frac{x}{2}\right] - \frac{2}{\pi}\sum_{k=1}^{\infty}\frac{(-1)^k}{(k!)^2}\left(1 + \frac{1}{2} + \cdots + \frac{1}{k}\right)\left(\frac{x}{2}\right)^{2k},$$

where $\gamma = 0.57721566\ldots$ is **Euler's constant**. Because of the presence of the logarithmic term, it is apparent that $Y_0(x)$ is discontinuous at $x = 0$.

■ **Numerical Values** The first five nonnegative zeros of $J_0(x)$, $J_1(x)$, $Y_0(x)$, and $Y_1(x)$ are given in Table 5.3.1. Some additional functional values of these four functions are given in Table 5.3.2.

TABLE 5.3.1 Zeros of J_0, J_1, Y_0, and Y_1

$J_0(x)$	$J_1(x)$	$Y_0(x)$	$Y_1(x)$
2.4048	0.0000	0.8936	2.1971
5.5201	3.8317	3.9577	5.4297
8.6537	7.0156	7.0861	8.5960
11.7915	10.1735	10.2223	11.7492
14.9309	13.3237	13.3611	14.8974

TABLE 5.3.2 Numerical Values of J_0, J_1, Y_0, and Y_1

x	$J_0(x)$	$J_1(x)$	$Y_0(x)$	$Y_1(x)$
0	1.0000	0.0000	—	—
1	0.7652	0.4401	0.0883	−0.7812
2	0.2239	0.5767	0.5104	−0.1070
3	−0.2601	0.3391	0.3769	0.3247
4	−0.3971	−0.0660	−0.0169	0.3979
5	−0.1776	−0.3276	−0.3085	0.1479
6	0.1506	−0.2767	−0.2882	−0.1750
7	0.3001	−0.0047	−0.0259	−0.3027
8	0.1717	0.2346	0.2235	−0.1581
9	−0.0903	0.2453	0.2499	0.1043
10	−0.2459	0.0435	0.0557	0.2490
11	−0.1712	−0.1768	−0.1688	0.1637
12	0.0477	−0.2234	−0.2252	−0.0571
13	0.2069	−0.0703	−0.0782	−0.2101
14	0.1711	0.1334	0.1272	−0.1666
15	−0.0142	0.2051	0.2055	0.0211

■ Differential Recurrence Relation

Recurrence formulas that relate Bessel functions of different orders are important in theory and in applications. In the next example we derive a **differential recurrence relation**.

EXAMPLE 6 **Derivation Using the Series Definition**

Derive the formula $xJ_\nu'(x) = \nu J_\nu(x) - xJ_{\nu+1}(x)$.

SOLUTION It follows from (7) that

$$xJ_\nu'(x) = \sum_{n=0}^{\infty} \frac{(-1)^n(2n + \nu)}{n!\,\Gamma(1 + \nu + n)} \left(\frac{x}{2}\right)^{2n+\nu}$$

$$= \nu \sum_{n=0}^{\infty} \frac{(-1)^n}{n!\,\Gamma(1 + \nu + n)} \left(\frac{x}{2}\right)^{2n+\nu} + 2\sum_{n=0}^{\infty} \frac{(-1)^n n}{n!\,\Gamma(1 + \nu + n)} \left(\frac{x}{2}\right)^{2n+\nu}$$

$$= \nu J_\nu(x) + x \underbrace{\sum_{n=1}^{\infty} \frac{(-1)^n}{(n-1)!\,\Gamma(1 + \nu + n)} \left(\frac{x}{2}\right)^{2n+\nu-1}}_{k = n - 1}$$

$$= \nu J_\nu(x) - x \sum_{k=0}^{\infty} \frac{(-1)^k}{k!\,\Gamma(2 + \nu + k)} \left(\frac{x}{2}\right)^{2k+\nu+1} = \nu J_\nu(x) - xJ_{\nu+1}(x). \qquad \equiv$$

The result in Example 6 can be written in an alternative form. Dividing $xJ_\nu'(x) - \nu J_\nu(x) = -xJ_{\nu+1}(x)$ by x gives

$$J_\nu'(x) - \frac{\nu}{x} J_\nu(x) = -J_{\nu+1}(x).$$

This last expression is recognized as a linear first-order differential equation in $J_\nu(x)$. Multiplying both sides of the equality by the integrating factor $x^{-\nu}$ then yields

$$\frac{d}{dx} [x^{-\nu}J_\nu(x)] = -x^{-\nu}J_{\nu+1}(x). \qquad (22)$$

It can be shown in a similar manner that

$$\frac{d}{dx} [x^\nu J_\nu(x)] = x^\nu J_{\nu-1}(x). \qquad (23)$$

See Problem 27 in Exercises 5.3. The differential recurrence relations (22) and (23) are also valid for the Bessel function of the second kind $Y_\nu(x)$. Observe that when $\nu = 0$ it follows from (22) that

$$\frac{d}{dx} J_0(x) = -J_1(x) \quad \text{and} \quad \frac{d}{dx} Y_0(x) = -Y_1(x). \qquad (24)$$

Results similar to (24) also hold for the modified Bessel functions of the first and second kind of order $\nu = 0$:

$$\frac{d}{dx} I_0(x) = I_1(x) \quad \text{and} \quad \frac{d}{dx} K_0(x) = -K_1(x). \qquad (25)$$

See Problems 43 and 45 in Exercises 5.3 for applications of these derivatives.

■ Bessel Functions of Half-Integral Order

When the order ν is half an odd integer, that is, $\pm\frac{1}{2}, \pm\frac{3}{2}, \pm\frac{5}{2}, \ldots$, Bessel functions of the first and second kinds can be expressed in terms of the elementary functions $\sin x$, $\cos x$, and powers of x. To see this let's consider the case when $\nu = \frac{1}{2}$. From (7) we have

$$J_{1/2}(x) = \sum_{n=0}^{\infty} \frac{(-1)^n}{n!\,\Gamma(1 + \frac{1}{2} + n)} \left(\frac{x}{2}\right)^{2n+1/2}.$$

See Appendix A. ▶ In view of the properties the gamma function, $\Gamma(1 + \alpha) = \alpha\Gamma(\alpha)$ and the fact that $\Gamma(\frac{1}{2}) = \sqrt{\pi}$ the values of $\Gamma(1 + \frac{1}{2} + n)$ for $n = 0$, $n = 1$, $n = 2$, and $n = 3$ are, respectively,

$$\Gamma(\tfrac{3}{2}) = \Gamma(1 + \tfrac{1}{2}) = \tfrac{1}{2}\Gamma(\tfrac{1}{2}) = \frac{1}{2}\sqrt{\pi}$$

$$\Gamma(\tfrac{5}{2}) = \Gamma(1 + \tfrac{3}{2}) = \tfrac{3}{2}\Gamma(\tfrac{3}{2}) = \frac{3}{2^2}\sqrt{\pi}$$

$$\Gamma(\tfrac{7}{2}) = \Gamma(1 + \tfrac{5}{2}) = \tfrac{5}{2}\Gamma(\tfrac{5}{2}) = \frac{5 \cdot 3}{2^3}\sqrt{\pi} = \frac{5 \cdot 4 \cdot 3 \cdot 2 \cdot 1}{2^3 4 \cdot 2}\sqrt{\pi} = \frac{5!}{2^5 2!}\sqrt{\pi}$$

$$\Gamma(\tfrac{9}{2}) = \Gamma(1 + \tfrac{7}{2}) = \tfrac{7}{2}\Gamma(\tfrac{7}{2}) = \frac{7 \cdot 5!}{2^6 2!}\sqrt{\pi} = \frac{7 \cdot 6 \cdot 5!}{2^6 \cdot 6 \cdot 2!}\sqrt{\pi} = \frac{7!}{2^7 3!}\sqrt{\pi}.$$

In general,
$$\Gamma(1 + \tfrac{1}{2} + n) = \frac{(2n + 1)!}{2^{2n+1}n!}\sqrt{\pi}.$$

Hence,
$$J_{1/2}(x) = \sum_{n=0}^{\infty} \frac{(-1)^n}{n!\dfrac{(2n + 1)!}{2^{2n+1}n!}\sqrt{\pi}} \left(\frac{x}{2}\right)^{2n+1/2} = \sqrt{\frac{2}{\pi x}} \sum_{n=0}^{\infty} \frac{(-1)^n}{(2n + 1)!} x^{2n+1}.$$

The infinite series in the last line is the Maclaurin series for $\sin x$, and so we have shown that

$$J_{1/2}(x) = \sqrt{\frac{2}{\pi x}} \sin x. \tag{26}$$

We leave it as an exercise to show that

$$J_{-1/2}(x) = \sqrt{\frac{2}{\pi x}} \cos x. \tag{27}$$

See Problems 31 and 32 in Exercises 5.3.

If n is an integer, then the order $\nu = n + \frac{1}{2}$ is half an odd integer. Because $\cos(n + \frac{1}{2})\pi = 0$ and $\sin(n + \frac{1}{2})\pi = \cos n\pi = (-1)^n$, we see from (10) that

$$Y_{n+1/2}(x) = (-1)^{n+1}J_{-(n+1/2)}(x). \tag{28}$$

For $n = 0$ and $n = -1$ in the last formula, we get, in turn, $Y_{1/2}(x) = -J_{-1/2}(x)$ and $Y_{-1/2}(x) = J_{1/2}(x)$. In view of (26) and (27) these results are the same as

$$Y_{1/2}(x) = -\sqrt{\frac{2}{\pi x}} \cos x \tag{29}$$

and

$$Y_{-1/2}(x) = \sqrt{\frac{2}{\pi x}} \sin x. \tag{30}$$

■ Spherical Bessel Functions

Bessel functions of half-integral order are used to define two more important functions:

$$j_n(x) = \sqrt{\frac{\pi}{2x}} J_{n+1/2}(x) \quad \text{and} \quad y_n(x) = \sqrt{\frac{\pi}{2x}} Y_{n+1/2}(x). \tag{31}$$

The function $j_n(x)$ is called the **spherical Bessel function of the first kind** and $y_n(x)$ is the **spherical Bessel function of the second kind**. For example, by using (26) and (29) we see that for $n = 0$ the expressions in (31) become

$$j_0(x) = \sqrt{\frac{\pi}{2x}} J_{1/2}(x) = \sqrt{\frac{\pi}{2x}} \sqrt{\frac{2}{\pi x}} \sin x = \frac{\sin x}{x}$$

and

$$y_0(x) = \sqrt{\frac{\pi}{2x}} Y_{1/2}(x) = -\sqrt{\frac{\pi}{2x}} \sqrt{\frac{2}{\pi x}} \cos x = -\frac{\cos x}{x}.$$

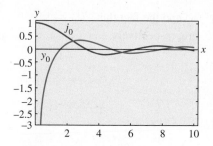

FIGURE 5.3.5 Spherical Bessel functions $j_0(x)$ and $y_0(x)$

The graphs of $j_n(x)$ and $y_n(x)$ for $n \geq 0$ are very similar to those given in Figures 5.3.1 and 5.3.2; that is, both functions are oscillatory, and $y_n(x)$ becomes unbounded as $x \to 0^+$. The graphs of $j_0(x)$ (blue) and $y_0(x)$ (red) are given in **FIGURE 5.3.5**. See Problems 39 and 40 in Exercises 5.3.

Spherical Bessel functions arise in the solution of a special partial differential equation expressed in spherical coordinates. See Problems 41 and 42 in Exercises 5.3 and Problem 14 in Exercises 14.3.

5.3.2 Legendre Functions

■ **The Solution** Since $x = 0$ is an ordinary point of Legendre's equation (2), we substitute the series $y = \sum_{k=0}^{\infty} c_k x^k$, shift summation indices, and combine series to get

$$(1 - x^2)y'' - 2xy' + n(n + 1)y = [n(n + 1)c_0 + 2c_2] + [(n - 1)(n + 2)c_1 + 6c_3]x$$

$$+ \sum_{j=2}^{\infty} [(j + 2)(j + 1)c_{j+2} + (n - j)(n + j + 1)c_j]x^j = 0,$$

which implies that

$$n(n + 1)c_0 + 2c_2 = 0$$

$$(n - 1)(n + 2)c_1 + 6c_3 = 0$$

$$(j + 2)(j + 1)c_{j+2} + (n - j)(n + j + 1)c_j = 0$$

or

$$c_2 = -\frac{n(n + 1)}{2!}c_0$$

$$c_3 = -\frac{(n - 1)(n + 2)}{3!}c_1$$

$$c_{j+2} = -\frac{(n - j)(n + j + 1)}{(j + 2)(j + 1)}c_j, \quad j = 2, 3, 4, \ldots. \tag{32}$$

Letting j take on the values 2, 3, 4, $\ldots$, recurrence relation (32) yields

$$c_4 = -\frac{(n - 2)(n + 3)}{4 \cdot 3}c_2 = \frac{(n - 2)n(n + 1)(n + 3)}{4!}c_0$$

$$c_5 = -\frac{(n - 3)(n + 4)}{5 \cdot 4}c_3 = \frac{(n - 3)(n - 1)(n + 2)(n + 4)}{5!}c_1$$

$$c_6 = -\frac{(n - 4)(n + 5)}{6 \cdot 5}c_4 = -\frac{(n - 4)(n - 2)n(n + 1)(n + 3)(n + 5)}{6!}c_0$$

$$c_7 = -\frac{(n - 5)(n + 6)}{7 \cdot 6}c_5 = -\frac{(n - 5)(n - 3)(n - 1)(n + 2)(n + 4)(n + 6)}{7!}c_1$$

and so on. Thus for at least $|x| < 1$ we obtain two linearly independent power series solutions:

$$y_1(x) = c_0 \left[1 - \frac{n(n + 1)}{2!}x^2 + \frac{(n - 2)n(n + 1)(n + 3)}{4!}x^4 \right.$$

$$\left. - \frac{(n - 4)(n - 2)n(n + 1)(n + 3)(n + 5)}{6!}x^6 + \cdots \right]$$

$$y_2(x) = c_1 \left[x - \frac{(n - 1)(n + 2)}{3!}x^3 + \frac{(n - 3)(n - 1)(n + 2)(n + 4)}{5!}x^5 \right.$$

$$\left. - \frac{(n - 5)(n - 3)(n - 1)(n + 2)(n + 4)(n + 6)}{7!}x^7 + \cdots \right]. \tag{33}$$

Notice that if n is an even integer, the first series terminates, whereas $y_2(x)$ is an infinite series. For example, if $n = 4$, then

$$y_1(x) = c_0\left[1 - \frac{4 \cdot 5}{2!}x^2 + \frac{2 \cdot 4 \cdot 5 \cdot 7}{4!}x^4\right] = c_0\left[1 - 10x^2 + \frac{35}{3}x^4\right].$$

Similarly, when n is an odd integer, the series for $y_2(x)$ terminates with x^n; that is, *when n is a nonnegative integer, we obtain an nth-degree polynomial solution* of Legendre's equation.

Since we know that a constant multiple of a solution of Legendre's equation is also a solution, it is traditional to choose specific values for c_0 or c_1, depending on whether n is an even or odd positive integer, respectively. For $n = 0$ we choose $c_0 = 1$, and for $n = 2, 4, 6, \ldots$

$$c_0 = (-1)^{n/2}\frac{1 \cdot 3 \cdots (n-1)}{2 \cdot 4 \cdots n};$$

whereas for $n = 1$ we choose $c_1 = 1$, and for $n = 3, 5, 7, \ldots$,

$$c_1 = (-1)^{(n-1)/2}\frac{1 \cdot 3 \cdots n}{2 \cdot 4 \cdots (n-1)}.$$

For example, when $n = 4$ we have

$$y_1(x) = (-1)^{4/2}\frac{1 \cdot 3}{2 \cdot 4}\left[1 - 10x^2 + \frac{35}{3}x^4\right] = \frac{1}{8}(35x^4 - 30x^2 + 3).$$

▇ **Legendre Polynomials** These specific nth-degree polynomial solutions are called **Legendre polynomials** and are denoted by $P_n(x)$. From the series for $y_1(x)$ and $y_2(x)$ and from the above choices of c_0 and c_1 we find that the first six Legendre polynomials are

$$P_0(x) = 1, \qquad\qquad P_1(x) = x,$$

$$P_2(x) = \frac{1}{2}(3x^2 - 1), \qquad P_3(x) = \frac{1}{2}(5x^3 - 3x), \qquad (34)$$

$$P_4(x) = \frac{1}{8}(35x^4 - 30x^2 + 3), \qquad P_5(x) = \frac{1}{8}(63x^5 - 70x^3 + 15x).$$

Remember, $P_0(x)$, $P_1(x)$, $P_2(x)$, $P_3(x)$, $\ldots$, are, in turn, particular solutions of the differential equations

$$
\begin{aligned}
n = 0: \quad & (1 - x^2)y'' - 2xy' = 0 \\
n = 1: \quad & (1 - x^2)y'' - 2xy' + 2y = 0 \\
n = 2: \quad & (1 - x^2)y'' - 2xy' + 6y = 0 \qquad (35) \\
n = 3: \quad & (1 - x^2)y'' - 2xy' + 12y = 0 \\
& \quad\vdots \qquad\qquad \vdots
\end{aligned}
$$

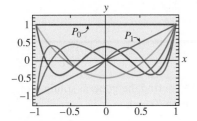

FIGURE 5.3.6 Legendre polynomials for $n = 0, 1, 2, 3, 4, 5$

The graphs, on the interval $[-1, 1]$, of the six Legendre polynomials in (34) are given in **FIGURE 5.3.6**.

▇ **Properties** You are encouraged to verify the following properties for the Legendre polynomials in (34):

$$(i) \; P_n(-x) = (-1)^n P_n(x)$$

$$(ii) \; P_n(1) = 1 \qquad\qquad (iii) \; P_n(-1) = (-1)^n$$

$$(iv) \; P_n(0) = 0, \; n \text{ odd} \qquad (v) \; P_n'(0) = 0, \; n \text{ even}.$$

Property (i) indicates, as is apparent in Figure 5.3.6, that $P_n(x)$ is an even or odd function according to whether n is even or odd.

Recurrence Relation Recurrence relations that relate Legendre polynomials of different degrees are also important in some aspects of their applications. We state, without proof, the following three-term recurrence relation

$$(k + 1)P_{k+1}(x) - (2k + 1)x P_k(x) + kP_{k-1}(x) = 0, \tag{36}$$

which is valid for $k = 1, 2, 3, \ldots$. In (34) we listed the first six Legendre polynomials. If, say, we wish to find $P_6(x)$, we can use (36) with $k = 5$. This relation expresses $P_6(x)$ in terms of the known $P_4(x)$ and $P_5(x)$. See Problem 51 in Exercises 5.3.

Another formula, although not a recurrence relation, can generate the Legendre polynomials by differentiation:

$$P_n(x) = \frac{1}{2^n n!} \frac{d^n}{dx^n}(x^2 - 1)^n, \quad n = 0, 1, 2, \ldots. \tag{37}$$

The formula given in (37) first appeared in 1815 in the doctoral dissertation of the French mathematician **Benjamin Olinde Rodrigues** (1795–1851). So naturally, (37) is called **Rodrigues' formula**. There are several such derivative formulas for generating polynomial solutions of differential equations with variable coefficients. See Problems 29 and 30 in Chapter 5 in Review.

REMARKS

Although we have assumed that the parameter n in Legendre's differential equation

$$(1 - x^2)y'' - 2xy' + n(n + 1)y = 0$$

represented a nonnegative integer, in a more general setting n can represent any real number. If n is *not* a nonnegative integer, then both Legendre functions $y_1(x)$ and $y_2(x)$ given in (33) are infinite series convergent on the open interval $(-1, 1)$ and divergent (unbounded) at $x = \pm 1$. If n is a nonnegative integer, then as we have just seen one of the Legendre functions in (33) is a polynomial and the other is an infinite series convergent for $-1 < x < 1$. You should be aware of the fact that Legendre's equation possesses solutions that are bounded on the *closed* interval $[-1, 1]$ only in the case when $n = 0, 1, 2, \ldots$. More to the point, the only Legendre functions that are bounded on the closed interval $[-1, 1]$ are the Legendre polynomials $P_n(x)$ or constant multiples of these polynomials. See Problem 53 in Exercises 5.3 and Problem 24 in Chapter 5 in Review.

5.3 Exercises Answers to selected odd-numbered problems begin on page ANS-13.

5.3.1 Bessel Functions

In Problems 1–6, use (1) to find the general solution of the given differential equation on $(0, \infty)$.

1. $x^2y'' + xy' + (x^2 - \frac{1}{9})y = 0$

2. $x^2y'' + xy' + (x^2 - 1)y = 0$

3. $4x^2y'' + 4xy' + (4x^2 - 25)y = 0$

4. $16x^2y'' + 16xy' + (16x^2 - 1)y = 0$

5. $xy'' + y' + xy = 0$

6. $\dfrac{d}{dx}[xy'] + \left(x - \dfrac{4}{x}\right)y = 0$

In Problems 7 and 8, use (12) to find the general solution of the given differential equation on the interval $(0, \infty)$.

7. $x^2y'' + xy' + (9x^2 - 4)y = 0$

8. $x^2y'' + xy' + (36x^2 - \frac{1}{4})y = 0$

In Problems 9 and 10, use (18) to find the general solution of the given differential equation on $(0, \infty)$.

9. $x^2y'' + xy' - (16x^2 + \frac{4}{9})y = 0$

10. $x^2y'' + xy' - (2x^2 + 64)y = 0$

In Problems 11 and 12, use the indicated change of variable to find the general solution of the given differential equation on the interval $(0, \infty)$.

11. $x^2y'' + 2xy' + \alpha^2x^2y = 0; \quad y = x^{-1/2}u(x)$

12. $x^2y'' + (\alpha^2x^2 - \nu^2 + \frac{1}{4})y = 0; \quad y = \sqrt{x}u(x)$

In Problems 13–20, use (20) to find the general solution of the given differential equation on the interval $(0, \infty)$.

13. $xy'' + 2y' + 4y = 0$ **14.** $xy'' + 3y' + xy = 0$

15. $xy'' - y' + xy = 0$ **16.** $xy'' - 5y' + xy = 0$

17. $x^2y'' + (x^2 - 2)y = 0$ **18.** $4x^2y'' + (16x^2 + 1)y = 0$

19. $xy'' + 3y' + x^3 y = 0$

20. $9x^2 y'' + 9xy' + (x^6 - 36)y = 0$

21. Use the series in (7) to verify that $I_\nu(x) = i^{-\nu} J_\nu(ix)$ is a real function.

22. Assume that b in equation (20) can be pure imaginary; that is, $b = \beta i, \beta > 0, i^2 = -1$. Use this assumption to express the general solution of the given differential equation in terms of the modified Bessel functions I_n and K_n.

 (a) $y'' - x^2 y = 0$

 (b) $xy'' + y' - 7x^3 y = 0$

In Problems 23–26, first use (20) to express the general solution of the given differential equation in terms of Bessel functions. Then use (26) and (27) to express the general solution in terms of elementary functions.

23. $y'' + y = 0$

24. $x^2 y'' + 4xy' + (x^2 + 2)y = 0$

25. $16x^2 y'' + 32xy' + (x^4 - 12)y = 0$

26. $4x^2 y'' - 4xy' + (16x^2 + 3)y = 0$

27. **(a)** Proceed as in Example 6 to show that

$$x J_\nu'(x) = -\nu J_\nu(x) + x J_{\nu-1}(x).$$

 [*Hint*: Write $2n + \nu = 2(n + \nu) - \nu$.]

 (b) Use the result in part (a) to derive (23).

28. Use the formula obtained in Example 6 along with part (a) of Problem 27 to derive the recurrence relation

$$2\nu J_\nu(x) = x J_{\nu+1}(x) + x J_{\nu-1}(x).$$

In Problems 29 and 30, use (22) or (23) to obtain the given result.

29. $\displaystyle\int_0^x r J_0(r)\, dr = x J_1(x)$ **30.** $J_0'(x) = J_{-1}(x) = -J_1(x)$

31. **(a)** Proceed as on pages 296–297 to derive the elementary form of $J_{-1/2}(x)$ given in (27).

 (b) Use $\nu = -\frac{1}{2}$ along with (26) and (27) in the recurrence relation in Problem 28 to express $J_{-3/2}(x)$ in terms of $\sin x$, $\cos x$, and powers of x.

 (c) Use a graphing utility to plot the graph of $J_{-3/2}(x)$.

32. **(a)** Use the recurrence relation in Problem 28 to express $J_{3/2}(x)$, $J_{5/2}(x)$, and $J_{7/2}(x)$ in terms of $\sin x$, $\cos x$, and powers of x.

 (b) Use a graphing utility to plot the graphs of $J_{3/2}(x)$, $J_{5/2}(x)$, and $J_{7/2}(x)$ in the same coordinate plane.

33. Use the solution of the aging spring equation $mx'' + ke^{-\alpha t} x = 0$, obtained in Example 5, to discuss the behavior of $x(t)$ as $t \to \infty$ in the three cases

 (a) $c_1 \neq 0, c_2 = 0$

 (b) $c_1 = 0, c_2 \neq 0$

 (c) $c_1 \neq 0, c_2 \neq 0$.

34. Show that $y = x^{1/2} w(\frac{2}{3}\alpha x^{3/2})$ is a solution of the given form of Airy's differential equation whenever w is a solution of the indicated Bessel's equation. [*Hint*: After differentiating, substituting, and simplifying, then let $t = \frac{2}{3}\alpha x^{3/2}$.]

(a) $y'' + \alpha^2 xy = 0, x > 0; t^2 w'' + tw' + (t^2 - \frac{1}{9})w = 0, t > 0$

(b) $y'' - \alpha^2 xy = 0, x > 0; t^2 w'' + tw' - (t^2 + \frac{1}{9})w = 0, t > 0$

35. Use the result in parts (a) and (b) of Problem 34 to express the general solution on $(0, \infty)$ of each of the two forms of Airy's equation in terms of Bessel functions.

36. Use Table 5.3.1 to find the first three positive eigenvalues and corresponding eigenfunctions of the boundary-value problem

$$xy'' + y' + \lambda xy = 0,$$
$$y(x), y'(x) \text{ bounded as } x \to 0^+, y(2) = 0.$$

[*Hint*: By identifying $\lambda = \alpha^2$, the DE is the parametric Bessel equation of order zero.]

37. **(a)** Use (20) to show that the general solution of the differential equation $xy'' + \lambda y = 0$ on the interval $(0, \infty)$ is

$$y = c_1 \sqrt{x} J_1(2\sqrt{\lambda x}) + c_2 \sqrt{x} Y_1(2\sqrt{\lambda x}).$$

 (b) Verify by direct substitution that $y = \sqrt{x} J_1(2\sqrt{x})$ is a particular solution of the DE in the case $\lambda = 1$.

38. Use $I_\nu(x) = i^{-\nu} J_\nu(ix)$ along with (7) and (8) to show that:

 (a) $I_{1/2}(x) = \sqrt{\dfrac{2}{\pi x}} \sinh x$

 (b) $I_{-1/2}(x) = \sqrt{\dfrac{2}{\pi x}} \cosh x$

 (c) Use (16) to express $K_{1/2}(x)$ in terms of elementary functions.

39. **(a)** Use the first formula in (31) to find the spherical Bessel functions $j_1(x), j_2(x),$ and $j_3(x)$.

 (b) Use a graphing utility to plot the graphs of $j_1(x), j_2(x)$, and $j_3(x)$ in the same coordinate plane.

40. **(a)** Use the second formula in (31) to find the spherical Bessel functions $y_1(x), y_2(x),$ and $y_3(x)$.

 (b) Use a graphing utility to plot the graphs of $y_1(x), y_2(x)$, and $y_3(x)$ in the same coordinate plane.

41. If n is an integer, use the substitution $R(x) = (\alpha x)^{-1/2} Z(x)$ to show that the differential equation

$$x^2 \frac{d^2 R}{dx^2} + 2x \frac{dR}{dx} + [\alpha^2 x^2 - n(n+1)] R = 0 \quad \text{(38)}$$

becomes

$$x^2 \frac{d^2 Z}{dx^2} + x \frac{dZ}{dx} + [\alpha^2 x^2 - (n + \tfrac{1}{2})^2] Z = 0. \quad \text{(39)}$$

42. **(a)** In Problem 41, find the general solution of the differential equation in (39) on the interval $(0, \infty)$.

 (b) Use part (a) to find the general solution of the differential equation in (38) on the interval $(0, \infty)$.

 (c) Use part (b) to express the general solution of (38) in terms of the spherical Bessel functions of the first and second kind defined in (31).

Mathematical Model

43. Cooling Fin A cooling fin is an outward projection from a mechanical or electronic device from which heat can be radiated away from the device into the surrounding medium (such as air). See the accompanying photo. An annular, or ring-shaped, cooling fin is normally used on cylindrical surfaces such as a circular heating pipe. See **FIGURE 5.3.7**. In the latter case, let r denote the radial distance measured from the center line of the pipe and $T(r)$ the temperature within the fin defined for $r_0 \leq r \leq r_1$. It can be shown that $T(r)$ satisfies the differential equation

$$\frac{d}{dr}\left(r\frac{dT}{dr}\right) = a^2 r(T - T_m),$$

where a^2 is a constant and T_m is the constant air temperature. Suppose $r_0 = 1$, $r_1 = 3$, and $T_m = 70$. Use the substitution $w(r) = T(r) - 70$ to show that the solution of the given differential equation subject to the boundary conditions

$$T(1) = 160, \quad T'(3) = 0$$

is $T(r) = 70 + 90\dfrac{K_1(3a)I_0(ar) + I_1(3a)K_0(ar)}{K_1(3a)I_0(a) + I_1(3a)K_0(a)},$

where $I_0(x)$ and $K_0(x)$ are the modified Bessel functions of the first and second kind. You will also have to use the derivatives given in (25) on page 296.

<image_placeholder><document_segment>© GDragan/iStock/Thinkstock
Cooling fins on a motorcycle engine</document_segment></image_placeholder>

FIGURE 5.3.7 Annular cooling fin

44. Solve the differential equation in Problem 43 when the boundary conditions are

$$T(1) = 160, \quad T(3) = 90.$$

Computer Lab Assignments

45. (a) Use the general solution given in Example 5 to solve the IVP

$$4x'' + e^{-0.1t}x = 0, \quad x(0) = 1, \quad x'(0) = -\tfrac{1}{2}.$$

Also use $J_0'(x) = -J_1(x)$ and $Y_0'(x) = -Y_1(x)$ along with Table 5.3.1 or a CAS to evaluate coefficients.

(b) Use a CAS to graph the solution obtained in part (a) for $0 \leq t < \infty$.

46. (a) Use the general solution obtained in Problem 35 to solve the IVP

$$4x'' + tx = 0, \quad x(0.1) = 1, \quad x'(0.1) = -\tfrac{1}{2}.$$

Use a CAS to evaluate coefficients.

(b) Use a CAS to graph the solution obtained in part (a) for $0 \leq t \leq 200$.

47. Column Bending Under Its Own Weight A uniform thin column of length L, positioned vertically with one end embedded in the ground, will deflect, or bend away, from the vertical under the influence of its own weight when its length or height exceeds a certain critical value. It can be shown that the angular deflection $\theta(x)$ of the column from the vertical at a point $P(x)$ is a solution of the boundary-value problem

$$EI\frac{d^2\theta}{dx^2} + \delta g(L - x)\theta = 0, \quad \theta(0) = 0, \quad \theta'(L) = 0,$$

where E is Young's modulus, I is the cross-sectional moment of inertia, δ is the constant linear density, and x is the distance along the column measured from its base. See **FIGURE 5.3.8**.

The column will bend only for those values of L for which the boundary-value problem has a nontrivial solution.

(a) Restate the boundary-value problem by making the change of variables $t = L - x$. Then use the results of a problem earlier in this exercise set to express the general solution of the differential equation in terms of Bessel functions.

(b) Use the general solution found in part (a) to find a solution of the BVP and an equation that defines the critical length L, that is, the smallest value of L for which the column will start to bend.

(c) With the aid of a CAS, find the critical length L of a solid steel rod of radius $r = 0.05$ in., $\delta g = 0.28\,A$ lb/in., $E = 2.6 \times 10^7$ lb/in.2, $A = \pi r^2$, and $I = \tfrac{1}{4}\pi r^4$.

FIGURE 5.3.8 Column in Problem 47

48. Buckling of a Thin Vertical Column In Example 4 of Section 3.9 we saw that when a constant vertical compressive force, or load, P was applied to a thin column of uniform cross section and hinged at both ends, the deflection $y(x)$ is a solution of the boundary-value problem:

$$EI\frac{d^2y}{dx^2} + Py = 0, \quad y(0) = 0, \quad y(L) = 0.$$

(a) If the bending stiffness factor EI is proportional to x, then $EI(x) = kx$, where k is a constant of proportionality. If $EI(L) = kL = M$ is the maximum stiffness factor, then $k = M/L$ and so $EI(x) = Mx/L$. Use the information in Problem 37 to find a solution of

$$M\frac{x}{L}\frac{d^2y}{dx^2} + Py = 0, \quad y(0) = 0, \quad y(L) = 0$$

if it is known that $\sqrt{x}\,Y_1(2\sqrt{\lambda x})$ is *not* zero at $x = 0$.

(b) Use Table 5.3.1 to find the Euler load P_1 for the column.

(c) Use a CAS to graph the first buckling mode $y_1(x)$ corresponding to the Euler load P_1. For simplicity assume that $c_1 = 1$ and $L = 1$.

49. Pendulum of Varying Length For the simple pendulum described on page 194 of Section 3.11, suppose that the rod holding the mass m at one end is replaced by a flexible wire and that the wire is fed by a pulley through the horizontal support at point O in **FIGURE 5.3.9**. In this manner, while it is in motion in a vertical plane, the mass m can be raised or lowered. In other words, the length $l(t)$ of the pendulum varies with time. Under the same assumptions leading to equation (6) in Section 3.11, it follows from (1) in Chapter 3 in Review that the differential equation for the displacement angle $\theta(t)$ is now

$$l\frac{d^2\theta}{dt^2} + 2\frac{dl}{dt}\frac{d\theta}{dt} + g\sin\theta = 0.$$

(a) If l increases at a constant rate v and $l(0) = l_0$, then show that a linearization of the foregoing differential equation is

$$(l_0 + vt)\frac{d^2\theta}{dt^2} + 2v\frac{d\theta}{dt} + g\theta = 0. \tag{40}$$

(b) Make the change of variables $x = (l_0 + vt)/v$ and show that (40) becomes

$$\frac{d^2\theta}{dx^2} + \frac{2}{x}\frac{d\theta}{dx} + \frac{g}{vx}\theta = 0.$$

(c) Use part (b) and (20) to express the general solution of equation (40) in terms of Bessel functions.

(d) Use the general solution obtained in part (c) to solve the initial-value problem consisting of equation (40) and the initial conditions $\theta(0) = \theta_0$, $\theta'(0) = 0$. [*Hints:* To simplify calculations use a further change of variable

$$u = \frac{2}{v}\sqrt{g(l_0 + vt)} = 2\sqrt{\frac{g}{v}}\,x^{1/2}.$$ Also, recall (22) holds for both $J_1(u)$ and $Y_1(u)$. Finally, the identity

$$J_1(u)Y_2(u) - J_2(u)Y_1(u) = -\frac{2}{\pi u}$$

will be helpful.]

(e) Use a CAS to graph the solution $\theta(t)$ of the IVP in part (d) when $l_0 = 1$ ft, $\theta_0 = \frac{1}{10}$ radian, and $v = \frac{1}{60}$ ft/s. Experiment with the graph using different time intervals such as $[0, 10]$, $[0, 30]$, and so on.

(f) What do the graphs indicate about the displacement angle $\theta(t)$ as the length l of the wire increases with time?

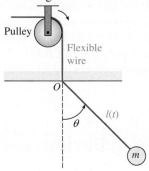

FIGURE 5.3.9 Pendulum of varying length in Problem 49

5.3.2 Legendre Functions

50. (a) Use the explicit solutions $y_1(x)$ and $y_2(x)$ of Legendre's equation given in (33) and the appropriate choice of c_0 and c_1 to find the Legendre polynomials $P_6(x)$ and $P_7(x)$.

(b) Write the differential equations for which $P_6(x)$ and $P_7(x)$ are particular solutions.

51. Use the recurrence relation (36) and $P_0(x) = 1$, $P_1(x) = x$, to generate the next six Legendre polynomials.

52. Show that the differential equation

$$\sin\theta\frac{d^2y}{d\theta^2} + \cos\theta\frac{dy}{d\theta} + n(n + 1)(\sin\theta)y = 0$$

can be transformed into Legendre's equation by means of the substitution $x = \cos\theta$.

53. Find the first three positive values of λ for which the problem

$$(1 - x^2)y'' - 2xy' + \lambda y = 0,$$
$$y(0) = 0, y(x), y'(x) \text{ bounded on } [-1, 1]$$

has nontrivial solutions.

54. The differential equation

$$(1 - x^2)y'' - 2xy' + \left[n(n + 1) - \frac{m^2}{1 - x^2}\right]y = 0,$$

is known as the **associated Legendre equation**. When $m = 0$ this equation reduces to Legendre's equation (2). A solution of the associated equation is

$$P_n^m(x) = (1 - x^2)^{m/2}\frac{d^m}{dx^m}P_n(x),$$

where $P_n(x)$, $n = 0, 1, 2, \ldots$ are the Legendre polynomials given in (34). The solutions $P_n^m(x)$ for $m = 0, 1, 2, \ldots$, are called **associated Legendre functions**.

(a) Find the associated Legendre functions $P_0^0(x)$, $P_1^0(x)$, $P_1^1(x)$, $P_2^1(x)$, $P_2^2(x)$, $P_3^1(x)$, $P_3^2(x)$, and $P_3^3(x)$.

(b) What can you say about $P_n^m(x)$ when m is an even nonnegative integer?

(c) What can you say about $P_n^m(x)$ when m is a nonnegative integer and $m > n$?

(d) Verify that $y = P_1^1(x)$ satisfies the associated Legendre equation when $n = 1$ and $m = 1$.

Computer Lab Assignments

55. For purposes of this problem, ignore the list of Legendre polynomials given on page 299 and the graphs given in Figure 5.3.6. Use Rodrigues' formula (37) to generate the Legendre polynomials $P_1(x)$, $P_2(x),\ldots,P_7(x)$. Use a CAS to carry out the differentiations and simplifications.

56. Use a CAS to graph $P_1(x)$, $P_2(x),\ldots,P_7(x)$ on the closed interval $[-1, 1]$.

57. Use a root-finding application to find the zeros of $P_1(x)$, $P_2(x),\ldots,P_7(x)$. If the Legendre polynomials are built-in functions of your CAS, find the zeros of Legendre polynomials of higher degree. Form a conjecture about the location of the zeros of any Legendre polynomial $P_n(x)$, and then investigate to see whether it is true.

5 Chapter in Review Answers to selected odd-numbered problems begin on page ANS-14.

In Problems 1 and 2, answer true or false without referring back to the text.

1. The general solution of $x^2y'' + xy' + (x^2 - 1)y = 0$ is $y = c_1J_1(x) + c_2J_{-1}(x)$. _____

2. Since $x = 0$ is an irregular singular point of $x^3y'' - xy' + y = 0$, the DE possesses no solution that is analytic at $x = 0$. _____

3. Both power series solutions of $y'' + \ln(x + 1)y' + y = 0$ centered at the ordinary point $x = 0$ are guaranteed to converge for all x in which *one* of the following intervals?

(a) $(-\infty, \infty)$ (b) $(-1, \infty)$

(c) $[-\frac{1}{2}, \frac{1}{2}]$ (d) $[-1, 1]$

4. $x = 0$ is an ordinary point of a certain linear differential equation. After the assumed solution $y = \sum_{n=0}^{\infty} c_n x^n$ is substituted into the DE, the following algebraic system is obtained by equating the coefficients of $x^0, x^1, x^2,$ and x^3 to zero:

$$2c_2 + 2c_1 + c_0 = 0$$

$$6c_3 + 4c_2 + c_1 = 0$$

$$12c_4 + 6c_3 + c_2 - \tfrac{1}{3}c_1 = 0$$

$$20c_5 + 8c_4 + c_3 - \tfrac{2}{3}c_2 = 0.$$

Bearing in mind that c_0 and c_1 are arbitrary, write down the first five terms of two power series solutions of the differential equation.

5. Suppose the powers series $\sum_{n=0}^{\infty} c_k(x - 4)^k$ is known to converge at -2 and diverge at 13. Discuss whether the series converges at $-7, 0, 7, 10,$ and 11. Possible answers are *does*, *does not*, or *might*.

6. Use the Maclaurin series for $\sin x$ and $\cos x$ along with long division to find the first three nonzero terms of a power series in x for the function $f(x) = \dfrac{\sin x}{\cos x}$.

In Problems 7 and 8, construct a linear second-order differential equation that has the given properties.

7. A regular singular point at $x = 1$ and an irregular singular point at $x = 0$.

8. Regular singular points at $x = 1$ and at $x = -3$.

In Problems 9–14, use an appropriate infinite series method about $x = 0$ to find two solutions of the given differential equation.

9. $2xy'' + y' + y = 0$ **10.** $y'' - xy' - y = 0$

11. $(x - 1)y'' + 3y = 0$ **12.** $y'' - x^2y' + xy = 0$

13. $xy'' - (x + 2)y' + 2y = 0$ **14.** $(\cos x)y'' + y = 0$

In Problems 15 and 16, solve the given initial-value problem.

15. $y'' + xy' + 2y = 0$, $y(0) = 3$, $y'(0) = -2$

16. $(x + 2)y'' + 3y = 0$, $y(0) = 0$, $y'(0) = 1$

17. Without actually solving the differential equation

$$(1 - 2\sin x)y'' + xy = 0$$

find a lower bound for the radius of convergence of power series solutions about the ordinary point $x = 0$.

18. Even though $x = 0$ is an ordinary point of the differential equation, explain why it is not a good idea to try to find a solution of the initial-value problem

$$y'' + xy' + y = 0,\quad y(1) = -6,\quad y'(1) = 3$$

of the form $\sum_{n=0}^{\infty} c_n x^n$. Using power series, find a better way to solve the problem.

In Problems 19 and 20, investigate whether $x = 0$ is an ordinary point, singular point, or irregular singular point of the given differential equation. [*Hint*: Recall the Maclaurin series for $\cos x$ and e^x.]

19. $xy'' + (1 - \cos x)y' + x^2y = 0$

20. $(e^x - 1 - x)y'' + xy = 0$

21. Note that $x = 0$ is an ordinary point of the differential equation

$$y'' + x^2y' + 2xy = 5 - 2x + 10x^3.$$

Use the assumption $y = \sum_{n=0}^{\infty} c_n x^n$ to find the general solution $y = y_c + y_p$ that consists of three power series centered at $x = 0$.

22. The first-order differential equation $dy/dx = x^2 + y^2$ cannot be solved in terms of elementary functions. However, a solution can be expressed in terms of Bessel functions.

(a) Show that the substitution $y = -\dfrac{1}{u}\dfrac{du}{dx}$ leads to the equation $u'' + x^2u = 0$.

(b) Use (20) in Section 5.3 to find the general solution of $u'' + x^2u = 0$.

(c) Use (22) and (23) in Section 5.3 in the forms

$$J_\nu'(x) = \frac{\nu}{x}J_\nu(x) - J_{\nu+1}(x)$$

and

$$J_\nu'(x) = -\frac{\nu}{x}J_\nu(x) + J_{\nu-1}(x)$$

as an aid in showing that a one-parameter family of solutions of $dy/dx = x^2 + y^2$ is given by

$$y = x\,\frac{J_{3/4}(\frac{1}{2}x^2) - cJ_{-3/4}(\frac{1}{2}x^2)}{cJ_{1/4}(\frac{1}{2}x^2) + J_{-1/4}(\frac{1}{2}x^2)}.$$

23. Express the general solution of the given differential equation on the interval $(0, \infty)$ in terms of Bessel functions.
 (a) $4x^2y'' + 4xy' + (64x^2 - 9)y = 0$
 (b) $x^2y'' + xy' - (36x^2 + 9)y = 0$

24. (a) From (34) and (35) of Section 5.3 we know that when $n = 0$, Legendre's differential equation $(1 - x^2)y'' - 2xy' = 0$ has the polynomial solution $y = P_0(x) = 1$. Use (5) of Section 3.2 to show that a second Legendre function satisfying the DE on the interval $(-1, 1)$ is

$$y = \frac{1}{2}\ln\left(\frac{1 + x}{1 - x}\right).$$

 (b) We also know from (34) and (35) of Section 5.3 that when $n = 1$, Legendre's differential equation $(1 - x^2)y'' - 2xy' + 2y = 0$ possesses the polynomial solution $y = P_1(x) = x$. Use (5) of Section 3.2 to show that a second Legendre function satisfying the DE on the interval $(-1, 1)$ is

$$y = \frac{x}{2}\ln\left(\frac{1 + x}{1 - x}\right) - 1.$$

 (c) Use a graphing utility to graph the logarithmic Legendre functions given in parts (a) and (b).

25. Use binomial series to formally show that

$$(1 - 2xt + t^2)^{-1/2} = \sum_{n=0}^{\infty} P_n(x)\, t^n.$$

26. Use the result obtained in Problem 25 to show that $P_n(1) = 1$ and $P_n(-1) = (-1)^n$. See Properties (*ii*) and (*iii*) on page 299.

27. The differential equation

$$y'' - 2xy' + 2\alpha y = 0$$

 is known as **Hermite's equation of order α** after the French mathematician **Charles Hermite** (1822–1901). Show that the general solution of the equation is $y(x) = c_0y_1(x) + c_1y_2(x)$, where

$$y_1(x) = 1 + \sum_{k=1}^{\infty}(-1)^k\frac{2^k\alpha(\alpha - 2)\cdots(\alpha - 2k + 2)}{(2k)!}x^{2k}$$

$$y_2(x) = x + \sum_{k=1}^{\infty}(-1)^k\frac{2^k(\alpha - 1)(\alpha - 3)\cdots(\alpha - 2k + 1)}{(2k + 1)!}x^{2k + 1}$$

 are power series solutions centered at the ordinary point 0.

28. (a) When $\alpha = n$ is a nonnegative integer Hermite's differential equation always possesses a polynomial solution of degree n. Use $y_1(x)$ given in Problem 27 to find polynomial solutions for $n = 0, n = 2$, and $n = 4$. Then use $y_2(x)$ in Problem 27 to find polynomial solutions for $n = 1, n = 3$, and $n = 5$.

(b) A **Hermite polynomial** $H_n(x)$ is defined to be an nth degree polynomial solution of Hermite's differential equation multiplied by an appropriate constant so that the coefficient of x^n in $H_n(x)$ is 2^n. Use the polynomial solutions found in part (a) to show that the first six Hermite polynomials are

$$H_0(x) = 1$$
$$H_1(x) = 2x$$
$$H_2(x) = 4x^2 - 2$$
$$H_3(x) = 8x^3 - 12x$$
$$H_4(x) = 16x^4 - 48x^2 + 12$$
$$H_5(x) = 32x^5 - 160x^3 + 120x.$$

29. Rodrigues' formula for the Hermite polynomials is

$$H_n(x) = (-1)^n e^{x^2}\frac{d^n}{dx^n}e^{-x^2}.$$

 Use this formula to find the six Hermite polynomials listed in part (b) of Problem 28.

30. The differential equation

$$xy'' + (1 - x)y' + ny = 0$$

 is known as **Laguerre's equation** after the French mathematician **Edmond Laguerre** (1834–1886). When n is a nonnegative integer, the DE is known to possess polynomial solutions. These solutions are naturally called **Laguerre polynomials** and are denoted by $L_n(t)$. Rodrigues' formula for the Laguerre polynomials is

$$L_n(x) = \frac{e^x}{n!}\frac{d^n}{dx^n}x^ne^{-x},$$

 Use this formula to find the Laguerre polynomials corresponding to $n = 0, 1, 2, 3, 4$.

31. The differential equation

$$(1 - x^2)y'' - xy' + \alpha^2 y = 0$$

 where α is a parameter, is known as **Chebyshev's equation** after the Russian mathematician **Pafnuty Chebyshev** (1821–1894). Find the general solution $y(x) = c_0y_1(x) + c_1y_2(x)$ of the equation, where $y_1(x)$ and $y_2(x)$ are power series solutions centered at the ordinary point 0 and containing only even powers of x and odd powers of x, respectively.

32. (a) When $\alpha = n$ is a nonnegative integer Chebyshev's differential equation always possesses a polynomial solution of degree n. Use $y_1(x)$ found in Problem 31 to find polynomial solutions for $n = 0, n = 2$, and $n = 4$. Then use $y_2(x)$ in Problem 31 to find polynomial solutions for $n = 1, n = 3$, and $n = 5$.

(b) A **Chebyshev polynomial** $T_n(x)$ is defined to be an nth degree polynomial solution of Chebyshev's equation multiplied by the constant $(-1)^{n/2}$ when n is even and by $(-1)^{(n-1)/2}n$ when n is odd. Use the solutions found in part (a) to obtain the first six Chebyshev polynomials $T_0(x), T_1(x), \ldots, T_5(x)$.

6

Numerical Solutions of Ordinary Differential Equations

6.1 Euler Methods and Error Analysis

6.2 Runge–Kutta Methods

6.3 Multistep Methods

6.4 Higher-Order Equations and Systems

6.5 Second-Order Boundary-Value Problems

Chapter 6 in Review

A differential equation need not possess a solution. But even when a solution exists, we may not be able to exhibit it in an explicit or implicit form; we may have to be content with an **approximation** to the solution. In this chapter we continue to explore the basic idea introduced in Section 2.6, that is, using the differential equation to construct an algorithm to approximate the y-coordinates of points on the actual solution curve.

6.1 Euler Methods and Error Analysis

INTRODUCTION In Section 2.6 we examined one of the simplest numerical methods for approximating solutions of first-order initial-value problems $y' = f(x, y)$, $y(x_0) = y_0$. Recall that the backbone of **Euler's method** was the formula

$$y_{n+1} = y_n + hf(x_n, y_n), \tag{1}$$

where f is the function obtained from the differential equation $y' = f(x, y)$. The recursive use of (1) for $n = 0, 1, 2, \ldots$ yields the y-coordinates $y_1, y_2, y_3, \ldots$ of points on successive "tangent lines" to the solution curve at $x_1, x_2, x_3, \ldots$ or $x_n = x_0 + nh$, where h is a constant and is the size of the step between x_n and x_{n+1}. The values $y_1, y_2, y_3, \ldots$ approximate the values of a solution $y(x)$ of the IVP at $x_1, x_2, x_3, \ldots$. But whatever advantage (1) has in its simplicity is lost in the crudeness of its approximations.

A Comparison In Problem 4 in Exercises 2.6 you were asked to use Euler's method to obtain the approximate value of $y(1.5)$ for the solution of the initial-value problem $y' = 2xy$, $y(1) = 1$. You should have obtained the analytic solution $y = e^{x^2-1}$ and results similar to those given in Tables 6.1.1 and 6.1.2.

TABLE 6.1.1 Euler's Method with $h = 0.1$				
x_n	y_n	Actual Value	Abs. Error	% Rel. Error
1.00	1.0000	1.0000	0.0000	0.00
1.10	1.2000	1.2337	0.0337	2.73
1.20	1.4640	1.5527	0.0887	5.71
1.30	1.8154	1.9937	0.1784	8.95
1.40	2.2874	2.6117	0.3244	12.42
1.50	2.9278	3.4903	0.5625	16.12

TABLE 6.1.2 Euler's Method with $h = 0.05$				
x_n	y_n	Actual Value	Abs. Error	% Rel. Error
1.00	1.0000	1.0000	0.0000	0.00
1.05	1.1000	1.1079	0.0079	0.72
1.10	1.2155	1.2337	0.0182	1.47
1.15	1.3492	1.3806	0.0314	2.27
1.20	1.5044	1.5527	0.0483	3.11
1.25	1.6849	1.7551	0.0702	4.00
1.30	1.8955	1.9937	0.0982	4.93
1.35	2.1419	2.2762	0.1343	5.90
1.40	2.4311	2.6117	0.1806	6.92
1.45	2.7714	3.0117	0.2403	7.98
1.50	3.1733	3.4903	0.3171	9.08

In this case, with a step size $h = 0.1$, a 16% relative error in the calculation of the approximation to $y(1.5)$ is totally unacceptable. At the expense of doubling the number of calculations, some improvement in accuracy is obtained by halving the step size to $h = 0.05$.

Errors in Numerical Methods In choosing and using a numerical method for the solution of an initial-value problem, we must be aware of the various sources of errors. For some kinds of computation the accumulation of errors might reduce the accuracy of an approximation to the point of making the computation useless. On the other hand, depending on the use to which a numerical solution may be put, extreme accuracy may not be worth the added expense and complication.

One source of error always present in calculations is **round-off error**. This error results from the fact that any calculator or computer can represent numbers using only a finite number of digits. Suppose, for the sake of illustration, that we have a calculator that uses base 10 arithmetic and carries four digits, so that $\frac{1}{3}$ is represented in the calculator as 0.3333 and $\frac{1}{9}$ is represented as 0.1111. If we use this calculator to compute $(x^2 - \frac{1}{9})/(x - \frac{1}{3})$ for $x = 0.3334$, we obtain

$$\frac{(0.3334)^2 - 0.1111}{0.3334 - 0.3333} = \frac{0.1112 - 0.1111}{0.3334 - 0.3333} = 1.$$

With the help of a little algebra, however, we see that

$$\frac{x^2 - \frac{1}{9}}{x - \frac{1}{3}} = \frac{(x - \frac{1}{3})(x + \frac{1}{3})}{x - \frac{1}{3}} = x + \frac{1}{3},$$

so that when $x = 0.3334$, $(x^2 - \frac{1}{9})/(x - \frac{1}{3}) \approx 0.3334 + 0.3333 = 0.6667$. This example shows that the effects of round-off error can be quite serious unless some care is taken. One way to reduce the effect of round-off error is to minimize the number of calculations. Another technique on a computer is to use double-precision arithmetic to check the results. In general, round-off error is unpredictable and difficult to analyze, and we will neglect it in the error analysis that follows. We will concentrate on investigating the error introduced by using a formula or algorithm to approximate the values of the solution.

■ **Truncation Errors for Euler's Method** In the sequence of values $y_1, y_2, y_3, \ldots$ generated from (1), usually the value of y_1 will not agree with the actual solution evaluated at x_1, namely, $y(x_1)$, because the algorithm gives only a straight-line approximation to the solution. See Figure 2.6.2. The error is called the **local truncation error**, **formula error**, or **discretization error**. It occurs at each step; that is, if we assume that y_n is accurate, then y_{n+1} will contain a local truncation error.

To derive a formula for the local truncation error for Euler's method, we use Taylor's formula with remainder. If a function $y(x)$ possesses $k + 1$ derivatives that are continuous on an open interval containing a and x, then

$$y(x) = y(a) + y'(a)\frac{x - a}{1!} + \cdots + y^{(k)}(a)\frac{(x - a)^k}{k!} + y^{(k+1)}(c)\frac{(x - a)^{k+1}}{(k + 1)!},$$

where c is some point between a and x. Setting $k = 1$, $a = x_n$, and $x = x_{n+1} = x_n + h$, we get

$$y(x_{n+1}) = y(x_n) + y'(x_n)\frac{h}{1!} + y''(c)\frac{h^2}{2!} \quad \text{or} \quad y(x_{n+1}) = \underbrace{y_n + hf(x_n, y_n)}_{y_{n+1}} + y''(c)\frac{h^2}{2!}.$$

Euler's method (1) is the last formula without the last term; hence the local truncation error in y_{n+1} is

$$y''(c)\frac{h^2}{2!} \quad \text{where} \quad x_n < c < x_{n+1}.$$

The value of c is usually unknown (it exists theoretically), and so the *exact* error cannot be calculated, but an upper bound on the absolute value of the error is

$$M\frac{h^2}{2!} \quad \text{where} \quad M = \max_{x_n < x < x_{n+1}} |y''(x)|.$$

In discussing errors arising from the use of numerical methods, it is helpful to use the notation $O(h^n)$. To define this concept we let $e(h)$ denote the error in a numerical calculation depending on h. Then $e(h)$ is said to be of order h^n, denoted by $O(h^n)$, if there is a constant C and a positive integer n such that $|e(h)| \leq Ch^n$ for h is sufficiently small. Thus the local truncation error for Euler's method is $O(h^2)$. We note that, in general, if $e(h)$ in a numerical method is of order h^n, and h is halved, the new error is approximately $C(h/2)^n = Ch^n/2^n$; that is, the error is reduced by a factor of $(\frac{1}{2})^n$.

EXAMPLE 1 **Bound for Local Truncation Error**

Find a bound for the local truncation errors for Euler's method applied to $y' = 2xy$, $y(1) = 1$.

SOLUTION From the solution $y = e^{x^2 - 1}$ we get $y'' = (2 + 4x^2)e^{x^2 - 1}$, and so the local truncation error is

$$y''(c)\frac{h^2}{2} = (2 + 4c^2)e^{(c^2 - 1)}\frac{h^2}{2},$$

where c is between x_n and $x_n + h$. In particular, for $h = 0.1$ we can get an upper bound on the local truncation error for y_1 by replacing c by 1.1:

$$[2 + (4)(1.1)^2]e^{((1.1)^2 - 1)}\frac{(0.1)^2}{2} = 0.0422.$$

From Table 6.1.1 we see that the error after the first step is 0.0337, less than the value given by the bound.

Similarly, we can get a bound for the local truncation error for any of the five steps given in Table 6.1.1 by replacing c by 1.5 (this value of c gives the largest value of $y''(c)$ for any of the steps and may be too generous for the first few steps). Doing this gives

$$[2 + (4)(1.5)^2]e^{((1.5)^2 - 1)}\frac{(0.1)^2}{2} = 0.1920 \tag{2}$$

as an upper bound for the local truncation error in each step. ≡

Note in Example 1 that if h is halved to 0.05, then the error bound is 0.0480, about one-fourth as much as shown in (2). This is expected because the local truncation error for Euler's method is $O(h^2)$.

In the above analysis we assumed that the value of y_n was exact in the calculation of y_{n+1}, but it is not because it contains local truncation errors from previous steps. The total error in y_{n+1} is an accumulation of the errors in each of the previous steps. This total error is called the **global truncation error**. A complete analysis of the global truncation error is beyond the scope of this text, but it can be shown that the global truncation error for Euler's method is $O(h)$.

We expect that, for Euler's method, if the step size is halved the error will be approximately halved as well. This is borne out in Tables 6.1.1 and 6.1.2, where the absolute error at $x = 1.50$ with $h = 0.1$ is 0.5625 and with $h = 0.05$ is 0.3171, approximately half as large.

In general it can be shown that if a method for the numerical solution of a differential equation has local truncation error $O(h^{\alpha + 1})$, then the global truncation error is $O(h^{\alpha})$.

For the remainder of this section and in the subsequent sections we study methods that give significantly greater accuracy than Euler's method.

FIGURE 6.1.1 Slope m_{ave} is average of m_0 and m_1

▮ Improved Euler's Method The numerical method defined by the formula

$$y_{n+1} = y_n + h\frac{f(x_n, y_n) + f(x_{n+1}, y_{n+1}^*)}{2}, \tag{3}$$

where

$$y_{n+1}^* = y_n + hf(x_n, y_n), \tag{4}$$

is commonly known as the **improved Euler's method**. In order to compute y_{n+1} for $n = 0$, 1, 2, ..., from (3) we must, at each step, first use Euler's method (4) to obtain an initial estimate y_{n+1}^*. For example, with $n = 0$, (4) would give $y_1^* = y_0 + hf(x_0, y_0)$, and then knowing this value we use (3) to get $y_1 = y_0 + h(f(x_0, y_0) + f(x_1, y_1^*))/2$, where $x_1 = x_0 + h$. These equations can be readily visualized. In **FIGURE 6.1.1** observe that $m_0 = f(x_0, y_0)$ and $m_1 = f(x_1, y_1^*)$ are slopes of the solid straight lines shown passing through the points (x_0, y_0) and (x_1, y_1^*), respectively. By taking an average of these slopes, that is, $m_{\text{ave}} = (f(x_0, y_0) + f(x_1, y_1^*))/2$, we obtain the slope of the parallel dashed skew lines. With the first step, rather than advancing along the line through (x_0, y_0) with slope $f(x_0, y_0)$ to the point with y-coordinate y_1^* obtained by Euler's method, we advance instead along the red dashed line through (x_0, y_0) with slope m_{ave} until we reach x_1. It seems plausible from inspection of the figure that y_1 is an improvement over y_1^*.

In general, the improved Euler's method is an example of a **predictor–corrector method**. The value of y_{n+1}^* given by (4) *predicts* a value of $y(x_n)$, whereas the value of y_{n+1} defined by formula (3) *corrects* this estimate.

EXAMPLE 2 Improved Euler's Method

Use the improved Euler's method to obtain the approximate value of $y(1.5)$ for the solution of the initial-value problem $y' = 2xy$, $y(1) = 1$. Compare the results for $h = 0.1$ and $h = 0.05$.

SOLUTION With $x_0 = 1, y_0 = 1, f(x_n, y_n) = 2x_ny_n, n = 0$, and $h = 0.1$ we first compute (4):

$$y_1^* = y_0 + (0.1)(2x_0y_0) = 1 + (0.1)2(1)(1) = 1.2.$$

We use this last value in (3) along with $x_1 = 1 + h = 1 + 0.1 = 1.1$:

$$y_1 = y_0 + (0.1)\frac{2x_0y_0 + 2x_1y_1^*}{2} = 1 + (0.1)\frac{2(1)(1) + 2(1.1)(1.2)}{2} = 1.232.$$

The comparative values of the calculations for $h = 0.1$ and $h = 0.05$ are given in Tables 6.1.3 and 6.1.4, respectively.

TABLE 6.1.3 Improved Euler's Method with $h = 0.1$

x_n	y_n	Actual Value	Abs. Error	% Rel. Error
1.00	1.0000	1.0000	0.0000	0.00
1.10	1.2320	1.2337	0.0017	0.14
1.20	1.5479	1.5527	0.0048	0.31
1.30	1.9832	1.9937	0.0106	0.53
1.40	2.5908	2.6117	0.0209	0.80
1.50	3.4509	3.4904	0.0394	1.13

TABLE 6.1.4 Improved Euler's Method with $h = 0.05$

x_n	y_n	Actual Value	Abs. Error	% Rel. Error
1.00	1.0000	1.0000	0.0000	0.00
1.05	1.1077	1.1079	0.0002	0.02
1.10	1.2332	1.2337	0.0004	0.04
1.15	1.3798	1.3806	0.0008	0.06
1.20	1.5514	1.5527	0.0013	0.08
1.25	1.7531	1.7551	0.0020	0.11
1.30	1.9909	1.9937	0.0029	0.14
1.35	2.2721	2.2762	0.0041	0.18
1.40	2.6060	2.6117	0.0057	0.22
1.45	3.0038	3.0117	0.0079	0.26
1.50	3.4795	3.4904	0.0108	0.31

Note that we can't compute all the y_n^* first. ▶ A brief word of caution is in order here. We cannot compute all the values of y_n^* first and then substitute these values into formula (3). In other words, we cannot use the data in Table 6.1.1 to help construct the values in Table 6.1.3. Why not?

■ **Truncation Errors for the Improved Euler's Method** The local truncation error for the improved Euler's method is $O(h^3)$. The derivation of this result is similar to the derivation of the local truncation error for Euler's method. Since the local truncation error for the improved Euler's method is $O(h^3)$, the global truncation error is $O(h^2)$. This can be seen in Example 2; when the step size is halved from $h = 0.1$ to $h = 0.05$, the absolute error at $x = 1.50$ is reduced from 0.0394 to 0.0108, a reduction of approximately $(\frac{1}{2})^2 = \frac{1}{4}$.

6.1 Exercises Answers to selected odd-numbered problems begin on page ANS-14.

Given the initial-value problems in Problems 1–10, use the improved Euler's method to obtain a four-decimal approximation to the indicated value. First use $h = 0.1$ and then use $h = 0.05$.

1. $y' = 2x - 3y + 1, \quad y(1) = 5; \quad y(1.5)$
2. $y' = 4x - 2y, \quad y(0) = 2; \quad y(0.5)$
3. $y' = 1 + y^2, \quad y(0) = 0; \quad y(0.5)$
4. $y' = x^2 + y^2, \quad y(0) = 1; \quad y(0.5)$
5. $y' = e^{-y}, \quad y(0) = 0; \quad y(0.5)$
6. $y' = x + y^2, \quad y(0) = 0; \quad y(0.5)$

7. $y' = (x - y)^2, \quad y(0) = 0.5; \quad y(0.5)$
8. $y' = xy + \sqrt{y}, \quad y(0) = 1; \quad y(0.5)$
9. $y' = xy^2 - \frac{y}{x}, \quad y(1) = 1; \quad y(1.5)$
10. $y' = y - y^2, \quad y(0) = 0.5; \quad y(0.5)$
11. Consider the initial-value problem $y' = (x + y - 1)^2$, $y(0) = 2$. Use the improved Euler's method with $h = 0.1$ and $h = 0.05$ to obtain approximate values of the solution at $x = 0.5$. At each step compare the approximate value with the exact value of the analytic solution.

12. Although it may not be obvious from the differential equation, its solution could "behave badly" near a point x at which we wish to approximate $y(x)$. Numerical procedures may give widely differing results near this point. Let $y(x)$ be the solution of the initial-value problem $y' = x^2 + y^3$, $y(1) = 1$.

 (a) Use a numerical solver to obtain the graph of the solution on the interval $[1, 1.4]$.

 (b) Using the step size $h = 0.1$, compare the results obtained from Euler's method with the results from the improved Euler's method in the approximation of $y(1.4)$.

13. Consider the initial-value problem $y' = 2y$, $y(0) = 1$. The analytic solution is $y = e^{2x}$.

 (a) Approximate $y(0.1)$ using one step and Euler's method.

 (b) Find a bound for the local truncation error in y_1.

 (c) Compare the actual error in y_1 with your error bound.

 (d) Approximate $y(0.1)$ using two steps and Euler's method.

 (e) Verify that the global truncation error for Euler's method is $O(h)$ by comparing the errors in parts (a) and (d).

14. Repeat Problem 13 using the improved Euler's method. Its global truncation error is $O(h^2)$.

15. Repeat Problem 13 using the initial-value problem $y' = x - 2y$, $y(0) = 1$. The analytic solution is

$$y = \tfrac{1}{2}x - \tfrac{1}{4} + \tfrac{5}{4}e^{-2x}.$$

16. Repeat Problem 15 using the improved Euler's method. Its global truncation error is $O(h^2)$.

17. Consider the initial-value problem $y' = 2x - 3y + 1$, $y(1) = 5$. The analytic solution is

$$y(x) = \tfrac{1}{9} + \tfrac{2}{3}x + \tfrac{38}{9}e^{-3(x-1)}.$$

 (a) Find a formula involving c and h for the local truncation error in the nth step if Euler's method is used.

 (b) Find a bound for the local truncation error in each step if $h = 0.1$ is used to approximate $y(1.5)$.

 (c) Approximate $y(1.5)$ using $h = 0.1$ and $h = 0.05$ with Euler's method. See Problem 1 in Exercises 2.6.

 (d) Calculate the errors in part (c) and verify that the global truncation error of Euler's method is $O(h)$.

18. Repeat Problem 17 using the improved Euler's method, which has a global truncation error $O(h^2)$. See Problem 1. You may need to keep more than four decimal places to see the effect of reducing the order of error.

19. Repeat Problem 17 for the initial-value problem $y' = e^{-y}$, $y(0) = 0$. The analytic solution is $y(x) = \ln(x + 1)$. Approximate $y(0.5)$. See Problem 5 in Exercises 2.6.

20. Repeat Problem 19 using the improved Euler's method, which has a global truncation error $O(h^2)$. See Problem 5. You may need to keep more than four decimal places to see the effect of reducing the order of error.

Discussion Problem

21. Answer the question: "Why not?" that follows the three sentences after Example 2 on page 310.

6.2 Runge–Kutta Methods

INTRODUCTION Probably one of the more popular, as well as most accurate, numerical procedures used in obtaining approximate solutions to a first-order initial-value problem $y' = f(x, y)$, $y(x_0) = y_0$ is the **fourth-order Runge–Kutta method** devised by the German applied mathematicians **Carl David Runge** (1856–1927) and **Martin Wilhelm Kutta** (1867–1944) in the late 1890s. As the words *fourth-order* suggest, there are Runge–Kutta methods of different orders.

■ Runge–Kutta Methods Fundamentally, all Runge–Kutta methods are generalizations of the basic Euler formula (1) of Section 6.1 in that the slope function f is replaced by a weighted average of slopes over the interval defined by $x_n \leq x \leq x_{n+1}$. That is,

$$\overbrace{}^{\text{weighted average}}$$
$$y_{n+1} = y_n + h(w_1 k_1 + w_2 k_2 + \cdots + w_m k_m). \tag{1}$$

Here the weights w_i, $i = 1, 2, \ldots, m$ are constants that generally satisfy $w_1 + w_2 + \cdots + w_m = 1$ and each k_i, $i = 1, 2, \ldots, m$ is the function f evaluated at a selected point (x, y) for which $x_n \leq x \leq x_{n+1}$. We shall see that the k_i are defined recursively. The number m is called the **order** of the method. Observe that by taking $m = 1$, $w_1 = 1$, and $k_1 = f(x_n, y_n)$ we get the familiar Euler formula $y_{n+1} = y_n + hf(x_n, y_n)$. Hence Euler's method is said to be a **first-order Runge–Kutta method**.

The average in (1) is not formed willy-nilly, but parameters are chosen so that (1) agrees with a Taylor polynomial of degree m. As we have seen in the last section, if a function $y(x)$ possesses $k + 1$ derivatives that are continuous on an open interval containing a and x, then we can write

$$y(x) = y(a) + y'(a)\frac{x - a}{1!} + y''(a)\frac{(x - a)^2}{2!} + \cdots + y^{(k+1)}(c)\frac{(x - a)^{k+1}}{(k + 1)!},$$

where c is some number between a and x. If we replace a by x_n and x by $x_{n+1} = x_n + h$, then the foregoing formula becomes

$$y(x_{n+1}) = y(x_n + h) = y(x_n) + hy'(x_n) + \frac{h^2}{2!} y''(x_n) + \cdots + \frac{h^{k+1}}{(k+1)!} y^{(k+1)}(c),$$

where c is now some number between x_n and x_{n+1}. When $y(x)$ is a solution of $y' = f(x, y)$, in the case $k = 1$ and the remainder $\frac{1}{2} h^2 y''(c)$ is small, we see that a Taylor polynomial $y(x_{n+1}) = y(x_n) + hy'(x_n)$ of degree 1 agrees with the approximation formula of Euler's method

$$y_{n+1} = y_n + hy_n' = y_n + hf(x_n, y_n).$$

■ A Second-Order Runge–Kutta Method

To further illustrate (1), we consider now a **second-order Runge–Kutta method**. This consists of finding constants, or parameters, w_1, w_2, α, and β so that the formula

$$y_{n+1} = y_n + h(w_1 k_1 + w_2 k_2), \tag{2}$$

where
$$k_1 = f(x_n, y_n),$$
$$k_2 = f(x_n + \alpha h, y_n + \beta h k_1)$$

agrees with a Taylor polynomial of degree 2. For our purposes it suffices to say that this can be done whenever the constants satisfy

$$w_1 + w_2 = 1, \quad w_2 \alpha = \frac{1}{2}, \quad \text{and} \quad w_2 \beta = \frac{1}{2}. \tag{3}$$

This is an algebraic system of three equations in four unknowns and has infinitely many solutions:

$$w_1 = 1 - w_2, \quad \alpha = \frac{1}{2w_2}, \quad \text{and} \quad \beta = \frac{1}{2w_2}, \tag{4}$$

where $w_2 \neq 0$. For example, the choice $w_2 = \frac{1}{2}$ yields $w_1 = \frac{1}{2}$, $\alpha = 1$, $\beta = 1$ and so (2) becomes

$$y_{n+1} = y_n + \frac{h}{2}(k_1 + k_2),$$

where
$$k_1 = f(x_n, y_n) \quad \text{and} \quad k_2 = f(x_n + h, y_n + h k_1).$$

Since $x_n + h = x_{n+1}$ and $y_n + h k_1 = y_n + hf(x_n, y_n)$ the foregoing result is recognized to be the improved Euler's method that is summarized in (3) and (4) of Section 6.1.

In view of the fact that $w_2 \neq 0$ can be chosen arbitrarily in (4), there are many possible second-order Runge–Kutta methods. See Problem 2 in Exercises 6.2.

We shall skip any discussion of third-order methods in order to come to the principal point of discussion in this section.

■ A Fourth-Order Runge–Kutta Method

A **fourth-order Runge–Kutta procedure** consists of finding parameters so that the formula

$$y_{n+1} = y_n + h(w_1 k_1 + w_2 k_2 + w_3 k_3 + w_4 k_4), \tag{5}$$

where
$$k_1 = f(x_n, y_n)$$
$$k_2 = f(x_n + \alpha_1 h, y_n + \beta_1 h k_1)$$
$$k_3 = f(x_n + \alpha_2 h, y_n + \beta_2 h k_1 + \beta_3 h k_2)$$
$$k_4 = f(x_n + \alpha_3 h, y_n + \beta_4 h k_1 + \beta_5 h k_2 + \beta_6 h k_3)$$

agrees with a Taylor polynomial of degree 4. This results in a system of 11 equations in 13 unknowns. The most commonly used set of values for the parameters yields the following result:

$$y_{n+1} = y_n + \frac{h}{6}(k_1 + 2k_2 + 2k_3 + k_4),$$
$$k_1 = f(x_n, y_n)$$
$$k_2 = f(x_n + \tfrac{1}{2}h, y_n + \tfrac{1}{2}hk_1) \tag{6}$$
$$k_3 = f(x_n + \tfrac{1}{2}h, y_n + \tfrac{1}{2}hk_2)$$
$$k_4 = f(x_n + h, y_n + hk_3).$$

While other fourth-order formulas are easily derived, the algorithm summarized in (6) is so widely used and recognized as a valuable computational tool it is often referred to as *the* fourth-order Runge–Kutta method or *the classical* Runge–Kutta method. It is (6) that we have in mind, hereafter, when we use the abbreviation "the RK4 method."

You are advised to look carefully at the formulas in (6); note that k_2 depends on k_1, k_3 depends on k_2, and k_4 depends on k_3. Also, k_2 and k_3 involve approximations to the slope at the midpoint $x_n + \tfrac{1}{2}h$ of the interval $[x_n, x_{n+1}]$.

EXAMPLE 1 **RK4 Method**

Use the RK4 method with $h = 0.1$ to obtain an approximation to $y(1.5)$ for the solution of $y' = 2xy$, $y(1) = 1$.

SOLUTION For the sake of illustration, let us compute the case when $n = 0$. From (6) we find

$$k_1 = f(x_0, y_0) = 2x_0 y_0 = 2$$
$$k_2 = f(x_0 + \tfrac{1}{2}(0.1), y_0 + \tfrac{1}{2}(0.1)2)$$
$$= 2(x_0 + \tfrac{1}{2}(0.1))(y_0 + \tfrac{1}{2}(0.2)) = 2.31$$
$$k_3 = f(x_0 + \tfrac{1}{2}(0.1), y_0 + \tfrac{1}{2}(0.1)2.31)$$
$$= 2(x_0 + \tfrac{1}{2}(0.1))(y_0 + \tfrac{1}{2}(0.231)) = 2.34255$$
$$k_4 = f(x_0 + (0.1), y_0 + (0.1)2.34255)$$
$$= 2(x_0 + 0.1)(y_0 + 0.234255) = 2.715361$$

and therefore

$$y_1 = y_0 + \frac{0.1}{6}(k_1 + 2k_2 + 2k_3 + k_4)$$

$$= 1 + \frac{0.1}{6}(2 + 2(2.31) + 2(2.34255) + 2.715361) = 1.23367435.$$

The remaining calculations are summarized in Table 6.2.1, whose entries are rounded to four decimal places. ≡

Inspection of Table 6.2.1 shows why the fourth-order Runge–Kutta method is so popular. If four-decimal-place accuracy is all that we desire, there is no need to use a smaller step size. Table 6.2.2 on page 314 compares the results of applying Euler's, the improved Euler's, and the fourth-order Runge–Kutta methods to the initial-value problem $y' = 2xy$, $y(1) = 1$. See Tables 6.1.1–6.1.4.

TABLE 6.2.1 RK4 Method with $h = 0.1$

x_n	y_n	Actual Value	Abs. Error	% Rel. Error
1.00	1.0000	1.0000	0.0000	0.00
1.10	1.2337	1.2337	0.0000	0.00
1.20	1.5527	1.5527	0.0000	0.00
1.30	1.9937	1.9937	0.0000	0.00
1.40	2.6116	2.6117	0.0001	0.00
1.50	3.4902	3.4904	0.0001	0.00

TABLE 6.2.2 $y' = 2xy,\ y(1) = 1$

	Comparison of Numerical Methods with $h = 0.1$				Comparison of Numerical Methods with $h = 0.05$				
x_n	Euler	Improved Euler	RK4	Actual Value	x_n	Euler	Improved Euler	RK4	Actual Value
1.00	1.0000	1.0000	1.0000	1.0000	1.00	1.0000	1.0000	1.0000	1.0000
1.10	1.2000	1.2320	1.2337	1.2337	1.05	1.1000	1.1077	1.1079	1.1079
1.20	1.4640	1.5479	1.5527	1.5527	1.10	1.2155	1.2332	1.2337	1.2337
1.30	1.8154	1.9832	1.9937	1.9937	1.15	1.3492	1.3798	1.3806	1.3806
1.40	2.2874	2.5908	2.6116	2.6117	1.20	1.5044	1.5514	1.5527	1.5527
1.50	2.9278	3.4509	3.4902	3.4904	1.25	1.6849	1.7531	1.7551	1.7551
					1.30	1.8955	1.9909	1.9937	1.9937
					1.35	2.1419	2.2721	2.2762	2.2762
					1.40	2.4311	2.6060	2.6117	2.6117
					1.45	2.7714	3.0038	3.0117	3.0117
					1.50	3.1733	3.4795	3.4903	3.4904

■ **Truncation Errors for the RK4 Method** In Section 6.1 we saw that global trunca-tion errors for Euler's method and for the improved Euler's method are, respectively, $O(h)$ and $O(h^2)$. Because the first equation in (6) agrees with a Taylor polynomial of degree 4, the local truncation error for this method is $y^{(5)}(c)h^5/5!$ or $O(h^5)$, and the global truncation error is thus $O(h^4)$. It is now obvious why Euler's method, the improved Euler's method, and (6) are *first-*, *second-*, and *fourth-order* Runge–Kutta methods, respectively.

EXAMPLE 2 **Bound for Local Truncation Errors**

Find a bound for the local truncation errors for the RK4 method applied to $y' = 2xy,\ y(1) = 1$.

SOLUTION By computing the fifth derivative of the known solution $y(x) = e^{x^2 - 1}$ we get

$$y^{(5)}(c)\,\frac{h^5}{5!} = (120c + 160c^3 + 32c^5)e^{c^2-1}\,\frac{h^5}{5!}. \tag{7}$$

Thus with $c = 1.5$, (7) yields a bound of 0.00028 on the local truncation error for each of the five steps when $h = 0.1$. Note that in Table 6.2.1 the error in y_1 is much less than this bound.

Table 6.2.3 gives the approximations to the solution of the initial-value problem at $x = 1.5$ that are obtained from the RK4 method. By computing the value of the analytic solution at $x = 1.5$ we can find the error in these approximations. Because the method is so accurate, many decimal places must be used in the numerical solution to see the effect of halving the step size. Note that when h is halved, from $h = 0.1$ to $h = 0.05$, the error is divided by a factor of about $2^4 = 16$, as expected. ≡

TABLE 6.2.3 RK4 Method

h	Approx.	Error
0.1	3.49021064	$1.32321089 \times 10^{-4}$
0.05	3.49033382	$9.13776090 \times 10^{-6}$

■ **Adaptive Methods** We have seen that the accuracy of a numerical method for approximating solutions of differential equations can be improved by decreasing the step size h. Of course, this enhanced accuracy is usually obtained at a cost—namely, increased compu-tation time and greater possibility of round-off error. In general, over the interval of approxi-mation there may be subintervals where a relatively large step size suffices and other subintervals where a smaller step size is necessary in order to keep the truncation error within a desired limit. Numerical methods that use a variable step size are called **adaptive methods**. One of the more popular of the adaptive routines is the **Runge–Kutta–Fehlberg method**. Because the German mathematician **Erwin Fehlberg** (1911–1990) employed two Runge–Kutta methods of differing orders, a fourth- and a fifth-order method, this algorithm is frequently denoted as the **RKF45 method**.*

*The Runge–Kutta method of order four used in RKF45 is *not* the same as that given in (6).

1. Use the RK4 method with $h = 0.1$ to approximate $y(0.5)$, where $y(x)$ is the solution of the initial-value problem $y' = (x + y - 1)^2, y(0) = 2$. Compare this approximate value with the actual value obtained in Problem 11 in Exercises 6.1.

2. Assume that $w_2 = \frac{3}{4}$ in (4). Use the resulting second-order Runge–Kutta method to approximate $y(0.5)$, where $y(x)$ is the solution of the initial-value problem in Problem 1. Compare this approximate value with the approximate value obtained in Problem 11 in Exercises 6.1.

In Problems 3–12, use the RK4 method with $h = 0.1$ to obtain a four-decimal approximation to the indicated value.

3. $y' = 2x - 3y + 1$, $y(1) = 5$; $y(1.5)$

4. $y' = 4x - 2y$, $y(0) = 2$; $y(0.5)$

5. $y' = 1 + y^2$, $y(0) = 0$; $y(0.5)$

6. $y' = x^2 + y^2$, $y(0) = 1$; $y(0.5)$

7. $y' = e^{-y}$, $y(0) = 0$; $y(0.5)$

8. $y' = x + y^2$, $y(0) = 0$; $y(0.5)$

9. $y' = (x - y)^2$, $y(0) = 0.5$; $y(0.5)$

10. $y' = xy + \sqrt{y}$, $y(0) = 1$; $y(0.5)$

11. $y' = xy^2 - \dfrac{y}{x}$, $y(1) = 1$; $y(1.5)$

12. $y' = y - y^2$, $y(0) = 0.5$; $y(0.5)$

13. If air resistance is proportional to the square of the instantaneous velocity, then the velocity v of a mass m dropped from a given height h is determined from

$$m \frac{dv}{dt} = mg - kv^2, \ k > 0.$$

Let $v(0) = 0$, $k = 0.125$, $m = 5$ slugs, and $g = 32 \text{ ft/s}^2$.
(a) Use the RK4 method with $h = 1$ to approximate the velocity $v(5)$.
(b) Use a numerical solver to graph the solution of the IVP on the interval $[0, 6]$.
(c) Use separation of variables to solve the IVP and then find the actual value $v(5)$.

14. A mathematical model for the area A (in cm^2) that a colony of bacteria (*B. dendroides*) occupies is given by

$$\frac{dA}{dt} = A(2.128 - 0.0432A).$$

Suppose that the initial area is 0.24 cm^2.
(a) Use the RK4 method with $h = 0.5$ to complete the following table.

t (days)	1	2	3	4	5
A (observed)	2.78	13.53	36.30	47.50	49.40
A (approximated)					

(b) Use a numerical solver to graph the solution of the initial-value problem. Estimate the values $A(1)$, $A(2)$, $A(3)$, $A(4)$, and $A(5)$ from the graph.

(c) Use separation of variables to solve the initial-value problem and compute the values $A(1), A(2), A(3), A(4)$, and $A(5)$.

15. Consider the initial-value problem $y' = x^2 + y^3$, $y(1) = 1$. See Problem 12 in Exercises 6.1.
(a) Compare the results obtained from using the RK4 method over the interval $[1, 1.4]$ with step sizes $h = 0.1$ and $h = 0.05$.
(b) Use a numerical solver to graph the solution of the initial-value problem on the interval $[1, 1.4]$.

16. Consider the initial-value problem $y' = 2y$, $y(0) = 1$. The analytic solution is $y(x) = e^{2x}$.
(a) Approximate $y(0.1)$ using one step and the fourth-order RK4 method.
(b) Find a bound for the local truncation error in y_1.
(c) Compare the actual error in y_1 with your error bound.
(d) Approximate $y(0.1)$ using two steps and the RK4 method.
(e) Verify that the global truncation error for the RK4 method is $O(h^4)$ by comparing the errors in parts (a) and (d).

17. Repeat Problem 16 using the initial-value problem $y' = -2y + x$, $y(0) = 1$. The analytic solution is

$$y(x) = \tfrac{1}{2}x - \tfrac{1}{4} + \tfrac{5}{4}e^{-2x}.$$

18. Consider the initial-value problem $y' = 2x - 3y + 1$, $y(1) = 5$. The analytic solution is

$$y(x) = \tfrac{1}{9} + \tfrac{2}{3}x + \tfrac{38}{9}e^{-3(x-1)}.$$

(a) Find a formula involving c and h for the local truncation error in the nth step if the RK4 method is used.
(b) Find a bound for the local truncation error in each step if $h = 0.1$ is used to approximate $y(1.5)$.
(c) Approximate $y(1.5)$ using the RK4 method with $h = 0.1$ and $h = 0.05$. See Problem 3. You will need to carry more than six decimal places to see the effect of reducing the step size.

19. Repeat Problem 18 for the initial-value problem $y' = e^{-y}$, $y(0) = 0$. The analytic solution is $y(x) = \ln(x + 1)$. Approximate $y(0.5)$. See Problem 7.

Discussion Problem

20. A count of the number of evaluations of the function f used in solving the initial-value problem $y' = f(x, y)$, $y(x_0) = y_0$ is used as a measure of the computational complexity of a numerical method. Determine the number of evaluations of f required for each step of Euler's, the improved Euler's, and the RK4 methods. By considering some specific examples, compare the accuracy of these methods when used with comparable computational complexities.

Computer Lab Assignment

21. The RK4 method for solving an initial-value problem over an interval $[a, b]$ results in a finite set of points that are

supposed to approximate points on the graph of the exact solution. In order to expand this set of discrete points to an approximate solution defined at *all* points on the interval $[a, b]$, we can use an **interpolating function**. This is a function, supported by most computer algebra systems, that agrees with the given data exactly and assumes a smooth transition between data points. These interpolating functions may be polynomials or sets of polynomials joined together smoothly. In *Mathematica* the command $\mathbf{y} = \mathbf{Interpolation[data]}$ can be used to obtain an interpolating function through the points

$\mathbf{data} = \{ \{x_0, y_0\}, \{x_1, y_1\}, \dots, \{x_n, y_n\} \}$. The interpolating function $\mathbf{y[x]}$ can now be treated like any other function built into the computer algebra system.

(a) Find the analytic solution of the initial-value problem $y' = -y + 10 \sin 3x$; $y(0) = 0$ on the interval $[0, 2]$. Graph this solution and find its positive roots.

(b) Use the RK4 method with $h = 0.1$ to approximate a solution of the initial-value problem in part (a). Obtain an interpolating function and graph it. Find the positive roots of the interpolating function on the interval $[0, 2]$.

6.3 Multistep Methods

INTRODUCTION Euler's, the improved Euler's, and the Runge–Kutta methods are examples of **single-step** or **starting methods**. In these methods each successive value y_{n+1} is computed based only on information about the immediately preceding value y_n. On the other hand, **multistep** or **continuing methods** use the values from several computed steps to obtain the value of y_{n+1}. There are a large number of multistep method formulas for approximating solutions of DEs, but since it is not our intention to survey the vast field of numerical procedures, we will consider only one such method here.

Adams–Bashforth–Moulton Method

The multistep method discussed in this section is called the Adams–Bashforth–Moulton method after the English astronomer/mathematician **John Couch Adams** (1819–1892) who co-discovered the planet Neptune, the English applied mathematician **Francis Bashforth** (1819–1912), and the American astronomer **Forest Ray Moulton** (1872–1952). Like the improved Euler's method it is a predictor–corrector method—that is, one formula is used to predict a value y_{n+1}^{*}, which in turn is used to obtain a corrected value y_{n+1}. The predictor in this method is the Adams–Bashforth formula

$$y_{n+1}^{*} = y_n + \frac{h}{24}(55y_n' - 59y_{n-1}' + 37y_{n-2}' - 9y_{n-3}'), \tag{1}$$

$$y_n' = f(x_n, y_n)$$

$$y_{n-1}' = f(x_{n-1}, y_{n-1})$$

$$y_{n-2}' = f(x_{n-2}, y_{n-2})$$

$$y_{n-3}' = f(x_{n-3}, y_{n-3})$$

for $n \geq 3$. The value of y_{n+1}^{*} is then substituted into the Adams–Moulton corrector

$$y_{n+1} = y_n + \frac{h}{24}(9y_{n+1}' + 19y_n' - 5y_{n-1}' + y_{n-2}'), \tag{2}$$

$$y_{n+1}' = f(x_{n+1}, y_{n+1}^{*}).$$

Notice that formula (1) requires that we know the values of y_0, y_1, y_2, and y_3 in order to obtain y_4. The value of y_0 is, of course, the given initial condition. Since the local truncation error of the Adams–Bashforth–Moulton method is $O(h^5)$, the values of y_1, y_2, and y_3 are generally computed by a method with the same error property, such as the fourth-order Runge–Kutta formula.

EXAMPLE 1 **Adams–Bashforth–Moulton Method**

Use the Adams–Bashforth–Moulton method with $h = 0.2$ to obtain an approximation to $y(0.8)$ for the solution of

$$y' = x + y - 1, \qquad y(0) = 1.$$

SOLUTION With a step size of $h = 0.2$, $y(0.8)$ will be approximated by y_4. To get started we use the RK4 method with $x_0 = 0$, $y_0 = 1$, and $h = 0.2$ to obtain

$$y_1 = 1.02140000, \qquad y_2 = 1.09181796, \qquad y_3 = 1.22210646.$$

Now with the identifications $x_0 = 0$, $x_1 = 0.2$, $x_2 = 0.4$, $x_3 = 0.6$, and $f(x, y) = x + y - 1$, we find

$$y_0' = f(x_0, y_0) = (0) + (1) - 1 = 0$$

$$y_1' = f(x_1, y_1) = (0.2) + (1.02140000) - 1 = 0.22140000$$

$$y_2' = f(x_2, y_2) = (0.4) + (1.09181796) - 1 = 0.49181796$$

$$y_3' = f(x_3, y_3) = (0.6) + (1.22210646) - 1 = 0.82210646.$$

With the foregoing values, the predictor (1) then gives

$$y_4^* = y_3 + \frac{0.2}{24}(55y_3' - 59y_2' + 37y_1' - 9y_0') = 1.42535975.$$

To use the corrector (2) we first need

$$y_4' = f(x_4, y_4^*) = 0.8 + 1.42535975 - 1 = 1.22535975.$$

Finally, (2) yields

$$y_4 = y_3 + \frac{0.2}{24}(9y_4' + 19y_3' - 5y_2' + y_1') = 1.42552788. \qquad \equiv$$

You should verify that the exact value of $y(0.8)$ in Example 1 is $y(0.8) = 1.42554093$.

▇ Stability of Numerical Methods

An important consideration in using numerical methods to approximate the solution of an initial-value problem is the stability of the method. Simply stated, a numerical method is **stable** if small changes in the initial condition result in only small changes in the computed solution. A numerical method is said to be **unstable** if it is not stable. The reason that stability considerations are important is that in each step after the first step of a numerical technique we are essentially starting over again with a new initial-value problem, where the initial condition is the approximate solution value computed in the preceding step. Because of the presence of round-off error, this value will almost certainly vary at least slightly from the true value of the solution. Besides round-off error, another common source of error occurs in the initial condition itself; in physical applications the data are often obtained by imprecise measurements.

One possible method for detecting instability in the numerical solution of a specific initial-value problem is to compare the approximate solutions obtained when decreasing step sizes are used. If the numerical method is unstable, the error may actually increase with smaller step sizes. Another way of checking stability is to observe what happens to solutions when the initial condition is slightly perturbed (for example, change $y(0) = 1$ to $y(0) = 0.999$).

For a more detailed and precise discussion of stability, consult a numerical analysis text. In general, all of the methods we have discussed in this chapter have good stability characteristics.

▇ Advantages/Disadvantages of Multistep Methods

Many considerations enter into the choice of a method to solve a differential equation numerically. Single-step methods, particularly the Runge–Kutta method, are often chosen because of their accuracy and the fact that they are easy to program. However, a major drawback is that the right-hand side of the differential equation must be evaluated many times at each step. For instance, the RK4 method requires four function evaluations for each step. On the other hand, if the function evaluations in the previous step have been calculated and stored, a multistep method requires only one new function evaluation for each step. This can lead to great savings in time and expense.

As an example, to solve $y' = f(x, y)$, $y(x_0) = y_0$ numerically using n steps by the RK4 method requires $4n$ function evaluations. The Adams–Bashforth multistep method requires 16 function evaluations for the Runge–Kutta fourth-order starter and $n - 4$ for the n Adams–Bashforth steps, giving a total of $n + 12$ function evaluations for this method. In general the Adams–Bashforth multistep method requires slightly more than a quarter of the number of function evaluations required for the RK4 method. If the evaluation of $f(x, y)$ is complicated, the multistep method will be more efficient.

Another issue involved with multistep methods is how many times the Adams–Moulton corrector formula should be repeated in each step. Each time the corrector is used, another function

evaluation is done, and so the accuracy is increased at the expense of losing an advantage of the multistep method. In practice, the corrector is calculated once, and if the value of y_{n+1} is changed by a large amount, the entire problem is restarted using a smaller step size. This is often the basis of the variable step size methods, whose discussion is beyond the scope of this text.

6.3 Exercises Answers to selected odd-numbered problems begin on page ANS-15.

1. Find the analytic solution of the initial-value problem in Example 1. Compare the actual values of $y(0.2)$, $y(0.4)$, $y(0.6)$, and $y(0.8)$ with the approximations y_1, y_2, y_3, and y_4.

2. Write a computer program to implement the Adams–Bashforth–Moulton method.

In Problems 3 and 4, use the Adams–Bashforth–Moulton method to approximate $y(0.8)$, where $y(x)$ is the solution of the given initial-value problem. Use $h = 0.2$ and the RK4 method to compute y_1, y_2, and y_3.

3. $y' = 2x - 3y + 1$, $y(0) = 1$
4. $y' = 4x - 2y$, $y(0) = 2$

In Problems 5–8, use the Adams–Bashforth–Moulton method to approximate $y(1.0)$, where $y(x)$ is the solution of the given initial-value problem. First use $h = 0.2$ and then use $h = 0.1$. Use the RK4 method to compute y_1, y_2, and y_3.

5. $y' = 1 + y^2$, $y(0) = 0$
6. $y' = y + \cos x$, $y(0) = 1$
7. $y' = (x - y)^2$, $y(0) = 0$
8. $y' = xy + \sqrt{y}$, $y(0) = 1$

6.4 Higher-Order Equations and Systems

INTRODUCTION So far we have focused on numerical techniques that can be used to approximate the solution of a first-order initial-value problem $y' = f(x, y)$, $y(x_0) = y_0$. In order to approximate the solution of a second-order initial-value problem we must express a second-order DE as a system of two first-order DEs. To do this we begin by writing the second-order DE in normal form by solving for y'' in terms of x, y, and y'.

▇ Second-Order IVPs A second-order initial-value problem

$$y'' = f(x, y, y'), \qquad y(x_0) = y_0, \qquad y'(x_0) = u_0, \tag{1}$$

can be expressed as an initial-value problem for a system of first-order differential equations. If we let $y' = u$, the differential equation in (1) becomes the system

$$\begin{aligned} y' &= u \\ u' &= f(x, y, u). \end{aligned} \tag{2}$$

Since $y'(x_0) = u(x_0)$, the corresponding initial conditions for (2) are then $y(x_0) = y_0$, $u(x_0) = u_0$. The system (2) can now be solved numerically by simply applying a particular numerical method to each first-order differential equation in the system. For example, **Euler's method** applied to the system (2) would be

$$\begin{aligned} y_{n+1} &= y_n + hu_n \\ u_{n+1} &= u_n + hf(x_n, y_n, u_n), \end{aligned} \tag{3}$$

whereas the **fourth-order Runge–Kutta method**, or **RK4 method**, would be

$$\begin{aligned} y_{n+1} &= y_n + \frac{h}{6}(m_1 + 2m_2 + 2m_3 + m_4) \\ u_{n+1} &= u_n + \frac{h}{6}(k_1 + 2k_2 + 2k_3 + k_4) \end{aligned} \tag{4}$$

where

$$\begin{aligned} m_1 &= u_n & k_1 &= f(x_n, y_n, u_n) \\ m_2 &= u_n + \tfrac{1}{2}hk_1 & k_2 &= f(x_n + \tfrac{1}{2}h, y_n + \tfrac{1}{2}hm_1, u_n + \tfrac{1}{2}hk_1) \\ m_3 &= u_n + \tfrac{1}{2}hk_2 & k_3 &= f(x_n + \tfrac{1}{2}h, y_n + \tfrac{1}{2}hm_2, u_n + \tfrac{1}{2}hk_2) \\ m_4 &= u_n + hk_3 & k_4 &= f(x_n + h, y_n + hm_3, u_n + hk_3). \end{aligned}$$

In general, we can express every nth-order differential equation

$$y^{(n)} = f(x, y, y', \ldots, y^{(n-1)})$$

as a system of n first-order equations using the substitutions $y = u_1$, $y' = u_2$, $y'' = u_3$, ..., $y^{(n-1)} = u_n$.

EXAMPLE 1 Euler's Method

Use Euler's method to obtain the approximate value of $y(0.2)$, where $y(x)$ is the solution of the initial-value problem

$$y'' + xy' + y = 0, \quad y(0) = 1, \quad y'(0) = 2. \tag{5}$$

SOLUTION In terms of the substitution $y' = u$, the equation is equivalent to the system

$$y' = u$$
$$u' = -xu - y.$$

Thus from (3) we obtain

$$y_{n+1} = y_n + hu_n$$
$$u_{n+1} = u_n + h[-x_n u_n - y_n].$$

Using the step size $h = 0.1$ and $y_0 = 1$, $u_0 = 2$, we find

$$y_1 = y_0 + (0.1)u_0 = 1 + (0.1)2 = 1.2$$
$$u_1 = u_0 + (0.1)[-x_0 u_0 - y_0] = 2 + (0.1)[-(0)(2) - 1] = 1.9$$
$$y_2 = y_1 + (0.1)u_1 = 1.2 + (0.1)(1.9) = 1.39$$
$$u_2 = u_1 + (0.1)[-x_1 u_1 - y_1] = 1.9 + (0.1)[-(0.1)(1.9) - 1.2] = 1.761.$$

In other words, $y(0.2) \approx 1.39$ and $y'(0.2) \approx 1.761$. $\equiv$

With the aid of the graphing feature of a numerical solver we have compared in **FIGURE 6.4.1(a)** the solution curve of (5) generated by Euler's method ($h = 0.1$) on the interval $[0, 3]$ with the solution curve generated by the RK4 method ($h = 0.1$). From Figure 6.4.1(b), it would appear that the solution $y(x)$ of (4) has the property that $y(x) \to 0$ as $x \to \infty$.

(a) Euler's method (blue)
Runge–Kutta method (red)

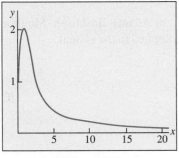

(b) Runge–Kutta

FIGURE 6.4.1 Numerical solution curves

■ Systems Reduced to First-Order Systems

Using a procedure similar to that just discussed, we can often reduce a system of higher-order differential equations to a system of first-order equations by first solving for the highest-order derivative of each dependent variable and then making appropriate substitutions for the lower-order derivatives.

EXAMPLE 2 A System Rewritten as a First-Order System

Write

$$x'' - x' + 5x + 2y'' = e^t$$
$$-2x + y'' + 2y = 3t^2$$

as a system of first-order differential equations.

SOLUTION Write the system as

$$x'' + 2y'' = e^t - 5x + x'$$
$$y'' = 3t^2 + 2x - 2y$$

and then eliminate y'' by multiplying the second equation by 2 and subtracting. This gives

$$x'' = -9x + 4y + x' + e^t - 6t^2.$$

Since the second equation of the system already expresses the highest-order derivative of y in terms of the remaining functions, we are now in a position to introduce new variables. If we let $x' = u$ and $y' = v$, the expressions for x'' and y'' become, respectively,

$$u' = x'' = -9x + 4y + u + e^t - 6t^2$$
$$v' = y'' = 2x - 2y + 3t^2.$$

The original system can then be written in the form

$$x' = u$$
$$y' = v$$
$$u' = -9x + 4y + u + e^t - 6t^2$$
$$v' = 2x - 2y + 3t^2.$$

$\equiv$

It may not always be possible to carry out the reductions illustrated in Example 2.

Numerical Solution of a System The solution of a system of the form

$$\frac{dx_1}{dt} = g_1(t, x_1, x_2, \ldots, x_n)$$

$$\frac{dx_2}{dt} = g_2(t, x_1, x_2, \ldots, x_n)$$

$$\vdots \qquad \vdots$$

$$\frac{dx_n}{dt} = g_n(t, x_1, x_2, \ldots, x_n)$$

can be approximated by a version of the Euler, Runge–Kutta, or Adams–Bashforth–Moulton method adapted to the system. For example, the **RK4 method** applied to the system

$$x' = f(t, x, y)$$
$$y' = g(t, x, y) \tag{6}$$
$$x(t_0) = x_0, \quad y(t_0) = y_0$$

looks like this:

$$x_{n+1} = x_n + \frac{h}{6}(m_1 + 2m_2 + 2m_3 + m_4)$$
$$y_{n+1} = y_n + \frac{h}{6}(k_1 + 2k_2 + 2k_3 + k_4), \tag{7}$$

where

$$m_1 = f(t_n, x_n, y_n) \qquad\qquad k_1 = g(t_n, x_n, y_n)$$
$$m_2 = f(t_n + \tfrac{1}{2}h, x_n + \tfrac{1}{2}hm_1, y_n + \tfrac{1}{2}hk_1) \qquad k_2 = g(t_n + \tfrac{1}{2}h, x_n + \tfrac{1}{2}hm_1, y_n + \tfrac{1}{2}hk_1)$$
$$m_3 = f(t_n + \tfrac{1}{2}h, x_n + \tfrac{1}{2}hm_2, y_n + \tfrac{1}{2}hk_2) \qquad k_3 = g(t_n + \tfrac{1}{2}h, x_n + \tfrac{1}{2}hm_2, y_n + \tfrac{1}{2}hk_2) \tag{8}$$
$$m_4 = f(t_n + h, x_n + hm_3, y_n + hk_3) \qquad k_4 = g(t_n + h, x_n + hm_3, y_n + hk_3).$$

EXAMPLE 3 RK4 Method

Consider the initial-value problem

$$x' = 2x + 4y$$
$$y' = -x + 6y$$
$$x(0) = -1, \quad y(0) = 6.$$

Use the RK4 method to approximate $x(0.6)$ and $y(0.6)$. Compare the results for $h = 0.2$ and $h = 0.1$.

SOLUTION We illustrate the computations of x_1 and y_1 with the step size $h = 0.2$. With the identifications $f(t, x, y) = 2x + 4y$, $g(t, x, y) = -x + 6y$, $t_0 = 0$, $x_0 = -1$, and $y_0 = 6$, we see from (8) that

$$m_1 = f(t_0, x_0, y_0) = f(0, -1, 6) = 2(-1) + 4(6) = 22$$

$$k_1 = g(t_0, x_0, y_0) = g(0, -1, 6) = -1(-1) + 6(6) = 37$$

$$m_2 = f(t_0 + \tfrac{1}{2}h, x_0 + \tfrac{1}{2}hm_1, y_0 + \tfrac{1}{2}hk_1) = f(0.1, 1.2, 9.7) = 41.2$$

$$k_2 = g(t_0 + \tfrac{1}{2}h, x_0 + \tfrac{1}{2}hm_1, y_0 + \tfrac{1}{2}hk_1) = g(0.1, 1.2, 9.7) = 57$$

$$m_3 = f(t_0 + \tfrac{1}{2}h, x_0 + \tfrac{1}{2}hm_2, y_0 + \tfrac{1}{2}hk_2) = f(0.1, 3.12, 11.7) = 53.04$$

$$k_3 = g(t_0 + \tfrac{1}{2}h, x_0 + \tfrac{1}{2}hm_2, y_0 + \tfrac{1}{2}hk_2) = g(0.1, 3.12, 11.7) = 67.08$$

$$m_4 = f(t_0 + h, x_0 + hm_3, y_0 + hk_3) = f(0.2, 9.608, 19.416) = 96.88$$

$$k_4 = g(t_0 + h, x_0 + hm_3, y_0 + hk_3) = g(0.2, 9.608, 19.416) = 106.888.$$

Therefore from (7) we get

$$x_1 = x_0 + \frac{0.2}{6}(m_1 + 2m_2 + 2m_3 + m_4)$$

$$= -1 + \frac{0.2}{6}(2.2 + 2(41.2) + 2(53.04) + 96.88) = 9.2453$$

$$y_1 = y_0 + \frac{0.2}{6}(k_1 + 2k_2 + 2k_3 + k_4)$$

$$= 6 + \frac{0.2}{6}(37 + 2(57) + 2(67.08) + 106.888) = 19.0683,$$

where, as usual, the computed values of x_1 and y_1 are rounded to four decimal places. These numbers give us the approximations $x_1 \approx x(0.2)$ and $y_1 \approx y(0.2)$. The subsequent values, obtained with the aid of a computer, are summarized in Tables 6.4.1 and 6.4.2.

x, y

FIGURE 6.4.2 Numerical solution curves for IVP in Example 3

TABLE 6.4.1 $h = 0.2$		
t_n	x_n	y_n
0.00	−1.0000	6.0000
0.20	9.2453	19.0683
0.40	46.0327	55.1203
0.60	158.9430	150.8192

TABLE 6.4.2 $h = 0.1$		
t_n	x_n	y_n
0.00	−1.0000	6.0000
0.10	2.3840	10.8883
0.20	9.3379	19.1332
0.30	22.5541	32.8539
0.40	46.5103	55.4420
0.50	88.5729	93.3006
0.60	160.7563	152.0025

You should verify that the solution of the initial-value problem in Example 3 is given by $x(t) = (26t - 1)e^{4t}$, $y(t) = (13t + 6)e^{4t}$. From these equations we see that the actual values $x(0.6) = 160.9384$ and $y(0.6) = 152.1198$ compare favorably with the entries in the last line of Table 6.4.2. The graph of the solution in a neighborhood of $t = 0$ is shown in **FIGURE 6.4.2**; the graph was obtained from a numerical solver using the RK4 method with $h = 0.1$.

In conclusion, we state **Euler's method** for the general system (6):

$$x_{n+1} = x_n + hf(t_n, x_n, y_n)$$
$$y_{n+1} = y_n + hg(t_n, x_n, y_n).$$

1. Use Euler's method to approximate $y(0.2)$, where $y(x)$ is the solution of the initial-value problem

$$y'' - 4y' + 4y = 0, \quad y(0) = -2, \ y'(0) = 1.$$

Use $h = 0.1$. Find the exact solution of the problem, and compare the actual value of $y(0.2)$ with y_2.

2. Use Euler's method to approximate $y(1.2)$, where $y(x)$ is the solution of the initial-value problem

$$x^2 y'' - 2xy' + 2y = 0, \quad y(1) = 4, \ y'(1) = 9,$$

where $x > 0$. Use $h = 0.1$. Find the analytic solution of the problem, and compare the actual value of $y(1.2)$ with y_2.

In Problems 3 and 4, repeat the indicated problem using the RK4 method. First use $h = 0.2$ and then use $h = 0.1$.

3. Problem 1 4. Problem 2

5. Use the RK4 method to approximate $y(0.2)$, where $y(x)$ is a solution of the initial-value problem

$$y'' - 2y' + 2y = e^t \cos t, \quad y(0) = 1, \ y'(0) = 2.$$

First use $h = 0.2$ and then $h = 0.1$.

6. When $E = 100$ V, $R = 10 \ \Omega$, and $L = 1$ h, the system of differential equations for the currents $i_1(t)$ and $i_3(t)$ in the electrical network given in **FIGURE 6.4.3** is

$$\frac{di_1}{dt} = -20i_1 + 10i_3 + 100$$

$$\frac{di_3}{dt} = 10i_1 - 20i_3,$$

where $i_1(0) = 0$ and $i_3(0) = 0$. Use the RK4 method to approximate $i_1(t)$ and $i_3(t)$ at $t = 0.1, 0.2, 0.3, 0.4,$ and 0.5. Use $h = 0.1$. Use a numerical solver to graph the solution for $0 \leq t \leq 5$. Use their graphs to predict the behavior of $i_1(t)$ and $i_3(t)$ as $t \to \infty$.

FIGURE 6.4.3 Network in Problem 6

In Problems 7–12, use the Runge–Kutta method to approximate $x(0.2)$ and $y(0.2)$. First use $h = 0.2$ and then use $h = 0.1$. Use a numerical solver and $h = 0.1$ to graph the solution in a neighborhood of $t = 0$.

7. $x' = 2x - y$
 $y' = x$
 $x(0) = 6, \ y(0) = 2$

8. $x' = x + 2y$
 $y' = 4x + 3y$
 $x(0) = 1, \ y(0) = 1$

9. $x' = -y + t$
 $y' = x - t$
 $x(0) = -3, \ y(0) = 5$

10. $x' = 6x + y + 6t$
 $y' = 4x + 3y - 10t + 4$
 $x(0) = 0.5, \ y(0) = 0.2$

11. $x' + 4x - y' = 7t$
 $x' + y' - 2y = 3t$
 $x(0) = 1, \ y(0) = -2$

12. $x' + y' = 4t$
 $-x' + y' + y = 6t^2 + 10$
 $x(0) = 3, \ y(0) = -1$

6.5 Second-Order Boundary-Value Problems

INTRODUCTION We just saw in Section 6.4 how to approximate the solution of a *second-order initial-value problem* $y'' = f(x, y, y')$, $y(x_0) = y_0$, $y'(x_0) = u_0$. In this section we are going to examine two methods for approximating a solution of a *second-order boundary-value problem* $y'' = f(x, y, y')$, $y(a) = \alpha$, $y(b) = \beta$. Unlike the procedures used with second-order initial-value problems, the methods of second-order boundary-value problems do not require rewriting the second-order DE as a system of first-order DEs.

Finite Difference Approximations The Taylor series expansion, centered at a point a, of a function $y(x)$ is

$$y(x) = y(a) + y'(a)\frac{x-a}{1!} + y''(a)\frac{(x-a)^2}{2!} + y'''(a)\frac{(x-a)^3}{3!} + \cdots.$$

If we set $h = x - a$, then the preceding line is the same as

$$y(x) = y(a) + y'(a)\frac{h}{1!} + y''(a)\frac{h^2}{2!} + y'''(a)\frac{h^3}{3!} + \cdots.$$

For the subsequent discussion it is convenient, then, to rewrite this last expression in two alternative forms:

$$y(x + h) = y(x) + y'(x)h + y''(x)\frac{h^2}{2} + y'''(x)\frac{h^3}{6} + \cdots \tag{1}$$

and

$$y(x - h) = y(x) - y'(x)h + y''(x)\frac{h^2}{2} - y'''(x)\frac{h^3}{6} + \cdots. \tag{2}$$

If h is small, we can ignore terms involving $h^4, h^5, \ldots$, since these values are negligible. Indeed, if we ignore all terms involving h^2 and higher, then solving (1) and (2), in turn, for $y'(x)$ yields the following approximations for the first derivative:

$$y'(x) \approx \frac{1}{h}[y(x + h) - y(x)] \tag{3}$$

$$y'(x) \approx \frac{1}{h}[y(x) - y(x - h)]. \tag{4}$$

Subtracting (1) and (2) also gives

$$y'(x) \approx \frac{1}{2h}[y(x + h) - y(x - h)]. \tag{5}$$

On the other hand, if we ignore terms involving h^3 and higher, then by adding (1) and (2) we obtain an approximation for the second derivative $y''(x)$:

$$y''(x) \approx \frac{1}{h^2}[y(x + h) - 2y(x) + y(x - h)]. \tag{6}$$

The right sides of (3), (4), (5), and (6) are called **difference quotients**. The expressions

$$y(x + h) - y(x), \quad y(x) - y(x - h), \quad y(x + h) - y(x - h)$$

and

$$y(x + h) - 2y(x) + y(x - h)$$

are called **finite differences**. Specifically, $y(x + h) - y(x)$ is called a **forward difference**, $y(x) - y(x - h)$ is a **backward difference**, and both $y(x + h) - y(x - h)$ and $y(x + h) - 2y(x) + y(x - h)$ are called **central differences**. The results given in (5) and (6) are referred to as **central difference approximations** for the derivatives y' and y''.

▨ Finite Difference Method

Consider now a linear second-order boundary-value problem

$$y'' + P(x)y' + Q(x)y = f(x), \quad y(a) = \alpha, \quad y(b) = \beta. \tag{7}$$

Suppose $a = x_0 < x_1 < x_2 < \cdots < x_{n-1} < x_n = b$ represents a regular partition of the interval $[a, b]$; that is, $x_i = a + ih$, where $i = 0, 1, 2, \ldots, n$ and $h = (b - a)/n$. The points

$$x_1 = a + h, \quad x_2 = a + 2h, \quad \ldots, \quad x_{n-1} = a + (n - 1)h,$$

are called **interior mesh points** of the interval $[a, b]$. If we let

$$y_i = y(x_i), \quad P_i = P(x_i), \quad Q_i = Q(x_i), \quad \text{and} \quad f_i = f(x_i)$$

and if y'' and y' in (7) are replaced by the central difference approximations (5) and (6), we get

$$\frac{y_{i+1} - 2y_i + y_{i-1}}{h^2} + P_i\frac{y_{i+1} - y_{i-1}}{2h} + Q_iy_i = f_i$$

or, after simplifying,

$$\left(1 + \frac{h}{2}P_i\right)y_{i+1} + (-2 + h^2Q_i)y_i + \left(1 - \frac{h}{2}P_i\right)y_{i-1} = h^2f_i. \tag{8}$$

The last equation, known as a **finite difference equation**, is an approximation to the differential equation. It enables us to approximate the solution $y(x)$ of (7) at the interior mesh points $x_1, x_2, \ldots, x_{n-1}$ of the interval $[a, b]$. By letting i take on the values $1, 2, \ldots, n-1$ in (8), we obtain $n-1$ equations in the $n-1$ unknowns $y_1, y_2, \ldots, y_{n-1}$. Bear in mind that we know y_0 and y_n, since these are the prescribed boundary conditions $y_0 = y(x_0) = y(a) = \alpha$ and $y_n = y(x_n) = y(b) = \beta$.

In Example 1 we consider a boundary-value problem for which we can compare the approximate values found with the exact values of an explicit solution.

EXAMPLE 1 Using the Finite Difference Method

Use the difference equation (8) with $n = 4$ to approximate the solution of the boundary-value problem

$$y'' - 4y = 0, \quad y(0) = 0, \quad y(1) = 5.$$

SOLUTION To use (8) we identify $P(x) = 0$, $Q(x) = -4$, $f(x) = 0$, and $h = (1 - 0)/4 = \frac{1}{4}$. Hence the difference equation is

$$y_{i+1} - 2.25y_i + y_{i-1} = 0. \tag{9}$$

Now the interior points are $x_1 = 0 + \frac{1}{4}$, $x_2 = 0 + \frac{2}{4}$, $x_3 = 0 + \frac{3}{4}$, and so for $i = 1, 2$, and 3, (9) yields the following system for the corresponding y_1, y_2, and y_3:

$$y_2 - 2.25y_1 + y_0 = 0$$
$$y_3 - 2.25y_2 + y_1 = 0$$
$$y_4 - 2.25y_3 + y_2 = 0.$$

With the boundary conditions $y_0 = 0$ and $y_4 = 5$, the foregoing system becomes

$$-2.25y_1 + \quad y_2 \qquad\qquad = 0$$
$$y_1 - 2.25y_2 + \quad y_3 = 0$$
$$y_2 - 2.25y_3 = -5.$$

Solving the system gives $y_1 = 0.7256$, $y_2 = 1.6327$, and $y_3 = 2.9479$.

Now the general solution of the given differential equation is $y = c_1 \cosh 2x + c_2 \sinh 2x$. The condition $y(0) = 0$ implies $c_1 = 0$. The other boundary condition gives c_2. In this way we see that an explicit solution of the boundary-value problem is $y(x) = (5 \sinh 2x)/\sinh 2$.

Thus the exact values (rounded to four decimal places) of this solution at the interior points are as follows: $y(0.25) = 0.7184$, $y(0.5) = 1.6201$, and $y(0.75) = 2.9354$. $\equiv$

The accuracy of the approximations in Example 1 can be improved by using a smaller value of h. Of course, the trade-off here is that a smaller value of h necessitates solving a larger system of equations. It is left as an exercise to show that with $h = \frac{1}{8}$, approximations to $y(0.25)$, $y(0.5)$, and $y(0.75)$ are 0.7202, 1.6233, and 2.9386, respectively. See Problem 11 in Exercises 6.5.

EXAMPLE 2 Using the Finite Difference Method

Use the difference equation (8) with $n = 10$ to approximate the solution of

$$y'' + 3y' + 2y = 4x^2, \quad y(1) = 1, \quad y(2) = 6.$$

SOLUTION In this case we identify $P(x) = 3$, $Q(x) = 2$, $f(x) = 4x^2$, and $h = (2 - 1)/10 = 0.1$, and so (8) becomes

$$1.15y_{i+1} - 1.98y_i + 0.85y_{i-1} = 0.04x_i^2. \tag{10}$$

Now the interior points are $x_1 = 1.1, x_2 = 1.2, x_3 = 1.3, x_4 = 1.4, x_5 = 1.5, x_6 = 1.6, x_7 = 1.7,$ $x_8 = 1.8,$ and $x_9 = 1.9$. For $i = 1, 2, \ldots, 9$ and $y_0 = 1, y_{10} = 6$, (10) gives a system of nine equations and nine unknowns:

$$1.15y_2 - 1.98y_1 \qquad\qquad = -0.8016$$
$$1.15y_3 - 1.98y_2 + 0.85y_1 = 0.0576$$
$$1.15y_4 - 1.98y_3 + 0.85y_2 = 0.0676$$
$$1.15y_5 - 1.98y_4 + 0.85y_3 = 0.0784$$
$$1.15y_6 - 1.98y_5 + 0.85y_4 = 0.0900$$
$$1.15y_7 - 1.98y_6 + 0.85y_5 = 0.1024$$
$$1.15y_8 - 1.98y_7 + 0.85y_6 = 0.1156$$
$$1.15y_9 - 1.98y_8 + 0.85y_7 = 0.1296$$
$$-1.98y_9 + 0.85y_8 = -6.7556.$$

We can solve this large system using Gaussian elimination or, with relative ease, by means of a computer algebra system. The result is found to be $y_1 = 2.4047$, $y_2 = 3.4432$, $y_3 = 4.2010$, $y_4 = 4.7469$, $y_5 = 5.1359$, $y_6 = 5.4124$, $y_7 = 5.6117$, $y_8 = 5.7620$, and $y_9 = 5.8855$. ≡

▨ Shooting Method

Another way of approximating a solution of a second-order boundary-value problem $y'' = f(x, y, y')$, $y(a) = \alpha$, $y(b) = \beta$, is called the **shooting method**. The starting point in this method is the replacement of the boundary-value problem by an initial-value problem

$$y'' = f(x, y, y'), \qquad y(a) = \alpha, \qquad y'(a) = m_1. \tag{11}$$

The number m_1 in (11) is simply a guess for the unknown slope of the solution curve at the known point $(a, y(a))$. We then apply one of the step-by-step numerical techniques to the second-order equation in (11) to find an approximation β_1 for the value of $y(b)$. If β_1 agrees with the given value $y(b) = \beta$ to some preassigned tolerance, we stop; otherwise the calculations are repeated, starting with a different guess $y'(a) = m_2$ to obtain a second approximation β_2 for $y(b)$. This method can be continued in a trial-and-error manner or the subsequent slopes $m_3, m_4, \ldots$, can be adjusted in some systematic way; linear interpolation is particularly successful when the differential equation in (11) is linear. The procedure is analogous to shooting (the "aim" is the choice of the initial slope) at a target until the bullseye $y(b)$ is hit. See Problem 14 in Exercises 6.5.

Of course, underlying the use of these numerical methods is the assumption, which we know is not always warranted, that a solution of the boundary-value problem exists.

REMARKS

The approximation method using finite differences can be extended to boundary-value problems in which the first derivative is specified at a boundary—for example, a problem such as $y'' = f(x, y, y')$, $y'(a) = \alpha$, $y(b) = \beta$. See Problem 13 in Exercises 6.5.

6.5 Exercises Answers to selected odd-numbered problems begin on page ANS-15.

In Problems 1–10, use the finite difference method and the indicated value of n to approximate the solution of the given boundary-value problem.

1. $y'' + 9y = 0$, $y(0) = 4$, $y(2) = 1$; $n = 4$
2. $y'' - y = x^2$, $y(0) = 0$, $y(1) = 0$; $n = 4$
3. $y'' + 2y' + y = 5x$, $y(0) = 0$, $y(1) = 0$; $n = 5$
4. $y'' - 10y' + 25y = 1$, $y(0) = 1$, $y(1) = 0$; $n = 5$
5. $y'' - 4y' + 4y = (x + 1)e^{2x}$, $y(0) = 3$, $y(1) = 0$; $n = 6$
6. $y'' + 5y' = 4\sqrt{x}$, $y(1) = 1$, $y(2) = -1$; $n = 6$

7. $x^2 y'' + 3xy' + 3y = 0$, $y(1) = 5$, $y(2) = 0$; $n = 8$

8. $x^2 y'' - xy' + y = \ln x$, $y(1) = 0$, $y(2) = -2$; $n = 8$

9. $y'' + (1 - x)y' + xy = x$, $y(0) = 0$, $y(1) = 2$; $n = 10$

10. $y'' + xy' + y = x$, $y(0) = 1$, $y(1) = 0$; $n = 10$

11. Rework Example 1 using $n = 8$.

12. The electrostatic potential u between two concentric spheres of radius $r = 1$ and $r = 4$ is determined from

$$\frac{d^2 u}{dr^2} + \frac{2}{r} \frac{du}{dr} = 0, \quad u(1) = 50, \quad u(4) = 100.$$

Use the method of this section with $n = 6$ to approximate the solution of this boundary-value problem.

13. Consider the boundary-value problem

$$y'' + xy = 0, \quad y'(0) = 1, \quad y(1) = -1.$$

(a) Find the difference equation corresponding to the differential equation. Show that for $i = 0, 1, 2, \ldots, n - 1$, the difference equation yields n equations in $n + 1$

unknowns $y_{-1}, y_0, y_1, y_2, \ldots, y_{n-1}$. Here y_{-1} and y_0 are unknowns since y_{-1} represents an approximation to y at the exterior point $x = -h$ and y_0 is not specified at $x = 0$.

(b) Use the central difference approximation (5) to show that $y_1 - y_{-1} = 2h$. Use this equation to eliminate y_{-1} from the system in part (a).

(c) Use $n = 5$ and the system of equations found in parts (a) and (b) to approximate the solution of the original boundary-value problem.

Computer Lab Assignment

14. Consider the boundary-value problem $y'' = y' - \sin(xy)$, $y(0) = 1$, $y(1) = 1.5$. Use the shooting method to approximate the solution of this problem. (The actual approximation can be obtained using a numerical technique, say, the fourth-order Runge–Kutta method with $h = 0.1$; even better, if you have access to a CAS, such as *Mathematica* or *Maple*, the **NDSolve** function can be used.)

6 Chapter in Review Answers to selected odd-numbered problems begin on page ANS-15.

In Problems 1–4, construct a table comparing the indicated values of $y(x)$ using Euler's method, the improved Euler's method, and the RK4 method. Compute to four rounded decimal places. Use $h = 0.1$ and then use $h = 0.05$.

1. $y' = 2 \ln xy$, $y(1) = 2$;

$y(1.1), y(1.2), y(1.3), y(1.4), y(1.5)$

2. $y' = \sin x^2 + \cos y^2$, $y(0) = 0$;

$y(0.1), y(0.2), y(0.3), y(0.4), y(0.5)$

3. $y' = \sqrt{x + y}$, $y(0.5) = 0.5$;

$y(0.6), y(0.7), y(0.8), y(0.9), y(1.0)$

4. $y' = xy + y^2$, $y(1) = 1$;

$y(1.1), y(1.2), y(1.3), y(1.4), y(1.5)$

5. Use Euler's method to approximate $y(0.2)$, where $y(x)$ is the solution of the initial-value problem $y'' - (2x + 1)y = 1$, $y(0) = 3$, $y'(0) = 1$. First use one step with $h = 0.2$, and then repeat the calculations using two steps with $h = 0.1$.

6. Use the Adams–Bashforth–Moulton method to approximate $y(0.4)$, where $y(x)$ is the solution of the initial-value problem $y' = 4x - 2y$, $y(0) = 2$. Use $h = 0.1$ and the RK4 method to compute y_1, y_2, and y_3.

7. Use Euler's method with $h = 0.1$ to approximate $x(0.2)$ and $y(0.2)$, where $x(t), y(t)$ is the solution of the initial-value problem

$$x' = x + y$$
$$y' = x - y$$
$$x(0) = 1, \quad y(0) = 2.$$

8. Use the finite difference method with $n = 10$ to approximate the solution of the boundary-value problem $y'' + 6.55(1 + x)y = 1$, $y(0) = 0$, $y(1) = 0$.

Vectors, Matrices, and Vector Calculus

CHAPTER 7

Vectors

You have undoubtedly encountered the notion of **vectors** in your study of calculus, as well as in physics and engineering. For most of you then, this chapter is a review of familiar topics such as the dot and cross products. However, in Section 7.6 we shall consider an **abstraction of the vector concept**.

7.1 Vectors in 2-Space

INTRODUCTION In science, mathematics, and engineering, we distinguish two important quantities: **scalars** and **vectors**. A scalar is simply a real number or a quantity that has *magnitude*. For example, length, temperature, and blood pressure are represented by numbers such as 80 m, 20°C, and the systolic/diastolic ratio 120/80, respectively. A vector, on the other hand, is usually described as a quantity that has both *magnitude* and *direction*.

■ **Geometric Vectors** Geometrically, a vector can be represented by a directed line segment—that is, by an arrow—and is denoted either by a boldface symbol or by a symbol with an arrow over it; for example, **v**, $\vec{u}$, or $\overrightarrow{AB}$. Examples of vector quantities shown in **FIGURE 7.1.1** are weight **w**, velocity **v**, and the retarding force of friction $\mathbf{F}_f$.

FIGURE 7.1.2 Vectors are equal

FIGURE 7.1.3 Parallel vectors

The question of what is the direction of **0** is usually answered by saying that the zero vector can be assigned *any* direction. More to the point, **0** is needed in order to have a vector algebra.

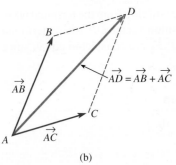

FIGURE 7.1.4 Vector $\overrightarrow{AD}$ is the sum of $\overrightarrow{AB}$ and $\overrightarrow{AC}$

FIGURE 7.1.1 Examples of vector quantities

■ **Notation and Terminology** A vector whose initial point (or end) is A and whose terminal point (or tip) is B is written $\overrightarrow{AB}$. The magnitude of a vector is written $\|\overrightarrow{AB}\|$. Two vectors that have the same magnitude and same direction are said to be **equal**. Thus, in **FIGURE 7.1.2**, we have $\overrightarrow{AB} = \overrightarrow{CD}$. Vectors are said to be **free**, which means that a vector can be moved from one position to another provided its magnitude and direction are not changed. The **negative** of a vector $\overrightarrow{AB}$, written $-\overrightarrow{AB}$, is a vector that has the same magnitude as $\overrightarrow{AB}$ but is opposite in direction. If $k \neq 0$ is a scalar, the **scalar multiple** of a vector, $k\overrightarrow{AB}$, is a vector that is $|k|$ times as long as $\overrightarrow{AB}$. If $k > 0$, then $k\overrightarrow{AB}$ has the same direction as the vector $\overrightarrow{AB}$; if $k < 0$, then $k\overrightarrow{AB}$ has the direction opposite that of $\overrightarrow{AB}$. When $k = 0$, we say $0\overrightarrow{AB} = \mathbf{0}$ is the **zero vector**. Two vectors are **parallel** if and only if they are nonzero scalar multiples of each other. See **FIGURE 7.1.3**.

■ **Addition and Subtraction** Two vectors can be considered as having a common initial point, such as A in **FIGURE 7.1.4(a)**. Thus, if nonparallel vectors $\overrightarrow{AB}$ and $\overrightarrow{AC}$ are the sides of a parallelogram as in Figure 7.1.4(b), we say the vector that is the main diagonal, or $\overrightarrow{AD}$, is the **sum** of $\overrightarrow{AB}$ and $\overrightarrow{AC}$. We write

$$\overrightarrow{AD} = \overrightarrow{AB} + \overrightarrow{AC}.$$

The **difference** of two vectors $\overrightarrow{AB}$ and $\overrightarrow{AC}$ is defined by

$$\overrightarrow{AB} - \overrightarrow{AC} = \overrightarrow{AB} + (-\overrightarrow{AC}).$$

As seen in **FIGURE 7.1.5(a)**, the difference $\overrightarrow{AB} - \overrightarrow{AC}$ can be interpreted as the main diagonal of the parallelogram with sides $\overrightarrow{AB}$ and $-\overrightarrow{AC}$. However, as shown in Figure 7.1.5(b), we can also interpret the same vector difference as the third side of a triangle with sides $\overrightarrow{AB}$ and $\overrightarrow{AC}$. In this second interpretation, observe that the vector difference $\overrightarrow{CB} = \overrightarrow{AB} - \overrightarrow{AC}$ points toward the terminal point of the vector *from* which we are subtracting the second vector. If $\overrightarrow{AB} = \overrightarrow{AC}$, then

$$\overrightarrow{AB} - \overrightarrow{AC} = \mathbf{0}.$$

$$\overrightarrow{AB} + (-\overrightarrow{AC})$$

(a)

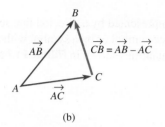

$$\overrightarrow{AB} \qquad \overrightarrow{CB} = \overrightarrow{AB} - \overrightarrow{AC}$$

(b)

FIGURE 7.1.5 Vector $\overrightarrow{CB}$ is the difference of $\overrightarrow{AB}$ and $\overrightarrow{AC}$

FIGURE 7.1.6 Position vector

(a)

(b)

FIGURE 7.1.7 Graphs of vectors in Example 1

Vectors in a Coordinate Plane

To describe a vector analytically, let us suppose for the remainder of this section that the vectors we are considering lie in a two-dimensional coordinate plane or **2-space**. We shall denote the set of all vectors in the plane by R^2. The vector shown in **FIGURE 7.1.6**, with initial point the origin O and terminal point $P(x_1, y_1)$, is called the **position vector** of the point P and is written

$$\overrightarrow{OP} = \langle x_1, y_1 \rangle.$$

Components

In general, a vector **a** in R^2 is any ordered pair of real numbers,

$$\mathbf{a} = \langle a_1, a_2 \rangle.$$

The numbers a_1 and a_2 are said to be the **components** of the vector **a**.

As we shall see in the first example, the vector **a** is not necessarily a position vector.

EXAMPLE 1 Position Vector

The displacement from the initial point $P_1(x, y)$ to the terminal point $P_2(x + 4, y + 3)$ in **FIGURE 7.1.7(a)** is 4 units to the right and 3 units up. As seen in Figure 7.1.7(b), the position vector of $\mathbf{a} = \langle 4, 3 \rangle$ emanating from the origin is equivalent to the displacement vector $\overrightarrow{P_1P_2}$ from $P_1(x, y)$ to $P_2(x + 4, y + 3)$. ≡

Addition and subtraction of vectors, multiplication of vectors by scalars, and so on, are defined in terms of their components.

DEFINITION 7.1.1 Addition, Scalar Multiplication, Equality

Let $\mathbf{a} = \langle a_1, a_2 \rangle$ and $\mathbf{b} = \langle b_1, b_2 \rangle$ be vectors in R^2.

(*i*) Addition: $\mathbf{a} + \mathbf{b} = \langle a_1 + b_1, a_2 + b_2 \rangle$ (1)
(*ii*) Scalar multiplication: $k\mathbf{a} = \langle ka_1, ka_2 \rangle$ (2)
(*iii*) Equality: $\mathbf{a} = \mathbf{b}$ if and only if $a_1 = b_1, a_2 = b_2$ (3)

Subtraction

Using (2), we define the **negative** of a vector **b** by

$$-\mathbf{b} = (-1)\mathbf{b} = \langle -b_1, -b_2 \rangle.$$

We can then define the **subtraction**, or the difference, of two vectors as

$$\mathbf{a} - \mathbf{b} = \mathbf{a} + (-\mathbf{b}) = \langle a_1 - b_1, a_2 - b_2 \rangle. \qquad (4)$$

In **FIGURE 7.1.8(a)** on page 331, we have illustrated the sum of two vectors $\overrightarrow{OP_1}$ and $\overrightarrow{OP_2}$. In Figure 7.1.8(b), the vector $\overrightarrow{P_1P_2}$, with initial point P_1 and terminal point P_2, is the difference of position vectors

$$\overrightarrow{P_1P_2} = \overrightarrow{OP_2} - \overrightarrow{OP_1} = \langle x_2 - x_1, y_2 - y_1 \rangle.$$

As shown in Figure 7.1.8(b), the vector $\overrightarrow{P_1P_2}$ can be drawn either starting from the terminal point of $\overrightarrow{OP_1}$ and ending at the terminal point of $\overrightarrow{OP_2}$, or as the position vector $\overrightarrow{OP}$ whose terminal point has coordinates $(x_2 - x_1, y_2 - y_1)$. Remember, $\overrightarrow{OP}$ and $\overrightarrow{P_1P_2}$ are considered equal, since they have the same magnitude and direction.

EXAMPLE 2 Addition and Subtraction of Two Vectors

If $\mathbf{a} = \langle 1, 4 \rangle$ and $\mathbf{b} = \langle -6, 3 \rangle$, find $\mathbf{a} + \mathbf{b}$, $\mathbf{a} - \mathbf{b}$, and $2\mathbf{a} + 3\mathbf{b}$.

SOLUTION We use (1), (2), and (4):

$$\mathbf{a} + \mathbf{b} = \langle 1 + (-6), 4 + 3 \rangle = \langle -5, 7 \rangle$$
$$\mathbf{a} - \mathbf{b} = \langle 1 - (-6), 4 - 3 \rangle = \langle 7, 1 \rangle$$
$$2\mathbf{a} + 3\mathbf{b} = \langle 2, 8 \rangle + \langle -18, 9 \rangle = \langle -16, 17 \rangle. \qquad ≡$$

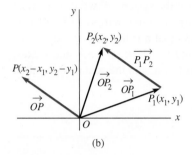

FIGURE 7.1.8 In (b), $\overrightarrow{OP}$ and $\overrightarrow{P_1P_2}$ are the same vector

FIGURE 7.1.9 A right triangle

Properties The component definition of a vector can be used to verify each of the following **properties** of vectors in R^2:

THEOREM 7.1.1 Properties of Vectors

(i)	$\mathbf{a} + \mathbf{b} = \mathbf{b} + \mathbf{a}$	← commutative law
(ii)	$\mathbf{a} + (\mathbf{b} + \mathbf{c}) = (\mathbf{a} + \mathbf{b}) + \mathbf{c}$	← associative law
(iii)	$\mathbf{a} + \mathbf{0} = \mathbf{a}$	← additive identity
(iv)	$\mathbf{a} + (-\mathbf{a}) = \mathbf{0}$	← additive inverse
(v)	$k(\mathbf{a} + \mathbf{b}) = k\mathbf{a} + k\mathbf{b}$, k a scalar	
(vi)	$(k_1 + k_2)\mathbf{a} = k_1\mathbf{a} + k_2\mathbf{a}$, k_1 and k_2 scalars	
(vii)	$k_1(k_2\mathbf{a}) = (k_1 k_2)\mathbf{a}$, k_1 and k_2 scalars	
(viii)	$1\mathbf{a} = \mathbf{a}$	
(ix)	$0\mathbf{a} = \mathbf{0}$	← zero vector

The **zero vector 0** in properties (iii), (iv), and (ix) is defined as

$$\mathbf{0} = \langle 0, 0 \rangle.$$

Magnitude The **magnitude, length**, or **norm** of a vector $\mathbf{a}$ is denoted by $\|\mathbf{a}\|$. Motivated by the Pythagorean theorem and **FIGURE 7.1.9**, we define the magnitude of a vector

$$\mathbf{a} = \langle a_1, a_2 \rangle \quad \text{to be} \quad \|\mathbf{a}\| = \sqrt{a_1^2 + a_2^2}.$$

Clearly, $\|\mathbf{a}\| \geq 0$ for any vector $\mathbf{a}$, and $\|\mathbf{a}\| = 0$ if and only if $\mathbf{a} = \mathbf{0}$. For example, if $\mathbf{a} = \langle 6, -2 \rangle$, then $\|\mathbf{a}\| = \sqrt{6^2 + (-2)^2} = \sqrt{40} = 2\sqrt{10}$.

Unit Vectors A vector that has magnitude 1 is called a **unit vector**. We can obtain a unit vector $\mathbf{u}$ in the same direction as a nonzero vector $\mathbf{a}$ by multiplying $\mathbf{a}$ by the positive scalar $k = 1/\|\mathbf{a}\|$ (reciprocal of its magnitude). In this case we say that $\mathbf{u} = (1/\|\mathbf{a}\|)\mathbf{a}$ is the **normalization** of the vector $\mathbf{a}$. The normalization of the vector $\mathbf{a}$ is a unit vector because

$$\|\mathbf{u}\| = \left\| \frac{1}{\|\mathbf{a}\|} \mathbf{a} \right\| = \frac{1}{\|\mathbf{a}\|} \|\mathbf{a}\| = 1.$$

Note: It is often convenient to write the scalar multiple $\mathbf{u} = (1/\|\mathbf{a}\|)\mathbf{a}$ as

$$\mathbf{u} = \frac{\mathbf{a}}{\|\mathbf{a}\|}.$$

EXAMPLE 3 **Unit Vectors**

Given $\mathbf{a} = \langle 2, -1 \rangle$, form a unit vector in the same direction as $\mathbf{a}$. In the opposite direction of $\mathbf{a}$.

SOLUTION The magnitude of the vector $\mathbf{a}$ is $\|\mathbf{a}\| = \sqrt{2^2 + (-1)^2} = \sqrt{5}$. Thus a unit vector in the same direction as $\mathbf{a}$ is the scalar multiple

$$\mathbf{u} = \frac{1}{\sqrt{5}}\mathbf{a} = \frac{1}{\sqrt{5}} \langle 2, -1 \rangle = \left\langle \frac{2}{\sqrt{5}}, \frac{-1}{\sqrt{5}} \right\rangle.$$

A unit vector in the opposite direction of $\mathbf{a}$ is the negative of $\mathbf{u}$:

$$-\mathbf{u} = \left\langle -\frac{2}{\sqrt{5}}, \frac{1}{\sqrt{5}} \right\rangle. \qquad \equiv$$

If $\mathbf{a}$ and $\mathbf{b}$ are vectors and c_1 and c_2 are scalars, then the expression $c_1\mathbf{a} + c_2\mathbf{b}$ is called a **linear combination** of $\mathbf{a}$ and $\mathbf{b}$. As we see next, any vector in R^2 can be written as a linear combination of two special vectors.

(a)

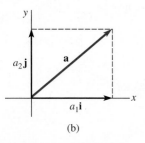

(b)

FIGURE 7.1.10 **i** and **j** form a basis for R^2

The i, j Vectors

In view of (1) and (2), any vector $\mathbf{a} = \langle a_1, a_2 \rangle$ can be written as a sum:

$$\langle a_1, a_2 \rangle = \langle a_1, 0 \rangle + \langle 0, a_2 \rangle = a_1 \langle 1, 0 \rangle + a_2 \langle 0, 1 \rangle. \tag{5}$$

The unit vectors $\langle 1, 0 \rangle$ and $\langle 0, 1 \rangle$ are usually given the special symbols **i** and **j**. See **FIGURE 7.1.10(a)**. Thus, if

$$\mathbf{i} = \langle 1, 0 \rangle \quad \text{and} \quad \mathbf{j} = \langle 0, 1 \rangle,$$

then (5) becomes

$$\mathbf{a} = a_1 \mathbf{i} + a_2 \mathbf{j}. \tag{6}$$

The unit vectors **i** and **j** are referred to as the **standard basis** for the system of two-dimensional vectors, since any vector **a** can be written uniquely as a linear combination of **i** and **j**. If $\mathbf{a} = a_1 \mathbf{i} + a_2 \mathbf{j}$ is a position vector, then Figure 7.1.10(b) shows that **a** is the sum of the vectors $a_1 \mathbf{i}$ and $a_2 \mathbf{j}$, which have the origin as a common initial point and which lie on the x- and y-axes, respectively. The scalar a_1 is called the **horizontal component** of **a**, and the scalar a_2 is called the **vertical component** of **a**.

EXAMPLE 4 Vector Operations Using i and j

(a) $\langle 4, 7 \rangle = 4\mathbf{i} + 7\mathbf{j}$

(b) $(2\mathbf{i} - 5\mathbf{j}) + (8\mathbf{i} + 13\mathbf{j}) = 10\mathbf{i} + 8\mathbf{j}$

(c) $\|\mathbf{i} + \mathbf{j}\| = \sqrt{2}$

(d) $10(3\mathbf{i} - \mathbf{j}) = 30\mathbf{i} - 10\mathbf{j}$

(e) $\mathbf{a} = 6\mathbf{i} + 4\mathbf{j}$ and $\mathbf{b} = 9\mathbf{i} + 6\mathbf{j}$ are parallel, since **b** is a scalar multiple of **a**. We see that $\mathbf{b} = \frac{3}{2}\mathbf{a}$. ≡

EXAMPLE 5 Graphs of Vector Sum/Vector Difference

Let $\mathbf{a} = 4\mathbf{i} + 2\mathbf{j}$ and $\mathbf{b} = -2\mathbf{i} + 5\mathbf{j}$. Graph $\mathbf{a} + \mathbf{b}$ and $\mathbf{a} - \mathbf{b}$.

SOLUTION The graphs of $\mathbf{a} + \mathbf{b} = 2\mathbf{i} + 7\mathbf{j}$ and $\mathbf{a} - \mathbf{b} = 6\mathbf{i} - 3\mathbf{j}$ are given in **FIGURE 7.1.11(a)** and 7.1.11(b), respectively.

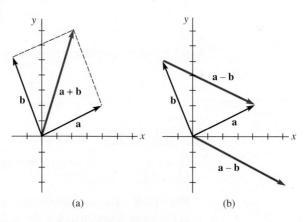

(a) (b)

FIGURE 7.1.11 Graphs of vectors in Example 5 ≡

7.1 Exercises
Answers to selected odd-numbered problems begin on page ANS-15.

In Problems 1–8, find (a) 3**a**, (b) **a** + **b**, (c) **a** − **b**, (d) $\|\mathbf{a} + \mathbf{b}\|$, and (e) $\|\mathbf{a} - \mathbf{b}\|$.

1. $\mathbf{a} = 2\mathbf{i} + 4\mathbf{j}$, $\mathbf{b} = -\mathbf{i} + 4\mathbf{j}$

2. $\mathbf{a} = \langle 1, 1 \rangle$, $\mathbf{b} = \langle 2, 3 \rangle$

3. $\mathbf{a} = \langle 4, 0 \rangle$, $\mathbf{b} = \langle 0, -5 \rangle$

4. $\mathbf{a} = \frac{1}{6}\mathbf{i} - \frac{1}{6}\mathbf{j}$, $\mathbf{b} = \frac{1}{2}\mathbf{i} + \frac{5}{6}\mathbf{j}$

5. $\mathbf{a} = -3\mathbf{i} + 2\mathbf{j}$, $\mathbf{b} = 7\mathbf{j}$

6. $\mathbf{a} = \langle 1, 3 \rangle$, $\mathbf{b} = -5\mathbf{a}$

7. $\mathbf{a} = -\mathbf{b}$, $\mathbf{b} = 2\mathbf{i} - 9\mathbf{j}$

8. $\mathbf{a} = \langle 7, 10 \rangle$, $\mathbf{b} = \langle 1, 2 \rangle$

In Problems 9–14, find (a) $4\mathbf{a} - 2\mathbf{b}$ and (b) $-3\mathbf{a} - 5\mathbf{b}$.

9. $\mathbf{a} = \langle 1, -3 \rangle$, $\mathbf{b} = \langle -1, 1 \rangle$

10. $\mathbf{a} = \mathbf{i} + \mathbf{j}$, $\mathbf{b} = 3\mathbf{i} - 2\mathbf{j}$ **11.** $\mathbf{a} = \mathbf{i} - \mathbf{j}$, $\mathbf{b} = -3\mathbf{i} + 4\mathbf{j}$

12. $\mathbf{a} = \langle 2, 0 \rangle$, $\mathbf{b} = \langle 0, -3 \rangle$ **13.** $\mathbf{a} = \langle 4, 10 \rangle$, $\mathbf{b} = -2\langle 1, 3 \rangle$

14. $\mathbf{a} = \langle 3, 1 \rangle + \langle -1, 2 \rangle$, $\mathbf{b} = \langle 6, 5 \rangle - \langle 1, 2 \rangle$

In Problems 15–18, find the vector $\overrightarrow{P_1P_2}$. Graph $\overrightarrow{P_1P_2}$ and its corresponding position vector.

15. $P_1(3, 2)$, $P_2(5, 7)$ **16.** $P_1(-2, -1)$, $P_2(4, -5)$

17. $P_1(3, 3)$, $P_2(5, 5)$ **18.** $P_1(0, 3)$, $P_2(2, 0)$

19. Find the terminal point of the vector $\overrightarrow{P_1P_2} = 4\mathbf{i} + 8\mathbf{j}$ if its initial point is $(-3, 10)$.

20. Find the initial point of the vector $\overrightarrow{P_1P_2} = \langle -5, -1 \rangle$ if its terminal point is $\langle 4, 7 \rangle$.

21. Determine which of the following vectors are parallel to $\mathbf{a} = 4\mathbf{i} + 6\mathbf{j}$.
 (a) $-4\mathbf{i} - 6\mathbf{j}$ (b) $-\mathbf{i} - \frac{3}{2}\mathbf{j}$
 (c) $10\mathbf{i} + 15\mathbf{j}$ (d) $2(\mathbf{i} - \mathbf{j}) - 3(\frac{1}{2}\mathbf{i} - \frac{5}{12}\mathbf{j})$
 (e) $8\mathbf{i} + 12\mathbf{j}$ (f) $(5\mathbf{i} + \mathbf{j}) - (7\mathbf{i} + 4\mathbf{j})$

22. Determine a scalar c so that $\mathbf{a} = 3\mathbf{i} + c\mathbf{j}$ and $\mathbf{b} = -\mathbf{i} + 9\mathbf{j}$ are parallel.

In Problems 23 and 24, find $\mathbf{a} + (\mathbf{b} + \mathbf{c})$ for the given vectors.

23. $\mathbf{a} = \langle 5, 1 \rangle$, $\mathbf{b} = \langle -2, 4 \rangle$, $\mathbf{c} = \langle 3, 10 \rangle$

24. $\mathbf{a} = \langle 1, 1 \rangle$, $\mathbf{b} = \langle 4, 3 \rangle$, $\mathbf{c} = \langle 0, -2 \rangle$

In Problems 25–28, find a unit vector (a) in the same direction as $\mathbf{a}$, and (b) in the opposite direction of $\mathbf{a}$.

25. $\mathbf{a} = \langle 2, 2 \rangle$ **26.** $\mathbf{a} = \langle -3, 4 \rangle$

27. $\mathbf{a} = \langle 0, -5 \rangle$ **28.** $\mathbf{a} = \langle 1, -\sqrt{3} \rangle$

In Problems 29 and 30, $\mathbf{a} = \langle 2, 8 \rangle$ and $\mathbf{b} = \langle 3, 4 \rangle$. Find a unit vector in the same direction as the given vector.

29. $\mathbf{a} + \mathbf{b}$ **30.** $2\mathbf{a} - 3\mathbf{b}$

In Problems 31 and 32, find a vector $\mathbf{b}$ that is parallel to the given vector and has the indicated magnitude.

31. $\mathbf{a} = 3\mathbf{i} + 7\mathbf{j}$, $\|\mathbf{b}\| = 2$ **32.** $\mathbf{a} = \frac{1}{2}\mathbf{i} - \frac{1}{2}\mathbf{j}$, $\|\mathbf{b}\| = 3$

33. Find a vector in the opposite direction of $\mathbf{a} = \langle 4, 10 \rangle$ but $\frac{3}{4}$ as long.

34. Given that $\mathbf{a} = \langle 1, 1 \rangle$ and $\mathbf{b} = \langle -1, 0 \rangle$, find a vector in the same direction as $\mathbf{a} + \mathbf{b}$ but five times as long.

In Problems 35 and 36, use the given figure to draw the indicated vector.

35. $3\mathbf{b} - \mathbf{a}$ **36.** $\mathbf{a} + (\mathbf{b} + \mathbf{c})$

FIGURE 7.1.12 Vectors for Problem 35

FIGURE 7.1.13 Vectors for Problem 36

In Problems 37 and 38, express the vector $\mathbf{x}$ in terms of the vectors $\mathbf{a}$ and $\mathbf{b}$.

37.

38.

FIGURE 7.1.14 Vector $\mathbf{x}$ in Problem 37 **FIGURE 7.1.15** Vector $\mathbf{x}$ in Problem 38

In Problems 39 and 40, use the given figure to prove the given result.

39. $\mathbf{a} + \mathbf{b} + \mathbf{c} = \mathbf{0}$ **40.** $\mathbf{a} + \mathbf{b} + \mathbf{c} + \mathbf{d} = \mathbf{0}$

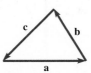

FIGURE 7.1.16 Vectors for Problem 39 **FIGURE 7.1.17** Vectors for Problem 40

In Problems 41 and 42, express the vector $\mathbf{a} = 2\mathbf{i} + 3\mathbf{j}$ as a linear combination of the given vectors $\mathbf{b}$ and $\mathbf{c}$.

41. $\mathbf{b} = \mathbf{i} + \mathbf{j}$, $\mathbf{c} = \mathbf{i} - \mathbf{j}$ **42.** $\mathbf{b} = -2\mathbf{i} + 4\mathbf{j}$, $\mathbf{c} = 5\mathbf{i} + 7\mathbf{j}$

A vector is said to be tangent to a curve at a point if it is parallel to the tangent line at the point. In Problems 43 and 44, find a unit tangent vector to the given curve at the indicated point.

43. $y = \frac{1}{4}x^2 + 1$, $(2, 2)$ **44.** $y = -x^2 + 3x$, $(0, 0)$

45. When walking, a person's foot strikes the ground with a force $\mathbf{F}$ at an angle θ from the vertical. In **FIGURE 7.1.18**, the vector $\mathbf{F}$ is resolved into vector components $\mathbf{F}_g$, which is parallel to the ground, and $\mathbf{F}_n$, which is perpendicular to the ground. In order that the foot does not slip, the force $\mathbf{F}_g$ must be offset by the opposing force $\mathbf{F}_f$ of friction; that is, $\mathbf{F}_f = -\mathbf{F}_g$.
 (a) Use the fact that $\|\mathbf{F}_f\| = \mu \|\mathbf{F}_n\|$, where μ is the coefficient of friction, to show that $\tan\theta = \mu$. The foot will not slip for angles less than or equal to θ.
 (b) Given that $\mu = 0.6$ for a rubber heel striking an asphalt sidewalk, find the "no-slip" angle.

FIGURE 7.1.18 Vector $\mathbf{F}$ in Problem 45

46. A 200-lb traffic light supported by two cables hangs in equilibrium. As shown in **FIGURE 7.1.19(b)**, let the weight of the light be represented by **w** and the forces in the two cables by $\mathbf{F}_1$ and $\mathbf{F}_2$. From Figure 7.1.19(c), we see that a condition of equilibrium is

$$\mathbf{w} + \mathbf{F}_1 + \mathbf{F}_2 = \mathbf{0}. \qquad (7)$$

See Problem 39. If

$$\mathbf{w} = -200\mathbf{j}$$
$$\mathbf{F}_1 = (\|\mathbf{F}_1\| \cos 20°)\mathbf{i} + (\|\mathbf{F}_1\| \sin 20°)\mathbf{j}$$
$$\mathbf{F}_2 = -(\|\mathbf{F}_2\| \cos 15°)\mathbf{i} + (\|\mathbf{F}_2\| \sin 15°)\mathbf{j},$$

use (7) to determine the magnitudes of $\mathbf{F}_1$ and $\mathbf{F}_2$. [*Hint:* Reread (*iii*) of Definition 7.1.1.]

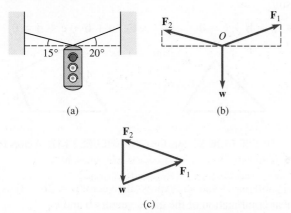

(a) (b)

(c)

FIGURE 7.1.19 Three force vectors in Problem 46

47. An electric charge Q is uniformly distributed along the y-axis between $y = -a$ and $y = a$. See **FIGURE 7.1.20**. The total force exerted on the charge q on the x-axis by the charge Q is $\mathbf{F} = F_x\mathbf{i} + F_y\mathbf{j}$, where

$$F_x = \frac{qQ}{4\pi\epsilon_0} \int_{-a}^{a} \frac{L\,dy}{2a(L^2 + y^2)^{3/2}}$$

and

$$F_y = -\frac{qQ}{4\pi\epsilon_0} \int_{-a}^{a} \frac{y\,dy}{2a(L^2 + y^2)^{3/2}}.$$

Determine **F**.

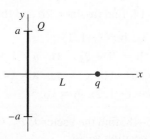

FIGURE 7.1.20 Charge on x-axis in Problem 47

48. Using vectors, show that the diagonals of a parallelogram bisect each other. [*Hint:* Let M be the midpoint of one diagonal and N the midpoint of the other.]

49. Using vectors, show that the line segment between the midpoints of two sides of a triangle is parallel to the third side and half as long.

50. An airplane starts from an airport located at the origin O and flies 150 miles in the direction 20° north of east to city A. From A, the airplane then flies 200 miles in the direction 23° west of north to city B. From B, the airplane flies 240 miles in the direction 10° south of west to city C. Express the location of city C as a vector **r** as shown in **FIGURE 7.1.21**. Find the distance from O to C.

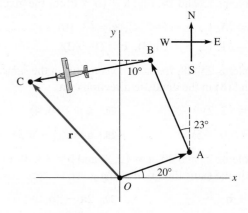

FIGURE 7.1.21 Airplane in Problem 50

Vectors in 3-Space

INTRODUCTION In the plane, or 2-space, one way of describing the position of a point P is to assign to it coordinates relative to two mutually orthogonal, or perpendicular, axes called the x- and y-axes. If P is the point of intersection of the line $x = a$ (perpendicular to the x-axis) and the line $y = b$ (perpendicular to the y-axis), then the **ordered pair** (a, b) is said to be the **rectangular** or **Cartesian coordinates** of the point. See **FIGURE 7.2.1**. In this section we extend the notions of Cartesian coordinates and vectors to three dimensions.

■ **Rectangular Coordinate System in 3-Space** In three dimensions, or **3-space**, a rectangular coordinate system is constructed using three mutually orthogonal axes. The point at which these axes intersect is called the origin O. These axes, shown in **FIGURE 7.2.2(a)**, are labeled in accordance with the so-called **right-hand rule**: If the fingers of the right hand, pointing in the direction of the positive x-axis, are curled toward the positive y-axis, then the thumb will point in the direction of a new axis perpendicular to the plane of the x- and y-axes. This new axis is labeled the z-axis.

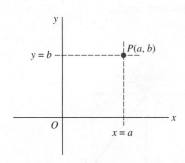

FIGURE 7.2.1 Rectangular coordinates in 2-space

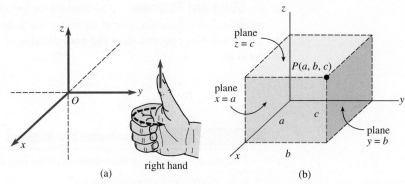

FIGURE 7.2.2 Rectangular coordinates in 3-space

The dashed lines in Figure 7.2.2(a) represent the negative axes. Now, if

$$x = a, \quad y = b, \quad z = c$$

are planes perpendicular to the x-axis, y-axis, and z-axis, respectively, then the point P at which these planes intersect can be represented by an **ordered triple** of numbers (a, b, c) said to be the **rectangular** or **Cartesian coordinates** of the point. The numbers a, b, and c are, in turn, called the x-, y-, and z-**coordinates** of $P(a, b, c)$. See Figure 7.2.2(b).

FIGURE 7.2.3 Octants

Octants Each pair of coordinate axes determines a **coordinate plane**. As shown in **FIGURE 7.2.3**, the x- and y-axes determine the xy-plane, the x- and z-axes determine the xz-plane, and so on. The coordinate planes divide 3-space into eight parts known as **octants**. The octant in which all three coordinates of a point are *positive* is called the **first octant**. There is no agreement for naming the other seven octants.

The following table summarizes the coordinates of a point either on a coordinate axis or in a coordinate plane. As seen in the table, we can also describe, say, the xy-plane by the simple equation $z = 0$. Similarly, the xz-plane is $y = 0$ and the yz-plane is $x = 0$.

Axes	Coordinates	Plane	Coordinates
x	$(a, 0, 0)$	xy	$(a, b, 0)$
y	$(0, b, 0)$	xz	$(a, 0, c)$
z	$(0, 0, c)$	yz	$(0, b, c)$

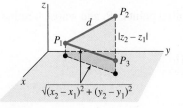

FIGURE 7.2.4 Points in Example 1

EXAMPLE 1 **Graphs of Three Points**

Graph the points $(4, 5, 6)$, $(3, -3, -1)$, and $(-2, -2, 0)$.

SOLUTION Of the three points shown in **FIGURE 7.2.4**, only $(4, 5, 6)$ is in the first octant. The point $(-2, -2, 0)$ is in the xy-plane. ≡

Distance Formula To find the **distance** between two points $P_1(x_1, y_1, z_1)$ and $P_2(x_2, y_2, z_2)$ in 3-space, let us first consider their projection onto the xy-plane. As seen in **FIGURE 7.2.5**, the distance between $(x_1, y_1, 0)$ and $(x_2, y_2, 0)$ follows from the usual distance formula in the plane and is $\sqrt{(x_2 - x_1)^2 + (y_2 - y_1)^2}$. If the coordinates of P_3 are (x_2, y_2, z_1), then the Pythagorean theorem applied to the right triangle $P_1P_2P_3$ yields

$$[d(P_1, P_2)]^2 = [\sqrt{(x_2 - x_1)^2 + (y_2 - y_1)^2}]^2 + |z_2 - z_1|^2$$

or

$$d(P_1, P_2) = \sqrt{(x_2 - x_1)^2 + (y_2 - y_1)^2 + (z_2 - z_1)^2}. \tag{1}$$

FIGURE 7.2.5 Distance d between two points in 3-space

EXAMPLE 2 **Distance Between Two Points**

Find the distance between $(2, -3, 6)$ and $(-1, -7, 4)$.

SOLUTION Choosing P_2 as $(2, -3, 6)$ and P_1 as $(-1, -7, 4)$, formula (1) gives

$$d = \sqrt{(2 - (-1))^2 + (-3 - (-7))^2 + (6 - 4)^2} = \sqrt{29}.$$

≡

Midpoint Formula The formula for finding the midpoint of a line segment between two points in 2-space carries over in an analogous fashion to 3-space. If $P_1(x_1, y_1, z_1)$ and $P_2(x_2, y_2, z_2)$ are two distinct points, then the **coordinates of the midpoint** of the line segment between them are

$$\left(\frac{x_1 + x_2}{2}, \frac{y_1 + y_2}{2}, \frac{z_1 + z_2}{2} \right). \tag{2}$$

EXAMPLE 3 **Coordinates of a Midpoint**

Find the coordinates of the midpoint of the line segment between the two points in Example 2.

SOLUTION From (2) we obtain

$$\left(\frac{2 + (-1)}{2}, \frac{-3 + (-7)}{2}, \frac{6 + 4}{2} \right) \quad \text{or} \quad (\tfrac{1}{2}, -5, 5). \qquad \equiv$$

Vectors in 3-Space A **vector a in 3-space** is any ordered triple of real numbers

$$\mathbf{a} = \langle a_1, a_2, a_3 \rangle,$$

FIGURE 7.2.6 Position vector

where a_1, a_2, and a_3 are the **components** of the vector. The set of all vectors in 3-space will be denoted by the symbol R^3. The **position vector** of a point $P(x_1, y_1, z_1)$ in space is the vector $\overrightarrow{OP} = \langle x_1, y_1, z_1 \rangle$ whose initial point is the origin O and whose terminal point is P. See **FIGURE 7.2.6**.

The component definitions of addition, subtraction, scalar multiplication, and so on are natural generalizations of those given for vectors in R^2. Moreover, the vectors in R^3 possess all the properties listed in Theorem 7.1.1.

DEFINITION 7.2.1 **Component Definitions in 3-Space**

Let $\mathbf{a} = \langle a_1, a_2, a_3 \rangle$ and $\mathbf{b} = \langle b_1, b_2, b_3 \rangle$ be vectors in R^3.
(*i*) Addition: $\mathbf{a} + \mathbf{b} = \langle a_1 + b_1, a_2 + b_2, a_3 + b_3 \rangle$
(*ii*) Scalar multiplication: $k\mathbf{a} = \langle ka_1, ka_2, ka_3 \rangle$
(*iii*) Equality: $\mathbf{a} = \mathbf{b}$ if and only if $a_1 = b_1, a_2 = b_2, a_3 = b_3$
(*iv*) Negative: $-\mathbf{b} = (-1)\mathbf{b} = \langle -b_1, -b_2, -b_3 \rangle$
(*v*) Subtraction: $\mathbf{a} - \mathbf{b} = \mathbf{a} + (-\mathbf{b}) = \langle a_1 - b_1, a_2 - b_2, a_3 - b_3 \rangle$
(*vi*) Zero vector: $\mathbf{0} = \langle 0, 0, 0 \rangle$
(*vii*) Magnitude: $\|\mathbf{a}\| = \sqrt{a_1^2 + a_2^2 + a_3^2}$

FIGURE 7.2.7 $\overrightarrow{OP}$ and $\overrightarrow{P_1P_2}$ are the same vector

If $\overrightarrow{OP_1}$ and $\overrightarrow{OP_2}$ are the position vectors of the points $P_1(x_1, y_1, z_1)$ and $P_2(x_2, y_2, z_2)$, then the vector $\overrightarrow{P_1P_2}$ is given by

$$\overrightarrow{P_1P_2} = \overrightarrow{OP_2} - \overrightarrow{OP_1} = \langle x_2 - x_1, y_2 - y_1, z_2 - z_1 \rangle. \tag{3}$$

As in 2-space, $\overrightarrow{P_1P_2}$ can be drawn either as a vector whose initial point is P_1 and whose terminal point is P_2 or as position vector $\overrightarrow{OP}$ with terminal point

$$P(x_2 - x_1, y_2 - y_1, z_2 - z_1).$$

See **FIGURE 7.2.7**.

EXAMPLE 4 **Vector Between Two Points**

Find the vector $\overrightarrow{P_1P_2}$ if the points P_1 and P_2 are given by $P_1(4, 6, -2)$ and $P_2(1, 8, 3)$, respectively.

SOLUTION If the position vectors of the points are $\overrightarrow{OP_1} = \langle 4, 6, -2 \rangle$ and $\overrightarrow{OP_2} = \langle 1, 8, 3 \rangle$, then from (3) we have

$$\overrightarrow{P_1P_2} = \overrightarrow{OP_2} - \overrightarrow{OP_1} = \langle 1 - 4, 8 - 6, 3 - (-2) \rangle = \langle -3, 2, 5 \rangle. \qquad \equiv$$

EXAMPLE 5 A Unit Vector

Find a unit vector in the direction of $\mathbf{a} = \langle -2, 3, 6 \rangle$.

SOLUTION Since a unit vector has length 1, we first find the magnitude of $\mathbf{a}$ and then use the fact that $\mathbf{a}/\|\mathbf{a}\|$ is a unit vector in the direction of $\mathbf{a}$. The magnitude of $\mathbf{a}$ is

$$\|\mathbf{a}\| = \sqrt{(-2)^2 + 3^2 + 6^2} = \sqrt{49} = 7.$$

A unit vector in the direction of $\mathbf{a}$ is

$$\frac{\mathbf{a}}{\|\mathbf{a}\|} = \frac{1}{7}\langle -2, 3, 6 \rangle = \left\langle -\frac{2}{7}, \frac{3}{7}, \frac{6}{7} \right\rangle.$$

≡

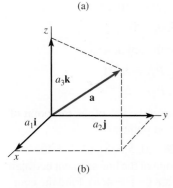

FIGURE 7.2.8 $\mathbf{i}$, $\mathbf{j}$, and $\mathbf{k}$ form a basis for R^3

�newline The i, j, k Vectors

The i, j, k Vectors We saw in the preceding section that the set of two unit vectors $\mathbf{i} = \langle 1, 0 \rangle$ and $\mathbf{j} = \langle 0, 1 \rangle$ constitute a basis for the system of two-dimensional vectors. That is, any vector $\mathbf{a}$ in 2-space can be written as a linear combination of $\mathbf{i}$ and $\mathbf{j}$: $\mathbf{a} = a_1\mathbf{i} + a_2\mathbf{j}$. Likewise any vector $\mathbf{a} = \langle a_1, a_2, a_3 \rangle$ in 3-space can be expressed as a linear combination of the unit vectors

$$\mathbf{i} = \langle 1, 0, 0 \rangle, \quad \mathbf{j} = \langle 0, 1, 0 \rangle, \quad \mathbf{k} = \langle 0, 0, 1 \rangle.$$

To see this we use (*i*) and (*ii*) of Definition 7.2.1 to write

$$\langle a_1, a_2, a_3 \rangle = \langle a_1, 0, 0 \rangle + \langle 0, a_2, 0 \rangle + \langle 0, 0, a_3 \rangle$$
$$= a_1\langle 1, 0, 0 \rangle + a_2\langle 0, 1, 0 \rangle + a_3\langle 0, 0, 1 \rangle;$$

that is, $\mathbf{a} = a_1\mathbf{i} + a_2\mathbf{j} + a_3\mathbf{k}.$

The vectors $\mathbf{i}$, $\mathbf{j}$, and $\mathbf{k}$ illustrated in **FIGURE 7.2.8(a)** are called the **standard basis** for the system of three-dimensional vectors. In Figure 7.2.8(b) we see that a position vector $\mathbf{a} = a_1\mathbf{i} + a_2\mathbf{j} + a_3\mathbf{k}$ is the sum of the vectors $a_1\mathbf{i}$, $a_2\mathbf{j}$, and $a_3\mathbf{k}$, which lie along the coordinate axes and have the origin as a common initial point.

EXAMPLE 6 **Using the i, j, k Vectors**

The vector $\mathbf{a} = \langle 7, -5, 13 \rangle$ is the same as $\mathbf{a} = 7\mathbf{i} - 5\mathbf{j} + 13\mathbf{k}$. ≡

When the third dimension is taken into consideration, any vector in the xy-plane is equivalently described as a three-dimensional vector that lies in the coordinate plane $z = 0$. Although the vectors $\langle a_1, a_2 \rangle$ and $\langle a_1, a_2, 0 \rangle$ are technically not equal, we shall ignore the distinction. That is why, for example, we denoted $\langle 1, 0 \rangle$ and $\langle 1, 0, 0 \rangle$ by the same symbol $\mathbf{i}$. But to avoid any possible confusion, hereafter we shall always consider a vector a three-dimensional vector, and the symbols $\mathbf{i}$ and $\mathbf{j}$ will represent only $\langle 1, 0, 0 \rangle$ and $\langle 0, 1, 0 \rangle$, respectively. Similarly, a vector in either the xy-plane or the xz-plane must have one zero component. In the yz-plane, a vector

$$\mathbf{b} = \langle 0, b_2, b_3 \rangle \quad \text{is written} \quad \mathbf{b} = b_2\mathbf{j} + b_3\mathbf{k}.$$

In the xz-plane, a vector

$$\mathbf{c} = \langle c_1, 0, c_3 \rangle \quad \text{is the same as} \quad \mathbf{c} = c_1\mathbf{i} + c_3\mathbf{k}.$$

EXAMPLE 7 **Vector in xz-Plane**

(a) The vector $\mathbf{a} = 5\mathbf{i} + 3\mathbf{k} = 5\mathbf{i} + 0\mathbf{j} + 3\mathbf{k}$ lies in the xz-plane and can also be written as $\mathbf{a} = \langle 5, 0, 3 \rangle$.

(b) $\|5\mathbf{i} + 3\mathbf{k}\| = \sqrt{5^2 + 0^2 + 3^2} = \sqrt{25 + 9} = \sqrt{34}$ ≡

EXAMPLE 8 **Linear Combination**

If $\mathbf{a} = 3\mathbf{i} - 4\mathbf{j} + 8\mathbf{k}$ and $\mathbf{b} = \mathbf{i} - 4\mathbf{k}$, find $5\mathbf{a} - 2\mathbf{b}$.

SOLUTION We treat $\mathbf{b}$ as a three-dimensional vector and write, for emphasis, $\mathbf{b} = \mathbf{i} + 0\mathbf{j} - 4\mathbf{k}$. From

$$5\mathbf{a} = 15\mathbf{i} - 20\mathbf{j} + 40\mathbf{k} \quad \text{and} \quad 2\mathbf{b} = 2\mathbf{i} + 0\mathbf{j} - 8\mathbf{k}$$

we get $5\mathbf{a} - 2\mathbf{b} = (15\mathbf{i} - 20\mathbf{j} + 40\mathbf{k}) - (2\mathbf{i} + 0\mathbf{j} - 8\mathbf{k})$
$$= 13\mathbf{i} - 20\mathbf{j} + 48\mathbf{k}.$$

≡

In Problems 1–6, graph the given point. Use the same coordinate axes.

1. $(1, 1, 5)$ **2.** $(0, 0, 4)$

3. $(3, 4, 0)$ **4.** $(6, 0, 0)$

5. $(6, -2, 0)$ **6.** $(5, -4, 3)$

In Problems 7–10, describe geometrically all points $P(x, y, z)$ that satisfy the given condition.

7. $z = 5$ **8.** $x = 1$

9. $x = 2,\ y = 3$ **10.** $x = 4,\ y = -1,\ z = 7$

11. Give the coordinates of the vertices of the rectangular parallelepiped whose sides are the coordinate planes and the planes $x = 2, y = 5, z = 8$.

12. In **FIGURE 7.2.9**, two vertices are shown of a rectangular parallelepiped having sides parallel to the coordinate planes. Find the coordinates of the remaining six vertices.

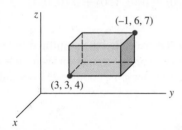

FIGURE 7.2.9 Rectangular parallelepiped in Problem 12

13. Consider the point $P(-2, 5, 4)$.
 (a) If lines are drawn from P perpendicular to the coordinate planes, what are the coordinates of the point at the base of each perpendicular?
 (b) If a line is drawn from P to the plane $z = -2$, what are the coordinates of the point at the base of the perpendicular?
 (c) Find the point in the plane $x = 3$ that is closest to P.

14. Determine an equation of a plane parallel to a coordinate plane that contains the given pair of points.
 (a) $(3, 4, -5), (-2, 8, -5)$
 (b) $(1, -1, 1), (1, -1, -1)$
 (c) $(-2, 1, 2), (2, 4, 2)$

In Problems 15–20, describe the locus of points $P(x, y, z)$ that satisfy the given equation(s).

15. $xyz = 0$ **16.** $x^2 + y^2 + z^2 = 0$

17. $(x + 1)^2 + (y - 2)^2 + (z + 3)^2 = 0$

18. $(x - 2)(z - 8) = 0$

19. $z^2 - 25 = 0$ **20.** $x = y = z$

In Problems 21 and 22, find the distance between the given points.

21. $(3, -1, 2), (6, 4, 8)$ **22.** $(-1, -3, 5), (0, 4, 3)$

23. Find the distance from the point $(7, -3, -4)$ to (a) the yz-plane and (b) the x-axis.

24. Find the distance from the point $(-6, 2, -3)$ to (a) the xz-plane and (b) the origin.

In Problems 25–28, the given three points form a triangle. Determine which triangles are isosceles and which are right triangles.

25. $(0, 0, 0), (3, 6, -6), (2, 1, 2)$

26. $(0, 0, 0), (1, 2, 4), (3, 2, 2\sqrt{2})$

27. $(1, 2, 3), (4, 1, 3), (4, 6, 4)$

28. $(1, 1, -1), (1, 1, 1), (0, -1, 1)$

In Problems 29 and 30, use the distance formula to prove that the given points are collinear.

29. $P_1(1, 2, 0), P_2(-2, -2, -3), P_3(7, 10, 6)$

30. $P_1(2, 3, 2), P_2(1, 4, 4), P_3(5, 0, -4)$

In Problems 31 and 32, solve for the unknown.

31. $P_1(x, 2, 3), P_2(2, 1, 1);\ d(P_1, P_2) = \sqrt{21}$

32. $P_1(x, x, 1), P_2(0, 3, 5);\ d(P_1, P_2) = 5$

In Problems 33 and 34, find the coordinates of the midpoint of the line segment between the given points.

33. $(1, 3, \frac{1}{2}), (7, -2, \frac{5}{2})$ **34.** $(0, 5, -8), (4, 1, -6)$

35. The coordinates of the midpoint of the line segment between $P_1(x_1, y_1, z_1)$ and $P_2(2, 3, 6)$ are $(-1, -4, 8)$. Find the coordinates of P_1.

36. Let P_3 be the midpoint of the line segment between $P_1(-3, 4, 1)$ and $P_2(-5, 8, 3)$. Find the coordinates of the midpoint of the line segment (a) between P_1 and P_3 and (b) between P_3 and P_2.

In Problems 37–40, find the vector $\overrightarrow{P_1P_2}$.

37. $P_1(3, 4, 5), P_2(0, -2, 6)$ **38.** $P_1(-2, 4, 0), P_2(6, \frac{3}{2}, 8)$

39. $P_1(0, -1, 0), P_2(2, 0, 1)$ **40.** $P_1(\frac{1}{2}, \frac{3}{4}, 5), P_2(-\frac{5}{2}, -\frac{9}{4}, 12)$

In Problems 41–48, $\mathbf{a} = \langle 1, -3, 2 \rangle$, $\mathbf{b} = \langle -1, 1, 1 \rangle$, and $\mathbf{c} = \langle 2, 6, 9 \rangle$. Find the indicated vector or scalar.

41. $\mathbf{a} + (\mathbf{b} + \mathbf{c})$ **42.** $2\mathbf{a} - (\mathbf{b} - \mathbf{c})$

43. $\mathbf{b} + 2(\mathbf{a} - 3\mathbf{c})$ **44.** $4(\mathbf{a} + 2\mathbf{c}) - 6\mathbf{b}$

45. $\|\mathbf{a} + \mathbf{c}\|$ **46.** $\|\mathbf{c}\|\ \|2\mathbf{b}\|$

47. $\left\| \dfrac{\mathbf{a}}{\|\mathbf{a}\|} \right\| + 5 \left\| \dfrac{\mathbf{b}}{\|\mathbf{b}\|} \right\|$ **48.** $\|\mathbf{b}\|\mathbf{a} + \|\mathbf{a}\|\mathbf{b}$

49. Find a unit vector in the opposite direction of $\mathbf{a} = \langle 10, -5, 10 \rangle$.

50. Find a unit vector in the same direction as $\mathbf{a} = \mathbf{i} - 3\mathbf{j} + 2\mathbf{k}$.

51. Find a vector $\mathbf{b}$ that is four times as long as $\mathbf{a} = \mathbf{i} - \mathbf{j} + \mathbf{k}$ in the same direction as $\mathbf{a}$.

52. Find a vector $\mathbf{b}$ for which $\|\mathbf{b}\| = \frac{1}{2}$ that is parallel to $\mathbf{a} = \langle -6, 3, -2 \rangle$ but has the opposite direction.

53. Using the vectors $\mathbf{a}$ and $\mathbf{b}$ shown in **FIGURE 7.2.10**, sketch the "average vector" $\frac{1}{2}(\mathbf{a} + \mathbf{b})$.

FIGURE 7.2.10 Vectors for Problem 53

In this and the following section, we shall consider two kinds of products between vectors that originated in the study of mechanics and electricity and magnetism. The first of these products, known as the **dot product**, is studied in this section.

Component Form of the Dot Product

The dot product, defined next, is also known as the **inner product**, or **scalar product**. The dot product of two vectors **a** and **b** is denoted by **a · b** and is a real number, or scalar, defined in terms of the components of the vectors.

DEFINITION 7.3.1 Dot Product of Two Vectors

In 2-space the **dot product** of two vectors $\mathbf{a} = \langle a_1, a_2 \rangle$ and $\mathbf{b} = \langle b_1, b_2 \rangle$ is the number

$$\mathbf{a} \cdot \mathbf{b} = a_1 b_1 + a_2 b_2. \tag{1}$$

In 3-space the **dot product** of two vectors $\mathbf{a} = \langle a_1, a_2, a_3 \rangle$ and $\mathbf{b} = \langle b_1, b_2, b_3 \rangle$ is the number

$$\mathbf{a} \cdot \mathbf{b} = a_1 b_1 + a_2 b_2 + a_3 b_3. \tag{2}$$

EXAMPLE 1 Dot Product Using (2)

If $\mathbf{a} = 10\mathbf{i} + 2\mathbf{j} - 6\mathbf{k}$ and $\mathbf{b} = -\frac{1}{2}\mathbf{i} + 4\mathbf{j} - 3\mathbf{k}$, then it follows from (2) that

$$\mathbf{a} \cdot \mathbf{b} = (10)(-\tfrac{1}{2}) + (2)(4) + (-6)(-3) = 21.$$

≡

EXAMPLE 2 Dot Products of the Basis Vectors

Since $\mathbf{i} = \langle 1, 0, 0 \rangle$, $\mathbf{j} = \langle 0, 1, 0 \rangle$, and $\mathbf{k} = \langle 0, 0, 1 \rangle$, we see from (2) that

$$\mathbf{i} \cdot \mathbf{j} = \mathbf{j} \cdot \mathbf{i} = 0, \quad \mathbf{j} \cdot \mathbf{k} = \mathbf{k} \cdot \mathbf{j} = 0, \quad \text{and} \quad \mathbf{k} \cdot \mathbf{i} = \mathbf{i} \cdot \mathbf{k} = 0. \tag{3}$$

Similarly, by (2)

$$\mathbf{i} \cdot \mathbf{i} = 1, \quad \mathbf{j} \cdot \mathbf{j} = 1, \quad \text{and} \quad \mathbf{k} \cdot \mathbf{k} = 1. \tag{4}$$ ≡

Properties

The dot product possesses the following properties.

THEOREM 7.3.1 Properties of the Dot Product

(*i*) $\mathbf{a} \cdot \mathbf{b} = 0$ if $\mathbf{a} = \mathbf{0}$ or $\mathbf{b} = \mathbf{0}$

(*ii*) $\mathbf{a} \cdot \mathbf{b} = \mathbf{b} \cdot \mathbf{a}$ ← commutative law

(*iii*) $\mathbf{a} \cdot (\mathbf{b} + \mathbf{c}) = \mathbf{a} \cdot \mathbf{b} + \mathbf{a} \cdot \mathbf{c}$ ← distributive law

(*iv*) $\mathbf{a} \cdot (k\mathbf{b}) = (k\mathbf{a}) \cdot \mathbf{b} = k(\mathbf{a} \cdot \mathbf{b})$, k a scalar

(*v*) $\mathbf{a} \cdot \mathbf{a} \geq 0$

(*vi*) $\mathbf{a} \cdot \mathbf{a} = \|\mathbf{a}\|^2$

PROOF: All the properties can be proved directly from (2). We illustrate by proving parts (*iii*) and (*vi*).

To prove part (*iii*) we let $\mathbf{a} = \langle a_1, a_2, a_3 \rangle$, $\mathbf{b} = \langle b_1, b_2, b_3 \rangle$, and $\mathbf{c} = \langle c_1, c_2, c_3 \rangle$. Then

$$
\begin{aligned}
\mathbf{a} \cdot (\mathbf{b} + \mathbf{c}) &= \langle a_1, a_2, a_3 \rangle \cdot (\langle b_1, b_2, b_3 \rangle + \langle c_1, c_2, c_3 \rangle) \\
&= \langle a_1, a_2, a_3 \rangle \cdot \langle b_1 + c_1, b_2 + c_2, b_3 + c_3 \rangle \\
&= a_1(b_1 + c_1) + a_2(b_2 + c_2) + a_3(b_3 + c_3) \\
&= a_1 b_1 + a_1 c_1 + a_2 b_2 + a_2 c_2 + a_3 b_3 + a_3 c_3 \quad \leftarrow \begin{cases} \text{since multiplication of real} \\ \text{numbers is distributive over} \\ \text{addition} \end{cases} \\
&= (a_1 b_1 + a_2 b_2 + a_3 b_3) + (a_1 c_1 + a_2 c_2 + a_3 c_3) \\
&= \mathbf{a} \cdot \mathbf{b} + \mathbf{a} \cdot \mathbf{c}.
\end{aligned}
$$

To prove part (*vi*) we note that

$$\mathbf{a} \cdot \mathbf{a} = \langle a_1, a_2, a_3 \rangle \cdot \langle a_1, a_2, a_3 \rangle = a_1^2 + a_2^2 + a_3^2 = \|\mathbf{a}\|^2. \qquad \equiv$$

Notice that (*vi*) of Theorem 7.3.1 states that the **magnitude** of a vector can be written in terms of the dot product:

$$\|\mathbf{a}\| = \sqrt{\mathbf{a} \cdot \mathbf{a}} = \sqrt{a_1^2 + a_2^2 + a_3^2}.$$

■ **Alternative Form** The dot product of two vectors **a** and **b** can also be expressed in terms of the lengths of the vectors and the angle between them. If the vectors **a** and **b** are positioned in such a manner that their initial points coincide, then we define the angle between **a** and **b** as the angle θ that satisfies $0 \le \theta \le \pi$.

THEOREM 7.3.2 Alternative Form of Dot Product

The dot product of two vectors **a** and **b** is

$$\mathbf{a} \cdot \mathbf{b} = \|\mathbf{a}\| \|\mathbf{b}\| \cos\theta \qquad (5)$$

where θ is the angle between the vectors.

> This more geometric form is what is generally used as the definition of the dot product in a physics course.

PROOF: Suppose θ is the angle between the vectors $\mathbf{a} = a_1\mathbf{i} + a_2\mathbf{j} + a_3\mathbf{k}$ and $\mathbf{b} = b_1\mathbf{i} + b_2\mathbf{j} + b_3\mathbf{k}$. Then the vector

$$\mathbf{c} = \mathbf{b} - \mathbf{a} = (b_1 - a_1)\mathbf{i} + (b_2 - a_2)\mathbf{j} + (b_3 - a_3)\mathbf{k}$$

is the third side of the triangle indicated in **FIGURE 7.3.1**. By the law of cosines we can write
$\|\mathbf{c}\|^2 = \|\mathbf{b}\|^2 + \|\mathbf{a}\|^2 - 2\|\mathbf{a}\| \|\mathbf{b}\| \cos\theta$ or

$$\|\mathbf{a}\| \|\mathbf{b}\| \cos\theta = \frac{1}{2}(\|\mathbf{b}\|^2 + \|\mathbf{a}\|^2 - \|\mathbf{c}\|^2). \qquad (6)$$

Using

$$\|\mathbf{a}\|^2 = a_1^2 + a_2^2 + a_3^2, \quad \|\mathbf{b}\|^2 = b_1^2 + b_2^2 + b_3^2,$$

and

$$\|\mathbf{c}\|^2 = \|\mathbf{b} - \mathbf{a}\|^2 = (b_1 - a_1)^2 + (b_2 - a_2)^2 + (b_3 - a_3)^2,$$

FIGURE 7.3.1 Vector **c** in the proof of Theorem 7.3.2

the right-hand side of equation (6) simplifies to $a_1b_1 + a_2b_2 + a_3b_3$. Since the last expression is (2) in Definition 7.3.1, we see that $\|\mathbf{a}\| \|\mathbf{b}\| \cos\theta = \mathbf{a} \cdot \mathbf{b}$. $\qquad \equiv$

■ **Orthogonal Vectors** If **a** and **b** are nonzero vectors, then $\|\mathbf{a}\|$ and $\|\mathbf{b}\|$ are both positive numbers. In this case it follows from (5) that the sign of the dot product $\mathbf{a} \cdot \mathbf{b}$ is the same as the sign of $\cos\theta$. In **FIGURE 7.3.2** we see various orientations of two vectors for angles that satisfy $0 \le \theta \le \pi$. Observe from Figure 7.3.2 and (5) that $\mathbf{a} \cdot \mathbf{b} > 0$ for $\theta = 0$, $\mathbf{a} \cdot \mathbf{b} > 0$ when θ is acute, $\mathbf{a} \cdot \mathbf{b} = 0$ for $\theta = \pi/2$, $\mathbf{a} \cdot \mathbf{b} < 0$ when θ is obtuse, and $\mathbf{a} \cdot \mathbf{b} < 0$ for $\theta = \pi$. In the special case $\theta = \pi/2$ we say that the vectors are **orthogonal** or **perpendicular**. Moreover, if we know that $\mathbf{a} \cdot \mathbf{b} = 0$, then the only angle for which this is true and satisfying $0 \le \theta \le \pi$ is $\theta = \pi/2$. This leads us to the next result.

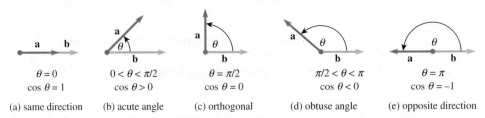

$\theta = 0$	$0 < \theta < \pi/2$	$\theta = \pi/2$	$\pi/2 < \theta < \pi$	$\theta = \pi$
$\cos\theta = 1$	$\cos\theta > 0$	$\cos\theta = 0$	$\cos\theta < 0$	$\cos\theta = -1$
(a) same direction	(b) acute angle	(c) orthogonal	(d) obtuse angle	(e) opposite direction

FIGURE 7.3.2 Angles between vectors **a** and **b**

THEOREM 7.3.3 Criterion for Orthogonal Vectors

Two nonzero vectors **a** and **b** are orthogonal if and only if $\mathbf{a} \cdot \mathbf{b} = 0$.

Since $\mathbf{0} \cdot \mathbf{b} = \langle 0, 0, 0 \rangle \cdot \langle b_1, b_2, b_3 \rangle = 0(b_1) + 0(b_2) + 0(b_3) = 0$ for every vector **b**, the zero vector **0** is considered to be orthogonal to every vector.

EXAMPLE 3 **Orthogonal Vectors**

If $\mathbf{a} = -3\mathbf{i} - \mathbf{j} + 4\mathbf{k}$ and $\mathbf{b} = 2\mathbf{i} + 14\mathbf{j} + 5\mathbf{k}$, then

$$\mathbf{a} \cdot \mathbf{b} = (-3)(2) + (-1)(14) + (4)(5) = 0.$$

From Theorem 7.3.3, we conclude that $\mathbf{a}$ and $\mathbf{b}$ are orthogonal. ≡

In Example 1, from $\mathbf{a} \cdot \mathbf{b} \neq 0$ we can conclude that the vectors $\mathbf{a} = 10\mathbf{i} + 2\mathbf{j} - 6\mathbf{k}$ and $\mathbf{b} = -\frac{1}{2}\mathbf{i} + 4\mathbf{j} - 3\mathbf{k}$ are *not* orthogonal. From (3) of Example 2, we see what is apparent in Figure 7.2.8, namely, the vectors $\mathbf{i}$, $\mathbf{j}$, and $\mathbf{k}$ constitute a set of mutually orthogonal unit vectors.

■ **Angle Between Two Vectors** By equating the two forms of the dot product, (2) and (5), we can determine **the angle between two vectors** from

$$\cos\theta = \frac{a_1b_1 + a_2b_2 + a_3b_3}{\|\mathbf{a}\| \, \|\mathbf{b}\|}. \tag{7}$$

EXAMPLE 4 **Angle Between Two Vectors**

Find the angle between $\mathbf{a} = 2\mathbf{i} + 3\mathbf{j} + \mathbf{k}$ and $\mathbf{b} = -\mathbf{i} + 5\mathbf{j} + \mathbf{k}$.

SOLUTION From $\|\mathbf{a}\| = \sqrt{14}$, $\|\mathbf{b}\| = \sqrt{27}$, $\mathbf{a} \cdot \mathbf{b} = 14$, we see from (7) that

$$\cos\theta = \frac{14}{\sqrt{14}\sqrt{27}} = \frac{\sqrt{42}}{9},$$

and so $\theta = \cos^{-1}(\sqrt{42}/9) \approx 0.77$ radian or $\theta \approx 43.9°$. ≡

FIGURE 7.3.3 Direction angles α, β, and γ

■ **Direction Cosines** For a nonzero vector $\mathbf{a} = a_1\mathbf{i} + a_2\mathbf{j} + a_3\mathbf{k}$ in 3-space, the angles α, β, and γ between $\mathbf{a}$ and the unit vectors $\mathbf{i}$, $\mathbf{j}$, and $\mathbf{k}$, respectively, are called **direction angles** of $\mathbf{a}$. See **FIGURE 7.3.3**. Now, by (7),

$$\cos\alpha = \frac{\mathbf{a} \cdot \mathbf{i}}{\|\mathbf{a}\| \, \|\mathbf{i}\|}, \quad \cos\beta = \frac{\mathbf{a} \cdot \mathbf{j}}{\|\mathbf{a}\| \, \|\mathbf{j}\|}, \quad \cos\gamma = \frac{\mathbf{a} \cdot \mathbf{k}}{\|\mathbf{a}\| \, \|\mathbf{k}\|},$$

which simplify to

$$\cos\alpha = \frac{a_1}{\|\mathbf{a}\|}, \quad \cos\beta = \frac{a_2}{\|\mathbf{a}\|}, \quad \cos\gamma = \frac{a_3}{\|\mathbf{a}\|}.$$

We say that $\cos\alpha$, $\cos\beta$, and $\cos\gamma$ are the **direction cosines** of $\mathbf{a}$. The direction cosines of a nonzero vector $\mathbf{a}$ are simply the components of the unit vector $(1/\|\mathbf{a}\|)\mathbf{a}$:

$$\frac{1}{\|\mathbf{a}\|}\mathbf{a} = \frac{a_1}{\|\mathbf{a}\|}\mathbf{i} + \frac{a_2}{\|\mathbf{a}\|}\mathbf{j} + \frac{a_3}{\|\mathbf{a}\|}\mathbf{k} = (\cos\alpha)\,\mathbf{i} + (\cos\beta)\,\mathbf{j} + (\cos\gamma)\,\mathbf{k}.$$

Since the magnitude of $(1/\|\mathbf{a}\|)\,\mathbf{a}$ is 1, it follows from the last equation that

$$\cos^2\alpha + \cos^2\beta + \cos^2\gamma = 1.$$

EXAMPLE 5 **Direction Cosines/Angles**

Find the direction cosines and direction angles of the vector $\mathbf{a} = 2\mathbf{i} + 5\mathbf{j} + 4\mathbf{k}$.

SOLUTION From $\|\mathbf{a}\| = \sqrt{2^2 + 5^2 + 4^2} = \sqrt{45} = 3\sqrt{5}$, we see that the direction cosines are

$$\cos\alpha = \frac{2}{3\sqrt{5}}, \quad \cos\beta = \frac{5}{3\sqrt{5}}, \quad \cos\gamma = \frac{4}{3\sqrt{5}}.$$

The direction angles are

$$\alpha = \cos^{-1}\left(\frac{2}{3\sqrt{5}}\right) \approx 1.27 \text{ radians} \quad \text{or} \quad \alpha \approx 72.7°$$

$$\beta = \cos^{-1}\left(\frac{5}{3\sqrt{5}}\right) \approx 0.73 \text{ radian} \quad \text{or} \quad \beta \approx 41.8°$$

$$\gamma = \cos^{-1}\left(\frac{4}{3\sqrt{5}}\right) \approx 0.93 \text{ radian} \quad \text{or} \quad \gamma \approx 53.4°. \quad ≡$$

(a)

(b)

FIGURE 7.3.4 Component of **a** on **b**

Observe in Example 5 that

$$\cos^2\alpha + \cos^2\beta + \cos^2\gamma = \frac{4}{45} + \frac{25}{45} + \frac{16}{45} = 1.$$

■ **Component of a on b** The distributive law and (3) enable us to express the components of a vector $\mathbf{a} = a_1\mathbf{i} + a_2\mathbf{j} + a_3\mathbf{k}$ in terms of the dot product:

$$a_1 = \mathbf{a} \cdot \mathbf{i}, \quad a_2 = \mathbf{a} \cdot \mathbf{j}, \quad a_3 = \mathbf{a} \cdot \mathbf{k}. \tag{8}$$

Symbolically, we write the components of **a** as

$$\text{comp}_\mathbf{i}\mathbf{a} = \mathbf{a} \cdot \mathbf{i}, \quad \text{comp}_\mathbf{j}\mathbf{a} = \mathbf{a} \cdot \mathbf{j}, \quad \text{comp}_\mathbf{k}\mathbf{a} = \mathbf{a} \cdot \mathbf{k}. \tag{9}$$

We shall now see that the results indicated in (9) carry over to finding the **component of a on an arbitrary vector b**. Note that in either of the two cases shown in **FIGURE 7.3.4**,

$$\text{comp}_\mathbf{b}\mathbf{a} = \|\mathbf{a}\| \cos \theta. \tag{10}$$

In Figure 7.3.4(b), $\text{comp}_\mathbf{b}\mathbf{a} < 0$ since $\pi/2 < \theta \leq \pi$. Now, by writing (10) as

$$\text{comp}_\mathbf{b}\mathbf{a} = \frac{\|\mathbf{a}\| \|\mathbf{b}\| \cos\theta}{\|\mathbf{b}\|} = \frac{\mathbf{a} \cdot \mathbf{b}}{\|\mathbf{b}\|},$$

we see that

$$\text{comp}_\mathbf{b}\mathbf{a} = \mathbf{a} \cdot \left(\frac{1}{\|\mathbf{b}\|}\mathbf{b}\right) = \frac{\mathbf{a} \cdot \mathbf{b}}{\|\mathbf{b}\|}. \tag{11}$$

In other words:

*To find the component of **a** on a vector **b**, we dot **a** with a unit vector in the direction of **b**.*

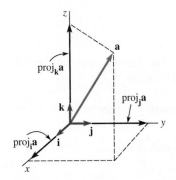

FIGURE 7.3.5 Projections of **a** onto **i**, **j**, and **k**

| EXAMPLE 6 | **Component of a Vector on Another Vector** |

Let $\mathbf{a} = 2\mathbf{i} + 3\mathbf{j} - 4\mathbf{k}$ and $\mathbf{b} = \mathbf{i} + \mathbf{j} + 2\mathbf{k}$. Find $\text{comp}_\mathbf{b}\mathbf{a}$ and $\text{comp}_\mathbf{a}\mathbf{b}$.

SOLUTION We first form a unit vector in the direction of **b**:

$$\|\mathbf{b}\| = \sqrt{6}, \quad \frac{1}{\|\mathbf{b}\|}\mathbf{b} = \frac{1}{\sqrt{6}}(\mathbf{i} + \mathbf{j} + 2\mathbf{k}).$$

Then from (11) we have

$$\text{comp}_\mathbf{b}\mathbf{a} = (2\mathbf{i} + 3\mathbf{j} - 4\mathbf{k}) \cdot \frac{1}{\sqrt{6}}(\mathbf{i} + \mathbf{j} + 2\mathbf{k}) = -\frac{3}{\sqrt{6}}.$$

By modifying (11) accordingly, we have

$$\text{comp}_\mathbf{a}\mathbf{b} = \mathbf{b} \cdot \left(\frac{1}{\|\mathbf{a}\|}\mathbf{a}\right).$$

Therefore, $\quad \|\mathbf{a}\| = \sqrt{29}, \quad \frac{1}{\|\mathbf{a}\|}\mathbf{a} = \frac{1}{\sqrt{29}}(2\mathbf{i} + 3\mathbf{j} - 4\mathbf{k})$

and $\quad \text{comp}_\mathbf{a}\mathbf{b} = (\mathbf{i} + \mathbf{j} + 2\mathbf{k}) \cdot \frac{1}{\sqrt{29}}(2\mathbf{i} + 3\mathbf{j} - 4\mathbf{k}) = -\frac{3}{\sqrt{29}}.$ ≡

FIGURE 7.3.6 Projection of **a** onto **b**

■ **Projection of a onto b** As illustrated in **FIGURE 7.3.5**, the **projection** of a vector **a** in any of the directions determined by **i**, **j**, **k** is simply the *vector* formed by multiplying the component of **a** in the specified direction with a unit vector in that direction; for example,

$$\text{proj}_\mathbf{i}\mathbf{a} = (\text{comp}_\mathbf{i}\mathbf{a})\mathbf{i} = (\mathbf{a} \cdot \mathbf{i})\mathbf{i} = a_1\mathbf{i}$$

and so on. **FIGURE 7.3.6** shows the general case of the **projection of a onto b**:

$$\text{proj}_\mathbf{b}\mathbf{a} = (\text{comp}_\mathbf{b}\mathbf{a})\left(\frac{1}{\|\mathbf{b}\|}\mathbf{b}\right) = \left(\frac{\mathbf{a} \cdot \mathbf{b}}{\mathbf{b} \cdot \mathbf{b}}\right)\mathbf{b}. \tag{12}$$

$$\frac{22}{13}\mathbf{i} + \frac{33}{13}\mathbf{j}$$

FIGURE 7.3.7 Projection of **a** onto **b** in Example 7

FIGURE 7.3.8 Work done by a force **F**

EXAMPLE 7 **Projection of a Vector on Another Vector**

Find the projection of $\mathbf{a} = 4\mathbf{i} + \mathbf{j}$ onto the vector $\mathbf{b} = 2\mathbf{i} + 3\mathbf{j}$. Graph.

SOLUTION First, we find the component of **a** and **b**. Since $\|\mathbf{b}\| = \sqrt{13}$, we find from (11) that

$$\mathrm{comp}_{\mathbf{b}}\mathbf{a} = (4\mathbf{i} + \mathbf{j}) \cdot \frac{1}{\sqrt{13}} (2\mathbf{i} + 3\mathbf{j}) = \frac{11}{\sqrt{13}}.$$

Thus, from (12),

$$\mathrm{proj}_{\mathbf{b}}\mathbf{a} = \left(\frac{11}{\sqrt{13}}\right)\left(\frac{1}{\sqrt{13}}\right)(2\mathbf{i} + 3\mathbf{j}) = \frac{22}{13}\mathbf{i} + \frac{33}{13}\mathbf{j}.$$

The graph of this vector is shown in **FIGURE 7.3.7**. ≡

■ **Physical Interpretation of the Dot Product** When a constant force of magnitude F moves an object a distance d in the same direction of the force, the work done is simply $W = Fd$. However, if a constant force $\mathbf{F}$ applied to a body acts at an angle θ to the direction of motion, then the work done by $\mathbf{F}$ is defined to be the product of the component of $\mathbf{F}$ in the direction of the displacement and the distance $\|\mathbf{d}\|$ that the body moves:

$$W = (\|\mathbf{F}\| \cos \theta) \|\mathbf{d}\| = \|\mathbf{F}\| \|\mathbf{d}\| \cos \theta.$$

See **FIGURE 7.3.8**. It follows from Theorem 7.3.2 that if $\mathbf{F}$ causes a displacement $\mathbf{d}$ of a body, then the work done is

$$W = \mathbf{F} \cdot \mathbf{d}. \tag{13}$$

EXAMPLE 8 **Work Done by a Constant Force**

Find the work done by a constant force $\mathbf{F} = 2\mathbf{i} + 4\mathbf{j}$ if its point of application to a block moves from $P_1(1, 1)$ to $P_2(4, 6)$. Assume that $\|\mathbf{F}\|$ is measured in newtons and $\|\mathbf{d}\|$ is measured in meters.

SOLUTION The displacement of the block is given by

$$\mathbf{d} = \overrightarrow{P_1P_2} = \overrightarrow{OP_2} - \overrightarrow{OP_1} = 3\mathbf{i} + 5\mathbf{j}.$$

It follows from (13) that the work done is

$$W = (2\mathbf{i} + 4\mathbf{j}) \cdot (3\mathbf{i} + 5\mathbf{j}) = 26 \text{ N-m}.$$ ≡

7.3 Exercises Answers to selected odd-numbered problems begin on page ANS-16.

In Problems 1–12, $\mathbf{a} = \langle 2, -3, 4 \rangle$, $\mathbf{b} = \langle -1, 2, 5 \rangle$, and $\mathbf{c} = \langle 3, 6, -1 \rangle$. Find the indicated scalar or vector.

1. $\mathbf{a} \cdot \mathbf{b}$ **2.** $\mathbf{b} \cdot \mathbf{c}$

3. $\mathbf{a} \cdot \mathbf{c}$ **4.** $\mathbf{a} \cdot (\mathbf{b} + \mathbf{c})$

5. $\mathbf{a} \cdot (4\mathbf{b})$ **6.** $\mathbf{b} \cdot (\mathbf{a} - \mathbf{c})$

7. $\mathbf{a} \cdot \mathbf{a}$ **8.** $(2\mathbf{b}) \cdot (3\mathbf{c})$

9. $\mathbf{a} \cdot (\mathbf{a} + \mathbf{b} + \mathbf{c})$ **10.** $(2\mathbf{a}) \cdot (\mathbf{a} - 2\mathbf{b})$

11. $\left(\dfrac{\mathbf{a} \cdot \mathbf{b}}{\mathbf{b} \cdot \mathbf{b}}\right)\mathbf{b}$ **12.** $(\mathbf{c} \cdot \mathbf{b})\mathbf{a}$

In Problems 13 and 14, find $\mathbf{a} \cdot \mathbf{b}$ if the smaller angle between **a** and **b** is as given.

13. $\|\mathbf{a}\| = 10$, $\|\mathbf{b}\| = 5$, $\theta = \pi/4$

14. $\|\mathbf{a}\| = 6$, $\|\mathbf{b}\| = 12$, $\theta = \pi/6$

15. Determine which pairs of the following vectors are orthogonal:
 (a) $\langle 2, 0, 1 \rangle$ (b) $3\mathbf{i} + 2\mathbf{j} - \mathbf{k}$
 (c) $2\mathbf{i} - \mathbf{j} - \mathbf{k}$ (d) $\mathbf{i} - 4\mathbf{j} + 6\mathbf{k}$
 (e) $\langle 1, -1, 1 \rangle$ (f) $\langle -4, 3, 8 \rangle$

16. Determine a scalar c so that the given vectors are orthogonal.
 (a) $\mathbf{a} = 2\mathbf{i} - c\mathbf{j} + 3\mathbf{k}$, $\mathbf{b} = 3\mathbf{i} + 2\mathbf{j} + 4\mathbf{k}$
 (b) $\mathbf{a} = \langle c, \frac{1}{2}, c \rangle$, $\mathbf{b} = \langle -3, 4, c \rangle$

17. Find a vector $\mathbf{v} = \langle x_1, y_1, 1 \rangle$ that is orthogonal to both $\mathbf{a} = \langle 3, 1, -1 \rangle$ and $\mathbf{b} = \langle -3, 2, 2 \rangle$.

18. A rhombus is an oblique-angled parallelogram with all four sides equal. Use the dot product to show that the diagonals of a rhombus are perpendicular.

19. Verify that the vector

$$c = b - \frac{a \cdot b}{\|a\|^2} a$$

is orthogonal to the vector **a**.

20. Determine a scalar c so that the angle between $\mathbf{a} = \mathbf{i} + c\mathbf{j}$ and $\mathbf{b} = \mathbf{i} + \mathbf{j}$ is $45°$.

In Problems 21–24, find the angle θ between the given vectors.

21. $\mathbf{a} = 3\mathbf{i} - \mathbf{k}, \ \mathbf{b} = 2\mathbf{i} + 2\mathbf{k}$
22. $\mathbf{a} = 2\mathbf{i} + \mathbf{j}, \ \mathbf{b} = -3\mathbf{i} - 4\mathbf{j}$
23. $\mathbf{a} = \langle 2, 4, 0 \rangle, \ \mathbf{b} = \langle -1, -1, 4 \rangle$
24. $\mathbf{a} = \langle \frac{1}{2}, \frac{1}{2}, \frac{3}{2} \rangle, \ \mathbf{b} = \langle 2, -4, 6 \rangle$

In Problems 25–28, find the direction cosines and direction angles of the given vector.

25. $\mathbf{a} = \mathbf{i} + 2\mathbf{j} + 3\mathbf{k}$ 26. $\mathbf{a} = 6\mathbf{i} + 6\mathbf{j} - 3\mathbf{k}$

27. $\mathbf{a} = \langle 1, 0, -\sqrt{3} \rangle$ 28. $\mathbf{a} = \langle 5, 7, 2 \rangle$

29. Find the angle between the diagonal $\overrightarrow{AD}$ of the cube shown in **FIGURE 7.3.9** and the edge AB. Find the angle between the diagonal $\overrightarrow{AD}$ of the cube and the diagonal $\overrightarrow{AC}$.

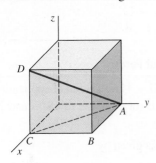

FIGURE 7.3.9 Diagonal in Problem 29

30. Show that if two nonzero vectors **a** and **b** are orthogonal, then their direction cosines satisfy

$$\cos \alpha_1 \cos \alpha_2 + \cos \beta_1 \cos \beta_2 + \cos \gamma_1 \cos \gamma_2 = 0.$$

31. An airplane is 4 km high, 5 km south, and 7 km east of an airport. See **FIGURE 7.3.10**. Find the direction angles of the plane.

FIGURE 7.3.10 Airplane in Problem 31

32. Determine a unit vector whose direction angles, relative to the three coordinate axes, are equal.

In Problems 33–36, $\mathbf{a} = \langle 1, -1, 3 \rangle$ and $\mathbf{b} = \langle 2, 6, 3 \rangle$. Find the indicated number.

33. $\text{comp}_b \mathbf{a}$ 34. $\text{comp}_a \mathbf{b}$
35. $\text{comp}_a(\mathbf{b} - \mathbf{a})$ 36. $\text{comp}_{2b}(\mathbf{a} + \mathbf{b})$

In Problems 37 and 38, find the component of the given vector in the direction from the origin to the indicated point.

37. $\mathbf{a} = 4\mathbf{i} + 6\mathbf{j}, \quad P(3, 10)$
38. $\mathbf{a} = \langle 2, 1, -1 \rangle, \quad P(1, -1, 1)$

In Problems 39–42, find $\text{proj}_b \mathbf{a}$.

39. $\mathbf{a} = -5\mathbf{i} + 5\mathbf{j}, \ \mathbf{b} = -3\mathbf{i} + 4\mathbf{j}$
40. $\mathbf{a} = 4\mathbf{i} + 2\mathbf{j}, \ \mathbf{b} = -3\mathbf{i} + \mathbf{j}$
41. $\mathbf{a} = -\mathbf{i} - 2\mathbf{j} + 7\mathbf{k}, \ \mathbf{b} = 6\mathbf{i} - 3\mathbf{j} - 2\mathbf{k}$
42. $\mathbf{a} = \langle 1, 1, 1 \rangle, \ \mathbf{b} = \langle -2, 2, -1 \rangle$

In Problems 43 and 44, $\mathbf{a} = 4\mathbf{i} + 3\mathbf{j}$ and $\mathbf{b} = -\mathbf{i} + \mathbf{j}$. Find the indicated vector.

43. $\text{proj}_{(a+b)} \mathbf{a}$ 44. $\text{proj}_{(a-b)} \mathbf{b}$

45. A sled is pulled horizontally over ice by a rope attached to the front of the sled. A 20-lb force acting at an angle of $60°$ with the horizontal moves the sled 100 ft. Find the work done.

46. Find the work done if the point at which the constant force $\mathbf{F} = 4\mathbf{i} + 3\mathbf{j} + 5\mathbf{k}$ is applied to an object moves from $P_1(3, 1, -2)$ to $P_2(2, 4, 6)$. Assume $\|\mathbf{F}\|$ is measured in newtons and $\|\mathbf{d}\|$ is measured in meters.

47. A block with weight **w** is pulled along a frictionless horizontal surface by a constant force **F** of magnitude 30 newtons in the direction given by a vector **d**. See **FIGURE 7.3.11**. Assume $\|\mathbf{d}\|$ is measured in meters.
 (a) What is the work done by the weight **w**?
 (b) What is the work done by the force **F** if $\mathbf{d} = 4\mathbf{i} + 3\mathbf{j}$?

FIGURE 7.3.11 Block in Problem 47

48. A constant force **F** of magnitude 3 lb is applied to the block shown in **FIGURE 7.3.12**. **F** has the same direction as the vector $\mathbf{a} = 3\mathbf{i} + 4\mathbf{j}$. Find the work done in the direction of motion if the block moves from $P_1(3, 1)$ to $P_2(9, 3)$. Assume distance is measured in feet.

FIGURE 7.3.12 Block in Problem 48

49. In the methane molecule CH_4, the hydrogen atoms are located at the four vertices of a regular tetrahedron. See **FIGURE 7.3.13**. The distance between the center of a hydrogen atom and the center of a carbon atom is 1.10 angstroms (1 angstrom $= 10^{-10}$ meter), and the hydrogen–carbon–hydrogen bond angle is $\theta = 109.5°$. Using vector methods only, find the distance between two hydrogen atoms.

FIGURE 7.3.13 Molecule in Problem 49

Discussion Problems

50. Use the dot product to prove the Cauchy–Schwarz inequality: $|\mathbf{a} \cdot \mathbf{b}| \leq \|\mathbf{a}\| \, \|\mathbf{b}\|$.

51. Use the dot product to prove the triangle inequality $\|\mathbf{a} + \mathbf{b}\| \leq \|\mathbf{a}\| + \|\mathbf{b}\|$. [*Hint:* Consider property (*vi*) of the dot product.]

52. Prove that the vector $\mathbf{n} = a\mathbf{i} + b\mathbf{j}$ is perpendicular to the line whose equation is $ax + by + c = 0$. [*Hint:* Let $P_1(x_1, y_1)$ and $P_2(x_2, y_2)$ be distinct points on the line.]

53. Use the result of Problem 52 and **FIGURE 7.3.14** to show that the distance d from a point $P_1(x_1, y_1)$ to a line $ax + by + c = 0$ is

$$d = \frac{|ax_1 + by_1 + c|}{\sqrt{a^2 + b^2}}.$$

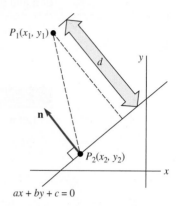

FIGURE 7.3.14 Distance d in Problem 53

7.4 Cross Product

INTRODUCTION The dot product, introduced in the preceding section, works in both two- and three-dimensional spaces and results in a *number*. On the other hand, the **cross product**, introduced in this section, is only defined for vectors in 3-space and results in another *vector* in 3-space.

Before proceeding we need the following two results from the theory of determinants.

See Sections 8.4 and 8.5 for a complete discussion of determinants.

▶ **Review of Determinants** A **determinant of order 2** is the number

$$\begin{vmatrix} a_1 & a_2 \\ b_1 & b_2 \end{vmatrix} = a_1 b_2 - a_2 b_1. \tag{1}$$

A **determinant of order 3** is the number defined in terms of three determinants of order 2:

$$\begin{vmatrix} a_1 & a_2 & a_3 \\ b_1 & b_2 & b_3 \\ c_1 & c_2 & c_3 \end{vmatrix} = a_1 \begin{vmatrix} b_2 & b_3 \\ c_2 & c_3 \end{vmatrix} - a_2 \begin{vmatrix} b_1 & b_3 \\ c_1 & c_3 \end{vmatrix} + a_3 \begin{vmatrix} b_1 & b_2 \\ c_1 & c_2 \end{vmatrix}. \tag{2}$$

This is called **expanding the determinant** along the first row. For example, from (1)

$$\begin{vmatrix} -4 & -2 \\ 5 & 3 \end{vmatrix} = (-4)3 - (-2)5 = -2$$

and from (2)

$$\begin{vmatrix} 8 & 5 & 4 \\ 2 & 4 & 6 \\ -1 & 2 & 3 \end{vmatrix} = 8 \begin{vmatrix} 4 & 6 \\ 2 & 3 \end{vmatrix} - 5 \begin{vmatrix} 2 & 6 \\ -1 & 3 \end{vmatrix} + 4 \begin{vmatrix} 2 & 4 \\ -1 & 2 \end{vmatrix} = 8(0) - 5(12) + 4(8) = -28.$$

▶ **Component Form of the Cross Product** As we did in the discussion of the dot product, we define the **cross product** of two vectors $\mathbf{a}$ and $\mathbf{b}$ in terms of the components of the vectors.

DEFINITION 7.4.1 Cross Product of Two Vectors

The **cross product** of two vectors $\mathbf{a} = \langle a_1, a_2, a_3 \rangle$ and $\mathbf{b} = \langle b_1, b_2, b_3 \rangle$ is the vector

$$\mathbf{a} \times \mathbf{b} = (a_2 b_3 - a_3 b_2)\mathbf{i} - (a_1 b_3 - a_3 b_1)\mathbf{j} + (a_1 b_2 - a_2 b_1)\mathbf{k}. \tag{3}$$

The coefficients of the basis vectors in (3) are recognized as determinants of order 2, so (3) can be written as

$$\mathbf{a} \times \mathbf{b} = \begin{vmatrix} a_2 & a_3 \\ b_2 & b_3 \end{vmatrix} \mathbf{i} - \begin{vmatrix} a_1 & a_3 \\ b_1 & b_3 \end{vmatrix} \mathbf{j} + \begin{vmatrix} a_1 & a_2 \\ b_1 & b_2 \end{vmatrix} \mathbf{k}. \tag{4}$$

By comparing (4) with (2), we see that the cross product can be formally expressed as a determinant of order 3 with the unit vectors $\mathbf{i}$, $\mathbf{j}$, and $\mathbf{k}$ as entries in the first row and the components of $\mathbf{a}$ and $\mathbf{b}$ entries of the second and third rows, respectively:

$$\mathbf{a} \times \mathbf{b} = \begin{vmatrix} \mathbf{i} & \mathbf{j} & \mathbf{k} \\ a_1 & a_2 & a_3 \\ b_1 & b_2 & b_3 \end{vmatrix}. \tag{5}$$

EXAMPLE 1 Cross Product Using (5)

Let $\mathbf{a} = 4\mathbf{i} - 2\mathbf{j} + 5\mathbf{k}$ and $\mathbf{b} = 3\mathbf{i} + \mathbf{j} - \mathbf{k}$. Find $\mathbf{a} \times \mathbf{b}$.

SOLUTION We use (5) and expand the determinant along the first row:

$$\mathbf{a} \times \mathbf{b} = \begin{vmatrix} \mathbf{i} & \mathbf{j} & \mathbf{k} \\ 4 & -2 & 5 \\ 3 & 1 & -1 \end{vmatrix} = \begin{vmatrix} -2 & 5 \\ 1 & -1 \end{vmatrix} \mathbf{i} - \begin{vmatrix} 4 & 5 \\ 3 & -1 \end{vmatrix} \mathbf{j} + \begin{vmatrix} 4 & -2 \\ 3 & 1 \end{vmatrix} \mathbf{k}$$

$$= -3\mathbf{i} + 19\mathbf{j} + 10\mathbf{k}. \qquad \equiv$$

EXAMPLE 2 Cross Product of Two Basis Vectors

If $\mathbf{i} = \langle 1, 0, 0 \rangle$ and $\mathbf{j} = \langle 0, 1, 0 \rangle$, then (5) gives

$$\mathbf{i} \times \mathbf{j} = \begin{vmatrix} \mathbf{i} & \mathbf{j} & \mathbf{k} \\ 1 & 0 & 0 \\ 0 & 1 & 0 \end{vmatrix} = \begin{vmatrix} 0 & 0 \\ 1 & 0 \end{vmatrix} \mathbf{i} - \begin{vmatrix} 1 & 0 \\ 0 & 0 \end{vmatrix} \mathbf{j} + \begin{vmatrix} 1 & 0 \\ 0 & 1 \end{vmatrix} \mathbf{k} = \mathbf{k}$$

and

$$\mathbf{i} \times \mathbf{i} = \begin{vmatrix} \mathbf{i} & \mathbf{j} & \mathbf{k} \\ 1 & 0 & 0 \\ 1 & 0 & 0 \end{vmatrix} = \begin{vmatrix} 0 & 0 \\ 0 & 0 \end{vmatrix} \mathbf{i} - \begin{vmatrix} 1 & 0 \\ 1 & 0 \end{vmatrix} \mathbf{j} + \begin{vmatrix} 1 & 0 \\ 1 & 0 \end{vmatrix} \mathbf{k} = \mathbf{0}. \qquad \equiv$$

Proceeding as in Example 2, it is readily shown that

$$\mathbf{i} \times \mathbf{j} = \mathbf{k}, \qquad \mathbf{j} \times \mathbf{k} = \mathbf{i}, \qquad \mathbf{k} \times \mathbf{i} = \mathbf{j}, \tag{6}$$

$$\mathbf{j} \times \mathbf{i} = -\mathbf{k}, \qquad \mathbf{k} \times \mathbf{j} = -\mathbf{i}, \qquad \mathbf{i} \times \mathbf{k} = -\mathbf{j}, \tag{7}$$

and

$$\mathbf{i} \times \mathbf{i} = \mathbf{0}, \qquad \mathbf{j} \times \mathbf{j} = \mathbf{0}, \qquad \mathbf{k} \times \mathbf{k} = \mathbf{0}. \tag{8}$$

FIGURE 7.4.1 A mnemonic for cross products involving $\mathbf{i}$, $\mathbf{j}$, and $\mathbf{k}$

The results in (6) can be obtained using the circular mnemonic illustrated in **FIGURE 7.4.1**. Notice in (7), if we cross two basis vectors in the direction opposite to that shown in Figure 7.4.1, then we get the negative of the corresponding vector in (6). The results in (6), (7), and (8) are special cases of (*ii*) and (*vi*) in Theorem 7.4.1.

Properties The next theorem summarizes some of the important properties of the cross product.

THEOREM 7.4.1 Properties of the Cross Product

(*i*) $\mathbf{a} \times \mathbf{b} = \mathbf{0}$ if $\mathbf{a} = \mathbf{0}$ or $\mathbf{b} = \mathbf{0}$

(*ii*) $\mathbf{a} \times \mathbf{b} = -\mathbf{b} \times \mathbf{a}$

(*iii*) $\mathbf{a} \times (\mathbf{b} + \mathbf{c}) = (\mathbf{a} \times \mathbf{b}) + (\mathbf{a} \times \mathbf{c})$ ← distributive law

(*iv*) $(\mathbf{a} + \mathbf{b}) \times \mathbf{c} = (\mathbf{a} \times \mathbf{c}) + (\mathbf{b} \times \mathbf{c})$ ← distributive law

(*v*) $\mathbf{a} \times (k\mathbf{b}) = (k\mathbf{a}) \times \mathbf{b} = k(\mathbf{a} \times \mathbf{b})$, k a scalar

(*vi*) $\mathbf{a} \times \mathbf{a} = \mathbf{0}$

(*vii*) $\mathbf{a} \cdot (\mathbf{a} \times \mathbf{b}) = 0$

(*viii*) $\mathbf{b} \cdot (\mathbf{a} \times \mathbf{b}) = 0$

PROOF: All the properties can be proved directly from (4) and (5). To prove part (*i*), we let $\mathbf{a} = \langle 0, 0, 0 \rangle = \mathbf{0}$ and $\mathbf{b} = \langle b_1, b_2, b_3 \rangle$. Then from (5)

$$\mathbf{a} \times \mathbf{b} = \begin{vmatrix} \mathbf{i} & \mathbf{j} & \mathbf{k} \\ 0 & 0 & 0 \\ b_1 & b_2 & b_3 \end{vmatrix} = \begin{vmatrix} 0 & 0 \\ b_2 & b_3 \end{vmatrix} \mathbf{i} - \begin{vmatrix} 0 & 0 \\ b_1 & b_3 \end{vmatrix} \mathbf{j} + \begin{vmatrix} 0 & 0 \\ b_1 & b_2 \end{vmatrix} \mathbf{k}$$

$$= [0(b_3) - 0(b_2)]\mathbf{i} - [0(b_3) - 0(b_1)]\mathbf{j} + [0(b_2) - 0(b_1)]\mathbf{k}$$

$$= 0\mathbf{i} + 0\mathbf{j} + 0\mathbf{k} = \mathbf{0}.$$

To prove part (*iii*) we let $\mathbf{a} = \langle a_1, a_2, a_3 \rangle$, $\mathbf{b} = \langle b_1, b_2, b_3 \rangle$, and $\mathbf{c} = \langle c_1, c_2, c_3 \rangle$. Then

$$\mathbf{a} \times (\mathbf{b} + \mathbf{c}) = \begin{vmatrix} a_2 & a_3 \\ b_2 + c_2 & b_3 + c_3 \end{vmatrix} \mathbf{i} - \begin{vmatrix} a_1 & a_3 \\ b_1 + c_1 & b_3 + c_3 \end{vmatrix} \mathbf{j} + \begin{vmatrix} a_1 & a_2 \\ b_1 + c_1 & b_2 + c_2 \end{vmatrix} \mathbf{k}$$

$$= [(a_2b_3 + a_2c_3) - (a_3b_2 + a_3c_2)]\mathbf{i} - [(a_1b_3 + a_1c_3) - (a_3b_1 + a_3c_1)]\mathbf{j}$$

$$+ [(a_1b_2 + a_1c_2) - (a_2b_1 + a_2c_1)]\mathbf{k}$$

$$= (a_2b_3 - a_3b_2)\mathbf{i} - (a_1b_3 - a_3b_1)\mathbf{j} + (a_1b_2 - a_2b_1)\mathbf{k}$$

$$+ (a_2c_3 - a_3c_2)\mathbf{i} - (a_1c_3 - a_3c_1)\mathbf{j} + (a_1c_2 - a_2c_1)\mathbf{k}$$

$$= (\mathbf{a} \times \mathbf{b}) + (\mathbf{a} \times \mathbf{c}). \qquad \equiv$$

Because part (*ii*) of Theorem 7.4.1 indicates that the cross product is not commutative, there are two distributive laws in parts (*iii*) and (*iv*) of the theorem.

Parts (*vii*) and (*viii*) of Theorem 7.4.1 deserve special attention. In view of Theorem 7.3.3 we see (*vii*) and (*viii*) imply that the vector $\mathbf{a} \times \mathbf{b}$ is orthogonal to $\mathbf{a}$ and that $\mathbf{a} \times \mathbf{b}$ is orthogonal to $\mathbf{b}$. In the case when $\mathbf{a}$ and $\mathbf{b}$ are nonzero vectors, then $\mathbf{a} \times \mathbf{b}$ is orthogonal to *every* vector in the plane containing $\mathbf{a}$ and $\mathbf{b}$. Put another way,

$$\mathbf{a} \times \mathbf{b} \text{ is orthogonal to the plane containing } \mathbf{a} \text{ and } \mathbf{b}. \qquad (9)$$

You can see in Figure 7.4.1 and from the results in (6) that $\mathbf{i} \times \mathbf{j}$ is orthogonal to the plane of $\mathbf{i}$ and $\mathbf{j}$; that is, the *xy*-plane, $\mathbf{j} \times \mathbf{k}$ is orthogonal to the *yz*-plane, and $\mathbf{k} \times \mathbf{i}$ is orthogonal to the *xz*-plane.

The results in (6) and (7) give us a clue as to the direction in which the vector $\mathbf{a} \times \mathbf{b}$ points.

■ **Right-Hand Rule** The vectors $\mathbf{a}$, $\mathbf{b}$, and $\mathbf{a} \times \mathbf{b}$ form a **right-handed system** or a **right-handed triple**. This means that $\mathbf{a} \times \mathbf{b}$ points in the direction given by the **right-hand rule**:

If the fingers of the right hand point along the vector $\mathbf{a}$ and then curl toward the vector $\mathbf{b}$, the thumb will give the direction of $\mathbf{a} \times \mathbf{b}$. (10)

See **FIGURE 7.4.2(a)**. In Figure 7.4.2(b), the right-hand rule shows the direction of $\mathbf{b} \times \mathbf{a}$.

FIGURE 7.4.2 Right-hand rule

Because $\mathbf{a} \times \mathbf{b}$ is a vector its magnitude $\|\mathbf{a} \times \mathbf{b}\|$ can be found from (*vii*) of Definition 7.2.1. As we see in the next theorem, $\|\mathbf{a} \times \mathbf{b}\|$ can also be expressed in terms of the angle θ between the vectors $\mathbf{a}$ and $\mathbf{b}$.

THEOREM 7.4.2 Magnitude of the Cross Product

For nonzero vectors $\mathbf{a}$ and $\mathbf{b}$, if θ is the angle between $\mathbf{a}$ and $\mathbf{b}$ ($0 \le \theta \le \pi$), then

$$\|\mathbf{a} \times \mathbf{b}\| = \|\mathbf{a}\| \, \|\mathbf{b}\| \sin \theta. \tag{11}$$

PROOF: We separately compute the squares of the left- and right-hand sides of equation (11) using the component forms of $\mathbf{a}$ and $\mathbf{b}$:

$$\|\mathbf{a} \times \mathbf{b}\|^2 = (a_2 b_3 - a_3 b_2)^2 + (a_1 b_3 - a_3 b_1)^2 + (a_1 b_2 - a_2 b_1)^2$$

$$= a_2^2 b_3^2 - 2a_2 b_3 a_3 b_2 + a_3^2 b_2^2 + a_1^2 b_3^2 - 2a_1 b_3 a_3 b_1 + a_3^2 b_1^2$$

$$+ \, a_1^2 b_2^2 - 2a_1 b_2 a_2 b_1 + a_2^2 b_1^2,$$

$$(\|\mathbf{a}\| \, \|\mathbf{b}\| \sin \theta)^2 = \|\mathbf{a}\|^2 \|\mathbf{b}\|^2 \sin^2 \theta = \|\mathbf{a}\|^2 \|\mathbf{b}\|^2 (1 - \cos^2 \theta)$$

$$= \|\mathbf{a}\|^2 \|\mathbf{b}\|^2 - \|\mathbf{a}\|^2 \|\mathbf{b}\|^2 \cos^2 \theta = \|\mathbf{a}\|^2 \|\mathbf{b}\|^2 - (\mathbf{a} \cdot \mathbf{b})^2$$

$$= (a_1^2 + a_2^2 + a_3^2)(b_1^2 + b_2^2 + b_3^2) - (a_1 b_1 + a_2 b_2 + a_3 b_3)^2$$

$$= a_2^2 b_3^2 - 2a_2 b_2 a_3 b_3 + a_3^2 b_2^2 + a_1^2 b_3^2 - 2a_1 b_1 a_3 b_3 + a_3^2 b_1^2$$

$$+ \, a_1^2 b_2^2 - 2a_1 b_1 a_2 b_2 + a_2^2 b_1^2.$$

Since both sides are equal to the same quantity, they must be equal to each other, so

$$\|\mathbf{a} \times \mathbf{b}\|^2 = (\|\mathbf{a}\| \, \|\mathbf{b}\| \sin \theta)^2.$$

Finally, taking the square root of both sides and using the fact that $\sqrt{\sin^2 \theta} = \sin \theta$ since $\sin \theta \ge 0$ for $0 \le \theta \le \pi$, we have $\|\mathbf{a} \times \mathbf{b}\| = \|\mathbf{a}\| \, \|\mathbf{b}\| \sin \theta$. ≡

■ **Alternative Form** Combining (9), (10), and Theorem 7.4.2 we see for *any* pair of vectors $\mathbf{a}$ and $\mathbf{b}$ in R^3 that the cross product has the alternative form

$$\mathbf{a} \times \mathbf{b} = (\|\mathbf{a}\| \, \|\mathbf{b}\| \sin \theta)\, \mathbf{n}, \tag{12}$$

where $\mathbf{n}$ is a unit vector given by the right-hand rule that is orthogonal to the plane of $\mathbf{a}$ and $\mathbf{b}$.

> This more geometric form is generally used as the definition of the cross product in a physics course.

■ **Parallel Vectors** We saw in Section 7.1 that two nonzero vectors are parallel if and only if one is a nonzero scalar multiple of the other. Thus, two vectors are parallel if they have the forms $\mathbf{a}$ and $k\mathbf{a}$, where $\mathbf{a}$ is any vector and k is a scalar. By properties (*v*) and (*vi*) in Theorem 7.4.1, the cross product of parallel vectors must be $\mathbf{0}$. This fact is summarized in the next theorem.

THEOREM 7.4.3 Criterion for Parallel Vectors

Two nonzero vectors $\mathbf{a}$ and $\mathbf{b}$ are parallel if and only if $\mathbf{a} \times \mathbf{b} = \mathbf{0}$.

Of course, Theorem 7.4.3 follows as well from (12) because the angle between parallel vectors $\mathbf{a}$ and $\mathbf{b}$ is either $\theta = 0$ or $\theta = \pi$.

EXAMPLE 3 Parallel Vectors

Determine whether $\mathbf{a} = 2\mathbf{i} + \mathbf{j} - \mathbf{k}$ and $\mathbf{b} = -6\mathbf{i} - 3\mathbf{j} + 3\mathbf{k}$ are parallel vectors.

SOLUTION From the cross product

$$\mathbf{a} \times \mathbf{b} = \begin{vmatrix} \mathbf{i} & \mathbf{j} & \mathbf{k} \\ 2 & 1 & -1 \\ -6 & -3 & 3 \end{vmatrix} = \begin{vmatrix} 1 & -1 \\ -3 & 3 \end{vmatrix} \mathbf{i} - \begin{vmatrix} 2 & -1 \\ -6 & 3 \end{vmatrix} \mathbf{j} + \begin{vmatrix} 2 & 1 \\ -6 & -3 \end{vmatrix} \mathbf{k}$$

$$= 0\mathbf{i} - 0\mathbf{j} + 0\mathbf{k} = \mathbf{0}$$

and Theorem 7.4.3 we conclude that $\mathbf{a}$ and $\mathbf{b}$ are parallel vectors. ≡

Special Products

The **scalar triple product** of vectors **a**, **b**, and **c** is $\mathbf{a} \cdot (\mathbf{b} \times \mathbf{c})$. Using the component forms of the definitions of the dot and cross products, we have

$$\mathbf{a} \cdot (\mathbf{b} \times \mathbf{c}) = (a_1\mathbf{i} + a_2\mathbf{j} + a_3\mathbf{k}) \cdot \left[\begin{vmatrix} b_2 & b_3 \\ c_2 & c_3 \end{vmatrix}\mathbf{i} - \begin{vmatrix} b_1 & b_3 \\ c_1 & c_3 \end{vmatrix}\mathbf{j} + \begin{vmatrix} b_1 & b_2 \\ c_1 & c_2 \end{vmatrix}\mathbf{k} \right]$$

$$= a_1 \begin{vmatrix} b_2 & b_3 \\ c_2 & c_3 \end{vmatrix} - a_2 \begin{vmatrix} b_1 & b_3 \\ c_1 & c_3 \end{vmatrix} + a_3 \begin{vmatrix} b_1 & b_2 \\ c_1 & c_2 \end{vmatrix}.$$

Thus, we see that the scalar triple product can be written as a determinant of order 3:

$$\mathbf{a} \cdot (\mathbf{b} \times \mathbf{c}) = \begin{vmatrix} a_1 & a_2 & a_3 \\ b_1 & b_2 & b_3 \\ c_1 & c_2 & c_3 \end{vmatrix}. \tag{13}$$

Using (4), (5), and (2) of Section 7.3 it can be shown that

$$\mathbf{a} \cdot (\mathbf{b} \times \mathbf{c}) = (\mathbf{a} \times \mathbf{b}) \cdot \mathbf{c}. \tag{14}$$

See Problem 61 in Exercises 7.4.

The **vector triple product** of three vectors **a**, **b**, and **c** is

$$\mathbf{a} \times (\mathbf{b} \times \mathbf{c}).$$

The vector triple product is related to the dot product by

$$\mathbf{a} \times (\mathbf{b} \times \mathbf{c}) = (\mathbf{a} \cdot \mathbf{c})\mathbf{b} - (\mathbf{a} \cdot \mathbf{b})\mathbf{c}. \tag{15}$$

See Problem 59 in Exercises 7.4.

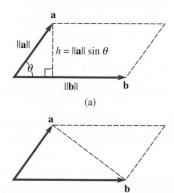

(a)

(b)

FIGURE 7.4.3 Areas of a parallelogram and a triangle

Areas

Two nonzero and nonparallel vectors **a** and **b** can be considered to be the sides of a parallelogram. The **area A of a parallelogram** is

$$A = (\text{base})(\text{altitude}).$$

From **FIGURE 7.4.3(a)** we see that $A = \|\mathbf{b}\| (\|\mathbf{a}\| \sin \theta) = \|\mathbf{a}\| \|\mathbf{b}\| \sin \theta$

or

$$A = \|\mathbf{a} \times \mathbf{b}\|. \tag{16}$$

Likewise from Figure 7.4.3(b), we see that the **area of a triangle** with sides **a** and **b** is

$$A = \frac{1}{2}\|\mathbf{a} \times \mathbf{b}\|. \tag{17}$$

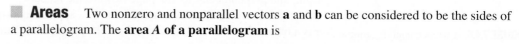

EXAMPLE 4 **Area of a Triangle**

Find the area of the triangle determined by the points $P_1(1, 1, 1)$, $P_2(2, 3, 4)$, and $P_3(3, 0, -1)$.

SOLUTION The vectors $\overrightarrow{P_1P_2}$ and $\overrightarrow{P_2P_3}$ can be taken as two sides of the triangle. Since $\overrightarrow{P_1P_2} = \mathbf{i} + 2\mathbf{j} + 3\mathbf{k}$ and $\overrightarrow{P_2P_3} = \mathbf{i} - 3\mathbf{j} - 5\mathbf{k}$, we have

$$\overrightarrow{P_1P_2} \times \overrightarrow{P_2P_3} = \begin{vmatrix} \mathbf{i} & \mathbf{j} & \mathbf{k} \\ 1 & 2 & 3 \\ 1 & -3 & -5 \end{vmatrix} = \begin{vmatrix} 2 & 3 \\ -3 & -5 \end{vmatrix}\mathbf{i} - \begin{vmatrix} 1 & 3 \\ 1 & -5 \end{vmatrix}\mathbf{j} + \begin{vmatrix} 1 & 2 \\ 1 & -3 \end{vmatrix}\mathbf{k}$$

$$= -\mathbf{i} + 8\mathbf{j} - 5\mathbf{k}.$$

From (17) we see that the area is

$$A = \frac{1}{2}\|-\mathbf{i} + 8\mathbf{j} - 5\mathbf{k}\| = \frac{3}{2}\sqrt{10}.$$

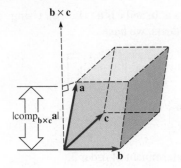

FIGURE 7.4.4 Parallelepiped formed by three vectors

FIGURE 7.4.5 A force acting at the end of a vector

FIGURE 7.4.6 A wrench applying torque to a bolt

Volume of a Parallelepiped
If the vectors **a**, **b**, and **c** do not lie in the same plane, then the **volume of the parallelepiped** with edges **a**, **b**, and **c** shown in **FIGURE 7.4.4** is

$$V = (\text{area of base})(\text{height})$$

$$= \|\mathbf{b} \times \mathbf{c}\| \,|\text{comp}_{\mathbf{b}\times\mathbf{c}}\mathbf{a}|$$

$$= \|\mathbf{b} \times \mathbf{c}\| \left|\mathbf{a} \cdot \left(\frac{1}{\|\mathbf{b} \times \mathbf{c}\|}\mathbf{b} \times \mathbf{c}\right)\right|$$

or

$$V = |\mathbf{a} \cdot (\mathbf{b} \times \mathbf{c})|. \qquad (18)$$

Thus, the volume of a parallelepiped determined by three vectors is the absolute value of the scalar triple product of the vectors.

Coplanar Vectors
Vectors that lie in the same plane are said to be **coplanar**. We have just seen that if the vectors **a**, **b**, and **c** are not coplanar, then necessarily $\mathbf{a} \cdot (\mathbf{b} \times \mathbf{c}) \neq 0$, since the volume of a parallelepiped with edges **a**, **b**, and **c** has nonzero volume. Equivalently stated, this means that if $\mathbf{a} \cdot (\mathbf{b} \times \mathbf{c}) = 0$, then the vectors **a**, **b**, and **c** are coplanar. Since the converse of this last statement is also true, we have:

$$\mathbf{a} \cdot (\mathbf{b} \times \mathbf{c}) = 0 \quad \textit{if and only if } \mathbf{a}, \mathbf{b}, \textit{ and } \mathbf{c} \textit{ are coplanar.}$$

Physical Interpretation of the Cross Product
In physics a force **F** acting at the end of a position vector **r**, as shown in **FIGURE 7.4.5**, is said to produce a **torque** $\boldsymbol{\tau}$ defined by $\boldsymbol{\tau} = \mathbf{r} \times \mathbf{F}$. For example, if $\|\mathbf{F}\| = 20$ N, $\|\mathbf{r}\| = 3.5$ m, and $\theta = 30°$, then from (11),

$$\|\boldsymbol{\tau}\| = (3.5)(20) \sin 30° = 35 \text{ N-m}.$$

If **F** and **r** are in the plane of the page, the right-hand rule implies that the direction of $\boldsymbol{\tau}$ is outward from, and perpendicular to, the page (toward the reader).

As we see in **FIGURE 7.4.6**, when a force **F** is applied to a wrench, the magnitude of the torque $\boldsymbol{\tau}$ is a measure of the turning effect about the pivot point P and the vector $\boldsymbol{\tau}$ is directed along the axis of the bolt. In this case $\boldsymbol{\tau}$ points inward from the page.

REMARKS

When working with vectors, one should be careful not to mix the symbols · and × with the symbols for ordinary multiplication, and to be especially careful in the use, or lack of use, of parentheses. For example, expressions such as

$$\mathbf{a} \times \mathbf{b} \times \mathbf{c} \qquad \mathbf{a} \cdot \mathbf{b} \times \mathbf{c} \qquad \mathbf{a} \cdot \mathbf{b} \cdot \mathbf{c} \qquad \mathbf{a} \cdot \mathbf{bc}$$

are not meaningful or well-defined.

7.4 Exercises
Answers to selected odd-numbered problems begin on page ANS-16.

In Problems 1–10, find $\mathbf{a} \times \mathbf{b}$.

1. $\mathbf{a} = \mathbf{i} - \mathbf{j}$, $\mathbf{b} = 3\mathbf{j} + 5\mathbf{k}$
2. $\mathbf{a} = 2\mathbf{i} + \mathbf{j}$, $\mathbf{b} = 4\mathbf{i} - \mathbf{k}$
3. $\mathbf{a} = \langle 1, -3, 1 \rangle$, $\mathbf{b} = \langle 2, 0, 4 \rangle$
4. $\mathbf{a} = \langle 1, 1, 1 \rangle$, $\mathbf{b} = \langle -5, 2, 3 \rangle$
5. $\mathbf{a} = 2\mathbf{i} - \mathbf{j} + 2\mathbf{k}$, $\mathbf{b} = -\mathbf{i} + 3\mathbf{j} - \mathbf{k}$
6. $\mathbf{a} = 4\mathbf{i} + \mathbf{j} - 5\mathbf{k}$, $\mathbf{b} = 2\mathbf{i} + 3\mathbf{j} - \mathbf{k}$
7. $\mathbf{a} = \langle \frac{1}{2}, 0, \frac{1}{2} \rangle$, $\mathbf{b} = \langle 4, 6, 0 \rangle$
8. $\mathbf{a} = \langle 0, 5, 0 \rangle$, $\mathbf{b} = \langle 2, -3, 4 \rangle$
9. $\mathbf{a} = \langle 2, 2, -4 \rangle$, $\mathbf{b} = \langle -3, -3, 6 \rangle$
10. $\mathbf{a} = \langle 8, 1, -6 \rangle$, $\mathbf{b} = \langle 1, -2, 10 \rangle$

In Problems 11 and 12, find $\overrightarrow{P_1P_2} \times \overrightarrow{P_1P_3}$.

11. $P_1(2, 1, 3)$, $P_2(0, 3, -1)$, $P_3(-1, 2, 4)$
12. $P_1(0, 0, 1)$, $P_2(0, 1, 2)$, $P_3(1, 2, 3)$

In Problems 13 and 14, find a vector that is perpendicular to both **a** and **b**.

13. $\mathbf{a} = 2\mathbf{i} + 7\mathbf{j} - 4\mathbf{k}$, $\mathbf{b} = \mathbf{i} + \mathbf{j} - \mathbf{k}$
14. $\mathbf{a} = \langle -1, -2, 4 \rangle$, $\mathbf{b} = \langle 4, -1, 0 \rangle$

In Problems 15 and 16, verify that $\mathbf{a} \cdot (\mathbf{a} \times \mathbf{b}) = 0$ and $\mathbf{b} \cdot (\mathbf{a} \times \mathbf{b}) = 0$.

15. $\mathbf{a} = \langle 5, -2, 1 \rangle$, $\mathbf{b} = \langle 2, 0, -7 \rangle$

16. $\mathbf{a} = \frac{1}{2}\mathbf{i} - \frac{1}{4}\mathbf{j}$, $\mathbf{b} = 2\mathbf{i} - 2\mathbf{j} + 6\mathbf{k}$

In Problems 17 and 18, **(a)** calculate $\mathbf{b} \times \mathbf{c}$ followed by $\mathbf{a} \times (\mathbf{b} \times \mathbf{c})$, and **(b)** verify the results in part (a) by (15) of this section.

17. $\mathbf{a} = \mathbf{i} - \mathbf{j} + 2\mathbf{k}$
$\quad\ \mathbf{b} = 2\mathbf{i} + \mathbf{j} + \mathbf{k}$
$\quad\ \mathbf{c} = 3\mathbf{i} + \mathbf{j} + \mathbf{k}$

18. $\mathbf{a} = 3\mathbf{i} - 4\mathbf{k}$
$\quad\ \mathbf{b} = \mathbf{i} + 2\mathbf{j} - \mathbf{k}$
$\quad\ \mathbf{c} = -\mathbf{i} + 5\mathbf{j} + 8\mathbf{k}$

In Problems 19–36, find the indicated scalar or vector *without* using (5), (13), or (15).

19. $(2\mathbf{i}) \times \mathbf{j}$
20. $\mathbf{i} \times (-3\mathbf{k})$
21. $\mathbf{k} \times (2\mathbf{i} - \mathbf{j})$
22. $\mathbf{i} \times (\mathbf{j} \times \mathbf{k})$
23. $[(2\mathbf{k}) \times (3\mathbf{j})] \times (4\mathbf{j})$
24. $(2\mathbf{i} - \mathbf{j} + 5\mathbf{k}) \times \mathbf{i}$
25. $(\mathbf{i} + \mathbf{j}) \times (\mathbf{i} + 5\mathbf{k})$
26. $\mathbf{i} \times \mathbf{k} - 2(\mathbf{j} \times \mathbf{i})$
27. $\mathbf{k} \cdot (\mathbf{j} \times \mathbf{k})$
28. $\mathbf{i} \cdot [\mathbf{j} \times (-\mathbf{k})]$
29. $\|4\mathbf{j} - 5(\mathbf{i} \times \mathbf{j})\|$
30. $(\mathbf{i} \times \mathbf{j}) \cdot (3\mathbf{j} \times \mathbf{i})$
31. $\mathbf{i} \times (\mathbf{i} \times \mathbf{j})$
32. $(\mathbf{i} \times \mathbf{j}) \times \mathbf{i}$
33. $(\mathbf{i} \times \mathbf{i}) \times \mathbf{j}$
34. $(\mathbf{i} \cdot \mathbf{i})(\mathbf{i} \times \mathbf{j})$
35. $2\mathbf{j} \cdot [\mathbf{i} \times (\mathbf{j} - 3\mathbf{k})]$
36. $(\mathbf{i} \times \mathbf{k}) \times (\mathbf{j} \times \mathbf{i})$

In Problems 37–44, $\mathbf{a} \times \mathbf{b} = 4\mathbf{i} - 3\mathbf{j} + 6\mathbf{k}$ and $\mathbf{c} = 2\mathbf{i} + 4\mathbf{j} - \mathbf{k}$. Find the indicated scalar or vector.

37. $\mathbf{a} \times (3\mathbf{b})$
38. $\mathbf{b} \times \mathbf{a}$
39. $(-\mathbf{a}) \times \mathbf{b}$
40. $\|\mathbf{a} \times \mathbf{b}\|$
41. $(\mathbf{a} \times \mathbf{b}) \times \mathbf{c}$
42. $(\mathbf{a} \times \mathbf{b}) \cdot \mathbf{c}$
43. $\mathbf{a} \cdot (\mathbf{b} \times \mathbf{c})$
44. $(4\mathbf{a}) \cdot (\mathbf{b} \times \mathbf{c})$

In Problems 45 and 46, **(a)** verify that the given quadrilateral is a parallelogram, and **(b)** find the area of the parallelogram.

45.

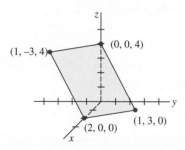

FIGURE 7.4.7 Parallelogram in Problem 45

46.

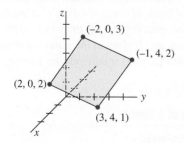

FIGURE 7.4.8 Parallelogram in Problem 46

In Problems 47–50, find the area of the triangle determined by the given points.

47. $P_1(1, 1, 1)$, $P_2(1, 2, 1)$, $P_3(1, 1, 2)$
48. $P_1(0, 0, 0)$, $P_2(0, 1, 2)$, $P_3(2, 2, 0)$
49. $P_1(1, 2, 4)$, $P_2(1, -1, 3)$, $P_3(-1, -1, 2)$
50. $P_1(1, 0, 3)$, $P_2(0, 0, 6)$, $P_3(2, 4, 5)$

In Problems 51 and 52, find the volume of the parallelepiped for which the given vectors are three edges.

51. $\mathbf{a} = \mathbf{i} + \mathbf{j}$, $\mathbf{b} = -\mathbf{i} + 4\mathbf{j}$, $\mathbf{c} = 2\mathbf{i} + 2\mathbf{j} + 2\mathbf{k}$
52. $\mathbf{a} = 3\mathbf{i} + \mathbf{j} + \mathbf{k}$, $\mathbf{b} = \mathbf{i} + 4\mathbf{j} + \mathbf{k}$, $\mathbf{c} = \mathbf{i} + \mathbf{j} + 5\mathbf{k}$

53. Determine whether the vectors $\mathbf{a} = 4\mathbf{i} + 6\mathbf{j}$, $\mathbf{b} = -2\mathbf{i} + 6\mathbf{j} - 6\mathbf{k}$, and $\mathbf{c} = \frac{5}{2}\mathbf{i} + 3\mathbf{j} + \frac{1}{2}\mathbf{k}$ are coplanar.

54. Determine whether the four points $P_1(1, 1, -2)$, $P_2(4, 0, -3)$, $P_3(1, -5, 10)$, and $P_4(-7, 2, 4)$ lie in the same plane.

55. As shown in **FIGURE 7.4.9**, the vector $\mathbf{a}$ lies in the xy-plane and the vector $\mathbf{b}$ lies along the positive z-axis. Their magnitudes are $\|\mathbf{a}\| = 6.4$ and $\|\mathbf{b}\| = 5$.
 (a) Use Theorem 7.4.2 to find $\|\mathbf{a} \times \mathbf{b}\|$.
 (b) Use the right-hand rule to find the direction of $\mathbf{a} \times \mathbf{b}$.
 (c) Use part (b) to express $\mathbf{a} \times \mathbf{b}$ in terms of the unit vectors $\mathbf{i}$, $\mathbf{j}$, and $\mathbf{k}$.

FIGURE 7.4.9 Vectors for Problem 55

56. Two vectors $\mathbf{a}$ and $\mathbf{b}$ lie in the xz-plane so that the angle between them is $120°$. If $\|\mathbf{a}\| = \sqrt{27}$ and $\|\mathbf{b}\| = 8$, find all possible values of $\mathbf{a} \times \mathbf{b}$.

57. A three-dimensional lattice is a collection of integer combinations of three noncoplanar basis vectors $\mathbf{a}$, $\mathbf{b}$, and $\mathbf{c}$. In crystallography, a lattice can specify the locations of atoms in a crystal. X-ray diffraction studies of crystals use the "reciprocal lattice" that has basis

$$A = \frac{\mathbf{b} \times \mathbf{c}}{\mathbf{a} \cdot (\mathbf{b} \times \mathbf{c})}, \quad B = \frac{\mathbf{c} \times \mathbf{a}}{\mathbf{b} \cdot (\mathbf{c} \times \mathbf{a})}, \quad C = \frac{\mathbf{a} \times \mathbf{b}}{\mathbf{c} \cdot (\mathbf{a} \times \mathbf{b})}.$$

 (a) A certain lattice has basis vectors $\mathbf{a} = \mathbf{i}$, $\mathbf{b} = \mathbf{j}$, and $\mathbf{c} = \frac{1}{2}(\mathbf{i} + \mathbf{j} + \mathbf{k})$. Find basis vectors for the reciprocal lattice.
 (b) The unit cell of the reciprocal lattice is the parallelepiped with edges $\mathbf{A}$, $\mathbf{B}$, and $\mathbf{C}$, while the unit cell of the original lattice is the parallelepiped with edges $\mathbf{a}$, $\mathbf{b}$, and $\mathbf{c}$. Show that the volume of the unit cell of the reciprocal lattice is the reciprocal of the volume of the unit cell of the original lattice. [*Hint*: Start with $\mathbf{B} \times \mathbf{C}$ and use (15).]

Discussion Problems

58. Use (4) to prove property (*iii*) of the cross product.
59. Prove $\mathbf{a} \times (\mathbf{b} \times \mathbf{c}) = (\mathbf{a} \cdot \mathbf{c})\mathbf{b} - (\mathbf{a} \cdot \mathbf{b})\mathbf{c}$.
60. Prove or disprove $\mathbf{a} \times (\mathbf{b} \times \mathbf{c}) = (\mathbf{a} \times \mathbf{b}) \times \mathbf{c}$.
61. Prove $\mathbf{a} \cdot (\mathbf{b} \times \mathbf{c}) = (\mathbf{a} \times \mathbf{b}) \cdot \mathbf{c}$.
62. Prove $\mathbf{a} \times (\mathbf{b} \times \mathbf{c}) + \mathbf{b} \times (\mathbf{c} \times \mathbf{a}) + \mathbf{c} \times (\mathbf{a} \times \mathbf{b}) = \mathbf{0}$.
63. Prove **Lagrange's identity**:

$$\|\mathbf{a} \times \mathbf{b}\|^2 = \|\mathbf{a}\|^2\|\mathbf{b}\|^2 - (\mathbf{a} \cdot \mathbf{b})^2.$$

64. Does $\mathbf{a} \times \mathbf{b} = \mathbf{a} \times \mathbf{c}$ imply that $\mathbf{b} = \mathbf{c}$?
65. Show that $(\mathbf{a} + \mathbf{b}) \times (\mathbf{a} - \mathbf{b}) = 2\mathbf{b} \times \mathbf{a}$.

INTRODUCTION In this section we discuss how to find various equations of lines and planes in 3-space.

Lines: Vector Equation

As in the plane, any two distinct points in 3-space determine only one line between them. To find an equation of the line through $P_1(x_1, y_1, z_1)$ and $P_2(x_2, y_2, z_2)$, let us assume that $P(x, y, z)$ is *any* point on the line. In **FIGURE 7.5.1**, if $\mathbf{r} = \overrightarrow{OP}$, $\mathbf{r}_1 = \overrightarrow{OP_1}$, and $\mathbf{r}_2 = \overrightarrow{OP_2}$, we see that vector $\mathbf{a} = \mathbf{r}_2 - \mathbf{r}_1$ is parallel to vector $\mathbf{r} - \mathbf{r}_2$. Thus,

$$\mathbf{r} - \mathbf{r}_2 = t(\mathbf{r}_2 - \mathbf{r}_1). \tag{1}$$

If we write

$$\mathbf{a} = \mathbf{r}_2 - \mathbf{r}_1 = \langle x_2 - x_1, y_2 - y_1, z_2 - z_1 \rangle = \langle a_1, a_2, a_3 \rangle, \tag{2}$$

then (1) implies that a **vector equation** for the line $\mathscr{L}_a$ is

$$\mathbf{r} = \mathbf{r}_2 + t\mathbf{a}.$$

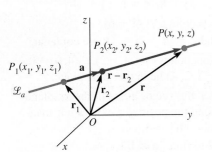

FIGURE 7.5.1 Line through distinct points in 3-space

The scalar t is called a **parameter** and the nonzero vector $\mathbf{a}$ is called a **direction vector**; the components a_1, a_2, and a_3 of the direction vector $\mathbf{a}$ are called **direction numbers** for the line.

Since $\mathbf{r} - \mathbf{r}_1$ is also parallel to $\mathscr{L}_a$, an alternative vector equation for the line is $\mathbf{r} = \mathbf{r}_1 + t\mathbf{a}$. Indeed, $\mathbf{r} = \mathbf{r}_1 + t(-\mathbf{a})$ and $\mathbf{r} = \mathbf{r}_1 + t(k\mathbf{a})$, k a nonzero scalar, are also equations for $\mathscr{L}_a$.

▶ Alternative forms of the vector equation.

EXAMPLE 1 **Vector Equation of a Line**

Find a vector equation for the line through $(2, -1, 8)$ and $(5, 6, -3)$.

SOLUTION Define $\mathbf{a} = \langle 2 - 5, -1 - 6, 8 - (-3) \rangle = \langle -3, -7, 11 \rangle$. The following are three different vector equations for the line:

$$\langle x, y, z \rangle = \langle 2, -1, 8 \rangle + t\langle -3, -7, 11 \rangle \tag{3}$$

$$\langle x, y, z \rangle = \langle 5, 6, -3 \rangle + t\langle -3, -7, 11 \rangle \tag{4}$$

$$\langle x, y, z \rangle = \langle 5, 6, -3 \rangle + t\langle 3, 7, -11 \rangle. \tag{5} \equiv$$

Parametric Equations

By writing (1) as

$$\langle x, y, z \rangle = \langle x_2 + t(x_2 - x_1), y_2 + t(y_2 - y_1), z_2 + t(z_2 - z_1) \rangle$$
$$= \langle x_2 + a_1 t, y_2 + a_2 t, z_2 + a_3 t \rangle$$

and equating components, we obtain

$$x = x_2 + a_1 t, \quad y = y_2 + a_2 t, \quad z = z_2 + a_3 t. \tag{6}$$

The equations in (6) are called **parametric equations** for the line through P_1 and P_2. As the parameter t increases from $-\infty$ to ∞, we can think of the point $P(x, y, z)$ tracing out the entire line. If the parameter t is restricted to a closed interval $[t_0, t_1]$, then $P(x, y, z)$ traces out a **line segment** starting at the point corresponding to t_0 and ending at the point corresponding to t_1. For example, in Figure 7.5.1, if $-1 \leq t \leq 0$, then $P(x, y, z)$ traces out the line segment starting at $P_1(x_1, y_1, z_1)$ and ending at $P_2(x_2, y_2, z_2)$.

EXAMPLE 2 **Parametric Equations of a Line**

Find parametric equations for the line in Example 1.

SOLUTION From (3), it follows that

$$x = 2 - 3t, \quad y = -1 - 7t, \quad z = 8 + 11t. \tag{7}$$

An alternative set of parametric equations can be obtained from (5):

$$x = 5 + 3t, \quad y = 6 + 7t, \quad z = -3 - 11t. \tag{8} \equiv$$

Note that the value $t = 0$ in (7) gives $(2, -1, 8)$, whereas in (8), $t = -1$ must be used to obtain the same point.

EXAMPLE 3 **Vector Parallel to a Line**

Find a vector $\mathbf{a}$ that is parallel to the line $\mathscr{L}_a$ whose parametric equations are $x = 4 + 9t$, $y = -14 + 5t$, $z = 1 - 3t$.

SOLUTION The coefficients (or a nonzero constant multiple of the coefficients) of the parameter in each equation are the components of a vector that is parallel to the line. Thus, $\mathbf{a} = 9\mathbf{i} + 5\mathbf{j} - 3\mathbf{k}$ is parallel to $\mathscr{L}_a$ and hence is a direction vector of the line. ≡

■ **Symmetric Equations** From (6), observe that we can clear the parameter by writing

$$t = \frac{x - x_2}{a_1} = \frac{y - y_2}{a_2} = \frac{z - z_2}{a_3}$$

provided that the three direction numbers a_1, a_2, and a_3 are nonzero. The resulting equations

$$\frac{x - x_2}{a_1} = \frac{y - y_2}{a_2} = \frac{z - z_2}{a_3} \tag{9}$$

are said to be **symmetric equations** for the line through P_1 and P_2.

EXAMPLE 4 **Symmetric Equations of a Line**

Find symmetric equations for the line through $(4, 10, -6)$ and $(7, 9, 2)$.

SOLUTION Define $a_1 = 7 - 4 = 3$, $a_2 = 9 - 10 = -1$, and $a_3 = 2 - (-6) = 8$. It follows from (9) that symmetric equations for the line are

$$\frac{x - 7}{3} = \frac{y - 9}{-1} = \frac{z - 2}{8}.$$ ≡

If one of the direction numbers a_1, a_2, or a_3 is zero in (6), we use the remaining two equations to eliminate the parameter t. For example, if $a_1 = 0$, $a_2 \neq 0$, $a_3 \neq 0$, then (6) yields

$$x = x_2 \quad \text{and} \quad t = \frac{y - y_2}{a_2} = \frac{z - z_2}{a_3}.$$

In this case, $$x = x_2, \quad \frac{y - y_2}{a_2} = \frac{z - z_2}{a_3}$$

are symmetric equations for the line.

EXAMPLE 5 **Symmetric Equations of a Line**

Find symmetric equations for the line through $(5, 3, 1)$ and $(2, 1, 1)$.

SOLUTION Define $a_1 = 5 - 2 = 3$, $a_2 = 3 - 1 = 2$, and $a_3 = 1 - 1 = 0$. From the preceding discussion, it follows that symmetric equations for the line are

$$\frac{x - 5}{3} = \frac{y - 3}{2}, \quad z = 1.$$

In other words, the symmetric equations describe a line in the plane $z = 1$. ≡

A line in space is also determined by specifying a point $P_1(x_1, y_1, z_1)$ and a nonzero direction vector $\mathbf{a}$. Through the point P_1, there passes only one line $\mathscr{L}_a$ parallel to the given vector. If $P(x, y, z)$ is a point on the line $\mathscr{L}_a$, shown in **FIGURE 7.5.2**, then, as before,

$$\overrightarrow{OP} - \overrightarrow{OP}_1 = t\mathbf{a} \quad \text{or} \quad \mathbf{r} = \mathbf{r}_1 + t\mathbf{a}.$$

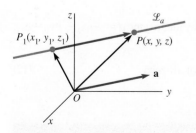

FIGURE 7.5.2 Line determined by a point P and vector $\mathbf{a}$

EXAMPLE 6 Line Parallel to a Vector

Write vector, parametric, and symmetric equations for the line through $(4, 6, -3)$ and parallel to $\mathbf{a} = 5\mathbf{i} - 10\mathbf{j} + 2\mathbf{k}$.

SOLUTION With $a_1 = 5$, $a_2 = -10$, and $a_3 = 2$ we have immediately

$$\textit{Vector:}\quad \langle x, y, z \rangle = \langle 4, 6, -3 \rangle + t\langle 5, -10, 2 \rangle$$

$$\textit{Parametric:}\quad x = 4 + 5t,\quad y = 6 - 10t,\quad z = -3 + 2t$$

$$\textit{Symmetric:}\quad \frac{x - 4}{5} = \frac{y - 6}{-10} = \frac{z + 3}{2}. \qquad \equiv$$

◼ Planes: Vector Equation

FIGURE 7.5.3(a) illustrates the fact that through a given point $P_1(x_1, y_1, z_1)$ there pass an infinite number of planes. However, as shown in Figure 7.5.3(b), if a point P_1 and a vector $\mathbf{n}$ are specified, there is only *one* plane $\mathcal{P}$ containing P_1 with $\mathbf{n}$ **normal**, or perpendicular, to the plane. Moreover, if $P(x, y, z)$ is any point on $\mathcal{P}$, and $\mathbf{r} = \overrightarrow{OP}$, $\mathbf{r}_1 = \overrightarrow{OP_1}$, then, as shown in Figure 7.5.3(c), $\mathbf{r} - \mathbf{r}_1$ is in the plane. It follows that a **vector equation** of the plane is

$$\mathbf{n} \cdot (\mathbf{r} - \mathbf{r}_1) = 0. \tag{10}$$

(a) (b) (c)

FIGURE 7.5.3 Vector $\mathbf{n}$ is perpendicular to a plane

◼ Cartesian Equation

Specifically, if the normal vector is $\mathbf{n} = a\mathbf{i} + b\mathbf{j} + c\mathbf{k}$, then (10) yields a **Cartesian equation** of the plane containing $P_1(x_1, y_1, z_1)$:

$$a(x - x_1) + b(y - y_1) + c(z - z_1) = 0. \tag{11}$$

Equation (11) is sometimes called the **point-normal form** of the equation of a plane.

EXAMPLE 7 Equation of a Plane

Find an equation of the plane with normal vector $\mathbf{n} = 2\mathbf{i} + 8\mathbf{j} - 5\mathbf{k}$ containing the point $(4, -1, 3)$.

SOLUTION It follows immediately from the point-normal form (11) that an equation of the plane is $2(x - 4) + 8(y + 1) - 5(z - 3) = 0$ or $2x + 8y - 5z + 15 = 0$. $\qquad \equiv$

Equation (11) can always be written as $ax + by + cz + d = 0$ by identifying $d = -ax_1 - by_1 - cz_1$. Conversely, we shall now prove that any linear equation

$$ax + by + cz + d = 0,\quad a, b, c \text{ not all zero} \tag{12}$$

is a plane.

THEOREM 7.5.1 Plane with Normal Vector

The graph of any equation $ax + by + cz + d = 0$, a, b, c not all zero, is a plane with the **normal vector $\mathbf{n} = a\mathbf{i} + b\mathbf{j} + c\mathbf{k}$.**

PROOF: Suppose x_0, y_0, and z_0 are numbers that satisfy the given equation. Then, $ax_0 + by_0 + cz_0 + d = 0$ implies that $d = -ax_0 - by_0 - cz_0$. Replacing this latter value of d in the original equation gives, after simplifying, $a(x - x_0) + b(y - y_0) + c(z - z_0) = 0$, or, in terms of vectors,

$$[a\mathbf{i} + b\mathbf{j} + c\mathbf{k}] \cdot [(x - x_0)\mathbf{i} + (y - y_0)\mathbf{j} + (z - z_0)\mathbf{k}] = 0.$$

This last equation implies that $a\mathbf{i} + b\mathbf{j} + c\mathbf{k}$ is normal to the plane containing the point (x_0, y_0, z_0) and to the vector $(x - x_0)\mathbf{i} + (y - y_0)\mathbf{j} + (z - z_0)\mathbf{k}$. ≡

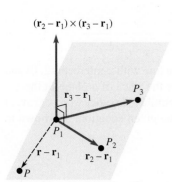

FIGURE 7.5.4 Vectors $\mathbf{r}_2 - \mathbf{r}_1$ and $\mathbf{r}_3 - \mathbf{r}_1$ are in the plane, and their cross product is normal to the plane

EXAMPLE 8 A Vector Normal to a Plane

A vector normal to the plane $3x - 4y + 10z - 8 = 0$ is $\mathbf{n} = 3\mathbf{i} - 4\mathbf{j} + 10\mathbf{k}$. ≡

Of course, a nonzero scalar multiple of a normal vector is still perpendicular to the plane. Three noncollinear points P_1, P_2, and P_3 also determine a plane.* To obtain an equation of the plane, we need only form two vectors between two pairs of points. As shown in **FIGURE 7.5.4**, their cross product is a vector normal to the plane containing these vectors. If $P(x, y, z)$ represents any point on the plane, and $\mathbf{r} = \overrightarrow{OP}$, $\mathbf{r}_1 = \overrightarrow{OP_1}$, $\mathbf{r}_2 = \overrightarrow{OP_2}$, $\mathbf{r}_3 = \overrightarrow{OP_3}$, then $\mathbf{r} - \mathbf{r}_1$ (or, for that matter, $\mathbf{r} - \mathbf{r}_2$ or $\mathbf{r} - \mathbf{r}_3$) is in the plane. Hence,

$$[(\mathbf{r}_2 - \mathbf{r}_1) \times (\mathbf{r}_3 - \mathbf{r}_1)] \cdot (\mathbf{r} - \mathbf{r}_1) = 0 \qquad (13)$$

is a vector equation of the plane. Do not memorize the last formula. The procedure is the same as (10) with the exception that the vector $\mathbf{n}$ normal to the plane is obtained by means of the cross product.

EXAMPLE 9 Three Points That Determine a Plane

Find an equation of the plane that contains $(1, 0, -1)$, $(3, 1, 4)$, and $(2, -2, 0)$.

SOLUTION We need three vectors. Pairing the points on the left as shown yields the vectors on the right. The order in which we subtract is irrelevant.

$$\left.\begin{matrix}(1, 0, -1)\\(3, 1, 4)\end{matrix}\right\}\mathbf{u} = 2\mathbf{i} + \mathbf{j} + 5\mathbf{k}, \quad \left.\begin{matrix}(3, 1, 4)\\(2, -2, 0)\end{matrix}\right\}\mathbf{v} = \mathbf{i} + 3\mathbf{j} + 4\mathbf{k}, \quad \left.\begin{matrix}(2, -2, 0)\\(x, y, z)\end{matrix}\right\}\mathbf{w} = (x - 2)\mathbf{i} + (y + 2)\mathbf{j} + z\mathbf{k}.$$

Now,
$$\mathbf{u} \times \mathbf{v} = \begin{vmatrix} \mathbf{i} & \mathbf{j} & \mathbf{k} \\ 2 & 1 & 5 \\ 1 & 3 & 4 \end{vmatrix} = -11\mathbf{i} - 3\mathbf{j} + 5\mathbf{k}$$

is a vector normal to the plane containing the given points. Hence, a vector equation of the plane is $(\mathbf{u} \times \mathbf{v}) \cdot \mathbf{w} = 0$. The latter equation yields

$$-11(x - 2) - 3(y + 2) + 5z = 0 \quad \text{or} \quad -11x - 3y + 5z + 16 = 0. \qquad ≡$$

◼ **Graphs** The graph of (12) with one or even two variables missing is still a plane. For example, we saw in Section 7.2 that the graphs of

$$x = x_0, \quad y = y_0, \quad z = z_0,$$

where x_0, y_0, z_0 are constants, are planes perpendicular to the x-, y-, and z-axes, respectively. In general, to graph a plane, we should try to find

 (*i*) the x-, y-, and z-intercepts and, if necessary,

 (*ii*) the trace of the plane in each coordinate plane.

A **trace** of a plane in a coordinate plane is the line of intersection of the plane with a coordinate plane.

*If you ever sit at a four-legged table that rocks, you might consider replacing it with a three-legged table.

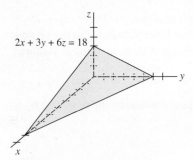

$2x + 3y + 6z = 18$

FIGURE 7.5.5 Plane in Example 10

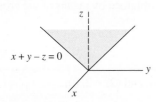

$6x + 4y = 12$

FIGURE 7.5.6 Plane in Example 11

$x + y - z = 0$

FIGURE 7.5.7 Plane in Example 12

FIGURE 7.5.8 Planes intersect in a line

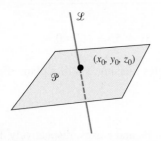

FIGURE 7.5.9 Point of intersection of a plane and a line

EXAMPLE 10 Graph of a Plane

Graph the equation $2x + 3y + 6z = 18$.

SOLUTION Setting: $y = 0, z = 0$ gives $x = 9$
$x = 0, z = 0$ gives $y = 6$
$x = 0, y = 0$ gives $z = 3$.

As shown in **FIGURE 7.5.5**, we use the x-, y-, and z-intercepts $(9, 0, 0)$, $(0, 6, 0)$, and $(0, 0, 3)$ to draw the graph of the plane in the first octant. ≡

EXAMPLE 11 Graph of a Plane

Graph the equation $6x + 4y = 12$.

SOLUTION In two dimensions, the graph of the equation is a line with x-intercept $(2, 0)$ and y-intercept $(3, 0)$. However, in three dimensions, this line is the trace of a plane in the xy-coordinate plane. Since z is not specified, it can be any real number. In other words, (x, y, z) is a point on the plane provided that x and y are related by the given equation. As shown in **FIGURE 7.5.6**, the graph is a plane parallel to the z-axis. ≡

EXAMPLE 12 Graph of a Plane

Graph the equation $x + y - z = 0$.

SOLUTION First observe that the plane passes through the origin $(0, 0, 0)$. Now, the trace of the plane in the xz-plane $(y = 0)$ is $z = x$, whereas its trace in the yz-plane $(x = 0)$ is $z = y$. Drawing these two lines leads to the graph given in **FIGURE 7.5.7**. ≡

Two planes $\mathcal{P}_1$ and $\mathcal{P}_2$ that are not parallel must intersect in a line $\mathcal{L}$. See **FIGURE 7.5.8**. Example 13 will illustrate one way of finding parametric equations for the line of intersection. In Example 14 we shall see how to find a point of intersection (x_0, y_0, z_0) of a plane $\mathcal{P}$ and a line $\mathcal{L}$. See **FIGURE 7.5.9**.

EXAMPLE 13 Line of Intersection of Two Planes

Find parametric equations for the line of intersection of

$$2x - 3y + 4z = 1$$
$$x - y - z = 5.$$

SOLUTION In a system of two equations and three unknowns, we choose one variable arbitrarily, say $z = t$, and solve for x and y from

$$2x - 3y = 1 - 4t$$
$$x - y = 5 + t.$$

Proceeding, we find $x = 14 + 7t$, $y = 9 + 6t$, $z = t$. These are parametric equations for the line of intersection of the given planes. ≡

EXAMPLE 14 Point of Intersection of a Line and a Plane

Find the point of intersection of the plane $3x - 2y + z = -5$ and the line $x = 1 + t$, $y = -2 + 2t$, $z = 4t$.

SOLUTION If (x_0, y_0, z_0) denotes the point of intersection, then we must have $3x_0 - 2y_0 + z_0 = -5$ and $x_0 = 1 + t_0$, $y_0 = -2 + 2t_0$, $z_0 = 4t_0$, for some number t_0. Substituting the latter equations into the equation of the plane gives

$$3(1 + t_0) - 2(-2 + 2t_0) + 4t_0 = -5 \quad \text{or} \quad t_0 = -4.$$

From the parametric equations for the line, we then obtain $x_0 = -3$, $y_0 = -10$, and $z_0 = -16$. The point of intersection is $(-3, -10, -16)$. ≡

In everyday speech, the words *orthogonal*, *perpendicular*, and *normal* are often used interchangeably in the sense that two objects touch, intersect, or abut at a 90° angle. But in recent years an unwritten convention has arisen to use these terms in specific mathematical contexts. As a general rule, we say that two vectors are *orthogonal*, two lines (or two planes) are *perpendicular*, and that a vector is *normal* to a plane.

7.5 Exercises Answers to selected odd-numbered problems begin on page ANS-16.

In Problems 1–6, find a vector equation for the line through the given points.

1. $(1, 2, 1), (3, 5, -2)$ **2.** $(0, 4, 5), (-2, 6, 3)$

3. $(\frac{1}{2}, -\frac{1}{2}, 1), (-\frac{3}{2}, \frac{5}{2}, -\frac{1}{2})$ **4.** $(10, 2, -10), (5, -3, 5)$

5. $(1, 1, -1), (-4, 1, -1)$ **6.** $(3, 2, 1), (\frac{5}{2}, 1, -2)$

In Problems 7–12, find parametric equations for the line through the given points.

7. $(2, 3, 5), (6, -1, 8)$ **8.** $(2, 0, 0), (0, 4, 9)$

9. $(1, 0, 0), (3, -2, -7)$ **10.** $(0, 0, 5), (-2, 4, 0)$

11. $(4, \frac{1}{2}, \frac{1}{3}), (-6, -\frac{1}{4}, \frac{1}{6})$ **12.** $(-3, 7, 9), (4, -8, -1)$

In Problems 13–18, find symmetric equations for the line through the given points.

13. $(1, 4, -9), (10, 14, -2)$ **14.** $(\frac{2}{3}, 0, -\frac{1}{4}), (1, 3, \frac{1}{4})$

15. $(4, 2, 1), (-7, 2, 5)$ **16.** $(-5, -2, -4), (1, 1, 2)$

17. $(5, 10, -2), (5, 1, -14)$ **18.** $(\frac{5}{6}, -\frac{1}{4}, \frac{1}{5}), (\frac{1}{3}, \frac{3}{8}, -\frac{1}{10})$

In Problems 19–22, find parametric and symmetric equations for the line through the given point parallel to the given vector.

19. $(4, 6, -7)$, $\mathbf{a} = \langle 3, \frac{1}{2}, -\frac{3}{2} \rangle$

20. $(1, 8, -2)$, $\mathbf{a} = -7\mathbf{i} - 8\mathbf{j}$

21. $(0, 0, 0)$, $\mathbf{a} = 5\mathbf{i} + 9\mathbf{j} + 4\mathbf{k}$

22. $(0, -3, 10)$, $\mathbf{a} = \langle 12, -5, -6 \rangle$

23. Find parametric equations for the line through $(6, 4, -2)$ that is parallel to the line $x/2 = (1 - y)/3 = (z - 5)/6$.

24. Find symmetric equations for the line through $(4, -11, -7)$ that is parallel to the line $x = 2 + 5t, y = -1 + \frac{1}{3}t, z = 9 - 2t$.

25. Find parametric equations for the line through $(2, -2, 15)$ that is parallel to the xz-plane and the xy-plane.

26. Find parametric equations for the line through $(1, 2, 8)$ that is **(a)** parallel to the y-axis, and **(b)** perpendicular to the xy-plane.

27. Show that the lines given by $\mathbf{r} = t\langle 1, 1, 1 \rangle$ and $\mathbf{r} = \langle 6, 6, 6 \rangle + t\langle -3, -3, -3 \rangle$ are the same.

28. Let $\mathcal{L}_a$ and $\mathcal{L}_b$ be lines with direction vectors $\mathbf{a}$ and $\mathbf{b}$, respectively. $\mathcal{L}_a$ and $\mathcal{L}_b$ are orthogonal if $\mathbf{a}$ and $\mathbf{b}$ are orthogonal and parallel if $\mathbf{a}$ and $\mathbf{b}$ are parallel. Determine which of the following lines are orthogonal and which are parallel.

 (a) $\mathbf{r} = \langle 1, 0, 2 \rangle + t\langle 9, -12, 6 \rangle$

 (b) $x = 1 + 9t, \ y = 12t, \ z = 2 - 6t$

 (c) $x = 2t, \ y = -3t, \ z = 4t$

 (d) $x = 5 + t, \ y = 4t, \ z = 3 + \frac{5}{2}t$

 (e) $x = 1 + t, \ y = \frac{3}{2}t, \ z = 2 - \frac{3}{2}t$

 (f) $\dfrac{x + 1}{-3} = \dfrac{y + 6}{4} = \dfrac{z - 3}{-2}$

In Problems 29 and 30, determine the points of intersection of the given line and the three coordinate planes.

29. $x = 4 - 2t, \ y = 1 + 2t, \ z = 9 + 3t$

30. $\dfrac{x - 1}{2} = \dfrac{y + 2}{3} = \dfrac{z - 4}{2}$

In Problems 31–34, determine whether the given lines intersect. If so, find the point of intersection.

31. $x = 4 + t, \ y = 5 + t, \ z = -1 + 2t$
$x = 6 + 2s, \ y = 11 + 4s, \ z = -3 + s$

32. $x = 1 + t, \ y = 2 - t, \ z = 3t$
$x = 2 - s, \ y = 1 + s, \ z = 6s$

33. $x = 2 - t, \ y = 3 + t, \ z = 1 + t$
$x = 4 + s, \ y = 1 + s, \ z = 1 - s$

34. $x = 3 - t, \ y = 2 + t, \ z = 8 + 2t$
$x = 2 + 2s, \ y = -2 + 3s, \ z = -2 + 8s$

The angle between two lines $\mathcal{L}_a$ and $\mathcal{L}_b$ is the angle between their direction vectors $\mathbf{a}$ and $\mathbf{b}$. In Problems 35 and 36, find the angle between the given lines.

35. $x = 4 - t, \ y = 3 + 2t, \ z = -2t$
$x = 5 + 2s, \ y = 1 + 3s, \ z = 5 - 6s$

36. $\dfrac{x - 1}{2} = \dfrac{y + 5}{7} = \dfrac{z - 1}{-1}; \quad \dfrac{x + 3}{-2} = y - 9 = \dfrac{z}{4}$

In Problems 37 and 38, the given lines lie in the same plane. Find parametric equations for the line through the indicated point that is perpendicular to this plane.

37. $x = 3 + t, \ y = -2 + t, \ z = 9 + t$
$x = 1 - 2s, \ y = 5 + s, \ z = -2 - 5s; \quad (4, 1, 6)$

38. $\dfrac{x - 1}{3} = \dfrac{y + 1}{2} = \dfrac{z}{4}$
$\dfrac{x + 4}{6} = \dfrac{y - 6}{4} = \dfrac{z - 10}{8}; \quad (1, -1, 0)$

In Problems 39–44, find an equation of the plane that contains the given point and is perpendicular to the indicated vector.

39. $(5, 1, 3); \ 2\mathbf{i} - 3\mathbf{j} + 4\mathbf{k}$

40. $(1, 2, 5); \ 4\mathbf{i} - 2\mathbf{j}$

41. $(6, 10, -7); \ -5\mathbf{i} + 3\mathbf{k}$

42. $(0, 0, 0); \ 6\mathbf{i} - \mathbf{j} + 3\mathbf{k}$

43. $(\frac{1}{2}, \frac{3}{4}, -\frac{1}{2})$; $6\mathbf{i} + 8\mathbf{j} - 4\mathbf{k}$

44. $(-1, 1, 0)$; $-\mathbf{i} + \mathbf{j} - \mathbf{k}$

In Problems 45–50, find, if possible, an equation of a plane that contains the given points.

45. $(3, 5, 2)$, $(2, 3, 1)$, $(-1, -1, 4)$

46. $(0, 1, 0)$, $(0, 1, 1)$, $(1, 3, -1)$

47. $(0, 0, 0)$, $(1, 1, 1)$, $(3, 2, -1)$

48. $(0, 0, 3)$, $(0, -1, 0)$, $(0, 0, 6)$

49. $(1, 2, -1)$, $(4, 3, 1)$, $(7, 4, 3)$

50. $(2, 1, 2)$, $(4, 1, 0)$, $(5, 0, -5)$

In Problems 51–60, find an equation of the plane that satisfies the given conditions.

51. Contains $(2, 3, -5)$ and is parallel to $x + y - 4z = 1$

52. Contains the origin and is parallel to $5x - y + z = 6$

53. Contains $(3, 6, 12)$ and is parallel to the xy-plane

54. Contains $(-7, -5, 18)$ and is perpendicular to the y-axis

55. Contains the lines $x = 1 + 3t$, $y = 1 - t$, $z = 2 + t$; $x = 4 + 4s$, $y = 2s$, $z = 3 + s$

56. Contains the lines $\dfrac{x - 1}{2} = \dfrac{y + 1}{-1} = \dfrac{z - 5}{6}$; $\mathbf{r} = \langle 1, -1, 5 \rangle + t\langle 1, 1, -3 \rangle$

57. Contains the parallel lines $x = 1 + t$, $y = 1 + 2t$, $z = 3 + t$; $x = 3 + s$, $y = 2s$, $z = -2 + s$

58. Contains the point $(4, 0, -6)$ and the line $x = 3t$, $y = 2t$, $z = -2t$

59. Contains $(2, 4, 8)$ and is perpendicular to the line $x = 10 - 3t$, $y = 5 + t$, $z = 6 - \frac{1}{2}t$

60. Contains $(1, 1, 1)$ and is perpendicular to the line through $(2, 6, -3)$ and $(1, 0, -2)$

61. Let $\mathcal{P}_1$ and $\mathcal{P}_2$ be planes with normal vectors $\mathbf{n}_1$ and $\mathbf{n}_2$, respectively. $\mathcal{P}_1$ and $\mathcal{P}_2$ are orthogonal if $\mathbf{n}_1$ and $\mathbf{n}_2$ are orthogonal and parallel if $\mathbf{n}_1$ and $\mathbf{n}_2$ are parallel. Determine which of the following planes are orthogonal and which are parallel.
 (a) $2x - y + 3z = 1$ (b) $x + 2y + 2z = 9$
 (c) $x + y - \frac{3}{2}z = 2$ (d) $-5x + 2y + 4z = 0$
 (e) $-8x - 8y + 12z = 1$ (f) $-2x + y - 3z = 5$

62. Find parametric equations for the line that contains $(-4, 1, 7)$ and is perpendicular to the plane $-7x + 2y + 3z = 1$.

63. Determine which of the following planes are perpendicular to the line $x = 4 - 6t$, $y = 1 + 9t$, $z = 2 + 3t$.
 (a) $4x + y + 2z = 1$ (b) $2x - 3y + z = 4$
 (c) $10x - 15y - 5z = 2$ (d) $-4x + 6y + 2z = 9$

64. Determine which of the following planes are parallel to the line $(1 - x)/2 = (y + 2)/4 = z - 5$.
 (a) $x - y + 3z = 1$ (b) $6x - 3y = 1$
 (c) $x - 2y + 5z = 0$ (d) $-2x + y - 2z = 7$

In Problems 65–68, find parametric equations for the line of intersection of the given planes.

65. $5x - 4y - 9z = 8$
 $x + 4y + 3z = 4$

66. $x + 2y - z = 2$
 $3x - y + 2z = 1$

67. $4x - 2y - z = 1$
 $x + y + 2z = 1$

68. $2x - 5y + z = 0$
 $y = 0$

In Problems 69–72, find the point of intersection of the given plane and line.

69. $2x - 3y + 2z = -7$; $x = 1 + 2t$, $y = 2 - t$, $z = -3t$

70. $x + y + 4z = 12$; $x = 3 - 2t$, $y = 1 + 6t$, $z = 2 - \frac{1}{2}t$

71. $x + y - z = 8$; $x = 1$, $y = 2$, $z = 1 + t$

72. $x - 3y + 2z = 0$; $x = 4 + t$, $y = 2 + t$, $z = 1 + 5t$

In Problems 73 and 74, find parametric equations for the line through the indicated point that is parallel to the given planes.

73. $x + y - 4z = 2$
 $2x - y + z = 10$; $(5, 6, -12)$

74. $2x + z = 0$
 $-x + 3y + z = 1$; $(-3, 5, -1)$

In Problems 75 and 76, find an equation of the plane that contains the given line and is orthogonal to the indicated plane.

75. $x = 4 + 3t$, $y = -t$, $z = 1 + 5t$; $x + y + z = 7$

76. $\dfrac{2 - x}{3} = \dfrac{y + 2}{5} = \dfrac{z - 8}{2}$; $2x - 4y - z + 16 = 0$

In Problems 77–82, graph the given equation.

77. $5x + 2y + z = 10$

78. $3x + 2z = 9$

79. $-y - 3z + 6 = 0$

80. $3x + 4y - 2z - 12 = 0$

81. $-x + 2y + z = 4$

82. $x - y - 1 = 0$

7.6 Vector Spaces

INTRODUCTION In the preceding sections we were dealing with points and vectors in 2- and 3-space. Mathematicians in the nineteenth century, notably the English mathematicians **Arthur Cayley** (1821–1895) and **James Joseph Sylvester** (1814–1897) and the Irish mathematician **William Rowan Hamilton** (1805–1865), realized that the concepts of *point* and *vector* could be generalized. A realization developed that vectors could be described, or defined, by analytic rather than geometric properties. This was a truly significant breakthrough in the history of mathematics. There is no need to stop with three dimensions; ordered quadruples $\langle a_1, a_2, a_3, a_4 \rangle$, quintuples $\langle a_1, a_2, a_3, a_4, a_5 \rangle$, and n-tuples $\langle a_1, a_2, \dots, a_n \rangle$ of real numbers can be thought of as vectors just as well as ordered pairs $\langle a_1, a_2 \rangle$ and ordered triples $\langle a_1, a_2, a_3 \rangle$, the only difference being that we lose our ability to visualize directed line segments or arrows in 4-dimensional, 5-dimensional, or n-dimensional space.

■ *n*-Space

In formal terms, a **vector in *n*-space** is any ordered *n*-tuple $\mathbf{a} = \langle a_1, a_2, \ldots, a_n \rangle$ of real numbers called the components of $\mathbf{a}$. The set of all vectors in *n*-space is denoted by R^n. The concepts of vector addition, scalar multiplication, equality, and so on listed in Definition 7.2.1 carry over to R^n in a natural way. For example, if $\mathbf{a} = \langle a_1, a_2, \ldots, a_n \rangle$ and $\mathbf{b} = \langle b_1, b_2, \ldots, b_n \rangle$, then addition and scalar multiplication in *n*-space are defined by

$$\mathbf{a} + \mathbf{b} = \langle a_1 + b_1, a_2 + b_2, \ldots, a_n + b_n \rangle \quad \text{and} \quad k\mathbf{a} = \langle ka_1, ka_2, \ldots, ka_n \rangle. \tag{1}$$

The **zero vector** in R^n is $\mathbf{0} = \langle 0, 0, \ldots, 0 \rangle$. The notion of **length** or **magnitude** of a vector $\mathbf{a} = \langle a_1, a_2, \ldots, a_n \rangle$ in *n*-space is just an extension of that concept in 2- and 3-space:

$$\|\mathbf{a}\| = \sqrt{a_1^2 + a_2^2 + \cdots + a_n^2}.$$

The length of a vector is also called its **norm**. A **unit vector** is one whose norm is 1. For a nonzero vector $\mathbf{a}$, the process of constructing a unit vector $\mathbf{u}$ by multiplying $\mathbf{a}$ by the reciprocal of its norm, that is, $\mathbf{u} = \dfrac{1}{\|\mathbf{a}\|}\mathbf{a}$, is referred to as **normalizing a**. For example, if $\mathbf{a} = \langle 3, 1, 2, -1 \rangle$, then $\|\mathbf{a}\| = \sqrt{3^2 + 1^2 + 2^2 + (-1)^2} = \sqrt{15}$ and a unit vector is

$$\mathbf{u} = \frac{1}{\sqrt{15}}\mathbf{a} = \left\langle \frac{3}{\sqrt{15}}, \frac{1}{\sqrt{15}}, \frac{2}{\sqrt{15}}, -\frac{1}{\sqrt{15}} \right\rangle.$$

The **standard inner product**, also known as the **Euclidean inner product** or **dot product**, of two *n*-vectors $\mathbf{a} = \langle a_1, a_2, \ldots, a_n \rangle$ and $\mathbf{b} = \langle b_1, b_2, \ldots, b_n \rangle$ is the real number defined by

$$\mathbf{a} \cdot \mathbf{b} = \langle a_1, a_2, \ldots, a_n \rangle \cdot \langle b_1, b_2, \ldots, b_n \rangle = a_1 b_1 + a_2 b_2 + \cdots + a_n b_n. \tag{2}$$

Two nonzero vectors $\mathbf{a}$ and $\mathbf{b}$ in R^n are said to be **orthogonal** if and only if $\mathbf{a} \cdot \mathbf{b} = 0$. For example, $\mathbf{a} = \langle 3, 4, 1, -6 \rangle$ and $\mathbf{b} = \langle 1, \frac{1}{2}, 1, 1 \rangle$ are orthogonal in R^4 since $\mathbf{a} \cdot \mathbf{b} = 3 \cdot 1 + 4 \cdot \frac{1}{2} + 1 \cdot 1 + (-6) \cdot 1 = 0$.

■ Vector Space

We can even go beyond the notion of a vector as an ordered *n*-tuple in R^n. A vector can be defined as anything we want it to be: an ordered *n*-tuple, a number, an array of numbers, or even a function. But we are particularly interested in vectors that are elements in a special kind of set called a **vector space**. Fundamental to the notion of vector space are two kinds of objects, **vectors** and **scalars**, and two algebraic operations analogous to those given in (1). For a set of vectors we want to be able to add two vectors in this set and get another vector in the same set, and we want to multiply a vector by a scalar and obtain a vector in the same set. To determine whether a set of objects is a vector space depends on whether the set possesses these two algebraic operations along with certain other properties. These properties, the *axioms* of a vector space, are given next.

DEFINITION 7.6.1 Vector Space

Let V be a set of elements on which two operations called **vector addition** and **scalar multiplication** are defined. Then V is said to be a **vector space** if the following 10 properties are satisfied.

Axioms for Vector Addition:
(i) If $\mathbf{x}$ and $\mathbf{y}$ are in V, then $\mathbf{x} + \mathbf{y}$ is in V.
(ii) For all $\mathbf{x}, \mathbf{y}$ in V, $\mathbf{x} + \mathbf{y} = \mathbf{y} + \mathbf{x}$. ← commutative law
(iii) For all $\mathbf{x}, \mathbf{y}, \mathbf{z}$ in V, $\mathbf{x} + (\mathbf{y} + \mathbf{z}) = (\mathbf{x} + \mathbf{y}) + \mathbf{z}$. ← associative law
(iv) There is a unique vector $\mathbf{0}$ in V such that
$$\mathbf{0} + \mathbf{x} = \mathbf{x} + \mathbf{0} = \mathbf{x}.$$ ← zero vector
(v) For each $\mathbf{x}$ in V, there exists a vector $-\mathbf{x}$ such that
$$\mathbf{x} + (-\mathbf{x}) = (-\mathbf{x}) + \mathbf{x} = \mathbf{0}.$$ ← negative of a vector

Axioms for Scalar Multiplication:
(vi) If k is any scalar and $\mathbf{x}$ is in V, then $k\mathbf{x}$ is in V.
(vii) $k(\mathbf{x} + \mathbf{y}) = k\mathbf{x} + k\mathbf{y}$ ← distributive law
(viii) $(k_1 + k_2)\mathbf{x} = k_1\mathbf{x} + k_2\mathbf{x}$ ← distributive law
(ix) $k_1(k_2\mathbf{x}) = (k_1 k_2)\mathbf{x}$
(x) $1\mathbf{x} = \mathbf{x}$

In this brief introduction to abstract vectors we shall take the scalars in Definition 7.6.1 to be real numbers. In this case V is referred to as a **real vector space**, although we shall not belabor this term. When the scalars are allowed to be complex numbers we obtain a **complex vector space**. Since properties (i)–$(viii)$ on page 331 are the prototypes for the axioms in Definition 7.6.1, it is clear that R^2 is a vector space. Moreover, since vectors in R^3 and R^n have these same properties, we conclude that R^3 and R^n are also vector spaces. Axioms (i) and (vi) are called the **closure axioms** and we say that a vector space V is closed under vector addition and scalar multiplication. Note, too, that concepts such as *length* and *inner product* are not part of the axiomatic structure of a vector space.

EXAMPLE 1 Checking the Closure Axioms

Determine whether the sets **(a)** $V = \{1\}$ and **(b)** $V = \{0\}$ under ordinary addition and multiplication by real numbers are vector spaces.

SOLUTION **(a)** For this system consisting of one element, many of the axioms given in Definition 7.6.1 are violated. In particular, axioms (i) and (vi) of closure are not satisfied. Neither the sum $1 + 1 = 2$ nor the scalar multiple $k \cdot 1 = k$, for $k \neq 1$, is in V. Hence V is not a vector space.

(b) In this case the closure axioms are satisfied since $0 + 0 = 0$ and $k \cdot 0 = 0$ for any real number k. The commutative and associative axioms are satisfied since $0 + 0 = 0 + 0$ and $0 + (0 + 0) = (0 + 0) + 0$. In this manner it is easy to verify that the remaining axioms are also satisfied. Hence V is a vector space. ≡

The vector space $V = \{0\}$ is often called the **trivial** or **zero vector space**.

If this is your first experience with the notion of an abstract vector, then you are cautioned to not take the names *vector addition* and *scalar multiplication* too literally. These operations are *defined* and as such you must accept them at face value even though these operations may not bear any resemblance to the usual understanding of ordinary addition and multiplication in, say, R, R^2, R^3, or R^n. For example, the addition of two vectors $\mathbf{x}$ and $\mathbf{y}$ could be $\mathbf{x} - \mathbf{y}$. With this forewarning, consider the next example.

EXAMPLE 2 An Example of a Vector Space

Consider the set V of positive real numbers. If x and y denote positive real numbers, then we write vectors in V as $\mathbf{x} = x$ and $\mathbf{y} = y$. Now, *addition* of vectors is defined by

$$\mathbf{x} + \mathbf{y} = xy,$$

and *scalar multiplication* is defined by

$$k\mathbf{x} = x^k.$$

Determine whether V is a vector space.

SOLUTION We shall go through all 10 axioms in Definition 7.6.1.
- (i) For $\mathbf{x} = x > 0$ and $\mathbf{y} = y > 0$, $\mathbf{x} + \mathbf{y} = xy > 0$. Thus, the sum $\mathbf{x} + \mathbf{y}$ is in V; V is closed under addition.
- (ii) Since multiplication of positive real numbers is commutative, we have for all $\mathbf{x} = x$ and $\mathbf{y} = y$ in V, $\mathbf{x} + \mathbf{y} = xy = yx = \mathbf{y} + \mathbf{x}$. Thus, addition is commutative.
- (iii) For all $\mathbf{x} = x$, $\mathbf{y} = y$, $\mathbf{z} = z$ in V,

$$\mathbf{x} + (\mathbf{y} + \mathbf{z}) = x(yz) = (xy)z = (\mathbf{x} + \mathbf{y}) + \mathbf{z}.$$

Thus, addition is associative.
- (iv) Since $1 + \mathbf{x} = 1x = x = \mathbf{x}$ and $\mathbf{x} + 1 = x1 = x = \mathbf{x}$, the zero vector $\mathbf{0}$ is $1 = 1$.
- (v) If we define $-\mathbf{x} = \dfrac{1}{x}$, then

$$\mathbf{x} + (-\mathbf{x}) = x\frac{1}{x} = 1 = \mathbf{1} = \mathbf{0} \text{ and } (-\mathbf{x}) + \mathbf{x} = \frac{1}{x}x = 1 = \mathbf{1} = \mathbf{0}.$$

Therefore, the negative of a vector is its reciprocal.

(vi) If k is any scalar and $\mathbf{x} = x > 0$ is any vector, then $k\mathbf{x} = x^k > 0$. Hence V is closed under scalar multiplication.

(vii) If k is any scalar, then

$$k(\mathbf{x} + \mathbf{y}) = (xy)^k = x^k y^k = k\mathbf{x} + k\mathbf{y}.$$

(viii) For scalars k_1 and k_2,

$$(k_1 + k_2)\mathbf{x} = x^{(k_1 + k_2)} = x^{k_1} x^{k_2} = k_1\mathbf{x} + k_2\mathbf{x}.$$

(ix) For scalars k_1 and k_2,

$$k_1(k_2\mathbf{x}) = (x^{k_2})^{k_1} = x^{k_1 k_2} = (k_1 k_2)\mathbf{x}.$$

(x) $1\mathbf{x} = x^1 = x = \mathbf{x}$.

Since all the axioms of Definition 7.6.1 are satisfied, we conclude that V is a vector space. ≡

Here are some important vector spaces—we have mentioned some of these previously. The operations of vector addition and scalar multiplication are the usual operations associated with the set.

- The set R of real numbers
- The set R^2 of ordered pairs
- The set R^3 of ordered triples
- The set R^n of ordered n-tuples
- The set P_n of polynomials of degree less than or equal to n
- The set P of all polynomials
- The set of real-valued functions f defined on the entire real line
- The set $C[a, b]$ of real-valued functions f continuous on the closed interval $[a, b]$
- The set $C(-\infty, \infty)$ of real-valued functions f continuous on the entire real line
- The set $C^n[a, b]$ of all real-valued functions f for which $f, f', f'', \ldots, f^{(n)}$ exist and are continuous on the closed interval $[a, b]$

■ **Subspace** It may happen that a subset of vectors W of a vector space V is itself a vector space.

DEFINITION 7.6.2 Subspace

If a subset W of a vector space V is itself a vector space under the operations of vector addition and scalar multiplication defined on V, then W is called a **subspace** of V.

Every vector space V has at least two subspaces: V itself and the zero subspace $\{\mathbf{0}\}$; $\{\mathbf{0}\}$ is a subspace since the zero vector must be an element in every vector space.

To show that a subset W of a vector space V is a subspace, it is not necessary to demonstrate that all 10 axioms of Definition 7.6.1 are satisfied. Since all the vectors in W are also in V, these vectors must satisfy axioms such as (ii) and (iii). In other words, W inherits most of the properties of a vector space from V. As the next theorem indicates, we need only check the two closure axioms to demonstrate that a subset W is a subspace of V.

THEOREM 7.6.1 Criteria for a Subspace

A nonempty subset W of a vector space V is a subspace of V if and only if W is closed under vector addition and scalar multiplication defined on V:

(i) If $\mathbf{x}$ and $\mathbf{y}$ are in W, then $\mathbf{x} + \mathbf{y}$ is in W.

(ii) If $\mathbf{x}$ is in W and k is any scalar, then $k\mathbf{x}$ is in W.

EXAMPLE 3 **A Subspace**

Suppose $\mathbf{f}$ and $\mathbf{g}$ are continuous real-valued functions defined on the entire real line. Then we know from calculus that $\mathbf{f} + \mathbf{g}$ and $k\mathbf{f}$, for any real number k, are continuous and real-valued functions. From this we can conclude that $C(-\infty, \infty)$ is a subspace of the vector space of real-valued functions defined on the entire real line. ≡

EXAMPLE 4 A Subspace

The set P_n of polynomials of degree less than or equal to n is a subspace of $C(-\infty, \infty)$, the set of real-valued functions continuous on the entire real line. ≡

It is always a good idea to have concrete visualizations of vector spaces and subspaces. The subspaces of the vector space R^3 of three-dimensional vectors can be easily visualized by thinking of a vector as a point (a_1, a_2, a_3). Of course, $\{0\}$ and R^3 itself are subspaces; other subspaces are all lines passing through the origin, and all planes passing through the origin. The lines and planes must pass through the origin since the zero vector $\mathbf{0} = (0, 0, 0)$ must be an element in each subspace.

Similar to Definition 3.1.1 we can define linearly independent vectors.

DEFINITION 7.6.3 Linear Independence

A set of vectors $\{\mathbf{x}_1, \mathbf{x}_2, \ldots, \mathbf{x}_n\}$ is said to be **linearly independent** if the only constants satisfying the equation

$$k_1\mathbf{x}_1 + k_2\mathbf{x}_2 + \cdots + k_n\mathbf{x}_n = \mathbf{0} \tag{3}$$

are $k_1 = k_2 = \cdots = k_n = 0$. If the set of vectors is not linearly independent, then it is said to be **linearly dependent**.

In R^3, the vectors $\mathbf{i} = \langle 1, 0, 0 \rangle$, $\mathbf{j} = \langle 0, 1, 0 \rangle$, and $\mathbf{k} = \langle 0, 0, 1 \rangle$ are linearly independent since the equation $k_1\mathbf{i} + k_2\mathbf{j} + k_3\mathbf{k} = \mathbf{0}$ is the same as

$$k_1\langle 1, 0, 0 \rangle + k_2\langle 0, 1, 0 \rangle + k_3\langle 0, 0, 1 \rangle = \langle 0, 0, 0 \rangle \quad \text{or} \quad \langle k_1, k_2, k_3 \rangle = \langle 0, 0, 0 \rangle.$$

By equality of vectors, (*iii*) of Definition 7.2.1, we conclude that $k_1 = 0$, $k_2 = 0$, and $k_3 = 0$. In Definition 7.6.3, linear dependence means that there are constants $k_1, k_2, \ldots, k_n$ *not all zero* such that $k_1\mathbf{x}_1 + k_2\mathbf{x}_2 + \cdots + k_n\mathbf{x}_n = \mathbf{0}$. For example, in R^3 the vectors $\mathbf{a} = \langle 1, 1, 1 \rangle$, $\mathbf{b} = \langle 2, -1, 4 \rangle$, and $\mathbf{c} = \langle 5, 2, 7 \rangle$ are linearly dependent since (3) is satisfied when $k_1 = 3$, $k_2 = 1$, and $k_3 = -1$:

$$3\langle 1, 1, 1 \rangle + \langle 2, -1, 4 \rangle - \langle 5, 2, 7 \rangle = \langle 0, 0, 0 \rangle \quad \text{or} \quad 3\mathbf{a} + \mathbf{b} - \mathbf{c} = \mathbf{0}.$$

We observe that two vectors are linearly independent if neither is a constant multiple of the other.

Basis Any vector in R^3 can be written as a linear combination of the linearly independent vectors $\mathbf{i}$, $\mathbf{j}$, and $\mathbf{k}$. In Section 7.2, we said that these vectors form a **basis** for the system of three-dimensional vectors.

DEFINITION 7.6.4 Basis for a Vector Space

Consider a set of vectors $B = \{\mathbf{x}_1, \mathbf{x}_2, \ldots, \mathbf{x}_n\}$ in a vector space V. If the set B is linearly independent and if every vector in V can be expressed as a linear combination of these vectors, then B is said to be a **basis** for V.

Standard Bases Although we cannot prove it in this course, every vector space has a basis. The vector space P_n of all polynomials of degree less than or equal to n has the basis $\{1, x, x^2, \ldots, x^n\}$ since any vector (polynomial) $p(x)$ of degree n or less can be written as the linear combination $p(x) = c_n x^n + \cdots + c_2 x^2 + c_1 x + c_0$. A vector space may have many bases. We mentioned previously the set of vectors $\{\mathbf{i}, \mathbf{j}, \mathbf{k}\}$ is a basis for R^3. But it can be proved that $\{\mathbf{u}_1, \mathbf{u}_2, \mathbf{u}_3\}$, where

$$\mathbf{u}_1 = \langle 1, 0, 0 \rangle, \quad \mathbf{u}_2 = \langle 1, 1, 0 \rangle, \quad \mathbf{u}_3 = \langle 1, 1, 1 \rangle$$

is a linearly independent set (see Problem 23 in Exercises 7.6) and, furthermore, every vector $\mathbf{a} = \langle a_1, a_2, a_3 \rangle$ can be expressed as a linear combination $\mathbf{a} = c_1\mathbf{u}_1 + c_2\mathbf{u}_2 + c_3\mathbf{u}_3$. Hence, the set of vectors $\{\mathbf{u}_1, \mathbf{u}_2, \mathbf{u}_3\}$ is another basis for R^3. Indeed, any set of *three* linearly independent vectors is a basis for that space. However, as mentioned in Section 7.2, the set $\{\mathbf{i}, \mathbf{j}, \mathbf{k}\}$ is referred to

as the **standard basis** for R^3. The standard basis for the space P_n is $\{1, x, x^2, \ldots, x^n\}$. For the vector space R^n, the **standard basis** consists of the n vectors

$$\mathbf{e}_1 = \langle 1, 0, 0, \ldots, 0 \rangle, \quad \mathbf{e}_2 = \langle 0, 1, 0, \ldots, 0 \rangle, \quad \ldots, \quad \mathbf{e}_n = \langle 0, 0, 0, \ldots, 1 \rangle. \tag{4}$$

If B is a basis for a vector space V, then for every vector $\mathbf{v}$ in V there exist scalars c_i, $i = 1, 2, \ldots, n$ such that

$$\mathbf{v} = c_1 \mathbf{x}_2 + c_2 \mathbf{x}_2 + \cdots + c_n \mathbf{x}_n. \tag{5}$$

Read the last sentence several times. ▶

The scalars c_i, $i = 1, 2, \ldots, n$, in the linear combination (5) are called the **coordinates of v relative to the basis B**. In R^n, the n-tuple notation $\langle a_1, a_2, \ldots, a_n \rangle$ for a vector $\mathbf{a}$ means that real numbers $a_1, a_2, \ldots, a_n$ are the coordinates of $\mathbf{a}$ *relative to the standard basis* with $\mathbf{e}_i$'s in the precise order given in (4).

▨ **Dimension** If a vector space V has a basis B consisting of n vectors, then it can be proved that *every* basis for that space must contain n vectors. This leads to the next definition.

DEFINITION 7.6.5 Dimension of a Vector Space

The number of vectors in a basis B for a vector space V is said to be the **dimension** of the space.

EXAMPLE 5 Dimensions of Some Vector Spaces

(a) In agreement with our intuition, the dimensions of the vector spaces R, R^2, R^3, and R^n are, in turn, 1, 2, 3, and n.

(b) Since there are $n + 1$ vectors in the standard basis $B = \{1, x, x^2, \ldots, x^n\}$, the dimension of the vector space P_n of polynomials of degree less than or equal to n is $n + 1$.

(c) The zero vector space $\{\mathbf{0}\}$ is given special consideration. This space contains only $\mathbf{0}$ and since $\{\mathbf{0}\}$ is a linearly dependent set, it is not a basis. In this case it is customary to take the empty set as the basis and to *define* the dimension of $\{\mathbf{0}\}$ as zero. ≡

If the basis of a vector space V contains a finite number of vectors, then we say that the vector space is **finite dimensional**; otherwise it is **infinite dimensional**. The function space $C^n(I)$ of n times continuously differentiable functions on an interval I is an example of an infinite-dimensional vector space.

▨ **Linear Differential Equations** Consider the homogeneous linear nth-order differential equation

$$a_n(x) \frac{d^n y}{dx^n} + a_{n-1}(x) \frac{d^{n-1} y}{dx^{n-1}} + \cdots + a_1(x) \frac{dy}{dx} + a_0(x)y = 0 \tag{6}$$

on an interval I on which the coefficients are continuous and $a_n(x) \neq 0$ for every x in the interval. A solution y_1 of (6) is necessarily a vector in the vector space $C^n(I)$. In addition, we know from the theory examined in Section 3.1 that if y_1 and y_2 are solutions of (6), then the sum $y_1 + y_2$ and any constant multiple ky_1 are also solutions. Since the solution set is closed under addition and scalar multiplication, it follows from Theorem 7.6.1 that the solution set of (6) is a subspace of $C^n(I)$. Hence the solution set of (6) deserves to be called the **solution space** of the differential equation. We also know that if $\{y_1, y_2, \ldots, y_n\}$ is a linearly independent set of solutions of (6), then its general solution of the differential equation is the linear combination

$$y = c_1 y_1(x) + c_2 y_2(x) + \cdots + c_n y_n(x).$$

Recall that any solution of the equation can be found from this general solution by specialization of the constants $c_1, c_2, \ldots, c_n$. Therefore, the linearly independent set of solutions $\{y_1, y_2, \ldots, y_n\}$ is a basis for the solution space. The dimension of this solution space is n.

EXAMPLE 6 Dimension of a Solution Space

The general solution of the homogeneous linear second-order differential equation $y'' + 25y = 0$ is $y = c_1 \cos 5x + c_2 \sin 5x$. A basis for the solution space consists of the linearly independent vectors $\{\cos 5x, \sin 5x\}$. The solution space is two-dimensional. ≡

The set of solutions of a nonhomogeneous linear differential equation is not a vector space. Several axioms of a vector space are not satisfied; most notably the set of solutions does not contain a zero vector. In other words, $y = 0$ is not a solution of a nonhomogeneous linear differential equation.

■ **Span** If S denotes any set of vectors $\{\mathbf{x}_1, \mathbf{x}_2, \ldots, \mathbf{x}_n\}$ in a vector space V, then the set of *all* linear combinations of the vectors $\mathbf{x}_1, \mathbf{x}_2, \ldots, \mathbf{x}_n$ in S,

$$\{k_1\mathbf{x}_1 + k_2\mathbf{x}_2 + \cdots + k_n\mathbf{x}_n\},$$

where the k_i, $i = 1, 2, \ldots, n$ are scalars, is called the **span** of the vectors and written Span(S) or Span($\mathbf{x}_1, \mathbf{x}_2, \ldots, \mathbf{x}_n$). It is left as an exercise to show that Span(S) is a subspace of the vector space V. See Problem 33 in Exercises 7.6. Span(S) is said to be a subspace *spanned* by the vectors $\mathbf{x}_1, \mathbf{x}_2, \ldots, \mathbf{x}_n$. If $V = $ Span(S), then we say that S is a **spanning set** for the vector space V, or that S **spans** V. For example, each of the three sets

$$\{\mathbf{i}, \mathbf{j}, \mathbf{k}\}, \quad \{\mathbf{i}, \mathbf{i} + \mathbf{j}, \mathbf{i} + \mathbf{j} + \mathbf{k}\}, \quad \text{and} \quad \{\mathbf{i}, \mathbf{j}, \mathbf{k}, \mathbf{i} + \mathbf{j}, \mathbf{i} + \mathbf{j} + \mathbf{k}\}$$

are spanning sets for the vector space R^3. But note that the first two sets are linearly independent, whereas the third set is dependent. With these new concepts we can rephrase Definitions 7.6.4 and 7.6.5 in the following manner:

*A set S of vectors $\{\mathbf{x}_1, \mathbf{x}_2, \ldots, \mathbf{x}_n\}$ in a vector space V is a **basis** for V if S is linearly independent and is a spanning set for V. The number of vectors in this spanning set S is the **dimension** of the space V.*

REMARKS

(*i*) Suppose V is an arbitrary real vector space. If there is an inner product defined on V it need not look at all like the standard or Euclidean inner product defined on R^n. In Chapter 12 we will work with an inner product that is a definite integral. We shall denote an inner product that is *not* the Euclidean inner product by the symbol $(\mathbf{u}, \mathbf{v})$. See Problems 30, 31, and 38(b) in Exercises 7.6.

(*ii*) A vector space V on which an inner product has been defined is called an **inner product space**. A vector space V can have more than one inner product defined on it. For example, a non-Euclidean inner product defined on R^2 is $(\mathbf{u}, \mathbf{v}) = u_1v_1 + 4u_2v_2$, where $\mathbf{u} = \langle u_1, u_2 \rangle$ and $\mathbf{v} = \langle v_1, v_2 \rangle$. See Problems 37 and 38(a) in Exercises 7.6.

(*iii*) A lot of our work in the later chapters in this text takes place in an **infinite-dimensional vector space**. As such, we need to extend the definition of linear independence of a finite set of vectors $S = \{\mathbf{x}_1, \mathbf{x}_2, \ldots, \mathbf{x}_n\}$ given in Definition 7.6.3 to an infinite set:

*An infinite set of vectors $S = \{\mathbf{x}_1, \mathbf{x}_2, \ldots\}$ is said to be **linearly independent** if every finite subset of the set S is linearly independent. If the set S is not linearly independent, then it is **linearly dependent**.*

We note that if S contains a linearly dependent subset, then the entire set S is linearly dependent. The vector space P of all polynomials has the standard basis $B = \{1, x, x^2, \ldots\}$. The infinite set B is linearly independent. P is another example of an infinite-dimensional vector space.

7.6 Exercises Answers to selected odd-numbered problems begin on page ANS-17.

In Problems 1–10, determine whether the given set is a vector space. If not, give at least one axiom that is not satisfied. Unless stated to the contrary, assume that vector addition and scalar multiplication are the ordinary operations defined on that set.

1. The set of vectors $\langle a_1, a_2 \rangle$, where $a_1 \geq 0$, $a_2 \geq 0$

2. The set of vectors $\langle a_1, a_2 \rangle$, where $a_2 = 3a_1 + 1$

3. The set of vectors $\langle a_1, a_2 \rangle$, scalar multiplication defined by $k\langle a_1, a_2 \rangle = \langle ka_1, 0 \rangle$

4. The set of vectors $\langle a_1, a_2 \rangle$, where $a_1 + a_2 = 0$

5. The set of vectors $\langle a_1, a_2, 0 \rangle$

6. The set of vectors $\langle a_1, a_2 \rangle$, addition and scalar multiplication defined by

$$\langle a_1, a_2 \rangle + \langle b_1, b_2 \rangle = \langle a_1 + b_1 + 1, a_2 + b_2 + 1 \rangle$$
$$k\langle a_1, a_2 \rangle = \langle ka_1 + k - 1, ka_2 + k - 1 \rangle$$

7. The set of real numbers, addition defined by $\mathbf{x} + \mathbf{y} = x - y$

8. The set of complex numbers $a + bi$, where $i^2 = -1$, addition and scalar multiplication defined by

$$(a_1 + b_1i) + (a_2 + b_2i) = (a_1 + a_2) + (b_1 + b_2)i$$

$$k(a + bi) = ka + kbi, \quad k \text{ a real number}$$

9. The set of arrays of real numbers $\begin{pmatrix} a_{11} & a_{12} \\ a_{21} & a_{22} \end{pmatrix}$, addition and scalar multiplication defined by

$$\begin{pmatrix} a_{11} & a_{12} \\ a_{21} & a_{22} \end{pmatrix} + \begin{pmatrix} b_{11} & b_{12} \\ b_{21} & b_{22} \end{pmatrix} = \begin{pmatrix} a_{12} + b_{12} & a_{11} + b_{11} \\ a_{22} + b_{22} & a_{21} + b_{21} \end{pmatrix}$$

$$k\begin{pmatrix} a_{11} & a_{12} \\ a_{21} & a_{22} \end{pmatrix} = \begin{pmatrix} ka_{11} & ka_{12} \\ ka_{21} & ka_{22} \end{pmatrix}$$

10. The set of all polynomials of degree 2

In Problems 11–16, determine whether the given set is a subspace of the vector space $C(-\infty, \infty)$.

11. All functions f such that $f(1) = 0$
12. All functions f such that $f(0) = 1$
13. All nonnegative functions f
14. All functions f such that $f(-x) = f(x)$
15. All differentiable functions f
16. All functions f of the form $f(x) = c_1e^x + c_2xe^x$

In Problems 17–20, determine whether the given set is a subspace of the indicated vector space.

17. Polynomials of the form $p(x) = c_3x^3 + c_1x$; P_3
18. Polynomials p that are divisible by $x - 2$; P_2
19. All unit vectors; R^3
20. Functions f such that $\int_a^b f(x)\, dx = 0$; $C[a, b]$
21. In 3-space, a line through the origin can be written as $S = \{(x, y, z) \,|\, x = at, y = bt, z = ct, a, b, c \text{ real numbers}\}$. With addition and scalar multiplication the same as for vectors $\langle x, y, z \rangle$, show that S is a subspace of R^3.
22. In 3-space, a plane through the origin can be written as $S = \{(x, y, z) \,|\, ax + by + cz = 0, a, b, c \text{ real numbers}\}$. Show that S is a subspace of R^3.
23. The vectors $\mathbf{u}_1 = \langle 1, 0, 0 \rangle$, $\mathbf{u}_2 = \langle 1, 1, 0 \rangle$, and $\mathbf{u}_3 = \langle 1, 1, 1 \rangle$ form a basis for the vector space R^3.
 (a) Show that $\mathbf{u}_1$, $\mathbf{u}_2$, and $\mathbf{u}_3$ are linearly independent.
 (b) Express the vector $\mathbf{a} = \langle 3, -4, 8 \rangle$ as a linear combination of $\mathbf{u}_1$, $\mathbf{u}_2$, and $\mathbf{u}_3$.
24. The vectors $p_1(x) = x + 1$, $p_2(x) = x - 1$ form a basis for the vector space P_1.
 (a) Show that $p_1(x)$ and $p_2(x)$ are linearly independent.
 (b) Express the vector $p(x) = 5x + 2$ as a linear combination of $p_1(x)$ and $p_2(x)$.

In Problems 25–28, determine whether the given vectors are linearly independent or linearly dependent.

25. $\langle 4, -8 \rangle$, $\langle -6, 12 \rangle$ in R^2
26. $\langle 1, 1 \rangle$, $\langle 0, 1 \rangle$, $\langle 2, 5 \rangle$ in R^2
27. 1, $(x + 1)$, $(x + 1)^2$ in P_2

28. 1, $(x + 1)$, $(x + 1)^2$, x^2 in P_2

29. Explain why $f(x) = \dfrac{x}{x^2 + 4x + 3}$ is a vector in $C[0, 3]$ but not a vector in $C[-3, 0]$.

30. A vector space V on which a dot or inner product has been defined is called an **inner product space**. An inner product for the vector space $C[a, b]$ is given by

$$(f, g) = \int_a^b f(x)g(x)\, dx.$$

In $C[0, 2\pi]$ compute $(x, \sin x)$.

31. The **norm** of a vector in an inner product space is defined in terms of the inner product. For the inner product given in Problem 30, the norm of a vector is given by $\|f\| = \sqrt{(f, f)}$. In $C[0, 2\pi]$ compute $\|x\|$ and $\|\sin x\|$.

32. Find a basis for the solution space of

$$\frac{d^4y}{dx^4} - 2\frac{d^3y}{dx^3} + 10\frac{d^2y}{dx^2} = 0.$$

33. Let $\{\mathbf{x}_1, \mathbf{x}_2, \dots, \mathbf{x}_n\}$ be any set of vectors in a vector space V. Show that $\text{Span}(\mathbf{x}_1, \mathbf{x}_2, \dots, \mathbf{x}_n)$ is a subspace of V.

Discussion Problems

34. Discuss: Is R^2 a subspace of R^3? Are R^2 and R^3 subspaces of R^4?

35. In Problem 9, you should have proved that the set M_{22} of 2×2 arrays of real numbers

$$M_{22} = \left\{ \begin{pmatrix} a_{11} & a_{12} \\ a_{21} & a_{22} \end{pmatrix} \right\},$$

or **matrices**, is a vector space with vector addition and scalar multiplication defined in that problem. Find a basis for M_{22}. What is the dimension of M_{22}?

36. Consider a finite orthogonal set of nonzero vectors $\{\mathbf{v}_1, \mathbf{v}_2, \dots, \mathbf{v}_k\}$ in R^n. Discuss: Is this set linearly independent or linearly dependent?

37. If $\mathbf{u}$, $\mathbf{v}$, and $\mathbf{w}$ are vectors in a vector space V, then the axioms of an inner product $(\mathbf{u}, \mathbf{v})$ are

 (i) $(\mathbf{u}, \mathbf{v}) = (\mathbf{v}, \mathbf{u})$
 (ii) $(k\mathbf{u}, \mathbf{v}) = k(\mathbf{u}, \mathbf{v})$, k a scalar
 (iii) $(\mathbf{u}, \mathbf{u}) = 0$ if $\mathbf{u} = \mathbf{0}$ and $(\mathbf{u}, \mathbf{u}) > 0$ if $\mathbf{u} \neq \mathbf{0}$
 (iv) $(\mathbf{u}, \mathbf{v} + \mathbf{w}) = (\mathbf{u}, \mathbf{v}) + (\mathbf{u}, \mathbf{w})$.

Show that $(\mathbf{u}, \mathbf{v}) = u_1v_1 + 4u_2v_2$, where $\mathbf{u} = \langle u_1, u_2 \rangle$ and $\mathbf{v} = \langle v_1, v_2 \rangle$, is an inner product on R^2.

38. **(a)** Find a pair of nonzero vectors $\mathbf{u}$ and $\mathbf{v}$ in R^2 that are not orthogonal with respect to the standard or Euclidean inner product $\mathbf{u} \cdot \mathbf{v}$, but are orthogonal with respect to the inner product $(\mathbf{u}, \mathbf{v})$ in Problem 37.
 (b) Find a pair of nonzero functions f and g in $C[0, 2\pi]$ that are orthogonal with respect to the inner product (f, g) given in Problem 30.

7.7 Gram–Schmidt Orthogonalization Process

INTRODUCTION In Section 7.6 we saw that a vector space V can have many different bases. Recall, the defining characteristics of any basis $B = \{\mathbf{x}_1, \mathbf{x}_2, \ldots, \mathbf{x}_n\}$ of a vector space V is that

- the set B is linearly independent, and
- the set B spans the space.

In this context the word *span* means that every vector in the space can be expressed as a linear combination of the vectors $\mathbf{x}_1, \mathbf{x}_2, \ldots, \mathbf{x}_n$. For example, every vector $\mathbf{u}$ in R^n can be written as a linear combination of the vectors in the standard basis $B = \{\mathbf{e}_1, \mathbf{e}_2, \ldots, \mathbf{e}_n\}$, where

$$\mathbf{e}_1 = \langle 1, 0, 0, \ldots, 0 \rangle, \quad \mathbf{e}_2 = \langle 0, 1, 0, \ldots, 0 \rangle, \quad \ldots, \quad \mathbf{e}_n = \langle 0, 0, 0, \ldots, 1 \rangle.$$

This standard basis $B = \{\mathbf{e}_1, \mathbf{e}_2, \ldots, \mathbf{e}_n\}$ is also an example of an **orthonormal basis**; that is, the $\mathbf{e}_i$, $i = 1, 2, \ldots, n$ are mutually orthogonal and are unit vectors; that is,

$$\mathbf{e}_i \cdot \mathbf{e}_j = 0, i \neq j \quad \text{and} \quad \|\mathbf{e}_i\| = 1, i = 1, 2, \ldots, n.$$

In this section we focus on orthonormal bases for R^n and examine a procedure whereby we can transform or convert any basis B of R^n into an orthonormal basis.

EXAMPLE 1 Orthonormal Basis for R^3

The set of three vectors

$$\mathbf{w}_1 = \left\langle \frac{1}{\sqrt{3}}, \frac{1}{\sqrt{3}}, \frac{1}{\sqrt{3}} \right\rangle, \mathbf{w}_2 = \left\langle -\frac{2}{\sqrt{6}}, \frac{1}{\sqrt{6}}, \frac{1}{\sqrt{6}} \right\rangle, \mathbf{w}_3 = \left\langle 0, \frac{1}{\sqrt{2}}, -\frac{1}{\sqrt{2}} \right\rangle \tag{1}$$

is linearly independent and spans the space R^3. Hence $B = \{\mathbf{w}_1, \mathbf{w}_2, \mathbf{w}_3\}$ is a basis for R^3. Using the standard inner product or dot product defined on R^3, observe

$$\mathbf{w}_1 \cdot \mathbf{w}_2 = 0, \mathbf{w}_1 \cdot \mathbf{w}_3 = 0, \mathbf{w}_2 \cdot \mathbf{w}_3 = 0, \quad \text{and} \quad \|\mathbf{w}_1\| = 1, \|\mathbf{w}_2\| = 1, \|\mathbf{w}_3\| = 1.$$

Hence B is an orthonormal basis. ≡

A basis B for R^n need not be orthogonal nor do the basis vectors need to be unit vectors. In fact, *any* linearly independent set of n vectors can serve as a basis for the n-dimensional vector space R^n. For example, it is a straightforward task to show that the vectors

$$\mathbf{u}_1 = \langle 1, 0, 0 \rangle, \quad \mathbf{u}_2 = \langle 1, 1, 0 \rangle, \quad \mathbf{u}_3 = \langle 1, 1, 1 \rangle$$

in R^3 are linearly independent, and hence $B = \{\mathbf{u}_1, \mathbf{u}_2, \mathbf{u}_3\}$ is a basis for R^3. Note that B is not an orthogonal basis.

Generally, an orthonormal basis for a vector space V turns out to be the most convenient basis for V. One of the advantages that an orthonormal basis has over any other basis for R^n is the comparative ease with which we can obtain the coordinates of a vector $\mathbf{u}$ relative to that basis.

THEOREM 7.7.1 Coordinates Relative to an Orthonormal Basis

Suppose $B = \{\mathbf{w}_1, \mathbf{w}_2, \ldots, \mathbf{w}_n\}$ is an orthonormal basis for R^n. If $\mathbf{u}$ is any vector in R^n, then

$$\mathbf{u} = (\mathbf{u} \cdot \mathbf{w}_1)\mathbf{w}_1 + (\mathbf{u} \cdot \mathbf{w}_2)\mathbf{w}_2 + \cdots + (\mathbf{u} \cdot \mathbf{w}_n)\mathbf{w}_n.$$

PROOF: The vector $\mathbf{u}$ is in R^n and so it is an element of the set Span(B). In other words, there exist real scalars k_i, $i = 1, 2, \ldots, n$ such that $\mathbf{u}$ can be expressed as the linear combination

$$\mathbf{u} = k_1\mathbf{w}_1 + k_2\mathbf{w}_2 + \cdots + k_n\mathbf{w}_n.$$

The scalars k_i are the coordinates of $\mathbf{u}$ relative to the basis B. These coordinates can be found by taking the dot product of $\mathbf{u}$ with each of the basis vectors:

$$\mathbf{u} \cdot \mathbf{w}_i = (k_1\mathbf{w}_1 + k_2\mathbf{w}_2 + \cdots + k_n\mathbf{w}_n) \cdot \mathbf{w}_i = k_1(\mathbf{w}_1 \cdot \mathbf{w}_i) + k_2(\mathbf{w}_2 \cdot \mathbf{w}_i) + \cdots + k_n(\mathbf{w}_n \cdot \mathbf{w}_i). \quad (2)$$

Since B is orthonormal, $\mathbf{w}_i$ is orthogonal to all vectors in B with the exception of $\mathbf{w}_i$ itself. That is, $\mathbf{w}_i \cdot \mathbf{w}_j = 0$, $i \neq j$ and $\mathbf{w}_i \cdot \mathbf{w}_i = \|\mathbf{w}_i\|^2 = 1$. Hence from (2), we obtain $k_i = (\mathbf{u} \cdot \mathbf{w}_i)$ for $i = 1, 2, \ldots, n$. $\equiv$

EXAMPLE 2 **Coordinates of a Vector in R^3**

Find the coordinates of the vector $\mathbf{u} = \langle 3, -2, 9 \rangle$ relative to the orthonormal basis B for R^3 given in (1) of Example 1. Write $\mathbf{u}$ in terms of the basis B.

SOLUTION From Theorem 7.7.1, the coordinates of $\mathbf{u}$ relative to the basis B in (1) of Example 1 are simply

$$\mathbf{u} \cdot \mathbf{w}_1 = \frac{10}{\sqrt{3}}, \quad \mathbf{u} \cdot \mathbf{w}_2 = \frac{1}{\sqrt{6}}, \quad \text{and} \quad \mathbf{u} \cdot \mathbf{w}_3 = -\frac{11}{\sqrt{2}}.$$

Hence we can write

$$\mathbf{u} = \frac{10}{\sqrt{3}}\mathbf{w}_1 + \frac{1}{\sqrt{6}}\mathbf{w}_2 - \frac{11}{\sqrt{2}}\mathbf{w}_3. \quad \equiv$$

■ **Gram–Schmidt Orthogonalization Process** The procedure known as the **Gram–Schmidt orthogonalization process** is a straightforward algorithm for generating an orthogonal basis $B' = \{\mathbf{v}_1, \mathbf{v}_2, \ldots, \mathbf{v}_n\}$ from any given basis $B = \{\mathbf{u}_1, \mathbf{u}_2, \ldots, \mathbf{u}_n\}$ for R^n. We then produce an orthonormal basis $B'' = \{\mathbf{w}_1, \mathbf{w}_2, \ldots, \mathbf{w}_n\}$ by normalizing the vectors in the orthogonal basis B'. The key idea in the orthogonalization process is vector projection, and so we suggest that you review that concept in Section 7.3. Also, for the sake of gaining some geometric insight into the process, we shall begin in R^2 and R^3.

■ **Constructing an Orthogonal Basis for R^2** The Gram–Schmidt orthogonalization process for R^n is a sequence of steps; at each step we construct a vector $\mathbf{v}_i$ that is orthogonal to the vector in the preceding step. The transformation of a basis $B = \{\mathbf{u}_1, \mathbf{u}_2\}$ for R^2 into an orthogonal basis $B' = \{\mathbf{v}_1, \mathbf{v}_2\}$ consists of two steps. See **FIGURE 7.7.1(a)**. The first step is simple, we merely choose one of the vectors in B, say, $\mathbf{u}_1$, and rename it $\mathbf{v}_1$. Next, as shown in Figure 7.7.1(b), we project the remaining vector $\mathbf{u}_2$ in B onto the vector $\mathbf{v}_1$ and define a second vector to be $\mathbf{v}_2 = \mathbf{u}_2 - \text{proj}_{\mathbf{v}_1}\mathbf{u}_2$. Recall from (12) of Section 7.3 that $\text{proj}_{\mathbf{v}_1}\mathbf{u}_2 = \left(\dfrac{\mathbf{u}_2 \cdot \mathbf{v}_1}{\mathbf{v}_1 \cdot \mathbf{v}_1}\right)\mathbf{v}_1$. As seen in Figure 7.7.1(c), the vectors

$$\mathbf{v}_1 = \mathbf{u}_1$$

$$\mathbf{v}_2 = \mathbf{u}_2 - \left(\frac{\mathbf{u}_2 \cdot \mathbf{v}_1}{\mathbf{v}_1 \cdot \mathbf{v}_1}\right)\mathbf{v}_1 \quad (3)$$

are orthogonal. If you are not convinced of this, we suggest you verify the orthogonality of $\mathbf{v}_1$ and $\mathbf{v}_2$ by demonstrating that $\mathbf{v}_1 \cdot \mathbf{v}_2 = 0$.

(a) Linearly independent vectors $\mathbf{u}_1$ and $\mathbf{u}_2$

(b) Projection of $\mathbf{u}_2$ onto $\mathbf{v}_1$

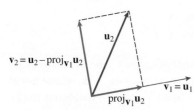

(c) $\mathbf{v}_1$ and $\mathbf{v}_2$ are orthogonal

FIGURE 7.7.1 The orthogonal vectors $\mathbf{v}_1$ and $\mathbf{v}_2$ are defined in terms of $\mathbf{u}_1$ and $\mathbf{u}_2$

EXAMPLE 3 **Gram–Schmidt Process in R^2**

The set $B = \{\mathbf{u}_1, \mathbf{u}_2\}$, where $\mathbf{u}_1 = \langle 3, 1 \rangle$, $\mathbf{u}_2 = \langle 1, 1 \rangle$, is a basis for R^2. Transform B into an orthonormal basis $B'' = \{\mathbf{w}_1, \mathbf{w}_2\}$.

SOLUTION We choose $\mathbf{v}_1$ as $\mathbf{u}_1$: $\mathbf{v}_1 = \langle 3, 1 \rangle$. Then from the second equation in (3), with $\mathbf{u}_2 \cdot \mathbf{v}_1 = 4$ and $\mathbf{v}_1 \cdot \mathbf{v}_1 = 10$, we obtain

$$\mathbf{v}_2 = \langle 1, 1 \rangle - \frac{4}{10}\langle 3, 1 \rangle = \left\langle -\frac{1}{5}, \frac{3}{5} \right\rangle.$$

(a) Basis B

(b) Basis B''

FIGURE 7.7.2 The two bases in Example 3

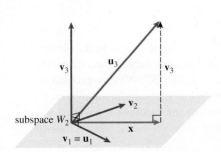

FIGURE 7.7.3 The vectors $\mathbf{v}_1$, $\mathbf{v}_2$, $\mathbf{v}_3$ (in blue) obtained by the Gram–Schmidt process

The set $B' = \{\mathbf{v}_1, \mathbf{v}_2\} = \{\langle 3, 1 \rangle, \langle -\frac{1}{5}, \frac{3}{5} \rangle\}$ is an orthogonal basis for R^2. We finish by normalizing the vectors $\mathbf{v}_1$ and $\mathbf{v}_2$:

$$\mathbf{w}_1 = \frac{1}{\|\mathbf{v}_1\|}\mathbf{v}_1 = \left\langle \frac{3}{\sqrt{10}}, \frac{1}{\sqrt{10}} \right\rangle \quad \text{and} \quad \mathbf{w}_2 = \frac{1}{\|\mathbf{v}_2\|}\mathbf{v}_2 = \left\langle -\frac{1}{\sqrt{10}}, \frac{3}{\sqrt{10}} \right\rangle.$$

The basis B is shown in **FIGURE 7.7.2(a)**, and the new orthonormal basis $B'' = \{\mathbf{w}_1, \mathbf{w}_2\}$ is shown in blue in Figure 7.7.2(b). ≡

In Example 3 we are free to choose either vector in $B = \{\mathbf{u}_1, \mathbf{u}_2\}$ as the vector $\mathbf{v}_1$. However, by choosing $\mathbf{v}_1 = \mathbf{u}_2 = \langle 1, 1 \rangle$, we obtain a different orthonormal basis, namely, $B'' = \{\mathbf{w}_1, \mathbf{w}_2\}$, where $\mathbf{w}_1 = \langle 1/\sqrt{2}, 1/\sqrt{2} \rangle$ and $\mathbf{w}_2 = \langle 1/\sqrt{2}, -1/\sqrt{2} \rangle$. See Problems 5–8 in Exercises 7.7.

Constructing an Orthogonal Basis for R^3

Now suppose $B = \{\mathbf{u}_1, \mathbf{u}_2, \mathbf{u}_3\}$ is a basis for R^3. Then the set $B' = \{\mathbf{v}_1, \mathbf{v}_2, \mathbf{v}_3\}$, where

$$\mathbf{v}_1 = \mathbf{u}_1$$

$$\mathbf{v}_2 = \mathbf{u}_2 - \left(\frac{\mathbf{u}_2 \cdot \mathbf{v}_1}{\mathbf{v}_1 \cdot \mathbf{v}_1} \right)\mathbf{v}_1 \tag{4}$$

$$\mathbf{v}_3 = \mathbf{u}_3 - \left(\frac{\mathbf{u}_3 \cdot \mathbf{v}_1}{\mathbf{v}_1 \cdot \mathbf{v}_1} \right)\mathbf{v}_1 - \left(\frac{\mathbf{u}_3 \cdot \mathbf{v}_2}{\mathbf{v}_2 \cdot \mathbf{v}_2} \right)\mathbf{v}_2$$

is an orthogonal basis for R^3. Again, if you do not see this, then compute $\mathbf{v}_1 \cdot \mathbf{v}_2$, $\mathbf{v}_1 \cdot \mathbf{v}_3$, and $\mathbf{v}_2 \cdot \mathbf{v}_3$.

Since the vectors $\mathbf{v}_1$ and $\mathbf{v}_2$ in the list (4) are by construction orthogonal, the set $\{\mathbf{v}_1, \mathbf{v}_2\}$ must be linearly independent. See Problem 36 in Exercises 7.6. Thus, $W_2 = \text{Span}(\mathbf{v}_1, \mathbf{v}_2)$ is necessarily a two-dimensional subspace of R^3. Now the vector $\mathbf{x} = \left(\frac{\mathbf{u}_3 \cdot \mathbf{v}_1}{\mathbf{v}_1 \cdot \mathbf{v}_1} \right)\mathbf{v}_1 + \left(\frac{\mathbf{u}_3 \cdot \mathbf{v}_2}{\mathbf{v}_2 \cdot \mathbf{v}_2} \right)\mathbf{v}_2$ is a vector in W_2 because it is a linear combination of $\mathbf{v}_1$ and $\mathbf{v}_2$. The vector $\mathbf{x}$ is called the **orthogonal projection** of $\mathbf{u}_3$ onto the subspace W_2 and is usually denoted by $\mathbf{x} = \text{proj}_{W_2}\mathbf{u}_3$. In **FIGURE 7.7.3**, $\mathbf{x}$ is the red vector. Notice, too, that $\mathbf{x}$ is the sum of two projections. Using (12) of Section 7.3, we can write

$$\mathbf{x} = \text{proj}_{W_2}\mathbf{u}_3 = \overbrace{\left(\frac{\mathbf{u}_3 \cdot \mathbf{v}_1}{\mathbf{v}_1 \cdot \mathbf{v}_1} \right)\mathbf{v}_1}^{\text{proj}_{\mathbf{v}_1}\mathbf{u}_3} + \overbrace{\left(\frac{\mathbf{u}_3 \cdot \mathbf{v}_2}{\mathbf{v}_2 \cdot \mathbf{v}_2} \right)\mathbf{v}_2}^{\text{proj}_{\mathbf{v}_2}\mathbf{u}_3}. \tag{5}$$

The difference $\mathbf{v}_3 = \mathbf{u}_3 - \mathbf{x}$ is orthogonal to $\mathbf{x}$. Indeed, $\mathbf{v}_3$ is orthogonal to $\mathbf{v}_1$ and $\mathbf{v}_2$ and to every vector in W_2. This is precisely the same idea in (3). In that context, $\mathbf{v}_2 = \mathbf{u}_2 - \mathbf{x}$, where $\mathbf{x}$ was the projection of $\mathbf{u}_2$ onto the one-dimensional subspace $W_1 = \text{Span}(\mathbf{v}_1)$ of R^2. Analogous to (5), we have

$$\mathbf{x} = \text{proj}_{W_1}\mathbf{u}_2 = \overbrace{\left(\frac{\mathbf{u}_2 \cdot \mathbf{v}_1}{\mathbf{v}_1 \cdot \mathbf{v}_1} \right)\mathbf{v}_1}^{\text{proj}_{\mathbf{v}_1}\mathbf{u}_2}. \tag{6}$$

■ **EXAMPLE 4** **Gram–Schmidt Process in R^3**

The set $B = \{\mathbf{u}_1, \mathbf{u}_2, \mathbf{u}_3\}$, where

$$\mathbf{u}_1 = \langle 1, 1, 1 \rangle, \, \mathbf{u}_2 = \langle 1, 2, 2 \rangle, \, \mathbf{u}_3 = \langle 1, 1, 0 \rangle$$

is a basis for R^3. Transform B into an orthonormal basis B''.

SOLUTION We choose $\mathbf{v}_1$ as $\mathbf{u}_1$: $\mathbf{v}_1 = \langle 1, 1, 1 \rangle$. Then from the second equation in (4), with $\mathbf{u}_2 \cdot \mathbf{v}_1 = 5$ and $\mathbf{v}_1 \cdot \mathbf{v}_1 = 3$, we obtain

$$\mathbf{v}_2 = \langle 1, 2, 2 \rangle - \frac{5}{3} \langle 1, 1, 1 \rangle = \left\langle -\frac{2}{3}, \frac{1}{3}, \frac{1}{3} \right\rangle.$$

Now with $\mathbf{u}_3 \cdot \mathbf{v}_1 = 2$, $\mathbf{v}_1 \cdot \mathbf{v}_1 = 3$, $\mathbf{u}_3 \cdot \mathbf{v}_2 = -\frac{1}{3}$, and $\mathbf{v}_2 \cdot \mathbf{v}_2 = \frac{2}{3}$, the third equation in (4) yields

$$\mathbf{v}_3 = \langle 1, 1, 0 \rangle - \frac{2}{3} \langle 1, 1, 1 \rangle + \frac{1}{2} \left\langle -\frac{2}{3}, \frac{1}{3}, \frac{1}{3} \right\rangle$$

$$= \langle 1, 1, 0 \rangle + \left\langle -\frac{2}{3}, -\frac{2}{3}, -\frac{2}{3} \right\rangle + \left\langle -\frac{1}{3}, \frac{1}{6}, \frac{1}{6} \right\rangle$$

$$= \left\langle 0, \frac{1}{2}, -\frac{1}{2} \right\rangle.$$

The set $B' = \{\mathbf{v}_1, \mathbf{v}_2, \mathbf{v}_3\} = \{\langle 1, 1, 1 \rangle, \langle -\frac{2}{3}, \frac{1}{3}, \frac{1}{3} \rangle, \langle 0, \frac{1}{2}, -\frac{1}{2} \rangle\}$ is an orthogonal basis for R^3. As in Example 3, we finish the job by normalizing each vector in B'. Using $\|\mathbf{v}_1\| = \sqrt{3}$, $\|\mathbf{v}_2\| = \sqrt{6}/3$, $\|\mathbf{v}_3\| = \sqrt{2}/2$, and $\mathbf{w}_i = \frac{1}{\|\mathbf{v}_i\|} \mathbf{v}_i$, $i = 1, 2, 3$, we find that an orthonormal basis for R^3 is $B'' = \{\mathbf{w}_1, \mathbf{w}_2, \mathbf{w}_3\}$, where

$$\mathbf{w}_1 = \left\langle \frac{1}{\sqrt{3}}, \frac{1}{\sqrt{3}}, \frac{1}{\sqrt{3}} \right\rangle, \quad \mathbf{w}_2 = \left\langle -\frac{2}{\sqrt{6}}, \frac{1}{\sqrt{6}}, \frac{1}{\sqrt{6}} \right\rangle, \quad \mathbf{w}_3 = \left\langle 0, \frac{1}{\sqrt{2}}, -\frac{1}{\sqrt{2}} \right\rangle.$$

The set B'' is recognized as the orthonormal basis for R^3 examined in Example 1. $\equiv$

We conclude this section with a theorem that summarizes the most general case of the Gram–Schmidt process for R^n. The orthogonalization process can be used on any linearly independent set S, and so we can use it to find orthonormal bases for subspaces of R^n.

THEOREM 7.7.2 Gram–Schmidt Orthogonalization Process

Let $B = \{\mathbf{u}_1, \mathbf{u}_2, \ldots, \mathbf{u}_m\}$, $m \leq n$, be a basis for a subspace W_m of R^n. Then $B' = \{\mathbf{v}_1, \mathbf{v}_2, \ldots, \mathbf{v}_m\}$, where

$$\mathbf{v}_1 = \mathbf{u}_1$$

$$\mathbf{v}_2 = \mathbf{u}_2 - \left(\frac{\mathbf{u}_2 \cdot \mathbf{v}_1}{\mathbf{v}_1 \cdot \mathbf{v}_1} \right) \mathbf{v}_1$$

$$\mathbf{v}_3 = \mathbf{u}_3 - \left(\frac{\mathbf{u}_3 \cdot \mathbf{v}_1}{\mathbf{v}_1 \cdot \mathbf{v}_1} \right) \mathbf{v}_1 - \left(\frac{\mathbf{u}_3 \cdot \mathbf{v}_2}{\mathbf{v}_2 \cdot \mathbf{v}_2} \right) \mathbf{v}_2 \tag{7}$$

$$\vdots$$

$$\mathbf{v}_m = \mathbf{u}_m - \left(\frac{\mathbf{u}_m \cdot \mathbf{v}_1}{\mathbf{v}_1 \cdot \mathbf{v}_1} \right) \mathbf{v}_1 - \left(\frac{\mathbf{u}_m \cdot \mathbf{v}_2}{\mathbf{v}_2 \cdot \mathbf{v}_2} \right) \mathbf{v}_2 - \cdots - \left(\frac{\mathbf{u}_m \cdot \mathbf{v}_{m-1}}{\mathbf{v}_{m-1} \cdot \mathbf{v}_{m-1}} \right) \mathbf{v}_{m-1},$$

is an orthogonal basis for W_m. An orthonormal basis for W_m is

$$B'' = \{\mathbf{w}_1, \mathbf{w}_2, \ldots, \mathbf{w}_m\} = \left\{ \frac{1}{\|\mathbf{v}_1\|} \mathbf{v}_1, \frac{1}{\|\mathbf{v}_2\|} \mathbf{v}_2, \ldots, \frac{1}{\|\mathbf{v}_m\|} \mathbf{v}_m \right\}.$$

7.7 Exercises Answers to selected odd-numbered problems begin on page ANS-17.

In Problems 1 and 2, verify that the basis B for the given vector space is orthonormal. Use Theorem 7.7.1 to find the coordinates of the vector $\mathbf{u}$ relative to the basis B. Then write $\mathbf{u}$ as a linear combination of the basis vectors.

1. $B = \left\{ \left\langle \dfrac{12}{13}, \dfrac{5}{13} \right\rangle, \left\langle \dfrac{5}{13}, -\dfrac{12}{13} \right\rangle \right\}, \quad R^2; \ \mathbf{u} = \langle 4, 2 \rangle$

2. $B = \left\{ \left\langle \dfrac{1}{\sqrt{3}}, \dfrac{1}{\sqrt{3}}, -\dfrac{1}{\sqrt{3}} \right\rangle, \left\langle 0, -\dfrac{1}{\sqrt{2}}, -\dfrac{1}{\sqrt{2}} \right\rangle, \right.$
$\left. \left\langle -\dfrac{2}{\sqrt{6}}, \dfrac{1}{\sqrt{6}}, -\dfrac{1}{\sqrt{6}} \right\rangle \right\}, \quad R^3; \ \mathbf{u} = \langle 5, -1, 6 \rangle$

In Problems 3 and 4, verify that the basis B for the given vector space is orthogonal. Use Theorem 7.7.1 as an aid in finding the coordinates of the vector $\mathbf{u}$ relative to the basis B. Then write $\mathbf{u}$ as a linear combination of the basis vectors.

3. $B = \{ \langle 1, 0, 1 \rangle, \langle 0, 1, 0 \rangle, \langle -1, 0, 1 \rangle, \ R^3;$
$\mathbf{u} = \langle 10, 7, -13 \rangle$

4. $B = \{ \langle 2, 1, -2, 0 \rangle, \langle 1, 2, 2, 1 \rangle, \langle 3, -4, 1, 3 \rangle, \langle 5, -2, 4, -9 \rangle \}, \ R^4;$
$\mathbf{u} = \langle 1, 2, 4, 3 \rangle$

In Problems 5–8, use the Gram–Schmidt orthogonalization process (3) to transform the given basis $B = \{\mathbf{u}_1, \mathbf{u}_2\}$ for R^2 into an orthogonal basis $B' = \{\mathbf{v}_1, \mathbf{v}_2\}$. Then form an orthonormal basis $B'' = \{\mathbf{w}_1, \mathbf{w}_2\}$.

 (a) First construct B'' using $\mathbf{v}_1, \mathbf{u}_1$.
 (b) Then construct B'' using $\mathbf{v}_1, \mathbf{u}_2$.
 (c) Sketch B and each basis B''.

5. $B = \{ \langle -3, 2 \rangle, \langle -1, -1 \rangle \}$ **6.** $B = \{ \langle -3, 4 \rangle, \langle -1, 0 \rangle \}$

7. $B = \{ \langle 1, 1 \rangle, \langle 1, 0 \rangle \}$ **8.** $B = \{ \langle 5, 7 \rangle, \langle 1, -2 \rangle \}$

In Problems 9–12, use the Gram–Schmidt orthogonalization process (4) to transform the given basis $B = \{\mathbf{u}_1, \mathbf{u}_2, \mathbf{u}_3\}$ for R^3 into an orthogonal basis $B' = \{\mathbf{v}_1, \mathbf{v}_2, \mathbf{v}_3\}$. Then form an orthonormal basis $B'' = \{\mathbf{w}_1, \mathbf{w}_2, \mathbf{w}_3\}$.

9. $B = \{ \langle 1, 1, 0 \rangle, \langle 1, 2, 2 \rangle, \langle 2, 2, 1 \rangle \}$

10. $B = \{ \langle -3, 1, 1 \rangle, \langle 1, 1, 0 \rangle, \langle -1, 4, 1 \rangle \}$

11. $B = \{ \langle \tfrac{1}{2}, \tfrac{1}{2}, 1 \rangle, \langle -1, 1, -\tfrac{1}{2} \rangle, \langle -1, \tfrac{1}{2}, 1 \rangle \}$

12. $B = \{ \langle 1, 1, 1 \rangle, \langle 9, -1, 1 \rangle, \langle -1, 4, -2 \rangle \}$

In Problems 13 and 14, the given vectors span a subspace W of R^3. Use the Gram–Schmidt orthogonalization process to construct an orthonormal basis for the subspace.

13. $\mathbf{u}_1 = \langle 1, 5, 2 \rangle, \mathbf{u}_2 = \langle -2, 1, 1 \rangle$

14. $\mathbf{u}_1 = \langle 1, 2, 3 \rangle, \mathbf{u}_2 = \langle 3, 4, 1 \rangle$

In Problems 15 and 16, the given vectors span a subspace W of R^4. Use the Gram–Schmidt orthogonalization process to construct an orthonormal basis for the subspace.

15. $\mathbf{u}_1 = \langle 1, -1, 1, -1 \rangle, \mathbf{u}_2 = \langle 1, 3, 0, 1 \rangle$

16. $\mathbf{u}_1 = \langle 4, 0, 2, -1 \rangle, \mathbf{u}_2 = \langle 2, 1, -1, 1 \rangle, \mathbf{u}_3 = \langle 1, 1, -1, 0 \rangle$

In Problems 17 and 18, an inner product defined on the vector space P_2 of all polynomials of degree less than or equal to 2, is given by

$$(p, q) = \int_{-1}^{1} p(x) q(x) \, dx.$$

Use the Gram–Schmidt orthogonalization process to transform the given basis B for P_2 into an orthogonal basis B'.

17. $B = \{1, x, x^2\}$

18. $B = \{x^2 - x, x^2 + 1, 1 - x^2\}$

For the inner product (p, q) defined on P_2 in Problems 17 and 18, the norm $\|p(x)\|$ of a polynomial p is defined by

$$\|p(x)\|^2 = (p, p) = \int_{-1}^{1} p^2(x) \, dx.$$

Use this norm in Problems 19 and 20.

19. Construct an orthonormal basis B'' from B' obtained in Problem 17.

20. Construct an orthonormal basis B'' from B' obtained in Problem 18.

In Problems 21 and 22, let $p(x) = 9x^2 - 6x + 5$ be a vector in P_2. Use Theorem 7.7.1 and the indicated orthonormal basis B'' to find the coordinates $p(x)$ relative to B''. Then write $p(x)$ as a linear combination of the basis vectors.

21. B'' in Problem 19 **22.** B'' in Problem 20

Discussion Problem

23. The set of vectors $\{\mathbf{u}_1, \mathbf{u}_2, \mathbf{u}_3\}$, where

$$\mathbf{u}_1 = \langle 1, 1, 3 \rangle, \mathbf{u}_2 = \langle 1, 4, 1 \rangle, \text{ and } \mathbf{u}_3 = \langle 1, 10, -3 \rangle,$$

is linearly dependent in R^3 since $\mathbf{u}_3 = -2\mathbf{u}_1 + 3\mathbf{u}_2$. Discuss what you would expect when the Gram–Schmidt process in (4) is applied to these vectors. Then carry out the orthogonalization process.

Answer Problems 1–30 without referring back to the text. Fill in the blank or answer true/false.

1. The vectors $\langle -4, -6, 10 \rangle$ and $\langle -10, -15, 25 \rangle$ are parallel. _____

2. In 3-space, any three distinct points determine a plane. _____

3. The line $x = 1 + 5t$, $y = 1 - 2t$, $z = 4 + t$ and the plane $2x + 3y - 4z = 1$ are perpendicular. _____

4. Nonzero vectors **a** and **b** are parallel if $\mathbf{a} \times \mathbf{b} = \mathbf{0}$. _____

5. If the angle between **a** and **b** is obtuse, $\mathbf{a} \cdot \mathbf{b} < 0$. _____

6. If **a** is a unit vector, then $\mathbf{a} \cdot \mathbf{a} = 1$. _____

7. The cross product of two vectors is not commutative. _____

8. The terminal point of the vector $\mathbf{a} - \mathbf{b}$ is at the terminal point of **a**. _____

9. $(\mathbf{a} \times \mathbf{b}) \cdot \mathbf{c} = \mathbf{a} \cdot (\mathbf{b} \times \mathbf{c})$. _____

10. If **a**, **b**, **c**, and **d** are nonzero coplanar vectors, then $(\mathbf{a} \times \mathbf{b}) \times (\mathbf{c} \times \mathbf{d}) = \mathbf{0}$. _____

11. The sum of $3\mathbf{i} + 4\mathbf{j} + 5\mathbf{k}$ and $6\mathbf{i} - 2\mathbf{j} - 3\mathbf{k}$ is _____.

12. If $\mathbf{a} \cdot \mathbf{b} = 0$, the nonzero vectors **a** and **b** are _____.

13. $(-\mathbf{k}) \times (5\mathbf{j}) = $ _____

14. $\mathbf{i} \cdot (\mathbf{i} \times \mathbf{j}) = $ _____

15. $\| -12\mathbf{i} + 4\mathbf{j} + 6\mathbf{k} \| = $ _____

16. $\begin{vmatrix} \mathbf{i} & \mathbf{j} & \mathbf{k} \\ 2 & 1 & 5 \\ 0 & 4 & -1 \end{vmatrix} = $ _____

17. A vector that is normal to the plane $-6x + y - 7z + 10 = 0$ is _____.

18. The plane $x + 3y - z = 5$ contains the point $(1, -2, ___)$.

19. The point of intersection of the line $x - 1 = (y + 2)/3 = (z + 1)/2$ and the plane $x + 2y - z = 13$ is _____.

20. A unit vector that has the opposite direction of $\mathbf{a} = 4\mathbf{i} + 3\mathbf{j} - 5\mathbf{k}$ is _____.

21. If $\overrightarrow{P_1P_2} = \langle 3, 5, -4 \rangle$ and P_1 has coordinates $(2, 1, 7)$, then the coordinates of P_2 are _____.

22. The midpoint of the line segment between $P_1(4, 3, 10)$ and $P_2(6, -2, -5)$ has coordinates _____.

23. If $\|\mathbf{a}\| = 7.2$, $\|\mathbf{b}\| = 10$, and the angle between **a** and **b** is $135°$, then $\mathbf{a} \cdot \mathbf{b} = $ _____.

24. If $\mathbf{a} = \langle 3, 1, 0 \rangle$, $\mathbf{b} = \langle -1, 2, 1 \rangle$, and $\mathbf{c} = \langle 0, -2, 2 \rangle$, then $\mathbf{a} \cdot (2\mathbf{b} + 4\mathbf{c}) = $ _____.

25. The x-, y-, and z-intercepts of the plane $2x - 3y + 4z = 24$ are, respectively, _____.

26. The angle θ between the vectors $\mathbf{a} = \mathbf{i} + \mathbf{j}$ and $\mathbf{b} = \mathbf{i} - \mathbf{k}$ is _____.

27. The area of a triangle with two sides given by $\mathbf{a} = \langle 1, 3, -1 \rangle$ and $\mathbf{b} = \langle 2, -1, 2 \rangle$ is _____.

28. An equation of the plane containing $(3, 6, -2)$ and with normal vector $\mathbf{n} = 3\mathbf{i} + \mathbf{k}$ is _____.

29. The distance from the plane $y = -5$ to the point $(4, -3, 1)$ is _____.

30. The vectors $\langle 1, 3, c \rangle$ and $\langle -2, -6, 5 \rangle$ are parallel for $c = $ _____ and orthogonal for $c = $ _____.

31. Find a unit vector that is perpendicular to both $\mathbf{a} = \mathbf{i} + \mathbf{j}$ and $\mathbf{b} = \mathbf{i} - 2\mathbf{j} + \mathbf{k}$.

32. Find the direction cosines and direction angles of the vector $\mathbf{a} = \frac{1}{2}\mathbf{i} + \frac{1}{2}\mathbf{j} - \frac{1}{4}\mathbf{k}$.

In Problems 33–36, let $\mathbf{a} = \langle 1, 2, -2 \rangle$ and $\mathbf{b} = \langle 4, 3, 0 \rangle$. Find the indicated number or vector.

33. $\text{comp}_{\mathbf{b}}\mathbf{a}$

34. $\text{proj}_{\mathbf{a}}\mathbf{b}$

35. $\text{proj}_{\mathbf{a}}(\mathbf{a} + \mathbf{b})$

36. $\text{proj}_{\mathbf{b}}(\mathbf{a} - \mathbf{b})$

37. Let **r** be the position vector of a variable point $P(x, y, z)$ in space and let **a** be a constant vector. Determine the surface described by **(a)** $(\mathbf{r} - \mathbf{a}) \cdot \mathbf{r} = 0$ and **(b)** $(\mathbf{r} - \mathbf{a}) \cdot \mathbf{a} = 0$.

38. Use the dot product to determine whether the points $(4, 2, -2)$, $(2, 4, -3)$, and $(6, 7, -5)$ are vertices of a right triangle.

39. Find symmetric equations for the line through the point $(7, 3, -5)$ that is parallel to $(x - 3)/4 = (y + 4)/(-2) = (z - 9)/6$.

40. Find parametric equations for the line through the point $(5, -9, 3)$ that is perpendicular to the plane $8x + 3y - 4z = 13$.

41. Show that the lines $x = 1 - 2t$, $y = 3t$, $z = 1 + t$ and $x = 1 + 2s$, $y = -4 + s$, $z = -1 + s$ intersect orthogonally.

42. Find an equation of the plane containing the points $(0, 0, 0)$, $(2, 3, 1)$, $(1, 0, 2)$.

43. Find an equation of the plane containing the lines $x = t$, $y = 4t$, $z = -2t$, and $x = 1 + t$, $y = 1 + 4t$, $z = 3 - 2t$.

44. Find an equation of the plane containing $(1, 7, -1)$ that is perpendicular to the line of intersection of $-x + y - 8z = 4$ and $3x - y + 2z = 0$.

45. A constant force of 10 N in the direction of $\mathbf{a} = \mathbf{i} + \mathbf{j}$ moves a block on a frictionless surface from $P_1(4, 1, 0)$ to $P_2(7, 4, 0)$. Suppose distance is measured in meters. Find the work done.

46. In Problem 45, find the work done in moving the block between the same points if another constant force of 50 N in the direction of $\mathbf{b} = \mathbf{i}$ acts simultaneously with the original force.

47. Water rushing from a fire hose exerts a horizontal force $\mathbf{F}_1$ of magnitude 200 lb. See **FIGURE 7.R.1**. What is the magnitude of the force $\mathbf{F}_3$ that a firefighter must exert to hold the hose at an angle of $45°$ from the horizontal?

FIGURE 7.R.1 Fire hose in Problem 47

48. A uniform ball of weight 50 lb is supported by two friction-less planes as shown in **FIGURE 7.R.2**. Let the force exerted by the supporting plane $\mathscr{P}_1$ on the ball be $\mathbf{F}_1$ and the force exerted by the plane $\mathscr{P}_2$ be $\mathbf{F}_2$. Since the ball is held in equilibrium, we must have $\mathbf{w} + \mathbf{F}_1 + \mathbf{F}_2 = \mathbf{0}$, where $\mathbf{w} = -50\mathbf{j}$. Find the magnitudes of the forces $\mathbf{F}_1$ and $\mathbf{F}_2$. [*Hint*: Assume the forces $\mathbf{F}_1$ and $\mathbf{F}_2$ are normal to the planes $\mathscr{P}_1$ and $\mathscr{P}_2$, respectively, and act along lines through the center C of the ball. Place the origin of a two-dimensional coordinate system at C.]

FIGURE 7.R.2 Supported ball in Problem 48

49. Determine whether the set of vectors $\langle a_1, 0, a_3 \rangle$ under addition and scalar multiplication defined by

$$\langle a_1, 0, a_3 \rangle + \langle b_1, 0, b_3 \rangle = \langle a_1 + b_1, 0, a_3 + b_3 \rangle$$

$$k\langle a_1, 0, a_3 \rangle = \langle ka_1, 0, a_3 \rangle$$

is a vector space.

50. Determine whether the vectors $\langle 1, 1, 2 \rangle$, $\langle 0, 2, 3 \rangle$, and $\langle 0, 1, -1 \rangle$ are linearly independent in R^3.

51. Determine whether the set of polynomials in P_n satisfying the condition $d^2p/dx^2 = 0$ is a subspace of P_n. If it is, find a basis for the subspace.

52. Recall that the intersection of two sets W_1 and W_2 is the set of all elements common to both sets, and the union of W_1 and W_2 is the set of elements that are in either W_1 or W_2. Suppose W_1 and W_2 are subspaces of a vector space V. Prove, or disprove by counterexample, the following propositions:

(a) $W_1 \cap W_2$ is a subspace of V.

(b) $W_1 \cup W_2$ is a subspace of V.

CHAPTER 8

Matrices

In mathematics one is often faced with the task of handling large arrays of numbers or functions. Such work is often expedited by using a **matrix**. The theory of matrices was the invention of the prolific English mathematician **Arthur Cayley** (1821–1895).

8.1 Matrix Algebra

INTRODUCTION In the last section of Chapter 7 we saw that a vector in R^n is an ordered n-tuple $(x_1, x_2, \ldots, x_n)$. Vectors are also written as a horizontal or vertical array with no commas:

$$(x_1 \;\; x_2 \;\; \cdots \;\; x_n) \quad \text{or} \quad \begin{pmatrix} x_1 \\ x_2 \\ \vdots \\ x_n \end{pmatrix}. \tag{1}$$

Each array in (1) is called a **matrix**. Our objective in this section is to consider the algebra of such arrays.

■ **A Definition** The arrays in (1) are just special cases of (2) in the definition that follows.

DEFINITION 8.1.1 Matrix

A **matrix** is any rectangular array of numbers or functions:

$$\begin{pmatrix} a_{11} & a_{12} & \cdots & a_{1n} \\ a_{21} & a_{22} & \cdots & a_{2n} \\ \vdots & & & \vdots \\ a_{m1} & a_{m2} & \cdots & a_{mn} \end{pmatrix}. \tag{2}$$

The numbers or functions in the array (2) are called **entries** or **elements** of the matrix. If a matrix has m rows and n columns we say that its **size** is m by n (written $m \times n$). An $n \times n$ matrix is called a **square** matrix or a matrix of **order n**. A 1×1 matrix is simply a constant or a single function. For example, $\mathbf{A} = \begin{pmatrix} 1 & 2 & 5 \\ 6 & 9 & 3 \end{pmatrix}$ is a 2×3 matrix, whereas

$$\mathbf{B} = \begin{pmatrix} 9 & 7 & 0 & 8 \\ \frac{1}{2} & -2 & 6 & 1 \\ 0 & 0 & -1 & 6 \\ 5 & \sqrt{3} & \pi & -4 \end{pmatrix} \tag{3}$$

is a 4×4 square matrix or a matrix of order 4. Throughout this text we shall denote a matrix by a capital boldfaced letter such as $\mathbf{A}$, $\mathbf{B}$, $\mathbf{C}$, or $\mathbf{X}$.

The entry in the ith row and jth column of an $m \times n$ matrix $\mathbf{A}$ is written a_{ij}. An $m \times n$ matrix $\mathbf{A}$ is then abbreviated as $\mathbf{A} = (a_{ij})_{m \times n}$. For an $n \times n$ square matrix, the entries $a_{11}, a_{22}, \ldots, a_{nn}$ are called the **main diagonal entries**. The main diagonal entries for the matrix $\mathbf{B}$ in (3) are 9, -2, -1, and -4.

DEFINITION 8.1.2 Column and Row Vectors

An $n \times 1$ matrix,

$$\begin{pmatrix} a_1 \\ a_2 \\ \vdots \\ a_n \end{pmatrix},$$

is called a **column vector**. A $1 \times n$ matrix,

$$(a_1 \;\; a_2 \;\; \cdots \;\; a_n),$$

is called a **row vector**.

Two $m \times n$ matrices **A** and **B** are **equal** if $a_{ij} = b_{ij}$ for each i and j.

In other words, two matrices are equal if and only if they have the same size and their corresponding entries are equal.

EXAMPLE 1 **Equality**

(a) The matrices $\begin{pmatrix} 1 & 1 \\ 1 & 1 \end{pmatrix}$ and $\begin{pmatrix} 1 & 1 & 1 \\ 1 & 1 & 1 \end{pmatrix}$ are not equal, since the size of the first matrix is 2×2 and the size of the second matrix is 2×3.

(b) The matrices $\begin{pmatrix} 1 & 2 \\ 3 & 4 \end{pmatrix}$ and $\begin{pmatrix} 1 & 2 \\ 4 & 3 \end{pmatrix}$ are not equal, since the corresponding entries in the second rows of the matrices are not the same. ≡

■ Matrix Addition When two matrices **A** and **B** are of the same size we can add them by adding their corresponding entries.

DEFINITION 8.1.4 Matrix Addition

If **A** and **B** are $m \times n$ matrices, then their **sum** is

$$\mathbf{A} + \mathbf{B} = (a_{ij} + b_{ij})_{m \times n}.$$

EXAMPLE 2 **Addition of Two Matrices**

(a) The sum of $\mathbf{A} = \begin{pmatrix} 2 & -1 & 3 \\ 0 & 4 & 6 \\ -6 & 10 & -5 \end{pmatrix}$ and $\mathbf{B} = \begin{pmatrix} 4 & 7 & -8 \\ 9 & 3 & 5 \\ 1 & -1 & 2 \end{pmatrix}$ is

$$\mathbf{A} + \mathbf{B} = \begin{pmatrix} 2+4 & -1+7 & 3+(-8) \\ 0+9 & 4+3 & 6+5 \\ -6+1 & 10+(-1) & -5+2 \end{pmatrix} = \begin{pmatrix} 6 & 6 & -5 \\ 9 & 7 & 11 \\ -5 & 9 & -3 \end{pmatrix}.$$

(b) The sum of $\mathbf{A} = \begin{pmatrix} 1 & 2 & 3 \\ 4 & 5 & 6 \end{pmatrix}$ and $\mathbf{B} = \begin{pmatrix} 1 & 0 \\ 1 & 0 \end{pmatrix}$ is not defined, since **A** and **B** are of different sizes. ≡

DEFINITION 8.1.5 Scalar Multiple of a Matrix

If k is a real number, then the **scalar multiple** of a matrix **A** is

$$k\mathbf{A} = \begin{pmatrix} ka_{11} & ka_{12} & \cdots & ka_{1n} \\ ka_{21} & ka_{22} & \cdots & ka_{2n} \\ \vdots & & & \vdots \\ ka_{m1} & ka_{m2} & \cdots & ka_{mn} \end{pmatrix} = (ka_{ij})_{m \times n}.$$

In other words, to compute $k\mathbf{A}$, we simply multiply each entry of **A** by k. For example, from Definition 8.1.5,

$$5\begin{pmatrix} 2 & -3 \\ 4 & -1 \end{pmatrix} = \begin{pmatrix} 5 \cdot 2 & 5 \cdot (-3) \\ 5 \cdot 4 & 5 \cdot (-1) \end{pmatrix} = \begin{pmatrix} 10 & -15 \\ 20 & -5 \end{pmatrix}.$$

We note in passing that for any matrix **A** the scalar multiple $k\mathbf{A}$ is the same as $\mathbf{A}k$.

The **difference** of two $m \times n$ matrices is defined in the usual manner: $\mathbf{A} - \mathbf{B} = \mathbf{A} + (-\mathbf{B})$ where $-\mathbf{B} = (-1)\mathbf{B}$.

The following theorem summarizes some of the properties of matrix addition and scalar multiplication. Each of the six parts of the theorem can be proved using Definitions 8.1.4 and 8.1.5.

THEOREM 8.1.1 Properties of Matrix Addition and Scalar Multiplication

Suppose $\mathbf{A}$, $\mathbf{B}$, and $\mathbf{C}$ are $m \times n$ matrices and k_1 and k_2 are scalars. Then

(i) $\mathbf{A} + \mathbf{B} = \mathbf{B} + \mathbf{A}$ ← commutative law of addition

(ii) $\mathbf{A} + (\mathbf{B} + \mathbf{C}) = (\mathbf{A} + \mathbf{B}) + \mathbf{C}$ ← associative law of addition

(iii) $(k_1 k_2)\mathbf{A} = k_1(k_2\mathbf{A})$

(iv) $1\mathbf{A} = \mathbf{A}$

(v) $k_1(\mathbf{A} + \mathbf{B}) = k_1\mathbf{A} + k_1\mathbf{B}$ ← distributive law

(vi) $(k_1 + k_2)\mathbf{A} = k_1\mathbf{A} + k_2\mathbf{A}.$ ← distributive law

■ **Matrix Multiplication** We have just seen that any matrix $\mathbf{A}$ can be multiplied by a scalar. But can two matrices be multiplied together? The next definition gives the answer.

DEFINITION 8.1.6 Matrix Multiplication

Let $\mathbf{A}$ be a matrix having m rows and p columns, and let $\mathbf{B}$ be a matrix having p rows and n columns. The **product $\mathbf{AB}$** is the $m \times n$ matrix

$$\mathbf{AB} = \begin{pmatrix} a_{11} & a_{12} & \cdots & a_{1p} \\ a_{21} & a_{22} & \cdots & a_{2p} \\ \vdots & & & \vdots \\ a_{m1} & a_{m2} & \cdots & a_{mp} \end{pmatrix} \begin{pmatrix} b_{11} & b_{12} & \cdots & b_{1n} \\ b_{21} & b_{22} & \cdots & b_{2n} \\ \vdots & & & \vdots \\ b_{p1} & b_{p2} & \cdots & b_{pn} \end{pmatrix}$$

$$= \begin{pmatrix} a_{11}b_{11} + a_{12}b_{21} + \cdots + a_{1p}b_{p1} & \cdots & a_{11}b_{1n} + a_{12}b_{2n} + \cdots + a_{1p}b_{pn} \\ a_{21}b_{11} + a_{22}b_{21} + \cdots + a_{2p}b_{p1} & \cdots & a_{21}b_{1n} + a_{22}b_{2n} + \cdots + a_{2p}b_{pn} \\ \vdots & & \vdots \\ a_{m1}b_{11} + a_{m2}b_{21} + \cdots + a_{mp}b_{p1} & \cdots & a_{m1}b_{1n} + a_{m2}b_{2n} + \cdots + a_{mp}b_{pn} \end{pmatrix}$$

$$= \left(\sum_{k=1}^{p} a_{ik}b_{kj} \right)_{m \times n}.$$

Definition 8.1.6 states that the product $\mathbf{C} = \mathbf{AB}$ is defined only when the number of columns in the matrix $\mathbf{A}$ is the same as the number of rows in $\mathbf{B}$. The dimension of the product can be determined from

$$\mathbf{A}_{m \times p}\mathbf{B}_{p \times n} = \mathbf{C}_{m \times n}.$$

▶ The number of columns of $\mathbf{A}$ must equal the number of rows of $\mathbf{B}$.

Also, you might recognize that the entries in, say, the ith row of the final matrix $\mathbf{C} = \mathbf{AB}$ are formed by using the component definition of the inner or dot product of the ith row (vector) of $\mathbf{A}$ with each of the columns (vectors) of $\mathbf{B}$.

EXAMPLE 3　**Matrix Multiplication**

Find the product $\mathbf{AB}$ for the following matrices:

(a) $\mathbf{A} = \begin{pmatrix} 4 & 7 \\ 3 & 5 \end{pmatrix}$, $\mathbf{B} = \begin{pmatrix} 9 & -2 \\ 6 & 8 \end{pmatrix}$ **(b)** $\mathbf{A} = \begin{pmatrix} 5 & 8 \\ 1 & 0 \\ 2 & 7 \end{pmatrix}$, $\mathbf{B} = \begin{pmatrix} -4 & -3 \\ 2 & 0 \end{pmatrix}.$

SOLUTION　From Definition 8.1.6 we have

(a) $\mathbf{AB} = \begin{pmatrix} 4 \cdot 9 + 7 \cdot 6 & 4 \cdot (-2) + 7 \cdot 8 \\ 3 \cdot 9 + 5 \cdot 6 & 3 \cdot (-2) + 5 \cdot 8 \end{pmatrix} = \begin{pmatrix} 78 & 48 \\ 57 & 34 \end{pmatrix}$

(b) $\mathbf{AB} = \begin{pmatrix} 5 \cdot (-4) + 8 \cdot 2 & 5 \cdot (-3) + 8 \cdot 0 \\ 1 \cdot (-4) + 0 \cdot 2 & 1 \cdot (-3) + 0 \cdot 0 \\ 2 \cdot (-4) + 7 \cdot 2 & 2 \cdot (-3) + 7 \cdot 0 \end{pmatrix} = \begin{pmatrix} -4 & -15 \\ -4 & -3 \\ 6 & -6 \end{pmatrix}.$ ≡

Unlike matrix addition, matrix multiplication is, in general, not commutative. That is, $\mathbf{BA} \neq \mathbf{AB}$. Observe that in part (a) of Example 3, $\mathbf{BA} = \begin{pmatrix} 30 & 53 \\ 48 & 82 \end{pmatrix}$, whereas in part (b) the product $\mathbf{BA}$ is *not defined* because the first matrix (in this case, $\mathbf{B}$) does not have the same number of columns as the second matrix has rows.

The product of an $m \times n$ matrix with an $n \times 1$ column vector is an $m \times 1$ column vector. For example,

$$\begin{pmatrix} -4 & 2 \\ 3 & 8 \end{pmatrix}\begin{pmatrix} x_1 \\ x_2 \end{pmatrix} = \begin{pmatrix} -4x_1 + 2x_2 \\ 3x_1 + 8x_2 \end{pmatrix}. \tag{4}$$

It is often convenient to write a column vector as a sum of two or more column vectors. In view of Definitions 8.1.4 and 8.1.5, the result in (4) can be written

$$\begin{pmatrix} -4x_1 + 2x_2 \\ 3x_1 + 8x_2 \end{pmatrix} = x_1 \begin{pmatrix} -4 \\ 3 \end{pmatrix} + x_2 \begin{pmatrix} 2 \\ 8 \end{pmatrix}.$$

■ **Associative Law** Although we shall not prove it, matrix multiplication is **associative**. If $\mathbf{A}$ is an $m \times p$ matrix, $\mathbf{B}$ a $p \times r$ matrix, and $\mathbf{C}$ an $r \times n$ matrix, then the product

$$\mathbf{A(BC)} = \mathbf{(AB)C}$$

is an $m \times n$ matrix.

■ **Distributive Law** If $\mathbf{B}$ and $\mathbf{C}$ are both $r \times n$ matrices and $\mathbf{A}$ is an $m \times r$ matrix, then the **distributive law** is

$$\mathbf{A(B + C)} = \mathbf{AB} + \mathbf{AC}.$$

Furthermore, if the product $\mathbf{(B + C)A}$ is defined, then

$$\mathbf{(B + C)A} = \mathbf{BA} + \mathbf{CA}.$$

DEFINITION 8.1.7 **Transpose of a Matrix**

The **transpose** of the $m \times n$ matrix (2) is the $n \times m$ matrix $\mathbf{A}^T$ given by

$$\mathbf{A}^T = \begin{pmatrix} a_{11} & a_{21} & \cdots & a_{m1} \\ a_{12} & a_{22} & \cdots & a_{m2} \\ \vdots & & & \vdots \\ a_{1n} & a_{2n} & \cdots & a_{mn} \end{pmatrix}.$$

In other words, the rows of a matrix $\mathbf{A}$ become the columns of its transpose $\mathbf{A}^T$.

EXAMPLE 4 **Using Definition 8.1.7**

(a) If $\mathbf{A} = \begin{pmatrix} 9 & 0 & 4 \\ 3 & 7 & 5 \end{pmatrix}$, then $\mathbf{A}^T = \begin{pmatrix} 9 & 3 \\ 0 & 7 \\ 4 & 5 \end{pmatrix}$.

(b) If $\mathbf{B} = \begin{pmatrix} 3 & 2 & -1 \\ 6 & 5 & 8 \\ 1 & -2 & 4 \end{pmatrix}$, then $\mathbf{B}^T = \begin{pmatrix} 3 & 6 & 1 \\ 2 & 5 & -2 \\ -1 & 8 & 4 \end{pmatrix}$.

(c) If $\mathbf{C} = (7 \quad 2)$, then $\mathbf{C}^T = \begin{pmatrix} 7 \\ 2 \end{pmatrix}$.

≡

In the next theorem we give some important properties of the transpose.

THEOREM 8.1.2 **Properties of Transpose**

Suppose $\mathbf{A}$ and $\mathbf{B}$ are matrices and k a scalar. Then

(i) $(\mathbf{A}^T)^T = \mathbf{A}$ ← transpose of a transpose
(ii) $(\mathbf{A} + \mathbf{B})^T = \mathbf{A}^T + \mathbf{B}^T$ ← transpose of a sum
(iii) $(\mathbf{AB})^T = \mathbf{B}^T\mathbf{A}^T$ ← transpose of a product
(iv) $(k\mathbf{A})^T = k\mathbf{A}^T$. ← transpose of a scalar multiple

Of course, in properties (ii) and (iii) of Theorem 8.1.2 we assume that the sum and product of $\mathbf{A}$ and $\mathbf{B}$ are defined. Note carefully that part (iii) of the theorem indicates that the transpose of the product is the product of the transposes *in reverse order*. Moreover, (ii) and (iii) extend to any finite sum or product of matrices. For example, in the case of three matrices we have

$$(\mathbf{A} + \mathbf{B} + \mathbf{C})^T = \mathbf{A}^T + \mathbf{B}^T + \mathbf{C}^T \quad \text{and} \quad (\mathbf{ABC})^T = \mathbf{C}^T\mathbf{B}^T\mathbf{A}^T.$$

■ Special Matrices

In matrix theory there are many special kinds of matrices that are important because they possess certain properties. The following is a list of some of these matrices:

- A matrix that consists of all zero entries is called a **zero matrix** and is denoted by $\mathbf{0}$. For example,

$$\mathbf{0} = \begin{pmatrix} 0 \\ 0 \end{pmatrix}, \quad \mathbf{0} = \begin{pmatrix} 0 & 0 \\ 0 & 0 \end{pmatrix}, \quad \mathbf{0} = \begin{pmatrix} 0 & 0 \\ 0 & 0 \\ 0 & 0 \end{pmatrix}$$

are zero matrices. If $\mathbf{A}$ and $\mathbf{0}$ are $m \times n$ matrices, then

$$\mathbf{A} + \mathbf{0} = \mathbf{A}. \tag{5}$$

In addition, $\mathbf{A} + (-\mathbf{A}) = \mathbf{0}.$ (6)

- An $n \times n$ matrix $\mathbf{A}$ is said to be a **triangular matrix** if all its entries *above* the main diagonal are zeros or if all its entries *below* the main diagonal are zeros. In other words, a square matrix $\mathbf{A} = (a_{ij})_{n \times n}$ is triangular if $a_{ij} = 0$ whenever $i < j$ or $a_{ij} = 0$ whenever $i > j$. More specifically, in the first case the matrix is called **lower triangular**, and in the second the matrix is called **upper triangular**. The following two matrices are triangular:

$$\begin{pmatrix} -5 & 0 & 0 & 0 & 0 \\ 1 & 6 & 0 & 0 & 0 \\ 8 & 9 & 3 & 0 & 0 \\ 1 & 1 & -1 & 2 & 0 \\ 13 & 0 & 4 & 7 & -2 \end{pmatrix} \qquad \begin{pmatrix} 1 & 2 & 3 & 1 \\ 0 & 2 & 1 & 5 \\ 0 & 0 & -1 & 4 \\ 0 & 0 & 0 & 8 \end{pmatrix}$$

$\underbrace{}_{\text{lower triangular matrix}}$ $\underbrace{}_{\text{upper triangular matrix}}$

- An $n \times n$ matrix $\mathbf{A}$ is said to be a **diagonal matrix** if all its entries not on the main diagonal are zeros. In terms of the symbolism $\mathbf{A} = (a_{ij})_{n \times n}$, $\mathbf{A}$ is a diagonal matrix if $a_{ij} = 0$ for $i \neq j$. The following is a diagonal matrix:

$$\begin{pmatrix} 7 & 0 & 0 \\ 0 & \frac{1}{2} & 0 \\ 0 & 0 & 1 \end{pmatrix}.$$

- When the entries a_{ii} of a diagonal matrix $\mathbf{A}$ are all equal, it is called a **scalar matrix**. For example, $\begin{pmatrix} 5 & 0 \\ 0 & 5 \end{pmatrix}$ is a scalar matrix. An $n \times n$ scalar matrix is simply a scalar multiple of a diagonal matrix in which the main diagonal entries are all equal to 1. For example, $\begin{pmatrix} 5 & 0 \\ 0 & 5 \end{pmatrix} = 5 \begin{pmatrix} 1 & 0 \\ 0 & 1 \end{pmatrix}$. In general, the $n \times n$ matrix

$$\begin{pmatrix} 1 & 0 & 0 & \cdots & 0 \\ 0 & 1 & 0 & \cdots & 0 \\ \vdots & & & & \vdots \\ 0 & 0 & 0 & \cdots & 1 \end{pmatrix}$$

 is denoted by the symbol $\mathbf{I}$ (or $\mathbf{I}_n$ when there is a need to emphasize the order of the matrix). For any $m \times n$ matrix $\mathbf{A}$ it is readily verified that $\mathbf{I}_m \mathbf{A} = \mathbf{A} \mathbf{I}_n = \mathbf{A}$. Because this last property is analogous to $1 \cdot a = a \cdot 1 = a$, for any real number a, the matrix $\mathbf{I}$ is called the **identity matrix**.

- An $n \times n$ matrix $\mathbf{A}$ is said to be **symmetric** if $\mathbf{A}^T = \mathbf{A}$; that is, $\mathbf{A}$ is symmetric if $a_{ij} = a_{ji}$ for all i and j. This means that the entries in a symmetric matrix are symmetric with respect to the main diagonal of the matrix. For example, a quick inspection of the matrix

$$\mathbf{A} = \begin{pmatrix} 1 & 2 & 7 \\ 2 & 5 & 6 \\ 7 & 6 & 4 \end{pmatrix}$$

 shows that it is symmetric. In addition, by taking the transpose of $\mathbf{A}$ we see that

$$\mathbf{A}^T = \begin{pmatrix} 1 & 2 & 7 \\ 2 & 5 & 6 \\ 7 & 6 & 4 \end{pmatrix} = \mathbf{A}.$$

REMARKS

Suppose the symbol $M_{m,n}$ denotes the set of all $m \times n$ matrices on which the operations of matrix addition and scalar multiplication are defined. Then

$$\mathbf{A} + \mathbf{B} \text{ is in } M_{m,n} \quad \text{and} \quad k\mathbf{A} \text{ is in } M_{m,n} \tag{7}$$

for every $\mathbf{A}$ and $\mathbf{B}$ in $M_{m,n}$ and for every scalar k. That is to say, $M_{m,n}$ is closed under matrix addition and scalar multiplication. When we combine (7) with properties (5) and (6) and the properties listed in Theorem 8.1.1, it follows immediately that $M_{m,n}$ is a **vector space**. For practical purposes, the vector spaces $M_{1,n}$ (row vectors) and $M_{n,1}$ (column vectors) are indistinguishable from the vector space R^n.

In Problems 1–6, state the size of the given matrix.

1. $\begin{pmatrix} 1 & 2 & 3 & 9 \\ 5 & 6 & 0 & 1 \end{pmatrix}$

2. $\begin{pmatrix} 0 & 2 \\ 8 & 4 \\ 5 & 6 \end{pmatrix}$

3. $\begin{pmatrix} 1 & 2 & -1 \\ 0 & 7 & -2 \\ 0 & 0 & 5 \end{pmatrix}$

4. $(5 \quad 7 \quad -15)$

5. $\begin{pmatrix} 1 & 5 & -6 & 0 \\ 7 & -10 & 2 & 12 \\ 0 & 9 & 2 & -1 \end{pmatrix}$

6. $\begin{pmatrix} 1 \\ 5 \\ -6 \\ 0 \\ 7 \\ -10 \\ 2 \\ 12 \end{pmatrix}$

In Problems 7–10, determine whether the given matrices are equal.

7. $\begin{pmatrix} 1 & 2 & 3 \\ 4 & 5 & 6 \end{pmatrix}$, $\begin{pmatrix} 1 & 2 \\ 3 & 4 \\ 5 & 6 \end{pmatrix}$

8. $\begin{pmatrix} 1 & 2 \\ 0 & 1 \end{pmatrix}$, $\begin{pmatrix} 1 & 0 \\ 2 & 1 \end{pmatrix}$

9. $\begin{pmatrix} \sqrt{(-2)^2} & 1 \\ 2 & \frac{2}{8} \end{pmatrix}$, $\begin{pmatrix} -2 & 1 \\ 2 & \frac{1}{4} \end{pmatrix}$

10. $\begin{pmatrix} \frac{1}{8} & \frac{1}{5} \\ \sqrt{2} & 1 \end{pmatrix}$, $\begin{pmatrix} 0.125 & 0.2 \\ 1.414 & 1 \end{pmatrix}$

In Problems 11 and 12, determine the values of x and y for which the matrices are equal.

11. $\begin{pmatrix} 1 & x \\ y & -3 \end{pmatrix}$, $\begin{pmatrix} 1 & y-2 \\ 3x-2 & -3 \end{pmatrix}$

12. $\begin{pmatrix} x^2 & 1 \\ y & 5 \end{pmatrix}$, $\begin{pmatrix} 9 & 1 \\ 4x & 5 \end{pmatrix}$

In Problems 13 and 14, find the entries c_{23} and c_{12} for the matrix $\mathbf{C} = 2\mathbf{A} - 3\mathbf{B}$.

13. $\mathbf{A} = \begin{pmatrix} 2 & 3 & -1 \\ -1 & 6 & 0 \end{pmatrix}$, $\mathbf{B} = \begin{pmatrix} 4 & -2 & 6 \\ 1 & 3 & -3 \end{pmatrix}$

14. $\mathbf{A} = \begin{pmatrix} 1 & -1 & 1 \\ 2 & 2 & 1 \\ 0 & -4 & 1 \end{pmatrix}$, $\mathbf{B} = \begin{pmatrix} 2 & 0 & 5 \\ 0 & 4 & 0 \\ 3 & 0 & 7 \end{pmatrix}$

15. If $\mathbf{A} = \begin{pmatrix} 4 & 5 \\ -6 & 9 \end{pmatrix}$ and $\mathbf{B} = \begin{pmatrix} -2 & 6 \\ 8 & -10 \end{pmatrix}$, find **(a)** $\mathbf{A} + \mathbf{B}$, **(b)** $\mathbf{B} - \mathbf{A}$, **(c)** $2\mathbf{A} + 3\mathbf{B}$.

16. If $\mathbf{A} = \begin{pmatrix} -2 & 0 \\ 4 & 1 \\ 7 & 3 \end{pmatrix}$ and $\mathbf{B} = \begin{pmatrix} 3 & -1 \\ 0 & 2 \\ -4 & -2 \end{pmatrix}$, find **(a)** $\mathbf{A} - \mathbf{B}$, **(b)** $\mathbf{B} - \mathbf{A}$, **(c)** $2(\mathbf{A} + \mathbf{B})$.

17. If $\mathbf{A} = \begin{pmatrix} 2 & -3 \\ -5 & 4 \end{pmatrix}$ and $\mathbf{B} = \begin{pmatrix} -1 & 6 \\ 3 & 2 \end{pmatrix}$, find **(a)** $\mathbf{AB}$, **(b)** $\mathbf{BA}$, **(c)** $\mathbf{A}^2 = \mathbf{AA}$, **(d)** $\mathbf{B}^2 = \mathbf{BB}$.

18. If $\mathbf{A} = \begin{pmatrix} 1 & 4 \\ 5 & 10 \\ 8 & 12 \end{pmatrix}$ and $\mathbf{B} = \begin{pmatrix} -4 & 6 & -3 \\ 1 & -3 & 2 \end{pmatrix}$, find **(a)** $\mathbf{AB}$, **(b)** $\mathbf{BA}$.

19. If $\mathbf{A} = \begin{pmatrix} 1 & -2 \\ -2 & 4 \end{pmatrix}$, $\mathbf{B} = \begin{pmatrix} 6 & 3 \\ 2 & 1 \end{pmatrix}$, and $\mathbf{C} = \begin{pmatrix} 0 & 2 \\ 3 & 4 \end{pmatrix}$, find **(a)** $\mathbf{BC}$, **(b)** $\mathbf{A}(\mathbf{BC})$, **(c)** $\mathbf{C}(\mathbf{BA})$, **(d)** $\mathbf{A}(\mathbf{B} + \mathbf{C})$.

20. If $\mathbf{A} = (5 \quad -6 \quad 7)$, $\mathbf{B} = \begin{pmatrix} 3 \\ 4 \\ -1 \end{pmatrix}$, and $\mathbf{C} = \begin{pmatrix} 1 & 2 & 4 \\ 0 & 1 & -1 \\ 3 & 2 & 1 \end{pmatrix}$, find **(a)** $\mathbf{AB}$, **(b)** $\mathbf{BA}$, **(c)** $(\mathbf{BA})\mathbf{C}$, **(d)** $(\mathbf{AB})\mathbf{C}$.

21. If $\mathbf{A} = \begin{pmatrix} 4 \\ 8 \\ -10 \end{pmatrix}$ and $\mathbf{B} = (2 \quad 4 \quad 5)$, find **(a)** $\mathbf{A}^T\mathbf{A}$, **(b)** $\mathbf{B}^T\mathbf{B}$, **(c)** $\mathbf{A} + \mathbf{B}^T$.

22. If $\mathbf{A} = \begin{pmatrix} 1 & 2 \\ 2 & 4 \end{pmatrix}$ and $\mathbf{B} = \begin{pmatrix} -2 & 3 \\ 5 & 7 \end{pmatrix}$, find **(a)** $\mathbf{A} + \mathbf{B}^T$, **(b)** $2\mathbf{A}^T - \mathbf{B}^T$, **(c)** $\mathbf{A}^T(\mathbf{A} - \mathbf{B})$.

23. If $\mathbf{A} = \begin{pmatrix} 3 & 4 \\ 8 & 1 \end{pmatrix}$ and $\mathbf{B} = \begin{pmatrix} 5 & 10 \\ -2 & -5 \end{pmatrix}$, find **(a)** $(\mathbf{AB})^T$, **(b)** $\mathbf{B}^T\mathbf{A}^T$.

24. If $\mathbf{A} = \begin{pmatrix} 5 & 9 \\ -4 & 6 \end{pmatrix}$ and $\mathbf{B} = \begin{pmatrix} -3 & 11 \\ -7 & 2 \end{pmatrix}$, find **(a)** $\mathbf{A}^T + \mathbf{B}$, **(b)** $2\mathbf{A} + \mathbf{B}^T$.

In Problems 25–28, write the given sum as a single-column matrix.

25. $4\begin{pmatrix} -1 \\ 2 \end{pmatrix} - 2\begin{pmatrix} 2 \\ 8 \end{pmatrix} + 3\begin{pmatrix} -2 \\ 3 \end{pmatrix}$

26. $3\begin{pmatrix} 2 \\ 1 \\ -1 \end{pmatrix} + 5\begin{pmatrix} -1 \\ -1 \\ 3 \end{pmatrix} - 2\begin{pmatrix} 3 \\ 4 \\ -5 \end{pmatrix}$

27. $\begin{pmatrix} 2 & -3 \\ 1 & 4 \end{pmatrix}\begin{pmatrix} -2 \\ 5 \end{pmatrix} - \begin{pmatrix} -1 & 6 \\ -2 & 3 \end{pmatrix}\begin{pmatrix} -7 \\ 2 \end{pmatrix}$

28. $\begin{pmatrix} 1 & -3 & 4 \\ 2 & 5 & -1 \\ 0 & -4 & -2 \end{pmatrix} \begin{pmatrix} 3 \\ 2 \\ -1 \end{pmatrix} + \begin{pmatrix} -1 \\ 1 \\ 4 \end{pmatrix} - \begin{pmatrix} 2 \\ 8 \\ -6 \end{pmatrix}$

In Problems 29 and 30, determine the size of the matrix **A** such that the given product is defined.

29. $\begin{pmatrix} 2 & 1 & 3 & 3 \\ 9 & 6 & 7 & 0 \end{pmatrix} \mathbf{A} \begin{pmatrix} 0 \\ 5 \\ 7 \\ 9 \\ 2 \end{pmatrix}$

30. $\begin{pmatrix} 2 & 1 & 3 \\ 3 & 9 & 6 \\ 7 & 0 & -1 \end{pmatrix} \mathbf{A} \begin{pmatrix} 0 & 1 \\ 7 & 4 \end{pmatrix}$

In Problems 31–34, suppose $\mathbf{A} = \begin{pmatrix} 2 & 4 \\ -3 & 2 \end{pmatrix}$ and $\mathbf{B} = \begin{pmatrix} 4 & 10 \\ 2 & 5 \end{pmatrix}$. Verify the given property by computing the left and right members of the given equality.

31. $(\mathbf{A}^T)^T = \mathbf{A}$ **32.** $(\mathbf{A} + \mathbf{B})^T = \mathbf{A}^T + \mathbf{B}^T$

33. $(\mathbf{AB})^T = \mathbf{B}^T\mathbf{A}^T$ **34.** $(6\mathbf{A})^T = 6\mathbf{A}^T$

35. Suppose $\mathbf{A} = \begin{pmatrix} 2 & 1 \\ 6 & 3 \\ 2 & 5 \end{pmatrix}$. Verify that the matrix $\mathbf{B} = \mathbf{AA}^T$ is symmetric.

36. Show that if **A** is an $m \times n$ matrix, then $\mathbf{AA}^T$ is symmetric.

37. In matrix theory, many of the familiar properties of the real number system are not valid. If a and b are real numbers, then $ab = 0$ implies that $a = 0$ or $b = 0$. Find two matrices such that $\mathbf{AB} = \mathbf{0}$ but $\mathbf{A} \neq \mathbf{0}$ and $\mathbf{B} \neq \mathbf{0}$.

38. If a, b, and c are real numbers and $c \neq 0$, then $ac = bc$ implies $a = b$. For matrices, $\mathbf{AC} = \mathbf{BC}$, $\mathbf{C} \neq \mathbf{0}$, does not necessarily imply $\mathbf{A} = \mathbf{B}$. Verify this for

$$\mathbf{A} = \begin{pmatrix} 2 & 1 & 4 \\ 3 & 2 & 1 \\ 1 & 3 & 2 \end{pmatrix}, \mathbf{B} = \begin{pmatrix} 5 & 1 & 6 \\ 9 & 2 & -3 \\ -1 & 3 & 7 \end{pmatrix},$$

and $$\mathbf{C} = \begin{pmatrix} 0 & 0 & 0 \\ 2 & 3 & 4 \\ 0 & 0 & 0 \end{pmatrix}.$$

In Problems 39 and 40, let **A** and **B** be $n \times n$ matrices. Explain why, in general, the given formula is *not* valid.

39. $(\mathbf{A} + \mathbf{B})^2 = \mathbf{A}^2 + 2\mathbf{AB} + \mathbf{B}^2$

40. $(\mathbf{A} + \mathbf{B})(\mathbf{A} - \mathbf{B}) = \mathbf{A}^2 - \mathbf{B}^2$

41. Write $\begin{pmatrix} a_{11} & a_{12} \\ a_{21} & a_{22} \end{pmatrix} \begin{pmatrix} x_1 \\ x_2 \end{pmatrix} = \begin{pmatrix} b_1 \\ b_2 \end{pmatrix}$ without matrices.

42. Write the system of equations

$$2x_1 + 6x_2 + x_3 = 7$$
$$x_1 + 2x_2 - x_3 = -1$$
$$5x_1 + 7x_2 - 4x_3 = 9$$

as a matrix equation $\mathbf{AX} = \mathbf{B}$, where **X** and **B** are column vectors.

43. Verify that the quadratic form $ax^2 + bxy + cy^2$ is the same as

$$(x \quad y)\begin{pmatrix} a & \frac{1}{2}b \\ \frac{1}{2}b & c \end{pmatrix}\begin{pmatrix} x \\ y \end{pmatrix}.$$

44. Verify that the curl of the vector field $\mathbf{F} = P\mathbf{i} + Q\mathbf{j} + R\mathbf{k}$ can be written

$$\text{curl } \mathbf{F} = \begin{pmatrix} 0 & -\partial/\partial x & \partial/\partial x \\ \partial/\partial x & 0 & -\partial/\partial x \\ -\partial/\partial y & \partial/\partial x & 0 \end{pmatrix}\begin{pmatrix} P \\ Q \\ R \end{pmatrix}.$$

(Readers who are not familiar with the concept of the curl of a vector field should see Section 9.7.)

If a vector $\mathbf{a} = \overrightarrow{OP} = \langle x, y \rangle$ in R^2 is rotated counterclockwise about the origin through an angle θ, then the components of the resulting vector $\mathbf{b} = \overrightarrow{OP_1} = \langle x_1, y_1 \rangle$ are given by $\mathbf{B} = \mathbf{MA}$, where

$$\mathbf{A} = \begin{pmatrix} x \\ y \end{pmatrix}, \mathbf{B} = \begin{pmatrix} x_1 \\ y_1 \end{pmatrix}, \text{ and } \mathbf{M} = \begin{pmatrix} \cos\theta & -\sin\theta \\ \sin\theta & \cos\theta \end{pmatrix}.$$

The 2×2 matrix **M** is called a **rotation matrix**. See **FIGURE 8.1.1**. In Problems 45–48, find the resulting vector **b** if the given vector $\mathbf{a} = \overrightarrow{OP} = \langle x, y \rangle$ is rotated through the indicated angle.

45. $\mathbf{a} = \langle 1, 1 \rangle$, $\theta = \pi/2$ **46.** $\mathbf{a} = \langle -2, 4 \rangle$, $\theta = \pi/6$

47. $\mathbf{a} = \langle \sqrt{2}, \sqrt{2} \rangle$, $\theta = 3\pi/4$ **48.** $\mathbf{a} = \langle 1, -1 \rangle$, $\theta = 2\pi/3$

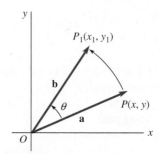

FIGURE 8.1.1 Rotated vector in Problems 45–48

In Problems 49 and 50, let $\mathbf{A} = \begin{pmatrix} x \\ y \end{pmatrix}$ and $\mathbf{B} = \begin{pmatrix} x_1 \\ y_1 \end{pmatrix}$. Find a 2×2 matrix **M** so that $\mathbf{B} = \mathbf{MA}$ represents a reflection of the vector $\mathbf{a} = \overrightarrow{OP} = \langle x, y \rangle$ in the given axis.

49. y-axis **50.** x-axis

51. As shown in **FIGURE 8.1.2(a)**, a spacecraft can perform rotations called **pitch**, **roll**, and **yaw** about three distinct axes. To describe the coordinates of a point P we use two coordinate systems: a fixed three-dimensional Cartesian coordinate system in which the coordinates of P are (x, y, z) and a spacecraft coordinate system that moves with the particular rotation. In Figure 8.1.2(b) we have illustrated a yaw—that is, a rotation around the z-axis (which is perpendicular to the plane of the paper). The coordinates

(x_Y, y_Y, z_Y) of the point P in the spacecraft system after the yaw are related to the coordinates (x, y, z) of P in the fixed coordinate system by the equations

$$x_Y = x \cos \gamma + y \sin \gamma$$
$$y_Y = -x \sin \gamma + y \cos \gamma$$
$$z_Y = z,$$

where γ is the angle of rotation.

(a) Verify that the foregoing system of equations can be written as the matrix equation

$$\begin{pmatrix} x_Y \\ y_Y \\ z_Y \end{pmatrix} = \mathbf{M}_Y \begin{pmatrix} x \\ y \\ z \end{pmatrix}$$

where $\quad \mathbf{M}_Y = \begin{pmatrix} \cos \gamma & \sin \gamma & 0 \\ -\sin \gamma & \cos \gamma & 0 \\ 0 & 0 & 1 \end{pmatrix}.$

(b) When the spacecraft performs a pitch, roll, and yaw *in sequence* through the angles α, β, and γ, respectively, the final coordinates of the point P in the spacecraft system (x_S, y_S, z_S) are obtained from the sequence of transformations

$$x_P = x \qquad\qquad x_R = x_P \cos \beta - z_P \sin \beta$$
$$y_P = y \cos \alpha + z \sin \alpha \qquad y_R = y_P$$
$$z_P = -y \sin \alpha + z \cos \alpha; \qquad z_R = x_P \sin \beta + z_P \cos \beta;$$
$$x_S = x_R \cos \gamma + y_R \sin \gamma$$
$$y_S = -x_R \sin \gamma + y_R \cos \gamma$$
$$z_S = z_R.$$

Write this sequence of transformations as a matrix equation

$$\begin{pmatrix} x_S \\ y_S \\ z_S \end{pmatrix} = \mathbf{M}_Y \mathbf{M}_R \mathbf{M}_P \begin{pmatrix} x \\ y \\ z \end{pmatrix}.$$

The matrix $\mathbf{M}_Y$ is the same as in part (a). Identify the matrices $\mathbf{M}_R$ and $\mathbf{M}_P$.

(c) Suppose the coordinates of a point are $(1, 1, 1)$ in the fixed coordinate system. Determine the coordinates of the point in the spacecraft system if the spacecraft performs a pitch, roll, and yaw in sequence through the angles $\alpha = 30°$, $\beta = 45°$, and $\gamma = 60°$.

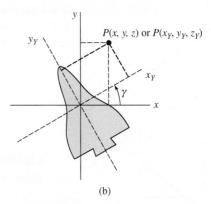

FIGURE 8.1.2 Spacecraft in Problem 51

52. **Project (a)** A matrix $\mathbf{A}$ can be **partitioned** into submatrices. For example, the 3×5 and 5×2 matrices

$$\mathbf{A} = \begin{pmatrix} 3 & 2 & -1 & 2 & 4 \\ 1 & 6 & 3 & -1 & 5 \\ 0 & 4 & 6 & -2 & 3 \end{pmatrix}, \quad \mathbf{B} = \begin{pmatrix} 3 & 4 \\ 0 & 7 \\ -4 & 1 \\ -2 & -1 \\ 2 & 5 \end{pmatrix}$$

can be written

$$\mathbf{A} = \begin{pmatrix} A_{11} & A_{12} \\ A_{21} & A_{22} \end{pmatrix}, \quad \mathbf{B} = \begin{pmatrix} B_1 \\ B_2 \end{pmatrix},$$

where A_{11} is the upper left-hand block, or submatrix, indicated in blue in $\mathbf{A}$, A_{12} is the upper right-hand block, and so on. Compute the product $\mathbf{AB}$ using the partitioned matrices.

(b) Investigate how partitioned matrices can be useful when using a computer to perform matrix calculations involving large matrices.

8.2 Systems of Linear Equations

INTRODUCTION Recall, any equation of the form $ax + by = c$, where a, b, and c are real numbers, is said to be a **linear equation** in the variables x and y. The graph of a linear equation in two variables is a straight line. For real numbers a, b, c, and d, $ax + by + cz = d$ is a linear equation in the variables x, y, and z and is the equation of a plane in 3-space. In general, an equation of the form

$$a_1 x_1 + a_2 x_2 + \cdots + a_n x_n = b_n,$$

where $a_1, a_2, \ldots, a_n$, and b_n are real numbers, is a linear equation in the n variables $x_1, x_2, \ldots, x_n$.

In this section we will study **systems of linear equations**. Systems of linear equations are also called **linear systems**.

General Form

A system of m linear equations in n variables, or unknowns, has the general form

$$a_{11}x_1 + a_{12}x_2 + \cdots + a_{1n}x_n = b_1$$
$$a_{21}x_1 + a_{22}x_2 + \cdots + a_{2n}x_n = b_2$$
$$\vdots \qquad\qquad \vdots$$
$$a_{m1}x_1 + a_{m2}x_2 + \cdots + a_{mn}x_n = b_m.$$

(1)

The **coefficients** of the variables in the linear system (1) can be abbreviated as a_{ij}, where i denotes the row and j denotes the column in which the coefficient appears. For example, a_{23} is the coefficient of the unknown in the second row and third column (that is, x_3). Thus, $i = 1, 2, 3, \ldots, m$ and $j = 1, 2, 3, \ldots, n$. The numbers $b_1, b_2, \ldots, b_m$ are called the **constants** of the system. If all the constants are zero, the system (1) is said to be **homogeneous**; otherwise it is **nonhomogeneous**. For example,

<center>this system is homogeneous</center>
<center>↓</center>

$$5x_1 - 9x_2 + x_3 = 0$$
$$x_1 + 3x_2 \qquad = 0$$
$$4x_1 + 6x_2 - x_3 = 0$$

<center>this system is nonhomogeneous</center>
<center>↓</center>

$$2x_1 + 5x_2 + 6x_3 = 1$$
$$4x_1 + 3x_2 - x_3 = 9.$$

Solution

A **solution** of a linear system (1) is a set of n numbers $x_1, x_2, \ldots, x_n$ that satisfies each equation in the system. For example, $x_1 = 3$, $x_2 = -1$ is a solution of the system

$$3x_1 + 6x_2 = 3$$
$$x_1 - 4x_2 = 7.$$

To see this we replace x_1 by 3 and x_2 by -1 in each equation:

$$3(3) + 6(-1) = 9 - 6 = 3 \quad \text{and} \quad 3 - 4(-1) = 3 + 4 = 7.$$

Solutions of linear systems are also written as an ordered n-tuple $(x_1, x_2, \ldots, x_n)$. The solution for the above system is then the ordered pair $(3, -1)$.

A linear system of equations is said to be **consistent** if it has at least one solution and **inconsistent** if it has no solutions. If a linear system is consistent, it has either

- a unique solution (that is, precisely one solution) or
- infinitely many solutions.

Thus, a system of linear equations cannot have, say, exactly three solutions.

For a linear system with two equations and two variables, the lines in the plane, or 2-space, intersect at one point as in **FIGURE 8.2.1(a)** (unique solution), are identical as in Figure 8.2.1(b) (infinitely many solutions), or are parallel as in Figure 8.2.1(c) (no solutions).

For a linear system with three equations and three variables, each equation in the system represents a plane in 3-space. **FIGURE 8.2.2** shows some of the many ways that this kind of linear system can be interpreted.

(a) Consistent

(b) Consistent

(c) Inconsistent

FIGURE 8.2.1 A linear system of two equations in two variables interpreted as lines in 2-space

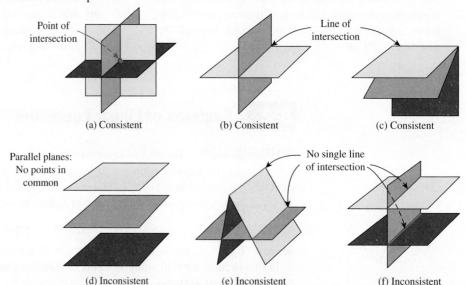

(a) Consistent (b) Consistent (c) Consistent

(d) Inconsistent (e) Inconsistent (f) Inconsistent

FIGURE 8.2.2 A linear system of three equations in three variables interpreted as planes in 3-space

EXAMPLE 1 Verification of a Solution

Verify that $x_1 = 14 + 7t$, $x_2 = 9 + 6t$, $x_3 = t$, where t is any real number, is a solution of the system

$$2x_1 - 3x_2 + 4x_3 = 1$$

$$x_1 - x_2 - x_3 = 5.$$

SOLUTION Replacing x_1, x_2, and x_3 in turn by $14 + 7t$, $9 + 6t$, and t, we have

$$2(14 + 7t) - 3(9 + 6t) + 4t = 1$$

$$14 + 7t - (9 + 6t) - t = 5.$$

For each real number t we obtain a different solution of the system; in other words, the system has an infinite number of solutions. For instance, $t = 0$, $t = 4$, and $t = -2$ give the three solutions

$$x_1 = 14, \quad x_2 = 9, \quad x_3 = 0,$$

$$x_1 = 42, \quad x_2 = 33, \quad x_3 = 4,$$

and $\quad\quad\quad\quad x_1 = 0, \quad x_2 = -3, \quad x_3 = -2,$

respectively. Geometrically, each equation in the system represents a plane in R^3. In this case, the planes intersect in a line as shown in Figure 8.2.2(b). Parametric equations of the line are $x_1 = 14 + 7t$, $x_2 = 9 + 6t$, $x_3 = t$. The solution can also be written as the ordered triple (x_1, x_2, x_3) or $(14 + 7t, 9 + 6t, t)$. ≡

■ **Solving Systems** We can transform a system of linear equations into an equivalent system (that is, one having the same solutions) using the following **elementary operations**:

(*i*) Multiply an equation by a nonzero constant.

(*ii*) Interchange the positions of equations in the system.

(*iii*) Add a nonzero multiple of one equation to any other equation.

As the next example will show, these elementary operations enable us to systematically eliminate variables from the equations of the system.

EXAMPLE 2 Solving a Linear System

Solve $\quad\quad\quad\quad\quad\quad 2x_1 + 6x_2 + x_3 = 7$

$$x_1 + 2x_2 - x_3 = -1$$

$$5x_1 + 7x_2 - 4x_3 = 9.$$

SOLUTION We begin by interchanging the first and second rows:

$$x_1 + 2x_2 - x_3 = -1$$

$$2x_1 + 6x_2 + x_3 = 7$$

$$5x_1 + 7x_2 - 4x_3 = 9.$$

Our goal now is to eliminate x_1 from the second and third equations. If we add to the second equation -2 times the first equation, we obtain the equivalent system

$$x_1 + 2x_2 - x_3 = -1$$

$$2x_2 + 3x_3 = 9$$

$$5x_1 + 7x_2 - 4x_3 = 9.$$

By adding to the third equation -5 times the first equation, we get a new equivalent system:

$$x_1 + 2x_2 - x_3 = -1$$
$$2x_2 + 3x_3 = 9$$
$$-3x_2 + x_3 = 14.$$

We are now going to use the second equation to eliminate the variable x_2 from the first and third equations. To simplify matters, let us multiply the second equation by $\frac{1}{2}$:

$$x_1 + 2x_2 - x_3 = -1$$
$$x_2 + \frac{3}{2}x_3 = \frac{9}{2}$$
$$-3x_2 + x_3 = 14.$$

Adding to the first equation -2 times the second equation yields

$$x_1 \qquad - 4x_3 = -10$$
$$x_2 + \frac{3}{2}x_3 = \frac{9}{2}$$
$$-3x_2 + x_3 = 14.$$

Next, by adding 3 times the second equation to the third equation we get

$$x_1 \qquad - 4x_3 = -10$$
$$x_2 + \frac{3}{2}x_3 = \frac{9}{2}$$
$$\frac{11}{2}x_3 = \frac{55}{2}.$$

We shall use the last equation to eliminate the variable x_3 from the first and second equations. To this end, we multiply the third equation by $\frac{2}{11}$:

$$x_1 \qquad - 4x_3 = -10$$
$$x_2 + \frac{3}{2}x_3 = \frac{9}{2}$$
$$x_3 = 5.$$

At this point we could use **back-substitution**; that is, substitute the value $x_3 = 5$ back into the remaining equations to determine x_1 and x_2. However, by continuing with our systematic elimination, we add to the second equation $-\frac{3}{2}$ times the third equation:

$$x_1 \qquad - 4x_3 = -10$$
$$x_2 \qquad = -3$$
$$x_3 = 5.$$

Finally, by adding to the first equation 4 times the third equation, we obtain

$$x_1 \qquad = 10$$
$$x_2 \qquad = -3$$
$$x_3 = 5.$$

It is now apparent that $x_1 = 10$, $x_2 = -3$, $x_3 = 5$ is the solution of the original system. The answer written as the ordered triple $(10, -3, 5)$ means that the planes represented by the three equations in the system intersect at a point as in Figure 8.2.2(a). ≡

▨ Augmented Matrix

Reflecting on the solution of the linear system in Example 2 should convince you that the solution of the system does not depend on what symbols are used as variables. Thus, the systems

$$
\begin{aligned}
2x + 6y + z &= 7 \\
x + 2y - z &= -1 \\
5x + 7y - 4z &= 9
\end{aligned}
\qquad \text{and} \qquad
\begin{aligned}
2u + 6v + w &= 7 \\
u + 2v - w &= -1 \\
5u + 7v - 4w &= 9
\end{aligned}
$$

have the same solution as the system in Example 2. In other words, in the solution of a linear system, the symbols used to denote the variables are immaterial; it is the coefficients of the variables and the constants that determine the solution of the system. In fact, we can solve a system of form (1) by dropping the variables entirely and performing operations on the rows of the array of coefficients and constants:

$$
\left(\begin{array}{cccc|c}
a_{11} & a_{12} & \cdots & a_{1n} & b_1 \\
a_{21} & a_{22} & \cdots & a_{2n} & b_2 \\
\vdots & & & \vdots & \vdots \\
a_{m1} & a_{m2} & \cdots & a_{mn} & b_m
\end{array} \right). \tag{2}
$$

This array is called the **augmented matrix** of the system or simply the **matrix of the system** (1).

EXAMPLE 3 Augmented Matrices

(a) The augmented matrix $\begin{pmatrix} 1 & -3 & 5 & | & 2 \\ 4 & 7 & -1 & | & 8 \end{pmatrix}$ represents the linear system

$$
\begin{aligned}
x_1 - 3x_2 + 5x_3 &= 2 \\
4x_1 + 7x_2 - x_3 &= 8.
\end{aligned}
$$

(b) The linear system

$$
\begin{aligned}
x_1 - 5x_3 &= -1 \\
2x_1 + 8x_2 &= 7 \\
x_2 + 9x_3 &= 1
\end{aligned}
\qquad \text{is the same as} \qquad
\begin{aligned}
x_1 + 0x_2 - 5x_3 &= -1 \\
2x_1 + 8x_2 + 0x_3 &= 7 \\
0x_1 + x_2 + 9x_3 &= 1.
\end{aligned}
$$

Thus the matrix of the system is

$$
\left(\begin{array}{ccc|c}
1 & 0 & -5 & -1 \\
2 & 8 & 0 & 7 \\
0 & 1 & 9 & 1
\end{array} \right).
$$

≡

▨ Elementary Row Operations

Since the rows of an augmented matrix represent the equations in a linear system, the three elementary operations on a linear system listed previously are equivalent to the following **elementary row operations** on a matrix:

(*i*) Multiply a row by a nonzero constant.
(*ii*) Interchange any two rows.
(*iii*) Add a nonzero multiple of one row to any other row.

Of course, when we add a multiple of one row to another, we add the corresponding entries in the rows. We say that two matrices are **row equivalent** if one can be obtained from the other through a sequence of elementary row operations. The procedure of carrying out elementary row operations on a matrix to obtain a row-equivalent matrix is called **row reduction**.

■ **Elimination Methods** To solve a system such as (1) using an augmented matrix, we shall use either **Gaussian elimination** or the **Gauss–Jordan elimination method**. In the former method, we row-reduce the augmented matrix of the system until we arrive at a row-equivalent augmented matrix in **row-echelon form**:

(*i*) The first nonzero entry in a nonzero row is a 1.

(*ii*) In consecutive nonzero rows, the first entry 1 in the lower row appears to the right of the 1 in the higher row.

(*iii*) Rows consisting of all zeros are at the bottom of the matrix.

In the Gauss–Jordan method, the row operations are continued until we obtain an augmented matrix that is in **reduced row-echelon form**. A reduced row-echelon matrix has the same three properties listed previously, but in addition:

(*iv*) A column containing a first entry 1 has zeros everywhere else.

EXAMPLE 4 **Echelon Forms**

(a) The augmented matrices

$$\begin{pmatrix} 1 & 5 & 0 & | & 2 \\ 0 & 1 & 0 & | & -1 \\ 0 & 0 & 0 & | & 0 \end{pmatrix} \quad \text{and} \quad \begin{pmatrix} 0 & 0 & 1 & -6 & 2 & | & 2 \\ 0 & 0 & 0 & 0 & 1 & | & 4 \end{pmatrix}$$

are in row-echelon form. The reader should verify that the three criteria for this form are satisfied.

(b) The augmented matrices

$$\begin{pmatrix} 1 & 0 & 0 & | & 7 \\ 0 & 1 & 0 & | & -1 \\ 0 & 0 & 0 & | & 0 \end{pmatrix} \quad \text{and} \quad \begin{pmatrix} 0 & 0 & 1 & -6 & 0 & | & -6 \\ 0 & 0 & 0 & 0 & 1 & | & 4 \end{pmatrix}$$

are in reduced row-echelon form. Note that the remaining entries in the columns that contain a leading entry 1 are all zeros. ≡

It should be noted that in Gaussian elimination, we stop when we have obtained *an* augmented matrix in row-echelon form. In other words, by using different sequences of row operations, we may arrive at different row-echelon forms. This method then requires the use of back-substitution. In Gauss–Jordan elimination, we stop when we have obtained *the* augmented matrix in reduced row-echelon form. Any sequence of row operations will lead to the same augmented matrix in reduced row-echelon form. This method does not require back-substitution; the solution of the system will be apparent by inspection of the final matrix. In terms of the equations of the original system, our goal in both methods is simply to make the coefficient of x_1 in the first equation* equal to one and then use multiples of that equation to eliminate x_1 from other equations. The process is repeated for the other variables.

To keep track of the row operations used on an augmented matrix, we shall utilize the following notation:

Note: Row operations can lead to different row-echelon forms. ▶

Symbol	Meaning
$R_i \leftrightarrow R_j$	Interchange rows i and j
cR_i	Multiply the ith row by the nonzero constant c
$cR_i + R_j$	Multiply the ith row by c and add to the jth row

*We can always interchange equations so that the first equation contains the variable x_1.

Solve the linear system in Example 2 using **(a)** Gaussian elimination and **(b)** Gauss–Jordan elimination.

SOLUTION **(a)** Using row operations on the augmented matrix of the system, we obtain:

$$\begin{pmatrix} 2 & 6 & 1 & | & 7 \\ 1 & 2 & -1 & | & -1 \\ 5 & 7 & -4 & | & 9 \end{pmatrix} \overset{R_1 \leftrightarrow R_2}{\Rightarrow} \begin{pmatrix} 1 & 2 & -1 & | & -1 \\ 2 & 6 & 1 & | & 7 \\ 5 & 7 & -4 & | & 9 \end{pmatrix}$$

$$\overset{\substack{-2R_1 + R_2 \\ -5R_1 + R_3}}{\Rightarrow} \begin{pmatrix} 1 & 2 & -1 & | & -1 \\ 0 & 2 & 3 & | & 9 \\ 0 & -3 & 1 & | & 14 \end{pmatrix} \overset{\frac{1}{2}R_2}{\Rightarrow} \begin{pmatrix} 1 & 2 & -1 & | & -1 \\ 0 & 1 & \frac{3}{2} & | & \frac{9}{2} \\ 0 & -3 & 1 & | & 14 \end{pmatrix}$$

$$\overset{3R_2 + R_3}{\Rightarrow} \begin{pmatrix} 1 & 2 & -1 & | & -1 \\ 0 & 1 & \frac{3}{2} & | & \frac{9}{2} \\ 0 & 0 & \frac{11}{2} & | & \frac{55}{2} \end{pmatrix} \overset{\frac{2}{11}R_3}{\Rightarrow} \begin{pmatrix} 1 & 2 & -1 & | & -1 \\ 0 & 1 & \frac{3}{2} & | & \frac{9}{2} \\ 0 & 0 & 1 & | & 5 \end{pmatrix}.$$

The last matrix is in row-echelon form and represents the system

$$x_1 + 2x_2 - x_3 = -1$$

$$x_2 + \frac{3}{2}x_3 = \frac{9}{2}$$

$$x_3 = 5.$$

Substituting $x_3 = 5$ into the second equation gives $x_2 = -3$. Substituting both these values back into the first equation finally yields $x_1 = 10$.

(b) We start with the last matrix above. Since the first entries in the second and third rows are ones, we must, in turn, make the remaining entries in the second and third columns zeros:

$$\begin{pmatrix} 1 & 2 & -1 & | & -1 \\ 0 & 1 & \frac{3}{2} & | & \frac{9}{2} \\ 0 & 0 & 1 & | & 5 \end{pmatrix} \overset{-2R_2 + R_1}{\Rightarrow} \begin{pmatrix} 1 & 0 & -4 & | & -10 \\ 0 & 1 & \frac{3}{2} & | & \frac{9}{2} \\ 0 & 0 & 1 & | & 5 \end{pmatrix} \overset{\substack{-4R_3 + R_1 \\ -\frac{3}{2}R_3 + R_2}}{\Rightarrow} \begin{pmatrix} 1 & 0 & 0 & | & 10 \\ 0 & 1 & 0 & | & -3 \\ 0 & 0 & 1 & | & 5 \end{pmatrix}$$

The last matrix is in reduced row-echelon form. Bearing in mind what the matrix means in terms of equations, we see that the solution of the system is $x_1 = 10$, $x_2 = -3$, $x_3 = 5$. ≡

EXAMPLE 6 **Gauss–Jordan Elimination**

Use Gauss–Jordan elimination to solve

$$x_1 + 3x_2 - 2x_3 = -7$$

$$4x_1 + x_2 + 3x_3 = 5$$

$$2x_1 - 5x_2 + 7x_3 = 19.$$

SOLUTION Row operations give

$$\begin{pmatrix} 1 & 3 & -2 & | & -7 \\ 4 & 1 & 3 & | & 5 \\ 2 & -5 & 7 & | & 19 \end{pmatrix} \overset{\substack{-4R_1 + R_2 \\ -2R_1 + R_3}}{\Rightarrow} \begin{pmatrix} 1 & 3 & -2 & | & -7 \\ 0 & -11 & 11 & | & 33 \\ 0 & -11 & 11 & | & 33 \end{pmatrix}$$

$$\overset{-\frac{1}{11}R_2}{\Rightarrow} \begin{pmatrix} 1 & 3 & -2 & | & -7 \\ 0 & 1 & -1 & | & -3 \\ 0 & 1 & -1 & | & -3 \end{pmatrix} \overset{\substack{-3R_2 + R_1 \\ -R_2 + R_3}}{\Rightarrow} \begin{pmatrix} 1 & 0 & 1 & | & 2 \\ 0 & 1 & -1 & | & -3 \\ 0 & 0 & 0 & | & 0 \end{pmatrix}.$$

In this case, the last matrix in reduced row-echelon form implies that the original system of three equations in three variables is really equivalent to two equations in the variables.

Since only x_3 is common to both equations (the nonzero rows), we can assign its values arbitrarily. If we let $x_3 = t$, where t represents any real number, then we see that the system has infinitely many solutions: $x_1 = 2 - t, x_2 = -3 + t, x_3 = t$. Geometrically, these equations are the parametric equations for the line of intersection of the planes $x_1 + 0x_2 + x_3 = 2$ and $0x_1 + x_2 - x_3 = -3$. ≡

EXAMPLE 7 **Inconsistent System**

Solve

$$x_1 + x_2 = 1$$
$$4x_1 - x_2 = -6$$
$$2x_1 - 3x_2 = 8.$$

SOLUTION In the process of applying Gauss–Jordan elimination to the matrix of the system, we stop at

$$\begin{pmatrix} 1 & 1 & | & 1 \\ 4 & -1 & | & -6 \\ 2 & -3 & | & 8 \end{pmatrix} \xrightarrow[\text{operations}]{\text{row}} \begin{pmatrix} 1 & 0 & | & -1 \\ 0 & 1 & | & 2 \\ 0 & 0 & | & 16 \end{pmatrix}.$$

The third row of the last matrix means $0x_1 + 0x_2 = 16$ (or $0 = 16$). Since no numbers x_1 and x_2 can satisfy this equation, we conclude that the system has no solution. ≡

Worth remembering. ▶ Inconsistent systems of m linear equations in n variables will always yield the situation illustrated in Example 7; that is, there will be a row in the reduced row-echelon form of the augmented matrix in which the first n entries are zero and the $(n + 1)$st entry is nonzero.

■ **Networks** The currents in the branches of an electrical network can be determined by using Kirchhoff's point and loop rules:

Point rule: The algebraic sum of the currents toward any branch point is 0.

Loop rule: The algebraic sum of the potential differences in any loop is 0.

When a loop is traversed in a chosen direction (clockwise or counterclockwise), an emf is taken to be *positive* when it is traversed from $-$ to $+$ and *negative* when traversed from $+$ to $-$. An iR product is taken to be *positive* if the chosen direction through the resistor is opposite that of the assumed current, and *negative* if the chosen direction is in the same direction as the assumed current.

In **FIGURE 8.2.3**, the branch points of the network are labeled A and B, the loops are labeled L_1 and L_2, and the chosen direction in each loop is clockwise. Now, applying the foregoing rules to the network yields the nonhomogeneous system of linear equations

$$i_1 - i_2 - i_3 = 0 \qquad\qquad i_1 - i_2 - i_3 = 0$$
$$E - i_1R_1 - i_2R_2 = 0 \quad \text{or} \quad i_1R_1 + i_2R_2 = E \qquad (3)$$
$$i_2R_2 - i_3R_3 = 0 \qquad\qquad i_2R_2 - i_3R_3 = 0.$$

FIGURE 8.2.3 Electrical network

EXAMPLE 8 **Currents in a Network**

Use Gauss–Jordan elimination to solve the system (3) when $R_1 = 10$ ohms, $R_2 = 20$ ohms, $R_3 = 10$ ohms, and $E = 12$ volts.

SOLUTION The system to be solved is

$$i_1 - i_2 - i_3 = 0$$
$$10i_1 + 20i_2 = 12$$
$$20i_2 - 10i_3 = 0.$$

In this case, Gauss–Jordan elimination yields

$$\begin{pmatrix} 1 & -1 & -1 & 0 \\ 10 & 20 & 0 & 12 \\ 0 & 20 & -10 & 0 \end{pmatrix} \overset{\text{row}}{\underset{\text{operations}}{\Rightarrow}} \begin{pmatrix} 1 & 0 & 0 & \frac{18}{25} \\ 0 & 1 & 0 & \frac{6}{25} \\ 0 & 0 & 1 & \frac{12}{25} \end{pmatrix}.$$

Hence, we see that the currents in the three branches are $i_1 = \frac{18}{25} = 0.72$ ampere, $i_2 = \frac{6}{25} = 0.24$ ampere, and $i_3 = \frac{12}{25} = 0.48$ ampere. $\equiv$

■ **Homogeneous Systems** All the systems in the preceding examples are nonhomogeneous systems. As we have seen, a nonhomogeneous system can be consistent or inconsistent. By contrast, a homogeneous system of linear equations

$$a_{11}x_1 + a_{12}x_2 + \cdots + a_{1n}x_n = 0$$
$$a_{12}x_1 + a_{22}x_2 + \cdots + a_{2n}x_n = 0$$
$$\vdots \qquad\qquad \vdots \qquad\qquad (4)$$
$$a_{m1}x_1 + a_{m2}x_2 + \cdots + a_{mn}x_n = 0$$

is *always consistent*, since $x_1 = 0$, $x_2 = 0$, ..., $x_n = 0$ will satisfy each equation in the system. The solution consisting of all zeros is called the **trivial solution**. But naturally we are interested in whether a system of form (4) has any solutions for which *some* of the x_i, $i = 1, 2, ..., n$, are not zero. Such a solution is called a **nontrivial solution**. A homogeneous system either possesses only the trivial solution or possesses the trivial solution along with infinitely many nontrivial solutions. The next theorem, presented without proof, will give us a sufficient condition for the existence of nontrivial solutions.

THEOREM 8.2.1 Existence of Nontrivial Solutions

A homogeneous system of form (4) possesses nontrivial solutions if the number m of equations is less than the number n of variables ($m < n$).

EXAMPLE 9 Solving a Homogeneous System

Solve
$$2x_1 - 4x_2 + 3x_3 = 0$$
$$x_1 + x_2 - 2x_3 = 0.$$

SOLUTION Since the number of equations is less than the number of variables, we know from Theorem 8.2.1 that the given system has nontrivial solutions. Using Gauss–Jordan elimination, we find

$$\begin{pmatrix} 2 & -4 & 3 & 0 \\ 1 & 1 & -2 & 0 \end{pmatrix} \overset{R_1 \leftrightarrow R_2}{\Rightarrow} \begin{pmatrix} 1 & 1 & -2 & 0 \\ 2 & -4 & 3 & 0 \end{pmatrix} \overset{-2R_1+R_2}{\Rightarrow} \begin{pmatrix} 1 & 1 & -2 & 0 \\ 0 & -6 & 7 & 0 \end{pmatrix}$$

$$\overset{-\frac{1}{6}R_2}{\Rightarrow} \begin{pmatrix} 1 & 1 & -2 & 0 \\ 0 & 1 & -\frac{7}{6} & 0 \end{pmatrix} \overset{-R_2+R_1}{\Rightarrow} \begin{pmatrix} 1 & 0 & -\frac{5}{6} & 0 \\ 0 & 1 & -\frac{7}{6} & 0 \end{pmatrix}.$$

As in Example 6, if $x_3 = t$, then the solution of the system is $x_1 = \frac{5}{6}t$, $x_2 = \frac{7}{6}t$, $x_3 = t$. Note we obtain the trivial solution $x_1 = 0$, $x_2 = 0$, $x_3 = 0$ for this system by choosing $t = 0$. For $t \neq 0$ we get nontrivial solutions. For example, the solutions corresponding to $t = 6$, $t = -12$, and $t = 3$ are, in turn, $x_1 = 5$, $x_2 = 7$, $x_3 = 6$; $x_1 = -10$, $x_2 = -14$, $x_3 = -12$; and $x_1 = \frac{5}{2}$, $x_2 = \frac{7}{2}$, $x_3 = 3$. $\equiv$

■ **Chemical Equations** The next example will give an application of homogeneous systems in chemistry.

EXAMPLE 10 **Balancing a Chemical Equation**

Balance the chemical equation $C_2H_6 + O_2 \rightarrow CO_2 + H_2O$.

SOLUTION We want to find positive integers x_1, x_2, x_3, and x_4 so that

$$x_1 C_2H_6 + x_2 O_2 \rightarrow x_3 CO_2 + x_4 H_2O.$$

Because the number of atoms of each element must be the same on each side of the last equation, we have

carbon (C): $2x_1 = x_3$	$2x_1 + 0x_2 - x_3 + 0x_4 = 0$
hydrogen (H): $6x_1 = 2x_4$ or	$6x_1 + 0x_2 + 0x_3 - 2x_4 = 0$
oxygen (O): $2x_2 = 2x_3 + x_4$	$0x_1 + 2x_2 - 2x_3 - x_4 = 0.$

Gauss–Jordan elimination gives

$$\begin{pmatrix} 2 & 0 & -1 & 0 & | & 0 \\ 6 & 0 & 0 & -2 & | & 0 \\ 0 & 2 & -2 & -1 & | & 0 \end{pmatrix} \overset{\underset{\text{operations}}{\text{row}}}{\Longrightarrow} \begin{pmatrix} 1 & 0 & 0 & -\frac{1}{3} & | & 0 \\ 0 & 1 & 0 & -\frac{7}{6} & | & 0 \\ 0 & 0 & 1 & -\frac{2}{3} & | & 0 \end{pmatrix}$$

and so $x_1 = \frac{1}{3}t$, $x_2 = \frac{7}{6}t$, $x_3 = \frac{2}{3}t$, $x_4 = t$. In this case t must be a positive integer chosen in such a manner so that x_1, x_2, and x_3 are positive integers. To accomplish this we pick $t = 6$. This gives $x_1 = 2$, $x_2 = 7$, $x_3 = 4$, $x_4 = 6$. The balanced chemical equation is then

$$2C_2H_6 + 7O_2 \rightarrow 4CO_2 + 6H_2O. \qquad \equiv$$

Notation In view of matrix multiplication and equality of matrices defined in Section 8.1, if we define matrices **A**, **X**, and **B** as

$$\mathbf{A} = \begin{pmatrix} a_{11} & a_{12} & \cdots & a_{1n} \\ a_{21} & a_{22} & \cdots & a_{2n} \\ \vdots & \vdots & & \vdots \\ a_{m1} & a_{m2} & \cdots & a_{mn} \end{pmatrix}, \quad \mathbf{X} = \begin{pmatrix} x_1 \\ x_2 \\ \vdots \\ x_n \end{pmatrix}, \quad \mathbf{B} = \begin{pmatrix} b_1 \\ b_2 \\ \vdots \\ b_m \end{pmatrix},$$

then observe that the linear system (1) can be written as the matrix equation

$$\begin{pmatrix} a_{11} & a_{12} & \cdots & a_{1n} \\ a_{21} & a_{22} & \cdots & a_{2n} \\ \vdots & \vdots & & \vdots \\ a_{m1} & a_{m2} & \cdots & a_{mn} \end{pmatrix} \begin{pmatrix} x_1 \\ x_2 \\ \vdots \\ x_n \end{pmatrix} = \begin{pmatrix} b_1 \\ b_2 \\ \vdots \\ b_m \end{pmatrix}$$

or more compactly as

$$\mathbf{AX} = \mathbf{B},$$

where the matrix **A** is, naturally, called the **coefficient matrix**. The augmented matrix of a system $\mathbf{AX} = \mathbf{B}$ is often denoted by $(\mathbf{A} | \mathbf{B})$. A homogeneous system of equations is written $\mathbf{AX} = \mathbf{0}$. This notation makes it easier to discuss linear systems and to prove theorems for such systems.

THEOREM 8.2.2 Two Properties of Homogeneous Systems

Let $\mathbf{AX} = \mathbf{0}$ denote a homogeneous system of linear equations.

(*i*) If $\mathbf{X}_1$ is a solution of $\mathbf{AX} = \mathbf{0}$, then so is $c\mathbf{X}_1$ for any constant c.
(*ii*) If $\mathbf{X}_1$ and $\mathbf{X}_2$ are solutions of $\mathbf{AX} = \mathbf{0}$, then so is $\mathbf{X}_1 + \mathbf{X}_2$.

PROOF: (*i*) Because $\mathbf{X}_1$ is a solution, then $\mathbf{AX}_1 = \mathbf{0}$. Now

$$\mathbf{A}(c\mathbf{X}_1) = c(\mathbf{AX}_1) = c\mathbf{0} = \mathbf{0}.$$

This shows that a constant multiple of a solution of a homogeneous linear system is also a solution.
(*ii*) Because $\mathbf{X}_1$ and $\mathbf{X}_2$ are solutions, then $\mathbf{AX}_1 = \mathbf{0}$ and $\mathbf{AX}_2 = \mathbf{0}$. Now

$$\mathbf{A}(\mathbf{X}_1 + \mathbf{X}_2) = \mathbf{AX}_1 + \mathbf{AX}_2 = \mathbf{0} + \mathbf{0} = \mathbf{0}.$$

This shows that $\mathbf{X}_1 + \mathbf{X}_2$ is a solution. ≡

By combining parts (*i*) and (*ii*) of Theorem 8.2.2 we can say that if $\mathbf{X}_1$ and $\mathbf{X}_2$ are solutions of $\mathbf{AX} = \mathbf{0}$, then so is the linear combination

$$c_1\mathbf{X}_1 + c_2\mathbf{X}_2,$$

where c_1 and c_2 are constants. Moreover, this **superposition principle** extends to three or more solutions of $\mathbf{AX} = \mathbf{0}$.

EXAMPLE 11 Example 9 Revisited

At the end of Example 9, we obtained three distinct solutions

$$\mathbf{X}_1 = \begin{pmatrix} 5 \\ 7 \\ 6 \end{pmatrix}, \quad \mathbf{X}_2 = \begin{pmatrix} -10 \\ -14 \\ -12 \end{pmatrix}, \quad \text{and} \quad \mathbf{X}_3 = \begin{pmatrix} \frac{5}{2} \\ \frac{7}{2} \\ 3 \end{pmatrix}$$

of the given homogeneous linear system. It follows from the preceding discussion that

$$\mathbf{X} = \mathbf{X}_1 + \mathbf{X}_2 + \mathbf{X}_3 = \begin{pmatrix} 5 \\ 7 \\ 6 \end{pmatrix} + \begin{pmatrix} -10 \\ -14 \\ -12 \end{pmatrix} + \begin{pmatrix} \frac{5}{2} \\ \frac{7}{2} \\ 3 \end{pmatrix} = \begin{pmatrix} -\frac{5}{2} \\ -\frac{7}{2} \\ -3 \end{pmatrix}$$

is also a solution of the system. Verify this. ≡

■ **Terminology** Suppose a linear system has m equations and n variables. If there are more equations than variables, that is, $m > n$, then the system is said to be **overdetermined**. If the system has fewer equations than variables, that is, $m < n$, then the system is called **underdetermined**. An overdetermined linear system may put too many constraints on the variables and so is usually—not always—inconsistent. The system in Example 7 is overdetermined and is inconsistent. On the other hand, an underdetermined system is usually—but not always—consistent. The systems in Examples 9 and 10 are underdetermined and are consistent. It should be noted that it is impossible for a consistent underdetermined system to possess a single or unique solution. To see this, suppose that $m < n$. If Gaussian elimination is used to solve the system, then the row-echelon form that is row equivalent to the matrix of the system will contain r nonzero rows where $r \leq m < n$. Thus we can solve for r of the variables in terms of $n - r > 0$ variables or parameters. If the underdetermined system is consistent, then those remaining $n - r$ variables can be chosen arbitrarily and so the system has an infinite number of solutions.

REMARKS

(*i*) To solve large systems of linear equations, we obviously need the help of a computer. Since the two methods in this section are so systematic, they can be programmed easily. However, the requirement that each nonzero row start with a one may at times demand division by a very small number. Problems can then occur. Large systems are often solved indirectly—that is, by an approximation technique such as **Gauss–Seidel iteration**. See Section 16.1.

(*ii*) Since Gauss–Jordan elimination avoids the necessity of back-substitution, it would appear that it is the more efficient of the two methods we have considered. Actually, this is not the case. It has been shown that for large systems, Gauss–Jordan elimination can require about 50% more operations than Gaussian elimination.

(*iii*) A consistent nonhomogeneous linear system $\mathbf{AX} = \mathbf{B}$, $\mathbf{B} \neq \mathbf{0}$, shares a property with nonhomogeneous linear differential equations. If $\mathbf{X}_h$ is a solution of the associated homogeneous system $\mathbf{AX} = \mathbf{0}$, and $\mathbf{X}_p$ is a particular solution of the nonhomogeneous system $\mathbf{AX} = \mathbf{B}$, then the superposition $\mathbf{X}_h + \mathbf{X}_p$ is also a solution of the nonhomogeneous system. This is readily verified by the distributive law of matrix multiplication:

$$\mathbf{A}(\mathbf{X}_h + \mathbf{X}_p) = \mathbf{AX}_h + \mathbf{AX}_p = \mathbf{0} + \mathbf{B} = \mathbf{B}.$$

(*iv*) The set S of all solution vectors $\mathbf{X}$ of a homogeneous linear system with m equations and n variables,

$$S = \{\mathbf{X} \mid \mathbf{AX} = \mathbf{0}\}$$

is called the **nullspace** of the matrix $\mathbf{A}$. If you have studied Section 7.6, then S is a *subspace* of the vector space R^n. See pages 361 and 362. This fact follows immediately from Theorems 7.6.1 and 8.2.2.

8.2 Exercises Answers to selected odd-numbered problems begin on page ANS-18.

In Problems 1–20, use either Gaussian elimination or Gauss–Jordan elimination to solve the given system or show that no solution exists.

1. $\begin{aligned} x_1 - x_2 &= 11 \\ 4x_1 + 3x_2 &= -5 \end{aligned}$

2. $\begin{aligned} 3x_1 - 2x_2 &= 4 \\ x_1 - x_2 &= -2 \end{aligned}$

3. $\begin{aligned} 9x_1 + 3x_2 &= -5 \\ 2x_1 + x_2 &= -1 \end{aligned}$

4. $\begin{aligned} 10x_1 + 15x_2 &= 1 \\ 3x_1 + 2x_2 &= -1 \end{aligned}$

5. $\begin{aligned} x_1 - x_2 - x_3 &= -3 \\ 2x_1 + 3x_2 + 5x_3 &= 7 \\ x_1 - 2x_2 + 3x_3 &= -11 \end{aligned}$

6. $\begin{aligned} x_1 + 2x_2 - x_3 &= 0 \\ 2x_1 + x_2 + 2x_3 &= 9 \\ x_1 - x_2 + x_3 &= 3 \end{aligned}$

7. $\begin{aligned} x_1 + x_2 + x_3 &= 0 \\ x_1 + x_2 + 3x_3 &= 0 \end{aligned}$

8. $\begin{aligned} x_1 + 2x_2 - 4x_3 &= 9 \\ 5x_1 - x_2 + 2x_3 &= 1 \end{aligned}$

9. $\begin{aligned} x_1 - x_2 - x_3 &= 8 \\ x_1 - x_2 + x_3 &= 3 \\ -x_1 + x_2 + x_3 &= 4 \end{aligned}$

10. $\begin{aligned} 3x_1 + x_2 &= 4 \\ 4x_1 + 3x_2 &= -3 \\ 2x_1 - x_2 &= 11 \end{aligned}$

11. $\begin{aligned} 2x_1 + 2x_2 &= 0 \\ -2x_1 + x_2 + x_3 &= 0 \\ 3x_1 + x_3 &= 0 \end{aligned}$

12. $\begin{aligned} x_1 - x_2 - 2x_3 &= 0 \\ 2x_1 + 4x_2 + 5x_3 &= 0 \\ 6x_1 - 3x_3 &= 0 \end{aligned}$

13. $\begin{aligned} x_1 + 2x_2 + 2x_3 &= 2 \\ x_1 + x_2 + x_3 &= 0 \\ x_1 - 3x_2 - x_3 &= 0 \end{aligned}$

14. $\begin{aligned} x_1 - 2x_2 + x_3 &= 2 \\ 3x_1 - x_2 + 2x_3 &= 5 \\ 2x_1 + x_2 + x_3 &= 1 \end{aligned}$

15. $\begin{aligned} x_1 + x_2 + x_3 &= 3 \\ x_1 - x_2 - x_3 &= -1 \\ 3x_1 + x_2 + x_3 &= 5 \end{aligned}$

16. $\begin{aligned} x_1 - x_2 - 2x_3 &= -1 \\ -3x_1 - 2x_2 + x_3 &= -7 \\ 2x_1 + 3x_2 + x_3 &= 8 \end{aligned}$

17. $\begin{aligned} x_1 + x_3 - x_4 &= 1 \\ 2x_2 + x_3 + x_4 &= 3 \\ x_1 - x_2 + x_4 &= -1 \\ x_1 + x_2 + x_3 + x_4 &= 2 \end{aligned}$

18. $\begin{aligned} 2x_1 + x_2 + x_3 &= 3 \\ 3x_1 + x_2 + x_3 + x_4 &= 4 \\ x_1 + 2x_2 + 2x_3 + 3x_4 &= 3 \\ 4x_1 + 5x_2 - 2x_3 + x_4 &= 16 \end{aligned}$

19. $\begin{aligned} x_2 + x_3 - x_4 &= 4 \\ x_1 + 3x_2 + 5x_3 - x_4 &= 1 \\ x_1 + 2x_2 + 5x_3 - 4x_4 &= -2 \\ x_1 + 4x_2 + 6x_3 - 2x_4 &= 6 \end{aligned}$

20. $\begin{aligned} x_1 + 2x_2 + x_4 &= 0 \\ 4x_1 + 9x_2 + x_3 + 12x_4 &= 0 \\ 3x_1 + 9x_2 + 6x_3 + 21x_4 &= 0 \\ x_1 + 3x_2 + x_3 + 9x_4 &= 0 \end{aligned}$

In Problems 21 and 22, use a calculator to solve the given system.

21.
$$x_1 + x_2 + x_3 = 4.280$$
$$0.2x_1 - 0.1x_2 - 0.5x_3 = -1.978$$
$$4.1x_1 + 0.3x_2 + 0.12x_3 = 1.686$$

22.
$$2.5x_1 + 1.4x_2 + 4.5x_3 = 2.6170$$
$$1.35x_1 + 0.95x_2 + 1.2x_3 = 0.7545$$
$$2.7x_1 + 3.05x_2 - 1.44x_3 = -1.4292$$

In Problems 23–28, use the procedures illustrated in Example 10 to balance the given chemical equation.

23. $Na + H_2O \rightarrow NaOH + H_2$

24. $KClO_3 \rightarrow KCl + O_2$

25. $Fe_3O_4 + C \rightarrow Fe + CO$

26. $C_5H_8 + O_2 \rightarrow CO_2 + H_2O$

27. $Cu + HNO_3 \rightarrow Cu(NO_3)_2 + H_2O + NO$

28. $Ca_3(PO_4)_2 + H_3PO_4 \rightarrow Ca(H_2PO_4)_2$

In Problems 29 and 30, set up and solve the system of equations for the currents in the branches of the given network.

29.

FIGURE 8.2.4 Network in Problem 29

30.

FIGURE 8.2.5 Network in Problem 30

In Problems 31 and 32, write the homogeneous system of linear equations in the form $AX = 0$. Then verify by matrix multiplication that the given matrix X is a solution of the system for any real number c_1.

31.
$$\begin{cases} x_1 + x_2 + x_3 = 0 \\ 5x_1 - 2x_2 + 2x_3 = 0, \\ 8x_1 + x_2 + 5x_3 = 0 \end{cases} \quad X = c_1 \begin{pmatrix} 4 \\ 3 \\ -7 \end{pmatrix}$$

32.
$$\begin{cases} x_1 + x_2 + x_3 + x_4 = 0 \\ -x_1 + x_2 - x_3 + x_4 = 0 \\ x_1 + x_2 - x_3 - x_4 = 0, \\ 3x_1 + x_2 + x_3 - x_4 = 0 \end{cases} \quad X = c_1 \begin{pmatrix} 1 \\ -1 \\ -1 \\ 1 \end{pmatrix}$$

In Problems 33 and 34, write the nonhomogeneous system of linear equations in the form $AX = B$. Then verify by matrix multiplication that the given matrix X is a solution of the system for any real numbers c_1 and c_2.

33.
$$\begin{cases} 2x_1 - 3x_2 + x_3 = -12 \\ x_1 + x_2 - x_3 = 1 \\ 4x_1 - x_2 - x_3 = -10 \end{cases}, \quad X = \begin{pmatrix} -1 \\ 4 \\ 2 \end{pmatrix} + c_1 \begin{pmatrix} 2 \\ 3 \\ 5 \end{pmatrix}$$

34.
$$\begin{cases} x_1 + 2x_2 - 3x_3 = 2 \\ 2x_1 + 4x_2 - 6x_3 = 4, \\ 3x_1 + 6x_2 - 9x_3 = 6 \end{cases} \quad X = \begin{pmatrix} 2 \\ 0 \\ 0 \end{pmatrix} + c_1 \begin{pmatrix} 1 \\ 1 \\ 1 \end{pmatrix} + c_2 \begin{pmatrix} -2 \\ 1 \\ 0 \end{pmatrix}$$

An **elementary matrix E** is one obtained by performing a single row operation on the identity matrix I. In Problems 35–38, verify that the given matrix is an elementary matrix.

35. $\begin{pmatrix} 0 & 1 & 0 \\ 1 & 0 & 0 \\ 0 & 0 & 1 \end{pmatrix}$

36. $\begin{pmatrix} 1 & 0 & 0 \\ 0 & 1 & 0 \\ 0 & 0 & c \end{pmatrix}$

37. $\begin{pmatrix} 1 & 0 & 0 \\ 0 & 1 & 0 \\ 0 & c & 1 \end{pmatrix}$

38. $\begin{pmatrix} 1 & 0 & 0 & 1 \\ 0 & 1 & 0 & 0 \\ 0 & 0 & 1 & 0 \\ 0 & 0 & 0 & 1 \end{pmatrix}$

If a matrix A is premultiplied by an elementary matrix E, the product EA will be that matrix obtained from A by performing the elementary row operation symbolized by E. In Problems 39–42, compute the given product for an arbitrary 3×3 matrix A.

39. $\begin{pmatrix} 0 & 1 & 0 \\ 1 & 0 & 0 \\ 0 & 0 & 1 \end{pmatrix} A$

40. $\begin{pmatrix} 1 & 0 & 0 \\ 0 & 1 & 0 \\ 0 & 0 & c \end{pmatrix} A$

41. $\begin{pmatrix} 1 & 0 & 0 \\ 0 & 1 & 0 \\ 0 & c & 1 \end{pmatrix} A$

42. $\begin{pmatrix} 0 & 1 & 0 \\ 1 & 0 & 0 \\ 0 & 0 & 1 \end{pmatrix} \begin{pmatrix} 1 & 0 & 0 \\ 0 & 1 & 0 \\ 0 & c & 1 \end{pmatrix} A$

Computer Lab Assignments

In Problems 43–46, use a CAS to solve the given system.

43.
$$1.567x_1 - 3.48x_2 + 5.22x_3 = 1.045$$
$$3.56x_1 + 4.118x_2 + 1.57x_3 = -1.625$$

44.
$$x_1 + 2x_2 - 2x_3 = 0$$
$$2x_1 - 2x_2 + x_3 = 0$$
$$3x_1 - 6x_2 + 4x_3 = 0$$
$$4x_1 + 14x_2 - 13x_3 = 0$$

45.
$$1.2x_1 + 3.5x_2 - 4.4x_3 + 3.1x_4 = 1.8$$
$$0.2x_1 - 6.1x_2 - 2.3x_3 + 5.4x_4 = -0.6$$
$$3.3x_1 - 3.5x_2 - 2.4x_3 - 0.1x_4 = 2.5$$
$$5.2x_1 + 8.5x_2 - 4.4x_3 - 2.9x_4 = 0$$

46.
$$x_1 - x_2 - x_3 + 2x_4 - x_5 = 5$$
$$6x_1 + 9x_2 - 6x_3 + 17x_4 - x_5 = 40$$
$$2x_1 + x_2 - 2x_3 + 5x_4 - x_5 = 12$$
$$x_1 + 2x_2 - x_3 + 3x_4 = 7$$
$$x_1 + 2x_2 + x_3 + 3x_4 = 1$$

8.3 Rank of a Matrix

INTRODUCTION In a general $m \times n$ matrix,

$$
\mathbf{A} = \begin{pmatrix}
a_{11} & a_{12} & \cdots & a_{1n} \\
a_{21} & a_{22} & \cdots & a_{2n} \\
\vdots & & & \vdots \\
a_{m1} & a_{m2} & \cdots & a_{mn}
\end{pmatrix},
$$

the rows

$$
\mathbf{u}_1 = (a_{11}\, a_{12}\, \ldots\, a_{1n}), \quad \mathbf{u}_2 = (a_{21}\, a_{22}\, \ldots\, a_{2n}), \quad \ldots, \quad \mathbf{u}_m = (a_{m1}\, a_{m2}\, \ldots\, a_{mn})
$$

and columns

$$
\mathbf{v}_1 = \begin{pmatrix} a_{11} \\ a_{21} \\ \vdots \\ a_{m1} \end{pmatrix}, \quad
\mathbf{v}_2 = \begin{pmatrix} a_{12} \\ a_{22} \\ \vdots \\ a_{m2} \end{pmatrix}, \ldots, \quad
\mathbf{v}_n = \begin{pmatrix} a_{1n} \\ a_{2n} \\ \vdots \\ a_{mn} \end{pmatrix}
$$

are called the **row vectors** of $\mathbf{A}$ and the **column vectors** of $\mathbf{A}$, respectively.

■ **A Definition** As vectors, the set $\mathbf{u}_1, \mathbf{u}_2, \ldots, \mathbf{u}_m$ is either linearly independent or linearly dependent. We have the following definition.

DEFINITION 8.3.1 Rank of a Matrix

The **rank** of an $m \times n$ matrix $\mathbf{A}$, denoted by rank($\mathbf{A}$), is the maximum number of linearly independent row vectors in $\mathbf{A}$.

EXAMPLE 1 Rank of a 3 × 4 Matrix

Find the rank of the 3×4 matrix

$$
\mathbf{A} = \begin{pmatrix}
1 & 1 & -1 & 3 \\
2 & -2 & 6 & 8 \\
3 & 5 & -7 & 8
\end{pmatrix}.
\tag{1}
$$

SOLUTION With $\mathbf{u}_1 = (1 \;\; 1 \;\; -1 \;\; 3)$, $\mathbf{u}_2 = (2 \;\; -2 \;\; 6 \;\; 8)$, and $\mathbf{u}_3 = (3 \;\; 5 \;\; -7 \;\; 8)$, we see that $4\mathbf{u}_1 - \frac{1}{2}\mathbf{u}_2 - \mathbf{u}_3 = \mathbf{0}$. In view of Definition 7.6.3 and the discussion following it, we conclude that the set $\mathbf{u}_1, \mathbf{u}_2, \mathbf{u}_3$ is linearly dependent. On the other hand, since neither $\mathbf{u}_1$ nor $\mathbf{u}_2$ is a constant multiple of the other, the set of row vectors $\mathbf{u}_1, \mathbf{u}_2$ is linearly independent. Hence by Definition 8.3.1, rank($\mathbf{A}$) $= 2$. ≡

■ **Row Space** In the terminology of the preceding chapter, the row vectors $\mathbf{u}_1, \mathbf{u}_2, \mathbf{u}_3$ of the matrix (1) are a set of vectors in the *vector space* R^4. Since $R_\mathbf{A} = \text{Span}(\mathbf{u}_1, \mathbf{u}_2, \mathbf{u}_3)$ (the set of all linear combinations of the vectors $\mathbf{u}_1, \mathbf{u}_2, \mathbf{u}_3$) is a subspace of R^4, we are justified in calling $R_\mathbf{A}$ the **row space** of the matrix $\mathbf{A}$. Now the set of vectors $\mathbf{u}_1, \mathbf{u}_2$ is linearly independent and also spans $R_\mathbf{A}$; in other words, the set $\mathbf{u}_1, \mathbf{u}_2$ is a *basis* for $R_\mathbf{A}$. The dimension (the number of vectors in a basis) of the row space $R_\mathbf{A}$ is 2, which is rank($\mathbf{A}$).

See pages 359 and 360 in Section 7.6.

■ **Rank by Row Reduction** Example 1 notwithstanding, it is generally not easy to determine the rank of a matrix by inspection. Although there are several mechanical ways of finding rank($\mathbf{A}$), we examine one way that uses the elementary row operations introduced in the preceding section. *Specifically, the rank of $\mathbf{A}$ can be found by row reducing $\mathbf{A}$ to a row-echelon matrix $\mathbf{B}$.* To understand this, first recall that an $m \times n$ matrix $\mathbf{B}$ is row equivalent to an $m \times n$ matrix $\mathbf{A}$ if the rows of $\mathbf{B}$ were obtained from the rows of $\mathbf{A}$ by applying elementary row operations. If we simply interchange two rows in $\mathbf{A}$ to obtain $\mathbf{B}$, then the row space $R_\mathbf{A}$ of $\mathbf{A}$ and the row space $R_\mathbf{B}$ of $\mathbf{B}$ are equal because the row vectors of $\mathbf{A}$ and $\mathbf{B}$ are the same.

When the row vectors of **B** are linear combinations of the rows of **A**, it follows that the row vectors of **B** are in the row space R_A, and so R_B is a subset of R_A (written $R_B \subseteq R_A$). Conversely, **A** is row equivalent to **B**, since we can obtain **A** by applying row operations to **B**. Hence the rows of **A** are linear combinations of the rows of **B**, and so it follows that R_A is a subset of R_B ($R_A \subseteq R_B$). From $R_B \subseteq R_A$ and $R_A \subseteq R_B$, we conclude that $R_A = R_B$. Finally, if we row-reduce **A** into a row-echelon matrix **B**, then the nonzero rows of **B** are linearly independent. (Why?) The nonzero rows of **B** form a basis for the row space R_A, and so we have the result that rank(**A**) = dimension of R_A.

We summarize these conclusions in the next theorem.

THEOREM 8.3.1 Rank of a Matrix by Row Reduction

If a matrix **A** is row equivalent to a row-echelon form **B**, then

(*i*) the row space of **A** = the row space of **B**,
(*ii*) the nonzero rows of **B** form a basis for the row space of **A**, and
(*iii*) rank(**A**) = the number of nonzero rows in **B**.

EXAMPLE 2 Rank by Row Reduction—Example 1 Revisited

We row-reduce a matrix **A** to a row-echelon form **B** in exactly the same manner as we row-reduced the augmented matrix of a system of linear equations to an echelon form when we solved the system using Gaussian elimination. Using the matrix (1) in Example 1, elementary row operations give

$$\mathbf{A} = \begin{pmatrix} 1 & 1 & -1 & 3 \\ 2 & -2 & 6 & 8 \\ 3 & 5 & -7 & 8 \end{pmatrix} \overset{-2R_1+R_2}{\underset{-3R_1+R_3}{\Rightarrow}} \begin{pmatrix} 1 & 1 & -1 & 3 \\ 0 & -4 & 8 & 2 \\ 0 & 2 & -4 & -1 \end{pmatrix} \overset{\frac{1}{2}R_2+R_3}{\underset{-\frac{1}{4}R_2}{\Rightarrow}} \begin{pmatrix} 1 & 1 & -1 & 3 \\ 0 & 1 & -2 & -\frac{1}{2} \\ 0 & 0 & 0 & 0 \end{pmatrix}.$$

Since the last matrix is in row-echelon form, and since the last matrix has two nonzero rows, we conclude from (*iii*) of Theorem 8.3.1 that rank(**A**) = 2. ≡

EXAMPLE 3 Linear Independence/Dependence

Determine whether the set of vectors $\mathbf{u}_1 = \langle 2, 1, 1 \rangle$, $\mathbf{u}_2 = \langle 0, 3, 0 \rangle$, $\mathbf{u}_3 = \langle 3, 1, 2 \rangle$, in R^3 is linearly dependent or linearly independent.

SOLUTION It should be clear from the discussion above that if we form a matrix **A** with the given vectors as rows, and if we row-reduce **A** to a row-echelon form **B** with rank 3, then the set of vectors is linearly independent. If rank(**A**) < 3, then the set of vectors is linearly dependent. In this case, it is easy to carry the row reduction all the way to a reduced row-echelon form:

$$\mathbf{A} = \begin{pmatrix} 2 & 1 & 1 \\ 0 & 3 & 0 \\ 3 & 1 & 2 \end{pmatrix} \overset{\text{row}}{\underset{\text{operations}}{\Rightarrow}} \begin{pmatrix} 1 & 0 & 0 \\ 0 & 1 & 0 \\ 0 & 0 & 1 \end{pmatrix}.$$

Thus rank(**A**) = 3 and the set of vectors $\mathbf{u}_1$, $\mathbf{u}_2$, $\mathbf{u}_3$ is linearly independent. ≡

As mentioned previously, the vectors in the row-echelon form of a matrix **A** can serve as a basis for the row space of the matrix **A**. In Example 3, we see that a basis for the row space of **A** is the standard basis $\langle 1, 0, 0 \rangle$, $\langle 0, 1, 0 \rangle$, $\langle 0, 0, 1 \rangle$ of R^3.

■ **Rank and Linear Systems** The concept of rank can be related back to solvability of linear systems of algebraic equations. Suppose $\mathbf{AX} = \mathbf{B}$ is a linear system and that $(\mathbf{A}|\mathbf{B})$ denotes the augmented matrix of the system. In Example 7 of Section 8.2, we saw that the system

$$x_1 + x_2 = 1$$
$$4x_1 - x_2 = -6$$
$$2x_1 - 3x_2 = 8$$

was inconsistent. The inconsistency of the system is seen in the fact that, after row reduction of the augmented matrix $(\mathbf{A}|\mathbf{B})$,

$$\begin{pmatrix} 1 & 1 & | & 1 \\ 4 & -1 & | & -6 \\ 2 & -3 & | & 8 \end{pmatrix} \xrightarrow[\text{operations}]{\text{row}} \begin{pmatrix} 1 & 0 & | & 1 \\ 0 & 1 & | & 2 \\ 0 & 0 & | & 16 \end{pmatrix} \xrightarrow[\text{operations}]{\text{row}} \begin{pmatrix} 1 & 0 & | & 0 \\ 0 & 1 & | & 0 \\ 0 & 0 & | & 1 \end{pmatrix},\qquad(2)$$

the last row of the reduced row-echelon form is nonzero. Of course, this reduction shows that $\text{rank}(\mathbf{A}|\mathbf{B}) = 3$. But note, too, that the result in (2) indicates that $\text{rank}(\mathbf{A}) = 2$ because

$$\begin{pmatrix} 1 & 1 \\ 4 & -1 \\ 2 & -3 \end{pmatrix} \xrightarrow[\text{operations}]{\text{row}} \begin{pmatrix} 1 & 0 \\ 0 & 1 \\ 0 & 0 \end{pmatrix}.$$

We have illustrated a special case of the next theorem.

THEOREM 8.3.2 Consistency of AX = B

A linear system of equations $\mathbf{AX} = \mathbf{B}$ is consistent if and only if the rank of the coefficient matrix $\mathbf{A}$ is the same as the rank of the augmented matrix of the system $(\mathbf{A}|\mathbf{B})$.

In Example 6 of Section 8.2, we saw that the system

$$\begin{aligned} x_1 + 3x_2 - 2x_3 &= -7 \\ 4x_1 + x_2 + 3x_3 &= 5 \\ 2x_1 - 5x_2 + 7x_3 &= 19 \end{aligned}\qquad(3)$$

was consistent and had an infinite number of solutions. We solved for two of the variables, x_1 and x_2, in terms of the remaining variable x_3, which we relabeled as a parameter t. The number of parameters in a solution of a system is related to the rank of the coefficient matrix $\mathbf{A}$.

THEOREM 8.3.3 Number of Parameters in a Solution

Suppose a linear system $\mathbf{AX} = \mathbf{B}$ with m equations and n variables is consistent. If the coefficient matrix $\mathbf{A}$ has rank r, then the solution of the system contains $n - r$ parameters.

For the system (3), we can see from the row reduction

$$\begin{pmatrix} 1 & 3 & -2 & | & -7 \\ 4 & 1 & 3 & | & 5 \\ 2 & -5 & 7 & | & 19 \end{pmatrix} \xrightarrow[\text{operations}]{\text{row}} \begin{pmatrix} 1 & 0 & 1 & | & 2 \\ 0 & 1 & -1 & | & -3 \\ 0 & 0 & 0 & | & 0 \end{pmatrix}$$

that $\text{rank}(\mathbf{A}) = \text{rank}(\mathbf{A}|\mathbf{B}) = 2$, and so the system is consistent by Theorem 8.3.2. With $n = 3$, we see from Theorem 8.3.3 the number of parameters in the solution is $3 - 2 = 1$.

FIGURE 8.3.1 outlines the connection between the concept of rank of a matrix and the solution of a linear system.

FIGURE 8.3.1 For m linear equations in n variables $\mathbf{AX} = \mathbf{B}$.
Two cases: $\mathbf{B} = \mathbf{0}$, $\mathbf{B} \neq \mathbf{0}$. Let rank$(\mathbf{A}) = r$.

REMARKS

We have not mentioned the connection between the columns of a matrix $\mathbf{A}$ and the rank of $\mathbf{A}$. It turns out that the maximum number of independent columns that a matrix $\mathbf{A}$ can have must equal the maximum number of independent rows. In the terminology of vector spaces, the row space $R_\mathbf{A}$ of matrix $\mathbf{A}$ has the same dimension as its column space $C_\mathbf{A}$. For example, if we take the transpose of the matrix in (1) and reduce it to a row-echelon form

$$\mathbf{A}^T = \begin{pmatrix} 1 & 2 & 3 \\ 1 & -2 & 5 \\ -1 & 6 & -7 \\ 3 & 8 & 8 \end{pmatrix} \overset{\text{row}}{\underset{\text{operations}}{\Rightarrow}} \begin{pmatrix} 1 & 2 & 3 \\ 0 & 1 & -\frac{1}{2} \\ 0 & 0 & 0 \\ 0 & 0 & 0 \end{pmatrix},$$

we see that the maximum number of rows of $\mathbf{A}^T$ is 2, and so the maximum number of linearly independent columns of $\mathbf{A}$ is 2.

8.3 Exercises Answers to selected odd-numbered problems begin on page ANS-18.

In Problems 1–10, use (*iii*) of Theorem 8.3.1 to find the rank of the given matrix.

1. $\begin{pmatrix} 3 & -1 \\ 1 & 3 \end{pmatrix}$

2. $\begin{pmatrix} 2 & -2 \\ 0 & 0 \end{pmatrix}$

3. $\begin{pmatrix} 2 & 1 & 3 \\ 6 & 3 & 9 \\ -1 & -\frac{1}{2} & -\frac{3}{2} \end{pmatrix}$

4. $\begin{pmatrix} 1 & 1 & 2 \\ -1 & 2 & 4 \\ -1 & 0 & 3 \end{pmatrix}$

5. $\begin{pmatrix} 1 & 1 & 1 \\ 1 & 0 & 4 \\ 1 & 4 & 1 \end{pmatrix}$

6. $\begin{pmatrix} 3 & -1 & 2 & 0 \\ 6 & 2 & 4 & 5 \end{pmatrix}$

7. $\begin{pmatrix} 1 & -2 \\ 3 & -6 \\ 7 & -1 \\ 4 & 5 \end{pmatrix}$

8. $\begin{pmatrix} 1 & -2 & 3 & 4 \\ 1 & 4 & 6 & 8 \\ 0 & 1 & 0 & 0 \\ 2 & 5 & 6 & 8 \end{pmatrix}$

9. $\begin{pmatrix} 0 & 2 & 4 & 2 & 2 \\ 4 & 1 & 0 & 5 & 1 \\ 2 & 1 & \frac{2}{3} & 3 & \frac{1}{3} \\ 6 & 6 & 6 & 12 & 0 \end{pmatrix}$

10. $\begin{pmatrix} 1 & -2 & 1 & 8 & -1 & 1 & 1 & 6 \\ 0 & 0 & 1 & 3 & -1 & 1 & 1 & 5 \\ 0 & 0 & 1 & 3 & -1 & 2 & 10 & 8 \\ 0 & 0 & 0 & 0 & 0 & 1 & 1 & 3 \\ 1 & -2 & 1 & 8 & -1 & 1 & 2 & 6 \end{pmatrix}$

In Problems 11–14, determine whether the given set of vectors is linearly dependent or linearly independent.

11. $\mathbf{u}_1 = \langle 1, 2, 3 \rangle$, $\mathbf{u}_2 = \langle 1, 0, 1 \rangle$, $\mathbf{u}_3 = \langle 1, -1, 5 \rangle$

12. $\mathbf{u}_1 = \langle 2, 6, 3 \rangle$, $\mathbf{u}_2 = \langle 1, -1, 4 \rangle$, $\mathbf{u}_3 = \langle 3, 2, 1 \rangle$, $\mathbf{u}_4 = \langle 2, 5, 4 \rangle$

13. $\mathbf{u}_1 = \langle 1, -1, 3, -1 \rangle$, $\mathbf{u}_2 = \langle 1, -1, 4, 2 \rangle$, $\mathbf{u}_3 = \langle 1, -1, 5, 7 \rangle$

14. $\mathbf{u}_1 = \langle 2, 1, 1, 5 \rangle$, $\mathbf{u}_2 = \langle 2, 2, 1, 1 \rangle$, $\mathbf{u}_3 = \langle 3, -1, 6, 1 \rangle$, $\mathbf{u}_4 = \langle 1, 1, 1, -1 \rangle$

15. Suppose the system $\mathbf{AX} = \mathbf{B}$ is consistent and $\mathbf{A}$ is a 5×8 matrix and rank$(\mathbf{A}) = 3$. How many parameters does the solution of the system have?

16. Let $\mathbf{A}$ be a nonzero 4×6 matrix.
 (a) What is the maximum rank that $\mathbf{A}$ can have?
 (b) If rank$(\mathbf{A}|\mathbf{B}) = 2$, then for what value(s) of rank$(\mathbf{A})$ is the system $\mathbf{AX} = \mathbf{B}$, $\mathbf{B} \neq \mathbf{0}$, inconsistent? Consistent?
 (c) If rank$(\mathbf{A}) = 3$, then how many parameters does the solution of the system $\mathbf{AX} = \mathbf{0}$ have?

17. Let $\mathbf{v}_1$, $\mathbf{v}_2$, and $\mathbf{v}_3$ be the first, second, and third column vectors, respectively, of the matrix

$$\mathbf{A} = \begin{pmatrix} 2 & 1 & 7 \\ 1 & 0 & 2 \\ -1 & 5 & 13 \end{pmatrix}.$$

What can we conclude about rank$(\mathbf{A})$ from the observation $2\mathbf{v}_1 + 3\mathbf{v}_2 - \mathbf{v}_3 = \mathbf{0}$? [*Hint*: Read the *Remarks* at the end of this section.]

Discussion Problems

18. Suppose the system $\mathbf{AX} = \mathbf{B}$ is consistent and $\mathbf{A}$ is a 6×3 matrix. Suppose the maximum number of linearly independent rows in $\mathbf{A}$ is 3. Discuss: Is the solution of the system unique?

19. Suppose we wish to determine whether the set of column vectors

$$\mathbf{v}_1 = \begin{pmatrix} 4 \\ 3 \\ 2 \\ 1 \end{pmatrix}, \quad \mathbf{v}_2 = \begin{pmatrix} 1 \\ 2 \\ 2 \\ 1 \end{pmatrix}, \quad \mathbf{v}_3 = \begin{pmatrix} -1 \\ 1 \\ 1 \\ 1 \end{pmatrix},$$

$$\mathbf{v}_4 = \begin{pmatrix} 2 \\ 3 \\ 4 \\ 1 \end{pmatrix}, \quad \mathbf{v}_5 = \begin{pmatrix} 1 \\ 7 \\ -5 \\ 1 \end{pmatrix}$$

is linearly dependent or linearly independent. By Definition 7.6.3, if

$$c_1\mathbf{v}_1 + c_2\mathbf{v}_2 + c_3\mathbf{v}_3 + c_4\mathbf{v}_4 + c_5\mathbf{v}_5 = \mathbf{0} \qquad (4)$$

only for $c_1 = 0$, $c_2 = 0$, $c_3 = 0$, $c_4 = 0$, $c_5 = 0$, then the set of vectors is linearly independent; otherwise the set is linearly dependent. But (4) is equivalent to the linear system

$$4c_1 + c_2 - c_3 + 2c_4 + c_5 = 0$$
$$3c_1 + 2c_2 + c_3 + 3c_4 + 7c_5 = 0$$
$$2c_1 + 2c_2 + c_3 + 4c_4 - 5c_5 = 0$$
$$c_1 + c_2 + c_3 + c_4 + c_5 = 0.$$

Without doing any further work, explain why we can now conclude that the set of vectors is linearly dependent.

Computer Lab Assignment

20. A CAS can be used to row-reduce a matrix to a row-echelon form. Use a CAS to determine the ranks of the augmented matrix $(\mathbf{A}|\mathbf{B})$ and the coefficient matrix $\mathbf{A}$ for

$$x_1 + 2x_2 - 6x_3 + x_4 + x_5 + x_6 = 2$$
$$5x_1 + 2x_2 - 2x_3 + 5x_4 + 4x_5 + 2x_6 = 3$$
$$6x_1 + 2x_2 - 2x_3 + x_4 + x_5 + 3x_6 = -1$$
$$-x_1 + 2x_2 + 3x_3 + x_4 - x_5 + 6x_6 = 0$$
$$9x_1 + 7x_2 - 2x_3 + x_4 + 4x_5 = 5.$$

Is the system consistent or inconsistent? If consistent, solve the system.

8.4 Determinants

INTRODUCTION Suppose $\mathbf{A}$ is an $n \times n$ matrix. Associated with $\mathbf{A}$ is a *number* called the **determinant of** $\mathbf{A}$ and is denoted by det $\mathbf{A}$. Symbolically, we distinguish a matrix $\mathbf{A}$ from the determinant of $\mathbf{A}$ by replacing the parentheses by vertical bars:

$$\mathbf{A} = \begin{pmatrix} a_{11} & a_{12} & \cdots & a_{1n} \\ a_{21} & a_{22} & \cdots & a_{2n} \\ \vdots & & & \vdots \\ a_{n1} & a_{n2} & \cdots & a_{nn} \end{pmatrix} \quad \text{and} \quad \det \mathbf{A} = \begin{vmatrix} a_{11} & a_{12} & \cdots & a_{1n} \\ a_{21} & a_{22} & \cdots & a_{2n} \\ \vdots & & & \vdots \\ a_{n1} & a_{n2} & \cdots & a_{nn} \end{vmatrix}.$$

A determinant of an $n \times n$ matrix is said to be a **determinant of order n**. We begin by defining the determinants of 1×1, 2×2, and 3×3 matrices.

■ **A Definition** For a 1×1 matrix $\mathbf{A} = (a)$, we have $\det \mathbf{A} = |a| = a$. For example, if $\mathbf{A} = (-5)$, then $\det \mathbf{A} = |-5| = -5$. In this case the vertical bars $|\,|$ around a number do *not* mean the absolute value of the number.

DEFINITION 8.4.1 Determinant of a 2 × 2 Matrix

The determinant of $\mathbf{A} = \begin{pmatrix} a_{11} & a_{12} \\ a_{21} & a_{22} \end{pmatrix}$ is the number

$$\det \mathbf{A} = \begin{vmatrix} a_{11} & a_{12} \\ a_{21} & a_{22} \end{vmatrix} = a_{11}a_{22} - a_{12}a_{21}. \tag{1}$$

As a mnemonic, a determinant of order 2 is thought to be the product of the main diagonal entries of $\mathbf{A}$ minus the product of the other diagonal entries:

$$\begin{matrix} \text{multiply} & \text{multiply} & \begin{matrix}\text{subtract}\\ \text{products}\end{matrix} \end{matrix}$$

$$\begin{vmatrix} a_{11} & a_{12} \\ a_{21} & a_{22} \end{vmatrix} = a_{11}a_{22} - a_{12}a_{21}. \tag{2}$$

For example, if $\mathbf{A} = \begin{pmatrix} 6 & -3 \\ 5 & 9 \end{pmatrix}$, then $\det \mathbf{A} = \begin{vmatrix} 6 & -3 \\ 5 & 9 \end{vmatrix} = 6(9) - (-3)(5) = 69$.

DEFINITION 8.4.2 Determinant of a 3 × 3 Matrix

The determinant of $\mathbf{A} = \begin{pmatrix} a_{11} & a_{12} & a_{13} \\ a_{21} & a_{22} & a_{23} \\ a_{31} & a_{32} & a_{33} \end{pmatrix}$ is the number

$$\det \mathbf{A} = \begin{vmatrix} a_{11} & a_{12} & a_{13} \\ a_{21} & a_{22} & a_{23} \\ a_{31} & a_{32} & a_{33} \end{vmatrix} = \begin{matrix} a_{11}a_{22}a_{33} + a_{12}a_{23}a_{31} + a_{13}a_{21}a_{32} - a_{13}a_{22}a_{31} \\ - a_{11}a_{23}a_{32} - a_{12}a_{21}a_{33}. \end{matrix} \tag{3}$$

The expression in (3) can be written in a more tractable form. By factoring, we have

$$\det \mathbf{A} = a_{11}(a_{22}a_{33} - a_{23}a_{32}) + a_{12}(-a_{21}a_{33} + a_{23}a_{31}) + a_{13}(a_{21}a_{32} - a_{22}a_{31}).$$

But in view of (1), each term in parentheses is recognized as the determinant of a 2×2 matrix:

$$\det \mathbf{A} = a_{11}\begin{vmatrix} a_{22} & a_{23} \\ a_{32} & a_{33} \end{vmatrix} + a_{12}\left(-\begin{vmatrix} a_{21} & a_{23} \\ a_{31} & a_{33} \end{vmatrix}\right) + a_{13}\begin{vmatrix} a_{21} & a_{22} \\ a_{31} & a_{32} \end{vmatrix}. \tag{4}$$

Observe that each determinant in (4) is a determinant of a *submatrix* of the matrix $\mathbf{A}$ and corresponds to its coefficient in the following manner: a_{11} is the coefficient of the determinant of the submatrix obtained by deleting the *first row and first column* of $\mathbf{A}$; a_{12} is the coefficient of the negative of the determinant of the submatrix obtained by deleting the *first row and second column* of $\mathbf{A}$; and finally, a_{13} is the coefficient of the determinant of the submatrix obtained by deleting the *first row and third column* of $\mathbf{A}$. In other words, the coefficients in (4) are simply the entries of the first row of $\mathbf{A}$. We say that $\det \mathbf{A}$ has been **expanded by cofactors along the first row**, with the cofactors of a_{11}, a_{12}, and a_{13} being the determinants

$$C_{11} = \begin{vmatrix} a_{22} & a_{23} \\ a_{32} & a_{33} \end{vmatrix} \qquad C_{12} = -\begin{vmatrix} a_{21} & a_{23} \\ a_{31} & a_{33} \end{vmatrix} \qquad C_{13} = \begin{vmatrix} a_{21} & a_{22} \\ a_{31} & a_{32} \end{vmatrix}.$$

Thus (4) is

$$\det \mathbf{A} = a_{11}C_{11} + a_{12}C_{12} + a_{13}C_{13}. \tag{5}$$

In general, the **cofactor of** a_{ij} is the determinant

$$C_{ij} = (-1)^{i+j}M_{ij}, \tag{6}$$

where M_{ij} is the determinant of the submatrix obtained by deleting the *i*th row and the *j*th column of **A**. The determinant M_{ij} is called a **minor determinant**. A cofactor is a *signed* minor determinant; that is, $C_{ij} = M_{ij}$ when $i + j$ is even and $C_{ij} = -M_{ij}$ when $i + j$ is odd.

A 3×3 matrix has nine cofactors:

$$
\begin{array}{ccc}
C_{11} = M_{11} & C_{12} = -M_{12} & C_{13} = M_{13} \\
C_{21} = -M_{21} & C_{22} = M_{22} & C_{23} = -M_{23} \\
C_{31} = M_{31} & C_{32} = -M_{32} & C_{33} = M_{33}.
\end{array}
$$

Inspection of the above array shows that the sign factor $+1$ or -1 associated with a cofactor can be obtained from the checkerboard pattern:

$$
\begin{array}{ccc}
+ & - & + \\
- & + & - \\
+ & - & +
\end{array} \tag{7}
$$

3×3 matrix

Now observe that (3) can be rearranged and factored again as

$$\det \mathbf{A} = -a_{12}(a_{21}a_{33} - a_{23}a_{31}) + a_{22}(a_{11}a_{33} - a_{13}a_{31}) - a_{32}(a_{11}a_{23} - a_{13}a_{21})$$

$$= a_{12}\left(-\begin{vmatrix} a_{21} & a_{23} \\ a_{31} & a_{33} \end{vmatrix}\right) + a_{22}\begin{vmatrix} a_{11} & a_{13} \\ a_{31} & a_{33} \end{vmatrix} + a_{32}\left(-\begin{vmatrix} a_{11} & a_{13} \\ a_{21} & a_{23} \end{vmatrix}\right) \tag{8}$$

$$= a_{12}C_{12} + a_{22}C_{22} + a_{32}C_{32},$$

which is the cofactor expansion of $\det \mathbf{A}$ along the second column. It is left as an exercise to show from (3) that $\det \mathbf{A}$ can also be expanded by cofactors along the third row:

$$\det \mathbf{A} = a_{31}C_{31} + a_{32}C_{32} + a_{33}C_{33}. \tag{9}$$

We are, of course, suggesting in (5), (8), and (9) the following general result:

The determinant of a 3×3 matrix can be evaluated by expanding $\det \mathbf{A}$ *by cofactors along any row or along any column.*

EXAMPLE 1 **Cofactor Expansion Along the First Row**

Evaluate the determinant of $\mathbf{A} = \begin{pmatrix} 2 & 4 & 7 \\ 6 & 0 & 3 \\ 1 & 5 & 3 \end{pmatrix}$.

SOLUTION Using cofactor expansion along the first row gives

$$\det \mathbf{A} = \begin{vmatrix} 2 & 4 & 7 \\ 6 & 0 & 3 \\ 1 & 5 & 3 \end{vmatrix} = 2C_{11} + 4C_{12} + 7C_{13}.$$

Now, the cofactors of the entries in the first row of $\mathbf{A}$ are

$$C_{11} = (-1)^{1+1} \begin{vmatrix} 2 & 4 & 7 \\ 6 & 0 & 3 \\ 1 & 5 & 3 \end{vmatrix} = (-1)^{1+1} \begin{vmatrix} 0 & 3 \\ 5 & 3 \end{vmatrix}$$

$$C_{12} = (-1)^{1+2} \begin{vmatrix} 2 & 4 & 7 \\ 6 & 0 & 3 \\ 1 & 5 & 3 \end{vmatrix} = (-1)^{1+2} \begin{vmatrix} 6 & 3 \\ 1 & 3 \end{vmatrix}$$

$$C_{13} = (-1)^{1+3} \begin{vmatrix} 2 & 4 & 7 \\ 6 & 0 & 3 \\ 1 & 5 & 3 \end{vmatrix} = (-1)^{1+3} \begin{vmatrix} 6 & 0 \\ 1 & 5 \end{vmatrix},$$

where the dashed lines indicate the row and column that are deleted. Thus,

$$\det \mathbf{A} = 2(-1)^{1+1} \begin{vmatrix} 0 & 3 \\ 5 & 3 \end{vmatrix} + 4(-1)^{1+2} \begin{vmatrix} 6 & 3 \\ 1 & 3 \end{vmatrix} + 7(-1)^{1+3} \begin{vmatrix} 6 & 0 \\ 1 & 5 \end{vmatrix}$$

$$= 2[0(3) - 3(5)] - 4[6(3) - 3(1)] + 7[6(5) - 0(1)] = 120. \qquad \equiv$$

If a matrix has a row (or a column) containing many zero entries, then wisdom dictates that we evaluate the determinant of the matrix using cofactor expansion along that row (or column). Thus, in Example 1, had we expanded the determinant of $\mathbf{A}$ using cofactors along, say, the second row, then

$$\det \mathbf{A} = 6C_{21} + 0C_{22} + 3C_{23} = 6C_{21} + 3C_{23}$$

$$= 6(-1)^{2+1} \begin{vmatrix} 4 & 7 \\ 5 & 3 \end{vmatrix} + 3(-1)^{2+3} \begin{vmatrix} 2 & 4 \\ 1 & 5 \end{vmatrix}$$

$$= -6(-23) - 3(6) = 120.$$

<div style="border-left: 4px solid; padding-left: 8px;">

EXAMPLE 2 **Cofactor Expansion Along the Third Column**

</div>

Evaluate the determinant of $\mathbf{A} = \begin{pmatrix} 6 & 5 & 0 \\ -1 & 8 & -7 \\ -2 & 4 & 0 \end{pmatrix}$.

SOLUTION Since there are two zeros in the third column, we expand by cofactors of that column:

$$\det \mathbf{A} = \begin{vmatrix} 6 & 5 & 0 \\ -1 & 8 & -7 \\ -2 & 4 & 0 \end{vmatrix} = 0C_{13} + (-7)C_{23} + 0C_{33}$$

$$= (-7)(-1)^{2+3} \begin{vmatrix} 6 & 5 & 0 \\ -1 & 8 & -7 \\ -2 & 4 & 0 \end{vmatrix} = (-7)(-1)^{2+3} \begin{vmatrix} 6 & 5 \\ -2 & 4 \end{vmatrix}$$

$$= 7[6(4) - 5(-2)] = 238. \qquad \equiv$$

Carrying the above ideas one step further, we can evaluate the determinant of a 4×4 matrix by multiplying the entries in a row (or column) by their corresponding cofactors and adding the

products. In this case, each cofactor is a signed minor determinant of an appropriate 3×3 submatrix. The following theorem, which we shall give without proof, states that the determinant of any $n \times n$ matrix $\mathbf{A}$ can be evaluated by means of cofactors.

THEOREM 8.4.1 Cofactor Expansion of a Determinant

Let $\mathbf{A} = (a_{ij})_{n \times n}$ be an $n \times n$ matrix. For each $1 \le i \le n$, **the cofactor expansion of det A along the ith row** is

$$\det \mathbf{A} = a_{i1}C_{i1} + a_{i2}C_{i2} + \cdots + a_{in}C_{in}.$$

For each $1 \le j \le n$, **the cofactor expansion of det A along the jth column** is

$$\det \mathbf{A} = a_{1j}C_{1j} + a_{2j}C_{2j} + \cdots + a_{nj}C_{nj}.$$

The sign factor pattern for the cofactors illustrated in (7) extends to matrices of order greater than 3:

$$
\begin{matrix}
 & & & & + & - & + & - & + & \cdots \\
+ & - & + & - & & - & + & - & + & - & \cdots \\
- & + & - & + & & + & - & + & - & + & \cdots \\
+ & - & + & - & & - & + & - & + & - & \cdots \\
- & + & - & + & & + & - & + & - & + & \cdots \\
 & & & & & \vdots & \vdots & \vdots & \vdots & \vdots
\end{matrix}
$$

$\quad\quad$ **4×4 matrix** $\quad\quad\quad\quad\quad\quad$ **$n \times n$ matrix**

EXAMPLE 3 **Cofactor Expansion Along the Fourth Row**

Evaluate the determinant of the matrix

$$\mathbf{A} = \begin{pmatrix} 5 & 1 & 2 & 4 \\ -1 & 0 & 2 & 3 \\ 1 & 1 & 6 & 1 \\ 1 & 0 & 0 & -4 \end{pmatrix}.$$

SOLUTION Since the matrix has two zero entries in its fourth row, we choose to expand det $\mathbf{A}$ by cofactors along that row:

$$\det \mathbf{A} = \begin{vmatrix} 5 & 1 & 2 & 4 \\ -1 & 0 & 2 & 3 \\ 1 & 1 & 6 & 1 \\ 1 & 0 & 0 & -4 \end{vmatrix} = (1)C_{41} + 0C_{42} + 0C_{43} + (-4)C_{44}, \quad\quad (10)$$

where $\quad C_{41} = (-1)^{4+1}\begin{vmatrix} 1 & 2 & 4 \\ 0 & 2 & 3 \\ 1 & 6 & 1 \end{vmatrix}$ and $\quad C_{44} = (-1)^{4+4}\begin{vmatrix} 5 & 1 & 2 \\ -1 & 0 & 2 \\ 1 & 1 & 6 \end{vmatrix}.$

We then expand both these determinants by cofactors along the second row:

$$C_{41} = (-1)\begin{vmatrix} 1 & 2 & 4 \\ 0 & 2 & 3 \\ 1 & 6 & 1 \end{vmatrix} = -\left(0(-1)^{2+1}\begin{vmatrix} 2 & 4 \\ 6 & 1 \end{vmatrix} + 2(-1)^{2+2}\begin{vmatrix} 1 & 4 \\ 1 & 1 \end{vmatrix} + 3(-1)^{2+3}\begin{vmatrix} 1 & 2 \\ 1 & 6 \end{vmatrix} \right) = 18$$

$$C_{44} = \begin{vmatrix} 5 & 1 & 2 \\ -1 & 0 & 2 \\ 1 & 1 & 6 \end{vmatrix} = (-1)(-1)^{2+1}\begin{vmatrix} 1 & 2 \\ 1 & 6 \end{vmatrix} + 0(-1)^{2+2}\begin{vmatrix} 5 & 2 \\ 1 & 6 \end{vmatrix} + 2(-1)^{2+3}\begin{vmatrix} 5 & 1 \\ 1 & 1 \end{vmatrix} = -4.$$

Therefore (10) becomes

$$\det \mathbf{A} = \begin{vmatrix} 5 & 1 & 2 & 4 \\ -1 & 0 & 2 & 3 \\ 1 & 1 & 6 & 1 \\ 1 & 0 & 0 & -4 \end{vmatrix} = (1)(18) + (-4)(-4) = 34.$$

You should verify this result by expanding $\det \mathbf{A}$ by cofactors along the second column. ≡

REMARKS

In previous mathematics courses you may have seen the following memory device, analogous to (2), for evaluating a determinant of order 3:

(11)

(*i*) Add the products of the entries on the arrows that go from left to right.
(*ii*) Subtract from the number in (*i*) the sum of the products of the entries on the arrows that go from right to left.

Note: Method illustrated in (11) does not work for determinants of order $n > 3$. ▶

A word of caution is in order here. The memory device given in (11), though easily adapted to matrices larger than 3×3, does not give the correct results. There are no mnemonic devices for evaluating the determinants of order 4 or greater.

8.4 Exercises Answers to selected odd-numbered problems begin on page ANS-18.

In Problems 1–4, suppose

$$\mathbf{A} = \begin{pmatrix} 2 & 3 & 4 \\ 1 & -1 & 2 \\ -2 & 3 & 5 \end{pmatrix}.$$

Evaluate the indicated minor determinant or cofactor.

1. M_{12} **2.** M_{32} **3.** C_{13} **4.** C_{22}

In Problems 5–8, suppose

$$\mathbf{A} = \begin{pmatrix} 0 & 2 & 4 & 0 \\ 1 & 2 & -2 & 3 \\ 5 & 1 & 0 & -1 \\ 1 & 1 & 1 & 2 \end{pmatrix}.$$

Evaluate the indicated minor determinant or cofactor.

5. M_{33} **6.** M_{41} **7.** C_{34} **8.** C_{23}

In Problems 9–14, evaluate the determinant of the given matrix.

9. (-7) **10.** (2)

11. $\begin{pmatrix} 3 & 5 \\ -1 & 4 \end{pmatrix}$ **12.** $\begin{pmatrix} \frac{1}{4} & \frac{1}{2} \\ \frac{1}{3} & -\frac{4}{3} \end{pmatrix}$

13. $\begin{pmatrix} 1 - \lambda & 3 \\ 2 & 2 - \lambda \end{pmatrix}$ **14.** $\begin{pmatrix} -3 - \lambda & -4 \\ -2 & 5 - \lambda \end{pmatrix}$

In Problems 15–28, evaluate the determinant of the given matrix by cofactor expansion.

15. $\begin{pmatrix} 0 & 2 & 0 \\ 3 & 0 & 1 \\ 0 & 5 & 8 \end{pmatrix}$ **16.** $\begin{pmatrix} 5 & 0 & 0 \\ 0 & -3 & 0 \\ 0 & 0 & 2 \end{pmatrix}$

17. $\begin{pmatrix} 3 & 0 & 2 \\ 2 & 7 & 1 \\ 2 & 6 & 4 \end{pmatrix}$

18. $\begin{pmatrix} 1 & -1 & -1 \\ 2 & 2 & -2 \\ 1 & 1 & 9 \end{pmatrix}$

19. $\begin{pmatrix} 4 & 5 & 3 \\ 1 & 2 & 3 \\ 1 & 2 & 3 \end{pmatrix}$

20. $\begin{pmatrix} \frac{1}{4} & 6 & 0 \\ \frac{1}{3} & 8 & 0 \\ \frac{1}{2} & 9 & 0 \end{pmatrix}$

21. $\begin{pmatrix} -2 & -1 & 4 \\ -3 & 6 & 1 \\ -3 & 4 & 8 \end{pmatrix}$

22. $\begin{pmatrix} 3 & 5 & 1 \\ -1 & 2 & 5 \\ 7 & -4 & 10 \end{pmatrix}$

23. $\begin{pmatrix} 1 & 1 & 1 \\ x & y & z \\ 2 & 3 & 4 \end{pmatrix}$

24. $\begin{pmatrix} 1 & 1 & 1 \\ x & y & z \\ 2+x & 3+y & 4+z \end{pmatrix}$

25. $\begin{pmatrix} 1 & 1 & -3 & 0 \\ 1 & 5 & 3 & 2 \\ 1 & -2 & 1 & 0 \\ 4 & 8 & 0 & 0 \end{pmatrix}$

26. $\begin{pmatrix} 2 & 1 & -2 & 1 \\ 0 & 5 & 0 & 4 \\ 1 & 6 & 1 & 0 \\ 5 & -1 & 1 & 1 \end{pmatrix}$

27. $\begin{pmatrix} 3 & 2 & 0 & 1 & -1 \\ 0 & 1 & 4 & 2 & 3 \\ 0 & 0 & 2 & -1 & 1 \\ 0 & 0 & 0 & 4 & 3 \\ 0 & 0 & 0 & 0 & 2 \end{pmatrix}$

28. $\begin{pmatrix} 2 & 2 & 0 & 0 & -2 \\ 1 & 1 & 6 & 0 & 5 \\ 1 & 0 & 2 & -1 & -1 \\ 2 & 0 & 1 & -2 & 3 \\ 0 & 1 & 0 & 0 & 1 \end{pmatrix}$

In Problems 29 and 30, find the values of λ that satisfy the given equation.

29. $\begin{vmatrix} -3-\lambda & 10 \\ 2 & 5-\lambda \end{vmatrix} = 0$

30. $\begin{vmatrix} 1-\lambda & 0 & -1 \\ 1 & 2-\lambda & 1 \\ 3 & 3 & -\lambda \end{vmatrix} = 0$

8.5 Properties of Determinants

INTRODUCTION In this section we are going to consider some of the many properties of determinants. Our goal in the discussion is to use these properties to develop a means of evaluating a determinant that is an alternative to cofactor expansion.

■ **Properties** The first property states that the determinant of an $n \times n$ matrix and its transpose are the same.

THEOREM 8.5.1 Determinant of a Transpose

If $\mathbf{A}^T$ is the transpose of the $n \times n$ matrix $\mathbf{A}$, then $\det \mathbf{A}^T = \det \mathbf{A}$.

For example, for the matrix $\mathbf{A} = \begin{pmatrix} 5 & 7 \\ 3 & -4 \end{pmatrix}$, we have $\mathbf{A}^T = \begin{pmatrix} 5 & 3 \\ 7 & -4 \end{pmatrix}$. Observe that

$$\det \mathbf{A} = \begin{vmatrix} 5 & 7 \\ 3 & -4 \end{vmatrix} = -41 \quad \text{and} \quad \det \mathbf{A}^T = \begin{vmatrix} 5 & 3 \\ 7 & -4 \end{vmatrix} = -41.$$

Since transposing a matrix interchanges its rows and columns, the significance of Theorem 8.5.1 is that statements concerning determinants and the rows of a matrix also hold when the word *row* is replaced by the word *column*.

THEOREM 8.5.2 Two Identical Rows

If any two rows (columns) of an $n \times n$ matrix $\mathbf{A}$ are the same, then $\det \mathbf{A} = 0$.

EXAMPLE 1 Matrix with Two Identical Rows

Since the second and third columns in the matrix $\mathbf{A} = \begin{pmatrix} 6 & 2 & 2 \\ 4 & 2 & 2 \\ 9 & 2 & 2 \end{pmatrix}$ are the same, it follows from Theorem 8.5.2 that

$$\det \mathbf{A} = \begin{vmatrix} 6 & 2 & 2 \\ 4 & 2 & 2 \\ 9 & 2 & 2 \end{vmatrix} = 0.$$

You should verify this by expanding the determinant by cofactors. ≡

If all the entries in a row (column) of an $n \times n$ matrix $\mathbf{A}$ are zero, then det $\mathbf{A} = 0$.

PROOF: Suppose the ith row of $\mathbf{A}$ consists of all zeros. Hence all the products in the expansion of det $\mathbf{A}$ by cofactors along the ith row are zero and consequently det $\mathbf{A} = 0$. ≡

For example, it follows immediately from Theorem 8.5.3 that

$$\text{zero row} \rightarrow \begin{vmatrix} 0 & 0 \\ 7 & -6 \end{vmatrix} = 0 \quad \text{and} \quad \begin{vmatrix} 4 & 6 & \overset{\text{zero column } \downarrow}{0} \\ 1 & 5 & 0 \\ 8 & -1 & 0 \end{vmatrix} = 0.$$

THEOREM 8.5.4 Interchanging Rows

If $\mathbf{B}$ is the matrix obtained by interchanging any two rows (columns) of an $n \times n$ matrix $\mathbf{A}$, then det $\mathbf{B} = -$det $\mathbf{A}$.

For example, if $\mathbf{B}$ is the matrix obtained by interchanging the first and third rows of $\mathbf{A} = \begin{pmatrix} 4 & -1 & 9 \\ 6 & 0 & 7 \\ 2 & 1 & 3 \end{pmatrix}$, then from Theorem 8.5.4 we have

$$\det \mathbf{B} = \begin{vmatrix} 2 & 1 & 3 \\ 6 & 0 & 7 \\ 4 & -1 & 9 \end{vmatrix} = -\begin{vmatrix} 4 & -1 & 9 \\ 6 & 0 & 7 \\ 2 & 1 & 3 \end{vmatrix} = -\det \mathbf{A}.$$

You should verify this by computing both determinants.

THEOREM 8.5.5 Constant Multiple of a Row

If $\mathbf{B}$ is the matrix obtained from an $n \times n$ matrix $\mathbf{A}$ by multiplying a row (column) by a nonzero real number k, then det $\mathbf{B} = k$ det $\mathbf{A}$.

PROOF: Suppose the entries in the ith row of $\mathbf{A}$ are multiplied by the number k. Call the resulting matrix $\mathbf{B}$. Expanding det $\mathbf{B}$ by cofactors along the ith row then gives

$$\det \mathbf{B} = ka_{i1}C_{i1} + ka_{i2}C_{i2} + \cdots + ka_{in}C_{in}$$
$$= k\underbrace{(a_{i1}C_{i1} + a_{i2}C_{i2} + \cdots + a_{in}C_{in})}_{\text{expansion of det } \mathbf{A} \text{ by cofactors along the } i\text{th row}} = k \det \mathbf{A}.$$

≡

EXAMPLE 2 Theorems 8.5.5 and 8.5.2

(a) $\begin{vmatrix} 5 & 8 \\ 20 & 16 \end{vmatrix} = 5\overset{\underset{\displaystyle\downarrow}{\text{from first column}}}{\begin{vmatrix} 1 & 8 \\ 4 & 16 \end{vmatrix}} = 5 \cdot 8 \overset{\underset{\displaystyle\downarrow}{\text{from second column}}}{\begin{vmatrix} 1 & 1 \\ 4 & 2 \end{vmatrix}} = 5 \cdot 8 \cdot 2 \overset{\underset{\displaystyle\downarrow}{\text{from second row}}}{\begin{vmatrix} 1 & 1 \\ 2 & 1 \end{vmatrix}} = 80(1 - 2) = -80$

$$\textbf{(b)} \quad \begin{vmatrix} 4 & 2 & -1 \\ 5 & -2 & 1 \\ 7 & 4 & -2 \end{vmatrix} = (-2) \begin{vmatrix} 4 & -1 & -1 \\ 5 & 1 & 1 \\ 7 & -2 & -2 \end{vmatrix} = (-2) \cdot 0 = 0 \qquad \equiv$$

from second column ↓ from Theorem 8.5.2 ↓

THEOREM 8.5.6 Determinant of a Matrix Product

If $\mathbf{A}$ and $\mathbf{B}$ are both $n \times n$ matrices, then $\det \mathbf{AB} = \det \mathbf{A} \cdot \det \mathbf{B}$.

In other words, the determinant of a product of two $n \times n$ matrices is the same as the product of the determinants.

EXAMPLE 3 Determinant of a Matrix Product

Suppose $\mathbf{A} = \begin{pmatrix} 2 & 6 \\ 1 & -1 \end{pmatrix}$ and $\mathbf{B} = \begin{pmatrix} 3 & -4 \\ -3 & 5 \end{pmatrix}$. Then $\mathbf{AB} = \begin{pmatrix} -12 & 22 \\ 6 & -9 \end{pmatrix}$. Now $\det \mathbf{AB} = -24$, $\det \mathbf{A} = -8$, $\det \mathbf{B} = 3$, and so we see that

$$\det \mathbf{A} \cdot \det \mathbf{B} = (-8)(3) = -24 = \det \mathbf{AB}. \qquad \equiv$$

THEOREM 8.5.7 Determinant Is Unchanged

Suppose $\mathbf{B}$ is the matrix obtained from an $n \times n$ matrix $\mathbf{A}$ by multiplying the entries in a row (column) by a nonzero real number k and adding the result to the corresponding entries in another row (column). Then $\det \mathbf{B} = \det \mathbf{A}$.

EXAMPLE 4 A Multiple of a Row Added to Another

Suppose $\mathbf{A} = \begin{pmatrix} 5 & 1 & 2 \\ 3 & 0 & 7 \\ 4 & -1 & 4 \end{pmatrix}$ and suppose the matrix $\mathbf{B}$ is defined as that matrix obtained from $\mathbf{A}$ by the elementary row operation

$$\mathbf{A} = \begin{pmatrix} 5 & 1 & 2 \\ 3 & 0 & 7 \\ 4 & -1 & 4 \end{pmatrix} \xrightarrow{-3R_1 + R_3} \begin{pmatrix} 5 & 1 & 2 \\ 3 & 0 & 7 \\ -11 & -4 & -2 \end{pmatrix} = \mathbf{B}.$$

Expanding by cofactors along, say, the second column, we find $\det \mathbf{A} = 45$ and $\det \mathbf{B} = 45$. You should verify this result. $\qquad \equiv$

THEOREM 8.5.8 Determinant of a Triangular Matrix

Suppose $\mathbf{A}$ is an $n \times n$ triangular matrix (upper or lower). Then

$$\det \mathbf{A} = a_{11}a_{22} \cdots a_{nn},$$

where $a_{11}, a_{22}, \ldots, a_{nn}$ are the entries on the main diagonal of $\mathbf{A}$.

PROOF: We prove the result for a 3×3 lower triangular matrix

$$\mathbf{A} = \begin{pmatrix} a_{11} & 0 & 0 \\ a_{21} & a_{22} & 0 \\ a_{31} & a_{32} & a_{33} \end{pmatrix}.$$

Expanding det **A** by cofactors along the first row gives

$$\det \mathbf{A} = a_{11} \begin{vmatrix} a_{22} & 0 \\ a_{32} & a_{33} \end{vmatrix} = a_{11}(a_{22}a_{33} - 0 \cdot a_{32}) = a_{11}a_{22}a_{33}.$$

≡

EXAMPLE 5 Determinant of a Triangular Matrix

(a) The determinant of the lower triangular matrix

$$\mathbf{A} = \begin{pmatrix} 3 & 0 & 0 & 0 \\ 2 & 6 & 0 & 0 \\ 5 & 9 & -4 & 0 \\ 7 & 2 & 4 & -2 \end{pmatrix}$$

is

$$\det \mathbf{A} = \begin{vmatrix} 3 & 0 & 0 & 0 \\ 2 & 6 & 0 & 0 \\ 5 & 9 & -4 & 0 \\ 7 & 2 & 4 & -2 \end{vmatrix} = 3 \cdot 6 \cdot (-4) \cdot (-2) = 144.$$

(b) The determinant of the diagonal matrix $\mathbf{A} = \begin{pmatrix} -3 & 0 & 0 \\ 0 & 6 & 0 \\ 0 & 0 & 4 \end{pmatrix}$ is

$$\det \mathbf{A} = \begin{vmatrix} -3 & 0 & 0 \\ 0 & 6 & 0 \\ 0 & 0 & 4 \end{vmatrix} = (-3) \cdot 6 \cdot 4 = -72.$$

≡

■ **Row Reduction** Evaluating the determinant of an $n \times n$ matrix by the method of cofactor expansion requires a Herculean effort when the order of the matrix is large. To expand the determinant of, say, a 5×5 matrix with nonzero entries requires evaluating five cofactors that are determinants of 4×4 submatrices; each of these in turn requires four additional cofactors that are determinants of 3×3 submatrices, and so on. There is a more practical (and programmable) method for evaluating the determinant of a matrix. This method is based on **reducing** the matrix to a triangular form by row operations and the fact that determinants of triangular matrices are easy to evaluate (see Theorem 8.5.8).

EXAMPLE 6 Reducing a Determinant to Triangular Form

Evaluate the determinant of $\mathbf{A} = \begin{pmatrix} 6 & 2 & 7 \\ -4 & -3 & 2 \\ 2 & 4 & 8 \end{pmatrix}$.

SOLUTION

$$\det \mathbf{A} = \begin{vmatrix} 6 & 2 & 7 \\ -4 & -3 & 2 \\ 2 & 4 & 8 \end{vmatrix}$$

$$= 2 \begin{vmatrix} 6 & 2 & 7 \\ -4 & -3 & 2 \\ 1 & 2 & 4 \end{vmatrix} \qquad \leftarrow 2 \text{ is a common factor in third row: Theorem 8.5.5}$$

$$= -2 \begin{vmatrix} 1 & 2 & 4 \\ -4 & -3 & 2 \\ 6 & 2 & 7 \end{vmatrix} \qquad \leftarrow \text{interchange first and third rows: Theorem 8.5.4}$$

$$= -2 \begin{vmatrix} 1 & 2 & 4 \\ 0 & 5 & 18 \\ 6 & 2 & 7 \end{vmatrix} \qquad \leftarrow \text{4 times first row added to second row: Theorem 8.5.7}$$

$$= -2 \begin{vmatrix} 1 & 2 & 4 \\ 0 & 5 & 18 \\ 0 & -10 & -17 \end{vmatrix} \qquad \leftarrow -6 \text{ times first row added to third row: Theorem 8.5.7}$$

$$= -2 \begin{vmatrix} 1 & 2 & 4 \\ 0 & 5 & 18 \\ 0 & 0 & 19 \end{vmatrix} \qquad \leftarrow \text{2 times second row added to third row: Theorem 8.5.7}$$

$$= (-2)(1)(5)(19) = -190 \qquad \leftarrow \text{Theorem 8.5.8} \qquad\qquad \equiv$$

Our final theorem concerns cofactors. We saw in Section 8.4 that a determinant det $\mathbf{A}$ of an $n \times n$ matrix $\mathbf{A}$ could be evaluated by cofactor expansion along any row (column). This means that the n entries a_{ij} of a row (column) are multiplied by the corresponding cofactors C_{ij} and the n products are added. If, however, the entries a_{ij} of a row (a_{ij} of a column) of $\mathbf{A}$ are multiplied by the corresponding cofactors C_{kj} of a different row (C_{ik} of a different column), the sum of the n products is zero.

THEOREM 8.5.9 A Property of Cofactors

Suppose $\mathbf{A}$ is an $n \times n$ matrix. If $a_{i1}, a_{i2}, \ldots, a_{in}$ are the entries in the ith row and $C_{k1}, C_{k2}, \ldots, C_{kn}$ are the cofactors of the entries in the kth row, then

$$a_{i1}C_{k1} + a_{i2}C_{k2} + \cdots + a_{in}C_{kn} = 0 \quad \text{for } i \neq k.$$

If $a_{1j}, a_{2j}, \ldots, a_{nj}$ are the entries in the jth column and $C_{1k}, C_{2k}, \ldots, C_{nk}$ are the cofactors of the entries in the kth column, then

$$a_{1j}C_{1k} + a_{2j}C_{2k} + \cdots + a_{nj}C_{nk} = 0 \quad \text{for } j \neq k.$$

PROOF: We shall prove the result for rows. Let $\mathbf{B}$ be the matrix obtained from $\mathbf{A}$ by letting the entries in the ith row of $\mathbf{A}$ be the same as the entries in the kth row—that is, $a_{i1} = a_{k1}, a_{i2} = a_{k2}, \ldots, a_{in} = a_{kn}$. Since $\mathbf{B}$ has two rows that are the same, it follows from Theorem 8.5.2 that det $\mathbf{B} = 0$. Cofactor expansion along the kth row then gives the desired result:

$$0 = \det \mathbf{B} = a_{k1}C_{k1} + a_{k2}C_{k2} + \cdots + a_{kn}C_{kn}$$
$$= a_{i1}C_{k1} + a_{i2}C_{k2} + \cdots + a_{in}C_{kn}. \qquad\qquad \equiv$$

EXAMPLE 7 Cofactors of Third Row/Entries of First Row

Consider the matrix $\mathbf{A} = \begin{pmatrix} 6 & 2 & 7 \\ -4 & -3 & 2 \\ 2 & 4 & 8 \end{pmatrix}$. Suppose we multiply the entries of the first row by the cofactors of the third row and add the results; that is,

$$a_{11}C_{31} + a_{12}C_{32} + a_{13}C_{33} = 6 \begin{vmatrix} 2 & 7 \\ -3 & 2 \end{vmatrix} + 2\left(-\begin{vmatrix} 6 & 7 \\ -4 & 2 \end{vmatrix}\right) + 7 \begin{vmatrix} 6 & 2 \\ -4 & -3 \end{vmatrix}$$

$$= 6(25) + 2(-40) + 7(-10) = 0. \qquad\qquad \equiv$$

In Problems 1–10, state the appropriate theorem(s) in this section that justifies the given equality. Do not expand the determinants by cofactors.

1. $\begin{vmatrix} 1 & 2 \\ 3 & 4 \end{vmatrix} = - \begin{vmatrix} 3 & 4 \\ 1 & 2 \end{vmatrix}$ **2.** $\begin{vmatrix} 1 & 2 \\ 3 & 4 \end{vmatrix} = \begin{vmatrix} 1 & 2 \\ 4 & 6 \end{vmatrix}$

3. $\begin{vmatrix} -5 & 6 \\ 2 & -8 \end{vmatrix} = \begin{vmatrix} 1 & 6 \\ -6 & -8 \end{vmatrix}$ **4.** $\begin{vmatrix} 1 & 0 & 0 \\ 0 & 0 & 2 \\ 0 & 1 & 0 \end{vmatrix} = -2 \begin{vmatrix} 1 & 0 & 0 \\ 0 & 1 & 0 \\ 0 & 0 & 1 \end{vmatrix}$

5. $\begin{vmatrix} 1 & 2 & 3 \\ 4 & 2 & 18 \\ 5 & 9 & -12 \end{vmatrix} = 6 \begin{vmatrix} 1 & 2 & 1 \\ 2 & 1 & 3 \\ 5 & 9 & -4 \end{vmatrix}$

6. $\begin{vmatrix} 1 & 2 & 3 \\ 4 & 2 & 18 \\ 5 & 9 & -12 \end{vmatrix} = \begin{vmatrix} 4 & 2 & 18 \\ 5 & 9 & -12 \\ 1 & 2 & 3 \end{vmatrix}$

7. $\begin{vmatrix} 0 & 5 & 0 & 6 \\ 2 & 1 & 0 & 8 \\ 0 & 2 & 0 & -9 \\ 0 & 6 & 0 & 4 \end{vmatrix} = 0$ **8.** $\begin{vmatrix} 3 & 2 & 1 \\ 2 & 6 & 3 \\ 5 & -8 & -4 \end{vmatrix} = 0$

9. $\begin{vmatrix} 1 & 2 & 3 \\ 4 & 5 & 6 \\ 7 & 8 & 9 \end{vmatrix} = \begin{vmatrix} 1 & 4 & 7 \\ 2 & 5 & 8 \\ 3 & 6 & 9 \end{vmatrix}$

10. $\begin{vmatrix} 1 & 0 & 0 & 0 \\ 0 & 2 & 0 & 0 \\ 0 & 0 & 3 & 0 \\ 0 & 0 & 0 & 4 \end{vmatrix} = \begin{vmatrix} 0 & 0 & 0 & 1 \\ 0 & 0 & 2 & 0 \\ 0 & 3 & 0 & 0 \\ 4 & 0 & 0 & 0 \end{vmatrix}$

In Problems 11–16, evaluate the determinant of the given matrix using the result

$$\begin{vmatrix} a_1 & a_2 & a_3 \\ b_1 & b_2 & b_3 \\ c_1 & c_2 & c_3 \end{vmatrix} = 5.$$

11. $\mathbf{A} = \begin{pmatrix} a_3 & a_2 & a_1 \\ b_3 & b_2 & b_1 \\ c_3 & c_2 & c_1 \end{pmatrix}$ **12.** $\mathbf{B} = \begin{pmatrix} 2a_1 & a_2 & a_3 \\ 6b_1 & 3b_2 & 3b_3 \\ 2c_1 & c_2 & c_3 \end{pmatrix}$

13. $\mathbf{C} = \begin{pmatrix} -a_1 & -a_2 & -a_3 \\ b_1 & b_2 & b_3 \\ c_1 - a_1 & c_2 - a_2 & c_3 - a_3 \end{pmatrix}$

14. $\mathbf{D} = \begin{pmatrix} a_1 & b_1 & c_1 \\ a_2 & b_2 & c_2 \\ a_3 & b_3 & c_3 \end{pmatrix}$

15. $\begin{vmatrix} a_1 - 2b_1 + 3c_1 & a_2 - 2b_2 + 3c_2 & a_3 - 2b_3 + 3c_3 \\ b_1 & b_2 & b_3 \\ c_1 & c_2 & c_3 \end{vmatrix}$

16. $\begin{vmatrix} 4a_1 - 2a_3 & a_2 & a_3 \\ 4b_1 - 2b_3 & b_2 & b_3 \\ 2c_1 - c_3 & \frac{1}{2}c_2 & \frac{1}{2}c_3 \end{vmatrix}$

In Problems 17–20, evaluate the determinant of the given matrix without expanding by cofactors.

17. $\mathbf{A} = \begin{pmatrix} 6 & 1 & 8 & 10 \\ 0 & \frac{2}{3} & 7 & 2 \\ 0 & 0 & -4 & 9 \\ 0 & 0 & 0 & -5 \end{pmatrix}$ **18.** $\mathbf{B} = \begin{pmatrix} 0 & 0 & a_{13} \\ 0 & a_{22} & a_{23} \\ a_{31} & a_{32} & a_{33} \end{pmatrix}$

19. $\mathbf{C} = \begin{pmatrix} -5 & 0 & 0 \\ 0 & 7 & 0 \\ 0 & 0 & 3 \end{pmatrix}$ **20.** $\mathbf{D} = \begin{pmatrix} 0 & 7 & 0 \\ 4 & 0 & 0 \\ 0 & 0 & -2 \end{pmatrix}$

In Problems 21 and 22, verify that $\det \mathbf{A} = \det \mathbf{A}^T$ for the given matrix $\mathbf{A}$.

21. $\mathbf{A} = \begin{pmatrix} 1 & 2 & 1 \\ 4 & 1 & -1 \\ 1 & 2 & -1 \end{pmatrix}$ **22.** $\mathbf{A} = \begin{pmatrix} 2 & 3 & 4 \\ 1 & 0 & 5 \\ 7 & 2 & -1 \end{pmatrix}$

23. Consider the matrices

$$\mathbf{A} = \begin{pmatrix} 2 & -1 & 1 \\ 3 & 1 & -1 \\ 0 & 2 & 2 \end{pmatrix} \quad \text{and} \quad \mathbf{B} = \begin{pmatrix} 2 & 1 & 5 \\ 4 & 3 & 8 \\ 0 & -1 & 0 \end{pmatrix}.$$

Verify that $\det \mathbf{AB} = \det \mathbf{A} \cdot \det \mathbf{B}$.

24. Suppose $\mathbf{A}$ is an $n \times n$ matrix such that $\mathbf{A}^2 = \mathbf{I}$. Then show that $\det \mathbf{A} = \pm 1$.

25. Suppose $\mathbf{A}$ is an $n \times n$ matrix such that $\mathbf{A}^2 = \mathbf{A}$. Then show that either $\det \mathbf{A} = 0$ or $\det \mathbf{A} = 1$.

26. If $\mathbf{A}$ and $\mathbf{B}$ are $n \times n$ matrices, then prove or disprove that $\det \mathbf{AB} = \det \mathbf{BA}$.

27. Consider the matrix

$$\mathbf{A} = \begin{pmatrix} a & a + 1 & a + 2 \\ b & b + 1 & b + 2 \\ c & c + 1 & c + 2 \end{pmatrix}.$$

Without expanding, evaluate $\det \mathbf{A}$.

28. Consider the matrix

$$\mathbf{A} = \begin{pmatrix} 1 & 1 & 1 \\ x & y & z \\ y + z & x + z & x + y \end{pmatrix}.$$

Without expanding, show that $\det \mathbf{A} = 0$.

In Problems 29–36, use the procedure illustrated in Example 6 to evaluate the determinant of the given matrix.

29. $\begin{pmatrix} 1 & 1 & 5 \\ 4 & 3 & 6 \\ 0 & -1 & 1 \end{pmatrix}$ **30.** $\begin{pmatrix} 2 & 4 & 5 \\ 4 & 2 & 0 \\ 8 & 7 & -2 \end{pmatrix}$

31. $\begin{pmatrix} -1 & 2 & 3 \\ 4 & -5 & -2 \\ 9 & -9 & 6 \end{pmatrix}$ **32.** $\begin{pmatrix} -2 & 2 & -6 \\ 5 & 0 & 1 \\ 1 & -2 & 2 \end{pmatrix}$

33. $\begin{pmatrix} 1 & -2 & 2 & 1 \\ 2 & 1 & -2 & 3 \\ 3 & 4 & -8 & 1 \\ 3 & -11 & 12 & 2 \end{pmatrix}$ **34.** $\begin{pmatrix} 0 & 1 & 4 & 5 \\ 2 & 5 & 0 & 1 \\ 1 & 2 & 2 & 0 \\ 3 & 1 & 3 & 2 \end{pmatrix}$

35. $\begin{pmatrix} 1 & 2 & 3 & 4 \\ 1 & 3 & 5 & 7 \\ 2 & 3 & 6 & 7 \\ 1 & 5 & 8 & 20 \end{pmatrix}$ **36.** $\begin{pmatrix} 2 & 9 & 1 & 8 \\ 1 & 3 & 7 & 4 \\ 0 & 1 & 6 & 5 \\ 3 & 1 & 4 & 2 \end{pmatrix}$

37. By proceeding as in Example 6, show that

$$\begin{vmatrix} 1 & 1 & 1 \\ a & b & c \\ a^2 & b^2 & c^2 \end{vmatrix} = (b - a)(c - a)(c - b).$$

38. Evaluate $\begin{vmatrix} 1 & 1 & 1 & 1 \\ a & b & c & d \\ a^2 & b^2 & c^2 & d^2 \\ a^3 & b^3 & c^3 & d^3 \end{vmatrix}$. [*Hint*: See Problem 37.]

In Problems 39 and 40, verify Theorem 8.5.9 by evaluating $a_{21}C_{11} + a_{22}C_{12} + a_{23}C_{13}$ and $a_{13}C_{12} + a_{23}C_{22} + a_{33}C_{32}$ for the given matrix.

39. $\mathbf{A} = \begin{pmatrix} 1 & 1 & 2 \\ -1 & 2 & 1 \\ 4 & -2 & 1 \end{pmatrix}$ **40.** $\mathbf{A} = \begin{pmatrix} 3 & 0 & 5 \\ -2 & 3 & -1 \\ 2 & 2 & -3 \end{pmatrix}$

41. Let $\mathbf{A} = \begin{pmatrix} 3 & -4 \\ 1 & 2 \end{pmatrix}$ and $\mathbf{B} = \begin{pmatrix} 7 & 4 \\ -1 & -5 \end{pmatrix}$. Verify that

$$\det(\mathbf{A} + \mathbf{B}) \neq \det \mathbf{A} + \det \mathbf{B}.$$

42. Suppose $\mathbf{A}$ is a 5×5 matrix for which $\det \mathbf{A} = -7$. What is the value of $\det(2\mathbf{A})$?

43. An $n \times n$ matrix $\mathbf{A}$ is said to be **skew-symmetric** if $\mathbf{A}^T = -\mathbf{A}$. If $\mathbf{A}$ is a 5×5 skew-symmetric matrix, show that $\det \mathbf{A} = 0$.

44. It takes about $n!$ multiplications to evaluate the determinant of an $n \times n$ matrix using expansion by cofactors, whereas it takes about $n^3/3$ arithmetic operations using the row-reduction method. Compare the number of operations for both methods using a 25×25 matrix.

8.6 Inverse of a Matrix

INTRODUCTION The concept of the determinant of an $n \times n$, or square, matrix will play an important role in this and the following section.

8.6.1 Finding the Inverse

In the real number system, if a is a nonzero number, then there exists a number b such that $ab = ba = 1$. The number b is called the multiplicative inverse of the number a and is denoted by a^{-1}. For a square matrix $\mathbf{A}$ it is also important to know whether we can find another square matrix $\mathbf{B}$ of the same order such that $\mathbf{AB} = \mathbf{BA} = \mathbf{I}$. We have the following definition.

DEFINITION 8.6.1 Inverse of a Matrix

Let $\mathbf{A}$ be an $n \times n$ matrix. If there exists an $n \times n$ matrix $\mathbf{B}$ such that

$$\mathbf{AB} = \mathbf{BA} = \mathbf{I}, \tag{1}$$

where $\mathbf{I}$ is the $n \times n$ identity, then the matrix $\mathbf{A}$ is said to be **nonsingular** or **invertible**. The matrix $\mathbf{B}$ is said to be the **inverse** of $\mathbf{A}$.

For example, the matrix $\mathbf{A} = \begin{pmatrix} 2 & 1 \\ 1 & 1 \end{pmatrix}$ is nonsingular or invertible since the matrix $\mathbf{B} = \begin{pmatrix} 1 & -1 \\ -1 & 2 \end{pmatrix}$ is its inverse. To verify this, observe that

$$\mathbf{AB} = \begin{pmatrix} 2 & 1 \\ 1 & 1 \end{pmatrix}\begin{pmatrix} 1 & -1 \\ -1 & 2 \end{pmatrix} = \begin{pmatrix} 1 & 0 \\ 0 & 1 \end{pmatrix} = \mathbf{I}$$

and

$$\mathbf{BA} = \begin{pmatrix} 1 & -1 \\ -1 & 2 \end{pmatrix}\begin{pmatrix} 2 & 1 \\ 1 & 1 \end{pmatrix} = \begin{pmatrix} 1 & 0 \\ 0 & 1 \end{pmatrix} = \mathbf{I}.$$

Unlike the real number system, where *every* nonzero number a has a multiplicative inverse, not every nonzero $n \times n$ matrix $\mathbf{A}$ has an inverse.

EXAMPLE 1 Matrix with No Inverse

The matrix $\mathbf{A} = \begin{pmatrix} 1 & 1 \\ 0 & 0 \end{pmatrix}$ has no multiplicative inverse. To see this suppose $\mathbf{B} = \begin{pmatrix} b_{11} & b_{12} \\ b_{21} & b_{22} \end{pmatrix}$. Then

$$\mathbf{AB} = \begin{pmatrix} 1 & 1 \\ 0 & 0 \end{pmatrix} \begin{pmatrix} b_{11} & b_{12} \\ b_{21} & b_{22} \end{pmatrix} = \begin{pmatrix} b_{11} + b_{21} & b_{12} + b_{22} \\ 0 & 0 \end{pmatrix}.$$

Inspection of the last matrix shows that it is impossible to obtain the 2×2 identity matrix $\mathbf{I}$, because there is no way of selecting $b_{11}, b_{12}, b_{21},$ and b_{22} to get 1 as the entry in the second row and second column. We conclude that the matrix $\mathbf{A} = \begin{pmatrix} 1 & 1 \\ 0 & 0 \end{pmatrix}$ has no inverse. ≡

An $n \times n$ matrix that has no inverse is called **singular**. If $\mathbf{A}$ is **nonsingular**, its inverse is denoted by $\mathbf{B} = \mathbf{A}^{-1}$.

Note that the symbol -1 in the notation $\mathbf{A}^{-1}$ is not an exponent; in other words, $\mathbf{A}^{-1}$ is not a reciprocal. Also, if $\mathbf{A}$ is nonsingular, its inverse is unique.

Important. ▶

■ **Properties** The following theorem lists some properties of the inverse of a matrix.

THEOREM 8.6.1 Properties of the Inverse

Let $\mathbf{A}$ and $\mathbf{B}$ be nonsingular matrices. Then

(*i*) $(\mathbf{A}^{-1})^{-1} = \mathbf{A}$
(*ii*) $(\mathbf{AB})^{-1} = \mathbf{B}^{-1}\mathbf{A}^{-1}$
(*iii*) $(\mathbf{A}^T)^{-1} = (\mathbf{A}^{-1})^T$

PROOF OF (*i*): This part of the theorem states that if $\mathbf{A}$ is nonsingular, then its inverse $\mathbf{A}^{-1}$ is also nonsingular and *its* inverse is $\mathbf{A}$. To prove that $\mathbf{A}^{-1}$ is nonsingular, we have to show that a matrix $\mathbf{B}$ can be found such that $\mathbf{A}^{-1}\mathbf{B} = \mathbf{BA}^{-1} = \mathbf{I}$. But since $\mathbf{A}$ is assumed to be nonsingular, we know from (1) that $\mathbf{AA}^{-1} = \mathbf{A}^{-1}\mathbf{A} = \mathbf{I}$ and, equivalently, $\mathbf{A}^{-1}\mathbf{A} = \mathbf{AA}^{-1} = \mathbf{I}$. The last matrix equation indicates that the required matrix, the inverse of $\mathbf{A}^{-1}$, is $\mathbf{B} = \mathbf{A}$. Consequently, $(\mathbf{A}^{-1})^{-1} = \mathbf{A}$. ≡

Theorem 8.6.1(*ii*) extends to any finite number of nonsingular matrices:

$$(\mathbf{A}_1\mathbf{A}_2 \cdots \mathbf{A}_k)^{-1} = \mathbf{A}_k^{-1}\mathbf{A}_{k-1}^{-1} \cdots \mathbf{A}_1^{-1};$$

that is, the inverse of a product of nonsingular matrices is the product of the inverses in reverse order.

In the discussion that follows we are going to consider two different ways of finding $\mathbf{A}^{-1}$ for a nonsingular matrix $\mathbf{A}$. The first method utilizes determinants, whereas the second employs the elementary row operations introduced in Section 8.2.

■ **Adjoint Method** Recall from (6) of Section 8.4 that the cofactor C_{ij} of the entry a_{ij} of an $n \times n$ matrix $\mathbf{A}$ is $C_{ij} = (-1)^{i+j}M_{ij}$, where M_{ij} is the minor of a_{ij}; that is, the determinant of the $(n-1) \times (n-1)$ submatrix obtained by deleting the ith row and the jth column of $\mathbf{A}$.

DEFINITION 8.6.2 Adjoint Matrix

Let $\mathbf{A}$ be an $n \times n$ matrix. The matrix that is the transpose of the matrix of cofactors corresponding to the entries of $\mathbf{A}$:

$$\begin{pmatrix} C_{11} & C_{12} & \cdots & C_{1n} \\ C_{21} & C_{22} & \cdots & C_{2n} \\ \vdots & & & \vdots \\ C_{n1} & C_{n2} & \cdots & C_{nn} \end{pmatrix}^T = \begin{pmatrix} C_{11} & C_{21} & \cdots & C_{n1} \\ C_{12} & C_{22} & \cdots & C_{n2} \\ \vdots & & & \vdots \\ C_{1n} & C_{2n} & \cdots & C_{nn} \end{pmatrix}$$

is called the **adjoint** of $\mathbf{A}$ and is denoted by adj $\mathbf{A}$.

The next theorem will give a compact formula for the inverse of a nonsingular matrix in terms of the adjoint of the matrix. However, because of the determinants involved, this method becomes unwieldy for $n \geq 4$.

THEOREM 8.6.2 Finding the Inverse

Let $\mathbf{A}$ be an $n \times n$ matrix. If $\det \mathbf{A} \neq 0$, then

$$\mathbf{A}^{-1} = \left(\frac{1}{\det \mathbf{A}} \right) \text{adj } \mathbf{A}. \tag{2}$$

PROOF: For brevity we prove the case when $n = 3$. Observe that

$$
\mathbf{A}(\text{adj } \mathbf{A}) = \begin{pmatrix} a_{11} & a_{12} & a_{13} \\ a_{21} & a_{22} & a_{23} \\ a_{31} & a_{32} & a_{33} \end{pmatrix} \begin{pmatrix} C_{11} & C_{21} & C_{31} \\ C_{12} & C_{22} & C_{32} \\ C_{13} & C_{23} & C_{33} \end{pmatrix}
$$
$$
= \begin{pmatrix} \det \mathbf{A} & 0 & 0 \\ 0 & \det \mathbf{A} & 0 \\ 0 & 0 & \det \mathbf{A} \end{pmatrix} \tag{3}
$$

since $\det \mathbf{A} = a_{i1}C_{i1} + a_{i2}C_{i2} + a_{i3}C_{i3}$, for $i = 1, 2, 3$, are the expansions of $\det \mathbf{A}$ by cofactors along the first, second, and third rows, and

$$a_{11}C_{21} + a_{12}C_{22} + a_{13}C_{23} = 0 \qquad a_{11}C_{31} + a_{12}C_{32} + a_{13}C_{33} = 0$$
$$a_{21}C_{11} + a_{22}C_{12} + a_{23}C_{13} = 0 \qquad a_{21}C_{31} + a_{22}C_{32} + a_{23}C_{33} = 0$$
$$a_{31}C_{11} + a_{32}C_{12} + a_{33}C_{13} = 0 \qquad a_{31}C_{21} + a_{32}C_{22} + a_{33}C_{23} = 0$$

in view of Theorem 8.5.9. Thus, (3) is the same as

$$\mathbf{A}(\text{adj } \mathbf{A}) = (\det \mathbf{A}) \begin{pmatrix} 1 & 0 & 0 \\ 0 & 1 & 0 \\ 0 & 0 & 1 \end{pmatrix} = (\det \mathbf{A}) \mathbf{I}$$

or $\mathbf{A}((1/\det \mathbf{A}) \text{adj } \mathbf{A}) = \mathbf{I}$. Similarly, it can be shown in exactly the same manner that $((1/\det \mathbf{A}) \text{adj } \mathbf{A})\mathbf{A} = \mathbf{I}$. Hence, by definition, $\mathbf{A}^{-1} = (1/\det \mathbf{A}) \text{adj } \mathbf{A}$. ≡

For future reference we observe in the case of a 2×2 nonsingular matrix

$$\mathbf{A} = \begin{pmatrix} a_{11} & a_{12} \\ a_{21} & a_{22} \end{pmatrix}$$

that the cofactors are $C_{11} = a_{22}$, $C_{12} = -a_{21}$, $C_{21} = -a_{12}$, and $C_{22} = a_{11}$. In this case,

$$\text{adj } \mathbf{A} = \begin{pmatrix} C_{11} & C_{12} \\ C_{21} & C_{22} \end{pmatrix}^T = \begin{pmatrix} a_{22} & -a_{21} \\ -a_{12} & a_{11} \end{pmatrix}^T = \begin{pmatrix} a_{22} & -a_{12} \\ -a_{21} & a_{11} \end{pmatrix}.$$

It follows from (2) that

$$\mathbf{A}^{-1} = \frac{1}{\det \mathbf{A}} \begin{pmatrix} a_{22} & -a_{12} \\ -a_{21} & a_{11} \end{pmatrix}. \tag{4}$$

For a 3×3 nonsingular matrix

$$\mathbf{A} = \begin{pmatrix} a_{11} & a_{12} & a_{13} \\ a_{21} & a_{22} & a_{23} \\ a_{31} & a_{32} & a_{33} \end{pmatrix}$$

$$C_{11} = \begin{vmatrix} a_{22} & a_{23} \\ a_{32} & a_{33} \end{vmatrix} \quad C_{12} = -\begin{vmatrix} a_{21} & a_{23} \\ a_{31} & a_{33} \end{vmatrix} \quad C_{13} = \begin{vmatrix} a_{21} & a_{22} \\ a_{31} & a_{32} \end{vmatrix},$$

and so on. After the adjoint of $\mathbf{A}$ is formed, (2) gives

$$\mathbf{A}^{-1} = \frac{1}{\det \mathbf{A}} \begin{pmatrix} C_{11} & C_{21} & C_{31} \\ C_{12} & C_{22} & C_{32} \\ C_{13} & C_{23} & C_{33} \end{pmatrix}. \tag{5}$$

EXAMPLE 2 **Inverse of a Matrix**

Find the inverse of $\mathbf{A} = \begin{pmatrix} 1 & 4 \\ 2 & 10 \end{pmatrix}$.

SOLUTION Since $\det \mathbf{A} = 10 - 8 = 2$, it follows from (4) that

$$\mathbf{A}^{-1} = \frac{1}{2} \begin{pmatrix} 10 & -4 \\ -2 & 1 \end{pmatrix} = \begin{pmatrix} 5 & -2 \\ -1 & \frac{1}{2} \end{pmatrix}.$$

Check: $\mathbf{A}\mathbf{A}^{-1} = \begin{pmatrix} 1 & 4 \\ 2 & 10 \end{pmatrix}\begin{pmatrix} 5 & -2 \\ -1 & \frac{1}{2} \end{pmatrix} = \begin{pmatrix} 5-4 & -2+2 \\ 10-10 & -4+5 \end{pmatrix} = \begin{pmatrix} 1 & 0 \\ 0 & 1 \end{pmatrix}$

$\mathbf{A}^{-1}\mathbf{A} = \begin{pmatrix} 5 & -2 \\ -1 & \frac{1}{2} \end{pmatrix}\begin{pmatrix} 1 & 4 \\ 2 & 10 \end{pmatrix} = \begin{pmatrix} 5-4 & 20-20 \\ -1+1 & -4+5 \end{pmatrix} = \begin{pmatrix} 1 & 0 \\ 0 & 1 \end{pmatrix}.$ ≡

EXAMPLE 3 **Inverse of a Matrix**

Find the inverse of $\mathbf{A} = \begin{pmatrix} 2 & 2 & 0 \\ -2 & 1 & 1 \\ 3 & 0 & 1 \end{pmatrix}$.

SOLUTION Since $\det \mathbf{A} = 12$, we can find $\mathbf{A}^{-1}$ from (5). The cofactors corresponding to the entries in $\mathbf{A}$ are

$$C_{11} = \begin{vmatrix} 1 & 1 \\ 0 & 1 \end{vmatrix} = 1 \quad C_{12} = -\begin{vmatrix} -2 & 1 \\ 3 & 1 \end{vmatrix} = 5 \quad C_{13} = \begin{vmatrix} -2 & 1 \\ 3 & 0 \end{vmatrix} = -3$$

$$C_{21} = -\begin{vmatrix} 2 & 0 \\ 0 & 1 \end{vmatrix} = -2 \quad C_{22} = \begin{vmatrix} 2 & 0 \\ 3 & 1 \end{vmatrix} = 2 \quad C_{23} = -\begin{vmatrix} 2 & 2 \\ 3 & 0 \end{vmatrix} = 6$$

$$C_{31} = \begin{vmatrix} 2 & 0 \\ 1 & 1 \end{vmatrix} = 2 \quad C_{32} = -\begin{vmatrix} 2 & 0 \\ -2 & 1 \end{vmatrix} = -2 \quad C_{33} = \begin{vmatrix} 2 & 2 \\ -2 & 1 \end{vmatrix} = 6.$$

From (5) we then obtain

$$\mathbf{A}^{-1} = \frac{1}{12} \begin{pmatrix} 1 & -2 & 2 \\ 5 & 2 & -2 \\ -3 & 6 & 6 \end{pmatrix} = \begin{pmatrix} \frac{1}{12} & -\frac{1}{6} & \frac{1}{6} \\ \frac{5}{12} & \frac{1}{6} & -\frac{1}{6} \\ -\frac{1}{4} & \frac{1}{2} & \frac{1}{2} \end{pmatrix}.$$

The reader is urged to verify that $\mathbf{A}\mathbf{A}^{-1} = \mathbf{A}^{-1}\mathbf{A} = \mathbf{I}$. ≡

We are now in a position to prove a necessary and sufficient condition for an $n \times n$ matrix $\mathbf{A}$ to have an inverse.

THEOREM 8.6.3 Nonsingular Matrices and det A

An $n \times n$ matrix $\mathbf{A}$ is nonsingular if and only if det $\mathbf{A} \neq 0$.

PROOF: We shall first prove the sufficiency. Assume det $\mathbf{A} \neq 0$. Then $\mathbf{A}$ is nonsingular, since $\mathbf{A}^{-1}$ can be found from Theorem 8.6.2.

To prove the necessity, we must assume that $\mathbf{A}$ is nonsingular and prove that det $\mathbf{A} \neq 0$. Now from Theorem 8.5.6, $\mathbf{A}\mathbf{A}^{-1} = \mathbf{A}^{-1}\mathbf{A} = \mathbf{I}$ implies

$$(\det \mathbf{A})(\det \mathbf{A}^{-1}) = (\det \mathbf{A}^{-1})(\det \mathbf{A}) = \det \mathbf{I}.$$

But since det $\mathbf{I} = 1$ (why?), the product $(\det \mathbf{A})(\det \mathbf{A}^{-1}) = 1 \neq 0$ shows that we must have det $\mathbf{A} \neq 0$. ≡

For emphasis we restate Theorem 8.6.3 in an alternative manner:

An $n \times n$ matrix $\mathbf{A}$ is singular if and only if det $\mathbf{A} = 0$. (6)

EXAMPLE 4 Using (6)

The 2×2 matrix $\mathbf{A} = \begin{pmatrix} 2 & 2 \\ 3 & 3 \end{pmatrix}$ has no inverse; that is, $\mathbf{A}$ is singular, because det $\mathbf{A} = 6 - 6 = 0$. ≡

Because of the number of determinants that must be evaluated, the foregoing method for calculating the inverse of a matrix is tedious when the order of the matrix is large. In the case of 3×3 or larger matrices, the next method is a particularly efficient means for finding $\mathbf{A}^{-1}$.

■ **Row Operations Method** Although it would take us beyond the scope of this book to prove it, we shall nonetheless use the following results:

THEOREM 8.6.4 Finding the Inverse

If an $n \times n$ matrix $\mathbf{A}$ can be transformed into the $n \times n$ identity $\mathbf{I}$ by a sequence of elementary row operations, then $\mathbf{A}$ is nonsingular. The same sequence of operations that transforms $\mathbf{A}$ into the identity $\mathbf{I}$ will also transform $\mathbf{I}$ into $\mathbf{A}^{-1}$.

It is convenient to carry out these row operations on $\mathbf{A}$ and $\mathbf{I}$ simultaneously by means of an $n \times 2n$ matrix obtained by augmenting $\mathbf{A}$ with the identity $\mathbf{I}$ as shown here:

$$(\mathbf{A}|\mathbf{I}) = \begin{pmatrix} a_{11} & a_{12} & \cdots & a_{1n} & 1 & 0 & \cdots & 0 \\ a_{21} & a_{22} & \cdots & a_{2n} & 0 & 1 & \cdots & 0 \\ \vdots & & & \vdots & \vdots & & & \vdots \\ a_{n1} & a_{n2} & \cdots & a_{nn} & 0 & 0 & \cdots & 1 \end{pmatrix}.$$

The procedure for finding $\mathbf{A}^{-1}$ is outlined in the following diagram:

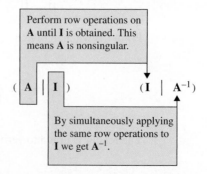

Perform row operations on $\mathbf{A}$ until $\mathbf{I}$ is obtained. This means $\mathbf{A}$ is nonsingular.

$(\mathbf{A} | \mathbf{I})$ $(\mathbf{I} | \mathbf{A}^{-1})$

By simultaneously applying the same row operations to $\mathbf{I}$ we get $\mathbf{A}^{-1}$.

EXAMPLE 5 **Inverse by Elementary Row Operations**

Find the inverse of $\mathbf{A} = \begin{pmatrix} 2 & 0 & 1 \\ -2 & 3 & 4 \\ -5 & 5 & 6 \end{pmatrix}$.

SOLUTION We shall use the same notation as we did in Section 8.2 when we reduced an augmented matrix to reduced row-echelon form:

$$\left(\begin{array}{rrr|rrr} 2 & 0 & 1 & 1 & 0 & 0 \\ -2 & 3 & 4 & 0 & 1 & 0 \\ -5 & 5 & 6 & 0 & 0 & 1 \end{array}\right) \overset{\frac{1}{2}R_1}{\Longrightarrow} \left(\begin{array}{rrr|rrr} 1 & 0 & \frac{1}{2} & \frac{1}{2} & 0 & 0 \\ -2 & 3 & 4 & 0 & 1 & 0 \\ -5 & 5 & 6 & 0 & 0 & 1 \end{array}\right)$$

$$\overset{\substack{2R_1+R_2 \\ 5R_1+R_3}}{\Longrightarrow} \left(\begin{array}{rrr|rrr} 1 & 0 & \frac{1}{2} & \frac{1}{2} & 0 & 0 \\ 0 & 3 & 5 & 1 & 1 & 0 \\ 0 & 5 & \frac{17}{2} & \frac{5}{2} & 0 & 1 \end{array}\right)$$

$$\overset{\substack{\frac{1}{3}R_2 \\ \frac{1}{5}R_3}}{\Longrightarrow} \left(\begin{array}{rrr|rrr} 1 & 0 & \frac{1}{2} & \frac{1}{2} & 0 & 0 \\ 0 & 1 & \frac{5}{3} & \frac{1}{3} & \frac{1}{3} & 0 \\ 0 & 1 & \frac{17}{10} & \frac{1}{2} & 0 & \frac{1}{5} \end{array}\right)$$

$$\overset{-R_2+R_3}{\Longrightarrow} \left(\begin{array}{rrr|rrr} 1 & 0 & \frac{1}{2} & \frac{1}{2} & 0 & 0 \\ 0 & 1 & \frac{5}{3} & \frac{1}{3} & \frac{1}{3} & 0 \\ 0 & 0 & \frac{1}{30} & \frac{1}{6} & -\frac{1}{3} & \frac{1}{5} \end{array}\right)$$

$$\overset{30R_3}{\Longrightarrow} \left(\begin{array}{rrr|rrr} 1 & 0 & \frac{1}{2} & \frac{1}{2} & 0 & 0 \\ 0 & 1 & \frac{5}{3} & \frac{1}{3} & \frac{1}{3} & 0 \\ 0 & 0 & 1 & 5 & -10 & 6 \end{array}\right)$$

$$\overset{\substack{-\frac{1}{2}R_3+R_1 \\ -\frac{5}{3}R_3+R_2}}{\Longrightarrow} \left(\begin{array}{rrr|rrr} 1 & 0 & 0 & -2 & 5 & -3 \\ 0 & 1 & 0 & -8 & 17 & -10 \\ 0 & 0 & 1 & 5 & -10 & 6 \end{array}\right).$$

Since $\mathbf{I}$ appears to the left of the vertical line, we conclude that the matrix to the right of the line is

$$\mathbf{A}^{-1} = \begin{pmatrix} -2 & 5 & -3 \\ -8 & 17 & -10 \\ 5 & -10 & 6 \end{pmatrix}.$$

≡

If row reduction of $(\mathbf{A}|\mathbf{I})$ leads to the situation

$$(\mathbf{A}|\mathbf{I}) \overset{\substack{\text{row} \\ \text{operations}}}{\Longrightarrow} (\mathbf{B}|\mathbf{C}),$$

where the matrix $\mathbf{B}$ contains a row of zeros, then necessarily $\mathbf{A}$ is singular. Since further reduction of $\mathbf{B}$ always yields another matrix with a row of zeros, we can never transform $\mathbf{A}$ into $\mathbf{I}$.

EXAMPLE 6 **A Singular Matrix**

The matrix $\mathbf{A} = \begin{pmatrix} 1 & -1 & -2 \\ 2 & 4 & 5 \\ 6 & 0 & -3 \end{pmatrix}$ has no inverse, since

$$\left(\begin{array}{rrr|rrr} 1 & -1 & -2 & 1 & 0 & 0 \\ 2 & 4 & 5 & 0 & 1 & 0 \\ 6 & 0 & -3 & 0 & 0 & 1 \end{array}\right) \overset{-2R_1+R_2}{\Longrightarrow} \left(\begin{array}{rrr|rrr} 1 & -1 & -2 & 1 & 0 & 0 \\ 0 & 6 & 9 & -2 & 1 & 0 \\ 6 & 0 & -3 & 0 & 0 & 1 \end{array}\right)$$

$$\overset{-6R_1+R_3}{\Longrightarrow} \left(\begin{array}{rrr|rrr} 1 & -1 & -2 & 1 & 0 & 0 \\ 0 & 6 & 9 & -2 & 1 & 0 \\ 0 & 6 & 9 & -6 & 0 & 1 \end{array}\right)$$

$$\overset{-R_2+R_3}{\Longrightarrow} \quad \begin{pmatrix} 1 & -1 & -2 & 1 & 0 & 0 \\ 0 & 6 & 9 & -2 & 1 & 0 \\ 0 & 0 & 0 & -4 & -1 & 1 \end{pmatrix}.$$

Since the matrix to the left of the vertical bar has a row of zeros, we can stop at this point and conclude that $\mathbf{A}$ is singular. $\equiv$

8.6.2 Using the Inverse to Solve Systems

A system of m linear equations in n variables $x_1, x_2, \ldots, x_n$,

$$
\begin{aligned}
a_{11}x_1 + a_{12}x_2 + \cdots + a_{1n}x_n &= b_1 \\
a_{21}x_1 + a_{22}x_2 + \cdots + a_{2n}x_n &= b_2 \\
&\vdots \\
a_{m1}x_1 + a_{m2}x_2 + \cdots + a_{mn}x_n &= b_m
\end{aligned}
\tag{7}
$$

can be written compactly as a matrix equation $\mathbf{AX} = \mathbf{B}$, where

$$\mathbf{A} = \begin{pmatrix} a_{11} & a_{12} & \cdots & a_{1n} \\ a_{21} & a_{22} & \cdots & a_{2n} \\ \vdots & & & \vdots \\ a_{m1} & a_{m2} & \cdots & a_{mn} \end{pmatrix}, \quad \mathbf{X} = \begin{pmatrix} x_1 \\ x_2 \\ \vdots \\ x_n \end{pmatrix}, \quad \mathbf{B} = \begin{pmatrix} b_1 \\ b_2 \\ \vdots \\ b_m \end{pmatrix}.$$

◼ **Special Case** Let us suppose now that $m = n$ in (7) so that the coefficient matrix $\mathbf{A}$ is $n \times n$. In particular, if $\mathbf{A}$ is nonsingular, then the system $\mathbf{AX} = \mathbf{B}$ can be solved by multiplying both of the equations by $\mathbf{A}^{-1}$. From $\mathbf{A}^{-1}(\mathbf{AX}) = \mathbf{A}^{-1}\mathbf{B}$, we get $(\mathbf{A}^{-1}\mathbf{A})\mathbf{X} = \mathbf{A}^{-1}\mathbf{B}$. Since $\mathbf{A}^{-1}\mathbf{A} = \mathbf{I}$ and $\mathbf{IX} = \mathbf{X}$, we have

$$\mathbf{X} = \mathbf{A}^{-1}\mathbf{B}. \tag{8}$$

EXAMPLE 7 **Using (8) to Solve a System**

Use the inverse of the coefficient matrix to solve the system

$$2x_1 - 9x_2 = 15$$
$$3x_1 + 6x_2 = 16.$$

SOLUTION The given system can be written as

$$\begin{pmatrix} 2 & -9 \\ 3 & 6 \end{pmatrix}\begin{pmatrix} x_1 \\ x_2 \end{pmatrix} = \begin{pmatrix} 15 \\ 16 \end{pmatrix}.$$

Since $\begin{vmatrix} 2 & -9 \\ 3 & 6 \end{vmatrix} = 39 \neq 0$, the coefficient matrix is nonsingular. Consequently, from (4) we get

$$\begin{pmatrix} 2 & -9 \\ 3 & 6 \end{pmatrix}^{-1} = \frac{1}{39}\begin{pmatrix} 6 & 9 \\ -3 & 2 \end{pmatrix}.$$

Using (8) it follows that

$$\begin{pmatrix} x_1 \\ x_2 \end{pmatrix} = \frac{1}{39}\begin{pmatrix} 6 & 9 \\ -3 & 2 \end{pmatrix}\begin{pmatrix} 15 \\ 16 \end{pmatrix} = \frac{1}{39}\begin{pmatrix} 234 \\ -13 \end{pmatrix} = \begin{pmatrix} 6 \\ -\frac{1}{3} \end{pmatrix},$$

and so $x_1 = 6$ and $x_2 = -\frac{1}{3}$. $\equiv$

EXAMPLE 8 **Using (8) to Solve a System**

Use the inverse of the coefficient matrix to solve the system

$$2x_1 + x_3 = 2$$
$$-2x_1 + 3x_2 + 4x_3 = 4$$
$$-5x_1 + 5x_2 + 6x_3 = -1$$

SOLUTION We found the inverse of the coefficient matrix

$$\mathbf{A} = \begin{pmatrix} 2 & 0 & 1 \\ -2 & 3 & 4 \\ -5 & 5 & 6 \end{pmatrix}$$

in Example 5. Thus, (8) gives

$$\begin{pmatrix} x_1 \\ x_2 \\ x_3 \end{pmatrix} = \begin{pmatrix} 2 & 0 & 1 \\ -2 & 3 & 4 \\ -5 & 5 & 6 \end{pmatrix}^{-1} \begin{pmatrix} 2 \\ 4 \\ -1 \end{pmatrix} = \begin{pmatrix} -2 & 5 & -3 \\ -8 & 17 & -10 \\ 5 & -10 & 6 \end{pmatrix} \begin{pmatrix} 2 \\ 4 \\ -1 \end{pmatrix} = \begin{pmatrix} 19 \\ 62 \\ -36 \end{pmatrix}.$$

Consequently, $x_1 = 19$, $x_2 = 62$, and $x_3 = -36$. $\equiv$

■ **Uniqueness** When det $\mathbf{A} \neq 0$ the solution of the system $\mathbf{AX} = \mathbf{B}$ is unique. Suppose not—that is, suppose det $\mathbf{A} \neq 0$ and $\mathbf{X}_1$ and $\mathbf{X}_2$ are two *different* solution vectors. Then $\mathbf{AX}_1 = \mathbf{B}$ and $\mathbf{AX}_2 = \mathbf{B}$ imply $\mathbf{AX}_1 = \mathbf{AX}_2$. Since $\mathbf{A}$ is nonsingular, $\mathbf{A}^{-1}$ exists, and so $\mathbf{A}^{-1}(\mathbf{AX}_1) = \mathbf{A}^{-1}(\mathbf{AX}_2)$ and $(\mathbf{A}^{-1}\mathbf{A})\mathbf{X}_1 = (\mathbf{A}^{-1}\mathbf{A})\mathbf{X}_2$. This gives $\mathbf{IX}_1 = \mathbf{IX}_2$ or $\mathbf{X}_1 = \mathbf{X}_2$, which contradicts our assumption that $\mathbf{X}_1$ and $\mathbf{X}_2$ were different solution vectors.

■ **Homogeneous Systems** A homogeneous system of equations can be written $\mathbf{AX} = \mathbf{0}$. Recall that a homogeneous system always possesses the trivial solution $\mathbf{X} = \mathbf{0}$ and possibly an infinite number of solutions. In the next theorem we shall see that a homogeneous system of n equations in n variables possesses *only* the trivial solution when $\mathbf{A}$ is nonsingular.

THEOREM 8.6.5 Trivial Solution Only

A homogeneous system of n linear equations in n variables $\mathbf{AX} = \mathbf{0}$ has only the trivial solution if and only if $\mathbf{A}$ is nonsingular.

PROOF: We prove the sufficiency part of the theorem. Suppose $\mathbf{A}$ is nonsingular. Then by (8), we have the unique solution $\mathbf{X} = \mathbf{A}^{-1}\mathbf{0} = \mathbf{0}$. $\equiv$

The next theorem will answer the question: When does a homogeneous system of n linear equations in n variables possess a nontrivial solution? Bear in mind that if a homogeneous system has one nontrivial solution, it must have an infinite number of solutions.

THEOREM 8.6.6 Existence of Nontrivial Solutions

A homogeneous system of n linear equations in n variables $\mathbf{AX} = \mathbf{0}$ has a nontrivial solution if and only if $\mathbf{A}$ is singular.

In view of Theorem 8.6.6, we can conclude that a homogeneous system of n linear equations in n variables $\mathbf{AX} = \mathbf{0}$ possesses

- only the trivial solution if and only if det $\mathbf{A} \neq 0$, and
- a nontrivial solution if and only if det $\mathbf{A} = 0$.

The last result will be put to use in Section 8.8.

REMARKS

(*i*) As a practical means of solving n linear equations in n variables, the use of an inverse matrix offers few advantages over the method of Section 8.2. However, in applications we sometimes need to solve a system $\mathbf{AX} = \mathbf{B}$ several times; that is, we need to examine the solutions of the system corresponding to the same coefficient matrix $\mathbf{A}$ but different input vectors $\mathbf{B}$. In this case, the single calculation of $\mathbf{A}^{-1}$ enables us to obtain these solutions quickly through the matrix multiplication $\mathbf{A}^{-1}\mathbf{B}$.

(ii) In Definition 8.6.1 we saw that if **A** is an $n \times n$ matrix and if there exists another $n \times n$ matrix **B** that commutes with **A** such that

$$\mathbf{AB} = \mathbf{I} \quad \text{and} \quad \mathbf{BA} = \mathbf{I}, \tag{9}$$

then **B** is the inverse of **A**. Although matrix multiplication is in general not commutative, the condition in (9) can be relaxed somewhat in this sense: If we find an $n \times n$ matrix **B** for which $\mathbf{AB} = \mathbf{I}$, then it can be proved that $\mathbf{BA} = \mathbf{I}$ as well, and so **B** is the inverse of **A**. As a consequence of this result, in the subsequent sections of this chapter if we wish to prove that a certain matrix **B** is the inverse of a given matrix **A**, it will suffice to show only that $\mathbf{AB} = \mathbf{I}$. We need not demonstrate that **B** commutes with **A** to give **I**.

8.6 Exercises Answers to selected odd-numbered problems begin on page ANS-18.

8.6.1 Finding the Inverse

In Problems 1 and 2, verify that the matrix **B** is the inverse of the matrix **A**.

1. $\mathbf{A} = \begin{pmatrix} 1 & \frac{1}{2} \\ 2 & \frac{3}{2} \end{pmatrix}$, $\mathbf{B} = \begin{pmatrix} 3 & -1 \\ -4 & 2 \end{pmatrix}$

2. $\mathbf{A} = \begin{pmatrix} 1 & -1 & 0 \\ 3 & 0 & 2 \\ 1 & 1 & 1 \end{pmatrix}$, $\mathbf{B} = \begin{pmatrix} 2 & -1 & 2 \\ 1 & -1 & 2 \\ -3 & 2 & -3 \end{pmatrix}$

In Problems 3–14, use Theorem 8.6.3 to determine whether the given matrix is singular or nonsingular. If it is nonsingular, use Theorem 8.6.2 to find the inverse.

3. $\begin{pmatrix} 5 & -1 \\ 4 & 1 \end{pmatrix}$

4. $\begin{pmatrix} \frac{1}{3} & -1 \\ 4 & 3 \end{pmatrix}$

5. $\begin{pmatrix} 6 & 0 \\ -3 & 2 \end{pmatrix}$

6. $\begin{pmatrix} -2\pi & -\pi \\ -\pi & \pi \end{pmatrix}$

7. $\begin{pmatrix} 1 & 3 & 5 \\ 2 & 4 & 4 \\ 1 & -1 & 1 \end{pmatrix}$

8. $\begin{pmatrix} 2 & 3 & 0 \\ 0 & 11 & 14 \\ -1 & 4 & 7 \end{pmatrix}$

9. $\begin{pmatrix} 1 & 2 & 3 \\ 0 & -4 & 2 \\ -1 & 5 & 1 \end{pmatrix}$

10. $\begin{pmatrix} 2 & -1 & 5 \\ 3 & 0 & -2 \\ 1 & 4 & 0 \end{pmatrix}$

11. $\begin{pmatrix} 3 & 0 & 0 \\ 0 & 6 & 0 \\ 0 & 0 & -2 \end{pmatrix}$

12. $\begin{pmatrix} 0 & 2 & 0 \\ 0 & 0 & 1 \\ 8 & 0 & 0 \end{pmatrix}$

13. $\begin{pmatrix} 0 & -1 & 1 & 4 \\ 3 & 2 & -2 & 1 \\ 0 & 4 & 0 & 1 \\ 1 & 0 & -1 & 1 \end{pmatrix}$

14. $\begin{pmatrix} 1 & 2 & 1 & 1 \\ 0 & 0 & 3 & 0 \\ 3 & 1 & 2 & 0 \\ 1 & 1 & 1 & 0 \end{pmatrix}$

In Problems 15–26, use Theorem 8.6.4 to find the inverse of the given matrix or show that no inverse exists.

15. $\begin{pmatrix} 6 & -2 \\ 0 & 4 \end{pmatrix}$

16. $\begin{pmatrix} 8 & 0 \\ 0 & \frac{1}{2} \end{pmatrix}$

17. $\begin{pmatrix} 1 & 3 \\ 5 & 3 \end{pmatrix}$

18. $\begin{pmatrix} 2 & -3 \\ -2 & 4 \end{pmatrix}$

19. $\begin{pmatrix} 1 & 2 & 3 \\ 4 & 5 & 6 \\ 7 & 8 & 9 \end{pmatrix}$

20. $\begin{pmatrix} 1 & 0 & -1 \\ 0 & -2 & 1 \\ 2 & -1 & 3 \end{pmatrix}$

21. $\begin{pmatrix} 4 & 2 & 3 \\ 2 & 1 & 0 \\ -1 & -2 & 0 \end{pmatrix}$

22. $\begin{pmatrix} 2 & 4 & -2 \\ 4 & 2 & -2 \\ 8 & 10 & -6 \end{pmatrix}$

23. $\begin{pmatrix} -1 & 3 & 0 \\ 1 & -2 & 1 \\ 0 & 1 & 2 \end{pmatrix}$

24. $\begin{pmatrix} 1 & 2 & 3 \\ 0 & 1 & 4 \\ 0 & 0 & 8 \end{pmatrix}$

25. $\begin{pmatrix} 1 & 2 & 3 & 1 \\ -1 & 0 & 2 & 1 \\ 2 & 1 & -3 & 0 \\ 1 & 1 & 2 & 1 \end{pmatrix}$

26. $\begin{pmatrix} 1 & 0 & 0 & 0 \\ 0 & 0 & 1 & 0 \\ 0 & 0 & 0 & 1 \\ 0 & 1 & 0 & 0 \end{pmatrix}$

In Problems 27 and 28, use the given matrices to find $(\mathbf{AB})^{-1}$.

27. $\mathbf{A}^{-1} = \begin{pmatrix} \frac{1}{2} & -\frac{5}{2} \\ -\frac{1}{2} & \frac{3}{2} \end{pmatrix}$, $\mathbf{B}^{-1} = \begin{pmatrix} \frac{2}{3} & \frac{4}{3} \\ -\frac{1}{3} & \frac{5}{2} \end{pmatrix}$

28. $\mathbf{A}^{-1} = \begin{pmatrix} 1 & 3 & -15 \\ 0 & -1 & 5 \\ -1 & -2 & 11 \end{pmatrix}$, $\mathbf{B}^{-1} = \begin{pmatrix} -1 & 1 & 0 \\ 2 & 0 & 0 \\ 1 & 1 & -2 \end{pmatrix}$

29. If $\mathbf{A}^{-1} = \begin{pmatrix} 4 & 3 \\ 3 & 2 \end{pmatrix}$, what is **A**?

30. If **A** is nonsingular, then $(\mathbf{A}^T)^{-1} = (\mathbf{A}^{-1})^T$. Verify this for $\mathbf{A} = \begin{pmatrix} 1 & 4 \\ 2 & 10 \end{pmatrix}$.

31. Find a value of x such that the matrix $\mathbf{A} = \begin{pmatrix} 4 & -3 \\ x & -4 \end{pmatrix}$ is its own inverse.

32. The rotation matrix

$$\mathbf{M} = \begin{pmatrix} \cos\theta & -\sin\theta \\ \sin\theta & \cos\theta \end{pmatrix}$$

played an important role in Problems 45–48 of Exercises 8.1. Find $\mathbf{M}^{-1}$. In the context of those earlier problems, what does $\mathbf{A} = \mathbf{M}^{-1}\mathbf{B}$ represent?

33. A nonsingular matrix **A** is said to be **orthogonal** if $\mathbf{A}^{-1} = \mathbf{A}^T$.
(a) Verify that the matrix in Problem 32 is orthogonal.
(b) Verify that $\mathbf{A} = \begin{pmatrix} 1/\sqrt{3} & 0 & -2/\sqrt{6} \\ 1/\sqrt{3} & 1/\sqrt{2} & 1/\sqrt{6} \\ 1/\sqrt{3} & -1/\sqrt{2} & 1/\sqrt{6} \end{pmatrix}$ is an orthogonal matrix.

34. Show that if **A** is an orthogonal matrix (see Problem 33), then $\det \mathbf{A} = \pm 1$.

35. Suppose **A** and **B** are nonsingular $n \times n$ matrices. Then show that **AB** is nonsingular.

36. Suppose **A** and **B** are $n \times n$ matrices and that either **A** or **B** is singular. Then show that **AB** is singular.

37. Suppose **A** is a nonsingular matrix. Then show that $\det \mathbf{A}^{-1} = 1/\det \mathbf{A}$.

38. Suppose $\mathbf{A}^2 = \mathbf{A}$. Then show that either $\mathbf{A} = \mathbf{I}$ or **A** is singular.

39. Suppose **A** and **B** are $n \times n$ matrices, **A** is nonsingular, and $\mathbf{AB} = \mathbf{0}$. Then show that $\mathbf{B} = \mathbf{0}$.

40. Suppose **A** and **B** are $n \times n$ matrices, **A** is nonsingular, and $\mathbf{AB} = \mathbf{AC}$. Then show that $\mathbf{B} = \mathbf{C}$.

41. Suppose **A** and **B** are nonsingular $n \times n$ matrices. Is $\mathbf{A} + \mathbf{B}$ necessarily nonsingular?

42. Suppose **A** is a nonsingular matrix. Then show that $\mathbf{A}^T$ is nonsingular.

43. Suppose **A** and **B** are $n \times n$ nonzero matrices and $\mathbf{AB} = \mathbf{0}$. Then show that both **A** and **B** are singular.

44. Consider the 3×3 diagonal matrix

$$\mathbf{A} = \begin{pmatrix} a_{11} & 0 & 0 \\ 0 & a_{22} & 0 \\ 0 & 0 & a_{33} \end{pmatrix}.$$

Determine conditions such that **A** is nonsingular. If **A** is nonsingular, find $\mathbf{A}^{-1}$. Generalize your results to an $n \times n$ diagonal matrix.

8.6.2 Using the Inverse to Solve Systems

In Problems 45–52, use an inverse matrix to solve the given system of equations.

45. $x_1 + x_2 = 4$
$2x_1 - x_2 = 14$

46. $x_1 - x_2 = 2$
$2x_1 + 4x_2 = -5$

47. $4x_1 - 6x_2 = 6$
$2x_1 + x_2 = 1$

48. $x_1 + 2x_2 = 4$
$3x_1 + 4x_2 = -3$

49. $x_1 + x_3 = -4$
$x_1 + x_2 + x_3 = 0$
$5x_1 - x_2 = 6$

50. $x_1 - x_2 + x_3 = 1$
$2x_1 + x_2 + 2x_3 = 2$
$3x_1 + 2x_2 - x_3 = -3$

51. $x_1 + 2x_2 + 2x_3 = 1$
$x_1 - 2x_2 + 2x_3 = -3$
$3x_1 - x_2 + 5x_3 = 7$

52. $x_1 - x_3 = 2$
$x_2 + x_3 = 1$
$-x_1 + x_2 + 2x_3 + x_4 = -5$
$x_3 - x_4 = 3$

In Problems 53 and 54, write the system in the form $\mathbf{AX} = \mathbf{B}$. Use $\mathbf{X} = \mathbf{A}^{-1}\mathbf{B}$ to solve the system for each matrix **B**.

53. $7x_1 - 2x_2 = b_1$
$3x_1 - 2x_2 = b_2$,

$$\mathbf{B} = \begin{pmatrix} 5 \\ 4 \end{pmatrix}, \quad \mathbf{B} = \begin{pmatrix} 10 \\ 50 \end{pmatrix}, \quad \mathbf{B} = \begin{pmatrix} 0 \\ -20 \end{pmatrix}$$

54. $x_1 + 2x_2 + 5x_3 = b_1$
$2x_1 + 3x_2 + 8x_3 = b_2$
$-x_1 + x_2 + 2x_3 = b_3$,

$$\mathbf{B} = \begin{pmatrix} -1 \\ 4 \\ 6 \end{pmatrix}, \quad \mathbf{B} = \begin{pmatrix} 3 \\ 3 \\ 3 \end{pmatrix}, \quad \mathbf{B} = \begin{pmatrix} 0 \\ -5 \\ 4 \end{pmatrix}$$

In Problems 55–58, without solving, determine whether the given homogeneous system of equations has only the trivial solution or a nontrivial solution.

55. $x_1 + 2x_2 - x_3 = 0$
$4x_1 - x_2 + x_3 = 0$
$5x_1 + x_2 - 2x_3 = 0$

56. $x_1 + x_2 + x_3 = 0$
$x_1 - 2x_2 + x_3 = 0$
$-2x_1 + x_2 - 2x_3 = 0$

57. $x_1 + x_2 - x_3 + x_4 = 0$
$5x_2 + 2x_4 = 0$
$x_1 + x_3 - x_4 = 0$
$3x_1 + 2x_2 - x_3 + x_4 = 0$

58. $x_1 + x_2 - x_3 + x_4 = 0$
$x_1 + x_2 + x_3 - x_4 = 0$
$2x_2 + x_3 + x_4 = 0$
$x_2 - x_3 - x_4 = 0$

59. The system of equations for the currents i_1, i_2, and i_3 in the network shown in **FIGURE 8.6.1** is

$$i_1 + i_2 + i_3 = 0$$
$$-R_1 i_1 + R_2 i_2 = E_2 - E_1$$
$$-R_2 i_2 + R_3 i_3 = E_3 - E_2$$

where R_k and E_k, $k = 1, 2, 3$, are constants.

(a) Express the system as a matrix equation $\mathbf{AX} = \mathbf{B}$.
(b) Show that the coefficient matrix **A** is nonsingular.
(c) Use $\mathbf{X} = \mathbf{A}^{-1}\mathbf{B}$ to solve for the currents.

FIGURE 8.6.1 Network in Problem 59

60. Consider the square plate shown in **FIGURE 8.6.2**, with the temperatures as indicated on each side. Under some circumstances it can be shown that the approximate temperatures u_1, u_2, u_3, and u_4 at the points P_1, P_2, P_3, and P_4, respectively, are given by

$$u_1 = \frac{u_2 + u_4 + 100 + 100}{4}$$

$$u_2 = \frac{200 + u_3 + u_1 + 100}{4}$$

$$u_3 = \frac{200 + 100 + u_4 + u_2}{4}$$

$$u_4 = \frac{u_3 + 100 + 100 + u_1}{4}.$$

(a) Show that the above system can be written as the matrix equation

$$\begin{pmatrix} -4 & 1 & 0 & 1 \\ 1 & -4 & 1 & 0 \\ 0 & 1 & -4 & 1 \\ 1 & 0 & 1 & -4 \end{pmatrix} \begin{pmatrix} u_1 \\ u_2 \\ u_3 \\ u_4 \end{pmatrix} = \begin{pmatrix} -200 \\ -300 \\ -300 \\ -200 \end{pmatrix}.$$

(b) Solve the system in part (a) by finding the inverse of the coefficient matrix.

FIGURE 8.6.2 Plate in Problem 60

8.7 Cramer's Rule

INTRODUCTION We saw at the end of the preceding section that a system of n linear equations in n variables $\mathbf{AX} = \mathbf{B}$ has precisely one solution when $\det \mathbf{A} \neq 0$. This solution, as we shall now see, can be expressed in terms of determinants. For example, the system of two equations in two variables

$$
\begin{aligned}
a_{11}x_1 + a_{12}x_2 &= b_1 \\
a_{21}x_1 + a_{22}x_2 &= b_2
\end{aligned}
\tag{1}
$$

possesses the solution

$$
x_1 = \frac{b_1a_{22} - a_{12}b_2}{a_{11}a_{22} - a_{12}a_{21}} \quad \text{and} \quad x_2 = \frac{a_{11}b_2 - b_1a_{21}}{a_{11}a_{22} - a_{12}a_{21}}
\tag{2}
$$

provided that $a_{11}a_{22} - a_{12}a_{21} \neq 0$. The numerators and denominators in (2) are recognized as determinants. That is, the system (1) has the unique solution

$$
x_1 = \frac{\begin{vmatrix} b_1 & a_{12} \\ b_2 & a_{22} \end{vmatrix}}{\begin{vmatrix} a_{11} & a_{12} \\ a_{21} & a_{22} \end{vmatrix}}, \quad x_2 = \frac{\begin{vmatrix} a_{11} & b_1 \\ a_{21} & b_2 \end{vmatrix}}{\begin{vmatrix} a_{11} & a_{12} \\ a_{21} & a_{22} \end{vmatrix}}
\tag{3}
$$

provided that the determinant of the matrix of coefficients $\begin{vmatrix} a_{11} & a_{12} \\ a_{21} & a_{22} \end{vmatrix} \neq 0$. In this section we generalize the result in (3).

Using Determinants to Solve Systems
For a system of n linear equations in n variables

$$
\begin{aligned}
a_{11}x_1 + a_{12}x_2 + \cdots + a_{1n}x_n &= b_1 \\
a_{21}x_1 + a_{22}x_2 + \cdots + a_{2n}x_n &= b_2 \\
&\;\;\vdots \\
a_{n1}x_1 + a_{n2}x_2 + \cdots + a_{nn}x_n &= b_n
\end{aligned}
\tag{4}
$$

it is convenient to define a special matrix

$$
\mathbf{A}_k = \begin{pmatrix}
a_{11} & a_{12} & \cdots & a_{1\,k-1} & b_1 & a_{1\,k+1} & \cdots & a_{1n} \\
a_{21} & a_{22} & \cdots & a_{2\,k-1} & b_2 & a_{2\,k+1} & \cdots & a_{2n} \\
\vdots & & & & \vdots & & & \vdots \\
a_{n1} & a_{n2} & \cdots & a_{n\,k-1} & b_n & a_{n\,k+1} & \cdots & a_{nn}
\end{pmatrix}.
\tag{5}
$$

where the kth column is replaced by b entries (labeled "kth column" above the matrix).

In other words, $\mathbf{A}_k$ is the same as the matrix $\mathbf{A}$ except that the kth column of $\mathbf{A}$ has been replaced by the entries of the column matrix

$$
\mathbf{B} = \begin{pmatrix} b_1 \\ b_2 \\ \vdots \\ b_n \end{pmatrix}.
$$

The generalization of (3), known as **Cramer's rule**, is given in the next theorem. The mathematician **Gabriel Cramer** (1704–1752), born in the Republic of Geneva, published this method of solving linear systems using determinants in 1750.

Let $\mathbf{A}$ be the coefficient matrix of system (4). If $\det \mathbf{A} \neq 0$, then the solution of (4) is given by

$$x_1 = \frac{\det \mathbf{A}_1}{\det \mathbf{A}}, \quad x_2 = \frac{\det \mathbf{A}_2}{\det \mathbf{A}}, \quad \dots, \quad x_n = \frac{\det \mathbf{A}_n}{\det \mathbf{A}}, \tag{6}$$

where $\mathbf{A}_k$, $k = 1, 2, \dots, n$, is defined in (5).

PROOF: We first write system (4) as $\mathbf{AX} = \mathbf{B}$. Since $\det \mathbf{A} \neq 0$, $\mathbf{A}^{-1}$ exists, and so

$$\mathbf{X} = \mathbf{A}^{-1}\mathbf{B} = \frac{1}{\det \mathbf{A}} \begin{pmatrix} C_{11} & C_{21} & \cdots & C_{n1} \\ C_{12} & C_{22} & \cdots & C_{n2} \\ \vdots & & & \vdots \\ C_{1n} & C_{2n} & \cdots & C_{nn} \end{pmatrix} \begin{pmatrix} b_1 \\ b_2 \\ \vdots \\ b_n \end{pmatrix}$$

$$= \frac{1}{\det \mathbf{A}} \begin{pmatrix} b_1 C_{11} + b_2 C_{21} + \cdots + b_n C_{n1} \\ b_1 C_{12} + b_2 C_{22} + \cdots + b_n C_{n2} \\ \vdots \\ b_1 C_{1n} + b_2 C_{2n} + \cdots + b_n C_{nn} \end{pmatrix}.$$

Now the entry in the kth row of the last matrix is

$$x_k = \frac{b_1 C_{1k} + b_2 C_{2k} + \cdots + b_n C_{nk}}{\det \mathbf{A}}. \tag{7}$$

But $b_1 C_{1k} + b_2 C_{2k} + \cdots + b_n C_{nk}$ is the cofactor expansion of $\det \mathbf{A}_k$, where $\mathbf{A}_k$ is the matrix given in (5), along the kth column. Hence, we have $x_k = \det \mathbf{A}_k / \det \mathbf{A}$ for $k = 1, 2, \dots, n$. ≡

EXAMPLE 1 Using Cramer's Rule to Solve a System

Use Cramer's rule to solve the system

$$3x_1 + 2x_2 + x_3 = 7$$
$$x_1 - x_2 + 3x_3 = 3$$
$$5x_1 + 4x_2 - 2x_3 = 1.$$

SOLUTION The solution requires the evaluation of four determinants:

$$\det \mathbf{A} = \begin{vmatrix} 3 & 2 & 1 \\ 1 & -1 & 3 \\ 5 & 4 & -2 \end{vmatrix} = 13, \qquad \det \mathbf{A}_1 = \begin{vmatrix} 7 & 2 & 1 \\ 3 & -1 & 3 \\ 1 & 4 & -2 \end{vmatrix} = -39,$$

$$\det \mathbf{A}_2 = \begin{vmatrix} 3 & 7 & 1 \\ 1 & 3 & 3 \\ 5 & 1 & -2 \end{vmatrix} = 78, \qquad \det \mathbf{A}_3 = \begin{vmatrix} 3 & 2 & 7 \\ 1 & -1 & 3 \\ 5 & 4 & 1 \end{vmatrix} = 52.$$

Thus, (6) gives

$$x_1 = \frac{\det \mathbf{A}_1}{\det \mathbf{A}} = -3, \quad x_2 = \frac{\det \mathbf{A}_2}{\det \mathbf{A}} = 6, \quad x_3 = \frac{\det \mathbf{A}_3}{\det \mathbf{A}} = 4.$$ ≡

REMARKS

Like the method of the preceding section, Cramer's rule is not a very practical means for solving systems of n linear equations in n variables. For $n \geq 4$, the work required to evaluate the determinants becomes formidable. Nonetheless, as we saw in Section 3.5, Cramer's rule is used *sometimes*, and it is of theoretical importance.

In the application of Cramer's rule, some shortcuts can be taken. In Example 1, for instance, we really did not have to compute det $\mathbf{A}_3$ since once we found the values of x_1 and x_2, the value of x_3 can be found by using one of the equations in the system.

8.7 Exercises Answers to selected odd-numbered problems begin on page ANS-19.

In Problems 1–10, solve the given system of equations by Cramer's rule.

1. $-3x_1 + x_2 = 3$
$2x_1 - 4x_2 = -6$

2. $x_1 + x_2 = 4$
$2x_1 - x_2 = 2$

3. $0.1x_1 - 0.4x_2 = 0.13$
$x_1 - x_2 = 0.4$

4. $0.21x_1 + 0.57x_2 = 0.369$
$0.1x_1 + 0.2x_2 = 0.135$

5. $2x + y = 1$
$3x + 2y = -2$

6. $5r + 4s = -1$
$10r - 6s = 5$

7. $x_1 - 2x_2 - 3x_3 = 3$
$x_1 + x_2 - x_3 = 5$
$3x_1 + 2x_2 = -4$

8. $x_1 - x_2 + 6x_3 = -2$
$-x_1 + 2x_2 + 4x_3 = 9$
$2x_1 + 3x_2 - x_3 = \frac{1}{2}$

9. $u + 2v + w = 8$
$2u - 2v + 2w = 7$
$u - 4v + 3w = 1$

10. $4x + 3y + 2z = 8$
$-x + 2z = 12$
$3x + 2y + z = 3$

11. Use Cramer's rule to determine the solution of the system

$$(2 - k)x_1 + kx_2 = 4$$
$$kx_1 + (3 - k)x_2 = 3.$$

For what value(s) of k is the system inconsistent?

12. Consider the system

$$x_1 + x_2 = 1$$
$$x_1 + \epsilon x_2 = 2.$$

When ϵ is close to 1, the lines that make up the system are almost parallel.

(a) Use Cramer's rule to show that a solution of the system is

$$x_1 = 1 - \frac{1}{\epsilon - 1}, \quad x_2 = \frac{1}{\epsilon - 1}.$$

(b) The system is said to be **ill-conditioned** since small changes in the input data (for example, the coefficients) causes a significant or large change in the output or solution. Verify this by finding the solution of the system for $\epsilon = 1.01$ and then for $\epsilon = 0.99$.

13. The magnitudes T_1 and T_2 of the tensions in the support wires shown in **FIGURE 8.7.1** satisfy the equations

$$(\cos 25°)T_1 - (\cos 15°)T_2 = 0$$
$$(\sin 25°)T_1 + (\sin 15°)T_2 = 300.$$

Use Cramer's rule to solve for T_1 and T_2.

FIGURE 8.7.1 Support wires in Problem 13

14. The 400-lb block shown in **FIGURE 8.7.2** is kept from sliding down the inclined plane by friction and a force of smallest magnitude F. If the coefficient of friction between the block and the inclined plane is 0.5, then the magnitude of the frictional force is $0.5N$, where N is the magnitude of the normal force exerted on the block by the plane. Use the fact that the system is in equilibrium to set up a system of equations for F and N. Use Cramer's rule to solve for F and N.

FIGURE 8.7.2 Inclined plane in Problem 14

15. As shown in **FIGURE 8.7.3**, a circuit consists of two batteries with internal resistances r_1 and r_2 connected in parallel with a resistor. Use Cramer's rule to show that the current i through the resistor is given by

$$i = \frac{r_1 E_2 + r_2 E_1}{r_1 R + r_2 R + r_1 r_2}.$$

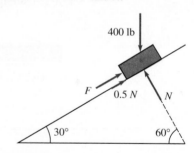

FIGURE 8.7.3 Circuit in Problem 15

8.8 Eigenvalue Problem

INTRODUCTION If $\mathbf{A}$ is an $n \times n$ matrix and $\mathbf{K}$ is an $n \times 1$ matrix (column vector), then the product $\mathbf{AK}$ is defined and is another $n \times 1$ matrix. It is important in many applications to determine whether there exist nonzero $n \times 1$ matrices $\mathbf{K}$ such that the product vector $\mathbf{AK}$ is a constant multiple λ of $\mathbf{K}$ itself. The problem of solving $\mathbf{AK} = \lambda\mathbf{K}$ for nonzero vectors $\mathbf{K}$ is called the **eigenvalue problem** for the matrix $\mathbf{A}$.

■ **A Definition** The foregoing introductory remarks are summarized in the next definition.

DEFINITION 8.8.1 Eignvalues and Eigenvectors

Let $\mathbf{A}$ be an $n \times n$ matrix. A number λ is said to be an **eigenvalue** of $\mathbf{A}$ if there exists a nonzero solution vector $\mathbf{K}$ of the linear system

$$\mathbf{AK} = \lambda\mathbf{K}. \tag{1}$$

The solution vector $\mathbf{K}$ is said to be an **eigenvector** corresponding to the eigenvalue λ.

The word *eigenvalue* is a combination of German and English terms adapted from the German word *Eigenwert*, which, translated literally, is *proper value*. Eigenvalues and eigenvectors are also called **characteristic values** and **characteristic vectors**, respectively.

Gauss–Jordan elimination introduced in Section 8.2 can be used to find the **eigenvectors** of a square matrix $\mathbf{A}$.

EXAMPLE 1 **Verification of an Eigenvector**

Verify that $\mathbf{K} = \begin{pmatrix} 1 \\ -1 \\ 1 \end{pmatrix}$ is an eigenvector of the 3×3 matrix

$$\mathbf{A} = \begin{pmatrix} 0 & -1 & -3 \\ 2 & 3 & 3 \\ -2 & 1 & 1 \end{pmatrix}.$$

SOLUTION By carrying out the multiplication $\mathbf{AK}$ we see

$$\mathbf{AK} = \begin{pmatrix} 0 & -1 & -3 \\ 2 & 3 & 3 \\ -2 & 1 & 1 \end{pmatrix}\begin{pmatrix} 1 \\ -1 \\ 1 \end{pmatrix} = \begin{pmatrix} -2 \\ 2 \\ -2 \end{pmatrix} = (-2)\begin{pmatrix} 1 \\ -1 \\ 1 \end{pmatrix} \overset{\text{eigenvalue}}{=} (-2)\mathbf{K}.$$

We see from the preceding line and Definition 8.8.1 that $\lambda = -2$ is an eigenvalue of $\mathbf{A}$. ≡

Using properties of matrix algebra, we can write (1) in the alternative form

$$(\mathbf{A} - \lambda\mathbf{I})\mathbf{K} = \mathbf{0}, \tag{2}$$

where $\mathbf{I}$ is the multiplicative identity. If we let

$$\mathbf{K} = \begin{pmatrix} k_1 \\ k_2 \\ \vdots \\ k_n \end{pmatrix},$$

then (2) is the same as

$$
\begin{aligned}
(a_{11} - \lambda)k_1 + & & a_{12}k_2 + \cdots + & & a_{1n}k_n = 0 \\
a_{21}k_1 + & (a_{22} - \lambda)k_2 + \cdots + & & & a_{2n}k_n = 0 \\
& \vdots & & & \vdots \\
a_{n1}k_1 + & & a_{n2}k_2 + \cdots + & (a_{nn} - \lambda)k_n = 0.
\end{aligned}
\tag{3}
$$

Although an obvious solution of (3) is $k_1 = 0, k_2 = 0, \ldots, k_n = 0$, we are seeking only nontrivial solutions. Now we know from Theorem 8.6.5 that a homogeneous system of n linear equations in n variables has a nontrivial solution if and only if the determinant of the coefficient matrix is equal to zero. Thus to find a nonzero solution $\mathbf{K}$ for (2), we must have

$$
\det(\mathbf{A} - \lambda\mathbf{I}) = 0.
\tag{4}
$$

Inspection of (4) shows that expansion of $\det(\mathbf{A} - \lambda\mathbf{I})$ by cofactors results in an nth-degree polynomial in λ. The equation (4) is called the **characteristic equation** of $\mathbf{A}$. Thus, *the eigenvalues of $\mathbf{A}$ are the roots of the characteristic equation.* To find an eigenvector corresponding to an eigenvalue λ, we simply solve the system of equations $(\mathbf{A} - \lambda\mathbf{I})\mathbf{K} = \mathbf{0}$ by applying Gauss–Jordan elimination to the augmented matrix $(\mathbf{A} - \lambda\mathbf{I} \,|\, \mathbf{0})$.

EXAMPLE 2 Finding Eigenvalues and Eigenvectors

Find the eigenvalues and eigenvectors of

$$
\mathbf{A} = \begin{pmatrix} 1 & 2 & 1 \\ 6 & -1 & 0 \\ -1 & -2 & -1 \end{pmatrix}.
\tag{5}
$$

SOLUTION To expand the determinant in the characteristic equation

$$
\det(\mathbf{A} - \lambda\mathbf{I}) = \begin{vmatrix} 1-\lambda & 2 & 1 \\ 6 & -1-\lambda & 0 \\ -1 & -2 & -1-\lambda \end{vmatrix} = 0,
$$

we use the cofactors of the second row. It follows that the characteristic equation is

$$
-\lambda^3 - \lambda^2 + 12\lambda = 0 \quad \text{or} \quad \lambda(\lambda + 4)(\lambda - 3) = 0.
$$

Hence the eigenvalues are $\lambda_1 = 0, \lambda_2 = -4, \lambda_3 = 3$. To find the eigenvectors, we must now reduce $(\mathbf{A} - \lambda\mathbf{I} \,|\, \mathbf{0})$ three times corresponding to the three distinct eigenvalues.

For $\lambda_1 = 0$, we have

$$
(\mathbf{A} - 0\mathbf{I} \,|\, \mathbf{0}) = \begin{pmatrix} 1 & 2 & 1 & | & 0 \\ 6 & -1 & 0 & | & 0 \\ -1 & -2 & -1 & | & 0 \end{pmatrix}
\overset{\substack{-6R_1 + R_2 \\ R_1 + R_3}}{\Longrightarrow}
\begin{pmatrix} 1 & 2 & 1 & | & 0 \\ 0 & -13 & -6 & | & 0 \\ 0 & 0 & 0 & | & 0 \end{pmatrix}
$$

$$
\overset{-\frac{1}{13}R_2}{\Longrightarrow}
\begin{pmatrix} 1 & 2 & 1 & | & 0 \\ 0 & 1 & \frac{6}{13} & | & 0 \\ 0 & 0 & 0 & | & 0 \end{pmatrix}
\overset{-2R_2 + R_1}{\Longrightarrow}
\begin{pmatrix} 1 & 0 & \frac{1}{13} & | & 0 \\ 0 & 1 & \frac{6}{13} & | & 0 \\ 0 & 0 & 0 & | & 0 \end{pmatrix}.
$$

Thus we see that $k_1 = -\frac{1}{13}k_3$ and $k_2 = -\frac{6}{13}k_3$. Choosing $k_3 = -13$ gives the eigenvector

$$
\mathbf{K}_1 = \begin{pmatrix} 1 \\ 6 \\ -13 \end{pmatrix}.
$$

For $\lambda_2 = -4$,

$$(\mathbf{A} + 4\mathbf{I}\,|\,\mathbf{0}) = \begin{pmatrix} 5 & 2 & 1 & 0 \\ 6 & 3 & 0 & 0 \\ -1 & -2 & 3 & 0 \end{pmatrix} \overset{-R_3}{\underset{R_1 \leftrightarrow R_3}{\Longrightarrow}} \begin{pmatrix} 1 & 2 & -3 & 0 \\ 6 & 3 & 0 & 0 \\ 5 & 2 & 1 & 0 \end{pmatrix}$$

$$\overset{-6R_1 + R_2}{\underset{-5R_1 + R_3}{\Longrightarrow}} \begin{pmatrix} 1 & 2 & -3 & 0 \\ 0 & -9 & 18 & 0 \\ 0 & -8 & 16 & 0 \end{pmatrix} \overset{-\frac{1}{9}R_2}{\underset{-\frac{1}{8}R_3}{\Longrightarrow}} \begin{pmatrix} 1 & 2 & -3 & 0 \\ 0 & 1 & -2 & 0 \\ 0 & 1 & -2 & 0 \end{pmatrix}$$

$$\overset{-2R_2 + R_1}{\underset{-R_2 + R_3}{\Longrightarrow}} \begin{pmatrix} 1 & 0 & 1 & 0 \\ 0 & 1 & -2 & 0 \\ 0 & 0 & 0 & 0 \end{pmatrix}$$

implies $k_1 = -k_3$ and $k_2 = 2k_3$. Choosing $k_3 = 1$ then yields a second eigenvector

$$\mathbf{K}_2 = \begin{pmatrix} -1 \\ 2 \\ 1 \end{pmatrix}.$$

Finally, for $\lambda_3 = 3$, Gauss–Jordan elimination gives

$$(\mathbf{A} - 3\mathbf{I}\,|\,\mathbf{0}) = \begin{pmatrix} -2 & 2 & 1 & 0 \\ 6 & -4 & 0 & 0 \\ -1 & -2 & -4 & 0 \end{pmatrix} \overset{\text{row}}{\underset{\text{operations}}{\Longrightarrow}} \begin{pmatrix} 1 & 0 & 1 & 0 \\ 0 & 1 & \frac{3}{2} & 0 \\ 0 & 0 & 0 & 0 \end{pmatrix},$$

and so $k_1 = -k_3$ and $k_2 = -\frac{3}{2}k_3$. The choice of $k_3 = -2$ leads to a third eigenvector,

$$\mathbf{K}_3 = \begin{pmatrix} 2 \\ 3 \\ -2 \end{pmatrix}.$$

≡

See Figure 8.3.1 on page 397 and reread Theorem 8.2.2(*i*). ▶ It should be obvious from Example 2 that eigenvectors are not uniquely determined. For example, in obtaining, say, $\mathbf{K}_2$, had we not chosen a specific value for k_3 then a solution of the homogeneous system $(\mathbf{A} + 4\mathbf{I})\mathbf{K} = \mathbf{0}$ can be written

$$\mathbf{K}_2 = \begin{pmatrix} -k_3 \\ 2k_3 \\ k_3 \end{pmatrix} = k_3 \begin{pmatrix} -1 \\ 2 \\ 1 \end{pmatrix}.$$

The point of showing $\mathbf{K}_2$ in this form is:

A nonzero constant multiple of an eigenvector is another eigenvector.

For example, $\lambda = 7$ and $\mathbf{K} = \begin{pmatrix} -\frac{1}{2} \\ \frac{1}{3} \end{pmatrix}$ are, in turn, an eigenvalue and corresponding eigenvector of the matrix $\mathbf{A} = \begin{pmatrix} 5 & -3 \\ -4 & 1 \end{pmatrix}$. In practice, it may be more convenient to work with an eigenvector with integer entries:

$$\mathbf{K}_1 = 6\mathbf{K} = 6\begin{pmatrix} -\frac{1}{2} \\ \frac{1}{3} \end{pmatrix} = \begin{pmatrix} -3 \\ 2 \end{pmatrix}.$$

Observe that

$$\mathbf{AK} = \begin{pmatrix} 5 & -3 \\ -4 & 1 \end{pmatrix} \begin{pmatrix} -\frac{1}{2} \\ \frac{1}{3} \end{pmatrix} = \begin{pmatrix} -\frac{7}{2} \\ \frac{7}{3} \end{pmatrix} = 7 \begin{pmatrix} -\frac{1}{2} \\ \frac{1}{3} \end{pmatrix} = \lambda \mathbf{K}$$

and

$$\mathbf{AK}_1 = \begin{pmatrix} 5 & -3 \\ -4 & 1 \end{pmatrix} \begin{pmatrix} -3 \\ 2 \end{pmatrix} = \begin{pmatrix} -21 \\ 14 \end{pmatrix} = 7 \begin{pmatrix} -3 \\ 2 \end{pmatrix} = \lambda \mathbf{K}_1.$$

When an $n \times n$ matrix $\mathbf{A}$ possesses n distinct eigenvalues $\lambda_1, \lambda_2, \ldots, \lambda_n$, it can be proved that a set of n linearly independent eigenvectors $\mathbf{K}_1, \mathbf{K}_2, \ldots, \mathbf{K}_n$ can be found. However, when the characteristic equation has repeated roots, it may not be possible to find n linearly independent eigenvectors for $\mathbf{A}$.

EXAMPLE 3 Finding Eigenvalues and Eigenvectors

Find the eigenvalues and eigenvectors of $\mathbf{A} = \begin{pmatrix} 3 & 4 \\ -1 & 7 \end{pmatrix}$.

SOLUTION From the characteristic equation

$$\det(\mathbf{A} - \lambda \mathbf{I}) = \begin{vmatrix} 3 - \lambda & 4 \\ -1 & 7 - \lambda \end{vmatrix} = (\lambda - 5)^2 = 0,$$

we see $\lambda_1 = \lambda_2 = 5$ is an eigenvalue of multiplicity 2. In the case of a 2×2 matrix, there is no need to use Gauss–Jordan elimination. To find the eigenvector(s) corresponding to $\lambda_1 = 5$, we resort to the system $(\mathbf{A} - 5\mathbf{I} \,|\, \mathbf{0})$ in its equivalent form

$$-2k_1 + 4k_2 = 0$$
$$-k_1 + 2k_2 = 0.$$

It is apparent from this system that $k_1 = 2k_2$. Thus, if we choose $k_2 = 1$, we find the single eigenvector $\mathbf{K}_1 = \begin{pmatrix} 2 \\ 1 \end{pmatrix}$.

≡

EXAMPLE 4 Finding Eigenvalues and Eigenvectors

Find the eigenvalues and eigenvectors of $\mathbf{A} = \begin{pmatrix} 9 & 1 & 1 \\ 1 & 9 & 1 \\ 1 & 1 & 9 \end{pmatrix}$.

SOLUTION The characteristic equation

$$\det(\mathbf{A} - \lambda \mathbf{I}) = \begin{vmatrix} 9 - \lambda & 1 & 1 \\ 1 & 9 - \lambda & 1 \\ 1 & 1 & 9 - \lambda \end{vmatrix} = -(\lambda - 11)(\lambda - 8)^2 = 0$$

shows that $\lambda_1 = 11$ and that $\lambda_2 = \lambda_3 = 8$ is an eigenvalue of multiplicity 2.
 For $\lambda_1 = 11$, Gauss–Jordan elimination gives

$$(\mathbf{A} - 11\mathbf{I} \,|\, \mathbf{0}) = \begin{pmatrix} -2 & 1 & 1 & | & 0 \\ 1 & -2 & 1 & | & 0 \\ 1 & 1 & -2 & | & 0 \end{pmatrix} \xrightarrow[\text{operations}]{\text{row}} \begin{pmatrix} 1 & 0 & -1 & | & 0 \\ 0 & 1 & -1 & | & 0 \\ 0 & 0 & 0 & | & 0 \end{pmatrix}.$$

Hence $k_1 = k_3$ and $k_2 = k_3$. If $k_3 = 1$, then

$$\mathbf{K}_1 = \begin{pmatrix} 1 \\ 1 \\ 1 \end{pmatrix}.$$

Now for $\lambda_2 = 8$ we have

$$(\mathbf{A} - 8\mathbf{I} \,|\, \mathbf{0}) = \begin{pmatrix} 1 & 1 & 1 & | & 0 \\ 1 & 1 & 1 & | & 0 \\ 1 & 1 & 1 & | & 0 \end{pmatrix} \xRightarrow[\text{operations}]{\text{row}} \begin{pmatrix} 1 & 1 & 1 & | & 0 \\ 0 & 0 & 0 & | & 0 \\ 0 & 0 & 0 & | & 0 \end{pmatrix}.$$

In the equation $k_1 + k_2 + k_3 = 0$ we are free to select two of the variables arbitrarily. Choosing, on the one hand, $k_2 = 1$, $k_3 = 0$, and on the other, $k_2 = 0$, $k_3 = 1$, we obtain two linearly independent eigenvectors:

$$\mathbf{K}_2 = \begin{pmatrix} -1 \\ 1 \\ 0 \end{pmatrix} \quad \text{and} \quad \mathbf{K}_3 = \begin{pmatrix} -1 \\ 0 \\ 1 \end{pmatrix}$$

corresponding to a single eigenvalue. $\equiv$

Complex Eigenvalues A matrix $\mathbf{A}$ may have **complex eigenvalues**.

PROOF: Since $\mathbf{A}$ is a matrix with real entries, the characteristic equation $\det(\mathbf{A} - \lambda\mathbf{I}) = 0$ is a polynomial equation with real coefficients. From algebra we know that complex roots of such equations appear in conjugate pairs. In other words, if $\lambda = \alpha + i\beta$ is a root, then $\overline{\lambda} = \alpha - i\beta$ is also a root. Now let $\mathbf{K}$ be an eigenvector of $\mathbf{A}$ corresponding to λ. By definition, $\mathbf{AK} = \lambda\mathbf{K}$. Taking complex conjugates of the latter equation gives

$$\overline{\mathbf{A}}\,\overline{\mathbf{K}} = \overline{\lambda}\,\overline{\mathbf{K}} \quad \text{or} \quad \mathbf{A}\overline{\mathbf{K}} = \overline{\lambda}\,\overline{\mathbf{K}},$$

since $\mathbf{A}$ is a real matrix. The last equation indicates that $\overline{\mathbf{K}}$ is an eigenvector corresponding to $\overline{\lambda}$. $\equiv$

EXAMPLE 5 Complex Eigenvalues and Eigenvectors

Find the eigenvalues and eigenvectors of $\mathbf{A} = \begin{pmatrix} 6 & -1 \\ 5 & 4 \end{pmatrix}$.

SOLUTION The characteristic equation is

$$\det(\mathbf{A} - \lambda\mathbf{I}) = \begin{vmatrix} 6 - \lambda & -1 \\ 5 & 4 - \lambda \end{vmatrix} = \lambda^2 - 10\lambda + 29 = 0.$$

From the quadratic formula, we find $\lambda_1 = 5 + 2i$ and $\lambda_2 = \overline{\lambda}_1 = 5 - 2i$.

Now for $\lambda_1 = 5 + 2i$, we must solve

$$(1 - 2i)k_1 - \qquad k_2 = 0$$
$$5k_1 - (1 + 2i)k_2 = 0.$$

Since $k_2 = (1 - 2i)k_1$,* it follows, after choosing $k_1 = 1$, that one eigenvector is

$$\mathbf{K}_1 = \begin{pmatrix} 1 \\ 1 - 2i \end{pmatrix}.$$

From Theorem 8.8.1, we see that an eigenvector corresponding to $\lambda_2 = \overline{\lambda}_1 = 5 - 2i$ is

$$\mathbf{K}_2 = \overline{\mathbf{K}}_1 = \begin{pmatrix} 1 \\ 1 + 2i \end{pmatrix}. \qquad \equiv$$

■ **Eigenvalues and Singular Matrices** If the number 0 is an eigenvalue of an $n \times n$ matrix $\mathbf{A}$, then by Definition 8.8.1 the homogeneous linear system

$$\begin{array}{c} \lambda = 0 \\ \downarrow \\ \mathbf{AK} = 0\mathbf{K} = \mathbf{0} \end{array}$$

of n equations in n variables must possess a *nontrivial* solution vector $\mathbf{K}$. But this fact tells us something important about the matrix $\mathbf{A}$. By Theorem 8.6.6 a homogeneous system of n equations in n variables possesses a nontrivial solution if and only if the coefficient matrix $\mathbf{A}$ is *singular*. We summarize this result in the following theorem.

THEOREM 8.8.2 Zero Eigenvalue and a Singular Matrix

Let $\mathbf{A}$ be an $n \times n$ matrix. Then the number $\lambda = 0$ is an eigenvalue of $\mathbf{A}$ if and only if $\mathbf{A}$ is singular.

Put a different way:

A matrix $\mathbf{A}$ is nonsingular if and only if the number 0 is not an eigenvalue of $\mathbf{A}$. (6)

Reexamination of Example 2 shows that $\lambda_1 = 0$ is an eigenvalue of the 3×3 matrix (5). So we can conclude from Theorem 8.8.2 that the matrix (5) is singular, that is, does not possess an inverse. On the other hand, we conclude from (6) that the matrices in Examples 3 and 4 are nonsingular because none of the eigenvalues of the matrices are 0.

The eigenvalues of an $n \times n$ matrix $\mathbf{A}$ are related to $\det \mathbf{A}$. Because the characteristic equation $\det(\mathbf{A} - \lambda\mathbf{I}) = 0$ is an nth degree polynomial equation, it must, counting multiplicities and complex numbers, have n roots $\lambda_1, \lambda_2, \lambda_3, \ldots \lambda_n$. By the Factor Theorem of algebra, the characteristic polynomial $\det(\mathbf{A} - \lambda\mathbf{I})$ can then be written

$$\det(\mathbf{A} - \lambda\mathbf{I}) = (\lambda_1 - \lambda)(\lambda_2 - \lambda)(\lambda_3 - \lambda)\cdots(\lambda_n - \lambda). \qquad (7)$$

By setting $\lambda = 0$ in (7) we see that

$$\det \mathbf{A} = \lambda_1\lambda_2\lambda_3\cdots\lambda_n, \qquad (8)$$

that is:

The determinant of $\mathbf{A}$ is the product of its eigenvalues.

The result in (8) provides an alternative proof of Theorem 8.8.2: If, say, $\lambda_1 = 0$ then $\det \mathbf{A} = 0$. Conversely, if $\det \mathbf{A} = 0$, then $\lambda_1\lambda_2\lambda_3\cdots\lambda_n = 0$ implies that at least one of the eigenvalues of $\mathbf{A}$ is 0.

*Note that the second equation is simply $1 + 2i$ times the first.

EXAMPLE 6 **Examples 4 and 5 Revisited**

(a) In Example 4 we saw that the eigenvalues of the matrix

$$\mathbf{A} = \begin{pmatrix} 9 & 1 & 1 \\ 1 & 9 & 1 \\ 1 & 1 & 9 \end{pmatrix}$$

are $\lambda_1 = 11$, $\lambda_2 = \lambda_3 = 8$. With no further work we see that $\det \mathbf{A} = 11 \cdot 8 \cdot 8 = 704$. Because $\det \mathbf{A} \neq 0$ the matrix $\mathbf{A}$ is nonsingular.

(b) In Example 5 it is easily seen that the determinant of the 2×2 matrix $\mathbf{A} = \begin{pmatrix} 6 & -1 \\ 5 & 4 \end{pmatrix}$ is

$$\det \mathbf{A} = \begin{vmatrix} 6 & -1 \\ 5 & 4 \end{vmatrix} = 29.$$

Also, using the two complex eigenvalues $\lambda_1 = 5 + 2i$, $\lambda_2 = 5 - 2i$ we see that $\det \mathbf{A} = \lambda_1 \lambda_2 = (5 + 2i)(5 - 2i) = 5^2 + 2^2 = 29$. ≡

▮ Eigenvalues and Eigenvectors of $\mathbf{A}^{-1}$

If an $n \times n$ matrix $\mathbf{A}$ is nonsingular, then $\mathbf{A}^{-1}$ exists and we now know that none of the eigenvalues of $\mathbf{A}$ are 0. Suppose that λ is an eigenvalue of $\mathbf{A}$ with corresponding eigenvector $\mathbf{K}$. By multiplying both sides of the equation $\mathbf{AK} = \lambda \mathbf{K}$ by $\mathbf{A}^{-1}$, we get

$$\mathbf{A}^{-1}(\mathbf{AK}) = \lambda \mathbf{A}^{-1} \mathbf{K}$$

$$(\mathbf{A}^{-1}\mathbf{A})\mathbf{K} = \lambda \mathbf{A}^{-1} \mathbf{K}$$

$$\mathbf{IK} = \lambda \mathbf{A}^{-1} \mathbf{K}$$

or

$$\mathbf{A}^{-1}\mathbf{K} = \frac{1}{\lambda} \mathbf{K}. \tag{9}$$

Comparing the result in (9) with (1) leads us to the following conclusion.

THEOREM 8.8.3 **Eigenvalues and Eigenvectors of $\mathbf{A}^{-1}$**

Let $\mathbf{A}$ be a nonsingular matrix. If λ is an eigenvalue of $\mathbf{A}$ with corresponding eigenvector $\mathbf{K}$, then $1/\lambda$ is an eigenvalue of $\mathbf{A}^{-1}$ with the same corresponding eigenvector $\mathbf{K}$.

EXAMPLE 7 **Eigenvalues of an Inverse**

The matrix $\mathbf{A} = \begin{pmatrix} 4 & 0 \\ 2 & 3 \end{pmatrix}$ has distinct eigenvalues $\lambda_1 = 4$ and $\lambda_2 = 3$ with corresponding eigenvectors $\mathbf{K}_1 = \begin{pmatrix} 1 \\ 2 \end{pmatrix}$ and $\mathbf{K}_2 = \begin{pmatrix} 0 \\ 1 \end{pmatrix}$. Since 0 is not an eigenvalue of $\mathbf{A}$ it is nonsingular. So in view of Theorem 8.8.3 we can also say that the reciprocals

$$\lambda_3 = \frac{1}{\lambda_1} = \frac{1}{4} \quad \text{and} \quad \lambda_4 = \frac{1}{\lambda_2} = \frac{1}{3}$$

are, respectively, eigenvalues of $\mathbf{A}^{-1}$ with corresponding eigenvectors $\mathbf{K}_1 = \begin{pmatrix} 1 \\ 2 \end{pmatrix}$ and $\mathbf{K}_2 = \begin{pmatrix} 0 \\ 1 \end{pmatrix}$. This is easily verified using $\mathbf{A}^{-1} = \begin{pmatrix} \frac{1}{4} & 0 \\ -\frac{1}{6} & \frac{1}{3} \end{pmatrix}$:

$$\mathbf{A}^{-1}\mathbf{K}_1 = \begin{pmatrix} \frac{1}{4} & 0 \\ -\frac{1}{6} & \frac{1}{3} \end{pmatrix}\begin{pmatrix} 1 \\ 2 \end{pmatrix} = \begin{pmatrix} \frac{1}{4} \\ \frac{1}{2} \end{pmatrix} = \frac{1}{4}\begin{pmatrix} 1 \\ 2 \end{pmatrix} = \frac{1}{\lambda_1} \mathbf{K}_1 = \lambda_3 \mathbf{K}_1$$

$$\mathbf{A}^{-1}\mathbf{K}_2 = \begin{pmatrix} \frac{1}{4} & 0 \\ -\frac{1}{6} & \frac{1}{3} \end{pmatrix}\begin{pmatrix} 0 \\ 1 \end{pmatrix} = \begin{pmatrix} 0 \\ \frac{1}{3} \end{pmatrix} = \frac{1}{3}\begin{pmatrix} 0 \\ 1 \end{pmatrix} = \frac{1}{\lambda_2} \mathbf{K}_2 = \lambda_4 \mathbf{K}_2. \quad ≡$$

Our last theorem follows immediately from the fact that the determinant of an upper triangular, lower triangular, and a diagonal matrix is the product of the main diagonal entries. See Theorem 8.5.8.

THEOREM 8.8.4 Triangular and Diagonal Matrices

The eigenvalues of an upper triangular, lower triangular, and diagonal matrix are the main diagonal entries.

Notice in Example 7 that the matrix $\mathbf{A} = \begin{pmatrix} 4 & 0 \\ 2 & 3 \end{pmatrix}$ is a lower triangular matrix and its eigenvalues $\lambda_1 = 4$ and $\lambda_2 = 3$ are the entries on the main diagonal.

8.8 Exercises Answers to selected odd-numbered problems begin on page ANS-19.

In Problems 1–6, determine which of the indicated column vectors are eigenvectors of the given matrix $\mathbf{A}$. Give the corresponding eigenvalue.

1. $\mathbf{A} = \begin{pmatrix} 4 & 2 \\ 5 & 1 \end{pmatrix};$ $\mathbf{K}_1 = \begin{pmatrix} 5 \\ -2 \end{pmatrix},$

 $\mathbf{K}_2 = \begin{pmatrix} 2 \\ 5 \end{pmatrix},$ $\mathbf{K}_3 = \begin{pmatrix} -2 \\ 5 \end{pmatrix}$

2. $\mathbf{A} = \begin{pmatrix} 2 & -1 \\ 2 & -2 \end{pmatrix};$ $\mathbf{K}_1 = \begin{pmatrix} 1 \\ 2 - \sqrt{2} \end{pmatrix},$

 $\mathbf{K}_2 = \begin{pmatrix} 2 + \sqrt{2} \\ 2 \end{pmatrix},$ $\mathbf{K}_3 = \begin{pmatrix} \sqrt{2} \\ -\sqrt{2} \end{pmatrix}$

3. $\mathbf{A} = \begin{pmatrix} 6 & 3 \\ 2 & 1 \end{pmatrix};$ $\mathbf{K}_1 = \begin{pmatrix} 3 \\ -2 \end{pmatrix},$

 $\mathbf{K}_2 = \begin{pmatrix} 1 \\ 0 \end{pmatrix},$ $\mathbf{K}_3 = \begin{pmatrix} -5 \\ 10 \end{pmatrix}$

4. $\mathbf{A} = \begin{pmatrix} 2 & 8 \\ -1 & -2 \end{pmatrix};$ $\mathbf{K}_1 = \begin{pmatrix} 0 \\ 0 \end{pmatrix},$

 $\mathbf{K}_2 = \begin{pmatrix} 2 + 2i \\ -1 \end{pmatrix},$ $\mathbf{K}_3 = \begin{pmatrix} 2 + 2i \\ 1 \end{pmatrix}$

5. $\mathbf{A} = \begin{pmatrix} 1 & -2 & 2 \\ -2 & 1 & -2 \\ 2 & 2 & 1 \end{pmatrix};$ $\mathbf{K}_1 = \begin{pmatrix} 0 \\ 1 \\ 1 \end{pmatrix},$

 $\mathbf{K}_2 = \begin{pmatrix} 4 \\ -4 \\ 0 \end{pmatrix},$ $\mathbf{K}_3 = \begin{pmatrix} -1 \\ 1 \\ 1 \end{pmatrix}$

6. $\mathbf{A} = \begin{pmatrix} -1 & 1 & 0 \\ 1 & 2 & 1 \\ 0 & 3 & -1 \end{pmatrix};$ $\mathbf{K}_1 = \begin{pmatrix} -1 \\ 4 \\ 3 \end{pmatrix},$

 $\mathbf{K}_2 = \begin{pmatrix} 1 \\ 4 \\ 3 \end{pmatrix},$ $\mathbf{K}_3 = \begin{pmatrix} 3 \\ 1 \\ 4 \end{pmatrix}$

In Problems 7–22, find the eigenvalues and eigenvectors of the given matrix. Using Theorem 8.8.2 or (6), state whether the matrix is singular or nonsingular.

7. $\begin{pmatrix} -1 & 2 \\ -7 & 8 \end{pmatrix}$

8. $\begin{pmatrix} 2 & 1 \\ 2 & 1 \end{pmatrix}$

9. $\begin{pmatrix} -8 & -1 \\ 16 & 0 \end{pmatrix}$

10. $\begin{pmatrix} 1 & 1 \\ \frac{1}{4} & 1 \end{pmatrix}$

11. $\begin{pmatrix} -1 & 2 \\ -5 & 1 \end{pmatrix}$

12. $\begin{pmatrix} 1 & -1 \\ 1 & 1 \end{pmatrix}$

13. $\begin{pmatrix} 4 & 8 \\ 0 & -5 \end{pmatrix}$

14. $\begin{pmatrix} 7 & 0 \\ 0 & 13 \end{pmatrix}$

15. $\begin{pmatrix} 5 & -1 & 0 \\ 0 & -5 & 9 \\ 5 & -1 & 0 \end{pmatrix}$

16. $\begin{pmatrix} 3 & 0 & 0 \\ 0 & 2 & 0 \\ 4 & 0 & 1 \end{pmatrix}$

17. $\begin{pmatrix} 0 & 4 & 0 \\ -1 & -4 & 0 \\ 0 & 0 & -2 \end{pmatrix}$

18. $\begin{pmatrix} 1 & 6 & 0 \\ 0 & 2 & 1 \\ 0 & 1 & 2 \end{pmatrix}$

19. $\begin{pmatrix} 0 & 0 & -1 \\ 1 & 0 & 0 \\ 1 & 1 & -1 \end{pmatrix}$

20. $\begin{pmatrix} 2 & -1 & 0 \\ 5 & 2 & 4 \\ 0 & 1 & 2 \end{pmatrix}$

21. $\begin{pmatrix} 1 & 2 & 3 \\ 0 & 5 & 6 \\ 0 & 0 & -7 \end{pmatrix}$

22. $\begin{pmatrix} 0 & 0 & 0 \\ 0 & 0 & 0 \\ 0 & 0 & 1 \end{pmatrix}$

In Problems 23–26, find the eigenvalues and eigenvectors of the given nonsingular matrix $\mathbf{A}$. Then without finding $\mathbf{A}^{-1}$, find its eigenvalues and corresponding eigenvectors.

23. $\mathbf{A} = \begin{pmatrix} 5 & 1 \\ 1 & 5 \end{pmatrix}$

24. $\mathbf{A} = \begin{pmatrix} 4 & 2 \\ 7 & -1 \end{pmatrix}$

25. $\mathbf{A} = \begin{pmatrix} 4 & 2 & -1 \\ 0 & 3 & -2 \\ 0 & 0 & 5 \end{pmatrix}$ **26.** $\mathbf{A} = \begin{pmatrix} 1 & 2 & -1 \\ 1 & 0 & 1 \\ 4 & -4 & 5 \end{pmatrix}$

Discussion Problems

27. Review the definitions of upper triangular, lower triangular, and diagonal matrices on page 379. Explain why the eigenvalues for those matrices are the main diagonal entries.

28. True or false: If λ is an eigenvalue of an $n \times n$ matrix $\mathbf{A}$, then the matrix $\mathbf{A} - \lambda\mathbf{I}$ is singular. Justify your answer.

29. Suppose λ is an eigenvalue with corresponding eigenvector $\mathbf{K}$ of an $n \times n$ matrix $\mathbf{A}$.
 (a) If $\mathbf{A}^2 = \mathbf{A}\mathbf{A}$, then show that $\mathbf{A}^2\mathbf{K} = \lambda^2\mathbf{K}$. Explain the meaning of the last equation.
 (b) Verify the result obtained in part (a) for the matrix $\mathbf{A} = \begin{pmatrix} 2 & 3 \\ 5 & 4 \end{pmatrix}$.
 (c) Generalize the result in part (a).

30. Let $\mathbf{A}$ and $\mathbf{B}$ be $n \times n$ matrices. The matrix $\mathbf{B}$ is said to be **similar** to the matrix $\mathbf{A}$ if there exists a nonsingular matrix $\mathbf{S}$ such that $\mathbf{B} = \mathbf{S}^{-1}\mathbf{A}\mathbf{S}$. If $\mathbf{B}$ is similar to $\mathbf{A}$, then show that $\mathbf{A}$ is similar to $\mathbf{B}$.

31. Suppose $\mathbf{A}$ and $\mathbf{B}$ are similar matrices. See Problem 30. Show that $\mathbf{A}$ and $\mathbf{B}$ have the same eigenvalues. [*Hint*: Review Theorem 8.5.6 and Problem 37 in Exercises 8.6.]

Computer Lab Assignment

32. An $n \times n$ matrix $\mathbf{A}$ is said to be a **stochastic matrix** if all its entries are nonnegative and the sum of the entries in each row (or the sum of the entries in each column) add up to 1. Stochastic matrices are important in probability theory.
 (a) Verify that

$$\mathbf{A} = \begin{pmatrix} p & 1-p \\ q & 1-q \end{pmatrix}, \quad 0 \le p \le 1, \ 0 \le q \le 1,$$

and $\mathbf{A} = \begin{pmatrix} \frac{1}{2} & \frac{1}{4} & \frac{1}{4} \\ \frac{1}{3} & \frac{1}{3} & \frac{1}{3} \\ \frac{1}{6} & \frac{1}{3} & \frac{1}{2} \end{pmatrix}$

are stochastic matrices.
 (b) Use a CAS or linear algebra software to find the eigenvalues and eigenvectors of the 3×3 matrix $\mathbf{A}$ in part (a). Make up at least six more stochastic matrices of various sizes, 2×2, 3×3, 4×4, and 5×5. Find the eigenvalues and eigenvectors of each matrix. If you discern a pattern, form a conjecture and then try to prove it.
 (c) For the 3×3 matrix $\mathbf{A}$ in part (a), use the software to find $\mathbf{A}^2, \mathbf{A}^3, \mathbf{A}^4, \dots$. Repeat for the matrices that you constructed in part (b). If you discern a pattern, form a conjecture and then try to prove it.

8.9 Powers of Matrices

INTRODUCTION It is sometimes important to be able to quickly compute a power $\mathbf{A}^m$, m a positive integer, of an $n \times n$ matrix $\mathbf{A}$:

$$\mathbf{A}^m = \underbrace{\mathbf{A}\mathbf{A}\mathbf{A} \cdots \mathbf{A}}_{m \text{ factors}}.$$

Of course, computation of $\mathbf{A}^m$ could be done with the appropriate software or by writing a short computer program, but even then, you should be aware that it is inefficient to simply use brute force to carry out repeated multiplications: $\mathbf{A}^2 = \mathbf{A}\mathbf{A}$, $\mathbf{A}^3 = \mathbf{A}\mathbf{A}^2$, $\mathbf{A}^4 = \mathbf{A}\mathbf{A}\mathbf{A}\mathbf{A} = \mathbf{A}(\mathbf{A}^3) = \mathbf{A}^2\mathbf{A}^2$, and so on.

Computation of $\mathbf{A}^m$ We are going to sketch an alternative method for computing $\mathbf{A}^m$ by means of the following theorem known as the Cayley–Hamilton theorem.

THEOREM 8.9.1 Cayley–Hamilton Theorem

An $n \times n$ matrix $\mathbf{A}$ satisfies its own characteristic equation.

If $(-1)^n\lambda^n + c_{n-1}\lambda^{n-1} + \cdots + c_1\lambda + c_0 = 0$ is the characteristic equation of $\mathbf{A}$, then Theorem 8.9.1 states that by replacing λ by $\mathbf{A}$ we have

$$(-1)^n\mathbf{A}^n + c_{n-1}\mathbf{A}^{n-1} + \cdots + c_1\mathbf{A} + c_0\mathbf{I} = \mathbf{0}. \tag{1}$$

Matrices of Order 2 The characteristic equation of the 2×2 matrix $\mathbf{A} = \begin{pmatrix} -2 & 4 \\ -1 & 3 \end{pmatrix}$ is $\lambda^2 - \lambda - 2 = 0$, and the eigenvalues of $\mathbf{A}$ are $\lambda_1 = -1$ and $\lambda_2 = 2$. Theorem 8.9.1 implies $\mathbf{A}^2 - \mathbf{A} - 2\mathbf{I} = \mathbf{0}$, or, solving for the highest power of $\mathbf{A}$,

$$\mathbf{A}^2 = 2\mathbf{I} + \mathbf{A}. \tag{2}$$

Now if we multiply (2) by $\mathbf{A}$, we get $\mathbf{A}^3 = 2\mathbf{A} + \mathbf{A}^2$, and if we use (2) again to eliminate $\mathbf{A}^2$ on the right side of this new equation, then

$$\mathbf{A}^3 = 2\mathbf{A} + \mathbf{A}^2 = 2\mathbf{A} + (2\mathbf{I} + \mathbf{A}) = 2\mathbf{I} + 3\mathbf{A}.$$

Continuing in this manner—in other words, multiplying the last result by $\mathbf{A}$ and using (2) to eliminate $\mathbf{A}^2$—we obtain in succession powers of $\mathbf{A}$ expressed solely in terms of the identity matrix $\mathbf{I}$ and $\mathbf{A}$:

$$\mathbf{A}^4 = 6\mathbf{I} + 5\mathbf{A}$$
$$\mathbf{A}^5 = 10\mathbf{I} + 11\mathbf{A} \tag{3}$$
$$\mathbf{A}^6 = 22\mathbf{I} + 21\mathbf{A}$$

and so on (verify). Thus, for example,

$$\mathbf{A}^6 = 22 \begin{pmatrix} 1 & 0 \\ 0 & 1 \end{pmatrix} + 21 \begin{pmatrix} -2 & 4 \\ -1 & 3 \end{pmatrix} = \begin{pmatrix} -20 & 84 \\ -21 & 85 \end{pmatrix}. \tag{4}$$

Now we can determine the c_k without actually carrying out the repeated multiplications and resubstitutions as we did in (3). First, note that since the characteristic equation of the matrix $\mathbf{A} = \begin{pmatrix} -2 & 4 \\ -1 & 3 \end{pmatrix}$ can be written $\lambda^2 = 2 + \lambda$, results analogous to (3) must also hold for the eigenvalues $\lambda_1 = -1$ and $\lambda_2 = 2$; that is, $\lambda^3 = 2 + 3\lambda$, $\lambda^4 = 6 + 5\lambda$, $\lambda^5 = 10 + 11\lambda$, $\lambda^6 = 22 + 21\lambda, \dots$. It follows then that the equations

$$\mathbf{A}^m = c_0\mathbf{I} + c_1\mathbf{A} \quad \text{and} \quad \lambda^m = c_0 + c_1\lambda \tag{5}$$

hold *for the same pair of constants* c_0 and c_1. We can determine the constants c_0 and c_1 by simply setting $\lambda = -1$ and $\lambda = 2$ in the last equation in (5) and solving the resulting system of two equations in two unknowns. The solution of the system

$$(-1)^m = c_0 + c_1(-1)$$
$$2^m = c_0 + c_1(2)$$

is $c_0 = \frac{1}{3}[2^m + 2(-1)^m]$, $c_1 = \frac{1}{3}[2^m - (-1)^m]$. Now by substituting these coefficients in the first equation in (5), adding the two matrices and simplifying each entry, we obtain

$$\mathbf{A}^m = \begin{pmatrix} \frac{1}{3}[-2^m + 4(-1)^m] & \frac{4}{3}[2^m - (-1)^m] \\ -\frac{1}{3}[2^m - (-1)^m] & \frac{1}{3}[2^{m+2} - (-1)^m] \end{pmatrix}. \tag{6}$$

You should verify the result in (4) by setting $m = 6$ in (6). Note that (5) and (6) are valid for $m \geq 0$ since $\mathbf{A}^0 = \mathbf{I}$ and $\mathbf{A}^1 = \mathbf{A}$.

Matrices of Order n If the matrix $\mathbf{A}$ were 3×3, then the characteristic equation (1) is a cubic polynomial equation, and the analogue of (2) would enable us to express $\mathbf{A}^3$ in terms of

I, A, and $\mathbf{A}^2$. We could proceed as just illustrated to write any power $\mathbf{A}^m$ in terms of $\mathbf{I}$, $\mathbf{A}$, and $\mathbf{A}^2$. In general, for an $n \times n$ matrix $\mathbf{A}$, we can write

$$\mathbf{A}^m = c_0\mathbf{I} + c_1\mathbf{A} + c_2\mathbf{A}^2 + \cdots + c_{n-1}\mathbf{A}^{n-1},$$

and where each of the coefficients c_k, $k = 0, 1, \ldots, n - 1$, depends on the value of m.

EXAMPLE 1 $\mathbf{A}^m$ for a 3 × 3 Matrix

Compute $\mathbf{A}^m$ for $\mathbf{A} = \begin{pmatrix} 1 & 1 & -2 \\ -1 & 2 & 1 \\ 0 & 1 & -1 \end{pmatrix}$.

SOLUTION The characteristic equation of $\mathbf{A}$ is $-\lambda^3 + 2\lambda^2 + \lambda - 2 = 0$ or $\lambda^3 = -2 + \lambda + 2\lambda^2$, and the eigenvalues are $\lambda_1 = -1$, $\lambda_2 = 1$, and $\lambda_3 = 2$. From the preceding discussion we know that the same coefficients hold in both of the following equations:

$$\mathbf{A}^m = c_0\mathbf{I} + c_1\mathbf{A} + c_2\mathbf{A}^2 \quad \text{and} \quad \lambda^m = c_0 + c_1\lambda + c_2\lambda^2. \tag{7}$$

Setting, in turn, $\lambda = -1$, $\lambda = 1$, $\lambda = 2$ in the last equation generates three equations in three unknowns:

$$\begin{aligned} (-1)^m &= c_0 - c_1 + c_2 \\ 1 &= c_0 + c_1 + c_2 \\ 2^m &= c_0 + 2c_1 + 4c_2. \end{aligned} \tag{8}$$

Solving (8) gives

$$\begin{aligned} c_0 &= \tfrac{1}{3}[3 + (-1)^m - 2^m], \\ c_1 &= 2[1 - (-1)^m], \\ c_2 &= \tfrac{1}{6}[-3 + (-1)^m + 2^{m+1}]. \end{aligned}$$

After computing $\mathbf{A}^2$, we substitute these coefficients into the first equation of (7) and simplify the entries of the resulting matrix. The result is

$$\mathbf{A}^m = \begin{pmatrix} \tfrac{1}{6}[9 - 2^{m+1} - (-1)^m] & \tfrac{1}{3}[2^m - (-1)^m] & \tfrac{1}{6}[-9 + 2^{m+1} + 7(-1)^m] \\ 1 - 2^m & 2^m & 2^m - 1 \\ \tfrac{1}{6}[3 - 2^{m+1} - (-1)^m] & \tfrac{1}{3}[2^m - (-1)^m] & \tfrac{1}{6}[-3 + 2^{m+1} + 7(-1)^m] \end{pmatrix}.$$

For example, with $m = 10$,

$$\mathbf{A}^{10} = \begin{pmatrix} -340 & 341 & 341 \\ -1023 & 1024 & 1023 \\ -341 & 341 & 342 \end{pmatrix}. \qquad \equiv$$

Finding the Inverse
Suppose $\mathbf{A}$ is a nonsingular matrix. The fact that $\mathbf{A}$ satisfies its own characteristic equation can be used to compute $\mathbf{A}^{-1}$ as a linear combination of powers of $\mathbf{A}$. For example, we have just seen that the nonsingular matrix $\mathbf{A} = \begin{pmatrix} -2 & 4 \\ -1 & 3 \end{pmatrix}$ satisfies $\mathbf{A}^2 - \mathbf{A} - 2\mathbf{I} = \mathbf{0}$. Solving for the identity matrix gives $\mathbf{I} = \tfrac{1}{2}\mathbf{A}^2 - \tfrac{1}{2}\mathbf{A}$. By multiplying the last result by $\mathbf{A}^{-1}$, we find $\mathbf{A}^{-1} = \tfrac{1}{2}\mathbf{A} - \tfrac{1}{2}\mathbf{I}$. In other words,

$$\begin{pmatrix} -2 & 4 \\ -1 & 3 \end{pmatrix}^{-1} = \frac{1}{2}\begin{pmatrix} -2 & 4 \\ -1 & 3 \end{pmatrix} - \frac{1}{2}\begin{pmatrix} 1 & 0 \\ 0 & 1 \end{pmatrix} = \begin{pmatrix} -\tfrac{3}{2} & 2 \\ -\tfrac{1}{2} & 1 \end{pmatrix}. \tag{9}$$

REMARKS

There are some obvious problems with the method just illustrated for finding $\mathbf{A}^m$. If, for example, the matrix in Example 1 had an eigenvalue of multiplicity two, then instead of three equations and three unknowns as in (8) we would have only two equations in three unknowns. How do we then find *unique* coefficients c_0, c_1, and c_2? See Problems 11–14 in Exercises 8.9. Also, for larger matrices, even with distinct eigenvalues, solving a large system of equations for $c_0, c_1, c_2, \ldots, c_{n-1}$ is far too tedious to do by hand.

8.9 Exercises Answers to selected odd-numbered problems begin on page ANS-19.

In Problems 1 and 2, verify that the given matrix satisfies its own characteristic equation.

1. $\mathbf{A} = \begin{pmatrix} 1 & -2 \\ 4 & 5 \end{pmatrix}$
2. $\mathbf{A} = \begin{pmatrix} 0 & 1 & 2 \\ 1 & 0 & 3 \\ 0 & 1 & 1 \end{pmatrix}$

In Problems 3–10, use the method of this section to compute $\mathbf{A}^m$. Use this result to compute the indicated power of the matrix $\mathbf{A}$.

3. $\mathbf{A} = \begin{pmatrix} -1 & 3 \\ 2 & 4 \end{pmatrix}$; $m = 3$ **4.** $\mathbf{A} = \begin{pmatrix} 5 & -3 \\ -3 & 5 \end{pmatrix}$; $m = 4$

5. $\mathbf{A} = \begin{pmatrix} 8 & 5 \\ 4 & 0 \end{pmatrix}$; $m = 5$ **6.** $\mathbf{A} = \begin{pmatrix} -1 & 2 \\ 0 & -3 \end{pmatrix}$; $m = 6$

7. $\mathbf{A} = \begin{pmatrix} 1 & 1 & 1 \\ 0 & 1 & 2 \\ 0 & 1 & 0 \end{pmatrix}$; $m = 10$

8. $\mathbf{A} = \begin{pmatrix} 0 & 1 & 1 \\ 0 & -1 & 1 \\ 1 & 1 & 0 \end{pmatrix}$; $m = 6$

9. $\mathbf{A} = \begin{pmatrix} 2 & 2 & 0 \\ 4 & 0 & 0 \\ 1 & 2 & 1 \end{pmatrix}$; $m = 10$

10. $\mathbf{A} = \begin{pmatrix} 0 & -\frac{1}{2} & 0 \\ -1 & \frac{1}{2} & 0 \\ 2 & -\frac{1}{2} & -2 \end{pmatrix}$; $m = 8$

In Problems 11 and 12, show that the given matrix has an eigenvalue λ_1 of multiplicity two. As a consequence, the equations $\lambda^m = c_0 + c_1\lambda$ (Problem 11) and $\lambda^m = c_0 + c_1\lambda + c_2\lambda^2$ (Problem 12) do not yield enough independent equations to form a system for determining the coefficients c_i. Use the derivative (with respect to λ) of each of these equations evaluated at λ_1 as the extra needed equation to form a system. Compute $\mathbf{A}^m$ and use this result to compute the indicated power of the matrix $\mathbf{A}$.

11. $\mathbf{A} = \begin{pmatrix} 7 & 3 \\ -3 & 1 \end{pmatrix}$; $m = 6$

12. $\mathbf{A} = \begin{pmatrix} -2 & 2 & -1 \\ 2 & 1 & -2 \\ -3 & -6 & 0 \end{pmatrix}$; $m = 5$

13. Show that $\lambda = 0$ is an eigenvalue of each matrix. In this case, the coefficient c_0 in the characteristic equation (1) is 0. Compute $\mathbf{A}^m$ in each case. In parts (a) and (b), explain why we do not have to solve any system for the coefficients c_i in determining $\mathbf{A}^m$.

(a) $\mathbf{A} = \begin{pmatrix} 1 & 1 \\ 3 & 3 \end{pmatrix}$ (b) $\mathbf{A} = \begin{pmatrix} 1 & -1 \\ 1 & -1 \end{pmatrix}$

(c) $\mathbf{A} = \begin{pmatrix} 2 & 1 & 1 \\ 1 & 0 & -2 \\ 1 & 1 & 3 \end{pmatrix}$

14. **Leonardo of Pisa** (1170–c.1250), also known as *Fibonacci*, was an Italian mathematician from the Republic of Pisa. His work *Liber Abaci* (Book of Calculation), first published in 1202, was a strong advocation of the value and practical use of the new Hindu-Arabic numeral system. In *Liber Abaci* Leonardo also speculated on the reproduction of rabbits:

How many pairs of rabbits will be produced in a year beginning with a single pair, if every month each pair bears a new pair which become productive from the second month on?

The answer to his question is contained in a sequence known as a Fibonacci sequence.

After Each Month													
Start $n =$	0	1	2	3	4	5	6	7	8	9	10	11	12
Adult pairs	1	1	2	3	5	8	13	21	...				
Baby pairs	0	1	1	2	3	5	8	13	...				
Total pairs	1	2	3	5	8	13	21	34	...				

Each of the three rows describing rabbit pairs is a Fibonacci sequence and can be defined recursively by a **second-order difference equation** $x_n = x_{n-2} + x_{n-1}$, $n = 2, 3, \ldots$, where x_0 and x_1 depend on the row. For example, for the first row designating adult pairs of rabbits, $x_0 = 1$, $x_1 = 1$.

(a) If we let $y_{n-1} = x_{n-2}$, then $y_n = x_{n-1}$, and the difference equation can be written as a system of first-order difference equations

$$x_n = x_{n-1} + y_{n-1}$$
$$y_n = x_{n-1}.$$

Write this system in the matrix form $\mathbf{X}_n = \mathbf{A}\mathbf{X}_{n-1}$, $n = 2, 3, \ldots$.

(b) Show that

$$\mathbf{A}^m = \begin{pmatrix} \dfrac{\lambda_2\lambda_1^m - \lambda_1\lambda_2^m + \lambda_2^m - \lambda_1^m}{\lambda_2 - \lambda_1} & \dfrac{\lambda_2^m - \lambda_1^m}{\lambda_2 - \lambda_1} \\ \dfrac{\lambda_2^m - \lambda_1^m}{\lambda_2 - \lambda_1} & \dfrac{\lambda_2\lambda_1^m - \lambda_1\lambda_2^m}{\lambda_2 - \lambda_1} \end{pmatrix}$$

or

$$\mathbf{A}^m = \frac{1}{2^{m+1}\sqrt{5}}\begin{pmatrix} (1+\sqrt{5})^{m+1} - (1-\sqrt{5})^{m+1} & 2(1+\sqrt{5})^m - 2(1-\sqrt{5})^m \\ 2(1+\sqrt{5})^m - 2(1-\sqrt{5})^m & (1+\sqrt{5})(1-\sqrt{5})^m - (1-\sqrt{5})(1+\sqrt{5})^m \end{pmatrix},$$

where $\lambda_1 = \frac{1}{2}(1 - \sqrt{5})$ and $\lambda_2 = \frac{1}{2}(1 + \sqrt{5})$ are the distinct eigenvalues of $\mathbf{A}$.

(c) Use the result in part (a) to show $\mathbf{X}_n = \mathbf{A}^{n-1}\mathbf{X}_1$. Use the last result and the result in part (b) to find the number of adult pairs, baby pairs, and total pairs of rabbits after the twelfth month.

In Problems 15 and 16, use the procedure illustrated in (9) to find $\mathbf{A}^{-1}$.

15. $\mathbf{A} = \begin{pmatrix} 2 & -4 \\ 1 & 3 \end{pmatrix}$

16. $\mathbf{A} = \begin{pmatrix} 1 & 1 & -2 \\ -1 & 2 & 1 \\ 0 & 1 & -1 \end{pmatrix}$

17. A nonzero $n \times n$ matrix $\mathbf{A}$ is said to be **nilpotent of index** m if m is the smallest positive integer for which $\mathbf{A}^m = \mathbf{0}$. Which of the following matrices are nilpotent? If nilpotent, what is its index?

(a) $\begin{pmatrix} 1 & 0 \\ -1 & 0 \end{pmatrix}$

(b) $\begin{pmatrix} 2 & 2 \\ -2 & -2 \end{pmatrix}$

(c) $\begin{pmatrix} 0 & 0 & 0 \\ 1 & 0 & 0 \\ 2 & 3 & 0 \end{pmatrix}$

(d) $\begin{pmatrix} 0 & 0 & 5 \\ 0 & 0 & 0 \\ 0 & 0 & 0 \end{pmatrix}$

(e) $\begin{pmatrix} 0 & 0 & 0 & 0 \\ -1 & 0 & 0 & 1 \\ 0 & 1 & 0 & 1 \\ -1 & 0 & 0 & 0 \end{pmatrix}$

(f) $\begin{pmatrix} 0 & 0 & 0 & 0 \\ 1 & 0 & 0 & 0 \\ 3 & 1 & 0 & 0 \\ 2 & 2 & 1 & 0 \end{pmatrix}$

18. (a) Explain why any nilpotent matrix $\mathbf{A}$ is singular. [*Hint*: Review Section 8.5.] **(b)** Show that all the eigenvalues of a nilpotent matrix $\mathbf{A}$ are 0. [*Hint*: Use (1) of Section 8.8.]

8.10 Orthogonal Matrices

INTRODUCTION In this section we are going to use some elementary properties of **complex numbers**. Suppose $z = a + ib$ denotes a complex number, where a and b are real and the symbol i is defined by $i^2 = -1$. If $\bar{z} = a - ib$ is the **conjugate** of z, then the equality $z = \bar{z}$ or $a + ib = a - ib$ implies that $b = 0$. In other words, if $z = \bar{z}$, then z is a real number. In addition, it is easily verified that the product of a complex number z and its conjugate $\bar{z}$ is a real number: $z\bar{z} = a^2 + b^2$. The **magnitude** of z is defined to be the real number $|z| = \sqrt{a^2 + b^2}$. The magnitude of z can be expressed in terms of the product $z\bar{z}$: $|z| = \sqrt{a^2 + b^2} = |z\bar{z}|$, or $|z|^2 = z\bar{z}$. A detailed discussion of complex numbers can be found in Section 17.1.

There are many types of special matrices, but two types occur again and again in applications: symmetric matrices (page 380) and orthogonal matrices (page 420). In this section we are going to examine both these matrices in further detail.

Symmetric Matrices We begin by recalling, in formal terms, the definition of a symmetric matrix.

DEFINITION 8.10.1 Symmetric Matrix

An $n \times n$ matrix $\mathbf{A}$ is **symmetric** if $\mathbf{A} = \mathbf{A}^T$, where $\mathbf{A}^T$ is the transpose of $\mathbf{A}$.

The proof of the next theorem depends on the properties of complex numbers discussed in the review at the start of this section.

THEOREM 8.10.1 Real Eigenvalues

Let $\mathbf{A}$ be a symmetric matrix with real entries. Then the eigenvalues of $\mathbf{A}$ are real.

PROOF: If $\mathbf{K}$ is an eigenvector corresponding to an eigenvalue λ of $\mathbf{A}$, then $\mathbf{AK} = \lambda\mathbf{K}$. The conjugate of the last equation is

$$\overline{\mathbf{A}}\,\overline{\mathbf{K}} = \overline{\lambda}\,\overline{\mathbf{K}}. \tag{1}$$

Since the entries of $\mathbf{A}$ are real, we have $\mathbf{A} = \overline{\mathbf{A}}$, and so (1) is

$$\mathbf{A}\overline{\mathbf{K}} = \overline{\lambda}\,\overline{\mathbf{K}}. \tag{2}$$

We now take the transpose of (2), use the fact that $\mathbf{A}$ is symmetric, and multiply the resulting equation on the right by $\mathbf{K}$:

$$\overline{\mathbf{K}}^{T}\mathbf{AK} = \overline{\lambda}\,\overline{\mathbf{K}}^{T}\mathbf{K}. \tag{3}$$

But if we multiply $\mathbf{AK} = \lambda\mathbf{K}$ on the left by $\overline{\mathbf{K}}^{T}$, we obtain

$$\overline{\mathbf{K}}^{T}\mathbf{AK} = \lambda\overline{\mathbf{K}}^{T}\mathbf{K}. \tag{4}$$

Subtracting (4) from (3) then gives

$$0 = (\overline{\lambda} - \lambda)\overline{\mathbf{K}}^{T}\mathbf{K}. \tag{5}$$

Now $\overline{\mathbf{K}}^{T}$ is a $1 \times n$ matrix and $\mathbf{K}$ is an $n \times 1$ matrix, so the product $\overline{\mathbf{K}}^{T}\mathbf{K}$ is the 1×1 matrix $\overline{\mathbf{K}}^{T}\mathbf{K} = (|k_1|^2 + |k_2|^2 + \cdots + |k_n|^2)$. Since by definition, $\mathbf{K} \neq \mathbf{0}$, the last expression is a positive quantity. Therefore we conclude from (5) that $\overline{\lambda} - \lambda = 0$ or $\overline{\lambda} = \lambda$. This implies that λ is a real number. ≡

■ Inner Product In R^n the **inner product** or **dot product** of two vectors $\mathbf{x} = (x_1, x_2, \ldots, x_n)$ and $\mathbf{y} = (y_1, y_2, \ldots, y_n)$ is given by

$$\mathbf{x} \cdot \mathbf{y} = x_1 y_1 + x_2 y_2 + \cdots + x_n y_n. \tag{6}$$

Now if $\mathbf{X}$ and $\mathbf{Y}$ are $n \times 1$ column vectors, $\mathbf{X} = \begin{pmatrix} x_1 \\ x_2 \\ \vdots \\ x_n \end{pmatrix}$ and $\mathbf{Y} = \begin{pmatrix} y_1 \\ y_2 \\ \vdots \\ y_n \end{pmatrix}$, then the matrix analogue of (6) is

$$\mathbf{X} \cdot \mathbf{Y} = \mathbf{X}^{T}\mathbf{Y} = (x_1 y_1 + x_2 y_2 + \cdots + x_n y_n).* \tag{7}$$

Of course, for the column vectors given, $\mathbf{Y}^{T}\mathbf{X} = \mathbf{X}^{T}\mathbf{Y}$. The **norm** of a column vector $\mathbf{X}$ is given by

$$\|\mathbf{X}\| = \sqrt{\mathbf{X} \cdot \mathbf{X}} = \sqrt{\mathbf{X}^{T}\mathbf{X}} = \sqrt{x_1^2 + x_2^2 + \cdots + x_n^2}.$$

THEOREM 8.10.2 Orthogonal Eigenvectors

Let $\mathbf{A}$ be an $n \times n$ symmetric matrix. Then eigenvectors corresponding to distinct (different) eigenvalues are orthogonal.

*Since a 1×1 matrix is simply a scalar, we will hereafter drop the parentheses and write $\mathbf{X}^{T}\mathbf{Y} = x_1 y_1 + x_2 y_2 + \cdots + x_n y_n$.

PROOF: Let λ_1 and λ_2 be two distinct eigenvalues of $\mathbf{A}$ corresponding to the eigenvectors $\mathbf{K}_1$ and $\mathbf{K}_2$, respectively. We wish to show that $\mathbf{K}_1 \cdot \mathbf{K}_2 = \mathbf{K}_1^T \mathbf{K}_2 = 0$.

Now by definition we must have

$$\mathbf{AK}_1 = \lambda_1 \mathbf{K}_1 \quad \text{and} \quad \mathbf{AK}_2 = \lambda_2 \mathbf{K}_2. \tag{8}$$

We form the transpose of the first of these equations, use $\mathbf{A}^T = \mathbf{A}$, and then multiply the result on the right by $\mathbf{K}_2$:

$$\mathbf{K}_1^T \mathbf{AK}_2 = \lambda_1 \mathbf{K}_1^T \mathbf{K}_2. \tag{9}$$

The second equation in (8) is multiplied on the left by $\mathbf{K}_1^T$:

$$\mathbf{K}_1^T \mathbf{AK}_2 = \lambda_2 \mathbf{K}_1^T \mathbf{K}_2. \tag{10}$$

Subtracting (10) from (9) yields

$$0 = \lambda_1 \mathbf{K}_1^T \mathbf{K}_2 - \lambda_2 \mathbf{K}_1^T \mathbf{K}_2 \quad \text{or} \quad 0 = (\lambda_1 - \lambda_2) \mathbf{K}_1^T \mathbf{K}_2.$$

Since $\lambda_1 \neq \lambda_2$, it follows that $\mathbf{K}_1^T \mathbf{K}_2 = 0$. ≡

EXAMPLE 1 **Orthogonal Eigenvectors**

The eigenvalues of the symmetric matrix $\mathbf{A} = \begin{pmatrix} 0 & -1 & 0 \\ -1 & -1 & 1 \\ 0 & 1 & 0 \end{pmatrix}$ are $\lambda_1 = 0$, $\lambda_2 = 1$, and $\lambda_3 = -2$. In turn, the corresponding eigenvectors are

$$\mathbf{K}_1 = \begin{pmatrix} 1 \\ 0 \\ 1 \end{pmatrix}, \quad \mathbf{K}_2 = \begin{pmatrix} -1 \\ 1 \\ 1 \end{pmatrix}, \quad \mathbf{K}_3 = \begin{pmatrix} 1 \\ 2 \\ -1 \end{pmatrix}.$$

Since all the eigenvalues are distinct, it follows from Theorem 8.10.2 that the eigenvectors are orthogonal, that is,

$$\mathbf{K}_1^T \mathbf{K}_2 = (1 \ 0 \ 1) \begin{pmatrix} -1 \\ 1 \\ 1 \end{pmatrix} = 1 \cdot (-1) + 0 \cdot 1 + 1 \cdot 1 = 0$$

$$\mathbf{K}_1^T \mathbf{K}_3 = (1 \ 0 \ 1) \begin{pmatrix} 1 \\ 2 \\ -1 \end{pmatrix} = 1 \cdot 1 + 0 \cdot 2 + 1 \cdot (-1) = 0$$

$$\mathbf{K}_2^T \mathbf{K}_3 = (-1 \ 1 \ 1) \begin{pmatrix} 1 \\ 2 \\ -1 \end{pmatrix} = (-1) \cdot 1 + 1 \cdot 2 + 1 \cdot (-1) = 0.$$ ≡

We saw in Example 3 of Section 8.8 that it may not be possible to find n linearly independent eigenvectors for an $n \times n$ matrix $\mathbf{A}$ when some of the eigenvalues are repeated. But a symmetric matrix is an exception. It can be proved that a set of n linearly independent eigenvectors can always be found for an $n \times n$ symmetric matrix $\mathbf{A}$ even when there is some repetition of the eigenvalues. (See Example 4 of Section 8.8.)

See (2) of Section 7.6 for the definition of the inner product in R^n.

A set of vectors $\mathbf{x}_1, \mathbf{x}_2, \ldots, \mathbf{x}_n$ in R^n is called **orthonormal** if every pair of distinct vectors is orthogonal and each vector in the set is a unit vector. In terms of the inner product of vectors, the set is orthonormal if

$$\mathbf{x}_i \cdot \mathbf{x}_j = 0, \quad i \neq j, \quad i, j = 1, 2, \ldots, n \quad \text{and} \quad \mathbf{x}_i \cdot \mathbf{x}_i = 1, \quad i = 1, 2, \ldots, n.$$

The last condition simply states that $\|\mathbf{x}_i\| = \sqrt{\mathbf{x}_i \cdot \mathbf{x}_i} = 1, i = 1, 2, \ldots, n$.

Orthogonal Matrix

The concept of an orthonormal set of vectors plays an important role in the consideration of the next type of matrix.

DEFINITION 8.10.2 Orthogonal Matrix

An $n \times n$ nonsingular matrix $\mathbf{A}$ is **orthogonal** if $\mathbf{A}^{-1} = \mathbf{A}^T$.

In other words, $\mathbf{A}$ is orthogonal if $\mathbf{A}^T \mathbf{A} = \mathbf{I}$.

EXAMPLE 2 Orthogonal Matrices

(a) The $n \times n$ identity matrix $\mathbf{I}$ is an orthogonal matrix. For example, in the case of the 3×3 identity

$$\mathbf{I} = \begin{pmatrix} 1 & 0 & 0 \\ 0 & 1 & 0 \\ 0 & 0 & 1 \end{pmatrix}$$

it is readily seen that $\mathbf{I}^T = \mathbf{I}$ and $\mathbf{I}^T \mathbf{I} = \mathbf{I}\mathbf{I} = \mathbf{I}$.

(b) The matrix

$$\mathbf{A} = \begin{pmatrix} \frac{1}{4} & -\frac{2}{3} & \frac{2}{3} \\ \frac{2}{3} & \frac{2}{3} & \frac{1}{3} \\ -\frac{2}{3} & \frac{1}{3} & \frac{2}{3} \end{pmatrix}$$

is orthogonal. To see this, we need only verify that $\mathbf{A}^T \mathbf{A} = \mathbf{I}$:

$$\mathbf{A}^T \mathbf{A} = \begin{pmatrix} \frac{1}{3} & \frac{2}{3} & -\frac{2}{3} \\ -\frac{2}{3} & \frac{2}{3} & \frac{1}{3} \\ \frac{2}{3} & \frac{1}{3} & \frac{2}{3} \end{pmatrix} \begin{pmatrix} \frac{1}{3} & -\frac{2}{3} & \frac{2}{3} \\ \frac{2}{3} & \frac{2}{3} & \frac{1}{3} \\ -\frac{2}{3} & \frac{1}{3} & \frac{2}{3} \end{pmatrix} = \begin{pmatrix} 1 & 0 & 0 \\ 0 & 1 & 0 \\ 0 & 0 & 1 \end{pmatrix}. \qquad \equiv$$

THEOREM 8.10.3 Criterion for an Orthogonal Matrix

An $n \times n$ matrix $\mathbf{A}$ is orthogonal if and only if its columns $\mathbf{X}_1, \mathbf{X}_2, \ldots, \mathbf{X}_n$ form an orthonormal set.

PARTIAL PROOF: Let us suppose that $\mathbf{A}$ is an $n \times n$ orthogonal matrix with columns $\mathbf{X}_1$, $\mathbf{X}_2, \ldots, \mathbf{X}_n$. Hence, the rows of $\mathbf{A}^T$ are $\mathbf{X}_1^T, \mathbf{X}_2^T, \ldots, \mathbf{X}_n^T$. But since $\mathbf{A}$ is orthogonal, $\mathbf{A}^T \mathbf{A} = \mathbf{I}$; that is,

$$\mathbf{A}^T \mathbf{A} = \begin{pmatrix} \mathbf{X}_1^T \mathbf{X}_1 & \mathbf{X}_1^T \mathbf{X}_2 & \cdots & \mathbf{X}_1^T \mathbf{X}_n \\ \mathbf{X}_2^T \mathbf{X}_1 & \mathbf{X}_2^T \mathbf{X}_2 & \cdots & \mathbf{X}_2^T \mathbf{X}_n \\ \vdots & & & \vdots \\ \mathbf{X}_n^T \mathbf{X}_1 & \mathbf{X}_n^T \mathbf{X}_2 & \cdots & \mathbf{X}_n^T \mathbf{X}_n \end{pmatrix} = \begin{pmatrix} 1 & 0 & \cdots & 0 \\ 0 & 1 & \cdots & 0 \\ \vdots & & & \vdots \\ 0 & 0 & \cdots & 1 \end{pmatrix}.$$

It follows from the definition of equality of matrices that

$$\mathbf{X}_i^T \mathbf{X}_j = 0, \quad i \neq j, \quad i, j = 1, 2, \ldots, n \quad \text{and} \quad \mathbf{X}_i^T \mathbf{X}_i = 1, \quad i = 1, 2, \ldots, n.$$

This means that the columns of the orthogonal matrix form an orthonormal set of n vectors. ≡

If we write the columns of the matrix in part (b) of Example 2 as

$$\mathbf{X}_1 = \begin{pmatrix} \frac{1}{3} \\ \frac{2}{3} \\ -\frac{2}{3} \end{pmatrix}, \quad \mathbf{X}_2 = \begin{pmatrix} -\frac{2}{3} \\ \frac{2}{3} \\ \frac{1}{3} \end{pmatrix}, \quad \mathbf{X}_3 = \begin{pmatrix} \frac{2}{3} \\ \frac{1}{3} \\ \frac{2}{3} \end{pmatrix},$$

then the vectors are orthogonal:

$$\mathbf{X}_1^T \mathbf{X}_2 = \begin{pmatrix} \frac{1}{3} & \frac{2}{3} & -\frac{2}{3} \end{pmatrix} \begin{pmatrix} -\frac{2}{3} \\ \frac{2}{3} \\ \frac{1}{3} \end{pmatrix} = -\frac{2}{9} + \frac{4}{9} - \frac{2}{9} = 0$$

$$\mathbf{X}_1^T \mathbf{X}_3 = \begin{pmatrix} \frac{1}{3} & \frac{2}{3} & -\frac{2}{3} \end{pmatrix} \begin{pmatrix} \frac{2}{3} \\ \frac{1}{3} \\ \frac{2}{3} \end{pmatrix} = \frac{2}{9} + \frac{2}{9} - \frac{4}{9} = 0$$

$$\mathbf{X}_2^T \mathbf{X}_3 = \begin{pmatrix} -\frac{2}{3} & \frac{2}{3} & \frac{1}{3} \end{pmatrix} \begin{pmatrix} \frac{2}{3} \\ \frac{1}{3} \\ \frac{2}{3} \end{pmatrix} = -\frac{4}{9} + \frac{2}{9} + \frac{2}{9} = 0$$

and are unit vectors:

$$\mathbf{X}_1^T \mathbf{X}_1 = \begin{pmatrix} \frac{1}{3} & \frac{2}{3} & -\frac{2}{3} \end{pmatrix} \begin{pmatrix} \frac{1}{3} \\ \frac{2}{3} \\ -\frac{2}{3} \end{pmatrix} = \frac{1}{9} + \frac{4}{9} + \frac{4}{9} = 1$$

$$\mathbf{X}_2^T \mathbf{X}_2 = \begin{pmatrix} -\frac{2}{3} & \frac{2}{3} & \frac{1}{3} \end{pmatrix} \begin{pmatrix} -\frac{2}{3} \\ \frac{2}{3} \\ \frac{1}{3} \end{pmatrix} = \frac{4}{9} + \frac{4}{9} + \frac{1}{9} = 1$$

$$\mathbf{X}_3^T \mathbf{X}_3 = \begin{pmatrix} \frac{2}{3} & \frac{1}{3} & \frac{2}{3} \end{pmatrix} \begin{pmatrix} \frac{2}{3} \\ \frac{1}{3} \\ \frac{2}{3} \end{pmatrix} = \frac{4}{9} + \frac{1}{9} + \frac{4}{9} = 1.$$

Constructing an Orthogonal Matrix If an $n \times n$ real symmetric matrix $\mathbf{A}$ possesses n distinct eigenvalues $\lambda_1, \lambda_2, \ldots, \lambda_n$, it follows from Theorem 8.10.2 that the corresponding eigenvectors $\mathbf{K}_1, \mathbf{K}_2, \ldots, \mathbf{K}_n$ are mutually orthogonal. By multiplying each vector by the reciprocal of its norm, we obtain a set of mutually orthogonal unit vectors—that is, an orthonormal set. We can then construct an orthogonal matrix by forming an $n \times n$ matrix $\mathbf{P}$ whose columns are these normalized eigenvectors of $\mathbf{A}$.

EXAMPLE 3 **Constructing an Orthogonal Matrix**

In Example 1, we verified that the eigenvectors

$$\mathbf{K}_1 = \begin{pmatrix} 1 \\ 0 \\ 1 \end{pmatrix}, \quad \mathbf{K}_2 = \begin{pmatrix} -1 \\ 1 \\ 1 \end{pmatrix}, \quad \mathbf{K}_3 = \begin{pmatrix} 1 \\ 2 \\ -1 \end{pmatrix}$$

of the given symmetric matrix $\mathbf{A}$ are orthogonal. Now the norms of the eigenvectors are

$$\|\mathbf{K}_1\| = \sqrt{\mathbf{K}_1^T \mathbf{K}_1} = \sqrt{2}, \quad \|\mathbf{K}_2\| = \sqrt{\mathbf{K}_2^T \mathbf{K}_2} = \sqrt{3}, \quad \|\mathbf{K}_3\| = \sqrt{\mathbf{K}_3^T \mathbf{K}_3} = \sqrt{6}.$$

Thus, an orthonormal set of vectors is

$$\begin{pmatrix} \dfrac{1}{\sqrt{2}} \\ 0 \\ \dfrac{1}{\sqrt{2}} \end{pmatrix}, \quad \begin{pmatrix} -\dfrac{1}{\sqrt{3}} \\ \dfrac{1}{\sqrt{3}} \\ \dfrac{1}{\sqrt{3}} \end{pmatrix}, \quad \begin{pmatrix} \dfrac{1}{\sqrt{6}} \\ \dfrac{2}{\sqrt{6}} \\ -\dfrac{1}{\sqrt{6}} \end{pmatrix}.$$

Using these vectors as columns, we obtain the orthogonal matrix

$$\mathbf{P} = \begin{pmatrix} \dfrac{1}{\sqrt{2}} & -\dfrac{1}{\sqrt{3}} & \dfrac{1}{\sqrt{6}} \\ 0 & \dfrac{1}{\sqrt{3}} & \dfrac{2}{\sqrt{6}} \\ \dfrac{1}{\sqrt{2}} & \dfrac{1}{\sqrt{3}} & -\dfrac{1}{\sqrt{6}} \end{pmatrix}.$$

You should verify that $\mathbf{P}^T = \mathbf{P}^{-1}$. ≡

We will use the technique of constructing an orthogonal matrix from the eigenvectors of a symmetric matrix in the next section.

Do not misinterpret Theorem 8.10.2. We can always find n linearly independent eigenvectors for an $n \times n$ real symmetric matrix $\mathbf{A}$. However, the theorem does not state that *all* the eigenvectors are mutually orthogonal. The set of eigenvectors corresponding to distinct eigenvalues are orthogonal, but the different eigenvectors corresponding to a repeated eigenvalue may not be orthogonal. Consider the symmetric matrix in the next example.

EXAMPLE 4 Using the Gram–Schmidt Process

For the symmetric matrix

$$\mathbf{A} = \begin{pmatrix} 7 & 4 & -4 \\ 4 & -8 & -1 \\ -4 & -1 & -8 \end{pmatrix}$$

we find that the eigenvalues are $\lambda_1 = \lambda_2 = -9$, and $\lambda_3 = 9$. Proceeding as in Section 8.8, for $\lambda_1 = \lambda_2 = -9$, we find

$$(\mathbf{A} + 9\mathbf{I}\,|\,\mathbf{0}) = \begin{pmatrix} 16 & 4 & -4 & | & 0 \\ 4 & 1 & -1 & | & 0 \\ -4 & -1 & 1 & | & 0 \end{pmatrix} \overset{\text{row operations}}{\Longrightarrow} \begin{pmatrix} 1 & \frac{1}{4} & -\frac{1}{4} & | & 0 \\ 0 & 0 & 0 & | & 0 \\ 0 & 0 & 0 & | & 0 \end{pmatrix}.$$

From the last matrix we see that $k_1 = -\frac{1}{4}k_2 + \frac{1}{4}k_3$. The choices $k_2 = 1$, $k_3 = 1$ followed by $k_2 = -4$, $k_3 = 0$ yield, in turn, the distinct eigenvectors

$$\mathbf{K}_1 = \begin{pmatrix} 0 \\ 1 \\ 1 \end{pmatrix} \quad \text{and} \quad \mathbf{K}_2 = \begin{pmatrix} 1 \\ -4 \\ 0 \end{pmatrix}.$$

Now for $\lambda_1 = 9$,

$$(\mathbf{A} - 9\mathbf{I}\,|\,\mathbf{0}) = \begin{pmatrix} -2 & 4 & -4 & | & 0 \\ 4 & -17 & -1 & | & 0 \\ -4 & -1 & -17 & | & 0 \end{pmatrix} \overset{\text{row operations}}{\Longrightarrow} \begin{pmatrix} 1 & 0 & 4 & | & 0 \\ 0 & 1 & 1 & | & 0 \\ 0 & 0 & 0 & | & 0 \end{pmatrix}$$

indicates that $\mathbf{K}_3 = \begin{pmatrix} 4 \\ 1 \\ -1 \end{pmatrix}$ is a third eigenvector.

Observe, in accordance with Theorem 8.10.2, the vector $\mathbf{K}_3$ is orthogonal to $\mathbf{K}_1$ and to $\mathbf{K}_2$, but $\mathbf{K}_1$ and $\mathbf{K}_2$, eigenvectors corresponding to the repeated eigenvalue $\lambda_1 = -9$, are not orthogonal since $\mathbf{K}_1 \cdot \mathbf{K}_2 = -4 \neq 0$.

We use the Gram–Schmidt orthogonalization process (see pages 366–369) to transform the set $\{\mathbf{K}_1, \mathbf{K}_2\}$ into an orthogonal set. Let $\mathbf{V}_1 = \mathbf{K}_1$ and then

$$\mathbf{V}_2 = \mathbf{K}_2 - \left(\frac{\mathbf{K}_2 \cdot \mathbf{V}_1}{\mathbf{V}_1 \cdot \mathbf{V}_1}\right)\mathbf{V}_1 = \begin{pmatrix} 1 \\ -2 \\ 2 \end{pmatrix}.$$

The set $\{\mathbf{V}_1, \mathbf{V}_2\}$ is an orthogonal set of vectors (verify). Moreover, the set $\{\mathbf{V}_1, \mathbf{V}_2, \mathbf{K}_3\}$ is an orthogonal set of eigenvectors. Using the norms $\|\mathbf{V}_1\| = \sqrt{2}$, $\|\mathbf{V}_2\| = 3$, and $\|\mathbf{K}_3\| = 3\sqrt{2}$, we obtain an orthonormal set of vectors

$$\begin{pmatrix} 0 \\ \dfrac{1}{\sqrt{2}} \\ \dfrac{1}{\sqrt{2}} \end{pmatrix}, \begin{pmatrix} \dfrac{1}{3} \\ -\dfrac{2}{3} \\ \dfrac{2}{3} \end{pmatrix}, \begin{pmatrix} \dfrac{4}{3\sqrt{2}} \\ \dfrac{1}{3\sqrt{2}} \\ -\dfrac{1}{3\sqrt{2}} \end{pmatrix},$$

and so the matrix

$$\mathbf{P} = \begin{pmatrix} 0 & \dfrac{1}{3} & \dfrac{4}{3\sqrt{2}} \\ \dfrac{1}{\sqrt{2}} & -\dfrac{2}{3} & \dfrac{1}{3\sqrt{2}} \\ \dfrac{1}{\sqrt{2}} & \dfrac{2}{3} & -\dfrac{1}{3\sqrt{2}} \end{pmatrix}$$

is orthogonal.

$\equiv$

REMARKS

For a real $n \times n$ symmetric matrix with repeated eigenvalues it is always possible to *find*, rather than *construct*, a set of n mutually orthogonal eigenvectors. In other words, the Gram–Schmidt process does not have to be used. See Problem 23 in Exercises 8.10.

8.10 Exercises Answers to selected odd-numbered problems begin on page ANS-20.

In Problems 1–4, (a) verify that the indicated column vectors are eigenvectors of the given symmetric matrix, (b) identify the corresponding eigenvalues, and (c) verify that the column vectors are orthogonal.

1. $\begin{pmatrix} 0 & 0 & -4 \\ 0 & -4 & 0 \\ -4 & 0 & 15 \end{pmatrix}$; $\begin{pmatrix} 0 \\ 1 \\ 0 \end{pmatrix}$, $\begin{pmatrix} 4 \\ 0 \\ 1 \end{pmatrix}$, $\begin{pmatrix} 1 \\ 0 \\ -4 \end{pmatrix}$

2. $\begin{pmatrix} 1 & -1 & -1 \\ -1 & 1 & -1 \\ -1 & -1 & 1 \end{pmatrix}$; $\begin{pmatrix} -2 \\ 1 \\ 1 \end{pmatrix}$, $\begin{pmatrix} 0 \\ 1 \\ -1 \end{pmatrix}$, $\begin{pmatrix} 1 \\ 1 \\ 1 \end{pmatrix}$

3. $\begin{pmatrix} 5 & 13 & 0 \\ 13 & 5 & 0 \\ 0 & 0 & -8 \end{pmatrix}$;

$\begin{pmatrix} \dfrac{\sqrt{2}}{2} \\ \dfrac{\sqrt{2}}{2} \\ 0 \end{pmatrix}$, $\begin{pmatrix} \dfrac{\sqrt{3}}{3} \\ -\dfrac{\sqrt{3}}{3} \\ \dfrac{\sqrt{3}}{3} \end{pmatrix}$, $\begin{pmatrix} \dfrac{\sqrt{6}}{6} \\ -\dfrac{\sqrt{6}}{6} \\ -\dfrac{\sqrt{6}}{3} \end{pmatrix}$

4. $\begin{pmatrix} 3 & 2 & 2 \\ 2 & 2 & 0 \\ 2 & 0 & 4 \end{pmatrix}$; $\begin{pmatrix} -2 \\ 2 \\ 1 \end{pmatrix}$, $\begin{pmatrix} 1 \\ 2 \\ -2 \end{pmatrix}$, $\begin{pmatrix} 2 \\ 1 \\ 2 \end{pmatrix}$

In Problems 5–10, determine whether the given matrix is orthogonal.

5. $\begin{pmatrix} 0 & 1 & 0 \\ 1 & 0 & 0 \\ 0 & 0 & 1 \end{pmatrix}$

6. $\begin{pmatrix} \frac{1}{2} & 0 & \frac{1}{2} \\ 0 & 1 & 0 \\ \frac{1}{2} & 0 & -\frac{1}{2} \end{pmatrix}$

7. $\begin{pmatrix} 0 & 0 & 1 \\ -\frac{12}{13} & \frac{5}{13} & 0 \\ \frac{5}{13} & \frac{12}{13} & 0 \end{pmatrix}$

8. $\begin{pmatrix} 0 & 0 & 0 \\ 0 & 0 & 0 \\ 0 & 0 & 0 \end{pmatrix}$

9. $\begin{pmatrix} 1 & -1 & 1 \\ 1 & -1 & -1 \\ 1 & 2 & 0 \end{pmatrix}$

10. $\begin{pmatrix} 0 & \frac{8}{17} & 0 & -\frac{15}{17} \\ 0 & 0 & 1 & 0 \\ 1 & 0 & 0 & 0 \\ 0 & \frac{15}{17} & 0 & \frac{8}{17} \end{pmatrix}$

In Problems 11–18, proceed as in Example 3 to construct an orthogonal matrix from the eigenvectors of the given symmetric matrix. (The answers are not unique.)

11. $\begin{pmatrix} 1 & 9 \\ 9 & 1 \end{pmatrix}$

12. $\begin{pmatrix} 7 & 0 \\ 0 & 4 \end{pmatrix}$

13. $\begin{pmatrix} 1 & 3 \\ 3 & 9 \end{pmatrix}$

14. $\begin{pmatrix} 1 & 1 \\ 1 & 0 \end{pmatrix}$

15. $\begin{pmatrix} 1 & 0 & 1 \\ 0 & 1 & 0 \\ 1 & 0 & 1 \end{pmatrix}$

16. $\begin{pmatrix} 0 & 1 & 1 \\ 1 & 1 & 1 \\ 1 & 1 & 0 \end{pmatrix}$

17. $\begin{pmatrix} -8 & 5 & 4 \\ 5 & 3 & 1 \\ 4 & 1 & 0 \end{pmatrix}$

18. $\begin{pmatrix} 2 & 8 & -2 \\ 8 & -4 & 10 \\ -2 & 10 & -7 \end{pmatrix}$

In Problems 19 and 20, use Theorem 8.10.3 to find values of a and b so that the given matrix is orthogonal.

19. $\begin{pmatrix} \frac{3}{5} & a \\ \frac{4}{5} & b \end{pmatrix}$

20. $\begin{pmatrix} 1/\sqrt{5} & b \\ a & 1/\sqrt{5} \end{pmatrix}$

In Problems 21 and 22, **(a)** verify that the indicated column vectors are eigenvectors of the given symmetric matrix and **(b)** identify the corresponding eigenvalues. **(c)** Proceed as in Example 4 and use the Gram–Schmidt process to construct an orthogonal matrix **P** from the eigenvectors.

21. $\mathbf{A} = \begin{pmatrix} 0 & 2 & 2 \\ 2 & 0 & 2 \\ 2 & 2 & 0 \end{pmatrix}$; $\mathbf{K}_1 = \begin{pmatrix} 1 \\ -1 \\ 0 \end{pmatrix}$,

$\mathbf{K}_2 = \begin{pmatrix} 1 \\ 0 \\ -1 \end{pmatrix}$, $\mathbf{K}_3 = \begin{pmatrix} 1 \\ 1 \\ 1 \end{pmatrix}$

22. $\mathbf{A} = \begin{pmatrix} 1 & 1 & 1 & 1 \\ 1 & 1 & 1 & 1 \\ 1 & 1 & 1 & 1 \\ 1 & 1 & 1 & 1 \end{pmatrix}$; $\mathbf{K}_1 = \begin{pmatrix} -1 \\ 0 \\ 0 \\ 1 \end{pmatrix}$,

$\mathbf{K}_2 = \begin{pmatrix} -1 \\ 0 \\ 1 \\ 0 \end{pmatrix}$, $\mathbf{K}_3 = \begin{pmatrix} -1 \\ 1 \\ 0 \\ 0 \end{pmatrix}$, $\mathbf{K}_4 = \begin{pmatrix} 1 \\ 1 \\ 1 \\ 1 \end{pmatrix}$

23. In Example 4, use the equation $k_1 = -\frac{1}{4}k_2 + \frac{1}{4}k_3$ and choose two different sets of values for k_2 and k_3 so that the corresponding eigenvectors $\mathbf{K}_1$ and $\mathbf{K}_2$ are orthogonal.

24. Construct an orthogonal matrix from the eigenvectors of

$$\mathbf{A} = \begin{pmatrix} 1 & 2 & 0 & 0 \\ 2 & 1 & 0 & 0 \\ 0 & 0 & 1 & 2 \\ 0 & 0 & 2 & 1 \end{pmatrix}.$$

25. Suppose **A** and **B** are $n \times n$ orthogonal matrices. Then show that **AB** is orthogonal.

26. Suppose **A** is an orthogonal matrix. Is $\mathbf{A}^2$ orthogonal?

27. Suppose **A** is an orthogonal matrix. Then show that $\mathbf{A}^{-1}$ is orthogonal.

28. Suppose **A** is an orthogonal matrix. Then show that $\det \mathbf{A} = \pm 1$.

29. Suppose **A** is an orthogonal matrix such that $\mathbf{A}^2 = \mathbf{I}$. Then show that $\mathbf{A}^T = \mathbf{A}$.

30. Show that the rotation matrix

$$\mathbf{A} = \begin{pmatrix} \cos\theta & -\sin\theta \\ \sin\theta & \cos\theta \end{pmatrix}$$

is orthogonal.

8.11 Approximation of Eigenvalues

INTRODUCTION Recall, to find the eigenvalues for a matrix **A** we must find the roots of the polynomial equation $p(\lambda) = \det(\mathbf{A} - \lambda\mathbf{I}) = 0$. If **A** is a large matrix, the computations involved in obtaining this characteristic equation can be overwhelming. Moreover, even if we can find the exact characteristic equation it is likely that we would have to use a numerical procedure to approximate its roots. There are alternative numerical procedures for approximating eigenvalues and the corresponding eigenvectors. The procedure that we shall consider in this section deals with matrices that possess a **dominant eigenvalue**.

A Definition A dominant eigenvalue of a square matrix **A** is one whose absolute value is greater than the absolute value of each of the remaining eigenvalues. We formalize the last sentence in the next definition.

Let $\lambda_1, \lambda_2, \ldots, \lambda_k, \ldots, \lambda_n$ denote the eigenvalues of an $n \times n$ matrix $\mathbf{A}$. The eigenvalue λ_k is said to be the **dominant eigenvalue** of $\mathbf{A}$ if

$$|\lambda_k| > \lambda_i, \quad i = 1, 2, \ldots, n, \quad \text{but } i \neq k.$$

An eigenvector corresponding to λ_k is called the **dominant eigenvector** of $\mathbf{A}$.

In Example 2 of Section 8.8, we saw that the eigenvalues of the matrix

$$\mathbf{A} = \begin{pmatrix} 1 & 2 & 1 \\ 6 & -1 & 0 \\ -1 & -2 & -1 \end{pmatrix}$$

are $\lambda_1 = 0$, $\lambda_2 = -4$, and $\lambda_3 = 3$. Since $|-4| > 0$ and $|-4| > 3$, we see that $\lambda_2 = -4$ is the dominant eigenvalue of $\mathbf{A}$.

EXAMPLE 1 Matrices with No Dominant Eigenvalue

(a) The matrix $\mathbf{A} = \begin{pmatrix} 2 & 0 \\ 0 & -2 \end{pmatrix}$ has eigenvalues $\lambda_1 = -2$ and $\lambda_2 = 2$. Since $|\lambda_1| = |\lambda_2| = 2$, it follows that there is no dominant eigenvalue.

(b) The eigenvalues of the matrix

$$\mathbf{A} = \begin{pmatrix} 2 & 0 & 0 \\ 0 & 5 & 1 \\ 0 & 0 & 5 \end{pmatrix}$$

are $\lambda_1 = 2$, $\lambda_2 = \lambda_3 = 5$. Again, the matrix has no dominant eigenvalue. ≡

Power Method Suppose the $n \times n$ matrix $\mathbf{A}$ has a dominant eigenvalue λ_1. The iterative technique for approximating a corresponding dominant eigenvector is due to the German mathematician **Richard von Mises** (1883–1953) and is called the **power method**. The basic idea of this procedure is first to compute an approximation to a dominant eigenvector by means of the sequence

$$\mathbf{X}_i = \mathbf{A}\mathbf{X}_{i-1}, \quad i = 1, 2, 3, \ldots, \tag{1}$$

where $\mathbf{X}_0$ represents a nonzero $n \times 1$ vector that is an initial guess or approximation for the eigenvector we seek. Iterating (1) gives

$$\mathbf{X}_1 = \mathbf{A}\mathbf{X}_0$$
$$\mathbf{X}_2 = \mathbf{A}\mathbf{X}_1 = \mathbf{A}^2\mathbf{X}_0$$
$$\vdots \tag{2}$$
$$\mathbf{X}_m = \mathbf{A}\mathbf{X}_{m-1} = \mathbf{A}^m\mathbf{X}_0.$$

Under certain circumstances, for large values of m the vector defined by $\mathbf{X}_m = \mathbf{A}^m\mathbf{X}_0$ is an approximation to a dominant eigenvector. To see this, let us make some further assumptions about the matrix $\mathbf{A}$. Let us assume that the eigenvalues of $\mathbf{A}$ are such that

$$|\lambda_1| > |\lambda_2| \geq |\lambda_3| \geq \cdots \geq |\lambda_n|$$

and that the corresponding n eigenvectors $\mathbf{K}_1, \mathbf{K}_2, \ldots, \mathbf{K}_n$ are linearly independent. Because of this last assumption, $\mathbf{K}_1, \mathbf{K}_2, \ldots, \mathbf{K}_n$ can serve as a basis for R^n (see Section 7.6). Thus, for any nonzero $n \times 1$ vector $\mathbf{X}_0$, constants $c_1, c_2, \ldots, c_n$ can be found such that

$$\mathbf{X}_0 = c_1\mathbf{K}_1 + c_2\mathbf{K}_2 + \cdots + c_n\mathbf{K}_n. \tag{3}$$

We shall also assume $\mathbf{X}_0$ is chosen so that $c_1 \neq 0$. Multiplying (3) by $\mathbf{A}$ then gives

$$\mathbf{A}\mathbf{X}_0 = c_1\mathbf{A}\mathbf{K}_1 + c_2\mathbf{A}\mathbf{K}_2 + \cdots + c_n\mathbf{A}\mathbf{K}_n.$$

Since $\mathbf{A}\mathbf{K}_1 = \lambda_1\mathbf{K}_1, \mathbf{A}\mathbf{K}_2 = \lambda_2\mathbf{K}_2, \ldots, \mathbf{A}\mathbf{K}_n = \lambda_n\mathbf{K}_n$, the last line becomes

$$\mathbf{A}\mathbf{X}_0 = c_1\lambda_1\mathbf{K}_1 + c_2\lambda_2\mathbf{K}_2 + \cdots + c_n\lambda_n\mathbf{K}_n. \tag{4}$$

Multiplying (4) by $\mathbf{A}$ gives

$$\mathbf{A}^2\mathbf{X}_0 = c_1\lambda_1\mathbf{A}\mathbf{K}_1 + c_2\lambda_2\mathbf{A}\mathbf{K}_2 + \cdots + c_n\lambda_n\mathbf{A}\mathbf{K}_n$$
$$= c_1\lambda_1^2\mathbf{K}_1 + c_2\lambda_2^2\mathbf{K}_2 + \cdots + c_n\lambda_n^2\mathbf{K}_n.$$

Continuing in this manner, we find that

$$\mathbf{A}^m\mathbf{X}_0 = c_1\lambda_1^m\mathbf{K}_1 + c_2\lambda_2^m\mathbf{K}_2 + \cdots + c_n\lambda_n^m\mathbf{K}_n \tag{5}$$

$$= \lambda_1^m\left(c_1\mathbf{K}_1 + c_2\left(\frac{\lambda_2}{\lambda_1}\right)^m\mathbf{K}_2 + \cdots + c_n\left(\frac{\lambda_n}{\lambda_1}\right)^m\mathbf{K}_n\right). \tag{6}$$

Since $|\lambda_1| > |\lambda_i|$ for $i = 2, 3, \ldots, n$, we have $|\lambda_i/\lambda_1| < 1$ and consequently $\lim_{m\to\infty}(\lambda_i/\lambda_1)^m = 0$. Therefore, as $m \to \infty$, we see from (6) that

$$\mathbf{A}^m\mathbf{X}_0 \approx \lambda_1^m c_1\mathbf{K}_1. \tag{7}$$

Since a nonzero constant multiple of an eigenvector is an eigenvector, it follows from (7) that for large values of m and under all the assumptions that were made, the $n \times 1$ matrix $\mathbf{X}_m = \mathbf{A}^m\mathbf{X}_0$ is an approximation to a dominant eigenvector associated with the dominant eigenvalue λ_1. The rate at which this method converges depends on the quotient λ_2/λ_1: If $|\lambda_2/\lambda_1|$ is very small, then convergence is fast, whereas if $|\lambda_2/\lambda_1|$ is close to one, convergence is slow. Of course, this information is not as useful as it seems because we generally do not know the eigenvalues in advance.

It remains, then, to approximate the dominant eigenvalue itself. This can be done by means of the inner product. If $\mathbf{K}$ is an eigenvector of a matrix $\mathbf{A}$ corresponding to the eigenvalue λ, we have $\mathbf{A}\mathbf{K} = \lambda\mathbf{K}$, and so we have $\mathbf{A}\mathbf{K} \cdot \mathbf{K} = \lambda\mathbf{K} \cdot \mathbf{K}$. Since $\mathbf{A}\mathbf{K} \cdot \mathbf{K}$ and $\mathbf{K} \cdot \mathbf{K}$ are scalars, we can solve this last equation for λ:

$$\lambda = \frac{\mathbf{A}\mathbf{K} \cdot \mathbf{K}}{\mathbf{K} \cdot \mathbf{K}}.$$

Hence, if $\mathbf{X}_m = \mathbf{A}^m\mathbf{X}_0$ is an approximation to a dominant eigenvector obtained by the iteration of (1), then the dominant eigenvalue λ_1 can be approximated by the quotient

$$\lambda_1 \approx \frac{\mathbf{A}\mathbf{X}_m \cdot \mathbf{X}_m}{\mathbf{X}_m \cdot \mathbf{X}_m}. \tag{8}$$

The quotient in (8) is sometimes referred to as the **Rayleigh quotient**.

EXAMPLE 2 Using the Power Method

Use the power method to approximate the dominant eigenvalue and a corresponding dominant eigenvector of $\mathbf{A} = \begin{pmatrix} 4 & 2 \\ 3 & -1 \end{pmatrix}$.

SOLUTION Since we do not know the eigenvalues and eigenvectors, we may as well take $\mathbf{X}_0 = \begin{pmatrix} 1 \\ 1 \end{pmatrix}$. Now the first two terms of the sequence of vectors defined by (1) are

$$\mathbf{X}_1 = \mathbf{A}\mathbf{X}_0 = \begin{pmatrix} 4 & 2 \\ 3 & -1 \end{pmatrix} \begin{pmatrix} 1 \\ 1 \end{pmatrix} = \begin{pmatrix} 6 \\ 2 \end{pmatrix}$$

$$\mathbf{X}_2 = \mathbf{A}\mathbf{X}_1 = \begin{pmatrix} 4 & 2 \\ 3 & -1 \end{pmatrix} \begin{pmatrix} 6 \\ 2 \end{pmatrix} = \begin{pmatrix} 28 \\ 16 \end{pmatrix}.$$

The next five vectors obtained in this manner are given in the following table:

i	3	4	5	6	7
$\mathbf{X}_i$	$\begin{pmatrix} 144 \\ 68 \end{pmatrix}$	$\begin{pmatrix} 712 \\ 364 \end{pmatrix}$	$\begin{pmatrix} 3576 \\ 1772 \end{pmatrix}$	$\begin{pmatrix} 17{,}848 \\ 8956 \end{pmatrix}$	$\begin{pmatrix} 89{,}304 \\ 44{,}588 \end{pmatrix}$

At this point it may not appear that we are getting anywhere, since the entries of the vectors in the table appear to be growing without bound. But bear in mind that (7) indicates we are obtaining a constant multiple of a vector. If the power method converges, then by factoring the entry with the largest absolute value from $\mathbf{X}_m$ (for a large value of m), we should obtain a reasonable approximation to a dominant eigenvector. From the table,

$$\mathbf{X}_7 = 89{,}304 \begin{pmatrix} 1 \\ 0.4933 \end{pmatrix}. \tag{9}$$

It appears then that the vectors are approaching scalar multiples of $\begin{pmatrix} 1 \\ 0.5 \end{pmatrix}$.

We now use (8) to approximate the dominant eigenvalue λ_1. First we have

$$\mathbf{A}\mathbf{X}_7 = \begin{pmatrix} 4 & 2 \\ 3 & -1 \end{pmatrix} \begin{pmatrix} 1 \\ 0.4993 \end{pmatrix} = \begin{pmatrix} 4.9986 \\ 2.5007 \end{pmatrix}$$

$$\mathbf{A}\mathbf{X}_7 \cdot \mathbf{X}_7 = \begin{pmatrix} 4.9986 \\ 2.5007 \end{pmatrix}^T \begin{pmatrix} 1 \\ 0.4993 \end{pmatrix} = 6.2472$$

$$\mathbf{X}_7 \cdot \mathbf{X}_7 = \begin{pmatrix} 1 \\ 0.4993 \end{pmatrix}^T \begin{pmatrix} 1 \\ 0.4993 \end{pmatrix} = 1.2493.$$

Finally, we have

$$\lambda_1 \approx \frac{\mathbf{A}\mathbf{X}_7 \cdot \mathbf{X}_7}{\mathbf{X}_7 \cdot \mathbf{X}_7} = \frac{6.2472}{1.2493} = 5.0006.$$

The reader should use the procedure of Section 8.7 to verify that the eigenvalues and corresponding eigenvectors of $\mathbf{A}$ are $\lambda_1 = 5$, $\lambda_2 = -2$, $\mathbf{K}_1 = \begin{pmatrix} 1 \\ 0.5 \end{pmatrix}$, and $\mathbf{K}_2 = \begin{pmatrix} 1 \\ -3 \end{pmatrix}$. ≡

■ **Scaling** As we have just seen, iteration of (1) often results in vectors whose entries become very large in absolute value. Large numbers can, of course, cause a problem if a computer is used to carry out a great number of iterations. The result in (9) suggests that one way around this difficulty is to use a **scaled-down** vector at each step of the iteration. To do this, we simply multiply the vector $\mathbf{A}\mathbf{X}_0$ by the reciprocal of the entry with the largest absolute value. That is, we multiply

$$\mathbf{A}\mathbf{X}_0 = \begin{pmatrix} x_1 \\ \vdots \\ x_n \end{pmatrix} \quad \text{by} \quad \frac{1}{\max\{|x_1|, |x_2|, \ldots, |x_n|\}}.$$

This resulting matrix, whose entries are now less than or equal to one, we label $\mathbf{X}_1$. We repeat the process with the vector $\mathbf{A}\mathbf{X}_1$ to obtain the scaled-down vector $\mathbf{X}_2$, and so on.

EXAMPLE 3 **Example 2 Revisited**

Repeat the iterations of Example 2 using scaled-down vectors.

SOLUTION From $\mathbf{AX}_0 = \begin{pmatrix} 4 & 2 \\ 3 & -1 \end{pmatrix} \begin{pmatrix} 1 \\ 1 \end{pmatrix} = \begin{pmatrix} 6 \\ 2 \end{pmatrix}$ we define

$$\mathbf{X}_1 = \frac{1}{6} \begin{pmatrix} 6 \\ 2 \end{pmatrix} = \begin{pmatrix} 1 \\ 0.3333 \end{pmatrix}.$$

From $\mathbf{AX}_1 = \begin{pmatrix} 4 & 2 \\ 3 & -1 \end{pmatrix} \begin{pmatrix} 1 \\ 0.3333 \end{pmatrix} = \begin{pmatrix} 4.6666 \\ 2.6667 \end{pmatrix}$, we define

$$\mathbf{X}_2 = \frac{1}{4.6666} \begin{pmatrix} 4.6666 \\ 2.6667 \end{pmatrix} = \begin{pmatrix} 1 \\ 0.5714 \end{pmatrix}.$$

We continue in this manner to construct the following table:

i	3	4	5	6	7
$\mathbf{X}_i$	$\begin{pmatrix} 1 \\ 0.4722 \end{pmatrix}$	$\begin{pmatrix} 1 \\ 0.5112 \end{pmatrix}$	$\begin{pmatrix} 1 \\ 0.4955 \end{pmatrix}$	$\begin{pmatrix} 1 \\ 0.5018 \end{pmatrix}$	$\begin{pmatrix} 1 \\ 0.4993 \end{pmatrix}$

In contrast to the table in Example 2, it is apparent from this table that the vectors are approaching $\begin{pmatrix} 1 \\ 0.5 \end{pmatrix}$. ≡

Method of Deflation
After we have found the dominant eigenvalue λ_1 of a matrix $\mathbf{A}$ it may still be necessary to find nondominant eigenvalues. The procedure we shall consider next is a modification of the power method and is called the **method of deflation**. We will limit the discussion to the case where $\mathbf{A}$ is a *symmetric* matrix.

Suppose λ_1 and $\mathbf{K}_1$ are, respectively, the dominant eigenvalue and a corresponding *normalized* eigenvector* (that is, $\|\mathbf{K}_1\| = 1$) of a symmetric matrix $\mathbf{A}$. Furthermore, suppose the eigenvalues of $\mathbf{A}$ are such that

$$|\lambda_1| > |\lambda_2| > |\lambda_3| \geq \cdots \geq |\lambda_n|.$$

It can be proved that the matrix

$$\mathbf{B} = \mathbf{A} - \lambda_1 \mathbf{K}_1 \mathbf{K}_1^T \tag{10}$$

has eigenvalues $0, \lambda_2, \lambda_3, \ldots, \lambda_n$ and that eigenvectors of $\mathbf{B}$ are also eigenvectors of $\mathbf{A}$. Note that λ_2 is now the dominant eigenvalue of $\mathbf{B}$. We apply the power method to $\mathbf{B}$ to approximate λ_2 and a corresponding eigenvector.

EXAMPLE 4 **Using the Method of Deflation**

Use the method of deflation to approximate the eigenvalues of

$$\mathbf{A} = \begin{pmatrix} 1 & 2 & -1 \\ 2 & 1 & 1 \\ -1 & 1 & 0 \end{pmatrix}.$$

SOLUTION We begin by using the power method with scaling to find the dominant eigenvalue and a corresponding eigenvector of $\mathbf{A}$. Choosing $\mathbf{X}_0 = \begin{pmatrix} 1 \\ 1 \\ 1 \end{pmatrix}$, we see that

*See Example 3 in Section 8.10.

$$\mathbf{AX}_0 = \begin{pmatrix} 1 & 2 & -1 \\ 2 & 1 & 1 \\ -1 & 1 & 0 \end{pmatrix} \begin{pmatrix} 1 \\ 1 \\ 1 \end{pmatrix} = \begin{pmatrix} 2 \\ 4 \\ 0 \end{pmatrix} \qquad \text{so} \qquad \mathbf{X}_1 = \frac{1}{4}\begin{pmatrix} 2 \\ 4 \\ 0 \end{pmatrix} = \begin{pmatrix} 0.5 \\ 1 \\ 0 \end{pmatrix}$$

$$\mathbf{AX}_1 = \begin{pmatrix} 1 & 2 & -1 \\ 2 & 1 & 1 \\ -1 & 1 & 0 \end{pmatrix} \begin{pmatrix} 0.5 \\ 1 \\ 0 \end{pmatrix} = \begin{pmatrix} 2.5 \\ 2 \\ 0.5 \end{pmatrix} \qquad \text{so} \qquad \mathbf{X}_2 = \frac{1}{2.5}\begin{pmatrix} 2.5 \\ 2 \\ 0.5 \end{pmatrix} = \begin{pmatrix} 1 \\ 0.8 \\ 0.2 \end{pmatrix}.$$

The scaled vectors $\mathbf{X}_3$ to $\mathbf{X}_{10}$ are given in the following table:

i	3	4	5	6	7	8	9	10
$\mathbf{X}_i$	$\begin{pmatrix} 0.8 \\ 1 \\ -0.0667 \end{pmatrix}$	$\begin{pmatrix} 1 \\ 0.8837 \\ 0.0698 \end{pmatrix}$	$\begin{pmatrix} 0.9134 \\ 1 \\ -0.0394 \end{pmatrix}$	$\begin{pmatrix} 1 \\ 0.9440 \\ 0.0293 \end{pmatrix}$	$\begin{pmatrix} 0.9614 \\ 1 \\ -0.0188 \end{pmatrix}$	$\begin{pmatrix} 1 \\ 0.9744 \\ 0.0129 \end{pmatrix}$	$\begin{pmatrix} 0.9828 \\ 1 \\ -0.0086 \end{pmatrix}$	$\begin{pmatrix} 1 \\ 0.9885 \\ 0.0058 \end{pmatrix}$

Utilizing $\mathbf{X}_{10}$ and (8), we find

$$\lambda_1 \approx \frac{\mathbf{AX}_{10} \cdot \mathbf{X}_{10}}{\mathbf{X}_{10} \cdot \mathbf{X}_{10}} = 2.9997.$$

It appears that the dominant eigenvalue and a corresponding eigenvector are, respectively,
$\lambda_1 = 3$ and $\mathbf{K} = \begin{pmatrix} 1 \\ 1 \\ 0 \end{pmatrix}$.

Our next task is to construct the matrix $\mathbf{B}$ defined by (10). With $\|\mathbf{K}\| = \sqrt{2}$, the normalized eigenvector is $\mathbf{K}_1 = \begin{pmatrix} 1/\sqrt{2} \\ 1/\sqrt{2} \\ 0 \end{pmatrix}$. Thus,

$$\mathbf{B} = \begin{pmatrix} 1 & 2 & -1 \\ 2 & 1 & 1 \\ -1 & 1 & 0 \end{pmatrix} - 3\begin{pmatrix} 1/\sqrt{2} \\ 1/\sqrt{2} \\ 0 \end{pmatrix}(1/\sqrt{2} \quad 1/\sqrt{2} \quad 0)$$

$$= \begin{pmatrix} 1 & 2 & -1 \\ 2 & 1 & 1 \\ -1 & 1 & 0 \end{pmatrix} - \begin{pmatrix} \frac{3}{2} & \frac{3}{2} & 0 \\ \frac{3}{2} & \frac{3}{2} & 0 \\ 0 & 0 & 0 \end{pmatrix} = \begin{pmatrix} -0.5 & 0.5 & -1 \\ 0.5 & -0.5 & 1 \\ -1 & 1 & 0 \end{pmatrix}.$$

We now use the power method with scaling to find the dominant eigenvalue of $\mathbf{B}$. With
$\mathbf{X}_0 = \begin{pmatrix} 1 \\ 1 \\ 1 \end{pmatrix}$ again, the results are summarized in the following table:

i	1	2	3	4	5	6	7
$\mathbf{X}_i$	$\begin{pmatrix} -1 \\ 1 \\ 0 \end{pmatrix}$	$\begin{pmatrix} 0.5 \\ -0.5 \\ 1 \end{pmatrix}$	$\begin{pmatrix} -1 \\ 1 \\ -0.6667 \end{pmatrix}$	$\begin{pmatrix} 0.8333 \\ -0.8333 \\ 1 \end{pmatrix}$	$\begin{pmatrix} -1 \\ 1 \\ -0.9091 \end{pmatrix}$	$\begin{pmatrix} 0.9545 \\ -0.9545 \\ 1 \end{pmatrix}$	$\begin{pmatrix} -1 \\ 1 \\ -0.9767 \end{pmatrix}$

Utilizing $\mathbf{X}_7$ and (8), we find

$$\lambda_2 \approx \frac{\mathbf{AX}_7 \cdot \mathbf{X}_7}{\mathbf{X}_7 \cdot \mathbf{X}_7} = -1.9996.$$

From these computations it seems apparent that the dominant eigenvalue of $\mathbf{B}$ and a corresponding eigenvector are $\lambda_2 = -2$ and $\mathbf{K} = \begin{pmatrix} -1 \\ 1 \\ -1 \end{pmatrix}$.

To find the last eigenvalue of $\mathbf{A}$, we repeat the deflation process to find the dominant eigenvalue and a corresponding eigenvector of the matrix

$$\mathbf{C} = \mathbf{B} - \lambda_2 \mathbf{K}_2 \mathbf{K}_2^T = \begin{pmatrix} 0.1667 & -0.1667 & -0.3333 \\ -0.1667 & 0.1667 & 0.3333 \\ -0.3333 & 0.3333 & 0.6667 \end{pmatrix},$$

where we have used $\mathbf{K}_2 = \begin{pmatrix} -1/\sqrt{3} \\ 1/\sqrt{3} \\ -1/\sqrt{3} \end{pmatrix}$. The student is encouraged to verify that $\lambda_3 = 1$.

≡

Example 4 is somewhat artificial since eigenvalues of a matrix need not be "nice" numbers such as 3, -2, and 1. Moreover, we used the exact values of the dominant eigenvalues λ_1 and λ_2 in the formation of the matrices $\mathbf{B}$ and $\mathbf{C}$. In practice, of course, we may have to be content with working with approximations to the dominant eigenvalue λ_1 and a corresponding normalized dominant eigenvector $\mathbf{K}_1$ of $\mathbf{A}$. If these approximations are used in (10), an error is introduced in the computation of the matrix $\mathbf{B}$, and so more errors may be introduced in the computation of its dominant eigenvalue λ_2 and dominant eigenvector $\mathbf{K}_2$. If λ_2 and $\mathbf{K}_2$ are now used to construct the matrix $\mathbf{C}$, it seems reasonable to conclude that errors are being compounded. In other words, the method of deflation can become increasingly inaccurate as more eigenvalues are computed.

■ **Inverse Power Method** In some applied problems we are more interested in approximating the eigenvalue of a matrix $\mathbf{A}$ of smallest absolute value than the dominant eigenvalue. If $\mathbf{A}$ is nonsingular, then the eigenvalues of $\mathbf{A}$ are nonzero, and if $\lambda_1, \lambda_2, \ldots, \lambda_n$ are the eigenvalues of $\mathbf{A}$, then $1/\lambda_1, 1/\lambda_2, \ldots, 1/\lambda_n$ are the eigenvalues of $\mathbf{A}^{-1}$. Now if the eigenvalues of $\mathbf{A}$ can be arranged in the order

Review (6) and Theorems 8.8.2 and 8.8.3 on pages 430–431.

$$|\lambda_1| \geq |\lambda_2| \geq |\lambda_3| \geq \cdots \geq |\lambda_{n-1}| > |\lambda_n|,$$

then we see that $1/\lambda_n$ is the dominant eigenvalue of $\mathbf{A}^{-1}$. By applying the power method to $\mathbf{A}^{-1}$, we approximate the eigenvalue of largest magnitude and, by taking its reciprocal, we find the eigenvalue of $\mathbf{A}$ of least magnitude. This is called the **inverse power method**. See Problems 11–13 in Exercises 8.11.

8.11 Exercises Answers to selected odd-numbered problems begin on page ANS-20.

To the Instructor/Student: A calculator with matrix capabilities or a CAS would be useful in the following problems.

Each matrix in Problems 1–10 has a dominant eigenvalue.

In Problems 1 and 2, use the power method as illustrated in Example 3 to find the dominant eigenvalue and a corresponding dominant eigenvector of the given matrix.

1. $\begin{pmatrix} 1 & 1 \\ 2 & 0 \end{pmatrix}$

2. $\begin{pmatrix} -7 & 2 \\ 8 & -1 \end{pmatrix}$

In Problems 3–6, use the power method with scaling to find the dominant eigenvalue and a corresponding eigenvector of the given matrix.

3. $\begin{pmatrix} 2 & 4 \\ 3 & 13 \end{pmatrix}$

4. $\begin{pmatrix} -1 & 2 \\ -2 & 7 \end{pmatrix}$

5. $\begin{pmatrix} 5 & 4 & 2 \\ 4 & 5 & 2 \\ 2 & 2 & 2 \end{pmatrix}$

6. $\begin{pmatrix} 3 & 1 & 1 \\ 0 & 1 & 1 \\ 0 & 0 & 2 \end{pmatrix}$

In Problems 7–10, use the method of deflation to find the eigenvalues of the given matrix.

7. $\begin{pmatrix} 3 & 2 \\ 2 & 6 \end{pmatrix}$ **8.** $\begin{pmatrix} 1 & 3 \\ 3 & 9 \end{pmatrix}$

9. $\begin{pmatrix} 3 & -1 & 0 \\ -1 & 2 & -1 \\ 0 & -1 & 3 \end{pmatrix}$ **10.** $\begin{pmatrix} 0 & 0 & -4 \\ 0 & -4 & 0 \\ -4 & 0 & 15 \end{pmatrix}$

In Problems 11 and 12, use the inverse power method to find the eigenvalue of least magnitude for the given matrix.

11. $\begin{pmatrix} 1 & 1 \\ 3 & 4 \end{pmatrix}$ **12.** $\begin{pmatrix} -0.2 & 0.3 \\ 0.4 & -0.1 \end{pmatrix}$

13. In Example 4 of Section 3.9 we saw that the deflection curve of a thin column under an applied load P was defined by the boundary-value problem

$$EI \frac{d^2y}{dx^2} + Py = 0, \quad y(0) = 0, \quad y(L) = 0.$$

In this problem we show how to apply matrix techniques to compute the smallest critical load.

Let the interval $[0, L]$ be divided into n subintervals of length $h = L/n$, and let $x_i = ih$, $i = 0, 1, \ldots, n$. For small values of h it follows from (6) of Section 6.5 that

$$\frac{d^2y}{dx^2} \approx \frac{y_{i+1} - 2y_i + y_{i-1}}{h^2},$$

where $y_i = y(x_i)$.

(a) Show that the differential equation can be replaced by the difference equation

$$EI(y_{i+1} - 2y_i + y_{i-1}) + Ph^2 y_i = 0, \quad i = 1, 2, \ldots, n - 1.$$

(b) Show that for $n = 4$ the difference equation in part (a) yields the system of linear equations

$$\begin{pmatrix} 2 & -1 & 0 \\ -1 & 2 & -1 \\ 0 & -1 & 2 \end{pmatrix} \begin{pmatrix} y_1 \\ y_2 \\ y_3 \end{pmatrix} = \frac{PL^2}{16EI} \begin{pmatrix} y_1 \\ y_2 \\ y_3 \end{pmatrix}.$$

Note that this system has the form of the eigenvalue problem $\mathbf{AY} = \lambda \mathbf{Y}$, where $\lambda = PL^2/16EI$.

(c) Find $\mathbf{A}^{-1}$.

(d) Use the inverse power method to find, to two decimal places, the eigenvalue of $\mathbf{A}$ of least magnitude.

(e) Use the result of part (d) to compute the approximate smallest critical load. Compare your answer with that given in Section 3.9.

14. Suppose the column in Problem 13 is tapered so that the moment of inertia of a cross-section I varies linearly from $I(0) = I_0 = 0.002$ to $I(L) = I_L = 0.001$.

(a) Use the difference equation in part (a) of Problem 13 with $n = 4$ to set up a system of equations analogous to that given in part (b).

(b) Proceed as in Problem 13 to find an approximation to the smallest critical load.

Computer Lab Assignment

15. In Section 8.9 we saw how to compute a power $\mathbf{A}^m$ for an $n \times n$ matrix $\mathbf{A}$. Consult the documentation for the CAS you have on hand for the command to compute the power $\mathbf{A}^m$. (In *Mathematica* the command is **MatrixPower[A, m]**.) The matrix

$$\mathbf{A} = \begin{pmatrix} 5 & -2 & 0 \\ -2 & 3 & -1 \\ 0 & -1 & 1 \end{pmatrix}$$

possesses a dominant eigenvalue.

(a) Use a CAS to compute $\mathbf{A}^{10}$.

(b) Now use (2), $\mathbf{X}_m = \mathbf{A}^m \mathbf{X}_0$, with $m = 10$ and

$$\mathbf{X}_0 = \begin{pmatrix} 1 \\ 0 \\ 0 \end{pmatrix}, \text{ to compute } \mathbf{X}_{10}. \text{ In the same manner compute}$$

$\mathbf{X}_{12}$. Then proceed as in (9) to find the approximate dominant eigenvector $\mathbf{K}$.

(c) If $\mathbf{K}$ is an eigenvector of $\mathbf{A}$, then $\mathbf{AK} = \lambda \mathbf{K}$. Use this definition and the result in part (b) to find the dominant eigenvalue.

8.12 Diagonalization

INTRODUCTION In Chapter 10 we shall see that eigenvalues, eigenvectors, orthogonal matrices, and the topic of this present section, diagonalization, are important tools in the solution of systems of linear first-order differential equations. The basic question that we shall consider in this section is

For an $n \times n$ matrix $\mathbf{A}$, can we find an $n \times n$ nonsingular matrix $\mathbf{P}$ such that $\mathbf{P}^{-1}\mathbf{AP} = \mathbf{D}$ is a diagonal matrix?

■ **A Special Notation** We begin by introducing a shorthand notation for the product of two $n \times n$ matrices. This notation will be useful in proving the principal theorem of this section. To illustrate, suppose $\mathbf{A}$ and $\mathbf{B}$ are 2×2 matrices. Then

$$\mathbf{AB} = \begin{pmatrix} a_{11} & a_{12} \\ a_{21} & a_{22} \end{pmatrix} \begin{pmatrix} b_{11} & b_{12} \\ b_{21} & b_{22} \end{pmatrix} = \begin{pmatrix} a_{11}b_{11} + a_{12}b_{21} & a_{11}b_{12} + a_{12}b_{22} \\ a_{21}b_{11} + a_{22}b_{21} & a_{21}b_{12} + a_{22}b_{22} \end{pmatrix}. \tag{1}$$

<div style="text-align:center">column 1 column 2</div>

If we write the columns of the matrix $\mathbf{B}$ as vectors $\mathbf{X}_1 = \begin{pmatrix} b_{11} \\ b_{21} \end{pmatrix}$ and $\mathbf{X}_2 = \begin{pmatrix} b_{12} \\ b_{22} \end{pmatrix}$, then column 1 and column 2 in the product (1) can be expressed as the products $\mathbf{AX}_1$ and $\mathbf{AX}_2$. That is,

$$\mathbf{AB} = (\underset{\text{column 1}}{\mathbf{AX}_1} \quad \underset{\text{column 2}}{\mathbf{AX}_2}).$$

In general, for two $n \times n$ matrices

$$\mathbf{AB} = \mathbf{A}(\mathbf{X}_1 \, \mathbf{X}_2 \dots \mathbf{X}_n) = (\mathbf{AX}_1 \, \mathbf{AX}_2 \dots \mathbf{AX}_n), \tag{2}$$

where $\mathbf{X}_1, \mathbf{X}_2, \dots, \mathbf{X}_n$, are the columns of $\mathbf{B}$.

■ Diagonalizable Matrix

If an $n \times n$ nonsingular matrix $\mathbf{P}$ can be found so that $\mathbf{P}^{-1}\mathbf{AP} = \mathbf{D}$ is a diagonal matrix, then we say that the $n \times n$ matrix $\mathbf{A}$ can be **diagonalized**, or is **diagonalizable**, and that $\mathbf{P}$ **diagonalizes** $\mathbf{A}$.

To discover how to diagonalize a matrix, let us assume for the sake of discussion that $\mathbf{A}$ is a 3×3 diagonalizable matrix. Then there exists a 3×3 nonsingular matrix $\mathbf{P}$ such that $\mathbf{P}^{-1}\mathbf{AP} = \mathbf{D}$ or $\mathbf{AP} = \mathbf{PD}$, where $\mathbf{D}$ is a diagonal matrix

$$\mathbf{D} = \begin{pmatrix} d_{11} & 0 & 0 \\ 0 & d_{22} & 0 \\ 0 & 0 & d_{33} \end{pmatrix}.$$

If $\mathbf{P}_1$, $\mathbf{P}_2$, and $\mathbf{P}_3$ denote the columns of $\mathbf{P}$, then it follows from (2) that the equation $\mathbf{AP} = \mathbf{PD}$ is the same as

$$(\mathbf{AP}_1 \, \mathbf{AP}_2 \, \mathbf{AP}_3) = (d_{11}\mathbf{P}_1 \, d_{22}\mathbf{P}_2 \, d_{33}\mathbf{P}_3)$$

or

$$\mathbf{AP}_1 = d_{11}\mathbf{P}_1, \quad \mathbf{AP}_2 = d_{22}\mathbf{P}_2, \quad \mathbf{AP}_3 = d_{33}\mathbf{P}_3.$$

But by Definition 8.8.1 we see that d_{11}, d_{22}, and d_{33} are eigenvalues of $\mathbf{A}$ associated with the eigenvectors $\mathbf{P}_1$, $\mathbf{P}_2$, and $\mathbf{P}_3$. These eigenvectors are linearly independent, since $\mathbf{P}$ was assumed to be nonsingular.

We have just discovered, in a particular case, that if $\mathbf{A}$ is diagonalizable, then the columns of the diagonalizing matrix $\mathbf{P}$ consist of linearly independent eigenvectors of $\mathbf{A}$. Since we wish to diagonalize a matrix, we are really concerned with the validity of the converse of the last sentence. In other words, if we can find n linearly independent eigenvectors of an $n \times n$ matrix $\mathbf{A}$ and form an $n \times n$ matrix $\mathbf{P}$ whose columns consist of these eigenvectors, then does $\mathbf{P}$ diagonalize $\mathbf{A}$? The answer is yes and will be proved in the next theorem.

THEOREM 8.12.1 Sufficient Condition for Diagonalizability

If an $n \times n$ matrix $\mathbf{A}$ has n linearly independent eigenvectors $\mathbf{K}_1, \mathbf{K}_2, \dots, \mathbf{K}_n$, then $\mathbf{A}$ is diagonalizable.

PROOF: We shall prove the theorem in the case when $\mathbf{A}$ is a 3×3 matrix. Let $\mathbf{K}_1$, $\mathbf{K}_2$, and $\mathbf{K}_3$ be linearly independent eigenvectors corresponding to eigenvalues λ_1, λ_2, and λ_3; that is,

$$\mathbf{AK}_1 = \lambda_1\mathbf{K}_1, \quad \mathbf{AK}_2 = \lambda_2\mathbf{K}_2, \quad \text{and} \quad \mathbf{AK}_3 = \lambda_3\mathbf{K}_3. \tag{3}$$

Next form the 3×3 matrix $\mathbf{P}$ with column vectors $\mathbf{K}_1$, $\mathbf{K}_2$, and $\mathbf{K}_3$: $\mathbf{P} = (\mathbf{K}_1 \, \mathbf{K}_2 \, \mathbf{K}_3)$. $\mathbf{P}$ is nonsingular since, by hypothesis, the eigenvectors are linearly independent. Now using (2) and (3),

we can write the product **AP** as

$$\mathbf{AP} = (\mathbf{AK}_1\ \mathbf{AK}_2\ \mathbf{AK}_3) = (\lambda_1\mathbf{K}_1\ \lambda_2\mathbf{K}_2\ \lambda_3\mathbf{K}_3)$$

$$= (\mathbf{K}_1\ \mathbf{K}_2\ \mathbf{K}_3)\begin{pmatrix} \lambda_1 & 0 & 0 \\ 0 & \lambda_2 & 0 \\ 0 & 0 & \lambda_3 \end{pmatrix} = \mathbf{PD}.$$

Multiplying the last equation on the left by $\mathbf{P}^{-1}$ then gives $\mathbf{P}^{-1}\mathbf{AP} = \mathbf{D}$. ≡

Note carefully in the proof of Theorem 8.12.1 that the entries in the diagonalized matrix are the eigenvalues of **A** and the order in which these numbers appear on the diagonal of **D** corresponds to the order in which the eigenvectors are used as columns in the matrix **P**.

In view of the motivational discussion preceding Theorem 8.12.1, we can state the general result:

THEOREM 8.12.2 Criterion for Diagonalizability

An $n \times n$ matrix **A** is diagonalizable if and only if **A** has n linearly independent eigenvectors.

We saw in Section 8.8 that an $n \times n$ matrix **A** has n linearly independent eigenvectors whenever it possesses n distinct eigenvalues.

THEOREM 8.12.3 Sufficient Condition for Diagonalizability

If an $n \times n$ matrix **A** has n distinct eigenvalues, it is diagonalizable.

EXAMPLE 1 **Diagonalizing a Matrix**

Diagonalize $\mathbf{A} = \begin{pmatrix} -5 & 9 \\ -6 & 10 \end{pmatrix}$ if possible.

SOLUTION First we find the eigenvalues of **A**. The characteristic equation is $\det(\mathbf{A} - \lambda\mathbf{I}) = \begin{vmatrix} -5 - \lambda & 9 \\ -6 & 10 - \lambda \end{vmatrix} = \lambda^2 - 5\lambda + 4 = (\lambda - 1)(\lambda - 4) = 0$. The eigenvalues are $\lambda_1 = 1$ and $\lambda_2 = 4$. Since the eigenvalues are distinct, we know from Theorem 8.12.3 that **A** is diagonalizable.

Next the eigenvectors of **A** corresponding to $\lambda_1 = 1$ and $\lambda_2 = 4$ are, respectively,

$$\mathbf{K}_1 = \begin{pmatrix} 3 \\ 2 \end{pmatrix} \quad \text{and} \quad \mathbf{K}_2 = \begin{pmatrix} 1 \\ 1 \end{pmatrix}.$$

Using these vectors as columns, we find that the nonsingular matrix **P** that diagonalizes **A** is

$$\mathbf{P} = (\mathbf{K}_1\ \mathbf{K}_2) = \begin{pmatrix} 3 & 1 \\ 2 & 1 \end{pmatrix}.$$

Now

$$\mathbf{P}^{-1} = \begin{pmatrix} 1 & -1 \\ -2 & 3 \end{pmatrix},$$

and so carrying out the multiplication gives

$$\mathbf{P}^{-1}\mathbf{AP} = \begin{pmatrix} 1 & -1 \\ -2 & 3 \end{pmatrix}\begin{pmatrix} -5 & 9 \\ -6 & 10 \end{pmatrix}\begin{pmatrix} 3 & 1 \\ 2 & 1 \end{pmatrix} = \begin{pmatrix} 1 & 0 \\ 0 & 4 \end{pmatrix} = \mathbf{D}. \quad ≡$$

In Example 1, had we reversed the columns in **P**, that is, $\mathbf{P} = \begin{pmatrix} 1 & 3 \\ 1 & 2 \end{pmatrix}$, then the diagonal matrix would have been $\mathbf{D} = \begin{pmatrix} 4 & 0 \\ 0 & 1 \end{pmatrix}$.

EXAMPLE 2 Diagonalizing a Matrix

Consider the matrix $\mathbf{A} = \begin{pmatrix} 1 & 2 & 1 \\ 6 & -1 & 0 \\ -1 & -2 & -1 \end{pmatrix}$. We saw in Example 2 of Section 8.8 that the eigenvalues and corresponding eigenvectors are

$$\lambda_1 = 0, \quad \lambda_2 = -4, \quad \lambda_3 = 3, \quad \mathbf{K}_1 = \begin{pmatrix} 1 \\ 6 \\ -13 \end{pmatrix}, \quad \mathbf{K}_2 = \begin{pmatrix} -1 \\ 2 \\ 1 \end{pmatrix}, \quad \mathbf{K}_3 = \begin{pmatrix} 2 \\ 3 \\ -2 \end{pmatrix}.$$

Since the eigenvalues are distinct, $\mathbf{A}$ is diagonalizable. We form the matrix

$$\mathbf{P} = (\mathbf{K}_1 \, \mathbf{K}_2 \, \mathbf{K}_3) = \begin{pmatrix} 1 & -1 & 2 \\ 6 & 2 & 3 \\ -13 & 1 & -2 \end{pmatrix}.$$

Matching the eigenvalues with the order in which the eigenvectors appear in $\mathbf{P}$, we know that the diagonal matrix will be

$$\mathbf{D} = \begin{pmatrix} 0 & 0 & 0 \\ 0 & -4 & 0 \\ 0 & 0 & 3 \end{pmatrix}.$$

Now from either of the methods of Section 8.6 we find

$$\mathbf{P}^{-1} = \begin{pmatrix} -\frac{1}{12} & 0 & -\frac{1}{12} \\ -\frac{9}{28} & \frac{2}{7} & \frac{3}{28} \\ \frac{8}{21} & \frac{1}{7} & \frac{2}{21} \end{pmatrix},$$

and so

$$\mathbf{P}^{-1}\mathbf{A}\mathbf{P} = \begin{pmatrix} -\frac{1}{12} & 0 & -\frac{1}{12} \\ -\frac{9}{28} & \frac{2}{7} & \frac{3}{28} \\ \frac{8}{21} & \frac{1}{7} & \frac{2}{21} \end{pmatrix} \begin{pmatrix} 1 & 2 & 1 \\ 6 & -1 & 0 \\ -1 & -2 & -1 \end{pmatrix} \begin{pmatrix} 1 & -1 & 2 \\ 6 & 2 & 3 \\ -13 & 1 & -2 \end{pmatrix}$$

$$= \begin{pmatrix} 0 & 0 & 0 \\ 0 & -4 & 0 \\ 0 & 0 & 3 \end{pmatrix} = \mathbf{D}. \qquad \equiv$$

A matrix with repeated eigenvalues ▶ could be diagonalizable.

The condition that an $n \times n$ matrix $\mathbf{A}$ have n distinct eigenvalues is sufficient—that is, a guarantee—that $\mathbf{A}$ is diagonalizable. The condition that there be n distinct eigenvalues is not a necessary condition for the diagonalization of $\mathbf{A}$. In other words, if the matrix $\mathbf{A}$ does not have n distinct eigenvalues, then it may or may not be diagonalizable.

EXAMPLE 3 A Matrix That Is Not Diagonalizable

In Example 3 of Section 8.8 we saw that the matrix $\mathbf{A} = \begin{pmatrix} 3 & 4 \\ -1 & 7 \end{pmatrix}$ has a repeated eigenvalue $\lambda_1 = \lambda_2 = 5$. Correspondingly we were able to find only a *single* eigenvector $\mathbf{K}_1 = \begin{pmatrix} 2 \\ 1 \end{pmatrix}$. We conclude from Theorem 8.12.2 that $\mathbf{A}$ is not diagonalizable. $\qquad \equiv$

EXAMPLE 4 Repeated Eigenvalues Yet Diagonalizable

The eigenvalues of the matrix $\mathbf{A} = \begin{pmatrix} 0 & 1 & 0 \\ 1 & 0 & 0 \\ 0 & 0 & 1 \end{pmatrix}$ are $\lambda_1 = -1$ and $\lambda_2 = \lambda_3 = 1$.

For $\lambda_1 = -1$ we find $\mathbf{K}_1 = \begin{pmatrix} 1 \\ -1 \\ 0 \end{pmatrix}$. For the repeated eigenvalue $\lambda_2 = \lambda_3 = 1$, Gauss–Jordan elimination gives

$$(\mathbf{A} - \mathbf{I}|\mathbf{0}) = \begin{pmatrix} -1 & 1 & 0 \\ 1 & -1 & 0 \\ 0 & 0 & 0 \end{pmatrix} \xrightarrow[\text{operations}]{\text{row}} \begin{pmatrix} 1 & -1 & 0 \\ 0 & 0 & 0 \\ 0 & 0 & 0 \end{pmatrix}.$$

From the last matrix we see that $k_1 - k_2 = 0$. Since k_3 is not determined from the last matrix, we can choose its value arbitrarily. The choice $k_2 = 1$ gives $k_1 = 1$. If we then pick $k_3 = 0$, we get the eigenvector

$$\mathbf{K}_2 = \begin{pmatrix} 1 \\ 1 \\ 0 \end{pmatrix}.$$

The alternative choice $k_2 = 0$ gives $k_1 = 0$. If $k_3 = 1$, we get another eigenvector corresponding to $\lambda_2 = \lambda_3 = 1$:

$$\mathbf{K}_3 = \begin{pmatrix} 0 \\ 0 \\ 1 \end{pmatrix}.$$

Since the eigenvectors $\mathbf{K}_1$, $\mathbf{K}_2$, and $\mathbf{K}_3$ are linearly independent, a matrix that diagonalizes $\mathbf{A}$ is

$$\mathbf{P} = \begin{pmatrix} 1 & 1 & 0 \\ -1 & 1 & 0 \\ 0 & 0 & 1 \end{pmatrix}.$$

Matching the eigenvalues with the eigenvectors in $\mathbf{P}$, we have $\mathbf{P}^{-1}\mathbf{A}\mathbf{P} = \mathbf{D}$, where

$$\mathbf{D} = \begin{pmatrix} -1 & 0 & 0 \\ 0 & 1 & 0 \\ 0 & 0 & 1 \end{pmatrix}. \qquad \equiv$$

■ **Symmetric Matrices** An $n \times n$ symmetric matrix $\mathbf{A}$ with real entries can always be diagonalized. This is a consequence of the fact that we can always find n linearly independent eigenvectors for such a matrix. Moreover, since we can find n mutually orthogonal eigenvectors, we can use an orthogonal matrix $\mathbf{P}$ to diagonalize $\mathbf{A}$. A symmetric matrix is said to be **orthogonally diagonalizable**.

THEOREM 8.12.4 **Criterion for Orthogonal Diagonalizability**

An $n \times n$ matrix $\mathbf{A}$ can be orthogonally diagonalized if and only if $\mathbf{A}$ is symmetric.

PARTIAL PROOF: We shall prove the necessity part (that is, the "only if" part) of the theorem. Assume an $n \times n$ matrix $\mathbf{A}$ is orthogonally diagonalizable. Then there exists an orthogonal matrix $\mathbf{P}$ such that $\mathbf{P}^{-1}\mathbf{A}\mathbf{P} = \mathbf{D}$ or $\mathbf{A} = \mathbf{P}\mathbf{D}\mathbf{P}^{-1}$. Since $\mathbf{P}$ is orthogonal, $\mathbf{P}^{-1} = \mathbf{P}^T$ and consequently $\mathbf{A} = \mathbf{P}\mathbf{D}\mathbf{P}^T$. But from (*i*) and (*iii*) of Theorem 8.1.2 and the fact that a diagonal matrix is symmetric, we have

$$\mathbf{A}^T = (\mathbf{P}\mathbf{D}\mathbf{P}^T)^T = (\mathbf{P}^T)^T\mathbf{D}^T\mathbf{P}^T = \mathbf{P}\mathbf{D}\mathbf{P}^T = \mathbf{A}.$$

Thus, $\mathbf{A}$ is symmetric. $\equiv$

EXAMPLE 5 **Diagonalizing a Symmetric Matrix**

Consider the symmetric matrix $\mathbf{A} = \begin{pmatrix} 9 & 1 & 1 \\ 1 & 9 & 1 \\ 1 & 1 & 9 \end{pmatrix}$. We saw in Example 4 of Section 8.8 that the eigenvalues and corresponding eigenvectors are

$$\lambda_1 = 11, \quad \lambda_2 = \lambda_3 = 8, \quad \mathbf{K}_1 = \begin{pmatrix} 1 \\ 1 \\ 1 \end{pmatrix}, \quad \mathbf{K}_2 = \begin{pmatrix} -1 \\ 1 \\ 0 \end{pmatrix}, \quad \mathbf{K}_3 = \begin{pmatrix} -1 \\ 0 \\ 1 \end{pmatrix}.$$

See the *Remarks* on page 458. ▶

The eigenvectors $\mathbf{K}_1$, $\mathbf{K}_2$, and $\mathbf{K}_3$ are linearly independent, but note that they are not mutually orthogonal since $\mathbf{K}_2$ and $\mathbf{K}_3$, the eigenvectors corresponding to the repeated eigenvalue $\lambda_2 = \lambda_3 = 8$, are not orthogonal. For $\lambda_2 = \lambda_3 = 8$, we found the eigenvectors from Gauss–Jordan elimination

$$(\mathbf{A} - 8\mathbf{I}\,|\,\mathbf{0}) = \begin{pmatrix} 1 & 1 & 1 \\ 1 & 1 & 1 \\ 1 & 1 & 1 \end{pmatrix} \overset{\text{row}}{\underset{\text{operations}}{\Rightarrow}} \begin{pmatrix} 1 & 1 & 1 \\ 0 & 0 & 0 \\ 0 & 0 & 0 \end{pmatrix},$$

which implies that $k_1 + k_2 + k_3 = 0$. Since two of the variables are arbitrary, we selected $k_2 = 1, k_3 = 0$ to obtain $\mathbf{K}_2$, and $k_2 = 0, k_3 = 1$ to obtain $\mathbf{K}_3$. Now if instead we choose $k_2 = 1$, $k_3 = 1$ and then $k_2 = 1, k_3 = -1$, we obtain, respectively, two entirely different but orthogonal eigenvectors:

$$\mathbf{K}_2 = \begin{pmatrix} -2 \\ 1 \\ 1 \end{pmatrix} \quad \text{and} \quad \mathbf{K}_3 = \begin{pmatrix} 0 \\ 1 \\ -1 \end{pmatrix}.$$

Thus, a new set of mutually orthogonal eigenvectors is

$$\mathbf{K}_1 = \begin{pmatrix} 1 \\ 1 \\ 1 \end{pmatrix}, \quad \mathbf{K}_2 = \begin{pmatrix} -2 \\ 1 \\ 1 \end{pmatrix}, \quad \mathbf{K}_3 = \begin{pmatrix} 0 \\ 1 \\ -1 \end{pmatrix}.$$

Multiplying these vectors, in turn, by the reciprocals of the norms $\|\mathbf{K}_1\| = \sqrt{3}$, $\|\mathbf{K}_2\| = \sqrt{6}$, and $\|\mathbf{K}_3\| = \sqrt{2}$, we obtain an orthonormal set

$$\begin{pmatrix} \dfrac{1}{\sqrt{3}} \\ \dfrac{1}{\sqrt{3}} \\ \dfrac{1}{\sqrt{3}} \end{pmatrix}, \begin{pmatrix} -\dfrac{2}{\sqrt{6}} \\ \dfrac{1}{\sqrt{6}} \\ \dfrac{1}{\sqrt{6}} \end{pmatrix}, \begin{pmatrix} 0 \\ \dfrac{1}{\sqrt{2}} \\ -\dfrac{1}{\sqrt{2}} \end{pmatrix}.$$

We then use these vectors as columns to construct an orthogonal matrix that diagonalizes $\mathbf{A}$:

$$\mathbf{P} = \begin{pmatrix} \dfrac{1}{\sqrt{3}} & -\dfrac{2}{\sqrt{6}} & 0 \\ \dfrac{1}{\sqrt{3}} & \dfrac{1}{\sqrt{6}} & \dfrac{1}{\sqrt{2}} \\ \dfrac{1}{\sqrt{3}} & \dfrac{1}{\sqrt{6}} & -\dfrac{1}{\sqrt{2}} \end{pmatrix}.$$

The diagonal matrix whose entries are the eigenvalues of $\mathbf{A}$ corresponding to the order in which the eigenvectors appear in $\mathbf{P}$ is then

$$\mathbf{D} = \begin{pmatrix} 11 & 0 & 0 \\ 0 & 8 & 0 \\ 0 & 0 & 8 \end{pmatrix}.$$

This is verified from

$$
\mathbf{P}^{-1}\mathbf{AP} = \mathbf{P}^T\mathbf{AP} = \begin{pmatrix} \dfrac{1}{\sqrt{3}} & \dfrac{1}{\sqrt{3}} & \dfrac{1}{\sqrt{3}} \\ -\dfrac{2}{\sqrt{6}} & \dfrac{1}{\sqrt{6}} & \dfrac{1}{\sqrt{6}} \\ 0 & \dfrac{1}{\sqrt{2}} & -\dfrac{1}{\sqrt{2}} \end{pmatrix} \begin{pmatrix} 9 & 1 & 1 \\ 1 & 9 & 1 \\ 1 & 1 & 9 \end{pmatrix} \begin{pmatrix} \dfrac{1}{\sqrt{3}} & -\dfrac{2}{\sqrt{6}} & 0 \\ \dfrac{1}{\sqrt{3}} & \dfrac{1}{\sqrt{6}} & \dfrac{1}{\sqrt{2}} \\ \dfrac{1}{\sqrt{3}} & \dfrac{1}{\sqrt{6}} & -\dfrac{1}{\sqrt{2}} \end{pmatrix}
$$

$$
= \begin{pmatrix} 11 & 0 & 0 \\ 0 & 8 & 0 \\ 0 & 0 & 8 \end{pmatrix} = \mathbf{D}. \qquad\qquad \equiv
$$

■ **Quadratic Forms** An algebraic expression of the form

$$ ax^2 + bxy + cy^2 \qquad\qquad (4) $$

is said to be a **quadratic form**. If we let $\mathbf{X} = \begin{pmatrix} x \\ y \end{pmatrix}$, then (4) can be written as the matrix product

$$ \mathbf{X}^T\mathbf{AX} = (x \ \ y)\begin{pmatrix} a & \frac{1}{2}b \\ \frac{1}{2}b & c \end{pmatrix}\begin{pmatrix} x \\ y \end{pmatrix}. \qquad\qquad (5) $$

Observe that the matrix $\mathbf{A} = \begin{pmatrix} a & \frac{1}{2}b \\ \frac{1}{2}b & c \end{pmatrix}$ is symmetric.

In calculus you may have seen that an appropriate rotation of axes enables us to eliminate the xy-term in an equation:

$$ ax^2 + bxy + cy^2 + dx + ey + f = 0. $$

As the next example will illustrate, we can eliminate the xy-term by means of an orthogonal matrix and diagonalization rather than by using trigonometry.

| EXAMPLE 6 | **Identifying a Conic Section** |

Identify the conic section whose equation is $2x^2 + 4xy - y^2 = 1$.

SOLUTION From (5) we can write the given equation as

$$ (x \ \ y)\begin{pmatrix} 2 & 2 \\ 2 & -1 \end{pmatrix}\begin{pmatrix} x \\ y \end{pmatrix} = 1 \quad \text{or} \quad \mathbf{X}^T\mathbf{AX} = 1, \qquad\qquad (6) $$

where $\mathbf{A} = \begin{pmatrix} 2 & 2 \\ 2 & -1 \end{pmatrix}$ and $\mathbf{X} = \begin{pmatrix} x \\ y \end{pmatrix}$. Now the eigenvalues and corresponding eigenvectors of $\mathbf{A}$ are found to be

$$ \lambda_1 = -2, \quad \lambda_2 = 3, \quad \mathbf{K}_1 = \begin{pmatrix} 1 \\ -2 \end{pmatrix}, \quad \mathbf{K}_2 = \begin{pmatrix} 2 \\ 1 \end{pmatrix}. $$

Observe that $\mathbf{K}_1$ and $\mathbf{K}_2$ are orthogonal. Moreover, $\|\mathbf{K}_1\| = \|\mathbf{K}_2\| = \sqrt{5}$, and so the vectors

$$ \begin{pmatrix} \dfrac{1}{\sqrt{5}} \\ -\dfrac{2}{\sqrt{5}} \end{pmatrix} \quad \text{and} \quad \begin{pmatrix} \dfrac{2}{\sqrt{5}} \\ \dfrac{1}{\sqrt{5}} \end{pmatrix} $$

are orthonormal. Hence, the matrix

$$\mathbf{P} = \begin{pmatrix} \dfrac{1}{\sqrt{5}} & \dfrac{2}{\sqrt{5}} \\ -\dfrac{2}{\sqrt{5}} & \dfrac{1}{\sqrt{5}} \end{pmatrix}$$

is orthogonal. If we define the change of variables $\mathbf{X} = \mathbf{PX}'$ where $\mathbf{X}' = \begin{pmatrix} X \\ Y \end{pmatrix}$, then the quadratic form $2x^2 + 4xy - y^2$ can be written

$$\mathbf{X}^T\mathbf{AX} = (\mathbf{X}')^T\mathbf{P}^T\mathbf{APX}' = (\mathbf{X}')^T(\mathbf{P}^T\mathbf{AP})\mathbf{X}'.$$

Since $\mathbf{P}$ orthogonally diagonalizes the symmetric matrix $\mathbf{A}$, the last equation is the same as

$$\mathbf{X}^T\mathbf{AX} = (\mathbf{X}')^T\mathbf{DX}'. \tag{7}$$

Using (7), we see that (6) becomes

$$(X \ \ Y)\begin{pmatrix} -2 & 0 \\ 0 & 3 \end{pmatrix}\begin{pmatrix} X \\ Y \end{pmatrix} = 1 \quad \text{or} \quad -2X^2 + 3Y^2 = 1.$$

This last equation is recognized as the standard form of a hyperbola. The xy-coordinates of the eigenvectors are $(1, -2)$ and $(2, 1)$. Using the substitution $\mathbf{X} = \mathbf{PX}'$ in the form $\mathbf{X}' = \mathbf{P}^{-1}\mathbf{X} = \mathbf{P}^T\mathbf{X}$, we find that the XY-coordinates of these two points are $(\sqrt{5}, 0)$ and $(0, \sqrt{5})$, respectively. From this we conclude that the X-axis and Y-axis are as shown in **FIGURE 8.12.1**. The eigenvectors, shown in red in the figure, lie along the new axes. The X- and Y-axes are called the **principal axes** of the conic. ≡

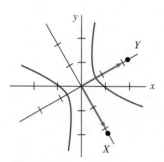

FIGURE 8.12.1 X- and Y-axes in Example 6

REMARKS

The matrix $\mathbf{A}$ in Example 5 is symmetric and as such eigenvectors corresponding to distinct eigenvalues are orthogonal. In the third line of the example, note that $\mathbf{K}_1$, an eigenvector for $\lambda_1 = 11$, is orthogonal to both $\mathbf{K}_2$ and $\mathbf{K}_3$. The eigenvectors $\mathbf{K}_2 = \begin{pmatrix} -1 \\ 1 \\ 0 \end{pmatrix}$ and $\mathbf{K}_3 = \begin{pmatrix} -1 \\ 0 \\ 1 \end{pmatrix}$ corresponding to $\lambda_2 = \lambda_3 = 8$ are not orthogonal. As an alternative to searching for orthogonal eigenvectors for this repeated eigenvalue by performing Gauss–Jordan elimination a second time, we could simply apply the Gram–Schmidt orthogonalization process and transform the set $\{\mathbf{K}_2, \mathbf{K}_3\}$ into an orthogonal set. See Section 7.7 and Example 4 in Section 8.10.

8.12 Exercises Answers to selected odd-numbered problems begin on page ANS-20.

In Problems 1–20, determine whether the given matrix $\mathbf{A}$ is diagonalizable. If so, find the matrix $\mathbf{P}$ that diagonalizes $\mathbf{A}$ and the diagonal matrix $\mathbf{D}$ such that $\mathbf{D} = \mathbf{P}^{-1}\mathbf{AP}$.

1. $\begin{pmatrix} 2 & 3 \\ 1 & 4 \end{pmatrix}$

2. $\begin{pmatrix} -4 & -5 \\ 8 & 10 \end{pmatrix}$

3. $\begin{pmatrix} 0 & 1 \\ -1 & 2 \end{pmatrix}$

4. $\begin{pmatrix} 0 & 5 \\ 1 & 0 \end{pmatrix}$

5. $\begin{pmatrix} -9 & 13 \\ -2 & 6 \end{pmatrix}$

6. $\begin{pmatrix} -5 & -3 \\ 5 & 11 \end{pmatrix}$

7. $\begin{pmatrix} \frac{1}{2} & \frac{1}{6} \\ \frac{1}{6} & \frac{1}{2} \end{pmatrix}$

8. $\begin{pmatrix} -2 & -1 \\ 1 & -4 \end{pmatrix}$

9. $\begin{pmatrix} 0 & 1 \\ -1 & 0 \end{pmatrix}$

10. $\begin{pmatrix} 1 & 2 \\ -\frac{1}{2} & 1 \end{pmatrix}$

11. $\begin{pmatrix} 1 & 0 & 1 \\ 0 & -1 & 3 \\ 0 & 0 & 2 \end{pmatrix}$

12. $\begin{pmatrix} 1 & 2 & 2 \\ 2 & 3 & -2 \\ -5 & 3 & 8 \end{pmatrix}$

13. $\begin{pmatrix} 1 & -1 & 1 \\ 0 & 1 & 0 \\ 1 & -1 & 1 \end{pmatrix}$

14. $\begin{pmatrix} 0 & -9 & 0 \\ 1 & 0 & 0 \\ 0 & 0 & 1 \end{pmatrix}$

15. $\begin{pmatrix} 1 & 3 & -1 \\ 0 & 2 & 4 \\ 0 & 0 & 1 \end{pmatrix}$

16. $\begin{pmatrix} 1 & 1 & 0 \\ 0 & 2 & 0 \\ 0 & 0 & 3 \end{pmatrix}$

17. $\begin{pmatrix} 1 & 2 & 0 \\ 2 & -1 & 0 \\ 0 & 0 & 1 \end{pmatrix}$ **18.** $\begin{pmatrix} 0 & 0 & 1 \\ 1 & 0 & -3 \\ 0 & 1 & 3 \end{pmatrix}$

19. $\begin{pmatrix} -8 & -10 & 7 & -9 \\ 0 & 2 & 0 & 0 \\ -9 & -9 & 8 & -9 \\ 1 & 1 & -1 & 2 \end{pmatrix}$

20. $\begin{pmatrix} 4 & 2 & -1 & 4 \\ 0 & 2 & 0 & 0 \\ 1 & 3 & 2 & 1 \\ 0 & 0 & 0 & 2 \end{pmatrix}$

In Problems 21–30, the given matrix **A** is symmetric. Find an orthogonal matrix **P** that diagonalizes **A** and the diagonal matrix **D** such that $\mathbf{D} = \mathbf{P}^T\mathbf{AP}$.

21. $\begin{pmatrix} 1 & 1 \\ 1 & 1 \end{pmatrix}$ **22.** $\begin{pmatrix} 3 & 2 \\ 2 & 0 \end{pmatrix}$

23. $\begin{pmatrix} 5 & \sqrt{10} \\ \sqrt{10} & 8 \end{pmatrix}$ **24.** $\begin{pmatrix} 1 & -2 \\ -2 & 1 \end{pmatrix}$

25. $\begin{pmatrix} 0 & 1 & 0 \\ 1 & 0 & 0 \\ 0 & 0 & 1 \end{pmatrix}$ **26.** $\begin{pmatrix} 1 & -2 & 2 \\ -2 & 1 & -2 \\ 2 & -2 & 1 \end{pmatrix}$

27. $\begin{pmatrix} 5 & -2 & 0 \\ -2 & 6 & -2 \\ 0 & -2 & 7 \end{pmatrix}$ **28.** $\begin{pmatrix} 3 & 0 & 1 \\ 0 & 1 & 0 \\ 1 & 0 & 1 \end{pmatrix}$

29. $\begin{pmatrix} 1 & 0 & 7 \\ 0 & 1 & 0 \\ 7 & 0 & 1 \end{pmatrix}$ **30.** $\begin{pmatrix} 0 & 1 & 0 & 1 \\ 1 & 0 & 1 & 0 \\ 0 & 1 & 0 & 1 \\ 1 & 0 & 1 & 0 \end{pmatrix}$

In Problems 31–34, use the procedure illustrated in Example 6 to identify the given conic section. Graph.

31. $5x^2 - 2xy + 5y^2 = 24$
32. $13x^2 - 10xy + 13y^2 = 288$
33. $-3x^2 + 8xy + 3y^2 = 20$
34. $16x^2 + 24xy + 9y^2 - 3x + 4y = 0$

35. Find a 2×2 matrix **A** that has eigenvalues $\lambda_1 = 2$ and $\lambda_2 = 3$ and corresponding eigenvectors

$$\mathbf{K}_1 = \begin{pmatrix} 1 \\ 2 \end{pmatrix} \quad \text{and} \quad \mathbf{K}_2 = \begin{pmatrix} 1 \\ 1 \end{pmatrix}.$$

36. Find a 3×3 symmetric matrix that has eigenvalues $\lambda_1 = 1$, $\lambda_2 = 3$, and $\lambda_3 = 5$ and corresponding eigenvectors

$$\mathbf{K}_1 = \begin{pmatrix} 1 \\ -1 \\ 1 \end{pmatrix}, \mathbf{K}_2 = \begin{pmatrix} 1 \\ 0 \\ -1 \end{pmatrix}, \text{ and } \mathbf{K}_3 = \begin{pmatrix} 1 \\ 2 \\ 1 \end{pmatrix}.$$

37. If **A** is an $n \times n$ diagonalizable matrix, then $\mathbf{D} = \mathbf{P}^{-1}\mathbf{AP}$, where **D** is a diagonal matrix. Show that if m is a positive integer, then $\mathbf{A}^m = \mathbf{PD}^m\mathbf{P}^{-1}$.

38. The mth power of a diagonal matrix

$$\mathbf{D} = \begin{pmatrix} a_{11} & 0 & \cdots & 0 \\ 0 & a_{22} & \cdots & 0 \\ \vdots & & & \vdots \\ 0 & 0 & \cdots & a_{nn} \end{pmatrix}$$

is

$$\mathbf{D}^m = \begin{pmatrix} a_{11}^m & 0 & \cdots & 0 \\ 0 & a_{22}^m & \cdots & 0 \\ \vdots & & & \vdots \\ 0 & 0 & \cdots & a_{nn}^m \end{pmatrix}.$$

Use this result to compute

$$\begin{pmatrix} 2 & 0 & 0 & 0 \\ 0 & 3 & 0 & 0 \\ 0 & 0 & -1 & 0 \\ 0 & 0 & 0 & 5 \end{pmatrix}^4.$$

In Problems 39 and 40, use the results of Problems 37 and 38 to find the indicated power of the given matrix.

39. $\mathbf{A} = \begin{pmatrix} 1 & 1 \\ 2 & 0 \end{pmatrix}$, $\mathbf{A}^5$ **40.** $\mathbf{A} = \begin{pmatrix} 6 & -10 \\ 3 & -5 \end{pmatrix}$, $\mathbf{A}^{10}$

41. Suppose **A** is a nonsingular diagonalizable matrix. Then show that $\mathbf{A}^{-1}$ is diagonalizable.

42. Suppose **A** is a diagonalizable matrix. Is the matrix **P** unique?

8.13 LU-Factorization

INTRODUCTION Just as positive integers and polynomials can be factored, so too a matrix can sometimes be factored into other matrices. For example, in the last section we saw that if an $n \times n$ matrix **A** is diagonalizable, then there exists a nonsingular matrix **P** and a diagonal matrix **D** such that $\mathbf{P}^{-1}\mathbf{AP} = \mathbf{D}$. When the last equation is written as $\mathbf{A} = \mathbf{PDP}^{-1}$ we say that **A** has been **factored** or **decomposed** into three matrices. There are many ways of factoring an $n \times n$ matrix **A**, but in this section we are interested in a special type of factorization that involves triangular matrices.

The notion of a triangular matrix was introduced in Section 8.1. See page 379. ▶

LU-Factorization

Recall, a triangular matrix is a square matrix that is either a **lower triangular matrix** whose entries above the main diagonal are all zero or an **upper triangular matrix** whose entries below the main diagonal are all zero:

$$\begin{pmatrix} l_{11} & 0 & 0 & \cdots & 0 \\ l_{21} & l_{22} & 0 & \cdots & 0 \\ l_{31} & l_{32} & l_{33} & \cdots & 0 \\ \vdots & \vdots & \vdots & & \vdots \\ l_{n1} & l_{n2} & l_{n3} & \cdots & l_{nn} \end{pmatrix} \qquad \begin{pmatrix} u_{11} & u_{12} & u_{13} & \cdots & u_{1n} \\ 0 & u_{22} & u_{23} & \cdots & u_{2n} \\ 0 & 0 & u_{23} & \cdots & u_{3n} \\ \vdots & \vdots & \vdots & & \vdots \\ 0 & 0 & 0 & \cdots & u_{nn} \end{pmatrix}. \qquad (1)$$

<div align="center">lower triangular matrix upper triangular matrix</div>

Lower and upper triangular matrices are very important in a variety of computational applications, but in this section we will confine our attention to their use in solving linear systems $\mathbf{AX} = \mathbf{B}$. We will denote a lower triangle matrix by the letter $\mathbf{L}$ and an upper triangular matrix by $\mathbf{U}$. This leads to the following definition.

DEFINITION 8.13.1 LU-Factorization

Let $\mathbf{A}$ be an $n \times n$ matrix. If $\mathbf{A}$ can be written as a product $\mathbf{A} = \mathbf{LU}$ where $\mathbf{L}$ and $\mathbf{U}$ are lower triangular and upper triangular matrices, respectively, then we say that $\mathbf{A} = \mathbf{LU}$ is an **LU-factorization** of $\mathbf{A}$.

An LU-factorization of a matrix $\mathbf{A}$ is also called an **LU-decomposition** of the matrix $\mathbf{A}$.

EXAMPLE 1 An LU-Factorization

Suppose $\mathbf{A} = \begin{pmatrix} 1 & 1 & 1 \\ 3 & 1 & 2 \\ 1 & -1 & 1 \end{pmatrix}$. Then for the lower triangular matrix $\mathbf{L} = \begin{pmatrix} 1 & 0 & 0 \\ 3 & 1 & 0 \\ 1 & 1 & 1 \end{pmatrix}$ and the upper triangular matrix $\mathbf{U} = \begin{pmatrix} 1 & 1 & 1 \\ 0 & -2 & -1 \\ 0 & 0 & 1 \end{pmatrix}$ it is easily verified by matrix multiplication that an LU-factorization of $\mathbf{A}$ is

$$\mathbf{LU} = \begin{pmatrix} 1 & 0 & 0 \\ 3 & 1 & 0 \\ 1 & 1 & 1 \end{pmatrix} \begin{pmatrix} 1 & 1 & 1 \\ 0 & -2 & -1 \\ 0 & 0 & 1 \end{pmatrix} = \begin{pmatrix} 1 & 1 & 1 \\ 3 & 1 & 2 \\ 1 & -1 & 1 \end{pmatrix} = \mathbf{A}. \qquad (2) \quad \equiv$$

Finding an LU-Factorization

Of course, the obvious question is: How do we find an LU-factorization for a given matrix $\mathbf{A}$? For the sake of discussion let's assume that $\mathbf{A}$ is a 3×3 matrix, although the procedures discussed in this section are applicable to square matrices of any order. We want to find two 3×3 matrices $\mathbf{L}$ and $\mathbf{U}$ such that $\mathbf{A} = \mathbf{LU}$ or

$$\overbrace{\begin{pmatrix} a_{11} & a_{12} & a_{13} \\ a_{21} & a_{22} & a_{23} \\ a_{31} & a_{32} & a_{33} \end{pmatrix}}^{\mathbf{A}} = \overbrace{\begin{pmatrix} l_{11} & 0 & 0 \\ l_{21} & l_{22} & 0 \\ l_{31} & l_{32} & l_{33} \end{pmatrix}}^{\mathbf{L}} \overbrace{\begin{pmatrix} u_{11} & u_{12} & u_{13} \\ 0 & u_{22} & u_{23} \\ 0 & 0 & u_{33} \end{pmatrix}}^{\mathbf{U}}. \qquad (3)$$

By carrying out the indicated multiplication on the right-hand side of (3) and equating its entries to the corresponding known entries a_{ij} on the left-hand side we obtain a system of 9 equations in 12 variables, the variables being the l_{ij} and u_{ij}. Recall, such a system is *underdetermined* and, if consistent, would have many solutions. This leads us immediately to the

See page 393 at the end of Section 8.2. ▶

conclusion that the matrices **L** and **U** in an LU-factorization are not uniquely determined. In other words:

• An $n \times n$ matrix **A** can have several different LU-factorizations.

The factorization depends on the method used.

Doolittle's Method

In order to determine unique matrices **L** and **U** we need to impose three conditions on the variables l_{ij} and u_{ij} in (3). The method that we examine in the examples that follow is called **Doolittle's method** after the American mathematician **Myrick H. Doolittle** (1830–1913). In this method the main diagonal entries l_{ii} of **L** in (3) are chosen to be all 1s. The next example illustrates a method for obtaining the remaining nine variables.

EXAMPLE 2 **Example 1 Revisited**

Use Doolittle's method to find an LU-factorization of

$$
\mathbf{A} = \begin{pmatrix} 1 & 1 & 1 \\ 3 & 1 & 2 \\ 1 & -1 & 1 \end{pmatrix}.
$$

SOLUTION The matrix **A** is the one considered in Example 1. We are going to show how the factorization given in (2) can be found. Using (3) with $l_{ii} = 1$, $i = 1, 2, 3$ in the lower triangular matrix **L** we write

$$
\begin{pmatrix} 1 & 1 & 1 \\ 3 & 1 & 2 \\ 1 & -1 & 1 \end{pmatrix} = \overbrace{\begin{pmatrix} 1 & 0 & 0 \\ l_{21} & 1 & 0 \\ l_{31} & l_{32} & 1 \end{pmatrix}}^{\mathbf{L}} \overbrace{\begin{pmatrix} u_{11} & u_{12} & u_{13} \\ 0 & u_{22} & u_{23} \\ 0 & 0 & u_{33} \end{pmatrix}}^{\mathbf{U}} = \begin{pmatrix} u_{11} & u_{12} & u_{13} \\ l_{21}u_{11} & l_{21}u_{12} + u_{22} & l_{21}u_{13} + u_{23} \\ l_{31}u_{11} & l_{31}u_{12} + l_{32}u_{22} & l_{31}u_{13} + l_{32}u_{23} + u_{33} \end{pmatrix}. \quad (4)
$$

Then by the definition of equality of matrices we must have

$$
\begin{array}{lll}
1 = u_{11}, & 1 = u_{12}, & 1 = u_{13}, \\
3 = l_{21}u_{11}, & 1 = l_{21}u_{12} + u_{22}, & 2 = l_{21}u_{13} + u_{23}, \\
1 = l_{31}u_{11}, & -1 = l_{31}u_{12} + l_{32}u_{22}, & 1 = l_{31}u_{13} + l_{32}u_{23} + u_{33}.
\end{array} \quad (5)
$$

At first glance (5) may look a bit formidable to solve, but it is not. Indeed, (5) is not so much a system of equations as it is a sequence of nine linear equations in *one* variable. The trick in solving (5) is to read each equation beginning with the first row moving left to right and use previously found values to determine the one variable in each equation. The variable determined by each equation is indicated in red in (5). Using the values $u_{11} = 1$, $u_{12} = 1$, $u_{13} = 1$ from the first row in the three equations in the second row yield, in turn, $l_{21} = 3$, $u_{22} = -2$, $u_{23} = -1$. Note that we also had to use $l_{21} = 3$ to find u_{22}, and so on. The three equations in the third row then give $l_{31} = 1$, $l_{32} = 1$, $u_{33} = 1$. Substituting these values in the appropriate triangular matrices in (4) gives LU-factorization,

$$
\mathbf{A} = \begin{pmatrix} 1 & 1 & 1 \\ 3 & 1 & 2 \\ 1 & -1 & 1 \end{pmatrix} = \overbrace{\begin{pmatrix} 1 & 0 & 0 \\ 3 & 1 & 0 \\ 1 & 1 & 1 \end{pmatrix}}^{\mathbf{L}} \overbrace{\begin{pmatrix} 1 & 1 & 1 \\ 0 & -2 & -1 \\ 0 & 0 & 1 \end{pmatrix}}^{\mathbf{U}}.
$$

≡

An Algorithm

Doolittle's method is very popular in that it is easily programmed in computer algebra systems such as *Mathematica* or *MATLAB*. In lieu of the brute force technique illustrated in Example 2, we can carry out Doolittle's method using one of the elementary row operations that we used in Gaussian elimination to create an upper triangular matrix **U** out of the matrix **A**. In particular, we use the notion of *row addition*, that is, adding a nonzero multiple of one row of a matrix **A** to another row. Recall from Section 8.2, the notation $cR_i + R_j$ denotes the multiplication of the entries in the ith row of a matrix **A** by a nonzero constant c and adding them

to the corresponding entries in the jth row. We state the following two-step algorithm without a formal proof. We assume that $\mathbf{A}$ has an LU-factorization.

(*i*) Create an upper triangular matrix $\mathbf{U}$ using row addition to *zero out* the entries below the main diagonal of $\mathbf{A}$. For a 3×3 matrix, this looks like

$$\mathbf{A} = \begin{pmatrix} a_{11} & a_{12} & a_{13} \\ a_{21} & a_{22} & a_{23} \\ a_{31} & a_{32} & a_{33} \end{pmatrix} \overset{\substack{\text{row} \\ \text{addition} \\ \text{operations}}}{\Rightarrow} \begin{pmatrix} a_{11} & a_{12} & a_{13} \\ 0 & a'_{23} & a'_{23} \\ 0 & 0 & a''_{33} \end{pmatrix} = \mathbf{U}. \tag{6}$$

(*ii*) Keep a record of the *negatives* of the multipliers of the rows used in step (*i*) by entering these numbers in an appropriate-sized identity matrix $\mathbf{I}$. If we perform the addition $cR_i + R_j$, then enter $-c$ in the jth row and ith column of $\mathbf{I}$. For example, if we perform $5R_1 + R_2$ on $\mathbf{A}$ in (6) to zero out the entry in the second row and first column, we then replace the 0 in the second row and first column of the 3×3 identity $\mathbf{I}$ by the number -5:

$$\begin{pmatrix} 1 & 0 & 0 \\ -5 & 1 & 0 \\ 0 & 0 & 1 \end{pmatrix}.$$

When finished with the reduction of $\mathbf{A}$ to $\mathbf{U}$, the matrix created by recording the negatives of all the multipliers of the rows we used is a lower triangular matrix $\mathbf{L}$ because these numbers are placed in $\mathbf{I}$ in exactly the same position as the 0 s in the matrix $\mathbf{U}$ (below the diagonal).

EXAMPLE 3 Adding Multiples of Rows to Rows

Find an LU-factorization of

$$\mathbf{A} = \begin{pmatrix} 2 & -8 \\ 3 & 0 \end{pmatrix}.$$

SOLUTION We need only perform one row-addition operation on $\mathbf{A}$ to obtain the desired upper triangular matrix $\mathbf{U}$. We simultaneously create $\mathbf{L}$ using the 2×2 identity $\mathbf{I}$:

$$\mathbf{A} = \begin{pmatrix} 2 & -8 \\ 3 & 0 \end{pmatrix} \overset{-\frac{3}{2}R_1 + R_2}{\Rightarrow} \begin{pmatrix} 2 & -8 \\ 0 & 12 \end{pmatrix}$$

$$\mathbf{I} = \begin{pmatrix} 1 & 0 \\ 0 & 1 \end{pmatrix} \overset{\text{record } \frac{3}{2}}{\Rightarrow} \begin{pmatrix} 1 & 0 \\ \frac{3}{2} & 1 \end{pmatrix}. \tag{7}$$

From (7) we see that

$$\mathbf{U} = \begin{pmatrix} 2 & -8 \\ 0 & 12 \end{pmatrix} \quad \text{and} \quad \mathbf{L} = \begin{pmatrix} 1 & 0 \\ \frac{3}{2} & 1 \end{pmatrix}.$$

An LU-factorization of $\mathbf{A}$ is given by

$$\mathbf{A} = \begin{pmatrix} 2 & -8 \\ 3 & 0 \end{pmatrix} = \begin{pmatrix} 1 & 0 \\ \frac{3}{2} & 1 \end{pmatrix} \begin{pmatrix} 2 & -8 \\ 0 & 12 \end{pmatrix}. \qquad \equiv$$

EXAMPLE 4 Adding Multiples of Rows to Rows

Find an LU-factorization of

$$\mathbf{A} = \begin{pmatrix} -1 & 2 & -4 \\ 2 & -5 & 10 \\ 3 & 1 & 6 \end{pmatrix}.$$

SOLUTION In this case the reduction of **A** to **U** requires three row-addition operations. We record the negatives of the multiples of the rows in the 3×3 identity **I**:

$$\mathbf{A} = \begin{pmatrix} -1 & 2 & -4 \\ 2 & -5 & 10 \\ 3 & 1 & 6 \end{pmatrix} \overset{\substack{2R_1 + R_2 \\ 3R_1 + R_3}}{\Rightarrow} \begin{pmatrix} -1 & 2 & -4 \\ 0 & -1 & 2 \\ 0 & 7 & -6 \end{pmatrix} \overset{7R_2 + R_3}{\Rightarrow} \begin{pmatrix} -1 & 2 & -4 \\ 0 & -1 & 2 \\ 0 & 0 & 8 \end{pmatrix}$$

$$\mathbf{I} = \begin{pmatrix} 1 & 0 & 0 \\ 0 & 1 & 0 \\ 0 & 0 & 1 \end{pmatrix} \overset{\substack{\text{record } -2 \\ \text{record } -3}}{\Rightarrow} \begin{pmatrix} 1 & 0 & 0 \\ -2 & 1 & 0 \\ -3 & 0 & 1 \end{pmatrix} \overset{\text{record } -7}{\Rightarrow} \begin{pmatrix} 1 & 0 & 0 \\ -2 & 1 & 0 \\ -3 & -7 & 1 \end{pmatrix}. \tag{8}$$

Inspection of (8) yields

$$\mathbf{U} = \begin{pmatrix} -1 & 2 & -4 \\ 0 & -1 & 2 \\ 0 & 0 & 8 \end{pmatrix} \quad \text{and} \quad \mathbf{L} = \begin{pmatrix} 1 & 0 & 0 \\ -2 & 1 & 0 \\ -3 & -7 & 1 \end{pmatrix}.$$

Thus an LU-factorization of **A** is

$$\mathbf{A} = \begin{pmatrix} -1 & 2 & -4 \\ 2 & -5 & 10 \\ 3 & 1 & 6 \end{pmatrix} = \begin{pmatrix} 1 & 0 & 0 \\ -2 & 1 & 0 \\ -3 & -7 & 1 \end{pmatrix} \begin{pmatrix} -1 & 2 & -4 \\ 0 & -1 & 2 \\ 0 & 0 & 8 \end{pmatrix}. \qquad \equiv$$

■ **Solving Linear Systems** Suppose $\mathbf{AX} = \mathbf{B}$ is a linear system of n equations in n variables. If an $n \times n$ coefficient matrix **A** in the system has an LU-factorization $\mathbf{A} = \mathbf{LU}$ then $\mathbf{AX} = \mathbf{B}$ is the same as $(\mathbf{LU})\mathbf{X} = \mathbf{B}$, *or* $\mathbf{L}(\mathbf{UX}) = \mathbf{B}$. This last system can be solved efficiently in two steps:

(*i*) First, let $\mathbf{Y} = \mathbf{UX}$ so that $\mathbf{L}(\mathbf{UX}) = \mathbf{B}$ becomes $\mathbf{LY} = \mathbf{B}$. Solve for **Y** by forward-substitution.

(*ii*) Then solve $\mathbf{UX} = \mathbf{Y}$ for **X** using back-substitution.

The next example illustrates the method.

EXAMPLE 5 **Example 1 Revisited**

Solve the system

$$\begin{aligned} x_1 + x_2 + x_3 &= 0 \\ 3x_1 + x_2 + 2x_3 &= 1 \\ x_1 - x_2 + x_3 &= 4. \end{aligned}$$

SOLUTION In terms of the matrix notation $\mathbf{AX} = \mathbf{B}$, the linear system is the same as

$$\begin{pmatrix} 1 & 1 & 1 \\ 3 & 1 & 2 \\ 1 & -1 & 1 \end{pmatrix} \begin{pmatrix} x_1 \\ x_2 \\ x_3 \end{pmatrix} = \begin{pmatrix} 0 \\ 1 \\ 4 \end{pmatrix}, \tag{9}$$

where the coefficient matrix **A** in (9) is the matrix in Example 1. Using the LU-factorization of **A** given in (2) of Example 1, (9) can be written

$$\overset{\mathbf{L}}{\overbrace{\begin{pmatrix} 1 & 0 & 0 \\ 3 & 1 & 0 \\ 1 & 1 & 1 \end{pmatrix}}} \overset{\mathbf{UX}}{\overbrace{\begin{pmatrix} 1 & 1 & 1 \\ 0 & -2 & -1 \\ 0 & 0 & 1 \end{pmatrix} \begin{pmatrix} x_1 \\ x_2 \\ x_3 \end{pmatrix}}} = \overset{\mathbf{B}}{\overbrace{\begin{pmatrix} 0 \\ 1 \\ 4 \end{pmatrix}}}. \tag{10}$$

If we let $\mathbf{UX} = \mathbf{Y}$ in (10), where $\mathbf{Y}$ is the column matrix

$$\mathbf{Y} = \begin{pmatrix} y_1 \\ y_2 \\ y_3 \end{pmatrix}, \tag{11}$$

then (10) is the same as $\mathbf{LY} = \mathbf{B}$:

$$\begin{pmatrix} 1 & 0 & 0 \\ 3 & 1 & 0 \\ 1 & 1 & 1 \end{pmatrix} \begin{pmatrix} y_1 \\ y_2 \\ y_3 \end{pmatrix} = \begin{pmatrix} 0 \\ 1 \\ 4 \end{pmatrix} \quad \text{or} \quad \begin{aligned} y_1 &&&= 0 \\ 3y_1 + y_2 &&&= 1 \\ y_1 + y_2 + y_3 &&&= 4. \end{aligned} \tag{12}$$

We now use forward-substitution in the second form of the system in (12). That is, we substitute $y_1 = 0$ obtained from the first equation into the second equation of (12). This immediately gives $y_2 = 1$. Then substituting both these values into the third equation of (12) yields $y_3 = 3$. Since the column matrix $\mathbf{Y}$ is now determined, we now solve $\mathbf{UX} = \mathbf{Y}$, that is

$$\begin{pmatrix} 1 & 1 & 1 \\ 0 & -2 & -1 \\ 0 & 0 & 1 \end{pmatrix} \begin{pmatrix} x_1 \\ x_2 \\ x_3 \end{pmatrix} = \begin{pmatrix} 0 \\ 1 \\ 3 \end{pmatrix} \quad \text{or} \quad \begin{aligned} x_1 + x_2 + x_3 &= 0 \\ -2x_2 - x_3 &= 1 \\ x_3 &= 3. \end{aligned} \tag{13}$$

We solve the second form of the system in (13) by back-substitution, that is, we substitute $x_3 = 3$ from the third equation into the second equation to obtain $x_2 = -2$. Substituting these two values into the first equation of (13) yields $x_1 = -1$. Hence the solution of the original system is $x_1 = -1$, $x_2 = -2$, $x_3 = 3$. ≡

See Theorem 8.5.6 on page 408.

■ **Relationship to Determinants** If an $n \times n$ matrix $\mathbf{A}$ has an LU-factorization $\mathbf{A} = \mathbf{LU}$, then $\det \mathbf{A}$ is easy to evaluate using the determinant property $\det \mathbf{A} = \det \mathbf{L} \cdot \det \mathbf{U}$.

EXAMPLE 6 **Determinant of a Factored Matrix**

Evaluate the determinant of the matrix $\mathbf{A}$ in Example 4.

SOLUTION From Example 4 an LU-factorization $\mathbf{A} = \mathbf{LU}$ is

$$\mathbf{A} = \begin{pmatrix} -1 & 2 & -4 \\ 2 & -5 & 10 \\ 3 & 1 & 6 \end{pmatrix} = \begin{pmatrix} 1 & 0 & 0 \\ -2 & 1 & 0 \\ -3 & -7 & 1 \end{pmatrix} \begin{pmatrix} -1 & 2 & -4 \\ 0 & -1 & 2 \\ 0 & 0 & 8 \end{pmatrix}$$

so that

$$\det \mathbf{A} = \det \begin{pmatrix} 1 & 0 & 0 \\ -2 & 1 & 0 \\ -3 & -7 & 1 \end{pmatrix} \det \begin{pmatrix} -1 & 2 & -4 \\ 0 & -1 & 2 \\ 0 & 0 & 8 \end{pmatrix}. \tag{14}$$

Now here is the beauty of this technique. Going back to Theorem 8.5.8 we know that the determinant of a triangular matrix is the product of its main diagonal entries. So with no additional work we have immediately that (14) is

$$\det \mathbf{A} = \overbrace{(1 \cdot 1 \cdot 1)}^{\substack{\text{diagonal} \\ \text{entries of } \mathbf{L}}} \overbrace{((-1) \cdot (-1) \cdot 8)}^{\substack{\text{diagonal} \\ \text{entries of } \mathbf{U}}} = 8.$$

≡

REMARKS

(*i*) Not every $n \times n$ matrix $\mathbf{A}$ has an LU-factorization. For example, the 2×2 matrix $\mathbf{A} = \begin{pmatrix} 0 & 2 \\ 1 & 1 \end{pmatrix}$ has no LU-factorization. Try the method in Example 2 and see what happens. In general, if interchanges of rows are required in the method illustrated in Examples 3 and 4 in order to obtain $\mathbf{U}$, then we say that an LU-factorization does not exist.

(*ii*) When solving a linear system $\mathbf{AX} = \mathbf{B}$ by Gaussian elimination, the steps of that method must be repeated if the coefficient matrix $\mathbf{A}$ is the same but the column vector $\mathbf{B}$ is changed. The method for solving systems of linear equations illustrated in Example 5 is especially useful if a system $\mathbf{AX} = \mathbf{B}$ has to be solved for different vectors $\mathbf{B}_i$, $i = 1, 2, \ldots, n$, since the matrices $\mathbf{L}$ and $\mathbf{U}$ remain the same in each system and so can be retained in a computer program. See Problems 31–34 in Exercises 8.13.

(*iii*) If a matrix $\mathbf{A}$ already has a 0 below its main diagonal such as

$$\mathbf{A} = \begin{pmatrix} 1 & 4 & 5 \\ 2 & 1 & 3 \\ 0 & 4 & 6 \end{pmatrix},$$

then we do not have to zero out the entry in the third row and first column by a row addition. So in the matrix in which we are recording the negatives of the multipliers we simply retain the 0 in the third row and first column.

(*iv*) Had we chosen the main diagonal entries u_{ii} of $\mathbf{U}$ in (3) to be all 1s, then an LU-factorization of a 3×3 matrix $\mathbf{A}$ looks like

$$\mathbf{A} = \overbrace{\begin{pmatrix} l_{11} & 0 & 0 \\ l_{21} & l_{22} & 0 \\ l_{31} & l_{32} & l_{33} \end{pmatrix}}^{\mathbf{L}} \overbrace{\begin{pmatrix} 1 & u_{12} & u_{13} \\ 0 & 1 & u_{23} \\ 0 & 0 & 1 \end{pmatrix}}^{\mathbf{U}} = \begin{pmatrix} l_{11} & l_{11}u_{12} & l_{11}u_{13} \\ l_{21} & l_{21}u_{12} + l_{22} & l_{21}u_{13} + l_{22}u_{23} \\ l_{31} & l_{31}u_{12} + l_{32} & l_{31}u_{13} + l_{32}u_{23} + l_{33} \end{pmatrix}. \text{(15)}$$

This is called **Crout's method** because it was devised by the American mathematician **Preston D. Crout** (1907–1984). Like the Doolittle method, Crout's method also gives a unique LU-factorization of $\mathbf{A}$. See Problems 41 and 42 in Exercises 8.13.

(*v*) **Cholesky's method** discovered by the French military officer and mathematician **André-Louis Cholesky** (1875–1918) is a way of finding an LU-factorization of a symmetric matrix, that is, a matrix $\mathbf{A}$ for which $\mathbf{A} = \mathbf{A}^T$. The form of the factorization for a 3×3 symmetric matrix is $\mathbf{A} = \mathbf{LU} = \mathbf{LL}^T$:

$$\mathbf{A} = \overbrace{\begin{pmatrix} l_{11} & 0 & 0 \\ l_{21} & l_{22} & 0 \\ l_{31} & l_{32} & l_{33} \end{pmatrix}}^{\mathbf{L}} \overbrace{\begin{pmatrix} l_{11} & l_{21} & l_{31} \\ 0 & l_{22} & l_{32} \\ 0 & 0 & l_{33} \end{pmatrix}}^{\mathbf{U} = \mathbf{L}^T} = \begin{pmatrix} l_{11}^2 & l_{11}l_{21} & l_{11}l_{31} \\ l_{21}l_{11} & l_{21}^2 + l_{22}^2 & l_{21}l_{31} + l_{22}l_{32} \\ l_{31}l_{11} & l_{31}l_{21} + l_{32}l_{22} & l_{31}^2 + l_{32}^2 + l_{33}^2 \end{pmatrix}. \text{(16)}$$

See Problems 43 and 44 in Exercises 8.13.

8.13 Exercises Answers to selected odd-numbered problems begin on page ANS-21.

In Problems 1–10, use the procedure illustrated in Example 2 to find the LU-factorization of the given matrix.

1. $\begin{pmatrix} 2 & -2 \\ 1 & 2 \end{pmatrix}$ **2.** $\begin{pmatrix} 6 & 2 \\ 4 & 1 \end{pmatrix}$ **3.** $\begin{pmatrix} -1 & 4 \\ 2 & 2 \end{pmatrix}$ **4.** $\begin{pmatrix} 5 & -4 \\ 15 & 2 \end{pmatrix}$

5. $\begin{pmatrix} 4 & -2 & 1 \\ -4 & 1 & 2 \\ 12 & 1 & 3 \end{pmatrix}$ **6.** $\begin{pmatrix} -3 & 2 & 1 \\ 9 & 3 & 2 \\ 3 & 1 & -1 \end{pmatrix}$

7. $\begin{pmatrix} 1 & 2 & 7 \\ 2 & 5 & 6 \\ 7 & 6 & 4 \end{pmatrix}$ **8.** $\begin{pmatrix} -4 & 2 & -10 \\ 4 & 1 & 4 \\ 6 & 6 & -8 \end{pmatrix}$

9. $\begin{pmatrix} 1 & -2 & 1 \\ 0 & 1 & 2 \\ 2 & 6 & 1 \end{pmatrix}$ **10.** $\begin{pmatrix} 1 & 0 & 1 \\ 1 & 9 & 1 \\ 1 & 0 & -1 \end{pmatrix}$

In Problems 11–20, use the procedure illustrated in Examples 3 and 4 to find the LU-factorization of the given matrix.

11. $\begin{pmatrix} 3 & 9 \\ 1 & 11 \end{pmatrix}$

12. $\begin{pmatrix} -2 & 10 \\ 1 & -4 \end{pmatrix}$

13. $\begin{pmatrix} -4 & -2 \\ -1 & -3 \end{pmatrix}$

14. $\begin{pmatrix} 0.2 & 0.3 \\ 0.8 & 3.5 \end{pmatrix}$

15. $\begin{pmatrix} 1 & 1 & 1 \\ 3 & 1 & 2 \\ 1 & -1 & 1 \end{pmatrix}$

16. $\begin{pmatrix} -1 & 2 & -4 \\ 2 & -5 & 10 \\ 3 & 1 & 6 \end{pmatrix}$

17. $\begin{pmatrix} 4 & 2 & 12 \\ 2 & 26 & -4 \\ 12 & -4 & 56 \end{pmatrix}$

18. $\begin{pmatrix} 16 & 4 & 20 \\ 4 & 5 & 3 \\ 20 & 3 & 29 \end{pmatrix}$

19. $\begin{pmatrix} 1 & -2 & 1 & 0 \\ -2 & 3 & -2 & 1 \\ 1 & 2 & 4 & 3 \\ 5 & 1 & -1 & 1 \end{pmatrix}$

20. $\begin{pmatrix} 1 & -1 & 3 & 2 \\ 3 & -5 & 19 & 3 \\ 2 & -2 & 3 & 10 \\ -1 & 5 & -5 & -2 \end{pmatrix}$

In Problems 21–30, proceed as in Example 5 and use the corresponding LU-factorization from Problems 1–10 to solve the given linear system of equations.

21. $\begin{pmatrix} 2 & -2 \\ 1 & 2 \end{pmatrix}\begin{pmatrix} x_1 \\ x_2 \end{pmatrix} = \begin{pmatrix} 1 \\ -2 \end{pmatrix}$

22. $\begin{pmatrix} 6 & 2 \\ 4 & 1 \end{pmatrix}\begin{pmatrix} x_1 \\ x_2 \end{pmatrix} = \begin{pmatrix} 12 \\ 2 \end{pmatrix}$

23. $\begin{pmatrix} -1 & 4 \\ 2 & 2 \end{pmatrix}\begin{pmatrix} x_1 \\ x_2 \end{pmatrix} = \begin{pmatrix} 15 \\ 5 \end{pmatrix}$

24. $\begin{pmatrix} 5 & -4 \\ 15 & 2 \end{pmatrix}\begin{pmatrix} x_1 \\ x_2 \end{pmatrix} = \begin{pmatrix} 1 \\ 7 \end{pmatrix}$

25. $\begin{pmatrix} 4 & -2 & 1 \\ -4 & 1 & 2 \\ 12 & 1 & 3 \end{pmatrix}\begin{pmatrix} x_1 \\ x_2 \\ x_3 \end{pmatrix} = \begin{pmatrix} 7 \\ 7 \\ 28 \end{pmatrix}$

26. $\begin{pmatrix} -3 & 2 & 1 \\ 9 & 3 & 2 \\ 3 & 1 & -1 \end{pmatrix}\begin{pmatrix} x_1 \\ x_2 \\ x_3 \end{pmatrix} = \begin{pmatrix} 15 \\ 3 \\ -24 \end{pmatrix}$

27. $\begin{pmatrix} 1 & 2 & 7 \\ 2 & 5 & 6 \\ 7 & 6 & 4 \end{pmatrix}\begin{pmatrix} x_1 \\ x_2 \\ x_3 \end{pmatrix} = \begin{pmatrix} 109 \\ 109 \\ 218 \end{pmatrix}$

28. $\begin{pmatrix} -4 & 2 & -10 \\ 4 & 1 & 4 \\ 6 & 6 & -8 \end{pmatrix}\begin{pmatrix} x_1 \\ x_2 \\ x_3 \end{pmatrix} = \begin{pmatrix} -20 \\ 14 \\ -18 \end{pmatrix}$

29. $\begin{pmatrix} 1 & -2 & 1 \\ 0 & 1 & 2 \\ 2 & 6 & 1 \end{pmatrix}\begin{pmatrix} x_1 \\ x_2 \\ x_3 \end{pmatrix} = \begin{pmatrix} 14 \\ -42 \\ 7 \end{pmatrix}$

30. $\begin{pmatrix} 1 & 0 & 1 \\ 1 & 9 & 1 \\ 1 & 0 & -1 \end{pmatrix}\begin{pmatrix} x_1 \\ x_2 \\ x_3 \end{pmatrix} = \begin{pmatrix} 18 \\ 27 \\ -12 \end{pmatrix}$

In Problems 31–34, use the given LU-factorization

$$\mathbf{A} = \begin{pmatrix} 1 & 1 & 1 \\ 1 & 2 & 2 \\ 1 & 2 & 3 \end{pmatrix} = \begin{pmatrix} 1 & 0 & 0 \\ 1 & 1 & 0 \\ 1 & 1 & 1 \end{pmatrix}\begin{pmatrix} 1 & 1 & 1 \\ 0 & 1 & 1 \\ 0 & 0 & 1 \end{pmatrix}$$

to solve the linear system $\mathbf{AX} = \mathbf{B}_i$ for the given column matrix $\mathbf{B}_i, i = 1, 2, 3, 4$.

31. $\mathbf{B}_1 = \begin{pmatrix} 2 \\ 4 \\ 1 \end{pmatrix}$

32. $\mathbf{B}_2 = \begin{pmatrix} -4 \\ 7 \\ 10 \end{pmatrix}$

33. $\mathbf{B}_3 = \begin{pmatrix} \frac{1}{2} \\ \frac{3}{4} \\ -\frac{1}{2} \end{pmatrix}$

34. $\mathbf{B}_4 = \begin{pmatrix} 30 \\ -42 \\ -18 \end{pmatrix}$

In Problems 35–40, proceed as in Example 6 and use an LU-factorization to evaluate the determinant det $\mathbf{A}$ of the indicated matrix $\mathbf{A}$.

35. $\mathbf{A}$ is the matrix in Problem 15

36. $\mathbf{A}$ is the matrix in Problem 16

37. $\mathbf{A}$ is the matrix in Problem 17

38. $\mathbf{A}$ is the matrix in Problem 18

39. $\mathbf{A}$ is the matrix in Problem 19

40. $\mathbf{A}$ is the matrix in Problem 20

In Problems 41 and 42, use Crout's method discussed in (*iv*) of the *Remarks* to obtain an LU-factorization of the indicated matrix $\mathbf{A}$.

41. $\mathbf{A}$ is the matrix in Problem 15

42. $\mathbf{A}$ is the matrix in Problem 16

In Problems 43 and 44, use Cholesky's method discussed in (*v*) of the *Remarks* to obtain an LU-factorization of the indicated symmetric matrix $\mathbf{A}$.

43. $\mathbf{A}$ is the matrix in Problem 17

44. $\mathbf{A}$ is the matrix in Problem 18

8.14 Cryptography

INTRODUCTION The word *cryptography* is a combination of two Greek words: *crypto,* meaning "hidden" or "secret," and *grapho,* which means "writing." Cryptography then is the study of making "secret writings" or **codes**.

In this section we will consider a system of encoding and decoding messages that requires both the sender of the message and the receiver of the message to know:

- a specified rule of correspondence between a set of symbols (such as letters of the alphabet and punctuation marks from which messages are composed) and a set of integers; and
- a specified *nonsingular* matrix $\mathbf{A}$.

A natural correspondence between the first 27 nonnegative integers and the letters of the alphabet and a blank space (to separate words) is given by

0	1	2	3	4	5	6	7	8	9	10	11	12	13	14	15	16	17	18	19	20	21	22	23	24	25	26
space	a	b	c	d	e	f	g	h	i	j	k	l	m	n	o	p	q	r	s	t	u	v	w	x	y	z

$$(1)$$

From (1) the numerical equivalent of the message

<div align="center">SEND THE DOCUMENT TODAY</div>

is 19 5 14 4 0 20 8 5 0 4 15 3 21 13 5 14 20 0 20 15 4 1 25. $$(2)$$

The sender will **encode** the message by means of the nonsingular matrix **A** and, as we shall see, the receiver of the encoded message will **decode** the message by means of the (unique) matrix $\mathbf{A}^{-1}$. The numerical message (2) is now written as a matrix. Since there are 23 symbols in the message, we need a matrix that will hold a minimum of 24 entries (an $m \times n$ matrix has mn entries). We choose to write (2) as the 3×8 matrix

$$\mathbf{M} = \begin{pmatrix} 19 & 5 & 14 & 4 & 0 & 20 & 8 & 5 \\ 0 & 4 & 15 & 3 & 21 & 13 & 5 & 14 \\ 20 & 0 & 20 & 15 & 4 & 1 & 25 & 0 \end{pmatrix}. \qquad (3)$$

Note that the last entry a_{38} in the message matrix **M** has been simply padded with a space represented by the number 0. Of course, we could have written (2) as a 6×4 or a 4×6 matrix but that would require a larger encoding matrix. A 3×8 matrix allows us to encode the message by means of a 3×3 matrix. The size of the matrices used is only a concern when the encoding and decoding are done by hand rather than by a computer.

The encoding matrix **A** is chosen, or rather constructed, so that

- **A** is nonsingular,
- **A** has only integer entries, and
- $\mathbf{A}^{-1}$ has only integer entries.

The last criterion is not particularly difficult to accomplish. We need only select the integer entries of **A** in such a manner that $\det \mathbf{A} = \pm 1$. For a 2×2 or a 3×3 matrix we can then find $\mathbf{A}^{-1}$ by the formulas in (4) and (5) of Section 8.6. If **A** has integer entries, then all the cofactors C_{11}, C_{12}, and so on, are also integers. For the discussion on hand we choose

$$\mathbf{A} = \begin{pmatrix} -1 & 0 & -1 \\ 2 & 3 & 4 \\ 2 & 4 & 5 \end{pmatrix}. \qquad (4)$$

You should verify that $\det \mathbf{A} = -1$.

The original message is encoded by premultiplying the message matrix **M** by **A**; that is, the message is sent as the matrix:

$$\mathbf{B} = \mathbf{AM} = \begin{pmatrix} -1 & 0 & -1 \\ 2 & 3 & 4 \\ 2 & 4 & 5 \end{pmatrix} \begin{pmatrix} 19 & 5 & 14 & 4 & 0 & 20 & 8 & 5 \\ 0 & 4 & 15 & 3 & 21 & 13 & 5 & 14 \\ 20 & 0 & 20 & 15 & 4 & 1 & 25 & 0 \end{pmatrix}$$

$$= \begin{pmatrix} -39 & -5 & -34 & -19 & -4 & -21 & -33 & -5 \\ 118 & 22 & 153 & 77 & 79 & 83 & 131 & 52 \\ 138 & 26 & 188 & 95 & 104 & 97 & 161 & 66 \end{pmatrix}. \qquad (5)$$

You should try to imagine the difficulty of decoding (5) without knowledge of **A**. But the receiver of the encoded message **B** knows **A** and its inverse, and so decoding is the straightforward computation of premultiplying **B** by $\mathbf{A}^{-1}$:

<div align="center">$\mathbf{AM} = \mathbf{B}$ implies $\mathbf{M} = \mathbf{A}^{-1}\mathbf{B}$.</div>

For the matrix (4) we find from (5) of Section 8.6 that

$$\mathbf{A}^{-1} = \begin{pmatrix} 1 & 4 & -3 \\ 2 & 3 & -2 \\ -2 & -4 & 3 \end{pmatrix}.$$

Thus, the decoded message is

$$\mathbf{M} = \begin{pmatrix} 1 & 4 & -3 \\ 2 & 3 & -2 \\ -2 & -4 & 3 \end{pmatrix} \begin{pmatrix} -39 & -5 & -34 & -19 & -4 & -21 & -33 & -5 \\ 118 & 22 & 153 & 77 & 79 & 83 & 131 & 52 \\ 138 & 26 & 188 & 95 & 104 & 97 & 161 & 66 \end{pmatrix}$$

$$= \begin{pmatrix} 19 & 5 & 14 & 4 & 0 & 20 & 8 & 5 \\ 0 & 4 & 15 & 3 & 21 & 13 & 5 & 14 \\ 20 & 0 & 20 & 15 & 4 & 1 & 25 & 0 \end{pmatrix}$$

or 19 5 14 4 0 20 8 5 0 4 15 3 21 13 5 14 20 0 20 15 4 1 25 0.

By also knowing the original correspondence (1), the receiver translates the numbers into

<p style="text-align:center">SEND_THE_DOCUMENT_TODAY</p>

where we have indicated the blank spaces by lines.

Some observations are in order. The correspondence or mapping (1) is one of many such correspondences that can be set up between the letters of the alphabet (we could even include punctuation symbols such as the comma and period) and integers. Using the 26 letters of the alphabet and the blank space, we can set up 27! of these correspondences. (Why?) Furthermore, we could have used a 2×2 matrix to encode (2). The size of the message matrix $\mathbf{M}$ would then have to be at least 2×12 in order to contain the 23 entries of the message. For example, if

$$\mathbf{A} = \begin{pmatrix} 1 & 2 \\ 0 & 1 \end{pmatrix} \quad \text{and} \quad \mathbf{M} = \begin{pmatrix} 19 & 5 & 14 & 4 & 0 & 20 & 8 & 5 & 0 & 4 & 15 & 3 \\ 21 & 13 & 5 & 14 & 20 & 0 & 20 & 15 & 4 & 1 & 25 & 0 \end{pmatrix},$$

then $\mathbf{B} = \mathbf{AM} = \begin{pmatrix} 61 & 31 & 24 & 32 & 40 & 20 & 48 & 35 & 8 & 6 & 65 & 3 \\ 21 & 13 & 5 & 14 & 20 & 0 & 20 & 15 & 4 & 1 & 25 & 0 \end{pmatrix}.$

Using $\mathbf{A}^{-1} = \begin{pmatrix} 1 & -2 \\ 0 & 1 \end{pmatrix}$, we obtain as before

$$\mathbf{M} = \mathbf{A}^{-1}\mathbf{B} = \begin{pmatrix} 1 & -2 \\ 0 & 1 \end{pmatrix} \begin{pmatrix} 61 & 31 & 24 & 32 & 40 & 20 & 48 & 35 & 8 & 6 & 65 & 3 \\ 21 & 13 & 5 & 14 & 20 & 0 & 20 & 15 & 4 & 1 & 25 & 0 \end{pmatrix}$$

$$= \begin{pmatrix} 19 & 5 & 14 & 4 & 0 & 20 & 8 & 5 & 0 & 4 & 15 & 3 \\ 21 & 13 & 5 & 14 & 20 & 0 & 20 & 15 & 4 & 1 & 25 & 0 \end{pmatrix}.$$

There is no particular reason why the numerical message (2) has to be broken down as rows (1×8 vectors) as in the matrix (3). Alternatively, (2) could be broken down as columns (3×1 vectors) as shown in the matrix

$$\begin{pmatrix} 19 & 4 & 8 & 4 & 21 & 14 & 20 & 1 \\ 5 & 0 & 5 & 15 & 13 & 20 & 15 & 25 \\ 14 & 20 & 0 & 3 & 5 & 0 & 4 & 0 \end{pmatrix}.$$

Finally, it may be desirable to send the encoded message as letters of the alphabet rather than as numbers. In Problem 13 in Exercises 8.14, we shall show how to transmit SEND THE DOCUMENT TODAY encoded as

OVTHWFUVJVRWYBWYCZZNWPZL.

8.14 Exercises Answers to selected odd-numbered problems begin on page ANS-21.

In Problems 1–6, use the matrix **A** and the correspondence (1) to encode the given message. Verify your work by decoding the encoded message.

1. $\mathbf{A} = \begin{pmatrix} 1 & 2 \\ 1 & 1 \end{pmatrix}$; SEND HELP

2. $\mathbf{A} = \begin{pmatrix} 3 & 5 \\ 1 & 2 \end{pmatrix}$; THE MONEY IS HERE

3. $\mathbf{A} = \begin{pmatrix} 3 & 5 \\ 2 & 3 \end{pmatrix}$; PHONE HOME

4. $\mathbf{A} = \begin{pmatrix} 1 & 2 & 3 \\ 1 & 1 & 2 \\ 0 & 1 & 2 \end{pmatrix}$; MADAME X HAS THE PLANS

5. $\mathbf{A} = \begin{pmatrix} 2 & 1 & 1 \\ 1 & 1 & 1 \\ -1 & 1 & 0 \end{pmatrix}$; GO NORTH ON MAIN ST

6. $\mathbf{A} = \begin{pmatrix} 5 & 3 & 0 \\ 4 & 3 & -1 \\ 5 & 2 & 2 \end{pmatrix}$; DR JOHN IS THE SPY

In Problems 7–10, use the matrix **A** and the correspondence (1) to decode the given message.

7. $\mathbf{A} = \begin{pmatrix} 8 & 3 \\ 5 & 2 \end{pmatrix}$; $\mathbf{B} = \begin{pmatrix} 152 & 184 & 171 & 86 & 212 \\ 95 & 116 & 107 & 56 & 133 \end{pmatrix}$

8. $\mathbf{A} = \begin{pmatrix} 2 & -1 \\ 1 & -1 \end{pmatrix}$;

$\mathbf{B} = \begin{pmatrix} 46 & -7 & -13 & 22 & -18 & 1 & 10 \\ 23 & -15 & -14 & 2 & -18 & -12 & 5 \end{pmatrix}$

9. $\mathbf{A} = \begin{pmatrix} 1 & 0 & 1 \\ 0 & 1 & 0 \\ 1 & 0 & 0 \end{pmatrix}$;

$\mathbf{B} = \begin{pmatrix} 31 & 21 & 21 & 22 & 20 & 9 \\ 19 & 0 & 9 & 13 & 16 & 15 \\ 13 & 1 & 20 & 8 & 0 & 9 \end{pmatrix}$

10. $\mathbf{A} = \begin{pmatrix} 2 & 1 & 1 \\ 0 & 0 & -1 \\ 1 & 1 & 1 \end{pmatrix}$;

$\mathbf{B} = \begin{pmatrix} 36 & 32 & 28 & 61 & 26 & 56 & 10 & 12 \\ -9 & -2 & -18 & -1 & -18 & -25 & 0 & 0 \\ 23 & 27 & 23 & 41 & 26 & 43 & 5 & 12 \end{pmatrix}$

11. Using the correspondence (1), we encoded the following message by a 2 × 2 matrix:

$\begin{pmatrix} 17 & 16 & 18 & 5 & 34 & 0 & 34 & 20 & 9 & 5 & 25 \\ -30 & -31 & -32 & -10 & -59 & 0 & -54 & -35 & -13 & -6 & -50 \end{pmatrix}$

Decode the message if its first two letters are DA and its last two letters are AY.

12. (a) Using the correspondence

1	2	3	4	5	6	7	8	9	10	11	12	13	14	15	16	17	18	19	20	21	22	23	24	25	26	27
j	k	l	n	m	s	t	u	w	x	g	h	i	o	p	q	r	v	y	z	a	b	c	d	e	f	space

find the numerical equivalent of the message

BUY ALL AVAILABLE STOCK AT MARKET.

(b) Encode the message by *postmultiplying* the message matrix **M** by

$$\mathbf{A} = \begin{pmatrix} 1 & 1 & 0 \\ 1 & 0 & 1 \\ 1 & 1 & -1 \end{pmatrix}.$$

(c) Verify your work by decoding the encoded message in part (b).

13. Consider the matrices **A** and **B** defined in (4) and (5), respectively.
(a) Rewrite **B** as **B′** using integers modulo 27.*
(b) Verify that the encoded message to be sent in letters is

OVTHWFUVJVRWYBWYCZZNWPZL.

(c) Decode the encoded message by computing $\mathbf{A}^{-1}\mathbf{B'}$ and rewriting the result using integers modulo 27.

*For integers a and b, we write $a = b \pmod{27}$ if b is the remainder ($0 \leq b < 27$) when a is divided by 27. For example, $33 = 6 \pmod{27}$, $28 = 1 \pmod{27}$, and so on. Negative integers are handled in the following manner: If $27 = 0 \pmod{27}$, then, for example, $25 + 2 = 0 \pmod{27}$ so that $-25 = 2 \pmod{27}$ and $-2 = 25 \pmod{27}$. Also, $-30 = 24 \pmod{27}$, since $30 + 24 (= 54) = 0 \pmod{27}$.

8.15 Error-Correcting Code

INTRODUCTION In contrast to the last section, there is no connotation of secretiveness to the word *code* as used in this section. We are going to examine briefly the concept of digital communication, say, communication between a satellite and a computer. As a consequence, we will deal only with matrices whose entries are binary digits, namely, 0s and 1s. When adding and multiplying such matrices, we will use **arithmetic modulo 2**. This arithmetic is defined by the addition and multiplication tables

$$
\begin{array}{c|cc}
+ & 0 & 1 \\
\hline
0 & 0 & 1 \\
1 & 1 & 0
\end{array}
\qquad
\begin{array}{c|cc}
\times & 0 & 1 \\
\hline
0 & 0 & 0 \\
1 & 0 & 1
\end{array}
$$

Fundamental properties such as the commutative and associative laws hold in this system. The only notable exception here is that $1 + 1 = 0$.

■ **Binary Strings** In digital communication the **messages** or **words** are binary n-tuples; that is, n-tuples consisting of only 0s and 1s or **bits**. An n-bit word is also said to be a **binary string of length** n.

EXAMPLE 1 Binary Strings

(a) The ordered 4-tuple $(0, 1, 0, 1)$ is a 4-bit word or a string of length four.
(b) The binary (that is, base 2) representation of the number 39 is 1 0 0 1 1 1 or as a 6-tuple $(1, 0, 0, 1, 1, 1)$.
(c) The ASCII* word for the letter Z is the string of length eight: $(1, 0, 0, 1, 1, 0, 1, 0)$. ≡

For convenience a word of length n will be written as a $1 \times n$ matrix, that is, as a row vector. For example, the 4-bit word in Example 1 would be written as the 1×4 matrix $\mathbf{W} = (0 \quad 1 \quad 0 \quad 1)$.

■ **Codes** By **encoding** a message, we mean a process whereby we transform a word $\mathbf{W}$ of length n into another word $\mathbf{C}$ of length $n + m$ by augmenting $\mathbf{W}$ with m additional bits, called **parity check bits**. An encoded word is said to be a **code word**. By **decoding** a received message we mean another process that gives either the decoded message or an indication that an error has occurred during transmission. An encoding/decoding scheme is called a **code**.

One of the simplest codes is the **parity check code**. In this code a word is encoded according to the rule:

$$
\text{If the number of ones in the word is} \begin{cases} \text{even: Add 0 to the word.} \\ \text{odd: Add 1 to the word.} \end{cases} \tag{1}
$$

The word *parity* refers to whether the number of ones in a word is even or odd. The encoding rule given in (1) makes the parity of the code word always *even*.

EXAMPLE 2 Encoding Words

Use the parity check code to encode the words:
(a) $\mathbf{W} = (1 \quad 0 \quad 0 \quad 0 \quad 1 \quad 1)$ (b) $\mathbf{W} = (1 \quad 1 \quad 1 \quad 0 \quad 0 \quad 1)$.

SOLUTION (a) Since the number of ones in $\mathbf{W}$ is odd, we add the extra bit 1 to the end of the word $\mathbf{W}$. The code word is then $\mathbf{C} = (1 \quad 0 \quad 0 \quad 0 \quad 1 \quad 1 \quad 1)$.
(b) In this case the number of ones is even, so the extra bit added to the word is 0. The encoded word is $\mathbf{C} = (1 \quad 1 \quad 1 \quad 0 \quad 0 \quad 1 \quad 0)$. ≡

*American Standard Code for Information Interchange.

In digital communication it is the encoded word **C** that is transmitted. But due to some kind of interference or noise in the transmission channel, one or more of the bits of **C** may be changed. Thus, the message transmitted is not always the message received. See **FIGURE 8.15.1**.

The parity check code enables the decoder to detect single errors. Suppose **R** is the message received. A single error in **R** means that one bit has been changed; either a zero has been changed to a one or a one has been changed to a zero. In either circumstance the parity of the word **R** is odd.

FIGURE 8.15.1 Bits of encoded word may be changed by interference

EXAMPLE 3 **Decoding Words**

Use the parity check code to decode the words:

(a) $\mathbf{R} = (1\ 1\ 0\ 0\ 1\ 0\ 1)$ **(b)** $\mathbf{R} = (1\ 0\ 1\ 1\ 0\ 0\ 0)$.

SOLUTION **(a)** The parity of **R** is even. We drop the last bit and take the decoded message to be $(1\ 1\ 0\ 0\ 1\ 0)$.
(b) The parity of **R** is odd. The decoding is simple: a parity error. ≡

For some types of digital communication, such as communication internal to a computer, the parity check code is adequate. But Example 2 clearly indicates one major pitfall of the code: If an error occurs, we do not know how to correct it since we do not know which bit is incorrect. Moreover, multiple errors could occur in transmission. If, say, two ones were changed to zeros in transmission, the message received still has even parity and the decoding takes place by dropping the last bit. In this case at least one of the bits in the decoded message is in error.

■ **Hamming Codes** The parity check code is an example of an *error-detecting*, but not an *error-correcting*, code. For the remainder of this discussion we will consider an error-detecting/correcting code called the **Hamming (7, 4) code**. This code, one of a widely used set of codes invented by the American mathematician **Richard W. Hamming** (1915–1998) at Bell Laboratories in the 1950s, is an encoding/decoding scheme that can detect the presence of a single error in a received message and can tell which bit must be corrected. In the (7, 4) code the encoding process consists of transforming a 4-bit word $\mathbf{W} = (w_1\ w_2\ w_3\ w_4)$ into a 7-bit code word

$$\mathbf{C} = (c_1\ c_2\ w_1\ c_3\ w_2\ w_3\ w_4),$$

where c_1, c_2, and c_3 denote the parity check bits. (Words longer than four bits can be broken up into sequences of words of length four.)

■ **Encoding** In the Hamming (7, 4) code, the parity check bits c_1, c_2, and c_3 are defined in terms of the information bits w_1, w_2, w_3, and w_4:

$$\begin{aligned}
c_1 &= w_1 + w_2 + w_4 \\
c_2 &= w_1 + w_3 + w_4 \\
c_3 &= w_2 + w_3 + w_4,
\end{aligned} \tag{2}$$

where the arithmetic is carried out modulo 2. Using matrices, we can write (2) as the product

$$\begin{pmatrix} c_1 \\ c_2 \\ c_3 \end{pmatrix} = \begin{pmatrix} 1 & 1 & 0 & 1 \\ 1 & 0 & 1 & 1 \\ 0 & 1 & 1 & 1 \end{pmatrix} \begin{pmatrix} w_1 \\ w_2 \\ w_3 \\ w_4 \end{pmatrix}. \tag{3}$$

EXAMPLE 4 **Encoding a Word**

Encode the word $\mathbf{W} = (1 \quad 0 \quad 1 \quad 1)$.

SOLUTION From (3) we have, with $w_1 = 1$, $w_2 = 0$, $w_3 = 1$, and $w_4 = 1$:

$$
\begin{pmatrix} c_1 \\ c_2 \\ c_3 \end{pmatrix} = \begin{pmatrix} 1 & 1 & 0 & 1 \\ 1 & 0 & 1 & 1 \\ 0 & 1 & 1 & 1 \end{pmatrix} \begin{pmatrix} 1 \\ 0 \\ 1 \\ 1 \end{pmatrix} = \begin{pmatrix} 1 \cdot 1 + 1 \cdot 0 + 0 \cdot 1 + 1 \cdot 1 \\ 1 \cdot 1 + 0 \cdot 0 + 1 \cdot 1 + 1 \cdot 1 \\ 0 \cdot 1 + 1 \cdot 0 + 1 \cdot 1 + 1 \cdot 1 \end{pmatrix} = \begin{pmatrix} 0 \\ 1 \\ 0 \end{pmatrix}.
$$

From this product we see that $c_1 = 0$, $c_2 = 1$, $c_3 = 0$, and so the corresponding code word is $\mathbf{C} = (0 \quad 1 \quad 1 \quad 0 \quad 0 \quad 1 \quad 1)$. ≡

Before launching into the details of how to decode a message we need to introduce a special matrix. We first observe that in modulo 2 arithmetic there are no negative numbers; the additive inverse of 1 is 1, not -1. With this in mind, we can write the system (2) in the equivalent form

$$
\begin{aligned}
c_3 + w_2 + w_3 + w_4 &= 0 \\
c_2 + w_1 + w_3 + w_4 &= 0 \\
c_1 + w_1 + w_2 + w_4 &= 0.
\end{aligned}
\tag{4}
$$

These are called **parity check equations**. This means that each c_i is a parity check on three of the digits in the original word. For instance, if the number of ones in the three digits w_2, w_3, and w_4 is odd, then, as in the parity check code discussed previously, we would take $c_1 = 1$ and so on. As a matrix product, (4) can be written

$$
\begin{pmatrix} 0 & 0 & 0 & 1 & 1 & 1 & 1 \\ 0 & 1 & 1 & 0 & 0 & 1 & 1 \\ 1 & 0 & 1 & 0 & 1 & 0 & 1 \end{pmatrix} \begin{pmatrix} c_1 \\ c_2 \\ w_1 \\ c_3 \\ w_2 \\ w_3 \\ w_4 \end{pmatrix} = \begin{pmatrix} 0 \\ 0 \\ 0 \end{pmatrix}.
\tag{5}
$$

The 3×7 matrix in (5),

$$
\mathbf{H} = \begin{pmatrix} 0 & 0 & 0 & 1 & 1 & 1 & 1 \\ 0 & 1 & 1 & 0 & 0 & 1 & 1 \\ 1 & 0 & 1 & 0 & 1 & 0 & 1 \end{pmatrix},
$$

is called the **parity check matrix**. We have shown in (5) that the binary digits of a code word $\mathbf{C} = (c_1 \quad c_2 \quad w_1 \quad c_3 \quad w_2 \quad w_3 \quad w_4)$ satisfy the matrix equation

$$
\mathbf{H}\mathbf{C}^T = \mathbf{0}.
\tag{6}
$$

A closer inspection of $\mathbf{H}$ shows a surprising fact: The columns of $\mathbf{H}$, left to right, are the numbers 1 through 7 written in binary. For example, by writing the column $\begin{pmatrix} 1 \\ 1 \\ 0 \end{pmatrix}$ as 1 1 0, we recognize the binary representation of the number 6.

Let $\mathbf{R}$ be a 1×7 matrix representing the received message. Since $\mathbf{H}$ is a 3×7 matrix and $\mathbf{R}^T$ is a 7×1 matrix, the product

$$
\mathbf{S} = \mathbf{H}\mathbf{R}^T
\tag{7}
$$

is a 3×1 matrix called the **syndrome** of $\mathbf{R}$.

Decoding If the syndrome of the received message $\mathbf{R}$ is

$$\mathbf{S} = \mathbf{HR}^T = \mathbf{0},$$

then, in view of the result in (6), we conclude that $\mathbf{R}$ is a code word, and it is assumed that the transmission is correct and that $\mathbf{R}$ is the same as the original encoded message $\mathbf{C}$. The decoding of the message is accomplished by simply dropping the three check bits in $\mathbf{R}$.

EXAMPLE 5 **Syndromes**

Compute the syndrome of:
(a) $\mathbf{R} = (1 \ \ 1 \ \ 0 \ \ 1 \ \ 0 \ \ 0 \ \ 1)$ **(b)** $\mathbf{R} = (1 \ \ 0 \ \ 0 \ \ 1 \ \ 0 \ \ 1 \ \ 0).$

SOLUTION **(a)** From (7) we have

$$\mathbf{S} = \begin{pmatrix} 0 & 0 & 0 & 1 & 1 & 1 & 1 \\ 0 & 1 & 1 & 0 & 0 & 1 & 1 \\ 1 & 0 & 1 & 0 & 1 & 0 & 1 \end{pmatrix} \begin{pmatrix} 1 \\ 1 \\ 0 \\ 1 \\ 0 \\ 0 \\ 1 \end{pmatrix} = \begin{pmatrix} 0 \\ 0 \\ 0 \end{pmatrix}.$$

We conclude that $\mathbf{R}$ is a code word. By dropping the blue check bits in $(1 \ \ 1 \ \ 0 \ \ 1 \ \ 0 \ \ 0 \ \ 1)$, we get the decoded message $(0 \ \ 0 \ \ 0 \ \ 1)$.

(b) From (7), $\quad \mathbf{S} = \begin{pmatrix} 0 & 0 & 0 & 1 & 1 & 1 & 1 \\ 0 & 1 & 1 & 0 & 0 & 1 & 1 \\ 1 & 0 & 1 & 0 & 1 & 0 & 1 \end{pmatrix} \begin{pmatrix} 1 \\ 0 \\ 0 \\ 1 \\ 0 \\ 1 \\ 0 \end{pmatrix} = \begin{pmatrix} 0 \\ 1 \\ 1 \end{pmatrix}.$

Since $\mathbf{S} \neq \mathbf{0}$, the received message $\mathbf{R}$ is not a code word. ≡

As mentioned earlier, the Hamming $(7, 4)$ code enables us to detect and also to correct a single error in the message $\mathbf{R}$. Now let $\mathbf{C}$ be a code word and let $\mathbf{E} = (e_1 \ e_2 \ e_3 \ e_4 \ e_5 \ e_6 \ e_7)$ be a single-error noise word added to $\mathbf{C}$ during its transmission. The entries of $\mathbf{E}$ are defined by

$$e_i = \begin{cases} 1, & \text{if noise changes the } i\text{th bit} \\ 0, & \text{if noise does not change the } i\text{th bit.} \end{cases}$$

The received message is then $\mathbf{R} = \mathbf{C} + \mathbf{E}$. From the property $\mathbf{R}^T = \mathbf{C}^T + \mathbf{E}^T$ and the distributive law, we see that the syndrome of $\mathbf{R}$ is the same as the syndrome of $\mathbf{E}$:

$$\mathbf{HR}^T = \mathbf{H}(\mathbf{C}^T + \mathbf{E}^T) = \mathbf{HC}^T + \mathbf{HE}^T = \mathbf{0} + \mathbf{HE}^T = \mathbf{HE}^T.$$

From the definition of matrix addition, it is a straightforward matter to verify that the syndrome of $\mathbf{E}$

$$\mathbf{HE}^T = \begin{pmatrix} e_4 + e_5 + e_6 + e_7 \\ e_2 + e_3 + e_6 + e_7 \\ e_1 + e_3 + e_5 + e_7 \end{pmatrix}$$

can be written as the sum of the column vectors of $\mathbf{H}$ with coefficients the symbols that denote the bits where the error might occur:

$$\mathbf{HE}^T = e_1\begin{pmatrix}0\\0\\1\end{pmatrix} + e_2\begin{pmatrix}0\\1\\0\end{pmatrix} + e_3\begin{pmatrix}0\\1\\1\end{pmatrix} + e_4\begin{pmatrix}1\\0\\0\end{pmatrix} + e_5\begin{pmatrix}1\\0\\1\end{pmatrix} + e_6\begin{pmatrix}1\\1\\0\end{pmatrix} + e_7\begin{pmatrix}1\\1\\1\end{pmatrix}. \quad (8)$$

Now consider the set of 3×1 column vectors whose entries are binary digits. Since there are only two ways to select each of the three entries, there are $2^3 = 8$ such vectors. The seven *nonzero* vectors are the columns of $\mathbf{H}$ or the column vectors displayed in (8). The syndrome $\mathbf{S}$ of the received message $\mathbf{R}$ is a 3×1 column vector with binary entries; hence, if $\mathbf{S} \neq \mathbf{0}$, then $\mathbf{S}$ must be one of the columns of $\mathbf{H}$. If $\mathbf{R}$ contains a single error, then $\mathbf{S} \neq \mathbf{0}$ and, since the entries of $\mathbf{E}$ are all zero except for one entry of one, we see from (8) that the syndrome itself indicates which bit is in error. In practice there is no need to write out (8); just compute the syndrome $\mathbf{S}$ of the received message $\mathbf{R}$. $\mathbf{S}$ is a column of $\mathbf{H}$ and consequently is the binary number of the bit that is in error.

EXAMPLE 6 Decoding a Word

In part (b) of Example 5 we saw that the syndrome of the message $\mathbf{R} = (1 \quad 0 \quad 0 \quad 1 \quad 0 \quad 1 \quad 0)$ was $\mathbf{S} = \begin{pmatrix}0\\1\\1\end{pmatrix}$. This is the third column of $\mathbf{H}$ (or the binary representation of the number 3) and so we conclude that the third bit in $\mathbf{R}$ is incorrect. Changing zero to one gives the code word $\mathbf{C} = (1 \quad 0 \quad 1 \quad 1 \quad 0 \quad 1 \quad 0)$. Hence by dropping the first, second, and fourth bits from $\mathbf{C}$ we arrive at the decoded message $(1 \quad 0 \quad 1 \quad 0)$. ≡

In these brief descriptions of cryptography and coding theory we have not even begun to scratch the surface of these fascinating subjects. Our goal was a modest one: to show how matrix theory is a natural working tool in various areas of mathematics and computer science.

REMARKS

The Hamming (7, 4) code can detect but not correct any pair of errors. Students interested in how this is done or in further details of coding theory should check their library for more specialized texts.

8.15 Exercises Answers to selected odd-numbered problems begin on page ANS-21.

In Problems 1–6, encode the given word using the parity check code.

1. $(0 \quad 1 \quad 1)$
2. $(1 \quad 1 \quad 1)$
3. $(0 \quad 0 \quad 0 \quad 1)$
4. $(1 \quad 0 \quad 1 \quad 0)$
5. $(1 \quad 0 \quad 1 \quad 0 \quad 1 \quad 0 \quad 0)$
6. $(0 \quad 1 \quad 1 \quad 0 \quad 1 \quad 0 \quad 1)$

In Problems 7–12, decode the given message using the parity check code.

7. $(1 \quad 0 \quad 0 \quad 1)$
8. $(0 \quad 0 \quad 1 \quad 1)$
9. $(1 \quad 1 \quad 1 \quad 0 \quad 0)$
10. $(1 \quad 0 \quad 1 \quad 0 \quad 0)$
11. $(1 \quad 0 \quad 0 \quad 1 \quad 1 \quad 1)$
12. $(1 \quad 0 \quad 0 \quad 1 \quad 0 \quad 1)$

In Problems 13–18, encode the given word using the Hamming (7, 4) code.

13. $(1 \quad 1 \quad 1 \quad 0)$
14. $(0 \quad 0 \quad 1 \quad 1)$
15. $(0 \quad 1 \quad 0 \quad 1)$
16. $(0 \quad 0 \quad 0 \quad 1)$
17. $(0 \quad 1 \quad 1 \quad 0)$
18. $(1 \quad 1 \quad 0 \quad 0)$

In Problems 19–28, determine whether the given message is a code word in the Hamming (7, 4) code. If it is, decode it. If it is not, correct the single error and decode the corrected message.

19. $(0 \quad 0 \quad 0 \quad 0 \quad 0 \quad 0 \quad 0)$
20. $(1 \quad 1 \quad 0 \quad 0 \quad 0 \quad 0 \quad 0)$
21. $(1 \quad 1 \quad 0 \quad 1 \quad 1 \quad 0 \quad 1)$
22. $(0 \quad 1 \quad 0 \quad 1 \quad 0 \quad 1 \quad 0)$
23. $(1 \quad 1 \quad 1 \quad 1 \quad 1 \quad 1 \quad 1)$
24. $(1 \quad 1 \quad 0 \quad 0 \quad 1 \quad 1 \quad 0)$
25. $(0 \quad 1 \quad 1 \quad 1 \quad 0 \quad 0 \quad 1)$
26. $(1 \quad 0 \quad 0 \quad 1 \quad 0 \quad 0 \quad 1)$
27. $(1 \quad 0 \quad 1 \quad 1 \quad 0 \quad 1 \quad 1)$
28. $(0 \quad 0 \quad 1 \quad 0 \quad 0 \quad 1 \quad 1)$

29. (a) Determine the total number of 7-tuples with binary entries.
 (b) How many 7-tuple code words are there in the Hamming (7, 4) code?
 (c) List all code words in the Hamming (7, 4) code.

30. (a) In the Hamming (8, 4) code a word

$$\mathbf{W} = (w_1 \; w_2 \; w_3 \; w_4)$$

of length four is transformed into a code word of length eight:

$$\mathbf{C} = (c_1 \; c_2 \; c_3 \; w_1 \; c_4 \; w_2 \; w_3 \; w_4),$$

where the parity check equations are

$$c_4 + w_2 + w_3 + w_4 = 0$$
$$c_3 + w_1 + w_3 + w_4 = 0$$
$$c_2 + w_1 + w_2 + w_4 = 0$$
$$c_1 + c_2 + c_3 + w_1 + c_4 + w_2 + w_3 + w_4 = 0.$$

Encode the word (0 1 1 0).

(b) From the system in part (a), determine the parity check matrix $\mathbf{H}$.

(c) Using the matrix $\mathbf{H}$ from part (b), compute the syndrome $\mathbf{S}$ of the received message

$$\mathbf{R} = (0 \; 0 \; 1 \; 1 \; 1 \; 1 \; 0 \; 0).$$

8.16 Method of Least Squares

INTRODUCTION When performing experiments we often tabulate data in the form of ordered pairs $(x_1, y_1), (x_2, y_2), \ldots, (x_n, y_n)$, with each x_i distinct. Given the data, it is then often desirable to be able to extrapolate or predict y from x by finding a mathematical model, that is, a function that approximates or "fits" the data. In other words, we want a function $f(x)$ such that

$$f(x_1) \approx y_1, \quad f(x_2) \approx y_2, \quad \ldots, \quad f(x_n) \approx y_n.$$

But naturally we do not want just *any* function but a function that fits the data as closely as possible.

In the discussion that follows we shall confine our attention to the problem of finding a linear polynomial $f(x) = ax + b$ or straight line that "best fits" the data $(x_1, y_1), (x_2, y_2), \ldots, (x_n, y_n)$. The procedure for finding this linear function is known as the **method of least squares**.

We begin with an example.

(a)

(b)

FIGURE 8.16.1 Data and line in Example 1

EXAMPLE 1 Line of Best Fit

Consider the data (1, 1), (2, 3), (3, 4), (4, 6), (5, 5) shown in **FIGURE 8.16.1(a)**. Visually, and the fact that the line $y = x + 1$, shown in Figure 8.16.1(b), passes through two of the data points, we might take this line as that best fitting the data. ≡

Obviously we need something better than a visual guess to determine the linear function $y = f(x)$, as in the last example. We need a criterion that defines the concept of "best fit" or, as it is sometimes called, "the goodness of fit."

If we try to match the data points with the function $f(x) = ax + b$, then we wish to find a and b that satisfy the system of equations

$$
\begin{aligned}
y_1 &= ax_1 + b \\
y_2 &= ax_2 + b \\
&\;\;\vdots \\
y_n &= ax_n + b
\end{aligned}
\tag{1}
$$

or $\qquad \mathbf{Y} = \mathbf{AX} \qquad$ where $\qquad \mathbf{Y} = \begin{pmatrix} y_1 \\ y_2 \\ \vdots \\ y_n \end{pmatrix}, \quad \mathbf{A} = \begin{pmatrix} x_1 & 1 \\ x_2 & 1 \\ \vdots & \vdots \\ x_n & 1 \end{pmatrix}, \quad \mathbf{X} = \begin{pmatrix} a \\ b \end{pmatrix}. \tag{2}$

Unfortunately (1) is an overdetermined system and, unless the data points lie on the same line, has no solution. Thus we shall be content to find a vector $\mathbf{X} = \begin{pmatrix} a \\ b \end{pmatrix}$ so that the right side $\mathbf{AX}$ is close to the left side $\mathbf{Y}$.

FIGURE 8.16.2 e_i is the error in approximating y_i with $f(x_i)$

Least Squares Line If the data points are (x_1, y_1), (x_2, y_2), ..., (x_n, y_n), then one way to determine how well the linear function $f(x) = ax + b$ fits the data is to measure the vertical distances between the points and the graph of f:

$$e_i = |y_i - f(x_i)|, \quad i = 1, 2, \ldots, n.$$

We can think of each e_i as the error in approximating the data value y_i by the functional value $f(x_i)$. See **FIGURE 8.16.2**. Intuitively, we know that the function f will fit the data well if the sum of all the e_i values is a minimum. Actually, a more convenient approach to the problem is to find a linear function f so that the sum of the *squares* of all the e_i values is a minimum. We shall define the solution of the system (1) to be those coefficients a and b that minimize the expression $E = e_1^2 + e_2^2 + \cdots + e_n^2$; that is,

$$E = [y_1 - f(x_1)]^2 + [y_2 - f(x_2)]^2 + \cdots + [y_n - f(x_n)]^2$$
$$= [y_1 - (ax_1 + b)]^2 + [y_2 - (ax_2 + b)]^2 + \cdots + [y_n - (ax_n + b)]^2$$

or

$$E = \sum_{i=1}^{n} [y_i - ax_i - b]^2. \tag{3}$$

The expression E is called the **sum of the square errors**. The line $y = ax + b$ that minimizes the sum of the square errors (3) is defined to be the **line of best fit** and is called the **least squares line** for the data (x_1, y_1), (x_2, y_2), ..., (x_n, y_n).

The problem remains: How does one find a and b so that (3) is a minimum? The answer can be found from calculus. If we think of (3) as a function of two variables a and b, then to find the minimum value of E we set the first partial derivatives equal to zero:

Partial differentiation is reviewed in Section 9.4.

$$\frac{\partial E}{\partial a} = 0 \quad \text{and} \quad \frac{\partial E}{\partial b} = 0.$$

The last two conditions yield, in turn,

$$-2 \sum_{i=1}^{n} x_i [y_i - ax_i - b] = 0$$
$$-2 \sum_{i=1}^{n} [y_i - ax_i - b] = 0. \tag{4}$$

Expanding the sums and using $\sum_{i=1}^{n} b = nb$, we find the system (4) is the same as

$$\left(\sum_{i=1}^{n} x_i^2 \right) a + \left(\sum_{i=1}^{n} x_i \right) b = \sum_{i=1}^{n} x_i y_i$$
$$\left(\sum_{i=1}^{n} x_i \right) a + \qquad nb = \sum_{i=1}^{n} y_i. \tag{5}$$

Although we shall not give the details, the values of a and b that satisfy the system (5) yield the minimum value of E.

In terms of matrices it can be shown that (5) is equivalent to

$$\mathbf{A}^T \mathbf{A} \mathbf{X} = \mathbf{A}^T \mathbf{Y}, \tag{6}$$

where $\mathbf{A}$, $\mathbf{Y}$, and $\mathbf{X}$ are defined in (2). Since $\mathbf{A}$ is an $n \times 2$ matrix and $\mathbf{A}^T$ is a $2 \times n$ matrix, the matrix $\mathbf{A}^T \mathbf{A}$ is a 2×2 matrix. Moreover, unless the data points all lie on the same vertical line,

the matrix $\mathbf{A}^T\mathbf{A}$ is nonsingular. Thus, (6) has the unique solution

$$\mathbf{X} = (\mathbf{A}^T\mathbf{A})^{-1}\mathbf{A}^T\mathbf{Y}. \tag{7}$$

We say that $\mathbf{X}$ is the **least squares solution** of the overdetermined system (1).

EXAMPLE 2 **Least Squares Line**

Find the least squares line for the data in Example 1. Calculate the sum of the square errors E for this line and the line $y = x + 1$.

SOLUTION For the function $f(x) = ax + b$ the data $(1, 1)$, $(2, 3)$, $(3, 4)$, $(4, 6)$, $(5, 5)$ lead to the overdetermined system

$$a + b = 1$$
$$2a + b = 3$$
$$3a + b = 4 \tag{8}$$
$$4a + b = 6$$
$$5a + b = 5.$$

Now by identifying

$$\mathbf{Y} = \begin{pmatrix} 1 \\ 3 \\ 4 \\ 6 \\ 5 \end{pmatrix} \quad \text{and} \quad \mathbf{A} = \begin{pmatrix} 1 & 1 \\ 2 & 1 \\ 3 & 1 \\ 4 & 1 \\ 5 & 1 \end{pmatrix} \quad \text{we have} \quad \mathbf{A}^T\mathbf{A} = \begin{pmatrix} 55 & 15 \\ 15 & 5 \end{pmatrix},$$

and so (7) gives

$$\mathbf{X} = \begin{pmatrix} 55 & 15 \\ 15 & 5 \end{pmatrix}^{-1} \begin{pmatrix} 1 & 1 \\ 2 & 1 \\ 3 & 1 \\ 4 & 1 \\ 5 & 1 \end{pmatrix}^T \begin{pmatrix} 1 \\ 3 \\ 4 \\ 6 \\ 5 \end{pmatrix} = \frac{1}{50}\begin{pmatrix} 5 & -15 \\ -15 & 55 \end{pmatrix}\begin{pmatrix} 1 & 2 & 3 & 4 & 5 \\ 1 & 1 & 1 & 1 & 1 \end{pmatrix}\begin{pmatrix} 1 \\ 3 \\ 4 \\ 6 \\ 5 \end{pmatrix}$$

$$= \frac{1}{50}\begin{pmatrix} 5 & -15 \\ -15 & 55 \end{pmatrix}\begin{pmatrix} 68 \\ 19 \end{pmatrix} = \begin{pmatrix} 1.1 \\ 0.5 \end{pmatrix}.$$

Thus the least squares solution of (8) is $a = 1.1$ and $b = 0.5$, and the least squares line is $y = 1.1x + 0.5$. For this line the sum of the square errors is

$$E = [1 - f(1)]^2 + [3 - f(2)]^2 + [4 - f(3)]^2 + [6 - f(4)]^2 + [5 - f(5)]^2$$
$$= [1 - 1.6]^2 + [3 - 2.7]^2 + [4 - 3.8]^2 + [6 - 4.9]^2 + [5 - 6]^2 = 2.7.$$

For the line $y = x + 1$ that we guessed and that also passed through two of the data points, we find $E = 3.0$.

By way of comparison, **FIGURE 8.16.3** shows the data points, the line $y = x + 1$ (green), and the least squares line $y = 1.1x + 0.5$ (blue). ≡

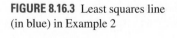

FIGURE 8.16.3 Least squares line (in blue) in Example 2

■ **Least Squares Parabola** The procedure illustrated in Example 2 is easily modified to find a least squares parabola.

EXAMPLE 3 **Least Squares Parabola**

Find the least squares parabola for the data (1, 1), (2, 4), (3, 7), (4, 5).

SOLUTION For the quadratic function $f(x) = ax^2 + bx + c$, the analogue of system (8) is

$$
\begin{aligned}
a + b + c &= 1 \\
4a + 2b + c &= 4 \\
9a + 3b + c &= 7 \\
16a + 4b + c &= 5.
\end{aligned}
$$

From this system we see the matrix **A** now has three columns. So with

$$
\mathbf{Y} = \begin{pmatrix} 1 \\ 4 \\ 7 \\ 5 \end{pmatrix}, \quad
\mathbf{A} = \begin{pmatrix} 1 & 1 & 1 \\ 4 & 2 & 1 \\ 9 & 3 & 1 \\ 16 & 4 & 1 \end{pmatrix} \quad \text{and} \quad
\mathbf{A}^T\mathbf{A} = \begin{pmatrix} 354 & 100 & 30 \\ 100 & 30 & 10 \\ 30 & 10 & 4 \end{pmatrix}
$$

equation (7) gives

$$
\mathbf{X} = \begin{pmatrix} 354 & 100 & 30 \\ 100 & 30 & 10 \\ 30 & 10 & 4 \end{pmatrix}^{-1}
\begin{pmatrix} 1 & 1 & 1 \\ 4 & 2 & 1 \\ 9 & 3 & 1 \\ 16 & 4 & 1 \end{pmatrix}^T
\begin{pmatrix} 1 \\ 4 \\ 7 \\ 5 \end{pmatrix}
$$

$$
= \frac{1}{20} \begin{pmatrix} 5 & -25 & 25 \\ -25 & 129 & -135 \\ 25 & -135 & 155 \end{pmatrix}
\begin{pmatrix} 1 & 4 & 9 & 16 \\ 1 & 2 & 3 & 4 \\ 1 & 1 & 1 & 1 \end{pmatrix}
\begin{pmatrix} 1 \\ 4 \\ 7 \\ 5 \end{pmatrix}
$$

$$
= \frac{1}{20} \begin{pmatrix} 5 & -25 & 25 \\ -25 & 129 & -135 \\ 25 & -135 & 155 \end{pmatrix}
\begin{pmatrix} 160 \\ 50 \\ 17 \end{pmatrix}
= \begin{pmatrix} -1.25 \\ 7.75 \\ -5.75 \end{pmatrix}
$$

FIGURE 8.16.4 Least squares parabola in Example 3

Therefore $a = -1.25$, $b = 7.75$, $c = -5.75$ and the equation of the least squares parabola is $f(x) = -1.25x^2 + 7.75x - 5.75$. The graphs of the data points and the quadratic function f are given in **FIGURE 8.16.4**. ≡

8.16 Exercises Answers to selected odd-numbered problems begin on page ANS-22.

In Problems 1–6, find the least squares line for the given data.

1. (2, 1), (3, 2), (4, 3), (5, 2)

2. (0, −1), (1, 3), (2, 5), (3, 7)

3. (1, 1), (2, 1.5), (3, 3), (4, 4.5), (5, 5)

4. (0, 0), (2, 1.5), (3, 3), (4, 4.5), (5, 5)

5. (0, 2), (1, 3), (2, 5), (3, 5), (4, 9), (5, 8), (6, 10)

6. (1, 2), (2, 2.5), (3, 1), (4, 1.5), (5, 2), (6, 3.2), (7, 5)

7. In an experiment, the following correspondence was found between temperature T (in °C) and kinematic viscosity v (in Centistokes) of an oil with a certain additive:

T	20	40	60	80	100	120
v	220	200	180	170	150	135

Find the least squares line $v = aT + b$. Use this line to estimate the viscosity of the oil at $T = 140$ and $T = 160$.

8. In an experiment the following correspondence was found between temperature T (in °C) and electrical resistance R (in MΩ):

T	400	450	500	550	600	650
R	0.47	0.90	2.0	3.7	7.5	15

Find the least squares line $R = aT + b$. Use this line to estimate the resistance at $T = 700$.

In Problems 9 and 10, proceed as in Example 3 and find the least squares parabola for the given data.

9. (1, 1), (2, 1), (3, 2), (4, 5)

10. (−2, 1), (−1, 1), (1, 2), (2, 3)

8.17 Discrete Compartmental Models

INTRODUCTION The construction of the mathematical model (3) in Section 2.9, describing the number of pounds of salt in two connected tanks in which brine is flowing into and out of the tanks, is an example of **compartmental analysis**. In the discussion in Section 2.9, the compartmental model was a system of differential equations. In this section we introduce the notion of a discrete mathematical model.

■ The General Two-Compartment Model

Suppose material flows between two tanks with volumes V_1 and V_2. In the diagram shown in **FIGURE 8.17.1**, F_{01}, F_{12}, F_{21}, F_{10}, and F_{20} denote flow rates. Note that the double-subscripted symbol F_{ij} denotes the flow rate from tank i to tank j. Next, suppose a second substance, called the **tracer**, is infused into compartment 1 at a known rate $I(t)$. As we did in Section 2.9, we will assume that the tracer is thoroughly mixed in both compartments at all times t. If $x(t)$ denotes the amount of tracer in compartment 1 and $y(t)$ the amount of tracer in compartment 2, then the concentrations are $c_1(t) = x(t)/V_1$ and $c_2(t) = y(t)/V_2$, respectively. It follows that the general two-compartment model is

$$\frac{dx}{dt} = -(F_{12} + F_{10})c_1(t) + F_{21}c_2(t) + I(t)$$

$$\frac{dy}{dt} = F_{21}c_1(t) - (F_{21} + F_{20})c_2(t). \tag{1}$$

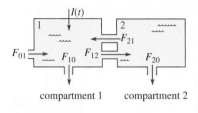

FIGURE 8.17.1 Material flowing between two compartments at specified rates

The model in (1) keeps track of the amount of tracer that flows between the compartments. The material consisting of, say, a fluid and a tracer is *continually interchanged*. We present next a model that keeps track of compartmental contents every Δt units of time and assumes that the system changes only at times $\Delta t, 2\Delta t, \ldots, n\Delta t, \ldots$. Of course, by selecting Δt very small, we can approximate the continuous case.

■ Discrete Compartmental Models

In constructing a compartmental model of a physical system, we conceptually separate the system into a distinct number of small components between which material is transported. Compartments need not be spatially distinct (like the tanks used in Section 2.9) but must be distinguishable on some basis. The following are a few examples:

- Acid rain (containing strontium 90, for example) is deposited onto pastureland. Compartments might be grasses, soil, streams, and litter.
- In studying the flow of energy through an aquatic ecosystem, we might separate the system into phytoplankton, zooplankton, plankton predators, seaweed, small carnivores, large carnivores, and decay organisms.
- A tracer is infused into the bloodstream and is lost to the body by the metabolism of a particular organ and by excretion. Appropriate compartments might be arterial blood, venous blood, the organ, and urine.

Suppose then that a system is divided into n compartments and, after each Δt units of time, material is interchanged between compartments. We will assume that a fixed fraction τ_{ij} of the contents of compartment j are passed to compartment i every Δt units of time, as depicted in **FIGURE 8.17.2**. This hypothesis is known as the **linear donor-controlled hypothesis**.

Let the entries x_i in the $n \times 1$ matrix **X**,

FIGURE 8.17.2 Interchange of material between compartments

$$\mathbf{X} = \begin{pmatrix} x_1 \\ x_2 \\ \vdots \\ x_n \end{pmatrix}, \quad \mathbf{Y} = \begin{pmatrix} y_1 \\ y_2 \\ \vdots \\ y_n \end{pmatrix}, \tag{2}$$

represent the amount of tracer in compartment i. We say that **X** specifies the **state of the system**. The $n \times 1$ matrix **Y** is the state of the system Δt units of time later. We will show that **X** and **Y**

are related by the matrix equation $\mathbf{Y} = \mathbf{TX}$, where $\mathbf{T}$ is an $n \times n$ matrix determined by the **transfer coefficients** τ_{ij}. To find $\mathbf{T}$, observe, for example, that

$$y_1 = x_1 + (\text{amount of tracer entering } 1) - (\text{amount of tracer leaving } 1)$$
$$= x_1 + (\tau_{12}x_2 + \tau_{13}x_3 + \cdots + \tau_{1n}x_n) - (\tau_{21} + \tau_{31} + \cdots + \tau_{n1})x_1$$
$$= (1 - \tau_{21} - \tau_{31} - \cdots - \tau_{n1})x_1 + \tau_{12}x_2 + \cdots + \tau_{1n}x_n.$$

If we let $\tau_{11} = 1 - \tau_{21} - \tau_{31} - \cdots - \tau_{n1}$, then τ_{11} is just the fraction of the contents of compartment 1 that remains in 1.

Letting $\tau_{ii} = 1 - \Sigma_{j \neq i}\tau_{ji}$ we have, in general,

$$
\begin{aligned}
y_1 &= \tau_{11}x_1 + \tau_{12}x_2 + \cdots + \tau_{1n}x_n \\
y_2 &= \tau_{21}x_1 + \tau_{22}x_2 + \cdots + \tau_{2n}x_n \\
&\vdots \qquad \vdots \\
y_n &= \tau_{n1}x_1 + \tau_{n2}x_2 + \cdots + \tau_{nn}x_n
\end{aligned}
\quad \text{or} \quad
\begin{pmatrix} y_1 \\ y_2 \\ \vdots \\ y_n \end{pmatrix}
=
\begin{pmatrix}
\tau_{11} & \tau_{12} & \cdots & \tau_{1n} \\
\tau_{21} & \tau_{22} & \cdots & \tau_{2n} \\
\vdots & & & \vdots \\
\tau_{n1} & \tau_{n2} & \cdots & \tau_{nn}
\end{pmatrix}
\begin{pmatrix} x_1 \\ x_2 \\ \vdots \\ x_n \end{pmatrix}. \quad (3)
$$

The matrix equation in (3) is the desired equation $\mathbf{Y} = \mathbf{TX}$. The matrix $\mathbf{T} = (\tau_{ij})_{n \times n}$ is called the **transfer matrix**. Note that the sum of the entries in any column, the transfer coefficients, is equal to 1.

Discrete compartmental models are illustrated in the next two examples.

Note: A transfer matrix is an example of a stochastic matrix. See Problem 32 in Exercises 8.8.

EXAMPLE 1 Transfer Matrix

In **FIGURE 8.17.3** the three boxes represent three compartments. The content of each compartment at time t is indicated in each box. The transfer coefficients are shown along the arrows connecting the compartments.

(a) Find the transfer matrix $\mathbf{T}$.
(b) Suppose $\Delta t = 1$ day. Find the state of the system $\mathbf{Y}$ 1 day later.

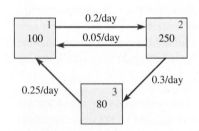

FIGURE 8.17.3 Compartments and transfer coefficients in Example 1

SOLUTION **(a)** The state of the system at time $t = 0$ is $\mathbf{X} = \begin{pmatrix} 100 \\ 250 \\ 80 \end{pmatrix}$. Remember that τ_{ij} specifies the rate of transfer to compartment i from compartment j. Hence we are given that $\tau_{21} = 0.2$, $\tau_{12} = 0.05$, $\tau_{32} = 0.3$, $\tau_{23} = 0$, $\tau_{13} = 0.25$, and $\tau_{31} = 0$. From these numbers we see that the matrix $\mathbf{T}$ is

$$
\mathbf{T} = \begin{pmatrix}
- & 0.05 & 0.25 \\
0.2 & - & 0 \\
0 & 0.3 & -
\end{pmatrix}. \quad (4)
$$

But since the column entries must sum to 1, we can fill in the blanks in (4):

$$
\mathbf{T} = \begin{pmatrix}
0.8 & 0.05 & 0.25 \\
0.2 & 0.65 & 0 \\
0 & 0.3 & 0.75
\end{pmatrix}
$$

(b) The state of the system 1 day later is then

$$
\mathbf{Y} = \mathbf{TX} = \begin{pmatrix}
0.8 & 0.05 & 0.25 \\
0.2 & 0.65 & 0 \\
0 & 0.3 & 0.75
\end{pmatrix}
\begin{pmatrix} 100 \\ 250 \\ 80 \end{pmatrix}
= \begin{pmatrix} 112.5 \\ 182.5 \\ 135 \end{pmatrix}. \qquad \equiv
$$

If $\mathbf{X}_0$ denotes the initial state of the system, and $\mathbf{X}_n$ is the state after $n(\Delta t)$ units of time, then

$$\mathbf{X}_1 = \mathbf{T}\mathbf{X}_0, \quad \mathbf{X}_2 = \mathbf{T}\mathbf{X}_1, \quad \mathbf{X}_3 = \mathbf{T}\mathbf{X}_2, \dots, \quad \mathbf{X}_{n+1} = \mathbf{T}\mathbf{X}_n.$$

Because

$$\mathbf{X}_2 = \mathbf{T}(\mathbf{T}\mathbf{X}_0) = \mathbf{T}^2\mathbf{X}_0, \quad \mathbf{X}_3 = \mathbf{T}(\mathbf{T}^2\mathbf{X}_0) = \mathbf{T}^3\mathbf{X}_0, \dots,$$

we have in general

$$\mathbf{X}_n = \mathbf{T}^n\mathbf{X}_0, \quad n = 1, 2, \dots. \tag{5}$$

We could, of course, use the method illustrated in Section 8.9 to compute $\mathbf{T}^n$, but with the aid of a calculator or a CAS it is just as easy to use the recursion formula $\mathbf{X}_{n+1} = \mathbf{T}\mathbf{X}_n$ by letting $n = 0, 1, \dots$.

EXAMPLE 2 States of an Ecosystem

Strontium-90 is deposited into pastureland by rainfall. To study how this material is cycled through the ecosystem, we divide the system into the compartments shown in **FIGURE 8.17.4**. Suppose that $\Delta t = 1$ month and the transfer coefficients (which have been estimated experimentally) shown in the figure are measured in fraction/month. (We will ignore that some strontium-90 is lost due to radioactive decay.) Suppose that rainfall has deposited the strontium-90 into the compartments so that $\mathbf{X}_0 = \begin{pmatrix} 20 \\ 60 \\ 15 \\ 20 \end{pmatrix}$. (Units might be grams per hectare.) Compute the states of the ecosystem over the next 12 months.

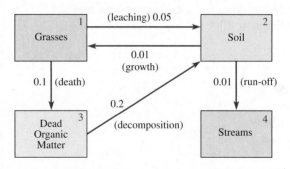

FIGURE 8.17.4 Ecosystem in Example 2

SOLUTION From the data in Figure 8.17.4 we see that transfer matrix $\mathbf{T}$ is

$$\mathbf{T} = \begin{pmatrix} 0.85 & 0.01 & 0 & 0 \\ 0.05 & 0.98 & 0.2 & 0 \\ 0.1 & 0 & 0.8 & 0 \\ 0 & 0.01 & 0 & 1 \end{pmatrix}.$$

We must compute $\mathbf{X}_1, \mathbf{X}_2, \dots, \mathbf{X}_{12}$. The state of the ecosystem after the first month is

$$\mathbf{X}_1 = \mathbf{T}\mathbf{X}_0 = \begin{pmatrix} 0.85 & 0.01 & 0 & 0 \\ 0.05 & 0.98 & 0.2 & 0 \\ 0.1 & 0 & 0.8 & 0 \\ 0 & 0.01 & 0 & 1 \end{pmatrix} \begin{pmatrix} 20 \\ 60 \\ 15 \\ 20 \end{pmatrix} = \begin{pmatrix} 17.6 \\ 62.8 \\ 14.0 \\ 20.6 \end{pmatrix}.$$

The remaining states, computed with the aid of a CAS and the recursion formula $\mathbf{X}_{n+1} = \mathbf{T}\mathbf{X}_n$ with $n = 1, 2, \dots, 11$, are given in Table 8.17.1. ≡

TABLE 8.17.1

Month	Grasses	Soil	Dead Organic Matter	Streams
0	20.00	60.00	15.00	20.00
1	17.60	62.80	14.00	20.60
2	15.59	65.22	12.96	21.23
3	13.90	67.29	11.93	21.88
4	12.49	69.03	10.93	22.55
5	11.31	70.46	9.99	23.24
6	10.32	71.61	9.13	23.95
7	9.48	72.52	8.33	24.66
8	8.79	73.21	7.61	25.39
9	8.20	73.71	6.97	26.12
10	7.71	74.04	6.40	26.86
11	7.29	74.22	5.89	27.60
12	6.94	74.28	5.44	28.34

1. (a) Use the data in the compartmental diagram in **FIGURE 8.17.5** to determine the appropriate transfer matrix **T** and the initial state of the system $\mathbf{X}_0$.

(b) Find the state of the system after 1 day. After 2 days.

(c) Eventually, the system will reach an equilibrium state $\hat{\mathbf{X}} = \begin{pmatrix} x_1 \\ x_2 \end{pmatrix}$ that satisfies $\mathbf{T}\hat{\mathbf{X}} = \hat{\mathbf{X}}$. Find $\hat{\mathbf{X}}$. [*Hint*: $x_1 + x_2 = 150$.]

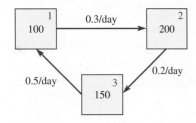

FIGURE 8.17.5 Compartments for Problem 1

2. (a) Use the data in the compartmental diagram in **FIGURE 8.17.6** to determine the appropriate transfer matrix **T** and the initial state of the system $\mathbf{X}_0$.

(b) Find the state of the system after 1 day. After 2 days.

(c) Find the equilibrium state $\hat{\mathbf{X}} = \begin{pmatrix} x_1 \\ x_2 \\ x_3 \end{pmatrix}$ that satisfies $\mathbf{T}\hat{\mathbf{X}} = \hat{\mathbf{X}}$. [*Hint*: What is the analogue of the hint in part (c) of Problem 1?]

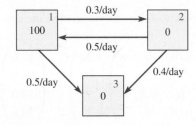

FIGURE 8.17.6 Compartments for Problem 2

3. (a) Use the data in the compartmental diagram in **FIGURE 8.17.7** to determine the appropriate transfer matrix **T** and the initial state of the system $\mathbf{X}_0$.

(b) Find the state of the system after 1 day. After 2 days.

(c) Find the equilibrium state $\hat{\mathbf{X}} = \begin{pmatrix} x_1 \\ x_2 \\ x_3 \end{pmatrix}$ that satisfies $\mathbf{T}\hat{\mathbf{X}} = \hat{\mathbf{X}}$.

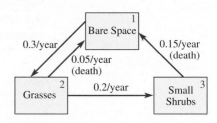

FIGURE 8.17.7 Compartments for Problem 3

4. A field has been completely devastated by fire. Two types of vegetation, grasses, and small shrubs will first begin to grow, but the small shrubs can take over an area only if preceded by the grasses. In **FIGURE 8.17.8**, the transfer coefficient of 0.3 indicates that, by the end of the summer, 30% of the prior bare space in the field becomes occupied by grasses.

(a) Find the transfer matrix **T**.

(b) Suppose $\mathbf{X} = \begin{pmatrix} 10 \\ 0 \\ 0 \end{pmatrix}$ and that area is measured in acres. Use the recursion formula $\mathbf{X}_{n+1} = \mathbf{T}\mathbf{X}_n$, along with a calculator or a CAS, to determine the ground cover in each of the next 6 years.

FIGURE 8.17.8 Compartments for Problem 4

Discussion Problem

5. Characterize the vector $\hat{\mathbf{X}}$ in part (c) of Problems 1–3 in terms of one of the principal concepts in Section 8.8.

Computer Lab Assignment

6. Radioisotopes (such as phosphorus-32 and carbon-14) have been used to study the transfer of nutrients in food chains. **FIGURE 8.17.9** is a compartmental representation of a simple aquatic food chain. One hundred units (for example, microcuries) of tracer are dissolved in the water of an aquarium containing a species of phytoplankton and a species of zooplankton.

(a) Find the transfer matrix **T** and the initial state of the system $\mathbf{X}_0$.

(b) Instead of the recursion formula, use $\mathbf{X}_n = \mathbf{T}^n\mathbf{X}_0$, $n = 1, 2, \ldots, 12$, to predict the state of the system for the next 12 hours. Use a CAS and the command to compute powers of matrices (in *Mathematica* it is **MatrixPower[T, n]**) to find $\mathbf{T}^2, \mathbf{T}^3, \ldots, \mathbf{T}^{12}$.

FIGURE 8.17.9 Aquatic food chain in Problem 6

In Problems 1–20, fill in the blanks or answer true/false.

1. A matrix $\mathbf{A} = (a_{ij})_{4\times3}$ such that $a_{ij} = i + j$ is given by _____.

2. If $\mathbf{A}$ is a 4×7 matrix and $\mathbf{B}$ is a 7×3 matrix, then the size of $\mathbf{AB}$ is _____.

3. If $\mathbf{A} = \begin{pmatrix} 1 \\ 2 \end{pmatrix}$ and $\mathbf{B} = (3 \quad 4)$, then $\mathbf{AB} =$ _____ and $\mathbf{BA} =$ _____.

4. If $\mathbf{A} = \begin{pmatrix} 1 & 2 \\ 3 & 4 \end{pmatrix}$, then $\mathbf{A}^{-1} =$ _____.

5. If $\mathbf{A}$ and $\mathbf{B}$ are $n \times n$ nonsingular matrices, then $\mathbf{A} + \mathbf{B}$ is necessarily nonsingular. _____

6. If $\mathbf{A}$ is a nonsingular matrix for which $\mathbf{AB} = \mathbf{AC}$, then $\mathbf{B} = \mathbf{C}$. _____

7. If $\mathbf{A}$ is a 3×3 matrix such that $\det \mathbf{A} = 5$, then $\det(\frac{1}{2}\mathbf{A}) =$ _____ and $\det(-\mathbf{A}^T) =$ _____.

8. If $\det \mathbf{A} = 6$ and $\det \mathbf{B} = 2$, then $\det \mathbf{AB}^{-1} =$ _____.

9. If $\mathbf{A}$ and $\mathbf{B}$ are $n \times n$ matrices whose corresponding entries in the third column are the same, then $\det(\mathbf{A} - \mathbf{B}) =$ _____.

10. Suppose $\mathbf{A}$ is a 3×3 matrix such that $\det \mathbf{A} = 2$. If $\mathbf{B} = 10\mathbf{A}$ and $\mathbf{C} = -\mathbf{B}^{-1}$, then $\det \mathbf{C} =$ _____.

11. Let $\mathbf{A}$ be an $n \times n$ matrix. The eigenvalues of $\mathbf{A}$ are the nonzero solutions of $\det(\mathbf{A} - \lambda\mathbf{I}) = 0$. _____

12. A nonzero scalar multiple of an eigenvector is also an eigenvector corresponding to the same eigenvalue. _____

13. An $n \times 1$ column vector $\mathbf{K}$ with all zero entries is never an eigenvector of an $n \times n$ matrix $\mathbf{A}$. _____

14. Let $\mathbf{A}$ be an $n \times n$ matrix with real entries. If λ is a complex eigenvalue, then $\bar{\lambda}$ is also an eigenvalue of $\mathbf{A}$. _____

15. An $n \times n$ matrix $\mathbf{A}$ always possesses n linearly independent eigenvectors. _____

16. The augmented matrix $\begin{pmatrix} 1 & 1 & 1 & | & 2 \\ 0 & 1 & 0 & | & 3 \\ 0 & 0 & 0 & | & 0 \end{pmatrix}$ is in reduced row-echelon form. _____

17. If a 3×3 matrix $\mathbf{A}$ is diagonalizable, then it possesses three linearly independent eigenvectors. _____

18. The only matrices that are orthogonally diagonalizable are symmetric matrices. _____

19. The symmetric matrix $\mathbf{A} = \begin{pmatrix} 1 & -1 \\ -1 & 1 \end{pmatrix}$ is orthogonal. _____

20. The eigenvalues of a symmetric matrix with real entries are always real numbers. _____

21. An $n \times n$ matrix $\mathbf{B}$ is symmetric if $\mathbf{B}^T = \mathbf{B}$, and an $n \times n$ matrix $\mathbf{C}$ is skew-symmetric if $\mathbf{C}^T = -\mathbf{C}$. By noting the identity $2\mathbf{A} = \mathbf{A} + \mathbf{A}^T + \mathbf{A} - \mathbf{A}^T$, show that any $n \times n$ matrix $\mathbf{A}$ can be written as the sum of a symmetric matrix and a skew-symmetric matrix.

22. Show that there exists no 2×2 matrix with real entries such that $\mathbf{A}^2 = \begin{pmatrix} 0 & 1 \\ 1 & 0 \end{pmatrix}$.

23. An $n \times n$ matrix $\mathbf{A}$ is said to be **nilpotent** if, for some positive integer m, $\mathbf{A}^m = \mathbf{0}$. Find a 2×2 nilpotent matrix $\mathbf{A} \neq \mathbf{0}$.

24. (a) Two $n \times n$ matrices $\mathbf{A}$ and $\mathbf{B}$ are said to **anticommute** if $\mathbf{AB} = -\mathbf{BA}$. Show that each of the **Pauli spin matrices**

$$\sigma_x = \begin{pmatrix} 0 & 1 \\ 1 & 0 \end{pmatrix} \quad \sigma_y = \begin{pmatrix} 0 & -i \\ i & 0 \end{pmatrix} \quad \sigma_z = \begin{pmatrix} 1 & 0 \\ 0 & -1 \end{pmatrix}$$

where $i^2 = -1$, anticommutes with the others. **Wolfgang Ernst Pauli** (1900–1958) was an Austrian-born theoretical physicist and one the pioneers of quantum physics. Pauli received the Nobel Prize in Physics in 1945.

(b) The matrix $\mathbf{C} = \mathbf{AB} - \mathbf{BA}$ is said to be the **commutator** of the $n \times n$ matrices $\mathbf{A}$ and $\mathbf{B}$. Find the commutators of σ_x and σ_y, σ_y and σ_z, and σ_z and σ_x.

In Problems 25 and 26, solve the given system of equations by Gauss–Jordan elimination.

25. $\begin{pmatrix} 5 & -1 & 1 \\ 2 & 4 & 0 \\ 1 & 1 & 5 \end{pmatrix}\mathbf{X} = \begin{pmatrix} -9 \\ 27 \\ 9 \end{pmatrix}$

26.
$$x_1 + x_2 + x_3 = 6$$
$$x_1 - 2x_2 + 3x_3 = 2$$
$$2x_1 - \quad\quad 3x_3 = 3$$

27. Without expanding, show that $\begin{vmatrix} 1 & 1 & 1 \\ \frac{1}{a} & \frac{1}{b} & \frac{1}{c} \\ bc & ac & ab \end{vmatrix} = 0$.

28. Show that $\begin{vmatrix} y & x^2 & x & 1 \\ 2 & 1 & 1 & 1 \\ 3 & 4 & 2 & 1 \\ 5 & 9 & 3 & 1 \end{vmatrix} = 0$ is the equation of a parabola passing through the three points $(1, 2)$, $(2, 3)$, and $(3, 5)$.

In Problems 29 and 30, evaluate the determinant of the given matrix by inspection.

29. $\begin{pmatrix} 4 & 0 & 0 & 0 & 0 & 0 \\ 0 & -2 & 0 & 0 & 0 & 0 \\ 0 & 0 & 3 & 0 & 0 & 0 \\ 0 & 0 & 0 & -1 & 0 & 0 \\ 0 & 0 & 0 & 0 & 2 & 0 \\ 0 & 0 & 0 & 0 & 0 & 5 \end{pmatrix}$

30. $\begin{pmatrix} -3 & 0 & 0 & 0 \\ 4 & 6 & 0 & 0 \\ 1 & 3 & 9 & 0 \\ 6 & 4 & 2 & 1 \end{pmatrix}$

In Problems 31 and 32, without solving, state whether the given homogeneous system has only the trivial solution or has infinitely many solutions.

31.
$$x_1 - x_2 + x_3 = 0$$
$$5x_1 + x_2 - x_3 = 0$$
$$x_1 + 2x_2 + x_3 = 0$$

32.
$$x_1 - x_2 - x_3 = 0$$
$$5x_1 + x_2 - x_3 = 0$$
$$x_1 + 2x_2 + x_3 = 0$$

In Problems 33 and 34, use Gauss–Jordan elimination to balance the given chemical equation.

33. $I_2 + HNO_3 \rightarrow HIO_3 + NO_2 + H_2O$

34. $Ca + H_3PO_4 \rightarrow Ca_3P_2O_8 + H_2$

In Problems 35 and 36, solve the given system of equations by Cramer's rule.

35. $x_1 + 2x_2 - 3x_3 = -2$

$2x_1 - 4x_2 + 3x_3 = 0$

$4x_2 + 6x_3 = 5$

36. $x_1 + \qquad x_3 = 4$

$2x_1 + 3x_2 + 4x_3 = 5$

$x_1 + 4x_2 + 5x_3 = 0$

37. Use Cramer's rule to solve the system

$$X = x \cos \theta + y \sin \theta$$
$$Y = -x \sin \theta + y \cos \theta$$

for x and y.

38. **(a)** Set up the system of equations for the currents in the branches of the network given in **FIGURE 8.R.1**.

(b) Use Cramer's rule to show that

$$i_1 = E\left(\frac{1}{R_1} + \frac{1}{R_2} + \frac{1}{R_3}\right).$$

FIGURE 8.R.1 Network in Problem 38

39. Solve the system

$$2x_1 + 3x_2 - x_3 = 6$$
$$x_1 - 2x_2 \qquad = -3$$
$$-2x_1 + \qquad x_3 = 9$$

by writing it as a matrix equation and finding the inverse of the coefficient matrix.

40. Use the inverse of the matrix **A** to solve the system $\mathbf{AX} = \mathbf{B}$, where

$$\mathbf{A} = \begin{pmatrix} 1 & 2 & 3 \\ 2 & 3 & 0 \\ 0 & 1 & 2 \end{pmatrix}$$

and the vector **B** is given by **(a)** $\begin{pmatrix} 1 \\ 1 \\ 1 \end{pmatrix}$ **(b)** $\begin{pmatrix} -2 \\ 1 \\ 3 \end{pmatrix}$.

In Problems 41–46, find the eigenvalues and corresponding eigenvectors of the given matrix.

41. $\begin{pmatrix} 1 & 2 \\ 4 & 3 \end{pmatrix}$

42. $\begin{pmatrix} 0 & 0 \\ 4 & 0 \end{pmatrix}$

43. $\begin{pmatrix} 3 & 2 & 4 \\ 2 & 0 & 2 \\ 4 & 2 & 3 \end{pmatrix}$

44. $\begin{pmatrix} 7 & -2 & 0 \\ -2 & 6 & 2 \\ 0 & 2 & 5 \end{pmatrix}$

45. $\begin{pmatrix} -2 & 2 & -3 \\ 2 & 1 & -6 \\ -1 & -2 & 0 \end{pmatrix}$

46. $\begin{pmatrix} 0 & 0 & 0 \\ 0 & 0 & 1 \\ 2 & 2 & 1 \end{pmatrix}$

47. Supply a first column so that the matrix is orthogonal:

$$\begin{pmatrix} \square & -\dfrac{1}{\sqrt{2}} & \dfrac{1}{\sqrt{3}} \\ \square & 0 & \dfrac{1}{\sqrt{3}} \\ \square & \dfrac{1}{\sqrt{2}} & \dfrac{1}{\sqrt{3}} \end{pmatrix}.$$

48. Consider the symmetric matrix $\mathbf{A} = \begin{pmatrix} 1 & 0 & -2 \\ 0 & 0 & 0 \\ -2 & 0 & 4 \end{pmatrix}$.

(a) Find matrices **P** and $\mathbf{P}^{-1}$ that orthogonally diagonalize the matrix **A**.

(b) Find the diagonal matrix **D** by actually carrying out the multiplication $\mathbf{P}^{-1}\mathbf{AP}$.

49. Identify the conic section $x^2 + 3xy + y^2 = 1$.

50. Consider the following population data:

Year	1890	1900	1910	1920	1930
Population (in millions)	63	76	92	106	123

The actual population in 1940 was 132 million. Compare this amount with the population predicted from the least squares line for the given data.

In Problems 51 and 52, use the matrix $\mathbf{A} = \begin{pmatrix} 10 & 1 \\ 9 & 1 \end{pmatrix}$ to encode the given message. Use the correspondence (1) of Section 8.14.

51. SATELLITE LAUNCHED ON FRI

52. SEC AGNT ARRVS TUES AM

In Problems 53 and 54, use the matrix $\mathbf{A} = \begin{pmatrix} 0 & 1 & 0 \\ 1 & 1 & 1 \\ 1 & -1 & 2 \end{pmatrix}$ to decode the given message. Use the correspondence (1) in Section 8.14.

53. $\mathbf{B} = \begin{pmatrix} 19 & 0 & 15 & 14 & 0 & 20 \\ 35 & 10 & 27 & 53 & 1 & 54 \\ 5 & 15 & -3 & 48 & 2 & 39 \end{pmatrix}$

54. $\mathbf{B} = \begin{pmatrix} 5 & 2 & 21 \\ 27 & 17 & 40 \\ 21 & 13 & -2 \end{pmatrix}$

55. Decode the following messages using the parity check code.

(a) $(1\,1\,0\,0\,1\,1)$ **(b)** $(0\,1\,1\,0\,1\,1\,1\,0)$

56. Encode the word $(1\,0\,0\,1)$ using the Hamming $(7, 4)$ code.

In Problems 57 and 58, solve the given system of equations using LU-factorization.

57. The system in Problem 26

58. The system in Problem 36

59. Find the least squares line for the data $(1, -2), (2, 0), (3, 5), (4, -1)$.

60. Find the least squares parabola for the data given in Problem 59.

Vector Calculus

In Chapter 7 we studied the properties of vectors in 2- and 3-space. In this chapter we combine vector concepts with those from differential and integral calculus.

9.1 Vector Functions

INTRODUCTION Recall that a **curve** C in the xy-plane is simply a set of ordered pairs (x, y). We say that C is a **parametric curve** if the x- and y-coordinates of a point on the curve are defined by a pair of functions $x = f(t)$, $y = g(t)$ that are continuous on some interval $a \leq t \leq b$. The notion of a parametric curve extends to 3-space as well. A **parametric curve in space**, or **space curve**, is a set of ordered triples (x, y, z) where

$$x = f(t), \quad y = g(t), \quad z = h(t), \tag{1}$$

are continuous on an interval defined by $a \leq t \leq b$. In this section we combine the concepts of parametric curves with vectors.

Vector-Valued Functions It is often convenient in science and engineering to introduce a vector $\mathbf{r}$ whose components are functions of a parameter t. We say that

$$\mathbf{r}(t) = \langle f(t), g(t) \rangle = f(t)\mathbf{i} + g(t)\mathbf{j}$$

and

$$\mathbf{r}(t) = \langle f(t), g(t), h(t) \rangle = f(t)\mathbf{i} + g(t)\mathbf{j} + h(t)\mathbf{k},$$

are **vector-valued functions** or simply **vector functions**. As shown in **FIGURE 9.1.1**, for a given value of the parameter, say t_0, the vector $\mathbf{r}(t_0)$ is the position vector of a point P on a curve C. In other words, as the parameter t varies, we can envision the curve C being traced out by the moving arrowhead of $\mathbf{r}(t)$.

We have already seen an example of parametric equations, as well as the vector function of a space curve, in Section 7.5, when we discussed the line in 3-space.

EXAMPLE 1 Circular Helix

Graph the curve traced by the vector function

$$\mathbf{r}(t) = 2\cos t\,\mathbf{i} + 2\sin t\,\mathbf{j} + t\mathbf{k}, \quad t \geq 0.$$

SOLUTION The parametric equations of the curve are $x = 2\cos t$, $y = 2\sin t$, $z = t$. By eliminating the parameter t from the first two equations,

$$x^2 + y^2 = (2\cos t)^2 + (2\sin t)^2 = 2^2,$$

we see that a point on the curve lies on the circular cylinder $x^2 + y^2 = 4$. As seen in **FIGURE 9.1.2** and the accompanying table, as the value of t increases, the curve winds upward in a spiral or **circular helix**.

(a) 2-space

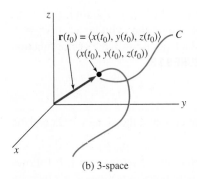

(b) 3-space

FIGURE 9.1.1 Curves defined by vector functions

t	x	y	z
0	2	0	0
$\pi/2$	0	2	$\pi/2$
π	-2	0	π
$3\pi/2$	0	-2	$3\pi/2$
2π	2	0	2π
$5\pi/2$	0	2	$5\pi/2$
3π	-2	0	3π
$7\pi/2$	0	-2	$7\pi/2$
4π	2	0	4π
$9\pi/2$	0	2	$9\pi/2$

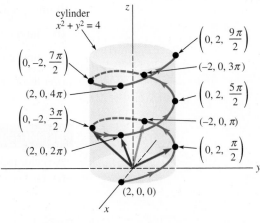

FIGURE 9.1.2 Circular helix in Example 1

The curve in Example 1 is a special case of the vector function

$$\mathbf{r}(t) = a \cos t\,\mathbf{i} + b \sin t\,\mathbf{j} + ct\,\mathbf{k}, \quad a > 0, \quad b > 0, \quad c > 0,$$

which describes an **elliptical helix**. When $a = b$, the helix is circular. The **pitch** of a helix is defined to be the number $2\pi c$. Problems 9 and 10 in Exercises 9.1 illustrate two other kinds of helixes.

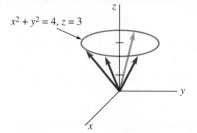

$x^2 + y^2 = 4, z = 3$

FIGURE 9.1.3 Curve in Example 2

EXAMPLE 2 **Circle in a Plane**

Graph the curve traced by the vector function

$$\mathbf{r}(t) = 2 \cos t\,\mathbf{i} + 2 \sin t\,\mathbf{j} + 3\,\mathbf{k}.$$

SOLUTION The parametric equations of this curve are $x = 2 \cos t$, $y = 2 \sin t$, $z = 3$. As in Example 1, we see that a point on the curve must also lie on the cylinder $x^2 + y^2 = 4$. However, since the z-coordinate of any point has the constant value $z = 3$, the vector function $\mathbf{r}(t)$ traces out a circle 3 units above the xy-plane. See **FIGURE 9.1.3**. ≡

EXAMPLE 3 **Curve of Intersection**

Find the vector function that describes the curve C of intersection of the plane $y = 2x$ and the paraboloid $z = 9 - x^2 - y^2$.

SOLUTION We first parameterize the curve C of intersection by letting $x = t$. It follows that $y = 2t$ and $z = 9 - t^2 - (2t)^2 = 9 - 5t^2$. From the parametric equations $x = t$, $y = 2t$, $z = 9 - 5t^2$, we see that a vector function describing the trace of the paraboloid in the plane $y = 2x$ is given by $\mathbf{r}(t) = t\,\mathbf{i} + 2t\,\mathbf{j} + (9 - 5t^2)\,\mathbf{k}$. See **FIGURE 9.1.4**. ≡

$z = 9 - x^2 - y^2$

C

$x^2 + y^2 = 9$ $y = 2x$

FIGURE 9.1.4 Curve in Example 3

■ **Limits, Continuity, and Derivatives** The fundamental notion of the **limit** of a vector function $\mathbf{r}(t) = \langle f(t), g(t), h(t) \rangle$ is defined in terms of the limits of the component functions.

DEFINITION 9.1.1 **Limit of a Vector Function**

If $\lim_{t \to a} f(t)$, $\lim_{t \to a} g(t)$, and $\lim_{t \to a} h(t)$ exist, then

$$\lim_{t \to a} \mathbf{r}(t) = \left\langle \lim_{t \to a} f(t), \lim_{t \to a} g(t), \lim_{t \to a} h(t) \right\rangle.$$

The notation $t \to a$ in Definition 9.1.1 can, of course, be replaced by $t \to a^+$, $t \to a^-$, $t \to \infty$, or $t \to -\infty$.

As an immediate consequence of Definition 9.1.1, we have the following result.

THEOREM 9.1.1 **Properties of Limits**

If $\lim_{t \to a} \mathbf{r}_1(t) = \mathbf{L}_1$ and $\lim_{t \to a} \mathbf{r}_2(t) = \mathbf{L}_2$, then

(*i*) $\lim_{t \to a} c\mathbf{r}_1(t) = c\mathbf{L}_1$, $\quad c$ a scalar

(*ii*) $\lim_{t \to a} [\mathbf{r}_1(t) + \mathbf{r}_2(t)] = \mathbf{L}_1 + \mathbf{L}_2$

(*iii*) $\lim_{t \to a} \mathbf{r}_1(t) \cdot \mathbf{r}_2(t) = \mathbf{L}_1 \cdot \mathbf{L}_2$.

DEFINITION 9.1.2 **Continuity of a Vector Function**

A vector function $\mathbf{r}$ is said to be **continuous** at $t = a$ if
(*i*) $\mathbf{r}(a)$ is defined, (*ii*) $\lim_{t \to a} \mathbf{r}(t)$ exists, and (*iii*) $\lim_{t \to a} \mathbf{r}(t) = \mathbf{r}(a)$.

Equivalently, $\mathbf{r}(t)$ is continuous at $t = a$ if and only if the component functions f, g, and h are continuous there.

DEFINITION 9.1.3 Derivative of Vector Function

The **derivative** of a vector function **r** is

$$\mathbf{r}'(t) = \lim_{\Delta t \to 0} \frac{1}{\Delta t}[\mathbf{r}(t + \Delta t) - \mathbf{r}(t)] \qquad (2)$$

for all t for which the limit exists.

The derivative of **r** is also written $d\mathbf{r}/dt$. The next theorem will show that on a practical level the derivative of a vector function is obtained by simply differentiating its component functions.

THEOREM 9.1.2 Differentiation of Components

If $\mathbf{r}(t) = \langle f(t), g(t), h(t) \rangle$, where f, g, and h are differentiable, then

$$\mathbf{r}'(t) = \langle f'(t), g'(t), h'(t) \rangle.$$

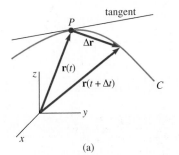

PROOF: From (2) we have

$$\mathbf{r}'(t) = \lim_{\Delta t \to 0} \frac{1}{\Delta t}\left[\langle f(t + \Delta t), g(t + \Delta t), h(t + \Delta t) \rangle - \langle f(t), g(t), h(t) \rangle \right]$$

$$= \lim_{\Delta t \to 0} \left\langle \frac{f(t + \Delta t) - f(t)}{\Delta t}, \frac{g(t + \Delta t) - g(t)}{\Delta t}, \frac{h(t + \Delta t) - h(t)}{\Delta t} \right\rangle.$$

Taking the limit of each component yields the desired result. ≡

■ **Smooth Curves** When the component functions of a vector function **r** have continuous first derivatives and $\mathbf{r}'(t) \neq \mathbf{0}$ for all t in the open interval (a, b), then **r** is said to be a **smooth function** and the curve C traced by **r** is called a **smooth curve**.

■ **Geometric Interpretation of r'(t)** If the vector $\mathbf{r}'(t)$ is not **0** at a point P, then it may be drawn *tangent to the curve* at P. As seen in **FIGURE 9.1.5**, the vectors

$$\Delta\mathbf{r} = \mathbf{r}(t + \Delta t) - \mathbf{r}(t) \qquad \text{and} \qquad \frac{\Delta\mathbf{r}}{\Delta t} = \frac{1}{\Delta t}[\mathbf{r}(t + \Delta t) - \mathbf{r}(t)]$$

are parallel. If we assume $\lim_{\Delta t \to 0} \Delta\mathbf{r}/\Delta t$ exists, it seems reasonable to conclude that as $\Delta t \to 0$, $\mathbf{r}(t)$ and $\mathbf{r}(t + \Delta t)$ become close and, as a consequence, the limiting position of the vector $\Delta\mathbf{r}/\Delta t$ is the tangent line at P. Indeed, the tangent line at P is *defined* as that line through P parallel to $\mathbf{r}'(t)$.

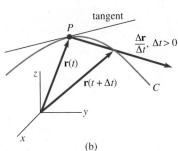

FIGURE 9.1.5 Vector $\mathbf{r}'(t)$ is tangent to curve C at P

EXAMPLE 4 Tangent Vectors

Graph the curve C that is traced by a point P whose position is given by $\mathbf{r}(t) = \cos 2t\,\mathbf{i} + \sin t\,\mathbf{j}$, $0 \le t \le 2\pi$. Graph $\mathbf{r}'(0)$ and $\mathbf{r}'(\pi/6)$.

SOLUTION By clearing the parameter from the parametric equations $x = \cos 2t$, $y = \sin t$, $0 \le t \le 2\pi$, we find that C is the parabola $x = 1 - 2y^2$, $-1 \le x \le 1$. From $\mathbf{r}'(t) = -2 \sin 2t\,\mathbf{i} + \cos t\,\mathbf{j}$ we find

$$\mathbf{r}'(0) = \mathbf{j} \qquad \text{and} \qquad \mathbf{r}'\left(\frac{\pi}{6}\right) = -\sqrt{3}\,\mathbf{i} + \frac{\sqrt{3}}{2}\,\mathbf{j}.$$

In **FIGURE 9.1.6** these vectors are drawn tangent to the curve C at $(1, 0)$ and $(\frac{1}{2}, \frac{1}{2})$, respectively. ≡

FIGURE 9.1.6 Tangent vectors in Example 4

EXAMPLE 5 **Tangent Line**

Find parametric equations of the tangent line to the graph of the curve C whose parametric equations are $x = t^2$, $y = t^2 - t$, $z = -7t$ at $t = 3$.

SOLUTION The vector function that gives the position of a point P on the curve is given by $\mathbf{r}(t) = t^2\mathbf{i} + (t^2 - t)\mathbf{j} - 7t\mathbf{k}$. Now,

$$\mathbf{r}'(t) = 2t\mathbf{i} + (2t - 1)\mathbf{j} - 7\mathbf{k} \quad \text{and so} \quad \mathbf{r}'(3) = 6\mathbf{i} + 5\mathbf{j} - 7\mathbf{k},$$

which is tangent to C at the point whose position vector is

$$\mathbf{r}(3) = 9\mathbf{i} + 6\mathbf{j} - 21\mathbf{k};$$

that is, $P(9, 6, -21)$. Using the components of $\mathbf{r}'(3)$, we see that parametric equations of the tangent line are $x = 9 + 6t$, $y = 6 + 5t$, $z = -21 - 7t$. ☰

Higher-Order Derivatives

Higher-order derivatives of a vector function are also obtained by differentiating its components. In the case of the **second derivative**, we have

$$\mathbf{r}''(t) = \langle f''(t), g''(t), h''(t)\rangle = f''(t)\mathbf{i} + g''(t)\mathbf{j} + h''(t)\mathbf{k}.$$

EXAMPLE 6 **Derivative of a Vector Function**

If $\mathbf{r}(t) = (t^3 - 2t^2)\mathbf{i} + 4t\mathbf{j} + e^{-t}\mathbf{k}$, then

$$\mathbf{r}'(t) = (3t^2 - 4t)\mathbf{i} + 4\mathbf{j} - e^{-t}\mathbf{k} \quad \text{and} \quad \mathbf{r}''(t) = (6t - 4)\mathbf{i} + e^{-t}\mathbf{k}.$$ ☰

THEOREM 9.1.3 Chain Rule

If $\mathbf{r}$ is a differentiable vector function and $s = u(t)$ is a differentiable scalar function, then the derivative of $\mathbf{r}(s)$ with respect to t is

$$\frac{d\mathbf{r}}{dt} = \frac{d\mathbf{r}}{ds}\frac{ds}{dt} = \mathbf{r}'(s)\,u'(t).$$

EXAMPLE 7 **Chain Rule**

If $\mathbf{r}(s) = \cos 2s\,\mathbf{i} + \sin 2s\,\mathbf{j} + e^{-3s}\mathbf{k}$, where $s = t^4$, then

$$\frac{d\mathbf{r}}{dt} = [-2\sin 2s\,\mathbf{i} + 2\cos 2s\,\mathbf{j} - 3e^{-3s}\mathbf{k}]\,4t^3$$

$$= -8t^3\sin(2t^4)\mathbf{i} + 8t^3\cos(2t^4)\mathbf{j} - 12t^3 e^{-3t^4}\mathbf{k}.$$ ☰

Details of the proof of the next theorem are left as exercises.

THEOREM 9.1.4 Rules of Differentiation

Let $\mathbf{r}_1$ and $\mathbf{r}_2$ be differentiable vector functions and $u(t)$ a differentiable scalar function.

(i) $\dfrac{d}{dt}\big[\mathbf{r}_1(t) + \mathbf{r}_2(t)\big] = \mathbf{r}_1'(t) + \mathbf{r}_2'(t)$

(ii) $\dfrac{d}{dt}\big[u(t)\mathbf{r}_1(t)\big] = u(t)\mathbf{r}_1'(t) + u'(t)\mathbf{r}_1(t)$

(iii) $\dfrac{d}{dt}\big[\mathbf{r}_1(t) \cdot \mathbf{r}_2(t)\big] = \mathbf{r}_1(t) \cdot \mathbf{r}_2'(t) + \mathbf{r}_1'(t) \cdot \mathbf{r}_2(t)$

(iv) $\dfrac{d}{dt}\big[\mathbf{r}_1(t) \times \mathbf{r}_2(t)\big] = \mathbf{r}_1(t) \times \mathbf{r}_2'(t) + \mathbf{r}_1'(t) \times \mathbf{r}_2(t)$.

Note of caution. ▶ Since the cross product of two vectors is not commutative, the order in which $\mathbf{r}_1$ and $\mathbf{r}_2$ appear in part (*iv*) of Theorem 9.1.4 must be strictly observed.

Integrals of Vector Functions If f, g, and h are integrable, then the indefinite and definite integrals of a vector function $\mathbf{r}(t) = f(t)\mathbf{i} + g(t)\mathbf{j} + h(t)\mathbf{k}$ are defined, respectively, by

$$\int \mathbf{r}(t)\, dt = \left[\int f(t)\, dt\right]\mathbf{i} + \left[\int g(t)\, dt\right]\mathbf{j} + \left[\int h(t)\, dt\right]\mathbf{k}$$

$$\int_a^b \mathbf{r}(t)\, dt = \left[\int_a^b f(t)\, dt\right]\mathbf{i} + \left[\int_a^b g(t)\, dt\right]\mathbf{j} + \left[\int_a^b h(t)\, dt\right]\mathbf{k}.$$

The indefinite integral of $\mathbf{r}$ is another vector function $\mathbf{R} + \mathbf{c}$ such that $\mathbf{R}'(t) = \mathbf{r}(t)$.

EXAMPLE 8 Integral of a Vector Function

If $\mathbf{r}(t) = 6t^2\mathbf{i} + 4e^{-2t}\mathbf{j} + 8\cos 4t\,\mathbf{k}$,

then

$$\int \mathbf{r}(t)\, dt = \left[\int 6t^2\, dt\right]\mathbf{i} + \left[\int 4e^{-2t}\, dt\right]\mathbf{j} + \left[\int 8\cos 4t\, dt\right]\mathbf{k}$$

$$= [2t^3 + c_1]\mathbf{i} + [-2e^{-2t} + c_2]\mathbf{j} + [2\sin 4t + c_3]\mathbf{k}$$

$$= 2t^3\mathbf{i} - 2e^{-2t}\mathbf{j} + 2\sin 4t\,\mathbf{k} + \mathbf{c},$$

where $\mathbf{c} = c_1\mathbf{i} + c_2\mathbf{j} + c_3\mathbf{k}$. ≡

Length of a Space Curve If $\mathbf{r}(t) = f(t)\mathbf{i} + g(t)\mathbf{j} + h(t)\mathbf{k}$ is a smooth function, then it can be shown that the **length** of the smooth curve traced by $\mathbf{r}$ is given by

$$s = \int_a^b \sqrt{[f'(t)]^2 + [g'(t)]^2 + [h'(t)]^2}\, dt = \int_a^b \|\mathbf{r}'(t)\|\, dt. \qquad (3)$$

Arc Length as a Parameter A curve in the plane or in space can be parameterized in terms of the arc length s.

EXAMPLE 9 Example 1 Revisited

Consider the helix of Example 1. Since $\|\mathbf{r}'(t)\| = \sqrt{5}$, it follows from (3) that the length of the curve from $\mathbf{r}(0)$ to an arbitrary point $\mathbf{r}(t)$ is

$$s = \int_0^t \sqrt{5}\, du = \sqrt{5}\,t,$$

where we have used u as a dummy variable of integration. Using $t = s/\sqrt{5}$, we obtain a vector equation of the helix as a function of arc length:

$$\mathbf{r}(s) = 2\cos\frac{s}{\sqrt{5}}\mathbf{i} + 2\sin\frac{s}{\sqrt{5}}\mathbf{j} + \frac{s}{\sqrt{5}}\mathbf{k}. \qquad (4)$$

Parametric equations of the helix are then

$$f(s) = 2\cos\frac{s}{\sqrt{5}}, \qquad g(s) = 2\sin\frac{s}{\sqrt{5}}, \qquad h(s) = \frac{s}{\sqrt{5}}. \qquad ≡$$

The derivative of a vector function $\mathbf{r}(t)$ with respect to the parameter t is a tangent vector to the curve traced by $\mathbf{r}$. However, if the curve is parameterized in terms of arc length s, then $\mathbf{r}'(s)$ *is a unit tangent vector*. To see this, let a curve be described by $\mathbf{r}(s)$, where s is arc length. From (3), the length of the curve from $\mathbf{r}(0)$ to $\mathbf{r}(s)$ is $s = \int_0^s \|\mathbf{r}'(u)\|\, du$. Differentiation of this last equation with respect to s then yields $\|\mathbf{r}'(s)\| = 1$.

In Problems 1–10, graph the curve traced by the given vector function.

1. $\mathbf{r}(t) = 2 \sin t \mathbf{i} + 4 \cos t \mathbf{j} + t \mathbf{k}; \ t \geq 0$
2. $\mathbf{r}(t) = \cos t \mathbf{i} + t \mathbf{j} + \sin t \mathbf{k}; \ t \geq 0$
3. $\mathbf{r}(t) = t \mathbf{i} + 2t \mathbf{j} + \cos t \mathbf{k}; \ t \geq 0$
4. $\mathbf{r}(t) = 4 \mathbf{i} + 2 \cos t \mathbf{j} + 3 \sin t \mathbf{k}$
5. $\mathbf{r}(t) = \langle e^t, e^{2t} \rangle$
6. $\mathbf{r}(t) = \cosh t \mathbf{i} + 3 \sinh t \mathbf{j}$
7. $\mathbf{r}(t) = \langle \sqrt{2} \sin t, \sqrt{2} \sin t, 2 \cos t \rangle; \ 0 \leq t \leq \pi/2$
8. $\mathbf{r}(t) = t \mathbf{i} + t^3 \mathbf{j} + t \mathbf{k}$
9. $\mathbf{r}(t) = e^t \cos t \mathbf{i} + e^t \sin t \mathbf{j} + e^t \mathbf{k}$
10. $\mathbf{r}(t) = \langle t \cos t, t \sin t, t^2 \rangle$

In Problems 11–14, find the vector function that describes the curve C of intersection between the given surfaces. Sketch the curve C. Use the indicated parameter.

11. $z = x^2 + y^2, \ y = x; \ x = t$
12. $x^2 + y^2 - z^2 = 1, \ y = 2x; \ x = t$
13. $x^2 + y^2 = 9, \ z = 9 - x^2; \ x = 3 \cos t$
14. $z = x^2 + y^2, \ z = 1; \ x = \sin t$

15. Given that $\mathbf{r}(t) = \dfrac{\sin 2t}{t} \mathbf{i} + (t - 2)^5 \mathbf{j} + t \ln t \mathbf{k}$, find $\lim_{t \to 0^+} \mathbf{r}(t)$.

16. Given that $\lim_{t \to a} \mathbf{r}_1(t) = \mathbf{i} - 2\mathbf{j} + \mathbf{k}$ and $\lim_{t \to a} \mathbf{r}_2(t) = 2\mathbf{i} + 5\mathbf{j} + 7\mathbf{k}$, find:
 (a) $\lim_{t \to a} [-4\mathbf{r}_1(t) + 3\mathbf{r}_2(t)]$
 (b) $\lim_{t \to a} \mathbf{r}_1(t) \cdot \mathbf{r}_2(t)$.

In Problems 17–20, find $\mathbf{r}'(t)$ and $\mathbf{r}''(t)$ for the given vector function.

17. $\mathbf{r}(t) = \ln t \mathbf{i} + \mathbf{j}, \ t > 0$
18. $\mathbf{r}(t) = \langle t \cos t - \sin t, t + \cos t \rangle$
19. $\mathbf{r}(t) = \langle t e^{2t}, t^3, 4t^2 - t \rangle$
20. $\mathbf{r}(t) = t^2 \mathbf{i} + t^3 \mathbf{j} + \tan^{-1} t \mathbf{k}$

In Problems 21–24, graph the curve C that is described by $\mathbf{r}$ and graph $\mathbf{r}'$ at the indicated value of t.

21. $\mathbf{r}(t) = 2 \cos t \mathbf{i} + 6 \sin t \mathbf{j}; \ t = \pi/6$
22. $\mathbf{r}(t) = t^3 \mathbf{i} + t^2 \mathbf{j}; \ t = -1$
23. $\mathbf{r}(t) = 2 \mathbf{i} + t \mathbf{j} + \dfrac{4}{1 + t^2} \mathbf{k}; \ t = 1$
24. $\mathbf{r}(t) = 3 \cos t \mathbf{i} + 3 \sin t \mathbf{j} + 2t \mathbf{k}; \ t = \pi/4$

In Problems 25 and 26, find parametric equations of the tangent line to the given curve at the indicated value of t.

25. $x = t, \ y = \frac{1}{2} t^2, \ z = \frac{1}{3} t^3; \ t = 2$
26. $x = t^3 - t, \ y = \dfrac{6t}{t + 1}, \ z = (2t + 1)^2; \ t = 1$

In Problems 27–32, find the indicated derivative. Assume that all vector functions are differentiable.

27. $\dfrac{d}{dt} [\mathbf{r}(t) \times \mathbf{r}'(t)]$

28. $\dfrac{d}{dt} [\mathbf{r}(t) \cdot (t \mathbf{r}(t))]$

29. $\dfrac{d}{dt} [\mathbf{r}(t) \cdot (\mathbf{r}'(t) \times \mathbf{r}''(t))]$

30. $\dfrac{d}{dt} [\mathbf{r}_1(t) \times (\mathbf{r}_2(t) \times \mathbf{r}_3(t))]$

31. $\dfrac{d}{dt} \left[\mathbf{r}_1(2t) + \mathbf{r}_2 \left(\dfrac{1}{t} \right) \right]$

32. $\dfrac{d}{dt} [t^3 \mathbf{r}(t^2)]$

In Problems 33–36, evaluate the given integral.

33. $\displaystyle \int_{-1}^{2} (t \mathbf{i} + 3t^2 \mathbf{j} + 4t^3 \mathbf{k}) \, dt$

34. $\displaystyle \int_{0}^{4} (\sqrt{2t + 1} \mathbf{i} - \sqrt{t} \mathbf{j} + \sin \pi t \mathbf{k}) \, dt$

35. $\displaystyle \int (t e^t \mathbf{i} - e^{-2t} \mathbf{j} + t e^{t^2} \mathbf{k}) \, dt$

36. $\displaystyle \int \dfrac{1}{1 + t^2} (\mathbf{i} + t \mathbf{j} + t^2 \mathbf{k}) \, dt$

In Problems 37–40, find a vector function $\mathbf{r}$ that satisfies the indicated conditions.

37. $\mathbf{r}'(t) = 6\mathbf{i} + 6t\mathbf{j} + 3t^2 \mathbf{k}; \ \mathbf{r}(0) = \mathbf{i} - 2\mathbf{j} + \mathbf{k}$
38. $\mathbf{r}'(t) = t \sin t^2 \mathbf{i} - \cos 2t \mathbf{j}; \ \mathbf{r}(0) = \frac{3}{2} \mathbf{i}$
39. $\mathbf{r}''(t) = 12t \mathbf{i} - 3t^{-1/2} \mathbf{j} + 2\mathbf{k}; \ \mathbf{r}'(1) = \mathbf{j}, \ \mathbf{r}(1) = 2\mathbf{i} - \mathbf{k}$
40. $\mathbf{r}''(t) = \sec^2 t \mathbf{i} + \cos t \mathbf{j} - \sin t \mathbf{k};$
 $\mathbf{r}'(0) = \mathbf{i} + \mathbf{j} + \mathbf{k}, \ \mathbf{r}(0) = -\mathbf{j} + 5\mathbf{k}$

In Problems 41–44, find the length of the curve traced by the given vector function on the indicated interval.

41. $\mathbf{r}(t) = a \cos t \mathbf{i} + a \sin t \mathbf{j} + ct \mathbf{k}; \ 0 \leq t \leq 2\pi$
42. $\mathbf{r}(t) = t \mathbf{i} + t \cos t \mathbf{j} + t \sin t \mathbf{k}; \ 0 \leq t \leq \pi$
43. $\mathbf{r}(t) = e^t \cos 2t \mathbf{i} + e^t \sin 2t \mathbf{j} + e^t \mathbf{k}; \ 0 \leq t \leq 3\pi$
44. $\mathbf{r}(t) = 3t \mathbf{i} + \sqrt{3} t^2 \mathbf{j} + \frac{2}{3} t^3 \mathbf{k}; \ 0 \leq t \leq 1$

45. Express the vector equation of a circle $\mathbf{r}(t) = a \cos t \mathbf{i} + a \sin t \mathbf{j}$ as a function of arc length s. Verify that $\mathbf{r}'(s)$ is a unit vector.

46. If $\mathbf{r}(s)$ is the vector function given in (4), verify that $\mathbf{r}'(s)$ is a unit vector.

47. Suppose $\mathbf{r}$ is a differentiable vector function for which $\| \mathbf{r}(t) \| = c$ for all t. Show that the tangent vector $\mathbf{r}'(t)$ is perpendicular to the position vector $\mathbf{r}(t)$ for all t.

48. In Problem 47, describe geometrically the kind of curve C for which $\| \mathbf{r}(t) \| = c$.

Miscellaneous Problems

49. Prove Theorem 9.1.4(*ii*).
50. Prove Theorem 9.1.4(*iii*).
51. Prove Theorem 9.1.4(*iv*).
52. If $\mathbf{v}$ is a constant vector and $\mathbf{r}$ is integrable on $[a, b]$, prove that $\int_a^b \mathbf{v} \cdot \mathbf{r}(t) \, dt = \mathbf{v} \cdot \int_a^b \mathbf{r}(t) \, dt$.

9.2 Motion on a Curve

INTRODUCTION In the preceding section we saw that the first and second derivatives of the vector function $\mathbf{r}(t) = \langle f(t), g(t), h(t) \rangle = f(t)\mathbf{i} + g(t)\mathbf{j} + h(t)\mathbf{k}$ can be obtained by differentiating the component functions f, g, and h. In this section we give a physical interpretation to the vectors $\mathbf{r}'(t)$ and $\mathbf{r}''(t)$.

Velocity and Acceleration Suppose a particle or body moves along a curve C so that its position at time t is given by the vector function $\mathbf{r}(t) = f(t)\mathbf{i} + g(t)\mathbf{j} + h(t)\mathbf{k}$. If the component functions f, g, and h have second derivatives, then the vectors

$$\mathbf{v}(t) = \mathbf{r}'(t) = f'(t)\mathbf{i} + g'(t)\mathbf{j} + h'(t)\mathbf{k}$$
$$\mathbf{a}(t) = \mathbf{r}''(t) = f''(t)\mathbf{i} + g''(t)\mathbf{j} + h''(t)\mathbf{k}$$

are called the **velocity** and **acceleration** of the particle, respectively. The scalar function $\|\mathbf{v}(t)\|$ is the **speed** of the particle. Since

$$\|\mathbf{v}(t)\| = \left\|\frac{d\mathbf{r}}{dt}\right\| = \sqrt{\left(\frac{dx}{dt}\right)^2 + \left(\frac{dy}{dt}\right)^2 + \left(\frac{dz}{dt}\right)^2}$$

speed is related to arc length s by $s'(t) = \|\mathbf{v}(t)\|$. In other words, arc length is given by $s = \int_{t_0}^{t_1} \|\mathbf{v}(t)\|\, dt$. It also follows from the discussion of Section 9.1 that if $P(x_1, y_1, z_1)$ is the position of the particle on C at time t_1, then we may draw $\mathbf{v}(t_1)$ *tangent to C at P*. Similar remarks hold for curves traced by the vector function $\mathbf{r}(t) = f(t)\mathbf{i} + g(t)\mathbf{j}$.

FIGURE 9.2.1 Velocity and acceleration vectors in Example 1

EXAMPLE 1 Velocity and Acceleration Vectors

The position of a moving particle is given by $\mathbf{r}(t) = t^2\mathbf{i} + t\mathbf{j} + \frac{5}{2}t\mathbf{k}$. Graph the curve defined by $\mathbf{r}(t)$ and the vectors $\mathbf{v}(2)$ and $\mathbf{a}(2)$.

SOLUTION Since $x = t^2$, $y = t$, the path of the particle is above the parabola $x = y^2$. When $t = 2$ the position vector $\mathbf{r}(2) = 4\mathbf{i} + 2\mathbf{j} + 5\mathbf{k}$ indicates that the particle is at the point $P(4, 2, 5)$. Now,

$$\mathbf{v}(t) = \mathbf{r}'(t) = 2t\mathbf{i} + \mathbf{j} + \frac{5}{2}\mathbf{k} \quad \text{and} \quad \mathbf{a}(t) = \mathbf{r}''(t) = 2\mathbf{i}$$

so that $\mathbf{v}(2) = 4\mathbf{i} + \mathbf{j} + \frac{5}{2}\mathbf{k}$ and $\mathbf{a}(2) = 2\mathbf{i}$. These vectors are shown in **FIGURE 9.2.1**. ≡

If a particle moves with a constant speed c, then its acceleration vector is perpendicular to the velocity vector $\mathbf{v}$. To see this, note that $\|\mathbf{v}\|^2 = c^2$ or $\mathbf{v} \cdot \mathbf{v} = c^2$. We differentiate both sides with respect to t and obtain, with the aid of Theorem 9.1.4(*iii*):

$$\frac{d}{dt}(\mathbf{v} \cdot \mathbf{v}) = \mathbf{v} \cdot \frac{d\mathbf{v}}{dt} + \frac{d\mathbf{v}}{dt} \cdot \mathbf{v} = 2\mathbf{v} \cdot \frac{d\mathbf{v}}{dt} = 0.$$

Thus,
$$\frac{d\mathbf{v}}{dt} \cdot \mathbf{v} = 0 \quad \text{or} \quad \mathbf{a}(t) \cdot \mathbf{v}(t) = 0 \text{ for all } t.$$

EXAMPLE 2 Velocity and Acceleration Vectors

Suppose the vector function in Example 2 of Section 9.1 represents the position of a particle moving in a circular orbit. Graph the velocity and acceleration vector at $t = \pi/4$.

SOLUTION Recall that $\mathbf{r}(t) = 2\cos t\mathbf{i} + 2\sin t\mathbf{j} + 3\mathbf{k}$ is the position vector of a particle moving in a circular orbit of radius 2 in the plane $z = 3$. When $t = \pi/4$ the particle is at the point $P(\sqrt{2}, \sqrt{2}, 3)$. Now,

$$\mathbf{v}(t) = \mathbf{r}'(t) = -2\sin t\mathbf{i} + 2\cos t\mathbf{j}$$
$$\mathbf{a}(t) = \mathbf{r}''(t) = -2\cos t\mathbf{i} - 2\sin t\mathbf{j}.$$

FIGURE 9.2.2 Velocity and acceleration vectors in Example 2

FIGURE 9.2.3 Circular motion

FIGURE 9.2.4 Trajectory of a projectile

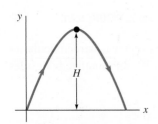

(a) **Maximum height H:**
Find t_1 for which $y'(t_1) = 0$;
$H = y_{max} = y(t_1)$

(b) **Range R:**
Find $t_1 > 0$ for which $y(t_1) = 0$;
$R = x_{max} = x(t_1)$

FIGURE 9.2.5 Maximum height and range of a projectile

Since the speed is $\|\mathbf{v}(t)\| = 2$ for all time t, it follows from the discussion preceding this example that $\mathbf{a}(t)$ is perpendicular to $\mathbf{v}(t)$. (Verify this.) As shown in **FIGURE 9.2.2**, the vectors

$$\mathbf{v}\left(\frac{\pi}{4}\right) = -2\sin\frac{\pi}{4}\mathbf{i} + 2\cos\frac{\pi}{4}\mathbf{j} = -\sqrt{2}\mathbf{i} + \sqrt{2}\mathbf{j}$$

$$\mathbf{a}\left(\frac{\pi}{4}\right) = -2\cos\frac{\pi}{4}\mathbf{i} - 2\sin\frac{\pi}{4}\mathbf{j} = -\sqrt{2}\mathbf{i} - \sqrt{2}\mathbf{j}$$

are drawn at the point P. The vector $\mathbf{v}(\pi/4)$ is tangent to the circular path and $\mathbf{a}(\pi/4)$ points along a radius toward the center of the circle. ≡

■ **Centripetal Acceleration** For circular motion in the plane, described by $\mathbf{r}(t) = r_0 \cos\omega t\,\mathbf{i} + r_0 \sin\omega t\,\mathbf{j}$, r_0 and ω constants, it is evident that $\mathbf{r}'' = -\omega^2\mathbf{r}$. This means that the acceleration vector $\mathbf{a}(t) = \mathbf{r}''(t)$ points in the direction opposite to that of the position vector $\mathbf{r}(t)$. We then say $\mathbf{a}(t)$ is **centripetal acceleration**. See **FIGURE 9.2.3**. If $v = \|\mathbf{v}(t)\|$ and $a = \|\mathbf{a}(t)\|$, we leave it as an exercise to show that $a = v^2/r_0$.

■ **Curvilinear Motion in the Plane** Many important applications of vector functions occur in regard to curvilinear motion in a plane. For example, planetary and projectile motion take place in a plane.

In analyzing the motion of short-range ballistic projectiles,* we begin with the acceleration of gravity written in vector form: $\mathbf{a}(t) = -g\mathbf{j}$.

If, as shown in **FIGURE 9.2.4**, a projectile is launched with an initial velocity $\mathbf{v}_0 = v_0 \cos\theta\mathbf{i} + v_0 \sin\theta\mathbf{j}$ from an initial height $\mathbf{s}_0 = s_0\mathbf{j}$, then

$$\mathbf{v}(t) = \int(-g\mathbf{j})\,dt = -gt\mathbf{j} + \mathbf{c}_1,$$

where $\mathbf{v}(0) = \mathbf{v}_0$ implies that $\mathbf{c}_1 = \mathbf{v}_0$. Therefore,

$$\mathbf{v}(t) = (v_0 \cos\theta)\mathbf{i} + (-gt + v_0 \sin\theta)\mathbf{j}.$$

Integrating again and using $\mathbf{r}(0) = \mathbf{s}_0$ yields

$$\mathbf{r}(t) = (v_0 \cos\theta)t\mathbf{i} + \left[-\frac{1}{2}gt^2 + (v_0 \sin\theta)t + s_0\right]\mathbf{j}. \tag{1}$$

Hence, parametric equations for the trajectory of the projectile are

$$x(t) = (v_0 \cos\theta)t, \qquad y(t) = -\frac{1}{2}gt^2 + (v_0 \sin\theta)t + s_0. \tag{2}$$

We are naturally interested in finding the maximum height H and the range R attained by a projectile. As shown in **FIGURE 9.2.5**, these quantities are the maximum values of $y(t)$ and $x(t)$, respectively.

EXAMPLE 3 **Trajectory of a Projectile**

A projectile is launched from ground level with an initial speed $v_0 = 768$ ft/s at an angle of elevation $\theta = 30°$. Find **(a)** the vector function and parametric equations of the projectile's trajectory, **(b)** the maximum altitude attained, **(c)** the range of the projectile, and **(d)** the impact speed.

SOLUTION **(a)** The initial height and velocity are, respectively, $\mathbf{s}_0 = \mathbf{0}$ and

$$\mathbf{r}'(0) = \mathbf{v}_0 = (768\cos 30°)\mathbf{i} + (768\sin 30°)\mathbf{j} = 384\sqrt{3}\mathbf{i} + 384\mathbf{j}. \tag{3}$$

Integrating $\mathbf{a}(t) = -32\mathbf{j}$ and using (3) give

$$\mathbf{v}(t) = (384\sqrt{3})\mathbf{i} + (-32t + 384)\mathbf{j}. \tag{4}$$

*A projectile is shot or hurled rather than self-propelled. In the analysis of *long-range* ballistic motion, the curvature of the Earth must be taken into consideration.

Integrating (4) and using $\mathbf{r}(0) = \mathbf{s}_0 = \mathbf{0}$ then give

$$\mathbf{r}(t) = (384\sqrt{3})t\,\mathbf{i} + (-16t^2 + 384t)\mathbf{j}.$$

The components of this vector function,

$$x(t) = (384\sqrt{3})t, \quad y(t) = -16t^2 + 384t, \tag{5}$$

are the parametric equations of the projectile's trajectory.

(b) From (5) we see that $y'(t) = 0$ when $-32t + 384 = 0$ or $t = 12$ s. The maximum height attained by the projectile is

$$H = y(12) = -16(12)^2 + 384(12) = 2304 \text{ ft}.$$

(c) From (5) we see that $y(t) = 0$ when $-16t^2 + 384t = -16t(t - 24) = 0$. The time that projectile hits the ground is $t = 24$ s and the corresponding range is

$$R = x(24) = 384\sqrt{3}(24) \approx 15{,}963 \text{ ft}.$$

(d) Finally, from (4) we see that $\mathbf{v}(24) = (384\sqrt{3})\mathbf{i} + (-384)\mathbf{j}$ and so the impact speed is

$$\|\mathbf{v}(24)\| = \sqrt{(384\sqrt{3})^2 + (-384)^2} = 768 \text{ ft/s}. \qquad \equiv$$

In Example 3, note that the impact speed is the same as the launch speed $v_0 = 768$ ft/s. Verify that this is still the case if we change the angle of elevation to, say, $\theta = 50°$. See Problem 16 in Exercises 9.2.

REMARKS

We have seen that the rate of change of arc length ds/dt is the same as the speed $\|\mathbf{v}(t)\| = \|\mathbf{r}'(t)\|$. However, as we shall see in the next section, it does *not* follow that the *scalar acceleration* d^2s/dt^2 is the same as $\|\mathbf{a}(t)\| = \|\mathbf{r}''(t)\|$. See Problem 24 in Exercises 9.2.

9.2 Exercises Answers to selected odd-numbered problems begin on page ANS-23.

In Problems 1–8, $\mathbf{r}(t)$ is the position vector of a moving particle. Graph the curve and the velocity and acceleration vectors at the indicated time. Find the speed at that time.

1. $\mathbf{r}(t) = t^2\mathbf{i} + \frac{1}{4}t^4\mathbf{j}$; $t = 1$

2. $\mathbf{r}(t) = t^2\mathbf{i} + \frac{1}{t^2}\mathbf{j}$; $t = 1$

3. $\mathbf{r}(t) = -\cosh 2t\,\mathbf{i} + \sinh 2t\,\mathbf{j}$; $t = 0$

4. $\mathbf{r}(t) = 2\cos t\,\mathbf{i} + (1 + \sin t)\mathbf{j}$; $t = \pi/3$

5. $\mathbf{r}(t) = 2\mathbf{i} + (t - 1)^2\mathbf{j} + t\mathbf{k}$; $t = 2$

6. $\mathbf{r}(t) = t\mathbf{i} + t\mathbf{j} + t^3\mathbf{k}$; $t = 2$

7. $\mathbf{r}(t) = t\mathbf{i} + t^2\mathbf{j} + t^3\mathbf{k}$; $t = 1$

8. $\mathbf{r}(t) = t\mathbf{i} + t^3\mathbf{j} + t\mathbf{k}$; $t = 1$

9. Suppose $\mathbf{r}(t) = t^2\mathbf{i} + (t^3 - 2t)\mathbf{j} + (t^2 - 5t)\mathbf{k}$ is the position vector of a moving particle. At what points does the particle

pass through the xy-plane? What are its velocity and acceleration at these points?

10. Suppose a particle moves in space so that $\mathbf{a}(t) = \mathbf{0}$ for all time t. Describe its path.

11. A shell is fired from ground level with an initial speed of 480 ft/s at an angle of elevation of 30°. Find:
 (a) a vector function and parametric equations of the shell's trajectory,
 (b) the maximum altitude attained,
 (c) the range of the shell, and
 (d) the speed at impact.

12. Rework Problem 11 if the shell is fired with the same initial speed and the same angle of elevation but from a cliff 1600 ft high.

13. A used car is pushed off an 81-ft-high sheer seaside cliff with a speed of 4 ft/s. Find the speed at which the car hits the water.

14. A small projectile is launched from ground level with an initial speed of 98 m/s. Find the possible angles of elevation so that its range is 490 m.

15. A football quarterback throws a 100-yd "bomb" at an angle of 45° from the horizontal. What is the initial speed of the football at the point of release?

16. If a projectile is launched from level ground, then the initial conditions are $\mathbf{r}(0) = \mathbf{0}$, $\mathbf{r}'(0) = \mathbf{v}_0 = (v_0 \cos \theta)\mathbf{i} + (v_0 \sin \theta)\mathbf{j}$ and the vector function (1) becomes

$$\mathbf{r}(t) = (v_0 \cos \theta)t\mathbf{i} + [-\tfrac{1}{2}gt^2 + (v_0 \sin \theta)t]\mathbf{j}. \qquad (6)$$

This function is equivalent to the solution $x(t)$, $y(t)$ of the system of linear differential equations (10) in Problem 21 of Exercises 4.6 subject to the same initial conditions. Underlying that earlier discussion, as well as in the derivation of (1), is the assumption that the projectile is not subject to air resistance. Use (6) to show that the impact speed of a projectile is the same as the initial speed $\|\mathbf{r}'(0)\| = \|\mathbf{v}_0\| = v_0$ for any angle of elevation θ, $0° < \theta < 90°$.

17. When air resistance is ignored, we saw in Problem 21 of Exercises 4.6 that a projectile launched at an angle θ, $0° < \theta < 90°$ from level ground ($\mathbf{r}(0) = \mathbf{0}$), then its horizontal range and maximum height are, respectively,

$$R = \frac{v_0^2}{g} \sin 2\theta \quad \text{and} \quad H = \frac{v_0^2}{2g} \sin^2 \theta.$$

From the first formula, it follows that when the projectile is launched at distinct complementary angles the horizontal range is the same and that the maximum range is attained when the angle of elevation is $\theta = 45°$. But if the projectile is launched from an initial height $s_0 > 0$ the foregoing formulas and statements are not valid.

(a) With $v_0 = 480$ ft/s, $s_0 = 1600$ ft rework Problem 12, but this time use the complementary angle of elevation $\theta = 60°$. Compare the maximum height, range, and impact speed of the projectile with the answers to parts (b), (c) and (d) of Problem 12.

(b) Use a graphing utility or CAS to plot the trajectory of the projectile defined by the parametric equations $x(t)$ and $y(t)$ in part (a) of Problem 12. Repeat for the parametric equations in part (a) of this problem. Superimpose both of these curves on the same coordinate system.

18. As mentioned in Problem 17, if a projectile is launched from an initial height $s_0 > 0$ the maximum range of the projectile is not attained using $\theta = 45°$ as the angle of elevation.

(a) A projectile is launched from an initial height of $v_0 = 480$ ft/s, $s_0 = 1600$ ft, at an angle of elevation of $\theta = 45°$. Use a calculator or CAS to find the time the projectile hits the ground and the corresponding range.

(b) If the angle of elevation in part (a) is changed to $\theta = 39.76°$, show that the range is greater than that in part (a).

(c) Use a graphing utility or CAS to plot the trajectories of the projectiles in parts (a) and (b). Superimpose both of these curves on the same coordinate system.

19. If a projectile is launched from level ground, the initial conditions are $\mathbf{r}(0) = \mathbf{0}$, $\mathbf{r}'(0) = \mathbf{v}_0 = (v_0 \cos \theta)\mathbf{i} + (v_0 \sin \theta)\mathbf{j}$, and if linear air resistance is taken into consideration, the vector function analogue of (1) is

$$\mathbf{r}(t) = \frac{mv_0 \cos \theta}{\beta}(1 - e^{-\beta t/m})\mathbf{i}$$

$$+ \left[\left(\frac{mv_0 \sin \theta}{\beta} + \frac{m^2 g}{\beta^2} \right)(1 - e^{-\beta t/m}) - \frac{mg}{\beta}t \right]\mathbf{j}.$$

This function is equivalent to the solution $x(t)$, $y(t)$ of the system of linear differential equations (13) in Problem 22 of Exercises 4.6 subject to the same initial conditions. Here $\beta > 0$ is a constant of proportionality related to the air resistance or drag. If $m = \tfrac{1}{4}$ slug, $g = 32$ ft/s², $\beta = 0.02$, $v_0 = 300$ ft/s, and $\theta = 38°$, use a calculator or CAS to find the impact speed of the projectile. See part (b) of Problem 22 in Exercises 4.6.

20. Suppose the angle of elevation of the projectile in Problem 19 is changed to $\theta = 52°$. Do you think that the impact speed of the projectile is the same, greater than, or less than the value found in that problem? Prove your assertion.

21. A projectile is fired from a cannon directly at a target that is dropped from rest simultaneously as the cannon is fired. Show that the projectile will strike the target in midair. See **FIGURE 9.2.6**. [*Hint*: Assume that the origin is at the muzzle of the cannon and that the angle of elevation is θ. If $\mathbf{r}_p$ and $\mathbf{r}_t$ are position vectors of the projectile and target, respectively, is there a time at which $\mathbf{r}_p = \mathbf{r}_t$?]

FIGURE 9.2.6 Cannon in Problem 21

22. In army field maneuvers sturdy equipment and supply packs are simply dropped from planes that fly horizontally at a slow speed and a low altitude. A supply plane flies horizontally over a target at an altitude of 1024 ft at a constant speed of 180 mi/h. Use (1) to determine the horizontal distance a supply pack travels relative to the point from which it was dropped. At what line-of-sight angle α should the supply pack be released in order to hit the target indicated in **FIGURE 9.2.7**?

FIGURE 9.2.7 Supply plane in Problem 22

23. Suppose that $\mathbf{r}(t) = r_0 \cos \omega t \mathbf{i} + r_0 \sin \omega t \mathbf{j}$ is the position vector of an object that is moving in a circle of radius r_0 in the xy-plane. If $\|\mathbf{v}(t)\| = v$, show that the magnitude of the centripetal acceleration is $a = \|\mathbf{a}(t)\| = v^2/r_0$.

24. The motion of a particle in space is described by

$$\mathbf{r}(t) = b \cos t \mathbf{i} + b \sin t \mathbf{j} + ct\mathbf{k}, \quad t \geq 0.$$

(a) Compute $\|\mathbf{v}(t)\|$.
(b) Compute $s = \int_0^t \|\mathbf{v}(t)\| \, dt$ and verify that ds/dt is the same as the result of part (a).
(c) Verify that $d^2s/dt^2 \neq \|\mathbf{a}(t)\|$.

25. The **effective weight** w_e of mass m at the equator of the Earth is defined by $w_e = mg - ma$, where a is the magnitude of the centripetal acceleration given in Problem 23. Determine the effective weight of a 192-lb person if the radius of the Earth is 4000 mi, $g = 32$ ft/s^2, and $v = 1530$ ft/s.

26. Consider a bicyclist riding on a flat circular track of radius r_0. If m is the combined mass of the rider and bicycle, fill in the blanks in **FIGURE 9.2.8**. [*Hint*: Use Problem 23 and force = mass × acceleration. Assume that the directions are upward and to the left.] The resultant vector $\mathbf{U}$ gives the direction the bicyclist must be tipped to avoid falling. Find the angle ϕ from the vertical at which the bicyclist must be tipped if her speed is 44 ft/s and the radius of the track is 60 ft.

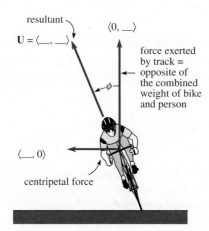

FIGURE 9.2.8 Bicyclist in Problem 26

27. The velocity of a particle moving in a fluid is described by means of a **velocity field** $\mathbf{v} = v_1\mathbf{i} + v_2\mathbf{j} + v_3\mathbf{k}$, where the components v_1, v_2, and v_3 are functions of x, y, z, and time t. If the velocity of the particle is $\mathbf{v}(t) = 6t^2x\mathbf{i} - 4ty^2\mathbf{j} + 2t(z + 1)\mathbf{k}$, find $\mathbf{r}(t)$. [*Hint*: Use separation of variables.]

28. Suppose m is the mass of a moving particle. Newton's second law of motion can be written in vector form as

$$\mathbf{F} = m\mathbf{a} = \frac{d}{dt}(m\mathbf{v}) = \frac{d\mathbf{p}}{dt},$$

where $\mathbf{p} = m\mathbf{v}$ is called **linear momentum**. The angular momentum of the particle with respect to the origin is defined to be $\mathbf{L} = \mathbf{r} \times \mathbf{p}$, where $\mathbf{r}$ is its position vector. If the torque of the particle about the origin is $\boldsymbol{\tau} = \mathbf{r} \times \mathbf{F} = \mathbf{r} \times d\mathbf{p}/dt$, show that $\boldsymbol{\tau}$ is the time rate of change of angular momentum.

29. Suppose the Sun is located at the origin. The gravitational force $\mathbf{F}$ exerted on a planet of mass m by the Sun of mass M is

$$\mathbf{F} = -k\frac{Mm}{r^2}\mathbf{u}.$$

$\mathbf{F}$ is a **central force**—that is, a force directed along the position vector $\mathbf{r}$ of the planet. Here k is the gravitational constant, $r = \|\mathbf{r}\|$, $\mathbf{u} = \mathbf{r}/r$ is a unit vector in the direction of $\mathbf{r}$, and the minus sign indicates that $\mathbf{F}$ is an attractive force—that is, a force directed toward the Sun. See **FIGURE 9.2.9**.

(a) Use Problem 28 to show that the torque acting on the planet due to this central force is $\mathbf{0}$.
(b) Explain why the angular momentum $\mathbf{L}$ of a planet is constant.

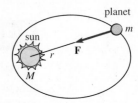

FIGURE 9.2.9 Force $\mathbf{F}$ in Problem 29

Discussion Problems

30. In this problem the student will use the properties in Sections 7.4 and 9.1 to prove **Kepler's first law of planetary motion**: The orbit of a planet is an ellipse with the Sun at one focus. We assume that the Sun has mass M and is located at the origin, $\mathbf{r}$ is the position vector of a body of mass m moving under the gravitational attraction of the Sun, and $\mathbf{u} = \mathbf{r}/r$ is a unit vector in the direction of $\mathbf{r}$.

(a) Use Problem 29 and Newton's second law of motion $\mathbf{F} = m\mathbf{a}$ to show that

$$\frac{d^2\mathbf{r}}{dt^2} = -kM\frac{\mathbf{u}}{r^2}.$$

(b) Use part (a) to show that $\mathbf{r} \times \mathbf{r}'' = \mathbf{0}$.
(c) Use part (b) to show that $\dfrac{d}{dt}(\mathbf{r} \times \mathbf{v}) = \mathbf{0}$.
(d) It follows from part (c) that $\mathbf{r} \times \mathbf{v} = \mathbf{c}$, where $\mathbf{c}$ is a constant vector. Show that $\mathbf{c} = r^2(\mathbf{u} \times \mathbf{u}')$.
(e) Show that $\dfrac{d}{dt}(\mathbf{u} \cdot \mathbf{u}) = 0$ and consequently $\mathbf{u} \cdot \mathbf{u}' = 0$.
(f) Use parts (a), (e), and (d) to show that

$$\frac{d}{dt}(\mathbf{v} \times \mathbf{c}) = kM\frac{d\mathbf{u}}{dt}.$$

(g) By integrating the result in part (f) with respect to t, we get $\mathbf{v} \times \mathbf{c} = kM\mathbf{u} + \mathbf{d}$, where $\mathbf{d}$ is another constant vector. Dot both sides of this last expression by the vector $\mathbf{r} = r\mathbf{u}$ and use Problem 61 in Exercises 7.4 to show that

$$r = \frac{c^2/kM}{1 + (d/kM)\cos\theta},$$

where $c = \|\mathbf{c}\|$, $d = \|\mathbf{d}\|$, and θ is the angle between $\mathbf{d}$ and $\mathbf{r}$.

(h) Explain why the result in part (g) proves Kepler's first law.

(i) At perihelion, the point in the orbit where the body is closest to the Sun, the vectors **r** and **v** are perpendicular and have magnitudes r_0 and v_0, respectively. Use this information and parts (d) and (g) to show that $c = r_0 v_0$ and $d = r_0 v_0^2 - kM$.

31. Suppose a projectile is launched from an initial height s_0 at an angle of elevation θ. If air resistance is ignored, show that the horizontal range is given by

$$R = \frac{v_0 \cos \theta}{g} (v_0 \sin \theta + \sqrt{v_0^2 \sin^2 \theta + 2 s_0 g}).$$

Note that when $s_0 = 0$ this formula reduces to the range R given in Problem 17.

32. Consider the formula in Problem 31 as a function of the single variable θ. Show that the range of a projectile launched from an initial height s_0 is a maximum for the angle of inclination

$$\theta = \cos^{-1} \sqrt{\frac{2 s_0 g + v_0^2}{2 s_0 g + 2 v_0^2}}.$$

Note that when $s_0 = 0$ this formula reduces to the value given in Problem 17, that is, $\theta = \pi/4$ or $45°$. Use this formula to verify the angle of inclination used in part (b) of Problem 18.

9.3 Curvature

INTRODUCTION Let C be a smooth curve in either 2- or 3-space traced out by a vector function $\mathbf{r}(t)$. In this section we are going to consider in greater detail the acceleration vector $\mathbf{a}(t) = \mathbf{r}''(t)$ introduced in the last section. But before doing this, we need to examine a scalar quantity called the **curvature** of a curve.

■ A Definition We know that $\mathbf{r}'(t)$ is a tangent vector to the curve C, and consequently

$$\mathbf{T}(t) = \frac{\mathbf{r}'(t)}{\|\mathbf{r}'(t)\|} \tag{1}$$

is a **unit tangent**. But recall from the end of Section 9.1 that if C is parameterized by arc length s, then a unit tangent to the curve is also given by $d\mathbf{r}/ds$. The quantity $\|\mathbf{r}'(t)\|$ in (1) is related to arc length s by $ds/dt = \|\mathbf{r}'(t)\|$. Since the curve C is smooth, we know from pages 489 and 491 that $ds/dt > 0$. Hence by the Chain Rule,

$$\frac{d\mathbf{r}}{dt} = \frac{d\mathbf{r}}{ds}\frac{ds}{dt} \quad \text{and so} \quad \frac{d\mathbf{r}}{ds} = \frac{d\mathbf{r}/dt}{ds/dt} = \frac{\mathbf{r}'(t)}{\|\mathbf{r}'(t)\|} = \mathbf{T}(t). \tag{2}$$

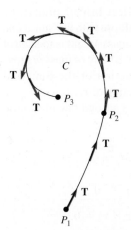

FIGURE 9.3.1 Unit tangents

Now suppose C is as shown in **FIGURE 9.3.1**. As s increases, **T** moves along C, changing direction but not length (it is always of unit length). Along the portion of the curve between P_1 and P_2 the vector **T** varies little in direction; along the curve between P_2 and P_3, where C obviously bends more sharply, the change in the direction of the tangent **T** is more pronounced. We use the *rate* at which the unit vector **T** changes direction with respect to arc length as an indicator of the curvature of a smooth curve C.

DEFINITION 9.3.1 Curvature

Let $\mathbf{r}(t)$ be a vector function defining a smooth curve C. If s is the arc length parameter and $\mathbf{T} = d\mathbf{r}/ds$ is the unit tangent vector, then the **curvature** of C at a point is

$$\kappa = \left\| \frac{d\mathbf{T}}{ds} \right\|. \tag{3}$$

The symbol κ in (3) is the Greek letter *kappa*. Now since curves are generally not parameterized by arc length, it is convenient to express (3) in terms of a general parameter t. Using the Chain Rule again, we can write

$$\frac{d\mathbf{T}}{dt} = \frac{d\mathbf{T}}{ds}\frac{ds}{dt} \quad \text{and consequently} \quad \frac{d\mathbf{T}}{ds} = \frac{d\mathbf{T}/dt}{ds/dt}.$$

In other words, curvature is given by

$$\kappa(t) = \frac{\|\mathbf{T}'(t)\|}{\|\mathbf{r}'(t)\|}. \tag{4}$$

EXAMPLE 1 **Curvature of a Circle**

Find the curvature of a circle of radius a.

SOLUTION A circle can be described by the vector function $\mathbf{r}(t) = a\cos t\,\mathbf{i} + a\sin t\,\mathbf{j}$. Now from $\mathbf{r}'(t) = -a\sin t\,\mathbf{i} + a\cos t\,\mathbf{j}$ and $\|\mathbf{r}'(t)\| = a$, we get

$$\mathbf{T}(t) = \frac{\mathbf{r}'(t)}{\|\mathbf{r}'(t)\|} = -\sin t\,\mathbf{i} + \cos t\,\mathbf{j} \quad \text{and} \quad \mathbf{T}'(t) = -\cos t\,\mathbf{i} - \sin t\,\mathbf{j}.$$

Hence, from (4) the curvature is

$$\kappa(t) = \frac{\|\mathbf{T}'(t)\|}{\|\mathbf{r}'(t)\|} = \frac{\sqrt{\cos^2 t + \sin^2 t}}{a} = \frac{1}{a}. \tag{5}$$

The result in (5) shows that the curvature at a point on a circle is the reciprocal of the radius of the circle and indicates a fact that is in keeping with our intuition: A circle with a small radius curves more than one with a large radius. See **FIGURE 9.3.2**. ≡

■ **Tangential and Normal Components of Acceleration** Suppose a particle moves in 2- or 3-space on a smooth curve C described by the vector function $\mathbf{r}(t)$. Then the velocity of the particle on C is $\mathbf{v}(t) = \mathbf{r}'(t)$, whereas its speed is $ds/dt = v = \|\mathbf{v}(t)\|$. Thus, (1) implies $\mathbf{v}(t) = v\mathbf{T}$. Differentiating this last expression with respect to t gives acceleration:

$$\mathbf{a}(t) = v\frac{d\mathbf{T}}{dt} + \frac{dv}{dt}\mathbf{T}. \tag{6}$$

Furthermore, with the help of Theorem 9.1.4(*iii*), it follows from the differentiation of $\mathbf{T} \cdot \mathbf{T} = 1$ that $\mathbf{T} \cdot d\mathbf{T}/dt = 0$. Hence, at a point P on C the vectors $\mathbf{T}$ and $d\mathbf{T}/dt$ are orthogonal. If $\|d\mathbf{T}/dt\| \neq 0$, the vector

$$\mathbf{N}(t) = \frac{d\mathbf{T}/dt}{\|d\mathbf{T}/dt\|} \tag{7}$$

is a unit normal to the curve C at P with direction given by $d\mathbf{T}/dt$. The vector $\mathbf{N}$ is also called the **principal normal**. But since curvature is $\kappa = \dfrac{\|d\mathbf{T}/dt\|}{v}$, it follows from (7) that $d\mathbf{T}/dt = \kappa v\mathbf{N}$. Thus, (6) becomes

$$\mathbf{a}(t) = \kappa v^2\mathbf{N} + \frac{dv}{dt}\mathbf{T}. \tag{8}$$

By writing (8) as $\qquad \mathbf{a}(t) = a_N\mathbf{N} + a_T\mathbf{T}, \tag{9}$

we see that the acceleration vector $\mathbf{a}$ of the moving particle is the sum of two orthogonal vectors $a_N\mathbf{N}$ and $a_T\mathbf{T}$. See **FIGURE 9.3.3**. The scalar functions $a_T = dv/dt$ and $a_N = \kappa v^2$ are called the **tangential** and **normal components of the acceleration**, respectively. Note that the tangential component of the acceleration results from a change in the *magnitude* of the velocity $\mathbf{v}$, whereas the normal component of the acceleration results from a change in the *direction* of $\mathbf{v}$.

■ **The Binormal** A third unit vector defined by

$$\mathbf{B}(t) = \mathbf{T}(t) \times \mathbf{N}(t)$$

is called the **binormal**. The three unit vectors $\mathbf{T}$, $\mathbf{N}$, and $\mathbf{B}$ form a right-handed set of mutually orthogonal vectors called the **moving trihedral**. The plane of $\mathbf{T}$ and $\mathbf{N}$ is called the **osculating plane**,* the plane of $\mathbf{N}$ and $\mathbf{B}$ is said to be the **normal plane**, and the plane of $\mathbf{T}$ and $\mathbf{B}$ is the **rectifying plane**. See **FIGURE 9.3.4**.

*Literally, this means the "kissing" plane.

FIGURE 9.3.2 Curvature of a circle in Example 1

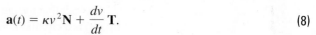

FIGURE 9.3.3 Components of acceleration

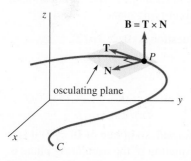

FIGURE 9.3.4 Osculating plane

The three mutually orthogonal unit vectors **T**, **N**, and **B** can be thought of as a movable right-handed coordinate system since

$$\mathbf{B}(t) = \mathbf{T}(t) \times \mathbf{N}(t), \qquad \mathbf{N}(t) = \mathbf{B}(t) \times \mathbf{T}(t), \qquad \mathbf{T}(t) = \mathbf{N}(t) \times \mathbf{B}(t).$$

This movable coordinate system is referred to as the **TNB-frame**.

EXAMPLE 2 **Tangent, Normal, and Binormal Vectors**

The position of a moving particle is given by $\mathbf{r}(t) = 2\cos t\,\mathbf{i} + 2\sin t\,\mathbf{j} + 3t\,\mathbf{k}$. Find the vectors **T**, **N**, and **B**. Find the curvature.

SOLUTION Since $\mathbf{r}'(t) = -2\sin t\,\mathbf{i} + 2\cos t\,\mathbf{j} + 3\,\mathbf{k}$, $\|\mathbf{r}'(t)\| = \sqrt{13}$, and so from (1) we see that a unit tangent is

$$\mathbf{T}(t) = -\frac{2}{\sqrt{13}}\sin t\,\mathbf{i} + \frac{2}{\sqrt{13}}\cos t\,\mathbf{j} + \frac{3}{\sqrt{13}}\,\mathbf{k}.$$

Next, we have

$$\frac{d\mathbf{T}}{dt} = -\frac{2}{\sqrt{13}}\cos t\,\mathbf{i} - \frac{2}{\sqrt{13}}\sin t\,\mathbf{j} \quad\text{and}\quad \left\|\frac{d\mathbf{T}}{dt}\right\| = \frac{2}{\sqrt{13}}.$$

Hence, (7) gives the principal normal

$$\mathbf{N}(t) = -\cos t\,\mathbf{i} - \sin t\,\mathbf{j}.$$

Now, the binormal is

$$\mathbf{B}(t) = \mathbf{T}(t) \times \mathbf{N}(t) = \begin{vmatrix} \mathbf{i} & \mathbf{j} & \mathbf{k} \\ -\dfrac{2}{\sqrt{13}}\sin t & \dfrac{2}{\sqrt{13}}\cos t & \dfrac{3}{\sqrt{13}} \\ -\cos t & -\sin t & 0 \end{vmatrix}$$

$$= \frac{3}{\sqrt{13}}\sin t\,\mathbf{i} - \frac{3}{\sqrt{13}}\cos t\,\mathbf{j} + \frac{2}{\sqrt{13}}\,\mathbf{k}.$$

Finally, using $\|d\mathbf{T}/dt\| = 2/\sqrt{13}$ and $\|\mathbf{r}'(t)\| = \sqrt{13}$, we obtain from (4) that the curvature at any point is the constant

$$\kappa = \frac{2/\sqrt{13}}{\sqrt{13}} = \frac{2}{13}.$$

≡

The fact that the curvature in Example 2 is constant is not surprising, since the curve defined by $\mathbf{r}(t)$ is a circular helix.

EXAMPLE 3 **Osculating, Normal, Rectifying Planes**

At the point corresponding to $t = \pi/2$ on the circular helix in Example 2, find an equation of (**a**) the osculating plane, (**b**) the normal plane, and (**c**) the rectifying plane.

SOLUTION From $\mathbf{r}(\pi/2) = \langle 0, 2, 3\pi/2 \rangle$ the point P in question is $(0, 2, 3\pi/2)$.

(**a**) A normal vector to the osculating plane at P is

$$\mathbf{B}(\pi/2) = \mathbf{T}(\pi/2) \times \mathbf{N}(\pi/2) = \frac{3}{\sqrt{13}}\,\mathbf{i} + \frac{2}{\sqrt{13}}\,\mathbf{k}.$$

To find an equation of a plane we do not require a *unit* normal, so in lieu of $\mathbf{B}(\pi/2)$ it is a bit simpler to use $\langle 3, 0, 2 \rangle$. From (11) of Section 7.5 an equation of the osculating plane is

$$3(x - 0) + 0(y - 2) + 2\left(z - \frac{3\pi}{2}\right) = 0 \quad\text{or}\quad 3x + 2z = 3\pi.$$

FIGURE 9.3.5 Helix and osculating plane in Example 3

(b) At the point P, the vector $\mathbf{T}(\pi/2) = \frac{1}{\sqrt{13}}\langle -2, 0, 3 \rangle$ or $\langle -2, 0, 3 \rangle$ is normal to the plane containing $\mathbf{N}(\pi/2)$ and $\mathbf{B}(\pi/2)$. Hence an equation of the normal plane is

$$-2(x - 0) + 0(y - 2) + 3\left(z - \frac{3\pi}{2}\right) = 0 \quad \text{or} \quad -4x + 6z = 9\pi.$$

(c) Finally, at the point P, the vector $\mathbf{N}(\pi/2) = \langle 0, -1, 0 \rangle$ is normal to the plane containing $\mathbf{T}(\pi/2)$ and $\mathbf{B}(\pi/2)$. An equation of the rectifying plane is

$$0(x - 0) + (-1)(y - 2) + 0\left(z - \frac{3\pi}{2}\right) = 0 \quad \text{or} \quad y = 2. \qquad \equiv$$

With the help of *Mathematica*, portions of the helix and the osculating plane in Example 3 are shown in **FIGURE 9.3.5**. The point $(0, 2, 3\pi/2)$ is indicated in the figure by the red dot.

▨ Formulas for a_T, a_N, and Curvature By dotting, and in turn crossing, the vector $\mathbf{v} = v\mathbf{T}$ with (9), it is possible to obtain explicit formulas involving $\mathbf{r}$, $\mathbf{r}'$, and $\mathbf{r}''$ for the tangential and normal components of the acceleration and the curvature. Observe that

$$\mathbf{v} \cdot \mathbf{a} = a_N \underbrace{(v\,\mathbf{T} \cdot \mathbf{N})}_{0} + a_T \underbrace{(v\,\mathbf{T} \cdot \mathbf{T})}_{1} = a_T v$$

yields the tangential component of acceleration

$$a_T = \frac{dv}{dt} = \frac{\mathbf{v} \cdot \mathbf{a}}{\|\mathbf{v}\|} = \frac{\mathbf{r}'(t) \cdot \mathbf{r}''(t)}{\|\mathbf{r}'(t)\|}. \tag{10}$$

On the other hand,

$$\mathbf{v} \times \mathbf{a} = a_N \underbrace{(v\,\mathbf{T} \times \mathbf{N})}_{\mathbf{B}} + a_T \underbrace{(v\,\mathbf{T} \times \mathbf{T})}_{0} = a_N v\,\mathbf{B}.$$

Since $\|\mathbf{B}\| = 1$, it follows that the normal component of acceleration is

$$a_N = \kappa v^2 = \frac{\|\mathbf{v} \times \mathbf{a}\|}{\|\mathbf{v}\|} = \frac{\|\mathbf{r}'(t) \times \mathbf{r}''(t)\|}{\|\mathbf{r}'(t)\|}. \tag{11}$$

Solving (11) for the curvature gives

$$\kappa(t) = \frac{\|\mathbf{v} \times \mathbf{a}\|}{\|\mathbf{v}\|^3} = \frac{\|\mathbf{r}'(t) \times \mathbf{r}''(t)\|}{\|\mathbf{r}'(t)\|^3}. \tag{12}$$

EXAMPLE 4 **Curvature of Twisted Cubic**

The curve traced by $\mathbf{r}(t) = t\mathbf{i} + \frac{1}{2}t^2\mathbf{j} + \frac{1}{3}t^3\mathbf{k}$ is said to be a "twisted cubic." If $\mathbf{r}(t)$ is the position vector of a moving particle, find the tangential and normal components of the acceleration at any t. Find the curvature.

SOLUTION $\mathbf{v}(t) = \mathbf{r}'(t) = \mathbf{i} + t\mathbf{j} + t^2\mathbf{k}$, $\mathbf{a}(t) = \mathbf{r}''(t) = \mathbf{j} + 2t\mathbf{k}$.

Since $\mathbf{v} \cdot \mathbf{a} = t + 2t^3$ and $\|\mathbf{v}\| = \sqrt{1 + t^2 + t^4}$, it follows from (10) that

$$a_T = \frac{dv}{dt} = \frac{t + 2t^3}{\sqrt{1 + t^2 + t^4}}.$$

Now,

$$\mathbf{v} \times \mathbf{a} = \begin{vmatrix} \mathbf{i} & \mathbf{j} & \mathbf{k} \\ 1 & t & t^2 \\ 0 & 1 & 2t \end{vmatrix} = t^2\mathbf{i} - 2t\mathbf{j} + \mathbf{k}$$

and $\|\mathbf{v} \times \mathbf{a}\| = \sqrt{t^4 + 4t^2 + 1}$. Thus, (11) gives

$$a_N = \kappa v^2 = \frac{\sqrt{t^4 + 4t^2 + 1}}{\sqrt{1 + t^2 + t^4}} = \sqrt{\frac{t^4 + 4t^2 + 1}{t^4 + t^2 + 1}}.$$

From (12) we find that the curvature of the twisted cubic is given by

$$\kappa(t) = \frac{(t^4 + 4t^2 + 1)^{1/2}}{(t^4 + t^2 + 1)^{3/2}}.$$

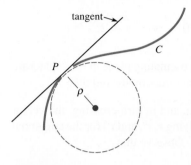

FIGURE 9.3.6 Radius of curvature

Radius of Curvature The reciprocal of the curvature, $\rho = 1/\kappa$, is called the **radius of curvature**. The radius of curvature at a point P on a curve C is the radius of a circle that "fits" the curve there better than any other circle. The circle at P is called the **circle of curvature** and its center is the **center of curvature**. The circle of curvature has the same tangent line at P as the curve C, and its center lies on the concave side of C. For example, a car moving on a curved track, as shown in FIGURE 9.3.6, can, at any instant, be thought to be moving on a circle of radius ρ. Hence, the normal component of its acceleration $a_N = \kappa v^2$ must be the same as the magnitude of its centripetal acceleration $a = v^2/\rho$. Therefore, $\kappa = 1/\rho$ and $\rho = 1/\kappa$. Knowing the radius of curvature, we can determine the speed v at which a car can negotiate a banked curve without skidding. (This is essentially the idea in Problem 26 in Exercises 9.2.)

REMARKS

By writing (6) as

$$\mathbf{a}(t) = \frac{ds}{dt}\frac{d\mathbf{T}}{dt} + \frac{d^2s}{dt^2}\mathbf{T},$$

we note that the so-called scalar acceleration d^2s/dt^2, referred to in the last remark, is now seen to be the tangential component of the acceleration a_T.

9.3 Exercises Answers to selected odd-numbered problems begin on page ANS-23.

In Problems 1 and 2, for the given position function, find the unit tangent.

1. $\mathbf{r}(t) = (t \cos t - \sin t)\mathbf{i} + (t \sin t + \cos t)\mathbf{j} + t^2\mathbf{k}, \ t > 0$

2. $\mathbf{r}(t) = e^t \cos t\,\mathbf{i} + e^t \sin t\,\mathbf{j} + \sqrt{2}e^t\mathbf{k}$

3. Use the procedure outlined in Example 2 to find $\mathbf{T}, \mathbf{N}, \mathbf{B}$, and κ for motion on a general circular helix that is described by $\mathbf{r}(t) = a \cos t\,\mathbf{i} + a \sin t\,\mathbf{j} + ct\,\mathbf{k}$.

4. Use the procedure outlined in Example 2 to show on the twisted cubic of Example 4 that at $t = 1$:

$$\mathbf{T} = \frac{1}{\sqrt{3}}(\mathbf{i} + \mathbf{j} + \mathbf{k}), \quad \mathbf{N} = -\frac{1}{\sqrt{2}}(\mathbf{i} - \mathbf{k}),$$

$$\mathbf{B} = -\frac{1}{\sqrt{6}}(-\mathbf{i} + 2\mathbf{j} - \mathbf{k}), \quad \kappa = \frac{\sqrt{2}}{3}.$$

In Problems 5 and 6, find an equation of the osculating plane to the given space curve at the point that corresponds to the indicated value of t.

5. The circular helix of Example 2; $t = \pi/4$

6. The twisted cubic of Example 4; $t = 1$

In Problems 7–16, $\mathbf{r}(t)$ is the position vector of a moving particle. Find the tangential and normal components of the acceleration at any t.

7. $\mathbf{r}(t) = \mathbf{i} + t\mathbf{j} + t^2\mathbf{k}$

8. $\mathbf{r}(t) = 3 \cos t\,\mathbf{i} + 2 \sin t\,\mathbf{j} + t\mathbf{k}$

9. $\mathbf{r}(t) = t^2\mathbf{i} + (t^2 - 1)\mathbf{j} + 2t^2\mathbf{k}$

10. $\mathbf{r}(t) = t^2\mathbf{i} - t^3\mathbf{j} + t^4\mathbf{k}$

11. $\mathbf{r}(t) = 2t\,\mathbf{i} + t^2\mathbf{j}$

12. $\mathbf{r}(t) = \tan^{-1}t\,\mathbf{i} + \frac{1}{2}\ln(1 + t^2)\mathbf{j}$

13. $\mathbf{r}(t) = 5 \cos t\,\mathbf{i} + 5 \sin t\,\mathbf{j}$

14. $\mathbf{r}(t) = \cosh t\,\mathbf{i} + \sinh t\,\mathbf{j}$

15. $\mathbf{r}(t) = e^{-t}(\mathbf{i} + \mathbf{j} + \mathbf{k})$

16. $\mathbf{r}(t) = t\mathbf{i} + (2t - 1)\mathbf{j} + (4t + 2)\mathbf{k}$

17. Find the curvature of an elliptical helix that is described by $\mathbf{r}(t) = a \cos t\,\mathbf{i} + b \sin t\,\mathbf{j} + ct\,\mathbf{k}, a > 0, b > 0, c > 0$.

18. (a) Find the curvature of an elliptical orbit that is described by $\mathbf{r}(t) = a \cos t\,\mathbf{i} + b \sin t\,\mathbf{j} + c\mathbf{k}, a > 0, b > 0, c > 0$.
 (b) Show that when $a = b$, the curvature of a circular orbit is the constant $\kappa = 1/a$.

19. Show that the curvature of a straight line is the constant $\kappa = 0$. [*Hint:* Use (2) in Section 7.5.]

20. Find the curvature at $t = \pi$ of the cycloid that is described by

$$\mathbf{r}(t) = a(t - \sin t)\mathbf{i} + a(1 - \cos t)\mathbf{j}, \quad a > 0.$$

21. Let C be a plane curve traced by $\mathbf{r}(t) = f(t)\mathbf{i} + g(t)\mathbf{j}$, where f and g have second derivatives. Show that the curvature at a point is given by

$$\kappa = \frac{|f'(t)g''(t) - g'(t)f''(t)|}{([f'(t)]^2 + [g'(t)]^2)^{3/2}}.$$

22. Show that if $y = F(x)$, the formula for κ in Problem 21 reduces to

$$\kappa = \frac{|F''(x)|}{[1 + (F'(x))^2]^{3/2}}.$$

In Problems 23 and 24, use the result of Problem 22 to find the curvature and radius of curvature of the curve at the indicated points. Decide at which point the curve is "sharper."

23. $y = x^2$; $(0, 0)$, $(1, 1)$

24. $y = x^3$; $(-1, -1)$, $(\frac{1}{2}, \frac{1}{8})$

Discussion Problems

25. Discuss the curvature near a point of inflection of $y = F(x)$.

26. Show that $\|\mathbf{a}(t)\|^2 = a_N^2 + a_T^2$.

INTRODUCTION In this section we consider functions of two or more variables and how to find the instantaneous rate of change—that is, the derivative—of such functions with respect to each variable.

■ **Functions of Two Variables** Recall from calculus that a **function of two variables** is a rule of correspondence that assigns to each ordered pair of real numbers (x, y) of a subset of the xy-plane one and only one number z in the set R of real numbers. The set of ordered pairs (x, y) is called the **domain** of the function and the set of corresponding values of z is called the **range**. A function of two variables is usually written $z = f(x, y)$. The variables x and y are called the **independent variables** of the function, and z is called the **dependent variable**. The **graph** of a function $z = f(x, y)$ is a surface in 3-space. See **FIGURE 9.4.1**.

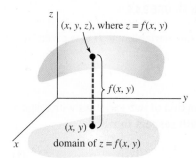

FIGURE 9.4.1 Function of two variables

■ **Level Curves** For a function $z = f(x, y)$, the curves defined by $f(x, y) = c$, for suitable c, are called the **level curves** of f. The word *level* arises from the fact that we can interpret the equation $f(x, y) = c$ as the projection onto the xy-plane of the curve of intersection, or trace, of $z = f(x, y)$ and the (horizontal or level) plane $z = c$. See **FIGURE 9.4.2**.

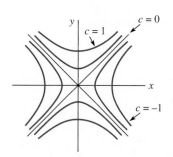

(b) Level curves

FIGURE 9.4.3 Surface and level curves in Example 1

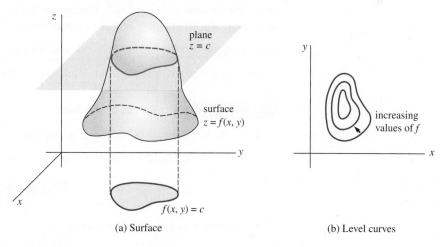

(a) Surface

(b) Level curves

FIGURE 9.4.2 Surface and level curves

| EXAMPLE 1 | Level Curves |

The level curves of the function $f(x, y) = y^2 - x^2$ are defined by $y^2 - x^2 = c$. As shown in **FIGURE 9.4.3**, when $c > 0$ or $c < 0$, a member of this family of curves is a hyperbola. For $c = 0$, we obtain the lines $y = x$ and $y = -x$. ≡

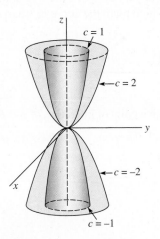

FIGURE 9.4.4 Level surfaces in Example 2

Functions of Three or More Variables
Functions of three or more variables are defined analogously to functions of two variables. For example, a **function of three variables** is a rule of correspondence that assigns to each ordered triple of real numbers (x, y, z) of a subset of 3-space one and only one number w in the set R of real numbers. We write $w = F(x, y, z)$.

Level Surfaces
Although we cannot draw a graph of a function of three variables $w = F(x, y, z)$, we can draw the surfaces defined by $F(x, y, z) = c$ for suitable values of c. These surfaces are called **level surfaces**. This is an unfortunate, though standard, choice of words, since *level* surfaces are usually not level.

EXAMPLE 2 Level Surfaces

Describe the level surfaces of the function $F(x, y, z) = (x^2 + y^2)/z$.

SOLUTION For $c \neq 0$ the level surfaces are given by

$$\frac{x^2 + y^2}{z} = c \quad \text{or} \quad x^2 + y^2 = cz.$$

A few members of this family of paraboloids are shown in **FIGURE 9.4.4**. ≡

Partial Derivatives
The derivative of a function of one variable $y = f(x)$ is given by

$$\frac{dy}{dx} = \lim_{\Delta x \to 0} \frac{f(x + \Delta x) - f(x)}{\Delta x}.$$

In exactly the same manner, we can define a derivative of a function of two variables with respect to *each* variable. If $z = f(x, y)$, then the **partial derivative with respect to x** is

$$\frac{\partial z}{\partial x} = \lim_{\Delta x \to 0} \frac{f(x + \Delta x, y) - f(x, y)}{\Delta x} \tag{1}$$

and the **partial derivative with respect to y** is

$$\frac{\partial z}{\partial y} = \lim_{\Delta y \to 0} \frac{f(x, y + \Delta y) - f(x, y)}{\Delta y}, \tag{2}$$

provided each limit exists.

In (1) the variable y does not change in the limiting process; that is, y is held fixed. Similarly, in (2) the variable x is held fixed. The two partial derivatives (1) and (2) then represent the *rates of change* of f with respect to x and y, respectively. On a practical level:

To compute $\partial z/\partial x$, use the laws of ordinary differentiation while treating y as a constant.

To compute $\partial z/\partial y$, use the laws of ordinary differentiation while treating x as a constant.

EXAMPLE 3 Partial Derivatives

If $z = 4x^3 y^2 - 4x^2 + y^6 + 1$, find $\partial z/\partial x$ and $\partial z/\partial y$.

SOLUTION We hold y fixed and treat constants in the usual manner. Thus,

$$\frac{\partial z}{\partial x} = 12x^2 y^2 - 8x.$$

By treating x as a constant, we obtain

$$\frac{\partial z}{\partial y} = 8x^3 y + 6y^5.$$ ≡

Alternative Symbols
The partial derivatives $\partial z/\partial x$ and $\partial z/\partial y$ are often represented by alternative symbols. If $z = f(x, y)$, then

$$\frac{\partial z}{\partial x} = \frac{\partial f}{\partial x} = z_x = f_x \quad \text{and} \quad \frac{\partial z}{\partial y} = \frac{\partial f}{\partial y} = z_y = f_y.$$

Higher-Order and Mixed Derivatives

For a function of two variables $z = f(x, y)$, the partial derivatives $\partial z / \partial x$ and $\partial z / \partial y$ are themselves functions of x and y. Consequently, we can compute **second** and higher **partial derivatives**. Indeed, we can find the partial derivative of $\partial z / \partial x$ with respect to y, and the partial derivative of $\partial z / \partial y$ with respect to x. The latter types of partial derivatives are called **mixed partial derivatives**. In summary, for $z = f(x, y)$:

Second-order partial derivatives:

$$\frac{\partial^2 z}{\partial x^2} = \frac{\partial}{\partial x}\left(\frac{\partial z}{\partial x}\right) \quad \text{and} \quad \frac{\partial^2 z}{\partial y^2} = \frac{\partial}{\partial y}\left(\frac{\partial z}{\partial y}\right).$$

Third-order partial derivatives:

$$\frac{\partial^3 z}{\partial x^3} = \frac{\partial}{\partial x}\left(\frac{\partial^2 z}{\partial x^2}\right) \quad \text{and} \quad \frac{\partial^3 z}{\partial y^3} = \frac{\partial}{\partial y}\left(\frac{\partial^2 z}{\partial y^2}\right).$$

Mixed second-order partial derivatives:

$$\frac{\partial^2 z}{\partial x \partial y} = \frac{\partial}{\partial x}\left(\frac{\partial z}{\partial y}\right) \quad \text{and} \quad \frac{\partial^2 z}{\partial y \partial x} = \frac{\partial}{\partial y}\left(\frac{\partial z}{\partial x}\right).$$

Alternative Symbols

The second- and third-order partial derivatives are denoted by f_{xx}, f_{yy}, f_{xxx}, and so on. The subscript notation for mixed second partial derivatives is f_{xy} or f_{yx}. Note that

$$f_{xy} = (f_x)_y = \frac{\partial}{\partial y}\left(\frac{\partial z}{\partial x}\right) = \frac{\partial^2 z}{\partial y \partial x} \quad \text{and} \quad f_{yx} = \frac{\partial^2 z}{\partial x \partial y}.$$

Although we shall not prove it, if a function f has continuous second partial derivatives, then the order in which a mixed second partial derivative is done is irrelevant; that is,

$$f_{xy} = f_{yx}. \tag{3}$$

Functions of Three or More Variables

The rates of change of a function of three variables $w = F(x, y, z)$ in the x-, y-, and z-directions are $\partial w / \partial x$, $\partial w / \partial y$, and $\partial w / \partial z$, respectively. To compute, say, $\partial w / \partial x$, we differentiate with respect to x in the usual manner while holding *both* y and z constant. In this manner we extend the process of partial differentiation to functions of any number of variables.

EXAMPLE 4 **Partial Derivatives**

If $F(x, y, t) = e^{-3\pi t} \cos 4x \sin 6y$, then the partial derivatives with respect to x, y, and t are, in turn,

$$F_x(x, y, t) = -4e^{-3\pi t} \sin 4x \sin 6y,$$
$$F_y(x, y, t) = 6e^{-3\pi t} \cos 4x \cos 6y,$$
$$F_t(x, y, t) = -3\pi e^{-3\pi t} \cos 4x \sin 6y. \qquad \equiv$$

Chain Rule

The Chain Rule for functions of one variable states that if $y = f(u)$ is a differentiable function of u, and $u = g(x)$ is a differentiable function of x, then the derivative of the composite function is

$$\frac{dy}{dx} = \frac{dy}{du}\frac{du}{dx}. \tag{4}$$

For a composite function of two variables $z = f(u, v)$, where $u = g(x, y)$ and $v = h(x, y)$, we would naturally expect *two* formulas analogous to (4), since we can compute both $\partial z / \partial x$ and $\partial z / \partial y$. The **Chain Rule** for functions of two variables is summarized as follows:

THEOREM 9.4.1 **Chain Rule**

If $z = f(u, v)$ is differentiable and $u = g(x, y)$ and $v = h(x, y)$ have continuous first partial derivatives, then

$$\frac{\partial z}{\partial x} = \frac{\partial z}{\partial u}\frac{\partial u}{\partial x} + \frac{\partial z}{\partial v}\frac{\partial v}{\partial x}, \qquad \frac{\partial z}{\partial y} = \frac{\partial z}{\partial u}\frac{\partial u}{\partial y} + \frac{\partial z}{\partial v}\frac{\partial v}{\partial y}. \tag{5}$$

EXAMPLE 5 **Chain Rule**

If $z = u^2 - v^3$ and $u = e^{2x-3y}$, $v = \sin(x^2 - y^2)$, find $\partial z/\partial x$ and $\partial z/\partial y$.

SOLUTION Since $\partial z/\partial u = 2u$ and $\partial z/\partial v = -3v^2$, it follows from (5) that

$$\frac{\partial z}{\partial x} = 2u(2e^{2x-3y}) - 3v^2[2x\cos(x^2 - y^2)] = 4ue^{2x-3y} - 6xv^2\cos(x^2 - y^2) \tag{6}$$

$$\frac{\partial z}{\partial y} = 2u(-3e^{2x-3y}) - 3v^2[(-2y)\cos(x^2 - y^2)] = -6ue^{2x-3y} + 6yv^2\cos(x^2 - y^2). \tag{7} \equiv$$

Of course, in Example 5 we could substitute the expressions for u and v in the original function and then find the partial derivatives directly. In the same manner, the answers (6) and (7) can be expressed in terms of x and y.

■ **Special Case** If $z = f(u, v)$ is differentiable and $u = g(t)$ and $v = h(t)$ are differentiable functions of a single variable t, then Theorem 9.4.1 implies that the ordinary derivative dz/dt is

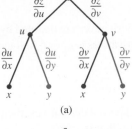

$$\frac{dz}{dt} = \frac{\partial z}{\partial u}\frac{du}{dt} + \frac{\partial z}{\partial v}\frac{dv}{dt}. \tag{8}$$

■ **Generalizations** The results given in (5) and (8) immediately generalize to any number of variables. If $z = f(u_1, u_2, \ldots, u_n)$ and each of the variables $u_1, u_2, u_3, \ldots, u_n$ are functions of $x_1, x_2, \ldots, x_k$, then under the same assumptions as in Theorem 9.4.1 we have

$$\frac{\partial z}{\partial x_i} = \frac{\partial z}{\partial u_1}\frac{\partial u_1}{\partial x_i} + \frac{\partial z}{\partial u_2}\frac{\partial u_2}{\partial x_i} + \cdots + \frac{\partial z}{\partial u_n}\frac{\partial u_n}{\partial x_i}, \tag{9}$$

where $i = 1, 2, \ldots, k$. Similarly, if the u_i, $i = 1, \ldots, n$, are differentiable functions of a single variable t, then

$$\frac{dz}{dt} = \frac{\partial z}{\partial u_1}\frac{du_1}{dt} + \frac{\partial z}{\partial u_2}\frac{du_2}{dt} + \cdots + \frac{\partial z}{\partial u_n}\frac{du_n}{dt}. \tag{10}$$

■ **Tree Diagrams** The results in (5) can be memorized in terms of a **tree diagram**. The dots in the first diagram in **FIGURE 9.4.5(a)** indicate the fact that z depends on u and v; u and v depend, in turn, on x and y. To compute $\partial z/\partial x$ for example, we read the diagram vertically downward starting from z and following the two blue polygonal paths leading to x, multiply the partial derivatives on each path, and then add the products. The result given in (8) is represented by the second tree diagram in Figure 9.4.5(b).

We shall use tree diagrams in the next two examples to illustrate special cases of (9) and (10).

EXAMPLE 6 **Using Tree Diagrams**

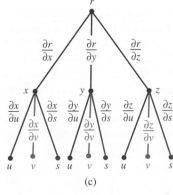

If $r = x^2 + y^5z^3$ and $x = uve^{2s}$, $y = u^2 - v^2s$, $z = \sin(uvs^2)$, find $\partial r/\partial s$.

SOLUTION From the blue and green paths in the tree diagram in Figure 9.4.5(c) we obtain

$$\frac{\partial r}{\partial s} = \frac{\partial r}{\partial x}\frac{\partial x}{\partial s} + \frac{\partial r}{\partial y}\frac{\partial y}{\partial s} + \frac{\partial r}{\partial z}\frac{\partial z}{\partial s}$$

$$= 2x(2uve^{2s}) + 5y^4z^3(-v^2) + 3y^5z^2(2uvs\cos(uvs^2)). \qquad \equiv$$

EXAMPLE 7 **Using Tree Diagrams**

If $z = u^2v^3w^4$ and $u = t^2$, $v = 5t - 8$, $w = t^3 + t$, find dz/dt.

SOLUTION In this case the tree diagram in Figure 9.4.5(d) indicates that

$$\frac{dz}{dt} = \frac{\partial z}{\partial u}\frac{du}{dt} + \frac{\partial z}{\partial v}\frac{dv}{dt} + \frac{\partial z}{\partial w}\frac{dw}{dt}$$

$$= 2uv^3w^4(2t) + 3u^2v^2w^4(5) + 4u^2v^3w^3(3t^2 + 1).$$

FIGURE 9.4.5 Tree diagrams

ALTERNATIVE SOLUTION Differentiate $z = t^4(5t - 8)^3(t^3 + t)^4$ by the Product Rule. $\equiv$

REMARKS

If $w = F(x, y, z)$ has continuous partial derivatives of any order, then analogous to (3), the mixed partial derivatives are equal:

$$F_{xyz} = F_{yzx} = F_{zyx}, \qquad F_{xxy} = F_{yxx} = F_{xyx},$$

and so on.

9.4 Exercises Answers to selected odd-numbered problems begin on page ANS-23.

In Problems 1–6, sketch some of the level curves associated with the given function.

1. $f(x, y) = x + 2y$ **2.** $f(x, y) = y^2 - x$

3. $f(x, y) = \sqrt{x^2 - y^2 - 1}$

4. $f(x, y) = \sqrt{36 - 4x^2 - 9y^2}$

5. $f(x, y) = e^{y - x^2}$ **6.** $f(x, y) = \tan^{-1}(y - x)$

In Problems 7–10, describe the level surfaces but do not graph.

7. $F(x, y, z) = \dfrac{x^2}{9} + \dfrac{z^2}{4}$

8. $F(x, y, z) = x^2 + y^2 + z^2$

9. $F(x, y, z) = x^2 + 3y^2 + 6z^2$

10. $F(x, y, z) = 4y - 2z + 1$

11. Graph some of the level surfaces associated with $F(x, y, z) = x^2 + y^2 - z^2$ for $c = 0$, $c > 0$, and $c < 0$.

12. Given that

$$F(x, y, z) = \frac{x^2}{16} + \frac{y^2}{4} + \frac{z^2}{9},$$

find the x-, y-, and z-intercepts of the level surface that passes through $(-4, 2, -3)$.

In Problems 13–32, find the first partial derivatives of the given function.

13. $z = x^2 - xy^2 + 4y^5$ **14.** $z = -x^3 + 6x^2y^3 + 5y^2$

15. $z = 5x^4y^3 - x^2y^6 + 6x^5 - 4y$

16. $z = \tan(x^3y^2)$

17. $z = \dfrac{4\sqrt{x}}{3y^2 + 1}$ **18.** $z = 4x^3 - 5x^2 + 8x$

19. $z = (x^3 - y^2)^{-1}$ **20.** $z = (-x^4 + 7y^2 + 3y)^6$

21. $z = \cos^2 5x + \sin^2 5y$ **22.** $z = e^{x^2 \tan^{-1} y^2}$

23. $f(x, y) = xe^{x^3y}$ **24.** $f(\theta, \phi) = \phi^2 \sin \dfrac{\theta}{\phi}$

25. $f(x, y) = \dfrac{3x - y}{x + 2y}$ **26.** $f(x, y) = \dfrac{xy}{(x^2 - y^2)^2}$

27. $g(u, v) = \ln(4u^2 + 5v^3)$ **28.** $h(r, s) = \dfrac{\sqrt{r}}{s} - \dfrac{\sqrt{s}}{r}$

29. $w = 2\sqrt{xy} - ye^{y/z}$ **30.** $w = xy \ln(xz)$

31. $F(u, v, x, t) = u^2w^2 - uv^3 + vw \cos(ut^2) + (2x^2t)^4$

32. $G(p, q, r, s) = (p^2q^3)^{r^4s^5}$

In Problems 33 and 34, verify that the given function satisfies **Laplace's equation**:

$$\frac{\partial^2 z}{\partial x^2} + \frac{\partial^2 z}{\partial y^2} = 0.$$

33. $z = \ln(x^2 + y^2)$ **34.** $z = e^{x^2 - y^2} \cos 2xy$

In Problems 35 and 36 verify that the given function satisfies the **wave equation**:

$$a^2 \frac{\partial^2 u}{\partial x^2} = \frac{\partial^2 u}{\partial t^2}.$$

35. $u = \cos at \sin x$ **36.** $u = \cos(x + at) + \sin(x - at)$

37. The molecular concentration $C(x, t)$ of a liquid is given by $C(x, t) = t^{-1/2} e^{-x^2/kt}$. Verify that this function satisfies the **diffusion equation**:

$$\frac{k}{4} \frac{\partial^2 C}{\partial x^2} = \frac{\partial C}{\partial t}.$$

38. The pressure P exerted by an enclosed ideal gas is given by $P = k(T/V)$, where k is a constant, T is temperature, and V is volume. Find:
 (a) the rate of change of P with respect to V,
 (b) the rate of change of V with respect to T, and
 (c) the rate of change of T with respect to P.

In Problems 39–48, use the Chain Rule to find the indicated partial derivatives.

39. $z = e^{uv^2}$; $u = x^3$, $v = x - y^2$; $\dfrac{\partial z}{\partial x}, \dfrac{\partial z}{\partial y}$

40. $z = u^2 \cos 4v$; $u = x^2y^3$, $v = x^3 + y^3$; $\dfrac{\partial z}{\partial x}, \dfrac{\partial z}{\partial y}$

41. $z = 4x - 5y^2$; $x = u^4 - 8v^3$, $y = (2u - v)^2$; $\dfrac{\partial z}{\partial u}, \dfrac{\partial z}{\partial v}$

42. $z = \dfrac{x - y}{x + y}$; $x = \dfrac{u}{v}$, $y = \dfrac{v^2}{u}$; $\dfrac{\partial z}{\partial u}, \dfrac{\partial z}{\partial v}$

43. $w = (u^2 + v^2)^{3/2}$; $u = e^{-t} \sin \theta$, $v = e^{-t} \cos \theta$; $\dfrac{\partial w}{\partial t}, \dfrac{\partial w}{\partial \theta}$

44. $w = \tan^{-1}\sqrt{uv}$; $u = r^2 - s^2$, $v = r^2s^2$; $\dfrac{\partial w}{\partial r}, \dfrac{\partial w}{\partial s}$

45. $R = rs^2t^4$; $r = ue^{v^2}$, $s = ve^{-u^2}$, $t = e^{u^2v^2}$; $\dfrac{\partial R}{\partial u}, \dfrac{\partial R}{\partial v}$

46. $Q = \ln(pqr)$; $p = t^2 \sin^{-1} x$, $q = \dfrac{x}{t^2}$, $r = \tan^{-1}\dfrac{x}{t}$; $\dfrac{\partial Q}{\partial x}, \dfrac{\partial Q}{\partial t}$

47. $w = \sqrt{x^2 + y^2}$; $x = \ln(rs + tu)$,

$y = \dfrac{t}{u}\cosh rs$; $\dfrac{\partial w}{\partial t}, \dfrac{\partial w}{\partial r}, \dfrac{\partial w}{\partial u}$

48. $s = p^2 + q^2 - r^2 + 4t$; $p = \phi e^{3\theta}, q = \cos(\phi + \theta), r = \phi\theta^2$,

$t = 2\phi + 8\theta$; $\dfrac{\partial s}{\partial \phi}, \dfrac{\partial s}{\partial \theta}$

In Problems 49–52, use (8) to find the indicated derivative.

49. $z = \ln(u^2 + v^2)$; $u = t^2, v = t^{-2}$; $\dfrac{dz}{dt}$

50. $z = u^3v - uv^4$; $u = e^{-5t}, v = \sec 5t$; $\dfrac{dz}{dt}$

51. $w = \cos(3u + 4v)$; $u = 2t + \dfrac{\pi}{2}, v = -t - \dfrac{\pi}{4}$; $\dfrac{dw}{dt}\bigg|_{t=\pi}$

52. $w = e^{xy}$; $x = \dfrac{4}{2t + 1}, y = 3t + 5$; $\dfrac{dw}{dt}\bigg|_{t=0}$

53. If $u = f(x, y)$ and $x = r\cos\theta, y = r\sin\theta$, show that **Laplace's equation** $\partial^2 u/\partial x^2 + \partial^2 u/\partial y^2 = 0$ becomes

$$\frac{\partial^2 u}{\partial r^2} + \frac{1}{r}\frac{\partial u}{\partial r} + \frac{1}{r^2}\frac{\partial^2 u}{\partial \theta^2} = 0.$$

54. Van der Waals' equation of state for the real gas CO_2 is

$$P = \frac{0.08T}{V - 0.0427} - \frac{3.6}{V^2}.$$

If dT/dt and dV/dt are rates at which the temperature and volume change, respectively, use the Chain Rule to find dP/dt.

55. The equation of state for a thermodynamic system is $F(P, V, T) = 0$, where P, V, and T are pressure, volume, and temperature, respectively. If the equation defines V as a function of P and T, and also defines T as a function of V and P, show that

$$\frac{\partial V}{\partial T} = -\frac{\dfrac{\partial F}{\partial T}}{\dfrac{\partial F}{\partial V}} = -\frac{1}{\dfrac{\partial T}{\partial V}}.$$

56. The voltage across a conductor is increasing at a rate of 2 volts/min and the resistance is decreasing at a rate of 1 ohm/min. Use $I = E/R$ and the Chain Rule to find the rate at which the current passing through the conductor is changing when $R = 50$ ohms and $E = 60$ volts.

57. The length of the side labeled x of the triangle in **FIGURE 9.4.6** increases at a rate of 0.3 cm/s, the side labeled y increases at a rate of 0.5 cm/s, and the included angle θ increases at a rate of 0.1 rad/s. Use the Chain Rule to find the rate at which the area of the triangle is changing at the instant $x = 10$ cm, $y = 8$ cm, and $\theta = \pi/6$.

FIGURE 9.4.6 Triangle in Problem 57

58. A particle moves in 3-space so that its coordinates at any time are $x = 4\cos t, y = 4\sin t, z = 5t, t \geq 0$. Use the Chain Rule to find the rate at which its distance

$$w = \sqrt{x^2 + y^2 + z^2}$$

from the origin is changing at $t = 5\pi/2$ seconds.

9.5 Directional Derivative

INTRODUCTION We saw in the last section that for a function f of two variables x and y, the partial derivatives $\partial z/\partial x$ and $\partial z/\partial y$ give the slope of the tangent to the trace, or curve of intersection of the surface defined by $z = f(x, y)$ and vertical planes that are, respectively, parallel to the x- and y-coordinates axes. Equivalently, we can think of the partial derivative $\partial z/\partial x$ as the rate of change of the function f in the direction given by the vector $\mathbf{i}$, and $\partial z/\partial y$ as the rate of change of the function f in the $\mathbf{j}$-direction. There is no reason to confine our attention to just two directions. In this section we shall see how to find the rate of change of a differentiable function in *any* direction. See **FIGURE 9.5.1**.

■ **The Gradient of a Function** In this and the next sections it is convenient to introduce a new vector based on partial differentiation. When the **vector differential operator**

$$\nabla = \mathbf{i}\frac{\partial}{\partial x} + \mathbf{j}\frac{\partial}{\partial y} \quad \text{or} \quad \nabla = \mathbf{i}\frac{\partial}{\partial x} + \mathbf{j}\frac{\partial}{\partial y} + \mathbf{k}\frac{\partial}{\partial z}$$

is applied to a differentiable function $z = f(x, y)$ or $w = F(x, y, z)$, we say that the vectors

$$\nabla f(x, y) = \frac{\partial f}{\partial x}\mathbf{i} + \frac{\partial f}{\partial y}\mathbf{j} \tag{1}$$

$$\nabla F(x, y, z) = \frac{\partial F}{\partial x}\mathbf{i} + \frac{\partial F}{\partial y}\mathbf{j} + \frac{\partial F}{\partial z}\mathbf{k} \tag{2}$$

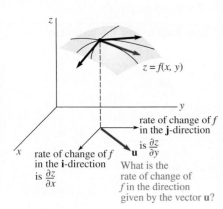

rate of change of f in the **j**-direction is $\frac{\partial z}{\partial y}$

$\mathbf{u}$

rate of change of f in the **i**-direction is $\frac{\partial z}{\partial x}$

What is the rate of change of f in the direction given by the vector $\mathbf{u}$?

FIGURE 9.5.1 An arbitrary direction is denoted by the vector $\mathbf{u}$

are the **gradients** of the respective functions. The symbol ∇, an inverted capital Greek delta, is called "del" or "nabla." The vector ∇f is usually read "grad f."

EXAMPLE 1 **Gradient**

Compute $\nabla f(x, y)$ for $f(x, y) = 5y - x^3 y^2$.

SOLUTION From (1), $\nabla f(x, y) = \dfrac{\partial}{\partial x}(5y - x^3 y^2)\mathbf{i} + \dfrac{\partial}{\partial y}(5y - x^3 y^2)\mathbf{j}$; therefore

$$\nabla f(x, y) = -3x^2 y^2 \mathbf{i} + (5 - 2x^3 y)\mathbf{j}.$$

≡

EXAMPLE 2 **Gradient at a Point**

If $F(x, y, z) = xy^2 + 3x^2 - z^3$, find $\nabla F(x, y, z)$ at $(2, -1, 4)$.

SOLUTION From (2), $\nabla F(x, y, z) = (y^2 + 6x)\mathbf{i} + 2xy\mathbf{j} - 3z^2\mathbf{k}$ and so

$$\nabla F(2, -1, 4) = 13\mathbf{i} - 4\mathbf{j} - 48\mathbf{k}.$$

≡

■ **A Generalization of Partial Differentiation** Suppose $\mathbf{u} = \cos\theta\,\mathbf{i} + \sin\theta\,\mathbf{j}$ is a unit vector in the xy-plane that makes an angle θ with the positive x-axis and is parallel to the vector $\mathbf{v}$ from $(x, y, 0)$ to $(x + \Delta x, y + \Delta y, 0)$. If $h = \sqrt{(\Delta x)^2 + (\Delta y)^2} > 0$, then $\mathbf{v} = h\mathbf{u}$. Furthermore, let the plane perpendicular to the xy-plane that contains these points slice the surface $z = f(x, y)$ in a curve C. We ask: What is the slope of the tangent line to C at a point P with coordinates $(x, y, f(x, y))$ in the direction given by $\mathbf{v}$? See **FIGURE 9.5.2**.

From the figure we see that $\Delta x = h\cos\theta$ and $\Delta y = h\sin\theta$ so that the slope of the indicated secant line is

$$\frac{f(x + \Delta x, y + \Delta y) - f(x, y)}{h} = \frac{f(x + h\cos\theta, y + h\sin\theta) - f(x, y)}{h}. \tag{3}$$

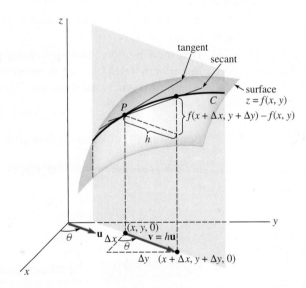

FIGURE 9.5.2 C is the curve of intersection of the surface and the plane determined by vector $\mathbf{v}$

We expect the slope of the tangent at P to be the limit of (3) as $h \to 0$. This slope is the rate of change of f at P in the direction specified by the unit vector $\mathbf{u}$. This leads us to the following definition:

DEFINITION 9.5.1 **Directional Derivative**

The **directional derivative** of $z = f(x, y)$ in the direction of a unit vector $\mathbf{u} = \cos\theta\,\mathbf{i} + \sin\theta\,\mathbf{j}$ is

$$D_{\mathbf{u}}f(x, y) = \lim_{h \to 0}\frac{f(x + h\cos\theta, y + h\sin\theta) - f(x, y)}{h} \tag{4}$$

provided the limit exists.

Observe that (4) is truly a generalization of partial differentiation, since

$$\theta = 0 \quad \text{implies that} \quad D_i f(x, y) = \lim_{h \to 0} \frac{f(x + h, y) - f(x, y)}{h} = \frac{\partial z}{\partial x}$$

and $\quad \theta = \dfrac{\pi}{2} \quad$ implies that $\quad D_j f(x, y) = \lim_{h \to 0} \dfrac{f(x, y + h) - f(x, y)}{h} = \dfrac{\partial z}{\partial y}.$

■ Method for Computing the Directional Derivative

While (4) could be used to find $D_u f(x, y)$ for a given function, as usual we seek a more efficient procedure. The next theorem will show how the concept of the gradient of a function plays a key role in computing a directional derivative.

THEOREM 9.5.1 Computing a Directional Derivative

If $z = f(x, y)$ is a differentiable function of x and y and $\mathbf{u} = \cos \theta \mathbf{i} + \sin \theta \mathbf{j}$, then

$$D_u f(x, y) = \nabla f(x, y) \cdot \mathbf{u}. \tag{5}$$

PROOF: Let x, y, and θ be fixed so that $g(t) = f(x + t \cos \theta, y + t \sin \theta)$ is a function of one variable. We wish to compare the value of $g'(0)$, which is found by two different methods. First, by the definition of a derivative,

$$g'(0) = \lim_{h \to 0} \frac{g(0 + h) - g(0)}{h} = \lim_{h \to 0} \frac{f(x + h \cos \theta, y + h \sin \theta) - f(x, y)}{h}. \tag{6}$$

Second, by the Chain Rule, (8) of Section 9.4,

$$g'(t) = f_1(x + t \cos\theta, y + t \sin\theta) \frac{d}{dt}(x + t \cos\theta) + f_2(x + t\cos\theta, y + t \sin\theta) \frac{d}{dt}(x + t \sin\theta) \tag{7}$$

$$= f_1(x + t\cos\theta, y + t\sin\theta)\cos\theta + f_2(x + t\cos\theta, y + t\sin\theta)\sin\theta.$$

Here the subscripts 1 and 2 refer to the partial derivatives of $f(x + t \cos\theta, y + t \sin\theta)$ with respect to $x + t \cos\theta$ and $y + t \sin\theta$, respectively. When $t = 0$, we note that $x + t\cos\theta$ and $y + t\sin\theta$ are simply x and y, and therefore (7) becomes

$$g'(0) = f_x(x, y)\cos\theta + f_y(x, y)\sin\theta. \tag{8}$$

Comparing (4), (6), and (8) then gives

$$D_u f(x, y) = f_x(x, y)\cos\theta + f_y(x, y)\sin\theta$$

$$= [f_x(x, y)\mathbf{i} + f_y(x, y)\mathbf{j}] \cdot (\cos\theta\mathbf{i} + \sin\theta\mathbf{j})$$

$$= \nabla f(x, y) \cdot \mathbf{u}. \qquad \equiv$$

EXAMPLE 3 Directional Derivative

Find the directional derivative of $f(x, y) = 2x^2 y^3 + 6xy$ at $(1, 1)$ in the direction of a unit vector whose angle with the positive x-axis is $\pi/6$.

SOLUTION Since $\dfrac{\partial f}{\partial x} = 4xy^3 + 6y$ and $\dfrac{\partial f}{\partial y} = 6x^2 y^2 + 6x$, we have

$$\nabla f(x, y) = (4xy^3 + 6y)\mathbf{i} + (6x^2 y^2 + 6x)\mathbf{j} \quad \text{and} \quad \nabla f(1, 1) = 10\mathbf{i} + 12\mathbf{j}.$$

Now, at $\theta = \pi/6$, $\mathbf{u} = \cos\theta\mathbf{i} + \sin\theta\mathbf{j}$ becomes $\mathbf{u} = \dfrac{\sqrt{3}}{2}\mathbf{i} + \dfrac{1}{2}\mathbf{j}$. Therefore,

$$D_u f(1, 1) = \nabla f(1, 1) \cdot \mathbf{u} = (10\mathbf{i} + 12\mathbf{j}) \cdot \left(\frac{\sqrt{3}}{2}\mathbf{i} + \frac{1}{2}\mathbf{j}\right) = 5\sqrt{3} + 6. \qquad \equiv$$

EXAMPLE 4 **Directional Derivative**

Consider the plane that is perpendicular to the xy-plane and passes through the points $P(2, 1)$ and $Q(3, 2)$. What is the slope of the tangent line to the curve of intersection of this plane with the surface $f(x, y) = 4x^2 + y^2$ at $(2, 1, 17)$ in the direction of Q?

SOLUTION We want $D_{\mathbf{u}}f(2, 1)$ in the direction given by the vector $\overrightarrow{PQ} = \mathbf{i} + \mathbf{j}$. But since $\overrightarrow{PQ}$ is not a unit vector, we form $\mathbf{u} = (1/\sqrt{2})\mathbf{i} + (1/\sqrt{2})\mathbf{j}$. Now,

$$\nabla f(x, y) = 8x\mathbf{i} + 2y\mathbf{j} \quad \text{and} \quad \nabla f(2, 1) = 16\mathbf{i} + 2\mathbf{j}.$$

Therefore, the required slope is

$$D_{\mathbf{u}}f(2, 1) = (16\mathbf{i} + 2\mathbf{j}) \cdot \left(\frac{1}{\sqrt{2}}\mathbf{i} + \frac{1}{\sqrt{2}}\mathbf{j} \right) = 9\sqrt{2}. \qquad \equiv$$

■ **Functions of Three Variables** For a function $w = F(x, y, z)$ the directional derivative is defined by

$$D_{\mathbf{u}}F(x, y, z) = \lim_{h \to 0} \frac{F(x + h\cos\alpha, y + h\cos\beta, z + h\cos\gamma) - F(x, y, z)}{h},$$

where α, β, and γ are the direction angles of the unit vector $\mathbf{u}$ measured relative to the positive x-, y-, and z-axes, respectively.* But in the same manner as before, we can show that

$$D_{\mathbf{u}}F(x, y, z) = \nabla F(x, y, z) \cdot \mathbf{u}. \qquad (9)$$

Notice that since $\mathbf{u}$ is a unit vector, it follows from (10) of Section 7.3 that

$$D_{\mathbf{u}}f(x, y) = \text{comp}_{\mathbf{u}} \nabla f(x, y) \quad \text{and} \quad D_{\mathbf{u}}F(x, y, z) = \text{comp}_{\mathbf{u}} \nabla F(x, y, z).$$

In addition, (9) reveals that

$$D_{\mathbf{k}}F(x, y, z) = \frac{\partial w}{\partial z}.$$

EXAMPLE 5 **Directional Derivative**

Find the directional derivative of $F(x, y, z) = xy^2 - 4x^2y + z^2$ at $(1, -1, 2)$ in the direction of $6\mathbf{i} + 2\mathbf{j} + 3\mathbf{k}$.

SOLUTION We have $\dfrac{\partial F}{\partial x} = y^2 - 8xy$, $\dfrac{\partial F}{\partial y} = 2xy - 4x^2$, and $\dfrac{\partial F}{\partial z} = 2z$ so that

$$\nabla F(x, y, z) = (y^2 - 8xy)\mathbf{i} + (2xy - 4x^2)\mathbf{j} + 2z\mathbf{k}$$

$$\nabla F(1, -1, 2) = 9\mathbf{i} - 6\mathbf{j} + 4\mathbf{k}.$$

Since $\| 6\mathbf{i} + 2\mathbf{j} + 3\mathbf{k} \| = 7$ then $\mathbf{u} = \frac{6}{7}\mathbf{i} + \frac{2}{7}\mathbf{j} + \frac{3}{7}\mathbf{k}$ is a unit vector in the indicated direction. It follows from (9) that

$$D_{\mathbf{u}}F(1, -1, 2) = (9\mathbf{i} - 6\mathbf{j} + 4\mathbf{k}) \cdot \left(\frac{6}{7}\mathbf{i} + \frac{2}{7}\mathbf{j} + \frac{3}{7}\mathbf{k} \right) = \frac{54}{7}. \qquad \equiv$$

■ **Maximum Value of the Directional Derivative** Let f represent a function of either two or three variables. Since (5) and (9) express the directional derivative as a dot product, we see from (5) of Theorem 7.3.2 that

$$D_{\mathbf{u}}f = \| \nabla f \| \, \| \mathbf{u} \| \cos\phi = \| \nabla f \| \cos\phi, \quad (\| \mathbf{u} \| = 1),$$

*Note that the numerator in (4) can be written $f(x + h\cos\alpha, y + h\cos\beta) - f(x, y)$, where $\beta = (\pi/2) - \alpha$.

where ϕ is the angle between ∇f and $\mathbf{u}$. Because $0 \leq \phi \leq \pi$, we have $-1 \leq \cos \phi \leq 1$ and, consequently, $-\|\nabla f\| \leq D_{\mathbf{u}} f \leq \|\nabla f\|$. In other words:

> *The maximum value of the directional derivative is $\|\nabla f\|$ and it occurs when $\mathbf{u}$ has the same direction as ∇f (when $\cos \phi = 1$).* (10)

> *The minimum value of the directional derivative is $-\|\nabla f\|$ and it occurs when $\mathbf{u}$ and ∇f have opposite directions (when $\cos \phi = -1$).* (11)

EXAMPLE 6 Max/Min of Directional Derivative

In Example 5 the maximum value of the directional derivative of F of $(1, -1, 2)$ is $\|\nabla F(1, -1, 2)\| = \sqrt{133}$. The minimum value of $D_{\mathbf{u}} F(1, -1, 2)$ is then $-\sqrt{133}$. ≡

▉ Gradient Points in Direction of Most Rapid Increase of f Put yet another way, (10) and (11) state:

> *The gradient vector ∇f points in the direction in which f increases most rapidly, whereas $-\nabla f$ points in the direction of the most rapid decrease of f.*

EXAMPLE 7 Direction of Steepest Ascent

FIGURE 9.5.3 Model of a hill in Example 7

Each year in Los Angeles there is a bicycle race up to the top of a hill by a road known to be the steepest in the city. To understand why a bicyclist with a modicum of sanity will zigzag up the road, let us suppose the graph of $f(x, y) = 4 - \frac{2}{3}\sqrt{x^2 + y^2}$, $0 \leq z \leq 4$, shown in **FIGURE 9.5.3(a)** is a mathematical model of the hill. The gradient of f is

$$\nabla f(x, y) = \frac{2}{3}\left[\frac{-x}{\sqrt{x^2 + y^2}}\mathbf{i} + \frac{-y}{\sqrt{x^2 + y^2}}\mathbf{j}\right] = \frac{\frac{2}{3}}{\sqrt{x^2 + y^2}}\mathbf{r},$$

where $\mathbf{r} = -x\mathbf{i} - y\mathbf{j}$ is a vector pointing to the center of the circular base.

Thus the steepest ascent up the hill is a straight road whose projection in the xy-plane is a radius of the circular base. Since $D_{\mathbf{u}} f = \text{comp}_{\mathbf{u}} \nabla f$, a bicyclist will zigzag, or seek a direction $\mathbf{u}$ other than ∇f, in order to reduce this component. ≡

EXAMPLE 8 Direction to Cool Off Fastest

The temperature in a rectangular box is approximated by

$$T(x, y, z) = xyz(1 - x)(2 - y)(3 - z), \quad 0 \leq x \leq 1, \quad 0 \leq y \leq 2, \quad 0 \leq z \leq 3.$$

If a mosquito is located at $(\frac{1}{2}, 1, 1)$, in which direction should it fly to cool off as rapidly as possible?

SOLUTION The gradient of T is

$$\nabla T(x, y, z) = yz(2 - y)(3 - z)(1 - 2x)\mathbf{i} + xz(1 - x)(3 - z)(2 - 2y)\mathbf{j} + xy(1 - x)(2 - y)(3 - 2z)\mathbf{k}.$$

Therefore, $\nabla T(\frac{1}{2}, 1, 1) = \frac{1}{4}\mathbf{k}$. To cool off most rapidly, the mosquito should fly in the direction of $-\frac{1}{4}\mathbf{k}$; that is, it should dive for the floor of the box, where the temperature is $T(x, y, 0) = 0$. ≡

9.5 Exercises Answers to selected odd-numbered problems begin on page ANS-24.

In Problems 1–4, compute the gradient for the given function.

1. $f(x, y) = x^2 - x^3y^2 + y^4$ **2.** $f(x, y) = y - e^{-2x^2y}$

3. $F(x, y, z) = \dfrac{xy^2}{z^3}$ **4.** $F(x, y, z) = xy \cos yz$

In Problems 5–8, find the gradient of the given function at the indicated point.

5. $f(x, y) = x^2 - 4y^2$; $(2, 4)$

6. $f(x, y) = \sqrt{x^3y - y^4}$; $(3, 2)$

7. $F(x, y, z) = x^2z^2 \sin 4y$; $(-2, \pi/3, 1)$
8. $F(x, y, z) = \ln(x^2 + y^2 + z^2)$; $(-4, 3, 5)$

In Problems 9 and 10, use Definition 9.5.1 to find $D_{\mathbf{u}} f(x, y)$ given that $\mathbf{u}$ makes the indicated angle with the positive x-axis.

9. $f(x, y) = x^2 + y^2$; $\theta = 30°$
10. $f(x, y) = 3x - y^2$; $\theta = 45°$

In Problems 11–20, find the directional derivative of the given function at the given point in the indicated direction.

11. $f(x, y) = 5x^3y^6$; $(-1, 1)$, $\theta = \pi/6$
12. $f(x, y) = 4x + xy^2 - 5y$; $(3, -1)$, $\theta = \pi/4$
13. $f(x, y) = \tan^{-1}\dfrac{y}{x}$; $(2, -2)$, $\mathbf{i} - 3\mathbf{j}$
14. $f(x, y) = \dfrac{xy}{x + y}$; $(2, -1)$, $6\mathbf{i} + 8\mathbf{j}$
15. $f(x, y) = (xy + 1)^2$; $(3, 2)$, in the direction of $(5, 3)$
16. $f(x, y) = x^2 \tan y$; $(\frac{1}{2}, \pi/3)$, in the direction of the negative x-axis
17. $F(x, y, z) = x^2y^2(2z + 1)^2$; $(1, -1, 1)$, $\langle 0, 3, 3 \rangle$
18. $F(x, y, z) = \dfrac{x^2 - y^2}{z^2}$; $(2, 4, -1)$, $\mathbf{i} - 2\mathbf{j} + \mathbf{k}$
19. $F(x, y, z) = \sqrt{x^2y + 2y^2z}$; $(-2, 2, 1)$, in the direction of the negative z-axis
20. $F(x, y, z) = 2x - y^2 + z^2$; $(4, -4, 2)$, in the direction of the origin

In Problems 21 and 22, consider the plane through the points P and Q that is perpendicular to the xy-plane. Find the slope of the tangent at the indicated point to the curve of intersection of this plane and the graph of the given function in the direction of Q.

21. $f(x, y) = (x - y)^2$; $P(4, 2)$, $Q(0, 1)$; $(4, 2, 4)$
22. $f(x, y) = x^3 - 5xy + y^2$; $P(1, 1)$, $Q(-1, 6)$; $(1, 1, -3)$

In Problems 23–26, find a vector that gives the direction in which the given function increases most rapidly at the indicated point. Find the maximum rate.

23. $f(x, y) = e^{2x} \sin y$; $(0, \pi/4)$
24. $f(x, y) = xye^{x-y}$; $(5, 5)$
25. $F(x, y, z) = x^2 + 4xz + 2yz^2$; $(1, 2, -1)$
26. $F(x, y, z) = xyz$; $(3, 1, -5)$

In Problems 27–30, find a vector that gives the direction in which the given function decreases most rapidly at the indicated point. Find the minimum rate.

27. $f(x, y) = \tan(x^2 + y^2)$; $(\sqrt{\pi/6}, \sqrt{\pi/6})$
28. $f(x, y) = x^3 - y^3$; $(2, -2)$
29. $F(x, y, z) = \sqrt{xz}e^y$; $(16, 0, 9)$
30. $F(x, y, z) = \ln\left(\dfrac{xy}{z}\right)$; $(\frac{1}{2}, \frac{1}{6}, \frac{1}{3})$

31. Find the directional derivative(s) of $f(x, y) = x + y^2$ at $(3, 4)$ in the direction of a tangent vector to the graph of $2x^2 + y^2 = 9$ at $(2, 1)$.

32. If $f(x, y) = x^2 + xy + y^2 - x$, find all points where $D_{\mathbf{u}} f(x, y)$ in the direction of $\mathbf{u} = (1/\sqrt{2})(\mathbf{i} + \mathbf{j})$ is zero.

33. Suppose $\nabla f(a, b) = 4\mathbf{i} + 3\mathbf{j}$. Find a unit vector $\mathbf{u}$ so that
 (a) $D_{\mathbf{u}} f(a, b) = 0$,
 (b) $D_{\mathbf{u}} f(a, b)$ is a maximum, and
 (c) $D_{\mathbf{u}} f(a, b)$ is a minimum.

34. Suppose $D_{\mathbf{u}} f(a, b) = 6$. What is the value of $D_{-\mathbf{u}} f(a, b)$?
35. (a) If $f(x, y) = x^3 - 3x^2y^2 + y^3$, find the directional derivative of f at a point (x, y) in the direction of $\mathbf{u} = (1/\sqrt{10})(3\mathbf{i} + \mathbf{j})$.
 (b) If $F(x, y) = D_{\mathbf{u}} f(x, y)$ in part (a), find $D_{\mathbf{u}} F(x, y)$.
36. Consider the gravitational potential

$$U(x, y) = \frac{-Gm}{\sqrt{x^2 + y^2}},$$

where G and m are constants. Show that U increases or decreases most rapidly along a line through the origin.

37. If $f(x, y) = x^3 - 12x + y^2 - 10y$, find all points at which $\|\nabla f\| = 0$.

38. Suppose

$$D_{\mathbf{u}} f(a, b) = 7, \quad D_{\mathbf{v}} f(a, b) = 3$$

$$\mathbf{u} = \frac{5}{13}\mathbf{i} - \frac{12}{13}\mathbf{j}, \quad \mathbf{v} = \frac{5}{13}\mathbf{i} + \frac{12}{13}\mathbf{j}.$$

Find $\nabla f(a, b)$.

39. Consider the rectangular plate shown in **FIGURE 9.5.4**. The temperature at a point (x, y) on the plate is given by $T(x, y) = 5 + 2x^2 + y^2$. Determine the direction an insect should take, starting at $(4, 2)$, in order to cool off as rapidly as possible.

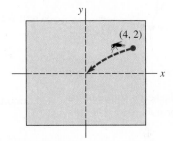

FIGURE 9.5.4 Insect in Problem 39

40. In Problem 39, observe that $(0, 0)$ is the coolest point of the plate. Find the path the cold-seeking insect, starting at $(4, 2)$, will take to the origin. If $\langle x(t), y(t) \rangle$ is the vector equation of the path, then use the fact that $-\nabla T(x, y) = \langle x'(t), y'(t) \rangle$. Why is this? [*Hint*: Remember separation of variables?]

41. The temperature at a point (x, y) on a rectangular metal plate is given by $T(x, y) = 100 - 2x^2 - y^2$. Find the path a heat-seeking particle will take, starting at $(3, 4)$, as it moves in the direction in which the temperature increases most rapidly.

42. The temperature T at a point (x, y, z) in space is inversely proportional to the square of the distance from (x, y, z) to the origin. It is known that $T(0, 0, 1) = 500$. Find the rate of change of T at $(2, 3, 3)$ in the direction of $(3, 1, 1)$. In which direction from $(2, 3, 3)$ does the temperature T increase most rapidly? At $(2, 3, 3)$ what is the maximum rate of change of T?

43. Find a function f such that

$$\nabla f = (3x^2 + y^3 + ye^{xy})\mathbf{i} + (-2y^2 + 3xy^2 + xe^{xy})\mathbf{j}.$$

44. Let f_x, f_y, f_{xy}, f_{yx} be continuous and $\mathbf{u}$ and $\mathbf{v}$ be unit vectors. Show that $D_{\mathbf{u}} D_{\mathbf{v}} f = D_{\mathbf{v}} D_{\mathbf{u}} f$.

In Problems 45–48, assume that f and g are differentiable functions of two variables. Prove the given identity.

45. $\nabla(cf) = c\,\nabla f$

46. $\nabla(f + g) = \nabla f + \nabla g$

47. $\nabla(fg) = f\,\nabla g + g\,\nabla f$

48. $\nabla\!\left(\dfrac{f}{g}\right) = \dfrac{g\nabla f - f\nabla g}{g^2}$

49. If $\mathbf{F}(x, y, z) = f_1(x, y, z)\mathbf{i} + f_2(x, y, z)\mathbf{j} + f_3(x, y, z)\mathbf{k}$ and

$$\nabla = \mathbf{i}\frac{\partial}{\partial x} + \mathbf{j}\frac{\partial}{\partial y} + \mathbf{k}\frac{\partial}{\partial z},$$

show that

$$\nabla \times \mathbf{F} = \left(\frac{\partial f_3}{\partial y} - \frac{\partial f_2}{\partial z}\right)\mathbf{i} + \left(\frac{\partial f_1}{\partial z} - \frac{\partial f_3}{\partial x}\right)\mathbf{j} + \left(\frac{\partial f_2}{\partial x} - \frac{\partial f_1}{\partial y}\right)\mathbf{k}.$$

9.6 Tangent Planes and Normal Lines

INTRODUCTION The notion of the gradient of a function of two or more variables was introduced in the preceding section as an aid in computing directional derivatives. In this section we give a geometric interpretation of the gradient vector.

Geometric Interpretation of the Gradient—Functions of Two Variables

Suppose $f(x, y) = c$ is the level curve of the differentiable function $z = f(x, y)$ that passes through a specified point $P(x_0, y_0)$; that is, $f(x_0, y_0) = c$. If this level curve is parameterized by the differentiable functions

$$x = g(t), \quad y = h(t) \quad \text{such that} \quad x_0 = g(t_0), \; y_0 = h(t_0),$$

then the derivative of $f(g(t), h(t)) = c$ with respect to t is

$$\frac{\partial f}{\partial x}\frac{dx}{dt} + \frac{\partial f}{\partial y}\frac{dy}{dt} = 0. \tag{1}$$

When we introduce the vectors

$$\nabla f(x, y) = \frac{\partial f}{\partial x}\mathbf{i} + \frac{\partial f}{\partial y}\mathbf{j} \quad \text{and} \quad \mathbf{r}'(t) = \frac{dx}{dt}\mathbf{i} + \frac{dy}{dt}\mathbf{j},$$

(1) becomes $\nabla f \cdot \mathbf{r}' = 0$. Specifically, at $t = t_0$, we have

$$\nabla f(x_0, y_0) \cdot \mathbf{r}'(t_0) = 0. \tag{2}$$

Thus, if $\mathbf{r}'(t_0) \neq 0$, the vector $\nabla f(x_0, y_0)$ is orthogonal to the tangent vector $\mathbf{r}'(t_0)$ at $P(x_0, y_0)$. We interpret this to mean:

∇f is orthogonal to the level curve at P.

See **FIGURE 9.6.1**.

curve $f(x, y) = c$

FIGURE 9.6.1 Gradient is perpendicular to tangent vector at P

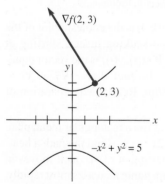

FIGURE 9.6.2 Gradient in Example 1

EXAMPLE 1 **Gradient at a Point**

Find the level curve of $f(x, y) = -x^2 + y^2$ passing through $(2, 3)$. Graph the gradient at the point.

SOLUTION Since $f(2, 3) = -4 + 9 = 5$, the level curve is the hyperbola $-x^2 + y^2 = 5$. Now,

$$\nabla f(x, y) = -2x\mathbf{i} + 2y\mathbf{j} \quad \text{and} \quad \nabla f(2, 3) = -4\mathbf{i} + 6\mathbf{j}.$$

FIGURE 9.6.2 shows the level curve and $\nabla f(2, 3)$. ≡

Geometric Interpretation of the Gradient—Functions of Three Variables

Proceeding as before, let $F(x, y, z) = c$ be the *level surface* of a differentiable function $w = F(x, y, z)$ that passes through $P(x_0, y_0, z_0)$. If the differentiable functions $x = f(t)$, $y = g(t)$, $z = h(t)$ are the parametric equations of a curve C on the surface for which $x_0 = f(t_0)$, $y_0 = g(t_0)$, $z_0 = h(t_0)$, then the derivative of $F(f(t), g(t), h(t)) = 0$ implies that

$$\frac{\partial F}{\partial x}\frac{dx}{dt} + \frac{\partial F}{\partial y}\frac{dy}{dt} + \frac{\partial F}{\partial z}\frac{dz}{dt} = 0$$

$$\left(\frac{\partial F}{\partial x}\mathbf{i} + \frac{\partial F}{\partial y}\mathbf{j} + \frac{\partial F}{\partial z}\mathbf{k}\right) \cdot \left(\frac{dx}{dt}\mathbf{i} + \frac{dy}{dt}\mathbf{j} + \frac{dz}{dt}\mathbf{k}\right) = 0. \tag{3}$$

In particular, at $t = t_0$, (3) is

$$\nabla F(x_0, y_0, z_0) \cdot \mathbf{r}'(t_0) = 0. \tag{4}$$

Thus, when $\mathbf{r}'(t_0) \neq 0$, the vector $\nabla F(x_0, y_0, z_0)$ is orthogonal to the tangent vector $\mathbf{r}'(t_0)$. Since this argument holds for any differentiable curve through $P(x_0, y_0, z_0)$ on the surface, we conclude that:

> ∇F is normal (*perpendicular*) to the level surface at P.

See **FIGURE 9.6.3**.

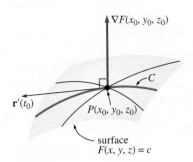

FIGURE 9.6.3 Gradient is normal to level surface at P

EXAMPLE 2 **Gradient at a Point**

Find the level surface of $F(x, y, z) = x^2 + y^2 + z^2$ passing through (1, 1, 1). Graph the gradient at the point.

SOLUTION Since $F(1, 1, 1) = 3$, the level surface passing through (1, 1, 1) is the sphere $x^2 + y^2 + z^2 = 3$. The gradient of the function is

$$\nabla F(x, y, z) = 2x\mathbf{i} + 2y\mathbf{j} + 2z\mathbf{k},$$

and so, at the given point, $\nabla F(1, 1, 1) = 2\mathbf{i} + 2\mathbf{j} + 2\mathbf{k}$. The level surface and $\nabla F(1, 1, 1)$ are illustrated in **FIGURE 9.6.4**. ≡

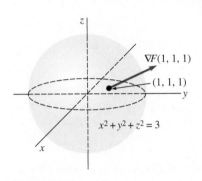

FIGURE 9.6.4 Gradient in Example 2

■ **Tangent Plane** In the study of differential calculus a basic problem was finding an equation of a tangent line to the graph of a function. In 3-space the analogous problem is finding an equation of a **tangent plane** to a surface. We assume again that $w = F(x, y, z)$ is a differentiable function and that a surface is given by $F(x, y, z) = c$.

DEFINITION 9.6.1 **Tangent Plane**

Let $P(x_0, y_0, z_0)$ be a point on the graph of $F(x, y, z) = c$, where ∇F is not $\mathbf{0}$. The **tangent plane** at P is that plane through P that is normal to ∇F evaluated at P.

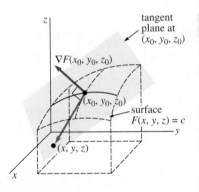

FIGURE 9.6.5 Tangent plane is normal to gradient at P

Thus, if $P(x, y, z)$ and $P(x_0, y_0, z_0)$ are points on the tangent plane and $\mathbf{r}$ and $\mathbf{r}_0$ are their respective position vectors, then a vector equation of the tangent plane is $\nabla F(x_0, y_0, z_0) \cdot (\mathbf{r} - \mathbf{r}_0) = 0$. See **FIGURE 9.6.5**. We summarize this last result:

THEOREM 9.6.1 **Equation of Tangent Plane**

Let $P(x_0, y_0, z_0)$ be a point on the graph of $F(x, y, z) = c$, where ∇F is not $\mathbf{0}$. Then an equation of the tangent plane at P is

$$F_x(x_0, y_0, z_0)(x - x_0) + F_y(x_0, y_0, z_0)(y - y_0) + F_z(x_0, y_0, z_0)(z - z_0) = 0. \tag{5}$$

EXAMPLE 3 **Equation of Tangent Plane**

Find an equation of the tangent plane to the graph of $x^2 - 4y^2 + z^2 = 16$ at (2, 1, 4).

SOLUTION By defining $F(x, y, z) = x^2 - 4y^2 + z^2$, the given surface is the level surface $F(x, y, z) = F(2, 1, 4) = 16$ passing through (2, 1, 4). Now, $F_x(x, y, z) = 2x$, $F_y(x, y, z) = -8y$, and $F_z(x, y, z) = 2z$, so that

$$\nabla F(x, y, z) = 2x\mathbf{i} - 8y\mathbf{j} + 2z\mathbf{k} \quad \text{and} \quad \nabla F(2, 1, 4) = 4\mathbf{i} - 8\mathbf{j} + 8\mathbf{k}.$$

It follows from (5) that an equation of the tangent plane is

$$4(x - 2) - 8(y - 1) + 8(z - 4) = 0 \quad \text{or} \quad x - 2y + 2z = 8.$$ ≡

▨ Surfaces Given by $z = f(x, y)$

For a surface given explicitly by a differentiable function $z = f(x, y)$, we define $F(x, y, z) = f(x, y) - z$ or $F(x, y, z) = z - f(x, y)$. Thus, a point (x_0, y_0, z_0) is on the graph of $z = f(x, y)$ if and only if it is also on the level surface $F(x, y, z) = 0$. This follows from $F(x_0, y_0, z_0) = f(x_0, y_0) - z_0 = 0$.

FIGURE 9.6.6 Tangent plane in Example 4

EXAMPLE 4 **Equation of Tangent Plane**

Find an equation of the tangent plane to the graph of $z = \frac{1}{2}x^2 + \frac{1}{2}y^2 + 4$ at $(1, -1, 5)$.

SOLUTION Define $F(x, y, z) = \frac{1}{2}x^2 + \frac{1}{2}y^2 - z + 4$ so that the level surface of F passing through the given point is $F(x, y, z) = F(1, -1, 5)$ or $F(x, y, z) = 0$. Now, $F_x = x$, $F_y = y$, and $F_z = -1$, so that

$$\nabla F(x, y, z) = x\mathbf{i} + y\mathbf{j} - \mathbf{k} \quad \text{and} \quad \nabla F(1, -1, 5) = \mathbf{i} - \mathbf{j} - \mathbf{k}.$$

Thus, from (5) the desired equation is

$$(x - 1) - (y + 1) - (z - 5) = 0 \quad \text{or} \quad -x + y + z = 3.$$

See **FIGURE 9.6.6**. ≡

▨ Normal Line

Let $P(x_0, y_0, z_0)$ be a point on the graph of $F(x, y, z) = c$, where ∇F is not **0**. The line containing $P(x_0, y_0, z_0)$ that is parallel to $\nabla F(x_0, y_0, z_0)$ is called the **normal line** to the surface at P. As indicated by its name, this line is normal to the tangent plane to the surface at P.

EXAMPLE 5 **Normal Line to a Surface**

Find parametric equations for the normal line to the surface in Example 4 at $(1, -1, 5)$.

SOLUTION A direction vector for the normal line at $(1, -1, 5)$ is $\nabla F(1, -1, 5) = \mathbf{i} - \mathbf{j} - \mathbf{k}$. It follows that parametric equations for the normal line are

$$x = 1 + t, \quad y = -1 - t, \quad z = 5 - t.$$ ≡

Expressed as symmetric equations the normal line to a surface $F(x, y, z) = c$ at $P(x_0, y_0, z_0)$ is given by

$$\frac{x - x_0}{F_x(x_0, y_0, z_0)} = \frac{y - y_0}{F_y(x_0, y_0, z_0)} = \frac{z - z_0}{F_z(x_0, y_0, z_0)}.$$

In Example 5, you should verify that symmetric equations of the normal line at $(1, -1, 5)$ are

$$x - 1 = \frac{y + 1}{-1} = \frac{z - 5}{-1}.$$

FIGURE 9.6.7 Stream is perpendicular to contours

REMARKS

Water flowing down a hill chooses a path in the direction of the greatest change in altitude. **FIGURE 9.6.7** shows the contours, or level curves, of a hill. As shown in the figure, a stream starting at point P will take a path that is perpendicular to the contours. After reading Sections 9.5 and 9.6, the student should be able to explain why.

In Problems 1–12, sketch the level curve or surface passing through the indicated point. Sketch the gradient at the point.

1. $f(x, y) = x - 2y$; $(6, 1)$
2. $f(x, y) = \dfrac{y + 2x}{x}$; $(1, 3)$

3. $f(x, y) = y - x^2$; $(2, 5)$
4. $f(x, y) = x^2 + y^2$; $(-1, 3)$

5. $f(x, y) = \dfrac{x^2}{4} + \dfrac{y^2}{9}$; $(-2, -3)$

6. $f(x, y) = \dfrac{y^2}{x}$; $(2, 2)$

7. $f(x, y) = (x - 1)^2 - y^2$; $(1, 1)$

8. $f(x, y) = \dfrac{y - 1}{\sin x}$; $(\pi/6, \frac{3}{2})$

9. $F(x, y, z) = y + z$; $(3, 1, 1)$
10. $F(x, y, z) = x^2 + y^2 - z$; $(1, 1, 3)$
11. $F(x, y, z) = \sqrt{x^2 + y^2 + z^2}$; $(3, 4, 0)$
12. $F(x, y, z) = x^2 - y^2 + z$; $(0, -1, 1)$

In Problems 13 and 14, find the points on the given surface at which the gradient is parallel to the indicated vector.

13. $z = x^2 + y^2$; $4\mathbf{i} + \mathbf{j} + \frac{1}{2}\mathbf{k}$
14. $x^3 + y^2 + z = 15$; $27\mathbf{i} + 8\mathbf{j} + \mathbf{k}$

In Problems 15–24, find an equation of the tangent plane to the graph of the given equation at the indicated point.

15. $x^2 + y^2 + z^2 = 9$; $(-2, 2, 1)$
16. $5x^2 - y^2 + 4z^2 = 8$; $(2, 4, 1)$
17. $x^2 - y^2 - 3z^2 = 5$; $(6, 2, 3)$
18. $xy + yz + zx = 7$; $(1, -3, -5)$
19. $z = 25 - x^2 - y^2$; $(3, -4, 0)$
20. $xz = 6$; $(2, 0, 3)$
21. $z = \cos(2x + y)$; $(\pi/2, \pi/4, -1/\sqrt{2})$
22. $x^2 y^3 + 6z = 10$; $(2, 1, 1)$
23. $z = \ln(x^2 + y^2)$; $(1/\sqrt{2}, 1/\sqrt{2}, 0)$
24. $z = 8e^{-2y} \sin 4x$; $(\pi/24, 0, 4)$

In Problems 25 and 26, find the points on the given surface at which the tangent plane is parallel to the indicated plane.

25. $x^2 + y^2 + z^2 = 7$; $2x + 4y + 6z = 1$

26. $x^2 - 2y^2 - 3z^2 = 33$; $8x + 4y + 6z = 5$
27. Find points on the surface $x^2 + 4x + y^2 + z^2 - 2z = 11$ at which the tangent plane is horizontal.
28. Find points on the surface $x^2 + 3y^2 + 4z^2 - 2xy = 16$ at which the tangent plane is parallel to (a) the xz-plane, (b) the yz-plane, and (c) the xy-plane.

In Problems 29 and 30, show that the second equation is an equation of the tangent plane to the graph of the first equation at (x_0, y_0, z_0).

29. $\dfrac{x^2}{a^2} + \dfrac{y^2}{b^2} + \dfrac{z^2}{c^2} = 1$; $\dfrac{xx_0}{a^2} + \dfrac{yy_0}{b^2} + \dfrac{zz_0}{c^2} = 1$

30. $\dfrac{x^2}{a^2} - \dfrac{y^2}{b^2} + \dfrac{z^2}{c^2} = 1$; $\dfrac{xx_0}{a^2} - \dfrac{yy_0}{b^2} + \dfrac{zz_0}{c^2} = 1$

31. Show that every tangent plane to the graph of $z^2 = x^2 + y^2$ passes through the origin.

32. Show that the sum of the x-, y-, and z-intercepts of every tangent plane to the graph of $\sqrt{x} + \sqrt{y} + \sqrt{z} = \sqrt{a}$, $a > 0$, is the number a.

In Problems 33 and 34, find parametric equations for the normal line at the indicated point. In Problems 35 and 36, find symmetric equations for the normal line.

33. $x^2 + 2y^2 + z^2 = 4$; $(1, -1, 1)$
34. $z = 2x^2 - 4y^2$; $(3, -2, 2)$
35. $z = 4x^2 + 9y^2 + 1$; $(\frac{1}{2}, \frac{1}{3}, 3)$
36. $x^2 + y^2 - z^2 = 0$; $(3, 4, 5)$
37. Show that every normal line to the graph $x^2 + y^2 + z^2 = a^2$ passes through the origin.
38. Two surfaces are said to be **orthogonal** at a point P of intersection if their normal lines at P are orthogonal. Prove that the surfaces given by $F(x, y, z) = 0$ and $G(x, y, z) = 0$ are orthogonal at P if and only if $F_x G_x + F_y G_y + F_z G_z = 0$.

In Problems 39 and 40, use the result of Problem 38 to show that the given surfaces are orthogonal at a point of intersection.

39. $x^2 + y^2 + z^2 = 25$; $-x^2 + y^2 + z^2 = 0$
40. $x^2 - y^2 + z^2 = 4$; $z = 1/xy^2$

9.7 Curl and Divergence

INTRODUCTION In Section 9.1 we introduced the concept of vector function of one variable. In this section we examine vector functions of two and three variables.

■ **Vector Fields** Vector functions of two and three variables,

$$\mathbf{F}(x, y) = P(x, y)\mathbf{i} + Q(x, y)\mathbf{j}$$

$$\mathbf{F}(x, y, z) = P(x, y, z)\mathbf{i} + Q(x, y, z)\mathbf{j} + R(x, y, z)\mathbf{k}$$

are also called **vector fields**. For example, the motion of a wind or a fluid can be described by means of a *velocity field* because a vector can be assigned at each point representing the velocity of a particle at the point. See **FIGURE 9.7.1(a)** and 9.7.1(b). The concept of a *force field* plays an important role in mechanics, electricity, and magnetism. See Figure 9.7.1(c) and 9.7.1(d).

(a) Airflow around an airplane wing

(b) Laminar flow of blood in an artery; cylindrical layers of blood flow faster near the center of the artery

(c) Inverse square force field; magnitude of the attractive force is large near the particle

(d) Lines of force around two equal positive charges

FIGURE 9.7.1 Various vector fields

EXAMPLE 1 **Two-Dimensional Vector Field**

Graph the two-dimensional vector field $\mathbf{F}(x, y) = -y\mathbf{i} + x\mathbf{j}$.

SOLUTION One manner of proceeding is simply to choose points in the *xy*-plane and then graph the vector $\mathbf{F}$ at that point. For example, at (1, 1) we would draw the vector $\mathbf{F}(1, 1) = -\mathbf{i} + \mathbf{j}$.

For the given vector field it is possible to systematically draw vectors of the same length. Observe that $\|\mathbf{F}\| = \sqrt{x^2 + y^2}$, and so vectors of the same length k must lie along the curve defined by $\sqrt{x^2 + y^2} = k$; that is, at any point on the circle $x^2 + y^2 = k^2$ a vector would have length k. For simplicity let us choose circles that have some points on them with integer coordinates. For example, for $k = 1$, $k = \sqrt{2}$, and $k = 2$, we have:

$x^2 + y^2 = 1$: At the points (1, 0), (0, 1), (−1, 0), (0, −1) the corresponding vectors $\mathbf{j}$, $-\mathbf{i}$, $-\mathbf{j}$, $\mathbf{i}$ have the same length 1.

$x^2 + y^2 = 2$: At the points (1, 1), (−1, 1), (−1, −1), (1, −1) the corresponding vectors $-\mathbf{i} + \mathbf{j}$, $-\mathbf{i} - \mathbf{j}$, $\mathbf{i} - \mathbf{j}$, $\mathbf{i} + \mathbf{j}$ have the same length $\sqrt{2}$.

$x^2 + y^2 = 4$: At the points (2, 0), (0, 2), (−2, 0), (0, −2) the corresponding vectors $2\mathbf{j}$, $-2\mathbf{i}$, $-2\mathbf{j}$, $2\mathbf{i}$ have the same length 2.

FIGURE 9.7.2 Vector field in Example 1

The vectors at these points are shown in **FIGURE 9.7.2**. ≡

In Section 9.5 we saw that the del operator

$$\nabla = \frac{\partial}{\partial x}\mathbf{i} + \frac{\partial}{\partial y}\mathbf{j} + \frac{\partial}{\partial z}\mathbf{k}$$

combined with a scalar function $\phi(x, y, z)$ produces a vector field

$$\mathbf{F}(x, y, z) = \nabla\phi = \frac{\partial\phi}{\partial x}\mathbf{i} + \frac{\partial\phi}{\partial y}\mathbf{j} + \frac{\partial\phi}{\partial z}\mathbf{k}$$

called the **gradient of ϕ** or a **gradient field**. The del operator can also be combined with a vector field $\mathbf{F}(x, y, z) = P(x, y, z)\mathbf{i} + Q(x, y, z)\mathbf{j} + R(x, y, z)\mathbf{k}$ in two different ways: in one case producing another vector field and in the other producing a scalar function. We will assume hereafter that P, Q, and R have continuous partial derivatives.

The **curl** of a vector field $\mathbf{F} = P\mathbf{i} + Q\mathbf{j} + R\mathbf{k}$ is the vector field

$$\text{curl } \mathbf{F} = \left(\frac{\partial R}{\partial y} - \frac{\partial Q}{\partial z}\right)\mathbf{i} + \left(\frac{\partial P}{\partial z} - \frac{\partial R}{\partial x}\right)\mathbf{j} + \left(\frac{\partial Q}{\partial x} - \frac{\partial P}{\partial y}\right)\mathbf{k}.$$

In practice, curl $\mathbf{F}$ can be computed from the cross product of the del operator and the vector $\mathbf{F}$:

$$\text{curl } \mathbf{F} = \nabla \times \mathbf{F} = \begin{vmatrix} \mathbf{i} & \mathbf{j} & \mathbf{k} \\ \dfrac{\partial}{\partial x} & \dfrac{\partial}{\partial y} & \dfrac{\partial}{\partial z} \\ P & Q & R \end{vmatrix}. \tag{1}$$

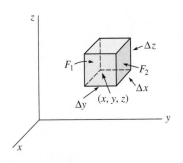

FIGURE 9.7.3 Fluid flow through element of area ΔS

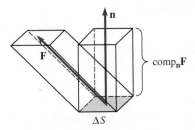

FIGURE 9.7.4 What is total flux of vector field through this element?

There is another combination of partial derivatives of the component functions of a vector field that occurs frequently in science and engineering. Before we state the next definition, consider the following motivation.

If $\mathbf{F}(x, y, z) = P(x, y, z)\mathbf{i} + Q(x, y, z)\mathbf{j} + R(x, y, z)\mathbf{k}$ is the velocity field of a fluid, then as shown in **FIGURE 9.7.3**, the volume of the fluid flowing through an element of surface area ΔS per unit time—that is, the **flux** of the vector field $\mathbf{F}$ through the area ΔS—is approximated by

$$(\text{height})(\text{area of base}) = (\text{comp}_\mathbf{n} \mathbf{F})\,\Delta S = (\mathbf{F} \cdot \mathbf{n})\,\Delta S, \tag{2}$$

where $\mathbf{n}$ is a unit vector normal to the surface. Now consider the rectangular parallelepiped shown in **FIGURE 9.7.4**. To compute the total flux of $\mathbf{F}$ through its six sides in the outward direction we first compute the total flux out of parallel faces. The area of face F_1 is $\Delta x\,\Delta z$ and its outward unit normal is $-\mathbf{j}$, and so by (2) the flux of $\mathbf{F}$ through F_1 is approximately

$$\mathbf{F} \cdot (-\mathbf{j})\,\Delta x\,\Delta z = -Q(x, y, z)\,\Delta x\,\Delta z.$$

The flux out of face F_2, whose outward normal is $\mathbf{j}$, is approximated by

$$(\mathbf{F} \cdot \mathbf{j})\,\Delta x\,\Delta z = Q(x, y + \Delta y, z)\,\Delta x\,\Delta z.$$

Consequently the total flux out of these parallel faces is

$$Q(x, y + \Delta y, z)\,\Delta x\,\Delta z + (-Q(x, y, z)\,\Delta x\,\Delta z) = [Q(x, y + \Delta y, z) - Q(x, y, z)]\,\Delta x\,\Delta z. \tag{3}$$

By multiplying (3) by $\Delta y/\Delta y$ and recalling the definition of a partial derivative, we get for Δy close to 0,

$$\frac{Q(x, y + \Delta y, z) - Q(x, y, z)}{\Delta y}\,\Delta x\,\Delta y\,\Delta z \approx \frac{\partial Q}{\partial y}\,\Delta x\,\Delta y\,\Delta z.$$

Arguing in exactly the same manner, we see that the contributions to the total flux out of the parallelepiped from the two faces parallel to the yz-plane, and from the two faces parallel to the xy-plane, are, in turn,

$$\frac{\partial P}{\partial x}\,\Delta x\,\Delta y\,\Delta z \quad \text{and} \quad \frac{\partial R}{\partial z}\,\Delta x\,\Delta y\,\Delta z.$$

Adding the results, we see that the net flux of $\mathbf{F}$ out of the parallelepiped is approximately

$$\left(\frac{\partial P}{\partial x} + \frac{\partial Q}{\partial y} + \frac{\partial R}{\partial z}\right)\Delta x\,\Delta y\,\Delta z.$$

By dividing the last expression by $\Delta x \, \Delta y \, \Delta z$, we get the outward flux of $\mathbf{F}$ per unit volume:

$$\frac{\partial P}{\partial x} + \frac{\partial Q}{\partial y} + \frac{\partial R}{\partial z}.$$

It is this combination of partial derivatives that is given a special name.

DEFINITION 9.7.2 Divergence

The **divergence** of a vector field $\mathbf{F} = P\mathbf{i} + Q\mathbf{j} + R\mathbf{k}$ is the scalar function

$$\text{div } \mathbf{F} = \frac{\partial P}{\partial x} + \frac{\partial Q}{\partial y} + \frac{\partial R}{\partial z}.$$

Observe that div $\mathbf{F}$ can also be written in terms of the del operator as

$$\text{div } \mathbf{F} = \nabla \cdot \mathbf{F} = \frac{\partial}{\partial x} P(x, y, z) + \frac{\partial}{\partial y} Q(x, y, z) + \frac{\partial}{\partial z} R(x, y, z). \tag{4}$$

EXAMPLE 2 **Curl and Divergence**

If $\mathbf{F} = (x^2y^3 - z^4)\mathbf{i} + 4x^5y^2z\,\mathbf{j} - y^4z^6\mathbf{k}$, find **(a)** curl $\mathbf{F}$, **(b)** div $\mathbf{F}$, and **(c)** div(curl $\mathbf{F}$).

SOLUTION **(a)** From (1),

$$\text{curl } \mathbf{F} = \nabla \times \mathbf{F} = \begin{vmatrix} \mathbf{i} & \mathbf{j} & \mathbf{k} \\ \dfrac{\partial}{\partial x} & \dfrac{\partial}{\partial y} & \dfrac{\partial}{\partial z} \\ x^2y^3 - z^4 & 4x^5y^2z & -y^4z^6 \end{vmatrix}$$

$$= (-4y^3z^6 - 4x^5y^2)\mathbf{i} - 4z^3\mathbf{j} + (20x^4y^2z - 3x^2y^2)\mathbf{k}.$$

(b) From (4),

$$\text{div } \mathbf{F} = \nabla \cdot \mathbf{F} = \frac{\partial}{\partial x}(x^2y^3 - z^4) + \frac{\partial}{\partial y}(4x^5y^2z) + \frac{\partial}{\partial z}(-y^4z^6)$$

$$= 2xy^3 + 8x^5yz - 6y^4z^5.$$

(c) From Definition 9.7.2 and part (a) we find

$$\text{div(curl } \mathbf{F}) = \frac{\partial}{\partial x}(-4y^3z^6 - 4x^5y^2) + \frac{\partial}{\partial y}(-4z^3) + \frac{\partial}{\partial z}(20x^4y^2z - 3x^2y^2)$$

$$= 0 - 20x^4y^2 + 0 + 20x^4y^2 = 0. \qquad \equiv$$

We ask you to prove the following two important properties. If f is a scalar function with continuous second partial derivatives, then

$$\text{curl(grad } f) = \nabla \times \nabla f = \mathbf{0}. \tag{5}$$

Also, if $\mathbf{F}$ is a vector field having continuous second partial derivatives, then

$$\text{div(curl } \mathbf{F}) = \nabla \cdot (\nabla \times \mathbf{F}) = 0. \tag{6}$$

See part (c) of Example 2 and Problems 29 and 30 in Exercises 9.7.

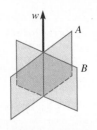

FIGURE 9.7.5 Paddle device

Physical Interpretations The word *curl* was introduced by **James Clerk Maxwell** (1831–1879), a Scottish physicist, in his studies of electromagnetic fields. However, the curl is easily understood in connection with the flow of fluids. If a paddle device, such as shown in **FIGURE 9.7.5**, is inserted in a flowing fluid, then the curl of the velocity field **F** is a measure of the tendency of the fluid to turn the device about its vertical axis *w*. If curl **F** = **0**, then the flow of the fluid is said to be **irrotational**, which means that it is free of vortices or whirlpools that would cause the paddle to rotate.* In **FIGURE 9.7.6** the axis *w* of the paddle points straight out of the page.

(a) Irrotational flow (b) Rotational flow

FIGURE 9.7.6 Irrotational flow in (a); rotational flow in (b)

(a) div **F**(*P*) > 0; *P* a source

(b) div **F**(*P*) < 0; *P* a sink

FIGURE 9.7.7 *P* a source in (a); *P* a sink in (b)

In the motivational discussion leading to Definition 9.7.2 we saw that the divergence of a velocity field **F** near a point *P*(*x*, *y*, *z*) is the flux per unit volume. If div **F**(*P*) > 0, then *P* is said to be a **source** for **F**, since there is a net outward flow of fluid near *P*; if div **F**(*P*) < 0, then *P* is said to be a **sink** for **F**, since there is a net inward flow of fluid near *P*; if div **F**(*P*) = 0, there are no sources or sinks near *P*. See **FIGURE 9.7.7**.

The divergence of a vector field can also be interpreted as a measure of the rate of change of the density of the fluid at a point. In other words, div **F** is a measure of the fluid's compressibility. If $\nabla \cdot \mathbf{F} = 0$, the fluid is said to be **incompressible**. In electromagnetic theory, if $\nabla \cdot \mathbf{F} = 0$, the vector field **F** is said to be **solenoidal**.

*In science texts the word *rotation* is sometimes used instead of *curl*. The symbol curl **F** is then replaced by rot **F**.

9.7 Exercises Answers to selected odd-numbered problems begin on page ANS-25.

In Problems 1–6, graph some representative vectors in the given vector field.

1. $\mathbf{F}(x, y) = x\mathbf{i} + y\mathbf{j}$
2. $\mathbf{F}(x, y) = -x\mathbf{i} + y\mathbf{j}$
3. $\mathbf{F}(x, y) = y\mathbf{i} + x\mathbf{j}$
4. $\mathbf{F}(x, y) = x\mathbf{i} + 2y\mathbf{j}$
5. $\mathbf{F}(x, y) = y\mathbf{j}$
6. $\mathbf{F}(x, y) = x\mathbf{j}$

In Problems 7–16, find the curl and the divergence of the given vector field.

7. $\mathbf{F}(x, y, z) = xz\mathbf{i} + yz\mathbf{j} + xy\mathbf{k}$
8. $\mathbf{F}(x, y, z) = 10yz\mathbf{i} + 2x^2z\mathbf{j} + 6x^3\mathbf{k}$
9. $\mathbf{F}(x, y, z) = 4xy\mathbf{i} + (2x^2 + 2yz)\mathbf{j} + (3z^2 + y^2)\mathbf{k}$
10. $\mathbf{F}(x, y, z) = (x - y)^3\mathbf{i} + e^{-yz}\mathbf{j} + xye^{2y}\mathbf{k}$
11. $\mathbf{F}(x, y, z) = 3x^2y\mathbf{i} + 2xz^3\mathbf{j} + y^4\mathbf{k}$

12. $\mathbf{F}(x, y, z) = 5y^3\mathbf{i} + (\frac{1}{2}x^3y^2 - xy)\mathbf{j} - (x^3yz - xz)\mathbf{k}$
13. $\mathbf{F}(x, y, z) = xe^{-z}\mathbf{i} + 4yz^2\mathbf{j} + 3ye^{-z}\mathbf{k}$
14. $\mathbf{F}(x, y, z) = yz \ln x\mathbf{i} + (2x - 3yz)\mathbf{j} + xy^2z^3\mathbf{k}$
15. $\mathbf{F}(x, y, z) = xye^x\mathbf{i} - x^3yze^z\mathbf{j} + xy^2e^y\mathbf{k}$
16. $\mathbf{F}(x, y, z) = x^2 \sin yz\mathbf{i} + z \cos xz^3\mathbf{j} + ye^{5xy}\mathbf{k}$

In Problems 17–24, let **a** be a constant vector and $\mathbf{r} = x\mathbf{i} + y\mathbf{j} + z\mathbf{k}$. Verify the given identity.

17. div **r** = 3
18. curl **r** = **0**
19. $(\mathbf{a} \times \nabla) \times \mathbf{r} = -2\mathbf{a}$
20. $\nabla \times (\mathbf{a} \times \mathbf{r}) = 2\mathbf{a}$
21. $\nabla \cdot (\mathbf{a} \times \mathbf{r}) = 0$
22. $\mathbf{a} \times (\nabla \times \mathbf{r}) = \mathbf{0}$
23. $\nabla \times [(\mathbf{r} \cdot \mathbf{r})\mathbf{a}] = 2(\mathbf{r} \times \mathbf{a})$
24. $\nabla \cdot [(\mathbf{r} \cdot \mathbf{r})\mathbf{a}] = 2(\mathbf{r} \cdot \mathbf{a})$

In Problems 25–32, verify the given identity. Assume continuity of all partial derivatives.

25. $\nabla \cdot (\mathbf{F} + \mathbf{G}) = \nabla \cdot \mathbf{F} + \nabla \cdot \mathbf{G}$

26. $\nabla \times (\mathbf{F} + \mathbf{G}) = \nabla \times \mathbf{F} + \nabla \times \mathbf{G}$

27. $\nabla \cdot (f\mathbf{F}) = f(\nabla \cdot \mathbf{F}) + \mathbf{F} \cdot \nabla f$

28. $\nabla \times (f\mathbf{F}) = f(\nabla \times \mathbf{F}) + (\nabla f) \times \mathbf{F}$

29. $\text{curl}(\text{grad } f) = \mathbf{0}$

30. $\text{div}(\text{curl } \mathbf{F}) = 0$

31. $\text{div}(\mathbf{F} \times \mathbf{G}) = \mathbf{G} \cdot \text{curl } \mathbf{F} - \mathbf{F} \cdot \text{curl } \mathbf{G}$

32. $\text{curl}(\text{curl } \mathbf{F} + \text{grad } f) = \text{curl}(\text{curl } \mathbf{F})$

33. Show that

$$\nabla \cdot \nabla f = \frac{\partial^2 f}{\partial x^2} + \frac{\partial^2 f}{\partial y^2} + \frac{\partial^2 f}{\partial z^2}.$$

This is known as the **Laplacian** and is also written $\nabla^2 f$.

34. Show that $\nabla \cdot (f \nabla f) = f \nabla^2 f + \|\nabla f\|^2$, where $\nabla^2 f$ is the Laplacian defined in Problem 33. [*Hint*: See Problem 27.]

35. Find curl(curl **F**) for the vector field $\mathbf{F} = xy\mathbf{i} + 4yz^2\mathbf{j} + 2xz\mathbf{k}$.

36. **(a)** Assuming continuity of all partial derivatives, show that curl(curl **F**) $= -\nabla^2 \mathbf{F} + \text{grad}(\text{div } \mathbf{F})$, where

$$\nabla^2 \mathbf{F} = \nabla^2(P\mathbf{i} + Q\mathbf{j} + R\mathbf{k}) = \nabla^2 P\mathbf{i} + \nabla^2 Q\mathbf{j} + \nabla^2 R\mathbf{k}.$$

(b) Use the identity in part (a) to obtain the result in Problem 35.

37. Any scalar function f for which $\nabla^2 f = 0$ is said to be **harmonic**. Verify that $f(x, y, z) = (x^2 + y^2 + z^2)^{-1/2}$ is harmonic except at the origin. $\nabla^2 f = 0$ is called **Laplace's equation**.

38. Verify that

$$f(x, y) = \arctan\left(\frac{2y}{x^2 + y^2 - 1}\right), \quad x^2 + y^2 \neq 1$$

satisfies Laplace's equation in two variables

$$\nabla^2 f = \frac{\partial^2 f}{\partial x^2} + \frac{\partial^2 f}{\partial y^2} = 0.$$

39. Let $\mathbf{r} = x\mathbf{i} + y\mathbf{j} + z\mathbf{k}$ be the position vector of a mass m_1 and let the mass m_2 be located at the origin. If the force of gravitational attraction is

$$\mathbf{F} = -\frac{Gm_1m_2}{\|\mathbf{r}\|^3}\,\mathbf{r},$$

verify that curl **F** $= \mathbf{0}$ and div **F** $= 0$, $\mathbf{r} \neq \mathbf{0}$.

40. Suppose a body rotates with a constant angular velocity $\boldsymbol{\omega}$ about an axis. If **r** is the position vector of a point P on the body measured from the origin, then the linear velocity vector **v** of rotation is $\mathbf{v} = \boldsymbol{\omega} \times \mathbf{r}$. See **FIGURE 9.7.8**. If $\mathbf{r} = x\mathbf{i} + y\mathbf{j} + z\mathbf{k}$ and $\boldsymbol{\omega} = \omega_1\mathbf{i} + \omega_2\mathbf{j} + \omega_3\mathbf{k}$, show that $\boldsymbol{\omega} = \frac{1}{2}\,\text{curl } \mathbf{v}$.

FIGURE 9.7.8 Rotating body in Problem 40

In Problems 41 and 42, assume that f and g have continuous second partial derivatives. Show that the given vector field is solenoidal. [*Hint*: See Problem 31.]

41. $\mathbf{F} = \nabla f \times \nabla g$

42. $\mathbf{F} = \nabla f \times (f\nabla g)$

43. The velocity vector field for the two-dimensional flow of an ideal fluid around a cylinder is given by

$$\mathbf{F}(x, y) = A\left[\left(1 - \frac{x^2 - y^2}{(x^2 + y^2)^2}\right)\mathbf{i} - \frac{2xy}{(x^2 + y^2)^2}\mathbf{j}\right]$$

for some positive constant A. See **FIGURE 9.7.9**.

(a) Show that when the point (x, y) is far from the origin, $\mathbf{F}(x, y) \approx A\mathbf{i}$.

(b) Show that **F** is irrotational.

(c) Show that **F** is incompressible.

FIGURE 9.7.9 Vector field in Problem 43

44. If $\mathbf{E} = \mathbf{E}(x, y, z, t)$ and $\mathbf{H} = \mathbf{H}(x, y, z, t)$ represent electric and magnetic fields in empty space, then Maxwell's equations are

$$\text{div } \mathbf{E} = 0, \quad \text{curl } \mathbf{E} = -\frac{1}{c}\frac{\partial \mathbf{H}}{\partial t},$$

$$\text{div } \mathbf{H} = 0, \quad \text{curl } \mathbf{H} = \frac{1}{c}\frac{\partial \mathbf{E}}{\partial t},$$

where c is the speed of light. Use the identity in Problem 36(a) to show that **E** and **H** satisfy

$$\nabla^2 \mathbf{E} = \frac{1}{c^2}\frac{\partial^2 \mathbf{E}}{\partial t^2}, \quad \nabla^2 \mathbf{H} = \frac{1}{c^2}\frac{\partial^2 \mathbf{H}}{\partial t^2}.$$

45. Consider the vector field $\mathbf{F} = x^2yz\mathbf{i} - xy^2z\mathbf{j} + (z + 5x)\mathbf{k}$. Explain why **F** is not the curl of another vector field **G**.

(a) Smooth curve (b) Piecewise-smooth curve

(c) Closed but not simple (d) Simple closed curve

FIGURE 9.8.1 Various curves

INTRODUCTION The notion of the definite integral $\int_a^b f(x)\,dx$; that is, *integration of a function defined over an interval,* can be generalized to *integration of a function defined along a curve.* To this end we need to introduce some terminology about curves.

▉ **Terminology** Suppose C is a curve parameterized by $x = f(t)$, $y = g(t)$, $a \le t \le b$, and A and B are the points $(f(a), g(a))$ and $(f(b), g(b))$, respectively. We say that

- (*i*) C is a **smooth curve** if f' and g' are continuous on the closed interval $[a, b]$ and not simultaneously zero on the open interval (a, b).
- (*ii*) C is **piecewise smooth** if it consists of a finite number of smooth curves $C_1, C_2, \ldots, C_n$ joined end to end—that is, $C = C_1 \cup C_2 \cup \cdots \cup C_n$.
- (*iii*) C is a **closed curve** if $A = B$.
- (*iv*) C is a **simple closed curve** if $A = B$ and the curve does not cross itself.
- (*v*) If C is not a closed curve, then the **positive direction** on C is the direction corresponding to increasing values of t.

FIGURE 9.8.1 illustrates each type of curve defined in (*i*)−(*iv*).

This same terminology carries over in a natural manner to curves in space. For example, a curve C defined by $x = f(t)$, $y = g(t)$, $z = h(t)$, $a \le t \le b$, is smooth if f', g', and h' are continuous on $[a, b]$ and not simultaneously zero on (a, b).

▉ **Definite Integral** Before defining integration along a curve, let us review the five steps leading to the definition of the definite integral.

1. Let $y = f(x)$ be defined on a closed interval $[a, b]$.
2. Partition the interval $[a, b]$ into n subintervals $[x_{k-1}, x_k]$ of lengths $\Delta x_k = x_k - x_{k-1}$. Let P denote the partition shown in **FIGURE 9.8.2(a)**.
3. Let $\lVert P \rVert$ be the length of the longest subinterval. The number $\lVert P \rVert$ is called the **norm** of the partition P.
4. Choose a sample point x_k^* in each subinterval. See Figure 9.8.2(b).
5. Form the sum $\displaystyle\sum_{k=1}^{n} f(x_k^*)\Delta x_k$.

$$a = x_0 < x_1 < x_2 < \cdots < x_{n-1} < x_n = b$$

(a)

(b)

FIGURE 9.8.2 Sample point in kth subinterval

The definite integral of a function of a single variable is given by the limit of a sum:

$$\int_a^b f(x)\,dx = \lim_{\lVert P \rVert \to 0} \sum_{k=1}^{n} f(x_k^*)\Delta x_k.$$

▉ **Line Integrals in the Plane** The following analogous five steps lead to the definitions of three **line integrals*** in the plane.

1. Let $z = G(x, y)$ be defined in some region that contains the smooth curve C defined by $x = f(t)$, $y = g(t)$, $a \le t \le b$.
2. Divide C into n subarcs of lengths Δs_k according to the partition $a = t_0 < t_1 < t_2 < \cdots < t_n = b$ of $[a, b]$. Let the projection of each subarc onto the x- and y-axes have lengths Δx_k and Δy_k, respectively.
3. Let $\lVert P \rVert$ be the **norm** of the partition or the length of the longest subarc.
4. Choose a sample point (x_k^*, y_k^*) on each subarc. See **FIGURE 9.8.3**.
5. Form the sums

FIGURE 9.8.3 Sample point in kth subarc

$$\sum_{k=1}^{n} G(x_k^*, y_k^*)\,\Delta x_k, \quad \sum_{k=1}^{n} G(x_k^*, y_k^*)\,\Delta y_k, \quad \sum_{k=1}^{n} G(x_k^*, y_k^*)\,\Delta s_k.$$

*An unfortunate choice of names. **Curve integrals** would be more appropriate.

Let G be a function of two variables x and y defined on a region of the plane containing a smooth curve C.

(*i*) The **line integral of G along C from A to B with respect to x** is

$$\int_C G(x, y)\, dx = \lim_{\|P\| \to 0} \sum_{k=1}^{n} G(x_k^*, y_k^*)\, \Delta x_k.$$

(*ii*) The **line integral of G along C from A to B with respect to y** is

$$\int_C G(x, y)\, dy = \lim_{\|P\| \to 0} \sum_{k=1}^{n} G(x_k^*, y_k^*)\, \Delta y_k.$$

(*iii*) The **line integral of G along C from A to B with respect to arc length** is

$$\int_C G(x, y)\, ds = \lim_{\|P\| \to 0} \sum_{k=1}^{n} G(x_k^*, y_k^*)\, \Delta s_k.$$

It can be proved that if $G(x, y)$ is continuous on C, then the integrals defined in (*i*), (*ii*), and (*iii*) exist. We shall assume continuity of G as a matter of course.

▉ Method of Evaluation—Curve Defined Parametrically

The line integrals in Definition 9.8.1 can be evaluated in two ways, depending on whether the curve C is defined parametrically or by an explicit function. In either case the basic idea is to convert the line integral to a definite integral in a single variable. If C is a smooth curve parameterized by $x = f(t)$, $y = g(t)$, $a \le t \le b$, then we simply replace x and y in the integral by the functions $f(t)$ and $g(t)$, and the appropriate differential dx, dy, or ds by $f'(t)\, dt$, $g'(t)\, dt$, or $\sqrt{[f'(t)]^2 + [g'(t)]^2}\, dt$. The expression $ds = \sqrt{[f'(t)]^2 + [g'(t)]^2}\, dt$ is called the **differential of arc length**. The integration is carried out with respect to the variable t in the usual manner:

$$\int_C G(x, y)\, dx = \int_a^b G(f(t), g(t))\, f'(t)\, dt \tag{1}$$

$$\int_C G(x, y)\, dy = \int_a^b G(f(t), g(t))\, g'(t)\, dt \tag{2}$$

$$\int_C G(x, y)\, ds = \int_a^b G(f(t), g(t))\sqrt{[f'(t)]^2 + [g'(t)]^2}\, dt. \tag{3}$$

EXAMPLE 1 Evaluation of Line Integrals

Evaluate (**a**) $\int_C xy^2\, dx$, (**b**) $\int_C xy^2\, dy$, and (**c**) $\int_C xy^2\, ds$ on the quarter-circle C defined by $x = 4 \cos t$, $y = 4 \sin t$, $0 \le t \le \pi/2$. See **FIGURE 9.8.4**.

SOLUTION (**a**) From (1),

$$\int_C xy^2\, dx = \int_0^{\pi/2} \overbrace{(4 \cos t)}^{x}\,\overbrace{(16 \sin^2 t)}^{y^2}\,\overbrace{(-4 \sin t\, dt)}^{dx}$$

$$= -256 \int_0^{\pi/2} \sin^3 t \cos t\, dt$$

$$= -256 \left[\frac{1}{4} \sin^4 t \right]_0^{\pi/2} = -64.$$

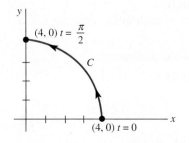

$(4, 0)\ t = \dfrac{\pi}{2}$

C

$(4, 0)\ t = 0$

FIGURE 9.8.4 Curve C in Example 1

(b) From (2),

$$\int_C xy^2\, dy = \int_0^{\pi/2} \overbrace{(4\cos t)}^{x}\,\overbrace{(16\sin^2 t)}^{y^2}\,\overbrace{(4\cos t\, dt)}^{dy}$$

$$= 256\int_0^{\pi/2} \sin^2 t\, \cos^2 t\, dt$$

$$= 256\int_0^{\pi/2} \frac{1}{4}\sin^2 2t\, dt \qquad \leftarrow \text{trigonometric identities}$$

$$= 64\int_0^{\pi/2} \frac{1}{2}(1 - \cos 4t)\, dt$$

$$= 32\left[t - \frac{1}{4}\sin 4t\right]_0^{\pi/2} = 16\pi.$$

(c) From (3),

$$\int_C xy^2\, ds = \int_0^{\pi/2} \overbrace{(4\cos t)}^{x}\,\overbrace{(16\sin^2 t)}^{y^2}\,\overbrace{\sqrt{16(\cos^2 t + \sin^2 t)}\, dt}^{ds}$$

$$= 256\int_0^{\pi/2} \sin^2 t\, \cos t\, dt$$

$$= 256\left[\frac{1}{3}\sin^3 t\right]_0^{\pi/2} = \frac{256}{3}. \qquad \equiv$$

■ **Method of Evaluation—Curve Defined by an Explicit Function** If the curve C is defined by an explicit function $y = f(x)$, $a \le x \le b$, we can use x as a parameter. With $dy = f'(x)\, dx$ and $ds = \sqrt{1 + [f'(x)]^2}\, dx$, the foregoing line integrals become, in turn,

$$\int_C G(x, y)\, dx = \int_a^b G(x, f(x))\, dx \tag{4}$$

$$\int_C G(x, y)\, dy = \int_a^b G(x, f(x)) f'(x)\, dx \tag{5}$$

$$\int_C G(x, y)\, ds = \int_a^b G(x, f(x))\sqrt{1 + [f'(x)]^2}\, dx. \tag{6}$$

A line integral along a *piecewise-smooth* curve C is defined as the *sum* of the integrals over the various smooth curves whose union comprises C. For example, if C is composed of smooth curves C_1 and C_2, then

$$\int_C G(x, y)\, ds = \int_{C_1} G(x, y)\, ds + \int_{C_2} G(x, y)\, ds.$$

■ **Notation** In many applications, line integrals appear as a sum

$$\int_C P(x, y)\, dx + \int_C Q(x, y)\, dy.$$

It is common practice to write this sum as one integral without parentheses as

$$\int_C P(x, y)\, dx + Q(x, y)\, dy \quad \text{or simply} \quad \int_C P\, dx + Q\, dy. \tag{7}$$

A line integral along a *closed* curve C is very often denoted by

$$\oint_C P\,dx + Q\,dy.$$

Curve Defined by an Explicit Function

Evaluate $\int_C xy\,dx + x^2\,dy$, where C is given by $y = x^3$, $-1 \le x \le 2$.

SOLUTION The curve C is illustrated in **FIGURE 9.8.5** and is defined by the explicit function $y = x^3$. Hence we can use x as the parameter. Using $dy = 3x^2\,dx$, we get

$$\int_C xy\,dx + x^2\,dy = \int_{-1}^{2} x(\overbrace{x^3}^{y})\,dx + x^2(\overbrace{3x^2\,dx}^{dy})$$

$$= \int_{-1}^{2} 4x^4\,dx$$

$$= \frac{4}{5}x^5\bigg]_{-1}^{2} = \frac{132}{5}. \qquad \equiv$$

FIGURE 9.8.5 Curve C in Example 2

Curve Defined Parametrically

Evaluate $\oint_C x\,dx$, where C is the circle $x = \cos t$, $y = \sin t$, $0 \le t \le 2\pi$.

SOLUTION From (1),

$$\oint_C x\,dx = \int_0^{2\pi} \cos t\,(-\sin t\,dt) = \frac{1}{2}\cos^2 t\bigg]_0^{2\pi} = \frac{1}{2}[1-1] = 0. \qquad \equiv$$

Closed Curve

Evaluate $\oint_C y^2\,dx - x^2\,dy$ on the closed curve C that is shown in **FIGURE 9.8.6(a)**.

SOLUTION Since C is piecewise smooth, we express the integral as a sum of integrals. Symbolically, we write

$$\oint_C = \int_{C_1} + \int_{C_2} + \int_{C_3}$$

where C_1, C_2, and C_3 are the curves shown in Figure 9.8.6(b). On C_1, we use x as a parameter. Since $y = 0$, $dy = 0$; therefore,

$$\int_{C_1} y^2\,dx - x^2\,dy = \int_0^2 0\,dx - x^2(0) = 0.$$

On C_2, we use y as a parameter. From $x = 2$, $dx = 0$, we have

$$\int_{C_2} y^2\,dx - x^2\,dy = \int_0^4 y^2(0) - 4\,dy$$

$$= -\int_0^4 4\,dy = -4y\bigg]_0^4 = -16.$$

Finally, on C_3, we again use x as a parameter. From $y = x^2$, we get $dy = 2x\,dx$ and so

$$\int_{C_3} y^2\,dx - x^2\,dy = \int_2^0 x^4\,dx - x^2(2x\,dx)$$

(a)

(b)

FIGURE 9.8.6 Curve C in Example 4

FIGURE 9.8.7 Curves with opposite orientation

$$= \int_2^0 (x^4 - 2x^3)\, dx$$

$$= \left(\frac{1}{5} x^5 - \frac{1}{2} x^4 \right) \Big]_2^0 = \frac{8}{5}.$$

Hence,

$$\int_C y^2\, dx - x^2\, dy = 0 - 16 + \frac{8}{5} = -\frac{72}{5}. \qquad \equiv$$

It is important to be aware that a line integral is independent of the parameterization of the curve C provided C is given the same orientation by all sets of parametric equations defining the curve. See Problem 37 in Exercises 9.8. Also, recall for definite integrals that $\int_b^a f(x)\, dx = -\int_a^b f(x)\, dx$. Line integrals possess a similar property. Suppose, as shown in **FIGURE 9.8.7**, that $-C$ denotes the curve having the opposite orientation of C. Then it can be shown that

$$\int_{-C} P\, dx + Q\, dy = -\int_C P\, dx + Q\, dy,$$

or equivalently,

$$\int_{-C} P\, dx + Q\, dy + \int_C P\, dx + Q\, dy = 0. \qquad (8)$$

For example, in part (*a*) of Example 1, $\int_{-C} xy^2\, dx = 64$.

■ **Line Integrals in Space** Line integrals of a function G of three variables, $\int_C G(x, y, z)\, dx$, $\int_C G(x, y, z)\, dy$, and $\int_C G(x, y, z)\, ds$, are defined in a manner analogous to Definition 9.8.1. However, to that list we add a fourth **line integral along a space curve C with respect to z**:

$$\int_C G(x, y, z)\, dz = \lim_{\|P\| \to 0} \sum_{k=1}^n G(x_k^*, y_k^*, z_k^*)\, \Delta z_k. \qquad (9)$$

■ **Method of Evaluation** If C is a smooth curve in 3-space defined by the parametric equations $x = f(t), y = g(t), z = h(t), a \le t \le b$, then the integral in (9) can be evaluated by using

$$\int_C G(x, y, z)\, dz = \int_a^b G(f(t), g(t), h(t))\, h'(t)\, dt.$$

The integrals $\int_C G(x, y, z)\, dx$ and $\int_C G(x, y, z)\, dy$ are evaluated in a similar fashion. The line integral with respect to arc length is

$$\int_C G(x, y, z)\, ds = \int_a^b G(f(t), g(t), h(t)) \sqrt{[f'(t)]^2 + [g'(t)]^2 + [h'(t)]^2}\, dt.$$

As in (7), in 3-space we are often concerned with line integrals in the form of a sum:

$$\int_C P(x, y, z)\, dx + Q(x, y, z)\, dy + R(x, y, z)\, dz.$$

EXAMPLE 5 **Line Integral on a Curve in 3-Space**

Evaluate $\int_C y\, dx + x\, dy + z\, dz$, where C is the helix $x = 2 \cos t, y = 2 \sin t, z = t, 0 \le t \le 2\pi$.

SOLUTION Substituting the expressions for x, y, and z along with $dx = -2 \sin t\, dt$, $dy = 2 \cos t\, dt, dz = dt$, we get

$$\int_C y\, dx + x\, dy + z\, dz = \int_0^{2\pi} \underbrace{(-4 \sin^2 t + 4 \cos^2 t)}_{\text{double angle formula}}\, dt + t\, dt$$

$$= \int_0^{2\pi} (4 \cos 2t + t)\, dt$$

$$= \left(2 \sin 2t + \frac{t^2}{2} \right) \Big]_0^{2\pi} = 2\pi^2. \qquad \equiv$$

We can use the concept of a vector function of several variables to write a general line integral in a compact fashion. For example, suppose the vector-valued function $\mathbf{F}(x, y) = P(x, y)\mathbf{i} + Q(x, y)\mathbf{j}$ is defined along a curve $C: x = f(t)$, $y = g(t)$, $a \leq t \leq b$, and suppose $\mathbf{r}(t) = f(t)\mathbf{i} + g(t)\mathbf{j}$ is the position vector of points on C. Then the derivative of $\mathbf{r}(t)$,

$$\frac{d\mathbf{r}}{dt} = f'(t)\mathbf{i} + g'(t)\mathbf{j} = \frac{dx}{dt}\mathbf{i} + \frac{dy}{dt}\mathbf{j},$$

prompts us to define $d\mathbf{r} = \dfrac{d\mathbf{r}}{dt}\, dt = dx\,\mathbf{i} + dy\,\mathbf{j}$. Since $\mathbf{F}(x, y) \cdot d\mathbf{r} = P(x, y)\, dx + Q(x, y)\, dy$ we can write

$$\int_C P(x, y)\, dx + Q(x, y)\, dy = \int_C \mathbf{F} \cdot d\mathbf{r}. \tag{10}$$

Similarly, for a line integral on a space curve,

$$\int_C P(x, y, z)\, dx + Q(x, y, z)\, dy + R(x, y, z)\, dz = \int_C \mathbf{F} \cdot d\mathbf{r}, \tag{11}$$

where $\mathbf{F}(x, y, z) = P(x, y, z)\mathbf{i} + Q(x, y, z)\mathbf{j} + R(x, y, z)\mathbf{k}$ and $d\mathbf{r} = dx\,\mathbf{i} + dy\,\mathbf{j} + dz\,\mathbf{k}$.

(a)

(b)

FIGURE 9.8.8 Force field $\mathbf{F}$ varies along curve C

■ **Work** In Section 7.3 we saw that the work W done by a constant force $\mathbf{F}$ that causes a straight-line displacement $\mathbf{d}$ of an object is $W = \mathbf{F} \cdot \mathbf{d}$. In beginning courses in calculus or physics it is then shown that the work done in moving an object from $x = a$ to $x = b$ by a force $F(x)$, which varies in magnitude but not in direction, is given by the definite integral $W = \int_a^b F(x)\, dx$. In general, a force field $\mathbf{F}(x, y) = P(x, y)\mathbf{i} + Q(x, y)\mathbf{j}$ acting at each point on a smooth curve $C: x = f(t)$, $y = g(t)$, $a \leq t \leq b$, varies in both magnitude and direction. See **FIGURE 9.8.8(a)**. If A and B are the points $(f(a), g(a))$ and $(f(b), g(b))$, respectively, we ask: What is the work done by $\mathbf{F}$ as its point of application moves along C from A to B? To answer this question, suppose C is divided into n subarcs of lengths Δs_k. On each subarc $\mathbf{F}(x_k^*, y_k^*)$ is a constant force. If, as shown in Figure 9.8.8(b), the length of the vector $\Delta\mathbf{r}_k = (x_k - x_{k-1})\mathbf{i} + (y_k - y_{k-1})\mathbf{j} = \Delta x_k\mathbf{i} + \Delta y_k\mathbf{j}$ is an approximation to the length of the kth subarc, then the approximate work done by $\mathbf{F}$ over the subarc is

$$\left(\|\mathbf{F}(x_k^*, y_k^*)\| \cos\theta\right)\|\Delta\mathbf{r}_k\| = \mathbf{F}(x_k^*, y_k^*) \cdot \Delta\mathbf{r}_k$$

$$= P(x_k^*, y_k^*)\, \Delta x_k + Q(x_k^*, y_k^*)\, \Delta y_k.$$

By summing these elements of work and passing to the limit, we naturally define the **work done by $\mathbf{F}$ along C** as the line integral

$$W = \int_C P(x, y)\, dx + Q(x, y)\, dy \qquad \text{or} \qquad W = \int_C \mathbf{F} \cdot d\mathbf{r}. \tag{12}$$

Of course, (12) extends to force fields acting at points on a space curve. In this case, work $\int_C \mathbf{F} \cdot d\mathbf{r}$ is defined as in (11).

Now, since

$$\frac{d\mathbf{r}}{dt} = \frac{d\mathbf{r}}{ds}\frac{ds}{dt},$$

we let $d\mathbf{r} = \mathbf{T}\, ds$, where $\mathbf{T} = d\mathbf{r}/ds$ is a unit tangent to C. Hence,

$$W = \int_C \mathbf{F} \cdot d\mathbf{r} = \int_C \mathbf{F} \cdot \mathbf{T}\, ds = \int_C \text{comp}_\mathbf{T}\, \mathbf{F}\, ds. \tag{13}$$

In other words:

The work done by a force $\mathbf{F}$ along a curve C is due entirely to the tangential component of $\mathbf{F}$.

FIGURE 9.8.9 Force field in (a) of Example 6

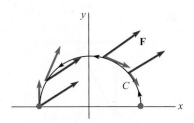

FIGURE 9.8.10 Force field in (b) of Example 6

flow of fluid

FIGURE 9.8.11 Does the velocity field turn the curve C?

(a) Vertical rectangle

(b) "Fence" or "curtain" of varying height $G(x, y)$ with base C

FIGURE 9.8.12 A geometric interpretation of a line integral

EXAMPLE 6 **Work Done by a Force**

Find the work done by **(a)** $\mathbf{F} = x\mathbf{i} + y\mathbf{j}$ and **(b)** $\mathbf{F} = \frac{3}{4}\mathbf{i} + \frac{1}{2}\mathbf{j}$ along the curve C traced by $\mathbf{r}(t) = \cos t\,\mathbf{i} + \sin t\,\mathbf{j}$ from $t = 0$ to $t = \pi$.

SOLUTION **(a)** The vector function $\mathbf{r}(t)$ gives the parametric equations $x = \cos t$, $y = \sin t$, $0 \le t \le \pi$, which we recognize as a half-circle. As seen in **FIGURE 9.8.9**, the force field $\mathbf{F}$ is perpendicular to C at every point. Since the tangential components of $\mathbf{F}$ are zero, the work done along C is zero. To see this we use (12):

$$W = \int_C \mathbf{F} \cdot d\mathbf{r} = \int_C (x\mathbf{i} + y\mathbf{j}) \cdot d\mathbf{r}$$

$$= \int_0^\pi (\cos t\,\mathbf{i} + \sin t\,\mathbf{j}) \cdot (-\sin t\,\mathbf{i} + \cos t\,\mathbf{j})\, dt$$

$$= \int_0^\pi (-\cos t \sin t + \sin t \cos t)\, dt = 0.$$

(b) In **FIGURE 9.8.10** the vectors in red are the projections of $\mathbf{F}$ on the unit tangent vectors. The work done by $\mathbf{F}$ is

$$W = \int_C \mathbf{F} \cdot d\mathbf{r} = \int_C \left(\frac{3}{4}\mathbf{i} + \frac{1}{2}\mathbf{j}\right) \cdot d\mathbf{r}$$

$$= \int_0^\pi \left(\frac{3}{4}\mathbf{i} + \frac{1}{2}\mathbf{j}\right) \cdot (-\sin t\,\mathbf{i} + \cos t\,\mathbf{j})\, dt$$

$$= \int_0^\pi \left(-\frac{3}{4}\sin t + \frac{1}{2}\cos t\right) dt$$

$$= \left(\frac{3}{4}\cos t + \frac{1}{2}\sin t\right)\Bigg]_0^\pi = -\frac{3}{2}.$$

The units of work depend on the units of $\|\mathbf{F}\|$ and on the units of distance. ≡

■ **Circulation** A line integral of a vector field $\mathbf{F}$ around a simple closed curve C is said to be the **circulation** of $\mathbf{F}$ around C; that is,

$$\text{circulation} = \oint_C \mathbf{F} \cdot d\mathbf{r} = \oint_C \mathbf{F} \cdot \mathbf{T}\, ds.$$

In particular, if $\mathbf{F}$ is the velocity field of a fluid, then the circulation is a measure of the amount by which the fluid tends to turn the curve C by rotating, or circulating, around it. For example, if $\mathbf{F}$ is perpendicular to $\mathbf{T}$ for every (x, y) on C, then $\int_C \mathbf{F} \cdot \mathbf{T}\, ds = 0$, and the curve does not move at all. On the other hand, $\int_C \mathbf{F} \cdot \mathbf{T}\, ds > 0$ and $\int_C \mathbf{F} \cdot \mathbf{T}\, ds < 0$ mean that the fluid tends to rotate C in the counterclockwise and clockwise directions, respectively. See **FIGURE 9.8.11**.

REMARKS

In the case of two variables, the line integral with respect to arc length $\int_C G(x, y)\, ds$ can be interpreted in a geometric manner when $G(x, y) \ge 0$ on C. In Definition 9.8.1 the symbol Δs_k represents the length of the kth subarc on the curve C. But from Figure 9.8.3 accompanying that definition, we have the approximation $\Delta s_k = \sqrt{(\Delta x_k)^2 + (\Delta y_k)^2}$. With this interpretation of Δs_k we see from **FIGURE 9.8.12(a)** that the product $G(x_k^*, y_k^*)\,\Delta s_k$ is the area of a vertical rectangle of height $G(x_k^*, y_k^*)$ and width Δs_k. The integral $\int_C G(x, y)\, ds$ then represents the area of one side of a "fence" or "curtain" extending from the curve C in the xy-plane up to the graph of $G(x, y)$ that corresponds to points (x, y) on C. See Figure 9.8.12(b).

In Problems 1–4, evaluate $\int_C G(x, y)\, dx$, $\int_C G(x, y)\, dy$, and $\int_C G(x, y)\, ds$ on the indicated curve C.

1. $G(x, y) = 2xy$; $x = 5\cos t$, $y = 5\sin t$, $0 \le t \le \pi/4$

2. $G(x, y) = x^3 + 2xy^2 + 2x$; $x = 2t$, $y = t^2$, $0 \le t \le 1$

3. $G(x, y) = 3x^2 + 6y^2$; $y = 2x + 1$, $-1 \le x \le 0$

4. $G(x, y) = x^2/y^3$; $2y = 3x^{2/3}$, $1 \le x \le 8$

In Problems 5 and 6, evaluate $\int_C G(x, y, z)\, dx$, $\int_C G(x, y, z)\, dy$, $\int_C G(x, y, z)\, dz$, and $\int_C G(x, y, z)\, ds$ on the indicated curve C.

5. $G(x, y, z) = z$; $x = \cos t$, $y = \sin t$, $z = t$, $0 \le t \le \pi/2$

6. $G(x, y, z) = 4xyz$; $x = \frac{1}{3}t^3$, $y = t^2$, $z = 2t$, $0 \le t \le 1$

In Problems 7–10, evaluate $\int_C (2x + y)\, dx + xy\, dy$ on the given curve C between $(-1, 2)$ and $(2, 5)$.

7. $y = x + 3$

8. $y = x^2 + 1$

9.

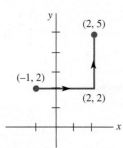

FIGURE 9.8.13 Curve C for Problem 9

10.

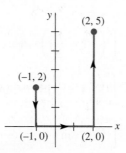

FIGURE 9.8.14 Curve C for Problem 10

In Problems 11–14, evaluate $\int_C y\, dx + x\, dy$ on the given curve C between $(0, 0)$ and $(1, 1)$.

11. $y = x^2$

12. $y = x$

13. C consists of the line segments from $(0, 0)$ to $(0, 1)$ and from $(0, 1)$ to $(1, 1)$.

14. C consists of the line segments from $(0, 0)$ to $(1, 0)$ and from $(1, 0)$ to $(1, 1)$.

15. Evaluate $\int_C (6x^2 + 2y^2)\, dx + 4xy\, dy$, where C is given by $x = \sqrt{t}$, $y = t$, $4 \le t \le 9$.

16. Evaluate $\int_C -y^2\, dx + xy\, dy$, where C is given by $x = 2t$, $y = t^3$, $0 \le t \le 2$.

17. Evaluate $\int_C 2x^3y\, dx + (3x + y)\, dy$, where C is given by $x = y^2$ from $(1, -1)$ to $(1, 1)$.

18. Evaluate $\int_C 4x\, dx + 2y\, dy$, where C is given by $x = y^3 + 1$ from $(0, -1)$ to $(9, 2)$.

In Problems 19 and 20, evaluate $\oint_C (x^2 + y^2)\, dx - 2xy\, dy$ on the given closed curve C.

19.

FIGURE 9.8.15 Closed curve C for Problem 19

20.

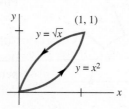

FIGURE 9.8.16 Closed curve C for Problem 20

In Problems 21 and 22, evaluate $\oint_C x^2y^3\, dx - xy^2\, dy$ on the given closed curve C.

21.

FIGURE 9.8.17 Closed curve C for Problem 21

22.

FIGURE 9.8.18 Closed curve C for Problem 22

23. Evaluate $\oint_C (x^2 - y^2)\, ds$, where C is given by
$$x = 5\cos t, \quad y = 5\sin t, \quad 0 \le t \le 2\pi.$$

24. Evaluate $\int_{-C} y\, dx - x\, dy$, where C is given by
$$x = 2\cos t, \quad y = 3\sin t, \quad 0 \le t \le \pi.$$

In Problems 25–28, evaluate $\int_C y\, dx + z\, dy + x\, dz$ on the given curve C between $(0, 0, 0)$ and $(6, 8, 5)$.

25. C consists of the line segments from $(0, 0, 0)$ to $(2, 3, 4)$ and from $(2, 3, 4)$ to $(6, 8, 5)$.

26. $x = 3t$, $y = t^3$, $z = \frac{5}{4}t^2$, $0 \le t \le 2$

27.

FIGURE 9.8.19 Curve C for Problem 27

28.

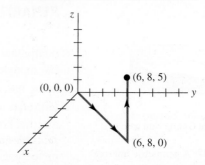

FIGURE 9.8.20 Curve C for Problem 28

In Problems 29 and 30, evaluate $\int_C \mathbf{F} \cdot d\mathbf{r}$.

29. $\mathbf{F}(x, y) = y^3\mathbf{i} - x^2 y\mathbf{j}$; $\mathbf{r}(t) = e^{-2t}\mathbf{i} + e^t\mathbf{j}$, $0 \le t \le \ln 2$

30. $\mathbf{F}(x, y, z) = e^x\mathbf{i} + xe^{xy}\mathbf{j} + xye^{xyz}\mathbf{k}$; $\mathbf{r}(t) = t\mathbf{i} + t^2\mathbf{j} + t^3\mathbf{k}$, $0 \le t \le 1$

31. Find the work done by the force $\mathbf{F}(x, y) = y\mathbf{i} + x\mathbf{j}$ acting along $y = \ln x$ from $(1, 0)$ to $(e, 1)$.

32. Find the work done by the force $\mathbf{F}(x, y) = 2xy\mathbf{i} + 4y^2\mathbf{j}$ acting along the piecewise-smooth curve consisting of the line segments from $(-2, 2)$ to $(0, 0)$ and from $(0, 0)$ to $(2, 3)$.

33. Find the work done by the force $\mathbf{F}(x, y) = (x + 2y)\mathbf{i} + (6y - 2x)\mathbf{j}$ acting counterclockwise once around the triangle with vertices $(1, 1)$, $(3, 1)$, and $(3, 2)$.

34. Find the work done by the force $\mathbf{F}(x, y, z) = yz\mathbf{i} + xz\mathbf{j} + xy\mathbf{k}$ acting along the curve given by $\mathbf{r}(t) = t^3\mathbf{i} + t^2\mathbf{j} + t\mathbf{k}$ from $t = 1$ to $t = 3$.

35. Find the work done by a constant force $\mathbf{F}(x, y) = a\mathbf{i} + b\mathbf{j}$ acting counterclockwise once around the circle $x^2 + y^2 = 9$.

36. In an inverse square force field $\mathbf{F} = c\mathbf{r}/\|\mathbf{r}\|^3$, where c is a constant and $\mathbf{r} = x\mathbf{i} + y\mathbf{j} + z\mathbf{k}$,* find the work done in moving a particle along the line from $(1, 1, 1)$ to $(3, 3, 3)$.

37. Verify that the line integral $\int_C y^2 \, dx + xy \, dy$ has the same value on C for each of the following parameterizations:

$C: x = 2t + 1$, $y = 4t + 2$, $0 \le t \le 1$
$C: x = t^2$, $y = 2t^2$, $1 \le t \le \sqrt{3}$
$C: x = \ln t$, $y = 2 \ln t$, $e \le t \le e^3$.

38. Consider the three curves between $(0, 0)$ and $(2, 4)$:

$C_1: x = t$, $y = 2t$, $0 \le t \le 2$
$C_2: x = t$, $y = t^2$, $0 \le t \le 2$
$C_3: x = 2t - 4$, $y = 4t - 8$, $2 \le t \le 3$.

Show that $\int_{C_1} xy \, ds = \int_{C_3} xy \, ds$, but $\int_{C_1} xy \, ds \ne \int_{C_2} xy \, ds$. Explain.

39. Assume a smooth curve C is described by the vector function $\mathbf{r}(t)$ for $a \le t \le b$. Let acceleration, velocity, and speed be given by $\mathbf{a} = d\mathbf{v}/dt$, $\mathbf{v} = d\mathbf{r}/dt$, and $v = \|\mathbf{v}\|$, respectively. Using Newton's second law $\mathbf{F} = m\mathbf{a}$, show that, in the absence of friction, the work done by $\mathbf{F}$ in moving a particle of constant mass m from point A at $t = a$ to point B at $t = b$ is the same as the change in kinetic energy:

*Note that the magnitude of $\mathbf{F}$ is inversely proportional to $\|\mathbf{r}\|^2$.

$$K(B) - K(A) = \frac{1}{2} m[v(b)]^2 - \frac{1}{2} m[v(a)]^2.$$

[*Hint*: Consider $\dfrac{d}{dt} v^2 = \dfrac{d}{dt} \mathbf{v} \cdot \mathbf{v}$.]

40. If $\rho(x, y)$ is the density of a wire (mass per unit length), then $m = \int_C \rho(x, y) \, ds$ is the mass of the wire. Find the mass of a wire having the shape of the semicircle $x = 1 + \cos t$, $y = \sin t$, $0 \le t \le \pi$, if the density at a point P is directly proportional to distance from the y-axis.

41. The coordinates of the center of mass of a wire with variable density are given by $\bar{x} = M_y/m$, $\bar{y} = M_x/m$, where

$$m = \int_C \rho(x, y) \, ds, \quad M_x = \int_C y \, \rho(x, y) \, ds$$

and
$$M_y = \int_C x \, \rho(x, y) \, ds.$$

Find the center of mass of the wire in Problem 40.

42. A force field $\mathbf{F}(x, y)$ acts at each point on the curve C, which is the union of C_1, C_2, and C_3 shown in red in **FIGURE 9.8.21**. $\|\mathbf{F}\|$ is measured in pounds and distance is measured in feet using the scale given in the figure. Use the representative vectors shown to approximate the work done by $\mathbf{F}$ along C. [*Hint*: Use $W = \int_C \mathbf{F} \cdot \mathbf{T} \, ds$.]

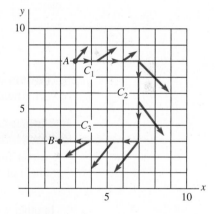

FIGURE 9.8.21 Force field in Problem 42

9.9 Independence of Path

INTRODUCTION In this section we refer to a piecewise-smooth curve C between an initial point A and a terminal point B as a **path of integration** or simply a **path**. We begin the discussion by considering line integrals in 2-space. Recall, in (10) of Section 9.8 we saw that if $\mathbf{F}(x, y) = P(x, y)\mathbf{i} + Q(x, y)\mathbf{j}$ is a vector field in 2-space and a path C is defined by the vector function $\mathbf{r}(t) = f(t)\mathbf{i} + g(t)\mathbf{j}$, $a \le t \le b$, then a line integral can be written as

$$\int_C P \, dx + Q \, dy = \int_C \mathbf{F} \cdot d\mathbf{r},$$

where $d\mathbf{r} = dx\,\mathbf{i} + dy\,\mathbf{j}$. Generally the value of a line integral $\int_C \mathbf{F} \cdot d\mathbf{r}$ depends on the path of integration. Stated another way, if C_1 and C_2 are two different paths between the same points A and B, then we expect that $\int_{C_1} \mathbf{F} \cdot d\mathbf{r} \neq \int_{C_2} \mathbf{F} \cdot d\mathbf{r}$. However, there is an important exception. As we see in this section, line integrals that involve a certain kind of vector field $\mathbf{F}$ depend not on the path C but only on the endpoints $A = (f(a), g(a))$ and $B = (f(b), g(b))$ of the path.

Note: To avoid needless repetition we assume throughout that the component functions of the vector field $\mathbf{F}$ are continuous in some region of 2- or 3-space, its component functions have continuous first partial derivatives in the region, and the path C of integration lies entirely in the region.

EXAMPLE 1 Path Independence

The integral $\int_C y\,dx + x\,dy$ has the same value on each path C between $(0, 0)$ and $(1, 1)$ shown in **FIGURE 9.9.1**. You may recall from Problems 11–14 in Exercises 9.8 that on these paths

$$\int_C y\,dx + x\,dy = 1.$$

You are also urged to verify $\int_C y\,dx + x\,dy = 1$ on the curves $y = x^3$, $y = x^4$, and $y = \sqrt{x}$ between $(0, 0)$ and $(1, 1)$. The relevance of all this is to suggest that the integral $\int_C y\,dx + x\,dy$ does not depend on the path joining these two points. We continue this discussion in Examples 2 and 3.

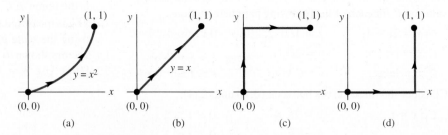

FIGURE 9.9.1 Line integral in Example 1 is the same on four paths

■ Conservative Vector Fields Before proceeding, we need to introduce a special kind of vector field $\mathbf{F}$ called a **conservative field**.

DEFINITION 9.9.1 Conservative Vector Field

A vector function $\mathbf{F}$ in 2- or 3-space is said to be **conservative** if $\mathbf{F}$ can be written as the gradient of a scalar function ϕ. The function ϕ is called a **potential function** for $\mathbf{F}$.

In other words, $\mathbf{F}$ is conservative if there exists a function ϕ such that $\mathbf{F} = \nabla\phi$. A conservative vector field is also called a **gradient vector field**.

EXAMPLE 2 Conservative Vector Field

The integral in Example 1 can be interpreted as a line integral of a vector field $\mathbf{F}$ along a path C. If $\mathbf{F}(x, y) = y\,\mathbf{i} + x\,\mathbf{j}$ and $d\mathbf{r} = dx\,\mathbf{i} + dy\,\mathbf{j}$, then $\int_C y\,dx + x\,dy = \int_C \mathbf{F} \cdot d\mathbf{r}$. Now consider the function $\phi(x, y) = xy$. The gradient of the scalar function ϕ is

$$\nabla\phi = \frac{\partial\phi}{\partial x}\,\mathbf{i} + \frac{\partial\phi}{\partial y}\,\mathbf{j} = y\,\mathbf{i} + x\,\mathbf{j}.$$

Because $\nabla\phi = \mathbf{F}(x, y)$ we conclude that $\mathbf{F}(x, y) = y\,\mathbf{i} + x\,\mathbf{j}$ is a conservative vector field and that ϕ is a potential function for $\mathbf{F}$.

Of course, not every vector field is a conservative field although many vector fields encountered in physics are conservative. For present purposes, the importance of conservative vector fields will become evident as we continue our study of line integrals.

■ **Path Independence** If the value of a line integral is the same for *every* path in a region connecting the initial point A and terminal point B, then the integral is said to be **independent of the path**. In other words, a line integral $\int_C \mathbf{F} \cdot d\mathbf{r}$ of $\mathbf{F}$ along C is independent of the path if $\int_{C_1} \mathbf{F} \cdot d\mathbf{r} = \int_{C_2} \mathbf{F} \cdot d\mathbf{r}$ for any two paths C_1 and C_2 between A and B.

■ **A Fundamental Theorem** The next theorem establishes an important relationship between the value of a line integral $\int_C \mathbf{F} \cdot d\mathbf{r}$ and the fact that its path of integration C lies within a conservative vector field $\mathbf{F}$. In addition, it provides a means of evaluating these line integrals in a manner that is akin to the Fundamental Theorem of Calculus:

$$\int_a^b f'(x)\, dx = f(b) - f(a),\tag{1}$$

where $f(x)$ is an antiderivative of $f'(x)$. In the next theorem, known as the **Fundamental Theorem for Line Integrals**, the gradient $\nabla\phi$ of a scalar function ϕ plays the part of the derivative $f'(x)$ in (1).

THEOREM 9.9.1 Fundamental Theorem

Suppose C is a path in an open region R of the xy-plane and is defined by $\mathbf{r}(t) = x(t)\mathbf{i} + y(t)\mathbf{j}$, $a \le t \le b$. If $\mathbf{F}(x, y) = P(x, y)\mathbf{i} + Q(x, y)\mathbf{j}$ is a conservative vector field in R and ϕ is a potential function for $\mathbf{F}$, then

$$\int_C \mathbf{F} \cdot d\mathbf{r} = \int_C \nabla\phi \cdot d\mathbf{r} = \phi(B) - \phi(A),\tag{2}$$

where $A = (x(a), y(a))$ and $B = (x(b), y(b))$.

PROOF: We will prove the theorem for a smooth path C. Because ϕ is a potential function for $\mathbf{F}$ we have

$$\mathbf{F} = \nabla\phi = \frac{\partial\phi}{\partial x}\mathbf{i} + \frac{\partial\phi}{\partial y}\mathbf{j}.$$

Then using $\mathbf{r}'(t) = (dx/dt)\mathbf{i} + (dy/dt)\mathbf{j}$ we can write the line integral of $\mathbf{F}$ along the path C as

$$\int_C \mathbf{F} \cdot d\mathbf{r} = \int_C \mathbf{F} \cdot \mathbf{r}'(t)\, dt = \int_a^b \left(\frac{\partial\phi}{\partial x}\frac{dx}{dt} + \frac{\partial\phi}{\partial y}\frac{dy}{dt} \right) dt.$$

In view of the Chain Rule (see (8) in Section 9.4),

$$\frac{d\phi}{dt} = \frac{\partial\phi}{\partial x}\frac{dx}{dt} + \frac{\partial\phi}{\partial y}\frac{dy}{dt},$$

it follows that

$$\int_C \mathbf{F} \cdot d\mathbf{r} = \int_a^b \frac{d\phi}{dt}\, dt = \phi(x(t), y(t)) \Big]_a^b$$

$$= \phi(x(b), y(b)) - \phi(x(a), y(a))$$

$$= \phi(B) - \phi(A).\qquad \equiv$$

For piecewise-smooth curves, the foregoing proof must be modified by considering each smooth arc of the curve C.

Theorem 9.9.1 shows that if $\mathbf{F}$ is a conservative vector field in an open region in 2- or 3-space, then $\int_C \mathbf{F} \cdot d\mathbf{r}$ depends only on the initial and terminal points A and B of the path C, and not on C itself. In other words, line integrals of conservative vector fields are independent of the path. Such integrals are often written

$$\int_A^B \mathbf{F} \cdot d\mathbf{r} = \int_A^B \nabla\phi \cdot d\mathbf{r}. \qquad (3)$$

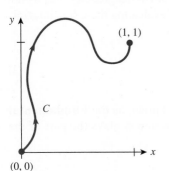

FIGURE 9.9.2 Piecewise-smooth curve in Example 3

EXAMPLE 3 **Example 1 Revisited Again**

Evaluate $\int_C y\,dx + x\,dy$, where C is a path with initial point $(0, 0)$ and terminal point $(1, 1)$.

SOLUTION The path C shown in **FIGURE 9.9.2** represents any piecewise-smooth curve with initial and terminal points $(0, 0)$ and $(1, 1)$. We have just seen that $\mathbf{F} = y\mathbf{i} + x\mathbf{j}$ is a conservative vector field defined at each point of the xy-plane and that $\phi(x, y) = xy$ is a potential function for $\mathbf{F}$. Thus, in view of (2) of Theorem 9.9.1 and (3), we can write

$$\int_C y\,dx + x\,dy = \int_{(0,0)}^{(1,1)} \mathbf{F} \cdot d\mathbf{r} = \int_{(0,0)}^{(1,1)} \nabla\phi \cdot d\mathbf{r}$$

$$= xy \Big]_{(0,0)}^{(1,1)}$$

$$= 1 \cdot 1 - 0 \cdot 0 = 1. \qquad \equiv$$

In using the Fundamental Theorem of Calculus (1), *any* antiderivative of $f'(x)$ can be used, such as $f(x) + K$, where K is a constant. Similarly, a potential function for the vector field in Example 2 is $\phi(x, y) = xy + K$ where K is a constant. We may disregard this constant when using (2) of Theorem 9.9.1 since

$$\int_A^B \mathbf{F} \cdot d\mathbf{r} = (\phi(B) + K) - (\phi(A) + K) = \phi(B) - \phi(A).$$

(a) Connected region R

(b) R is not connected

(c) Multiply connected region R

FIGURE 9.9.3 Regions in the plane

Some Terminology We say that a region (in the plane or in space) is **connected** if every pair of points A and B in the region can be joined by a piecewise-smooth curve that lies entirely in the region. A region R in the plane is **simply connected** if it is connected and every simple closed curve C lying entirely within the region can be shrunk, or contracted, to a point without leaving R. The last condition means that if C is any simple closed curve lying entirely in R, then the region in the interior of C also lies entirely in R. Roughly put, a simply connected region has no holes in it. The region R in **FIGURE 9.9.3(a)** is a simply connected region. In Figure 9.9.3(b) the region R shown is not connected, or **disconnected**, since A and B cannot be joined by a piecewise-smooth curve C that lies in R. The region in Figure 9.9.3(c) is connected but not simply connected because it has three holes in it. The representative curve C in the figure surrounds one of the holes, and so cannot be shrunk to a point without leaving the region. This last region is said to be **multiply connected**. Lastly, a region R is said to be **open** if it contains no boundary points.

In an open connected region R, the notions of path independence and a conservative vector field are equivalent. This means: If $\mathbf{F}$ is conservative in R, then $\int_C \mathbf{F} \cdot d\mathbf{r}$ is independent of the path C, and conversely, if $\int_C \mathbf{F} \cdot d\mathbf{r}$ is independent of the path, then $\mathbf{F}$ is conservative.

We state this formally in the next theorem.

THEOREM 9.9.2 Equivalent Concepts

In an open connected region R, $\int_C \mathbf{F} \cdot d\mathbf{r}$ is independent of the path C if and only if the vector field $\mathbf{F}$ is conservative in R.

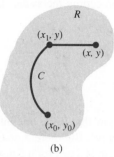

FIGURE 9.9.4 Region R in the proof of Theorem 9.9.2

PROOF: If $\mathbf{F}$ is conservative in R, then we have already seen that $\int_C \mathbf{F} \cdot d\mathbf{r}$ is independent of the path C as a consequence of Theorem 9.9.1.

For convenience we prove the converse for a region R in the plane. Assume that $\int_C \mathbf{F} \cdot d\mathbf{r}$ is independent of the path in R and that (x_0, y_0) and (x, y) are arbitrary points in the region R. Let the function $\phi(x, y)$ be defined as

$$\phi(x, y) = \int_{(x_0, y_0)}^{(x, y)} \mathbf{F} \cdot d\mathbf{r},$$

where C is an arbitrary path in R from (x_0, y_0) to (x, y) and $\mathbf{F} = P\mathbf{i} + Q\mathbf{j}$. See **FIGURE 9.9.4(a)**. Now choose a point (x_1, y), $x_1 \neq x$, so that the line segment from (x_1, y) to (x, y) is in R. See Figure 9.9.4(b). Then by path independence we can write

$$\phi(x, y) = \int_{(x_0, y_0)}^{(x_1, y)} \mathbf{F} \cdot d\mathbf{r} + \int_{(x_1, y)}^{(x, y)} \mathbf{F} \cdot d\mathbf{r}.$$

Now,

$$\frac{\partial \phi}{\partial x} = 0 + \frac{\partial}{\partial x} \int_{(x_1, y)}^{(x, y)} P\, dx + Q\, dy$$

since the first integral does not depend on x. But on the line segment between (x_1, y) and (x, y), y is constant so that $dy = 0$. Hence, $\int_{(x_1, y)}^{(x, y)} P\, dx + Q\, dy = \int_{(x_1, y)}^{(x, y)} P\, dx$. By the derivative form of the Fundamental Theorem of Calculus we then have

$$\frac{\partial \phi}{\partial x} = \frac{\partial}{\partial x} \int_{(x_1, y)}^{(x, y)} P(x, y)\, dx = P(x, y).$$

Likewise we can show that $\partial \phi / \partial y = Q(x, y)$. Hence, from

$$\nabla \phi = \frac{\partial \phi}{\partial x} \mathbf{i} + \frac{\partial \phi}{\partial y} \mathbf{j} = P\mathbf{i} + Q\mathbf{j} = \mathbf{F}(x, y)$$

we conclude that $\mathbf{F}$ is conservative. ≡

▨ Integrals Around Closed Paths Recall from Section 9.8 that a path, or curve, C is said to be closed when its initial point A is the same as the terminal point B. If C is a parametric curve defined by a vector function $\mathbf{r}(t), a \leq t \leq b$, then C is **closed** when $A = B$; that is, $\mathbf{r}(a) = \mathbf{r}(b)$. The next theorem is an immediate consequence of Theorem 9.9.1.

THEOREM 9.9.3 Equivalent Concepts

In an open connected region R, $\int_C \mathbf{F} \cdot d\mathbf{r}$ is independent of the path if and only if $\int_C \mathbf{F} \cdot d\mathbf{r} = 0$ for every closed path C in R.

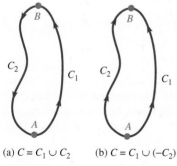

(a) $C = C_1 \cup C_2$ (b) $C = C_1 \cup (-C_2)$

FIGURE 9.9.5 Paths in the proof of Theorem 9.9.3

PROOF: First, we show that if $\int_C \mathbf{F} \cdot d\mathbf{r}$ is independent of the path, then $\int_C \mathbf{F} \cdot d\mathbf{r} = 0$ for every closed path C in R. To see this let us suppose A and B are any two points on C and that $C = C_1 \cup C_2$, where C_1 is a path from A to B and C_2 is a path from B to A. See **FIGURE 9.9.5(a)**. Then

$$\int_C \mathbf{F} \cdot d\mathbf{r} = \int_{C_1} \mathbf{F} \cdot d\mathbf{r} + \int_{C_2} \mathbf{F} \cdot d\mathbf{r} = \int_{C_1} \mathbf{F} \cdot d\mathbf{r} - \int_{-C_2} \mathbf{F} \cdot d\mathbf{r}, \qquad (4)$$

where $-C_2$ is now a path from A to B. Because of path independence $\int_{C_1} \mathbf{F} \cdot d\mathbf{r} = \int_{-C_2} \mathbf{F} \cdot d\mathbf{r}$. Thus (4) implies that $\int_C \mathbf{F} \cdot d\mathbf{r} = 0$.

Next, we prove the converse that if $\int_C \mathbf{F} \cdot d\mathbf{r} = 0$ for every closed path C in R, then $\int_C \mathbf{F} \cdot d\mathbf{r}$ is independent of the path. Let C_1 and C_2 represent any two paths from A to B and so $C = C_1 \cup (-C_2)$ is a closed path. See Figure 9.9.5(b). It follows from $\int_C \mathbf{F} \cdot d\mathbf{r} = 0$ or

$$0 = \int_C \mathbf{F} \cdot d\mathbf{r} = \int_{C_1} \mathbf{F} \cdot d\mathbf{r} + \int_{-C_2} \mathbf{F} \cdot d\mathbf{r} = \int_{C_1} \mathbf{F} \cdot d\mathbf{r} - \int_{C_2} \mathbf{F} \cdot d\mathbf{r}$$

that $\int_{C_1} \mathbf{F} \cdot d\mathbf{r} = \int_{C_2} \mathbf{F} \cdot d\mathbf{r}$. Hence $\int_C \mathbf{F} \cdot d\mathbf{r}$ is independent of the path. ≡

Suppose $\mathbf{F}$ is a conservative vector field defined over an open connected region and C is a closed path lying entirely in the region. When the results of the preceding theorems are put together we conclude that:

$$\mathbf{F} \text{ conservative} \Leftrightarrow \text{path independence} \Leftrightarrow \int_C \mathbf{F} \cdot d\mathbf{r} = 0. \qquad (5)$$

The symbol $\Leftrightarrow$ in (5) is read "equivalent to" or "if and only if."

■ **Test for a Conservative Field** The implications in (5) show that if the line integral $\int_C \mathbf{F} \cdot d\mathbf{r}$ is not path independent, then the vector field is not conservative. But there is an easier way of determining whether $\mathbf{F}$ is conservative. The following theorem is a test for a conservative vector field that uses the partial derivatives of the component functions of $\mathbf{F} = P\mathbf{i} + Q\mathbf{j}$.

THEOREM 9.9.4 Test for a Conservative Field

Suppose $\mathbf{F}(x, y) = P(x, y)\mathbf{i} + Q(x, y)\mathbf{j}$ is a conservative vector field in an open region R, and that P and Q are continuous and have continuous first partial derivatives in R. Then

$$\frac{\partial P}{\partial y} = \frac{\partial Q}{\partial x} \qquad (6)$$

for all (x, y) in R. Conversely, if the equality (6) holds for all (x, y) in a simply connected region R, then $\mathbf{F} = P\mathbf{i} + Q\mathbf{j}$ is conservative in R.

PARTIAL PROOF: We prove the first half of the theorem. Assume that the component functions of the conservative vector field $\mathbf{F} = P\mathbf{i} + Q\mathbf{j}$ are continuous and have continuous first partial derivatives in an open region R. Since $\mathbf{F}$ is conservative there exists a potential function ϕ such that

$$\mathbf{F} = P\mathbf{i} + Q\mathbf{j} = \nabla\phi = \frac{\partial\phi}{\partial x}\mathbf{i} + \frac{\partial\phi}{\partial y}\mathbf{j}.$$

Thus $P = \partial\phi/\partial x$ and $Q = \partial\phi/\partial y$. Now

$$\frac{\partial P}{\partial y} = \frac{\partial}{\partial y}\left(\frac{\partial\phi}{\partial x}\right) = \frac{\partial^2\phi}{\partial y\partial x} \quad \text{and} \quad \frac{\partial Q}{\partial x} = \frac{\partial}{\partial x}\left(\frac{\partial\phi}{\partial y}\right) = \frac{\partial^2\phi}{\partial x\partial y}.$$

From continuity of the partial derivatives we have $\partial P/\partial y = \partial Q/\partial x$ as was to be shown. ≡

EXAMPLE 4	Using Theorem 9.9.4

The conservative vector field $\mathbf{F}(x, y) = y\mathbf{i} + x\mathbf{j}$ in Example 2 is continuous and has component functions that have continuous first partial derivatives throughout the open region R consisting of the entire xy-plane. With the identifications $P = y$ and $Q = x$ it follows from (6) of Theorem 9.9.4,

$$\frac{\partial P}{\partial y} = 1 = \frac{\partial Q}{\partial x}.$$

≡

EXAMPLE 5 **Using Theorem 9.9.4**

Determine whether the vector field $\mathbf{F}(x, y) = (x^2 - 2y^3)\mathbf{i} + (x + 5y)\mathbf{j}$ is conservative.

SOLUTION With $P = x^2 - 2y^3$ and $Q = x + 5y$, we find

$$\frac{\partial P}{\partial y} = -6y^2 \quad \text{and} \quad \frac{\partial Q}{\partial x} = 1.$$

Because $\partial P/\partial y \neq \partial Q/\partial x$ for all points in the plane, it follows from Theorem 9.9.4 that $\mathbf{F}$ is not conservative. ≡

EXAMPLE 6 **Using Theorem 9.9.4**

Determine whether the vector field $\mathbf{F}(x, y) = -ye^{-xy}\mathbf{i} - xe^{-xy}\mathbf{j}$ is conservative.

SOLUTION With $P = -ye^{-xy}$ and $Q = -xe^{-xy}$, we find

$$\frac{\partial P}{\partial y} = xye^{-xy} - e^{-xy} = \frac{\partial Q}{\partial x}.$$

The components of $\mathbf{F}$ are continuous and have continuous partial derivatives. Thus (6) holds throughout the xy-plane, which is a simply connected region. From the converse in Theorem 9.9.4 we can conclude that $\mathbf{F}$ is conservative. ≡

We have one more important question to answer in this section:

> *If $\mathbf{F}$ is a conservative vector field, how does one find a potential function ϕ for $\mathbf{F}$?* (7)

In the next example we give the answer to the question posed in (7).

EXAMPLE 7 **Integral That Is Path Independent**

(a) Show that $\int_C \mathbf{F} \cdot d\mathbf{r}$, where $\mathbf{F}(x, y) = (y^2 - 6xy + 6)\mathbf{i} + (2xy - 3x^2 - 2y)\mathbf{j}$, is independent of the path C between $(-1, 0)$ and $(3, 4)$.

(b) Find a potential function ϕ for $\mathbf{F}$.

(c) Evaluate $\displaystyle\int_{(-1,0)}^{(3,4)} \mathbf{F} \cdot d\mathbf{r}$.

SOLUTION **(a)** Identifying $P = y^2 - 6xy + 6$ and $Q = 2xy - 3x^2 - 2y$ yields

$$\frac{\partial P}{\partial y} = 2y - 6x = \frac{\partial Q}{\partial x}.$$

The vector field $\mathbf{F}$ is conservative because (6) holds throughout the xy-plane and as a consequence the integral $\int_C \mathbf{F} \cdot d\mathbf{r}$ is independent of the path between any two points A and B in the plane.

(b) Because $\mathbf{F}$ is conservative there is a potential function ϕ such that

$$\frac{\partial \phi}{\partial x} = y^2 - 6xy + 6 \quad \text{and} \quad \frac{\partial \phi}{\partial y} = 2xy - 3x^2 - 2y.$$ (8)

> In partial integration with respect to x, the variable y is treated as a constant. Similarly, in partial integration with respect to y we treat x as a constant.

Employing partial integration on the first expression in (8) gives

$$\phi = \int (y^2 - 6xy + 6)\, dx = xy^2 - 3x^2y + 6x + g(y),$$ (9)

where $g(y)$ is the "constant" of integration. Now we take the partial derivative of (9) with respect to y and equate it to the second expression in (8):

$$\frac{\partial \phi}{\partial y} = 2xy - 3x^2 + g'(y) = 2xy - 3x^2 - 2y.$$

From the last equality we find $g'(y) = -2y$. Integrating again gives $g(y) = -y^2 + C$, where C is a constant. Thus

$$\phi = xy^2 - 3x^2y + 6x - y^2 + C. \qquad (10)$$

(c) We can now use Theorem 9.9.2 and the potential function (10) (without the constant) to evaluate the line integral:

$$\int_C \mathbf{F} \cdot d\mathbf{r} = \int_{(-1,0)}^{(3,4)} \mathbf{F} \cdot d\mathbf{r} = (xy^2 - 3x^2y + 6x - y^2) \Big]_{(-1,0)}^{(3,4)}$$

$$= (48 - 108 + 18 - 16) - (-6) = -52. \qquad \equiv$$

Note: Since the integral in Example 7 was shown to be independent of the path in part (a), we can evaluate it without finding a potential function. We can integrate along *any convenient curve C* connecting the given points. In particular, the line $y = x + 1$ is such a curve. Using x as a parameter then gives

$$\int_C \mathbf{F} \cdot d\mathbf{r} = \int_C (y^2 - 6xy + 6)\,dx + (2xy - 3x^2 - 2y)\,dy$$

$$= \int_{-1}^{3} [(x+1)^2 - 6x(x+1) + 6]\,dx + [2x(x+1) - 3x^2 - 2(x+1)]\,dx$$

$$= \int_{-1}^{3} (-6x^2 - 4x + 5)\,dx = -52.$$

■ **Conservative Vector Fields in 3-Space** For a three-dimensional conservative vector field

$$\mathbf{F}(x, y, z) = P(x, y, z)\mathbf{i} + Q(x, y, z)\mathbf{j} + R(x, y, z)\mathbf{k}$$

and a piecewise-smooth space curve $\mathbf{r}(t) = x(t)\mathbf{i} + y(t)\mathbf{j} + z(t)\mathbf{k}$, $a \le t \le b$, the basic form of (2) is the same:

$$\int_C \mathbf{F} \cdot d\mathbf{r} = \int_C \nabla\phi \cdot d\mathbf{r} = \phi(x(b), y(b), z(b)) - \phi(x(a), y(a), z(a)) = \phi(B) - \phi(A). \qquad (11)$$

If C is a space curve, a line integral $\int_C \mathbf{F} \cdot d\mathbf{r}$ is independent of the path whenever the three-dimensional vector field

$$\mathbf{F}(x, y, z) = P(x, y, z)\mathbf{i} + Q(x, y, z)\mathbf{j} + R(x, y, z)\mathbf{k}$$

is conservative. The three-dimensional analogue of Theorem 9.9.4 goes like this: If $\mathbf{F}$ is conservative and P, Q, and R are continuous and have continuous first partial derivatives in some open region of 3-space, then

$$\frac{\partial P}{\partial y} = \frac{\partial Q}{\partial x}, \quad \frac{\partial P}{\partial z} = \frac{\partial R}{\partial x}, \quad \frac{\partial Q}{\partial z} = \frac{\partial R}{\partial y}. \qquad (12)$$

Conversely, if (12) holds throughout an appropriate region of 3-space, then $\mathbf{F}$ is conservative.

The necessity of (12) can be seen from (5) of Section 9.7. If $\mathbf{F}$ is conservative then $\mathbf{F} = \nabla\phi$ and curl $\nabla\phi = $ curl $\mathbf{F} = \mathbf{0}$; that is,

$$\text{curl } \mathbf{F} = \left(\frac{\partial R}{\partial y} - \frac{\partial Q}{\partial z}\right)\mathbf{i} + \left(\frac{\partial P}{\partial z} - \frac{\partial R}{\partial x}\right)\mathbf{j} + \left(\frac{\partial Q}{\partial x} - \frac{\partial P}{\partial y}\right)\mathbf{k} = \mathbf{0}.$$

Setting the three components of the vector curl $\mathbf{F}$ equal to 0 yields (12).

EXAMPLE 8 **Integral That Is Path Independent**

(a) Show that the line integral

$$\int_C (y + yz)\,dx + (x + 3z^3 + xz)\,dy + (9yz^2 + xy - 1)\,dz$$

is independent of the path C between $(1, 1, 1)$ to $(2, 1, 4)$.

(b) Evaluate $\displaystyle\int_{(1,1,1)}^{(2,1,4)} \mathbf{F} \cdot d\mathbf{r}$.

SOLUTION **(a)** With the identifications

$$\mathbf{F}(x, y, z) = (y + yz)\mathbf{i} + (x + 3z^3 + xz)\mathbf{j} + (9yz^2 + xy - 1)\mathbf{k},$$

$$P = y + yz, \quad Q = x + 3z^3 + xz, \quad \text{and} \quad R = 9yz^2 + xy - 1,$$

we see that the equalities

$$\frac{\partial P}{\partial y} = 1 + z = \frac{\partial Q}{\partial x}, \quad \frac{\partial P}{\partial z} = y = \frac{\partial R}{\partial x}, \quad \text{and} \quad \frac{\partial Q}{\partial z} = 9z^2 + x = \frac{\partial R}{\partial y}$$

hold throughout 3-space. From (12) we conclude that $\mathbf{F}$ is conservative and therefore the integral is independent of the path.

(b) The path C shown in **FIGURE 9.9.6** represents any path with initial and terminal points $(1, 1, 1)$ and $(2, 1, 4)$. To evaluate the integral we again illustrate how to find a potential function $\phi(x, y, z)$ for $\mathbf{F}$ using partial integration.

First we know that

$$\frac{\partial \phi}{\partial x} = P, \quad \frac{\partial \phi}{\partial y} = Q, \quad \text{and} \quad \frac{\partial \phi}{\partial z} = R.$$

Integrating the first of these three equations with respect to x gives

$$\phi = xy + xyz + g(y, z).$$

The derivative of this last expression with respect to y must then be equal to Q:

$$\frac{\partial \phi}{\partial y} = x + xz + \frac{\partial g}{\partial y} = x + 3z^3 + xz.$$

Hence,

$$\frac{\partial g}{\partial y} = 3z^3 \quad \text{implies} \quad g = 3yz^3 + h(z).$$

Consequently, $\phi = xy + xyz + 3yz^3 + h(z).$

The partial derivative of this last expression with respect to z must now be equal to the function R:

$$\frac{\partial \phi}{\partial z} = xy + 9yz^2 + h'(z) = 9yz^2 + xy - 1.$$

From this we get $h'(z) = -1$ and $h(z) = -z + K$. Disregarding K, we can write

$$\phi = xy + xyz + 3yz^3 - z. \tag{13}$$

Finally, from (11) and the potential function (13) we obtain

$$\int_{(1,1,1)}^{(2,1,4)} (y + yz)\,dx + (x + 3z^3 + xz)\,dy + (9yz^2 + xy - 1)\,dz$$

$$= (xy + xyz + 3yz^3 - z)\Big]_{(1,1,1)}^{(2,1,4)} = 198 - 4 = 194. \qquad \equiv$$

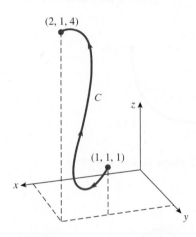

FIGURE 9.9.6 Representative path C in Example 8

Conservation of Energy In a conservative force field **F**, the work done by the force on a particle moving from position A to position B is the same for all paths between these points. Moreover, the work done by the force along a closed path is *zero*. See Problem 29 in Exercises 9.9. For this reason, such a force field is also said to be **conservative**. In a conservative field **F** the **law of conservation of mechanical energy** holds:

> *For a particle moving along a path in a conservative field,*
> *kinetic energy + potential energy = constant.*

See Problem 31 in Exercises 9.9.

A frictional force such as air resistance is **nonconservative**. Nonconservative forces are *dissipative* in that their action reduces kinetic energy without a corresponding increase in potential energy. Stated in another way, if the work done $\int_C \mathbf{F} \cdot d\mathbf{r}$ depends on the path, then **F** is nonconservative.

9.9 Exercises Answers to selected odd-numbered problems begin on page ANS-25.

In Problems 1–10, show that the given line integral is independent of the path. Evaluate in two ways: **(a)** Find a potential function ϕ and then use Theorem 9.9.1, and **(b)** Use any convenient path between the endpoints of the path.

1. $\displaystyle\int_{(0,0)}^{(2,2)} x^2\, dx + y^2\, dy$

2. $\displaystyle\int_{(1,1)}^{(2,4)} 2xy\, dx + x^2\, dy$

3. $\displaystyle\int_{(1,0)}^{(3,2)} (x + 2y)\, dx + (2x - y)\, dy$

4. $\displaystyle\int_{(0,0)}^{(\pi/2,0)} \cos x \cos y\, dx + (1 - \sin x \sin y)\, dy$

5. $\displaystyle\int_{(4,1)}^{(4,4)} \frac{-y\, dx + x\, dy}{y^2}$ on any path not crossing the x-axis

6. $\displaystyle\int_{(1,0)}^{(3,4)} \frac{x\, dx + y\, dy}{\sqrt{x^2 + y^2}}$ on any path not through the origin

7. $\displaystyle\int_{(1,2)}^{(3,6)} (2y^2x - 3)\, dx + (2yx^2 + 4)\, dy$

8. $\displaystyle\int_{(-1,1)}^{(0,0)} (5x + 4y)\, dx + (4x - 8y^3)\, dy$

9. $\displaystyle\int_{(0,0)}^{(2,8)} (y^3 + 3x^2y)\, dx + (x^3 + 3y^2x + 1)\, dy$

10. $\displaystyle\int_{(-2,0)}^{(1,0)} (2x - y \sin xy - 5y^4)\, dx - (20xy^3 + x \sin xy)\, dy$

In Problems 11–16, determine whether the given vector field is a conservative field. If so, find a potential function ϕ for **F**.

11. $\mathbf{F}(x, y) = (4x^3y^3 + 3)\mathbf{i} + (3x^4y^2 + 1)\mathbf{j}$

12. $\mathbf{F}(x, y) = 2xy^3\mathbf{i} + 3y^2(x^2 + 1)\mathbf{j}$

13. $\mathbf{F}(x, y) = y^2 \cos xy^2\mathbf{i} - 2xy \sin xy^2\mathbf{j}$

14. $\mathbf{F}(x, y) = (x^2 + y^2 + 1)^{-2}(x\mathbf{i} + y\mathbf{j})$

15. $\mathbf{F}(x, y) = (x^3 + y)\mathbf{i} + (x + y^3)\mathbf{j}$

16. $\mathbf{F}(x, y) = 2e^{2y}\mathbf{i} + xe^{2y}\mathbf{j}$

In Problems 17 and 18, find the work done by the force $\mathbf{F}(x, y) = (2x + e^{-y})\mathbf{i} + (4y - xe^{-y})\mathbf{j}$ along the indicated curve.

17.

FIGURE 9.9.7 Curve for Problem 17

18.

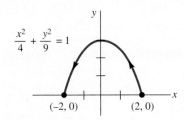

FIGURE 9.9.8 Curve for Problem 18

In Problems 19–24, show that the given integral is independent of the path. Evaluate.

19. $\displaystyle\int_{(1,1,1)}^{(2,4,8)} yz\, dx + xz\, dy + xy\, dz$

20. $\displaystyle\int_{(0,0,0)}^{(1,1,1)} 2x\, dx + 3y^2\, dy + 4z^3\, dz$

21. $\displaystyle\int_{(1,0,0)}^{(2,\pi/2,1)} (2x \sin y + e^{3z})\, dx + x^2 \cos y\, dy + (3xe^{3z} + 5)\, dz$

22. $\displaystyle\int_{(1,2,1)}^{(3,4,1)} (2x + 1)\, dx + 3y^2\, dy + \frac{1}{z}\, dz$

23. $\displaystyle\int_{(1,1,\ln 3)}^{(2,2,\ln 3)} e^{2z}\, dx + 3y^2\, dy + 2xe^{2z}\, dz$

24. $\displaystyle\int_{(-2,\,3,\,1)}^{(0,\,0,\,0)} 2xz\,dx + 2yz\,dy + (x^2 + y^2)\,dz$

In Problems 25 and 26, evaluate $\int_C \mathbf{F} \cdot d\mathbf{r}$.

25. $\mathbf{F}(x, y, z) = (y - yz\sin x)\mathbf{i} + (x + z\cos x)\mathbf{j} + y\cos x\,\mathbf{k}$;
$\mathbf{r}(t) = 2t\mathbf{i} + (1 + \cos t)^2\mathbf{j} + 4\sin^3 t\,\mathbf{k}, \ 0 \le t \le \pi/2$

26. $\mathbf{F}(x, y, z) = (2 - e^z)\mathbf{i} + (2y - 1)\mathbf{j} + (2 - xe^z)\mathbf{k}$;
$\mathbf{r}(t) = t\mathbf{i} + t^2\mathbf{j} + t^3\mathbf{k}, \ (-1, 1, -1)$ to $(2, 4, 8)$

27. The inverse square law of gravitational attraction between two masses m_1 and m_2 is given by $\mathbf{F} = -Gm_1m_2\,\mathbf{r}/\|\mathbf{r}\|^3$, where $\mathbf{r} = x\mathbf{i} + y\mathbf{j} + z\mathbf{k}$. Show that $\mathbf{F}$ is conservative. Find a potential function for $\mathbf{F}$.

28. Find the work done by the force $\mathbf{F}(x, y, z) = 8xy^3z\mathbf{i} + 12x^2y^2z\mathbf{j} + 4x^2y^3\mathbf{k}$ acting along the helix $\mathbf{r}(t) = 2\cos t\mathbf{i} + 2\sin t\mathbf{j} + t\mathbf{k}$ from $(2, 0, 0)$ to $(1, \sqrt{3}, \pi/3)$. From $(2, 0, 0)$ to $(0, 2, \pi/2)$. [*Hint:* Show that $\mathbf{F}$ is conservative.]

29. If $\mathbf{F}$ is a conservative force field, show that the work done along any simple closed path is zero.

30. A particle in the plane is attracted to the origin with a force $\mathbf{F} = \|\mathbf{r}\|^n\mathbf{r}$, where n is a positive integer and $\mathbf{r} = x\mathbf{i} + y\mathbf{j}$ is the position vector of the particle. Show that $\mathbf{F}$ is conservative. Find the work done in moving the particle between (x_1, y_1) and (x_2, y_2).

31. Suppose $\mathbf{F}$ is a conservative force field with potential function ϕ. In physics the function $p = -\phi$ is called **potential energy**. Since $\mathbf{F} = -\nabla p$, Newton's second law becomes

$$m\mathbf{r}'' = -\nabla p \quad \text{or} \quad m\frac{d\mathbf{v}}{dt} + \nabla p = \mathbf{0}.$$

By integrating $m\dfrac{d\mathbf{v}}{dt}\cdot\dfrac{d\mathbf{r}}{dt} + \nabla p\cdot\dfrac{d\mathbf{r}}{dt} = 0$ with respect to t, derive the law of conservation of mechanical energy: $\frac{1}{2}mv^2 + p = \text{constant}$. [*Hint:* See Problem 39 in Exercises 9.8.]

32. Suppose that C is a smooth curve between points A (at $t = a$) and B (at $t = b$) and that p is potential energy, defined in Problem 31. If $\mathbf{F}$ is a conservative force field and $K = \frac{1}{2}mv^2$ is kinetic energy, show that $p(B) + K(B) = p(A) + K(A)$.

9.10 Double Integrals

INTRODUCTION In Section 9.8 we gave the five steps leading to the definition of the definite integral $\int_a^b f(x)\,dx$. The analogous steps that lead to the concept of the *two-dimensional definite integral*, known simply as the **double integral** of a function f of two variables, will be given next.

1. Let $z = f(x, y)$ be defined in a closed and bounded region R of 2-space.
2. By means of a grid of vertical and horizontal lines parallel to the coordinate axes, form a partition P of R into n rectangular subregions R_k of areas ΔA_k that lie entirely in R.
3. Let $\|P\|$ be the **norm** of the partition or the length of the longest diagonal of the R_k.
4. Choose a sample point (x_k^*, y_k^*) in each subregion R_k. See **FIGURE 9.10.1**.
5. Form the sum $\displaystyle\sum_{k=1}^{n} f(x_k^*, y_k^*)\,\Delta A_k$.

Thus we have the following definition:

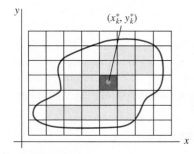

FIGURE 9.10.1 Sample point in kth rectangle

DEFINITION 9.10.1 The Double Integral

Let f be a function of two variables defined on a closed region R of 2-space. Then the **double integral of f over R** is given by

$$\iint_R f(x, y)\,dA = \lim_{\|P\|\to 0}\sum_{k=1}^{n} f(x_k^*, y_k^*)\,\Delta A_k. \tag{1}$$

■ **Integrability** If the limit in (1) exists, we say that f is **integrable** over R and that R is the **region of integration**. When f is continuous on R, then f is necessarily integrable over R.

■ **Area** When $f(x, y) = 1$ on R, then $\lim_{\|P\|\to 0}\sum_{k=1}^{n}\Delta A_k$ simply gives the **area** A of the region; that is,

$$A = \iint_R dA. \tag{2}$$

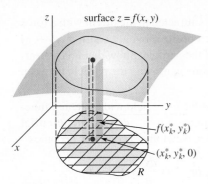

FIGURE 9.10.2 Volume under a surface

Volume If $f(x, y) \geq 0$ on R, then, as shown in **FIGURE 9.10.2**, the product $f(x_k^*, y_k^*)\, \Delta A_k$ can be interpreted as the volume of a rectangular prism of height $f(x_k^*, y_k^*)$ and base of area ΔA_k. The summation of volumes $\sum_{k=1}^{n} f(x_k^*, y_k^*)\, \Delta A_k$ is an approximation to the **volume** V of the solid *above* the region R and *below* the surface $z = f(x, y)$. The limit of this sum as $\|P\| \to 0$, if it exists, gives the exact volume of this solid; that is, if f is nonnegative on R, then

$$V = \iint\limits_{R} f(x, y)\, dA. \tag{3}$$

Properties The double integral possesses the following properties:

> **THEOREM 9.10.1** Properties of Double Integrals
>
> Let f and g be functions of two variables that are integrable over a region R. Then
>
> (*i*) $\displaystyle\iint\limits_{R} kf(x, y)\, dA = k \iint\limits_{R} f(x, y)\, dA$, where k is any constant
>
> (*ii*) $\displaystyle\iint\limits_{R} [f(x, y) \pm g(x, y)]\, dA = \iint\limits_{R} f(x, y)\, dA \pm \iint\limits_{R} g(x, y)\, dA$
>
> (*iii*) $\displaystyle\iint\limits_{R} f(x, y)\, dA = \iint\limits_{R_1} f(x, y)\, dA + \iint\limits_{R_2} f(x, y)\, dA$, where R_1 and R_2 are subregions of R
>
> that do not overlap and $R = R_1 \cup R_2$.

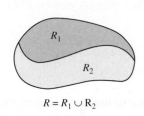

$R = R_1 \cup R_2$

FIGURE 9.10.3 R is the union of two regions

Part (*iii*) of Theorem 9.10.1 is the two-dimensional equivalent of $\int_a^b f(x)\, dx = \int_a^c f(x)\, dx + \int_c^b f(x)\, dx$. **FIGURE 9.10.3** illustrates the division of a region into subregions R_1 and R_2 for which $R = R_1 \cup R_2$. R_1 and R_2 can have no points in common except possibly on their common border. Furthermore, Theorem 9.10.1(*iii*) extends to any finite number of nonoverlapping subregions whose union is R.

Regions of Type I and II The region shown in **FIGURE 9.10.4(a)**,

$$R: a \leq x \leq b, \quad g_1(x) \leq y \leq g_2(x),$$

where the boundary functions g_1 and g_2 are continuous, is called a **region of Type I**. In Figure 9.10.4(b), the region

$$R: c \leq y \leq d, \quad h_1(y) \leq x \leq h_2(y),$$

where the functions h_1 and h_2 are continuous, is called a **region of Type II**.

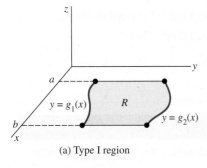

(a) Type I region

Iterated Integrals Since the partial integral $\int_{g_1(x)}^{g_2(x)} f(x, y)\, dy$ is a function of x alone, we may in turn integrate the resulting function now with respect to x. If f is continuous on a region of Type I, we define an **iterated integral** of f over the region by

$$\int_a^b \int_{g_1(x)}^{g_2(x)} f(x, y)\, dy\, dx = \int_a^b \left[\int_{g_1(x)}^{g_2(x)} f(x, y)\, dy \right] dx. \tag{4}$$

The basic idea in (4) is to carry out *successive integrations*. The partial integral with respect to y gives a function of x, which is then integrated in the usual manner from $x = a$ to $x = b$. The end result of both integrations will be a real number. In a similar manner, we define an iterated integral of a continuous function f over a region of Type II by

(b) Type II region

FIGURE 9.10.4 Regions of integration

$$\int_c^d \int_{h_1(y)}^{h_2(y)} f(x, y)\, dx\, dy = \int_c^d \left[\int_{h_1(y)}^{h_2(y)} f(x, y)\, dx \right] dy. \tag{5}$$

■ **Evaluation of Double Integrals** Iterated integrals provide the means for evaluating a double integral $\iint_R f(x, y)\, dA$ over a region of Type I or Type II or a region that can be expressed as a union of a finite number of these regions. The following result is due to the Italian mathematician **Guido Fubini** (1879–1943).

THEOREM 9.10.2 Fubini's Theorem

Let f be continuous on a region R.
(*i*) If R is of Type I, then

$$\iint_R f(x, y)\, dA = \int_a^b \int_{g_1(x)}^{g_2(x)} f(x, y)\, dy\, dx. \tag{6}$$

(*ii*) If R is of Type II, then

$$\iint_R f(x, y)\, dA = \int_c^d \int_{h_1(y)}^{h_2(y)} f(x, y)\, dx\, dy. \tag{7}$$

Theorem 9.10.2 is the double integral analogue of the Fundamental Theorem of Calculus. While Theorem 9.10.2 is difficult to prove, we can get some intuitive feeling for its significance by considering volumes. Let R be a Type I region and $z = f(x, y)$ be continuous and nonnegative on R. The area A of the vertical plane, as shown in **FIGURE 9.10.5**, is the area under the trace of the surface $z = f(x, y)$ in the plane $x = $ constant and hence is given by the partial integral

$$A(x) = \int_{g_1(x)}^{g_2(x)} f(x, y)\, dy.$$

By summing all these areas from $x = a$ to $x = b$, we obtain the volume V of the solid above R and below the surface:

$$V = \int_a^b A(x)\, dx = \int_a^b \int_{g_1(x)}^{g_2(x)} f(x, y)\, dy\, dx.$$

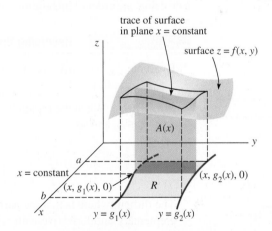

FIGURE 9.10.5 Geometric interpretation of (6)

But, as we have already seen in (3), this volume is also given by the double integral

$$V = \iint_R f(x, y)\, dA.$$

EXAMPLE 1 **Evaluation of a Double Integral**

Evaluate the double integral $\iint_R e^{x+3y}\, dA$ over the region bounded by the graphs of $y = 1$, $y = 2$, $y = x$, and $y = -x + 5$. See **FIGURE 9.10.6**.

FIGURE 9.10.6 Region of integration in Example 1

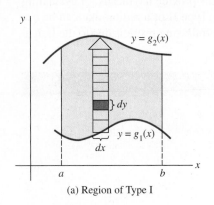

(a) Region of Type I

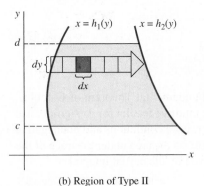

(b) Region of Type II

FIGURE 9.10.7 Summation in y-direction in (a); summation in x-direction in (b)

(a) Type I region

(b) Type II region

FIGURE 9.10.8 Reversing order of integration in Example 2

SOLUTION As seen in the figure, the region is of Type II; hence, by (7) we integrate first with respect to x from the left boundary $x = y$ to the right boundary $x = 5 - y$:

$$\iint\limits_{R} e^{x+3y}\, dA = \int_{1}^{2}\int_{y}^{5-y} e^{x+3y}\, dx\, dy$$

$$= \int_{1}^{2} e^{x+3y}\Big]_{y}^{5-y}\, dy$$

$$= \int_{1}^{2} (e^{5+2y} - e^{4y})\, dy = \left[\frac{1}{2}e^{5+2y} - \frac{1}{4}e^{4y}\right]_{1}^{2}$$

$$= \frac{1}{2}e^{9} - \frac{1}{4}e^{8} - \frac{1}{2}e^{7} + \frac{1}{4}e^{4} \approx 2771.64. \qquad \equiv$$

As an aid in reducing a double integral to an iterated integral with correct limits of integration, it is useful to visualize, as suggested in the foregoing discussion, the double integral as a double summation process. Over a Type I region the iterated integral $\int_{a}^{b}\int_{g_{1}(x)}^{g_{2}(x)} f(x, y)\, dy\, dx$ is first a summation in the y-direction. Pictorially, this is indicated by the vertical arrow in **FIGURE 9.10.7(a)**; the typical rectangle in the arrow has area $dy\, dx$. The dy placed before the dx signifies that the "volumes" $f(x, y)\, dy\, dx$ of prisms built up on the rectangles are summed vertically with respect to y from the lower boundary curve g_{1} to the upper boundary curve g_{2}. The dx following the dy signifies that the result of each vertical summation is then summed horizontally with respect to x from left ($x = a$) to right ($x = b$). Similar remarks hold for double integrals over regions of Type II. See Figure 9.10.7(b). Recall from (2) that when $f(x, y) = 1$, the double integral $A = \iint_{R} dA$ gives the area of the region. Thus, Figure 9.10.7(a) shows that $\int_{a}^{b}\int_{g_{1}(x)}^{g_{2}(x)} dy\, dx$ adds the rectangular areas vertically and then horizontally, whereas Figure 9.10.7(b) shows that $\int_{c}^{d}\int_{h_{1}(y)}^{h_{2}(y)} dx\, dy$ adds the rectangular areas horizontally and then vertically.

■ **Reversing the Order of Integration** A problem may become easier when the order of integration is **changed** or **reversed**. Also, some iterated integrals that may be impossible to evaluate using one order of integration can perhaps be evaluated using the reverse order of integration.

EXAMPLE 2 Reversing the Order of Integration

Evaluate $\iint_{R} xe^{y^{2}}\, dA$ over the region R in the first quadrant bounded by the graphs of $y = x^{2}, x = 0, y = 4$.

SOLUTION When the region is viewed as Type I, we have from **FIGURE 9.10.8(a)**, $0 \leq x \leq 2$, $x^{2} \leq y \leq 4$, and so

$$\iint\limits_{R} xe^{y^{2}}\, dA = \int_{0}^{2}\int_{x^{2}}^{4} xe^{y^{2}}\, dy\, dx.$$

The difficulty here is that the partial integral $\int_{x^{2}}^{4} xe^{y^{2}}\, dy$ cannot be evaluated, since $e^{y^{2}}$ has no elementary antiderivative with respect to y. However, as we see in Figure 9.10.8(b), we can interpret the same region as a Type II region defined by $0 \leq y \leq 4, 0 \leq x \leq \sqrt{y}$. Hence, from (7)

$$\iint\limits_{R} xe^{y^{2}}\, dA = \int_{0}^{4}\int_{0}^{\sqrt{y}} xe^{y^{2}}\, dx\, dy$$

$$= \int_{0}^{4} \frac{x^{2}}{2}e^{y^{2}}\Big]_{0}^{\sqrt{y}}\, dy$$

$$= \int_{0}^{4} \frac{1}{2} ye^{y^{2}}\, dy = \frac{1}{4}e^{y^{2}}\Big]_{0}^{4} = \frac{1}{4}(e^{16} - 1). \qquad \equiv$$

Laminas with Variable Density—Center of Mass

If ρ is a constant density (mass per unit area), then the mass of the lamina coinciding with a region bounded by the graphs of $y = f(x)$, the x-axis, and the lines $x = a$ and $x = b$ is

$$m = \lim_{\|P\| \to 0} \sum_{k=1}^{n} \rho\, f(x_k^*)\, \Delta x_k = \int_a^b \rho\, f(x)\, dx. \tag{8}$$

If a lamina corresponding to a region R has a *variable density* $\rho(x, y)$, where ρ is nonnegative and continuous on R, then analogous to (8) we define its **mass** m by the double integral

$$m = \lim_{\|P\| \to 0} \sum_{k=1}^{n} \rho(x_k^*, y_k^*)\, \Delta A_k = \iint\limits_R \rho(x, y)\, dA. \tag{9}$$

The coordinates of the **center of mass** of the lamina are then

$$\bar{x} = \frac{M_y}{m}, \quad \bar{y} = \frac{M_x}{m}, \tag{10}$$

where

$$M_y = \iint\limits_R x\rho(x, y)\, dA \quad \text{and} \quad M_x = \iint\limits_R y\rho(x, y)\, dA \tag{11}$$

are the **moments** of the lamina about the y- and x-axes, respectively. The center of mass is the point where we consider all the mass of the lamina to be concentrated. If $\rho(x, y)$ is a constant, the center of mass is called the **centroid** of the lamina.

EXAMPLE 3 Center of Mass

A lamina has the shape of the region in the first quadrant that is bounded by the graphs of $y = \sin x$, $y = \cos x$, between $x = 0$ and $x = \pi/4$. Find its center of mass if the density is $\rho(x, y) = y$.

SOLUTION From **FIGURE 9.10.9** we see that

$$
\begin{aligned}
m &= \iint\limits_R y\, dA = \int_0^{\pi/4} \int_{\sin x}^{\cos x} y\, dy\, dx \\
&= \int_0^{\pi/4} \frac{y^2}{2} \bigg]_{\sin x}^{\cos x} dx \\
&= \frac{1}{2} \int_0^{\pi/4} (\cos^2 x - \sin^2 x)\, dx \quad \leftarrow \text{double angle formula} \\
&= \frac{1}{2} \int_0^{\pi/4} \cos 2x\, dx = \frac{1}{4} \sin 2x \bigg]_0^{\pi/4} = \frac{1}{4}.
\end{aligned}
$$

Now,

$$
\begin{aligned}
M_y &= \iint\limits_R xy\, dA = \int_0^{\pi/4} \int_{\sin x}^{\cos x} xy\, dy\, dx \\
&= \int_0^{\pi/4} \frac{1}{2} xy^2 \bigg]_{\sin x}^{\cos x} dx \\
&= \frac{1}{2} \int_0^{\pi/4} x \cos 2x\, dx \quad \leftarrow \text{integration by parts} \\
&= \left[\frac{1}{4} x \sin 2x + \frac{1}{8} \cos 2x \right]_0^{\pi/4} = \frac{\pi - 2}{16}.
\end{aligned}
$$

FIGURE 9.10.9 Region in Example 3

Similarly,

$$M_x = \iint\limits_R y^2 \, dA = \int_0^{\pi/4} \int_{\sin x}^{\cos x} y^2 \, dy \, dx$$

$$= \frac{1}{3} \int_0^{\pi/4} (\cos^3 x - \sin^3 x) \, dx$$

$$= \frac{1}{3} \int_0^{\pi/4} [\cos x (1 - \sin^2 x) - \sin x (1 - \cos^2 x)] \, dx$$

$$= \frac{1}{3} \left[\sin x - \frac{1}{3} \sin^3 x + \cos x - \frac{1}{3} \cos^3 x \right]_0^{\pi/4} = \frac{5\sqrt{2} - 4}{18}.$$

Hence, from (10),

$$\bar{x} = \frac{M_y}{m} = \frac{(\pi - 2)/16}{1/4} \approx 0.29 \quad \text{and} \quad \bar{y} = \frac{M_x}{m} = \frac{(5\sqrt{2} - 4)/18}{1/4} \approx 0.68.$$

Thus the center of mass has the approximate coordinates $(0.29, 0.68)$. ≡

■ **Moments of Inertia** The integrals M_x and M_y in (11) are also called the **first moments** of a lamina about the x-axis and y-axis, respectively. The so-called **second moments** of a lamina or **moments of inertia** about the x- and y-axes are, in turn, defined by the double integrals

$$I_x = \iint\limits_R y^2 \rho(x, y) \, dA \quad \text{and} \quad I_y = \iint\limits_R x^2 \rho(x, y) \, dA. \tag{12}$$

A moment of inertia is the rotational equivalent of mass. For translational motion, kinetic energy is given by $K = \frac{1}{2} mv^2$, where m is mass and v is linear speed. The kinetic energy of a particle of mass m rotating at a distance r from an axis is $K = \frac{1}{2} mv^2 = \frac{1}{2} m(r\omega)^2 = \frac{1}{2} (mr^2)\omega^2 = \frac{1}{2} I\omega^2$, where $I = mr^2$ is its moment of inertia about the axis of rotation and ω is angular speed.

EXAMPLE 4 **Moment of Inertia**

Find the moment of inertia about the y-axis of the thin homogeneous disk of mass m shown in **FIGURE 9.10.10**.

SOLUTION Since the disk is homogeneous, its density is the constant $\rho(x, y) = m/\pi r^2$. Hence, from (12),

FIGURE 9.10.10 Disk in Example 4

$$I_y = \iint\limits_R x^2 \left(\frac{m}{\pi r^2} \right) dA = \frac{m}{\pi r^2} \int_{-r}^{r} \int_{-\sqrt{r^2 - x^2}}^{\sqrt{r^2 - x^2}} x^2 \, dy \, dx$$

$$= \frac{2m}{\pi r^2} \int_{-r}^{r} x^2 \sqrt{r^2 - x^2} \, dx \quad \leftarrow \text{trigonometric substitution } x = r \sin \theta$$

$$= \frac{2mr^2}{\pi} \int_{-\pi/2}^{\pi/2} \sin^2 \theta \cos^2 \theta \, d\theta$$

$$= \frac{mr^2}{2\pi} \int_{-\pi/2}^{\pi/2} \sin^2 2\theta \, d\theta$$

$$= \frac{mr^2}{4\pi} \int_{-\pi/2}^{\pi/2} (1 - \cos 4\theta) \, d\theta = \frac{1}{4} mr^2.$$ ≡

Radius of Gyration The **radius of gyration** of a lamina of mass m and the moment of inertia I about an axis is defined by

$$R_g = \sqrt{\frac{I}{m}}. \tag{13}$$

Since (13) implies that $I = mR_g^2$, the radius of gyration is interpreted as the radial distance the lamina, considered as a point mass, can rotate about the axis without changing the rotational inertia of the body. In Example 4 the radius of gyration is $R_g = \sqrt{I_y/m} = \sqrt{(mr^2/4)/m} = r/2$.

9.10 Exercises Answers to selected odd-numbered problems begin on page ANS-25.

In Problems 1–8, evaluate the given partial integral.

1. $\displaystyle\int_{-1}^{3} (6xy - 5e^y)\, dx$

2. $\displaystyle\int_{1}^{2} \tan xy\, dy$

3. $\displaystyle\int_{1}^{3x} x^3 e^{xy}\, dy$

4. $\displaystyle\int_{\sqrt{y}}^{y^3} (8x^3 y - 4xy^2)\, dx$

5. $\displaystyle\int_{0}^{2x} \frac{xy}{x^2 + y^2}\, dy$

6. $\displaystyle\int_{x^3}^{x} e^{2y/x}\, dy$

7. $\displaystyle\int_{\tan y}^{\sec y} (2x + \cos y)\, dx$

8. $\displaystyle\int_{\sqrt{y}}^{1} y \ln x\, dx$

In Problems 9–12, sketch the region of integration for the given iterated integral.

9. $\displaystyle\int_{0}^{2}\int_{1}^{2x+1} f(x, y)\, dy\, dx$

10. $\displaystyle\int_{1}^{4}\int_{-\sqrt{y}}^{\sqrt{y}} f(x, y)\, dx\, dy$

11. $\displaystyle\int_{-1}^{3}\int_{0}^{\sqrt{16-y^2}} f(x, y)\, dx\, dy$

12. $\displaystyle\int_{-1}^{2}\int_{-x^2}^{x^2+1} f(x, y)\, dy\, dx$

In Problems 13–22, evaluate the double integral over the region R that is bounded by the graphs of the given equations. Choose the most convenient order of integration.

13. $\displaystyle\iint_R x^3 y^2\, dA; \quad y = x, \ y = 0, \ x = 1$

14. $\displaystyle\iint_R (x + 1)\, dA; \quad y = x, \ x + y = 4, \ x = 0$

15. $\displaystyle\iint_R (2x + 4y + 1)\, dA; \quad y = x^2, \ y = x^3$

16. $\displaystyle\iint_R xe^y\, dA; \quad R \text{ the same as in Problem 13}$

17. $\displaystyle\iint_R 2xy\, dA; \quad y = x^3, \ y = 8, \ x = 0$

18. $\displaystyle\iint_R \frac{x}{\sqrt{y}}\, dA; \quad y = x^2 + 1, \ y = 3 - x^2$

19. $\displaystyle\iint_R \frac{y}{1 + xy}\, dA; \quad y = 0, \ y = 1, \ x = 0, \ x = 1$

20. $\displaystyle\iint_R \sin\frac{\pi x}{y}\, dA; \quad x = y^2, \ x = 0, \ y = 1, \ y = 2$

21. $\displaystyle\iint_R \sqrt{x^2 + 1}\, dA; \quad x = y, \ x = -y, \ x = \sqrt{3}$

22. $\displaystyle\iint_R x\, dA; \quad y = \tan^{-1}x, \ y = 0, \ x = 1$

23. Consider the solid bounded by the graphs of $x^2 + y^2 = 4$, $z = 4 - y$, and $z = 0$ shown in **FIGURE 9.10.11**. Choose and evaluate the correct integral representing the volume V of the solid.

(a) $\displaystyle 4\int_{0}^{2}\int_{0}^{\sqrt{4-x^2}} (4 - y)\, dy\, dx$

(b) $\displaystyle 2\int_{-2}^{2}\int_{0}^{\sqrt{4-x^2}} (4 - y)\, dy\, dx$

(c) $\displaystyle 2\int_{-2}^{2}\int_{0}^{\sqrt{4-y^2}} (4 - y)\, dx\, dy$

FIGURE 9.10.11 Solid for Problem 23

24. Consider the solid bounded by the graphs of $x^2 + y^2 = 4$ and $y^2 + z^2 = 4$. An eighth of the solid is shown in **FIGURE 9.10.12**. Choose and evaluate the correct integral representing the volume V of the solid.

(a) $\displaystyle 4\int_{-2}^{2}\int_{-\sqrt{4-x^2}}^{\sqrt{4-x^2}} (4 - y^2)^{1/2}\, dy\, dx$

(b) $\displaystyle 8\int_{0}^{2}\int_{0}^{\sqrt{4-y^2}} (4 - y^2)^{1/2}\, dx\, dy$

(c) $\displaystyle 8\int_{0}^{2}\int_{0}^{\sqrt{4-x^2}} (4 - x^2)^{1/2}\, dy\, dx$

FIGURE 9.10.12 Solid for Problem 24

In Problems 25–34, find the volume of the solid bounded by the graphs of the given equations.

25. $2x + y + z = 6, x = 0, y = 0, z = 0$, first octant
26. $z = 4 - y^2, x = 3, x = 0, y = 0, z = 0$, first octant
27. $x^2 + y^2 = 4, x - y + 2z = 4, x = 0, y = 0, z = 0$, first octant
28. $y = x^2, y + z = 3, z = 0$
29. $z = 1 + x^2 + y^2, 3x + y = 3, x = 0, y = 0, z = 0$, first octant
30. $z = x + y, x^2 + y^2 = 9, x = 0, y = 0, z = 0$, first octant
31. $yz = 6, x = 0, x = 5, y = 1, y = 6, z = 0$
32. $z = 4 - x^2 - \frac{1}{4}y^2, z = 0$
33. $z = 4 - y^2, x^2 + y^2 = 2x, z = 0$
34. $z = 1 - x^2, z = 1 - y^2, x = 0, y = 0, z = 0$, first octant

In Problems 35–40, evaluate the given iterated integral by reversing the order of integration.

35. $\displaystyle\int_0^1 \int_x^1 x^2 \sqrt{1 + y^4}\, dy\, dx$ **36.** $\displaystyle\int_0^1 \int_{2y}^2 e^{-y/x}\, dx\, dy$

37. $\displaystyle\int_0^2 \int_{y^2}^4 \cos \sqrt{x^3}\, dx\, dy$

38. $\displaystyle\int_{-1}^1 \int_{-\sqrt{1-x^2}}^{\sqrt{1-x^2}} x\sqrt{1 - x^2 - y^2}\, dy\, dx$

39. $\displaystyle\int_0^1 \int_x^1 \frac{1}{1 + y^4}\, dy\, dx$ **40.** $\displaystyle\int_0^4 \int_{\sqrt{y}}^2 \sqrt{x^3 + 1}\, dx\, dy$

In Problems 41–50, find the center of mass of the lamina that has the given shape and density.

41. $x = 0, \; x = 4, \; y = 0, \; y = 3; \; \rho(x, y) = xy$
42. $x = 0, \; y = 0, \; 2x + y = 4; \; \rho(x, y) = x^2$
43. $y = x, \; x + y = 6, \; y = 0; \; \rho(x, y) = 2y$
44. $y = |x|, \; y = 3; \; \rho(x, y) = x^2 + y^2$
45. $y = x^2, \; x = 1, \; y = 0; \; \rho(x, y) = x + y$
46. $x = y^2, \; x = 4; \; \rho(x, y) = y + 5$
47. $y = 1 - x^2, \; y = 0$; density at a point P directly proportional to the distance from the x-axis
48. $y = \sin x, \; 0 \le x \le \pi, \; y = 0$; density at a point P directly proportional to the distance from the y-axis
49. $y = e^x, \; x = 0, \; x = 1, \; y = 0; \; \rho(x, y) = y^3$
50. $y = \sqrt{9 - x^2}, \; y = 0; \; \rho(x, y) = x^2$

In Problems 51–54, find the moment of inertia about the x-axis of the lamina that has the given shape and density.

51. $x = y - y^2, \; x = 0; \; \rho(x, y) = 2x$
52. $y = x^2, \; y = \sqrt{x}; \; \rho(x, y) = x^2$

53. $y = \cos x, \; -\pi/2 \le x \le \pi/2, \; y = 0; \; \rho(x, y) = k$ (constant)
54. $y = \sqrt{4 - x^2}, x = 0, y = 0$, first quadrant; $\rho(x, y) = y$

In Problems 55–58, find the moment of inertia about the y-axis of the lamina that has the given shape and density.

55. $y = x^2, \; x = 0, \; y = 4$, first quadrant; $\rho(x, y) = y$
56. $y = x^2, \; y = \sqrt{x}; \; \rho(x, y) = x^2$
57. $y = x, \; y = 0, \; y = 1, \; x = 3; \; \rho(x, y) = 4x + 3y$
58. Same R and density as in Problem 47

In Problems 59 and 60, find the radius of gyration about the indicated axis of the lamina that has the given shape and density.

59. $x = \sqrt{a^2 - y^2}, \; x = 0; \rho(x, y) = x; \; y$-axis
60. $x + y = a, \; a > 0, \; x = 0, \; y = 0; \; \rho(x, y) = k$ (constant); x-axis

61. A lamina has the shape of the region bounded by the graph of the ellipse $x^2/a^2 + y^2/b^2 = 1$. If its density is $\rho(x, y) = 1$, find:
 (a) the moment of inertia about the x-axis of the lamina,
 (b) the moment of inertia about the y-axis of the lamina,
 (c) the radius of gyration about the x-axis [*Hint*: The area of the ellipse is πab], and
 (d) the radius of gyration about the y-axis.

62. A cross section of an experimental airfoil is the lamina shown in **FIGURE 9.10.13**. The arc ABC is elliptical, whereas the two arcs AD and CD are parabolic. Find the moment of inertia about the x-axis of the lamina under the assumption that the density is $\rho(x, y) = 1$.

FIGURE 9.10.13 Airfoil in Problem 62

The **polar moment of inertia** of a lamina with respect to the origin is defined to be

$$I_0 = \iint\limits_R (x^2 + y^2)\rho(x, y)\, dA = I_x + I_y.$$

In Problems 63–66, find the polar moment of inertia of the lamina that has the given shape and density.

63. $x + y = a, \; a > 0, \; x = 0, \; y = 0; \; \rho(x, y) = k$ (constant)
64. $y = x^2, \; y = \sqrt{x}; \; \rho(x, y) = x^2$ [*Hint*: See Problems 52 and 56.]
65. $x = y^2 + 2, \; x = 6 - y^2$; density at a point P inversely proportional to the square of the distance from the origin
66. $y = x, \; y = 0, \; y = 3, \; x = 4; \; \rho(x, y) = k$ (constant)
67. Find the radius of gyration in Problem 63.
68. Show that the polar moment of inertia about the center of a thin homogeneous rectangular plate of mass m, width w, and length l is $I_0 = m(l^2 + w^2)/12$.

9.11 Double Integrals in Polar Coordinates

INTRODUCTION A double integral, which may be difficult or even impossible to evaluate in rectangular xy-coordinates, may become more tractable when expressed in a different coordinate system. In this section we examine double integrals in polar $r\theta$-coordinates.

■ Polar Rectangles

Suppose R is a region in the plane bounded by the graphs of the polar equations $r = g_1(\theta)$, $r = g_2(\theta)$, and the rays $\theta = \alpha$, $\theta = \beta$, and f is a function of r and θ that is continuous on R. In order to define the double integral of f over R, we use rays and concentric circles to partition the region into a grid of **polar rectangles** or subregions R_k. See **FIGURE 9.11.1(a)** and 9.11.1(b). The area ΔA_k of typical subregion R_k, shown in Figure 9.11.1(c), is the difference of the areas of two circular sectors: $\Delta A_k = \frac{1}{2}r_{k+1}^2 \Delta\theta_k - \frac{1}{2}r_k^2 \Delta\theta_k$. Now ΔA_k can be written as

$$\Delta A_k = \tfrac{1}{2}(r_{k+1}^2 - r_k^2)\Delta\theta_k = \tfrac{1}{2}(r_{k+1} + r_k)(r_{k+1} - r_k)\Delta\theta_k = r_k^* \Delta r_k \Delta\theta_k$$

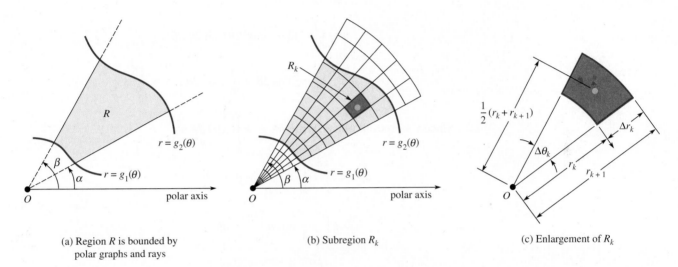

(a) Region R is bounded by polar graphs and rays

(b) Subregion R_k

(c) Enlargement of R_k

FIGURE 9.11.1 R_k in (b) and (c) is called a polar rectangle

where $\Delta r_k = r_{k+1} - r_k$ and $r_k^* = \frac{1}{2}(r_{k+1} + r_k)$ denotes the average radius. By choosing (r_k^*, θ_k^*) on each R_k, the double integral of f over R is

$$\lim_{\|P\| \to 0} \sum_{k=1}^{n} f(r_k^*, \theta_k^*) r_k^* \, \Delta r_k \, \Delta\theta_k = \iint_R f(r, \theta) \, dA.$$

The double integral is then evaluated by means of the iterated integral:

$$\iint_R f(r, \theta) \, dA = \int_\alpha^\beta \int_{g_1(\theta)}^{g_2(\theta)} f(r, \theta) \, r \, dr \, d\theta. \qquad (1)$$

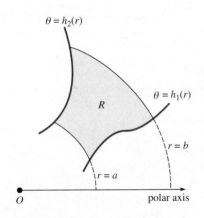

FIGURE 9.11.2 R bounded by polar graphs and circular arcs

On the other hand, if the region R is as given in **FIGURE 9.11.2**, the double integral of f over R is then

$$\iint_R f(r, \theta) \, dA = \int_a^b \int_{h_1(r)}^{h_2(r)} f(r, \theta) \, r \, d\theta \, dr. \qquad (2)$$

EXAMPLE 1	Center of Mass

Find the center of mass of the lamina that corresponds to the region bounded by one leaf of the rose $r = 2 \sin 2\theta$ in the first quadrant if the density at a point P in the lamina is directly proportional to the distance from the pole.

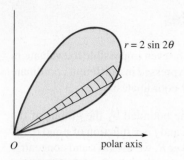

$r = 2 \sin 2\theta$

O polar axis

FIGURE 9.11.3 Lamina in Example 1

SOLUTION By varying θ from 0 to $\pi/2$, we obtain the graph in **FIGURE 9.11.3**. Now, $d(O, P) = |r|$. Hence, $\rho(r, \theta) = k|r|$, where k is a constant of proportionality. From (9) of Section 9.10, we have

$$m = \iint_R k|r|\, dA$$

$$= k \int_0^{\pi/2} \int_0^{2\sin 2\theta} (r) r\, dr\, d\theta$$

$$= k \int_0^{\pi/2} \left. \frac{r^3}{3} \right]_0^{2\sin 2\theta} d\theta$$

$$= \frac{8}{3} k \int_0^{\pi/2} \sin^3 2\theta\, d\theta \qquad \leftarrow \sin^2 2\theta = 1 - \cos^2 2\theta$$

$$= \frac{8}{3} k \int_0^{\pi/2} (1 - \cos^2 2\theta) \sin 2\theta\, d\theta$$

$$= \frac{8}{3} k \left[-\frac{1}{2} \cos 2\theta + \frac{1}{6} \cos^3 2\theta \right]_0^{\pi/2} = \frac{16}{9} k.$$

Since $x = r \cos \theta$, we can write $M_y = k \iint_R x|r|\, dA$ as

$$M_y = k \int_0^{\pi/2} \int_0^{2\sin 2\theta} r^3 \cos \theta\, dr\, d\theta$$

$$= k \int_0^{\pi/2} \left. \frac{r^4}{4} \cos \theta \right]_0^{2\sin 2\theta} d\theta$$

$$= 4k \int_0^{\pi/2} \sin^4 2\theta \cos \theta\, d\theta \qquad \leftarrow \text{double-angle formula}$$

$$= 4k \int_0^{\pi/2} 16 \sin^4 \theta \cos^4 \theta \cos \theta\, d\theta$$

$$= 64k \int_0^{\pi/2} \sin^4 \theta \cos^5 \theta\, d\theta$$

$$= 64k \int_0^{\pi/2} \sin^4 \theta (1 - \sin^2 \theta)^2 \cos \theta\, d\theta$$

$$= 64k \int_0^{\pi/2} (\sin^4 \theta - 2 \sin^6 \theta + \sin^8 \theta) \cos \theta\, d\theta$$

$$= 64k \left[\frac{1}{5} \sin^5 \theta - \frac{2}{7} \sin^7 \theta + \frac{1}{9} \sin^9 \theta \right]_0^{\pi/2} = \frac{512}{315} k.$$

Similarly, by using $x = r \sin \theta$, we find*

$$M_x = k \int_0^{\pi/2} \int_0^{2\sin 2\theta} r^3 \sin \theta\, dr\, d\theta = \frac{512}{315} k.$$

*We could have argued to the fact that $M_x = M_y$ and hence $\bar{x} = \bar{y}$ from the fact that the lamina and the density function are symmetric about the ray $\theta = \pi/4$.

Here the rectangular coordinates of the center of mass are

$$\bar{x} = \bar{y} = \frac{512k/315}{16k/9} = \frac{32}{35}.$$

Change of Variables: Rectangular to Polar Coordinates

In some instances a double integral $\iint_R f(x, y)\, dA$ that is difficult or even impossible to evaluate using rectangular coordinates may be readily evaluated when a change of variables is used. If we assume that f is continuous on the region R and if R can be described in polar coordinates as $0 \leq g_1(\theta) \leq r \leq g_2(\theta)$, $\alpha \leq \theta \leq \beta$, $0 < \beta - \alpha \leq 2\pi$, then

$$\iint_R f(x, y)\, dA = \int_\alpha^\beta \int_{g_1(\theta)}^{g_2(\theta)} f(r\cos\theta, r\sin\theta) r\, dr\, d\theta. \tag{3}$$

Equation (3) is particularly useful when f contains the expression $x^2 + y^2$ since, in polar coordinates, we can now write

$$x^2 + y^2 = r^2 \quad \text{and} \quad \sqrt{x^2 + y^2} = r.$$

EXAMPLE 2 **Changing an Integral to Polar Coordinates**

Use polar coordinates to evaluate

$$\int_0^2 \int_x^{\sqrt{8-x^2}} \frac{1}{5 + x^2 + y^2}\, dy\, dx.$$

SOLUTION From $x \leq y \leq \sqrt{8 - x^2}$, $0 \leq x \leq 2$, we have sketched the region R of integration in **FIGURE 9.11.4**. Since $x^2 + y^2 = r^2$, the polar description of the circle $x^2 + y^2 = 8$ is $r = \sqrt{8}$. Hence, in polar coordinates, the region of R is given by $0 \leq r \leq \sqrt{8}$, $\pi/4 \leq \theta \leq \pi/2$. With $1/(5 + x^2 + y^2) = 1/(5 + r^2)$, the original integral becomes

$$\int_0^2 \int_x^{\sqrt{8-x^2}} \frac{1}{5 + x^2 + y^2}\, dy\, dx = \int_{\pi/4}^{\pi/2} \int_0^{\sqrt{8}} \frac{1}{5 + r^2} r\, dr\, d\theta$$

$$= \frac{1}{2} \int_{\pi/4}^{\pi/2} \int_0^{\sqrt{8}} \frac{2r\, dr}{5 + r^2}\, d\theta$$

$$= \frac{1}{2} \int_{\pi/4}^{\pi/2} \ln(5 + r^2) \Big]_0^{\sqrt{8}}\, d\theta$$

$$= \frac{1}{2}(\ln 13 - \ln 5) \int_{\pi/4}^{\pi/2} d\theta$$

$$= \frac{1}{2}(\ln 13 - \ln 5)\left(\frac{\pi}{2} - \frac{\pi}{4}\right) = \frac{\pi}{8} \ln \frac{13}{5}.$$

FIGURE 9.11.4 Region R in Example 2

EXAMPLE 3 **Volume**

Find the volume of the solid that is under the hemisphere $z = \sqrt{1 - x^2 - y^2}$ and above the region bounded by the graph of the circle $x^2 + y^2 - y = 0$.

SOLUTION From **FIGURE 9.11.5**, we see that $V = \iint_R \sqrt{1 - x^2 - y^2}\, dA$. In polar coordinates the equations of the hemisphere and the circle become, respectively, $z = \sqrt{1 - r^2}$ and $r = \sin\theta$. Now, using symmetry, we have

$$V = \iint_R \sqrt{1 - r^2}\, dA = 2\int_0^{\pi/2} \int_0^{\sin\theta} (1 - r^2)^{1/2} r\, dr\, d\theta$$

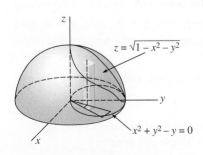

FIGURE 9.11.5 Solid in Example 3

$$= 2 \int_0^{\pi/2} \left[-\frac{1}{3} (1 - r^2)^{3/2} \right]_0^{\sin \theta} d\theta$$

$$= \frac{2}{3} \int_0^{\pi/2} [1 - (1 - \sin^2 \theta)^{3/2}] \, d\theta$$

$$= \frac{2}{3} \int_0^{\pi/2} [1 - (\cos^2 \theta)^{3/2}] \, d\theta$$

$$= \frac{2}{3} \int_0^{\pi/2} [1 - \cos^3 \theta] \, d\theta$$

$$= \frac{2}{3} \int_0^{\pi/2} [1 - (1 - \sin^2 \theta) \cos \theta] \, d\theta$$

$$= \frac{2}{3} \left[\theta - \sin \theta + \frac{1}{3} \sin^3 \theta \right]_0^{\pi/2} = \frac{\pi}{3} - \frac{4}{9} \approx 0.60. \qquad \equiv$$

■ **Area** Note that in (1) if $f(r, \theta) = 1$, then the **area** of the region R in Figure 9.11.1(a) is given by

$$A = \iint\limits_R dA = \int_\alpha^\beta \int_{g_1(\theta)}^{g_2(\theta)} r \, dr \, d\theta.$$

The same observation holds for (2) and Figure 9.11.2 when $f(r, \theta) = 1$.

REMARKS

The reader is invited to reexamine Example 3. The graph of the circle $r = \sin \theta$ is obtained by varying θ from 0 to π. However, carry out the iterated integration

$$V = \int_0^\pi \int_0^{\sin \theta} (1 - r^2)^{1/2} r \, dr \, d\theta$$

and see if the result is the *incorrect* answer $\pi/3$. What goes wrong?

9.11 Exercises Answers to selected odd-numbered problems begin on page ANS-25.

In Problems 1–4, use a double integral in polar coordinates to find the area of the region bounded by the graphs of the given polar equations.

1. $r = 3 + 3 \sin \theta$

2. $r = 2 + \cos \theta$

3. $r = 2 \sin \theta$, $r = 1$, common area

4. $r = 8 \sin 4\theta$, one petal

In Problems 5–10, find the volume of the solid bounded by the graphs of the given equations.

5. One petal of $r = 5 \cos 3\theta$, $z = 0$, $z = 4$

6. $x^2 + y^2 = 4$, $z = \sqrt{9 - x^2 - y^2}$, $z = 0$

7. Between $x^2 + y^2 = 1$ and $x^2 + y^2 = 9$,
 $z = \sqrt{16 - x^2 - y^2}$, $z = 0$

8. $z = \sqrt{x^2 + y^2}$, $x^2 + y^2 = 25$, $z = 0$

9. $r = 1 + \cos \theta$, $z = y$, $z = 0$, first octant

10. $r = \cos \theta$, $z = 2 + x^2 + y^2$, $z = 0$

In Problems 11–16, find the center of mass of the lamina that has the given shape and density.

11. $r = 1$, $r = 3$, $x = 0$, $y = 0$, first quadrant; $\rho(r, \theta) = k$
 (constant)

12. $r = \cos \theta$; density at point P directly proportional to the distance from the pole

13. $y = \sqrt{3}x$, $y = 0$, $x = 3$; $\rho(r, \theta) = r^2$

14. $r = 4 \cos 2\theta$, petal on the polar axis; $\rho(r, \theta) = k$ (constant)

15. Outside $r = 2$ and inside $r = 2 + 2 \cos \theta$, $y = 0$, first quadrant; density at a point P inversely proportional to the distance from the pole

16. $r = 2 + 2 \cos \theta$, $y = 0$, first and second quadrants; $\rho(r, \theta) = k$ (constant)

In Problems 17–20, find the indicated moment of inertia of the lamina that has the given shape and density.

17. $r = a$; $\rho(r, \theta) = k$ (constant); I_x

18. $r = a$; $\rho(r, \theta) = \dfrac{1}{1 + r^4}$; I_x

19. Outside $r = a$ and inside $r = 2a \cos \theta$; density at a point P inversely proportional to the cube of the distance from the pole; I_y

20. Outside $r = 1$ and inside $r = 2 \sin 2\theta$, first quadrant; $\rho(r, \theta) = \sec^2 \theta$; I_y

In Problems 21–24, find the **polar moment of inertia**

$$I_0 = \iint_R r^2 \rho(r, \theta) \, dA = I_x + I_y$$

of the lamina that has the given shape and density.

21. $r = a$; $\rho(r, \theta) = k$ (constant) [*Hint*: Use Problem 17 and the fact that $I_x = I_y$.]

22. $r = \theta$, $0 \leq \theta \leq \pi$, $y = 0$; density at a point P proportional to the distance from the pole

23. $r\theta = 1$, $\frac{1}{3} \leq \theta \leq 1$, $r = 1$, $r = 3$, $y = 0$; density at a point P inversely proportional to the distance from the pole [*Hint*: Integrate first with respect to θ.]

24. $r = 2a \cos \theta$; $\rho(r, \theta) = k$ (constant)

In Problems 25–32, evaluate the given iterated integral by changing to polar coordinates.

25. $\displaystyle\int_{-3}^{3} \int_{0}^{\sqrt{9 - x^2}} \sqrt{x^2 + y^2} \, dy \, dx$

26. $\displaystyle\int_{0}^{\sqrt{2}/2} \int_{y}^{\sqrt{1 - y^2}} \frac{y^2}{\sqrt{x^2 + y^2}} \, dx \, dy$

27. $\displaystyle\int_{0}^{1} \int_{0}^{\sqrt{1 - y^2}} e^{x^2 + y^2} \, dx \, dy$

28. $\displaystyle\int_{-\sqrt{\pi}}^{\sqrt{\pi}} \int_{0}^{\sqrt{\pi - x^2}} \sin(x^2 + y^2) \, dy \, dx$

29. $\displaystyle\int_{0}^{1} \int_{\sqrt{1 - x^2}}^{\sqrt{4 - x^2}} \frac{x^2}{x^2 + y^2} \, dy \, dx + \int_{1}^{2} \int_{0}^{\sqrt{4 - x^2}} \frac{x^2}{x^2 + y^2} \, dy \, dx$

30. $\displaystyle\int_{0}^{1} \int_{0}^{\sqrt{2y - y^2}} (1 - x^2 - y^2) \, dx \, dy$

31. $\displaystyle\int_{-5}^{5} \int_{0}^{\sqrt{25 - x^2}} (4x + 3y) \, dy \, dx$

32. $\displaystyle\int_{0}^{1} \int_{0}^{\sqrt{1 - y^2}} \frac{1}{1 + \sqrt{x^2 + y^2}} \, dx \, dy$

33. The liquid hydrogen tank in the space shuttle has the form of a right circular cylinder with a semi-ellipsoidal cap at each end. The radius of the cylindrical part of the tank is 4.2 m. Find the volume of the tank shown in **FIGURE 9.11.6**.

5.15 m

19.3 m

5.15 m

FIGURE 9.11.6 Fuel tank in Problem 33

34. Evaluate $\iint_R (x + y) \, dA$ over the region shown in **FIGURE 9.11.7**.

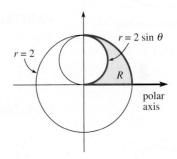

$r = 2$

$r = 2 \sin \theta$

R

polar axis

FIGURE 9.11.7 Region R for Problem 34

Discussion Problem

35. The improper integral $\int_0^\infty e^{-x^2} \, dx$ is important in the theory of probability, statistics, and other areas of applied mathematics. If I denotes the integral, then

$$I = \int_0^\infty e^{-x^2} \, dx \quad \text{and} \quad I = \int_0^\infty e^{-y^2} \, dy,$$

and consequently

$$I^2 = \left(\int_0^\infty e^{-x^2} dx \right)\left(\int_0^\infty e^{-y^2} dy \right) = \int_0^\infty \int_0^\infty e^{-(x^2 + y^2)} \, dx \, dy.$$

Discuss how to use polar coordinates to evaluate the last integral. Find the value of I.

9.12 Green's Theorem

INTRODUCTION One of the most important theorems in vector integral calculus relates a line integral around a piecewise-smooth simple closed curve C to a double integral over the region R bounded by the curve.

■ **Line Integrals Along Simple Closed Curves** We say the **positive direction** around a simple closed curve C is that direction a point on the curve must move, or the direction a person

(a) Positive direction

(b) Positive direction

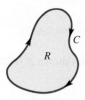

(c) Negative direction

FIGURE 9.12.1 Directions around curve C

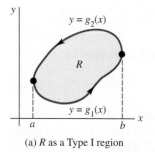

(a) R as a Type I region

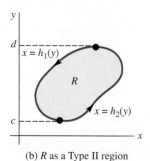

(b) R as a Type II region

FIGURE 9.12.2 Region R in Theorem 9.12.1

must walk on C, in order to keep the region R bounded by C to the left. See **FIGURE 9.12.1(a)**. Roughly, as shown in Figure 9.12.1(b) and 9.12.1(c), the *positive* and *negative* directions correspond to the *counterclockwise* and *clockwise* directions, respectively. Line integrals on simple closed curves are written

$$\oint_C P(x, y)\, dx + Q(x, y)\, dy, \quad \oint_C P(x, y)\, dx + Q(x, y)\, dy, \quad \oint_C F(x, y)\, ds,$$

and so on. The symbols $\oint_C$ and $\oint_C$ refer, in turn, to integrations in the positive and negative directions.

THEOREM 9.12.1 Green's Theorem in the Plane

Suppose that C is a piecewise-smooth simple closed curve bounding a simply connected region R. If P, Q, $\partial P/\partial y$, and $\partial Q/\partial x$ are continuous on R, then

$$\oint_C P\, dx + Q\, dy = \iint_R \left(\frac{\partial Q}{\partial x} - \frac{\partial P}{\partial y} \right) dA. \tag{1}$$

PARTIAL PROOF: We shall prove (1) only for a region R that is simultaneously of Type I and Type II:

$$R: g_1(x) \le y \le g_2(x), \quad a \le x \le b$$
$$R: h_1(y) \le x \le h_2(y), \quad c \le y \le d.$$

Using **FIGURE 9.12.2(a)**, we have

$$-\iint_R \frac{\partial P}{\partial y}\, dA = -\int_a^b \int_{g_1(x)}^{g_2(x)} \frac{\partial P}{\partial y}\, dy\, dx$$

$$= -\int_a^b [P(x, g_2(x)) - P(x, g_1(x))]\, dx$$

$$= \int_a^b P(x, g_1(x))\, dx + \int_b^a P(x, g_2(x))\, dx \tag{2}$$

$$= \oint_C P(x, y)\, dx.$$

Similarly, from Figure 9.12.2(b),

$$\iint_R \frac{\partial Q}{\partial x}\, dA = \int_c^d \int_{h_1(y)}^{h_2(y)} \frac{\partial Q}{\partial x}\, dx\, dy$$

$$= \int_c^d [Q(h_2(y), y) - Q(h_1(y), y)]\, dy$$

$$= \int_c^d Q(h_2(y), y)\, dy + \int_d^c Q(h_1(y), y)\, dy \tag{3}$$

$$= \oint_C Q(x, y)\, dy.$$

Adding the results in (2) and (3) yields (1). ≡

The result given in Theorem 9.12.1 is named after George Green, the same English mathematical physicist who developed the Green's functions discussed in Section 3.10. The words *in the plane* suggest that Theorem 9.12.1 generalizes to 3-space. It does—read on.

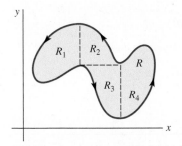

FIGURE 9.12.3 Subregions of R

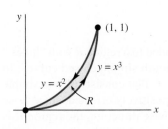

FIGURE 9.12.4 Curve C in Example 1

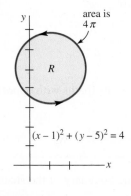

FIGURE 9.12.5 Curve C in Example 2

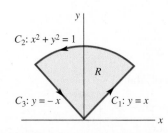

FIGURE 9.12.6 Curve C in Example 3

Although the foregoing proof is not valid, the theorem is applicable to more complicated regions, such as those shown in **FIGURE 9.12.3**. The proof consists of decomposing R into a finite number of subregions to which (1) can be applied and then adding the results.

EXAMPLE 1 **Using Green's Theorem**

Evaluate $\oint_C (x^2 - y^2)\, dx + (2y - x)\, dy$, where C consists of the boundary of the region in the first quadrant that is bounded by the graphs of $y = x^2$ and $y = x^3$.

SOLUTION If $P(x, y) = x^2 - y^2$ and $Q(x, y) = 2y - x$, then $\partial P/\partial y = -2y$ and $\partial Q/\partial x = -1$. From (1) and **FIGURE 9.12.4**, we have

$$
\oint_C (x^2 - y^2)\, dx + (2y - x)\, dy = \iint_R (-1 + 2y)\, dA
$$

$$
= \int_0^1 \int_{x^3}^{x^2} (-1 + 2y)\, dy\, dx
$$

$$
= \int_0^1 (-y + y^2) \Big]_{x^3}^{x^2} dx
$$

$$
= \int_0^1 (-x^6 + x^4 + x^3 - x^2)\, dx = -\frac{11}{420}. \qquad \equiv
$$

We note that the line integral in Example 1 could have been evaluated in a straightforward manner using the variable x as a parameter. However, in the next example, ponder the problem of evaluating the given line integral in the usual manner.

EXAMPLE 2 **Using Green's Theorem**

Evaluate $\oint_C (x^5 + 3y)\, dx + (2x - e^{y^3})\, dy$, where C is the circle $(x - 1)^2 + (y - 5)^2 = 4$.

SOLUTION Identifying $P(x, y) = x^5 + 3y$ and $Q(x, y) = 2x - e^{y^3}$, we have $\partial P/\partial y = 3$ and $\partial Q/\partial x = 2$. Hence, (1) gives

$$
\oint_C (x^5 + 3y)\, dx + (2x - e^{y^3})\, dy = \iint_R (2 - 3)\, dA = -\iint_R dA.
$$

Now the double integral $\iint_R dA$ gives the area of the region R bounded by the circle of radius 2 shown in **FIGURE 9.12.5**. Since the area of the circle is $\pi 2^2 = 4\pi$, it follows that

$$
\oint_C (x^5 + 3y)\, dx + (2x - e^{y^3})\, dy = -4\pi. \qquad \equiv
$$

EXAMPLE 3 **Work Done by a Force**

Find the work done by the force $\mathbf{F} = (-16y + \sin x^2)\mathbf{i} + (4e^y + 3x^2)\mathbf{j}$ acting along the simple closed curve C shown in **FIGURE 9.12.6**.

SOLUTION From (12) of Section 9.8 the work done by $\mathbf{F}$ is given by

$$
W = \oint_C \mathbf{F} \cdot d\mathbf{r} = \oint_C (-16y + \sin x^2)\, dx + (4e^y + 3x^2)\, dy
$$

and so by Green's theorem $W = \iint_R (6x + 16)\, dA$. In view of the region R, the last integral is best handled in polar coordinates. Since R is defined by $0 \le r \le 1$, $\pi/4 \le \theta \le 3\pi/4$,

$$
W = \int_{\pi/4}^{3\pi/4} \int_0^1 (6r \cos \theta + 16)\, r\, dr\, d\theta
$$

FIGURE 9.12.7 Curve C in Example 4

(a)

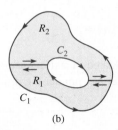

(b)

FIGURE 9.12.8 Boundary of R is $C = C_1 \cup C_2$

FIGURE 9.12.9 Boundary C in Example 5

FIGURE 9.12.10 Curves C_1 and C_2 in (5)

$$= \int_{\pi/4}^{3\pi/4} (2r^3 \cos\theta + 8r^2) \Big]_0^1 d\theta$$

$$= \int_{\pi/4}^{3\pi/4} (2\cos\theta + 8)\, d\theta = 4\pi.$$

<div style="text-align:right">≡</div>

EXAMPLE 4 **Green's Theorem Not Applicable**

Let C be the closed curve consisting of the four straight line segments C_1, C_2, C_3, C_4 shown in **FIGURE 9.12.7**. Green's theorem is *not* applicable to the line integral

$$\oint_C \frac{-y}{x^2 + y^2}\, dx + \frac{x}{x^2 + y^2}\, dy$$

since P, Q, $\partial P/\partial y$, and $\partial Q/\partial x$ are not continuous at the origin.

<div style="text-align:right">≡</div>

■ **Region with Holes** Green's theorem can also be extended to a region R with "holes," that is, a region bounded between two or more piecewise-smooth simple closed curves. In **FIGURE 9.12.8(a)** we have shown a region R bounded by a curve C that consists of two simple closed curves C_1 and C_2; that is, $C = C_1 \cup C_2$. The curve C is positively oriented, since if we traverse C_1 in a counterclockwise direction and C_2 in a clockwise direction, the region R is always to the left. If we now introduce crosscuts as shown in Figure 9.12.8(b), the region R is divided into two subregions, R_1 and R_2. By applying Green's theorem to R_1 and R_2, we obtain

$$\iint_R \left(\frac{\partial Q}{\partial x} - \frac{\partial P}{\partial y} \right) dA = \iint_{R_1} \left(\frac{\partial Q}{\partial x} - \frac{\partial P}{\partial y} \right) dA + \iint_{R_2} \left(\frac{\partial Q}{\partial x} - \frac{\partial P}{\partial y} \right) dA$$

$$= \oint_{C_1} P\, dx + Q\, dy + \oint_{C_2} P\, dx + Q\, dy \qquad (4)$$

$$= \oint_C P\, dx + Q\, dy.$$

The last result follows from the fact that the line integrals on the crosscuts (paths with opposite orientations) cancel each other. See (8) of Section 9.8.

EXAMPLE 5 **Region with a Hole**

Evaluate $\oint_C \dfrac{-y}{x^2 + y^2}\, dx + \dfrac{x}{x^2 + y^2}\, dy$, where $C = C_1 \cup C_2$ is the boundary of the shaded region R shown in **FIGURE 9.12.9**.

SOLUTION Because $P(x, y) = \dfrac{-y}{x^2 + y^2}$, $Q(x, y) = \dfrac{x}{x^2 + y^2}$, and the partial derivatives

$$\frac{\partial P}{\partial y} = \frac{y^2 - x^2}{(x^2 + y^2)^2}, \qquad \frac{\partial Q}{\partial x} = \frac{y^2 - x^2}{(x^2 + y^2)^2},$$

are continuous on the region R bounded by C, it follows from (4) that

$$\oint_C \frac{-y}{x^2 + y^2}\, dx + \frac{x}{x^2 + y^2}\, dy = \iint_R \left[\frac{y^2 - x^2}{(x^2 + y^2)^2} - \frac{y^2 - x^2}{(x^2 + y^2)^2} \right] dA = 0.$$

<div style="text-align:right">≡</div>

As a consequence of the discussion preceding Example 5 we can establish a result for line integrals that enables us, under certain circumstances, to replace a complicated closed path with a path that is simpler. Suppose, as shown in **FIGURE 9.12.10**, that C_1 and C_2 are two nonintersecting

piecewise-smooth simple closed paths that have the same counterclockwise orientation. Suppose further that P and Q have continuous first partial derivatives such that

$$\frac{\partial P}{\partial y} = \frac{\partial Q}{\partial x}$$

in the region R bounded between C_1 and C_2. Then from (4) above and (8) of Section 9.8 we have

$$\oint_{C_1} P\,dx + Q\,dy + \oint_{-C_2} P\,dx + Q\,dy = 0$$

or

$$\oint_{C_1} P\,dx + Q\,dy = \oint_{C_2} P\,dx + Q\,dy. \tag{5}$$

EXAMPLE 6 **Example 4 Revisited**

Evaluate the line integral in Example 4.

SOLUTION One method of evaluating the line integral is to write

$$\oint_C = \int_{C_1} + \int_{C_2} + \int_{C_3} + \int_{C_4}$$

and then evaluate the four integrals on the line segments C_1, C_2, C_3, and C_4. Alternatively, if we note that the circle C': $x^2 + y^2 = 1$ lies entirely within C (see **FIGURE 9.12.11**), then from Example 5 it is apparent that $P = -y/(x^2 + y^2)$ and $Q = x/(x^2 + y^2)$ have continuous first partial derivatives in the region R bounded between C and C'. Moreover,

$$\frac{\partial P}{\partial y} = \frac{y^2 - x^2}{(x^2 + y^2)^2} = \frac{\partial Q}{\partial x}$$

in R. Hence, it follows from (5) that

$$\oint_C \frac{-y}{x^2 + y^2}\,dx + \frac{x}{x^2 + y^2}\,dy = \oint_{C'} \frac{-y}{x^2 + y^2}\,dx + \frac{x}{x^2 + y^2}\,dy.$$

Using the parameterization $x = \cos t$, $y = \sin t$, $0 \le t \le 2\pi$ for C' we obtain

$$\oint_C \frac{-y}{x^2 + y^2}\,dx + \frac{x}{x^2 + y^2}\,dy = \int_0^{2\pi} [-\sin t(-\sin t) + \cos t(\cos t)]\,dt$$

$$= \int_0^{2\pi} (\sin^2 t + \cos^2 t)\,dt \tag{6}$$

$$= \int_0^{2\pi} dt = 2\pi. \qquad \equiv$$

It is interesting to note that the result in (6),

$$\oint_C \frac{-y}{x^2 + y^2}\,dx + \frac{x}{x^2 + y^2}\,dy = 2\pi,$$

is true for every piecewise-smooth simple closed curve C with the origin in its interior. We need only choose C' to be $x^2 + y^2 = a^2$, where a is small enough so that the circle lies entirely within C.

FIGURE 9.12.11 Curves C and C' in Example 6

9.12 Exercises Answers to selected odd-numbered problems begin on page ANS-25.

In Problems 1–4, verify Green's theorem by evaluating both integrals.

1. $\oint_C (x - y)\,dx + xy\,dy = \iint_R (y + 1)\,dA$, where C is the triangle with vertices $(0, 0)$, $(1, 0)$, $(1, 3)$

2. $\oint_C 3x^2 y\,dx + (x^2 - 5y)\,dy = \iint_R (2x - 3x^2)\,dA$, where C is the rectangle with vertices $(-1, 0)$, $(1, 0)$, $(1, 1)$, $(-1, 1)$

3. $\oint_C -y^2\,dx + x^2\,dy = \iint_R (2x + 2y)\,dA$, where C is the circle $x = 3 \cos t$, $y = 3 \sin t$, $0 \le t \le 2\pi$

4. $\oint_C -2y^2\,dx + 4xy\,dy = \iint_R 8y\,dA$, where C is the boundary of the region in the first quadrant determined by the graphs of $y = 0$, $y = \sqrt{x}$, $y = -x + 2$

In Problems 5–14, use Green's theorem to evaluate the given line integral.

5. $\oint_C 2y\,dx + 5x\,dy$, where C is the circle $(x - 1)^2 + (y + 3)^2 = 25$

6. $\oint_C (x + y^2)\,dx + (2x^2 - y)\,dy$, where C is the boundary of the region determined by the graphs of $y = x^2$, $y = 4$

7. $\oint_C (x^4 - 2y^3)\,dx + (2x^3 - y^4)\,dy$, where C is the circle $x^2 + y^2 = 4$

8. $\oint_C (x - 3y)\,dx + (4x + y)\,dy$, where C is the rectangle with vertices $(-2, 0)$, $(3, 0)$, $(3, 2)$, $(-2, 2)$

9. $\oint_C 2xy\,dx + 3xy^2\,dy$, where C is the triangle with vertices $(1, 2)$, $(2, 2)$, $(2, 4)$

10. $\oint_C e^{2x}\sin 2y\,dx + e^{2x}\cos 2y\,dy$, where C is the ellipse $9(x - 1)^2 + 4(y - 3)^2 = 36$

11. $\oint_C xy\,dx + x^2\,dy$, where C is the boundary of the region determined by the graphs of $x = 0$, $x^2 + y^2 = 1$, $x \geq 0$

12. $\oint_C e^{x^2}\,dx + 2\tan^{-1} x\,dy$, where C is the triangle with vertices $(0, 0)$, $(0, 1)$, $(-1, 1)$

13. $\oint_C \frac{1}{3}y^3\,dx + (xy + xy^2)\,dy$, where C is the boundary of the region in the first quadrant determined by the graphs of $y = 0$, $x = y^2$, $x = 1 - y^2$

14. $\oint_C xy^2\,dx + 3\cos y\,dy$, where C is the boundary of the region in the first quadrant determined by the graphs of $y = x^2$, $y = x^3$

In Problems 15 and 16, evaluate the given integral on any piecewise-smooth simple closed curve C.

15. $\oint_C ay\,dx + bx\,dy$ **16.** $\oint_C P(x)\,dx + Q(y)\,dy$

In Problems 17 and 18, let R be the region bounded by a piecewise-smooth simple closed curve C. Prove the given result.

17. $\oint_C x\,dy = -\oint_C y\,dx = $ area of R

18. $\frac{1}{2}\oint_C -y\,dx + x\,dy = $ area of R

In Problems 19 and 20, use the results of Problems 17 and 18 to find the area of the region bounded by the given closed curve.

19. The hypocycloid $x = a\cos^3 t$, $y = a\sin^3 t$, $a > 0$, $0 \leq t \leq 2\pi$

20. The ellipse $x = a\cos t$, $y = b\sin t$, $a > 0$, $b > 0$, $0 \leq t \leq 2\pi$

21. (a) Show that

$$\oint_C -y\,dx + x\,dy = x_1 y_2 - x_2 y_1,$$

where C is the line segment from the point (x_1, y_1) to (x_2, y_2).

(b) Use part (a) and Problem 18 to show that the area A of a polygon with vertices (x_1, y_1), (x_2, y_2), …, (x_n, y_n), labeled counterclockwise, is

$$A = \frac{1}{2}(x_1 y_2 - x_2 y_1) + \frac{1}{2}(x_2 y_3 - x_3 y_2) + \cdots$$
$$+ \frac{1}{2}(x_{n-1}y_n - x_n y_{n-1}) + \frac{1}{2}(x_n y_1 - x_1 y_n).$$

22. Use part (b) of Problem 21 to find the area of the quadrilateral with vertices $(-1, 3)$, $(1, 1)$, $(4, 2)$, and $(3, 5)$.

In Problems 23 and 24, evaluate the given line integral where $C = C_1 \cup C_2$ is the boundary of the shaded region R.

23. $\oint_C (4x^2 - y^3)\,dx + (x^3 + y^2)\,dy$

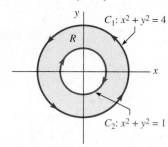

FIGURE 9.12.12 Boundary C for Problem 23

24. $\oint_C (\cos x^2 - y)\,dx + \sqrt{y^2 + 1}\,dy$

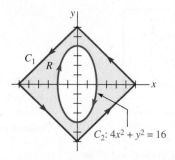

FIGURE 9.12.13 Boundary C for Problem 24

In Problems 25 and 26, proceed as in Example 6 to evaluate the given line integral.

25. $\oint_C \frac{-y^3\,dx + xy^2\,dy}{(x^2 + y^2)^2}$, where C is the ellipse $x^2 + 4y^2 = 4$

26. $\oint_C \frac{-y}{(x + 1)^2 + 4y^2}\,dx + \frac{x + 1}{(x + 1)^2 + 4y^2}\,dy$, where C is the circle $x^2 + y^2 = 16$

In Problems 27 and 28, use Green's theorem to evaluate the given double integral by means of a line integral. [*Hint*: Find appropriate functions P and Q.]

27. $\iint_R x^2\,dA$; R is the region bounded by the ellipse $x^2/9 + y^2/4 = 1$

28. $\iint_R [1 - 2(y - 1)]\,dA$; R is the region in the first quadrant bounded by the circle $x^2 + (y - 1)^2 = 1$ and $x = 0$

In Problems 29 and 30, use Green's theorem to find the work done by the given force $\mathbf{F}$ around the closed curve in **FIGURE 9.12.14**.

29. $\mathbf{F} = (x - y)\mathbf{i} + (x + y)\mathbf{j}$ **30.** $\mathbf{F} = -xy^2\mathbf{i} + x^2 y\mathbf{j}$

FIGURE 9.12.14 Curve for Problems 29 and 30

31. Let P and Q be continuous and have continuous first partial derivatives in a simply connected region of the xy-plane. If $\int_A^B P\,dx + Q\,dy$ is independent of the path, show that $\oint_C P\,dx + Q\,dy = 0$ on every piecewise-smooth simple closed curve C in the region.

32. Let R be a region bounded by a piecewise-smooth simple closed curve C. Show that the coordinates of the centroid of the region are given by

$$\bar{x} = \frac{1}{2A}\oint_C x^2\,dy, \quad \bar{y} = -\frac{1}{2A}\oint_C y^2\,dx.$$

33. Find the work done by the force $\mathbf{F} = -y\mathbf{i} + x\mathbf{j}$ acting along the cardioid $r = 1 + \cos\theta$.

9.13 Surface Integrals

INTRODUCTION In the xy-plane, the length of an arc of the graph of $y = f(x)$ from $x = a$ to $x = b$ is given by the definite integral

$$s = \int_a^b \sqrt{1 + \left(\frac{dy}{dx}\right)^2}\,dx. \tag{1}$$

The problem in three dimensions, which is the counterpart of the arc length problem, is to find the area $A(s)$ of that portion of the surface S given by a function $z = f(x, y)$ having continuous first partial derivatives on a closed region R in the xy-plane. Such a surface is said to be **smooth**.

Surface Area Suppose, as shown in **FIGURE 9.13.1(a)**, that an inner partition P of R is formed using lines parallel to the x- and y-axes. P then consists of n rectangular elements R_k of area $\Delta A_k = \Delta x_k \Delta y_k$ that lie entirely within R. Let $(x_k, y_k, 0)$ denote any point in an R_k. As we see in Figure 9.13.1(a), by projecting the sides of R_k upward, we determine two quantities: a portion or patch S_k of the surface and a portion T_k of a tangent plane at $(x_k, y_k, f(x_k, y_k))$. It seems reasonable to assume that when R_k is small, the area ΔT_k of T_k is approximately the same as the area ΔS_k of S_k.

(a)

(b) Enlargement of R_k, S_k, and T_k

FIGURE 9.13.1 What is the area of the surface above R?

To find the area of T_k let us choose $(x_k, y_k, 0)$ at a corner of R_k as shown in Figure 9.13.1(b). The indicated vectors $\mathbf{u}$ and $\mathbf{v}$, which form two sides of T_k, are given by

$$\mathbf{u} = \Delta x_k\mathbf{i} + f_x(x_k, y_k)\,\Delta x_k\mathbf{k},$$

$$\mathbf{v} = \Delta y_k\mathbf{j} + f_y(x_k, y_k)\,\Delta y_k\mathbf{k},$$

where $f_x(x_k, y_k)$ and $f_y(x_k, y_k)$ are the slopes of the lines containing **u** and **v**, respectively. Now from (16) of Section 7.4 we know that $\Delta T_k = \|\mathbf{u} \times \mathbf{v}\|$, where

$$\mathbf{u} \times \mathbf{v} = \begin{vmatrix} \mathbf{i} & \mathbf{j} & \mathbf{k} \\ \Delta x_k & 0 & f_x(x_k, y_k)\Delta x_k \\ 0 & \Delta y_k & f_y(x_k, y_k)\Delta y_k \end{vmatrix} = [-f_x(x_k, y_k)\mathbf{i} - f_y(x_k, y_k)\mathbf{j} + \mathbf{k}]\Delta x_k\, \Delta y_k.$$

In other words,

$$\Delta T_k = \sqrt{[f_x(x_k, y_k)]^2 + [f_y(x_k, y_k)]^2 + 1}\; \Delta x_k\, \Delta y_k$$
$$= \sqrt{1 + [f_x(x_k, y_k)]^2 + [f_y(x_k, y_k)]^2}\; \Delta A_k.$$

Consequently, the area A is approximately

$$\sum_{k=1}^{n} \sqrt{1 + [f_x(x_k, y_k)]^2 + [f_y(x_k, y_k)]^2}\; \Delta A_k.$$

Taking the limit of the foregoing sum as $\|P\| \to 0$ leads us to the next definition.

DEFINITION 9.13.1 Surface Area

Let f be a function for which the first partial derivatives f_x and f_y are continuous on a closed region R. Then the **area of the surface over R** is given by

$$A(S) = \iint_R \sqrt{1 + [f_x(x, y)]^2 + [f_y(x, y)]^2}\; dA. \tag{2}$$

One could have almost guessed the form of (2) by naturally extending the one-variable structure of (1) to two variables.

EXAMPLE 1 Surface Area

Find the surface area of that portion of the sphere $x^2 + y^2 + z^2 = a^2$ that is above the xy-plane and within the cylinder $x^2 + y^2 = b^2$, $0 < b < a$.

SOLUTION If we define $z = f(x, y)$ by $f(x, y) = \sqrt{a^2 - x^2 - y^2}$, then

$$f_x(x, y) = \frac{-x}{\sqrt{a^2 - x^2 - y^2}} \quad \text{and} \quad f_y(x, y) = \frac{-y}{\sqrt{a^2 - x^2 - y^2}}$$

and so

$$1 + [f_x(x, y)]^2 + [f_y(x, y)]^2 = \frac{a^2}{a^2 - x^2 - y^2}.$$

Hence, (2) is

$$A(S) = \iint_R \frac{a}{\sqrt{a^2 - x^2 - y^2}}\; dA,$$

where R is indicated in **FIGURE 9.13.2**. To evaluate this double integral, we change to polar coordinates:

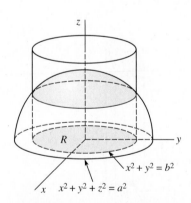

FIGURE 9.13.2 Portion of a sphere in Example 1

$$A(S) = a \int_0^{2\pi} \int_0^b (a^2 - r^2)^{-1/2} r\, dr\, d\theta$$
$$= a \int_0^{2\pi} \left[-(a^2 - r^2)^{1/2} \right]_0^b d\theta = a(a - \sqrt{a^2 - b^2}) \int_0^{2\pi} d\theta$$
$$= 2\pi a(a - \sqrt{a^2 - b^2}) \text{ square units.} \qquad \equiv$$

Differential of Surface Area The function

$$dS = \sqrt{1 + [f_x(x, y)]^2 + [f_y(x, y)]^2} \, dA \qquad (3)$$

is called the **differential of the surface area**. We will use this function in the discussion that follows.

Surface Integral As we have seen, a double integral

$$\iint\limits_R f(x, y) \, dA$$

is a generalization of the definite integral $\int_a^b f(x) \, dx$. The surface area integral (2) is a generalization of the arc length integral (1). We are now going to consider a generalization of the line integral $\int_C G(x, y) \, ds$. This generalization is called a **surface integral**.

1. Let $w = G(x, y, z)$ be defined in a region of 3-space that contains a surface S, which is the graph of a function $z = f(x, y)$. Let the projection R of the surface onto the xy-plane be either a Type I or a Type II region.

2. Divide the surface S into n patches S_k with areas ΔS_k that correspond to a partition P of R into n rectangles R_k with areas ΔA_k.

3. Let $\|P\|$ be the **norm** of the partition or the length of the longest diagonal of the R_k.

4. Choose a sample point (x_k^*, y_k^*, z_k^*) on each patch S_k of surface area. See **FIGURE 9.13.3**.

5. Form the sum $\displaystyle\sum_{k=1}^{n} G(x_k^*, y_k^*, z_k^*) \, \Delta S_k$.

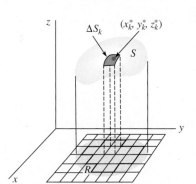

FIGURE 9.13.3 Sample point on kth patch

DEFINITION 9.13.2 Surface Integral

Let G be a function of three variables defined over a region of 3-space containing the surface S. Then the **surface integral of G over S** is given by

$$\iint\limits_S G(x, y, z) \, dS = \lim_{\|P\| \to 0} \sum_{k=1}^{n} G(x_k^*, y_k^*, z_k^*) \, \Delta S_k. \qquad (4)$$

Method of Evaluation If G, f, f_x, and f_y are continuous throughout a region containing S, we can evaluate (4) by means of a double integral. From (3) the left side of (4) becomes

$$\iint\limits_S G(x, y, z) \, dS = \iint\limits_R G(x, y, f(x, y))\sqrt{1 + [f_x(x, y)]^2 + [f_y(x, y)]^2} \, dA. \qquad (5)$$

Note that when $G = 1$, (5) reduces to formula (2) for surface area.

Projection of S into Other Planes If $y = g(x, z)$ is the equation of a surface S that projects onto a region R of the xz-plane, then

$$\iint\limits_S G(x, y, z) \, dS = \iint\limits_R G(x, g(x, z), z)\sqrt{1 + [g_x(x, z)]^2 + [g_z(x, z)]^2} \, dA. \qquad (6)$$

Similarly, if $x = h(y, z)$ is the equation of a surface that projects onto the yz-plane, then the analogue of (5) is

$$\iint\limits_S G(x, y, z) \, dS = \iint\limits_R G(h(y, z), y, z)\sqrt{1 + [h_y(y, z)]^2 + [h_z(y, z)]^2} \, dA. \qquad (7)$$

Mass of a Surface Suppose $\rho(x, y, z)$ represents the density of a surface at any point, or mass per unit surface area; then the **mass m** of the surface is

$$m = \iint\limits_S \rho(x, y, z) \, dS. \qquad (8)$$

FIGURE 9.13.4 Surface in Example 2

FIGURE 9.13.5 Surface in Example 3

(a) Two-sided surface

(b) One-sided surface

FIGURE 9.13.6 Oriented surface in (a); non-oriented surface in (b)

EXAMPLE 2 **Mass of a Surface**

Find the mass of the surface of the paraboloid $z = 1 + x^2 + y^2$ in the first octant for $1 \le z \le 5$ if the density at a point P on the surface is directly proportional to its distance from the xy-plane.

SOLUTION The surface in question and its projection onto the xy-plane are shown in **FIGURE 9.13.4**. Now, since $\rho(x, y, z) = kz$ and $z = 1 + x^2 + y^2$, (8) and (5) give

$$m = \iint_S kz \, dS = k \iint_R (1 + x^2 + y^2)\sqrt{1 + 4x^2 + 4y^2} \, dA.$$

By changing to polar coordinates, we obtain

$$m = k \int_0^{\pi/2} \int_0^2 (1 + r^2)\sqrt{1 + 4r^2} \, r \, dr \, d\theta$$

$$= k \int_0^{\pi/2} \int_0^2 [r(1 + 4r^2)^{1/2} + r^3(1 + 4r^2)^{1/2}] \, dr \, d\theta \quad \leftarrow \text{integration by parts}$$

$$= k \int_0^{\pi/2} \left[\frac{1}{12}(1 + 4r^2)^{3/2} + \frac{1}{12}r^2(1 + 4r^2)^{3/2} - \frac{1}{120}(1 + 4r^2)^{5/2} \right]_0^2 d\theta$$

$$= \frac{k\pi}{2} \left[\frac{5(17)^{3/2}}{12} - \frac{17^{5/2}}{120} - \frac{3}{40} \right] \approx 30.16 \, k. \qquad \equiv$$

EXAMPLE 3 **Evaluating a Surface Integral**

Evaluate $\iint_S xz^2 \, dS$, where S is that portion of the cylinder $y = 2x^2 + 1$ in the first octant bounded by $x = 0$, $x = 2$, $z = 4$, and $z = 8$.

SOLUTION We shall use (6) with $g(x, z) = 2x^2 + 1$ and R the rectangular region in the xz-plane shown in **FIGURE 9.13.5**. Since $g_x(x, z) = 4x$ and $g_z(x, z) = 0$, it follows that

$$\iint_S xz^2 \, dS = \int_0^2 \int_4^8 xz^2\sqrt{1 + 16x^2} \, dz \, dx$$

$$= \int_0^2 \frac{z^3}{3} x\sqrt{1 + 16x^2} \Big]_4^8 \, dx = \frac{448}{3} \int_0^2 x(1 + 16x^2)^{1/2} \, dx$$

$$= \frac{28}{9}(1 + 16x^2)^{3/2} \Big]_0^2 = \frac{28}{9}[65^{3/2} - 1] \approx 1627.3. \qquad \equiv$$

■ **Orientable Surfaces** In Example 5, we are going to evaluate a surface integral of a vector field. In order to do this we need to examine the concept of an **orientable surface**. Roughly, an orientable surface S, such as that given in **FIGURE 9.13.6(a)**, has two sides that could be painted different colors. The Möbius strip* shown in Figure 9.13.6(b) is not an orientable surface and is one-sided. A person who starts to paint the surface of a Möbius strip at a point will paint the entire surface and return to the starting point.

Specifically, we say a smooth surface S is **orientable** or is an **oriented surface** if there exists a continuous unit normal vector function $\mathbf{n}$ defined at each point (x, y, z) on the surface. The vector field $\mathbf{n}(x, y, z)$ is called the **orientation** of S. But since a unit normal to the surface S at (x, y, z) can be either $\mathbf{n}(x, y, z)$ or $-\mathbf{n}(x, y, z)$, an orientable surface has two orientations. See **FIGURE 9.13.7**(a)–(c). The Möbius strip shown again in Figure 9.13.7(d) is not an oriented surface, since if a unit normal $\mathbf{n}$ starts at P on the surface and moves *once* around the strip on the curve C, it ends up on the "opposite side" of the strip at P and so points in the opposite direction.

*To construct a Möbius strip cut out a long strip of paper, give one end a half-twist, and then attach the ends with tape.

A surface S defined by $z = f(x, y)$ has an **upward orientation** (Figure 9.13.7(b)) when the unit normals are directed upward—that is, have positive **k** components, and it has a **downward orientation** (Figure 9.13.7(c)) when the unit normals are directed downward—that is, have negative **k** components.

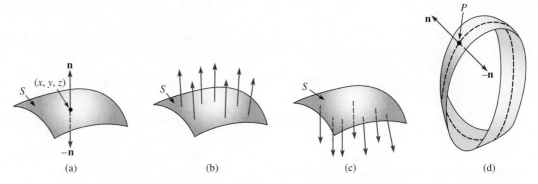

FIGURE 9.13.7 Upward orientation in (b); downward orientation in (c)

If a smooth surface S is defined by $g(x, y, z) = 0$, then recall that a unit normal is

$$\mathbf{n} = \frac{1}{\|\nabla g\|} \nabla g, \tag{9}$$

where $\nabla g = \dfrac{\partial g}{\partial x} \mathbf{i} + \dfrac{\partial g}{\partial y} \mathbf{j} + \dfrac{\partial g}{\partial z} \mathbf{k}$ is the gradient of g. If S is defined by $z = f(x, y)$, then we can use $g(x, y, z) = z - f(x, y) = 0$ or $g(x, y, z) = f(x, y) - z = 0$ depending on the orientation of S.

As we shall see in the next example, the two orientations of an orientable closed surface are **outward** and **inward**.

(a)

(b)

FIGURE 9.13.8 Sphere in Example 4

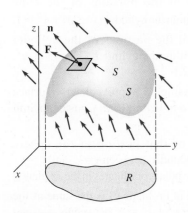

FIGURE 9.13.9 Surface S in (10)

EXAMPLE 4 **Orientations of a Surface**

Consider the sphere of radius $a > 0$: $x^2 + y^2 + z^2 = a^2$. If we define $g(x, y, z) = x^2 + y^2 + z^2 - a^2$, then

$$\nabla g = 2x\mathbf{i} + 2y\mathbf{j} + 2z\mathbf{k} \quad \text{and} \quad \|\nabla g\| = \sqrt{4x^2 + 4y^2 + 4z^2} = 2a.$$

Then the two orientations of the surface are

$$\mathbf{n} = \frac{x}{a}\mathbf{i} + \frac{y}{a}\mathbf{j} + \frac{z}{a}\mathbf{k} \quad \text{and} \quad \mathbf{n}_1 = -\mathbf{n} = -\frac{x}{a}\mathbf{i} - \frac{y}{a}\mathbf{j} - \frac{z}{a}\mathbf{k}.$$

The vector field $\mathbf{n}$ defines an outward orientation, whereas $\mathbf{n}_1 = -\mathbf{n}$ defines an inward orientation. See **FIGURE 9.13.8**. ≡

Integrals of Vector Fields If $\mathbf{F}(x, y, z) = P(x, y, z)\mathbf{i} + Q(x, y, z)\mathbf{j} + R(x, y, z)\mathbf{k}$ is the velocity field of a fluid, then, as we saw in Figure 9.7.3, the volume of the fluid flowing through an element of surface area ΔS per unit time is approximated by

$$(\text{height})(\text{area of base}) = (\text{comp}_{\mathbf{n}} \mathbf{F}) \, \Delta S = (\mathbf{F} \cdot \mathbf{n}) \, \Delta S,$$

where $\mathbf{n}$ is a unit normal to the surface. See **FIGURE 9.13.9**. The total volume of a fluid passing through S per unit time is called the **flux of F through S** and is given by

$$\text{flux} = \iint_S (\mathbf{F} \cdot \mathbf{n}) \, dS. \tag{10}$$

In the case of a closed surface S, if $\mathbf{n}$ is the outer (inner) normal, then (10) gives the volume of fluid flowing out (in) through S per unit time.

FIGURE 9.13.10 Surface in Example 5

FIGURE 9.13.11 Piecewise-defined surface

EXAMPLE 5 **Flux Through a Surface**

Let $\mathbf{F}(x, y, z) = z\mathbf{j} + z\mathbf{k}$ represent the flow of a liquid. Find the flux of $\mathbf{F}$ through the surface S given by that portion of the plane $z = 6 - 3x - 2y$ in the first octant oriented upward.

SOLUTION The vector field and the surface are illustrated in **FIGURE 9.13.10**. By defining the plane by $g(x, y, z) = 3x + 2y + z - 6 = 0$, we see that a unit normal with a positive $\mathbf{k}$ component is

$$\mathbf{n} = \frac{\nabla g}{\|\nabla g\|} = \frac{3}{\sqrt{14}}\mathbf{i} + \frac{2}{\sqrt{14}}\mathbf{j} + \frac{1}{\sqrt{14}}\mathbf{k}.$$

Hence,

$$\text{flux} = \iint_S (\mathbf{F} \cdot \mathbf{n}) \, dS = \frac{1}{\sqrt{14}} \iint_S 3z \, dS.$$

With R the projection of the surface onto the xy-plane, we find from (10) that

$$\text{flux} = \frac{1}{\sqrt{14}} \iint_R 3(6 - 3x - 2y)(\sqrt{14} \, dA)$$

$$= 3 \int_0^2 \int_0^{3 - 3x/2} (6 - 3x - 2y) \, dy \, dx = 18. \qquad \equiv$$

Depending on the nature of the vector field, the integral in (10) can represent other kinds of flux. For example, (10) could also give electric flux, magnetic flux, flux of heat, and so on.

REMARKS

If the surface S is piecewise defined, we express a surface integral over S as the sum of the surface integrals over the various pieces of the surface. For example, suppose S is the orientable piecewise-smooth closed surface bounded by the paraboloid $z = x^2 + y^2$ (S_1) and the plane $z = 1$ (S_2). Then the flux of a vector field $\mathbf{F}$ out of the surface S is

$$\iint_S \mathbf{F} \cdot \mathbf{n} \, dS = \iint_{S_1} \mathbf{F} \cdot \mathbf{n} \, dS + \iint_{S_2} \mathbf{F} \cdot \mathbf{n} \, dS,$$

where we take S_1 oriented downward and S_2 oriented upward. See **FIGURE 9.13.11** and Problem 35 in Exercises 9.13.

9.13 Exercises Answers to selected odd-numbered problems begin on page ANS-26.

1. Find the surface area of that portion of the plane $2x + 3y + 4z = 12$ that is bounded by the coordinate planes in the first octant.

2. Find the surface area of that portion of the plane $2x + 3y + 4z = 12$ that is above the region in the first quadrant bounded by the graph $r = \sin 2\theta$.

3. Find the surface area of that portion of the cylinder $x^2 + z^2 = 16$ that is above the region in the first quadrant bounded on the graphs of $x = 0$, $x = 2$, $y = 0$, $y = 5$.

4. Find the surface area of that portion of the paraboloid $z = x^2 + y^2$ that is below the plane $z = 2$.

5. Find the surface area of that portion of the paraboloid $z = 4 - x^2 - y^2$ that is above the xy-plane.

6. Find the surface area of those portions of the sphere $x^2 + y^2 + z^2 = 2$ that are within the cone $z^2 = x^2 + y^2$.

7. Find the surface area of the portion of the sphere $x^2 + y^2 + z^2 = 25$ that is above the region in the first quadrant bounded by the graphs of $x = 0$, $y = 0$, $4x^2 + y^2 = 25$. [*Hint*: Integrate first with respect to x.]

8. Find the surface area of that portion of the graph of $z = x^2 - y^2$ that is in the first octant within the cylinder $x^2 + y^2 = 4$.

9. Find the surface area of the portions of the sphere $x^2 + y^2 + z^2 = a^2$ that are within the cylinder $x^2 + y^2 = ay$.

10. Find the surface area of the portions of the cone $z^2 = \frac{1}{4}(x^2 + y^2)$ that are within the cylinder $(x - 1)^2 + y^2 = 1$.

11. Find the surface area of the portions of the cylinder $y^2 + z^2 = a^2$ that are within the cylinder $x^2 + y^2 = a^2$. [*Hint*: See Figure 9.10.12.]

12. Use the result given in Example 1 to prove that the surface area of a sphere of radius a is $4\pi a^2$. [*Hint*: Consider a limit as $b \to a$.]

13. Find the surface area of that portion of the sphere $x^2 + y^2 + z^2 = a^2$ that is bounded between $y = c_1$ and $y = c_2$, $0 < c_1 < c_2 < a$. [*Hint*: Use polar coordinates in the xz-plane.]

14. Show that the area found in Problem 13 is the same as the surface area of the cylinder $x^2 + z^2 = a^2$ between $y = c_1$ and $y = c_2$.

In Problems 15–24, evaluate the surface integral $\iint_S G(x, y, z)\, dS$.

15. $G(x, y, z) = x$; S the portion of the cylinder $z = 2 - x^2$ in the first octant bounded by $x = 0$, $y = 0$, $y = 4$, $z = 0$

16. $G(x, y, z) = xy(9 - 4z)$; same surface as in Problem 15

17. $G(x, y, z) = xz^3$; S the cone $z = \sqrt{x^2 + y^2}$ inside the cylinder $x^2 + y^2 = 1$

18. $G(x, y, z) = x + y + z$; S the cone $z = \sqrt{x^2 + y^2}$ between $z = 1$ and $z = 4$

19. $G(x, y, z) = (x^2 + y^2)z$; S that portion of the sphere $x^2 + y^2 + z^2 = 36$ in the first octant

20. $G(x, y, z) = z^2$; S that portion of the plane $z = x + 1$ within the cylinder $y = 1 - x^2$, $0 \le y \le 1$

21. $G(x, y, z) = xy$; S that portion of the paraboloid $2z = 4 - x^2 - y^2$ within $0 \le x \le 1$, $0 \le y \le 1$

22. $G(x, y, z) = 2z$; S that portion of the paraboloid $2z = 1 + x^2 + y^2$ in the first octant bounded by $x = 0$, $y = \sqrt{3}x$, $z = 1$

23. $G(x, y, z) = 24\sqrt{yz}$; S that portion of the cylinder $y = x^2$ in the first octant bounded by $y = 0$, $y = 4$, $z = 0$, $z = 3$

24. $G(x, y, z) = (1 + 4y^2 + 4z^2)^{1/2}$; S that portion of the paraboloid $x = 4 - y^2 - z^2$ in the first octant outside the cylinder $y^2 + z^2 = 1$

In Problems 25 and 26, evaluate $\iint_S (3z^2 + 4yz)\, dS$, where S is that portion of the plane $x + 2y + 3z = 6$ in the first octant. Use the projection of S onto the coordinate plane indicated in the given figure.

25.

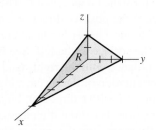

FIGURE 9.13.12 Region R for Problem 25

26.

FIGURE 9.13.13 Region R for Problem 26

In Problems 27 and 28, find the mass of the given surface with the indicated density function.

27. S that portion of the plane $x + y + z = 1$ in the first octant; density at a point P directly proportional to the square of the distance from the yz-plane

28. S the hemisphere $z = \sqrt{4 - x^2 - y^2}$; $\rho(x, y, z) = |xy|$

In Problems 29–34, let $\mathbf{F}$ be a vector field. Find the flux of $\mathbf{F}$ through the given surface. Assume the surface S is oriented upward.

29. $\mathbf{F} = x\mathbf{i} + 2z\mathbf{j} + y\mathbf{k}$; S that portion of the cylinder $y^2 + z^2 = 4$ in the first octant bounded by $x = 0$, $x = 3$, $y = 0$, $z = 0$

30. $\mathbf{F} = z\mathbf{k}$; S that part of the paraboloid $z = 5 - x^2 - y^2$ inside the cylinder $x^2 + y^2 = 4$

31. $\mathbf{F} = x\mathbf{i} + y\mathbf{j} + z\mathbf{k}$; same surface S as in Problem 30

32. $\mathbf{F} = -x^3y\mathbf{i} + yz^3\mathbf{j} + xy^3\mathbf{k}$; S that portion of the plane $z = x + 3$ in the first octant within the cylinder $x^2 + y^2 = 2x$

33. $\mathbf{F} = \frac{1}{2}x^2\mathbf{i} + \frac{1}{2}y^2\mathbf{j} + z\mathbf{k}$; S that portion of the paraboloid $z = 4 - x^2 - y^2$ for $0 \le z \le 4$

34. $\mathbf{F} = e^y\mathbf{i} + e^x\mathbf{j} + 18y\mathbf{k}$; S that portion of the plane $x + y + z = 6$ in the first octant

35. Find the flux of $\mathbf{F} = y^2\mathbf{i} + x^2\mathbf{j} + 5z\mathbf{k}$ out of the closed surface S given in Figure 9.13.11.

36. Find the flux of $\mathbf{F} = -y\mathbf{i} + x\mathbf{j} + 6z^2\mathbf{k}$ out of the closed surface S bounded by the paraboloids $z = 4 - x^2 - y^2$ and $z = x^2 + y^2$.

37. Let $T(x, y, z) = x^2 + y^2 + z^2$ represent temperature and let the "flow" of heat be given by the vector field $\mathbf{F} = -\nabla T$. Find the flux of heat out of the sphere $x^2 + y^2 + z^2 = a^2$. [*Hint*: The surface area of a sphere of radius a is $4\pi a^2$.]

38. Find the flux of $\mathbf{F} = x\mathbf{i} + y\mathbf{j} + z\mathbf{k}$ out of the unit cube $0 \le x \le 1$, $0 \le y \le 1$, $0 \le z \le 1$. See **FIGURE 9.13.14**. Use the fact that the flux out of the cube is the sum of the fluxes out of the sides.

FIGURE 9.13.14 Cube in Problem 38

39. Coulomb's law states that the electric field $\mathbf{E}$ due to a point charge q at the origin is given by $\mathbf{E} = kq\mathbf{r}/\|\mathbf{r}\|^3$, where k is a constant and $\mathbf{r} = x\mathbf{i} + y\mathbf{j} + z\mathbf{k}$. Determine the flux out of a sphere $x^2 + y^2 + z^2 = a^2$.

40. If $\sigma(x, y, z)$ is charge density in an electrostatic field, then the total charge on a surface S is $Q = \iint_S \sigma(x, y, z)\, dS$. Find the total charge on that part of the hemisphere $z = \sqrt{16 - x^2 - y^2}$ that is inside the cylinder $x^2 + y^2 = 9$ if the charge density at a point P on the surface is directly proportional to distance from the xy-plane.

41. The coordinates of the centroid of a surface are given by

$$\bar{x} = \frac{\iint_S x\, dS}{A(S)}, \qquad \bar{y} = \frac{\iint_S y\, dS}{A(S)}, \qquad \bar{z} = \frac{\iint_S z\, dS}{A(S)},$$

where $A(S)$ is the area of the surface. Find the centroid of that portion of the plane $2x + 3y + z = 6$ in the first octant.

42. Use the information in Problem 41 to find the centroid of the hemisphere $z = \sqrt{a^2 - x^2 - y^2}$.

43. Let $z = f(x, y)$ be the equation of a surface S and $\mathbf{F}$ be the vector field $\mathbf{F}(x, y, z) = P(x, y, z)\mathbf{i} + Q(x, y, z)\mathbf{j} + R(x, y, z)\mathbf{k}$. Show that $\iint_S (\mathbf{F} \cdot \mathbf{n})\, dS$ equals

$$\iint_R \left[-P(x, y, z)\frac{\partial z}{\partial x} - Q(x, y, z)\frac{\partial z}{\partial y} + R(x, y, z) \right] dA.$$

9.14 Stokes' Theorem

INTRODUCTION Green's theorem of the preceding section has two vector forms. In this and in Section 9.16 we shall generalize these forms to three dimensions.

■ **Vector Form of Green's Theorem** If $\mathbf{F}(x, y) = P(x, y)\mathbf{i} + Q(x, y)\mathbf{j}$ is a two-dimensional vector field, then

$$\text{curl } \mathbf{F} = \nabla \times \mathbf{F} = \begin{vmatrix} \mathbf{i} & \mathbf{j} & \mathbf{k} \\ \dfrac{\partial}{\partial x} & \dfrac{\partial}{\partial y} & \dfrac{\partial}{\partial z} \\ P & Q & 0 \end{vmatrix} = \left(\frac{\partial Q}{\partial x} - \frac{\partial P}{\partial y} \right) \mathbf{k}.$$

From (12) and (13) of Section 9.8, Green's theorem can be written in vector notation as

$$\oint_C \mathbf{F} \cdot d\mathbf{r} = \oint_C \mathbf{F} \cdot \mathbf{T} \, ds = \iint_R (\text{curl } \mathbf{F}) \cdot \mathbf{k} \, dA; \tag{1}$$

that is, the line integral of the tangential component of $\mathbf{F}$ is the double integral of the normal component of curl $\mathbf{F}$.

■ **Green's Theorem in 3-Space** The vector form of Green's theorem given in (1) relates a line integral around a piecewise-smooth simple closed curve C forming the boundary of a plane region R to a double integral over R. Green's theorem in 3-space relates a line integral around a piecewise-smooth simple closed curve C forming the boundary of a surface S with a surface integral over S. Suppose $z = f(x, y)$ is a continuous function whose graph is a piecewise-smooth orientable surface over a region R on the xy-plane. Let C form the boundary of S and let the projection of C onto the xy-plane form the boundary of R. The positive direction on C is induced by the orientation of the surface S; the positive direction on C corresponds to the direction a person would have to walk on C to have his or her head point in the direction of the orientation of the surface while keeping the surface to the left. See **FIGURE 9.14.1**. More precisely, the positive orientation of C is in accordance with the right-hand rule: If the thumb of the right hand points in the direction of the orientation of the surface, then roughly the fingers of the right hand wrap around the surface in the positive direction. Finally, let $\mathbf{T}$ be a unit tangent vector to C that points in the positive direction. The three-dimensional form of Green's theorem, which we now give, is called **Stokes' theorem** after the Irish mathematical physicist **George G. Stokes** (1819–1903).

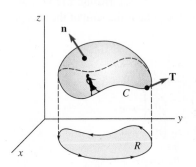

FIGURE 9.14.1 Boundary C of surface S has positive orientation

THEOREM 9.14.1 Stokes' Theorem

Let S be a piecewise-smooth orientable surface bounded by a piecewise-smooth simple closed curve C. Let $\mathbf{F}(x, y, z) = P(x, y, z)\mathbf{i} + Q(x, y, z)\mathbf{j} + R(x, y, z)\mathbf{k}$ be a vector field for which P, Q, and R are continuous and have continuous first partial derivatives in a region of 3-space containing S. If C is traversed in the positive direction, then

$$\oint_C \mathbf{F} \cdot d\mathbf{r} = \oint_C (\mathbf{F} \cdot \mathbf{T}) \, dS = \iint_S (\text{curl } \mathbf{F}) \cdot \mathbf{n} \, dS, \tag{2}$$

where $\mathbf{n}$ is a unit normal to S in the direction of the orientation of S.

PARTIAL PROOF: Suppose the surface S is oriented upward and is defined by a function $z = f(x, y)$ that has continuous second partial derivatives. From Definition 9.7.1 we have

$$\text{curl } \mathbf{F} = \left(\frac{\partial R}{\partial y} - \frac{\partial Q}{\partial z} \right)\mathbf{i} + \left(\frac{\partial P}{\partial z} - \frac{\partial R}{\partial x} \right)\mathbf{j} + \left(\frac{\partial Q}{\partial x} - \frac{\partial P}{\partial y} \right)\mathbf{k}.$$

Furthermore, if we write $g(x, y, z) = z - f(x, y) = 0$, then

$$\mathbf{n} = \frac{\nabla g}{\|\nabla g\|} = \frac{-\dfrac{\partial f}{\partial x}\mathbf{i} - \dfrac{\partial f}{\partial y}\mathbf{j} + \mathbf{k}}{\sqrt{1 + \left(\dfrac{\partial f}{\partial x}\right)^2 + \left(\dfrac{\partial f}{\partial y}\right)^2}}.$$

Hence,

$$\iint\limits_{S} (\text{curl } \mathbf{F}) \cdot \mathbf{n}\, dS = \iint\limits_{R} \left[-\left(\frac{\partial R}{\partial y} - \frac{\partial Q}{\partial z}\right)\frac{\partial f}{\partial x} - \left(\frac{\partial P}{\partial z} - \frac{\partial R}{\partial x}\right)\frac{\partial f}{\partial y} + \left(\frac{\partial Q}{\partial x} - \frac{\partial P}{\partial y}\right) \right] dA. \quad (3)$$

Our goal is now to show that $\oint_C \mathbf{F} \cdot d\mathbf{r}$ reduces to (3).

If C_{xy} is the projection of C onto the xy-plane and has the parametric equations $x = x(t), y = y(t)$, $a \leq t \leq b$, then parametric equations for C are $x = x(t), y = y(t), z = f(x(t), y(t)), a \leq t \leq b$. Thus,

$$\oint_C \mathbf{F} \cdot d\mathbf{r} = \oint_C P\, dx + Q\, dy + R\, dz$$

$$= \int_a^b \left[P\frac{dx}{dt} + Q\frac{dy}{dt} + R\left(\frac{\partial f}{\partial x}\frac{dx}{dt} + \frac{\partial f}{\partial y}\frac{dy}{dt}\right) \right] dt \quad \leftarrow \text{Chain Rule}$$

$$= \oint_{C_{xy}} \left(P + R\frac{\partial f}{\partial x} \right) dx + \left(Q + R\frac{\partial f}{\partial y} \right) dy \quad (4)$$

$$= \iint\limits_{R} \left[\frac{\partial}{\partial x}\left(Q + R\frac{\partial f}{\partial y} \right) - \frac{\partial}{\partial y}\left(P + R\frac{\partial f}{\partial x} \right) \right] dA. \quad \leftarrow \text{Green's theorem}$$

Now,

$$\frac{\partial}{\partial x}\left(Q + R\frac{\partial f}{\partial y} \right) = \frac{\partial}{\partial x}\left[Q(x, y, f(x, y)) + R(x, y, f(x, y))\frac{\partial f}{\partial y} \right]$$

$$= \frac{\partial Q}{\partial x} + \frac{\partial Q}{\partial z}\frac{\partial f}{\partial x} + R\frac{\partial^2 f}{\partial x\, \partial y} + \frac{\partial f}{\partial y}\left(\frac{\partial R}{\partial x} + \frac{\partial R}{\partial z}\frac{\partial f}{\partial x} \right) \quad \begin{array}{l}\leftarrow \text{Chain and} \\ \text{Product Rules}\end{array} \quad (5)$$

$$= \frac{\partial Q}{\partial x} + \frac{\partial Q}{\partial z}\frac{\partial f}{\partial x} + R\frac{\partial^2 f}{\partial x\, \partial y} + \frac{\partial R}{\partial x}\frac{\partial f}{\partial y} + \frac{\partial R}{\partial z}\frac{\partial f}{\partial y}\frac{\partial f}{\partial x}.$$

Similarly,

$$\frac{\partial}{\partial y}\left(P + R\frac{\partial f}{\partial x} \right) = \frac{\partial P}{\partial y} + \frac{\partial P}{\partial z}\frac{\partial f}{\partial y} + R\frac{\partial^2 f}{\partial y\, \partial x} + \frac{\partial R}{\partial y}\frac{\partial f}{\partial x} + \frac{\partial R}{\partial z}\frac{\partial f}{\partial x}\frac{\partial f}{\partial y}. \quad (6)$$

Subtracting (6) from (5) and using the fact that $\partial^2 f / \partial x\, \partial y = \partial^2 f / \partial y\, \partial x$, we see that (4) becomes, after rearranging,

$$\iint\limits_{R} \left[-\left(\frac{\partial R}{\partial y} - \frac{\partial Q}{\partial z}\right)\frac{\partial f}{\partial x} - \left(\frac{\partial P}{\partial z} - \frac{\partial R}{\partial x}\right)\frac{\partial f}{\partial y} + \left(\frac{\partial Q}{\partial x} - \frac{\partial P}{\partial y}\right) \right] dA.$$

This last expression is the same as the right side of (3), which was to be shown. ≡

EXAMPLE 1 Verifying Stokes' Theorem

Let S be the part of the cylinder $z = 1 - x^2$ for $0 \leq x \leq 1, -2 \leq y \leq 2$. Verify Stokes' theorem for the vector field $\mathbf{F} = xy\mathbf{i} + yz\mathbf{j} + xz\mathbf{k}$. Assume S is oriented upward.

SOLUTION The surface S, the curve C (which is composed of the union of C_1, C_2, C_3, and C_4), and the region R are shown in **FIGURE 9.14.2**.

Surface Integral: From $\mathbf{F} = xy\mathbf{i} + yz\mathbf{j} + xz\mathbf{k}$, we find

$$\operatorname{curl} \mathbf{F} = \begin{vmatrix} \mathbf{i} & \mathbf{j} & \mathbf{k} \\ \dfrac{\partial}{\partial x} & \dfrac{\partial}{\partial y} & \dfrac{\partial}{\partial z} \\ xy & yz & xz \end{vmatrix} = -y\mathbf{i} - z\mathbf{j} - x\mathbf{k}.$$

FIGURE 9.14.2 Surface S and region R in Example 1

Now, if $g(x, y, z) = z + x^2 - 1 = 0$ defines the cylinder, then the upper normal is

$$\mathbf{n} = \frac{\nabla g}{\|\nabla g\|} = \frac{2x\mathbf{i} + \mathbf{k}}{\sqrt{4x^2 + 1}}.$$

Therefore,
$$\iint\limits_{S} (\operatorname{curl} \mathbf{F} \cdot \mathbf{n}) \, dS = \iint\limits_{S} \frac{-2xy - x}{\sqrt{4x^2 + 1}} \, dS.$$

To evaluate the latter surface integral, we use (5) of Section 9.13:

$$
\begin{aligned}
\iint\limits_{S} \frac{-2xy - x}{\sqrt{4x^2 + 1}} \, dS &= \iint\limits_{R} (-2xy - x) \, dA \\
&= \int_0^1 \int_{-2}^{2} (-2xy - x) \, dy \, dx \\
&= \int_0^1 \left[-xy^2 - xy \right]_{-2}^{2} \, dx \\
&= \int_0^1 (-4x) \, dx = -2.
\end{aligned}
$$

(7)

Line Integral: We write $\oint_C = \int_{C_1} + \int_{C_2} + \int_{C_3} + \int_{C_4}$.

On C_1: $x = 1, z = 0, dx = 0, dz = 0$, so

$$\int_{C_1} y(0) + y(0) \, dy + 0 = 0.$$

On C_2: $y = 2, z = 1 - x^2, dy = 0, dz = -2x \, dx$, so

$$\int_{C_2} 2x \, dx + 2(1 - x^2)0 + x(1 - x^2)(-2x \, dx) = \int_1^0 (2x - 2x^2 + 2x^4) \, dx = -\frac{11}{15}.$$

On C_3: $x = 0$, $z = 1$, $dx = 0$, $dz = 0$, so

$$\int_{C_3} 0 + y \, dy + 0 = \int_2^{-2} y \, dy = 0.$$

On C_4: $y = -2$, $z = 1 - x^2$, $dy = 0$, $dz = -2x \, dx$, so

$$\int_{C_4} -2x \, dx - 2(1 - x^2)0 + x(1 - x^2)(-2x \, dx) = \int_0^1 (-2x - 2x^2 + 2x^4) \, dx = -\frac{19}{15}.$$

Hence, $\qquad \oint_C xy \, dx + yz \, dy + xz \, dz = 0 - \frac{11}{15} + 0 - \frac{19}{15} = -2,$

which, of course, agrees with (7). $\qquad\qquad\qquad\qquad\qquad\qquad\qquad\equiv$

EXAMPLE 2 **Using Stokes' Theorem**

Evaluate $\oint_C z \, dx + x \, dy + y \, dz$, where C is the trace of the cylinder $x^2 + y^2 = 1$ in the plane $y + z = 2$. Orient C counterclockwise as viewed from above. See **FIGURE 9.14.3**.

SOLUTION If $\mathbf{F} = z\mathbf{i} + x\mathbf{j} + y\mathbf{k}$, then

$$\text{curl } \mathbf{F} = \begin{vmatrix} \mathbf{i} & \mathbf{j} & \mathbf{k} \\ \dfrac{\partial}{\partial x} & \dfrac{\partial}{\partial y} & \dfrac{\partial}{\partial z} \\ z & x & y \end{vmatrix} = \mathbf{i} + \mathbf{j} + \mathbf{k}.$$

The given orientation of C corresponds to an upward orientation of the surface S. Thus, if $g(x, y, z) = y + z - 2 = 0$ defines the plane, then the upper normal is

$$\mathbf{n} = \frac{\nabla g}{\|\nabla g\|} = \frac{1}{\sqrt{2}}\mathbf{j} + \frac{1}{\sqrt{2}}\mathbf{k}.$$

Hence, from (2),

$$\oint_C \mathbf{F} \cdot d\mathbf{r} = \iint_S \left[(\mathbf{i} + \mathbf{j} + \mathbf{k}) \cdot \left(\frac{1}{\sqrt{2}}\mathbf{j} + \frac{1}{\sqrt{2}}\mathbf{k} \right) \right] dS$$

$$= \sqrt{2} \iint_S dS = \sqrt{2} \iint_R \sqrt{2} \, dA = 2\pi. \qquad\qquad \equiv$$

Note that if $\mathbf{F}$ is the gradient of a scalar function, then, in view of (5) in Section 9.7, (2) implies that the circulation $\oint_C \mathbf{F} \cdot d\mathbf{r}$ is zero. Conversely, it can be shown that if the circulation is zero for every simple closed curve, then $\mathbf{F}$ is the gradient of a scalar function. In other words, $\mathbf{F}$ is irrotational if and only if $\mathbf{F} = \nabla\phi$, where ϕ is a potential for $\mathbf{F}$. Equivalently, this gives a test for a conservative vector field:

$$\mathbf{F} \text{ is a conservative vector field if and only if curl } \mathbf{F} = \mathbf{0}.$$

Physical Interpretation of Curl In Section 9.8 we saw that if $\mathbf{F}$ is a velocity field of a fluid, then the circulation $\oint_C \mathbf{F} \cdot d\mathbf{r}$ of $\mathbf{F}$ around C is a measure of the amount by which the fluid tends to turn the curve C by circulating around it. The circulation of $\mathbf{F}$ is closely related to the curl of $\mathbf{F}$. To see this, suppose $P_0(x_0, y_0, z_0)$ is any point in the fluid and C_r is a small circle of radius r centered at P_0. See **FIGURE 9.14.4**. Then by Stokes' theorem,

$$\oint_{C_r} \mathbf{F} \cdot d\mathbf{r} = \iint_{S_r} (\text{curl } \mathbf{F}) \cdot \mathbf{n} \, dS. \qquad\qquad (8)$$

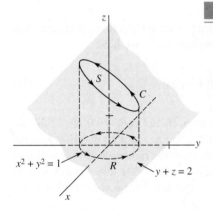

FIGURE 9.14.3 Curve C in Example 2

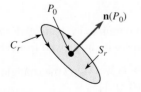

FIGURE 9.14.4 Curve C_r and surface S_r in (8)

Now at all points $P(x, y, z)$ within the small circle C_r, if we take curl $\mathbf{F}(P) \approx$ curl $\mathbf{F}(P_0)$, then (8) gives the approximation

$$\oint_{C_r} \mathbf{F} \cdot d\mathbf{r} \approx \iint_{S_r} (\text{curl } \mathbf{F}(P_0)) \cdot \mathbf{n}(P_0) \, dS$$

$$= (\text{curl } \mathbf{F}(P_0)) \cdot \mathbf{n}(P_0) \iint_{S_r} dS \qquad (9)$$

$$= (\text{curl } \mathbf{F}(P_0)) \cdot \mathbf{n}(P_0) A_r,$$

where A_r is the area (πr^2) of the circular surface S_r. As we let $r \to 0$, the approximation curl $\mathbf{F}(P) \approx$ curl $\mathbf{F}(P_0)$ becomes better, and so (9) yields

$$(\text{curl } \mathbf{F}(P_0)) \cdot \mathbf{n}(P_0) = \lim_{r \to 0} \frac{1}{A_r} \oint_{C_r} \mathbf{F} \cdot d\mathbf{r}. \qquad (10)$$

Thus we see that the normal component of curl $\mathbf{F}$ is the limiting value of the ratio of the circulation of $\mathbf{F}$ to the area of the circular surface. For a small but fixed value of r, we have

$$(\text{curl } \mathbf{F}(P_0)) \cdot \mathbf{n}(P_0) \approx \frac{1}{A_r} \oint_{C_r} \mathbf{F} \cdot d\mathbf{r}. \qquad (11)$$

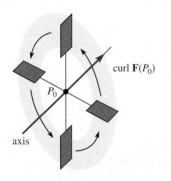

curl $\mathbf{F}(P_0)$

P_0

axis

FIGURE 9.14.5 Paddle wheel

Roughly then, the curl of $\mathbf{F}$ is the circulation of $\mathbf{F}$ per unit area. If curl $\mathbf{F}(P_0) \neq \mathbf{0}$, then the left-hand side of (11) is a maximum when the circle C_r is situated in a manner so that $\mathbf{n}(P_0)$ points in the same direction as curl $\mathbf{F}(P_0)$. In this case, the circulation on the right side of (11) will also be a maximum. Thus, a paddle wheel inserted into the fluid at P_0 will rotate fastest when its axis points in the direction of curl $\mathbf{F}(P_0)$. See **FIGURE 9.14.5**. Note, too, that the paddle wheel will not rotate if its axis is perpendicular to curl $\mathbf{F}(P_0)$.

S_1

C

(a)

C

S_2

(b)

FIGURE 9.14.6 Two surfaces with same boundary C

REMARKS

The value of the surface integral in (2) is determined solely by the integral around its boundary C. This basically means that the shape of the surface S is irrelevant. Assuming that the hypotheses of Theorem 9.14.1 are satisfied, then for two different surfaces S_1 and S_2 with the same orientation and with the same boundary C, we have

$$\oint_C \mathbf{F} \cdot d\mathbf{r} = \iint_{S_1} (\text{curl } \mathbf{F}) \cdot \mathbf{n} \, dS = \iint_{S_2} (\text{curl } \mathbf{F}) \cdot \mathbf{n} \, dS.$$

See **FIGURE 9.14.6** and Problems 17 and 18 in Exercises 9.14.

9.14 Exercises Answers to selected odd-numbered problems begin on page ANS-26.

In Problems 1–4, verify Stokes' theorem. Assume that the surface S is oriented upward.

1. $\mathbf{F} = 5y\mathbf{i} - 5x\mathbf{j} + 3\mathbf{k}$; S that portion of the plane $z = 1$ within the cylinder $x^2 + y^2 = 4$

2. $\mathbf{F} = 2z\mathbf{i} - 3x\mathbf{j} + 4y\mathbf{k}$; S that portion of the paraboloid $z = 16 - x^2 - y^2$ for $z \geq 0$

3. $\mathbf{F} = z\mathbf{i} + x\mathbf{j} + y\mathbf{k}$; S that portion of the plane $2x + y + 2z = 6$ in the first octant

4. $\mathbf{F} = x\mathbf{i} + y\mathbf{j} + z\mathbf{k}$; S that portion of the sphere $x^2 + y^2 + z^2 = 1$ for $z \geq 0$

In Problems 5–12, use Stokes' theorem to evaluate $\oint_C \mathbf{F} \cdot d\mathbf{r}$. Assume C is oriented counterclockwise as viewed from above.

5. $\mathbf{F} = (2z + x)\mathbf{i} + (y - z)\mathbf{j} + (x + y)\mathbf{k}$; C the triangle with vertices $(1, 0, 0)$, $(0, 1, 0)$, $(0, 0, 1)$

6. $\mathbf{F} = z^2 y \cos xy\,\mathbf{i} + z^2 x(1 + \cos xy)\mathbf{j} + 2z \sin xy\,\mathbf{k}$; C the boundary of the plane $z = 1 - y$ shown in **FIGURE 9.14.7** on page 571.

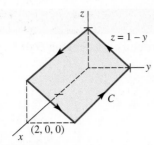

FIGURE 9.14.7 Curve C for Problem 6

7. $\mathbf{F} = xy\mathbf{i} + 2yz\mathbf{j} + xz\mathbf{k}$; C the boundary given in Problem 6

8. $\mathbf{F} = (x + 2z)\mathbf{i} + (3x + y)\mathbf{j} + (2y - z)\mathbf{k}$; C the curve of intersection of the plane $x + 2y + z = 4$ with the coordinate planes

9. $\mathbf{F} = y^3\mathbf{i} - x^3\mathbf{j} + z^3\mathbf{k}$; C the trace of the cylinder $x^2 + y^2 = 1$ in the plane $x + y + z = 1$

10. $\mathbf{F} = x^2y\mathbf{i} + (x + y^2)\mathbf{j} + xy^2z\mathbf{k}$; C the boundary of the surface shown in **FIGURE 9.14.8**

FIGURE 9.14.8 Curve C for Problem 10

11. $\mathbf{F} = x\mathbf{i} + x^3y^2\mathbf{j} + z\mathbf{k}$; C the boundary of the semi-ellipsoid $z = \sqrt{4 - 4x^2 - y^2}$ in the plane $z = 0$

12. $\mathbf{F} = z\mathbf{i} + x\mathbf{j} + y\mathbf{k}$; C the curve of intersection of the plane $x + y + z = 0$ and the sphere $x^2 + y^2 + z^2 = 1$ [*Hint*: Recall that the area of an ellipse $x^2/a^2 + y^2/b^2 = 1$ is πab.]

In Problems 13–16, use Stokes' theorem to evaluate $\iint_S (\text{curl } \mathbf{F}) \cdot \mathbf{n} \, dS$. Assume that the surface S is oriented upward.

13. $\mathbf{F} = 6yz\mathbf{i} + 5x\mathbf{j} + yze^{x^2}\mathbf{k}$; S that portion of the paraboloid $z = \frac{1}{4}x^2 + y^2$ for $0 \le z \le 4$

14. $\mathbf{F} = y\mathbf{i} + (y - x)\mathbf{j} + z^2\mathbf{k}$; S that portion of the sphere $x^2 + y^2 + (z - 4)^2 = 25$ for $z \ge 0$

15. $\mathbf{F} = 3x^2\mathbf{i} + 8x^3y\mathbf{j} + 3x^2y\mathbf{k}$; S that portion of the plane $z = x$ that lies inside the rectangular cylinder defined by the planes $x = 0, y = 0, x = 2, y = 2$

16. $\mathbf{F} = 2xy^2z\mathbf{i} + 2x^2yz\mathbf{j} + (x^2y^2 - 6x)\mathbf{k}$; S that portion of the plane $z = y$ that lies inside the cylinder $x^2 + y^2 = 1$

17. Use Stokes' theorem to evaluate

$$\oint_C z^2 e^{x^2} dx + xy^2 \, dy + \tan^{-1} y \, dz,$$

where C is the circle $x^2 + y^2 = 9$, by finding a surface S with C as its boundary and such that the orientation of C is counterclockwise as viewed from above.

18. Consider the surface integral $\iint_S (\text{curl } \mathbf{F}) \cdot \mathbf{n} \, dS$, where $\mathbf{F} = xyz\mathbf{k}$ and S is that portion of the paraboloid $z = 1 - x^2 - y^2$ for $z \ge 0$ oriented upward.
 (a) Evaluate the surface integral by the method of Section 9.13; that is, do not use Stokes' theorem.
 (b) Evaluate the surface integral by finding a simpler surface that is oriented upward and has the same boundary as the paraboloid.
 (c) Use Stokes' theorem to verify the result in part (b).

9.15 Triple Integrals

INTRODUCTION The steps leading to the definition of the *three-dimensional definite integral* or **triple integral** are quite similar to the steps leading to the definition of the double integral. Obvious differences: Instead of a function of two variables we are integrating a function f of three variables, not over a region R in a coordinate plane, but over a region D of 3-space.

1. Let $w = F(x, y, z)$ be defined over a closed and bounded region D of space.
2. By means of a three-dimensional grid of vertical and horizontal planes parallel to the coordinate planes, form a partition P of D into n subregions (boxes) D_k of volumes ΔV_k that lie entirely in D.
3. Let $\|P\|$ be the **norm** of the partition or the length of the longest diagonal of the D_k.
4. Choose a sample point (x_k^*, y_k^*, z_k^*) in each subregion D_k. See **FIGURE 9.15.1**.
5. Form the sum $\sum_{k=1}^{n} F(x_k^*, y_k^*, z_k^*) \, \Delta V_k$.

FIGURE 9.15.1 Sample point in kth subregion

The type of partition used in step 2, where all the D_k lie completely within D, is called an **inner partition** of D. A sum $\sum_{k=1}^{n} F(x_k^*, y_k^*, z_k^*) \, \Delta V_k$, where (x_k^*, y_k^*, z_k^*) is an arbitrary point within each D_k and ΔV_k denotes the volume of each D_k, is called a **Riemann sum** after the German mathematician **Georg Friedrich Bernhard Riemann** (1826–1866). Although Riemann made significant contributions to complex analysis, he is famous mostly for the first rigorous definition of the integral of a real function on an interval, known to every student of calculus as a definite integral.

Let F be a function of three variables defined over a closed region D of 3-space. Then the **triple integral of F over D** is given by

$$\iiint\limits_{D} F(x, y, z)\, dV = \lim_{\|P\| \to 0} \sum_{k=1}^{n} F(x_k^*, y_k^*, z_k^*)\, \Delta V_k. \tag{1}$$

As in our previous discussions on the integral, when F is continuous over D, the limit in (1) exists; that is, F is **integrable** over D.

▧ Evaluation by Iterated Integrals

If the region D is bounded above by the graph of $z = f_2(x, y)$ and bounded below by the graph of $z = f_1(x, y)$, then it can be shown that the triple integral (1) can be expressed as a double integral of the partial integral $\int_{f_1(x, y)}^{f_2(x, y)} F(x, y, z)\, dz$; that is,

$$\iiint\limits_{D} F(x, y, z)\, dV = \iint\limits_{R} \left[\int_{f_1(x, y)}^{f_2(x, y)} F(x, y, z)\, dz \right] dA,$$

where R is the orthogonal projection of D onto the xy-plane. In particular, if R is a Type I region, then, as shown in **FIGURE 9.15.2**, the triple integral of F over D can be written as an iterated integral:

$$\iiint\limits_{D} F(x, y, z)\, dV = \int_{a}^{b} \int_{g_1(x)}^{g_2(x)} \int_{f_1(x, y)}^{f_2(x, y)} F(x, y, z)\, dz\, dy\, dx. \tag{2}$$

FIGURE 9.15.2 Geometric interpretation of (2)

To evaluate the iterated integral in (2) we begin by evaluating the partial integral

$$\int_{f_1(x, y)}^{f_2(x, y)} F(x, y, z)\, dz,$$

in which *both* x and y are held fixed.

In a double integral there are only two possible orders of integration: $dy\, dx$ and $dx\, dy$. The triple integral in (2) illustrates one of *six* possible orders of integration:

$$dz\, dy\, dx, \quad dz\, dx\, dy, \quad dy\, dx\, dz,$$
$$dx\, dy\, dz, \quad dx\, dz\, dy, \quad dy\, dz\, dx.$$

The last two differentials tell the coordinate plane in which the region R is situated. For example, the iterated integral corresponding to the order of integration $dx\, dz\, dy$ must have

the form

$$\iiint\limits_{D} F(x, y, z)\, dV = \int_c^d \int_{k_1(y)}^{k_2(y)} \int_{h_1(y,z)}^{h_2(y,z)} F(x, y, z)\, dx\, dz\, dy.$$

The geometric interpretation of this integral and the region R of integration in the yz-plane are shown in **FIGURE 9.15.3**.

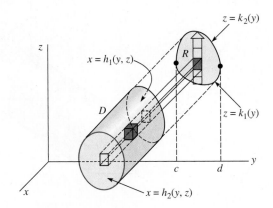

FIGURE 9.15.3 Integration: first x, then z, then y

■ **Applications** A list of some of the standard applications of the triple integral follows:

Volume: If $F(x, y, z) = 1$, then the **volume** of the solid D is

$$V = \iiint\limits_{D} dV.$$

Mass: If $\rho(x, y, z)$ is density, then the **mass** of the solid D is given by

$$m = \iiint\limits_{D} \rho(x, y, z)\, dV.$$

First Moments: The **first moments** of the solid about the coordinate planes indicated by the subscripts are given by

$$M_{xy} = \iiint\limits_{D} z\rho(x, y, z)\, dV, \qquad M_{xz} = \iiint\limits_{D} y\rho(x, y, z)\, dV,$$

$$M_{yz} = \iiint\limits_{D} x\rho(x, y, z)\, dV.$$

Center of Mass: The coordinates of the **center of mass** of D are given by

$$\bar{x} = \frac{M_{yz}}{m}, \quad \bar{y} = \frac{M_{xz}}{m}, \quad \bar{z} = \frac{M_{xy}}{m}.$$

Centroid: If $\rho(x, y, z) = $ a constant, the center of mass is called the **centroid** of the solid.

Second Moments: The **second moments**, or **moments of inertia** of D about the coordinate axes indicated by the subscripts, are given by

$$I_x = \iiint\limits_{D} (y^2 + z^2)\rho(x, y, z)\, dV, \quad I_y = \iiint\limits_{D} (x^2 + z^2)\rho(x, y, z)\, dV, \quad I_z = \iiint\limits_{D} (x^2 + y^2)\rho(x, y, z)\, dV.$$

Radius of Gyration: As in Section 9.10, if I is a moment of inertia of the solid about a given axis, then the **radius of gyration** is

$$R_g = \sqrt{\frac{I}{m}}.$$

EXAMPLE **1** Volume of a Solid

Find the volume of the solid in the first octant bounded by the graphs of $z = 1 - y^2$, $y = 2x$, and $x = 3$.

SOLUTION As indicated in **FIGURE 9.15.4(a)**, the first integration with respect to z is from 0 to $1 - y^2$. Furthermore, from Figure 9.15.4(b) we see that the projection of the solid D in the

(a) (b)

FIGURE 9.15.4 Solid D and region R of integration in Example 1

xy-plane is a region of Type II. Hence, we next integrate, with respect to x, from $y/2$ to 3. The last integration is with respect to y from 0 to 1. Thus,

$$V = \iiint\limits_{D} dV = \int_0^1 \int_{y/2}^3 \int_0^{1-y^2} dz \, dx \, dy$$

$$= \int_0^1 \int_{y/2}^3 (1 - y^2) \, dx \, dy$$

$$= \int_0^1 \left[x - xy^2 \right]_{y/2}^3 dy$$

$$= \int_0^1 \left(3 - 3y^2 - \frac{1}{2}y + \frac{1}{2}y^3 \right) dy$$

$$= \left[3y - y^3 - \frac{1}{4}y^2 + \frac{1}{8}y^4 \right]_0^1 = \frac{15}{8}.$$ $\equiv$

EXAMPLE **2** Changing the Order of Integration

Change the order of integration in

$$\int_0^6 \int_0^{4-2x/3} \int_0^{3-x/2-3y/4} F(x, y, z) \, dz \, dy \, dx$$

to $dy \, dx \, dz$.

SOLUTION As seen in **FIGURE 9.15.5(a)**, the region D is the solid in the first octant bounded by the three coordinate planes and the plane $2x + 3y + 4z = 12$. Referring to Figure 9.15.5(b) and the table, we conclude that

$$\int_0^6 \int_0^{4-2x/3} \int_0^{3-x/2-3y/4} F(x, y, z) \, dz \, dy \, dx = \int_0^3 \int_0^{6-2z} \int_0^{4-2x/3-4z/3} F(x, y, z) \, dy \, dx \, dz.$$

Order of Integration	First Integration	Second Integration	Third Integration
$dz \, dy \, dx$	0 to $3 - x/2 - 3y/4$	0 to $4 - 2x/3$	0 to 6
$dy \, dx \, dz$	0 to $4 - 2x/3 - 4z/3$	0 to $6 - 2z$	0 to 3

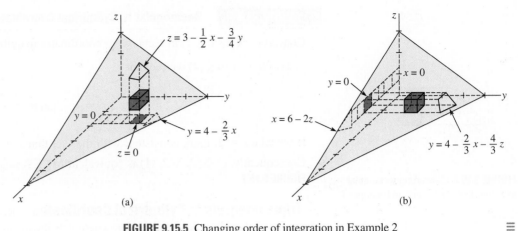

(a) (b)

FIGURE 9.15.5 Changing order of integration in Example 2

Depending on the geometry of a region in 3-space, the evaluation of a triple integral over that region may be made easier by utilizing a new coordinate system.

Cylindrical Coordinates

The **cylindrical coordinate system** combines the polar description of a point in the plane with the rectangular description of the z-component of a point in space. As seen in **FIGURE 9.15.6(a)**, the cylindrical coordinates of a point P are denoted by the ordered triple (r, θ, z). The word *cylindrical* arises from the fact that a point P in space is determined by the intersection of the planes $z = $ constant and $\theta = $ constant with a cylinder $r = $ constant. See Figure 9.15.6(b).

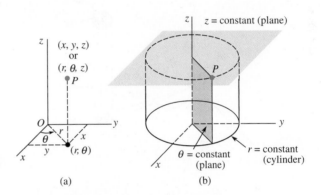

(a) (b)

FIGURE 9.15.6 Cylindrical coordinates

Conversion of Cylindrical Coordinates to Rectangular Coordinates

From Figure 9.15.6(a) we also see that the rectangular coordinates (x, y, z) of a point can be obtained from the cylindrical coordinates (r, θ, z) by means of

$$x = r \cos \theta, \quad y = r \sin \theta, \quad z = z. \tag{3}$$

EXAMPLE 3 **Cylindrical to Rectangular Coordinates**

Convert $(8, \pi/3, 7)$ in cylindrical coordinates to rectangular coordinates.

SOLUTION From (3),

$$x = 8 \cos \frac{\pi}{3} = 4, \quad y = 8 \sin \frac{\pi}{3} = 4\sqrt{3}, \quad z = 7.$$

Thus, $(8, \pi/3, 7)$ is equivalent to $(4, 4\sqrt{3}, 7)$ in rectangular coordinates.

Conversion of Rectangular Coordinates to Cylindrical Coordinates

To express rectangular coordinates (x, y, z) as cylindrical coordinates, we use

$$r^2 = x^2 + y^2, \quad \tan \theta = \frac{y}{x}, \quad z = z. \tag{4}$$

$(-\sqrt{2}, \sqrt{2}, 1)$
or
$(2, 3\pi/4, 1)$

FIGURE 9.15.7 Converting rectangular to cylindrical coordinates in Example 4

EXAMPLE 4 **Rectangular to Cylindrical Coordinates**

Convert $(-\sqrt{2}, \sqrt{2}, 1)$ in rectangular coordinates to cylindrical coordinates.

SOLUTION From (4) we see that

$$r^2 = (-\sqrt{2})^2 + (\sqrt{2})^2 = 4, \quad \tan\theta = \frac{\sqrt{2}}{-\sqrt{2}} = -1, \quad z = 1.$$

If we take $r = 2$, then, consistent with the fact that $x < 0$ and $y > 0$, we take $\theta = 3\pi/4$.*
Consequently, $(-\sqrt{2}, \sqrt{2}, 1)$ is equivalent to $(2, 3\pi/4, 1)$ in cylindrical coordinates. See
FIGURE 9.15.7. ≡

■ **Triple Integrals in Cylindrical Coordinates** Recall from Section 9.11 that the area
of a **polar rectangle** is $\Delta A = r^* \Delta r \Delta\theta$, where r^* is the average radius. From **FIGURE 9.15.8(a)** we
see that the volume of a **cylindrical wedge** is simply $\Delta V = $ (area of base)(height) $= r^* \Delta r \Delta\theta$
Δz. Thus, if $F(r, \theta, z)$ is a continuous function over the region D, as shown in Figure 9.15.8(b),
then the triple integral of F over D is given by

$$\iiint\limits_{D} F(r, \theta, z)\, dV = \iint\limits_{R}\left[\int_{f_1(r,\theta)}^{f_2(r,\theta)} F(r, \theta, z)\, dz\right] dA = \int_{\alpha}^{\beta}\int_{g_1(\theta)}^{g_2(\theta)}\int_{f_1(r,\theta)}^{f_2(r,\theta)} F(r, \theta, z)\, r\, dz\, dr\, d\theta.$$

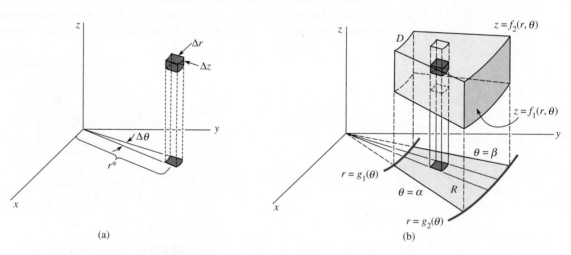

(a) (b)

FIGURE 9.15.8 Cylindrical wedge in (a); region D in (b)

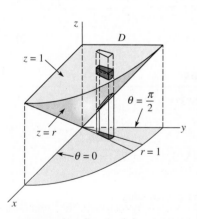

FIGURE 9.15.9 Solid in Example 5

EXAMPLE 5 **Center of Mass**

A solid in the first octant has the shape determined by the graph of the cone $z = \sqrt{x^2 + y^2}$
and the planes $z = 1$, $x = 0$, and $y = 0$. Find the center of mass if the density is given by
$\rho(r, \theta, z) = r$.

SOLUTION In view of (4), the equation of the cone is $z = r$. Hence, we see from **FIGURE 9.15.9**
that

$$m = \iiint\limits_{D} r\, dV = \int_{0}^{\pi/2}\int_{0}^{1}\int_{r}^{1} r(r\, dz\, dr\, d\theta)$$

$$= \int_{0}^{\pi/2}\int_{0}^{1} r^2 z\Big]_{r}^{1} dr\, d\theta$$

$$= \int_{0}^{\pi/2}\int_{0}^{1} (r^2 - r^3)\, dr\, d\theta = \frac{\pi}{24}$$

*If we use $\theta = \tan^{-1}(-1) = -\pi/4$, then we can use $r = -2$. Notice that the combinations $r = 2$,
$\theta = -\pi/4$ and $r = -2$, $\theta = 3\pi/4$ are inconsistent.

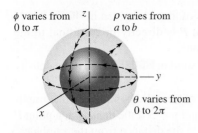

ϕ varies from 0 to π

ρ varies from a to b

θ varies from 0 to 2π

FIGURE 9.15.12 Limits of integration in Example 7

From (5) we find $x^2 + y^2 = \rho^2 \sin^2 \phi$ and $x^2 + y^2 + z^2 = \rho^2$. Thus the equations of the spheres are simply $\rho = a$ and $\rho = b$. See **FIGURE 9.15.12**. Consequently, in spherical coordinates the foregoing integral becomes

$$I_z = k \int_0^{2\pi} \int_0^{\pi} \int_a^b \rho^2 \sin^2\phi (\rho^2 \sin \phi \, d\rho \, d\phi \, d\theta)$$

$$= k \int_0^{2\pi} \int_0^{\pi} \int_a^b \rho^4 \sin^3\phi \, d\rho \, d\phi \, d\theta$$

$$= k \int_0^{2\pi} \int_0^{\pi} \frac{\rho^5}{5} \sin^3\phi \Big]_a^b \, d\phi \, d\theta$$

$$= \frac{k}{5}(b^5 - a^5) \int_0^{2\pi} \int_0^{\pi} (1 - \cos^2 \phi) \sin \phi \, d\phi \, d\theta$$

$$= \frac{k}{5}(b^5 - a^5) \int_0^{2\pi} \left[-\cos \phi + \frac{1}{3} \cos^3 \phi \right]_0^{\pi} d\theta$$

$$= \frac{4k}{15}(b^5 - a^5) \int_0^{2\pi} d\theta = \frac{8\pi k}{15}(b^5 - a^5).$$

≡

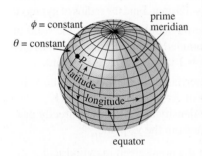

$\phi = $ constant

$\theta = $ constant

prime meridian

P

latitude

longitude

equator

FIGURE 9.15.13 Parallels and great circles

REMARKS

Spherical coordinates are used in navigation. If we think of the Earth as a sphere of fixed radius centered at the origin, then a point P can be located by specifying two angles θ and ϕ. As shown in **FIGURE 9.15.13**, when ϕ is held constant, the resulting curve is called a **parallel**. Fixed values of θ result in curves called **great circles**. Half of one of these great circles joining the north and south poles is called a **meridian**. The intersection of a parallel and a meridian gives the position of a point P. If $0° \leq \phi \leq 180°$ and $-180° \leq \theta \leq 180°$, the angles $90° - \phi$ and θ are said to be the **latitude** and **longitude** of P, respectively. The **prime meridian** corresponds to a longitude of $0°$. The latitude of the equator is $0°$; the latitudes of the north and south poles are, in turn, $+90°$ (or 90° North) and $-90°$ (or 90° South).

9.15 Exercises Answers to selected odd-numbered problems begin on page ANS-26.

In Problems 1–8, evaluate the given iterated integral.

1. $\int_2^4 \int_{-2}^2 \int_{-1}^1 (x + y + z) \, dx \, dy \, dz$

2. $\int_1^3 \int_1^x \int_2^{xy} 24xy \, dz \, dy \, dx$

3. $\int_0^6 \int_0^{6-x} \int_0^{6-x-z} dy \, dz \, dx$

4. $\int_0^1 \int_0^{1-x} \int_0^{\sqrt{y}} 4x^2z^3 \, dz \, dy \, dx$

5. $\int_0^{\pi/2} \int_0^{y^2} \int_0^y \cos\left(\frac{x}{y}\right) dz \, dx \, dy$

6. $\int_0^{\sqrt{2}} \int_{\sqrt{y}}^2 \int_0^{e^{x^2}} x \, dz \, dx \, dy$

7. $\int_0^1 \int_0^1 \int_0^{2-x^2-y^2} xye^z \, dz \, dx \, dy$

8. $\int_0^4 \int_0^{1/2} \int_0^{x^2} \frac{1}{\sqrt{x^2 - y^2}} \, dy \, dx \, dz$

9. Evaluate $\iiint_D z \, dV$, where D is the region in the first octant bounded by the graphs of $y = x$, $y = x - 2$, $y = 1$, $y = 3$, $z = 0$, and $z = 5$.

10. Evaluate $\iiint_D (x^2 + y^2) \, dV$, where D is the region bounded by the graphs of $y = x^2$, $z = 4 - y$, and $z = 0$.

In Problems 11 and 12, change the indicated order of integration to each of the other five orders.

11. $\int_0^2 \int_0^{4-2y} \int_{x+2y}^4 F(x, y, z) \, dz \, dx \, dy$

12. $\displaystyle\int_0^2 \int_0^{\sqrt{36-9x^2}/2} \int_1^3 F(x, y, z)\, dz\, dy\, dx$

In Problems 13 and 14, consider the solid given in the figure. Set up, but do not evaluate, the integrals giving the volume V of the solid using the indicated orders of integration.

13.

FIGURE 9.15.14 Solid for Problem 13

 (a) $dz\, dy\, dx$ **(b)** $dx\, dz\, dy$ **(c)** $dy\, dx\, dz$

14.

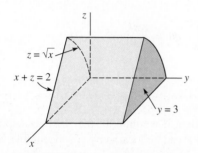

FIGURE 9.15.15 Solid for Problem 14

 (a) $dx\, dz\, dy$ **(b)** $dy\, dx\, dz$ **(c)** $dz\, dx\, dy$

[*Hint*: Part (c) will require two integrals.]

In Problems 15–20, sketch the region D whose volume V is given by the iterated integral.

15. $\displaystyle\int_0^4 \int_0^3 \int_0^{2-2z/3} dx\, dz\, dy$

16. $\displaystyle 4\int_0^3 \int_0^{\sqrt{9-y^2}} \int_4^{\sqrt{25-x^2-y^2}} dz\, dx\, dy$

17. $\displaystyle\int_{-1}^1 \int_{-\sqrt{1-x^2}}^{\sqrt{1-x^2}} \int_0^5 dz\, dy\, dx$

18. $\displaystyle\int_0^2 \int_0^{\sqrt{4-x^2}} \int_{x^2+y^2}^4 dz\, dy\, dx$

19. $\displaystyle\int_0^2 \int_0^{2-y} \int_{-\sqrt{y}}^{\sqrt{y}} dx\, dz\, dy$

20. $\displaystyle\int_1^3 \int_0^{1/x} \int_0^3 dy\, dz\, dx$

In Problems 21–24, find the volume of the solid bounded by the graphs of the given equations.

21. $x = y^2,\ 4 - x = y^2,\ z = 0,\ z = 3$
22. $x^2 + y^2 = 4,\ z = x + y$, the coordinate planes, first octant
23. $y = x^2 + z^2,\ y = 8 - x^2 - z^2$
24. $x = 2,\ y = x,\ y = 0,\ z = x^2 + y^2,\ z = 0$

25. Find the center of mass of the solid given in **FIGURE 9.15.14** if the density at a point P is directly proportional to the distance from the xy-plane.

26. Find the centroid of the solid in **FIGURE 9.15.15** if the density is constant.

27. Find the center of mass of the solid bounded by the graphs of $x^2 + z^2 = 4,\ y = 0$, and $y = 3$ if the density at a point P is directly proportional to the distance from the xz-plane.

28. Find the center of mass of the solid bounded by the graphs of $y = x^2,\ y = x,\ z = y + 2$, and $z = 0$ if the density at a point P is directly proportional to the distance from the xy-plane.

In Problems 29 and 30, set up, but do not evaluate, the iterated integrals giving the mass of the solid that has the given shape and density.

29. $x^2 + y^2 = 1,\ z + y = 8,\ z - 2y = 2;\ \rho(x, y, z) = x + y + 4$
30. $x^2 + y^2 - z^2 = 1,\ z = -1,\ z = 2;\ \rho(x, y, z) = z^2$ [*Hint*: Do not use $dz\, dy\, dx$.]

31. Find the moment of inertia of the solid in Figure 9.15.14 about the y-axis if the density is as given in Problem 25. Find the radius of gyration.

32. Find the moment of inertia of the solid in Figure 9.15.15 about the x-axis if the density is constant. Find the radius of gyration.

33. Find the moment of inertia about the z-axis of the solid in the first octant that is bounded by the coordinate planes and the graph of $x + y + z = 1$ if the density is constant.

34. Find the moment of inertia about the y-axis of the solid bounded by the graphs of $z = y,\ z = 4 - y,\ z = 1,\ z = 0$, $x = 2$, and $x = 0$ if the density at a point P is directly proportional to the distance from the yz-plane.

In Problems 35–38, convert the point given in cylindrical coordinates to rectangular coordinates.

35. $\left(10, \dfrac{3\pi}{4}, 5\right)$ **36.** $\left(2, \dfrac{5\pi}{6}, -3\right)$

37. $\left(\sqrt{3}, \dfrac{\pi}{3}, -4\right)$ **38.** $\left(4, \dfrac{7\pi}{4}, 0\right)$

In Problems 39–42, convert the point given in rectangular coordinates to cylindrical coordinates.

39. $(1, -1, -9)$ **40.** $(2\sqrt{3}, 2, 17)$
41. $(-\sqrt{2}, \sqrt{6}, 2)$ **42.** $(1, 2, 7)$

In Problems 43–46, convert the given equation to cylindrical coordinates.

43. $x^2 + y^2 + z^2 = 25$ **44.** $x + y - z = 1$
45. $x^2 + y^2 - z^2 = 1$ **46.** $x^2 + z^2 = 16$

In Problems 47–50, convert the given equation to rectangular coordinates.

47. $z = r^2$ **48.** $z = 2r \sin \theta$
49. $r = 5 \sec \theta$ **50.** $\theta = \pi/6$

In Problems 51–58, use triple integrals and cylindrical coordinates. In Problems 51–54, find the volume of the solid that is bounded by the graphs of the given equations.

51. $x^2 + y^2 = 4,\ x^2 + y^2 + z^2 = 16,\ z = 0$

52. $z = 10 - x^2 - y^2$, $z = 1$

53. $z = x^2 + y^2$, $x^2 + y^2 = 25$, $z = 0$

54. $y = x^2 + z^2$, $2y = x^2 + z^2 + 4$

55. Find the centroid of the homogeneous solid that is bounded by the hemisphere $z = \sqrt{a^2 - x^2 - y^2}$ and the plane $z = 0$.

56. Find the center of mass of the solid that is bounded by the graphs of $y^2 + z^2 = 16$, $x = 0$, and $x = 5$ if the density at a point P is directly proportional to distance from the yz-plane.

57. Find the moment of inertia about the z-axis of the solid that is bounded above by the hemisphere $z = \sqrt{9 - x^2 - y^2}$ and below by the plane $z = 2$ if the density at a point P is inversely proportional to the square of the distance from the z-axis.

58. Find the moment of inertia about the x-axis of the solid that is bounded by the cone $z = \sqrt{x^2 - y^2}$ and the plane $z = 1$ if the density at a point P is directly proportional to the distance from the z-axis.

In Problems 59–62, convert the point given in spherical coordinates to **(a)** rectangular coordinates and **(b)** cylindrical coordinates.

59. $\left(\dfrac{2}{3}, \dfrac{\pi}{2}, \dfrac{\pi}{6} \right)$

60. $\left(5, \dfrac{5\pi}{4}, \dfrac{2\pi}{3} \right)$

61. $\left(8, \dfrac{\pi}{4}, \dfrac{3\pi}{4} \right)$

62. $\left(\dfrac{1}{3}, \dfrac{5\pi}{3}, \dfrac{\pi}{6} \right)$

In Problems 63–66, convert the points given in rectangular coordinates to spherical coordinates.

63. $(-5, -5, 0)$

64. $(1, -\sqrt{3}, 1)$

65. $\left(\dfrac{\sqrt{3}}{2}, \dfrac{1}{2}, 1 \right)$

66. $\left(-\dfrac{\sqrt{3}}{2}, 0, -\dfrac{1}{2} \right)$

In Problems 67–70, convert the given equation to spherical coordinates.

67. $x^2 + y^2 + z^2 = 64$

68. $x^2 + y^2 + z^2 = 4z$

69. $z^2 = 3x^2 + 3y^2$

70. $-x^2 - y^2 + z^2 = 1$

In Problems 71–74, convert the given equation to rectangular coordinates.

71. $\rho = 10$

72. $\phi = \pi/3$

73. $\rho = 2 \sec \phi$

74. $\rho \sin^2 \phi = \cos \phi$

In Problems 75–82, use triple integrals and spherical coordinates. In Problems 75–78, find the volume of the solid that is bounded by the graphs of the given equations.

75. $z = \sqrt{x^2 + y^2}$, $x^2 + y^2 + z^2 = 9$

76. $x^2 + y^2 + z^2 = 4$, $y = x$, $y = \sqrt{3}x$, $z = 0$, first octant

77. $z^2 = 3x^2 + 3y^2$, $x = 0$, $y = 0$, $z = 2$, first octant

78. Inside $x^2 + y^2 + z^2 = 1$ and outside $z^2 = x^2 + y^2$

79. Find the centroid of the homogeneous solid that is bounded by the cone $z = \sqrt{x^2 + y^2}$ and the sphere $x^2 + y^2 + z^2 = 2z$.

80. Find the center of mass of the solid that is bounded by the hemisphere $z = \sqrt{1 - x^2 - y^2}$ and the plane $z = 0$ if the density at a point P is directly proportional to the distance from the xy-plane.

81. Find the mass of the solid that is bounded above by the hemisphere $z = \sqrt{25 - x^2 - y^2}$ and below by the plane $z = 4$ if the density at a point P is inversely proportional to the distance from the origin. [*Hint*: Express the upper ϕ limit of integration as an inverse cosine.]

82. Find the moment of inertia about the z-axis of the solid that is bounded by the sphere $x^2 + y^2 + z^2 = a^2$ if the density at a point P is directly proportional to the distance from the origin.

9.16 Divergence Theorem

INTRODUCTION In Section 9.14 we saw that Stokes' theorem was a three-dimensional generalization of a vector form of Green's theorem. In this section we present a second vector form of Green's theorem and its three-dimensional analogue.

Another Vector Form of Green's Theorem Let $\mathbf{F}(x, y) = P(x, y)\mathbf{i} + Q(x, y)\mathbf{j}$ be a two-dimensional vector field, and let $\mathbf{T} = (dx/ds)\mathbf{i} + (dy/ds)\mathbf{j}$ be a *unit tangent* to a simple closed plane curve C. In (1) of Section 9.14 we saw that $\oint_C (\mathbf{F} \cdot \mathbf{T})\, ds$ can be evaluated by a double integral involving curl $\mathbf{F}$. Similarly, if $\mathbf{n} = (dy/ds)\mathbf{i} - (dx/ds)\mathbf{j}$ is a *unit normal* to C (check $\mathbf{T} \cdot \mathbf{n}$), then $\oint_C (\mathbf{F} \cdot \mathbf{n})\, ds$ can be expressed in terms of a double integral of div $\mathbf{F}$. From Green's theorem,

$$\oint_C (\mathbf{F} \cdot \mathbf{n})\, ds = \oint_C P\, dy - Q\, dx = \iint_R \left[\frac{\partial P}{\partial x} - \left(-\frac{\partial Q}{\partial y} \right) \right] dA = \iint_R \left[\frac{\partial P}{\partial x} + \frac{\partial Q}{\partial y} \right] dA;$$

that is,

$$\oint_C (\mathbf{F} \cdot \mathbf{n})\, ds = \iint_R \text{div } \mathbf{F}\, dA. \tag{1}$$

The result in (1) is a special case of the **divergence** or **Gauss' theorem**. The following is a generalization of (1) to 3-space:

THEOREM 9.16.1 Divergence Theorem

Let D be a closed and bounded region in 3-space with a piecewise-smooth boundary S that is oriented outward. Let $\mathbf{F}(x, y, z) = P(x, y, z)\mathbf{i} + Q(x, y, z)\mathbf{j} + R(x, y, z)\mathbf{k}$ be a vector field for which P, Q, and R are continuous and have continuous first partial derivatives in a region of 3-space containing D. Then

$$\iint_S (\mathbf{F} \cdot \mathbf{n})\, dS = \iiint_D \operatorname{div} \mathbf{F}\, dV. \tag{2}$$

PARTIAL PROOF: We will prove (2) for the special region D shown in **FIGURE 9.16.1** whose surface S consists of three pieces:

$$\text{(bottom)} \quad S_1: z = f_1(x, y), \quad (x, y) \text{ in } R$$
$$\text{(top)} \quad S_2: z = f_2(x, y), \quad (x, y) \text{ in } R$$
$$\text{(side)} \quad S_3: f_1(x, y) \le z \le f_2(x, y), \quad (x, y) \text{ on } C,$$

where R is the projection of D onto the xy-plane and C is the boundary of R. Since

$$\operatorname{div} \mathbf{F} = \frac{\partial P}{\partial x} + \frac{\partial Q}{\partial y} + \frac{\partial R}{\partial z} \quad \text{and} \quad \mathbf{F} \cdot \mathbf{n} = P(\mathbf{i} \cdot \mathbf{n}) + Q(\mathbf{j} \cdot \mathbf{n}) + R(\mathbf{k} \cdot \mathbf{n}),$$

we can write

$$\iint_S (\mathbf{F} \cdot \mathbf{n})\, dS = \iint_S P(\mathbf{i} \cdot \mathbf{n})\, dS + \iint_S Q(\mathbf{j} \cdot \mathbf{n})\, dS + \iint_S R(\mathbf{k} \cdot \mathbf{n})\, dS$$

and

$$\iiint_D \operatorname{div} \mathbf{F}\, dV = \iiint_D \frac{\partial P}{\partial x}\, dV + \iiint_D \frac{\partial Q}{\partial y}\, dV + \iiint_D \frac{\partial R}{\partial z}\, dV.$$

To prove (2) we need only establish that

$$\iint_S P(\mathbf{i} \cdot \mathbf{n})\, dS = \iiint_D \frac{\partial P}{\partial x}\, dV \tag{3}$$

$$\iint_S Q(\mathbf{j} \cdot \mathbf{n})\, dS = \iiint_D \frac{\partial Q}{\partial y}\, dV \tag{4}$$

$$\iint_S R(\mathbf{k} \cdot \mathbf{n})\, dS = \iiint_D \frac{\partial R}{\partial z}\, dV. \tag{5}$$

Indeed, we shall prove only (5), since the proofs of (3) and (4) follow in a similar manner. Now,

$$\iiint_D \frac{\partial R}{\partial z}\, dV = \iint_R \left[\int_{f_1(x, y)}^{f_2(x, y)} \frac{\partial R}{\partial z}\, dz \right] dA = \iint_R [R(x, y, f_2(x, y)) - R(x, y, f_1(x, y))]\, dA. \tag{6}$$

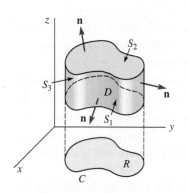

FIGURE 9.16.1 Region D used in proof of Theorem 9.16.1

Next we write

$$\iint\limits_{S} R(\mathbf{k} \cdot \mathbf{n}) \; dS = \iint\limits_{S_1} R(\mathbf{k} \cdot \mathbf{n}) \; dS + \iint\limits_{S_2} R(\mathbf{k} \cdot \mathbf{n}) \; dS + \iint\limits_{S_3} R(\mathbf{k} \cdot \mathbf{n}) \; dS.$$

On S_1: Since the outward normal points downward, we describe the surface as $g(x, y, z) = f_1(x, y) - z = 0$. Thus,

$$\mathbf{n} = \frac{\dfrac{\partial f_1}{\partial x}\mathbf{i} + \dfrac{\partial f_1}{\partial y}\mathbf{j} - \mathbf{k}}{\sqrt{1 + \left(\dfrac{\partial f_1}{\partial x}\right)^2 + \left(\dfrac{\partial f_1}{\partial y}\right)^2}} \quad \text{so that} \quad \mathbf{k} \cdot \mathbf{n} = \frac{-1}{\sqrt{1 + \left(\dfrac{\partial f_1}{\partial x}\right)^2 + \left(\dfrac{\partial f_1}{\partial y}\right)^2}}.$$

From the definition of dS we then have

$$\iint\limits_{S_1} R(\mathbf{k} \cdot \mathbf{n}) \; dS = -\iint\limits_{R} R(x, y, f_1(x, y)) \; dA. \tag{7}$$

On S_2: The outward normal points upward, so

$$\mathbf{n} = \frac{\dfrac{\partial f_2}{\partial x}\mathbf{i} - \dfrac{\partial f_2}{\partial y}\mathbf{j} + \mathbf{k}}{\sqrt{1 + \left(\dfrac{\partial f_2}{\partial x}\right)^2 + \left(\dfrac{\partial f_2}{\partial y}\right)^2}} \quad \text{so that} \quad \mathbf{k} \cdot \mathbf{n} = \frac{1}{\sqrt{1 + \left(\dfrac{\partial f_2}{\partial x}\right)^2 + \left(\dfrac{\partial f_2}{\partial y}\right)^2}}$$

from which we find

$$\iint\limits_{S_2} R(\mathbf{k} \cdot \mathbf{n}) \; dS = \iint\limits_{R} R(x, y, f_2(x, y)) \; dA. \tag{8}$$

On S_3: Since this side is vertical, $\mathbf{k}$ is perpendicular to $\mathbf{n}$. Consequently, $\mathbf{k} \cdot \mathbf{n} = 0$ and

$$\iint\limits_{S_3} R(\mathbf{k} \cdot \mathbf{n}) \; dS = 0. \tag{9}$$

Finally, adding (7), (8), and (9), we get

$$\iint\limits_{R} [R(x, y, f_2(x, y)) - R(x, y, f_1(x, y))] \; dA,$$

which is the same as (6). $\equiv$

FIGURE 9.16.2 Region D with no vertical side

FIGURE 9.16.3 Region D is bounded between two concentric spheres

Although we proved (2) for a special region D that has a vertical side, we note that this type of region is not required in Theorem 9.16.1. A region D with no vertical side is illustrated in **FIGURE 9.16.2**; a region bounded by a sphere or an ellipsoid also does not have a vertical side. The divergence theorem also holds for the region D bounded between two closed surfaces, such as the concentric spheres S_a and S_b shown in **FIGURE 9.16.3**; the boundary surface S of D is the union of S_a and S_b. In this case $\iint_S (\mathbf{F} \cdot \mathbf{n}) \; dS = \iiint_D \text{div } \mathbf{F} \; dV$ becomes

$$\iint\limits_{S_b} (\mathbf{F} \cdot \mathbf{n}) \; dS + \iint\limits_{S_a} (\mathbf{F} \cdot \mathbf{n}) \; dS = \iiint\limits_{D} \text{div } \mathbf{F} \; dV,$$

where $\mathbf{n}$ points outward from D; that is, $\mathbf{n}$ points away from the origin on S_b and $\mathbf{n}$ points toward the origin on S_a.

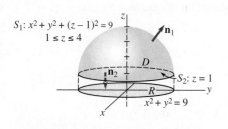

$S_1: x^2 + y^2 + (z-1)^2 = 9$
$1 \le z \le 4$

FIGURE 9.16.4 Hemispherical region D in Example 1

EXAMPLE 1 **Verifying Divergence Theorem**

Let D be the region bounded by the hemisphere $x^2 + y^2 + (z-1)^2 = 9$, $1 \le z \le 4$, and the plane $z = 1$. Verify the divergence theorem if $\mathbf{F} = x\mathbf{i} + y\mathbf{j} + (z-1)\mathbf{k}$.

SOLUTION The closed region is shown in **FIGURE 9.16.4**.

Triple Integral: Since $\mathbf{F} = x\mathbf{i} + y\mathbf{j} + (z-1)\mathbf{k}$, we see div $\mathbf{F} = 3$. Hence,

$$\iiint_D \text{div } \mathbf{F} \, dV = \iiint_D 3 \, dV = 3\iiint_D dV = 54\pi. \tag{10}$$

In the last calculation, we used the fact that $\iiint_D dV$ gives the volume of the hemisphere ($\frac{2}{3}\pi 3^3$).

Surface Integral: We write $\iint_S = \iint_{S_1} + \iint_{S_2}$, where S_1 is the hemisphere and S_2 is the plane $z = 1$. If S_1 is a level surface of $g(x, y, z) = x^2 + y^2 + (z-1)^2$, then a unit outer normal is

$$\mathbf{n} = \frac{\nabla g}{\|\nabla g\|} = \frac{x\mathbf{i} + y\mathbf{j} + (z-1)\mathbf{k}}{\sqrt{x^2 + y^2 + (z-1)^2}} = \frac{x}{3}\mathbf{i} + \frac{y}{3}\mathbf{j} + \frac{z-1}{3}\mathbf{k}.$$

Now
$$\mathbf{F} \cdot \mathbf{n} = \frac{x^2}{3} + \frac{y^2}{3} + \frac{(z-1)^2}{3} = \frac{1}{3}(x^2 + y^2 + (z-1)^2) = \frac{1}{3} \cdot 9 = 3$$

and so
$$\iint_{S_1} (\mathbf{F} \cdot \mathbf{n}) \, dS = \iint_R (3)\left(\frac{3}{\sqrt{9 - x^2 - y^2}} \, dA\right)$$

$$= 9\int_0^{2\pi} \int_0^3 (9 - r^2)^{-1/2} r \, dr \, d\theta = 54\pi. \quad \leftarrow \text{polar coordinates}$$

On S_2, we take $\mathbf{n} = -\mathbf{k}$ so that $\mathbf{F} \cdot \mathbf{n} = -z + 1$. But since $z = 1$, $\iint_{S_2}(-z + 1) \, dS = 0$. Hence, we see that $\iint_S (\mathbf{F} \cdot \mathbf{n}) \, dS = 54\pi + 0 = 54\pi$ agrees with (10). $\equiv$

EXAMPLE 2 **Using Divergence Theorem**

If $\mathbf{F} = xy\mathbf{i} + y^2z\mathbf{j} + z^3\mathbf{k}$, evaluate $\iint_S(\mathbf{F} \cdot \mathbf{n}) \, dS$, where S is the unit cube defined by $0 \le x \le 1$, $0 \le y \le 1$, $0 \le z \le 1$.

SOLUTION See Figure 9.13.14 and Problem 38 in Exercises 9.13. Rather than evaluate six surface integrals, we apply the divergence theorem. Since div $\mathbf{F} = \nabla \cdot \mathbf{F} = y + 2yz + 3z^2$, we have from (2)

$$\iint_S (\mathbf{F} \cdot \mathbf{n}) \, dS = \iiint_D (y + 2yz + 3z^2) \, dV$$

$$= \int_0^1 \int_0^1 \int_0^1 (y + 2yz + 3z^2) \, dx \, dy \, dz$$

$$= \int_0^1 \int_0^1 (y + 2yz + 3z^2) \, dy \, dz$$

$$= \int_0^1 \left(\frac{y^2}{2} + y^2 z + 3yz^2 \right) \Bigg]_0^1 dz$$

$$= \int_0^1 \left(\frac{1}{2} + z + 3z^2 \right) dz = \left(\frac{1}{2}z + \frac{1}{2}z^2 + z^3 \right) \Bigg]_0^1 = 2. \qquad \equiv$$

■ Physical Interpretation of Divergence

In Section 9.14 we saw that we could express the normal component of the curl of a vector field $\mathbf{F}$ at a point as a limit involving the circulation of $\mathbf{F}$. In view of (2), it is possible to interpret the divergence of $\mathbf{F}$ at a point as a limit involving the flux of $\mathbf{F}$. Recall from Section 9.7 that the flux of the velocity field $\mathbf{F}$ of a fluid is the rate of fluid flow—that is, the volume of fluid flowing through a surface per unit time. In Section 9.7 we saw that the divergence of $\mathbf{F}$ is the flux per unit volume. To reinforce this last idea let us suppose $P_0(x_0, y_0, z_0)$ is any point in the fluid and S_r is a small sphere of radius r centered at P_0. See **FIGURE 9.16.5**. If D_r is the sphere S_r and its interior, then the divergence theorem gives

FIGURE 9.16.5 Region D_r in (11)

$$\iint_{S_r} (\mathbf{F} \cdot \mathbf{n}) \, dS = \iint_{D_r} \text{div } \mathbf{F} \, dV. \tag{11}$$

If we take the approximation div $\mathbf{F}(P) \approx$ div $\mathbf{F}(P_0)$ at all points $P(x, y, z)$ within the small sphere, then (11) gives

$$\iint_{S_r} (\mathbf{F} \cdot \mathbf{n}) \, dS \approx \iiint_{D_r} \text{div } \mathbf{F}(P_0) \, dV$$

$$= \text{div } \mathbf{F}(P_0) \iiint_{D_r} dV \tag{12}$$

$$= \text{div } \mathbf{F}(P_0) V_r,$$

where V_r is the volume $\left(\frac{4}{3}\pi r^3 \right)$ of the spherical region D_r. By letting $r \to 0$, we see from (12) that the divergence of $\mathbf{F}$ is the limiting value of the ratio of the flux of $\mathbf{F}$ to the volume of the spherical region:

$$\text{div } \mathbf{F}(P_0) = \lim_{r \to 0} \frac{1}{V_r} \iint_{S_r} (\mathbf{F} \cdot \mathbf{n}) \, dS.$$

Hence, divergence $\mathbf{F}$ is flux per unit volume.

The divergence theorem is extremely useful in the derivation of some of the famous equations in electricity and magnetism and hydrodynamics. In the discussion that follows we shall consider an example from the study of fluids.

■ Continuity Equation

At the end of Section 9.7 we mentioned that one interpretation of div $\mathbf{F}$ was a measure of the rate of change of the density of a fluid at a point. To see why this is so, let us suppose that $\mathbf{F}$ is a velocity field of a fluid and that $\rho(x, y, z, t)$ is the density of the fluid at a point $P(x, y, z)$ at a time t. Let D be the closed region consisting of a sphere S and its interior. We know from Section 9.15 that the total mass m of the fluid in D is given by $m = \iiint_D \rho(x, y, z, t) \, dV$. The rate at which the mass increases in D is given by

$$\frac{dm}{dt} = \frac{d}{dt} \iiint_D \rho(x, y, z, t) \, dV = \iiint_D \frac{\partial \rho}{\partial t} \, dV. \tag{13}$$

Now from Figure 9.7.3 we saw that the volume of fluid flowing through an element of surface area ΔS per unit time is approximated by $(\mathbf{F} \cdot \mathbf{n}) \, \Delta S$. The mass of the fluid flowing through an element of surface area ΔS per unit time is then $(\rho \mathbf{F} \cdot \mathbf{n}) \, \Delta S$. If we assume that the change in

mass in D is due only to the flow in and out of D, then the *volume of fluid* flowing out of D per unit time is given by (10) of Section 9.13, $\iint_S (\mathbf{F} \cdot \mathbf{n}) \, dS$, whereas the *mass of the fluid* flowing out of D per unit time is $\iint_S (\rho \mathbf{F} \cdot \mathbf{n}) \, dS$. Hence, an alternative expression for the rate at which the mass increases in D is

$$-\iint_S (\rho \mathbf{F} \cdot \mathbf{n}) \, dS. \tag{14}$$

By the divergence theorem, (14) is the same as

$$-\iiint_D \operatorname{div}(\rho \mathbf{F}) \, dV. \tag{15}$$

Equating (13) and (15) then yields

$$\iiint_D \frac{\partial \rho}{\partial t} \, dV = -\iiint_D \operatorname{div}(\rho \mathbf{F}) \, dV \quad \text{or} \quad \iiint_D \left(\frac{\partial \rho}{\partial t} + \operatorname{div}(\rho \mathbf{F}) \right) dV = 0.$$

Since this last result is to hold for every sphere, we obtain the **equation of continuity** for fluid flows:

$$\frac{\partial \rho}{\partial t} + \operatorname{div}(\rho \mathbf{F}) = 0. \tag{16}$$

On page 521 we stated that if div $\mathbf{F} = \nabla \cdot \mathbf{F} = 0$, then a fluid is incompressible. This fact follows immediately from (16). If a fluid is incompressible (such as water), then ρ is constant, so consequently $\nabla \cdot (\rho \mathbf{F}) = \rho \nabla \cdot \mathbf{F}$. But in addition $\partial \rho / \partial t = 0$ and so (16) implies $\nabla \cdot \mathbf{F} = 0$.

9.16 Exercises Answers to selected odd-numbered problems begin on page ANS-26.

In Problems 1 and 2, verify the divergence theorem.

1. $\mathbf{F} = xy\mathbf{i} + yz\mathbf{j} + xz\mathbf{k}$; D the region bounded by the unit cube defined by $0 \le x \le 1, 0 \le y \le 1, 0 \le z \le 1$

2. $\mathbf{F} = 6xy\mathbf{i} + 4yz\mathbf{j} + xe^{-y}\mathbf{k}$; D the region bounded by the three coordinate planes and the plane $x + y + z = 1$

In Problems 3–14, use the divergence theorem to find the outward flux $\iint_S (\mathbf{F} \cdot \mathbf{n}) \, dS$ of the given vector field $\mathbf{F}$.

3. $\mathbf{F} = x^3\mathbf{i} + y^3\mathbf{j} + z^3\mathbf{k}$; D the region bounded by the sphere $x^2 + y^2 + z^2 = a^2$

4. $\mathbf{F} = 4x\mathbf{i} + y\mathbf{j} + 4z\mathbf{k}$; D the region bounded by the sphere $x^2 + y^2 + z^2 = 4$

5. $\mathbf{F} = y^2\mathbf{i} + xz^3\mathbf{j} + (z-1)^2\mathbf{k}$; D the region bounded by the cylinder $x^2 + y^2 = 16$ and the planes $z = 1, z = 5$

6. $\mathbf{F} = x^2\mathbf{i} + 2yz\mathbf{j} + 4z^3\mathbf{k}$; D the region bounded by the parallelepiped defined by $0 \le x \le 1, 0 \le y \le 2, 0 \le z \le 3$

7. $\mathbf{F} = y^3\mathbf{i} + x^3\mathbf{j} + z^3\mathbf{k}$; D the region bounded within by $z = \sqrt{4 - x^2 - y^2}, x^2 + y^2 = 3, z = 0$

8. $\mathbf{F} = (x^2 + \sin y)\mathbf{i} + z^2\mathbf{j} + xy^3\mathbf{k}$; D the region bounded by $y = x^2, z = 9 - y, z = 0$

9. $\mathbf{F} = (x\mathbf{i} + y\mathbf{j} + z\mathbf{k})/(x^2 + y^2 + z^2)$; D the region bounded by the concentric spheres $x^2 + y^2 + z^2 = a^2, x^2 + y^2 + z^2 = b^2$, $b > a$

10. $\mathbf{F} = 2yz\mathbf{i} + x^3\mathbf{j} + xy^2\mathbf{k}$; D the region bounded by the ellipsoid $x^2/a^2 + y^2/b^2 + z^2/c^2 = 1$

11. $\mathbf{F} = 2xz\mathbf{i} + 5y^2\mathbf{j} - z^2\mathbf{k}$; D the region bounded by $z = y$, $z = 4 - y, z = 2 - \frac{1}{2}x^2, x = 0, z = 0$. See **FIGURE 9.16.6**.

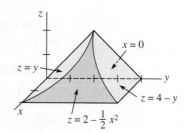

FIGURE 9.16.6 Region D for Problem 11

12. $\mathbf{F} = 15x^2y\mathbf{i} + x^2z\mathbf{j} + y^4\mathbf{k}$; D the region bounded by $x + y = 2, z = x + y, z = 3, x = 0, y = 0$

13. $\mathbf{F} = 3x^2y^2\mathbf{i} + y\mathbf{j} - 6zxy^2\mathbf{k}$; D the region bounded by the paraboloid $z = x^2 + y^2$ and the plane $z = 2y$

14. $\mathbf{F} = xy^2\mathbf{i} + x^2y\mathbf{j} + 6 \sin x\mathbf{k}$; D the region bounded by the cone $z = \sqrt{x^2 + y^2}$ and the planes $z = 2, z = 4$

15. The electric field at a point $P(x, y, z)$ due to a point charge q located at the origin is given by the inverse square field

$$E = q \frac{\mathbf{r}}{\|\mathbf{r}\|^3},$$

where $r = x\mathbf{i} + y\mathbf{j} + z\mathbf{k}$.

(a) Suppose S is a closed surface, S_a is a sphere $x^2 + y^2 + z^2 = a^2$ lying completely within S, and D is the region bounded between S and S_a. See **FIGURE 9.16.7**. Show that the outward flux of $\mathbf{E}$ for the region D is zero.

(b) Use the result of part (a) to prove **Gauss' law**:

$$\iint_S (\mathbf{E} \cdot \mathbf{n}) \, dS = 4\pi q;$$

that is, the outward flux of the electric field $\mathbf{E}$ through *any* closed surface (for which the divergence theorem applies) containing the origin is $4\pi q$.

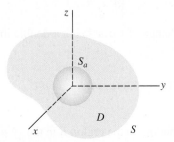

FIGURE 9.16.7 Region D for Problem 15(a)

16. Suppose there is a continuous distribution of charge throughout a closed and bounded region D enclosed by a surface S. Then the natural extension of Gauss' law is given by

$$\iint_S (\mathbf{E} \cdot \mathbf{n}) \, dS = \iiint_D 4\pi \rho \, dV,$$

where $\rho(x, y, z)$ is the charge density or charge per unit volume.

(a) Proceed as in the derivation of the continuity equation (16) to show that div $\mathbf{E} = 4\pi\rho$.

(b) Given that $\mathbf{E}$ is an irrotational vector field, show that the potential ϕ for $\mathbf{E}$ satisfies Poisson's equation $\nabla^2 \phi = 4\pi\rho$.

In Problems 17–21, assume that S forms the boundary of a closed and bounded region D.

17. If $\mathbf{a}$ is a constant vector, show that $\iint_S (\mathbf{a} \cdot \mathbf{n}) \, dS = 0$.

18. If $\mathbf{F} = P\mathbf{i} + Q\mathbf{j} + R\mathbf{k}$ and P, Q, and R have continuous second partial derivatives, prove that

$$\iint_S (\text{curl } \mathbf{F} \cdot \mathbf{n}) \, dS = 0.$$

In Problems 19 and 20, assume that f and g are scalar functions with continuous second partial derivatives. Use the divergence theorem to establish **Green's identities**.

19. $\iint_S (f\nabla g) \cdot \mathbf{n} \, dS = \iiint_D (f\nabla^2 g + \nabla f \cdot \nabla g) \, dV$

20. $\iint_S (f\nabla g - g\nabla f) \cdot \mathbf{n} \, dS = \iiint_D (f\nabla^2 g - g\nabla^2 f) \, dV$

21. If f is a scalar function with continuous first partial derivatives, prove that

$$\iint_S f\mathbf{n} \, dS = \iiint_D \nabla f \, dV.$$

[*Hint*: Use (2) on $f\mathbf{a}$, where $\mathbf{a}$ is a constant vector, and Problem 27 in Exercises 9.7.]

22. The buoyancy force on a floating object is $\mathbf{B} = -\iint_S p\mathbf{n} \, dS$, where p is the fluid pressure. The pressure p is related to the density of the fluid $\rho(x, y, z)$ by a law of hydrostatics: $\nabla p = \rho(x, y, z)\mathbf{g}$, where $\mathbf{g}$ is the constant acceleration of gravity. If the weight of the object is $\mathbf{W} = m\mathbf{g}$, use the result of Problem 21 to prove Archimedes' principle, $\mathbf{B} + \mathbf{W} = \mathbf{0}$. See **FIGURE 9.16.8**.

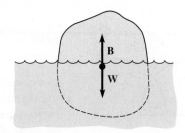

FIGURE 9.16.8 Floating object in Problem 22

9.17 Change of Variables in Multiple Integrals

INTRODUCTION In many instances it is either a matter of convenience or of necessity to make a substitution, or change of variable, in a definite integral $\int_a^b f(x) \, dx$ in order to evaluate it. If f is continuous on $[a, b]$, $x = g(u)$ has a continuous derivative, and $dx = g'(u) \, du$, then

$$\int_a^b f(x) \, dx = \int_c^d f(g(u)) \, g'(u) \, du, \tag{1}$$

If the function g is one-to-one, then it has an inverse and so $c = g^{-1}(a)$ and $d = g^{-1}(b)$. ▶

where the u-limits of integration c and d are defined by $a = g(c)$ and $b = g(d)$. There are three things that bear emphasizing in (1). To change the variable in a definite integral we replace x where it appears in the integrand by $g(u)$, we change the interval of integration $[a, b]$ on the x-axis to the corresponding interval $[c, d]$ on the u-axis, and we replace dx by a function multiple (namely, the derivative of g) of du. If we write $J(u) = dx/du$, then (1) has the form

$$\int_a^b f(x)\, dx = \int_c^d f(g(u))\, J(u)\, du. \tag{2}$$

For example, using $x = 2 \sin \theta$, $-\pi/2 \le \theta \le \pi/2$, we get

$$\int_0^2 \overbrace{\sqrt{4 - x^2}}^{f(x)}\, dx = \int_0^{\pi/2} \overbrace{2\cos\theta}^{f(2\sin\theta)}\, \overbrace{(2\cos\theta)}^{J(\theta)}\, d\theta = 4\int_0^{\pi/2} \cos^2\theta\, d\theta = \pi.$$

where the x-limits $\downarrow$ θ-limits $\downarrow$

■ **Double Integrals** Although changing variables in a multiple integral is not as straightforward as the procedure in (1), the basic idea illustrated in (2) carries over. To change variables in a double integral we need two equations such as

$$x = f(u, v), \qquad y = g(u, v). \tag{3}$$

To be analogous with (2), we expect that a change of variables in a double integral would take the form

$$\iint_R F(x, y)\, dA = \iint_S F(f(u, v), g(u, v))\, J(u, v)\, dA', \tag{4}$$

where S is the region in the uv-plane corresponding to the region R in the xy-plane and $J(u, v)$ is some function that depends on the partial derivatives of the equations in (3). The symbol dA' on the right side of (4) represents either $du\, dv$ or $dv\, du$.

In Section 9.11 we briefly discussed how to change a double integral $\iint_R F(x, y)\, dA$ from rectangular coordinates to polar coordinates. Recall that in Example 2 of that section the substitutions

$$x = r \cos \theta, \qquad y = r \sin \theta \tag{5}$$

led to

$$\int_0^2 \int_x^{\sqrt{8-x^2}} \frac{1}{5 + x^2 + y^2}\, dy\, dx = \int_{\pi/4}^{\pi/2} \int_0^{\sqrt{8}} \frac{1}{5 + r^2}\, r\, dr\, d\theta. \tag{6}$$

As we see in **FIGURE 9.17.1**, the introduction of polar coordinates changes the original region of integration R in the xy-plane to the more convenient rectangular region of integration S in the $r\theta$-plane. We note, too, that by comparing (4) with (6), we can identify $J(r, \theta) = r$ and $dA' = dr\, d\theta$.

The change-of-variable equations in (3) define a **transformation** or **mapping** T from the uv-plane to the xy-plane. A point (x_0, y_0) in the xy-plane determined from $x_0 = f(u_0, v_0)$, $y_0 = g(u_0, v_0)$ is said to be an **image** of (u_0, v_0).

EXAMPLE 1 **Image of a Region**

Find the image of the region S shown in **FIGURE 9.17.2(a)** under the transformation $x = u^2 + v^2$, $y = u^2 - v^2$.

SOLUTION We begin by finding the images of the sides of S that we have indicated by S_1, S_2, and S_3.

S_1: On this side $v = 0$ so that $x = u^2$, $y = u^2$. Eliminating u then gives $y = x$. Now imagine moving along the boundary from $(1, 0)$ to $(2, 0)$ (that is, $1 \le u \le 2$). The equations $x = u^2$, $y = u^2$ then indicate that x ranges from $x = 1$ to $x = 4$ and y ranges

(b) Region S in $r\theta$-plane

FIGURE 9.17.1 Region S is used to evaluate (6)

(a)

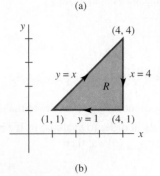

(b)

FIGURE 9.17.2 Region R is the image of region S in Example 1

simultaneously from $y = 1$ to $y = 4$. In other words, in the xy-plane the image of S_1 is the line segment $y = x$ from $(1, 1)$ to $(4, 4)$.

S_2: On this boundary $u^2 + v^2 = 4$ and so $x = 4$. Now as we move from the point $(2, 0)$ to $(\sqrt{\frac{5}{2}}, \sqrt{\frac{3}{2}})$, the remaining equation $y = u^2 - v^2$ indicates that y ranges from $y = 2^2 - 0^2 = 4$ to $y = (\sqrt{\frac{5}{2}})^2 - (\sqrt{\frac{3}{2}})^2 = 1$. In this case the image of S_2 is the vertical line segment $x = 4$ starting at $(4, 4)$ and going down to $(4, 1)$.

S_3: Since $u^2 - v^2 = 1$, we get $y = 1$. But as we move on this boundary from $(\sqrt{\frac{5}{2}}, \sqrt{\frac{3}{2}})$, to $(1, 0)$, the equation $x = u^2 + v^2$ indicates that x ranges from $x = 4$ to $x = 1$. The image of S_3 is the horizontal line segment $y = 1$ starting at $(4, 1)$ and ending at $(1, 1)$.

The image of S is the region R given in Figure 9.17.2(b). ≡

Observe in Example 1 that as we traverse the boundary of S in the counterclockwise direction, the boundary of R is traversed in a clockwise manner. We say that the transformation of the boundary of S has *induced* an orientation on the boundary of R.

While a proof of the formula for changing variables in a multiple integral is beyond the level of this text, we will give *some* of the underlying assumptions that are made about the equations (3) and the regions R and S. We assume that

- The functions f and g have continuous first partial derivatives on S.
- The transformation is one-to-one.
- Each of the regions R and S consists of a piecewise-smooth simple closed curve and its interior.
- The determinant

$$\begin{vmatrix} \dfrac{\partial x}{\partial u} & \dfrac{\partial x}{\partial v} \\ \dfrac{\partial y}{\partial u} & \dfrac{\partial y}{\partial v} \end{vmatrix} = \dfrac{\partial x}{\partial u}\dfrac{\partial y}{\partial v} - \dfrac{\partial x}{\partial v}\dfrac{\partial y}{\partial u} \tag{7}$$

is not zero on S.

A transformation T is said to be **one-to-one** if each point (x_0, y_0) in R is the image under T of a unique point (u_0, v_0) in S. Put another way, no two points in S have the same image in R. With the restrictions that $r \geq 0$ and $0 \leq \theta \leq 2\pi$, the equations in (5) define a one-to-one transformation from the $r\theta$-plane to the xy-plane. The determinant in (7) is called the **Jacobian determinant**, or simply **Jacobian**, of the transformation T and is the key to changing variables in a multiple integral. The Jacobian of the transformation defined by the equations in (3) is denoted by the symbol

We will encounter another Jacobian in Section 11.3.

$$\dfrac{\partial(x, y)}{\partial(u, v)}.$$

Similar to the notion of a one-to-one function, a one-to-one transformation T has an **inverse transformation** T^{-1} such that (u_0, v_0) is the image under T^{-1} of (x_0, y_0). See **FIGURE 9.17.3**. If it is possible to solve (3) for u and v in terms of x and y, then the inverse transformation is defined by a pair of equations

$$u = h(x, y), \qquad v = k(x, y). \tag{8}$$

The Jacobian of the inverse transformation T^{-1} is

$$\dfrac{\partial(u, v)}{\partial(x, y)} = \begin{vmatrix} \dfrac{\partial u}{\partial x} & \dfrac{\partial u}{\partial y} \\ \dfrac{\partial v}{\partial x} & \dfrac{\partial v}{\partial y} \end{vmatrix} \tag{9}$$

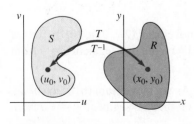

FIGURE 9.17.3 Transformation T and its inverse

and is related to the Jacobian of the transformation T by

$$\frac{\partial(x, y)}{\partial(u, v)} \frac{\partial(u, v)}{\partial(x, y)} = 1. \tag{10}$$

EXAMPLE 2 **Jacobian**

The Jacobian of the transformation $x = r \cos \theta$, $y = r \sin \theta$ is

$$\frac{\partial(x, y)}{\partial(r, \theta)} = \begin{vmatrix} \dfrac{\partial x}{\partial r} & \dfrac{\partial x}{\partial \theta} \\[2mm] \dfrac{\partial y}{\partial r} & \dfrac{\partial y}{\partial \theta} \end{vmatrix} = \begin{vmatrix} \cos \theta & -r \sin \theta \\ \sin \theta & r \cos \theta \end{vmatrix} = r(\cos^2 \theta + \sin^2 \theta) = r. \qquad \equiv$$

We now turn our attention to the main point of this discussion: how to change variables in a multiple integral. The idea expressed in (4) is valid; the function $J(u, v)$ turns out to be $|\partial(x, y)/\partial(u, v)|$. Under the assumptions made above, we have the following result:

THEOREM 9.17.1 **Change of Variables in a Double Integral**

If F is continuous on R, then

$$\iint\limits_{R} F(x, y) \, dA = \iint\limits_{S} F(f(u, v), g(u, v)) \left| \frac{\partial(x, y)}{\partial(u, v)} \right| dA'. \tag{11}$$

Formula (3) of Section 9.11 for changing a double integral to polar coordinates is just a special case of (11) with

$$\left| \frac{\partial(x, y)}{\partial(r, \theta)} \right| = |r| = r,$$

since $r \geq 0$. In (6) then we have $J(r, \theta) = |\partial(x, y)/\partial(r, \theta)| = r$.

A change of variables in a multiple integral can be used for either a simplification of the integrand or a simplification of the region of integration. The actual change of variables used is often inspired by the structure of the integrand $F(x, y)$ or by equations that define the region R. As a consequence, the transformation is then defined by equations of the form given in (8); that is, we are dealing with the inverse transformation. The next two examples will illustrate these ideas.

EXAMPLE 3 **Changing Variables in a Double Integral**

Evaluate $\iint_R \sin(x + 2y) \cos(x - 2y) \, dA$ over the region R shown in **FIGURE 9.17.4(a)**.

SOLUTION The difficulty in evaluating this double integral is clearly the integrand. The presence of the terms $x + 2y$ and $x - 2y$ prompts us to define the change of variables $u = x + 2y$, $v = x - 2y$. These equations will map R onto a region S in the uv-plane. As in Example 1, we transform the sides of the region.

S_1: $y = 0$ implies $u = x$ and $v = x$ or $v = u$. As we move from $(2\pi, 0)$ to $(0, 0)$, we see that the corresponding image points in the uv-plane lie on the line segment $v = u$ from $(2\pi, 2\pi)$ to $(0, 0)$.

S_2: $x = 0$ implies $u = 2y$ and $v = -2y$, or $v = -u$. As we move from $(0, 0)$ to $(0, \pi)$, the corresponding image points in the uv-plane lie on the line segment $v = -u$ from $(0, 0)$ to $(2\pi, -2\pi)$.

S_3: $x + 2y = 2\pi$ implies $u = 2\pi$. As we move from $(0, \pi)$ to $(2\pi, 0)$, the equation $v = x - 2y$ shows that v ranges from $v = -2\pi$ to $v = 2\pi$. Thus, the image of S_3 is the

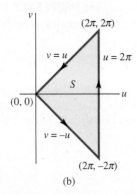

FIGURE 9.17.4 Region S is the image of region R in Example 3

vertical line segment $u = 2\pi$ starting at $(-2\pi, -2\pi)$ and going up to $(2\pi, 2\pi)$. See Figure 9.17.4(b).

Now, solving for x and y in terms of u and v gives

$$x = \frac{1}{2}(u + v), \quad y = \frac{1}{4}(u - v).$$

Therefore,

$$\frac{\partial(x, y)}{\partial(u, v)} = \begin{vmatrix} \dfrac{\partial x}{\partial u} & \dfrac{\partial x}{\partial v} \\ \dfrac{\partial y}{\partial u} & \dfrac{\partial y}{\partial v} \end{vmatrix} = \begin{vmatrix} \dfrac{1}{2} & \dfrac{1}{2} \\ \dfrac{1}{4} & -\dfrac{1}{4} \end{vmatrix} = -\frac{1}{4}.$$

Hence, from (11) we find that

$$\iint_R \sin(x + 2y)\cos(x - 2y)\, dA = \iint_S \sin u \cos v \left| -\frac{1}{4} \right| dA'$$

$$= \frac{1}{4} \int_0^{2\pi} \int_{-u}^u \sin u \cos v \, dv \, du$$

$$= \frac{1}{4} \int_0^{2\pi} \sin u \sin v \Big]_{-u}^{u} du$$

$$= \frac{1}{2} \int_0^{2\pi} \sin^2 u \, du$$

$$= \frac{1}{4} \int_0^{2\pi} (1 - \cos 2u) \, du$$

$$= \frac{1}{4}\left[u - \frac{1}{2}\sin 2u \right]_0^{2\pi} = \frac{\pi}{2}. \qquad \equiv$$

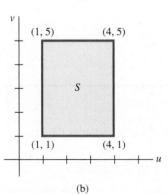

FIGURE 9.17.5 Region S is image of region R in Example 4

EXAMPLE 4 Changing Variables in a Double Integral

Evaluate $\iint_R xy \, dA$ over the region R shown in **FIGURE 9.17.5(a)**.

SOLUTION In this case the integrand is fairly simple, but integration over the region R would be tedious since we would have to express $\iint_R xy \, dA$ as the sum of three integrals. (Verify this.) The equations of the boundaries of R suggest the change of variables

$$u = \frac{y}{x^2}, \quad v = xy. \tag{12}$$

Obtaining the image of R is a straightforward matter in this case, since the images of the curves that make up the four boundaries are simply $u = 1$, $u = 4$, $v = 1$, and $v = 5$. In other words, the image of the region R is the rectangular region S: $1 \le u \le 4$, $1 \le v \le 5$. See Figure 9.17.5(b). Now instead of trying to solve the equations in (12) for x and y in terms of u and v, we can compute the Jacobian $\partial(x, y)/\partial(u, v)$ by computing $\partial(u, v)/\partial(x, y)$ and using (10). We have

$$\frac{\partial(u, v)}{\partial(x, y)} = \begin{vmatrix} \dfrac{\partial u}{\partial x} & \dfrac{\partial u}{\partial y} \\ \dfrac{\partial v}{\partial x} & \dfrac{\partial v}{\partial y} \end{vmatrix} = \begin{vmatrix} -\dfrac{2y}{x^3} & \dfrac{1}{x^2} \\ y & x \end{vmatrix} = -\frac{3y}{x^2},$$

and so from (10),

$$\frac{\partial(x, y)}{\partial(u, v)} = \frac{1}{\dfrac{\partial(u, v)}{\partial(x, y)}} = -\frac{x^2}{3y} = -\frac{1}{3u}.$$

Hence,

$$\iint\limits_{R} xy \, dA = \iint\limits_{S} v \left| -\frac{1}{3u} \right| dA'$$

$$= \frac{1}{3} \int_{1}^{4} \int_{1}^{5} \frac{v}{u} \, dv \, du$$

$$= \frac{1}{3} \int_{1}^{4} \left. \frac{v^2}{2u} \right]_{1}^{5} du$$

$$= 4 \int_{1}^{4} \frac{1}{u} \, du = 4 \ln u \Big]_{1}^{4} = 4 \ln 4. \qquad \equiv$$

■ Triple Integrals To change variables in a triple integral, let

$$x = f(u, v, w), \qquad y = g(u, v, w), \qquad z = h(u, v, w),$$

be a one-to-one transformation T from a region E in uvw-space to a region D in xyz-space. If F is continuous on D, then

$$\iiint\limits_{D} F(x, y, z) \, dV = \iiint\limits_{E} F(f(u, v, w), g(u, v, w), h(u, v, w)) \left| \frac{\partial(x, y, z)}{\partial(u, v, w)} \right| dV',$$

where

$$\frac{\partial(x, y, z)}{\partial(u, v, w)} = \begin{vmatrix} \dfrac{\partial x}{\partial u} & \dfrac{\partial x}{\partial v} & \dfrac{\partial x}{\partial w} \\[2mm] \dfrac{\partial y}{\partial u} & \dfrac{\partial y}{\partial v} & \dfrac{\partial y}{\partial w} \\[2mm] \dfrac{\partial z}{\partial u} & \dfrac{\partial z}{\partial v} & \dfrac{\partial z}{\partial w} \end{vmatrix}.$$

We leave it as an exercise for the reader to show that if T is the transformation from spherical to rectangular coordinates defined by

$$x = \rho \sin \phi \cos \theta, \qquad y = \rho \sin \phi \sin \theta, \qquad z = \rho \cos \phi, \qquad (13)$$

then

$$\frac{\partial(x, y, z)}{\partial(\rho, \phi, \theta)} = \rho^2 \sin \phi.$$

9.17 Exercises Answers to selected odd-numbered problems begin on page ANS-26.

1. Consider a transformation T defined by $x = 4u - v$, $y = 5u + 4v$. Find the images of the points $(0, 0), (0, 2), (4, 0)$, and $(4, 2)$ in the uv-plane under T.

2. Consider a transformation T defined by $x = \sqrt{v - u}$, $y = v + u$. Find the images of the points $(1, 1), (1, 3)$, and $(\sqrt{2}, 2)$ in the xy-plane under T^{-1}.

In Problems 3–6, find the image of the set S under the given transformation.

3. $S: 0 \le u \le 2, 0 \le v \le u; x = 2u + v, y = u - 3v$

4. $S: -1 \le u \le 4, 1 \le v \le 5; u = x - y, v = x + 2y$

5. $S: 0 \le u \le 1, 0 \le v \le 2; x = u^2 - v^2, y = uv$

6. $S: 1 \le u \le 2, 1 \le v \le 2; x = uv, y = v^2$

In Problems 7–10, find the Jacobian of the transformation T from the uv-plane to the xy-plane.

7. $x = ve^{-u}, y = ve^u$ **8.** $x = e^{3u} \sin v, y = e^{3u} \cos v$

9. $u = \dfrac{y}{x^2}, v = \dfrac{y^2}{x}$ **10.** $u = \dfrac{2x}{x^2 + y^2}, v = \dfrac{-2y}{x^2 + y^2}$

11. (a) Find the image of the region $S: 0 \le u \le 1, 0 \le v \le 1$ under the transformation $x = u - uv, y = uv$.

 (b) Explain why the transformation is not one-to-one on the boundary of S.

12. Determine where the Jacobian $\partial(x, y)/\partial(u, v)$ of the transformation in Problem 11 is zero.

In Problems 13–22, evaluate the given integral by means of the indicated change of variables.

13. $\iint_R (x + y)\, dA$, where R is the region bounded by the graphs of $x - 2y = -6, x - 2y = 6, x + y = -1, x + y = 3$; $u = x - 2y, v = x + y$

14. $\iint_R \dfrac{\cos\frac{1}{2}(x - y)}{3x + y}\, dA$, where R is the region bounded by the graphs of $y = x, y = x - \pi, y = -3x + 3, y = -3x + 6$; $u = x - y, v = 3x + y$

15. $\iint_R \dfrac{y^2}{x}\, dA$, where R is the region bounded by the graphs $y = x^2, y = \frac{1}{2}x^2, x = y^2, x = \frac{1}{2}y^2; u = \dfrac{x^2}{y}, v = \dfrac{y^2}{x}$

16. $\iint_R (x^2 + y^2)^{-3}\, dA$, where R is the region bounded by the circles $x^2 + y^2 = 2x, x^2 + y^2 = 4x, x^2 + y^2 = 2y, x^2 + y^2 = 6y$; $u = \dfrac{2x}{x^2 + y^2}, v = \dfrac{2y}{x^2 + y^2}$ [*Hint:* Form $u^2 + v^2$.]

17. $\iint_R (x^2 + y^2)\, dA$, where R is the region in the first quadrant bounded by the graphs of $x^2 - y^2 = a, x^2 - y^2 = b, 2xy = c, 2xy = d, 0 < a < b, 0 < c < d; u = x^2 - y^2, v = 2xy$

18. $\iint_R (x^2 + y^2) \sin xy\, dA$, where R is the region bounded by the graphs of $x^2 - y^2 = 1, x^2 - y^2 = 9, xy = 2, xy = -2$; $u = x^2 - y^2, v = xy$

19. $\iint_R \dfrac{x}{y + x^2}\, dA$, where R is the region in the first quadrant bounded by the graphs of $x = 1, y = x^2, y = 4 - x^2$; $x = \sqrt{v - u}, y = v + u$

20. $\iint_R y\, dA$, where R is the triangular region with vertices $(0, 0), (2, 3)$ and $(-4, 1); x = 2u - 4v, y = 3u + v$

21. $\iint_R y^4\, dA$, where R is the region in the first quadrant bounded by the graphs of $xy = 1, xy = 4, y = x, y = 4x; u = xy, v = y/x$

22. $\iiint_D (4z + 2x - 2y)\, dV$, where D is the parallelepiped $1 \le y + z \le 3, -1 \le -y + z \le 1, 0 \le x - y \le 3; u = y + z, v = -y + z, w = x - y$

In Problems 23–26, evaluate the given double integral by means of an appropriate change of variables.

23. $\int_0^1 \int_0^{1-x} e^{(y-x)/(y+x)}\, dy\, dx$ **24.** $\int_{-2}^0 \int_0^{x+2} e^{y^2 - 2xy + x^2}\, dy\, dx$

25. $\iint_R (6x + 3y)\, dA$, where R is the trapezoidal region in the first quadrant with vertices $(1, 0), (4, 0), (2, 4)$, and $(\frac{1}{2}, 1)$

26. $\iint_R (x + y)^4\, e^{x-y}\, dA$, where R is the square region with vertices $(1, 0), (0, 1), (1, 2)$, and $(2, 1)$

27. A problem in thermodynamics is to find the work done by an ideal Carnot engine. This work is defined to be the area of the region R in the first quadrant bounded by the isothermals $xy = a, xy = b, 0 < a < b$, and the adiabatics $xy^{1.4} = c, xy^{1.4} = d, 0 < c < d$. Use $A = \iint_R dA$ and an appropriate substitution to find the area shown in **FIGURE 9.17.6**.

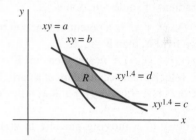

FIGURE 9.17.6 Region R for Problem 27

28. Use $V = \iiint_D dV$ and the substitutions $u = x/a, v = y/b, w = z/c$ to show that the volume of the ellipsoid $x^2/a^2 + y^2/b^2 + z^2/c^2 = 1$ is $V = \frac{4}{3}\pi abc$.

29. Evaluate the double integral $\iint_R \left(\dfrac{x^2}{25} + \dfrac{y^2}{9}\right) dA$, where R is the elliptical region whose boundary is the graph of $x^2/25 + y^2/9 = 1$. Use the substitutions $u = x/5, v = y/3$, and polar coordinates.

30. Verify that the Jacobian of the transformation given in (13) is $\partial(x, y, z)/\partial(\rho, \phi, \theta) = \rho^2 \sin \phi$.

9 Chapter in Review Answers to selected odd-numbered problems begin on page ANS-26.

Answer Problems 1–20 without referring back to the text. Fill in the blank or answer true/false. Where appropriate, assume continuity of P, Q, and their first partial derivatives.

1. A particle whose position vector is $\mathbf{r}(t) = \cos t\mathbf{i} + \cos t\mathbf{j} + \sqrt{2} \sin t\mathbf{k}$ moves with constant speed. _____

2. The path of a moving particle whose position vector is $\mathbf{r}(t) = (t^2 + 1)\mathbf{i} + 4\mathbf{j} + t^4\mathbf{k}$ lies in a plane. _____

3. The binormal vector is perpendicular to the osculating plane. _____

4. If $\mathbf{r}(t)$ is the position vector of a moving particle, then the velocity vector $\mathbf{v}(t) = \mathbf{r}'(t)$ and the acceleration vector $\mathbf{a}(t) = \mathbf{r}''(t)$ are orthogonal. _____

5. ∇z is perpendicular to the graph of $z = f(x, y)$. _____

6. If $\nabla f = \mathbf{0}$, then $f = $ constant. _____

7. The integral $\int_C (x^2 + y^2)\, dx + 2xy\, dy$, where C is given by $y = x^3$ from $(0, 0)$ to $(1, 1)$, has the same value on the curve $y = x^6$ from $(0, 0)$ to $(1, 1)$. _____

8. The value of the integral $\int_C 2xy\, dx - x^2\, dy$ between two points A and B depends on the path C. _____

9. If C_1 and C_2 are two smooth curves such that $\int_{C_1} P\,dx + Q\,dy = \int_{C_2} P\,dx + Q\,dy$, then $\int_C P\,dx + Q\,dy$ is independent of the path. _____

10. If the work $\int_C \mathbf{F} \cdot d\mathbf{r}$ depends on the curve C, then $\mathbf{F}$ is nonconservative. _____

11. If $\partial P/\partial x = \partial Q/\partial y$, then $\int_C P\,dx + Q\,dy$ is independent of the path. _____

12. In a conservative force field $\mathbf{F}$, the work done by $\mathbf{F}$ around a simple closed curve is zero. _____

13. Assuming continuity of all partial derivatives, $\nabla \times \nabla f = \mathbf{0}$. _____

14. The surface integral of the normal component of the curl of a conservative vector field $\mathbf{F}$ over a surface S is equal to zero. _____

15. Work done by a force $\mathbf{F}$ along a curve C is due entirely to the tangential component of $\mathbf{F}$. _____

16. For a two-dimensional vector field $\mathbf{F}$ in the plane $z = 0$, Stokes' theorem is the same as Green's theorem. _____

17. If $\mathbf{F}$ is a conservative force field, then the sum of the potential and kinetic energies of an object is constant. _____

18. If $\int_C P\,dx + Q\,dy$ is independent of the path, then $\mathbf{F} = P\mathbf{i} + Q\mathbf{j}$ is the gradient of some function ϕ. _____

19. If $\phi = 1/\sqrt{x^2 + y^2}$ is a potential function for a conservative force field $\mathbf{F}$, then $\mathbf{F} = $ _____.

20. If $\mathbf{F} = f(x)\mathbf{i} + g(y)\mathbf{j} + h(z)\mathbf{k}$, then curl $\mathbf{F} = $ _____.

21. Find the velocity and acceleration of a particle whose position vector is $\mathbf{r}(t) = 6t\mathbf{i} + t\mathbf{j} + t^2\mathbf{k}$ as it passes through the plane $-x + y + z = -4$.

22. The velocity of a moving particle is $\mathbf{v}(t) = -10t\mathbf{i} + (3t^2 - 4t)\mathbf{j} + \mathbf{k}$. If the particle starts at $t = 0$ at $(1, 2, 3)$, what is its position at $t = 2$?

23. The acceleration of a moving particle is $\mathbf{a}(t) = \sqrt{2}\sin t\,\mathbf{i} + \sqrt{2}\cos t\,\mathbf{j}$. Given that the velocity and position of the particle at $t = \pi/4$ are $\mathbf{v}(\pi/4) = -\mathbf{i} + \mathbf{j} + \mathbf{k}$ and $\mathbf{r}(\pi/4) = \mathbf{i} + 2\mathbf{j} + (\pi/4)\mathbf{k}$, respectively, what was the position of the particle at $t = 3\pi/4$?

24. Given that $\mathbf{r}(t) = \frac{1}{2}t^2\mathbf{i} + \frac{1}{3}t^3\mathbf{j} - \frac{1}{2}t^2\mathbf{k}$ is the position vector of a moving particle, find the tangential and normal components of the acceleration at any t. Find the curvature.

25. Sketch the curve traced by $\mathbf{r}(t) = \cosh t\,\mathbf{i} + \sinh t\,\mathbf{j} + t\mathbf{k}$.

26. Suppose that the vector function of Problem 25 is the position vector of a moving particle. Find the vectors $\mathbf{T}$, $\mathbf{N}$, and $\mathbf{B}$ at $t = 1$. Find the curvature at that point.

In Problems 27 and 28, find the directional derivative of the given function in the indicated direction.

27. $f(x, y) = x^2 y - y^2 x$; $D_{\mathbf{u}}f$ in the direction of $2\mathbf{i} + 6\mathbf{j}$

28. $F(x, y, z) = \ln(x^2 + y^2 + z^2)$; $D_{\mathbf{u}}F$ in the direction of $-2\mathbf{i} + \mathbf{j} + 2\mathbf{k}$

29. Consider the function $f(x, y) = x^2 y^4$. At $(1, 1)$ what is
 (a) The rate of change of f in the direction of $\mathbf{i}$?
 (b) The rate of change of f in the direction of $\mathbf{i} - \mathbf{j}$?
 (c) The rate of change of f in the direction of $\mathbf{j}$?

30. Let $w = \sqrt{x^2 + y^2 + z^2}$.
 (a) If $x = 3\sin 2t$, $y = 4\cos 2t$, and $z = 5t^3$, find dw/dt.
 (b) If $x = 3\sin 2\dfrac{t}{r}$, $y = 4\cos 2\dfrac{r}{t}$, and $z = 5t^3 r^3$, find $\partial w/\partial t$.

31. Find an equation of the tangent plane to the graph of
$$z = \sin(xy) \text{ at } (\tfrac{1}{2}, \tfrac{2}{3}\pi, \tfrac{1}{2}\sqrt{3}).$$

32. Determine whether there are any points on the surface $z^2 + xy - 2x - y^2 = 1$ at which the tangent plane is parallel to $z = 2$.

33. Express the volume of the solid shown in **FIGURE 9.R.1** as one or more iterated integrals using the order of integration **(a)** $dy\,dx$ and **(b)** $dx\,dy$. Choose either part **(a)** or part **(b)** to find the volume.

FIGURE 9.R.1 Solid for Problem 33

34. A lamina has the shape of the region in the first quadrant bounded by the graphs of $y = x^2$ and $y = x^3$. Find the center of mass if the density at a point P is directly proportional to the square of the distance from the origin.

35. Find the moment of inertia of the lamina described in Problem 34 about the y-axis.

36. Find the volume of the sphere $x^2 + y^2 + z^2 = a^2$ using a triple integral in **(a)** rectangular coordinates, **(b)** cylindrical coordinates, and **(c)** spherical coordinates.

37. Find the volume of the solid that is bounded between the cones $z = \sqrt{x^2 + y^2}$ and $z = \sqrt{9x^2 + 9y^2}$ and the plane $z = 3$.

38. Find the volume of the solid shown in **FIGURE 9.R.2**.

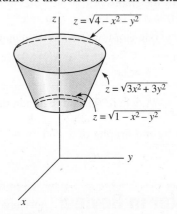

FIGURE 9.R.2 Solid for Problem 38

In Problems 39–42, find the indicated expression for the vector field $\mathbf{F} = x^2 y\mathbf{i} + xy^2\mathbf{j} + 2xyz\mathbf{k}$.

39. $\nabla \cdot \mathbf{F}$ **40.** $\nabla \times \mathbf{F}$

41. $\nabla \cdot (\nabla \times \mathbf{F})$ **42.** $\nabla(\nabla \cdot \mathbf{F})$

43. Evaluate $\displaystyle\int_C \frac{z^2}{x^2 + y^2}\,ds$, where C is given by
$$x = \cos 2t, \quad y = \sin 2t, \quad z = 2t, \quad \pi \le t \le 2\pi.$$

44. Evaluate $\int_C (xy + 4x)\,ds$, where C is given by $2x + y = 2$ from $(1, 0)$ to $(0, 2)$.

45. Evaluate $\int_C 3x^2y^2\,dx + (2x^3y - 3y^2)\,dy$, where C is given by $y = 5x^4 + 7x^2 - 14x$ from $(0, 0)$ to $(1, -2)$.

46. Show that $\oint_C \dfrac{-y\,dx + x\,dy}{x^2 + y^2} = 2\pi$, where C is the circle $x^2 + y^2 = a^2$.

47. Evaluate $\int_C y \sin \pi z\,dx + x^2e^y\,dy + 3xyz\,dz$, where C is given by $x = t$, $y = t^2$, $z = t^3$ from $(0, 0, 0)$ to $(1, 1, 1)$.

48. If $\mathbf{F} = 4y\mathbf{i} + 6x\mathbf{j}$ and C is given by $x^2 + y^2 = 1$, evaluate $\oint_C \mathbf{F}\cdot d\mathbf{r}$ in two different ways.

49. Find the work done by the force $\mathbf{F} = x \sin y\mathbf{i} + y \sin x\mathbf{j}$ acting along the line segments from $(0, 0)$ to $(\pi/2, 0)$ and from $(\pi/2, 0)$ to $(\pi/2, \pi)$.

50. Find the work done by $\mathbf{F} = \dfrac{2}{x^2 + y^2}\mathbf{i} + \dfrac{1}{x^2 + y^2}\mathbf{j}$ from $(-\frac{1}{2}, \frac{1}{2})$ to $(1, \sqrt{3})$ acting on the path shown in **FIGURE 9.R.3**.

FIGURE 9.R.3 Path for Problem 50

51. Evaluate $\iint_S (z/xy)\,dS$, where S is that portion of the cylinder $z = x^2$ in the first octant that is bounded by $y = 1$, $y = 3$, $z = 1$, $z = 4$.

52. If $\mathbf{F} = \mathbf{i} + 2\mathbf{j} + 3\mathbf{k}$, find the flux of $\mathbf{F}$ through the square defined by $0 \le x \le 1$, $0 \le y \le 1$, $z = 2$.

53. If $\mathbf{F} = c\nabla(1/r)$, where c is constant and $\|\mathbf{r}\| = r$, $\mathbf{r} = x\mathbf{i} + y\mathbf{j} + z\mathbf{k}$, find the flux of $\mathbf{F}$ through the sphere $x^2 + y^2 + z^2 = a^2$.

54. Explain why the divergence theorem is not applicable in Problem 53.

55. Find the flux of $\mathbf{F} = c\nabla(1/r)$ through any surface S that forms the boundary of a closed bounded region of space not containing the origin.

56. If $\mathbf{F} = 6x\mathbf{i} + 7z\mathbf{j} + 8y\mathbf{k}$, use Stokes' theorem to evaluate $\iint_S (\text{curl } \mathbf{F}\cdot\mathbf{n})\,dS$, where S is that portion of the paraboloid $z = 9 - x^2 - y^2$ within the cylinder $x^2 + y^2 = 4$.

57. Use Stokes' theorem to evaluate $\oint_C -2y\,dx + 3x\,dy + 10z\,dz$, where C is the circle $(x - 1)^2 + (y - 3)^2 = 25$, $z = 3$.

58. Find the work $\oint_C \mathbf{F}\cdot d\mathbf{r}$ done by the force $\mathbf{F} = x^2\mathbf{i} + y^2\mathbf{j} + z^2\mathbf{k}$ around the curve C that is formed by the intersection of the plane $z = 2 - y$ and the sphere $x^2 + y^2 + z^2 = 4z$.

59. If $\mathbf{F} = x\mathbf{i} + y\mathbf{j} + z\mathbf{k}$, use the divergence theorem to evaluate $\iint_S (\mathbf{F}\cdot\mathbf{n})\,dS$, where S is the surface of the region bounded by $x^2 + y^2 = 1$, $z = 0$, $z = 1$.

60. Repeat Problem 59 for $\mathbf{F} = \frac{1}{3}x^3\mathbf{i} + \frac{1}{3}y^3\mathbf{j} + \frac{1}{3}z^3\mathbf{k}$.

61. If $\mathbf{F} = (x^2 - e^y \tan^{-1}z)\mathbf{i} + (x + y)^2\mathbf{j} - (2yz + x^{10})\mathbf{k}$, use the divergence theorem to evaluate $\iint_S (\mathbf{F}\cdot\mathbf{n})\,dS$, where S is the surface of the region in the first octant bounded by $z = 1 - x^2$, $z = 0$, $z = 2 - y$, $y = 0$.

62. Suppose $\mathbf{F} = x\mathbf{i} + y\mathbf{j} + (z^2 + 1)\mathbf{k}$ and S is the surface of the region bounded by $x^2 + y^2 = a^2$, $z = 0$, $z = c$. Evaluate $\iint_S (\mathbf{F}\cdot\mathbf{n})\,dS$ without the aid of the divergence theorem. [*Hint*: The lateral surface area of the cylinder is $2\pi ac$.]

63. Evaluate the integral $\iint_R (x^2 + y^2)\sqrt[3]{x^2 - y^2}\,dA$, where R is the region bounded by the graphs of $x = 0$, $x = 1$, $y = 0$, and $y = 1$ by means of the change of variables $u = 2xy$, $v = x^2 - y^2$.

64. Evaluate the integral

$$\iint_R \frac{1}{\sqrt{(x - y)^2 + 2(x + y) + 1}}\,dA,$$

where R is the region bounded by the graphs of $y = x$, $x = 2$, and $y = 0$ by means of the change of variables $x = u + uv$, $y = v + uv$.

65. As shown in **FIGURE 9.R.4**, a sphere of radius 1 has its center on the surface of a sphere of radius $a > 1$. Find the surface area of that portion of the larger sphere cut out by the smaller sphere.

FIGURE 9.R.4 Spheres in Problem 65

66. On the surface of a globe or, more precisely, on the surface of the Earth, the boundaries of the states of Colorado and Wyoming are both "spherical rectangles." (In this problem we assume that the Earth is a perfect sphere.) Colorado is bounded by the lines of longitude 102°W and 109°W and the lines of latitude 37°N and 41°N. Wyoming is bounded by longitudes 104°W and 111°W and latitudes 41°N and 45°N. See **FIGURE 9.R.5**.

(a) Without explicitly computing their areas, determine which state is larger and explain why.

(b) By what percentage is Wyoming larger (or smaller) than Colorado? [*Hint*: Suppose the radius of the Earth is R. Project a spherical rectangle in the Northern Hemisphere that is determined by latitudes θ_1 and θ_2 and longitudes ϕ_1 and ϕ_2 onto the xy-plane.]

(c) One reference book gives the areas of the two states as 104,247 and 97,914 mi^2. How does this answer compare with the answer in part (b)?

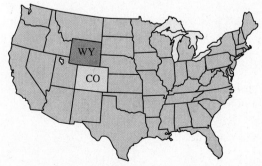

FIGURE 9.R.5 States WY and CO are spherical rectangles in Problem 66

PART 3

Systems of Differential Equations

Systems of Linear Differential Equations

We encountered systems of ordinary differential equations in Sections 2.9, 3.12, and 4.6 and were able to solve some systems using either systematic elimination or the Laplace transform. In this chapter we focus only on systems of linear first-order differential equations. We will see that the general theory of systems of linear DEs and the solution procedure is similar to that of linear higher-order equations considered in Chapter 3. This material is fundamental to the analysis of systems of nonlinear first-order equations in Chapter 11. Matrix notation and properties are used extensively throughout this chapter. If you are unfamiliar with these concepts, a review of Chapter 8 is recommended.

10.1 Theory of Linear Systems

INTRODUCTION Recall that in Section 3.12 we illustrated how to solve systems of n linear differential equations in n unknowns of the form

$$
\begin{aligned}
P_{11}(D)x_1 + P_{12}(D)x_2 + \cdots + P_{1n}(D)x_n &= b_1(t) \\
P_{21}(D)x_1 + P_{22}(D)x_2 + \cdots + P_{2n}(D)x_n &= b_2(t) \\
&\vdots \qquad\qquad\qquad \vdots \\
P_{n1}(D)x_1 + P_{n2}(D)x_2 + \cdots + P_{nn}(D)x_n &= b_n(t),
\end{aligned}
\tag{1}
$$

where the P_{ij} were polynomials of various degrees in the differential operator D. In this chapter we confine our study to systems of first-order DEs that are special cases of systems that have the normal form

$$
\begin{aligned}
\frac{dx_1}{dt} &= g_1(t, x_1, x_2, \ldots, x_n) \\
\frac{dx_2}{dt} &= g_2(t, x_1, x_2, \ldots, x_n) \\
&\vdots \qquad\qquad \vdots \\
\frac{dx_n}{dt} &= g_n(t, x_1, x_2, \ldots, x_n).
\end{aligned}
\tag{2}
$$

A system such as (2) of n first-order equations is called a **first-order system**.

Linear Systems

When each of the functions $g_1, g_2, \ldots, g_n$ in (2) is linear in the dependent variables $x_1, x_2, \ldots, x_n$, we get the **normal form** of a first-order system of linear equations:

$$
\begin{aligned}
\frac{dx_1}{dt} &= a_{11}(t)x_1 + a_{12}(t)x_2 + \cdots + a_{1n}(t)x_n + f_1(t) \\
\frac{dx_2}{dt} &= a_{21}(t)x_1 + a_{22}(t)x_2 + \cdots + a_{2n}(t)x_n + f_2(t) \\
&\vdots \qquad\qquad\qquad\qquad \vdots \\
\frac{dx_n}{dt} &= a_{n1}(t)x_1 + a_{n2}(t)x_2 + \cdots + a_{nn}(t)x_n + f_n(t).
\end{aligned}
\tag{3}
$$

We refer to a system of the form given in (3) simply as a **linear system**. We assume that the coefficients $a_{ij}(t)$ as well as the functions $f_i(t)$ are continuous on a common interval I. When $f_i(t) = 0$, $i = 1, 2, \ldots, n$, the linear system is said to be **homogeneous**; otherwise it is **nonhomogeneous**.

Matrix Form of a Linear System

If $\mathbf{X}, \mathbf{X}', \mathbf{A}(t), \mathbf{F}(t)$ denote the respective matrices

$$
\mathbf{X} = \begin{pmatrix} x_1(t) \\ x_2(t) \\ \vdots \\ x_n(t) \end{pmatrix}, \quad
\mathbf{X}' = \frac{d}{dt}\begin{pmatrix} x_1(t) \\ x_2(t) \\ \vdots \\ x_n(t) \end{pmatrix} = \begin{pmatrix} x_1'(t) \\ x_2'(t) \\ \vdots \\ x_n'(t) \end{pmatrix}, \quad
\mathbf{A}(t) = \begin{pmatrix} a_{11}(t) & a_{12}(t) & \cdots & a_{1n}(t) \\ a_{21}(t) & a_{22}(t) & \cdots & a_{2n}(t) \\ \vdots & & & \vdots \\ a_{n1}(t) & a_{n2}(t) & \cdots & a_{nn}(t) \end{pmatrix}, \quad
\mathbf{F}(t) = \begin{pmatrix} f_1(t) \\ f_2(t) \\ \vdots \\ f_n(t) \end{pmatrix},
$$

then the system of linear first-order equations (3) can be written

$$
\begin{pmatrix} x_1'(t) \\ x_2'(t) \\ \vdots \\ x_n'(t) \end{pmatrix} = \begin{pmatrix} a_{11}(t) & a_{12}(t) & \cdots & a_{1n}(t) \\ a_{21}(t) & a_{22}(t) & \cdots & a_{2n}(t) \\ \vdots & & & \vdots \\ a_{n1}(t) & a_{n2}(t) & \cdots & a_{nn}(t) \end{pmatrix}\begin{pmatrix} x_1(t) \\ x_2(t) \\ \vdots \\ x_n(t) \end{pmatrix} + \begin{pmatrix} f_1(t) \\ f_2(t) \\ \vdots \\ f_n(t) \end{pmatrix}
$$

or simply

$$
\mathbf{X}' = \mathbf{AX} + \mathbf{F}.
\tag{4}
$$

If the system is homogeneous, then the matrix form is

$$
\mathbf{X}' = \mathbf{AX}.
\tag{5}
$$

EXAMPLE 1 **Systems Written in Matrix Notation**

(a) If $\mathbf{X} = \begin{pmatrix} x \\ y \end{pmatrix}$, then the matrix form of the homogeneous linear system

$$\frac{dx}{dt} = 3x + 4y$$
$$\frac{dy}{dt} = 5x - 7y$$

is $\mathbf{X}' = \begin{pmatrix} 3 & 4 \\ 5 & -7 \end{pmatrix} \mathbf{X}.$

(b) If $\mathbf{X} = \begin{pmatrix} x \\ y \\ z \end{pmatrix}$, then the matrix form of the nonhomogeneous linear system

$$\frac{dx}{dt} = 6x + y + z + t$$
$$\frac{dy}{dt} = 8x + 7y - z + 10t$$
$$\frac{dz}{dt} = 2x + 9y - z + 6t$$

is $\mathbf{X}' = \begin{pmatrix} 6 & 1 & 1 \\ 8 & 7 & -1 \\ 2 & 9 & -1 \end{pmatrix} \mathbf{X} + \begin{pmatrix} t \\ 10t \\ 6t \end{pmatrix}.$

≡

EXAMPLE 2 **Writing a Linear System Without Matrices**

Write the linear system

$$\mathbf{X}' = \begin{pmatrix} 1 & 7 \\ 2 & -9 \end{pmatrix} \mathbf{X} + \begin{pmatrix} -4 \\ 1 \end{pmatrix} \cos t + \begin{pmatrix} 1 \\ -1 \end{pmatrix}$$

without matrices.

SOLUTION First, let $\mathbf{X} = \begin{pmatrix} x \\ y \end{pmatrix}$ and so the left-hand side of the given system is

$$\mathbf{X}' = \begin{pmatrix} \dfrac{dx}{dt} \\ \dfrac{dy}{dt} \end{pmatrix}.$$

Then, using matrix multiplication and addition, the terms on the right-hand side of the given linear system can be written as a single matrix:

$$\begin{pmatrix} 1 & 7 \\ 2 & -9 \end{pmatrix} \mathbf{X} + \begin{pmatrix} -4 \\ 1 \end{pmatrix} \cos t + \begin{pmatrix} 1 \\ -1 \end{pmatrix} = \begin{pmatrix} 1 & 7 \\ 2 & -9 \end{pmatrix} \begin{pmatrix} x \\ y \end{pmatrix} + \begin{pmatrix} -4 \\ 1 \end{pmatrix} \cos t + \begin{pmatrix} 1 \\ -1 \end{pmatrix}$$

$$= \begin{pmatrix} x + 7y \\ 2x - 9y \end{pmatrix} + \begin{pmatrix} -4\cos t \\ \cos t \end{pmatrix} + \begin{pmatrix} 1 \\ -1 \end{pmatrix}$$

$$= \begin{pmatrix} x + 7y - 4\cos t + 1 \\ 2x - 9y + \cos t - 1 \end{pmatrix}.$$

Now two matrices are equal if their entries in the same position are equal. Therefore

$$\begin{pmatrix} \dfrac{dx}{dt} \\ \dfrac{dy}{dt} \end{pmatrix} = \begin{pmatrix} x + 7y - 4\cos t + 1 \\ 2x - 9y + \cos t - 1 \end{pmatrix} \quad \text{implies} \quad \begin{aligned} \frac{dx}{dt} &= x + 7y - 4\cos t + 1 \\ \frac{dy}{dt} &= 2x - 9y + \cos t - 1. \end{aligned}$$

≡

An Equation Written as a System

At times it is to your advantage to work with a system of equations rather than with a differential equation. Suppose

$$y^{(n)} = f(t, y, y', \ldots, y^{(n-1)}) \tag{6}$$

is a linear nth-order differential equation. If we let

$$y = x_1, \ y' = x_2, \ y'' = x_3, \ \ldots, \ y^{(n-2)} = x_{n-1}, \ y^{(n-1)} = x_n, \tag{7}$$

then observe by taking the derivative of each of the terms in (7) we get

$$y' = x_1' = x_2, \ y'' = x_2' = x_3, \ldots, \ y^{(n-1)} = x_{n-1}' = x_n, \ y^{(n)} = x_n' = f(t, x_1, x_2, \ldots, x_n).$$

Thus a first-order system of n equations is

$$
\begin{aligned}
x_1' &= x_2 \\
x_2' &= x_3 \\
&\ \ \vdots \\
x_{n-1}' &= x_n \\
x_n' &= f(t, x_1, x_2, \ldots, x_n).
\end{aligned}
\tag{8}
$$

EXAMPLE 3 **Writing an Equation as a First-Order System**

Write the third-order differential equation

$$y''' - 3y'' + 5y' - y = 2e^{-2t}$$

as a first-order system of equations using matrix notation.

SOLUTION First, we write the differential equation as $y''' = y - 5y' + 3y'' + 2e^{-2t}$ to agree with the form given in (6). Then, using the substitutions $y = x_1, y' = x_2, y'' = x_3$, we see that

$$y' = x_1' = x_2, \quad y'' = x_2' = x_3, \quad y''' = x_3' = x_1 - 5x_2 + 3x_3 + 2e^{-2t}.$$

The result is a nonhomogeneous linear system

$$
\begin{aligned}
x_1' &= x_2 \\
x_2' &= x_3 \\
x_3' &= x_1 - 5x_2 + 3x_3 + 2e^{-2t}.
\end{aligned}
$$

Using matrices, the foregoing system has the form $\mathbf{X}' = \mathbf{AX} + \mathbf{F}(t)$ or

$$
\begin{pmatrix} x_1' \\ x_2' \\ x_3' \end{pmatrix} =
\begin{pmatrix} 0 & 1 & 0 \\ 0 & 0 & 1 \\ 1 & -5 & 3 \end{pmatrix}
\begin{pmatrix} x_1 \\ x_2 \\ x_3 \end{pmatrix} +
\begin{pmatrix} 0 \\ 0 \\ 2e^{-2t} \end{pmatrix}.
$$

≡

Solution of a Linear System

A **solution** of a linear system (4) is a set of n differentiable functions $x_1(t), x_2(t), \ldots, x_n(t)$ defined on a common interval that satisfy each of the differential equations in (3). Equivalently, we have the following definition in terms of matrices.

DEFINITION 10.1.1 **Solution Vector**

A **solution vector** on an interval I is any column matrix

$$
\mathbf{X} = \begin{pmatrix} x_1(t) \\ x_2(t) \\ \vdots \\ x_n(t) \end{pmatrix}
$$

whose entries are differentiable functions satisfying the system (4) on the interval.

A solution vector of (4) is equivalent to n scalar equations $x_1 = \phi_1(t)$, $x_2 = \phi_2(t), \ldots,$ $x_n = \phi_n(t)$, and can be interpreted geometrically as a set of parametric equations of a space curve. In the cases $n = 2$ and $n = 3$, the equations $x_1 = \phi_1(t)$, $x_2 = \phi_2(t)$, and $x_1 = \phi_1(t)$, $x_2 = \phi_2(t)$, $x_3 = \phi_3(t)$ represent curves in 2-space and 3-space, respectively. It is common practice to call such a solution curve a **trajectory**. The plane is also called the **phase plane**. We will illustrate these concepts in the section that follows, as well as in Chapter 11.

EXAMPLE 4 **Verification of Solutions**

Verify that on the interval $(-\infty, \infty)$

$$\mathbf{X}_1 = \begin{pmatrix} 1 \\ -1 \end{pmatrix} e^{-2t} = \begin{pmatrix} e^{-2t} \\ -e^{-2t} \end{pmatrix} \quad \text{and} \quad \mathbf{X}_2 = \begin{pmatrix} 3 \\ 5 \end{pmatrix} e^{6t} = \begin{pmatrix} 3e^{6t} \\ 5e^{6t} \end{pmatrix}$$

are solutions of
$$\mathbf{X}' = \begin{pmatrix} 1 & 3 \\ 5 & 3 \end{pmatrix} \mathbf{X}. \tag{9}$$

SOLUTION From $\mathbf{X}_1' = \begin{pmatrix} -2e^{-2t} \\ 2e^{-2t} \end{pmatrix}$ and $\mathbf{X}_2' = \begin{pmatrix} 18e^{6t} \\ 30e^{6t} \end{pmatrix}$ we see that

$$\mathbf{AX}_1 = \begin{pmatrix} 1 & 3 \\ 5 & 3 \end{pmatrix} \begin{pmatrix} e^{-2t} \\ -e^{-2t} \end{pmatrix} = \begin{pmatrix} e^{-2t} - 3e^{-2t} \\ 5e^{-2t} - 3e^{-2t} \end{pmatrix} = \begin{pmatrix} -2e^{-2t} \\ 2e^{-2t} \end{pmatrix} = \mathbf{X}_1'$$

and $$\mathbf{AX}_2 = \begin{pmatrix} 1 & 3 \\ 5 & 3 \end{pmatrix} \begin{pmatrix} 3e^{6t} \\ 5e^{6t} \end{pmatrix} = \begin{pmatrix} 3e^{6t} + 15e^{6t} \\ 15e^{6t} + 15e^{6t} \end{pmatrix} = \begin{pmatrix} 18e^{6t} \\ 30e^{6t} \end{pmatrix} = \mathbf{X}_2'. \qquad \equiv$$

Much of the theory of systems of n linear first-order differential equations is similar to that of linear nth-order differential equations.

■ **Initial-Value Problem** Let t_0 denote a point on an interval I and

$$\mathbf{X}(t_0) = \begin{pmatrix} x_1(t_0) \\ x_2(t_0) \\ \vdots \\ x_n(t_0) \end{pmatrix} \quad \text{and} \quad \mathbf{X}_0 = \begin{pmatrix} \gamma_1 \\ \gamma_2 \\ \vdots \\ \gamma_n \end{pmatrix},$$

where the γ_i, $i = 1, 2, \ldots, n$ are given constants. Then the problem

> **Solve:** $\mathbf{X}' = \mathbf{A}(t)\mathbf{X} + \mathbf{F}(t)$
>
> **Subject to:** $\mathbf{X}(t_0) = \mathbf{X}_0$
$$\tag{10}$$

is an **initial-value problem** on the interval.

THEOREM 10.1.1 **Existence of a Unique Solution**

Let the entries of the matrices $\mathbf{A}(t)$ and $\mathbf{F}(t)$ be functions continuous on a common interval I that contains the point t_0. Then there exists a unique solution of the initial-value problem (10) on the interval.

■ **Homogeneous Systems** In the next several definitions and theorems we are concerned only with homogeneous systems. Without stating it, we shall always assume that the a_{ij} and the f_i are continuous functions of t on some common interval I.

■ **Superposition Principle** The following result is a **superposition principle** for solutions of linear systems.

Let $\mathbf{X}_1, \mathbf{X}_2, \ldots, \mathbf{X}_k$ be a set of solution vectors of the homogeneous system (5) on an interval I. Then the linear combination

$$\mathbf{X} = c_1\mathbf{X}_1 + c_2\mathbf{X}_2 + \cdots + c_k\mathbf{X}_k,$$

where the c_i, $i = 1, 2, \ldots, k$ are arbitrary constants, is also a solution of the system on the interval.

It follows from Theorem 10.1.2 that a constant multiple of any solution vector of a homogeneous system of linear first-order differential equations is also a solution.

EXAMPLE 5 Using the Superposition Principle

You should practice by verifying that the two vectors

$$\mathbf{X}_1 = \begin{pmatrix} \cos t \\ -\frac{1}{2}\cos t + \frac{1}{2}\sin t \\ -\cos t - \sin t \end{pmatrix} \quad \text{and} \quad \mathbf{X}_2 = \begin{pmatrix} 0 \\ e^t \\ 0 \end{pmatrix}$$

are solutions of the homogeneous system

$$\mathbf{X}' = \begin{pmatrix} 1 & 0 & 1 \\ 1 & 1 & 0 \\ -2 & 0 & -1 \end{pmatrix}\mathbf{X}. \tag{11}$$

By the superposition principle, Theorem 10.1.2, the linear combination

$$\mathbf{X} = c_1\mathbf{X}_1 + c_2\mathbf{X}_2 = c_1\begin{pmatrix} \cos t \\ -\frac{1}{2}\cos t + \frac{1}{2}\sin t \\ -\cos t - \sin t \end{pmatrix} + c_2\begin{pmatrix} 0 \\ e^t \\ 0 \end{pmatrix}$$

is yet another solution of the system. ≡

Linear Dependence and Linear Independence We are primarily interested in linearly independent solutions of the homogeneous system (5).

Let $\mathbf{X}_1, \mathbf{X}_2, \ldots, \mathbf{X}_k$ be a set of solution vectors of the homogeneous system (5) on an interval I. We say that the set is **linearly dependent** on the interval if there exist constants $c_1, c_2, \ldots, c_k$, not all zero, such that

$$c_1\mathbf{X}_1 + c_2\mathbf{X}_2 + \cdots + c_k\mathbf{X}_k = \mathbf{0}$$

for every t in the interval. If the set of vectors is not linearly dependent on the interval, it is said to be **linearly independent**.

The case when $k = 2$ should be clear; two solution vectors $\mathbf{X}_1$ and $\mathbf{X}_2$ are linearly dependent if one is a constant multiple of the other, and conversely. For $k > 2$, a set of solution vectors is linearly dependent if we can express at least one solution vector as a linear combination of the remaining vectors.

Wronskian As in our earlier consideration of the theory of a single ordinary differential equation, we can introduce the concept of the **Wronskian** determinant as a test for linear independence. We state the following theorem without proof.

Let
$$\mathbf{X}_1 = \begin{pmatrix} x_{11} \\ x_{21} \\ \vdots \\ x_{n1} \end{pmatrix}, \quad \mathbf{X}_2 = \begin{pmatrix} x_{12} \\ x_{22} \\ \vdots \\ x_{n2} \end{pmatrix}, \quad \ldots, \quad \mathbf{X}_n = \begin{pmatrix} x_{1n} \\ x_{2n} \\ \vdots \\ x_{nn} \end{pmatrix}$$

be n solution vectors of the homogeneous system (5) on an interval I. Then the set of solution vectors is linearly independent on I if and only if the **Wronskian**

$$W(\mathbf{X}_1, \mathbf{X}_2, \ldots, \mathbf{X}_n) = \begin{vmatrix} x_{11} & x_{12} & \cdots & x_{1n} \\ x_{21} & x_{22} & \cdots & x_{2n} \\ \vdots & & & \vdots \\ x_{n1} & x_{n2} & \cdots & x_{nn} \end{vmatrix} \neq 0 \tag{12}$$

for every t in the interval.

It can be shown that if $\mathbf{X}_1, \mathbf{X}_2, \ldots, \mathbf{X}_n$ are solution vectors of (5), then for every t in I, either $W(\mathbf{X}_1, \mathbf{X}_2, \ldots, \mathbf{X}_n) \neq 0$ or $W(\mathbf{X}_1, \mathbf{X}_2, \ldots, \mathbf{X}_n) = 0$. Thus if we can show that $W \neq 0$ for some t_0 in I, then $W \neq 0$ for every t; hence the set of solutions is linearly independent on the interval.

Notice that, unlike our definition of the Wronskian in Section 3.1, here the definition of the determinant (12) does not involve differentiation.

EXAMPLE 6 **Linearly Independent Solutions**

In Example 4 we saw that $\mathbf{X}_1 = \begin{pmatrix} 1 \\ -1 \end{pmatrix} e^{-2t}$ and $\mathbf{X}_2 = \begin{pmatrix} 3 \\ 5 \end{pmatrix} e^{6t}$ are solutions of system (9). Clearly $\mathbf{X}_1$ and $\mathbf{X}_2$ are linearly independent on the interval $(-\infty, \infty)$ since neither vector is a constant multiple of the other. In addition, we have

$$W(\mathbf{X}_1, \mathbf{X}_2) = \begin{vmatrix} e^{-2t} & 3e^{6t} \\ -e^{-2t} & 5e^{6t} \end{vmatrix} = 8e^{4t} \neq 0$$

for all real values of t. ≡

DEFINITION 10.1.3 **Fundamental Set of Solutions**

Any set $\mathbf{X}_1, \mathbf{X}_2, \ldots, \mathbf{X}_n$ of n linearly independent solution vectors of the homogeneous system (5) on an interval I is said to be a **fundamental set of solutions** on the interval.

THEOREM 10.1.4 **Existence of a Fundamental Set**

There exists a fundamental set of solutions for the homogeneous system (5) on an interval I.

The next two theorems are the linear system equivalents of Theorems 3.1.5 and 3.1.6.

THEOREM 10.1.5 **General Solution—Homogeneous Systems**

Let $\mathbf{X}_1, \mathbf{X}_2, \ldots, \mathbf{X}_n$ be a fundamental set of solutions of the homogeneous system (5) on an interval I. Then the **general solution** of the system on the interval is

$$\mathbf{X} = c_1 \mathbf{X}_1 + c_2 \mathbf{X}_2 + \cdots + c_n \mathbf{X}_n,$$

where the c_i, $i = 1, 2, \ldots, n$ are arbitrary constants.

EXAMPLE 7　　General Solution of System (9)

From Example 4 we know that $\mathbf{X}_1 = \begin{pmatrix} 1 \\ -1 \end{pmatrix} e^{-2t}$ and $\mathbf{X}_2 = \begin{pmatrix} 3 \\ 5 \end{pmatrix} e^{6t}$ are linearly independent solutions of (9) on $(-\infty, \infty)$. Hence $\mathbf{X}_1$ and $\mathbf{X}_2$ form a fundamental set of solutions on the interval. The general solution of the homogeneous system on the interval is then

$$\mathbf{X} = c_1 \mathbf{X}_1 + c_2 \mathbf{X}_2 = c_1 \begin{pmatrix} 1 \\ -1 \end{pmatrix} e^{-2t} + c_2 \begin{pmatrix} 3 \\ 5 \end{pmatrix} e^{6t}. \tag{13} \equiv$$

EXAMPLE 8　　General Solution of System (11)

The vectors

$$\mathbf{X}_1 = \begin{pmatrix} \cos t \\ -\frac{1}{2}\cos t + \frac{1}{2}\sin t \\ -\cos t - \sin t \end{pmatrix}, \quad \mathbf{X}_2 = \begin{pmatrix} 0 \\ 1 \\ 0 \end{pmatrix} e^t, \quad \mathbf{X}_3 = \begin{pmatrix} \sin t \\ -\frac{1}{2}\sin t - \frac{1}{2}\cos t \\ -\sin t + \cos t \end{pmatrix}$$

are solutions of the system (11) in Example 5 (see Problem 22 in Exercises 10.1). Now

$$W(\mathbf{X}_1, \mathbf{X}_2, \mathbf{X}_3) = \begin{vmatrix} \cos t & 0 & \sin t \\ -\frac{1}{2}\cos t + \frac{1}{2}\sin t & e^t & -\frac{1}{2}\sin t - \frac{1}{2}\cos t \\ -\cos t - \sin t & 0 & -\sin t + \cos t \end{vmatrix} = e^t \neq 0$$

for all real values of t. We conclude that $\mathbf{X}_1$, $\mathbf{X}_2$, and $\mathbf{X}_3$ form a fundamental set of solutions on $(-\infty, \infty)$. Thus the general solution of the homogeneous system on the interval is the linear combination $\mathbf{X} = c_1 \mathbf{X}_1 + c_2 \mathbf{X}_2 + c_3 \mathbf{X}_3$; that is,

$$\mathbf{X} = c_1 \begin{pmatrix} \cos t \\ -\frac{1}{2}\cos t + \frac{1}{2}\sin t \\ -\cos t - \sin t \end{pmatrix} + c_2 \begin{pmatrix} 0 \\ 1 \\ 0 \end{pmatrix} e^t + c_3 \begin{pmatrix} \sin t \\ -\frac{1}{2}\sin t - \frac{1}{2}\cos t \\ -\sin t + \cos t \end{pmatrix}. \qquad \equiv$$

Nonhomogeneous Systems　　For nonhomogeneous systems, a **particular solution** $\mathbf{X}_p$ on an interval I is any vector, free of arbitrary parameters, whose entries are functions that satisfy system (4).

THEOREM 10.1.6　**General Solution—Nonhomogeneous Systems**

Let $\mathbf{X}_p$ be a given solution of the nonhomogeneous system (4) on an interval I, and let

$$\mathbf{X}_c = c_1 \mathbf{X}_1 + c_2 \mathbf{X}_2 + \cdots + c_n \mathbf{X}_n$$

denote the general solution on the same interval of the associated homogeneous system (5). Then the **general solution** of the nonhomogeneous system on the interval is

$$\mathbf{X} = \mathbf{X}_c + \mathbf{X}_p.$$

The general solution $\mathbf{X}_c$ of the associated homogeneous system (5) is called the **complementary function** of the nonhomogeneous system (4).

EXAMPLE 9　　General Solution—Nonhomogeneous System

The vector $\mathbf{X}_p = \begin{pmatrix} 3t - 4 \\ -5t + 6 \end{pmatrix}$ is a particular solution of the nonhomogeneous system

$$\mathbf{X}' = \begin{pmatrix} 1 & 3 \\ 5 & 3 \end{pmatrix} \mathbf{X} + \begin{pmatrix} 12t - 11 \\ -3 \end{pmatrix} \tag{14}$$

on the interval $(-\infty, \infty)$. (Verify this.) The complementary function of (14) on the same interval, or the general solution of $\mathbf{X}' = \begin{pmatrix} 1 & 3 \\ 5 & 3 \end{pmatrix}\mathbf{X}$, was seen in (13) of Example 7 to be $\mathbf{X}_c = c_1\begin{pmatrix} 1 \\ -1 \end{pmatrix}e^{-2t} + c_2\begin{pmatrix} 3 \\ 5 \end{pmatrix}e^{6t}$. Hence by Theorem 10.1.6

$$\mathbf{X} = \mathbf{X}_c + \mathbf{X}_p = c_1\begin{pmatrix} 1 \\ -1 \end{pmatrix}e^{-2t} + c_2\begin{pmatrix} 3 \\ 5 \end{pmatrix}e^{6t} + \begin{pmatrix} 3t - 4 \\ -5t + 6 \end{pmatrix}$$

is the general solution of the nonhomogeneous system (14) on the interval $(-\infty, \infty)$. ≡

10.1 Exercises Answers to selected odd-numbered problems begin on page ANS-27.

In Problems 1–8, write the given linear system in matrix form.

1. $\dfrac{dx}{dt} = 3x - 5y$

$\dfrac{dy}{dt} = 4x + 8y$

2. $\dfrac{dx}{dt} = 4x - 7y$

$\dfrac{dy}{dt} = 5x$

3. $2\dfrac{dx}{dt} = 5x - 14y + 9t$

$4\dfrac{dy}{dt} = -x + 6y - 2$

4. $6\dfrac{dx}{dt} = -2x + 12y - 3\sin 4t$

$\dfrac{dy}{dt} = 8x - 3y + 10\cos 4t$

5. $\dfrac{dx}{dt} = -3x + 4y - 9z$

$\dfrac{dy}{dt} = 6x - y$

$\dfrac{dz}{dt} = 10x + 4y + 3z$

6. $\dfrac{dx}{dt} = x - y$

$\dfrac{dy}{dt} = x + 2z$

$\dfrac{dz}{dt} = -x + z$

7. $\dfrac{dx}{dt} = x - y + z + t - 1$

$\dfrac{dy}{dt} = 2x + y - z - 3t^2$

$\dfrac{dz}{dt} = x + y + z + t^2 - t + 2$

8. $\dfrac{dx}{dt} = -3x + 4y + e^{-t}\sin 2t$

$\dfrac{dy}{dt} = 5x + 9z + 4e^{-t}\cos 2t$

$\dfrac{dz}{dt} = y + 6z - e^{-t}$

In Problems 9–12, write the given linear system without the use of matrices.

9. $\mathbf{X}' = \begin{pmatrix} 4 & 2 \\ -1 & 3 \end{pmatrix}\mathbf{X} + \begin{pmatrix} 1 \\ -1 \end{pmatrix}e^t$

10. $\mathbf{X}' = \begin{pmatrix} 7 & 5 & -9 \\ 4 & 1 & 1 \\ 0 & -2 & 3 \end{pmatrix}\mathbf{X} + \begin{pmatrix} 0 \\ 2 \\ 1 \end{pmatrix}e^{5t} - \begin{pmatrix} 8 \\ 0 \\ 3 \end{pmatrix}e^{-2t}$

11. $\begin{pmatrix} x' \\ y' \\ z' \end{pmatrix} = \begin{pmatrix} 1 & -1 & 2 \\ 3 & -4 & 1 \\ -2 & 5 & 6 \end{pmatrix}\begin{pmatrix} x \\ y \\ z \end{pmatrix} + \begin{pmatrix} 1 \\ 2 \\ 2 \end{pmatrix}e^{-t} - \begin{pmatrix} 3 \\ -1 \\ 1 \end{pmatrix}t$

12. $\begin{pmatrix} x' \\ y' \end{pmatrix} = \begin{pmatrix} 3 & -7 \\ 1 & 1 \end{pmatrix}\begin{pmatrix} x \\ y \end{pmatrix} + \begin{pmatrix} 4 \\ 8 \end{pmatrix}\sin t + \begin{pmatrix} t - 4 \\ 2t + 1 \end{pmatrix}e^{4t}$

In Problems 13–16, write the given differential equation as a first-order system using matrix notation.

13. $y'' + 4y = t\cos t$

14. $y'' + 2y' - 2y = 3t - 9$

15. $y''' + 6y'' + 10y' - y = t^2$

16. $y^{(4)} + 2y'' + y = e^t$

In Problems 17–22, verify that the vector $\mathbf{X}$ is a solution of the given homogeneous linear system.

17. $\dfrac{dx}{dt} = 3x - 4y$

$\dfrac{dy}{dt} = 4x - 7y;$ $\mathbf{X} = \begin{pmatrix} 1 \\ 2 \end{pmatrix}e^{-5t}$

18. $\dfrac{dx}{dt} = -2x + 5y$

$\dfrac{dy}{dt} = -2x + 4y;$ $\mathbf{X} = \begin{pmatrix} 5\cos t \\ 3\cos t - \sin t \end{pmatrix}e^t$

19. $\mathbf{X}' = \begin{pmatrix} -1 & \frac{1}{4} \\ 1 & -1 \end{pmatrix}\mathbf{X};$ $\mathbf{X} = \begin{pmatrix} -1 \\ 2 \end{pmatrix}e^{-3t/2}$

20. $\mathbf{X}' = \begin{pmatrix} 2 & 1 \\ -1 & 0 \end{pmatrix}\mathbf{X};$ $\mathbf{X} = \begin{pmatrix} 1 \\ 3 \end{pmatrix}e^t + \begin{pmatrix} 4 \\ -4 \end{pmatrix}te^t$

21. $\mathbf{X}' = \begin{pmatrix} 1 & 2 & 1 \\ 6 & -1 & 0 \\ -1 & -2 & -1 \end{pmatrix}\mathbf{X};$ $\mathbf{X} = \begin{pmatrix} 1 \\ 6 \\ -13 \end{pmatrix}$

22. $\mathbf{X}' = \begin{pmatrix} 1 & 0 & 1 \\ 1 & 1 & 0 \\ -2 & 0 & -1 \end{pmatrix}\mathbf{X};$ $\mathbf{X} = \begin{pmatrix} \sin t \\ -\frac{1}{2}\sin t - \frac{1}{2}\cos t \\ -\sin t + \cos t \end{pmatrix}$

In Problems 23–26, the given vectors are solutions of a system $\mathbf{X}' = \mathbf{AX}$. Determine whether the vectors form a fundamental set on the interval $(-\infty, \infty)$.

23. $\mathbf{X}_1 = \begin{pmatrix} 1 \\ 1 \end{pmatrix}e^{-2t}, \mathbf{X}_2 = \begin{pmatrix} 1 \\ -1 \end{pmatrix}e^{-6t}$

24. $\mathbf{X}_1 = \begin{pmatrix} 1 \\ -1 \end{pmatrix}e^t, \mathbf{X}_2 = \begin{pmatrix} 2 \\ 6 \end{pmatrix}e^t + \begin{pmatrix} 8 \\ -8 \end{pmatrix}te^t$

25. $\mathbf{X}_1 = \begin{pmatrix} 1 \\ -2 \\ 4 \end{pmatrix} + t\begin{pmatrix} 1 \\ 2 \\ 2 \end{pmatrix}, \mathbf{X}_2 = \begin{pmatrix} 1 \\ -2 \\ 4 \end{pmatrix},$

$\mathbf{X}_3 = \begin{pmatrix} 3 \\ -6 \\ 12 \end{pmatrix} + t\begin{pmatrix} 2 \\ 4 \\ 4 \end{pmatrix}$

26. $\mathbf{X}_1 = \begin{pmatrix} 1 \\ 6 \\ -13 \end{pmatrix}, \mathbf{X}_2 = \begin{pmatrix} 1 \\ -2 \\ -1 \end{pmatrix}e^{-4t}, \mathbf{X}_3 = \begin{pmatrix} 2 \\ 3 \\ -2 \end{pmatrix}e^{3t}$

In Problems 27–30, verify that the vector $\mathbf{X}_p$ is a particular solution of the given nonhomogeneous linear system.

27. $\dfrac{dx}{dt} = x + 4y + 2t - 7$

$\dfrac{dy}{dt} = 3x + 2y - 4t - 18; \quad \mathbf{X}_p = \begin{pmatrix} 2 \\ -1 \end{pmatrix}t + \begin{pmatrix} 5 \\ 1 \end{pmatrix}$

28. $\mathbf{X}' = \begin{pmatrix} 2 & 1 \\ 1 & -1 \end{pmatrix}\mathbf{X} + \begin{pmatrix} -5 \\ 2 \end{pmatrix}; \quad \mathbf{X}_p = \begin{pmatrix} 1 \\ 3 \end{pmatrix}$

29. $\mathbf{X}' = \begin{pmatrix} 2 & 1 \\ 3 & 4 \end{pmatrix}\mathbf{X} - \begin{pmatrix} 1 \\ 7 \end{pmatrix}e^t;$

$\mathbf{X}_p = \begin{pmatrix} 1 \\ 1 \end{pmatrix}e^t + \begin{pmatrix} 1 \\ -1 \end{pmatrix}te^t$

30. $\mathbf{X}' = \begin{pmatrix} 1 & 2 & 3 \\ -4 & 2 & 0 \\ -6 & 1 & 0 \end{pmatrix}\mathbf{X} + \begin{pmatrix} -1 \\ 4 \\ 3 \end{pmatrix}\sin 3t;$

$\mathbf{X}_p = \begin{pmatrix} \sin 3t \\ 0 \\ \cos 3t \end{pmatrix}$

31. Prove that the general solution of the homogeneous linear system

$$\mathbf{X}' = \begin{pmatrix} 0 & 6 & 0 \\ 1 & 0 & 1 \\ 1 & 1 & 0 \end{pmatrix}\mathbf{X}$$

on the interval $(-\infty, \infty)$ is

$$\mathbf{X} = c_1\begin{pmatrix} 6 \\ -1 \\ -5 \end{pmatrix}e^{-t} + c_2\begin{pmatrix} -3 \\ 1 \\ 1 \end{pmatrix}e^{-2t} + c_3\begin{pmatrix} 2 \\ 1 \\ 1 \end{pmatrix}e^{3t}.$$

32. Prove that the general solution of the nonhomogeneous linear system

$$\mathbf{X}' = \begin{pmatrix} -1 & -1 \\ -1 & 1 \end{pmatrix}\mathbf{X} + \begin{pmatrix} 1 \\ 1 \end{pmatrix}t^2 + \begin{pmatrix} 4 \\ -6 \end{pmatrix}t + \begin{pmatrix} -1 \\ 5 \end{pmatrix}$$

on the interval $(-\infty, \infty)$ is

$$\mathbf{X} = c_1\begin{pmatrix} 1 \\ -1 - \sqrt{2} \end{pmatrix}e^{\sqrt{2}t} + c_2\begin{pmatrix} 1 \\ -1 + \sqrt{2} \end{pmatrix}e^{-\sqrt{2}t}$$

$$+ \begin{pmatrix} 1 \\ 0 \end{pmatrix}t^2 + \begin{pmatrix} -2 \\ 4 \end{pmatrix}t + \begin{pmatrix} 1 \\ 0 \end{pmatrix}.$$

10.2 Homogeneous Linear Systems

INTRODUCTION In Example 7 of Section 10.1 we saw that the general solution of the homogeneous system $\mathbf{X}' = \begin{pmatrix} 1 & 3 \\ 5 & 3 \end{pmatrix}\mathbf{X}$ is $\mathbf{X} = c_1\mathbf{X}_1 + c_2\mathbf{X}_2 = c_1\begin{pmatrix} 1 \\ -1 \end{pmatrix}e^{-2t} + c_2\begin{pmatrix} 3 \\ 5 \end{pmatrix}e^{6t}$. Since both solution vectors have the form $\mathbf{X}_i = \begin{pmatrix} k_1 \\ k_2 \end{pmatrix}e^{\lambda_i t}$, $i = 1, 2$, where $k_1, k_2, \lambda_1,$ and λ_2 are constants, we are prompted to ask whether we can always find a solution of the form

$$\mathbf{X} = \begin{pmatrix} k_1 \\ k_2 \\ \vdots \\ k_n \end{pmatrix}e^{\lambda t} = \mathbf{K}e^{\lambda t} \tag{1}$$

for the general homogeneous linear first-order system

$$\mathbf{X}' = \mathbf{AX}, \tag{2}$$

We will be dealing only with linear systems with constant coefficients. ▶

where the coefficient matrix $\mathbf{A}$ is an $n \times n$ matrix of constants.

■ **Eigenvalues and Eigenvectors** If (1) is to be a solution vector of the system, then $\mathbf{X}' = \mathbf{K}\lambda e^{\lambda t}$ so that (2) becomes $\mathbf{K}\lambda e^{\lambda t} = \mathbf{AK}e^{\lambda t}$. After dividing out $e^{\lambda t}$ and rearranging, we obtain $\mathbf{AK} = \lambda\mathbf{K}$ or $\mathbf{AK} - \lambda\mathbf{K} = \mathbf{0}$. Since $\mathbf{K} = \mathbf{IK}$, the last equation is the same as

$$(\mathbf{A} - \lambda\mathbf{I})\mathbf{K} = \mathbf{0}. \tag{3}$$

The matrix equation (3) is equivalent to the system of linear algebraic equations

$$
\begin{aligned}
(a_{11} - \lambda)k_1 + \quad & a_{12}k_2 + \cdots + \quad & a_{1n}k_n &= 0 \\
a_{21}k_1 + (a_{22} - \lambda)k_2 + \cdots + \quad & a_{2n}k_n &= 0 \\
& \vdots \qquad\qquad\qquad\qquad \vdots \\
a_{n1}k_1 + \quad & a_{n2}k_2 + \cdots + (a_{nn} - \lambda)k_n &= 0.
\end{aligned}
$$

Thus to find a nontrivial solution $\mathbf{X}$ of (2) we must first find a nontrivial solution of the foregoing system; in other words, we must find a nontrivial vector $\mathbf{K}$ that satisfies (3). But in order for (3) to have solutions other than the obvious trivial solution $k_1 = k_2 = \cdots = k_n = 0$, we must have

See Theorem 8.6.6 on page 419. ▶

$$
\det(\mathbf{A} - \lambda \mathbf{I}) = 0.
$$

This polynomial equation in λ is called the **characteristic equation** of the matrix $\mathbf{A}$; its solutions are the **eigenvalues** of $\mathbf{A}$. A solution $\mathbf{K} \neq \mathbf{0}$ of (3) corresponding to an eigenvalue λ is called an **eigenvector** of $\mathbf{A}$. A solution of the homogeneous system (2) is then $\mathbf{X} = \mathbf{K}e^{\lambda t}$.

In the following discussion we examine three cases: all the eigenvalues are real and distinct (that is, no eigenvalues are equal), repeated eigenvalues, and finally, complex eigenvalues.

10.2.1 Distinct Real Eigenvalues

When the $n \times n$ matrix $\mathbf{A}$ possesses n distinct real eigenvalues $\lambda_1, \lambda_2, \ldots, \lambda_n$, then a set of n linearly independent eigenvectors $\mathbf{K}_1, \mathbf{K}_2, \ldots, \mathbf{K}_n$ can always be found and

$$
\mathbf{X}_1 = \mathbf{K}_1 e^{\lambda_1 t}, \quad \mathbf{X}_2 = \mathbf{K}_2 e^{\lambda_2 t}, \quad \ldots, \quad \mathbf{X}_n = \mathbf{K}_n e^{\lambda_n t}
$$

is a fundamental set of solutions of (2) on the interval $(-\infty, \infty)$.

THEOREM 10.2.1 **General Solution—Homogeneous Systems**

Let $\lambda_1, \lambda_2, \ldots, \lambda_n$ be n distinct real eigenvalues of the coefficient matrix $\mathbf{A}$ of the homogeneous system (2), and let $\mathbf{K}_1, \mathbf{K}_2, \ldots, \mathbf{K}_n$ be the corresponding eigenvectors. Then the **general solution** of (2) on the interval $(-\infty, \infty)$ is given by

$$
\mathbf{X} = c_1 \mathbf{K}_1 e^{\lambda_1 t} + c_2 \mathbf{K}_2 e^{\lambda_2 t} + \cdots + c_n \mathbf{K}_n e^{\lambda_n t}.
$$

EXAMPLE 1 **Distinct Eigenvalues**

Solve

$$
\begin{aligned}
\frac{dx}{dt} &= 2x + 3y \\
\frac{dy}{dt} &= 2x + y.
\end{aligned}
\tag{4}
$$

SOLUTION We first find the eigenvalues and eigenvectors of the matrix of coefficients. From the characteristic equation

$$
\det(\mathbf{A} - \lambda\mathbf{I}) = \begin{vmatrix} 2 - \lambda & 3 \\ 2 & 1 - \lambda \end{vmatrix} = \lambda^2 - 3\lambda - 4 = (\lambda + 1)(\lambda - 4) = 0
$$

we see that the eigenvalues are $\lambda_1 = -1$ and $\lambda_2 = 4$.

Now for $\lambda_1 = -1$, (3) is equivalent to

$$
\begin{aligned}
3k_1 + 3k_2 &= 0 \\
2k_1 + 2k_2 &= 0.
\end{aligned}
$$

Thus $k_1 = -k_2$. When $k_2 = -1$, the related eigenvector is

$$
\mathbf{K}_1 = \begin{pmatrix} 1 \\ -1 \end{pmatrix}.
$$

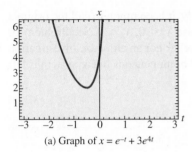

(a) Graph of $x = e^{-t} + 3e^{4t}$

(b) Graph of $y = -e^{-t} + 2e^{4t}$

(c) Trajectory defined by $x = e^{-t} + 3e^{4t}$, $y = -e^{-t} + 2e^{4t}$ in the phase plane

FIGURE 10.2.1 A particular solution of (5) gives three different curves in three different coordinate planes

FIGURE 10.2.2 A phase portrait of system (4)

For $\lambda_2 = 4$, we have

$$-2k_1 + 3k_2 = 0$$
$$2k_1 - 3k_2 = 0$$

so that $k_1 = \tfrac{3}{2}k_2$, and therefore with $k_2 = 2$, the corresponding eigenvector is

$$\mathbf{K}_2 = \begin{pmatrix} 3 \\ 2 \end{pmatrix}.$$

Since the matrix of coefficients $\mathbf{A}$ is a 2×2 matrix, and since we have found two linearly independent solutions of (4),

$$\mathbf{X}_1 = \begin{pmatrix} 1 \\ -1 \end{pmatrix} e^{-t} \quad \text{and} \quad \mathbf{X}_2 = \begin{pmatrix} 3 \\ 2 \end{pmatrix} e^{4t},$$

we conclude that the general solution of the system is

$$\mathbf{X} = c_1\mathbf{X}_1 + c_2\mathbf{X}_2 = c_1 \begin{pmatrix} 1 \\ -1 \end{pmatrix} e^{-t} + c_2 \begin{pmatrix} 3 \\ 2 \end{pmatrix} e^{4t}. \qquad (5) \equiv$$

■ **Phase Portrait** For the sake of review, you should keep firmly in mind that a solution of a system of linear first-order differential equations, when written in terms of matrices, is simply an alternative to the method that we employed in Section 3.12—namely, listing the individual functions and the relationship between the constants. If we add the vectors on the right side of (5) and then equate the entries with the corresponding entries in the vector of the left, we obtain the more familiar statement

$$x = c_1 e^{-t} + 3c_2 e^{4t}, \qquad y = -c_1 e^{-t} + 2c_2 e^{4t}.$$

As pointed out in Section 10.1, we can interpret these equations as parametric equations of a curve or **trajectory** in the xy-plane or **phase plane**. The three graphs shown in **FIGURE 10.2.1**, $x(t)$ in the tx-plane, $y(t)$ in the ty-plane, and the trajectory in the phase plane, correspond to the choice of constants $c_1 = c_2 = 1$ in the solution. A collection of trajectories in the phase plane as shown in **FIGURE 10.2.2** is said to be a **phase portrait** of the given linear system. What appears to be two red lines in Figure 10.2.2 are actually four half-lines defined parametrically in the first, second, third, and fourth quadrants by the solutions $\mathbf{X}_2$, $-\mathbf{X}_1$, $-\mathbf{X}_2$, and $\mathbf{X}_1$, respectively. For example, the Cartesian equations $y = \tfrac{2}{3}x$, $x > 0$, and $y = -x$, $x > 0$, of the half-lines in the first and fourth quadrants were obtained by eliminating the parameter t in the solutions $x = 3e^{4t}$, $y = 2e^{4t}$, and $x = e^{-t}$, $y = -e^{-t}$, respectively. Moreover, each eigenvector can be visualized as a two-dimensional vector lying along one of these half-lines. The eigenvector $\mathbf{K}_2 = \begin{pmatrix} 3 \\ 2 \end{pmatrix}$ lies along $y = \tfrac{2}{3}x$ in the first quadrant and $\mathbf{K}_1 = \begin{pmatrix} 1 \\ -1 \end{pmatrix}$ lies along $y = -x$ in the fourth quadrant; each vector starts at the origin with $\mathbf{K}_2$ terminating at the point $(2, 3)$ and $\mathbf{K}_1$ terminating at $(1, -1)$.

The origin is not only a constant solution, $x = 0$, $y = 0$, of every 2×2 homogeneous linear system $\mathbf{X}' = \mathbf{AX}$ but is as well an important point in the qualitative study of such systems. If we think in physical terms, the arrowheads on a trajectory in Figure 10.2.2 indicate the direction that a particle with coordinates $(x(t), y(t))$ on a trajectory at time t would move as time increases. Observe that the arrowheads, with only those on the half-lines in the second and fourth quadrants being the exception, indicate that a particle would move away from the origin as time t increases. If we imagine time ranging from $-\infty$ to ∞, then inspection of the solution $x = c_1 e^{-t} + 3c_2 e^{4t}$, $y = -c_1 e^{-t} + 2c_2 e^{4t}$, $c_1 \neq 0$, $c_2 \neq 0$, shows that a trajectory, or moving particle, "starts" asymptotic to one of the half-lines defined by $\mathbf{X}_1$ or $-\mathbf{X}_1$ (since e^{4t} is negligible for $t \to -\infty$) and "finishes" asymptotic to one of the half-lines defined by $\mathbf{X}_2$ and $-\mathbf{X}_2$ (since e^{-t} is negligible for $t \to \infty$).

We note in passing that Figure 10.2.2 represents a phase portrait that is typical of *all* 2×2 homogeneous linear systems $\mathbf{X}' = \mathbf{AX}$ with real eigenvalues of opposite signs. See Problem 19 in Exercises 10.2. Moreover, phase portraits in the two cases when distinct real eigenvalues have the same algebraic sign would be typical portraits of all such 2×2 linear systems; the only difference is that the arrowheads would indicate that a particle would move away from the origin on any trajectory as $t \to \infty$ when both λ_1 and λ_2 are positive and would move toward the origin on

any trajectory when both λ_1 and λ_2 are negative. Consequently, it is common to call the origin a **repeller** in the case $\lambda_1 > 0$, $\lambda_2 > 0$, and an **attractor** in the case $\lambda_1 < 0$, $\lambda_2 < 0$. See Problem 20 in Exercises 10.2. The origin in Figure 10.2.2 is neither a repeller nor an attractor. Investigation of the remaining case when $\lambda = 0$ is an eigenvalue of a 2×2 homogeneous linear system is left as an exercise. See Problem 52 in Exercises 10.2.

EXAMPLE 2 **Distinct Eigenvalues**

Solve

$$\frac{dx}{dt} = -4x + y + z$$

$$\frac{dy}{dt} = x + 5y - z \qquad (6)$$

$$\frac{dz}{dt} = y - 3z.$$

SOLUTION Using the cofactors of the third row, we find

$$\det(\mathbf{A} - \lambda \mathbf{I}) = \begin{vmatrix} -4 - \lambda & 1 & 1 \\ 1 & 5 - \lambda & -1 \\ 0 & 1 & -3 - \lambda \end{vmatrix} = -(\lambda + 3)(\lambda + 4)(\lambda - 5) = 0,$$

and so the eigenvalues are $\lambda_1 = -3$, $\lambda_2 = -4$, $\lambda_3 = 5$.

For $\lambda_1 = -3$, Gauss–Jordan elimination gives

$$(\mathbf{A} + 3\mathbf{I}|\mathbf{0}) = \begin{pmatrix} -1 & 1 & 1 & | & 0 \\ 1 & 8 & -1 & | & 0 \\ 0 & 1 & 0 & | & 0 \end{pmatrix} \overset{\text{row}}{\underset{\text{operations}}{\Longrightarrow}} \begin{pmatrix} 1 & 0 & -1 & | & 0 \\ 0 & 1 & 0 & | & 0 \\ 0 & 0 & 0 & | & 0 \end{pmatrix}.$$

Therefore $k_1 = k_3$ and $k_2 = 0$. The choice $k_3 = 1$ gives an eigenvector and corresponding solution vector

$$\mathbf{K}_1 = \begin{pmatrix} 1 \\ 0 \\ 1 \end{pmatrix}, \quad \mathbf{X}_1 = \begin{pmatrix} 1 \\ 0 \\ 1 \end{pmatrix} e^{-3t}. \qquad (7)$$

Similarly, for $\lambda_2 = -4$,

$$(\mathbf{A} + 4\mathbf{I}|\mathbf{0}) = \begin{pmatrix} 0 & 1 & 1 & | & 0 \\ 1 & 9 & -1 & | & 0 \\ 0 & 1 & 1 & | & 0 \end{pmatrix} \overset{\text{row}}{\underset{\text{operations}}{\Longrightarrow}} \begin{pmatrix} 1 & 0 & -10 & | & 0 \\ 0 & 1 & 1 & | & 0 \\ 0 & 0 & 0 & | & 0 \end{pmatrix}$$

implies $k_1 = 10k_3$ and $k_2 = -k_3$. Choosing $k_3 = 1$, we get a second eigenvector and solution vector

$$\mathbf{K}_2 = \begin{pmatrix} 10 \\ -1 \\ 1 \end{pmatrix}, \quad \mathbf{X}_2 = \begin{pmatrix} 10 \\ -1 \\ 1 \end{pmatrix} e^{-4t}. \qquad (8)$$

Finally, when $\lambda_3 = 5$, the augmented matrices

$$(\mathbf{A} - 5\mathbf{I}|\mathbf{0}) = \begin{pmatrix} -9 & 1 & 1 & | & 0 \\ 1 & 0 & -1 & | & 0 \\ 0 & 1 & -8 & | & 0 \end{pmatrix} \overset{\text{row}}{\underset{\text{operations}}{\Longrightarrow}} \begin{pmatrix} 1 & 0 & -1 & | & 0 \\ 0 & 1 & -8 & | & 0 \\ 0 & 0 & 0 & | & 0 \end{pmatrix}$$

yield

$$\mathbf{K}_3 = \begin{pmatrix} 1 \\ 8 \\ 1 \end{pmatrix}, \quad \mathbf{X}_3 = \begin{pmatrix} 1 \\ 8 \\ 1 \end{pmatrix} e^{5t}. \qquad (9)$$

The general solution of (6) is a linear combination of the solution vectors in (7), (8), and (9):

$$\mathbf{X} = c_1 \begin{pmatrix} 1 \\ 0 \\ 1 \end{pmatrix} e^{-3t} + c_2 \begin{pmatrix} 10 \\ -1 \\ 1 \end{pmatrix} e^{-4t} + c_3 \begin{pmatrix} 1 \\ 8 \\ 1 \end{pmatrix} e^{5t}. \qquad \equiv$$

■ **Use of Computers** Software packages such as MATLAB, *Mathematica*, and *Maple* can be real time savers in finding eigenvalues and eigenvectors of a matrix. For example, to find the eigenvalues and eigenvectors of the matrix of coefficients in (6) using *Mathematica*, we first input the definition of the matrix by rows:

$$\mathbf{m} = \{\,\{-4, 1, 1\}, \{\,1, 5, -1\}, \{\,0, 1, -3\}\,\}.$$

The commands **Eigenvalues[m]** and **Eigenvectors[m]** given in sequence yield

$$\{5, -4, -3\} \quad \text{and} \quad \{\,\{\,1, 8, 1\}, \{\,10, -1, 1\}, \{\,1, 0, 1\}\,\},$$

respectively. In *Mathematica*, eigenvalues and eigenvectors can also be obtained at the same time by using **Eigensystem[m]**.

10.2.2 Repeated Eigenvalues

Of course, not all of the n eigenvalues $\lambda_1, \lambda_2, \ldots, \lambda_n$ of an $n \times n$ matrix $\mathbf{A}$ need be distinct; that is, some of the eigenvalues may be repeated. For example, the characteristic equation of the coefficient matrix in the system

$$\mathbf{X}' = \begin{pmatrix} 3 & -18 \\ 2 & -9 \end{pmatrix} \mathbf{X} \qquad (10)$$

is readily shown to be $(\lambda + 3)^2 = 0$, and therefore $\lambda_1 = \lambda_2 = -3$ is a root of *multiplicity two*. For this value we find the single eigenvector

$$\mathbf{K}_1 = \begin{pmatrix} 3 \\ 1 \end{pmatrix}, \quad \text{so} \quad \mathbf{X}_1 = \begin{pmatrix} 3 \\ 1 \end{pmatrix} e^{-3t} \qquad (11)$$

is one solution of (10). But since we are obviously interested in forming the general solution of the system, we need to pursue the question of finding a second solution.

In general, if m is a positive integer and $(\lambda - \lambda_1)^m$ is a factor of the characteristic equation while $(\lambda - \lambda_1)^{m+1}$ is not a factor, then λ_1 is said to be an **eigenvalue of multiplicity** m. The next three examples illustrate the following cases:

(*i*) For some $n \times n$ matrices $\mathbf{A}$ it may be possible to find m linearly independent eigenvectors $\mathbf{K}_1, \mathbf{K}_2, \ldots, \mathbf{K}_m$ corresponding to an eigenvalue λ_1 of multiplicity $m \leq n$. In this case, the general solution of the system contains the linear combination

$$c_1 \mathbf{K}_1 e^{\lambda_1 t} + c_2 \mathbf{K}_2 e^{\lambda_1 t} + \cdots + c_m \mathbf{K}_m e^{\lambda_1 t}.$$

(*ii*) If there is only one eigenvector corresponding to the eigenvalue λ_1 of multiplicity m, then m linearly independent solutions of the form

$$\mathbf{X}_1 = \mathbf{K}_{11} e^{\lambda_1 t}$$

$$\mathbf{X}_2 = \mathbf{K}_{21} t e^{\lambda_1 t} + \mathbf{K}_{22} e^{\lambda_1 t}$$

$$\vdots$$

$$\mathbf{X}_m = \mathbf{K}_{m1} \frac{t^{m-1}}{(m-1)!} e^{\lambda_1 t} + \mathbf{K}_{m2} \frac{t^{m-2}}{(m-2)!} e^{\lambda_1 t} + \cdots + \mathbf{K}_{mm} e^{\lambda_1 t},$$

where $\mathbf{K}_{ij}$ are column vectors, can always be found.

■ **Eigenvalue of Multiplicity Two** We begin by considering eigenvalues of multiplicity two. In the first example we illustrate a matrix for which we can find two distinct eigenvectors corresponding to a repeated eigenvalue.

EXAMPLE 3 **Repeated Eigenvalues**

Solve $\mathbf{X}' = \begin{pmatrix} 1 & -2 & 2 \\ -2 & 1 & -2 \\ 2 & -2 & 1 \end{pmatrix} \mathbf{X}$.

SOLUTION Expanding the determinant in the characteristic equation

$$\det(\mathbf{A} - \lambda\mathbf{I}) = \begin{vmatrix} 1 - \lambda & -2 & 2 \\ -2 & 1 - \lambda & -2 \\ 2 & -2 & 1 - \lambda \end{vmatrix} = 0$$

yields $-(\lambda + 1)^2(\lambda - 5) = 0$. We see that $\lambda_1 = \lambda_2 = -1$ and $\lambda_3 = 5$.

For $\lambda_1 = -1$, Gauss–Jordan elimination immediately gives

$$(\mathbf{A} + \mathbf{I}|\mathbf{0}) = \begin{pmatrix} 2 & -2 & 2 & | & 0 \\ -2 & 2 & -2 & | & 0 \\ 2 & -2 & 2 & | & 0 \end{pmatrix} \xrightarrow[\text{operations}]{\text{row}} \begin{pmatrix} 1 & -1 & 1 & | & 0 \\ 0 & 0 & 0 & | & 0 \\ 0 & 0 & 0 & | & 0 \end{pmatrix}.$$

The first row of the last matrix means $k_1 - k_2 + k_3 = 0$ or $k_1 = k_2 - k_3$. The choices $k_2 = 1$, $k_3 = 0$ and $k_2 = 1, k_3 = 1$ yield, in turn, $k_1 = 1$ and $k_1 = 0$. Thus two eigenvectors corresponding to $\lambda_1 = -1$ are

$$\mathbf{K}_1 = \begin{pmatrix} 1 \\ 1 \\ 0 \end{pmatrix} \quad \text{and} \quad \mathbf{K}_2 = \begin{pmatrix} 0 \\ 1 \\ 1 \end{pmatrix}.$$

Since neither eigenvector is a constant multiple of the other, we have found, corresponding to the same eigenvalue, two linearly independent solutions

$$\mathbf{X}_1 = \begin{pmatrix} 1 \\ 1 \\ 0 \end{pmatrix} e^{-t} \quad \text{and} \quad \mathbf{X}_2 = \begin{pmatrix} 0 \\ 1 \\ 1 \end{pmatrix} e^{-t}.$$

Lastly, for $\lambda_3 = 5$, the reduction

$$(\mathbf{A} - 5\mathbf{I}|\mathbf{0}) = \begin{pmatrix} -4 & -2 & 2 & | & 0 \\ -2 & -4 & -2 & | & 0 \\ 2 & -2 & -4 & | & 0 \end{pmatrix} \xrightarrow[\text{operations}]{\text{row}} \begin{pmatrix} 1 & 0 & -1 & | & 0 \\ 0 & 1 & 1 & | & 0 \\ 0 & 0 & 0 & | & 0 \end{pmatrix}$$

implies $k_1 = k_3$ and $k_2 = -k_3$. Picking $k_3 = 1$ gives $k_1 = 1, k_2 = -1$, and thus a third eigenvector is

$$\mathbf{K}_3 = \begin{pmatrix} 1 \\ -1 \\ 1 \end{pmatrix}.$$

We conclude that the general solution of the system is

$$\mathbf{X} = c_1 \begin{pmatrix} 1 \\ 1 \\ 0 \end{pmatrix} e^{-t} + c_2 \begin{pmatrix} 0 \\ 1 \\ 1 \end{pmatrix} e^{-t} + c_3 \begin{pmatrix} 1 \\ -1 \\ 1 \end{pmatrix} e^{5t}. \qquad \equiv$$

The matrix of coefficients $\mathbf{A}$ in Example 3 is a special kind of matrix known as a symmetric matrix. An $n \times n$ matrix $\mathbf{A}$ is said to be **symmetric** if its transpose $\mathbf{A}^T$ (where the rows and columns are interchanged) is the same as $\mathbf{A}$; that is, if $\mathbf{A}^T = \mathbf{A}$. It can be proved that if the matrix $\mathbf{A}$ in the system $\mathbf{X}' = \mathbf{A}\mathbf{X}$ is symmetric and has real entries, then we can always find n linearly independent eigenvectors $\mathbf{K}_1, \mathbf{K}_2, \ldots, \mathbf{K}_n$, and the general solution of such a system is as given in Theorem 10.2.1. As illustrated in Example 3, this result holds even when some of the eigenvalues are repeated.

■ **Second Solution** Now suppose that λ_1 is an eigenvalue of multiplicity two and that there is only one eigenvector associated with this value. A second solution can be found of the form

$$\mathbf{X}_2 = \mathbf{K}te^{\lambda_1 t} + \mathbf{P}e^{\lambda_1 t},\tag{12}$$

where
$$\mathbf{K} = \begin{pmatrix} k_1 \\ k_2 \\ \vdots \\ k_n \end{pmatrix} \quad \text{and} \quad \mathbf{P} = \begin{pmatrix} p_1 \\ p_2 \\ \vdots \\ p_n \end{pmatrix}.$$

To see this we substitute (12) into the system $\mathbf{X}' = \mathbf{AX}$ and simplify:

$$(\mathbf{AK} - \lambda_1\mathbf{K})te^{\lambda_1 t} + (\mathbf{AP} - \lambda_1\mathbf{P} - \mathbf{K})e^{\lambda_1 t} = \mathbf{0}.$$

Since this last equation is to hold for all values of t, we must have

$$(\mathbf{A} - \lambda_1\mathbf{I})\mathbf{K} = \mathbf{0}\tag{13}$$

and
$$(\mathbf{A} - \lambda_1\mathbf{I})\mathbf{P} = \mathbf{K}.\tag{14}$$

Equation (13) simply states that $\mathbf{K}$ must be an eigenvector of $\mathbf{A}$ associated with λ_1. By solving (13), we find one solution $\mathbf{X}_1 = \mathbf{K}e^{\lambda_1 t}$. To find the second solution $\mathbf{X}_2$ we need only solve the additional system (14) for the vector $\mathbf{P}$.

EXAMPLE 4 **Repeated Eigenvalues**

Find the general solution of the system given in (10).

SOLUTION From (11) we know that $\lambda_1 = -3$ and that one solution is $\mathbf{X}_1 = \begin{pmatrix} 3 \\ 1 \end{pmatrix}e^{-3t}$. Identifying $\mathbf{K} = \begin{pmatrix} 3 \\ 1 \end{pmatrix}$ and $\mathbf{P} = \begin{pmatrix} p_1 \\ p_2 \end{pmatrix}$, we find from (14) that we must now solve

$$(\mathbf{A} + 3\mathbf{I})\mathbf{P} = \mathbf{K} \quad \text{or} \quad \begin{array}{r} 6p_1 - 18p_2 = 3 \\ 2p_1 - 6p_2 = 1. \end{array}$$

Since this system is obviously equivalent to one equation, we have an infinite number of choices for p_1 and p_2. For example, by choosing $p_1 = 1$ we find $p_2 = \frac{1}{6}$. However, for simplicity, we shall choose $p_1 = \frac{1}{2}$ so that $p_2 = 0$. Hence $\mathbf{P} = \begin{pmatrix} \frac{1}{2} \\ 0 \end{pmatrix}$. Thus from (12),

$$\mathbf{X}_2 = \begin{pmatrix} 3 \\ 1 \end{pmatrix}te^{-3t} + \begin{pmatrix} \frac{1}{2} \\ 0 \end{pmatrix}e^{-3t}.$$

The general solution of (10) is then

$$\mathbf{X} = c_1\begin{pmatrix} 3 \\ 1 \end{pmatrix}e^{-3t} + c_2\left[\begin{pmatrix} 3 \\ 1 \end{pmatrix}te^{-3t} + \begin{pmatrix} \frac{1}{2} \\ 0 \end{pmatrix}e^{-3t}\right]. \qquad \equiv$$

By assigning various values to c_1 and c_2 in the solution in Example 4, we can plot trajectories of the system in (10). A phase portrait of (10) is given in **FIGURE 10.2.3**. The solutions $\mathbf{X}_1$ and $-\mathbf{X}_1$ determine two half-lines $y = \frac{1}{3}x$, $x > 0$, and $y = \frac{1}{3}x$, $x < 0$, respectively, that are shown in red in Figure 10.2.3. Because the single eigenvalue is negative and $e^{-3t} \to 0$ as $t \to \infty$ on *every* trajectory, we have $(x(t), y(t)) \to (0, 0)$ as $t \to \infty$. This is why the arrowheads in Figure 10.2.3 indicate that a particle on any trajectory would move toward the origin as time increases and why the origin is an attractor in this case. Moreover, a moving particle on a trajectory $x = 3c_1e^{-3t} + c_2(te^{-3t} + \frac{1}{2}e^{-3t})$, $y = c_1e^{-3t} + c_2te^{-3t}$, $c_2 \neq 0$, approaches $(0, 0)$ tangentially to one of the half-lines as $t \to \infty$. In contrast, when the repeated eigenvalue is positive the situation is reversed and the origin is a repeller. See Problem 23 in Exercises 10.2. Analogous to Figure 10.2.2, Figure 10.2.3 is typical of all 2×2 homogeneous linear systems $\mathbf{X}' = \mathbf{AX}$ that have repeated negative eigenvalues. See Problem 34 in Exercises 10.2.

FIGURE 10.2.3 A phase portrait of system (10)

Eigenvalue of Multiplicity Three

When the coefficient matrix $\mathbf{A}$ has only one eigenvector associated with an eigenvalue λ_1 of multiplicity three, we can find a solution of the form (12) and a third solution of the form

$$\mathbf{X}_3 = \mathbf{K}\frac{t^2}{2}e^{\lambda_1 t} + \mathbf{P}te^{\lambda_1 t} + \mathbf{Q}e^{\lambda_1 t}, \tag{15}$$

where

$$\mathbf{K} = \begin{pmatrix} k_1 \\ k_2 \\ \vdots \\ k_n \end{pmatrix}, \quad \mathbf{P} = \begin{pmatrix} p_1 \\ p_2 \\ \vdots \\ p_n \end{pmatrix}, \quad \text{and} \quad \mathbf{Q} = \begin{pmatrix} q_1 \\ q_2 \\ \vdots \\ q_n \end{pmatrix}.$$

By substituting (15) into the system $\mathbf{X}' = \mathbf{AX}$, we find that the column vectors $\mathbf{K}$, $\mathbf{P}$, and $\mathbf{Q}$ must satisfy

$$(\mathbf{A} - \lambda_1\mathbf{I})\mathbf{K} = \mathbf{0} \tag{16}$$

$$(\mathbf{A} - \lambda_1\mathbf{I})\mathbf{P} = \mathbf{K} \tag{17}$$

and

$$(\mathbf{A} - \lambda_1\mathbf{I})\mathbf{Q} = \mathbf{P}. \tag{18}$$

Of course, the solutions of (16) and (17) can be used in forming the solutions $\mathbf{X}_1$ and $\mathbf{X}_2$.

EXAMPLE 5 Repeated Eigenvalues

Solve $\mathbf{X}' = \begin{pmatrix} 2 & 1 & 6 \\ 0 & 2 & 5 \\ 0 & 0 & 2 \end{pmatrix}\mathbf{X}$.

SOLUTION The characteristic equation $(\lambda - 2)^3 = 0$ shows that $\lambda_1 = 2$ is an eigenvalue of multiplicity three. By solving $(\mathbf{A} - 2\mathbf{I})\mathbf{K} = \mathbf{0}$ we find the single eigenvector

$$\mathbf{K} = \begin{pmatrix} 1 \\ 0 \\ 0 \end{pmatrix}.$$

We next solve the systems $(\mathbf{A} - 2\mathbf{I})\mathbf{P} = \mathbf{K}$ and $(\mathbf{A} - 2\mathbf{I})\mathbf{Q} = \mathbf{P}$ in succession and find that

$$\mathbf{P} = \begin{pmatrix} 0 \\ 1 \\ 0 \end{pmatrix} \quad \text{and} \quad \mathbf{Q} = \begin{pmatrix} 0 \\ -\frac{6}{5} \\ \frac{1}{5} \end{pmatrix}.$$

Using (12) and (15), we see that the general solution of the system is

$$\mathbf{X} = c_1\begin{pmatrix} 1 \\ 0 \\ 0 \end{pmatrix}e^{2t} + c_2\left[\begin{pmatrix} 1 \\ 0 \\ 0 \end{pmatrix}te^{2t} + \begin{pmatrix} 0 \\ 1 \\ 0 \end{pmatrix}e^{2t}\right] + c_3\left[\begin{pmatrix} 1 \\ 0 \\ 0 \end{pmatrix}\frac{t^2}{2}e^{2t} + \begin{pmatrix} 0 \\ 1 \\ 0 \end{pmatrix}te^{2t} + \begin{pmatrix} 0 \\ -\frac{6}{5} \\ \frac{1}{5} \end{pmatrix}e^{2t}\right]. \quad \equiv$$

REMARKS

When an eigenvalue λ_1 has multiplicity m, then we can either find m linearly independent eigenvectors or the number of corresponding eigenvectors is less than m. Hence the two cases listed on page 611 are not all the possibilities under which a repeated eigenvalue can occur. It could happen, say, that a 5×5 matrix has an eigenvalue of multiplicity 5 and there exist three corresponding linearly independent eigenvectors. See Problems 33 and 53 in Exercises 10.2.

10.2.3 Complex Eigenvalues

If $\lambda_1 = \alpha + i\beta$ and $\lambda_2 = \alpha - i\beta$, $\beta > 0$, $i^2 = -1$, are complex eigenvalues of the coefficient matrix $\mathbf{A}$, we can then certainly expect their corresponding eigenvectors to also have complex entries.*

For example, the characteristic equation of the system

$$\frac{dx}{dt} = 6x - y$$

$$\frac{dy}{dt} = 5x + 4y$$

(19)

is

$$\det(\mathbf{A} - \lambda\mathbf{I}) = \begin{vmatrix} 6 - \lambda & -1 \\ 5 & 4 - \lambda \end{vmatrix} = \lambda^2 - 10\lambda + 29 = 0.$$

From the quadratic formula we find $\lambda_1 = 5 + 2i$, $\lambda_2 = 5 - 2i$.

Now for $\lambda_1 = 5 + 2i$ we must solve

$$(1 - 2i)k_1 - \qquad k_2 = 0$$

$$5k_1 - (1 + 2i)k_2 = 0.$$

Since $k_2 = (1 - 2i)k_1$,[†] the choice $k_1 = 1$ gives the following eigenvector and a solution vector:

$$\mathbf{K}_1 = \begin{pmatrix} 1 \\ 1 - 2i \end{pmatrix}, \quad \mathbf{X}_1 = \begin{pmatrix} 1 \\ 1 - 2i \end{pmatrix} e^{(5 + 2i)t}.$$

In like manner, for $\lambda_2 = 5 - 2i$ we find

$$\mathbf{K}_2 = \begin{pmatrix} 1 \\ 1 + 2i \end{pmatrix}, \quad \mathbf{X}_2 = \begin{pmatrix} 1 \\ 1 + 2i \end{pmatrix} e^{(5 - 2i)t}.$$

We can verify by means of the Wronskian that these solution vectors are linearly independent, and so the general solution of (19) is

$$\mathbf{X} = c_1 \begin{pmatrix} 1 \\ 1 - 2i \end{pmatrix} e^{(5 + 2i)t} + c_2 \begin{pmatrix} 1 \\ 1 + 2i \end{pmatrix} e^{(5 - 2i)t}.$$

(20)

Note that the entries in $\mathbf{K}_2$ corresponding to λ_2 are the conjugates of the entries in $\mathbf{K}_1$ corresponding to λ_1. The conjugate of λ_1 is, of course, λ_2. We write this as $\lambda_2 = \overline{\lambda}_1$ and $\mathbf{K}_2 = \overline{\mathbf{K}}_1$. We have illustrated the following general result.

THEOREM 10.2.2 Solutions Corresponding to a Complex Eigenvalue

Let $\mathbf{A}$ be the coefficient matrix having real entries of the homogeneous system (2), and let $\mathbf{K}_1$ be an eigenvector corresponding to the complex eigenvalue $\lambda_1 = \alpha + i\beta$, α and β real. Then

$$\mathbf{K}_1 e^{\lambda_1 t} \quad \text{and} \quad \overline{\mathbf{K}}_1 e^{\overline{\lambda}_1 t}$$

are solutions of (2).

It is desirable and relatively easy to rewrite a solution such as (20) in terms of real functions. To this end we first use Euler's formula to write

$$e^{(5 + 2i)t} = e^{5t} e^{2ti} = e^{5t}(\cos 2t + i \sin 2t)$$

$$e^{(5 - 2i)t} = e^{5t} e^{-2ti} = e^{5t}(\cos 2t - i \sin 2t).$$

*When the characteristic equation has real coefficients, complex eigenvalues always appear in conjugate pairs.

[†]Note that the second equation in the system is simply $1 + 2i$ times the first equation.

Then (20) becomes

$$\mathbf{X} = c_1 e^{5t} \binom{1}{1-2i}(\cos 2t + i\sin 2t) + c_2 e^{5t} \binom{1}{1+2i}(\cos 2t - i\sin 2t)$$

$$= c_1 e^{5t} \binom{\cos 2t + i\sin 2t}{(1-2i)(\cos 2t + i\sin 2t)} + c_2 e^{5t} \binom{\cos 2t - i\sin 2t}{(1+2i)(\cos 2t - i\sin 2t)}$$

$$= c_1 e^{5t} \binom{\cos 2t + i\sin 2t}{\cos 2t + 2\sin 2t \ + i(-2\cos 2t + \sin 2t)} + c_2 e^{5t} \binom{\cos 2t - i\sin 2t}{\cos 2t + 2\sin 2t \ + i(2\cos 2t - \sin 2t)}$$

$$= (c_1 + c_2) \binom{\cos 2t}{\cos 2t + 2\sin 2t} e^{5t} + (c_1 - c_2)i \binom{\sin 2t}{-2\cos 2t + \sin 2t} e^{5t}$$

If we let $C_1 = c_1 + c_2$ and $C_2 = (c_1 - c_2)i$, then the last line can be written

$$\mathbf{X} = C_1 \mathbf{X}_1 + C_2 \mathbf{X}_2, \tag{21}$$

where

$$\mathbf{X}_1 = \binom{\cos 2t}{\cos 2t + 2\sin 2t} e^{5t} = \left[\binom{1}{1}\cos 2t - \binom{0}{-2}\sin 2t \right] e^{5t}$$

and

$$\mathbf{X}_2 = \binom{\sin 2t}{-2\cos 2t + \sin 2t} e^{5t} = \left[\binom{0}{-2}\cos 2t + \binom{1}{1}\sin 2t \right] e^{5t}.$$

It is now important to realize that the two vectors $\mathbf{X}_1$ and $\mathbf{X}_2$ in (21) are themselves linearly independent *real* solutions of the original system. Consequently, we are justified in ignoring the relationship between C_1, C_2 and c_1, c_2, and we can regard C_1 and C_2 as completely arbitrary and real. In other words, the linear combination (21) is an alternative general solution of (19).

The foregoing process can be generalized. Let $\mathbf{K}_1$ be an eigenvector of the coefficient matrix $\mathbf{A}$ (with real entries) corresponding to the complex eigenvalue $\lambda_1 = \alpha + i\beta$. Then the two solution vectors in Theorem 10.2.2 can be written as

$$\mathbf{K}_1 e^{\lambda_1 t} = \mathbf{K}_1 e^{\alpha t} e^{i\beta t} = \mathbf{K}_1 e^{\alpha t}(\cos \beta t + i\sin \beta t)$$

$$\overline{\mathbf{K}}_1 e^{\overline{\lambda}_1 t} = \overline{\mathbf{K}}_1 e^{\alpha t} e^{-i\beta t} = \overline{\mathbf{K}}_1 e^{\alpha t}(\cos \beta t - i\sin \beta t).$$

By the superposition principle, Theorem 10.1.2, the following vectors are also solutions:

$$\mathbf{X}_1 = \frac{1}{2}(\mathbf{K}_1 e^{\lambda_1 t} + \overline{\mathbf{K}}_1 e^{\overline{\lambda}_1 t}) = \frac{1}{2}(\mathbf{K}_1 + \overline{\mathbf{K}}_1) e^{\alpha t}\cos \beta t - \frac{i}{2}(-\mathbf{K}_1 + \overline{\mathbf{K}}_1) e^{\alpha t}\sin \beta t$$

$$\mathbf{X}_2 = \frac{i}{2}(-\mathbf{K}_1 e^{\lambda_1 t} + \overline{\mathbf{K}}_1 e^{\overline{\lambda}_1 t}) = \frac{i}{2}(-\mathbf{K}_1 + \overline{\mathbf{K}}_1) e^{\alpha t}\cos \beta t + \frac{1}{2}(\mathbf{K}_1 + \overline{\mathbf{K}}_1) e^{\alpha t}\sin \beta t.$$

For *any* complex number $z = a + ib$, both $\frac{1}{2}(z + \overline{z}) = a$ and $\frac{i}{2}(-z + \overline{z}) = b$ are *real* numbers. Therefore, the entries in the column vectors $\frac{1}{2}(\mathbf{K}_1 + \overline{\mathbf{K}}_1)$ and $\frac{i}{2}(-\mathbf{K}_1 + \overline{\mathbf{K}}_1)$ are real numbers. By defining

$$\mathbf{B}_1 = \frac{1}{2}(\mathbf{K}_1 + \overline{\mathbf{K}}_1) \quad \text{and} \quad \mathbf{B}_2 = \frac{i}{2}(-\mathbf{K}_1 + \overline{\mathbf{K}}_1), \tag{22}$$

we are led to the following theorem.

Let $\lambda_1 = \alpha + i\beta$ be a complex eigenvalue of the coefficient matrix $\mathbf{A}$ in the homogeneous system (2), and let $\mathbf{B}_1$ and $\mathbf{B}_2$ denote the column vectors defined in (22). Then

$$\mathbf{X}_1 = [\mathbf{B}_1 \cos \beta t - \mathbf{B}_2 \sin \beta t]e^{\alpha t}$$
$$\mathbf{X}_2 = [\mathbf{B}_2 \cos \beta t + \mathbf{B}_1 \sin \beta t]e^{\alpha t} \tag{23}$$

are linearly independent solutions of (2) on $(-\infty, \infty)$.

The matrices $\mathbf{B}_1$ and $\mathbf{B}_2$ in (22) are often denoted by

$$\mathbf{B}_1 = \text{Re}(\mathbf{K}_1) \quad \text{and} \quad \mathbf{B}_2 = \text{Im}(\mathbf{K}_1) \tag{24}$$

since these vectors are, respectively, the *real* and *imaginary* parts of the eigenvector $\mathbf{K}_1$. For example, (21) follows from (23) with

$$\mathbf{K}_1 = \begin{pmatrix} 1 \\ 1 - 2i \end{pmatrix} = \begin{pmatrix} 1 \\ 1 \end{pmatrix} + i\begin{pmatrix} 0 \\ -2 \end{pmatrix}$$

$$\mathbf{B}_1 = \text{Re}(\mathbf{K}_1) = \begin{pmatrix} 1 \\ 1 \end{pmatrix} \quad \text{and} \quad \mathbf{B}_2 = \text{Im}(\mathbf{K}_1) = \begin{pmatrix} 0 \\ -2 \end{pmatrix}.$$

EXAMPLE 6 Complex Eigenvalues

Solve the initial-value problem

$$\mathbf{X}' = \begin{pmatrix} 2 & 8 \\ -1 & -2 \end{pmatrix}\mathbf{X}, \quad \mathbf{X}(0) = \begin{pmatrix} 2 \\ -1 \end{pmatrix}. \tag{25}$$

SOLUTION First we obtain the eigenvalues from

$$\det(\mathbf{A} - \lambda\mathbf{I}) = \begin{vmatrix} 2 - \lambda & 8 \\ -1 & -2 - \lambda \end{vmatrix} = \lambda^2 + 4 = 0.$$

The eigenvalues are $\lambda_1 = 2i$ and $\lambda_2 = \bar{\lambda}_1 = -2i$. For λ_1 the system

$$(2 - 2i)k_1 + \qquad 8k_2 = 0$$
$$-k_1 + (-2 - 2i)k_2 = 0$$

gives $k_1 = -(2 + 2i)k_2$. By choosing $k_2 = -1$ we get

$$\mathbf{K}_1 = \begin{pmatrix} 2 + 2i \\ -1 \end{pmatrix} = \begin{pmatrix} 2 \\ -1 \end{pmatrix} + i\begin{pmatrix} 2 \\ 0 \end{pmatrix}.$$

Now from (24) we form

$$\mathbf{B}_1 = \text{Re}(\mathbf{K}_1) = \begin{pmatrix} 2 \\ -1 \end{pmatrix} \quad \text{and} \quad \mathbf{B}_2 = \text{Im}(\mathbf{K}_1) = \begin{pmatrix} 2 \\ 0 \end{pmatrix}.$$

Since $\alpha = 0$, it follows from (23) that the general solution of the system is

$$\mathbf{X} = c_1\left[\begin{pmatrix} 2 \\ -1 \end{pmatrix}\cos 2t - \begin{pmatrix} 2 \\ 0 \end{pmatrix}\sin 2t\right] + c_2\left[\begin{pmatrix} 2 \\ 0 \end{pmatrix}\cos 2t + \begin{pmatrix} 2 \\ -1 \end{pmatrix}\sin 2t\right]$$

$$= c_1\begin{pmatrix} 2\cos 2t - 2\sin 2t \\ -\cos 2t \end{pmatrix} + c_2\begin{pmatrix} 2\cos 2t + 2\sin 2t \\ -\sin 2t \end{pmatrix}. \tag{26}$$

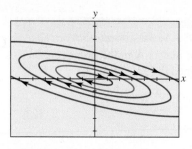

FIGURE 10.2.4 A phase portrait of system (25) in Example 6

Some graphs of the curves or trajectories defined by the solution (26) of the system are illustrated in the phase portrait in **FIGURE 10.2.4**. Now the initial condition $\mathbf{X}(0) = \begin{pmatrix} 2 \\ -1 \end{pmatrix}$, or equivalently $x(0) = 2$, and $y(0) = -1$, yields the algebraic system $2c_1 + 2c_2 = 2$, $-c_1 = -1$ whose solution is $c_1 = 1$, $c_2 = 0$. Thus the solution to the problem is $\mathbf{X} = \begin{pmatrix} 2\cos 2t - 2\sin 2t \\ -\cos 2t \end{pmatrix}$. The specific trajectory defined parametrically by the particular solution $x = 2\cos 2t - 2\sin 2t$, $y = -\cos 2t$ is the red curve in Figure 10.2.4. Note that this curve passes through $(2, -1)$. $\equiv$

10.2 Exercises Answers to selected odd-numbered problems begin on page ANS-27.

10.2.1 Distinct Real Eigenvalues

In Problems 1–12, find the general solution of the given system.

1. $\dfrac{dx}{dt} = x + 2y$

$\dfrac{dy}{dt} = 4x + 3y$

2. $\dfrac{dx}{dt} = 2x + 2y$

$\dfrac{dy}{dt} = x + 3y$

3. $\dfrac{dx}{dt} = -4x + 2y$

$\dfrac{dy}{dt} = -\tfrac{5}{2}x + 2y$

4. $\dfrac{dx}{dt} = -\tfrac{5}{2}x + 2y$

$\dfrac{dy}{dt} = \tfrac{3}{4}x - 2y$

5. $\mathbf{X}' = \begin{pmatrix} 10 & -5 \\ 8 & -12 \end{pmatrix}\mathbf{X}$

6. $\mathbf{X}' = \begin{pmatrix} -6 & 2 \\ -3 & 1 \end{pmatrix}\mathbf{X}$

7. $\dfrac{dx}{dt} = x + y - z$

$\dfrac{dy}{dt} = 2y$

$\dfrac{dz}{dt} = y - z$

8. $\dfrac{dx}{dt} = 2x - 7y$

$\dfrac{dy}{dt} = 5x + 10y + 4z$

$\dfrac{dz}{dt} = 5y + 2z$

9. $\mathbf{X}' = \begin{pmatrix} -1 & 1 & 0 \\ 1 & 2 & 1 \\ 0 & 3 & -1 \end{pmatrix}\mathbf{X}$ **10.** $\mathbf{X}' = \begin{pmatrix} 1 & 0 & 1 \\ 0 & 1 & 0 \\ 1 & 0 & 1 \end{pmatrix}\mathbf{X}$

11. $\mathbf{X}' = \begin{pmatrix} -1 & -1 & 0 \\ \tfrac{3}{4} & -\tfrac{3}{2} & 3 \\ \tfrac{1}{8} & \tfrac{1}{4} & -\tfrac{1}{2} \end{pmatrix}\mathbf{X}$ **12.** $\mathbf{X}' = \begin{pmatrix} -1 & 5 & 2 \\ 4 & -1 & -2 \\ 0 & 0 & 6 \end{pmatrix}\mathbf{X}$

In Problems 13 and 14, solve the given initial-value problem.

13. $\mathbf{X}' = \begin{pmatrix} \tfrac{1}{2} & 0 \\ 1 & -\tfrac{1}{2} \end{pmatrix}\mathbf{X}, \quad \mathbf{X}(0) = \begin{pmatrix} 3 \\ 5 \end{pmatrix}$

14. $\mathbf{X}' = \begin{pmatrix} 1 & 1 & 4 \\ 0 & 2 & 0 \\ 1 & 1 & 1 \end{pmatrix}\mathbf{X}, \quad \mathbf{X}(0) = \begin{pmatrix} 1 \\ 3 \\ 0 \end{pmatrix}$

15. Consider the large mixing tanks shown in **FIGURE 10.2.5**. Suppose that both tanks A and B initially contain 100 gallons of brine. Liquid is pumped in and out of the tanks as indicated in the figure; the mixture pumped between and out of the tanks is assumed to be well-stirred.
 (a) Construct a mathematical model in the form of a linear system of first-order differential equations for

the number of pounds $x_1(t)$ and $x_2(t)$ of salt in tanks A and B, respectively, at time t. Write the system in matrix form. [*Hint*: See Section 2.9 and Problem 52, Chapter 2 in Review.]
 (b) Use the eigenvalue method of this section to solve the linear system in part (a) subject to $x_1(0) = 20$, $x_2(0) = 5$.
 (c) Use a graphing utility or CAS to plot the graphs of $x_1(t)$ and $x_2(t)$ in the same coordinate plane.
 (d) Suppose the system of mixing tanks is to be turned off when the number of pounds of salt in tank B equals that in tank A. Use a root-finding application of a CAS or calculator to approximate that time.

pure water
2 gal/min

mixture
1 gal/min

mixture
1 gal/min

mixture
2 gal/min

mixture
1 gal/min

FIGURE 10.2.5 Mixing tanks in Problem 15

16. In Problem 26 of Exercises 3.12 you were asked to solve the following linear system

$$\dfrac{dx_1}{dt} = -\dfrac{1}{50}x_1$$

$$\dfrac{dx_2}{dt} = \dfrac{1}{50}x_1 - \dfrac{2}{75}x_2$$

$$\dfrac{dx_3}{dt} = \dfrac{2}{75}x_2 - \dfrac{1}{25}x_3$$

using elimination techniques. Recall, this linear system is a mathematical model for the number of pounds of salt $x_1(t)$, $x_2(t)$, and $x_3(t)$ in the connected mixing tanks A, B, and C shown in Figure 2.9.7 on page 102.
 (a) Use the eigenvalue method of this section to solve the system subject to $x_1(0) = 15$, $x_2(0) = 10$, $x_3(0) = 5$.
 (b) What are $\lim_{t\to\infty}x_1(t)$, $\lim_{t\to\infty}x_2(t)$, and $\lim_{t\to\infty}x_3(t)$? Interpret this result.

Computer Lab Assignments

In Problems 17 and 18, use a CAS or linear algebra software as an aid in finding the general solution of the given system.

17. $\mathbf{X}' = \begin{pmatrix} 0.9 & 2.1 & 3.2 \\ 0.7 & 6.5 & 4.2 \\ 1.1 & 1.7 & 3.4 \end{pmatrix} \mathbf{X}$

18. $\mathbf{X}' = \begin{pmatrix} 1 & 0 & 2 & -1.8 & 0 \\ 0 & 5.1 & 0 & -1 & 3 \\ 1 & 2 & -3 & 0 & 0 \\ 0 & 1 & -3.1 & 4 & 0 \\ -2.8 & 0 & 0 & 1.5 & 1 \end{pmatrix} \mathbf{X}$

19. (a) Use computer software to obtain the phase portrait of the system in Problem 5. If possible, include the arrowheads as in Figure 10.2.2. Also, include four half-lines in your phase portrait.
 (b) Obtain the Cartesian equations of each of the four half-lines in part (a).
 (c) Draw the eigenvectors on your phase portrait of the system.

20. Find phase portraits for the systems in Problems 2 and 4. For each system, find any half-line trajectories and include these lines in your phase portrait.

10.2.2 Repeated Eigenvalues

In Problems 21–30, find the general solution of the given system.

21. $\dfrac{dx}{dt} = 3x - y$
$\dfrac{dy}{dt} = 9x - 3y$

22. $\dfrac{dx}{dt} = -6x + 5y$
$\dfrac{dy}{dt} = -5x + 4y$

23. $\mathbf{X}' = \begin{pmatrix} -1 & 3 \\ -3 & 5 \end{pmatrix} \mathbf{X}$

24. $\mathbf{X}' = \begin{pmatrix} 12 & -9 \\ 4 & 0 \end{pmatrix} \mathbf{X}$

25. $\dfrac{dx}{dt} = 3x - y - z$
$\dfrac{dy}{dt} = x + y - z$
$\dfrac{dz}{dt} = x - y + z$

26. $\dfrac{dx}{dt} = 3x + 2y + 4z$
$\dfrac{dy}{dt} = 2x + 2z$
$\dfrac{dz}{dt} = 4x + 2y + 3z$

27. $\mathbf{X}' = \begin{pmatrix} 5 & -4 & 0 \\ 1 & 0 & 2 \\ 0 & 2 & 5 \end{pmatrix} \mathbf{X}$

28. $\mathbf{X}' = \begin{pmatrix} 1 & 0 & 0 \\ 0 & 3 & 1 \\ 0 & -1 & 1 \end{pmatrix} \mathbf{X}$

29. $\mathbf{X}' = \begin{pmatrix} 1 & 0 & 0 \\ 2 & 2 & -1 \\ 0 & 1 & 0 \end{pmatrix} \mathbf{X}$

30. $\mathbf{X}' = \begin{pmatrix} 4 & 1 & 0 \\ 0 & 4 & 1 \\ 0 & 0 & 4 \end{pmatrix} \mathbf{X}$

In Problems 31 and 32, solve the given initial-value problem.

31. $\mathbf{X}' = \begin{pmatrix} 2 & 4 \\ -1 & 6 \end{pmatrix} \mathbf{X}, \quad \mathbf{X}(0) = \begin{pmatrix} -1 \\ 6 \end{pmatrix}$

32. $\mathbf{X}' = \begin{pmatrix} 0 & 0 & 1 \\ 0 & 1 & 0 \\ 1 & 0 & 0 \end{pmatrix} \mathbf{X}, \quad \mathbf{X}(0) = \begin{pmatrix} 1 \\ 2 \\ 5 \end{pmatrix}$

33. Show that the 5×5 matrix

$$\mathbf{A} = \begin{pmatrix} 2 & 1 & 0 & 0 & 0 \\ 0 & 2 & 0 & 0 & 0 \\ 0 & 0 & 2 & 0 & 0 \\ 0 & 0 & 0 & 2 & 1 \\ 0 & 0 & 0 & 0 & 2 \end{pmatrix}$$

has an eigenvalue λ_1 of multiplicity 5. Show that three linearly independent eigenvectors corresponding to λ_1 can be found.

Computer Lab Assignment

34. Find phase portraits for the systems in Problems 22 and 23. For each system, find any half-line trajectories and include these lines in your phase portrait.

10.2.3 Complex Eigenvalues

In Problems 35–46, find the general solution of the given system.

35. $\dfrac{dx}{dt} = 6x - y$
$\dfrac{dy}{dt} = 5x + 2y$

36. $\dfrac{dx}{dt} = x + y$
$\dfrac{dy}{dt} = -2x - y$

37. $\dfrac{dx}{dt} = 5x + y$
$\dfrac{dy}{dt} = -2x + 3y$

38. $\dfrac{dx}{dt} = 4x + 5y$
$\dfrac{dy}{dt} = -2x + 6y$

39. $\mathbf{X}' = \begin{pmatrix} 4 & -5 \\ 5 & -4 \end{pmatrix} \mathbf{X}$

40. $\mathbf{X}' = \begin{pmatrix} 1 & -8 \\ 1 & -3 \end{pmatrix} \mathbf{X}$

41. $\dfrac{dx}{dt} = z$
$\dfrac{dy}{dt} = -z$
$\dfrac{dz}{dt} = y$

42. $\dfrac{dx}{dt} = 2x + y + 2z$
$\dfrac{dy}{dt} = 3x + 6z$
$\dfrac{dz}{dt} = -4x - 3z$

43. $\mathbf{X}' = \begin{pmatrix} 1 & -1 & 2 \\ -1 & 1 & 0 \\ -1 & 0 & 1 \end{pmatrix} \mathbf{X}$

44. $\mathbf{X}' = \begin{pmatrix} 4 & 0 & 1 \\ 0 & 6 & 0 \\ -4 & 0 & 4 \end{pmatrix} \mathbf{X}$

45. $\mathbf{X}' = \begin{pmatrix} 2 & 5 & 1 \\ -5 & -6 & 4 \\ 0 & 0 & 2 \end{pmatrix} \mathbf{X}$

46. $\mathbf{X}' = \begin{pmatrix} 2 & 4 & 4 \\ -1 & -2 & 0 \\ -1 & 0 & -2 \end{pmatrix} \mathbf{X}$

In Problems 47 and 48, solve the given initial-value problem.

47. $\mathbf{X}' = \begin{pmatrix} 1 & -12 & -14 \\ 1 & 2 & -3 \\ 1 & 1 & -2 \end{pmatrix} \mathbf{X}, \quad \mathbf{X}(0) = \begin{pmatrix} 4 \\ 6 \\ -7 \end{pmatrix}$

48. $\mathbf{X}' = \begin{pmatrix} 6 & -1 \\ 5 & 4 \end{pmatrix} \mathbf{X}, \quad \mathbf{X}(0) = \begin{pmatrix} -2 \\ 8 \end{pmatrix}$

49. (a) In the closed system shown in **FIGURE 10.2.6** the three large tanks A, B, and C initially contain the number of gallons of brine indicated. Construct a mathematical

model for the number of pounds of salt $x_1(t)$, $x_2(t)$, and $x_3(t)$ in the tanks A, B, and C at time t, respectively.

(b) Use the eigenvalue method of this section to solve the system in part (a) subject to $x_1(0) = 30$, $x_2(0) = 20$, $x_3(0) = 5$.

mixture
5 gal/min

mixture
5 gal/min

mixture
5 gal/min

FIGURE 10.2.6 Mixing tanks in Problem 49

50. For the system in Problem 49:

(a) Show that $x_1(t) + x_2(t) + x_3(t) = 55$. Interpret this result.

(b) What are $\lim_{t\to\infty} x_1(t)$, $\lim_{t\to\infty} x_2(t)$, and $\lim_{t\to\infty} x_3(t)$? Interpret this result.

Computer Lab Assignments

51. Find phase portraits for the systems in Problems 38–40.

52. Solve each of the following linear systems.

(a) $X' = \begin{pmatrix} 1 & 1 \\ 1 & 1 \end{pmatrix} X$

(b) $X' = \begin{pmatrix} 1 & 1 \\ -1 & -1 \end{pmatrix} X$

Find a phase portrait of each system. What is the geometric significance of the line $y = -x$ in each portrait?

Discussion Problems

53. Consider the 5×5 matrix given in Problem 33. Solve the system $X' = AX$ without the aid of matrix methods, but write the general solution using the matrix notation. Use the general solution as a basis for a discussion on how the system can be solved using the matrix methods of this section. Carry out your ideas.

54. Obtain a Cartesian equation of the curve defined parametrically by the solution of the linear system in Example 6. Identify the curve passing through $(2, -1)$ in Figure 10.2.4. [*Hint:* Compute x^2, y^2, and xy.]

55. Examine your phase portraits in Problem 51. Under what conditions will the phase portrait of a 2×2 homogeneous linear system with complex eigenvalues consist of a family of closed curves? Consist of a family of spirals? Under what conditions is the origin $(0, 0)$ a repeller? An attractor?

56. The system of linear second-order differential equations

$$m_1 x_1'' = -k_1 x_1 + k_2(x_2 - x_1)$$
$$m_2 x_2'' = -k_2(x_2 - x_1) \tag{27}$$

describes the motion of two coupled spring/mass systems (see Figure 3.12.1). We have already solved a special case of this system in Sections 3.12 and 4.6. In this problem we describe yet another method for solving the system.

(a) Show that the system in (27) can be written as the matrix equation $X'' = AX$, where

$$X = \begin{pmatrix} x_1 \\ x_2 \end{pmatrix} \quad \text{and} \quad A = \begin{pmatrix} -\dfrac{k_1 + k_2}{m_1} & \dfrac{k_2}{m_1} \\ \dfrac{k_2}{m_2} & -\dfrac{k_2}{m_2} \end{pmatrix}.$$

(b) If a solution is assumed of the form $X = K e^{\omega t}$, show that $X'' = AX$ yields

$$(A - \lambda I)K = 0 \quad \text{where} \quad \lambda = \omega^2.$$

(c) Show that if $m_1 = 1$, $m_2 = 1$, $k_1 = 3$, and $k_2 = 2$, a solution of the system is

$$X = c_1 \begin{pmatrix} 1 \\ 2 \end{pmatrix} e^{it} + c_2 \begin{pmatrix} 1 \\ 2 \end{pmatrix} e^{-it} + c_3 \begin{pmatrix} -2 \\ 1 \end{pmatrix} e^{\sqrt{6} it} + c_4 \begin{pmatrix} -2 \\ 1 \end{pmatrix} e^{-\sqrt{6} it}.$$

(d) Show that the solution in part (c) can be written as

$$X = b_1 \begin{pmatrix} 1 \\ 2 \end{pmatrix} \cos t + b_2 \begin{pmatrix} 1 \\ 2 \end{pmatrix} \sin t$$
$$+ b_3 \begin{pmatrix} -2 \\ 1 \end{pmatrix} \cos \sqrt{6}t + b_4 \begin{pmatrix} -2 \\ 1 \end{pmatrix} \sin \sqrt{6}t.$$

10.3 Solution by Diagonalization

INTRODUCTION In this section we are going to consider an alternative method for solving a homogeneous system of linear first-order differential equations. This method is applicable to

Review Section 8.12. ▶ such a system $X' = AX$ whenever the coefficient matrix A is diagonalizable.

■ **Coupled Systems** A homogeneous linear system $X' = AX$,

$$\begin{pmatrix} x_1' \\ x_2' \\ \vdots \\ x_n' \end{pmatrix} = \begin{pmatrix} a_{11} & a_{12} & \cdots & a_{1n} \\ a_{21} & a_{22} & \cdots & a_{2n} \\ \vdots & & & \vdots \\ a_{n1} & a_{n2} & \cdots & a_{nn} \end{pmatrix} \begin{pmatrix} x_1 \\ x_2 \\ \vdots \\ x_n \end{pmatrix}, \tag{1}$$

in which each x_i' is expressed as a linear combination of $x_1, x_2, \ldots, x_n$ is said to be **coupled**. If the coefficient matrix A is diagonalizable, then the system can be **uncoupled** in that each x_i' can be expressed solely in terms of x_i.

If the matrix A has n linearly independent eigenvectors then we know from Theorem 8.12.2 that we can find a matrix P such that $P^{-1}AP = D$, where D is a diagonal matrix. If we make the substitution $X = PY$ in the system $X' = AX$, then

$$PY' = APY \quad \text{or} \quad Y' = P^{-1}APY \quad \text{or} \quad Y' = DY. \tag{2}$$

The last equation in (2) is the same as

$$\begin{pmatrix} y_1' \\ y_2' \\ \vdots \\ y_n' \end{pmatrix} = \begin{pmatrix} \lambda_1 & 0 & 0 & \cdots & 0 \\ 0 & \lambda_2 & 0 & \cdots & 0 \\ \vdots & & & & \vdots \\ 0 & 0 & 0 & \cdots & \lambda_n \end{pmatrix} \begin{pmatrix} y_1 \\ y_2 \\ \vdots \\ y_n \end{pmatrix}. \tag{3}$$

Since D is diagonal, an inspection of (3) reveals that this new system is uncoupled: Each differential equation in the system is of the form $y_i' = \lambda_i y_i$, $i = 1, 2, \ldots, n$. The solution of each of these linear equations is $y_i = c_i e^{\lambda_i t}$, $i = 1, 2, \ldots, n$. Hence the general solution of (3) can be written as the column vector

$$Y = \begin{pmatrix} c_1 e^{\lambda_1 t} \\ c_2 e^{\lambda_2 t} \\ \vdots \\ c_n e^{\lambda_n t} \end{pmatrix}. \tag{4}$$

Since we now know Y and since the matrix P can be constructed from the eigenvectors of A, the general solution of the original system $X' = AX$ is obtained from $X = PY$.

EXAMPLE 1 **Uncoupling a Linear System**

Solve $X' = \begin{pmatrix} -2 & -1 & 8 \\ 0 & -3 & 8 \\ 0 & -4 & 9 \end{pmatrix} X$ by diagonalization.

SOLUTION We begin by finding the eigenvalues and corresponding eigenvectors of the coefficient matrix.

From $\det(A - \lambda I) = -(\lambda + 2)(\lambda - 1)(\lambda - 5)$, we get $\lambda_1 = -2$, $\lambda_2 = 1$, and $\lambda_3 = 5$. Since the eigenvalues are distinct, the eigenvectors are linearly independent. Solving $(A - \lambda_i I)K = 0$ for $i = 1, 2,$ and 3 gives, respectively,

$$K_1 = \begin{pmatrix} 1 \\ 0 \\ 0 \end{pmatrix}, \quad K_2 = \begin{pmatrix} 2 \\ 2 \\ 1 \end{pmatrix}, \quad K_3 = \begin{pmatrix} 1 \\ 1 \\ 1 \end{pmatrix}. \tag{5}$$

Thus a matrix that diagonalizes the coefficient matrix is

$$P = \begin{pmatrix} 1 & 2 & 1 \\ 0 & 2 & 1 \\ 0 & 1 & 1 \end{pmatrix}.$$

The entries on the main diagonal of D are the eigenvalues of A corresponding to the order in which the eigenvectors appear in P:

$$D = \begin{pmatrix} -2 & 0 & 0 \\ 0 & 1 & 0 \\ 0 & 0 & 5 \end{pmatrix}.$$

As we have shown previously, the substitution $\mathbf{X} = \mathbf{PY}$ in $\mathbf{X}' = \mathbf{AX}$ gives the uncoupled system $\mathbf{Y}' = \mathbf{DY}$. The general solution of this last system is immediate:

$$\mathbf{Y} = \begin{pmatrix} c_1 e^{-2t} \\ c_2 e^{t} \\ c_3 e^{5t} \end{pmatrix}.$$

Hence the solution of the given system is

$$\mathbf{X} = \mathbf{PY} = \begin{pmatrix} 1 & 2 & 1 \\ 0 & 2 & 1 \\ 0 & 1 & 1 \end{pmatrix}\begin{pmatrix} c_1 e^{-2t} \\ c_2 e^{t} \\ c_3 e^{5t} \end{pmatrix} = \begin{pmatrix} c_1 e^{-2t} + 2c_2 e^{t} + c_3 e^{5t} \\ 2c_2 e^{t} + c_3 e^{5t} \\ c_2 e^{t} + c_3 e^{5t} \end{pmatrix}. \qquad (6) \equiv$$

Note that (6) can be written in the usual manner by expressing the last matrix as a sum of column matrices:

$$\mathbf{X} = c_1 \begin{pmatrix} 1 \\ 0 \\ 0 \end{pmatrix} e^{-2t} + c_2 \begin{pmatrix} 2 \\ 2 \\ 1 \end{pmatrix} e^{t} + c_3 \begin{pmatrix} 1 \\ 1 \\ 1 \end{pmatrix} e^{5t}.$$

Solution by diagonalization will always work provided we can find n linearly independent eigenvectors of the $n \times n$ matrix $\mathbf{A}$; the eigenvalues of $\mathbf{A}$ could be real and distinct, complex, or repeated. The method fails when $\mathbf{A}$ has repeated eigenvalues and n linearly independent eigenvectors cannot be found. Of course, in this last situation $\mathbf{A}$ is not diagonalizable.

Since we have to find eigenvalues and eigenvectors of $\mathbf{A}$, this method is essentially equivalent to the procedure presented in the last section.

In the next section we shall see that diagonalization can also be used to solve nonhomogeneous linear systems $\mathbf{X}' = \mathbf{AX} + \mathbf{F}(t)$.

10.3 Exercises Answers to selected odd-numbered problems begin on page ANS-28.

In Problems 1–10, use diagonalization to solve the given system.

1. $\mathbf{X}' = \begin{pmatrix} 5 & 6 \\ 3 & -2 \end{pmatrix}\mathbf{X}$

2. $\mathbf{X}' = \begin{pmatrix} \frac{1}{2} & \frac{1}{2} \\ \frac{1}{2} & \frac{1}{2} \end{pmatrix}\mathbf{X}$

3. $\mathbf{X}' = \begin{pmatrix} 1 & \frac{1}{4} \\ 1 & 1 \end{pmatrix}\mathbf{X}$

4. $\mathbf{X}' = \begin{pmatrix} 1 & 1 \\ 1 & -1 \end{pmatrix}\mathbf{X}$

5. $\mathbf{X}' = \begin{pmatrix} -1 & 3 & 0 \\ 3 & -1 & 0 \\ -2 & -2 & 6 \end{pmatrix}\mathbf{X}$

6. $\mathbf{X}' = \begin{pmatrix} 1 & 1 & 2 \\ 1 & 2 & 1 \\ 2 & 1 & 1 \end{pmatrix}\mathbf{X}$

7. $\mathbf{X}' = \begin{pmatrix} 1 & -1 & -1 \\ -1 & 1 & -1 \\ -1 & -1 & 1 \end{pmatrix}\mathbf{X}$

8. $\mathbf{X}' = \begin{pmatrix} 1 & 1 & 1 & 1 \\ 1 & 1 & 1 & 1 \\ 1 & 1 & 1 & 1 \\ 1 & 1 & 1 & 1 \end{pmatrix}\mathbf{X}$

9. $\mathbf{X}' = \begin{pmatrix} -3 & 2 & 2 \\ -6 & 5 & 2 \\ -7 & 4 & 4 \end{pmatrix}\mathbf{X}$

10. $\mathbf{X}' = \begin{pmatrix} 0 & 2 & 0 \\ 2 & 0 & 2 \\ 0 & 2 & 0 \end{pmatrix}\mathbf{X}$

11. We have already shown how to solve the system of linear second-order differential equations that describes the motion of the coupled spring/mass system in Figure 3.12.1,

$$\begin{aligned} m_1 x_1'' &= -k_1 x_1 + k_2(x_2 - x_1) \\ m_2 x_2'' &= -k_2(x_2 - x_1), \end{aligned} \qquad (7)$$

in three different ways (see Example 4 in Section 3.12, Problem 56 in Exercises 10.2, and Example 1 in Section 4.6). In this problem you are led through the steps of how (7) can also be solved using diagonalization.

(a) Express (7) in the form $\mathbf{MX}'' + \mathbf{KX} = \mathbf{0}$, where $\mathbf{X} = \begin{pmatrix} x_1 \\ x_2 \end{pmatrix}$. Identify the 2×2 matrices $\mathbf{M}$ and $\mathbf{K}$. Explain why the matrix $\mathbf{M}$ has an inverse.

(b) Express the system in part (a) as

$$\mathbf{X}'' + \mathbf{BX} = \mathbf{0}. \qquad (8)$$

Identify the matrix $\mathbf{B}$.

(c) Solve system (7) in the special case where $m_1 = 1$, $m_2 = 1$, $k_1 = 3$, and $k_2 = 2$ by solving (8) using the diagonalization method. In other words, let $\mathbf{X} = \mathbf{PY}$, where $\mathbf{P}$ is a matrix whose columns are the eigenvectors of $\mathbf{B}$.

(d) Show that your solution $\mathbf{X}$ in part (c) is the same as that given in part (d) of Problem 56 in Exercises 10.2.

10.4 Nonhomogeneous Linear Systems

INTRODUCTION The methods of **undetermined coefficients** and **variation of parameters** used in Chapter 3 to find particular solutions of nonhomogeneous linear ODEs can both be adapted to the solution of nonhomogeneous linear systems. Of the two methods, variation of parameters is the more powerful technique. However, there are instances when the method of undetermined coefficients provides a quick means of finding a particular solution.

In Section 10.1, we saw that the general solution of a nonhomogeneous linear system $\mathbf{X}' = \mathbf{AX} + \mathbf{F}(t)$ on an interval I is $\mathbf{X} = \mathbf{X}_c + \mathbf{X}_p$ where $\mathbf{X}_c = c_1\mathbf{X}_1 + c_2\mathbf{X}_2 + \cdots + c_n\mathbf{X}_n$ is the complementary function or general solution of the associated homogeneous linear system $\mathbf{X}' = \mathbf{AX}$ and $\mathbf{X}_p$ is any particular solution of the nonhomogeneous system. We just saw how to obtain $\mathbf{X}_c$ in Section 10.2 when $\mathbf{A}$ was an $n \times n$ matrix of constants; we now consider three methods for obtaining $\mathbf{X}_p$.

10.4.1 Undetermined Coefficients

■ **The Assumptions** As in Section 3.4, the **method of undetermined coefficients** consists of making an educated guess about the form of a particular solution vector $\mathbf{X}_p$; the guess is motivated by the types of functions that comprise the entries of the column matrix $\mathbf{F}(t)$. Not surprisingly, the matrix version of undetermined coefficients is only applicable to $\mathbf{X}' = \mathbf{AX} + \mathbf{F}(t)$ when the entries of $\mathbf{A}$ are constants and the entries of $\mathbf{F}(t)$ are constants, polynomials, exponential functions, sines and cosines, or finite sums and products of these functions.

EXAMPLE 1 Undetermined Coefficients

Solve the system $\mathbf{X}' = \begin{pmatrix} -1 & 2 \\ -1 & 1 \end{pmatrix}\mathbf{X} + \begin{pmatrix} -8 \\ 3 \end{pmatrix}$ on the interval $(-\infty, \infty)$.

SOLUTION We first solve the associated homogeneous system

$$\mathbf{X}' = \begin{pmatrix} -1 & 2 \\ -1 & 1 \end{pmatrix}\mathbf{X}.$$

The characteristic equation of the coefficient matrix $\mathbf{A}$,

$$\det(\mathbf{A} - \lambda\mathbf{I}) = \begin{vmatrix} -1 - \lambda & 2 \\ -1 & 1 - \lambda \end{vmatrix} = \lambda^2 + 1 = 0,$$

yields the complex eigenvalues $\lambda_1 = i$ and $\lambda_2 = \overline{\lambda}_1 = -i$. By the procedures of Section 10.2, we find

$$\mathbf{X}_c = c_1\begin{pmatrix} \cos t + \sin t \\ \cos t \end{pmatrix} + c_2\begin{pmatrix} \cos t - \sin t \\ -\sin t \end{pmatrix}.$$

Now since $\mathbf{F}(t)$ is a constant vector, we assume a constant particular solution vector $\mathbf{X}_p = \begin{pmatrix} a_1 \\ b_1 \end{pmatrix}$. Substituting this latter assumption into the original system and equating entries leads to

$$0 = -a_1 + 2b_1 - 8$$
$$0 = -a_1 + b_1 + 3.$$

Solving this algebraic system gives $a_1 = 14$ and $b_1 = 11$, and so a particular solution is $\mathbf{X}_p = \begin{pmatrix} 14 \\ 11 \end{pmatrix}$. The general solution of the original system of differential equations on the interval $(-\infty, \infty)$ is then $\mathbf{X} = \mathbf{X}_c + \mathbf{X}_p$ or

$$\mathbf{X} = c_1\begin{pmatrix} \cos t + \sin t \\ \cos t \end{pmatrix} + c_2\begin{pmatrix} \cos t - \sin t \\ -\sin t \end{pmatrix} + \begin{pmatrix} 14 \\ 11 \end{pmatrix}.$$

≡

EXAMPLE 2 **Undetermined Coefficients**

Solve the system $\mathbf{X}' = \begin{pmatrix} 6 & 1 \\ 4 & 3 \end{pmatrix} \mathbf{X} + \begin{pmatrix} 6t \\ -10t + 4 \end{pmatrix}$ on the interval $(-\infty, \infty)$.

SOLUTION The eigenvalues and corresponding eigenvectors of the associated homogeneous system $\mathbf{X}' = \begin{pmatrix} 6 & 1 \\ 4 & 3 \end{pmatrix} \mathbf{X}$ are found to be $\lambda_1 = 2$, $\lambda_2 = 7$, $\mathbf{K}_1 = \begin{pmatrix} 1 \\ -4 \end{pmatrix}$, and $\mathbf{K}_2 = \begin{pmatrix} 1 \\ 1 \end{pmatrix}$. Hence the complementary function is

$$\mathbf{X}_c = c_1 \begin{pmatrix} 1 \\ -4 \end{pmatrix} e^{2t} + c_2 \begin{pmatrix} 1 \\ 1 \end{pmatrix} e^{7t}.$$

Now because $\mathbf{F}(t)$ can be written $\mathbf{F}(t) = \begin{pmatrix} 6 \\ -10 \end{pmatrix} t + \begin{pmatrix} 0 \\ 4 \end{pmatrix}$ we shall try to find a particular solution of the system that possesses the *same* form:

$$\mathbf{X}_p = \begin{pmatrix} a_2 \\ b_2 \end{pmatrix} t + \begin{pmatrix} a_1 \\ b_1 \end{pmatrix}.$$

Substituting this last assumption into the given system yields

$$\begin{pmatrix} a_2 \\ b_2 \end{pmatrix} = \begin{pmatrix} 6 & 1 \\ 4 & 3 \end{pmatrix} \left[\begin{pmatrix} a_2 \\ b_2 \end{pmatrix} t + \begin{pmatrix} a_1 \\ b_1 \end{pmatrix} \right] + \begin{pmatrix} 6 \\ -10 \end{pmatrix} t + \begin{pmatrix} 0 \\ 4 \end{pmatrix}$$

or $\begin{pmatrix} 0 \\ 0 \end{pmatrix} = \begin{pmatrix} (6a_2 + b_2 + 6)t + 6a_1 + b_1 - a_2 \\ (4a_2 + 3b_2 - 10)t + 4a_1 + 3b_1 - b_2 + 4 \end{pmatrix}.$

From the last identity we obtain four algebraic equations in four unknowns:

$$\begin{array}{rr} 6a_2 + b_2 + 6 = 0 \\ 4a_2 + 3b_2 - 10 = 0 \end{array} \quad \text{and} \quad \begin{array}{rr} 6a_1 + b_1 - a_2 = 0 \\ 4a_1 + 3b_1 - b_2 + 4 = 0. \end{array}$$

Solving the first two equations simultaneously yields $a_2 = -2$ and $b_2 = 6$. We then substitute these values into the last two equations and solve for a_1 and b_1. The results are $a_1 = -\frac{4}{7}$, $b_1 = \frac{10}{7}$. It follows, therefore, that a particular solution vector is

$$\mathbf{X}_p = \begin{pmatrix} -2 \\ 6 \end{pmatrix} t + \begin{pmatrix} -\frac{4}{7} \\ \frac{10}{7} \end{pmatrix}.$$

The general solution of the system on the interval $(-\infty, \infty)$ is $\mathbf{X} = \mathbf{X}_c + \mathbf{X}_p$ or

$$\mathbf{X} = c_1 \begin{pmatrix} 1 \\ -4 \end{pmatrix} e^{2t} + c_2 \begin{pmatrix} 1 \\ 1 \end{pmatrix} e^{7t} + \begin{pmatrix} -2 \\ 6 \end{pmatrix} t + \begin{pmatrix} -\frac{4}{7} \\ \frac{10}{7} \end{pmatrix}. \qquad \equiv$$

EXAMPLE 3 **Form of $\mathbf{X}_p$**

Determine the form of a particular solution vector $\mathbf{X}_p$ for the system

$$\frac{dx}{dt} = 5x + 3y - 2e^{-t} + 1$$

$$\frac{dy}{dt} = -x + y + e^{-t} - 5t + 7.$$

SOLUTION Because $\mathbf{F}(t)$ can be written in matrix terms as

$$\mathbf{F}(t) = \begin{pmatrix} -2 \\ 1 \end{pmatrix} e^{-t} + \begin{pmatrix} 0 \\ -5 \end{pmatrix} t + \begin{pmatrix} 1 \\ 7 \end{pmatrix}$$

a natural assumption for a particular solution would be

$$\mathbf{X}_p = \begin{pmatrix} a_3 \\ b_3 \end{pmatrix} e^{-t} + \begin{pmatrix} a_2 \\ b_2 \end{pmatrix} t + \begin{pmatrix} a_1 \\ b_1 \end{pmatrix}.$$

REMARKS

The method of undetermined coefficients for linear systems is not as straightforward as the last three examples would seem to indicate. In Section 3.4, the form of a particular solution y_p was predicated on prior knowledge of the complementary function y_c. The same is true for the formation of $\mathbf{X}_p$. But there are further difficulties; the special rules governing the form of y_p in Section 3.4 do not *quite* carry to the formation of $\mathbf{X}_p$. For example, if $\mathbf{F}(t)$ is a constant vector as in Example 1 and $\lambda = 0$ is an eigenvalue of multiplicity one, then $\mathbf{X}_c$ contains a constant vector. Under the "multiplication rule" on page 137, we would ordinarily try a particular solution of the form $\mathbf{X}_p = \begin{pmatrix} a_1 \\ b_1 \end{pmatrix} t$. This is not the proper assumption for linear systems; it should be $\mathbf{X}_p = \begin{pmatrix} a_2 \\ b_2 \end{pmatrix} t + \begin{pmatrix} a_1 \\ b_1 \end{pmatrix}$. Similarly, in Example 3, if we replace e^{-t} in $\mathbf{F}(t)$ by e^{2t} ($\lambda = 2$ is an eigenvalue), then the correct form of the particular solution vector is

$$\mathbf{X}_p = \begin{pmatrix} a_4 \\ b_4 \end{pmatrix} t e^{2t} + \begin{pmatrix} a_3 \\ b_3 \end{pmatrix} e^{2t} + \begin{pmatrix} a_2 \\ b_2 \end{pmatrix} t + \begin{pmatrix} a_1 \\ b_1 \end{pmatrix}.$$

Rather than delve into these difficulties, we turn instead to the method of variation of parameters.

10.4.2 Variation of Parameters

■ **A Fundamental Matrix** If $\mathbf{X}_1, \mathbf{X}_2, \ldots, \mathbf{X}_n$ is a fundamental set of solutions of the homogeneous system $\mathbf{X}' = \mathbf{A}\mathbf{X}$ on an interval I, then its general solution on the interval is the linear combination $\mathbf{X} = c_1\mathbf{X}_1 + c_2\mathbf{X}_2 + \cdots + c_n\mathbf{X}_n$, or

$$\mathbf{X} = c_1 \begin{pmatrix} x_{11} \\ x_{21} \\ \vdots \\ x_{n1} \end{pmatrix} + c_2 \begin{pmatrix} x_{12} \\ x_{22} \\ \vdots \\ x_{n2} \end{pmatrix} + \cdots + c_n \begin{pmatrix} x_{1n} \\ x_{2n} \\ \vdots \\ x_{nn} \end{pmatrix} = \begin{pmatrix} c_1 x_{11} + c_2 x_{12} + \cdots + c_n x_{1n} \\ c_1 x_{21} + c_2 x_{22} + \cdots + c_n x_{2n} \\ \vdots \\ c_1 x_{n1} + c_2 x_{n2} + \cdots + c_n x_{nn} \end{pmatrix}. \tag{1}$$

The last matrix in (1) is recognized as the product of an $n \times n$ matrix with an $n \times 1$ matrix. In other words, the general solution (1) can be written as the product

$$\mathbf{X} = \mathbf{\Phi}(t)\mathbf{C}, \tag{2}$$

where $\mathbf{C}$ is the $n \times 1$ column vector of arbitrary constants $c_1, c_2, \ldots, c_n$, and $\mathbf{\Phi}(t)$ is the $n \times n$ matrix whose columns consist of the entries of the solution vectors of the system $\mathbf{X}' = \mathbf{A}\mathbf{X}$:

$$\mathbf{\Phi}(t) = \begin{pmatrix} x_{11} & x_{12} & \cdots & x_{1n} \\ x_{21} & x_{22} & \cdots & x_{2n} \\ \vdots & & & \vdots \\ x_{n1} & x_{n2} & \cdots & x_{nn} \end{pmatrix}.$$

The matrix $\mathbf{\Phi}(t)$ is called a **fundamental matrix** of the system on the interval.
 In the discussion that follows, we need to use two properties of a fundamental matrix:

- A fundamental matrix $\mathbf{\Phi}(t)$ is nonsingular.
- If $\mathbf{\Phi}(t)$ is a fundamental matrix of the system $\mathbf{X}' = \mathbf{A}\mathbf{X}$, then

$$\mathbf{\Phi}'(t) = \mathbf{A}\mathbf{\Phi}(t). \tag{3}$$

A reexamination of (12) of Theorem 10.1.3 shows that det $\boldsymbol{\Phi}(t)$ is the same as the Wronskian $W(\mathbf{X}_1, \mathbf{X}_2, \ldots, \mathbf{X}_n)$. Hence the linear independence of the columns of $\boldsymbol{\Phi}(t)$ on the interval I guarantees that det $\boldsymbol{\Phi}(t) \neq 0$ for every t in the interval. Since $\boldsymbol{\Phi}(t)$ is nonsingular, the multiplicative inverse $\boldsymbol{\Phi}^{-1}(t)$ exists for every t in the interval. The result given in (3) follows immediately from the fact that every column of $\boldsymbol{\Phi}(t)$ is a solution vector of $\mathbf{X}' = \mathbf{AX}$.

■ **Variation of Parameters** Analogous to the procedure in Section 3.5, we ask whether it is possible to replace the matrix of constants $\mathbf{C}$ in (2) by a column matrix of functions

$$\mathbf{U}(t) = \begin{pmatrix} u_1(t) \\ u_2(t) \\ \vdots \\ u_n(t) \end{pmatrix} \qquad \text{so that} \qquad \mathbf{X}_p = \boldsymbol{\Phi}(t)\mathbf{U}(t) \tag{4}$$

is a particular solution of the nonhomogeneous system

$$\mathbf{X}' = \mathbf{AX} + \mathbf{F}(t). \tag{5}$$

By the Product Rule, the derivative of the last expression in (4) is

$$\mathbf{X}'_p = \boldsymbol{\Phi}(t)\mathbf{U}'(t) + \boldsymbol{\Phi}'(t)\mathbf{U}(t). \tag{6}$$

Note that the order of the products in (6) is very important. Since $\mathbf{U}(t)$ is a column matrix, the products $\mathbf{U}'(t)\boldsymbol{\Phi}(t)$ and $\mathbf{U}(t)\boldsymbol{\Phi}'(t)$ are not defined. Substituting (4) and (6) into (5) gives

$$\boldsymbol{\Phi}(t)\mathbf{U}'(t) + \boldsymbol{\Phi}'(t)\mathbf{U}(t) = \mathbf{A}\boldsymbol{\Phi}(t)\mathbf{U}(t) + \mathbf{F}(t). \tag{7}$$

Now if we use (3) to replace $\boldsymbol{\Phi}'(t)$, (7) becomes

$$\boldsymbol{\Phi}(t)\mathbf{U}'(t) + \mathbf{A}\boldsymbol{\Phi}(t)\mathbf{U}(t) = \mathbf{A}\boldsymbol{\Phi}(t)\mathbf{U}(t) + \mathbf{F}(t)$$

or

$$\boldsymbol{\Phi}(t)\mathbf{U}'(t) = \mathbf{F}(t). \tag{8}$$

Multiplying both sides of equation (8) by $\boldsymbol{\Phi}^{-1}(t)$ gives

$$\mathbf{U}'(t) = \boldsymbol{\Phi}^{-1}(t)\mathbf{F}(t) \quad \text{and so} \quad \mathbf{U}(t) = \int \boldsymbol{\Phi}^{-1}(t)\mathbf{F}(t)\, dt.$$

Since $\mathbf{X}_p = \boldsymbol{\Phi}(t)\mathbf{U}(t)$, we conclude that a particular solution of (5) is

$$\mathbf{X}_p = \boldsymbol{\Phi}(t) \int \boldsymbol{\Phi}^{-1}(t)\mathbf{F}(t)\, dt. \tag{9}$$

To calculate the indefinite integral of the column matrix $\boldsymbol{\Phi}^{-1}(t)\mathbf{F}(t)$ in (9), we integrate each entry. Thus the general solution of the nonhomogeneous system (5) is $\mathbf{X} = \mathbf{X}_c + \mathbf{X}_p$ or

$$\mathbf{X} = \boldsymbol{\Phi}(t)\mathbf{C} + \boldsymbol{\Phi}(t) \int \boldsymbol{\Phi}^{-1}(t)\mathbf{F}(t)\, dt. \tag{10}$$

EXAMPLE 4 **Variation of Parameters**

Find the general solution of the nonhomogeneous system

$$\mathbf{X}' = \begin{pmatrix} -3 & 1 \\ 2 & -4 \end{pmatrix} \mathbf{X} + \begin{pmatrix} 3t \\ e^{-t} \end{pmatrix} \tag{11}$$

on the interval $(-\infty, \infty)$.

SOLUTION We first solve the associated homogeneous system

$$\mathbf{X}' = \begin{pmatrix} -3 & 1 \\ 2 & -4 \end{pmatrix} \mathbf{X}. \tag{12}$$

The characteristic equation of the coefficient matrix is

$$\det(\mathbf{A} - \lambda\mathbf{I}) = \begin{vmatrix} -3 - \lambda & 1 \\ 2 & -4 - \lambda \end{vmatrix} = (\lambda + 2)(\lambda + 5) = 0,$$

so the eigenvalues are $\lambda_1 = -2$ and $\lambda_2 = -5$. By the usual method we find that the eigenvectors corresponding to λ_1 and λ_2 are, respectively,

$$\begin{pmatrix} 1 \\ 1 \end{pmatrix} \quad \text{and} \quad \begin{pmatrix} 1 \\ -2 \end{pmatrix}.$$

The solution vectors of the homogeneous system (12) are then

$$\mathbf{X}_1 = \begin{pmatrix} 1 \\ 1 \end{pmatrix} e^{-2t} = \begin{pmatrix} e^{-2t} \\ e^{-2t} \end{pmatrix} \quad \text{and} \quad \mathbf{X}_2 = \begin{pmatrix} 1 \\ -2 \end{pmatrix} e^{-5t} = \begin{pmatrix} e^{-5t} \\ -2e^{-5t} \end{pmatrix}.$$

The entries in $\mathbf{X}_1$ form the first column of $\mathbf{\Phi}(t)$, and the entries in $\mathbf{X}_2$ form the second column of $\mathbf{\Phi}(t)$. Hence

$$\mathbf{\Phi}(t) = \begin{pmatrix} e^{-2t} & e^{-5t} \\ e^{-2t} & -2e^{-5t} \end{pmatrix} \quad \text{and} \quad \mathbf{\Phi}^{-1}(t) = \begin{pmatrix} \frac{2}{3}e^{2t} & \frac{1}{3}e^{2t} \\ \frac{1}{3}e^{5t} & -\frac{1}{3}e^{5t} \end{pmatrix}.$$

From (9) we obtain

$$\mathbf{X}_p = \mathbf{\Phi}(t) \int \mathbf{\Phi}^{-1}(t)\,\mathbf{F}(t)\,dt = \begin{pmatrix} e^{-2t} & e^{-5t} \\ e^{-2t} & -2e^{-5t} \end{pmatrix} \int \begin{pmatrix} \frac{2}{3}e^{2t} & \frac{1}{3}e^{2t} \\ \frac{1}{3}e^{5t} & -\frac{1}{3}e^{5t} \end{pmatrix} \begin{pmatrix} 3t \\ e^{-t} \end{pmatrix} dt$$

$$= \begin{pmatrix} e^{-2t} & e^{-5t} \\ e^{-2t} & -2e^{-5t} \end{pmatrix} \int \begin{pmatrix} 2te^{2t} + \frac{1}{3}e^{t} \\ te^{5t} - \frac{1}{3}e^{4t} \end{pmatrix} dt$$

$$= \begin{pmatrix} e^{-2t} & e^{-5t} \\ e^{-2t} & -2e^{-5t} \end{pmatrix} \begin{pmatrix} te^{2t} - \frac{1}{2}e^{2t} + \frac{1}{3}e^{t} \\ \frac{1}{5}te^{5t} - \frac{1}{25}e^{5t} - \frac{1}{12}e^{4t} \end{pmatrix}$$

$$= \begin{pmatrix} \frac{6}{5}t - \frac{27}{50} + \frac{1}{4}e^{-t} \\ \frac{3}{5}t - \frac{21}{50} + \frac{1}{2}e^{-t} \end{pmatrix}.$$

Hence from (10), the general solution of (11) on the interval $(-\infty, \infty)$ is

$$\mathbf{X} = \begin{pmatrix} e^{-2t} & e^{-5t} \\ e^{-2t} & -2e^{-5t} \end{pmatrix} \begin{pmatrix} c_1 \\ c_2 \end{pmatrix} + \begin{pmatrix} \frac{6}{5}t - \frac{27}{50} + \frac{1}{4}e^{-t} \\ \frac{3}{5}t - \frac{21}{50} + \frac{1}{2}e^{-t} \end{pmatrix}$$

$$= c_1 \begin{pmatrix} 1 \\ 1 \end{pmatrix} e^{-2t} + c_2 \begin{pmatrix} 1 \\ -2 \end{pmatrix} e^{-5t} + \begin{pmatrix} \frac{6}{5} \\ \frac{3}{5} \end{pmatrix} t - \begin{pmatrix} \frac{27}{50} \\ \frac{21}{50} \end{pmatrix} + \begin{pmatrix} \frac{1}{4} \\ \frac{1}{2} \end{pmatrix} e^{-t}. \qquad \equiv$$

Initial-Value Problem The general solution of the nonhomogeneous system (5) on an interval can be written in an alternative manner

$$\mathbf{X} = \mathbf{\Phi}(t)\mathbf{C} + \mathbf{\Phi}(t) \int_{t_0}^{t} \mathbf{\Phi}^{-1}(s)\mathbf{F}(s)\,ds, \qquad (13)$$

where t and t_0 are points in the interval I. The last form is useful in solving (5) subject to an initial condition $\mathbf{X}(t_0) = \mathbf{X}_0$, because the limits of integration are chosen so that the particular solution vanishes at $t = t_0$. Substituting $t = t_0$ in (13) yields $\mathbf{X}_0 = \mathbf{\Phi}(t_0)\mathbf{C}$, from which we get $\mathbf{C} = \mathbf{\Phi}^{-1}(t_0)\mathbf{X}_0$. Substituting this last result in (13) gives the following solution of the initial-value problem:

$$\mathbf{X} = \mathbf{\Phi}(t)\,\mathbf{\Phi}^{-1}(t_0)\,\mathbf{X}_0 + \mathbf{\Phi}(t) \int_{t_0}^{t} \mathbf{\Phi}^{-1}(s)\mathbf{F}(s)\,ds. \qquad (14)$$

10.4.3 Diagonalization

The Assumptions As in Section 10.3, if the coefficient matrix $\mathbf{A}$ possesses n linearly independent eigenvectors, then we can use diagonalization to uncouple the system $\mathbf{X}' = \mathbf{AX} + \mathbf{F}(t)$. Suppose $\mathbf{P}$ is the matrix such that $\mathbf{P}^{-1}\mathbf{AP} = \mathbf{D}$, where $\mathbf{D}$ is a diagonal matrix. Substituting $\mathbf{X} = \mathbf{PY}$ into the nonhomogeneous system $\mathbf{X}' = \mathbf{AX} + \mathbf{F}(t)$ gives

$$\mathbf{PY}' = \mathbf{APY} + \mathbf{F} \quad \text{or} \quad \mathbf{Y}' = \mathbf{P}^{-1}\mathbf{APY} + \mathbf{P}^{-1}\mathbf{F} \quad \text{or} \quad \mathbf{Y}' = \mathbf{DY} + \mathbf{G}. \quad (15)$$

In the last equation in (15), $\mathbf{G} = \mathbf{P}^{-1}\mathbf{F}$ is a column vector. So each differential equation in this new system has the form $y_i' = \lambda_i y_i + g_i(t)$, $i = 1, 2, \ldots, n$. But notice that, unlike the procedure for solving a homogeneous system $\mathbf{X}' = \mathbf{AX}$, we now are required to compute the inverse of the matrix $\mathbf{P}$.

EXAMPLE 5 **Diagonalization**

Solve $\mathbf{X}' = \begin{pmatrix} 4 & 2 \\ 2 & 1 \end{pmatrix}\mathbf{X} + \begin{pmatrix} 3e^t \\ e^t \end{pmatrix}$ by diagonalization.

SOLUTION The eigenvalues and corresponding eigenvectors of the coefficient matrix are found to be $\lambda_1 = 0$, $\lambda_2 = 5$, $\mathbf{K}_1 = \begin{pmatrix} 1 \\ -2 \end{pmatrix}$, $\mathbf{K}_2 = \begin{pmatrix} 2 \\ 1 \end{pmatrix}$. Thus we find $\mathbf{P} = \begin{pmatrix} 1 & 2 \\ -2 & 1 \end{pmatrix}$ and $\mathbf{P}^{-1} = \begin{pmatrix} \frac{1}{5} & -\frac{2}{5} \\ \frac{2}{5} & \frac{1}{5} \end{pmatrix}$. Using the substitution $\mathbf{X} = \mathbf{PY}$ and

$$\mathbf{P}^{-1}\mathbf{F} = \begin{pmatrix} \frac{1}{5} & -\frac{2}{5} \\ \frac{2}{5} & \frac{1}{5} \end{pmatrix}\begin{pmatrix} 3e^t \\ e^t \end{pmatrix} = \begin{pmatrix} \frac{1}{5}e^t \\ \frac{7}{5}e^t \end{pmatrix}$$

the uncoupled system is

$$\mathbf{Y}' = \begin{pmatrix} 0 & 0 \\ 0 & 5 \end{pmatrix}\mathbf{Y} + \begin{pmatrix} \frac{1}{5}e^t \\ \frac{7}{5}e^t \end{pmatrix}.$$

The solutions of the two linear first-order differential equations

$$y_1' = \tfrac{1}{5}e^t \quad \text{and} \quad y_2' = 5y_2 + \tfrac{7}{5}e^t$$

are, respectively, $y_1 = \frac{1}{5}e^t + c_1$ and $y_2 = -\frac{7}{20}e^t + c_2 e^{5t}$. Hence, the solution of the original system is

$$\mathbf{X} = \mathbf{PY} = \begin{pmatrix} 1 & 2 \\ -2 & 1 \end{pmatrix}\begin{pmatrix} \frac{1}{5}e^t + c_1 \\ -\frac{7}{20}e^t + c_2 e^{5t} \end{pmatrix} = \begin{pmatrix} -\frac{1}{2}e^t + c_1 + 2c_2 e^{5t} \\ -\frac{3}{4}e^t - 2c_1 + c_2 e^{5t} \end{pmatrix}. \quad (16)$$

Written in the usual manner using column vectors, (16) is

$$\mathbf{X} = c_1\begin{pmatrix} 1 \\ -2 \end{pmatrix} + c_2\begin{pmatrix} 2 \\ 1 \end{pmatrix}e^{5t} - \begin{pmatrix} \frac{1}{2} \\ \frac{3}{4} \end{pmatrix}e^t. \qquad \equiv$$

10.4 Exercises Answers to selected odd-numbered problems begin on page ANS-28.

10.4.1 Undetermined Coefficients

In Problems 1–8, use the method of undetermined coefficients to solve the given system.

1. $\dfrac{dx}{dt} = 2x + 3y - 7$

$\dfrac{dy}{dt} = -x - 2y + 5$

2. $\dfrac{dx}{dt} = 5x + 9y + 2$

$\dfrac{dy}{dt} = -x + 11y + 6$

3. $\mathbf{X}' = \begin{pmatrix} 1 & 3 \\ 3 & 1 \end{pmatrix}\mathbf{X} + \begin{pmatrix} -2t^2 \\ t + 5 \end{pmatrix}$

4. $\mathbf{X}' = \begin{pmatrix} 1 & -4 \\ 4 & 1 \end{pmatrix}\mathbf{X} + \begin{pmatrix} 4t + 9e^{6t} \\ -t + e^{6t} \end{pmatrix}$

5. $\mathbf{X}' = \begin{pmatrix} 4 & \frac{1}{3} \\ 9 & 6 \end{pmatrix}\mathbf{X} + \begin{pmatrix} -3 \\ 10 \end{pmatrix}e^t$

6. $\mathbf{X}' = \begin{pmatrix} -1 & 5 \\ -1 & 1 \end{pmatrix}\mathbf{X} + \begin{pmatrix} \sin t \\ -2\cos t \end{pmatrix}$

7. $\mathbf{X}' = \begin{pmatrix} 1 & 1 & 1 \\ 0 & 2 & 3 \\ 0 & 0 & 5 \end{pmatrix}\mathbf{X} + \begin{pmatrix} 1 \\ -1 \\ 2 \end{pmatrix}e^{4t}$

8. $\mathbf{X}' = \begin{pmatrix} 0 & 0 & 5 \\ 0 & 5 & 0 \\ 5 & 0 & 0 \end{pmatrix} \mathbf{X} + \begin{pmatrix} 5 \\ -10 \\ 40 \end{pmatrix}$

In Problems 9 and 10, solve the given initial-value problem.

9. $\mathbf{X}' = \begin{pmatrix} -1 & -2 \\ 3 & 4 \end{pmatrix} \mathbf{X} + \begin{pmatrix} 3 \\ 3 \end{pmatrix}, \quad \mathbf{X}(0) = \begin{pmatrix} -4 \\ 5 \end{pmatrix}$

10. $\mathbf{X}' = \begin{pmatrix} 1 & -1 \\ 1 & 3 \end{pmatrix} \mathbf{X} + \begin{pmatrix} t \\ t+1 \end{pmatrix}, \quad \mathbf{X}(0) = \begin{pmatrix} 0 \\ 2 \end{pmatrix}$

11. Consider the large mixing tanks shown in **FIGURE 10.4.1**. Suppose that both tanks A and B initially contain 100 gallons of brine. Liquid is pumped in and out of the tanks as indicated in the figure; the mixture pumped between and out of the tanks is assumed to be well-stirred.
 (a) Construct a mathematical model in the form of a linear system of first-order differential equations for the number of pounds $x_1(t)$ and $x_2(t)$ of salt in tanks A and B, respectively, at time t. Write the system in matrix form. [*Hint*: See Section 2.9 and Problem 52, Chapter 2 in Review.]
 (b) Use the method of undetermined coefficients to solve the linear system in part (a) subject to $x_1(0) = 60, x_2(0) = 10$.
 (c) What are $\lim_{t \to \infty} x_1(t)$ and $\lim_{t \to \infty} x_2(t)$? Interpret this result.
 (d) Use a graphing utility or CAS to plot the graphs of $x_1(t)$ and $x_2(t)$ in the same coordinate plane.

pure water mixture ½ lb/gal
2 gal/min 1 gal/min 2 gal/min

mixture mixture mixture
1 gal/min 2 gal/min 3 gal/min

FIGURE 10.4.1 Mixing tanks in Problem 11

12. (a) The system of differential equations for the currents $i_2(t)$ and $i_3(t)$ in the electrical network shown in **FIGURE 10.4.2** is

$$\frac{d}{dt}\begin{pmatrix} i_2 \\ i_3 \end{pmatrix} = \begin{pmatrix} -R_1/L_1 & -R_1/L_1 \\ -R_1/L_2 & -(R_1 + R_2)/L_2 \end{pmatrix}\begin{pmatrix} i_2 \\ i_3 \end{pmatrix} + \begin{pmatrix} E/L_1 \\ E/L_2 \end{pmatrix}.$$

Use the method of undetermined coefficients to solve the system if $R_1 = 2\,\Omega, R_2 = 3\,\Omega, L_1 = 1$ h, $L_2 = 1$ h, $E = 60$ V, $i_2(0) = 0$, and $i_3(0) = 0$.
 (b) Determine the current $i_1(t)$.

FIGURE 10.4.2 Network in Problem 12

10.4.2 Variation of Parameters

In Problems 13–32, use variation of parameters to solve the given system.

13. $\dfrac{dx}{dt} = 3x - 3y + 4$

$\dfrac{dy}{dt} = 2x - 2y - 1$

14. $\dfrac{dx}{dt} = 2x - y$

$\dfrac{dy}{dt} = 3x - 2y + 4t$

15. $\mathbf{X}' = \begin{pmatrix} 3 & -5 \\ \frac{3}{4} & -1 \end{pmatrix} \mathbf{X} + \begin{pmatrix} 1 \\ -1 \end{pmatrix} e^{t/2}$

16. $\mathbf{X}' = \begin{pmatrix} 2 & -1 \\ 4 & 2 \end{pmatrix} \mathbf{X} + \begin{pmatrix} \sin 2t \\ 2\cos 2t \end{pmatrix} e^{2t}$

17. $\mathbf{X}' = \begin{pmatrix} 0 & 2 \\ -1 & 3 \end{pmatrix} \mathbf{X} + \begin{pmatrix} 1 \\ -1 \end{pmatrix} e^{t}$

18. $\mathbf{X}' = \begin{pmatrix} 0 & 2 \\ -1 & 3 \end{pmatrix} \mathbf{X} + \begin{pmatrix} 2 \\ e^{-3t} \end{pmatrix}$

19. $\mathbf{X}' = \begin{pmatrix} 1 & 8 \\ 1 & -1 \end{pmatrix} \mathbf{X} + \begin{pmatrix} 12 \\ 12 \end{pmatrix} t$

20. $\mathbf{X}' = \begin{pmatrix} 1 & 8 \\ 1 & -1 \end{pmatrix} \mathbf{X} + \begin{pmatrix} e^{-t} \\ te^{t} \end{pmatrix}$

21. $\mathbf{X}' = \begin{pmatrix} 3 & 2 \\ -2 & -1 \end{pmatrix} \mathbf{X} + \begin{pmatrix} 2e^{-t} \\ e^{-t} \end{pmatrix}$

22. $\mathbf{X}' = \begin{pmatrix} 3 & 2 \\ -2 & -1 \end{pmatrix} \mathbf{X} + \begin{pmatrix} 1 \\ 1 \end{pmatrix}$

23. $\mathbf{X}' = \begin{pmatrix} 0 & -1 \\ 1 & 0 \end{pmatrix} \mathbf{X} + \begin{pmatrix} \sec t \\ 0 \end{pmatrix}$

24. $\mathbf{X}' = \begin{pmatrix} 1 & -1 \\ 1 & 1 \end{pmatrix} \mathbf{X} + \begin{pmatrix} 3 \\ 3 \end{pmatrix} e^{t}$

25. $\mathbf{X}' = \begin{pmatrix} 1 & -1 \\ 1 & 1 \end{pmatrix} \mathbf{X} + \begin{pmatrix} \cos t \\ \sin t \end{pmatrix} e^{t}$

26. $\mathbf{X}' = \begin{pmatrix} 2 & -2 \\ 8 & -6 \end{pmatrix} \mathbf{X} + \begin{pmatrix} 1 \\ 3 \end{pmatrix} \dfrac{e^{-2t}}{t}$

27. $\mathbf{X}' = \begin{pmatrix} 0 & 1 \\ -1 & 0 \end{pmatrix} \mathbf{X} + \begin{pmatrix} 0 \\ \sec t \tan t \end{pmatrix}$

28. $\mathbf{X}' = \begin{pmatrix} 0 & 1 \\ -1 & 0 \end{pmatrix} \mathbf{X} + \begin{pmatrix} 1 \\ \cot t \end{pmatrix}$

29. $\mathbf{X}' = \begin{pmatrix} 1 & 2 \\ -\frac{1}{2} & 1 \end{pmatrix} \mathbf{X} + \begin{pmatrix} \csc t \\ \sec t \end{pmatrix} e^{t}$

30. $\mathbf{X}' = \begin{pmatrix} 1 & -2 \\ 1 & -1 \end{pmatrix} \mathbf{X} + \begin{pmatrix} \tan t \\ 1 \end{pmatrix}$

31. $\mathbf{X}' = \begin{pmatrix} 1 & 1 & 0 \\ 1 & 1 & 0 \\ 0 & 0 & 3 \end{pmatrix} \mathbf{X} + \begin{pmatrix} e^{t} \\ e^{2t} \\ te^{3t} \end{pmatrix}$

32. $\mathbf{X}' = \begin{pmatrix} 3 & -1 & -1 \\ 1 & 1 & -1 \\ 1 & -1 & 1 \end{pmatrix} \mathbf{X} + \begin{pmatrix} 0 \\ t \\ 2e^{t} \end{pmatrix}$

In Problems 33 and 34, use (14) to solve the given initial-value problem.

33. $\mathbf{X}' = \begin{pmatrix} 3 & -1 \\ -1 & 3 \end{pmatrix} \mathbf{X} + \begin{pmatrix} 4e^{2t} \\ 4e^{4t} \end{pmatrix}, \quad \mathbf{X}(0) = \begin{pmatrix} 1 \\ 1 \end{pmatrix}$

34. $\mathbf{X}' = \begin{pmatrix} 1 & -1 \\ 1 & -1 \end{pmatrix} \mathbf{X} + \begin{pmatrix} 1/t \\ 1/t \end{pmatrix}$, $\mathbf{X}(1) = \begin{pmatrix} 2 \\ -1 \end{pmatrix}$

35. The system of differential equations for the currents $i_1(t)$ and $i_2(t)$ in the electrical network shown in **FIGURE 10.4.3** is

$$\frac{d}{dt} \begin{pmatrix} i_1 \\ i_2 \end{pmatrix} = \begin{pmatrix} -(R_1 + R_2)/L_2 & R_2/L_2 \\ R_2/L_1 & -R_2/L_1 \end{pmatrix} \begin{pmatrix} i_1 \\ i_2 \end{pmatrix} + \begin{pmatrix} E/L_2 \\ 0 \end{pmatrix}.$$

Use variation of parameters to solve the system if $R_1 = 8\,\Omega$, $R_2 = 3\,\Omega$, $L_1 = 1$ h, $L_2 = 1$ h, $E(t) = 100 \sin t$ V, $i_1(0) = 0$, and $i_2(0) = 0$.

FIGURE 10.4.3 Network in Problem 35

Computer Lab Assignment

36. Solving a nonhomogeneous linear system $\mathbf{X}' = \mathbf{AX} + \mathbf{F}(t)$ by variation of parameters when $\mathbf{A}$ is a 3×3 (or larger) matrix is almost an impossible task to do by hand. Consider the system

$$\mathbf{X}' = \begin{pmatrix} 2 & -2 & 2 & 1 \\ -1 & 3 & 0 & 3 \\ 0 & 0 & 4 & -2 \\ 0 & 0 & 2 & -1 \end{pmatrix} \mathbf{X} + \begin{pmatrix} te^t \\ e^{-t} \\ e^{2t} \\ 1 \end{pmatrix}.$$

(a) Use a CAS or linear algebra software to find the eigenvalues and eigenvectors of the coefficient matrix.

(b) Form a fundamental matrix $\mathbf{\Phi}(t)$ and use the computer to find $\mathbf{\Phi}^{-1}(t)$.

(c) Use the computer to carry out the computations of
$$\mathbf{\Phi}^{-1}(t)\mathbf{F}(t), \int \mathbf{\Phi}^{-1}(t)\mathbf{F}(t)\,dt, \mathbf{\Phi}(t) \int \mathbf{\Phi}^{-1}(t)\mathbf{F}(t)\,dt, \mathbf{\Phi}(t)\mathbf{C},$$
and $\mathbf{\Phi}(t)\mathbf{C} + \int \mathbf{\Phi}^{-1}(t)\mathbf{F}(t)\,dt$, where $\mathbf{C}$ is a column matrix of constants c_1, c_2, c_3, and c_4.

(d) Rewrite the computer output for the general solution of the system in the form $\mathbf{X} = \mathbf{X}_c + \mathbf{X}_p$, where $\mathbf{X}_c = c_1\mathbf{X}_1 + c_2\mathbf{X}_2 + c_3\mathbf{X}_3 + c_4\mathbf{X}_4$.

10.4.3 Diagonalization

In Problems 37–40, use diagonalization to solve the given system.

37. $\mathbf{X}' = \begin{pmatrix} 5 & -2 \\ 21 & -8 \end{pmatrix} \mathbf{X} + \begin{pmatrix} 6 \\ 4 \end{pmatrix}$

38. $\mathbf{X}' = \begin{pmatrix} 1 & 3 \\ 2 & 2 \end{pmatrix} \mathbf{X} + \begin{pmatrix} e^t \\ e^t \end{pmatrix}$

39. $\mathbf{X}' = \begin{pmatrix} 5 & 5 \\ 5 & 5 \end{pmatrix} \mathbf{X} + \begin{pmatrix} 2t \\ 8 \end{pmatrix}$

40. $\mathbf{X}' = \begin{pmatrix} 0 & 1 \\ 1 & 0 \end{pmatrix} \mathbf{X} + \begin{pmatrix} 4 \\ 8e^{-2t} \end{pmatrix}$

10.5 Matrix Exponential

INTRODUCTION Matrices can be used in an entirely different manner to solve a system of linear first-order differential equations. Recall that the simple linear first-order differential equation $x' = ax$, where a is a constant, has the general solution $x = ce^{at}$. It seems natural, then, to ask whether we can define a matrix exponential $e^{\mathbf{A}t}$, where $\mathbf{A}$ is a matrix of constants, so that $e^{\mathbf{A}t}$ is a solution of the system $\mathbf{X}' = \mathbf{AX}$.

Homogeneous Systems We shall now see that it is possible to define a **matrix exponential** $e^{\mathbf{A}t}$ so that the homogeneous system $\mathbf{X}' = \mathbf{AX}$, where $\mathbf{A}$ is an $n \times n$ matrix of constants, has a solution

$$\mathbf{X} = e^{\mathbf{A}t}\mathbf{C}. \tag{1}$$

Since $\mathbf{C}$ is to be an $n \times 1$ column matrix of arbitrary constants, we want $e^{\mathbf{A}t}$ to be an $n \times n$ matrix. While the complete development of the meaning and theory of the matrix exponential would require a thorough knowledge of matrix algebra, one way of defining $e^{\mathbf{A}t}$ is inspired by the power series representation of the scalar exponential function e^{at}:

$$e^{at} = 1 + at + a^2 \frac{t^2}{2!} + \cdots + a^m \frac{t^m}{m!} + \cdots = \sum_{k=0}^{\infty} a^k \frac{t^k}{k!}. \tag{2}$$

The series in (2) converges for all t. Using this series, with 1 replaced by the identity matrix $\mathbf{I}$ and the constant a replaced with an $n \times n$ matrix $\mathbf{A}$ of constants, we arrive at a definition for the $n \times n$ matrix $e^{\mathbf{A}t}$.

For any $n \times n$ matrix $\mathbf{A}$,

$$e^{\mathbf{A}t} = \mathbf{I} + \mathbf{A}t + \mathbf{A}^2 \frac{t^2}{2!} + \cdots + \mathbf{A}^m \frac{t^m}{m!} + \cdots = \sum_{k=0}^{\infty} \mathbf{A}^k \frac{t^k}{k!}. \tag{3}$$

It can be shown that the series given in (3) converges to an $n \times n$ matrix for every value of t. Also, in (3), $\mathbf{A}^0 = \mathbf{I}, \mathbf{A}^2 = \mathbf{A}\mathbf{A}, \mathbf{A}^3 = \mathbf{A}(\mathbf{A}^2)$, and so on.

EXAMPLE 1 Matrix Exponential Using (3)

Compute $e^{\mathbf{A}t}$ for the matrix

$$\mathbf{A} = \begin{pmatrix} 2 & 0 \\ 0 & 3 \end{pmatrix}.$$

SOLUTION From the various powers

$$\mathbf{A}^2 = \begin{pmatrix} 2^2 & 0 \\ 0 & 3^2 \end{pmatrix}, \mathbf{A}^3 = \begin{pmatrix} 2^3 & 0 \\ 0 & 3^3 \end{pmatrix}, \mathbf{A}^4 = \begin{pmatrix} 2^4 & 0 \\ 0 & 3^4 \end{pmatrix}, \ldots, \mathbf{A}^n = \begin{pmatrix} 2^n & 0 \\ 0 & 3^n \end{pmatrix}, \ldots,$$

we see from (3) that

$$e^{\mathbf{A}t} = \mathbf{I} + \mathbf{A}t + \mathbf{A}^2 \frac{t^2}{2!} + \cdots$$

$$= \begin{pmatrix} 1 & 0 \\ 0 & 1 \end{pmatrix} + \begin{pmatrix} 2 & 0 \\ 0 & 3 \end{pmatrix}t + \begin{pmatrix} 2^2 & 0 \\ 0 & 3^2 \end{pmatrix}\frac{t^2}{2!} + \cdots + \begin{pmatrix} 2^n & 0 \\ 0 & 3^n \end{pmatrix}\frac{t^n}{n!} + \cdots$$

$$= \begin{pmatrix} 1 + 2t + 2^2\dfrac{t^2}{2!} + \cdots & 0 \\ 0 & 1 + 3t + 3^2\dfrac{t^2}{2!} + \cdots \end{pmatrix}.$$

In view of (2) and the identifications $a = 2$ and $a = 3$, the power series in the first and second rows of the last matrix represent, respectively, e^{2t} and e^{3t} and so we have

$$e^{\mathbf{A}t} = \begin{pmatrix} e^{2t} & 0 \\ 0 & e^{3t} \end{pmatrix}. \qquad \equiv$$

The matrix in Example 1 is an example of a 2×2 diagonal matrix. In general, an $n \times n$ matrix $\mathbf{A}$ is a **diagonal matrix** if all its entries off the main diagonal are zero; that is,

$$\mathbf{A} = \begin{pmatrix} a_{11} & 0 & \ldots & 0 \\ 0 & a_{22} & \ldots & 0 \\ \vdots & & & \vdots \\ 0 & 0 & \ldots & a_{nn} \end{pmatrix}.$$

Hence if $\mathbf{A}$ is any $n \times n$ diagonal matrix it follows from Example 1 that

$$e^{\mathbf{A}t} = \begin{pmatrix} e^{a_{11}t} & 0 & \ldots & 0 \\ 0 & e^{a_{22}t} & \ldots & 0 \\ \vdots & & & \vdots \\ 0 & 0 & \ldots & e^{a_{nn}t} \end{pmatrix}.$$

▌ Derivative of $e^{\mathbf{A}t}$

The derivative of the matrix exponential $e^{\mathbf{A}t}$ is analogous to that of the scalar exponential; that is, $d/dt\, e^{at} = ae^{at}$. To justify

$$\frac{d}{dt}e^{\mathbf{A}t} = \mathbf{A}e^{\mathbf{A}t} \tag{4}$$

we differentiate (3) term-by-term:

$$\frac{d}{dt} e^{\mathbf{A}t} = \frac{d}{dt} \left[\mathbf{I} + \mathbf{A}t + \mathbf{A}^2 \frac{t^2}{2!} + \cdots + \mathbf{A}^m \frac{t^m}{m!} + \cdots \right] = \mathbf{A} + \mathbf{A}^2 t + \mathbf{A}^3 \frac{t^2}{2!} + \cdots$$

$$= \mathbf{A} \left[\mathbf{I} + \mathbf{A}t + \mathbf{A}^2 \frac{t^2}{2!} + \cdots \right] = \mathbf{A}e^{\mathbf{A}t}.$$

Because of (4), we can now prove that (1) is a solution of $\mathbf{X}' = \mathbf{A}\mathbf{X}$ for every $n \times 1$ vector $\mathbf{C}$ of constants:

$$\mathbf{X}' = \frac{d}{dt} e^{\mathbf{A}t}\mathbf{C} = \mathbf{A}e^{\mathbf{A}t}\mathbf{C} = \mathbf{A}(e^{\mathbf{A}t}\mathbf{C}) = \mathbf{A}\mathbf{X}. \tag{5}$$

▨ $e^{\mathbf{A}t}$ Is a Fundamental Matrix

If we denote the matrix exponential $e^{\mathbf{A}t}$ by the symbol $\boldsymbol{\Psi}(t)$, then (4) is equivalent to the matrix differential equation $\boldsymbol{\Psi}'(t) = \mathbf{A}\boldsymbol{\Psi}$ (see (3) of Section 10.4). In addition, it follows immediately from Definition 10.5.1 that $\boldsymbol{\Psi}(0) = e^{\mathbf{A}0} = \mathbf{I}$ and so det $\boldsymbol{\Psi}(0) \neq 0$. It turns out that these two properties are sufficient for us to conclude that $\boldsymbol{\Psi}(t)$ is a fundamental matrix of the system $\mathbf{X}' = \mathbf{A}\mathbf{X}$.

▨ Nonhomogeneous Systems

We have seen in (4) of Section 2.3 that the general solution of the single linear first-order differential equation $x' = ax + f(t)$, where a is a constant, can be expressed as

$$x = x_c + x_p = ce^{at} + e^{at} \int_{t_0}^{t} e^{-as} f(s) \, ds.$$

For a nonhomogeneous system of linear first-order differential equations, it can be shown that the general solution of $\mathbf{X}' = \mathbf{A}\mathbf{X} + \mathbf{F}(t)$, where $\mathbf{A}$ is an $n \times n$ matrix of constants, is

$$\mathbf{X} = \mathbf{X}_c + \mathbf{X}_p = e^{\mathbf{A}t}\mathbf{C} + e^{\mathbf{A}t} \int_{t_0}^{t} e^{-\mathbf{A}s} \mathbf{F}(s) \, ds. \tag{6}$$

Since the matrix exponential $e^{\mathbf{A}t}$ is a fundamental matrix, it is always nonsingular and $e^{-\mathbf{A}s} = (e^{\mathbf{A}s})^{-1}$. Note that $e^{-\mathbf{A}s}$ can be obtained from $e^{\mathbf{A}t}$ by replacing t by $-s$.

▨ Computation of $e^{\mathbf{A}t}$

The definition of $e^{\mathbf{A}t}$ given in (3) can, of course, always be used to compute $e^{\mathbf{A}t}$. However, the practical utility of (3) is limited by the fact that the entries in $e^{\mathbf{A}t}$ are power series in t. With one's natural desire to work with simple and familiar things, we then try to recognize whether these series define a closed form function. See Problems 1–4 in Exercises 10.5. Fortunately there are many alternative ways of computing $e^{\mathbf{A}t}$. We sketch two of these methods in the discussion that follows.

▨ Using the Laplace Transform

We saw in (5) that $\mathbf{X} = e^{\mathbf{A}t}$ is a solution of $\mathbf{X}' = \mathbf{A}\mathbf{X}$. Indeed, since $e^{\mathbf{A}0} = \mathbf{I}$, $\mathbf{X} = e^{\mathbf{A}t}$ is a solution of the initial-value problem

$$\mathbf{X}' = \mathbf{A}\mathbf{X}, \quad \mathbf{X}(0) = \mathbf{I}. \tag{7}$$

If $\mathbf{x}(s) = \mathscr{L}\{\mathbf{X}(t)\} = \mathscr{L}\{e^{\mathbf{A}t}\}$, then the Laplace transform of (7) is

$$s\mathbf{x}(s) - \mathbf{X}(0) = \mathbf{A}\mathbf{x}(s) \quad \text{or} \quad (s\mathbf{I} - \mathbf{A})\mathbf{x}(s) = \mathbf{I}.$$

Multiplying the last equation by $(s\mathbf{I} - \mathbf{A})^{-1}$ implies $\mathbf{x}(s) = (s\mathbf{I} - \mathbf{A})^{-1}\mathbf{I} = (s\mathbf{I} - \mathbf{A})^{-1}$. In other words, $\mathscr{L}\{e^{\mathbf{A}t}\} = (s\mathbf{I} - \mathbf{A})^{-1}$ or

$$e^{\mathbf{A}t} = \mathscr{L}^{-1}\{(s\mathbf{I} - \mathbf{A})^{-1}\}. \tag{8}$$

EXAMPLE 2 Matrix Exponential

Use the Laplace transform to compute $e^{\mathbf{A}t}$ for $\mathbf{A} = \begin{pmatrix} 1 & -1 \\ 2 & -2 \end{pmatrix}$.

SOLUTION First we compute the matrix $s\mathbf{I} - \mathbf{A}$ and then find its inverse:

$$s\mathbf{I} - \mathbf{A} = \begin{pmatrix} s-1 & 1 \\ -2 & s+2 \end{pmatrix}$$

$$(s\mathbf{I} - \mathbf{A})^{-1} = \begin{pmatrix} s-1 & 1 \\ -2 & s+2 \end{pmatrix}^{-1} = \begin{pmatrix} \dfrac{s+2}{s(s+1)} & \dfrac{-1}{s(s+1)} \\ \dfrac{2}{s(s+1)} & \dfrac{s-1}{s(s+1)} \end{pmatrix}.$$

Then we decompose the entries of the last matrix into partial fractions:

$$(s\mathbf{I} - \mathbf{A})^{-1} = \begin{pmatrix} \dfrac{2}{s} - \dfrac{1}{s+1} & -\dfrac{1}{s} + \dfrac{1}{s+1} \\ \dfrac{2}{s} - \dfrac{2}{s+1} & -\dfrac{1}{s} + \dfrac{2}{s+1} \end{pmatrix}. \tag{9}$$

Taking the inverse Laplace transform of (9) gives the desired result,

$$e^{\mathbf{A}t} = \begin{pmatrix} 2 - e^{-t} & -1 + e^{-t} \\ 2 - 2e^{-t} & -1 + 2e^{-t} \end{pmatrix}. \qquad \equiv$$

■ Using Powers $\mathbf{A}^m$

In Section 8.9, we developed a method for computing an arbitrary power $\mathbf{A}^k$, k a nonnegative integer, of an $n \times n$ matrix $\mathbf{A}$. Recall from Section 8.9 that we can write

$$\mathbf{A}^k = \sum_{j=0}^{n-1} c_j \mathbf{A}^j \quad \text{and} \quad \lambda^k = \sum_{j=0}^{n-1} c_j \lambda^j, \tag{10}$$

where the coefficients c_j are the same in each and the last expression is valid for the eigenvalues $\lambda_1, \lambda_2, \ldots, \lambda_n$ of $\mathbf{A}$. We assume here that the eigenvalues of $\mathbf{A}$ are distinct. By setting $\lambda = \lambda_1, \lambda_2, \ldots, \lambda_n$ in the second expression in (10), we were able to find the c_j in the first expression by solving n equations in n unknowns. It will be convenient in the development that follows to emphasize the fact that the coefficients c_j in (10) depend on the power k by replacing c_j by $c_j(k)$. From (3) and (2), we have

$$e^{\mathbf{A}t} = \sum_{k=0}^{\infty} \mathbf{A}^k \frac{t^k}{k!} \quad \text{and} \quad e^{\lambda t} = \sum_{k=0}^{\infty} \lambda^k \frac{t^k}{k!}. \tag{11}$$

We next use (10) in (11) to replace $\mathbf{A}^k$ and λ^k as finite sums followed by an interchange of the order of summations

$$e^{\mathbf{A}t} = \sum_{k=0}^{\infty} \frac{t^k}{k!} \left(\sum_{j=0}^{n-1} c_j(k) \mathbf{A}^j \right) = \sum_{j=0}^{n-1} \mathbf{A}^j \left(\sum_{k=0}^{\infty} \frac{t^k}{k!} c_j(k) \right) = \sum_{j=0}^{n-1} \mathbf{A}^j b_j(t) \tag{12}$$

$$e^{\lambda t} = \sum_{k=0}^{\infty} \frac{t^k}{k!} \left(\sum_{j=0}^{n-1} c_j(k) \lambda^j \right) = \sum_{j=0}^{n-1} \lambda^j \left(\sum_{k=0}^{\infty} \frac{t^k}{k!} c_j(k) \right) = \sum_{j=0}^{n-1} \lambda^j b_j(t), \tag{13}$$

where $b_j(t) = \sum_{k=0}^{\infty} (t^k/k!) c_j(k)$. Analogous to how we used the eigenvalues of $\mathbf{A}$ in (10) to determine the c_j, we again use the eigenvalues, but this time in the finite sum (13) to determine a system of equations to determine the b_j; these coefficients, in turn, are used in (12) to determine $e^{\mathbf{A}t}$.

EXAMPLE 3 **Matrix Exponential**

Compute $e^{\mathbf{A}t}$ for $\mathbf{A} = \begin{pmatrix} -2 & 4 \\ -1 & 3 \end{pmatrix}$.

SOLUTION We have already seen the matrix $\mathbf{A}$ in Section 8.9, and there we found its eigenvalues to be $\lambda_1 = -1$ and $\lambda_2 = 2$. Now since $\mathbf{A}$ is a 2×2 matrix, we have from (12) and (13)

$$e^{\mathbf{A}t} = b_0\mathbf{I} + b_1\mathbf{A} \quad \text{and} \quad e^{\lambda t} = b_0 + b_1\lambda. \tag{14}$$

Setting $\lambda = -1$ and $\lambda = 2$ in the second equation of (14) gives two equations in the two unknowns b_0 and b_1. Solving the system

$$e^{-t} = b_0 - b_1$$

$$e^{2t} = b_0 + 2b_1$$

yields $b_0 = \frac{1}{3}[e^{2t} + 2e^{-t}]$, $b_1 = \frac{1}{3}[e^{2t} - e^{-t}]$. Substituting these values in the first equation of (14) and simplifying the entries yields

$$e^{\mathbf{A}t} = \begin{pmatrix} -\frac{1}{3}e^{2t} + \frac{4}{3}e^{-t} & \frac{4}{3}e^{2t} - \frac{4}{3}e^{-t} \\ -\frac{1}{3}e^{2t} + \frac{1}{3}e^{-t} & \frac{4}{3}e^{2t} - \frac{1}{3}e^{-t} \end{pmatrix}. \tag{15} \equiv$$

In Problems 23–26 in Exercises 10.5, we show how to compute the matrix exponential $e^{\mathbf{A}t}$ when the matrix $\mathbf{A}$ is diagonalizable (see Section 8.12).

■ **Use of Computers** For those willing to momentarily trade understanding for speed of solution, $e^{\mathbf{A}t}$ can be computed with the aid of computer software; for example, in *Mathematica*, the function **MatrixExp[A t]** computes the matrix exponential for a square matrix $\mathbf{A}$; in *Maple*, the command is **exponential(A, t)**; in MATLAB the function is `expm(At)`. See Problems 27 and 28 in Exercises 10.5.

10.5 Exercises Answers to selected odd-numbered problems begin on page ANS-29.

In Problems 1 and 2, use (3) to compute $e^{\mathbf{A}t}$ and $e^{-\mathbf{A}t}$.

1. $\mathbf{A} = \begin{pmatrix} 1 & 0 \\ 0 & 2 \end{pmatrix}$ **2.** $\mathbf{A} = \begin{pmatrix} 0 & 1 \\ 1 & 0 \end{pmatrix}$

In Problems 3 and 4, use (3) to compute $e^{\mathbf{A}t}$.

3. $\mathbf{A} = \begin{pmatrix} 1 & 1 & 1 \\ 1 & 1 & 1 \\ -2 & -2 & -2 \end{pmatrix}$ **4.** $\mathbf{A} = \begin{pmatrix} 0 & 0 & 0 \\ 3 & 0 & 0 \\ 5 & 1 & 0 \end{pmatrix}$

In Problems 5–8, use (1) and the results in Problems 1–4 to find the general solution of the given system.

5. $\mathbf{X}' = \begin{pmatrix} 1 & 0 \\ 0 & 2 \end{pmatrix}\mathbf{X}$ **6.** $\mathbf{X}' = \begin{pmatrix} 0 & 1 \\ 1 & 0 \end{pmatrix}\mathbf{X}$

7. $\mathbf{X}' = \begin{pmatrix} 1 & 1 & 1 \\ 1 & 1 & 1 \\ -2 & -2 & -2 \end{pmatrix}\mathbf{X}$ **8.** $\mathbf{X}' = \begin{pmatrix} 0 & 0 & 0 \\ 3 & 0 & 0 \\ 5 & 1 & 0 \end{pmatrix}\mathbf{X}$

In Problems 9–12, use (6) to find the general solution of the given system.

9. $\mathbf{X}' = \begin{pmatrix} 1 & 0 \\ 0 & 2 \end{pmatrix}\mathbf{X} + \begin{pmatrix} 3 \\ -1 \end{pmatrix}$ **10.** $\mathbf{X}' = \begin{pmatrix} 1 & 0 \\ 0 & 2 \end{pmatrix}\mathbf{X} + \begin{pmatrix} t \\ e^{4t} \end{pmatrix}$

11. $\mathbf{X}' = \begin{pmatrix} 0 & 1 \\ 1 & 0 \end{pmatrix}\mathbf{X} + \begin{pmatrix} 1 \\ 1 \end{pmatrix}$ **12.** $\mathbf{X}' = \begin{pmatrix} 0 & 1 \\ 1 & 0 \end{pmatrix}\mathbf{X} + \begin{pmatrix} \cosh t \\ \sinh t \end{pmatrix}$

13. Solve the system in Problem 7 subject to the initial condition

$$\mathbf{X}(0) = \begin{pmatrix} 1 \\ -4 \\ 6 \end{pmatrix}.$$

14. Solve the system in Problem 9 subject to the initial condition

$$\mathbf{X}(0) = \begin{pmatrix} 4 \\ 3 \end{pmatrix}.$$

In Problems 15–18, use the method of Example 2 to compute $e^{\mathbf{A}t}$ for the coefficient matrix. Use (1) to find the general solution of the given system.

15. $\mathbf{X}' = \begin{pmatrix} 4 & 3 \\ -4 & -4 \end{pmatrix}\mathbf{X}$ **16.** $\mathbf{X}' = \begin{pmatrix} 4 & -2 \\ 1 & 1 \end{pmatrix}\mathbf{X}$

17. $\mathbf{X}' = \begin{pmatrix} 5 & -9 \\ 1 & -1 \end{pmatrix}\mathbf{X}$ **18.** $\mathbf{X}' = \begin{pmatrix} 0 & 1 \\ -2 & -2 \end{pmatrix}\mathbf{X}$

In Problems 19–22, use the method of Example 3 to compute $e^{\mathbf{A}t}$ for the coefficient matrix. Use (1) to find the general solution of the given system.

19. $\mathbf{X}' = \begin{pmatrix} 2 & -2 \\ -2 & 5 \end{pmatrix}\mathbf{X}$ **20.** $\mathbf{X}' = \begin{pmatrix} 1 & 2 \\ -1 & 4 \end{pmatrix}\mathbf{X}$

21. $\mathbf{X}' = \begin{pmatrix} 3 & 8 \\ 0 & -1 \end{pmatrix}\mathbf{X}$ **22.** $\mathbf{X}' = \begin{pmatrix} 1 & -\frac{3}{2} \\ \frac{1}{4} & -\frac{1}{4} \end{pmatrix}\mathbf{X}$

23. If the matrix $\mathbf{A}$ can be diagonalized, then $\mathbf{P}^{-1}\mathbf{AP} = \mathbf{D}$ or $\mathbf{A} = \mathbf{PDP}^{-1}$. Use this last result and (3) to show that $e^{\mathbf{A}t} = \mathbf{P}e^{\mathbf{D}t}\mathbf{P}^{-1}$.

24. Use $\mathbf{D} = \begin{pmatrix} \lambda_1 & 0 & \cdots & 0 \\ 0 & \lambda_2 & \cdots & 0 \\ \vdots & & & \vdots \\ 0 & 0 & \cdots & \lambda_n \end{pmatrix}$ and (3) to show that

$$e^{\mathbf{D}t} = \begin{pmatrix} e^{\lambda_1 t} & 0 & \cdots & 0 \\ 0 & e^{\lambda_2 t} & \cdots & 0 \\ \vdots & & & \vdots \\ 0 & 0 & \cdots & e^{\lambda_n t} \end{pmatrix}.$$

In Problems 25 and 26, use the results of Problems 23 and 24 to solve the given system.

25. $\mathbf{X}' = \begin{pmatrix} 2 & 1 \\ -3 & 6 \end{pmatrix} \mathbf{X}$ **26.** $\mathbf{X}' = \begin{pmatrix} 2 & 1 \\ 1 & 2 \end{pmatrix} \mathbf{X}$

Computer Lab Assignments

27. (a) Use (1) to find the general solution of $\mathbf{X}' = \begin{pmatrix} 4 & 2 \\ 3 & 3 \end{pmatrix} \mathbf{X}$. Use a CAS to find $e^{\mathbf{A}t}$. Then use the computer to find eigenvalues and eigenvectors of the coefficient matrix $\mathbf{A} = \begin{pmatrix} 4 & 2 \\ 3 & 3 \end{pmatrix}$ and form the general solution in the manner of Section 10.2. Finally, reconcile the two forms of the general solution of the system.

(b) Use (1) to find the general solution of $\mathbf{X}' = \begin{pmatrix} -3 & -1 \\ 2 & -1 \end{pmatrix} \mathbf{X}$. Use a CAS to find $e^{\mathbf{A}t}$. In the case of complex output, utilize the software to do the simplification; for example, in *Mathematica*, if **m=MatrixExp[A t]** has complex entries, then try the command **Simplify [ComplexExpand[m]]**.

28. Use (1) to find the general solution of

$$\mathbf{X}' = \begin{pmatrix} -4 & 0 & 6 & 0 \\ 0 & -5 & 0 & -4 \\ -1 & 0 & 1 & 0 \\ 0 & 3 & 0 & 2 \end{pmatrix} \mathbf{X}.$$

Use a CAS to find $e^{\mathbf{A}t}$.

Discussion Problems

29. Reread the discussion leading to the result given in (8). Does the matrix $s\mathbf{I} - \mathbf{A}$ always have an inverse? Discuss.

30. In Exercises 8.9 we saw that a nonzero $n \times n$ matrix $\mathbf{A}$ is **nilpotent** if m is the smallest positive integer such that $\mathbf{A}^m = \mathbf{0}$. Verify that $\mathbf{A} = \begin{pmatrix} -1 & 1 & 1 \\ -1 & 0 & 1 \\ -1 & 1 & 1 \end{pmatrix}$ is nilpotent. Discuss why it is relatively easy to compute $e^{\mathbf{A}t}$ when $\mathbf{A}$ is nilpotent. Compute $e^{\mathbf{A}t}$ for the given matrix and then use (2) to solve the system $\mathbf{X}' = \mathbf{A}\mathbf{X}$.

10 Chapter in Review Answers to selected odd-numbered problems begin on page ANS-30.

In Problems 1 and 2, fill in the blanks.

1. The column vector $\mathbf{X} = k \begin{pmatrix} 4 \\ 5 \end{pmatrix}$ is a solution of the linear system $\mathbf{X}' = \begin{pmatrix} 1 & 4 \\ 2 & -1 \end{pmatrix} \mathbf{X} - \begin{pmatrix} 8 \\ 1 \end{pmatrix}$ for $k = $ _____.

2. The vector $\mathbf{X} = c_1 \begin{pmatrix} -1 \\ 1 \end{pmatrix} e^{-9t} + c_2 \begin{pmatrix} 5 \\ 3 \end{pmatrix} e^{7t}$ is a solution of the initial-value problem $\mathbf{X}' = \begin{pmatrix} 1 & 10 \\ 6 & -3 \end{pmatrix} \mathbf{X}, \mathbf{X}(0) = \begin{pmatrix} 2 \\ 0 \end{pmatrix}$ for $c_1 = $ _____ and $c_2 = $ _____.

3. Consider the linear system $\mathbf{X}' = \begin{pmatrix} 4 & 6 & 6 \\ 1 & 3 & 2 \\ -1 & -4 & -3 \end{pmatrix} \mathbf{X}$. Without attempting to solve the system, which one of the following vectors,

$$\mathbf{K}_1 = \begin{pmatrix} 0 \\ 1 \\ 1 \end{pmatrix}, \quad \mathbf{K}_2 = \begin{pmatrix} 1 \\ 1 \\ -1 \end{pmatrix}, \quad \mathbf{K}_3 = \begin{pmatrix} 3 \\ 1 \\ -1 \end{pmatrix}, \quad \mathbf{K}_4 = \begin{pmatrix} 6 \\ 2 \\ -5 \end{pmatrix},$$

is an eigenvector of the coefficient matrix? What is the solution of the system corresponding to this eigenvector?

4. Consider the linear system $\mathbf{X}' = \mathbf{A}\mathbf{X}$ of two differential equations where $\mathbf{A}$ is a real coefficient matrix. What is the general solution of the system if it is known that $\lambda_1 = 1 + 2i$ is an eigenvalue and $\mathbf{K}_1 = \begin{pmatrix} 1 \\ i \end{pmatrix}$ is a corresponding eigenvector?

In Problems 5–14, solve the given linear system by the methods of this chapter.

5. $\dfrac{dx}{dt} = 2x + y$
$\dfrac{dy}{dt} = -x$

6. $\dfrac{dx}{dt} = -4x + 2y$
$\dfrac{dy}{dt} = 2x - 4y$

7. $\mathbf{X}' = \begin{pmatrix} 1 & 2 \\ -2 & 1 \end{pmatrix} \mathbf{X}$ **8.** $\mathbf{X}' = \begin{pmatrix} -2 & 5 \\ -2 & 4 \end{pmatrix} \mathbf{X}$

9. $\mathbf{X}' = \begin{pmatrix} 1 & -1 & 1 \\ 0 & 1 & 3 \\ 4 & 3 & 1 \end{pmatrix} \mathbf{X}$ **10.** $\mathbf{X}' = \begin{pmatrix} 0 & 2 & 1 \\ 1 & 1 & -2 \\ 2 & 2 & -1 \end{pmatrix} \mathbf{X}$

11. $\mathbf{X}' = \begin{pmatrix} 2 & 8 \\ 0 & 4 \end{pmatrix} \mathbf{X} + \begin{pmatrix} 2 \\ 16t \end{pmatrix}$

12. $\mathbf{X}' = \begin{pmatrix} 1 & 2 \\ -\frac{1}{2} & 1 \end{pmatrix} \mathbf{X} + \begin{pmatrix} 0 \\ e^t \tan t \end{pmatrix}$

13. $\mathbf{X}' = \begin{pmatrix} -1 & 1 \\ -2 & 1 \end{pmatrix} \mathbf{X} + \begin{pmatrix} 1 \\ \cot t \end{pmatrix}$

14. $\mathbf{X}' = \begin{pmatrix} 3 & 1 \\ -1 & 1 \end{pmatrix} \mathbf{X} + \begin{pmatrix} -2 \\ 1 \end{pmatrix} e^{2t}$

15. (a) Consider the linear system $\mathbf{X}' = \mathbf{AX}$ of three first-order differential equations where the coefficient matrix is

$$\mathbf{A} = \begin{pmatrix} 5 & 3 & 3 \\ 3 & 5 & 3 \\ -5 & -5 & -3 \end{pmatrix},$$

and $\lambda = 2$ is known to be an eigenvalue of multiplicity two. Find two different solutions of the system corresponding to this eigenvalue without using any special formula (such as (12) of Section 10.2).

(b) Use the procedure in part (a) to solve

$$\mathbf{X}' = \begin{pmatrix} 1 & 1 & 1 \\ 1 & 1 & 1 \\ 1 & 1 & 1 \end{pmatrix} \mathbf{X}.$$

16. Verify that $\mathbf{X} = \begin{pmatrix} c_1 \\ c_2 \end{pmatrix} e^t$ is a solution of the linear system

$$\mathbf{X}' = \begin{pmatrix} 1 & 0 \\ 0 & 1 \end{pmatrix} \mathbf{X}$$

for arbitrary constants c_1 and c_2. By hand, draw a phase portrait of the system.

CHAPTER 11

Systems of Nonlinear Differential Equations

In Chapter 10 we used matrix techniques to solve systems of linear first-order differential equations. When a system of differential equations is not linear, it is usually not possible to find solutions that can be expressed in terms of elementary functions. In this chapter we will see that information on the geometric nature of the solutions can be acquired by first analyzing special constant solutions of the system and then searching for periodic solutions. The concept of **stability** will be introduced and illustrated with mathematical models from physics and ecology.

11.1 Autonomous Systems

INTRODUCTION We introduced the notions of autonomous first-order DEs, critical points of an autonomous DE, and the stability of a critical point in Section 2.1. This earlier consideration of stability was purposely kept at a fairly intuitive level; it is now time to give the precise definition of this concept. To do this we need to examine autonomous *systems* of first-order DEs. In this section we define critical points of autonomous systems of two first-order DEs; the autonomous systems can be linear or nonlinear.

◼ **Autonomous Systems** A system of first-order differential equations is called **autonomous** when the system can be written in the form

$$\frac{dx_1}{dt} = g_1(x_1, x_2, \ldots, x_n)$$

$$\frac{dx_2}{dt} = g_2(x_1, x_2, \ldots, x_n)$$

$$\vdots \qquad \vdots$$

$$\frac{dx_n}{dt} = g_n(x_1, x_2, \ldots, x_n). \tag{1}$$

Notice that the independent variable t does not appear explicitly on the right-hand side of each differential equation. Compare (1) with the general system given in (2) of Section 10.1.

EXAMPLE 1 A Nonautonomous System

The system of nonlinear first-order differential equations

$$\overset{t \text{ dependence}}{\underset{\downarrow}{}}$$

$$\frac{dx_1}{dt} = x_1 - 3x_2 + t^2$$

$$\frac{dx_2}{dt} = tx_1 (\sin x_2 t)$$

$$\underset{t \text{ dependence}}{\uparrow}$$

is *not* autonomous because of the presence of t on the right-hand side of both DEs. ≡

When $n = 1$ in (1) a single first-order differential equation takes on the form $dx/dt = g(x)$. This last equation is equivalent to (1) of Section 2.1 with the symbols x and t playing the parts of y and x, respectively. Explicit solutions can be constructed since the differential equation $dx/dt = g(x)$ is separable, and we will make use of this fact to give illustrations of the concepts in this chapter.

See pages 154 and 155. ▶

◼ **Second-Order DE as a System** Any autonomous second-order differential equation $x'' = g(x, x')$ can be written as an autonomous system. As we did in Section 3.7, if we let $y = x'$, then $x'' = g(x, x')$ becomes $y' = g(x, y)$. Thus the second-order differential equation becomes the system of two first-order equations

$$x' = y$$

$$y' = g(x, y).$$

EXAMPLE 2 The Pendulum DE as an Autonomous System

In (6) of Section 3.11 we showed that the displacement angle θ for a pendulum satisfies the nonlinear second-order differential equation

$$\frac{d^2\theta}{dt^2} + \frac{g}{l}\sin\theta = 0.$$

If we let $x = \theta$ and $y = \theta'$, this second-order differential equation can be rewritten as the autonomous system

$$x' = y$$

$$y' = -\frac{g}{l}\sin x. \qquad\qquad \equiv$$

Matrix Forms of (1) If $\mathbf{X}(t)$ and $\mathbf{g}(\mathbf{X})$ denote the respective column vectors

$$\mathbf{X}(t) = \begin{pmatrix} x_1(t) \\ x_2(t) \\ \vdots \\ x_n(t) \end{pmatrix}, \quad \mathbf{g}(\mathbf{X}) = \begin{pmatrix} g_1(x_1, x_2, \ldots, x_n) \\ g_2(x_1, x_2, \ldots, x_n) \\ \vdots \\ g_n(x_1, x_2, \ldots, x_n) \end{pmatrix},$$

then the autonomous system (1) may be written in the compact **column vector form** $\mathbf{X}' = \mathbf{g}(\mathbf{X})$. The homogeneous linear system $\mathbf{X}' = \mathbf{A}\mathbf{X}$ studied in Section 10.2 is an important special case. In this chapter it is also convenient to write (1) using row vectors. If we let

$$\mathbf{X}(t) = (x_1(t), x_2(t), \ldots, x_n(t))$$

and
$$\mathbf{g}(\mathbf{X}) = (g_1(x_1, x_2, \ldots, x_n), g_2(x_1, x_2, \ldots, x_n), \ldots, g_n(x_1, x_2, \ldots, x_n)),$$

then the autonomous system (1) can also be written in the compact **row vector form** $\mathbf{X}' = \mathbf{g}(\mathbf{X})$. It should be clear from the context whether we are using column or row vector form, and therefore we will not distinguish between $\mathbf{X}$ and $\mathbf{X}^T$, the transpose of $\mathbf{X}$. In particular, when $n = 2$ it is convenient to use row vector form and write an initial condition as $\mathbf{X}(0) = (x_0, y_0)$.

When the variable t is interpreted as time, we can refer to the system of differential equations in (1) as a **dynamical system** and a solution $\mathbf{X}(t)$ as the **state of the system** or the **response of the system** at time t. With this terminology, a dynamical system is autonomous when the rate $\mathbf{X}'(t)$ at which the system changes depends only on the system's present state $\mathbf{X}(t)$. The linear system $\mathbf{X}' = \mathbf{A}\mathbf{X} + \mathbf{F}(t)$ studied in Chapter 10 is then autonomous when $\mathbf{F}(t)$ is constant. In the case $n = 2$ or 3 we call a solution a **path** or **trajectory** since we may think of $x = x_1(t)$, $y = x_2(t)$, $z = x_3(t)$ as parametric equations of a curve.

Vector Field Interpretation When $n = 2$ the system in (1) is called a **plane autonomous system**, and we write the system as

$$\frac{dx}{dt} = P(x, y)$$

$$\frac{dy}{dt} = Q(x, y).$$

The vector $\mathbf{V}(x, y) = (P(x, y), Q(x, y))$ defines a **vector field** in a region of the plane, and a solution to the system may be interpreted as the resulting path of a particle as it moves through the region. To be more specific, let $\mathbf{V}(x, y) = (P(x, y), Q(x, y))$ denote the velocity of a stream at position (x, y), and suppose that a small particle (such as a cork) is released at a position (x_0, y_0) in the stream. If $\mathbf{X}(t) = (x(t), y(t))$ denotes the position of the particle at time t, then $\mathbf{X}'(t) = (x'(t), y'(t))$ is the velocity vector $\mathbf{v}$. When external forces are not present and frictional forces are neglected, the velocity of the particle at time t is the velocity of the stream at position $\mathbf{X}(t)$; that is, $\mathbf{X}'(t) = \mathbf{V}(x(t), y(t))$, or

$$\frac{dx}{dt} = P(x(t), y(t))$$

$$\frac{dy}{dt} = Q(x(t), y(t)).$$

Thus the path of the particle is a solution to the system that satisfies the initial condition $\mathbf{X}(0) = (x_0, y_0)$. We will frequently call on this simple interpretation of a plane autonomous system to illustrate new concepts.

EXAMPLE 3 Plane Autonomous System of a Vector Field

A vector field for the steady-state flow of a fluid around a cylinder of radius 1 is given by

$$\mathbf{V}(x, y) = V_0\left(1 - \frac{x^2 - y^2}{(x^2 + y^2)^2}, \frac{-2xy}{(x^2 + y^2)^2}\right),$$

where V_0 is the speed of the fluid far from the cylinder. If a small cork is released at $(-3, 1)$, the path $\mathbf{X}(t) = (x(t), y(t))$ of the cork satisfies the plane autonomous system

$$\frac{dx}{dt} = V_0\left(1 - \frac{x^2 - y^2}{(x^2 + y^2)^2}\right)$$

$$\frac{dy}{dt} = V_0\frac{-2xy}{(x^2 + y^2)^2}$$

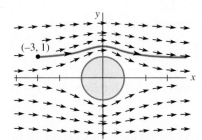

FIGURE 11.1.1 Vector field of a fluid flow around a cylinder in Example 3

subject to the initial condition $\mathbf{X}(0) = (-3, 1)$. See **FIGURE 11.1.1**. ≡

■ **Types of Solutions** If $P(x, y)$, $Q(x, y)$, and the first-order partial derivatives $\partial P/\partial x$, $\partial P/\partial y$, $\partial Q/\partial x$, and $\partial Q/\partial y$ are continuous in a region R of the plane, then a solution to the plane autonomous system

$$\frac{dx}{dt} = P(x, y)$$

$$\frac{dy}{dt} = Q(x, y)$$

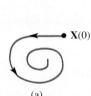

(a)

that satisfies $\mathbf{X}(0) = \mathbf{X}_0$ is unique and one of three basic types:

(*i*) A **constant solution** $x(t) = x_0$, $y(t) = y_0$ (or $\mathbf{X}(t) = \mathbf{X}_0$ for all t). A constant solution is called a **critical** or **stationary point**. When the particle is placed at a critical point $\mathbf{X}_0$ (that is, $\mathbf{X}(0) = \mathbf{X}_0$), it remains there indefinitely. For this reason a constant solution is also called an **equilibrium solution**. Note that since $\mathbf{X}'(t) = \mathbf{0}$, a critical point is a solution of the system of algebraic equations

$$P(x, y) = 0$$

$$Q(x, y) = 0.$$

(b)

FIGURE 11.1.2 Curve in (a) is called an arc

(*ii*) A solution $x = x(t)$, $y = y(t)$ defines an **arc**—a plane curve that does *not* cross itself. Thus the curve in **FIGURE 11.1.2(a)** can be a solution to a plane autonomous system, whereas the curve in Figure 11.1.2(b) cannot be a solution. There would be *two solutions* that start from the point **P** of intersection.

(*iii*) A **periodic solution** $x = x(t)$, $y = y(t)$. A periodic solution is called a **cycle**. If p is the period of the solution, then $\mathbf{X}(t + p) = \mathbf{X}(t)$ and a particle placed on the curve at $\mathbf{X}_0$ will cycle around the curve and return to $\mathbf{X}_0$ in p units of time. See **FIGURE 11.1.3**.

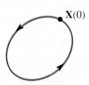

FIGURE 11.1.3 Periodic solution or a cycle

EXAMPLE 4 Finding Critical Points

Find all critical points of each of the following plane autonomous systems.

(a) $x' = -x + y$ **(b)** $x' = x^2 + y^2 - 6$ **(c)** $x' = 0.01x(100 - x - y)$
 $y' = x - y$ $y' = x^2 - y$ $y' = 0.05y(60 - y - 0.2x)$

SOLUTION We find the critical points by setting the right-hand side of the differential equations equal to zero.
(a) The solution to the system

$$-x + y = 0$$
$$x - y = 0$$

consists of all points on the line $y = x$. Thus there are infinitely many critical points.
(b) To solve the system

$$x^2 + y^2 - 6 = 0$$
$$x^2 - y = 0$$

we substitute the second equation, $x^2 = y$, into the first equation to obtain $y^2 + y - 6 = (y + 3)(y - 2) = 0$. If $y = -3$, then $x^2 = -3$, and so there are no real solutions. If $y = 2$, then $x = \pm\sqrt{2}$, and so the critical points are $(\sqrt{2}, 2)$ and $(-\sqrt{2}, 2)$.
(c) Finding the critical points requires a careful consideration of cases. The equation $0.01x(100 - x - y) = 0$ implies $x = 0$ or $x + y = 100$.
 If $x = 0$, then substituting in $0.05y(60 - y - 0.2x) = 0$ we have $y(60 - y) = 0$. Thus $y = 0$ or 60, and so $(0, 0)$ and $(0, 60)$ are critical points.
 If $x + y = 100$, then $0 = y(60 - y - 0.2(100 - y)) = y(40 - 0.8y)$. It follows that $y = 0$ or 50, and so $(100, 0)$ and $(50, 50)$ are critical points. $\equiv$

When the plane autonomous system is linear, we can use the methods in Chapter 10 to investigate solutions.

EXAMPLE 5 **Discovering Periodic Solutions**

Determine whether the given linear dynamical system possesses a periodic solution.
(a) $x' = 2x + 8y$ **(b)** $x' = x + 2y$
 $y' = -x - 2y$ $y' = -\frac{1}{2}x + y$
In each case sketch the graph of the solution that satisfies $\mathbf{X}(0) = (2, 0)$.

SOLUTION **(a)** In Example 6 of Section 10.2 we used the eigenvalue–eigenvector method to show that

$$x = c_1(2 \cos 2t - 2 \sin 2t) + c_2(2 \cos 2t + 2 \sin 2t)$$
$$y = c_1(-\cos 2t) - c_2 \sin 2t.$$

Thus every solution is periodic with period $p = \pi$. The solution satisfying $\mathbf{X}(0) = (2, 0)$ is $x = 2 \cos 2t + 2 \sin 2t$, $y = -\sin 2t$. This solution generates the ellipse shown in
FIGURE 11.1.4(a).
(b) Using the eigenvalue–eigenvector method, we can show that

$$x = c_1(2e^t \cos t) + c_2(2e^t \sin t), \quad y = c_1(-e^t \sin t) + c_2(e^t \cos t).$$

Because of the presence of e^t in the general solution, there are no periodic solutions (that is, cycles). The solution satisfying $\mathbf{X}(0) = (2, 0)$ is $x = 2e^t \cos t$, $y = -e^t \sin t$, and the resulting curve is shown in Figure 11.1.4(b). $\equiv$

(a)

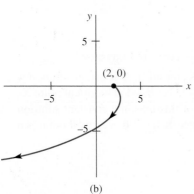

(b)

FIGURE 11.1.4 Solution curves in Example 5

■ **Changing to Polar Coordinates** Except for the case of constant solutions, it is usually not possible to find explicit expressions for the solutions of a *nonlinear* autonomous system. We can solve some nonlinear systems, however, by changing to polar coordinates. From the formulas $r^2 = x^2 + y^2$ and $\theta = \tan^{-1}(y/x)$ we obtain

$$\frac{dr}{dt} = \frac{1}{r}\left(x\frac{dx}{dt} + y\frac{dy}{dt}\right), \quad \frac{d\theta}{dt} = \frac{1}{r^2}\left(-y\frac{dx}{dt} + x\frac{dy}{dt}\right). \tag{2}$$

We can sometimes use (2) to convert a plane autonomous system in rectangular coordinates to a simpler system in polar coordinates.

EXAMPLE 6 **Changing to Polar Coordinates**

Find the solution of the nonlinear plane autonomous system

$$x' = -y - x\sqrt{x^2 + y^2}$$

$$y' = x - y\sqrt{x^2 + y^2}$$

satisfying the initial condition $\mathbf{X}(0) = (3, 3)$.

SOLUTION Substituting for dx/dt and dy/dt in the expressions for dr/dt and $d\theta/dt$ in (2), we obtain

$$\frac{dr}{dt} = \frac{1}{r}[x(-y - xr) + y(x - yr)] = -r^2$$

$$\frac{d\theta}{dt} = \frac{1}{r^2}[-y(-y - xr) + x(x - yr)] = 1.$$

Since $(3, 3)$ is $(3\sqrt{2}, \pi/4)$ in polar coordinates, the initial condition $\mathbf{X}(0) = (3, 3)$ becomes $r(0) = \sqrt{2}$ and $\theta(0) = \pi/4$. Using separation of variables, we see that the solution of the system is

$$r = \frac{1}{t + c_1}, \quad \theta = t + c_2,$$

for $r \neq 0$. (Check this!) Applying the initial conditions then gives

$$r = \frac{1}{t + \sqrt{2}/6}, \quad \theta = t + \frac{\pi}{4}.$$

FIGURE 11.1.5 Solution curve in Example 6

The spiral $r = \dfrac{1}{\theta + \sqrt{2}/6 - \pi/4}$ is sketched in **FIGURE 11.1.5**. ≡

EXAMPLE 7 **Solutions in Polar Coordinates**

When expressed in polar coordinates, a plane autonomous system takes the form

$$\frac{dr}{dt} = 0.5(3 - r)$$

$$\frac{d\theta}{dt} = 1.$$

Find and sketch the solutions satisfying $\mathbf{X}(0) = (0, 1)$ and $\mathbf{X}(0) = (3, 0)$ in rectangular coordinates.

SOLUTION Applying separation of variables to $dr/dt = 0.5(3 - r)$ and integrating $d\theta/dt$ leads to the solution

$$r = 3 + c_1 e^{-0.5t}, \quad \theta = t + c_2.$$

FIGURE 11.1.6 Solution curves in Example 7

If $\mathbf{X}(0) = (0, 1)$, then $r(0) = 1$ and $\theta(0) = \pi/2$, and so $c_1 = -2$ and $c_2 = \pi/2$. The solution curve is the spiral $r = 3 - 2e^{-0.5(\theta - \pi/2)}$. Note that as $t \to \infty$, θ increases without bound and r approaches 3.

If $\mathbf{X}(0) = (3, 0)$, then $r(0) = 3$ and $\theta(0) = 0$. It follows that $c_1 = c_2 = 0$, and so $r = 3$ and $\theta = t$. Hence $x = r \cos \theta = 3 \cos t$ and $y = r \sin \theta = 3 \sin t$, and so the solution is periodic. The solution generates a circle of radius 3 about $(0, 0)$. Both solutions are shown in **FIGURE 11.1.6**. ≡

In Problems 1–6, write the given nonlinear second-order differential equation as a plane autonomous system. Find all critical points of the resulting system.

1. $x'' + 9 \sin x = 0$ **2.** $x'' + (x')^2 + 2x = 0$

3. $x'' + x'(1 - x^3) - x^2 = 0$ **4.** $x'' + 4 \dfrac{x}{1 + x^2} + 2x' = 0$

5. $x'' + x = \epsilon x^3$ for $\epsilon > 0$

6. $x'' + x - \epsilon x|x| = 0$ for $\epsilon > 0$

In Problems 7–16, find all critical points of the given plane autonomous system.

7. $x' = x + xy$
 $y' = -y - xy$

8. $x' = y^2 - x$
 $y' = x^2 - y$

9. $x' = 3x^2 - 4y$
 $y' = x - y$

10. $x' = x^3 - y$
 $y' = x - y^3$

11. $x' = x(10 - x - \tfrac{1}{2}y)$
 $y' = y(16 - y - x)$

12. $x' = -2x + y + 10$
 $y' = 2x - y - 15 \dfrac{y}{y + 5}$

13. $x' = x^2 e^y$
 $y' = y(e^x - 1)$

14. $x' = \sin y$
 $y' = e^{x-y} - 1$

15. $x' = x(1 - x^2 - 3y^2)$
 $y' = y(3 - x^2 - 3y^2)$

16. $x' = -x(4 - y^2)$
 $y' = 4y(1 - x^2)$

In Problems 17–22, for the given linear dynamical system (taken from Exercises 10.2)

(a) find the general solution and determine whether there are periodic solutions,

(b) find the solution satisfying the given initial condition, and,

(c) with the aid of a graphing utility, plot the solution in part (b) and indicate the direction in which the curve is traversed.

17. $x' = x + 2y$
 $y' = 4x + 3y$, $\mathbf{X}(0) = (2, -2)$ (Problem 1, Exercises 10.2)

18. $x' = -6x + 2y$
 $y' = -3x + y$, $\mathbf{X}(0) = (3, 4)$ (Problem 6, Exercises 10.2)

19. $x' = 4x - 5y$
 $y' = 5x - 4y$, $\mathbf{X}(0) = (4, 5)$ (Problem 39, Exercises 10.2)

20. $x' = x + y$
 $y' = -2x - y$, $\mathbf{X}(0) = (-2, 2)$ (Problem 36, Exercises 10.2)

21. $x' = 5x + y$
 $y' = -2x + 3y$, $\mathbf{X}(0) = (-1, 2)$ (Problem 37, Exercises 10.2)

22. $x' = x - 8y$
 $y' = x - 3y$, $\mathbf{X}(0) = (2, 1)$ (Problem 40, Exercises 10.2)

In Problems 23–26, solve the given nonlinear plane autonomous system by changing to polar coordinates. Describe the geometric behavior of the solution that satisfies the given initial condition(s).

23. $x' = -y - x(x^2 + y^2)^2$
 $y' = x - y(x^2 + y^2)^2$, $\mathbf{X}(0) = (4, 0)$

24. $x' = y + x(x^2 + y^2)$
 $y' = -x + y(x^2 + y^2)$, $\mathbf{X}(0) = (4, 0)$

25. $x' = -y + x(1 - x^2 - y^2)$
 $y' = x + y(1 - x^2 - y^2)$, $\mathbf{X}(0) = (1, 0)$; $\mathbf{X}(0) = (2, 0)$

 [*Hint*: The resulting differential equation for r is a Bernoulli differential equation. See Section 2.5.]

26. $x' = y - \dfrac{x}{\sqrt{x^2 + y^2}}(4 - x^2 - y^2)$

 $y' = -x - \dfrac{x}{\sqrt{x^2 + y^2}}(4 - x^2 - y^2)$,

 $\mathbf{X}(0) = (1, 0)$; $\mathbf{X}(0) = (2, 0)$

 [*Hint*: See Example 4 in Section 2.2.]

If a plane autonomous system has a periodic solution, then there must be at least one critical point inside the curve generated by the solution. In Problems 27–30, use this fact together with a numerical solver to investigate the possibility of periodic solutions.

27. $x' = -x + 6y$
 $y' = xy + 12$

28. $x' = -x + 6xy$
 $y' = -8xy + 2y$

29. $x' = y$
 $y' = y(1 - 3x^2 - 2y^2) - x$

30. $x' = xy$
 $y' = -1 - x^2 - y^2$

31. If $z = f(x, y)$ is a function with continuous first partial derivatives in a region R, then a flow $\mathbf{V}(x, y) = (P(x, y), Q(x, y))$ in R may be defined by letting $P(x, y) = -\dfrac{\partial f}{\partial y}(x, y)$ and $Q(x, y) = \dfrac{\partial f}{\partial x}(x, y)$. Show that if $\mathbf{X}(t) = (x(t), y(t))$ is a solution of the plane autonomous system

$$x' = P(x, y)$$
$$y' = Q(x, y),$$

then $f(x(t), y(t)) = c$ for some constant c. Thus a solution curve lies on the level curves of f. [*Hint*: Use the Chain Rule to compute $\dfrac{d}{dt} f(x(t), y(t))$.]

11.2 Stableity of Linear Systems

INTRODUCTION We have seen that the plane autonomous system

$$\frac{dx}{dt} = P(x, y)$$

$$\frac{dy}{dt} = Q(x, y)$$

gives rise to a vector field $\mathbf{V}(x, y) = (P(x, y), Q(x, y))$, and a solution $\mathbf{X} = \mathbf{X}(t)$ of the system may be interpreted as the resulting path of a particle that is initially placed at position $\mathbf{X}(0) = \mathbf{X}_0$. If $\mathbf{X}_0$ is a critical point of the system, then the particle remains stationary. In this section we examine the behavior of solutions when $\mathbf{X}_0$ is chosen *close* to a critical point of the system.

critical point
(a)

critical point
(b)

critical point

critical point
(c)

FIGURE 11.2.1 Critical points

■ **Some Fundamental Questions** Suppose that $\mathbf{X}_1$ is a critical point of a plane autonomous system and $\mathbf{X} = \mathbf{X}(t)$ is a solution of the system that satisfies $\mathbf{X}(0) = \mathbf{X}_0$. If the solution is interpreted as a path of a moving particle, we are interested in the answers to the following questions when $\mathbf{X}_0$ is placed near $\mathbf{X}_1$:

(*i*) Will the particle return to the critical point? More precisely, if $\mathbf{X} = \mathbf{X}(t)$ is the solution that satisfies $\mathbf{X}(0) = \mathbf{X}_0$, is $\lim_{t \to \infty} \mathbf{X}(t) = \mathbf{X}_1$?

(*ii*) If the particle does *not* return to the critical point, does it remain close to the critical point or move away from the critical point? It is conceivable, for example, that the particle may simply circle the critical point, or it may even return to a different critical point or to no critical point at all. See **FIGURE 11.2.1**.

If in some neighborhood of the critical point case (a) or (b) in Figure 11.2.1 *always* occurs, we call the critical point **locally stable**. If, however, an initial value $\mathbf{X}_0$ that results in behavior similar to (c) can be found in *any* given neighborhood, we call the critical point **unstable**. These concepts will be made more precise in Section 11.3, where questions (*i*) and (*ii*) will be investigated for nonlinear systems.

■ **Stability Analysis** We will first investigate these two stability questions for *linear* plane autonomous systems and lay the foundation for Section 11.3. The solution methods of Chapter 10 enable us to give a careful geometric analysis of the solutions to

$$x' = ax + by$$
$$y' = cx + dy$$

$$\tag{1}$$

in terms of the eigenvalues and eigenvectors of the coefficient matrix

$$\mathbf{A} = \begin{pmatrix} a & b \\ c & d \end{pmatrix}.$$

Here a, b, c, and d are constants. To ensure that $\mathbf{X}_0 = (0, 0)$ is the only critical point, we will assume that the determinant $\Delta = ad - bc \neq 0$. If $\tau = a + d$ is the trace* of matrix $\mathbf{A}$, then the characteristic equation $\det(\mathbf{A} - \lambda \mathbf{I}) = 0$ can be rewritten as

$$\lambda^2 - \tau\lambda + \Delta = 0.$$

Therefore the eigenvalues of $\mathbf{A}$ are $\lambda = (\tau \pm \sqrt{\tau^2 - 4\Delta})/2$, and the usual three cases for these roots occur according to whether $\tau^2 - 4\Delta$ is positive, negative, or zero. In the next example we use a numerical solver to discover the nature of the solutions corresponding to these cases.

*In general, if $\mathbf{A}$ is an $n \times n$ matrix, the **trace** of $\mathbf{A}$ is the sum of the main diagonal entries.

EXAMPLE 1 **Eigenvalues and the Shape of Solutions**

Find the eigenvalues of the linear system

$$x' = -x + y$$
$$y' = cx - y$$

in terms of c, and use a numerical solver to discover the shapes of solutions corresponding to the cases $c = \frac{1}{4}, 4, 0,$ and -9.

SOLUTION The coefficient matrix $\begin{pmatrix} -1 & 1 \\ c & -1 \end{pmatrix}$ has trace $\tau = -2$ and determinant $\Delta = 1 - c$, and so the eigenvalues are

$$\lambda = \frac{\tau \pm \sqrt{\tau^2 - 4\Delta}}{2} = \frac{-2 \pm \sqrt{4 - 4(1 - c)}}{2} = -1 \pm \sqrt{c}.$$

The nature of the eigenvalues is therefore determined by the sign of c.

If $c = \frac{1}{4}$, then the eigenvalues are negative and distinct: $\lambda = -\frac{1}{2}$ and $-\frac{3}{2}$. In **FIGURE 11.2.2(a)** we have used a numerical solver to generate solution curves or trajectories that correspond to various initial conditions. Note that, except for the trajectories drawn in red in the figure, the trajectories all appear to approach **0** from a fixed direction. Recall from Chapter 10 that a collection of trajectories in the xy-plane or **phase plane** is called a **phase portrait** of the system.

When $c = 4$ the eigenvalues have opposite signs, $\lambda = 1$ and -3, and an interesting phenomenon occurs. All trajectories move away from the origin in a fixed direction except for solutions that start along the single line drawn in red in Figure 11.2.2(b). We have already seen behavior like this in the phase portrait given in Figure 10.2.2. Experiment with your numerical solver and verify these observations.

The selection $c = 0$ leads to a single real eigenvalue $\lambda = -1$. This case is very similar to the case $c = \frac{1}{4}$ with one notable exception. All solution curves in Figure 11.2.2(c) appear to approach **0** from a fixed direction as t increases.

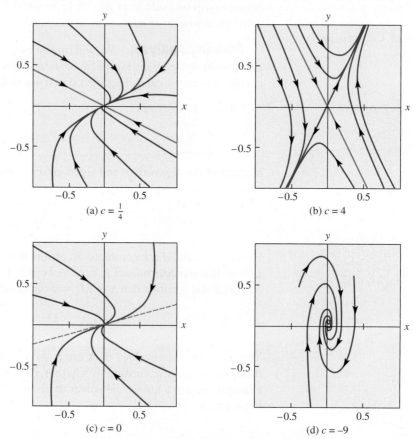

(a) $c = \frac{1}{4}$

(b) $c = 4$

(c) $c = 0$

(d) $c = -9$

FIGURE 11.2.2 Phase portraits of linear system in Example 1

FIGURE 11.2.3 Stable node

FIGURE 11.2.4 Unstable node

FIGURE 11.2.5 Saddle point

Finally, when $c = -9$, $\lambda = -1 \pm \sqrt{-9} = -1 \pm 3i$. Thus the eigenvalues are conjugate complex numbers with negative real part -1. Figure 11.2.2(d) shows that solution curves spiral in toward the origin **0** as t increases.

The behaviors of the trajectories observed in the four phase portraits in Figure 11.2.2 in Example 1 can be explained using the eigenvalue–eigenvector solution results from Chapter 10.

Case I: *Real Distinct Eigenvalues* $(\tau^2 - 4\Delta > 0)$

According to Theorem 10.2.1 in Section 10.2, the general solution of (1) is given by

$$\mathbf{X}(t) = c_1\mathbf{K}_1 e^{\lambda_1 t} + c_2\mathbf{K}_2 e^{\lambda_2 t}, \tag{2}$$

where λ_1 and λ_2 are the eigenvalues and $\mathbf{K}_1$ and $\mathbf{K}_2$ are the corresponding eigenvectors. Note that $\mathbf{X}(t)$ can also be written as

$$\mathbf{X}(t) = e^{\lambda_1 t}[c_1\mathbf{K}_1 + c_2\mathbf{K}_2 e^{(\lambda_2 - \lambda_1)t}]. \tag{3}$$

(a) *Both eigenvalues negative* $(\tau^2 - 4\Delta > 0, \tau < 0, \text{ and } \Delta > 0)$

Stable Node $(\lambda_2 < \lambda_1 < 0)$: Since both eigenvalues are negative, it follows from (2) that $\lim_{t\to\infty}\mathbf{X}(t) = \mathbf{0}$. If we assume that $\lambda_2 < \lambda_1$, then $\lambda_2 - \lambda_1 < 0$ and so $e^{(\lambda_2 - \lambda_1)t}$ is an exponential decay function. We may therefore conclude from (3) that $\mathbf{X}(t) \approx c_1\mathbf{K}_1 e^{\lambda_1 t}$ for large values of t. When $c_1 \neq 0$, $\mathbf{X}(t)$ will approach **0** from one of the two directions determined by the eigenvector $\mathbf{K}_1$ corresponding to λ_1. If $c_1 = 0$, $\mathbf{X}(t) = c_2\mathbf{K}_2 e^{\lambda_2 t}$ and $\mathbf{X}(t)$ approaches **0** along the line determined by the eigenvector $\mathbf{K}_2$. **FIGURE 11.2.3** shows a collection of solution curves around the origin. A critical point is called a **stable node** when both eigenvalues are negative.

(b) *Both eigenvalues positive* $(\tau^2 - 4\Delta > 0, \tau > 0, \text{ and } \Delta > 0)$

Unstable Node $(0 < \lambda_2 < \lambda_1)$: The analysis for this case is similar to (a). Again from (2), $\mathbf{X}(t)$ becomes unbounded as t increases. Moreover, again assuming $\lambda_2 < \lambda_1$ and using (3), we see that $\mathbf{X}(t)$ becomes unbounded in one of the directions determined by the eigenvector $\mathbf{K}_1$ (when $c_1 \neq 0$) or along the line determined by the eigenvector $\mathbf{K}_2$ (when $c_1 = 0$). **FIGURE 11.2.4** shows a typical collection of solution curves. This type of critical point, corresponding to the case when both eigenvalues are positive, is called an **unstable node**.

(c) *Eigenvalues have opposite signs* $(\tau^2 - 4\Delta > 0 \text{ and } \Delta < 0)$

Saddle Point $(\lambda_2 < 0 < \lambda_1)$: The analysis of the solutions is identical to (b) with one exception. When $c_1 = 0$, $\mathbf{X}(t) = c_2\mathbf{K}_2 e^{\lambda_2 t}$ and, since $\lambda_2 < 0$, $\mathbf{X}(t)$ will approach **0** along the line determined by the eigenvector $\mathbf{K}_2$. If $\mathbf{X}(0)$ does not lie on the line determined by $\mathbf{K}_2$, the line determined by $\mathbf{K}_1$ serves as an asymptote for $\mathbf{X}(t)$. Thus the critical point is unstable even though some solutions approach **0** as t increases. This unstable critical point is called a **saddle point**. See **FIGURE 11.2.5**.

EXAMPLE 2 **Real Distinct Eigenvalues**

Classify the critical point $(0, 0)$ of each of the following linear systems $\mathbf{X}' = \mathbf{A}\mathbf{X}$ as either a stable node, an unstable node, or a saddle point.

(a) $\mathbf{A} = \begin{pmatrix} 2 & 3 \\ 2 & 1 \end{pmatrix}$ **(b)** $\mathbf{A} = \begin{pmatrix} -10 & 6 \\ 15 & -19 \end{pmatrix}$

In each case discuss the nature of the solutions in a neighborhood of $(0, 0)$.

SOLUTION **(a)** Since the trace $\tau = 3$ and the determinant $\Delta = -4$, the eigenvalues are

$$\lambda = \frac{\tau \pm \sqrt{\tau^2 - 4\Delta}}{2} = \frac{3 \pm \sqrt{3^2 - 4(-4)}}{2} = \frac{3 \pm 5}{2} = 4, -1.$$

FIGURE 11.2.6 Saddle point

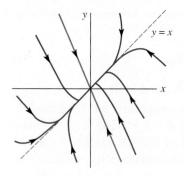

FIGURE 11.2.7 Stable node

The eigenvalues have opposite signs, and so $(0, 0)$ is a saddle point. It is not hard to show (see Example 1, Section 10.2) that eigenvectors corresponding to $\lambda_1 = 4$ and $\lambda_2 = -1$ are

$$\mathbf{K}_1 = \begin{pmatrix} 3 \\ 2 \end{pmatrix} \quad \text{and} \quad \mathbf{K}_2 = \begin{pmatrix} 1 \\ -1 \end{pmatrix},$$

respectively. If $\mathbf{X}(0) = \mathbf{X}_0$ lies on the line $y = -x$, then $\mathbf{X}(t)$ approaches $\mathbf{0}$. For any other initial condition $\mathbf{X}(t)$ will become unbounded in the directions determined by $\mathbf{K}_1$. In other words, the line $y = \frac{2}{3}x$ serves as an asymptote for all these solution curves. See **FIGURE 11.2.6**.

(b) From $\tau = -29$ and $\Delta = 100$ it follows that the eigenvalues of $\mathbf{A}$ are $\lambda_1 = -4$ and $\lambda_2 = -25$. Both eigenvalues are negative, and so $(0, 0)$ is in this case a stable node. Since eigenvectors corresponding to $\lambda_1 = -4$ and $\lambda_2 = -25$ are

$$\mathbf{K}_1 = \begin{pmatrix} 1 \\ 1 \end{pmatrix} \quad \text{and} \quad \mathbf{K}_2 = \begin{pmatrix} 2 \\ -5 \end{pmatrix},$$

respectively, it follows that all solutions approach $\mathbf{0}$ from the direction defined by $\mathbf{K}_1$ except those solutions for which $\mathbf{X}(0) = \mathbf{X}_0$ lies on the line $y = -\frac{5}{2}x$ determined by $\mathbf{K}_2$. These solutions approach $\mathbf{0}$ along $y = -\frac{5}{2}x$. See **FIGURE 11.2.7**. ≡

Case II: *A Repeated Real Eigenvalue* $(\tau^2 - 4\Delta = 0)$

Degenerate Nodes: Recall from Section 10.2 that the general solution takes on one of two different forms depending on whether one or two linearly independent eigenvectors can be found for the repeated eigenvalue λ_1.

(a) *Two linearly independent eigenvectors*

If $\mathbf{K}_1$ and $\mathbf{K}_2$ are two linearly independent eigenvectors corresponding to λ_1, then the general solution is given by

$$\mathbf{X}(t) = c_1 \mathbf{K}_1 e^{\lambda_1 t} + c_2 \mathbf{K}_2 e^{\lambda_1 t} = (c_1 \mathbf{K}_1 + c_2 \mathbf{K}_2) e^{\lambda_1 t}.$$

If $\lambda_1 < 0$, then $\mathbf{X}(t)$ approaches $\mathbf{0}$ along the line determined by the vector $c_1 \mathbf{K}_1 + c_2 \mathbf{K}_2$ and the critical point is called a **degenerate stable node**. See **FIGURE 11.2.8(a)**. The arrows in Figure 11.2.8(a) are reversed when $\lambda_1 > 0$, and we have a **degenerate unstable node**.

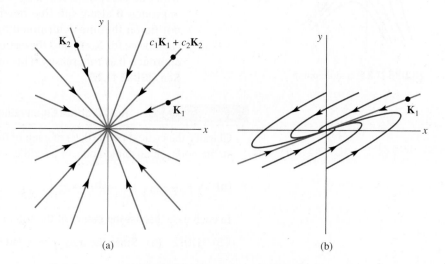

(a) (b)

FIGURE 11.2.8 Degenerate stable nodes

FIGURE 11.2.9 Center

(a)

(b)

FIGURE 11.2.10 Stable and unstable spiral points

(b) *A single linearly independent eigenvector*

When only a single linearly independent eigenvector $\mathbf{K}_1$ exists, the general solution is given by

$$\mathbf{X}(t) = c_1\mathbf{K}_1 e^{\lambda_1 t} + c_2(\mathbf{K}_1 t e^{\lambda_1 t} + \mathbf{P}e^{\lambda_1 t}),$$

where $(\mathbf{A} - \lambda_1 \mathbf{I})\mathbf{P} = \mathbf{K}_1$ (see Section 10.2, (12)−(14)), and the solution may be rewritten as

$$\mathbf{X}(t) = te^{\lambda_1 t}\left[c_2\mathbf{K}_1 + \frac{c_1}{t}\mathbf{K}_1 + \frac{c_2}{t}\mathbf{P}\right].$$

If $\lambda_1 < 0$, then $\lim_{t\to\infty} te^{\lambda_1 t} = 0$ and it follows that $\mathbf{X}(t)$ approaches **0** in one of the directions determined by the vector $\mathbf{K}_1$ (see Figure 11.2.8(b)). The critical point is again called a **degenerate stable node**. When $\lambda_1 > 0$, the solutions look like those in Figure 11.2.8(b) with the arrows reversed. The line determined by $\mathbf{K}_1$ is an asymptote for *all* solutions. The critical point is again called a **degenerate unstable node**.

Case III: *Complex Eigenvalues* $(\tau^2 - 4\Delta < 0)$

If $\lambda_1 = \alpha + i\beta$ and $\overline{\lambda}_1 = \alpha - i\beta$ are the complex eigenvalues and $\mathbf{K}_1 = \mathbf{B}_1 + i\mathbf{B}_2$ is a complex eigenvector corresponding to λ_1, the general solution can be written as $\mathbf{X}(t) = c_1\mathbf{X}_1(t) + c_2\mathbf{X}_2(t)$, where

$$\mathbf{X}_1(t) = (\mathbf{B}_1 \cos \beta t - \mathbf{B}_2 \sin \beta t)e^{\alpha t} \quad \text{and} \quad \mathbf{X}_2(t) = (\mathbf{B}_2 \cos \beta t + \mathbf{B}_1 \sin \beta t)e^{\alpha t}.$$

See equations (23) and (24) in Section 10.2. A solution can therefore be written in the form

$$x(t) = e^{\alpha t}(c_{11} \cos \beta t + c_{12} \sin \beta t), \quad y(t) = e^{\alpha t}(c_{21} \cos \beta t + c_{22} \sin \beta t), \quad (4)$$

and when $\alpha = 0$ we have

$$x(t) = c_{11} \cos \beta t + c_{12} \sin \beta t, \quad y(t) = c_{21} \cos \beta t + c_{22} \sin \beta t. \quad (5)$$

(a) *Pure imaginary roots* $(\tau^2 - 4\Delta < 0, \tau = 0)$

Center: When $\alpha = 0$, the eigenvalues are pure imaginary and, from (5), all solutions are periodic with period $p = 2\pi/\beta$. Notice that if both c_{12} and c_{21} happened to be 0, then (5) would reduce to

$$x(t) = c_{11} \cos \beta t, \quad y(t) = c_{22} \sin \beta t,$$

which is a standard parametric representation for the ellipse $x^2/c_{11}^2 + y^2/c_{22}^2 = 1$. By solving the system of equations in (4) for $\cos \beta t$ and $\sin \beta t$ and using the identity $\sin^2 \beta t + \cos^2 \beta t = 1$, it is possible to show that *all solutions are ellipses* with center at the origin. The critical point $(0, 0)$ is called a **center**, and **FIGURE 11.2.9** shows a typical collection of solution curves. The ellipses are either *all* traversed in the clockwise direction or *all* traversed in the counterclockwise direction.

(b) *Nonzero real part* $(\tau^2 - 4\Delta < 0, \tau \neq 0)$

Spiral Points: When $\alpha \neq 0$, the effect of the term $e^{\alpha t}$ in (4) is similar to the effect of the exponential term in the analysis of damped motion given in Section 3.8. When $\alpha < 0$, $e^{\alpha t} \to 0$, and the elliptical-like solution spirals closer and closer to the origin. The critical point is called a **stable spiral point**. When $\alpha > 0$, the effect is the opposite. An elliptical-like solution is driven farther and farther from the origin, and the critical point is now called an **unstable spiral point**. See **FIGURE 11.2.10**.

EXAMPLE 3 **Repeated and Complex Eigenvalues**

Classify the critical point $(0, 0)$ of each of the following linear systems $\mathbf{X}' = \mathbf{AX}$.

(a) $\mathbf{A} = \begin{pmatrix} 3 & -18 \\ 2 & -9 \end{pmatrix}$ **(b)** $\mathbf{A} = \begin{pmatrix} -1 & 2 \\ -1 & 1 \end{pmatrix}$

In each case discuss the nature of the solution that satisfies $\mathbf{X}(0) = (1, 0)$. Determine parametric equations for each solution.

SOLUTION **(a)** Since $\tau = -6$ and $\Delta = 9$, the characteristic polynomial is $\lambda^2 + 6\lambda + 9 = (\lambda + 3)^2$, and so $(0, 0)$ is a degenerate stable node. For the repeated eigenvalue $\lambda = -3$ we find a single eigenvector $\mathbf{K}_1 = \begin{pmatrix} 3 \\ 1 \end{pmatrix}$, and so the solution $\mathbf{X}(t)$ that satisfies $\mathbf{X}(0) = (1, 0)$ approaches $(0, 0)$ from the direction specified by the line $y = x/3$.

(b) Since $\tau = 0$ and $\Delta = 1$, the eigenvalues are $\lambda = \pm i$, and so $(0, 0)$ is a center. The solution $\mathbf{X}(t)$ that satisfies $\mathbf{X}(0) = (1, 0)$ is an ellipse that circles the origin every 2π units of time.

From Example 4 of Section 10.2 the general solution of the system in part (a) is

$$\mathbf{X}(t) = c_1 \begin{pmatrix} 3 \\ 1 \end{pmatrix} e^{-3t} + c_2 \left[\begin{pmatrix} 3 \\ 1 \end{pmatrix} t e^{-3t} + \begin{pmatrix} \frac{1}{2} \\ 0 \end{pmatrix} e^{-3t} \right].$$

The initial condition gives $c_1 = 0$ and $c_2 = 2$, and so $x = (6t + 1)e^{-3t}$, $y = 2te^{-3t}$ are parametric equations for the solution.

The general solution of the system in part (b) is

$$\mathbf{X}(t) = c_1 \begin{pmatrix} \cos t + \sin t \\ \cos t \end{pmatrix} + c_2 \begin{pmatrix} \cos t - \sin t \\ -\sin t \end{pmatrix}.$$

The initial condition gives $c_1 = 0$ and $c_2 = 1$, and so $x = \cos t - \sin t$, $y = -\sin t$ are parametric equations for the ellipse. Note that $y < 0$ for small positive values of t, and therefore the ellipse is traversed in the clockwise direction.

The solutions of parts (a) and (b) are shown in **FIGURES 11.2.11(a)** and 11.2.11(b), respectively.

$\equiv$

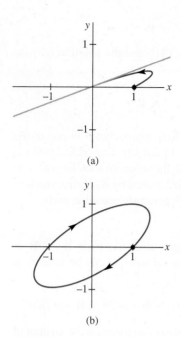

(a)

(b)

FIGURE 11.2.11 Solution curves in Example 3

Classifying Critical Points
FIGURE 11.2.12 conveniently summarizes the results of this section. The general geometric nature of the solutions can be determined by computing the trace and determinant of $\mathbf{A}$. In practice, graphs of the solutions are most easily obtained *not* by constructing explicit eigenvalue–eigenvector solutions but rather by generating the solutions numerically using a numerical solver and a method such as the Runge–Kutta method for first-order systems (Section 6.4).

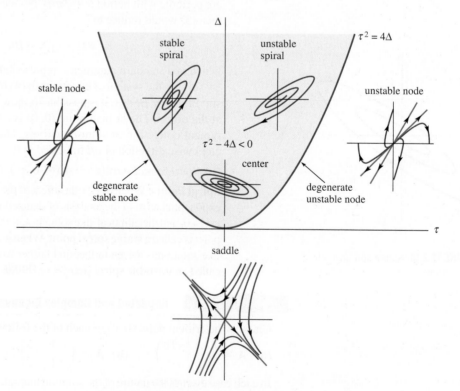

FIGURE 11.2.12 Geometric summary of Cases I, II, and III

EXAMPLE 4 **Classifying Critical Points**

Classify the critical point $(0, 0)$ of each of the following linear systems $\mathbf{X}' = \mathbf{AX}$

(a) $\mathbf{A} = \begin{pmatrix} 1.01 & 3.10 \\ -1.10 & -1.02 \end{pmatrix}$ (b) $\mathbf{A} = \begin{pmatrix} -a\hat{x} & -ab\hat{x} \\ -cd\hat{y} & -d\hat{y} \end{pmatrix}$

for positive constants $a, b, c, d, \hat{x},$ and $\hat{y}$.

SOLUTION (a) For this matrix $\tau = -0.01$, $\Delta = 2.3798$, and so $\tau^2 - 4\Delta < 0$. Using Figure 11.2.12, we see that $(0, 0)$ is a stable spiral point.

(b) This matrix arises from the Lotka–Volterra competition model, which we will study in Section 11.4. Since $\tau = -(a\hat{x} + d\hat{y})$ and all constants in the matrix are positive, $\tau < 0$. The determinant may be written as $\Delta = ad\hat{x}\hat{y}(1 - bc)$. If $bc > 1$, then $\Delta < 0$ and the critical point is a saddle point. If $bc < 1$, then $\Delta > 0$ and the critical point is either a stable node, a degenerate stable node, or a stable spiral point. In all three of these cases $\lim_{t\to\infty}\mathbf{X}(t) = 0$. ≡

We can now give answers to each of the questions posed at the beginning of this section for the linear plane autonomous system

$$x' = ax + by$$
$$y' = cx + dy$$

with $ad - bc \neq 0$. The answers are summarized in the theorem below.

THEOREM 11.2.1 Stability Criteria for Linear Systems

For a linear plane autonomous system $\mathbf{X}' = \mathbf{AX}$ with $\det \mathbf{A} \neq 0$, let $\mathbf{X} = \mathbf{X}(t)$ denote the solution that satisfies the initial condition $\mathbf{X}(0) = \mathbf{X}_0$, where $\mathbf{X}_0 \neq \mathbf{0}$.

(a) $\lim_{t\to\infty}\mathbf{X}(t) = \mathbf{0}$ if and only if the eigenvalues of $\mathbf{A}$ have negative real parts. This will occur when $\Delta > 0$ and $\tau < 0$.

(b) $\mathbf{X}(t)$ is periodic if and only if the eigenvalues of $\mathbf{A}$ are pure imaginary. This will occur when $\Delta > 0$ and $\tau = 0$.

(c) In all other cases, given any neighborhood of the origin, there is at least one $\mathbf{X}_0$ in the neighborhood for which $\mathbf{X}(t)$ becomes unbounded as t increases.

REMARKS

The terminology used to describe the types of critical points varies from text to text. The following table lists many of the alternative terms that you may encounter in your reading.

Term	Alternative Terms
critical point	equilibrium point, singular point, stationary point, rest point
spiral point	focus, focal point, vortex point
stable node or spiral point	attractor, sink
unstable node or spiral point	repeller, source

11.2 Exercises Answers to selected odd-numbered problems begin on page ANS-30.

In Problems 1–8, the general solution of the linear system $\mathbf{X}' = \mathbf{AX}$ is given.

(a) In each case discuss the nature of the solution in a neighborhood of $(0, 0)$.

(b) With the aid of a graphing utility plot the solution that satisfies $\mathbf{X}(0) = (1, 1)$.

1. $\mathbf{A} = \begin{pmatrix} -2 & -2 \\ -2 & -5 \end{pmatrix}$, $\mathbf{X}(t) = c_1\begin{pmatrix} 2 \\ -1 \end{pmatrix}e^{-t} + c_2\begin{pmatrix} 1 \\ 2 \end{pmatrix}e^{-6t}$

2. $\mathbf{A} = \begin{pmatrix} -1 & -2 \\ 3 & 4 \end{pmatrix}$, $\mathbf{X}(t) = c_1\begin{pmatrix} 1 \\ -1 \end{pmatrix}e^{t} + c_2\begin{pmatrix} -4 \\ 6 \end{pmatrix}e^{2t}$

3. $\mathbf{A} = \begin{pmatrix} 1 & -1 \\ 1 & 1 \end{pmatrix}$, $\mathbf{X}(t) = e^{t}\left[c_1\begin{pmatrix} -\sin t \\ \cos t \end{pmatrix} + c_2\begin{pmatrix} \cos t \\ \sin t \end{pmatrix}\right]$

4. $A = \begin{pmatrix} -1 & -4 \\ 1 & -1 \end{pmatrix}$,

$$X(t) = e^{-t}\left[c_1 \begin{pmatrix} 2\cos 2t \\ \sin 2t \end{pmatrix} + c_2 \begin{pmatrix} -2\sin 2t \\ \cos 2t \end{pmatrix} \right]$$

5. $A = \begin{pmatrix} -6 & 5 \\ -5 & 4 \end{pmatrix}$,

$$X(t) = c_1 \begin{pmatrix} 1 \\ 1 \end{pmatrix} e^{-t} + c_2 \left[\begin{pmatrix} 1 \\ 1 \end{pmatrix} t e^{-t} + \begin{pmatrix} 0 \\ \frac{1}{5} \end{pmatrix} e^{-t} \right]$$

6. $A = \begin{pmatrix} 2 & 4 \\ -1 & 6 \end{pmatrix}$,

$$X(t) = c_1 \begin{pmatrix} 2 \\ 1 \end{pmatrix} e^{4t} + c_2 \left[\begin{pmatrix} 2 \\ 1 \end{pmatrix} t e^{4t} + \begin{pmatrix} 1 \\ 1 \end{pmatrix} e^{4t} \right]$$

7. $A = \begin{pmatrix} 2 & -1 \\ 3 & -2 \end{pmatrix}$, $\quad X(t) = c_1 \begin{pmatrix} 1 \\ 1 \end{pmatrix} e^{t} + c_2 \begin{pmatrix} 1 \\ 3 \end{pmatrix} e^{-t}$

8. $A = \begin{pmatrix} -1 & 5 \\ -1 & 1 \end{pmatrix}$,

$$X(t) = c_1 \begin{pmatrix} 5\cos 2t \\ \cos 2t - 2\sin 2t \end{pmatrix} + c_2 \begin{pmatrix} 5\sin 2t \\ 2\cos 2t + \sin 2t \end{pmatrix}$$

In Problems 9–16, classify the critical point (0, 0) of the given linear system by computing the trace τ and determinant Δ and using Figure 11.2.12.

9. $x' = -5x + 3y$
$y' = 2x + 7y$

10. $x' = -5x + 3y$
$y' = 2x - 7y$

11. $x' = -5x + 3y$
$y' = -2x + 5y$

12. $x' = -5x + 3y$
$y' = -7x + 4y$

13. $x' = -\frac{3}{2}x + \frac{1}{4}y$
$y' = -x - \frac{1}{2}y$

14. $x' = \frac{3}{2}x + \frac{1}{4}y$
$y' = -x + \frac{1}{2}y$

15. $x' = 0.02x - 0.11y$
$y' = 0.10x - 0.05y$

16. $x' = 0.03x + 0.01y$
$y' = -0.01x + 0.05y$

17. Determine conditions on the real constant μ so that (0, 0) is a center for the linear system

$$x' = -\mu x + y$$
$$y' = -x + \mu y.$$

18. Determine a condition on the real constant μ so that (0, 0) is a stable spiral point of the linear system

$$x' = y$$
$$y' = -x + \mu y.$$

19. Show that (0, 0) is always an unstable critical point of the linear system

$$x' = \mu x + y$$
$$y' = -x + y,$$

where μ is a real constant and $\mu \neq -1$. When is (0, 0) an unstable saddle point? When is (0, 0) an unstable spiral point?

20. Let $X = X(t)$ be the response of the linear dynamical system

$$x' = \alpha x - \beta y$$
$$y' = \beta x + \alpha y$$

that satisfies the initial condition $X(0) = X_0$. Determine conditions on the real constants α and β that will ensure $\lim_{t \to \infty} X(t) = (0, 0)$. Can (0, 0) be a node or saddle point?

21. Show that the nonhomogeneous linear system $X' = AX + F$ has a unique critical point X_1 when $\Delta = \det A \neq 0$. Conclude that if $X = X(t)$ is a solution to the nonhomogeneous system, $\tau < 0$ and $\Delta > 0$, then $\lim_{t \to \infty} X(t) = X_1$. [*Hint:* $X(t) = X_c(t) + X_1$.]

22. In Example 4(b) show that (0, 0) is a stable node when $bc < 1$.

In Problems 23–26, a nonhomogeneous linear system $X' = AX + F$ is given.
(a) In each case determine the unique critical point X_1.
(b) Use a numerical solver to determine the nature of the critical point in part (a).
(c) Investigate the relationship between X_1 and the critical point (0, 0) of the homogeneous linear system $X' = AX$.

23. $x' = 2x + 3y - 6$
$y' = -x - 2y + 5$

24. $x' = -5x + 9y + 13$
$y' = -x - 11y - 23$

25. $x' = 0.1x - 0.2y + 0.35$
$y' = 0.1x + 0.1y - 0.25$

26. $x' = 3x - 2y - 1$
$y' = 5x - 3y - 2$

11.3 Linearization and Local Stability

INTRODUCTION The key idea in this section is that of **linearization**. Recall from calculus and Section 2.6, a linearization of a differentiable function $f(x)$ at a number x_1 is the equation of the **tangent line** to the graph of f at the point:

$$y = f(x_1) + f'(x_1)(x - x_1).$$

For x close to x_1 the points on the graph of f are close to the points on the tangent line so values $y(x)$ obtained from the equation of the tangent line are said to be *local linear approximations* to the corresponding function values $f(x)$. Similarly, a linearization of a function of two variables $f(x, y)$ that is differentiable at a point (x_1, y_1) is the equation of the **tangent plane** to the graph of f at the point:

$$z = f(x_1, y_1) + f_x(x_1, y_1)(x - x_1) + f_y(x_1, y_1)(y - y_1),$$

where f_x and f_y are partial derivatives with respect to x and y, respectively. In this section we will use linearization as a means of analyzing nonlinear DEs and nonlinear systems; the idea is to replace them by linear DEs and linear systems.

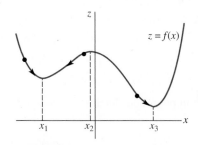

FIGURE 11.3.1 Bead sliding on graph of $z = f(x)$

■ Sliding Bead
We start this section by refining the stability concepts introduced in Section 11.2 in such a way that they will apply to nonlinear autonomous systems as well. Although the linear system $\mathbf{X}' = \mathbf{AX}$ had only one critical point when $\det \mathbf{A} \neq 0$, we saw in Section 11.1 that a nonlinear system may have many critical points. We therefore cannot expect that a particle placed initially at $\mathbf{X}_0$ will remain near a given critical point $\mathbf{X}_1$ unless $\mathbf{X}_0$ has been placed sufficiently close to $\mathbf{X}_1$ to begin with. The particle might well be driven to a second critical point $\mathbf{X}_2$. To emphasize this idea we will consider the physical system shown in **FIGURE 11.3.1** in which a bead slides along the curve $z = f(x)$ under the influence of gravity alone. We will show in Section 11.4 that the x-coordinate of the bead satisfies a nonlinear second-order differential equation $x'' = g(x, x')$, and therefore letting $y = x'$ satisfies the nonlinear autonomous system

$$x' = y$$
$$y' = g(x, y).$$

If the bead is positioned at $P = (x, f(x))$ and given zero initial velocity, the bead will remain at P provided $f'(x) = 0$. If the bead is placed near the critical point located at $x = x_1$, it will remain near $x = x_1$ only if its initial velocity does not drive it over the "hump" at $x = x_2$ toward the critical point located at $x = x_3$. Therefore $\mathbf{X}(0) = (x(0), x'(0))$ must be near $(x_1, 0)$.

In the next definition we will denote the distance between two points $\mathbf{X}$ and $\mathbf{Y}$ by $|\mathbf{X} - \mathbf{Y}|$. Recall that if $\mathbf{X} = (x_1, x_2, \ldots, x_n)$ and $\mathbf{Y} = (y_1, y_2 \ldots, y_n)$, then

$$|\mathbf{X} - \mathbf{Y}| = \sqrt{(x_1 - y_1)^2 + (x_2 - y_2)^2 + \cdots + (x_n - y_n)^2}.$$

DEFINITION 11.3.1 Stable Critical Points

Let $\mathbf{X}_1$ be a critical point of an autonomous system, and let $\mathbf{X} = \mathbf{X}(t)$ denote the solution that satisfies the initial condition $\mathbf{X}(0) = \mathbf{X}_0$, where $\mathbf{X}_0 \neq \mathbf{X}_1$. We say that $\mathbf{X}_1$ is a **stable critical point** when, given any radius $\rho > 0$, there is a corresponding radius $r > 0$ such that if the initial position $\mathbf{X}_0$ satisfies $|\mathbf{X}_0 - \mathbf{X}_1| < r$, then the corresponding solution $\mathbf{X}(t)$ satisfies $|\mathbf{X}(t) - \mathbf{X}_1| < \rho$ for all $t > 0$. If, in addition, $\lim_{t \to \infty} \mathbf{X}(t) = \mathbf{X}_1$ whenever $|\mathbf{X}_0 - \mathbf{X}_1| < r$, we call $\mathbf{X}_1$ an **asymptotically stable critical point**.

(a) stable

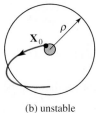

(b) unstable

FIGURE 11.3.2 Critical points

This definition is illustrated in **FIGURE 11.3.2(a)**. Given any disk of radius ρ about the critical point $\mathbf{X}_1$, a solution will remain inside this disk provided $\mathbf{X}(0) = \mathbf{X}_0$ is selected sufficiently close to $\mathbf{X}_1$. It is *not* necessary that a solution approach the critical point in order for $\mathbf{X}_1$ to be stable. Stable nodes, stable spiral points, and centers are all examples of stable critical points for linear systems. To emphasize that $\mathbf{X}_0$ must be selected close to $\mathbf{X}_1$ the terminology **locally stable critical point** is also used.

By negating Definition 11.3.1 we obtain the definition of an unstable critical point.

DEFINITION 11.3.2 Unstable Critical Point

Let $\mathbf{X}_1$ be a critical point of an autonomous system, and let $\mathbf{X} = \mathbf{X}(t)$ denote the solution that satisfies the initial condition $\mathbf{X}(0) = \mathbf{X}_0$, where $\mathbf{X}_0 \neq \mathbf{X}_1$. We say that $\mathbf{X}_1$ is an **unstable critical point** if there is a disk of radius $\rho > 0$ with the property that, for any $r > 0$, there is at least one initial position $\mathbf{X}_0$ that satisfies $|\mathbf{X}_0 - \mathbf{X}_1| < r$, yet the corresponding solution $\mathbf{X}(t)$ satisfies $|\mathbf{X}(t) - \mathbf{X}_1| \geq \rho$ for at least one $t > 0$.

If a critical point $\mathbf{X}_1$ is unstable, no matter how small the neighborhood about $\mathbf{X}_1$, an initial position $\mathbf{X}_0$ can always be found that results in the solution leaving some disk of radius ρ at some future time t. See Figure 11.3.2(b). Therefore unstable nodes, unstable spiral points, and saddle points are all examples of unstable critical points for linear systems. In Figure 11.3.1 the critical point $(x_2, 0)$ is unstable. The slightest displacement or initial velocity results in the bead sliding away from the point $(x_2, f(x_2))$.

EXAMPLE 1 A Stable Critical Point

Show that $(0, 0)$ is a stable critical point of the nonlinear plane autonomous system

$$x' = -y - x\sqrt{x^2 + y^2}$$
$$y' = x - y\sqrt{x^2 + y^2}$$

considered in Example 6 of Section 11.1.

SOLUTION In Example 6 of Section 11.1 we showed that in polar coordinates

$$r = \frac{1}{t + c_1}, \qquad \theta = t + c_2,$$

is the solution of the system. If $\mathbf{X}(0) = (r_0, \theta_0)$ is the initial condition in polar coordinates, then

$$r = \frac{r_0}{r_0 t + 1}, \qquad \theta = t + \theta_0.$$

Note that $r \leq r_0$ for $t \geq 0$, and r approaches $(0, 0)$ as t increases. Therefore, given $\rho > 0$, a solution that starts less than ρ units from $(0, 0)$ remains within ρ units of the origin for all $t \geq 0$. Hence the critical point $(0, 0)$ is stable and is in fact asymptotically stable. A typical solution is shown in **FIGURE 11.3.3**. ≡

FIGURE 11.3.3 Asymptotically stable critical point in Example 1

EXAMPLE 2 An Unstable Critical Point

When expressed in polar coordinates, a plane autonomous system takes the form

$$\frac{dr}{dt} = 0.05r(3 - r)$$

$$\frac{d\theta}{dt} = -1.$$

Show that $(x, y) = (0, 0)$ is an unstable critical point.

SOLUTION Since $x = r\cos\theta$ and $y = r\sin\theta$, we have

$$\frac{dx}{dt} = -r\sin\theta\,\frac{d\theta}{dt} + \frac{dr}{dt}\cos\theta$$

$$\frac{dy}{dt} = r\cos\theta\,\frac{d\theta}{dt} + \frac{dr}{dt}\sin\theta.$$

From $dr/dt = 0.05r(3 - r)$ we see that $dr/dt = 0$ when $r = 0$ and can conclude that $(x, y) = (0, 0)$ is a critical point by substituting $r = 0$ into the new system.

The differential equation $dr/dt = 0.05r(3 - r)$ is a logistic equation that can be solved using separation of variables. If $r(0) = r_0$ and $r_0 \neq 0$, then

$$r = \frac{3}{1 + c_0 e^{-0.15t}},$$

where $c_0 = (3 - r_0)/r_0$. Since

$$\lim_{t \to \infty} \frac{3}{1 + c_0 e^{-0.15t}} = 3,$$

it follows that no matter how close to $(0, 0)$ a solution starts, the solution will leave a disk of radius 1 about the origin. Therefore $(0, 0)$ is an unstable critical point. A typical solution that starts near $(0, 0)$ is shown in **FIGURE 11.3.4**. ≡

FIGURE 11.3.4 Unstable critical point in Example 2

■ **Linearization** It is rarely possible to determine the stability of a critical point of a nonlinear system by finding explicit solutions, as in Examples 1 and 2. Instead we replace the term

$g(\mathbf{X})$ in the original autonomous system $\mathbf{X}' = g(\mathbf{X})$ by a linear term $\mathbf{A}(\mathbf{X} - \mathbf{X}_1)$ that most closely approximates $g(\mathbf{X})$ in a neighborhood of $\mathbf{X}_1$. This replacement process, called **linearization**, will be illustrated first for the first-order differential equation $x' = g(x)$.

An equation of the tangent line to the curve $y = g(x)$ at $x = x_1$ is $y = g(x_1) + g'(x_1)(x - x_1)$, and if x_1 is a critical point of $x' = g(x)$, we have

$$x' = g(x) \approx g'(x_1)(x - x_1).$$

The general solution to the linear differential equation

$$x' = g'(x_1)(x - x_1)$$

is $x = x_1 + ce^{\lambda_1 t}$, where $\lambda_1 = g'(x_1)$. Thus if $g'(x_1) < 0$, then $x(t)$ approaches x_1. Theorem 11.3.1 asserts that the same behavior occurs in the original differential equation provided $x(0) = x_0$ is selected close enough to x_1.

Autonomous first-order DEs were discussed in Section 2.1.

THEOREM 11.3.1 Stability Criteria for $x' = g(x)$

Let x_1 be a critical point of the autonomous first-order differential equation $x' = g(x)$, where g is differentiable at x_1.

(a) If $g'(x_1) < 0$, then x_1 is an asymptotically stable critical point.
(b) If $g'(x_1) > 0$, then x_1 is an unstable critical point.

EXAMPLE 3 Stability in a Nonlinear First-Order DE

Both $x = \pi/4$ and $x = 5\pi/4$ are critical points of the autonomous differential equation $x' = \cos x - \sin x$. This differential equation is difficult to solve explicitly, but we can use Theorem 11.3.1 to predict the behavior of solutions near these two critical points.

Since $g'(x) = -\sin x - \cos x$, $g'(\pi/4) = -\sqrt{2} < 0$ and $g'(5\pi/4) = \sqrt{2} > 0$. Therefore $x = \pi/4$ is an asymptotically stable critical point, but $x = 5\pi/4$ is unstable. In **FIGURE 11.3.5** we used a numerical solver to investigate solutions that start near $(0, \pi/4)$ and $(0, 5\pi/4)$. Observe that solution curves that start close to $(0, 5\pi/4)$ quickly move away from the line $x = 5\pi/4$, as predicted. ≡

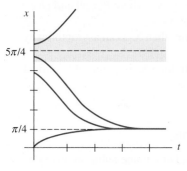

FIGURE 11.3.5 In Example 3, $\pi/4$ is asymptotically stable; $5\pi/4$ is unstable

EXAMPLE 4 Stability Analysis of the Logistic DE

Without solving explicitly, analyze the critical points of the logistic differential equation (see Section 2.8) $x' = \dfrac{r}{K} x(K - x)$, where r and K are positive constants.

SOLUTION The two critical points are $x = 0$ and $x = K$. Since

$$g'(x) = \frac{r}{K}(K - 2x),$$

$g'(0) = r$ and $g'(K) = -r$ and it follows from Theorem 11.3.1 that $x = 0$ is an unstable critical point and $x = K$ is an asymptotically stable critical point. ≡

Jacobian Matrix A similar analysis may be carried out for a plane autonomous system. An equation of the tangent plane to the surface $z = g(x, y)$ at $\mathbf{X}_1 = (x_1, y_1)$ is

$$z = g(x_1, y_1) + \left.\frac{\partial g}{\partial x}\right|_{(x_1, y_1)} (x - x_1) + \left.\frac{\partial g}{\partial y}\right|_{(x_1, y_1)} (y - y_1),$$

and $g(x, y)$ may be approximated by its tangent plane in a neighborhood of $\mathbf{X}_1$.

When $\mathbf{X}_1$ is a critical point of a plane autonomous system, $P(x_1, y_1) = Q(x_1, y_1) = 0$ and we have

$$x' = P(x, y) \approx \left.\frac{\partial P}{\partial x}\right|_{(x_1, y_1)} (x - x_1) + \left.\frac{\partial P}{\partial y}\right|_{(x_1, y_1)} (y - y_1)$$

$$y' = Q(x, y) \approx \left.\frac{\partial Q}{\partial x}\right|_{(x_1, y_1)} (x - x_1) + \left.\frac{\partial Q}{\partial y}\right|_{(x_1, y_1)} (y - y_1).$$

The original system $X' = g(X)$ may be approximated in a neighborhood of the critical point X_1 by the linear system $X' = A(X - X_1)$, where

$$A = \begin{pmatrix} \dfrac{\partial P}{\partial x}\bigg|_{(x_1, y_1)} & \dfrac{\partial P}{\partial y}\bigg|_{(x_1, y_1)} \\ \dfrac{\partial Q}{\partial x}\bigg|_{(x_1, y_1)} & \dfrac{\partial Q}{\partial y}\bigg|_{(x_1, y_1)} \end{pmatrix}.$$

This matrix is called the **Jacobian matrix** at X_1 after the German mathematician **Carl Gustav Jacob Jacobi** (1804–1851). The Jacobian matrix at X_1 is also denoted by $g'(X_1)$. If we let $H = X - X_1$, then the linear system $X' = A(X - X_1)$ becomes $H' = AH$, which is the form of the linear system analyzed in Section 11.2. The critical point $X = X_1$ for $X' = A(X - X_1)$ now corresponds to the critical point $H = 0$ for $H' = AH$. If the eigenvalues of A have negative real parts, then, by Theorem 11.2.1, 0 is an asymptotically stable critical point for $H' = AH$. If there is an eigenvalue with a positive real part, $H = 0$ is an unstable critical point. Theorem 11.3.2 asserts that the same conclusions can be made for the critical point X_1 of the original system.

THEOREM 11.3.2 Stability Criteria for Plane Autonomous Systems

Let X_1 be a crtical point of the plane autonomous sytem $X' = g(X)$, where $P(x, y)$ and $Q(x, y)$ have continuous first partial derivatives in a neighborhood of X_1.
 (a) If the eigenvalues of $A = g'(X_1)$ have negative real part, then X_1 is an asymptotically stable critical point.
 (b) If $A = g'(X_1)$ has an eigenvalue with positive real part, then X_1 is an unstable critical point.

EXAMPLE 5 **Stability Analysis of Nonlinear Systems**

Classify (if possible) the critical points of each of the following plane autonomous systems as stable or unstable.
 (a) $x' = x^2 + y^2 - 6$ (b) $x' = 0.01x(100 - x - y)$
 $y' = x^2 - y$ $y' = 0.05y(60 - y - 0.2x)$

SOLUTION The critical points of each system were determined in Example 4 of Section 11.1.

(a) The critical points are $(\sqrt{2}, 2)$ and $(-\sqrt{2}, 2)$, the Jacobian matrix is

$$g'(X) = \begin{pmatrix} 2x & 2y \\ 2x & -1 \end{pmatrix},$$

and so

$$A_1 = g'((\sqrt{2}, 2)) = \begin{pmatrix} 2\sqrt{2} & 4 \\ 2\sqrt{2} & -1 \end{pmatrix} \quad \text{and} \quad A_2 = g'((-\sqrt{2}, 2)) = \begin{pmatrix} -2\sqrt{2} & 4 \\ -2\sqrt{2} & -1 \end{pmatrix}.$$

Since the determinant of A_1 is negative, A_1 has a positive real eigenvalue. Therefore $(\sqrt{2}, 2)$ is an unstable critical point. Matrix A_2 has a positive determinant and a negative trace, and so both eigenvalues have negative real parts. It follows that $(-\sqrt{2}, 2)$ is a stable critical point.

(b) The critical points are $(0, 0)$, $(0, 60)$, $(100, 0)$, and $(50, 50)$, the Jacobian matrix is

$$g'(X) = \begin{pmatrix} 0.01(100 - 2x - y) & -0.01x \\ -0.01y & 0.05(60 - 2y - 0.2x) \end{pmatrix},$$

and so

$$A_1 = g'((0, 0)) = \begin{pmatrix} 1 & 0 \\ 0 & 3 \end{pmatrix} \qquad\qquad A_2 = g'((0, 60)) = \begin{pmatrix} 0.4 & 0 \\ -0.6 & -3 \end{pmatrix}$$

$$A_3 = g'((100, 0)) = \begin{pmatrix} -1 & -1 \\ 0 & 2 \end{pmatrix} \qquad A_4 = g'((50, 50)) = \begin{pmatrix} -0.5 & -0.5 \\ -0.5 & -2.5 \end{pmatrix}.$$

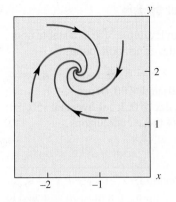

FIGURE 11.3.6 Point appears to be a stable spiral point

Since the matrix $\mathbf{A}_1$ has a positive determinant and a positive trace, both eigenvalues have positive real parts. Therefore $(0, 0)$ is an unstable critical point. The determinants of matrices $\mathbf{A}_2$ and $\mathbf{A}_3$ are negative, and so in each case one of the eigenvalues is positive. Therefore both $(0, 60)$ and $(100, 0)$ are unstable critical points. Since the matrix $\mathbf{A}_4$ has a positive determinant and a negative trace, $(50, 50)$ is a stable critical point. ≡

In Example 5 we did not compute $\tau^2 - 4\Delta$ (as in Section 11.2) and attempt to further classify the critical points as stable nodes, stable spiral points, saddle points, and so on. For example, for $\mathbf{X}_1 = (-\sqrt{2}, 2)$ in Example 4(a), $\tau^2 - 4\Delta < 0$, and if the system were linear, we would be able to conclude that $\mathbf{X}_1$ was a stable spiral point. **FIGURE 11.3.6** shows several solution curves near $\mathbf{X}_1$ that were obtained with a numerical solver, and each solution does *appear* to spiral in toward the critical point.

▦ **Classifying Critical Points** It is natural to ask whether we can infer more geometric information about the solutions near a critical point $\mathbf{X}_1$ of a nonlinear autonomous system from an analysis of the critical point of the corresponding linear system. The answer is summarized in **FIGURE 11.3.7**, but you should note the comments following the figure.

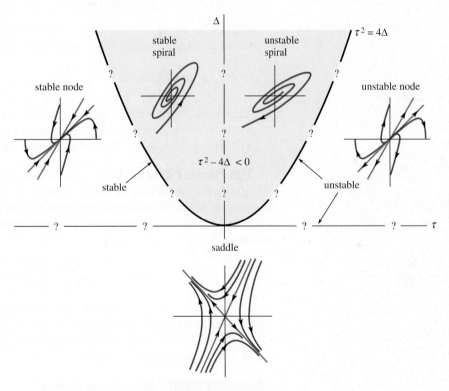

FIGURE 11.3.7 Geometric summary

(*i*) In five separate cases (stable node, stable spiral point, unstable spiral point, unstable node, and saddle) the critical point may be categorized like the critical point in the corresponding linear system. The solutions have *the same general geometric features* as the solutions to the linear system, and the smaller the neighborhood about $\mathbf{X}_1$, the closer the resemblance.

(*ii*) If $\tau^2 = 4\Delta$ and $\tau > 0$, the critical point $\mathbf{X}_1$ is unstable, but in this borderline case *we are not yet able to decide whether $\mathbf{X}_1$ is an unstable spiral, unstable node, or degenerate unstable node.* Likewise, if $\tau^2 = 4\Delta$ and $\tau < 0$, the critical point $\mathbf{X}_1$ is stable but may be either a stable spiral, a stable node, or a degenerate stable node.

(*iii*) If $\tau = 0$ and $\Delta > 0$, the eigenvalues of $\mathbf{A} = \mathbf{g}'(\mathbf{X})$ are pure imaginary, and in this borderline case $\mathbf{X}_1$ may be either a stable spiral, an unstable spiral, or a center. *It is therefore not yet possible to determine whether $\mathbf{X}_1$ is stable or unstable.*

EXAMPLE 6 **Classifying Critical Points of a Nonlinear System**

Classify each critical point of the plane autonomous system in Example 5(b) as a stable node, a stable spiral point, an unstable spiral point, an unstable node, or a saddle point.

SOLUTION For the matrix $\mathbf{A}_1$ corresponding to $(0, 0)$, $\Delta = 3$, $\tau = 4$, and so $\tau^2 - 4\Delta = 4$. Therefore $(0, 0)$ is an unstable node. The critical points $(0, 60)$ and $(100, 0)$ are saddles since $\Delta < 0$ in both cases. For matrix $\mathbf{A}_4$, $\Delta > 0$, $\tau < 0$, and $\tau^2 - 4\Delta > 0$. It follows that $(50, 50)$ is a stable node. Experiment with a numerical solver to verify these conclusions. ≡

EXAMPLE 7 **Stability Analysis for a Soft Spring**

Recall from Section 3.11 that the second-order differential equation $mx'' + kx + k_1x^3 = 0$, for $k > 0$, represents a general model for the free, undamped oscillations of a mass m attached to a nonlinear spring. If $k = 1$ and $k_1 = -1$, the spring is called *soft* and the plane autonomous system corresponding to the nonlinear second-order differential equation $x'' + x - x^3 = 0$ is

$$x' = y$$
$$y' = x^3 - x.$$

Find and classify (if possible) the critical points.

SOLUTION Since $x^3 - x = x(x^2 - 1)$, the critical points are $(0, 0)$, $(1, 0)$, and $(-1, 0)$. The corresponding Jacobian matrices are

$$\mathbf{A}_1 = \mathbf{g}'((0, 0)) = \begin{pmatrix} 0 & 1 \\ -1 & 0 \end{pmatrix}, \quad \mathbf{A}_2 = \mathbf{g}'((1, 0)) = \mathbf{g}'((-1, 0)) = \begin{pmatrix} 0 & 1 \\ 2 & 0 \end{pmatrix}.$$

Since det $\mathbf{A}_2 < 0$, critical points $(1, 0)$ and $(-1, 0)$ are both saddle points. The eigenvalues of matrix $\mathbf{A}_1$ are $\pm i$, and according to comment (*iii*) on page 657, the status of the critical point at $(0, 0)$ remains in doubt. It may be either a stable spiral, an unstable spiral, or a center. ≡

■ The Phase-Plane Method The linearization method, when successful, can provide useful information on the local behavior of solutions near critical points. It is of little help if we are interested in solutions whose initial position $\mathbf{X}(0) = \mathbf{X}_0$ is not close to a critical point or if we wish to obtain a global view of the family of solution curves. The **phase-plane method** is based on the fact that

$$\frac{dy}{dx} = \frac{dy/dt}{dx/dt} = \frac{Q(x, y)}{P(x, y)}$$

and it attempts to find y as a function of x using one of the methods available for solving first-order differential equations (Chapter 2). As we show in Examples 8 and 9, the method can sometimes be used to decide whether a critical point such as $(0, 0)$ in Example 7 is a stable spiral, an unstable spiral, or a center.

EXAMPLE 8 **Phase-Plane Method**

Use the phase-plane method to classify the sole critical point $(0, 0)$ of the plane autonomous system

$$x' = y^2$$
$$y' = x^2.$$

SOLUTION The determinant of the Jacobian matrix

$$\mathbf{g}'(\mathbf{X}) = \begin{pmatrix} 0 & 2y \\ 2x & 0 \end{pmatrix}$$

is 0 at $(0, 0)$, and so the nature of the critical point $(0, 0)$ remains in doubt. Using the phase-plane method, we obtain the first-order differential equation

$$\frac{dy}{dx} = \frac{dy/dt}{dx/dt} = \frac{x^2}{y^2},$$

FIGURE 11.3.8 Phase portrait of nonlinear system in Example 8

which can be easily solved by separation of variables:

$$\int y^2 \, dy = \int x^2 \, dx \quad \text{or} \quad y^3 = x^3 + c.$$

If $\mathbf{X}(0) = (0, y_0)$, it follows that $y^3 = x^3 + y_0^3$ or $y = \sqrt[3]{x^3 + y_0^3}$. **FIGURE 11.3.8** shows a collection of solution curves corresponding to various choices for y_0, and the nature of the critical point is clear. No matter how close to $(0, 0)$ the solution starts, $\mathbf{X}(t)$ moves away from the origin as t increases. The critical point at $(0, 0)$ is therefore unstable. $\equiv$

EXAMPLE 9 **Phase-Plane Analysis of a Soft Spring**

Use the phase-plane method to determine the nature of the solutions to $x'' + x - x^3 = 0$ in a neighborhood of $(0, 0)$.

SOLUTION If we let $dx/dt = y$, then $dy/dt = x^3 - x$. From this we obtain the first-order differential equation

$$\frac{dy}{dx} = \frac{dy/dt}{dx/dt} = \frac{x^3 - x}{y},$$

which can be solved by separation of variables. Integrating

$$\int y \, dy = \int (x^3 - x) \, dx \quad \text{gives} \quad \frac{y^2}{2} = \frac{x^4}{4} - \frac{x^2}{2} + c.$$

After completing the square, we can write the solution as $y^2 = (x^2 - 1)^2/2 + c_0$. If $\mathbf{X}(0) = (x_0, 0)$, where $0 < x_0 < 1$, then $c_0 = -(x_0^2 - 1)^2/2$, and so

$$y^2 = \frac{(x^2 - 1)^2}{2} - \frac{(x_0^2 - 1)^2}{2} = \frac{(2 - x^2 - x_0^2)(x_0^2 - x^2)}{2}.$$

Note that $y = 0$ when $x = -x_0$. In addition, the right-hand side is positive when $-x_0 < x < x_0$, and so each x has *two* corresponding values of y. The solution $\mathbf{X} = \mathbf{X}(t)$ that satisfies $\mathbf{X}(0) = (x_0, 0)$ is therefore periodic, and so $(0, 0)$ is a center.

FIGURE 11.3.9 shows a family of solution curves or phase portrait of the original system. We used the original plane autonomous system to determine the directions indicated on each trajectory. $\equiv$

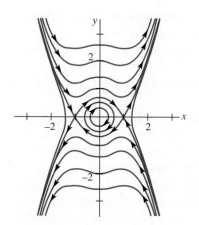

FIGURE 11.3.9 Phase portrait of nonlinear system in Example 9

11.3 Exercises Answers to selected odd-numbered problems begin on page ANS-30.

1. Show that $(0, 0)$ is an asymptotically stable critical point of the nonlinear autonomous system

$$x' = \alpha x - \beta y + y^2$$
$$y' = \beta x + \alpha y - xy$$

when $\alpha < 0$ and an unstable critical point when $\alpha > 0$. [*Hint*: Switch to polar coordinates.]

2. When expressed in polar coordinates, a plane autonomous system takes the form

$$\frac{dr}{dt} = \alpha r(5 - r)$$

$$\frac{d\theta}{dt} = -1.$$

Show that $(0, 0)$ is an asymptotically stable critical point if and only if $\alpha < 0$.

In Problems 3–10, without solving explicitly, classify the critical points of the given first-order autonomous differential equation as either asymptotically stable or unstable. All constants are assumed to be positive.

3. $\dfrac{dx}{dt} = kx(n + 1 - x)$

4. $\dfrac{dx}{dt} = -kx \ln \dfrac{x}{k}, \ x > 0$

5. $\dfrac{dT}{dt} = k(T - T_0)$

6. $m \dfrac{dv}{dt} = mg - kv$

7. $\dfrac{dx}{dt} = k(\alpha - x)(\beta - x), \ \alpha > \beta$

8. $\dfrac{dx}{dt} = k(\alpha - x)(\beta - x)(\gamma - x), \ \alpha > \beta > \gamma$

9. $\dfrac{dP}{dt} = P(a - bP)(1 - cP^{-1}), \ P > 0, a < bc$

10. $\dfrac{dA}{dt} = k\sqrt{A}(K - \sqrt{A}), \ A > 0$

In Problems 11–20, classify (if possible) each critical point of the given plane autonomous system as a stable node, a stable spiral point, an unstable spiral point, an unstable node, or a saddle point.

11. $x' = 1 - 2xy$
$y' = 2xy - y$

12. $x' = x^2 - y^2 - 1$
$y' = 2y$

13. $x' = y - x^2 + 2$
$y' = 2xy - y$

14. $x' = 2x - y^2$
$y' = -y + xy$

15. $x' = -3x + y^2 + 2$
$y' = x^2 - y^2$

16. $x' = xy - 3y - 4$
$y' = y^2 - x^2$

17. $x' = -2xy$
$y' = y - x + xy - y^3$

18. $x' = x(1 - x^2 - 3y^2)$
$y' = y(3 - x^2 - 3y^2)$

19. $x' = x(10 - x - \frac{1}{2}y)$
$y' = y(16 - y - x)$

20. $x' = -2x + y + 10$
$y' = 2x - y - 15\dfrac{y}{y + 5}$

In Problems 21–26, classify (if possible) each critical point of the given second-order differential equation as a stable node, a stable spiral point, an unstable spiral point, an unstable node, or a saddle point.

21. $\theta'' = (\cos\theta - 0.5)\sin\theta, \quad |\theta| < \pi$

22. $x'' + x = (\frac{1}{2} - 3(x')^2)x' - x^2$

23. $x'' + x'(1 - x^3) - x^2 = 0$

24. $x'' + 4\dfrac{x}{1 + x^2} + 2x' = 0$

25. $x'' + x = \epsilon x^3$ for $\epsilon > 0$

26. $x'' + x - \epsilon x|x| = 0$ for $\epsilon > 0$ [*Hint:* $\dfrac{d}{dx}\, x|x| = 2|x|.$]

27. Show that the nonlinear second-order differential equation
$$(1 + \alpha^2 x^2)x'' + (\beta + \alpha^2(x')^2)x = 0$$
has a saddle point at $(0, 0)$ when $\beta < 0$.

28. Show that the dynamical system
$$x' = -\alpha x + xy$$
$$y' = 1 - \beta y - x^2$$
has a unique critical point when $\alpha\beta > 1$ and that this critical point is stable when $\beta > 0$.

29. (a) Show that the plane autonomous system
$$x' = -x + y - x^3$$
$$y' = -x - y + y^2$$
has two critical points by sketching the graphs of $-x + y - x^3 = 0$ and $-x - y + y^2 = 0$. Classify the critical point at $(0, 0)$.

(b) Show that the second critical point $\mathbf{X}_1 = (0.88054, 1.56327)$ is a saddle point.

30. (a) The equation
$$x'' + \epsilon\left(\frac{1}{3}(x')^3 - x'\right) + x = 0$$
is called **Rayleigh's differential equation** after the English physicist **John William Strutt, 3rd Baron Rayleigh** (1842–1919). Strutt was the co-discoverer of argon gas and was awarded the Nobel Prize in physics in 1904. Show that $(0, 0)$ is the only critical point of this differential equation.

(b) Show that $(0, 0)$ is unstable when $\epsilon > 0$. When is $(0, 0)$ an unstable spiral point?

(c) Show that $(0, 0)$ is stable when $\epsilon < 0$. When is $(0, 0)$ a stable spiral point?

(d) Show that $(0, 0)$ is a center when $\epsilon = 0$.

31. Use the phase-plane method to show that $(0, 0)$ is a center of the nonlinear second-order differential equation $x'' + 2x^3 = 0$.

32. Use the phase-plane method to show that the solution to the nonlinear second-order differential equation $x'' + 2x - x^2 = 0$ that satisfies $x(0) = 1$ and $x'(0) = 0$ is periodic.

33. (a) Find the critical points of the plane autonomous system
$$x' = 2xy$$
$$y' = 1 - x^2 + y^2,$$
and show that linearization gives no information about the nature of these critical points.

(b) Use the phase-plane method to show that the critical points in part (a) are both centers. [*Hint:* Let $u = y^2/x$, and show that $(x - c)^2 + y^2 = c^2 - 1.$]

34. The origin is the only critical point of the nonlinear second-order differential equation $x'' + (x')^2 + x = 0$.

(a) Show that the phase-plane method leads to the Bernoulli differential equation $dy/dx = -y - xy^{-1}$.

(b) Show that the solution satisfying $x(0) = \frac{1}{2}$ and $x'(0) = 0$ is not periodic.

35. A solution of the nonlinear second-order differential equation $x'' + x - x^3 = 0$ satisfies $x(0) = 0$ and $x'(0) = v_0$. Use the phase-plane method to determine when the resulting solution is periodic. [*Hint:* See Example 9.]

36. The nonlinear differential equation $x'' + x = 1 + \epsilon x^2$ arises in the analysis of planetary motion using relativity theory. Classify (if possible) all critical points of the corresponding plane autonomous system.

37. When a nonlinear capacitor is present in an *LRC*-series circuit, the voltage drop is no longer given by q/C but is more accurately described by $\alpha q + \beta q^3$, where α and β are constants and $\alpha > 0$. Differential equation (34) of Section 3.8 for the free circuit is then replaced by
$$L\frac{d^2q}{dt^2} + R\frac{dq}{dt} + \alpha q + \beta q^3 = 0.$$
Find and classify all critical points of this nonlinear differential equation. [*Hint:* Divide into the two cases $\beta > 0$ and $\beta < 0$.]

38. The nonlinear second-order differential equation
$$mx'' + kx + k_1 x^3 = 0,$$
for $k > 0$, represents a general model for the free, undamped oscillations of a mass m attached to a spring. If $k_1 > 0$, the spring is called *hard* (see Example 1 in Section 3.11). Determine the nature of the solutions to $x'' + x + x^3 = 0$ in a neighborhood of $(0, 0)$.

39. The nonlinear differential equation
$$\frac{d^2\theta}{dt^2} + \sin\theta = \frac{1}{2}$$

can be interpreted as a model for a certain pendulum with a constant driving function.

(a) Show that $(\pi/6, 0)$ and $(5\pi/6, 0)$ are critical points of the corresponding plane autonomous system.

(b) Classify the critical point $(5\pi/6, 0)$ using linearization.

(c) Use the phase-plane method to classify the critical point $(\pi/6, 0)$.

40. (a) Show that $(0, 0)$ is an isolated critical point of the plane autonomous system

$$x' = x^4 - 2xy^3$$
$$y' = 2x^3y - y^4$$

but that linearization gives no useful information about the nature of this critical point.

(b) Use the phase-plane method to show that a family of solutions of the system in part (a) is

$$x^3 + y^3 = 3cxy.$$

The graphs of the members of this one-parameter family are called **folia of Descartes**. [*Hint*: The differential equation in x and y is homogeneous.]

(c) Parametric equations for the folia in part (b) are

$$x = \frac{3ct}{1 + t^3}, \quad y = \frac{3ct^2}{1 + t^3}.$$

Use graphing software or a numerical solver to graph various solution curves defined by these equations. Based on your graphs, would you classify the critical point $(0, 0)$ as stable or unstable? Would you classify the critical point as a node, saddle point, center, or spiral point?

11.4 Autonomous Systems as Mathematical Models

INTRODUCTION Many applications from physics give rise to nonlinear autonomous second-order differential equations—that is, DEs of the form $x'' = g(x, x')$. For example, in the analysis of free, damped motion of a spring/mass system in Section 3.8 we assumed that the damping force was proportional to the velocity x' and the resulting model $mx'' = -\beta x' - kx$ is a linear differential equation. But if the magnitude of the damping force is proportional to the square of the velocity, the new differential equation $mx'' = -\beta x'|x'| - kx$ is nonlinear. The corresponding plane autonomous system is nonlinear:

$$x' = y$$
$$y' = -\frac{\beta}{m} y|y| - \frac{k}{m} x.$$

In this section we will also analyze the nonlinear pendulum, motion of a bead on a curve, the Lotka–Volterra predator–prey models, and the Lotka–Volterra competition model. Additional models are presented in these exercises.

Nonlinear Pendulum In (6) of Section 3.11 we showed that the displacement angle θ for a simple pendulum satisfies the nonlinear second-order differential equation

$$\frac{d^2\theta}{dt^2} + \frac{g}{l} \sin \theta = 0.$$

When we let $x = \theta$ and $y = \theta'$, this second-order differential equation may be rewritten as the dynamical system

$$x' = y$$
$$y' = -\frac{g}{l} \sin x.$$

The critical points are $(\pm k\pi, 0)$, and the Jacobian matrix is easily shown to be

$$\mathbf{g}'((\pm k\pi, 0)) = \begin{pmatrix} 0 & 1 \\ (-1)^{k+1}\dfrac{g}{l} & 0 \end{pmatrix}.$$

(a) $\theta = 0, \theta' = 0$

(b) $\theta = \pi, \theta' = 0$

FIGURE 11.4.1 $(0, 0)$ is stable; $(\pi, 0)$ is unstable

If $k = 2n + 1$, $\Delta < 0$, and so all critical points $(\pm(2n + 1)\pi, 0)$ are saddle points. In particular, the critical point at $(\pi, 0)$ is unstable as expected. See **FIGURE 11.4.1**. When $k = 2n$, the eigenvalues are pure imaginary, and so the nature of these critical points remains in doubt. Since we have assumed that there are no damping forces acting on the pendulum, we expect that all the critical points $(\pm 2n\pi, 0)$ are centers. This can be verified using the phase-plane method. From

$$\frac{dy}{dx} = \frac{dy/dt}{dx/dt} = -\frac{g}{l} \frac{\sin x}{y}$$

FIGURE 11.4.2 Phase portrait of nonlinear pendulum

it follows that

$$y^2 = \frac{2g}{l}\cos x + c.$$

If $\mathbf{X}(0) = (x_0, 0)$, then

$$y^2 = \frac{2g}{l}(\cos x - \cos x_0).$$

Note that $y = 0$ when $x = -x_0$, and that $(2g/l)(\cos x - \cos x_0) > 0$ for $|x| < |x_0| < \pi$. Thus each such x has two corresponding values of y, and so the solution $\mathbf{X} = \mathbf{X}(t)$ that satisfies $\mathbf{X}(0) = (x_0, 0)$ is periodic. We may conclude that $(0, 0)$ is a center. Observe that $x = \theta$ increases for solutions, such as the one drawn in red in **FIGURE 11.4.2**, that correspond to large initial velocities. In this case the pendulum spins in complete circles about its pivot.

EXAMPLE 1 **Periodic Solutions of the Pendulum DE**

A pendulum in an equilibrium position with $\theta = 0$ is given an initial angular velocity of ω_0 rad/s. Determine under what conditions the resulting motion is periodic.

SOLUTION We are asked to examine the solution of the plane autonomous system that satisfies $\mathbf{X}(0) = (0, \omega_0)$. From $y^2 = (2g/l)\cos x + c$ it follows that

$$y^2 = \frac{2g}{l}\left(\cos x - 1 + \frac{l}{2g}\omega_0^2\right).$$

To establish that the solution $\mathbf{X}(t)$ is periodic it is sufficient to show that there are two x-intercepts $x = \pm x_0$ between $-\pi$ and π and that the right-hand side is positive for $|x| < |x_0|$. Each such x then has two corresponding values of y.

If $y = 0$, $\cos x = 1 - (l/2g)\omega_0^2$, and this equation has two solutions $x = \pm x_0$ between $-\pi$ and π, provided $1 - (l/2g)\omega_0^2 > -1$. Note that $(2g/l)(\cos x - \cos x_0)$ is then positive for $|x| < |x_0|$. This restriction on the initial angular velocity can be written as

$$|\omega_0| < 2\sqrt{\frac{g}{l}}.$$

$\equiv$

Nonlinear Oscillations: The Sliding Bead Suppose, as shown in **FIGURE 11.4.3**, a bead with mass m slides along a thin wire whose shape is described by the function $z = f(x)$. A wide variety of nonlinear oscillations can be obtained by changing the shape of the wire and by making different assumptions about the forces acting on the bead.

The tangential force $\mathbf{F}$ due to the weight $W = mg$ has magnitude $mg \sin \theta$, and therefore the x-component of $\mathbf{F}$ is $F_x = -mg \sin \theta \cos \theta$. Since $\tan \theta = f'(x)$, we can use the identities $1 + \tan^2 \theta = \sec^2 \theta$ and $\sin^2 \theta = 1 - \cos^2 \theta$ to conclude that

$$F_x = -mg \sin \theta \cos \theta = -mg \frac{f'(x)}{1 + [f'(x)]^2}.$$

FIGURE 11.4.3 Forces acting on sliding bead

We assume (as in Section 3.8) that a damping force $\mathbf{D}$, acting in the direction opposite to the motion, is a constant multiple of the velocity of the bead. The x-component of $\mathbf{D}$ is therefore

$$D_x = -\beta \frac{dx}{dt}.$$

If we ignore the frictional force between the wire and the bead and assume that no other external forces are impressed on the system, it follows from Newton's second law that

$$mx'' = -mg \frac{f'(x)}{1 + [f'(x)]^2} - \beta x',$$

and the corresponding plane autonomous system is

$$x' = y$$

$$y' = -g \frac{f'(x)}{1 + [f'(x)]^2} - \frac{\beta}{m} y.$$

If $\mathbf{X}_1 = (x_1, y_1)$ is a critical point of the system, $y_1 = 0$ and therefore $f'(x_1) = 0$. The bead must therefore be at rest at a point on the wire where the tangent line is horizontal. When f is twice differentiable, the Jacobian matrix at $\mathbf{X}_1$ is

$$\mathbf{g}'(\mathbf{X}_1) = \begin{pmatrix} 0 & 1 \\ -gf''(x_1) & -\beta/m \end{pmatrix},$$

and so $\tau = -\beta/m$, $\Delta = gf''(x_1)$, and $\tau^2 - 4\Delta = \beta^2/m^2 - 4gf''(x_1)$. Using the results of Section 11.3, we can make the following conclusions:

(*i*) $f''(x_1) < 0$:

A relative maximum therefore occurs at $x = x_1$, and since $\Delta < 0$, an *unstable saddle point* occurs at $\mathbf{X}_1 = (x_1, 0)$.

(*ii*) $f''(x_1) > 0$ and $\beta > 0$:

A relative minimum therefore occurs at $x = x_1$, and since $\tau < 0$ and $\Delta > 0$, $\mathbf{X}_1 = (x_1, 0)$ is a *stable critical point*. If $\beta^2 > 4gm^2f''(x_1)$, the system is **overdamped** and the critical point is a *stable node*. If $\beta^2 < 4gm^2f''(x_1)$, the system is **underdamped** and the critical point is a *stable spiral point*. The exact nature of the stable critical point is still in doubt if $\beta^2 = 4gm^2f''(x_1)$.

(*iii*) $f''(x_1) > 0$ and the system is undamped ($\beta = 0$):

In this case the eigenvalues are pure imaginary, but the phase-plane method can be used to show that the critical point is a *center*. Therefore solutions with $\mathbf{X}(0) = (x(0), x'(0))$ near $\mathbf{X}_1 = (x_1, 0)$ are periodic.

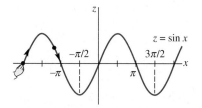

FIGURE 11.4.4 $-\pi/2$ and $3\pi/2$ are stable in Example 2

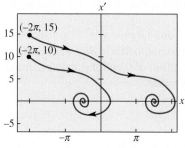

FIGURE 11.4.5 $\beta = 0.01$ in Example 2

FIGURE 11.4.6 $\beta = 0$ in Example 2

EXAMPLE 2 **Bead Sliding Along a Sine Wave**

A 10-gram bead slides along the graph of $z = \sin x$. According to conclusion (*ii*), the relative minima at $x_1 = -\pi/2$ and $3\pi/2$ give rise to stable critical points. See **FIGURE 11.4.4**. Since $f''(-\pi/2) = f''(3\pi/2) = 1$, the system will be underdamped provided $\beta^2 < 4gm^2$. If we use SI units, $m = 0.01$ kg and $g = 9.8$ m/s^2, and so the condition for an underdamped system becomes $\beta^2 < 3.92 \times 10^{-3}$.

If $\beta = 0.01$ is the damping constant, both of these critical points are stable spiral points. The two solutions corresponding to initial conditions $\mathbf{X}(0) = (x(0), x'(0)) = (-2\pi, 10)$ and $\mathbf{X}(0) = (-2\pi, 15)$, respectively, were obtained from a numerical solver and are shown in **FIGURE 11.4.5**. When $x'(0) = 10$, the bead has enough momentum to make it over the hill at $x = -3\pi/2$ but not over the hill at $x = \pi/2$. The bead then approaches the relative minimum based at $x = -\pi/2$. If $x'(0) = 15$, the bead has the momentum to make it over both hills, but then it rocks back and forth in the valley based at $x = 3\pi/2$ and approaches the point $(3\pi/2, -1)$ on the wire. Experiment with other initial conditions using your numerical solver.

FIGURE 11.4.6 shows a collection of solution curves obtained from a numerical solver for the undamped case. Since $\beta = 0$, the critical points corresponding to $x_1 = -\pi/2$ and $3\pi/2$ are now centers. When $\mathbf{X}(0) = (-2\pi, 10)$, the bead has sufficient momentum to move over *all* hills. The figure also indicates that when the bead is released from rest at a position on the wire between $x = -3\pi/2$ and $x = \pi/2$, the resulting motion is periodic. $\equiv$

▣ Lotka–Volterra Predator–Prey Model

A **predator–prey interaction** between two species occurs when one species (the predator) feeds on a second species (the prey). For example, the snowy owl feeds almost exclusively on a common arctic rodent called a lemming, while a lemming uses arctic tundra plants as its food supply. Interest in using mathematics to help explain predator–prey interactions has been stimulated by the observation of population cycles in many arctic mammals. In the MacKenzie River district of Canada, for example, the principal prey of the lynx is the snowshoe hare, and both populations cycle with a period of about 10 years.

There are many predator–prey models that lead to plane autonomous systems with at least one periodic solution. The first such model was constructed independently by the American mathematician and biophysicist **Alfred James Lotka** (1880–1949) and the Italian mathematician and physicist **Vito Volterra** (1860–1940). If x denotes the number of predators and y denotes the number of prey, then the Lotka–Volterra model takes the form

$$x' = -ax + bxy = x(-a + by)$$
$$y' = -cxy + dy = y(-cx + d),$$

where a, b, c, and d are positive constants.

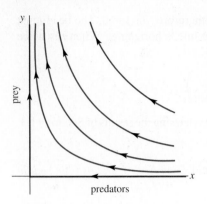

FIGURE 11.4.7 Solutions near (0, 0)

Note that in the absence of predators ($x = 0$), $y' = dy$, and so the number of prey grows exponentially. In the absence of prey, $x' = -ax$, and so the predator population becomes extinct. The term $-cxy$ represents the death rate due to predation. The model therefore assumes that this death rate is directly proportional to the number of *possible encounters* xy between predator and prey at a particular time t, and the term bxy represents the resulting positive contribution to the predator population.

The critical points of this plane autonomous system are $(0, 0)$ and $(d/c, a/b)$, and the corresponding Jacobian matrices are

$$\mathbf{A}_1 = \mathbf{g}'((0, 0)) = \begin{pmatrix} -a & 0 \\ 0 & d \end{pmatrix} \quad \text{and} \quad \mathbf{A}_2 = \mathbf{g}'((d/c, a/b)) = \begin{pmatrix} 0 & bd/c \\ -ac/b & 0 \end{pmatrix}.$$

The critical point at $(0, 0)$ is a saddle point, and **FIGURE 11.4.7** shows a typical profile of solutions that are in the first quadrant and near $(0, 0)$.

Since the matrix $\mathbf{A}_2$ has pure imaginary eigenvalues $\lambda = \pm\sqrt{adi}$, the critical point $(d/c, a/b)$ *may* be a center. This possibility can be investigated using the phase-plane method. Since

$$\frac{dy}{dx} = \frac{y(-cx + d)}{x(-a + by)},$$

we may separate variables and obtain

$$\int \frac{-a + by}{y}\, dy = \int \frac{-cx + d}{x}\, dx$$

so that

$$-a \ln y + by = -cx + d \ln x + c_1 \quad \text{or} \quad (x^d e^{-cx})(y^a e^{-by}) = c_0.$$

The following argument establishes that all solution curves that originate in the first quadrant are periodic.

Typical graphs of the nonnegative functions $F(x) = x^d e^{-cx}$ and $G(y) = y^a e^{-by}$ are shown in **FIGURE 11.4.8**. It is not hard to show that $F(x)$ has an absolute maximum at $x = d/c$, whereas $G(y)$ has an absolute maximum at $y = a/b$. Note that, with the exception of 0 and the absolute maximum, F and G each take on all values in their range precisely twice.

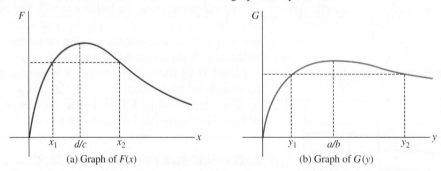

(a) Graph of $F(x)$ (b) Graph of $G(y)$

FIGURE 11.4.8 Graphs of F and G help to establish properties 1–3

These graphs can be used to establish the following properties of a solution curve that originates at a noncritical point (x_0, y_0) in the first quadrant.

1. If $y = a/b$, the equation $F(x)G(y) = c_0$ has exactly two solutions x_m and x_M that satisfy $x_m < d/c < x_M$.
2. If $x_m < x_1 < x_M$ and $x = x_1$, then $F(x)G(y) = c_0$ has exactly two solutions y_1 and y_2 that satisfy $y_1 < a/b < y_2$.
3. If x is outside the interval $[x_m, x_M]$, then $F(x)G(y) = c_0$ has no solutions.

We will give the demonstration of 1. and outline parts 2. and 3. in the exercises. Since $(x_0, y_0) \neq (d/c, a/b)$, $F(x_0)G(y_0) < F(d/c)G(a/b)$. If $y = a/b$, then

$$0 < \frac{c_0}{G(a/b)} = \frac{F(x_0)G(y_0)}{G(a/b)} < \frac{F(d/c)G(a/b)}{G(a/b)} = F(d/c).$$

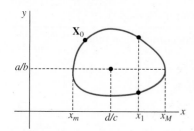

FIGURE 11.4.9 Periodic solution of Lotka–Volterra model

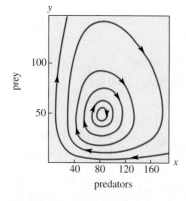

FIGURE 11.4.10 Phase portrait of Lotka–Volterra model in Example 3

(a) $\alpha_{12}\alpha_{21} < 1$

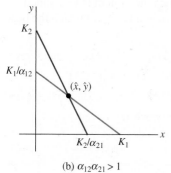

(b) $\alpha_{12}\alpha_{21} > 1$

FIGURE 11.4.11 Two conditions when critical point is in first quadrant

Therefore $F(x) = c_0/G(a/b)$ has precisely two solutions x_m and x_M that satisfy $x_m < d/c < x_M$. The graph of a typical periodic solution is shown in **FIGURE 11.4.9**.

EXAMPLE 3 **Predator–Prey Population Cycles**

If we let $a = 0.1$, $b = 0.002$, $c = 0.0025$, and $d = 0.2$ in the Lotka–Volterra predator–prey model, the critical point in the first quadrant is $(d/c, a/b) = (80, 50)$, and we know that this critical point is a center. See **FIGURE 11.4.10** in which we have used a numerical solver to generate these cycles. The closer the initial condition $\mathbf{X}_0$ is to $(80, 50)$, the more the periodic solutions resemble the elliptical solutions to the corresponding linear system. The eigenvalues of $\mathbf{g}'((80, 50))$ are $\lambda = \pm\sqrt{ad}\,i = \pm(\sqrt{2}/10)i$, and so the solutions near the critical point have period $p \approx 10\sqrt{2}\pi$, or about 44.4. ≡

▣ Lotka–Volterra Competition Model

A **competitive interaction** occurs when two or more species compete for the food, water, light, and space resources of an ecosystem. The use of one of these resources by one population therefore inhibits the ability of another population to survive and grow. Under what conditions can two competing species coexist? A number of mathematical models have been constructed that offer insights into conditions that permit *coexistence*. If x denotes the number in species I and y denotes the number in species II, then the Lotka–Volterra model takes the form

$$x' = \frac{r_1}{K_1}x(K_1 - x - \alpha_{12}y)$$

$$y' = \frac{r_2}{K_2}y(K_2 - y - \alpha_{21}x). \tag{1}$$

Note that in the absence of species II ($y = 0$), $x' = (r_1/K_1)x(K_1 - x)$, and so the first population grows logistically and approaches the steady-state population K_1 (see Section 2.8 and Example 4 in Section 11.3). A similar statement holds for species II growing in the absence of species I. The term $-\alpha_{21}xy$ in the second equation stems from the competitive effect of species I on species II. The model therefore assumes that this rate of inhibition is directly proportional to the number of possible competitive pairs xy at a particular time t.

This plane autonomous system has critical points at $(0, 0)$, $(K_1, 0)$, and $(0, K_2)$. When $\alpha_{12}\alpha_{21} \neq 0$, the lines $K_1 - x - \alpha_{12}y = 0$ and $K_2 - y - \alpha_{21}x = 0$ intersect to produce a fourth critical point $\hat{\mathbf{X}} = (\hat{x}, \hat{y})$. **FIGURE 11.4.11** shows the two conditions under which $(\hat{x}, \hat{y})$ is in the first quadrant. The trace and determinant of the Jacobian matrix at $(\hat{x}, \hat{y})$ are, respectively,

$$\tau = -\hat{x}\frac{r_1}{K_1} - \hat{y}\frac{r_2}{K_2} \quad \text{and} \quad \Delta = (1 - \alpha_{12}\alpha_{21})\hat{x}\hat{y}\frac{r_1 r_2}{K_1 K_2}.$$

In case (a), $K_1/\alpha_{12} > K_2$ and $K_2/\alpha_{21} > K_1$. It follows that $\alpha_{12}\alpha_{21} < 1$, $\tau < 0$, and $\Delta > 0$. Since

$$\tau^2 - 4\Delta = \left(\hat{x}\frac{r_1}{K_1} + \hat{y}\frac{r_2}{K_2}\right)^2 + 4(\alpha_{12}\alpha_{21} - 1)\hat{x}\hat{y}\frac{r_1 r_2}{K_1 K_2}$$

$$= \left[\left(\hat{x}\frac{r_1}{K_1} - \hat{y}\frac{r_2}{K_2}\right)^2 - 4\hat{x}\hat{y}\frac{r_1 r_2}{K_1 K_2}\right] + 4\alpha_{12}\alpha_{21}\hat{x}\hat{y}\frac{r_1 r_2}{K_1 K_2}$$

$$= \left(\hat{x}\frac{r_1}{K_1} - \hat{y}\frac{r_2}{K_2}\right)^2 + 4\alpha_{12}\alpha_{21}\hat{x}\hat{y}\frac{r_1 r_2}{K_1 K_2},$$

$\tau^2 - 4\Delta > 0$, and so $(\hat{x}, \hat{y})$ is a stable node. Therefore if $\mathbf{X}(0) = \mathbf{X}_0$ is sufficiently close to $\hat{\mathbf{X}} = (\hat{x}, \hat{y})$, $\lim_{t\to\infty} \mathbf{X}(t) = \hat{\mathbf{X}}$, and we may conclude that coexistence is possible. The demonstration that case (b) leads to a saddle point and the investigation of the nature of critical points at $(0, 0)$, $(K_1, 0)$, and $(0, K_2)$ are left to the exercises.

When the competitive interactions between two species are weak, both of the coefficients α_{12} and α_{21} will be small, and so the conditions $K_1/\alpha_{12} > K_2$ and $K_2/\alpha_{21} > K_1$ may be satisfied. This might occur when there is a small overlap in the ranges of two predator species that hunt for a common prey.

EXAMPLE 4 A Lotka–Volterra Competition Model

A competitive interaction is described by the Lotka–Volterra competition model

$$x' = 0.004x(50 - x - 0.75y)$$
$$y' = 0.001y(100 - y - 3.0x).$$

Find and classify all critical points of the system.

SOLUTION Critical points occur at $(0, 0)$, $(50, 0)$, $(0, 100)$, and at the solution $(20, 40)$ of the system

$$x + 0.75y = 50$$
$$3.0x + y = 100.$$

Since $\alpha_{12}\alpha_{21} = 2.25 > 1$, we have case (b) in Figure 11.4.11, and so the critical point at $(20, 40)$ is a saddle point. The Jacobian matrix is

$$\mathbf{g}'(\mathbf{X}) = \begin{pmatrix} 0.2 - 0.008x - 0.003y & -0.003x \\ -0.003y & 0.1 - 0.002y - 0.003x \end{pmatrix},$$

and we obtain

$$\mathbf{A}_1 = \mathbf{g}'((0, 0)) = \begin{pmatrix} 0.2 & 0 \\ 0 & 0.1 \end{pmatrix} \qquad \mathbf{A}_2 = \mathbf{g}'((50, 0)) = \begin{pmatrix} -0.2 & -0.15 \\ 0 & -0.05 \end{pmatrix}$$

$$\mathbf{A}_3 = \mathbf{g}'((20, 40)) = \begin{pmatrix} -0.08 & -0.12 \\ -0.06 & -0.04 \end{pmatrix} \qquad \mathbf{A}_4 = \mathbf{g}'((0, 100)) = \begin{pmatrix} -0.1 & 0 \\ -0.3 & -0.1 \end{pmatrix}.$$

Therefore $(0, 0)$ is an unstable node, whereas both $(50, 0)$ and $(0, 100)$ are stable nodes. (Check this!) Since $\det \mathbf{A}_3 < 0$, we have a second demonstration that $(20, 40)$ is a saddle point. ≡

Coexistence can also occur in the Lotka–Volterra competition model if there is at least one periodic solution lying entirely in the first quadrant. It is possible to show, however, that this model has no periodic solutions.

11.4 Exercises Answers to selected odd-numbered problems begin on page ANS-31.

The Nonlinear Pendulum

1. A pendulum is released at $\theta = \pi/3$ and is given an initial angular velocity of ω_0 rad/s. Determine under what conditions the resulting motion is periodic.

2. (a) If a pendulum is released from rest at $\theta = \theta_0$, show that the angular velocity is again 0 when $\theta = -\theta_0$.
 (b) The period T of the pendulum is the amount of time needed for θ to change from θ_0 to $-\theta_0$ and back to θ_0. Show that

$$T = \sqrt{\frac{2L}{g}} \int_{-\theta_0}^{\theta_0} \frac{1}{\sqrt{\cos\theta - \cos\theta_0}} \, d\theta.$$

The Sliding Bead

3. A bead with mass m slides along a thin wire whose shape is described by the function $z = f(x)$. If $\mathbf{X}_1 = (x_1, y_1)$ is a critical point of the plane autonomous system associated with the sliding bead, verify that the Jacobian matrix at $\mathbf{X}_1$ is

$$\mathbf{g}'(\mathbf{X}_1) = \begin{pmatrix} 0 & 1 \\ -gf''(x_1) & -\beta/m \end{pmatrix}.$$

4. A bead with mass m slides along a thin wire whose shape is described by the function $z = f(x)$. When $f'(x_1) = 0$, $f''(x_1) > 0$, and the system is undamped, the critical point $\mathbf{X}_1 = (x_1, 0)$ is a center. Estimate the period of the bead when $x(0)$ is near x_1 and $x'(0) = 0$.

5. A bead is released from the position $x(0) = x_0$ on the curve $z = \frac{1}{2}x^2$ with initial velocity $x'(0) = v_0$ cm/s.
 (a) Use the phase-plane method to show that the resulting solution is periodic when the system is undamped.
 (b) Show that the maximum height z_{max} to which the bead rises is given by

$$z_{max} = \frac{1}{2}\left[e^{v_0^2/g}(1 + x_0^2) - 1\right].$$

6. Rework Problem 5 with $z = \cosh x$.

Interaction Models

7. Refer to Figure 11.4.9. If $x_m < x_1 < x_M$ and $x = x_1$, show that $F(x)G(y) = c_0$ has exactly two solutions y_1 and y_2 that satisfy $y_1 < a/b < y_2$. [*Hint:* First show that $G(y) = c_0/F(x_1) < G(a/b)$.]

8. From **1.** and **3.** on page 664, conclude that the maximum number of predators occurs when $y = a/b$.

9. In many fishery science models the rate at which a species is caught is assumed to be directly proportional to its abundance. If both predator and prey are being exploited in this manner, the Lotka–Volterra differential equations take the form

$$x' = -ax + bxy - \epsilon_1 x$$
$$y' = -cxy + dy - \epsilon_2 y,$$

where ϵ_1 and ϵ_2 are positive constants.

(a) When $\epsilon_2 < d$, show that there is a new critical point in the first quadrant that is a center.

(b) **Volterra's principle** states that a moderate amount of exploitation increases the average number of prey and decreases the average number of predators. Is this fisheries model consistent with Volterra's principle?

10. A predator–prey interaction is described by the Lotka–Volterra model

$$x' = -0.1x + 0.02xy$$
$$y' = 0.2y - 0.025xy.$$

(a) Find the critical point in the first quadrant, and use a numerical solver to sketch some population cycles.

(b) Estimate the period of the periodic solutions that are close to the critical point in part (a).

11. A competitive interaction is described by the Lotka–Volterra competition model

$$x' = 0.08x(20 - 0.4x - 0.3y)$$
$$y' = 0.06y(10 - 0.1y - 0.3x).$$

Find and classify all critical points of the system.

12. In (1) show that $(0, 0)$ is always an unstable node.

13. In (1) show that $(K_1, 0)$ is a stable node when $K_1 > K_2/\alpha_{21}$ and a saddle point when $K_1 < K_2/\alpha_{21}$.

14. Use Problems 12 and 13 to establish that $(0, 0)$, $(K_1, 0)$, and $(0, K_2)$ are unstable when $\hat{\mathbf{X}} = (\hat{x}, \hat{y})$ is a stable node.

15. In (1) show that $\hat{\mathbf{X}} = (\hat{x}, \hat{y})$ is a saddle point when

$$\frac{K_1}{\alpha_{12}} < K_2 \quad \text{and} \quad \frac{K_2}{\alpha_{21}} < K_1.$$

Miscellaneous Nonlinear Models

16. If we assume that a damping force acts in a direction opposite to the motion of a pendulum and with a magnitude directly proportional to the angular velocity $d\theta/dt$, the displacement angle θ for the pendulum satisfies the nonlinear second-order differential equation

$$ml\frac{d^2\theta}{dt^2} = -mg \sin\theta - \beta\frac{d\theta}{dt}.$$

(a) Write the second-order differential equation as a plane autonomous system, and find all critical points.

(b) Find a condition on m, l, and β that will make $(0, 0)$ a stable spiral point.

17. In the analysis of free, damped motion in Section 3.8 we assumed that the damping force was proportional to the velocity x'. Frequently the magnitude of this damping force is proportional to the square of the velocity, and the new differential equation becomes

$$x'' = -\frac{\beta}{m}x'|x'| - \frac{k}{m}x.$$

(a) Write the second-order differential equation as a plane autonomous system, and find all critical points.

(b) The system is called *overdamped* when $(0, 0)$ is a stable node and is called *underdamped* when $(0, 0)$ is a stable spiral point. Physical considerations suggest that $(0, 0)$ must be an asymptotically stable critical point. Show that the system is necessarily underdamped. [*Hint*: $d/dy\,(y|y|) = 2|y|$.]

18. A bead with mass m slides along a thin wire whose shape may be described by the function $z = f(x)$. Small stretches of the wire act like an inclined plane, and in mechanics it is assumed that the magnitude of the frictional force between the bead and wire is directly proportional to $mg \cos\theta$. See Figure 11.4.3.

(a) Explain why the new differential equation for the x-coordinate of the bead is

$$x'' = g\frac{\mu - f'(x)}{1 + [f'(x)]^2} - \frac{\beta}{m}x'$$

for some positive constant μ.

(b) Investigate the critical points of the corresponding plane autonomous system. Under what conditions is a critical point a saddle point? A stable spiral point?

19. An undamped oscillation satisfies a nonlinear second-order differential equation of the form $x'' + f(x) = 0$, where $f(0) = 0$ and $xf(x) > 0$ for $x \neq 0$ and $-d < x < d$. Use the phase-plane method to investigate whether it is possible for the critical point $(0, 0)$ to be a stable spiral point. [*Hint*: Let $F(x) = \int_0^x f(u)\,du$, and show that $y^2 + 2F(x) = c$.]

20. The Lotka–Volterra predator–prey model assumes that, in the absence of predators, the number of prey grows exponentially. If we make the alternative assumption that the prey population grows logistically, the new system is

$$x' = -ax + bxy$$
$$y' = -cxy + \frac{r}{K}y(K - y),$$

where a, b, c, r, and K are positive and $K > a/b$.

(a) Show that the system has critical points at $(0, 0)$, $(0, K)$, and $(\hat{x}, \hat{y})$, where $\hat{y} = a/b$ and $c\hat{x} = \frac{r}{K}(K - \hat{y})$.

(b) Show that the critical points at $(0, 0)$ and $(0, K)$ are saddle points, whereas the critical point at $(\hat{x}, \hat{y})$ is either a stable node or a stable spiral point.

(c) Show that $(\hat{x}, \hat{y})$ is a stable spiral point if $\hat{y} < \frac{4bK^2}{r + 4bK}$. Explain why this case will occur when the carrying capacity K of the prey is large.

21. The nonlinear system

$$x' = \alpha \frac{y}{1+y} x - x$$

$$y' = -\frac{y}{1+y} x - y + \beta$$

arises in a model for the growth of microorganisms in a *chemostat*, a simple laboratory device in which a nutrient from a supply source flows into a growth chamber. In the system, x denotes the concentration of the microorganisms in the growth chamber, y denotes the concentration

of nutrients, and $\alpha > 1$ and $\beta > 0$ are constants that can be adjusted by the experimenter. Find conditions on α and β that ensure that the system has a single critical point $(\hat{x}, \hat{y})$ in the first quadrant, and investigate the stability of this critical point.

22. Use the methods of this chapter together with a numerical solver to investigate stability in the nonlinear spring/mass system modeled by

$$x'' + 8x - 6x^3 + x^5 = 0.$$

See Problem 8 in Exercises 3.11.

11.5 Periodic Solutions, Limit Cycles, and Global Stability

INTRODUCTION In this section we will investigate the existence of periodic solutions of nonlinear plane autonomous systems and introduce special periodic solutions called limit cycles.

We saw in Sections 11.3 and 11.4 that an analysis of critical points using linearization can provide valuable information on the behavior of solutions near critical points and insight into a variety of biological and physical phenomena. There are, however, some inherent limitations to this approach. When the eigenvalues of the Jacobian matrix are pure imaginary, we cannot conclude that there are periodic solutions near the critical point. In some cases we were able to solve $dy/dx = Q(x, y)/P(x, y)$, obtain an implicit representation $f(x, y) = c$ of the solution curves, and investigate whether any of these solutions formed closed curves. More often than not, this differential equation will not possess closed-form solutions. For example, the Lotka–Volterra competition model cannot be handled by this procedure. The first goal of this section is *to determine conditions under which we can either exclude the possibility of periodic solutions or assert their existence.*

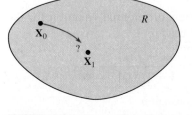

FIGURE 11.5.1 For any $\mathbf{X}_0$ in R, does $\mathbf{X}$ approach $\mathbf{X}_1$ as $t \to \infty$?

We encountered an additional problem in studying the models in Section 11.3. **FIGURE 11.5.1** illustrates the common situation in which a region R contains a *single* asymptotically stable critical point $\mathbf{X}_1$. We can assert that $\lim_{t \to \infty} \mathbf{X}(t) = \mathbf{X}_1$ when the initial position $\mathbf{X}(0) = \mathbf{X}_0$ is "near" $\mathbf{X}_1$, but under what conditions is $\lim_{t \to \infty} \mathbf{X}(t) = \mathbf{X}_1$ for *all initial positions* in R? Such a critical point is called **globally stable** in R. A second goal is *to determine conditions under which an asymptotically stable critical point is globally stable.*

In motivating and discussing the methods in this section, we will use the fact that the vector field $\mathbf{V}(x, y) = (P(x, y), Q(x, y))$ can be interpreted as defining a fluid flow in a region of the plane, and a solution to the autonomous system may be interpreted as the resulting path of a particle as it moves through the region.

Negative Criteria A number of results can sometimes be used to establish that there are no periodic solutions in a given region R of the plane. We will assume that $P(x, y)$ and $Q(x, y)$ have continuous first partial derivatives in R and that R is simply connected. Recall that in a simply connected region, any simple closed curve C in R encloses only points in R. Therefore, if there is a periodic solution $\mathbf{X} = \mathbf{X}(t)$ in R, then R will contain all points in the interior of the resulting curve.

> **THEOREM 11.5.1 Cycles and Critical Points**
>
> If a plane autonomous system has a periodic solution $\mathbf{X} = \mathbf{X}(t)$ in a simply connected region R, then the system has at least one critical point inside the corresponding simple closed curve C. If there is a single critical point inside C, then that critical point cannot be a saddle point.

> **Corollary to Theorem 11.5.1**
>
> If a simply connected region R either contains no critical points of a plane autonomous system or contains a single saddle point, then there are no periodic solutions in R.

EXAMPLE 1 No Periodic Solutions

Show that the plane autonomous system

$$x' = xy$$
$$y' = -1 - x^2 - y^2$$

has no periodic solutions.

SOLUTION If (x, y) is a critical point, then, from the first equation, either $x = 0$ or $y = 0$. If $x = 0$, then $-1 - y^2 = 0$ or $y^2 = -1$. Likewise, $y = 0$ implies $x^2 = -1$. Therefore, this plane autonomous system has no critical points, and by the corollary possesses no periodic solutions in the plane. ≡

EXAMPLE 2 Lotka–Volterra Competition Model

Show that the Lotka–Volterra competition model

$$x' = 0.004x(50 - x - 0.75y)$$
$$y' = 0.001y(100 - y - 3.0x)$$

has no periodic solutions in the first quadrant.

SOLUTION In Example 4 in Section 11.4 we showed that this system has critical points at $(0, 0)$, $(50, 0)$, $(0, 100)$, and $(20, 40)$, and that $(20, 40)$ is a saddle point. Since only $(20, 40)$ lies in the first quadrant, by the corollary there are no periodic solutions in the first quadrant. ≡

Another sometimes useful result can be formulated in terms of the divergence of the vector field $\mathbf{V}(x, y) = (P(x, y), Q(x, y))$:

THEOREM 11.5.2 Bendixson Negative Criterion

If div $\mathbf{V} = \partial P/\partial x + \partial Q/\partial y$ does not change sign in a simply connected region R, then the plane autonomous system has no periodic solutions in R.

PROOF: Suppose, to the contrary, that there is a periodic solution $\mathbf{X} = \mathbf{X}(t)$ lying in R, and let C be the resulting simple closed curve and R_1 the region bounded by C. Green's theorem states that

$$\int_C M(x, y)\, dx + N(x, y)\, dy = \iint_{R_1} \left(\frac{\partial N}{\partial x} - \frac{\partial M}{\partial y} \right) dx\, dy$$

whenever $M(x, y)$ and $N(x, y)$ have continuous first partials in R. If we let $N = P$ and $M = -Q$, we obtain

$$\int_C -Q(x, y)\, dx + P(x, y)\, dy = \iint_{R_1} \left(\frac{\partial P}{\partial x} + \frac{\partial Q}{\partial y} \right) dx\, dy.$$

Since $\mathbf{X} = \mathbf{X}(t)$ is a solution with period p, we have $x'(t) = P(x(t), y(t))$ and $y'(t) = Q(x(t), y(t))$, and so

$$\int_C -Q(x, y)\, dx + P(x, y)\, dy = \int_0^P [-Q(x(t), y(t))\, x'(t) + P(x(t), y(t))\, y'(t)]\, dt$$

$$= \int_0^P [-QP + PQ]\, dt = 0.$$

Since div $\mathbf{V} = \partial P/\partial x + \partial Q/\partial y$ is continuous and does not change sign in R, it follows that either div $\mathbf{V} \geq 0$ in R or div $\mathbf{V} \leq 0$ in R, and so

$$\iint_{R_1} \left(\frac{\partial P}{\partial x} + \frac{\partial Q}{\partial y} \right) dx\, dy \neq 0.$$

This contradiction establishes that there are no periodic solutions in R. ≡

EXAMPLE 3 **Bendixson Negative Criterion**

Investigate possible periodic solutions of each system.

(a) $x' = x + 2y + 4x^3 - y^2$
 $y' = -x + 2y + yx^2 + y^3$

(b) $x' = y + x(2 - x^2 - y^2)$
 $y' = -x + y(2 - x^2 - y^2)$

SOLUTION (a) We have div $\mathbf{V} = \partial P/\partial x + \partial Q/\partial y = 1 + 12x^2 + 2 + x^2 + 3y^2 \geq 3$, and so there are no periodic solutions in the plane.

(b) For this system div $\mathbf{V} = (2 - 3x^2 - y^2) + (2 - x^2 - 3y^2) = 4 - 4(x^2 + y^2)$. Therefore, if R is the interior of the circle $x^2 + y^2 = 1$, div $\mathbf{V} > 0$, and so there are no periodic solutions inside this disk. Note that div $\mathbf{V} < 0$ on the exterior of the circle. If R is any simply connected subset of the exterior, then there are no periodic solutions in R. It follows that if there is a periodic solution in the exterior, it must enclose the circle $x^2 + y^2 = 1$. In fact, the reader can verify that $\mathbf{X}(t) = (\sqrt{2}\sin t, \sqrt{2}\cos t)$ is a periodic solution that generates the circle $x^2 + y^2 = 2$. ≡

EXAMPLE 4 **Sliding Bead and Periodic Solutions**

The sliding bead discussed in Section 11.4 satisfies the differential equation

$$mx'' = -mg\frac{f'(x)}{1 + [f'(x)]^2} - \beta x'.$$

Show that there are no periodic solutions.

SOLUTION The corresponding plane autonomous system is

$$x' = y$$

$$y' = -g\frac{f'(x)}{1 + [f'(x)]^2} - \frac{\beta}{m}y$$

and so div $\mathbf{V} = \dfrac{\partial P}{\partial x} + \dfrac{\partial Q}{\partial y} = -\dfrac{\beta}{m} < 0$. By Theorem 11.5.2 we conclude the system has no periodic solutions. ≡

The following theorem is a generalization of the Bendixson negative criterion, which leaves it to the reader to construct an appropriate function $\delta(x, y)$.

THEOREM **11.5.3** **Dulac Negative Criterion**

If $\delta(x, y)$ has continuous first partial derivatives in a simply connected region R and $\dfrac{\partial(\delta P)}{\partial x} + \dfrac{\partial(\delta Q)}{\partial y}$ does not change sign in R, then the plane autonomous system has no periodic solutions in R.

There are no general techniques for constructing an appropriate function $\delta(x, y)$. Instead, we experiment with simple functions of the form $ax^2 + by^2$, e^{ax+by}, $x^a y^b$, and so on, and try to determine constants for which $\partial(\delta P)/\partial x + \partial(\delta Q)/\partial y$ is nonzero in a given region.

EXAMPLE 5 **Dulac Negative Criterion**

Show that the second-order nonlinear differential equation

$$x'' = x^2 + (x')^2 - x - x'$$

has no periodic solutions.

SOLUTION The corresponding plane autonomous system is

$$x' = y$$
$$y' = x^2 + y^2 - x - y.$$

If we let $\delta(x, y) = e^{ax+by}$, then

$$\frac{\partial(\delta P)}{\partial x} + \frac{\partial(\delta Q)}{\partial y} = e^{ax+by}(ay + 2y - 1) + e^{ax+by}b(x^2 + y^2 - x - y).$$

If we set $a = -2$ and $b = 0$, then $\partial(\delta P)/\partial x + \partial(\delta Q)/\partial y = -e^{ax+by}$, which is always negative. Therefore, by the Dulac negative criterion, the second-order differential equation has no periodic solutions. ≡

EXAMPLE 6 **Dulac Negative Criterion**

Use $\delta(x, y) = 1/(xy)$ to show that the Lotka–Volterra competition equations

$$x' = \frac{r_1}{K_1} x(K_1 - x - \alpha_{12}y)$$

$$y' = \frac{r_2}{K_2} y(K_2 - y - \alpha_{21}x)$$

have no periodic solutions in the first quadrant.

SOLUTION If $\delta(x, y) = 1/(xy)$, then

$$\delta P = \frac{r_1}{K_1}\left[\frac{K_1}{y} - \frac{x}{y} - \alpha_{12}\right] \quad \text{and} \quad \delta Q = \frac{r_2}{K_2}\left[\frac{K_2}{x} - \frac{y}{x} - \alpha_{21}\right]$$

and so

$$\frac{\partial(\delta P)}{\partial x} + \frac{\partial(\delta Q)}{\partial y} = \frac{r_1}{K_1}\left(-\frac{1}{y}\right) + \frac{r_2}{K_2}\left(-\frac{1}{x}\right).$$

For (x, y) in the first quadrant, the latter expression is always negative. Therefore, there are no periodic solutions. ≡

Positive Criteria: Poincaré–Bendixson Theory The Poincaré–Bendixson theorem is an advanced result that describes the long-range behavior of a *bounded* solution to a plane autonomous system. Rather than present the result in its full generality, we will concentrate on a number of special cases that occur frequently in applications. One of these cases will lead to a new type of periodic solution called a *limit cycle*.

DEFINITION 11.5.1 **Invariant Region**

A region R is called an **invariant region** for a plane autonomous system if, whenever X_0 is in R, the solution $X = X(t)$ satisfying $X(0) = X_0$ remains in R.

(a) Type I invariant region

(b) Type II invariant region

FIGURE 11.5.2 Two types of invariant regions

FIGURE 11.5.2 shows two standard types of invariant regions. A **Type I invariant region** R is bounded by a simple closed curve C, and the flow at the boundary defined by the vector field $V(x, y) = (P(x, y), Q(x, y))$ is always directed into the region. This prevents a particle from crossing the boundary. A **Type II invariant region** is an annular region bounded by simple closed curves C_1 and C_2, and the flow at the boundary is again directed toward the interior of R. The following theorem provides a method for verifying that a given region is invariant.

THEOREM 11.5.4 **Normal Vectors and Invariant Regions**

If $n(x, y)$ denotes a normal vector on the boundary that points inside the region, then R will be an invariant region for the plane autonomous system provided $V(x, y) \cdot n(x, y) \geq 0$ for all points (x, y) on the boundary.

PROOF: If θ is the angle between $V(x, y)$ and $n(x, y)$, then from $V \cdot n = \|V\| \|n\| \cos\theta$, we may conclude that $\cos\theta \geq 0$ and so θ is between $0°$ and $90°$. The flow is therefore directed into the region (or at worst along the boundary) for any boundary point (x, y). This prevents a solution that starts in R from leaving R. Therefore, R is an invariant region for the plane autonomous system. ≡

The problem of finding an invariant region for a given nonlinear system is an extremely difficult one. An excellent first step is to use software that plots the vector field $\mathbf{V}(x, y) = (P(x, y), Q(x, y))$ together with the curves $P(x, y) = 0$ (along which the vectors are vertical) and $Q(x, y) = 0$ (along which the vectors are horizontal). This can lead to choices for R. In the following examples, we will construct invariant regions bounded by lines and circles. In more complicated cases, we will be content to offer empirical evidence that an invariant region exists.

EXAMPLE 7 Circular Invariant Region

Find a circular region with center at $(0, 0)$ that serves as an invariant region for the plane autonomous system

$$x' = -y - x^3$$
$$y' = x - y^3.$$

SOLUTION For the circle $x^2 + y^2 = r^2$, $\mathbf{n} = (-2x, -2y)$ is a normal vector that points toward the interior of the circle. Since

$$\mathbf{V} \cdot \mathbf{n} = (-y - x^3, x - y^3) \cdot (-2x, -2y) = 2(x^4 + y^4),$$

we may conclude that $\mathbf{V} \cdot \mathbf{n} \geq 0$ on the circle $x^2 + y^2 = r^2$. Therefore, by Theorem 11.5.4 the circular region defined by $x^2 + y^2 \leq r^2$ serves as an invariant region for the system for any $r > 0$. ≡

EXAMPLE 8 Annular Invariant Region

Find an annular region bounded by circles that serves as an invariant region for the plane autonomous system

$$x' = x - y - 5x(x^2 + y^2) + x^5$$
$$y' = x + y - 5y(x^2 + y^2) + y^5.$$

SOLUTION As in Example 7, the normal vector $\mathbf{n}_1 = (-2x, -2y)$ points inside the circle $x^2 + y^2 = r^2$, while the normal vector $\mathbf{n}_2 = -\mathbf{n}_1$ is directed toward the exterior. Computing $\mathbf{V} \cdot \mathbf{n}_1$ and simplifying, we obtain

$$\mathbf{V} \cdot \mathbf{n}_1 = -2(r^2 - 5r^4 + x^6 + y^6).$$

Note that $r^2 - 5r^4 = r^2(1 - 5r^2)$ takes on both positive and negative values.

If $r = 1$, $\mathbf{V} \cdot \mathbf{n}_1 = 8 - 2(x^6 + y^6) \geq 0$, since the maximum value of $x^6 + y^6$ on the circle $x^2 + y^2 = 1$ is 1. The flow is therefore directed toward the interior of the circular region $x^2 + y^2 \leq 1$.

If $r = \frac{1}{4}$, $\mathbf{V} \cdot \mathbf{n}_1 \leq -2(r^2 - 5r^4) < 0$, and so $\mathbf{V} \cdot \mathbf{n}_2 = -\mathbf{V} \cdot \mathbf{n}_1 > 0$. The flow is therefore directed toward the exterior of the circle $x^2 + y^2 = \frac{1}{16}$, and so the annular region R defined by $\frac{1}{16} \leq x^2 + y^2 \leq 1$ is an invariant region for the system. ≡

EXAMPLE 9 The Van der Pol Equation

The Van der Pol equation is a nonlinear second-order differential equation that arises in electronics, and as a plane autonomous system it takes the form

$$x' = y$$
$$y' = -\mu(x^2 - 1)y - x.$$

FIGURE 11.5.3 shows the corresponding vector field for $\mu = 1$, together with the curves $y = 0$ and $(x^2 - 1)y = -x$ along which the vectors are vertical and horizontal, respectively. (For convenience we have sketched the normalized vector field $\mathbf{V}/\|\mathbf{V}\|$.) It is not possible to find a simple invariant region whose boundary consists of lines or circles. The figure does offer empirical evidence that an invariant region R, with $(0, 0)$ in its interior, does exist. Advanced methods are required to demonstrate this mathematically.* ≡

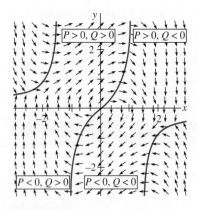

FIGURE 11.5.3 Vector field for Van der Pol's equation in Example 9

*See M. Hirsch and S. Smale, *Differential Equations, Dynamical Systems, and Linear Algebra* (New York: Academic Press, 1974).

We next present two important special cases of the Poincaré–Bendixson theorem that guarantee the existence of periodic solutions.

THEOREM 11.5.5 **Poincaré–Bendixson I**

Let R be an invariant region for a plane autonomous system and suppose that R has no critical points on its boundary.
 (a) If R is a Type I region that has a single unstable node or an unstable spiral point in its interior, then there is at least one periodic solution in R.
 (b) If R is a Type II region that contains no critical points of the system, then there is at least one periodic solution in R.
In either of the two cases, if $\mathbf{X} = \mathbf{X}(t)$ is a nonperiodic solution in R, then $\mathbf{X}(t)$ spirals toward a cycle that is a solution to the system. This periodic solution is called a **limit cycle**.

The flow interpretation portrayed in Figure 11.5.2 can be used to make the result plausible. If a particle is released at a point $\mathbf{X}_0$ in a Type II invariant region R, then, with no escape from the region and with no resting points, the particle will begin to rotate around the boundary C_2 and settle into a periodic orbit. It is not possible for the particle to return to an earlier position unless the solution is itself periodic.

EXAMPLE 10 **Existence of a Periodic Solution**

Use Theorem 11.5.5 to show that the system

$$x' = -y + x(1 - x^2 - y^2) - y(x^2 + y^2)$$
$$y' = x + y(1 - x^2 - y^2) + x(x^2 + y^2)$$

has at least one periodic solution.

SOLUTION We first construct an invariant region that is bounded by circles. If $\mathbf{n}_1 = (-2x, -2y)$, then $\mathbf{V} \cdot \mathbf{n}_1 = -2r^2(1 - r^2)$. If we let $r = 2$ and then $r = \frac{1}{2}$, we may conclude that the annular region R defined by $\frac{1}{4} \le x^2 + y^2 \le 4$ is invariant. If (x_1, y_1) is a critical point of the system, then $\mathbf{V} \cdot \mathbf{n}_1 = (0, 0) \cdot \mathbf{n}_1 = 0$. Therefore, $r = 0$ or $r = 1$. If $r = 0$, then $(x_1, y_1) = (0, 0)$ is a critical point. If $r = 1$, the system reduces to $-2y = 0$, $2x = 0$, and we have reached a contradiction. Therefore, $(0, 0)$ is the only critical point, and this critical point is not in R. By part (b) of Theorem 11.5.5, the system has at least one periodic solution in R.

The reader can verify that $\mathbf{X}(t) = (\cos 2t, \sin 2t)$ is a periodic solution. $\equiv$

EXAMPLE 11 **Limit Cycle in the Van der Pol Equation**

Show that the Van der Pol differential equation

$$x'' + \mu(x^2 - 1)x' + x = 0$$

has a periodic solution when $\mu > 0$.

SOLUTION We will assume that there is a Type I invariant region R for the corresponding plane autonomous system and that this region contains $(0, 0)$ in its interior (see Example 9 and Figure 11.5.3). The only critical point is $(0, 0)$, and the Jacobian matrix is given by

$$\mathbf{g}'((0, 0)) = \begin{pmatrix} 0 & 1 \\ -1 & \mu \end{pmatrix}.$$

Therefore, $\tau = \mu$, $\Delta = 1$, and $\tau^2 - 4\Delta = \mu^2 - 4$. Since $\mu > 0$, the critical point is either an unstable spiral point or an unstable node. By part (a) of Theorem 11.5.5, the system has at least one periodic solution in R. **FIGURE 11.5.4** shows solutions corresponding to $\mathbf{X}(0) = (0.5, 0.5)$ and $\mathbf{X}(0) = (3, 3)$ for $\mu = 1$. Each of these solutions spirals around the origin and approaches a limit cycle. It can be shown that the Van der Pol differential equation has a unique limit cycle for all values of the parameter μ. $\equiv$

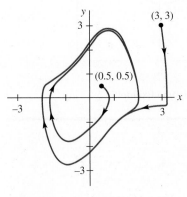

FIGURE 11.5.4 Two solutions of Van der Pol's equation approaching the same limit cycle in Example 11

Global Stability

Another version of the Poincaré–Bendixson theorem can be used to show that a locally stable critical point is globally stable:

THEOREM 11.5.6 Poincaré–Bendixson II

Let R be a Type I invariant region for a plane autonomous system that has *no* periodic solutions in R.

 (a) If R has a finite number of nodes or spiral points, then given any initial position $\mathbf{X}_0$ in R, $\lim_{t \to \infty} \mathbf{X}(t) = \mathbf{X}_1$ for some critical point $\mathbf{X}_1$.

 (b) If R has a *single* stable node or stable spiral point $\mathbf{X}_1$ in its interior and no critical points on its boundary, then $\lim_{t \to \infty} \mathbf{X}(t) = \mathbf{X}_1$ for all initial positions $\mathbf{X}_0$ in R.

In Theorem 11.5.6 the particle cannot escape from R, cannot return to any of its prior positions, and therefore, in the absence of cycles, must be attracted to some stable critical point $\mathbf{X}_1$.

EXAMPLE 12 A Globally Stable Critical Point

Investigate global stability for the system from Example 7:

$$x' = -y - x^3$$
$$y' = x - y^3.$$

SOLUTION In Example 7 we showed that the circular region defined by $x^2 + y^2 \leq r^2$ serves as an invariant region for the system for any $r > 0$. Since $\partial P/\partial x + \partial Q/\partial y = -3x^2 - 3y^2$ does not change sign, there are no periodic solutions by the Bendixson negative criterion. It is not hard to show that $(0, 0)$ is the only critical point and that the Jacobian matrix is

$$\mathbf{g}'((0, 0)) = \begin{pmatrix} 0 & -1 \\ 1 & 0 \end{pmatrix}.$$

Since $\tau = 0$ and $\Delta = 1$, $(0, 0)$ may be either a stable or an unstable spiral (it cannot be a center). Theorem 11.5.6, however, guarantees that $\lim_{t \to \infty} \mathbf{X}(t) = \mathbf{X}_1$ for some critical point $\mathbf{X}_1$. Since $(0, 0)$ is the only critical point, we must have $\lim_{t \to \infty} \mathbf{X}(t) = (0, 0)$ for *any* initial position $\mathbf{X}_0$ in the plane. The critical point is therefore a globally stable spiral point. **FIGURE 11.5.5** shows two views of the solution satisfying $\mathbf{X}(0) = (4, 4)$. Figure 11.5.5(b) is an enlarged view of the curve around $(0, 0)$. Notice how slowly the solution spirals toward $(0, 0)$.

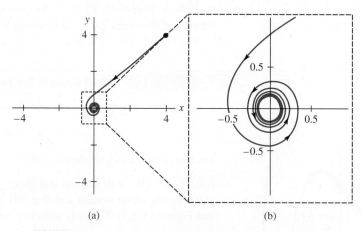

FIGURE 11.5.5 Zoom-in of a region around $(0, 0)$ in Example 12

11.5 Exercises Answers to selected odd-numbered problems begin on page ANS-31.

In Problems 1–8, show that the given plane autonomous system (or second-order differential equation) has no periodic solutions.

1. $x' = 2 + xy$
 $y' = x - y$

2. $x' = 2x - xy$
 $y' = -1 - x^2 + 2x - y^2$

3. $x' = -x + y^2$
 $y' = x - y$

4. $x' = xy^2 - x^2y$
 $y' = x^2y - 1$

5. $x' = -\mu x - y$
 $y' = x + y^3$
 for $\mu < 0$

6. $x' = 2x + y^2$
 $y' = xy - y$

7. $x'' - 2x + (x')^4 = 0$

8. $x'' + x = [\frac{1}{2} + 3(x')^2]x' - x^2$

In Problems 9 and 10, use the Dulac negative criterion to show that the given plane autonomous system has no periodic solutions. Experiment with simple functions of the form $\delta(x, y) = ax^2 + by^2$, e^{ax+by}, or $x^a y^b$.

9. $x' = -2x + xy$
 $y' = 2y - x^2$

10. $x' = -x^3 + 4xy$
 $y' = -5x^2 - y^2$

11. Show that the plane autonomous system

$$x' = x(1 - x^2 - 3y^2)$$

$$y' = y(3 - x^2 - 3y^2)$$

has no periodic solutions in an elliptical region about the origin.

12. If $\partial g/\partial x' \neq 0$ in a region R, prove that $x'' = g(x, x')$ has no periodic solutions in R.

13. Show that the predator–prey model

$$x' = -ax + bxy$$

$$y' = -cxy + \frac{r}{K}y(K - y)$$

in Problem 20 of Exercises 11.4 has no periodic solutions in the first quadrant.

In Problems 14 and 15, find a circular invariant region for the given plane autonomous system.

14. $x' = -y - xe^{x+y}$
 $y' = x - ye^{x+y}$

15. $x' = -x + y + xy$
 $y' = x - y - x^2 - y^3$

16. Verify that the region bounded by the closed curve $x^6 + 3y^2 = 1$ is an invariant region for the second-order nonlinear differential equation $x'' + x' = -(x')^3 - x^5$. See **FIGURE 11.5.6**.

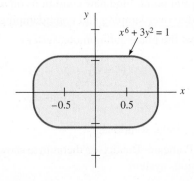

FIGURE 11.5.6 Invariant region in Problem 16

17. The plane autonomous system in Example 8 has only one critical point. Can we conclude that this system has at least one periodic solution?

18. Use the Poincaré–Bendixson theorem to show that the second-order nonlinear differential equation

$$x'' = x'[1 - 3x^2 - 2(x')^2] - x$$

has at least one periodic solution. [*Hint*: Find an invariant annular region for the corresponding plane autonomous system.]

19. Let $\mathbf{X} = \mathbf{X}(t)$ be the solution of the plane autonomous system

$$x' = y$$

$$y' = -x - (1 - x^2)y$$

that satisfies $\mathbf{X}(0) = (x_0, y_0)$. Show that if $x_0^2 + y_0^2 < 1$, then $\lim_{t \to \infty} \mathbf{X}(t) = (0, 0)$. [*Hint*: Select $r < 1$ with $x_0^2 + y_0^2 < r^2$ and first show that the circular region R defined by $x^2 + y^2 \leq r^2$ is an invariant region.]

20. Investigate global stability for the system

$$x' = y - x$$

$$y' = -x - y^3.$$

21. Empirical evidence suggests that the plane autonomous system

$$x' = x^2 y - x + 1$$

$$y' = -x^2 y + \frac{1}{2}$$

has a Type I invariant region R that lies inside the rectangle defined by $0 \leq x \leq 2, 0 \leq y \leq 1$.

(a) Use the Bendixson negative criterion to show that there are no periodic solutions in R.

(b) If $\mathbf{X}_0$ is in R and $\mathbf{X} = \mathbf{X}(t)$ is the solution satisfying $\mathbf{X}(t) = \mathbf{X}_0$, use Theorem 11.5.6 to find $\lim_{t \to \infty} \mathbf{X}(t)$.

22. (a) Find and classify all critical points of the plane autonomous system

$$x' = x\left(\frac{2y}{y + 2} - 1\right)$$

$$y' = y\left(1 - \frac{2x}{y + 2} - \frac{y}{8}\right).$$

(b) **FIGURE 11.5.7** shows the vector field $\mathbf{V}/\|\mathbf{V}\|$ and offers empirical evidence that there is an invariant region R in the first quadrant with a critical point in its interior. Assuming that such a region exists, prove that there is at least one periodic solution.

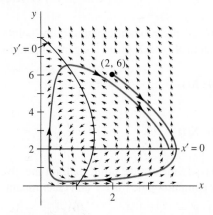

FIGURE 11.5.7 Vector field in Problem 22

Answer Problems 1–10 without referring back to the text. Fill in the blank or answer true/false.

1. The second-order differential equation $x'' + f(x') + g(x) = 0$ can be written as a plane autonomous system. _____

2. If $\mathbf{X} = \mathbf{X}(t)$ is a solution of a plane autonomous system and $\mathbf{X}(t_1) = \mathbf{X}(t_2)$ for $t_1 \neq t_2$, then $\mathbf{X}(t)$ is a periodic solution. _____

3. If the trace of the matrix $\mathbf{A}$ is 0 and det $\mathbf{A} \neq 0$, then the critical point $(0, 0)$ of the linear system $\mathbf{X}' = \mathbf{AX}$ may be classified as _____.

4. If the critical point $(0, 0)$ of the linear system $\mathbf{X}' = \mathbf{AX}$ is a stable spiral point, then the eigenvalues of $\mathbf{A}$ are _____.

5. If the critical point $(0, 0)$ of the linear system $\mathbf{X}' = \mathbf{AX}$ is a saddle point and $\mathbf{X} = \mathbf{X}(t)$ is a solution, then $\lim_{t\to\infty} \mathbf{X}(t)$ does not exist. _____

6. If the Jacobian matrix $\mathbf{A} = \mathbf{g}'(\mathbf{X}_1)$ at a critical point of a plane autonomous system has positive trace and determinant, then the critical point $\mathbf{X}_1$ is unstable. _____

7. It is possible to show that a nonlinear plane autonomous system has periodic solutions using linearization. _____

8. All solutions to the pendulum equation $\dfrac{d^2\theta}{dt^2} + \dfrac{g}{l}\sin\theta = 0$ are periodic. _____

9. If a simply connected region R contains no critical points of a plane autonomous system, then there are no periodic solutions in R. _____

10. If a plane autonomous system has no critical points in an annular invariant region R, then there is at least one periodic solution in R. _____

11. Solve the following nonlinear plane autonomous system by switching to polar coordinates, and describe the geometric behavior of the solution that satisfies the given initial condition.

$$x' = -y - x\left(\sqrt{x^2 + y^2}\right)^3$$
$$y' = x - y\left(\sqrt{x^2 + y^2}\right)^3, \quad \mathbf{X}(0) = (1, 0)$$

12. Discuss the geometric nature of the solutions to the linear system $\mathbf{X}' = \mathbf{AX}$ given the general solution.

(a) $\mathbf{X}(t) = c_1 \begin{pmatrix} 1 \\ 1 \end{pmatrix} e^{-t} + c_2 \begin{pmatrix} 1 \\ -2 \end{pmatrix} e^{-2t}$

(b) $\mathbf{X}(t) = c_1 \begin{pmatrix} 1 \\ -1 \end{pmatrix} e^{-t} + c_2 \begin{pmatrix} 1 \\ 2 \end{pmatrix} e^{2t}$

13. Classify the critical point $(0, 0)$ of the given linear system by computing the trace τ and determinant Δ.

(a) $x' = -3x + 4y$
$\quad\; y' = -5x + 3y$

(b) $x' = -3x + 2y$
$\quad\; y' = -2x + y$

14. Find and classify (if possible) the critical points of the plane autonomous system

$$x' = x + xy - 3x^2$$
$$y' = 4y - 2xy - y^2.$$

Does this system have any periodic solutions in the first quadrant?

15. Classify the critical point $(0, 0)$ of the plane autonomous system corresponding to the nonlinear second-order differential equation

$$x'' + \mu(x^2 - 1)x' + x = 0$$

where μ is a real constant.

16. Without solving explicitly, classify (if possible) the critical points of the autonomous first-order differential equation $x' = (x^2 - 1)e^{-x/2}$ as asymptotically stable or unstable.

17. Use the phase-plane method to show that the solutions of the nonlinear second-order differential equation

$$x'' = -2x\sqrt{(x')^2 + 1}$$

that satisfy $x(0) = x_0$ and $x'(0) = 0$ are periodic.

18. In Section 3.8 we assumed that the restoring force F of the spring satisfied Hooke's law $F = ks$, where s is the elongation of the spring and k is a positive constant of proportionality. If we replace this assumption with the nonlinear law $F = ks^3$, then the new differential equation for damped motion becomes $mx'' = -\beta x' - k(s + x)^3 + mg$, where $ks^3 = mg$. The system is called overdamped when $(0, 0)$ is a stable node and is called underdamped when $(0, 0)$ is a stable spiral point. Find new conditions on m, k, and β that will lead to overdamping and underdamping.

19. Show that the plane autonomous system

$$x' = 4x + 2y - 2x^2$$
$$y' = 4x - 3y + 4xy$$

has no periodic solutions.

20. Use the Poincaré–Bendixson theorem to show that the plane autonomous system

$$x' = \epsilon x + y - x(x^2 + y^2)$$
$$y' = -x + \epsilon y - y(x^2 + y^2)$$

has at least one periodic solution when $\epsilon > 0$. What occurs when $\epsilon < 0$?

21. The rod of a pendulum is attached to a movable joint at point P and rotates at an angular speed of ω (radians/s) in the plane perpendicular to the rod. See **FIGURE 11.R.1**. As a

result, the bob of the pendulum experiences an additional centripetal force and the new differential equation for θ becomes

$$ml\frac{d^2\theta}{dt^2} = \omega^2 ml \sin\theta\cos\theta - mg\sin\theta - \beta\frac{d\theta}{dt}.$$

(a) Establish that there are no periodic solutions.

(b) If $\omega^2 < g/l$, show that $(0, 0)$ is a stable critical point and is the only critical point in the domain $-\pi < \theta < \pi$. Describe what occurs physically when $\theta(0) = \theta_0$, $\theta'(0) = 0$, and θ_0 is small.

(c) If $\omega^2 > g/l$, show that $(0, 0)$ is unstable and there are two additional stable critical points $(\pm\hat{\theta}, 0)$ in the domain $-\pi < \theta < \pi$. Describe what occurs physically when $\theta(0) = \theta_0$, $\theta'(0) = 0$, and θ_0 is small.

(d) Determine under what conditions the critical points in parts (a) and (b) are stable spiral points.

FIGURE 11.R.1 Rotating pendulum in Problem 21

22. The nonlinear second-order differential equation

$$x'' - 2kx' + c(x')^3 + \omega^2 x = 0$$

arises in modeling the motion of an electrically driven tuning fork. See **FIGURE 11.R.2**, where $k = c = 0.1$ and $\omega = 1$. Assume that this differential equation possesses a Type I invariant region that contains $(0, 0)$. Show that there is at least one periodic solution.

FIGURE 11.R.2 Solution curve in Problem 22

Partial Differential Equations

Fourier Series

In calculus you saw that a sufficiently differentiable function f could be expanded in a Taylor series, which is essentially an infinite series consisting of powers of x. The principal concept examined in this chapter also involves expanding a function in an infinite series. In the early 1800s, mathematicians developed the idea of expanding a function in a series of trigonometric functions. This idea is the linchpin of the procedures discussed in the two chapters that follow.

12.1 Orthogonal Functions

INTRODUCTION In certain areas of advanced mathematics, a function is considered to be a generalization of a vector. In this section we shall see how the two vector concepts of inner, or dot, product and orthogonality of vectors can be extended to functions. The remainder of the chapter is a practical application of this discussion.

■ **Inner Product** Recall, if $\mathbf{u} = u_1\mathbf{i} + u_2\mathbf{j} + u_3\mathbf{k}$ and $\mathbf{v} = v_1\mathbf{i} + v_2\mathbf{j} + v_3\mathbf{k}$ are two vectors in R^3 or 3-space, then the inner product or dot product of $\mathbf{u}$ and $\mathbf{v}$ is a real number, called a **scalar**, defined as the sum of the products of their corresponding components:

In Chapter 7, the inner product was denoted by $\mathbf{u} \cdot \mathbf{v}$. ▶

$$(\mathbf{u}, \mathbf{v}) = u_1v_1 + u_2v_2 + u_3v_3 = \sum_{k=1}^{3} u_kv_k.$$

The inner product $(\mathbf{u}, \mathbf{v})$ possesses the following properties:

(*i*) $(\mathbf{u}, \mathbf{v}) = (\mathbf{v}, \mathbf{u})$
(*ii*) $(k\mathbf{u}, \mathbf{v}) = k(\mathbf{u}, \mathbf{v})$, k a scalar
(*iii*) $(\mathbf{u}, \mathbf{u}) = 0$ if $\mathbf{u} = \mathbf{0}$ and $(\mathbf{u}, \mathbf{u}) > 0$ if $\mathbf{u} \neq \mathbf{0}$
(*iv*) $(\mathbf{u} + \mathbf{v}, \mathbf{w}) = (\mathbf{u}, \mathbf{w}) + (\mathbf{v}, \mathbf{w})$.

We expect any generalization of the inner product to possess these same properties.

Suppose that f_1 and f_2 are piecewise-continuous functions defined on an interval $[a, b]$.* Since a definite integral on the interval of the product $f_1(x)f_2(x)$ possesses properties (*i*)−(*iv*) of the inner product of vectors, whenever the integral exists we are prompted to make the following definition.

DEFINITION 12.1.1 Inner Product of Functions

The **inner product** of two functions f_1 and f_2 on an interval $[a, b]$ is the number

$$(f_1, f_2) = \int_a^b f_1(x)f_2(x)\, dx.$$

■ **Orthogonal Functions** Motivated by the fact that two vectors $\mathbf{u}$ and $\mathbf{v}$ are orthogonal whenever their inner product is zero, we define **orthogonal functions** in a similar manner.

DEFINITION 12.1.2 Orthogonal Functions

Two functions f_1 and f_2 are said to be **orthogonal** on an interval $[a, b]$ if

$$(f_1, f_2) = \int_a^b f_1(x)f_2(x)\, dx = 0. \tag{1}$$

EXAMPLE 1 Orthogonal Functions

The functions $f_1(x) = x^2$ and $f_2(x) = x^3$ are orthogonal on the interval $[-1, 1]$. This fact follows from (1):

$$(f_1, f_2) = \int_{-1}^{1} x^2 \cdot x^3\, dx = \int_{-1}^{1} x^5\, dx = \frac{1}{6}x^6 \Big]_{-1}^{1} = 0. \qquad \equiv$$

Unlike vector analysis, where the word *orthogonal* is a synonym for *perpendicular*, in this present context the term *orthogonal* and condition (1) have no geometric significance.

■ **Orthogonal Sets** We are primarily interested in infinite sets of orthogonal functions.

*The interval could also be $(-\infty, \infty)$, $[0, \infty)$, and so on.

DEFINITION 12.1.3 Orthogonal Set

A set of real-valued functions $\{\phi_0(x), \phi_1(x), \phi_2(x), \dots\}$ is said to be **orthogonal** on an interval $[a, b]$ if

$$(\phi_m, \phi_n) = \int_a^b \phi_m(x)\phi_n(x)\,dx = 0, \quad m \neq n. \qquad (2)$$

■ Orthonormal Sets The norm, or length $\|\mathbf{u}\|$, of a vector $\mathbf{u}$ can be expressed in terms of the inner product. The expression $(\mathbf{u}, \mathbf{u}) = \|\mathbf{u}\|^2$ is called the square norm, and so the norm is $\|\mathbf{u}\| = \sqrt{(\mathbf{u}, \mathbf{u})}$. Similarly, the **square norm** of a function ϕ_n is $\|\phi_n(x)\|^2 = (\phi_n, \phi_n)$, and so the **norm**, or its generalized length, is $\|\phi_n(x)\| = \sqrt{(\phi_n, \phi_n)}$. In other words, the square norm and norm of a function ϕ_n in an orthogonal set $\{\phi_n(x)\}$ are, respectively,

$$\|\phi_n(x)\|^2 = \int_a^b \phi_n^2(x)\,dx \quad \text{and} \quad \|\phi_n(x)\| = \sqrt{\int_a^b \phi_n^2(x)\,dx}. \qquad (3)$$

If $\{\phi_n(x)\}$ is an orthogonal set of functions on the interval $[a, b]$ with the property that $\|\phi_n(x)\| = 1$ for $n = 0, 1, 2, \dots$, then $\{\phi_n(x)\}$ is said to be an **orthonormal set** on the interval.

EXAMPLE 2 Orthogonal Set of Functions

Show that the set $\{1, \cos x, \cos 2x, \dots\}$ is orthogonal on the interval $[-\pi, \pi]$.

SOLUTION If we make the identification $\phi_0(x) = 1$ and $\phi_n(x) = \cos nx$, we must then show that $\int_{-\pi}^{\pi} \phi_0(x)\phi_n(x)\,dx = 0$, $n \neq 0$, and $\int_{-\pi}^{\pi} \phi_m(x)\phi_n(x)\,dx = 0$, $m \neq n$. We have, in the first case, for $n \neq 0$,

$$(\phi_0, \phi_n) = \int_{-\pi}^{\pi} \phi_0(x)\phi_n(x)\,dx = \int_{-\pi}^{\pi} \cos nx\,dx$$

$$= \frac{1}{n} \sin nx \Big]_{-\pi}^{\pi} = \frac{1}{n}[\sin n\pi - \sin(-n\pi)] = 0,$$

and in the second, for $m \neq n$,

$$(\phi_m, \phi_n) = \int_{-\pi}^{\pi} \phi_m(x)\phi_n(x)\,dx = \int_{-\pi}^{\pi} \cos mx \cos nx\,dx$$

$$= \frac{1}{2} \int_{-\pi}^{\pi} [\cos(m+n)x + \cos(m-n)x]\,dx \quad \leftarrow \text{trigonometric identity}$$

$$= \frac{1}{2} \left[\frac{\sin(m+n)x}{m+n} + \frac{\sin(m-n)x}{m-n} \right]_{-\pi}^{\pi} = 0. \qquad \equiv$$

EXAMPLE 3 Norms

Find the norms of each function in the orthogonal set given in Example 2.

SOLUTION For $\phi_0(x) = 1$ we have from (3)

$$\|\phi_0(x)\|^2 = \int_{-\pi}^{\pi} dx = 2\pi$$

so that $\|\phi_0(x)\| = \sqrt{2\pi}$. For $\phi_n(x) = \cos nx$, $n > 0$, it follows that

$$\|\phi_n(x)\|^2 = \int_{-\pi}^{\pi} \cos^2 nx\,dx = \frac{1}{2} \int_{-\pi}^{\pi} [1 + \cos 2nx]\,dx = \pi.$$

Thus for $n > 0$, $\|\phi_n(x)\| = \sqrt{\pi}$. $\qquad \equiv$

An orthogonal set can be made into an orthonormal set. ▶

Any orthogonal set of nonzero functions $\{\phi_n(x)\}$, $n = 0, 1, 2, \ldots$, can be *normalized*—that is, made into an orthonormal set—by dividing each function by its norm. It follows from Examples 2 and 3 that the set

$$\left\{ \frac{1}{\sqrt{2\pi}}, \frac{\cos x}{\sqrt{\pi}}, \frac{\cos 2x}{\sqrt{\pi}}, \ldots \right\}$$

is orthonormal on the interval $[-\pi, \pi]$.

■ **Vector Analogy** We shall make one more analogy between vectors and functions. Suppose $\mathbf{v}_1$, $\mathbf{v}_2$, and $\mathbf{v}_3$ are three mutually orthogonal nonzero vectors in 3-space. Such an orthogonal set can be used as a basis for 3-space; that is, any three-dimensional vector can be written as a linear combination

$$\mathbf{u} = c_1\mathbf{v}_1 + c_2\mathbf{v}_2 + c_3\mathbf{v}_3, \tag{4}$$

where the c_i, $i = 1, 2, 3$, are scalars called the components of the vector. Each component c_i can be expressed in terms of $\mathbf{u}$ and the corresponding vector $\mathbf{v}_i$. To see this we take the inner product of (4) with $\mathbf{v}_1$:

$$(\mathbf{u}, \mathbf{v}_1) = c_1(\mathbf{v}_1, \mathbf{v}_1) + c_2(\mathbf{v}_2, \mathbf{v}_1) + c_3(\mathbf{v}_3, \mathbf{v}_1) = c_1\|\mathbf{v}_1\|^2 + c_2 \cdot 0 + c_3 \cdot 0.$$

Hence
$$c_1 = \frac{(\mathbf{u}, \mathbf{v}_1)}{\|\mathbf{v}_1\|^2}.$$

In like manner we find that the components c_2 and c_3 are given by

$$c_2 = \frac{(\mathbf{u}, \mathbf{v}_2)}{\|\mathbf{v}_2\|^2} \quad \text{and} \quad c_3 = \frac{(\mathbf{u}, \mathbf{v}_3)}{\|\mathbf{v}_3\|^2}.$$

Hence (4) can be expressed as

$$\mathbf{u} = \frac{(\mathbf{u}, \mathbf{v}_1)}{\|\mathbf{v}_1\|^2}\mathbf{v}_1 + \frac{(\mathbf{u}, \mathbf{v}_2)}{\|\mathbf{v}_2\|^2}\mathbf{v}_2 + \frac{(\mathbf{u}, \mathbf{v}_3)}{\|\mathbf{v}_3\|^2}\mathbf{v}_3 = \sum_{n=1}^{3} \frac{(\mathbf{u}, \mathbf{v}_n)}{\|\mathbf{v}_n\|^2}\mathbf{v}_n. \tag{5}$$

■ **Orthogonal Series Expansion** Suppose $\{\phi_n(x)\}$ is an infinite orthogonal set of functions on an interval $[a, b]$. We ask: If $y = f(x)$ is a function defined on the interval $[a, b]$, is it possible to determine a set of coefficients c_n, $n = 0, 1, 2, \ldots$, for which

$$f(x) = c_0\phi_0(x) + c_1\phi_1(x) + \cdots + c_n\phi_n(x) + \cdots? \tag{6}$$

As in the foregoing discussion on finding components of a vector, we can find the coefficients c_n by utilizing the inner product. Multiplying (6) by $\phi_m(x)$ and integrating over the interval $[a, b]$ gives

$$\int_a^b f(x)\phi_m(x)\,dx = c_0\int_a^b \phi_0(x)\phi_m(x)\,dx + c_1\int_a^b \phi_1(x)\phi_m(x)\,dx + \cdots + c_n\int_a^b \phi_n(x)\phi_m(x)\,dx + \cdots$$

$$= c_0(\phi_0, \phi_m) + c_1(\phi_1, \phi_m) + \cdots + c_n(\phi_n, \phi_m) + \cdots.$$

By orthogonality, each term on the right-hand side of the last equation is zero *except* when $m = n$. In this case we have

$$\int_a^b f(x)\phi_n(x)\,dx = c_n\int_a^b \phi_n^2(x)\,dx.$$

It follows that the required coefficients c_n are given by

$$c_n = \frac{\int_a^b f(x)\phi_n(x)\,dx}{\int_a^b \phi_n^2(x)\,dx}, \quad n = 0, 1, 2, \ldots.$$

In other words,
$$f(x) = \sum_{n=0}^{\infty} c_n \phi_n(x), \tag{7}$$

where
$$c_n = \frac{\int_a^b f(x)\phi_n(x)\,dx}{\|\phi_n(x)\|^2}. \tag{8}$$

With inner product notation, (7) becomes

$$f(x) = \sum_{n=0}^{\infty} \frac{(f, \phi_n)}{\|\phi_n(x)\|^2}\,\phi_n(x). \tag{9}$$

Thus (9) is seen to be the function analogue of the vector result given in (5).

DEFINITION 12.1.4 Orthogonal Set/Weight Function

A set of real-valued functions $\{\phi_0(x), \phi_1(x), \phi_2(x), \ldots\}$ is said to be **orthogonal with respect to a weight function** $w(x)$ on an interval $[a, b]$ if

$$\int_a^b w(x)\phi_m(x)\phi_n(x)\,dx = 0, \quad m \neq n.$$

The usual assumption is that $w(x) > 0$ on the interval of orthogonality $[a, b]$. The set $\{1, \cos x, \cos 2x, \ldots\}$ in Example 2 is orthogonal with respect to the weight function $w(x) = 1$ on the interval $[-\pi, \pi]$.

If $\{\phi_n(x)\}$ is orthogonal with respect to a weight function $w(x)$ on the interval $[a, b]$, then multiplying (6) by $w(x)\phi_n(x)$ and integrating yields

$$c_n = \frac{\int_a^b f(x)w(x)\phi_n(x)\,dx}{\|\phi_n(x)\|^2}, \tag{10}$$

where
$$\|\phi_n(x)\|^2 = \int_a^b w(x)\phi_n^2(x)\,dx. \tag{11}$$

The series (7) with coefficients c_n given by either (8) or (10) is said to be an **orthogonal series expansion** of f or a **generalized Fourier series**.

■ **Complete Sets** The procedure outlined for determining the coefficients c_n was *formal*; that is, basic questions on whether an orthogonal series expansion such as (7) is actually possible were ignored. Also, to expand f in a series of orthogonal functions, it is certainly necessary that f not be orthogonal to each ϕ_n of the orthogonal set $\{\phi_n(x)\}$. (If f were orthogonal to every ϕ_n, then $c_n = 0$, $n = 0, 1, 2, \ldots$.) To avoid the latter problem we shall assume, for the remainder of the discussion, that an orthogonal set is **complete**. This means that the only continuous function orthogonal to each member of the set is the zero function.

REMARKS

Suppose that $\{f_0(x), f_1(x), f_2(x), \ldots\}$ is an infinite set of real-valued functions that are continuous on an interval $[a, b]$. If this set is *linearly independent* on $[a, b]$ (see page 364 for the definition of an infinite linearly independent set), then it can always be made into an orthogonal set and, as described earlier in this section, can be made into an orthonormal set. See Problem 27 in Exercises 12.1.

In Problems 1–6, show that the given functions are orthogonal on the indicated interval.

1. $f_1(x) = x, f_2(x) = x^2$; $[-2, 2]$

2. $f_1(x) = x^3, f_2(x) = x^2 + 1$; $[-1, 1]$

3. $f_1(x) = e^x, f_2(x) = xe^{-x} - e^{-x}$; $[0, 2]$

4. $f_1(x) = \cos x, f_2(x) = \sin^2 x$; $[0, \pi]$

5. $f_1(x) = x, f_2(x) = \cos 2x$; $[-\pi/2, \pi/2]$

6. $f_1(x) = e^x, f_2(x) = \sin x$; $[\pi/4, 5\pi/4]$

In Problems 7–12, show that the given set of functions is orthogonal on the indicated interval. Find the norm of each function in the set.

7. $\{\sin x, \ \sin 3x, \ \sin 5x, \ ...\}$; $[0, \pi/2]$

8. $\{\cos x, \ \cos 3x, \ \cos 5x, \ ...\}$; $[0, \pi/2]$

9. $\{\sin nx\}$, $n = 1, 2, 3, ...$; $[0, \pi]$

10. $\left\{\sin \dfrac{n\pi}{p} x\right\}$, $n = 1, 2, 3, ...$; $[0, p]$

11. $\left\{1, \cos \dfrac{n\pi}{p} x\right\}$, $n = 1, 2, 3, ...$; $[0, p]$

12. $\left\{1, \cos \dfrac{n\pi}{p} x, \sin \dfrac{m\pi}{p} x\right\}$, $n = 1, 2, 3, ...$,

 $m = 1, 2, 3, ...$; $[-p, p]$

In Problems 13 and 14, verify by direct integration that the functions are orthogonal with respect to the indicated weight function on the given interval.

13. $H_0(x) = 1, H_1(x) = 2x, H_2(x) = 4x^2 - 2$; $w(x) = e^{-x^2}$, $(-\infty, \infty)$

14. $L_0(x) = 1, L_1(x) = -x + 1, L_2(x) = \frac{1}{2}x^2 - 2x + 1$; $w(x) = e^{-x}$, $[0, \infty)$

15. Let $\{\phi_n(x)\}$ be an orthogonal set of functions on $[a, b]$ such that $\phi_0(x) = 1$. Show that $\int_a^b \phi_n(x)\, dx = 0$ for $n = 1, 2, ...$.

16. Let $\{\phi_n(x)\}$ be an orthogonal set of functions on $[a, b]$ such that $\phi_0(x) = 1$ and $\phi_1(x) = x$. Show that $\int_a^b (\alpha x + \beta)\phi_n(x)\, dx = 0$ for $n = 2, 3, ...$ and any constants α and β.

17. Let $\{\phi_n(x)\}$ be an orthogonal set of functions on $[a, b]$. Show that $\|\phi_m(x) + \phi_n(x)\|^2 = \|\phi_m(x)\|^2 + \|\phi_n(x)\|^2$, $m \ne n$.

18. From Problem 1 we know that $f_1(x) = x$ and $f_2(x) = x^2$ are orthogonal on $[-2, 2]$. Find constants c_1 and c_2 such that $f_3(x) = x + c_1 x^2 + c_2 x^3$ is orthogonal to both f_1 and f_2 on the same interval.

19. The set of functions $\{\sin nx\}, n = 1, 2, 3, ...,$ is orthogonal on the interval $[-\pi, \pi]$. Show that the set is not complete.

20. Suppose $f_1, f_2,$ and f_3 are functions continuous on the interval $[a, b]$. Show that $(f_1 + f_2, f_3) = (f_1, f_3) + (f_2, f_3)$.

A real-valued function is said to be **periodic** with period $T \ne 0$ if $f(x + T) = f(x)$ for all x in the domain of f. If T is the smallest positive value for which $f(x + T) = f(x)$ holds, then T is called the **fundamental period** of f. In Problems 21–26, determine the fundamental period T of the given function.

21. $f(x) = \cos 2\pi x$

22. $f(x) = \sin \dfrac{4}{L} x, L > 0$

23. $f(x) = \sin x + \sin 2x$

24. $f(x) = \sin 2x + \cos 4x$

25. $f(x) = \sin 3x + \cos 2x$

26. $f(x) = \sin^2 \pi x$

Discussion Problems

27. The **Gram–Schmidt process** for constructing an orthogonal set that was discussed in Section 7.7 carries over to a linearly independent set $\{f_0(x), f_1(x), f_2(x), ...\}$ of real-valued functions continuous on an interval $[a, b]$. With the inner product $(f_n, \phi_n) = \int_a^b f_n(x)\phi_n(x)\, dx$, define the functions in the set $B' = \{\phi_0(x), \phi_1(x), \phi_2(x), ...\}$ to be

$$\phi_0(x) = f_0(x)$$

$$\phi_1(x) = f_1(x) - \frac{(f_1, \phi_0)}{(\phi_0, \phi_0)} \phi_0(x)$$

$$\phi_2(x) = f_2(x) - \frac{(f_2, \phi_0)}{(\phi_0, \phi_0)} \phi_0(x) - \frac{(f_2, \phi_1)}{(\phi_1, \phi_1)} \phi_1(x)$$

$$\vdots \qquad\qquad \vdots$$

and so on.

(a) Write out $\phi_3(x)$ in the set.

(b) By construction, the set $B' = \{\phi_0(x), \phi_1(x), \phi_2(x), ...\}$ is orthogonal on $[a, b]$. Demonstrate that $\phi_0(x), \phi_1(x),$ and $\phi_2(x)$ are mutually orthogonal.

28. (a) Consider the set of functions $\{1, x, x^2, x^3, ...\}$ defined on the interval $[-1, 1]$. Apply the Gram–Schmidt process given in Problem 27 to this set and find $\phi_0(x),$ $\phi_1(x), \phi_2(x),$ and $\phi_3(x)$ of the orthogonal set B'.

(b) Discuss: Do you recognize the orthogonal set?

29. Verify that the inner product (f_1, f_2) in Definition 12.1.1 satisfies properties $(i)-(iv)$ given on page 681.

30. In R^3, give an example of a set of orthogonal vectors that is not complete. Give a set of orthogonal vectors that is complete.

31. The function

$$f(x) = A_0 + \sum_{n=1}^{\infty}\left(A_n \cos \frac{n\pi}{p} x + B_n \sin \frac{n\pi}{p} x\right),$$

where the coefficients A_n and B_n depend only on n, is periodic. Find the period T of f.

12.2 Fourier Series

INTRODUCTION We have just seen in the preceding section that if $\{\phi_0(x), \phi_1(x), \phi_2(x), \ldots\}$ is a set of real-valued functions that is orthogonal on an interval $[a, b]$ and if f is a function defined on the same interval, then we can formally expand f in an orthogonal series $c_0\phi_0(x) + c_1\phi_1(x) + c_2\phi_2(x) + \cdots$. In this section we shall expand functions in terms of a special orthogonal set of trigonometric functions.

■ Trigonometric Series

In Problem 12 in Exercises 12.1, you were asked to show that the set of trigonometric functions

$$\left\{ 1, \cos\frac{\pi}{p}x, \cos\frac{2\pi}{p}x, \cos\frac{3\pi}{p}x, \ldots, \sin\frac{\pi}{p}x, \sin\frac{2\pi}{p}x, \sin\frac{3\pi}{p}x, \ldots \right\} \tag{1}$$

is orthogonal on the interval $[-p, p]$. This set will be of special importance later on in the solution of certain kinds of boundary-value problems involving linear partial differential equations. In those applications we will need to expand a function f defined on $[-p, p]$ in an orthogonal series consisting of the trigonometric functions in (1); that is,

$$f(x) = \frac{a_0}{2} + \sum_{n=1}^{\infty}\left(a_n\cos\frac{n\pi}{p}x + b_n\sin\frac{n\pi}{p}x \right). \tag{2}$$

The coefficients $a_0, a_1, a_2, \ldots, b_1, b_2, \ldots$, can be determined in exactly the same manner as in the general discussion of orthogonal series expansions on pages 683 and 684. Before proceeding, note that we have chosen to write the coefficient of 1 in the set (1) as $\frac{1}{2}a_0$ rather than a_0; this is for convenience only because the formula of a_n will then reduce to a_0 for $n = 0$.

This is why $\frac{1}{2}a_0$ is used instead of a_0. ▶

Now integrating both sides of (2) from $-p$ to p gives

$$\int_{-p}^{p} f(x)\,dx = \frac{a_0}{2}\int_{-p}^{p} dx + \sum_{n=1}^{\infty}\left(a_n\int_{-p}^{p}\cos\frac{n\pi}{p}x\,dx + b_n\int_{-p}^{p}\sin\frac{n\pi}{p}x\,dx \right). \tag{3}$$

Since $\cos(n\pi x/p)$ and $\sin(n\pi x/p)$, $n \geq 1$, are orthogonal to 1 on the interval, the right side of (3) reduces to a single term:

$$\int_{-p}^{p} f(x)\,dx = \frac{a_0}{2}\int_{-p}^{p} dx = \frac{a_0}{2}x\,\Big]_{-p}^{p} = pa_0.$$

Solving for a_0 yields

$$a_0 = \frac{1}{p}\int_{-p}^{p} f(x)\,dx. \tag{4}$$

Now we multiply (2) by $\cos(m\pi x/p)$ and integrate:

$$\int_{-p}^{p} f(x)\cos\frac{m\pi}{p}x\,dx = \frac{a_0}{2}\int_{-p}^{p}\cos\frac{m\pi}{p}x\,dx$$

$$+ \sum_{n=1}^{\infty}\left(a_n\int_{-p}^{p}\cos\frac{m\pi}{p}x\cos\frac{n\pi}{p}x\,dx + b_n\int_{-p}^{p}\cos\frac{m\pi}{p}x\sin\frac{n\pi}{p}x\,dx \right). \tag{5}$$

By orthogonality we have

$$\int_{-p}^{p}\cos\frac{m\pi}{p}x\,dx = 0, \quad m > 0, \qquad \int_{-p}^{p}\cos\frac{m\pi}{p}x\sin\frac{n\pi}{p}x\,dx = 0$$

and

$$\int_{-p}^{p}\cos\frac{m\pi}{p}x\cos\frac{n\pi}{p}x\,dx = \begin{cases} 0, & m \neq n \\ p, & m = n. \end{cases}$$

Thus (5) reduces to

$$\int_{-p}^{p} f(x) \cos \frac{n\pi}{p} x \, dx = a_n p,$$

and so

$$a_n = \frac{1}{p} \int_{-p}^{p} f(x) \cos \frac{n\pi}{p} x \, dx. \qquad (6)$$

Finally, if we multiply (2) by $\sin(m\pi x/p)$, integrate, and make use of the results

$$\int_{-p}^{p} \sin \frac{m\pi}{p} x \, dx = 0, \quad m > 0, \qquad \int_{-p}^{p} \sin \frac{m\pi}{p} x \cos \frac{n\pi}{p} x \, dx = 0$$

and

$$\int_{-p}^{p} \sin \frac{m\pi}{p} x \sin \frac{n\pi}{p} x \, dx = \begin{cases} 0, & m \neq n \\ p, & m = n, \end{cases}$$

we find that

$$b_n = \frac{1}{p} \int_{-p}^{p} f(x) \sin \frac{n\pi}{p} x \, dx. \qquad (7)$$

The trigonometric series (2) with coefficients $a_0, a_n,$ and b_n defined by (4), (6), and (7), respectively, is said to be the **Fourier series** of the function f. The coefficients obtained from (4), (6), and (7) are referred to as **Fourier coefficients** of f. If the Fourier series (2) converges, it is sometimes called a **representation** of f.

Although the French mathematical physicist and scientific advisor to Napoleon Bonaparte, **Jean-Baptiste Joseph Fourier** (1768–1830), did not invent the series that bears his name, he was at least responsible for sparking the interest of mathematicians in trigonometric series by his less-than-rigorous use of them in his research on the conduction of heat.

In finding the coefficients $a_0, a_n,$ and b_n, we assumed that f was integrable on the interval and that (2), as well as the series obtained by multiplying (2) by $\cos(m\pi x/p)$, converged in such a manner as to permit term-by-term integration. Until (2) is shown to be convergent for a given function f, the equality sign is not to be taken in a strict or literal sense. Some texts use the symbol $\sim$ in place of $=$. In view of the fact that most functions in applications are of a type that guarantees convergence of the series, we shall use the equality symbol. We summarize the results:

DEFINITION 12.2.1 Fourier Series

The **Fourier series** of a function f defined on the interval $(-p, p)$ is given by

$$f(x) = \frac{a_0}{2} + \sum_{n=1}^{\infty} \left(a_n \cos \frac{n\pi}{p} x + b_n \sin \frac{n\pi}{p} x \right), \qquad (8)$$

where

$$a_0 = \frac{1}{p} \int_{-p}^{p} f(x) \, dx \qquad (9)$$

$$a_n = \frac{1}{p} \int_{-p}^{p} f(x) \cos \frac{n\pi}{p} x \, dx \qquad (10)$$

$$b_n = \frac{1}{p} \int_{-p}^{p} f(x) \sin \frac{n\pi}{p} x \, dx. \qquad (11)$$

EXAMPLE 1 **Expansion in a Fourier Series**

Expand

$$f(x) = \begin{cases} 0, & -\pi < x < 0 \\ \pi - x, & 0 \leq x < \pi \end{cases} \qquad (12)$$

in a Fourier series.

SOLUTION The graph of f is given in **FIGURE 12.2.1**. With $p = \pi$ we have from (9) and (10) that

$$a_0 = \frac{1}{\pi} \int_{-\pi}^{\pi} f(x) \, dx = \frac{1}{\pi} \left[\int_{-\pi}^{0} 0 \, dx + \int_{0}^{\pi} (\pi - x) \, dx \right] = \frac{1}{\pi} \left[\pi x - \frac{x^2}{2} \right]_{0}^{\pi} = \frac{\pi}{2}$$

FIGURE 12.2.1 Function f in Example 1

$$a_n = \frac{1}{\pi} \int_{-\pi}^{\pi} f(x) \cos nx \, dx = \frac{1}{\pi} \left[\int_{-\pi}^{0} 0 \, dx + \int_{0}^{\pi} (\pi - x) \cos nx \, dx \right]$$

$$= \frac{1}{\pi} \left[(\pi - x) \frac{\sin nx}{n} \bigg|_{0}^{\pi} + \frac{1}{n} \int_{0}^{\pi} \sin nx \, dx \right]$$

$$= -\frac{1}{n\pi} \frac{\cos nx}{n} \bigg|_{0}^{\pi}$$

$$= \frac{-\cos n\pi + 1}{n^2 \pi} \qquad \leftarrow \cos n\pi = (-1)^n$$

$$= \frac{1 - (-1)^n}{n^2 \pi}.$$

In like manner we find from (11) that

$$b_n = \frac{1}{\pi} \int_{0}^{\pi} (\pi - x) \sin nx \, dx = \frac{1}{n}.$$

Note that
$$1 - (-1)^n = \begin{cases} 0, & n \text{ even} \\ 2, & n \text{ odd.} \end{cases}$$

▶ Therefore $\quad f(x) = \dfrac{\pi}{4} + \displaystyle\sum_{n=1}^{\infty} \left\{ \dfrac{1 - (-1)^n}{n^2 \pi} \cos nx + \dfrac{1}{n} \sin nx \right\}.$ (13) ≡

Note that a_n defined by (10) reduces to a_0 given by (9) when we set $n = 0$. But as Example 1 shows, this may not be the case *after* the integral for a_n is evaluated.

■ Convergence of a Fourier Series The following theorem gives sufficient conditions for convergence of a Fourier series at a point.

THEOREM 12.2.1 Conditions for Convergence

Let f and f' be piecewise continuous on the interval $[-p, p]$; that is, let f and f' be continuous except at a finite number of points in the interval and have only finite discontinuities at these points. Then for all x in the interval $(-p, p)$ the Fourier series of f converges to $f(x)$ at a point of continuity. At a point of discontinuity, the Fourier series converges to the average

$$\frac{f(x+) + f(x-)}{2},$$

where $f(x+)$ and $f(x-)$ denote the limit of f at x from the right and from the left, respectively.*

For a proof of this theorem you are referred to the classic text by Churchill and Brown.[†]

EXAMPLE 2 Convergence of a Point of Discontinuity

The function (12) in Example 1 satisfies the conditions of Theorem 12.2.1. Thus for every x in the interval $(-\pi, \pi)$, except at $x = 0$, the series (13) will converge to $f(x)$. At $x = 0$ the function is discontinuous, and so the series (13) will converge to

$$\frac{f(0+) + f(0-)}{2} = \frac{\pi + 0}{2} = \frac{\pi}{2}.$$

≡

*In other words, for x a point in the interval and $h > 0$,

$$f(x+) = \lim_{h \to 0} f(x + h), \quad f(x-) = \lim_{h \to 0} f(x - h).$$

[†]Ruel V. Churchill and James Ward Brown, *Fourier Series and Boundary Value Problems* (New York: McGraw-Hill, 2011).

Periodic Extension

Periodic Extension Observe that each of the functions in the basic set (1) has a different fundamental period,* namely, $2p/n$, $n \geq 1$, but since a positive integer multiple of a period is also a period we see that all of the functions have in common the period $2p$ (verify). Hence the right-hand side of (2) is $2p$-periodic; indeed, $2p$ is the fundamental period of the sum. We conclude that a Fourier series not only represents the function on the interval $(-p, p)$ but also gives the **periodic extension** of f outside this interval. We can now apply Theorem 12.2.1 to the periodic extension of f, or we may assume from the outset that the given function is periodic with period $T = 2p$; that is, $f(x + T) = f(x)$. When f is piecewise continuous and the right- and left-hand derivatives exist at $x = -p$ and $x = p$, respectively, then the series (8) converges to the average $[f(p-) + f(-p+)]/2$ at these endpoints and to this value extended periodically to $\pm 3p$, $\pm 5p$, $\pm 7p$, and so on. The Fourier series in (13) converges to the periodic extension of (12) on the entire x-axis. At 0, $\pm 2\pi$, $\pm 4\pi$, ..., and at $\pm\pi$, $\pm 3\pi$, $\pm 5\pi$, ..., the series converges to the values

$$\frac{f(0+) + f(0-)}{2} = \frac{\pi}{2} \quad \text{and} \quad \frac{f(\pi-) + f(-\pi+)}{2} = 0,$$

respectively. The solid dots in **FIGURE 12.2.2** represent the value $\pi/2$.

FIGURE 12.2.2 Periodic extension of the function f shown in Figure 12.2.1

Sequence of Partial Sums

Sequence of Partial Sums It is interesting to see how the sequence of partial sums $\{S_N(x)\}$ of a Fourier series approximates a function. For example, the first three partial sums of (13) are

$$S_1(x) = \frac{\pi}{4}, \quad S_2(x) = \frac{\pi}{4} + \frac{2}{\pi}\cos x + \sin x, \quad S_3(x) = \frac{\pi}{4} + \frac{2}{\pi}\cos x + \sin x + \frac{1}{2}\sin 2x.$$

In **FIGURE 12.2.3** we have used a CAS to graph the partial sums $S_5(x)$, $S_8(x)$, and $S_{15}(x)$ of (13) on the interval $(-\pi, \pi)$. Figure 12.2.3(d) shows the periodic extension using $S_{15}(x)$ on $(-4\pi, 4\pi)$.

(a) $S_5(x)$ on $(-\pi, \pi)$

(b) $S_8(x)$ on $(-\pi, \pi)$

(c) $S_{15}(x)$ on $(-\pi, \pi)$

(d) $S_{15}(x)$ on $(-4\pi, 4\pi)$

FIGURE 12.2.3 Partial sums of a Fourier series

*See Problems 21–26 in Exercises 12.1.

In Problems 1–18, find the Fourier series of the function f on the given interval. Give the number to which the Fourier series converges at a point of discontinuity of f.

1. $f(x) = \begin{cases} 0, & -\pi < x < 0 \\ 1, & 0 \le x < \pi \end{cases}$

2. $f(x) = \begin{cases} -1, & -\pi < x < 0 \\ 2, & 0 \le x < \pi \end{cases}$

3. $f(x) = \begin{cases} 1, & -1 < x < 0 \\ x, & 0 \le x < 1 \end{cases}$

4. $f(x) = \begin{cases} 0, & -1 < x < 0 \\ x, & 0 \le x < 1 \end{cases}$

5. $f(x) = \begin{cases} 0, & -\pi < x < 0 \\ x^2, & 0 \le x < \pi \end{cases}$

6. $f(x) = \begin{cases} \pi^2, & -\pi < x < 0 \\ \pi^2 - x^2, & 0 \le x < \pi \end{cases}$

7. $f(x) = x + \pi, \quad -\pi < x < \pi$

8. $f(x) = 3 - 2x, \quad -\pi < x < \pi$

9. $f(x) = \begin{cases} 0, & -\pi < x < 0 \\ \sin x, & 0 \le x < \pi \end{cases}$

10. $f(x) = \begin{cases} 0, & -\pi/2 < x < 0 \\ \cos x, & 0 \le x < \pi/2 \end{cases}$

11. $f(x) = \begin{cases} 0, & -2 < x < -1 \\ -2, & -1 \le x < 0 \\ 1, & 0 \le x < 1 \\ 0, & 1 \le x < 2 \end{cases}$

12. $f(x) = \begin{cases} 0, & -2 < x < 0 \\ x, & 0 \le x < 1 \\ 1, & 1 \le x < 2 \end{cases}$

13. $f(x) = \begin{cases} 1, & -5 < x < 0 \\ 1 + x, & 0 \le x < 5 \end{cases}$

14. $f(x) = \begin{cases} 2 + x, & -2 < x < 0 \\ 2, & 0 \le x < 2 \end{cases}$

15. $f(x) = e^x, \quad -\pi < x < \pi$

16. $f(x) = \begin{cases} 0, & -\pi < x < 0 \\ e^x - 1, & 0 \le x < \pi \end{cases}$

17. $f(x) = 2x, \quad -\pi < x < \pi$

18. $f(x) = \pi^2 - x^2, \quad -\pi < x < \pi$

In Problems 19 and 20, sketch the periodic extension of the indicated function.

19. The function f in Problem 9

20. The function f in Problem 14

21. Use the result of Problem 5 to show

$$\frac{\pi^2}{6} = 1 + \frac{1}{2^2} + \frac{1}{3^2} + \frac{1}{4^2} + \cdots$$

and

$$\frac{\pi^2}{12} = 1 - \frac{1}{2^2} + \frac{1}{3^2} - \frac{1}{4^2} + \cdots.$$

22. Use Problem 21 to find a series that gives the numerical value of $\pi^2/8$.

23. Use the result of Problem 7 to show

$$\frac{\pi}{4} = 1 - \frac{1}{3} + \frac{1}{5} - \frac{1}{7} + \cdots.$$

24. Use the result of Problem 9 to show

$$\frac{\pi}{4} = \frac{1}{2} + \frac{1}{1\cdot3} - \frac{1}{3\cdot5} + \frac{1}{5\cdot7} - \frac{1}{7\cdot9} + \cdots.$$

25. The **root–mean–square value** of a function $f(x)$ defined over an interval (a, b) is given by

$$\text{RMS}(f) = \sqrt{\frac{\int_a^b f^2(x)\, dx}{b - a}}.$$

If the Fourier series expansion of f is given by (8), show that the RMS value of f over the interval $(-p, p)$ is given by

$$\text{RMS}(f) = \sqrt{\frac{1}{4}a_0^2 + \frac{1}{2}\sum_{n=1}^{\infty}(a_n^2 + b_n^2)},$$

where a_0, a_n, and b_n are the Fourier coefficients in (9), (10), and (11), respectively.

12.3 Fourier Cosine and Sine Series

INTRODUCTION The effort expended in the evaluation of coefficients a_0, a_n, and b_n in expanding a function f in a Fourier series is reduced significantly when f is either an even or an odd function. A function f is said to be

even if $f(-x) = f(x)$ and **odd** if $f(-x) = -f(x)$.

On a symmetric interval such as $(-p, p)$, the graph of an even function possesses symmetry with respect to the y-axis, whereas the graph of an odd function possesses symmetry with respect to the origin.

FIGURE 12.3.1 Even function

FIGURE 12.3.2 Odd function

Even and Odd Functions It is likely the origin of the words *even* and *odd* derives from the fact that the graphs of polynomial functions that consist of all even powers of x are symmetric with respect to the y-axis, whereas graphs of polynomials that consist of all odd powers of x are symmetric with respect to the origin. For example,

$$\overset{\underset{\text{even integer}}{\downarrow}}{f(x) = x^2} \text{ is even since } f(-x) = (-x)^2 = x^2 = f(x)$$

$$\overset{\underset{\text{odd integer}}{\downarrow}}{f(x) = x^3} \text{ is odd since } f(-x) = (-x)^3 = -x^3 = -f(x).$$

See **FIGURES 12.3.1** and **12.3.2**. The trigonometric cosine and sine functions are even and odd functions, respectively, since $\cos(-x) = \cos x$ and $\sin(-x) = -\sin x$. The exponential functions $f(x) = e^x$ and $f(x) = e^{-x}$ are neither even nor odd.

Properties The following theorem lists some properties of even and odd functions.

THEOREM 12.3.1 Properties of Even/Odd Functions

 (a) The product of two even functions is even.
 (b) The product of two odd functions is even.
 (c) The product of an even function and an odd function is odd.
 (d) The sum (difference) of two even functions is even.
 (e) The sum (difference) of two odd functions is odd.
 (f) If f is even, then $\int_{-a}^{a} f(x)\, dx = 2\int_{0}^{a} f(x)\, dx$.
 (g) If f is odd, then $\int_{-a}^{a} f(x)\, dx = 0$.

PROOF OF (b): Let us suppose that f and g are odd functions. Then we have $f(-x) = -f(x)$ and $g(-x) = -g(x)$. If we define the product of f and g as $F(x) = f(x)g(x)$, then

$$F(-x) = f(-x)g(-x) = (-f(x))(-g(x)) = f(x)g(x) = F(x).$$

This shows that the product F of two odd functions is an even function. The proofs of the remaining properties are left as exercises. See Problem 56 in Exercises 12.3. ≡

Cosine and Sine Series If f is an even function on the interval $(-p, p)$, then in view of the foregoing properties, the coefficients (9), (10), and (11) of Section 12.2 become

$$a_0 = \frac{1}{p}\int_{-p}^{p} f(x)\, dx = \frac{2}{p}\int_{0}^{p} f(x)\, dx$$

$$a_n = \frac{1}{p}\int_{-p}^{p} \underbrace{f(x)\cos\frac{n\pi}{p}x}_{\text{even}}\, dx = \frac{2}{p}\int_{0}^{p} f(x)\cos\frac{n\pi}{p}x\, dx$$

$$b_n = \frac{1}{p}\int_{-p}^{p} \underbrace{f(x)\sin\frac{n\pi}{p}x}_{\text{odd}}\, dx = 0.$$

Similarly, when f is odd on the interval $(-p, p)$,

$$a_n = 0, \quad n = 0, 1, 2, \ldots, \qquad b_n = \frac{2}{p}\int_{0}^{p} f(x)\sin\frac{n\pi}{p}x\, dx.$$

We summarize the results in the following definition.

DEFINITION 12.3.1 Fourier Cosine and Sine Series

(*i*) The Fourier series of an even function on the interval $(-p, p)$ is the **cosine series**

$$f(x) = \frac{a_0}{2} + \sum_{n=1}^{\infty} a_n \cos \frac{n\pi}{p} x, \tag{1}$$

where

$$a_0 = \frac{2}{p} \int_0^p f(x)\, dx \tag{2}$$

$$a_n = \frac{2}{p} \int_0^p f(x) \cos \frac{n\pi}{p} x\, dx. \tag{3}$$

(*ii*) The Fourier series of an odd function on the interval $(-p, p)$ is the **sine series**

$$f(x) = \sum_{n=1}^{\infty} b_n \sin \frac{n\pi}{p} x, \tag{4}$$

where

$$b_n = \frac{2}{p} \int_0^p f(x) \sin \frac{n\pi}{p} x\, dx. \tag{5}$$

EXAMPLE 1 Expansion in a Sine Series

Expand $f(x) = x$, $-2 < x < 2$, in a Fourier series.

SOLUTION Inspection of **FIGURE 12.3.3** shows that the given function is odd on the interval $(-2, 2)$, and so we expand f in a sine series. With the identification $2p = 4$, we have $p = 2$. Thus (5), after integration by parts, is

$$b_n = \int_0^2 x \sin \frac{n\pi}{2} x\, dx = \frac{4(-1)^{n+1}}{n\pi}.$$

Therefore

$$f(x) = \frac{4}{\pi} \sum_{n=1}^{\infty} \frac{(-1)^{n+1}}{n} \sin \frac{n\pi}{2} x. \tag{6} \equiv$$

The function in Example 1 satisfies the conditions of Theorem 12.2.1. Hence the series (6) converges to the function on $(-2, 2)$ and the periodic extension (of period 4) given in **FIGURE 12.3.4**.

FIGURE 12.3.4 Periodic extension of the function f shown in Figure 12.3.3

EXAMPLE 2 Expansion in a Sine Series

The function $f(x) = \begin{cases} -1, & -\pi < x < 0 \\ 1, & 0 \le x < \pi \end{cases}$ shown in **FIGURE 12.3.5** is odd on the interval $(-\pi, \pi)$. With $p = \pi$ we have from (5)

$$b_n = \frac{2}{\pi} \int_0^\pi (1) \sin nx\, dx = \frac{2}{\pi} \frac{1 - (-1)^n}{n},$$

and so

$$f(x) = \frac{2}{\pi} \sum_{n=1}^{\infty} \frac{1 - (-1)^n}{n} \sin nx. \tag{7} \equiv$$

Margin figures:

FIGURE 12.3.3 Odd function f in Example 1

$y = x$, $-2 < x < 2$

FIGURE 12.3.5 Odd function f in Example 2

Gibbs Phenomenon With the aid of a CAS we have plotted in **FIGURE 12.3.6** the graphs $S_1(x)$, $S_2(x)$, $S_3(x)$, $S_{15}(x)$ of the partial sums of nonzero terms of (7). As seen in Figure 12.3.6(d), the graph of $S_{15}(x)$ has pronounced spikes near the discontinuities at $x = 0$, $x = \pi$, $x = -\pi$, and so on. This "overshooting" by the partial sums S_N from the function values near a point of discontinuity does not smooth out but remains fairly constant, even when the value N is taken to be large. This behavior of a Fourier series near a point at which f is discontinuous is called the **Gibbs phenomenon** after the American mathematician/scientist **Josiah Willard Gibbs** (1839–1903). Gibbs was the co-inventor of modern vector analysis.

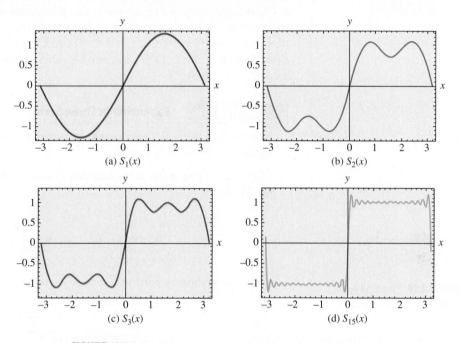

(a) $S_1(x)$ (b) $S_2(x)$

(c) $S_3(x)$ (d) $S_{15}(x)$

FIGURE 12.3.6 Partial sums of sine series (7) on the interval $(-\pi, \pi)$

The periodic extension of f in Example 2 onto the entire x-axis is a meander function (see page 255).

Half-Range Expansions Throughout the preceding discussion it was understood that a function f was defined on an interval with the origin as midpoint, that is, $(-p, p)$. However, in many instances we are interested in representing a function that is defined on an interval $(0, L)$ by a trigonometric series. This can be done in many different ways by supplying an arbitrary *definition* of the function on the interval $(-L, 0)$. For brevity we consider the three most important cases. If $y = f(x)$ is defined on the interval $(0, L)$, then:

(*i*) reflect the graph of the function about the y-axis onto $(-L, 0)$; the function is now even on the interval $(-L, L)$ (see **FIGURE 12.3.7**); or

(*ii*) reflect the graph of the function through the origin onto $(-L, 0)$; the function is now odd on the interval $(-L, L)$ (see **FIGURE 12.3.8**); or

(*iii*) define f on $(-L, 0)$ by $f(x) = f(x + L)$ (see **FIGURE 12.3.9**).

FIGURE 12.3.7 Even reflection **FIGURE 12.3.8** Odd reflection **FIGURE 12.3.9** Identity reflection

Note that the coefficients of the series (1) and (4) utilize only the definition of the function on $(0, p)$, that is, for half of the interval $(-p, p)$. Hence in practice there is no actual need to make the reflections described in (*i*) and (*ii*). If f is defined on $(0, L)$, we simply identify the half-period as the length of the interval $p = L$. The coefficient formulas (2), (3), and (5) and the corresponding series yield either an even or an odd periodic extension of period $2L$ of the original function. The cosine and sine series obtained in this manner are known as **half-range expansions**. Last, in case (*iii*) we are defining the function values on the interval $(-L, 0)$ to be the same as the values on $(0, L)$. As in the previous two cases, there is no real need to do this. It can be shown that the set of functions in (1) of Section 12.2 is orthogonal on $[a, a + 2p]$ for any real number a. Choosing $a = -p$, we obtain the limits of integration in (9), (10), and (11) of that section. But for $a = 0$ the limits of integration are from $x = 0$ to $x = 2p$. Thus if f is defined over the interval $(0, L)$, we identify $2p = L$ or $p = L/2$. The resulting Fourier series will give the periodic extension of f with period L. In this manner the values to which the series converges will be the same on $(-L, 0)$ as on $(0, L)$.

FIGURE 12.3.10 Function f in Example 3

EXAMPLE 3 **Expansion in Three Series**

Expand $f(x) = x^2$, $0 < x < L$, **(a)** in a cosine series, **(b)** in a sine series, **(c)** in a Fourier series.

SOLUTION The graph of the function is given in **FIGURE 12.3.10**.
(a) We have

$$a_0 = \frac{2}{L} \int_0^L x^2 \, dx = \frac{2}{3} L^2, \quad a_n = \frac{2}{L} \int_0^L x^2 \cos \frac{n\pi}{L} x \, dx = \frac{4L^2(-1)^n}{n^2\pi^2},$$

where integration by parts was used twice in the evaluation of a_n. Thus

$$f(x) = \frac{L^2}{3} + \frac{4L^2}{\pi^2} \sum_{n=1}^{\infty} \frac{(-1)^n}{n^2} \cos \frac{n\pi}{L} x. \tag{8}$$

(b) In this case we must again integrate by parts twice:

$$b_n = \frac{2}{L} \int_0^L x^2 \sin \frac{n\pi}{L} x \, dx = \frac{2L^2(-1)^{n+1}}{n\pi} + \frac{4L^2}{n^3\pi^3} [(-1)^n - 1].$$

Hence $$f(x) = \frac{2L^2}{\pi} \sum_{n=1}^{\infty} \left\{ \frac{(-1)^{n+1}}{n} + \frac{2}{n^3\pi^2} [(-1)^n - 1] \right\} \sin \frac{n\pi}{L} x. \tag{9}$$

(c) With $p = L/2$, $1/p = 2/L$, and $n\pi/p = 2n\pi/L$, we have

$$a_0 = \frac{2}{L} \int_0^L x^2 \, dx = \frac{2}{3} L^2, \quad a_n = \frac{2}{L} \int_0^L x^2 \cos \frac{2n\pi}{L} x \, dx = \frac{L^2}{n^2\pi^2}$$

and $$b_n = \frac{2}{L} \int_0^L x^2 \sin \frac{2n\pi}{L} x \, dx = -\frac{L^2}{n\pi}.$$

Therefore $$f(x) = \frac{L^2}{3} + \frac{L^2}{\pi} \sum_{n=1}^{\infty} \left\{ \frac{1}{n^2\pi} \cos \frac{2n\pi}{L} x - \frac{1}{n} \sin \frac{2n\pi}{L} x \right\}. \tag{10} \equiv$$

The series (8), (9), and (10) converge to the $2L$-periodic even extension of f, the $2L$-periodic odd extension of f, and the L-periodic extension of f, respectively. The graphs of these periodic extensions are shown in **FIGURE 12.3.11**.

(a) Cosine series

(b) Sine series

(c) Fourier series

FIGURE 12.3.11 Different periodic extensions of the function f in Example 3

■ **Periodic Driving Force** Fourier series are sometimes useful in determining a particular solution of a differential equation describing a physical system in which the input or driving force $f(t)$ is periodic. In the next example we find a particular solution of the differential equation

$$m \frac{d^2x}{dt^2} + kx = f(t) \tag{11}$$

by first representing f by a half-range sine expansion and then assuming a particular solution of the form

$$x_p(t) = \sum_{n=1}^{\infty} B_n \sin \frac{n\pi}{p} t. \tag{12}$$

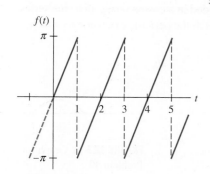

FIGURE 12.3.12 Periodic forcing function f in Example 4

> **EXAMPLE 4** **Particular Solution of a DE**
>
> An undamped spring/mass system, in which the mass $m = \frac{1}{16}$ slug and the spring constant $k = 4$ lb/ft, is driven by the 2-periodic external force $f(t)$ shown in **FIGURE 12.3.12**. Although the force $f(t)$ acts on the system for $t > 0$, note that if we extend the graph of the function in a 2-periodic manner to the negative t-axis, we obtain an odd function. In practical terms this means that we need only find the half-range sine expansion of $f(t) = \pi t$, $0 < t < 1$. With $p = 1$ it follows from (5) and integration by parts that
>
> $$b_n = 2 \int_0^1 \pi t \sin n\pi t \, dt = \frac{2(-1)^{n+1}}{n}.$$
>
> From (11) the differential equation of motion is seen to be
>
> $$\frac{1}{16} \frac{d^2x}{dt^2} + 4x = \sum_{n=1}^{\infty} \frac{2(-1)^{n+1}}{n} \sin n\pi t. \tag{13}$$
>
> To find a particular solution $x_p(t)$ of (13), we substitute the series (12) into the differential equation and equate coefficients of $\sin n\pi t$. This yields
>
> $$\left(-\frac{1}{16} n^2\pi^2 + 4 \right) B_n = \frac{2(-1)^{n+1}}{n} \quad \text{or} \quad B_n = \frac{32(-1)^{n+1}}{n(64 - n^2\pi^2)}.$$

Thus

$$x_p(t) = \sum_{n=1}^{\infty} \frac{32(-1)^{n+1}}{n(64 - n^2\pi^2)} \sin n\pi t. \qquad (14) \equiv$$

Observe in the solution (14) that there is no integer $n \geq 1$ for which the denominator $64 - n^2\pi^2$ of B_n is zero. In general, if there *is* a value of n, say, N, for which $N\pi/p = \omega$, where $\omega = \sqrt{k/m}$, then the system described by (11) is in a state of pure resonance. In other words, we have pure resonance if the Fourier series expansion of the driving force $f(t)$ contains a term $\sin(N\pi/L)t$ (or $\cos(N\pi/L)t$) that has the same frequency as the free vibrations.

Of course, if the $2p$-periodic extension of the driving force f onto the negative t-axis yields an even function, then we expand f in a cosine series.

12.3 Exercises Answers to selected odd-numbered problems begin on page ANS-32.

In Problems 1–10, determine whether the given function is even, odd, or neither.

1. $f(x) = \sin 3x$
2. $f(x) = x \cos x$
3. $f(x) = x^2 + x$
4. $f(x) = x^3 - 4x$
5. $f(x) = e^{|x|}$
6. $f(x) = e^x - e^{-x}$

7. $f(x) = \begin{cases} x^2, & -1 < x < 0 \\ -x^2, & 0 \leq x < 1 \end{cases}$

8. $f(x) = \begin{cases} x + 5, & -2 < x < 0 \\ -x + 5, & 0 \leq x < 2 \end{cases}$

9. $f(x) = x^3, \ 0 \leq x \leq 2$
10. $f(x) = |x^5|$

In Problems 11–24, expand the given function in an appropriate cosine or sine series.

11. $f(x) = \begin{cases} \pi, & -1 < x < 0 \\ -\pi, & 0 \leq x < 1 \end{cases}$

12. $f(x) = \begin{cases} 1, & -2 < x < -1 \\ 0, & -1 < x < 1 \\ 1, & 1 < x < 2 \end{cases}$

13. $f(x) = |x|, \ -\pi < x < \pi$
14. $f(x) = x, \ -\pi < x < \pi$
15. $f(x) = x^2, \ -1 < x < 1$
16. $f(x) = x|x|, \ -1 < x < 1$
17. $f(x) = \pi^2 - x^2, \ -\pi < x < \pi$
18. $f(x) = x^3, \ -\pi < x < \pi$

19. $f(x) = \begin{cases} x - 1, & -\pi < x < 0 \\ x + 1, & 0 \leq x < \pi \end{cases}$

20. $f(x) = \begin{cases} x + 1, & -1 < x < 0 \\ x - 1, & 0 \leq x < 1 \end{cases}$

21. $f(x) = \begin{cases} 1, & -2 < x < -1 \\ -x, & -1 \leq x < 0 \\ x, & 0 \leq x < 1 \\ 1, & 1 \leq x < 2 \end{cases}$

22. $f(x) = \begin{cases} -\pi, & -2\pi < x < -\pi \\ x, & -\pi \leq x < \pi \\ \pi, & \pi \leq x < 2\pi \end{cases}$

23. $f(x) = |\sin x|, \ -\pi < x < \pi$
24. $f(x) = \cos x, \ -\pi/2 < x < \pi/2$

In Problems 25–34, find the half-range cosine and sine expansions of the given function.

25. $f(x) = \begin{cases} 1, & 0 < x < \frac{1}{2} \\ 0, & \frac{1}{2} \leq x < 1 \end{cases}$
26. $f(x) = \begin{cases} 0, & 0 < x < \frac{1}{2} \\ 1, & \frac{1}{2} \leq x < 1 \end{cases}$

27. $f(x) = \cos x, \ 0 < x < \pi/2$
28. $f(x) = \sin x, \ 0 < x < \pi$

29. $f(x) = \begin{cases} x, & 0 < x < \pi/2 \\ \pi - x, & \pi/2 \leq x < \pi \end{cases}$

30. $f(x) = \begin{cases} 0, & 0 < x < \pi \\ x - \pi, & \pi \leq x < 2\pi \end{cases}$

31. $f(x) = \begin{cases} x, & 0 < x < 1 \\ 1, & 1 \leq x < 2 \end{cases}$

32. $f(x) = \begin{cases} 1, & 0 < x < 1 \\ 2 - x, & 1 \leq x < 2 \end{cases}$

33. $f(x) = x^2 + x, \ 0 < x < 1$
34. $f(x) = x(2 - x), \ 0 < x < 2$

In Problems 35–38, expand the given function in a Fourier series.

35. $f(x) = x^2, \ 0 < x < 2\pi$
36. $f(x) = x, \ 0 < x < \pi$
37. $f(x) = x + 1, \ 0 < x < 1$
38. $f(x) = 2 - x, \ 0 < x < 2$

In Problems 39–42, suppose the function $y = f(x), 0 < x < L$, given in the figure is expanded in a cosine series, in a sine series, and in a Fourier series. Sketch the periodic extension to which each series converges.

39.

FIGURE 12.3.13 Graph for Problem 39

40.

FIGURE 12.3.14 Graph for Problem 40

41.

FIGURE 12.3.15 Graph for Problem 41

42.

FIGURE 12.3.16 Graph for Problem 42

In Problems 43 and 44, proceed as in Example 4 to find a particular solution $x_p(t)$ of equation (11) when $m = 1$, $k = 10$, and the driving force $f(t)$ is as given. Assume that when $f(t)$ is extended to the negative t-axis in a periodic manner, the resulting function is odd.

43. $f(t) = \begin{cases} 5, & 0 < t < \pi \\ -5, & \pi < t < 2\pi \end{cases}$; $\quad f(t + 2\pi) = f(t)$

44. $f(t) = 1 - t$, $0 < t < 2$; $\quad f(t + 2) = f(t)$

In Problems 45 and 46, proceed as in Example 4 to find a particular solution $x_p(t)$ of equation (11) when $m = \frac{1}{4}$, $k = 12$, and the driving force $f(t)$ is as given. Assume that when $f(t)$ is extended to the negative t-axis in a periodic manner, the resulting function is even.

45. $f(t) = 2\pi t - t^2$, $0 < t < 2\pi$; $\quad f(t + 2\pi) = f(t)$

46. $f(x) = \begin{cases} t, & 0 < t < \frac{1}{2} \\ 1 - t, & \frac{1}{2} < t < 1 \end{cases}$; $\quad f(t + 1) = f(t)$

47. (a) Solve the differential equation in Problem 43, $x'' + 10x = f(t)$, subject to the initial conditions $x(0) = 0$, $x'(0) = 0$.
(b) Use a CAS to plot the graph of the solution $x(t)$ in part (a).

48. (a) Solve the differential equation in Problem 45, $\frac{1}{4}x'' + 12x = f(t)$, subject to the initial conditions $x(0) = 1$, $x'(0) = 0$.
(b) Use a CAS to plot the graph of the solution $x(t)$ in part (a).

49. Suppose a uniform beam of length L is simply supported at $x = 0$ and at $x = L$. If the load per unit length is given by $w(x) = w_0 x/L$, $0 < x < L$, then the differential equation for the deflection $y(x)$ is

$$EI \frac{d^4 y}{dx^4} = \frac{w_0 x}{L},$$

where E, I, and w_0 are constants. See (4) in Section 3.9.
(a) Expand $w(x)$ in a half-range sine series.
(b) Use the method of Example 4 to find a particular solution $y(x)$ of the differential equation.

50. Proceed as in Problem 49 to find a particular solution $y(x)$ when the load per unit length is as given in **FIGURE 12.3.17**.

FIGURE 12.3.17 Graph for Problem 50

Computer Lab Assignments

In Problems 51 and 52, use a CAS to graph the partial sums $\{S_N(x)\}$ of the given trigonometric series. Experiment with different values of N and graphs on different intervals of the x-axis. Use your graphs to conjecture a closed-form expression for a function f defined for $0 < x < L$ that is represented by the series.

51. $f(x) = -\frac{\pi}{4} + \sum_{n=1}^{\infty} \left[\frac{(-1)^n - 1}{n^2 \pi} \cos nx + \frac{1 - 2(-1)^n}{n} \sin nx \right]$

52. $f(x) = -\frac{1}{4} + \frac{4}{\pi^2} \sum_{n=1}^{\infty} \frac{1}{n^2} \left(1 - \cos \frac{n\pi}{2} \right) \cos \frac{n\pi}{2} x$

Discussion Problems

53. Is your answer in Problem 51 or in Problem 52 unique? Give a function f defined on a symmetric interval about the origin $(-a, a)$ that has the same trigonometric series as in Problem 51; as in Problem 52.

54. Discuss why the Fourier cosine series expansion of $f(x) = e^x$, $0 < x < \pi$ converges to e^{-x} on the interval $(-\pi, 0)$.

55. Suppose $f(x) = e^x$, $0 < x < \pi$ is expanded in a cosine series, and then $f(x) = e^x$, $0 < x < \pi$ is expanded in a sine series. If the two series are added and then divided by 2 (that is, the average of the two series) we get a series with cosines and sines that also represents $f(x) = e^x$ on the interval $(0, \pi)$. Is this a full Fourier series of f? [*Hint*: What does the averaging of the cosine and sine series represent on the interval $(-\pi, 0)$?]

56. Prove properties (a), (c), (d), (e), (f), and (g) in Theorem 12.3.1.

12.4 Complex Fourier Series

INTRODUCTION As we have seen in the preceding two sections, a real function f can be represented by a series of sines and cosines. The functions $\cos nx$, $n = 0, 1, 2, \ldots$ and $\sin nx$, $n = 1, 2, \ldots$ are *real-valued* functions of a real variable x. The three different real forms of Fourier series given in Definitions 12.2.1 and 12.3.1 will be exceedingly important in Chapters 13 and 14 when we set about to solve linear partial differential equations. However, in certain applications, for example, the analysis of periodic signals in electrical engineering, it is actually more convenient to represent a function f in an infinite series of *complex-valued* functions of a real variable x such as the exponential functions e^{inx}, $n = 0, 1, 2, \ldots$, and where i is the imaginary unit defined by $i^2 = -1$. Recall for x a real number, Euler's formula

$$e^{ix} = \cos x + i \sin x \quad \text{gives} \quad e^{-ix} = \cos x - i \sin x. \tag{1}$$

In this section we are going to use the results in (1) to recast the Fourier series in Definition 12.2.1 into a **complex form** or **exponential form**. We will see that we can represent a real function by

a complex series: a series in which the coefficients are complex numbers. To that end, recall that a complex number is a number $z = a + ib$, where a and b are real numbers, and $i^2 = -1$. The number $\bar{z} = a - ib$ is called the conjugate of z.

Complex Fourier Series

If we first add the two expressions in (1) and solve for $\cos x$ and then subtract the two expressions and solve for $\sin x$, we arrive at

$$\cos x = \frac{e^{ix} + e^{-ix}}{2} \quad \text{and} \quad \sin x = \frac{e^{ix} - e^{-ix}}{2i}. \tag{2}$$

Using (2) to replace $\cos(n\pi x/p)$ and $\sin(n\pi x/p)$ in (8) of Section 12.2, the Fourier series of a function f can be written

$$\frac{a_0}{2} + \sum_{n=1}^{\infty} \left[a_n \frac{e^{in\pi x/p} + e^{-in\pi x/p}}{2} + b_n \frac{e^{in\pi x/p} - e^{-in\pi x/p}}{2i} \right]$$

$$= \frac{a_0}{2} + \sum_{n=1}^{\infty} \left[\frac{1}{2}(a_n - ib_n)e^{in\pi x/p} + \frac{1}{2}(a_n + ib_n)e^{-in\pi x/p} \right]$$

$$= c_0 + \sum_{n=1}^{\infty} c_n e^{in\pi x/p} + \sum_{n=1}^{\infty} c_{-n} e^{-in\pi x/p}, \tag{3}$$

where $c_0 = \frac{1}{2}a_0$, $c_n = \frac{1}{2}(a_n - ib_n)$, and $c_{-n} = \frac{1}{2}(a_n + ib_n)$. The symbols a_0, a_n, and b_n are the coefficients (9), (10), and (11) respectively, in Definition 12.2.1. When the function f is real, c_n and c_{-n} are complex conjugates and can also be written in terms of complex exponential functions:

$$c_0 = \frac{1}{2} \cdot \frac{1}{p} \int_{-p}^{p} f(x) \, dx, \tag{4}$$

$$c_n = \frac{1}{2}(a_n - ib_n) = \frac{1}{2} \left(\frac{1}{p} \int_{-p}^{p} f(x) \cos \frac{n\pi}{p} x \, dx - i \frac{1}{p} \int_{-p}^{p} f(x) \sin \frac{n\pi}{p} x \, dx \right)$$

$$= \frac{1}{2p} \int_{-p}^{p} f(x) \left[\cos \frac{n\pi}{p} x - i \sin \frac{n\pi}{p} x \right] dx$$

$$= \frac{1}{2p} \int_{-p}^{p} f(x) \, e^{-in\pi x/p} \, dx, \tag{5}$$

$$c_{-n} = \frac{1}{2}(a_n + ib_n) = \frac{1}{2} \left(\frac{1}{p} \int_{-p}^{p} f(x) \cos \frac{n\pi}{p} x \, dx + i \frac{1}{p} \int_{-p}^{p} f(x) \sin \frac{n\pi}{p} x \, dx \right)$$

$$= \frac{1}{2p} \int_{-p}^{p} f(x) \left[\cos \frac{n\pi}{p} x + i \sin \frac{n\pi}{p} x \right] dx$$

$$= \frac{1}{2p} \int_{-p}^{p} f(x) \, e^{in\pi x/p} \, dx. \tag{6}$$

Since the subscripts of the coefficients and exponents range over the entire set of integers... -3, -2, -1, 0, 1, 2, 3, ..., we can write the results in (3), (4), (5), and (6) in a more compact manner by summing over both the negative and nonnegative integers. In other words, we can use *one* summation and *one* integral that defines all three coefficients c_0, c_n, and c_{-n}.

DEFINITION 12.4.1 Complex Fourier Series

The **complex Fourier series** of functions f defined on an interval $(-p, p)$ is given by

$$f(x) = \sum_{n=-\infty}^{\infty} c_n e^{in\pi x/p}, \tag{7}$$

where

$$c_n = \frac{1}{2p} \int_{-p}^{p} f(x) \, e^{-in\pi x/p} \, dx, \quad n = 0, \pm 1, \pm 2, \dots. \tag{8}$$

If f satisfies the hypotheses of Theorem 12.2.1, a complex Fourier series converges to $f(x)$ at a point of continuity and to the average

$$\frac{f(x+) + f(x-)}{2}$$

at a point of discontinuity.

EXAMPLE 1 Complex Fourier Series

Expand $f(x) = e^{-x}$, $-\pi < x < \pi$, in a complex Fourier series.

SOLUTION With $p = \pi$, (8) gives

$$c_n = \frac{1}{2\pi}\int_{-\pi}^{\pi} e^{-x}e^{-inx}\,dx = \frac{1}{2\pi}\int_{-\pi}^{\pi} e^{-(in+1)x}\,dx$$

$$= -\frac{1}{2\pi(in+1)}\left[e^{-(in+1)\pi} - e^{(in+1)\pi}\right].$$

We can simplify the coefficients c_n somewhat using Euler's formula:

$$e^{-(in+1)\pi} = e^{-\pi}(\cos n\pi - i\sin n\pi) = (-1)^n e^{-\pi}$$

and

$$e^{(in+1)\pi} = e^{\pi}(\cos n\pi + i\sin n\pi) = (-1)^n e^{\pi},$$

since $\cos n\pi = (-1)^n$ and $\sin n\pi = 0$. Hence

$$c_n = (-1)^n \frac{(e^{\pi} - e^{-\pi})}{2(in+1)\pi} = (-1)^n \frac{\sinh \pi}{\pi}\frac{1 - in}{n^2 + 1}. \qquad (9)$$

The complex Fourier series is then

$$f(x) = \frac{\sinh \pi}{\pi}\sum_{n=-\infty}^{\infty} (-1)^n \frac{1 - in}{n^2 + 1} e^{inx}. \qquad (10) \equiv$$

The series (10) converges to the 2π-periodic extension of f.

You may get the impression that we have just made life more complicated by introducing a complex version of a Fourier series. The reality of the situation is that in areas of engineering, the form (7) given in Definition 12.4.1 is sometimes more useful than that given in (8) of Definition 12.2.1.

▨ Fundamental Frequency The Fourier series in Definitions 12.2.1 and 12.4.1 define a periodic function and the **fundamental period** of that function (that is, the periodic extension of f) is $T = 2p$. Since $p = T/2$, (8) of Section 12.2 and (7) become, respectively,

$$\frac{a_0}{2} + \sum_{n=1}^{\infty}(a_n \cos n\omega x + b_n \sin n\omega x) \qquad \text{and} \qquad \sum_{n=-\infty}^{\infty} c_n e^{-in\omega x}, \qquad (11)$$

where number $\omega = 2\pi/T$ is called the **fundamental angular frequency**. In Example 1 the periodic extension of the function has period $T = 2\pi$; the fundamental angular frequency is $\omega = 2\pi/2\pi = 1$.

▨ Frequency Spectrum In the study of time-periodic signals, electrical engineers find it informative to examine various spectra of a wave form. If f is periodic and has fundamental period T, the plot of the points $(n\omega, |c_n|)$, where ω is the fundamental angular frequency and the c_n are the coefficients defined in (8), is called the **frequency spectrum** of f.

EXAMPLE 2 Frequency Spectrum

In Example 1, $\omega = 1$ so that $n\omega$ takes on the values $0, \pm 1, \pm 2, \dots$. Using $|\alpha + i\beta| = \sqrt{\alpha^2 + \beta^2}$, we see from (9) that

$$|c_n| = \frac{\sinh \pi}{\pi}\frac{1}{\sqrt{n^2 + 1}}.$$

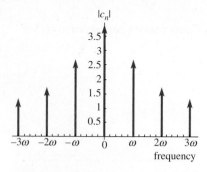

FIGURE 12.4.1 Frequency spectrum of f in Example 2

FIGURE 12.4.2 Periodic pulse in Example 3

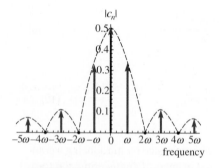

FIGURE 12.4.3 Frequency spectrum of f in Example 3

The following table shows some values of n and corresponding values of c_n.

n	-3	-2	-1	0	1	2	3		
$	c_n	$	1.162	1.644	2.599	3.676	2.599	1.644	1.162

The graph in **FIGURE 12.4.1**, lines with arrowheads terminating at the points, is a portion of the frequency spectrum of f.

EXAMPLE 3 Frequency Spectrum

Find the frequency spectrum of the periodic square wave or periodic pulse shown in **FIGURE 12.4.2**. The wave is the periodic extension of the function f:

$$f(x) = \begin{cases} 0, & -\frac{1}{2} < x < -\frac{1}{4} \\ 1, & -\frac{1}{4} < x < \frac{1}{4} \\ 0, & \frac{1}{4} < x < \frac{1}{2}. \end{cases}$$

SOLUTION Here $T = 1 = 2p$ so $p = \frac{1}{2}$. Since f is 0 on the intervals $(-\frac{1}{2}, -\frac{1}{4})$ and $(\frac{1}{4}, \frac{1}{2})$, (8) becomes

$$c_n = \int_{-1/2}^{1/2} f(x)\, e^{2in\pi x}\, dx = \int_{-1/4}^{1/4} 1 \cdot e^{2in\pi x}\, dx$$

$$= \frac{e^{2in\pi x}}{2in\pi} \Bigg]_{-1/4}^{1/4}$$

$$= \frac{1}{n\pi} \frac{e^{in\pi/2} - e^{-in\pi/2}}{2i}.$$

That is,
$$c_n = \frac{1}{n\pi} \sin \frac{n\pi}{2}. \quad \leftarrow \text{by (2)}$$

Since the last result is not valid at $n = 0$, we compute that term separately:

$$c_0 = \int_{-1/4}^{1/4} dx = \frac{1}{2}.$$

The following table shows some of the values of $|c_n|$, and **FIGURE 12.4.3** shows the frequency spectrum of f.

n	-5	-4	-3	-2	-1	0	1	2	3	4	5		
$	c_n	$	$\frac{1}{5\pi}$	0	$\frac{1}{3\pi}$	0	$\frac{1}{\pi}$	$\frac{1}{2}$	$\frac{1}{\pi}$	0	$\frac{1}{3\pi}$	0	$\frac{1}{5\pi}$

Since the fundamental frequency is $\omega = 2\pi/T = 2\pi$, the units $n\omega$ on the horizontal scale are $\pm 2\pi, \pm 4\pi, \pm 6\pi, \dots$. The curved dashed lines were added in Figure 12.4.3 to emphasize the presence of the zero values of $|c_n|$ when n is an even nonzero integer.

12.4 Exercises Answers to selected odd-numbered problems begin on page ANS-32.

In Problems 1–6, find the complex Fourier series of f on the given interval.

1. $f(x) = \begin{cases} -1, & -2 < x < 0 \\ 1, & 0 < x < 2 \end{cases}$

2. $f(x) = \begin{cases} 0, & 0 < x < 1 \\ 1, & 1 < x < 2 \end{cases}$

3. $f(x) = \begin{cases} 0, & -\frac{1}{2} < x < 0 \\ 1, & 0 < x < \frac{1}{4} \\ 0, & \frac{1}{4} < x < \frac{1}{2} \end{cases}$

4. $f(x) = \begin{cases} 0, & -\pi < x < 0 \\ x, & 0 < x < \pi \end{cases}$

5. $f(x) = x,\ 0 < x < 2\pi$

6. $f(x) = e^{-|x|},\ -1 < x < 1$

7. Find the frequency spectrum of the periodic wave that is the periodic extension of the function f in Problem 1.

8. Find the frequency spectrum of the periodic wave that is the periodic extension of the function f in Problem 3.

In Problems 9 and 10, sketch the given periodic wave. Find the frequency spectrum of f.

9. $f(x) = 4 \sin x, 0 < x < \pi; f(x + \pi) = f(x)$ [*Hint*: Use (2).]

10. $f(x) = \begin{cases} \cos x, & 0 < x < \pi/2 \\ 0, & \pi/2 < x < \pi \end{cases}; \ f(x + \pi) = f(x)$

11. **(a)** Show that $a_n = c_n + c_{-n}$ and $b_n = i(c_n - c_{-n})$.
 (b) Use the results in part (a) and the complex Fourier series in Example 1 to obtain the Fourier series expansion of f.

12. The function f in Problem 1 is odd. Use the complex Fourier series to obtain the Fourier sine series expansion of f.

12.5 Sturm–Liouville Problem

INTRODUCTION For convenience we present here a brief review of some of the ordinary differential equations that will be of importance in the sections and chapters that follow.

Linear equations	General solutions
$y' + \alpha y = 0,$	$y = c_1 e^{-\alpha x}$
$y'' + \alpha^2 y = 0, \quad \alpha > 0$	$y = c_1 \cos \alpha x + c_2 \sin \alpha x$
$y'' - \alpha^2 y = 0, \quad \alpha > 0$	$\begin{cases} y = c_1 e^{-\alpha x} + c_2 e^{\alpha x}, & \text{or} \\ y = c_1 \cosh \alpha x + c_2 \sinh \alpha x \end{cases}$

Cauchy–Euler equation	General solutions, $x > 0$
$x^2 y'' + xy' - \alpha^2 y = 0, \quad \alpha \geq 0$	$\begin{cases} y = c_1 x^{-\alpha} + c_2 x^{\alpha}, & \alpha \neq 0 \\ y = c_1 + c_2 \ln x, & \alpha = 0 \end{cases}$

Parametric Bessel equation ($\nu = 0$)	General solution, $x > 0$
$xy'' + y' + \alpha^2 xy = 0$	$y = c_1 J_0(\alpha x) + c_2 Y_0(\alpha x)$

Legendre's equation ($n = 0, 1, 2, \ldots$)	Particular solutions are polynomials
$(1 - x^2)y'' - 2xy' + n(n+1)y = 0$	$y = P_0(x) = 1,$
	$y = P_1(x) = x,$
	$y = P_2(x) = \frac{1}{2}(3x^2 - 1), \ldots$

Regarding the two forms of the general solution of $y'' - \alpha^2 y = 0$, we will, in the future, employ the following informal rule:

This rule will be useful in Chapters 13 and 14. ▶

> *Use the exponential form $y = c_1 e^{-\alpha x} + c_2 e^{\alpha x}$ when the domain of x is an infinite or semi-infinite interval; use the hyperbolic form $y = c_1 \cosh \alpha x + c_2 \sinh \alpha x$ when the domain of x is a finite interval.*

Eigenvalues and Eigenfunctions
Orthogonal functions arise in the solution of differential equations. More to the point, an orthogonal set of functions can be generated by solving a two-point boundary-value problem involving a linear second-order differential equation containing a parameter λ. In Example 2 of Section 3.9 we saw that the boundary-value problem

$$y'' + \lambda y = 0, \quad y(0) = 0, \quad y(L) = 0, \tag{1}$$

possessed nontrivial solutions only when the parameter λ took on the values $\lambda_n = n^2 \pi^2 / L^2$, $n = 1, 2, 3, \ldots$ called **eigenvalues**. The corresponding nontrivial solutions $y = c_2 \sin(n\pi x/L)$ or simply $y = \sin(n\pi x/L)$ are called the **eigenfunctions** of the problem. For example, for (1) we have

$$\overset{\text{not an eigenvalue}}{\underset{\downarrow}{}}$$

BVP: $\quad y'' + 5y = 0, \quad y(0) = 0, \quad y(L) = 0$

Solution is trivial: $\quad y = 0.$

$$\overset{\text{is an eigenvalue } (n = 2)}{\underset{\downarrow}{}}$$

BVP: $\quad y'' + \dfrac{4\pi^2}{L^2} y = 0, \quad y(0) = 0, \quad y(L) = 0$

Solution is nontrivial: $\quad y = \sin(2\pi x/L).$

For our purposes in this chapter it is important to recognize the set of functions generated by this BVP; that is, $\{\sin(n\pi x/L)\}$, $n = 1, 2, 3, \ldots$, is the orthogonal set of functions on the interval $[0, L]$ used as the basis for the Fourier sine series.

EXAMPLE 1 **Eigenvalues and Eigenfunctions**

It is left as an exercise to show, by considering the three possible cases for the parameter λ (zero, negative, or positive; that is, $\lambda = 0$, $\lambda = -\alpha^2 < 0$, $\alpha > 0$, and $\lambda = \alpha^2 > 0$, $\alpha > 0$), that the eigenvalues and eigenfunctions for the boundary-value problem

$$y'' + \lambda y = 0, \qquad y'(0) = 0, \qquad y'(L) = 0 \qquad (2)$$

are, respectively, $\lambda_n = \alpha_n^2 = n^2\pi^2/L^2$, $n = 0, 1, 2, \ldots$, and $y = c_1 \cos(n\pi x/L)$, $c_1 \neq 0$. In contrast to (1), $\lambda_0 = 0$ is an eigenvalue for this BVP and $y = 1$ is the corresponding eigenfunction. The latter comes from solving $y'' = 0$ subject to the same boundary conditions $y'(0) = 0$, $y'(L) = 0$. Note also that $y = 1$ can be incorporated into the family $y = \cos(n\pi x/L)$ by permitting $n = 0$. The set $\{\cos(n\pi x/L)\}$, $n = 0, 1, 2, 3, \ldots$, is orthogonal on the interval $[0, L]$. See Problem 3 in Exercises 12.5. ≡

Regular Sturm–Liouville Problem
The problems in (1) and (2) are special cases of an important general two-point boundary-value problem. Let p, q, r, and r' be real-valued functions continuous on an interval $[a, b]$, and let $r(x) > 0$ and $p(x) > 0$ for every x in the interval. Then

$$\textbf{Solve:} \qquad \frac{d}{dx}[r(x)y'] + (q(x) + \lambda p(x))y = 0 \qquad (3)$$

$$\textbf{Subject to:} \qquad A_1 y(a) + B_1 y'(a) = 0 \qquad (4)$$

$$A_2 y(b) + B_2 y'(b) = 0 \qquad (5)$$

is said to be a **regular Sturm–Liouville problem**. **Jacques Charles François Sturm** (1803–1855) was born in Switzerland but spent his adult life as a professor of mathematics and mechanics in Paris, France. Sturm's name is one of 72 names of scientists, engineers, and mathematicians engraved on the Eiffel Tower. The French mathematician **Joseph Liouville** (1809–1882) was also a professor in Paris. In addition to his work with Sturm on boundary-value problems, an important theorem in complex analysis bears his name (Liouville's theorem). The small impact crater *Liouville* on the moon is named after him.

The coefficients in the boundary-conditions (4) and (5) are assumed to be real and independent of λ. In addition, A_1 and B_1 are not both zero, and A_2 and B_2 are not both zero. The boundary-value problems in (1) and (2) are regular Sturm–Liouville problems. From (1) we can identify $r(x) = 1$, $q(x) = 0$, and $p(x) = 1$ in the differential equation (3); in boundary condition (4) we identify $a = 0$, $A_1 = 1$, $B_1 = 0$, and in (5), $b = L$, $A_2 = 1$, $B_2 = 0$. From (2) the identifications would be $a = 0$, $A_1 = 0$, $B_1 = 1$ in (4), and $b = L$, $A_2 = 0$, $B_2 = 1$ in (5).

The differential equation (3) is linear and homogeneous. The boundary conditions in (4) and (5), both a linear combination of y and y' *equal to zero at a point*, are also called **homogeneous**. A boundary condition such as $A_2 y(b) + B_2 y'(b) = C_2$, where C_2 is a nonzero constant, is **nonhomogeneous**. Naturally, a boundary-value problem that consists of a homogeneous linear differential equation and homogeneous boundary conditions is said to be homogeneous; otherwise it is nonhomogeneous. The boundary conditions (4) and (5) are said to be **separated** because each condition involves only a single boundary point. Boundary conditions are referred to as **mixed** if each condition involves both boundary points $x = a$ and $x = b$. For example, the periodic boundary conditions $y(a) = y(b)$, $y'(a) = y'(b)$ are mixed boundary conditions.

Because a regular Sturm–Liouville problem is a homogeneous BVP, it always possesses the trivial solution $y = 0$. However, this solution is of no interest to us. As in Example 1, in solving such a problem we seek numbers λ (eigenvalues) and nontrivial solutions y that depend on λ (eigenfunctions).

Properties
Theorem 12.5.1 is a list of some of the more important of the many properties of the regular Sturm–Liouville problem. We shall prove only the last property.

THEOREM 12.5.1 Properties of the Regular Sturm–Liouville Problem

(a) There exist an infinite number of real eigenvalues that can be arranged in increasing order $\lambda_1 < \lambda_2 < \lambda_3 < \cdots < \lambda_n < \cdots$ such that $\lambda_n \to \infty$ as $n \to \infty$.

(b) For each eigenvalue there is only one eigenfunction (except for nonzero constant multiples).

(c) Eigenfunctions corresponding to different eigenvalues are linearly independent.

(d) The set of eigenfunctions corresponding to the set of eigenvalues is orthogonal with respect to the weight function $p(x)$ on the interval $[a, b]$.

PROOF OF (d): Let y_m and y_n be eigenfunctions corresponding to eigenvalues λ_m and λ_n, respectively. Then

$$\frac{d}{dx}[r(x)y_m'] + (q(x) + \lambda_m p(x))y_m = 0 \tag{6}$$

$$\frac{d}{dx}[r(x)y_n'] + (q(x) + \lambda_n p(x))y_n = 0. \tag{7}$$

Multiplying (6) by y_n and (7) by y_m and subtracting the two equations gives

$$(\lambda_m - \lambda_n)p(x)y_m y_n = y_m \frac{d}{dx}[r(x)y_n'] - y_n \frac{d}{dx}[r(x)y_m'].$$

Integrating this last result by parts from $x = a$ to $x = b$ then yields

$$(\lambda_m - \lambda_n)\int_a^b p(x)y_m y_n \, dx = r(b)[y_m(b)y_n'(b) - y_n(b)y_m'(b)] \\ - r(a)[y_m(a)y_n'(a) - y_n(a)y_m'(a)]. \tag{8}$$

Now the eigenfunctions y_m and y_n must both satisfy the boundary conditions (4) and (5). In particular, from (4) we have

$$A_1 y_m(a) + B_1 y_m'(a) = 0$$
$$A_1 y_n(a) + B_1 y_n'(a) = 0.$$

For this system to be satisfied by A_1 and B_1, not both zero, the determinant of the coefficients must be zero:

$$y_m(a)y_n'(a) - y_n(a)y_m'(a) = 0.$$

A similar argument applied to (5) also gives

$$y_m(b)y_n'(b) - y_n(b)y_m'(b) = 0.$$

Using these last two results in (8) shows that both members of the right-hand side are zero. Hence we have established the orthogonality relation

$$\int_a^b p(x)y_m(x)y_n(x) \, dx = 0, \quad \lambda_m \neq \lambda_n. \tag{9} \equiv$$

It can also be proved that the orthogonal set of eigenfunctions $\{y_1(x), y_2(x), y_3(x), \ldots\}$ of a regular Sturm–Liouville problem is complete on $[a, b]$. See page 684.

EXAMPLE 2 **A Regular Sturm–Liouville Problem**

Solve the boundary-value problem

$$y'' + \lambda y = 0, \quad y(0) = 0, \quad y(1) + y'(1) = 0. \tag{10}$$

SOLUTION You should verify that for $\lambda = 0$ and for $\lambda = -\alpha^2 < 0$, where $\alpha > 0$, the BVP in (10) possesses only the trivial solution $y = 0$. For $\lambda = \alpha^2 > 0$, $\alpha > 0$, the general solution of the differential equation $y'' + \alpha^2 y = 0$ is $y = c_1 \cos \alpha x + c_2 \sin \alpha x$. Now the condition $y(0) = 0$ implies $c_1 = 0$ in this solution and so we are left with $y = c_2 \sin \alpha x$. The second boundary condition $y(1) + y'(1) = 0$ is satisfied if

$$c_2 \sin \alpha + c_2 \alpha \cos \alpha = 0.$$

Choosing $c_2 \neq 0$, we see that the last equation is equivalent to

$$\tan \alpha = -\alpha. \tag{11}$$

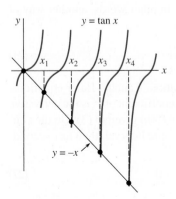

FIGURE 12.5.1 Positive roots of $\tan x = -x$ in Example 2

If we let $x = \alpha$ in (11), then **FIGURE 12.5.1** shows the plausibility that there exists an infinite number of roots of the equation $\tan x = -x$, namely, the x-coordinates of the points where

the graph of $y = -x$ intersects the branches of the graph of $y = \tan x$. The eigenvalues of problem (10) are then $\lambda_n = \alpha_n^2$, where α_n, $n = 1, 2, 3, \ldots$, are the consecutive *positive* roots $\alpha_1, \alpha_2, \alpha_3, \ldots$ of (11). With the aid of a CAS it is easily shown that, to four rounded decimal places, $\alpha_1 = 2.0288$, $\alpha_2 = 4.9132$, $\alpha_3 = 7.9787$, and $\alpha_4 = 11.0855$, and the corresponding solutions are $y_1 = \sin 2.0288x$, $y_2 = \sin 4.9132x$, $y_3 = \sin 7.9787x$, and $y_4 = \sin 11.0855x$. In general, the eigenfunctions of the problem are $\{\sin \alpha_n x\}$, $n = 1, 2, 3, \ldots$.

With identifications $r(x) = 1$, $q(x) = 0$, $p(x) = 1$, $A_1 = 1$, $B_1 = 0$, $A_2 = 1$, and $B_2 = 1$ we see that (10) is a regular Sturm–Liouville problem. Thus $\{\sin \alpha_n x\}$, $n = 1, 2, 3, \ldots$ is an orthogonal set with respect to the weight function $p(x) = 1$ on the interval $[0, 1]$. $\equiv$

In some circumstances we can prove the orthogonality of the solutions of (3) without the necessity of specifying a boundary condition at $x = a$ and at $x = b$.

Singular Sturm–Liouville Problem

There are several other important conditions under which we seek nontrivial solutions of the differential equation (3):

- $r(a) = 0$ and a boundary condition of the type given in (5) is specified at $x = b$; (12)
- $r(b) = 0$ and a boundary condition of the type given in (4) is specified at $x = a$; (13)
- $r(a) = r(b) = 0$ and no boundary condition is specified at either $x = a$ or at $x = b$; (14)
- $r(a) = r(b)$ and boundary conditions $y(a) = y(b)$, $y'(a) = y'(b)$. (15)

The differential equation (3) along with one of conditions (12) or (13) is said to be a **singular boundary-value problem**. Equation (3) with the conditions specified in (15) is said to be a **periodic boundary-value problem** because the boundary conditions are periodic. Observe that if, say, $r(a) = 0$, then $x = a$ *may* be a singular point of the differential equation, and consequently a solution of (3) may become unbounded as $x \to a$. However, we see from (8) that if $r(a) = 0$, then no boundary condition is required at $x = a$ to prove orthogonality of the eigenfunctions *provided* these solutions are bounded at that point. This latter requirement guarantees the existence of the integrals involved. By assuming the solutions of (3) are bounded on the closed interval $[a, b]$ we can see from inspection of (8) that

- If $r(a) = 0$, then the orthogonality relation (9) holds with no boundary condition at $x = a$; (16)
- If $r(b) = 0$, then the orthogonality relation (9) holds with no boundary condition at $x = b^*$; (17)
- If $r(a) = r(b) = 0$, then the orthogonality relation (9) holds with no boundary conditions specified at either $x = a$ or $x = b$; (18)
- If $r(a) = r(b)$, then the orthogonality relation (9) holds with the periodic boundary conditions $y(a) = y(b)$, $y'(a) = y'(b)$. (19)

Self-Adjoint Form

If we carry out the differentiation $\dfrac{d}{dx}[r(x)y']$, the differential equation in (3) is the same as

$$r(x)y'' + r'(x)y' + (q(x) + \lambda p(x))y = 0. \tag{20}$$

For example, Legendre's differential equation $(1 - x^2)y'' - 2xy' + n(n + 1)y = 0$ is exactly of the form given in (20) with $r(x) = 1 - x^2$ and $r'(x) = -2x$. In other words, another way of writing Legendre's DE is

$$\frac{d}{dx}[(1 - x^2)y'] + n(n + 1)y = 0. \tag{21}$$

But if you compare other second-order DEs (say, Bessel's equation, Cauchy–Euler equations, and DEs with constant coefficients) you might believe, given the coefficient of y' is the derivative of the coefficient of y'', that few other second-order DEs have the form given in (3). On the contrary, if the coefficients are continuous and $a(x) \neq 0$ for all x in some interval, then *any* second-order differential equation

$$a(x)y'' + b(x)y' + (c(x) + \lambda d(x))y = 0 \tag{22}$$

*Conditions (16) and (17) are equivalent to choosing $A_1 = 0$, $B_1 = 0$ in (4), and $A_2 = 0$, $B_2 = 0$ in (5), respectively.

can be recast into the so-called **self-adjoint form** (3). To see this, we proceed as in Section 2.3 where we rewrote a linear first-order equation $a_1(x)y' + a_0(x)y = 0$ in the form $\dfrac{d}{dx}[\mu y] = 0$ by dividing the equation by $a_1(x)$ and then multiplying by the integrating factor $\mu = e^{\int P(x)\,dx}$ where, assuming no common factors, $P(x) = a_0(x)/a_1(x)$. So first, we divide (22) by $a(x)$. The first two terms are then $Y' + \dfrac{b(x)}{a(x)}Y + \cdots$ where, for emphasis, we have written $Y = y'$. Second, we multiply this equation by the integrating factor $e^{\int (b(x)/a(x))\,dx}$, where $a(x)$ and $b(x)$ are assumed to have no common factors

$$\underbrace{e^{\int(b(x)/a(x))\,dx}Y' + \frac{b(x)}{a(x)}e^{\int(b(x)/a(x))\,dx}Y} + \cdots = \frac{d}{dx}[e^{\int(b(x)/a(x))\,dx}Y] + \cdots = \frac{d}{dx}[e^{\int(b(x)/a(x))\,dx}y'] + \cdots.$$

derivative of a product

In summary, by dividing (22) by $a(x)$ and then multiplying by $e^{\int(b(x)/a(x))\,dx}$ we get

$$e^{\int(b/a)\,dx}y'' + \frac{b(x)}{a(x)}e^{\int(b/a)\,dx}y' + \left(\frac{c(x)}{a(x)}e^{\int(b/a)\,dx} + \lambda\frac{d(x)}{a(x)}e^{\int(b/a)\,dx}\right)y = 0. \tag{23}$$

Equation (23) is the desired form given in (20) and is the same as (3):

$$\frac{d}{dx}\left[\underbrace{e^{\int(b/a)\,dx}}_{r(x)}y'\right] + \left(\underbrace{\frac{c(x)}{a(x)}e^{\int(b/a)\,dx}}_{q(x)} + \lambda\underbrace{\frac{d(x)}{a(x)}e^{\int(b/a)\,dx}}_{p(x)}\right)y = 0.$$

For example, to express $3y'' + 6y' + \lambda y = 0$ in self-adjoint form, we write $y'' + 2y' + \lambda\frac{1}{3}y = 0$ and then multiply by $e^{\int 2\,dx} = e^{2x}$. The resulting equation is

$$\overset{\underset{r(x)}{\downarrow}}{e^{2x}}y'' + \overset{\underset{r'(x)}{\downarrow}}{2e^{2x}}y' + \lambda\frac{1}{3}\overset{\underset{p(x)}{\downarrow}}{e^{2x}}y = 0 \qquad \text{or} \qquad \frac{d}{dx}[e^{2x}y'] + \lambda\frac{1}{3}e^{2x}y = 0.$$

Note. ▶ It is certainly not necessary to put a second-order differential equation (22) into the self-adjoint form (3) in order to *solve* the DE. For our purposes we use the form given in (3) to determine the weight function $p(x)$ needed in the orthogonality relation (9). The next two examples illustrate orthogonality relations for Bessel functions and for Legendre polynomials.

Parametric Bessel Equation

In Section 5.3 we saw that the general solution of the parametric Bessel differential equation $x^2 y'' + xy' + (\alpha^2 x^2 - n^2)y = 0$, $n = 0, 1, 2, \ldots$ is $y = c_1 J_n(\alpha x) + c_2 Y_n(\alpha x)$. After dividing the parametric Bessel equation by the lead coefficient x^2 and multiplying the resulting equation by the integrating factor $e^{\int(1/x)\,dx} = e^{\ln x} = x$, $x > 0$, we obtain the self-adjoint form

$$xy'' + y' + \left(\alpha^2 x - \frac{n^2}{x}\right)y = 0 \qquad \text{or} \qquad \frac{d}{dx}[xy'] + \left(\alpha^2 x - \frac{n^2}{x}\right)y = 0,$$

where we identify $r(x) = x$, $q(x) = -n^2/x$, $p(x) = x$, and $\lambda = \alpha^2$. Now $r(0) = 0$, and of the two solutions $J_n(\alpha x)$ and $Y_n(\alpha x)$ only $J_n(\alpha x)$ is bounded at $x = 0$. Thus in view of (16) above, the set $\{J_n(\alpha_i x)\}$, $i = 1, 2, 3, \ldots$, is orthogonal with respect to the weight function $p(x) = x$ on an interval $[0, b]$. The orthogonality relation is

$$\int_0^b x J_n(\alpha_i x)J_n(\alpha_j x)\,dx = 0, \qquad \lambda_i \neq \lambda_j, \tag{24}$$

provided the α_i, and hence the eigenvalues $\lambda_i = \alpha_i^2$, $i = 1, 2, 3, \ldots$, are defined by means of a boundary condition at $x = b$ of the type given in (5):

$$A_2 J_n(\lambda b) + B_2 \alpha J_n'(\alpha b) = 0. \tag{25}$$

The extra factor of α in (25) comes from the Chain Rule:

$$\frac{d}{dx} J_n(\alpha x) = J_n'(\alpha x) \frac{d}{dx} \alpha x = \alpha J_n'(\alpha x).$$

≡

For any choice of A_2 and B_2, not both zero, it is known that (25) has an infinite number of roots $x_i = \alpha_i b$. The eigenvalues are then $\lambda_i = \alpha_i^2 = (x_i/b)^2$. More will be said about eigenvalues in the next chapter.

EXAMPLE 4 **Legendre's Equation**

From the result given in (21) we can identify $q(x) = 0$, $p(x) = 1$, and $\lambda = n(n + 1)$. Recall from Section 5.3 when $n = 0, 1, 2, \ldots$ Legendre's DE possesses polynomial solutions $P_n(x)$. Now we can put the observation that $r(-1) = r(1) = 0$ together with the fact that the Legendre polynomials $P_n(x)$ are the only solutions of (21) that are bounded on the closed interval $[-1, 1]$, to conclude from (18) that the set $\{P_n(x)\}$, $n = 0, 1, 2, \ldots$, is orthogonal with respect to the weight function $p(x) = 1$ on $[-1, 1]$. The orthogonality relation is

$$\int_{-1}^{1} P_m(x)P_n(x)\, dx = 0, \quad m \neq n.$$

≡

REMARKS

(*i*) A Sturm–Liouville problem is also considered to be singular when the interval under consideration is infinite. See Problems 9 and 10 in Exercises 12.5.
(*ii*) Even when the conditions on the coefficients p, q, r, and r' are as assumed in the regular Sturm–Liouville problem, if the boundary conditions are periodic, then property (b) of Theorem 12.5.1 does not hold. You are asked to show in Problem 4 of Exercises 12.5 that corresponding to each eigenvalue of the BVP

$$y'' + \lambda y = 0, \quad y(-L) = y(L), \quad y'(-L) = y'(L)$$

there exist two linearly independent eigenfunctions.

12.5 Exercises Answers to selected odd-numbered problems begin on page ANS-32.

In Problems 1 and 2, find the eigenfunctions and the equation that defines the eigenvalues for the given boundary-value problem. Use a CAS to approximate the first four eigenvalues λ_1, λ_2, λ_3, and λ_4. Give the eigenfunctions corresponding to these approximations.

1. $y'' + \lambda y = 0$, $y'(0) = 0$, $y(1) + y'(1) = 0$
2. $y'' + \lambda y = 0$, $y(0) + y'(0) = 0$, $y(1) = 0$
3. Consider $y'' + \lambda y = 0$ subject to $y'(0) = 0$, $y'(L) = 0$. Show that the eigenfunctions are

$$\left\{ 1, \cos\frac{\pi}{L}x, \cos\frac{2\pi}{L}x, \ldots \right\}.$$

This set, which is orthogonal on $[0, L]$, is the basis for the Fourier cosine series.

4. Consider $y'' + \lambda y = 0$ subject to the periodic boundary conditions $y(-L) = y(L)$, $y'(-L) = y'(L)$. Show that the eigenfunctions are

$$\left\{ 1, \cos\frac{\pi}{L}x, \cos\frac{2\pi}{L}x, \ldots, \sin\frac{\pi}{L}x, \sin\frac{2\pi}{L}x, \sin\frac{3\pi}{L}x, \ldots \right\}.$$

This set, which is orthogonal on $[-L, L]$, is the basis for the Fourier series.

5. Find the square norm of each eigenfunction in Problem 1.
6. Show that for the eigenfunctions in Example 2,

$$\| \sin \alpha_n x \|^2 = \frac{1}{2}[1 + \cos^2\alpha_n].$$

7. **(a)** Find the eigenvalues and eigenfunctions of the boundary-value problem

$$x^2 y'' + xy' + \lambda y = 0, \quad y(1) = 0, \quad y(5) = 0.$$

 (b) Put the differential equation in self-adjoint form.
 (c) Give an orthogonality relation.

8. **(a)** Find the eigenvalues and eigenfunctions of the boundary-value problem

$$y'' + y' + \lambda y = 0, \quad y(0) = 0, \quad y(2) = 0.$$

 (b) Put the differential equation in self-adjoint form.
 (c) Give an orthogonality relation.

9. **Laguerre's differential equation**

$$xy'' + (1 - x)y' + ny = 0, \quad n = 0, 1, 2, \ldots,$$

has polynomial solutions $L_n(x)$. Put the equation in self-adjoint form and give an orthogonality relation on $(0, \infty)$.

10. Hermite's differential equation

$$y'' - 2xy' + 2ny = 0, \quad n = 0, 1, 2, \ldots,$$

has polynomial solutions $H_n(x)$. Put the equation in self-adjoint form and give an orthogonality relation on $(-\infty, \infty)$.

11. Consider the regular Sturm–Liouville problem:

$$\frac{d}{dx}[(1 + x^2)y'] + \frac{\lambda}{1 + x^2} y = 0, \quad y(0) = 0, \quad y(1) = 0.$$

(a) Find the eigenvalues and eigenfunctions of the boundary-value problem. [*Hint*: Let $x = \tan\theta$ and then use the Chain Rule.]

(b) Give an orthogonality relation.

12. (a) Find the eigenfunctions and the equation that defines the eigenvalues for the boundary-value problem

$$x^2 y'' + xy' + (\lambda x^2 - 1)y = 0,$$

y is bounded at $x = 0, \quad y(3) = 0.$

(b) Use Table 5.3.1 of Section 5.3 to find the approximate values of the first four eigenvalues $\lambda_1, \lambda_2, \lambda_3,$ and λ_4.

Discussion Problem

13. Consider the special case of the regular Sturm–Liouville problem on the interval $[a, b]$:

$$\frac{d}{dx}[r(x)y'] + \lambda p(x)y = 0, \quad y'(a) = 0, \quad y'(b) = 0.$$

Is $\lambda = 0$ an eigenvalue of the problem? Defend your answer.

Computer Lab Assignments

14. (a) Give an orthogonality relation for the Sturm–Liouville problem in Problem 1.

(b) Use a CAS as an aid in verifying the orthogonality relation for the eigenfunctions y_1 and y_2 that correspond to the first two eigenvalues λ_1 and λ_2, respectively.

15. (a) Give an orthogonality relation for the Sturm–Liouville problem in Problem 2.

(b) Use a CAS as an aid in verifying the orthogonality relation for the eigenfunctions y_1 and y_2 that correspond to the first two eigenvalues λ_1 and λ_2, respectively.

12.6 Bessel and Legendre Series

INTRODUCTION Fourier series, Fourier cosine series, and Fourier sine series are three ways of expanding a function in terms of an orthogonal set of functions. But such expansions are by no means limited to orthogonal sets of trigonometric functions. We saw in Section 12.1 that a function f defined on an interval (a, b) could be expanded, at least in a formal manner, in terms of any set of functions $\{\phi_n(x)\}$ that is orthogonal with respect to a weight function on $[a, b]$. Many of these orthogonal series expansions or generalized Fourier series derive from Sturm–Liouville problems that, in turn, arise from attempts to solve linear partial differential equations serving as models for physical systems. Fourier series and orthogonal series expansions (the latter includes the two series considered in this section) will appear in the subsequent consideration of these applications in Chapters 13 and 14.

12.6.1 Fourier–Bessel Series

We saw in Example 3 of Section 12.5 that for a fixed value of n the set of Bessel functions $\{J_n(\alpha_i x)\}$, $i = 1, 2, 3, \ldots$, is orthogonal with respect to the weight function $p(x) = x$ on an interval $[0, b]$ when the α_i are defined by means of a boundary condition of the form

$$A_2 J_n(\alpha b) + B_2 \alpha J_n'(\alpha b) = 0. \tag{1}$$

The eigenvalues of the corresponding Sturm–Liouville problem are $\lambda_i = \alpha_i^2$. From (7) and (8) of Section 12.1 the orthogonal series expansion or generalized Fourier series of a function f defined on the interval $(0, b)$ in terms of this orthogonal set is

$$f(x) = \sum_{i=1}^{\infty} c_i J_n(\alpha_i x), \tag{2}$$

where

$$c_i = \frac{\int_0^b x J_n(\alpha_i x) f(x)\, dx}{\|J_n(\alpha_i x)\|^2}. \tag{3}$$

The square norm of the function $J_n(\alpha_i x)$ is defined by (11) of Section 12.1:

$$\|J_n(\alpha_i x)\|^2 = \int_0^b x J_n^2(\alpha_i x)\, dx. \tag{4}$$

The series (2) with coefficients (3) is called a **Fourier–Bessel series**.

Differential Recurrence Relations

The differential recurrence relations that were given in (22) and (23) of Section 5.3 are often useful in the evaluation of the coefficients (3). For convenience we reproduce those relations here:

$$\frac{d}{dx}[x^n J_n(x)] = x^n J_{n-1}(x) \tag{5}$$

$$\frac{d}{dx}[x^{-n} J_n(x)] = -x^{-n} J_{n+1}(x). \tag{6}$$

Square Norm

The value of the square norm (4) depends on how the eigenvalues $\lambda_i = \alpha_i^2$ are defined. If $y = J_n(\alpha x)$, then we know from Example 3 of Section 12.5 that

$$\frac{d}{dx}[xy'] + \left(\alpha^2 x - \frac{n^2}{x}\right)y = 0.$$

After we multiply by $2xy'$, this equation can be written as

$$\frac{d}{dx}[xy']^2 + (\alpha^2 x^2 - n^2)\frac{d}{dx}[y]^2 = 0.$$

Integrating the last result by parts on $[0, b]$ then gives

$$2\alpha^2 \int_0^b xy^2\, dx = \left([xy']^2 + (\alpha^2 x^2 - n^2)y^2\right)\Big]_0^b.$$

Since $y = J_n(\alpha x)$, the lower limit is zero for $n > 0$ because $J_n(0) = 0$. For $n = 0$, the quantity $[xy']^2 + \alpha^2 x^2 y^2$ is zero at $x = 0$. Thus

$$2\alpha^2 \int_0^b x J_n^2(\alpha x)\, dx = \alpha^2 b^2 [J_n'(ab)]^2 + (\alpha^2 b^2 - n^2)[J_n(\alpha b)]^2, \tag{7}$$

where we have used the Chain Rule to write $y' = \alpha J_n'(\alpha x)$.

We now consider three cases of the boundary condition (1).

Case I: If we choose $A_2 = 1$ and $B_2 = 0$, then (1) is

$$J_n(\alpha b) = 0. \tag{8}$$

There are an infinite number of positive roots $x_i = \alpha_i b$ of (8) (see Figure 5.3.1) that define the α_i as $\alpha_i = x_i/b$. The eigenvalues are positive and are then $\lambda_i = \alpha_i^2 = x_i^2/b^2$. No new eigenvalues result from the negative roots of (8) since $J_n(-x) = (-1)^n J_n(x)$. (See page 295.) The number 0 is not an eigenvalue for any n since $J_n(0) = 0$ for $n = 1$, $2, 3, \ldots$ and $J_0(0) = 1$. In other words, if $\lambda = 0$, we get the trivial function (which is never an eigenfunction) for $n = 1, 2, 3, \ldots$, and for $n = 0$, $\lambda = 0$ (or equivalently, $\alpha = 0$) does not satisfy the equation in (8). When (6) is written in the form $xJ_n'(x) = nJ_n(x) - xJ_{n+1}(x)$, it follows from (7) and (8) that the square norm of $J_n(\alpha_i x)$ is

$$\|J_n(\alpha_i x)\|^2 = \frac{b^2}{2} J_{n+1}^2(\alpha_i b). \tag{9}$$

Case II: If we choose $A_2 = h \geq 0$, $B_2 = b$, then (1) is

$$hJ_n(\alpha b) + \alpha b J_n'(\alpha b) = 0. \tag{10}$$

Equation (10) has an infinite number of positive roots $x_i = \alpha_i b$ for each positive integer $n = 1, 2, 3, \ldots$. As before, the eigenvalues are obtained from $\lambda_i = \alpha_i^2 = x_i^2/b^2$. $\lambda = 0$ is not an eigenvalue for $n = 1, 2, 3, \ldots$. Substituting $\alpha_i b J_n'(\alpha_i b) = -hJ_n(\alpha_i b)$ into (7), we find that the square norm of $J_n(\alpha_i x)$ is now

$$\|J_n(\alpha_i x)\|^2 = \frac{\alpha_i^2 b^2 - n^2 + h^2}{2\alpha_i^2} J_n^2(\alpha_i b). \tag{11}$$

Case III: If $h = 0$ and $n = 0$ in (10), the α_i are defined from the roots of

$$J_0'(\alpha b) = 0. \tag{12}$$

Even though (12) is just a special case of (10), it is the only situation for which $\lambda = 0$ is an eigenvalue. To see this, observe that for $n = 0$, the result in (6) implies that $J_0'(\alpha b) = 0$ is equivalent to $J_1(\alpha b) = 0$. Since $x_1 = \alpha_i b = 0$ is a root of the last equation, $\alpha_1 = 0$, and because $J_0(0) = 1$ is nontrivial, we conclude from $\lambda_1 = \alpha_1^2 = x_1^2/b^2$ that $\lambda_1 = 0$ is an eigenvalue. But obviously we cannot use (11) when $\alpha_1 = 0$, $h = 0$, and $n = 0$. However, from the square norm (4) we have

$$\|1\|^2 = \int_0^b x\,dx = \frac{b^2}{2}. \tag{13}$$

For $\alpha_i > 0$ we can use (11) with $h = 0$ and $n = 0$:

$$\|J_0(\alpha_i x)\|^2 = \frac{b^2}{2} J_0^2(\alpha_i b). \tag{14}$$

The following definition summarizes three forms of the series (2) corresponding to the square norms in the three cases.

DEFINITION 12.6.1 Fourier–Bessel Series

The **Fourier–Bessel series** of a function f defined on the interval $(0, b)$ is given by

(i)
$$f(x) = \sum_{i=1}^{\infty} c_i J_n(\alpha_i x) \tag{15}$$

$$c_i = \frac{2}{b^2 J_{n+1}^2(\alpha_i b)} \int_0^b x J_n(\alpha_i x) f(x)\,dx, \tag{16}$$

where the α_i are defined by $J_n(\alpha b) = 0$.

(ii)
$$f(x) = \sum_{i=1}^{\infty} c_i J_n(\alpha_i x) \tag{17}$$

$$c_i = \frac{2\alpha_i^2}{(\alpha_i^2 b^2 - n^2 + h^2) J_n^2(\alpha_i b)} \int_0^b x J_n(\alpha_i x) f(x)\,dx, \tag{18}$$

where the α_i are defined by $h J_n(\alpha b) + \alpha b J_n'(\alpha b) = 0$.

(iii)
$$f(x) = c_1 + \sum_{i=2}^{\infty} c_i J_0(\alpha_i x) \tag{19}$$

$$c_1 = \frac{2}{b^2} \int_0^b x f(x)\,dx, \quad c_i = \frac{2}{b^2 J_0^2(\alpha_i b)} \int_0^b x J_0(\alpha_i x) f(x)\,dx, \tag{20}$$

where the α_i are defined by $J_0'(\alpha b) = 0$.

■ **Convergence of a Fourier–Bessel Series** Sufficient conditions for the convergence of a Fourier–Bessel series are not particularly restrictive.

THEOREM 12.6.1 Conditions for Convergence

Let f and f' be piecewise continuous on the interval $[0, b]$. Then for all x in the interval $(0, b)$, the Fourier–Bessel series of f converges to $f(x)$ at a point where f is continuous and to the average

$$\frac{f(x+) + f(x-)}{2}$$

at a point where f is discontinuous.

EXAMPLE 1 **Expansion in a Fourier–Bessel Series**

Expand $f(x) = x$, $0 < x < 3$, in a Fourier–Bessel series, using Bessel functions of order one that satisfy the boundary condition $J_1(3\alpha) = 0$.

SOLUTION We use (15) where the coefficients c_i are given by (16) with $b = 3$:

$$c_i = \frac{2}{3^2 J_2^2(3\alpha_i)} \int_0^3 x^2 J_1(\alpha_i x) \, dx.$$

To evaluate this integral we let $t = \alpha_i x$, $dx = dt/\alpha_i$, $x^2 = t^2/\alpha_i^2$, and use (5) in the form $\frac{d}{dt}[t^2 J_2(t)] = t^2 J_1(t)$:

$$c_i = \frac{2}{9\alpha_i^3 J_2^2(3\alpha_i)} \int_0^{3\alpha_i} \frac{d}{dt}[t^2 J_2(t)] \, dt = \frac{2}{\alpha_i J_2(3\alpha_i)}.$$

Therefore the desired expansion is

$$f(x) = 2\sum_{i=1}^{\infty} \frac{1}{\alpha_i J_2(3\alpha_i)} J_1(\alpha_i x). \qquad \equiv$$

You are asked to find the first four values of the α_i for the foregoing Bessel series in Problem 1 in Exercises 12.6.

EXAMPLE 2 **Expansion in a Fourier–Bessel Series**

If the α_i in Example 1 are defined by $J_1(3\alpha) + \alpha J_1'(3\alpha) = 0$, then the only thing that changes in the expansion is the value of the square norm. Multiplying the boundary condition by 3 gives $3J_1(3\alpha) + 3\alpha J_1'(3\alpha) = 0$, which now matches (10) when $h = 3$, $b = 3$, and $n = 1$. Thus (18) and (17) yield, in turn,

$$c_i = \frac{18\alpha_i J_2(3\alpha_i)}{(9\alpha_i^2 + 8)J_1^2(3\alpha_i)}$$

and

$$f(x) = 18\sum_{i=1}^{\infty} \frac{\alpha_i J_2(3\alpha_i)}{(9\alpha_i^2 + 8)J_1^2(3\alpha_i)} J_1(\alpha_i x). \qquad \equiv$$

Use of Computers Since Bessel functions are "built-in functions" in a CAS, it is a straightforward task to find the approximate values of the α_i and the coefficients c_i in a Fourier–Bessel series. For example, in (9) we can think of $x_i = \alpha_i b$ as a positive root of the equation $hJ_n(x) + xJ_n'(x) = 0$. Thus in Example 2 we have used a CAS to find the first five positive roots x_i of $3J_1(x) + xJ_1'(x) = 0$ and from these roots we obtain the first five values of α_i: $\alpha_1 = x_1/3 = 0.98320$, $\alpha_2 = x_2/3 = 1.94704$, $\alpha_3 = x_3/3 = 2.95758$, $\alpha_4 = x_4/3 = 3.98538$, and $\alpha_5 = x_5/3 = 5.02078$. Knowing the roots $x_i = 3\alpha_i$ and the α_i, we again use a CAS to calculate the numerical values of $J_2(3\alpha_i)$, $J_1^2(3\alpha_i)$, and finally the coefficients c_i. In this manner we find that the fifth partial sum $S_5(x)$ for the Fourier–Bessel series representation of $f(x) = x$, $0 < x < 3$ in Example 2 is

$$S_5(x) = 4.01844 \, J_1(0.98320x) - 1.86937 \, J_1(1.94704x)$$
$$+ 1.07106 \, J_1(2.95758x) - 0.70306 \, J_1(3.98538x) + 0.50343 \, J_1(5.02078x).$$

The graph of $S_5(x)$ on the interval $(0, 3)$ is shown in **FIGURE 12.6.1(a)**. In Figure 12.6.1(b) we have graphed $S_{10}(x)$ on the interval $(0, 50)$. Notice that outside the interval of definition $(0, 3)$ the series does not converge to a periodic extension of f because Bessel functions are not periodic functions. See Problems 11 and 12 in Exercises 12.6.

(a) $S_5(x)$, $0 < x < 3$

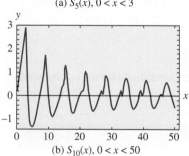

(b) $S_{10}(x)$, $0 < x < 50$

FIGURE 12.6.1 Partial sums of a Fourier–Bessel series

12.6.2 Fourier–Legendre Series

From Example 4 of Section 12.5 we know that the set of Legendre polynomials $\{P_n(x)\}$, $n = 0, 1, 2, \ldots$, is orthogonal with respect to the weight function $p(x) = 1$ on the interval $[-1, 1]$.

Furthermore, it can be proved that the square norm of a polynomial $P_n(x)$ depends on n in the following manner:

$$\|P_n(x)\|^2 = \int_{-1}^{1} P_n^2(x)\, dx = \frac{2}{2n + 1}.$$

The orthogonal series expansion of a function in terms of the Legendre polynomials is summarized in the next definition.

DEFINITION 12.6.2 Fourier–Legendre Series

The **Fourier–Legendre series** of a function f defined on the interval $(-1, 1)$ is given by

$$f(x) = \sum_{n=0}^{\infty} c_n P_n(x), \tag{21}$$

where

$$c_n = \frac{2n + 1}{2} \int_{-1}^{1} f(x) P_n(x)\, dx. \tag{22}$$

■ **Convergence of a Fourier–Legendre Series** Sufficient conditions for convergence of a Fourier–Legendre series are given in the next theorem.

THEOREM 12.6.2 Conditions for Convergence

Let f and f' be piecewise continuous on the interval $[-1, 1]$. Then for all x in the interval $(-1, 1)$, the Fourier–Legendre series of f converges to $f(x)$ at a point where f is continuous and to the average

$$\frac{f(x+) + f(x-)}{2}$$

at a point where f is discontinuous.

EXAMPLE 3 **Expansion in a Fourier–Legendre Series**

Write out the first four nonzero terms in the Fourier–Legendre expansion of

$$f(x) = \begin{cases} 0, & -1 < x < 0 \\ 1, & 0 \le x < 1. \end{cases}$$

SOLUTION The first several Legendre polynomials are listed on page 299. From these and (22) we find

$$c_0 = \frac{1}{2} \int_{-1}^{1} f(x) P_0(x)\, dx = \frac{1}{2} \int_{0}^{1} 1 \cdot 1\, dx = \frac{1}{2}$$

$$c_1 = \frac{3}{2} \int_{-1}^{1} f(x) P_1(x)\, dx = \frac{3}{2} \int_{0}^{1} 1 \cdot x\, dx = \frac{3}{4}$$

$$c_2 = \frac{5}{2} \int_{-1}^{1} f(x) P_2(x)\, dx = \frac{5}{2} \int_{0}^{1} 1 \cdot \frac{1}{2} (3x^2 - 1)\, dx = 0$$

$$c_3 = \frac{7}{2} \int_{-1}^{1} f(x) P_3(x)\, dx = \frac{7}{2} \int_{0}^{1} 1 \cdot \frac{1}{2} (5x^3 - 3x)\, dx = -\frac{7}{16}$$

$$c_4 = \frac{9}{2} \int_{-1}^{1} f(x) P_4(x)\, dx = \frac{9}{2} \int_{0}^{1} 1 \cdot \frac{1}{8} (35x^4 - 30x^2 + 3)\, dx = 0$$

$$c_5 = \frac{11}{2} \int_{-1}^{1} f(x) P_5(x)\, dx = \frac{11}{2} \int_{0}^{1} 1 \cdot \frac{1}{8} (63x^5 - 70x^3 + 15x)\, dx = \frac{11}{32}.$$

Hence

$$f(x) = \frac{1}{2} P_0(x) + \frac{3}{4} P_1(x) - \frac{7}{16} P_3(x) + \frac{11}{32} P_5(x) + \cdots.$$

≡

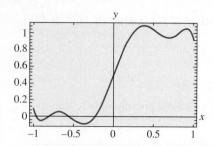

FIGURE 12.6.2 Partial sum $S_5(x)$ of Fourier–Legendre series in Example 3

Like the Bessel functions, Legendre polynomials are built-in functions in computer algebra systems such as *Mathematica* and *Maple*, and so each of the coefficients just listed can be found using the integration application of such a program. Indeed, using a CAS, we further find that $c_6 = 0$ and $c_7 = -\frac{65}{256}$. The fifth partial sum of the Fourier–Legendre series representation of the function f defined in Example 3 is then

$$S_5(x) = \frac{1}{2}P_0(x) + \frac{3}{4}P_1(x) - \frac{7}{16}P_3(x) + \frac{11}{32}P_5(x) - \frac{65}{256}P_7(x).$$

The graph of $S_5(x)$ on the interval $(-1, 1)$ is given in **FIGURE 12.6.2**.

■ **Alternative Form of Series** In applications, the Fourier–Legendre series appears in an alternative form. If we let $x = \cos\theta$, then $x = 1$ implies $\theta = 0$, whereas $x = -1$ implies $\theta = \pi$. Since $dx = -\sin\theta\,d\theta$, (21) and (22) become, respectively,

$$F(\theta) = \sum_{n=0}^{\infty} c_n P_n(\cos\theta) \qquad (23)$$

$$c_n = \frac{2n+1}{2}\int_0^{\pi} F(\theta)P_n(\cos\theta)\sin\theta\,d\theta, \qquad (24)$$

where $f(\cos\theta)$ has been replaced by $F(\theta)$.

12.6 Exercises Answers to selected odd-numbered problems begin on page ANS-33.

12.6.1 Fourier–Bessel Series

In Problems 1 and 2, use Table 5.3.1 in Section 5.3.

1. Find the first four $\alpha_i > 0$ defined by $J_1(3\alpha) = 0$.

2. Find the first four $\alpha_i \geq 0$ defined by $J_0'(2\alpha) = 0$.

In Problems 3–6, expand $f(x) = 1, 0 < x < 2$, in a Fourier–Bessel series using Bessel functions of order zero that satisfy the given boundary condition.

3. $J_0(2\alpha) = 0$ **4.** $J_0'(2\alpha) = 0$

5. $J_0(2\alpha) + 2\alpha J_0'(2\alpha) = 0$ **6.** $J_0(2\alpha) + \alpha J_0'(2\alpha) = 0$

In Problems 7–10, expand the given function in a Fourier–Bessel series using Bessel functions of the same order as in the indicated boundary condition.

7. $f(x) = 5x, \; 0 < x < 4$ **8.** $f(x) = x^2, \; 0 < x < 1$
$3J_1(4\alpha) + 4\alpha J_1'(4\alpha) = 0$ $J_2(\alpha) = 0$

9. $f(x) = x^2, \; 0 < x < 3$ **10.** $f(x) = 1 - x^2, \; 0 < x < 1$
$J_0'(3\alpha) = 0$ $J_0(\alpha) = 0$
[*Hint:* $t^3 = t^2 \cdot t$.]

Computer Lab Assignments

11. (a) Use a CAS to graph $y = 3J_1(x) + xJ_1'(x)$ on an interval so that the first five positive x-intercepts of the graph are shown.

(b) Use the root-finding capability of your CAS to approximate the first five roots x_i of the equation

$$3J_1(x) + xJ_1'(x) = 0.$$

(c) Use the data obtained in part (b) to find the first five positive values of α_i that satisfy

$$3J_1(4\alpha) + 4\alpha J_1'(4\alpha) = 0.$$

See Problem 7.

(d) If instructed, find the first 10 positive values of α_i.

12. (a) Use the values of α_i in part (c) of Problem 11 and a CAS to approximate the values of the first five coefficients c_i of the Fourier–Bessel series obtained in Problem 7.

(b) Use a CAS to graph the partial sums $S_N(x)$, $N = 1, 2, 3, 4, 5$, of the Fourier–Bessel series in Problem 7.

(c) If instructed, graph the partial sum $S_{10}(x)$ for $0 < x < 4$ and for $0 < x < 50$.

Discussion Problems

13. If the partial sums in Problem 12 are plotted on a symmetric interval such as $(-30, 30)$, would the graphs possess any symmetry? Explain.

14. (a) Sketch, by hand, a graph of what you think the Fourier–Bessel series in Problem 3 converges to on the interval $(-2, 2)$.

(b) Sketch, by hand, a graph of what you think the Fourier–Bessel series would converge to on the interval $(-4, 4)$ if the values α_i in Problem 7 were defined by $3J_2(4\alpha) + 4\alpha J_2'(4\alpha) = 0$.

12.6.2 Fourier–Legendre Series

In Problems 15 and 16, write out the first five nonzero terms in the Fourier–Legendre expansion of the given function. If instructed, use a CAS as an aid in evaluating the coefficients. Use a CAS to graph the partial sum $S_5(x)$.

15. $f(x) = \begin{cases} 0, & -1 < x < 0 \\ x, & 0 < x < 1 \end{cases}$

16. $f(x) = e^x, \; -1 < x < 1$

17. The first three Legendre polynomials are $P_0(x) = 1, P_1(x) = x,$ and $P_2(x) = \frac{1}{2}(3x^2 - 1)$. If $x = \cos\theta$, then $P_0(\cos\theta) = 1$ and $P_1(\cos\theta) = \cos\theta$. Show that $P_2(\cos\theta) = \frac{1}{4}(3\cos 2\theta + 1)$.

18. Use the results of Problem 17 to find a Fourier–Legendre expansion (23) of $F(\theta) = 1 - \cos 2\theta$.

19. A Legendre polynomial $P_n(x)$ is an even or odd function, depending on whether n is even or odd. Show that if f is an even function on the interval $(-1, 1)$, then (21) and (22) become, respectively,

$$f(x) = \sum_{n=0}^{\infty} c_{2n} P_{2n}(x) \qquad (25)$$

$$c_{2n} = (4n + 1) \int_0^1 f(x) P_{2n}(x) \, dx. \qquad (26)$$

20. Show that if f is an odd function on the interval $(-1, 1)$, then (21) and (22) become, respectively,

$$f(x) = \sum_{n=0}^{\infty} c_{2n+1} P_{2n+1}(x) \qquad (27)$$

$$c_{2n+1} = (4n + 3) \int_0^1 f(x) P_{2n+1}(x) \, dx. \qquad (28)$$

The series (25) and (27) can also be used when f is defined on only the interval $(0, 1)$. Both series represent f on $(0, 1)$; but on the interval $(-1, 0)$, (25) represents an even extension, whereas (27) represents an odd extension. In Problems 21 and 22, write out the first four nonzero terms in the indicated expansion of the given function. What function does the series represent on the interval $(-1, 1)$? Use a CAS to graph the partial sum $S_4(x)$.

21. $f(x) = x, \ 0 < x < 1$; (25)

22. $f(x) = 1, \ 0 < x < 1$; (27)

Discussion Problems

23. Why is a Fourier–Legendre expansion of a polynomial function that is defined on the interval $(-1, 1)$ necessarily a finite series?

24. Use your conclusion from Problem 23 to find the finite Fourier–Legendre series of $f(x) = x^2$. The series of $f(x) = x^3$. Do *not* use (21) and (22).

12 Chapter in Review Answers to selected odd-numbered problems begin on page ANS-33.

In Problems 1–10, fill in the blank or answer true/false without referring back to the text.

1. The functions $f(x) = x^2 - 1$ and $g(x) = x^5$ are orthogonal on the interval $[-\pi, \pi]$. _____

2. The product of an odd function f with an odd function g is an _____ function.

3. To expand $f(x) = |x| + 1, -\pi < x < \pi$, in an appropriate trigonometric series we would use a _____ series.

4. $y = 0$ is never an eigenfunction of a Sturm–Liouville problem. _____

5. $\lambda = 0$ is never an eigenvalue of a Sturm–Liouville problem. _____

6. If the function

$$f(x) = \begin{cases} x + 1, & -1 < x < 0 \\ -x, & 0 < x < 1 \end{cases}$$

is expanded in a Fourier series, the series will converge to _____ at $x = -1$, to _____ at $x = 0$, and to _____ at $x = 1$.

7. Suppose the function $f(x) = x^2 + 1, 0 < x < 3$, is expanded in a Fourier series, a cosine series, and a sine series. At $x = 0$, the Fourier series will converge to _____, the cosine series will converge to _____, and the sine series will converge to _____.

8. The corresponding eigenfunction for the boundary-value problem

$$y'' + \lambda y = 0, \quad y'(0) = 0, \quad y(\pi/2) = 0$$

for $\lambda = 25$ is _____.

9. The set $\{P_{2n}(x)\}, n = 0, 1, 2, \ldots$ of Legendre polynomials of even degree is orthogonal with respect to the weight function $p(x) = 1$ on the interval $[0, 1]$. _____

10. The set $\{P_n(x)\}, n = 0, 1, 2, \ldots$ of Legendre polynomials is orthogonal with respect to the weight function $p(x) = 1$ on the interval $[-1, 1]$. Hence, for $n > 0$, $\int_{-1}^{1} P_n(x) \, dx = $ _____.

11. Without doing any work, explain why the cosine series of $f(x) = \cos^2 x, 0 < x < \pi$, is the finite series

$$f(x) = \frac{1}{2} + \frac{1}{2} \cos 2x.$$

12. **(a)** Show that the set

$$\left\{ \sin \frac{\pi}{2L}x, \ \sin \frac{3\pi}{2L}x, \ \sin \frac{5\pi}{2L}x, \ \ldots \right\}$$

is orthogonal on the interval $[0, L]$.

(b) Find the norm of each function in part (a). Construct an orthonormal set.

13. Expand $f(x) = |x| - x, -1 < x < 1$, in a Fourier series.

14. Expand $f(x) = 2x^2 - 1, -1 < x < 1$, in a Fourier series.

15. Expand $f(x) = e^{-x}, 0 < x < 1$, in a cosine series. In a sine series.

16. In Problems 13, 14, and 15, sketch the periodic extension of f to which each series converges.

17. Find the eigenvalues and eigenfunctions of the boundary-value problem

$$x^2 y'' + xy' + 9\lambda y = 0, \quad y'(1) = 0, \quad y(e) = 0.$$

18. Give an orthogonality relation for the eigenfunctions in Problem 17.

19. **Chebyshev's differential equation**

$$(1 - x^2)y'' - xy' + n^2 y = 0$$

has a polynomial solution $y = T_n(x)$ for $n = 0, 1, 2, \ldots$. Specify the weight function $p(x)$ and the interval over which the set of Chebyshev polynomials $\{T_n(x)\}$ is orthogonal. Give an orthogonality relation.

20. Expand the periodic function shown in **FIGURE 12.R.1** in an appropriate Fourier series.

FIGURE 12.R.1 Graph for Problem 20

21. Expand

$$f(x) = \begin{cases} 1, & 0 < x < 2 \\ 0, & 2 < x < 4 \end{cases}$$

in a Fourier–Bessel series, using Bessel functions of order zero that satisfy the boundary condition $J_0(4\alpha) = 0$.

22. Expand $f(x) = x^4$, $-1 < x < 1$, in a Fourier–Legendre series.

23. Suppose the function $y = f(x)$ is defined on the interval $(-\infty, \infty)$.

(a) Verify the identity $f(x) = f_e(x) + f_o(x)$, where

$$f_e(x) = \frac{f(x) + f(-x)}{2} \quad \text{and} \quad f_o(x) = \frac{f(x) - f(-x)}{2}.$$

(b) Show that f_e is an even function and f_o an odd function.

24. The function $f(x) = e^x$ is neither even nor odd. Use Problem 23 to write f as the sum of an even function and an odd function. Identify f_e and f_o.

25. Suppose f is an integrable $2p$-periodic function. Prove that for any real number a,

$$\int_0^{2p} f(x)\, dx = \int_a^{a+2p} f(x)\, dx.$$

Boundary-Value Problems in Rectangular Coordinates

In this and the next two chapters the emphasis will be on two procedures used in solving partial differential equations that occur in problems involving temperature distributions, vibrations, and potentials. These problems, called **boundary-value problems** (BVPs), are described by relatively simple linear second-order partial differential equations (PDEs). The thrust of these procedures is to find particular solutions of a linear PDE by reducing it to one or more linear ordinary differential equations.

13.1 Separable Partial Differential Equations

INTRODUCTION Partial differential equations (PDEs), like ordinary differential equations (ODEs), are classified as *linear* or *nonlinear*. Analogous to a linear ODE (see (6) of Section 1.1), the dependent variable and its partial derivatives appear only to the first power in a linear PDE. In this and the chapters that follow, we are concerned only with linear partial differential equations.

◼ Linear Partial Differential Equation
If we let u denote the dependent variable and x and y the independent variables, then the general form of a **linear second-order partial differential equation** is given by

$$A\frac{\partial^2 u}{\partial x^2} + B\frac{\partial^2 u}{\partial x \partial y} + C\frac{\partial^2 u}{\partial y^2} + D\frac{\partial u}{\partial x} + E\frac{\partial u}{\partial y} + Fu = G, \tag{1}$$

where the coefficients $A, B, C, \ldots, G$ are constants or functions of x and y. When $G(x, y) = 0$, equation (1) is said to be **homogeneous**; otherwise, it is **nonhomogeneous**.

EXAMPLE 1 Linear Second-Order PDEs

The equations

$$\frac{\partial^2 u}{\partial x^2} + \frac{\partial^2 u}{\partial y^2} = 0 \quad \text{and} \quad \frac{\partial^2 u}{\partial x^2} - \frac{\partial u}{\partial y} = xy$$

are examples of linear second-order PDEs. The first equation is homogeneous and the second is nonhomogeneous. ≡

◼ Solution of a PDE
A **solution** of a linear partial differential equation (1) is a function $u(x, y)$ of two independent variables that possesses all partial derivatives occurring in the equation and that satisfies the equation in some region of the xy-plane.

It is not our intention to examine procedures for finding *general solutions* of linear partial differential equations. Not only is it often difficult to obtain a general solution of a linear second-order PDE, but a general solution is usually not all that useful in applications. Thus our focus throughout will be on finding *particular solutions* of some of the important linear PDEs, that is, equations that appear in many applications.

We are interested only in particular solutions of PDEs. ▶

◼ Separation of Variables
Although there are several methods that can be tried to find particular solutions of a linear PDE, the one we are interested in at the moment is called the method of **separation of variables**. In this method if we are seeking a particular solution of, say, a linear second-order PDE in which the independent variables are x and y, then we seek to find a particular solution in the form of *product* of a function x and a function of y:

$$u(x, y) = X(x)Y(y).$$

With this assumption, it is *sometimes* possible to reduce a linear PDE in two variables to two ODEs. To this end we observe that

$$\frac{\partial u}{\partial x} = X'Y, \quad \frac{\partial u}{\partial y} = XY', \quad \frac{\partial^2 u}{\partial x^2} = X''Y, \quad \frac{\partial^2 u}{\partial y^2} = XY'',$$

where the primes denote ordinary differentiation.

EXAMPLE 2 Using Separation of Variables

Find product solutions of $\dfrac{\partial^2 u}{\partial x^2} = 4\dfrac{\partial u}{\partial y}$.

SOLUTION Substituting $u(x, y) = X(x)Y(y)$ into the partial differential equation yields

$$X''Y = 4XY'.$$

After dividing both sides by $4XY$, we have separated the variables:

$$\frac{X''}{4X} = \frac{Y'}{Y}.$$

Since the left-hand side of the last equation is independent of y and is equal to the right-hand side, which is independent of x, we conclude that both sides of the equation are independent of x *and* y. In other words, each side of the equation must be a constant. As a practical matter it is convenient to write this real **separation constant** as $-\lambda$. From the two equalities,

$$\frac{X''}{4X} = \frac{Y'}{Y} = -\lambda,$$

we obtain the two linear ordinary differential equations

$$X'' + 4\lambda X = 0 \quad \text{and} \quad Y' + \lambda Y = 0. \tag{2}$$

See Example 2, Section 3.9 and Example 1, Section 12.5.

For the three cases for λ: zero, negative, or positive; that is, $\lambda = 0$, $\lambda = -\alpha^2 < 0$, and $\lambda = \alpha^2 > 0$, where $\alpha > 0$, the ODEs in (2) are, in turn,

$$X'' = 0 \qquad\qquad \text{and} \quad Y' = 0, \tag{3}$$

$$X'' - 4\alpha^2 X = 0 \quad \text{and} \quad Y' - \alpha^2 Y = 0, \tag{4}$$

$$X'' + 4\alpha^2 X = 0 \quad \text{and} \quad Y' + \alpha^2 Y = 0. \tag{5}$$

Case I ($\lambda = 0$): The DEs in (3) can be solved by integration. The solutions are $X = c_1 + c_2 x$ and $Y = c_3$. Thus a particular product solution of the given PDE is

$$u = XY = (c_1 + c_2 x)c_3 = A_1 + B_1 x, \tag{6}$$

where we have replaced $c_1 c_3$ and $c_2 c_3$ by A_1 and B_1, respectively.

Case II ($\lambda = -\alpha^2$): The general solutions of the DEs in (4) are

$$X = c_4 \cosh 2\alpha x + c_5 \sinh 2\alpha x \quad \text{and} \quad Y = c_6 e^{\alpha^2 y},$$

respectively. Thus, another particular product solution of the PDE is

$$u = XY = (c_4 \cosh 2\alpha x + c_5 \sinh 2\alpha x)c_6 e^{\alpha^2 y}$$

or

$$u = A_2 e^{\alpha^2 y} \cosh 2\alpha x + B_2 e^{\alpha^2 y} \sinh 2\alpha x, \tag{7}$$

where $A_2 = c_4 c_6$ and $B_2 = c_5 c_6$.

Case III ($\lambda = \alpha^2$): Finally, the general solutions of the DEs in (5) are

$$X = c_7 \cos 2\alpha x + c_8 \sin 2\alpha x \quad \text{and} \quad Y = c_9 e^{-\alpha^2 y},$$

respectively. These results give yet another particular solution

$$u = A_3 e^{-\alpha^2 y} \cos 2\alpha x + B_3 e^{-\alpha^2 y} \sin 2\alpha x, \tag{8}$$

where $A_3 = c_7 c_9$ and $B_3 = c_8 c_9$.

It is left as an exercise to verify that (6), (7), and (8) satisfy the given partial differential equation $u_{xx} = 4u_y$. See Problem 29 in Exercises 13.1.

Separation of variables is not a general method for finding particular solutions; some linear partial differential equations are simply not separable. You should verify that the assumption $u = XY$ does not lead to a solution for $\partial^2 u/\partial x^2 - \partial u/\partial y = x$.

■ Superposition Principle
 The following theorem is analogous to Theorem 3.1.2 and is known as the **superposition principle**.

THEOREM 13.1.1 Superposition Principle

If $u_1, u_2, \dots, u_k$ are solutions of a homogeneous linear partial differential equation, then the linear combination

$$u = c_1 u_1 + c_2 u_2 + \cdots + c_k u_k,$$

where the c_i, $i = 1, 2, \dots, k$ are constants, is also a solution.

Throughout the remainder of the chapter we shall assume that whenever we have an infinite set $u_1, u_2, u_3, \ldots$ of solutions of a homogeneous linear equation, we can construct yet another solution u by forming the infinite series

$$u = \sum_{k=1}^{\infty} c_k u_k,$$

where the $c_k, k = 1, 2, \ldots$, are constants.

Classification of Equations A linear second-order partial differential equation in two independent variables with constant coefficients can be classified as one of three types. This classification depends only on the coefficients of the second-order derivatives. Of course, we assume that at least one of the coefficients A, B, and C is not zero.

DEFINITION 13.1.1 **Classification of Equations**

The linear second-order partial differential equation

$$A\frac{\partial^2 u}{\partial x^2} + B\frac{\partial^2 u}{\partial x \partial y} + C\frac{\partial^2 u}{\partial y^2} + D\frac{\partial u}{\partial x} + E\frac{\partial u}{\partial y} + Fu = G,$$

where A, B, C, D, E, F, and G are real constants, is said to be

> **hyperbolic** if $B^2 - 4AC > 0,$
> **parabolic** if $B^2 - 4AC = 0,$
> **elliptic** if $B^2 - 4AC < 0.$

EXAMPLE 3 **Classifying Linear Second-Order PDEs**

Classify the following equations:

(a) $3\dfrac{\partial^2 u}{\partial x^2} = \dfrac{\partial u}{\partial y}$ **(b)** $\dfrac{\partial^2 u}{\partial x^2} = \dfrac{\partial^2 u}{\partial y^2}$ **(c)** $\dfrac{\partial^2 u}{\partial x^2} + \dfrac{\partial^2 u}{\partial y^2} = 0.$

SOLUTION **(a)** By rewriting the given equation as

$$3\frac{\partial^2 u}{\partial x^2} - \frac{\partial u}{\partial y} = 0$$

we can make the identifications $A = 3$, $B = 0$, and $C = 0$. Since $B^2 - 4AC = 0$, the equation is parabolic.

(b) By rewriting the equation as

$$\frac{\partial^2 u}{\partial x^2} - \frac{\partial^2 u}{\partial y^2} = 0,$$

we see that $A = 1$, $B = 0$, $C = -1$, and $B^2 - 4AC = -4(1)(-1) > 0$. The equation is hyperbolic.

(c) With $A = 1$, $B = 0$, $C = 1$, and $B^2 - 4AC = -4(1)(1) < 0$, the equation is elliptic. ≡

REMARKS

(*i*) Separation of variables is not a general method for finding particular solutions of linear partial differential equations. Some equations are simply not separable. You are encouraged to verify that the assumption $u(x, y) = X(x)Y(y)$ does not lead to a solution of the linear second-order PDE $\partial^2 u/\partial x^2 + \partial u/\partial x = y$.

(*ii*) A detailed explanation of why the classifications given in Definition 13.1.1 are important is beyond the scope of this text. But you should at least be aware that these classifications do have a practical importance. Beginning in Section 13.3 we are going to solve some PDEs subject to both boundary and initial conditions. The kinds of side conditions appropriate for a given equation depend on whether the equation is hyperbolic, parabolic, or elliptic. Also, we shall see in Chapter 16 that numerical solution methods for linear second-order PDEs differ in conformity with the classification of the equation.

In Problems 1–16, use separation of variables to find, if possible, product solutions for the given partial differential equation.

1. $\dfrac{\partial u}{\partial x} = \dfrac{\partial u}{\partial y}$

2. $\dfrac{\partial u}{\partial x} + 3\dfrac{\partial u}{\partial y} = 0$

3. $u_x + u_y = u$

4. $u_x = u_y + u$

5. $x\dfrac{\partial u}{\partial x} = y\dfrac{\partial u}{\partial y}$

6. $y\dfrac{\partial u}{\partial x} + x\dfrac{\partial u}{\partial y} = 0$

7. $\dfrac{\partial^2 u}{\partial x^2} + \dfrac{\partial^2 u}{\partial x\partial y} + \dfrac{\partial^2 u}{\partial y^2} = 0$

8. $y\dfrac{\partial^2 u}{\partial x\partial y} + u = 0$

9. $k\dfrac{\partial^2 u}{\partial x^2} - u = \dfrac{\partial u}{\partial t}$, $k > 0$

10. $k\dfrac{\partial^2 u}{\partial x^2} = \dfrac{\partial u}{\partial t}$, $k > 0$

11. $a^2\dfrac{\partial^2 u}{\partial x^2} = \dfrac{\partial^2 u}{\partial t^2}$

12. $a^2\dfrac{\partial^2 u}{\partial x^2} = \dfrac{\partial^2 u}{\partial t^2} + 2k\dfrac{\partial u}{\partial t}$, $k > 0$

13. $\dfrac{\partial^2 u}{\partial x^2} + \dfrac{\partial^2 u}{\partial y^2} + 2k\dfrac{\partial u}{\partial t}$, $k > 0$

14. $x^2\dfrac{\partial^2 u}{\partial x^2} + \dfrac{\partial^2 u}{\partial y^2} = 0$

15. $u_{xx} + u_{yy} = u$

16. $a^2 u_{xx} - g = u_{tt}$, g a constant

In Problems 17–26, classify the given partial differential equation as hyperbolic, parabolic, or elliptic.

17. $\dfrac{\partial^2 u}{\partial x^2} + \dfrac{\partial^2 u}{\partial x\partial y} + \dfrac{\partial^2 u}{\partial y^2} = 0$

18. $3\dfrac{\partial^2 u}{\partial x^2} + 5\dfrac{\partial^2 u}{\partial x\partial y} + \dfrac{\partial^2 u}{\partial y^2} = 0$

19. $\dfrac{\partial^2 u}{\partial x^2} + 6\dfrac{\partial^2 u}{\partial x\partial y} + 9\dfrac{\partial^2 u}{\partial y^2} = 0$

20. $\dfrac{\partial^2 u}{\partial x^2} - \dfrac{\partial^2 u}{\partial x\partial y} - 3\dfrac{\partial^2 u}{\partial y^2} = 0$

21. $\dfrac{\partial^2 u}{\partial x^2} = 9\dfrac{\partial^2 u}{\partial x\partial y}$

22. $\dfrac{\partial^2 u}{\partial x\partial y} - \dfrac{\partial^2 u}{\partial y^2} + 2\dfrac{\partial u}{\partial x} = 0$

23. $\dfrac{\partial^2 u}{\partial x^2} + 2\dfrac{\partial^2 u}{\partial x\partial y} + \dfrac{\partial^2 u}{\partial y^2} + \dfrac{\partial u}{\partial x} - 6\dfrac{\partial u}{\partial y} = 0$

24. $\dfrac{\partial^2 u}{\partial x^2} + \dfrac{\partial^2 u}{\partial y^2} = u$

25. $a^2\dfrac{\partial^2 u}{\partial x^2} = \dfrac{\partial^2 u}{\partial t^2}$

26. $k\dfrac{\partial^2 u}{\partial x^2} = \dfrac{\partial u}{\partial t}$, $k > 0$

In Problems 27 and 28, show that the given partial differential equation possesses the indicated product solution.

27. $k\left(\dfrac{\partial^2 u}{\partial r^2} + \dfrac{1}{r}\dfrac{\partial u}{\partial r}\right) = \dfrac{\partial u}{\partial t}$;

$u = e^{-k\alpha^2 t}(c_1 J_0(\alpha r) + c_2 Y_0(\alpha r))$

28. $\dfrac{\partial^2 u}{\partial r^2} + \dfrac{1}{r}\dfrac{\partial u}{\partial r} + \dfrac{1}{r^2}\dfrac{\partial^2 u}{\partial \theta^2} = 0$;

$u = (c_1 \cos \alpha\theta + c_2 \sin \alpha\theta)(c_3 r^\alpha + c_4 r^{-\alpha})$

29. Verify that each of the products $u = X(x)Y(y)$ in (6), (7), and (8) satisfies the second-order PDE in Example 2.

30. Definition 13.1.1 generalizes to linear PDEs with coefficients that are functions of x and y. Determine the regions in the xy-plane for which the equation

$$(xy + 1)\dfrac{\partial^2 u}{\partial x^2} + (x + 2y)\dfrac{\partial^2 u}{\partial x\partial y} + \dfrac{\partial^2 u}{\partial y^2} + xy^2 u = 0$$

is hyperbolic, parabolic, or elliptic.

Discussion Problems

In Problems 31 and 32, discuss whether product solutions $u = X(x)Y(y)$ can be found for the given partial differential equation. [*Hint*: Use the superposition principle.]

31. $\dfrac{\partial^2 u}{\partial x^2} - u = 0$

32. $\dfrac{\partial^2 u}{\partial x\partial y} + \dfrac{\partial u}{\partial x} = 0$

13.2 Classical PDEs and Boundary-Value Problems

INTRODUCTION For the remainder of this and the next chapter we shall be concerned with finding product solutions of the second-order partial differential equations

$$k\dfrac{\partial^2 u}{\partial x^2} = \dfrac{\partial u}{\partial t}, \quad k > 0 \tag{1}$$

$$a^2\dfrac{\partial^2 u}{\partial x^2} = \dfrac{\partial^2 u}{\partial t^2} \tag{2}$$

$$\dfrac{\partial^2 u}{\partial x^2} + \dfrac{\partial^2 u}{\partial y^2} = 0 \tag{3}$$

or slight variations of these equations. These classical equations of mathematical physics are known, respectively, as the **one-dimensional heat equation**, the **one-dimensional wave equation**, and **Laplace's equation in two dimensions**. "One-dimensional" refers to the fact that x denotes a spatial dimension whereas t represents time; "two dimensional" in (3) means that x and y are both spatial dimensions. Laplace's equation is abbreviated $\nabla^2 u = 0$, where

$$\nabla^2 u = \frac{\partial^2 u}{\partial x^2} + \frac{\partial^2 u}{\partial y^2}$$

is called the **two-dimensional Laplacian** of the function u. In three dimensions the **Laplacian** of u is

$$\nabla^2 u = \frac{\partial^2 u}{\partial x^2} + \frac{\partial^2 u}{\partial y^2} + \frac{\partial^2 u}{\partial z^2}.$$

By comparing equations (1)−(3) with the linear second-order PDE given in Definition 13.1.1, with t playing the part of y, we see that the heat equation (1) is parabolic, the wave equation (2) is hyperbolic, and Laplace's equation (3) is elliptic. This classification is important in Chapter 16.

▉ Heat Equation

Equation (1) occurs in the theory of heat flow—that is, heat transferred by conduction in a rod or thin wire. The function $u(x, t)$ is temperature. Problems in mechanical vibrations often lead to the wave equation (2). For purposes of discussion, a solution $u(x, t)$ of (2) will represent the displacement of an idealized string. Finally, a solution $u(x, y)$ of Laplace's equation (3) can be interpreted as the steady-state (that is, time-independent) temperature distribution throughout a thin, two-dimensional plate.

Even though we have to make many simplifying assumptions, it is worthwhile to see how equations such as (1) and (2) arise.

Suppose a thin rod of length L has a circular cross-sectional area A and coincides with the x-axis on the interval $[0, L]$. See **FIGURE 13.2.1**. Let us suppose:

- The flow of heat within the rod takes place only in the x-direction.
- The lateral, or curved, surface of the rod is insulated; that is, no heat escapes from this surface.
- No heat is being generated within the rod by either chemical or electrical means.
- The rod is homogeneous; that is, its mass per unit volume ρ is a constant.
- The specific heat γ and thermal conductivity K of the material of the rod are constants.

To derive the partial differential equation satisfied by the temperature $u(x, t)$, we need two empirical laws of heat conduction:

(*i*) *The quantity of heat Q in an element of mass m is*

$$Q = \gamma m u, \tag{4}$$

 where u is the temperature of the element.

(*ii*) *The rate of heat flow Q_t through the cross section indicated in Figure 13.2.1 is proportional to the area A of the cross section and the partial derivative with respect to x of the temperature:*

$$Q_t = -K A u_x. \tag{5}$$

Since heat flows in the direction of decreasing temperature, the minus sign in (5) is used to ensure that Q_t is positive for $u_x < 0$ (heat flow to the right) and negative for $u_x > 0$ (heat flow to the left). If the circular slice of the rod shown in Figure 13.2.1 between x and $x + \Delta x$ is very thin, then $u(x, t)$ can be taken as the approximate temperature at each point in the interval. Now the mass of the slice is $m = \rho(A \, \Delta x)$, and so it follows from (4) that the quantity of heat in it is

$$Q = \gamma \rho A \, \Delta x \, u. \tag{6}$$

Furthermore, when heat flows in the positive x-direction, we see from (5) that heat builds up in the slice at the net rate

$$-K A u_x(x, t) - [-K A u_x(x + \Delta x, t)] = K A[u_x(x + \Delta x, t) - u_x(x, t)]. \tag{7}$$

cross section of area A

FIGURE 13.2.1 One-dimensional flow of heat

By differentiating (6) with respect to t we see that this net rate is also given by

$$Q_t = \gamma \rho A \, \Delta x \, u_t. \tag{8}$$

Equating (7) and (8) gives

$$\frac{K}{\gamma \rho} \frac{u_x(x + \Delta x, t) - u_x(x, t)}{\Delta x} = u_t. \tag{9}$$

Taking the limit of (9) as $\Delta x \to 0$ finally yields (1) in the form*

$$\frac{K}{\gamma \rho} u_{xx} = u_t.$$

It is customary to let $k = K/\gamma \rho$ and call this positive constant the **thermal diffusivity**.

Wave Equation

Consider a string of length L, such as a guitar string, stretched taut between two points on the x-axis—say, $x = 0$ and $x = L$. When the string starts to vibrate, assume that the motion takes place in the xy-plane in such a manner that each point on the string moves in a direction perpendicular to the x-axis (transverse vibrations). As shown in **FIGURE 13.2.2(a)**, let $u(x, t)$ denote the vertical displacement of any point on the string measured from the x-axis for $t > 0$. We further assume:

(a) Segment of string

(b) Enlargement of segment

FIGURE 13.2.2 Taut string anchored at two points on the x-axis

- The string is perfectly flexible.
- The string is homogeneous; that is, its mass per unit length ρ is a constant.
- The displacements u are small compared to the length of the string.
- The slope of the curve is small at all points.
- The tension $\mathbf{T}$ acts tangent to the string, and its magnitude T is the same at all points.
- The tension is large compared with the force of gravity.
- No other external forces act on the string.

Now in Figure 13.2.2(b) the tensions $\mathbf{T}_1$ and $\mathbf{T}_2$ are tangent to the ends of the curve on the interval $[x, x + \Delta x]$. For small values of θ_1 and θ_2 the net vertical force acting on the corresponding element Δs of the string is then

$$T \sin \theta_2 - T \sin \theta_1 \approx T \tan \theta_2 - T \tan \theta_1$$

$$= T[u_x(x + \Delta x, t) - u_x(x, t)],^{\dagger}$$

where $T = |\mathbf{T}_1| = |\mathbf{T}_2|$. Now $\rho \, \Delta s \approx \rho \, \Delta x$ is the mass of the string on $[x, x + \Delta x]$, and so Newton's second law gives

$$T[u_x(x + \Delta x, t) - u_x(x, t)] = \rho \, \Delta x \, u_{tt}$$

or

$$\frac{u_x(x + \Delta x, t) - u_x(x, t)}{\Delta x} = \frac{\rho}{T} u_{tt}.$$

If the limit is taken as $\Delta x \to 0$, the last equation becomes $u_{xx} = (\rho/T)u_{tt}$. This of course is (2) with $a^2 = T/\rho$.

temperature as a function of position on the hot plate thermometer

FIGURE 13.2.3 Steady-state temperatures in a rectangular plate

Laplace's Equation

Although we shall not present its derivation, Laplace's equation in two and three dimensions occurs in time-independent problems involving potentials such as electrostatic, gravitational, and velocity in fluid mechanics. Moreover, a solution of Laplace's equation can also be interpreted as a steady-state temperature distribution. As illustrated in **FIGURE 13.2.3**, a solution $u(x, y)$ of (3) could represent the temperature that varies from point to point—but not with time—of a rectangular plate.

We often wish to find solutions of equations (1), (2), and (3) that satisfy certain side conditions.

*Recall from calculus that $u_{xx} = \lim\limits_{\Delta x \to 0} \dfrac{u_x(x + \Delta x, t) - u_x(x, t)}{\Delta x}$.

$^{\dagger}\tan \theta_2 = u_x(x + \Delta x, t)$ and $\tan \theta_1 = u_x(x, t)$ are equivalent expressions for slope.

FIGURE 13.2.4 Plucked string

Initial Conditions Since solutions of (1) and (2) depend on time t, we can prescribe what happens at $t = 0$; that is, we can give **initial conditions** (**IC**). If $f(x)$ denotes the initial temperature distribution throughout the rod in Figure 13.2.1, then a solution $u(x, t)$ of (1) must satisfy the single initial condition $u(x, 0) = f(x)$, $0 < x < L$. On the other hand, for a vibrating string, we can specify its initial displacement (or shape) $f(x)$ as well as its initial velocity $g(x)$. In mathematical terms we seek a function $u(x, t)$ satisfying (2) and the two initial conditions:

$$u(x, 0) = f(x), \qquad \frac{\partial u}{\partial t}\bigg|_{t=0} = g(x), \quad 0 < x < L. \tag{10}$$

For example, the string could be plucked, as shown in **FIGURE 13.2.4**, and released from rest ($g(x) = 0$).

Boundary Conditions The string in Figure 13.2.4 is secured to the x-axis at $x = 0$ and $x = L$ for all time. We interpret this by the two **boundary conditions** (**BC**):

$$u(0, t) = 0, \qquad u(L, t) = 0, \qquad t > 0.$$

Note that in this context the function f in (10) is continuous, and consequently $f(0) = 0$ and $f(L) = 0$. In general, there are three types of boundary conditions associated with equations (1), (2), and (3). On a boundary we can specify the values of *one* of the following:

$$(i)\ u, \qquad (ii)\ \frac{\partial u}{\partial n}, \qquad \text{or} \qquad (iii)\ \frac{\partial u}{\partial n} + hu, \quad h \text{ a constant.}$$

Here $\partial u/\partial n$ denotes the normal derivative of u (the directional derivative of u in the direction perpendicular to the boundary). A boundary condition of the first type (i) is called a **Dirichlet condition**, a boundary condition of the second type (ii) is called a **Neumann condition**, and a boundary condition of the third type (iii) is known as a **Robin condition**. For example, for $t > 0$ a typical condition at the right-hand end of the rod in Figure 13.2.1 can be

$$(i)'\quad u(L, t) = u_0, \quad u_0 \text{ a constant,}$$

$$(ii)'\quad \frac{\partial u}{\partial x}\bigg|_{x=L} = 0, \quad \text{or}$$

$$(iii)'\quad \frac{\partial u}{\partial x}\bigg|_{x=L} = -h(u(L, t) - u_m), \quad h > 0 \text{ and } u_m \text{ constants.}$$

Condition $(i)'$ simply states that the boundary $x = L$ is held by some means at a constant *temperature* u_0 for all time $t > 0$. Condition $(ii)'$ indicates that the boundary $x = L$ is *insulated*. From the empirical law of heat transfer, the flux of heat across a boundary (that is, the amount of heat per unit area per unit time conducted across the boundary) is proportional to the value of the normal derivative $\partial u/\partial n$ of the temperature u. Thus when the boundary $x = L$ is thermally insulated, no heat flows into or out of the rod and so

$$\frac{\partial u}{\partial x}\bigg|_{x=L} = 0.$$

We can interpret $(iii)'$ to mean that *heat is lost* from the right-hand end of the rod by being in contact with a medium, such as air or water, that is held at a constant temperature. From Newton's law of cooling, the outward flux of heat from the rod is proportional to the difference between the temperature $u(L, t)$ at the boundary and the temperature u_m of the surrounding medium. We note that if heat is lost from the left-hand end of the rod, the boundary condition is

$$\frac{\partial u}{\partial x}\bigg|_{x=0} = h(u(0, t) - u_m).$$

The change in algebraic sign is consistent with the assumption that the rod is at a higher temperature than the medium surrounding the ends so that $u(0, t) > u_m$ and $u(L, t) > u_m$. At $x = 0$ and $x = L$, the slopes $u_x(0, t)$ and $u_x(L, t)$ must be positive and negative, respectively.

Of course, at the ends of the rod we can specify different conditions at the same time. For example, we could have

$$\left.\frac{\partial u}{\partial x}\right|_{x=0} = 0 \quad \text{and} \quad u(L, t) = u_0, \ t > 0.$$

We note that the boundary condition in $(i)'$ is homogeneous if $u_0 = 0$; if $u_0 \neq 0$, the boundary condition is nonhomogeneous. The boundary condition $(ii)'$ is homogeneous; $(iii)'$ is homogeneous if $u_m = 0$ and nonhomogeneous if $u_m \neq 0$.

▧ Boundary-Value Problems Problems such as

Solve: $a^2 \dfrac{\partial^2 u}{\partial x^2} = \dfrac{\partial^2 u}{\partial t^2}, \ 0 < x < L, \ t > 0$

Subject to: (BC) $u(0, t) = 0, \ u(L, t) = 0, \ t > 0$ (11)

(IC) $u(x, 0) = f(x), \ \left.\dfrac{\partial u}{\partial t}\right|_{t=0} = g(x), \ 0 < x < L$

and

Solve: $\dfrac{\partial^2 u}{\partial x^2} + \dfrac{\partial^2 u}{\partial y^2} = 0, \ 0 < x < a, \ 0 < y < b$

Subject to: (BC) $\begin{cases} \left.\dfrac{\partial u}{\partial x}\right|_{x=0} = 0, \ \left.\dfrac{\partial u}{\partial x}\right|_{x=a} = 0, \ 0 < y < b \\ u(x, 0) = 0, \ u(x, b) = f(x), \ 0 < x < a \end{cases}$ (12)

are called **boundary-value problems**. The problem in (11) is classified as a **homogeneous BVP** since the partial differential equation and the boundary conditions are homogeneous.

▧ Variations The partial differential equations (1), (2), and (3) must be modified to take into consideration internal or external influences acting on the physical system. More general forms of the one-dimensional heat and wave equations are, respectively,

$$k \frac{\partial^2 u}{\partial x^2} + F(x, t, u, u_x) = \frac{\partial u}{\partial t} \tag{13}$$

and

$$a^2 \frac{\partial^2 u}{\partial x^2} + F(x, t, u, u_t) = \frac{\partial^2 u}{\partial t^2}. \tag{14}$$

For example, if there is heat transfer from the lateral surface of a rod into a surrounding medium that is held at a constant temperature u_m, then the heat equation (13) is

$$k \frac{\partial^2 u}{\partial x^2} - h(u - u_m) = \frac{\partial u}{\partial t},$$

where h is a constant. In (14) the function F could represent the various forces acting on the string. For example, when external, damping, and elastic restoring forces are taken into account, (14) assumes the form

$$a^2 \frac{\partial^2 u}{\partial x^2} + \underbrace{\overset{\text{external force}}{f(x, t)} - \overset{\text{damping}}{c \frac{\partial u}{\partial t}} - \overset{\text{restoring force}}{ku}}_{F(x, t, u, u_t)} = \frac{\partial^2 u}{\partial t^2}. \tag{15}$$

REMARKS

The analysis of a wide variety of diverse phenomena yields the mathematical models (1), (2), or (3) or their generalizations involving a greater number of spatial variables. For example, (1) is sometimes called the **diffusion equation** since the diffusion of dissolved substances in solution is analogous to the flow of heat in a solid. The function $c(x, t)$ satisfying the partial differential equation in this case represents the concentration of the dissolved substance. Similarly, equation (2) and its generalization (15) arise in the analysis of the flow of electricity in a long cable or transmission line. In this setting (2) is known as the **telegraph equation**. It can be shown that under certain assumptions the current $i(x, t)$ and the voltage $v(x, t)$ in the line satisfy two partial differential equations identical to (2) (or (15)). The wave equation (2) also appears in fluid mechanics, acoustics, and elasticity. Laplace's equation (3) is encountered in determining the static displacement of membranes.

13.2 Exercises Answers to selected odd-numbered problems begin on page ANS-33.

In Problems 1–6, a rod of length L coincides with the interval $[0, L]$ on the x-axis. Set up the boundary-value problem for the temperature $u(x, t)$.

1. The left end is held at temperature zero, and the right end is insulated. The initial temperature is $f(x)$ throughout.

2. The left end is held at temperature u_0, and the right end is held at temperature u_1. The initial temperature is zero throughout.

3. The left end is held at temperature 100°, and there is heat transfer from the right end into the surrounding medium at temperature zero. The initial temperature is $f(x)$ throughout.

4. There is heat transfer from the left end into a surrounding medium at temperature 20°, and the right end is insulated. The initial temperature is $f(x)$ throughout.

5. The left end is at temperature $\sin(\pi t/L)$, the right end is held at zero, and there is heat transfer from the lateral surface of the rod into the surrounding medium held at temperature zero. The initial temperature is $f(x)$ throughout.

6. The ends are insulated, and there is heat transfer from the lateral surface of the rod into the surrounding medium held at temperature 50°. The initial temperature is 100° throughout.

In Problems 7–10, a string of length L coincides with the interval $[0, L]$ on the x-axis. Set up the boundary-value problem for the displacement $u(x, t)$.

7. The ends are secured to the x-axis. The string is released from rest from the initial displacement $x(L - x)$.

8. The ends are secured to the x-axis. Initially the string is undisplaced but has the initial velocity $\sin(\pi x/L)$.

9. The left end is secured to the x-axis, but the right end moves in a transverse manner according to $\sin \pi t$. The string is released from rest from the initial displacement $f(x)$. For $t > 0$ the transverse vibrations are damped with a force proportional to the instantaneous velocity.

10. The ends are secured to the x-axis, and the string is initially at rest on that axis. An external vertical force proportional to the horizontal distance from the left end acts on the string for $t > 0$.

In Problems 11 and 12, set up the boundary-value problem for the steady-state temperature $u(x, y)$.

11. A thin rectangular plate coincides with the region in the xy-plane defined by $0 \le x \le 4, 0 \le y \le 2$. The left end and the bottom of the plate are insulated. The top of the plate is held at temperature zero, and the right end of the plate is held at temperature $f(y)$.

12. A semi-infinite plate coincides with the region defined by $0 \le x \le \pi, y \ge 0$. The left end is held at temperature e^{-y}, and the right end is held at temperature 100° for $0 < y \le 1$ and temperature zero for $y > 1$. The bottom of the plate is held at temperature $f(x)$.

13.3 Heat Equation

INTRODUCTION Consider a thin rod of length L with an initial temperature $f(x)$ throughout and whose ends are held at temperature zero for all time $t > 0$. If the rod shown in **FIGURE 13.3.1** satisfies the assumptions given on page 721, then the temperature $u(x, t)$ in the rod is determined from the boundary-value problem

$$k \frac{\partial^2 u}{\partial x^2} = \frac{\partial u}{\partial t}, \quad 0 < x < L, \ t > 0 \tag{1}$$

$$u(0, t) = 0, \quad u(L, t) = 0, \quad t > 0 \tag{2}$$

$$u(x, 0) = f(x), \quad 0 < x < L. \tag{3}$$

FIGURE 13.3.1 Find the temperature u in a finite rod

In the discussion that follows next we show how to solve this BVP using the method of separation of variables introduced in Section 13.1.

Solution of the BVP Using the product $u(x, t) = X(x)T(t)$, and $-\lambda$ as the separation constant, leads to

$$\frac{X''}{X} = \frac{T'}{kT} = -\lambda \qquad (4)$$

and

$$X'' + \lambda X = 0 \qquad (5)$$

$$T' + k\lambda T = 0. \qquad (6)$$

Now the boundary conditions in (2) become

$$u(0, t) = X(0)T(t) = 0 \quad \text{and} \quad u(L, t) = X(L)T(t) = 0.$$

Since the last equalities must hold for all time t, we must have $X(0) = 0$ and $X(L) = 0$. These homogeneous boundary conditions together with the homogeneous ODE (5) constitute a regular Sturm–Liouville problem:

$$X'' + \lambda X = 0, \quad X(0) = 0, \quad X(L) = 0. \qquad (7)$$

The solution of this BVP was discussed in detail in Example 2 of Section 3.9 and on page 701 of Section 12.5. In that example, we considered three possible cases for the parameter λ: zero, negative, and positive. The corresponding general solutions of the DEs are

$$X(x) = c_1 + c_2 x, \qquad\qquad \lambda = 0 \qquad (8)$$

$$X(x) = c_1 \cosh \alpha x + c_2 \sinh \alpha x, \quad \lambda = -\alpha^2 < 0 \qquad (9)$$

$$X(x) = c_1 \cos \alpha x + c_2 \sin \alpha x, \quad \lambda = \alpha^2 > 0. \qquad (10)$$

Recall, when the boundary conditions $X(0) = 0$ and $X(L) = 0$ are applied to (8) and (9) these solutions yield only $X(x) = 0$ and so we are left with the unusable result $u = 0$. Applying the first boundary condition $X(0) = 0$ to the solution in (10) gives $c_1 = 0$. Therefore $X(x) = c_2 \sin \alpha x$. The second boundary condition $X(L) = 0$ now implies

$$X(L) = c_2 \sin \alpha L = 0. \qquad (11)$$

If $c_2 = 0$, then $X = 0$ so that $u = 0$. But (11) can be satisfied for $c_2 \neq 0$ when $\sin \alpha L = 0$. This last equation implies that $\alpha L = n\pi$ or $\alpha = n\pi/L$, where $n = 1, 2, 3, \ldots$. Hence (7) possesses nontrivial solutions when $\lambda_n = \alpha_n^2 = n^2\pi^2/L^2, n = 1, 2, 3, \ldots$. The values λ_n and the corresponding solutions

$$X(x) = c_2 \sin \frac{n\pi}{L} x, \quad n = 1, 2, 3, \ldots \qquad (12)$$

are the **eigenvalues** and **eigenfunctions**, respectively, of the problem in (7).

The general solution of (6) is $T(t) = c_3 e^{-k(n^2\pi^2/L^2)t}$, and so

$$u_n = X(x)T(t) = A_n e^{-k(n^2\pi^2/L^2)t} \sin \frac{n\pi}{L} x, \qquad (13)$$

where we have replaced the constant $c_2 c_3$ by A_n. The products $u_n(x, t)$ given in (13) satisfy the partial differential equation (1) as well as the boundary conditions (2) for each value of the positive integer n. However, in order for the functions in (13) to satisfy the initial condition (3), we would have to choose the coefficient A_n in such a manner that

$$u_n(x, 0) = f(x) = A_n \sin \frac{n\pi}{L} x. \qquad (14)$$

In general, we would not expect condition (14) to be satisfied for an arbitrary, but reasonable, choice of f. Therefore we are forced to admit that $u_n(x, t)$ *is not a solution of the problem given in* (1)–(3). Now by the superposition principle the function

$$u(x, t) = \sum_{n=1}^{\infty} u_n = \sum_{n=1}^{\infty} A_n e^{-k(n^2\pi^2/L^2)t} \sin \frac{n\pi}{L} x \qquad (15)$$

must also, although formally, satisfy equation (1) and the conditions in (2). If we substitute $t = 0$ into (15), then

$$u(x, 0) = f(x) = \sum_{n=1}^{\infty} A_n \sin \frac{n\pi}{L} x.$$

This last expression is recognized as the half-range expansion of f in a sine series. If we make the identification $A_n = b_n$, $n = 1, 2, 3, \ldots$, it follows from (5) of Section 12.3 that

$$A_n = \frac{2}{L} \int_0^L f(x) \sin \frac{n\pi}{L} x \, dx. \tag{16}$$

We conclude that a solution of the boundary-value problem described in (1), (2), and (3) is given by the infinite series

$$u(x, t) = \frac{2}{L} \sum_{n=1}^{\infty} \left(\int_0^L f(x) \sin \frac{n\pi}{L} x \, dx \right) e^{-k(n^2\pi^2/L^2)t} \sin \frac{n\pi}{L} x. \tag{17}$$

In the special case when the initial temperature is $u(x, 0) = 100$, $L = \pi$, and $k = 1$, you should verify that the coefficients (16) are given by

$$A_n = \frac{200}{\pi} \left[\frac{1 - (-1)^n}{n} \right]$$

and that the series (17) is

$$u(x, t) = \frac{200}{\pi} \sum_{n=1}^{\infty} \left[\frac{1 - (-1)^n}{n} \right] e^{-n^2 t} \sin nx. \tag{18}$$

■ Insulated Boundaries

Another boundary-value problem associated with the heat equation is

$$k \frac{\partial^2 u}{\partial x^2} = \frac{\partial u}{\partial t}, \quad 0 < x < L, \ t > 0$$

$$\left. \frac{\partial u}{\partial x} \right|_{x=0} = 0, \quad \left. \frac{\partial u}{\partial x} \right|_{x=L} = 0, \ t > 0$$

$$u(x, 0) = f(x), \quad 0 < x < L.$$

The boundary conditions in this problem indicate that both ends of the rod in Figure 13.3.1 are insulated. See the discussion on page 723 of Section 13.2. The major change in the solution of this boundary-value problem by separation of variables is that the analogue of (7) is now

$$X'' + \lambda X = 0, \quad X'(0) = 0, \ X'(L) = 0.$$

In solving this type of Sturm–Liouville problem, you are cautioned to pay close attention to the case $\lambda = 0$.

■ Use of Computers

The solution u in (18) is a function of two variables and as such its graph is a surface in 3-space. We could use the 3D-plot application of a computer algebra system to approximate this surface by graphing partial sums $S_n(x, t)$ over a rectangular region defined by $0 \leq x \leq \pi$, $0 \leq t \leq T$. Alternatively, with the aid of the 2D-plot application of a CAS we plot the solution $u(x, t)$ on the x-interval $[0, \pi]$ for increasing values of time t. See **FIGURE 13.3.2(a)**. In Figure 13.3.2(b) the solution $u(x, t)$ is graphed on the t-interval $[0, 6]$ for increasing values of x ($x = 0$ is the left end and $x = \pi/2$ is the midpoint of the rod of length $L = \pi$). Both sets of graphs verify that which is apparent in (18)—namely, $u(x, t) \to 0$ as $t \to \infty$.

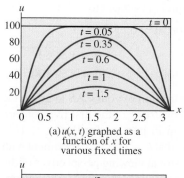

(a) $u(x, t)$ graphed as a function of x for various fixed times

(b) $u(x, t)$ graphed as a function of t for various fixed positions

FIGURE 13.3.2 Graphs obtained using partial sums of (18)

13.3 Exercises Answers to selected odd-numbered problems begin on page ANS-34.

In Problems 1–6, solve the heat equation (1) subject to the given conditions.

1. $u(0, t) = 0$, $u(L, t) = 0$, $t > 0$
 $u(x, 0) = 100$, $0 < x < L$

2. $u(0, t) = 0$, $u(L, t) = 0$, $t > 0$
 $u(x, 0) = x(L - x)$, $0 < x < L$

3. $u(0, t) = 0$, $u(L, t) = 0$, $t > 0$
 $u(x, 0) = \begin{cases} 1, & 0 < x < L/2 \\ 0, & L/2 < x < L \end{cases}$

4. $u(0, t) = 0$, $u(\pi, t) = 0$, $t > 0$
 $u(x, 0) = \begin{cases} x, & 0 < x < \pi/2 \\ \pi - x, & \pi/2 < x < \pi \end{cases}$

5. $\left.\dfrac{\partial u}{\partial x}\right|_{x=0} = 0, \quad \left.\dfrac{\partial u}{\partial x}\right|_{x=L} = 0, \quad t > 0$

$u(x, 0) = f(x), \quad 0 < x < L$

6. $\left.\dfrac{\partial u}{\partial x}\right|_{x=0} = 0, \quad \left.\dfrac{\partial u}{\partial x}\right|_{x=2} = 0, \quad t > 0$

$u(x, 0) = \begin{cases} x, & 0 < x < 1 \\ 0, & 1 < x < 2 \end{cases}$

In Problems 7 and 8, solve the heat equation (1) subject to the given conditions. [*Hint*: A solution of the boundary-value problem (1)–(3) need not be an infinite series.]

7. $u(0, t) = 0, \quad u(\pi, t) = 0, \quad t > 0$

$u(x, 0) = 10 \sin 5x, \quad 0 < x < \pi$

8. $u(0, t) = 0, \quad u(1, t) = 0, \quad t > 0$

$u(x, 0) = 4 \sin 3\pi x + 8 \sin 6\pi x, \quad 0 < x < 1$

9. Suppose heat is lost from the lateral surface of a thin rod of length L into a surrounding medium at temperature zero. If the linear law of heat transfer applies, then the heat equation takes on the form

$$k \dfrac{\partial^2 u}{\partial x^2} - hu = \dfrac{\partial u}{\partial t}, \quad 0 < x < L, \, t > 0,$$

h a constant. Find the temperature $u(x, t)$ if the initial temperature is $f(x)$ throughout and the ends $x = 0$ and $x = L$ are insulated. See **FIGURE 13.3.3**.

insulated $0°$ insulated

$0°$

heat transfer from
lateral surface of
the rod

FIGURE 13.3.3 Rod in Problem 9

10. Solve Problem 9 if the ends $x = 0$ and $x = L$ are held at temperature zero.

11. A thin rod (or wire) coinciding with the x-axis on the interval $[-L, L]$ is bent into a circular ring by joining the two ends $x = -L$ and $x = L$. See **FIGURE 13.3.4**. If the temperature u and the flux of heat $\partial u/\partial x$ are continuous at the point of juncture, then the temperature $u(x, t)$ in the rod satisfies the boundary-value problem

$k \dfrac{\partial^2 u}{\partial x^2} = \dfrac{\partial u}{\partial t}, \quad -L < x < L, \quad t > 0,$

$u(-L, t) = u(L, t), \quad t > 0$

$\left.\dfrac{\partial u}{\partial x}\right|_{x=-L} = \left.\dfrac{\partial u}{\partial x}\right|_{x=L}, \quad t > 0$

$u(x, t) = f(x), \quad -L < x < L.$

See (15) in Section 12.5. Find the temperature $u(x, t)$. [*Hint*: Think full Fourier series.]

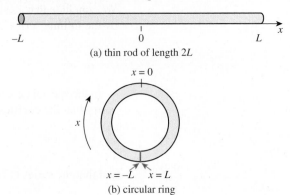

(a) thin rod of length $2L$

(b) circular ring

FIGURE 13.3.4 Thin rod bent into a circular ring in Problem 11

Computer Lab Assignments

12. (a) Solve the heat equation (1) subject to

$u(0, t) = 0, \quad u(100, t) = 0, \quad t > 0$

$u(x, 0) = \begin{cases} 0.8x, & 0 \le x \le 50 \\ 0.8(100 - x), & 50 < x \le 100. \end{cases}$

(b) Use the 3D-plot application of your CAS to graph the partial sum $S_5(x, t)$ consisting of the first five nonzero terms of the solution in part (a) for $0 \le x \le 100$, $0 \le t \le 200$. Assume that $k = 1.6352$. Experiment with various three-dimensional viewing perspectives of the surface (called the **ViewPoint** option in *Mathematica*).

Discussion Problems

13. In Figure 13.3.2(b) we have the graphs of $u(x, t)$ on the interval [0, 6] for $x = 0, x = \pi/12, x = \pi/6, x = \pi/4$, and $x = \pi/2$. Describe or sketch the graphs of $u(x, t)$ on the same time interval but for the fixed values $x = 3\pi/4, x = 5\pi/6$, $x = 11\pi/12$, and $x = \pi$.

13.4 Wave Equation

INTRODUCTION We are now in a position to solve the boundary-value problem (11) discussed in Section 13.2. The vertical displacement $u(x, t)$ of a string of length L that is freely vibrating in the vertical plane shown in Figure 13.2.2(a) is determined from

$$a^2 \dfrac{\partial^2 u}{\partial x^2} = \dfrac{\partial^2 u}{\partial t^2}, \quad 0 < x < L, \quad t > 0 \tag{1}$$

$$u(0, t) = 0, \quad u(L, t) = 0, \quad t > 0 \tag{2}$$

$$u(x, 0) = f(x), \quad \left.\dfrac{\partial u}{\partial t}\right|_{t=0} = g(x), \quad 0 < x < L. \tag{3}$$

Solution of the BVP With the usual assumption that $u(x, t) = X(x)T(t)$, separating variables in (1) gives

$$\frac{X''}{X} = \frac{T''}{a^2 T} = -\lambda$$

so that

$$X'' + \lambda X = 0 \tag{4}$$

$$T'' + a^2 \lambda T = 0. \tag{5}$$

As in Section 13.3, the boundary conditions (2) translate into $X(0) = 0$ and $X(L) = 0$. The ODE in (4) along with these boundary conditions is the regular Sturm–Liouville problem

$$X'' + \lambda X = 0, \quad X(0) = 0, \quad X(L) = 0. \tag{6}$$

Of the usual three possibilities for the parameter λ: $\lambda = 0$, $\lambda = -\alpha^2 < 0$, and $\lambda = \alpha^2 > 0$, only the last choice leads to nontrivial solutions. Corresponding to $\lambda = \alpha^2$, $\alpha > 0$, the general solution of (4) is

$$X(x) = c_1 \cos \alpha x + c_2 \sin \alpha x.$$

$X(0) = 0$ and $X(L) = 0$ indicate that $c_1 = 0$ and $c_2 \sin \alpha L = 0$. The last equation again implies that $\alpha L = n\pi$ or $\alpha = n\pi/L$. The eigenvalues and corresponding eigenfunctions of (6) are $\lambda_n = n^2 \pi^2/L^2$ and $X(x) = c_2 \sin \dfrac{n\pi}{L} x$, $n = 1, 2, 3, \ldots$. The general solution of the second-order equation (5) is then

$$T(t) = c_3 \cos \frac{n\pi a}{L} t + c_4 \sin \frac{n\pi a}{L} t.$$

By rewriting $c_2 c_3$ as A_n and $c_2 c_4$ as B_n, solutions that satisfy both the wave equation (1) and boundary conditions (2) are

$$u_n = \left(A_n \cos \frac{n\pi a}{L} t + B_n \sin \frac{n\pi a}{L} t \right) \sin \frac{n\pi}{L} x \tag{7}$$

and

$$u(x, t) = \sum_{n=1}^{\infty} \left(A_n \cos \frac{n\pi a}{L} t + B_n \sin \frac{n\pi a}{L} t \right) \sin \frac{n\pi}{L} x. \tag{8}$$

Setting $t = 0$ in (8) and using the initial condition $u(x, 0) = f(x)$ gives

$$u(x, 0) = f(x) = \sum_{n=1}^{\infty} A_n \sin \frac{n\pi}{L} x.$$

Since the last series is a half-range expansion for f in a sine series, we can write $A_n = b_n$:

$$A_n = \frac{2}{L} \int_0^L f(x) \sin \frac{n\pi}{L} x \, dx. \tag{9}$$

To determine B_n we differentiate (8) with respect to t and then set $t = 0$:

$$\frac{\partial u}{\partial t} = \sum_{n=1}^{\infty} \left(-A_n \frac{n\pi a}{L} \sin \frac{n\pi a}{L} t + B_n \frac{n\pi a}{L} \cos \frac{n\pi a}{L} t \right) \sin \frac{n\pi}{L} x$$

$$\left. \frac{\partial u}{\partial t} \right|_{t=0} = g(x) = \sum_{n=1}^{\infty} \left(B_n \frac{n\pi a}{L} \right) \sin \frac{n\pi}{L} x.$$

In order for this last series to be the half-range sine expansion of the initial velocity g on the interval, the *total* coefficient $B_n n\pi a/L$ must be given by the form b_n in (5) of Section 12.3—that is,

$$B_n \frac{n\pi a}{L} = \frac{2}{L} \int_0^L g(x) \sin \frac{n\pi}{L} x \, dx,$$

from which we obtain

$$B_n = \frac{2}{n\pi a} \int_0^L g(x) \sin \frac{n\pi}{L} x \, dx. \tag{10}$$

The solution of the boundary-value problem (1)–(3) consists of the series (8) with coefficients A_n and B_n defined by (9) and (10), respectively.

We note that when the string is released from *rest*, then $g(x) = 0$ for every x in the interval $[0, L]$ and consequently $B_n = 0$.

Plucked String

A special case of the boundary-value problem in (1)–(3) is the model of a **plucked string**. When a string is plucked, a point on the string at $x = x_0$ is raised to a height h above the x-axis so that its initial shape is given by the graph of a function of the form

$$
f(x) = \begin{cases} \dfrac{h}{x_0} x, & 0 < x < x_0 \\[2mm] \dfrac{h}{L - x_0}(L - x), & x_0 < x < L. \end{cases}
$$

The point is released from rest at $t = 0$. See Figure 13.2.4 and Problems 7 ($x_0 = L/2$) and 8 ($x_0 = L/3$) in Exercises 13.4. We can see the motion of the string by plotting the solution or displacement $u(x, t)$ for increasing values of time and using the animation feature of a CAS. Some frames of a "movie" generated in this manner are given in **FIGURE 13.4.1**; the initial shape of the string is given in Figure 13.4.1(a). You are asked to emulate the results given in the figure by plotting a sequence of partial sums of (8). See Problem 27 in Exercises 13.4.

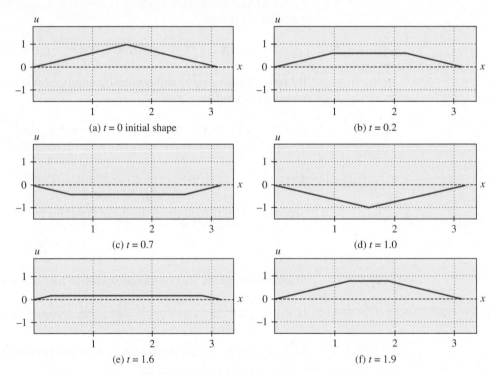

(a) $t = 0$ initial shape

(b) $t = 0.2$

(c) $t = 0.7$

(d) $t = 1.0$

(e) $t = 1.6$

(f) $t = 1.9$

FIGURE 13.4.1 Frames of plucked-string "movie"

Standing Waves

Recall from the derivation of the wave equation in Section 13.2 that the constant a appearing in the solution of the boundary-value problem in (1)–(3) is given by $\sqrt{T/\rho}$, where ρ is mass per unit length and T is the magnitude of the tension in the string. When T is large enough, the vibrating string produces a musical sound. This sound is the result of standing waves. The solution (8) is a superposition of product solutions called **standing waves** or **normal modes**:

$$
u(x, t) = u_1(x, t) + u_2(x, t) + u_3(x, t) + \cdots .
$$

In view of (6) and (7) of Section 3.8, the product solutions (7) can be written as

$$
u_n(x, t) = C_n \sin\left(\frac{n\pi a}{L} t + \phi_n\right) \sin \frac{n\pi}{L} x, \tag{11}
$$

(a) First standing wave

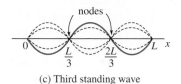

node

(b) Second standing wave

nodes

(c) Third standing wave

FIGURE 13.4.2 First three standing waves

where $C_n = \sqrt{A_n^2 + B_n^2}$ and ϕ_n is defined by $\sin \phi_n = A_n/C_n$ and $\cos \phi_n = B_n/C_n$. For $n = 1, 2, 3, \ldots$ the standing waves are essentially the graphs of $\sin(n\pi x/L)$, with a time-varying amplitude given by

$$C_n \sin\left(\frac{n\pi a}{L}t + \phi_n\right).$$

Alternatively, we see from (11) that at a fixed value of x each product function $u_n(x, t)$ represents simple harmonic motion with amplitude $C_n|\sin(n\pi x/L)|$ and frequency $f_n = na/2L$. In other words, each point on a standing wave vibrates with a different amplitude but with the same frequency. When $n = 1$,

$$u_1(x, t) = C_1 \sin\left(\frac{\pi a}{L}t + \phi_1\right)\sin\frac{\pi}{L}x$$

is called the **first standing wave**, the **first normal mode**, or the **fundamental mode of vibration**. The first three standing waves, or normal modes, are shown in **FIGURE 13.4.2**. The dashed graphs represent the standing waves at various values of time. The points in the interval $(0, L)$, for which $\sin(n\pi/L)x = 0$, correspond to points on a standing wave where there is no motion. These points are called **nodes**. For example, in Figures 13.4.2(b) and (c) we see that the second standing wave has one node at $L/2$ and the third standing wave has two nodes at $L/3$ and $2L/3$. In general, the nth normal mode of vibration has $n - 1$ nodes.

The frequency

$$f_1 = \frac{a}{2L} = \frac{1}{2L}\sqrt{\frac{T}{\rho}}$$

of the first normal mode is called the **fundamental frequency** or **first harmonic** and is directly related to the pitch produced by a stringed instrument. It is apparent that the greater the tension on the string, the higher the pitch of the sound. The frequencies f_n of the other normal modes, which are integer multiples of the fundamental frequency, are called **overtones**. The second harmonic is the first overtone, and so on.

■ **Superposition Principle** The superposition principle, Theorem 13.1.1, is the key in making the method of separation of variables an effective means of solving certain kinds of boundary-value problems involving linear partial differential equations. Sometimes a problem can also be solved by using a superposition of solutions of two easier problems. If we can solve each of the problems,

<table>
<tr><td align="center">Problem 1</td><td align="center">Problem 2</td></tr>
</table>

$$a^2\frac{\partial^2 u_1}{\partial x^2} = \frac{\partial^2 u_1}{\partial t^2}, \quad 0 < x < L, \quad t > 0 \qquad a^2\frac{\partial^2 u_2}{\partial x^2} = \frac{\partial^2 u_2}{\partial t^2}, \quad 0 < x < L, \quad t > 0$$

$$u_1(0, t) = 0, \quad u_1(L, t) = 0, \quad t > 0 \qquad u_2(0, t) = 0, \quad u_2(L, t) = 0, \quad t > 0 \qquad (12)$$

$$u_1(x, 0) = f(x), \quad \left.\frac{\partial u_1}{\partial t}\right|_{t=0} = 0, \quad 0 < x < L \qquad u_2(x, 0) = 0, \quad \left.\frac{\partial u_2}{\partial t}\right|_{t=0} = g(x), \quad 0 < x < L$$

then a solution of (1)–(3) is given by $u(x, t) = u_1(x, t) + u_2(x, t)$. To see this we know that $u(x, t) = u_1(x, t) + u_2(x, t)$ is a solution of the homogeneous equation in (1) because of Theorem 13.1.1. Moreover, $u(x, t)$ satisfies the boundary condition (2) and the initial conditions (3) because, in turn,

$$\text{BC}\begin{cases} u(0, t) = u_1(0, t) + u_2(0, t) = 0 + 0 = 0 \\ u(L, t) = u_1(L, t) + u_2(L, t) = 0 + 0 = 0, \end{cases}$$

and

$$\text{IC}\begin{cases} u(x, 0) = u_1(x, 0) + u_2(x, 0) = f(x) + 0 = f(x) \\ \left.\dfrac{\partial u}{\partial t}\right|_{t=0} = \left.\dfrac{\partial u_1}{\partial t}\right|_{t=0} + \left.\dfrac{\partial u_2}{\partial t}\right|_{t=0} = 0 + g(x) = g(x). \end{cases}$$

You are encouraged to try this method to obtain (8), (9), and (10). See Problems 5 and 14 in Exercises 13.4.

REMARKS

Perhaps you are aware that a troop of soldiers is instructed to break stride, that is, not move in unison, when crossing a long suspension bridge. On April 12, 1831, the Broughton Suspension Bridge in England collapsed while a company of soldiers marched in step across it. On April 16, 1850, the suspension bridge in Angers, France, collapsed as a battalion of soldiers was marching in step across it. There were many injuries but no fatalities in the Broughton bridge collapse, whereas there were 226 persons killed in the Angers bridge collapse. Both collapses were blamed on mechanical resonance between the frequency of the natural vibrations of the bridge and the frequency of the vibrational force of the many feet applied to the bridge in unison.

For a mathematical model for the oscillations of a suspension bridge under an external force, similar to the boundary-value problem in (1), (2), and (3), see Chapter 14 in the text *Topics in Mathematical Modeling* by K. K. Tung (Princeton University Press, 2007).

© goja1/Shutterstock
Soldiers marching in step

Salford Local History Library, retrieved from: https://commons.wikimedia.org/wiki/File:Broughton-suspension-bridge.jpg
Rebuilt Broughton Suspension Bridge

Retrieved from: https://commons.wikimedia.org/wiki/File:Pont1839.jpg
Angers Suspension Bridge

13.4 Exercises Answers to selected odd-numbered problems begin on page ANS-34.

In Problems 1–6, solve the wave equation (1) subject to the given conditions.

1. $u(0, t) = 0$, $u(L, t) = 0$, $t > 0$

$u(x, 0) = \dfrac{1}{4}x(L - x)$, $\dfrac{\partial u}{\partial t}\bigg|_{t=0} = 0$, $0 < x < L$

2. $u(0, t) = 0$, $u(L, t) = 0$, $t > 0$

$u(x, 0) = 0$, $\dfrac{\partial u}{\partial t}\bigg|_{t=0} = x(L - x)$, $0 < x < L$

3. $u(0, t) = 0$, $u(\pi, t) = 0$, $t > 0$

$u(x, 0) = 0$, $\dfrac{\partial u}{\partial t}\bigg|_{t=0} = \sin x$, $0 < x < \pi$

4. $u(0, t) = 0$, $u(\pi, t) = 0$, $t > 0$

$u(x, 0) = \dfrac{1}{6}x(\pi^2 - x^2)$, $\dfrac{\partial u}{\partial t}\bigg|_{t=0} = 0$, $0 < x < \pi$

5. $u(0, t) = 0$, $u(1, t) = 0$, $t > 0$

$u(x, 0) = x(1 - x)$, $\dfrac{\partial u}{\partial t}\bigg|_{t=0} = x(1 - x)$, $0 < x < 1$

6. $u(0, t) = 0$, $u(1, t) = 0$, $t > 0$

$u(x, 0) = 0.01 \sin 3\pi x$, $\dfrac{\partial u}{\partial t}\bigg|_{t=0} = 0$, $0 < x < 1$

In Problems 7–10, a string is secured to the *x*-axis at $x = 0$ and at $x = L$ and its initial displacement $u(x, 0) = f(x)$, $0 < x < L$, is shown in the figure. Find $u(x, t)$ if the string is released from rest.

7.

FIGURE 13.4.3 Initial displacement for Problem 7

8.

FIGURE 13.4.4 Initial displacement for Problem 8

9.

FIGURE 13.4.5 Initial displacement for Problem 9

10.

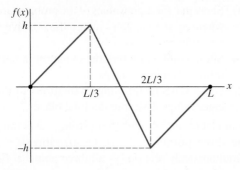

FIGURE 13.4.6 Initial displacement for Problem 10

11. The longitudinal displacement of a vibrating elastic bar shown in **FIGURE 13.4.7** satisfies the wave equation (1) and the conditions

$$\frac{\partial u}{\partial x}\bigg|_{x=0} = 0, \quad \frac{\partial u}{\partial x}\bigg|_{x=L} = 0, \ t > 0$$

$$u(x, 0) = x, \quad \frac{\partial u}{\partial t}\bigg|_{t=0} = 0, \ 0 < x < L.$$

The boundary conditions at $x = 0$ and $x = L$ are called **free-end conditions**. Find the displacement $u(x, t)$.

FIGURE 13.4.7 Elastic bar in Problem 11

12. A model for the motion of a vibrating string whose ends are allowed to slide on frictionless sleeves attached to the vertical axes $x = 0$ and $x = L$ is given by the wave equation (1) and the conditions

$$\frac{\partial u}{\partial x}\bigg|_{x=0} = 0, \quad \frac{\partial u}{\partial x}\bigg|_{x=L} = 0, \ t > 0$$

$$u(x, 0) = f(x), \quad \frac{\partial u}{\partial t}\bigg|_{t=0} = g(x), \ 0 < x < L.$$

See **FIGURE 13.4.8**. The boundary conditions indicate that the motion is such that the slope of the curve is zero at its ends for $t > 0$. Find the displacement $u(x, t)$.

FIGURE 13.4.8 String whose ends are attached to frictionless sleeves in Problem 12

13. In Problem 10, determine the value of $u(L/2, t)$ for $t \geq 0$.

14. Rederive the results given in (8), (9), and (10), but this time use the superposition principle discussed on page 731.

15. A string stretched along the x-axis and secured at $x = 0$ and $x = \pi$ for $t > 0$ is given an initial displacement and released. If the transverse vibrations take place in a medium that imparts a **linear damping** proportional to the instantaneous velocity, then the wave equation takes on the form

$$\frac{\partial^2 u}{\partial x^2} = \frac{\partial^2 u}{\partial t^2} + 2\beta \frac{\partial u}{\partial t}, \quad 0 < x < \pi, \quad t > 0,$$

where β is a constant satisfying $0 < \beta < 1$. Find the displacement $u(x, t)$ if the boundary and initial conditions are

$$u(0, t) = 0, \quad u(\pi, t) = 0, \quad t > 0$$

$$u(x, 0) = f(x), \quad \frac{\partial u}{\partial t}\bigg|_{t=0} = 0, \quad 0 < x < \pi.$$

16. Show that a solution of the boundary-value problem

$$\frac{\partial^2 u}{\partial x^2} = \frac{\partial^2 u}{\partial t^2} + u, \quad 0 < x < \pi, \ t > 0$$

$$u(0, t) = 0, \quad u(\pi, t) = 0, \ t > 0$$

$$u(x, 0) = \begin{cases} x, & 0 < x < \pi/2 \\ \pi - x, & \pi/2 \leq x < \pi \end{cases}$$

$$\frac{\partial u}{\partial t}\bigg|_{t=0} = 0, \quad 0 < x < \pi$$

is

$$u(x, t) = \frac{4}{\pi} \sum_{k=1}^{\infty} \frac{(-1)^{k+1}}{(2k-1)^2} \sin(2k-1)x \cos\sqrt{(2k-1)^2 + 1}\,t.$$

17. Consider the boundary-value problem given in (1)–(3) of this section. If $g(x) = 0$ on $0 < x < L$, show that the solution of the problem can be written as

$$u(x, t) = \frac{1}{2}[f(x + at) + f(x - at)].$$

[*Hint*: Use the identity

$$2 \sin \theta_1 \cos \theta_2 = \sin(\theta_1 + \theta_2) + \sin(\theta_1 - \theta_2).]$$

18. The vertical displacement $u(x, t)$ of an infinitely long string is determined from the initial-value problem

$$a^2 \frac{\partial^2 u}{\partial x^2} = \frac{\partial^2 u}{\partial t^2}, \quad -\infty < x < \infty, \ t > 0 \tag{13}$$

$$u(x, 0) = f(x), \quad \frac{\partial u}{\partial t}\bigg|_{t=0} = g(x).$$

This problem can be solved without separating variables.

(a) Show that the wave equation can be put into the form $\partial^2 u/\partial \eta \, \partial \xi = 0$ by means of the substitutions $\xi = x + at$ and $\eta = x - at$.

(b) Integrate the partial differential equation in part (a), first with respect to η and then with respect to ξ, to show that $u(x, t) = F(x + at) + G(x - at)$, where F and G are arbitrary twice differentiable functions, is a

solution of the wave equation. Use this solution and the given initial conditions to show that

$$F(x) = \frac{1}{2} f(x) + \frac{1}{2a} \int_{x_0}^{x} g(s)\, ds + c$$

and $\quad G(x) = \frac{1}{2} f(x) - \frac{1}{2a} \int_{x_0}^{x} g(s)\, ds - c,$

where x_0 is arbitrary and c is a constant of integration.

(c) Use the results in part (b) to show that

$$u(x, t) = \frac{1}{2}\left[f(x + at) + f(x - at)\right] + \frac{1}{2a} \int_{x-at}^{x+at} g(s)\, ds. \quad (14)$$

Note that when the initial velocity $g(x) = 0$ we obtain

$$u(x, t) = \frac{1}{2}\left[f(x + at) + f(x - at)\right], \quad -\infty < x < \infty.$$

The last solution can be interpreted as a superposition of two **traveling waves**, one moving to the right (that is, $\frac{1}{2} f(x - at)$) and one moving to the left ($\frac{1}{2} f(x + at)$). Both waves travel with speed a and have the same basic shape as the initial displacement $f(x)$. The form of $u(x, t)$ given in (14) is called **d'Alembert's solution** after the French mathematician, physicist, philosopher, and music theorist **Jean-Baptiste le Rond d'Alembert** (1717–1783).

In Problems 19–21, use d'Alembert's solution (14) to solve the initial-value problem in Problem 18 subject to the given initial conditions.

19. $f(x) = \sin x, g(x) = 1$

20. $f(x) = \sin x, g(x) = \cos x$

21. $f(x) = 0, g(x) = \sin 2x$

22. Suppose $f(x) = 1/(1 + x^2), g(x) = 0$, and $a = 1$ for the initial-value problem given in Problem 18. Graph d'Alembert's solution in this case at the time $t = 0, t = 1$, and $t = 3$.

23. The transverse displacement $u(x, t)$ of a vibrating beam of length L is determined from a fourth-order partial differential equation

$$a^2 \frac{\partial^4 u}{\partial x^4} + \frac{\partial^2 u}{\partial t^2} = 0, \ 0 < x < L, \ t > 0.$$

If the beam is **simply supported**, as shown in **FIGURE 13.4.9**, the boundary and initial conditions are

$$u(0, t) = 0, \qquad u(L, t) = 0, \ t > 0$$

$$\left.\frac{\partial^2 u}{\partial x^2}\right|_{x=0} = 0, \qquad \left.\frac{\partial^2 u}{\partial x^2}\right|_{x=L} = 0, \ t > 0$$

$$u(x, 0) = f(x), \qquad \left.\frac{\partial u}{\partial t}\right|_{t=0} = g(x), \ 0 < x < L.$$

Solve for $u(x, t)$. [*Hint*: For convenience use $\lambda = \alpha^4$ when separating variables.]

FIGURE 13.4.9 Simply supported beam in Problem 23

24. If the ends of the beam in Problem 23 are **embedded** at $x = 0$ and $x = L$, the boundary conditions become, for $t > 0$,

$$u(0, t) = 0, \qquad u(L, t) = 0$$

$$\left.\frac{\partial u}{\partial x}\right|_{x=0} = 0, \qquad \left.\frac{\partial u}{\partial x}\right|_{x=L} = 0.$$

(a) Show that the eigenvalues of the problem are $\lambda = x_n^2/L^2$ where $x_n, n = 1, 2, 3, \ldots$, are the positive roots of the equation $\cosh x \cos x = 1$.

(b) Show graphically that the equation in part (a) has an infinite number of roots.

(c) Use a CAS to find approximations to the first four eigenvalues. Use four decimal places.

25. A model for an infinitely long string that is initially held at the three points $(-1, 0), (1, 0)$, and $(0, 1)$ and then simultaneously released at all three points at time $t = 0$ is given by (13) with

$$f(x) = \begin{cases} 1 - |x|, & |x| \le 1 \\ 0, & |x| > 1 \end{cases} \quad \text{and} \quad g(x) = 0.$$

(a) Plot the initial position of the string on the interval $[-6, 6]$.

(b) Use a CAS to plot d'Alembert's solution (14) on $[-6, 6]$ for $t = 0.2k, k = 0, 1, 2, \ldots, 25$. Assume that $a = 1$.

(c) Use the animation feature of your computer algebra system to make a movie of the solution. Describe the motion of the string over time.

26. An infinitely long string coinciding with the x-axis is struck at the origin with a hammer whose head is 0.2 inch in diameter. A model for the motion of the string is given by (13) with

$$f(x) = 0 \quad \text{and} \quad g(x) = \begin{cases} 1, & |x| \le 0.1 \\ 0, & |x| > 0.1. \end{cases}$$

(a) Use a CAS to plot d'Alembert's solution (14) on $[-6, 6]$ for $t = 0.2k, k = 0, 1, 2, \ldots, 25$. Assume that $a = 1$.

(b) Use the animation feature of your computer algebra system to make a movie of the solution. Describe the motion of the string over time.

27. The model of the vibrating string in Problem 7 is a **plucked string**.

(a) Use a CAS to plot the partial sum $S_6(x, t)$; that is, the first six nonzero terms of your solution $u(x, t)$, for $t = 0.1k, k = 0, 1, 2, \ldots, 20$. Assume that $a = 1, h = 1$, and $L = \pi$.

(b) Use the animation feature of your computer algebra system to make a movie of the solution to Problem 7.

28. Consider the vibrating string in Problem 10. Use a CAS to plot the partial sum $S_6(x, t)$; that is, the first six nonzero terms of your solution $u(x, t)$ for $t = 0.25k, k = 0, 2, 3, 4, 6, 8, 10, 14$. Assume that $a = 1, h = 1$, and $L = \pi$. Then superimpose the eight graphs on the same coordinate system.

13.5 Laplace's Equation

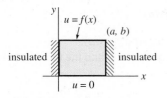

FIGURE 13.5.1 Find the temperature u in a rectangular plate

INTRODUCTION Suppose we wish to find the steady-state temperature $u(x, y)$ in a rectangular plate whose vertical edges $x = 0$ and $x = a$ are insulated, and whose upper and lower edges $y = b$ and $y = 0$ are maintained at temperatures $f(x)$ and 0, respectively. See **FIGURE 13.5.1**. When no heat escapes from the lateral faces of the plate, we solve the following boundary-value problem:

$$\frac{\partial^2 u}{\partial x^2} + \frac{\partial^2 u}{\partial y^2} = 0, \quad 0 < x < a, \quad 0 < y < b \tag{1}$$

$$\left.\frac{\partial u}{\partial x}\right|_{x=0} = 0, \quad \left.\frac{\partial u}{\partial x}\right|_{x=a} = 0, \quad 0 < y < b \tag{2}$$

$$u(x, 0) = 0, \quad u(x, b) = f(x), \quad 0 < x < a. \tag{3}$$

■ **Solution of the BVP** With $u(x, y) = X(x)Y(y)$, separation of variables in (1) leads to

$$\frac{X''}{X} = -\frac{Y''}{Y} = -\lambda$$

$$X'' + \lambda X = 0 \tag{4}$$

$$Y'' - \lambda Y = 0. \tag{5}$$

The three homogeneous boundary conditions in (2) and (3) translate into $X'(0) = 0$, $X'(a) = 0$, and $Y(0) = 0$. The Sturm–Liouville problem associated with the equation in (4) is then

$$X'' + \lambda X = 0, \quad X'(0) = 0, \quad X'(a) = 0. \tag{6}$$

Examination of the cases corresponding to $\lambda = 0$, $\lambda = -\alpha^2 < 0$, and $\lambda = \alpha^2 > 0$, where $\alpha > 0$, has already been carried out in Example 1 in Section 12.5. For convenience a shortened version of that analysis follows.

For $\lambda = 0$, (6) becomes

$$X'' = 0, \quad X'(0) = 0, \quad X'(a) = 0.$$

The solution of the ODE is $X = c_1 + c_2 x$. The boundary condition $X'(0) = 0$ then implies $c_2 = 0$, and so $X = c_1$. Note that for any c_1, this constant solution satisfies the second boundary condition $X'(a) = 0$. By imposing $c_1 \neq 0$, $X = c_1$ is a nontrivial solution of the BVP (6). For $\lambda = -\alpha^2 < 0$, (6) possesses no nontrivial solution. For $\lambda = \alpha^2 > 0$, (6) becomes

$$X'' + \alpha^2 X = 0, \quad X'(0) = 0, \quad X'(a) = 0.$$

Applying the boundary condition $X'(0) = 0$ the solution $X = c_1 \cos \alpha x + c_2 \sin \alpha x$ implies $c_2 = 0$ and so $X = c_1 \cos \alpha x$. The second boundary condition $X'(a) = 0$ applied to this last expression then gives $-c_1 \alpha \sin \alpha a = 0$. Because $\alpha > 0$, the last equation is satisfied when $\alpha a = n\pi$ or $\alpha = n\pi/a, n = 1, 2, \ldots$. The eigenvalues of (6) are then λ_0 and $\lambda_n = \alpha_n^2 = n^2 \pi^2/a^2, n = 1, 2, \ldots$. By corresponding $\lambda_0 = 0$ with $n = 0$, the eigenfunctions of (6) are

$$X = c_1, \, n = 0, \quad \text{and} \quad X = c_1 \cos \frac{n\pi}{a} x, \quad n = 1, 2, \ldots.$$

We must now solve equation (5) subject to the single homogeneous boundary condition $Y(0) = 0$. First, for $\lambda_0 = 0$ the DE in (5) is simply $Y'' = 0$, and thus its solution is $Y = c_3 + c_4 y$. But $Y(0) = 0$ implies $c_3 = 0$ so $Y = c_4 y$. Second, for $\lambda_n = n^2 \pi^2/a^2$, the DE in (5) is $Y'' - \dfrac{n^2 \pi^2}{a^2} Y = 0$.

Because $0 < y < b$ is a finite interval, we write the general solution in terms of hyperbolic functions:

$$Y(y) = c_3 \cosh(n\pi y/a) + c_4 \sinh(n\pi y/a).$$

Why hyperbolic functions?
See pages 127 and 701.

From this solution we see $Y(0) = 0$ again implies $c_3 = 0$ so $Y = c_4 \sinh(n\pi y/a)$.

Thus product solutions $u_n = X(x)Y(y)$ that satisfy Laplace's equation (1) and the three homogeneous boundary conditions in (2) and (3) are

$$A_0 y, \quad n = 0, \quad \text{and} \quad A_n \sinh \frac{n\pi}{a} y \cos \frac{n\pi}{a} x, \quad n = 1, 2, \ldots,$$

where we have rewritten $c_1 c_4$ as A_0 for $n = 0$ and as A_n for $n = 1, 2, \ldots$.

The superposition principle yields another solution

$$u(x, y) = A_0 y + \sum_{n=1}^{\infty} A_n \sinh \frac{n\pi}{a} y \cos \frac{n\pi}{a} x. \tag{7}$$

Finally, by substituting $y = b$ in (7) we see

$$u(x, b) = f(x) = A_0 b + \sum_{n=1}^{\infty} \left(A_n \sinh \frac{n\pi}{a} b \right) \cos \frac{n\pi}{a} x$$

is a half-range expansion of f in a Fourier cosine series. If we make the identifications $A_0 b = a_0/2$ and $A_n \sinh(n\pi b/a) = a_n$, $n = 1, 2, \ldots$, it follows from (2) and (3) of Section 12.3 that

$$2A_0 b = \frac{2}{a} \int_0^a f(x) \, dx$$

$$A_0 = \frac{1}{ab} \int_0^a f(x) \, dx \tag{8}$$

and

$$A_n \sinh \frac{n\pi}{a} b = \frac{2}{a} \int_0^a f(x) \cos \frac{n\pi}{a} x \, dx$$

$$A_n = \frac{2}{a \sinh \dfrac{n\pi}{a} b} \int_0^a f(x) \cos \frac{n\pi}{a} x \, dx. \tag{9}$$

The solution of the boundary-value problem (1)–(3) consists of the series in (7), with coefficients A_0 and A_n defined in (8) and (9), respectively.

Dirichlet Problem

A boundary-value problem in which we seek a solution to an elliptic partial differential equation such as Laplace's equation $\nabla^2 u = 0$ within a region R (in the plane or 3-space) such that u takes on prescribed values on the entire boundary of the region is called a **Dirichlet problem**. This problem was named after the German mathematician **Johann Peter Gustav Lejeune Dirichlet** (1805–1859). In addition to being remembered for his contributions to the theory of Fourier series, Dirichlet is credited with being one of the first mathematicians to give the formal definition of a function as we know it today. In Problem 1 in Exercises 13.5 you are asked to show that the solution of the Dirichlet problem for a rectangular region

$$\frac{\partial^2 u}{\partial x^2} + \frac{\partial^2 u}{\partial y^2} = 0, \quad 0 < x < a, \quad 0 < y < b$$

$$u(0, y) = 0, \quad u(a, y) = 0$$

$$u(x, 0) = 0, \quad u(x, b) = f(x)$$

is

$$u(x, y) = \sum_{n=1}^{\infty} A_n \sinh \frac{n\pi}{a} y \sin \frac{n\pi}{a} x \quad \text{where} \quad A_n = \frac{2}{a \sinh \dfrac{n\pi b}{a}} \int_0^a f(x) \sin \frac{n\pi}{a} x \, dx. \tag{10}$$

(a) Surface

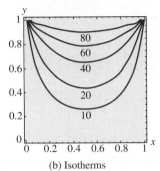
(b) Isotherms

FIGURE 13.5.2 Surface is graph of partial sums when $f(x) = 100$ and $a = b = 1$ in (10)

In the special case when $f(x) = 100$, $a = 1$, $b = 1$, the coefficients A_n are given by $A_n = 200 \dfrac{1 - (-1)^n}{n\pi \sinh n\pi}$. With the help of a CAS the plot of the surface defined by $u(x, y)$ over the region R: $0 \leq x \leq 1$, $0 \leq y \leq 1$ is given in **FIGURE 13.5.2(a)**. You can see in the figure that boundary conditions are satisfied; especially note that along $y = 1$, $u = 100$ for $0 \leq x \leq 1$. The isotherms, or curves, in the rectangular region along which the temperature $u(x, y)$ is constant can be obtained using the contour plotting capabilities of a CAS and are illustrated in Figure 13.5.2(b). The isotherms can also be visualized as the curves of intersection (projected into the xy-plane) of horizontal planes $u = 80$, $u = 60$, and so on, with the surface in Figure 13.5.2(a). Notice that throughout the region the maximum temperature is $u = 100$ and occurs on the portion of the boundary corresponding to $y = 1$. This is no coincidence. There is a **maximum principle** that states a solution u of Laplace's equation within a bounded region R with boundary B (such as a rectangle, circle, sphere, and so on) takes on its maximum and minimum values on B. In addition, it can be proved that u can have no relative extrema (maxima or minima) in the interior of R. This last statement is clearly borne out by the surface shown in Figure 13.5.2(a).

■ **Superposition Principle** A Dirichlet problem for a rectangle can be readily solved by separation of variables when homogeneous boundary conditions are specified on two *parallel* boundaries. However, the method of separation of variables is not applicable to a Dirichlet problem when the boundary conditions on all four sides of the rectangle are nonhomogeneous. To get around this difficulty we break the boundary-value problem

$$\frac{\partial^2 u}{\partial x^2} + \frac{\partial^2 u}{\partial y^2} = 0, \quad 0 < x < a, \ 0 < y < b$$

$$u(0, y) = F(y), \quad u(a, y) = G(y), \ 0 < y < b \tag{11}$$

$$u(x, 0) = f(x), \quad u(x, b) = g(x), \ 0 < x < a$$

into two problems, each of which has homogeneous boundary conditions on parallel boundaries, as shown.

Problem 1

$$\frac{\partial^2 u_1}{\partial x^2} + \frac{\partial^2 u_1}{\partial y^2} = 0, \quad 0 < x < a, \ 0 < y < b$$

$$u_1(0, y) = 0, \quad u_1(a, y) = 0, \ 0 < y < b$$

$$u_1(x, 0) = f(x), \quad u_1(x, b) = g(x), \ 0 < x < a$$

Problem 2

$$\frac{\partial^2 u_2}{\partial x^2} + \frac{\partial^2 u_2}{\partial y^2} = 0, \quad 0 < x < a, \ 0 < y < b$$

$$u_2(0, y) = F(y), \quad u_2(a, y) = G(y), \ 0 < y < b$$

$$u_2(x, 0) = 0, \quad u_2(x, b) = 0, \ 0 < x < a$$

Suppose u_1 and u_2 are the solutions of Problems 1 and 2, respectively. If we define $u(x, y) = u_1(x, y) + u_2(x, y)$, it is seen that u satisfies all boundary conditions in the original problem (11). For example,

$$u(0, y) = u_1(0, y) + u_2(0, y) = 0 + F(y) = F(y)$$

$$u(x, b) = u_1(x, b) + u_2(x, b) = g(x) + 0 = g(x)$$

and so on. Furthermore, u is a solution of Laplace's equation by Theorem 13.1.1. In other words, by solving Problems 1 and 2 and adding their solutions we have solved the original problem. This additive property of solutions is known as the superposition principle. See **FIGURE 13.5.3**.

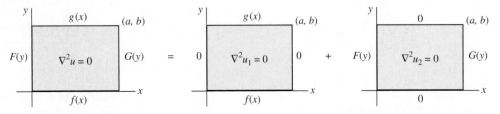

FIGURE 13.5.3 Solution u = Solution u_1 of Problem 1 + Solution u_2 of Problem 2

We leave as exercises (see Problems 13 and 14 in Exercises 13.5) to show that a solution of Problem 1 is

$$u_1(x, y) = \sum_{n=1}^{\infty}\left[A_n \cosh\frac{n\pi}{a}y + B_n \sinh\frac{n\pi}{a}y\right]\sin\frac{n\pi}{a}x,$$

where

$$A_n = \frac{2}{a}\int_0^a f(x)\sin\frac{n\pi}{a}x\,dx$$

$$B_n = \frac{1}{\sinh\dfrac{n\pi}{a}b}\left(\frac{2}{a}\int_0^a g(x)\sin\frac{n\pi}{a}x\,dx - A_n\cosh\frac{n\pi}{a}b\right),$$

and that a solution of Problem 2 is

$$u_2(x, y) = \sum_{n=1}^{\infty}\left[A_n\cosh\frac{n\pi}{b}x + B_n\sinh\frac{n\pi}{b}x\right]\sin\frac{n\pi}{b}y,$$

where

$$A_n = \frac{2}{b}\int_0^b F(y)\sin\frac{n\pi}{b}y\,dy$$

$$B_n = \frac{1}{\sinh\dfrac{n\pi}{b}a}\left(\frac{2}{b}\int_0^b G(y)\sin\frac{n\pi}{b}y\,dy - A_n\cosh\frac{n\pi}{b}a\right).$$

13.5 Exercises Answers to selected odd-numbered problems begin on page ANS-34.

In Problems 1–10, solve Laplace's equation (1) for a rectangular plate subject to the given boundary conditions.

1. $u(0, y) = 0$, $u(a, y) = 0$

$u(x, 0) = 0$, $u(x, b) = f(x)$

2. $u(0, y) = 0$, $u(a, y) = 0$

$\dfrac{\partial u}{\partial y}\bigg|_{y=0} = 0$, $u(x, b) = f(x)$

3. $u(0, y) = 0$, $u(a, y) = 0$

$u(x, 0) = f(x)$, $u(x, b) = 0$

4. $\dfrac{\partial u}{\partial x}\bigg|_{x=0} = 0$, $\dfrac{\partial u}{\partial x}\bigg|_{x=a} = 0$

$u(x, 0) = x$, $u(x, b) = 0$

5. $u(0, y) = 0$, $u(1, y) = 1 - y$

$\dfrac{\partial u}{\partial y}\bigg|_{y=0} = 0$, $\dfrac{\partial u}{\partial y}\bigg|_{y=1} = 0$

6. $u(0, y) = g(y)$, $\dfrac{\partial u}{\partial x}\bigg|_{x=1} = 0$

$\dfrac{\partial u}{\partial y}\bigg|_{y=0} = 0$, $\dfrac{\partial u}{\partial y}\bigg|_{y=\pi} = 0$

7. $\dfrac{\partial u}{\partial x}\bigg|_{x=0} = u(0, y)$, $u(\pi, y) = 1$

$u(x, 0) = 0$, $u(x, \pi) = 0$

8. $u(0, y) = 0$, $u(1, y) = 0$

$\dfrac{\partial u}{\partial y}\bigg|_{y=0} = u(x, 0)$, $u(x, 1) = f(x)$

9. $u(0, y) = 0$, $u(1, y) = 0$

$u(x, 0) = 100$, $u(x, 1) = 200$

10. $u(0, y) = 10y$, $\dfrac{\partial u}{\partial x}\bigg|_{x=1} = -1$

$u(x, 0) = 0$, $u(x, 1) = 0$

In Problems 11 and 12, solve Laplace's equation (1) for the semi-infinite plate extending in the positive y-direction. In each case assume that $u(x, y)$ is bounded at $y \to \infty$.

11.

12.

FIGURE 13.5.4 Semi-infinite plate in Problem 11

FIGURE 13.5.5 Semi-infinite plate in Problem 12

In Problems 13 and 14, solve Laplace's equation (1) for a rectangular plate subject to the given boundary conditions.

13. $u(0, y) = 0$, $u(a, y) = 0$

$u(x, 0) = f(x)$, $u(x, b) = g(x)$

14. $u(0, y) = F(y)$, $u(a, y) = G(y)$

$u(x, 0) = 0$, $u(x, b) = 0$

In Problems 15 and 16, use the superposition principle to solve Laplace's equation (1) for a square plate subject to the given boundary conditions.

15. $u(0, y) = 1$, $u(\pi, y) = 1$

$u(x, 0) = 0$, $u(x, \pi) = 1$

16. $u(0, y) = 0$, $u(2, y) = y(2 - y)$

$u(x, 0) = 0$, $u(x, 2) = \begin{cases} x, & 0 < x < 1 \\ 2 - x, & 1 \le x < 2 \end{cases}$

17. In Problem 16, what is the maximum value of the temperature u for $0 \le x \le 2, 0 \le y \le 2$?

Computer Lab Assignments

18. (a) In Problem 1 suppose $a = b = \pi$ and $f(x) = 100x(\pi - x)$. Without using the solution $u(x, y)$ sketch, by hand, what the surface would look like over the rectangular region defined by $0 \le x \le \pi, 0 \le y \le \pi$.

(b) What is the maximum value of the temperature u for $0 \le x \le \pi, 0 \le y \le \pi$?

(c) Use the information in part (a) to compute the coefficients for your answer in Problem 1. Then use the 3D-plot application of your CAS to graph the partial sum $S_5(x, y)$ consisting of the first five nonzero terms of the solution in part (a) for $0 \le x \le \pi, 0 \le y \le \pi$. Use different perspectives and then compare with part (a).

19. (a) Use the contour-plot application of your CAS to graph the isotherms $u = 170, 140, 110, 80, 60, 30$ for the solution of Problem 9. Use the partial sum $S_5(x, y)$ consisting of the first five nonzero terms of the solution.

(b) Use the 3D-plot application of your CAS to graph the partial sum $S_5(x, y)$.

20. Use the contour-plot application of your CAS to graph the isotherms $u = 2, 1, 0.5, 0.2, 0.1, 0.05, 0, -0.05$ for the solution of Problem 10. Use the partial sum $S_5(x, y)$ consisting of the first five nonzero terms of the solution.

Discussion Problems

21. Solve the **Neumann problem** for a rectangle:

$$\frac{\partial^2 u}{\partial x^2} + \frac{\partial^2 u}{\partial y^2} = 0, \ 0 < x < a, \ 0 < y < b$$

$$\left.\frac{\partial u}{\partial y}\right|_{y=0} = 0, \ \left.\frac{\partial u}{\partial y}\right|_{y=b} = 0, \ 0 < x < a$$

$$\left.\frac{\partial u}{\partial x}\right|_{x=0} = 0, \ \left.\frac{\partial u}{\partial x}\right|_{x=a} = g(y), \ 0 < y < b.$$

(a) Explain why a necessary condition for a solution u to exist is that g satisfy

$$\int_0^b g(y)\,dy = 0.$$

This is sometimes called a **compatibility condition**. Do some extra reading and explain the compatibility condition on physical grounds.

(b) If u is a solution of the BVP, explain why $u + c$, where c is an arbitrary constant, is also a solution.

22. Consider the boundary-value problem

$$\frac{\partial^2 u}{\partial x^2} + \frac{\partial^2 u}{\partial y^2} = 0, \ 0 < x < 1, \ 0 < y < \pi$$

$$u(0, y) = u_0 \cos y, \ u(1, y) = u_0(1 + \cos 2y)$$

$$\left.\frac{\partial u}{\partial y}\right|_{y=0} = 0, \ \left.\frac{\partial u}{\partial y}\right|_{y=\pi} = 0.$$

Discuss how the following answer was obtained:

$$u(x, y) = u_0 x + \frac{u_0}{\sinh 1}\sinh(1 - x)\cos y + \frac{u_0}{\sinh 2}\sinh 2x \cos 2y.$$

Carry out your ideas.

13.6 Nonhomogeneous Boundary-Value Problems

INTRODUCTION A boundary-value problem is said to be **nonhomogeneous** if either the partial differential equation or the boundary conditions are nonhomogeneous. The method of separation of variables employed in the preceding three sections may not be applicable to a nonhomogeneous boundary-value problem *directly*. In the first of the two techniques examined in this section we employ a change of dependent variable $u = v + \psi$ that transforms a nonhomogeneous boundary-value problem into two BVPs: one involving an ODE and the other involving a PDE. The latter problem is homogeneous and solvable by separation of variables. The second technique may also start with a change of a dependent variable, but is basically a frontal attack on the BVP using orthogonal series expansions.

The two solution methods that follow are distinguished by different types of nonhomogeneous boundary-value problems.

Time-Independent PDE and BCs We first consider a BVP involving a *time-independent nonhomogeneous equation* and *time-independent boundary conditions*. An example of such a problem is

$$k\frac{\partial^2 u}{\partial x^2} + F(x) = \frac{\partial u}{\partial t}, \quad 0 < x < L, \ t > 0$$

$$u(0, t) = u_0, \quad u(L, t) = u_1, \ t > 0 \tag{1}$$

$$u(x, 0) = f(x), \quad 0 < x < L,$$

where $k > 0$ is a constant. We can interpret (1) as a model for the temperature distribution $u(x, t)$ within a rod of length L where heat is being generated internally throughout the rod by either electrical or chemical means at a rate $F(x)$ and the boundaries $x = 0$ and $x = L$ are held at constant temperatures u_0 and u_1, respectively. When heat is generated at a constant rate r within the rod, the heat equation in (1) takes on the form

$$k\frac{\partial^2 u}{\partial x^2} + r = \frac{\partial u}{\partial t}. \tag{2}$$

Now equation (2) is readily shown not to be separable. On the other hand, suppose we wish to solve the usual homogeneous heat equation $ku_{xx} = u_t$ when the boundaries $x = 0$ and $x = L$ are held at, say, nonzero temperatures. Even though the substitution $u(x, t) = X(x)T(t)$ separates the PDE, we quickly find ourselves at an impasse in determining eigenvalues and eigenfunctions because no conclusion about the values of $X(0)$ and $X(L)$ can be drawn from $u(x, 0) = X(0)T(t) = u_0$ and $u(x, L) = X(L)T(t) = u_1$.

By changing the dependent variable u to a new dependent variable v by the substitution $u(x, t) = v(x, t) + \psi(x)$, (1) can be reduced to two problems:

Problem A: $\{k\psi'' + F(x) = 0, \quad \psi(0) = u_0, \quad \psi(L) = u_1$

Problem B: $\begin{cases} k\dfrac{\partial^2 v}{\partial x^2} = \dfrac{\partial v}{\partial t}, \\ v(0, t) = 0, \quad v(L, t) = 0 \\ v(x, 0) = f(x) - \psi(x) \end{cases}$

Observe in Problem A that the simple ODE $k\psi'' + F(x) = 0$ can be solved by integration. Moreover, Problem B is a homogeneous BVP that can be solved straight away by the method of separation of variables. A solution of the given nonhomogeneous problem is then a superposition of solutions:

Solution $u =$ Solution ψ of Problem A + Solution v of Problem B.

There is nothing in the above discussion that should be memorized; you should work through the substitution $u(x, t) = v(x, t) + \psi(x)$ each time as outlined in the next example.

EXAMPLE 1 **Time-Independent PDE and BCs**

Solve

$$k\frac{\partial^2 u}{\partial x^2} + r = \frac{\partial u}{\partial t}, \quad 0 < x < 1, \ t > 0$$

$$u(0, t) = 0, \quad u(1, t) = u_1, \ t > 0 \tag{3}$$

$$u(x, 0) = f(x), \quad 0 < x < 1,$$

where r and u_1 are nonzero constants.

SOLUTION Both the partial differential equation and the boundary condition at $x = 1$ are nonhomogeneous. If we let $u(x, t) = v(x, t) + \psi(x)$, then

$$\frac{\partial^2 u}{\partial x^2} = \frac{\partial^2 v}{\partial x^2} + \psi''(x) \qquad \text{and} \qquad \frac{\partial u}{\partial t} = \frac{\partial v}{\partial t}.$$

After substituting these results, the PDE in (3) then becomes

$$k \frac{\partial^2 v}{\partial x^2} + k\psi''(x) + r = \frac{\partial v}{\partial t}. \tag{4}$$

Equation (4) reduces to a homogeneous equation if we demand that ψ satisfy

$$k\psi''(x) + r = 0 \quad \text{or} \quad \psi''(x) = -\frac{r}{k}.$$

Integrating the last equation twice reveals that

$$\psi(x) = -\frac{r}{2k}x^2 + c_1 x + c_2. \tag{5}$$

Furthermore, $\qquad u(0, t) = v(0, t) + \psi(0) = 0$

$$u(1, t) = v(1, t) + \psi(1) = u_1.$$

We have $v(0, t) = 0$ and $v(1, t) = 0$ provided that ψ also satisfies

$$\psi(0) = 0 \quad \text{and} \quad \psi(1) = u_1.$$

By applying the latter two conditions to (5) we obtain, in turn, $c_2 = 0$ and $c_1 = r/2k + u_1$. Consequently

$$\psi(x) = -\frac{r}{2k}x^2 + \left(\frac{r}{2k} + u_1\right)x.$$

Finally, the initial condition in (3) implies $v(x, 0) = u(x, 0) - \psi(x) = f(x) - \psi(x)$. Thus to determine $v(x, t)$ we solve the new boundary-value problem

$$k \frac{\partial^2 v}{\partial x^2} = \frac{\partial v}{\partial t}, \ 0 < x < 1, \ t > 0$$

$$v(0, t) = 0, \quad v(1, t) = 0, \ t > 0$$

$$v(x, 0) = f(x) + \frac{r}{2k}x^2 - \left(\frac{r}{2k} + u_1\right)x, \ 0 < x < 1$$

by separation of variables. In the usual manner we find

$$v(x, t) = \sum_{n=1}^{\infty} A_n e^{-kn^2\pi^2 t} \sin n\pi x, \tag{6}$$

where $\qquad A_n = 2 \int_0^1 \left[f(x) + \frac{r}{2k}x^2 - \left(\frac{r}{2k} + u_1\right)x \right] \sin n\pi x \, dx. \tag{7}$

A solution of the original problem is $u(x, t) = v(x, t) + \psi(x)$, or

$$u(x, t) = -\frac{r}{2k}x^2 + \left(\frac{r}{2k} + u_1\right)x + \sum_{n=1}^{\infty} A_n e^{-kn^2\pi^2 t} \sin n\pi x, \tag{8}$$

where the coefficients A_n are defined in (7). $\qquad\qquad\equiv$

Inspection of (6) shows that $v(x, t) \to 0$ as $t \to \infty$, and so $v(x, t)$ is called a **transient solution**. But observe in (8) that $u(x, t) \to \psi(x)$ as $t \to \infty$. In the context of solving forms of the heat equation, $\psi(x)$ is called a **steady-state solution**.

■ **Time-Dependent PDE and BCs** We turn now to a method for solving some kinds of BVPs that involve a *time-dependent nonhomogeneous equation* and *time-dependent boundary conditions*. A problem similar to (1),

$$k \frac{\partial^2 u}{\partial x^2} + F(x, t) = \frac{\partial u}{\partial t}, \quad 0 < x < L, \ t > 0$$

$$u(0, t) = u_0(t), \quad u(L, t) = u_1(t), \ t > 0 \tag{9}$$

$$u(x, 0) = f(x), \quad 0 < x < L,$$

describes the temperatures of a rod of length L, but in this case the heat-source term F and the temperatures at the two ends of the rod can vary with time t. Intuitively one might expect that the line of attack for this problem would be a natural extension of the procedure that worked in Example 1, namely, seek a solution of the form $u(x, t) = v(x, t) + \psi(x, t)$. While this form of the solution is correct in some instances, it is usually not possible to find a function of two variables $\psi(x, t)$ that reduces a problem for $v(x, t)$ to a homogeneous one. To understand why this is so, let's see what happens when $u(x, t) = v(x, t) + \psi(x, t)$ is substituted into the PDE in (9). Because

$$\frac{\partial^2 u}{\partial x^2} = \frac{\partial^2 v}{\partial x^2} + \frac{\partial^2 \psi}{\partial x^2} \quad \text{and} \quad \frac{\partial u}{\partial t} = \frac{\partial v}{\partial t} + \frac{\partial \psi}{\partial t}, \tag{10}$$

the BVP (9) becomes

$$k \frac{\partial^2 v}{\partial x^2} + k \frac{\partial^2 \psi}{\partial x^2} + F(x, t) = \frac{\partial v}{\partial t} + \frac{\partial \psi}{\partial t}$$

$$v(0, t) + \psi(0, t) = u_0(t), \quad v(L, t) + \psi(L, t) = u_1(t) \tag{11}$$

$$v(x, 0) = f(x) - \psi(x, 0).$$

The boundary conditions on v in (11) will be homogeneous if we demand that

$$\psi(0, t) = u_0(t), \quad \psi(L, t) = u_1(t). \tag{12}$$

Were we, at this point, to follow the same steps in the method used in Example 1, we would try to force the problem in (11) to be homogeneous by requiring $k\psi_{xx} + F(x, t) = \psi_t$ and then imposing the conditions in (12) on the solution ψ. But in view of the fact that the defining equation for ψ is itself a nonhomogeneous PDE, this is an unrealistic expectation. So we try an entirely different tack by simply *constructing* a function ψ that satisfies both conditions given in (12). One such function is given by

$$\psi(x, t) = u_0(t) + \frac{x}{L}[u_1(t) - u_0(t)]. \tag{13}$$

Reinspection of (11) shows that we have gained some additional simplification with this choice of ψ because $\psi_{xx} = 0$. We now start over. This time if we substitute

$$u(x, t) = v(x, t) + u_0(t) + \frac{x}{L}[u_1(t) - u_0(t)] \tag{14}$$

the boundary-value problem (11) then becomes

$$k \frac{\partial^2 v}{\partial x^2} + G(x, t) = \frac{\partial v}{\partial t}, \quad 0 < x < L, \ t > 0$$

$$v(0, t) = 0, \quad v(L, t) = 0, \ t > 0 \tag{15}$$

$$v(x, 0) = f(x) - \psi(x, 0), \quad 0 < x < L,$$

where $G(x, t) = F(x, t) - \psi_t$. While the problem (15) is still nonhomogeneous (the boundary conditions are homogeneous but the partial differential equation is nonhomogeneous) it is a problem that we can solve.

■ The Basic Strategy

The solution method for (15) is a bit involved, so before illustrating with a specific example, we first outline the basic strategy:

Make the assumption that time-dependent coefficients $v_n(t)$ and $G_n(t)$ can be found such that both $v(x, t)$ and $G(x, t)$ in (15) can be expanded in the series

$$v(x, t) = \sum_{n=1}^{\infty} v_n(t) \sin\frac{n\pi}{L}x \quad \text{and} \quad G(x, t) = \sum_{n=1}^{\infty} G_n(t) \sin\frac{n\pi}{L}x, \tag{16}$$

where $\sin(n\pi x/L)$, $n = 1, 2, 3, \ldots$, are the eigenfunctions of $X'' + \lambda X = 0$, $X(0) = 0$, $X(L) = 0$ corresponding to the eigenvalues $\lambda_n = \alpha_n^2 = n^2\pi^2/L^2$. This Sturm–Liouville problem would have been obtained had separation of variables been applied to the associated homogeneous BVP of (15). In (16), observe that the assumed series $v(x, t) = \sum_{n=1}^{\infty} v_n(t) \sin\frac{n\pi}{L}x$ already satisfies the boundary conditions in (15). Now substitute this series for $v(x, t)$ into the nonhomogeneous PDE in (15), collect terms, and equate the resulting series with the actual series expansion found for $G(x, t)$.

In the next example we illustrate this method by solving a special case of (9).

EXAMPLE 2 **Time-Dependent Boundary Condition**

Solve
$$\frac{\partial^2 u}{\partial x^2} = \frac{\partial u}{\partial t}, \quad 0 < x < 1, \ t > 0$$

$$u(0, t) = \cos t, \quad u(1, t) = 0, \ t > 0$$

$$u(x, 0) = 0, \ 0 < x < 1.$$

SOLUTION We match this problem with (9) by identifying $k = 1$, $L = 1$, $F(x, t) = 0$, $u_0(t) = \cos t$, $u_1(t) = 0$, and $f(x) = 0$. We begin with the construction of ψ. From (13) we get

$$\psi(x, t) = \cos t + x[0 - \cos t] = (1 - x)\cos t,$$

and then, as indicated in (14), we use

$$u(x, t) = v(x, t) + (1 - x)\cos t \tag{17}$$

and substitute the quantities

$$\frac{\partial^2 u}{\partial x^2} = \frac{\partial^2 v}{\partial x^2}, \quad \frac{\partial u}{\partial t} = \frac{\partial v}{\partial t} + (1 - x)(-\sin t),$$

$$u(0, t) = v(0, t) + \cos t, \quad u(1, t) = v(1, t) \text{ and } u(x, 0) = v(x, 0) + 1 - x$$

into the given problem to obtain the BVP for $v(x, t)$:

$$\frac{\partial^2 v}{\partial x^2} + (1 - x)\sin t = \frac{\partial v}{\partial t}, \quad 0 < x < 1, \ t > 0$$

$$v(0, t) = 0, \quad v(1, t) = 0, \ t > 0 \tag{18}$$

$$v(x, 0) = x - 1, \ 0 < x < 1.$$

The eigenvalues and eigenfunctions of the Sturm–Liouville problem

$$X'' + \lambda X = 0, \quad X(0) = 0, \quad X(1) = 0$$

are found to be $\lambda_n = \alpha_n^2 = n^2\pi^2$ and $\sin n\pi x$, $n = 1, 2, 3, \ldots$. With $G(x, t) = (1 - x)\sin t$ we assume from (18) that for fixed t, v and G can be written as Fourier sine series:

$$v(x, t) = \sum_{n=1}^{\infty} v_n(t) \sin n\pi x \tag{19}$$

and
$$(1 - x)\sin t = \sum_{n=1}^{\infty} G_n(t) \sin n\pi x. \tag{20}$$

By treating t as a parameter, the coefficients G_n in (20) can be computed:

$$G_n(t) = \frac{2}{1}\int_0^1 (1 - x)\sin t \sin n\pi x \, dx = 2\sin t \int_0^1 (1 - x)\sin n\pi x \, dx = \frac{2}{n\pi}\sin t.$$

Hence,
$$(1 - x)\sin t = \sum_{n=1}^{\infty} \frac{2}{n\pi}\sin t \sin n\pi x. \tag{21}$$

We can determine the coefficients $v_n(t)$ by substituting (20) and (21) back into the PDE in (18). To that end, the partial derivatives of v are

$$\frac{\partial^2 v}{\partial x^2} = \sum_{n=1}^{\infty} v_n(t)(-n^2\pi^2)\sin n\pi x \quad \text{and} \quad \frac{\partial v}{\partial t} = \sum_{n=1}^{\infty} v_n'(t)\sin n\pi x. \tag{22}$$

Writing the PDE in (18) as $v_t - v_{xx} = (1 - x)\sin t$ and using (21) and (22) we get

$$\sum_{n=1}^{\infty} [v_n'(t) + n^2\pi^2 v_n(t)]\sin n\pi x = \sum_{n=1}^{\infty} \frac{2\sin t}{n\pi}\sin n\pi x.$$

We then equate the coefficients of $\sin n\pi x$ on each side of the equality to get the linear first-order ODE

$$v_n'(t) + n^2\pi^2 v_n(t) = \frac{2\sin t}{n\pi}.$$

Proceeding as in Section 2.3, we multiply the last equation by the integrating factor $e^{n^2\pi^2 t}$ and rewrite it as

$$\frac{d}{dt}[e^{n^2\pi^2 t}v_n(t)] = \frac{2}{n\pi}e^{n^2\pi^2 t}\sin t.$$

Integrating both sides, we find that the general solution of this equation is

$$v_n(t) = 2\frac{n^2\pi^2\sin t - \cos t}{n\pi(n^4\pi^4 + 1)} + C_n e^{-n^2\pi^2 t},$$

where C_n denotes the arbitrary constant. Therefore the assumed form of $v(x, t)$ in (19) can be written

$$v(x, t) = \sum_{n=1}^{\infty} \left[2\frac{n^2\pi^2 \sin t - \cos t}{n\pi(n^4\pi^4 + 1)} + C_n e^{-n^2\pi^2 t}\right]\sin n\pi x. \tag{23}$$

The coefficients C_n can be found by applying the initial condition $v(x, 0)$ to (23). From the Fourier sine series,

$$x - 1 = \sum_{n=1}^{\infty} \left[\frac{-2}{n\pi(n^4\pi^4 + 1)} + C_n\right]\sin n\pi x, \tag{24}$$

we see that the quantity in the brackets represents the Fourier sine coefficients b_n for $x - 1$. That is,

$$\frac{-2}{n\pi(n^4\pi^4 + 1)} + C_n = 2\int_0^1 (x - 1)\sin n\pi x \, dx \quad \text{or} \quad \frac{-2}{n\pi(n^4\pi^4 + 1)} + C_n = \frac{-2}{n\pi}.$$

Therefore,
$$C_n = \frac{2}{n\pi(n^4\pi^4 + 1)} - \frac{2}{n\pi}.$$

By substituting the last result into (23) we obtain a solution of (18):

$$v(x, t) = \frac{2}{\pi}\sum_{n=1}^{\infty} \left[\frac{n^2\pi^2\sin t - \cos t + e^{-n^2\pi^2 t}}{n(n^4\pi^4 + 1)} - \frac{e^{-n^2\pi^2 t}}{n}\right]\sin n\pi x.$$

At long last, then, it follows from (17) that the desired solution $u(x, t)$ is

$$u(x, t) = (1 - x)\cos t + \frac{2}{\pi}\sum_{n=1}^{\infty}\left[\frac{n^2\pi^2\sin t - \cos t + e^{-n^2\pi^2 t}}{n(n^4\pi^4 + 1)} - \frac{e^{-n^2\pi^2 t}}{n}\right]\sin n\pi x. \quad \equiv$$

If the boundary-value problem has homogeneous boundary conditions and a time-dependent term $F(x, t)$ in the PDE, then there is no actual need to change the dependent variable through the substitution $u(x, t) = v(x, t) + \psi(x, t)$. For example, if both u_0 and u_1 are 0 in a problem such as (9), then it follows from (13) that $\psi(x, t) = 0$. The method of solution then begins by assuming appropriate orthogonal series expansions for $u(x, t)$ and $F(x, t)$ as in (16), where the symbols v and G in (16) are naturally replaced by u and F, respectively.

<table>
<tr><td>EXAMPLE 3</td><td>Time-Dependent PDE and Homogeneous BCs</td></tr>
</table>

Solve
$$\frac{\partial^2 u}{\partial x^2} + (1 - x)\sin t = \frac{\partial u}{\partial t}, \quad 0 < x < 1, \ t > 0$$

$$u(0, t) = 0, \quad u(1, t) = 0, \ t > 0 \qquad (25)$$

$$u(x, 0) = 0, \ 0 < x < 1.$$

SOLUTION Except for the initial condition, the BVP (25) is basically (18). As pointed out in the paragraph preceding this example, because the boundary conditions are both homogeneous we have $\psi(x, t) = 0$. Thus all steps in Example 2 used in the solution of (18) are the same *except* the initial condition $u(x, 0) = 0$ indicates that the analogue of (24) is then

$$0 = \sum_{n=1}^{\infty}\left[\frac{-2}{n\pi(n^4\pi^4 + 1)} + C_n\right]\sin n\pi x.$$

We conclude from this identity that the coefficient of $\sin n\pi x$ must be 0 and so

$$C_n = \frac{2}{n\pi(n^4\pi^4 + 1)}.$$

Hence a solution of (25) is

$$u(x, t) = \frac{2}{\pi}\sum_{n=1}^{\infty}\left[\frac{n^2\pi^2\sin t - \cos t}{n(n^4\pi^4 + 1)} + \frac{e^{-n^2\pi^2 t}}{n(n^4\pi^4 + 1)}\right]\sin n\pi x. \quad \equiv$$

In Problems 13–16 in Exercises 13.6 you are asked to construct $\psi(x, t)$ as illustrated in Example 2. In Problems 17–20 of Exercises 13.6 the given boundary conditions are homogenous and so you can start as we did in Example 3 with the assumption that $\psi(x, t) = 0$.

REMARKS

Don't put any special emphasis on the fact that we used the heat equation throughout the foregoing discussion. The method outlined in Example 1 can be applied to the wave equation and Laplace's equation as well. See Problems 1–12 in Exercises 13.6. The method outlined in Example 2 is predicated on time dependence in the problem and so is not applicable to BVPs involving Laplace's equation.

13.6 Exercises Answers to selected odd-numbered problems begin on page ANS-35.

In Problems 1–12, proceed as in Example 1 of this section to solve the given boundary-value problem.

In Problems 1 and 2, solve the heat equation $ku_{xx} = u_t$, $0 < x < 1$, $t > 0$ subject to the given conditions.

1. $u(0, t) = 100, \quad u(1, t) = 100$
$u(x, 0) = 0$

2. $u(0, t) = u_0, \quad u(1, t) = 0$
$u(x, 0) = f(x)$

In Problems 3 and 4, solve the heat equation (2) subject to the given conditions.

3. $u(0, t) = u_0, \quad u(1, t) = u_0$
$u(x, 0) = 0$

4. $u(0, t) = u_0, \quad u(1, t) = u_1$

$u(x, 0) = f(x)$

5. Solve the boundary-value problem

$$k\frac{\partial^2 u}{\partial x^2} + Ae^{-\beta x} = \frac{\partial u}{\partial t}, \quad \beta > 0, 0 < x < 1, \ t > 0$$

$$u(0, t) = 0, \quad u(1, t) = 0, \ t > 0$$

$$u(x, 0) = f(x), \ 0 < x < 1,$$

where A is a constant. The PDE is a form of the heat equation when heat is generated within a thin rod due to radioactive decay of the material.

6. Solve the boundary-value problem

$$k\frac{\partial^2 u}{\partial x^2} - hu = \frac{\partial u}{\partial t}, \quad 0 < x < \pi, \ t > 0$$

$$u(0, t) = 0, \quad u(\pi, t) = u_0, \ t > 0$$

$$u(x, 0) = 0, \ 0 < x < \pi.$$

The PDE is a form of the heat equation when heat is lost by radiation from the lateral surface of a thin rod into a medium at temperature zero.

7. Find a steady-state solution $\psi(x)$ of the boundary-value problem

$$k\frac{\partial^2 u}{\partial x^2} - h(u - u_0) = \frac{\partial u}{\partial t}, \quad 0 < x < 1, \ t > 0$$

$$u(0, t) = u_0, \quad u(1, t) = 0, \ t > 0$$

$$u(x, 0) = f(x), \ 0 < x < 1.$$

8. Find a steady-state solution $\psi(x)$ if the rod in Problem 7 is semi-infinite extending in the positive x-direction, radiates from its lateral surface into a medium at temperature zero, and

$$u(0, t) = u_0, \quad \lim_{x \to \infty} u(x, t) = 0, \ t > 0$$

$$u(x, 0) = f(x), \ x > 0.$$

9. When a vibrating string is subjected to an external vertical force that varies with the horizontal distance from the left end, the wave equation takes on the form

$$a^2 \frac{\partial^2 u}{\partial x^2} + Ax = \frac{\partial^2 u}{\partial t^2},$$

where A is constant. Solve this partial differential equation subject to

$$u(0, t) = 0, \quad u(1, t) = 0, \ t > 0$$

$$u(x, 0) = 0, \quad \left.\frac{\partial u}{\partial t}\right|_{t=0} = 0, \ 0 < x < 1.$$

10. A string initially at rest on the x-axis is secured on the x-axis at $x = 0$ and $x = 1$. If the string is allowed to fall under its own weight for $t > 0$, the displacement $u(x, t)$ satisfies

$$a^2 \frac{\partial^2 u}{\partial x^2} - g = \frac{\partial^2 u}{\partial t^2}, \quad 0 < x < 1, \ t > 0,$$

where g is the acceleration of gravity. Solve for $u(x, t)$.

11. Find the steady-state temperature $u(x, y)$ in the semi-infinite plate shown in **FIGURE 13.6.1**. Assume that the temperature is bounded as $x \to \infty$. [*Hint:* Use $u(x, y) = v(x, y) + \psi(y)$.]

FIGURE 13.6.1 Semi-infinite plate in Problem 11

12. The partial differential equation

$$\frac{\partial^2 u}{\partial x^2} + \frac{\partial^2 u}{\partial y^2} = -h,$$

where $h > 0$ is a constant, is known as **Poisson's equation** after the French mathematician, engineer, and physicist **Siméon Denis Poisson** (1781–1840). Poisson is remembered for his scientific and mathematical advances, as well as for his wrong-headed opposition to the wave theory of light (subsequently proved by Augustin-Jean Fresnel). Solve the equation above subject to the conditions

$$u(0, y) = 0, \quad u(\pi, y) = 1, \ y > 0$$

$$u(x, 0) = 0, \ 0 < x < \pi.$$

In Problems 13–16, proceed as in Example 2 to solve the given boundary-value problem.

13. $\dfrac{\partial^2 u}{\partial x^2} = \dfrac{\partial u}{\partial t}, \quad 0 < x < 1, \ t > 0$

$u(0, t) = \sin t, \quad u(1, t) = 0, \ t > 0$

$u(x, 0) = 0, \ 0 < x < 1$

14. $\dfrac{\partial^2 u}{\partial x^2} + 2t + 3tx = \dfrac{\partial u}{\partial t}, \quad 0 < x < 1, \ t > 0$

$u(0, t) = t^2, \quad u(1, t) = 1, \ t > 0$

$u(x, 0) = x^2, \ 0 < x < 1$

15. $\dfrac{\partial^2 u}{\partial x^2} = \dfrac{\partial^2 u}{\partial t^2}, \quad 0 < x < 1, \ t > 0$

$u(0, t) = 0, \quad u(1, t) = \sin t, \ t > 0$

$u(x, 0) = 0, \quad \left.\dfrac{\partial u}{\partial t}\right|_{t=0} = 0, \ 0 < x < 1$

16. $\dfrac{\partial^2 u}{\partial x^2} = \dfrac{\partial u}{\partial t}, \quad 0 < x < 1, \ t > 0$

$u(0, t) = 1 - e^{-t}, \quad u(1, t) = 1 - e^{-t}, \ t > 0$

$u(x, 0) = 0, \ 0 < x < 1$

In Problems 17–20, proceed as in Example 3 to solve the given boundary-value problem.

17. $\dfrac{\partial^2 u}{\partial x^2} + xe^{-3t} = \dfrac{\partial u}{\partial t}$, $0 < x < \pi$, $t > 0$

$u(0, t) = 0$, $u(\pi, t) = 0$, $t > 0$

$u(x, 0) = 0$, $0 < x < \pi$

18. $\dfrac{\partial^2 u}{\partial x^2} + xe^{-3t} = \dfrac{\partial u}{\partial t}$, $0 < x < \pi$, $t > 0$

$\dfrac{\partial u}{\partial x}\bigg|_{x=0} = 0$, $\dfrac{\partial u}{\partial x}\bigg|_{x=\pi} = 0$, $t > 0$

$u(x, 0) = 0$, $0 < x < \pi$

19. $\dfrac{\partial^2 u}{\partial x^2} - 1 + x - x\cos t = \dfrac{\partial u}{\partial t}$, $0 < x < 1$, $t > 0$

$u(0, t) = 0$, $u(1, t) = 0$, $t > 0$

$u(x, 0) = x(1 - x)$, $0 < x < 1$

20. $\dfrac{\partial^2 u}{\partial x^2} + \sin x \cos t = \dfrac{\partial^2 u}{\partial t^2}$, $0 < x < \pi$, $t > 0$

$u(0, t) = 0$, $u(\pi, t) = 0$, $t > 0$

$u(x, 0) = 0$, $\dfrac{\partial u}{\partial t}\bigg|_{t=0} = 0$, $0 < x < \pi$

13.7 Orthogonal Series Expansions

INTRODUCTION For certain types of boundary conditions, the method of separation of variables and the superposition principle lead to an expansion of a function in an infinite series that is *not* a Fourier series. To solve the problems in this section we shall utilize the concept of orthogonal series expansions or generalized Fourier series developed in Section 12.1.

EXAMPLE 1 Using Orthogonal Series Expansions

The temperature in a rod of unit length in which there is heat transfer from its right boundary into a surrounding medium kept at a constant temperature zero is determined from

$$k\dfrac{\partial^2 u}{\partial x^2} = \dfrac{\partial u}{\partial t}, \quad 0 < x < 1, \; t > 0$$

$$u(0, t) = 0, \quad \dfrac{\partial u}{\partial x}\bigg|_{x=1} = -hu(1, t), \; h > 0, \; t > 0$$

$$u(x, 0) = 1, \; 0 < x < 1.$$

Solve for $u(x, t)$.

SOLUTION Proceeding exactly as we did in Section 13.3, with $u(x, t) = X(x)T(t)$ and $-\lambda$ as the separation constant, we find the separated ODEs and boundary conditions to be, respectively,

$$X'' + \lambda X = 0 \tag{1}$$

$$T' + k\lambda T = 0 \tag{2}$$

$$X(0) = 0 \quad \text{and} \quad X'(1) = -hX(1). \tag{3}$$

Equation (1) along with the homogeneous boundary conditions (3) comprise the regular Sturm–Liouville problem:

$$X'' + \lambda X = 0, \quad X(0) = 0, \quad X'(1) + hX(1) = 0. \tag{4}$$

Except for the presence of the symbol h, the BVP in (4) is essentially the problem solved in Example 2 of Section 12.5. As in that example, (4) possesses nontrivial solutions only in the case $\lambda = \alpha^2 > 0$, $\alpha > 0$. The general solution of the DE in (4) is $X(x) = c_1 \cos \alpha x + c_2 \sin \alpha x$. The first boundary condition in (4) immediately gives $c_1 = 0$. Applying the second boundary condition in (4) to $X(x) = c_2 \sin \alpha x$ yields

$$\alpha \cos \alpha + h \sin \alpha = 0 \quad \text{or} \quad \tan \alpha = -\dfrac{\alpha}{h}. \tag{5}$$

Because the graphs of $y = \tan x$ and $y = -x/h$, $h > 0$, have an infinite number of points of intersection for $x > 0$ (Figure 12.5.1 illustrates the case $h = 1$), the last equation in (5) has an infinite number of roots. Of course, these roots depend on the value of h. If the consecutive positive roots are denoted α_n, $n = 1, 2, 3, \ldots$, then the eigenvalues of the problem are $\lambda_n = \alpha_n^2$, and the corresponding eigenfunctions are $X(x) = c_2 \sin \alpha_n x$, $n = 1, 2, 3, \ldots$. The solution of the first-order DE (2) is $T(t) = c_3 e^{-k\alpha_n^2 t}$ and so

$$u_n = X(x)T(t) = A_n e^{-k\alpha_n^2 t} \sin \alpha_n x \quad \text{and} \quad u(x, t) = \sum_{n=1}^{\infty} A_n e^{-k\alpha_n^2 t} \sin \alpha_n x.$$

Now at $t = 0$, $u(x, 0) = 1$, $0 < x < 1$, so that

$$1 = \sum_{n=1}^{\infty} A_n \sin \alpha_n x. \tag{6}$$

The series in (6) is not a Fourier sine series; rather, it is an expansion of $u(x, 0) = 1$ in terms of the orthogonal functions arising from the Sturm–Liouville problem (4). It follows that the set of eigenfunctions $\{\sin \alpha_n x\}$, $n = 1, 2, 3, \ldots$, where the αs are defined by $\tan \alpha = -\alpha/h$ is orthogonal with respect to the weight function $p(x) = 1$ on the interval $[0, 1]$. With $f(x) = 1$ and $\phi_n(x) = \sin \alpha_n x$, it follows from (8) of Section 12.1 that the coefficients A_n in (6) are

$$A_n = \frac{\int_0^1 \sin \alpha_n x \, dx}{\int_0^1 \sin^2 \alpha_n x \, dx}. \tag{7}$$

To evaluate the square norm of each of the eigenfunctions we use a trigonometric identity:

$$\int_0^1 \sin^2 \alpha_n x \, dx = \frac{1}{2} \int_0^1 (1 - \cos 2\alpha_n x) \, dx = \frac{1}{2}\left(1 - \frac{1}{2\alpha_n} \sin 2\alpha_n\right). \tag{8}$$

With the aid of the double angle formula $\sin 2\alpha_n = 2 \sin \alpha_n \cos \alpha_n$ and the first equation in (5) in the form $\alpha_n \cos \alpha_n = -h \sin \alpha_n$, we can simplify (8) to

$$\int_0^1 \sin^2 \alpha_n x \, dx = \frac{1}{2h}(h + \cos^2 \alpha_n).$$

Also

$$\int_0^1 \sin \alpha_n x \, dx = -\frac{1}{\alpha_n} \cos \alpha_n x \Big]_0^1 = \frac{1}{\alpha_n}(1 - \cos \alpha_n).$$

Consequently (7) becomes

$$A_n = \frac{2h(1 - \cos \alpha_n)}{\alpha_n(h + \cos^2 \alpha_n)}.$$

Finally, a solution of the boundary-value problem is

$$u(x, t) = 2h \sum_{n=1}^{\infty} \frac{1 - \cos \alpha_n}{\alpha_n(h + \cos^2 \alpha_n)} e^{-k\alpha_n^2 t} \sin \alpha_n x. \qquad \equiv$$

EXAMPLE 2 **Using Orthogonal Series Expansions**

The twist angle $\theta(x, t)$ of a torsionally vibrating shaft of unit length is determined from

$$a^2 \frac{\partial^2 \theta}{\partial x^2} = \frac{\partial^2 \theta}{\partial t^2}, \quad 0 < x < 1, \ t > 0$$

$$\theta(0, t) = 0, \quad \frac{\partial \theta}{\partial x}\Big|_{x=1} = 0, \ t > 0$$

$$\theta(x, 0) = x, \quad \frac{\partial \theta}{\partial t}\Big|_{t=0} = 0, \ 0 < x < 1.$$

See **FIGURE 13.7.1**. The boundary condition at $x = 1$ is called a free-end condition. Solve for $\theta(x, t)$.

FIGURE 13.7.1 The twist angle θ in Example 2

SOLUTION Proceeding as in Section 13.4 with $\theta(x, t) = X(x)T(t)$ and using $-\lambda$ once again as the separation constant, the separated equations and boundary conditions are

$$X'' + \lambda X = 0 \tag{9}$$

$$T'' + a^2\lambda T = 0. \tag{10}$$

$$X(0) = 0 \quad \text{and} \quad X'(1) = 0. \tag{11}$$

Equation (9) together with the homogeneous boundary conditions in (11),

$$X'' + \lambda X = 0, \quad X(0) = 0, \quad X'(1) = 0, \tag{12}$$

is a regular Sturm–Liouville problem. You are encouraged to verify that for $\lambda = 0$ and for $\lambda = -\alpha^2$, $\alpha > 0$, the only solution of (12) is $X = 0$. For $\lambda = \alpha^2 > 0$, $\alpha > 0$, the boundary conditions $X(0) = 0$ and $X'(1) = 0$ applied to the general solution $X(x) = c_1 \cos \alpha x + c_2 \sin \alpha x$ give, in turn, $c_1 = 0$ and $c_2 \cos \alpha = 0$. Since $\cos \alpha$ is zero only when α is an odd integer multiple of $\pi/2$ we write $\alpha_n = (2n - 1)\pi/2$. The eigenvalues of (12) are $\lambda_n = \alpha_n^2 = (2n - 1)^2\pi^2/4$, and the corresponding eigenfunctions are $X(x) = c_2 \sin \alpha_n x = c_2 \sin\left(\dfrac{2n - 1}{2}\right)\pi x$, $n = 1, 2, 3, \ldots$.

Since the rod is released from rest, the initial condition $\theta_t(x, 0) = 0$ translates into $X(x)T'(0) = 0$ or $T'(0) = 0$. When applied to the general solution $T(t) = c_3 \cos a\alpha_n t + c_4 \sin a\alpha_n t$ of the second-order DE (10), $T'(0) = 0$ implies $c_4 = 0$ leaving $T(t) = c_3 \cos a\alpha_n t = c_3 \cos a\left(\dfrac{2n - 1}{2}\right)\pi t$. Therefore,

$$\theta_n = X(x)T(t) = A_n \cos a\left(\frac{2n - 1}{2}\right)\pi t \sin\left(\frac{2n - 1}{2}\right)\pi x.$$

In order to satisfy the remaining initial condition we form the superposition of the θ_n,

$$\theta(x, t) = \sum_{n=1}^{\infty} A_n \cos a\left(\frac{2n - 1}{2}\right)\pi t \sin\left(\frac{2n - 1}{2}\right)\pi x. \tag{13}$$

When $t = 0$, we must have, for $0 < x < 1$,

$$\theta(x, 0) = x = \sum_{n=1}^{\infty} A_n \sin\left(\frac{2n - 1}{2}\right)\pi x. \tag{14}$$

As in Example 1, the set of eigenfunctions $\left\{ \sin\left(\dfrac{2n - 1}{2}\right)\pi x \right\}$, $n = 1, 2, 3, \ldots$, is orthogonal with respect to the weight function $p(x) = 1$ on the interval $[0, 1]$. Even though the trigonometric series in (14) looks more like a Fourier series than (6), it is not a Fourier sine series because the argument of the sine function is not an integer multiple of $\pi x/L$ (here $L = 1$). The series is again an orthogonal series expansion or generalized Fourier series. Hence from (8) of Section 12.1 the coefficients A_n in (14) are given by

$$A_n = \frac{\displaystyle\int_0^1 x \sin\left(\frac{2n - 1}{2}\right)\pi x \, dx}{\displaystyle\int_0^1 \sin^2\left(\frac{2n - 1}{2}\right)\pi x \, dx}.$$

Carrying out the two integrations, we arrive at

$$A_n = \frac{8(-1)^{n+1}}{(2n - 1)^2\pi^2}.$$

The twist angle is then

$$\theta(x, t) = \frac{8}{\pi^2} \sum_{n=1}^{\infty} \frac{(-1)^{n+1}}{(2n - 1)^2} \cos a\left(\frac{2n - 1}{2}\right)\pi t \sin\left(\frac{2n - 1}{2}\right)\pi x. \qquad \equiv$$

1. In Example 1 find the temperature $u(x, t)$ when the left end of the rod is insulated.

2. Solve the boundary-value problem

$$k\frac{\partial^2 u}{\partial x^2} = \frac{\partial u}{\partial t}, \ 0 < x < 1, \ t > 0$$

$$u(0, t) = 0, \ \frac{\partial u}{\partial x}\bigg|_{x=1} = -h(u(1, t) - u_0), \ h > 0, \ t > 0$$

$$u(x, 0) = f(x), \ 0 < x < 1.$$

3. Find the steady-state temperature $u(x, y)$ for a rectangular plate for which the boundary conditions are

$$u(0, y) = 0, \ \frac{\partial u}{\partial x}\bigg|_{x=a} = -hu(a, y), \ h > 0, \ 0 < y < b,$$

$$u(x, 0) = 0, \quad u(x, b) = f(x), \ 0 < x < a.$$

4. Solve the boundary-value problem

$$\frac{\partial^2 u}{\partial x^2} + \frac{\partial^2 u}{\partial y^2} = 0, \ x > 0, \ 0 < y < 1$$

$$u(0, y) = u_0, \ \lim_{x \to \infty} u(x, y) = 0, \ 0 < y < 1$$

$$\frac{\partial u}{\partial y}\bigg|_{y=0} = 0, \ \frac{\partial u}{\partial y}\bigg|_{y=1} = -hu(x, 1), \ h > 0, \ x > 0.$$

5. Find the temperature $u(x, t)$ in a rod of length L if the initial temperature is $f(x)$ throughout and if the end $x = 0$ is maintained at temperature zero and the end $x = L$ is insulated.

6. Solve the boundary-value problem

$$a^2\frac{\partial^2 u}{\partial x^2} = \frac{\partial^2 u}{\partial t^2}, \ 0 < x < L, \ t > 0$$

$$u(0, t) = 0, \ E\frac{\partial u}{\partial x}\bigg|_{x=L} = F_0, \ t > 0$$

$$u(x, 0) = 0, \ \frac{\partial u}{\partial t}\bigg|_{t=0} = g(x), \ 0 < x < L.$$

The solution $u(x, t)$ represents the longitudinal displacement of a vibrating elastic bar that is anchored at its left end and is subjected to a constant force F_0 at its right end. See Figure 13.4.7 on page 733. E is called the modulus of elasticity.

7. Solve the boundary-value problem

$$\frac{\partial^2 u}{\partial x^2} + \frac{\partial^2 u}{\partial y^2} = 0, \ 0 < x < 1, \ 0 < y < 1$$

$$\frac{\partial u}{\partial x}\bigg|_{x=0} = 0, \ u(1, y) = u_0, \ 0 < y < 1$$

$$u(x, 0) = 0, \ \frac{\partial u}{\partial y}\bigg|_{y=1} = 0, \ 0 < x < 1.$$

8. Solve the boundary-value problem

$$\frac{\partial^2 u}{\partial x^2} - u = \frac{\partial u}{\partial t}, \ 0 < x < 1, \ t > 0$$

$$\frac{\partial u}{\partial x}\bigg|_{x=0} = 0, \ u(1, t) = 0, \ t > 0$$

$$u(x, 0) = 1 - x^2, \ 0 < x < 1.$$

9. The initial temperature in a rod of unit length is $f(x)$ throughout. There is heat transfer from both ends, $x = 0$ and $x = 1$, into a surrounding medium kept at a constant temperature zero. Show that

$$u(x, t) = \sum_{n=1}^{\infty} A_n e^{-k\alpha_n^2 t}(\alpha_n \cos \alpha_n x + h \sin \alpha_n x),$$

where

$$A_n = \frac{2}{(\alpha_n^2 + 2h + h^2)} \int_0^1 f(x)(\alpha_n \cos \alpha_n x + h \sin \alpha_n x)\, dx.$$

The eigenvalues are $\lambda_n = \alpha_n^2, n = 1, 2, 3, \ldots$, where the α_n are the consecutive positive roots of $\tan \alpha = 2\alpha h/(\alpha^2 - h^2)$.

10. Use the method discussed in Example 3 of Section 13.6 to solve the nonhomogeneous boundary-value problem

$$k\frac{\partial^2 u}{\partial x^2} + xe^{-2t} = \frac{\partial u}{\partial t}, \ 0 < x < 1, \ t > 0$$

$$u(0, t) = 0, \ \frac{\partial u}{\partial x}\bigg|_{x=1} = -u(1, t), \ t > 0$$

$$u(x, 0) = 0, \ 0 < x < 1.$$

Computer Lab Assignments

11. A vibrating cantilever beam is embedded at its left end ($x = 0$) and free at its right end ($x = 1$). See **FIGURE 13.7.2**. The transverse displacement $u(x, t)$ of the beam is determined from

$$\frac{\partial^4 u}{\partial x^4} + \frac{\partial^2 u}{\partial t^2} = 0, \ 0 < x < 1, \ t > 0$$

$$u(0, t) = 0, \ \frac{\partial u}{\partial x}\bigg|_{x=0} = 0, \ t > 0$$

$$\frac{\partial^2 u}{\partial x^2}\bigg|_{x=1} = 0, \ \frac{\partial^3 u}{\partial x^3}\bigg|_{x=1} = 0, \ t > 0$$

$$u(x, 0) = f(x), \ \frac{\partial u}{\partial t}\bigg|_{t=0} = g(x), \ 0 < x < 1.$$

This boundary-value problem could serve as a model for the displacements of a vibrating airplane wing.

(a) Show that the eigenvalues of the problem are determined from the equation $\cos \alpha \cosh \alpha = -1$.

(b) Use a CAS to find approximations to the first two positive eigenvalues of the problem. [*Hint*: See Problem 23 in Exercises 13.4.]

FIGURE 13.7.2 Cantilever beam in Problem 11

12. (a) Find an equation that defines the eigenvalues when the ends of the beam in Problem 11 are embedded at $x = 0$ and $x = 1$.

(b) Use a CAS to find approximations to the first two eigenvalues of the problem.

13.8 Higher-Dimensional Problems

INTRODUCTION In the preceding sections we solved one-dimensional forms of the heat and wave equations. In this section we extend the method of separation of variables to certain problems involving the two-dimensional heat and wave equations.

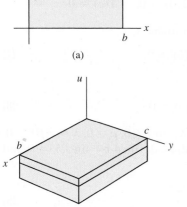

FIGURE 13.8.1 (a) Rectangular plate (b) Rectangular membrane

■ **Heat and Wave Equations in Two Dimensions** Suppose the rectangular region in **FIGURE 13.8.1(a)** is a thin plate in which the temperature u is a function of time t and position (x, y). Then, under suitable conditions, $u(x, y, t)$ can be shown to satisfy the **two-dimensional heat equation**

$$k\left(\frac{\partial^2 u}{\partial x^2} + \frac{\partial^2 u}{\partial y^2}\right) = \frac{\partial u}{\partial t}. \tag{1}$$

On the other hand, suppose Figure 13.8.1(b) represents a rectangular frame over which a thin flexible membrane has been stretched (a rectangular drum). If the membrane is set in motion, then its displacement u, measured from the xy-plane (transverse vibrations), is also a function of time t and position (x, y). When the displacements are small, free, and undamped, $u(x, y, t)$ satisfies the **two-dimensional wave equation**

$$a^2\left(\frac{\partial^2 u}{\partial x^2} + \frac{\partial^2 u}{\partial y^2}\right) = \frac{\partial^2 u}{\partial t^2}. \tag{2}$$

As the next example will show, solutions of boundary-value problems involving (1) and (2) lead to the concept of a Fourier series in two variables. Because the analysis of problems involving (1) and (2) are quite similar, we illustrate a solution only of the heat equation (1).

EXAMPLE 1 **Temperatures in a Plate**

Find the temperature $u(x, y, t)$ in the plate shown in Figure 13.8.1(a) if the initial temperature is $f(x, y)$ throughout and if the boundaries are held at temperature zero for time $t > 0$.

SOLUTION We must solve

$$k\left(\frac{\partial^2 u}{\partial x^2} + \frac{\partial^2 u}{\partial y^2}\right) = \frac{\partial u}{\partial t}, \quad 0 < x < b, \ 0 < y < c, \ t > 0$$

subject to

$$u(0, y, t) = 0, \qquad u(b, y, t) = 0, \quad 0 < y < c, \ t > 0$$

$$u(x, 0, t) = 0, \qquad u(x, c, t) = 0, \quad 0 < x < b, \ t > 0$$

$$u(x, y, 0) = f(x, y), \quad 0 < x < b, \ 0 < y < c.$$

To separate variables for the PDE in three independent variables x, y, and t we try to find a product solution $u(x, y, t) = X(x)Y(y)T(t)$. Substituting, we get

$$k(X''YT + XY''T) = XYT' \quad \text{or} \quad \frac{X''}{X} = -\frac{Y''}{Y} + \frac{T'}{kT}. \tag{3}$$

Since the left side of the last equation in (3) depends only on x and the right side depends only on y and t, we must have both sides equal to a constant $-\lambda$:

$$\frac{X''}{X} = -\frac{Y''}{Y} + \frac{T'}{kT} = -\lambda$$

and so

$$X'' + \lambda X = 0 \tag{4}$$

$$\frac{Y''}{Y} = \frac{T'}{kT} + \lambda. \tag{5}$$

By the same reasoning, if we introduce another separation constant $-\mu$ in (5), then

$$\frac{Y''}{Y} = -\mu \quad \text{and} \quad \frac{T'}{kT} + \lambda = -\mu$$

$$Y'' + \mu Y = 0 \quad \text{and} \quad T' + k(\lambda + \mu)T = 0. \tag{6}$$

Now the homogeneous boundary conditions

$$\left.\begin{array}{ll} u(0, y, t) = 0, & u(b, y, t) = 0 \\ u(x, 0, t) = 0, & u(x, c, t) = 0 \end{array}\right\} \quad \text{imply} \quad \begin{cases} X(0) = 0, & X(b) = 0 \\ Y(0) = 0, & Y(c) = 0. \end{cases}$$

Thus we have two Sturm–Liouville problems, one in the variable x,

$$X'' + \lambda X = 0, \quad X(0) = 0, \; X(b) = 0, \tag{7}$$

and the other in the variable y,

$$Y'' + \mu Y = 0, \quad Y(0) = 0, \; Y(c) = 0. \tag{8}$$

The usual consideration of cases ($\lambda = 0, \lambda = -\alpha^2 < 0, \lambda = \alpha^2 > 0, \mu = 0, \lambda = -\beta^2 < 0$, and so on) leads to two independent sets of eigenvalues defined by $\sin \lambda b = 0$ and $\sin \mu c = 0$. These equations in turn imply

$$\lambda_m = \frac{m^2\pi^2}{b^2}, \quad \text{and} \quad \mu_n = \frac{n^2\pi^2}{c^2}. \tag{9}$$

The corresponding eigenfunctions are

$$X(x) = c_2 \sin \frac{m\pi}{b} x, \; m = 1, 2, 3, \dots \quad \text{and} \quad Y(y) = c_4 \sin \frac{n\pi}{c} y, \; n = 1, 2, 3, \dots. \tag{10}$$

After substituting the values in (9) into the first-order DE in (6), its general solution is $T(t) = c_5 e^{-k[(m\pi/b)^2 + (n\pi/c)^2]t}$. A product solution of the two-dimensional heat equation that satisfies the four homogeneous boundary conditions is then

$$u_{mn}(x, y, t) = A_{mn} e^{-k[(m\pi/b)^2 + (n\pi/c)^2]t} \sin \frac{m\pi}{b} x \sin \frac{n\pi}{c} y,$$

where A_{mn} is an arbitrary constant. Because we have two sets of eigenvalues, we are prompted to try the superposition principle in the form of a double sum

$$u(x, y, t) = \sum_{m=1}^{\infty} \sum_{n=1}^{\infty} A_{mn} e^{-k[(m\pi/b)^2 + (n\pi/c)^2]t} \sin \frac{m\pi}{b} x \sin \frac{n\pi}{c} y. \tag{11}$$

At $t = 0$ we want the temperature $f(x, y)$ to be represented by

$$u(x, y, 0) = f(x, y) = \sum_{m=1}^{\infty} \sum_{n=1}^{\infty} A_{mn} \sin \frac{m\pi}{b} x \sin \frac{n\pi}{c} y. \tag{12}$$

Finding the coefficients A_{mn} in (12) really does not pose a problem; we simply multiply the double sum (12) by the product $\sin (m\pi x/b) \sin (n\pi y/c)$ and integrate over the rectangle defined by $0 \le x \le b, 0 \le y \le c$. It follows that

$$A_{mn} = \frac{4}{bc} \int_0^c \int_0^b f(x, y) \sin \frac{m\pi}{b} x \sin \frac{n\pi}{c} y \, dx \, dy. \tag{13}$$

Thus, the solution of the boundary-value problem consists of (11) with the A_{mn} defined by (13). $\equiv$

The series (11) with coefficients (13) is called a **sine series in two variables**, or a **double sine series**. The **cosine series in two variables** of a function $f(x, y)$ is a little more complicated. If the function f is defined over a rectangular region defined by $0 \le x \le b$, $0 \le y \le c$, then the **double cosine series** is given by

$$f(x, y) = A_{00} + \sum_{m=1}^{\infty} A_{m0} \cos \frac{m\pi}{b} x + \sum_{n=1}^{\infty} A_{0n} \cos \frac{n\pi}{c} y$$

$$+ \sum_{m=1}^{\infty} \sum_{n=1}^{\infty} A_{mn} \cos \frac{m\pi}{b} x \cos \frac{n\pi}{c} y,$$

where

$$A_{00} = \frac{1}{bc} \int_0^c \int_0^b f(x, y)\, dx\, dy$$

$$A_{m0} = \frac{2}{bc} \int_0^c \int_0^b f(x, y) \cos \frac{m\pi}{b} x\, dx\, dy$$

$$A_{0n} = \frac{2}{bc} \int_0^c \int_0^b f(x, y) \cos \frac{n\pi}{c} y\, dx\, dy$$

$$A_{mn} = \frac{4}{bc} \int_0^c \int_0^b f(x, y) \cos \frac{m\pi}{b} x \cos \frac{n\pi}{c} y\, dx\, dy.$$

See Problem 2 in Exercises 13.8 for a boundary-value problem leading to a double cosine series.

13.8 Exercises Answers to selected odd-numbered problems begin on page ANS-36.

In Problems 1 and 2, solve the heat equation (1) subject to the given conditions.

1. $u(0, y, t) = 0$, $u(\pi, y, t) = 0$
$u(x, 0, t) = 0$, $u(x, \pi, t) = 0$
$u(x, y, 0) = u_0$

2. $\dfrac{\partial u}{\partial x}\Big|_{x=0} = 0$, $\quad \dfrac{\partial u}{\partial x}\Big|_{x=1} = 0$

$\dfrac{\partial u}{\partial y}\Big|_{y=0} = 0$, $\quad \dfrac{\partial u}{\partial y}\Big|_{y=1} = 0$

$u(x, y, 0) = xy$

In Problems 3 and 4, solve the wave equation (2) subject to the given conditions.

3. $u(0, y, t) = 0$, $u(\pi, y, t) = 0$
$u(x, 0, t) = 0$, $u(x, \pi, t) = 0$
$u(x, y, 0) = xy(x - \pi)(y - \pi)$

$\dfrac{\partial u}{\partial t}\Big|_{t=0} = 0$

4. $u(0, y, t) = 0$, $u(b, y, t) = 0$
$u(x, 0, t) = 0$, $u(x, c, t) = 0$
$u(x, y, 0) = f(x, y)$

$\dfrac{\partial u}{\partial t}\Big|_{t=0} = g(x, y)$

In Problems 5–7, solve Laplace's equation

$$\frac{\partial^2 u}{\partial x^2} + \frac{\partial^2 u}{\partial y^2} + \frac{\partial^2 u}{\partial z^2} = 0 \qquad (14)$$

for the steady-state temperature $u(x, y, z)$ in the rectangular parallelepiped shown in **FIGURE 13.8.2**.

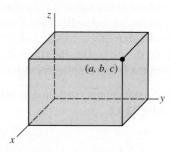

FIGURE 13.8.2 Rectangular parallelepiped in Problems 5–7

5. The top ($z = c$) of the parallelepiped is kept at temperature $f(x, y)$ and the remaining sides are kept at temperature zero.

6. The bottom ($z = 0$) of the parallelepiped is kept at temperature $f(x, y)$ and the remaining sides are kept at temperature zero.

7. The parallelepiped is a unit cube ($a = b = c = 1$) with the top ($z = 1$) and bottom ($z = 0$) kept at constant temperatures u_0 and $-u_0$, respectively, and the remaining sides kept at temperature zero.

In Problems 1 and 2, use separation of variables to find product solutions $u = X(x)Y(y)$ of the given partial differential equation.

1. $\dfrac{\partial^2 u}{\partial x \partial y} = u$

2. $\dfrac{\partial^2 u}{\partial x^2} + \dfrac{\partial^2 u}{\partial y^2} + 2\dfrac{\partial u}{\partial x} + 2\dfrac{\partial u}{\partial y} = 0$

3. Find a steady-state solution $\psi(x)$ of the boundary-value problem

$$k\dfrac{\partial^2 u}{\partial x^2} = \dfrac{\partial u}{\partial t}, \quad 0 < x < \pi, \ t > 0$$

$$u(0, t) = u_0, \quad -\dfrac{\partial u}{\partial x}\Big|_{x=\pi} = u(\pi, t) - u_1, \ t > 0$$

$$u(x, 0) = 0, \ 0 < x < \pi.$$

4. Give a physical interpretation for the boundary conditions in Problem 3.

5. At $t = 0$ a string of unit length is stretched on the positive x-axis. The ends of the string $x = 0$ and $x = 1$ are secured on the x-axis for $t > 0$. Find the displacement $u(x, t)$ if the initial velocity $g(x)$ is as given in **FIGURE 13.R.1**.

FIGURE 13.R.1 Initial velocity in Problem 5

6. The partial differential equation

$$\dfrac{\partial^2 u}{\partial x^2} + x^2 = \dfrac{\partial^2 u}{\partial t^2}$$

is a form of the wave equation when an external vertical force proportional to the square of the horizontal distance from the left end is applied to the string. The string is secured at $x = 0$ one unit above the x-axis and on the x-axis at $x = 1$ for $t > 0$. Find the displacement $u(x, t)$ if the string starts from rest from the initial displacement $f(x)$.

7. Find the steady-state temperature $u(x, y)$ in the square plate shown in **FIGURE 13.R.2**.

FIGURE 13.R.2 Square plate in Problem 7

8. Find the steady-state temperature $u(x, y)$ in the semi-infinite plate shown in **FIGURE 13.R.3**.

FIGURE 13.R.3 Semi-infinite plate in Problem 8

9. Solve Problem 8 if the boundaries $y = 0$ and $y = \pi$ are held at temperature zero for all time.

10. Find the temperature $u(x, t)$ in the infinite plate of width $2L$ shown in **FIGURE 13.R.4** if the initial temperature is u_0 throughout. [*Hint*: $u(x, 0) = u_0, -L < x < L$ is an even function of x.]

FIGURE 13.R.4 Infinite plate in Problem 10

11. (a) Solve the boundary-value problem

$$\dfrac{\partial^2 u}{\partial x^2} = \dfrac{\partial u}{\partial t}, \quad 0 < x < \pi, \ t > 0$$

$$u(0, t) = 0, \quad u(\pi, t) = 0, \ t > 0$$

$$u(x, 0) = \sin x, \ 0 < x < \pi.$$

(b) What is the solution of the BVP in part (a) if the initial temperature is

$$u(x, 0) = 100 \sin 3x - 30 \sin 5x?$$

12. Solve the boundary-value problem

$$\dfrac{\partial^2 u}{\partial x^2} + \sin x = \dfrac{\partial u}{\partial t}, \quad 0 < x < \pi, \ t > 0$$

$$u(0, t) = 400, \quad u(\pi, t) = 200, \ t > 0$$

$$u(x, 0) = 400 + \sin x, \ 0 < x < \pi.$$

13. Find a series solution of the problem

$$\frac{\partial^2 u}{\partial x^2} + 2\frac{\partial u}{\partial x} = \frac{\partial^2 u}{\partial t^2} + 2\frac{\partial u}{\partial t} + u, \ 0 < x < \pi, \ t > 0$$

$$u(0, t) = 0, \quad u(\pi, t) = 0, \ t > 0$$

$$\left. \frac{\partial u}{\partial t} \right|_{t=0} = 0, \ 0 < x < \pi.$$

Do not attempt to evaluate the coefficients in the series.

14. The concentration $c(x, t)$ of a substance that both diffuses in a medium and is convected by the currents in the medium satisfies the partial differential equation

$$k\frac{\partial^2 c}{\partial x^2} - h\frac{\partial c}{\partial x} = \frac{\partial c}{\partial t}, \ 0 < x < 1, \ t > 0,$$

where k and h are constants. Solve the PDE subject to

$$c(0, t) = 0, \quad c(1, t) = 0, \ t > 0$$

$$c(x, 0) = c_0, \ 0 < x < 1,$$

where c_0 is a constant.

15. Solve the boundary-value problem

$$\frac{\partial^2 u}{\partial x^2} = \frac{\partial u}{\partial t}, \ 0 < x < 1, \ t > 0$$

$$u(0, t) = u_0, \quad \left. \frac{\partial u}{\partial x} \right|_{x=1} = -u(1, t) + u_1, \ t > 0$$

$$u(x, 0) = u_0, \ 0 < x < 1,$$

where u_0 and u_1 are constants.

16. Solve Laplace's equation for a rectangular plate subject to the boundary-value conditions

$$u(0, y) = 0, \quad u(\pi, y) = 0, \ 0 < y < \pi$$

$$u(x, 0) = x^2 - \pi x, \quad u(x, \pi) = x^2 - \pi x, \ 0 < x < \pi.$$

17. Use the substitution $u(x, y) = v(x, y) + \psi(x)$ and the result of Problem 16 to solve the boundary-value problem

$$\frac{\partial^2 u}{\partial x^2} + \frac{\partial^2 u}{\partial y^2} = -2, \quad 0 < x < \pi, \ 0 < y < \pi$$

$$u(0, y) = 0, \quad u(\pi, y) = 0, \quad 0 < y < \pi$$

$$u(x, 0) = 0, \quad u(x, \pi) = 0, \quad 0 < x < \pi.$$

18. Solve the boundary-value problem

$$\frac{\partial^2 u}{\partial x^2} + e^x = \frac{\partial u}{\partial t}, \quad 0 < x < \pi, \ t > 0$$

$$u(0, t) = 0, \quad \left. \frac{\partial u}{\partial x} \right|_{x=\pi} = 0, \ t > 0$$

$$u(x, 0) = f(x), \ 0 < x < \pi.$$

19. A rectangular plate is described by the region in the xy-plane defined by $0 \le x \le a, 0 \le y \le b$. In the analysis of the deflection $w(x, y)$ of the plate under a sinusoidal load, the following linear fourth-order partial differential equation is encountered:

$$\frac{\partial^4 w}{\partial x^4} + 2\frac{\partial^4 w}{\partial x^2 \partial y^2} + \frac{\partial^4 w}{\partial y^4} = \frac{q_0}{D}\sin\frac{\pi x}{a}\sin\frac{\pi y}{b},$$

where q_0 and D are constants. Find a constant C so that the product

$$w(x, y) = C\sin\frac{\pi x}{a}\sin\frac{\pi y}{b}$$

is a particular solution of the PDE.

20. If the four edges of the rectangular plate in Problem 19 are simply supported, then show that the given particular solution satisfies the boundary conditions

$$w(0, y) = 0, \quad w(a, y) = 0, \quad 0 < y < b$$

$$w(x, 0) = 0, \quad w(x, b) = 0, \quad 0 < x < a$$

$$\left. \frac{\partial^2 w}{\partial x^2} \right|_{x=0} = 0, \quad \left. \frac{\partial^2 w}{\partial x^2} \right|_{x=a} = 0, \quad 0 < y < b$$

$$\left. \frac{\partial^2 w}{\partial y^2} \right|_{y=0} = 0, \quad \left. \frac{\partial^2 w}{\partial y^2} \right|_{y=b} = 0, \quad 0 < x < a.$$

Boundary-Value Problems in Other Coordinate Systems

In the previous chapter we utilized Fourier series to solve boundary-value problems that were described in rectangular coordinates. But if we wish to find, say, temperatures in a circular plate, circular cylinder, or sphere, we would naturally describe the problem in polar coordinates, cylindrical coordinates, or spherical coordinates, respectively. We will see that the method of separation of variables in the last two coordinate systems leads to Fourier–Bessel series and Fourier–Legendre series.

14.1 Polar Coordinates

INTRODUCTION All the boundary-value problems that have been considered so far have been expressed in terms of rectangular coordinates. If, however, we wish to find temperatures in a circular disk, in a circular cylinder, or in a sphere, we would naturally try to describe the problems in polar coordinates, cylindrical coordinates, or spherical coordinates, respectively.

Because we consider only steady-state temperature problems in polar coordinates in this section, the first thing that must be done is to convert the familiar Laplace's equation in rectangular coordinates to polar coordinates.

Laplacian in Polar Coordinates The relationships between polar coordinates in the plane and rectangular coordinates are given by

$$x = r\cos\theta, \qquad y = r\sin\theta \qquad \text{and} \qquad r^2 = x^2 + y^2, \qquad \tan\theta = \frac{y}{x}.$$

FIGURE 14.1.1 Polar coordinates of a point (x, y) are (r, θ)

See **FIGURE 14.1.1**. The first pair of equations transform polar coordinates (r, θ) into rectangular coordinates (x, y); the second pair of equations enable us to transform rectangular coordinates into polar coordinates. These equations also make it possible to convert the two-dimensional Laplacian of the function u, $\nabla^2 u = \partial^2 u/\partial x^2 + \partial^2 u/\partial y^2$, into polar coordinates. You are encouraged to work through the details of the Chain Rule and show that

$$\frac{\partial u}{\partial x} = \frac{\partial u}{\partial r}\frac{\partial r}{\partial x} + \frac{\partial u}{\partial \theta}\frac{\partial \theta}{\partial x} = \cos\theta\,\frac{\partial u}{\partial r} - \frac{\sin\theta}{r}\frac{\partial u}{\partial \theta}$$

$$\frac{\partial u}{\partial y} = \frac{\partial u}{\partial r}\frac{\partial r}{\partial y} + \frac{\partial u}{\partial \theta}\frac{\partial \theta}{\partial y} = \sin\theta\,\frac{\partial u}{\partial r} + \frac{\cos\theta}{r}\frac{\partial u}{\partial \theta}$$

$$\frac{\partial^2 u}{\partial x^2} = \cos^2\theta\,\frac{\partial^2 u}{\partial r^2} - \frac{2\sin\theta\cos\theta}{r}\frac{\partial^2 u}{\partial r\partial\theta} + \frac{\sin^2\theta}{r^2}\frac{\partial^2 u}{\partial\theta^2} + \frac{\sin^2\theta}{r}\frac{\partial u}{\partial r} + \frac{2\sin\theta\cos\theta}{r^2}\frac{\partial u}{\partial\theta} \qquad (1)$$

$$\frac{\partial^2 u}{\partial y^2} = \sin^2\theta\,\frac{\partial^2 u}{\partial r^2} + \frac{2\sin\theta\cos\theta}{r}\frac{\partial^2 u}{\partial r\partial\theta} + \frac{\cos^2\theta}{r^2}\frac{\partial^2 u}{\partial\theta^2} + \frac{\cos^2\theta}{r}\frac{\partial u}{\partial r} - \frac{2\sin\theta\cos\theta}{r^2}\frac{\partial u}{\partial\theta}. \qquad (2)$$

Adding (1) and (2) and simplifying yields the Laplacian of u in polar coordinates:

$$\nabla^2 u = \frac{\partial^2 u}{\partial r^2} + \frac{1}{r}\frac{\partial u}{\partial r} + \frac{1}{r^2}\frac{\partial^2 u}{\partial\theta^2}.$$

In this section we shall concentrate only on boundary-value problems involving Laplace's equation in polar coordinates:

$$\frac{\partial^2 u}{\partial r^2} + \frac{1}{r}\frac{\partial u}{\partial r} + \frac{1}{r^2}\frac{\partial^2 u}{\partial\theta^2} = 0. \qquad (3)$$

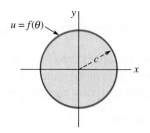

FIGURE 14.1.2 Dirichlet problem for a circular plate

Our first example is the Dirichlet problem for a circular disk. We wish to solve Laplace's equation (3) for the steady-state temperature $u(r, \theta)$ in a circular disk or plate of radius c when the temperature of the circumference is $u(c, \theta) = f(\theta)$, $0 < \theta < 2\pi$. See **FIGURE 14.1.2**. It is assumed that the two faces of the plate are insulated. This seemingly simple problem is unlike any we have encountered in the previous chapter.

EXAMPLE 1 Steady Temperatures in a Circular Plate

Solve Laplace's equation (3) subject to $u(c, \theta) = f(\theta)$, $0 < \theta < 2\pi$.

SOLUTION Before attempting separation of variables we note that the single boundary condition is nonhomogeneous. In other words, there are no explicit conditions in the statement of the problem that enable us to determine either the coefficients in the solutions of the separated ODEs or the required eigenvalues. However, there are some implicit conditions.

First, our physical intuition leads us to expect that the temperature $u(r, \theta)$ should be continuous and therefore bounded inside the circle $r = c$. In addition, the temperature $u(r, \theta)$

should be single-valued; this means that the value of u should be the same at a specified point in the plate regardless of the polar description of that point. Since $(r, \theta + 2\pi)$ is an equivalent description of the point (r, θ), we must have $u(r, \theta) = u(r, \theta + 2\pi)$. That is, $u(r, \theta)$ must be periodic in θ with period 2π. If we seek a product solution $u = R(r)\Theta(\theta)$, then $\Theta(\theta)$ needs to be 2π-periodic.

With all this in mind, we choose to write the separation constant in the separation of variables as λ:

$$\frac{r^2 R'' + rR'}{R} = -\frac{\Theta''}{\Theta} = \lambda.$$

The separated equations are then

$$r^2 R'' + rR' - \lambda R = 0 \tag{4}$$

$$\Theta'' + \lambda \Theta = 0 \tag{5}$$

We are seeking a solution of the problem

$$\Theta'' + \lambda \Theta = 0, \quad \Theta(\theta) = \Theta(\theta + 2\pi). \tag{6}$$

Although (6) is not a regular Sturm–Liouville problem, nonetheless the problem generates eigenvalues and eigenfunctions. The latter form an orthogonal set on the interval $[0, 2\pi]$.

Of the three possible general solutions of (5),

$$\Theta(\theta) = c_1 + c_2\theta, \qquad \qquad \lambda = 0 \tag{7}$$

$$\Theta(\theta) = c_1 \cosh \alpha\theta + c_2 \sinh \alpha\theta, \qquad \lambda = -\alpha^2 < 0 \tag{8}$$

$$\Theta(\theta) = c_1 \cos \alpha\theta + c_2 \sin \alpha\theta, \qquad \lambda = \alpha^2 > 0, \tag{9}$$

we can dismiss (8) as inherently nonperiodic unless $c_1 = c_2 = 0$. Similarly, solution (7) is nonperiodic unless we define $c_2 = 0$. The remaining constant solution $\Theta(\theta) = c_1, c_1 \neq 0$, can be assigned any period and so $\lambda = 0$ is an eigenvalue. Finally, solution (9) will be 2π-periodic if we take $\alpha = n$, where $n = 1, 2, \ldots$.* The eigenvalues of (6) are then $\lambda_0 = 0$ and $\lambda_n = n^2$, $n = 1, 2, \ldots$. If we correspond $\lambda_0 = 0$ with $n = 0$, the eigenfunctions of (6) are

$$\Theta(\theta) = c_1, \; n = 0, \quad \text{and} \quad \Theta(\theta) = c_1 \cos n\theta + c_2 \sin n\theta, \; n = 1, 2, \ldots.$$

When $\lambda_n = n^2, n = 0, 1, 2, \ldots$ the solutions of the Cauchy–Euler DE (4) are

$$R(r) = c_3 + c_4 \ln r, \; n = 0, \tag{10}$$

$$R(r) = c_3 r^n + c_4 r^{-n}, \; n = 1, 2, \ldots. \tag{11}$$

Now observe in (11) that $r^{-n} = 1/r^n$. In either of the solutions (10) or (11), we must define $c_4 = 0$ in order to guarantee that the solution u is bounded at the center of the plate (which is $r = 0$). Thus product solutions $u_n = R(r)\Theta(\theta)$ for Laplace's equation in polar coordinates are

$$u_0 = A_0, \; n = 0, \quad \text{and} \quad u_n = r^n(A_n \cos n\theta + B_n \sin n\theta), \; n = 1, 2, \ldots,$$

where we have replaced $c_3 c_1$ by A_0 for $n = 0$ and by A_n for $n = 1, 2, \ldots$; the product $c_3 c_2$ has been replaced by B_n. The superposition principle then gives

$$u(r, \theta) = A_0 + \sum_{n=1}^{\infty} r^n(A_n \cos n\theta + B_n \sin n\theta). \tag{12}$$

By applying the boundary condition at $r = c$ to the result in (12) we recognize

$$f(\theta) = A_0 + \sum_{n=1}^{\infty} c^n(A_n \cos n\theta + B_n \sin n\theta)$$

*For example, note that $\cos n(\theta + 2\pi) = \cos(n\theta + 2n\pi) = \cos n\theta$.

as an expansion of f in a full Fourier series. Consequently we can make the identifications

$$A_0 = \frac{a_0}{2}, \quad c^n A_n = a_n, \quad \text{and} \quad c^n B_n = b_n.$$

That is,

$$A_0 = \frac{1}{2\pi} \int_0^{2\pi} f(\theta)\, d\theta \tag{13}$$

$$A_n = \frac{1}{c^n \pi} \int_0^{2\pi} f(\theta) \cos n\theta\, d\theta \tag{14}$$

$$B_n = \frac{1}{c^n \pi} \int_0^{2\pi} f(\theta) \sin n\theta\, d\theta. \tag{15}$$

The solution of the problem consists of the series given in (12), where the coefficients A_0, A_n, and B_n are defined in (13), (14), and (15), respectively. ≡

Observe in Example 1 that corresponding to each *positive* eigenvalue, $\lambda_n = n^2$, $n = 1, 2, \ldots$, there are two different eigenfunctions—namely, $\cos n\theta$ and $\sin n\theta$. In this situation the eigenvalues are sometimes called **double eigenvalues**.

EXAMPLE 2　Steady Temperatures in a Semicircular Plate

Find the steady-state temperature $u(r, \theta)$ in the semicircular plate shown in **FIGURE 14.1.3**.

SOLUTION　The boundary-value problem is

$$\frac{\partial^2 u}{\partial r^2} + \frac{1}{r}\frac{\partial u}{\partial r} + \frac{1}{r^2}\frac{\partial^2 u}{\partial \theta^2} = 0, \quad 0 < \theta < \pi, \quad 0 < r < c$$

$$u(c, \theta) = u_0, \quad 0 < \theta < \pi$$

$$u(r, 0) = 0, \quad u(r, \pi) = 0, \quad 0 < r < c.$$

FIGURE 14.1.3 Semicircular plate in Example 2

Defining $u = R(r)\Theta(\theta)$ and separating variables gives

$$\frac{r^2 R'' + rR'}{R} = -\frac{\Theta''}{\Theta} = \lambda$$

and

$$r^2 R'' + rR' - \lambda R = 0 \tag{16}$$

$$\Theta'' + \lambda\Theta = 0. \tag{17}$$

The homogeneous conditions stipulated at the boundaries $\theta = 0$ and $\theta = \pi$ translate into $\Theta(0) = 0$ and $\Theta(\pi) = 0$. These conditions together with equation (17) constitute a regular Sturm–Liouville problem:

$$\Theta'' + \lambda\Theta = 0, \quad \Theta(0) = 0, \quad \Theta(\pi) = 0. \tag{18}$$

This familiar problem* possesses eigenvalues $\lambda_n = n^2$ and eigenfunctions $\Theta(\theta) = c_2 \sin n\theta$, $n = 1, 2, \ldots$. Also, by replacing λ by n^2 the solution of (16) is $R(r) = c_3 r^n + c_4 r^{-n}$. The reasoning used in Example 1, namely, that we expect a solution u of the problem to be bounded at $r = 0$, prompts us to define $c_4 = 0$. Therefore $u_n = R(r)\Theta(\theta) = A_n r^n \sin n\theta$ and

$$u(r, \theta) = \sum_{n=1}^{\infty} A_n r^n \sin n\theta.$$

The remaining boundary condition at $r = c$ gives the Fourier sine series

$$u_0 = \sum_{n=1}^{\infty} A_n c^n \sin n\theta.$$

Consequently

$$A_n c^n = \frac{2}{\pi} \int_0^{\pi} u_0 \sin n\theta\, d\theta,$$

*The problem in (18) is Example 2 of Section 3.9 with $L = \pi$.

and so
$$A_n = \frac{2u_0}{\pi c^n} \frac{1 - (-1)^n}{n}.$$

Hence the solution of the problem is given by

$$u(r, \theta) = \frac{2u_0}{\pi} \sum_{n=1}^{\infty} \frac{1 - (-1)^n}{n} \left(\frac{r}{c}\right)^n \sin n\theta.$$

≡

14.1 Exercises Answers to selected odd-numbered problems begin on page ANS-36.

In Problems 1–4, find the steady-state temperature $u(r, \theta)$ in a circular plate of radius 1 if the temperature on the circumference is as given.

1. $u(1, \theta) = \begin{cases} u_0, & 0 < \theta < \pi \\ 0, & \pi < \theta < 2\pi \end{cases}$

2. $u(1, \theta) = \begin{cases} \theta, & 0 < \theta < \pi \\ \pi - \theta, & \pi < \theta < 2\pi \end{cases}$

3. $u(1, \theta) = 2\pi\theta - \theta^2, \quad 0 < \theta < 2\pi$

4. $u(1, \theta) = \theta, \quad 0 < \theta < 2\pi$

5. If the boundaries $\theta = 0$ and $\theta = \pi$ of a semicircular plate of radius 2 are insulated, we then have

$$\frac{\partial u}{\partial \theta}\bigg|_{\theta=0} = 0, \quad \frac{\partial u}{\partial \theta}\bigg|_{\theta=\pi} = 0, \quad 0 < r < 2.$$

Find the steady-state temperature $u(r, \theta)$ if

$$u(2, \theta) = \begin{cases} u_0, & 0 < \theta < \pi/2 \\ 0, & \pi/2 < \theta < \pi, \end{cases}$$

where u_0 is a constant.

6. Find the steady-state temperature $u(r, \theta)$ in a semicircular plate of radius 1 if the boundary conditions are

$$u(1, \theta) = u_0, \quad 0 < \theta < \pi$$
$$u(r, 0) = 0, \quad u(r, \pi) = u_0, \quad 0 < r < 1,$$

where u_0 is a constant.

7. Find the steady-state temperature $u(r, \theta)$ in the quarter-circular plate shown in **FIGURE 14.1.4**.

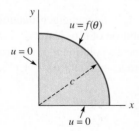

FIGURE 14.1.4 Quarter-circular plate in Problem 7

8. Find the steady-state temperature $u(r, \theta)$ in the quarter-circular plate shown in Figure 14.1.4 if the boundaries $\theta = 0$ and $\theta = \pi/2$ are insulated, and

$$u(c, \theta) = \begin{cases} 1, & 0 < \theta < \pi/4 \\ 0, & \pi/4 < \theta < \pi. \end{cases}$$

9. Find the steady-state temperature $u(r, \theta)$ in the portion of a circular plate shown in **FIGURE 14.1.5**.

FIGURE 14.1.5 Portion of a circular plate in Problem 9

10. Find the steady-state temperature $u(r, \theta)$ in the infinite wedge-shaped plate shown in **FIGURE 14.1.6**. [*Hint*: Assume that the temperature is bounded as $r \to 0$ and as $r \to \infty$.]

FIGURE 14.1.6 Wedge-shaped plate in Problem 10

11. Find the steady-state temperature $u(r, \theta)$ in the plate in the form of an annulus bounded between two concentric circles of radius a and b, $a < b$, shown in **FIGURE 14.1.7**. [*Hint*: Proceed as in Example 1.]

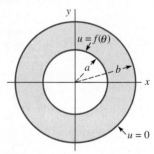

FIGURE 14.1.7 Annular plate in Problem 11

12. If the boundary conditions for the annular plate in Figure 14.1.7 are

$$u(a, \theta) = u_0, \quad u(b, \theta) = u_1, \quad 0 < \theta < 2\pi,$$

where u_0 and u_1 are constants, show that the steady-state temperature is given by

$$u(r, \theta) = \frac{u_0 \ln(r/b) - u_1 \ln(r/a)}{\ln(a/b)}.$$

[*Hint*: Try a solution of the form $u(r, \theta) = v(r, \theta) + \psi(r)$.]

13. Find the steady-state temperature $u(r, \theta)$ in the annular plate shown in Figure 14.1.7 if the boundary conditions are

$$\left.\frac{\partial u}{\partial r}\right|_{r=a} = 0, \quad u(b, \theta) = f(\theta), \quad 0 < \theta < 2\pi.$$

14. Find the steady-state temperature $u(r, \theta)$ in the annular plate shown in Figure 14.1.7 if $a = 1$, $b = 2$, and

$$u(1, \theta) = 75 \sin \theta, \quad u(2, \theta) = 60 \cos \theta, \quad 0 < \theta < 2\pi.$$

15. Find the steady-state temperature $u(r, \theta)$ in the semiannular plate shown in **FIGURE 14.1.8** if the boundary conditions are

$$u(a, \theta) = \theta(\pi - \theta), \quad u(b, \theta) = 0, \quad 0 < \theta < \pi$$
$$u(r, 0) = 0, \quad u(r, \pi) = 0, \quad a < r < b.$$

FIGURE 14.1.8 Semiannular plate in Problem 15

16. Find the steady-state temperature $u(r, \theta)$ in the semiannular plate shown in Figure 14.1.8 if $a = 1$, $b = 2$, and

$$u(1, \theta) = 0, \quad u(2, \theta) = u_0, \quad 0 < \theta < \pi$$
$$u(r, 0) = 0, \quad u(r, \pi) = 0, \quad 1 < r < 2,$$

where u_0 is a constant.

17. Find the steady-state temperature $u(r, \theta)$ in the quarter-annular plate shown in **FIGURE 14.1.9**.

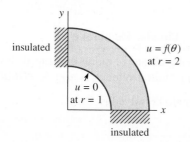

FIGURE 14.1.9 Quarter-annular plate in Problem 17

18. The plate in the first quadrant shown in **FIGURE 14.1.10** is one-eighth of the annular plate in Figure 14.1.7. Find the steady-state temperature $u(r, \theta)$.

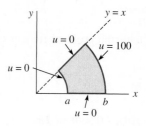

FIGURE 14.1.10 One-eighth annular plate in Problem 18

19. Solve the *exterior* Dirichlet problem for a circular disk of radius c shown in **FIGURE 14.1.11**. In other words, find the steady-state temperature $u(r, \theta)$ in an infinite plate that coincides with the entire *xy*-plane in which a circular hole of radius c has been cut out around the origin and the temperature on the circumference of the hole is $f(\theta)$. [*Hint*: Assume that the temperature u is bounded as $r \to \infty$.]

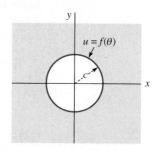

FIGURE 14.1.11 Infinite plate in Problem 19

20. Consider the steady-state temperature $u(r, \theta)$ in the semiannular plate shown in Figure 14.1.8 with $a = 1$, $b = 2$, and boundary conditions

$$u(1, \theta) = 0, \quad u(2, \theta) = 0, \quad 0 < \theta < \pi$$
$$u(r, 0) = 0, \quad u(r, \pi) = r, \quad 1 < r < 2.$$

Show that in this case the choice of $\lambda = -\alpha^2$ in (4) and (5) leads to eigenvalues and eigenfunctions. Find the steady-state temperature $u(r, \theta)$.

Computer Lab Assignments

21. (a) Find the series solution for $u(r, \theta)$ in Example 1 when

$$u(1, \theta) = \begin{cases} 100, & 0 < \theta < \pi \\ 0, & \pi < \theta < 2\pi. \end{cases}$$

See Problem 1.

(b) Use a CAS or a graphing utility to plot the partial sum $S_5(r, \theta)$ consisting of the first five nonzero terms of the solution in part (a) for $r = 0.9$, $r = 0.7$, $r = 0.5$, $r = 0.3$, $r = 0.1$. Superimpose the graphs on the same coordinate axes.

(c) Approximate the temperatures $u(0.9, 1.3)$, $u(0.7, 2)$, $u(0.5, 3.5)$, $u(0.3, 4)$, $u(0.1, 5.5)$. Then approximate $u(0.9, 2\pi - 1.3)$, $u(0.7, 2\pi - 2)$, $u(0.5, 2\pi - 3.5)$, $u(0.3, 2\pi - 4)$, $u(0.1, 2\pi - 5.5)$.

(d) What is the temperature at the center of the circular plate? Why is it appropriate to call this value the *average temperature* in the plate? [*Hint*: Look at the graphs in part (b) and look at the numbers in part (c).]

Discussion Problems

22. Solve the Neumann problem for a circular plate:

$$\frac{\partial^2 u}{\partial r^2} + \frac{1}{r}\frac{\partial u}{\partial r} + \frac{1}{r^2}\frac{\partial^2 u}{\partial \theta^2} = 0, \quad 0 < \theta < 2\pi, \quad 0 < r < c$$

$$\left.\frac{\partial u}{\partial r}\right|_{r=c} = f(\theta), \quad 0 < \theta < 2\pi.$$

Give the compatibility condition. [*Hint*: See Problem 21 of Exercises 13.5.]

23. Consider the annular plate shown in Figure 14.1.7. Discuss how the steady-state temperature $u(r, \theta)$ can be found when the boundary conditions are

$$u(a, \theta) = f(\theta), \quad u(b, \theta) = g(\theta), \quad 0 \le \theta \le 2\pi.$$

24. Verify that $u(r, \theta) = \frac{3}{4}r\sin\theta - \frac{1}{4}r^3\sin 3\theta$ is a solution of the boundary-value problem

$$\frac{\partial^2 u}{\partial r^2} + \frac{1}{r}\frac{\partial u}{\partial r} + \frac{1}{r^2}\frac{\partial^2 u}{\partial \theta^2} = 0, \quad 0 < \theta < 2\pi, \quad 0 < r < 1$$

$$u(1, \theta) = \sin^3\theta, \quad 0 < \theta < 2\pi.$$

14.2 Cylindrical Coordinates

INTRODUCTION In this section we are going to consider boundary-value problems involving forms of the heat and wave equation in polar coordinates and a form of Laplace's equation in cylindrical coordinates. There is a commonality throughout the examples and most of the exercises—the boundary-value problem possesses radial symmetry.

■ **Radial Symmetry** The two-dimensional heat and wave equations

$$k\left(\frac{\partial^2 u}{\partial x^2} + \frac{\partial^2 u}{\partial y^2}\right) = \frac{\partial u}{\partial t} \quad \text{and} \quad a^2\left(\frac{\partial^2 u}{\partial x^2} + \frac{\partial^2 u}{\partial y^2}\right) = \frac{\partial^2 u}{\partial t^2}$$

expressed in polar coordinates are, in turn,

$$k\left(\frac{\partial^2 u}{\partial r^2} + \frac{1}{r}\frac{\partial u}{\partial r} + \frac{1}{r^2}\frac{\partial^2 u}{\partial \theta^2}\right) = \frac{\partial u}{\partial t} \quad \text{and} \quad a^2\left(\frac{\partial^2 u}{\partial r^2} + \frac{1}{r}\frac{\partial u}{\partial r} + \frac{1}{r^2}\frac{\partial^2 u}{\partial \theta^2}\right) = \frac{\partial^2 u}{\partial t^2}, \tag{1}$$

where $u = u(r, \theta, t)$. To solve a boundary-value problem involving either of these equations by separation of variables we must define $u = R(r)\Theta(\theta)T(t)$. As in Section 13.8, this assumption leads to multiple infinite series. See Problem 18 in Exercises 14.2. In the discussion that follows we shall consider the simpler, but still important, problems that possess **radial symmetry**—that is, problems in which the unknown function u is independent of the angular coordinate θ. In this case the heat and wave equations in (1) take, in turn, the forms

$$k\left(\frac{\partial^2 u}{\partial r^2} + \frac{1}{r}\frac{\partial u}{\partial r}\right) = \frac{\partial u}{\partial t} \quad \text{and} \quad a^2\left(\frac{\partial^2 u}{\partial r^2} + \frac{1}{r}\frac{\partial u}{\partial r}\right) = \frac{\partial^2 u}{\partial t^2}, \tag{2}$$

where $u = u(r, t)$. Vibrations described by the second equation in (2) are said to be **radial vibrations**.

The first example deals with the free undamped radial vibrations of a thin circular membrane. We assume that the displacements are small and that the motion is such that each point on the membrane moves in a direction perpendicular to the *xy*-plane (transverse vibrations)—that is, the *u*-axis is perpendicular to the *xy*-plane. A physical model to keep in mind while studying this example is a vibrating drumhead.

EXAMPLE 1 Radial Vibrations of a Circular Membrane

Find the displacement $u(r, t)$ of a circular membrane of radius c clamped along its circumference if its initial displacement is $f(r)$ and its initial velocity is $g(r)$. See **FIGURE 14.2.1**.

SOLUTION The boundary-value problem to be solved is

$$a^2\left(\frac{\partial^2 u}{\partial r^2} + \frac{1}{r}\frac{\partial u}{\partial r}\right) = \frac{\partial^2 u}{\partial t^2}, \quad 0 < r < c, \quad t > 0$$

$$u(c, t) = 0, \quad t > 0$$

$$u(r, 0) = f(r), \quad \left.\frac{\partial u}{\partial t}\right|_{t=0} = g(r), \quad 0 < r < c.$$

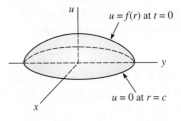

FIGURE 14.2.1 Initial displacement of circular membrane in Example 1

Substituting $u = R(r)T(t)$ into the partial differential equation and separating variables gives

$$\frac{R'' + \frac{1}{r}R'}{R} = \frac{T''}{a^2 T} = -\lambda. \tag{3}$$

Note in (3) we have returned to our usual separation constant $-\lambda$. The two equations obtained from (3) are

$$rR'' + R' + \lambda rR = 0 \tag{4}$$

and

$$T'' + a^2 \lambda T = 0. \tag{5}$$

Because of the vibrational nature of the problem, equation (5) suggests that we use only $\lambda = \alpha^2 > 0$, $\alpha > 0$. Now (4) is not a Cauchy–Euler equation but is the parametric Bessel differential equation of order $\nu = 0$; that is, $rR'' + R' + \alpha^2 rR = 0$. From (13) of Section 5.3 the general solution of the last equation is

$$R(r) = c_1 J_0(\alpha r) + c_2 Y_0(\alpha r). \tag{6}$$

The general solution of the familiar equation (5) is

$$T(t) = c_3 \cos a\alpha t + c_4 \sin a\alpha t.$$

Now recall, the Bessel function of the second kind of order zero has the property that $Y_0(\alpha r) \to -\infty$ as $r \to 0^+$, and so the implicit assumption that the displacement $u(r, t)$ should be bounded at $r = 0$ forces us to define $c_2 = 0$ in (6). Thus $R(r) = c_1 J_0(\alpha r)$.

Since the boundary condition $u(c, t) = 0$ is equivalent to $R(c) = 0$, we must have $c_1 J_0(\alpha c) = 0$. We rule out $c_1 = 0$ (this would lead to a trivial solution of the PDE), so consequently

$$J_0(\alpha c) = 0. \tag{7}$$

If $x_n = \alpha_n c$ are the positive roots of (7), then $\alpha_n = x_n/c$ and so the eigenvalues of the problem are $\lambda_n = \alpha_n^2 = x_n^2/c^2$ and the eigenfunctions are $c_1 J_0(\alpha_n r)$. Product solutions that satisfy the partial differential equation and the boundary condition are

$$u_n = R(r)T(t) = (A_n \cos a\alpha_n t + B_n \sin a\alpha_n t) J_0(\alpha_n r), \tag{8}$$

where we have done the usual relabeling of the constants. The superposition principle then gives

$$u(r, t) = \sum_{n=1}^{\infty} (A_n \cos a\alpha_n t + B_n \sin a\alpha_n t) J_0(\alpha_n r). \tag{9}$$

The given initial conditions determine the coefficients A_n and B_n.

Setting $t = 0$ in (9) and using $u(r, 0) = f(r)$ gives

$$f(r) = \sum_{n=1}^{\infty} A_n J_0(\alpha_n r). \tag{10}$$

This last result is recognized as the Fourier–Bessel expansion of the function f on the interval $(0, c)$. Hence by a direct comparison of (7) and (10) with (8) and (15) of Section 12.6, we can identify the coefficients A_n with those given in (16) of Section 12.6:

$$A_n = \frac{2}{c^2 J_1^2(\alpha_n c)} \int_0^c r J_0(\alpha_n r) f(r)\, dr. \tag{11}$$

Next, we differentiate (9) with respect to t, set $t = 0$, and use $u_t(r, 0) = g(r)$:

$$g(r) = \sum_{n=1}^{\infty} a\alpha_n B_n J_0(\alpha_n r).$$

This is now a Fourier–Bessel expansion of the function g. By identifying the total coefficient $a\alpha_n B_n$ with (16) of Section 12.6 we can write

$$B_n = \frac{2}{a\alpha_n c^2 J_1^2(\alpha_n c)} \int_0^c r J_0(\alpha_n r) g(r)\, dr. \tag{12}$$

Finally, the solution of the given boundary-value problem is the series (9) with coefficients A_n and B_n defined in (11) and (12), respectively. ≡

$n = 1$
(a)

$n = 2$
(b)

$n = 3$
(c)

FIGURE 14.2.2 Standing waves

Standing Waves

Analogous to (11) of Section 13.4, the product solutions (8) are called **standing waves**. For $n = 1, 2, 3, \ldots$, the standing waves are basically the graph of $J_0(\alpha_n r)$ with the time-varying amplitude

$$A_n \cos a\alpha_n t + B_n \sin a\alpha_n t.$$

The standing waves at different values of time are represented by the dashed graphs in **FIGURE 14.2.2**. The zeros of each standing wave in the interval $(0, c)$ are the roots of $J_0(\alpha_n r) = 0$ and correspond to the set of points on a standing wave where there is no motion. This set of points is called a **nodal line**. If (as in Example 1) the positive roots of $J_0(\alpha_n c) = 0$ are denoted by x_n, then $x_n = \alpha_n c$ implies $\alpha_n = x_n/c$ and consequently the zeros of the standing waves are determined from

$$J_0(\alpha_n r) = J_0\left(\frac{x_n}{c} r\right) = 0.$$

Now from Table 5.3.1, the first three positive zeros of J_0 are (approximately) $x_1 = 2.4$, $x_2 = 5.5$, and $x_3 = 8.7$. Thus for $n = 1$, the first positive root of

$$J_0\left(\frac{x_1}{c} r\right) = 0 \quad \text{is} \quad \frac{2.4}{c} r = 2.4 \quad \text{or} \quad r = c.$$

Since we are seeking zeros of the standing waves in the open interval $(0, c)$, the last result means that the first standing wave has no nodal line. For $n = 2$, the first two positive roots of

$$J_0\left(\frac{x_2}{c} r\right) = 0 \quad \text{are determined from} \quad \frac{5.5}{c} r = 2.4 \quad \text{and} \quad \frac{5.5}{c} r = 5.5.$$

Thus the second standing wave has one nodal line defined by $r = x_1 c/x_2 = 2.4c/5.5$. Note that $r \approx 0.44c < c$. For $n = 3$, a similar analysis shows that there are two nodal lines defined by $r = x_1 c/x_3 = 2.4c/8.7$ and $r = x_2 c/x_3 = 5.5c/8.7$. In general, the nth standing wave has $n - 1$ nodal lines $r = x_1 c/x_n$, $r = x_2 c/x_n$, $\ldots$, $r = x_{n-1} c/x_n$. Since $r = constant$ is an equation of a circle in polar coordinates, we see in Figure 14.2.2 that the nodal lines of a standing wave are concentric circles.

Use of Computers

It is possible to see the effect of a single drumbeat for the model solved in Example 1 by means of the animation capabilities of a computer algebra system. In Problem 21 in Exercises 14.2 you are asked to find the solution given in (9) when

$$c = 1, \quad f(r) = 0, \quad \text{and} \quad g(r) = \begin{cases} -v_0, & 0 \le r < b \\ 0, & b \le r < 1. \end{cases}$$

Some frames of a "movie" of the vibrating drumhead are given in **FIGURE 14.2.3**.

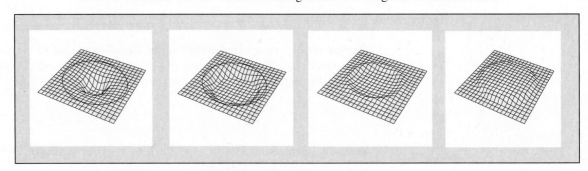

FIGURE 14.2.3 Frames of a CAS "movie"

Laplacian in Cylindrical Coordinates

From **FIGURE 14.2.4** we can see that the relationship between the cylindrical coordinates of a point in space and its rectangular coordinates is given by

$$x = r \cos \theta, \quad y = r \sin \theta, \quad z = z.$$

We saw in Section 14.1 that the Laplacian of a function u in polar coordinates (r, θ) is

$$\nabla^2 u = \frac{\partial^2 u}{\partial r^2} + \frac{1}{r} \frac{\partial u}{\partial r} + \frac{1}{r^2} \frac{\partial^2 u}{\partial \theta^2}.$$

(x, y, z) or
(r, θ, z)

FIGURE 14.2.4 Cylindrical coordinates of a point (x, y, z) are (r, θ, z)

It follows with no extra work that the Laplacian of a function u in cylindrical coordinates (r, θ, z) is then

$$\nabla^2 u = \frac{\partial^2 u}{\partial r^2} + \frac{1}{r}\frac{\partial u}{\partial r} + \frac{1}{r^2}\frac{\partial^2 u}{\partial \theta^2} + \frac{\partial^2 u}{\partial z^2}.$$

Laplace's equation is then

$$\frac{\partial^2 u}{\partial r^2} + \frac{1}{r}\frac{\partial u}{\partial r} + \frac{1}{r^2}\frac{\partial^2 u}{\partial \theta^2} + \frac{\partial^2 u}{\partial z^2} = 0. \tag{13}$$

If u does not depend on the polar angle θ, then the radially symmetric form of Laplace's equation is

$$\frac{\partial^2 u}{\partial r^2} + \frac{1}{r}\frac{\partial u}{\partial r} + \frac{\partial^2 u}{\partial z^2} = 0. \tag{14}$$

EXAMPLE 2 Steady Temperatures in a Circular Cylinder

Find the steady-state temperature in the circular cylinder shown in **FIGURE 14.2.5**.

SOLUTION The boundary conditions suggest that the temperature u has radial symmetry. Accordingly, $u(r, z)$ is determined from

$$\frac{\partial^2 u}{\partial r^2} + \frac{1}{r}\frac{\partial u}{\partial r} + \frac{\partial^2 u}{\partial z^2} = 0, \quad 0 < r < 2, \quad 0 < z < 4$$

$$u(2, z) = 0, \quad 0 < z < 4$$

$$u(r, 0) = 0, \quad u(r, 4) = u_0, \quad 0 < r < 2.$$

Using $u = R(r)Z(z)$ and separating variables gives

$$\frac{R'' + \dfrac{1}{r}R'}{R} = -\frac{Z''}{Z} = -\lambda \tag{15}$$

and

$$rR'' + R' + \lambda rR = 0 \tag{16}$$

$$Z'' - \lambda Z = 0. \tag{17}$$

For the choice $\lambda = \alpha^2 > 0$, $\alpha > 0$, the general solution of (16) is

$$R(r) = c_1 J_0(\alpha r) + c_2 Y_0(\alpha r),$$

and since a solution of (17) is defined on the finite interval $[0, 2]$, we write its general solution as

$$Z(z) = c_3 \cosh \alpha z + c_4 \sinh \alpha z.$$

As in Example 1, the assumption that the temperature u is bounded at $r = 0$ demands that $c_2 = 0$. The condition $u(2, z) = 0$ implies $R(2) = 0$. This equation,

$$J_0(2\alpha) = 0, \tag{18}$$

defines the positive eigenvalues $\lambda_n = \alpha_n^2$ of the problem. Last, $Z(0) = 0$ implies $c_3 = 0$. Hence we have $R = c_1 J_0(\alpha_n r)$, $Z = c_4 \sinh \alpha_n z$,

$$u_n = R(r)Z(z) = A_n \sinh \alpha_n z J_0(\alpha_n r)$$

and

$$u(r, z) = \sum_{n=1}^{\infty} A_n \sinh \alpha_n z J_0(\alpha_n r).$$

The remaining boundary condition at $z = 4$ then yields the Fourier–Bessel series

$$u_0 = \sum_{n=1}^{\infty} A_n \sinh 4\alpha_n J_0(\alpha_n r),$$

so that in view of (18) the coefficients are defined by (16) of Section 12.6,

$$A_n \sinh 4\alpha_n = \frac{2u_0}{2^2 J_1^2(2\alpha_n)} \int_0^2 r J_0(\alpha_n r)\, dr.$$

$u = u_0$ at $z = 4$

$u = 0$ at $r = 2$

$u = 0$ at $z = 0$

FIGURE 14.2.5 Finite cylinder in Example 2

To evaluate the last integral we first use the substitution $t = \alpha_n r$, followed by $\dfrac{d}{dt}[tJ_1(t)] = tJ_0(t)$. From

$$A_n \sinh 4\alpha_n = \frac{u_0}{2\alpha_n^2 J_1^2(2\alpha_n)} \int_0^{2\alpha_n} \frac{d}{dt}[tJ_1(t)]\, dt = \frac{u_0}{\alpha_n J_1(2\alpha_n)},$$

we obtain

$$A_n = \frac{u_0}{\alpha_n \sinh 4\alpha_n J_1(2\alpha_n)}.$$

Finally, the temperature in the cylinder is

$$u(r, z) = u_0 \sum_{n=1}^{\infty} \frac{\sinh \alpha_n z}{\alpha_n \sinh 4\alpha_n \, J_1(2\alpha_n)} J_0(\alpha_n r).$$

≡

Do not conclude from two examples that every boundary-value problem in cylindrical coordinates gives rise to a Fourier–Bessel series.

EXAMPLE 3 **Steady Temperatures in a Circular Cylinder**

Find the steady-state temperatures $u(r, z)$ in the circular cylinder defined by $0 \le r \le 1$, $0 \le z \le 1$ if the boundary conditions are

$$u(1, z) = 1 - z, \quad 0 < z < 1$$
$$u(r, 0) = 0, \quad u(r, 1) = 0, \quad 0 < r < 1.$$

SOLUTION Because of the nonhomogeneous condition specified at $r = 1$ we do not expect the eigenvalues of the problem to be defined in terms of zeros of a Bessel function of the first kind. As we did in Section 14.1 it is convenient in this problem to use λ as the separation constant. Thus from (15) of Example 2 we see that separation of variables now gives the two ordinary differential equations

$$rR'' + R' - \lambda rR = 0 \quad \text{and} \quad Z'' + \lambda Z = 0.$$

You should verify that the two cases $\lambda = 0$ and $\lambda = -\alpha^2 < 0$ lead only to the trivial solution $u = 0$. In the case $\lambda = \alpha^2 > 0$ the DEs are

$$rR'' + R' - \alpha^2 rR = 0 \quad \text{and} \quad Z'' + \alpha^2 Z = 0.$$

Review pages 292–293 of Section 5.3. See also Figures 5.3.3 and 5.3.4. ▶

The first equation is the parametric form of Bessel's modified DE of order $\nu = 0$. The solution of this equation is $R(r) = c_1 I_0(\alpha r) + c_2 K_0(\alpha r)$. We immediately define $c_2 = 0$ because the modified Bessel function of the second kind $K_0(\alpha r)$ is unbounded at $r = 0$. Therefore, $R(r) = c_1 I_0(\alpha r)$.

Now the eigenvalues and eigenfunctions of the Sturm–Liouville problem

$$Z'' + \alpha^2 Z = 0, \quad Z(0) = 0, \quad Z(1) = 0$$

are $\lambda_n = n^2 \pi^2, n = 1, 2, 3, \ldots$ and $Z(z) = c_3 \sin n\pi z$. Thus product solutions that satisfy the PDE and the homogeneous boundary conditions are

$$u_n = R(r)Z(z) = A_n I_0(n\pi r) \sin n\pi z.$$

Next we form

$$u(r, z) = \sum_{n=1}^{\infty} A_n I_0(n\pi r) \sin n\pi z.$$

The remaining condition at $r = 1$ yields the Fourier sine series

$$u(1, z) = 1 - z = \sum_{n=1}^{\infty} A_n I_0(n\pi) \sin n\pi z.$$

From (5) of Section 12.3 we can write

$$A_n I_0(n\pi) = 2 \int_0^1 (1 - z) \sin n\pi z \, dz = \frac{2}{n\pi} \quad \leftarrow \text{integration by parts}$$

and so
$$A_n = \frac{2}{n\pi I_0(n\pi)}.$$

The steady-state temperature is then

$$u(r, z) = \frac{2}{\pi} \sum_{n=1}^{\infty} \frac{I_0(n\pi r)}{n I_0(n\pi)} \sin n\pi z.$$

REMARKS

Because Bessel functions appear so frequently in the solutions of boundary-value problems expressed in cylindrical coordinates, they are also referred to as **cylinder functions**.

14.2 Exercises

Answers to selected odd-numbered problems begin on page ANS-37.

1. Find the displacement $u(r, t)$ in Example 1 if $f(r) = 0$ and the circular membrane is given an initial unit velocity in the upward direction.

2. A circular membrane of radius 1 is clamped along its circumference. Find the displacement $u(r, t)$ if the membrane starts from rest from the initial displacement $f(r) = 1 - r^2$, $0 < r < 1$. [*Hint*: See Problem 10 in Exercises 12.6.]

3. Find the steady-state temperature $u(r, z)$ in the cylinder in Example 2 if the boundary conditions are $u(2, z) = 0$, $0 < z < 4$, $u(r, 0) = u_0$, $u(r, 4) = 0$, $0 < r < 2$.

4. If the lateral side of the cylinder in Example 2 is insulated, then
$$\frac{\partial u}{\partial r}\bigg|_{r=2} = 0, \quad 0 < z < 4.$$

 (a) Find the steady-state temperature $u(r, z)$ when $u(r, 4) = f(r)$, $0 < r < 2$.

 (b) Show that the steady-state temperature in part (a) reduces to $u(r, z) = u_0 z/4$ when $f(r) = u_0$. [*Hint*: Use (12) of Section 12.6.]

In Problems 5–8, find the steady-state temperature $u(r, z)$ in a finite cylinder defined by $0 \le r \le 1$, $0 \le z \le 1$ if the boundary conditions are as given and u_0 is a constant.

5. $u(1, z) = z, \quad 0 < z < 1$
$$\frac{\partial u}{\partial z}\bigg|_{z=0} = 0, \quad 0 < r < 1$$
$$\frac{\partial u}{\partial z}\bigg|_{z=1} = 0, \quad 0 < r < 1$$

6. $u(1, z) = z, \quad 0 < z < 1$
$$\frac{\partial u}{\partial z}\bigg|_{z=0} = 0, \quad 0 < r < 1$$
$$u(r, 1) = 0, \quad 0 < r < 1$$

7. $u(1, z) = u_0, \quad 0 < z < 1$
$$u(r, 0) = 0, \quad 0 < r < 1$$
$$\frac{\partial u}{\partial z}\bigg|_{z=1} = 0, \quad 0 < r < 1$$

8. $u(1, z) = 0, \quad 0 < z < 1$
$$\frac{\partial u}{\partial z}\bigg|_{z=0} = 0, \quad 0 < r < 1$$
$$u(r, 1) = u_0, \quad 0 < r < 1$$

9. The temperature in a circular plate of radius c is determined from the boundary-value problem
$$k\left(\frac{\partial^2 u}{\partial r^2} + \frac{1}{r}\frac{\partial u}{\partial r}\right) = \frac{\partial u}{\partial t}, \quad 0 < r < c, \quad t > 0$$
$$u(c, t) = 0, \quad t > 0$$
$$u(r, 0) = f(r), \quad 0 < r < c.$$

Solve for $u(r, t)$.

10. Solve Problem 9 if the edge $r = c$ of the plate is insulated.

11. When there is heat transfer from the lateral side of an infinite circular cylinder of radius 1 (see **FIGURE 14.2.6**) into a surrounding medium at temperature zero, the temperature inside the cylinder is determined from
$$k\left(\frac{\partial^2 u}{\partial r^2} + \frac{1}{r}\frac{\partial u}{\partial r}\right) = \frac{\partial u}{\partial t}, \quad 0 < r < 1, \quad t > 0$$
$$\frac{\partial u}{\partial r}\bigg|_{r=1} = -hu(1, t), \quad h > 0, t > 0$$
$$u(r, 0) = f(r), \quad 0 < r < 1.$$

Solve for $u(r, t)$.

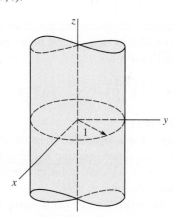

FIGURE 14.2.6 Infinite cylinder in Problem 11

12. When there is heat transfer from the lateral side of a semi-infinite ($z \ge 0$) circular cylinder of unit radius into a surrounding medium at temperature zero, the steady-state temperature $u(r, z)$ inside the cylinder is determined from the boundary-value problem
$$\frac{\partial^2 u}{\partial r^2} + \frac{1}{r}\frac{\partial u}{\partial r} + \frac{\partial^2 u}{\partial z^2} = 0, \quad 0 < r < 1, \quad z > 0$$
$$\frac{\partial u}{\partial r}\bigg|_{r=1} = -hu(1, z), \quad h > 0, \quad z > 0$$
$$u(r, 0) = u_0, \quad 0 < r < 1,$$

where h and u_0 are constants. Solve for $u(r, z)$.

In Problems 13 and 14, use the substitution $u(r, t) = v(r, t) + \psi(r)$ to solve the given boundary-value problem. [*Hint*: Review Section 13.6.]

13. A circular plate is a composite of two different materials in the form of concentric circles. See **FIGURE 14.2.7**. The temperature $u(r, t)$ in the plate is determined from the boundary-value problem

$$\frac{\partial^2 u}{\partial r^2} + \frac{1}{r}\frac{\partial u}{\partial r} = \frac{\partial u}{\partial t}, \quad 0 < r < 2, \quad t > 0$$

$$u(2, t) = 100, \quad t > 0$$

$$u(r, 0) = \begin{cases} 200, & 0 < r < 1 \\ 100, & 1 < r < 2. \end{cases}$$

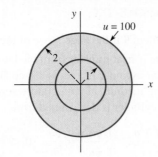

FIGURE 14.2.7 Circular plate in Problem 13

14. $\dfrac{\partial^2 u}{\partial r^2} + \dfrac{1}{r}\dfrac{\partial u}{\partial r} + \beta = \dfrac{\partial u}{\partial t}, \quad 0 < r < 1, \quad t > 0, \beta \text{ a constant}$

$$u(1, t) = 0, \quad t > 0$$

$$u(r, 0) = 0, \quad 0 < r < 1.$$

15. The horizontal displacement $u(x, t)$ of a heavy chain of length L oscillating in a vertical plane satisfies the partial differential equation

$$g\frac{\partial}{\partial x}\left(x\frac{\partial u}{\partial x}\right) = \frac{\partial^2 u}{\partial t^2}, \quad 0 < x < L, \quad t > 0.$$

See **FIGURE 14.2.8**.

(a) Using $-\lambda$ as a separation constant, show that the ordinary differential equation in the spatial variable x is $xX'' + X' + \lambda X = 0$. Solve this equation by means of the substitution $x = \tau^2/4$.

(b) Use the result of part (a) to solve the given partial differential equation subject to

$$u(L, t) = 0, \quad t > 0$$

$$u(x, 0) = f(x), \quad \frac{\partial u}{\partial t}\Big|_{t=0} = 0, \quad 0 < x < L.$$

[*Hint*: Assume the oscillations at the free end $x = 0$ are finite.]

FIGURE 14.2.8 Oscillating chain in Problem 15

16. Consider the boundary-value problem

$$\frac{\partial^2 u}{\partial r^2} + \frac{1}{r}\frac{\partial u}{\partial r} = \frac{\partial u}{\partial t}, \quad 0 < r < 1, \quad t > 0$$

$$\frac{\partial u}{\partial r}\Big|_{r=1} = 1, \quad t > 0$$

$$u(r, 0) = 0, \quad 0 < r < 1.$$

(a) Use the substitution $u(r, t) = v(r, t) + Bt$ in the preceding problem to show that $v(r, t)$ satisfies

$$\frac{\partial^2 v}{\partial r^2} + \frac{1}{r}\frac{\partial v}{\partial r} = \frac{\partial v}{\partial t} + B, \quad 0 < r < 1, \quad t > 0$$

$$\frac{\partial v}{\partial r}\Big|_{r=1} = 1, \quad t > 0$$

$$v(r, 0) = 0, \quad 0 < r < 1.$$

Here B is a constant to be determined.

(b) Now use the substitution $v(r, t) = w(r, t) + \psi(r)$ to solve the boundary-value problem in part (a). [*Hint*: You may need to review Section 3.5.]

(c) What is the solution $u(r, t)$ of the first problem?

17. Suppose heat is lost from the flat surfaces of a very thin circular plate into a surrounding medium at temperature zero. If the linear law of heat transfer applies, the heat equation assumes the form

$$\frac{\partial^2 u}{\partial r^2} + \frac{1}{r}\frac{\partial u}{\partial r} - hu = \frac{\partial u}{\partial t}, \quad 0 < r < 1, \quad t > 0,$$

where h is a positive constant. See **FIGURE 14.2.9**. Find the temperature $u(r, t)$ if the edge $r = 1$ is kept at temperature zero and if initially the temperature of the plate is unity throughout.

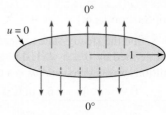

FIGURE 14.2.9 Circular plate in Problem 17

18. In this problem we consider the general case—that is, with θ dependence—of the vibrating circular membrane of radius c:

$$a^2\left(\frac{\partial^2 u}{\partial r^2} + \frac{1}{r}\frac{\partial u}{\partial r} + \frac{1}{r^2}\frac{\partial^2 u}{\partial \theta^2}\right) = \frac{\partial^2 u}{\partial t^2}, \quad 0 < r < c, \quad t > 0$$

$$u(c, \theta, t) = 0, \quad 0 < \theta < 2\pi, t > 0$$

$$u(r, \theta, 0) = f(r, \theta), \quad 0 < r < c, 0 < \theta < 2\pi$$

$$\frac{\partial u}{\partial t}\Big|_{t=0} = g(r, \theta), \quad 0 < r < c, 0 < \theta < 2\pi.$$

(a) Assume that $u = R(r)\Theta(\theta)T(t)$ and the separation constants are $-\lambda$ and $-\nu$. Show that the separated differential equations are

$$T'' + a^2\lambda T = 0, \quad \Theta'' + \nu\Theta = 0$$

$$r^2R'' + rR' + (\lambda r^2 - \nu)R = 0.$$

(b) Let $\lambda = \alpha^2$ and $\nu = \beta^2$ and solve the separated equations in part (a).

(c) Determine the eigenvalues and eigenfunctions of the problem.

(d) Use the superposition principle to determine a multiple series solution. Do not attempt to evaluate the coefficients.

Discussion Problem

19. Discuss how to solve the partial differential equation

$$\frac{\partial^2 u}{\partial r^2} + \frac{1}{r}\frac{\partial u}{\partial r} + \frac{\partial^2 u}{\partial z^2} = 0, \quad 0 < r < b, \quad 0 < z < L$$

subject to the boundary conditions indicated in **FIGURE 14.2.10**.

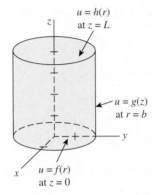

$u = h(r)$
at $z = L$

$u = g(z)$
at $r = b$

$u = f(r)$
at $z = 0$

FIGURE 14.2.10 Circular cylinder in Problem 19

Computer Lab Assignments

20. (a) Consider Example 1 with $a = 1$, $c = 10$, $g(r) = 0$, and $f(r) = 1 - r/10$, $0 < r < 10$. Use a CAS as an aid in finding the numerical values of the first three eigenvalues $\lambda_1, \lambda_2, \lambda_3$ of the boundary-value problem and the first three coefficients A_1, A_2, A_3 of the solution $u(r, t)$ given in (9). Write the third partial sum $S_3(r, t)$ of the series solution.

(b) Use a CAS to plot the graph of $S_3(r, t)$ for $t = 0, 4, 10, 12, 20$.

21. Solve Problem 9 with boundary conditions $u(c, t) = 200$, $u(r, 0) = 0$. With these imposed conditions, one would expect intuitively that at any interior point of the plate, $u(r, t) \to 200$

as $t \to \infty$. Assume that $c = 10$ and that the plate is cast iron so that $k = 0.1$ (approximately). Use a CAS as an aid in finding the numerical values of the first five eigenvalues $\lambda_1, \lambda_2, \lambda_3, \lambda_4, \lambda_5$ of the boundary-value problem and the five coefficients A_1, A_2, A_3, A_4, A_5 in the solution $u(r, t)$. Let the corresponding approximate solution be denoted by $S_5(r, t)$. Plot $S_5(5, t)$ and $S_5(0, t)$ on a sufficiently large time interval $[0, T]$. Use the plots of $S_5(5, t)$ and $S_5(0, t)$ to estimate the times (in seconds) for which $u(5, t) \approx 100$ and $u(0, t) \approx 100$. Repeat for $u(5, t) \approx 200$ and $u(0, t) \approx 200$.

22. Consider an idealized drum consisting of a thin membrane stretched over a circular frame of radius 1. When such a drum is struck at its center, one hears a sound that is frequently described as a dull thud rather than a melodic tone. We can model a single drumbeat using the boundary-value problem solved in Example 1.

(a) Find the solution $u(r, t)$ given in (9) when $c = 1$, $f(r) = 0$, and

$$g(r) = \begin{cases} -v_0, & 0 \le r < b \\ 0, & b \le r < 1. \end{cases}$$

(b) Show that the frequency of the standing wave $u_n(r, t)$ is $f_n = a\lambda_n/2\pi$, where λ_n is the nth positive zero of $J_0(x)$. Unlike the solution of the one-dimensional wave equation in Section 13.4, the frequencies are not integer multiples of the fundamental frequency f_1. Show that $f_2 \approx 2.295f_1$ and $f_3 \approx 3.598f_1$. We say that the drumbeat produces **anharmonic overtones**. As a result the displacement function $u(r, t)$ is not periodic, and so our ideal drum cannot produce a sustained tone.

(c) Let $a = 1$, $b = \frac{1}{4}$, and $v_0 = 1$ in your solution in part (a). Use a CAS to graph the fifth partial sum $S_5(r, t)$ at the times $t = 0, 0.1, 0.2, 0.3, \ldots, 5.9, 6.0$ on the interval $[-1, 1]$. Use the animation capabilities of your CAS to produce a movie of these vibrations.

(d) For a greater challenge, use the 3D plotting capabilities of your CAS to make a movie of the motion of the circular drumhead that is shown in cross section in part (c). [*Hint*: There are several ways of proceeding. For a fixed time, either graph u as a function of x and y using $r = \sqrt{x^2 + y^2}$ or use the equivalent of *Mathematica*'s **RevolutionPlot3D**.]

14.3 Spherical Coordinates

INTRODUCTION In this section we continue our examination of boundary-value problems in different coordinate systems. This time we are going to consider problems involving the heat, wave, and Laplace's equation in spherical coordinates.

■ **Laplacian in Spherical Coordinates** As shown in **FIGURE 14.3.1**, a point in 3-space is described in terms of rectangular coordinates and in spherical coordinates. The rectangular coordinates x, y, and z of the point are related to its spherical coordinates r, θ, and ϕ through the equations

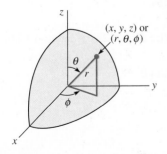

(x, y, z) or (r, θ, ϕ)

FIGURE 14.3.1 Spherical coordinates of a point (x, y, z) are (r, θ, ϕ)

$$x = r\sin\theta\cos\phi, \quad y = r\sin\theta\sin\phi, \quad z = r\cos\theta. \tag{1}$$

By using the equations in (1) it can be shown that the Laplacian $\nabla^2 u$ in the spherical coordinate system is

$$\nabla^2 u = \frac{\partial^2 u}{\partial r^2} + \frac{2}{r}\frac{\partial u}{\partial r} + \frac{1}{r^2 \sin^2\theta}\frac{\partial^2 u}{\partial\phi^2} + \frac{1}{r^2}\frac{\partial^2 u}{\partial\theta^2} + \frac{\cot\theta}{r^2}\frac{\partial u}{\partial\theta}. \tag{2}$$

As you might imagine, problems involving (1) can be quite formidable. Consequently we shall consider only a few of the simpler problems that are independent of the azimuthal angle ϕ.

Our first example is the Dirichlet problem for a sphere.

<div style="border:1px solid;">EXAMPLE 1</div> **Steady Temperatures in a Sphere**

Find the steady-state temperature $u(r, \theta)$ in the sphere shown in **FIGURE 14.3.2**.

SOLUTION The temperature is determined from

$$\frac{\partial^2 u}{\partial r^2} + \frac{2}{r}\frac{\partial u}{\partial r} + \frac{1}{r^2}\frac{\partial^2 u}{\partial\theta^2} + \frac{\cot\theta}{r^2}\frac{\partial u}{\partial\theta} = 0, \quad 0 < r < c, \quad 0 < \theta < \pi$$

$$u(c, \theta) = f(\theta), \quad 0 < \theta < \pi.$$

If $u = R(r)\Theta(\theta)$, the partial differential equation separates as

$$\frac{r^2 R'' + 2rR'}{R} = -\frac{\Theta'' + \cot\theta\,\Theta'}{\Theta} = \lambda,$$

and so

$$r^2 R'' + 2rR' - \lambda R = 0 \tag{3}$$

$$\sin\theta\,\Theta'' + \cos\theta\,\Theta' + \lambda\sin\theta\,\Theta = 0. \tag{4}$$

After we substitute $x = \cos\theta$, $0 \le \theta \le \pi$, (4) becomes

$$(1 - x^2)\frac{d^2\Theta}{dx^2} - 2x\frac{d\Theta}{dx} + \lambda\Theta = 0, \quad -1 \le x \le 1. \tag{5}$$

The latter equation is a form of Legendre's equation (see Problems 52 and 53 in Exercises 5.3). Now the only solutions of (5) that are continuous and have continuous derivatives on the closed interval $[-1, 1]$ are the Legendre polynomials $P_n(x)$ corresponding to $\lambda = n(n + 1)$, $n = 0$, $1, 2, \ldots$ Thus we take the solutions of (4) to be

$$\Theta(\theta) = P_n(\cos\theta).$$

Furthermore, when $\lambda = n(n + 1)$, the general solution of the Cauchy–Euler equation (3) is

$$R(r) = c_1 r^n + c_2 r^{-(n+1)}.$$

Since we again expect $u(r, \theta)$ to be bounded at $r = 0$, we define $c_2 = 0$. Hence $u_n = A_n r^n P_n(\cos\theta)$, and

$$u(r, \theta) = \sum_{n=0}^{\infty} A_n r^n P_n(\cos\theta).$$

At $r = c$,

$$f(\theta) = \sum_{n=0}^{\infty} A_n c^n P_n(\cos\theta).$$

Therefore $A_n c^n$ are the coefficients of the Fourier–Legendre series (23) of Section 12.6:

$$A_n = \frac{2n + 1}{2c^n}\int_0^\pi f(\theta)P_n(\cos\theta)\sin\theta\,d\theta.$$

It follows that the solution is

$$u(r, \theta) = \sum_{n=0}^{\infty}\left(\frac{2n + 1}{2}\int_0^\pi f(\theta)P_n(\cos\theta)\sin\theta\,d\theta\right)\left(\frac{r}{c}\right)^n P_n(\cos\theta). \qquad \equiv$$

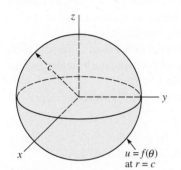

FIGURE 14.3.2 Dirichlet problem for a sphere in Example 1

Spherical Symmetry When u depends on r and t but is independent of both angles θ and ϕ, then the problem is said to possess **spherical symmetry**. The heat and wave equations are then

$$k\left(\frac{\partial^2 u}{\partial r^2} + \frac{2}{r}\frac{\partial u}{\partial r}\right) = \frac{\partial u}{\partial t} \quad \text{and} \quad a^2\left(\frac{\partial^2 u}{\partial r^2} + \frac{2}{r}\frac{\partial u}{\partial r}\right) = \frac{\partial^2 u}{\partial t^2}. \tag{6}$$

See Problems 9–11 in Exercises 14.3.

14.3 Exercises Answers to selected odd-numbered problems begin on page ANS-37.

1. Solve the problem in Example 1 if

 $$f(\theta) = \begin{cases} 50, & 0 < \theta < \pi/2 \\ 0, & \pi/2 < \theta < \pi. \end{cases}$$

 Write out the first four nonzero terms of the series solution. [*Hint*: See Example 3, Section 12.6.] The solution $u(r, \theta)$ can be interpreted as the potential due to a charge of 50 volts on the upper hemisphere and a grounded lower hemisphere.

2. The solution $u(r, \theta)$ in Example 1 could also be interpreted as the potential inside the sphere due to a charge distribution $f(\theta)$ on its surface. Find the potential outside the sphere.

3. Find the solution of the problem in Example 1 if $f(\theta) = \cos\theta$, $0 < \theta < \pi$. [*Hint*: $P_1(\cos\theta) = \cos\theta$. Use orthogonality.]

4. Find the solution of the problem in Example 1 if $f(\theta) = 1 - \cos 2\theta, 0 < \theta < \pi$. [*Hint*: See Problem 18, Exercises 12.6.]

5. Find the steady-state temperature $u(r, \theta)$ within a hollow sphere $a < r < b$ if its inner surface $r = a$ is kept at temperature $f(\theta)$ and its outer surface $r = b$ is kept at temperature zero. The sphere in the first octant is shown in **FIGURE 14.3.3**.

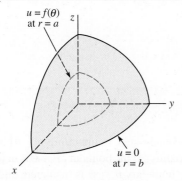

$u = f(\theta)$
at $r = a$

$u = 0$
at $r = b$

FIGURE 14.3.3 Hollow sphere in Problem 5

6. The steady-state temperature in a hemisphere of radius c is determined from

 $$\frac{\partial^2 u}{\partial r^2} + \frac{2}{r}\frac{\partial u}{\partial r} + \frac{1}{r^2}\frac{\partial^2 u}{\partial \theta^2} + \frac{\cot\theta}{r^2}\frac{\partial u}{\partial \theta} = 0, 0 < r < c, \quad 0 < \theta < \pi/2$$

 $$u(r, \pi/2) = 0, \quad 0 < r < c$$

 $$u(c, \theta) = f(\theta), \quad 0 < \theta < \pi/2.$$

 Solve for $u(r, \theta)$. [*Hint*: $P_n(0) = 0$ only if n is odd. Also see Problem 20, Exercises 12.6.]

7. Solve Problem 6 when the base of the hemisphere is insulated; that is,

 $$\frac{\partial u}{\partial \theta}\bigg|_{\theta = \pi/2} = 0, \quad 0 < r < c.$$

8. Solve Problem 6 for $r > c$.

9. The time-dependent temperature within a sphere of radius 1 is determined from

 $$\frac{\partial^2 u}{\partial r^2} + \frac{2}{r}\frac{\partial u}{\partial r} = \frac{\partial u}{\partial t}, \quad 0 < r < 1, \quad t > 0$$

 $$u(1, t) = 100, \quad t > 0$$

 $$u(r, 0) = 0, \quad 0 < r < 1.$$

 Solve for $u(r, t)$. [*Hint*: Verify that the left side of the partial differential equation can be written as $\dfrac{1}{r}\dfrac{\partial^2}{\partial r^2}(ru)$. Let $ru(r, t) = v(r, t) + \psi(r)$. Use only functions that are bounded as $r \to 0$.]

10. A uniform solid sphere of radius 1 at an initial constant temperature u_0 throughout is dropped into a large container of fluid that is kept at a constant temperature u_1 ($u_1 > u_0$) for all time. See **FIGURE 14.3.4**. Since there is heat transfer across the boundary $r = 1$, the temperature $u(r, t)$ in the sphere is determined from the boundary-value problem

 $$\frac{\partial^2 u}{\partial r^2} + \frac{2}{r}\frac{\partial u}{\partial r} = \frac{\partial u}{\partial t}, \quad 0 < r < 1, \quad t > 0$$

 $$\frac{\partial u}{\partial r}\bigg|_{r=1} = -h(u(1, t) - u_1), \quad 0 < h < 1$$

 $$u(r, 0) = u_0, \quad 0 < r < 1.$$

 Solve for $u(r, t)$. [*Hint*: Proceed as in Problem 9.]

u_1

FIGURE 14.3.4 Container in Problem 10

11. Solve the boundary-value problem involving spherical vibrations:

 $$a^2\left(\frac{\partial^2 u}{\partial r^2} + \frac{2}{r}\frac{\partial u}{\partial r}\right) = \frac{\partial^2 u}{\partial t^2}, \quad 0 < r < c, \quad t > 0$$

 $$u(c, t) = 0, \quad t > 0$$

 $$u(r, 0) = f(r), \quad \frac{\partial u}{\partial t}\bigg|_{t=0} = g(r), \quad 0 < r < c.$$

 [*Hint*: Write the left side of the partial differential equation as $a^2\dfrac{1}{r}\dfrac{\partial^2}{\partial r^2}(ru)$. Let $v(r, t) = ru(r, t)$.]

12. A conducting sphere of radius c is grounded and placed in a uniform electric field that has intensity E in the z-direction. The potential $u(r, \theta)$ outside the sphere is determined from the boundary-value problem

$$\frac{\partial^2 u}{\partial r^2} + \frac{2}{r}\frac{\partial u}{\partial r} + \frac{1}{r^2}\frac{\partial^2 u}{\partial \theta^2} + \frac{\cot\theta}{r^2}\frac{\partial u}{\partial \theta} = 0, \quad r > c, \quad 0 < \theta < \pi$$

$$u(c, \theta) = 0, \quad 0 < \theta < \pi$$

$$\lim_{r \to \infty} u(r, \theta) = -Ez = -Er\cos\theta.$$

Show that

$$u(r, \theta) = -Er\cos\theta + E\frac{c^3}{r^2}\cos\theta.$$

[*Hint*: Explain why $\int_0^\pi \cos\theta\, P_n(\cos\theta)\sin\theta\, d\theta = 0$ for all nonnegative integers except $n = 1$. See (24) of Section 12.6.]

In Problems 13 and 14, the time-independent partial differential equation

$$\nabla^2 u + k^2 u = 0$$

that arises in physics is called the **Helmholtz's equation**. This equation is named after the German physician and physicist **Hermann Ludwig Ferdinand von Helmholtz** (1821–1894). The three-dimensional Laplacian $\nabla^2 u$ is the spherical coordinates form given in (2) of this section.

13. (a) Proceed as in Example 1 but using $u(r, \theta, \phi) = R(r)\Theta(\theta)\Phi(\phi)$ and the separation constant $n(n + 1)$ to

show that the radial dependence of the solution u is defined by the equation

$$r^2\frac{d^2 R}{dr^2} + 2r\frac{dR}{dr} + [k^2 r^2 - n(n + 1)]R = 0.$$

(b) Now use the second separation constant m^2 to show that the remaining separated equations are

$$\frac{d^2\Phi}{d\phi^2} + m^2\Phi = 0$$

$$\frac{d^2\Theta}{d\theta^2} + \frac{\cos\theta}{\sin\theta}\frac{d\Theta}{d\theta} + \left[n(n + 1) - \frac{m^2}{\sin^2\theta}\right]\Theta = 0.$$

(c) Use the substitution $x = \cos\theta$ to show that the second differential equation in part (b) becomes

$$(1 - x^2)\frac{d^2\Theta}{dx^2} - 2x\frac{d\Theta}{dx} + \left[n(n + 1) - \frac{m^2}{1 - x^2}\right]\Theta = 0.$$

14. (a) Assume that m and n are nonnegative integers. Then find a product solution $u(r, \theta, \phi) = R(r)\Theta(\theta)\Phi(\phi)$ of Helmholtz's PDE using the general solution of the ODE in part (a), the general solution of the first ODE in part (b), and a particular solution of the second ODE in part (b) of Problem 13. [*Hint*: See Problems 41, 42(c), and 54 in Exercises 5.3.]

(b) What product solution in part (a) would be bounded at the origin?

14 Chapter in Review Answers to selected odd-numbered problems begin on page ANS-37.

In Problems 1 and 2, find the steady-state temperature $u(r, \theta)$ in a circular plate of radius c if the temperature on the circumference is as given.

1. $u(c, \theta) = \begin{cases} u_0, & 0 < \theta < \pi \\ -u_0, & \pi < \theta < 2\pi \end{cases}$

2. $u(c, \theta) = \begin{cases} 1, & 0 < \theta < \pi/2 \\ 0, & \pi/2 < \theta < 3\pi/2 \\ 1, & 3\pi/2 < \theta < 2\pi \end{cases}$

In Problems 3 and 4, find the steady-state temperature $u(r, \theta)$ in a semicircular plate of radius 1 if boundary conditions are as given.

3. $u(1, \theta) = u_0(\pi\theta - \theta^2), \quad 0 < \theta < \pi$
 $u(r, 0) = 0, \quad u(r, \pi) = 0, \quad 0 < r < 1$

4. $u(1, \theta) = \sin\theta, \quad 0 < \theta < \pi$
 $u(r, 0) = 0, \quad u(r, \pi) = 0, \quad 0 < r < 1$

5. Find the steady-state temperature $u(r, \theta)$ in a semicircular plate of radius c if the boundaries $\theta = 0$ and $\theta = \pi$ are insulated and $u(c, \theta) = f(\theta), \quad 0 < \theta < \pi$.

6. Find the steady-state temperature $u(r, \theta)$ in a semicircular plate of radius c if the boundary $\theta = 0$ is held at temperature zero, the boundary $\theta = \pi$ is insulated, and $u(c, \theta) = f(\theta)$, $0 < \theta < \pi$.

In Problems 7 and 8, find the steady-state temperature $u(r, \theta)$ in the plate shown in the figure.

7.

FIGURE 14.R.1 Plate in Problem 7

8.

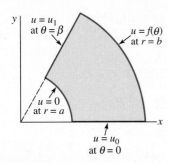

FIGURE 14.R.2 Plate in Problem 8

9. If the boundary conditions for an annular plate defined by $1 < r < 2$ are

$$u(1, \theta) = \sin^2\theta, \quad \frac{\partial u}{\partial r}\bigg|_{r=2} = 0, \quad 0 < \theta < 2\pi,$$

show that the steady-state temperature is

$$u(r, \theta) = \frac{1}{2} - \left(\frac{1}{34}r^2 + \frac{8}{17}r^{-2}\right)\cos 2\theta.$$

[*Hint*: See Figure 14.1.4. Also, use the identity $\sin^2\theta = \frac{1}{2}(1 - \cos 2\theta)$.]

10. Find the steady-state temperature $u(r, \theta)$ in the infinite plate shown in **FIGURE 14.R.3**.

FIGURE 14.R.3 Infinite plate in Problem 10

11. Solve the boundary-value problem

$$\frac{\partial^2 u}{\partial r^2} + \frac{1}{r}\frac{\partial u}{\partial r} - \frac{1}{r^2}u + \frac{\partial^2 u}{\partial z^2} = 0, \quad 0 < r < 1, \quad 0 < z < 1$$

$$u(1, z) = 0, \quad 0 < z < 1$$

$$u(r, 0) = 0, \quad u(r, 1) = r - r^3, \quad 0 < r < 1.$$

[*Hint*: See equation (12) in Section 5.3.]

12. Suppose x_k is a positive zero of J_0. Show that a solution of the boundary-value problem

$$a^2\left(\frac{\partial^2 u}{\partial r^2} + \frac{1}{r}\frac{\partial u}{\partial r}\right) = \frac{\partial^2 u}{\partial t^2}, \quad 0 < r < 1, \quad t > 0$$

$$u(1, t) = 0, \quad t > 0$$

$$u(r, 0) = u_0 J_0(x_k r), \quad \frac{\partial u}{\partial t}\bigg|_{t=0} = 0, \quad 0 < r < 1$$

is $u(r, t) = u_0 J_0(x_k r) \cos ax_k t$.

13. Find the steady-state temperature $u(r, z)$ in the cylinder in Figure 14.2.5 if the lateral side is kept at temperature zero, the top $z = 4$ is kept at temperature 50, and the base $z = 0$ is insulated.

14. Solve the boundary-value problem

$$\frac{\partial^2 u}{\partial r^2} + \frac{1}{r}\frac{\partial u}{\partial r} + \frac{\partial^2 u}{\partial z^2} = 0, \quad 0 < r < 1, \quad 0 < z < 1$$

$$\frac{\partial u}{\partial r}\bigg|_{r=1} = 0, \quad 0 < z < 1$$

$$u(r, 0) = f(r), \quad u(r, 1) = g(r), \quad 0 < r < 1.$$

15. Find the steady-state temperature $u(r, \theta)$ in a sphere of unit radius if the surface is kept at

$$u(1, \theta) = \begin{cases} 100, & 0 < \theta < \pi/2 \\ -100, & \pi/2 < \theta < \pi. \end{cases}$$

[*Hint*: See Problem 22 in Exercises 12.6.]

16. Solve the boundary-value problem

$$\frac{\partial^2 u}{\partial r^2} + \frac{2}{r}\frac{\partial u}{\partial r} = \frac{\partial^2 u}{\partial t^2}, \quad 0 < r < 1, \quad t > 0$$

$$\frac{\partial u}{\partial r}\bigg|_{r=1} = 0, \quad t > 0$$

$$u(r, 0) = f(r), \quad \frac{\partial u}{\partial t}\bigg|_{t=0} = g(r), \quad 0 < r < 1.$$

[*Hint*: Proceed as in Problems 9 and 10 in Exercises 14.3, but let $v(r, t) = ru(r, t)$. See Section 13.7.]

17. The function $u(x) = Y_0(\alpha a)J_0(\alpha x) - J_0(\alpha a)Y_0(\alpha x)$, $a > 0$ is a solution of the parametric Bessel equation

$$x^2\frac{d^2 u}{dx^2} + x\frac{du}{dx} + \alpha^2 x^2 u = 0$$

on the interval $[a, b]$. If the eigenvalues $\lambda_n = \alpha_n^2$ are defined by the positive roots of the equation

$$Y_0(\alpha a)J_0(\alpha b) - J_0(\alpha a)Y_0(\alpha b) = 0,$$

show that the functions

$$u_m(x) = Y_0(\alpha_m a)J_0(\alpha_m x) - J_0(\alpha_m a)Y_0(\alpha_m x)$$
$$u_n(x) = Y_0(\alpha_n a)J_0(\alpha_n x) - J_0(\alpha_n a)Y_0(\alpha_n x)$$

are orthogonal with respect to the weight function $p(x) = x$ on the interval $[a, b]$; that is,

$$\int_a^b x u_m(x)u_n(x)\, dx = 0, \quad m \neq n.$$

[*Hint*: Follow the procedure on pages 702 and 703.]

18. Use the results of Problem 17 to solve the following boundary-value problem for the temperature $u(r, t)$ in an annular plate:

$$\frac{\partial^2 u}{\partial r^2} + \frac{1}{r}\frac{\partial u}{\partial r} = \frac{\partial u}{\partial t}, \quad a < r < b, \quad t > 0$$

$$u(a, t) = 0, \quad u(b, t) = 0, \quad t > 0$$

$$u(r, 0) = f(r), \quad a < r < b.$$

In Problems 19–22, solve the given boundary-value problem.

19. $\dfrac{\partial^2 u}{\partial r^2} + \dfrac{1}{r}\dfrac{\partial u}{\partial r} + \dfrac{\partial^2 u}{\partial z^2} = 0, \quad 0 < r < 1, \quad 0 < z < 1$

$u(1, z) = 100, \quad 0 < z < 1$

$\dfrac{\partial u}{\partial z}\bigg|_{z=0} = 0, \quad 0 < r < 1$

$u(r, 1) = 200, \quad 0 < r < 1.$

20. $\dfrac{\partial^2 u}{\partial r^2} + \dfrac{1}{r}\dfrac{\partial u}{\partial r} + \dfrac{\partial^2 u}{\partial z^2} = 0, \quad 0 < r < 3, \quad 0 < z < 1$

$u(3, z) = u_0, \quad 0 < z < 1$

$u(r, 0) = 0, \quad 0 < r < 3$

$u(r, 1) = 0, \quad 0 < r < 3.$

21. $\dfrac{\partial^2 u}{\partial r^2} + \dfrac{1}{r}\dfrac{\partial u}{\partial r} + \dfrac{\partial^2 u}{\partial z^2} = 0, \quad 0 < r < 1, \quad z > 0$

$u(1, z) = 0, \quad z > 0$

$u(r, 0) = 100, \quad 0 < r < 1$

22. $\dfrac{\partial^2 u}{\partial r^2} + \dfrac{1}{r}\dfrac{\partial u}{\partial r} + \dfrac{\partial^2 u}{\partial z^2} = 0, \quad 0 < r < 1, \quad z > 0$

$u(1, z) = 0, \quad z > 0$

$u(r, 0) = u_0(1 - r^2), \quad 0 < r < 1$

Integral Transforms

The method of separation of variables is a powerful, but not universally applicable, method for solving boundary-value problems (BVPs). If the partial differential equation (PDE) is nonhomogeneous, or the boundary conditions are time dependent, or the domain of the spatial variable is the infinite interval $(-\infty, \infty)$ or a semi-infinite interval (a, ∞), we may be able to use an **integral transform** to solve the problem. In Section 15.2 we will solve BVPs that involve the heat and wave equations by means of the Laplace transform. In Sections 15.4 and 15.5 we introduce five new integral transforms: three **Fourier transforms** and two **finite Fourier transforms**.

15.1 Error Function

INTRODUCTION There are many functions in mathematics that are defined by means of an integral. For example, in many traditional calculus texts the natural logarithm is defined in the following manner: $\ln x = \int_1^x dt/t, x > 0$. In earlier chapters we have already seen, albeit briefly, the error function erf(x), the complementary error function erfc(x), the sine integral function Si(x), the Fresnel sine integral $S(x)$, and the gamma function $\Gamma(\alpha)$; all of these functions are defined in terms of an integral. Before applying the Laplace transform to boundary-value problems, we need to know a little more about the error function and the complementary error function. In this section we examine the graphs and a few of the more obvious properties of erf(x) and erfc(x).

See Appendix A. ▶

■ **Properties and Graphs** Recall from (14) of Section 2.3 that the definitions of the **error function** erf(x) and **complementary error function** erfc(x) are, respectively,

$$\text{erf}(x) = \frac{2}{\sqrt{\pi}} \int_0^x e^{-u^2} \, du \quad \text{and} \quad \text{erfc}(x) = \frac{2}{\sqrt{\pi}} \int_x^\infty e^{-u^2} \, du. \tag{1}$$

With the aid of polar coordinates, it can be demonstrated that

$$\int_0^\infty e^{-u^2} \, du = \frac{\sqrt{\pi}}{2} \quad \text{or} \quad \frac{2}{\sqrt{\pi}} \int_0^\infty e^{-u^2} \, du = 1. \tag{2}$$

We have already seen in (15) of Section 2.3 that when the second integral in (2) is written as $\int_0^\infty = \int_0^x + \int_x^\infty$ we obtain an identity that relates the error function and the complementary error function:

$$\text{erf}(x) + \text{erfc}(x) = 1. \tag{3}$$

For $x > 0$ it is seen in **FIGURE 15.1.1** that erf(x) can be interpreted as the area of the blue region under the graph of $f(t) = (2/\sqrt{\pi})e^{-t^2}$ on the interval $[0, x]$ and erfc(x) is the area of the red region on $[x, \infty)$. The graph of the function f is often referred to as a **bell curve**.

Because of the importance of erf(x) and erfc(x) in the solution of partial differential equations and in the theory of probability and statistics, these functions are built into computer algebra systems. So with the aid of *Mathematica* we get the graphs of erf(x) (in blue) and erfc(x) (in red) given in **FIGURE 15.1.2**. The y-intercepts of the two graphs give the values

- erf$(0) = 0$, erfc$(0) = 1$.

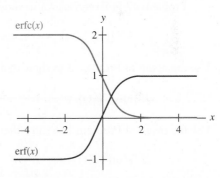

FIGURE 15.1.1 Bell curve

FIGURE 15.1.2 Graphs of erf(x) and erfc(x)

Other numerical values of erf(x) and erfc(x) can be obtained directly from a CAS. Further inspection of the two graphs in Figure 15.1.2 shows that:

- the domains of erf(x) and erfc(x) are $(-\infty, \infty)$,
- erf(x) and erfc(x) are continuous functions,
- $\lim_{x \to \infty} \text{erf}(x) = 1$, $\lim_{x \to -\infty} \text{erf}(x) \to -1$,
- $\lim_{x \to \infty} \text{erfc}(x) \to 0$, $\lim_{x \to -\infty} \text{erfc}(x) \to 2$.

It should also be apparent that the graph of the error function is symmetric with respect to the origin and so erf(x) is an odd function:

$$\text{erf}(-x) = -\text{erf}(x). \tag{4}$$

You are asked to prove (4) in Problem 14 of Exercises 15.1.

Table 15.1.1 contains Laplace transforms of some functions involving the error and complementary error functions. These results will be useful in the exercises in the next section.

TABLE 15.1.1

$f(t), a > 0$	$\mathscr{L}\{f(t)\} = F(s)$	$f(t), a > 0$	$\mathscr{L}\{f(t)\} = F(s)$
1. $\dfrac{1}{\sqrt{\pi t}} e^{-a^2/4t}$	$\dfrac{e^{-a\sqrt{s}}}{\sqrt{s}}$	**4.** $2\sqrt{\dfrac{t}{\pi}} e^{-a^2/4t} - a\,\mathrm{erfc}\left(\dfrac{a}{2\sqrt{t}}\right)$	$\dfrac{e^{-a\sqrt{s}}}{s\sqrt{s}}$
2. $\dfrac{a}{2\sqrt{\pi t^3}} e^{-a^2/4t}$	$e^{-a\sqrt{s}}$	**5.** $e^{ab}\,e^{b^2 t}\,\mathrm{erfc}\left(b\sqrt{t} + \dfrac{a}{2\sqrt{t}}\right)$	$\dfrac{e^{-a\sqrt{s}}}{\sqrt{s}(\sqrt{s}+b)}$
3. $\mathrm{erfc}\left(\dfrac{a}{2\sqrt{t}}\right)$	$\dfrac{e^{-a\sqrt{s}}}{s}$	**6.** $-e^{ab}\,e^{b^2 t}\,\mathrm{erfc}\left(b\sqrt{t} + \dfrac{a}{2\sqrt{t}}\right) + \mathrm{erfc}\left(\dfrac{a}{2\sqrt{t}}\right)$	$\dfrac{be^{-a\sqrt{s}}}{s(\sqrt{s}+b)}$

REMARKS

The proofs of the results in Table 15.1.1 will not be given because they are long and somewhat complicated. For example, the proofs of entries 2 and 3 of the table require several changes of variables and the use of the convolution theorem. For those who are curious, see *Introduction to the Laplace Transform*, by Holl, Maple, and Vinograde (Appleton-Century-Crofts, 1959). A flavor of these kinds of proofs can be gotten by working Problem 1 in Exercises 15.1.

15.1 Exercises Answers to selected odd-numbered problems begin on page ANS-38.

1. (a) Show that
$$\mathrm{erf}(\sqrt{t}) = \frac{1}{\sqrt{\pi}} \int_0^t \frac{e^{-\tau}}{\sqrt{\tau}}\,d\tau.$$

(b) Use part (a), the convolution theorem, and the result of Problem 47 in Exercises 4.1 to show that
$$\mathscr{L}\{\mathrm{erf}(\sqrt{t})\} = \frac{1}{s\sqrt{s+1}}.$$

2. Use the result of Problem 1 to show that
$$\mathscr{L}\{\mathrm{erfc}(\sqrt{t})\} = \frac{1}{s}\left[1 - \frac{1}{\sqrt{s+1}}\right].$$

3. Use the result of Problem 1 to show that
$$\mathscr{L}\{e^t \mathrm{erf}(\sqrt{t})\} = \frac{1}{\sqrt{s}(s-1)}.$$

4. Use the result of Problem 2 to show that
$$\mathscr{L}\{e^t \mathrm{erfc}(\sqrt{t})\} = \frac{1}{\sqrt{s}(\sqrt{s}+1)}.$$

5. Use the result of Problem 4 to show that
$$\mathscr{L}\left\{\frac{1}{\sqrt{\pi t}} - e^t \mathrm{erfc}(\sqrt{t})\right\} = \frac{1}{\sqrt{s}+1}.$$

6. Find the inverse transform
$$\mathscr{L}^{-1}\left\{\frac{1}{1+\sqrt{s+1}}\right\}.$$

[*Hint*: Rationalize a denominator followed by a rationalization of a numerator.]

7. Let C, G, R, and x be constants. Use Table 15.1.1 to show that
$$\mathscr{L}^{-1}\left\{\frac{C}{Cs+G}\left(1 - e^{-x\sqrt{RCs+RG}}\right)\right\} = e^{-Gt/C}\mathrm{erf}\left(\frac{x}{2}\sqrt{\frac{RC}{t}}\right).$$

8. Let a be a constant. Show that
$$\mathscr{L}^{-1}\left\{\frac{\sinh a\sqrt{s}}{s\sinh \sqrt{s}}\right\} =$$
$$\sum_{n=0}^{\infty}\left[\mathrm{erf}\left(\frac{2n+1+a}{2\sqrt{t}}\right) - \mathrm{erf}\left(\frac{2n+1-a}{2\sqrt{t}}\right)\right].$$
[*Hint*: Use the exponential definition of the hyperbolic sine. Expand $1/(1 - e^{-2\sqrt{s}})$ in a geometric series.]

9. Use the Laplace transform and Table 15.1.1 to solve the integral equation
$$y(t) = 1 - \int_0^t \frac{y(\tau)}{\sqrt{t-\tau}}\,d\tau.$$

10. Use the third and fifth entries in Table 15.1.1 to derive the sixth entry.

11. Show that $\int_a^b e^{-u^2}\,du = \dfrac{\sqrt{\pi}}{2}[\mathrm{erf}(b) - \mathrm{erf}(a)]$.

12. Show that $\int_{-a}^a e^{-u^2}\,du = \sqrt{\pi}\,\mathrm{erf}(a)$.

13. Show that $\lim\limits_{x\to 0} \dfrac{\mathrm{erf}(x)}{x} = \dfrac{2}{\sqrt{\pi}}$.

14. Prove that $\mathrm{erf}(x)$ is an odd function.

15. Show that $\mathrm{erfc}(-x) = 1 + \mathrm{erf}(x)$.

15.2 Laplace Transform

INTRODUCTION In Chapter 4 we defined the Laplace transform of a function $f(t)$, $t \geq 0$, to be

$$\mathcal{L}\{f(t)\} = \int_0^\infty e^{-st} f(t)\, dt,$$

whenever the improper integral converges. This integral transforms a function $f(t)$ into another function F of the transform parameter s, that is, $\mathcal{L}\{f(t)\} = F(s)$. The main application of the Laplace transform in Chapter 4 was to the solution of certain types of initial-value problems involving linear ordinary differential equations with constant coefficients. Recall, the Laplace transform of such an equation reduces the ODE to an algebraic equation. In this section we are going to apply the Laplace transform to linear partial differential equations. We will see that this transform reduces a PDE to an ODE.

Transform of Partial Derivatives

The boundary-value problems considered in this section will involve either the one-dimensional wave and heat equations or slight variations of these equations. These PDEs involve an unknown function of two independent variables $u(x, t)$, where the variable t represents time $t \geq 0$. We define the Laplace transform of $u(x, t)$ with respect to t by

$$\mathcal{L}\{u(x, t)\} = \int_0^\infty e^{-st} u(x, t)\, dt = U(x, s),$$

where x is treated as a parameter. Throughout this section we shall assume that all the operational properties of Sections 4.2, 4.3, and 4.4 apply to functions of two variables. For example, by Theorem 4.2.2, the transform of the partial derivative $\partial u / \partial t$ is

$$\mathcal{L}\left\{\frac{\partial u}{\partial t}\right\} = s\mathcal{L}\{u(x, t)\} - u(x, 0);$$

that is,
$$\mathcal{L}\left\{\frac{\partial u}{\partial t}\right\} = sU(x, s) - u(x, 0). \tag{1}$$

Similarly,
$$\mathcal{L}\left\{\frac{\partial^2 u}{\partial t^2}\right\} = s^2 U(x, s) - su(x, 0) - u_t(x, 0). \tag{2}$$

Since we are transforming with respect to t, we further suppose that it is legitimate to interchange integration and differentiation in the transform of $\partial^2 u / \partial x^2$:

$$\mathcal{L}\left\{\frac{\partial^2 u}{\partial x^2}\right\} = \int_0^\infty e^{-st} \frac{\partial^2 u}{\partial x^2}\, dt = \int_0^\infty \frac{\partial^2}{\partial x^2}[e^{-st} u(x, t)]\, dt$$

$$= \frac{d^2}{dx^2} \int_0^\infty e^{-st} u(x, t)\, dt = \frac{d^2}{dx^2}\mathcal{L}\{u(x, t)\};$$

that is,
$$\mathcal{L}\left\{\frac{\partial^2 u}{\partial x^2}\right\} = \frac{d^2 U}{dx^2}. \tag{3}$$

In view of (1) and (2) we see that the Laplace transform is suited to problems with initial conditions—namely, those problems associated with the heat equation or the wave equation. We will see in Section 15.4 that boundary-value problems involving Laplace's equation in which one (or both) of the spatial variables is defined on an unbounded interval can often be solved using different integral transforms.

EXAMPLE 1 Laplace Transform of a PDE

Find the Laplace transform of the wave equation $a^2 \dfrac{\partial^2 u}{\partial x^2} = \dfrac{\partial^2 u}{\partial t^2}$, $t > 0$.

SOLUTION From (2) and (3),

$$\mathcal{L}\left\{a^2 \frac{\partial^2 u}{\partial x^2}\right\} = \mathcal{L}\left\{\frac{\partial^2 u}{\partial t^2}\right\}$$

becomes $\quad a^2 \dfrac{d^2}{dx^2} \mathscr{L}\{u(x, t)\} = s^2 \mathscr{L}\{u(x, t)\} - su(x, 0) - u_t(x, 0)$

or $\quad\quad a^2 \dfrac{d^2 U}{dx^2} - s^2 U = -su(x, 0) - u_t(x, 0).$ $\quad\quad\quad\quad$ (4) $\equiv$

The Laplace transform with respect to t of either the wave equation or the heat equation eliminates that variable, and for the one-dimensional equations the transformed equations are then *ordinary differential equations* in the spatial variable x. In solving a transformed equation, we treat s as a parameter.

EXAMPLE 2 Using the Laplace Transform to Solve a BVP

Solve $\quad \dfrac{\partial^2 u}{\partial x^2} = \dfrac{\partial^2 u}{\partial t^2}, \quad 0 < x < 1, \ t > 0$

subject to $\quad u(0, t) = 0, \quad u(1, t) = 0, \ t > 0$

$$u(x, 0) = 0, \quad \left. \dfrac{\partial u}{\partial t} \right|_{t=0} = \sin \pi x, \ 0 < x < 1.$$

SOLUTION The partial differential equation is recognized as the wave equation with $a = 1$. From (4) and the given initial conditions, the transformed equation is

$$\dfrac{d^2 U}{dx^2} - s^2 U = -\sin \pi x, \quad\quad\quad\quad (5)$$

where $U(x, s) = \mathscr{L}\{u(x, t)\}$. Since the boundary conditions are functions of t, we must also find their Laplace transforms:

$$\mathscr{L}\{u(0, t)\} = U(0, s) = 0 \quad\quad \text{and} \quad\quad \mathscr{L}\{u(1, t)\} = U(1, s) = 0. \quad\quad (6)$$

The results in (6) are boundary conditions for the ordinary differential equation (5). Since (5) is defined over a finite interval, its complementary function is

$$U_c(x, s) = c_1 \cosh sx + c_2 \sinh sx.$$

The method of undetermined coefficients yields a particular solution

$$U_p(x, s) = \dfrac{1}{s^2 + \pi^2} \sin \pi x.$$

Hence $\quad\quad U(x, s) = c_1 \cosh sx + c_2 \sinh sx + \dfrac{1}{s^2 + \pi^2} \sin \pi x.$

But the conditions $U(0, s) = 0$ and $U(1, s) = 0$ yield, in turn, $c_1 = 0$ and $c_2 = 0$. We conclude that

$$U(x, s) = \dfrac{1}{s^2 + \pi^2} \sin \pi x$$

$$u(x, t) = \mathscr{L}^{-1}\left\{ \dfrac{1}{s^2 + \pi^2} \sin \pi x \right\} = \dfrac{1}{\pi} \sin \pi x \, \mathscr{L}^{-1}\left\{ \dfrac{\pi}{s^2 + \pi^2} \right\}.$$

Therefore $\quad\quad\quad\quad u(x, t) = \dfrac{1}{\pi} \sin \pi x \sin \pi t.$ $\quad\quad\quad\quad\quad\quad\quad\quad \equiv$

EXAMPLE 3 Using the Laplace Transform to Solve a BVP

A very long string is initially at rest on the nonnegative x-axis. The string is secured at $x = 0$, and its distant right end slides down a frictionless vertical support. The string is set in motion by letting it fall under its own weight. Find the displacement $u(x, t)$.

SOLUTION Since the force of gravity is taken into consideration, it can be shown that the wave equation has the form

$$a^2 \dfrac{\partial^2 u}{\partial x^2} - g = \dfrac{\partial^2 u}{\partial t^2}, \quad x > 0, \ t > 0,$$

where g is the acceleration due to gravity. The boundary and initial conditions are, respectively,

$$u(0, t) = 0, \quad \lim_{x \to \infty} \frac{\partial u}{\partial x} = 0, \ t > 0$$

$$u(x, 0) = 0, \quad \left.\frac{\partial u}{\partial t}\right|_{t=0} = 0, \ x > 0.$$

The second boundary condition $\lim_{x \to \infty} \partial u / \partial x = 0$ indicates that the string is horizontal at a great distance from the left end. Now from (2) and (3),

$$\mathscr{L}\left\{a^2 \frac{\partial^2 u}{\partial x^2}\right\} - \mathscr{L}\{g\} = \mathscr{L}\left\{\frac{\partial^2 u}{\partial t^2}\right\}$$

becomes $\quad\quad a^2 \dfrac{d^2 U}{dx^2} - \dfrac{g}{s} = s^2 U - su(x, 0) - u_t(x, 0)$

or, in view of the initial conditions,

$$\frac{d^2 U}{dx^2} - \frac{s^2}{a^2} U = \frac{g}{a^2 s}.$$

The transforms of the boundary conditions are

$$\mathscr{L}\{u(0, t)\} = U(0, s) = 0 \quad \text{and} \quad \mathscr{L}\left\{\lim_{x \to \infty} \frac{\partial u}{\partial x}\right\} = \lim_{x \to \infty} \frac{dU}{dx} = 0.$$

With the aid of undetermined coefficients, the general solution of the transformed equation is found to be

$$U(x, s) = c_1 e^{-(x/a)s} + c_2 e^{(x/a)s} - \frac{g}{s^3}.$$

The boundary condition $\lim_{x \to \infty} dU/dx = 0$ implies $c_2 = 0$, and $U(0, s) = 0$ gives $c_1 = g/s^3$. Therefore

$$U(x, s) = \frac{g}{s^3} e^{-(x/a)s} - \frac{g}{s^3}.$$

Now by the second translation theorem we have

$$u(x, t) = \mathscr{L}^{-1}\left\{\frac{g}{s^3} e^{-(x/a)s} - \frac{g}{s^3}\right\} = \frac{1}{2} g\left(t - \frac{x}{a}\right)^2 \mathscr{U}\left(t - \frac{x}{a}\right) - \frac{1}{2} gt^2$$

$(at, -\frac{1}{2} gt^2)$

FIGURE 15.2.1 A long string falling under its own weight in Example 3

or $\quad\quad u(x, t) = \begin{cases} -\dfrac{1}{2} gt^2, & 0 \le t < \dfrac{x}{a} \\[2mm] -\dfrac{g}{2a^2}(2axt - x^2), & t \ge \dfrac{x}{a}. \end{cases}$

To interpret the solution, let us suppose $t > 0$ is fixed. For $0 \le x \le at$, the string is the shape of a parabola passing through the points $(0, 0)$ and $(at, -\frac{1}{2} gt^2)$. For $x > at$, the string is described by the horizontal line $u = -\frac{1}{2} gt^2$. See **FIGURE 15.2.1**. $\equiv$

Observe that the problem in the next example could be solved by the procedure in Section 13.6. The Laplace transform provides an alternative solution.

EXAMPLE 4 A Solution in Terms of erf(x)

Solve the heat equation

$$\frac{\partial^2 u}{\partial x^2} = \frac{\partial u}{\partial t}, \quad 0 < x < 1, \ t > 0$$

subject to $\quad\quad u(0, t) = 0, \quad u(1, t) = u_0, \ t > 0$

$$u(x, 0) = 0, \quad 0 < x < 1.$$

SOLUTION From (1) and (3) and the given initial condition,

$$\mathscr{L}\left\{\frac{\partial^2 u}{\partial x^2}\right\} = \mathscr{L}\left\{\frac{\partial u}{\partial t}\right\}$$

becomes

$$\frac{d^2 U}{dx^2} - sU = 0. \qquad (7)$$

The transforms of the boundary conditions are

$$U(0, s) = 0 \quad \text{and} \quad U(1, s) = \frac{u_0}{s}. \qquad (8)$$

Since we are concerned with a finite interval on the x-axis, we choose to write the general solution of (7) as

$$U(x, s) = c_1 \cosh(\sqrt{s}\,x) + c_2 \sinh(\sqrt{s}\,x).$$

Applying the two boundary conditions in (8) yields, respectively, $c_1 = 0$ and $c_2 = u_0/(s \sinh \sqrt{s})$. Thus

$$U(x, s) = u_0 \frac{\sinh(\sqrt{s}\,x)}{s \sinh \sqrt{s}}.$$

Now the inverse transform of the latter function cannot be found in most tables. However, by writing

$$\frac{\sinh(\sqrt{s}\,x)}{s \sinh \sqrt{s}} = \frac{e^{\sqrt{s}\,x} - e^{-\sqrt{s}\,x}}{s(e^{\sqrt{s}} - e^{-\sqrt{s}})} = \frac{e^{(x-1)\sqrt{s}} - e^{-(x+1)\sqrt{s}}}{s(1 - e^{-2\sqrt{s}})}$$

and using the geometric series

$$\frac{1}{1 - e^{-2\sqrt{s}}} = \sum_{n=0}^{\infty} e^{-2n\sqrt{s}},$$

we find

$$\frac{\sinh(\sqrt{s}\,x)}{s \sinh \sqrt{s}} = \sum_{n=0}^{\infty}\left[\frac{e^{-(2n+1-x)\sqrt{s}}}{s} - \frac{e^{-(2n+1+x)\sqrt{s}}}{s}\right].$$

If we assume that the inverse Laplace transform can be done term by term, it follows from entry 3 of Table 15.1.1 that

Also see Problem 8 in Exercises 15.1

$$u(x, t) = u_0 \mathscr{L}^{-1}\left\{\frac{\sinh(\sqrt{s}\,x)}{s \sinh \sqrt{s}}\right\}$$

$$= u_0 \sum_{n=0}^{\infty}\left[\mathscr{L}^{-1}\left\{\frac{e^{-(2n+1-x)\sqrt{s}}}{s}\right\} - \mathscr{L}^{-1}\left\{\frac{e^{-(2n+1+x)\sqrt{s}}}{s}\right\}\right]$$

$$= u_0 \sum_{n=0}^{\infty}\left[\text{erfc}\left(\frac{2n + 1 - x}{2\sqrt{t}}\right) - \text{erfc}\left(\frac{2n + 1 + x}{2\sqrt{t}}\right)\right]. \qquad (9)$$

The solution (9) can be rewritten in terms of the error function using $\text{erfc}(x) = 1 - \text{erf}(x)$:

$$u(x, t) = u_0 \sum_{n=0}^{\infty}\left[\text{erf}\left(\frac{2n + 1 + x}{2\sqrt{t}}\right) - \text{erf}\left(\frac{2n + 1 - x}{2\sqrt{t}}\right)\right]. \qquad (10) \equiv$$

FIGURE 15.2.2(a), obtained with the aid of the 3D plot function in a CAS, shows the surface over the rectangular region $0 \leq x \leq 1$, $0 \leq t \leq 6$ defined by the partial sum $S_{10}(x, t)$ of the solution (10). It is apparent from the surface and the accompanying two-dimensional graphs that at a fixed value of x (the curve of intersection of a plane slicing the surface perpendicular to the x-axis) on the interval [0, 1], the temperature $u(x, t)$ increases rapidly to a constant value as time increases. See Figure 15.2.2(b) and 15.2.2(c). For a fixed time (the curve of intersection of a plane slicing

782 | **CHAPTER 15** Integral Transforms

the surface perpendicular to the t-axis), the temperature $u(x, t)$ naturally increases from 0 to 100. See Figure 15.2.2(d) and 15.2.2(e).

(a) $u_0 = 100$

$u(0.2, t)$

(b) $x = 0.2$

$u(0.7, t)$

(c) $x = 0.7$

$u(x, 0.1)$

(d) $t = 0.1$

$u(x, 4)$

(e) $t = 4$

FIGURE 15.2.2 Graph of solution given in (10). In (b) and (c), x is held constant. In (d) and (e), t is held constant.

15.2 Exercises Answers to selected odd-numbered problems begin on page ANS-38.

In the following problems use tables as necessary.

1. A string is secured to the x-axis at $(0, 0)$ and $(L, 0)$. Find the displacement $u(x, t)$ if the string starts from rest in the initial position $A \sin(\pi x/L)$.

2. Solve the boundary-value problem

$$\frac{\partial^2 u}{\partial x^2} = \frac{\partial^2 u}{\partial t^2}, \quad 0 < x < 1, \ t > 0$$

$$u(0, t) = 0, \quad u(1, t) = 0, \ t > 0$$

$$u(x, 0) = 0, \quad \left.\frac{\partial u}{\partial t}\right|_{t=0} = 2 \sin \pi x + 4 \sin 3\pi x.$$

3. The displacement of a semi-infinite elastic string is determined from

$$a^2 \frac{\partial^2 u}{\partial x^2} = \frac{\partial^2 u}{\partial t^2}, \quad x > 0, \ t > 0$$

$$u(0, t) = f(t), \quad \lim_{x \to \infty} u(x, t) = 0, \ t > 0$$

$$u(x, 0) = 0, \quad \left.\frac{\partial u}{\partial t}\right|_{t=0} = 0, \ x > 0.$$

Solve for $u(x, t)$.

4. Solve the boundary-value problem in Problem 3 when

$$f(t) = \begin{cases} \sin \pi t, & 0 \le t \le 1 \\ 0, & t > 1. \end{cases}$$

Sketch the displacement $u(x, t)$ for $t > 1$.

5. In Example 3, find the displacement $u(x, t)$ when the left end of the string at $x = 0$ is given an oscillatory motion described by $f(t) = A \sin \omega t$.

6. The displacement $u(x, t)$ of a string that is driven by an external force is determined from

$$\frac{\partial^2 u}{\partial x^2} + \sin \pi x \sin \omega t = \frac{\partial^2 u}{\partial t^2}, \quad 0 < x < 1, \ t > 0$$

$$u(0, t) = 0, \quad u(1, t) = 0, \ t > 0$$

$$u(x, 0) = 0, \quad \left.\frac{\partial u}{\partial t}\right|_{t=0} = 0, \ 0 < x < 1.$$

Solve for $u(x, t)$.

7. (a) A uniform elastic bar is clamped at $x = 0$ and is initially at rest. If a constant unit force is applied to the free end at $x = 1$, the longitudinal displacement $u(x, t)$ of a cross section is determined from

$$\frac{\partial^2 u}{\partial x^2} = \frac{\partial^2 u}{\partial t^2}, \quad 0 < x < 1, \ t > 0$$

$$u(0, t) = 0, \quad \left.\frac{\partial u}{\partial x}\right|_{x=1} = 1, \ t > 0$$

$$u(x, 0) = 0, \quad \left.\frac{\partial u}{\partial t}\right|_{t=0} = 0, \ 0 < x < 1.$$

Solve for $u(x, t)$. [*Hint*: Expand $1/(1 + e^{-2s})$ in a geometric series. See Example 4.]

(b) For a fixed value of x satisfying $0 < x < 1$, sketch the graph of $u(x, t)$ as a function of t for $0 \le t \le 5 - x$.

(c) Sketch the graph of $u(1, t)$ as a function of t for $0 \le t \le 12$.

8. A uniform semi-infinite elastic beam moving along the x-axis with a constant velocity $-v_0$ is brought to a stop by hitting a wall at time $t = 0$. See **FIGURE 15.2.3**. The longitudinal displacement $u(x, t)$ is determined from

$$a^2 \frac{\partial^2 u}{\partial x^2} = \frac{\partial^2 u}{\partial t^2}, \quad x > 0, \ t > 0$$

$$u(0, t) = 0, \quad \lim_{x \to \infty} \frac{\partial u}{\partial x} = 0, \ t > 0$$

$$u(x, 0) = 0, \quad \left.\frac{\partial u}{\partial t}\right|_{t=0} = -v_0, \ x > 0.$$

Solve for $u(x, t)$.

FIGURE 15.2.3 Moving elastic beam in Problem 8

9. Solve the boundary-value problem

$$\frac{\partial^2 u}{\partial x^2} = \frac{\partial^2 u}{\partial t^2}, \quad x > 0, \ t > 0$$

$$u(0, t) = 0, \quad \lim_{x \to \infty} u(x, t) = 0, \ t > 0$$

$$u(x, 0) = xe^{-x}, \quad \left.\frac{\partial u}{\partial t}\right|_{t=0} = 0, \ x > 0.$$

10. Solve the boundary-value problem

$$\frac{\partial^2 u}{\partial x^2} = \frac{\partial^2 u}{\partial t^2}, \quad x > 0, \ t > 0$$

$$u(0, t) = 1, \quad \lim_{x \to \infty} u(x, t) = 0, \ t > 0$$

$$u(x, 0) = e^{-x}, \quad \left.\frac{\partial u}{\partial t}\right|_{t=0} = 0, \ x > 0.$$

In Problems 11–18, use the Laplace transform to solve the heat equation $u_{xx} = u_t$, $x > 0$, $t > 0$ subject to the given conditions.

11. $u(0, t) = u_0, \quad \lim_{x \to \infty} u(x, t) = u_1, \quad u(x, 0) = u_1$

12. $u(0, t) = u_0, \quad \lim_{x \to \infty} \dfrac{u(x, t)}{x} = u_1, \quad u(x, 0) = u_1 x$

13. $\left.\dfrac{\partial u}{\partial x}\right|_{x=0} = u(0, t), \quad \lim_{x \to \infty} u(x, t) = u_0, \quad u(x, 0) = u_0$

14. $\left.\dfrac{\partial u}{\partial x}\right|_{x=0} = u(0, t) - 50, \quad \lim_{x \to \infty} u(x, t) = 0, \quad u(x, 0) = 0$

15. $u(0, t) = f(t), \quad \lim_{x \to \infty} u(x, t) = 0, \quad u(x, 0) = 0$

 [*Hint*: Use the convolution theorem.]

16. $\left.\dfrac{\partial u}{\partial x}\right|_{x=0} = -f(t), \quad \lim_{x \to \infty} u(x, t) = 0, \quad u(x, 0) = 0$

17. $u(0, t) = 60 + 40 \,\mathcal{U}(t - 2), \quad \lim_{x \to \infty} u(x, t) = 60,$

 $u(x, 0) = 60$

18. $u(0, t) = \begin{cases} 20, & 0 < t < 1 \\ 0, & t \ge 1 \end{cases}, \quad \lim_{x \to \infty} u(x, t) = 100,$

 $u(x, 0) = 100$

19. Solve the boundary-value problem

$$\frac{\partial^2 u}{\partial x^2} = \frac{\partial u}{\partial t}, \quad -\infty < x < 1, \ t > 0$$

$$\left.\frac{\partial u}{\partial x}\right|_{x=1} = 100 - u(1, t), \quad \lim_{x \to -\infty} u(x, t) = 0, \ t > 0$$

$$u(x, 0) = 0, \quad -\infty < x < 1.$$

20. Show that a solution of the boundary-value problem

$$k \frac{\partial^2 u}{\partial x^2} + r = \frac{\partial u}{\partial t}, \quad x > 0, \ t > 0$$

$$u(0, t) = 0, \quad \lim_{x \to \infty} \frac{\partial u}{\partial x} = 0, \ t > 0$$

$$u(x, 0) = 0, \quad x > 0,$$

where r is a constant, is given by

$$u(x, t) = rt - r \int_0^t \text{erfc}\left(\frac{x}{2\sqrt{k\tau}}\right) d\tau.$$

21. Solve the boundary-value problem

$$\frac{\partial^2 u}{\partial x^2} = \frac{\partial^2 u}{\partial t^2} + 2\frac{\partial u}{\partial t} + u, \quad x > 0, \ t > 0$$

$$u(0, t) = \sin 3t, \ t > 0$$

$$u(x, 0) = 0, \quad \left.\frac{\partial u}{\partial t}\right|_{t=0} = 0, \ x > 0.$$

22. Solve the boundary-value problem

$$\frac{\partial^2 u}{\partial x^2} = \frac{\partial^2 u}{\partial t^2} + 1, \quad x > 0, \ t > 0$$

$$u(0, t) = 1, \ t > 0$$

$$u(x, 0) = 1, \quad \left.\frac{\partial u}{\partial t}\right|_{t=0} = 1, \ x > 0.$$

23. A rod of length L is held at a constant temperature u_0 at its ends $x = 0$ and $x = L$. If the rod's initial temperature is $u_0 + u_0 \sin(x\pi/L)$, solve the heat equation $u_{xx} = u_t, 0 < x < L$, $t > 0$ for the temperature $u(x, t)$.

24. If there is a heat transfer from the lateral surface of a thin wire of length L into a medium at constant temperature u_m, then the heat equation takes on the form

$$k\frac{\partial^2 u}{\partial x^2} - h(u - u_m) = \frac{\partial u}{\partial t}, \quad 0 < x < L, \ t > 0,$$

where h is a constant. Find the temperature $u(x, t)$ if the initial temperature is a constant u_0 throughout and the ends $x = 0$ and $x = L$ are insulated.

25. A rod of unit length is insulated at $x = 0$ and is kept at temperature zero at $x = 1$. If the initial temperature of the rod is a constant u_0, solve $ku_{xx} = u_t, 0 < x < 1, t > 0$ for the temperature $u(x, t)$. [*Hint*: Expand $1/(1 + e^{-2\sqrt{s/k}})$ in a geometric series. See Example 4.]

26. An infinite porous slab of unit width is immersed in a solution of constant concentration c_0. A dissolved substance in the solution diffuses into the slab. The concentration $c(x, t)$ in the slab is determined from

$$D\frac{\partial^2 c}{\partial x^2} = \frac{\partial c}{\partial t}, \quad 0 < x < 1, \ t > 0$$

$$c(0, t) = c_0, \quad c(1, t) = c_0, \ t > 0$$

$$c(x, 0) = 0, \quad 0 < x < 1,$$

where D is a constant. Solve for $c(x, t)$.

27. A very long telephone transmission line is initially at a constant potential u_0. If the line is grounded at $x = 0$ and insulated at the distant right end, then the potential $u(x, t)$ at a point x along the line at time t is determined from

$$\frac{\partial^2 u}{\partial x^2} - RC\frac{\partial u}{\partial t} - RGu = 0, \quad x > 0, \ t > 0$$

$$u(0, t) = 0, \quad \lim_{x \to \infty} \frac{\partial u}{\partial x} = 0, \ t > 0$$

$$u(x, 0) = u_0, \ x > 0,$$

where R, C, and G are constants known as resistance, capacitance, and conductance, respectively. Solve for $u(x, t)$. [*Hint*: See Problem 7 in Exercises 15.1.]

28. Show that a solution of the boundary-value problem

$$\frac{\partial^2 u}{\partial x^2} - hu = \frac{\partial u}{\partial t}, \quad x > 0, \ t > 0, \ h \text{ constant}$$

$$u(0, t) = u_0, \quad \lim_{x \to \infty} u(x, t) = 0, \ t > 0$$

$$u(x, 0) = 0, \ x > 0$$

is

$$u(x, t) = \frac{u_0 x}{2\sqrt{\pi}} \int_0^t \frac{e^{-h\tau - x^2/4\tau}}{\tau^{3/2}} d\tau.$$

29. Starting at $t = 0$, a concentrated load of magnitude F_0 moves with a constant velocity v_0 along a semi-infinite string. In this case, the wave equation becomes

$$a^2\frac{\partial^2 u}{\partial x^2} = \frac{\partial^2 u}{\partial t^2} + F_0\delta\left(t - \frac{x}{v_0}\right),$$

where $\delta(t - x/v_0)$ is the Dirac delta function. Solve the above PDE subject to

$$u(0, t) = 0, \quad \lim_{x \to \infty} u(x, t) = 0, \ t > 0$$

$$u(x, 0) = 0, \quad \left.\frac{\partial u}{\partial t}\right|_{t=0} = 0, \ x > 0$$

when $v_0 \neq a$.

30. Solve the boundary-value problem in Problem 29 when $v_0 = a$.

31. The temperature in a semi-infinite solid is modeled by the boundary-value problem

$$k\frac{\partial^2 u}{\partial x^2} = \frac{\partial u}{\partial t}, \quad x > 0, \ t > 0$$

$$u(0, t) = u_0, \quad \lim_{x \to \infty} u(x, t) = 0, \ t > 0$$

$$u(x, 0) = 0, \ x > 0,$$

where u_0 is a nonzero constant. Solve for $u(x, t)$. Use the solution to determine analytically the value of $\lim_{t \to \infty} u(x, t)$, $x > 0$.

32. In Problem 31, if there is a constant flux of heat into the solid at its left-hand boundary, then the boundary condition is

$$\left.\frac{\partial u}{\partial x}\right|_{x=0} = -u_0, \ t > 0,$$

where u_0 is a positive constant. Solve for $u(x, t)$. Use the solution to determine analytically the value of $\lim_{t \to \infty} u(x, t)$, $x > 0$.

In Problems 33–36, the partial differential equation is the heat equation in spherical coordinates

$$\frac{\partial^2 u}{\partial r^2} + \frac{2}{r}\frac{\partial u}{\partial r} = \frac{\partial u}{\partial t}.$$

Take the Laplace transform of this PDE with respect to the variable t. To solve the transformed equation, use the substitution $v(r, s) = rU(r, s)$, where $U(r, s) = \mathscr{L}\{u(r, t)\}$.

33. In Problem 9 of Exercises 14.3 you were asked to find the time-dependent temperature $u(r, t)$ inside a unit sphere centered at the origin. Find the temperature $u(r, t)$ *outside* a unit sphere if the boundary and initial conditions are

$$u(1, t) = 100, \quad \lim_{r \to \infty} u(r, t) = 0, \ t > 0$$

$$u(r, 0) = 0, \ r > 1.$$

34. Find the temperature $u(r, t)$ outside a unit sphere if the boundary and initial conditions are

$$u(1, t) = u_0, \quad \lim_{r \to \infty} u(r, t) = u_1, \ t > 0$$

$$u(r, 0) = u_1, \ r > 1,$$

where u_0 and u_1 are nonzero constants.

35. Find the temperature $u(r, t)$ inside a unit sphere if the boundary and initial conditions are

$$u(1, t) = u_0, \quad u(r, t) \text{ is finite as } r \to 0, \ t > 0$$

$$u(r, 0) = u_1, \quad 0 < r < 1,$$

where u_0 and u_1 are nonzero constants.

36. Find the temperature $u(r, t)$ outside a unit sphere if the boundary and initial conditions are

$$\left.\frac{\partial u}{\partial r}\right|_{r=1} = u(1, t), \quad \lim_{r \to \infty} u(r, t) = u_0, \ t > 0$$

$$u(r, 0) = u_0, \ r > 1,$$

where u_0 is a nonzero constant.

Mathematical Model

37. **Pulsating Sphere** A sphere of radius $r = a$ centered at the origin is submerged in an ideal compressible fluid. Starting at $t = 0$ the surface of the sphere is caused to pulsate with radial velocity $u_0 \cos \omega t$ where u_0 is a nonzero constant. The pulsation continues indefinitely. A cross section of the sphere is shown in **FIGURE 15.2.4**. A simplified model for the velocity potential $u(r, t)$ of points in the surrounding fluid is given by the boundary-value problem

$$\frac{\partial^2 u}{\partial r^2} + \frac{2}{r}\frac{\partial u}{\partial r} = \frac{\partial^2 u}{\partial t^2}, \quad r > a, \ t > 0$$

$$\left.\frac{\partial u}{\partial r}\right|_{r=a} = u_0 \cos \omega t, \quad \frac{\partial u}{\partial r} \text{ is finite as } r \to \infty, \ t > 0$$

$$u(r, 0) = 0, \quad \left.\frac{\partial u}{\partial t}\right|_{t=0} = 0, \ r > a.$$

Solve for $u(r, t)$. Then determine the radial velocity $\partial u/\partial r$ of points in the fluid.

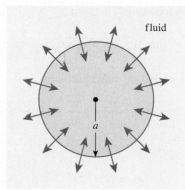

FIGURE 15.2.4 Pulsating sphere in Problem 37

38. Find the radial velocity $\partial u/\partial r$ of points in the fluid surrounding the sphere of radius $r = a$ in Problem 37 if the boundary and initial conditions are

$$\left.\frac{\partial u}{\partial r}\right|_{r=a} = u_0[1 - \mathscr{U}(t - t_0)], \quad \frac{\partial u}{\partial r} \text{ is finite as } r \to \infty, \ t > 0$$

$$u(r, 0) = 0, \quad \left.\frac{\partial u}{\partial t}\right|_{t=0} = 0, \ r > a.$$

In this case, the sphere is expanding for a period of time but is not pulsating.

Computer Lab Assignments

39. Use a CAS to obtain the graph of $u(x, t)$ in Problem 31 over the rectangular region defined by $0 \le x \le 10, 0 \le t \le 15$. Assume $u_0 = 100$ and $k = 1$. Indicate the two boundary conditions and initial condition on your graph. Use 2D and 3D plots of $u(x, t)$ to verify the value of $\lim_{t \to \infty} u(x, t)$.

40. Use a CAS to obtain the graph of $u(x, t)$ in Problem 32 over the rectangular region defined by $0 \le x \le 10, 0 \le t \le 15$. Assume $u_0 = 100$ and $k = 1$. Use 2D and 3D plots of $u(x, t)$ to verify the value of $\lim_{t \to \infty} u(x, t)$.

41. **Chemical Messages** Humans gather most of their information on the outside world through sight and sound. But many creatures use chemical signals as their primary means of communication; for example, honeybees, when alarmed, emit a substance and fan their wings feverishly to relay the warning signal to the bees that attend to the queen. These molecular messages between members of the same species are called pheromones. The signals may be carried by moving air or water or by a diffusion process in which the random movement of gas molecules transports the chemical away from its source. **FIGURE 15.2.5** shows an ant emitting an alarm chemical into the still air of a tunnel. If $c(x, t)$ denotes the concentration of the chemical x centimeters from the source at time t, then $c(x, t)$ satisfies

$$k\frac{\partial^2 c}{\partial x^2} = \frac{\partial c}{\partial t}, \quad x > 0, \ t > 0,$$

and k is a positive constant. The emission of pheromones as a discrete pulse gives rise to a boundary condition of the form

$$\frac{\partial c}{\partial x}\bigg|_{x=0} = -A\delta(t),$$

where $\delta(t)$ is the Dirac delta function.

(a) Solve the boundary-value problem if it is further known that $c(x, 0) = 0$, $x > 0$, and $\lim_{x\to\infty} c(x, t) = 0$, $t > 0$.

(b) Use a CAS to plot the graph of the solution in part (a) for $x > 0$ at the fixed times $t = 0.1$, $t = 0.5$, $t = 1$, $t = 2$, $t = 5$.

(c) For a fixed time t, show that $\int_0^\infty c(x, t)\, dx = Ak$. Thus Ak represents the total amount of chemical discharged.

FIGURE 15.2.5 Ants in Problem 41

15.3 Fourier Integral

INTRODUCTION In preceding chapters, Fourier series were used to represent a function f defined on a *finite* interval $(-p, p)$ or $(0, L)$. When f and f' are piecewise continuous on such an interval, a Fourier series represents the function on the interval and converges to the periodic extension of f outside the interval. In this way we are justified in saying that Fourier series are associated only with *periodic* functions. We shall now derive, in a nonrigorous fashion, a means of representing certain kinds of *nonperiodic* functions that are defined on either an infinite interval $(-\infty, \infty)$ or a semi-infinite interval $(0, \infty)$.

From Fourier Series to Fourier Integral Suppose a function f is defined on $(-p, p)$. If we use the integral definitions of the coefficients (9), (10), and (11) of Section 12.2 in (8) of that section, then the Fourier series of f on the interval is

$$f(x) = \frac{1}{2p}\int_{-p}^{p} f(t)\, dt + \frac{1}{p}\sum_{n=1}^{\infty}\left[\left(\int_{-p}^{p} f(t)\cos\frac{n\pi}{p}t\, dt\right)\cos\frac{n\pi}{p}x + \left(\int_{-p}^{p} f(t)\sin\frac{n\pi}{p}t\, dt\right)\sin\frac{n\pi}{p}x\right]. \quad (1)$$

If we let $\alpha_n = n\pi/p$, $\Delta\alpha = \alpha_{n+1} - \alpha_n = \pi/p$, then (1) becomes

$$f(x) = \frac{1}{2\pi}\left(\int_{-p}^{p} f(t)\, dt\right)\Delta\alpha + \frac{1}{\pi}\sum_{n=1}^{\infty}\left[\left(\int_{-p}^{p} f(t)\cos\alpha_n t\, dt\right)\cos\alpha_n x + \left(\int_{-p}^{p} f(t)\sin\alpha_n t\, dt\right)\sin\alpha_n x\right]\Delta\alpha. \quad (2)$$

We now expand the interval $(-p, p)$ by letting $p \to \infty$. Since $p \to \infty$ implies that $\Delta\alpha \to 0$, the limit of (2) has the form $\lim_{\Delta\alpha\to 0}\sum_{n=1}^{\infty}F(\alpha_n)\Delta\alpha$, which is suggestive of the definition of the integral $\int_0^\infty F(\alpha)\, d\alpha$. Thus if $\int_{-\infty}^{\infty} f(t)\, dt$ exists, the limit of the first term in (2) is zero and the limit of the sum becomes

$$f(x) = \frac{1}{\pi}\int_0^\infty\left[\left(\int_{-\infty}^{\infty} f(t)\cos\alpha t\, dt\right)\cos\alpha x + \left(\int_{-\infty}^{\infty} f(t)\sin\alpha t\, dt\right)\sin\alpha x\right]d\alpha. \quad (3)$$

The result given in (3) is called the **Fourier integral** of f on the interval $(-\infty, \infty)$. As the following summary shows, the basic structure of the Fourier integral is reminiscent of that of a Fourier series.

DEFINITION 15.3.1 Fourier Integral

The **Fourier integral** of a function f defined on the interval $(-\infty, \infty)$ is given by

$$f(x) = \frac{1}{\pi}\int_0^\infty [A(\alpha)\cos\alpha x + B(\alpha)\sin\alpha x]\, d\alpha, \quad (4)$$

where

$$A(\alpha) = \int_{-\infty}^{\infty} f(x)\cos\alpha x\, dx \quad (5)$$

$$B(\alpha) = \int_{-\infty}^{\infty} f(x)\sin\alpha x\, dx. \quad (6)$$

Convergence of a Fourier Integral Sufficient conditions under which a Fourier integral converges to $f(x)$ are similar to, but slightly more restrictive than, the conditions for a Fourier series.

THEOREM 15.3.1 **Conditions for Convergence**

Let f and f' be piecewise continuous on every finite interval, and let f be absolutely integrable on $(-\infty, \infty)$.* Then the Fourier integral of f on the interval converges to $f(x)$ at a point of continuity. At a point of discontinuity, the Fourier integral will converge to the average

$$\frac{f(x+) + f(x-)}{2},$$

where $f(x+)$ and $f(x-)$ denote the limit of f at x from the right and from the left, respectively.

EXAMPLE 1 **Fourier Integral Representation**

Find the Fourier integral representation of the piecewise-continuous function

$$f(x) = \begin{cases} 0, & x < 0 \\ 1, & 0 < x < 2 \\ 0, & x > 2. \end{cases}$$

FIGURE 15.3.1 Function f in Example 1

SOLUTION The function, whose graph is shown in **FIGURE 15.3.1**, satisfies the hypotheses of Theorem 15.3.1. Hence from (5) and (6) we have at once

$$A(\alpha) = \int_{-\infty}^{\infty} f(x) \cos \alpha x \, dx$$

$$= \int_{-\infty}^{0} f(x) \cos \alpha x \, dx + \int_{0}^{2} f(x) \cos \alpha x \, dx + \int_{2}^{\infty} f(x) \cos \alpha x \, dx$$

$$= \int_{0}^{2} \cos \alpha x \, dx = \frac{\sin 2\alpha}{\alpha}$$

$$B(\alpha) = \int_{-\infty}^{\infty} f(x) \sin \alpha x \, dx = \int_{0}^{2} \sin \alpha x \, dx = \frac{1 - \cos 2\alpha}{\alpha}.$$

Substituting these coefficients into (4) then gives

$$f(x) = \frac{1}{\pi} \int_{0}^{\infty} \left[\left(\frac{\sin 2\alpha}{\alpha} \right) \cos \alpha x + \left(\frac{1 - \cos 2\alpha}{\alpha} \right) \sin \alpha x \right] d\alpha.$$

When we use trigonometric identities, the last integral simplifies to

$$f(x) = \frac{2}{\pi} \int_{0}^{\infty} \frac{\sin \alpha \cos \alpha(x - 1)}{\alpha} \, d\alpha. \tag{7} \equiv$$

The Fourier integral can be used to evaluate integrals. For example, at $x = 1$ it follows from Theorem 15.3.1 that (7) converges to $f(1)$. From this we find

$$\int_{0}^{\infty} \frac{\sin \alpha}{\alpha} \, d\alpha = \frac{\pi}{2}.$$

The latter result is worthy of special note since it cannot be obtained in the "usual" manner; the integrand $(\sin x)/x$ does not possess an antiderivative that is an elementary function.

Cosine and Sine Integrals When f is an even function on the interval $(-\infty, \infty)$, then the product $f(x) \cos \alpha x$ is also an even function, whereas $f(x) \sin \alpha x$ is an odd function. As a consequence of property (g) of Theorem 12.3.1, $B(\alpha) = 0$, and so (4) becomes

*This means that the integral $\int_{-\infty}^{\infty} |f(x)| \, dx$ converges.

$$f(x) = \frac{2}{\pi} \int_0^\infty \left(\int_0^\infty f(t) \cos \alpha t \, dt \right) \cos \alpha x \, d\alpha.$$

Here we have also used property (f) of Theorem 12.3.1 to write

$$\int_{-\infty}^\infty f(t) \cos \alpha t \, dt = 2 \int_0^\infty f(t) \cos \alpha t \, dt.$$

Similarly, when f is an odd function on $(-\infty, \infty)$ the products $f(x) \cos \alpha x$ and $f(x) \sin \alpha x$ are odd and even functions, respectively. Therefore $A(\alpha) = 0$ and

$$f(x) = \frac{2}{\pi} \int_0^\infty \left(\int_0^\infty f(t) \sin \alpha t \, dt \right) \sin \alpha x \, d\alpha.$$

We summarize in the following definition.

DEFINITION 15.3.2 Fourier Cosine and Sine Integrals

(*i*) The Fourier integral of an even function on the interval $(-\infty, \infty)$ is the **cosine integral**

$$f(x) = \frac{2}{\pi} \int_0^\infty A(\alpha) \cos \alpha x \, d\alpha, \tag{8}$$

where

$$A(\alpha) = \int_0^\infty f(x) \cos \alpha x \, dx. \tag{9}$$

(*ii*) The Fourier integral of an odd function on the interval $(-\infty, \infty)$ is the **sine integral**

$$f(x) = \frac{2}{\pi} \int_0^\infty B(\alpha) \sin \alpha x \, d\alpha, \tag{10}$$

where

$$B(\alpha) = \int_0^\infty f(x) \sin \alpha x \, dx. \tag{11}$$

EXAMPLE 2 **Cosine Integral Representation**

Find the Fourier integral representation of the function

$$f(x) = \begin{cases} 1, & |x| < a \\ 0, & |x| > a. \end{cases}$$

SOLUTION It is apparent from **FIGURE 15.3.2** that f is an even function. Hence we represent f by the Fourier cosine integral (8). From (9) we obtain

$$A(\alpha) = \int_0^\infty f(x) \cos \alpha x \, dx = \int_0^a f(x) \cos \alpha x \, dx + \int_a^\infty f(x) \cos \alpha x \, dx$$

$$= \int_0^a \cos \alpha x \, dx = \frac{\sin a\alpha}{\alpha},$$

FIGURE 15.3.2 Function f in Example 2

and so

$$f(x) = \frac{2}{\pi} \int_0^\infty \frac{\sin a\alpha \cos \alpha x}{\alpha} \, d\alpha. \tag{12} \equiv$$

The integrals (8) and (10) can be used when f is neither odd nor even and defined only on the half-line $(0, \infty)$. In this case (8) represents f on the interval $(0, \infty)$ and its even (but not periodic) extension to $(-\infty, 0)$, whereas (10) represents f on $(0, \infty)$ and its odd extension to the interval $(-\infty, 0)$. The next example illustrates this concept.

FIGURE 15.3.3 Function *f* in Example 3

(a) Cosine integral

(b) Sine integral

FIGURE 15.3.4 In Example 3, (a) is the even extension of *f*; (b) is the odd extension of *f*

EXAMPLE 3 **Cosine and Sine Integral Representations**

Represent $f(x) = e^{-x}$, $x > 0$ **(a)** by a cosine integral; **(b)** by a sine integral.

SOLUTION The graph of the function is given in **FIGURE 15.3.3**.

(a) Using integration by parts, we find

$$A(\alpha) = \int_0^\infty e^{-x} \cos \alpha x \, dx = \frac{1}{1 + \alpha^2}.$$

Therefore from (8) the cosine integral of *f* is

$$f(x) = \frac{2}{\pi} \int_0^\infty \frac{\cos \alpha x}{1 + \alpha^2} \, d\alpha. \tag{13}$$

(b) Similarly, we have

$$B(\alpha) = \int_0^\infty e^{-x} \sin \alpha x \, dx = \frac{\alpha}{1 + \alpha^2}.$$

From (10) the sine integral of *f* is then

$$f(x) = \frac{2}{\pi} \int_0^\infty \frac{\alpha \sin \alpha x}{1 + \alpha^2} \, d\alpha. \tag{14}$$

FIGURE 15.3.4 shows the graphs of the functions and their extensions represented by the two integrals. ≡

▇ **Complex Form** The Fourier integral (4) also possesses an equivalent **complex form**, or **exponential form**, that is analogous to the complex form of a Fourier series (see Section 12.4). If (5) and (6) are substituted into (4), then

$$f(x) = \frac{1}{\pi} \int_0^\infty \int_{-\infty}^\infty f(t)[\cos \alpha t \cos \alpha x + \sin \alpha t \sin \alpha x] \, dt \, d\alpha$$

$$= \frac{1}{\pi} \int_0^\infty \int_{-\infty}^\infty f(t) \cos \alpha(t - x) \, dt \, d\alpha$$

$$= \frac{1}{2\pi} \int_{-\infty}^\infty \int_{-\infty}^\infty f(t) \cos \alpha(t - x) \, dt \, d\alpha \tag{15}$$

$$= \frac{1}{2\pi} \int_{-\infty}^\infty \int_{-\infty}^\infty f(t)[\cos \alpha(t - x) + i \sin \alpha(t - x)] \, dt \, d\alpha \tag{16}$$

$$= \frac{1}{2\pi} \int_{-\infty}^\infty \int_{-\infty}^\infty f(t) \, e^{i\alpha(t - x)} \, dt \, d\alpha$$

$$= \frac{1}{2\pi} \int_{-\infty}^\infty \left(\int_{-\infty}^\infty f(t) e^{i\alpha t} \, dt \right) e^{-i\alpha x} \, d\alpha. \tag{17}$$

We note that (15) follows from the fact that the integrand is an even function of α. In (16) we have simply added zero to the integrand,

$$i \int_{-\infty}^\infty \int_{-\infty}^\infty f(t) \sin \alpha(t - x) \, dt \, d\alpha = 0,$$

because the integrand is an odd function of α. The integral in (17) can be expressed as

$$f(x) = \frac{1}{2\pi} \int_{-\infty}^\infty C(\alpha) e^{-i\alpha x} \, d\alpha, \tag{18}$$

where

$$C(\alpha) = \int_{-\infty}^\infty f(x) e^{i\alpha x} \, dx. \tag{19}$$

This latter form of the Fourier integral will be put to use in the next section when we return to the solution of boundary-value problems.

Use of Computers

The convergence of a Fourier integral can be examined in a manner that is similar to graphing partial sums of a Fourier series. To illustrate, let's use the results in parts (a) and (b) of Example 3. By definition of an improper integral, the Fourier cosine integral representation of $f(x) = e^{-x}$, $x > 0$ in (13) can be written as $f(x) = \lim_{b \to \infty} F_b(x)$, where

$$F_b(x) = \frac{2}{\pi} \int_0^b \frac{\cos \alpha x}{1 + \alpha^2}\, d\alpha,$$

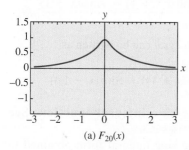

(a) $F_{20}(x)$

and x is treated as a parameter. Similarly, the Fourier sine integral representation of $f(x) = e^{-x}$, $x > 0$ in (14) can be written as $f(x) = \lim_{b \to \infty} G_b(x)$, where

$$G_b(x) = \frac{2}{\pi} \int_0^b \frac{\alpha \sin \alpha x}{1 + \alpha^2}\, d\alpha.$$

(b) $G_{20}(x)$

FIGURE 15.3.5 Graphs of partial integrals

Because the Fourier integrals (13) and (14) converge, the graphs of the partial integrals $F_b(x)$ and $G_b(x)$ for a specified value of $b > 0$ will be an approximation to the graph of f and its even and odd extensions shown in Figure 15.3.4(a) and 15.3.4(b), respectively. The graphs of $F_b(x)$ and $G_b(x)$ for $b = 20$ given in **FIGURE 15.3.5** were obtained using *Mathematica* and its **NIntegrate** application. See Problem 21 in Exercises 15.3.

15.3 Exercises Answers to selected odd-numbered problems begin on page ANS-38.

In Problems 1–6, find the Fourier integral representation of the given function.

1. $f(x) = \begin{cases} 0, & x < -1 \\ -1, & -1 < x < 0 \\ 2, & 0 < x < 1 \\ 0, & x > 1 \end{cases}$

2. $f(x) = \begin{cases} 0, & x < \pi \\ 4, & \pi < x < 2\pi \\ 0, & x > 2\pi \end{cases}$

3. $f(x) = \begin{cases} 0, & x < 0 \\ x, & 0 < x < 3 \\ 0, & x > 3 \end{cases}$

4. $f(x) = \begin{cases} 0, & x < 0 \\ \sin x, & 0 \le x \le \pi \\ 0, & x > \pi \end{cases}$

5. $f(x) = \begin{cases} 0, & x < 0 \\ e^{-x}, & x > 0 \end{cases}$

6. $f(x) = \begin{cases} e^x, & |x| < 1 \\ 0, & |x| > 1 \end{cases}$

In Problems 7–12, represent the given function by an appropriate cosine or sine integral.

7. $f(x) = \begin{cases} 0, & x < -1 \\ -5, & -1 < x < 0 \\ 5, & 0 < x < 1 \\ 0, & x > 1 \end{cases}$

8. $f(x) = \begin{cases} 0, & |x| < 1 \\ \pi, & 1 < |x| < 2 \\ 0, & |x| > 2 \end{cases}$

9. $f(x) = \begin{cases} |x|, & |x| < \pi \\ 0, & |x| > \pi \end{cases}$

10. $f(x) = \begin{cases} x, & |x| < \pi \\ 0, & |x| > \pi \end{cases}$

11. $f(x) = e^{-|x|} \sin x$

12. $f(x) = xe^{-|x|}$

In Problems 13–16, find the cosine and sine integral representations of the given function.

13. $f(x) = e^{-kx}$, $k > 0$, $x > 0$

14. $f(x) = e^{-x} - e^{-3x}$, $x > 0$

15. $f(x) = xe^{-2x}$, $x > 0$

16. $f(x) = e^{-x} \cos x$, $x > 0$

In Problems 17 and 18, solve the given integral equation for the function f.

17. $\displaystyle\int_0^\infty f(x) \cos \alpha x\, dx = e^{-\alpha}$

18. $\displaystyle\int_0^\infty f(x) \sin \alpha x\, dx = \begin{cases} 1, & 0 < \alpha < 1 \\ 0, & \alpha > 1 \end{cases}$

19. **(a)** Use (7) to show that $\displaystyle\int_0^\infty \frac{\sin 2x}{x}\, dx = \frac{\pi}{2}$.

 [*Hint:* α is a dummy variable of integration.]

 (b) Show in general that, for $k > 0$, $\displaystyle\int_0^\infty \frac{\sin kx}{x}\, dx = \frac{\pi}{2}$.

20. Use the complex form (19) to find the Fourier integral representation of $f(x) = e^{-|x|}$. Show that the result is the same as that obtained from (8) and (9).

Computer Lab Assignment

21. While the integral (12) can be graphed in the same manner discussed on page 791 to obtain Figure 15.3.5, it can also be expressed in terms of a special function that is built into a CAS.

(a) Use a trigonometric identity to show that an alternative form of the Fourier integral representation (12) of the function f in Example 2 (with $a = 1$) is

$$f(x) = \frac{1}{\pi} \int_0^{\infty} \frac{\sin \alpha(x + 1) - \sin \alpha(x - 1)}{\alpha} \, d\alpha.$$

(b) As a consequence of part (a), $f(x) = \lim_{b \to \infty} F_b(x)$, where

$$F_b(x) = \frac{1}{\pi} \int_0^b \frac{\sin \alpha(x + 1) - \sin \alpha(x - 1)}{\alpha} \, d\alpha.$$

Show that the last integral can be written as

$$F_b(x) = \frac{1}{\pi} [\text{Si}(b(x + 1)) - \text{Si}(b(x - 1))],$$

where $\text{Si}(x)$ is the **sine integral function**. See Problem 47 in Exercises 2.3.

(c) Use a CAS and the sine integral form obtained in part (b) to graph $F_b(x)$ on the interval $[-3, 3]$ for $b = 4, 6$, and 15. Then graph $F_b(x)$ for larger values of $b > 0$.

15.4 Fourier Transforms

INTRODUCTION Up to now we have studied and used only one integral transform: the Laplace transform. But in Section 15.3 we saw that the Fourier integral had three alternative forms: the cosine integral, the sine integral, and the complex or exponential form. In the present section we shall take these three forms of the Fourier integral and develop them into three new integral transforms naturally called **Fourier transforms**. In addition, we shall expand on the concept of a transform pair, that is, an integral transform and its inverse. We shall also see that the inverse of an integral transform is itself another integral transform.

■ **Transform Pairs** The Laplace transform $F(s)$ of a function $f(t)$ is defined by an integral, but up to now we have been using the symbolic representation $f(t) = \mathscr{L}^{-1}\{F(s)\}$ to denote the inverse Laplace transform of $F(s)$. Actually, the inverse Laplace transform is also an *integral* transform. If

$$\mathscr{L}\{f(t)\} = \int_0^{\infty} e^{-st} f(t) \, dt = F(s), \tag{1}$$

then the **inverse Laplace transform** is

$$\mathscr{L}^{-1}\{F(s)\} = \frac{1}{2\pi i} \int_{\gamma - i\infty}^{\gamma + i\infty} e^{st} F(s) \, ds = f(t). \tag{2}$$

The last integral is called a **contour integral**; its evaluation requires the use of complex variables and is beyond the scope of this discussion. The point here is this: Integral transforms appear in **transform pairs**. If $f(x)$ is transformed into $F(\alpha)$ by an **integral transform**

$$F(\alpha) = \int_a^b f(x) K(\alpha, x) \, dx, \tag{3}$$

then the function f can be recovered by another integral transform

$$f(x) = \int_c^b F(\alpha) H(\alpha, x) \, d\alpha, \tag{4}$$

called the **inverse transform**. The functions K and H in the integrands of (3) and (4) are called the **kernels** of their respective transforms. We identify $K(s, t) = e^{-st}$ as the kernel of the Laplace transform and $H(s, t) = e^{st}/2\pi i$ as the kernel of the inverse Laplace transform.

■ **Fourier Transform Pairs** The Fourier integral is the source of three integral transforms. From (19) and (18), (9) and (8), and (11) and (10) of Section 15.3, we are prompted to define the following **Fourier transform pairs**.

DEFINITION 15.4.1 Fourier Transform Pairs

(i) Fourier transform:
$$\mathscr{F}\{f(x)\} = \int_{-\infty}^{\infty} f(x)e^{i\alpha x}\, dx = F(\alpha) \tag{5}$$

Inverse Fourier transform:
$$\mathscr{F}^{-1}\{F(\alpha)\} = \frac{1}{2\pi}\int_{-\infty}^{\infty} F(\alpha)e^{-i\alpha x}\, d\alpha = f(x) \tag{6}$$

(ii) Fourier cosine transform:
$$\mathscr{F}_c\{f(x)\} = \int_{0}^{\infty} f(x)\cos\alpha x\, dx = F(\alpha) \tag{7}$$

Inverse Fourier cosine transform:
$$\mathscr{F}_c^{-1}\{F(\alpha)\} = \frac{2}{\pi}\int_{0}^{\infty} F(\alpha)\cos\alpha x\, d\alpha = f(x) \tag{8}$$

(iii) Fourier sine transform:
$$\mathscr{F}_s\{f(x)\} = \int_{0}^{\infty} f(x)\sin\alpha x\, dx = F(\alpha) \tag{9}$$

Inverse Fourier sine transform:
$$\mathscr{F}_s^{-1}\{F(\alpha)\} = \frac{2}{\pi}\int_{0}^{\infty} F(\alpha)\sin\alpha x\, d\alpha = f(x) \tag{10}$$

■ **Existence** The conditions under which (5), (7), and (9) exist are more stringent than those for the Laplace transform. For example, $\mathscr{F}\{1\}$, $\mathscr{F}_c\{1\}$, and $\mathscr{F}_s\{1\}$ do not exist. Sufficient conditions for existence are that f be absolutely integrable on the appropriate interval and that f and f' be piecewise continuous on every finite interval.

■ **Linearity of the Fourier Transforms** Like the Laplace transform, each of the six transforms given in Definition 15.4.1 are **linear transforms**. For example, in the case of the Fourier transform (5), under conditions such that the Fourier transform of each of the functions f and g exists, then for constants c_1 and c_2

$$\mathscr{F}\{c_1 f(x) + c_2 g(x)\} = c_1 \mathscr{F}\{f(x)\} + c_2 \mathscr{F}\{g(x)\}.$$

■ **Operational Properties** Since our immediate goal is to apply these new transforms to boundary-value problems, we need to examine the transforms of derivatives.

■ **Fourier Transform** Suppose that f is continuous and absolutely integrable on the interval $(-\infty, \infty)$ and f' is piecewise continuous on every finite interval. If $f(x) \to 0$ as $x \to \pm\infty$, then integration by parts gives

$$\mathscr{F}\{f'(x)\} = \int_{-\infty}^{\infty} f'(x)e^{i\alpha x}\, dx$$

$$= f(x)e^{i\alpha x}\Big|_{-\infty}^{\infty} - i\alpha \int_{-\infty}^{\infty} f(x)e^{i\alpha x}\, dx$$

$$= -i\alpha \int_{-\infty}^{\infty} f(x)e^{i\alpha x}\, dx;$$

that is,
$$\mathscr{F}\{f'(x)\} = -i\alpha F(\alpha). \tag{11}$$

Similarly, under the added assumptions that f' is continuous on $(-\infty, \infty)$, $f''(x)$ is piecewise continuous on every finite interval, and $f'(x) \to 0$ as $x \to \pm\infty$, we have

$$\mathscr{F}\{f''(x)\} = (-i\alpha)^2 \mathscr{F}\{f(x)\} = -\alpha^2 F(\alpha). \tag{12}$$

In general, under conditions analogous to those leading to (12), we have

$$\mathscr{F}\{f^{(n)}(x)\} = (-i\alpha)^n \mathscr{F}\{f(x)\} = (-i\alpha)^n F(\alpha),$$

where $n = 0, 1, 2, \ldots$.

It is important to be aware that the cosine and sine transforms are not suitable for transforming the first derivative of a function (or for that matter, any derivative of *odd* order). For example, it is readily shown that

$$\mathscr{F}_c\{f'(x)\} = \alpha\mathscr{F}_s\{f(x)\} - f(0) \quad \text{and} \quad \mathscr{F}_s\{f'(x)\} = -\alpha\mathscr{F}_c\{f(x)\}.$$

The difficulty is apparent: The transform of $f'(x)$ is not expressed in terms of the original integral transform.

■ Fourier Cosine Transform

Suppose that f and f' are continuous, f is absolutely integrable on the interval $[0, \infty)$, and f'' is piecewise continuous on every finite interval. If $f \to 0$ and $f' \to 0$ as $x \to \infty$, then the Fourier cosine transform of f'' is

$$\mathscr{F}_c\{f''(x)\} = \int_0^\infty f''(x)\cos\alpha x\, dx$$

$$= f'(x)\cos\alpha x\Big|_0^\infty - \alpha\int_0^\infty f'(x)\sin\alpha x\, dx$$

$$= -f'(0) + \alpha\left[f(x)\sin\alpha x\Big|_0^\infty - \alpha\int_0^\infty f(x)\cos\alpha x\, dx\right]$$

$$= -f'(0) - \alpha^2\mathscr{F}_c\{f(x)\},$$

that is,
$$\mathscr{F}_c\{f''(x)\} = -\alpha^2 F(\alpha) - f'(0). \tag{13}$$

■ Fourier Sine Transform

Under the same assumptions that lead to (13), we find the Fourier sine transform of f'' to be

$$\mathscr{F}_s\{f''(x)\} = -\alpha^2 F(\alpha) + \alpha f(0). \tag{14}$$

■ Choosing the Transform

The transform properties (12), (13), and (14) indicate, in contrast to the Laplace transform, that Fourier transforms are suitable for problems in which the spatial variable x (or y) is defined on an infinite or semi-infinite interval. But a natural question then arises:

How do we know which transform to use for a given boundary-value problem?

Clearly, to use the Fourier transform (5), the domain of the variable to be eliminated must be $(-\infty, \infty)$. To utilize a cosine or a sine transform, the domain of one of the spatial variables in the problem must be $[0, \infty)$. However, the determining factor in choosing between the cosine transform (7) and the sine transform (9) is the type of boundary condition specified at $x = 0$ (or $y = 0$), that is, whether a first partial derivative or u is given at this boundary.

In solving boundary-value problems using integral transforms most solutions are *formal*. In the language of mathematics, this means assumptions about the solution u and its partial derivatives go unstated. But one assumption should be kept in the back of your mind. In the examples that follow, it will be assumed without further mention that u and $\partial u/\partial x$ (or $\partial u/\partial y$) approach 0 as $x \to \pm\infty$ (or $y \to \pm\infty$). These are not major restrictions since these conditions hold in most applications.

These assumptions are sometimes used during the actual solution process. See Problems 13, 14, and 24 in Exercises 15.4.

EXAMPLE 1 Using the Fourier Transform

Solve the heat equation $k\dfrac{\partial^2 u}{\partial x^2} = \dfrac{\partial u}{\partial t}$, $-\infty < x < \infty$, $t > 0$, subject to

$$u(x, 0) = f(x) \quad \text{where} \quad f(x) = \begin{cases} u_0, & |x| < 1, \\ 0, & |x| > 1. \end{cases}$$

SOLUTION The problem can be interpreted as finding the temperature $u(x, t)$ in an infinite rod. Because the domain of x is the infinite interval $(-\infty, \infty)$ we use the Fourier transform (5) and define the transform of $u(x, t)$ to be

$$\mathscr{F}\{u(x, t)\} = \int_{-\infty}^\infty u(x, t)e^{i\alpha x}\, dx = U(\alpha, t).$$

If we write

$$\mathscr{F}\left\{\frac{\partial^2 u}{\partial x^2}\right\} = -\alpha^2 U(\alpha, t) \qquad \text{and} \qquad \mathscr{F}\left\{\frac{\partial u}{\partial t}\right\} = \frac{d}{dt}\mathscr{F}\{u(x, t)\} = \frac{dU}{dt},$$

then the Fourier transform of the partial differential equation,

$$\mathscr{F}\left\{k\frac{\partial^2 u}{\partial x^2}\right\} = \mathscr{F}\left\{\frac{\partial u}{\partial t}\right\},$$

becomes the ordinary differential equation

$$-k\alpha^2 U(\alpha, t) = \frac{dU}{dt} \qquad \text{or} \qquad \frac{dU}{dt} + k\alpha^2 U(\alpha, t) = 0.$$

FIGURE 15.4.1 Initial temperature f in Example 1

Solving the last equation by the method of Section 2.3 gives $U(\alpha, t) = ce^{-k\alpha^2 t}$. The initial temperature $u(x, 0) = f(x)$ in the rod is shown in **FIGURE 15.4.1** and its Fourier transform is

$$\mathscr{F}\{u(x, 0)\} = U(\alpha, 0) = \int_{-\infty}^{\infty} f(x)e^{i\alpha x}\,dx = \int_{-1}^{1} u_0 e^{i\alpha x}\,dx = u_0\frac{e^{i\alpha} - e^{-i\alpha}}{i\alpha}.$$

By Euler's formula

$$e^{i\alpha} = \cos\alpha + i\sin\alpha$$

$$e^{-i\alpha} = \cos\alpha - i\sin\alpha.$$

Subtracting these two results and solving for $\sin\alpha$ gives $\sin\alpha = \dfrac{e^{i\alpha} - e^{-i\alpha}}{2i}$. Hence we can rewrite the transform of the initial condition as $U(\alpha, 0) = 2u_0\dfrac{\sin\alpha}{\alpha}$. Applying this condition to the solution $U(\alpha, t) = ce^{-k\alpha^2 t}$ gives $U(\alpha, 0) = c = 2u_0\dfrac{\sin\alpha}{\alpha}$ and so

$$U(\alpha, t) = 2u_0\frac{\sin\alpha}{\alpha}e^{-k\alpha^2 t}.$$

It then follows from the inverse Fourier transform (6) that

$$u(x, t) = \frac{u_0}{\pi}\int_{-\infty}^{\infty} \frac{\sin\alpha}{\alpha}e^{-k\alpha^2 t}e^{-i\alpha x}\,d\alpha.$$

This integral can be simplified somewhat by using Euler's formula again as $e^{-i\alpha x} = \cos\alpha x - i\sin\alpha x$ and noting that

$$\int_{-\infty}^{\infty} \frac{\sin\alpha}{\alpha}e^{-k\alpha^2 t}\sin\alpha x\,d\alpha = 0$$

because the integrand is an odd function of α. Hence we finally have the solution

$$u(x, t) = \frac{u_0}{\pi}\int_{-\infty}^{\infty} \frac{\sin\alpha\cos\alpha x}{\alpha}e^{-k\alpha^2 t}\,d\alpha. \tag{15} \equiv$$

It is left to the reader to show that the solution (15) in Example 1 can be expressed in terms of the error function. See Problem 26 in Exercises 15.4.

EXAMPLE 2 **Two Useful Fourier Transforms**

It is a straightforward exercise in integration by parts to show that the Fourier cosine and sine transforms of $f(x) = e^{-bx}$, $x > 0$, $b > 0$, are, in turn,

$$\mathscr{F}_c\{e^{-bx}\} = \int_0^\infty e^{-bx} \cos \alpha x \, dx = \frac{b}{b^2 + \alpha^2}, \tag{16}$$

$$\mathscr{F}_s\{e^{-bx}\} = \int_0^\infty e^{-bx} \sin \alpha x \, dx = \frac{\alpha}{b^2 + \alpha^2}. \tag{17}$$

Another way to quickly obtain (and remember) these two results is to identify the two integrals with the more familiar Laplace transform in (2) of Section 4.1. With the symbols x, b, and α playing the part of t, s, and k, respectively, it follows that (16) and (17) are identical to (e) and (d) in Theorem 4.1.1. $\equiv$

EXAMPLE 3 **Using the Cosine Transform**

The steady-state temperature in the semi-infinite plate shown in **FIGURE 15.4.2** is determined from

$$\frac{\partial^2 u}{\partial x^2} + \frac{\partial^2 u}{\partial y^2} = 0, \quad 0 < x < \pi, \quad y > 0$$

$$u(0, y) = 0, \quad u(\pi, y) = e^{-y}, \quad y > 0$$

$$\left.\frac{\partial u}{\partial y}\right|_{y=0} = 0, \quad 0 < x < \pi.$$

$u = 0$ $u = e^{-y}$

insulated

FIGURE 15.4.2 Plate in Example 3

Solve for $u(x, y)$.

SOLUTION The domain of the variable y and the prescribed condition at $y = 0$ indicate that the Fourier cosine transform is suitable for the problem. We define

$$\mathscr{F}_c\{u(x, y)\} = \int_0^\infty u(x, y) \cos \alpha y \, dy = U(x, \alpha).$$

In view of (13), $$\mathscr{F}_c\left\{\frac{\partial^2 u}{\partial x^2}\right\} + \mathscr{F}_c\left\{\frac{\partial^2 u}{\partial y^2}\right\} = \mathscr{F}_c\{0\}$$

becomes

$$\frac{d^2 U}{dx^2} - \alpha^2 U(x, \alpha) - u_y(x, 0) = 0 \quad \text{or} \quad \frac{d^2 U}{dx^2} - \alpha^2 U = 0.$$

Since the domain of x is a finite interval, we choose to write the solution of the ordinary differential equation as

$$U(x, \alpha) = c_1 \cosh \alpha x + c_2 \sinh \alpha x. \tag{18}$$

Now $\mathscr{F}_c\{u(0, y)\} = \mathscr{F}_c\{0\}$ and $\mathscr{F}_c\{u(\pi, y)\} = \mathscr{F}_c\{e^{-y}\}$ are in turn equivalent to

$$U(0, \alpha) = 0 \quad \text{and} \quad U(\pi, \alpha) = \frac{1}{1 + \alpha^2}. \tag{19}$$

Note that the value $U(\pi, \alpha)$ in (19) is (16) of Example 2 with $b = 1$. When we apply the two conditions in (19) to the solution (18) we obtain $c_1 = 0$ and $c_2 = 1/[(1 + \alpha^2) \sinh \alpha\pi]$. Therefore,

$$U(x, \alpha) = \frac{\sinh \alpha x}{(1 + \alpha^2) \sinh \alpha\pi},$$

and so from (8) we arrive at

$$u(x, y) = \frac{2}{\pi} \int_0^\infty \frac{\sinh \alpha x}{(1 + \alpha^2) \sinh \alpha\pi} \cos \alpha y \, d\alpha. \tag{20} \equiv$$

Had $u(x, 0)$ been given in Example 3 rather than $u_y(x, 0)$, then the sine transform would have been appropriate.

In Problems 1–24, use a Fourier transform in this section to solve the given boundary-value problem. Make assumptions about boundedness where necessary.

1. $k \dfrac{\partial^2 u}{\partial x^2} = \dfrac{\partial u}{\partial t}$, $-\infty < x < \infty$, $t > 0$

$u(x, 0) = e^{-|x|}$, $-\infty < x < \infty$

2. $k \dfrac{\partial^2 u}{\partial x^2} = \dfrac{\partial u}{\partial t}$, $-\infty < x < \infty$, $t > 0$

$u(x, 0) = \begin{cases} 0, & x < -1 \\ -100, & -1 < x < 0 \\ 100, & 0 < x < 1 \\ 0, & x > 1 \end{cases}$

3. Find the temperature $u(x, t)$ in a semi-infinite rod if $u(0, t) = u_0$, $t > 0$ and $u(x, 0) = 0$, $x > 0$.

4. Use the result $\displaystyle\int_0^\infty \dfrac{\sin \alpha x}{\alpha} \, d\alpha = \dfrac{\pi}{2}$, $x > 0$, to show that the solution in Problem 3 can be written as

$$u(x, t) = u_0 - \dfrac{2u_0}{\pi} \int_0^\infty \dfrac{\sin \alpha x}{\alpha} e^{-k\alpha^2 t} \, d\alpha.$$

5. Find the temperature $u(x, t)$ in a semi-infinite rod if $u(0, t) = 0$, $t > 0$, and

$$u(x, 0) = \begin{cases} 1, & 0 < x < 1 \\ 0, & x > 1. \end{cases}$$

6. Solve Problem 3 if the condition at the left boundary is

$$\dfrac{\partial u}{\partial x}\bigg|_{x=0} = -A, \ t > 0.$$

7. Solve Problem 5 if the end $x = 0$ is insulated.

8. Find the temperature $u(x, t)$ in a semi-infinite rod if $u(0, t) = 1$, $t > 0$, and $u(x, 0) = e^{-x}$, $x > 0$.

9. (a) $a^2 \dfrac{\partial^2 u}{\partial x^2} = \dfrac{\partial u}{\partial t}$, $-\infty < x < \infty$, $t > 0$

$u(x, 0) = f(x)$, $\dfrac{\partial u}{\partial t}\bigg|_{t=0} = g(x)$, $-\infty < x < \infty$

(b) If $g(x) = 0$, show that the solution of part (a) can be written as $u(x, t) = \frac{1}{2}[f(x + at) + f(x - at)]$.

10. Find the displacement $u(x, t)$ of a semi-infinite string if

$$u(0, t) = 0, \quad t > 0$$

$$u(x, 0) = xe^{-x}, \quad \dfrac{\partial u}{\partial t}\bigg|_{t=0} = 0, \ x > 0.$$

11. Solve the problem in Example 3 if the boundary conditions at $x = 0$ and $x = \pi$ are reversed:

$$u(0, y) = e^{-y}, \quad u(\pi, y) = 0, \ y > 0.$$

12. Solve the problem in Example 3 if the boundary condition at $y = 0$ is $u(x, 0) = 1$, $0 < x < \pi$.

13. Find the steady-state temperature $u(x, y)$ in a plate defined by $x \geq 0$, $y \geq 0$ if the boundary $x = 0$ is insulated and, at $y = 0$,

$$u(x, 0) = \begin{cases} 50, & 0 < x < 1 \\ 0, & x > 1. \end{cases}$$

14. Solve Problem 13 if the boundary condition at $x = 0$ is $u(0, y) = 0$, $y > 0$.

15. $\dfrac{\partial^2 u}{\partial x^2} + \dfrac{\partial^2 u}{\partial y^2} = 0$, $x > 0$, $0 < y < 2$

$u(0, y) = 0$, $0 < y < 2$

$u(x, 0) = f(x)$, $u(x, 2) = 0$, $x > 0$

16. $\dfrac{\partial^2 u}{\partial x^2} + \dfrac{\partial^2 u}{\partial y^2} = 0$, $0 < x < \pi$, $y > 0$

$u(0, y) = f(y)$, $\dfrac{\partial u}{\partial x}\bigg|_{x=\pi} = 0$, $y > 0$

$\dfrac{\partial u}{\partial y}\bigg|_{y=0} = 0$, $0 < x < \pi$

In Problems 17 and 18, find the steady-state temperature $u(x, y)$ in the plate given in the figure. [*Hint*: One way of proceeding is to express Problems 17 and 18 as two and three boundary-value problems, respectively. Use the superposition principle (see Section 13.5).]

17.

$u = e^{-y}$

$u = e^{-x}$

FIGURE 15.4.3 Infinite plate in Problem 17

18.

$u = 0$

$u = e^{-y}$

$u = 100$

$u = f(x)$

FIGURE 15.4.4 Semi-infinite plate in Problem 18

19. Use the result $\mathscr{F}\{e^{-x^2/4p^2}\} = 2\sqrt{\pi} p e^{-p^2 \alpha^2}$ to solve the boundary-value problem

$$k \dfrac{\partial^2 u}{\partial x^2} = \dfrac{\partial u}{\partial t}, \ -\infty < x < \infty, \ t > 0$$

$$u(x, 0) = e^{-x^2}, \ -\infty < x < \infty.$$

20. If $\mathscr{F}\{f(x)\} = F(\alpha)$ and $\mathscr{F}\{g(x)\} = G(\alpha)$, then the **convolution theorem** for the Fourier transform is given by

$$\int_{-\infty}^{\infty} f(\tau)g(x - \tau)\,d\tau = \mathscr{F}^{-1}\{F(\alpha)G(\alpha)\}.$$

Use this result and the transform $\mathscr{F}\{e^{-x^2/4p^2}\}$ given in Problem 19 to show that a solution of the boundary-value problem

$$k\frac{\partial^2 u}{\partial x^2} = \frac{\partial u}{\partial t}, \quad -\infty < x < \infty, \; t > 0$$

$$u(x, 0) = f(x), \quad -\infty < x < \infty$$

is

$$u(x, t) = \frac{1}{2\sqrt{k\pi t}}\int_{-\infty}^{\infty} f(\tau)e^{-(x-\tau)^2/4kt}\,d\tau.$$

21. Use the transform $\mathscr{F}\{e^{-x^2/4p^2}\}$ given in Problem 19 to find the steady-state temperature $u(x, y)$ in the infinite strip shown in **FIGURE 15.4.5**.

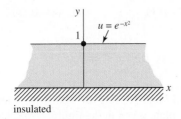

FIGURE 15.4.5 Infinite plate in Problem 21

22. Find the steady-state temperature $u(r, z)$ in a semi-infinite cylinder described by the boundary-value problem

$$\frac{\partial^2 u}{\partial r^2} + \frac{1}{r}\frac{\partial u}{\partial r} + \frac{\partial^2 u}{\partial z^2} = 0, \quad 0 < r < 1, \quad z > 0$$

$$u(1, z) = 0, \qquad z > 0$$

$$u(r, 0) = u_0, \qquad 0 < r < 1.$$

[*Hint*: Use the integral in Problem 4 and the parametric form of the modified Bessel equation on page 293.]

23. Find the steady-state temperature $u(r, z)$ in the semi-infinite cylinder in Problem 22 if the base of the cylinder is insulated and

$$u(1, z) = \begin{cases} u_0, & 0 < z < 1 \\ 0, & z > 1. \end{cases}$$

24. Find the steady-state temperature $u(x, y)$ in the infinite plate defined by $-\infty < x < \infty$, $y > 0$ if the boundary condition at $y = 0$ is

$$u(x, 0) = \begin{cases} u_0, & |x| < 1 \\ 0, & |x| > 1. \end{cases}$$

[*Hint*: Consider the two cases $\alpha > 0$ and $\alpha < 0$ when you solve the resulting ordinary differential equation.]

25. The solution of Problem 14 can be integrated. Use entries 50 and 51 of the table in Appendix C to show that

$$u(x, y) = \frac{100}{\pi}\left[\arctan\frac{x}{y} - \frac{1}{2}\arctan\frac{x-1}{y} - \frac{1}{2}\arctan\frac{x-1}{y}\right].$$

26. Use Problem 20, the change of variables $v = (x - \tau)/2\sqrt{kt}$, and Problem 11 in Exercises 15.1 to show that the solution of Example 1 can be expressed as

$$u(x, t) = \frac{u_0}{2}\left[\operatorname{erf}\left(\frac{x+1}{2\sqrt{kt}}\right) - \operatorname{erf}\left(\frac{x-1}{2\sqrt{kt}}\right)\right].$$

Computer Lab Assignment

27. Assume $u_0 = 100$ and $k = 1$ in the solution of Problem 26. Use a CAS to graph $u(x, t)$ over the rectangular region $-4 \le x \le 4, 0 \le t \le 6$. Use a 2D plot to superimpose the graphs of $u(x, t)$ for $t = 0.05, 0.125, 0.5, 1, 2, 4, 6$, and 15 for $-4 \le x \le 4$. Use the graphs to conjecture the values of $\lim_{t\to\infty} u(x, t)$ and $\lim_{x\to\infty} u(x, t)$. Then prove these results analytically using the properties of $\operatorname{erf}(x)$.

Discussion Problem

28. (a) Suppose

$$\int_0^{\infty} f(x)\cos \alpha x\,dx = F(\alpha),$$

where

$$F(\alpha) = \begin{cases} 1 - \alpha, & 0 \le \alpha \le 1 \\ 0, & \alpha > 1. \end{cases}$$

Find $f(x)$.

(b) Use part (a) to show that

$$\int_0^{\infty} \frac{\sin^2 x}{x^2}\,dx = \frac{\pi}{2}.$$

15.5 Finite Fourier Transforms

INTRODUCTION The two integral transform pairs considered in this section are a bit different from those considered in Section 15.4. The *finite* Fourier cosine and sine transforms are based not on the Fourier integral, but rather on the Fourier cosine and sine series discussed in Section 12.3. The transform of a function f is an integral, whereas the inverse transform is a Fourier series.

■ **Finite Transforms** We begin with the definitions of the finite Fourier cosine and sine transforms. Before reading Definition 15.5.1, we suggest that you review Definition 12.3.1. Let f be a function defined on the interval $[0, p]$.

(*i*) Finite cosine transform:
$$\mathscr{F}_{cn}\{f(x)\} = \int_0^p f(x) \cos \frac{n\pi}{p}\, dx = F(n) \tag{1}$$

Inverse transform:
$$\mathscr{F}_{cn}^{-1}\{F(n)\} = \frac{2}{p}\left[\frac{1}{2}F(0) + \sum_{n=1}^{\infty} F(n) \cos \frac{n\pi}{p}x\right] = f(x) \tag{2}$$

(*ii*) Finite sine transform:
$$\mathscr{F}_{sn}\{f(x)\} = \int_0^p f(x) \sin \frac{n\pi}{p}x\, dx = F(n) \tag{3}$$

Inverse transform:
$$\mathscr{F}_{sn}^{-1}\{F(n)\} = \frac{2}{p}\sum_{n=1}^{\infty} F(n) \sin \frac{n\pi}{p}x = f(x) \tag{4}$$

■ **Operational Properties** Like the Fourier cosine and sine transforms discussed in the preceding section, the finite Fourier cosine and sine transforms are not appropriate for transforming an odd-order derivative of a function. So if f and f' are continuous and f'' is piecewise continuous on the interval $[0, p]$, then the finite cosine transform of f'' follows from integration by parts:

$$\mathscr{F}_{cn}\{f''(x)\} = \int_0^p f''(x) \cos \frac{n\pi}{p}x\, dx$$

$$= f'(x) \cos \frac{n\pi}{p}x \Big|_0^p + \frac{n\pi}{p}\int_0^p f'(x) \sin \frac{n\pi}{p}x\, dx$$

$$= f'(p)(-1)^n - f'(0) + \frac{n\pi}{p}\left[f(x) \sin \frac{n\pi}{p}x \Big|_0^p - \frac{n\pi}{p}\int_0^p f(x) \cos \frac{n\pi}{p}x\, dx\right]$$

$$= -f'(0) + (-1)^n f'(p) - \frac{n^2\pi^2}{p^2}\mathscr{F}_{cn}\{f(x)\}.$$

Using the notation $\mathscr{F}_{cn}\{f(x)\} = F(n)$, we have

$$\mathscr{F}_{cn}\{f''(x)\} = -\frac{n^2\pi^2}{p^2}F(n) - f'(0) + (-1)^n f'(p). \tag{5}$$

Similarly, for the finite sine transform we have

$$\mathscr{F}_{sn}\{f''(x)\} = -\frac{n^2\pi^2}{p^2}F(n) + \frac{n\pi}{p}[f(0) - (-1)^n f(p)], \tag{6}$$

where $\mathscr{F}_{sn}\{f(x)\} = F(n)$.

The transforms given in (5) and (6) are the determining condition in deciding which transform to use in solving a boundary-value problem. If the value of the derivative u_x is given at 0 and p, we use the finite cosine transform, whereas if the function u is specified at 0 and p, we use the finite sine transform.

A word of caution when using the finite cosine transform. Consider the cases $n > 0$ and $n = 0$ separately. The solution of the ODE obtained by transforming the partial differential equation for $n > 0$ *may* not include the solution of that ODE when $n = 0$. See Problem 4 in Exercises 15.5. Also, in Problem 7, consider the cases $n = 1$ and $n > 1$.

EXAMPLE **1** **Finite Sine Transform**

Find the finite sine transform of $f(x) = x$ on $[0, 1]$. Find the inverse transform.

SOLUTION With $p = 1$ and integration by parts we see from (3) that the finite sine transform of f is

$$\mathscr{F}_{sn}\{x\} = \int_0^1 x \sin n\pi x \, dx$$

$$= x \frac{-\cos n\pi x}{n\pi} \Big|_0^1 + \frac{1}{n\pi} \int_0^1 \cos n\pi x \, dx$$

$$= \frac{(-1)^{n+1}}{n\pi} + \frac{\sin n\pi x}{n^2 \pi^2} \Big|_0^1$$

$$= \frac{(-1)^{n+1}}{n\pi} = F(n).$$

Then from (4) the inverse transform is

$$\mathscr{F}_{sn}^{-1}\{F(n)\} = 2 \sum_{n=1}^{\infty} \frac{(-1)^{n+1}}{n\pi} \sin n\pi x = x.$$

≡

EXAMPLE **2** **Using the Finite Cosine Transform**

Solve

$$\frac{\partial^2 u}{\partial x^2} - u = \frac{\partial u}{\partial t}, \quad 0 < x < \pi, \, t > 0$$

$$\frac{\partial u}{\partial x} \Big|_{x=0} = 0, \quad \frac{\partial u}{\partial x} \Big|_{x=\pi} = 0, \, t > 0$$

$$u(x, 0) = x, \quad 0 < x < \pi.$$

SOLUTION In view of the fact that the two boundary conditions involve the first partial derivative of $u(x, t)$ we use the finite cosine transform with $p = \pi$. From (1) we have

$$\mathscr{F}_{cn}\{u(x, t)\} = \int_0^{\pi} u(x, t) \cos nx \, dx = U(n, t), \tag{7}$$

and so, in view of (5),

$$\mathscr{F}_{cn}\left\{\frac{\partial^2 u}{\partial x^2}\right\} = -n^2 U(n, t).$$

As usual, we assume that we can interchange integration and differentiation:

$$\mathscr{F}_{cn}\left\{\frac{\partial u}{\partial t}\right\} = \frac{d}{dt}\mathscr{F}_{cn}\{u(x, t)\} = \frac{dU}{dt}.$$

Now in the case $n > 0$, the transform of the partial differential equation is

$$-n^2 U - U = \frac{dU}{dt} \quad \text{or} \quad \frac{dU}{dt} + (n^2 + 1)U = 0.$$

This ordinary first-order differential equation is linear and has the solution $U(n, t) = c_1 e^{-(n^2+1)t}$. Using integration by parts, the transform of the initial condition is

$$\mathscr{F}_{cn}\{u(x, 0)\} = U(n, 0) = c_1 = \int_0^{\pi} x \cos nx \, dx = \frac{(-1)^n - 1}{n^2}.$$

Hence

$$U(n, t) = \frac{(-1)^n - 1}{n^2} e^{-(n^2+1)t}. \tag{8}$$

Now in the case $n = 0$, the transform of the partial differential equation is

$$-U = \frac{dU}{dt} \quad \text{and has the solution} \quad U(0, t) = c_1 e^{-t}.$$

From (7) we see that the transform of the initial condition is

$$\mathcal{F}_{cn}\{u(x, 0)\} = U(0, 0) = c_1 = \int_0^{\pi} x \, dx = \frac{\pi^2}{2}.$$

Hence

$$U(0, t) = \frac{\pi^2}{2} e^{-t}. \tag{9}$$

From (2) the form of the solution is

$$u(x, t) = \mathcal{F}_{cn}^{-1}\{U(n, t)\} = \frac{2}{\pi}\left[\frac{1}{2}U(0, t) + \sum_{n=1}^{\infty} U(n, t) \cos nx\right].$$

Finally, using (8) and (9) the inverse transform is

$$u(x, t) = \frac{\pi}{2} e^{-t} + \frac{2}{\pi}\sum_{n=1}^{\infty}\frac{(-1)^n - 1}{n^2} e^{-(n^2+1)t} \cos nx.$$

$\equiv$

15.5 Exercises Answers to selected odd-numbered problems begin on page ANS-39.

In Problems 1 and 2, find the finite sine transform of the function f. Find the inverse transform.

1. $f(x) = 1$, $[0, 1]$

2. $f(x) = \pi - x$, $[0, \pi]$

In Problems 3 and 4, find the finite cosine transform of the function f. Find the inverse transform.

3. $f(x) = e^{-2x}$, $[0, \pi]$

4. $f(x) = x$, $[0, 1]$

In Problems 5–14, use a finite Fourier transform to solve the given boundary-value problem. The symbol u_0 is a constant.

5. $\dfrac{\partial^2 u}{\partial x^2} = \dfrac{\partial u}{\partial t}$, $0 < x < 1$, $t > 0$

$u(0, t) = 0$, $u(1, t) = 0$, $t > 0$
$u(x, 0) = f(x)$, $0 < x < 1$

6. $\dfrac{\partial^2 u}{\partial x^2} + \dfrac{\partial^2 u}{\partial y^2} = 0$, $0 < x < \pi$, $0 < y < \pi$

$u(0, y) = 0$, $u(\pi, y) = u_0$, $0 < y < \pi$
$u(x, 0) = 0$, $u(x, \pi) = 0$, $0 < x < \pi$

7. $\dfrac{\partial^2 u}{\partial x^2} = \dfrac{\partial u}{\partial t}$, $0 < x < \pi$, $t > 0$

$\dfrac{\partial u}{\partial x}\Big|_{x=0} = 0$, $\dfrac{\partial u}{\partial x}\Big|_{x=\pi} = 0$, $t > 0$
$u(x, 0) = f(x)$, $0 < x < \pi$

8. $\dfrac{\partial^2 u}{\partial x^2} + \dfrac{\partial^2 u}{\partial y^2} = 0$, $0 < x < 1$, $0 < y < 1$

$\dfrac{\partial u}{\partial x}\Big|_{x=0} = 0$, $\dfrac{\partial u}{\partial x}\Big|_{x=1} = 0$, $0 < y < 1$
$u(x, 0) = 0$, $u(x, 1) = f(x)$, $0 < x < 1$

9. $\dfrac{\partial^2 u}{\partial x^2} = \dfrac{\partial^2 u}{\partial t^2}$, $0 < x < \pi$, $t > 0$

$u(0, t) = 0$, $u(\pi, t) = 0$, $t > 0$
$u(x, 0) = f(x)$, $\dfrac{\partial u}{\partial t}\Big|_{t=0} = g(x)$, $0 < x < \pi$

10. $\dfrac{\partial^2 u}{\partial x^2} - \beta u = \dfrac{\partial u}{\partial t}$, $0 < x < \pi$, $t > 0$, β a constant

$\dfrac{\partial u}{\partial x}\Big|_{x=0} = 0$, $\dfrac{\partial u}{\partial x}\Big|_{x=\pi} = 0$, $t > 0$
$u(x, 0) = f(x)$, $0 < x < \pi$

11. $\dfrac{\partial^2 u}{\partial x^2} + \sin x = \dfrac{\partial^2 u}{\partial t^2}$, $0 < x < \pi$, $t > 0$

$u(0, t) = 0$, $u(\pi, t) = 0$, $t > 0$
$u(x, 0) = u_0$, $\dfrac{\partial u}{\partial t}\Big|_{t=0} = 0$, $0 < x < \pi$

12. $\dfrac{\partial^2 u}{\partial x^2} + \dfrac{\partial^2 u}{\partial y^2} = 0$, $0 < x < \pi$, $0 < y < \pi$

$u(0, y) = 0$, $u(\pi, y) = 0$, $0 < y < \pi$
$u(x, 0) = u_0$, $u(x, \pi) = u_0$, $0 < x < \pi$

13. $\dfrac{\partial^2 u}{\partial x^2} + \dfrac{\partial^2 u}{\partial y^2} = \beta u$, $0 < x < 1$, $0 < y < 1$, β a constant

$u(0, y) = 0$, $u(1, y) = 0$, $0 < y < 1$
$\dfrac{\partial u}{\partial y}\Big|_{y=0} = 0$, $\dfrac{\partial u}{\partial y}\Big|_{y=1} = u_0$, $0 < x < 1$

14. $\dfrac{\partial^2 u}{\partial x^2} + \dfrac{\partial^2 u}{\partial y^2} = 0, \quad 0 < x < \pi, \ y > 0$

$\left.\dfrac{\partial u}{\partial x}\right|_{x=0} = 0, \quad \left.\dfrac{\partial u}{\partial x}\right|_{x=\pi} = 0, \ y > 0$

$u(x, 0) = 10 \cos x, \ 0 < x < \pi$

$u(x, y)$ bounded for $0 < x < \pi, \ y > 0$

15. Use a finite Fourier transform to find the steady-state temperature $u(x, y)$ in the plate shown in **FIGURE 15.5.1**.

FIGURE 15.5.1 Plate in Problem 15

16. Use the substitution $u(x, y) = v(x, y) + \dfrac{x}{\pi}$ to change the boundary-value problem in Problem 15 into a boundary-value problem for $v(x, y)$. Then use a finite Fourier transform to solve this second problem for $v(x, y)$. Finally, determine $u(x, y)$. Resolve any differences between the solution $u(x, y)$ of Problem 15 with the solution $u(x, y)$ of this problem.

17. Use the finite cosine transform of the fourth derivative $f^{(4)}(x)$,

$$\mathcal{F}_{cn}\{f^{(4)}(x)\} = \dfrac{n^4\pi^4}{p^4} F(n) + \dfrac{n^2\pi^2}{p^2}[f'(0) - (-1)^n f'(p)]$$
$$- f'''(0) + (-1)^n f'''(p),$$

to show that a solution of the boundary-value problem

$$\dfrac{\partial^4 u}{\partial x^4} + \dfrac{\partial^4 u}{\partial t^4} = 0, \quad 0 < x < \pi, \ \ 0 < t < 2, \ \ u_0 \text{ constant}$$

$\left.\dfrac{\partial u}{\partial x}\right|_{x=0} = 0, \quad \left.\dfrac{\partial u}{\partial x}\right|_{x=\pi} = 0, \ \ 0 < t < 2$

$\left.\dfrac{\partial^3 u}{\partial x^3}\right|_{x=0} = 0, \quad \left.\dfrac{\partial^3 u}{\partial x^3}\right|_{x=\pi} = 0, \ \ 0 < t < 2$

$u(x, 0) = 0, \quad u(x, 2) = 0, \quad 0 < x < \pi$

$\left.\dfrac{\partial u}{\partial t}\right|_{t=0} = 0 \quad \left.\dfrac{\partial u}{\partial t}\right|_{t=2} = u_0, \quad 0 < x < \pi$

is $u(x, t) = \frac{1}{4} u_0(t^3 - 2t^2)$.

18. Use the finite sine transform of the fourth derivative $f^{(4)}(x)$,

$$\mathcal{F}_{sn}\{f^{(4)}(x)\} = \dfrac{n^4\pi^4}{p^4} F(n) - \dfrac{n^3\pi^3}{p^3}[f(0) - (-1)^n f(p)]$$
$$+ \dfrac{n\pi}{p}[f''(0) - (-1)^n f''(p)],$$

to solve the boundary-value problem

$$\dfrac{\partial^4 u}{\partial x^4} + \alpha^2 \dfrac{\partial^2 u}{\partial t^2} = \beta x, \quad 0 < x < \pi, \ t > 0, \ \alpha, \ \beta \text{ constants}$$

$u(0, t) = 0, \quad u(\pi, t) = 0, \ t > 0$

$\left.\dfrac{\partial^2 u}{\partial x^2}\right|_{x=0} = 0, \quad \left.\dfrac{\partial^2 u}{\partial x^2}\right|_{x=\pi} = 0, \ t > 0$

$u(x, 0) = 0, \quad \left.\dfrac{\partial u}{\partial t}\right|_{t=0} = 0, \quad 0 < x < \pi.$

15.6 Fast Fourier Transform

INTRODUCTION Consider a function f that is defined and continuous on the interval $[0, 2p]$. If $x_0, x_1, x_2, \ldots, x_n, \ldots$ are equally spaced points in the interval, then the corresponding function values $f_0, f_1, f_2, \ldots, f_n, \ldots$ shown in **FIGURE 15.6.1** are said to represent a discrete *sampling* of the function f. The notion of discrete samplings of a function is important in the analysis of continuous signals.

In this section, the complex or exponential form of a Fourier series plays an important role in the discussion. A review of Section 12.4 is recommended.

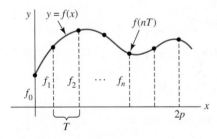

FIGURE 15.6.1 Sampling of a continuous function

Discrete Fourier Transform Consider a function f defined on the interval $[0, 2p]$. From (11) of Section 12.4 we saw that f can be written in a complex Fourier series,

$$f(x) = \sum_{n=-\infty}^{\infty} c_n e^{in\omega x} \quad \text{where} \quad c_n = \dfrac{1}{2p} \int_0^{2p} f(x) e^{-in\omega x}\, dx, \tag{1}$$

where the $\omega = 2\pi/2p = \pi/p$ is the fundamental angular frequency and $2p$ is the fundamental period. In the *discrete case*, however, the input is $f_0, f_1, f_2, \ldots$, which are the values of the function f at equally spaced points $x = nT, n = 0, 1, 2, \ldots$. The number T is called the **sampling rate** or the length of the sampling interval.* If f is continuous at T, then the **sample** of f at T is defined to be the product $f(x)\delta(x - T)$, where $\delta(x - T)$ is the Dirac delta function (see Section 4.5). We can

*Note that the symbol T used here does not have the same meaning as in Section 12.4.

then represent this discrete version of f, or **discrete signal**, as the sum of unit impulses acting on the function at $x = nT$:

$$\sum_{n=-\infty}^{\infty} f(x)\delta(x - nT). \tag{2}$$

If we apply the Fourier transform to the discrete signal (2), we have

$$\int_{-\infty}^{\infty} \sum_{n=-\infty}^{\infty} f(x)\delta(x - nT)e^{i\alpha x}\, dx. \tag{3}$$

By the sifting property of the Dirac delta function (see Section 4.5), (3) is the same as

$$F(\alpha) = \sum_{n=-\infty}^{\infty} f(nT)e^{i\alpha nT}. \tag{4}$$

The expression $F(\alpha)$ in (4) is called the **discrete Fourier transform** (DFT) of the function f. We often write the coefficients $f(nT)$ in (4) as $f(n)$ or f_n. It is also worth noting that since $e^{i\alpha x}$ is periodic in α and $e^{i\alpha T} = e^{i(\alpha T + 2\pi)} = e^{i(\alpha + 2\pi/T)T}$, we only need to consider the function for α in $[0, 2\pi/T]$. Let $N = 2\pi/T$. This places x in the interval $[0, 2\pi]$. So, because we sample over one period, the sum in (4) is actually finite.

Now consider the function values $f(x)$ at N equally spaced points, $x = nT$, $n = 0, 1, 2, \ldots, N - 1$, in the interval $[0, 2\pi]$; that is, $f_0, f_1, f_2, \ldots, f_{N-1}$. The (finite) discrete Fourier series $f(x) = \sum_{n=-\infty}^{\infty} c_n e^{inx}$ using these N terms gives us

$$
\begin{aligned}
f_0 &= c_0 + c_1 e^{i1 \cdot 0} & + c_2 e^{i2 \cdot 0} + \cdots & + c_{N-1} e^{i(N-1)\cdot 0} \\
f_1 &= c_0 + c_1 e^{i2\pi/N} & + c_2 e^{i4\pi/N} + \cdots & + c_{N-1} e^{i2(N-1)\pi/N} \\
f_2 &= c_0 + c_1 e^{i4\pi/N} & + c_2 e^{i8\pi/N} + \cdots & + c_{N-1} e^{i4(N-1)\pi/N} \\
&\;\vdots & & \qquad\vdots \\
f_{N-1} &= c_0 + c_1 e^{i2(N-1)\pi/N} & + c_2 e^{i4(N-1)\pi/N} + \cdots & + c_{N-1} e^{i2(N-1)2\pi/N}.
\end{aligned}
$$

If we let $\omega_n = e^{i2\pi/n} = \cos\dfrac{2\pi}{n} + i\sin\dfrac{2\pi}{n}$ and use the usual laws of exponents, this system of equations is the same as

$$
\begin{aligned}
f_0 &= c_0 + c_1 & + c_2 + \cdots & + c_{N-1} \\
f_1 &= c_0 + c_1 \omega_N & + c_2 \omega_N^2 + \cdots & + c_{N-1} \omega_N^{N-1} \\
f_2 &= c_0 + c_1 \omega_N^2 & + c_2 \omega_N^4 + \cdots & + c_{N-1} \omega_N^{2(N-1)} \\
&\;\vdots & & \qquad\vdots \\
f_{N-1} &= c_0 + c_1 \omega_N^{N-1} & + c_2 \omega_N^{2(N-1)} + \cdots & + c_{N-1} \omega_N^{(N-1)^2}.
\end{aligned}
\tag{5}
$$

If we use matrix notation (see Sections 8.1 and 8.2), then (5) is

$$
\begin{pmatrix} f_0 \\ f_1 \\ f_2 \\ \vdots \\ f_{N-1} \end{pmatrix}
=
\begin{pmatrix}
1 & 1 & 1 & \cdots & 1 \\
1 & \omega_N & \omega_N^2 & & \omega_N^{N-1} \\
1 & \omega_N^2 & \omega_N^4 & & \omega_N^{2(N-1)} \\
\vdots & & & & \vdots \\
1 & \omega_N^{N-1} & \omega_N^{2(N-1)} & \cdots & \omega_N^{(N-1)^2}
\end{pmatrix}
\begin{pmatrix} c_0 \\ c_1 \\ c_2 \\ \vdots \\ c_{N-1} \end{pmatrix}.
\tag{6}
$$

Let the $N \times N$ matrix in (6) be denoted by the symbol $\mathbf{F}_N$. Given the inputs $f_0, f_1, f_2, \ldots, f_{N-1}$, is there an easy way to find the Fourier coefficients $c_0, c_1, c_2, \ldots, c_{N-1}$? If $\overline{\mathbf{F}}_N$ is the matrix consisting of the complex conjugates of the entries of $\mathbf{F}_N$ and if $\mathbf{I}$ denotes the $N \times N$ identity matrix, then we have

$$\mathbf{F}_N \overline{\mathbf{F}}_N = \overline{\mathbf{F}}_N \mathbf{F}_N = N\mathbf{I} \qquad \text{and so} \qquad \mathbf{F}_N^{-1} = \frac{1}{N}\overline{\mathbf{F}}_N.$$

It follows from (6) and the last equation that

$$\begin{pmatrix} c_0 \\ c_1 \\ c_2 \\ \vdots \\ c_{N-1} \end{pmatrix} = \frac{1}{N} \overline{\mathbf{F}}_N \begin{pmatrix} f_0 \\ f_1 \\ f_2 \\ \vdots \\ f_{N-1} \end{pmatrix}.$$

■ **Discrete Transform Pair** Recall from Section 15.4 that in the Fourier transform pair we use a function $f(x)$ as input and compute the coefficients that give the amplitude for each frequency k (c_k in the case of periodic functions of period 2π) or we compute the coefficients that give the amplitude for each frequency α ($F(\alpha)$ in the case of nonperiodic functions).

Also, given these frequencies and coefficients, we could reconstruct the original function $f(x)$. In the discrete case, we use a sample of N values of the function $f(x)$ as input and compute the coefficients that give the amplitude for each sampled frequency. Given these frequencies and coefficients, it is possible to reconstruct the n sampled values of $f(x)$. The transform pair, the **discrete Fourier transform pair**, is given by

$$\mathbf{c} = \frac{1}{N} \overline{\mathbf{F}}_N \mathbf{f} \quad \text{and} \quad \mathbf{f} = \mathbf{F}_N \mathbf{c} \tag{7}$$

where

$$\mathbf{c} = \begin{pmatrix} c_0 \\ c_1 \\ c_2 \\ \vdots \\ c_{N-1} \end{pmatrix} \quad \text{and} \quad \mathbf{f} = \begin{pmatrix} f_0 \\ f_1 \\ f_2 \\ \vdots \\ f_{N-1} \end{pmatrix}.$$

EXAMPLE 1 **Discrete Fourier Transform**

Let $N = 4$ so that the input is f_0, f_1, f_2, f_3 at the four points $x = 0, \pi/2, \pi, 3\pi/2$. Since $\omega_4 = e^{i\pi/2} = \cos(\pi/2) + i\sin(\pi/2) = i$, the matrix $\mathbf{F}_4$ is

$$\mathbf{F}_4 = \begin{pmatrix} 1 & 1 & 1 & 1 \\ 1 & i & -1 & -i \\ 1 & -1 & 1 & -1 \\ 1 & -i & -1 & i \end{pmatrix}.$$

Hence from (7), the Fourier coefficients are given by $\mathbf{c} = \frac{1}{4} \overline{\mathbf{F}}_4 \mathbf{f}$:

$$\begin{pmatrix} c_0 \\ c_1 \\ c_2 \\ c_3 \end{pmatrix} = \frac{1}{4} \begin{pmatrix} 1 & 1 & 1 & 1 \\ 1 & -i & -1 & i \\ 1 & -1 & 1 & -1 \\ 1 & i & -1 & -i \end{pmatrix} \begin{pmatrix} f_0 \\ f_1 \\ f_2 \\ f_3 \end{pmatrix}.$$

If we let f_0, f_1, f_2, f_3 be 0, 2, 4, 6, respectively, we find from the preceding matrix product that

$$\mathbf{c} = \begin{pmatrix} c_0 \\ c_1 \\ c_2 \\ c_3 \end{pmatrix} = \begin{pmatrix} 3 \\ -1+i \\ -1 \\ -1-i \end{pmatrix}.$$

Note that we obtain the same result using (4); that is, $F(\alpha) = \sum_{n=0}^{3} f(nT)e^{i\alpha nT}$, with $T = \pi/2$, $\alpha = 0, 1, 2, 3$. The graphs of $|c_n|$, $n = 0, 1, 2, 3$, or equivalently $|F(\alpha)|$ for $\alpha = 0, 1, 2, 3$, are given in **FIGURE 15.6.2**. ≡

FIGURE 15.6.2 Graph of $|F(\alpha)|$ in Example 1

Finding the coefficients involves multiplying by matrices $\mathbf{F}_n$ and $\overline{\mathbf{F}}_n$. Because of the nature of these matrices, these multiplications can be done in a very computationally efficient manner, using the **Fast Fourier Transform** (**FFT**), which is discussed later in this section.

Heat Equation and Discrete Fourier Series

If the function f in the initial-value problem

$$k \frac{\partial^2 u}{\partial x^2} = \frac{\partial u}{\partial t}, \quad -\infty < x < \infty, \quad t > 0$$

$$(8)$$

$$u(x, 0) = f(x), \quad -\infty < x < \infty$$

is periodic with period 2π, the solution can be written in terms of a Fourier series for $f(x)$. We can also approximate this solution with a finite sum

$$u(x, t) = \sum_{k=0}^{n-1} c_k(t) e^{ikx}.$$

If we examine both sides of the one-dimensional heat equation in (8), we see that

$$\frac{\partial u}{\partial t} = \sum_{j=0}^{n-1} \frac{dc_j}{dt} e^{ijx}$$

and

$$k \frac{\partial^2 u}{\partial x^2} = k \sum_{j=0}^{n-1} c_j(t)(ij)^2 e^{ijx},$$

since $\dfrac{d^2 e^{ijx}}{dx^2} = (ij)^2 e^{ijx}$.

Equating these last two expressions, we have the first-order DE

$$\frac{dc_j}{dt} = -kj^2 c_j(t) \quad \text{with solution} \quad c_j(t) = c_j(0) e^{-kj^2 t}.$$

The final task is to find the values $c_j(0)$. However, recall that $u(x, t) = \sum_{k=0}^{n-1} c_k(t) e^{ikx}$ and $u(x, 0) = f(x)$, so $c_j(0)$ are the coefficients of the discrete Fourier series of $f(x)$. Compare this with Section 13.3.

Heat Equation and Discrete Fourier Transform

The initial-value problem (8) can be interpreted as the mathematical model for the temperature $u(x, t)$ in an infinitely long bar. In Section 15.4 we saw that we can solve (8) using the Fourier transform and that the solution $u(x, t)$ depends on the Fourier transform $F(\alpha)$ of $f(x)$ (see pages 794–795). We can approximate $F(\alpha)$ by taking a different look at the discrete Fourier transform.

First we approximate values of the transform by discretizing the integral $\mathscr{F}\{f(x)\} = F(\alpha) = \int_{-\infty}^{\infty} f(x) e^{i\alpha x} \, dx$. Consider an interval $[a, b]$. Let $f(x)$ be given at n equally spaced points

$$x_j = a + \frac{b-a}{n} j, \quad j = 0, 1, 2, \ldots, \quad n - 1.$$

Now approximate:

$$F(\alpha) \approx \frac{b-a}{n} \sum_{j=0}^{n-1} f(x_j) e^{i\alpha x_j}$$

$$= \frac{b-a}{n} \sum_{j=0}^{n-1} f\left(a + \frac{b-a}{n} j\right) e^{i\alpha x_j}$$

$$= \frac{b-a}{n} \sum_{j=0}^{n-1} f\left(a + \frac{b-a}{n} j\right) e^{i\alpha a} e^{i\alpha \frac{b-a}{n} j}$$

$$= \frac{b-a}{n} e^{i\alpha a} \sum_{j=0}^{n-1} f\left(a + \frac{b-a}{n} j\right) e^{i\alpha \frac{b-a}{n} j}.$$

If we now choose a convenient value for α, say $\dfrac{2\pi M}{b-a}$ with M an integer, we have

$$F\left(\frac{2\pi M}{b-a}\right) \approx \frac{b-a}{n}\, e^{i\frac{2\pi Ma}{b-a}} \sum_{j=0}^{n-1} f\left(a + \frac{b-a}{n}\,j\right) e^{i\frac{2\pi jM}{n}}$$

$$= \frac{b-a}{n}\, e^{i\frac{2\pi Ma}{b-a}} \sum_{j=0}^{n-1} f\left(a + \frac{b-a}{n}\,j\right) \omega_n^{jM},$$

where recall that $\omega_n = e^{i2\pi/n}$. This is a numerical approximation to the Fourier transform of $f(x)$ evaluated at points $\dfrac{2\pi M}{b-a}$ with M an integer.

EXAMPLE 2 Example 1, Section 15.4—Revisited

Recall from Example 1 in Section 15.4 (with $u_0 = 1$) that the Fourier transform of a rectangular pulse defined by

$$f(x) = \begin{cases} 1, & |x| < 1 \\ 0, & |x| > 1 \end{cases}$$

is

$$F(\alpha) = \frac{2\sin\alpha}{\alpha}.$$

The frequency spectrum is the graph of $|F(\alpha)|$ versus α given in **FIGURE 15.6.3(a)**. Using $n = 16$ equally spaced points between $a = -2$ and $b = 2$, and M running from -6 to 6, we get the discrete Fourier transform of $f(x)$, superimposed over the graph of $|F(\alpha)|$, in Figure 15.6.3(b).

FIGURE 15.6.3 In Example 2 (a) is the graph of $|F(\alpha)|$; (b) is the discrete Fourier transform of f

▇ Aliasing

A problem known as aliasing may appear whenever one is sampling data at equally spaced intervals. If you have ever seen a motion picture where rotating wheels appear to be rotating slowly (or even backwards!), you have experienced aliasing. The wheels may rotate at a high rate, but because the frames in a motion picture are "sampled" at equally spaced intervals, we see a low rate of rotation.

Graphing calculators also suffer from aliasing due to the way they sample points to create graphs. For example, plot the trigonometric function $y = \sin 20\pi x$ with frequency 10 on a Texas Instruments TI-92 and you get the nice graph in **FIGURE 15.6.4(a)**. At higher frequencies, say $y = \sin 100\pi x$ with frequency 50, you get the correct amount of cycles, but the amplitudes of the graph in Figure 15.6.4(b) are clearly not 1.

On a calculator such as the Texas Instruments TI-83, the graphs in **FIGURE 15.6.5** show aliasing much more clearly.

The problem lies in the fact that $e^{2n\pi i} = \cos 2n\pi + i \sin 2n\pi = 1$ for all integer values of n. The discrete Fourier series cannot distinguish e^{inx} from 1 as these functions are equal at sampled points $x = 2k\pi/n$. The higher frequency is seen as the lower one. Consider the functions $\cos(\pi n/2)$ and $\cos(7\pi n/2)$. If we sample at the points $n = 0, 1, 2, \ldots$, these two functions appear the same,

(a) $y = \sin 20\pi x$; x range: [0, 1]; y range: [−1, 1] (b) $y = \sin 100\pi x$; x range: [0, 1]; y range: [−1, 1]

FIGURE 15.6.4 TI-92

(a) $y = \sin 20\pi x$; x range: [0, 1]; y range: [−1, 1] (b) $y = \sin 100\pi x$; x range: [0, 1]; y range: [−1, 1]

FIGURE 15.6.5 TI-83

the lower frequency is assumed, and the amplitudes (Fourier coefficients) associated with the higher frequencies are added in with the amplitude of the low frequency. If these Fourier coefficients at large frequencies are small, however, we do not have a big problem. In the Sampling Theorem below, we will see what can take care of this problem.

■ **Signal Processing** Beyond solving PDEs as we have done earlier, the ideas of this section are useful in **signal processing**. Consider the functions we have been dealing with as **signals** from a source. We would like to reconstruct a signal transmitted by sampling it at discrete points. The problem of calculating an infinite number of Fourier coefficients and summing an infinite series to reconstruct a signal (function) is not practical. A finite sum could be a decent approximation, but certain signals can be reconstructed by a finite number of samples.

THEOREM **15.6.1** **Sampling Theorem**

If a signal $f(x)$ is **band-limited**; that is, if the range of frequencies of the signal lie in a band $-A < k < A$, then the signal can be reconstructed by sampling two times for every cycle of the highest frequency present; in fact,

$$f(x) = \sum_{n=-\infty}^{\infty} f\left(\frac{n\pi}{A}\right) \frac{\sin(Ax - n\pi)}{Ax - n\pi}.$$

To justify Theorem 15.6.1, consider the Fourier transform $F(\alpha)$ of $f(x)$ as a periodic extension so that $F(\alpha)$ is defined for all values of α, not just those in $-A < k < A$. Using the Fourier transform, we have

$$F(\alpha) = \int_{-\infty}^{\infty} f(x)e^{i\alpha x}\, dx \tag{9}$$

$$f(x) = \frac{1}{2\pi} \int_{-\infty}^{\infty} F(\alpha)e^{-i\alpha x}\, d\alpha = \frac{1}{2\pi} \int_{-A}^{A} F(\alpha)e^{-i\alpha x}\, d\alpha. \tag{10}$$

Treating $F(\alpha)$ as a periodic extension, the Fourier series for $F(\alpha)$ is

$$F(\alpha) = \sum_{n=-\infty}^{\infty} c_n e^{in\pi\alpha/A}, \tag{11}$$

where

$$c_n = \frac{1}{2A} \int_{-A}^{A} F(\alpha) e^{-in\pi\alpha/A} \, d\alpha. \tag{12}$$

Using (10), note that

$$\frac{\pi}{A} f\left(\frac{n\pi}{A}\right) = \frac{\pi}{A} \frac{1}{2\pi} \int_{-A}^{A} F(\alpha) e^{-in\pi\alpha/A} \, d\alpha,$$

which by (12) is equal to c_n. Substituting $c_n = \dfrac{\pi}{A} f\left(\dfrac{n\pi}{A}\right)$ into (11) yields

$$F(\alpha) = \sum_{n=-\infty}^{\infty} \frac{\pi}{A} f\left(\frac{n\pi}{A}\right) e^{in\pi\alpha/A}.$$

Substituting this expression for $F(\alpha)$ back into (10), we have

$$
\begin{aligned}
f(x) &= \frac{1}{2\pi} \int_{-A}^{A} \left(\sum_{n=-\infty}^{\infty} \frac{\pi}{A} f\left(\frac{n\pi}{A}\right) e^{in\pi\alpha/A} \right) e^{-i\alpha x} \, d\alpha \\
&= \frac{1}{2A} \sum_{n=-\infty}^{\infty} f\left(\frac{n\pi}{A}\right) \int_{-A}^{A} e^{in\pi\alpha/A} e^{-i\alpha x} \, d\alpha \\
&= \frac{1}{2A} \sum_{n=-\infty}^{\infty} f\left(\frac{n\pi}{A}\right) \int_{-A}^{A} e^{i\alpha\left(\frac{n\pi}{A} - x\right)} \, d\alpha \\
&= \frac{1}{2A} \sum_{n=-\infty}^{\infty} f\left(\frac{n\pi}{A}\right) \frac{1}{i\left(\frac{n\pi}{A} - x\right)} \left(e^{iA\left(\frac{n\pi}{A} - x\right)} - e^{-iA\left(\frac{n\pi}{A} - x\right)} \right) \\
&= \frac{1}{2A} \sum_{n=-\infty}^{\infty} f\left(\frac{n\pi}{A}\right) \frac{1}{i\left(\frac{n\pi}{A} - x\right)} 2i \sin(n\pi - Ax) \\
&= \sum_{n=-\infty}^{\infty} f\left(\frac{n\pi}{A}\right) \frac{\sin(n\pi - Ax)}{n\pi - Ax} \\
&= \sum_{n=-\infty}^{\infty} f\left(\frac{n\pi}{A}\right) \frac{\sin(Ax - n\pi)}{Ax - n\pi}.
\end{aligned}
$$

Note that we used, in succession, an interchange of summation and integration (not always allowed, but is acceptable here), integration of an exponential function, $\sin\theta = \dfrac{e^{i\theta} - e^{-i\theta}}{2i}$, and the fact that $\sin(-\theta) = -\sin\theta$.

So, from samples at intervals of π/A, all values of f can be reconstructed. Note that if we allow e^{iAx} (in other words, we allow $k = A$), then the Sampling Theorem will fail. If, for example, $f(x) = \sin Ax$, then all samples will be 0 and f cannot be reconstructed, as aliasing appears again.

▧ Band-Limited Signals

A signal that contains many frequencies can be **filtered** so that only frequencies in an interval survive, and it becomes a band-limited signal. Consider the signal $f(x)$. Multiply the Fourier transform $F(\alpha)$ of f by a function $G(\alpha)$ that is 1 on the interval containing the frequencies α you wish to keep, and 0 elsewhere. This multiplication of two Fourier transforms in the frequency domain is a convolution of $f(x)$ and $g(x)$ in the time domain. Recall that Problem 20 in Exercises 15.4 states that

$$\mathcal{F}^{-1}\{F(\alpha)G(\alpha)\} = \int_{-\infty}^{\infty} f(\tau) g(x - \tau) \, d\tau.$$

The integral on the right-hand side is called the **convolution of f and g** and is written $f*g$. The last statement can be written more compactly as

$$\mathscr{F}\{f*g\} = F(\alpha)G(\alpha).$$

The analogous idea for Laplace transforms is in Section 4.4. The function $g(x) = \dfrac{\sin Ax}{\pi x}$ has as its Fourier transform the pulse function

$$G(\alpha) = \begin{cases} 1, & -A < \alpha < A \\ 0, & \text{elsewhere.} \end{cases}$$

This implies that the function $(f*g)(x)$ is band-limited, with frequencies between $-A$ and A.

▨ Computing with the Fast Fourier Transform

Return to the discrete Fourier transform of $f(x)$, where we have f sampled at n equally spaced points a distance of T apart, namely, $0, T, 2T, 3T, \ldots, (n-1)T$. (We used $T = \pi/n$ at the beginning of this section.) Substituting this, the discrete Fourier transform

$$F\left(\frac{2\pi M}{b-a}\right) = \frac{b-a}{n}\, e^{i\frac{2\pi Ma}{b-a}} \sum_{j=0}^{n-1} f\left(a + \frac{b-a}{n}j\right)\omega_n^{jM}$$

becomes

$$F\left(\frac{2\pi k}{nT}\right) = T\sum_{j=0}^{n-1} f(jT)\omega_n^{kj}, \quad k = 0, 1, 2, \ldots, n-1.$$

For simplicity of notation, write this instead as

$$c_k = \sum_{j=0}^{n-1} f_j\omega_n^{kj}, \quad k = 0, 1, 2, \ldots, n-1.$$

This should remind you of (6), where we had

$$\begin{pmatrix} f_0 \\ f_1 \\ f_2 \\ \vdots \\ f_{n-1} \end{pmatrix} = \begin{pmatrix} 1 & 1 & 1 & \cdots & 1 \\ 1 & \omega_n & \omega_n^2 & \cdots & \omega_n^{n-1} \\ 1 & \omega_n^2 & \omega_n^4 & \cdots & \omega_n^{2(n-1)} \\ \vdots & \vdots & \vdots & & \vdots \\ 1 & \omega_n^{n-1} & \omega_n^{2(n-1)} & \cdots & \omega_n^{(n-1)^2} \end{pmatrix} \begin{pmatrix} c_0 \\ c_1 \\ c_2 \\ \vdots \\ c_{n-1} \end{pmatrix},$$

or $\mathbf{f} = \mathbf{F}_n\mathbf{c}$. The key to the FFT is properties of ω_n and matrix factorization. If $n = 2^N$, we can write $\mathbf{F}_n$ in the following way (which we will not prove):

$$\mathbf{F}_{2^N} = \begin{pmatrix} \mathbf{I}_{2^{N-1}} & \mathbf{D}_{2^{N-1}} \\ \mathbf{I}_{2^{N-1}} & -\mathbf{D}_{2^{N-1}} \end{pmatrix}\begin{pmatrix} \mathbf{F}_{2^{N-1}} & 0 \\ 0 & \mathbf{F}_{2^{N-1}} \end{pmatrix}\mathbf{P}, \tag{13}$$

where $\mathbf{I}_k$ is the $k \times k$ identity matrix and $\mathbf{P}$ is the permutation matrix that rearranges the matrix $\mathbf{c}$ so that the even subscripts are ordered on the top and the odd ones are ordered on the bottom. The matrix $\mathbf{D}$ is a diagonal matrix defined by

$$\mathbf{D}_{2^{N-1}} = \begin{pmatrix} 1 & & & & \\ & \omega_{2^N} & & & \\ & & (\omega_{2^N})^2 & & \\ & & & \ddots & \\ & & & & (\omega_{2^N})^{2^{N-1}-1} \end{pmatrix}.$$

Note that each of the $\mathbf{F}_{2^{N-1}}$ matrices can, in turn, be factored. In the end, the matrix $\mathbf{F}_n$ with n^2 nonzero entries is factored into the product of n simpler matrices at a great savings to the number of computations needed on the computer.

EXAMPLE 3 The FFT

Let $n = 2^2 = 4$ and let $\mathbf{F}_4$ be the matrix in Example 1:

$$\mathbf{F}_4 = \begin{pmatrix} 1 & 1 & 1 & 1 \\ 1 & i & -1 & -i \\ 1 & -1 & 1 & -1 \\ 1 & -i & -1 & i \end{pmatrix}.$$

From (13), the desired factorization of $\mathbf{F}_4$ is

$$\mathbf{F}_4 = \underbrace{\begin{pmatrix} 1 & 0 & 1 & 0 \\ 0 & 1 & 0 & i \\ 1 & 0 & -1 & 0 \\ 0 & 1 & 0 & -i \end{pmatrix}}_{\mathbf{A}} \underbrace{\begin{pmatrix} 1 & 1 & 0 & 0 \\ 1 & -1 & 0 & 0 \\ 0 & 0 & 1 & 1 \\ 0 & 0 & 1 & -1 \end{pmatrix}}_{\mathbf{B}} \underbrace{\begin{pmatrix} 1 & 0 & 0 & 0 \\ 0 & 0 & 1 & 0 \\ 0 & 1 & 0 & 0 \\ 0 & 0 & 0 & 1 \end{pmatrix}}_{\mathbf{P}}. \qquad (14)$$

We have inserted dashed lines in the matrices marked $\mathbf{A}$ and $\mathbf{B}$ so that you can identify the submatrices $\mathbf{I}_2$, $\mathbf{D}_2$, $-\mathbf{D}_2$, and $\mathbf{F}_2$ by comparing (14) directly with (13). You are also encouraged to multiply out the right side of (14) and verify that you get $\mathbf{F}_4$. Now if $\mathbf{c} = \begin{pmatrix} 3 \\ 5 \\ 8 \\ 20 \end{pmatrix}$, then

$$\mathbf{F}_4 \mathbf{c} = \begin{pmatrix} 1 & 0 & 1 & 0 \\ 0 & 1 & 0 & i \\ 1 & 0 & -1 & 0 \\ 0 & 1 & 0 & -i \end{pmatrix} \begin{pmatrix} 1 & 1 & 0 & 0 \\ 1 & -1 & 0 & 0 \\ 0 & 0 & 1 & 1 \\ 0 & 0 & 1 & -1 \end{pmatrix} \begin{pmatrix} 1 & 0 & 0 & 0 \\ 0 & 0 & 1 & 0 \\ 0 & 1 & 0 & 0 \\ 0 & 0 & 0 & 1 \end{pmatrix} \begin{pmatrix} 3 \\ 5 \\ 8 \\ 20 \end{pmatrix}$$

$$= \begin{pmatrix} 1 & 0 & 1 & 0 \\ 0 & 1 & 0 & i \\ 1 & 0 & -1 & 0 \\ 0 & 1 & 0 & -i \end{pmatrix} \begin{pmatrix} 1 & 1 & 0 & 0 \\ 1 & -1 & 0 & 0 \\ 0 & 0 & 1 & 1 \\ 0 & 0 & 1 & -1 \end{pmatrix} \begin{pmatrix} 3 \\ 8 \\ 5 \\ 20 \end{pmatrix}$$

$$= \begin{pmatrix} 1 & 0 & 1 & 0 \\ 0 & 1 & 0 & i \\ 1 & 0 & -1 & 0 \\ 0 & 1 & 0 & -i \end{pmatrix} \begin{pmatrix} 11 \\ -5 \\ 25 \\ -15 \end{pmatrix} = \begin{pmatrix} 36 \\ -5 - 15i \\ -14 \\ -5 + 15i \end{pmatrix} = \mathbf{f}. \qquad \equiv$$

Without going into details, a computation of $\mathbf{F}_n$ requires n^2 computations, while using the matrix factorization (the FFT) means the number of computations is reduced to one proportional to $n \ln n$. Try a few larger values of n and you will see substantial savings.

15.6 Exercises Answers to selected odd-numbered problems begin on page ANS-39.

1. Show that $\mathbf{F}_4^{-1} = \frac{1}{4}\overline{\mathbf{F}}_4$.

2. Prove the sifting property of the Dirac delta function:

$$\int_{-\infty}^{\infty} f(x)\,\delta(x - a)\,dx = f(a).$$

[*Hint*: Consider the function

$$\delta_\epsilon(x - a) = \begin{cases} \dfrac{1}{2\epsilon}, & |x - a| < \epsilon \\ 0, & \text{elsewhere.} \end{cases}$$

Use the mean value theorem for integrals and then let $\epsilon \to 0$.]

3. Find the Fourier transform of the Dirac delta function $\delta(x)$.

4. Show that the Dirac delta function is the identity under the convolution operation; that is, show $f * \delta = \delta * f = f$. [*Hint*: Use Fourier transforms and Problem 3.]

5. Show that the derivative of the Dirac delta function $\delta'(x - a)$ has the property that it sifts out the derivative of a function f at a. [*Hint*: Use integration by parts.]

6. Use a CAS to show that the Fourier transform of the function $g(x) = \dfrac{\sin Ax}{\pi x}$ is the pulse function

$$G(\alpha) = \begin{cases} 1, & -A < \alpha < A \\ 0, & \text{elsewhere.} \end{cases}$$

7. Write the matrix $\mathbf{F}_8$ and then write it in factored form (13). Verify that the product of the factors is $\mathbf{F}_8$. If instructed, use a CAS to verify the result.

8. Let $\omega_n = e^{i2\pi/n} = \cos(2\pi/n) + i \sin(2\pi/n)$. Since $e^{i2\pi k} = 1$, the numbers ω_n^k, $k = 0, 1, 2, \ldots, n-1$, all have the property that $(\omega_n^k)^n = 1$. Because of this, ω_n^k, $k = 0, 1, 2, \ldots, n-1$, are called the **nth roots of unity** and are solutions of the equation $z^n - 1 = 0$. Find the eighth roots of unity and plot them in the xy-plane where a complex number is written $z = x + iy$. What do you notice?

Computer Lab Assignments

9. Use a CAS to verify that the function $f * g$, where $f(x) = e^{-5x^2}$ and $g(x) = \dfrac{\sin 2x}{\pi x}$, is band-limited. If your CAS can handle it, plot the graphs of $\mathscr{F}\{f * g\}$ and $F(\alpha)G(\alpha)$ to verify the result.

10. If your CAS has a discrete Fourier transform command, choose any six points and compare the result obtained using this command with that obtained from $\mathbf{c} = \frac{1}{6}\,\overline{\mathbf{F}}_6 \mathbf{f}$.

15 Chapter in Review Answers to selected odd-numbered problems begin on page ANS-39.

In Problems 1–20, use either the Laplace transform or a Fourier transform to solve the given boundary-value problem. Make assumptions about boundedness where necessary. The symbols u_0 and u_1 are constants.

1. $\dfrac{\partial^2 u}{\partial x^2} + \dfrac{\partial^2 u}{\partial y^2} = 0$, $x > 0$, $0 < y < \pi$

 $\dfrac{\partial u}{\partial x}\bigg|_{x=0} = 0$, $0 < y < \pi$

 $u(x, 0) = 0$, $\dfrac{\partial u}{\partial y}\bigg|_{y=\pi} = e^{-x}$, $x > 0$

2. $\dfrac{\partial^2 u}{\partial x^2} = \dfrac{\partial u}{\partial t}$, $0 < x < 1$, $t > 0$

 $u(0, t) = 0$, $u(1, t) = 0$, $t > 0$

 $u(x, 0) = 50 \sin 2\pi x$, $0 < x < 1$

3. $\dfrac{\partial^2 u}{\partial x^2} - hu = \dfrac{\partial u}{\partial t}$, $h > 0$, $x > 0$, $t > 0$

 $u(0, t) = 0$, $\displaystyle\lim_{x\to\infty} \dfrac{\partial u}{\partial x} = 0$, $t > 0$

 $u(x, 0) = u_0$, $x > 0$

4. $\dfrac{\partial u}{\partial t} - \dfrac{\partial^2 u}{\partial x^2} = e^{-|x|}$, $-\infty < x < \infty$, $t > 0$

 $u(x, 0) = u_0$, $-\infty < x < \infty$

5. $\dfrac{\partial^2 u}{\partial x^2} = \dfrac{\partial u}{\partial t}$, $x > 1$, $t > 0$

 $u(0, t) = t$, $\displaystyle\lim_{x\to\infty} u(x, t) = 0$

 $u(x, 0) = 0$, $x > 0$ [*Hint*: Use Theorem 4.4.2.]

6. $\dfrac{\partial^2 u}{\partial x^2} = \dfrac{\partial^2 u}{\partial t^2}$, $0 < x < 1$, $t > 0$

 $u(0, t) = 0$, $u(1, t) = 0$, $t > 0$

 $u(x, 0) = \sin \pi x$, $\dfrac{\partial u}{\partial t}\bigg|_{t=0} = -\sin \pi x$, $0 < x < 1$

7. $k\dfrac{\partial^2 u}{\partial x^2} = \dfrac{\partial u}{\partial t}$, $-\infty < x < \infty$, $t > 0$

 $u(x, 0) = \begin{cases} 0, & x < 0 \\ u_0, & 0 < x < \pi \\ 0, & x > \pi \end{cases}$

8. $\dfrac{\partial^2 u}{\partial x^2} + \dfrac{\partial^2 u}{\partial y^2} = 0$, $0 < x < \pi$, $y > 0$

 $u(0, y) = 0$, $u(\pi, y) = \begin{cases} 0, & 0 < y < 1 \\ 1, & 1 < y < 2 \\ 0, & y > 2 \end{cases}$

 $\dfrac{\partial u}{\partial y}\bigg|_{y=0} = 0$, $0 < x < \pi$

9. $\dfrac{\partial^2 u}{\partial x^2} + \dfrac{\partial^2 u}{\partial y^2} = 0$, $x > 0$, $y > 0$

 $u(0, y) = \begin{cases} 50, & 0 < y < 1 \\ 0, & y > 1 \end{cases}$

 $u(x, 0) = \begin{cases} 100, & 0 < x < 1 \\ 0, & x > 1 \end{cases}$

10. $\dfrac{\partial^2 u}{\partial x^2} + r = \dfrac{\partial u}{\partial t}$, $0 < x < 1$, $t > 0$

 $\dfrac{\partial u}{\partial x}\bigg|_{x=0} = 0$, $u(1, t) = 0$, $t > 0$

 $u(x, 0) = 0$, $0 < x < 1$

11. $\dfrac{\partial^2 u}{\partial x^2} + \dfrac{\partial^2 u}{\partial y^2} = 0$, $x > 0$, $0 < y < \pi$

 $u(0, y) = A$, $0 < y < \pi$

 $\dfrac{\partial u}{\partial y}\bigg|_{y=0} = 0$, $\dfrac{\partial u}{\partial y}\bigg|_{y=\pi} = Be^{-x}$, $x > 0$

12. $\dfrac{\partial^2 u}{\partial x^2} = \dfrac{\partial u}{\partial t}$, $0 < x < 1$, $t > 0$

$u(0, t) = u_0$, $u(1, t) = u_0$, $t > 0$

$u(x, 0) = 0$, $0 < x < 1$

[*Hint*: Use the identity

$$\sinh(x - y) = \sinh x \cosh y - \cosh x \sinh y,$$

and then use Problem 8 in Exercises 15.1.]

13. $k\dfrac{\partial^2 u}{\partial x^2} = \dfrac{\partial u}{\partial t}$, $-\infty < x < \infty$, $t > 0$

$u(x, 0) = \begin{cases} 0, & x < 0 \\ e^{-x}, & x > 0 \end{cases}$

14. $\dfrac{\partial^2 u}{\partial x^2} = \dfrac{\partial u}{\partial t}$, $x > 0$, $t > 0$

$\dfrac{\partial u}{\partial x}\bigg|_{x=0} = -50$, $\lim\limits_{x\to\infty} u(x, t) = 100$, $t > 0$

$u(x, 0) = 100$, $x > 0$

15. Show that a solution of the BVP

$$\dfrac{\partial^2 u}{\partial x^2} + \dfrac{\partial^2 u}{\partial y^2} = 0, \quad -\infty < x < \infty, \ 0 < y < 1$$

$$\dfrac{\partial u}{\partial y}\bigg|_{y=0} = 0, \ u(x, 1) = f(x), \ -\infty < x < \infty$$

is $\quad u(x, y) = \dfrac{1}{\pi} \displaystyle\int_0^\infty \int_{-\infty}^\infty f(t) \, \dfrac{\cosh \alpha y \, \cos \alpha(t - x)}{\cosh \alpha} \, dt \, d\alpha$.

16. $0.5\dfrac{\partial^2 u}{\partial x^2} = \dfrac{\partial u}{\partial t}$, $x > 0$, $t > 0$

$\dfrac{\partial u}{\partial x}\bigg|_{x=0} = 0$, $t > 0$

$u(x, 0) = e^{-2x}$, $x > 0$

17. $\dfrac{\partial^2 u}{\partial x^2} = \dfrac{\partial u}{\partial t}$, $0 < x < \pi$, $t > 0$

$u(0, t) = 1$, $u(\pi, t) = 1$, $t > 0$

$u(x, 0) = 1 + \sin 2x$, $0 < x < \pi$

18. $\dfrac{\partial^2 u}{\partial x^2} = \dfrac{\partial u}{\partial t}$, $x > 0$, $t > 0$

$u(0, t) = 100 \, [\mathcal{U}(t - 5) - \mathcal{U}(t - 10)]$, $t > 0$

$\lim\limits_{x\to\infty} u(x, t) = 50$, $t > 0$

$u(x, 0) = 50$, $x > 0$

19. $\dfrac{\partial^2 u}{\partial x^2} + \dfrac{\partial^2 u}{\partial y^2} = hu$, h constant, $x > 0$, $0 < y < \pi$

$u(0, y) = 0$, $0 < y < \pi$

$u(x, 0) = 0$, $u(x, \pi) = f(x)$, $x > 0$

20. $\dfrac{\partial^2 u}{\partial x^2} - A\dfrac{\partial u}{\partial t} - Bu = 0$, A, B constants, $x > 0$, $t > 0$

$u(0, t) = u_1 e^{-10t}$, $\lim\limits_{x\to\infty} \dfrac{\partial u}{\partial x} = 0$, $t > 0$

$u(x, 0) = u_0$, $x > 0$

21. Solve the boundary-value problem

$$\dfrac{\partial^2 u}{\partial x^2} = \dfrac{\partial u}{\partial t}, \quad x > 0, \ t > 0$$

$$\dfrac{\partial u}{\partial x}\bigg|_{x=0} = -1, \quad \lim\limits_{x\to\infty} u(x, t) = 0, \ t > 0$$

$$u(x, 0) = 0, \quad x > 0$$

using the Laplace transform. Give two different forms of the solution $u(x, t)$.

22. (a) Solve the BVP in Problem 21 using a Fourier transform.
(b) Use a CAS to carry out an integration to show that the answer in part (a) is equivalent to one of the answers in Problem 21.

In Problems 23–26, use a finite Fourier transform to solve the given boundary-value problem. The symbols u_0 and u_1 are constants.

23. $\dfrac{\partial u}{\partial t} - \dfrac{\partial^2 u}{\partial x^2} = t \sin x$, $0 < x < \pi$, $t > 0$

$u(0, t) = 0$, $u(\pi, t) = 0$, $t > 0$

$u(x, 0) = 0$, $0 < x < \pi$

24. $\dfrac{\partial^2 u}{\partial x^2} = \dfrac{\partial^2 u}{\partial t^2} + u$, $0 < x < 1$, $t > 0$

$\dfrac{\partial u}{\partial x}\bigg|_{x=0} = 0$, $\dfrac{\partial u}{\partial x}\bigg|_{x=1} = 0$, $t > 0$

$u(x, 0) = 0$, $\dfrac{\partial u}{\partial t}\bigg|_{x=0} = x$, $0 < x < 1$

25. $\dfrac{\partial^2 u}{\partial x^2} = \dfrac{\partial u}{\partial t}$, $0 < x < \pi$, $t > 0$

$u(0, t) = u_0$, $u(\pi, t) = u_1$, $t > 0$

$u(x, 0) = 0$, $0 < x < \pi$

26. $\dfrac{\partial^2 u}{\partial x^2} + \dfrac{\partial^2 u}{\partial y^2} = u - u_0$, $0 < x < 1$, $0 < y < 1$

$u(0, y) = 0$, $u(1, y) = 0$, $0 < y < 1$

$u(x, 0) = 0$, $u(x, 1) = u_1$, $0 < x < 1$

CHAPTER 16

Numerical Solutions of Partial Differential Equations

In Section 6.5 we saw that one way of approximating a solution of a second-order boundary-value problem is to work with a finite difference equation replacement of the linear ordinary differential equation. The difference equation is constructed by replacing the ordinary derivatives dy/dx and d^2y/dx^2 with difference quotients. We will see in this chapter that the same idea carries over to boundary-value problems involving linear partial differential equations.

16.1 Laplace's Equation

INTRODUCTION Recall from Section 13.1 that linear second-order PDEs in two independent variables are classified as *elliptic*, *parabolic*, and *hyperbolic*. Roughly, elliptic PDEs involve partial derivatives with respect to spatial variables only and as a consequence solutions of such equations are determined by boundary conditions alone. Parabolic and hyperbolic equations involve partial derivatives with respect to both spatial and time variables, and so solutions of such equations generally are determined from boundary and initial conditions. A solution of an elliptic PDE (such as Laplace's equation) can describe a physical system whose state is in equilibrium (steady state), a solution of a parabolic PDE (such as the heat equation) can describe a diffusional state, whereas a hyperbolic PDE (such as the wave equation) can describe a vibrational state.

In this section we begin our discussion with approximation methods appropriate for elliptic equations. Our focus will be on the simplest but probably the most important PDE of the elliptic type: Laplace's equation.

■ Difference Equation Replacement Suppose that we are seeking a solution $u(x, y)$ of Laplace's equation

$$\frac{\partial^2 u}{\partial x^2} + \frac{\partial^2 u}{\partial y^2} = 0$$

in a planar region R that is bounded by some curve C. See **FIGURE 16.1.1**. Analogous to (6) of Section 6.5, using the central differences

$$u(x + h, y) - 2u(x, y) + u(x - h, y) \qquad \text{and} \qquad u(x, y + h) - 2u(x, y) + u(x, y - h),$$

approximations for the second partial derivatives u_{xx} and u_{yy} can be obtained using the difference quotients

$$\frac{\partial^2 u}{\partial x^2} \approx \frac{1}{h^2}\left[u(x + h, y) - 2u(x, y) + u(x - h, y)\right] \tag{1}$$

$$\frac{\partial^2 u}{\partial y^2} \approx \frac{1}{h^2}\left[u(x, y + h) - 2u(x, y) + u(x, y - h)\right]. \tag{2}$$

Now by adding (1) and (2) we obtain a **five-point approximation** to the Laplacian:

$$\frac{\partial^2 u}{\partial x^2} + \frac{\partial^2 u}{\partial y^2} \approx \frac{1}{h^2}\left[u(x + h, y) + u(x, y + h) + u(x - h, y) + u(x, y - h) - 4u(x, y)\right].$$

Hence we can replace Laplace's equation by the difference equation

$$u(x + h, y) + u(x, y + h) + u(x - h, y) + u(x, y - h) - 4u(x, y) = 0. \tag{3}$$

If we adopt the notation $u(x, y) = u_{ij}$ and

$$u(x + h, y) = u_{i+1, j}, \qquad u(x, y + h) = u_{i, j+1}$$

$$u(x - h, y) = u_{i-1, j}, \qquad u(x, y - h) = u_{i, j-1},$$

then (3) becomes

$$u_{i+1, j} + u_{i, j+1} + u_{i-1, j} + u_{i, j-1} - 4u_{ij} = 0. \tag{4}$$

To understand (4) a little better, suppose a rectangular grid consisting of horizontal lines spaced h units apart and vertical lines spaced h units apart is placed over the region R. The number h is called the **mesh size**. See **FIGURE 16.1.2(a)**. The points $P_{ij} = P(ih, jh)$, i and j integers, of intersection of the horizontal and vertical lines, are called **mesh points** or **lattice points**. A mesh point is an **interior point** if its four nearest neighboring mesh points are points of R. Points in R or on C that are not interior points are called **boundary points**. For example, in Figure 16.1.2(a) we have

$$P_{20} = P(2h, 0), \quad P_{11} = P(h, h), \quad P_{21} = P(2h, h), \quad P_{22} = P(2h, 2h),$$

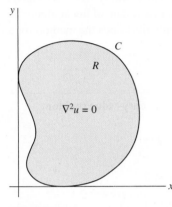

FIGURE 16.1.1 Planar region R with boundary C

(a)

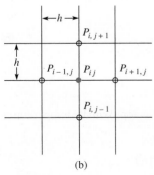

(b)

FIGURE 16.1.2 Region R overlaid with rectangular grid

and so on. Of the points listed, P_{21} and P_{22} are interior points, whereas P_{20} and P_{11} are boundary points. In Figure 16.1.2(a) interior points are the dots shown in red and the boundary points are shown in black. Now from (4) we see that

$$u_{ij} = \tfrac{1}{4} \left[u_{i+1,\,j} + u_{i,\,j+1} + u_{i-1,\,j} + u_{i,\,j-1} \right], \tag{5}$$

and so, as shown in Figure 16.1.2(b), the value of u_{ij} at an interior mesh point of R is the average of the values of u at four neighboring mesh points. The neighboring points $P_{i+1,\,j}$, $P_{i,\,j+1}$, $P_{i-1,\,j}$, and $P_{i,\,j-1}$ correspond, respectively, to the four points on a compass: E, N, W, and S.

▉ Dirichlet Problem

Recall that in the **Dirichlet problem** for Laplace's equation $\nabla^2 u = 0$, the values of $u(x, y)$ are prescribed on the boundary C of a region R. The basic idea is to find an approximate solution to Laplace's equation at interior mesh points by replacing the partial differential equation at these points by the difference equation (4). Hence the approximate values of u at the mesh points—namely, the u_{ij}—are related to each other and, possibly, to known values of u if a mesh point lies on the boundary C. In this manner we obtain a system of linear algebraic equations that we solve for the unknown u_{ij}. The following example illustrates the method for a square region.

| EXAMPLE 1 | A Boundary-Value Problem Revisited |

In Problem 16 of Exercises 13.5 you were asked to solve the boundary-value problem

$$\frac{\partial^2 u}{\partial x^2} + \frac{\partial^2 u}{\partial y^2} = 0, \quad 0 < x < 2, \quad 0 < y < 2$$

$$u(0, y) = 0, \quad u(2, y) = y(2 - y), \quad 0 < y < 2$$

$$u(x, 0) = 0, \quad u(x, 2) = \begin{cases} x, & 0 < x < 1 \\ 2 - x, & 1 \le x < 2 \end{cases}$$

FIGURE 16.1.3 Square region R for Example 1

utilizing the superposition principle. To apply the present numerical method, let us start with a mesh size of $h = \tfrac{2}{3}$. As we see in **FIGURE 16.1.3**, that choice yields four interior points and eight boundary points. The numbers listed next to the boundary points are the exact values of u obtained from the specified condition along that boundary. For example, at $P_{31} = P(3h, h) = P(2, \tfrac{2}{3})$ we have $x = 2$ and $y = \tfrac{2}{3}$, and so the condition $u(2, y)$ gives $u(2, \tfrac{2}{3}) = \tfrac{2}{3}(2 - \tfrac{2}{3}) = \tfrac{8}{9}$. Similarly, at $P_{13} = P(\tfrac{2}{3}, 2)$, the condition $u(x, 2)$ gives $u(\tfrac{2}{3}, 2) = \tfrac{2}{3}$. We now apply (4) at each interior point. For example, at P_{11} we have $i = 1$ and $j = 1$, so (4) becomes

$$u_{21} + u_{12} + u_{01} + u_{10} - 4u_{11} = 0.$$

Since $u_{01} = u(0, \tfrac{2}{3}) = 0$ and $u_{10} = u(\tfrac{2}{3}, 0) = 0$, the foregoing equation becomes $-4u_{11} + u_{21} + u_{12} = 0$. Repeating this, in turn, at P_{21}, P_{12}, and P_{22}, we get three additional equations:

$$
\begin{aligned}
-4u_{11} + u_{21} + u_{12} \quad\quad\quad &= 0 \\
u_{11} - 4u_{21} + \quad\quad u_{22} &= -\tfrac{8}{9} \\
u_{11} \quad\quad - 4u_{12} + u_{22} &= -\tfrac{2}{3} \\
u_{21} + u_{12} - 4u_{22} &= -\tfrac{14}{9}.
\end{aligned}
\tag{6}
$$

Using a computer algebra system to solve the system, we find the approximate temperatures at the four interior points to be

$$u_{11} = \tfrac{7}{36} = 0.1944, \qquad u_{21} = \tfrac{5}{12} = 0.4167,$$

$$u_{12} = \tfrac{13}{36} = 0.3611, \qquad u_{22} = \tfrac{7}{12} = 0.5833. \qquad\qquad \equiv$$

As in the discussion of ordinary differential equations, we expect that a smaller value of h will improve the accuracy of the approximation. However, using a smaller mesh size means, of course, that there are more interior mesh points, and correspondingly there is a much larger system of equations to be solved. For a *square* region whose length of side is L, a mesh size of $h = L/n$ will yield a total of $(n-1)^2$ interior mesh points. In Example 1, for $n = 8$, the mesh size is a reasonable $h = \frac{2}{8} = \frac{1}{4}$, but the number of interior points is $(8-1)^2 = 49$. Thus we have 49 equations in 49 unknowns. In the next example we use a mesh size of $h = \frac{1}{2}$.

EXAMPLE 2 **Example 1 with More Mesh Points**

FIGURE 16.1.4 Region R in Example 1 with additional mesh points

As we see in **FIGURE 16.1.4**, with $n = 4$, a mesh size $h = \frac{2}{4} = \frac{1}{2}$ for the square in Example 1 gives $3^2 = 9$ interior mesh points. Applying (4) at these points and using the indicated boundary conditions, we get nine equations in nine unknowns. So that you can verify the results, we give the system in an unsimplified form:

$$u_{21} + u_{12} + \quad 0 + \quad 0 - 4u_{11} = 0$$

$$u_{31} + u_{22} + u_{11} + \quad 0 - 4u_{21} = 0$$

$$\tfrac{3}{4} + u_{32} + u_{21} + \quad 0 - 4u_{31} = 0$$

$$u_{22} + u_{13} + u_{11} + \quad 0 - 4u_{12} = 0$$

$$u_{32} + u_{23} + u_{12} + u_{21} - 4u_{22} = 0 \tag{7}$$

$$1 + u_{33} + u_{22} + u_{31} - 4u_{32} = 0$$

$$u_{23} + \tfrac{1}{2} + \quad 0 + u_{12} - 4u_{13} = 0$$

$$u_{33} + \quad 1 + u_{13} + u_{22} - 4u_{23} = 0$$

$$\tfrac{3}{4} + \tfrac{1}{2} + u_{23} + u_{32} - 4u_{33} = 0.$$

In this case, a CAS yields

$$u_{11} = \tfrac{7}{64} = 0.1094, \qquad u_{21} = \tfrac{51}{224} = 0.2277, \qquad u_{31} = \tfrac{177}{448} = 0.3951$$

$$u_{12} = \tfrac{47}{224} = 0.2098, \qquad u_{22} = \tfrac{13}{32} = 0.4063, \qquad u_{32} = \tfrac{135}{224} = 0.6027$$

$$u_{13} = \tfrac{145}{448} = 0.3237, \qquad u_{23} = \tfrac{131}{224} = 0.5848, \qquad u_{33} = \tfrac{39}{64} = 0.6094. \qquad \equiv$$

After we simplify (7), it is interesting to note that the 9×9 matrix of coefficients is

$$
\begin{pmatrix}
-4 & 1 & 0 & 1 & 0 & 0 & 0 & 0 & 0 \\
1 & -4 & 1 & 0 & 1 & 0 & 0 & 0 & 0 \\
0 & 1 & -4 & 0 & 0 & 1 & 0 & 0 & 0 \\
1 & 0 & 0 & -4 & 1 & 0 & 1 & 0 & 0 \\
0 & 1 & 0 & 1 & -4 & 1 & 0 & 1 & 0 \\
0 & 0 & 1 & 0 & 1 & -4 & 0 & 0 & 1 \\
0 & 0 & 0 & 1 & 0 & 0 & -4 & 1 & 0 \\
0 & 0 & 0 & 0 & 1 & 0 & 1 & -4 & 1 \\
0 & 0 & 0 & 0 & 0 & 1 & 0 & 1 & -4
\end{pmatrix}. \tag{8}
$$

This is an example of a **sparse matrix** in that a large percentage of the entries are zeros. The matrix (8) is also an example of a **banded matrix**. These kinds of matrices are characterized by the properties that the entries on the main diagonal and on diagonals (or bands) parallel to the main diagonal are all nonzero. The bands shown in red in (8) are separated by diagonals consisting of all zeros or not.

■ **Gauss–Seidel Iteration** Problems requiring approximations to solutions of partial differential equations invariably lead to large systems of linear algebraic equations. It is not uncommon to have to solve systems involving hundreds of equations. Although a direct method of solution such as Gaussian elimination leaves unchanged the zero entries outside the bands in a matrix such as (8), it does fill in the positions between the bands with nonzeros. Since storing very large matrices uses up a large portion of computer memory, it is usual practice to solve a large system in an indirect manner. One popular indirect method is called **Gauss–Seidel iteration**. This method is named after the German mathematicians **Johann Carl Friedrich Gauss** (1777–1855) and **Philipp Ludwig von Seidel** (1821–1896).

We shall illustrate this method for the system in (6). For the sake of simplicity we replace the double-subscripted variables u_{11}, u_{21}, u_{12}, and u_{22} by x_1, x_2, x_3, and x_4, respectively.

EXAMPLE 3 Gauss–Seidel Iteration

Step 1: *Solve each equation for the variables on the main diagonal of the system.* That is, in (6), solve the first equation for x_1, the second equation for x_2, and so on:

$$x_1 = 0.25x_2 + 0.25x_3$$

$$x_2 = 0.25x_1 + 0.25x_4 + 0.2222$$

$$x_3 = 0.25x_1 + 0.25x_4 + 0.1667 \tag{9}$$

$$x_4 = 0.25x_2 + 0.25x_3 + 0.3889.$$

These equations can be obtained directly by using (5) rather than (4) at the interior points.

Step 2: *Iterations.* We start by making an initial guess for the values of x_1, x_2, x_3, and x_4. If this were simply a system of linear equations and we knew nothing about the solution, we could start with $x_1 = 0$, $x_2 = 0$, $x_3 = 0$, $x_4 = 0$. But since the solution of (9) represents approximations to a solution of a boundary-value problem, it would seem reasonable to use as the initial guess for the values of $x_1 = u_{11}$, $x_2 = u_{21}$, $x_3 = u_{12}$, and $x_4 = u_{22}$ the average of all the boundary conditions. In this case the average of the numbers at the eight boundary points shown in Figure 16.1.2 is approximately 0.4. Thus our initial guess is $x_1 = 0.4$, $x_2 = 0.4$, $x_3 = 0.4$, and $x_4 = 0.4$. Iterations of the Gauss–Seidel method use the x values as soon as they are computed. Note that the first equation in (9) depends only on x_2 and x_3; thus substituting $x_2 = 0.4$ and $x_3 = 0.4$ gives $x_1 = 0.2$. Since the second and third equations depend on x_1 and x_4, we use the newly calculated values $x_1 = 0.2$ and $x_4 = 0.4$ to obtain $x_2 = 0.3722$ and $x_3 = 0.3167$. The fourth equation depends on x_2 and x_3, so we use the new values $x_2 = 0.3722$ and $x_3 = 0.3167$ to get $x_4 = 0.5611$. In summary, the first iteration has given the values

$$x_1 = 0.2, \quad x_2 = 0.3722, \quad x_3 = 0.3167, \quad x_4 = 0.5611.$$

Note how close these numbers are already to the actual values given at the end of Example 1.

The second iteration starts with substituting $x_2 = 0.3722$ and $x_3 = 0.3167$ into the first equation. This gives $x_1 = 0.1722$. From $x_1 = 0.1722$ and the last computed value of x_4 (namely, $x_4 = 0.5611$), the second and third equations give, in turn, $x_2 = 0.4055$ and $x_3 = 0.3500$. Using these two values, we find from the fourth equation that $x_4 = 0.5678$. At the end of the second iteration we have

$$x_1 = 0.1722, \quad x_2 = 0.4055, \quad x_3 = 0.3500, \quad x_4 = 0.5678.$$

The third through seventh iterations are summarized in Table 16.1.1.

TABLE 16.1.1	Iteration				
	3rd	4th	5th	6th	7th
x_1	0.1889	0.1931	0.1941	0.1944	0.1944
x_2	0.4139	0.4160	0.4165	0.4166	0.4166
x_3	0.3584	0.3605	0.3610	0.3611	0.3611
x_4	0.5820	0.5830	0.5833	0.5833	0.5833

To apply Gauss–Seidel iteration to a general system of n linear equations in n unknowns, the variable x_i must actually appear in the ith equation of the system. Moreover, after each equation is solved for x_i, $i = 1, 2, \ldots, n$, the resulting system has the form $\mathbf{X} = \mathbf{AX} + \mathbf{B}$, where all the entries on the main diagonal of $\mathbf{A}$ are zero.

REMARKS

(*i*) In the examples given in this section, the values of u_{ij} were determined using known values of u at boundary points. But what do we do if the region is such that boundary points do not coincide with the actual boundary C of the region R? In this case the required values can be obtained by interpolation.

(*ii*) It is sometimes possible to cut down the number of equations to solve by using symmetry. Consider the rectangular region defined by $0 \le x \le 2, 0 \le y \le 1$, shown in **FIGURE 16.1.5**. The boundary conditions are $u = 0$ along the boundaries $x = 0$, $x = 2$, $y = 1$, and $u = 100$ along $y = 0$. The region is symmetric about the lines $x = 1$ and $y = \frac{1}{2}$, and the interior points P_{11} and P_{31} are equidistant from the neighboring boundary points at which the specified values of u are the same. Consequently we assume that $u_{11} = u_{31}$, and so the system of three equations in three unknowns reduces to two equations in two unknowns. See Problem 2 in Exercises 16.1.

(*iii*) It may not be noticeable on a computer, but convergence of Gauss–Seidel iteration (or Liebman's method) may not be particularly fast. Also, in a more general setting, Gauss–Seidel iteration may not converge at all. For conditions that are sufficient to guarantee convergence of Gauss–Seidel iteration you are encouraged to consult texts on numerical analysis.

FIGURE 16.1.5 Rectangular region R

16.1 Exercises Answers to selected odd-numbered problems begin on page ANS-40.

In Problems 1–8, use a computer as a computational aid.

In Problems 1–4, use (4) to approximate the solution of Laplace's equation at the interior points of the given region. Use symmetry when possible.

1. $u(0, y) = 0$, $u(3, y) = y(2 - y)$, $0 < y < 2$

$u(x, 0) = 0$, $u(x, 2) = x(3 - x)$, $0 < x < 3$

mesh size: $h = 1$

2. $u(0, y) = 0$, $u(2, y) = 0$, $0 < y < 1$

$u(x, 0) = 100$, $u(x, 1) = 0$, $0 < x < 2$

mesh size: $h = \frac{1}{2}$

3. $u(0, y) = 0$, $u(1, y) = 0$, $0 < y < 1$

$u(x, 0) = 0$, $u(x, 1) = \sin \pi x$, $0 < x < 1$

mesh size: $h = \frac{1}{3}$

4. $u(0, y) = 108y^2(1 - y)$, $u(1, y) = 0$, $0 < y < 1$

$u(x, 0) = 0$, $u(x, 1) = 0$, $0 < x < 1$

mesh size: $h = \frac{1}{3}$

In Problems 5 and 6, use (5) and Gauss–Seidel iteration to approximate the solution of Laplace's equation at the interior points of a unit square. Use the mesh size $h = \frac{1}{4}$. In Problem 5, the boundary conditions are given; in Problem 6, the values of u at boundary points are given in **FIGURE 16.1.6**.

5. $u(0, y) = 0$, $u(1, y) = 100y$, $0 < y < 1$

$u(x, 0) = 0$, $u(x, 1) = 100x$, $0 < x < 1$

6.

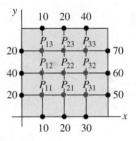

FIGURE 16.1.6 Region in Problem 6

7. (a) In Problem 12 of Exercises 13.6, you solved a potential problem using a special form of Poisson's equation

$$\frac{\partial^2 u}{\partial x^2} + \frac{\partial^2 u}{\partial y^2} = f(x, y).$$ Show that the difference equation replacement for Poisson's equation is

$$u_{i+1,j} + u_{i,j+1} + u_{i-1,j} + u_{i,j-1} - 4u_{ij} = h^2 f(x, y).$$

(b) Use the result in part (a) to approximate the solution of the Poisson equation $\dfrac{\partial^2 u}{\partial x^2} + \dfrac{\partial^2 u}{\partial y^2} = -2$ at the interior points of the region in **FIGURE 16.1.7**. The mesh size is $h = \frac{1}{2}$, $u = 1$ at every point along $ABCD$, and $u = 0$ at every point along $DEFGA$. Use symmetry and, if necessary, Gauss–Seidel iteration.

interior points of the region in **FIGURE 16.1.8**. The mesh is $h = \frac{1}{8}$, and $u = 0$ at every point on the boundary of the region. If necessary, use Gauss–Seidel iteration.

FIGURE 16.1.7 Region in Problem 7

8. Use the result in part (a) of Problem 7 to approximate the solution of the Poisson equation $\dfrac{\partial^2 u}{\partial x^2} + \dfrac{\partial^2 u}{\partial y^2} = -64$ at the

FIGURE 16.1.8 Region in Problem 8

16.2 Heat Equation

INTRODUCTION The basic idea in the following discussion is the same as in Section 16.1; we approximate a solution of a PDE—this time a parabolic PDE—by replacing the equation with a finite difference equation. But unlike the preceding section, we shall consider *two* finite-difference approximation methods for parabolic partial differential equations: one called an *explicit method* and the other an *implicit method*. For the sake of definiteness, we consider only the one-dimensional heat equation.

■ **Difference Equation Replacement** To approximate the solution $u(x, t)$ of the one-dimensional heat equation

$$c\,\frac{\partial^2 u}{\partial x^2} = \frac{\partial u}{\partial t} \tag{1}$$

we again replace the derivatives by difference quotients. Using the central difference approximation (2) of Section 16.1,

$$\frac{\partial^2 u}{\partial x^2} \approx \frac{1}{h^2}\,[u(x + h, t) - 2u(x, t) + u(x - h, t)]$$

and the forward difference approximation (3) of Section 6.5,

$$\frac{\partial u}{\partial t} \approx \frac{1}{h}\,[u(x, t + h) - u(x, t)]$$

equation (1) becomes

$$\frac{c}{h^2}\,[u(x + h, t) - 2u(x, t) + u(x - h, t)] = \frac{1}{k}\,[u(x, t + k) - u(x, t)]. \tag{2}$$

If we let $\lambda = ck/h^2$ and

$$u(x, t) = u_{ij}, \ \ u(x + h, t) = u_{i+1,\, j}, \ \ u(x - h, t) = u_{i-1,\, j}, \ \ u(x, t + k) = u_{i,\, j+1},$$

then, after simplifying, (2) is

$$u_{i,\, j+1} = \lambda u_{i+1,\, j} + (1 - 2\lambda)u_{ij} + \lambda u_{i-1,\, j}. \tag{3}$$

In the case of the heat equation (1), typical boundary conditions are $u(0, t) = u_1$, $u(a, t) = u_2$, $t > 0$, and an initial condition is $u(x, 0) = f(x)$, $0 < x < a$. The function f can be interpreted as the initial temperature distribution in a homogeneous rod extending from $x = 0$ to $x = a$; u_1 and u_2 can be interpreted as constant temperatures at the endpoints of the rod. Although we shall not

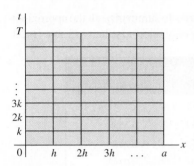

FIGURE 16.2.1 Rectangular region in xt-plane

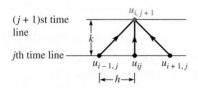

FIGURE 16.2.2 u at $t = j + 1$ is determined from three values of u at $t = j$

prove it, the boundary-value problem consisting of (1) and these two boundary conditions and one initial condition has a unique solution when f is continuous on the closed interval $[0, a]$. This latter condition will be assumed, and so we replace the initial condition by $u(x, 0) = f(x)$, $0 \leq x \leq a$. Moreover, instead of working with the semi-infinite region in the xt-plane defined by the inequalities $0 \leq x \leq a$, $t \geq 0$, we use a rectangular region defined by $0 \leq x \leq a$, $0 \leq t \leq T$, where T is some specified value of time. Over this region we place a rectangular grid consisting of vertical lines h units apart and horizontal lines k units apart. See **FIGURE 16.2.1**. If we choose two positive integers n and m and define

$$h = \frac{a}{n} \quad \text{and} \quad k = \frac{T}{m},$$

then the vertical and horizontal grid lines are defined by

$$x_i = ih, \quad i = 0, 1, 2, \ldots, n \quad \text{and} \quad t_j = jk, \quad j = 0, 1, 2, \ldots, m.$$

As illustrated in **FIGURE 16.2.2**, the plan here is to use formula (3) to estimate the values of the solution $u(x, t)$ at the points on the $(j + 1)$st time line using only values from the jth time line. For example, the values on the first time line $(j = 1)$ depend on the initial condition $u_{i, 0} = u(x_i, 0) = f(x_i)$ given on the zeroth time line $(j = 0)$. This kind of numerical procedure is called an **explicit finite difference method**.

EXAMPLE 1 **Using the Finite Difference Method**

Consider the boundary-value problem

$$\frac{\partial^2 u}{\partial x^2} = \frac{\partial u}{\partial t}, \quad 0 < x < 1, \quad 0 < t < 0.5$$

$$u(0, t) = 0, \quad u(1, t) = 0, \quad 0 \leq t \leq 0.5$$

$$u(x, 0) = \sin \pi x, \quad 0 \leq x \leq 1.$$

First we identify $c = 1$, $a = 1$, and $T = 0.5$. If we choose, say, $n = 5$ and $m = 50$, then $h = \frac{1}{5} = 0.2$, $k = \frac{0.5}{50} = 0.01$, $\lambda = 0.25$,

$$x_i = i\frac{1}{5}, \quad i = 0, 1, 2, 3, 4, 5, \quad \text{and} \quad t_j = j\frac{1}{100}, \quad j = 0, 1, 2, \ldots, 50.$$

Thus (3) becomes

$$u_{i, j+1} = 0.25(u_{i+1, j} + 2u_{ij} + u_{i-1, j}).$$

By setting $j = 0$ in this formula, we get a formula for the approximations to the temperature u on the first time line:

$$u_{i, 1} = 0.25(u_{i+1, 0} + 2u_{i, 0} + u_{i-1, 0}).$$

If we then let $i = 1, \ldots,$ 4 in the last equation, we obtain, in turn,

$$u_{11} = 0.25(u_{20} + 2u_{10} + u_{00})$$

$$u_{21} = 0.25(u_{30} + 2u_{20} + u_{10})$$

$$u_{31} = 0.25(u_{40} + 2u_{30} + u_{20})$$

$$u_{41} = 0.25(u_{50} + 2u_{40} + u_{30}).$$

The first equation in this list is interpreted as

$$u_{11} = 0.25(u(x_2, 0) + 2u(x_1, 0) + u(0, 0))$$

$$= 0.25(u(0.4, 0) + 2u(0.2, 0) + u(0, 0)).$$

From the initial condition $u(x, 0) = \sin \pi x$, the last line becomes

$$u_{11} = 0.25(0.951056516 + 2(0.587785252) + 0) = 0.531656755.$$

This number represents an approximation to the temperature $u(0.2, 0.01)$.

Since it would require a rather large table of over 200 entries to summarize all the approximations over the rectangular grid determined by h and k, we give only selected values in Table 16.2.1.

TABLE 16.2.1 Explicit Difference Equation Approximation with $h = 0.2$, $k = 0.01$, $\lambda = 0.25$

Time	$x = 0.20$	$x = 0.40$	$x = 0.60$	$x = 0.80$
0.00	0.5878	0.9511	0.9511	0.5878
0.10	0.2154	0.3486	0.3486	0.2154
0.20	0.0790	0.1278	0.1278	0.0790
0.30	0.0289	0.0468	0.0468	0.0289
0.40	0.0106	0.0172	0.0172	0.0106
0.50	0.0039	0.0063	0.0063	0.0039

TABLE 16.2.2

Exact	Approximation
$u(0.4, 0.05) = 0.5806$	$u_{25} = 0.5758$
$u(0.6, 0.06) = 0.5261$	$u_{36} = 0.5208$
$u(0.2, 0.10) = 0.2191$	$u_{1, 10} = 0.2154$
$u(0.8, 0.14) = 0.1476$	$u_{4, 14} = 0.1442$

You should verify, using the methods of Chapter 13, that an exact solution of the boundary-value problem in Example 1 is given by $u(x, t) = e^{-\pi^2 t} \sin \pi x$. Using this solution, we compare in Table 16.2.2 a sample of exact values with their corresponding approximations.

Stability These approximations are comparable to the exact values and accurate enough for some purposes. But there is a problem with the foregoing method. Recall that a numerical method is **unstable** if round-off errors or any other errors grow too rapidly as the computations proceed. The numerical procedure in Example 1 can exhibit this kind of behavior. It can be proved that the procedure is stable if λ is less than or equal to 0.5, but unstable otherwise. To obtain $\lambda = 0.25 \leq 0.5$ in Example 1, we had to choose the value $k = 0.01$; the necessity of using very small step sizes in the time direction is the principal fault of this method. You are urged to work Problem 12 in Exercises 16.2 and witness the predictable instability when $\lambda = 1$.

Crank–Nicholson Method There are **implicit finite difference methods** for solving parabolic partial differential equations. These methods require that we solve a system of equations to determine the approximate values of u on the $(j + 1)$st time line. However, implicit methods do not suffer from instability problems.

The finite difference algorithm introduced by the English mathematicians **John Crank** (1916–2006) and **Phyllis Nicolson** (1917–1968) in 1947 is used mostly for solving the heat equation or similar parabolic PDEs. The algorithm consists of replacing the second partial derivative in $c\dfrac{\partial^2 u}{\partial x^2} = \dfrac{\partial u}{\partial t}$ by an average of two central difference quotients, one evaluated at t and the other at $t + k$:

$$\frac{c}{2}\left[\frac{u(x + h, t) - 2u(x, t) + u(x - h, t)}{h^2} + \frac{u(x + h, t + k) - 2u(x, t + k) + u(x - h, t + k)}{h^2}\right] = \frac{1}{k}[u(x, t + k) - u(x, t)]. \quad (4)$$

If we again define $\lambda = ck/h^2$, then after rearranging terms we can write (4) as

$$-u_{i-1, j+1} + \alpha u_{i, j+1} - u_{i+1, j+1} = u_{i+1, j} - \beta u_{ij} + u_{i-1, j}, \quad (5)$$

where $\alpha = 2(1 + 1/\lambda)$ and $\beta = 2(1 - 1/\lambda)$, $j = 0, 1, \ldots, m - 1$, and $i = 1, 2, \ldots, n - 1$.

For each choice of j, the difference equation (5) for $i = 1, 2, \ldots, n - 1$ gives $n - 1$ equations in $n - 1$ unknowns $u_{i, j+1}$. Because of the prescribed boundary conditions, the values of $u_{i, j+1}$ are known for $i = 0$ and for $i = n$. For example, in the case $n = 4$, the system of equations for determining the approximate values of u on the $(j + 1)$st time line is

$$-u_{0, j+1} + \alpha u_{1, j+1} - u_{2, j+1} = u_{2, j} - \beta u_{1, j} + u_{0, j}$$
$$-u_{1, j+1} + \alpha u_{2, j+1} - u_{3, j+1} = u_{3, j} - \beta u_{2, j} + u_{1, j}$$
$$-u_{2, j+1} + \alpha u_{3, j+1} - u_{4, j+1} = u_{4, j} - \beta u_{3, j} + u_{2, j}$$

or

$$\alpha u_{1, j+1} - u_{2, j+1} \qquad\qquad = b_1$$
$$-u_{1, j+1} + \alpha u_{2, j+1} - u_{3, j+1} = b_2 \qquad (6)$$
$$-u_{2, j+1} + \alpha u_{3, j+1} = b_3,$$

where
$$b_1 = u_{2,j} - \beta u_{1,j} + u_{0,j} + u_{0,j+1}$$
$$b_2 = u_{3,j} - \beta u_{2,j} + u_{1,j}$$
$$b_3 = u_{4,j} - \beta u_{3,j} + u_{2,j} + u_{4,j+1}.$$

In general, if we use the difference equation (5) to determine values of u on the $(j+1)$st time line, we need to solve a linear system $\mathbf{AX} = \mathbf{B}$, where the coefficient matrix $\mathbf{A}$ is a **tridiagonal matrix**,

$$\mathbf{A} = \begin{pmatrix} \alpha & -1 & 0 & 0 & 0 & \cdots & & 0 \\ -1 & \alpha & -1 & 0 & 0 & & & 0 \\ 0 & -1 & \alpha & -1 & 0 & & & 0 \\ 0 & 0 & -1 & \alpha & -1 & & & 0 \\ \vdots & & & & & \ddots & & \vdots \\ 0 & 0 & 0 & 0 & 0 & & \alpha & -1 \\ 0 & 0 & 0 & 0 & 0 & \cdots & -1 & \alpha \end{pmatrix},$$

and the entries of the column matrix $\mathbf{B}$ are

$$b_1 = u_{2,j} - \beta u_{1,j} + u_{0,j} + u_{0,j+1}$$
$$b_2 = u_{3,j} - \beta u_{2,j} + u_{1,j}$$
$$b_3 = u_{4,j} - \beta u_{3,j} + u_{2,j}$$
$$\vdots$$
$$b_{n-1} = u_{n,j} - \beta u_{n-1,j} + u_{n-2,j} + u_{n,j+1}.$$

EXAMPLE 2 · Using the Crank–Nicholson Method

Use the Crank–Nicholson method to approximate the solution of the boundary-value problem

$$0.25 \frac{\partial^2 u}{\partial x^2} = \frac{\partial u}{\partial t}, \quad 0 < x < 2, \quad 0 < t < 0.3$$

$$u(0,t) = 0, \quad u(2,t) = 0, \quad 0 \le t \le 0.3$$

$$u(x,0) = \sin \pi x, \quad 0 \le x \le 2,$$

using $n = 8$ and $m = 30$.

SOLUTION From the identifications $a = 2$, $T = 0.3$, $h = \frac{1}{4} = 0.25$, $k = \frac{1}{100} = 0.01$, and $c = 0.25$, we get $\lambda = 0.04$. With the aid of a computer we get the results in Table 16.2.3. As in Example 1, the entries in this table represent only a selected number from the 210 approximations over the rectangular grid determined by h and k.

TABLE 16.2.3 Crank–Nicholson Method with $h = 0.25$, $k = 0.01$, $\lambda = 0.25$

Time	$x = 0.25$	$x = 0.50$	$x = 0.75$	$x = 1.00$	$x = 1.25$	$x = 1.50$	$x = 1.75$
0.00	0.7071	1.0000	0.7071	0.0000	−0.7071	−1.0000	−0.7071
0.05	0.6289	0.8894	0.6289	0.0000	−0.6289	−0.8894	−0.6289
0.10	0.5594	0.7911	0.5594	0.0000	−0.5594	−0.7911	−0.5594
0.15	0.4975	0.7036	0.4975	0.0000	−0.4975	−0.7036	−0.4975
0.20	0.4425	0.6258	0.4425	0.0000	−0.4425	−0.6258	−0.4425
0.25	0.3936	0.5567	0.3936	0.0000	−0.3936	−0.5567	−0.3936
0.30	0.3501	0.4951	0.3501	0.0000	−0.3501	−0.4951	−0.3501

TABLE 16.2.4

Exact	Approx.
$u(0.75, 0.05) = 0.6250$	$u_{35} = 0.6289$
$u(0.50, 0.20) = 0.6105$	$u_{2,20} = 0.6259$
$u(0.25, 0.10) = 0.5525$	$u_{1,10} = 0.5594$

Like Example 1, the boundary-value problem in Example 2 also possesses an exact solution given by $u(x,t) = e^{-\pi^2 t/4} \sin \pi x$. The sample comparisons listed in Table 16.2.4 show that the absolute errors are of the order 10^{-2} or 10^{-3}. Smaller errors can be obtained by decreasing either h or k.

In Problems 1–12, use a computer as a computational aid.

1. Use the difference equation (3) to approximate the solution of the boundary-value problem

$$\frac{\partial^2 u}{\partial x^2} = \frac{\partial u}{\partial t}, \quad 0 < x < 2, \quad 0 < t < 1$$

$$u(0, t) = 0, \quad u(2, t) = 0, \quad 0 \le t \le 1$$

$$u(x, 0) = \begin{cases} 1, & 0 \le x \le 1 \\ 0, & 1 < x \le 2. \end{cases}$$

Use $n = 8$ and $m = 40$.

2. Using the Fourier series solution obtained in Problem 3 of Exercises 13.3 with $L = 2$, one can sum the first 20 terms to estimate the values for $u(0.25, 0.1)$, $u(1, 0.5)$, and $u(1.5, 0.8)$ for the solution $u(x, t)$ of Problem 1 above. A student wrote a computer program to do this and obtained the results $u(0.25, 0.1) = 0.3794$, $u(1, 0.5) = 0.1854$, and $u(1.5, 0.8) = 0.0623$. Assume these results are accurate for all digits given. Compare these values with the approximations obtained in Problem 1 above. Find the absolute errors in each case.

3. Solve Problem 1 by the Crank–Nicholson method with $n = 8$ and $m = 40$. Use the values for $u(0.25, 0.1)$, $u(1, 0.5)$, and $u(1.5, 0.8)$ given in Problem 2 to compute the absolute errors.

4. Repeat Problem 1 using $n = 8$ and $m = 20$. Use the values for $u(0.25, 0.1)$, $u(1, 0.5)$, and $u(1.5, 0.8)$ given in Problem 2 to compute the absolute errors. Why are the approximations so inaccurate in this case?

5. Solve Problem 1 by the Crank–Nicholson method with $n = 8$ and $m = 20$. Use the values for $u(0.25, 0.1)$, $u(1, 0.5)$, and $u(1.5, 0.8)$ given in Problem 2 to compute the absolute errors. Compare the absolute errors with those obtained in Problem 4.

6. It was shown in Section 13.2 that if a rod of length L is made of a material with thermal conductivity K, specific heat γ, and density ρ, the temperature $u(x, t)$ satisfies the partial differential equation

$$\frac{K}{\gamma\rho}\frac{\partial^2 u}{\partial x^2} = \frac{\partial u}{\partial t}, \quad 0 < x < L.$$

Consider the boundary-value problem consisting of the foregoing equation and conditions

$$u(0, t) = 0, \quad u(L, t) = 0, \quad 0 \le t \le 10$$

$$u(x, 0) = f(x), \quad 0 \le x \le L.$$

Use the difference equation (3) in this section with $n = 10$ and $m = 10$ to approximate the solution of the boundary-value problem when

(a) $L = 20, K = 0.15, \rho = 8.0, \gamma = 0.11, f(x) = 30$
(b) $L = 50, K = 0.15, \rho = 8.0, \gamma = 0.11, f(x) = 30$
(c) $L = 20, K = 1.10, \rho = 2.7, \gamma = 0.22, f(x) = 0.5x(20 - x)$
(d) $L = 100, K = 1.04, \rho = 10.6, \gamma = 0.06,$

$$f(x) = \begin{cases} 0.8x, & 0 \le x \le 50 \\ 0.8(100 - x), & 50 < x \le 100. \end{cases}$$

7. Solve Problem 6 by the Crank–Nicholson method with $n = 10$ and $m = 10$.

8. Repeat Problem 6 if the endpoint temperatures are $u(0, t) = 0, u(L, t) = 20, 0 \le t \le 10$.

9. Solve Problem 8 by the Crank–Nicholson method.

10. Consider the boundary-value problem in Example 2. Assume that $n = 4$.
(a) Find the new value of λ.
(b) Use the Crank–Nicholson difference equation (5) to find the system of equations for u_{11}, u_{21}, and u_{31}, that is, the approximate values of u on the first time line. [*Hint*: Set $j = 0$ in (5), and let i take on the values 1, 2, 3.]
(c) Solve the system of three equations without the aid of a computer program. Compare your results with the corresponding entries in Table 16.2.3.

11. Consider a rod whose length is $L = 20$ for which $K = 1.05$, $\rho = 10.6$, and $\gamma = 0.056$. Suppose

$$u(0, t) = 20, \quad u(20, t) = 30$$

$$u(x, 0) = 50.$$

(a) Use the method outlined in Section 13.6 to find the steady-state solution $\psi(x)$.
(b) Use the Crank–Nicholson method to approximate the temperatures $u(x, t)$ for $0 \le t \le T_{max}$. Select T_{max} large enough to allow the temperatures to approach the steady-state values. Compare the approximations for $t = T_{max}$ with the values of $\psi(x)$ found in part (a).

12. Use the difference equation (3) to approximate the solution of the boundary-value problem

$$\frac{\partial^2 u}{\partial x^2} = \frac{\partial u}{\partial t}, \quad 0 < x < 1, \quad 0 < t < 1$$

$$u(0, t) = 0, \quad u(1, t) = 0, \quad 0 \le t \le 1$$

$$u(x, 0) = \sin \pi x, \quad 0 \le x \le 1.$$

Use $n = 5$ and $m = 25$.

16.3 Wave Equation

INTRODUCTION In this section we approximate a solution of the one-dimensional wave equation using the finite difference method used in the preceding two sections. The one-dimensional wave equation is the prototype hyperbolic partial differential equation.

■ **Difference Equation Replacement** Suppose $u(x, t)$ represents a solution of the one-dimensional wave equation

$$c^2 \frac{\partial^2 u}{\partial x^2} = \frac{\partial^2 u}{\partial t^2}. \tag{1}$$

Using two central differences,

$$\frac{\partial^2 u}{\partial x^2} \approx \frac{1}{h^2} [u(x + h, t) - 2u(x, t) + u(x - h, t)]$$

$$\frac{\partial^2 u}{\partial t^2} \approx \frac{1}{k^2} [u(x, t + k) - 2u(x, t) + u(x, t - k)],$$

we replace equation (1) by

$$\frac{c^2}{h^2} [u(x + h, t) - 2u(x, t) + u(x - h, t)] = \frac{1}{k^2} [u(x, t + k) - 2u(x, t) + u(x, t - k)]. \tag{2}$$

We solve (2) for $u(x, t + k)$, which is $u_{i, j+1}$. If $\lambda = ck/h$, then (2) yields

$$u_{i, j+1} = \lambda^2 u_{i+1, j} + 2(1 - \lambda^2)u_{ij} + \lambda^2 u_{i-1, j} - u_{i, j-1} \tag{3}$$

for $i = 1, 2, \ldots, n - 1$ and $j = 1, 2, \ldots, m - 1$.

In the case when the wave equation (1) is a model for the vertical displacements $u(x, t)$ of a vibrating string, typical boundary conditions are $u(0, t) = 0$, $u(a, t) = 0$, $t > 0$, and initial conditions are $u(x, 0) = f(x)$, $\partial u/\partial t|_{t=0} = g(x)$, $0 < x < a$. The functions f and g can be interpreted as the initial position and the initial velocity of the string. The numerical method based on equation (3), like the first method considered in Section 16.2, is an explicit finite difference method. As before, we use the difference equation (3) to approximate the solution $u(x, t)$ of (1), using the boundary and initial conditions, over a rectangular region in the xt-plane defined by the inequalities $0 \le x \le a$, $0 \le t \le T$, where T is some specified value of time. If n and m are positive integers and

$$h = \frac{a}{n} \quad \text{and} \quad k = \frac{T}{m},$$

the vertical and horizontal grid lines on this region are defined by

$$x_i = ih, \quad i = 0, 1, 2, \ldots, n \quad \text{and} \quad t_j = jk, \quad j = 0, 1, 2, \ldots, m.$$

$(j + 1)$st time line

jth time line

$(j - 1)$st time line

FIGURE 16.3.1 u at $t = j + 1$ is determined from three values of u at $t = j$ and one value at $t = j - 1$

As shown in **FIGURE 16.3.1**, (3) enables us to obtain the approximation $u_{i, j+1}$ on the $(j + 1)$st time line from the values indicated on the jth and $(j - 1)$st time lines. Moreover, we use

$$u_{0, j} = u(0, jk) = 0, \quad u_{n, j} = u(a, jk) = 0 \qquad \leftarrow \text{boundary conditions}$$

and

$$u_{i, 0} = u(x_i, 0) = f(x_i). \qquad \leftarrow \text{initial condition}$$

There is one minor problem in getting started. You can see from (3) that for $j = 1$ we need to know the values of $u_{i, 1}$ (that is, the estimates of u on the first time line) in order to find $u_{i, 2}$. But from Figure 16.3.1, with $j = 0$, we see that the values of $u_{i, 1}$ on the first time line depend on the values of $u_{i, 0}$ on the zeroth time line and on the values of $u_{i, -1}$. To compute these latter values we make use of the initial-velocity condition $u_t(x, 0) = g(x)$. At $t = 0$, it follows from (5) of Section 6.5 that

$$g(x_i) = u_t(x_i, 0) \approx \frac{u(x_i, k) - u(x_i, -k)}{2k}. \tag{4}$$

In order to make sense of the term $u(x_i, -k) = u_{i, -1}$ in (4), we have to imagine $u(x, t)$ extended backward in time. It follows from (4) that

$$u(x_i, -k) \approx u(x_i, k) - 2kg(x_i).$$

This last result suggests that we define

$$u_{i, -1} = u_{i, 1} - 2kg(x_i) \tag{5}$$

in the iteration of (3). By substituting (5) into (3) when $j = 0$, we get the special case

$$u_{i, 1} = \frac{\lambda^2}{2}(u_{i+1, 0} + u_{i-1, 0}) + (1 - \lambda^2)u_{i, 0} + kg(x_i). \tag{6}$$

EXAMPLE 1 **Using the Finite Difference Method**

Approximate the solution of the boundary-value problem

$$4\frac{\partial^2 u}{\partial x^2} = \frac{\partial^2 u}{\partial t^2}, \quad 0 < x < 1, \quad 0 < t < 1$$

$$u(0, t) = 0, \quad u(1, t) = 0, \quad 0 \le t \le 1$$

$$u(x, 0) = \sin \pi x, \quad \left.\frac{\partial u}{\partial t}\right|_{t=0} = 0, \quad 0 \le x \le 1,$$

using (3) with $n = 5$ and $m = 20$.

SOLUTION We make the identifications $c = 2$, $a = 1$, and $T = 1$. With $n = 5$ and $m = 20$ we get $h = \frac{1}{5} = 0.2$, $k = \frac{1}{20} = 0.05$, and $\lambda = 0.5$. Thus, with $g(x) = 0$, equations (6) and (3) become, respectively,

$$u_{i, 1} = 0.125(u_{i+1, 0} + u_{i-1, 0}) + 0.75u_{i, 0} \tag{7}$$

$$u_{i, j+1} = 0.25u_{i+1, j} + 1.5u_{ij} + 0.25u_{i-1, j} - u_{i, j-1}. \tag{8}$$

For $i = 1, 2, 3, 4$, equation (7) yields the following values for the $u_{i, 1}$ on the first time line:

$$u_{11} = 0.125(u_{20} + u_{00}) + 0.75u_{10} = 0.55972100$$

$$u_{21} = 0.125(u_{30} + u_{10}) + 0.75u_{20} = 0.90564761$$

$$u_{31} = 0.125(u_{40} + u_{20}) + 0.75u_{30} = 0.90564761 \tag{9}$$

$$u_{41} = 0.125(u_{50} + u_{30}) + 0.75u_{40} = 0.55972100.$$

Note that the results given in (9) were obtained from the initial condition $u(x, 0) = \sin \pi x$. For example, $u_{20} = \sin(0.2\pi)$, and so on. Now $j = 1$ in (8) gives

$$u_{i, 2} = 0.25u_{i+1, 1} + 1.5u_{i, 1} + 0.25u_{i-1, 1} - u_{i, 0},$$

and so for $i = 1, 2, 3, 4$ we get

$$u_{12} = 0.25u_{21} + 1.5u_{11} + 0.25u_{01} - u_{10}$$

$$u_{22} = 0.25u_{31} + 1.5u_{21} + 0.25u_{11} - u_{20}$$

$$u_{32} = 0.25u_{41} + 1.5u_{31} + 0.25u_{21} - u_{30}$$

$$u_{42} = 0.25u_{51} + 1.5u_{41} + 0.25u_{31} - u_{40}.$$

Using the boundary conditions, the initial conditions, and the data obtained in (9), we get from these equations the approximations for u on the second time line. These last results and an abbreviation of the remaining calculations are summarized in Table 16.3.1 on page 827.

TABLE 16.3.1	Explicit Difference Equation Approximation with $h = 0.2$, $k = 0.05$, $\lambda = 0.5$			
Time	$x = 0.20$	$x = 0.40$	$x = 0.60$	$x = 0.80$
0.00	0.5878	0.9511	0.9511	0.5878
0.10	0.4782	0.7738	0.7738	0.4782
0.20	0.1903	0.3080	0.3080	0.1903
0.30	−0.1685	−0.2727	−0.2727	−0.1685
0.40	−0.4645	−0.7516	−0.7516	−0.4645
0.50	−0.5873	−0.9503	−0.9503	−0.5873
0.60	−0.4912	−0.7947	−0.7947	−0.4912
0.70	−0.2119	−0.3428	−0.3428	−0.2119
0.80	0.1464	0.2369	0.2369	0.1464
0.90	0.4501	0.7283	0.7283	0.4501
1.00	0.5860	0.9482	0.9482	0.5860

TABLE 16.3.2	
Exact	Approx.
$u(0.4, 0.25) = 0$	$u_{25} = 0.0185$
$u(0.6, 0.3) = -0.2939$	$u_{36} = -0.2727$
$u(0.2, 0.5) = -0.5878$	$u_{1, 10} = -0.5873$
$u(0.8, 0.7) = -0.1816$	$u_{4, 14} = -0.2119$

TABLE 16.3.3	
Exact	Approx.
$u(0.25, 0.3125) = -0.2706$	$u_{25} = -0.2706$
$u(0.375, 0.375) = -0.6533$	$u_{36} = -0.6533$
$u(0.125, 0.625) = -0.2706$	$u_{1, 10} = -0.2706$

It is readily verified that the exact solution of the BVP in Example 1 is $u(x, t) = \sin \pi x \cos 2\pi t$. With this function we can compare the exact results with the approximations. For example, some selected comparisons are given in Table 16.3.2. As you can see in the table, the approximations are in the same "ball park" as the exact values, but the accuracy is not particularly impressive. We can, however, obtain more accurate results. The accuracy of this algorithm varies with the choice of λ. Of course, λ is determined by the choice of the integers n and m, which in turn determine the values of the step sizes h and k. It can be proved that the best accuracy is always obtained from this method when the ratio $\lambda = kc/h$ is equal to one—in other words, when the step in the time direction is $k = h/c$. For example, the choice $n = 8$ and $m = 16$ yields $h = \frac{1}{8}$, $k = \frac{1}{16}$, and $\lambda = 1$. The sample values listed in Table 16.3.3 clearly show the improved accuracy.

■ **Stability** We note in conclusion that this explicit finite difference method for the wave equation is stable when $\lambda \leq 1$ and unstable when $\lambda > 1$.

16.3 Exercises
Answers to selected odd-numbered problems begin on page ANS-43.

In Problems 1, 3, 5, and 6, use a computer as a computational aid.

1. Use the difference equation (3) to approximate the solution of the boundary-value problem

$$c^2 \frac{\partial^2 u}{\partial x^2} = \frac{\partial^2 u}{\partial t^2}, \quad 0 < x < a, \quad 0 < t < T$$

$$u(0, t) = 0, \quad u(a, t) = 0, \quad 0 \leq t \leq T$$

$$u(x, 0) = f(x), \quad \left.\frac{\partial u}{\partial t}\right|_{t=0} = 0, \quad 0 \leq x \leq a,$$

when
 (a) $c = 1$, $a = 1$, $T = 1$, $f(x) = x(1 - x)$; $n = 4$ and $m = 10$
 (b) $c = 1$, $a = 2$, $T = 1$, $f(x) = e^{-16(x-1)^2}$; $n = 5$ and $m = 10$
 (c) $c = \sqrt{2}$, $a = 1$, $T = 1$,

$$f(x) = \begin{cases} 0, & 0 \leq x \leq 0.5 \\ 0.5, & 0.5 < x \leq 1; \end{cases}$$

 $n = 10$ and $m = 25$.

2. Consider the boundary-value problem

$$\frac{\partial^2 u}{\partial x^2} = \frac{\partial^2 u}{\partial t^2}, \quad 0 < x < 1, \quad 0 < t < 0.5$$

$$u(0, t) = 0, \quad u(1, t) = 0, \quad 0 \leq t \leq 0.5$$

$$u(x, 0) = \sin \pi x, \quad \left.\frac{\partial u}{\partial t}\right|_{t=0} = 0, \quad 0 \leq x \leq 1.$$

 (a) Use the methods of Chapter 13 to verify that the solution of the problem is $u(x, t) = \sin \pi x \cos \pi t$.
 (b) Use the method of this section to approximate the solution of the problem without the aid of a computer program. Use $n = 4$ and $m = 5$.
 (c) Compute the absolute error at each interior grid point.

3. Approximate the solution of the boundary-value problem in Problem 2 using a computer program with
 (a) $n = 5$, $m = 10$ (b) $n = 5$, $m = 20$.

4. Given the boundary-value problem

$$\frac{\partial^2 u}{\partial x^2} = \frac{\partial^2 u}{\partial t^2}, \quad 0 < x < 1, \quad 0 < t < 1$$

$$u(0, t) = 0, \quad u(1, t) = 0, \quad 0 \le t \le 1$$

$$u(x, 0) = x(1 - x), \quad \left.\frac{\partial u}{\partial t}\right|_{t=0} = 0, \quad 0 \le x \le 1,$$

use $h = k = \frac{1}{5}$ in equation (6) to compute the values of $u_{i,1}$ by hand.

5. It was shown in Section 13.2 that the equation of a vibrating string is

$$\frac{T}{\rho}\frac{\partial^2 u}{\partial x^2} = \frac{\partial^2 u}{\partial t^2},$$

where T is the constant magnitude of the tension in the string and ρ is its mass per unit length. Suppose a string of length

60 centimeters is secured to the x-axis at its ends and is released from rest from the initial displacement

$$f(x) = \begin{cases} 0.01x, & 0 \le x \le 30 \\ 0.30 - \dfrac{x - 30}{100}, & 30 < x \le 60. \end{cases}$$

Use the difference equation (3) in this section to approximate the solution of the boundary-value problem when $h = 10$, $k = 5\sqrt{\rho/T}$ and where $\rho = 0.0225$ g/cm, $T = 1.4 \times 10^7$ dynes. Use $m = 50$.

6. Repeat Problem 5 using

$$f(x) = \begin{cases} 0.02x, & 0 \le x \le 15 \\ 0.30 - \dfrac{x - 15}{150}, & 15 < x \le 60 \end{cases}$$

and $h = 10, k = 2.5\sqrt{\rho/T}$. Use $m = 50$.

16 Chapter in Review Answers to selected odd-numbered problems begin on page ANS-44.

1. Consider the boundary-value problem

$$\frac{\partial^2 u}{\partial x^2} + \frac{\partial^2 u}{\partial y^2} = 0, \quad 0 < x < 2, \quad 0 < y < 1$$

$$u(0, y) = 0, \quad u(2, y) = 50, \quad 0 < y < 1$$

$$u(x, 0) = 0, \quad u(x, 1) = 0, \quad 0 < x < 2.$$

Approximate the solution of the differential equation at the interior points of the region with mesh size $h = \frac{1}{2}$. Use Gaussian elimination or Gauss–Seidel iteration.

2. Solve Problem 1 using mesh size $h = \frac{1}{4}$. Use Gauss–Seidel iteration.

3. Consider the boundary-value problem

$$\frac{\partial^2 u}{\partial x^2} = \frac{\partial u}{\partial t}, \quad 0 < x < 1, \quad 0 < t < 0.05$$

$$u(0, t) = 0, \quad u(1, t) = 0, \quad t > 0$$

$$u(x, 0) = x, \quad 0 < x < 1.$$

(a) Note that the initial temperature $u(x, 0) = x$ indicates that the temperature at the right boundary $x = 1$ should be $u(1, 0) = 1$, whereas the boundary conditions imply that $u(1, 0) = 0$. Write a computer problem for the explicit finite difference method so that the boundary conditions prevail for all times considered, including $t = 0$. Use the program to complete Table 16.R.1.

(b) Modify your computer program so that the initial condition prevails at the boundaries at $t = 0$. Use this program to complete Table 16.R.2.

(c) Are Tables 16.R.1 and 16.R.2 related in any way? Use a larger time interval if necessary.

TABLE 16.R.1

Time	$x = 0.00$	$x = 0.20$	$x = 0.40$	$x = 0.60$	$x = 0.80$	$x = 1.00$
0.00	0.0000	0.2000	0.4000	0.6000	0.8000	0.0000
0.01	0.0000					0.0000
0.02	0.0000					0.0000
0.03	0.0000					0.0000
0.04	0.0000					0.0000
0.05	0.0000					0.0000

TABLE 16.R.2

Time	$x = 0.00$	$x = 0.20$	$x = 0.40$	$x = 0.60$	$x = 0.80$	$x = 1.00$
0.00	0.0000	0.2000	0.4000	0.6000	0.8000	1.0000
0.01	0.0000					0.0000
0.02	0.0000					0.0000
0.03	0.0000					0.0000
0.04	0.0000					0.0000
0.05	0.0000					0.0000

Complex Analysis

CHAPTER

17

Functions of a Complex Variable

In elementary courses you learned about the existence and properties of complex numbers. But in courses such as calculus, it is likely that you did not even see a complex number. So, in the next four chapters we are going to introduce you to **complex analysis**; that is, the study functions of a complex variable.

17.1 Complex Numbers

INTRODUCTION You have undoubtedly encountered complex numbers in your earlier courses in mathematics. When you first learned to solve a quadratic equation $ax^2 + bx + c = 0$ by the quadratic formula, you saw that the roots of the equation are not real, that is, complex, whenever the discriminant $b^2 - 4ac$ is negative. So, for example, simple equations such as $x^2 + 5 = 0$ and $x^2 + x + 1 = 0$ have no real solutions. For example, the roots of the last equation are $-\frac{1}{2} + \frac{\sqrt{-3}}{2}$ and $-\frac{1}{2} - \frac{\sqrt{-3}}{2}$. If it is assumed that $\sqrt{-3} = \sqrt{3}\sqrt{-1}$, then the roots are written $-\frac{1}{2} + \frac{\sqrt{3}}{2}\sqrt{-1}$ and $-\frac{1}{2} - \frac{\sqrt{3}}{2}\sqrt{-1}$.

■ **A Definition** Two hundred years ago, around the time that complex numbers were gaining some respectability in the mathematical community, the symbol i was originally used as a disguise for the embarrassing symbol $\sqrt{-1}$. We now simply say that i is the **imaginary unit** and define it by the property $i^2 = -1$. Using the imaginary unit, we build a general complex number out of two real numbers.

DEFINITION 17.1.1 Complex Number

A **complex number** is any number of the form $z = a + ib$, where a and b are real numbers and i is the imaginary unit.

The notations $a + ib$ and $a + bi$ are used interchangeably.

■ **Terminology** The number i in Definition 17.1.1 is called the **imaginary unit**. The real number x in $z = x + iy$ is called the **real part** of z; the real number y is called the **imaginary part** of z. The real and imaginary parts of a complex number z are abbreviated $\text{Re}(z)$ and $\text{Im}(z)$, respectively. For example, if $z = 4 - 9i$, then $\text{Re}(z) = 4$ and $\text{Im}(z) = -9$. A real constant multiple of the imaginary unit is called a **pure imaginary number**. For example, $z = 6i$ is a pure imaginary number. Two complex numbers are **equal** if their real and imaginary parts are equal. Since this simple concept is sometimes useful, we formalize the last statement in the next definition.

Note: The imaginary part of $z = 4 - 9i$ is -9, not $-9i$. ▶

DEFINITION 17.1.2 Equality

Complex numbers $z_1 = x_1 + iy_1$ and $z_2 = x_2 + iy_2$ are **equal**, $z_1 = z_2$, if

$$\text{Re}(z_1) = \text{Re}(z_2) \quad \text{and} \quad \text{Im}(z_1) = \text{Im}(z_2).$$

A complex number $x + iy = 0$ if $x = 0$ and $y = 0$.

■ **Arithmetic Operations** Complex numbers can be added, subtracted, multiplied, and divided. If $z_1 = x_1 + iy_1$ and $z_2 = x_2 + iy_2$, these operations are defined as follows.

Addition: $\quad z_1 + z_2 = (x_1 + iy_1) + (x_2 + iy_2) = (x_1 + x_2) + i(y_1 + y_2)$

Subtraction: $\quad z_1 - z_2 = (x_1 + iy_1) - (x_2 + iy_2) = (x_1 - x_2) + i(y_1 - y_2)$

Multiplication: $\quad z_1 \cdot z_2 = (x_1 + iy_1)(x_2 + iy_2)$

$$= x_1 x_2 - y_1 y_2 + i(y_1 x_2 + x_1 y_2)$$

Division: $\quad \dfrac{z_1}{z_2} = \dfrac{x_1 + iy_1}{x_2 + iy_2}$

$$= \frac{x_1 x_2 + y_1 y_2}{x_2^2 + y_2^2} + i\,\frac{y_1 x_2 - x_1 y_2}{x_2^2 + y_2^2}$$

The familiar commutative, associative, and distributive laws hold for complex numbers.

Commutative laws: $\begin{cases} z_1 + z_2 = z_2 + z_1 \\ \quad z_1 z_2 = z_2 z_1 \end{cases}$

Associative laws: $\begin{cases} z_1 + (z_2 + z_3) = (z_1 + z_2) + z_3 \\ \quad z_1(z_2 z_3) = (z_1 z_2) z_3 \end{cases}$

Distributive law: $\quad z_1(z_2 + z_3) = z_1 z_2 + z_1 z_3$

In view of these laws, there is no need to memorize the definitions of addition, subtraction, and multiplication. To add (subtract) two complex numbers, we simply add (subtract) the corresponding real and imaginary parts. To multiply two complex numbers, we use the distributive law and the fact that $i^2 = -1$.

EXAMPLE 1 **Addition and Multiplication**

If $z_1 = 2 + 4i$ and $z_2 = -3 + 8i$, find **(a)** $z_1 + z_2$ and **(b)** $z_1 z_2$.

SOLUTION **(a)** By adding the real and imaginary parts of the two numbers, we get

$$(2 + 4i) + (-3 + 8i) = (2 - 3) + (4 + 8)i = -1 + 12i.$$

(b) Using the distributive law, we have

$$(2 + 4i)(-3 + 8i) = (2 + 4i)(-3) + (2 + 4i)(8i)$$

$$= -6 - 12i + 16i + 32i^2$$

$$= (-6 - 32) + (16 - 12)i = -38 + 4i. \qquad \equiv$$

There is also no need to memorize the definition of division, but before discussing that we need to introduce another concept.

Conjugate If z is a complex number, then the number obtained by changing the sign of its imaginary part is called the **complex conjugate** or, simply, the **conjugate** of z. If $z = x + iy$, then its conjugate is

$$\bar{z} = x - iy.$$

For example, if $z = 6 + 3i$, then $\bar{z} = 6 - 3i$; if $z = -5 - i$, then $\bar{z} = -5 + i$. If z is a real number, say $z = 7$, then $\bar{z} = 7$. From the definition of addition it can be readily shown that the conjugate of a sum of two complex numbers is the sum of the conjugates:

$$\overline{z_1 + z_2} = \bar{z}_1 + \bar{z}_2.$$

Moreover, we have the additional three properties

$$\overline{z_1 - z_2} = \bar{z}_1 - \bar{z}_2, \quad \overline{z_1 z_2} = \bar{z}_1 \bar{z}_2, \quad \overline{\left(\frac{z_1}{z_2}\right)} = \frac{\bar{z}_1}{\bar{z}_2}.$$

The definitions of addition and multiplication show that the sum and product of a complex number z and its conjugate $\bar{z}$ are also real numbers:

$$z + \bar{z} = (x + iy) + (x - iy) = 2x \tag{1}$$

$$z\bar{z} = (x + iy)(x - iy) = x^2 - i^2 y^2 = x^2 + y^2. \tag{2}$$

The difference between a complex number z and its conjugate $\bar{z}$ is a pure imaginary number:

$$z - \bar{z} = (x + iy) - (x - iy) = 2iy. \tag{3}$$

Since $x = \text{Re}(z)$ and $y = \text{Im}(z)$, (1) and (3) yield two useful formulas:

$$\text{Re}(z) = \frac{z + \bar{z}}{2} \quad \text{and} \quad \text{Im}(z) = \frac{z - \bar{z}}{2i}.$$

However, (2) is the important relationship that enables us to approach division in a more practical manner: To divide z_1 by z_2, we multiply both numerator and denominator of z_1/z_2 by the conjugate of z_2. This procedure is illustrated in the next example.

EXAMPLE 2 **Division**

If $z_1 = 2 - 3i$ and $z_2 = 4 + 6i$, find **(a)** $\dfrac{z_1}{z_2}$ and **(b)** $\dfrac{1}{z_1}$.

SOLUTION In both parts of this example we shall multiply both numerator and denominator by the conjugate of the denominator and then use (2).

(a)
$$\frac{2-3i}{4+6i} = \frac{2-3i}{4+6i}\frac{4-6i}{4-6i} = \frac{8-12i-12i+18i^2}{16+36}$$

$$= \frac{-10-24i}{52} = -\frac{5}{26} - \frac{6}{13}i.$$

(b)
$$\frac{1}{2-3i} = \frac{1}{2-3i}\frac{2+3i}{2+3i} = \frac{2+3i}{4+9} = \frac{2}{13} + \frac{3}{13}i. \qquad \equiv$$

■ **Geometric Interpretation** A complex number $z = x + iy$ is uniquely determined by an *ordered pair* of real numbers (x, y). The first and second entries of the ordered pairs correspond, in turn, with the real and imaginary parts of the complex number. For example, the ordered pair $(2, -3)$ corresponds to the complex number $z = 2 - 3i$. Conversely, $z = 2 - 3i$ determines the ordered pair $(2, -3)$. In this manner we are able to associate a complex number $z = x + iy$ with a point (x, y) in a coordinate plane. But, as we saw in Section 7.1, an ordered pair of real numbers can be interpreted as the components of a vector. Thus, a complex number $z = x + iy$ can also be viewed as a *vector* whose initial point is the origin and whose terminal point is (x, y). The coordinate plane illustrated in **FIGURE 17.1.1** is called the **complex plane** or simply the **z-plane**. The horizontal or *x*-axis is called the **real axis** and the vertical or *y*-axis is called the **imaginary axis**.

The length of a vector z, or the distance from the origin to the point (x, y), is clearly $\sqrt{x^2 + y^2}$. This real number is given a special name.

FIGURE 17.1.1 z as a position vector

DEFINITION **17.1.3** **Modulus or Absolute Value**

The **modulus** or **absolute value** of $z = x + iy$, denoted by $|z|$, is the real number

$$|z| = \sqrt{x^2 + y^2} = \sqrt{z\bar{z}}. \qquad (4)$$

EXAMPLE 3 **Modulus of a Complex Number**

If $z = 2 - 3i$, then $|z| = \sqrt{2^2 + (-3)^2} = \sqrt{13}$. $\qquad \equiv$

As **FIGURE 17.1.2** shows, the sum of the vectors z_1 and z_2 is the vector $z_1 + z_2$. For the triangle given in the figure, we know that the length of the side of the triangle corresponding to the vector $z_1 + z_2$ cannot be longer than the sum of the remaining two sides. In symbols this is

$$|z_1 + z_2| \le |z_1| + |z_2|. \qquad (5)$$

The result in (5) is known as the **triangle inequality** and extends to any finite sum:

$$|z_1 + z_2 + z_3 + \cdots + z_n| \le |z_1| + |z_2| + |z_3| + \cdots + |z_n|. \qquad (6)$$

Using (5) on $z_1 + z_2 + (-z_2)$, we obtain another important inequality:

$$|z_1 + z_2| \ge |z_1| - |z_2|. \qquad (7)$$

FIGURE 17.1.2 Sum of vectors

REMARKS

Many of the properties of the real system hold in the complex number system, but there are some remarkable differences as well. For example, we cannot compare two complex numbers $z_1 = x_1 + iy_1$, $y_1 \neq 0$, and $z_2 = x_2 + iy_2$, $y_2 \neq 0$, by means of inequalities. In other words, statements such as $z_1 < z_2$ and $z_2 \geq z_1$ have no meaning except in the case when the two numbers z_1 and z_2 are real. We can, however, compare the absolute values of two complex numbers. Thus, if $z_1 = 3 + 4i$ and $z_2 = 5 - i$, then $|z_1| = 5$ and $|z_2| = \sqrt{26}$, and consequently $|z_1| < |z_2|$. This last inequality means that the point $(3, 4)$ is closer to the origin than is the point $(5, -1)$.

17.1 Exercises Answers to selected odd-numbered problems begin on page ANS-44.

In Problems 1–4, evaluate the given power of i.

1. i^8

2. i^{11}

3. i^{42}

4. i^{105}

In Problems 5–32, write the given number in the form $a + ib$.

5. $2i^3 - 3i^2 + 5i$

6. $3i^5 - i^4 + 7i^3 - 10i^2 - 9$

7. $4i^{11} - 5i^3 + 10i^{-2}$

8. $6i^{-4} + 3i^{15} - i^6 + 7i^2$

9. $(5 - 9i) + (2 - 4i)$

10. $3(4 - i) - 3(5 + 2i)$

11. $i(5 + 7i)$

12. $i(4 - i) + 4i(1 + 2i)$

13. $(2 - 3i)(4 + i)$

14. $(\frac{1}{2} - \frac{1}{4}i)(\frac{2}{3} + \frac{5}{3}i)$

15. $(2 + 3i)^2$

16. $(1 - i)^3$

17. $\dfrac{2}{i}$

18. $\dfrac{i}{1 + i}$

19. $\dfrac{2 - 4i}{3 + 5i}$

20. $\dfrac{10 - 5i}{6 + 2i}$

21. $\dfrac{(3 - i)(2 + 3i)}{1 + i}$

22. $\dfrac{(1 + i)(1 - 2i)}{(2 + i)(4 - 3i)}$

23. $\dfrac{(5 - 4i) - (3 + 7i)}{(4 + 2i) + (2 - 3i)}$

24. $\dfrac{(4 + 5i) + 2i^3}{(2 + i)^2}$

25. $i(1 - i)(2 - i)(2 + 6i)$

26. $(1 + i)^2(1 - i)^3$

27. $(3 + 6i) + (4 - i)(3 + 5i) + \dfrac{1}{2 - i}$

28. $(2 + 3i)\left(\dfrac{2 - i}{1 + 2i}\right)^2$

29. $\left(\dfrac{i}{3 - i}\right)\left(\dfrac{1}{2 + 3i}\right)$

30. $\dfrac{1}{(1 + i)(1 - 2i)(1 + 3i)}$

31. $\dfrac{1}{2 - i} + \dfrac{1}{1 + 2i}$

32. $\dfrac{i}{1 + 3i} - \dfrac{1 - i}{2 + i}$

In Problems 33–38, let $z = x + iy$. Find the indicated expression.

33. $\text{Re}(1/z)$

34. $\text{Re}(z^2)$

35. $\text{Im}(2z + 4\bar{z} - 4i)$

36. $\text{Im}(\bar{z}^2 + z^2)$

37. $|z - 1 - 3i|$

38. $|z + 5\bar{z}|$

In Problems 39–44, use Definition 17.1.2 to find a complex number z satisfying the given equation.

39. $2z = i(2 + 9i)$

40. $z - 2\bar{z} + 7 - 6i = 0$

41. $z^2 = i$

42. $\bar{z}^2 = 4z$

43. $z + 2\bar{z} = \dfrac{2 - i}{1 + 3i}$

44. $\dfrac{z}{1 + \bar{z}} = 3 + 4i$

In Problems 45 and 46, determine which complex number is closer to the origin.

45. $10 + 8i$, $11 - 6i$

46. $\frac{1}{2} - \frac{1}{4}i$, $\frac{2}{3} + \frac{1}{6}i$

47. Prove that $|z_1 - z_2|$ is the distance between the points z_1 and z_2 in the complex plane.

48. Show for all complex numbers z on the circle $x^2 + y^2 = 4$ that $|z + 6 + 8i| \leq 12$.

Discussion Problems

49. For n a nonnegative integer, i^n can be one of four values: $i, -1, -i$, and 1. In each of the following four cases express the integer exponent n in terms of the symbol k, where $k = 0, 1, 2, \ldots$.
(a) $i^n = i$ **(b)** $i^n = -1$ **(c)** $i^n = -i$ **(d)** $i^n = 1$

50. (a) Without doing any significant work, such as multiplying out or using the binomial theorem, think of an easy way of evaluating $(1 + i)^8$.
(b) Use your method in part (a) to evaluate $(1 + i)^{64}$.

17.2 Powers and Roots

INTRODUCTION Recall from calculus that a point (x, y) in rectangular coordinates can also be expressed in terms of polar coordinates (r, θ). We shall see in this section that the ability to express a complex number z in terms of r and θ greatly facilitates finding powers and roots of z.

Polar Form Rectangular coordinates (x, y) and polar coordinates (r, θ) are related by the equations $x = r \cos \theta$ and $y = r \sin \theta$ (see Section 14.1). Thus a nonzero complex number $z = x + iy$ can be written as $z = (r \cos \theta) + i(r \sin \theta)$ or

$$z = r(\cos \theta + i \sin \theta). \tag{1}$$

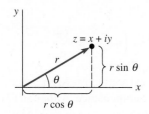

FIGURE 17.2.1 Polar coordinates

We say that (1) is the **polar form** of the complex number z. We see from **FIGURE 17.2.1** that the polar coordinate r can be interpreted as the distance from the origin to the point (x, y). In other words, we adopt the convention that r is never negative so that we can take r to be the modulus of z; that is, $r = |z|$. The angle θ of inclination of the vector z measured in radians from the positive real axis is positive when measured counterclockwise and negative when measured clockwise. The angle θ is called an **argument** of z and is written $\theta = \arg z$. From Figure 17.2.1 we see that an argument of a complex number must satisfy the equation $\tan \theta = y/x$. The solutions of this equation are not unique, since if θ_0 is an argument of z, then necessarily the angles $\theta_0 \pm 2\pi$, $\theta_0 \pm 4\pi, \ldots$, are also arguments. The argument of a complex number in the interval $-\pi < \theta \le \pi$ is called the **principal argument** of z and is denoted by Arg z. For example, $\text{Arg}(i) = \pi/2$.

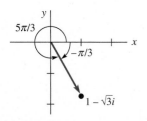

FIGURE 17.2.2 Two arguments of $z = 1 - \sqrt{3}i$ in Example 1

EXAMPLE 1 **A Complex Number in Polar Form**

Express $1 - \sqrt{3}i$ in polar form.

SOLUTION With $x = 1$ and $y = -\sqrt{3}$, we obtain $r = |z| = \sqrt{(1)^2 + (-\sqrt{3})^2} = 2$. Now since the point $(1, -\sqrt{3})$ lies in the fourth quadrant, we can take the solution of $\tan \theta = -\sqrt{3}/1 = -\sqrt{3}$ to be $\theta = \arg z = 5\pi/3$. It follows from (1) that a polar form of the number is

$$z = 2\left(\cos \frac{5\pi}{3} + i\sin \frac{5\pi}{3}\right).$$

As we see in **FIGURE 17.2.2**, the argument of $1 - \sqrt{3}i$ that lies in the interval $(-\pi, \pi]$, the principal argument of z, is $\text{Arg } z = -\pi/3$. Thus, an alternative polar form of the complex number is

$$z = 2\left[\cos\left(-\frac{\pi}{3}\right) + i\sin\left(-\frac{\pi}{3}\right)\right]. \qquad \equiv$$

■ **Multiplication and Division** The polar form of a complex number is especially convenient to use when multiplying or dividing two complex numbers. Suppose

$$z_1 = r_1(\cos \theta_1 + i \sin \theta_1) \quad \text{and} \quad z_2 = r_2(\cos \theta_2 + i \sin \theta_2),$$

where θ_1 and θ_2 are any arguments of z_1 and z_2, respectively. Then

$$z_1 z_2 = r_1 r_2[(\cos \theta_1 \cos \theta_2 - \sin \theta_1 \sin \theta_2) + i(\sin \theta_1 \cos \theta_2 + \cos \theta_1 \sin \theta_2)] \qquad (2)$$

and for $z_2 \ne 0$,

$$\frac{z_1}{z_2} = \frac{r_1}{r_2}[(\cos \theta_1 \cos \theta_2 + \sin \theta_1 \sin \theta_2) + i(\sin \theta_1 \cos \theta_2 - \cos \theta_1 \sin \theta_2)]. \qquad (3)$$

From the addition formulas from trigonometry, (2) and (3) can be rewritten, in turn, as

$$z_1 z_2 = r_1 r_2[\cos(\theta_1 + \theta_2) + i \sin(\theta_1 + \theta_2)] \qquad (4)$$

and

$$\frac{z_1}{z_2} = \frac{r_1}{r_2}[\cos(\theta_1 - \theta_2) + i \sin(\theta_1 - \theta_2)]. \qquad (5)$$

Inspection of (4) and (5) shows that

$$|z_1 z_2| = |z_1||z_2|, \qquad \left|\frac{z_1}{z_2}\right| = \frac{|z_1|}{|z_2|}, \qquad (6)$$

and

$$\arg(z_1 z_2) = \arg z_1 + \arg z_2, \qquad \arg\left(\frac{z_1}{z_2}\right) = \arg z_1 - \arg z_2. \qquad (7)$$

EXAMPLE 2 **Argument of a Product and of a Quotient**

We have seen that $\text{Arg } z_1 = \pi/2$ for $z_1 = i$. In Example 1 we saw that $\text{Arg } z_2 = -\pi/3$ for $z_2 = 1 - \sqrt{3}i$. Thus, for

$$z_1 z_2 = i(1 - \sqrt{3}i) = \sqrt{3} + i \quad \text{and} \quad \frac{z_1}{z_2} = \frac{i}{1 - \sqrt{3}i} = -\frac{\sqrt{3}}{4} + \frac{1}{4}i$$

it follows from (7) that

$$\arg(z_1 z_2) = \frac{\pi}{2} + \left(-\frac{\pi}{3}\right) = \frac{\pi}{6} \qquad \text{and} \qquad \arg\left(\frac{z_1}{z_2}\right) = \frac{\pi}{2} - \left(-\frac{\pi}{3}\right) = \frac{5\pi}{6}. \qquad \equiv$$

In Example 2 we used the principal arguments of z_1 and z_2 and obtained $\arg(z_1 z_2) = \text{Arg}(z_1 z_2)$ and $\arg(z_1/z_2) = \text{Arg}(z_1/z_2)$. It should be observed, however, that this was a coincidence. Although (7) is true for any arguments of z_1 and z_2, *it is not true*, in general, that $\text{Arg}(z_1 z_2) = \text{Arg } z_1 + \text{Arg } z_2$ and $\text{Arg}(z_1/z_2) = \text{Arg } z_1 - \text{Arg } z_2$. See Problem 39 in Exercises 17.2.

◾ **Integer Powers of z** We can find integer powers of the complex number z from the results in (4) and (5). For example, if $z = r(\cos \theta + i \sin \theta)$, then with $z_1 = z$ and $z_2 = z$, (4) gives

$$z^2 = r^2[\cos(\theta + \theta) + i \sin(\theta + \theta)] = r^2(\cos 2\theta + i \sin 2\theta).$$

Since $z^3 = z^2 z$, it follows that

$$z^3 = r^3(\cos 3\theta + i \sin 3\theta).$$

Moreover, since $\arg(1) = 0$, it follows from (5) that

$$\frac{1}{z^2} = z^{-2} = r^{-2}[\cos(-2\theta) + i \sin(-2\theta)].$$

Continuing in this manner, we obtain a formula for the *n*th power of z for any integer n:

$$z^n = r^n(\cos n\theta + i \sin n\theta). \tag{8}$$

> EXAMPLE 3 **Power of a Complex Number**

Compute z^3 for $z = 1 - \sqrt{3}i$.

SOLUTION In Example 1 we saw that

$$z = 2\left[\cos\left(-\frac{\pi}{3}\right) + i \sin\left(-\frac{\pi}{3}\right)\right].$$

Hence from (8) with $r = 2$, $\theta = -\pi/3$, and $n = 3$, we get

$$(1 - \sqrt{3}i)^3 = 2^3\left[\cos\left(3\left(-\frac{\pi}{3}\right)\right) + i \sin\left(3\left(-\frac{\pi}{3}\right)\right)\right]$$

$$= 8[\cos(-\pi) + i \sin(-\pi)] = -8. \qquad \equiv$$

◾ **De Moivre's Formula** When $z = \cos \theta + i \sin \theta$, we have $|z| = r = 1$ and so (8) yields

$$(\cos \theta + i \sin \theta)^n = \cos n\theta + i \sin n\theta. \tag{9}$$

The last result that links complex numbers and trigonometry is called **de Moivre's formula** after the French mathematician **Abraham de Moivre** (1667–1754). This formula is useful in deriving certain trigonometric identities. See Problems 37 and 38 in Exercises 17.2.

◾ **Roots** A number w is said to be an ***n*th root** of a nonzero complex number z if $w^n = z$. If we let $w = \rho(\cos \phi + i \sin \phi)$ and $z = r(\cos \theta + i \sin \theta)$ be the polar forms of w and z, then in view of (8), $w^n = z$ becomes

$$\rho^n(\cos n\phi + i \sin n\phi) = r(\cos \theta + i \sin \theta).$$

From this we conclude that $\rho^n = r$ or $\rho = r^{1/n}$ and

$$\cos n\phi + i \sin n\phi = \cos \theta + i \sin \theta.$$

By equating the real and imaginary parts, we get from this equation

$$\cos n\phi = \cos \theta \qquad \text{and} \qquad \sin n\phi = \sin \theta.$$

These equalities imply that $n\phi = \theta + 2k\pi$, where k is an integer. Thus,

$$\phi = \frac{\theta + 2k\pi}{n}.$$

As k takes on the successive integer values $k = 0, 1, 2,\ldots, n - 1$, we obtain n *distinct* roots with the same modulus but different arguments. But for $k \geq n$ we obtain the same roots because the sine and cosine are 2π-periodic. To see this, suppose $k = n + m$, where $m = 0, 1, 2,\ldots$. Then

$$\phi = \frac{\theta + 2(n + m)\pi}{n} = \frac{\theta + 2m\pi}{n} + 2\pi$$

and so

$$\sin\phi = \sin\left(\frac{\theta + 2m\pi}{n}\right), \qquad \cos\phi = \cos\left(\frac{\theta + 2m\pi}{n}\right).$$

We summarize this result. The n nth roots of a nonzero complex number $z = r(\cos\theta + i\sin\theta)$ are given by

$$w_k = r^{1/n}\left[\cos\left(\frac{\theta + 2k\pi}{n}\right) + i\sin\left(\frac{\theta + 2k\pi}{n}\right)\right], \tag{10}$$

where $k = 0, 1, 2,\ldots, n - 1$.

EXAMPLE 4 Roots of a Complex Number

Find the three cube roots of $z = i$.

SOLUTION With $r = 1$, $\theta = \arg z = \pi/2$, the polar form of the given number is $z = \cos(\pi/2) + i\sin(\pi/2)$. From (10) with $n = 3$ we obtain

$$w_k = (1)^{1/3}\left[\cos\left(\frac{\pi/2 + 2k\pi}{3}\right) + i\sin\left(\frac{\pi/2 + 2k\pi}{3}\right)\right], \quad k = 0, 1, 2.$$

Hence, the three roots are

$$k = 0, \quad w_0 = \cos\frac{\pi}{6} + i\sin\frac{\pi}{6} = \frac{\sqrt{3}}{2} + \frac{1}{2}i$$

$$k = 1, \quad w_1 = \cos\frac{5\pi}{6} + i\sin\frac{5\pi}{6} = -\frac{\sqrt{3}}{2} + \frac{1}{2}i$$

$$k = 2, \quad w_2 = \cos\frac{3\pi}{2} + i\sin\frac{3\pi}{2} = -i. \qquad \equiv$$

FIGURE 17.2.3 Three cube roots of i

The root w of a complex number z obtained by using the principal argument of z with $k = 0$ is sometimes called the **principal nth root** of z. In Example 4, since $\text{Arg}(i) = \pi/2$, $w_0 = \sqrt{3}/2 + (1/2)i$ is the principal third root of i.

Since the roots given by (8) have the same modulus, the n roots of a nonzero complex number z lie on a circle of radius $r^{1/n}$ centered at the origin in the complex plane. Moreover, since the difference between the arguments of any two successive roots is $2\pi/n$, the nth roots of z are equally spaced on this circle. **FIGURE 17.2.3** shows the three cube roots of i equally spaced on a unit circle; the angle between roots (vectors) w_k and w_{k+1} is $2\pi/3$.

As the next example will show, the roots of a complex number do not have to be "nice" numbers as in Example 3.

EXAMPLE 5 Roots of a Complex Number

Find the four fourth roots of $z = 1 + i$.

SOLUTION In this case, $r = \sqrt{2}$ and $\theta = \arg z = \pi/4$. From (10) with $n = 4$, we obtain

$$w_k = (\sqrt{2})^{1/4}\left[\cos\left(\frac{\pi/4 + 2k\pi}{4}\right) + i\sin\left(\frac{\pi/4 + 2k\pi}{4}\right)\right], \quad k = 0, 1, 2, 3.$$

The roots, rounded to four decimal places, are

$$k = 0, \quad w_0 = (\sqrt{2})^{1/4}\left[\cos\frac{\pi}{16} + i\sin\frac{\pi}{16}\right] = 1.0696 + 0.2127i$$

$$k = 1, \quad w_1 = (\sqrt{2})^{1/4}\left[\cos\frac{9\pi}{16} + i\sin\frac{9\pi}{16}\right] = -0.2127 + 1.0696i$$

$$k = 2, \quad w_2 = (\sqrt{2})^{1/4}\left[\cos\frac{17\pi}{16} + i\sin\frac{17\pi}{16}\right] = -1.0696 - 0.2127i$$

$$k = 3, \quad w_3 = (\sqrt{2})^{1/4}\left[\cos\frac{25\pi}{16} + i\sin\frac{25\pi}{16}\right] = 0.2127 - 1.0696i.$$

≡

17.2 Exercises Answers to selected odd-numbered problems begin on page ANS-44.

In Problems 1–10, write the given complex number in polar form.

1. 2

2. -10

3. $-3i$

4. $6i$

5. $1 + i$

6. $5 - 5i$

7. $-\sqrt{3} + i$

8. $-2 - 2\sqrt{3}i$

9. $\dfrac{3}{-1 + i}$

10. $\dfrac{12}{\sqrt{3} + i}$

In Problems 11–14, write the number given in polar form in the form $a + ib$.

11. $z = 5\left(\cos\dfrac{7\pi}{6} + i\sin\dfrac{7\pi}{6}\right)$

12. $z = 8\sqrt{2}\left(\cos\dfrac{11\pi}{4} + i\sin\dfrac{11\pi}{4}\right)$

13. $z = 6\left(\cos\dfrac{\pi}{8} + i\sin\dfrac{\pi}{8}\right)$

14. $z = 10\left(\cos\dfrac{\pi}{5} + i\sin\dfrac{\pi}{5}\right)$

In Problems 15 and 16, find $z_1 z_2$ and z_1/z_2. Write the number in the form $a + ib$.

15. $z_1 = 2\left(\cos\dfrac{\pi}{8} + i\sin\dfrac{\pi}{8}\right)$, $z_2 = 4\left(\cos\dfrac{3\pi}{8} + i\sin\dfrac{3\pi}{8}\right)$

16. $z_1 = \sqrt{2}\left(\cos\dfrac{\pi}{4} + i\sin\dfrac{\pi}{4}\right)$,

$z_2 = \sqrt{3}\left(\cos\dfrac{\pi}{12} + i\sin\dfrac{\pi}{12}\right)$

In Problems 17–20, write each complex number in polar form. Then use either (4) or (5) to obtain a polar form of the given number. Write the polar form in the form $a + ib$.

17. $(3 - 3i)(5 + 5\sqrt{3}i)$

18. $(4 + 4i)(-1 + i)$

19. $\dfrac{-i}{2 - 2i}$

20. $\dfrac{\sqrt{2} + \sqrt{6}i}{-1 + \sqrt{3}i}$

In Problems 21–26, use (8) to compute the indicated power.

21. $(1 + \sqrt{3}i)^9$

22. $(2 - 2i)^5$

23. $(\frac{1}{2} + \frac{1}{2}i)^{10}$

24. $(-\sqrt{2} + \sqrt{6}i)^4$

25. $\left(\cos\dfrac{\pi}{8} + i\sin\dfrac{\pi}{8}\right)^{12}$

26. $\left[\sqrt{3}\left(\cos\dfrac{2\pi}{9} + i\sin\dfrac{2\pi}{9}\right)\right]^6$

In Problems 27–32, use (10) to compute all roots. Sketch these roots on an appropriate circle centered at the origin.

27. $(8)^{1/3}$

28. $(1)^{1/8}$

29. $(i)^{1/2}$

30. $(-1 + i)^{1/3}$

31. $(-1 + \sqrt{3}i)^{1/2}$

32. $(-1 - \sqrt{3}i)^{1/4}$

In Problems 33 and 34, find all solutions of the given equation.

33. $z^4 + 1 = 0$

34. $z^8 - 2z^4 + 1 = 0$

In Problems 35 and 36, express the given complex number first in polar form and then in the form $a + ib$.

35. $\left(\cos\dfrac{\pi}{9} + i\sin\dfrac{\pi}{9}\right)^{12}\left[2\left(\cos\dfrac{\pi}{6} + i\sin\dfrac{\pi}{6}\right)\right]^5$

36. $\dfrac{\left[8\left(\cos\dfrac{3\pi}{8} + i\sin\dfrac{3\pi}{8}\right)\right]^3}{\left[2\left(\cos\dfrac{\pi}{16} + i\sin\dfrac{\pi}{16}\right)\right]^{10}}$

37. Use the result $(\cos\theta + i\sin\theta)^2 = \cos 2\theta + i\sin 2\theta$ to find trigonometric identities for $\cos 2\theta$ and $\sin 2\theta$.

38. Use the result $(\cos\theta + i\sin\theta)^3 = \cos 3\theta + i\sin 3\theta$ to find trigonometric identities for $\cos 3\theta$ and $\sin 3\theta$.

39. (a) If $z_1 = -1$ and $z_2 = 5i$, verify that

$$\text{Arg}(z_1 z_2) \neq \text{Arg}(z_1) + \text{Arg}(z_2).$$

(b) If $z_1 = -1$ and $z_2 = -5i$, verify that

$$\text{Arg}(z_1/z_2) \neq \text{Arg}(z_1) - \text{Arg}(z_2).$$

40. For the complex numbers given in Problem 39, verify in both parts (a) and (b) that

$$\arg(z_1 z_2) = \arg(z_1) + \arg(z_2)$$

and

$$\arg\left(\frac{z_1}{z_2}\right) = \arg(z_1) - \arg(z_2).$$

17.3 Sets in the Complex Plane

INTRODUCTION In the preceding sections we examined some rudiments of the algebra and geometry of complex numbers. But we have barely scratched the surface of the subject known as complex analysis; the main thrust of our study lies ahead. Our goal in the sections and chapters that follow is to examine functions of a single complex variable $z = x + iy$ and the calculus of these functions.

Before introducing the notion of a function of a complex variable, we need to state some essential definitions and terminology about sets in the complex plane.

■ **Terminology** Before discussing the concept of functions of a complex variable, we need to introduce some essential terminology about sets in the complex plane.

Suppose $z_0 = x_0 + iy_0$. Since $|z - z_0| = \sqrt{(x - x_0)^2 + (y - y_0)^2}$ is the distance between the points $z = x + iy$ and $z_0 = x_0 + iy_0$, the points $z = x + iy$ that satisfy the equation

$$|z - z_0| = \rho,$$

$|z - z_0| = \rho$

FIGURE 17.3.1 Circle of radius ρ

$\rho > 0$, lie on a **circle** of radius ρ centered at the point z_0. See **FIGURE 17.3.1**.

EXAMPLE 1 Circles

(a) $|z| = 1$ is the equation of a unit circle centered at the origin.
(b) $|z - 1 - 2i| = 5$ is the equation of a circle of radius 5 centered at $1 + 2i$. ≡

FIGURE 17.3.2 Open set

The points z satisfying the inequality $|z - z_0| < \rho$, $\rho > 0$, lie within, but not on, a circle of radius ρ centered at the point z_0. This set is called a **neighborhood** of z_0 or an **open disk**. A point z_0 is said to be an **interior point** of a set S of the complex plane if there exists some neighborhood of z_0 that lies entirely within S. If every point z of a set S is an interior point, then S is said to be an **open set**. See **FIGURE 17.3.2**. For example, the inequality $\text{Re}(z) > 1$ defines a *right half-plane*, which is an open set. All complex numbers $z = x + iy$ for which $x > 1$ are in this set. If we choose, for example, $z_0 = 1.1 + 2i$, then a neighborhood of z_0 lying entirely in the set is defined by $|z - (1.1 + 2i)| < 0.05$. See **FIGURE 17.3.3**. On the other hand, the set S of points in the complex plane defined by $\text{Re}(z) \geq 1$ is *not* open, since every neighborhood of a point on the line $x = 1$ must contain points in S and points not in S. See **FIGURE 17.3.4**.

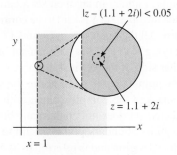

FIGURE 17.3.3 Open set magnified view of a point near $x = 1$

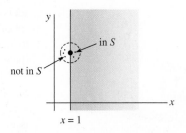

FIGURE 17.3.4 Set S is not open

EXAMPLE 2 **Open Sets**

FIGURE 17.3.5 illustrates some additional open sets.

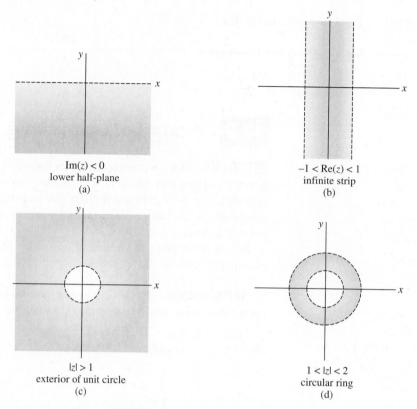

Im(z) < 0
lower half-plane
(a)

−1 < Re(z) < 1
infinite strip
(b)

$|z|$ > 1
exterior of unit circle
(c)

1 < $|z|$ < 2
circular ring
(d)

FIGURE 17.3.5 Four examples of open sets

The set of numbers satisfying the inequality

$$\rho_1 < |z - z_0| < \rho_2,$$

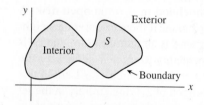

FIGURE 17.3.6 Interior, boundary, and exterior of set S

such as illustrated in Figure 17.3.5(d), is called an **open annulus**.

If *every* neighborhood of a point z_0 of a set S contains at least one point of S and at least one point not in S, then z_0 is said to be a **boundary point** of S. For the set of points defined by Re(z) ≥ 1, the points on the vertical line $x = 1$ are boundary points. The points that lie on the circle $|z - i| = 2$ are boundary points for the disk $|z - i| \le 2$, as well as for the neighborhood $|z - i| < 2$ of $z = i$. The collection of boundary points of a set S is called the **boundary** of S. The circle $|z - i| = 2$ is the boundary for both the disk $|z - i| \le 2$ and the neighborhood $|z - i| < 2$ of $z = i$. A point z_0 that is neither an interior point nor a boundary point of a set S is said to be an **exterior point** of S; in other words, z_0 is an exterior point of a set S if there exists *some* neighborhood of z_0 that contains no points of S. **FIGURE 17.3.6** shows a typical set S with interior, boundary, and exterior.

If any pair of points z_1 and z_2 in an open set S can be connected by a polygonal line that lies entirely in the set, then the open set S is said to be **connected**. See **FIGURE 17.3.7**. An open connected set is called a **domain**. All the open sets in Figure 17.3.5 are connected and so are domains. The set of numbers satisfying Re(z) ≠ 4 is an open set but is not connected, since it is not possible to join points on either side of the vertical line $x = 4$ by a polygonal line without leaving the set (bear in mind that the points on $x = 4$ are not in the set).

A **region** is a domain in the complex plane with all, some, or none of its boundary points. Since an open connected set does not contain any boundary points, it is automatically a region. A region containing all its boundary points is said to be **closed**. The disk defined by $|z - i| \le 2$ is an example of a closed region and is referred to as a **closed disk**. A region may be neither open nor closed; the annular region defined by $1 \le |z - 5| < 3$ contains only some of its boundary points and so is neither open nor closed.

FIGURE 17.3.7 Connected set

17.3 Exercises Answers to selected odd-numbered problems begin on page ANS-45.

In Problems 1–8, sketch the graph of the given equation.

1. $\text{Re}(z) = 5$
2. $\text{Im}(z) = -2$
3. $\text{Im}(\bar{z} + 3i) = 6$
4. $\text{Im}(z - i) = \text{Re}(z + 4 - 3i)$
5. $|z - 3i| = 2$
6. $|2z + 1| = 4$
7. $|z - 4 + 3i| = 5$
8. $|z + 2 + 2i| = 2$

In Problems 9–22, sketch the set of points in the complex plane satisfying the given inequality. Determine whether the set is a domain.

9. $\text{Re}(z) < -1$
10. $|\text{Re}(z)| > 2$
11. $\text{Im}(z) > 3$
12. $\text{Im}(z - i) < 5$
13. $2 < \text{Re}(z - 1) < 4$
14. $-1 \leq \text{Im}(z) < 4$

15. $\text{Re}(z^2) > 0$
16. $\text{Im}(1/z) < \frac{1}{2}$
17. $0 \leq \arg(z) \leq 2\pi/3$
18. $|\arg(z)| < \pi/4$
19. $|z - i| > 1$
20. $|z - i| > 0$
21. $2 < |z - i| < 3$
22. $1 \leq |z - 1 - i| < 2$

23. Describe the set of points in the complex plane that satisfies $|z + 1| = |z - i|$.

24. Describe the set of points in the complex plane that satisfies $|\text{Re}(z)| \leq |z|$.

25. Describe the set of points in the complex plane that satisfies $z^2 + \bar{z}^2 = 2$.

26. Describe the set of points in the complex plane that satisfies $|z - i| + |z + i| = 1$.

17.4 Functions of a Complex Variable

INTRODUCTION One of the most important concepts in mathematics is that of a function. You may recall from previous courses that a function is a certain kind of correspondence between two sets; more specifically: *A function f from a set A to a set B is a rule of correspondence that assigns to each element in A one and only one element in B.* If *b* is the element in the set *B* assigned to the element *a* in the set *A* by *f*, we say that *b* is the **image** of *a* and write $b = f(a)$. The set *A* is called the **domain** of the function *f* (but is not necessarily a domain in the sense defined in Section 17.3). The set of all images in *B* is called the **range** of the function. For example, suppose the set *A* is a set of real numbers defined by $3 \leq x < \infty$ and the function is given by $f(x) = \sqrt{x - 3}$; then $f(3) = 0$, $f(4) = 1, f(8) = \sqrt{5}$, and so on. In other words, the range of *f* is the set given by $0 \leq y < \infty$. Since *A* is a set of real numbers, we say *f* is a **function of a real variable** *x*.

Functions of a Complex Variable When the domain *A* in the foregoing definition of a function is a set of complex numbers *z*, we naturally say that *f* is a **function of a complex variable** *z* or a **complex function** for short. The image *w* of a complex number *z* will be some complex number $u + iv$; that is,

$$w = f(z) = u(x, y) + iv(x, y), \tag{1}$$

where *u* and *v* are the real and imaginary parts of *w* and are real-valued functions. Inherent in the mathematical statement (1) is the fact that *we cannot draw a graph of a complex function $w = f(z)$* since a graph would require four axes in a four-dimensional coordinate system.

Some examples of functions of a complex variable are

$$f(z) = z^2 - 4z, \qquad z \text{ any complex number}$$

$$f(z) = \frac{z}{z^2 + 1}, \qquad z \neq i \text{ and } z \neq -i$$

$$f(z) = z + \text{Re}(z), \qquad z \text{ any complex number}.$$

Each of these functions can be expressed in form (1). For example,

$$f(z) = z^2 - 4z = (x + iy)^2 - 4(x + iy) = (x^2 - y^2 - 4x) + i(2xy - 4y).$$

Thus, $u(x, y) = x^2 - y^2 - 4x$, and $v(x, y) = 2xy - 4y$.

Although we cannot draw a graph, a complex function $w = f(z)$ can be interpreted as a **mapping** or **transformation** from the *z*-plane to the *w*-plane. See **FIGURE 17.4.1**.

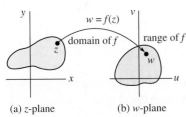

FIGURE 17.4.1 Mapping from *z*-plane to *w*-plane

(a) *z*-plane

(b) *w*-plane

(a) z-plane (b) w-plane

FIGURE 17.4.2 Image of $x = 1$ is a parabola

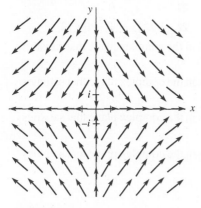

FIGURE 17.4.3 $f_1(z) = \bar{z}$ (normalized)

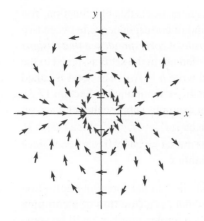

FIGURE 17.4.4 $f_2(z) = z^2$ (normalized)

EXAMPLE 1 **Image of a Vertical Line**

Find the image of the line $\text{Re}(z) = 1$ under the mapping $f(z) = z^2$.

SOLUTION For the function $f(z) = z^2$ we have $u(x, y) = x^2 - y^2$ and $v(x, y) = 2xy$. Now, $\text{Re}(z) = x$ and so by substituting $x = 1$ into the functions u and v, we obtain $u = 1 - y^2$ and $v = 2y$. These are parametric equations of a curve in the w-plane. Substituting $y = v/2$ into the first equation eliminates the parameter y to give $u = 1 - v^2/4$. In other words, the image of the line in **FIGURE 17.4.2(a)** is the parabola shown in Figure 17.4.2(b). ≡

We shall pursue the idea of $f(z)$ as a mapping in greater detail in Chapter 20.

It should be noted that a complex function is completely determined by the real-valued functions u and v. This means a complex function $w = f(z)$ can be defined by arbitrarily specifying $u(x, y)$ and $v(x, y)$, even though $u + iv$ may not be obtainable through the familiar operations on the symbol z alone. For example, if $u(x, y) = xy^2$ and $v(x, y) = x^2 - 4y^3$, then $f(z) = xy^2 + i(x^2 - 4y^3)$ is a function of a complex variable. To compute, say, $f(3 + 2i)$, we substitute $x = 3$ and $y = 2$ into u and v to obtain $f(3 + 2i) = 12 - 23i$.

■ **Complex Functions as Flows** We also may interpret a complex function $w = f(z)$ as a **two-dimensional fluid flow** by considering the complex number $f(z)$ as a vector based at the point z. The vector $f(z)$ specifies the speed and direction of the flow at a given point z. **FIGURES 17.4.3** and **17.4.4** show the flows corresponding to the complex functions $f_1(z) = \bar{z}$ and $f_2(z) = z^2$, respectively.

If $x(t) + iy(t)$ is a parametric representation for the path of a particle in the flow, the tangent vector $\mathbf{T} = x'(t) + iy'(t)$ must coincide with $f(x(t) + iy(t))$. When $f(z) = u(x, y) + iv(x, y)$, it follows that the path of the particle must satisfy the system of differential equations

$$\frac{dx}{dt} = u(x, y)$$

$$\frac{dy}{dt} = v(x, y).$$

We call the family of solutions of this system the **streamlines** of the flow associated with $f(z)$.

EXAMPLE 2 **Streamlines**

Find the streamlines of the flows associated with the complex functions **(a)** $f_1(z) = \bar{z}$ and **(b)** $f_2(z) = z^2$.

SOLUTION **(a)** The streamlines corresponding to $f_1(z) = x - iy$ satisfy the system

$$\frac{dx}{dt} = x$$

$$\frac{dy}{dt} = -y$$

and so $x(t) = c_1 e^t$ and $y(t) = c_2 e^{-t}$. By multiplying these two parametric equations, we see that the point $x(t) + iy(t)$ lies on the hyperbola $xy = c_1 c_2$.

(b) To find the streamlines corresponding to $f_2(z) = (x^2 - y^2) + i2xy$, note that $dx/dt = x^2 - y^2$, $dy/dt = 2xy$, and so

$$\frac{dy}{dx} = \frac{2xy}{x^2 - y^2}.$$

This homogeneous differential equation has the solution $x^2 + y^2 = c_2 y$, which represents a family of circles that have centers on the y-axis and pass through the origin. ≡

■ **Limits and Continuity** The definition of a limit of a complex function $f(z)$ as $z \to z_0$ has the same outward appearance as the limit in real variables.

DEFINITION 17.4.1 Limit of a Function

Suppose the function f is defined in some neighborhood of z_0, except possibly at z_0 itself. Then f is said to possess a **limit** at z_0, written

$$\lim_{z \to z_0} f(z) = L$$

if, for each $\epsilon > 0$, there exists a $\delta > 0$ such that $|f(z) - L| < \epsilon$ whenever $0 < |z - z_0| < \delta$.

In words, $\lim_{z \to z_0} f(z) = L$ means that the points $f(z)$ can be made arbitrarily close to the point L if we choose the point z sufficiently close to, but not equal to, the point z_0. As shown in **FIGURE 17.4.5**, for each ϵ-neighborhood of L (defined by $|f(z) - L| < \epsilon$) there is a δ-neighborhood of z_0 (defined by $|z - z_0| < \delta$) so that the images of all points $z \neq z_0$ in this neighborhood lie in the ϵ-neighborhood of L.

The fundamental difference between this definition and the limit concept in real variables lies in the understanding of $z \to z_0$. For a function f of a single real variable x, $\lim_{x \to x_0} f(x) = L$ means $f(x)$ approaches L as x approaches x_0 either from the right of x_0 or from the left of x_0 on the real number line. But since z and z_0 are points in the complex plane, when we say that $\lim_{z \to z_0} f(z)$ exists, we mean that $f(z)$ approaches L as the point z approaches z_0 from *any* direction.

The following theorem summarizes some properties of limits:

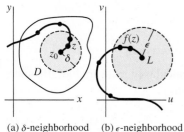

(a) δ-neighborhood (b) ϵ-neighborhood

FIGURE 17.4.5 Geometric meaning of a complex limit

THEOREM 17.4.1 Limit of Sum, Product, Quotient

Suppose $\lim_{z \to z_0} f(z) = L_1$ and $\lim_{z \to z_0} g(z) = L_2$. Then

(*i*) $\displaystyle\lim_{z \to z_0} [f(z) + g(z)] = L_1 + L_2$

(*ii*) $\displaystyle\lim_{z \to z_0} f(z)g(z) = L_1 L_2$

(*iii*) $\displaystyle\lim_{z \to z_0} \frac{f(z)}{g(z)} = \frac{L_1}{L_2}, \quad L_2 \neq 0.$

DEFINITION 17.4.2 Continuity at a Point

A function f is **continuous** at a point z_0 if

$$\lim_{z \to z_0} f(z) = f(z_0).$$

As a consequence of Theorem 17.4.1, it follows that if two functions f and g are continuous at a point z_0, then their sum and product are continuous at z_0. The quotient of the two functions is continuous at z_0 provided $g(z_0) \neq 0$.

A function f defined by

$$f(z) = a_n z^n + a_{n-1} z^{n-1} + \cdots + a_2 z^2 + a_1 z + a_0, \quad a_n \neq 0, \tag{2}$$

where n is a nonnegative integer and the coefficients a_i, $i = 0, 1, \ldots, n$, are complex constants, is called a **polynomial** of degree n. Although we shall not prove it, the limit result $\lim_{z \to z_0} z = z_0$ indicates that the simple polynomial function $f(z) = z$ is continuous everywhere—that is, on the entire z-plane. With this result in mind and with repeated applications of Theorem 17.4.1 (*i*) and (*ii*), it follows that a polynomial function (2) is continuous everywhere. A **rational function**

$$f(z) = \frac{g(z)}{h(z)},$$

where g and h are polynomial functions, is continuous except at those points at which $h(z)$ is zero.

■ **Derivative** The derivative of a complex function is defined in terms of a limit. The symbol Δz used in the following definition is the complex number $\Delta x + i \Delta y$.

Suppose the complex function f is defined in a neighborhood of a point z_0. The **derivative** of f at z_0 is

$$f'(z_0) = \lim_{\Delta z \to 0} \frac{f(z_0 + \Delta z) - f(z_0)}{\Delta z}, \tag{3}$$

provided this limit exists.

If the limit in (3) exists, the function f is said to be **differentiable** at z_0. The derivative of a function $w = f(z)$ is also written dw/dz.

As in real variables, differentiability implies continuity:

If f is differentiable at z_0, then f is continuous at z_0.

Moreover, the **rules of differentiation** are the same as in the calculus of real variables. If f and g are differentiable at a point z, and c is a complex constant, then

Constant Rules: $\quad \dfrac{d}{dz} c = 0, \quad \dfrac{d}{dz} cf(z) = cf'(z)$ $\hfill (4)$

Sum Rule: $\quad \dfrac{d}{dz}[f(z) + g(z)] = f'(z) + g'(z)$ $\hfill (5)$

Product Rule: $\quad \dfrac{d}{dz}[f(z)g(z)] = f(z)g'(z) + g(z)f'(z)$ $\hfill (6)$

Quotient Rule: $\quad \dfrac{d}{dz}\left[\dfrac{f(z)}{g(z)}\right] = \dfrac{g(z)f'(z) - f(z)g'(z)}{[g(z)]^2}$ $\hfill (7)$

Chain Rule: $\quad \dfrac{d}{dz} f(g(z)) = f'(g(z))g'(z).$ $\hfill (8)$

The usual **Power Rule** for differentiation of powers of z is also valid:

$$\frac{d}{dz} z^n = nz^{n-1}, \quad n \text{ an integer.} \tag{9}$$

Combining (9) and (8) gives the **Power Rule for Functions:**

$$\frac{d}{dz} [g(z)]^n = n[g(z)]^{n-1} g'(z), \, n \text{ an integer.} \tag{10}$$

Differentiate **(a)** $f(z) = 3z^4 - 5z^3 + 2z$ **(b)** $f(z) = \dfrac{z^2}{4z + 1}$ **(c)** $f(z) = (iz^2 + 3z)^5$.

SOLUTION **(a)** Using the Power Rule (9) along with the Sum Rule (5), we obtain

$$f'(z) = 3 \cdot 4z^3 - 5 \cdot 3z^2 + 2 = 12z^3 - 15z^2 + 2.$$

(b) From the Quotient Rule (7),

$$f'(z) = \frac{(4z + 1) \cdot 2z - z^2 \cdot 4}{(4z + 1)^2} = \frac{4z^2 + 2z}{(4z + 1)^2}.$$

(c) In the Power Rule for Functions (10), we identify $n = 5$, $g(z) = iz^2 + 3z$, and $g'(z) = 2iz + 3$, so that

$$f'(z) = 5(iz^2 + 3z)^4 (2zi + 3). \qquad \equiv$$

In order for a complex function f to be differentiable at a point z_0, $\lim\limits_{\Delta z \to 0} \dfrac{f(z_0 + \Delta z) - f(z_0)}{\Delta z}$ must approach the same complex number from any direction. Thus in the study of complex

variables, to require the differentiability of a function is a greater demand than in real variables. If a complex function is made up, such as $f(z) = x + 4iy$, there is a good chance that it is not differentiable.

EXAMPLE 4 A Function That Is Nowhere Differentiable

Show that the function $f(z) = x + 4iy$ is nowhere differentiable.

SOLUTION With $\Delta z = \Delta x + i\Delta y$, we have

$$f(z + \Delta z) - f(z) = (x + \Delta x) + 4i(y + \Delta y) - x - 4iy = \Delta x + 4i\Delta y$$

and so

$$\lim_{\Delta z \to 0} \frac{f(z + \Delta z) - f(z)}{\Delta z} = \lim_{\Delta z \to 0} \frac{\Delta x + 4i\Delta y}{\Delta x + i\Delta y}. \tag{11}$$

Now, if we let $\Delta z \to 0$ along a line parallel to the x-axis, then $\Delta y = 0$ and the value of (11) is 1. On the other hand, if we let $\Delta z \to 0$ along a line parallel to the y-axis, then $\Delta x = 0$ and the value of (11) is seen to be 4. Therefore, $f(z) = x + 4iy$ is not differentiable at any point z. ≡

Analytic Functions While the requirement of differentiability is a stringent demand, there is a class of functions that is of great importance whose members satisfy even more severe requirements. These functions are called **analytic functions**.

DEFINITION 17.4.4 Analyticity at a Point

A complex function $w = f(z)$ is said to be **analytic at a point** z_0 if f is differentiable at z_0 and at every point in some neighborhood of z_0.

A function f is analytic in a domain D if it is analytic at every point in D.

The student should read Definition 17.4.4 carefully. Analyticity *at a point* is a neighborhood property. Analyticity at a point is, therefore, not the same as differentiability at a point. It is left as an exercise to show that the function $f(z) = |z|^2$ is differentiable at $z = 0$ but is differentiable nowhere else. Hence, $f(z) = |z|^2$ is nowhere analytic. In contrast, the simple polynomial $f(z) = z^2$ is differentiable at every point z in the complex plane. Hence, $f(z) = z^2$ is analytic everywhere. A function that is analytic at every point z is said to be an **entire function**. Polynomial functions are differentiable at every point z and so are entire functions.

REMARKS

Recall from algebra that a number c is a zero of a polynomial function if and only if $x - c$ is a factor of $f(x)$. The same result holds in complex analysis. For example, since $f(z) = z^4 + 5z^2 + 4 = (z^2 + 1)(z^2 + 4)$, the zeros of f are $-i$, i, $-2i$, and $2i$. Hence, $f(z) = (z + i)(z - i)(z + 2i)(z - 2i)$. Moreover, the quadratic formula is also valid. For example, using this formula, we can write

$$f(z) = z^2 - 2z + 2 = (z - (1 + i))(z - (1 - i))$$

$$= (z - 1 - i)(z - 1 + i).$$

See Problems 21 and 22 in Exercises 17.4.

17.4 Exercises Answers to selected odd-numbered problems begin on page ANS-45.

In Problems 1–6, find the image of the given line under the mapping $f(z) = z^2$.

1. $y = 2$

2. $x = -3$

3. $x = 0$

4. $y = 0$

5. $y = x$

6. $y = -x$

In Problems 7–14, express the given function in the form $f(z) = u + iv$.

7. $f(z) = 6z - 5 + 9i$

8. $f(z) = 7z - 9i\bar{z} - 3 + 2i$

9. $f(z) = z^2 - 3z + 4i$

10. $f(z) = 3\bar{z}^2 + 2z$

11. $f(z) = z^3 - 4z$

12. $f(z) = z^4$

13. $f(z) = z + 1/z$

14. $f(z) = \dfrac{z}{z + 1}$

In Problems 15–18, evaluate the given function at the indicated points.

15. $f(z) = 2x - y^2 + i(xy^3 - 2x^2 + 1)$
 (a) $2i$ **(b)** $2 - i$ **(c)** $5 + 3i$

16. $f(z) = (x + 1 + 1/x) + i(4x^2 - 2y^2 - 4)$
 (a) $1 + i$ **(b)** $2 - i$ **(c)** $1 + 4i$

17. $f(z) = 4z + i\bar{z} + \text{Re}(z)$
 (a) $4 - 6i$ **(b)** $-5 + 12i$ **(c)** $2 - 7i$

18. $f(z) = e^x \cos y + ie^x \sin y$
 (a) $\pi i/4$ **(b)** $-1 - \pi i$ **(c)** $3 + \pi i/3$

In Problems 19–22, the given limit exists. Find its value.

19. $\lim\limits_{z \to i}(4z^3 - 5z^2 + 4z + 1 - 5i)$

20. $\lim\limits_{z \to 1-i} \dfrac{5z^2 - 2z + 2}{z + 1}$

21. $\lim\limits_{z \to i} \dfrac{z^4 - 1}{z - i}$ **22.** $\lim\limits_{z \to 1+i} \dfrac{z^2 - 2z + 2}{z^2 - 2i}$

In Problems 23 and 24, show that the given limit does not exist.

23. $\lim\limits_{z \to 0} \dfrac{z}{\bar{z}}$ **24.** $\lim\limits_{z \to 1} \dfrac{x + y - 1}{z - 1}$

In Problems 25 and 26, use (3) to obtain the indicated derivative of the given function.

25. $f(z) = z^2,\ f'(z) = 2z$
26. $f(z) = 1/z,\ f'(z) = -1/z^2$

In Problems 27–34, use (4)–(10) to find the derivative $f'(z)$ for the given function.

27. $f(z) = 4z^3 - (3 + i)z^2 - 5z + 4$
28. $f(z) = 5z^4 - iz^3 + (8 - i)z^2 - 6i$

29. $f(z) = (2z + 1)(z^2 - 4z + 8i)$
30. $f(z) = (z^5 + 3iz^3)(z^4 + iz^3 + 2z^2 - 6iz)$
31. $f(z) = (z^2 - 4i)^3$ **32.** $f(z) = (2z - 1/z)^6$
33. $f(z) = \dfrac{3z - 4 + 8i}{2z + i}$ **34.** $f(z) = \dfrac{5z^2 - z}{z^3 + 1}$

In Problems 35–38, give the points at which the given function will not be analytic.

35. $f(z) = \dfrac{z}{z - 3i}$ **36.** $f(z) = \dfrac{2i}{z^2 - 2z + 5iz}$

37. $f(z) = \dfrac{z^3 + z}{z^2 + 4}$ **38.** $f(z) = \dfrac{z - 4 + 3i}{z^2 - 6z + 25}$

39. Show that the function $f(z) = \bar{z}$ is nowhere differentiable.

40. The function $f(z) = |z|^2$ is continuous throughout the entire complex plane. Show, however, that f is differentiable only at the point $z = 0$. [*Hint:* Use (3) and consider two cases: $z = 0$ and $z \neq 0$. In the second case let Δz approach zero along a line parallel to the x-axis and then let Δz approach zero along a line parallel to the y-axis.]

In Problems 41–44, find the streamlines of the flow associated with the given complex function.

41. $f(z) = 2z$ **42.** $f(z) = iz$
43. $f(z) = 1/\bar{z}$ **44.** $f(z) = x^2 - iy^2$

In Problems 45 and 46, use a graphics calculator or computer to obtain the image of the given parabola under the mapping $f(z) = z^2$.

45. $y = \tfrac{1}{2}x^2$ **46.** $y = (x - 1)^2$

17.5 Cauchy–Riemann Equations

INTRODUCTION In the preceding section we saw that a function f of a complex variable z is *analytic at a point z* when f is differentiable at z and differentiable at every point in *some* neighborhood of z. This requirement is more stringent than simply *differentiability at a point* because a complex function can be differentiable at a point z but yet be differentiable nowhere else. A function f is *analytic in a domain D* if f is differentiable at all points in D. We shall now develop a test for analyticity of a complex function $f(z) = u(x, y) + iv(x, y)$.

A Necessary Condition for Analyticity In the next theorem we see that if a function $f(z) = u(x, y) + iv(x, y)$ is differentiable at a point z, then the functions u and v must satisfy a pair of equations that relate their first-order partial derivatives. This result is a necessary condition for analyticity.

THEOREM 17.5.1 Cauchy–Riemann Equations

Suppose $f(z) = u(x, y) + iv(x, y)$ is differentiable at a point $z = x + iy$. Then at z the first-order partial derivatives of u and v exist and satisfy the **Cauchy–Riemann equations**

$$\frac{\partial u}{\partial x} = \frac{\partial v}{\partial y} \quad \text{and} \quad \frac{\partial u}{\partial y} = -\frac{\partial v}{\partial x}. \tag{1}$$

PROOF: Since $f'(z)$ exists, we know that

$$f'(z) = \lim_{\Delta z \to 0} \frac{f(z + \Delta z) - f(z)}{\Delta z}. \tag{2}$$

By writing $f(z) = u(x, y) + iv(x, y)$ and $\Delta z = \Delta x + i\Delta y$, we get from (2)

$$f'(z) = \lim_{\Delta z \to 0} \frac{u(x + \Delta x, y + \Delta y) + iv(x + \Delta x, y + \Delta y) - u(x, y) - iv(x, y)}{\Delta x + i\Delta y}. \tag{3}$$

Since this limit exists, Δz can approach zero from any convenient direction. In particular, if $\Delta z \to 0$ horizontally, then $\Delta z = \Delta x$, and so (3) becomes

$$f'(z) = \lim_{\Delta x \to 0} \frac{u(x + \Delta x, y) - u(x, y)}{\Delta x} + i \lim_{\Delta x \to 0} \frac{v(x + \Delta x, y) - v(x, y)}{\Delta x}. \tag{4}$$

Since $f'(z)$ exists, the two limits in (4) exist. But by definition the limits in (4) are the first partial derivatives of u and v with respect to x. Thus, we have shown that

$$f'(z) = \frac{\partial u}{\partial x} + i \frac{\partial v}{\partial x}. \tag{5}$$

Now if we let $\Delta z \to 0$ vertically, then $\Delta z = i\Delta y$ and (3) becomes

$$f'(z) = \lim_{\Delta y \to 0} \frac{u(x, y + \Delta y) - u(x, y)}{i\Delta y} + i \lim_{\Delta y \to 0} \frac{v(x, y + \Delta y) - v(x, y)}{i\Delta y}, \tag{6}$$

which is the same as

$$f'(z) = -i \frac{\partial u}{\partial y} + \frac{\partial v}{\partial y}. \tag{7}$$

Equating the real and imaginary parts of (5) and (7) yields the pair of equations in (1). ≡

If a complex function $f(z) = u(x, y) + iv(x, y)$ is analytic throughout a domain D, then the real functions u and v must satisfy the Cauchy–Riemann equations (1) at every point in D.

EXAMPLE 1 Using the Cauchy–Riemann Equations

The polynomial $f(z) = z^2 + z$ is analytic for all z and $f(z) = x^2 - y^2 + x + i(2xy + y)$. Thus, $u(x, y) = x^2 - y^2 + x$ and $v(x, y) = 2xy + y$. For any point (x, y), we see that the Cauchy–Riemann equations are satisfied:

$$\frac{\partial u}{\partial x} = 2x + 1 = \frac{\partial v}{\partial y} \quad \text{and} \quad \frac{\partial u}{\partial y} = -2y = -\frac{\partial v}{\partial x}. \qquad \equiv$$

EXAMPLE 2 Using the Cauchy–Riemann Equations

Show that the function $f(z) = (2x^2 + y) + i(y^2 - x)$ is not analytic at any point.

SOLUTION We identify $u(x, y) = 2x^2 + y$ and $v(x, y) = y^2 - x$. Now from

$$\frac{\partial u}{\partial x} = 4x \quad \text{and} \quad \frac{\partial v}{\partial y} = 2y$$

$$\frac{\partial u}{\partial y} = 1 \quad \text{and} \quad \frac{\partial v}{\partial x} = -1$$

we see that $\partial u/\partial y = -\partial v/\partial x$ but that the equality $\partial u/\partial x = \partial v/\partial y$ is satisfied only on the line $y = 2x$. However, for any point z on the line, there is no neighborhood or open disk about z in which f is differentiable. We conclude that f is nowhere analytic. ≡

Important. ▶ By themselves, the Cauchy–Riemann equations are not sufficient to ensure analyticity. However, when we add the condition of continuity to u and v *and* the four partial derivatives, the Cauchy–Riemann equations can be shown to imply analyticity. The proof is long and complicated and so we state only the result.

THEOREM 17.5.2 Criterion for Analyticity

Suppose the real-valued functions $u(x, y)$ and $v(x, y)$ are continuous and have continuous first-order partial derivatives in a domain D. If u and v satisfy the Cauchy–Riemann equations at all points of D, then the complex function $f(z) = u(x, y) + iv(x, y)$ is analytic in D.

EXAMPLE 3 Using Theorem 17.5.2

For the function $f(z) = \dfrac{x}{x^2 + y^2} - i\dfrac{y}{x^2 + y^2}$ we have

$$\frac{\partial u}{\partial x} = \frac{y^2 - x^2}{(x^2 + y^2)^2} = \frac{\partial v}{\partial y} \quad \text{and} \quad \frac{\partial u}{\partial y} = -\frac{2xy}{(x^2 + y^2)^2} = -\frac{\partial v}{\partial x}.$$

In other words, the Cauchy–Riemann equations are satisfied except at the point where $x^2 + y^2 = 0$; that is, at $z = 0$. We conclude from Theorem 17.5.2 that f is analytic in any domain not containing the point $z = 0$. ≡

The results in (5) and (7) were obtained under the basic assumption that f was differentiable at the point z. In other words, (5) and (7) give us a formula for computing $f'(z)$:

$$f'(z) = \frac{\partial u}{\partial x} + i\frac{\partial v}{\partial x} = \frac{\partial v}{\partial y} - i\frac{\partial u}{\partial y}. \tag{8}$$

For example, we know that $f(z) = z^2$ is differentiable for all z. With $u(x, y) = x^2 - y^2$, $\partial u/\partial x = 2x$, $v(x, y) = 2xy$, and $\partial v/\partial x = 2y$, we see that

$$f'(z) = 2x + i2y = 2(x + iy) = 2z.$$

Recall that analyticity implies differentiability but not vice versa. Theorem 17.5.2 has an analogue that gives sufficient conditions for differentiability:

> *If the real-valued functions $u(x, y)$ and $v(x, y)$ are continuous and have continuous first-order partial derivatives in a neighborhood of z, and if u and v satisfy the Cauchy–Riemann equations at the point z, then the complex function $f(z) = u(x, y) + iv(x, y)$ is differentiable at z and $f'(z)$ is given by* (8).

The function $f(z) = x^2 - y^2 i$ is nowhere analytic. With the identifications $u(x, y) = x^2$ and $v(x, y) = -y^2$, we see from

$$\frac{\partial u}{\partial x} = 2x, \frac{\partial v}{\partial y} = -2y \quad \text{and} \quad \frac{\partial u}{\partial y} = 0, \frac{\partial v}{\partial x} = 0$$

that the Cauchy–Riemann equations are satisfied only when $y = -x$. But since the functions u, $\partial u/\partial x$, $\partial u/\partial y$, v, $\partial v/\partial x$, and $\partial v/\partial y$ are continuous at every point, it follows that f is differentiable on the line $y = -x$ and on that line (8) gives the derivative $f'(z) = 2x = -2y$.

▪ **Harmonic Functions** We saw in Chapter 13 that Laplace's equation $\partial^2 u/\partial x^2 + \partial^2 u/\partial y^2 = 0$ occurs in certain problems involving steady-state temperatures. This partial differential equation also plays an important role in many areas of applied mathematics. Indeed, as we now see, the real and imaginary parts of an analytic function cannot be chosen arbitrarily, since both u and v must satisfy Laplace's equation. It is this link between analytic functions and Laplace's equation that makes complex variables so essential in the serious study of applied mathematics.

DEFINITION 17.5.1 Harmonic Functions

A real-valued function $\phi(x, y)$ that has continuous second-order partial derivatives in a domain D and satisfies Laplace's equation is said to be **harmonic** in D.

Suppose $f(z) = u(x, y) + iv(x, y)$ is analytic in a domain D. Then the functions $u(x, y)$ and $v(x, y)$ are harmonic functions.

We will see in Chapter 18 that an analytic function possesses derivatives of all orders.

▶ **PROOF:** In this proof we shall assume that u and v have continuous second-order partial derivatives. Since f is analytic, the Cauchy–Riemann equations are satisfied. Differentiating both sides of $\partial u/\partial x = \partial v/\partial y$ with respect to x and differentiating both sides of $\partial u/\partial y = -\partial v/\partial x$ with respect to y then give

$$\frac{\partial^2 u}{\partial x^2} = \frac{\partial^2 v}{\partial x\,\partial y} \qquad \text{and} \qquad \frac{\partial^2 u}{\partial y^2} = -\frac{\partial^2 v}{\partial y\,\partial x}.$$

With the assumption of continuity, the mixed partials are equal. Hence, adding these two equations gives

$$\frac{\partial^2 u}{\partial x^2} + \frac{\partial^2 u}{\partial y^2} = 0.$$

This shows that $u(x, y)$ is harmonic.

Now differentiating both sides of $\partial u/\partial x = \partial v/\partial y$ with respect to y and differentiating both sides of $\partial u/\partial y = -\partial v/\partial x$ with respect to x and subtracting yield

$$\frac{\partial^2 v}{\partial x^2} + \frac{\partial^2 v}{\partial y^2} = 0.$$ ≡

■ **Harmonic Conjugate Functions** If $f(z) = u(x, y) + iv(x, y)$ is analytic in a domain D, then u and v are harmonic in D. Now suppose $u(x, y)$ is a given function that is harmonic in D. It is then sometimes possible to find another function $v(x, y)$ that is harmonic in D so that $u(x, y) + iv(x, y)$ is an analytic function in D. The function v is called a **harmonic conjugate function** of u.

EXAMPLE 4 **Harmonic Function/Harmonic Conjugate Function**

(a) Verify that the function $u(x, y) = x^3 - 3xy^2 - 5y$ is harmonic in the entire complex plane.
(b) Find the harmonic conjugate function of u.

SOLUTION **(a)** From the partial derivatives

$$\frac{\partial u}{\partial x} = 3x^2 - 3y^2, \quad \frac{\partial^2 u}{\partial x^2} = 6x, \quad \frac{\partial u}{\partial y} = -6xy - 5, \quad \frac{\partial^2 u}{\partial y^2} = -6x$$

we see that u satisfies Laplace's equation:

$$\frac{\partial^2 u}{\partial x^2} + \frac{\partial^2 u}{\partial y^2} = 6x - 6x = 0.$$

(b) Since the harmonic conjugate function v must satisfy the Cauchy–Riemann equations, we must have

$$\frac{\partial v}{\partial y} = \frac{\partial u}{\partial x} = 3x^2 - 3y^2 \qquad \text{and} \qquad \frac{\partial v}{\partial x} = -\frac{\partial u}{\partial y} = 6xy + 5. \tag{9}$$

Partial integration of the first equation in (9) with respect to y gives $v(x, y) = 3x^2 y - y^3 + h(x)$. From this we get

$$\frac{\partial v}{\partial x} = 6xy + h'(x).$$

Substituting this result into the second equation in (9) gives $h'(x) = 5$, and so $h(x) = 5x + C$. Therefore, the harmonic conjugate function of u is $v(x, y) = 3x^2 y - y^3 + 5x + C$. The analytic function is $f(z) = x^3 - 3xy^2 - 5y + i(3x^2 y - y^3 + 5x + C)$. ≡

REMARKS

Suppose u and v are the harmonic functions that comprise the real and imaginary parts of an analytic function $f(z)$. The level curves $u(x, y) = c_1$ and $v(x, y) = c_2$ defined by these functions form two **orthogonal families of curves**. See Problem 32 in Exercises 17.5. For example, the level curves generated by the simple analytic function $f(z) = z = x + iy$ are $x = c_1$ and $y = c_2$. The family of vertical lines defined by $x = c_1$ is clearly orthogonal to the family of horizontal lines defined by $y = c_2$. In electrostatics, if $u(x, y) = c_1$ defines the **equipotential curves**, then the other, and orthogonal, family $v(x, y) = c_2$ defines the lines of force.

17.5 Exercises Answers to selected odd-numbered problems begin on page ANS-45.

In Problems 1 and 2, the given function is analytic for all z. Show that the Cauchy–Riemann equations are satisfied at every point.

1. $f(z) = z^3$

2. $f(z) = 3z^2 + 5z - 6i$

In Problems 3–8, show that the given function is not analytic at any point.

3. $f(z) = \text{Re}(z)$

4. $f(z) = y + ix$

5. $f(z) = 4z - 6\bar{z} + 3$

6. $f(z) = \bar{z}^2$

7. $f(z) = x^2 + y^2$

8. $f(x) = \dfrac{x}{x^2 + y^2} + i\dfrac{y}{x^2 + y^2}$

In Problems 9–14, use Theorem 17.5.2 to show that the given function is analytic in an appropriate domain.

9. $f(z) = e^x \cos y + ie^x \sin y$

10. $f(z) = x + \sin x \cosh y + i(y + \cos x \sinh y)$

11. $f(z) = e^{x^2 - y^2} \cos 2xy + ie^{x^2 - y^2} \sin 2xy$

12. $f(z) = 4x^2 + 5x - 4y^2 + 9 + i(8xy + 5y - 1)$

13. $f(z) = \dfrac{x - 1}{(x - 1)^2 + y^2} - i\dfrac{y}{(x - 1)^2 + y^2}$

14. $f(x) = \dfrac{x^3 + xy^2 + x}{x^2 + y^2} + i\dfrac{x^2y + y^3 - y}{x^2 + y^2}$

In Problems 15 and 16, find real constants a, b, c, and d so that the given function is analytic.

15. $f(z) = 3x - y + 5 + i(ax + by - 3)$

16. $f(z) = x^2 + axy + by^2 + i(cx^2 + dxy + y^2)$

In Problems 17–20, show that the given function is not analytic at any point but is differentiable along the indicated curve(s).

17. $f(z) = x^2 + y^2 + 2xyi$; x-axis

18. $f(z) = 3x^2y^2 - 6x^2y^2i$; coordinate axes

19. $f(z) = x^3 + 3xy^2 - x + i(y^3 + 3x^2y - y)$; coordinate axes

20. $f(z) = x^2 - x + y + i(y^2 - 5y - x)$; $y = x + 2$

21. Use (8) to find the derivative of the function in Problem 9.

22. Use (8) to find the derivative of the function in Problem 11.

In Problems 23–28, verify that the given function u is harmonic. Find v, the harmonic conjugate function of u. Form the corresponding analytic function $f(z) = u + iv$.

23. $u(x, y) = x$

24. $u(x, y) = 2x - 2xy$

25. $u(x, y) = x^2 - y^2$

26. $u(x, y) = 4xy^3 - 4x^3y + x$

27. $u(x, y) = \log_e(x^2 + y^2)$

28. $u(x, y) = e^x(x \cos y - y \sin y)$

29. Sketch the level curves $u(x, y) = c_1$ and $v(x, y) = c_2$ of the analytic function $f(z) = z^2$.

30. Consider the function $f(z) = 1/z$. Describe the level curves.

31. Consider the function $f(z) = z + 1/z$. Describe the level curve $v(x, y) = 0$.

32. Suppose u and v are the harmonic functions forming the real and imaginary parts of an analytic function. Show that the level curves $u(x, y) = c_1$ and $v(x, y) = c_2$ are orthogonal. [*Hint*: Consider the gradient of u and the gradient of v. Ignore the case where a gradient vector is the zero vector.]

17.6 Exponential and Logarithmic Functions

INTRODUCTION In this and the next section, we shall examine the exponential, logarithmic, trigonometric, and hyperbolic functions of a complex variable z. Although the definitions of these complex functions are motivated by their real variable analogues, the properties of these complex functions will yield some surprises.

■ **Exponential Function** Recall that in real variables the exponential function $f(x) = e^x$ has the properties

$$f'(x) = f(x) \quad \text{and} \quad f(x_1 + x_2) = f(x_1)f(x_2). \tag{1}$$

We certainly want the definition of the complex function $f(z) = e^z$, where $z = x + iy$, to reduce e^x for $y = 0$ and to possess the same properties as in (1).

We have already used an exponential function with a pure imaginary exponent. Euler's formula,

$$e^{iy} = \cos y + i \sin y, \quad y \text{ a real number,} \tag{2}$$

played an important role in Section 3.3. We can formally establish the result in (2) by using the Maclaurin series for e^x and replacing x by iy and rearranging terms:

$$e^{iy} = \sum_{k=0}^{\infty} \frac{(iy)^k}{k!} = 1 + iy + \frac{(iy)^2}{2!} + \frac{(iy)^3}{3!} + \frac{(iy)^4}{4!} + \cdots$$

Maclaurin series for $\cos y$ and $\sin y$ ▶

$$= \left(1 - \frac{y^2}{2!} + \frac{y^4}{4!} - \frac{y^6}{6!} + \cdots\right) + i\left(y - \frac{y^3}{3!} + \frac{y^5}{5!} - \frac{y^7}{7!} + \cdots\right)$$

$$= \cos y + i \sin y.$$

For $z = x + iy$, it is natural to expect that

$$e^{x+iy} = e^x e^{iy}$$

and so by (2),

$$e^{x+iy} = e^x(\cos y + i \sin y).$$

Inspired by this formal result, we make the following definition.

DEFINITION 17.6.1 Exponential Function

$$e^z = e^{x+iy} = e^x(\cos y + i \sin y). \tag{3}$$

The exponential function e^z is also denoted by the symbol $\exp z$. Note that (3) reduces to e^x when $y = 0$.

EXAMPLE 1 Complex Value of the Exponential Function

Evaluate $e^{1.7+4.2i}$.

SOLUTION With the identifications $x = 1.7$ and $y = 4.2$ and the aid of a calculator, we have, to four rounded decimal places,

$$e^{1.7} \cos 4.2 = -2.6837 \quad \text{and} \quad e^{1.7} \sin 4.2 = -4.7710.$$

It follows from (3) that $e^{1.7+4.2i} = -2.6837 - 4.7710i$. ≡

The real and imaginary parts of e^z, $u(x, y) = e^x \cos y$ and $v(x, y) = e^x \sin y$, are continuous and have continuous first partial derivatives at every point z of the complex plane. Moreover, the Cauchy–Riemann equations are satisfied at all points of the complex plane:

$$\frac{\partial u}{\partial x} = e^x \cos y = \frac{\partial v}{\partial y} \quad \text{and} \quad \frac{\partial u}{\partial y} = -e^x \sin y = -\frac{\partial v}{\partial x}.$$

It follows from Theorem 17.5.2 that $f(z) = e^z$ is analytic for all z; in other words, f is an entire function.

Properties We shall now demonstrate that e^z possesses the two desired properties given in (1). First, the derivative of f is given by (5) of Section 17.5:

$$f'(z) = e^x \cos y + i(e^x \sin y) = e^x(\cos y + i \sin y) = f(z).$$

As desired, we have established that

$$\frac{d}{dz} e^z = e^z.$$

FIGURE 17.6.1 Values of $f(z) = e^z$ at the four points are the same

FIGURE 17.6.2 Flow over the fundamental region

Second, if $z_1 = x_1 + iy_1$ and $z_2 = x_2 + iy_2$, then by multiplication of complex numbers and the addition formulas of trigonometry, we obtain

$$f(z_1)f(z_2) = e^{x_1}(\cos y_1 + i \sin y_1)e^{x_2}(\cos y_2 + i \sin y_2)$$

$$= e^{x_1 + x_2}[(\cos y_1 \cos y_2 - \sin y_1 \sin y_2) + i(\sin y_1 \cos y_2 + \cos y_1 \sin y_2)]$$

$$= e^{x_1 + x_2}[(\cos(y_1 + y_2) + i \sin(y_1 + y_2)] = f(z_1 + z_2).$$

In other words, $\qquad\qquad e^{z_1}e^{z_2} = e^{z_1 + z_2}.$ (4)

It is left as an exercise to prove that

$$\frac{e^{z_1}}{e^{z_2}} = e^{z_1 - z_2}.$$

■ **Periodicity** Unlike the real function e^x, the complex function $f(z) = e^z$ is **periodic** with the complex period $2\pi i$. Since $e^{2\pi i} = \cos 2\pi + i \sin 2\pi = 1$ and, in view of (4), $e^{z+2\pi i} = e^z e^{2\pi i} = e^z$ for all z, it follows that $f(z + 2\pi i) = f(z)$. Because of this complex periodicity, all possible functional values of $f(z) = e^z$ are assumed in any infinite horizontal strip of width 2π. Thus, if we divide the complex plane into horizontal strips defined by $(2n - 1)\pi < y \le (2n + 1)\pi$, $n = 0, \pm 1, \pm 2, \ldots$, then, as shown in **FIGURE 17.6.1**, for any point z in the strip $-\pi < y \le \pi$, the values $f(z), f(z + 2\pi i), f(z - 2\pi i), f(z + 4\pi i)$, and so on, are the same. The strip $-\pi < y \le \pi$ is called the **fundamental region** for the exponential function $f(z) = e^z$. The corresponding flow over the fundamental region is shown in **FIGURE 17.6.2**.

■ **Polar Form of a Complex Number** In Section 17.2, we saw that the complex number z could be written in polar form as $z = r(\cos \theta + i \sin \theta)$. Since $e^{i\theta} = \cos \theta + i \sin \theta$, we can now write the polar form of a complex number as

$$z = re^{i\theta}.$$

For example, in polar form $z = 1 + i$ is $z = \sqrt{2}e^{\pi i/4}$.

■ **Circuits** In applying mathematics, mathematicians and engineers often approach the same problem in completely different ways. Consider, for example, the solution of Example 10 in Section 3.8. In this example we used strictly real analysis to find the steady-state current $i_p(t)$ in an *LRC*-series circuit described by the differential equation

$$L\frac{d^2q}{dt^2} + R\frac{dq}{dt} + \frac{1}{C}q = E_0 \sin \gamma t.$$

Electrical engineers often solve circuit problems such as this using complex analysis. To illustrate, let us first denote the imaginary unit $\sqrt{-1}$ by the symbol j to avoid confusion with the current i. Since current i is related to charge q by $i = dq/dt$, the differential equation is the same as

$$L\frac{di}{dt} + Ri + \frac{1}{C}q = E_0 \sin \gamma t.$$

Moreover, the impressed voltage $E_0 \sin \gamma t$ can be replaced by $\text{Im}(E_0 e^{j\gamma t})$, where Im means the "imaginary part of." Because of this last form, the method of undetermined coefficients suggests that we assume a solution in the form of a constant multiple of complex exponential—that is, $i_p(t) = \text{Im}(Ae^{j\gamma t})$. We substitute this expression into the last differential equation, use the fact that q is an antiderivative of i, and equate coefficients of $e^{j\gamma t}$:

$$\left(jL\gamma + R + \frac{1}{jC\gamma}\right)A = E_0 \quad \text{gives} \quad A = \frac{E_0}{R + j\left(L\gamma - \dfrac{1}{C\gamma}\right)}.$$

The quantity $Z = R + j(L\gamma - 1/C\gamma)$ is called the **complex impedance** of the circuit. Note that the modulus of the complex impedance, $|Z| = \sqrt{R^2 + (L\gamma - 1/C\gamma)^2}$, was denoted in Example 10 of Section 3.8 by the letter Z and called the impedance.

Now, in polar form the complex impedance is

$$Z = |Z|e^{j\theta} \quad \text{where} \quad \tan\theta = \frac{L\gamma - \dfrac{1}{C\gamma}}{R}.$$

Hence, $A = E_0/Z = E_0/(|Z|e^{j\theta})$, and so the steady-state current can be written as

$$i_p(t) = \operatorname{Im}\frac{E_0}{|Z|}e^{-j\theta}e^{j\gamma t}.$$

The reader is encouraged to verify that this last expression is the same as (35) in Section 3.8.

■ Logarithmic Function

The logarithm of a complex number $z = x + iy$, $z \neq 0$, is defined as the inverse of the exponential function—that is,

$$w = \ln z \quad \text{if} \quad z = e^w. \tag{5}$$

In (5) we note that $\ln z$ is not defined for $z = 0$, since there is no value of w for which $e^w = 0$. To find the real and imaginary parts of $\ln z$, we write $w = u + iv$ and use (3) and (5):

$$x + iy = e^{u+iv} = e^u(\cos v + i \sin v) = e^u \cos v + ie^u \sin v.$$

The last equality implies $x = e^u \cos v$ and $y = e^u \sin v$. We can solve these two equations for u and v. First, by squaring and adding the equations, we find

$$e^{2u} = x^2 + y^2 = r^2 = |z|^2 \quad \text{and so} \quad u = \log_e|z|,$$

where $\log_e|z|$ denotes the *real natural logarithm* of the modulus of z. Second, to solve for v, we divide the two equations to obtain

$$\tan v = \frac{y}{x}.$$

This last equation means that v is an argument of z; that is, $v = \theta = \arg z$. But since there is no unique argument of a given complex number $z = x + iy$, if θ is an argument of z, then so is $\theta + 2n\pi$, $n = 0, \pm1, \pm2, \dots$.

DEFINITION 17.6.2 **Logarithm of a Complex Number**

For $z \neq 0$, and $\theta = \arg z$,

$$\ln z = \log_e|z| + i(\theta + 2n\pi), \quad n = 0, \pm1, \pm2, \dots. \tag{6}$$

As is clearly indicated in (6), there are infinitely many values of the logarithm of a complex number z. This should not be any great surprise since the exponential function is periodic.

In real calculus, logarithms of negative numbers are not defined. As the next example will show, this is not the case in complex calculus.

EXAMPLE 2 Complex Values of the Logarithmic Function

Find the values of **(a)** $\ln(-2)$, **(b)** $\ln i$, and **(c)** $\ln(-1 - i)$.

SOLUTION **(a)** With $\theta = \arg(-2) = \pi$ and $\log_e|-2| = 0.6932$, we have from (6)

$$\ln(-2) = 0.6932 + i(\pi + 2n\pi).$$

(b) With $\theta = \arg(i) = \pi/2$ and $\log_e|i| = \log_e 1 = 0$, we have from (6)

$$\ln i = i\left(\frac{\pi}{2} + 2n\pi\right).$$

In other words, $\ln i = \pi i/2, -3\pi i/2, 5\pi i/2, -7\pi i/2$, and so on.

(c) With $\theta = \arg(-1-i) = 5\pi/4$ and $\log_e|-1-i| = \log_e\sqrt{2} = 0.3466$, we have from (6)

$$\ln(-1-i) = 0.3466 + i\left(\frac{5\pi}{4} + 2n\pi\right). \qquad \equiv$$

EXAMPLE 3 **Solving an Exponential Equation**

Find all values of z such that $e^z = \sqrt{3} + i$.

SOLUTION From (5), with the symbol w replaced by z, we have $z = \ln(\sqrt{3} + i)$. Now $|\sqrt{3} + i| = 2$ and $\tan\theta = 1/\sqrt{3}$ imply that $\arg(\sqrt{3} + i) = \pi/6$, and so (6) gives

$$z = \log_e 2 + i\left(\frac{\pi}{6} + 2n\pi\right) \quad \text{or} \quad z = 0.6931 + i\left(\frac{\pi}{6} + 2n\pi\right). \qquad \equiv$$

▨ **Principal Value** It is interesting to note that as a consequence of (6), the logarithm of a positive real number has many values. For example, in real calculus, $\log_e 5$ has only one value: $\log_e 5 = 1.6094$, whereas in complex calculus, $\ln 5 = 1.6094 + 2n\pi i$. The value of $\ln 5$ corresponding to $n = 0$ is the same as the real logarithm $\log_e 5$ and is called the **principal value** of $\ln 5$. Recall that in Section 17.2 we stipulated that the principal argument of a complex number, written Arg z, lies in the interval $(-\pi, \pi]$. In general, we define the **principal value** of $\ln z$ as that complex logarithm corresponding to $n = 0$ and $\theta = \text{Arg } z$. To emphasize the principal value of the logarithm, we shall adopt the notation Ln z. In other words,

$$\text{Ln } z = \log_e|z| + i \text{ Arg } z. \qquad (7)$$

Since Arg z is unique, there is only one value of Ln z for each $z \neq 0$.

EXAMPLE 4 **Principal Values**

The principal values of the logarithms in Example 2 are as follows:

(a) Since $\text{Arg}(-2) = \pi$, we need only set $n = 0$ in the result given in part (a) of Example 2:

$$\text{Ln}(-2) = 0.6932 + \pi i.$$

(b) Similarly, since $\text{Arg}(i) = \pi/2$, we set $n = 0$ in the result in part (b) of Example 2 to obtain

$$\text{Ln } i = \frac{\pi}{2} i.$$

(c) In part (c) of Example 2, $\arg(-1-i) = 5\pi/4$ is not the principal argument of $z = -1 - i$. The argument of z that lies in the interval $(-\pi, \pi]$ is $\text{Arg}(-1-i) = -3\pi/4$. Hence, it follows from (7) that

$$\text{Ln}(-1-i) = 0.3466 - \frac{3\pi}{4} i. \qquad \equiv$$

Up to this point we have avoided the use of the word *function* for the obvious reason that $\ln z$ defined in (6) is not a function in the strictest interpretation of that word. Nonetheless, it is customary to write $f(z) = \ln z$ and to refer to $f(z) = \ln z$ by the seemingly contradictory phrase *multiple-valued function*. Although we shall not pursue the details, (6) can be interpreted as an infinite collection of logarithmic functions (standard meaning of the word). Each function in the collection is called a **branch** of $\ln z$. The function $f(z) = \text{Ln } z$ is then called the **principal branch** of $\ln z$, or the **principal logarithmic function**. To minimize the confusion, we shall hereafter simply use the words *logarithmic function* when referring to either $f(z) = \ln z$ or $f(z) = \text{Ln } z$.

Some familiar properties of the logarithmic function hold in the complex case:

$$\ln(z_1 z_2) = \ln z_1 + \ln z_2 \qquad \text{and} \qquad \ln\left(\frac{z_1}{z_2}\right) = \ln z_1 - \ln z_2. \qquad (8)$$

Equations (8) and (9) are to be interpreted in the sense that if values are assigned to two of the terms, then a correct value is assigned to the third term.

EXAMPLE 5 **Properties of Logarithms**

Suppose $z_1 = 1$ and $z_2 = -1$. Then if we take $\ln z_1 = 2\pi i$ and $\ln z_2 = \pi i$, we get

$$\ln(z_1 z_2) = \ln(-1) = \ln z_1 + \ln z_2 = 2\pi i + \pi i = 3\pi i$$

$$\ln\left(\frac{z_1}{z_2}\right) = \ln(-1) = \ln z_1 - \ln z_2 = 2\pi i - \pi i = \pi i. \qquad \equiv$$

Just as (7) of Section 17.2 was not valid when arg z was replaced with Arg z, so too (8) is not true, in general, when $\ln z$ is replaced by Ln z. See Problems 45 and 46 in Exercises 17.6.

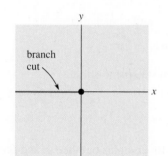

■ Analyticity The logarithmic function $f(z) = \text{Ln } z$ is not continuous at $z = 0$ since $f(0)$ is not defined. Moreover, $f(z) = \text{Ln } z$ is discontinuous at all points of the negative real axis. This is because the imaginary part of the function, $v = \text{Arg } z$, is discontinuous only at these points. To see this, suppose x_0 is a point on the negative real axis. As $z \to x_0$ from the upper half-plane, Arg $z \to \pi$, whereas if $z \to x_0$ from the lower half-plane, then Arg $z \to -\pi$. This means that $f(z) = \text{Ln } z$ is not analytic on the nonpositive real axis. However, $f(z) = \text{Ln } z$ is analytic throughout the domain D consisting of all the points in the complex plane except those on the nonpositive real axis. It is convenient to think of D as the complex plane from which the nonpositive real axis has been cut out. Since $f(z) = \text{Ln } z$ is the principal branch of $\ln z$, the nonpositive real axis is referred to as a **branch cut** for the function. See **FIGURE 17.6.3**. It is left as exercises to show that the Cauchy–Riemann equations are satisfied throughout this cut plane and that the derivative of Ln z is given by

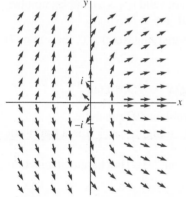

FIGURE 17.6.3 Branch cut for Ln z

$$\frac{d}{dz} \text{Ln } z = \frac{1}{z} \qquad (9)$$

for all z in D.

FIGURE 17.6.4 shows $w = \text{Ln } z$ as a flow. Note that the vector field is not continuous along the branch cut.

■ Complex Powers Inspired by the identity $x^a = e^{a \ln x}$ in real variables, we can define complex powers of a complex number. If α is a complex number and $z = x + iy$, then z^α is defined by

$$z^\alpha = e^{\alpha \ln z}, \quad z \neq 0. \qquad (10)$$

In general, z^α is multiple-valued since $\ln z$ is multiple-valued. However, in the special case when $\alpha = n, n = 0, \pm 1, \pm 2, \ldots$, (10) is single-valued since there is only one value for z^2, z^3, z^{-1}, and so on. To see that this is so, suppose $\alpha = 2$ and $z = re^{i\theta}$, where θ is any argument of z. Then

FIGURE 17.6.4 $w = \text{Ln } z$ as a flow

$$e^{2 \ln z} = e^{2(\log_e r + i\theta)} = e^{2 \log_e r + 2i\theta} = e^{2 \log_e r} e^{2i\theta} = r^2 e^{i\theta} e^{i\theta} = (re^{i\theta})(re^{i\theta}) = z^2.$$

If we use Ln z in place of $\ln z$, then (10) gives the **principal value** of z^α.

EXAMPLE 6 **Complex Power**

Find the value of i^{2i}.

SOLUTION With $z = i$, arg $z = \pi/2$, and $\alpha = 2i$, it follows from (10) that

$$i^{2i} = e^{2i[\log_e 1 + i(\pi/2 + 2n\pi)]} = e^{-(1 + 4n)\pi}$$

where $n = 0, \pm 1, \pm 2, \ldots$. Inspection of the equation shows that i^{2i} is real for every value of n. Since $\pi/2$ is the principal argument of $z = i$, we obtain the principal value of i^{2i} for $n = 0$. To four rounded decimal places, this principal value is $i^{2i} = e^{-\pi} = 0.0432$. $\equiv$

In Problems 1–10, express e^z in the form $a + ib$.

1. $z = \dfrac{\pi}{6}i$

2. $z = -\dfrac{\pi}{3}i$

3. $z = -1 + \dfrac{\pi}{4}i$

4. $z = 2 - \dfrac{\pi}{2}i$

5. $z = \pi + \pi i$

6. $z = -\pi + \dfrac{3\pi}{2}i$

7. $z = 1.5 + 2i$

8. $z = -0.3 + 0.5i$

9. $z = 5i$

10. $z = -0.23 - i$

In Problems 11 and 12, express the given number in the form $a + ib$.

11. $e^{1+5\pi i/4}e^{-1-\pi i/3}$

12. $\dfrac{e^{2+3\pi i}}{e^{-3+\pi i/2}}$

In Problems 13–16, use Definition 17.6.1 to express the given function in the form $f(z) = u + iv$.

13. $f(z) = e^{-iz}$

14. $f(z) = e^{2\bar{z}}$

15. $f(z) = e^{z^2}$

16. $f(z) = e^{1/z}$

In Problems 17–20, verify the given result.

17. $|e^z| = e^x$

18. $\dfrac{e^{z_1}}{e^{z_2}} = e^{z_1 - z_2}$

19. $e^{z+\pi i} = e^{z-\pi i}$

20. $(e^z)^n = e^{nz}$, n an integer

21. Show that $f(z) = e^{\bar{z}}$ is nowhere analytic.

22. (a) Use the result in Problem 15 to show that $f(z) = e^{z^2}$ is an entire function.

 (b) Verify that $u(x, y) = \text{Re}(e^{z^2})$ is a harmonic function.

In Problems 23–28, express $\ln z$ in the form $a + ib$.

23. $z = -5$

24. $z = -ei$

25. $z = -2 + 2i$

26. $z = 1 + i$

27. $z = \sqrt{2} + \sqrt{6}i$

28. $z = -\sqrt{3} + i$

In Problems 29–34, express $\text{Ln } z$ in the form $a + ib$.

29. $z = 6 - 6i$

30. $z = -e^3$

31. $z = -12 + 5i$

32. $z = 3 - 4i$

33. $z = (1 + \sqrt{3}i)^5$

34. $z = (1 + i)^4$

In Problems 35–38, find all values of z satisfying the given equation.

35. $e^z = 4i$

36. $e^{1/z} = -1$

37. $e^{z-1} = -ie^2$

38. $e^{2z} + e^z + 1 = 0$

In Problems 39–42, find all values of the given quantity.

39. $(-i)^{4i}$

40. $3^{i/\pi}$

41. $(1 + i)^{(1 + i)}$

42. $(1 + \sqrt{3}i)^{3i}$

In Problems 43 and 44, find the principal value of the given quantity. Express answers in the form $a + ib$.

43. $(-1)^{(-2i/\pi)}$

44. $(1 - i)^{2i}$

45. If $z_1 = i$ and $z_2 = -1 + i$, verify that

$$\text{Ln}(z_1 z_2) \neq \text{Ln } z_1 + \text{Ln } z_2.$$

46. Find two complex numbers z_1 and z_2 such that

$$\text{Ln}(z_1/z_2) \neq \text{Ln } z_1 - \text{Ln } z_2.$$

47. Determine whether the given statement is true.

 (a) $\text{Ln}(-1 + i)^2 = 2\,\text{Ln}(-1 + i)$

 (b) $\text{Ln } i^3 = 3\,\text{Ln } i$

 (c) $\ln i^3 = 3 \ln i$

48. The laws of exponents hold for complex numbers α and β:

$$z^{\alpha}z^{\beta} = z^{\alpha+\beta}, \quad \frac{z^{\alpha}}{z^{\beta}} = z^{\alpha-\beta}, \quad (z^{\alpha})^n = z^{n\alpha}, \quad n \text{ an integer.}$$

However, the last law is not valid if n is a complex number. Verify that $(i^i)^2 = i^{2i}$, but $(i^2)^i \neq i^{2i}$.

49. For complex numbers z satisfying $\text{Re}(z) > 0$, show that (7) can be written as

$$\text{Ln } z = \frac{1}{2}\log_e(x^2 + y^2) + i\tan^{-1}\frac{y}{x}.$$

50. The function given in Problem 49 is analytic.

 (a) Verify that $u(x, y) = \log_e(x^2 + y^2)$ is a harmonic function.

 (b) Verify that $v(x, y) = \tan^{-1}(y/x)$ is a harmonic function.

17.7 Trigonometric and Hyperbolic Functions

INTRODUCTION In this section we define the complex trigonometric and hyperbolic functions. Analogous to the complex functions e^z and $\text{Ln } z$ defined in the previous section, these functions will agree with their real counterparts for real values of z. In addition, we will show that the complex trigonometric and hyperbolic functions have the same derivatives and satisfy many of the same identities as the real trigonometric and hyperbolic functions.

Trigonometric Functions If x is a real variable, then Euler's formula gives

$$e^{ix} = \cos x + i \sin x \quad \text{and} \quad e^{-ix} = \cos x - i \sin x.$$

By subtracting and then adding these equations, we see that the real functions $\sin x$ and $\cos x$ can be expressed as a combination of exponential functions:

$$\sin x = \frac{e^{ix} - e^{-ix}}{2i}, \quad \cos x = \frac{e^{ix} + e^{-ix}}{2}. \tag{1}$$

Using (1) as a model, we now define the sine and cosine of a complex variable:

DEFINITION 17.7.1 Trigonometric Sine and Cosine

For any complex number $z = x + iy$,

$$\sin z = \frac{e^{iz} - e^{-iz}}{2i} \quad \text{and} \quad \cos z = \frac{e^{iz} + e^{-iz}}{2}. \tag{2}$$

As in trigonometry, we define four additional trigonometric functions in terms of $\sin z$ and $\cos z$:

$$\tan z = \frac{\sin z}{\cos z}, \quad \cot z = \frac{1}{\tan z}, \quad \sec z = \frac{1}{\cos z}, \quad \csc z = \frac{1}{\sin z}. \tag{3}$$

When $y = 0$, each function in (2) and (3) reduces to its real counterpart.

Analyticity Since the exponential functions e^{iz} and e^{-iz} are entire functions, it follows that $\sin z$ and $\cos z$ are entire functions. Now, as we shall see shortly, $\sin z = 0$ only for the real numbers $z = n\pi$, n an integer, and $\cos z = 0$ only for the real numbers $z = (2n + 1)\pi/2$, n an integer. Thus, $\tan z$ and $\sec z$ are analytic except at the points $z = (2n + 1)\pi/2$, and $\cot z$ and $\csc z$ are analytic except at the points $z = n\pi$.

Derivatives Since $(d/dz)e^z = e^z$, it follows from the Chain Rule that $(d/dz)e^{iz} = ie^{iz}$ and $(d/dz)e^{-iz} = -ie^{-iz}$. Hence,

$$\frac{d}{dz} \sin z = \frac{d}{dz} \frac{e^{iz} - e^{-iz}}{2i} = \frac{e^{iz} + e^{-iz}}{2} = \cos z.$$

In fact, it is readily shown that the forms of the derivatives of the complex trigonometric functions are the same as the real functions. We summarize the results:

$$\frac{d}{dz} \sin z = \cos z \qquad\qquad \frac{d}{dz} \cos z = -\sin z$$

$$\frac{d}{dz} \tan z = \sec^2 z \qquad\qquad \frac{d}{dz} \cot z = -\csc^2 z \tag{4}$$

$$\frac{d}{dz} \sec z = \sec z \tan z \qquad\qquad \frac{d}{dz} \csc z = -\csc z \cot z.$$

Identities The familiar trigonometric identities are also the same in the complex case:

$$\sin(-z) = -\sin z \quad \cos(-z) = \cos z$$

$$\cos^2 z + \sin^2 z = 1$$

$$\sin(z_1 \pm z_2) = \sin z_1 \cos z_2 \pm \cos z_1 \sin z_2$$

$$\cos(z_1 \pm z_2) = \cos z_1 \cos z_2 \mp \sin z_1 \sin z_2$$

$$\sin 2z = 2 \sin z \cos z \quad \cos 2z = \cos^2 z - \sin^2 z.$$

Zeros To find the zeros of $\sin z$ and $\cos z$ we need to express both functions in the form $u + iv$. Before proceeding, recall from calculus that if y is real, then the hyperbolic sine and hyperbolic cosine are defined in terms of the real exponential functions e^y and e^{-y}:

$$\sinh y = \frac{e^y - e^{-y}}{2} \quad \text{and} \quad \cosh y = \frac{e^y + e^{-y}}{2}. \tag{5}$$

Now from Definition 17.7.1 and Euler's formula we find, after simplifying,

$$\sin z = \frac{e^{i(x+iy)} - e^{-i(x+iy)}}{2i} = \sin x \left(\frac{e^y + e^{-y}}{2} \right) + i \cos x \left(\frac{e^y - e^{-y}}{2} \right).$$

Thus from (5) we have

$$\sin z = \sin x \cosh y + i \cos x \sinh y. \tag{6}$$

It is left as an exercise to show that

$$\cos z = \cos x \cosh y - i \sin x \sinh y. \tag{7}$$

From (6), (7), and $\cosh^2 y = 1 + \sinh^2 y$, we find

$$|\sin z|^2 = \sin^2 x + \sinh^2 y \tag{8}$$

and

$$|\cos z|^2 = \cos^2 x + \sinh^2 y. \tag{9}$$

Now a complex number z is zero if and only if $|z|^2 = 0$. Thus, if $\sin z = 0$, then from (8) we must have $\sin^2 x + \sinh^2 y = 0$. This implies that $\sin x = 0$ and $\sinh y = 0$, and so $x = n\pi$ and $y = 0$. Thus the only **zeros** of $\sin z$ are the real numbers $z = n\pi + 0i = n\pi, n = 0, \pm 1, \pm 2, \ldots$. Similarly, it follows from (9) that $\cos z = 0$ only when $z = (2n + 1)\pi/2, n = 0, \pm 1, \pm 2, \ldots$.

EXAMPLE 1 Complex Value of the Sine Function

From (6) we have, with the aid of a calculator,

$$\sin(2 + i) = \sin 2 \cosh 1 + i \cos 2 \sinh 1 = 1.4031 - 0.4891i. \qquad \equiv$$

In ordinary trigonometry we are accustomed to the fact that $|\sin x| \le 1$ and $|\cos x| \le 1$. Inspection of (8) and (9) shows that these inequalities do not hold for the complex sine and cosine, since $\sinh y$ can range from $-\infty$ to ∞. In other words, it is perfectly feasible to have solutions for equations such as $\cos z = 10$.

EXAMPLE 2 Solving a Trigonometric Equation

Solve the equation $\cos z = 10$.

SOLUTION From (2), $\cos z = 10$ is equivalent to $(e^{iz} + e^{-iz})/2 = 10$. Multiplying the last equation by e^{iz} then gives the quadratic equation in e^{iz}:

$$e^{2iz} - 20e^{iz} + 1 = 0.$$

From the quadratic formula we find $e^{iz} = 10 \pm 3\sqrt{11}$. Thus, for $n = 0, \pm 1, \pm 2, \ldots$, we have $iz = \log_e(10 \pm 3\sqrt{11}) + 2n\pi i$. Dividing by i and utilizing $\log_e(10 - 3\sqrt{11}) = -\log_e(10 + 3\sqrt{11})$, we can express the solutions of the given equation as $z = 2n\pi \pm i \log_e(10 + 3\sqrt{11})$. $\qquad \equiv$

Hyperbolic Functions We define the complex hyperbolic sine and cosine in a manner analogous to the real definitions given in (5).

DEFINITION 17.7.2 Hyperbolic Sine and Cosine

For any complex number $z = x + iy$,

$$\sinh z = \frac{e^z - e^{-z}}{2} \quad \text{and} \quad \cosh z = \frac{e^z + e^{-z}}{2}. \tag{10}$$

The hyperbolic tangent, cotangent, secant, and cosecant functions are defined in terms of $\sinh z$ and $\cosh z$:

$$\tanh z = \frac{\sinh z}{\cosh z}, \quad \coth z = \frac{1}{\tanh z}, \quad \operatorname{sech} z = \frac{1}{\cosh z}, \quad \operatorname{csch} z = \frac{1}{\sinh z}. \tag{11}$$

The hyperbolic sine and cosine are entire functions, and the functions defined in (11) are analytic except at points where the denominators are zero. It is also easy to see from (10) that

$$\frac{d}{dz}\sinh z = \cosh z \quad \text{and} \quad \frac{d}{dz}\cosh z = \sinh z. \tag{12}$$

It is interesting to observe that, in contrast to real calculus, the trigonometric and hyperbolic functions are related in complex calculus. If we replace z by iz everywhere in (10) and compare the results with (2), we see that $\sinh(iz) = i \sin z$ and $\cosh(iz) = \cos z$. These equations enable us to express

$\sin z$ and $\cos z$ in terms of $\sinh(iz)$ and $\cosh(iz)$, respectively. Similarly, by replacing z by iz in (2) we can express, in turn, $\sinh z$ and $\cosh z$ in terms of $\sin(iz)$ and $\cos(iz)$. We summarize the results:

$$\sin z = -i \sinh(iz), \quad \cos z = \cosh(iz) \tag{13}$$

$$\sinh z = -i \sin(iz), \quad \cosh z = \cos(iz). \tag{14}$$

■ **Zeros** The relationships given in (14) enable us to derive identities for the hyperbolic functions utilizing results for the trigonometric functions. For example, to express $\sinh z$ in the form $u + iv$ we write $\sinh z = -i \sin(iz)$ in the form $\sinh z = -i \sin(-y + ix)$ and use (6):

$$\sinh z = -i[\sin(-y) \cosh x + i \cos(-y) \sinh x].$$

Since $\sin(-y) = -\sin y$ and $\cos(-y) = \cos y$, the foregoing expression simplifies to

$$\sinh z = \sinh x \cos y + i \cosh x \sin y. \tag{15}$$

Similarly, $\qquad \cosh z = \cosh x \cos y + i \sinh x \sin y. \tag{16}$

It also follows directly from (14) that the **zeros** of $\sinh z$ and $\cosh z$ are pure imaginary and are, respectively,

$$z = n\pi i \quad \text{and} \quad z = (2n + 1)\frac{\pi i}{2}, \quad n = 0, \pm1, \pm2, \dots.$$

■ **Periodicity** Since $\sin x$ and $\cos x$ are 2π-periodic, we can easily demonstrate that $\sin z$ and $\cos z$ are also periodic with the same real period 2π. For example, from (6), note that

$$\sin(z + 2\pi) = \sin(x + 2\pi + iy)$$
$$= \sin(x + 2\pi) \cosh y + i \cos(x + 2\pi) \sinh y$$
$$= \sin x \cosh y + i \cos x \sinh y;$$

that is, $\sin(z + 2\pi) = \sin z$. In exactly the same manner, it follows from (7) that $\cos(z + 2\pi) = \cos z$. In addition, the hyperbolic functions $\sinh z$ and $\cosh z$ have the imaginary period $2\pi i$. This last result follows from either Definition 17.7.2 and the fact that e^z is periodic with period $2\pi i$, or from (15) and (16) and replacing z by $z + 2\pi i$.

17.7 Exercises Answers to selected odd-numbered problems begin on page ANS-46.

In Problems 1–12, express the given quantity in the form $a + ib$.

1. $\cos(3i)$

2. $\sin(-2i)$

3. $\sin\left(\dfrac{\pi}{4} + i\right)$

4. $\cos(2 - 4i)$

5. $\tan(i)$

6. $\cot\left(\dfrac{\pi}{2} + 3i\right)$

7. $\sec(\pi + i)$

8. $\csc(1 + i)$

9. $\cosh(\pi i)$

10. $\sinh\left(\dfrac{3\pi}{2} i\right)$

11. $\sinh\left(1 + \dfrac{\pi}{3} i\right)$

12. $\cosh(2 + 3i)$

In Problems 13 and 14, verify the given result.

13. $\sin\left(\dfrac{\pi}{2} + i \ln 2\right) = \dfrac{5}{4}$

14. $\cos\left(\dfrac{\pi}{2} + i \ln 2\right) = -\dfrac{3}{4} i$

In Problems 15–20, find all values of z satisfying the given equation.

15. $\sin z = 2$

16. $\cos z = -3i$

17. $\sinh z = -i$

18. $\sinh z = -1$

19. $\cos z = \sin z$

20. $\cos z = i \sin z$

In Problems 21 and 22, use the definition of equality of complex numbers to find all values of z satisfying the given equation.

21. $\cos z = \cosh 2$

22. $\sin z = i \sinh 2$

23. Prove that $\cos z = \cos x \cosh y - i \sin x \sinh y$.

24. Prove that $\sinh z = \sinh x \cos y + i \cosh x \sin y$.

25. Prove that $\cosh z = \cosh x \cos y + i \sinh x \sin y$.

26. Prove that $|\sinh z|^2 = \sin^2 y + \sinh^2 x$.

27. Prove that $|\cosh z|^2 = \cos^2 y + \sinh^2 x$.

28. Prove that $\cos^2 z + \sin^2 z = 1$.

29. Prove that $\cosh^2 z - \sinh^2 z = 1$.

30. Show that $\tan z = u + iv$, where

$$u = \frac{\sin 2x}{\cos 2x + \cosh 2y} \quad \text{and} \quad v = \frac{\sinh 2y}{\cos 2x + \cosh 2y}.$$

31. Prove that $\tanh z$ is periodic with period πi.

32. Prove that **(a)** $\overline{\sin z} = \sin \bar{z}$ and **(b)** $\overline{\cos z} = \cos \bar{z}$.

17.8 Inverse Trigonometric and Hyperbolic Functions

INTRODUCTION As functions of a complex variable z, we have seen that both the trigonometric and hyperbolic functions are periodic. Consequently, these functions do not possess inverses that are functions in the strictest interpretation of that word. The inverses of these analytic functions are multiple-valued functions. As we did in Section 17.6, in the examination of the logarithmic function, we shall drop the adjective *multiple-valued* throughout the discussion that follows.

Inverse Sine The **inverse sine** function, written as $\sin^{-1}z$ or arcsin z, is defined by

$$w = \sin^{-1}z \quad \text{if} \quad z = \sin w. \tag{1}$$

The inverse sine can be expressed in terms of the logarithmic function. To see this we use (1) and the definition of the sine function:

$$\frac{e^{iw} - e^{-iw}}{2i} = z \quad \text{or} \quad e^{2iw} - 2ize^{iw} - 1 = 0.$$

From the last equation and the quadratic formula, we then obtain

$$e^{iw} = iz + (1 - z^2)^{1/2}. \tag{2}$$

Note in (2) we did not use the customary symbolism $\pm\sqrt{1 - z^2}$, since we know from Section 17.2 that $(1 - z^2)^{1/2}$ is two-valued. Solving (2) for w then gives

$$\sin^{-1}z = -i \ln[iz + (1 - z^2)^{1/2}]. \tag{3}$$

Proceeding in a similar manner, we find the inverses of the cosine and tangent to be

$$\cos^{-1}z = -i \ln[z + i(1 - z^2)^{1/2}] \tag{4}$$

$$\tan^{-1}z = \frac{i}{2} \ln \frac{i + z}{i - z}. \tag{5}$$

EXAMPLE 1 **Values of an Inverse Sine**

Find all values of $\sin^{-1}\sqrt{5}$.

SOLUTION From (3) we have

$$\sin^{-1}\sqrt{5} = -i \ln [\sqrt{5}i + (1 - (\sqrt{5})^2)^{1/2}].$$

With $(1 - (\sqrt{5})^2)^{1/2} = (-4)^{1/2} = \pm 2i$, the preceding expression becomes

$$\sin^{-1}\sqrt{5} = -i \ln [(\sqrt{5} \pm 2)i]$$

$$= -i\left[\log_e(\sqrt{5} \pm 2) + \left(\frac{\pi}{2} + 2n\pi\right)i\right], \quad n = 0, \pm 1, \pm 2, \ldots.$$

The foregoing result can be simplified a little by noting that $\log_e(\sqrt{5} - 2) = \log_e(1/(\sqrt{5} + 2)) = -\log_e(\sqrt{5} + 2)$. Thus for $n = 0, \pm 1, \pm 2, \ldots,$

$$\sin^{-1}\sqrt{5} = \frac{\pi}{2} + 2n\pi \pm i \log_e(\sqrt{5} + 2). \tag{6} \equiv$$

To obtain particular values of, say, $\sin^{-1}z$, we must choose a specific root of $1 - z^2$ and a specific branch of the logarithm. For example, if we choose $(1 - (\sqrt{5})^2)^{1/2} = (-4)^{1/2} = 2i$ and the principal branch of the logarithm, then (6) gives the single value

$$\sin^{-1}\sqrt{5} = \frac{\pi}{2} - i \log_e(\sqrt{5} + 2).$$

Derivatives The derivatives of the three inverse trigonometric functions considered above can be found by implicit differentiation. To find the derivative of the inverse sine function $w = \sin^{-1}z$, we begin by differentiating $z = \sin w$:

$$\frac{d}{dz}z = \frac{d}{dz}\sin w \quad \text{gives} \quad \frac{dw}{dz} = \frac{1}{\cos w}.$$

Using the trigonometric identity $\cos^2 w + \sin^2 w = 1$ (see Problem 28 in Exercises 17.7) in the form $\cos w = (1 - \sin^2 w)^{1/2} = (1 - z^2)^{1/2}$, we obtain

$$\frac{d}{dz}\sin^{-1}z = \frac{1}{(1 - z^2)^{1/2}}. \tag{7}$$

Similarly, we find that

$$\frac{d}{dz}\cos^{-1}z = \frac{-1}{(1 - z^2)^{1/2}} \tag{8}$$

$$\frac{d}{dz}\tan^{-1}z = \frac{1}{1 + z^2}. \tag{9}$$

It should be noted that the square roots used in (7) and (8) must be consistent with the square roots used in (3) and (4).

EXAMPLE 2 **Evaluating a Derivative**

Find the derivative of $w = \sin^{-1}z$ at $z = \sqrt{5}$.

SOLUTION In Example 1, if we use $(1 - (\sqrt{5})^2)^{1/2} = (-4)^{1/2} = 2i$, then that same root must be used in (7). The value of the derivative consistent with this choice is given by

$$\frac{dw}{dz}\bigg|_{z=\sqrt{5}} = \frac{1}{(1 - (\sqrt{5})^2)^{1/2}} = \frac{1}{(-4)^{1/2}} = \frac{1}{2i} = -\frac{1}{2}i. \qquad \equiv$$

Inverse Hyperbolic Functions The inverse hyperbolic functions can also be expressed in terms of the logarithm. We summarize these results for the inverse hyperbolic sine, cosine, and tangent along with their derivatives:

$$\sinh^{-1}z = \ln[z + (z^2 + 1)^{1/2}] \tag{10}$$

$$\cosh^{-1}z = \ln[z + (z^2 - 1)^{1/2}] \tag{11}$$

$$\tanh^{-1}z = \frac{1}{2}\ln\frac{1 + z}{1 - z} \tag{12}$$

$$\frac{d}{dz}\sinh^{-1}z = \frac{1}{(z^2 + 1)^{1/2}} \tag{13}$$

$$\frac{d}{dz}\cosh^{-1}z = \frac{1}{(z^2 - 1)^{1/2}} \tag{14}$$

$$\frac{d}{dz}\tanh^{-1}z = \frac{1}{1 - z^2}. \tag{15}$$

EXAMPLE 3 **Values of an Inverse Hyperbolic Cosine**

Find all values of $\cosh^{-1}(-1)$.

SOLUTION From (11) with $z = -1$, we get

$$\cosh^{-1}(-1) = \ln(-1) = \log_e 1 + (\pi + 2n\pi)i.$$

Since $\log_e 1 = 0$ we have for $n = 0, \pm 1, \pm 2, \ldots,$

$$\cosh^{-1}(-1) = (2n + 1)\pi i. \qquad \equiv$$

17.8 Exercises Answers to selected odd-numbered problems begin on page ANS-46.

In Problems 1–14, find all values of the given quantity.

1. $\sin^{-1}(-i)$

2. $\sin^{-1}\sqrt{2}$

3. $\sin^{-1}0$

4. $\sin^{-1}\frac{13}{5}$

5. $\cos^{-1}2$

6. $\cos^{-1}2i$

7. $\cos^{-1}\frac{1}{2}$

8. $\cos^{-1}\frac{5}{3}$

9. $\tan^{-1}1$

10. $\tan^{-1}3i$

11. $\sinh^{-1}\frac{4}{3}$

12. $\cosh^{-1}i$

13. $\tanh^{-1}(1+2i)$

14. $\tanh^{-1}(-\sqrt{3}i)$

17 Chapter in Review Answers to selected odd-numbered problems begin on page ANS-46.

Answer Problems 1–16 without referring back to the text. Fill in the blank or answer true/false.

1. $\operatorname{Re}(1+i)^{10} = $ _____ and $\operatorname{Im}(1+i)^{10} = $ _____.

2. If z is a point in the third quadrant, then $i\bar{z}$ is in the _____ quadrant.

3. If $z = 3 + 4i$, then $\operatorname{Re}\left(\dfrac{z}{\bar{z}}\right) = $ _____.

4. $i^{127} - 5i^9 + 2i^{-1} = $ _____

5. If $z = \dfrac{4i}{-3 - 4i}$, then $|z| = $ _____.

6. Describe the region defined by $1 \le |z + 2| \le 3$. _____

7. $\operatorname{Arg}(z + \bar{z}) = 0$ _____

8. If $z = \dfrac{5}{-\sqrt{3} + i}$, then $\operatorname{Arg} z = $ _____.

9. If $e^z = 2i$, then $z = $ _____.

10. If $|e^z| = 1$, then z is a pure imaginary number. _____

11. The principal value of $(1 + i)^{(2+i)}$ is _____.

12. If $f(z) = x^2 - 3xy - 5y^3 + i(4x^2y - 4x + 7y)$, then $f(-1 + 2i) = $ _____.

13. If the Cauchy–Riemann equations are satisfied at a point, then the function is necessarily analytic there. _____

14. $f(z) = e^z$ is periodic with period _____.

15. $\operatorname{Ln}(-ie^3) = $ _____

16. $f(z) = \sin(x - iy)$ is nowhere analytic. _____

In Problems 17–20, write the given number in the form $a + ib$.

17. $i(2 - 3i)^2(4 + 2i)$

18. $\dfrac{3 - i}{2 + 3i} + \dfrac{2 - 2i}{1 + 5i}$

19. $\dfrac{(1 - i)^{10}}{(1 + i)^3}$

20. $4e^{\pi i/3}e^{-\pi i/4}$

In Problems 21–24, sketch the set of points in the complex plane satisfying the given inequality.

21. $\operatorname{Im}(z^2) \le 2$

22. $\operatorname{Im}(z + 5i) > 3$

23. $\dfrac{1}{|z|} \le 1$

24. $\operatorname{Im}(z) < \operatorname{Re}(z)$

25. Look up the definitions of conic sections in a calculus text. Now describe the set of points in the complex plane that satisfy the equation $|z - 2i| + |z + 2i| = 5$.

26. Let z and w be complex numbers such that $|z| = 1$ and $|w| \ne 1$. Prove that

$$\left|\frac{z - w}{1 - \bar{z}w}\right| = 1.$$

In Problems 27 and 28, find all solutions of the given equation.

27. $z^4 = 1 - i$

28. $z^{3/2} = \dfrac{1}{2 - i}$

29. If $f(z) = z^{24} - 3z^{20} + 4z^{12} - 5z^6$, find $f\left(\dfrac{1 + i}{\sqrt{2}}\right)$.

30. Write $f(z) = \operatorname{Im}(z - 3\bar{z}) + z\operatorname{Re}(z^2) - 5z$ in the form $f(z) = u(x, y) + iv(x, y)$.

In Problems 31 and 32, find the image of the line $x = 1$ in the w-plane under the given mapping.

31. $f(z) = x^2 - y + i(y^2 - x)$

32. $f(z) = \dfrac{1}{z}$

In Problems 33–36, find all complex numbers for which the given statement is true.

33. $z = z^{-1}$

34. $\bar{z} = \dfrac{1}{z}$

35. $\bar{z} = -z$

36. $z^2 = (\bar{z})^2$

37. Show that the function $f(z) = -(2xy + 5x) + i(x^2 - 5y - y^2)$ is analytic for all z. Find $f'(z)$.

38. Determine whether the function

$$f(z) = x^3 + xy^2 - 4x + i(4y - y^3 - x^2y)$$

is differentiable. Is it analytic?

In Problems 39 and 40, verify the given equality.

39. $\operatorname{Ln}[(1 + i)(1 - i)] = \operatorname{Ln}(1 + i) + \operatorname{Ln}(1 - i)$

40. $\operatorname{Ln}\left(\dfrac{1 + i}{1 - i}\right) = \operatorname{Ln}(1 + i) - \operatorname{Ln}(1 - i)$

CHAPTER 18

Integration in the Complex Plane

To define an integral of a complex function f, we start with f defined along some curve C or **contour** in the complex plane. We will see in this chapter that the definition of a complex integral, its properties, and the method of evaluation are quite similar to those of a real line integral in the plane (Section 9.8).

18.1 Contour Integrals

INTRODUCTION In Section 9.8 we saw that the definition of the definite integral $\int_a^b f(x)\,dx$ starts with a real function $y = f(x)$ that is defined on an interval $[a, b]$ on the x-axis. Because a planar curve is the two-dimensional analogue of an interval, we then generalized the definition of the definite integral to integrals of real functions of *two* variables defined on a curve C in the Cartesian plane. We shall see in this section that a complex integral is defined in a manner that is quite similar to that of a line integral in the Cartesian plane. In case you have not studied Sections 9.8 and 9.9, a review of those sections is recommended.

■ A Definition

Integration in the complex plane is defined in a manner similar to that of a line integral in the plane. In other words, we shall be dealing with an integral of a complex function $f(z)$ that is defined along a curve C in the complex plane. These curves are defined in terms of parametric equations $x = x(t)$, $y = y(t)$, $a \le t \le b$, where t is a real parameter. By using $x(t)$ and $y(t)$ as real and imaginary parts, we can also describe a curve C in the complex plane by means of a complex-valued function of a real variable t: $z(t) = x(t) + iy(t)$, $a \le t \le b$. For example, $x = \cos t$, $y = \sin t$, $0 \le t \le 2\pi$, describes a unit circle centered at the origin. This circle can also be described by $z(t) = \cos t + i \sin t$, or even more compactly by $z(t) = e^{it}$, $0 \le t \le 2\pi$. The same definitions of smooth curve, piecewise-smooth curve, closed curve, and simple closed curve given in Section 9.8 carry over to this discussion. As before, we shall assume that the positive direction on C corresponds to increasing values of t. In complex variables, a piecewise-smooth curve C is also called a **contour** or **path**. An integral of $f(z)$ on C is denoted by $\int_C f(z)\,dz$ or $\oint_C f(z)\,dz$ if the contour C is closed; it is referred to as a **contour integral** or simply as a **complex integral**.

1. Let $f(z) = u(x, y) + iv(x, y)$ be defined at all points on a smooth curve C defined by $x = x(t)$, $y = y(t)$, $a \le t \le b$.

2. Divide C into n subarcs according to the partition $a = t_0 < t_1 < \ldots < t_n = b$ of $[a, b]$. The corresponding points on the curve C are $z_0 = x_0 + iy_0 = x(t_0) + iy(t_0)$, $z_1 = x_1 + iy_1 = x(t_1) + iy(t_1), \ldots, z_n = x_n + iy_n = x(t_n) + iy(t_n)$. Let $\Delta z_k = z_k - z_{k-1}$, $k = 1, 2, \ldots, n$.

3. Let $\|P\|$ be the **norm** of the partition, that is, the maximum value of $|\Delta z_k|$.

4. Choose a sample point $z_k^* = x_k^* + iy_k^*$ on each subarc. See **FIGURE 18.1.1**.

5. Form the sum $\displaystyle\sum_{k=1}^{n} f(z_k^*)\,\Delta z_k$.

FIGURE 18.1.1 Sample points are the red dots

DEFINITION 18.1.1 Contour Integral

Let f be defined at points of a smooth curve C defined by $x = x(t)$, $y = y(t)$, $a \le t \le b$. The **contour integral** of f along C is

$$\int_C f(z)\,dz = \lim_{\|P\| \to 0} \sum_{k=1}^{n} f(z_k^*)\,\Delta z_k. \tag{1}$$

The limit in (1) exists if f is continuous at all points on C and C is either smooth or piecewise smooth. Consequently we shall, hereafter, assume these conditions as a matter of course.

■ Method of Evaluation

We shall turn now to the question of evaluating a contour integral. To facilitate the discussion, let us suppress the subscripts and write (1) in the abbreviated form

$$\int_C f(z)\,dz = \lim \Sigma(u + iv)(\Delta x + i\,\Delta y)$$

$$= \lim [\Sigma(u\Delta x - v\Delta y) + i\,\Sigma(v\Delta x + u\Delta y)].$$

This means

$$\int_C f(z)\,dz = \int_C u\,dx - v\,dy + i\int_C v\,dx + u\,dy. \tag{2}$$

In other words, a contour integral $\int_C f(z)\,dz$ is a combination of two real-line integrals $\int_C u\,dx - v\,dy$ and $\int_C v\,dx + u\,dy$. Now, since $x = x(t)$ and $y = y(t)$, $a \le t \le b$, the right side of (2) is the same as

$$\int_a^b [u(x(t), y(t))x'(t) - v(x(t), y(t))y'(t)]\,dt + i \int_a^b [v(x(t), y(t))x'(t) + u(x(t), y(t))y'(t)]\,dt.$$

But if we use $z(t) = x(t) + iy(t)$ to describe C, the last result is the same as $\int_a^b f(z(t))z'(t)\,dt$ when separated into two integrals. Thus we arrive at a practical means of evaluating a contour integral:

THEOREM 18.1.1 **Evaluation of a Contour Integral**

If f is continuous on a smooth curve C given by $z(t) = x(t) + iy(t)$, $a \le t \le b$, then

$$\int_C f(z)\,dz = \int_a^b f(z(t))z'(t)\,dt. \tag{3}$$

If f is expressed in terms of the symbol z, then to evaluate $f(z(t))$ we simply replace the symbol z by $z(t)$. If f is not expressed in terms of z, then to evaluate $f(z(t))$, we replace x and y wherever they appear by $x(t)$ and $y(t)$, respectively.

EXAMPLE 1 **Evaluating a Contour Integral**

Evaluate $\displaystyle\int_C \bar{z}\,dz$, where C is given by $x = 3t$, $y = t^2$, $-1 \le t \le 4$.

SOLUTION We write $z(t) = 3t + it^2$ so that $z'(t) = 3 + 2it$ and $f(z(t)) = \overline{3t + it^2} = 3t - it^2$. Thus,

$$\int_C \bar{z}\,dz = \int_{-1}^4 (3t - it^2)(3 + 2it)\,dt$$

$$= \int_{-1}^4 (2t^3 + 9t)\,dt + i \int_{-1}^4 3t^2\,dt = 195 + 65i. \qquad \equiv$$

EXAMPLE 2 **Evaluating a Contour Integral**

Evaluate $\displaystyle\oint_C \frac{1}{z}\,dz$, where C is the circle $x = \cos t$, $y = \sin t$, $0 \le t \le 2\pi$.

SOLUTION In this case $z(t) = \cos t + i \sin t = e^{it}$, $z'(t) = ie^{it}$, and $f(z) = 1/z = e^{-it}$. Hence,

$$\oint_C \frac{1}{z}\,dz = \int_0^{2\pi} (e^{-it})ie^{it}\,dt = i\int_0^{2\pi} dt = 2\pi i. \qquad \equiv$$

For some curves, the real variable x itself can be used as the parameter. For example, to evaluate $\int_C (8x^2 - iy)\,dz$ on the line segment $y = 5x$, $0 \le x \le 2$, we write $z = x + 5xi$, $dz = (1 + 5i)\,dx$,

$$\int_C (8x^2 - iy)\,dz = \int_0^2 (8x^2 - 5xi)(1 + 5i)\,dx$$

and integrate in the usual manner

$$\int_C (8x^2 - iy)\,dz = (1 + 5i)\int_0^2 (8x^2 - 5xi)\,dx$$

$$= (1 + 5i)\left[\frac{8}{3}x^3\right]_0^2 - (1 + 5i)i\left[\frac{5}{2}x^2\right]_0^2$$

$$= \frac{214}{3} + \frac{290}{3}\,i.$$

Properties The following properties of contour integrals are analogous to the properties of line integrals.

THEOREM 18.1.2 Properties of Contour Integrals

Suppose f and g are continuous in a domain D and C is a smooth curve lying entirely in D. Then

(i) $\displaystyle\int_C kf(z)\,dz = k\int_C f(z)\,dz,\ k$ a constant

(ii) $\displaystyle\int_C [f(z) + g(z)]\,dz = \int_C f(z)\,dz + \int_C g(z)\,dz$

(iii) $\displaystyle\int_C f(z)\,dz = \int_{C_1} f(z)\,dz + \int_{C_2} f(z)\,dz$, where C is the union of the smooth curves C_1 and C_2

(iv) $\displaystyle\int_{-C} f(z)\,dz = -\int_C f(z)\,dz$, where $-C$ denotes the curve having the opposite orientation of C.

The four parts of Theorem 18.1.2 also hold when C is a piecewise-smooth curve in D.

EXAMPLE 3 Evaluating a Contour Integral

Evaluate $\int_C (x^2 + iy^2)\,dz$, where C is the contour shown in **FIGURE 18.1.2**.

SOLUTION In view of Theorem 18.1.2(*iii*) we write

$$\int_C (x^2 + iy^2)\,dz = \int_{C_1} (x^2 + iy^2)\,dz + \int_{C_2} (x^2 + iy^2)\,dz.$$

Since the curve C_1 is defined by $y = x$, it makes sense to use x as a parameter. Therefore, $z(x) = x + ix$, $z'(x) = 1 + i$, $f(z(x)) = x^2 + ix^2$, and

$$\int_{C_1} (x^2 + iy^2)\,dz = \int_0^1 (x^2 + ix^2)(1 + i)\,dx$$

$$= (1 + i)^2\int_0^1 x^2\,dx = \frac{(1 + i)^2}{3} = \frac{2}{3}i.$$

The curve C_2 is defined by $x = 1$, $1 \le y \le 2$. Using y as a parameter, we have $z(y) = 1 + iy$, $z'(y) = i$, and $f(z(y)) = 1 + iy^2$. Thus,

$$\int_{C_2} (x^2 + iy^2)\,dz = \int_1^2 (1 + iy^2)\,i\,dy = -\int_1^2 y^2\,dy + i\int_1^2 dy = -\frac{7}{3} + i.$$

Finally, we have $\int_C (x^2 + iy^2)\,dz = \frac{2}{3}i + (-\frac{7}{3} + i) = -\frac{7}{3} + \frac{5}{3}i$. ≡

There are times in the application of complex integration that it is useful to find an upper bound for the absolute value of a contour integral. In the next theorem we shall use the fact that the length of a plane curve is $s = \int_a^b \sqrt{[x'(t)]^2 + [y'(t)]^2}\,dt$. But if $z'(t) = x'(t) + iy'(t)$, then $|z'(t)| = \sqrt{[x'(t)]^2 + [y'(t)]^2}$ and consequently $s = \int_a^b |z'(t)|\,dt$.

THEOREM 18.1.3 A Bounding Theorem

If f is continuous on a smooth curve C and if $|f(z)| \le M$ for all z on C, then $\left|\int_C f(z)\,dz\right| \le ML$, where L is the length of C.

PROOF: From the triangle inequality (6) of Section 17.1 we can write

$$\left|\sum_{k=1}^n f(z_k^*)\Delta z_k\right| \le \sum_{k=1}^n |f(z_k^*)||\Delta z_k| \le M\sum_{k=1}^n |\Delta z_k|. \qquad (4)$$

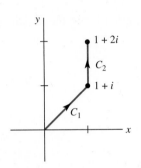

FIGURE 18.1.2 Piecewise-smooth contour in Example 3

Now, $|\Delta z_k|$ can be interpreted as the length of the chord joining the points z_k and z_{k-1}. Since the sum of the lengths of the chords cannot be greater than the length of C, (4) becomes $|\sum_{k=1}^{n} f(z_k^*) \, \Delta z_k| \le ML$. Hence, as $\|P\| \to 0$, the last inequality yields $|\int_C f(z) \, dz| \le ML$. ≡

Theorem 18.1.3 is used often in the theory of complex integration and is sometimes referred to as the *ML*-inequality.

EXAMPLE 4 **A Bound for a Contour Integral**

Find an upper bound for the absolute value of $\oint_C \dfrac{e^z}{z+1} \, dz$, where C is the circle $|z| = 4$.

SOLUTION First, the length s of the circle of radius 4 is 8π. Next, from the inequality (7) of Section 17.1, it follows that $|z + 1| \ge |z| - 1 = 4 - 1 = 3$, and so

$$\left| \frac{e^z}{z+1} \right| \le \frac{|e^z|}{|z| - 1} = \frac{|e^z|}{3}. \tag{5}$$

In addition, $|e^z| = |e^x(\cos y + i \sin y)| = e^x$. For points on the circle $|z| = 4$, the maximum that x can be is 4, and so (5) becomes

$$\left| \frac{e^z}{z+1} \right| \le \frac{e^4}{3}.$$

Hence from Theorem 18.1.3 we have

$$\left| \oint_C \frac{e^z}{z+1} \, dz \right| \le \frac{8\pi e^4}{3}.$$ ≡

Circulation and Net Flux Let $\mathbf{T}$ and $\mathbf{N}$ denote the unit tangent vector and the unit normal vector to a positively oriented simple closed contour C. When we interpret the complex function $f(z) = u(x, y) + iv(x, y)$ as a vector, the line integrals

$$\oint_C f \cdot \mathbf{T} \, ds = \oint_C u \, dx + v \, dy \tag{6}$$

and

$$\oint_C f \cdot \mathbf{N} \, ds = \oint_C u \, dy - v \, dx \tag{7}$$

have special interpretations. The line integral in (6) is called the **circulation** around C and measures the tendency of the flow to rotate the curve C. See Section 9.8 for the derivation. The **net flux** across C is the difference between the rate at which fluid enters and the rate at which fluid leaves the region bounded by C. The net flux across C is given by the line integral in (7), and a nonzero value for $\oint_C f \cdot \mathbf{N} \, ds$ indicates the presence of sources or sinks for the fluid inside the curve C. Note that

$$\left(\oint_C f \cdot \mathbf{T} \, ds \right) + i \left(\oint_C f \cdot \mathbf{N} \, ds \right) = \oint_C (u - iv)(dx + i \, dy) = \oint_C \overline{f(z)} \, dz$$

and so

$$\text{circulation} = \text{Re} \left(\oint_C \overline{f(z)} \, dz \right) \tag{8}$$

$$\text{net flux} = \text{Im} \left(\oint_C \overline{f(z)} \, dz \right). \tag{9}$$

Thus, both of these key quantities may be found by computing a single complex integral.

EXAMPLE 5 **Net Flux**

Given the flow $f(z) = (1 + i)z$, compute the circulation around, and the net flux across, the circle $C: |z| = 1$.

SOLUTION Since $\overline{f(z)} = (1 - i)\overline{z}$ and $z(t) = e^{it}, 0 \le t \le 2\pi$, we have

$$\oint_C \overline{f(z)}\, dz = \int_0^{2\pi} (1 - i)e^{-it}ie^{it}\, dt = (1 + i)\int_0^{2\pi} dt = 2\pi(1 + i) = 2\pi + 2\pi i.$$

Using (8) and (9), the circulation around C is 2π and the net flux across C is 2π. See **FIGURE 18.1.3**. ≡

FIGURE 18.1.3 Flow $f(z) + (1 + i)z$

18.1 Exercises Answers to selected odd-numbered problems begin on page ANS-46.

In Problems 1–16, evaluate the given integral along the indicated contour.

1. $\int_C(z + 3)\, dz$, where C is $x = 2t, y = 4t - 1, 1 \le t \le 3$

2. $\int_C(2\overline{z} - z)\, dz$, where C is $x = -t, y = t^2 + 2, 0 \le t \le 2$

3. $\int_C z^2\, dz$, where C is $z(t) = 3t + 2it, -2 \le t \le 2$

4. $\int_C(3z^2 - 2z)\, dz$, where C is $z(t) = t + it^2, 0 \le t \le 1$

5. $\int_C \dfrac{1 + z}{z}\, dz$, where C is the right half of the circle $|z| = 1$ from $z = -i$ to $z = i$

6. $\int_C |z|^2\, dz$, where C is $x = t^2, y = 1/t, 1 \le t \le 2$

7. $\oint_C \operatorname{Re}(z)\, dz$, where C is the circle $|z| = 1$

8. $\oint_C\left(\dfrac{1}{(z + i)^3} - \dfrac{5}{z + i} + 8\right)\, dz$, where C is the circle $|z + i| = 1, 0 \le t \le 2\pi$

9. $\int_C(x^2 + iy^3)\, dz$, where C is the straight line from $z = 1$ to $z = i$

10. $\int_C(x^3 - iy^3)\, dz$, where C is the lower half of the circle $|z| = 1$ from $z = -1$ to $z = 1$

11. $\int_C e^z\, dz$, where C is the polygonal path consisting of the line segments from $z = 0$ to $z = 2$ and from $z = 2$ to $z = 1 + \pi i$

12. $\int_C \sin z\, dz$, where C is the polygonal path consisting of the line segments from $z = 0$ to $z = 1$ and from $z = 1$ to $z = 1 + i$

13. $\int_C \operatorname{Im}(z - i)\, dz$, where C is the polygonal path consisting of the circular arc along $|z| = 1$ from $z = 1$ to $z = i$ and the line segment from $z = i$ to $z = -1$

14. $\int_C dz$, where C is the left half of the ellipse $x^2/36 + y^2/4 = 1$ from $z = 2i$ to $z = -2i$

15. $\oint_C ze^z\, dz$, where C is the square with vertices $z = 0, z = 1, z = 1 + i$, and $z = i$

16. $\int_C f(z)\, dz$, where $f(z) = \begin{cases} 2, & x < 0 \\ 6x, & x > 0 \end{cases}$ and C is the parabola $y = x^2$ from $z = -1 + i$ to $z = 1 + i$

In Problems 17–20, evaluate the given integral along the contour C given in **FIGURE 18.1.4**.

17. $\oint_C x\, dz$

18. $\oint_C(2z - 1)\, dz$

19. $\oint_C z^2\, dz$

20. $\oint_C \overline{z}^2\, dz$

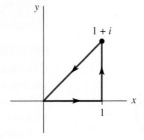

FIGURE 18.1.4 Contour in Problems 17–20

In Problems 21–24, evaluate $\int_C(z^2 - z + 2)\, dz$ from i to 1 along the indicated contours.

21.

FIGURE 18.1.5 Contour in Problem 21

22.

FIGURE 18.1.6 Contour in Problem 22

23.

FIGURE 18.1.7 Contour in Problem 23

24.

FIGURE 18.1.8 Contour in Problem 24

In Problems 25–28, find an upper bound for the absolute value of the given integral along the indicated contour.

25. $\oint_C \dfrac{e^z}{z^2+1}\, dz$, where C is the circle $|z|=5$

26. $\displaystyle\int_C \dfrac{1}{z^2-2i}\, dz$, where C is the right half of the circle $|z|=6$ from $z=-6i$ to $z=6i$

27. $\int_C (z^2+4)\, dz$, where C is the line segment from $z=0$ to $z=1+i$

28. $\displaystyle\int_C \dfrac{1}{z^3}\, dz$, where C is one quarter of the circle $|z|=4$ from $z=4i$ to $z=4$

29. (a) Use Definition 18.1.1 to show for any smooth curve C between z_0 and z_n that $\int_C dz = z_n - z_0$.
(b) Use the result in part (a) to verify the answer to Problem 14.

30. Use Definition 18.1.1 to show for any smooth curve C between z_0 and z_n that $\int_C z\, dz = \frac{1}{2}(z_n^2 - z_0^2)$. [*Hint:* The integral exists, so choose $z_k^* = z_k$ and $z_k^* = z_{k-1}$.]

31. Use the results of Problems 29 and 30 to evaluate $\oint_C (6z+4)\, dz$ where C is
(a) the straight line from $1+i$ to $2+3i$, and
(b) the closed contour $x^4 + y^4 = 4$.

In Problems 32–35, compute the circulation and net flux for the given flow and the indicated closed contour.

32. $f(z) = 1/z$, where C is the circle $|z|=2$
33. $f(z) = 2z$, where C is the circle $|z|=1$
34. $f(z) = 1/(z-1)$, where C is the circle $|z-1|=2$
35. $f(z) = \bar{z}$, where C is the square with vertices $z=0$, $z=1$, $z=1+i$, $z=i$

18.2 Cauchy–Goursat Theorem

INTRODUCTION In this section we shall concentrate on contour integrals where the contour C is a **simple closed curve** with a positive (counterclockwise) orientation. Specifically, we shall see that when f is analytic in a special kind of domain D, the value of the contour integral $\oint_C f(z)\, dz$ is the same for *any* simple closed curve C that lies entirely within D. This theorem, called the **Cauchy–Goursat theorem**, is one of *the* fundamental results in complex analysis. Preliminary to discussing the Cauchy–Goursat theorem and some of its ramifications, we first need to distinguish two kinds of domains in the complex plane: simply connected and multiply connected.

(a) Simply connected domain

(b) Multiply connected domain

FIGURE 18.2.1 Two kinds of domains

■ **Simply and Multiply Connected Domains** In the discussion that follows, we shall concentrate on contour integrals where the contour C is a **simple closed curve** with a positive (counterclockwise) orientation. Before doing this, we need to distinguish two kinds of domains. A domain D is said to be **simply connected** if every simple closed contour C lying entirely in D can be shrunk to a point without leaving D. In other words, in a simply connected domain, every simple closed contour C lying entirely within it encloses only points of the domain D. Expressed yet another way, a simply connected domain has no "holes" in it. The entire complex plane is an example of a simply connected domain. A domain that is not simply connected is called a **multiply connected domain**; that is, a multiply connected domain has "holes" in it. See **FIGURE 18.2.1**. As in Section 9.9, we call a domain with one "hole" **doubly connected**, a domain with two "holes" **triply connected**, and so on.

■ **Cauchy's Theorem** In 1825, the French mathematician Louis-Augustin Cauchy proved one of the most important theorems in complex analysis. **Cauchy's theorem** says:

Suppose that a function f is analytic in a simply connected domain D and that f′ is continuous in D. Then for every simple closed contour C in D, $\oint_C f(z)\, dz = 0$.

The proof of this theorem is an immediate consequence of Green's theorem and the Cauchy–Riemann equations. Since f' is continuous throughout D, the real and imaginary parts of $f(z) = u + iv$ and their first partial derivatives are continuous throughout D. By (2) of Section 18.1 we write $\oint_C f(z)\, dz$ in terms of real-line integrals and use Green's theorem on each line integral:

$$\oint_C f(z)\, dz = \oint_C u(x,y)\, dx - v(x,y)\, dy + i\oint_C v(x,y)\, dx + u(x,y)\, dy$$

$$= \iint_D \left(-\frac{\partial v}{\partial x} - \frac{\partial u}{\partial y}\right) dA + i\iint_D \left(\frac{\partial u}{\partial x} - \frac{\partial v}{\partial y}\right) dA. \tag{1}$$

Now since f is analytic, the Cauchy–Riemann equations, $\partial u/\partial x = \partial v/\partial y$ and $\partial u/\partial y = -\partial v/\partial x$, imply that the integrands in (1) are identically zero. Hence, we have $\oint_C f(z)\,dz = 0$.

In 1883, the French mathematician **Edouard Goursat** (1858–1936) proved Cauchy's theorem without the assumption of continuity of f'. The resulting modified version of Cauchy's theorem is known as the **Cauchy–Goursat theorem**.

THEOREM 18.2.1 Cauchy–Goursat Theorem

Suppose a function f is analytic in a simply connected domain D. Then for every simple closed contour C in D, $\oint_C f(z)\,dz = 0$.

Since the interior of a simple closed contour is a simply connected domain, the Cauchy–Goursat theorem can be stated in the slightly more practical manner:

If f is analytic at all points within and on a simple closed contour C, then $\oint_C f(z)\,dz = 0$. $\qquad(2)$

FIGURE 18.2.2 Contour in Example 1

EXAMPLE 1 Applying the Cauchy–Goursat Theorem

Evaluate $\oint_C e^z\,dz$, where C is the curve shown in **FIGURE 18.2.2**.

SOLUTION The function $f(z) = e^z$ is entire and C is a simple closed contour. It follows from the form of the Cauchy–Goursat theorem given in (2) that $\oint_C e^z\,dz = 0$. ≡

EXAMPLE 2 Applying the Cauchy–Goursat Theorem

Evaluate $\oint_C \dfrac{dz}{z^2}$, where C is the ellipse $(x-2)^2 + \dfrac{(y-5)^2}{4} = 1$.

SOLUTION The rational function $f(z) = 1/z^2$ is analytic everywhere except at $z = 0$. But $z = 0$ is not a point interior to or on the contour C. Thus, from (2) we have $\oint_C dz/z^2 = 0$. ≡

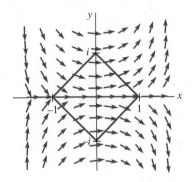

FIGURE 18.2.3 Flow $f(z) = \overline{\cos z}$

EXAMPLE 3 Applying the Cauchy–Goursat Theorem

Given the flow $f(z) = \overline{\cos z}$, compute the circulation around and net flux across C, where C is the square with vertices $z = 1$, $z = i$, $z = -1$, and $z = -i$.

SOLUTION We must compute $\oint_C \overline{f(z)}\,dz = \oint_C \cos z\,dz$ and then take the real and imaginary parts of the integral to find the circulation and net flux, respectively. The function $\cos z$ is analytic everywhere, and so $\oint_C \overline{f(z)}\,dz = 0$ from (2). The circulation and net flux are therefore both 0. **FIGURE 18.2.3** shows the flow $f(z) = \overline{\cos z}$ and the contour C. ≡

(a)

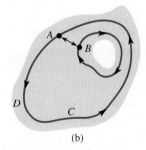

(b)

FIGURE 18.2.4 Doubly connected domain D

■ **Cauchy–Goursat Theorem for Multiply Connected Domains** If f is analytic in a multiply connected domain D, then we cannot conclude that $\oint_C f(z)\,dz = 0$ for every simple closed contour C in D. To begin, suppose D is a doubly connected domain and C and C_1 are simple closed contours such that C_1 surrounds the "hole" in the domain and is interior to C. See **FIGURE 18.2.4(a)**. Suppose, also, that f is analytic on each contour and at each point interior to C but exterior to C_1. When we introduce the cut AB shown in Figure 18.2.4(b), the region bounded by the curves is simply connected. Now the integral from A to B has the opposite value of the integral from B to A, and so from (2) we have $\oint_C f(z)\,dz + \oint_{C_1} f(z)\,dz = 0$ or

$$\oint_C f(z)\,dz = \oint_{C_1} f(z)\,dz. \qquad(3)$$

The last result is sometimes called the principle of **deformation of contours**, since we can think of the contour C_1 as a continuous deformation of the contour C. Under this deformation of contours, the value of the integral does not change. Thus, on a practical level, (3) allows us to evaluate an integral over a complicated simple closed contour by replacing that contour with one that is more convenient.

FIGURE 18.2.5 We use the simpler contour C_1 in Example 4

EXAMPLE 4 **Applying Deformation of Contours**

Evaluate $\oint_C \dfrac{dz}{z - i}$, where C is the outer contour shown in **FIGURE 18.2.5**.

SOLUTION In view of (3), we choose the more convenient circular contour C_1 in the figure. By taking the radius of the circle to be $r = 1$, we are guaranteed that C_1 lies within C. In other words, C_1 is the circle $|z - i| = 1$, which can be parameterized by $x = \cos t, y = 1 + \sin t, 0 \le t \le 2\pi$, or equivalently by $z = i + e^{it}, 0 \le t \le 2\pi$. From $z - i = e^{it}$ and $dz = ie^{it}\, dt$ we obtain

$$\oint_C \frac{dz}{z - i} = \oint_{C_1} \frac{dz}{z - i} = \int_0^{2\pi} \frac{ie^{it}}{e^{it}}\, dt = i\int_0^{2\pi} dt = 2\pi i.$$ ≡

The result in Example 4 can be generalized. Using the principle of deformation of contours (3) and proceeding as in the example, we can show that if z_0 is any constant complex number interior to *any* simple closed contour C, then

$$\oint_C \frac{dz}{(z - z_0)^n} = \begin{cases} 2\pi i, & n = 1 \\ 0, & n \text{ an integer} \ne 1. \end{cases} \tag{4}$$

The fact that the integral in (4) is zero when n is an integer $\ne 1$ follows only partially from the Cauchy–Goursat theorem. When n is zero or a negative integer, $1/(z - z_0)^n$ is a polynomial (for example, $n = -3, 1/(z - z_0)^{-3} = (z - z_0)^3$) and therefore entire. Theorem 18.2.1 then implies $\oint_C dz/(z - z_0)^n = 0$. It is left as an exercise to show that the integral is still zero when n is a positive integer different from one. See Problem 22 in Exercises 18.2.

EXAMPLE 5 **Applying Formula (4)**

Evaluate $\oint_C \dfrac{5z + 7}{z^2 + 2z - 3}\, dz$, where C is the circle $|z - 2| = 2$.

SOLUTION Since the denominator factors as $z^2 + 2z - 3 = (z - 1)(z + 3)$, the integrand fails to be analytic at $z = 1$ and $z = -3$. Of these two points, only $z = 1$ lies within the contour C, which is a circle centered at $z = 2$ of radius $r = 2$. Now by partial fractions,

$$\frac{5z + 7}{z^2 + 2z - 3} = \frac{3}{z - 1} + \frac{2}{z + 3}$$

and so

$$\oint_C \frac{5z + 7}{z^2 + 2z - 3}\, dz = 3\oint_C \frac{dz}{z - 1} + 2\oint_C \frac{dz}{z + 3}. \tag{5}$$

In view of the result given in (4), the first integral in (5) has the value $2\pi i$. By the Cauchy–Goursat theorem, the value of the second integral is zero. Hence, (5) becomes

$$\oint_C \frac{5z + 7}{z^2 + 2z - 3}\, dz = 3(2\pi i) + 2(0) = 6\pi i.$$ ≡

If C, C_1, and C_2 are the simple closed contours shown in **FIGURE 18.2.6** and if f is analytic on each of the three contours as well as at each point interior to C but exterior to both C_1 and C_2, then by introducing cuts, we get from Theorem 18.2.1 that $\oint_C f(z)\, dz + \oint_{C_1} f(z)\, dz + \oint_{C_2} f(z)\, dz = 0$. Hence,

$$\oint_C f(z)\, dz = \oint_{C_1} f(z)\, dz + \oint_{C_2} f(z)\, dz.$$

The next theorem summarizes the general result for a multiply connected domain with n "holes":

> **THEOREM 18.2.2 Cauchy–Goursat Theorem for Multiply Connected Domains**
>
> Suppose $C, C_1, \ldots, C_n$ are simple closed curves with a positive orientation such that $C_1, C_2, \ldots, C_n$ are interior to C but the regions interior to each $C_k, k = 1, 2, \ldots, n$, have no points in common. If f is analytic on each contour and at each point interior to C but exterior to all the $C_k, k = 1, 2, \ldots, n$, then
> $$\oint_C f(z)\, dz = \sum_{k=1}^{n} \oint_{C_k} f(z)\, dz. \tag{6}$$

FIGURE 18.2.6 Triply connected domain D

EXAMPLE 6 **Applying Theorem 18.2.2**

Evaluate $\displaystyle\oint_C \frac{dz}{z^2 + 1}$, where C is the circle $|z| = 3$.

SOLUTION In this case the denominator of the integrand factors as $z^2 + 1 = (z - i)(z + i)$. Consequently, the integrand $1/(z^2 + 1)$ is not analytic at $z = i$ and $z = -i$. Both of these points lie within the contour C. Using partial fraction decomposition once more, we have

$$\frac{1}{z^2 + 1} = \frac{1/2i}{z - i} - \frac{1/2i}{z + i}$$

and

$$\oint_C \frac{dz}{z^2 + 1} = \frac{1}{2i}\oint_C \left[\frac{1}{z - i} - \frac{1}{z + i}\right] dz.$$

We now surround the points $z = i$ and $z = -i$ by circular contours C_1 and C_2, respectively, that lie entirely within C. Specifically, the choice $|z - i| = \frac{1}{2}$ for C_1 and $|z + i| = \frac{1}{2}$ for C_2 will suffice. See **FIGURE 18.2.7**. From Theorem 18.2.2 we can then write

$$\oint_C \frac{dz}{z^2 + 1} = \frac{1}{2i}\oint_{C_1}\left[\frac{1}{z - i} - \frac{1}{z + i}\right] dz + \frac{1}{2i}\oint_{C_2}\left[\frac{1}{z - i} - \frac{1}{z + i}\right] dz$$

$$= \frac{1}{2i}\oint_{C_1}\frac{dz}{z - i} - \frac{1}{2i}\oint_{C_1}\frac{dz}{z + i} + \frac{1}{2i}\oint_{C_2}\frac{dz}{z - i} - \frac{1}{2i}\oint_{C_2}\frac{dz}{z + i}. \qquad (7)$$

Because $1/(z + i)$ is analytic on C_1 and at each point in its interior and because $1/(z - i)$ is analytic on C_2 and at each point in its interior, it follows from (4) that the second and third integrals in (7) are zero. Moreover, it follows from (4), with $n = 1$, that

$$\oint_{C_1}\frac{dz}{z - i} = 2\pi i \quad\text{and}\quad \oint_{C_2}\frac{dz}{z + i} = 2\pi i.$$

Thus (7) becomes

$$\oint_C \frac{dz}{z^2 + 1} = \pi - \pi = 0.$$

≡

FIGURE 18.2.7 Contour in Example 6

REMARKS

Throughout the foregoing discussion we assumed that C was a simple closed contour; in other words, C did not intersect itself. Although we shall not give the proof, it can be shown that the Cauchy–Goursat theorem is valid for any closed contour C in a simply connected domain D. As shown in **FIGURE 18.2.8**, the contour C is closed but not simple. Nevertheless, if f is analytic in D, then $\oint_C f(z)\, dz = 0$.

FIGURE 18.2.8 Contour C is closed but not simple

18.2 Exercises Answers to selected odd-numbered problems begin on page ANS-46.

In Problems 1–8, prove that $\oint_C f(z)\, dz = 0$, where f is the given function and C is the unit circle $|z| = 1$.

1. $f(z) = z^3 - 1 + 3i$

2. $f(z) = z^2 + \dfrac{1}{z - 4}$

3. $f(z) = \dfrac{z}{2z + 3}$

4. $f(z) = \dfrac{z - 3}{z^2 + 2z + 2}$

5. $f(z) = \dfrac{\sin z}{(z^2 - 25)(z^2 + 9)}$

6. $f(z) = \dfrac{e^z}{2z^2 + 11z + 15}$

7. $f(z) = \tan z$

8. $f(z) = \dfrac{z^2 - 9}{\cosh z}$

9. Evaluate $\oint_C \dfrac{1}{z}\,dz$, where C is the contour shown in **FIGURE 18.2.9**.

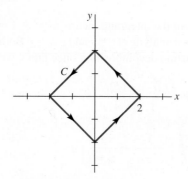

FIGURE 18.2.9 Contour in Problem 9

10. Evaluate $\oint_C \dfrac{5}{z+1+i}\,dz$, where C is the contour shown in **FIGURE 18.2.10**.

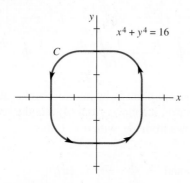

FIGURE 18.2.10 Contour in Problem 10

In Problems 11–20, use any of the results in this section to evaluate the given integral along the indicated closed contour(s).

11. $\oint_C \left(z + \dfrac{1}{z}\right) dz;\ |z| = 2$

12. $\oint_C \left(z + \dfrac{1}{z^2}\right) dz;\ |z| = 2$

13. $\oint_C \dfrac{z}{z^2 - \pi^2}\,dz;\ |z| = 3$

14. $\oint_C \dfrac{10}{(z+i)^4}\,dz;\ |z+i| = 1$

15. $\oint_C \dfrac{2z+1}{z^2 + z}\,dz;$ **(a)** $|z| = \frac{1}{2}$, **(b)** $|z| = 2$, **(c)** $|z - 3i| = 1$

16. $\oint_C \dfrac{2z}{z^2 + 3}\,dz;$ **(a)** $|z| = 1$, **(b)** $|z - 2i| = 1$, **(c)** $|z| = 4$

17. $\oint_C \dfrac{-3z + 2}{z^2 - 8z + 12}\,dz;$ **(a)** $|z - 5| = 2$, **(b)** $|z| = 9$

18. $\oint_C \left(\dfrac{3}{z+2} - \dfrac{1}{z - 2i}\right) dz;$ **(a)** $|z| = 5$, **(b)** $|z - 2i| = \frac{1}{2}$

19. $\oint_C \dfrac{z - 1}{z(z - i)(z - 3i)}\,dz;\ |z - i| = \frac{1}{2}$

20. $\oint_C \dfrac{1}{z^3 + 2iz^2}\,dz;\ |z| = 1$

21. Evaluate $\oint_C \dfrac{8z - 3}{z^2 - z}\,dz$, where C is the closed contour shown in **FIGURE 18.2.11**. [*Hint*: Express C as the union of two closed curves C_1 and C_2.]

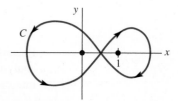

FIGURE 18.2.11 Contour in Problem 21

22. Suppose z_0 is any constant complex number interior to any simple closed contour C. Show that

$$\oint_C \dfrac{dz}{(z - z_0)^n} = \begin{cases} 2\pi i, & n = 1 \\ 0, & n \text{ a positive integer } \neq 1. \end{cases}$$

In Problems 23 and 24, evaluate the given integral by any means.

23. $\oint_C \left(\dfrac{e^z}{z + 3} - 3\bar{z}\right) dz$, C is the unit circle $|z| = 1$

24. $\oint_C (z^3 + z^2 + \text{Re}(z))\,dz$, C is the triangle with vertices $z = 0, z = 1 + 2i, z = 1$

18.3 Independence of Path

INTRODUCTION In real calculus when a function f possesses an elementary antiderivative, that is, a function F for which $F'(x) = f(x)$, a definite integral can be evaluated by the Fundamental Theorem of Calculus:

$$\int_a^b f(x)\,dx = F(b) - F(a). \tag{1}$$

Note that $\int_a^b f(x)\,dx$ depends only on the numbers a and b at the initial and terminal points of the interval of integration. In contrast, the value of a real-line integral $\int_C P\,dx + Q\,dy$ generally depends on the curve C. However, we saw in Section 9.9 that there exist line integrals whose value depends only on the initial point A and terminal point B of the curve C, and not on C itself. In this case we say that the line integral is **independent of the path**. These integrals can

be evaluated by the Fundamental Theorem for Line Integrals (Theorem 9.9.1). It seems natural then to ask:

Is there a complex version of the Fundamental Theorem of Calculus?

Can a contour integral $\int_C f(z)\, dz$ be independent of the path?

In this section we will see that the answer to both of these questions is "yes."

■ **Path Independence** As the next definition shows, the definition of **path independence** for a contour integral $\int_C f(z)\, dz$ is essentially the same as for a real-line integral $\int_C P\, dx + Q\, dy$.

DEFINITION 18.3.1 Independence of Path

Let z_0 and z_1 be points in a domain D. A contour integral $\int_C f(z)\, dz$ is said to be **independent of the path** if its value is the same for all contours C in D with an initial point z_0 and a terminal point z_1.

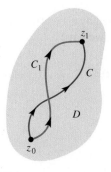

FIGURE 18.3.1 If f is analytic in D, integrals on C and C_1 are equal

At the end of the preceding section we noted that the Cauchy–Goursat theorem also holds for closed contours, not just simple closed contours, in a simply connected domain D. Now suppose, as shown in **FIGURE 18.3.1**, that C and C_1 are two contours in a simply connected domain D, both with initial point z_0 and terminal point z_1. Note that C and $-C_1$ form a closed contour. Thus, if f is analytic in D, it follows from the Cauchy–Goursat theorem that

$$\int_C f(z)\, dz + \int_{-C_1} f(z)\, dz = 0. \tag{2}$$

But (2) is equivalent to

$$\int_C f(z)\, dz = \int_{C_1} f(z)\, dz. \tag{3}$$

The result in (3) is also an example of the principle of deformation of contours introduced in (3) of Section 18.2. We summarize the last result as a theorem.

THEOREM 18.3.1 Analyticity Implies Path Independence

If f is an analytic function in a simply connected domain D, then $\int_C f(z)\, dz$ is independent of the path C.

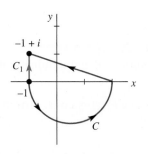

FIGURE 18.3.2 Contour in Example 1

EXAMPLE 1 **Choosing a Different Path**

Evaluate $\int_C 2z\, dz$, where C is the contour with initial point $z = -1$ and terminal point $z = -1 + i$ shown in **FIGURE 18.3.2**.

SOLUTION Since the function $f(z) = 2z$ is entire, we can replace the path C by any convenient contour C_1 joining $z = -1$ and $z = -1 + i$. In particular, by choosing C_1 to be the straight line segment $x = -1$, $0 \le y \le 1$, shown in red in Figure 18.3.2, we have $z = -1 + iy$, $dz = i\, dy$. Therefore,

$$\int_C 2z\, dz = \int_{C_1} 2z\, dz = -2 \int_0^1 y\, dy - 2i \int_0^1 dy = -1 - 2i. \qquad \equiv$$

A contour integral $\int_C f(z)\, dz$ that is independent of the path C is usually written $\int_{z_0}^{z_1} f(z)\, dz$, where z_0 and z_1 are the initial and terminal points of C. Hence in Example 1 we can write $\int_{-1}^{-1+i} 2z\, dz$.

There is an easier way to evaluate the contour integral in Example 1, but before proceeding we need another definition.

DEFINITION 18.3.2 Antiderivative

Suppose f is continuous in a domain D. If there exists a function F such that $F'(z) = f(z)$ for each z in D, then F is called an **antiderivative** of f.

For example, the function $F(z) = -\cos z$ is an antiderivative of $f(z) = \sin z$, since $F'(z) = \sin z$. As in real calculus, the most general antiderivative, or **indefinite integral**, of a function $f(z)$ is written $\int f(z)\, dz = F(z) + C$, where $F'(z) = f(z)$ and C is some complex constant.

Since an antiderivative F of a function f has a derivative at each point in a domain D, it is necessarily analytic and hence continuous in D (recall that differentiability implies continuity).

We are now in a position to prove the complex analogue of (1).

> ### THEOREM 18.3.2 Fundamental Theorem for Contour Integrals
>
> Suppose f is continuous in a domain D and F is an antiderivative of f in D. Then for any contour C in D with initial point z_0 and terminal point z_1,
>
> $$\int_C f(z)\, dz = F(z_1) - F(z_0). \tag{4}$$

PROOF: We will prove (4) in the case when C is a smooth curve defined by $z = z(t)$, $a \le t \le b$. Using (3) of Section 18.1 and the fact that $F'(z) = f(z)$ for each z in D, we have

$$\int_C f(z)\, dz = \int_a^b f(z(t)) z'(t)\, dt = \int_a^b F'(z(t)) z'(t)\, dt$$

$$= \int_a^b \frac{d}{dt} F(z(t))\, dt \quad \leftarrow \text{Chain Rule}$$

$$= F(z(t)) \Big|_a^b$$

$$= F(z(b)) - F(z(a)) = F(z_1) - F(z_0). \qquad \equiv$$

EXAMPLE 2 Using an Antiderivative

In Example 1 we saw that the integral $\int_C 2z\, dz$, where C is shown in Figure 18.3.2, is independent of the path. Now since $f(z) = 2z$ is an entire function, it is continuous. Moreover, $F(z) = z^2$ is an antiderivative of f, since $F'(z) = 2z$. Hence by (4) we have

$$\int_{-1}^{-1+i} 2z\, dz = z^2 \Big|_{-1}^{-1+i} = (-1 + i)^2 - (-1)^2 = -1 - 2i. \qquad \equiv$$

EXAMPLE 3 Using an Antiderivative

Evaluate $\int_C \cos z\, dz$, where C is any contour with initial point $z = 0$ and terminal point $z = 2 + i$.

SOLUTION $F(z) = \sin z$ is an antiderivative of $f(z) = \cos z$, since $F'(z) = \cos z$. Therefore from (4) we have

$$\int_C \cos z\, dz = \int_0^{2+i} \cos z\, dz = \sin z \Big|_0^{2+i} = \sin(2 + i) - \sin 0 = \sin(2 + i).$$

If we desired a complex number of the form $a + bi$ for an answer, we can use $\sin(2 + i) = 1.4031 - 0.4891i$ (see Example 1 in Section 17.7). Hence,

$$\int_C \cos z\, dz = 1.4031 - 0.4891i. \qquad \equiv$$

We can draw several immediate conclusions from Theorem 18.3.2. First, observe that if the contour C is closed, then $z_0 = z_1$ and consequently

$$\oint_C f(z)\, dz = 0. \tag{5}$$

Next, since the value of $\int_C f(z)\,dz$ depends on only the points z_0 and z_1, this value is the same for any contour C in D connecting these points. In other words:

> *If a continuous function f has an antiderivative F in D, then $\int_C f(z)\,dz$ is independent of the path.* (6)

In addition we have the following sufficient condition for the existence of an antiderivative:

> *If f is continuous and $\int_C f(z)\,dz$ is independent of the path in a domain D, then f has an antiderivative everywhere in D.* (7)

The last statement is important and deserves a proof. Assume that f is continuous, $\int_C f(z)\,dz$ is independent of the path in a domain D, and F is a function defined by $F(z) = \int_{z_0}^{z} f(s)\,ds$ where s denotes a complex variable, z_0 is a fixed point in D, and z represents any point in D. We wish to show that $F'(z) = f(z)$; that is, F is an antiderivative of f in D. Now,

$$F(z + \Delta z) - F(z) = \int_{z_0}^{z+\Delta z} f(s)\,ds - \int_{z_0}^{z} f(s)\,ds = \int_{z}^{z+\Delta z} f(s)\,ds. \quad (8)$$

Because D is a domain we can choose Δz so that $z + \Delta z$ is in D. Moreover, z and $z + \Delta z$ can be joined by a straight segment lying in D, as shown in **FIGURE 18.3.3**. This is the contour we use in the last integral in (8). With z fixed, we can write*

$$f(z)\,\Delta z = f(z)\int_{z}^{z+\Delta z} ds = \int_{z}^{z+\Delta z} f(z)\,ds \quad \text{and} \quad f(z) = \frac{1}{\Delta z}\int_{z}^{z+\Delta z} f(z)\,ds. \quad (9)$$

From (8) and (9) it follows that

$$\frac{F(z + \Delta z) - F(z)}{\Delta z} - f(z) = \frac{1}{\Delta z}\int_{z}^{z+\Delta z} [f(s) - f(z)]\,ds.$$

Now f is continuous at the point z. This means that for any $\epsilon > 0$ there exists a $\delta > 0$ so that $|f(s) - f(z)| < \epsilon$ whenever $|s - z| < \delta$. Consequently, if we choose Δz so that $|\Delta z| < \delta$, we have

$$\left|\frac{F(z + \Delta z) - F(z)}{\Delta z} - f(z)\right| = \left|\frac{1}{\Delta z}\int_{z}^{z+\Delta z}[f(s) - f(z)]\,ds\right|$$

$$= \left|\frac{1}{\Delta z}\right|\left|\int_{z}^{z+\Delta z}[f(s) - f(z)]\,ds\right| \leq \frac{1}{|\Delta z|}\,\epsilon\,|\Delta z| = \epsilon.$$

Hence, we have shown that

$$\lim_{\Delta z \to 0} \frac{F(z + \Delta z) - F(z)}{\Delta z} = f(z) \quad \text{or} \quad F'(z) = f(z).$$

If f is an analytic function in a simply connected domain D, it is necessarily continuous throughout D. This fact, when put together with the results in Theorem 18.3.1 and (7), leads to a theorem that states that an analytic function possesses an analytic antiderivative.

THEOREM 18.3.3 Existence of an Antiderivative

If f is analytic in a simply connected domain D, then f has an antiderivative in D; that is, there exists a function F such that $F'(z) = f(z)$ for all z in D.

In (9) of Section 17.6 we saw that $1/z$ is the derivative of $\text{Ln }z$. This means that under some circumstances $\text{Ln }z$ is an antiderivative of $1/z$. Care must be exercised in using this result. For example, suppose D is the entire complex plane without the origin. The function $1/z$ is analytic in this *multiply*

*See Problem 29 in Exercises 18.1.

connected domain. If C is any simple closed contour containing the origin, it does *not* follow from (5) that $\oint_C dz/z = 0$. In fact, from (4) of Section 18.2 with the identification $z_0 = 0$, we see that

$$\oint_C \frac{1}{z}\, dz = 2\pi i.$$

In this case, Ln z is not an antiderivative of $1/z$ in D, since Ln z is not analytic in D. Recall that Ln z fails to be analytic on the nonpositive real axis (the branch cut off the principal branch of the logarithm).

Using the Logarithmic Function

Evaluate $\displaystyle\int_C \frac{1}{z}\, dz$, where C is the contour shown in **FIGURE 18.3.4**.

SOLUTION Suppose that D is the simply connected domain defined by $x = \mathrm{Re}(z) > 0$, $y = \mathrm{Im}(z) > 0$. In this case, Ln z is an antiderivative of $1/z$, since both these functions are analytic in D. Hence by (4),

$$\int_3^{2i} \frac{1}{z}\, dz = \mathrm{Ln}\, z \Big|_3^{2i} = \mathrm{Ln}\, 2i - \mathrm{Ln}\, 3.$$

From (7) of Section 17.6, we have

$$\mathrm{Ln}\, 2i = \log_e 2 + \frac{\pi}{2}i \quad \text{and} \quad \mathrm{Ln}\, 3 = \log_e 3$$

and so

$$\int_3^{2i} \frac{1}{z}\, dz = \log_e \frac{2}{3} + \frac{\pi}{2}i = -0.4055 + 1.5708i. \qquad \equiv$$

FIGURE 18.3.4 Contour in Example 4

REMARKS

Suppose f and g are analytic in a simply connected domain D that contains the contour C. If z_0 and z_1 are the initial and terminal points of C, then the **integration by parts** formula is valid in D:

$$\int_{z_0}^{z_1} f(z) g'(z)\, dz = f(z) g(z) \Big|_{z_0}^{z_1} - \int_{z_0}^{z_1} f'(z) g(z)\, dz.$$

This can be proved in a straightforward manner using Theorem 18.3.2 on the function $(d/dz)(fg)$. See Problems 21–24 in Exercises 18.3.

18.3 Exercises Answers to selected odd-numbered problems begin on page ANS-46.

In Problems 1 and 2, evaluate the given integral, where C is the contour given in the figure, by **(a)** finding an alternative path of integration and **(b)** using Theorem 18.3.2.

1. $\displaystyle\int_C (4z - 1)\, dz$

FIGURE 18.3.5 Contour in Problem 1

2. $\displaystyle\int_C e^z\, dz$

FIGURE 18.3.6 Contour in Problem 2

In Problems 3 and 4, evaluate the given integral along the indicated contour C.

3. $\displaystyle\int_C 2z\, dz$, where C is $z(t) = 2t^3 + i(t^4 - 4t^3 + 2)$, $-1 \le t \le 1$

4. $\displaystyle\int_C 6z^2\, dz$, where C is $z(t) = 2\cos^3 \pi t - i \sin^2 \frac{\pi}{4} t$, $0 \le t \le 2$

In Problems 5–24, use Theorem 18.3.2 to evaluate the given integral. Write each answer in the form $a + ib$.

5. $\displaystyle\int_0^{3+i} z^2\, dz$

6. $\displaystyle\int_{-2i}^{1} (3z^2 - 4z + 5i)\, dz$

7. $\displaystyle\int_{1-i}^{1+i} z^3\, dz$

8. $\displaystyle\int_{-3i}^{2i} (z^3 - z)\, dz$

9. $\displaystyle\int_{-i/2}^{1-i} (2z + 1)^2\, dz$ **10.** $\displaystyle\int_{1}^{i} (iz + 1)^3\, dz$

11. $\displaystyle\int_{i/2}^{i} e^{\pi z}\, dz$ **12.** $\displaystyle\int_{1-i}^{1+2i} ze^{z^2}\, dz$

13. $\displaystyle\int_{\pi}^{\pi+2i} \sin\frac{z}{2}\, dz$ **14.** $\displaystyle\int_{1-2i}^{\pi i} \cos z\, dz$

15. $\displaystyle\int_{\pi i}^{2\pi i} \cosh z\, dz$ **16.** $\displaystyle\int_{i}^{1+(\pi/2)i} \sinh 3z\, dz$

17. $\displaystyle\int_{C} \frac{1}{z}\, dz$, C is the arc of the circle $z = 4e^{it}$, $-\pi/2 \le t \le \pi/2$

18. $\displaystyle\int_{C} \frac{1}{z}\, dz$, C is the straight line segment between $z = 1 + i$ and $z = 4 + 4i$

19. $\displaystyle\int_{-4i}^{4i} \frac{1}{z^2}\, dz$, C is any contour not passing through the origin

20. $\displaystyle\int_{1-i}^{1+\sqrt{3}i} \left(\frac{1}{z} + \frac{1}{z^2}\right) dz$, C is any contour in the right half-plane $\operatorname{Re}(z) > 0$

21. $\displaystyle\int_{\pi}^{i} e^z \cos z\, dz$ **22.** $\displaystyle\int_{0}^{i} z \sin z\, dz$

23. $\displaystyle\int_{i}^{1+i} ze^z\, dz$ **24.** $\displaystyle\int_{0}^{\pi i} z^2 e^z\, dz$

18.4 **Cauchy's Integral Formulas**

INTRODUCTION In the last two sections we saw the importance of the Cauchy–Goursat theorem in the evaluation of contour integrals. In this section we are going to examine several more consequences of the Cauchy–Goursat theorem. Unquestionably, the most significant of these is the following result:

> *The value of an analytic function f at any point z_0 in a simply connected domain can be represented by a contour integral.*

After establishing this proposition we shall use it to further show that

> *An analytic function f in a simply connected domain possesses derivatives of all orders.*

The ramifications of these two results alone will keep us busy not only for the remainder of this section but in the next chapter as well.

■ **First Formula** We begin with the Cauchy integral formula. The idea in the next theorem is this: If f is analytic in a simply connected domain and z_0 is any point D, then the quotient $f(z)/(z - z_0)$ is *not* analytic in D. As a consequence, the integral of $f(z)/(z - z_0)$ around a simple closed contour C that contains z_0 is not necessarily zero but has, as we shall now see, the value $2\pi i\, f(z_0)$. This remarkable result indicates that the values of an analytic function f at points *inside* a simple closed contour C are determined by the values of f *on* the contour C.

THEOREM 18.4.1 **Cauchy's Integral Formula**

Let f be analytic in a simply connected domain D, and let C be a simple closed contour lying entirely within D. If z_0 is any point within C, then

$$f(z_0) = \frac{1}{2\pi i} \oint_C \frac{f(z)}{z - z_0}\, dz. \tag{1}$$

PROOF: Let D be a simply connected domain, C a simple closed contour in D, and z_0 an interior point of C. In addition, let C_1 be a circle centered at z_0 with radius small enough that it is interior to C. By the principle of deformation of contours, we can write

$$\oint_C \frac{f(z)}{z - z_0}\, dz = \oint_{C_1} \frac{f(z)}{z - z_0}\, dz. \tag{2}$$

We wish to show that the value of the integral on the right is $2\pi i\, f(z_0)$. To this end we add and subtract the constant $f(z_0)$ in the numerator:

$$\oint_{C_1}\frac{f(z)}{z-z_0}\,dz = \oint_{C_1}\frac{f(z_0)-f(z_0)+f(z)}{z-z_0}\,dz$$

$$= f(z_0)\oint_{C_1}\frac{dz}{z-z_0}+\oint_{C_1}\frac{f(z)-f(z_0)}{z-z_0}\,dz. \qquad (3)$$

Now from (4) of Section 18.2 we know that

$$\oint_{C_1}\frac{dz}{z-z_0}=2\pi i.$$

Thus, (3) becomes

$$\oint_{C_1}\frac{f(z)}{z-z_0}\,dz = 2\pi i f(z_0)+\oint_{C_1}\frac{f(z)-f(z_0)}{z-z_0}\,dz. \qquad (4)$$

Since f is continuous at z_0 for any arbitrarily small $\epsilon > 0$, there exists a $\delta > 0$ such that $|f(z)-f(z_0)| < \epsilon$ whenever $|z-z_0| < \delta$. In particular, if we choose the circle C_1 to be $|z-z_0| = \delta/2 < \delta$, then by the *ML*-inequality (Theorem 18.1.3) the absolute value of the integral on the right side of (4) satisfies

$$\left|\oint_{C_1}\frac{f(z)-f(z_0)}{z-z_0}\,dz\right| \le \frac{\epsilon}{\delta/2}\,2\pi\left(\frac{\delta}{2}\right)=2\pi\epsilon.$$

In other words, the absolute value of the integral can be made arbitrarily small by taking the radius of the circle C_1 to be sufficiently small. This can happen only if the integral is zero. The Cauchy integral formula (1) follows from (4) by dividing both sides by $2\pi i$. ≡

The Cauchy integral formula (1) can be used to evaluate contour integrals. Since we often work problems without a simply connected domain explicitly defined, a more practical restatement of Theorem 18.4.1 is

If f is analytic at all points within and on a simple closed contour C, and z_0 is
any point interior to C, then $f(z_0) = \dfrac{1}{2\pi i}\oint_C\dfrac{f(z)}{z-z_0}\,dz.$ $\qquad (5)$

EXAMPLE 1 **Using Cauchy's Integral Formula**

Evaluate $\displaystyle\oint_C\frac{z^2-4z+4}{z+i}\,dz$, where C is the circle $|z| = 2$.

SOLUTION First, we identify $f(z) = z^2-4z+4$ and $z_0 = -i$ as a point within the circle C. Next, we observe that f is analytic at all points within and on the contour C. Thus by the Cauchy integral formula we obtain

$$\oint_C\frac{z^2-4z+4}{z+i}\,dz = 2\pi i f(-i)=2\pi i(3+4i)=2\pi(-4+3i).$$ ≡

EXAMPLE 2 **Using Cauchy's Integral Formula**

Evaluate $\displaystyle\oint_C\frac{z}{z^2+9}\,dz$, where C is the circle $|z-2i| = 4$.

SOLUTION By factoring the denominator as $z^2+9 = (z-3i)(z+3i)$, we see that $3i$ is the only point within the closed contour at which the integrand fails to be analytic. See **FIGURE 18.4.1**.

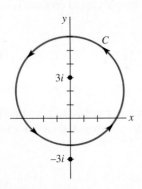

FIGURE 18.4.1 Contour in Example 2

Now by writing

$$\frac{z}{z^2 + 9} = \frac{\dfrac{z}{z + 3i}}{z - 3i},$$

we can identify $f(z) = z/(z + 3i)$. This function is analytic at all points within and on the contour C. From the Cauchy integral formula we then have

$$\oint_C \frac{z}{z^2 + 9}\, dz = \oint_C \frac{\dfrac{z}{z + 3i}}{z - 3i}\, dz = 2\pi i\, f(3i) = 2\pi i\, \frac{3i}{6i} = \pi i.$$

$\equiv$

EXAMPLE 3 Flux and Cauchy's Integral Formula

The complex function $f(z) = k/(\bar{z} - \bar{z}_1)$, where $k = a + ib$ and z_1 are complex numbers, gives rise to a flow in the domain $z \neq z_1$. If C is a simple closed contour containing $z = z_1$ in its interior, then from the Cauchy integral formula we have

$$\oint_C \overline{f(z)}\, dz = \oint_C \frac{a - ib}{z - z_1}\, dz = 2\pi i(a - ib).$$

Thus the circulation around C is $2\pi b$ and the net flux across C is $2\pi a$. If z_1 were in the exterior of C, both the circulation and net flux would be zero by Cauchy's theorem.

Note that when k is real, the circulation around C is zero but the net flux across C is $2\pi k$. The complex number z_1 is called a **source** for the flow when $k > 0$ and a **sink** when $k < 0$. Vector fields corresponding to these two cases are shown in **FIGURE 18.4.2(a)** and 18.4.2(b).

$\equiv$

(a) Source: $k > 0$

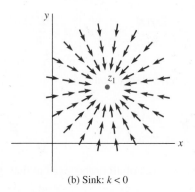

(b) Sink: $k < 0$

FIGURE 18.4.2 Vector fields in Example 3

■ **Second Formula** We can now use Theorem 18.4.1 to prove that an analytic function possesses derivatives of all orders; that is, if f is analytic at a point z_0, then f', f'', f''', and so on, are also analytic at z_0. Moreover, the values of the derivatives $f^{(n)}(z_0)$, $n = 1, 2, 3, \ldots$, are given by a formula similar to (1).

THEOREM 18.4.2 Cauchy's Integral Formula for Derivatives

Let f be analytic in a simply connected domain D, and let C be a simple closed contour lying entirely within D. If z_0 is any point interior to C, then

$$f^{(n)}(z_0) = \frac{n!}{2\pi i} \oint_C \frac{f(z)}{(z - z_0)^{n+1}}\, dz. \qquad (6)$$

PARTIAL PROOF: We will prove (6) only for the case $n = 1$. The remainder of the proof can be completed using the principle of mathematical induction.

We begin with the definition of the derivative and (1):

$$f'(z_0) = \lim_{\Delta z \to 0} \frac{f(z_0 + \Delta z) - f(z_0)}{\Delta z}$$

$$= \lim_{\Delta z \to 0} \frac{1}{2\pi i\, \Delta z} \left[\oint_C \frac{f(z)}{z - (z_0 + \Delta z)}\, dz - \oint_C \frac{f(z)}{z - z_0}\, dz \right]$$

$$= \lim_{\Delta z \to 0} \frac{1}{2\pi i} \oint_C \frac{f(z)}{(z - z_0 - \Delta z)(z - z_0)}\, dz.$$

Before proceeding, let us set up some preliminaries. Since f is continuous on C, it is bounded; that is, there exists a real number M such that $|f(z)| \leq M$ for all points z on C. In addition, let L be the length of C and let δ denote the shortest distance between points on C and the point z_0. Thus for all points z on C, we have

$$|z - z_0| \geq \delta \quad \text{or} \quad \frac{1}{|z - z_0|^2} \leq \frac{1}{\delta^2}.$$

Furthermore, if we choose $|\Delta z| \leq \delta/2$, then

$$|z - z_0 - \Delta z| \geq \left||z - z_0| - |\Delta z|\right| \geq \delta - |\Delta z| \geq \frac{\delta}{2} \quad \text{and so} \quad \frac{1}{|z - z_0 - \Delta z|} \leq \frac{2}{\delta}.$$

Now,

$$\left| \oint_C \frac{f(z)}{(z - z_0)^2} \, dz - \oint_C \frac{f(z)}{(z - z_0 - \Delta z)(z - z_0)} \, dz \right| = \left| \oint_C \frac{-\Delta z f(z)}{(z - z_0)^2 (z - z_0 - \Delta z)} \, dz \right| \leq \frac{2ML|\Delta z|}{\delta^3}.$$

Because the last expression approaches zero as $\Delta z \to 0$, we have shown that

$$f'(z_0) = \lim_{\Delta z \to 0} \frac{f(z_0 + \Delta z) - f(z_0)}{\Delta z} = \frac{1}{2\pi i} \oint_C \frac{f(z)}{(z - z_0)^2} \, dz. \qquad \equiv$$

If $f(z) = u(x, y) + iv(x, y)$ is analytic at a point, then its derivatives of all orders exist at that point and are continuous. Consequently, from

$$f'(z) = \frac{\partial u}{\partial x} + i\frac{\partial v}{\partial x} = \frac{\partial v}{\partial y} - i\frac{\partial u}{\partial y}$$

$$f''(z) = \frac{\partial^2 u}{\partial x^2} + i\frac{\partial^2 v}{\partial x^2} = \frac{\partial^2 v}{\partial y \, \partial x} - i\frac{\partial^2 u}{\partial y \, \partial x}$$

we can conclude that the real functions u and v have continuous partial derivatives of all orders at a point of analyticity.

Like (1), (6) can sometimes be used to evaluate integrals.

EXAMPLE 4 **Using Cauchy's Integral Formula for Derivatives**

Evaluate $\displaystyle\oint_C \frac{z + 1}{z^4 + 4z^3} \, dz$, where C is the circle $|z| = 1$.

SOLUTION Inspection of the integrand shows that it is not analytic at $z = 0$ and $z = -4$, but only $z = 0$ lies within the closed contour. By writing the integrand as

$$\frac{z + 1}{z^4 + 4z^3} = \frac{\dfrac{z + 1}{z + 4}}{z^3},$$

we can identify $z_0 = 0$, $n = 2$, and $f(z) = (z + 1)/(z + 4)$. By the Quotient Rule, $f''(z) = -6/(z + 4)^3$ and so by (6) we have

$$\oint_C \frac{z + 1}{z^4 + 4z^3} \, dz = \frac{2\pi i}{2!} f''(0) = -\frac{3\pi}{32} i. \qquad \equiv$$

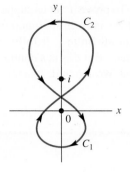

FIGURE 18.4.3 Contour in Example 5

EXAMPLE 5 Using Cauchy's Integral Formula for Derivatives

Evaluate $\oint_C \dfrac{z^3 + 3}{z(z - i)^2}\, dz$, where C is the contour shown in **FIGURE 18.4.3**.

SOLUTION Although C is not a simple closed contour, we can think of it as the union of two simple closed contours C_1 and C_2 as indicated in Figure 18.4.3. By writing

$$\oint_C \frac{z^3 + 3}{z(z - i)^2}\, dz = \oint_{C_1} \frac{z^3 + 3}{z(z - i)^2}\, dz + \oint_{C_2} \frac{z^3 + 3}{z(z - i)^2}\, dz$$

$$= -\oint_{C_1} \frac{\dfrac{z^3 + 3}{(z - i)^2}}{z}\, dz + \oint_{C_2} \frac{\dfrac{z^3 + 3}{z}}{(z - i)^2}\, dz = -I_1 + I_2,$$

we are in a position to use both (1) and (6).

To evaluate I_1, we identify $z_0 = 0$ and $f(z) = (z^3 + 3)/(z - i)^2$. By (1) it follows that

$$I_1 = \oint_{C_1} \frac{\dfrac{z^3 + 3}{(z - i)^2}}{z}\, dz = 2\pi i f(0) = -6\pi i.$$

To evaluate I_2 we identify $z_0 = i$, $n = 1$, $f(z) = (z^3 + 3)/z$, and $f'(z) = (2z^3 - 3)/z^2$. From (6) we obtain

$$I_2 = \oint_{C_2} \frac{\dfrac{z^3 + 3}{z}}{(z - i)^2}\, dz = \frac{2\pi i}{1!} f'(i) = 2\pi i(3 + 2i) = 2\pi(-2 + 3i).$$

Finally we get

$$\oint_C \frac{z^3 + 3}{z(z - i)^2}\, dz = -I_1 + I_2 = 6\pi i + 2\pi(-2 + 3i) = 4\pi(-1 + 3i). \qquad \equiv$$

Liouville's Theorem If we take the contour C to be the circle $|z - z_0| = r$, it follows from (6) and the *ML*-inequality that

$$|f^{(n)}(z_0)| = \frac{n!}{2\pi} \left| \oint_C \frac{f(z)}{(z - z_0)^{n+1}}\, dz \right| \leq \frac{n!}{2\pi} M \frac{1}{r^{n+1}} 2\pi r = \frac{n!M}{r^n}, \tag{7}$$

where M is a real number such that $|f(z)| \leq M$ for all points z on C. The result in (7), called **Cauchy's inequality**, is used to prove the next result.

THEOREM 18.4.3 Liouville's Theorem

The only bounded entire functions are constants.

PROOF: Suppose f is an entire function and is bounded; that is, $|f(z)| \leq M$ for all z. Then for any point z_0, (7) gives $|f'(z_0)| \leq M/r$. By making r arbitrarily large, we can make $|f'(z_0)|$ as small as we wish. This means $f'(z_0) = 0$ for all points z_0 in the complex plane. Hence f must be a constant. $\equiv$

Fundamental Theorem of Algebra Liouville's theorem enables us to prove, in turn, a result that is learned in elementary algebra:

If $P(z)$ is a nonconstant polynomial, then the equation $P(z) = 0$ has at least one root.

This result is known as the **Fundamental Theorem of Algebra**. To prove it, let us suppose that $P(z) \neq 0$ for all z. This implies that the reciprocal of P, $f(z) = 1/P(z)$, is an entire function. Now since $|f(z)| \to 0$ as $|z| \to \infty$, the function f must be bounded for all finite z. It follows from Liouville's theorem that f is a constant and therefore P is a constant. But this is a contradiction to our underlying assumption that P was *not* a constant polynomial. We conclude that there must exist at least one number z for which $P(z) = 0$.

18.4 Exercises
Answers to selected odd-numbered problems begin on page ANS-46.

In Problems 1–24, use Theorems 18.4.1 and 18.4.2, when appropriate, to evaluate the given integral along the indicated closed contour(s).

1. $\oint_C \dfrac{4}{z - 3i}\, dz;\quad |z| = 5$

2. $\oint_C \dfrac{z^2}{(z - 3i)^2}\, dz;\quad |z| = 5$

3. $\oint_C \dfrac{e^z}{z - \pi i}\, dz;\quad |z| = 4$

4. $\oint_C \dfrac{1 + 2e^z}{z}\, dz;\quad |z| = 1$

5. $\oint_C \dfrac{z^2 - 3z + 4i}{z + 2i}\, dz;\quad |z| = 3$

6. $\oint_C \dfrac{\cos z}{3z - \pi}\, dz;\quad |z| = 1.1$

7. $\oint_C \dfrac{z^2}{z^2 + 4}\, dz;\quad$ **(a)** $|z - i| = 2,$ **(b)** $|z + 2i| = 1$

8. $\oint_C \dfrac{z^2 + 3z + 2i}{z^2 + 3z - 4}\, dz;\quad$ **(a)** $|z| = 2,$ **(b)** $|z + 5| = \frac{3}{2}$

9. $\oint_C \dfrac{z^2 + 4}{z^2 - 5iz - 4}\, dz;\quad |z - 3i| = 1.3$

10. $\oint_C \dfrac{\sin z}{z^2 + \pi^2}\, dz;\quad |z - 2i| = 2$

11. $\oint_C \dfrac{e^{z^2}}{(z - i)^3}\, dz;\quad |z - i| = 1$

12. $\oint_C \dfrac{z}{(z + i)^4}\, dz;\quad |z| = 2$

13. $\oint_C \dfrac{\cos 2z}{z^5}\, dz;\quad |z| = 1$

14. $\oint_C \dfrac{e^{-z}\sin z}{z^3}\, dz;\quad |z - 1| = 3$

15. $\oint_C \dfrac{2z + 5}{z^2 - 2z}\, dz;\quad$ **(a)** $|z| = \frac{1}{2},$ **(b)** $|z + 1| = 2,$
 (c) $|z - 3| = 2,$ **(d)** $|z + 2i| = 1$

16. $\oint_C \dfrac{z}{(z - 1)(z - 2)}\, dz;\quad$ **(a)** $|z| = \frac{1}{2},$ **(b)** $|z + 1| = 1,$
 (c) $|z - 1| = \frac{1}{2},$ **(d)** $|z| = 4$

17. $\oint_C \dfrac{z + 2}{z^2(z - 1 - i)}\, dz;\quad$ **(a)** $|z| = 1,$ **(b)** $|z - 1 - i| = 1$

18. $\oint_C \dfrac{1}{z^3(z - 4)}\, dz;\quad$ **(a)** $|z| = 1,$ **(b)** $|z - 2| = 1$

19. $\oint_C \left(\dfrac{e^{2iz}}{z^4} - \dfrac{z^4}{(z - i)^3} \right) dz;\quad |z| = 6$

20. $\oint_C \left(\dfrac{\cosh z}{(z - \pi)^3} - \dfrac{\sin^2 z}{(2z - \pi)^3} \right) dz;\quad |z| = 3$

21. $\oint_C \dfrac{1}{z^3(z - 1)^2}\, dz;\quad |z - 2| = 5$

22. $\oint_C \dfrac{1}{z^2(z^2 + 1)}\, dz;\quad |z - i| = \frac{3}{2}$

23. $\oint_C \dfrac{3z + 1}{z(z - 2)^2}\, dz;\quad C$ is given in **FIGURE 18.4.4**

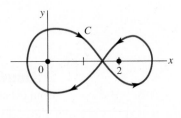

FIGURE 18.4.4 Contour in Problem 23

24. $\oint_C \dfrac{e^{iz}}{(z^2 + 1)^2}\, dz;\quad C$ is given in **FIGURE 18.4.5**

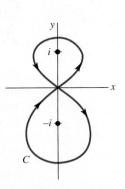

FIGURE 18.4.5 Contour in Problem 24

Answer Problems 1–12 without referring back to the text. Fill in the blank or answer true/false.

1. The sector defined by $-\pi/6 < \arg z < \pi/6$ is a simply connected domain. _____

2. If $\oint_C f(z)\, dz = 0$ for every simple closed contour C, then f is analytic within and on C. _____

3. The value of $\displaystyle\int_C \frac{z-2}{z}\, dz$ is the same for any path C in the right half-plane $\mathrm{Re}(z) > 0$ between $z = 1 + i$ and $z = 10 + 8i$. _____

4. If g is entire, then $\displaystyle\oint_C \frac{g(z)}{z-i}\, dz = \oint_{C_1} \frac{g(z)}{z-i}\, dz$, where C is the circle $|z| = 3$ and C_1 is the ellipse $x^2 + y^2/9 = 1$. _____

5. If f is a polynomial and C is a simple closed curve, then $\displaystyle\oint_C f(z)\, dz =$ _____.

6. If $f(z) = \displaystyle\oint_C \frac{\xi^2 + 6\xi - 2}{\xi - z}\, d\xi$, where C is $|z| = 3$, then $f(1 + i) =$ _____.

7. If $f(z) = z^3 + e^z$ and C is the contour $z = 8e^{it}, 0 \le t \le 2\pi$, then $\displaystyle\oint_C \frac{f(z)}{(z + \pi i)^3}\, dz =$ _____.

8. If f is entire and $|f(z)| \le 10$ for all z, then $f(z) =$ _____.

9. $\displaystyle\oint_C \frac{1}{(z - z_0)(z - z_1)}\, dz = 0$ for every simple closed contour C that encloses the points z_0 and z_1. _____

10. If f is analytic within and on the simple closed contour C and z_0 is a point within C, then
$$\oint_C \frac{f'(z)}{z - z_0}\, dz = \oint_C \frac{f(z)}{(z - z_0)^2}\, dz.$$ _____

11. $\displaystyle\oint_C z^n dz = \begin{cases} 0, & \text{if } n\ \underline{\quad} \\ 2\pi i, & \text{if } n\ \underline{\quad} \end{cases}$
where n is an integer and C is $|z| = 1$.

12. If $|f(z)| \le 2$ on $|z| = 3$, then $\left| \displaystyle\oint_C f(z)\, dz \right| \le$ _____.

In Problems 13–28, evaluate the given integral using the techniques considered in this chapter.

13. $\displaystyle\int_C (x + iy)\, dz$; C is the contour shown in **FIGURE 18.R.1**

14. $\displaystyle\int_C (x - iy)\, dz$; C is the contour shown in Figure 18.R.1

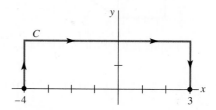

FIGURE 18.R.1 Contour in Problems 13 and 14

15. $\displaystyle\int_C |z^2|\, dz$; C is $z(t) = t + it^2, 0 \le t \le 2$

16. $\displaystyle\int_C e^{\pi z}\, dz$; C is the line segment from $z = i$ to $z = 1 + i$

17. $\displaystyle\oint_C e^{\pi z}\, dz$; C is the ellipse $x^2/100 + y^2/64 = 1$

18. $\displaystyle\int_{3i}^{1-i} (4z - 6)\, dz$

19. $\displaystyle\int_C \sin z\, dz$; C is $z(t) = t^4 + i(1 + t^3)^2, -1 \le t < 1$

20. $\displaystyle\int_C (4z^3 + 3z^2 + 2z + 1)\, dz$; C is the line segment from 0 to $2i$

21. $\displaystyle\oint_C (z^{-2} + z^{-1} + z + z^2)\, dz$; C is the circle $|z| = 1$

22. $\displaystyle\oint_C \frac{3z + 4}{z^2 - 1}\, dz$; C is the circle $|z| = 2$

23. $\displaystyle\oint_C \frac{e^{-2z}}{z^4}\, dz$; C is the circle $|z - 1| = 3$

24. $\displaystyle\oint_C \frac{\cos z}{z^3 - z^2}\, dz$; C is the circle $|z| = \frac{1}{2}$

25. $\displaystyle\oint_C \frac{1}{2z^2 + 7z + 3}\, dz$; C is the ellipse $x^2/4 + y^2 = 1$

26. $\displaystyle\oint_C z \csc z\, dz$; C is the rectangle with vertices $1 + i, 1 - i, 2 + i, 2 - i$

27. $\oint_C \dfrac{z}{z+i}\, dz$; C is the contour shown in **FIGURE 18.R.2**

FIGURE 18.R.2 Contour in Problem 27

28. $\oint_C \dfrac{e^{i\pi z}}{2z^2 - 5z + 2}\, dz$; C is **(a)** $|z| = 1$, **(b)** $|z - 3| = 2$, **(c)** $|z + 3| = 2$

29. Let $f(z) = z^n g(z)$, where n is a positive integer, $g(z)$ is entire, and $g(z) \neq 0$ for all z. Let C be a circle with center at the origin. Evaluate $\oint_C \dfrac{f'(z)}{f(z)}\, dz$.

30. Let C be the straight line segment from i to $2 + i$. Show that

$$\left| \int_C \mathrm{Ln}\,(z + 1)\, dz \right| \leq \log_e 10 + \frac{\pi}{2}.$$

Cauchy's integral formula for derivatives indicates that if a function f is analytic at a point z_0, then it possesses derivatives of all orders at that point. As a consequence, we will see that f can always be expanded in a power series at that point. On the other hand, if f fails to be analytic at z_0, then we may still be able to expand it in a different kind of series known as a **Laurent series**. The notion of Laurent series leads to the concept of a **residue**, and this, in turn, leads to yet another way of evaluating complex integrals.

19.1 Sequences and Series

INTRODUCTION Much of the theory of complex sequences and series is analogous to that encountered in real calculus. In this section we explore the definitions of convergence and divergence for complex sequences and complex infinite series. In addition, we give some tests for convergence of infinite series. You are urged to pay special attention to what is said about *geometric series* since this type of series will be important in the later sections of this chapter.

■ **Sequences** A **sequence** $\{z_n\}$ is a function whose domain is the set of positive integers; in other words, to each integer $n = 1, 2, 3, \ldots$, we assign a complex number z_n. For example, the sequence $\{1 + i^n\}$ is

$$
\begin{array}{ccccc}
1 + i, & 0, & 1 - i, & 2, & 1 + i, \ldots \\
\uparrow & \uparrow & \uparrow & \uparrow & \uparrow \\
n = 1, & n = 2, & n = 3, & n = 4, & n = 5, \ldots.
\end{array}
\tag{1}
$$

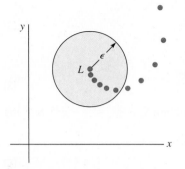

FIGURE 19.1.1 If $\{z_n\}$ converges to L, all but a finite number of terms are in any ϵ-neighborhood of L

If $\lim_{n \to \infty} z_n = L$ we say the sequence $\{z_n\}$ is **convergent**. In other words, $\{z_n\}$ **converges** to the number L if, for each positive number ϵ, an N can be found such that $|z_n - L| < \epsilon$ whenever $n > N$. As shown in **FIGURE 19.1.1**, when a sequence $\{z_n\}$ converges to L, all but a finite number of the terms of the sequence are within every ϵ-neighborhood of L. The sequence $\{1 + i^n\}$ illustrated in (1) is divergent, since the general term $z_n = 1 + i^n$ does not approach a fixed complex number as $n \to \infty$. Indeed, the first four terms of this sequence repeat endlessly as n increases.

EXAMPLE 1	A Convergent Sequence

The sequence $\left\{ \dfrac{i^{n+1}}{n} \right\}$ converges, since

$$
\lim_{n \to \infty} \frac{i^{n+1}}{n} = 0.
$$

FIGURE 19.1.2 The terms of the sequence spiral toward 0 in Example 1

As we see from

$$
-1, \, -\frac{i}{2}, \, \frac{1}{3}, \, \frac{i}{4}, \, -\frac{1}{5}, \ldots,
$$

and **FIGURE 19.1.2**, the terms of the sequence spiral toward the point $z = 0$. ≡

The following theorem should make intuitive sense.

THEOREM 19.1.1	Criterion for Convergence

A sequence $\{z_n\}$ converges to a complex number L if and only if $\text{Re}(z_n)$ converges to $\text{Re}(L)$ and $\text{Im}(z_n)$ converges to $\text{Im}(L)$.

EXAMPLE 2	Illustrating Theorem 19.1.1

The sequence $\left\{ \dfrac{ni}{n + 2i} \right\}$ converges to i. Note that $\text{Re}(i) = 0$ and $\text{Im}(i) = 1$. Then from

$$
z_n = \frac{ni}{n + 2i} = \frac{2n}{n^2 + 4} + i \frac{n^2}{n^2 + 4},
$$

we see that $\text{Re}(z_n) = 2n/(n^2 + 4) \to 0$ and $\text{Im}(z_n) = n^2/(n^2 + 4) \to 1$ as $n \to \infty$. ≡

■ **Series** An **infinite series** of complex numbers

$$
\sum_{k=1}^{\infty} z_k = z_1 + z_2 + z_3 + \cdots + z_n + \cdots
$$

is **convergent** if the sequence of partial sums $\{S_n\}$, where

$$S_n = z_1 + z_2 + z_3 + \cdots + z_n,$$

converges. If $S_n \to L$ as $n \to \infty$, we say that the **sum** of the series is L.

■ **Geometric Series** For the geometric series

$$\sum_{k=1}^{\infty} az^{k-1} = a + az + az^2 + \cdots + az^{n-1} + \cdots \tag{2}$$

the nth term of the sequence of partial sums is

$$S_n = a + az + az^2 + \cdots + az^{n-1}. \tag{3}$$

By multiplying S_n by z and subtracting this result from S_n, we obtain $S_n - zS_n = a - az^n$. Solving for S_n gives

$$S_n = \frac{a(1 - z^n)}{1 - z}. \tag{4}$$

Since $z^n \to 0$ as $n \to \infty$, whenever $|z| < 1$ we conclude from (4) that (2) converges to

$$\frac{a}{1 - z}$$

when $|z| < 1$; the series diverges when $|z| \geq 1$. The special geometric series

$$\frac{1}{1 - z} = 1 + z + z^2 + z^3 + \cdots \tag{5}$$

$$\frac{1}{1 + z} = 1 - z + z^2 - z^3 + \cdots \tag{6}$$

valid for $|z| < 1$, will be of particular usefulness in the next two sections. In addition, we shall use

$$\frac{1 - z^n}{1 - z} = 1 + z + z^2 + z^3 + \cdots + z^{n-1} \tag{7}$$

in the alternative form

$$\frac{1}{1 - z} = 1 + z + z^2 + z^3 + \cdots + z^{n-1} + \frac{z^n}{1 - z} \tag{8}$$

in the proofs of the two principal theorems of this chapter.

EXAMPLE 3 **Convergent Geometric Series**

The series

$$\sum_{k=1}^{\infty} \frac{(1 + 2i)^k}{5^k} = \frac{1 + 2i}{5} + \frac{(1 + 2i)^2}{5^2} + \frac{(1 + 2i)^3}{5^3} + \cdots$$

is a geometric series with $a = (1 + 2i)/5$ and $z = (1 + 2i)/5$. Since $|z| = \sqrt{5}/5 < 1$, the series converges and we can write

$$\sum_{k=1}^{\infty} \frac{(1 + 2i)^k}{5^k} = \frac{\dfrac{1 + 2i}{5}}{1 - \dfrac{1 + 2i}{5}} = \frac{i}{2}.$$

≡

THEOREM 19.1.2 **Necessary Condition for Convergence**

If $\sum_{k=1}^{\infty} z_k$ converges, then $\lim_{n \to \infty} z_n = 0$.

An equivalent form of Theorem 19.1.2 is the familiar nth term test for divergence of an infinite series.

THEOREM 19.1.3 The *n*th Term Test for Divergence

If $\lim_{n\to\infty} z_n \neq 0$, then the series $\sum_{k=1}^{\infty} z_k$ diverges.

For example, the series $\sum_{k=1}^{\infty} (k + 5i)/k$ diverges since $z_n = (n + 5i)/n \to 1$ as $n \to \infty$. The geometric series (2) diverges when $|z| \geq 1$, since, in this case, $\lim_{n\to\infty} |z^n|$ does not exist.

DEFINITION 19.1.1 Absolute Convergence

An infinite series $\sum_{k=1}^{\infty} z_k$ is said to be **absolutely convergent** if $\sum_{k=1}^{\infty} |z_k|$ converges.

EXAMPLE 4 Absolute Convergence

The series $\sum_{k=1}^{\infty} (i^k/k^2)$ is absolutely convergent since $|i^k/k^2| = 1/k^2$ and the real series $\sum_{k=1}^{\infty} (1/k^2)$ converges. Recall from calculus that a real series of the form $\sum_{k=1}^{\infty} (1/k^p)$ is called a p-series, p a real number, and converges for $p > 1$ and diverges for $p \leq 1$. $\equiv$

As in real calculus,

Absolute convergence implies convergence.

Thus in Example 4, because the series

$$\sum_{k=1}^{\infty} \frac{i^k}{k^2} = i - \frac{1}{2^2} - \frac{i}{3^2} + \cdots$$

converges absolutely, it is also convergent.

The following two tests are the complex versions of the ratio and root tests that are encountered in calculus:

THEOREM 19.1.4 Ratio Test

Suppose $\sum_{k=1}^{\infty} z_k$ is a series of nonzero complex terms such that

$$\lim_{n\to\infty} \left| \frac{z_{n+1}}{z_n} \right| = L. \tag{9}$$

(*i*) If $L < 1$, then the series converges absolutely.
(*ii*) If $L > 1$ or $L = \infty$, then the series diverges.
(*iii*) If $L = 1$, the test is inconclusive.

THEOREM 19.1.5 Root Test

Suppose $\sum_{k=1}^{\infty} z_k$ is a series of complex terms such that

$$\lim_{n\to\infty} \sqrt[n]{|z_n|} = L. \tag{10}$$

(*i*) If $L < 1$, then the series converges absolutely.
(*ii*) If $L > 1$ or $L = \infty$, then the series diverges.
(*iii*) If $L = 1$, the test is inconclusive.

We are interested primarily in applying these tests to power series.

Power Series

The notion of a power series is important in the study of analytic functions. An infinite series of the form

$$\sum_{k=0}^{\infty} a_k(z - z_0)^k = a_0 + a_1(z - z_0) + a_2(z - z_0)^2 + \cdots, \quad (11)$$

where the coefficients a_k are complex constants, is called a **power series** in $z - z_0$. The power series (11) is said to be **centered at z_0**, and the complex point z_0 is referred to as the **center** of the series. In (11), it is also convenient to define $(z - z_0)^0 = 1$ even when $z = z_0$.

Circle of Convergence

Every complex power series has **radius of convergence R**, where R is a real number. Analogous to the concept of an *interval* of convergence in real calculus, when $0 < R < \infty$, a complex power series (11) has a **circle of convergence** defined by $|z - z_0| = R$. The power series converges absolutely for all z satisfying $|z - z_0| < R$ and diverges for $|z - z_0| > R$. See **FIGURE 19.1.3**. The radius R of convergence can be

 (*i*) zero (in which case (11) converges at only $z = z_0$),

 (*ii*) a finite number (in which case (11) converges at all interior points of the circle $|z - z_0| = R$), or

 (*iii*) ∞ (in which case (11) converges for all z).

A power series may converge at some, all, or none of the points *on* the actual circle of convergence.

FIGURE 19.1.3 A power series converges at all points within the circle of convergence

EXAMPLE 5 Circle of Convergence

Consider the power series $\displaystyle\sum_{k=1}^{\infty} \frac{z^{k+1}}{k}$. By the ratio test (9),

$$\lim_{n\to\infty} \left| \frac{\dfrac{z^{n+2}}{n+1}}{\dfrac{z^{n+1}}{n}} \right| = \lim_{n\to\infty} \frac{n}{n+1} |z| = |z|.$$

Thus the series converges absolutely for $|z| < 1$. The circle of convergence is $|z| = 1$ and the radius of convergence is $R = 1$. Note that on the circle of convergence, the series does not converge absolutely, since the series of absolute values is the well-known divergent harmonic series $\sum_{k=1}^{\infty} (1/k)$. Bear in mind this does not say, however, that the series diverges on the circle of convergence. In fact, at $z = -1$, $\sum_{k=1}^{\infty} ((-1)^{k+1}/k)$ is the convergent alternating harmonic series. Indeed, it can be shown that the series converges at all points on the circle $|z| = 1$ except at $z = 1$. $\equiv$

It should be clear from Theorem 19.1.4 and Example 5 that for a power series $\sum_{k=0}^{\infty} a_k(z - z_0)^k$, the limit (9) depends on only the coefficients a_k. Thus, if

 (*i*) $\displaystyle\lim_{n\to\infty} \left| \frac{a_{n+1}}{a_n} \right| = L \neq 0$, the radius of convergence is $R = 1/L$;

 (*ii*) $\displaystyle\lim_{n\to\infty} \left| \frac{a_{n+1}}{a_n} \right| = 0$, the radius of convergence is ∞;

 (*iii*) $\displaystyle\lim_{n\to\infty} \left| \frac{a_{n+1}}{a_n} \right| = \infty$, the radius of convergence is $R = 0$.

Similar remarks can be made for the root test (10) by utilizing $\lim_{n\to\infty} \sqrt[n]{|a_n|}$.

EXAMPLE 6 Radius of Convergence

Consider the power series $\displaystyle\sum_{k=1}^{\infty} \frac{(-1)^{k+1}(z - 1 - i)^k}{k!}$. Identifying $a_n = (-1)^{n+1}/n!$, we have

$$\lim_{n\to\infty} \left| \frac{\dfrac{(-1)^{n+2}}{(n+1)!}}{\dfrac{(-1)^{n+1}}{n!}} \right| = \lim_{n\to\infty} \frac{1}{n+1} = 0.$$

Thus the radius of convergence is ∞; the power series with center $1 + i$ converges absolutely for all z.

≡

EXAMPLE 7 **Radius of Convergence**

Consider the power series $\displaystyle\sum_{k=1}^{\infty}\left(\frac{6k + 1}{2k + 5}\right)^{k}(z - 2i)^{k}$. With $a_n = \left(\dfrac{6n + 1}{2n + 5}\right)^{n}$, the root test in the form

$$\lim_{n\to\infty}\sqrt[n]{|a_n|} = \lim_{n\to\infty}\frac{6n + 1}{2n + 5} = 3$$

shows that the radius of convergence of the series is $R = \frac{1}{3}$. The circle of convergence is $|z - 2i| = \frac{1}{3}$; the series converges absolutely for $|z - 2i| < \frac{1}{3}$.

≡

19.1 Exercises Answers to selected odd-numbered problems begin on page ANS-47.

In Problems 1–4, write out the first five terms of the given sequence.

1. $\{5i^n\}$

2. $\{2 + (-i)^n\}$

3. $\{1 + e^{n\pi i}\}$

4. $\{(1 + i)^n\}$ [*Hint*: Write in polar form.]

In Problems 5–10, determine whether the given sequence converges or diverges.

5. $\left\{\dfrac{3ni + 2}{n + ni}\right\}$

6. $\left\{\dfrac{ni + 2^n}{3ni + 5^n}\right\}$

7. $\left\{\dfrac{(ni + 2)^2}{n^2 i}\right\}$

8. $\left\{\dfrac{n(1 + i^n)}{n + 1}\right\}$

9. $\left\{\dfrac{n + i^n}{\sqrt{n}}\right\}$

10. $\{e^{1/n} + 2(\tan^{-1}n)i\}$

In Problems 11 and 12, show that the given sequence $\{z_n\}$ converges to a complex number L by computing $\lim_{n\to\infty}\mathrm{Re}(z_n)$ and $\lim_{n\to\infty}\mathrm{Im}(z_n)$.

11. $\left\{\dfrac{4n + 3ni}{2n + i}\right\}$

12. $\left\{\left(\dfrac{1 + i}{4}\right)^n\right\}$

In Problems 13 and 14, use the sequence of partial sums to show that the given series is convergent.

13. $\displaystyle\sum_{k=1}^{\infty}\left[\frac{1}{k + 2i} - \frac{1}{k + 1 + 2i}\right]$

14. $\displaystyle\sum_{k=2}^{\infty}\frac{i}{k(k + 1)}$

In Problems 15–20, determine whether the given geometric series is convergent or divergent. If convergent, find its sum.

15. $\displaystyle\sum_{k=0}^{\infty}(1 - i)^k$

16. $\displaystyle\sum_{k=1}^{\infty}4i\left(\frac{1}{3}\right)^{k-1}$

17. $\displaystyle\sum_{k=1}^{\infty}\left(\frac{i}{2}\right)^{k}$

18. $\displaystyle\sum_{k=0}^{\infty}\frac{1}{2}i^k$

19. $\displaystyle\sum_{k=0}^{\infty}3\left(\frac{2}{1 + 2i}\right)^{k}$

20. $\displaystyle\sum_{k=2}^{\infty}\frac{i^k}{(1 + i)^{k-1}}$

In Problems 21–28, find the circle and radius of convergence of the given power series.

21. $\displaystyle\sum_{k=0}^{\infty}\frac{1}{(1 - 2i)^{k+1}}(z - 2i)^k$

22. $\displaystyle\sum_{k=1}^{\infty}\frac{1}{k}\left(\frac{i}{1 + i}\right)^{k}z^k$

23. $\displaystyle\sum_{k=1}^{\infty}\frac{(-1)^k}{k2^k}(z - 1 - i)^k$

24. $\displaystyle\sum_{k=1}^{\infty}\frac{1}{k^2(3 + 4i)^k}(z + 3i)^k$

25. $\displaystyle\sum_{k=0}^{\infty}(1 + 3i)^k(z - i)^k$

26. $\displaystyle\sum_{k=1}^{\infty}\frac{z^k}{k^k}$

27. $\displaystyle\sum_{k=0}^{\infty}\frac{(z - 4 - 3i)^k}{5^{2k}}$

28. $\displaystyle\sum_{k=0}^{\infty}(-1)^k\left(\frac{1 + 2i}{2}\right)^{k}(z + 2i)^k$

29. Show that the power series $\displaystyle\sum_{k=1}^{\infty}\frac{(z - i)^k}{k2^k}$ is not absolutely convergent on its circle of convergence. Determine at least one point on the circle of convergence at which the power series converges.

30. (a) Show that the power series $\displaystyle\sum_{k=1}^{\infty}\frac{z^k}{k^2}$ converges at every point on its circle of convergence.

(b) Show that the power series $\displaystyle\sum_{k=1}^{\infty}kz^k$ diverges at every point on its circle of convergence.

19.2 Taylor Series

INTRODUCTION The correspondence between a complex number z within the circle of convergence and the number to which the series $\sum_{k=1}^{\infty}a_k(z - z_0)^k$ converges is single-valued. In this sense, a power series *defines* or *represents* a function f; for a specified z within the circle of convergence, the number L to which the power series converges is defined to be the

value of f at z; that is, $f(z) = L$. In this section we present some important facts about the nature of this function f.

In the preceding section we saw that every power series has a radius of convergence R. Throughout the discussion in this section, we will assume that a power series $\sum_{k=1}^{\infty} a_k(z - z_0)^k$ has either a *positive* or an *infinite* radius R of convergence. The next three theorems will give some important facts about the nature of a power series within its circle of convergence $|z - z_0| = R$, $R \neq 0$.

THEOREM 19.2.1 Continuity

A power series $\sum_{k=0}^{\infty} a_k(z - z_0)^k$ represents a continuous function f within its circle of convergence $|z - z_0| = R$, $R \neq 0$.

THEOREM 19.2.2 Term-by-Term Integration

A power series $\sum_{k=0}^{\infty} a_k(z - z_0)^k$ can be integrated term by term within its circle of convergence $|z - z_0| = R$, $R \neq 0$, for every contour C lying entirely within the circle of convergence.

THEOREM 19.2.3 Term-by-Term Differentiation

A power series $\sum_{k=0}^{\infty} a_k(z - z_0)^k$ can be differentiated term by term within its circle of convergence $|z - z_0| = R$, $R \neq 0$.

■ **Taylor Series** Suppose a power series represents a function f for $|z - z_0| < R$, $R \neq 0$; that is,

$$f(z) = \sum_{k=0}^{\infty} a_k(z - z_0)^k = a_0 + a_1(z - z_0) + a_2(z - z_0)^2 + a_3(z - z_0)^3 + \cdots. \tag{1}$$

It follows from Theorem 19.2.3 that the derivatives of f are

$$f'(z) = \sum_{k=1}^{\infty} k a_k(z - z_0)^{k-1} = a_1 + 2a_2(z - z_0) + 3a_3(z - z_0)^2 + \cdots \tag{2}$$

$$f''(z) = \sum_{k=2}^{\infty} k(k-1) a_k(z - z_0)^{k-2} = 2 \cdot 1 a_2 + 3 \cdot 2 a_3(z - z_0) + \cdots \tag{3}$$

$$f'''(z) = \sum_{k=3}^{\infty} k(k-1)(k-2) a_k(z - z_0)^{k-3} = 3 \cdot 2 \cdot 1 a_3 + \cdots \tag{4}$$

and so on. Each of the differentiated series has the same radius of convergence as the original series. Moreover, since the original power series represents a differentiable function f within its circle of convergence, we conclude that when $R \neq 0$:

A power series represents an analytic function within its circle of convergence.

There is a relationship between the coefficients a_k and the derivatives of f. Evaluating (1), (2), (3), and (4) at $z = z_0$ gives

$$f(z_0) = a_0, \quad f'(z_0) = 1! a_1, \quad f''(z_0) = 2! a_2, \quad \text{and} \quad f'''(z_0) = 3! a_3,$$

respectively. In general, $f^{(n)}(z_0) = n! a_n$ or

$$a_n = \frac{f^{(n)}(z_0)}{n!}, \quad n \geq 0. \tag{5}$$

When $n = 0$, we interpret the zeroth derivative as $f(z_0)$ and $0! = 1$. Substituting (5) into (1) yields

$$f(z) = f(z_0) + \frac{f'(z_0)}{1!}(z - z_0) + \frac{f''(z_0)}{2!}(z - z_0)^2 + \cdots = \sum_{k=0}^{\infty} \frac{f^{(k)}(z_0)}{k!}(z - z_0)^k. \tag{6}$$

This series is called the **Taylor series** for f centered at z_0. Although the English mathematician **Brook Taylor** (1685–1731) did not invent this series, he did use it extensively in his research on finite differences. Because the Scottish mathematician **Colin Maclaurin** (1698–1746) frequently used Taylor series centered at 0 in his research, the series

$$f(z) = f(0) + \frac{f'(0)}{1!} z + \frac{f''(0)}{2!} z^2 + \cdots = \sum_{k=0}^{\infty} \frac{f^{(k)}(0)}{k!} z^k \tag{7}$$

was named the **Maclaurin series** in his honor.

We have just seen that a power series with a nonzero radius of convergence represents an analytic function. On the other hand, if we are given a function f that is analytic in some domain D, can we represent it by a power series of the form (6) and (7)? Since a power series converges in a circular domain, and a domain D is generally not circular, the question becomes: Can we expand f in *one or more* power series that are valid in circular domains that are all contained in D? The question will be answered in the affirmative by the next theorem.

> ### THEOREM 19.2.4 Taylor's Theorem
>
> Let f be analytic within a domain D and let z_0 be a point in D. Then f has the series representation
>
> $$f(z) = \sum_{k=0}^{\infty} \frac{f^{(k)}(z_0)}{k!} (z - z_0)^k \tag{8}$$
>
> valid for the largest circle C with center at z_0 and radius R that lies entirely within D.

PROOF: Let z be a fixed point within the circle C and let s denote the variable of integration. The circle C is then described by $|s - z_0| = R$. See **FIGURE 19.2.1**. To begin, we use the Cauchy integral formula to obtain the value of f at z:

$$f(z) = \frac{1}{2\pi i} \oint_C \frac{f(s)}{s - z} \, ds = \frac{1}{2\pi i} \oint_C \frac{f(s)}{(s - z_0) - (z - z_0)} \, ds$$

$$= \frac{1}{2\pi i} \oint_C \frac{f(s)}{s - z_0} \left\{ \frac{1}{1 - \dfrac{z - z_0}{s - z_0}} \right\} ds. \tag{9}$$

By replacing z by $(z - z_0)/(s - z_0)$ in (8) of Section 19.1, we have

$$\frac{1}{1 - \dfrac{z - z_0}{s - z_0}} = 1 + \frac{z - z_0}{s - z_0} + \left(\frac{z - z_0}{s - z_0} \right)^2 + \cdots + \left(\frac{z - z_0}{s - z_0} \right)^{n-1} + \frac{(z - z_0)^n}{(s - z)(s - z_0)^{n-1}},$$

and so (9) becomes

$$f(z) = \frac{1}{2\pi i} \oint_C \frac{f(s)}{s - z_0} \, ds + \frac{z - z_0}{2\pi i} \oint_C \frac{f(s)}{(s - z_0)^2} \, ds + \frac{(z - z_0)^2}{2\pi i} \oint_C \frac{f(s)}{(s - z_0)^3} \, ds + \cdots$$

$$+ \frac{(z - z_0)^{n-1}}{2\pi i} \oint_C \frac{f(s)}{(s - z_0)^n} \, ds + \frac{(z - z_0)^n}{2\pi i} \oint_C \frac{f(s)}{(s - z)(s - z_0)^n} \, ds. \tag{10}$$

Utilizing Cauchy's integral formula for derivatives, we can write (10) as

$$f(z) = f(z_0) + \frac{f'(z_0)}{1!} (z - z_0) + \frac{f''(z_0)}{2!} (z - z_0)^2 + \cdots$$

$$+ \frac{f^{(n-1)}(z_0)}{(n - 1)!} (z - z_0)^{n-1} + R_n(z), \tag{11}$$

where

$$R_n(z) = \frac{(z - z_0)^n}{2\pi i} \oint_C \frac{f(s)}{(s - z)(s - z_0)^n} \, ds.$$

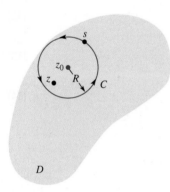

FIGURE 19.2.1 Circular contour C used in proof of Theorem 19.2.4

Equation (11) is called Taylor's formula with remainder R_n. We now wish to show that $R_n(z) \to 0$ as $n \to \infty$. Since f is analytic in D, $|f(z)|$ has a maximum value M on the contour C. In addition, since z is inside C, we have $|z - z_0| < R$, and, consequently,

$$|s - z| = |s - z_0 - (z - z_0)| \geq |s - z_0| - |z - z_0| = R - d,$$

where $d = |z - z_0|$ is the distance from z and z_0. The ML-inequality then gives

$$|R_n(z)| = \left| \frac{(z - z_0)^n}{2\pi i} \oint_C \frac{f(s)}{(s - z)(s - z_0)^n} \, ds \right| \leq \frac{d^n}{2\pi} \cdot \frac{M}{(R - d)R^n} \cdot 2\pi R = \frac{MR}{R - d} \left(\frac{d}{R} \right)^n.$$

Because $d < R$, $(d/R)^n \to 0$ as $n \to \infty$, we conclude that $|R_n(z)| \to 0$ as $n \to \infty$. It follows that the infinite series

$$f(z_0) + \frac{f'(z_0)}{1!} (z - z_0) + \frac{f''(z_0)}{2!} (z - z_0)^2 + \cdots$$

converges to $f(z)$. In other words, the result in (8) is valid for any point z interior to C. $\equiv$

We can find the radius of convergence of a Taylor series in exactly the same manner illustrated in Examples 5–7 of Section 19.1. However, we can simplify matters even further by noting that the radius of convergence is the *distance* from the center z_0 of the series to the nearest *isolated singularity* of f. We shall elaborate more on this concept in the next section, but an isolated singularity is a point at which f fails to be analytic but is, nonetheless, analytic at all other points throughout *some* neighborhood of the point. For example, $z = 5i$ is an isolated singularity of $f(z) = 1/(z - 5i)$. If the function f is entire, then the radius of convergence of a Taylor series centered at any point z_0 is necessarily infinite. Using (8) and the last fact, we can say that the Maclaurin series representations

$$e^z = 1 + \frac{z}{1!} + \frac{z^2}{2!} + \cdots = \sum_{k=0}^{\infty} \frac{z^k}{k!} \tag{12}$$

$$\sin z = z - \frac{z^3}{3!} + \frac{z^5}{5!} - \cdots = \sum_{k=0}^{\infty} (-1)^k \frac{z^{2k+1}}{(2k+1)!} \tag{13}$$

$$\cos z = 1 - \frac{z^2}{2!} + \frac{z^4}{4!} - \cdots = \sum_{k=0}^{\infty} (-1)^k \frac{z^{2k}}{(2k)!} \tag{14}$$

are valid for all z.

If two power series with center z_0,

$$\sum_{k=0}^{\infty} a_k (z - z_0)^k \quad \text{and} \quad \sum_{k=0}^{\infty} b_k (z - z_0)^k,$$

represent the same function and have the same nonzero radius of convergence, then $a_k = b_k$, $k = 0, 1, 2, \ldots$. Stated in another way, the power series expansion of a function with center z_0 is *unique*. On a practical level, this means that a power series expansion of an analytic function f centered at z_0, irrespective of the method used to obtain it, is the Taylor series expansion of the function. For example, we can obtain (14) by simply differentiating (13) term by term. The Maclaurin series for e^{z^2} can be obtained by replacing the symbol z in (12) by z^2.

<hr>

EXAMPLE 1 **Maclaurin Series**

Find the Maclaurin expansion of $f(z) = \dfrac{1}{(1 - z)^2}$.

SOLUTION We could, of course, begin by computing the coefficients using (8). However, we know from (5) of Section 19.1 that for $|z| < 1$,

$$\frac{1}{1 - z} = 1 + z + z^2 + z^3 + \cdots. \tag{15}$$

Differentiating both sides of the last result with respect to z then yields

$$\frac{1}{(1 - z)^2} = 1 + 2z + 3z^2 + \cdots = \sum_{k=1}^{\infty} k z^{k-1}.$$

Since we are using Theorem 19.2.3, the radius of convergence of this last series is the same as the original series, $R = 1$. $\equiv$

EXAMPLE 2 **Taylor Series**

Expand $f(z) = \dfrac{1}{1 - z}$ in a Taylor series with center $z_0 = 2i$.

SOLUTION We shall solve this problem in two ways. We begin by using (8). From the first several derivatives,

$$f'(z) = \frac{1}{(1 - z)^2}, \quad f''(z) = \frac{2 \cdot 1}{(1 - z)^3}, \quad f'''(z) = \frac{3 \cdot 2}{(1 - z)^4},$$

we conclude that $f^{(n)}(z) = n!/(1 - z)^{n+1}$ and so $f^{(n)}(2i) = n!/(1 - 2i)^{n+1}$. Thus from (8) we obtain the Taylor series

$$\frac{1}{1 - z} = \sum_{k=0}^{\infty} \frac{1}{(1 - 2i)^{k+1}} (z - 2i)^k. \tag{16}$$

Since the distance from the center $z_0 = 2i$ to the nearest singularity $z = 1$ is $\sqrt{5}$, we conclude that the circle of convergence for the power series in (16) is $|z - 2i| = \sqrt{5}$. This can be verified by the ratio test of the preceding section.

ALTERNATIVE SOLUTION In this solution we again use the geometric series (15). By adding and subtracting $2i$ in the denominator of $1/(1 - z)$, we can write

$$\frac{1}{1 - z} = \frac{1}{1 - z + 2i - 2i} = \frac{1}{1 - 2i - (z - 2i)} = \frac{1}{1 - 2i} \frac{1}{1 - \dfrac{z - 2i}{1 - 2i}}.$$

We now write $\dfrac{1}{1 - \dfrac{z - 2i}{1 - 2i}}$ as a power series by using (15) with the symbol z replaced by $(z - 2i)/(1 - 2i)$:

$$\frac{1}{1 - z} = \frac{1}{1 - 2i} \left[1 + \frac{z - 2i}{1 - 2i} + \left(\frac{z - 2i}{1 - 2i} \right)^2 + \left(\frac{z - 2i}{1 - 2i} \right)^3 + \cdots \right]$$

$$= \frac{1}{1 - 2i} + \frac{1}{(1 - 2i)^2} (z - 2i) + \frac{1}{(1 - 2i)^3} (z - 2i)^2 + \frac{1}{(1 - 2i)^4} (z - 2i)^3 + \cdots.$$

The reader should verify that this last series is exactly the same as that in (16). ≡

In (15) and (16) we represented the same function $1/(1 - z)$ by two different power series. The first series

$$\frac{1}{1 - z} = 1 + z + z^2 + z^3 + \cdots$$

has center zero and radius of convergence one. The second series

$$\frac{1}{1 - z} = \frac{1}{1 - 2i} + \frac{1}{(1 - 2i)^2} (z - 2i) + \frac{1}{(1 - 2i)^3} (z - 2i)^2 + \frac{1}{(1 - 2i)^4} (z - 2i)^3 + \cdots$$

has center $2i$ and radius of convergence $\sqrt{5}$. The two different circles of convergence are illustrated in **FIGURE 19.2.2**. The interior of the intersection of the two circles (shaded) is the region where *both* series converge; in other words, at a specified point $z*$ in this region, both series converge to the same value $f(z*) = 1/(1 - z*)$. Outside the shaded region, at least one of the two series must diverge.

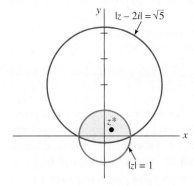

FIGURE 19.2.2 Series (15) and (16) both converge within the shaded region

In Problems 1–12, expand the given function in a Maclaurin series. Give the radius of convergence of each series.

1. $f(z) = \dfrac{z}{1 + z}$

2. $f(z) = \dfrac{1}{4 - 2z}$

3. $f(z) = \dfrac{1}{(1 + 2z)^2}$

4. $f(z) = \dfrac{z}{(1 - z)^3}$

5. $f(z) = e^{-2z}$

6. $f(z) = ze^{-z^2}$

7. $f(z) = \sinh z$

8. $f(z) = \cosh z$

9. $f(z) = \cos \dfrac{z}{2}$

10. $f(z) = \sin 3z$

11. $f(z) = \sin z^2$

12. $f(z) = \cos^2 z$ [*Hint*: Use a trigonometric identity.]

In Problems 13–22, expand the given function in a Taylor series centered at the indicated point. Give the radius of convergence of each series.

13. $f(z) = 1/z$, $z_0 = 1$

14. $f(z) = 1/z$, $z_0 = 1 + i$

15. $f(z) = \dfrac{1}{3 - z}$, $z_0 = 2i$

16. $f(z) = \dfrac{1}{1 + z}$, $z_0 = -i$

17. $f(z) = \dfrac{z - 1}{3 - z}$, $z_0 = 1$

18. $f(z) = \dfrac{1 + z}{1 - z}$, $z_0 = i$

19. $f(z) = \cos z$, $z_0 = \pi/4$

20. $f(z) = \sin z$, $z_0 = \pi/2$

21. $f(z) = e^z$, $z_0 = 3i$

22. $f(z) = (z - 1)e^{-2z}$, $z_0 = 1$

In Problems 23 and 24, use (7) to find the first three nonzero terms of the Maclaurin series of the given function.

23. $f(z) = \tan z$

24. $f(z) = e^{1/(1+z)}$

In Problems 25 and 26, use partial fractions as an aid in obtaining the Maclaurin series for the given function. Give the radius of convergence of the series.

25. $f(z) = \dfrac{i}{(z - i)(z - 2i)}$

26. $f(z) = \dfrac{z - 7}{z^2 - 2z - 3}$

In Problems 27 and 28, without actually expanding, determine the radius of convergence of the Taylor series of the given function centered at the indicated point.

27. $f(z) = \dfrac{4 + 5z}{1 + z^2}$, $z_0 = 2 + 5i$

28. $f(z) = \cot z$, $z_0 = \pi i$

In Problems 29 and 30, expand the given function in the Taylor series centered at the indicated points. Give the radius of convergence of each series. Sketch the region within which both series converge.

29. $f(z) = \dfrac{1}{2 + z}$, $z_0 = -1$, $z_0 = i$

30. $f(z) = \dfrac{1}{z}$, $z_0 = 1 + i$, $z_0 = 3$

31. (a) Suppose the principal branch of the logarithm $f(z) = \text{Ln } z = \log_e|z| + i \text{ Arg } z$ is expanded in a Taylor series with center $z_0 = -1 + i$. Explain why $R = 1$ is the radius of the largest circle centered at $z_0 = -1 + i$ within which f is analytic.

(b) Show that within the circle $|z - (-1 + i)| = 1$ the Taylor series for f is

$$\text{Ln } z = \frac{1}{2}\log_e 2 + \frac{3\pi}{4}i - \sum_{k=1}^{\infty} \frac{1}{k}\left(\frac{1 + i}{2}\right)^k (z + 1 - i)^k.$$

(c) Show that the radius of convergence for the power series in part (b) is $R = \sqrt{2}$. Explain why this does not contradict the result in part (a).

32. (a) Consider the function $f(z) = \text{Ln}(1 + z)$. What is the radius of the largest circle centered at the origin within which f is analytic?

(b) Expand f in a Maclaurin series. What is the radius of convergence of this series?

(c) Use the result in part (b) to find a Maclaurin series for $\text{Ln}(1 - z)$.

(d) Find a Maclaurin series for $\text{Ln}\left(\dfrac{1 + z}{1 - z}\right)$.

In Problems 33 and 34, approximate the value of the given expression using the indicated number of terms of a Maclaurin series.

33. $e^{(1+i)/10}$, three terms

34. $\sin\left(\dfrac{1 + i}{10}\right)$, two terms

35. In Section 15.1 we defined the error function as

$$\text{erf}(z) = \frac{2}{\sqrt{\pi}}\int_0^z e^{-t^2}dt.$$

Find a Maclaurin series for $\text{erf}(z)$.

36. Use the Maclaurin series for e^{iz} to prove Euler's formula for complex z:

$$e^{iz} = \cos z + i \sin z.$$

19.3 Laurent Series

INTRODUCTION If a complex function f fails to be analytic at a point $z = z_0$, then this point is said to be a **singularity** or a **singular point** of the function. For example, the complex numbers $z = 2i$ and $z = -2i$ are singularities of the function $f(z) = z/(z^2 + 4)$ because f is discontinuous at each of these points. Recall from Section 17.6 that the principal value of the logarithm, Ln z, is analytic at all points except those points on the branch cut consisting of the nonpositive x-axis; that is, the branch point $z = 0$ as well as all negative real numbers are singular points of Ln z. In this section we will be concerned with a new kind of "power series" expansion of f about an **isolated singularity** z_0. This new series will involve negative as well as nonnegative integer powers of $z - z_0$.

Isolated Singularities Suppose that $z = z_0$ is a singularity of a complex function f. The point $z = z_0$ is said to be an **isolated singularity** of the function f if there exists *some* deleted neighborhood, or punctured open disk, $0 < |z - z_0| < R$ of z_0 throughout which f is analytic. For example, we have just seen that $z = 2i$ and $z = -2i$ are singularities of $f(z) = z/(z^2 + 4)$. Both $2i$ and $-2i$ are isolated singularities since f is analytic at every point in the neighborhood defined by $|z - 2i| < 1$ except at $z = 2i$ and at every point in the neighborhood defined by $|z - (-2i)| < 1$ except at $z = -2i$. In other words, if f is analytic in the deleted neighborhood, $0 < |z - 2i| < 1$

and $0 < |z + 2i| < 1$. On the other hand, the branch point $z = 0$ is *not* an isolated singularity of Ln z since every neighborhood of $z = 0$ must contain points on the negative x-axis. We say that a singular point $z = z_0$ of a function f is **nonisolated** if *every* neighborhood of z_0 contains at least one singularity of f other than z_0. For example, the branch point $z = 0$ is a nonisolated singularity of Ln z since every neighborhood of $z = 0$ contains points on the negative real axis.

■ **A New Kind of Series** If $z = z_0$ is a singularity of a function f, then certainly f cannot be expanded in a power series with z_0 as its center. However, about an isolated singularity $z = z_0$ it is possible to represent f by a new kind of series involving both negative and nonnegative integer powers of $z - z_0$; that is,

$$f(z) = \cdots + \frac{a_{-2}}{(z - z_0)^2} + \frac{a_{-1}}{z - z_0} + a_0 + a_1(z - z_0) + a_2(z - z_0)^2 + \cdots.$$

Using summation notation, the last expression can be written as the sum of two series

$$f(z) = \sum_{k=1}^{\infty} a_{-k}(z - z_0)^{-k} + \sum_{k=0}^{\infty} a_k(z - z_0)^k. \tag{1}$$

The two series on the right-hand side in (1) are given special names. The part with negative powers of $z - z_0$, that is,

$$\sum_{k=1}^{\infty} a_{-k}(z - z_0)^{-k} = \sum_{k=1}^{\infty} \frac{a_{-k}}{(z - z_0)^k},$$

is called the **principal part** of the series (1) and will converge for $|1/(z - z_0)| < r^*$ or equivalently for $|z - z_0| > 1/r^* = r$. The part consisting of the nonnegative powers of $z - z_0$,

$$\sum_{k=0}^{\infty} a_k(z - z_0)^k,$$

is called the **analytic part** of the series (1) and will converge for $|z - z_0| < R$. Hence, the sum of these parts converges when z both $|z - z_0| > r$ *and* $|z - z_0| < R$; that is, when z is a point in an **annular domain** defined by $r < |z - z_0| < R$.

By summing over negative and nonnegative integers, (1) can be written compactly as

$$f(z) = \sum_{k=-\infty}^{\infty} a_k(z - z_0)^k.$$

The next example illustrates a series of the form (1) in which the principal part of the series consists of a finite number of nonzero terms, but the analytic part consists of an infinite number of nonzero terms.

A New Kind of Series

The function $f(z) = (\sin z)/z^3$ is not analytic at $z = 0$ and hence cannot be expanded in a Maclaurin series. However, $\sin z$ is an entire function, and from (13) of Section 19.2 we know that its Maclaurin series,

$$\sin z = z - \frac{z^3}{3!} + \frac{z^5}{5!} - \frac{z^7}{7!} + \cdots,$$

converges for all z. Dividing this power series by z^3 gives the following series with negative and nonnegative integer powers of z:

$$f(z) = \frac{\sin z}{z^3} = \frac{1}{z^2} - \frac{1}{3!} + \frac{z^2}{5!} - \frac{z^4}{7!} + \cdots. \tag{2}$$

This series converges for all z except $z = 0$; that is, for $0 < |z|$.

A series representation of a function f that has the form given in (1)—and (2) is such an example—is called a **Laurent series** or a **Laurent expansion** of f after the French mathematician and military engineer **Pierre Alphonse Laurent** (1813–1854). Basically, (1) is a generalization

of a Taylor series expansion, but Laurent's 1843 paper in which this series was discussed was not published in his lifetime.

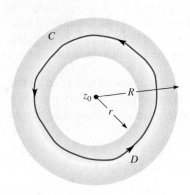

FIGURE 19.3.1 Contour in Theorem 19.3.1

THEOREM 19.3.1 Laurent's Theorem

Let f be analytic within the annular domain D defined by $r < |z - z_0| < R$. Then f has the series representation

$$f(z) = \sum_{k=-\infty}^{\infty} a_k(z - z_0)^k \qquad (3)$$

valid for $r < |z - z_0| < R$. The coefficients a_k are given by

$$a_k = \frac{1}{2\pi i} \oint_C \frac{f(s)}{(s - z_0)^{k+1}}\,ds, \quad k = 0, \pm 1, \pm 2, \ldots, \qquad (4)$$

where C is a simple closed curve that lies entirely within D and has z_0 in its interior (see **FIGURE 19.3.1**).

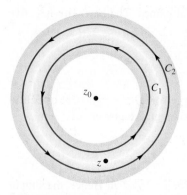

FIGURE 19.3.2 C_1 and C_2 are concentric circles

PROOF: Let C_1 and C_2 be concentric circles with center z_0 and radii r_1 and R_2, where $r < r_1 < R_2 < R$. Let z be a fixed point in D that also satisfies $r_1 < |z - z_0| < R_2$. See **FIGURE 19.3.2**. By introducing a cross cut between C_2 and C_1, we find from Cauchy's integral formula that

$$f(z) = \frac{1}{2\pi i} \oint_{C_2} \frac{f(s)}{s - z}\,ds - \frac{1}{2\pi i} \oint_{C_1} \frac{f(s)}{s - z}\,ds. \qquad (5)$$

Proceeding as in the proof of Theorem 19.2.4, we can write

$$\frac{1}{2\pi i} \oint_{C_2} \frac{f(s)}{s - z}\,ds = \sum_{k=0}^{\infty} a_k(z - z_0)^k, \qquad (6)$$

where

$$a_k = \frac{1}{2\pi i} \oint_{C_2} \frac{f(s)}{(s - z_0)^{k+1}}\,ds, \quad k = 0, 1, 2, \ldots. \qquad (7)$$

Now using (5) and (8) of Section 19.1, we have

$$-\frac{1}{2\pi i} \oint_{C_1} \frac{f(s)}{s - z}\,ds = \frac{1}{2\pi i} \oint_{C_1} \frac{f(s)}{(z - z_0) - (s - z_0)}\,ds$$

$$= \frac{1}{2\pi i} \oint_{C_1} \frac{f(s)}{z - z_0} \left\{ \frac{1}{1 - \dfrac{s - z_0}{z - z_0}} \right\} ds$$

$$= \frac{1}{2\pi i} \oint_{C_1} \frac{f(s)}{z - z_0} \left\{ 1 + \frac{s - z_0}{z - z_0} + \left(\frac{s - z_0}{z - z_0} \right)^2 + \cdots \right.$$

$$\left. + \left(\frac{s - z_0}{z - z_0} \right)^{n-1} + \frac{(s - z_0)^n}{(z - s)(z - z_0)^{n-1}} \right\} ds \qquad (8)$$

$$= \sum_{k=1}^{n} \frac{a_{-k}}{(z - z_0)^k} + R_n(z),$$

where

$$a_{-k} = \frac{1}{2\pi i} \oint_{C_1} \frac{f(s)}{(s - z_0)^{-k+1}}\,ds, \quad k = 1, 2, 3, \ldots, \qquad (9)$$

and

$$R_n(z) = \frac{1}{2\pi i(z - z_0)^n} \oint_{C_1} \frac{f(s)(s - z_0)^n}{z - s}\,ds.$$

Now let d denote the distance from z to z_0; that is, $|z - z_0| = d$, and let M denote the maximum value of $|f(z)|$ on the contour C_1. Since $|s - z_0| = r_1$,

$$|z - s| = |z - z_0 - (s - z_0)| \geq |z - z_0| - |s - z_0| = d - r_1.$$

The *ML*-inequality then gives

$$|R_n(z)| = \left| \frac{1}{2\pi i(z - z_0)^n} \oint_{C_1} \frac{f(s)(s - z_0)^n}{z - s} \, ds \right| \le \frac{1}{2\pi d^n} \cdot \frac{Mr_1^n}{d - r_1} \cdot 2\pi r_1 = \frac{Mr_1}{d - r_1} \left(\frac{r_1}{d} \right)^n.$$

Because $r_1 < d$, $(r_1/d)^n \to 0$ as $n \to \infty$ and so $|R_n(z)| \to 0$ as $n \to \infty$. Thus we have shown that

$$-\frac{1}{2\pi i} \oint_{C_1} \frac{f(s)}{s - z} \, ds = \sum_{k=1}^{\infty} \frac{a_{-k}}{(z - z_0)^k}, \tag{10}$$

where the coefficients a_{-k} are given in (9). Combining (6) and (10), we see that (5) yields

$$f(z) = \sum_{k=0}^{\infty} a_k(z - z_0)^k + \sum_{k=1}^{\infty} \frac{a_{-k}}{(z - z_0)^k}. \tag{11}$$

Finally, by summing over nonnegative and negative integers, we can write (11) as $f(z) = \sum_{k=-\infty}^{\infty} a_k(z - z_0)^k$. However, (7) and (9) can be written as a single integral:

$$a_k = \frac{1}{2\pi i} \oint_C \frac{f(s)}{(s - z_0)^{k+1}} \, ds, \quad k = 0, \pm 1, \pm 2, \ldots,$$

where, in view of (3) of Section 18.2, we have replaced the contours C_1 and C_2 by any simple closed contour C in D with z_0 in its interior. ≡

In the case when $a_{-k} = 0$ for $k = 1, 2, 3, \ldots$, the Laurent series (3) is a Taylor series. Because of this, a Laurent expansion is a generalization of a Taylor series.

The annular domain in Theorem 19.3.1 defined by $r < |z - z_0| < R$ need not have the "ring" shape illustrated in Figure 19.3.2. Some other possible annular domains are (*i*) $r = 0$, R finite; (*ii*) $r \ne 0$, $R \to \infty$; and (*iii*) $r = 0$, $R \to \infty$. In the first case, the series converges in the annular domain defined by $0 < |z - z_0| < R$. This is the interior of the circle $|z - z_0| = R$ except the point z_0. In the second case, the annular domain is defined by $r < |z - z_0|$; in other words, the domain consists of all points exterior to the circle $|z - z_0| = r$. In the third case, the domain is defined by $0 < |z - z_0|$. This represents the entire complex plane except the point z_0. The series we obtained in (2) is valid on this last type of domain.

In actual practice, the formula in (4) for the coefficients of a Laurent series is seldom used. As a consequence, finding the Laurent series of a function in a specified annular domain is generally not an easy task. We often use the geometric series (5) and (6) of Section 19.1 or, as we did in Example 1, other known series. But regardless of how a Laurent expansion of a function f is obtained in a specified annular domain, it is *the* Laurent series; that is, the series we obtain is unique.

EXAMPLE 2 **Laurent Expansions**

Expand $f(z) = \dfrac{1}{z(z - 1)}$ in a Laurent series valid for the following annular domains.

(a) $0 < |z| < 1$ **(b)** $1 < |z|$ **(c)** $0 < |z - 1| < 1$ **(d)** $1 < |z - 1|$

SOLUTION The four specified annular domains are shown in **FIGURE 19.3.3**. In parts (a) and (b), we want to represent f in a series involving only negative and nonnegative integer powers of z, whereas in parts (c) and (d) we want to represent f in a series involving negative and nonnegative integer powers of $z - 1$.

(a) By writing

$$f(z) = -\frac{1}{z} \frac{1}{1 - z},$$

we can use (5) of Section 19.1:

$$f(z) = -\frac{1}{z}[1 + z + z^2 + z^3 + \cdots].$$

The series in the brackets converges for $|z| < 1$, but after this expression is multiplied by $1/z$, the resulting series

$$f(z) = -\frac{1}{z} - 1 - z - z^2 - \cdots$$

converges for $0 < |z| < 1$.

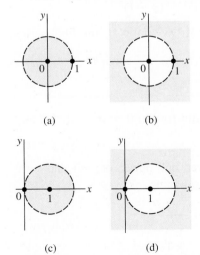

FIGURE 19.3.3 Annular domains for Example 2

(a) (b) (c) (d)

(b) To obtain a series that converges for $1 < |z|$ we start by constructing a series that converges for $|1/z| < 1$. To this end we write the given function f as

$$f(z) = \frac{1}{z^2} \frac{1}{1 - \frac{1}{z}}$$

and again use (5) of Section 19.1 with z replaced by $1/z$:

$$f(z) = \frac{1}{z^2}\left[1 + \frac{1}{z} + \frac{1}{z^2} + \frac{1}{z^3} + \cdots \right].$$

The series in the brackets converges for $|1/z| < 1$ or equivalently for $1 < |z|$. Thus the required Laurent series is

$$f(z) = \frac{1}{z^2} + \frac{1}{z^3} + \frac{1}{z^4} + \frac{1}{z^5} + \cdots.$$

(c) This is basically the same problem as part (a) except that we want all powers of $z - 1$. To that end we add and subtract 1 in the denominator and use (6) of Section 19.1 with z replaced by $z - 1$:

$$f(z) = \frac{1}{(1 - 1 + z)(z - 1)}$$

$$= \frac{1}{z - 1}\frac{1}{1 + (z - 1)}$$

$$= \frac{1}{z - 1}[1 - (z - 1) + (z - 1)^2 - (z - 1)^3 + \cdots]$$

$$= \frac{1}{z - 1} - 1 + (z - 1) - (z - 1)^2 + \cdots.$$

The series in brackets converges for $|z - 1| < 1$, and so the last series converges for $0 < |z - 1| < 1$.

(d) Proceeding as in part (b), we write

$$f(z) = \frac{1}{z - 1}\frac{1}{1 + (z - 1)} = \frac{1}{(z - 1)^2}\frac{1}{1 + \frac{1}{z - 1}}$$

$$= \frac{1}{(z - 1)^2}\left[1 - \frac{1}{z - 1} + \frac{1}{(z - 1)^2} - \frac{1}{(z - 1)^3} + \cdots \right]$$

$$= \frac{1}{(z - 1)^2} - \frac{1}{(z - 1)^3} + \frac{1}{(z - 1)^4} - \frac{1}{(z - 1)^5} + \cdots.$$

Because the series within the brackets converges for $|1/(z - 1)| < 1$, the final series converges for $1 < |z - 1|$. $\equiv$

EXAMPLE 3　Laurent Expansions

Expand $f(z) = \dfrac{1}{(z - 1)^2(z - 3)}$ in a Laurent series valid for **(a)** $0 < |z - 1| < 2$ and **(b)** $0 < |z - 3| < 2$.

SOLUTION **(a)** As in parts (c) and (d) of Example 2, we want only powers of $z - 1$, and so we need to express $z - 3$ in terms of $z - 1$. This can be done by writing

$$f(z) = \frac{1}{(z - 1)^2(z - 3)} = \frac{1}{(z - 1)^2}\frac{1}{-2 + (z - 1)}$$

$$= \frac{-1}{2(z - 1)^2}\frac{1}{1 - \frac{z - 1}{2}}$$

and then using (5) of Section 19.1 with z replaced by $(z-1)/2$:

$$f(z) = \frac{-1}{2(z-1)^2}\left[1 + \frac{z-1}{2} + \frac{(z-1)^2}{2^2} + \frac{(z-1)^3}{2^3} + \cdots\right]$$

$$= -\frac{1}{2(z-1)^2} - \frac{1}{4(z-1)} - \frac{1}{8} - \frac{1}{16}(z-1) - \cdots. \qquad (12)$$

(b) To obtain powers of $z-3$ we write $z-1 = 2 + (z-3)$ and

$$f(z) = \frac{1}{(z-1)^2(z-3)} = \frac{1}{z-3}[2 + (z-3)]^{-2} = \frac{1}{4(z-3)}\left[1 + \frac{z-3}{2}\right]^{-2}.$$

At this point we can expand $\left[1 + \dfrac{z-3}{2}\right]^{-2}$ in a power series using the general binomial theorem:

$$f(z) = \frac{1}{4(z-3)}\left[1 + \frac{(-2)}{1!}\left(\frac{z-3}{2}\right) + \frac{(-2)(-3)}{2!}\left(\frac{z-3}{2}\right)^2 + \frac{(-2)(-3)(-4)}{3!}\left(\frac{z-3}{2}\right)^3 + \cdots\right].$$

The binomial series is valid for $|(z-3)/2| < 1$ or $|z-3| < 2$. Multiplying this series by $1/4(z-3)$ gives a Laurent series that is valid for $0 < |z-3| < 2$:

$$f(z) = \frac{1}{4(z-3)} - \frac{1}{4} + \frac{3}{16}(z-3) - \frac{1}{8}(z-3)^2 + \cdots. \qquad \equiv$$

EXAMPLE 4 **A Laurent Expansion**

Expand $f(z) = \dfrac{8z+1}{z(1-z)}$ in a Laurent series valid for $0 < |z| < 1$.

SOLUTION By (5) of Section 19.1 we can write

$$f(z) = \frac{8z+1}{z(1-z)} = \frac{8z+1}{z}\frac{1}{1-z} = \left(8 + \frac{1}{z}\right)(1 + z + z^2 + z^3 + \cdots).$$

We then multiply the series by $8 + 1/z$ and collect like terms:

$$f(z) = \frac{1}{z} + 9 + 9z + 9z^2 + \cdots.$$

The geometric series converges for $|z| < 1$. After multiplying by $8 + 1/z$, the resulting Laurent series is valid for $0 < |z| < 1$. $\qquad \equiv$

In the preceding examples, the point at the center of the annular domain of validity for each Laurent series was an isolated singularity of the function f. A reexamination of Theorem 19.3.1 shows that this need not be the case.

EXAMPLE 5 **A Laurent Expansion**

Expand $f(z) = \dfrac{1}{z(z-1)}$ in a Laurent series valid for $1 < |z-2| < 2$.

SOLUTION The specified annular domain is shown in **FIGURE 19.3.4**. The center of this domain, $z = 2$, is a point of analyticity of the function f. Our goal now is to find two series involving integer powers of $z-2$: one converging for $1 < |z-2|$ and the other converging for $|z-2| < 2$. To accomplish this, we start with the decomposition of f into partial fractions:

$$f(z) = -\frac{1}{z} + \frac{1}{z-1} = f_1(z) + f_2(z). \qquad (13)$$

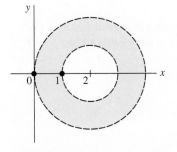

FIGURE 19.3.4 Annular domain for Example 5

Now,
$$f_1(z) = -\frac{1}{z} = -\frac{1}{2 + z - 2}$$
$$= -\frac{1}{2} \frac{1}{1 + \dfrac{z-2}{2}}$$
$$= -\frac{1}{2}\left[1 - \frac{z-2}{2} + \frac{(z-2)^2}{2^2} - \frac{(z-2)^3}{2^3} + \cdots\right]$$
$$= -\frac{1}{2} + \frac{z-2}{2^2} - \frac{(z-2)^2}{2^3} + \frac{(z-2)^3}{2^4} - \cdots.$$

This series converges for $|(z-2)/2| < 1$ or $|z - 2| < 2$. Furthermore,

$$f_2(z) = \frac{1}{z-1} = \frac{1}{1+z-2} = \frac{1}{z-2}\frac{1}{1 + \dfrac{1}{z-2}}$$

$$= \frac{1}{z-2}\left[1 - \frac{1}{z-2} + \frac{1}{(z-2)^2} - \frac{1}{(z-2)^3} + \cdots\right]$$

$$= \frac{1}{z-2} - \frac{1}{(z-2)^2} + \frac{1}{(z-2)^3} - \frac{1}{(z-2)^4} + \cdots$$

converges for $|1/(z-2)| < 1$ or $1 < |z - 2|$. Substituting these two results in (13) then gives

$$f(z) = \cdots - \frac{1}{(z-2)^4} + \frac{1}{(z-2)^3} - \frac{1}{(z-2)^2} + \frac{1}{z-2} - \frac{1}{2} + \frac{z-2}{2^2} - \frac{(z-2)^2}{2^3} + \frac{(z-2)^3}{2^4} - \cdots.$$

This representation is valid for $1 < |z - 2| < 2$. ≡

EXAMPLE 6 A Laurent Expansion

Expand $f(z) = e^{3/z}$ in a Laurent series valid for $0 < |z|$.

SOLUTION From (12) of Section 19.2 we know that for all finite z,

$$e^z = 1 + z + \frac{z^2}{2!} + \frac{z^3}{3!} + \cdots. \tag{14}$$

By replacing z in (14) by $3/z$, $z \neq 0$, we obtain the Laurent series

$$e^{3/z} = 1 + \frac{3}{z} + \frac{3^2}{2!z^2} + \frac{3^3}{3!z^3} + \cdots.$$

This series is valid for $0 < |z|$. ≡

REMARKS

In conclusion, we point out a result that will be of particular importance to us in Sections 19.5 and 19.6. Replacing the complex variable s with the usual symbol z, we see that when $k = -1$, (4) for the Laurent series coefficients yields

$$a_{-1} = \frac{1}{2\pi i} \oint_C f(z)\, dz$$

or, more importantly, the integral can be written as

$$\oint_C f(z)\, dz = 2\pi i a_{-1}. \tag{15}$$

In Problems 1–6, expand the given function in a Laurent series valid for the indicated annular domain.

1. $f(z) = \dfrac{\cos z}{z}$, $0 < |z|$

2. $f(z) = \dfrac{z - \sin z}{z^5}$, $0 < |z|$

3. $f(z) = e^{-1/z^2}$, $0 < |z|$

4. $f(z) = \dfrac{1 - e^z}{z^2}$, $0 < |z|$

5. $f(z) = \dfrac{e^z}{z - 1}$, $0 < |z - 1|$

6. $f(z) = z \cos \dfrac{1}{z}$, $0 < |z|$

In Problems 7–12, expand $f(z) = \dfrac{1}{z(z - 3)}$ in a Laurent series valid for the indicated annular domain.

7. $0 < |z| < 3$ **8.** $|z| > 3$

9. $0 < |z - 3| < 3$ **10.** $|z - 3| > 3$

11. $1 < |z - 4| < 4$ **12.** $1 < |z + 1| < 4$

In Problems 13–16, expand $f(z) = \dfrac{1}{(z - 1)(z - 2)}$ in a Laurent series valid for the indicated annular domain.

13. $1 < |z| < 2$ **14.** $|z| > 2$

15. $0 < |z - 1| < 1$ **16.** $0 < |z - 2| < 1$

In Problems 17–20, expand $f(z) = \dfrac{z}{(z + 1)(z - 2)}$ in a Laurent series valid for the indicated annular domain.

17. $0 < |z + 1| < 3$ **18.** $|z + 1| > 3$
19. $1 < |z| < 2$ **20.** $0 < |z - 2| < 3$

In Problems 21 and 22, expand $f(z) = \dfrac{1}{z(1 - z)^2}$ in a Laurent series valid for the indicated annular domain.

21. $0 < |z| < 1$ **22.** $|z| > 1$

In Problems 23 and 24, expand $f(z) = \dfrac{1}{(z - 2)(z - 1)^3}$ in a Laurent series valid for the indicated annular domain.

23. $0 < |z - 2| < 1$ **24.** $0 < |z - 1| < 1$

In Problems 25 and 26, expand $f(z) = \dfrac{7z - 3}{z(z - 1)}$ in a Laurent series valid for the indicated annular domain.

25. $0 < |z| < 1$ **26.** $0 < |z - 1| < 1$

In Problems 27 and 28, expand $f(z) = \dfrac{z^2 - 2z + 2}{z - 2}$ in a Laurent series valid for the indicated annular domain.

27. $1 < |z - 1|$ **28.** $0 < |z - 2|$

19.4 Zeros and Poles

INTRODUCTION Suppose that $z = z_0$ is an isolated singularity of a function f and that

$$f(z) = \sum_{k=-\infty}^{\infty} a_k(z - z_0)^k = \sum_{k=1}^{\infty} a_{-k}(z - z_0)^{-k} + \sum_{k=0}^{\infty} a_k(z - z_0)^k \tag{1}$$

is the Laurent series representation of f valid for the punctured open disk $0 < |z - z_0| < R$. We saw in the preceding section that a Laurent series (1) consists of two parts. That part of the series in (1) with negative powers of $z - z_0$, namely,

$$\sum_{k=1}^{\infty} a_{-k}(z - z_0)^{-k} = \sum_{k=1}^{\infty} \frac{a_{-k}}{(z - z_0)^k}, \tag{2}$$

is the **principal part** of the series. In the discussion that follows we will assign different names to the isolated singularity $z = z_0$ according to the number of terms in the principal part.

Classification of Isolated Singular Points An isolated singular point $z = z_0$ of a complex function f is given a classification depending on whether the principal part (2) of its Laurent expansion (1) contains zero, a finite number, or an infinite number of terms.

 (*i*) If the principal part is zero, that is, *all* the coefficients a_{-k} in (2) are zero, then $z = z_0$ is called a **removable singularity**.

 (*ii*) If the principal part contains a finite number of nonzero terms, then $z = z_0$ is called a **pole**. If, in this case, the last nonzero coefficient in (2) is a_{-n}, $n \geq 1$, then we say that $z = z_0$ is a **pole of order** n. If $z = z_0$ is a pole of order 1, then the principal part (2) contains exactly one term with coefficient a_{-1}. A pole of order 1 is commonly called a **simple pole**.

(*iii*) If the principal part (2) contains infinitely many nonzero terms, then $z = z_0$ is called an **essential singularity**.

The following table summarizes the form of the Laurent series for a function f when $z = z_0$ is one of the above types of isolated singularities. Of course, R in the table could be ∞.

$z = z_0$	Laurent Series
Removable singularity	$a_0 + a_1(z - z_0) + a_2(z - z_0)^2 + \cdots$
Pole of order n	$\dfrac{a_{-n}}{(z - z_0)^n} + \dfrac{a_{-(n-1)}}{(z - z_0)^{n-1}} + \cdots + \dfrac{a_{-1}}{z - z_0} + a_0 + a_1(z - z_0) + \cdots$
Simple pole	$\dfrac{a_{-1}}{z - z_0} + a_0 + a_1(z - z_0) + a_2(z - z_0)^2 + \cdots$
Essential singularity	$\cdots + \dfrac{a_{-2}}{(z - z_0)^2} + \dfrac{a_{-1}}{z - z_0} + a_0 + a_1(z - z_0) + a_2(z - z_0)^2 + \cdots$

EXAMPLE 1 Removable Discontinuity

Proceeding as we did in (2) of Section 19.3, we see from

$$\frac{\sin z}{z} = 1 - \frac{z^2}{3!} + \frac{z^4}{5!} - \cdots \tag{3}$$

that $z = 0$ is a removable singularity of the function $f(z) = (\sin z)/z$. ≡

If a function f has a removable singularity at the point $z = z_0$, then we can always supply an appropriate definition for the value of $f(z_0)$ so that f becomes analytic at the point. For instance, since the right side of (3) is 1 at $z = 0$, it makes sense to define $f(0) = 1$. With this definition, the function $f(z) = (\sin z)/z$ in Example 1 is now analytic at $z = 0$.

EXAMPLE 2 Poles and Essential Singularity

(a) From

principal part
↓

$$\frac{\sin z}{z^2} = \frac{1}{z} - \frac{z}{3!} + \frac{z^3}{5!} - \cdots,$$

$0 < |z|$, we see that $a_{-1} \neq 0$, and so $z = 0$ is a simple pole of the function $f(z) = (\sin z)/z^2$. The function $f(z) = (\sin z)/z^3$ represented by the series in (2) of Section 19.3 has a pole of order 2 at $z = 0$.

(b) In Example 3 of Section 19.3 we showed that the Laurent expansion of $f(z) = 1/(z - 1)^2(z - 3)$ valid for $0 < |z - 1| < 2$ was

principal part

$$f(z) = -\frac{1}{2(z - 1)^2} - \frac{1}{4(z - 1)} - \frac{1}{8} - \frac{z - 1}{16} - \cdots.$$

Since $a_{-2} \neq 0$, we conclude that $z = 1$ is a pole of order 2.

(c) From Example 6 of Section 19.3 we see from the Laurent series that the principal part of the function $f(z) = e^{3/z}$ contains an infinite number of terms. Thus $z = 0$ is an essential singularity. ≡

In part (b) of Example 2 in Section 19.3, we showed that the Laurent series representation of $f(z) = 1/z(z - 1)$ valid for $1 < |z|$ is

$$f(z) = \frac{1}{z^2} + \frac{1}{z^3} + \frac{1}{z^4} + \frac{1}{z^5} + \cdots.$$

The point $z = 0$ is an isolated singularity of f and the Laurent series contains an infinite number of terms involving negative integer powers of z. Does this mean that $z = 0$ is an essential singularity of f? The answer is "no." For this particular function, a reexamination of (1) shows that the Laurent series we are interested in is the one with the annular domain $0 < |z| < 1$. From part (a) of that same example we saw that

$$f(z) = -\frac{1}{z} - 1 - z - z^2 - \cdots$$

was valid for $0 < |z| < 1$. Thus we see that $z = 0$ is a simple pole.

■ **Zeros** Recall that z_0 is a **zero** of a function f if $f(z_0) = 0$. An analytic function f has a **zero of order n** at $z = z_0$ if

$$\overbrace{f(z_0) = 0, \; f'(z_0) = 0, \; f''(z_0) = 0, \ldots, f^{(n-1)}(z_0) = 0}^{z_0 \text{ is a zero of } f \text{ and its first } n-1 \text{ derivatives}} \quad \text{but} \quad f^{(n)}(z_0) \neq 0. \quad \textbf{(4)}$$

A zero of order n is also referred to as a **zero of multiplicity n**. For example, for $f(z) = (z - 5)^3$ we see that $f(5) = 0, f'(5) = 0, f''(5) = 0$ but $f'''(5) = 6 \neq 0$. Thus f has a zero of order (or multiplicity) 3 at $z_0 = 5$. A zero of order 1 is called a **simple zero**.

If an analytic function f has a zero of order n at $z = z_0$, it follows from (4) that the Taylor series expansion of f centered at z_0 must have the form

$$\begin{aligned}
f(z) &= a_n(z - z_0)^n + a_{n+1}(z - z_0)^{n+1} + a_{n+2}(z - z_0)^{n+2} + \cdots \\
&= (z - z_0)^n[a_n + a_{n+1}(z - z_0) + a_{n+2}(z - z_0)^2 + \cdots],
\end{aligned} \quad \textbf{(5)}$$

where $a_n \neq 0$.

EXAMPLE 3 Order of a Zero

The analytic function $f(z) = z \sin z^2$ has a zero at $z = 0$. By replacing z by z^2 in (13) of Section 19.2, we obtain

$$\sin z^2 = z^2 - \frac{z^6}{3!} + \frac{z^{10}}{5!} - \cdots$$

and so

$$f(z) = z \sin z^2 = z^3 \left[1 - \frac{z^4}{3!} + \frac{z^8}{5!} - \cdots \right].$$

Comparing the last result with (5) we see that $z = 0$ is a zero of order 3. ≡

A zero z_0 of a nontrivial analytic function f is *isolated* in the sense that there exists some neighborhood of z_0 for which $f(z) \neq 0$ at every point z in that neighborhood except at $z = z_0$. As a consequence, if z_0 is a zero of a nontrivial analytic function f, then the function $1/f(z)$ has an isolated singularity at the point $z = z_0$. The following result enables us, in some circumstances, to determine the poles of a function by inspection.

THEOREM 19.4.1 Pole of Order n

If the functions f and g are analytic at $z = z_0$ and f has a zero of order n at $z = z_0$ and $g(z_0) \neq 0$, then the function $F(z) = g(z)/f(z)$ has a pole of order n at $z = z_0$.

EXAMPLE 4 Order of Poles

(a) Inspection of the rational function

$$F(z) = \frac{2z + 5}{(z - 1)(z + 5)(z - 2)^4}$$

shows that the denominator has zeros of order 1 at $z = 1$ and $z = -5$, and a zero of order 4 at $z = 2$. Since the numerator is not zero at these points, it follows from Theorem 19.4.1 that F has simple poles at $z = 1$ and $z = -5$, and a pole of order 4 at $z = 2$.

(b) In Example 3 we saw that $z = 0$ is a zero of order 3 of $f(z) = z \sin z^2$. From Theorem 19.4.1, we conclude that the function $F(z) = 1/(z \sin z^2)$ has a pole of order 3 at $z = 0$. ≡

From the preceding discussion, it should be intuitively clear that if a function has a pole at $z = z_0$, then $|f(z)| \to \infty$ as $z \to z_0$ from any direction.

19.4 Exercises Answers to selected odd-numbered problems begin on page ANS-48.

In Problems 1 and 2, show that $z = 0$ is a removable singularity of the given function. Supply a definition of $f(0)$ so that f is analytic at $z = 0$.

1. $f(z) = \dfrac{e^{2z} - 1}{z}$

2. $f(z) = \dfrac{\sin 4z - 4z}{z^2}$

In Problems 3–8, determine the zeros and their orders for the given function.

3. $f(z) = (z + 2 - i)^2$

4. $f(z) = z^4 - 16$

5. $f(z) = z^4 + z^2$

6. $f(z) = z + \dfrac{9}{z}$

7. $f(z) = e^{2z} - e^z$

8. $f(z) = \sin^2 z$

In Problems 9–12, the indicated number is a zero of the given function. Use a Maclaurin or Taylor series to determine the order of the zero.

9. $f(z) = z(1 - \cos z^2); \ z = 0$

10. $f(z) = z - \sin z; \ z = 0$

11. $f(z) = 1 - e^{z-1}; \ z = 1$

12. $f(z) = 1 - \pi i + z + e^z; \ z = \pi i$

In Problems 13–24, determine the order of the poles for the given function.

13. $f(z) = \dfrac{3z - 1}{z^2 + 2z + 5}$

14. $f(z) = 5 - \dfrac{6}{z^2}$

15. $f(z) = \dfrac{1 + 4i}{(z + 2)(z + i)^4}$

16. $f(z) = \dfrac{z - 1}{(z + 1)(z^3 + 1)}$

17. $f(z) = \tan z$

18. $f(z) = \dfrac{\cot \pi z}{z^2}$

19. $f(z) = \dfrac{1 - \cosh z}{z^4}$

20. $f(z) = \dfrac{e^z}{z^2}$

21. $f(z) = \dfrac{1}{1 - e^z}$

22. $f(z) = \dfrac{e^z - 1}{z^4}$

23. $f(z) = \dfrac{\sin z}{z^2 - z}$

24. $f(z) = \dfrac{\cos z - \cos 2z}{z^6}$

25. Determine whether $z = 0$ is an isolated or nonisolated singularity of $f(z) = \tan(1/z)$.

26. Show that $z = 0$ is an essential singularity of $f(z) = z^3 \sin(1/z)$.

19.5 Residue Theorem

INTRODUCTION We saw in the last section that if the complex function f has an *isolated singularity* at the point z_0, then f has a Laurent series representation

$$f(z) = \sum_{k=-\infty}^{\infty} a_k(z - z_0)^k = \cdots + \frac{a_{-2}}{(z - z_0)^2} + \frac{a_{-1}}{z - z_0} + a_0 + a_1(z - z_0) + \cdots,$$

which converges for all z *near* z_0. More precisely, the representation is valid in some deleted neighborhood of z_0, or punctured open disk, $0 < |z - z_0| < R$. In this section our entire focus will be on the coefficient a_{-1} and its importance in the evaluation of contour integrals.

▨ **Residue** The coefficient a_{-1} of $1/(z - z_0)$ in the Laurent series given above is called the **residue** of the function f at the isolated singularity z_0. We shall use the notation

$$a_{-1} = \text{Res}\,(f(z), z_0)$$

to denote the residue of f at z_0. Recall, if the principal part of the Laurent series valid for $0 < |z - z_0| < R$ contains a finite number of terms with a_{-n} the last nonzero coefficient, then z_0 is a pole of order n; if the principal part of the series contains an infinite number of terms with nonzero coefficients, then z_0 is an essential singularity.

> **EXAMPLE 1** Residues

(a) In Example 2 of Section 19.4 we saw that $z = 1$ is a pole of order 2 of the function $f(z) = 1/(z - 1)^2(z - 3)$. From the Laurent series given in that example we see that the coefficient of $1/(z - 1)$ is $a_{-1} = \text{Res}\,(f(z), 1) = -\frac{1}{4}$.

(b) Example 6 of Section 19.3 showed that $z = 0$ is an essential singularity of $f(z) = e^{3/z}$. From the Laurent series given in that example we see that the coefficient of $1/z$ is $a_{-1} = \text{Res}\,(f(z), 0) = 3$. ≡

Later on in this section we will see why the coefficient a_{-1} is so important. In the meantime we are going to examine ways of obtaining this complex number when z_0 is a *pole* of a function f without the necessity of expanding f in a Laurent series at z_0. We begin with the residue at a simple pole.

THEOREM 19.5.1 Residue at a Simple Pole

If f has a simple pole at $z = z_0$, then

$$\operatorname{Res}(f(z), z_0) = \lim_{z \to z_0} (z - z_0) f(z). \tag{1}$$

PROOF: Since $z = z_0$ is a simple pole, the Laurent expansion of f about that point has the form

$$f(z) = \frac{a_{-1}}{z - z_0} + a_0 + a_1(z - z_0) + a_2(z - z_0)^2 + \cdots.$$

By multiplying both sides by $z - z_0$ and then taking the limit as $z \to z_0$, we obtain

$$\lim_{z \to z_0} (z - z_0) f(z) = \lim_{z \to z_0} [a_{-1} + a_0(z - z_0) + a_1(z - z_0)^2 + \cdots] = a_{-1} = \operatorname{Res}(f(z), z_0). \qquad \equiv$$

THEOREM 19.5.2 Residue at a Pole of Order n

If f has a pole of order n at $z = z_0$, then

$$\operatorname{Res}(f(z), z_0) = \frac{1}{(n-1)!} \lim_{z \to z_0} \frac{d^{n-1}}{dz^{n-1}} (z - z_0)^n f(z). \tag{2}$$

PROOF: Since f is assumed to have a pole of order n, its Laurent expansion for $0 < |z - z_0| < R$ must have the form

$$f(z) = \frac{a_{-n}}{(z - z_0)^n} + \cdots + \frac{a_{-2}}{(z - z_0)^2} + \frac{a_{-1}}{z - z_0} + a_0 + a_1(z - z_0) + \cdots.$$

We multiply the last expression by $(z - z_0)^n$:

$$(z - z_0)^n f(z) = a_{-n} + \cdots + a_{-2}(z - z_0)^{n-2} + a_{-1}(z - z_0)^{n-1} + a_0(z - z_0)^n + a_1(z - z_0)^{n+1} + \cdots$$

and then differentiate $n - 1$ times:

$$\frac{d^{n-1}}{dz^{n-1}} (z - z_0)^n f(z) = (n-1)! a_{-1} + n! a_0(z - z_0) + \cdots. \tag{3}$$

Since all the terms on the right side after the first involve positive integer powers of $z - z_0$, the limit of (3) as $z \to z_0$ is

$$\lim_{z \to z_0} \frac{d^{n-1}}{dz^{n-1}} (z - z_0)^n f(z) = (n-1)! a_{-1}.$$

Solving the last equation for a_{-1} gives (2). $\equiv$

Note that (2) reduces to (1) when $n = 1$.

EXAMPLE 2 Residue at a Pole

The function $f(z) = \dfrac{1}{(z-1)^2(z-3)}$ has a simple pole at $z = 3$ and a pole of order 2 at $z = 1$. Use Theorems 19.5.1 and 19.5.2 to find the residue at each pole.

SOLUTION Since $z = 3$ is a simple pole, we use (1):

$$\text{Res}(f(z), 3) = \lim_{z \to 3}(z - 3)f(z) = \lim_{z \to 3}\frac{1}{(z - 1)^2} = \frac{1}{4}.$$

Now at the pole of order 2 it follows from (2) that

$$\text{Res}(f(z), 1) = \frac{1}{1!}\lim_{z \to 1}\frac{d}{dz}(z - 1)^2 f(z)$$

$$= \lim_{z \to 1}\frac{d}{dz}\frac{1}{z - 3}$$

$$= \lim_{z \to 1}\frac{-1}{(z - 3)^2} = -\frac{1}{4}. \qquad \equiv$$

When f is not a rational function, calculating residues by means of (1) can sometimes be tedious. It is possible to devise alternative residue formulas. In particular, suppose a function f can be written as a quotient $f(z) = g(z)/h(z)$, where g and h are analytic at $z = z_0$. If $g(z_0) \neq 0$ and if the function h has a zero of order 1 at z_0, then f has a simple pole at $z = z_0$ and

An alternative method for computing a residue at a *simple* pole. ▶

$$\text{Res}(f(z), z_0) = \frac{g(z_0)}{h'(z_0)}. \qquad (4)$$

To see this last result, we use (1) and the facts that $h(z_0) = 0$ and that $\lim_{z \to z_0}(h(z) - h(z_0))/(z - z_0)$ is a definition of the derivative $h'(z_0)$:

$$\text{Res}(f(z), z_0) = \lim_{z \to z_0}(z - z_0)\frac{g(z)}{h(z)} = \lim_{z \to z_0}\frac{g(z)}{\dfrac{h(z) - h(z_0)}{z - z_0}} = \frac{g(z_0)}{h'(z_0)}.$$

Analogous formulas for residues at poles of order greater than 1 are complicated and will not be given.

EXAMPLE 3 **Using (4) to Compute a Residue**

The polynomial $z^4 + 1$ can be factored as $(z - z_1)(z - z_2)(z - z_3)(z - z_4)$, where $z_1, z_2, z_3,$ and z_4 are the four distinct roots of the equation $z^4 + 1 = 0$. It follows from Theorem 19.4.1 that the function

$$f(z) = \frac{1}{z^4 + 1}$$

has four simple poles. Now from (10) of Section 17.2 we have $z_1 = e^{\pi i/4}, z_2 = e^{3\pi i/4}, z_3 = e^{5\pi i/4}, z_4 = e^{7\pi i/4}$. To compute the residues, we use (4) and Euler's formula:

$$\text{Res}(f(z), z_1) = \frac{1}{4z_1^3} = \frac{1}{4}e^{-3\pi i/4} = -\frac{1}{4\sqrt{2}} - \frac{1}{4\sqrt{2}}i$$

$$\text{Res}(f(z), z_2) = \frac{1}{4z_2^3} = \frac{1}{4}e^{-9\pi i/4} = \frac{1}{4\sqrt{2}} - \frac{1}{4\sqrt{2}}i$$

$$\text{Res}(f(z), z_3) = \frac{1}{4z_3^3} = \frac{1}{4}e^{-15\pi i/4} = \frac{1}{4\sqrt{2}} + \frac{1}{4\sqrt{2}}i$$

$$\text{Res}(f(z), z_4) = \frac{1}{4z_4^3} = \frac{1}{4}e^{-21\pi i/4} = -\frac{1}{4\sqrt{2}} + \frac{1}{4\sqrt{2}}i. \qquad \equiv$$

Residue Theorem We come now to the reason for the importance of the residue concept. The next theorem states that under some circumstances, we can evaluate complex integrals $\oint_C f(z)\,dz$ by summing the residues at the isolated singularities of f within the closed contour C.

FIGURE 19.5.1 n singular points within contour C

Let D be a simply connected domain and C a simple closed contour lying entirely within D. If a function f is analytic on and within C, except at a finite number of singular points $z_1, z_2, \ldots, z_n$ within C, then

$$\oint_C f(z)\,dz = 2\pi i \sum_{k=1}^{n} \operatorname{Res}(f(z), z_k). \tag{5}$$

PROOF: Suppose $C_1, C_2, \ldots, C_n$ are circles centered at $z_1, z_2, \ldots, z_n$, respectively. Suppose further that each circle C_k has a radius r_k small enough so that $C_1, C_2, \ldots, C_n$ are mutually disjoint and are interior to the simple closed curve C. See **FIGURE 19.5.1**. Recall that (15) of Section 19.3 implies $\oint_{C_k} f(z)\,dz = 2\pi i \operatorname{Res}(f(z), z_k)$, and so Theorem 18.2.2 gives

$$\oint_C f(z)\,dz = \sum_{k=1}^{n} \oint_{C_k} f(z)\,dz = 2\pi i \sum_{k=1}^{n} \operatorname{Res}(f(z), z_k). \qquad \equiv$$

EXAMPLE 4 Evaluation by the Residue Theorem

Evaluate $\displaystyle\oint_C \frac{1}{(z-1)^2(z-3)}\,dz$, where

(a) the contour C is the rectangle defined by $x = 0$, $x = 4$, $y = -1$, $y = 1$,

(b) the contour C is the circle $|z| = 2$.

SOLUTION **(a)** Since both poles $z = 1$ and $z = 3$ lie within the square, we have from (5) that

$$\oint_C \frac{1}{(z-1)^2(z-3)}\,dz = 2\pi i[\operatorname{Res}(f(z), 1) + \operatorname{Res}(f(z), 3)].$$

We found these residues in Examples 2 and 3, and so

$$\oint_C \frac{1}{(z-1)^2(z-3)}\,dz = 2\pi i\left[-\frac{1}{4} + \frac{1}{4}\right] = 0.$$

(b) Since only the pole $z = 1$ lies within the circle $|z| = 2$, we have from (5) that

$$\oint_C \frac{1}{(z-1)^2(z-3)}\,dz = 2\pi i \operatorname{Res}(f(z), 1) = 2\pi i\left(-\frac{1}{4}\right) = -\frac{\pi}{2}i. \qquad \equiv$$

EXAMPLE 5 Evaluation by the Residue Theorem

Evaluate $\displaystyle\oint_C \frac{2z+6}{z^2+4}\,dz$, where the contour C is the circle $|z - i| = 2$.

SOLUTION By writing $z^2 + 4 = (z - 2i)(z + 2i)$, we see that the integrand has simple poles at $-2i$ and $2i$. Now since only $2i$ lies within the contour C, it follows from (5) that

$$\oint_C \frac{2z+6}{z^2+4}\,dz = 2\pi i \operatorname{Res}(f(z), 2i).$$

But

$$\operatorname{Res}(f(z), 2i) = \lim_{z \to 2i}(z - 2i)\frac{2z+6}{(z-2i)(z+2i)}$$

$$= \frac{6+4i}{4i} = \frac{3+2i}{2i}.$$

Hence,

$$\oint_C \frac{2z+6}{z^2+4}\,dz = 2\pi i\left(\frac{3+2i}{2i}\right) = \pi(3 + 2i). \qquad \equiv$$

EXAMPLE 6 **Evaluation by the Residue Theorem**

Evaluate $\displaystyle\oint_C \frac{e^z}{z^4 + 5z^3}\, dz$, where the contour C is the circle $|z| = 2$.

SOLUTION Since $z^4 + 5z^3 = z^3(z + 5)$ we see that the integrand has a pole of order 3 at $z = 0$ and a simple pole at $z = -5$. Since only $z = 0$ lies within the given contour, we have from (5) and (2)

$$\oint_C \frac{e^z}{z^4 + 5z^3}\, dz = 2\pi i\, \mathrm{Res}\,(f(z), 0)$$

$$= 2\pi i\, \frac{1}{2!}\lim_{z\to 0}\frac{d^2}{dz^2} z^3 \frac{e^z}{z^3(z + 5)}$$

$$= \pi i \lim_{z\to 0}\frac{(z^2 + 8z + 17)e^z}{(z + 5)^3} = \frac{17\pi}{125}\, i.$$ ≡

EXAMPLE 7 **Evaluation by the Residue Theorem**

Evaluate $\displaystyle\oint_C \tan z\, dz$, where the contour C is the circle $|z| = 2$.

SOLUTION The integrand $\tan z = \sin z/\cos z$ has simple poles at the points where $\cos z = 0$. We saw in Section 17.7 that the only zeros for $\cos z$ are the real numbers $z = (2n + 1)\pi/2$, $n = 0, \pm 1, \pm 2, \ldots$. Since only $-\pi/2$ and $\pi/2$ are within the circle $|z| = 2$, we have

$$\oint_C \tan z\, dz = 2\pi i\left[\mathrm{Res}\left(f(z), -\frac{\pi}{2}\right) + \mathrm{Res}\left(f(z), \frac{\pi}{2}\right)\right].$$

Now from (4) with $g(z) = \sin z$, $h(z) = \cos z$, and $h'(z) = -\sin z$, we see that

$$\mathrm{Res}\left(f(z), -\frac{\pi}{2}\right) = \frac{\sin(-\pi/2)}{-\sin(-\pi/2)} = -1 \quad \text{and} \quad \mathrm{Res}\left(f(z), \frac{\pi}{2}\right) = \frac{\sin(\pi/2)}{-\sin(\pi/2)} = -1.$$

Therefore, $\displaystyle\oint_C \tan z\, dz = 2\pi i[-1 - 1] = -4\pi i.$ ≡

EXAMPLE 8 **Evaluation by the Residue Theorem**

Evaluate $\displaystyle\oint_C e^{3/z}\, dz$, where the contour C is the circle $|z| = 1$.

SOLUTION As we have seen, $z = 0$ is an essential singularity of the integrand $f(z) = e^{3/z}$ and so neither formula (1) nor (2) is applicable to find the residue of f at that point. Nevertheless, we saw in part (b) of Example 1 that the Laurent series of f at $z = 0$ gives $\mathrm{Res}\,(f(z), 0) = 3$. Hence from (5) we have

$$\oint_C e^{3/z}\, dz = 2\pi i\, \mathrm{Res}\,(f(z), 0) = 6\pi i.$$ ≡

REMARKS

In the application of the limit formulas (1) and (2) for computing residues, the indeterminate form 0/0 may result. Although we are not going to prove it, it should be pointed out that **L'Hôpital's rule** is valid in complex analysis. If $f(z) = g(z)/h(z)$, where g and h are analytic at $z = z_0$, $g(z_0) = 0$, $h(z_0) = 0$, and $h'(z_0) \neq 0$, then

$$\lim_{z\to z_0}\frac{g(z)}{h(z)} = \frac{g'(z_0)}{h'(z_0)}.$$

In Problems 1–6, use a Laurent series to find the indicated residue.

1. $f(z) = \dfrac{2}{(z-1)(z+4)}$; Res $(f(z), 1)$

2. $f(z) = \dfrac{1}{z^3(1-z)^3}$; Res $(f(z), 0)$

3. $f(z) = \dfrac{4z-6}{z(2-z)}$; Res $(f(z), 0)$

4. $f(z) = (z+3)^2 \sin \dfrac{2}{z+3}$; Res $(f(z), -3)$

5. $f(z) = e^{-2/z^2}$; Res $(f(z), 0)$

6. $f(z) = \dfrac{e^{-z}}{(z-2)^2}$; Res $(f(z), 2)$

In Problems 7–16, use (1), (2), or (4) to find the residue at each pole of the given function.

7. $f(z) = \dfrac{z}{z^2 + 16}$

8. $f(z) = \dfrac{4z+8}{2z-1}$

9. $f(z) = \dfrac{1}{z^4 + z^3 - 2z^2}$

10. $f(z) = \dfrac{1}{(z^2 - 2z + 2)^2}$

11. $f(z) = \dfrac{5z^2 - 4z + 3}{(z+1)(z+2)(z+3)}$

12. $f(z) = \dfrac{2z-1}{(z-1)^4(z+3)}$

13. $f(z) = \dfrac{\cos z}{z^2(z-\pi)^3}$

14. $f(z) = \dfrac{e^z}{e^z - 1}$

15. $f(z) = \sec z$

16. $f(z) = \dfrac{1}{z \sin z}$

In Problems 17–20, use Cauchy's residue theorem, where appropriate, to evaluate the given integral along the indicated contours.

17. $\displaystyle\oint_C \dfrac{1}{(z-1)(z+2)^2}\, dz$

 (a) $|z| = \frac{1}{2}$ (b) $|z| = \frac{3}{2}$ (c) $|z| = 3$

18. $\displaystyle\oint_C \dfrac{z+1}{z^2(z-2i)}\, dz$

 (a) $|z| = 1$ (b) $|z - 2i| = 1$ (c) $|z - 2i| = 4$

19. $\displaystyle\oint_C z^3 e^{-1/z^2}\, dz$

 (a) $|z| = 5$ (b) $|z + i| = 2$ (c) $|z - 3| = 1$

20. $\displaystyle\oint_C \dfrac{1}{z \sin z}\, dz$

 (a) $|z - 2i| = 1$ (b) $|z - 2i| = 3$ (c) $|z| = 5$

In Problems 21–32, use Cauchy's residue theorem to evaluate the given integral along the indicated contour.

21. $\displaystyle\oint_C \dfrac{1}{z^2 + 4z + 13}\, dz$, $C: |z - 3i| = 3$

22. $\displaystyle\oint_C \dfrac{1}{z^3(z-1)^4}\, dz$, $C: |z - 2| = \frac{3}{2}$

23. $\displaystyle\oint_C \dfrac{z}{z^4 - 1}\, dz$, $C: |z| = 2$

24. $\displaystyle\oint_C \dfrac{z}{(z+1)(z^2+1)}\, dz$, C is the ellipse $16x^2 + y^2 = 4$

25. $\displaystyle\oint_C \dfrac{ze^z}{z^2 - 1}\, dz$, $C: |z| = 2$

26. $\displaystyle\oint_C \dfrac{e^z}{z^3 + 2z^2}\, dz$, $C: |z| = 3$

27. $\displaystyle\oint_C \dfrac{\tan z}{z}\, dz$, $C: |z - 1| = 2$

28. $\displaystyle\oint_C \dfrac{\cot \pi z}{z^2}\, dz$, $C: |z| = \frac{1}{2}$

29. $\displaystyle\oint_C \cot \pi z\, dz$, C is the rectangle defined by $x = \frac{1}{2}$, $x = \pi$, $y = -1$, $y = 1$

30. $\displaystyle\oint_C \dfrac{2z-1}{z^2(z^3+1)}\, dz$, C is the rectangle defined by $x = -2$, $x = 1$, $y = -\frac{1}{2}$, $y = 1$

31. $\displaystyle\oint_C \dfrac{e^{iz} + \sin z}{(z-\pi)^4}\, dz$, $C: |z - 3| = 1$

32. $\displaystyle\oint_C \dfrac{\cos z}{(z-1)^2(z^2+9)}\, dz$, $C: |z - 1| = 1$

19.6 Evaluation of Real Integrals

INTRODUCTION In this section we shall see how residue theory can be used to evaluate *real* integrals of the forms

$$\int_0^{2\pi} F(\cos\theta, \sin\theta)\, d\theta, \tag{1}$$

$$\int_{-\infty}^{\infty} f(x)\, dx, \tag{2}$$

$$\int_{-\infty}^{\infty} f(x) \cos \alpha x \, dx \quad \text{and} \quad \int_{-\infty}^{\infty} f(x) \sin \alpha x \, dx, \tag{3}$$

where F in (1) and f in (2) and (3) are rational functions. For the rational function $f(x) = p(x)/q(x)$ in (2) and (3), we will assume that the polynomials p and q have no common factors.

Integrals of the Form $\int_0^{2\pi} F(\cos \theta, \sin \theta) \, d\theta$

The basic idea here is to convert an integral of form (1) into a complex integral where the contour C is the unit circle centered at the origin. This contour can be parameterized by $z = \cos \theta + i \sin \theta = e^{i\theta}, 0 \le \theta \le 2\pi$. Using

$$dz = ie^{i\theta} \, d\theta, \quad \cos \theta = \frac{e^{i\theta} + e^{-i\theta}}{2}, \quad \sin \theta = \frac{e^{i\theta} - e^{-i\theta}}{2i},$$

we replace, in turn, $d\theta$, $\cos \theta$, and $\sin \theta$ by

$$d\theta = \frac{dz}{iz}, \quad \cos \theta = \frac{1}{2}(z + z^{-1}), \quad \sin \theta = \frac{1}{2i}(z - z^{-1}). \tag{4}$$

The integral in (1) then becomes

$$\oint_C F\left(\frac{1}{2}(z + z^{-1}), \frac{1}{2i}(z - z^{-1})\right) \frac{dz}{iz},$$

where C is $|z| = 1$.

A Real Trigonometric Integral

Evaluate $\displaystyle\int_0^{2\pi} \frac{1}{(2 + \cos \theta)^2} \, d\theta$.

SOLUTION Using the substitutions in (4) and simplifying yield the contour integral

$$\frac{4}{i} \oint_C \frac{z}{(z^2 + 4z + 1)^2} \, dz.$$

With the aid of the quadratic formula we can write

$$f(z) = \frac{z}{(z^2 + 4z + 1)^2} = \frac{z}{(z - z_0)^2 (z - z_1)^2},$$

where $z_0 = -2 - \sqrt{3}$ and $z_1 = -2 + \sqrt{3}$. Since only z_1 is inside the unit circle C, we have

$$\oint_C \frac{z}{(z^2 + 4z + 1)^2} \, dz = 2\pi i \operatorname{Res}(f(z), z_1).$$

Now z_1 is a pole of order 2 and so from (2) of Section 19.5,

$$\operatorname{Res}(f(z), z_1) = \lim_{z \to z_1} \frac{d}{dz}(z - z_1)^2 f(z) = \lim_{z \to z_1} \frac{d}{dz} \frac{z}{(z - z_0)^2}$$

$$= -\lim_{z \to z_1} \frac{z + z_0}{(z - z_0)^3} = \frac{1}{6\sqrt{3}}.$$

Hence,

$$\frac{4}{i} \oint_C \frac{z}{(z^2 + 4z + 1)^2} \, dz = \frac{4}{i} \cdot 2\pi i \operatorname{Res}(f(z), z_1) = \frac{4}{i} \cdot 2\pi i \cdot \frac{1}{6\sqrt{3}}$$

and finally

$$\int_0^{2\pi} \frac{1}{(2 + \cos \theta)^2} \, d\theta = \frac{4\pi}{3\sqrt{3}}. \qquad \equiv$$

Integrals of the Form $\int_{-\infty}^{\infty} f(x) \, dx$

When f is continuous on $(-\infty, \infty)$, recall from calculus that the improper integral $\int_{-\infty}^{\infty} f(x) \, dx$ is defined in terms of two distinct limits:

$$\int_{-\infty}^{\infty} f(x) \, dx = \lim_{r \to \infty} \int_{-r}^{0} f(x) \, dx + \lim_{R \to \infty} \int_{0}^{R} f(x) \, dx. \tag{5}$$

If both limits in (5) exist, the integral is said to be **convergent**; if one or both of the limits fail to exist, the integral is **divergent**. In the event that we know (a priori) that an integral $\int_{-\infty}^{\infty} f(x)\,dx$ converges, we can evaluate it by means of a single limiting process:

$$\int_{-\infty}^{\infty} f(x)\,dx = \lim_{R \to \infty} \int_{-R}^{R} f(x)\,dx. \tag{6}$$

It is important to note that the symmetric limit in (6) may exist even though the improper integral is divergent. For example, the integral $\int_{-\infty}^{\infty} x\,dx$ is divergent since $\lim_{R \to \infty} \int_0^R x\,dx = \lim_{R \to \infty} \frac{1}{2}R^2 = \infty$. However, using (6), we obtain

$$\lim_{R \to \infty} \int_{-R}^{R} x\,dx = \lim_{R \to \infty} \left[\frac{R^2}{2} - \frac{(-R)^2}{2} \right] = 0. \tag{7}$$

The limit in (6) is called the **Cauchy principal value** of the integral and is written

$$\text{P.V.} \int_{-\infty}^{\infty} f(x)\,dx = \lim_{R \to \infty} \int_{-R}^{R} f(x)\,dx.$$

In (7) we have shown that P.V. $\int_{-\infty}^{\infty} x\,dx = 0$. To summarize, when an integral of form (2) converges, its Cauchy principal value is the same as the value of the integral. If the integral diverges, it may still possess a Cauchy principal value.

To evaluate an integral $\int_{-\infty}^{\infty} f(x)\,dx$, where $f(x) = P(x)/Q(x)$ is continuous on $(-\infty, \infty)$, by residue theory we replace x by the complex variable z and integrate the complex function f over a closed contour C that consists of the interval $[-R, R]$ on the real axis and a semicircle C_R of radius large enough to enclose all the poles of $f(z) = P(z)/Q(z)$ in the upper half-plane $\text{Re}(z) > 0$. See **FIGURE 19.6.1**. By Theorem 19.5.3 we have

$$\oint_C f(z)\,dz = \int_{C_R} f(z)\,dz + \int_{-R}^{R} f(x)\,dx = 2\pi i \sum_{k=1}^{n} \text{Res}(f(z), z_k),$$

where z_k, $k = 1, 2, \ldots, n$, denotes poles in the upper half-plane. If we can show that the integral $\int_{C_R} f(z)\,dz \to 0$ as $R \to \infty$, then we have

$$\text{P.V.} \int_{-\infty}^{\infty} f(x)\,dx = \lim_{R \to \infty} \int_{-R}^{R} f(x)\,dx = 2\pi i \sum_{k=1}^{n} \text{Res}(f(z), z_k). \tag{8}$$

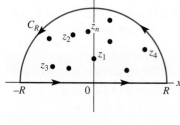

FIGURE 19.6.1 Closed contour C consists of a semicircle C_R and the interval $[-R, R]$

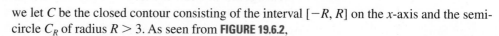

EXAMPLE 2 **Cauchy P.V. of an Improper Integral**

Evaluate the Cauchy principal value of

$$\int_{-\infty}^{\infty} \frac{1}{(x^2 + 1)(x^2 + 9)}\,dx.$$

SOLUTION Let $f(z) = 1/(z^2 + 1)(z^2 + 9)$. Since

$$(z^2 + 1)(z^2 + 9) = (z - i)(z + i)(z - 3i)(z + 3i),$$

we let C be the closed contour consisting of the interval $[-R, R]$ on the x-axis and the semicircle C_R of radius $R > 3$. As seen from **FIGURE 19.6.2**,

$$\oint_C \frac{1}{(z^2 + 1)(z^2 + 9)}\,dz = \int_{-R}^{R} \frac{1}{(x^2 + 1)(x^2 + 9)}\,dx + \int_{C_R} \frac{1}{(z^2 + 1)(z^2 + 9)}\,dz$$

$$= I_1 + I_2$$

and

$$I_1 + I_2 = 2\pi i [\text{Res}(f(z), i) + \text{Res}(f(z), 3i)].$$

At the simple poles $z = i$ and $z = 3i$ we find, respectively,

$$\text{Res}(f(z), i) = \frac{1}{16i} \quad \text{and} \quad \text{Res}(f(z), 3i) = -\frac{1}{48i},$$

FIGURE 19.6.2 Closed contour C for Example 2

so that

$$I_1 + I_2 = 2\pi i \left[\frac{1}{16i} - \frac{1}{48i} \right] = \frac{\pi}{12}. \tag{9}$$

We now want to let $R \to \infty$ in (9). Before doing this, we note that on C_R,

$$|(z^2 + 1)(z^2 + 9)| = |z^2 + 1| \, |z^2 + 9| \ge ||z|^2 - 1| \, ||z|^2 - 9| = (R^2 - 1)(R^2 - 9),$$

and so from the *ML*-inequality of Section 18.1 we can write

$$|I_2| = \left| \int_{C_R} \frac{1}{(z^2 + 1)(z^2 + 9)} \, dz \right| \le \frac{\pi R}{(R^2 - 1)(R^2 - 9)}.$$

This last result shows that $|I_2| \to 0$ as $R \to \infty$, and so we conclude that $\lim_{R \to \infty} I_2 = 0$. It follows from (9) that $\lim_{R \to \infty} I_1 = \pi/12$; in other words,

$$\lim_{R \to \infty} \int_{-R}^{R} \frac{1}{(x^2 + 1)(x^2 + 9)} \, dx = \frac{\pi}{12} \quad \text{or} \quad \text{P.V.} \int_{-\infty}^{\infty} \frac{1}{(x^2 + 1)(x^2 + 9)} \, dx = \frac{\pi}{12}. \qquad \equiv$$

It is often tedious to have to show that the contour integral along C_R approaches zero as $R \to \infty$. Sufficient conditions under which this is always true are given in the next theorem.

THEOREM 19.6.1 Behavior of Integral as $R \to \infty$

Suppose $f(z) = P(z)/Q(z)$, where the degree of $P(z)$ is n and the degree of $Q(z)$ is $m \ge n + 2$. If C_R is a semicircular contour $z = Re^{i\theta}, 0 \le \theta \le \pi$, then $\int_{C_R} f(z) \, dz \to 0$ as $R \to \infty$.

In other words, the integral along C_R approaches zero as $R \to \infty$ when the denominator of f is of a power at least 2 more than its numerator. The proof of this fact follows in the same manner as in Example 2. Notice in that example that the conditions stipulated in Theorem 19.6.1 are satisfied, since the degree of $P(z) = 1$ is 0 and the degree of $Q(z) = (z^2 + 1)(z^2 + 9)$ is 4.

EXAMPLE 3 **Cauchy P.V. of an Improper Integral**

Evaluate the Cauchy principal value of $\displaystyle\int_{-\infty}^{\infty} \frac{1}{x^4 + 1} \, dx$.

SOLUTION By inspection of the integrand, we see that the conditions given in Theorem 19.6.1 are satisfied. Moreover, we know from Example 3 of Section 19.5 that f has simple poles in the upper half-plane at $z_1 = e^{\pi i/4}$ and $z_2 = e^{3\pi i/4}$. We also saw in that example that the residues at these poles are

$$\text{Res}(f(z), z_1) = -\frac{1}{4\sqrt{2}} - \frac{1}{4\sqrt{2}} i \quad \text{and} \quad \text{Res}(f(z), z_2) = \frac{1}{4\sqrt{2}} - \frac{1}{4\sqrt{2}} i.$$

Thus, by (8),

$$\text{P.V.} \int_{-\infty}^{\infty} \frac{1}{x^4 + 1} \, dx = 2\pi i [\text{Res}(f(z), z_1) + \text{Res}(f(z), z_2)] = \frac{\pi}{\sqrt{2}}. \qquad \equiv$$

Integrals of the Forms $\int_{-\infty}^{\infty} f(x) \cos \alpha x \, dx$ or $\int_{-\infty}^{\infty} f(x) \sin \alpha x \, dx$

We encountered integrals of this type in Section 15.4 in the study of Fourier transforms. Accordingly, $\int_{-\infty}^{\infty} f(x) \cos \alpha x \, dx$ and $\int_{-\infty}^{\infty} f(x) \sin \alpha x \, dx, \alpha > 0$, are referred to as **Fourier integrals**. Fourier integrals appear as the real and imaginary parts in the improper integral $\int_{-\infty}^{\infty} f(x) e^{i\alpha x} \, dx$. Using Euler's formula $e^{i\alpha x} = \cos \alpha x + i \sin \alpha x$, we get

$$\int_{-\infty}^{\infty} f(x) e^{i\alpha x} \, dx = \int_{-\infty}^{\infty} f(x) \cos \alpha x \, dx + i \int_{-\infty}^{\infty} f(x) \sin \alpha x \, dx \tag{10}$$

whenever both integrals on the right side converge. When $f(x) = P(x)/Q(x)$ is continuous on $(-\infty, \infty)$ we can evaluate both Fourier integrals at the same time by considering the integral $\int_C f(z) e^{i\alpha z} \, dz$, where $\alpha > 0$ and the contour C again consists of the interval $[-R, R]$ on the real axis and a semicircular contour C_R with radius large enough to enclose the poles of $f(z)$ in the upper half-plane.

Before proceeding we give, without proof, sufficient conditions under which the contour integral along C_R approaches zero as $R \to \infty$:

THEOREM 19.6.2 **Behavior of Integral as $R \to \infty$**

Suppose $f(z) = P(z)/Q(z)$, where the degree of $P(z)$ is n and the degree of $Q(z)$ is $m \geq n + 1$. If C_R is a semicircular contour $z = Re^{i\theta}$, $0 \leq \theta \leq \pi$, and $\alpha > 0$, then $\int_{C_R} (P(z)/Q(z))e^{i\alpha z}\, dz \to 0$ as $R \to \infty$.

EXAMPLE 4 **Using Symmetry**

Evaluate the Cauchy principal value of $\displaystyle\int_0^\infty \frac{x \sin x}{x^2 + 9}\, dx$.

SOLUTION First note that the limits of integration are not from $-\infty$ to ∞ as required by the method. This can be rectified by observing that since the integrand is an even function of x, we can write

$$\int_0^\infty \frac{x \sin x}{x^2 + 9}\, dx = \frac{1}{2} \int_{-\infty}^\infty \frac{x \sin x}{x^2 + 9}\, dx. \tag{11}$$

With $\alpha = 1$, we now form the contour integral

$$\oint_C \frac{z}{z^2 + 9} e^{iz}\, dz,$$

where C is the same contour shown in Figure 19.6.2. By Theorem 19.5.3,

$$\int_{C_R} \frac{z}{z^2 + 9} e^{iz}\, dz + \int_{-R}^R \frac{x}{x^2 + 9} e^{ix}\, dx = 2\pi i \operatorname{Res}(f(z)e^{iz}, 3i),$$

where $f(z) = z/(z^2 + 9)$. From (4) of Section 19.5,

$$\operatorname{Res}(f(z)e^{iz}, 3i) = \frac{ze^{iz}}{2z}\bigg|_{z=3i} = \frac{e^{-3}}{2}.$$

Hence, in view of Theorem 19.6.2 we conclude $\int_{C_R} f(z)e^{iz}\, dz \to 0$ as $R \to \infty$ and so

$$\text{P.V.} \int_{-\infty}^\infty \frac{x}{x^2 + 9} e^{ix}\, dx = 2\pi i\left(\frac{e^{-3}}{2}\right) = \frac{\pi}{e^3} i.$$

But by (10),

$$\int_{-\infty}^\infty \frac{x}{x^2 + 9} e^{ix}\, dx = \int_{-\infty}^\infty \frac{x \cos x}{x^2 + 9}\, dx + i \int_{-\infty}^\infty \frac{x \sin x}{x^2 + 9}\, dx = \frac{\pi}{e^3} i.$$

Equating real and imaginary parts in the last line gives the bonus result

$$\text{P.V.} \int_{-\infty}^\infty \frac{x \cos x}{x^2 + 9}\, dx = 0 \quad \text{along with} \quad \text{P.V.} \int_{-\infty}^\infty \frac{x \sin x}{x^2 + 9}\, dx = \frac{\pi}{e^3}.$$

Finally, in view of (11) we obtain the value of the prescribed integral:

$$\int_0^\infty \frac{x \sin x}{x^2 + 9}\, dx = \frac{1}{2}\text{P.V.} \int_{-\infty}^\infty \frac{x \sin x}{x^2 + 9}\, dx = \frac{\pi}{2e^3}. \qquad \equiv$$

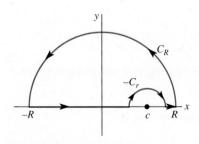

FIGURE 19.6.3 Indented contour

■ **Indented Contours** The improper integrals of form (2) and (3) that we have considered up to this point were continuous on the interval $(-\infty, \infty)$. In other words, the complex function $f(z) = P(z)/Q(z)$ did not have poles on the real axis. In the event f has poles on the real axis, we must modify the procedure used in Examples 2–4. For example, to evaluate $\int_{-\infty}^\infty f(x)\, dx$ by residues when $f(z)$ has a pole at $z = c$, where c is a real number, we use an **indented contour** as illustrated in **FIGURE 19.6.3**. The symbol C_r denotes a semicircular contour centered at $z = c$ oriented in the *positive* direction. The next theorem is important to this discussion.

Suppose f has a simple pole $z = c$ on the real axis. If C_r is the contour defined by $z = c + re^{i\theta}$, $0 \le \theta \le \pi$, then

$$\lim_{r \to 0} \int_{C_r} f(z)\, dz = \pi i \operatorname{Res}(f(z), c).$$

PROOF: Since f has a simple pole at $z = c$, its Laurent series is

$$f(z) = \frac{a_{-1}}{z - c} + g(z),$$

where $a_{-1} = \operatorname{Res}(f(z), c)$ and g is analytic at c. Using the Laurent series and the parameterization of C_r, we have

$$\int_{C_r} f(z)\, dz = a_{-1} \int_0^\pi \frac{ire^{i\theta}}{re^{i\theta}}\, d\theta + ir \int_0^\pi g(c + re^{i\theta})e^{i\theta}\, d\theta = I_1 + I_2. \tag{12}$$

First, we see that

$$I_1 = a_{-1} \int_0^\pi \frac{ire^{i\theta}}{re^{i\theta}}\, d\theta = a_{-1} \int_0^\pi i\, d\theta = \pi i a_{-1} = \pi i \operatorname{Res}(f(z), c).$$

Next, g is analytic at c and so it is continuous at this point and bounded in a neighborhood of the point; that is, there exists an $M > 0$ for which $|g(c + re^{i\theta})| \le M$. Hence,

$$|I_2| = \left| ir \int_0^\pi g(c + re^{i\theta})e^{i\theta}\, d\theta \right| \le r \int_0^\pi M\, d\theta = \pi r M.$$

It follows from this last inequality that $\lim_{r \to 0} |I_2| = 0$ and consequently $\lim_{r \to 0} I_2 = 0$. By taking the limit of (12) as $r \to 0$, we have proved the theorem. ≡

EXAMPLE 5 Using an Indented Contour

Evaluate the Cauchy principal value of $\displaystyle\int_{-\infty}^{\infty} \frac{\sin x}{x(x^2 - 2x + 2)}\, dx$.

SOLUTION Since the integral is of form (3), we consider the contour integral $\oint_C e^{iz}\, dz/z(z^2 - 2z + 2)$. The function $f(z) = 1/z(z^2 - 2z + 2)$ has simple poles at $z = 0$ and at $z = 1 + i$ in the upper half-plane. The contour C shown in **FIGURE 19.6.4** is indented at the origin. Adopting an obvious notation, we have

$$\oint_C = \int_{C_R} + \int_{-R}^{-r} + \int_{-C_r} + \int_r^R = 2\pi i \operatorname{Res}(f(z)e^{iz}, 1 + i), \tag{13}$$

where $\int_{-C_r} = -\int_{C_r}$. Taking the limits of (13) as $R \to \infty$ and as $r \to 0$, we find from Theorems 19.6.2 and 19.6.3 that

$$\text{P.V.} \int_{-\infty}^{\infty} \frac{e^{ix}}{x(x^2 - 2x + 2)}\, dx - \pi i \operatorname{Res}(f(z)e^{iz}, 0) = 2\pi i \operatorname{Res}(f(z)e^{iz}, 1 + i).$$

Now,

$$\operatorname{Res}(f(z)e^{iz}, 0) = \frac{1}{2} \quad \text{and} \quad \operatorname{Res}(f(z)e^{iz}, 1 + i) = -\frac{e^{-1+i}}{4}(1 + i).$$

Therefore,

$$\text{P.V.} \int_{-\infty}^{\infty} \frac{e^{ix}}{x(x^2 - 2x + 2)}\, dx = \pi i \left(\frac{1}{2} \right) + 2\pi i \left(-\frac{e^{-1+i}}{4}(1 + i) \right).$$

Using $e^{-1+i} = e^{-1}(\cos 1 + i \sin 1)$, simplifying, and then equating real and imaginary parts, we get from the last equality

$$\text{P.V.} \int_{-\infty}^{\infty} \frac{\cos x}{x(x^2 - 2x + 2)}\, dx = \frac{\pi}{2} e^{-1}(\sin 1 + \cos 1)$$

and

$$\text{P.V.} \int_{-\infty}^{\infty} \frac{\sin x}{x(x^2 - 2x + 2)}\, dx = \frac{\pi}{2}[1 + e^{-1}(\sin 1 - \cos 1)].$$

≡

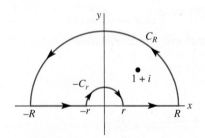

FIGURE 19.6.4 Indented contour C for Example 5

In Problems 1–10, evaluate the given trigonometric integral.

1. $\displaystyle\int_0^{2\pi} \frac{1}{1 + 0.5\sin\theta}\, d\theta$

2. $\displaystyle\int_0^{2\pi} \frac{1}{10 - 6\cos\theta}\, d\theta$

3. $\displaystyle\int_0^{2\pi} \frac{\cos\theta}{3 + \sin\theta}\, d\theta$

4. $\displaystyle\int_0^{2\pi} \frac{1}{1 + 3\cos^2\theta}\, d\theta$

5. $\displaystyle\int_0^{\pi} \frac{1}{2 - \cos\theta}\, d\theta$ [Hint: Let $t = 2\pi - \theta$.]

6. $\displaystyle\int_0^{\pi} \frac{1}{1 + \sin^2\theta}\, d\theta$

7. $\displaystyle\int_0^{2\pi} \frac{\sin^2\theta}{5 + 4\cos\theta}\, d\theta$

8. $\displaystyle\int_0^{2\pi} \frac{\cos^2\theta}{3 - \sin\theta}\, d\theta$

9. $\displaystyle\int_0^{2\pi} \frac{\cos 2\theta}{5 - 4\cos\theta}\, d\theta$

10. $\displaystyle\int_0^{2\pi} \frac{1}{\cos\theta + 2\sin\theta + 3}\, d\theta$

In Problems 11–30, evaluate the Cauchy principal value of the given improper integral.

11. $\displaystyle\int_{-\infty}^{\infty} \frac{1}{x^2 - 2x + 2}\, dx$

12. $\displaystyle\int_{-\infty}^{\infty} \frac{1}{x^2 - 6x + 25}\, dx$

13. $\displaystyle\int_{-\infty}^{\infty} \frac{1}{(x^2 + 4)^2}\, dx$

14. $\displaystyle\int_{-\infty}^{\infty} \frac{x^2}{(x^2 + 1)^2}\, dx$

15. $\displaystyle\int_{-\infty}^{\infty} \frac{1}{(x^2 + 1)^3}\, dx$

16. $\displaystyle\int_{-\infty}^{\infty} \frac{x}{(x^2 + 4)^3}\, dx$

17. $\displaystyle\int_{-\infty}^{\infty} \frac{2x^2 - 1}{x^4 + 5x^2 + 4}\, dx$

18. $\displaystyle\int_{-\infty}^{\infty} \frac{dx}{(x^2 + 1)^2(x^2 + 9)}$

19. $\displaystyle\int_0^{\infty} \frac{x^2 + 1}{x^4 + 1}\, dx$

20. $\displaystyle\int_0^{\infty} \frac{1}{x^6 + 1}\, dx$

21. $\displaystyle\int_{-\infty}^{\infty} \frac{\cos x}{x^2 + 1}\, dx$

22. $\displaystyle\int_{-\infty}^{\infty} \frac{\cos 2x}{x^2 + 1}\, dx$

23. $\displaystyle\int_{-\infty}^{\infty} \frac{x\sin x}{x^2 + 1}\, dx$

24. $\displaystyle\int_0^{\infty} \frac{\cos x}{(x^2 + 4)^2}\, dx$

25. $\displaystyle\int_0^{\infty} \frac{\cos 3x}{(x^2 + 1)^2}\, dx$

26. $\displaystyle\int_{-\infty}^{\infty} \frac{\sin x}{x^2 + 4x + 5}\, dx$

27. $\displaystyle\int_0^{\infty} \frac{\cos 2x}{x^4 + 1}\, dx$

28. $\displaystyle\int_0^{\infty} \frac{x\sin x}{x^4 + 1}\, dx$

29. $\displaystyle\int_{-\infty}^{\infty} \frac{\cos x}{(x^2 + 1)(x^2 + 9)}\, dx$

30. $\displaystyle\int_0^{\infty} \frac{x\sin x}{(x^2 + 1)(x^2 + 4)}\, dx$

In Problems 31 and 32, use an indented contour and residues to establish the given result.

31. P.V. $\displaystyle\int_{-\infty}^{\infty} \frac{\sin x}{x}\, dx = \pi$

32. P.V. $\displaystyle\int_{-\infty}^{\infty} \frac{\sin x}{x(x^2 + 1)}\, dx = \pi(1 - e^{-1})$

33. Establish the general result

$$\int_0^{\pi} \frac{d\theta}{(a + \cos\theta)^2} = \frac{a\pi}{(\sqrt{a^2 - 1})^3}, \quad a > 1$$

and use this formula to verify the answer in Example 1.

34. Establish the general result

$$\int_0^{2\pi} \frac{\sin^2\theta}{a + b\cos\theta}\, d\theta = \frac{2\pi}{b^2}(a - \sqrt{a^2 - b^2}), \quad a > b > 0$$

and use this formula to verify the answer to Problem 7.

35. Use the contour shown in **FIGURE 19.6.5** to show that

$$\text{P.V.} \int_{-\infty}^{\infty} \frac{e^{ax}}{1 + e^x}\, dx = \frac{\pi}{\sin a\pi}, \quad 0 < a < 1.$$

FIGURE 19.6.5 Contour in Problem 35

36. The steady-state temperature $u(x, y)$ in a semi-infinite plate is determined from

$$\frac{\partial^2 u}{\partial x^2} + \frac{\partial^2 u}{\partial y^2} = 0, \quad 0 < x < \pi, \ y > 0$$

$$u(0, y) = 0, \quad u(\pi, y) = \frac{2y}{y^4 + 4}, \ y > 0$$

$$u(x, 0) = 0, \quad 0 < x < \pi.$$

Use a Fourier transform and the residue method to show that

$$u(x, y) = \int_0^{\infty} \frac{e^{-\alpha}\sin\alpha\,\sinh\alpha x}{\sinh\alpha\pi}\sin\alpha y\, d\alpha.$$

19 Chapter in Review

Answer Problems 1–12 without referring back to the text. Fill in the blank or answer true/false.

1. A function f is analytic at a point z_0 if f can be expanded in a convergent power series centered at z_0. _____

2. A power series represents a continuous function at every point within and on its circle of convergence. _____

3. For $f(z) = 1/(z - 3)$, the Laurent series valid for $|z| > 3$ is $z^{-1} + 3z^{-2} + 9z^{-3} + \cdots$. Since there are an infinite number of negative powers of $z = z - 0$, $z = 0$ is an essential singularity. _____

4. The only possible singularities of a rational function are poles. _____

5. The function $f(z) = e^{1/(z-1)}$ has an essential singularity at $z = 1$. _____

6. The function $f(z) = z/(e^z - 1)$ has a removable singularity at $z = 0$. _____

7. The function $f(z) = z(e^z - 1)$ possesses a zero of order 2 at $z = 0$. _____

8. The function $f(z) = (z + 5)/(z^3 \sin^2 z)$ has a pole of order _____ at $z = 0$.

9. If $f(z) = \cot \pi z$, then $\text{Res}(f(z), 0) =$ _____.

10. The Laurent series of f valid for $0 < |z - 1|$ is given by
$$(z - 1)^{-3} - \frac{(z - 1)^{-1}}{3!} + \frac{(z - 1)}{5!} - \frac{(z - 1)^3}{7!} + \cdots.$$
From this series we see that f has a pole of order _____ at $z = 1$ and $\text{Res}(f(z), 1) =$ _____.

11. The circle of convergence of the power series $\sum_{k=1}^{\infty} \frac{(z - i)^k}{(2 + i)^{k+1}}$ is _____.

12. The power series $\sum_{k=1}^{\infty} \frac{z^k}{2^{k+1}}$ converges at $z = 2i$. _____

13. Find a Maclaurin expansion of $f(z) = e^z \cos z$. [*Hint:* Use the identity $\cos z = (e^{iz} + e^{-iz})/2$.]

14. Show that the function $f(z) = 1/\sin(\pi/z)$ has an infinite number of singular points. Are any of these isolated singular points?

In Problems 15–18, use known results as an aid in expanding the given function in a Laurent series valid for the indicated annular region.

15. $f(z) = \dfrac{1 - e^{iz}}{z^4}$, $0 < |z|$ 16. $f(z) = e^{z/(z-2)}$, $0 < |z - 2|$

17. $f(z) = (z - i)^2 \sin \dfrac{1}{z - i}$, $0 < |z - i|$

18. $f(z) = \dfrac{1 - \cos z^2}{z^5}$, $0 < |z|$

19. Expand $f(z) = \dfrac{2}{z^2 - 4z + 3}$ in an appropriate series valid for
 (a) $|z| < 1$ (b) $1 < |z| < 3$
 (c) $|z| > 3$ (d) $0 < |z - 1| < 2$.

20. Expand $f(z) = \dfrac{1}{(z - 5)^2}$ in an appropriate series valid for
 (a) $|z| < 5$ (b) $|z| > 5$ (c) $0 < |z - 5|$.

In Problems 21–30, use Cauchy's residue theorem to evaluate the given integral along the indicated contour.

21. $\displaystyle\oint_C \frac{2z + 5}{z(z + 2)(z - 1)^4} \, dz$, $C: |z + 2| = \frac{5}{2}$

22. $\displaystyle\oint_C \frac{z^2}{(z - 1)^3(z^2 + 4)} \, dz$, C is the ellipse $x^2/4 + y^2 = 1$

23. $\displaystyle\oint_C \frac{1}{2 \sin z - 1} \, dz$, $C: |z - \frac{1}{2}| = \frac{1}{3}$

24. $\displaystyle\oint_C \frac{z + 1}{\sinh z} \, dz$, C is the rectangle defined by $x = -1$, $x = 1$, $y = 4$, $y = -1$

25. $\displaystyle\oint_C \frac{e^{2z}}{z^4 + 2z^3 + 2z^2} \, dz$, $C: |z| = 4$

26. $\displaystyle\oint_C \frac{1}{z^4 - 2z^2 + 4} \, dz$, C is the square defined by $x = -2$, $x = 2$, $y = 0$, $y = 1$

27. $\displaystyle\oint_C \frac{1}{z(e^z - 1)} \, dz$, $C: |z| = 1$ [*Hint:* Use the Maclaurin series for $z(e^z - 1)$.]

28. $\displaystyle\oint_C \frac{z}{(z + 1)(z - 1)^{10}} \, dz$, $C: |z - 1| = 3$

29. $\displaystyle\oint_C \left[ze^{3/z} + \frac{\sin z}{z^2(z - \pi)^3} \right] dz$, $C: |z| = 6$

30. $\displaystyle\oint_C \csc \pi z \, dz$, C is the rectangle defined by $x = -\frac{1}{2}$, $x = \frac{5}{2}$, $y = -1$, $y = 1$

In Problems 31 and 32, evaluate the Cauchy principal value of the given improper integral.

31. $\displaystyle\int_{-\infty}^{\infty} \frac{x^2}{(x^2 + 2x + 2)(x^2 + 1)^2} \, dx$

32. $\displaystyle\int_{-\infty}^{\infty} \frac{a \cos x + x \sin x}{x^2 + a^2} \, dx$, $a > 0$ [*Hint:* Consider $e^{iz}/(z - ai)$.]

In Problems 33 and 34, evaluate the given trigonometric integral.

33. $\displaystyle\int_0^{2\pi} \frac{\cos^2 \theta}{2 + \sin \theta} \, d\theta$ 34. $\displaystyle\int_0^{2\pi} \frac{\cos 3\theta}{5 - 4\cos \theta} \, d\theta$

35. Use an indented contour to show that
$$\text{P.V.} \int_0^{\infty} \frac{1 - \cos x}{x^2} \, dx = \frac{\pi}{2}.$$

36. Show that $\int_0^{\infty} e^{-a^2 x^2} \cos bx \, dx = e^{-b^2/4a^2} \sqrt{\pi}/2a$ by considering the complex integral $\oint_C e^{-a^2 z^2} e^{ibz} \, dz$ along the contour C shown in **FIGURE 19.R.1**. Use the known result $\int_{-\infty}^{\infty} e^{-a^2 x^2} \, dx = \sqrt{\pi}/a$.

FIGURE 19.R.1 Contour in Problem 36

37. The Laurent expansion of $f(z) = e^{(u/2)(z - 1/z)}$ valid for $0 < |z|$ can be shown to be $f(z) = \sum_{k=-\infty}^{\infty} J_k(u) z^k$, where $J_k(u)$ is the Bessel function of the first kind of order k. Use (4) of Section 19.3 and the contour $C: |z| = 1$ to show that the coefficients $J_k(u)$ are given by
$$J_k(u) = \frac{1}{2\pi} \int_0^{2\pi} \cos(kt - u \sin t) \, dt.$$

20

Conformal Mappings

In earlier chapters we used Fourier series and integral transforms to solve boundary-value problems involving Laplace's equation. **Conformal mapping** methods discussed in this chapter can be used to transfer known solutions of Laplace's equation from one region to another. In addition, fluid flows around obstacles and through channels can be determined using conformal mappings.

INTRODUCTION In Chapter 17 we emphasized the algebraic definitions and properties of complex functions. In order to give a geometric interpretation of a complex function $w = f(z)$, we place a z-plane and a w-plane side by side and imagine that a point $z = x + iy$ in the domain of the definition of f has **mapped** (or transformed) to the point $w = f(z)$ in the second plane. Thus the complex function $w = f(z) = u(x, y) + iv(x, y)$ may be considered as the **planar transformation**

$$u = u(x, y), \quad v = v(x, y)$$

and $w = f(z)$ is called the **image** of z under f.

FIGURE 20.1.1 indicates the images of a finite number of complex numbers in the region R. More useful information is obtained by finding the image of the region R together with the images of a family of curves lying inside R. Common choices for the curves are families of lines, families of circles, and the system of level curves for the real and imaginary parts of f.

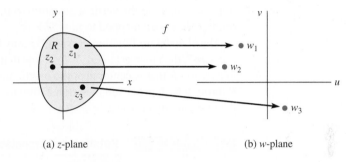

(a) z-plane (b) w-plane

FIGURE 20.1.1 w_1, w_2, w_3 are images of z_1, z_2, z_3

■ **Images of Curves** Note that if $z(t) = x(t) + iy(t)$, $a \leq t \leq b$, describes a curve C in the region, then $w = f(z(t))$, $a \leq t \leq b$, is a parametric representation of the corresponding curve C' in the w-plane. In addition, a point z on the level curve $u(x, y) = a$ will be mapped to a point w that lies on the vertical line $u = a$, and a point z on the level curve $v(x, y) = b$ will be mapped to a point w that lies on the horizontal line $v = b$.

EXAMPLE 1 **The Mapping $f(z) = e^z$**

The horizontal strip $0 \leq y \leq \pi$ lies in the fundamental region of the exponential function $f(z) = e^z$. A vertical line segment $x = a$ in this region can be described by $z(t) = a + it$, $0 \leq t \leq \pi$, and so $w = f(z(t)) = e^a e^{it}$. Thus the image is a semicircle with center at $w = 0$ and with radius $r = e^a$. Similarly, a horizontal line $y = b$ can be parametrized by $z(t) = t + ib$, $-\infty < t < \infty$, and so $w = f(z(t)) = e^t e^{ib}$. Since Arg $w = b$ and $|w| = e^t$, the image is a ray emanating from the origin, and since $0 \leq \text{Arg } w \leq \pi$, the image of the entire horizontal strip is the upper half-plane $v \geq 0$. Note that the horizontal lines $y = 0$ and $y = \pi$ are mapped onto the positive and negative u-axis, respectively. See **FIGURE 20.1.2** for the mapping by $f(z) = e^z$.

From $w = e^x e^{iy}$, we can conclude that $|w| = e^x$ and $y = \text{Arg } w$. Hence, $z = x + iy = \log_e |w| + i \text{ Arg } w = \text{Ln } w$. The inverse function $f^{-1}(w) = \text{Ln } w$ therefore maps the upper half-plane $v \geq 0$ back to the horizontal strip $0 \leq y \leq \pi$. ≡

EXAMPLE 2 **The Mapping $f(z) = 1/z$**

The complex function $f(z) = 1/z$ has domain $z \neq 0$ and real and imaginary parts $u(x, y) = x/(x^2 + y^2)$ and $v(x, y) = -y/(x^2 + y^2)$, respectively. When $a \neq 0$, a level curve $u(x, y) = a$ can be written as

$$x^2 - \frac{1}{a}x + y^2 = 0 \quad \text{or} \quad \left(x - \frac{1}{2a}\right)^2 + y^2 = \left(\frac{1}{2a}\right)^2.$$

(a)

(b)

Arg $w = \pi$ Arg $w = 0$

FIGURE 20.1.2 Images of vertical and horizontal lines in Example 1

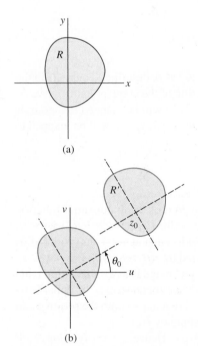

FIGURE 20.1.3 Images of circles in Example 2

FIGURE 20.1.4 Translation and rotation

The level curve is therefore a circle with its center on the *x*-axis and passing through the origin. A point *z* on this circle other than zero is mapped to a point *w* on the line $u = a$. Likewise, the level curve $v(x, y) = b$, $b \neq 0$, can be written as

$$x^2 + \left(y + \frac{1}{2b}\right)^2 = \left(\frac{1}{2b}\right)^2,$$

and a point *z* on this circle is mapped to a point *w* on the line $v = b$. **FIGURE 20.1.3** shows the mapping by $f(z) = 1/z$. Figure 20.1.3(a) shows the two collections of circular level curves, and Figure 20.1.3(b) shows their corresponding images in the *w*-plane.

Since $w = 1/z$, we have $z = 1/w$. Thus $f^{-1}(w) = 1/w$, and so $f = f^{-1}$. We can therefore conclude that *f* maps the horizontal line $y = b$ to the circle $u^2 + (v + \frac{1}{2}b)^2 = (\frac{1}{2}b)^2$, and *f* maps the vertical line $x = a$ to the circle $(u - \frac{1}{2}a)^2 + v^2 = (\frac{1}{2}a)^2$. ≡

■ **Translation and Rotation** The elementary linear function $f(z) = z + z_0$ may be interpreted as a **translation** in the *z*-plane. To see this, we let $z = x + iy$ and $z_0 = h + ik$. Since $w = f(z) = (x + h) + i(y + k)$, the point (x, y) has been translated *h* units in the horizontal direction and *k* units in the vertical direction to the new position at $(x + h, y + k)$. In particular, the origin *O* has been mapped to $z_0 = h + ik$.

The elementary function $g(z) = e^{i\theta_0}z$ may be interpreted as a **rotation** through θ_0 degrees, for if $z = re^{i\theta}$, then $w = g(z) = re^{i(\theta + \theta_0)}$. Note that if the complex mapping $h(z) = e^{i\theta_0}z + z_0$ is applied to a region *R* that is centered at the origin, the image region *R'* can be obtained by first rotating *R* through θ_0 degrees and then translating the center to the new position z_0. See **FIGURE 20.1.4** for the mapping by $h(z) = e^{i\theta_0}z + z_0$.

EXAMPLE 3 **Rotation and Translation**

Find a complex function that maps the horizontal strip $-1 \leq y \leq 1$ onto the vertical strip $2 \leq x \leq 4$.

SOLUTION If the horizontal strip $-1 \leq y \leq 1$ is rotated through 90°, the vertical strip $-1 \leq x \leq 1$ results, and the vertical strip $2 \leq x \leq 4$ can be obtained by shifting this vertical strip 3 units to the right. See **FIGURE 20.1.5**. Since $e^{i\pi/2} = i$, we obtain $h(z) = iz + 3$ as the desired complex mapping.

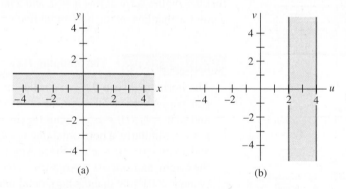

FIGURE 20.1.5 Image of horizontal strip in Example 3 ≡

■ **Magnification** A **magnification** is a complex function of the form $f(z) = \alpha z$, where α is a fixed positive real number. Note that $|w| = |\alpha z| = \alpha|z|$, and so *f* changes the length (but not the direction) of the complex number *z* by a fixed factor α. If $g(z) = az + b$ and $a = r_0 e^{i\theta_0}$, then the vector *z* is rotated through θ_0 degrees, magnified by a factor of r_0, and then translated using *b*.

EXAMPLE 4 **Contraction and Translation**

Find a complex function that maps the disk $|z| \leq 1$ onto the disk $|w - (1 + i)| \leq \frac{1}{2}$.

SOLUTION We must first contract the radius of the disk by a factor of $\frac{1}{2}$ and then translate its center to the point $1 + i$. Therefore, $w = f(z) = \frac{1}{2}z + (1 + i)$ maps $|z| \leq 1$ to the disk $|w - (1 + i)| \leq \frac{1}{2}$. ≡

(a)

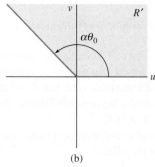

(b)

FIGURE 20.1.6 R' is the image of the angular wedge R

■ **Power Functions** A complex function of the form $f(z) = z^\alpha$, where α is a fixed positive real number, is called a **real power function**. **FIGURE 20.1.6** shows the effect of the complex function $f(z) = z^\alpha$ on the angular wedge $0 \leq \text{Arg } z \leq \theta_0$. If $z = re^{i\theta}$, then $w = f(z) = r^\alpha e^{i\alpha\theta}$. Hence, $0 \leq \text{Arg } w \leq \alpha\theta_0$ and the opening of the wedge is changed by a factor of α. It is not difficult to show that a circular arc with center at the origin is mapped to a similar circular arc, and rays emanating from the origin are mapped to similar rays.

EXAMPLE 5 **The Power Function $f(z) = z^{1/4}$**

Find a complex function that maps the upper half-plane $y \geq 0$ onto the wedge $0 \leq \text{Arg } w \leq \pi/4$.

SOLUTION The upper half-plane $y \geq 0$ can also be described by the inequality $0 \leq \text{Arg } z \leq \pi$. We must therefore find a complex mapping that reduces the angle $\theta_0 = \pi$ by a factor of $\alpha = \frac{1}{4}$. Hence, $f(z) = z^{1/4}$. ≡

■ **Successive Mappings** To find a complex mapping between two regions R and R', it is often convenient to first map R onto a third region R'' and then find a complex mapping from R'' onto R'. More precisely, if $\zeta = f(z)$ maps R onto R'', and $w = g(\zeta)$ maps R'' onto R', then the composite function $w = g(f(z))$ maps R onto R'. See **FIGURE 20.1.7** for a diagram of successive mappings.

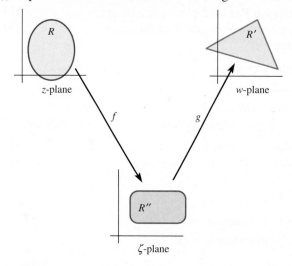

FIGURE 20.1.7 R' is image of R under successive mappings

EXAMPLE 6 **Successive Mappings**

Find a complex function that maps the horizontal strip $0 \leq y \leq \pi$ onto the wedge $0 \leq \text{Arg } w \leq \pi/4$.

SOLUTION We saw in Example 1 that the complex function $f(z) = e^z$ mapped the horizontal strip $0 \leq y \leq \pi$ onto the upper half-plane $0 \leq \text{Arg } \zeta \leq \pi$. From Example 5, the upper half-plane $0 \leq \text{Arg } \zeta \leq \pi$ is mapped onto the wedge $0 \leq \text{Arg } w \leq \pi/4$ by $g(\zeta) = \zeta^{1/4}$. It therefore follows that the composite function $w = g(f(z)) = g(e^z) = e^{z/4}$ maps the horizontal strip $0 \leq y \leq \pi$ onto the wedge $0 \leq \text{Arg } w \leq \pi/4$. ≡

EXAMPLE 7 **Successive Mappings**

Find a complex function that maps the wedge $\pi/4 \leq \text{Arg } z \leq 3\pi/4$ onto the upper half-plane $v \geq 0$.

SOLUTION We first rotate the wedge $\pi/4 \leq \text{Arg } z \leq 3\pi/4$ so that it is in the standard position shown in Figure 20.1.6. If $\zeta = f(z) = e^{-i\pi/4}z$, then the image of this wedge is the wedge R'' defined by $0 \leq \text{Arg } \zeta \leq \pi/2$. The real power function $w = g(\zeta) = \zeta^2$ expands the opening of R'' by a factor of two to give the upper half-plane $0 \leq \text{Arg } w \leq \pi$ as its image. Therefore, $w = g(f(z)) = (e^{-i\pi/4}z)^2 = -iz^2$ is the desired mapping. ≡

In Sections 20.2–20.4, we will expand our knowledge of complex mappings and show how they can be used to solve Laplace's equation in the plane.

In Problems 1–10, a curve in the z-plane and a complex mapping $w = f(z)$ are given. In each case, find the image curve in the w-plane.

1. $y = x$ under $w = 1/z$

2. $y = 1$ under $w = 1/z$

3. Hyperbola $xy = 1$ under $w = z^2$

4. Hyperbola $x^2 - y^2 = 4$ under $w = z^2$

5. Semicircle $|z| = 1$, $y > 0$, under $w = \text{Ln } z$

6. Ray $\theta = \pi/4$ under $w = \text{Ln } z$

7. Ray $\theta = \theta_0$ under $w = z^{1/2}$

8. Circular arc $r = 2$, $0 \le \theta \le \pi/2$, under $w = z^{1/2}$

9. Curve $e^x \cos y = 1$ under $w = e^z$

10. Circle $|z| = 1$ under $w = z + 1/z$

In Problems 11–20, a region R in the z-plane and a complex mapping $w = f(z)$ are given. In each case, find the image region R' in the w-plane.

11. First quadrant under $w = 1/z$

12. Strip $0 \le y \le 1$ under $w = 1/z$

13. Strip $\pi/4 \le y \le \pi/2$ under $w = e^z$

14. Rectangle $0 \le x \le 1$, $0 \le y \le \pi$, under $w = e^z$

15. Circle $|z| = 1$ under $w = z + 4i$

16. Circle $|z| = 1$ under $w = 2z - 1$

17. Strip $0 \le y \le 1$ under $w = iz$

18. First quadrant under $w = (1 + i)z$

19. Wedge $0 \le \text{Arg } z \le \pi/4$ under $w = z^3$

20. Wedge $0 \le \text{Arg } z \le \pi/4$ under $w = z^{1/2}$

In Problems 21–30, find a complex mapping from the given region R in the z-plane to the image region R' in the w-plane.

21. Strip $1 \le y \le 4$ to the strip $0 \le u \le 3$

22. Strip $1 \le y \le 4$ to the strip $0 \le v \le 3$

23. Disk $|z - 1| \le 1$ to the disk $|w| \le 2$

24. Strip $-1 \le x \le 1$ to the strip $-1 \le v \le 1$

25. Wedge $\pi/4 \le \text{Arg } z \le \pi/2$ to the upper half-plane $v \ge 0$

26. Strip $0 \le y \le 4$ to the upper half-plane $v \ge 0$

27. Strip $0 \le y \le \pi$ to the wedge $0 \le \text{Arg } w \le 3\pi/2$

28. Wedge $0 \le \text{Arg } z \le 3\pi/2$ to the half-plane $u \ge 2$

29.

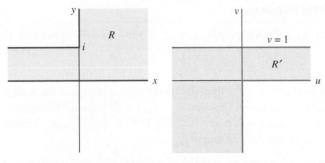

FIGURE 20.1.8 Regions R and R' for Problem 29

30.

FIGURE 20.1.9 Regions R and R' for Problem 30

31. Project The mapping in Problem 10 is a special case of the mapping $w = z + k^2/z$, where k is a positive constant, called the **Joukowski transformation** after the Russian mathematician and engineer **Nikolai Yegorovich Zhukovsky** (1847–1921) who published it in 1910.

(a) Show that the Joukowski transformation maps any circle $x^2 + y^2 = R^2$ into the ellipse

$$\frac{u^2}{\left(1 + \dfrac{k^2}{R^2}\right)^2} + \frac{v^2}{\left(1 - \dfrac{k^2}{R^2}\right)^2} = R^2, k \ne R.$$

(b) What is the image of the circle when $R = k$?

(c) The importance of the transformation $w = z + k^2/z$ does not lie in its effect on circles $|z| = R$ centered at the origin but on off-centered circles with center on the real axis. Show that the Joukowski transformation can be written

$$\frac{w - 2k}{w + 2k} = \left(\frac{z - k}{z + k}\right)^2.$$

With $k = 1$, this particular transformation maps a circle passing through $z = -1$ and containing the point $z = 1$ into a closed curve with a sharply pointed trailing edge. This kind of curve, which resembles a cross section of an airplane wing, is known as a **Joukowski airfoil**.

Write a report on the use of the Joukowski transformation in the study of the flow of air around an airfoil. There is a lot of information on this topic on the Internet.

20.2 Conformal Mappings

INTRODUCTION In Section 20.1 we saw that a nonconstant linear mapping $f(z) = az + b$, a and b complex numbers, acts by rotating, magnifying, and translating points in the complex plane. As a result, it is easily shown that the angle between any two intersecting curves in the z-plane is equal to the angle between the images of the arcs in the w-plane under a linear mapping. Other complex mappings that have this angle-preserving property are the subject of our study in this section.

■ **Angle-Preserving Mappings** A complex mapping $w = f(z)$ defined on a domain D is called **conformal at $z = z_0$** in D when f preserves the angles between any two curves in D that intersect at z_0. More precisely, if C_1 and C_2 intersect in D at z_0, and C_1' and C_2' are the corresponding images in the w-plane, we require that the angle θ between C_1 and C_2 equal the angle ϕ between C_1' and C_2'. See **FIGURE 20.2.1**.

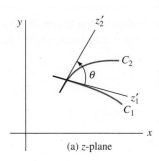

These angles can be computed in terms of tangent vectors to the curves. If z_1' and z_2' denote tangent vectors to curves C_1 and C_2, respectively, then, applying the law of cosines to the triangle determined by z_1' and z_2', we have

$$|z_1' - z_2'|^2 = |z_1'|^2 + |z_2'|^2 - 2|z_1'|\,|z_2'| \cos \theta$$

or

$$\theta = \cos^{-1}\left(\frac{|z_1'|^2 + |z_2'|^2 - |z_1' - z_2'|^2}{2|z_1'|\|z_2'|}\right). \tag{1}$$

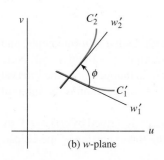

(a) z-plane

(b) w-plane

FIGURE 20.2.1 Conformal mapping if $\theta = \phi$

Likewise, if w_1' and w_2' denote tangent vectors to curves C_1' and C_2', respectively, then

$$\phi = \cos^{-1}\left(\frac{|w_1'|^2 + |w_2'|^2 - |w_1' - w_2'|^2}{2|w_1'|\|w_2'|}\right). \tag{2}$$

The next theorem gives a simple condition that guarantees that $\theta = \phi$.

THEOREM 20.2.1* Conformal Mapping

If $f(z)$ is analytic in the domain D and $f'(z_0) \neq 0$, then f is conformal at $z = z_0$.

PROOF: If a curve C in D is parameterized by $z = z(t)$, then $w = f(z(t))$ describes the image curve in the w-plane. Applying the Chain Rule to $w = f(z(t))$ gives $w' = f'(z(t))z'(t)$. If curves C_1 and C_2 intersect in D at z_0, then $w_1' = f'(z_0)z_1'$ and $w_2' = f'(z_0)z_2'$. Since $f'(z_0) \neq 0$, we can use (2) to obtain

$$\phi = \cos^{-1}\left(\frac{|f'(z_0)z_1'|^2 + |f'(z_0)z_2'|^2 - |f'(z_0)z_1' - f'(z_0)z_2'|^2}{2|f'(z_0)z_1'|\|f'(z_0)z_2'|}\right).$$

We can apply the laws of absolute value to factor out $|f'(z_0)|^2$ in the numerator and denominator and obtain

$$\phi = \cos^{-1}\left(\frac{|z_1'|^2 + |z_2'|^2 - |z_1' - z_2'|^2}{2|z_1'|\|z_2'|}\right).$$

Therefore, from (1), $\phi = \theta$. ≡

EXAMPLE 1 Conformal Mappings

(a) The analytic function $f(z) = e^z$ is conformal at all points in the z-plane, since $f'(z) = e^z$ is never zero.

(b) The analytic function $g(z) = z^2$ is conformal at all points except $z = 0$ since $g'(z) = 2z \neq 0$ for $z \neq 0$. From Figure 20.1.6 we see that $g(z)$ doubles the angle formed by the two rays at the origin. ≡

* It is also possible to prove that f preserves the *sense of direction* between the tangent vectors.

If $f'(z_0) = 0$ but $f''(z_0) \neq 0$, it is possible to show that f *doubles* the angle between any two curves in D that intersect at $z = z_0$. The next two examples will introduce two important complex mappings that are conformal at all but a finite number of points in their domains.

EXAMPLE 2 $f(z) = \sin z$ as a Conformal Mapping

The vertical strip $-\pi/2 \leq x \leq \pi/2$ is called the fundamental region of the trigonometric function $w = \sin z$. A vertical line $x = a$ in the interior of this region can be described by $z(t) = a + it$, $-\infty < t < \infty$. From (6) in Section 17.7, we have

$$\sin z = \sin x \cosh y + i \cos x \sinh y$$

and so
$$u + iv = \sin(a + it) = \sin a \cosh t + i \cos a \sinh t.$$

From the identity $\cosh^2 t - \sinh^2 t = 1$, it follows that

$$\frac{u^2}{\sin^2 a} - \frac{v^2}{\cos^2 a} = 1.$$

The image of the vertical line $x = a$ is therefore a hyperbola with $\pm \sin a$ as u-intercepts, and since $-\pi/2 < a < \pi/2$, the hyperbola crosses the u-axis between $u = -1$ and $u = 1$. Note that if $a = -\pi/2$, then $w = -\cosh t$, and so the line $x = -\pi/2$ is mapped onto the interval $(-\infty, -1]$ on the negative u-axis. Likewise, the line $x = \pi/2$ is mapped onto the interval $[1, \infty)$ on the positive u-axis.

A similar argument establishes that the horizontal line segment described by $z(t) = t + ib$, $-\pi/2 < t < \pi/2$, is mapped onto either the upper portion or the lower portion of the ellipse

$$\frac{u^2}{\cosh^2 b} + \frac{v^2}{\sinh^2 b} = 1$$

according to whether $b > 0$ or $b < 0$. These results are summarized in **FIGURE 20.2.2(b)**, which shows the mapping by $f(z) = \sin z$. Note that we have carefully used capital letters to indicate where portions of the boundary are mapped. Thus, for example, boundary segment AB is transformed to $A'B'$.

Since $f'(z) = \cos z$, f is conformal at all points in the region except $z = \pm\pi/2$. The hyperbolas and ellipses are therefore orthogonal since they are images of the orthogonal families of horizontal segments and vertical lines. Note that the $180°$ angle at $z = -\pi/2$ formed by segments AB and AC is doubled to form a single line segment at $w = -1$. ≡

FIGURE 20.2.2 Image of vertical strip in Example 2

EXAMPLE 3 $f(z) = z + 1/z$ as a Conformal Mapping

The complex mapping $f(z) = z + 1/z$ is conformal at all values of z except $z = \pm 1$ and $z = 0$. In particular, the function is conformal for all values of z in the upper half-plane that satisfy $|z| > 1$. If $z = re^{i\theta}$, then $w = re^{i\theta} + (1/r)e^{-i\theta}$, and so

$$u = \left(r + \frac{1}{r}\right)\cos\theta, \quad v = \left(r - \frac{1}{r}\right)\sin\theta. \tag{3}$$

Note that if $r = 1$, then $v = 0$ and $u = 2\cos\theta$. Therefore, the semicircle $z = e^{it}$, $0 \leq t \leq \pi$, is mapped to the segment $[-2, 2]$ on the u-axis. It follows from (3) that if $r > 1$, then the semicircle $z = re^{it}$, $0 \leq t \leq \pi$, is mapped onto the upper half of the ellipse $u^2/a^2 + v^2/b^2 = 1$, where $a = r + 1/r$ and $b = r - 1/r$. See **FIGURE 20.2.3** for the mapping by $f(z) = z + 1/z$.

For a fixed value of θ, the ray $z = te^{i\theta}$, for $t \geq 1$, is mapped to the portion of the hyperbola $u^2/\cos^2\theta - v^2/\sin^2\theta = 4$ in the upper half-plane $v \geq 0$. This follows from (3), since

$$\frac{u^2}{\cos^2\theta} - \frac{v^2}{\sin^2\theta} = \left(t + \frac{1}{t}\right)^2 - \left(t - \frac{1}{t}\right)^2 = 4.$$

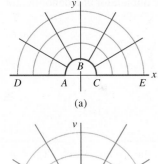

FIGURE 20.2.3 Images of rays and circles in Example 3

Since f is conformal for $|z| > 1$ and a ray $\theta = \theta_0$ intersects a circle $|z| = r$ at a right angle, the hyperbolas and ellipses in the w-plane are orthogonal. ≡

Conformal Mappings Using Tables
Conformal mappings are given in Appendix D. The mappings have been categorized as elementary mappings (E-1 to E-9), mappings to half-planes (H-1 to H-6), mappings to circular regions (C-1 to C-5), and miscellaneous mappings (M-1 to M-10). Some of these complex mappings will be derived in Sections 20.3 and 20.4.

The entries indicate not only the images of the region R but also the images of various portions of the boundary of R. This will be especially useful when we attempt to solve boundary-value problems using conformal maps. You should use the appendix much like you use integral tables to find antiderivatives. In some cases, a single entry can be used to find a conformal mapping between two given regions R and R'. In other cases, successive transformations may be required to map R to R'.

EXAMPLE 4 **Using a Table of Conformal Mappings**

Use the conformal mappings in Appendix D to find a conformal mapping between the strip $0 \leq y \leq 2$ and the upper half-plane $v \geq 0$. What is the image of the negative x-axis?

SOLUTION An appropriate mapping may be obtained directly from entry H-2. Letting $a = 2$ then $f(z) = e^{\pi z/2}$ and noting the positions of E, D, E', and D' in the figure, we can map the negative x-axis onto the interval $(0, 1)$ on the u-axis. ≡

EXAMPLE 5 **Using a Table of Conformal Mappings**

Use the conformal mappings in Appendix D to find a conformal mapping between the strip $0 \leq y \leq 2$ and the disk $|w| \leq 1$. What is the image of the negative x-axis?

SOLUTION Appendix D does not have an entry that maps the strip $0 \leq y \leq 2$ directly onto the disk. In Example 4, the strip was mapped by $f(z) = e^{\pi z/2}$ onto the upper half-plane and, from entry C-4, the complex mapping $w = \dfrac{i - \zeta}{i + \zeta}$ maps the upper half-plane to the disk $|w| \leq 1$. Therefore, $w = g(f(z)) = \dfrac{i - e^{\pi z/2}}{i + e^{\pi z/2}}$ maps the strip $0 \leq y \leq 2$ onto the disk $|w| \leq 1$.

The negative x-axis is first mapped to the interval $(0, 1)$ in the ζ-plane, and from the position of points C and C' in C-4, the interval $(0, 1)$ is mapped to the circular arc $w = e^{i\theta}$, $0 < \theta < \pi/2$, in the w-plane. ≡

Harmonic Functions and the Dirichlet Problem
A bounded harmonic function $u = u(x, y)$ that takes on prescribed values on the entire boundary of a region R is called a solution to a **Dirichlet problem** on R. In Chapters 13–15 we introduced a number of techniques for solving Laplace's equation in the plane, and we interpreted the solution to a Dirichlet problem as the steady-state temperature distribution in the interior of R that results from the fixed temperatures on the boundary.

There are at least two disadvantages to the Fourier series and integral transform methods presented in Chapters 13–15. The methods work only for simple regions in the plane and the solutions typically take the form of either infinite series or improper integrals. As such, they are difficult to evaluate. In Section 17.5 we saw that the real and imaginary parts of an analytic function are both harmonic. Since we have a large stockpile of analytic functions, we can find closed-form solutions to many Dirichlet problems and use these solutions to sketch the isotherms and lines of flow of the temperature distribution.

We will next show how conformal mappings can be used to solve a Dirichlet problem in a region R once the solution to the corresponding Dirichlet problem in the image region R' is known. The method depends on the following theorem.

THEOREM 20.2.2 Transformation Theorem for Harmonic Functions

Let f be an analytic function that maps a domain D onto a domain D'. If U is harmonic in D', then the real-valued function $u(x, y) = U(f(z))$ is harmonic in D.

PROOF: We will give a proof for the special case in which D' is simply connected. If U has a harmonic conjugate V in D', then $H = U + iV$ is analytic in D', and so the composite function

$H(f(z)) = U(f(z)) + iV(f(z))$ is analytic in D. By Theorem 17.5.3, it follows that the real part $U(f(z))$ is harmonic in D, and the proof is complete.

To establish that U has a harmonic conjugate, let $g(z) = \partial U/\partial x - i\,\partial U/\partial y$. The first Cauchy–Riemann equation $(\partial/\partial x)(\partial U/\partial x) = (\partial/\partial y)(-\partial U/\partial y)$ is equivalent to Laplace's equation $\partial^2 U/\partial x^2 + \partial^2 U/\partial y^2 = 0$, which is satisfied because U is harmonic in D'. The second Cauchy–Riemann equation $(\partial/\partial y)(\partial U/\partial x) = -(\partial/\partial x)(-\partial U/\partial y)$ is equivalent to the equality of the second-order mixed partial derivatives. Therefore, $g(z)$ is analytic in the simply connected domain D' and so, by Theorem 18.3.3, has an antiderivative $G(z)$. If $G(z) = U_1 + iV_1$, then $g(z) = G'(z) = \partial U_1/\partial x - i\,\partial U_1/\partial y$. Since $g(z) = \partial U/\partial x - i\,\partial U/\partial y$, it follows that U and U_1 have equal first partial derivatives. Therefore, $H = U + iV_1$ is analytic in D', and so U has a harmonic conjugate in D'. ≡

Theorem 20.2.2 can be used to solve a Dirichlet problem in a region R by transforming the problem to a region R' in which the solution U either is apparent or has been found by prior methods (including the Fourier series and integral transform methods of Chapters 13–15). The key steps are summarized next.

Solving Dirichlet Problems Using Conformal Mapping

1. Find a conformal mapping $w = f(z)$ that transforms the original region R onto the image region R'. The region R' may be a region for which many explicit solutions to Dirichlet problems are known.
2. Transfer the boundary conditions from the boundary of R to the boundary of R'. The value of u at a boundary point ξ of R is assigned as the value of U at the corresponding boundary point $f(\xi)$. See **FIGURE 20.2.4** for an illustration of transferring boundary conditions.

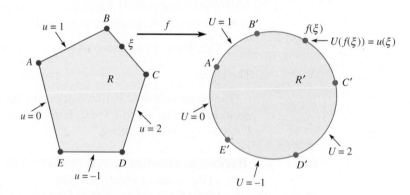

FIGURE 20.2.4 R' is image of R under a conformal mapping f

3. Solve the corresponding Dirichlet problem in R'. The solution U may be apparent from the simplicity of the problem in R' or may be found using Fourier or integral transform methods. (Additional methods will be presented in Sections 20.3 and 20.5.)
4. The solution to the original Dirichlet problem is $u(x, y) = U(f(z))$.

FIGURE 20.2.5 Image of semi-infinite vertical strip in Example 6

EXAMPLE 6 **Solving a Dirichlet Problem**

The function $U(u, v) = (1/\pi)\text{Arg } w$ is harmonic in the upper half-plane $v > 0$ since it is the imaginary part of the analytic function $g(w) = (1/\pi)\text{Ln } w$. Use this function to solve the Dirichlet problem in **FIGURE 20.2.5(a)**.

SOLUTION The analytic function $f(z) = \sin z$ maps the original region to the upper half-plane $v \geq 0$ and maps the boundary segments to the segments shown in Figure 20.2.5(b). The harmonic function $U(u, v) = (1/\pi)\text{Arg } w$ satisfies the transferred boundary conditions $U(u, 0) = 0$ for $u > 0$ and $U(u, 0) = 1$ for $u < 0$. Therefore, $u(x, y) = U(\sin z) = (1/\pi)\text{Arg}(\sin z)$ is the solution to the original problem. If $\tan^{-1}(v/u)$ is chosen to lie between 0 and π, the solution can also be written as

$$u(x, y) = \frac{1}{\pi}\tan^{-1}\left(\frac{\cos x \sinh y}{\sin x \cosh y}\right).$$ ≡

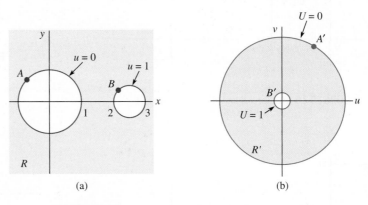

EXAMPLE 7 **Solving a Dirichlet Problem**

From C-1 in Appendix D of conformal mappings, the analytic function $f(z) = (z - a)/(az - 1)$, where $a = (7 + 2\sqrt{6})/5$, maps the region outside the two open disks $|z| < 1$ and $|z - \frac{5}{2}| < \frac{1}{2}$ onto the annular region $r_0 \leq |w| \leq 1$, where $r_0 = 5 - 2\sqrt{6}$. **FIGURE 20.2.6(a)** shows the original Dirichlet problem, and Figure 20.2.6(b) shows the transferred boundary conditions.

(a) (b)

FIGURE 20.2.6 Image of Dirichlet problem in Example 7

In Problem 12 in Exercises 14.1, we discovered that $U(r, \theta) = (\log_e r)/(\log_e r_0)$ is the solution to the new Dirichlet problem. From Theorem 20.2.2 we can conclude that the solution to the original boundary-value problem is

FIGURE 20.2.7 Image of a Dirichlet problem

$$u(x, y) = \frac{1}{\log_e(5 - 2\sqrt{6})} \log_e \left| \frac{z - (7 + 2\sqrt{6})/5}{(7 + 2\sqrt{6})z/5 - 1} \right|. \qquad \equiv$$

A favorite image region R' for a simply connected region R is the upper half-plane $y \geq 0$. For any real number a, the complex function $\text{Ln}(z - a) = \log_e|z - a| + i\,\text{Arg}(z - a)$ is analytic in R'. Therefore, $\text{Arg}(z - a)$ is harmonic in R' and is a solution to the Dirichlet problem shown in **FIGURE 20.2.7**.

It follows that the solution in R' to the Dirichlet problem with

$$U(x, 0) = \begin{cases} c_0, & a < x < b \\ 0, & \text{otherwise} \end{cases}$$

is the harmonic function $U(x, y) = (c_0/\pi)(\text{Arg}(z - b) - \text{Arg}(z - a))$. A large number of Dirichlet problems in the upper half-plane $y \geq 0$ can be solved by adding together harmonic functions of this form.

20.2 Exercises Answers to selected odd-numbered problems begin on page ANS-48.

In Problems 1–6, determine where the given complex mapping is conformal.

1. $f(z) = z^3 - 3z + 1$ **2.** $f(z) = \cos z$

3. $f(z) = z + e^z + 1$ **4.** $f(z) = z + \text{Ln}\,z + 1$

5. $f(z) = (z^2 - 1)^{1/2}$

6. $f(z) = \pi i - \frac{1}{2}[\text{Ln}(z + 1) + \text{Ln}(z - 1)]$

In Problems 7–10, use the results in Examples 2 and 3.

7. Use the identity $\cos z = \sin(\pi/2 - z)$ to find the image of the strip $0 \leq x \leq \pi$ under the complex mapping $w = \cos z$. What is the image of a horizontal line in the strip?

8. Use the identity $\sinh z = -i \sin(iz)$ to find the image of the strip $-\pi/2 \leq y \leq \pi/2$, $x \geq 0$, under the complex mapping $w = \sinh z$. What is the image of a vertical line segment in the strip?

9. Find the image of the region defined by $-\pi/2 \leq x \leq \pi/2$, $y \geq 0$, under the complex mapping $w = (\sin z)^{1/4}$. What is the image of the line segment $[-\pi/2, \pi/2]$ on the x-axis?

10. Find the image of the region $|z| \leq 1$ in the upper half-plane under the complex mapping $w = z + 1/z$. What is the image of the line segment $[-1, 1]$ on the x-axis?

In Problems 11–18, use the conformal mappings in Appendix D to find a conformal mapping from the given region R in the z-plane onto the target region R' in the w-plane, and find the image of the given boundary curve.

11.

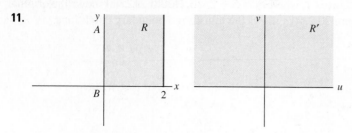

FIGURE 20.2.8 Regions R and R' for Problem 11

12.

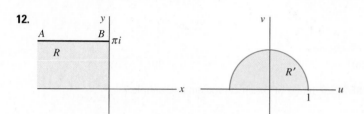

FIGURE 20.2.9 Regions R and R' for Problem 12

13.

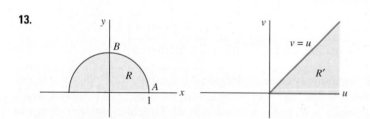

FIGURE 20.2.10 Regions R and R' for Problem 13

14.

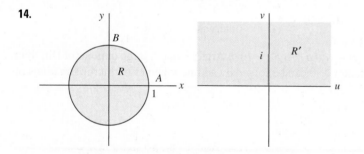

FIGURE 20.2.11 Regions R and R' for Problem 14

15.

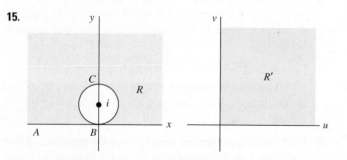

FIGURE 20.2.12 Regions R and R' for Problem 15

16.

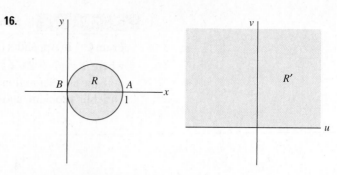

FIGURE 20.2.13 Regions R and R' for Problem 16

17.

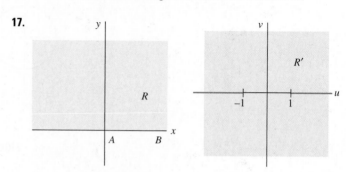

FIGURE 20.2.14 Regions R and R' for Problem 17

18.

FIGURE 20.2.15 Regions R and R' for Problem 18

In Problems 19–22, use an appropriate conformal mapping and the harmonic function $U = (1/\pi)\text{Arg } w$ to solve the given Dirichlet problem.

19.

FIGURE 20.2.16 Dirichlet problem in Problem 19

20.

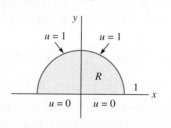

FIGURE 20.2.17 Dirichlet problem in Problem 20

21.

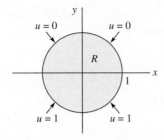

FIGURE 20.2.18 Dirichlet problem in Problem 21

22.

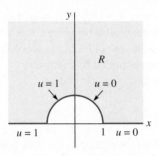

FIGURE 20.2.19 Dirichlet problem in Problem 22

In Problems 23–26, use an appropriate conformal mapping and the harmonic function $U = (c_0/\pi)[\text{Arg}(w - 1) - \text{Arg}(w + 1)]$ to solve the given Dirichlet problem.

23.

FIGURE 20.2.20 Dirichlet problem in Problem 23

24.

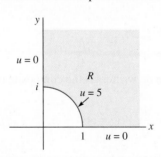

FIGURE 20.2.21 Dirichlet problem in Problem 24

25.

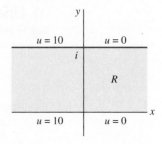

FIGURE 20.2.22 Dirichlet problem in Problem 25

26.

FIGURE 20.2.23 Dirichlet problem in Problem 26

27. A real-valued function $\phi(x, y)$ is called **biharmonic** in a domain D when the fourth-order differential equation

$$\frac{\partial^4 \phi}{\partial x^4} + 2 \frac{\partial^4 \phi}{\partial x^2 \partial y^2} + \frac{\partial^4 \phi}{\partial y^4} = 0$$

at all points in D. Examples of biharmonic functions are the Airy stress function in the mechanics of solids and velocity potentials in the analysis of viscous fluid flow.
 (a) Show that if ϕ is biharmonic in D, then $u = \partial^2 \phi/\partial x^2 + \partial^2 \phi/\partial y^2$ is harmonic in D.
 (b) If $g(z)$ is analytic in D and $\phi(x, y) = \text{Re}(\bar{z} g(z))$, show that ϕ is biharmonic in D.

20.3 Linear Fractional Transformations

INTRODUCTION In many applications that involve boundary-value problems associated with Laplace's equation, it is necessary to find a conformal mapping that maps a disk onto the half-plane $v \geq 0$. Such a mapping would have to map the circular boundary of the disk to the boundary line of the half-plane. An important class of elementary conformal mappings that map circles to lines (and vice versa) are the fractional transformations. In this section we will define and study this special class of mappings.

Linear Fractional Transformation

If a, b, c, and d are complex constants with $ad - bc \neq 0$, then the complex function defined by

$$T(z) = \frac{az + b}{cz + d}$$

is called a **linear fractional transformation**. Since

$$T'(z) = \frac{ad - bc}{(cz + d)^2},$$

T is conformal at z provided $\Delta = ad - bc \neq 0$ and $z \neq -d/c$. (If $\Delta = 0$, then $T'(z) = 0$ and $T(z)$ would be a constant function.) Linear fractional transformations are circle preserving in a sense that we will make precise in this section, and, as we saw in Example 7 of Section 20.2, they can be useful in solving Dirichlet problems in regions bounded by circles.

Note that when $c \neq 0$, $T(z)$ has a simple pole at $z_0 = -d/c$ and so

$$\lim_{z \to z_0} |T(z)| = \infty.$$

We will write $T(z_0) = \infty$ as shorthand for this limit. In addition, if $c \neq 0$, then

$$\lim_{|z| \to \infty} T(z) = \lim_{|z| \to \infty} \frac{a + b/z}{c + d/z} = \frac{a}{c},$$

and we write $T(\infty) = a/c$.

> **EXAMPLE 1** A Linear Fractional Transformation

If $T(z) = (2z + 1)/(z - i)$, compute $T(0)$, $T(\infty)$, and $T(i)$.

SOLUTION Note that $T(0) = 1/(-i) = i$ and $T(\infty) = \lim_{|z| \to \infty} T(z) = 2$. Since $z = i$ is a simple pole for $T(z)$, we have $\lim_{z \to i} |T(z)| = \infty$ and we write $T(i) = \infty$. ≡

Circle-Preserving Property

If $c = 0$, the linear fractional transformation reduces to a linear function $T(z) = Az + B$. We saw in Section 20.1 that such a complex mapping can be considered as the composite of a rotation, magnification, and translation. As such, a linear function will map a circle in the z-plane to a circle in the w-plane. When $c \neq 0$, we can divide $cz + d$ into $az + b$ to obtain

$$w = \frac{az + b}{cz + d} = \frac{bc - ad}{c} \frac{1}{cz + d} + \frac{a}{c}. \tag{1}$$

If we let $A = (bc - ad)/c$ and $B = a/c$, $T(z)$ can be written as the composite of transformations:

$$z_1 = cz + d, \quad z_2 = \frac{1}{z_1}, \quad w = Az_2 + B. \tag{2}$$

A general linear fractional transformation can therefore be written as the composite of two linear functions and the inversion $w = 1/z$. Note that if $|z - z_1| = r$ and $w = 1/z$, then

$$\left| \frac{1}{w} - \frac{1}{w_1} \right| = \frac{|w - w_1|}{|w||w_1|} = r \quad \text{or} \quad |w - w_1| = (r|w_1|)|w - 0|. \tag{3}$$

It is not hard to show that the set of all points w that satisfy

$$|w - w_1| = \lambda |w - w_2| \tag{4}$$

is a line when $\lambda = 1$ and is a circle when $\lambda > 0$ and $\lambda \neq 1$. It follows from (3) that the image of the circle $|z - z_1| = r$ under the inversion $w = 1/z$ is a circle except when $r = 1/|w_1| = |z_1|$. In the

latter case, the original circle passes through the origin and the image is a line. See Figure 20.1.3. From (2), we can deduce the following theorem.

THEOREM 20.3.1 Circle-Preserving Property

A linear fractional transformation maps a circle in the z-plane to either a line or a circle in the w-plane. The image is a line if and only if the original circle passes through a pole of the linear fractional transformation.

PROOF: We have shown that a linear function maps a circle to a circle, whereas an inversion maps a circle to a circle or a line. It follows from (2) that a circle in the z-plane will be mapped to either a circle or a line in the w-plane. If the original circle passes through a pole z_0, then $T(z_0) = \infty$, and so the image is unbounded. Therefore, the image of such a circle must be a line. If the original circle does not pass through z_0, then the image is bounded and must be a circle. ≡

EXAMPLE 2 Images of Circles

Find the images of the circles $|z| = 1$ and $|z| = 2$ under $T(z) = (z + 2)/(z - 1)$. What are the images of the interiors of these circles?

SOLUTION The circle $|z| = 1$ passes through the pole $z_0 = 1$ of the linear fractional transformation and so the image is a line. Since $T(-1) = -\frac{1}{2}$ and $T(i) = -\frac{1}{2} - \frac{3}{2}i$, we can conclude that the image is the line $u = -\frac{1}{2}$. The image of the interior $|z| < 1$ is either the half-plane $u < -\frac{1}{2}$ or the half-plane $u > -\frac{1}{2}$. Using $z = 0$ as a *test point*, $T(0) = -2$, and so the image is the half-plane $u < -\frac{1}{2}$.

The circle $|z| = 2$ does not pass through the pole and so the image is a circle. For $|z| = 2$,

$$|\bar{z}| = 2 \quad \text{and} \quad \overline{T(z)} = \overline{\frac{z + 2}{z - 1}} = \frac{\bar{z} + 2}{\bar{z} - 1} = T(\bar{z}).$$

Therefore, $\overline{T(z)}$ is a point on the image circle and so the image circle is symmetric with respect to the u-axis. Since $T(-2) = 0$ and $T(2) = 4$, the center of the circle is $w = 2$ and the image is the circle $|w - 2| = 2$. See **FIGURE 20.3.1**. The image of the interior $|z| < 2$ is either the interior or the exterior of the image circle $|w - 2| = 2$. Since $T(0) = -2$, we can conclude that the image is $|w - 2| > 2$.

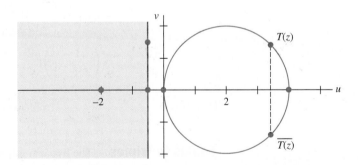

FIGURE 20.3.1 Images of test points in Example 2 ≡

Constructing Special Mappings In order to use linear fractional transformations to solve Dirichlet problems, we must construct special functions that map a given circular region R to a target region R' in which the corresponding Dirichlet problem is solvable. Since a circular boundary is determined by three of its points, we must find a linear fractional transformation $w = T(z)$ that maps three given points z_1, z_2, and z_3 on the boundary of R to three points w_1, w_2, and w_3 on the boundary of R'. In addition, the interior of R' must be the image of the interior of R. See **FIGURE 20.3.2**.

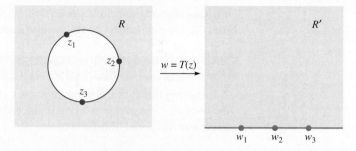

FIGURE 20.3.2 R' is image of R under T

Matrix Methods

Matrix methods can be used to simplify many of the computations. We can associate the matrix

$$\mathbf{A} = \begin{pmatrix} a & b \\ c & d \end{pmatrix}$$

with $T(z) = (az + b)/(cz + d)$.* If $T_1(z) = (a_1z + b_1)/(c_1z + d_1)$ and $T_2(z) = (a_2z + b_2)/(c_2z + d_2)$, then the composite function $T(z) = T_2(T_1(z))$ is given by $T(z) = (az + b)/(cz + d)$, where

$$\begin{pmatrix} a & b \\ c & d \end{pmatrix} = \begin{pmatrix} a_2 & b_2 \\ c_2 & d_2 \end{pmatrix}\begin{pmatrix} a_1 & b_1 \\ c_1 & d_1 \end{pmatrix}. \tag{5}$$

If $w = T(z) = (az + b)/(cz + d)$, we can solve for z to obtain $z = (dw - b)/(-cw + a)$. Therefore the inverse of the linear fractional transformation T is $T^{-1}(w) = (dw - b)/(-cw + a)$ and we associate the adjoint of the matrix $\mathbf{A}$,

$$\text{adj } \mathbf{A} = \begin{pmatrix} d & -b \\ -c & a \end{pmatrix}, \tag{6}$$

with T^{-1}. The matrix adj $\mathbf{A}$ is the adjoint matrix of $\mathbf{A}$ (see Section 8.6), the matrix for T.

EXAMPLE 3 Using Matrices to Find an Inverse Transform

If $T(z) = \dfrac{2z - 1}{z + 2}$ and $S(z) = \dfrac{z - i}{iz - 1}$, find $S^{-1}(T(z))$.

SOLUTION From (5) and (6), we have $S^{-1}(T(z)) = (az + b)/(cz + d)$, where

$$\begin{pmatrix} a & b \\ c & d \end{pmatrix} = \text{adj}\begin{pmatrix} 1 & -i \\ i & -1 \end{pmatrix}\begin{pmatrix} 2 & -1 \\ 1 & 2 \end{pmatrix}$$

$$= \begin{pmatrix} -1 & i \\ -i & 1 \end{pmatrix}\begin{pmatrix} 2 & -1 \\ 1 & 2 \end{pmatrix} = \begin{pmatrix} -2 + i & 1 + 2i \\ 1 - 2i & 2 + i \end{pmatrix}.$$

Therefore, $$S^{-1}(T(z)) = \frac{(-2 + i)z + 1 + 2i}{(1 - 2i)z + 2 + i}. \qquad\qquad \equiv$$

Triples to Triples

The linear fractional transformation

$$T(z) = \frac{z - z_1}{z - z_3}\frac{z_2 - z_3}{z_2 - z_1}$$

has a zero at $z = z_1$, a pole at $z = z_3$, and $T(z_2) = 1$. Therefore, $T(z)$ maps three *distinct* complex numbers z_1, z_2, and z_3 to 0, 1, and ∞, respectively. The term $\dfrac{z - z_1}{z - z_3}\dfrac{z_2 - z_3}{z_2 - z_1}$ is called the **cross-ratio** of the complex numbers z, z_1, z_2, and z_3.

* The matrix $\mathbf{A}$ is not unique since the numerator and denominator in $T(z)$ can be multiplied by a nonzero constant.

Likewise, the complex mapping $S(w) = \dfrac{w - w_1}{w - w_3} \dfrac{w_2 - w_3}{w_2 - w_1}$ sends w_1, w_2, and w_3 to 0, 1, and ∞, and so S^{-1} maps 0, 1, and ∞ to w_1, w_2, and w_3. It follows that the linear fractional transformation $w = S^{-1}(T(z))$ maps the triple z_1, z_2, and z_3 to w_1, w_2, and w_3. From $w = S^{-1}(T(z))$, we have $S(w) = T(z)$ and we can conclude that

$$\frac{w - w_1}{w - w_3} \frac{w_2 - w_3}{w_2 - w_1} = \frac{z - z_1}{z - z_3} \frac{z_2 - z_3}{z_2 - z_1}. \tag{7}$$

In constructing a linear fractional transformation that maps the triple z_1, z_2, and z_3 to w_1, w_2, and w_3, we can use matrix methods to compute $w = S^{-1}(T(z))$. Alternatively, we can substitute into (7) and solve the resulting equation for w.

EXAMPLE 4 **Constructing a Linear Fractional Transformation**

Construct a linear fractional transformation that maps the points 1, i, and -1 on the circle $|z| = 1$ to the points -1, 0, and 1 on the real axis.

SOLUTION Substituting into (7), we have

$$\frac{w + 1}{w - 1} \frac{0 - 1}{0 - (-1)} = \frac{z - 1}{z + 1} \frac{i + 1}{i - 1}$$

or

$$-\frac{w + 1}{w - 1} = -i \frac{z - 1}{z + 1}.$$

Solving for w, we obtain $w = -i(z - i)/(z + i)$. Alternatively, we could use the matrix method to compute $w = S^{-1}(T(z))$. ≡

When $z_k = \infty$ plays the role of one of the points in a triple, the definition of the cross-ratio is changed by replacing each factor that contains z_k by 1. For example, if $z_2 = \infty$, both $z_2 - z_3$ and $z_2 - z_1$ are replaced by 1, giving $(z - z_1)/(z - z_3)$ as the cross-ratio.

EXAMPLE 5 **Constructing a Linear Fractional Transformation**

Construct a linear fractional transformation that maps the points ∞, 0, and 1 on the real axis to the points 1, i, and -1 on the circle $|w| = 1$.

SOLUTION Since $z_1 = \infty$, the terms $z - z_1$ and $z_2 - z_1$ in the cross product are replaced by 1. It follows that

$$\frac{w - 1}{w + 1} \frac{i + 1}{i - 1} = \frac{1}{z - 1} \frac{0 - 1}{1}$$

or

$$S(w) = -i \frac{w - 1}{w + 1} = \frac{-1}{z - 1} = T(z).$$

If we use the matrix method to find $w = S^{-1}(T(z))$, then

$$\begin{pmatrix} a & b \\ c & d \end{pmatrix} = \text{adj} \begin{pmatrix} -i & i \\ 1 & 1 \end{pmatrix} \begin{pmatrix} 0 & -1 \\ 1 & -1 \end{pmatrix} = \begin{pmatrix} -i & -1 + i \\ -i & 1 + i \end{pmatrix}$$

and so $w = \dfrac{-iz - 1 + i}{-iz + 1 + i} = \dfrac{z - 1 - i}{z - 1 + i}.$ ≡

EXAMPLE 6 **Solving a Dirichlet Problem**

Solve the Dirichlet problem in **FIGURE 20.3.3(a)** using conformal mapping by constructing a linear fractional transformation that maps the given region into the upper half-plane.

SOLUTION The boundary circles $|z| = 1$ and $|z - \frac{1}{2}| = \frac{1}{2}$ each pass through $z = 1$. We can therefore map each boundary circle to a line by selecting a linear fractional transformation that has $z = 1$ as a pole. If we further require that $T(i) = 0$ and $T(-1) = 1$, then

$$T(z) = \frac{z - i}{z - 1} \frac{-1 - 1}{-1 - i} = (1 - i) \frac{z - i}{z - 1}.$$

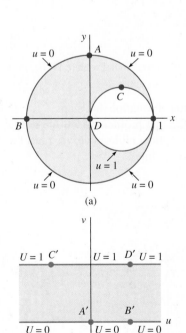

FIGURE 20.3.3 Image of Dirichlet problem in Example 6

Since $T(0) = 1 + i$ and $T(\frac{1}{2} + \frac{1}{2}i) = -1 + i$, T maps the interior of the circle $|z| = 1$ onto the upper half-plane and maps the circle $|z - \frac{1}{2}| = \frac{1}{2}$ onto the line $v = 1$. Figure 20.3.3(b) shows the transferred boundary conditions.

The harmonic function $U(u, v) = v$ is the solution to the simplified Dirichlet problem in the w-plane, and so, by Theorem 20.2.2, $u(x, y) = U(T(z))$ is the solution to the original Dirichlet problem in the z-plane.

Since the imaginary part of $T(z) = (1 - i)\dfrac{z - i}{z - 1}$ is $\dfrac{1 - x^2 - y^2}{(x - 1)^2 + y^2}$, the solution is given by

$$u(x, y) = \frac{1 - x^2 - y^2}{(x - 1)^2 + y^2}.$$

The level curves $u(x, y) = c$ can be written as

$$\left(x - \frac{c}{1 + c}\right)^2 + y^2 = \left(\frac{1}{1 + c}\right)^2$$

and are therefore circles that pass through $z = 1$. See **FIGURE 20.3.4**. These level curves may be interpreted as the isotherms of the steady-state temperature distribution induced by the boundary temperatures.

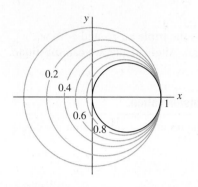

FIGURE 20.3.4 Circles are level curves in Example 6

20.3 Exercises Answers to selected odd-numbered problems begin on page ANS-49.

In Problems 1–4, a linear fractional transformation is given.

(a) Compute $T(0)$, $T(1)$, and $T(\infty)$.
(b) Find the images of the circles $|z| = 1$ and $|z - 1| = 1$.
(c) Find the image of the disk $|z| \le 1$.

1. $T(z) = \dfrac{i}{z}$

2. $T(z) = \dfrac{1}{z - 1}$

3. $T(z) = \dfrac{z + 1}{z - 1}$

4. $T(z) = \dfrac{z - i}{z}$

In Problems 5–8, use the matrix method to compute $S^{-1}(w)$ and $S^{-1}(T(z))$ for each pair of linear fractional transformations.

5. $T(z) = \dfrac{z}{iz - 1}$ and $S(z) = \dfrac{iz + 1}{z - 1}$

6. $T(z) = \dfrac{iz}{z - 2i}$ and $S(z) = \dfrac{2z + 1}{z + 1}$

7. $T(z) = \dfrac{2z - 3}{z - 3}$ and $S(z) = \dfrac{z - 2}{z - 1}$

8. $T(z) = \dfrac{z - 1 + i}{iz - 2}$ and $S(z) = \dfrac{(2 - i)z}{z - 1 - i}$

In Problems 9–16, construct a linear fractional transformation that maps the given triple z_1, z_2, and z_3 to the triple w_1, w_2, and w_3.

9. $-1, 0, 2$ to $0, 1, \infty$

10. $i, 0, -i$ to $0, 1, \infty$

11. $0, 1, \infty$ to $0, i, 2$

12. $0, 1, \infty$ to $1 + i, 0, 1 - i$

13. $-1, 0, 1$ to $i, \infty, 0$

14. $-1, 0, 1$ to $\infty, -i, 1$

15. $1, i, -i$ to $-1, 0, 3$

16. $1, i, -i$ to $i, -i, 1$

17. Use the results in Example 2 and the harmonic function $U = (\log_e r)/(\log_e r_0)$ to solve the Dirichlet problem in **FIGURE 20.3.5**. Explain why the level curves must be circles.

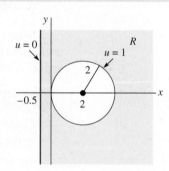

FIGURE 20.3.5 Dirichlet problem in Problem 17

18. Use the linear fractional transformation that maps $-1, 1, 0$ to $0, 1, \infty$ to solve the Dirichlet problem in **FIGURE 20.3.6**. Explain why, with one exception, all level curves must be circles. Which level curve is a line?

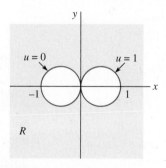

FIGURE 20.3.6 Dirichlet problem in Problem 18

19. Derive the conformal mapping H-1 in the conformal mappings in Appendix D.

20. Derive the conformal mapping H-5 in the conformal mappings in Appendix D by first mapping $1, i, -1$ to $\infty, i, 0$.

21. Show that the composite of two linear fractional transformations is a linear fractional transformation and verify (5).

22. If $w_1 \neq w_2$ and $\lambda > 0$, show that the set of all points w that satisfy $|w - w_1| = \lambda|w - w_2|$ is a line when $\lambda = 1$ and is a circle when $\lambda \neq 1$. [*Hint:* Write as $|w - w_1|^2 = \lambda^2|w - w_2|^2$ and expand.]

20.4 Schwarz–Christoffel Transformations

INTRODUCTION If D' is a simply connected domain with at least one boundary point, then the famous **Riemann mapping theorem** asserts the existence of an analytic function g that conformally maps the unit open disk $|z| < 1$ onto D'. The Riemann mapping theorem is a pure existence theorem that does not specify a formula for the conformal mapping. Since the upper half-plane $y > 0$ can be conformally mapped onto this disk using a linear fractional transformation, it follows that there exists a conformal mapping f between the upper half-plane and D'. In particular, there are analytic functions that map the upper half-plane onto polygonal regions of the types shown in **FIGURE 20.4.1**. Unlike the Riemann mapping theorem, the **Schwarz–Christoffel formula** specifies a form for the derivative $f'(z)$ of a conformal mapping from the upper half-plane to a bounded or unbounded polygonal region. This formula is named for the Polish mathematician **Karl Hermann Amandus Schwarz** (1843–1921) and the German mathematician/physicist **Elwin Bruno Christoffel** (1829–1900), who discovered it independently in the late 1860s.

(a) Bounded region

(b) Unbounded region

FIGURE 20.4.1 Polygonal regions

■ Special Cases To motivate the general Schwarz–Christoffel formula, we first examine the effect of the mapping $f(z) = (z - x_1)^{\alpha/\pi}$, $0 < \alpha < 2\pi$, on the upper half-plane $y \geq 0$ shown in **FIGURE 20.4.2(a)**. This mapping is the composite of the translation $\zeta = z - x_1$ and the real power function $w = \zeta^{\alpha/\pi}$. Since $w = \zeta^{\alpha/\pi}$ changes the angle in a wedge by a factor of α/π, the interior angle in the image region is $(\alpha/\pi)\pi = \alpha$. See Figure 20.4.2(b).

Note that $f'(z) = A(z - x_1)^{(\alpha/\pi)-1}$ for $A = \alpha/\pi$. Next assume that $f(z)$ is a function that is analytic in the upper half-plane and that has the derivative

$$f'(z) = A(z - x_1)^{(\alpha_1/\pi)-1}(z - x_2)^{(\alpha_2/\pi)-1}, \tag{1}$$

where $x_1 < x_2$. In determining the images of line segments on the x-axis, we will use the fact that a curve $w = w(t)$ in the w-plane is a line segment when the argument of its tangent vector $w'(t)$ is constant. From (1), an argument of $f'(t)$ is given by

$$\arg f'(t) = \operatorname{Arg} A + \left(\frac{\alpha_1}{\pi} - 1\right)\operatorname{Arg}(t - x_1) + \left(\frac{\alpha_2}{\pi} - 1\right)\operatorname{Arg}(t - x_2). \tag{2}$$

Since $\operatorname{Arg}(t - x) = \pi$ for $t < x$, we can find the variation of $\arg f'(t)$ along the x-axis. The results are shown in the following table.

(a)

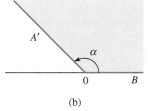

(b)

FIGURE 20.4.2 Image of upper half-plane

Interval	$\arg f'(t)$	Change in Argument
$(-\infty, x_1)$	$\operatorname{Arg} A + (\alpha_1 - \pi) + (\alpha_2 - \pi)$	0
(x_1, x_2)	$\operatorname{Arg} A + (\alpha_2 - \pi)$	$\pi - \alpha_1$
(x_2, ∞)	$\operatorname{Arg} A$	$\pi - \alpha_2$

Since $\arg f'(t)$ is constant on the intervals in the table, the images are line segments, and **FIGURE 20.4.3** shows the image of the upper half-plane. Note that the interior angles of the polygonal image region are α_1 and α_2. This discussion generalizes to produce the Schwarz–Christoffel formula.

THEOREM 20.4.1 Schwarz–Christoffel Formula

Let $f(z)$ be a function that is analytic in the upper half-plane $y > 0$ and that has the derivative

$$f'(z) = A(z - x_1)^{(\alpha_1/\pi)-1}(z - x_2)^{(\alpha_2/\pi)-1} \cdots (z - x_n)^{(\alpha_n/\pi)-1}, \tag{3}$$

where $x_1 < x_2 < \cdots < x_n$ and each α_i satisfies $0 < \alpha_i < 2\pi$. Then $f(z)$ maps the upper half-plane $y \geq 0$ to a polygonal region with interior angles $\alpha_1, \alpha_2, \ldots, \alpha_n$.

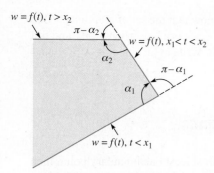

$w = f(t), t > x_2$

$\pi - \alpha_2$

$w = f(t), x_1 < t < x_2$

α_2

$\pi - \alpha_1$

α_1

$w = f(t), t < x_1$

FIGURE 20.4.3 Image of upper half-plane

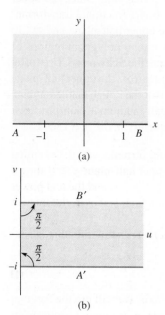

(a)

(b)

FIGURE 20.4.4 Image of upper half-plane in Example 1

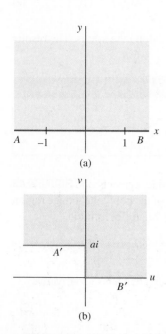

(a)

(b)

FIGURE 20.4.5 Image of upper half-plane in Example 2

In applying formula (3) to a particular polygonal target region, the reader should carefully note the following comments:

(*i*) One can select the location of three of the points x_k on the *x*-axis. A judicious choice can simplify the computation of $f(z)$. The selection of the remaining points depends on the shape of the target polygon.

(*ii*) A general formula for $f(z)$ is

$$f(z) = A\left(\int (z - x_1)^{(\alpha_1/\pi) - 1} (z - x_2)^{(\alpha_2/\pi) - 1} \cdots (z - x_n)^{(\alpha_n/\pi) - 1}\, dz \right) + B,$$

and therefore $f(z)$ may be considered as the composite of the conformal mapping

$$g(z) = \int (z - x_1)^{(\alpha_1/\pi) - 1} (z - x_2)^{(\alpha_2/\pi) - 1} \cdots (z - x_n)^{(\alpha_n/\pi) - 1}\, dz$$

and the linear function $w = Az + B$. The linear function $w = Az + B$ allows us to magnify, rotate, and translate the image polygon produced by $g(z)$. (See Section 20.1.)

(*iii*) If the polygonal region is bounded, only $n - 1$ of the n interior angles should be included in the Schwarz–Christoffel formula. As an illustration, the interior angles α_1, α_2, α_3, and α_4 are sufficient to determine the Schwarz–Christoffel formula for the pentagon shown in Figure 20.4.1(a).

EXAMPLE 1 **Constructing a Conformal Mapping**

Use the Schwarz–Christoffel formula to construct a conformal mapping from the upper half-plane to the strip $|v| \leq 1$, $u \geq 0$.

SOLUTION We may select $x_1 = -1$ and $x_2 = 1$ on the *x*-axis, and we will construct a conformal mapping f with $f(-1) = -i$ and $f(1) = i$. See **FIGURE 20.4.4**. Since $\alpha_1 = \alpha_2 = \pi/2$, the Schwarz–Christoffel formula (3) gives

$$f'(z) = A(z + 1)^{-1/2}(z - 1)^{-1/2} = A\frac{1}{(z^2 - 1)^{1/2}} = \frac{A}{i}\frac{1}{(1 - z^2)^{1/2}}.$$

Therefore, $f(z) = -Ai \sin^{-1} z + B$. Since $f(-1) = -i$ and $f(1) = i$, we obtain, respectively,

$$-i = Ai\frac{\pi}{2} + B \qquad \text{and} \qquad i = -Ai\frac{\pi}{2} + B$$

and conclude that $B = 0$ and $A = -2/\pi$. Thus, $f(z) = (2/\pi)i \sin^{-1} z$. ≡

EXAMPLE 2 **Constructing a Conformal Mapping**

Use the Schwarz–Christoffel formula to construct a conformal mapping from the upper half-plane to the region shown in **FIGURE 20.4.5(b)**.

SOLUTION We again select $x_1 = -1$ and $x_2 = 1$, and we will require that $f(-1) = ai$ and $f(1) = 0$. Since $\alpha_1 = 3\pi/2$ and $\alpha_2 = \pi/2$, the Schwarz–Christoffel formula (3) gives

$$f'(z) = A(z + 1)^{1/2}(z - 1)^{-1/2}.$$

If we write $f'(z)$ as $A(z/(z^2 - 1)^{1/2} + 1/(z^2 - 1)^{1/2})$, it follows that

$$f(z) = A[(z^2 - 1)^{1/2} + \cosh^{-1} z] + B.$$

Note that $\cosh^{-1}(-1) = \pi i$ and $\cosh^{-1} 1 = 0$, and so $ai = f(-1) = A(\pi i) + B$ and $0 = f(1) = B$. Therefore, $A = a/\pi$ and $f(z) = (a/\pi)[(z^2 - 1)^{1/2} + \cosh^{-1} z]$. ≡

The next example will show that it may not always be possible to find $f(z)$ in terms of elementary functions.

EXAMPLE 3 **Constructing a Conformal Mapping**

Use the Schwarz–Christoffel formula to construct a conformal mapping from the upper half-plane to the interior of the equilateral triangle shown in **FIGURE 20.4.6(b)**.

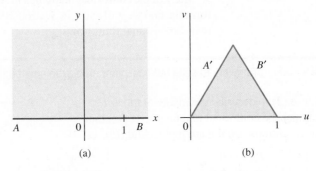

FIGURE 20.4.6 Image of upper half-plane in Example 3

SOLUTION Since the polygonal region is bounded, only two of the three 60° interior angles should be included in the Schwarz–Christoffel formula. If $x_1 = 0$ and $x_2 = 1$, we obtain $f'(z) = Az^{-2/3}(z - 1)^{-2/3}$. It is not possible to evaluate $f(z)$ in terms of elementary functions; however, we can use Theorem 18.3.3 to construct the antiderivative

$$f(z) = A \int_0^z \frac{1}{s^{2/3}(s - 1)^{2/3}} \, ds + B.$$

If we require that $f(0) = 0$ and $f(1) = 1$, it follows that $B = 0$ and

$$1 = A \int_0^1 \frac{1}{s^{2/3}(s - 1)^{2/3}} \, ds.$$

It can be shown that this last integral is $\Gamma(\tfrac{1}{3})$, where Γ denotes the gamma function. Therefore, the required conformal mapping is

$$f(z) = \frac{1}{\Gamma(\tfrac{1}{3})} \int_0^z \frac{1}{s^{2/3}(s - 1)^{2/3}} \, ds. \qquad \equiv$$

The Schwarz–Christoffel formula can sometimes be used to suggest a possible conformal mapping from the upper half-plane onto a nonpolygonal region R'. A key first step is to approximate R' by polygonal regions. This will be illustrated in the final example.

EXAMPLE 4 **Constructing a Conformal Mapping**

Use the Schwarz–Christoffel formula to construct a conformal mapping from the upper half-plane to the upper half-plane with the horizontal line $v = \pi$, $u \leq 0$, deleted.

SOLUTION The nonpolygonal target region can be approximated by a polygonal region by adjoining a line segment from $w = \pi i$ to a point u_0 on the negative u-axis. See **FIGURE 20.4.7(b)**. If we require that $f(-1) = \pi i$ and $f(0) = u_0$, the Schwarz–Christoffel transformation satisfies

$$f'(z) = A(z + 1)^{(\alpha_1/\pi) - 1} z^{(\alpha_2/\pi) - 1}.$$

Note that as u_0 approaches $-\infty$, the interior angles α_1 and α_2 approach 2π and 0, respectively. This suggests we examine conformal mappings that satisfy $w' = A(z + 1)^1 z^{-1} = A(1 + 1/z)$ or $w = A(z + \text{Ln } z) + B$.

We will first determine the image of the upper half-plane under $g(z) = z + \text{Ln } z$ and then translate the image region if needed. For t real,

$$g(t) = t + \log_e|t| + i \,\text{Arg } t.$$

If $t < 0$, $\text{Arg } t = \pi$ and $u(t) = t + \log_e|t|$ varies from $-\infty$ to -1. It follows that $w = g(t)$ moves along the line $v = \pi$ from $-\infty$ to -1. When $t > 0$, $\text{Arg } t = 0$ and $u(t)$ varies from $-\infty$ to ∞. Therefore, g maps the positive x-axis onto the u-axis. We can conclude that $g(z) = z + \text{Ln } z$ maps

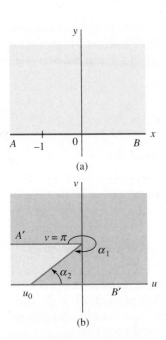

FIGURE 20.4.7 Image of upper half-plane in Example 4

the upper half-plane onto the upper half-plane with the horizontal line $v = \pi$, $u \leq -1$, deleted. Therefore, $w = z + \text{Ln } z + 1$ maps the upper half-plane onto the original target region. ≡

Many of the conformal mappings in Appendix D can be derived using the Schwarz–Christoffel formula, and we will show in Section 20.6 that these mappings are especially useful in analyzing two-dimensional fluid flows.

20.4 Exercises Answers to selected odd-numbered problems begin on page ANS-49.

In Problems 1–4, use (2) to describe the image of the upper half-plane $y \geq 0$ under the conformal mapping $w = f(z)$ that satisfies the given conditions. Do *not* attempt to find $f(z)$.

1. $f'(z) = (z - 1)^{-1/2}$, $f(1) = 0$
2. $f'(z) = (z + 1)^{-1/3}$, $f(-1) = 0$
3. $f'(z) = (z + 1)^{-1/2}(z - 1)^{1/2}$, $f(-1) = 0$
4. $f'(z) = (z + 1)^{-1/2}(z - 1)^{-3/4}$, $f(-1) = 0$

In Problems 5–8, find $f'(z)$ for the given polygonal region using $x_1 = -1$, $x_2 = 0$, $x_3 = 1$, $x_4 = 2$, and so on. Do *not* attempt to find $f(z)$.

5. $f(-1) = 0$, $f(0) = 1$

FIGURE 20.4.8 Polygonal region for Problem 5

6. $f(-1) = -1$, $f(0) = 0$

FIGURE 20.4.9 Polygonal region for Problem 6

7. $f(-1) = -1$, $f(0) = 1$

FIGURE 20.4.10 Polygonal region for Problem 7

8. $f(-1) = i$, $f(0) = 0$

FIGURE 20.4.11 Polygonal region for Problem 8

9. Use the Schwarz–Christoffel formula to construct a conformal mapping from the upper half-plane $y \geq 0$ to the region in **FIGURE 20.4.12**. Require that $f(-1) = \pi i$ and $f(1) = 0$.

FIGURE 20.4.12 Image of upper half-plane in Problem 9

10. Use the Schwarz–Christoffel formula to construct a conformal mapping from the upper half-plane $y \geq 0$ to the region in **FIGURE 20.4.13**. Require that $f(-1) = -ai$ and $f(1) = ai$.

FIGURE 20.4.13 Image of upper half-plane in Problem 10

11. Use the Schwarz–Christoffel formula to construct a conformal mapping from the upper half-plane $y \geq 0$ to the horizontal strip $0 \leq v \leq \pi$ by first approximating the strip by the polygonal region shown in **FIGURE 20.4.14**. Require that $f(-1) = \pi i$, $f(0) = w_2 = -\overline{w}_1$, and $f(1) = 0$, and let $w_1 \to \infty$ in the horizontal direction.

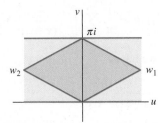

FIGURE 20.4.14 Image of upper half-plane in Problem 11

12. Use the Schwarz–Christoffel formula to construct a conformal mapping from the upper half-plane $y \geq 0$ to the wedge $0 \leq \operatorname{Arg} w \leq \pi/4$ by first approximating the wedge by the region shown in **FIGURE 20.4.15**. Require that $f(0) = 0$ and $f(1) = 1$ and let $\theta \to 0$.

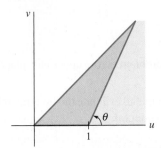

FIGURE 20.4.15 Image of upper half-plane in Problem 12

13. Verify M-4 in Appendix D by first approximating the region R' by the polygonal region shown in **FIGURE 20.4.16**. Require that $f(-1) = -u_1, f(0) = ai$, and $f(1) = u_1$ and let $u_1 \to 0$ along the u-axis.

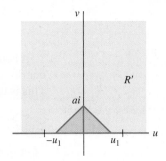

FIGURE 20.4.16 Image of upper half-plane in Problem 13

14. Show that if a curve in the w-plane is parameterized by $w = w(t)$, $a \leq t \leq b$, and $\arg w'(t)$ is constant, then the curve is a line segment. [*Hint*: If $w(t) = u(t) + iv(t)$, then $\tan(\arg w'(t)) = dv/du$.]

20.5 Poisson Integral Formulas

INTRODUCTION The success of the conformal mapping method depends on the recognition of the solution to the new Dirichlet problem in the image region R'. It would therefore be helpful if a general solution could be found for Dirichlet problems in either the upper half-plane $y \geq 0$ or the unit disk $|z| \leq 1$. The **Poisson integral formula** for the upper half-plane provides such a solution by expressing the value of a harmonic function $u(x, y)$ at a point in the interior of the upper half-plane in terms of its values on the boundary $y = 0$.

Formulas for the Upper Half-Plane
To develop the formula, we first assume that the boundary function is given by $u(x, 0) = f(x)$, where $f(x)$ is the step function indicated in **FIGURE 20.5.1**. The solution of the corresponding Dirichlet problem in the upper half-plane is

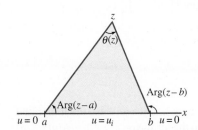

FIGURE 20.5.1 Boundary conditions on $y = 0$

$$u(x, y) = \frac{u_i}{\pi}\left[\operatorname{Arg}(z - b) - \operatorname{Arg}(z - a)\right]. \tag{1}$$

Since $\operatorname{Arg}(z - b)$ is an exterior angle in the triangle formed by z, a, and b, $\operatorname{Arg}(z - b) = \theta(z) + \operatorname{Arg}(z - a)$, where $0 < \theta(z) < \pi$, and we can write

$$u(x, y) = \frac{u_i}{\pi}\theta(z) = \frac{u_i}{\pi}\operatorname{Arg}\left(\frac{z - b}{z - a}\right). \tag{2}$$

The superposition principle can be used to solve the more general Dirichlet problem in **FIGURE 20.5.2**. If $u(x, 0) = u_i$ for $x_{i-1} \leq x \leq x_i$ and $u(x, 0) = 0$ outside the interval $[a, b]$, then from (1),

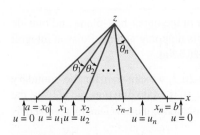

FIGURE 20.5.2 General boundary conditions on $y = 0$

$$u(x, y) = \sum_{i=1}^{n} \frac{u_i}{\pi}\left[\operatorname{Arg}(z - x_i) - \operatorname{Arg}(z - x_{i-1})\right] = \frac{1}{\pi}\sum_{i=1}^{n} u_i\theta_i(z). \tag{3}$$

Note that $\operatorname{Arg}(z - t) = \tan^{-1}(y/(x - t))$, where $\tan^{-1}$ is selected between 0 and π, and therefore $d/dt\operatorname{Arg}(z - t) = y/((x - t)^2 + y^2)$. From (3),

$$u(x, y) = \frac{1}{\pi} \sum_{i=1}^{n} \int_{x_{i-1}}^{x_i} u_i \frac{d}{dt} \text{Arg}(z - t) \, dt = \frac{1}{\pi} \sum_{i=1}^{n} \int_{x_{i-1}}^{x_i} \frac{u_i y}{(x - t)^2 + y^2} \, dt.$$

Since $u(x, 0) = 0$ outside of the interval $[a, b]$, we have

$$u(x, y) = \frac{y}{\pi} \int_{-\infty}^{\infty} \frac{u(t, 0)}{(x - t)^2 + y^2} \, dt. \tag{4}$$

A bounded piecewise-continuous function can be approximated by step functions, and therefore our discussion suggests that (4) is the solution to the Dirichlet problem in the upper half-plane. This is the content of Theorem 20.5.1.

THEOREM 20.5.1 **Poisson Integral Formula for the Upper Half-Plane**

Let $u(x, 0)$ be a piecewise-continuous function on every finite interval and bounded on $-\infty < x < \infty$. Then the function defined by

$$u(x, y) = \frac{y}{\pi} \int_{-\infty}^{\infty} \frac{u(t, 0)}{(x - t)^2 + y^2} \, dt$$

is the solution of the corresponding Dirichlet problem on the upper half-plane $y \geq 0$.

There are a few functions for which it is possible to evaluate the integral in (4), but in general, numerical methods are required to evaluate the integral.

EXAMPLE 1 **Solving a Dirichlet Problem**

Find the solution of the Dirichlet problem in the upper half-plane that satisfies the boundary condition $u(x, 0) = x$ when $|x| < 1$, and $u(x, 0) = 0$ otherwise.

SOLUTION By the Poisson integral formula,

$$u(x, y) = \frac{y}{\pi} \int_{-1}^{1} \frac{t}{(x - t)^2 + y^2} \, dt.$$

Using the substitution $s = x - t$, we can show that

$$u(x, y) = \frac{1}{\pi} \left[\frac{y}{2} \log_e((x - t)^2 + y^2) - x \tan^{-1}\left(\frac{x - t}{y}\right) \right]_{-1}^{1},$$

which can be simplified to

$$u(x, y) = \frac{y}{2\pi} \log_e\left[\frac{(x - 1)^2 + y^2}{(x + 1)^2 + y^2}\right] + \frac{x}{\pi}\left[\tan^{-1}\left(\frac{x + 1}{y}\right) - \tan^{-1}\left(\frac{x - 1}{y}\right)\right]. \qquad \equiv$$

In most of the examples and exercises $u(x, 0)$ is a step function, and we will use the integrated solution (3) rather than (4). If the first interval is $(-\infty, x_1)$, then the term $\text{Arg}(z - x_1) - \text{Arg}(z - a)$ in the sum should be replaced by $\text{Arg}(z - x_1)$. Likewise, if the last interval is (x_{n-1}, ∞), then $\text{Arg}(z - b) - \text{Arg}(z - x_{n-1})$ should be replaced by $\pi - \text{Arg}(z - x_{n-1})$.

EXAMPLE 2 **Solving a Dirichlet Problem**

The conformal mapping $f(z) = z + 1/z$ maps the region in the upper half-plane and outside the circle $|z| = 1$ onto the upper half-plane $v \geq 0$. Use this mapping and the Poisson integral formula to solve the Dirichlet problem shown in **FIGURE 20.5.3(a)**.

SOLUTION Using the results of Example 3 in Section 20.2, we can transfer the boundary conditions to the w-plane. See Figure 20.5.3(b). Since $U(u, 0)$ is a step function, we will use the integrated solution (3) rather than the Poisson integral. The solution to the new Dirichlet problem is

$$U(u, v) = \frac{1}{\pi} \text{Arg}(w + 2) + \frac{1}{\pi}[\pi - \text{Arg}(w - 2)] = 1 + \frac{1}{\pi} \text{Arg}\left(\frac{w + 2}{w - 2}\right),$$

FIGURE 20.5.3 Image of Dirichlet problem in Example 2

and therefore

$$u(x, y) = U\left(z + \frac{1}{z}\right) = 1 + \frac{1}{\pi} \operatorname{Arg}\left(\frac{z + 1/z + 2}{z + 1/z - 2}\right),$$

which can be simplified to $u(x, y) = 1 + \dfrac{1}{\pi} \operatorname{Arg}\left(\dfrac{z + 1}{z - 1}\right)^2.$ ≡

▪ Formula for the Unit Disk

A Poisson integral formula can also be developed to solve the general Dirichlet problem for the unit disk.

▪ Geometric Interpretation

FIGURE 20.5.4 shows a thin membrane (such as a soap film) that has been stretched across a frame defined by $u = u(e^{i\theta})$. The displacement u in the direction perpendicular to the z-plane satisfies the **two-dimensional wave equation**

$$a^2\left(\frac{\partial^2 u}{\partial x^2} + \frac{\partial^2 u}{\partial y^2}\right) = \frac{\partial^2 u}{\partial t^2},$$

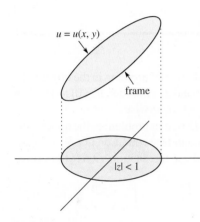

$u = u(x, y)$
frame
$|z| < 1$

FIGURE 20.5.4 Thin membrane on a frame

and so at equilibrium, the displacement function $u = u(x, y)$ is harmonic. Formula (5) provides an explicit solution for the displacement u and has the advantage that the integral is over the finite interval $[-\pi, \pi]$. When the integral cannot be evaluated, standard numerical integration procedures can be used to estimate $u(x, y)$ at a fixed point $z = x + iy$ with $|z| < 1$.

EXAMPLE 3 **Displacement of a Membrane**

A frame for a membrane is defined by $u(e^{i\theta}) = |\theta|$ for $-\pi \leq \theta \leq \pi$. Estimate the equilibrium displacement of the membrane at $(-0.5, 0)$, $(0, 0)$, and $(0.5, 0)$.

SOLUTION From (5), we get $u(x, y) = \dfrac{1}{2\pi} \displaystyle\int_{-\pi}^{\pi} |t| \dfrac{1 - |z|^2}{|e^{it} - z|^2} \, dt.$ When $(x, y) = (0, 0)$, we get

$$u(0, 0) = \frac{1}{2\pi} \int_{-\pi}^{\pi} |t| \, dt = \frac{\pi}{2}.$$

For the other two values of (x, y), the integral is not elementary and must be estimated using a numerical integration procedure. Using Simpson's rule, we obtain (to four decimal places) $u(-0.5, 0) = 2.2269$ and $u(0.5, 0) = 0.9147$. ≡

▪ Fourier Series Form

The Poisson integral formula for the unit disk is actually a compact way of writing the Fourier series solution to Laplace's equation that we developed in Chapter 14. To see this, first note that $u_n(r, \theta) = r^n \cos n\theta$ and $v_n(r, \theta) = r^n \sin n\theta$ are each harmonic, since these functions are the real and imaginary parts of z^n. If a_0, a_n, and b_n are chosen to be the Fourier coefficients of $u(e^{i\theta})$ for $-\pi < \theta < \pi$, then, by the superposition principle,

$$u(r, \theta) = \frac{a_0}{2} + \sum_{n=1}^{\infty} (a_n r^n \cos n\theta + b_n r^n \sin n\theta) \tag{6}$$

frame $u(e^{i\theta}) = \sin 4\theta$

z-plane

level curves

FIGURE 20.5.5 Level curves in Example 4

is harmonic and $u(1, \theta) = (a_0/2) + \sum_{n=1}^{\infty} (a_n \cos n\theta + b_n \sin n\theta) = u(e^{i\theta})$. Since the solution of the Dirichlet problem is also given by (5), we have

$$u(r, \theta) = \frac{1}{2\pi} \int_{-\pi}^{\pi} u(e^{it}) \frac{1 - r^2}{|e^{it} - re^{i\theta}|^2} \, dt = \frac{a_0}{2} + \sum_{n=1}^{\infty} (a_n r^n \cos n\theta + b_n r^n \sin n\theta).$$

EXAMPLE 4 **Solving a Dirichlet Problem**

Find the solution of the Dirichlet problem in the unit disk satisfying the boundary condition $u(e^{i\theta}) = \sin 4\theta$. Sketch the level curve $u = 0$.

SOLUTION Rather than working with the Poisson integral (5), we will use the Fourier series solution (6), which reduces to $u(r, \theta) = r^4 \sin 4\theta$. Therefore, $u = 0$ if and only if $\sin 4\theta = 0$. This implies $u = 0$ on the lines $x = 0$, $y = 0$, and $y = \pm x$.

If we switch to rectangular coordinates, $u(x, y) = 4xy(x^2 - y^2)$. The surface $u(x, y) = 4xy(x^2 - y^2)$, the frame $u(e^{i\theta}) = \sin 4\theta$, and the system of level curves were sketched using graphics software and are shown in **FIGURE 20.5.5**. ≡

20.5 Exercises Answers to selected odd-numbered problems begin on page ANS-49.

In Problems 1–4, use the integrated solution (3) to the Poisson integral formula to solve the given Dirichlet problem in the upper half-plane.

1.

Wait, figure 1:

Hmm.

FIGURE 20.5.6 Dirichlet problem in Problem 1

2.

FIGURE 20.5.7 Dirichlet problem in Problem 2

3.

FIGURE 20.5.8 Dirichlet problem in Problem 3

4.

FIGURE 20.5.9 Dirichlet problem in Problem 4

5. Find the solution of the Dirichlet problem in the upper half-plane that satisfies the boundary condition $u(x, 0) = x^2$ when $0 < x < 1$, and $u(x, 0) = 0$ otherwise.

6. Find the solution of the Dirichlet problem in the upper half-plane that satisfies the boundary condition $u(x, 0) = \cos x$. [*Hint*: Let $s = t - x$ and use the Section 19.6 formulas

$$\int_{-\infty}^{\infty} \frac{\cos s}{s^2 + a^2} \, ds = \frac{\pi e^{-a}}{a}, \qquad \int_{-\infty}^{\infty} \frac{\sin s}{s^2 + a^2} \, ds = 0$$

for $a > 0$.]

In Problems 7–10, solve the given Dirichlet problem by finding a conformal mapping from the given region R onto the upper half-plane $v \ge 0$.

7.

FIGURE 20.5.10 Dirichlet problem in Problem 7

8.

FIGURE 20.5.11 Dirichlet problem in Problem 8

9.

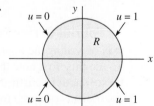

FIGURE 20.5.12 Dirichlet problem in Problem 9

10.

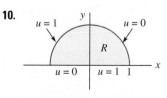

FIGURE 20.5.13 Dirichlet problem in Problem 10

11. A frame for a membrane is defined by $u(e^{i\theta}) = \theta^2/\pi^2$ for $-\pi \le \theta \le \pi$. Use the Poisson integral formula for the unit disk to estimate the equilibrium displacement of the membrane at $(-0.5, 0)$, $(0, 0)$, and $(0.5, 0)$.

12. A frame for a membrane is defined by $u(e^{i\theta}) = e^{-|\theta|}$ for $-\pi \le \theta \le \pi$. Use the Poisson integral formula for the unit disk to estimate the equilibrium displacement of the membrane at $(-0.5, 0)$, $(0, 0)$, and $(0.5, 0)$.

13. Use the Poisson integral formula for the unit disk to show that $u(0, 0)$ is the average value of the function $u = u(e^{i\theta})$ on the boundary $|z| = 1$.

In Problems 14 and 15, solve the given Dirichlet problem for the unit disk using the Fourier series form of the Poisson integral formula, and sketch the system of level curves.

14. $u(e^{i\theta}) = \cos 2\theta$ **15.** $u(e^{i\theta}) = \sin \theta + \cos \theta$

20.6 Applications

INTRODUCTION In Sections 20.2, 20.3, and 20.5 we demonstrated how Laplace's partial differential equation can be solved with conformal mapping methods, and we interpreted a solution $u = u(x, y)$ of the Dirichlet problem as either the steady-state temperature at the point (x, y) or the equilibrium displacement of a membrane at the point (x, y). Laplace's equation is a fundamental partial differential equation that arises in a variety of contexts. In this section we will establish a general relationship between vector fields and analytic functions and use our conformal mapping techniques to solve problems involving electrostatic force fields and two-dimensional fluid flows.

■ **Vector Fields** A vector field $\mathbf{F}(x, y) = P(x, y)\mathbf{i} + Q(x, y)\mathbf{j}$ in a domain D can also be expressed in the complex form

$$\mathbf{F}(x, y) = P(x, y) + iQ(x, y)$$

and thought of as a complex function. Recall from Chapter 9 that div $\mathbf{F} = \partial P/\partial x + \partial Q/\partial y$ and curl $\mathbf{F} = (\partial Q/\partial x - \partial P/\partial y)\mathbf{k}$. If we require that both div $\mathbf{F} = 0$ and curl $\mathbf{F} = \mathbf{0}$, then

$$\frac{\partial P}{\partial x} = -\frac{\partial Q}{\partial y} \quad \text{and} \quad \frac{\partial P}{\partial y} = \frac{\partial Q}{\partial x}. \tag{1}$$

This set of equations is reminiscent of the Cauchy–Riemann criterion for analyticity presented in Theorem 17.5.2 and suggests that we examine the complex function $g(z) = P(x, y) - iQ(x, y)$.

THEOREM 20.6.1 **Vector Fields and Analyticity**

(*i*) Suppose that $\mathbf{F}(x, y) = P(x, y) + iQ(x, y)$ is a vector field in a domain D and $P(x, y)$ and $Q(x, y)$ are continuous and have continuous first partial derivatives in D. If div $\mathbf{F} = 0$ and curl $\mathbf{F} = \mathbf{0}$, then the complex function

$$g(z) = P(x, y) - iQ(x, y)$$

is analytic in D.

(*ii*) Conversely, if $g(z)$ is analytic in D, then $\mathbf{F}(x, y) = \overline{g(z)}$ defines a vector field in D for which div $\mathbf{F} = 0$ and curl $\mathbf{F} = \mathbf{0}$.

PROOF: If $u(x, y)$ and $v(x, y)$ denote the real and imaginary parts of $g(z)$, then $u = P$ and $v = -Q$. Therefore the equations in (1) are equivalent to the equations

$$\frac{\partial u}{\partial x} = -\frac{\partial(-v)}{\partial y} \quad \text{and} \quad \frac{\partial u}{\partial y} = \frac{\partial(-v)}{\partial x};$$

that is,

$$\frac{\partial u}{\partial x} = \frac{\partial v}{\partial y} \quad \text{and} \quad \frac{\partial u}{\partial y} = -\frac{\partial v}{\partial x}. \tag{2}$$

The equations in (2) are the Cauchy–Riemann equations for analyticity. ≡

EXAMPLE 1 **Vector Field Gives an Analytic Function**

The vector field defined by $\mathbf{F}(x, y) = (-kq/|z - z_0|^2)(z - z_0)$ may be interpreted as the electric field produced by a wire that is perpendicular to the z-plane at $z = z_0$ and carries a charge of q coulombs per unit length. The corresponding complex function is

$$g(z) = \frac{-kq}{|z - z_0|^2} \overline{(z - z_0)} = \frac{-kq}{z - z_0}.$$

Since $g(z)$ is analytic for $z \neq z_0$, div $\mathbf{F} = 0$ and curl $\mathbf{F} = \mathbf{0}$. ≡

EXAMPLE 2 **Analytic Function Gives a Vector Field**

The complex function $g(z) = Az, A > 0$, is analytic in the first quadrant and therefore gives rise to the vector field $\mathbf{V}(x, y) = \overline{g(z)} = Ax - iAy$, which satisfies div $\mathbf{V} = 0$ and curl $\mathbf{V} = \mathbf{0}$. We will show toward the end of this section that $\mathbf{V}(x, y)$ may be interpreted as the velocity of a fluid that moves around the corner produced by the boundary of the first quadrant. ≡

The physical interpretation of the conditions div $\mathbf{F} = 0$ and curl $\mathbf{F} = \mathbf{0}$ depends on the setting. If $\mathbf{F}(x, y)$ represents the force in an electric field that acts on a unit test charge placed at (x, y), then, by Theorem 9.9.2, curl $\mathbf{F} = \mathbf{0}$ if and only if the field is **conservative**. The work done in transporting a test charge between two points in a domain D must be independent of the path. If C is a simple closed contour that lies in D, **Gauss's law** asserts that the line integral $\oint_C (\mathbf{F} \cdot \mathbf{n}) \, ds$ is proportional to the total charge enclosed by the curve C. If D is simply connected and all the electric charge is distributed on the boundary of D, then $\oint_C (\mathbf{F} \cdot \mathbf{n}) \, ds = 0$ for any simple closed contour in D. By the divergence theorem in the form (1) of Section 9.16,

$$\oint_C (\mathbf{F} \cdot \mathbf{n}) \, ds = \iint_R \text{div } \mathbf{F} \, dA, \tag{3}$$

where R is the region enclosed by C, and we can conclude that div $\mathbf{F} = 0$ in D. Conversely, if div $\mathbf{F} = 0$ in D, the double integral is zero and therefore the domain D contains no charge.

■ **Potential Functions** Suppose that $\mathbf{F}(x, y)$ is a vector field in a *simply connected* domain D with both div $\mathbf{F} = 0$ and curl $\mathbf{F} = \mathbf{0}$. By Theorem 18.3.3, the analytic function $g(z) = P(x, y) - iQ(x, y)$ has an antiderivative

$$G(z) = \phi(x, y) + i\psi(x, y) \tag{4}$$

in D, which is called a **complex potential** for the vector field $\mathbf{F}$. Note that

$$g(z) = G'(z) = \frac{\partial \phi}{\partial x} + i \frac{\partial \psi}{\partial x} = \frac{\partial \phi}{\partial x} - i \frac{\partial \phi}{\partial y}$$

and so

$$\frac{\partial \phi}{\partial x} = P \quad \text{and} \quad \frac{\partial \phi}{\partial y} = Q. \tag{5}$$

Therefore, $\mathbf{F} = \nabla \phi$ and, as in Section 9.9, the harmonic function ϕ is called a (real) **potential function** for $\mathbf{F}$.* When the potential ϕ is specified on the boundary of a region R, we can use conformal mapping techniques to solve the resulting Dirichlet problem. The equipotential lines $\phi(x, y) = c$ can be sketched and the vector field $\mathbf{F}$ can be determined using (5).

*If $\mathbf{F}$ is an electric field, the electric potential function Φ is defined to be $-\phi$ and $\mathbf{F} = -\nabla\Phi$.

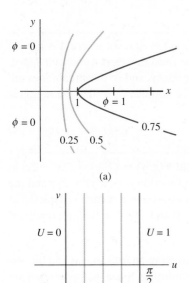

(a)

(b)

FIGURE 20.6.1 Images of boundary conditions in Example 3

EXAMPLE 3　　**Complex Potential**

The potential ϕ in the half-plane $x \geq 0$ satisfies the boundary conditions $\phi(0, y) = 0$ and $\phi(x, 0) = 1$ for $x \geq 1$. See **FIGURE 20.6.1(a)**. Determine a complex potential, the equipotential lines, and the force field **F**.

SOLUTION　We saw in Example 2 of Section 20.2 that the analytic function $z = \sin w$ maps the strip $0 \leq u \leq \pi/2$ in the w-plane to the region R in question. Therefore, $f(z) = \sin^{-1}z$ maps R onto the strip, and Figure 20.6.1(b) shows the transferred boundary conditions. The simplified Dirichlet problem has the solution $U(u, v) = (2/\pi)u$, and so $\phi(x, y) = U(\sin^{-1}z) = \text{Re}\,((2/\pi)\sin^{-1}z)$ is the potential function on D, and $G(z) = (2/\pi)u \sin^{-1}z$ is a complex potential for the force field **F**.

Note that the equipotential lines $\phi = c$ are the images of the equipotential lines $U = c$ in the w-plane under the inverse mapping $z = \sin w$. In Example 2 of Section 20.2 we showed that the vertical line $u = a$ is mapped onto a branch of the hyperbola

$$\frac{x^2}{\sin^2 a} - \frac{y^2}{\cos^2 a} = 1.$$

Since the equipotential line $U = c,\, 0 < c < 1$, is the vertical line $u = \pi/2c$, it follows that the equipotential line $\phi = c$ is the right branch of the hyperbola

$$\frac{x^2}{\sin^2(\pi c/2)} - \frac{y^2}{\cos^2(\pi c/2)} = 1.$$

Since $\mathbf{F} = \overline{G'(z)}$ and $d/dz\,\sin^{-1}z = 1/(1 - z^2)^{1/2}$, the force field is given by

$$\mathbf{F} = \frac{2}{\pi}\overline{\frac{1}{(1 - z^2)^{1/2}}} = \frac{2}{\pi}\frac{1}{(1 - \overline{z}^2)^{1/2}}. \qquad \equiv$$

▨ Steady-State Fluid Flow

The vector $\mathbf{V}(x, y) = P(x, y) + iQ(x, y)$ may also be interpreted as the velocity vector of a two-dimensional steady-state fluid flow at a point (x, y) in a domain D. The velocity at all points in the domain is therefore independent of time, and all movement takes place in planes that are parallel to a z-plane.

The physical interpretation of the conditions div $\mathbf{V} = 0$ and curl $\mathbf{V} = 0$ was discussed in Section 9.7. Recall that if curl $\mathbf{V} = 0$ in D, the flow is called **irrotational**. If a small circular paddle wheel is placed in the fluid, the net angular velocity on the boundary of the wheel is zero, and so the wheel will not rotate. If div $\mathbf{V} = 0$ in D, the flow is called **incompressible**. In a simply connected domain D, an incompressible flow has the special property that the amount of fluid in the interior of any simple closed contour C is independent of time. The rate at which fluid enters the interior of C matches the rate at which it leaves, and consequently there can be no fluid sources or sinks at points in D.

If div $\mathbf{V} = 0$ and curl $\mathbf{V} = 0$, $\mathbf{V}$ has a **complex velocity potential**

$$G(z) = \phi(x, y) + i\psi(x, y)$$

that satisfies $\overline{G'(z)} = \mathbf{V}$. In this setting, special importance is placed on the level curves $\psi(x, y) = c$. If $z(t) = x(t) + iy(t)$ is the path of a particle (such as a small cork) that has been placed in the fluid, then

$$\frac{dx}{dt} = P(x, y)$$

$$\frac{dy}{dt} = Q(x, y). \qquad (6)$$

Hence, $dy/dx = Q(x, y)/P(x, y)$ or $-Q(x, y)\,dx + P(x, y)\,dy = 0$. This differential equation is exact, since div $\mathbf{V} = 0$ implies $\partial(-Q)/\partial y = \partial P/\partial x$. By the Cauchy–Riemann equations, $\partial\psi/\partial x = -\partial\phi/\partial y = -Q$ and $\partial\psi/\partial y = \partial\phi/\partial x = P$, and therefore all solutions of (6) satisfy $\psi(x, y) = c$. The function $\psi(x, y)$ is therefore called a **stream function** and the level curves $\psi(x, y) = c$ are **streamlines** for the flow.

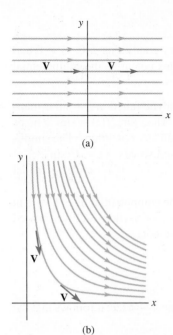

FIGURE 20.6.2 (a) Uniform flow in Example 4; (b) Flow around a corner in Example 5

FIGURE 20.6.3 Flow around a cylinder in Example 6

FIGURE 20.6.4 Flow in Example 7

EXAMPLE 4 Uniform Flow

The uniform flow in the upper half-plane is defined by $\mathbf{V}(x, y) = A(1, 0)$, where A is a fixed positive constant. Note that $|\mathbf{V}| = A$, and so a particle in the fluid moves at a constant speed. A complex potential for the vector field is $G(z) = Az = Ax + iAy$, and so the streamlines are the horizontal lines $Ay = c$. See **FIGURE 20.6.2(a)**. Note that the boundary $y = 0$ of the region is itself a streamline. ≡

EXAMPLE 5 Flow Around a Corner

The analytic function $G(z) = z^2$ gives rise to the vector field $\mathbf{V}(x, y) = \overline{G'(z)} = (2x, -2y)$ in the first quadrant. Since $z^2 = x^2 - y^2 + i(2xy)$, the stream function is $\psi(x, y) = 2xy$ and the streamlines are the hyperbolas $2xy = c$. This flow, called flow around a corner, is depicted in Figure 20.6.2(b). As in Example 4, the boundary lines $x = 0$ and $y = 0$ in the first quadrant are themselves streamlines. ≡

■ **Constructing Special Flows** The process of constructing an irrotational and incompressible flow that remains inside a given region R is called **streamlining**. Since the streamlines are described by $\psi(x, y) = c$, two distinct streamlines do not intersect. Therefore, if the boundary is itself a streamline, a particle that starts inside R cannot leave R. This is the content of the following theorem:

THEOREM 20.6.2 Streamlining

Suppose that $G(z) = \phi(x, y) + i\psi(x, y)$ is analytic in a region R and $\psi(x, y)$ is constant on the boundary of R. Then $\mathbf{V}(x, y) = \overline{G'(z)}$ defines an irrotational and incompressible fluid flow in R. Moreover, if a particle is placed inside R, its path $z = z(t)$ remains in R.

EXAMPLE 6 Flow Around a Cylinder

The analytic function $G(z) = z + 1/z$ maps the region R in the upper half-plane and outside the circle $|z| = 1$ onto the upper half-plane $v \geq 0$. The boundary of R is mapped onto the u-axis, and so $v = \psi(x, y) = y - y/(x^2 + y^2)$ is zero on the boundary of R. **FIGURE 20.6.3** shows the streamlines of the resulting flow. The velocity field is given by $\overline{G'(z)} = 1 - 1/\bar{z}^2$, and so

$$\overline{G'(re^{i\theta})} = 1 - \frac{1}{r^2}e^{2i\theta}.$$

It follows that $\mathbf{V} \approx (1, 0)$ for large values of r, and so the flow is approximately uniform at large distances from the circle $|z| = 1$. The resulting flow in the region R is called flow around a cylinder. The mirror image of the flow can be adjoined to give a flow around a complete cylinder. ≡

If R is a polygonal region, we can use the Schwarz–Christoffel formula to find a conformal mapping $z = f(w)$ from the upper half-plane R' onto R. The inverse function $G(z) = f^{-1}(z)$ maps the boundary of R onto the u-axis. Therefore, if $G(z) = \phi(x, y) + i\psi(x, y)$, then $\psi(x, y) = 0$ on the boundary of R. Note that the streamlines $\psi(x, y) = c$ in the z-plane are the images of the horizontal lines $v = c$ in the w-plane under $z = f(w)$.

EXAMPLE 7 Streamlines Defined Parametrically

The analytic function $f(w) = w + \text{Ln } w + 1$ maps the upper half-plane $v \geq 0$ to the upper half-plane $y \geq 0$ with the horizontal line $y = \pi$, $x \leq 0$, deleted. See Example 4 in Section 20.4. If $G(z) = f^{-1}(z) = \phi(x, y) + i\psi(x, y)$, then $G(z)$ maps R onto the upper half-plane and maps the boundary of R onto the u-axis. Therefore, $\psi(x, y) = 0$ on the boundary of R.

It is not possible to find an explicit formula for the stream function $\psi(x, y)$. The streamlines, however, are the images of the horizontal lines $v = c$ under $z = f(w)$. If we write $w = t + ic$, $c > 0$, then the streamlines can be represented in the parametric form

$$z = f(t + ic) = t + ic + \text{Ln}(t + ic) + 1;$$

that is, $$x = t + 1 + \frac{1}{2}\log_e(t^2 + c^2), \quad y = c + \text{Arg}(t + ic).$$

Graphing software was used to generate the streamlines in **FIGURE 20.6.4**. ≡

A stream function $\psi(x, y)$ is harmonic but, unlike a solution to a Dirichlet problem, we do *not* require $\psi(x, y)$ to be bounded (see Examples 4–6) or to assume a fixed set of constants on the boundary. Therefore, there may be many different stream functions for a given region that satisfy Theorem 20.6.2. This will be illustrated in the final example.

<table><tr><td>EXAMPLE 8</td><td>**Streamlines Defined Parametrically**</td></tr></table>

The analytic function $f(w) = w + e^w + 1$ maps the horizontal strip $0 \le v \le \pi$ onto the region R shown in Figure 20.6.4. Therefore, $G(z) = f^{-1}(z) = \phi(x, y) + i\psi(x, y)$ maps R back to the strip and, from M-1 in the conformal mappings in Appendix D, maps the boundary line $y = 0$ onto the u-axis and maps the boundary line $y = \pi$, $x \le 0$, onto the horizontal line $v = \pi$. Therefore, $\psi(x, y)$ is constant on the boundary of R.

The streamlines are the images of the horizontal lines $v = c$, $0 < c < \pi$, under $z = f(w)$. As in Example 7, a parametric representation of the streamlines is

$$z = f(t + ic) = t + ic + e^{t+ic} + 1$$

or
$$x = t + 1 + e^t \cos c, \quad y = c + e^t \sin c.$$

The streamlines are shown in **FIGURE 20.6.5**. Unlike the flow in Example 7, the fluid appears to emerge from the strip $0 \le y \le \pi$, $x \le 0$. ≡

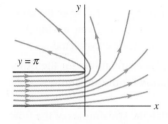

FIGURE 20.6.5 Flow in Example 8

$y = \pi$

20.6 Exercises Answers to selected odd-numbered problems begin on page ANS-50.

In Problems 1–4, verify that div $\mathbf{F} = 0$ and curl $\mathbf{F} = \mathbf{0}$ for the given vector field $\mathbf{F}(x, y)$ by examining the corresponding complex function $g(z) = P(x, y) - iQ(x, y)$. Find a complex potential for the vector field and sketch the equipotential lines.

1. $\mathbf{F}(x, y) = (\cos \theta_0)\mathbf{i} + (\sin \theta_0)\mathbf{j}$

2. $\mathbf{F}(x, y) = -y\mathbf{i} - x\mathbf{j}$

3. $\mathbf{F}(x, y) = \dfrac{x}{x^2 + y^2}\mathbf{i} + \dfrac{y}{x^2 + y^2}\mathbf{j}$

4. $\mathbf{F}(x, y) = \dfrac{x^2 - y^2}{(x^2 + y^2)^2}\mathbf{i} + \dfrac{2xy}{(x^2 + y^2)^2}\mathbf{j}$

5. The potential ϕ on the wedge $0 \le \text{Arg } z \le \pi/4$ satisfies the boundary conditions $\phi(x, 0) = 0$ and $\phi(x, x) = 1$ for $x > 0$. Determine a complex potential, the equipotential lines, and the corresponding force field $\mathbf{F}$.

6. Use the conformal mapping $f(z) = 1/z$ to determine a complex potential, the equipotential lines, and the corresponding force field $\mathbf{F}$ for the potential ϕ that satisfies the boundary conditions shown in **FIGURE 20.6.6**.

7. The potential ϕ on the semicircle $|z| \le 1$, $y \ge 0$, satisfies the boundary conditions $\phi(x, 0) = 0$, $-1 < x < 1$, and $\phi(e^{i\theta}) = 1$, $0 < \theta < \pi$. Show that

$$\phi(x, y) = \frac{1}{\pi} \text{Arg}\left(\frac{z-1}{z+1}\right)^2,$$

and use the mapping properties of linear fractional transformations to explain why the equipotential lines are arcs of circles.

8. Use the conformal mapping C-1 in Appendix D to find the potential ϕ in the region outside the two circles $|z| = 1$ and $|z - 3| = 1$ if the potential is kept at zero on $|z| = 1$ and one on $|z - 3| = 1$. Use the mapping properties of linear fractional transformations to explain why the equipotential lines are, with one exception, circles.

In Problems 9–14, a complex velocity potential $G(z)$ is defined on a region R.

(a) Find the stream function and verify that the boundary of R is a streamline.

(b) Find the corresponding velocity vector field $\mathbf{V}(x, y)$.

(c) Use a graphing utility to sketch the streamlines of the flow.

9. $G(z) = z^4$

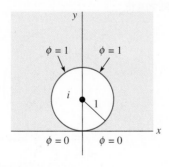

FIGURE 20.6.6 Boundary conditions in Problem 6

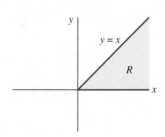

FIGURE 20.6.7 Region R for Problem 9

10. $G(z) = z^{2/3}$

FIGURE 20.6.8 Region R for Problem 10

11. $G(z) = \sin z$

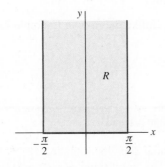

FIGURE 20.6.9 Region R for Problem 11

12. $G(z) = i \sin^{-1} z$

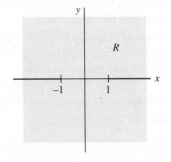

FIGURE 20.6.10 Region R for Problem 12

13. $G(z) = z^2 + 1/z^2$

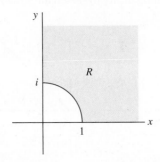

FIGURE 20.6.11 Region R for Problem 13

14. $G(z) = e^z$

FIGURE 20.6.12 Region R for Problem 14

In Problems 15–18, a conformal mapping $z = f(w)$ from the upper half-plane $v \geq 0$ to a region R in the z-plane is given and the flow in R with complex potential $G(z) = f^{-1}(z)$ is constructed.

(a) Verify that the boundary of R is a streamline for the flow.

(b) Find a parametric representation for the streamlines of the flow.

(c) Use a graphing utility to sketch the streamlines of the flow.

15. M-9 in Appendix D

16. M-4 in Appendix D; use $a = 1$

17. M-2 in Appendix D; use $a = 1$

18. M-5 in Appendix D

19. A stagnation point in a flow is a point at which $\mathbf{V} = \mathbf{0}$. Find all stagnation points for the flows in Examples 5 and 6.

20. For any two real numbers k and x_1, the function $G(z) = k \operatorname{Ln}(z - x_1)$ is analytic in the upper half-plane and therefore is a complex potential for a flow. The real number x_1 is called a *sink* when $k < 0$ and a *source* for the flow when $k > 0$.

(a) Show that the streamlines are rays emanating from x_1.

(b) Show that $\mathbf{V} = (k/|z - x_1|^2)(z - x_1)$ and conclude that the flow is directed toward x_1 precisely when $k < 0$.

21. If $f(z)$ is a conformal mapping from a domain D onto the upper half-plane, a flow with a source at a point ξ_0 on the boundary of D is defined by the complex potential $G(z) = k \operatorname{Ln}(f(z) - f(\xi_0))$, where $k > 0$. Determine the streamlines for a flow in the first quadrant with a source at $\xi_0 = 1$ and $k = 1$.

22. (a) Construct a flow on the horizontal strip $0 < y < \pi$ with a sink at the boundary point $\xi_0 = 0$. [*Hint*: See Problem 21.]

(b) Use a graphing utility to sketch the streamlines of the flow.

23. The complex potential $G(z) = k \operatorname{Ln}(z - 1) - k \operatorname{Ln}(z + 1)$ with $k > 0$ gives rise to a flow on the upper half-plane with a single source at $z = 1$ and a single sink at $z = -1$. Show that the streamlines are the family of circles $x^2 + (y - c)^2 = 1 + c^2$. See **FIGURE 20.6.13**.

FIGURE 20.6.13 Streamlines in Problem 23

24. The flow with velocity vector $\mathbf{V} = (a + ib)/\bar{z}$ is called a **vortex** at $z = 0$, and the geometric nature of the streamlines depends on the choice of a and b.

(a) Show that if $z = x(t) + iy(t)$ is the path of a particle, then

$$\frac{dx}{dt} = \frac{ax - by}{x^2 + y^2}$$

$$\frac{dy}{dt} = \frac{bx + ay}{x^2 + y^2}.$$

(b) Change to polar coordinates to establish that $dr/dt = a/r$ and $d\theta/dt = b/r^2$, and conclude that $r = ce^{a\theta/b}$ for $b \neq 0$. [*Hint*: See (2) of Section 11.1.]

(c) Conclude that the logarithmic spirals in part (b) spiral inward if and only if $a < 0$, and the curves are traversed clockwise if and only if $b < 0$. See **FIGURE 20.6.14**.

FIGURE 20.6.14 Logarithmic spiral in Problem 24

20 Chapter in Review Answers to selected odd-numbered problems begin on page ANS-51.

Answer Problems 1–10 without referring back to the text. Fill in the blank or answer true/false.

1. Under the complex mapping $f(z) = z^2$, the curve $xy = 2$ is mapped onto the line _____.

2. The complex mapping $f(z) = -iz$ is a rotation through _____ degrees.

3. The image of the upper half-plane $y \geq 0$ under the complex mapping $f(z) = z^{2/3}$ is _____.

4. The analytic function $f(z) = \cosh z$ is conformal except at $z = $ _____.

5. If $w = f(z)$ is an analytic function that maps a domain D onto the upper half-plane $v > 0$, then the function $u = \text{Arg}(f(z))$ is harmonic in D. _____

6. Is the image of the circle $|z - 1| = 1$ under the complex mapping $T(z) = (z - 1)/(z - 2)$ a circle or a line? _____

7. The linear fractional transformation $T(z) = \dfrac{z - z_1}{z - z_3} \dfrac{z_2 - z_3}{z_2 - z_1}$ maps the triple z_1, z_2, and z_3 to _____.

8. If $f'(z) = z^{-1/2}(z + 1)^{-1/2}(z - 1)^{-1/2}$, then $f(z)$ maps the upper half-plane $y > 0$ onto the interior of a rectangle. _____

9. If $\mathbf{F}(x, y) = P(x, y)\mathbf{i} + Q(x, y)\mathbf{j}$ is a vector field in a domain D with div $\mathbf{F} = 0$ and curl $\mathbf{F} = \mathbf{0}$, then the complex function $g(z) = P(x, y) + iQ(x, y)$ is analytic in D. _____

10. If $G(z) = \phi(x, y) + i\psi(x, y)$ is analytic in a region R and $\mathbf{V}(x, y) = \overline{iG'(z)}$, then the streamlines of the corresponding flow are described by $\phi(x, y) = c$. _____

11. Find the image of the first quadrant under the complex mapping $w = \text{Ln } z = \log_e|z| + i \text{Arg } z$. What are images of the rays $\theta = \theta_0$ that lie in the first quadrant?

In Problems 12 and 13, use the conformal mappings in Appendix D to find a conformal mapping from the given region R in the z-plane onto the target region R' in the w-plane, and find the image of the given boundary curve.

12.

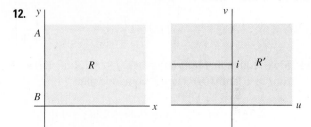

FIGURE 20.R.1 Regions R and R' for Problem 12

13.

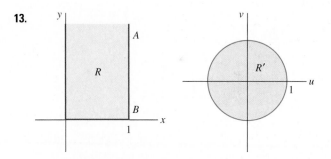

FIGURE 20.R.2 Regions R and R' for Problem 13

In Problems 14 and 15, use an appropriate conformal mapping to solve the given Dirichlet problem.

14.

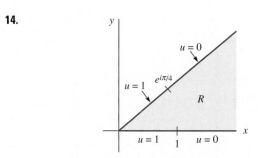

FIGURE 20.R.3 Dirichlet problem in Problem 14

15.

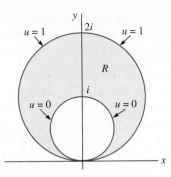

FIGURE 20.R.4 Dirichlet problem in Problem 15

16. Derive conformal mapping C-4 in Appendix D by constructing the linear fractional transformation that maps $1, -1, \infty$ to $i, -i, -1$.

17. (a) Approximate the region R' in M-9 in Appendix D by the polygonal region shown in **FIGURE 20.R.5**. Require that $f(-1) = u_1, f(0) = \pi i/2$, and $f(1) = u_1 + \pi i$.

(b) Show that when $u_1 \to \infty$,

$$f'(z) = Az(z + 1)^{-1}(z - 1)^{-1} = \frac{1}{2} A\left[\frac{1}{z + 1} + \frac{1}{z - 1}\right].$$

(c) If we require that $\mathrm{Im}(f(t)) = 0$ for $t < -1$, $\mathrm{Im}(f(t)) = \pi$ for $t > 1$, and $f(0) = \pi i/2$, conclude that

$$f(z) = \pi i - \frac{1}{2}[\mathrm{Ln}(z + 1) + \mathrm{Ln}(z - 1)].$$

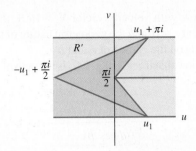

FIGURE 20.R.5 Image of upper half-plane in Problem 17

18. (a) Find the solution $u(x, y)$ of the Dirichlet problem in the upper half-plane $y \geq 0$ that satisfies the boundary condition $u(x, 0) = \sin x$. [*Hint*: See Problem 6 in Exercises 20.5.]

(b) Find the solution $u(x, y)$ of the Dirichlet problem in the unit disk $|z| \leq 1$ that satisfies the boundary condition $u(e^{i\theta}) = \sin \theta$.

19. Explain why the streamlines in Figure 20.6.5 may also be interpreted as the equipotential lines of the potential ϕ that satisfies $\phi(x, 0) = 0$ for $-\infty < x < \infty$ and $\phi(x, \pi) = 1$ for $x < 0$.

20. Verify that the boundary of the region R defined by $y^2 \geq 4(1 - x)$ is a streamline for the fluid flow with complex potential $G(z) = i(z^{1/2} - 1)$. Sketch the streamlines of the flow.

APPENDICES

APPENDIX

Integral-Defined Functions

INTRODUCTION Many important functions encountered in applied mathematics are defined in terms of an integral. There are several ways of doing this. In Sections 1.1, 2.3, and 3.2, we have already seen that a solution of a differential equation may lead to a function of the form

$$F(x) = \int_a^x g(t)\, dt. \tag{1}$$

In calculus, you may have seen that one definition of the natural logarithm is given in terms of an integral of the form given in (1):

$$\ln x = \int_1^x \frac{1}{t}\, dt.$$

An integral-defined function can also be of the form

$$F(x) = \int_a^b g(x, t)\, dt, \tag{2}$$

where it is understood that x is a parameter; that is, x is treated as a constant in the t-integration. In many cases, especially for functions with a name, the defining integral is nonelementary. The integrand g in (2) could also depend on several parameters, say, $g(x, y, t)$. In this case, the integral (2) defines a function of two variables x and y. Moreover, (1) and (2) can also be improper integrals. For example, in (1) the lower limit of integration a could be $-\infty$, or the interval of integration could be chosen to be $[x, \infty)$. In (2), one or both of the limits of integration could be infinite, or the integrand $g(x, t)$ could have an infinite discontinuity in the interval of integration $[a, b]$. The convergence of an improper integral will, in general, depend on values of the parameter x. See Examples 1, 2, and 3 in Section 4.1.

■ **The Error Function** In Sections 2.3 and 15.1, we saw that the **error function** and **complementary error function**,

$$\operatorname{erf}(x) = \frac{2}{\sqrt{\pi}} \int_0^x e^{-t^2}\, dt \quad \text{and} \quad \operatorname{erfc}(x) = \frac{2}{\sqrt{\pi}} \int_x^\infty e^{-t^2}\, dt, \tag{3}$$

are both of the type given in (1) with the identification $g(t) = (2/\sqrt{\pi})e^{-t^2}$. As we have seen in Chapter 4, the Laplace transform of a function $f(t)$ is the function $F(s) = \int_0^\infty e^{-st} f(t)\, dt$. The last integral is of the type given in (2) with the symbol s playing the part of x so that $g(s, t) = e^{-st} f(t)$. In Chapter 15, the Fourier transform $F(\alpha) = \int_{-\infty}^\infty f(x) e^{i\alpha x}\, dx$ is also of the type given in (2), with the symbols x and α replacing t and x, respectively, and $g(\alpha, x) = f(x) e^{i\alpha x}$.

We begin our discussion of specific integral-defined functions with one of the form given in (2).

■ **The Gamma Function** On a short list of important functions in the study of special functions (such as the Bessel function), the **gamma function** would appear near the top. The integral definition of this function,

$$\Gamma(x) = \int_0^\infty t^{x-1} e^{-t}\, dt, \tag{4}$$

was first given by the Swiss mathematician **Leonhard Euler** in his text *Institutiones calculi integralis*, published in 1768.

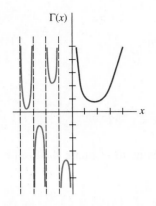

FIGURE A.1 Graph of the gamma function

■ **Graphs and Properties** Convergence of the improper integral (4) requires that $x - 1 > -1$ or $x > 0$. Although the integral (4) does not converge for $x \le 0$, it can be shown by alternative definitions that the domain of the gamma function can be expanded to the set of real numbers *except* the nonpositive integers: $x = -n$, $n = 0, 1, 2, \ldots$. See Problem 28 in Exercises for Appendix A. Considered as a function of a real variable x, the graph of $\Gamma(x)$ is given in **FIGURE A.1**. As seen in the figure, the dashed lines $x = 0$, $x = -1$, $x = -2, \ldots$ are vertical asymptotes of the graph.

There is a remarkably simple and extremely useful connection between the value of the gamma function at a number x and its value at $x + 1$. To obtain this relationship, we replace x by $x + 1$ in (4) and integrate by parts:

$$\Gamma(x + 1) = \int_0^\infty t^{(x+1)-1}e^{-t}\,dt = \int_0^\infty t^x e^{-t}\,dt$$

$$= \lim_{b\to\infty}\left[-t^x e^{-t}\Big|_0^b + x\int_0^b t^{x-1}e^{-t}\,dt \right] \tag{5}$$

$$= x\lim_{b\to\infty}\int_0^b t^{x-1}e^{-t}\,dt. \tag{6}$$

In (5), $\lim\limits_{b\to\infty} t^x e^{-t}\Big|_0^b = 0$ follows from the assumption that $x > 0$ and repeated applications of L'Hôpital's rule. The limit in (6), $\lim\limits_{b\to\infty}\int_0^b t^{x-1}e^{-t}\,dt$, is the definition of the improper integral (4). Hence we have shown that

$$\Gamma(x + 1) = x\Gamma(x). \tag{7}$$

Equation (7) is called a **recursion formula**. It is easy to find the value of $\Gamma(x)$ when x is a positive integer. For example, when $x = 1$ we see that (4) is an elementary integral:

$$\Gamma(1) = \int_0^\infty e^{-t}\,dt = \lim_{b\to\infty}\int_0^b e^{-t}\,dt = \lim_{b\to\infty}(-e^{-b} + 1) = 1.$$

From this single numerical value, the repeated application of formula (7) gives:

$$\Gamma(2) = \Gamma(1 + 1) = 1\cdot\Gamma(1) = 1$$

$$\Gamma(3) = \Gamma(2 + 1) = 2\cdot\Gamma(2) = 2\cdot 1$$

$$\Gamma(4) = \Gamma(3 + 1) = 3\cdot\Gamma(3) = 3\cdot 2\cdot 1 \tag{8}$$

$$\Gamma(5) = \Gamma(4 + 1) = 4\cdot\Gamma(4) = 4\cdot 3\cdot 2\cdot 1$$

and so on. In this manner, it is seen that when n is a positive integer,

$$\Gamma(n + 1) = n!. \tag{9}$$

Recall, $n!$ (read "n factorial") is defined as the product of consecutive integers from 1 to n; that is, $n! = n(n - 1)\cdots 3\cdot 2\cdot 1$ and is called the **factorial function**. For convenience, it is customary to define $0! = 1$. Because $n!$ is a special case of (4), the gamma function is sometimes referred to as the **generalized factorial function**.

In the discussion of the Bessel functions of half-integral order in Section 5.3, we used the value of $\Gamma(\frac{1}{2}) = \sqrt{\pi}$. With $x = \frac{1}{2}$ we see that (4) becomes

$$\Gamma(\tfrac{1}{2}) = \int_0^\infty t^{-1/2}e^{-t}\,dt. \tag{10}$$

Note that the foregoing integral is improper for two reasons: an infinite limit of integration, and the integrand has an infinite discontinuity at 0. Nonetheless, (10) can be explicitly evaluated. To that end, we begin with the substitutions $t = u^2$ and $dt = 2u \, du$:

$$\Gamma(\tfrac{1}{2}) = \int_0^\infty u^{-1} e^{-u^2} (2u \, du) = 2 \int_0^\infty e^{-u^2} \, du. \tag{11}$$

But $\int_0^\infty e^{-u^2} du = \int_0^\infty e^{-v^2} dv$, so

$$[\Gamma(\tfrac{1}{2})]^2 = \left(2 \int_0^\infty e^{-u^2} \, du \right)\left(2 \int_0^\infty e^{-v^2} \, dv \right) = 4 \int_0^\infty \int_0^\infty e^{-(u^2 + v^2)} \, du \, dv.$$

Switching to polar coordinates $u = r \cos \theta$, $v = r \sin \theta$ enables us to evaluate the double integral:

$$4 \int_0^\infty \int_0^\infty e^{-(u^2 + v^2)} \, du \, dv = 4 \int_0^{\pi/2} \int_0^\infty e^{-r^2} r \, dr \, d\theta = \pi.$$

Hence

$$[\Gamma(\tfrac{1}{2})]^2 = \pi \quad \text{or} \quad \Gamma(\tfrac{1}{2}) = \sqrt{\pi}. \tag{12}$$

We used (11), $2\int_0^\infty e^{-u^2} du = \sqrt{\pi}$, in the form $(2/\sqrt{\pi})\int_0^\infty e^{-u^2} du = 1$ to derive the fundamental identity that relates the error function and complementary error function: $\text{erf}(x) + \text{erfc}(x) = 1$. See (16) in Section 2.3.

Because of (7) and (12), the value of the gamma function for x equal to one half an odd integer, that is, $x = n + \tfrac{1}{2}$ where $n = 0, \pm 1, \pm 2, \ldots$, can be expressed in terms of $\Gamma(\tfrac{1}{2}) = \sqrt{\pi}$. The recursion formula (7) written as

$$\Gamma(x) = \frac{\Gamma(x + 1)}{x} \tag{13}$$

is one way to extend the definition of the gamma function to negative real numbers with the negative integers being the only exception. Part (b) of the next example illustrates this idea.

EXAMPLE 1 Using (7), (12), and (13)

Evaluate (a) $\Gamma(\tfrac{3}{2})$ (b) $\Gamma(-\tfrac{1}{2})$.

SOLUTION (a) With $x = \tfrac{1}{2}$ and the result in (12), the recursion formula (7) yields

$$\Gamma(\tfrac{3}{2}) = \Gamma(\tfrac{1}{2} + 1) = \tfrac{1}{2}\Gamma(\tfrac{1}{2}) = \tfrac{1}{2}\sqrt{\pi}.$$

(b) If we choose $x = -\tfrac{1}{2}$, then formula (13) gives

$$\Gamma(-\tfrac{1}{2}) = \frac{\Gamma(-\tfrac{1}{2} + 1)}{-\tfrac{1}{2}} = -2\Gamma(\tfrac{1}{2}) = -2\sqrt{\pi}. \qquad \equiv$$

The Beta Function Analogous to (2), an integral of the type $\int_a^b g(x, y, t) \, dt$ defines a function of two variables x and y. An example of this kind of integral is

$$B(x, y) = \int_0^1 t^{x-1}(1 - t)^{y-1} \, dt. \tag{14}$$

The foregoing integral is known as the **beta function** and is closely related to the gamma function. The integrand of (14) can be discontinuous at 0 and at 1 for particular choices of x and y, but it can be shown that the integral exists for $x > 0$, $y > 0$. Although we are not going to prove it, the beta function can be expressed in terms of the gamma function:

$$B(x, y) = \frac{\Gamma(x)\Gamma(y)}{\Gamma(x + y)}, \quad x > 0, y > 0. \tag{15}$$

EXAMPLE 2 Using (12) and (15)

Note that the integrand has an infinite discontinuity at $t = 1$. ▶

Evaluate $\displaystyle\int_0^1 \sqrt{\frac{t}{1-t}}\, dt.$

SOLUTION Even though the given integral is improper, we can evaluate it straight away using (15). By rewriting the integral in the equivalent form

$$\int_0^1 t^{1/2}(1-t)^{-1/2}\, dt$$

we recognize it as (14) in the case $x = \frac{3}{2}$ and $y = \frac{1}{2}$. From (15) it follows that

$$\int_0^1 t^{1/2}(1-t)^{-1/2}\, dt = B(\tfrac{3}{2}, \tfrac{1}{2}) = \frac{\Gamma(\frac{3}{2})\Gamma(\frac{1}{2})}{\Gamma(2)} = \frac{\pi}{2},$$

where we have used the previously obtained values $\Gamma(\frac{3}{2}) = \frac{1}{2}\sqrt{\pi}$, $\Gamma(\frac{1}{2}) = \sqrt{\pi}$, and $\Gamma(2) = 1$. ≡

▆ Other Integral-Defined Functions We have barely touched the subject of integral-defined functions. Here are a few more functions that occur in mathematics, physics, and engineering:

Sine and cosine integral functions:

$$\text{Si}(x) = \int_0^x \frac{\sin t}{t}\, dt, \quad \text{Ci}(x) = -\int_x^\infty \frac{\cos t}{t}\, dt \tag{16}$$

Fresnel sine and cosine integral functions:

$$S(x) = \int_0^x \sin\left(\tfrac{1}{2}\pi t^2\right) dt, \quad C(x) = \int_0^x \cos\left(\tfrac{1}{2}\pi t^2\right) dt \tag{17}$$

Exponential integral function:

$$\text{Ei}(x) = \int_{-\infty}^x \frac{e^t}{t}\, dt \tag{18}$$

Logarithmic integral function:

$$\text{Li}(x) = \int_2^x \frac{1}{\ln t}\, dt \tag{19}$$

Airy functions of first and second kind:

$$\text{Ai}(x) = \frac{1}{\pi}\int_0^\infty \cos\left(xt + \tfrac{1}{3}t^3\right) dt, \quad \text{Bi}(x) = \frac{1}{\pi}\int_0^\infty \left[e^{xt - \frac{1}{3}t^3} + \sin\left(xt + \tfrac{1}{3}t^3\right)\right] dt \tag{20}$$

The definitions of the functions in (16)–(20) vary slightly throughout the literature and websites. For example, the Fresnel integrals, which first appeared in the study of optics, are often defined without the factor $\frac{1}{2}\pi$ in the integrands. In the computer algebra system *Mathematica*, the exponential integral is written $\text{Ei}(x) = -\int_{-x}^\infty (e^{-t}/t)\, dt$. Using the substitution $t = -u$ it is easily shown that this latter form is equivalent to (18). Finally, it should be noted that the sine integral function $\text{Si}(x) = \int_0^x \frac{\sin t}{t}\, dt$ is not regarded as an improper integral because the integrand $f(t) = (\sin t)/t$ has a removable discontinuity at $t = 0$; that is, the integrand should be interpreted as

$$f(t) = \begin{cases} \dfrac{\sin t}{t}, & t > 0 \\[2mm] 1, & t = 0. \end{cases}$$

The graphs of the functions (16)–(20) are given in **FIGURE A.2**.

(a) Sine and cosine integral functions

(b) Fresnel sine and cosine integral functions

(c) Exponential integral function

(d) Logarithmic integral function

(e) Airy functions of first and second kind

FIGURE A.2 Graphs of Si(x), Ci(x), S(x), C(x), Ei(x), Li(x), Ai(x), and Bi(x)

▓ Derivative of an Integral-Defined Function

The integral (1) and its derivative are encountered in the first semester of a course in calculus. Recall, one form of the Fundamental Theorem of Calculus states that if g is continuous on an interval and a is a number in the interval, then $F(x) = \int_a^x g(t)\,dt$ is a differentiable function and its derivative is

$$F'(x) = \frac{d}{dx}\int_a^x g(t)\,dt = g(x). \tag{21}$$

In the most general case of (2) where the limits of integration are functions of x,

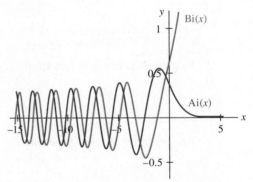

$$F(x) = \int_{u(x)}^{v(x)} g(x, t)\,dt,$$

the derivative is given by **Leibniz's rule**:

$$F'(x) = \frac{d}{dx}\int_{u(x)}^{v(x)} g(x, t)\, dt = g(x, v(x))\frac{dv}{dx} - g(x, u(x))\frac{du}{dx} + \int_{u(x)}^{v(x)} \frac{\partial}{\partial x} g(x, t)\, dt. \tag{22}$$

Without going into any details, it is usually assumed in (22) that $u(x)$, $v(x)$, $u'(x)$, and $v'(x)$ are continuous on an interval and $g(x, t)$, and $g_x(x, t)$ are continuous on some region of the xt-plane.

Two special cases of (22) are of particular interest. First, if the integrand $g(x, t)$ is replaced by $g(t)$, then $(\partial/\partial x)g(t) = 0$ and (22) reduces to

This result also follows directly from (21) and the Chain Rule.

$$F'(x) = \frac{d}{dx}\int_{u(x)}^{v(x)} g(t)\, dt = g(v(x))\frac{dv}{dx} - g(u(x))\frac{du}{dx}. \tag{23}$$

Second, if $u = a$ and $v = b$ denote constants, then $u' = 0$ and $v' = 0$ and (22) gives a rule for differentiating under an integral sign:

$$F'(x) = \frac{d}{dx}\int_{a}^{b} g(x, t)\, dt = \int_{a}^{b} \frac{\partial}{\partial x} g(x, t)\, dt. \tag{24}$$

Under slightly more stringent conditions, the result in (24) is also valid when the interval of integration is $[a, \infty)$. The last result leads us back to differential equations. See Problems 49–52 in Exercises for Appendix A.

EXAMPLE 3 Using (22), (23), and (24)

Compute the derivative of the functions

(a) $F(x) = \int_{x^2}^{x^3} \sin(4xt^2)\, dt$ **(b)** $F(x) = \int_{x^2}^{x^3} \sin(4t^2)\, dt$ **(c)** $F(x) = \int_{0}^{1} \sin(4xt^2)\, dt.$

SOLUTION **(a)** From (22) with $u = x^2$ and $v = x^3$ we have

$$F'(x) = \frac{d}{dx}\int_{x^2}^{x^3} \sin(4xt^2)\, dt$$

$$= \sin(4x(x^3)^2)\cdot 3x^2 - \sin(4x(x^2)^2)\cdot 2x + 4\int_{x^2}^{x^3} t^2\cos(4xt^2)\, dt$$

$$= 3x^2 \sin(4x^7) - 2x\sin(4x^5) + 4\int_{x^2}^{x^3} t^2\cos(4xt^2)\, dt.$$

(b) From (23),

$$F'(x) = \frac{d}{dx}\int_{x^2}^{x^3} \sin(4t^2)\, dt$$

$$= \sin(4(x^3)^2)\cdot 3x^2 - \sin(4(x^2)^2)\cdot 2x$$

$$= 3x^2 \sin(4x^6) - 2x\sin(4x^4).$$

(c) From (24),

$$F'(x) = \frac{d}{dx}\int_{0}^{1} \sin(4xt^2)\, dt = \int_{0}^{1} \frac{\partial}{\partial x} \sin(4xt^2)\, dt = 4\int_{0}^{1} t^2\cos(4xt^2)\, dt. \qquad \equiv$$

The result in (24) can sometimes be an aid in evaluating an integral that at first glance seems intractable. But the procedure usually involves formal manipulations and cleverness, which cannot be taught. Problems 53–56 in Exercises for Appendix A give you a small sample of the idea.

The Gamma Function

In Problems 1 and 2, evaluate the given quantity.

1. $\Gamma(6)$

2. $\Gamma(8)$

In Problems 3–6, use (7) and the known value $\Gamma(\frac{1}{2}) = \sqrt{\pi}$ to evaluate the given quantity.

3. $\Gamma(\frac{5}{2})$

4. $\Gamma(\frac{7}{2})$

5. $\Gamma(\frac{9}{2})$

6. $\Gamma(\frac{11}{2})$

In Problems 7–10, use (13) and the known value $\Gamma(-\frac{1}{2}) = -2\sqrt{\pi}$ to evaluate the given quantity.

7. $\Gamma(-\frac{3}{2})$

8. $\Gamma(-\frac{5}{2})$

9. $\Gamma(-\frac{7}{2})$

10. $\Gamma(-\frac{9}{2})$

In Problems 11–14, use (7) and the given numerical value to evaluate the indicated quantity.

11. $\Gamma(0.2) = 4.5908$; $\Gamma(1.2)$

12. $\Gamma(-0.1) = -10.6863$; $\Gamma(0.9)$

13. $\Gamma(-0.3) = -4.3269$; $\Gamma(1.7)$

14. $\Gamma(1.5) = 0.8862$; $\Gamma(3.5)$

15. Express $\Gamma(x + 4)$ in terms of $\Gamma(x)$.

16. For $c > 0, x > 0$, use a substitution to show that

$$\int_0^\infty t^{x-1} e^{-ct} \, dt = c^{-x} \Gamma(x).$$

In Problems 17–20, use the result in Problem 16 to express the given improper integral as a gamma function. Then evaluate the integral.

17. $\displaystyle\int_0^\infty \frac{e^{-2t}}{\sqrt{t}} \, dt$

18. $\displaystyle\int_0^\infty t^{3/2} e^{-4t} \, dt$

19. $\displaystyle\int_0^\infty t^4 e^{-6t} \, dt$

20. $\displaystyle\int_0^\infty t^3 e^{-t/3} \, dt$

In Problems 21–24, use the substitution $t = -\ln x$ and the result in Problem 16 to express the given improper integral as a gamma function. Then evaluate the integral.

21. $\displaystyle\int_0^1 x^3 \left(\ln \frac{1}{x}\right)^3 dx$

22. $\displaystyle\int_0^1 x^4 \left(\ln \frac{1}{x}\right)^2 dx$

23. $\displaystyle\int_0^1 \left(\frac{1}{x} \ln \frac{1}{x}\right)^{1/2} dx$

24. $\displaystyle\int_0^1 x^{1/2} \left(\ln \frac{1}{x}\right)^{-1/2} dx$

In Problems 25 and 26, use the indicated substitution to express the given improper integral as a gamma function. Then evaluate the integral using the indicated numerical value.

25. $\displaystyle\int_0^\infty x^5 e^{-x^5} dx$; $t = x^5$, $\Gamma(\frac{6}{5}) = 0.9182$

26. $\displaystyle\int_0^\infty x^4 e^{-x^3} dx$; $t = x^3$, $\Gamma(\frac{5}{3}) = 0.9027$

27. **(a)** For $x > 0$ the gamma function possesses derivatives of all orders. Use Leibniz's rule in the form given in (24) to show that

$$\Gamma'(x) = \int_0^\infty t^{x-1} e^{-t} \ln t \, dt$$

$$\Gamma''(x) = \int_0^\infty t^{x-1} e^{-t} (\ln t)^2 \, dt.$$

(b) For $x > 0$, $\Gamma(x) > 0$ and $\Gamma''(x) > 0$. Without looking back at Figure A.1, what does this say about the graph of $\Gamma(x)$?

28. A definition of the gamma function given by Carl Friedrich Gauss in 1811 that is valid for all real numbers, except $x = 0, x = -1, x = -2, \ldots$, is

$$\Gamma(x) = \lim_{n \to \infty} \frac{n! \, n^x}{x(x + 1)(x + 2) \cdots (x + n)}.$$

Use this definition to give an alternative derivation of the recursion formula (7).

The Beta Function

In Problems 29 and 30, express the given improper integral as a beta function. Then evaluate the integral using (15) and known values of the gamma function.

29. $\displaystyle\int_0^1 \frac{1}{\sqrt{t(1 - t)}} \, dt$

30. $\displaystyle\int_0^1 \sqrt{t} \sqrt{(1 - t)^3} \, dt$

In Problems 31 and 32, use the indicated substitution to express the given improper integral as a beta function. Then evaluate the integral using (15) and known values of the gamma function.

31. $\displaystyle\int_0^1 \frac{x^2}{\sqrt{1 - x^2}} \, dx$; $t = x^2$

32. $\displaystyle\int_0^1 \frac{x^5}{\sqrt{1 - x^4}} \, dx$; $t = x^4$

In Problems 33 and 34, use the substitution $t = x/2$ to express the given improper integral as a beta function. Then evaluate the integral using (15) and known values of the gamma function.

33. $\displaystyle\int_0^2 x^{-1/2} (2 - x)^{5/2} \, dx$

34. $\displaystyle\int_0^2 \frac{x^2}{\sqrt{2 - x}} \, dx$

35. Show that the beta function is symmetric in x and y; that is, $B(x, y) = B(y, x)$.

36. Show that if $m + 1$ and $n + 1$ are positive integers, then (15) becomes

$$B(m + 1, n + 1) = \frac{m! \, n!}{(m + n + 1)!}.$$

37. Use the substitution $t = \sin^2\theta$ to show that the beta function (14) can be expressed as

$$B(x, y) = 2 \int_0^{\pi/2} \sin^{2x-1}\theta \, \cos^{2y-1}\theta \, d\theta.$$

38. If $m = 2x - 1$ and $n = 2y - 1$, then show that the integral in Problem 37 can be written

$$\int_0^{\pi/2} \sin^m \theta \cos^n \theta \, d\theta = \frac{\Gamma\left(\dfrac{m+1}{2}\right)\Gamma\left(\dfrac{n+1}{2}\right)}{2\Gamma\left(\dfrac{m+n+2}{2}\right)}, \quad m > -1, n > -1.$$

In Problems 39–42, use the result in Problem 38 to evaluate the given integral.

39. $\displaystyle\int_0^{\pi/2} \sin^3 \theta \cos^3 \theta \, d\theta$

40. $\displaystyle\int_0^{\pi/2} \sin^6 \theta \cos^5 \theta \, d\theta$

41. $\displaystyle\int_0^{\pi/2} \cos^5 \theta \, d\theta$

42. $\displaystyle\int_0^{\pi/2} \sin^7 \theta \, d\theta$

Leibniz's Rule

In Problems 43–48, use Leibniz's rule in the forms given in (22) and (23) to find the derivative of the given function.

43. $F(x) = \displaystyle\int_x^{2x^3} e^{-x^3 t^2} dt$

44. $F(x) = \displaystyle\int_0^{x^2} \frac{\sin(x\sqrt{t})}{t} dt$

45. $F(x) = \displaystyle\int_{x^2}^{\cos x} (x^2 + t^2)^{10} dt$

46. $F(x) = \displaystyle\int_{\sqrt{3x}}^{\sin 2x} \ln(x^2 + t^2) dt$

47. $F(x) = \displaystyle\int_{\pi/2}^{\pi - x} e^{\sin t} \sin(\cos t) dt$

48. $F(x) = \displaystyle\int_{\sqrt{x}}^{4\sqrt{x}} e^{t^2} dt$

In Problems 49 and 50, use Leibniz's rule in the form given in (24) to show that the indicated integral-defined function is a solution of the given second-order differential equation. [*Hint*: After computing the first and second derivatives find a term in y' that can be evaluated using integration by parts.]

49. $xy'' + y' + xy = 0; \ y = \displaystyle\int_0^{\pi} \cos(x \cos t) dt$

50. $xy'' + y' - xy = 0; \ y = \displaystyle\int_0^{\pi} e^{x \cos t} dt$

51. Assume that Leibniz's rule (24) is valid for the improper integral

$$y = e^{x^2} \int_0^{\infty} t^2 e^{-t^2} \cos 2xt \, dt.$$

Show that this integral-defined function is a solution of the second-order differential equation $y'' + 2xy' + 4y = 0$. [*Hint*: After computing the first and second derivatives find a term in y'' that can be evaluated using integration by parts.]

52. (a) Find the derivative of the error function erf(x). Then use the identity erf(x) + erfc(x) = 1 to find the derivative of the complementary error function erfc(x).

(b) Use the last result in part (a) to show that

$$y = e^{-x^2} - 2x \int_x^{\infty} e^{-t^2} dt$$

is a solution of the differential equation $y'' + 2xy' - 2y = 0$.

53. (a) Show that for $x > 0$,

$$\int_0^{\infty} \frac{1}{x^2 + t^2} dt = \frac{\pi}{2x}.$$

[*Hint*: See formula 36 in the table of integrals given on the right inside page of the front cover.]

(b) Use the result in part (a) and (24) to show that

$$\int_0^{\infty} \frac{1}{(x^2 + t^2)^2} dt = \frac{\pi}{4x^3}.$$

54. Consider the function

$$F(x) = \int_0^1 \frac{t^x - 1}{\ln t} dt, \quad x \geq 0.$$

(a) Use (24) to find $F'(x)$ as an integral. Evaluate this integral.

(b) Use the result in part (a) to find $F(x) = \int F'(x) \, dx + C$, where C is a constant.

(c) Use the result in part (b) and $F(0) = 0$ to find an explicit function $F(x)$.

55. Consider the function

$$F(x) = \int_0^{\infty} \frac{1 - e^{-xt^2}}{t^2} dt, \quad x \geq 0.$$

(a) Use (24) to find $F'(x)$ as an integral. Evaluate this integral using a substitution and the known result $\int_0^{\infty} e^{-u^2} du = \sqrt{\pi}/2$. (See page APP-4.)

(b) Use the result in part (a) to find $F(x) = \int F'(x) \, dx + C$, where C is a constant.

(c) Use the result in part (b) and $F(0) = 0$ to find an explicit function $F(x)$.

56. Consider the function

$$F(x) = \int_0^{\infty} e^{-tx} \frac{\sin t}{t} dt, \quad x \geq 0.$$

(a) Use (24) to find $F'(x)$ as an integral. Evaluate this integral using integration by parts.

(b) Use the result in part (a) to find $F(x) = \int F'(x) \, dx + C$, where C is a constant.

(c) Use the result in part (b) and $F(0) = \pi/2$ to find an explicit function $F(x)$.

57. Express the function $F(x) = \sqrt{\dfrac{2}{\pi}} \displaystyle\int_0^x e^{-t^2/2} dt$ in terms of erf(x).

58. Evaluate the integral $\sqrt{\dfrac{2}{\pi}} \displaystyle\int_{x_0}^{x_1} e^{-t^2/2} dt$ in terms of erf(x).

Derivative and Integral Formulas

Differentiation Rules

1. Constant: $\dfrac{d}{dx}c = 0$

2. Constant Multiple: $\dfrac{d}{dx}cf(x) = cf'(x)$

3. Sum: $\dfrac{d}{dx}[f(x) \pm g(x)] = f'(x) \pm g'(x)$

4. Product: $\dfrac{d}{dx}f(x)g(x) = f(x)g'(x) + g(x)f'(x)$

5. Quotient: $\dfrac{d}{dx}\dfrac{f(x)}{g(x)} = \dfrac{g(x)f'(x) - f(x)g'(x)}{[g(x)]^2}$

6. Chain: $\dfrac{d}{dx}f(g(x)) = f'(g(x))g'(x)$

7. Power: $\dfrac{d}{dx}x^n = nx^{n-1}$

8. Power: $\dfrac{d}{dx}[g(x)]^n = n[g(x)]^{n-1}g'(x)$

Derivatives of Functions

Trigonometric:

9. $\dfrac{d}{dx}\sin x = \cos x$

10. $\dfrac{d}{dx}\cos x = -\sin x$

11. $\dfrac{d}{dx}\tan x = \sec^2 x$

12. $\dfrac{d}{dx}\cot x = -\csc^2 x$

13. $\dfrac{d}{dx}\sec x = \sec x \tan x$

14. $\dfrac{d}{dx}\csc x = -\csc x\cot x$

Inverse trigonometric:

15. $\dfrac{d}{dx}\sin^{-1}x = \dfrac{1}{\sqrt{1 - x^2}}$

16. $\dfrac{d}{dx}\cos^{-1}x = -\dfrac{1}{\sqrt{1 - x^2}}$

17. $\dfrac{d}{dx}\tan^{-1}x = \dfrac{1}{1 + x^2}$

18. $\dfrac{d}{dx}\cot^{-1}x = -\dfrac{1}{1 + x^2}$

19. $\dfrac{d}{dx}\sec^{-1}x = \dfrac{1}{|x|\sqrt{x^2 - 1}}$

20. $\dfrac{d}{dx}\csc^{-1}x = -\dfrac{1}{|x|\sqrt{x^2 - 1}}$

Hyperbolic:

21. $\dfrac{d}{dx}\sinh x = \cosh x$

22. $\dfrac{d}{dx}\cosh x = \sinh x$

23. $\dfrac{d}{dx}\tanh x = \text{sech}^2 x$

24. $\dfrac{d}{dx}\coth x = -\text{csch}^2 x$

25. $\dfrac{d}{dx}\text{sech}\,x = -\text{sech}\,x \tanh x$

26. $\dfrac{d}{dx}\text{csch}\,x = -\text{csch}\,x \coth x$

Inverse hyperbolic:

27. $\dfrac{d}{dx}\sinh^{-1}x = \dfrac{1}{\sqrt{x^2 + 1}}$

28. $\dfrac{d}{dx}\cosh^{-1}x = \dfrac{1}{\sqrt{x^2 - 1}}$

29. $\dfrac{d}{dx}\tanh^{-1}x = \dfrac{1}{1 - x^2},\ |x| < 1$

30. $\dfrac{d}{dx}\coth^{-1}x = \dfrac{1}{1 - x^2},\ |x| > 1$

31. $\dfrac{d}{dx}\text{sech}^{-1}x = -\dfrac{1}{x\sqrt{1 - x^2}}$

32. $\dfrac{d}{dx}\text{csch}^{-1}x = -\dfrac{1}{|x|\sqrt{x^2 + 1}}$

Exponential:

33. $\dfrac{d}{dx} e^x = e^x$ **34.** $\dfrac{d}{dx} b^x = b^x(\ln b)$

Logarithmic:

35. $\dfrac{d}{dx} \ln|x| = \dfrac{1}{x}$ **36.** $\dfrac{d}{dx} \log_b x = \dfrac{1}{x(\ln b)}$

Of an integral:

37. $\dfrac{d}{dx} \displaystyle\int_a^x g(t)\,dt = g(x)$ **38.** $\dfrac{d}{dx} \displaystyle\int_a^b g(x, t)\,dt = \int_a^b \dfrac{\partial}{\partial x} g(x, t)\,dt$

Integration Formulas

1. $\displaystyle\int u^n\,du = \dfrac{u^{n+1}}{n+1} + C,\, n \neq -1$ **2.** $\displaystyle\int \dfrac{1}{u}\,du = \ln|u| + C$ **3.** $\displaystyle\int e^u\,du = e^u + C$

4. $\displaystyle\int b^u\,du = \dfrac{1}{\ln b} b^u + C$ **5.** $\displaystyle\int \sin u\,du = -\cos u + C$ **6.** $\displaystyle\int \cos u\,du = \sin u + C$

7. $\displaystyle\int \sec^2 u\,du = \tan u + C$ **8.** $\displaystyle\int \csc^2 u\,du = -\cot u + C$ **9.** $\displaystyle\int \sec u \tan u\,du = \sec u + C$

10. $\displaystyle\int \csc u \cot u\,du = -\csc u + C$ **11.** $\displaystyle\int \tan u\,du = -\ln|\cos u| + C$ **12.** $\displaystyle\int \cot u\,du = \ln|\sin u| + C$

13. $\displaystyle\int \sec u\,du = \ln|\sec u + \tan u| + C$ **14.** $\displaystyle\int \csc u\,du = \ln|\csc u - \cot u| + C$ **15.** $\displaystyle\int u \sin u\,du = \sin u - u\cos u + C$

16. $\displaystyle\int u\cos u\,du = \cos u + u \sin u + C$ **17.** $\displaystyle\int \sin^2 u\,du = \tfrac{1}{2}u - \tfrac{1}{4}\sin 2u + C$ **18.** $\displaystyle\int \cos^2 u\,du = \tfrac{1}{2}u + \tfrac{1}{4}\sin 2u + C$

19. $\displaystyle\int \sin au \sin bu\,du = \dfrac{\sin(a-b)u}{2(a-b)} - \dfrac{\sin(a+b)u}{2(a+b)} + C$ **20.** $\displaystyle\int \cos au \cos bu\,du = \dfrac{\sin(a-b)u}{2(a-b)} + \dfrac{\sin(a+b)u}{2(a+b)} + C$

21. $\displaystyle\int e^{au} \sin bu\,du = \dfrac{e^{au}}{a^2+b^2}(a \sin bu - b \cos bu) + C$ **22.** $\displaystyle\int e^{au} \cos bu\,du = \dfrac{e^{au}}{a^2+b^2}(a \cos bu + b \sin bu) + C$

23. $\displaystyle\int \sinh u\,du = \cosh u + C$ **24.** $\displaystyle\int \cosh u\,du = \sinh u + C$

25. $\displaystyle\int \operatorname{sech}^2 u\,du = \tanh u + C$ **26.** $\displaystyle\int \operatorname{csch}^2 u\,du = -\coth u + C$

27. $\displaystyle\int \tanh u\,du = \ln(\cosh u) + C$ **28.** $\displaystyle\int \coth u\,du = \ln|\sinh u| + C$

29. $\displaystyle\int \ln u\,du = u \ln u - u + C$ **30.** $\displaystyle\int u \ln u\,du = \tfrac{1}{2}u^2 \ln u - \tfrac{1}{4}u^2 + C$

31. $\displaystyle\int \dfrac{1}{\sqrt{a^2-u^2}}\,du = \sin^{-1}\dfrac{u}{a} + C$ **32.** $\displaystyle\int \dfrac{1}{\sqrt{a^2+u^2}}\,du = \ln\left|u + \sqrt{a^2+u^2}\right| + C$

33. $\displaystyle\int \sqrt{a^2-u^2}\,du = \dfrac{u}{2}\sqrt{a^2-u^2} + \dfrac{a^2}{2}\sin^{-1}\dfrac{u}{a} + C$ **34.** $\displaystyle\int \sqrt{a^2+u^2}\,du = \dfrac{u}{2}\sqrt{a^2+u^2} + \dfrac{a^2}{2}\ln\left|u + \sqrt{a^2+u^2}\right| + C$

35. $\displaystyle\int \dfrac{1}{a^2-u^2}\,du = \dfrac{1}{2a}\ln\left|\dfrac{a+u}{a-u}\right| + C$ **36.** $\displaystyle\int \dfrac{1}{a^2+u^2}\,du = \dfrac{1}{a}\tan^{-1}\dfrac{u}{a} + C$

37. $\displaystyle\int \dfrac{1}{\sqrt{u^2-a^2}}\,du = \ln\left|u + \sqrt{u^2-a^2}\right| + C$ **38.** $\displaystyle\int \sqrt{u^2-a^2}\,du = \dfrac{u}{2}\sqrt{u^2-a^2} - \dfrac{a^2}{2}\ln\left|u + \sqrt{u^2-a^2}\right| + C$

Laplace Transforms

$f(t)$	$\mathscr{L}\{f(t)\} = F(s)$
1. 1	$\dfrac{1}{s}$
2. t	$\dfrac{1}{s^2}$
3. t^n	$\dfrac{n!}{s^{n+1}}$, n a positive integer
4. $t^{-1/2}$	$\dfrac{\sqrt{\pi}}{s^{1/2}}$
5. $t^{1/2}$	$\dfrac{\sqrt{\pi}}{2s^{3/2}}$
6. t^{α}	$\dfrac{\Gamma(\alpha+1)}{s^{\alpha+1}}$, $\alpha > -1$
7. $\sin kt$	$\dfrac{k}{s^2+k^2}$
8. $\cos kt$	$\dfrac{s}{s^2+k^2}$
9. $\sin^2 kt$	$\dfrac{2k^2}{s(s^2+4k^2)}$
10. $\cos^2 kt$	$\dfrac{s^2+2k^2}{s(s^2+4k^2)}$
11. e^{at}	$\dfrac{1}{s-a}$
12. $\sinh kt$	$\dfrac{k}{s^2-k^2}$
13. $\cosh kt$	$\dfrac{s}{s^2-k^2}$
14. $\sinh^2 kt$	$\dfrac{2k^2}{s(s^2-4k^2)}$
15. $\cosh^2 kt$	$\dfrac{s^2-2k^2}{s(s^2-4k^2)}$
16. $e^{at}t$	$\dfrac{1}{(s-a)^2}$

17. $e^{at}t^n$ $\qquad\qquad \dfrac{n!}{(s-a)^{n+1}}$, $\;n$ a positive integer

18. $e^{at}\sin kt$ $\qquad\qquad \dfrac{k}{(s-a)^2+k^2}$

19. $e^{at}\cos kt$ $\qquad\qquad \dfrac{s-a}{(s-a)^2+k^2}$

20. $e^{at}\sinh kt$ $\qquad\qquad \dfrac{k}{(s-a)^2-k^2}$

21. $e^{at}\cosh kt$ $\qquad\qquad \dfrac{s-a}{(s-a)^2-k^2}$

22. $t\sin kt$ $\qquad\qquad \dfrac{2ks}{(s^2+k^2)^2}$

23. $t\cos kt$ $\qquad\qquad \dfrac{s^2-k^2}{(s^2+k^2)^2}$

24. $\sin kt + kt\cos kt$ $\qquad\qquad \dfrac{2ks^2}{(s^2+k^2)^2}$

25. $\sin kt - kt\cos kt$ $\qquad\qquad \dfrac{2k^3}{(s^2+k^2)^2}$

26. $t\sinh kt$ $\qquad\qquad \dfrac{2ks}{(s^2-k^2)^2}$

27. $t\cosh kt$ $\qquad\qquad \dfrac{s^2+k^2}{(s^2-k^2)^2}$

28. $\dfrac{e^{at}-e^{bt}}{a-b}$ $\qquad\qquad \dfrac{1}{(s-a)(s-b)}$

29. $\dfrac{ae^{at}-be^{bt}}{a-b}$ $\qquad\qquad \dfrac{s}{(s-a)(s-b)}$

30. $1-\cos kt$ $\qquad\qquad \dfrac{k^2}{s(s^2+k^2)}$

31. $\cosh kt - 1$ $\qquad\qquad \dfrac{k^2}{s(s^2-k^2)}$

32. $kt-\sin kt$ $\qquad\qquad \dfrac{k^3}{s^2(s^2+k^2)}$

33. $a\sin bt - b\sin at$ $\qquad\qquad \dfrac{ab(a^2-b^2)}{(s^2+a^2)(s^2+b^2)}$

34. $\cos at - \cos bt$ $\qquad\qquad \dfrac{s(b^2-a^2)}{(s^2+a^2)(s^2+b^2)}$

35. $\sin kt\sinh kt$ $\qquad\qquad \dfrac{2k^2s}{s^4+4k^4}$

36. $\sin kt\cosh kt$ $\qquad\qquad \dfrac{k(s^2+2k^2)}{s^4+4k^4}$

37. $\cos kt\sinh kt$ $\qquad\qquad \dfrac{k(s^2-2k^2)}{s^4+4k^4}$

38. $\cos kt \cosh kt$

$$\frac{s^3}{s^4 + 4k^4}$$

39. $\sin kt \cosh kt + \cos kt \sinh kt$

$$\frac{2ks^2}{s^4 + 4k^4}$$

40. $\sin kt \cosh kt - \cos kt \sinh kt$

$$\frac{4k^3}{s^4 + 4k^4}$$

41. $\sinh kt + \sin kt$

$$\frac{2ks^2}{s^4 - k^4}$$

42. $\sinh kt - \sin kt$

$$\frac{2k^3}{s^4 - k^4}$$

43. $\cosh kt + \cos kt$

$$\frac{2s^3}{s^4 - k^4}$$

44. $\cosh kt - \cos kt$

$$\frac{2k^2 s}{s^4 - k^4}$$

45. $J_0(kt)$

$$\frac{1}{\sqrt{s^2 + k^2}}$$

46. $I_0(kt)$

$$\frac{1}{\sqrt{s^2 - k^2}}$$

47. $\dfrac{e^{bt} - e^{at}}{t}$

$$\ln\left(\frac{s - a}{s - b}\right)$$

48. $\dfrac{2(1 - \cos kt)}{t}$

$$\ln\left(\frac{s^2 + k^2}{s^2}\right)$$

49. $\dfrac{2(1 - \cosh kt)}{t}$

$$\ln\left(\frac{s^2 - k^2}{s^2}\right)$$

50. $\dfrac{\sin kt}{t}$

$$\arctan\left(\frac{k}{s}\right)$$

51. $\dfrac{\sin at \cos bt}{t}$

$$\frac{1}{2}\arctan\left(\frac{a + b}{s}\right)$$
$$+ \frac{1}{2}\arctan\left(\frac{a - b}{s}\right)$$

52. $\dfrac{\sin (2\sqrt{kt})}{\sqrt{\pi k}}$

$$\frac{e^{-k/s}}{s\sqrt{s}}$$

53. $\dfrac{\cos (2\sqrt{kt})}{\sqrt{\pi t}}$

$$\frac{e^{-k/s}}{\sqrt{s}}$$

54. $\dfrac{\sinh(2\sqrt{kt})}{\sqrt{\pi k}}$

$$\frac{e^{k/s}}{s\sqrt{s}}$$

55. $\dfrac{\cosh(2\sqrt{kt})}{\sqrt{\pi t}}$ $\dfrac{e^{k/s}}{\sqrt{s}}$

56. $\dfrac{1}{\sqrt{\pi t}}\, e^{-a^2/4t}$ $\dfrac{e^{-a\sqrt{s}}}{\sqrt{s}}$

57. $\dfrac{a}{2\sqrt{\pi t^3}}\, e^{-a^2/4t}$ $e^{-a\sqrt{s}}$

58. $\mathrm{erfc}\left(\dfrac{a}{2\sqrt{t}}\right)$ $\dfrac{e^{-a\sqrt{s}}}{s}$

59. $\mathrm{erf}(at)$ $\dfrac{e^{s^2/4a^2}}{s}\,\mathrm{erfc}\left(\dfrac{s}{2a}\right)$

60. $\mathrm{erf}(\sqrt{at})$ $\dfrac{\sqrt{a}}{s\sqrt{s+a}}$

61. $e^{a^2 t}\mathrm{erf}(a\sqrt{t})$ $\dfrac{a}{\sqrt{s}(s-a^2)}$

62. $\mathrm{erfc}(\sqrt{at})$ $\dfrac{\sqrt{s+a}-\sqrt{a}}{s\sqrt{s+a}}$

63. $e^{a^2 t}\mathrm{erfc}(a\sqrt{t})$ $\dfrac{1}{s+a\sqrt{s}}$

64. $\dfrac{1}{\sqrt{\pi t}} + ae^{a^2 t}\mathrm{erf}(a\sqrt{t})$ $\dfrac{\sqrt{s}}{s-a^2}$

65. $\dfrac{1}{\sqrt{\pi t}} - ae^{a^2 t}\mathrm{erfc}(a\sqrt{t})$ $\dfrac{1}{\sqrt{s}+a}$

66. $2\sqrt{\dfrac{t}{\pi}}\, e^{-a^2/4t} - a\,\mathrm{erfc}\left(\dfrac{a}{2\sqrt{t}}\right)$ $\dfrac{e^{-a\sqrt{s}}}{s\sqrt{s}}$

67. $-e^{ab}e^{b^2 t}\mathrm{erfc}\left(b\sqrt{t}+\dfrac{a}{2\sqrt{t}}\right) + \mathrm{erfc}\left(\dfrac{a}{2\sqrt{t}}\right)$ $\dfrac{be^{-a\sqrt{s}}}{s(\sqrt{s}+b)}$

68. $e^{ab}e^{b^2 t}\mathrm{erfc}\left(b\sqrt{t}+\dfrac{a}{2\sqrt{t}}\right)$ $\dfrac{e^{-a\sqrt{s}}}{s+b\sqrt{s}}$

69. $f(at)$ $\dfrac{1}{a}F\left(\dfrac{s}{a}\right)$

70. $e^{at}f(t)$ $F(s-a)$

71. $\mathcal{U}(t - a)$

$$\frac{e^{-as}}{s}$$

72. $f(t - a)\mathcal{U}(t - a)$

$$e^{-as}F(s)$$

73. $g(t)\mathcal{U}(t - a)$

$$e^{-as}\mathcal{L}\{g(t + a)\}$$

74. $f^{(n)}(t)$

$$s^nF(s) - s^{(n-1)}f(0) - \cdots - f^{(n-1)}(0)$$

75. $t^nf(t)$

$$(-1)^n \frac{d^n}{ds^n} F(s)$$

76. $\displaystyle\int_0^t f(\tau)g(t - \tau)\,d\tau$

$$F(s)G(s)$$

77. $\displaystyle\int_0^t f(\tau)\,d\tau$

$$\frac{F(s)}{s}$$

78. $\dfrac{f(t)}{t}$

$$\int_s^\infty F(s)\,ds$$

79. $f(t) = f(t + T)$

$$\frac{1}{1 - e^{-sT}}\int_0^T e^{-st}f(t)\,dt$$

80. $\delta(t - t_0)$

$$e^{-st_0}$$

APPENDIX **D**

Conformal Mappings

Elementary Mappings

E-1

$$w = z + z_0$$

E-2

$$w = e^{i\theta}z$$

E-3

$$w = \alpha z, \alpha > 0$$

E-4

$$w = z^\alpha, \alpha > 0$$

E-5

$$w = e^z$$
$$z = \operatorname{Ln} w$$

E-6

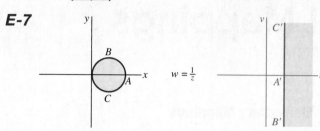

$$w = \sin z$$
$$z = \sin^{-1} w$$

E-7

$$w = \frac{1}{z}$$

E-8

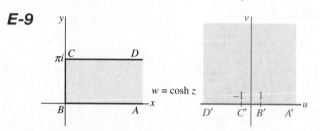

$$w = \log_e |z| + i \,\mathrm{Arg}\, z$$
$$a > 1$$

E-9

$$w = \cosh z$$

Mappings to Half-Planes

H-1

$$w = i\frac{1-z}{1+z}$$

H-2

width = a

$$w = e^{\pi z/a}$$

H-3

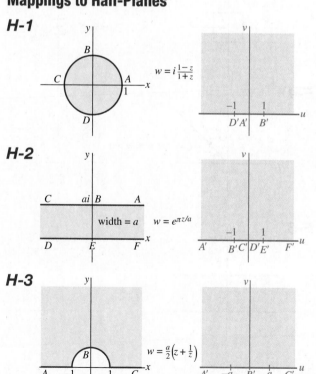

$$w = \frac{a}{2}\left(z + \frac{1}{z}\right)$$

H-4

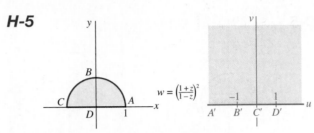

$$w = \cos\left(\frac{\pi z}{a}\right)$$

H-5

$$w = \left(\frac{1+z}{1-z}\right)^2$$

H-6

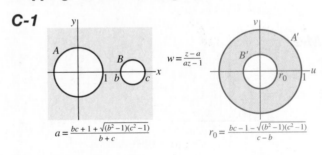

$$w = \frac{e^{\pi/z} + e^{-\pi/z}}{e^{\pi/z} - e^{-\pi/z}}$$

Mappings to Circular Regions

C-1

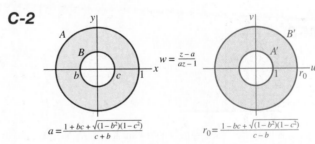

$$w = \frac{z-a}{az-1}$$

$$a = \frac{bc + 1 + \sqrt{(b^2-1)(c^2-1)}}{b+c} \qquad r_0 = \frac{bc - 1 - \sqrt{(b^2-1)(c^2-1)}}{c-b}$$

C-2

$$w = \frac{z-a}{az-1}$$

$$a = \frac{1 + bc + \sqrt{(1-b^2)(1-c^2)}}{c+b} \qquad r_0 = \frac{1 - bc + \sqrt{(1-b^2)(1-c^2)}}{c-b}$$

C-3

$$w = e^z$$

C-4

$$w = \frac{i-z}{i+z}$$

C-5

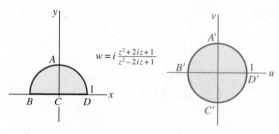

$$w = i\,\frac{z^2+2iz+1}{z^2-2iz+1}$$

Miscellaneous Mappings

M-1

$$w = z + e^z + 1$$

M-2

$$w = \frac{a}{\pi}\left[(z^2-1)^{1/2} + \cosh^{-1} z\right]$$

M-3

$$w = \frac{2a}{\pi}\left[(z^2-1)^{1/2} + \sin^{-1}(1/z)\right]$$

M-4

$$w = a(z^2-1)^{1/2}$$

M-5

$$w = 2\zeta + \mathrm{Ln}\!\left(\frac{\zeta-1}{\zeta+1}\right)$$
$$\zeta = (z+1)^{1/2}$$

M-6

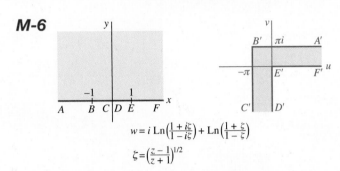

$$w = i\,\mathrm{Ln}\!\left(\frac{1+i\zeta}{1-i\zeta}\right) + \mathrm{Ln}\!\left(\frac{1+\zeta}{1-\zeta}\right)$$
$$\zeta = \left(\frac{z-1}{z+1}\right)^{1/2}$$

M-7

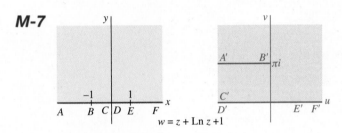

$$w = z + \mathrm{Ln}\,z + 1$$

M-8

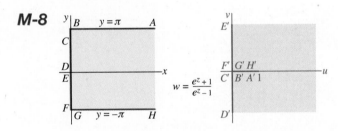

$$w = \frac{e^z + 1}{e^z - 1}$$

M-9

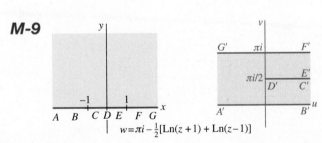

$$w = \pi i - \tfrac{1}{2}[\mathrm{Ln}(z+1) + \mathrm{Ln}(z-1)]$$

M-10

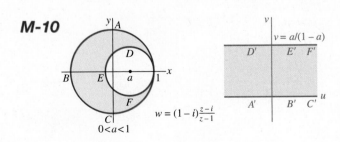

$$0 < a < 1$$

$$w = (1 - i)\frac{z - i}{z - 1}$$

Exercises 1.1, Page 11

1. linear, second order
3. linear, fourth order
5. nonlinear, second order
7. linear, third order
9. nonlinear, first order
11. linear in x but nonlinear in y
17. Domain of function is $[-2, \infty)$; largest interval of definition for solution is $(-2, \infty)$.
19. Domain of function is the set of real numbers except $x = 2$ and $x = -2$; largest intervals of definition for solution are $(-\infty, -2)$ $(-2, 2)$ or $(2, \infty)$.
21. $X = \dfrac{e^t - 1}{e^t - 2}$ defined on $(-\infty, \ln 2)$ or on $(\ln 2, \infty)$
33. $m = -2$
35. $m = 2, m = 3$
37. $m = 0, m = -1$
39. $m = 3, m = 5$
41. $y = 2$
43. no constant solutions
45. $y = 0$
47. $y = -4, y = 4$

Exercises 1.2, Page 17

1. $y = 1/(1 - 4e^{-x})$
3. $y = 1/(x^2 - 1); (1, \infty)$
5. $y = 1/(x^2 + 1); (-\infty, \infty)$
7. $x = -\cos t + 8 \sin t$
9. $x = \frac{\sqrt{3}}{4} \cos t + \frac{1}{4} \sin t$
11. $y = \frac{3}{2} e^x - \frac{1}{2} e^{-x}$
13. $y = 5e^{-x-1}$
15. $y = 0, y = x^3$
17. half-planes defined by either $y > 0$ or $y < 0$
19. half-planes defined by either $x > 0$ or $x < 0$
21. the regions defined by $y > 2$, $y < -2$, or $-2 < y < 2$
23. any region not containing $(0, 0)$
25. yes
27. no
29. (a) $y = cx$
 (b) any rectangular region not touching the y-axis
 (c) No, the function is not differentiable at $x = 0$.

31. (b) $y = 1/(1 - x)$ on $(-\infty, 1)$;
 $y = -1/(x + 1)$ on $(-1, \infty)$
39. $y = -\sin 3x$
41. $y = 0$
43. no solution

Exercises 1.3, Page 25

1. $\dfrac{dP}{dt} = kP + r; \quad \dfrac{dP}{dt} = kP - r$
3. $\dfrac{dP}{dt} = k_1 P - k_2 P^2$
7. $\dfrac{dx}{dt} = kx(1000 - x)$
9. $\dfrac{dA}{dt} + \dfrac{1}{100} A = 0; \quad A(0) = 50$
11. $\dfrac{dA}{dt} + \dfrac{7}{600 - t} A = 6$
13. $\dfrac{dh}{dt} = -\dfrac{c\pi}{450} \sqrt{h}$
15. $L \dfrac{di}{dt} + Ri = E(t)$
17. $m \dfrac{dv}{dt} = mg - kv^2$
19. $m \dfrac{d^2 x}{dt^2} = -kx$
21. $m \dfrac{dv}{dt} + v \dfrac{dm}{dt} + kv = -mg + R$
23. $\dfrac{d^2 r}{dt^2} + \dfrac{gR^2}{r^2} = 0$
25. $\dfrac{dA}{dt} = k(M - A), k > 0$
27. $\dfrac{dx}{dt} + kx = r, k > 0$
29. $\dfrac{dy}{dx} = \dfrac{-x + \sqrt{x^2 + y^2}}{y}$

Chapter 1 in Review, Page 30

1. $\dfrac{dy}{dx} = ky$
3. $y'' + k^2 y = 0$
5. $y'' - 2y' + y = 0$
7. (a), (d)
9. (b)
11. (b)
13. $y = c_1$ and $y = c_2 e^x$, c_1 and c_2 constants

15. $y' = x^2 + y^2$

17. (a) The domain is the set of all real numbers.
(b) either $(-\infty, 0)$ or $(0, \infty)$

19. For $x_0 = -1$, the interval is $(-\infty, 0)$, and for $x_0 = 2$, the interval is $(0, \infty)$.

21. (c) $y = \begin{cases} -x^2, & x < 0 \\ x^2, & x \geq 0 \end{cases}$ **23.** $(-\infty, \infty)$

25. $(0, \infty)$ **35.** $y = \frac{5}{9} + \cos 3x$

37. $y' + y = 4x - 2$ **39.** $y = e^{-3x} - e^x + 4x$

41. $y = -\frac{3}{2}e^{-3x+3} - \frac{9}{2}e^{x-1} + 4x$

45. $y_0 = -3, y_1 = 0$

47. $x\dfrac{d^2x}{dt^2} + \left(\dfrac{dx}{dt}\right)^2 + 32x = 160$

Exercises 2.1, Page 41

21. 0 is asymptotically stable (attractor); 3 is unstable (repeller).

23. 2 is semi-stable.

25. -2 is unstable (repeller); 0 is semi-stable; 2 is asymptotically stable (attractor).

27. -1 is asymptotically stable (attractor); 0 is unstable (repeller).

39. $0 < P_0 < h/k$ **41.** $\sqrt{mg/k}$

Exercises 2.2, Page 49

1. $y = -\frac{1}{5}\cos 5x + c$ **3.** $y = \frac{1}{3}e^{-3x} + c$
5. $y = cx^4$ **7.** $-3e^{-2y} = 2e^{3x} + c$
9. $\frac{1}{3}x^3 \ln x - \frac{1}{9}x^3 = \frac{1}{2}y^2 + 2y + \ln|y| + c$
11. $4\cos y = 2x + \sin 2x + c$
13. $(e^x + 1)^{-2} + 2(e^y + 1)^{-1} = c$

15. $S = ce^{kr}$ **17.** $P = \dfrac{ce^t}{1 + ce^t}$

19. $(y + 3)^5 e^x = c(x + 4)^5 e^y$

21. $y = \sin(\frac{1}{2}x^2 + c)$ **23.** $x = \tan(4t - \frac{3}{4}\pi)$

25. $y = \dfrac{e^{-(1+1/x)}}{x}$

27. $y = \frac{1}{2}x + \frac{1}{2}\sqrt{3}\sqrt{1 - x^2}$

29. $y = e^{e^{-x}}$

31. $y(x) = e^{\int_4^x e^{-t^2} dt}$

33. $y(x) = \tan\left(\pi/4 + \displaystyle\int_1^x \sqrt{1 + \cos t^3}\, dt\right)$

35. $y = -\sqrt{x^2 + x - 1}, (-\infty, -\frac{1}{2} - \frac{\sqrt{5}}{2})$
37. $y = -\ln(2 - e^x), (-\infty, \ln 2)$

39. (a) $y = 2, y = -2, y = 2\,\dfrac{3 - e^{4x-1}}{3 + e^{4x-1}}$

43. $y = 1$ **45.** $y = 1 + \frac{1}{10}\tan\left(\frac{1}{10}x\right)$

49. $y = \tan x - \sec x + c$

51. $y = [c(1 + \sqrt{x}) - 1]^2$

53. $y = 2\sqrt{\sqrt{x}e^{\sqrt{x}} - e^{\sqrt{x}} + 4}$

61. $m(t) = \sqrt[3]{m_0^3 - 3kt}$

63. $x = \dfrac{a}{2}(2\phi + \sin 2\phi), y = \dfrac{a}{2}(1 - \cos 2\phi)$

65. $y(x) = (4h/L^2)x^2 + a$

Exercises 2.3, Page 59

1. $y = ce^{5x}, (-\infty, \infty)$

3. $y = \frac{1}{4}e^{3x} + ce^{-x}, (-\infty, \infty)$

5. $y = \frac{1}{3} + ce^{-x^3}, (-\infty, \infty)$

7. $y = x^{-1}\ln x + cx^{-1}, (0, \infty)$

9. $y = cx - x\cos x, (0, \infty)$

11. $y = \frac{1}{7}x^3 - \frac{1}{5}x + cx^{-4}, (0, \infty)$

13. $y = \frac{1}{2}x^{-2}e^x + cx^{-2}e^{-x}, (0, \infty)$

15. $x = 2y^6 + cy^4, (0, \infty)$

17. $y = \sin x + c\cos x, (-\pi/2, \pi/2)$

19. $(x + 1)e^x y = x^2 + c, (-1, \infty)$

21. $(\sec \theta + \tan \theta)r = \theta - \cos \theta + c, (-\pi/2, \pi/2)$

23. $y = e^{-3x} + cx^{-1}e^{-3x}, (0, \infty)$

25. $y = x^{-1}e^x + (2 - e)x^{-1}, (0, \infty)$

27. $i = \dfrac{E}{R} + \left(i_0 - \dfrac{E}{R}\right)e^{-Rt/L}, (-\infty, \infty)$

29. $(x + 1)y = x\ln x - x + 21, (0, \infty)$

31. $y = (2\sqrt{x} + e^2 - 2)e^{-2\sqrt{x}}, (0, \infty)$

33. $y = \begin{cases} \frac{1}{2}(1 - e^{-2x}), & 0 \leq x \leq 3 \\ \frac{1}{2}(e^6 - 1)e^{-2x}, & x > 3 \end{cases}$

35. $y = \begin{cases} \frac{1}{2} + \frac{3}{2}e^{-x^2}, & 0 \leq x < 1 \\ (\frac{1}{2}e + \frac{3}{2})e^{-x^2}, & x \geq 1 \end{cases}$

37. $y = \begin{cases} 2x - 1 + 4e^{-2x}, & 0 \leq x \leq 1 \\ 4x^2 \ln x + (1 + 4e^{-2})x^2, & x > 1 \end{cases}$

39. $x = -y^2 - 2y - 2 + ce^y$

41. $y = e^{x^2-1} + \frac{1}{2}\sqrt{\pi}e^{x^2}[\text{erf}(x) - \text{erf}(1)]$

43. $y = e^{1-e^x} + e^{-e^x}\displaystyle\int_0^x e^{e^t}\, dt$

45. $y = 2e^{-x^2/2} + e^{-x^2/2}\displaystyle\int_0^x e^{t^2/2}\cos t\, dt$

47. $y = 10x^{-2}[\text{Si}(x) - \text{Si}(1)]$

57. $E(t) = E_0 e^{-(t-4)/RC}$

Exercises 2.4, Page 66

1. $x^2 - x + \frac{3}{2}y^2 + 7y = c$ **3.** $\frac{5}{2}x^2 + 4xy - 2y^4 = c$
5. $x^2y^2 - 3x + 4y = c$ **7.** not exact
9. $xy^3 + y^2\cos x - \frac{1}{2}x^2 = c$ **11.** not exact
13. $xy - 2xe^x + 2e^x - 2x^3 = c$ **15.** $x^3y^3 - \tan^{-1}3x = c$
17. $-\ln|\cos x| + \cos x \sin y = c$
19. $t^4y - 5t^3 - ty + y^3 = c$

21. $\frac{1}{3}x^3 + x^2y + xy^2 - y = \frac{4}{3}$

23. $4ty + t^2 - 5t + 3y^2 - y = 8$

25. $y^2 \sin x - x^3 y - x^2 + y \ln y - y = 0$

27. $k = 10$ **29.** $x^2 y^2 \cos x = c$

31. $\frac{1}{2}x^2 - 1/(xy) = c$

33. $x^2 y^2 + x^3 = c$ **35.** $3x^2 y^3 + y^4 = c$

37. $-2ye^{3x} + \frac{10}{3}e^{3x} + x = c$ **39.** $e^{y^2}(x^2 + 4) = 20$

41. **(c)** $y_1(x) = -x^2 - \sqrt{x^4 - x^3 + 4}$,

$\qquad y_2(x) = -x^2 + \sqrt{x^4 - x^3 + 4}$

47. **(a)** $v(x) = 8\sqrt{\dfrac{x}{3} - \dfrac{9}{x^2}}$ **(b)** 12.7 ft/s

Exercises 2.5, Page 71

1. $y + x \ln|x| = cx$

3. $(x - y) \ln|x - y| = y + c(x - y)$

5. $x + y \ln|x| = cy$

7. $\ln(x^2 + y^2) + 2 \tan^{-1}(y/x) = c$

9. $4x = y(\ln|y| - c)^2$ **11.** $y^3 + 3x^3 \ln|x| = 8x^3$

13. $\ln|x| = e^{y/x} - 1$ **15.** $y^3 = 1 + cx^{-3}$

17. $y^{-3} = x + \frac{1}{3} + ce^{3x}$ **19.** $e^{t/y} = ct$

21. $y^{-3} = -\frac{9}{5}x^{-1} + \frac{49}{5}x^{-6}$

23. $y = -x - 1 + \tan(x + c)$

25. $2y - 2x + \sin 2(x + y) = c$

27. $4(y - 2x + 3) = (x + c)^2$

29. $-\cot(x + y) + \csc(x + y) = x + \sqrt{2} - 1$

37. **(b)** $y = \dfrac{2}{x} + (-\frac{1}{4}x + cx^{-3})^{-1}$

39. $P = \dfrac{a}{b + c_1 e^{-at}}$

Exercises 2.6, Page 76

1. $y_2 = 2.9800$, $y_4 = 3.1151$

3. $y_{10} = 2.5937$, $y_{20} = 2.6533$; $y = e^x$

5. $y_5 = 0.4198$, $y_{10} = 0.4124$

7. $y_5 = 0.5639$, $y_{10} = 0.5565$

9. $y_5 = 1.2194$, $y_{10} = 1.2696$

13. Euler: $y_{10} = 3.8191$, $y_{20} = 5.9363$

 RK4: $y_{10} = 42.9931$, $y_{20} = 84.0132$

Exercises 2.7, Page 83

1. 7.9 years; 10 years

3. 760; approximately 11 persons/yr

5. 11 h **7.** 136.5 h

11. $I(15) = 0.00098 I_0$ or approximately 0.1% of I_0

13. 15,963 years

15. $T(1) = 36.76°$F; approximately 3.06 min

17. approximately 82.1 s; approximately 145.7 s

19. 390°F **21.** approximately 1.6 h

23. $A(t) = 200 - 170e^{-t/50}$

25. $A(t) = 1000 - 1000e^{-t/100}$

27. $A(t) = 1000 - 10t - \frac{1}{10}(100 - t)^2$; 100 min

29. 64.38 lb

31. $i(t) = \frac{3}{5} - \frac{3}{5}e^{-500t}$; $i \to \frac{3}{5}$ as $t \to \infty$

33. $q(t) = \frac{1}{100} - \frac{1}{100}e^{-50t}$; $i(t) = \frac{1}{2}e^{-50t}$

35. $i(t) = \begin{cases} 60 - 60e^{-t/10}, & 0 \le t \le 20 \\ 60(e^2 - 1)e^{-t/10}, & t > 20 \end{cases}$

37. **(a)** $v(t) = \dfrac{mg}{k} + \left(v_0 - \dfrac{mg}{k}\right)e^{-kt/m}$

 (b) $v(t) \to \dfrac{mg}{k}$ as $t \to \infty$

 (c)

$$s(t) = \dfrac{mg}{k}t - \dfrac{m}{k}\left(v_0 - \dfrac{mg}{k}\right)e^{-kt/m} + \dfrac{m}{k}\left(v_0 - \dfrac{mg}{k}\right) + s_0$$

41. **(a)** $v(t) = \dfrac{\rho g}{4k}\left(\dfrac{k}{\rho}t + r_0\right) - \dfrac{\rho g r_0}{4k}\left(\dfrac{r_0}{\frac{k}{\rho}t + r_0}\right)^3$

 (b) $33\frac{1}{3}$ min

43. **(a)** $P(t) = P_0 e^{(k_1 - k_2)t}$

45. **(a)** As $t \to \infty$, $x(t) \to r/k$.

 (b) $x(t) = r/k - (r/k)e^{-kt}$; $(\ln 2)/k$

47. **(a)** $t_b = 50$ s **(b)** 70 m/s

 (c) 1250 m **(e)** $\dfrac{dv}{dt} + \dfrac{1}{50}v = -9.8$

51. **(a)** $v(0) = 5$ m³, $v(t) = 0.8 + 4.2e^{-0.2t}$, approximately 0.117%

 (b) in approximately 11.757 min or at approximately 9:12 A.M.

 (c) approximately 829.114 m³/min

Exercises 2.8, Page 92

1. **(a)** $N = 2000$

 (b) $N(t) = \dfrac{2000e^t}{1999 + e^t}$; $N(10) = 1834$

3. 1,000,000; 52.9 mo

7. For $0 < P_0 < 1$, time of extinction is

$$t = -\frac{1}{3}\ln\frac{4(P_0 - 1)}{P_0 - 4}.$$

9. $P(t) = \dfrac{5}{2} + \dfrac{\sqrt{3}}{2}\tan\left[-\dfrac{\sqrt{3}}{2}t + \tan^{-1}\left(\dfrac{2P_0 - 5}{\sqrt{3}}\right)\right]$;

 time of extinction is

$$t = \frac{2}{\sqrt{3}}\left[\tan^{-1}\frac{5}{\sqrt{3}} + \tan^{-1}\left(\frac{2P_0 - 5}{\sqrt{3}}\right)\right]$$

11. $P(t) = e^{a/b}e^{-ce^{-bt}}$, where $c = (a/b) - \ln P_0$

13. 29.3 g; $X \to 60$ as $t \to \infty$; 0 g of A and 30 g of B

15. (a) $h(t) = \left(\sqrt{H} - \dfrac{4A_h}{A_w}t\right)^2$; I is $[0, \sqrt{H}A_w/4A_h]$

(b) $576\sqrt{10}$ s or 30.36 min

17. (a) approximately 858.65 s or 14.31 min

(b) 243 s or 4.05 min

19. (a) $v(t) = \sqrt{\dfrac{mg}{k}}\tanh\left(\sqrt{\dfrac{kg}{m}}\,t + c_1\right)$

where $c_1 = \tanh^{-1}\left(\sqrt{\dfrac{k}{mg}}\,v_0\right)$

(b) $\sqrt{\dfrac{mg}{k}}$

(c) $s(t) = \dfrac{m}{k}\ln\cosh\left(\sqrt{\dfrac{kg}{m}}\,t + c_1\right) + c_2$

where $c_2 = -(m/k)\ln(\cosh c_1)$

21. (a) $m\dfrac{dv}{dt} = mg - kv^2 - \rho V$, where ρ is the weight density of water

(b) $v(t) = \sqrt{\dfrac{mg - \rho V}{k}}\tanh\left(\dfrac{\sqrt{kmg - k\rho V}}{m}t + c_1\right)$

(c) $\sqrt{\dfrac{mg - \rho V}{k}}$

23. (a) $W = 0$ and $W = 2$

(b) $W(x) = 2\,\text{sech}^2(x - c_1)$

(c) $W(x) = 2\,\text{sech}^2 x$

25. $x(t) = \sqrt{2t}$, $t = \frac{1}{2}$

27. $x(t) = \begin{cases}\sqrt{2t}, & 0 \le t \le \frac{1}{4}\\ \sqrt{2} - \sqrt{1 - 2t}, & \frac{1}{4} < t \le \frac{1}{2}\end{cases}$, $t = \frac{1}{2}$

Exercises 2.9, Page 101

1. $x(t) = x_0 e^{-\lambda_1 t}$

$y(t) = \dfrac{x_0 \lambda_1}{\lambda_2 - \lambda_1}(e^{-\lambda_1 t} - e^{-\lambda_2 t})$

$z(t) = x_0\left(1 - \dfrac{\lambda_2}{\lambda_2 - \lambda_1}e^{-\lambda_1 t} + \dfrac{\lambda_1}{\lambda_2 - \lambda_1}e^{-\lambda_2 t}\right)$

3. 5, 20, 147 days. The time when $y(t)$ and $z(t)$ are the same makes sense because most of A and half of B are gone, so half of C should have been formed.

5. (a) $P(t) = P_0 e^{-(\lambda_A + \lambda_C)t}$

(b) approximately 1.25×10^9 years

(c) $A(t) = \dfrac{\lambda_A}{\lambda_A + \lambda_C}P_0[1 - e^{-(\lambda_A + \lambda_C)t}]$

$C(t) = \dfrac{\lambda_C}{\lambda_A + \lambda_C}P_0[1 - e^{-(\lambda_A + \lambda_C)t}]$

(d) 10.5% of P_0, 89.5% of P_0

7. $\dfrac{dx_1}{dt} = 6 - \frac{2}{25}x_1 + \frac{1}{50}x_2$

$\dfrac{dx_2}{dt} = \frac{2}{25}x_1 - \frac{2}{25}x_2$

9. (a) $\dfrac{dx_1}{dt} = 3\dfrac{x_2}{100 - t} - 2\dfrac{x_1}{100 + t}$

$\dfrac{dx_2}{dt} = 2\dfrac{x_1}{100 + t} - 3\dfrac{x_2}{100 - t}$

(b) $x_1(t) + x_2(t) = 150$; $x_2(30) \approx 47.4$ lb

15. $L_1\dfrac{di_2}{dt} + (R_1 + R_2)i_2 + R_1 i_3 = E(t)$

$L_2\dfrac{di_3}{dt} + R_1 i_2 + (R_1 + R_3)i_3 = E(t)$

17. $i(0) = i_0$, $s(0) = n - i_0$, $r(0) = 0$; because the population is assumed to be constant

Chapter 2 in Review, Page 104

1. $-A/k$, a repeller for $k > 0$, an attractor for $k < 0$

3. $y = 5$, $y = -5$ **5.** true

7. $y = c_1 e^{e^x}$ **9.** false

11. $\dfrac{dy}{dx} = (y - 1)^2(y - 3)^2$

13. semi-stable for n even and unstable for n odd; semi-stable for n even and asymptotically stable for n odd

17. $2x + \sin 2x = 2\ln(y^2 + 1) + c$

19. $(6x + 1)y^3 = -3x^3 + c$

21. $Q = ct^{-1} + \frac{1}{25}t^4(-1 + 5\ln t)$

23. $y = \frac{1}{4} + c(x^2 + 4)^{-4}$

25. $y = e^{-2\sin x} + \frac{1}{2}e^{-2\sin x}\displaystyle\int_0^x te^{2\sin t}\,dt$

27. $y = \dfrac{3}{x^2} + \dfrac{1}{x^2}\displaystyle\int_1^x t^2 e^{t^2}\,dt$

29. $y = \begin{cases}xe^{-x} + 5e^{-x}, & 0 \le x < 1\\ 6e^{-x}, & x \ge 1\end{cases}$

31. $y = \csc x$, $(\pi, 2\pi)$

33. (b) $y = \frac{1}{4}(x + 2\sqrt{y_0} - x_0)^2$, $[x_0 - 2\sqrt{y_0}, \infty]$

37. $P(45) = 8.99$ billion

39. (b) approximately 3257 BC

41. $x = 10\ln\left(\dfrac{10 + \sqrt{100 - y^2}}{y}\right) - \sqrt{100 - y^2}$

43. (a) $\dfrac{BT_1 + T_2}{1 + B}, \dfrac{BT_1 + T_2}{1 + B}$

(b) $T(t) = \dfrac{BT_1 + T_2}{1 + B} + \dfrac{T_1 - T_2}{1 + B}e^{k(1 + B)t}$

45. $q = E_0 C + (q_0 - E_0 C)\left(\dfrac{k_1}{k_1 + k_2 t}\right)^{1/Ck_2}$

47. $h(t) = (\sqrt{2} - 0.00000163t)^2$

49. no

51. $x(t) = \dfrac{\alpha c_1 e^{\alpha k_1 t}}{1 + c_1 e^{\alpha k_1 t}}, \ y(t) = c_2(1 + c_1 e^{\alpha k_1 t})^{k_2/k_1}$

53. $P(t) = 450 e^{-2.4204 e^{-0.02948 t}}$; 166

55. $x^2 + y^2 = c_2$

57. $x = -y + 1 + c_2 e^{-y}$

59. $(q - p, 2\sqrt{pq}), (q - p, -2\sqrt{pq})$

61. (a) $k = 0.083$ seems to work well;
$\quad\quad k = 0.1063$ and $k = 0.0823$

Exercises 3.1, Page 120

1. $y = \frac{1}{2}e^x - \frac{1}{2}e^{-x}$ **3.** $y = 3x - 4x \ln x$

9. $(-\infty, 2)$

11. (a) $y = \dfrac{e}{e^2 - 1}(e^x - e^{-x})$ **(b)** $y = \dfrac{\sinh x}{\sinh 1}$

13. (a) $y = e^x \cos x - e^x \sin x$
 (b) no solution
 (c) $y = e^x \cos x + e^{-\pi/2} e^x \sin x$
 (d) $y = c_2 e^x \sin x$, where c_2 is arbitrary

15. dependent **17.** dependent

19. dependent **21.** independent

23. The functions satisfy the DE and are linearly independent on the interval since $W(e^{-3x}, e^{4x}) = 7e^x \neq 0$; $y = c_1 e^{-3x} + c_2 e^{4x}$.

25. The functions satisfy the DE and are linearly independent on the interval since $W(e^x \cos 2x, e^x \sin 2x) = 2e^{2x} \neq 0$; $y = c_1 e^x \cos 2x + c_2 e^x \sin 2x$.

27. The functions satisfy the DE and are linearly independent on the interval since $W(x^3, x^4) = x^6 \neq 0$; $y = c_1 x^3 + c_2 x^4$.

29. The functions satisfy the DE and are linearly independent on the interval since $W(x, x^{-2}, x^{-2} \ln x) = 9x^{-6} \neq 0$; $y = c_1 x + c_2 x^{-2} + c_3 x^{-2} \ln x$.

35. (b) $y_p = x^2 + 3x + 3e^{2x}$; $y_p = -2x^2 - 6x - \frac{1}{3}e^{2x}$

Exercises 3.2, Page 124

1. $y_2 = xe^{2x}$ **3.** $y_2 = \sin 4x$

5. $y_2 = \sinh x$ **7.** $y_2 = xe^{2x/3}$

9. $y_2 = x^4 \ln |x|$ **11.** $y_2 = 1$

13. $y_2 = x \cos(\ln x)$ **15.** $y_2 = x^2 + x + 2$

17. $y_2 = x^2 e^x$ **19.** $y_2 = e^{2x}, y_p = -\frac{1}{2}$

21. $y_2 = e^{2x}, y_p = \frac{5}{2}e^{3x}$ **23.** $y_2 = x \displaystyle\int_{x_0}^{x} \dfrac{e^{-t}}{t}\, dt, x_0 > 0$

Exercises 3.3, Page 130

1. $y = c_1 + c_2 e^{-x/4}$ **3.** $y = c_1 e^{3x} + c_2 e^{-2x}$

5. $y = c_1 e^{-4x} + c_2 xe^{-4x}$ **7.** $y = c_1 e^{2x/3} + c_2 e^{-x/4}$

9. $y = c_1 \cos 3x + c_2 \sin 3x$

11. $y = e^{2x}(c_1 \cos x + c_2 \sin x)$

13. $y = e^{-x/3}(c_1 \cos \frac{1}{3}\sqrt{2}x + c_2 \sin \frac{1}{3}\sqrt{2}x)$

15. $y = c_1 + c_2 e^{-x} + c_3 e^{5x}$

17. $y = c_1 e^{-x} + c_2 e^{3x} + c_3 xe^{3x}$

19. $u = c_1 e^t + e^{-t}(c_2 \cos t + c_3 \sin t)$

21. $y = c_1 e^{-x} + c_2 xe^{-x} + c_3 x^2 e^{-x}$

23. $y = c_1 + c_2 x + e^{-x/2}(c_3 \cos \frac{1}{2}\sqrt{3}x + c_4 \sin \frac{1}{2}\sqrt{3}x)$

25. $y = c_1 \cos \frac{1}{2}\sqrt{3}x + c_2 \sin \frac{1}{2}\sqrt{3}x + c_3 x \cos \frac{1}{2}\sqrt{3}x$
$\quad\quad + c_4 x \sin \frac{1}{2}\sqrt{3}x$

27. $u = c_1 e^r + c_2 re^r + c_3 e^{-r} + c_4 re^{-r} + c_5 e^{-5r}$

29. $y = c_1 + c_2 x + c_3 x^2 + c_4 x^3 + c_5 e^x \cos 2x + c_6 e^x \sin 2x$

31. $y = 2 \cos 4x - \frac{1}{2} \sin 4x$

33. $y = -\frac{1}{3}e^{-(t-1)} + \frac{1}{3}e^{5(t-1)}$

35. $y = 0$

37. $y = \frac{5}{36} - \frac{5}{36}e^{-6x} + \frac{1}{6}xe^{-6x}$

39. $y = e^{5x} - xe^{5x}$ **41.** $y = 0$

43. $y = \frac{1}{2}(1 - \frac{5}{\sqrt{3}})e^{-\sqrt{3}x} + \frac{1}{2}(1 + \frac{5}{\sqrt{3}})e^{\sqrt{3}x}$;
$\quad\quad y = \cosh \sqrt{3}x + \frac{5}{\sqrt{3}} \sinh \sqrt{3}x$

45. $\alpha = n\pi/2, n = 1, 2, 3, \ldots$

53. $y'' - 7y' + 6y = 0$ **55.** $y'' - 3y' = 0$

57. $y'' + 64y = 0$ **59.** $y'' - 2y' + 2y = 0$

61. $y''' - 7y'' = 0$

Exercises 3.4, Page 139

1. $y = c_1 e^{-x} + c_2 e^{-2x} + 3$

3. $y = c_1 e^{5x} + c_2 xe^{5x} + \frac{6}{5}x + \frac{3}{5}$

5. $y = c_1 e^{-2x} + c_2 xe^{-2x} + x^2 - 4x + \frac{7}{2}$

7. $y = c_1 \cos \sqrt{3}x + c_2 \sin \sqrt{3}x$
$\quad\quad + (-4x^2 + 4x - \frac{4}{3})e^{3x}$

9. $y = c_1 + c_2 e^x + 3x$

11. $y = c_1 e^{x/2} + c_2 xe^{x/2} + 12 + \frac{1}{2}x^2 e^{x/2}$

13. $y = c_1 \cos 2x + c_2 \sin 2x - \frac{3}{4}x \cos 2x$

15. $y = c_1 \cos x + c_2 \sin x - \frac{1}{2}x^2 \cos x + \frac{1}{2}x \sin x$

17. $y = c_1 e^x \cos 2x + c_2 e^x \sin 2x + \frac{1}{4}xe^x \sin 2x$

19. $y = c_1 e^{-x} + c_2 xe^{-x} - \frac{1}{2} \cos x + \frac{12}{25} \sin 2x - \frac{9}{25} \cos 2x$

21. $y = c_1 + c_2 x + c_3 e^{6x} - \frac{1}{4}x^2 - \frac{6}{37} \cos x + \frac{1}{37} \sin x$

23. $y = c_1 e^x + c_2 xe^x + c_3 x^2 e^x - x - 3 - \frac{2}{3}x^3 e^x$

25. $y = c_1 \cos x + c_2 \sin x + c_3 x \cos x + c_4 x \sin x$
$\quad\quad + x^2 - 2x - 3$

27. $y = \sqrt{2} \sin 2x - \frac{1}{2}$

29. $y = -200 + 200e^{-x/5} - 3x^2 + 30x$

31. $y = -10e^{-2x} \cos x + 9e^{-2x} \sin x + 7e^{-4x}$

33. $x = \dfrac{F_0}{2\omega^2} \sin \omega t - \dfrac{F_0}{2\omega} t \cos \omega t$

35. $y = 11 - 11e^x + 9xe^x + 2x - 12x^2 e^x + \frac{1}{2}e^{5x}$

37. $y = 6 \cos x - 6(\cot 1) \sin x + x^2 - 1$

39. $y = \dfrac{-4 \sin \sqrt{3}x}{\sin \sqrt{3} + \sqrt{3} \cos \sqrt{3}} + 2x$

41. $y = \begin{cases} \cos 2x + \frac{5}{6}\sin 2x + \frac{1}{3}\sin x, & 0 \le x \le \pi/2 \\ \frac{2}{3}\cos 2x + \frac{5}{6}\sin 2x, & x > \pi/2 \end{cases}$

Exercises 3.5, Page 144

1. $y = c_1 \cos x + c_2 \sin x + x \sin x + \cos x \ln |\cos x|$

3. $y = c_1 \cos x + c_2 \sin x - \frac{1}{2}x \cos x$

5. $y = c_1 \cos x + c_2 \sin x + \frac{1}{2} - \frac{1}{6}\cos 2x$

7. $y = c_1 e^x + c_2 e^{-x} + \frac{1}{2}x \sinh x$

9. $y = c_1 e^{-3x} + c_3 e^{3x} - \frac{1}{4}xe^{-3x}(1 + 3x)$

11. $y = c_3 + c_2 e^{-x} + 3x$

13. $y = c_1 e^{-x} + c_2 e^{-2x} + (e^{-x} + e^{-2x})\ln(1 + e^x)$

15. $y = c_1 e^{-2x} + c_2 e^{-x} - e^{-2x}\sin e^x$

17. $y = c_1 e^{-t} + c_2 t e^{-t} + \frac{1}{2}t^2 e^{-t}\ln t - \frac{3}{4}t^2 e^{-t}$

19. $y = c_1 e^x \sin x + c_2 e^x \cos x + \frac{1}{3}xe^x \sin x$
$\qquad + \frac{1}{3}e^x \cos x \ln |\cos x|$

21. $y = \frac{1}{4}e^{-x/2} + \frac{3}{4}e^{x/2} + \frac{1}{8}x^2 e^{x/2} - \frac{1}{4}xe^{x/2}$

23. $y = e^x - xe^x - e^x \ln |\cos x|$

25. $y = \frac{4}{9}e^{-4x} + \frac{25}{36}e^{2x} - \frac{1}{4}e^{-2x} + \frac{1}{9}e^{-x}$

27. $y = c_1 \cos x + c_2 \sin x$
$\qquad - \cos x \int_{x_0}^x e^{t^2}\sin t\, dt + \sin x \int_{x_0}^x e^{t^2}\cos t\, dt$

29. $y = c_1 e^{-2x} + c_2 e^x$
$\qquad - \frac{1}{3}e^{-2x}\int_{x_0}^x e^{2t}\ln t\, dt + \frac{1}{3}e^x\int_{x_0}^x e^{-t}\ln t\, dt,\ x_0 > 0$

31. $y = c_1 x^{-1/2}\cos x + c_2 x^{-1/2}\sin x + x^{-1/2}$

33. $y = c_1 + c_2 \cos x + c_3 \sin x$
$\qquad - \ln |\cos x| - \sin x \ln |\sec x + \tan x|$

35. $y = c_1 e^x + c_2 e^{-x} + c_3 e^{2x} + \frac{1}{30}e^{4x}$

Exercises 3.6, Page 150

1. $y = c_1 x^{-1} + c_2 x^2$

3. $y = c_1 + c_2 \ln x$

5. $y = c_1 \cos(2 \ln x) + c_2 \sin(2 \ln x)$

7. $y = c_1 x^{(2-\sqrt{6})} + c_2 x^{(2+\sqrt{6})}$

9. $y = c_1 \cos(\frac{1}{5}\ln x) + c_2 \sin(\frac{1}{5}\ln x)$

11. $y = c_1 x^{-2} + c_2 x^{-2}\ln x$

13. $y = x^{-1/2}[c_1 \cos(\frac{1}{6}\sqrt{3}\ln x) + c_2 \sin(\frac{1}{6}\sqrt{3}\ln x)]$

15. $y = c_1 x^3 + c_2 \cos(\sqrt{2}\ln x) + c_3 \sin(\sqrt{2}\ln x)$

17. $y = c_1 + c_2 x + c_3 x^2 + c_4 x^{-3}$

19. $y = c_1 + c_2 x^5 + \frac{1}{5}x^5 \ln x$

21. $y = c_1 x + c_2 x \ln x + x(\ln x)^2$

23. $y = c_1 x^{-1} + c_2 x - \ln x$

25. $y = 2 - 2x^{-2}$ **27.** $y = \cos(\ln x) + 2\sin(\ln x)$

29. $y = \frac{3}{4} - \ln x + \frac{1}{4}x^2$

31. $y = c_1(x^2 - x^6),\ c_1$ any real constant

33. $x^2 y'' - xy' - 8y = 0$ **35.** $x^2 y'' + 7xy' + 9y = 0$

37. $x^2 y'' + xy' + y = 0$

39. $y = c_1(x + 3)^2 + c_2(x + 3)^7$

41. $y = c_1(x + 2) + c_2(x + 2)\ln(x + 2)$

43. $y = c_1 x^{-10} + c_2 x^2$

45. $y = c_1 x^{-1} + c_2 x^{-8} + \frac{1}{30}x^2$

47. $y = x^2[c_1 \cos(3 \ln x) + c_2 \sin(3 \ln x)] + \frac{4}{13} + \frac{3}{10}x$

49. $y = 2(-x)^{1/2} - 5(-x)^{1/2}\ln(-x),\ x < 0$

51. $T(r) = \frac{2}{5}T_0(r + r^{-1})$

53. (a) $w(r) = c_1 + c_2 \ln r + c_3 r^2 + \dfrac{q}{64D}r^4$

 (b) $w(r) = \dfrac{q}{64D}(a^2 - r^2)^2$

Exercises 3.7, Page 155

3. $y = \ln |\cos(c_1 - x)| + c_2$

5. $y = \dfrac{1}{c_1^2}\ln|c_1 x + 1| - \dfrac{1}{c_1}x + c_2$

7. $(x + c_2)^2 + y^2 = c_1^2$ **9.** $\frac{1}{3}y^3 - c_1 y = x + c_2$

11. $y = \frac{2}{3}(x + 1)^{3/2} + \frac{4}{3}$

13. (b) $y = \tan(\frac{1}{4}\pi - \frac{1}{2}x)$ **(c)** $(-\pi/2, 3\pi/2)$

15. $y = -\dfrac{1}{c_1}\sqrt{1 - c_1^2 x^2} + c_2$

17. $y = 1 + x + \frac{1}{2}x^2 + \frac{1}{2}x^3 + \frac{1}{6}x^4 + \frac{1}{10}x^5 + \cdots$

19. $y = 1 + x - \frac{1}{2}x^2 + \frac{2}{3}x^3 - \frac{1}{4}x^4 + \frac{7}{60}x^5 + \cdots$

21. $y = -\sqrt{1 - x^2}$

Exercises 3.8, Page 169

1. $\dfrac{\sqrt{2}\pi}{8}$ **3.** $x(t) = -\frac{1}{4}\cos 4\sqrt{6}t$

5. (a) $x(\pi/12) = -\frac{1}{4};\ x(\pi/8) = -\frac{1}{2};\ x(\pi/6) = -\frac{1}{4};$
$\qquad x(\pi/4) = \frac{1}{2};\ x(9\pi/32) = \frac{\sqrt{2}}{4}$

 (b) 4 ft/s; downward

 (c) $t = \dfrac{(2n + 1)\pi}{16},\ n = 0, 1, 2, \ldots$

7. (a) the 20-kg mass

 (b) the 20-kg mass; the 50-kg mass

 (c) $t = n\pi, n = 0, 1, 2, \ldots$; at the equilibrium posi-
tion; the 50-kg mass is moving upward whereas
the 20-kg mass is moving upward when n is even
and downward when n is odd.

9. (a) $x(t) = \frac{1}{2}\cos 2t + \frac{3}{4}\sin 2t$

 (b) $x(t) = \dfrac{\sqrt{13}}{4}\sin(2t + 0.588)$

 (c) $x(t) = \dfrac{\sqrt{13}}{4}\cos(2t - 0.983)$

11. (a) $x(t) = -\frac{2}{3}\cos 10t + \frac{1}{2}\sin 10t$
$\qquad = \frac{5}{6}\sin(10t - 0.927)$

 (b) $\frac{5}{6}$ ft; $\frac{\pi}{5}$

 (c) 15 cycles

(d) 0.721 s

(e) $\dfrac{(2n+1)\pi}{20} + 0.0927$, $n = 0, 1, 2, \ldots$

(f) $x(3) = -0.597$ ft

(g) $x'(3) = -5.814$ ft/s

(h) $x''(3) = 59.702$ ft/s^2

(i) $\pm 8\frac{1}{3}$ ft/s

(j) $0.1451 + \dfrac{n\pi}{5}$; $0.3545 + \dfrac{n\pi}{5}$, $n = 0, 1, 2, \ldots$

(k) $0.3545 + \dfrac{n\pi}{5}$, $n = 0, 1, 2, \ldots$

13. $k_{\text{eff}} = 160$ lb/ft; $x(t) = \frac{1}{8}\sin 16t$

15. $k_{\text{eff}} = 30$ lb/ft; $x(t) = \frac{\sqrt{3}}{6}\sin 4\sqrt{3}t$

17. Compared to a single-spring system with spring constant k, the parallel-spring system is more stiff.

21. (a) above **(b)** heading upward

23. (a) below **(b)** heading upward

25. $\frac{1}{4}$ s; $\frac{1}{2}$ s, $x(\frac{1}{2}) = e^{-2}$; that is, the weight is approximately 0.14 ft below the equilibrium position.

27. (a) $x(t) = \frac{4}{3}e^{-2t} - \frac{1}{3}e^{-8t}$

(b) $x(t) = -\frac{2}{3}e^{-2t} + \frac{5}{3}e^{-8t}$

29. (a) $x(t) = e^{-2t}(-\cos 4t - \frac{1}{2}\sin 4t)$

(b) $x(t) = \frac{\sqrt{5}}{2}e^{-2t}\sin(4t + 4.249)$

(c) $t = 1.294$ s

31. (a) $\beta > \frac{5}{2}$ **(b)** $\beta = \frac{5}{2}$ **(c)** $0 < \beta < \frac{5}{2}$

33. $x(t) = e^{-t/2}(-\frac{4}{3}\cos\frac{\sqrt{47}}{2}t - \frac{64}{3\sqrt{47}}\sin\frac{\sqrt{47}}{2}t)$
$+ \frac{10}{3}(\cos 3t + \sin 3t)$

35. $x(t) = \frac{1}{4}e^{-4t} + te^{-4t} - \frac{1}{4}\cos 4t$

37. $x(t) = -\frac{1}{2}\cos 4t + \frac{9}{4}\sin 4t + \frac{1}{2}e^{-2t}\cos 4t - 2e^{-2t}\sin 4t$

39. (a) $m\dfrac{d^2x}{dt^2} = -k(x - h) - \beta\dfrac{dx}{dt}$ or

$\dfrac{d^2x}{dt^2} + 2\lambda\dfrac{dx}{dt} + \omega^2 x = \omega^2 h(t)$,

where $2\lambda = \beta/m$ and $\omega^2 = k/m$

(b) $x(t) = e^{-2t}(-\frac{56}{13}\cos 2t - \frac{72}{13}\sin 2t) + \frac{56}{13}\cos t$
$+ \frac{32}{13}\sin t$

43. $x(t) = -\cos 2t - \frac{1}{8}\sin 2t + \frac{3}{4}t\sin 2t + \frac{5}{4}t\cos 2t$

45. (b) $\dfrac{F_0}{2\omega}t\sin\omega t$

51. 4.568 C; 0.0509 s

53. $q(t) = 10 - 10e^{-3t}(\cos 3t + \sin 3t)$

$i(t) = 60e^{-3t}\sin 3t$; 10.432 C

55. $q_p = \frac{100}{13}\sin t + \frac{150}{13}\cos t$, $i_p = \frac{100}{13}\cos t - \frac{150}{13}\sin t$

59. $q(t) = -\frac{1}{2}e^{-10t}(\cos 10t + \sin 10t) + \frac{3}{2}$; $\frac{3}{2}$ C

63. $q(t) = \left(q_0 - \dfrac{E_0 C}{1 - \gamma^2 LC}\right)\cos\dfrac{t}{\sqrt{LC}}$

$+ \sqrt{LC}i_0\sin\dfrac{t}{\sqrt{LC}} + \dfrac{E_0 C}{1 - \gamma^2 LC}\cos\gamma t$

$i(t) = i_0\cos\dfrac{t}{\sqrt{LC}} - \dfrac{1}{\sqrt{LC}}\left(q_0 - \dfrac{E_0 C}{1 - \gamma^2 LC}\right)\sin\dfrac{t}{\sqrt{LC}}$

$- \dfrac{E_0 C\gamma}{1 - \gamma^2 LC}\sin\gamma t$

Exercises 3.9, Page 178

1. (a) $y(x) = \dfrac{w_0}{24EI}(6L^2x^2 - 4Lx^3 + x^4)$

3. (a) $y(x) = \dfrac{w_0}{48EI}(3L^2x^2 - 5Lx^3 + 2x^4)$

5. (a) $y(x) = \dfrac{w_0}{360EI}(7L^4x - 10L^2x^3 + 3x^5)$

(c) $x \approx 0.51933$, $y_{\max} \approx 0.234799$

7. $y(x) = -\dfrac{w_0 EI}{P^2}\cosh\sqrt{\dfrac{P}{EI}}\,x$

$+ \left(\dfrac{w_0 EI}{P^2}\sinh\sqrt{\dfrac{P}{EI}}L - \dfrac{w_0 L\sqrt{EI}}{P\sqrt{P}}\right)\dfrac{\sinh\sqrt{\dfrac{P}{EI}}x}{\cosh\sqrt{\dfrac{P}{EI}}L}$

$+ \dfrac{w_0}{2P}x^2 + \dfrac{w_0 EI}{P^2}$

9. (a) $EIy'''(x) = \begin{cases} -\frac{1}{2}w_0 L, & 0 \le x < L/2 \\ w_0(x - L), & L/2 \le x \le L \end{cases}$

(b)

$y(x) = \begin{cases} \dfrac{w_0}{48EI}(-4Lx^3 + 9L^2x^2), & 0 \le x < L/2 \\ \dfrac{w_0}{384EI}(16x^4 - 64Lx^3 + 96L^2x^2 - 8L^3x + L^4), \\ \qquad\qquad\qquad\qquad L/2 \le x \le L \end{cases}$

(c) $y(L) = \dfrac{41w_0 L^4}{384EI}$

11. $\lambda_n = n^2$, $n = 1, 2, 3, \ldots$; $y_n = \sin nx$

13. $\lambda_n = \dfrac{(2n-1)^2\pi^2}{4L^2}$, $n = 1, 2, 3, \ldots$;

$y = \cos\dfrac{(2n-1)\pi x}{2L}$

15. $\lambda_n = n^2$, $n = 0, 1, 2, \ldots$; $y_n = \cos nx$

17. $\lambda_n = \dfrac{n^2\pi^2}{25}$, $n = 1, 2, 3, \ldots$; $y_n = e^{-x}\sin\dfrac{n\pi x}{5}$

19. $\lambda_n = n^2$, $n = 1, 2, 3, \ldots$; $y_n = \sin(n\ln x)$

21. $\lambda_n = n^4\pi^4$, $n = 1, 2, 3, \ldots$; $y_n = \sin n\pi x$

23. $x = L/4$, $x = L/2$, $x = 3L/4$

27. $\omega_n = \dfrac{n\pi\sqrt{T}}{L\sqrt{\rho}}$, $n = 1, 2, 3, \ldots$; $y_n = \sin\dfrac{n\pi x}{L}$

29. $u(r) = \left(\dfrac{u_0 - u_1}{b - a}\right)\dfrac{ab}{r} + \dfrac{u_1 b - u_0 a}{b - a}$

31. (a) $\lambda_n = \dfrac{n^4\pi^4}{L^4}$, $n = 1, 2, 3, \ldots$; $y_n(x) = \sin\left(\dfrac{n\pi x}{L}\right)$

 (b) $\omega_n = \dfrac{n^2\pi^2}{L^2}\sqrt{\dfrac{EI}{\rho}}$, $n = 1, 2, 3, \ldots$

Exercises 3.10, Page 192

1. $y_p(x) = \frac{1}{4}\int_{x_0}^{x}\sinh 4(x - t)f(t)\,dt$

3. $y_p(x) = \int_{x_0}^{x}(x - t)e^{-(x-t)}f(t)\,dt$

5. $y_p(x) = \frac{1}{3}\int_{x_0}^{x}\sin 3(x - t)f(t)\,dt$

7. $y = c_1 e^{-4x} + c_2 e^{4x} + \frac{1}{4}\int_{x_0}^{x}\sinh 4(x - t)te^{-2t}\,dt$

9. $y = c_1 e^{-x} + c_2 x e^{-x} + \int_{x_0}^{x}(x - t)e^{-(x-t)}e^{-t}\,dt$

11. $y = c_1\cos 3x + c_2\sin 3x + \frac{1}{3}\int_{x_0}^{x}\sin 3(x - t)(t + \sin t)\,dt$

13. $y_p(x) = \frac{1}{4}xe^{2x} - \frac{1}{16}e^{2x} + \frac{1}{16}e^{-2x}$

15. $y_p(x) = \frac{1}{2}x^2 e^{5x}$

17. $y_p(x) = -\cos x + \dfrac{\pi}{2}\sin x - x\sin x - \cos x\ln|\sin x|$

19. $y = \frac{25}{16}e^{-2x} - \frac{9}{16}e^{2x} + \frac{1}{4}xe^{2x}$

21. $y = -e^{5x} + 6xe^{5x} + \frac{1}{2}x^2 e^{5x}$

23. $y = -x\sin x - \cos x\ln|\sin x|$

25. $y = (\cos 1 - 2)e^{-x} + (1 + \sin 1 - \cos 1)e^{-2x}$
 $- e^{-2x}\sin e^{x}$

27. $y = 4x - 2x^2 - x\ln x$

29. $y = \frac{46}{45}x^3 - \frac{1}{20}x^{-2} + \frac{1}{36} - \frac{1}{6}\ln x$

31. $y(x) = 5e^{x} + 3e^{-x} + y_p(x)$,

 where $y_p(x) = \begin{cases} 1 - \cosh x, & x < 0 \\ -1 + \cosh x, & x \geq 0 \end{cases}$

33. $y = \cos x - \sin x + y_p(x)$,

 where $y_p(x) = \begin{cases} 0, & x < 0 \\ 10 - 10\cos x, & 0 \leq x \leq 3\pi \\ -20\cos x, & x > 3\pi \end{cases}$

35. $y_p(x) = (x - 1)\int_{0}^{x}tf(t)\,dt + x\int_{x}^{1}(t - 1)f(t)\,dt$

37. $y_p(x) = \frac{1}{2}x^2 - \frac{1}{2}x$

39. $y_p(x) = \dfrac{\sin(x - 1)}{\sin 1} - \dfrac{\sin x}{\sin 1} + 1$

41. $y_p(x) = -e^{x}\cos x - e^{x}\sin x + e^{x}$

43. $y_p(x) = \frac{1}{2}(\ln x)^2 + \frac{1}{2}\ln x$

Exercises 3.11, Page 199

7. $\dfrac{d^2 x}{dt^2} + x = 0$

15. (a) $x(t) = 5$　　　(c) $4\sqrt{10} \approx 12.65$ ft/s

17. (a) $\theta(t) = \omega_0\sqrt{l/g}\,\sin\sqrt{g/l}\,t$

19. (a) $xy'' = r\sqrt{1 + (y')^2}$. When $t = 0$, $x = a$, $y = 0$, $dy/dx = 0$.

 (b) When $r \neq 1$,

 $y(x) = \dfrac{a}{2}\left[\dfrac{1}{1 + r}\left(\dfrac{x}{a}\right)^{1+r} - \dfrac{1}{1 - r}\left(\dfrac{x}{a}\right)^{1-r}\right] + \dfrac{ar}{1 - r^2}$.

 When $r = 1$,

 $y(x) = \dfrac{1}{2}\left[\dfrac{1}{2a}(x^2 - a^2) + \dfrac{1}{a}\ln\dfrac{a}{x}\right]$.

 (c) The paths intersect when $r < 1$.

Exercises 3.12, Page 207

1. $x = c_1 e^{t} + c_2 te^{t}$
 $y = (c_1 - c_2)e^{t} + c_2 te^{t}$

3. $x = c_1\cos t + c_2\sin t + t + 1$
 $y = c_1\sin t - c_2\cos t + t - 1$

5. $x = \frac{1}{2}c_1\sin t + \frac{1}{2}c_2\cos t - 2c_3\sin\sqrt{6}t - 2c_4\cos\sqrt{6}t$
 $y = c_1\sin t + c_2\cos t + c_3\sin\sqrt{6}t + c_4\cos\sqrt{6}t$

7. $x = c_1 e^{2t} + c_2 e^{-2t} + c_3\sin 2t + c_4\cos 2t + \frac{1}{5}e^{t}$
 $y = c_1 e^{2t} + c_2 e^{-2t} - c_3\sin 2t - c_4\cos 2t - \frac{1}{5}e^{t}$

9. $x = c_1 - c_2\cos t + c_3\sin t + \frac{17}{15}e^{3t}$
 $y = c_1 + c_2\sin t + c_3\cos t - \frac{4}{15}e^{3t}$

11. $x = c_1 e^{t} + c_2 e^{-t/2}\cos\frac{1}{2}\sqrt{3}t + c_3 e^{-t/2}\sin\frac{1}{2}\sqrt{3}t$
 $y = (-\frac{3}{2}c_2 - \frac{1}{2}\sqrt{3}c_3)e^{-t/2}\cos\frac{1}{2}\sqrt{3}t$
 $\quad + (\frac{1}{2}\sqrt{3}c_2 - \frac{3}{2}c_3)e^{-t/2}\sin\frac{1}{2}\sqrt{3}t$

13. $x = c_1 e^{4t} + \frac{4}{3}e^{t}$
 $y = -\frac{3}{4}c_1 e^{4t} + c_2 + 5e^{t}$

15. $x = c_1 + c_2 t + c_3 e^{t} + c_4 e^{-t} - \frac{1}{2}t^2$
 $y = (c_1 - c_2 + 2) + (c_2 + 1)t + c_4 e^{-t} - \frac{1}{2}t^2$

17. $x = c_1 e^{t} + c_2 e^{-t/2}\sin\frac{1}{2}\sqrt{3}t + c_3 e^{-t/2}\cos\frac{1}{2}\sqrt{3}t$
 $y = c_1 e^{t} + (-\frac{1}{2}c_2 - \frac{1}{2}\sqrt{3}c_3)\,e^{-t/2}\sin\frac{1}{2}\sqrt{3}t$
 $\quad + (\frac{1}{2}\sqrt{3}c_2 - \frac{1}{2}c_3)\,e^{-t/2}\cos\frac{1}{2}\sqrt{3}t$
 $z = c_1 e^{t} + (-\frac{1}{2}c_2 + \frac{1}{2}\sqrt{3}c_3)\,e^{-t/2}\sin\frac{1}{2}\sqrt{3}t$
 $\quad + (-\frac{1}{2}\sqrt{3}c_2 - \frac{1}{2}c_3)\,e^{-t/2}\cos\frac{1}{2}\sqrt{3}t$

19. $x = -6c_1 e^{-t} - 3c_2 e^{-2t} + 2c_3 e^{3t}$
 $y = c_1 e^{-t} + c_2 e^{-2t} + c_3 e^{3t}$
 $z = 5c_1 e^{-t} + c_2 e^{-2t} + c_3 e^{3t}$

21. $x = e^{-3t+3} - te^{-3t+3}$
 $y = -e^{-3t+3} + 2te^{-3t+3}$

23. $mx'' = 0$
 $my'' = -mg$;
 $x = c_1 t + c_2$
 $y = -\frac{1}{2}gt^2 + c_3 t + c_4$

Chapter 3 in Review, Page 208

1. $y = 0$　　　3. false　　　5. 8 ft

7. $\frac{5}{4}$　　　9. $(-\infty, \infty)$; $(0, \infty)$

11. $y = c_1 e^{3x} + c_2 e^{-5x} + c_3 x e^{-5x} + c_4 e^x + c_5 x e^x + c_6 x^2 e^x$,

$y = c_1 x^3 + c_2 x^{-5} + c_3 x^{-5} \ln x + c_4 x + c_5 x \ln x$

$\quad + c_6 x (\ln x)^2$

13. $y = c_1 x + c_2 \int \dfrac{e^{x^2}}{x^2} \, dx$

15. $y = c_1 e^{(1+\sqrt{3})x} + c_2 e^{(1-\sqrt{3})x}$

17. $y = c_1 + c_2 e^{-5x} + c_3 x e^{-5x}$

19. $y = c_1 e^{-x/3} + e^{-3x/2}\left(c_2 \cos \frac{\sqrt{7}}{2}x + c_3 \sin \frac{\sqrt{7}}{2}x\right)$

21. $y = e^{3x/2}\left(c_2 \cos \frac{\sqrt{11}}{2}x + c_3 \sin \frac{\sqrt{11}}{2}x\right) + \frac{4}{5}x^3$

$\quad + \frac{36}{25}x^2 + \frac{46}{125}x - \frac{222}{625}$

23. $y = c_1 + c_2 e^{2x} + c_3 e^{3x} + \frac{1}{5}\sin x - \frac{1}{5}\cos x + \frac{4}{3}x$

25. $y = e^x(c_1 \cos x + c_2 \sin x) - e^x \cos x \ln |\sec x + \tan x|$

27. $y = c_1 x^{-1/3} + c_2 x^{1/2}$

29. $y = c_1 x^2 + c_2 x^3 + x^4 - x^2 \ln x$

31. (a) $y = c_1 \cos \omega x + c_2 \sin \omega x + A \cos \alpha x + B \sin \alpha x$,

$\quad \omega \neq \alpha, \; y = c_1 \cos \omega x + c_2 \sin \omega x + Ax \cos \omega x$

$\quad\quad + Bx \sin \omega x, \; \omega = \alpha$

(b) $y = c_1 e^{-\omega x} + c_2 e^{\omega x} + A e^{\alpha x}, \; \omega \neq \alpha$,

$\quad y = c_1 e^{-\omega x} + c_2 e^{\omega x} + Axe^{\omega x}, \; \omega = \alpha$

33. (a) $y = c_1 \cosh x + c_2 \sinh x + c_3 x \cosh x + c_4 x \sinh x$

(b) $y_p = Ax^2 \cosh x + Bx^2 \sinh x$

35. $y = e^{x-\pi} \cos x$

37. $y = \frac{13}{4}e^x - \frac{5}{4}e^{-x} - x - \frac{1}{2}\sin x$

39. $y = x^2 + 4$

43. $x = -c_1 e^t - \frac{3}{2}c_2 e^{2t} + \frac{5}{2}$

$\quad y = c_1 e^t + c_2 e^{2t} - 3$

45. $x = c_1 e^t + c_2 e^{5t} + te^t$

$\quad y = -c_1 e^t + 3c_2 e^{5t} - te^t + 2e^t$

47. 14.4 lb **49.** $0 < m \leq 2$

51. (a) $q(t) = -\frac{1}{150}\sin 100t + \frac{1}{75}\sin 50t$

(b) $i(t) = -\frac{2}{3}\cos 100t + \frac{2}{3}\cos 50t$

(c) $t = \dfrac{n\pi}{50}, \; n = 0, 1, 2, \ldots$

55. $m\dfrac{d^2 x}{dt^2} + kx = 0$

57. $y(x) = 2\cos x - 5\sin x + \int_0^x \sin(x-t)\tan t \, dt$

$\quad = 2\cos x - 4\sin x - \cos x \ln |\sec x + \tan x|$

59. (a) $\theta_1(t) = \theta_0 \cos \omega_1 t, \; \omega_1 = \sqrt{g/l}$

(b) $T = \pi/2\omega_1 = (\pi/2)\sqrt{l/g}$

(c) $\theta_2(t) = \frac{1}{2}\theta_0 \sin \omega_2 t, \; \omega_2 = 2\sqrt{g/l}$; the amplitude and period of the shorter pendulum are half that of the longer pendulum

63. (a) $x = (v_0 \cos \theta)t, \; y = -\frac{1}{2}gt^2 + (v_0 \sin \theta)t$;

$\quad y = -\frac{1}{2}\dfrac{g}{v_0^2 \cos^2 \theta}x^2 + \dfrac{\sin \theta}{\cos \theta}x$

Exercises 4.1, Page 223

1. $\dfrac{2}{s}e^{-s} - \dfrac{1}{s}$

3. $\dfrac{1}{s^2} - \dfrac{1}{s^2}e^{-s}$

5. $\dfrac{1 + e^{-\pi s}}{s^2 + 1}$

7. $\dfrac{1}{s}e^{-s} + \dfrac{1}{s^2}e^{-s}$

9. $\dfrac{1}{s} - \dfrac{1}{s^2} + \dfrac{1}{s^2}e^{-s}$

11. $\dfrac{e^7}{s - 1}$

13. $\dfrac{1}{(s - 4)^2}$

15. $\dfrac{1}{s^2 + 2s + 2}$

17. $\dfrac{s^2 - 1}{(s^2 + 1)^2}$

19. $\dfrac{48}{s^5}$

21. $\dfrac{4}{s^2} - \dfrac{10}{s}$

23. $\dfrac{2}{s^3} + \dfrac{6}{s^2} - \dfrac{3}{s}$

25. $\dfrac{6}{s^4} + \dfrac{6}{s^3} + \dfrac{3}{s^2} + \dfrac{1}{s}$

27. $\dfrac{1}{s} + \dfrac{1}{s - 4}$

29. $\dfrac{1}{s} + \dfrac{2}{s - 2} + \dfrac{1}{s - 4}$

31. $\dfrac{8}{s^3} - \dfrac{15}{s^2 + 9}$

33. Use $\sinh kt = \dfrac{e^{kt} - e^{-kt}}{2}$ to show that

$$\mathcal{L}\{\sinh kt\} = \dfrac{k}{s^2 - k^2}.$$

35. $\dfrac{1}{2(s - 2)} - \dfrac{1}{2s}$

37. $\dfrac{s^2 - 2k^2}{s(s^2 - 4k^2)}$

39. $\dfrac{s}{s^2 + 4}$

41. $\dfrac{3}{s^2 + 36}$

43. $\dfrac{s + 2}{s^2 + 4}$

47. $\dfrac{\sqrt{\pi}}{s^{1/2}}$

49. $\dfrac{3\sqrt{\pi}}{4s^{5/2}}$

Exercises 4.2, Page 231

1. $\frac{1}{2}t^2$

3. $t - 2t^4$

5. $1 + 3t + \frac{3}{2}t^2 + \frac{1}{6}t^3$

7. $t - 1 + e^{2t}$

9. $\frac{1}{4}e^{-t/4}$

11. $\frac{5}{7}\sin 7t$

13. $\cos \dfrac{t}{2}$

15. $2 \cos 3t - 2 \sin 3t$

17. $\frac{1}{3} - \frac{1}{3}e^{-3t}$

19. $\frac{3}{4}e^{-3t} + \frac{1}{4}e^t$

21. $0.3e^{0.1t} + 0.6e^{-0.2t}$

23. $\frac{1}{2}e^{2t} - e^{3t} + \frac{1}{2}e^{6t}$

25. $\frac{1}{5} - \frac{1}{5}\cos \sqrt{5}t$

27. $-4 + 3e^{-t} + \cos t + 3 \sin t$

29. $\frac{1}{3}\sin t - \frac{1}{6}\sin 2t$

31. $\dfrac{a \sin bt - b \sin at}{ab(a^2 - b^2)}$

33. $y = -1 + e^t$

35. $y = \frac{1}{10}e^{4t} + \frac{19}{10}e^{-6t}$

37. $y = \frac{4}{3}e^{-t} - \frac{1}{3}e^{-4t}$

39. $y = 10 \cos t + 2 \sin t - \sqrt{2} \sin \sqrt{2}t$

41. $y = -\frac{8}{9}e^{-t/2} + \frac{1}{9}e^{-2t} + \frac{5}{18}e^t + \frac{1}{2}e^{-t}$

43. $y = \sinh t - \sin t$

45. $y(t) = \frac{1}{2}\cos t + \frac{1}{2}\cosh t$

47. $y(t) = \frac{10}{21}\cos 2t - \frac{10}{21}\cos 5t$

49. $y = \frac{1}{4}e^{-t} - \frac{1}{4}e^{-3t}\cos 2t + \frac{1}{4}e^{-3t}\sin 2t$

Exercises 4.3, Page 240

1. $\dfrac{1}{(s-10)^2}$ **3.** $\dfrac{6}{(s+2)^4}$

5. $\dfrac{1}{(s-2)^2} + \dfrac{2}{(s-3)^2} + \dfrac{1}{(s-4)^2}$

7. $\dfrac{3}{(s-1)^2 + 9}$

9. $\dfrac{s}{s^2+25} - \dfrac{s-1}{(s-1)^2+25} + 3\dfrac{s+4}{(s+4)^2+25}$

11. $\frac{1}{2}t^2 e^{-2t}$ **13.** $e^{3t}\sin t$

15. $e^{-2t}\cos t - 2e^{-2t}\sin t$ **17.** $e^{-t} - te^{-t}$

19. $5 - t - 5e^{-t} - 4te^{-t} - \frac{3}{2}t^2 e^{-t}$

21. $\dfrac{2k^2 s}{s^4 + 4k^4}$ **23.** $\dfrac{k(s^2 - 2k^2)}{s^4 + 4k^4}$

25. $y = te^{-4t} + 2e^{-4t}$ **27.** $y = e^{-t} + 2te^{-t}$

29. $y = \frac{1}{9}t + \frac{2}{27} - \frac{2}{27}e^{3t} + \frac{10}{9}te^{3t}$

31. $y = -\frac{3}{2}e^{3t}\sin 2t$

33. $y = \frac{1}{2} - \frac{1}{2}e^t\cos t + \frac{1}{2}e^t\sin t$

35. $y = (e+1)te^{-t} + (e-1)e^{-t}$

37. $x(t) = -\frac{3}{2}e^{-7t/2}\cos\frac{\sqrt{15}}{2}t - \frac{7\sqrt{15}}{10}e^{-7t/2}\sin\frac{\sqrt{15}}{2}t$

41. $\dfrac{e^{-s}}{s^2}$ **43.** $\dfrac{e^{-2s}}{s^2} + 2\dfrac{e^{-2s}}{s}$

45. $\dfrac{s}{s^2+4}e^{-\pi s}$ **47.** $\frac{1}{2}(t-2)^2 \mathcal{U}(t-2)$

49. $-\sin t\,\mathcal{U}(t-\pi)$

51. $\mathcal{U}(t-1) - e^{-(t-1)}\mathcal{U}(t-1)$

53. (c) **55.** (f)

57. (a)

59. $f(t) = 2 - 4\,\mathcal{U}(t-3);\ \mathcal{L}\{f(t)\} = \dfrac{2}{s} - \dfrac{4}{s}e^{-3s}$

61. $f(t) = t^2\,\mathcal{U}(t-1);\ \mathcal{L}\{f(t)\} = 2\dfrac{e^{-s}}{s^3} + 2\dfrac{e^{-s}}{s^2} + \dfrac{e^{-s}}{s}$

63. $f(t) = t - t\,\mathcal{U}(t-2);\ \mathcal{L}\{f(t)\} = \dfrac{1}{s^2} - \dfrac{e^{-2s}}{s^2} - 2\dfrac{e^{-2s}}{s}$

65. $\dfrac{1}{s^2} - \dfrac{e^{-s}}{s^2} - \dfrac{e^{-s}}{s} + \dfrac{\pi e^{-2s}}{s^2 + \pi^2}$

67. $f(t) = \mathcal{U}(t-a) - \mathcal{U}(t-b);\ \mathcal{L}\{f(t)\} = \dfrac{e^{-as}}{s} - \dfrac{e^{-bs}}{s}$

69. $-\dfrac{1}{s^2} + \dfrac{2}{s} + \dfrac{e^{-s}}{s^2} - \dfrac{e^{-2s}}{s}$

71. $y = [5 - 5e^{-(t-1)}]\,\mathcal{U}(t-1)$

73. $y = -\frac{1}{4} + \frac{1}{2}t + \frac{1}{4}e^{-2t} - \frac{1}{4}\mathcal{U}(t-1)$
$- \frac{1}{2}(t-1)\,\mathcal{U}(t-1) + \frac{1}{4}e^{-2(t-1)}\,\mathcal{U}(t-1)$

75. $y = \cos 2t - \frac{1}{6}\sin 2(t-2\pi)\,\mathcal{U}(t-2\pi)$
$+ \frac{1}{3}\sin(t-2\pi)\,\mathcal{U}(t-2\pi)$

77. $y = \sin t + [1 - \cos(t-\pi)]\,\mathcal{U}(t-\pi)$
$- [1 - \cos(t-2\pi)]\,\mathcal{U}(t-2\pi)$

79. $x(t) = \frac{5}{4}t - \frac{5}{16}\sin 4t - \frac{5}{4}(t-5)\,\mathcal{U}(t-5)$
$+ \frac{5}{16}\sin 4(t-5)\,\mathcal{U}(t-5) - \frac{25}{4}\mathcal{U}(t-5)$
$+ \frac{25}{4}\cos 4(t-5)\,\mathcal{U}(t-5)$

81. $q(t) = \frac{2}{5}\mathcal{U}(t-3) - \frac{2}{5}e^{-5(t-3)}\mathcal{U}(t-3)$

83. (a) $i(t) = \frac{1}{101}e^{-10t} - \frac{1}{101}\cos t + \frac{10}{101}\sin t$
$- \frac{1}{101}e^{-10(t-3\pi/2)}\,\mathcal{U}(t-3\pi/2)$
$+ \frac{10}{101}\cos(t-3\pi/2)\,\mathcal{U}(t-3\pi/2)$
$+ \frac{1}{101}\sin(t-3\pi/2)$

(b) $i_{max} \approx 0.1$ at $t \approx 1.6$, $i_{min} \approx -0.1$ at $t \approx 4.7$

85. $y(x) = \dfrac{w_0 L^2}{16EI}x^2 - \dfrac{w_0 L}{12EI}x^3$
$+ \dfrac{w_0}{24EI}x^4 - \dfrac{w_0}{24EI}(x - \frac{1}{2}L)^4\,\mathcal{U}(x - \frac{1}{2}L)$

87. $y(x) = \dfrac{w_0 L^2}{48EI}x^2 - \dfrac{w_0 L}{24EI}x^3$
$+ \dfrac{w_0}{60EI}[\frac{5}{2}Lx^4 - x^5 + (x - \frac{1}{2}L)^5\,\mathcal{U}(x - \frac{1}{2}L)]$

89. (a) $\dfrac{dT}{dt} = k(T - 70 - 57.5t - (230 - 57.5t)\,\mathcal{U}(t-4))$

Exercises 4.4, Page 253

1. $\dfrac{1}{(s+10)^2}$ **3.** $\dfrac{s^2 - 4}{(s^2 + 4)^2}$

5. $\dfrac{6s^2 + 2}{(s^2 - 1)^3}$ **7.** $\dfrac{12s - 24}{[(s-2)^2 + 36]^2}$

9. $y = -\frac{1}{2}e^{-t} + \frac{1}{2}\cos t - \frac{1}{2}t\cos t + \frac{1}{2}t\sin t$

11. $y = 2\cos 3t + \frac{5}{3}\sin 3t + \frac{1}{6}t\sin 3t$

13. $y = \frac{1}{4}\sin 4t + \frac{1}{8}t\sin 4t$
$- \frac{1}{8}(t-\pi)\sin 4(t-\pi)\,\mathcal{U}(t-\pi)$

17. $y = \frac{2}{3}t^3 + ct^2$ **19.** $y = c(\sin t - t\cos t)$

21. $y = \frac{5}{2}t^2$ **23.** $\dfrac{24}{s^5}$ **25.** $\dfrac{1}{s^2 - 1}$ **27.** $\dfrac{6}{s^5}$

29. $\dfrac{s-1}{(s+1)[(s-1)^2 + 1]}$ **31.** $\dfrac{1}{s(s-1)}$

33. $\dfrac{s+1}{s[(s+1)^2 + 1]}$ **35.** $\dfrac{1}{s^2(s-1)}$

37. $\dfrac{3s^2 + 1}{s^2(s^2 + 1)^2}$ **39.** $e^t - 1$

41. $e^t - \frac{1}{2}t^2 - t - 1$

43. $y(t) = \cos t + \sin t + \displaystyle\int_0^t f(\tau)\sin(t-\tau)\,d\tau$

47. $f(t) = \sin t$

49. $f(t) = -\frac{1}{8}e^{-t} + \frac{1}{8}e^t + \frac{3}{4}te^t + \frac{1}{4}t^2 e^t$

51. $f(t) = e^{-t}$

53. $f(t) = \frac{3}{8}e^{2t} + \frac{1}{8}e^{-2t} + \frac{1}{2}\cos 2t + \frac{1}{4}\sin 2t$

55. $y(t) = \sin t - \tfrac{1}{2}t \sin t$ **57.** $y(t) = 10 - 5e^{-2t}$

59. $i(t) = 2e^{-2(t-1)}\mathcal{U}(t-1) - 2e^{-2(t-3)}\mathcal{U}(t-3)$

61. $i(t) = 100[e^{-10(t-1)} - e^{-20(t-1)}]\,\mathcal{U}(t-1)$
$\qquad - 100[e^{-10(t-2)} - e^{-20(t-2)}]\,\mathcal{U}(t-2)$

63. $y(t) = \displaystyle\int_0^t e^{-\tau^2}\sin 3(t-\tau)\,d\tau$

65. $\dfrac{1 - e^{-as}}{s(1 + e^{-as})}$ **67.** $\dfrac{a}{s}\left(\dfrac{1}{bs} - \dfrac{1}{e^{bs} - 1}\right)$

69. $\dfrac{\coth(\pi s/2)}{s^2 + 1}$

71. $i(t) = \dfrac{1}{R}(1 - e^{-Rt/L})$
$\qquad + \dfrac{2}{R}\displaystyle\sum_{n=1}^{\infty}(-1)^n(1 - e^{-R(t-n)/L})\,\mathcal{U}(t-n)$

73. $x(t) = 2(1 - e^{-t}\cos 3t - \tfrac{1}{3}e^{-t}\sin 3t)$
$\qquad + 4\displaystyle\sum_{n=1}^{\infty}(-1)^n\,[1 - e^{-(t-n\pi)}\cos 3(t-n\pi)$
$\qquad - \tfrac{1}{3}e^{-(t-n\pi)}\sin 3(t-n\pi)]\,\mathcal{U}(t-n\pi)$

Exercises 4.5, Page 259

1. $y = e^{3(t-2)}\mathcal{U}(t-2)$

3. $y = \sin t + \sin t\,\mathcal{U}(t-2\pi)$

5. $y = -\cos t\,\mathcal{U}(t-\pi/2) + \cos t\,\mathcal{U}(t-3\pi/2)$

7. $y = \tfrac{1}{2} - \tfrac{1}{2}e^{-2t} + [\tfrac{1}{2} - \tfrac{1}{2}e^{-2(t-1)}]\,\mathcal{U}(t-1)$

9. $y = e^{-2(t-2\pi)}\sin t\,\mathcal{U}(t-2\pi)$

11. $y = e^{-2t}\cos 3t + \tfrac{2}{3}e^{-2t}\sin 3t$
$\qquad + \tfrac{1}{3}e^{-2(t-\pi)}\sin 3(t-\pi)\,\mathcal{U}(t-\pi)$
$\qquad + \tfrac{1}{3}e^{-2(t-3\pi)}\sin 3(t-3\pi)\,\mathcal{U}(t-3\pi)$

13. $y(t) = \mathcal{U}(t-1) - \cos(t-1)\,\mathcal{U}(t-1)$
$\qquad + \sin(t-2)\,\mathcal{U}(t-2)$
$\qquad - \mathcal{U}(t-3) + \cos(t-3)\,\mathcal{U}(t-3)$

15. $y = \sin t + \sin t\displaystyle\sum_{k=1}^{\infty}(-1)^k\,\mathcal{U}(t-k\pi)$

17. $y(x) = \begin{cases} \dfrac{w_0}{EI}\left(\tfrac{1}{4}Lx^2 - \tfrac{1}{6}x^3\right), & 0 \le x < L/2 \\[2mm] \dfrac{w_0 L^2}{4EI}\left(\tfrac{1}{2}x - \tfrac{1}{12}L\right), & L/2 \le x \le L \end{cases}$

19. $y(t) = \cos t + 2\sin t\,\mathcal{U}(t-\pi) + 6\sin t\,\mathcal{U}(t-2\pi)$

Exercises 4.6, Page 263

1. $x = -\tfrac{1}{3}e^{-2t} + \tfrac{1}{3}e^t$
$\quad y = \tfrac{1}{3}e^{-2t} + \tfrac{2}{3}e^t$

3. $x = -\cos 3t - \tfrac{5}{3}\sin 3t$
$\quad y = 2\cos 3t - \tfrac{7}{3}\sin 3t$

5. $x = -2e^{3t} + \tfrac{5}{2}e^{2t} - \tfrac{1}{2}$
$\quad y = \tfrac{8}{3}e^{3t} - \tfrac{5}{2}e^{2t} - \tfrac{1}{6}$

7. $x = -\tfrac{1}{2}t - \tfrac{3}{4}\sqrt{2}\sin\sqrt{2}\,t$
$\quad y = -\tfrac{1}{2}t + \tfrac{3}{4}\sqrt{2}\sin\sqrt{2}\,t$

9. $x = 8 + \dfrac{2}{3!}t^3 + \dfrac{1}{4!}t^4$
$\qquad y = -\dfrac{2}{3!}t^3 + \dfrac{1}{4!}t^4$

11. $x = \tfrac{1}{2}t^2 + t + 1 - e^{-t}$
$\qquad y = -\tfrac{1}{3} + \tfrac{1}{3}e^{-t} + \tfrac{1}{3}te^{-t}$

13. $x_1 = \tfrac{1}{5}\sin t + \tfrac{2\sqrt{6}}{15}\sin\sqrt{6}\,t + \tfrac{2}{5}\cos t - \tfrac{2}{5}\cos\sqrt{6}\,t$

$\quad x_2 = \tfrac{2}{5}\sin t - \tfrac{\sqrt{6}}{15}\sin\sqrt{6}\,t + \tfrac{4}{5}\cos t + \tfrac{1}{5}\cos\sqrt{6}\,t$

15. (b) $i_2 = \tfrac{100}{9} - \tfrac{100}{9}e^{-900t}$
$\qquad i_3 = \tfrac{80}{9} - \tfrac{80}{9}e^{-900t}$
$\quad$ (c) $i_1 = 20 - 20e^{-900t}$

17. $i_2 = -\tfrac{20}{13}e^{-2t} + \tfrac{375}{1469}e^{-15t} + \tfrac{145}{113}\cos t + \tfrac{85}{113}\sin t$

$\quad i_3 = \tfrac{30}{13}e^{-2t} + \tfrac{250}{1469}e^{-15t} - \tfrac{280}{113}\cos t + \tfrac{810}{113}\sin t$

19. $i_1 = \tfrac{6}{5} - \tfrac{6}{5}e^{-100t}\cosh 50\sqrt{2}\,t - \tfrac{9\sqrt{2}}{5}e^{-100t}\sinh 50\sqrt{2}\,t$

$\quad i_2 = \tfrac{6}{5} - \tfrac{6}{5}e^{-100t}\cosh 50\sqrt{2}\,t - \tfrac{6\sqrt{2}}{5}e^{-100t}\sinh 50\sqrt{2}\,t$

21. (a) $x(t) = (v_0\cos\theta)t,\ y(t) = -\tfrac{1}{2}gt^2 + (v_0\sin\theta)t$

$\quad$ (b) $y(x) = -\dfrac{g}{2v_0^2\cos^2\theta}x^2 + (\tan\theta)x$ is a quadratic function, for a fixed value of θ its graph is a parabola.

$\quad$ (c) Solve $y(x) = 0$ to find the range R. To prove the complementary-angle property, show that $R(\theta) = R(\pi/2 - \theta)$.

$\quad$ (d) Solve $y'(x) = 0$ and find the corresponding value of $y(x)$.

$\quad$ (e) For $\theta = 38°$: range is 2728.96 ft, max. height is 533.02 ft
For $\theta = 52°$: range is 2728.96 ft, max. height is 873.23 ft

$\quad$ (f) For $\theta = 38°$: time to hit the ground is $t \approx 11.5437$ s, max. height occurs at $t \approx 5.7718$ s
For $\theta = 52°$: time to hit the ground is $t \approx 14.7752$ s, max. height occurs at $t \approx 7.3876$ s

$\quad$ (g)

23. (a) $\theta_1(t) = \dfrac{\theta_0 + \psi_0}{2}\cos\omega t + \dfrac{\theta_0 - \psi_0}{2}\cos\sqrt{\omega^2 + 2K}\,t$

$\quad \theta_2(t) = \dfrac{\theta_0 + \psi_0}{2}\cos\omega t - \dfrac{\theta_0 - \psi_0}{2}\cos\sqrt{\omega^2 + 2K}\,t$

Chapter 4 in Review, Page 266

1. $\dfrac{1}{s^2} - \dfrac{2}{s^2}e^{-s}$ **3.** false

5. true **7.** $\dfrac{1}{s+7}$

9. $\dfrac{2}{s^2 + 4}$

11. $\dfrac{4s}{(s^2 + 4)^2}$

13. $\dfrac{2}{s(s^2 + 4)}$

15. $\dfrac{1}{(s - 1)(s^2 + 1)}$

17. $\dfrac{e^{-4s}}{s - 3}$

19. $\frac{1}{6}t^5$

21. $\frac{1}{2}t^2 e^{5t}$

23. $1 - e^{-t}$

25. $e^{5t} \cos 2t + \frac{5}{2} e^{5t} \sin 2t$

27. $4e^{-t} \cos 10t - 9e^{-t} \sin 10t$

29. $\cos \pi(t - 1)\, \mathcal{U}(t - 1) + \sin \pi(t - 1)\, \mathcal{U}(t - 1)$

31. -5

33. $e^{-k(s - a)} F(s - a)$

35. t

37. $e^{-3t/2}\left(\cos \dfrac{t}{2} - 3 \sin \dfrac{t}{2}\right)$

39. $f(t)\, \mathcal{U}(t - t_0)$

41. $f(t - t_0)\, \mathcal{U}(t - t_0)$

43. $f(t) = t - (t - 1)\, \mathcal{U}(t - 1) - \mathcal{U}(t - 4);$

$$\mathcal{L}\{f(t)\} = \dfrac{1}{s^2} - \dfrac{1}{s^2} e^{-s} - \dfrac{1}{s} e^{-4s};$$

$$\mathcal{L}\{e^t f(t)\} = \dfrac{1}{(s - 1)^2} - \dfrac{1}{(s - 1)^2} e^{-(s-1)}$$

$$- \dfrac{1}{s - 1} e^{-4(s-1)}$$

45. $f(t) = 2 + (t - 2)\, \mathcal{U}(t - 2);$

$$\mathcal{L}\{f(t)\} = \dfrac{2}{s} + \dfrac{1}{s^2} e^{-2s};$$

$$\mathcal{L}\{e^t f(t)\} = \dfrac{2}{s - 1} + \dfrac{1}{(s - 1)^2} e^{-2(s-1)}$$

47. $\mathcal{L}\{f(t)\} = \dfrac{e^{-s} - 1}{s(1 + e^{-s})}$

49. $y = 5te^t + \frac{1}{2}t^2 e^t$

51. $y = -\frac{6}{25} + \frac{1}{5}t + \frac{3}{2}e^{-t} - \frac{13}{50}e^{-5t} - \frac{4}{25}\, \mathcal{U}(t - 2)$

$\qquad - \frac{1}{5}(t - 2)\, \mathcal{U}(t - 2) + \frac{1}{4}e^{-(t-2)}\, \mathcal{U}(t - 2)$

$\qquad - \frac{9}{100}e^{-5(t-2)}\, \mathcal{U}(t - 2)$

53. $y(t) = e^{-2t} + [-\frac{1}{4} + \frac{1}{2}(t - 1) + \frac{1}{4}e^{-2(t-1)}]\, \mathcal{U}(t - 1)$

$\qquad - 2[-\frac{1}{4} + \frac{1}{2}(t - 2) + \frac{1}{4}e^{-2(t-2)}]\, \mathcal{U}(t - 2)$

$\qquad + [-\frac{1}{4} + \frac{1}{2}(t - 3) + \frac{1}{4}e^{-2(t-3)}]\, \mathcal{U}(t - 3)$

55. $y = 1 + t + \frac{1}{2}t^2$

57. $y(t) = -\frac{1}{4} - \frac{1}{2}t + \frac{5}{4}e^{2t}$

59. $x = -\frac{1}{4} + \frac{9}{8}e^{-2t} + \frac{1}{8}e^{2t}$

$\qquad y = t + \frac{9}{4}e^{-2t} - \frac{1}{4}e^{2t}$

61. $i(t) = -9 + 2t + 9e^{-t/5}$

63. $y(x) = \dfrac{w_0}{12EIL}[-\frac{1}{5}x^5 + \frac{1}{2}Lx^4 - \frac{1}{2}L^2 x^3 + \frac{1}{4}L^3 x^2$

$\qquad + \frac{1}{5}(x - \frac{1}{2}L)^5\, \mathcal{U}(x - \frac{1}{2}L)]$

65. $y(x) = \dfrac{w_0}{2EI} \dfrac{\sinh \frac{\pi}{2}}{\sinh \pi} \sin x \sinh x$

$\qquad - \dfrac{w_0}{4EI} \dfrac{\cosh \frac{\pi}{2}}{\sinh \pi} (\sin x \cosh x - \cos x \sinh x)$

$\qquad + \dfrac{w_0}{4EI}\left[\sin\left(x - \dfrac{\pi}{2}\right)\cosh\left(x - \dfrac{\pi}{2}\right)\right.$

$\qquad \left. - \cos\left(x - \dfrac{\pi}{2}\right)\sinh\left(x - \dfrac{\pi}{2}\right)\right]\mathcal{U}\left(x - \dfrac{\pi}{2}\right)$

Exercises 5.1, Page 280

1. $R = \frac{1}{2},\ [-\frac{1}{2}, \frac{1}{2})$

3. $R = 10,\ (-5, 15)$

5. $x - \frac{2}{3}x^3 + \frac{2}{15}x^5 - \frac{4}{315}x^7 + \cdots$

7. $1 + \frac{1}{2}x^2 + \frac{5}{24}x^4 + \frac{61}{720}x^6 + \cdots,\ (-\pi/2, \pi/2)$

9. $\displaystyle\sum_{k=3}^{\infty} (k - 2)c_{k-2} x^k$

11. $2c_1 + \displaystyle\sum_{k=1}^{\infty} [2(k + 1)c_{k+1} + 6c_{k-1}]x^k$

17. $5;\ 4$

19. $y_1(x) = c_0\left[1 + \dfrac{3}{2\cdot 3}x^3 + \dfrac{3^2}{2\cdot 3\cdot 5\cdot 6}x^6\right.$

$\qquad \left. + \dfrac{3^3}{2\cdot 3\cdot 5\cdot 6\cdot 8\cdot 9}x^9 + \cdots\right]$

$\quad y_2(x) = c_1\left[x + \dfrac{3}{3\cdot 4}x^4 + \dfrac{3^2}{3\cdot 4\cdot 6\cdot 7}x^7\right.$

$\qquad \left. + \dfrac{3^3}{3\cdot 4\cdot 6\cdot 7\cdot 9\cdot 10}x^{10} + \cdots\right]$

21. $y_1(x) = c_0\left[1 - \dfrac{1}{2!}x^2 - \dfrac{3}{4!}x^4 - \dfrac{21}{6!}x^6 - \cdots\right]$

$\quad y_2(x) = c_1\left[x + \dfrac{1}{3!}x^3 + \dfrac{5}{5!}x^5 + \dfrac{45}{7!}x^7 + \cdots\right]$

23. $y_1(x) = c_0\left[1 - \dfrac{1}{3!}x^3 + \dfrac{4^2}{6!}x^6 - \dfrac{7^2\cdot 4^2}{9!}x^9 + \cdots\right]$

$\quad y_2(x) = c_1\left[x - \dfrac{2^2}{4!}x^4 + \dfrac{5^2\cdot 2^2}{7!}x^7\right.$

$\qquad \left. - \dfrac{8^2\cdot 5^2\cdot 2^2}{10!}x^{10} + \cdots\right]$

25. $y_1(x) = c_0;\ y_2(x) = c_1\displaystyle\sum_{n=1}^{\infty} \dfrac{1}{n}x^n$

27. $y_1(x) = c_0\left[1 + \dfrac{1}{2}x^2 + \dfrac{1}{6}x^3 + \dfrac{1}{6}x^4 + \cdots\right]$

$\quad y_2(x) = c_1\left[x + \dfrac{1}{2}x^2 + \dfrac{1}{2}x^3 + \dfrac{1}{4}x^4 + \cdots\right]$

29. $y_1(x) = c_0\left[1 + \dfrac{1}{4}x^2 - \dfrac{7}{4\cdot 4!}x^4 + \dfrac{23\cdot 7}{8\cdot 6!}x^6 - \cdots\right]$

$\quad y_2(x) = c_1\left[x - \dfrac{1}{6}x^3 + \dfrac{14}{2\cdot 5!}x^5 + \dfrac{34\cdot 14}{4\cdot 7!}x^7 - \cdots\right]$

31. $y(x) = -2\left[1 + \dfrac{1}{2!}x^2 + \dfrac{1}{3!}x^3 + \dfrac{1}{4!}x^4 + \cdots\right] + 6x$

$\quad = 8x - 2e^x$

33. $y(x) = 3 - 12x^2 + 4x^4$

35. $y_1(x) = c_0\left[1 - \dfrac{1}{6}x^3 + \dfrac{1}{120}x^5 + \cdots\right]$

$\quad y_2(x) = c_1\left[x - \dfrac{1}{12}x^4 + \dfrac{1}{180}x^6 + \cdots\right]$

Exercises 5.2, Page 288

1. $x = 0$, irregular singular point

3. $x = -3$, regular singular point; $x = 3$, irregular singular point

5. $x = 0, 2i, -2i$, regular singular points

7. $x = -3, 2$, regular singular points

9. $x = 0$, irregular singular point; $x = -5, 5, 2$, regular singular points

11. for $x = 1$: $p(x) = 5$, $q(x) = \dfrac{x(x-1)^2}{x+1}$

$\quad$ for $x = -1$: $p(x) = \dfrac{5(x+1)}{x-1}$, $q(x) = x^2 + x$

13. $r_1 = \frac{1}{3}, r_2 = -1$

15. $r_1 = \frac{3}{2}, r_2 = 0$

$\quad y(x) = C_1 x^{3/2}\left[1 - \dfrac{2}{5}x + \dfrac{2^2}{7\cdot5\cdot2}x^2\right.$

$\qquad\qquad \left. - \dfrac{2^3}{9\cdot7\cdot5\cdot3!}x^3 + \cdots\right]$

$\qquad + C_2\left[1 + 2x - 2x^2 + \dfrac{2^3}{3\cdot3!}x^3 - \cdots\right]$

17. $r_1 = \frac{7}{8}, r_2 = 0$

$\quad y(x) = C_1 x^{7/8}\left[1 - \dfrac{2}{15}x + \dfrac{2^2}{23\cdot15\cdot2}x^2\right.$

$\qquad\qquad \left. - \dfrac{2^3}{31\cdot23\cdot15\cdot3!}x^3 + \cdots\right]$

$\qquad + C_2\left[1 - 2x + \dfrac{2^2}{9\cdot2}x^2\right.$

$\qquad\qquad \left. - \dfrac{2^3}{17\cdot9\cdot3!}x^3 + \cdots\right]$

19. $r_1 = \frac{1}{3}, r_2 = 0$

$\quad y(x) = C_1 x^{1/3}\left[1 + \dfrac{1}{3}x + \dfrac{1}{3^2\cdot2}x^2 + \dfrac{1}{3^3\cdot3!}x^3 + \cdots\right]$

$\qquad + C_2\left[1 + \dfrac{1}{2}x + \dfrac{1}{5\cdot2}x^2 + \dfrac{1}{8\cdot5\cdot2}x^3 + \cdots\right]$

21. $r_1 = \frac{5}{2}, r_2 = 0$

$\quad y(x) = C_1 x^{5/2}\left[1 + \dfrac{2\cdot2}{7}x + \dfrac{2^2\cdot3}{9\cdot7}x^2\right.$

$\qquad\qquad \left. + \dfrac{2^3\cdot4}{11\cdot9\cdot7}x^3 + \cdots\right]$

23. $r_1 = \frac{2}{3}, r_2 = \frac{1}{3}$

$\quad y(x) = C_1 x^{2/3}\left[1 - \dfrac{1}{2}x + \dfrac{5}{28}x^2 - \dfrac{1}{21}x^3 + \cdots\right]$

$\qquad + C_2 x^{1/3}\left[1 - \dfrac{1}{2}x + \dfrac{1}{5}x^2 - \dfrac{7}{120}x^3 + \cdots\right]$

25. $r_1 = 0, r_2 = -1$

$\quad y(x) = C_1 \displaystyle\sum_{n=0}^{\infty} \dfrac{1}{(2n+1)!}x^{2n} + C_2 x^{-1}\sum_{n=0}^{\infty}\dfrac{1}{(2n)!}x^{2n}$

$\quad = C_1 x^{-1}\displaystyle\sum_{n=0}^{\infty}\dfrac{1}{(2n+1)!}x^{2n+1} + C_2 x^{-1}\sum_{n=0}^{\infty}\dfrac{1}{(2n)!}x^{2n}$

$\quad = \dfrac{1}{x}[C_1 \sinh x + C_2 \cosh x]$

27. $r_1 = 1, r_2 = 0$

$\quad y(x) = C_1 x + C_2\left[x\ln x - 1 + \frac{1}{2}x^2\right.$

$\qquad\qquad \left. + \frac{1}{12}x^3 + \frac{1}{72}x^4 + \cdots\right]$

29. $r_1 = r_2 = 0$

$\quad y(x) = C_1 y_1(x) + C_2\left[y_1(x)\ln x + y_1(x)\left(-x + \dfrac{1}{4}x^2\right.\right.$

$\qquad\qquad \left.\left. - \dfrac{1}{3\cdot3!}x^3 + \dfrac{1}{4\cdot4!}x^4 - \cdots\right)\right]$

$\quad$ where $y_1(x) = \displaystyle\sum_{n=0}^{\infty}\dfrac{1}{n!}x^n = e^x$

33. (b) $y_1(t) = \displaystyle\sum_{n=0}^{\infty}\dfrac{(-1)^n}{(2n+1)!}(\sqrt{\lambda}t)^{2n} = \dfrac{\sin(\sqrt{\lambda}t)}{\sqrt{\lambda}t}$

$\qquad y_2(t) = t^{-1}\displaystyle\sum_{n=0}^{\infty}\dfrac{(-1)^n}{(2n)!}(\sqrt{\lambda}t)^{2n} = \dfrac{\cos(\sqrt{\lambda}t)}{t}$

(c) $y = c_1 x \sin\left(\dfrac{\sqrt{\lambda}}{x}\right) + c_2 x \cos\left(\dfrac{\sqrt{\lambda}}{x}\right)$

Exercises 5.3, Page 300

1. $y = c_1 J_{1/3}(x) + c_2 J_{-1/3}(x)$

3. $y = c_1 J_{5/2}(x) + c_2 J_{-5/2}(x)$

5. $y = c_1 J_0(x) + c_2 Y_0(x)$

7. $y = c_1 J_2(3x) + c_2 Y_2(3x)$

9. $y = c_1 I_{2/3}(4x) + c_2 K_{2/3}(4x)$

11. $y = c_1 x^{-1/2}J_{1/2}(\alpha x) + c_2 x^{-1/2}J_{-1/2}(\alpha x)$

13. $y = x^{-1/2}[c_1 J_1(4x^{1/2}) + c_2 Y_1(4x^{1/2})]$

15. $y = x[c_1 J_1(x) + c_2 Y_1(x)]$

17. $y = x^{1/2}[c_1 J_{3/2}(x) + c_2 Y_{3/2}(x)]$

19. $y = x^{-1}[c_1 J_{1/2}(\frac{1}{2}x^2) + c_2 J_{-1/2}(\frac{1}{2}x^2)]$

23. $y = x^{1/2}[c_1 J_{1/2}(x) + c_2 J_{-1/2}(x)]$

$\quad = C_1 \sin x + C_2 \cos x$

25. $y = x^{-1/2}[c_1 J_{1/2}(\frac{1}{8}x^2) + c_2 J_{-1/2}(\frac{1}{8}x^2)]$

$\quad = C_1 x^{-3/2}\sin(\frac{1}{8}x^2) + C_2 x^{-3/2}\cos(\frac{1}{8}x^2)$

31. (b) $J_{-3/2}(x) = -\sqrt{\dfrac{2}{\pi x}}\left(\dfrac{\cos x}{x} + \sin x\right)$

35. $y = c_1 x^{1/2} J_{1/3}(\tfrac{2}{3}\alpha x^{3/2}) + c_2 x^{1/2} Y_{1/3}(\tfrac{2}{3}\alpha x^{3/2})$;

$y = c_1 x^{1/2} I_{1/3}(\tfrac{2}{3}\alpha x^{3/2}) + c_2 x^{1/2} K_{1/3}(\tfrac{2}{3}\alpha x^{3/2})$

39. (a) $j_1(x) = \dfrac{\sin x}{x^2} - \dfrac{\cos x}{x}$

$j_2(x) = \left(\dfrac{3}{x^3} - \dfrac{1}{x}\right)\sin x - \dfrac{3\cos x}{x^2}$

$j_3(x) = \left(\dfrac{15}{x^4} - \dfrac{6}{x^2}\right)\sin x - \left(\dfrac{15}{x^3} - \dfrac{1}{x}\right)\cos x$

51. $P_2(x), P_3(x), P_4(x),$ and $P_5(x)$ are given in the text,

$P_6(x) = \tfrac{1}{16}(231x^6 - 315x^4 + 105x^2 - 5)$,

$P_7(x) = \tfrac{1}{16}(429x^7 - 693x^5 + 315x^3 - 35x)$

53. $\lambda_1 = 2, \lambda_2 = 12, \lambda_3 = 30$

Chapter 5 in Review, Page 304

1. false **3.** $[-\tfrac{1}{2}, \tfrac{1}{2}]$

7. $x^2(x - 1)y'' + y' + y = 0$

9. $r_1 = \tfrac{1}{2}, r_2 = 0$

$y_1(x) = C_1 x^{1/2}[1 - \tfrac{1}{3}x + \tfrac{1}{30}x^2 - \tfrac{1}{630}x^3 + \cdots]$

$y_2(x) = C_2[1 - x + \tfrac{1}{6}x^2 - \tfrac{1}{90}x^3 + \cdots]$

11. $y_1(x) = c_0[1 + \tfrac{3}{2}x^2 + \tfrac{1}{2}x^3 + \tfrac{5}{8}x^4 + \cdots]$

$y_2(x) = c_1[x + \tfrac{1}{2}x^3 + \tfrac{1}{4}x^4 + \cdots]$

13. $r_1 = 3, r_2 = 0$

$y_1(x) = C_1 x^3[1 + \tfrac{1}{4}x + \tfrac{1}{20}x^2 + \tfrac{1}{120}x^3 + \cdots]$

$y_2(x) = C_2[1 + x + \tfrac{1}{2}x^2]$

15. $y(x) = 3[1 - x^2 + \tfrac{1}{3}x^4 - \tfrac{1}{15}x^6 + \cdots] - 2[x - \tfrac{1}{2}x^3$

$+ \tfrac{1}{8}x^5 - \tfrac{1}{48}x^7 + \cdots]$

17. $\tfrac{1}{6}\pi$ **19.** $x = 0$ is an ordinary point

21. $y(x) = c_0\left[1 - \dfrac{1}{3}x^3 + \dfrac{1}{3^2 \cdot 2!}x^6 - \dfrac{1}{3^3 \cdot 3!}x^9 + \cdots\right]$

$+ c_1\left[x - \dfrac{1}{4}x^4 + \dfrac{1}{4 \cdot 7}x^7\right.$

$- \dfrac{1}{4 \cdot 7 \cdot 10}x^{10} + \cdots\left.\right] + \left[\dfrac{5}{2}x^2 - \dfrac{1}{3}x^3\right.$

$+ \dfrac{1}{3^2 \cdot 2!}x^6 - \dfrac{1}{3^3 \cdot 3!}x^9 + \cdots\left.\right]$

23. (a) $y = c_1 J_{3/2}(4x) + c_2 Y_{3/2}(4x)$

(b) $y = c_1 I_3(6x) + c_2 K_3(6x)$

31. $y(x) = c_0 y_1(x) + c_1 y_2(x)$, where

$y_1(x) = 1 - \dfrac{\alpha^2}{2!}x^2 - \dfrac{(2^2 - \alpha^2)\alpha^2}{4!}x^4$

$- \dfrac{(4^2 - \alpha^2)(2^2 - \alpha^2)\alpha^2}{6!}x^6 - \cdots$

$y_2(x) = x + \dfrac{1 - \alpha^2}{3!}x^3 + \dfrac{(3^2 - \alpha^2)(1 - \alpha^2)}{5!}x^5$

$+ \dfrac{(5^2 - \alpha^2)(3^2 - \alpha^2)(1 - \alpha^2)}{7!}x^7 + \cdots$

Exercises 6.1, Page 310

1. for $h = 0.1$, $y_5 = 2.0801$; for $h = 0.05$, $y_{10} = 2.0592$

3. for $h = 0.1$, $y_5 = 0.5470$; for $h = 0.05$, $y_{10} = 0.5465$

5. for $h = 0.1$, $y_5 = 0.4053$; for $h = 0.05$, $y_{10} = 0.4054$

7. for $h = 0.1$, $y_5 = 0.5503$; for $h = 0.05$, $y_{10} = 0.5495$

9. for $h = 0.1$, $y_5 = 1.3260$; for $h = 0.05$, $y_{10} = 1.3315$

11. for $h = 0.1$, $y_5 = 3.8254$; for $h = 0.05$, $y_{10} = 3.8840$;

at $x = 0.5$ the actual value is $y(0.5) = 3.9082$

13. (a) $y_1 = 1.2$

(b) $y''(c)\tfrac{1}{2}h^2 = 4e^{2c}\tfrac{1}{2}(0.1)^2 = 0.02e^{2c} \le 0.02e^{0.2}$

$= 0.0244$

(c) Actual value is $y(0.1) = 1.2214$. Error is 0.0214.

(d) If $h = 0.05$, $y_2 = 1.21$.

(e) Error with $h = 0.1$ is 0.0214. Error with $h = 0.05$ is 0.0114.

15. (a) $y_1 = 0.8$

(b) $y''(c)\tfrac{1}{2}h^2 = 5e^{-2c}\tfrac{1}{2}(0.1)^2 = 0.025e^{-2c} \le 0.025$

for $0 \le c \le 0.1$

(c) Actual value is $y(0.1) = 0.8234$. Error is 0.0234.

(d) If $h = 0.05$, $y_2 = 0.8125$.

(e) Error with $h = 0.1$ is 0.0234. Error with $h = 0.05$ is 0.0109.

17. (a) Error is $19h^2 e^{-3(c-1)}$.

(b) $y''(c)\tfrac{1}{2}h^2 \le 19(0.1)^2(1) = 0.19$

(c) If $h = 0.1$, $y_5 = 1.8207$. If $h = 0.05$, $y_{10} = 1.9424$.

(d) Error with $h = 0.1$ is 0.2325. Error with $h = 0.05$ is 0.1109.

19. (a) Error is $\dfrac{1}{(c + 1)^2}\tfrac{1}{2}h^2$.

(b) $|y''(c)\tfrac{1}{2}h^2| \le (1)\tfrac{1}{2}(0.1)^2 = 0.005$

(c) If $h = 0.1$, $y_5 = 0.4198$. If $h = 0.05$, $y_{10} = 0.4124$.

(d) Error with $h = 0.1$ is 0.0143. Error with $h = 0.05$ is 0.0069.

Exercises 6.2, Page 315

1. $y_5 = 3.9078$; actual value is $y(0.5) = 3.9082$

3. $y_5 = 2.0533$ **5.** $y_5 = 0.5463$

7. $y_5 = 0.4055$ **9.** $y_5 = 0.5493$

11. $y_5 = 1.3333$

13. (a) 35.7678

(c) $v(t) = \sqrt{\dfrac{mg}{k}}\tanh\sqrt{\dfrac{kg}{m}}\,t$; $v(5) = 35.7678$

15. (a) $h = 0.1$, $y_4 = 903.0282$;
$h = 0.05$, $y_8 = 1.1 \times 10^{15}$

17. (a) $y_1 = 0.82341667$

(b) $y^{(5)}(c) \dfrac{h^5}{5!} = 40e^{-2c} \dfrac{h^5}{5!} \le 40e^{2(0)} \dfrac{(0.1)^5}{5!}$
$= 3.333 \times 10^{-6}$

(c) Actual value is $y(0.1) = 0.8234134413$. Error is
$3.225 \times 10^{-6} \le 3.333 \times 10^{-6}$.

(d) If $h = 0.05$, $y_2 = 0.82341363$.

(e) Error with $h = 0.1$ is 3.225×10^{-6}. Error with
$h = 0.05$ is 1.854×10^{-7}.

19. (a) $y^{(5)}(c) \dfrac{h^5}{5!} = \dfrac{24}{(c+1)^5} \dfrac{h^5}{5!}$

(b) $\dfrac{24}{(c+1)^5} \dfrac{h^5}{5!} \le 24 \dfrac{(0.1)^5}{5!} = 2.0000 \times 10^{-6}$

(c) From calculation with $h = 0.1$, $y_5 = 0.40546517$.
From calculation with $h = 0.05$, $y_{10} = 0.40546511$.

Exercises 6.3, Page 318

1. $y(x) = -x + e^x$; actual values are $y(0.2) = 1.0214$,
$y(0.4) = 1.0918$, $y(0.6) = 1.2221$, $y(0.8) = 1.4255$;
approximations are given in Example 1

3. $y_4 = 0.7232$

5. for $h = 0.2$, $y_5 = 1.5569$; for $h = 0.1$, $y_{10} = 1.5576$

7. for $h = 0.2$, $y_5 = 0.2385$; for $h = 0.1$, $y_{10} = 0.2384$

Exercises 6.4, Page 322

1. $y(x) = -2e^{2x} + 5xe^{2x}$; $y(0.2) = -1.4918$, $y_2 = -1.6800$

3. $y_1 = -1.4928$, $y_2 = -1.4919$

5. $y_1 = 1.4640$, $y_2 = 1.4640$

7. $x_1 = 8.3055$, $y_1 = 3.4199$; $x_2 = 8.3055$, $y_2 = 3.4199$

9. $x_1 = -3.9123$, $y_1 = 4.2857$; $x_2 = -3.9123$, $y_2 = 4.2857$

11. $x_1 = 0.4179$, $y_1 = -2.1824$; $x_2 = 0.4173$, $y_2 = -2.1821$

Exercises 6.5, Page 325

1. $y_1 = -5.6774$, $y_2 = -2.5807$, $y_3 = 6.3226$

3. $y_1 = -0.2259$, $y_2 = -0.3356$, $y_3 = -0.3308$,
$y_4 = -0.2167$

5. $y_1 = 3.3751$, $y_2 = 3.6306$, $y_3 = 3.6448$, $y_4 = 3.2355$,
$y_5 = 2.1411$

7. $y_1 = 3.8842$, $y_2 = 2.9640$, $y_3 = 2.2064$, $y_4 = 1.5826$,
$y_5 = 1.0681$, $y_6 = 0.6430$, $y_7 = 0.2913$

9. $y_1 = 0.2660$, $y_2 = 0.5097$, $y_3 = 0.7357$, $y_4 = 0.9471$,
$y_5 = 1.1465$, $y_6 = 1.3353$, $y_7 = 1.5149$, $y_8 = 1.6855$,
$y_9 = 1.8474$

11. $y_1 = 0.3492$, $y_2 = 0.7202$, $y_3 = 1.1363$, $y_4 = 1.6233$,
$y_5 = 2.2118$, $y_6 = 2.9386$, $y_7 = 3.8490$

13. (c) $y_0 = -2.2755$, $y_1 = -2.0755$, $y_2 = -1.8589$,
$y_3 = -1.6126$, $y_4 = -1.3275$

Chapter 6 in Review, Page 326

1. Comparison of Numerical Methods with $h = 0.1$

x_n	Euler	Improved Euler	RK4
1.10	2.1386	2.1549	2.1556
1.20	2.3097	2.3439	2.3454
1.30	2.5136	2.5672	2.5695
1.40	2.7504	2.8246	2.8278
1.50	3.0201	3.1157	3.1197

Comparison of Numerical Methods with $h = 0.05$

x_n	Euler	Improved Euler	RK4
1.10	2.1469	2.1554	2.1556
1.20	2.3272	2.3450	2.3454
1.30	2.5410	2.5689	2.5695
1.40	2.7883	2.8269	2.8278
1.50	3.0690	3.1187	3.1197

3. Comparison of Numerical Methods with $h = 0.1$

x_n	Euler	Improved Euler	RK4
0.60	0.6000	0.6048	0.6049
0.70	0.7095	0.7191	0.7194
0.80	0.8283	0.8427	0.8431
0.90	0.9559	0.9752	0.9757
1.00	1.0921	1.1163	1.1169

Comparison of Numerical Methods with $h = 0.05$

x_n	Euler	Improved Euler	RK4
0.60	0.6024	0.6049	0.6049
0.70	0.7144	0.7194	0.7194
0.80	0.8356	0.8431	0.8431
0.90	0.9657	0.9757	0.9757
1.00	1.1044	1.1170	1.1169

5. $h = 0.2$: $y(0.2) \approx 3.2$; $h = 0.1$: $y(0.2) \approx 3.23$

7. $x(0.2) \approx 1.62$, $y(0.2) \approx 1.84$

Exercises 7.1, Page 332

1. $6\mathbf{i} + 12\mathbf{j}$; $\mathbf{i} + 8\mathbf{j}$; $3\mathbf{i}$; $\sqrt{65}$; 3

3. $\langle 12, 0 \rangle$; $\langle 4, -5 \rangle$; $\langle 4, 5 \rangle$; $\sqrt{41}$; $\sqrt{41}$

5. $-9\mathbf{i} + 6\mathbf{j}$; $-3\mathbf{i} + 9\mathbf{j}$; $-3\mathbf{i} - 5\mathbf{j}$; $3\sqrt{10}$; $\sqrt{34}$

7. $-6\mathbf{i} + 27\mathbf{j}$; 0; $-4\mathbf{i} + 18\mathbf{j}$; 0; $2\sqrt{85}$

9. $\langle 6, -14 \rangle$; $\langle 2, 4 \rangle$ **11.** $10\mathbf{i} - 12\mathbf{j}$; $12\mathbf{i} - 17\mathbf{j}$

13. $\langle 20, 52 \rangle$; $\langle -2, 0 \rangle$

15. $2\mathbf{i} + 5\mathbf{j}$

17. $2\mathbf{i} + 2\mathbf{j}$

19. $(1, 18)$

21. (a), (b), (c), (e), (f)

23. $\langle 6, 15 \rangle$

25. $\langle \frac{1}{\sqrt{2}}, \frac{1}{\sqrt{2}} \rangle$; $\langle -\frac{1}{\sqrt{2}}, -\frac{1}{\sqrt{2}} \rangle$

27. $\langle 0, -1 \rangle$; $\langle 0, 1 \rangle$

29. $\langle \frac{5}{13}, \frac{12}{13} \rangle$

31. $\frac{6}{\sqrt{58}} \mathbf{i} + \frac{14}{\sqrt{58}} \mathbf{j}$

33. $\langle -3, -\frac{15}{2} \rangle$

35.

37. $-(\mathbf{a} + \mathbf{b})$

41. $\mathbf{a} = \frac{5}{2}\mathbf{b} - \frac{1}{2}\mathbf{c}$

43. $\pm \frac{1}{\sqrt{2}} (\mathbf{i} + \mathbf{j})$

45. (b) approximately $31°$

47. $F = \dfrac{qQ}{4\pi\epsilon_0} \dfrac{1}{L\sqrt{L^2 + a^2}} \mathbf{i}$

Exercises 7.2, Page 338

1. -5

7. The set $\{(x, y, 5) | x, y \text{ real numbers}\}$ is a plane perpendicular to the z-axis, 5 units above the xy-plane.

9. The set $\{(2, 3, z) | z \text{ a real number}\}$ is a line perpendicular to the xy-plane at $(2, 3, 0)$.

11. $(0, 0, 0)$, $(2, 0, 0)$, $(2, 5, 0)$, $(0, 5, 0)$, $(0, 0, 8)$, $(2, 0, 8)$, $(2, 5, 8)$, $(0, 5, 8)$

13. $(-2, 5, 0)$, $(-2, 0, 4)$, $(0, 5, 4)$; $(-2, 5, -2)$; $(3, 5, 4)$

15. the union of the coordinate planes

17. the point $(-1, 2, -3)$

19. the union of the planes $z = -5$ and $z = 5$

21. $\sqrt{70}$

23. $7; 5$

25. right triangle

27. isosceles

29. $d(P_1, P_2) + d(P_1, P_3) = d(P_2, P_3)$

31. 6 or -2

33. $(4, \frac{1}{2}, \frac{3}{2})$

35. $P_1(-4, -11, 10)$

37. $\langle -3, -6, 1 \rangle$

39. $\langle 2, 1, 1 \rangle$

41. $\langle 2, 4, 12 \rangle$

43. $\langle -11, -41, -49 \rangle$

45. $\sqrt{139}$

47. 6

49. $\langle -\frac{2}{3}, \frac{1}{3}, -\frac{2}{3} \rangle$

51. $4\mathbf{i} - 4\mathbf{j} + 4\mathbf{k}$

53.

Exercises 7.3, Page 343

1. 12

3. -16

5. 48

7. 29

9. 25

11. $\langle -\frac{2}{5}, \frac{4}{5}, 2 \rangle$

13. $25\sqrt{2}$

15. (a) and (f), (c) and (d), (b) and (e)

17. $\langle \frac{4}{9}, -\frac{1}{3}, 1 \rangle$

21. 1.11 radians or $63.43°$

23. 1.89 radians or $108.43°$

25. $\cos \alpha = 1/\sqrt{14}, \cos \beta = 2/\sqrt{14}, \cos \gamma = 3/\sqrt{14}$; $\alpha = 74.5°, \beta = 57.69°, \gamma = 36.7°$

27. $\cos \alpha = \frac{1}{2}, \cos \beta = 0, \cos \gamma = -\sqrt{3}/2$; $\alpha = 60°$, $\beta = 90°, \gamma = 150°$

29. 0.955 radian or $54.74°$; 0.616 radian or $35.26°$

31. $\alpha = 58.19°, \beta = 42.45°, \gamma = 65.06°$

33. $\frac{5}{7}$

35. $-6\sqrt{11}/11$

37. $72\sqrt{109}/109$

39. $\langle -\frac{21}{5}, \frac{28}{5} \rangle$

41. $\langle -\frac{12}{7}, \frac{6}{7}, \frac{4}{7} \rangle$

43. $\langle \frac{72}{25}, \frac{96}{25} \rangle$

45. 1000 ft-lb

47. 0; 150 N-m

49. approximately 1.80 angstroms

Exercises 7.4, Page 350

1. $-5\mathbf{i} - 5\mathbf{j} + 3\mathbf{k}$

3. $\langle -12, -2, 6 \rangle$

5. $-5\mathbf{i} + 5\mathbf{k}$

7. $\langle -3, 2, 3 \rangle$

9. $\mathbf{0}$

11. $6\mathbf{i} + 14\mathbf{j} + 4\mathbf{k}$

13. $-3\mathbf{i} - 2\mathbf{j} - 5\mathbf{k}$

17. $-\mathbf{i} + \mathbf{j} + \mathbf{k}$

19. $2\mathbf{k}$

21. $\mathbf{i} + 2\mathbf{j}$

23. $-24\mathbf{k}$

25. $5\mathbf{i} - 5\mathbf{j} - \mathbf{k}$

27. 0

29. $\sqrt{41}$

31. $-\mathbf{j}$

33. 0

35. 6

37. $12\mathbf{i} - 9\mathbf{j} + 18\mathbf{k}$

39. $-4\mathbf{i} + 3\mathbf{j} - 6\mathbf{k}$

41. $-21\mathbf{i} + 16\mathbf{j} + 22\mathbf{k}$

43. -10

45. 14 square units

47. $\frac{1}{2}$ square unit

49. $\frac{7}{2}$ square units

51. 10 cubic units

53. coplanar

55. 32; in the xy-plane, $30°$ from the positive x-axis in the direction of the negative y-axis; $16\sqrt{3}\mathbf{i} - 16\mathbf{j}$

57. $\mathbf{A} = \mathbf{i} - \mathbf{k}, \mathbf{B} = \mathbf{j} - \mathbf{k}, \mathbf{C} = 2\mathbf{k}$

Exercises 7.5, Page 357

1. $\langle x, y, z \rangle = \langle 1, 2, 1 \rangle + t\langle 2, 3, -3 \rangle$

3. $\langle x, y, z \rangle = \langle \frac{1}{2}, -\frac{1}{2}, 1 \rangle + t\langle -2, 3, -\frac{3}{2} \rangle$

5. $\langle x, y, z \rangle = \langle 1, 1, -1 \rangle + t\langle 5, 0, 0 \rangle$

7. $x = 2 + 4t, y = 3 - 4t, z = 5 + 3t$

9. $x = 1 + 2t, y = -2t, z = -7t$

11. $x = 4 + 10t, y = \frac{1}{2} + \frac{3}{4}t, z = \frac{1}{3} + \frac{1}{6}t$

13. $\dfrac{x-1}{9} = \dfrac{y-4}{10} = \dfrac{z+9}{7}$

15. $\dfrac{x+7}{11} = \dfrac{z-5}{-4}, y = 2$

17. $x = 5, \dfrac{y-10}{9} = \dfrac{z+2}{12}$

19. $x = 4 + 3t, y = 6 + \frac{1}{2}t, z = -7 - \frac{3}{2}t;$

$\dfrac{x-4}{3} = 2y - 12 = \dfrac{2z+14}{-3}$

21. $x = 5t, y = 9t, z = 4t; \dfrac{x}{5} = \dfrac{y}{9} = \dfrac{z}{4}$

23. $x = 6 + 2t, y = 4 - 3t, z = -2 + 6t$

25. $x = 2 + t, y = -2, z = 15$

27. Both lines pass through the origin and have parallel direction vectors.

29. $(0, 5, 15), (5, 0, \frac{15}{2}), (10, -5, 0)$

31. $(2, 3, -5)$ **33.** Lines do not intersect.

35. $40.37°$ **37.** $x = 4 - 6t, y = 1 + 3t, z = 6 + 3t$

39. $2x - 3y + 4z = 19$ **41.** $5x - 3z = 51$

43. $6x + 8y - 4z = 11$ **45.** $5x - 3y + z = 2$

47. $3x - 4y + z = 0$ **49.** The points are collinear.

51. $x + y - 4z = 25$ **53.** $z = 12$

55. $-3x + y + 10z = 18$ **57.** $9x - 7y + 5z = 17$

59. $6x - 2y + z = 12$

61. orthogonal: (a) and (d), (b) and (c), (d) and (f), (b) and (e); parallel: (a) and (f), (c) and (e)

63. (c) and (d) **65.** $x = 2 + t, y = \frac{1}{2} - t, z = t$

67. $x = \frac{1}{2} - \frac{1}{2}t, y = \frac{1}{2} - \frac{3}{2}t, z = t$

69. $(-5, 5, 9)$ **71.** $(1, 2, -5)$

73. $x = 5 + t, y = 6 + 3t, z = -12 + t$

75. $3x - y - 2z = 10$

77.

79.

81.

Exercises 7.6, Page 364

1. not a vector space, axiom (*vi*) is not satisfied

3. not a vector space, axiom (*x*) is not satisfied

5. vector space

7. not a vector space, axiom (*ii*) is not satisfied

9. vector space **11.** a subspace

13. not a subspace **15.** a subspace

17. a subspace **19.** not a subspace

23. (b) $\mathbf{a} = 7\mathbf{u}_1 - 12\mathbf{u}_2 + 8\mathbf{u}_3$

25. linearly dependent

27. linearly independent

29. f is discontinuous at $x = -1$ and at $x = -3$.

31. $2\sqrt{\frac{2}{3}}\,\pi^{2/3}, \sqrt{\pi}$

Exercises 7.7, Page 370

1. $\mathbf{u} = \frac{58}{13}\mathbf{w}_1 - \frac{4}{13}\mathbf{w}_2$, where

$\mathbf{w}_1 = \langle \frac{12}{13}, \frac{5}{13}\rangle, \mathbf{w}_2 = \langle \frac{5}{13}, -\frac{12}{13}\rangle$

3. $\mathbf{u} = -\frac{3}{2}\mathbf{v}_1 + 7\mathbf{v}_2 - \frac{23}{2}\mathbf{v}_3$, where

$\mathbf{v}_1 = \langle 1, 0, 1\rangle, \mathbf{v}_2 = \langle 0, 1, 0\rangle, \mathbf{v}_3 = \langle -1, 0, 1\rangle$

5. (a) $B'' = \{\langle -\frac{3}{\sqrt{13}}, \frac{2}{\sqrt{13}}\rangle, \langle -\frac{2}{\sqrt{13}}, -\frac{3}{\sqrt{13}}\rangle\}$

 (b) $B'' = \{\langle -\frac{1}{\sqrt{2}}, -\frac{1}{\sqrt{2}}\rangle, \langle -\frac{1}{\sqrt{2}}, \frac{1}{\sqrt{2}}\rangle\}$

7. (a) $B'' = \{\langle \frac{1}{\sqrt{2}}, \frac{1}{\sqrt{2}}\rangle, \langle -\frac{1}{\sqrt{2}}, \frac{1}{\sqrt{2}}\rangle\}$

 (b) $B'' = \{\langle 1, 0\rangle, \langle 0, 1\rangle\}$

9. $B'' = \{\langle \frac{1}{\sqrt{2}}, \frac{1}{\sqrt{2}}, 0\rangle, \langle -\frac{1}{3\sqrt{2}}, \frac{1}{3\sqrt{2}}, \frac{4}{3\sqrt{2}}\rangle, \langle \frac{2}{3}, -\frac{2}{3}, \frac{1}{3}\rangle\}$

11. $B'' = \{\langle \frac{1}{\sqrt{6}}, \frac{1}{\sqrt{6}}, \frac{2}{\sqrt{6}}\rangle, \langle -\frac{1}{\sqrt{3}}, \frac{7}{5\sqrt{3}}, -\frac{1}{5\sqrt{3}}\rangle,$

$\langle -\frac{1}{\sqrt{2}}, -\frac{3}{5\sqrt{2}}, \frac{4}{5\sqrt{2}}\rangle\}$

13. $B'' = \{\langle \frac{1}{\sqrt{30}}, \frac{5}{\sqrt{30}}, \frac{2}{\sqrt{30}}\rangle, \langle -\frac{13}{\sqrt{186}}, \frac{1}{\sqrt{186}}, \frac{4}{\sqrt{186}}\rangle\}$

15. $B'' = \{\langle \frac{1}{2}, -\frac{1}{2}, \frac{1}{2}, -\frac{1}{2}\rangle, \langle \frac{7}{2\sqrt{35}}, \frac{9}{2\sqrt{35}}, \frac{3}{2\sqrt{35}}, \frac{1}{2\sqrt{35}}\rangle\}$

17. $B' = \{1, x, \frac{1}{2}(-1 + 3x^2)\}$

19. $B'' = \{\frac{1}{\sqrt{2}}, \frac{3}{\sqrt{6}}x, \frac{5}{2\sqrt{10}}(3x^2 - 1)\}$

21. $p(x) = \frac{41}{\sqrt{15}}q_1(x) + 3\sqrt{6}\,q_2(x) + \frac{1}{\sqrt{3}}q_3(x)$, where

$q_1(x) = \frac{1}{\sqrt{2}}, q_2(x) = \frac{3}{\sqrt{6}}x, q_2(x) = \frac{5}{2\sqrt{10}}(3x^2 - 1)$

Chapter 7 in Review, Page 371

1. true **3.** false **5.** true **7.** true

9. true **11.** $9\mathbf{i} + 2\mathbf{j} + 2\mathbf{k}$ **13.** $5\mathbf{i}$

15. 14 **17.** $-6\mathbf{i} + \mathbf{j} - 7\mathbf{k}$ **19.** $(4, 7, 5)$

21. $(5, 6, 3)$ **23.** $-36\sqrt{2}$ **25.** $12, -8,$ and 6

27. $3\sqrt{10}/2$ **29.** 2 units

31. $(\mathbf{i} - \mathbf{j} - 3\mathbf{k})/\sqrt{11}$ **33.** 2

35. $\frac{26}{9}\mathbf{i} + \frac{7}{9}\mathbf{j} + \frac{20}{9}\mathbf{k}$ **37.** sphere; plane

39. $\dfrac{x-7}{4} = \dfrac{y-3}{-2} = \dfrac{z+5}{6}$

41. The direction vectors are orthogonal and the point of intersection is $(3, -3, 0)$.

43. $14x - 5y - 3z = 0$ **45.** $30\sqrt{2}$ N-m
47. approximately 153 lb
49. not a vector space **51.** a subspace; $1, x$

Exercises 8.1, Page 381

1. 2×4 **3.** 3×3 **5.** 3×4
7. not equal **9.** not equal
11. $x = 2, y = 4$ **13.** $c_{23} = 9, c_{12} = 12$

15. $\begin{pmatrix} 2 & 11 \\ 2 & -1 \end{pmatrix}; \begin{pmatrix} -6 & 1 \\ 14 & -19 \end{pmatrix}; \begin{pmatrix} 2 & 28 \\ 12 & -12 \end{pmatrix}$

17. $\begin{pmatrix} -11 & 6 \\ 17 & -22 \end{pmatrix}; \begin{pmatrix} -32 & 27 \\ -4 & -1 \end{pmatrix};$
$\begin{pmatrix} 19 & -18 \\ -30 & 31 \end{pmatrix}; \begin{pmatrix} 19 & 6 \\ 3 & 22 \end{pmatrix}$

19. $\begin{pmatrix} 9 & 24 \\ 3 & 8 \end{pmatrix}; \begin{pmatrix} 3 & 8 \\ -6 & -16 \end{pmatrix}; \begin{pmatrix} 0 & 0 \\ 0 & 0 \end{pmatrix}; \begin{pmatrix} -4 & -5 \\ 8 & 10 \end{pmatrix}$

21. $180; \begin{pmatrix} 4 & 8 & 10 \\ 8 & 16 & 20 \\ 10 & 20 & 25 \end{pmatrix}; \begin{pmatrix} 6 \\ 12 \\ -5 \end{pmatrix}$

23. $\begin{pmatrix} 7 & 38 \\ 10 & 75 \end{pmatrix}; \begin{pmatrix} 7 & 38 \\ 10 & 75 \end{pmatrix}$ **25.** $\begin{pmatrix} -14 \\ 1 \end{pmatrix}$

27. $\begin{pmatrix} -38 \\ -2 \end{pmatrix}$ **29.** 4×5

37. $\mathbf{A} = \begin{pmatrix} 1 & 2 \\ 0 & 0 \end{pmatrix}, \mathbf{B} = \begin{pmatrix} 2 & -4 \\ -1 & 2 \end{pmatrix}$

39. $\mathbf{AB}$ is not necessarily the same as $\mathbf{BA}$.
41. $a_{11}x_1 + a_{12}x_2 = b_1$
$a_{21}x_1 + a_{22}x_2 = b_2$
45. $\mathbf{b} = \langle -1, 1 \rangle$ **47.** $\mathbf{b} = \langle -2, 0 \rangle$

49. $\mathbf{M} = \begin{pmatrix} 1 & 0 \\ 0 & -1 \end{pmatrix}$

51. (b) $\mathbf{M}_R = \begin{pmatrix} \cos\beta & 0 & -\sin\beta \\ 0 & 1 & 0 \\ \sin\beta & 0 & \cos\beta \end{pmatrix},$

$\mathbf{M}_P = \begin{pmatrix} 1 & 0 & 0 \\ 0 & \cos\alpha & \sin\alpha \\ 0 & -\sin\alpha & \cos\alpha \end{pmatrix}$

(c) $x_S = \dfrac{3\sqrt{2} + 2\sqrt{3} - \sqrt{6} + 6}{8} \approx 1.4072,$

$y_S = \dfrac{3\sqrt{2} + 2\sqrt{3} - 3\sqrt{6} + 2}{8} \approx 0.2948,$

$z_S = \dfrac{\sqrt{2} + \sqrt{6}}{8} \approx 0.9659$

Exercises 8.2, Page 394

1. $x_1 = 4, x_2 = -7$ **3.** $x_1 = -\frac{2}{3}, x_2 = \frac{1}{3}$
5. $x_1 = 0, x_2 = 4, x_3 = -1$ **7.** $x_1 = -t, x_2 = t, x_3 = 0$

9. inconsistent **11.** $x_1 = 0, x_2 = 0, x_3 = 0$
13. $x_1 = -2, x_2 = -2, x_3 = 4$ **15.** $x_1 = 1, x_2 = 2 - t, x_3 = t$
17. $x_1 = 0, x_2 = 1, x_3 = 1, x_4 = 0$
19. inconsistent
21. $x_1 = 0.3, x_2 = -0.12, x_3 = 4.1$
23. $2\text{Na} + 2\text{H}_2\text{O} \rightarrow 2\text{NaOH} + \text{H}_2$
25. $\text{Fe}_3\text{O}_4 + 4\text{C} \rightarrow 3\text{Fe} + 4\text{CO}$
27. $3\text{Cu} + 8\text{HNO}_3 \rightarrow 3\text{Cu(NO}_3)_2 + 4\text{H}_2\text{O} + 2\text{NO}$
29. $i_1 = \frac{35}{9}, i_2 = \frac{38}{9}, i_3 = \frac{1}{3}$

31. $\begin{pmatrix} 1 & 1 & 1 \\ 5 & -2 & 2 \\ 8 & 1 & 5 \end{pmatrix} \begin{pmatrix} x_1 \\ x_2 \\ x_3 \end{pmatrix} = \begin{pmatrix} 0 \\ 0 \\ 0 \end{pmatrix}$

33. $\begin{pmatrix} 2 & -3 & 1 \\ 1 & 1 & -1 \\ 4 & -1 & -1 \end{pmatrix} \begin{pmatrix} x_1 \\ x_2 \\ x_3 \end{pmatrix} = \begin{pmatrix} -12 \\ 1 \\ -10 \end{pmatrix}$

35. Interchange row 1 and row 2 in $\mathbf{I}_3$.
37. Multiply the second row of $\mathbf{I}_3$ by c and add to the third row.

39. $\mathbf{EA} = \begin{pmatrix} a_{21} & a_{22} & a_{23} \\ a_{11} & a_{12} & a_{13} \\ a_{31} & a_{32} & a_{33} \end{pmatrix}$

41. $\mathbf{EA} = \begin{pmatrix} a_{11} & a_{12} & a_{13} \\ a_{21} & a_{22} & a_{23} \\ ca_{21} + a_{31} & ca_{22} + a_{32} & ca_{23} + a_{33} \end{pmatrix}$

Exercises 8.3, Page 399

1. 2 **3.** 1 **5.** 3
7. 2 **9.** 3 **11.** linearly independent
13. linearly independent **15.** 5
17. $\text{rank}(\mathbf{A}) = 2$

Exercises 8.4, Page 405

1. 9 **3.** 1 **5.** 2 **7.** 10 **9.** -7
11. 17 **13.** $\lambda^2 - 3\lambda - 4$ **15.** -48
17. 62 **19.** 0 **21.** -85
23. $-x + 2y - z$ **25.** -104 **27.** 48
29. $\lambda_1 = -5, \lambda_2 = 7$

Exercises 8.5, Page 411

1. Theorem 8.5.4 **3.** Theorem 8.5.7
5. Theorem 8.5.5 **7.** Theorem 8.5.3
9. Theorem 8.5.1 **11.** -5 **13.** -5
15. 5 **17.** 80 **19.** -105 **27.** 0
29. -15 **31.** -9 **33.** 0 **35.** 16

Exercises 8.6, Page 420

3. $\begin{pmatrix} \frac{1}{9} & \frac{1}{9} \\ -\frac{4}{9} & \frac{5}{9} \end{pmatrix}$ **5.** $\begin{pmatrix} \frac{1}{6} & 0 \\ \frac{1}{4} & \frac{1}{2} \end{pmatrix}$

7. $\begin{pmatrix} -\frac{1}{2} & \frac{1}{2} & \frac{1}{2} \\ -\frac{1}{8} & \frac{1}{4} & -\frac{3}{8} \\ \frac{3}{8} & -\frac{1}{4} & \frac{1}{8} \end{pmatrix}$

9. $\begin{pmatrix} \frac{7}{15} & -\frac{13}{30} & -\frac{8}{15} \\ \frac{1}{15} & -\frac{2}{15} & \frac{1}{15} \\ \frac{2}{15} & \frac{7}{30} & \frac{2}{15} \end{pmatrix}$

11. $\begin{pmatrix} \frac{1}{3} & 0 & 0 \\ 0 & \frac{1}{6} & 0 \\ 0 & 0 & -\frac{1}{2} \end{pmatrix}$

13. $\begin{pmatrix} \frac{2}{9} & \frac{7}{9} & -\frac{1}{3} & -\frac{4}{3} \\ -\frac{1}{27} & \frac{1}{27} & \frac{2}{9} & -\frac{1}{9} \\ \frac{10}{27} & \frac{17}{27} & -\frac{2}{9} & -\frac{17}{9} \\ \frac{4}{27} & -\frac{4}{27} & \frac{1}{9} & \frac{4}{9} \end{pmatrix}$

15. $\begin{pmatrix} \frac{1}{6} & \frac{1}{12} \\ 0 & \frac{1}{4} \end{pmatrix}$

17. $\begin{pmatrix} -\frac{1}{4} & \frac{1}{4} \\ \frac{5}{12} & -\frac{1}{12} \end{pmatrix}$

19. singular matrix

21. $\begin{pmatrix} 0 & \frac{2}{3} & \frac{1}{3} \\ 0 & -\frac{1}{3} & -\frac{2}{3} \\ \frac{1}{3} & -\frac{2}{3} & 0 \end{pmatrix}$

23. $\begin{pmatrix} 5 & 6 & -3 \\ 2 & 2 & -1 \\ -1 & -1 & 1 \end{pmatrix}$

25. $\begin{pmatrix} -\frac{1}{2} & -\frac{2}{3} & -\frac{1}{6} & \frac{7}{6} \\ 1 & \frac{1}{3} & \frac{1}{3} & -\frac{4}{3} \\ 0 & -\frac{1}{3} & -\frac{1}{3} & \frac{1}{3} \\ -\frac{1}{2} & 1 & \frac{1}{2} & \frac{1}{2} \end{pmatrix}$

27. $\begin{pmatrix} -\frac{1}{3} & \frac{1}{3} \\ -1 & \frac{10}{3} \end{pmatrix}$ **29.** $\begin{pmatrix} -2 & 3 \\ 3 & 4 \end{pmatrix}$ **31.** $x = 5$

45. $x_1 = 6,\ x_2 = -2$ **47.** $x_1 = \frac{3}{4},\ x_2 = -\frac{1}{2}$

49. $x_1 = 2,\ x_2 = 4,\ x_3 = -6$ **51.** $x_1 = 21,\ x_2 = 1,\ x_3 = -11$

53. $x_1 = \frac{9}{10},\ x_2 = \frac{13}{20};\ x_1 = 6,\ x_2 = 16;\ x_1 = -2,\ x_2 = -7$

55. System has only the trivial solution.

57. System has nontrivial solutions.

59. (c) $i_1 = \dfrac{-R_3E_2 + R_3E_1 + R_2E_1 - R_2E_3}{R_3R_1 + R_3R_2 + R_1R_2},$

$i_2 = \dfrac{R_3E_2 - R_3E_1 - R_1E_3 + R_1E_2}{R_3R_1 + R_3R_2 + R_1R_2},$

$i_3 = \dfrac{-R_2E_1 + R_1E_3 + R_2E_3 - R_1E_2}{R_3R_1 + R_3R_2 + R_1R_2}$

Exercises 8.7, Page 424

1. $x_1 = -\frac{3}{5},\ x_2 = \frac{6}{5}$ **3.** $x_1 = 0.1,\ x_2 = -0.3$

5. $x = 4,\ y = -7$ **7.** $x_1 = -4,\ x_2 = 4,\ x_3 = -5$

9. $u = 4,\ v = \frac{3}{2},\ w = 1$ **11.** $k = \frac{6}{5}$

13. $T_1 \approx 450.8$ lb, $T_2 \approx 423$ lb

Exercises 8.8, Page 432

1. $\mathbf{K}_3,\ \lambda = -1$

3. $\mathbf{K}_3,\ \lambda = 0$

5. $\mathbf{K}_2,\ \lambda = 3;\ \mathbf{K}_3,\ \lambda = 1$

7. $\lambda_1 = 6,\ \lambda_2 = 1,\ \mathbf{K}_1 = \begin{pmatrix} 2 \\ 7 \end{pmatrix},\ \mathbf{K}_2 = \begin{pmatrix} 1 \\ 1 \end{pmatrix}$; nonsingular

9. $\lambda_1 = \lambda_2 = -4,\ \mathbf{K}_1 = \begin{pmatrix} 1 \\ -4 \end{pmatrix}$; nonsingular

11. $\lambda_1 = 3i,\ \lambda_2 = -3i$

$\mathbf{K}_1 = \begin{pmatrix} 1 - 3i \\ 5 \end{pmatrix},\ \mathbf{K}_2 = \begin{pmatrix} 1 + 3i \\ 5 \end{pmatrix}$; nonsingular

13. $\lambda_1 = 4,\ \lambda_2 = -5,$

$\mathbf{K}_1 = \begin{pmatrix} 1 \\ 0 \end{pmatrix},\ \mathbf{K}_2 = \begin{pmatrix} -8 \\ 9 \end{pmatrix}$; nonsingular

15. $\lambda_1 = 0,\ \lambda_2 = 4,\ \lambda_3 = -4,$

$\mathbf{K}_1 = \begin{pmatrix} 9 \\ 45 \\ 25 \end{pmatrix},\ \mathbf{K}_2 = \begin{pmatrix} 1 \\ 1 \\ 1 \end{pmatrix},\ \mathbf{K}_3 = \begin{pmatrix} 1 \\ 9 \\ 1 \end{pmatrix}$; singular

17. $\lambda_1 = \lambda_2 = \lambda_3 = -2,$

$\mathbf{K}_1 = \begin{pmatrix} 2 \\ -1 \\ 0 \end{pmatrix},\ \mathbf{K}_2 = \begin{pmatrix} 0 \\ 0 \\ 1 \end{pmatrix}$; nonsingular

19. $\lambda_1 = -1,\ \lambda_2 = i,\ \lambda_3 = -i,$

$\mathbf{K}_1 = \begin{pmatrix} 1 \\ -1 \\ 1 \end{pmatrix},\ \mathbf{K}_2 = \begin{pmatrix} -1 \\ i \\ i \end{pmatrix},\ \mathbf{K}_3 = \begin{pmatrix} -1 \\ -i \\ -i \end{pmatrix}$; nonsingular

21. $\lambda_1 = 1,\ \lambda_2 = 5,\ \lambda_3 = -7,$

$\mathbf{K}_1 = \begin{pmatrix} 1 \\ 0 \\ 0 \end{pmatrix},\ \mathbf{K}_2 = \begin{pmatrix} 1 \\ 2 \\ 0 \end{pmatrix},\ \mathbf{K}_3 = \begin{pmatrix} -1 \\ -2 \\ 4 \end{pmatrix}$; nonsingular

23. Eigenvalues of $\mathbf{A}$ are $\lambda_1 = 6,\ \lambda_2 = 4$; eigenvalues of $\mathbf{A}^{-1}$ are $\lambda_1 = \frac{1}{6},\ \lambda_2 = \frac{1}{4}$; corresponding eigenvectors for both $\mathbf{A}$ and $\mathbf{A}^{-1}$ are $\mathbf{K}_1 = \begin{pmatrix} 1 \\ 1 \end{pmatrix},\ \mathbf{K}_2 = \begin{pmatrix} -1 \\ 1 \end{pmatrix}$.

25. Eigenvalues of $\mathbf{A}$ are $\lambda_1 = 5,\ \lambda_2 = 4,\ \lambda_3 = 3$, eigenvalues of $\mathbf{A}^{-1}$ are $\lambda_1 = \frac{1}{5},\ \lambda_2 = \frac{1}{4},\ \lambda_3 = \frac{1}{3}$; corresponding eigenvectors for both $\mathbf{A}$ and $\mathbf{A}^{-1}$ are $\mathbf{K}_1 = \begin{pmatrix} -3 \\ -1 \\ 1 \end{pmatrix},$ $\mathbf{K}_2 = \begin{pmatrix} 1 \\ 0 \\ 0 \end{pmatrix},\ \mathbf{K}_3 = \begin{pmatrix} -2 \\ 1 \\ 0 \end{pmatrix}$.

Exercises 8.9, Page 436

3. $\begin{pmatrix} \frac{1}{7}[3(-1)^m 2^{m+1} + 5^m] & -\frac{3}{7}[(-2)^m - 5^m] \\ -\frac{2}{7}[(-2)^m - 5^m] & \frac{1}{7}[(-2)^m - 6(5)^m] \end{pmatrix};$

$\begin{pmatrix} 11 & 57 \\ 38 & 106 \end{pmatrix}$

5. $\begin{pmatrix} \frac{1}{3}2^{m-1}[5^m + (-1)^m] & \frac{5}{3}2^{m-2}[5^m - (-1)^m] \\ \frac{1}{3}[10^m - (-2)^m] & \frac{1}{3}2^{m-1}[5^m + 5(-1)^m] \end{pmatrix};$

$\begin{pmatrix} 83328 & 41680 \\ 33344 & 16640 \end{pmatrix}$

7. $\begin{pmatrix} 1 & 2^m - 1 & 2^m - 1 \\ 0 & \frac{1}{3}[2^{m+1} + (-1)^m] & \frac{2}{3}[2^m - (-1)^m] \\ 0 & \frac{1}{3}[2^m - (-1)^m] & \frac{1}{3}[2^m + 2(-1)^m] \end{pmatrix}$;

$\begin{pmatrix} 1 & 1023 & 1023 \\ 0 & 683 & 682 \\ 0 & 341 & 342 \end{pmatrix}$

9. $\begin{pmatrix} \frac{1}{3}[2^{2m+1} + (-2)^m] & \frac{1}{3}[4^m - (-2)^m] & 0 \\ \frac{1}{3}[2^{2m+1} - (-1)^m 2^{m+1}] & \frac{1}{3}[4^m + (-1)^m 2^{m+1}] & 0 \\ \frac{1}{3}[2^{2m+1} + (-2)^m - 3] & \frac{1}{3}[4^m - (-2)^m] & 1 \end{pmatrix}$;

$\begin{pmatrix} 699392 & 349184 & 0 \\ 698368 & 350208 & 0 \\ 699391 & 349184 & 1 \end{pmatrix}$

11. $\begin{pmatrix} 7m4^{m+1} + (1-m)4^m & 3m4^{m-1} \\ -3m4^{m-1} & m4^{m-1} + (1-m)4^m \end{pmatrix}$;

$\begin{pmatrix} 22528 & 18432 \\ -18432 & -14336 \end{pmatrix}$

13. (a) $4^m \begin{pmatrix} 1 & 1 \\ 3 & 3 \end{pmatrix}, m > 1$

(b) $A^m = 0, m > 1$

(c)
$\begin{pmatrix} 2(3)^{m-1} & 3^{m-1} & 3^{m-1} \\ \frac{1}{6}[9(2)^m - 4(3)^m] & \frac{1}{6}[3(2)^m - 2(3)^m] & \frac{1}{6}[-3(2)^m - 2(3)^m] \\ \frac{1}{6}[-9(2)^m + 8(3)^m] & \frac{1}{6}[-3(2)^m + 4(3)^m] & \frac{1}{6}[3(2)^m + 4(3)^m] \end{pmatrix}$

15. $\begin{pmatrix} \frac{3}{10} & \frac{2}{5} \\ -\frac{1}{10} & \frac{1}{5} \end{pmatrix}$

17. (b), (c), (d), (e), (f)

Exercises 8.10, Page 443

1. (b) $\lambda_1 = -4, \lambda_2 = -1, \lambda_3 = 16$

3. (b) $\lambda_1 = 18, \lambda_2 = \lambda_3 = -8$ **5.** orthogonal

7. orthogonal **9.** not orthogonal

11. $\begin{pmatrix} \frac{1}{\sqrt{2}} & \frac{1}{\sqrt{2}} \\ -\frac{1}{\sqrt{2}} & \frac{1}{\sqrt{2}} \end{pmatrix}$ **13.** $\begin{pmatrix} \frac{3}{\sqrt{10}} & \frac{1}{\sqrt{10}} \\ -\frac{1}{\sqrt{10}} & -\frac{3}{\sqrt{10}} \end{pmatrix}$

15. $\begin{pmatrix} -\frac{1}{\sqrt{2}} & \frac{1}{\sqrt{2}} & 0 \\ 0 & 0 & 1 \\ \frac{1}{\sqrt{2}} & \frac{1}{\sqrt{2}} & 0 \end{pmatrix}$

17. $\begin{pmatrix} -\frac{3}{\sqrt{11}} & \frac{1}{\sqrt{66}} & \frac{1}{\sqrt{6}} \\ \frac{1}{\sqrt{11}} & -\frac{4}{\sqrt{66}} & \frac{2}{\sqrt{6}} \\ \frac{1}{\sqrt{11}} & \frac{7}{\sqrt{66}} & \frac{1}{\sqrt{6}} \end{pmatrix}$

19. $a = -\frac{4}{5}, b = \frac{3}{5}$

21. (b) $\lambda_1 = \lambda_2 = 2, \lambda_3 = 4$

(c) $P = \begin{pmatrix} \frac{1}{\sqrt{2}} & \frac{1}{\sqrt{6}} & \frac{1}{\sqrt{3}} \\ -\frac{1}{\sqrt{2}} & \frac{1}{\sqrt{6}} & \frac{1}{\sqrt{3}} \\ 0 & -\frac{2}{\sqrt{6}} & \frac{1}{\sqrt{3}} \end{pmatrix}$

Exercises 8.11, Page 450

1. $2, \begin{pmatrix} 1 \\ 1 \end{pmatrix}$ **3.** $14, \begin{pmatrix} 1 \\ 3 \end{pmatrix}$ **5.** $10, \begin{pmatrix} 1 \\ 1 \\ 0.5 \end{pmatrix}$

7. 7 and 2 **9.** 4, 3, and 1

11. approximately 0.2087

13. (c) $\frac{1}{4} \begin{pmatrix} 3 & 2 & 1 \\ 2 & 4 & 2 \\ 1 & 2 & 3 \end{pmatrix}$

(d) 0.59

(e) approximately $9.44EI/L^2$

Exercises 8.12, Page 458

1. $P = \begin{pmatrix} -3 & 1 \\ 1 & 1 \end{pmatrix}, D = \begin{pmatrix} 1 & 0 \\ 0 & 5 \end{pmatrix}$

3. not diagonalizable

5. $P = \begin{pmatrix} 13 & 1 \\ 2 & 1 \end{pmatrix}, D = \begin{pmatrix} -7 & 0 \\ 0 & 4 \end{pmatrix}$

7. $P = \begin{pmatrix} 1 & 1 \\ -1 & 1 \end{pmatrix}, D = \begin{pmatrix} \frac{1}{3} & 0 \\ 0 & \frac{2}{3} \end{pmatrix}$

9. $P = \begin{pmatrix} 1 & 1 \\ -i & i \end{pmatrix}, D = \begin{pmatrix} -i & 0 \\ 0 & i \end{pmatrix}$

11. $P = \begin{pmatrix} 1 & 0 & 1 \\ 0 & 1 & 1 \\ 0 & 0 & 1 \end{pmatrix}, D = \begin{pmatrix} 1 & 0 & 0 \\ 0 & -1 & 0 \\ 0 & 0 & 2 \end{pmatrix}$

13. $P = \begin{pmatrix} 1 & 1 & 1 \\ 0 & 1 & 0 \\ -1 & 1 & 1 \end{pmatrix}, D = \begin{pmatrix} 0 & 0 & 0 \\ 0 & 1 & 0 \\ 0 & 0 & 2 \end{pmatrix}$

15. not diagonalizable

17. $P = \begin{pmatrix} 0 & 1+\sqrt{5} & 1-\sqrt{5} \\ 0 & 2 & 2 \\ 1 & 0 & 0 \end{pmatrix}$,

$D = \begin{pmatrix} 1 & 0 & 0 \\ 0 & \sqrt{5} & 0 \\ 0 & 0 & -\sqrt{5} \end{pmatrix}$

19. $P = \begin{pmatrix} -3 & -1 & -1 & 1 \\ 0 & 1 & 0 & 0 \\ -3 & 0 & 0 & 1 \\ 1 & 0 & 1 & 0 \end{pmatrix}$,

$D = \begin{pmatrix} 2 & 0 & 0 & 0 \\ 0 & 2 & 0 & 0 \\ 0 & 0 & 1 & 0 \\ 0 & 0 & 0 & -1 \end{pmatrix}$

21. $P = \begin{pmatrix} \frac{1}{\sqrt{2}} & \frac{1}{\sqrt{2}} \\ -\frac{1}{\sqrt{2}} & \frac{1}{\sqrt{2}} \end{pmatrix}, D = \begin{pmatrix} 0 & 0 \\ 0 & 2 \end{pmatrix}$

23. $P = \begin{pmatrix} -\frac{\sqrt{10}}{\sqrt{14}} & \frac{\sqrt{10}}{\sqrt{35}} \\ \frac{2}{\sqrt{14}} & \frac{5}{\sqrt{35}} \end{pmatrix}, D = \begin{pmatrix} 3 & 0 \\ 0 & 10 \end{pmatrix}$

25. $P = \begin{pmatrix} -\frac{1}{\sqrt{2}} & \frac{1}{\sqrt{2}} & 0 \\ \frac{1}{\sqrt{2}} & \frac{1}{\sqrt{2}} & 0 \\ 0 & 0 & 1 \end{pmatrix}, D = \begin{pmatrix} -1 & 0 & 0 \\ 0 & 1 & 0 \\ 0 & 0 & 1 \end{pmatrix}$

27. $P = \begin{pmatrix} \frac{2}{3} & \frac{2}{3} & \frac{1}{3} \\ \frac{2}{3} & -\frac{1}{3} & -\frac{2}{3} \\ \frac{1}{3} & -\frac{2}{3} & \frac{2}{3} \end{pmatrix}, D = \begin{pmatrix} 3 & 0 & 0 \\ 0 & 6 & 0 \\ 0 & 0 & 9 \end{pmatrix}$

29. $P = \begin{pmatrix} 0 & \frac{1}{\sqrt{2}} & \frac{1}{\sqrt{2}} \\ 1 & 0 & 0 \\ 0 & -\frac{1}{\sqrt{2}} & \frac{1}{\sqrt{2}} \end{pmatrix}, D = \begin{pmatrix} 1 & 0 & 0 \\ 0 & -6 & 0 \\ 0 & 0 & 8 \end{pmatrix}$

31. Ellipse; using

$$X = \begin{pmatrix} \frac{1}{\sqrt{2}} & \frac{1}{\sqrt{2}} \\ -\frac{1}{\sqrt{2}} & \frac{1}{\sqrt{2}} \end{pmatrix} X'$$

we get $X^2/4 + Y^2/6 = 1$.

33. Hyperbola; using

$$X = \begin{pmatrix} \frac{1}{\sqrt{5}} & -\frac{2}{\sqrt{5}} \\ \frac{2}{\sqrt{5}} & \frac{1}{\sqrt{5}} \end{pmatrix} X'$$

we get $X^2/4 - Y^2/4 = 1$.

35. $A = \begin{pmatrix} 4 & -1 \\ 2 & 1 \end{pmatrix}$ **39.** $A^5 = \begin{pmatrix} 21 & 11 \\ 22 & 10 \end{pmatrix}$

Exercises 8.13, Page 465

1. $LU = \begin{pmatrix} 1 & 0 \\ \frac{1}{2} & 1 \end{pmatrix}\begin{pmatrix} 2 & -2 \\ 0 & 3 \end{pmatrix}$

3. $LU = \begin{pmatrix} 1 & 0 \\ -2 & 1 \end{pmatrix}\begin{pmatrix} -1 & 4 \\ 0 & 10 \end{pmatrix}$

5. $LU = \begin{pmatrix} 1 & 0 & 0 \\ -1 & 1 & 0 \\ 3 & -7 & 1 \end{pmatrix}\begin{pmatrix} 4 & -2 & 1 \\ 0 & -1 & 3 \\ 0 & 0 & 21 \end{pmatrix}$

7. $LU = \begin{pmatrix} 1 & 0 & 0 \\ 2 & 1 & 0 \\ 7 & -8 & 1 \end{pmatrix}\begin{pmatrix} 1 & 2 & 7 \\ 0 & 1 & -8 \\ 0 & 0 & -109 \end{pmatrix}$

9. $LU = \begin{pmatrix} 1 & 0 & 0 \\ 0 & 1 & 0 \\ 2 & 10 & 1 \end{pmatrix}\begin{pmatrix} 1 & -2 & 1 \\ 0 & 1 & 2 \\ 0 & 0 & -21 \end{pmatrix}$

11. $LU = \begin{pmatrix} 1 & 0 \\ \frac{1}{3} & 1 \end{pmatrix}\begin{pmatrix} 3 & 9 \\ 0 & 8 \end{pmatrix}$

13. $LU = \begin{pmatrix} 1 & 0 \\ \frac{1}{4} & 1 \end{pmatrix}\begin{pmatrix} -4 & -2 \\ 0 & -\frac{5}{2} \end{pmatrix}$

15. $LU = \begin{pmatrix} 1 & 0 & 0 \\ 3 & 1 & 0 \\ 1 & 1 & 1 \end{pmatrix}\begin{pmatrix} 1 & 1 & 1 \\ 0 & -2 & -1 \\ 0 & 0 & 1 \end{pmatrix}$

17. $LU = \begin{pmatrix} 1 & 0 & 0 \\ \frac{1}{2} & 1 & 0 \\ 3 & -\frac{2}{5} & 1 \end{pmatrix}\begin{pmatrix} 4 & 2 & 12 \\ 0 & 25 & -10 \\ 0 & 0 & 16 \end{pmatrix}$

19. $LU = \begin{pmatrix} 1 & 0 & 0 & 0 \\ -2 & 1 & 0 & 0 \\ 1 & -4 & 1 & 0 \\ 5 & -11 & -2 & 1 \end{pmatrix}\begin{pmatrix} 1 & -2 & 1 & 0 \\ 0 & -1 & 0 & 1 \\ 0 & 0 & 3 & 7 \\ 0 & 0 & 0 & 26 \end{pmatrix}$

21. $x_1 = -\frac{1}{3}, x_2 = -\frac{5}{6}$ **23.** $x_1 = -1, x_2 = \frac{7}{2}$

25. $x_1 = 1, x_2 = 1, x_3 = 5$

27. $x_1 = 28, x_2 = -5, x_3 = 13$

29. $x_1 = 25, x_2 = -4, x_3 = -19$

31. $X = \begin{pmatrix} 0 \\ 5 \\ -3 \end{pmatrix}$ **33.** $X = \begin{pmatrix} \frac{1}{4} \\ \frac{3}{2} \\ -\frac{5}{4} \end{pmatrix}$

35. -2 **37.** 1600 **39.** -78

41. $LU = \begin{pmatrix} 1 & 0 & 0 \\ 3 & -2 & 0 \\ 1 & -2 & 1 \end{pmatrix}\begin{pmatrix} 1 & 1 & 1 \\ 0 & 1 & \frac{1}{2} \\ 0 & 0 & 1 \end{pmatrix}$

43. $LL^T = \begin{pmatrix} 2 & 0 & 0 \\ 1 & 5 & 0 \\ 6 & -2 & 4 \end{pmatrix}\begin{pmatrix} 2 & 1 & 6 \\ 0 & 5 & -2 \\ 0 & 0 & 4 \end{pmatrix}$

Exercises 8.14, Page 469

1. (a) $\begin{pmatrix} 35 & 15 & 38 & 36 & 0 \\ 27 & 10 & 26 & 20 & 0 \end{pmatrix}$

3. (a) $\begin{pmatrix} 48 & 64 & 120 & 107 & 40 \\ 32 & 40 & 75 & 67 & 25 \end{pmatrix}$

5. (a) $\begin{pmatrix} 31 & 44 & 15 & 61 & 50 & 49 & 41 \\ 24 & 29 & 15 & 47 & 35 & 31 & 21 \\ 1 & -15 & 15 & 0 & -15 & -5 & -19 \end{pmatrix}$

7. STUDY_HARD

9. MATH_IS_IMPORTANT_

11. DAD_I_NEED_MONEY_TODAY

13. (a) $B' = \begin{pmatrix} 15 & 22 & 20 & 8 & 23 & 6 & 21 & 22 \\ 10 & 22 & 18 & 23 & 25 & 2 & 23 & 25 \\ 3 & 26 & 26 & 14 & 23 & 16 & 26 & 12 \end{pmatrix}$

Exercises 8.15, Page 474

1. $(0\ 1\ 1\ 0)$ **3.** $(0\ 0\ 0\ 1\ 1)$

5. $(1\ 0\ 1\ 0\ 1\ 0\ 0\ 1)$ **7.** $(1\ 0\ 0)$

9. parity error **11.** $(1\ 0\ 0\ 1\ 1)$

13. $(0\ 0\ 1\ 0\ 1\ 1\ 0)$ **15.** $(0\ 1\ 0\ 0\ 1\ 0\ 1)$

17. $(1\ 1\ 0\ 0\ 1\ 1\ 0)$ **19.** code word; $(0\ 0\ 0\ 0)$

21. $(0\ 0\ 0\ 1)$ **23.** code word; $(1\ 1\ 1\ 1)$

25. $(1\ 0\ 0\ 1)$ **27.** $(1\ 0\ 1\ 0)$

29. (a) $2^7 = 128$ **(b)** $2^4 = 16$
 (c) (0 0 0 0 0 0 0), (0 1 0 0 1 0 1),
 (0 1 1 0 0 1 1), (0 1 0 1 0 1 0),
 (0 1 1 1 1 0 0), (0 0 1 0 1 1 0),
 (0 0 1 1 0 0 1), (0 0 0 1 1 1 1),
 (1 0 0 0 0 1 1), (1 1 0 0 1 1 0),
 (1 0 1 0 1 0 1), (1 0 0 1 1 0 0),
 (1 1 1 0 0 0 0), (1 1 0 1 0 0 1),
 (1 0 1 1 0 1 0), (1 1 1 1 1 1 1)

Exercises 8.16, Page 478

1. $y = 0.4x + 0.6$ **3.** $y = 1.1x - 0.3$

5. $y = 1.3571x + 1.9286$

7. $v = -0.84T + 234$, 116.4, 99.6

9. $f(x) = 0.75x^2 - 2.45x + 2.75$

Exercises 8.17, Page 482

1. (a) $T = \begin{pmatrix} 0.8 & 0.4 \\ 0.2 & 0.6 \end{pmatrix}$, $X_0 = \begin{pmatrix} 90 \\ 60 \end{pmatrix}$

 (b) $\begin{pmatrix} 96 \\ 54 \end{pmatrix}$, $\begin{pmatrix} 98.4 \\ 51.6 \end{pmatrix}$ **(c)** $\hat{X} = \begin{pmatrix} 100 \\ 50 \end{pmatrix}$

3. (a) $T = \begin{pmatrix} 0.2 & 0.5 & 0 \\ 0.3 & 0.1 & 0 \\ 0.5 & 0.4 & 1 \end{pmatrix}$, $X_0 = \begin{pmatrix} 100 \\ 0 \\ 0 \end{pmatrix}$

 (b) $\begin{pmatrix} 20 \\ 30 \\ 50 \end{pmatrix}$, $\begin{pmatrix} 19 \\ 9 \\ 72 \end{pmatrix}$ **(c)** $\hat{X} = \begin{pmatrix} 0 \\ 0 \\ 100 \end{pmatrix}$

Chapter 8 in Review, Page 483

1. $\begin{pmatrix} 2 & 3 & 4 \\ 3 & 4 & 5 \\ 4 & 5 & 6 \\ 5 & 6 & 7 \end{pmatrix}$ **3.** $\begin{pmatrix} 3 & 4 \\ 6 & 8 \end{pmatrix}$, (11) **5.** false

7. $\frac{5}{8}, -5$ **9.** 0 **11.** false **13.** true

15. false **17.** true **19.** false

23. (a) $\begin{pmatrix} 1 & 1 \\ -1 & -1 \end{pmatrix}$ **25.** $x_1 = -\frac{1}{2}, x_2 = 7, x_3 = \frac{1}{2}$

29. 240 **31.** trivial solution only

33. $I_2 + 10HNO_3 \rightarrow 2HIO_3 + 10NO_2 + 4H_2O$

35. $x_1 = -\frac{1}{2}, x_2 = \frac{1}{4}, x_3 = \frac{2}{3}$

37. $x = X \cos \theta - Y \sin \theta$,
 $y = X \sin \theta + Y \cos \theta$

39. $x_1 = 7, x_2 = 5, x_3 = 23$

41. $\lambda_1 = 5, \lambda_2 = -1, K_1 = \begin{pmatrix} 1 \\ 2 \end{pmatrix}, K_2 = \begin{pmatrix} 1 \\ -1 \end{pmatrix}$

43. $\lambda_1 = \lambda_2 = -1, \lambda_3 = 8$,

 $K_1 = \begin{pmatrix} 1 \\ -2 \\ 0 \end{pmatrix}, K_2 = \begin{pmatrix} 1 \\ 0 \\ -1 \end{pmatrix}, K_3 = \begin{pmatrix} 2 \\ 1 \\ 2 \end{pmatrix}$

45. $\lambda_1 = \lambda_2 = -3, \lambda_3 = 5$,

$$K_1 = \begin{pmatrix} -2 \\ 1 \\ 0 \end{pmatrix}, K_2 = \begin{pmatrix} 3 \\ 0 \\ 1 \end{pmatrix}, K_3 = \begin{pmatrix} 1 \\ 2 \\ -1 \end{pmatrix}$$

47. $\begin{pmatrix} \frac{1}{\sqrt{6}} \\ -\frac{2}{\sqrt{6}} \\ \frac{1}{\sqrt{6}} \end{pmatrix}$ **49.** hyperbola

51.

$$\begin{pmatrix} 204 & 13 & 208 & 55 & 124 & 120 & 105 & 214 & 50 & 6 & 138 & 19 & 210 \\ 185 & 12 & 188 & 50 & 112 & 108 & 96 & 194 & 45 & 6 & 126 & 18 & 189 \end{pmatrix}$$

53. HELP_IS_ON_THE_WAY

55. (a) (1 1 0 0 1) **(b)** parity error

57. $x_1 = 3, x_2 = 2, x_3 = 1$

59. $f(x) = \frac{4}{5}x - \frac{3}{2}$

Exercises 9.1, Page 492

1. **3.**

5. **7.**

9. **11.**

$r(t) = t\mathbf{i} + t\mathbf{j} + 2t^2\mathbf{k}$

13.

$r(t) = 3 \cos t\mathbf{i} + 3 \sin t\mathbf{j} + 9 \sin^2 t\mathbf{k}$

15. $2\mathbf{i} - 32\mathbf{j}$

17. $(1/t)\mathbf{i} - (1/t^2)\mathbf{j}; \; -(1/t^2)\mathbf{i} + (2/t^3)\mathbf{j}$

19. $\langle e^{2t}(2t + 1), 3t^2, 8t - 1 \rangle; \; \langle 4e^{2t}(t + 1), 6t, 8 \rangle$

21.

23.

25. $x = 2 + t, y = 2 + 2t, z = \frac{8}{3} + 4t$

27. $\mathbf{r}(t) \times \mathbf{r}''(t)$

29. $\mathbf{r}(t) \cdot [\mathbf{r}'(t) \times \mathbf{r}'''(t)]$

31. $2\mathbf{r}_1'(2t) - (1/t^2)\mathbf{r}_2'(1/t)$

33. $\frac{3}{2}\mathbf{i} + 9\mathbf{j} + 15\mathbf{k}$

35. $e^t(t - 1)\mathbf{i} + \frac{1}{2}e^{-2t}\mathbf{j} + \frac{1}{2}e^{t^2}\mathbf{k} + \mathbf{c}$

37. $(6t + 1)\mathbf{i} + (3t^2 - 2)\mathbf{j} + (t^3 + 1)\mathbf{k}$

39. $(2t^3 - 6t + 6)\mathbf{i} + (7t - 4t^{3/2} - 3)\mathbf{j} + (t^2 - 2t)\mathbf{k}$

41. $2\sqrt{a^2 + c^2}\,\pi$

43. $6(e^{3\pi} - 1)$

45. $a \cos(s/a)\mathbf{i} + a \sin(s/a)\mathbf{j}$

47. Differentiate $\mathbf{r}(t) \cdot \mathbf{r}(t) = c^2$.

Exercises 9.2, Page 495

1. Speed is $\sqrt{5}$.

3. Speed is 2.

5. Speed is $\sqrt{5}$.

7. Speed is $\sqrt{14}$.

9. $(0, 0, 0)$ and $(25, 115, 0)$;
$\mathbf{v}(0) = -2\mathbf{j} - 5\mathbf{k}, \mathbf{a}(0) = 2\mathbf{i} + 2\mathbf{k}$,
$\mathbf{v}(5) = 10\mathbf{i} + 73\mathbf{j} + 5\mathbf{k}, \mathbf{a}(5) = 2\mathbf{i} + 30\mathbf{j} + 2\mathbf{k}$

11. (a) $\mathbf{r}(t) = (-16t^2 + 240t)\mathbf{j} + 240\sqrt{3}\,t\mathbf{i}$ and
$x(t) = 240\sqrt{3}\,t, y(t) = -16t^2 + 240t$
(b) 900 ft **(c)** 6235 ft **(d)** 480 ft/s

13. 72.11 ft/s

15. 97.98 ft/s

17. (a) 4300 ft, approximately 7052.15 ft, approximately 576.89 ft/s

(b)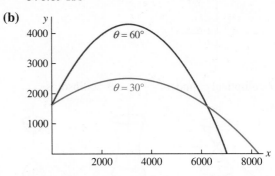

19. approximately 175.62 ft/s

21. Assume that (x_0, y_0) are the coordinates of the center of the target at $t = 0$. Then $\mathbf{r}_p = \mathbf{r}_t$ when $t = x_0/(v_0 \cos \theta) = y_0/(v_0 \sin \theta)$. This implies $\tan \theta = y_0/x_0$. In other words, aim directly at the target at $t = 0$.

25. 191.33 lb

27. $\mathbf{r}(t) = k_1 e^{2t^3}\mathbf{i} + \dfrac{1}{2t^2 + k_2}\mathbf{j} + (k_3 e^{t^2} - 1)\mathbf{k}$

29. (b) Since $\mathbf{F}$ is directed along $\mathbf{r}$, we must have $\mathbf{F} = c\mathbf{r}$ for some constant c. Hence $\boldsymbol{\tau} = \mathbf{r} \times (c\mathbf{r}) = c(\mathbf{r} \times \mathbf{r}) = \mathbf{0}$. If $\boldsymbol{\tau} = \mathbf{0}$, then $d\mathbf{L}/dt = \mathbf{0}$. This implies that $\mathbf{L}$ is a constant.

Exercises 9.3, Page 502

1. $\mathbf{T} = (\sqrt{5}/5)(-\sin t\mathbf{i} + \cos t\mathbf{j} + 2\mathbf{k})$

3. $\mathbf{T} = (a^2 + c^2)^{-1/2}(-a \sin t\mathbf{i} + a \cos t\mathbf{j} + c\mathbf{k})$,
$\mathbf{N} = -\cos t\mathbf{i} - \sin t\mathbf{j}$,
$\mathbf{B} = (a^2 + c^2)^{-1/2}(c \sin t\mathbf{i} - c \cos t\mathbf{j} + a\mathbf{k})$,
$\kappa = a/(a^2 + c^2)$

5. $3\sqrt{2}x - 3\sqrt{2}y + 4z = 3\pi$

7. $4t/\sqrt{1 + 4t^2}, \quad 2/\sqrt{1 + 4t^2}$

9. $2\sqrt{6}, 0, t > 0$ **11.** $2t/\sqrt{1 + t^2}, \quad 2/\sqrt{1 + t^2}$

13. $0, 5$ **15.** $-\sqrt{3}e^{-t}, 0$

17. $\kappa = \dfrac{\sqrt{b^2 c^2 \sin^2 t + a^2 c^2 \cos^2 t + a^2 b^2}}{(a^2 \sin^2 t + b^2 \cos^2 t + c^2)^{3/2}}$

23. $\kappa = 2, \rho = \frac{1}{2}; \kappa = 2/\sqrt{125} \approx 0.18$,
$\rho = \sqrt{125}/2 \approx 5.59$; the curve is sharper at $(0, 0)$.

Exercises 9.4, Page 507

1.

3.

5.

7. elliptical cylinders **9.** ellipsoids

11.

$c = 0$ $c > 0$ $c < 0$

13. $\partial z/\partial x = 2x - y^2, \partial z/\partial y = -2xy + 20y^4$

15. $\partial z/\partial x = 20x^3y^3 - 2xy^6 + 30x^4,$
$\partial z/\partial y = 15x^4y^2 - 6x^2y^5 - 4$

17. $\partial z/\partial x = 2x^{-1/2}/(3y^2 + 1),$
$\partial z/\partial y = -24\sqrt{xy}/(3y^2 + 1)^2$

19. $\partial z/\partial x = -3x^2(x^3 - y^2)^{-2},$
$\partial z/\partial y = 2y(x^3 - y^2)^{-2}$

21. $\partial z/\partial x = -10\cos 5x \sin 5x,$
$\partial z/\partial y = 10 \sin 5y \cos 5y$

23. $f_x = e^{x^3y}(3x^3y + 1), f_y = x^4e^{x^3y}$

25. $f_x = 7y/(x + 2y)^2, f_y = -7x/(x + 2y)^2$

27. $g_u = 8u/(4u^2 + 5v^3), g_v = 15v^2/(4u^2 + 5v^3)$

29. $w_x = x^{-1/2}y, w_y = 2\sqrt{x} - (y/z)e^{y/z} - e^{y/z},$
$w_z = (y^2/z^2)e^{y/z}$

31. $F_u = 2uw^2 - v^3 - vwt^2 \sin(ut^2),$
$F_v = -3uv^2 + w\cos(ut^2), F_x = 128x^7t^4,$
$F_t = -2uvwt \sin(ut^2) + 64x^8t^3$

39. $\partial z/\partial x = 3x^2v^2e^{uv^2} + 2uve^{uv^2}, \partial z/\partial y = -4yuve^{uv^2}$

41. $\partial z/\partial u = 16u^3 - 40y(2u - v),$
$\partial z/\partial v = -96v^2 + 20y(2u - v)$

43. $\partial w/\partial t = -3u(u^2 + v^2)^{1/2}e^{-t} \sin \theta$
$\qquad -3v(u^2 + v^2)^{1/2}e^{-t} \cos \theta,$
$\partial w/\partial \theta = 3u(u^2 + v^2)^{1/2}e^{-t} \cos \theta$
$\qquad -3v(u^2 + v^2)^{1/2}e^{-t} \sin \theta$

45. $\partial R/\partial u = s^2t^4e^{v^2} - 4rst^4uve^{-u^2} + 8rs^2t^3uv^2e^{u^2v^2},$
$\partial R/\partial v = 2s^2t^4uve^{v^2} + 2rst^4e^{-u^2} + 8rs^2t^3u^2ve^{u^2v^2}$

47. $\dfrac{\partial w}{\partial t} = \dfrac{xu}{(x^2 + y^2)^{1/2}(rs + tu)} + \dfrac{y\cosh rs}{u(x^2 + y^2)^{1/2}},$

$\dfrac{\partial w}{\partial r} = \dfrac{xs}{(x^2 + y^2)^{1/2}(rs + tu)} + \dfrac{sty\cosh rs}{u(x^2 + y^2)^{1/2}},$

$\dfrac{\partial w}{\partial u} = \dfrac{xt}{(x^2 + y^2)^{1/2}(rs + tu)} + \dfrac{ty\cosh rs}{u^2(x^2 + y^2)^{1/2}}$

49. $dz/dt = (4ut - 4vt^{-3})/(u^2 + v^2)$

51. $dw/dt|_{t=\pi} = -2$ **57.** $5.31 \text{ cm}^2/\text{s}$

Exercises 9.5, Page 512

1. $(2x - 3x^2y^2)\mathbf{i} + (-2x^3y + 4y^3)\mathbf{j}$

3. $(y^2/z^3)\mathbf{i} + (2xy/z^3)\mathbf{j} - (3xy^2/z^4)\mathbf{k}$

5. $4\mathbf{i} - 32\mathbf{j}$ **7.** $2\sqrt{3}\mathbf{i} - 8\mathbf{j} - 4\sqrt{3}\mathbf{k}$

9. $\sqrt{3}x + y$ **11.** $\frac{15}{2}(\sqrt{3} - 2)$

13. $-\frac{1}{2}\sqrt{10}$ **15.** $98/\sqrt{5}$ **17.** $-3\sqrt{2}$

19. -1 **21.** $-12/\sqrt{17}$

23. $\sqrt{2}\mathbf{i} + (\sqrt{2}/2)\mathbf{j}, \sqrt{5/2}$

25. $-2\mathbf{i} + 2\mathbf{j} - 4\mathbf{k}, 2\sqrt{6}$

27. $-8\sqrt{\pi/6}\mathbf{i} - 8\sqrt{\pi/6}\mathbf{j}, -8\sqrt{\pi/3}$

29. $-\frac{3}{8}\mathbf{i} - 12\mathbf{j} - \frac{2}{3}\mathbf{k}, -\sqrt{83,281}/24$

31. $\pm 31\sqrt{17}$

33. $\mathbf{u} = \frac{3}{5}\mathbf{i} - \frac{4}{5}\mathbf{j}; \mathbf{u} = \frac{4}{5}\mathbf{i} + \frac{3}{5}\mathbf{j}; \mathbf{u} = -\frac{4}{5}\mathbf{i} - \frac{3}{5}\mathbf{j}$

35. $D_{\mathbf{u}}f = (9x^2 + 3y^2 - 18xy^2 - 6x^2y)/\sqrt{10};$
$D_{\mathbf{u}}F = (-6x^2 - 54y^2 + 54x + 6y - 72xy)/10$

37. $(2, 5), (-2, 5)$

39. $-16\mathbf{i} - 4\mathbf{j}$

41. $x = 3e^{-4t}, y = 4e^{-2t}$

43. One possible function is $f(x, y) = x^3 - \frac{2}{3}y^3 + xy^3 + e^{xy}$.

Exercises 9.6, Page 517

1.

3.

5.

7.

9.

11.

13. $(-4, -1, 17)$

15. $-2x + 2y + z = 9$

17. $6x - 2y - 9z = 5$

19. $6x - 8y + z = 50$

21. $2x + y - \sqrt{2}z = (4 + 5\pi)/4$

23. $\sqrt{2}x + \sqrt{2}y - z = 2$

25. $(1/\sqrt{2}, \sqrt{2}, 3/\sqrt{2}), (-1/\sqrt{2}, -\sqrt{2}, -3/\sqrt{2})$

27. $(-2, 0, 5), (-2, 0, -3)$

33. $x = 1 + 2t, y = -1 - 4t, z = 1 + 2t$

35. $(x - \frac{1}{2})/4 = (y - \frac{1}{3})/6 = -(z - 3)$

Exercises 9.7, Page 521

1.

3.

5.
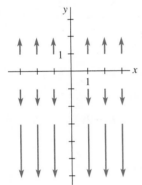

7. $(x - y)\mathbf{i} + (x - y)\mathbf{j}; 2z$ **9.** $\mathbf{0}; 4y + 8z$

11. $(4y^3 - 6xz^2)\mathbf{i} + (2z^3 - 3x^2)\mathbf{k}; 6xy$

13. $(3e^{-z} - 8yz)\mathbf{i} - xe^{-z}\mathbf{j}; e^{-z} + 4z^2 - 3ye^{-z}$

15. $(xy^2e^y + 2xye^y + x^3yze^z + x^3ye^z)\mathbf{i} - y^2e^y\mathbf{j}$
$+ (-3x^2yze^z - xe^x)\mathbf{k}; xye^x + ye^x - x^3ze^z$

35. $2\mathbf{i} + (1 - 8y)\mathbf{j} + 8z\mathbf{k}$

45. div $\mathbf{F} = 1 \neq 0$. If there existed a vector field $\mathbf{G}$ such that $\mathbf{F} = $ curl $\mathbf{G}$, then necessarily div $\mathbf{F} = $ div (curl $\mathbf{G}$) $= 0$.

Exercises 9.8, Page 530

1. $-125/3\sqrt{2}; -250(\sqrt{2} - 4)/12; \frac{125}{2}$

3. $3; 6; 3\sqrt{5}$ **5.** $-1; (\pi - 2)/2; \pi^2/8; \sqrt{2}\pi^2/8$

7. 21 **9.** 30 **11.** 1 **13.** 1

15. 460 **17.** $\frac{26}{9}$ **19.** $-\frac{64}{3}$ **21.** $-\frac{8}{3}$

23. 0 **25.** $\frac{123}{2}$ **27.** 70 **29.** $-\frac{19}{8}$

31. e **33.** -4 **35.** 0

37. On each curve the line integral has the value $\frac{208}{3}$.

41. $\bar{x} = \frac{3}{2}, \bar{y} = 2/\pi$

Exercises 9.9, Page 540

1. $\frac{16}{3}$ **3.** 14 **5.** 3 **7.** 330

9. 1096 **11.** $\phi = x^4y^3 + 3x + y$

13. not a conservative field **15.** $\phi = \frac{1}{4}x^4 + xy + \frac{1}{4}y^4$

17. $3 + e^{-1}$ **19.** 63 **21.** $8 + 2e^3$

23. 16 **25.** $\pi - 4$ **27.** $\phi = (Gm_1m_2)/|\mathbf{r}|$

Exercises 9.10, Page 547

1. $24y - 20e^y$ **3.** $x^2e^{3x^2} - x^2e^x$ **5.** $\frac{x}{2}\ln 5$

7. $2 - \sin y$

9.

11.

13. $\frac{1}{21}$ **15.** $\frac{25}{84}$ **17.** 96

19. $2\ln 2 - 1$ **21.** $\frac{14}{3}$ **23.** (c), 16π

25. 18 **27.** 2π **29.** 4 **31.** $30\ln 6$

33. $15\pi/4$ **35.** $(2^{3/2} - 1)/18$ **37.** $\frac{2}{3}\sin 8$

39. $\pi/8$ **41.** $\bar{x} = \frac{8}{3}, \bar{y} = 2$ **43.** $\bar{x} = 3, \bar{y} = \frac{3}{2}$

45. $\bar{x} = \frac{17}{21}, \bar{y} = \frac{55}{147}$ **47.** $\bar{x} = 0, \bar{y} = \frac{4}{7}$

49. $\bar{x} = (3e^4 + 1)/[4(e^4 - 1)],$
$\bar{y} = 16(e^5 - 1)/[25(e^4 - 1)]$

51. $\frac{1}{105}$ **53.** $4k/9$ **55.** $\frac{256}{21}$ **57.** $\frac{941}{10}$

59. $a\sqrt{10}/5$ **61.** $ab^3\pi/4; a^3b\pi/4; b/2; a/2$

63. $ka^4/6$ **65.** $16\sqrt{2}k/3$ **67.** $a\sqrt{3}/3$

Exercises 9.11, Page 552

1. $27\pi/2$ **3.** $(4\pi - 3\sqrt{3})/6$ **5.** $25\pi/3$

7. $(2\pi/3)(15^{3/2} - 7^{3/2})$ **9.** $\frac{5}{4}$

11. $\bar{x} = 13/3\pi, \bar{y} = 13/3\pi$ **13.** $\bar{x} = \frac{12}{5}, \bar{y} = 3\sqrt{3}/2$

15. $\bar{x} = (4 + 3\pi)/6, \bar{y} = \frac{4}{3}$ **17.** $\pi a^4k/4$

19. $(ka/12)(15\sqrt{3} - 4\pi)$ **21.** $\pi a^4k/2$

23. $4k$ **25.** 9π **27.** $(\pi/4)(e - 1)$

29. $3\pi/8$ **31.** 250

33. approximately 1450 m^3

Exercises 9.12, Page 557

1. 3 **3.** 0 **5.** 75π **7.** 48π

9. $\frac{56}{3}$ **11.** $\frac{2}{3}$ **13.** $\frac{1}{8}$

15. $(b - a) \times$ (area of region bounded by C)

19. $3a^2\pi/8$ **23.** $45\pi/2$ **25.** π

27. $27\pi/2$ **29.** $3\pi/2$ **33.** 3π

Exercises 9.13, Page 564

1. $3\sqrt{29}$ **3.** $10\pi/3$ **5.** $(\pi/6)(17^{3/2} - 1)$
7. $25\pi/6$ **9.** $2a^2(\pi - 2)$ **11.** $8a^2$
13. $2\pi a(c_2 - c_1)$ **15.** $\frac{26}{3}$ **17.** 0
19. 972π **21.** $(3^{5/2} - 2^{7/2} + 1)/15$
23. $9(17^{3/2} - 1)$ **25.** $12\sqrt{14}$ **27.** $k\sqrt{3}/12$
29. 18 **31.** 28π **33.** 8π **35.** $5\pi/2$
37. $-8\pi a^3$ **39.** $4\pi kq$ **41.** $(1, \frac{2}{3}, 2)$

Exercises 9.14, Page 570

1. -40π **3.** $\frac{45}{2}$ **5.** $\frac{3}{2}$ **7.** -3
9. $-3\pi/2$ **11.** π **13.** -152π
15. 112 **17.** Take the surface to be $z = 0$; $81\pi/4$.

Exercises 9.15, Page 579

1. 48 **3.** 36 **5.** $\pi - 2$
7. $\frac{1}{4}e^2 - \frac{1}{2}e$ **9.** 50

11. $\displaystyle\int_0^4 \int_0^{2-(x/2)} \int_{x+2y}^4 F(x, y, z)\, dz\, dy\, dx;$

$\displaystyle\int_0^2 \int_{2y}^4 \int_0^{z-2y} F(x, y, z)\, dx\, dz\, dy;$

$\displaystyle\int_0^4 \int_0^{z/2} \int_0^{z-2y} F(x, y, z)\, dx\, dy\, dz;$

$\displaystyle\int_0^4 \int_x^4 \int_0^{(z-x)/2} F(x, y, z)\, dy\, dz\, dx;$

$\displaystyle\int_0^4 \int_0^z \int_0^{(z-x)/2} F(x, y, z)\, dy\, dx\, dz$

13. (a) $\displaystyle\int_0^2 \int_{x^3}^8 \int_0^4 dz\, dy\, dx$ (b) $\displaystyle\int_0^8 \int_0^4 \int_0^{\sqrt[3]{y}} dx\, dz\, dy$

(c) $\displaystyle\int_0^4 \int_0^2 \int_{x^3}^8 dy\, dx\, dz$

15.

17. **19.**

21. $16\sqrt{2}$ **23.** 16π
25. $\bar{x} = \frac{4}{5}, \bar{y} = \frac{32}{7}, \bar{z} = \frac{8}{3}$ **27.** $\bar{x} = 0, \bar{y} = 2, \bar{z} = 0$
29. $\displaystyle\int_{-1}^1 \int_{-\sqrt{1-x^2}}^{\sqrt{1-x^2}} \int_{2y+2}^{8-y} (x + y + 4)\, dz\, dy\, dx$
31. $2560k/3$; $\sqrt{80/9}$ **33.** $k/30$

35. $(-10/\sqrt{2}, 10/\sqrt{2}, 5)$ **37.** $(\sqrt{3}/2, \frac{3}{2}, -4)$
39. $(\sqrt{2}, -\pi/4, -9)$ **41.** $(2\sqrt{2}, 2\pi/3, 2)$
43. $r^2 + z^2 = 25$ **45.** $r^2 - z^2 = 1$
47. $z = x^2 + y^2$ **49.** $x = 5$
51. $(2\pi/3)(64 - 12^{3/2})$ **53.** $625\pi/2$
55. $(0, 0, 3a/8)$ **57.** $8\pi k/3$
59. $(\sqrt{3}/3, \frac{1}{3}, 0)$; $(\frac{2}{3}, \pi/6, 0)$
61. $(-4, 4, 4\sqrt{2})$; $(4\sqrt{2}, 3\pi/4, 4\sqrt{2})$
63. $(5\sqrt{2}, \pi/2, 5\pi/4)$ **65.** $(\sqrt{2}, \pi/4, \pi/6)$
67. $\rho = 8$ **69.** $\phi = \pi/6, \phi = 5\pi/6$
71. $x^2 + y^2 + z^2 = 100$ **73.** $z = 2$
75. $9\pi(2 - \sqrt{2})$ **77.** $2\pi/9$
79. $(0, 0, \frac{7}{6})$ **81.** πk

Exercises 9.16, Page 586

1. $\frac{3}{2}$ **3.** $12a^5\pi/5$ **5.** 256π **7.** $62\pi/5$
9. $4\pi(b - a)$ **11.** 128 **13.** $\pi/2$

Exercises 9.17, Page 592

1. $(0, 0), (-2, 8), (16, 20), (14, 28)$
3.

5. **7.** $-2v$ **9.** $-\frac{1}{3}u^{-2}$

11.

(0, 0) is the image of every point on the boundary $u = 0$.
13. 16 **15.** $\frac{1}{2}$ **17.** $\frac{1}{4}(b - a)(d - c)$
19. $\frac{1}{2}(1 - \ln 2)$ **21.** $\frac{315}{4}$ **23.** $\frac{1}{4}(e - e^{-1})$
25. 126 **27.** $\frac{5}{2}(b - a) \ln\dfrac{d}{c}$ **29.** $15\pi/2$

Chapter 9 in Review, Page 593

1. true **3.** true **5.** false **7.** true
9. false **11.** false **13.** true
15. true **17.** true

19. $\nabla\phi = -\dfrac{x}{(x^2 + y^2)^{3/2}}\, \mathbf{i} - \dfrac{y}{(x^2 + y^2)^{3/2}}\, \mathbf{j}$

21. $\mathbf{v}(1) = 6\mathbf{i} + \mathbf{j} + 2\mathbf{k}$, $\mathbf{v}(4) = 6\mathbf{i} + \mathbf{j} + 8\mathbf{k}$, $\mathbf{a}(t) = 2\mathbf{k}$
for all t

23. $\mathbf{i} + 4\mathbf{j} + (3\pi/4)\mathbf{k}$

25.

27. $(6x^2 - 2y^2 - 8xy)/\sqrt{40}$ **29.** $2; -2/\sqrt{2}; 4$

31. $4\pi x + 3y - 12z = 4\pi - 6\sqrt{3}$

33. $\displaystyle\int_0^1 \int_x^{2x} \sqrt{1 - x^2}\, dy\, dx;$

$\displaystyle\int_0^1 \int_{y/2}^{y} \sqrt{1 - x^2}\, dx\, dy + \int_1^2 \int_{y/2}^1 \sqrt{1 - x^2}\, dx\, dy; \tfrac{1}{3}$

35. $41k/1512$ **37.** 8π

39. $6xy$ **41.** 0

43. $56\sqrt{2}\,\pi^3/3$ **45.** 12

47. $2 + 2/3\pi$ **49.** $\pi^2/2$

51. $(\ln 3)(17^{3/2} - 5^{3/2})/12$

53. $-4\pi c$ **55.** 0

57. 125π **59.** 3π

61. $\tfrac{5}{3}$ **63.** 0

65. π

Exercises 10.1, Page 606

1. $\mathbf{X}' = \begin{pmatrix} 3 & -5 \\ 4 & 8 \end{pmatrix}\mathbf{X}$, where $\mathbf{X} = \begin{pmatrix} x \\ y \end{pmatrix}$

3. $\mathbf{X}' = \begin{pmatrix} \tfrac{5}{2} & -7 \\ -\tfrac{1}{4} & \tfrac{3}{2} \end{pmatrix}\mathbf{X} + \begin{pmatrix} \tfrac{9}{2}t \\ -\tfrac{1}{2} \end{pmatrix}$

5. $\mathbf{X}' = \begin{pmatrix} -3 & 4 & -9 \\ 6 & -1 & 0 \\ 10 & 4 & 3 \end{pmatrix}\mathbf{X}$, where $\mathbf{X} = \begin{pmatrix} x \\ y \\ z \end{pmatrix}$

7. $\mathbf{X}' = \begin{pmatrix} 1 & -1 & 1 \\ 2 & 1 & -1 \\ 1 & 1 & 1 \end{pmatrix}\mathbf{X}$

$+ \begin{pmatrix} 0 \\ -3t^2 \\ t^2 \end{pmatrix} + \begin{pmatrix} t \\ 0 \\ -t \end{pmatrix} + \begin{pmatrix} -1 \\ 0 \\ 2 \end{pmatrix}$

9. $\dfrac{dx}{dt} = 4x + 2y + e^t$

$\dfrac{dy}{dt} = -x + 3y - e^t$

11. $\dfrac{dx}{dt} = x - y + 2z + e^{-t} - 3t$

$\dfrac{dy}{dt} = 3x - 4y + z + 2e^{-t} + t$

$\dfrac{dz}{dt} = -2x + 5y + 6z + 2e^{-t} - t$

13. $\begin{pmatrix} x_1' \\ x_2' \end{pmatrix} = \begin{pmatrix} 0 & 1 \\ -4 & 0 \end{pmatrix}\begin{pmatrix} x_1 \\ x_2 \end{pmatrix} + \begin{pmatrix} 0 \\ t\cos t \end{pmatrix}$

15. $\begin{pmatrix} x_1' \\ x_2' \\ x_3' \end{pmatrix} = \begin{pmatrix} 0 & 1 & 0 \\ 0 & 0 & 1 \\ 1 & -10 & -6 \end{pmatrix}\begin{pmatrix} x_1 \\ x_2 \\ x_3 \end{pmatrix} + \begin{pmatrix} 0 \\ 0 \\ t^2 \end{pmatrix}$

23. Yes; $W(\mathbf{X}_1, \mathbf{X}_2) = -2e^{-8t} \neq 0$ implies that $\mathbf{X}_1$ and $\mathbf{X}_2$ are linearly independent on $(-\infty, \infty)$.

25. No; $W(\mathbf{X}_1, \mathbf{X}_2, \mathbf{X}_3) = 0$ for every t. The solution vectors are linearly dependent on $(-\infty, \infty)$. Note that $\mathbf{X}_3 = 2\mathbf{X}_1 + \mathbf{X}_2$.

Exercises 10.2, Page 618

1. $\mathbf{X} = c_1 \begin{pmatrix} 1 \\ 2 \end{pmatrix}e^{5t} + c_2 \begin{pmatrix} 1 \\ -1 \end{pmatrix}e^{-t}$

3. $\mathbf{X} = c_1 \begin{pmatrix} 2 \\ 1 \end{pmatrix}e^{-3t} + c_2 \begin{pmatrix} 2 \\ 5 \end{pmatrix}e^t$

5. $\mathbf{X} = c_1 \begin{pmatrix} 5 \\ 2 \end{pmatrix}e^{8t} + c_2 \begin{pmatrix} 1 \\ 4 \end{pmatrix}e^{-10t}$

7. $\mathbf{X} = c_1 \begin{pmatrix} 1 \\ 0 \\ 0 \end{pmatrix}e^t + c_2 \begin{pmatrix} 2 \\ 3 \\ 1 \end{pmatrix}e^{2t} + c_3 \begin{pmatrix} 1 \\ 0 \\ 2 \end{pmatrix}e^{-t}$

9. $\mathbf{X} = c_1 \begin{pmatrix} -1 \\ 0 \\ 1 \end{pmatrix}e^{-t} + c_2 \begin{pmatrix} 1 \\ 4 \\ 3 \end{pmatrix}e^{3t} + c_3 \begin{pmatrix} 1 \\ -1 \\ 3 \end{pmatrix}e^{-2t}$

11. $\mathbf{X} = c_1 \begin{pmatrix} 4 \\ 0 \\ -1 \end{pmatrix}e^{-t} + c_2 \begin{pmatrix} -12 \\ 6 \\ 5 \end{pmatrix}e^{-t/2}$

$+ c_3 \begin{pmatrix} 4 \\ 2 \\ -1 \end{pmatrix}e^{-3t/2}$

13. $\mathbf{X} = 3 \begin{pmatrix} 1 \\ 1 \end{pmatrix}e^{t/2} + 2 \begin{pmatrix} 0 \\ 1 \end{pmatrix}e^{-t/2}$

15. **(a)** $\mathbf{X}' = \begin{pmatrix} -\tfrac{3}{100} & \tfrac{1}{100} \\ \tfrac{1}{50} & -\tfrac{1}{50} \end{pmatrix}\mathbf{X}$

(b) $\mathbf{X} = -\dfrac{35}{3}\begin{pmatrix} -1 \\ 1 \end{pmatrix}e^{-t/25} + \dfrac{25}{3}\begin{pmatrix} 1 \\ 2 \end{pmatrix}e^{-t/100}$

(d) approximately 34.3 minutes

21. $\mathbf{X} = c_1 \begin{pmatrix} 1 \\ 3 \end{pmatrix} + c_2 \left[\begin{pmatrix} 1 \\ 3 \end{pmatrix}t + \begin{pmatrix} \tfrac{1}{4} \\ -\tfrac{1}{4} \end{pmatrix} \right]$

23. $\mathbf{X} = c_1 \begin{pmatrix} 1 \\ 1 \end{pmatrix}e^{2t} + c_2 \left[\begin{pmatrix} 1 \\ 1 \end{pmatrix}te^{2t} + \begin{pmatrix} -\tfrac{1}{3} \\ 0 \end{pmatrix}e^{2t} \right]$

25. $\mathbf{X} = c_1 \begin{pmatrix} 1 \\ 1 \\ 1 \end{pmatrix}e^t + c_2 \begin{pmatrix} 1 \\ 1 \\ 0 \end{pmatrix}e^{2t} + c_3 \begin{pmatrix} 1 \\ 0 \\ 1 \end{pmatrix}e^{2t}$

27. $\mathbf{X} = c_1 \begin{pmatrix} -4 \\ -5 \\ 2 \end{pmatrix} + c_2 \begin{pmatrix} 2 \\ 0 \\ -1 \end{pmatrix} e^{5t}$

$+ c_3 \left[\begin{pmatrix} 2 \\ 0 \\ -1 \end{pmatrix} t e^{5t} + \begin{pmatrix} -\frac{1}{2} \\ -\frac{1}{2} \\ -1 \end{pmatrix} e^{5t} \right]$

29. $\mathbf{X} = c_1 \begin{pmatrix} 0 \\ 1 \\ 1 \end{pmatrix} e^t + c_2 \left[\begin{pmatrix} 0 \\ 1 \\ 1 \end{pmatrix} t e^t + \begin{pmatrix} 0 \\ 1 \\ 0 \end{pmatrix} e^t \right]$

$+ c_3 \left[\begin{pmatrix} 0 \\ 1 \\ 1 \end{pmatrix} \frac{t^2}{2} e^t + \begin{pmatrix} 0 \\ 1 \\ 0 \end{pmatrix} t e^t + \begin{pmatrix} \frac{1}{2} \\ 0 \\ 0 \end{pmatrix} e^t \right]$

31. $\mathbf{X} = -7 \begin{pmatrix} 2 \\ 1 \end{pmatrix} e^{4t} + 13 \begin{pmatrix} 2t + 1 \\ t + 1 \end{pmatrix} e^{4t}$

33. Corresponding to the eigenvalue $\lambda_1 = 2$ of multiplicity five, eigenvectors are

$$\mathbf{K}_1 = \begin{pmatrix} 1 \\ 0 \\ 0 \\ 0 \\ 0 \end{pmatrix}, \mathbf{K}_2 = \begin{pmatrix} 0 \\ 0 \\ 1 \\ 0 \\ 0 \end{pmatrix}, \mathbf{K}_3 = \begin{pmatrix} 0 \\ 0 \\ 0 \\ 1 \\ 0 \end{pmatrix}.$$

35. $\mathbf{X} = c_1 \begin{pmatrix} \cos t \\ 2\cos t + \sin t \end{pmatrix} e^{4t} + c_2 \begin{pmatrix} \sin t \\ 2\sin t - \cos t \end{pmatrix} e^{4t}$

37. $\mathbf{X} = c_1 \begin{pmatrix} \cos t \\ -\cos t - \sin t \end{pmatrix} e^{4t} + c_2 \begin{pmatrix} \sin t \\ -\sin t + \cos t \end{pmatrix} e^{4t}$

39. $\mathbf{X} = c_1 \begin{pmatrix} 5\cos 3t \\ 4\cos 3t + 3\sin 3t \end{pmatrix} + c_2 \begin{pmatrix} 5\sin 3t \\ 4\sin 3t - 3\cos 3t \end{pmatrix}$

41. $\mathbf{X} = c_1 \begin{pmatrix} 1 \\ 0 \\ 0 \end{pmatrix} + c_2 \begin{pmatrix} -\cos t \\ \cos t \\ \sin t \end{pmatrix} + c_3 \begin{pmatrix} \sin t \\ -\sin t \\ \cos t \end{pmatrix}$

43. $\mathbf{X} = c_1 \begin{pmatrix} 0 \\ 2 \\ 1 \end{pmatrix} e^t + c_2 \begin{pmatrix} \sin t \\ \cos t \\ \cos t \end{pmatrix} e^t + c_3 \begin{pmatrix} \cos t \\ -\sin t \\ -\sin t \end{pmatrix} e^t$

45. $\mathbf{X} = \begin{pmatrix} 28 \\ -5 \\ 25 \end{pmatrix} e^{2t} + c_2 \begin{pmatrix} 5\cos 3t \\ -4\cos 3t - 3\sin 3t \\ 0 \end{pmatrix} e^{-2t}$

$+ c_3 \begin{pmatrix} 5\sin 3t \\ -4\sin 3t + 3\cos 3t \\ 0 \end{pmatrix} e^{-2t}$

47. $\mathbf{X} = -\begin{pmatrix} 25 \\ -7 \\ 6 \end{pmatrix} e^t - \begin{pmatrix} \cos 5t - 5\sin 5t \\ \cos 5t \\ \cos 5t \end{pmatrix}$

$+ 6 \begin{pmatrix} 5\cos 5t + \sin 5t \\ \sin 5t \\ \sin 5t \end{pmatrix}$

49. (a) $\dfrac{dx_1}{dt} = -\dfrac{1}{20} x_1 + \dfrac{1}{10} x_3$

$\dfrac{dx_2}{dt} = \dfrac{1}{20} x_1 - \dfrac{1}{20} x_2$

$\dfrac{dx_3}{dt} = \dfrac{1}{20} x_2 - \dfrac{1}{10} x_3$

(b) $\mathbf{X} = -6 \begin{pmatrix} -\cos \frac{1}{20}t + \sin \frac{1}{20}t \\ -\sin \frac{1}{20}t \\ \cos \frac{1}{20}t \end{pmatrix} e^{-t/10}$

$- 2 \begin{pmatrix} -\cos \frac{1}{20}t - \sin \frac{1}{20}t \\ \cos \frac{1}{20}t \\ \sin \frac{1}{20}t \end{pmatrix} e^{-t/10} + 11 \begin{pmatrix} 2 \\ 2 \\ 1 \end{pmatrix}$

Exercises 10.3, Page 622

1. $\mathbf{X} = \begin{pmatrix} 3c_1 e^{7t} - 2c_2 e^{-4t} \\ c_1 e^{7t} + 3c_2 e^{-4t} \end{pmatrix}$

3. $\mathbf{X} = \begin{pmatrix} c_1 e^{t/2} + c_2 e^{3t/2} \\ -2c_1 e^{t/2} + 2c_2 e^{3t/2} \end{pmatrix}$

5. $\mathbf{X} = \begin{pmatrix} -c_1 e^{-4t} + c_2 e^{2t} \\ c_1 e^{-4t} + c_2 e^{2t} \\ c_2 e^{2t} + c_3 e^{6t} \end{pmatrix}$

7. $\mathbf{X} = \begin{pmatrix} c_1 e^{-t} - c_2 e^{2t} - c_3 e^{2t} \\ c_1 e^{-t} + c_2 e^{2t} \\ c_1 e^{-t} + c_3 e^{2t} \end{pmatrix}$

9. $\mathbf{X} = \begin{pmatrix} c_1 e^t + 2c_2 e^{2t} + 3c_3 e^{3t} \\ c_1 e^t + 2c_2 e^{2t} + 4c_3 e^{3t} \\ c_1 e^t + 2c_2 e^{2t} + 5c_3 e^{3t} \end{pmatrix}$

11. (a) $\mathbf{M} = \begin{pmatrix} m_1 & 0 \\ 0 & m_2 \end{pmatrix}$ and $\mathbf{K} = \begin{pmatrix} k_1 + k_2 & -k_2 \\ -k_2 & k_2 \end{pmatrix}$

Since $\mathbf{M}$ is a diagonal matrix with m_1 and m_2 nonzero, it has an inverse.

(b) $\mathbf{B} = \mathbf{M}^{-1}\mathbf{K} = \begin{pmatrix} \dfrac{k_1 + k_2}{m_1} & -\dfrac{k_2}{m_1} \\ -\dfrac{k_2}{m_2} & \dfrac{k_2}{m_2} \end{pmatrix}$

(c) $\mathbf{X} = c_1 \begin{pmatrix} 1 \\ 2 \end{pmatrix} \cos t + c_2 \begin{pmatrix} 1 \\ 2 \end{pmatrix} \sin t$

$+ c_3 \begin{pmatrix} -2 \\ 1 \end{pmatrix} \cos \sqrt{6}t + c_4 \begin{pmatrix} -2 \\ 1 \end{pmatrix} \sin \sqrt{6}t$

Exercises 10.4, Page 628

1. $\mathbf{X} = c_1 \begin{pmatrix} -1 \\ 1 \end{pmatrix} e^{-t} + c_2 \begin{pmatrix} -3 \\ 1 \end{pmatrix} e^t + \begin{pmatrix} -1 \\ 3 \end{pmatrix}$

3. $\mathbf{X} = c_1 \begin{pmatrix} 1 \\ -1 \end{pmatrix} e^{-2t} + c_2 \begin{pmatrix} 1 \\ 1 \end{pmatrix} e^{4t} + \begin{pmatrix} -\frac{1}{4} \\ \frac{3}{4} \end{pmatrix} t^2$

$+ \begin{pmatrix} \frac{1}{4} \\ -\frac{1}{4} \end{pmatrix} t + \begin{pmatrix} -2 \\ \frac{3}{4} \end{pmatrix}$

5. $\mathbf{X} = c_1 \begin{pmatrix} 1 \\ -3 \end{pmatrix} e^{3t} + c_2 \begin{pmatrix} 1 \\ 9 \end{pmatrix} e^{7t} + \begin{pmatrix} \frac{55}{36} \\ -\frac{19}{4} \end{pmatrix} e^{t}$

7. $\mathbf{X} = c_1 \begin{pmatrix} 1 \\ 0 \\ 0 \end{pmatrix} e^{t} + c_2 \begin{pmatrix} 1 \\ 1 \\ 0 \end{pmatrix} e^{2t} + c_3 \begin{pmatrix} 1 \\ 2 \\ 2 \end{pmatrix} e^{5t} - \begin{pmatrix} \frac{3}{2} \\ \frac{7}{2} \\ 2 \end{pmatrix} e^{4t}$

9. $\mathbf{X} = 13 \begin{pmatrix} 1 \\ -1 \end{pmatrix} e^{t} + 2 \begin{pmatrix} -4 \\ 6 \end{pmatrix} e^{2t} + \begin{pmatrix} -9 \\ 6 \end{pmatrix}$

11. (a) $\mathbf{X}' = \begin{pmatrix} -\frac{3}{100} & \frac{1}{100} \\ \frac{1}{50} & -\frac{1}{25} \end{pmatrix} \mathbf{X} + \begin{pmatrix} 0 \\ 1 \end{pmatrix}$

(b) $\mathbf{X} = -\frac{70}{3} \begin{pmatrix} -1 \\ 2 \end{pmatrix} e^{-t/20} + \frac{80}{3} \begin{pmatrix} 1 \\ 1 \end{pmatrix} e^{-t/50} + \begin{pmatrix} 10 \\ 30 \end{pmatrix}$

(c) 10; 30; as $t \to \infty$ the total amount of salt in the system of mixing tanks approaches a constant 40 lb

13. $\mathbf{X} = c_1 \begin{pmatrix} 1 \\ 1 \end{pmatrix} + c_2 \begin{pmatrix} 3 \\ 2 \end{pmatrix} e^{t} - \begin{pmatrix} 11 \\ 11 \end{pmatrix} t - \begin{pmatrix} 15 \\ 10 \end{pmatrix}$

15. $\mathbf{X} = c_1 \begin{pmatrix} 2 \\ 1 \end{pmatrix} e^{t/2} + c_2 \begin{pmatrix} 10 \\ 3 \end{pmatrix} e^{3t/2} - \begin{pmatrix} \frac{13}{2} \\ \frac{13}{4} \end{pmatrix} te^{t/2} - \begin{pmatrix} \frac{15}{2} \\ \frac{9}{4} \end{pmatrix} e^{t/2}$

17. $\mathbf{X} = c_1 \begin{pmatrix} 2 \\ 1 \end{pmatrix} e^{t} + c_2 \begin{pmatrix} 1 \\ 1 \end{pmatrix} e^{2t} + \begin{pmatrix} 3 \\ 3 \end{pmatrix} e^{t} + \begin{pmatrix} 4 \\ 2 \end{pmatrix} te^{t}$

19. $\mathbf{X} = c_1 \begin{pmatrix} 4 \\ 1 \end{pmatrix} e^{3t} + c_2 \begin{pmatrix} -2 \\ 1 \end{pmatrix} e^{-3t} + \begin{pmatrix} -12 \\ 0 \end{pmatrix} t - \begin{pmatrix} \frac{4}{3} \\ \frac{4}{3} \end{pmatrix}$

21. $\mathbf{X} = c_1 \begin{pmatrix} 1 \\ -1 \end{pmatrix} e^{t} + c_2 \begin{pmatrix} -t \\ \frac{1}{2} - t \end{pmatrix} e^{t} + \begin{pmatrix} \frac{1}{2} \\ -2 \end{pmatrix} e^{-t}$

23. $\mathbf{X} = c_1 \begin{pmatrix} \cos t \\ \sin t \end{pmatrix} + c_2 \begin{pmatrix} \sin t \\ -\cos t \end{pmatrix} + \begin{pmatrix} \cos t \\ \sin t \end{pmatrix} t$
$+ \begin{pmatrix} -\sin t \\ \cos t \end{pmatrix} \ln|\cos t|$

25. $\mathbf{X} = c_1 \begin{pmatrix} \cos t \\ \sin t \end{pmatrix} e^{t} + c_2 \begin{pmatrix} \sin t \\ -\cos t \end{pmatrix} e^{t} + \begin{pmatrix} \cos t \\ \sin t \end{pmatrix} te^{t}$

27. $\mathbf{X} = c_1 \begin{pmatrix} \cos t \\ -\sin t \end{pmatrix} + c_2 \begin{pmatrix} \sin t \\ \cos t \end{pmatrix} + \begin{pmatrix} \cos t \\ -\sin t \end{pmatrix} t$
$+ \begin{pmatrix} -\sin t \\ \sin t \tan t \end{pmatrix} - \begin{pmatrix} \sin t \\ \cos t \end{pmatrix} \ln|\cos t|$

29. $\mathbf{X} = c_1 \begin{pmatrix} 2\sin t \\ \cos t \end{pmatrix} e^{t} + c_2 \begin{pmatrix} 2\cos t \\ -\sin t \end{pmatrix} e^{t} + \begin{pmatrix} 3\sin t \\ \frac{3}{2}\cos t \end{pmatrix} te^{t}$
$+ \begin{pmatrix} \cos t \\ -\frac{1}{2}\sin t \end{pmatrix} e^{t} \ln|\sin t| + \begin{pmatrix} 2\cos t \\ -\sin t \end{pmatrix} e^{t} \ln|\cos t|$

31. $\mathbf{X} = c_1 \begin{pmatrix} 1 \\ -1 \\ 0 \end{pmatrix} + c_2 \begin{pmatrix} 1 \\ 1 \\ 0 \end{pmatrix} e^{2t} + c_3 \begin{pmatrix} 0 \\ 0 \\ 1 \end{pmatrix} e^{3t}$
$+ \begin{pmatrix} -\frac{1}{4}e^{2t} + \frac{1}{2}te^{2t} \\ -e^{t} + \frac{1}{4}e^{2t} + \frac{1}{2}te^{2t} \\ \frac{1}{2}t^2 e^{3t} \end{pmatrix}$

33. $\mathbf{X} = \begin{pmatrix} 2 \\ 2 \end{pmatrix} te^{2t} + \begin{pmatrix} -1 \\ 1 \end{pmatrix} e^{2t} + \begin{pmatrix} -2 \\ 2 \end{pmatrix} te^{4t} + \begin{pmatrix} 2 \\ 0 \end{pmatrix} e^{4t}$

35. $\begin{pmatrix} i_1 \\ i_2 \end{pmatrix} = 2 \begin{pmatrix} 1 \\ 3 \end{pmatrix} e^{-2t} + \frac{6}{29} \begin{pmatrix} 3 \\ -1 \end{pmatrix} e^{-12t} - \frac{4}{29} \begin{pmatrix} 19 \\ 42 \end{pmatrix} \cos t$
$+ \frac{4}{29} \begin{pmatrix} 83 \\ 69 \end{pmatrix} \sin t$

37. $\mathbf{X} = c_1 \begin{pmatrix} 1 \\ 3 \end{pmatrix} e^{-t} + c_2 \begin{pmatrix} 2 \\ 7 \end{pmatrix} e^{-2t} + \begin{pmatrix} 20 \\ 53 \end{pmatrix}$

39. $\mathbf{X} = c_1 \begin{pmatrix} 1 \\ -1 \end{pmatrix} + c_2 \begin{pmatrix} 1 \\ 1 \end{pmatrix} e^{10t} + \frac{1}{2} \begin{pmatrix} 1 \\ -1 \end{pmatrix} t^2$
$+ \frac{1}{10} \begin{pmatrix} -41 \\ 39 \end{pmatrix} t - \frac{41}{100} \begin{pmatrix} 1 \\ 1 \end{pmatrix}$

Exercises 10.5, Page 634

1. $e^{\mathbf{A}t} = \begin{pmatrix} e^{t} & 0 \\ 0 & e^{2t} \end{pmatrix}$; $e^{-\mathbf{A}t} = \begin{pmatrix} e^{-t} & 0 \\ 0 & e^{-2t} \end{pmatrix}$

3. $e^{\mathbf{A}t} = \begin{pmatrix} t+1 & t & t \\ t & t+1 & t \\ -2t & -2t & -2t+1 \end{pmatrix}$

5. $\mathbf{X} = c_1 \begin{pmatrix} 1 \\ 0 \end{pmatrix} e^{t} + c_2 \begin{pmatrix} 0 \\ 1 \end{pmatrix} e^{2t}$

7. $\mathbf{X} = c_1 \begin{pmatrix} t+1 \\ t \\ -2t \end{pmatrix} + c_2 \begin{pmatrix} t \\ t+1 \\ -2t \end{pmatrix} + c_3 \begin{pmatrix} t \\ t \\ -2t+1 \end{pmatrix}$

9. $\mathbf{X} = c_3 \begin{pmatrix} 1 \\ 0 \end{pmatrix} e^{t} + c_4 \begin{pmatrix} 0 \\ 1 \end{pmatrix} e^{2t} + \begin{pmatrix} -3 \\ \frac{1}{2} \end{pmatrix}$

11. $\mathbf{X} = c_1 \begin{pmatrix} \cosh t \\ \sinh t \end{pmatrix} + c_2 \begin{pmatrix} \sinh t \\ \cosh t \end{pmatrix} - \begin{pmatrix} 1 \\ 1 \end{pmatrix}$

13. $\mathbf{X} = \begin{pmatrix} t+1 \\ t \\ -2t \end{pmatrix} - 4 \begin{pmatrix} t \\ t+1 \\ -2t \end{pmatrix} + 6 \begin{pmatrix} t \\ t \\ -2t+1 \end{pmatrix}$

15. $e^{\mathbf{A}t} = \begin{pmatrix} \frac{3}{2}e^{2t} - \frac{1}{2}e^{-2t} & \frac{3}{4}e^{2t} - \frac{3}{4}e^{-2t} \\ -e^{2t} + e^{-2t} & -\frac{1}{2}e^{2t} + \frac{3}{2}e^{-2t} \end{pmatrix}$;
$\mathbf{X} = c_1 \begin{pmatrix} \frac{3}{2}e^{2t} - \frac{1}{2}e^{-2t} \\ -e^{2t} + e^{-2t} \end{pmatrix} + c_2 \begin{pmatrix} \frac{3}{4}e^{2t} - \frac{3}{4}e^{-2t} \\ -\frac{1}{2}e^{2t} + \frac{3}{2}e^{-2t} \end{pmatrix}$ or
$\mathbf{X} = c_3 \begin{pmatrix} 3 \\ -2 \end{pmatrix} e^{2t} + c_4 \begin{pmatrix} 1 \\ -2 \end{pmatrix} e^{-2t}$

17. $e^{\mathbf{A}t} = \begin{pmatrix} e^{2t} + 3te^{2t} & -9te^{2t} \\ te^{2t} & e^{2t} - 3te^{2t} \end{pmatrix}$;
$\mathbf{X} = c_1 \begin{pmatrix} 1+3t \\ t \end{pmatrix} e^{2t} + c_2 \begin{pmatrix} -9t \\ 1-3t \end{pmatrix} e^{2t}$

19. $e^{\mathbf{A}t} = \begin{pmatrix} \frac{4}{5}e^{t} + \frac{1}{5}e^{6t} & \frac{2}{5}e^{t} - \frac{2}{5}e^{6t} \\ \frac{2}{5}e^{t} - \frac{2}{5}e^{6t} & \frac{1}{5}e^{t} + \frac{4}{5}e^{6t} \end{pmatrix}$;
$\mathbf{X} = c_1 \begin{pmatrix} \frac{4}{5}e^{t} + \frac{1}{5}e^{6t} \\ \frac{2}{5}e^{t} - \frac{2}{5}e^{6t} \end{pmatrix} + c_2 \begin{pmatrix} \frac{2}{5}e^{t} - \frac{2}{5}e^{6t} \\ \frac{1}{5}e^{t} + \frac{4}{5}e^{6t} \end{pmatrix}$ or
$\mathbf{X} = c_3 \begin{pmatrix} 2 \\ 1 \end{pmatrix} e^{t} + c_4 \begin{pmatrix} 1 \\ -2 \end{pmatrix} e^{6t}$

21. $e^{\mathbf{A}t} = \begin{pmatrix} e^{3t} & -2e^{-t} + 2e^{3t} \\ 0 & e^{-t} \end{pmatrix}$;

$\mathbf{X} = c_1 \begin{pmatrix} 1 \\ 0 \end{pmatrix} e^{3t} + c_2 \begin{pmatrix} -2e^{-t} + 2e^{3t} \\ e^{-t} \end{pmatrix}$ or

$\mathbf{X} = c_2 \begin{pmatrix} -2 \\ 1 \end{pmatrix} e^{-t} + c_4 \begin{pmatrix} 1 \\ 0 \end{pmatrix} e^{3t}$

25. $\mathbf{X} = c_1 \begin{pmatrix} \frac{3}{2}e^{3t} - \frac{1}{2}e^{5t} \\ \frac{3}{2}e^{3t} - \frac{3}{2}e^{5t} \end{pmatrix} + c_2 \begin{pmatrix} -\frac{1}{2}e^{3t} + \frac{1}{2}e^{5t} \\ -\frac{1}{2}e^{3t} + \frac{3}{2}e^{5t} \end{pmatrix}$ or

$\mathbf{X} = c_3 \begin{pmatrix} 1 \\ 1 \end{pmatrix} e^{3t} + c_4 \begin{pmatrix} 1 \\ 3 \end{pmatrix} e^{5t}$

Chapter 10 in Review, Page 635

1. $k = \frac{1}{3}$ **5.** $\mathbf{X} = c_1 \begin{pmatrix} 1 \\ -1 \end{pmatrix} e^t + c_2 \left[\begin{pmatrix} 1 \\ -1 \end{pmatrix} te^t + \begin{pmatrix} 0 \\ 1 \end{pmatrix} e^t \right]$

7. $\mathbf{X} = c_1 \begin{pmatrix} \cos 2t \\ -\sin 2t \end{pmatrix} e^t + c_2 \begin{pmatrix} \sin 2t \\ \cos 2t \end{pmatrix} e^t$

9. $\mathbf{X} = c_1 \begin{pmatrix} -2 \\ 3 \\ 1 \end{pmatrix} e^{2t} + c_2 \begin{pmatrix} 0 \\ 1 \\ 1 \end{pmatrix} e^{4t} + c_3 \begin{pmatrix} 7 \\ 12 \\ -16 \end{pmatrix} e^{-3t}$

11. $\mathbf{X} = c_1 \begin{pmatrix} 1 \\ 0 \end{pmatrix} e^{2t} + c_2 \begin{pmatrix} 4 \\ 1 \end{pmatrix} e^{4t} + \begin{pmatrix} 16 \\ -4 \end{pmatrix} t + \begin{pmatrix} 11 \\ -1 \end{pmatrix}$

13. $\mathbf{X} = c_1 \begin{pmatrix} \cos t + \sin t \\ 2 \cos t \end{pmatrix} + c_2 \begin{pmatrix} \sin t - \cos t \\ 2 \sin t \end{pmatrix}$

$\quad - \begin{pmatrix} 1 \\ 1 \end{pmatrix} + \begin{pmatrix} \sin t \\ \sin t + \cos t \end{pmatrix} \ln|\csc t - \cot t|$

15. (b) $\mathbf{X} = c_1 \begin{pmatrix} -1 \\ 1 \\ 0 \end{pmatrix} + c_2 \begin{pmatrix} -1 \\ 0 \\ 1 \end{pmatrix} + c_3 \begin{pmatrix} 1 \\ 1 \\ 1 \end{pmatrix} e^{3t}$

Exercises 11.1, Page 644

1. $x' = y$

$y' = -9 \sin x$; critical points at $(\pm n\pi, 0)$

3. $x' = y$

$y' = x^2 + y(x^3 - 1)$; critical point at $(0, 0)$

5. $x' = y$

$y' = \epsilon x^3 - x$;

critical points at $(0, 0)$, $\left(\frac{1}{\sqrt{\epsilon}}, 0 \right)$, $\left(-\frac{1}{\sqrt{\epsilon}}, 0 \right)$

7. $(0, 0)$ and $(-1, -1)$

9. $(0, 0)$ and $(\frac{4}{3}, \frac{4}{3})$

11. $(0, 0)$, $(10, 0)$, $(0, 16)$, and $(4, 12)$

13. $(0, y)$, y arbitrary

15. $(0, 0)$, $(0, 1)$, $(0, -1)$, $(1, 0)$, $(-1, 0)$

17. (a) $x = c_1 e^{5t} - c_2 e^{-t}$

$y = 2c_1 e^{5t} + c_2 e^{-t}$

(b) $x = -2e^{-t}$

$y = 2e^{-t}$

19. (a) $x = c_1(4 \cos 3t - 3 \sin 3t) + c_2(4 \sin 3t + 3 \cos 3t)$

$y = c_1(5 \cos 3t) + c_2(5 \sin 3t)$

(b) $x = 4 \cos 3t - 3 \sin 3t$

$y = 5 \cos 3t$

21. (a) $x = c_1(\sin t - \cos t)e^{4t} + c_2(-\sin t - \cos t)e^{4t}$

$y = 2c_1(\cos t)e^{4t} + 2c_2(\sin t)e^{4t}$

(b) $x = (\sin t - \cos t)e^{4t}$

$y = 2(\cos t)e^{4t}$

23. $r = \dfrac{1}{\sqrt[4]{4t + c_1}}, \theta = t + c_2; r = 4\dfrac{1}{\sqrt[4]{1024t + 1}}, \theta = t$;

the solution spirals toward the origin as t increases.

25. $r = \dfrac{1}{\sqrt{1 + c_1 e^{-2t}}}, \theta = t + c_2; r = 1, \theta = t$ (or $x = \cos t$

and $y = \sin t$) is the solution that satisfies $\mathbf{X}(0) = (1, 0)$;

$r = \dfrac{1}{\sqrt{1 - \frac{3}{4}e^{-2t}}}, \theta = t$ is the solution that satisfies

$\mathbf{X}(0) = (2, 0)$. This solution spirals toward the circle

$r = 1$ as t increases.

27. There are no critical points and therefore no periodic solutions.

29. There appears to be a periodic solution enclosing the critical point $(0, 0)$.

Exercises 11.2, Page 651

1. (a) If $\mathbf{X}(0) = \mathbf{X}_0$ lies on the line $y = 2x$, then $\mathbf{X}(t)$ approaches $(0, 0)$ along this line. For all other initial conditions, $\mathbf{X}(t)$ approaches $(0, 0)$ from the direction determined by the line $y = -x/2$.

3. (a) All solutions are unstable spirals that become unbounded as t increases.

5. (a) All solutions approach $(0, 0)$ from the direction specified by the line $y = x$.

7. (a) If $\mathbf{X}(0) = \mathbf{X}_0$ lies on the line $y = 3x$, then $\mathbf{X}(t)$ approaches $(0, 0)$ along this line. For all other initial conditions, $\mathbf{X}(t)$ becomes unbounded and $y = x$ serves as the asymptote.

9. saddle point **11.** saddle point

13. degenerate stable node **15.** stable spiral

17. $|\mu| < 1$

19. $\mu < -1$ for a saddle point; $-1 < \mu < 3$ for an unstable spiral point

23. (a) $(-3, 4)$

(b) unstable node or saddle point

(c) $(0, 0)$ is a saddle point.

25. (a) $(\frac{1}{2}, 2)$

(b) unstable spiral point

(c) $(0, 0)$ is an unstable spiral point.

Exercises 11.3, Page 659

1. $r = r_0 e^{\alpha t}$

3. $x = 0$ is unstable; $x = n + 1$ is asymptotically stable.

5. $T = T_0$ is unstable.

7. $x = \alpha$ is unstable; $x = \beta$ is asymptotically stable.

9. $P = a/b$ is asymptotically stable; $P = c$ is unstable.

11. $(\frac{1}{2}, 1)$ is a stable spiral point.

13. $(\sqrt{2}, 0)$ and $(-\sqrt{2}, 0)$ are saddle points; $(\frac{1}{2}, -\frac{7}{4})$ is a stable spiral point.

15. $(1, 1)$ is a stable node; $(1, -1)$ is a saddle point; $(2, 2)$ is a saddle point; $(2, -2)$ is an unstable spiral point.

17. $(0, -1)$ is a saddle point; $(0, 0)$ is unclassified; $(0, 1)$ is stable but we are unable to classify further.

19. $(0, 0)$ is an unstable node; $(10, 0)$ is a saddle point; $(0, 16)$ is a saddle point; $(4, 12)$ is a stable node.

21. $\theta = 0$ is a saddle point. It is not possible to classify either $\theta = \pi/3$ or $\theta = -\pi/3$.

23. It is not possible to classify $x = 0$.

25. It is not possible to classify $x = 0$, but $x = 1/\sqrt{\epsilon}$ and $x = -1/\sqrt{\epsilon}$ are each saddle points.

29. (a) $(0, 0)$ is a stable spiral point.

33. (a) $(1, 0), (-1, 0)$ **35.** $|v_0| < \frac{1}{2}\sqrt{2}$

37. If $\beta > 0$, $(0, 0)$ is the only critical point and is stable. If $\beta < 0$, $(0, 0)$, $(\hat{x}, 0)$, and $(-\hat{x}, 0)$, where $\hat{x}^2 = -\alpha/\beta$, are critical points. $(0, 0)$ is stable, while $(\hat{x}, 0)$ and $(-\hat{x}, 0)$ are each saddle points.

39. (b) $(5\pi/6, 0)$ is a saddle point.

 (c) $(\pi/6, 0)$ is a center.

Exercises 11.4, Page 666

1. $|\omega_0| < \sqrt{3g/L}$

5. (a) First show that $y^2 = v_0^2 + g \ln\left(\dfrac{1 + x^2}{1 + x_0^2}\right)$.

9. (a) The new critical point is $(d/c - \epsilon_2/c, a/b + \epsilon_1/b)$.

 (b) yes

11. $(0, 0)$ is an unstable node, $(0, 100)$ is a stable node, $(50, 0)$ is a stable node, and $(20, 40)$ is a saddle point.

17. (a) $(0, 0)$ is the only critical point.

Exercises 11.5, Page 674

1. The system has no critical points.

3. $\dfrac{\partial P}{\partial x} + \dfrac{\partial Q}{\partial y} = -2 < 0$

5. $\dfrac{\partial P}{\partial x} + \dfrac{\partial Q}{\partial y} = -\mu + 9y^2 > 0$ if $\mu < 0$

7. The single critical point $(0, 0)$ is a saddle point.

9. $\delta(x, y) = e^{-y/2}$

11. $\dfrac{\partial P}{\partial x} + \dfrac{\partial Q}{\partial y} = 4(1 - x^2 - 3y^2) > 0$ for $x^2 + 3y^2 < 1$

13. Use $\delta(x, y) = 1/(xy)$ and show that

$$\frac{\partial(\delta P)}{\partial x} + \frac{\partial(\delta Q)}{\partial y} = -\frac{r}{Kx}.$$

15. If $\mathbf{n} = (-2x, -2y)$, show that $\mathbf{V} \cdot \mathbf{n} = 2(x - y)^2 + 2y^4$.

17. Yes; the sole critical point $(0, 0)$ lies outside the invariant region $\frac{1}{16} \le x^2 + y^2 \le 1$, and so Theorem 11.5.5(*ii*) applies.

19. $\mathbf{V} \cdot \mathbf{n} = 2y^2(1 - x^2) \ge 2y^2(1 - r^2)$ and $\partial P/\partial x + \partial Q/\partial y = x^2 - 1 < 0$. The sole critical point is $(0, 0)$ and this critical point is a stable spiral point. Therefore, Theorem 11.5.6(*ii*) applies.

21. (a) $\dfrac{\partial P}{\partial x} + \dfrac{\partial Q}{\partial y} = 2xy - 1 - x^2 \le 2x - 1 - x^2$

$$= -(x - 1)^2 \le 0$$

 (b) $\lim_{t \to \infty} \mathbf{X}(t) = (\frac{3}{2}, \frac{2}{9})$, a stable spiral point

Chapter 11 in Review, Page 676

1. true **3.** a center or a saddle point

5. false **7.** false **9.** true

11. $r = \dfrac{1}{\sqrt[3]{3t + 1}}$, $\theta = t$; the solution curve spirals toward the origin.

13. center; degenerate stable node

15. stable node for $\mu < -2$; stable spiral point for $-2 < \mu < 0$; unstable spiral point for $0 < \mu < 2$; unstable node for $\mu > 2$

17. Show that $y^2 = (1 + x_0^2 - x^2)^2 - 1$.

19. $\dfrac{\partial P}{\partial x} + \dfrac{\partial Q}{\partial y} = 1$

21. (a) *Hint*: Use the Bendixson negative criterion.

 (d) In (b), $(0, 0)$ is a stable spiral point when $\beta < 2ml \times \sqrt{g/l - \omega^2}$. In (c), $(\hat{x}, 0)$ and $(-\hat{x}, 0)$ are stable spiral points when $\beta < 2ml \sqrt{\omega^2 - g^2/(\omega^2 l^2)}$.

Exercises 12.1, Page 685

7. $\sqrt{\pi}/2$ **9.** $\sqrt{\pi/2}$

11. $\|1\| = \sqrt{p}$; $\left\| \cos \dfrac{n\pi}{p} x \right\| = \sqrt{\dfrac{p}{2}}$

21. $T = 1$ **23.** $T = 2\pi$

25. $T = 2\pi$

Exercises 12.2, Page 690

1. $f(x) = \dfrac{1}{2} + \dfrac{1}{\pi} \sum_{n=1}^{\infty} \dfrac{1 - (-1)^n}{n} \sin nx$;

 converges to $\frac{1}{2}$ at $x = 0$

3. $f(x) = \dfrac{3}{4} + \sum_{n=1}^{\infty} \left\{ \dfrac{(-1)^n - 1}{n^2 \pi^2} \cos n\pi x - \dfrac{1}{n\pi} \sin n\pi x \right\}$;

 converges to $\frac{1}{2}$ at $x = 0$

5. $f(x) = \dfrac{\pi^2}{6} + \sum_{n=1}^{\infty} \left\{ \dfrac{2(-1)^n}{n^2} \cos nx \right.$

$$\left. + \left(\dfrac{(-1)^{n+1}\pi}{n} + \dfrac{2}{\pi n^3}[(-1)^n - 1] \right) \sin nx \right\}$$

7. $f(x) = \pi + 2 \sum_{n=1}^{\infty} \frac{(-1)^{n+1}}{n} \sin nx$

9. $f(x) = \frac{1}{\pi} + \frac{1}{2} \sin x + \frac{1}{\pi} \sum_{n=2}^{\infty} \frac{(-1)^n + 1}{1 - n^2} \cos nx$

11. $f(x) = -\frac{1}{4} + \frac{1}{\pi} \sum_{n=1}^{\infty} \left\{ -\frac{1}{n} \sin \frac{n\pi}{2} \cos \frac{n\pi}{2} x \right.$

$\left. + \frac{3}{n} \left(1 - \cos \frac{n\pi}{2} \right) \sin \frac{n\pi}{2} x \right\};$

converges to -1 at $x = -1$, $-\frac{1}{2}$ at $x = 0$,

and $\frac{1}{2}$ at $x = 1$

13. $f(x) = \frac{9}{4} + 5 \sum_{n=1}^{\infty} \left\{ \frac{(-1)^n - 1}{n^2 \pi^2} \cos \frac{n\pi}{5} x \right.$

$\left. + \frac{(-1)^{n+1}}{n\pi} \sin \frac{n\pi}{5} x \right\}$

15. $f(x) = \frac{2 \sinh \pi}{\pi} \left[\frac{1}{2} + \sum_{n=1}^{\infty} \frac{(-1)^n}{1 + n^2} (\cos nx - n \sin nx) \right]$

17. $f(x) = 4 \sum_{n=1}^{\infty} \frac{(-1)^{n+1}}{n} \sin nx$

Exercises 12.3, Page 696

1. odd **3.** neither even nor odd

5. even **7.** odd

9. neither even nor odd

11. $f(x) = 2 \sum_{n=1}^{\infty} \frac{(-1)^n - 1}{n} \sin n\pi x$

13. $f(x) = \frac{\pi}{2} + \frac{2}{\pi} \sum_{n=1}^{\infty} \frac{(-1)^n - 1}{n^2} \cos nx$

15. $f(x) = \frac{1}{3} + \frac{4}{\pi^2} \sum_{n=1}^{\infty} \frac{(-1)^n}{n^2} \cos n\pi x$

17. $f(x) = \frac{2\pi^2}{3} + 4 \sum_{n=1}^{\infty} \frac{(-1)^{n+1}}{n^2} \cos nx$

19. $f(x) = \frac{2}{\pi} \sum_{n=1}^{\infty} \frac{1 - (-1)^n(1 + \pi)}{n} \sin nx$

21. $f(x) = \frac{3}{4} + \frac{4}{\pi^2} \sum_{n=1}^{\infty} \frac{\cos \frac{n\pi}{2} - 1}{n^2} \cos \frac{n\pi}{2} x$

23. $f(x) = \frac{2}{\pi} + \frac{2}{\pi} \sum_{n=2}^{\infty} \frac{1 + (-1)^n}{1 - n^2} \cos nx$

25. $f(x) = \frac{1}{2} + \frac{2}{\pi} \sum_{n=1}^{\infty} \frac{\sin \frac{n\pi}{2}}{n} \cos n\pi x$

$f(x) = \frac{2}{\pi} \sum_{n=1}^{\infty} \frac{1 - \cos \frac{n\pi}{2}}{n} \sin n\pi x$

27. $f(x) = \frac{2}{\pi} + \frac{4}{\pi} \sum_{n=1}^{\infty} \frac{(-1)^n}{1 - 4n^2} \cos 2nx$

$f(x) = \frac{8}{\pi} \sum_{n=1}^{\infty} \frac{n}{4n^2 - 1} \sin 2nx$

29. $f(x) = \frac{\pi}{4} + \frac{2}{\pi} \sum_{n=1}^{\infty} \frac{2 \cos \frac{n\pi}{2} - (-1)^n - 1}{n^2} \cos nx$

$f(x) = \frac{4}{\pi} \sum_{n=1}^{\infty} \frac{\sin \frac{n\pi}{2}}{n^2} \sin nx$

31. $f(x) = \frac{3}{4} + \frac{4}{\pi^2} \sum_{n=1}^{\infty} \frac{\cos \frac{n\pi}{2} - 1}{n^2} \cos \frac{n\pi}{2} x$

$f(x) = \sum_{n=1}^{\infty} \left\{ \frac{4}{n^2 \pi^2} \sin \frac{n\pi}{2} - \frac{2}{n\pi} (-1)^n \right\} \sin \frac{n\pi}{2} x$

33. $f(x) = \frac{5}{6} + \frac{2}{\pi^2} \sum_{n=1}^{\infty} \frac{3(-1)^n - 1}{n^2} \cos n\pi x$

$f(x) = 4 \sum_{n=1}^{\infty} \left\{ \frac{(-1)^{n+1}}{n\pi} + \frac{(-1)^n - 1}{n^3 \pi^3} \right\} \sin n\pi x$

35. $f(x) = \frac{4\pi^2}{3} + 4 \sum_{n=1}^{\infty} \left\{ \frac{1}{n^2} \cos nx - \frac{\pi}{n} \sin nx \right\}$

37. $f(x) = \frac{3}{2} - \frac{1}{\pi} \sum_{n=1}^{\infty} \frac{1}{n} \sin 2n\pi x$

43. $x_p(t) = \frac{10}{\pi} \sum_{n=1}^{\infty} \frac{1 - (-1)^n}{n(10 - n^2)} \sin nt$

45. $x_p(t) = \frac{\pi^2}{18} + 16 \sum_{n=1}^{\infty} \frac{1}{n^2(n^2 - 48)} \cos nt$

47. **(a)** $x(t) = \frac{10}{\pi} \sum_{n=1}^{\infty} \frac{1 - (-1)^n}{10 - n^2} \left\{ \frac{1}{n} \sin nt - \frac{1}{\sqrt{10}} \sin \sqrt{10} t \right\}$

49. **(b)** $y(x) = \frac{2w_0 L^4}{EI\pi^5} \sum_{n=1}^{\infty} \frac{(-1)^{n+1}}{n^5} \sin \frac{n\pi}{L} x$

Exercises 12.4, Page 700

1. $f(x) = \sum_{n=-\infty, n\neq 0}^{\infty} \frac{1 - (-1)^n}{n\pi i} e^{in\pi x/2}$

3. $f(x) = \frac{1}{4} + \sum_{n=-\infty, n\neq 0}^{\infty} \frac{1 - e^{-in\pi/2}}{2n\pi i} e^{2in\pi x}$

5. $f(x) = \pi + \sum_{n=-\infty, n\neq 0}^{\infty} \frac{i}{n} e^{inx}$

Exercises 12.5, Page 706

1. $y = \cos \alpha_n x$; α defined by $\cot \alpha = \alpha$;

$\lambda_1 = 0.7402, \lambda_2 = 11.7349,$

$\lambda_3 = 41.4388, \lambda_4 = 90.8082$

$y_1 = \cos 0.8603x, y_2 = \cos 3.4256x,$

$y_3 = \cos 6.4373x, y_4 = \cos 9.5293x$

5. $\frac{1}{2}[1 + \sin^2\alpha_n]$

7. (a) $\lambda_n = \left(\dfrac{n\pi}{\ln 5}\right)^2,\ y_n = \sin\left(\dfrac{n\pi}{\ln 5}\ln x\right),\ n = 1, 2, 3, \ldots$

(b) $\dfrac{d}{dx}[xy'] + \dfrac{\lambda}{x}y = 0$

(c) $\displaystyle\int_1^5 \dfrac{1}{x}\sin\left(\dfrac{m\pi}{\ln 5}\ln x\right)\sin\left(\dfrac{n\pi}{\ln 5}\ln x\right)dx = 0,\ m \neq n$

9. $\dfrac{d}{dx}[xe^{-x}y'] + ne^{-x}y = 0$;

$\displaystyle\int_0^\infty e^{-x}L_m(x)L_n(x)\,dx = 0,\ m \neq n$

11. (a) $\lambda_n = 16n^2,\ y_n = \sin(4n\tan^{-1}x),\ n = 1, 2, 3, \ldots$

(b) $\displaystyle\int_0^1 \dfrac{1}{1+x^2}\sin(4m\tan^{-1}x)\sin(4n\tan^{-1}x)\,dx = 0,$
$m \neq n$

Exercises 12.6, Page 712

1. $\alpha_1 = 1.277,\ \alpha_2 = 2.339,\ \alpha_3 = 3.391,\ \alpha_4 = 4.441$

3. $f(x) = \displaystyle\sum_{i=1}^\infty \dfrac{1}{\alpha_i J_1(2\alpha_i)} J_0(\alpha_i x)$

5. $f(x) = 4\displaystyle\sum_{i=1}^\infty \dfrac{\alpha_i J_1(2\alpha_i)}{(4\alpha_i^2 + 1)J_0^2(2\alpha_i)} J_0(\alpha_i x)$

7. $f(x) = 20\displaystyle\sum_{i=1}^\infty \dfrac{\alpha_i J_2(4\alpha_i)}{(2\alpha_i^2 + 1)J_1^2(4\alpha_i)} J_1(\alpha_i x)$

9. $f(x) = \dfrac{9}{2} - 4\displaystyle\sum_{i=1}^\infty \dfrac{J_2(3\alpha_i)}{\alpha_i^2 J_0^2(3\alpha_i)} J_0(\alpha_i x)$

15. $f(x) = \frac{1}{4}P_0(x) + \frac{1}{2}P_1(x) + \frac{5}{16}P_2(x) - \frac{3}{32}P_4(x) + \cdots$

21. $f(x) = \frac{1}{2}P_0(x) + \frac{5}{8}P_2(x) - \frac{3}{16}P_4(x) + \cdots,$
$f(x) = |x|$ on $(-1, 1)$

Chapter 12 in Review, Page 713

1. true
3. cosine
5. false
7. 5.5, 1, 0
9. true

13. $f(x) = \dfrac{1}{2} + \dfrac{2}{\pi}\displaystyle\sum_{n=1}^\infty \left\{\dfrac{1}{n^2\pi}[(-1)^n - 1]\cos n\pi x\right.$

$\left. + \dfrac{2}{n}(-1)^n\sin n\pi x\right\}$

15. $f(x) = 1 - e^{-1} + 2\displaystyle\sum_{n=1}^\infty \dfrac{1 - (-1)^n e^{-1}}{1 + n^2\pi^2}\cos n\pi x;$

$f(x) = \displaystyle\sum_{n=1}^\infty \dfrac{2n\pi[1 - (-1)^n e^{-1}]}{1 + n^2\pi^2}\sin n\pi x$

17. $\lambda_n = \dfrac{(2n-1)^2\pi^2}{36},\ n = 1, 2, 3, \ldots,$

$y_n = \cos\left(\dfrac{2n-1}{2}\pi \ln x\right)$

19. $p(x) = \dfrac{1}{\sqrt{1 - x^2}}$, the interval $(-1, 1)$,

$\displaystyle\int_{-1}^1 \dfrac{1}{\sqrt{1 - x^2}} T_m(x)T_n(x)\,dx = 0,\ m \neq n$

21. $f(x) = \dfrac{1}{4}\displaystyle\sum_{i=1}^\infty \dfrac{J_1(2\alpha_i)}{\alpha_i J_1^2(4\alpha_i)} J_0(\alpha_i x)$

Exercises 13.1, Page 720

1. The possible cases can be summarized in one form
$u = c_1 e^{c_2(x+y)}$, where c_1 and c_2 are constants.

3. $u = c_1 e^{y + c_2(x - y)}$ **5.** $u = c_1(xy)^{c_2}$

7. not separable

9. $u = e^{-t}(A_1 e^{k\alpha^2 t}\cosh\alpha x + B_1 e^{k\alpha^2 t}\sinh\alpha x)$
$u = e^{-t}(A_2 e^{-k\alpha^2 t}\cos\alpha x + B_2 e^{-k\alpha^2 t}\sin\alpha x)$
$u = e^{-t}(A_3 x + B_3)$

11. $u = (c_1\cosh\alpha x + c_2\sinh\alpha x)(c_3\cosh\alpha at + c_4\sinh\alpha at)$
$u = (c_5\cos\alpha x + c_6\sin\alpha x)(c_7\cos\alpha at + c_8\sin\alpha at)$
$u = (c_9 x + c_{10})(c_{11} t + c_{12})$

13. $u = (c_1\cosh\alpha x + c_2\sinh\alpha x)(c_3\cos\alpha y + c_4\sin\alpha y)$
$u = (c_5\cos\alpha x + c_6\sin\alpha x)(c_7\cosh\alpha y + c_8\sinh\alpha y)$
$u = (c_9 x + c_{10})(c_{11} y + c_{12})$

15. For $\lambda = \alpha^2 > 0$ there are three possibilities:

(i) For $0 < \alpha^2 < 1$,

$u = (c_1\cosh\alpha x + c_2\sinh\alpha x)(c_3\cosh\sqrt{1 - \alpha^2}\,y$
$+ c_4\sinh\sqrt{1 - \alpha^2}\,y).$

(ii) For $\alpha^2 > 1$,

$u = (c_1\cosh\alpha x + c_2\sinh\alpha x)(c_3\cos\sqrt{\alpha^2 - 1}\,y$
$+ c_4\sin\sqrt{\alpha^2 - 1}\,y).$

(iii) For $\alpha^2 = 1$,

$u = (c_1\cosh x + c_2\sinh x)(c_3 y + c_4).$

The results for the case $\lambda = -\alpha^2$ are similar.
For $\lambda = 0$,

$u = (c_1 x + c_2)(c_3\cosh y + c_4\sinh y).$

17. elliptic **19.** parabolic
21. hyperbolic **23.** parabolic
25. hyperbolic

Exercises 13.2, Page 725

1. $k\dfrac{\partial^2 u}{\partial x^2} = \dfrac{\partial u}{\partial t},\ 0 < x < L, t > 0$

$u(0, t) = 0,\ \left.\dfrac{\partial u}{\partial x}\right|_{x=L} = 0, t > 0$

$u(x, 0) = f(x),\ 0 < x < L$

3. $k\dfrac{\partial^2 u}{\partial x^2} = \dfrac{\partial u}{\partial t},\ 0 < x < L, t > 0$

$u(0, t) = 100,\ \left.\dfrac{\partial u}{\partial x}\right|_{x=L} = -hu(L, t), t > 0$

$u(x, 0) = f(x),\ 0 < x < L$

5. $k\dfrac{\partial^2 u}{\partial x^2} - hu = \dfrac{\partial u}{\partial t}, 0 < x < L, t > 0,$ h a constant

$u(0, t) = \sin(\pi t/L), u(L, t) = 0, t > 0$

$u(x, 0) = f(x), 0 < x < L$

7. $a^2\dfrac{\partial^2 u}{\partial x^2} = \dfrac{\partial^2 u}{\partial t^2}, 0 < x < L, t > 0$

$u(0, t) = 0, u(L, t) = 0, t > 0$

$u(x, 0) = x(L - x), \dfrac{\partial u}{\partial t}\bigg|_{t=0} = 0, 0 < x < L$

9. $a^2\dfrac{\partial^2 u}{\partial x^2} - 2\beta\dfrac{\partial u}{\partial t} = \dfrac{\partial^2 u}{\partial t^2}, 0 < x < L, t > 0$

$u(0, t) = 0, u(L, t) = \sin \pi t, t > 0$

$u(x, 0) = f(x), \dfrac{\partial u}{\partial t}\bigg|_{t=0} = 0, 0 < x < L$

11. $\dfrac{\partial^2 u}{\partial x^2} + \dfrac{\partial^2 u}{\partial y^2} = 0, 0 < x < 4, 0 < y < 2$

$\dfrac{\partial u}{\partial x}\bigg|_{x=0} = 0, u(4, y) = f(y), 0 < y < 2$

$\dfrac{\partial u}{\partial y}\bigg|_{y=0} = 0, u(x, 2) = 0, 0 < x < 4$

Exercises 13.3, Page 727

1. $u(x, t) = \dfrac{200}{\pi}\displaystyle\sum_{n=1}^{\infty}\dfrac{1 - (-1)^n}{n}e^{-k(n^2\pi^2/L^2)t}\sin\dfrac{n\pi}{L}x$

3. $u(x, t) = \dfrac{2}{\pi}\displaystyle\sum_{n=1}^{\infty}\dfrac{\left(-\cos\dfrac{n\pi}{2} + 1\right)}{n}e^{-k(n^2\pi^2/L^2)t}\sin\dfrac{n\pi}{L}x$

5. $u(x, t) = \dfrac{1}{L}\displaystyle\int_0^L f(x)\, dx$

$+ \dfrac{2}{L}\displaystyle\sum_{n=1}^{\infty}\left(\displaystyle\int_0^L f(x)\cos\dfrac{n\pi}{L}x\, dx\right)e^{-k(n^2\pi^2/L^2)t}\cos\dfrac{n\pi}{L}x$

7. $u(x, t) = 10e^{-25kt}\sin 5x$

9. $u(x, t) = e^{-ht}\left[\dfrac{1}{L}\displaystyle\int_0^L f(x)\, dx\right.$

$\left. + \dfrac{2}{L}\displaystyle\sum_{n=1}^{\infty}\left(\displaystyle\int_0^L f(x)\cos\dfrac{n\pi}{L}x\, dx\right)e^{-k(n^2\pi^2/L^2)t}\cos\dfrac{n\pi}{L}x\right]$

11. $u(x, t) = A_0 + \displaystyle\sum_{k=1}^{\infty}e^{-k(n\pi/L)^2 t}\left(A_n\cos\dfrac{n\pi}{L}x + B_n\sin\dfrac{n\pi}{L}x\right),$

where $A_0 = \dfrac{1}{2L}\displaystyle\int_{-L}^L f(x)\, dx,$

$A_n = \dfrac{1}{L}\displaystyle\int_{-L}^L f(x)\cos\dfrac{n\pi}{L}\, dx, B_n = \dfrac{1}{L}\displaystyle\int_{-L}^L f(x)\sin\dfrac{n\pi}{L}\, dx$

Exercises 13.4, Page 732

1. $u(x, t) = \dfrac{L^2}{\pi^3}\displaystyle\sum_{n=1}^{\infty}\dfrac{1 - (-1)^n}{n^3}\cos\dfrac{n\pi a}{L}t\sin\dfrac{n\pi}{L}x$

3. $u(x, t) = \dfrac{1}{a}\sin at\sin x$

5. $u(x, t) = \dfrac{4}{\pi^3}\displaystyle\sum_{n=1}^{\infty}\left[\dfrac{1 - (-1)^n}{n^3}\cos n\pi at\right.$

$\left. + \dfrac{1 - (-1)^n}{n^4\pi a}\sin n\pi at\right]\sin n\pi x$

7. $u(x, t) = \dfrac{8h}{\pi^2}\displaystyle\sum_{n=1}^{\infty}\dfrac{\sin\dfrac{n\pi}{2}}{n^2}\cos\dfrac{n\pi a}{L}t\sin\dfrac{n\pi}{L}x$

9. $u(x, t) = \dfrac{6h}{\pi^2}\displaystyle\sum_{n=1}^{\infty}\dfrac{1 - (-1)^n}{n^2}\sin\dfrac{n\pi}{3}\cos\dfrac{n\pi a}{L}t\sin\dfrac{n\pi}{L}x$

11. $u(x, t) = \dfrac{L}{2} + \dfrac{2L}{\pi^2}\displaystyle\sum_{n=1}^{\infty}\dfrac{(-1)^n - 1}{n^2}\cos\dfrac{n\pi a}{L}t\cos\dfrac{n\pi}{L}x$

13. $u(L/2, t) = 0$ for $t \geq 0$

15. $u(x, t) = e^{-\beta t}\displaystyle\sum_{n=1}^{\infty}A_n\left\{\cos q_n t + \dfrac{\beta}{q_n}\sin q_n t\right\}\sin nx,$

where $A_n = \dfrac{2}{\pi}\displaystyle\int_0^{\pi}f(x)\sin nx\, dx$ and $q_n = \sqrt{n^2 - \beta^2}$

19. $u(x, t) = t + \sin x\cos at$

21. $u(x, t) = \dfrac{1}{2a}\sin 2x\sin 2at$

23. $u(x, t) = \displaystyle\sum_{n=1}^{\infty}\left(A_n\cos\dfrac{n^2\pi^2}{L^2}at\right.$

$\left. + B_n\sin\dfrac{n^2\pi^2}{L^2}at\right)\sin\dfrac{n\pi}{L}x,$

where $A_n = \dfrac{2}{L}\displaystyle\int_0^L f(x)\sin\dfrac{n\pi}{L}x\, dx$

$B_n = \dfrac{2L}{n^2\pi^2 a}\displaystyle\int_0^L g(x)\sin\dfrac{n\pi}{L}x\, dx$

Exercises 13.5, Page 738

1. $u(x, y) = \dfrac{2}{a}\displaystyle\sum_{n=1}^{\infty}\left(\dfrac{1}{\sinh\dfrac{n\pi}{a}b}\displaystyle\int_0^a f(x)\sin\dfrac{n\pi}{a}x\, dx\right)$

$\times\sinh\dfrac{n\pi}{a}y\sin\dfrac{n\pi}{a}x$

3. $u(x, y) = \dfrac{2}{a}\displaystyle\sum_{n=1}^{\infty}\left(\dfrac{1}{\sinh\dfrac{n\pi}{a}b}\displaystyle\int_0^a f(x)\sin\dfrac{n\pi}{a}x\, dx\right)$

$\times\sinh\dfrac{n\pi}{a}(b - y)\sin\dfrac{n\pi}{a}x$

5. $u(x, y) = \dfrac{1}{2}x + \dfrac{2}{\pi^2}\displaystyle\sum_{n=1}^{\infty}\dfrac{1 - (-1)^n}{n^2\sinh n\pi}\sinh n\pi x\cos n\pi y$

7. $u(x, y) = \dfrac{2}{\pi} \displaystyle\sum_{n=1}^{\infty} \dfrac{[1 - (-1)^n]}{n}$

$\qquad \times \dfrac{n \cosh nx + \sinh nx}{n \cosh n\pi + \sinh n\pi} \sin ny$

9. $u(x, y) = \displaystyle\sum_{n=1}^{\infty} (A_n \cosh n\pi y + B_n \sinh n\pi y) \sin n\pi x$,

$\qquad$ where $A_n = 200 \dfrac{1 - (-1)^n}{n\pi}$

$\qquad B_n = 200 \dfrac{[1 - (-1)^n]}{n\pi} \dfrac{[2 - \cosh n\pi]}{\sinh n\pi}$

11. $u(x, y) = \dfrac{2}{\pi} \displaystyle\sum_{n=1}^{\infty} \left(\int_0^\pi f(x) \sin nx \, dx \right) e^{-ny} \sin nx$

13. $u(x, y) = \displaystyle\sum_{n=1}^{\infty} \left(A_n \cosh \dfrac{n\pi}{a} y + B_n \sinh \dfrac{n\pi}{a} y \right) \sin \dfrac{n\pi}{a} x$,

$\qquad$ where $A_n = \dfrac{2}{a} \displaystyle\int_0^a f(x) \sin \dfrac{n\pi}{a} x \, dx$

$\qquad B_n = \dfrac{1}{\sinh \dfrac{n\pi}{a} b} \left(\dfrac{2}{a} \displaystyle\int_0^a g(x) \sin \dfrac{n\pi}{a} x \, dx \right.$

$\qquad \left. - A_n \cosh \dfrac{n\pi}{a} b \right)$

15. $u = u_1 + u_2$ where

$\qquad u_1(x, y) = \dfrac{2}{\pi} \displaystyle\sum_{n=1}^{\infty} \dfrac{1 - (-1)^n}{n \sinh n\pi} \sinh ny \sin nx$

$\qquad u_2(x, y) = \dfrac{2}{\pi} \displaystyle\sum_{n=1}^{\infty} \dfrac{1 - (-1)^n}{n}$

$\qquad \times \dfrac{\sinh nx + \sinh n(\pi - x)}{\sinh n\pi} \sin ny$

17. max temperature is $u = 1$

Exercises 13.6, Page 745

1. $u(x, t) = 100 + \dfrac{200}{\pi} \displaystyle\sum_{n=1}^{\infty} \dfrac{(-1)^n - 1}{n} e^{-kn^2\pi^2 t} \sin n\pi x$

3. $u(x, t) = u_0 - \dfrac{r}{2k} x(x - 1) + 2 \displaystyle\sum_{n=1}^{\infty} \left[\dfrac{u_0}{n\pi} + \dfrac{r}{kn^3\pi^3} \right]$

$\qquad \times [(-1)^n - 1] e^{-kn^2\pi^2 t} \sin n\pi x$

5. $u(x, t) = \psi(x) + \displaystyle\sum_{n=1}^{\infty} A_n e^{-kn^2\pi^2 t} \sin n\pi x$,

$\qquad$ where $\psi(x) = \dfrac{A}{k\beta^2} [-e^{-\beta x} + (e^{-\beta} - 1)x + 1]$

$\qquad A_n = 2 \displaystyle\int_0^1 [f(x) - \psi(x)] \sin n\pi x \, dx$

7. $\psi(x) = u_0 \left[1 - \dfrac{\sinh \sqrt{h/k} \, x}{\sinh \sqrt{h/k}} \right]$

9. $u(x, t) = \dfrac{A}{6a^2} (x - x^3)$

$\qquad + \dfrac{2A}{a^2\pi^3} \displaystyle\sum_{n=1}^{\infty} \dfrac{(-1)^n}{n^3} \cos n\pi a t \sin n\pi x$

11. $u(x, y) = (u_0 - u_1) y + u_1$

$\qquad + \dfrac{2}{\pi} \displaystyle\sum_{n=1}^{\infty} \dfrac{u_0(-1)^n - u_1}{n} e^{-n\pi x} \sin n\pi y$

13. $u(x, t) = (1 - x) \sin t$

$\qquad + \dfrac{2}{\pi} \displaystyle\sum_{n=1}^{\infty} \left[\dfrac{n^2\pi^2 e^{-n^2\pi^2 t} - n^2\pi^2 \cos t - \sin t}{n(n^4\pi^4 + 1)} \right] \sin n\pi x$

15. $u(x, t) = x \sin t + 2 \displaystyle\sum_{n=1}^{\infty} \left[\left(\dfrac{(-1)^n}{n^2\pi^2} + \dfrac{(-1)^n}{n^2\pi^2(n^2\pi^2 - 1)} \right) \right.$

$\qquad \left. \times \sin n\pi t - \dfrac{(-1)^n}{n\pi(n^2\pi^2 - 1)} \sin t \right] \sin n\pi x$

17. $u(x, t) = 2 \displaystyle\sum_{n=1}^{\infty} \dfrac{(-1)^{n+1}}{n(n^2 - 3)} e^{-3t} \sin nx$

$\qquad + 2 \displaystyle\sum_{n=1}^{\infty} \dfrac{(-1)^n}{n(n^2 - 3)} e^{-n^2 t} \sin nx$

19. $u(x, t) = \displaystyle\sum_{n=1}^{\infty} \dfrac{2}{n\pi} \left[-\dfrac{1}{n^2\pi^2} + (-1)^n \dfrac{n^2\pi^2 \cos t + \sin t}{n^4\pi^4 + 1} \right]$

$\qquad \times \sin n\pi x + \displaystyle\sum_{n=1}^{\infty} \left[\dfrac{4 - 2(-1)^n}{n^3\pi^3} - (-1)^n \dfrac{2n\pi}{n^4\pi^4 + 1} \right]$

$\qquad \times e^{-n^2\pi^2 t} \sin n\pi x$

Exercises 13.7, Page 750

1. $u(x, t) = 2h \displaystyle\sum_{n=1}^{\infty} \dfrac{\sin \alpha_n}{(h + \sin^2\alpha_n)} e^{-k\alpha_n^2 t} \cos \alpha_n x$, where

$\qquad$ the α_n are the consecutive positive roots of $\cot \alpha = \alpha/h$

3. $u(x, y) = \displaystyle\sum_{n=1}^{\infty} A_n \sinh \alpha_n y \sin \alpha_n x$, where

$\qquad A_n = \dfrac{2h}{\sinh \alpha_n b (ah + \cos^2\alpha_n a)} \displaystyle\int_0^a f(x) \sin \alpha_n x \, dx$

$\qquad$ and the α_n are the consecutive positive roots of

$\qquad \tan \alpha a = -\alpha/h$

5. $u(x, t) = \displaystyle\sum_{n=1}^{\infty} A_n e^{-k(2n-1)^2\pi^2 t/4L^2} \sin \left(\dfrac{2n - 1}{2L} \right) \pi x$,

$\qquad$ where $A_n = \dfrac{2}{L} \displaystyle\int_0^L f(x) \sin \left(\dfrac{2n - 1}{2L} \right) \pi x \, dx$

7. $u(x, y) = \dfrac{4u_0}{\pi} \displaystyle\sum_{n=1}^{\infty} \dfrac{1}{(2n - 1) \cosh \left(\dfrac{2n - 1}{2} \right) \pi}$

$\qquad \times \cosh \left(\dfrac{2n - 1}{2} \right) \pi x \sin \left(\dfrac{2n - 1}{2} \right) \pi y$

1. $u(x, y, t) = \displaystyle\sum_{m=1}^{\infty}\sum_{n=1}^{\infty} A_{mn} e^{-k(m^2+n^2)t} \sin mx \sin ny,$

where $A_{mn} = \dfrac{4u_0}{mn\pi^2}[1-(-1)^m][1-(-1)^n]$

3. $u(x, y, t) = \displaystyle\sum_{m=1}^{\infty}\sum_{n=1}^{\infty} A_{mn} \sin mx \sin ny \cos a\sqrt{m^2+n^2}\,t,$

where $A_{mn} = \dfrac{16}{m^3 n^3 \pi^2}[(-1)^m-1][(-1)^n-1]$

5. $u(x, y, z) = \displaystyle\sum_{m=1}^{\infty}\sum_{n=1}^{\infty} A_{mn} \sinh \omega_{mn} z \sin \dfrac{m\pi}{a} x \sin \dfrac{n\pi}{b} y,$

where $\omega_{mn} = \sqrt{(m\pi/a)^2 + (n\pi/b)^2}$ and

$A_{mn} = \dfrac{4}{ab \sinh(c\omega_{mn})}\displaystyle\int_0^b\int_0^a f(x, y) \sin \dfrac{m\pi}{a} x \sin \dfrac{n\pi}{b} y\, dx\, dy$

7. Use $a = b = c = 1$ with $f(x, y) = u_0$ in Problem 5 and $f(x, y) = -u_0$ in Problem 6. Add the two solutions.

Chapter 13 in Review, Page 754

1. $u = c_1 e^{(c_2 x + y/c_2)}$

3. $\psi(x) = u_0 + \dfrac{(u_1 - u_0)}{1+\pi} x$

5. $u(x, t) = \dfrac{2h}{\pi^2 a} \displaystyle\sum_{n=1}^{\infty} \left(\dfrac{\cos \dfrac{n\pi}{4} - \cos \dfrac{3n\pi}{4}}{n^2} \right)$

$\times \sin n\pi a t \sin n\pi x$

7. $u(x, y) = \dfrac{100}{\pi} \displaystyle\sum_{n=1}^{\infty} \dfrac{1-(-1)^n}{n \sinh n\pi} \sinh nx \sin ny$

9. $u(x, y) = \dfrac{100}{\pi} \displaystyle\sum_{n=1}^{\infty} \dfrac{1-(-1)^n}{n} e^{-nx} \sin ny$

11. (a) $u(x, t) = e^{-t} \sin x$

13. $u(x, t) = e^{-(x+t)} \displaystyle\sum_{n=1}^{\infty} A_n[\sqrt{n^2+1} \cos \sqrt{n^2+1}\,t$

$+ \sin \sqrt{n^2+1}\,t] \sin nx$

15. $u(x, t) = u_0 + \tfrac{1}{2}(u_1 - u_0)x$

$+ 2(u_1 - u_0)\displaystyle\sum_{n=1}^{\infty} \dfrac{\cos \alpha_n}{\alpha_n(1+\cos^2 \alpha_n)} e^{-\alpha_n^2 t} \sin \alpha_n x$

17. $u(x, y) = -x^2 + \pi x + \dfrac{4}{\pi} \displaystyle\sum_{n=1}^{\infty} \dfrac{[(-1)^n - 1]}{n^3}$

$\times \left(\dfrac{\sinh n(\pi - y) + \sinh ny}{\sinh n\pi} \right) \sin nx$

19. $w(x, y) = \dfrac{q_0 a^4 b^4}{\pi^4 D(a^2 + b^2)^2} \sin \dfrac{\pi x}{a} \sin \dfrac{\pi y}{b}$

1. $u(r, \theta) = \dfrac{u_0}{2} + \dfrac{u_0}{\pi} \displaystyle\sum_{n=1}^{\infty} \dfrac{1-(-1)^n}{n} r^n \sin n\theta$

3. $u(r, \theta) = \dfrac{2\pi^2}{3} - 4\displaystyle\sum_{n=1}^{\infty} \dfrac{1}{n^2} r^n \cos n\theta$

5. $u(r, \theta) = \dfrac{u_0}{2} + \dfrac{2u_0}{\pi} \displaystyle\sum_{n=1}^{\infty} \dfrac{1}{n} \sin\left(\dfrac{n\pi}{2}\right)\left(\dfrac{r}{2}\right)^n \cos n\theta$

7. $u(r, \theta) = \displaystyle\sum_{n=1}^{\infty} A_n r^{2n} \sin 2n\theta,$

where $A_n = \dfrac{2}{\pi c^n} \displaystyle\int_0^{\pi/2} f(\theta) \sin 2n\theta\, d\theta$

9. $u(r, \theta) = \displaystyle\sum_{n=1}^{\infty} A_n r^{n\pi/\beta} \sin \dfrac{n\pi}{\beta} \theta,$

where $A_n = \dfrac{2}{\beta c^{n\pi/\beta}} \displaystyle\int_0^\beta f(\theta) \sin \dfrac{n\pi}{\beta} \theta\, d\theta$

11. $u(r, \theta) = A_0 \ln\left(\dfrac{r}{b}\right)$

$+ \displaystyle\sum_{n=1}^{\infty} \left[\left(\dfrac{b}{r}\right)^n - \left(\dfrac{r}{b}\right)^n \right] (A_n \cos n\theta + B_n \sin n\theta),$

where $A_0 \ln\left(\dfrac{a}{b}\right) = \dfrac{1}{2\pi} \displaystyle\int_0^{2\pi} f(\theta)\, d\theta$

$\left[\left(\dfrac{b}{a}\right)^n - \left(\dfrac{a}{b}\right)^n \right] A_n = \dfrac{1}{\pi} \displaystyle\int_0^{2\pi} f(\theta) \cos n\theta\, d\theta$

$\left[\left(\dfrac{b}{a}\right)^n - \left(\dfrac{a}{b}\right)^n \right] B_n = \dfrac{1}{\pi} \displaystyle\int_0^{2\pi} f(\theta) \sin n\theta\, d\theta$

13. $u(r, \theta) = A_0 + \displaystyle\sum_{n=1}^{\infty} \dfrac{(r^{2n} + a^{2n})}{r^n}(A_n \cos n\theta + B_n \sin n\theta),$

where $A_0 = \dfrac{1}{2\pi} \displaystyle\int_0^{2\pi} f(\theta)\, d\theta$

$\dfrac{(b^{2n} + a^{2n})}{b^n} A_n = \dfrac{1}{\pi} \displaystyle\int_0^{2\pi} f(\theta) \cos n\theta\, d\theta$

$\dfrac{(b^{2n} + a^{2n})}{b^n} B_n = \dfrac{1}{\pi} \displaystyle\int_0^{2\pi} f(\theta) \sin n\theta\, d\theta$

15. $u(r, \theta) = \dfrac{4}{\pi} \displaystyle\sum_{n=1}^{\infty} \dfrac{1-(-1)^n}{n^3} \dfrac{r^{2n} - b^{2n}}{a^{2n} - b^{2n}}\left(\dfrac{a}{r}\right)^n \sin n\theta$

17. $u(r, \theta) = A_0 \ln r + \displaystyle\sum_{n=1}^{\infty} A_n(r^{2n} - r^{-2n}) \cos 2n\theta,$

where $A_0 = \dfrac{2}{\pi \ln 2} \displaystyle\int_0^{\pi/2} f(\theta)\, d\theta,$

$A_n = \dfrac{4}{\pi(2^{2n} - 2^{-2n})} \displaystyle\int_0^{\pi/2} f(\theta) \cos 2n\theta\, d\theta$

19. $u(r, \theta) = A_0 + \sum_{n=1}^{\infty} r^{-n}(A_n \cos n\theta + B_n \sin n\theta)$,

where $A_0 = \dfrac{1}{2\pi} \displaystyle\int_0^{2\pi} f(\theta)\, d\theta$

$A_n = \dfrac{c^n}{\pi} \displaystyle\int_0^{2\pi} f(\theta) \cos n\theta\, d\theta$

$B_n = \dfrac{c^n}{\pi} \displaystyle\int_0^{2\pi} f(\theta) \sin n\theta\, d\theta$

Exercises 14.2, Page 767

1. $u(r, t) = \dfrac{2}{ac} \displaystyle\sum_{n=1}^{\infty} \dfrac{\sin \alpha_n at}{\alpha_n^2 J_1(\alpha_n c)} J_0(\alpha_n r)$

3. $u(r, z) = u_0 \displaystyle\sum_{n=1}^{\infty} \dfrac{\sinh \alpha_n(4 - z)}{\alpha_n \sinh 4\alpha_n J_1(2\alpha_n)} J_0(\alpha_n r)$

5. $u(r, z) = \frac{1}{2} + 2 \displaystyle\sum_{n=1}^{\infty} \dfrac{(-1)^n - 1}{n^2 \pi^2 I_0(n\pi)} I_0(n\pi r) \cos n\pi z$

7. $u(r, z) = \dfrac{4u_0}{\pi} \displaystyle\sum_{n=1}^{\infty} \dfrac{I_0\left(\dfrac{2n - 1}{2}\pi r\right)}{(2n - 1)I_0\left(\dfrac{2n - 1}{2}\pi\right)} \sin \dfrac{2n - 1}{2}\pi z$

9. $u(r, t) = \displaystyle\sum_{n=1}^{\infty} A_n J_0(\alpha_n r) e^{-k\alpha_n^2 t}$,

where $A_n = \dfrac{2}{c^2 J_1^2(\alpha_n c)} \displaystyle\int_0^c r J_0(\alpha_n r) f(r)\, dr$

11. $u(r, t) = \displaystyle\sum_{n=1}^{\infty} A_n J_0(\alpha_n r) e^{-k\alpha_n^2 t}$,

where $A_n = \dfrac{2\alpha_n^2}{(\alpha_n^2 + h^2)J_0^2(\alpha_n)} \displaystyle\int_0^1 r J_0(\alpha_n r) f(r)\, dr$

13. $u(r, t) = 100 + 50 \displaystyle\sum_{n=1}^{\infty} \dfrac{J_1(\alpha_n)}{\alpha_n J_1^2(2\alpha_n)} J_0(\alpha_n r) e^{-\alpha_n^2 t}$

15. (b) $u(x, t) = \displaystyle\sum_{n=1}^{\infty} A_n \cos(\alpha_n \sqrt{gt}) J_0(2\alpha_n \sqrt{x})$,

where $A_n = \dfrac{2}{L J_1^2(2\alpha_n \sqrt{L})} \displaystyle\int_0^{\sqrt{L}} v J_0(2\alpha_n v) f(v^2)\, dv$

17. $u(r, t) = 2e^{-ht} \displaystyle\sum_{n=1}^{\infty} \dfrac{J_0(\alpha_n r)}{\alpha_n J_1(\alpha_n)} e^{-\alpha_n^2 t}$

Exercises 14.3, Page 771

1. $u(r, \theta) = 50\left[\dfrac{1}{2} P_0(\cos \theta) + \dfrac{3}{4}\left(\dfrac{r}{c}\right)P_1(\cos \theta)\right.$

$\left. - \dfrac{7}{16}\left(\dfrac{r}{c}\right)^3 P_3(\cos \theta) + \dfrac{11}{32}\left(\dfrac{r}{c}\right)^5 P_5(\cos \theta) + \cdots\right]$

3. $u(r, \theta) = \dfrac{r}{c} \cos \theta$

5. $u(r, \theta) = \displaystyle\sum_{n=1}^{\infty} A_n \dfrac{b^{2n+1} - r^{2n+1}}{b^{2n+1} r^{n+1}} P_n(\cos \theta)$,

where $\dfrac{b^{2n+1} - a^{2n+1}}{b^{2n+1} a^{n+1}} A_n$

$= \dfrac{2n + 1}{2} \displaystyle\int_0^{\pi} f(\theta) P_n(\cos \theta) \sin \theta\, d\theta$

7. $u(r, \theta) = \displaystyle\sum_{n=0}^{\infty} A_{2n} r^{2n} P_{2n}(\cos \theta)$,

where $A_{2n} = \dfrac{4n + 1}{c^{2n}} \displaystyle\int_0^{\pi/2} f(\theta) P_{2n}(\cos \theta) \sin \theta\, d\theta$

9. $u(r, t) = 100 + \dfrac{200}{\pi r} \displaystyle\sum_{n=1}^{\infty} \dfrac{(-1)^n}{n} e^{-n^2 \pi^2 t} \sin n\pi r$

11. $u(r, t) = \dfrac{1}{r} \displaystyle\sum_{n=1}^{\infty} \left(A_n \cos \dfrac{n\pi a}{c} t + B_n \sin \dfrac{n\pi a}{c} t\right) \sin \dfrac{n\pi}{c} r$,

where $A_n = \dfrac{2}{c} \displaystyle\int_0^c r f(r) \sin \dfrac{n\pi}{c} r\, dr$,

$B_n = \dfrac{2}{n\pi a} \displaystyle\int_0^c r g(r) \sin \dfrac{n\pi}{c} r\, dr$

Chapter 14 in Review, Page 772

1. $u(r, \theta) = \dfrac{2u_0}{\pi} \displaystyle\sum_{n=1}^{\infty} \dfrac{1 - (-1)^n}{n} \left(\dfrac{r}{c}\right)^n \sin n\theta$

3. $u(r, \theta) = \dfrac{4u_0}{\pi} \displaystyle\sum_{n=1}^{\infty} \dfrac{1 - (-1)^n}{n^3} r^n \sin n\theta$

5. $u(r, \theta) = A_0 + \displaystyle\sum_{n=1}^{\infty} A_n r^n \cos n\theta$,

where $A_0 = \dfrac{1}{\pi} \displaystyle\int_0^{\pi} f(\theta)\, d\theta$

$A_n = \dfrac{2}{\pi c^n} \displaystyle\int_0^{\pi} f(\theta) \cos n\theta\, d\theta$

7. $u(r, \theta) = \dfrac{2u_0}{\pi} \displaystyle\sum_{n=1}^{\infty} \dfrac{r^{4n} + r^{-4n}}{2^{4n} + 2^{-4n}} \dfrac{1 - (-1)^n}{n} \sin 4n\theta$

11. $u(r, z) = 16 \displaystyle\sum_{n=1}^{\infty} \dfrac{\sinh \alpha_n z}{\alpha_n^3 J_2(\alpha_n)\sinh \alpha_n} J_1(\alpha_n r)$

13. $u(r, z) = 50 \displaystyle\sum_{n=1}^{\infty} \dfrac{\cosh \alpha_n z}{\alpha_n \cosh 4\alpha_n J_1(2\alpha_n)} J_0(\alpha_n r)$

15. $u(r, \theta) = 100\left[\dfrac{3}{2} r P_1(\cos \theta) - \dfrac{7}{8} r^3 P_3(\cos \theta)\right.$

$\left. + \dfrac{11}{16} r^5 P_5(\cos \theta) + \cdots\right]$

19. $u(r, z) = 100 + 200 \displaystyle\sum_{n=1}^{\infty} \dfrac{\cosh \alpha_n z}{\alpha_n \cosh \alpha_n J_1(\alpha_n)} J_0(\alpha_n r)$

21. $u(r, z) = 200 \displaystyle\sum_{n=1}^{\infty} \dfrac{e^{-\alpha_n z}}{\alpha_n J_1(\alpha_n)} J_0(\alpha_n r)$

Exercises 15.1, Page 778

1. (a) Let $\tau = u^2$ in the integral $\text{erf}(\sqrt{t})$.

9. $y(t) = e^{\pi t}\,\text{erf}(\sqrt{\pi t})$

11. Use the property $\displaystyle\int_0^b - \int_0^a = \int_0^b + \int_a^0$

Exercises 15.2, Page 783

1. $u(x, t) = A\cos\dfrac{\alpha\pi t}{L}\sin\dfrac{\pi x}{L}$

3. $u(x, t) = f\left(t - \dfrac{x}{a}\right)\mathcal{U}\left(t - \dfrac{x}{a}\right)$

5. $u(x, t) = \left[\dfrac{1}{2}g\left(t - \dfrac{x}{a}\right)^2 + A\sin\omega\left(t - \dfrac{x}{a}\right)\right]$
$$\times\,\mathcal{U}\left(t - \dfrac{x}{a}\right) - \dfrac{1}{2}gt^2$$

7. (a) $u(x, t) =$
$$\sum_{n=0}^{\infty}(-1)^n(t - (2n + 1 - x))\mathcal{U}(t - (2n + 1 - x))$$
$$-\sum_{n=0}^{\infty}(-1)^n(t - (2n + 1 + x))\mathcal{U}(t - (2n + 1 + x))$$

9. $u(x, t) = (t - x)\sinh(t - x)\,\mathcal{U}(t - x)$
$$+\, xe^{-x}\cosh t - e^{-x}t\sinh t$$

11. $u(x, t) = u_1 + (u_0 - u_1)\,\text{erfc}\left(\dfrac{x}{2\sqrt{t}}\right)$

13. $u(x, t) = u_0\left[1 - \left\{\text{erfc}\left(\dfrac{x}{2\sqrt{t}}\right)\right.\right.$
$$\left.\left. -\, e^{x+t}\text{erfc}\left(\sqrt{t} + \dfrac{x}{2\sqrt{t}}\right)\right\}\right]$$

15. $u(x, t) = \dfrac{x}{2\sqrt{\pi}}\displaystyle\int_0^t \dfrac{f(t - \tau)}{\tau^{3/2}}e^{-x^2/4\tau}\,d\tau$

17. $u(x, t) = 60 + 40\,\text{erfc}\left(\dfrac{x}{2\sqrt{t - 2}}\right)\mathcal{U}(t - 2)$

19. $u(x, t) = 100\left[-e^{1-x+t}\text{erfc}\left(\sqrt{t} + \dfrac{1 - x}{2\sqrt{t}}\right)\right.$
$$\left. +\,\text{erfc}\left(\dfrac{1 - x}{2\sqrt{t}}\right)\right]$$

21. $u(x, t) = e^{-x}\sin 3(t - x)\,\mathcal{U}(t - x)$

23. $u(x, t) = u_0 + u_0 e^{-(\pi^2/L^2)t}\sin\left(\dfrac{\pi}{L}x\right)$

25. $u(x, t) = u_0 - u_0\displaystyle\sum_{n=0}^{\infty}(-1)^n\left[\text{erfc}\left(\dfrac{2n + 1 - x}{2\sqrt{kt}}\right)\right.$
$$\left. +\,\text{erfc}\left(\dfrac{2n + 1 + x}{2\sqrt{kt}}\right)\right]$$

27. $u(x, t) = u_0 e^{-Gt/C}\text{erf}\left(\dfrac{x}{2}\sqrt{\dfrac{RC}{t}}\right)$

29. $u(x, t) = \dfrac{v_0^2 F_0}{a^2 - v_0^2}\left[\left(t - \dfrac{x}{v_0}\right)\mathcal{U}\left(t - \dfrac{x}{v_0}\right)\right.$
$$\left. -\left(t - \dfrac{x}{a}\right)\mathcal{U}\left(t - \dfrac{x}{a}\right)\right]$$

31. $u(x, t) = u_0\,\text{erfc}\left(\dfrac{x}{2\sqrt{kt}}\right)$; u_0

33. $u(r, t) = \dfrac{100}{r}\text{erfc}\left(\dfrac{r - 1}{2\sqrt{t}}\right)$

35. $u(r, t) = u_1 + \dfrac{u_0 - u_1}{r}\displaystyle\sum_{n=0}^{\infty}\left[\text{erfc}\left(\dfrac{2n + 1 - r}{2\sqrt{t}}\right)\right.$
$$\left. -\,\text{erfc}\left(\dfrac{2n + 1 + r}{2\sqrt{t}}\right)\right]$$

37. $u(r, t) = \dfrac{u_0 a^2}{r(a^2\omega^2 + 1)}[-a\omega\sin\omega(t - (r - a))$
$$-\cos\omega(t - (r - a)) + e^{-(t-(r-a))/a}]\,\mathcal{U}(t - (r - a)),$$
$$\dfrac{\partial u}{\partial r} = \dfrac{u_0 a^2}{r^2(a^2\omega^2 + 1)}\left[(ar\omega^2 + 1)\cos\omega(t - (r - a))\right.$$
$$-\,\omega(r - a)\sin\omega(t - (r - a))$$
$$\left. +\,\dfrac{1}{a}(r - a)e^{-(t-(r-a))/a}\right]\mathcal{U}(t - (r - a))$$

Exercises 15.3, Page 791

1. $f(x) = \dfrac{1}{\pi}\displaystyle\int_0^{\infty}\dfrac{\sin\alpha\cos\alpha x + 3(1 - \cos\alpha)\sin\alpha x}{\alpha}\,d\alpha$

3. $f(x) = \dfrac{1}{\pi}\displaystyle\int_0^{\infty}[A(\alpha)\cos\alpha x + B(\alpha)\sin\alpha x]\,d\alpha$,

where $A(\alpha) = \dfrac{3\alpha\sin 3\alpha + \cos 3\alpha - 1}{\alpha^2}$

$B(\alpha) = \dfrac{\sin 3\alpha - 3\alpha\cos 3\alpha}{\alpha^2}$

5. $f(x) = \dfrac{1}{\pi}\displaystyle\int_0^{\infty}\dfrac{\cos\alpha x + \alpha\sin\alpha x}{1 + \alpha^2}\,d\alpha$

7. $f(x) = \dfrac{10}{\pi}\displaystyle\int_0^{\infty}\dfrac{(1 - \cos\alpha)\sin\alpha x}{\alpha}\,d\alpha$

9. $f(x) = \dfrac{2}{\pi}\displaystyle\int_0^{\infty}\dfrac{(\pi\alpha\sin\pi\alpha + \cos\pi\alpha - 1)\cos\alpha x}{\alpha^2}\,d\alpha$

11. $f(x) = \dfrac{4}{\pi}\displaystyle\int_0^{\infty}\dfrac{\alpha\sin\alpha x}{4 + \alpha^4}\,d\alpha$

13. $f(x) = \dfrac{2k}{\pi}\displaystyle\int_0^{\infty}\dfrac{\cos\alpha x}{k^2 + \alpha^2}\,d\alpha$

$f(x) = \dfrac{2}{\pi}\displaystyle\int_0^{\infty}\dfrac{\alpha\sin\alpha x}{k^2 + \alpha^2}\,d\alpha$

15. $f(x) = \dfrac{2}{\pi} \displaystyle\int_0^\infty \dfrac{(4 - \alpha^2)\cos \alpha x}{(4 + \alpha^2)^2}\, d\alpha$

$f(x) = \dfrac{8}{\pi} \displaystyle\int_0^\infty \dfrac{\alpha \sin \alpha x}{(4 + \alpha^2)^2}\, d\alpha$

17. $f(x) = \dfrac{2}{\pi} \dfrac{1}{1 + x^2}, \quad x > 0$

19. Let $x = 2$ in (7). Use a trigonometric identity and replace α by x. In part (b) make the change of variable $2x = kt$.

Exercises 15.4, Page 797

1. $u(x, t) = \dfrac{1}{\pi} \displaystyle\int_{-\infty}^\infty \dfrac{e^{-k\alpha^2 t}}{1 + \alpha^2}\, e^{-i\alpha x}\, d\alpha$

$= \dfrac{1}{\pi} \displaystyle\int_{-\infty}^\infty \dfrac{\cos \alpha x}{1 + \alpha^2}\, e^{-k\alpha^2 t}\, d\alpha$

3. $u(x, t) = \dfrac{2u_0}{\pi} \displaystyle\int_{-\infty}^\infty \dfrac{1 - e^{-k\alpha^2 t}}{\alpha} \sin \alpha x\, d\alpha$

5. $u(x, t) = \dfrac{2}{\pi} \displaystyle\int_0^\infty \dfrac{1 - \cos \alpha}{\alpha} e^{-k\alpha^2 t}\sin \alpha x\, d\alpha$

7. $u(x, t) = \dfrac{2}{\pi} \displaystyle\int_0^\infty \dfrac{\sin \alpha}{\alpha} e^{-k\alpha^2 t}\cos \alpha x\, d\alpha$

9. (a) $u(x, t) =$

$\dfrac{1}{2\pi} \displaystyle\int_{-\infty}^\infty \left(F(\alpha)\cos \alpha at + G(\alpha)\dfrac{\sin \alpha at}{\alpha a}\right) e^{-i\alpha x}\, d\alpha$

11. $u(x, y) = \dfrac{2}{\pi} \displaystyle\int_0^\infty \dfrac{\sinh \alpha(\pi - x)}{(1 + \alpha^2)\sinh \alpha \pi}\cos \alpha y\, d\alpha$

13. $u(x, y) = \dfrac{100}{\pi} \displaystyle\int_0^\infty \dfrac{\sin \alpha}{\alpha} e^{-\alpha y}\cos \alpha x\, d\alpha$

15. $u(x, y) = \dfrac{2}{\pi} \displaystyle\int_0^\infty F(\alpha)\dfrac{\sinh \alpha(2 - y)}{\sinh 2\alpha}\sin \alpha x\, d\alpha$

17. $u(x, y) = \dfrac{2}{\pi} \displaystyle\int_0^\infty \dfrac{\alpha}{1 + \alpha^2}[e^{-\alpha x}\sin \alpha y + e^{-\alpha y}\sin \alpha x]\, d\alpha$

19. $u(x, t) = \dfrac{1}{\sqrt{1 + 4kt}}\, e^{-x^2/(1 + 4kt)}$

21. $u(x, y) = \dfrac{1}{2\sqrt\pi} \displaystyle\int_{-\infty}^\infty \dfrac{e^{-\alpha^2/4}\cosh \alpha y}{\cosh \alpha}\, e^{-i\alpha x}\, d\alpha$

$= \dfrac{1}{2\sqrt\pi} \displaystyle\int_{-\infty}^\infty \dfrac{e^{-\alpha^2/4}\cosh \alpha y}{\cosh \alpha}\cos \alpha x\, d\alpha$

23. $u(r, z) = \dfrac{2u_0}{\pi} \displaystyle\int_0^\infty \dfrac{I_0(\alpha r)}{\alpha I_0(\alpha)}\sin \alpha \cos \alpha z\, d\alpha$

Exercises 15.5, Page 801

1. $\mathscr{F}_{sn}\{f(x)\} = \dfrac{1 - (-1)^n}{n\pi} = F(n)$,

$\mathscr{F}_{sn}^{-1}\{F(n)\} = 2 \displaystyle\sum_{n=1}^\infty \dfrac{1 - (-1)^n}{n\pi}\sin n\pi x = f(x)$

3. $\mathscr{F}_{cn}\{f(x)\} = \dfrac{2 - 2(-1)^n e^{-2\pi}}{n^2 + 4} = F(n)$,

$\mathscr{F}_{cn}^{-1}\{F(n)\} = \dfrac{1}{2\pi}(1 - e^{-2\pi})$

$+ \dfrac{4}{\pi} \displaystyle\sum_{n=1}^\infty \dfrac{1 - (-1)^n e^{-2\pi}}{n^2 + 4}\cos nx = f(x)$

5. $u(x, t) = 2 \displaystyle\sum_{n=1}^\infty F(n)e^{-n^2 \pi^2 t}\sin n\pi x$

7. $u(x, t) = \dfrac{1}{\pi} \displaystyle\int_0^\pi f(x)\, dx + \dfrac{2}{\pi} \displaystyle\sum_{n=1}^\infty F(n)e^{-n^2 t}\cos nx$

9. $u(x, t) = \dfrac{2}{\pi} \displaystyle\sum_{n=1}^\infty \left(F(n)\cos nt + \dfrac{G(n)}{n}\sin nt\right)\sin nx$

11. $u(x, t) = \left(\dfrac{4}{\pi}u_0 - 1\right)\cos t \sin x + \sin x$

$+ \dfrac{2u_0}{\pi} \displaystyle\sum_{n=2}^\infty \dfrac{1 - (-1)^n}{n}\cos nt \sin nx$

13. $u(x, y) = \dfrac{2u_0}{\pi} \displaystyle\sum_{n=1}^\infty \dfrac{1 - (-1)^n}{n\sqrt{n^2\pi^2 + \beta}} \dfrac{\cosh\sqrt{n^2\pi^2 + \beta}\,y}{\sinh\sqrt{n^2\pi^2 + \beta}}\sin n\pi x$

15. $u(x, y) = \dfrac{2}{\pi} \displaystyle\sum_{n=1}^\infty \left[\dfrac{(-1)^n\cosh ny}{n\cosh n} - \dfrac{(-1)^n}{n}\right]\sin nx$

Exercises 15.6, Page 810

3. 1

7. $F_8 =$

$$\begin{pmatrix} 1 & 1 & 1 & 1 & 1 & 1 & 1 & 1 \\ 1 & \frac{\sqrt2}{2}+i\frac{\sqrt2}{2} & i & -\frac{\sqrt2}{2}+i\frac{\sqrt2}{2} & -1 & -\frac{\sqrt2}{2}-i\frac{\sqrt2}{2} & -i & \frac{\sqrt2}{2}-i\frac{\sqrt2}{2} \\ 1 & i & -1 & -i & 1 & i & -1 & -i \\ 1 & -\frac{\sqrt2}{2}+i\frac{\sqrt2}{2} & -i & \frac{\sqrt2}{2}+i\frac{\sqrt2}{2} & -1 & \frac{\sqrt2}{2}-i\frac{\sqrt2}{2} & i & -\frac{\sqrt2}{2}-i\frac{\sqrt2}{2} \\ 1 & -1 & 1 & -1 & 1 & -1 & 1 & -1 \\ 1 & -\frac{\sqrt2}{2}-i\frac{\sqrt2}{2} & i & \frac{\sqrt2}{2}-i\frac{\sqrt2}{2} & -1 & \frac{\sqrt2}{2}+i\frac{\sqrt2}{2} & -i & -\frac{\sqrt2}{2}+i\frac{\sqrt2}{2} \\ 1 & -i & -1 & i & 1 & -i & -1 & i \\ 1 & \frac{\sqrt2}{2}-i\frac{\sqrt2}{2} & -i & -\frac{\sqrt2}{2}-i\frac{\sqrt2}{2} & -1 & -\frac{\sqrt2}{2}+i\frac{\sqrt2}{2} & i & \frac{\sqrt2}{2}+i\frac{\sqrt2}{2} \end{pmatrix}$$

Chapter 15 in Review, Page 811

1. $u(x, y) = \dfrac{2}{\pi} \displaystyle\int_0^\infty \dfrac{\sinh \alpha y}{\alpha(1 + \alpha^2)\cosh \alpha \pi}\cos \alpha x\, d\alpha$

3. $u(x, t) = u_0 e^{-ht}\text{erf}\left(\dfrac{x}{2\sqrt{t}}\right)$

5. $u(x, t) = \displaystyle\int_0^t \text{erfc}\left(\dfrac{x}{2\sqrt{\tau}}\right) d\tau$

7. $u(x, t) = \dfrac{u_0}{2\pi} \displaystyle\int_{-\infty}^\infty \dfrac{\sin \alpha(\pi - x) + \sin \alpha x}{\alpha}\, e^{-k\alpha^2 t}\, d\alpha$

9. $u(x, y) =$

$$\frac{100}{\pi} \int_0^\infty \left(\frac{1 - \cos \alpha}{\alpha}\right) [e^{-\alpha x} \sin \alpha y + 2e^{-\alpha y} \sin \alpha x] \, d\alpha$$

11. $u(x, y) = \dfrac{2}{\pi} \int_0^\infty \left(\dfrac{B \cosh \alpha y}{(1 + \alpha^2) \sinh \alpha \pi} + \dfrac{A}{\alpha}\right) \sin \alpha x \, d\alpha$

13. $u(x, t) = \dfrac{1}{2\pi} \int_{-\infty}^\infty \dfrac{\cos \alpha x + \alpha \sin \alpha x}{1 + \alpha^2} e^{-k\alpha^2 t} \, d\alpha$

17. $u(x, t) = 1 + e^{-4t} \sin 2x$

19. $u(x, y) = \dfrac{2}{\pi} \int_0^\infty \dfrac{F(\alpha) \sinh \sqrt{\alpha^2 + h}\, y}{\sinh \sqrt{\alpha^2 + h}\, \pi} \sin \alpha x \, d\alpha,$

where $F(\alpha) = \displaystyle\int_0^\infty f(x) \sin \alpha x \, dx$

21. $u(x, t) = 2\sqrt{\dfrac{t}{\pi}} e^{-x^2/4t} - x \, \mathrm{erfc}\left(\dfrac{x}{2\sqrt{t}}\right)$

or $\quad u(x, t) = \dfrac{1}{\sqrt{\pi}} \displaystyle\int_0^t \dfrac{e^{-x^2/4(t-\tau)}}{\sqrt{t - \tau}} \, d\tau$

23. $u(x, t) = (t - 1 + e^{-t}) \sin x$

25. $u(x, t) = \dfrac{2}{\pi} \displaystyle\sum_{n=1}^\infty \dfrac{u_0 - (-1)^n u_1}{n} (1 - e^{-n^2 t}) \sin nx$

or $u(x, t) = u_0 + (u_1 - u_0)\dfrac{x}{\pi} - \dfrac{2}{\pi} \displaystyle\sum_{n=1}^\infty \dfrac{u_0 - (-1)^n u_1}{n} e^{-n^2 t} \sin nx$

Exercises 16.1, Page 819

1. $u_{11} = \frac{11}{15}, u_{21} = \frac{14}{15}$

3. $u_{11} = u_{21} = \sqrt{3}/16, u_{22} = u_{12} = 3\sqrt{3}/16$

5. $u_{21} = u_{12} = 12.50, u_{31} = u_{13} = 18.75, u_{32} = u_{23} = 37.50,$
$u_{11} = 6.25, u_{22} = 25.00, u_{33} = 56.25$

7. (b) $u_{14} = u_{41} = 0.5427, u_{24} = u_{42} = 0.6707,$
$u_{34} = u_{43} = 0.6402, u_{33} = 0.9451, u_{44} = 0.4451$

Exercises 16.2, Page 824
The tables in this section give a selection of the total number of approximations.

1.

Time	x = 0.25	x = 0.50	x = 0.75	x = 1.00	x = 1.25	x = 1.50	x = 1.75
0.000	1.0000	1.0000	1.0000	1.0000	0.0000	0.0000	0.0000
0.100	0.3728	0.6288	0.6800	0.5904	0.3840	0.2176	0.0768
0.200	0.2248	0.3942	0.4708	0.4562	0.3699	0.2517	0.1239
0.300	0.1530	0.2752	0.3448	0.3545	0.3101	0.2262	0.1183
0.400	0.1115	0.2034	0.2607	0.2757	0.2488	0.1865	0.0996
0.500	0.0841	0.1545	0.2002	0.2144	0.1961	0.1487	0.0800
0.600	0.0645	0.1189	0.1548	0.1668	0.1534	0.1169	0.0631
0.700	0.0499	0.0921	0.1201	0.1297	0.1196	0.0914	0.0494
0.800	0.0387	0.0715	0.0933	0.1009	0.0931	0.0712	0.0385
0.900	0.0301	0.0555	0.0725	0.0785	0.0725	0.0554	0.0300
1.000	0.0234	0.0432	0.0564	0.0610	0.0564	0.0431	0.0233

3.

Time	x = 0.25	x = 0.50	x = 0.75	x = 1.00	x = 1.25	x = 1.50	x = 1.75
0.000	1.0000	1.0000	1.0000	1.0000	0.0000	0.0000	0.0000
0.100	0.4015	0.6577	0.7084	0.5837	0.3753	0.1871	0.0684
0.200	0.2430	0.4198	0.4921	0.4617	0.3622	0.2362	0.1132
0.300	0.1643	0.2924	0.3604	0.3626	0.3097	0.2208	0.1136
0.400	0.1187	0.2150	0.2725	0.2843	0.2528	0.1871	0.0989
0.500	0.0891	0.1630	0.2097	0.2228	0.2020	0.1521	0.0814
0.600	0.0683	0.1256	0.1628	0.1746	0.1598	0.1214	0.0653
0.700	0.0530	0.0976	0.1270	0.1369	0.1259	0.0959	0.0518
0.800	0.0413	0.0762	0.0993	0.1073	0.0989	0.0755	0.0408
0.900	0.0323	0.0596	0.0778	0.0841	0.0776	0.0593	0.0321
1.000	0.0253	0.0466	0.0609	0.0659	0.0608	0.0465	0.0252

Absolute errors are approximately $2.2 \times 10^{-2}, 3.7 \times 10^{-2}, 1.3 \times 10^{-2}$.

5.

Time	$x = 0.25$	$x = 0.50$	$x = 0.75$	$x = 1.00$	$x = 1.25$	$x = 1.50$	$x = 1.75$
0.00	1.0000	1.0000	1.0000	1.0000	0.0000	0.0000	0.0000
0.10	0.3972	0.6551	0.7043	0.5883	0.3723	0.1955	0.0653
0.20	0.2409	0.4171	0.4901	0.4620	0.3636	0.2385	0.1145
0.30	0.1631	0.2908	0.3592	0.3624	0.3105	0.2220	0.1145
0.40	0.1181	0.2141	0.2718	0.2840	0.2530	0.1876	0.0993
0.50	0.0888	0.1625	0.2092	0.2226	0.2020	0.1523	0.0816
0.60	0.0681	0.1253	0.1625	0.1744	0.1597	0.1214	0.0654
0.70	0.0528	0.0974	0.1268	0.1366	0.1257	0.0959	0.0518
0.80	0.0412	0.0760	0.0991	0.1071	0.0987	0.0754	0.0408
0.90	0.0322	0.0594	0.0776	0.0839	0.0774	0.0592	0.0320
1.00	0.0252	0.0465	0.0608	0.0657	0.0607	0.0464	0.0251

Absolute errors are approximately 1.8×10^{-2}, 3.7×10^{-2}, 1.3×10^{-2}.

7. (a)

Time	$x = 2.00$	$x = 4.00$	$x = 6.00$	$x = 8.00$	$x = 10.00$	$x = 12.00$	$x = 14.00$	$x = 16.00$	$x = 18.00$
0.00	30.0000	30.0000	30.0000	30.0000	30.0000	30.0000	30.0000	30.0000	30.0000
2.00	27.6450	29.9037	29.9970	29.9999	30.0000	29.9999	29.9970	29.9037	27.6450
4.00	25.6452	29.6517	29.9805	29.9991	29.9999	29.9991	29.9805	29.6517	25.6452
6.00	23.9347	29.2922	29.9421	29.9963	29.9996	29.9963	29.9421	29.2922	23.9347
8.00	22.4612	28.8606	29.8782	29.9898	29.9986	29.9898	29.8782	28.8606	22.4612
10.00	21.1829	28.3831	29.7878	29.9782	29.9964	29.9782	29.7878	28.3831	21.1829

(b)

Time	$x = 5.00$	$x = 10.00$	$x = 15.00$	$x = 20.00$	$x = 25.00$	$x = 30.00$	$x = 35.00$	$x = 40.00$	$x = 45.00$
0.00	30.0000	30.0000	30.0000	30.0000	30.0000	30.0000	30.0000	30.0000	30.0000
2.00	29.5964	29.9973	30.0000	30.0000	30.0000	30.0000	30.0000	29.9973	29.5964
4.00	29.2036	29.9893	29.9999	30.0000	30.0000	30.0000	29.9999	29.9893	29.2036
6.00	28.8212	29.9762	29.9997	30.0000	30.0000	30.0000	29.9997	29.9762	28.8213
8.00	28.4490	29.9585	29.9992	30.0000	30.0000	30.0000	29.9993	29.9585	28.4490
10.00	28.0864	29.9363	29.9986	30.0000	30.0000	30.0000	29.9986	29.9363	28.0864

(c)

Time	$x = 2.00$	$x = 4.00$	$x = 6.00$	$x = 8.00$	$x = 10.00$	$x = 12.00$	$x = 14.00$	$x = 16.00$	$x = 18.00$
0.00	18.0000	32.0000	42.0000	48.0000	50.0000	48.0000	42.0000	32.0000	18.0000
2.00	15.3312	28.5348	38.3465	44.3067	46.3001	44.3067	38.3465	28.5348	15.3312
4.00	13.6371	25.6867	34.9416	40.6988	42.6453	40.6988	34.9416	25.6867	13.6371
6.00	12.3012	23.2863	31.8624	37.2794	39.1273	37.2794	31.8624	23.2863	12.3012
8.00	11.1659	21.1877	29.0757	34.0984	35.8202	34.0984	29.0757	21.1877	11.1659
10.00	10.1665	19.3143	26.5439	31.1662	32.7549	31.1662	26.5439	19.3143	10.1665

(d)

Time	x = 10.00	x = 20.00	x = 30.00	x = 40.00	x = 50.00	x = 60.00	x = 70.00	x = 80.00	x = 90.00
0.00	8.0000	16.0000	24.0000	32.0000	40.0000	32.0000	24.0000	16.0000	8.0000
2.00	8.0000	16.0000	23.9999	31.9918	39.4932	31.9918	23.9999	16.0000	8.0000
4.00	8.0000	16.0000	23.9993	31.9686	39.0175	31.9686	23.9993	16.0000	8.0000
6.00	8.0000	15.9999	23.9978	31.9323	38.5701	31.9323	23.9978	15.9999	8.0000
8.00	8.0000	15.9998	23.9950	31.8844	38.1483	31.8844	23.9950	15.9998	8.0000
10.00	8.0000	15.9996	23.9908	31.8265	37.7498	31.8265	23.9908	15.9996	8.0000

9. (a)

Time	x = 2.00	x = 4.00	x = 6.00	x = 8.00	x = 10.00	x = 12.00	x = 14.00	x = 16.00	x = 18.00
0.00	30.0000	30.0000	30.0000	30.0000	30.0000	30.0000	30.0000	30.0000	30.0000
2.00	27.6450	29.9037	29.9970	29.9999	30.0000	30.0000	29.9990	29.9679	29.2150
4.00	25.6452	29.6517	29.9805	29.9991	30.0000	29.9997	29.9935	29.8839	28.5484
6.00	23.9347	29.2922	29.9421	29.9963	29.9997	29.9988	29.9807	29.7641	27.9782
8.00	22.4612	28.8606	29.8782	29.9899	29.9991	29.9966	29.9594	29.6202	27.4870
10.00	21.1829	28.3831	29.7878	29.9783	29.9976	29.9927	29.9293	29.4610	27.0610

(b)

Time	x = 5.00	x = 10.00	x = 15.00	x = 20.00	x = 25.00	x = 30.00	x = 35.00	x = 40.00	x = 45.00
0.00	30.0000	30.0000	30.0000	30.0000	30.0000	30.0000	30.0000	30.0000	30.0000
2.00	29.5964	29.9973	30.0000	30.0000	30.0000	30.0000	30.0000	29.9991	29.8655
4.00	29.2036	29.9893	29.9999	30.0000	30.0000	30.0000	30.0000	29.9964	29.7345
6.00	28.8212	29.9762	29.9997	30.0000	30.0000	30.0000	29.9999	29.9921	29.6071
8.00	28.4490	29.9585	29.9992	30.0000	30.0000	30.0000	29.9997	29.9862	29.4830
10.00	28.0864	29.9363	29.9986	30.0000	30.0000	30.0000	29.9995	29.9788	29.3621

(c)

Time	x = 2.00	x = 4.00	x = 6.00	x = 8.00	x = 10.00	x = 12.00	x = 14.00	x = 16.00	x = 18.00
0.00	18.0000	32.0000	42.0000	48.0000	50.0000	48.0000	42.0000	32.0000	18.0000
2.00	15.3312	28.5350	38.3477	44.3130	46.3327	44.4671	39.0872	31.5755	24.6930
4.00	13.6381	25.6913	34.9606	40.7728	42.9127	41.5716	37.4340	31.7086	25.6986
6.00	12.3088	23.3146	31.9546	37.5566	39.8880	39.1565	36.9745	31.2134	25.7128
8.00	11.1946	21.2785	29.3217	34.7092	37.2109	36.9834	34.5032	30.4279	25.4167
10.00	10.2377	19.5150	27.0178	32.1929	34.8117	34.9710	33.0338	29.5224	25.0019

(d)

Time	x = 10.00	x = 20.00	x = 30.00	x = 40.00	x = 50.00	x = 60.00	x = 70.00	x = 80.00	x = 90.00
0.00	8.0000	16.0000	24.0000	32.0000	40.0000	32.0000	24.0000	16.0000	8.0000
2.00	8.0000	16.0000	23.9999	31.9918	39.4932	31.9918	24.0000	16.0102	8.6333
4.00	8.0000	16.0000	23.9993	31.9686	39.0175	31.9687	24.0002	16.0391	9.2272
6.00	8.0000	15.9999	23.9978	31.9323	38.5701	31.9324	24.0005	16.0845	9.7846
8.00	8.0000	15.9998	23.9950	31.8844	38.1483	31.8846	24.0012	16.1441	10.3084
10.00	8.0000	15.9996	23.9908	31.8265	37.7499	31.8269	24.0023	16.2160	10.8012

11. (a) $\psi(x) = \frac{1}{2}x + 20$

(b)

Time	$x = 4.00$	$x = 8.00$	$x = 12.00$	$x = 16.00$
0.00	50.0000	50.0000	50.0000	50.0000
10.00	32.7433	44.2679	45.4228	38.2971
20.00	29.9946	36.2354	38.3148	35.8160
30.00	26.9487	32.1409	34.0874	32.9644
50.00	24.1178	27.4348	29.4296	30.1207
70.00	22.8995	25.4560	27.4554	28.8998
90.00	22.3817	24.6176	26.6175	28.3817
110.00	22.1619	24.2620	26.2620	28.1619
130.00	22.0687	24.1112	26.1112	28.0687
150.00	22.0291	24.0472	26.0472	28.0291
170.00	22.0124	24.0200	26.0200	28.0124
190.00	22.0052	24.0085	26.0085	28.0052
210.00	22.0022	24.0036	26.0036	28.0022
230.00	22.0009	24.0015	26.0015	28.0009
250.00	22.0004	24.0007	26.0007	28.0004
270.00	22.0002	24.0003	26.0003	28.0002
290.00	22.0001	24.0001	26.0001	28.0001
310.00	22.0000	24.0001	26.0001	28.0000
330.00	22.0000	24.0000	26.0000	28.0000
350.00	22.0000	24.0000	26.0000	28.0000

1. (a)

Time	$x = 0.25$	$x = 0.50$	$x = 0.75$
0.00	0.1875	0.2500	0.1875
0.20	0.1491	0.2100	0.1491
0.40	0.0556	0.0938	0.0556
0.60	−0.0501	−0.0682	−0.0501
0.80	−0.1361	−0.2072	−0.1361
1.00	−0.1802	−0.2591	−0.1802

(b)

Time	$x = 0.4$	$x = 0.8$	$x = 1.2$	$x = 1.6$
0.00	0.0032	0.5273	0.5273	0.0032
0.20	0.0652	0.4638	0.4638	0.0652
0.40	0.2065	0.3035	0.3035	0.2065
0.60	0.3208	0.1190	0.1190	0.3208
0.80	0.3094	−0.0180	−0.0180	0.3094
1.00	0.1450	−0.0768	−0.0768	0.1450

(c)

Time	$x = 0.1$	$x = 0.2$	$x = 0.3$	$x = 0.4$	$x = 0.5$	$x = 0.6$	$x = 0.7$	$x = 0.8$	$x = 0.9$
0.00	0.0000	0.0000	0.0000	0.0000	0.0000	0.5000	0.5000	0.5000	0.5000
0.12	0.0000	0.0000	0.0082	0.1126	0.3411	0.1589	0.3792	0.3710	0.0462
0.24	0.0071	0.0657	0.2447	0.3159	0.1735	0.2463	−0.1266	−0.3056	−0.0625
0.36	0.1623	0.3197	0.2458	0.1657	0.0877	−0.2853	−0.2843	−0.2104	−0.2887
0.48	0.1965	0.1410	0.1149	−0.1216	−0.3593	−0.2381	−0.1977	−0.1715	0.0800
0.60	−0.2194	−0.2069	−0.3875	−0.3411	−0.1901	−0.1662	−0.0666	0.1140	−0.0446
0.72	−0.3003	−0.6865	−0.5097	−0.3230	−0.1585	0.0156	0.0893	−0.0874	0.0384
0.84	−0.2647	−0.1633	−0.3546	−0.3214	−0.1763	−0.0954	−0.1249	0.0665	−0.0386
0.96	0.3012	0.1081	0.1380	−0.0487	−0.2974	−0.3407	−0.1250	−0.1548	0.0092

3. (a)

Time	$x = 0.2$	$x = 0.4$	$x = 0.6$	$x = 0.8$
0.00	0.5878	0.9511	0.9511	0.5878
0.10	0.5599	0.9059	0.9059	0.5599
0.20	0.4788	0.7748	0.7748	0.4788
0.30	0.3524	0.5701	0.5701	0.3524
0.40	0.1924	0.3113	0.3113	0.1924
0.50	0.0142	0.0230	0.0230	0.0142

(b)

Time	$x = 0.2$	$x = 0.4$	$x = 0.6$	$x = 0.8$
0.00	0.5878	0.9511	0.9511	0.5878
0.05	0.5808	0.9397	0.9397	0.5808
0.10	0.5599	0.9060	0.9060	0.5599
0.15	0.5257	0.8507	0.8507	0.5257
0.20	0.4790	0.7750	0.7750	0.4790
0.25	0.4209	0.6810	0.6810	0.4209
0.30	0.3527	0.5706	0.5706	0.3527
0.35	0.2761	0.4467	0.4467	0.2761
0.40	0.1929	0.3122	0.3122	0.1929
0.45	0.1052	0.1701	0.1701	0.1052
0.50	0.0149	0.0241	0.0241	0.0149

5.

Time	$x = 10$	$x = 20$	$x = 30$	$x = 40$	$x = 50$
0.00000	0.1000	0.2000	0.3000	0.2000	0.1000
0.60134	0.0984	0.1688	0.1406	0.1688	0.0984
1.20268	0.0226	−0.0121	0.0085	−0.0121	0.0226
1.80401	−0.1271	−0.1347	−0.1566	−0.1347	−0.1271
2.40535	−0.0920	−0.2292	−0.2571	−0.2292	−0.0920
3.00669	−0.0932	−0.1445	−0.2018	−0.1445	−0.0932
3.60803	−0.0284	−0.0205	0.0336	−0.0205	−0.0284
4.20936	0.1064	0.1555	0.1265	0.1555	0.1064
4.81070	0.1273	0.2060	0.2612	0.2060	0.1273
5.41204	0.0625	0.1689	0.2038	0.1689	0.0625
6.01338	0.0436	0.0086	−0.0080	0.0086	0.0436
6.61472	−0.0931	−0.1364	−0.1578	−0.1364	−0.0931
7.21605	−0.1436	−0.2173	−0.2240	−0.2173	−0.1436
7.81739	−0.0625	−0.1644	−0.2247	−0.1644	−0.0625
8.41873	−0.0287	−0.0192	−0.0085	−0.0192	−0.0287
9.02007	0.0654	0.1332	0.1755	0.1332	0.0654
9.62140	0.1540	0.2189	0.2089	0.2189	0.1540

Note: Time is expressed in milliseconds.

Chapter 16 in Review, Page 828

1. $u_{11} = 0.8929$, $u_{21} = 3.5714$, $u_{31} = 13.3929$

3. (a)

$x = 0.20$	$x = 0.40$	$x = 0.60$	$x = 0.80$
0.2000	0.4000	0.6000	0.8000
0.2000	0.4000	0.6000	0.5500
0.2000	0.4000	0.5375	0.4250
0.2000	0.3844	0.4750	0.3469
0.1961	0.3609	0.4203	0.2922
0.1883	0.3346	0.3734	0.2512

(b)

$x = 0.20$	$x = 0.40$	$x = 0.60$	$x = 0.80$
0.2000	0.4000	0.6000	0.8000
0.2000	0.4000	0.6000	0.8000
0.2000	0.4000	0.6000	0.5500
0.2000	0.4000	0.5375	0.4250
0.2000	0.3844	0.4750	0.3469
0.1961	0.3609	0.4203	0.2922

(c) Yes; the table in part (b) is the table in part (a) shifted downward.

Exercises 17.1, Page 834

1. 1 **3.** -1 **5.** $3 + 3i$

7. $-10 + i$ **9.** $7 - 13i$ **11.** $-7 + 5i$

13. $11 - 10i$ **15.** $-5 + 12i$ **17.** $-2i$

19. $-\frac{7}{17} - \frac{11}{17}i$ **21.** $8 - i$ **23.** $\frac{23}{37} - \frac{64}{37}i$

25. $20i$ **27.** $\frac{102}{5} + \frac{116}{5}i$ **29.** $\frac{7}{130} + \frac{9}{130}i$

31. $\frac{3}{5} - \frac{1}{5}i$ **33.** $x/(x^2 + y^2)$

35. $-2y - 4$ **37.** $\sqrt{(x-1)^2 + (y-3)^2}$

39. $x = -\frac{9}{2}, y = 1$ **41.** $\frac{\sqrt{2}}{2} + \frac{\sqrt{2}}{2}i, -\frac{\sqrt{2}}{2} - \frac{\sqrt{2}}{2}i$

43. $z = -\frac{1}{30} + \frac{7}{10}i$ **45.** $11 - 6i$

Exercises 17.2, Page 838

1. $2(\cos 0 + i \sin 0)$ or $2(\cos 2\pi + i \sin 2\pi)$

3. $3\left(\cos \dfrac{3\pi}{2} + i \sin \dfrac{3\pi}{2}\right)$ **5.** $\sqrt{2}\left(\cos \dfrac{\pi}{4} + i \sin \dfrac{\pi}{4}\right)$

7. $2\left(\cos \dfrac{5\pi}{6} + i \sin \dfrac{5\pi}{6}\right)$

9. $\dfrac{3\sqrt{2}}{2}\left(\cos \dfrac{5\pi}{4} + i \sin \dfrac{5\pi}{4}\right)$

11. $-\frac{53}{2} - \frac{5}{2}i$ **13.** $5.5433 + 2.2961i$

15. $8i$; $\dfrac{\sqrt{2}}{4} - \dfrac{\sqrt{2}}{4}i$

17. $30\sqrt{2}\,[\cos(25\pi/12) + i \sin(25\pi/12)]$;
$40.9808 + 10.9808i$

19. $\frac{1}{2\sqrt{2}}[\cos(-\pi/4) + i \sin(-\pi/4)]$; $\frac{1}{4} - \frac{1}{4}i$

21. -512 **23.** $\frac{1}{32}i$ **25.** $-i$

27. $w_0 = 2$, $w_1 = -1 + \sqrt{3}i$, $w_2 = -1 - \sqrt{3}i$

29. $w_0 = \frac{\sqrt{2}}{2} + \frac{\sqrt{2}}{2}i$, $w_1 = -\frac{\sqrt{2}}{2} - \frac{\sqrt{2}}{2}i$

31. $w_0 = \frac{\sqrt{2}}{2} + \frac{\sqrt{6}}{2}i$, $w_1 = -\frac{\sqrt{2}}{2} - \frac{\sqrt{6}}{2}i$

33. $\pm \frac{\sqrt{2}}{2}(1 + i)$, $\pm \frac{\sqrt{2}}{2}(1 - i)$

35. $32\left(\cos \dfrac{13\pi}{6} + i \sin \dfrac{13\pi}{6}\right)$, $16\sqrt{3} + 16i$

37. $\cos 2\theta = \cos^2\theta - \sin^2\theta$, $\sin 2\theta = 2 \sin\theta \cos\theta$

1.

3.

5.

7.

9. domain

11. domain

13. domain

15. not a domain

17. not a domain

19. domain

21. domain

23. the line $y = -x$

25. the hyperbola $x^2 - y^2 = 1$

1.
$u = \dfrac{v^2}{16} - 4$

3.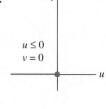
$u \le 0$
$v = 0$

5.
$v \ge 0$
$u = 0$

7. $f(z) = (6x - 5) + i(6y + 9)$

9. $f(z) = (x^2 - y^2 - 3x) + i(2xy - 3y + 4)$

11. $f(z) = (x^3 - 3xy^2 - 4x) + i(3x^2y - y^3 - 4y)$

13. $f(z) = \left(x + \dfrac{x}{x^2 + y^2}\right) + i\left(y - \dfrac{y}{x^2 + y^2}\right)$

15. $-4 + i;\ 3 - 9i;\ 1 + 86i$

17. $14 - 20i;\ -13 + 43i;\ 3 - 26i$ **19.** $6 - 5i$

21. $-4i$ **27.** $12z^2 - (6 + 2i)z - 5$

29. $6z^2 - 14z - 4 + 16i$ **31.** $6z(z^2 - 4i)^2$

33. $\dfrac{8 - 13i}{(2z + i)^2}$ **35.** $3i$ **37.** $2i, -2i$

41. $x(t) = c_1 e^{2t}$ and $y(t) = c_2 e^{2t}$; the streamlines lie on lines through the origin.

43. $y = cx$; the streamlines are lines through the origin.

45.

15. $a = 1, b = 3$

21. $f'(z) = e^x \cos y + i e^x \sin y$

23. $f(z) = x + i(y + C)$

25. $f(z) = x^2 - y^2 + i(2xy + C)$

27. $f(z) = \log_e(x^2 + y^2) + i\left(2\tan^{-1}\dfrac{y}{x} + C\right)$

29.

$y = c_2$

$u = c_1$

31. the x-axis and the circle $|z| = 1$

Exercises 17.6, Page 856

1. $\dfrac{\sqrt{3}}{2} + \dfrac{1}{2}i$ **3.** $e^{-1}\left(\dfrac{\sqrt{2}}{2} + \dfrac{\sqrt{2}}{2}i\right)$

5. $-e^{\pi}$ **7.** $-1.8650 + 4.0752i$

9. $0.2837 - 0.9589i$ **11.** $-0.9659 + 0.2588i$

13. $e^y(\cos x - i\sin x)$

15. $e^{x^2 - y^2}(\cos 2xy + i\sin 2xy)$

23. $1.6094 + i(\pi + 2n\pi)$

25. $1.0397 + i(3\pi/4 + 2n\pi)$

27. $1.0397 + i(\pi/3 + 2n\pi)$

29. $2.1383 - (\pi/4)i$ **31.** $2.5649 + 2.7468i$

33. $3.4657 - (\pi/3)i$ **35.** $1.3863 + i(\pi/2 + 2n\pi)$

37. $3 + i(-\pi/2 + 2n\pi)$ **39.** $e^{(2-8n)\pi}$

41. $e^{-2n\pi}(0.2740 + 0.5837i)$ **43.** e^2

47. no; no; no

Exercises 17.7, Page 859

1. 10.0677 **3.** $1.0911 + 0.8310i$

5. $0.7616i$ **7.** -0.6481 **9.** -1

11. $0.5876 + 1.3363i$

15. $\pi/2 + 2n\pi - i\log_e(2 \pm \sqrt{3})$

17. $(-\pi/2 + 2n\pi)i$ **19.** $\pi/4 + n\pi$

21. $2n\pi \pm 2i$

Exercises 17.8, Page 862

1. $n\pi + (-1)^{n+1}i\log_e(1 + \sqrt{2})$ **3.** $n\pi$

5. $2n\pi \pm i\log_e(2 + \sqrt{3})$ **7.** $\pm\pi/3 + 2n\pi$

9. $\pi/4 + n\pi$ **11.** $(-1)^n \log_e 3 + n\pi i$

Chapter 17 in Review, Page 862

1. $0; 32$ **3.** $-\dfrac{7}{25}$ **5.** $\dfrac{4}{5}$ **7.** false

9. $0.6931 + i(\pi/2 + 2n\pi)$ **11.** $-0.3097 + 0.8577i$

13. false **15.** $3 - \dfrac{\pi}{2}i$ **17.** $58 - 4i$

19. $-8 + 8i$

21.

23.

25. an ellipse with foci $(0, -2)$ and $(0, 2)$

27. $1.0696 - 0.2127i$, $0.2127 + 1.0696i$,
$-1.0696 + 0.2127i$, $-0.2127 - 1.0696i$

29. $5i$ **31.** the parabola $v = u^2 - 2u$

33. $1, -1$ **35.** pure imaginary numbers

37. $f'(z) = (-2y - 5) + 2xi$

Exercises 18.1, Page 869

1. $-28 + 84i$ **3.** $-48 + \dfrac{736}{3}i$ **5.** $(2 + \pi)i$

7. πi **9.** $-\dfrac{7}{12} + \dfrac{1}{12}i$ **11.** $-e - 1$

13. $\dfrac{3}{2} - \dfrac{\pi}{4}$ **15.** 0 **17.** $\dfrac{1}{2}i$ **19.** 0

21. $\dfrac{4}{3} - \dfrac{5}{3}i$ **23.** $\dfrac{4}{3} - \dfrac{5}{3}i$ **25.** $\dfrac{5\pi e^5}{12}$

27. $6\sqrt{2}$ **31.** $-11 + 38i; 0$

33. circulation $= 0$, net flux $= 4\pi$

35. circulation $= 0$, net flux $= 0$

Exercises 18.2, Page 873

9. $2\pi i$ **11.** $2\pi i$ **13.** 0

15. $2\pi i; 4\pi i; 0$ **17.** $-8\pi i; -6\pi i$

19. $-\pi(1 + i)$ **21.** $-4\pi i$ **23.** $-6\pi i$

Exercises 18.3, Page 878

1. $2i$ **3.** $48 + 24i$ **5.** $6 + \dfrac{26}{3}i$ **7.** 0

9. $-\dfrac{7}{16} - \dfrac{22}{3}i$ **11.** $-\dfrac{1}{\pi} - \dfrac{1}{\pi}i$ **13.** $2.3504i$

15. 0 **17.** πi **19.** $\dfrac{1}{2}i$

21. $11.4928 + 0.9667i$ **23.** $-0.9056 + 1.7699i$

Exercises 18.4, Page 884

1. $8\pi i$ **3.** $-2\pi i$ **5.** $-\pi(20 + 8i)$

7. $-2\pi; 2\pi$ **9.** -8π **11.** $-2\pi e^{-1}i$

13. $\dfrac{4}{3}\pi i$ **15.** $-5\pi i; -5\pi i; 9\pi i; 0$

17. $-\pi(3 + i); \pi(3 + i)$ **19.** $\pi\left(\dfrac{8}{3} + 12i\right)$

21. 0 **23.** $-\pi i$

Chapter 18 in Review, Page 885

1. true **3.** true **5.** 0 **7.** $\pi(6\pi - i)$

9. true **11.** 0 if $n \neq -1$, $2\pi i$ if $n = -1$

13. $-\frac{7}{2}$ **15.** $\frac{136}{15} + \frac{88}{3}i$ **17.** 0

19. $-14.2144 + 22.9637i$ **21.** $2\pi i$

23. $-\frac{8}{3}\pi i$ **25.** $\frac{2}{5}\pi i$ **27.** 2π **29.** $2n\pi i$

Exercises 19.1, Page 893

1. $5i, -5, -5i, 5, 5i$ **3.** $0, 2, 0, 2, 0$

5. converges **7.** converges **9.** diverges

11. $\lim_{n\to\infty} \text{Re}(z_n) = 2$ and $\lim_{n\to\infty} \text{Im}(z_n) = \frac{3}{2}$

13. The series converges to $1/(1 + 2i)$.

15. divergent

17. convergent, $-\frac{1}{5} + \frac{2}{5}i$

19. convergent, $\frac{9}{5} - \frac{12}{5}i$

21. $|z - 2i| = \sqrt{5}, R = \sqrt{5}$

23. $|z - 1 - i| = 2, R = 2$

25. $|z - i| = 1/\sqrt{10}, R = 1/\sqrt{10}$

27. $|z - 4 - 3i| = 25, R = 25$

29. The series converges at $z = -2 + i$.

Exercises 19.2, Page 897

1. $\displaystyle\sum_{k=1}^{\infty}(-1)^{k+1}z^k, R = 1$

3. $\displaystyle\sum_{k=1}^{\infty}(-1)^{k-1}k(2z)^{k-1}, R = \frac{1}{2}$

5. $\displaystyle\sum_{k=0}^{\infty}\frac{(-1)^k}{k!}(2z)^k, R = \infty$

7. $\displaystyle\sum_{k=0}^{\infty}\frac{z^{2k+1}}{(2k+1)!}, R = \infty$

9. $\displaystyle\sum_{k=0}^{\infty}\frac{(-1)^k}{(2k)!}\left(\frac{z}{2}\right)^{2k}, R = \infty$

11. $\displaystyle\sum_{k=0}^{\infty}\frac{(-1)^k}{(2k+1)!}z^{4k+2}, R = \infty$

13. $\displaystyle\sum_{k=0}^{\infty}(-1)^k(z-1)^k, R = 1$

15. $\displaystyle\sum_{k=0}^{\infty}\frac{(z-2i)^k}{(3-2i)^{k+1}}, R = \sqrt{13}$

17. $\displaystyle\sum_{k=1}^{\infty}\frac{(z-1)^k}{2^k}, R = 2$

19. $\dfrac{\sqrt{2}}{2} - \dfrac{\sqrt{2}}{2\cdot 1!}\left(z - \dfrac{\pi}{4}\right) - \dfrac{\sqrt{2}}{2\cdot 2!}\left(z - \dfrac{\pi}{4}\right)^2$

$\quad + \dfrac{\sqrt{2}}{2\cdot 3!}\left(z - \dfrac{\pi}{4}\right)^3 + \cdots, R = \infty$

21. $e^{3i}\displaystyle\sum_{k=0}^{\infty}\frac{(z-3i)^k}{k!}, R = \infty$

23. $z + \frac{1}{3}z^3 + \frac{2}{15}z^5 + \cdots$

25. $\dfrac{1}{2i} + \dfrac{3}{(2i)^2}z + \dfrac{7}{(2i)^3}z^2 + \dfrac{15}{(2i)^4}z^3 + \cdots, R = 1$

27. $2\sqrt{5}$

29. $\displaystyle\sum_{k=0}^{\infty}(-1)^k(z+1)^k, R = 1;$

$\displaystyle\sum_{k=0}^{\infty}\frac{(-1)^k}{(2+i)^{k+1}}(z-i)^k, R = \sqrt{5}$

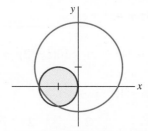

31. (a) The distance from z_0 to the branch cut is 1 unit.

(c) The series converges within the circle

$|z + 1 - i| = \sqrt{2}$. Although the series converges in the shaded region, it does not converge to (or represent) Ln z in this region.

33. $1.1 + 0.12i$ **35.** $\dfrac{2}{\sqrt{\pi}}\displaystyle\sum_{k=0}^{\infty}\frac{(-1)^k}{(2k+1)k!}z^{2k+1}$

Exercises 19.3, Page 905

1. $\dfrac{1}{z} - \dfrac{z}{2!} + \dfrac{z^3}{4!} - \dfrac{z^5}{6!} + \cdots$

3. $1 - \dfrac{1}{1!\cdot z^2} + \dfrac{1}{2!\cdot z^4} - \dfrac{1}{3!\cdot z^6} + \cdots$

5. $\dfrac{e}{z-1} + e + \dfrac{e(z-1)}{2!} + \dfrac{e(z-1)^2}{3!} + \cdots$

7. $-\dfrac{1}{3z} - \dfrac{1}{3^2} - \dfrac{z}{3^3} - \dfrac{z^2}{3^4} - \cdots$

9. $\dfrac{1}{3(z-3)} - \dfrac{1}{3^2} + \dfrac{z-3}{3^3} - \dfrac{(z-3)^2}{3^4} + \cdots$

11. $\cdots - \dfrac{1}{3(z-4)^2} + \dfrac{1}{3(z-4)} - \dfrac{1}{12} + \dfrac{z-4}{3\cdot 4^2}$

$\quad - \dfrac{(z-4)^2}{3\cdot 4^3} + \cdots$

13. $\cdots - \dfrac{1}{z^2} - \dfrac{1}{z} - \dfrac{1}{2} - \dfrac{z}{2^2} - \dfrac{z^2}{2^3} - \cdots$

15. $\dfrac{-1}{z-1} - 1 - (z-1) - (z-1)^2 - \cdots$

17. $\dfrac{1}{3(z+1)} - \dfrac{2}{3^2} - \dfrac{2(z+1)}{3^3} - \dfrac{2(z+1)^2}{3^4} - \cdots$

19. $\cdots - \dfrac{1}{3z^2} + \dfrac{1}{3z} - \dfrac{1}{3} - \dfrac{z}{3\cdot 2} - \dfrac{z^2}{3\cdot 2^2} - \cdots$

21. $\dfrac{1}{z} + 2 + 3z + 4z^2 + \cdots$

23. $\dfrac{1}{z-2} - 3 + 6(z-2) - 10(z-2)^2 + \cdots$

25. $\dfrac{3}{z} - 4 - 4z - 4z^2 - \cdots$

27. $\cdots + \dfrac{2}{(z-1)^3} + \dfrac{2}{(z-1)^2} + \dfrac{2}{z-1} + 1 + (z-1)$

Exercises 19.4, Page 908

1. Define $f(0) = 2$.

3. $-2 + i$ is a zero of order 2.

5. $-i$ and i are zeros of order 1; 0 is a zero of order 2.

7. $2n\pi i$, $n = 0, \pm 1, \ldots$, are zeros of order 1.

9. order 5 **11.** order 1

13. $-1 \pm 2i$ are simple poles.

15. -2 is a simple pole; $-i$ is a pole of order 4.

17. $(2n+1)\pi/2$, $n = 0, \pm 1, \ldots$, are simple poles.

19. 0 is a pole of order 2.

21. $2n\pi i$, $n = 0, \pm 1, \ldots$, are simple poles.

23. 0 is a removable singularity; 1 is a simple pole.

25. nonisolated

Exercises 19.5, Page 913

1. $\frac{2}{5}$ **3.** -3 **5.** 0

7. $\operatorname{Res}(f(z), -4i) = \frac{1}{2}$, $\operatorname{Res}(f(z), 4i) = \frac{1}{2}$

9. $\operatorname{Res}(f(z), 1) = \frac{1}{3}$, $\operatorname{Res}(f(z), -2) = -\frac{1}{12}$,
$\operatorname{Res}(f(z), 0) = -\frac{1}{4}$

11. $\operatorname{Res}(f(z), -1) = 6$, $\operatorname{Res}(f(z), -2) = -31$,
$\operatorname{Res}(f(z), -3) = 30$

13. $\operatorname{Res}(f(z), 0) = -3/\pi^4$, $\operatorname{Res}(f(z), \pi) = (\pi^2 - 6)/2\pi^4$

15. $\operatorname{Res}(f(z), (2n+1)\pi/2) = (-1)^{n+1}$, $n = 0, \pm 1, \pm 2, \ldots$

17. 0; $2\pi i/9$; 0 **19.** πi; πi; 0 **21.** $\pi/3$

23. 0 **25.** $2\pi i \cosh 1$ **27.** $-4i$ **29.** $6i$

31. $-\dfrac{\pi}{3} + \dfrac{\pi}{3} i$

Exercises 19.6, Page 919

1. $4\pi/\sqrt{3}$ **3.** 0 **5.** $\pi/\sqrt{3}$ **7.** $\pi/4$

9. $\pi/6$ **11.** π **13.** $\pi/16$ **15.** $3\pi/8$

17. $\pi/2$ **19.** $\pi/\sqrt{2}$ **21.** πe^{-1} **23.** πe^{-1}

25. πe^{-3} **27.** $\dfrac{\pi e^{-\sqrt{2}}}{2\sqrt{2}}(\cos\sqrt{2} + \sin\sqrt{2})$

29. $-\dfrac{\pi}{8}\left(\dfrac{e^{-3}}{3} - e^{-1}\right)$

Chapter 19 in Review, Page 919

1. true **3.** false **5.** true **7.** true

9. $\dfrac{1}{\pi}$ **11.** $|z - i| = \sqrt{5}$

13. $1 + \displaystyle\sum_{k=1}^{\infty} \dfrac{(\sqrt{2})^k \cos(k\pi/4)}{k!} z^k$

15. $-\dfrac{i}{z^3} + \dfrac{1}{2!z^2} + \dfrac{i}{3!z} - \dfrac{1}{4!} - \dfrac{i}{5!}z + \cdots$

17. $\cdots + \dfrac{1}{5!(z-i)^3} - \dfrac{1}{3!(z-i)} + (z-i)$

19. $\dfrac{2}{3} + \dfrac{8}{9}z + \dfrac{26}{27}z^2 + \cdots$;

$\cdots - \dfrac{1}{z^3} - \dfrac{1}{z^2} - \dfrac{1}{z} - \dfrac{1}{3} - \dfrac{z}{3^2} - \dfrac{z^2}{3^3} - \cdots$;

$\dfrac{2}{z^2} + \dfrac{8}{z^3} + \dfrac{26}{z^4} + \cdots$;

$-\dfrac{1}{z-1} - \dfrac{1}{2} - \dfrac{z-1}{2^2} - \dfrac{(z-1)^2}{2^3} - \cdots$

21. $\dfrac{404\pi}{81} i$ **23.** $\dfrac{2\pi}{\sqrt{3}} i$

25. $(\pi + \pi e^{-2}\cos 2)i$ **27.** $-\pi i$

29. $\dfrac{9\pi^3 + 2}{\pi^2} i$ **31.** $7\pi/50$

33. $\pi\left(\dfrac{90 - 52\sqrt{3}}{12 - 7\sqrt{3}}\right)$

Exercises 20.1, Page 926

1. the line $v = -u$ **3.** the line $v = 2$

5. open line segment from 0 to πi

7. the ray $\theta = \frac{1}{2}\theta_0$ **9.** the line $u = 1$

11. the fourth quadrant

13. the wedge $\pi/4 \le \operatorname{Arg} w \le \pi/2$

15. the circle with center $w = 4i$ and radius $r = 1$

17. the strip $-1 \le u \le 0$

19. the wedge $0 \le \operatorname{Arg} w \le 3\pi/4$

21. $w = -i(z - i) = -iz - 1$

23. $w = 2(z - 1)$ **25.** $w = -z^4$

27. $w = e^{3z/2}$ **29.** $w = -z + i$

Exercises 20.2, Page 931

1. conformal at all points except $z = \pm 1$

3. conformal at all points except $z = \pi i \pm 2n\pi i$

5. conformal at all points outside the interval $[-1, 1]$ on the x-axis

7. The image is the region shown in Figure 20.2.2(b). A horizontal segment $z(t) = t + ib, 0 < t < \pi$, is mapped onto the lower or upper portion of the ellipse

$$\frac{u^2}{\cosh^2 b} + \frac{v^2}{\sinh^2 b} = 1$$

according to whether $b > 0$ or $b < 0$.

9. The image of the region is the wedge $0 \le \text{Arg } w \le \pi/4$. The image of the line segment $[-\pi/2, \pi/2]$ is the union of the line segments joining $e^{i\pi/4}$ to 0 and 0 to 1.

11. $w = \cos(\pi z/2)$ using H-4

13. $w = \left(\dfrac{1+z}{1-z}\right)^{1/2}$ using H-5 and $w = z^{1/4}$

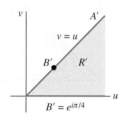

15. $w = \left(\dfrac{e^{\pi/z} + e^{-\pi/z}}{e^{\pi/z} - e^{-\pi/z}}\right)^{1/2}$ using H-6 and $w = z^{1/2}$

17. $w = \sin(-i\text{Ln } z - \pi/2)$; $A'B'$ is the real interval $(-\infty, -1]$.

19. $u = \dfrac{1}{\pi}\text{Arg}(z^4)$ or $u(r, \theta) = \dfrac{4}{\pi}\theta$

21. $u = \dfrac{1}{\pi}\text{Arg}\left(i\dfrac{1-z}{1+z}\right) = \dfrac{1}{\pi}\tan^{-1}\left(\dfrac{1-x^2-y^2}{2y}\right)$

23. $u = \dfrac{1}{\pi}[\text{Arg}(z^2 - 1) - \text{Arg}(z^2 + 1)]$

25. $u = \dfrac{10}{\pi}[\text{Arg}(e^{\pi z} - 1) - \text{Arg}(e^{\pi z} + 1)]$

Exercises 20.3, Page 938

1. $T(0) = \infty$, $T(1) = i$, $T(\infty) = 0$; $|w| = 1$ and the line $v = \frac{1}{2}$; $|w| \ge 1$

3. $T(0) = -1$, $T(1) = \infty$, $T(\infty) = 1$; the line $u = 0$ and the circle $|w - 1| = 2$; the half-plane $u \le 0$

5. $S^{-1}(w) = \dfrac{-w - 1}{-w + i} = \dfrac{w + 1}{w - i}$,

$S^{-1}(T(z)) = \dfrac{(1 + i)z - 1}{2z + i}$

7. $S^{-1}(w) = \dfrac{-w + 2}{-w + 1} = \dfrac{w - 2}{w - 1}$, $S^{-1}(T(z)) = \dfrac{3}{z}$

9. $w = -2\dfrac{z+1}{z-2}$ 11. $w = \dfrac{2z}{z - 1 - 2i}$

13. $w = \dfrac{i}{2}\dfrac{z-1}{z}$

15. $w = 3\dfrac{(1+i)z + 1 - i}{(-3+5i)z - 3 - 5i}$

17. $u = \dfrac{1}{\log_e 2}\log_e\left|\dfrac{z+2}{z-1}\right|$. The level curves are the images of the circles $|w| = r$, $1 < r < 2$, under the linear fractional transformation $T(w) = (w + 2)/(w - 1)$. Since the circles do not pass through the pole at $w = 1$, the images are circles.

19. Construct the linear fractional transformation that sends $1, i, -i$ to $0, 1, -1$.

21. Simplify $T_2(T_1(z)) = \dfrac{a_2\left(\dfrac{a_1 z + b_1}{c_1 z + d_1}\right) + b_2}{c_2\left(\dfrac{a_1 z + b_1}{c_1 z + d_1}\right) + d_2}$.

Exercises 20.4, Page 942

1. first quadrant

3.

5. $f'(z) = A(z + 1)^{-1/2}z^{-1/2}(z - 1)^{-1/2}$ for some constant A

7. $f'(z) = A(z + 1)^{-1/3}z^{-1/3}$ for some constant A

9. Show that $f'(z) = \dfrac{A}{(z^2 - 1)^{1/2}}$ and conclude that $f(z) = \cosh^{-1}z$.

11. Show that $f'(z) \to A/z$ as $w_1 \to \infty$ and conclude that $f(z) = \text{Ln } z$.

13. Show that $f'(z) \to A(z + 1)^{-1/2}z(z - 1)^{-1/2} = Az/(z^2 - 1)^{1/2}$ as $u_1 \to 0$.

Exercises 20.5, Page 946

1. $u = \dfrac{1}{\pi}\text{Arg}\left(\dfrac{z-1}{z}\right) - \dfrac{1}{\pi}\text{Arg}\left(\dfrac{z}{z+1}\right)$

3. $u = \dfrac{5}{\pi}[\pi - \text{Arg}(z - 1)] + \dfrac{1}{\pi}\text{Arg}\left(\dfrac{z}{z+1}\right)$

$- \dfrac{1}{\pi}\text{Arg}\left(\dfrac{z+1}{z+2}\right)$

5. $u = \dfrac{y}{\pi}\left\{1 + \dfrac{y^2 - x^2}{y}\left[\tan^{-1}\left(\dfrac{x-1}{y}\right) - \tan^{-1}\left(\dfrac{x}{y}\right)\right]\right.$

$\left. + x\log_e\left[\dfrac{(x-1)^2 + y^2}{x^2 + y^2}\right]\right\}$

7. $u = \dfrac{1}{\pi}\text{Arg}\left(\dfrac{z^2 - 1}{z^2}\right) + \dfrac{5}{\pi}\text{Arg}(z^2 + 1)$

9. $u = \dfrac{1}{\pi}\text{Arg}\left(\dfrac{(1-i)z - (1+i)}{1 - z}\right)$

$\quad - \dfrac{1}{\pi}\text{Arg}\left(\dfrac{1-z}{-(1+i)z + 1 - i}\right)$

11. $u(0,0) = \frac{1}{3}$, $u(-0.5, 0) = 0.5693$, $u(0.5, 0) = 0.1516$

13. Show that $u(0,0) = \dfrac{1}{2\pi}\displaystyle\int_{-\pi}^{\pi} u(e^{it})\, dt$.

15. $u(r, \theta) = r\sin\theta + r\cos\theta \quad$ or $\quad u(x, y) = x + y$

Exercises 20.6, Page 951

1. $g(z) = e^{-i\theta_0}$ is analytic everywhere and $G(z) = e^{-i\theta_0}z$ is a complex potential. The equipotential lines are the lines $x\cos\theta_0 + y\sin\theta_0 = c$.

3. $g(z) = 1/z$ is analytic for $z \neq 0$ and $G(z) = \text{Ln}\,z$ is analytic except for $z = x \leq 0$. The equipotential lines are the circles $x^2 + y^2 = e^{2c}$.

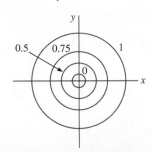

5. $\phi = \dfrac{4}{\pi}\text{Arg}\,z$ or $\phi(r, \theta) = \dfrac{4}{\pi}\theta$, and $G(z) = \dfrac{4}{\pi}\text{Ln}\,z$ is a complex potential. The equipotential lines are the rays

$\theta = \dfrac{\pi}{4}c$ and $\mathbf{F} = \dfrac{4}{\pi}\left(\dfrac{x}{x^2 + y^2}, \dfrac{y}{x^2 + y^2}\right).$

7. The equipotential lines are the images of the rays $\theta = \theta_0$ under the successive transformations $\zeta = w^{1/2}$ and $z = (\zeta + 1)/(-\zeta + 1)$. The transformation $\zeta = w^{1/2}$ maps the ray $\theta = \theta_0$ to the ray $\theta = \theta_0/2$ in the ζ-plane, and $z = (\zeta + 1)/(-\zeta + 1)$ maps this ray onto an arc of a circle that passes through $z = -1$ and $z = 1$.

9. (a) $\psi(x, y) = 4xy(x^2 - y^2)$ or, in polar coordinates, $\psi(r, \theta) = r^4\sin 4\theta$. Note that $\psi = 0$ on the boundary of R.

(b) $\mathbf{V} = \overline{4z^3} = 4(x^3 - 3xy^2, y^3 - 3x^2y)$

(c)

11. (a) $\psi(x, y) = \cos x \sinh y$ and $\psi = 0$ on the boundary of R.

(b) $\mathbf{V} = \overline{\cos z} = (\cos x \cosh y, \sin x \sinh y)$

(c)

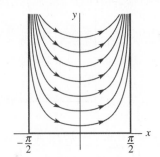

13. (a) $\psi(x, y) = 2xy - \dfrac{2xy}{(x^2 + y^2)^2}$ or, in polar coordinates,

$\psi(r, \theta) = (r^2 - 1/r^2)\sin 2\theta$. Note that $\psi = 0$ on the boundary of R.

(b) $\mathbf{V} = 2\bar{z} - 2/\bar{z}^3$

(c)

15. (a) $f(t) = \pi i - \dfrac{1}{2}[\log_e|t + 1| + \log_e|t - 1|$

$\qquad\qquad + i\operatorname{Arg}(t + 1) + i\operatorname{Arg}(t - 1)]$ and so

$$\operatorname{Im}(f(t)) = \begin{cases} 0, & t < -1 \\ \pi/2, & -1 < t < 1 \\ \pi, & t > 1. \end{cases}$$

Hence, $\operatorname{Im}(G(z)) = \psi(x, y) = 0$ on the boundary of R.

(b) $x = -\frac{1}{2}[\log_e|t + 1 + ic| + \log_e|t - 1 + ic|]$

$y = \pi - \frac{1}{2}[\operatorname{Arg}(t + 1 + ic) + \operatorname{Arg}(t - 1 + ic)]$,
for $c > 0$

(c)

17. (a) $f(t) = \dfrac{1}{\pi}((t^2 - 1)^{1/2} + \cosh^{-1}t) = \dfrac{1}{\pi}((t^2 - 1)^{1/2}$

$\qquad\qquad + \operatorname{Ln}(t + (t^2 - 1)^{1/2}))$ and so $\operatorname{Im}(f(t)) =$

$$\begin{cases} 1, & t < -1 \\ 0, & t > 1 \end{cases}$$ and $\operatorname{Re}(f(t)) = 0$ for $-1 < t < 1$.

Hence, $\operatorname{Im}(G(z)) = \psi(x, y) = 0$ on the boundary of R.

(b) $x = \operatorname{Re}\left[\dfrac{1}{\pi}\left(((t + ic)^2 - 1)^{1/2} + \cosh^{-1}(t + ic)\right)\right]$

$y = \operatorname{Im}\left[\dfrac{1}{\pi}\left(((t + ic)^2 - 1)^{1/2} + \cosh^{-1}(t + ic)\right)\right]$

for $c > 0$

(c)

19. $z = 0$ in Example 5; $z = 1$, $z = -1$ in Example 6

21. The streamlines are the branches of the family of hyperbolas $x^2 + Bxy - y^2 - 1 = 0$ that lie in the first quadrant. Each member of the family passes through $(1, 0)$.

23. *Hint:* For z in the upper half-plane,

$$k[\operatorname{Arg}(z - 1) - \operatorname{Arg}(z + 1)] = k\operatorname{Arg}\left(\dfrac{z - 1}{z + 1}\right).$$

Chapter 20 in Review, Page 953

1. $v = 4$ **3.** the wedge $0 \le \operatorname{Arg} w \le 2\pi/3$

5. true **7.** $0, 1, \infty$ **9.** false

11. The image of the first quadrant is the strip $0 < v < \pi/2$. Rays $\theta = \theta_0$ are mapped onto horizontal lines $v = \theta_0$ in the w-plane.

13. $w = \dfrac{i - \cos \pi z}{i + \cos \pi z}$

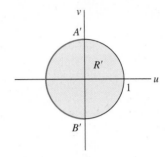

15. $u = 2 - 2y/(x^2 + y^2)$

17. (a) Note that $\alpha_1 \to 0$, $\alpha_2 \to 2\pi$, and $\alpha_3 \to 0$ as $u_1 \to \infty$.

(b) *Hint:* Write $f(t) = \frac{1}{2}A[\log_e|t + 1| + \log_e|t - 1| + i\operatorname{Arg}(t + 1) + i\operatorname{Arg}(t - 1)] + B$.

19. $G(z) = f^{-1}(z)$ maps R to the strip $0 \le v \le \pi$, and $U(u, v) = v/\pi$ is the solution to the transferred boundary problem. Hence, $\phi(x, y) = (1/\pi)\operatorname{Im}(G(z)) = (1/\pi)\psi(x, y)$, and so the equipotential lines $\phi(x, y) = c$ are the streamlines $\psi(x, y) = c\pi$.

Exercises Appendix A, Page APP-8

1. 120 **3.** $\frac{3}{4}\sqrt{\pi}$

5. $\frac{105}{16}\sqrt{\pi}$ **7.** $\frac{4}{3}\sqrt{\pi}$

9. $\frac{16}{105}\sqrt{\pi}$ **11.** 0.9182

13. 0.9086 **15.** $(x + 3)(x + 2)(x + 1)x\Gamma(x)$

17. $\sqrt{\pi/2}$ **19.** $\frac{1}{324}$

21. $\frac{3}{128}$ **23.** $\sqrt{2\pi}$

25. 0.1836 **29.** π

31. $\frac{1}{4}\pi$ **33.** $\frac{5}{2}\pi$

39. $\frac{1}{12}$ **41.** $\frac{8}{15}$

43. $F'(x) = 6x^2 e^{-4x^9} - e^{-x^5} - 3x^2 \displaystyle\int_x^{2x^3} t^2 e^{-x^3 t^2} dt$

45. $F'(x) = -\sin x(x^2 + \cos^2 x)^{10} - 2x(x^2 + x^4)^{10}$

$\qquad\qquad + 20x \displaystyle\int_{x^2}^{\cos x} (x^2 + t^2)^9 dt$

47. $F'(x) = e^{\sin x}\sin(\cos x)$

55. (a) $\dfrac{\sqrt{\pi}}{2\sqrt{x}}$ **(b)** $\sqrt{\pi x} + C$ **(c)** $\sqrt{\pi x}$

57. $\operatorname{erf}(x/\sqrt{2})$

INDEX

INDEX

INDEX

INDEX

Laplace Transforms

$f(t)$	$\mathscr{L}\{f(t)\} = F(s)$		$f(t)$	$\mathscr{L}\{f(t)\} = F(s)$
1. 1	$\dfrac{1}{s}$		**21.** $e^{at}\cosh kt$	$\dfrac{s-a}{(s-a)^2 - k^2}$
2. t	$\dfrac{1}{s^2}$		**22.** $t \sin kt$	$\dfrac{2ks}{(s^2+k^2)^2}$
3. t^n	$\dfrac{n!}{s^{n+1}}$, n a positive integer		**23.** $t \cos kt$	$\dfrac{s^2-k^2}{(s^2+k^2)^2}$
4. $t^{-1/2}$	$\dfrac{\sqrt{\pi}}{s^{1/2}}$		**24.** $\sin kt + kt \cos kt$	$\dfrac{2ks^2}{(s^2+k^2)^2}$
5. $t^{1/2}$	$\dfrac{\sqrt{\pi}}{2s^{3/2}}$		**25.** $\sin kt - kt \cos kt$	$\dfrac{2k^3}{(s^2+k^2)^2}$
6. t^{α}	$\dfrac{\Gamma(\alpha+1)}{s^{\alpha+1}}$, $\alpha > -1$		**26.** $t \sinh kt$	$\dfrac{2ks}{(s^2-k^2)^2}$
7. $\sin kt$	$\dfrac{k}{s^2+k^2}$		**27.** $t \cosh kt$	$\dfrac{s^2+k^2}{(s^2-k^2)^2}$
8. $\cos kt$	$\dfrac{s}{s^2+k^2}$		**28.** $\dfrac{e^{at}-e^{bt}}{a-b}$	$\dfrac{1}{(s-a)(s-b)}$
9. $\sin^2 kt$	$\dfrac{2k^2}{s(s^2+4k^2)}$		**29.** $\dfrac{ae^{at}-be^{bt}}{a-b}$	$\dfrac{s}{(s-a)(s-b)}$
10. $\cos^2 kt$	$\dfrac{s^2+2k^2}{s(s^2+4k^2)}$		**30.** $1 - \cos kt$	$\dfrac{k^2}{s(s^2+k^2)}$
11. e^{at}	$\dfrac{1}{s-a}$		**31.** $\cosh kt - 1$	$\dfrac{k^2}{s(s^2-k^2)}$
12. $\sinh kt$	$\dfrac{k}{s^2-k^2}$		**32.** $kt - \sin kt$	$\dfrac{k^3}{s^2(s^2+k^2)}$
13. $\cosh kt$	$\dfrac{s}{s^2-k^2}$		**33.** $a \sin bt - b \sin at$	$\dfrac{ab(a^2-b^2)}{(s^2+a^2)(s^2+b^2)}$
14. $\sinh^2 kt$	$\dfrac{2k^2}{s(s^2-4k^2)}$		**34.** $\cos at - \cos bt$	$\dfrac{s(b^2-a^2)}{(s^2+a^2)(s^2+b^2)}$
15. $\cosh^2 kt$	$\dfrac{s^2-2k^2}{s(s^2-4k^2)}$		**35.** $\sin kt \sinh kt$	$\dfrac{2k^2s}{s^4+4k^4}$
16. $e^{at}t$	$\dfrac{1}{(s-a)^2}$		**36.** $\sin kt \cosh kt$	$\dfrac{k(s^2+2k^2)}{s^4+4k^4}$
17. $e^{at}t^n$	$\dfrac{n!}{(s-a)^{n+1}}$, n a positive integer		**37.** $\cos kt \sinh kt$	$\dfrac{k(s^2-2k^2)}{s^4+4k^4}$
18. $e^{at}\sin kt$	$\dfrac{k}{(s-a)^2+k^2}$		**38.** $\cos kt \cosh kt$	$\dfrac{s^3}{s^4+4k^4}$
19. $e^{at}\cos kt$	$\dfrac{s-a}{(s-a)^2+k^2}$		**39.** $\sin kt \cosh kt + \cos kt \sinh kt$	$\dfrac{2ks^2}{s^4+4k^4}$
20. $e^{at}\sinh kt$	$\dfrac{k}{(s-a)^2-k^2}$			

40. $\sin kt \cosh kt - \cos kt \sinh kt$ $\qquad$ $\dfrac{4k^3}{s^4 + 4k^4}$

41. $\sinh kt + \sin kt$ $\qquad$ $\dfrac{2ks^2}{s^4 - k^4}$

42. $\sinh kt - \sin kt$ $\qquad$ $\dfrac{2k^3}{s^4 - k^4}$

43. $\cosh kt + \cos kt$ $\qquad$ $\dfrac{2s^3}{s^4 - k^4}$

44. $\cosh kt - \cos kt$ $\qquad$ $\dfrac{2k^2 s}{s^4 - k^4}$

45. $J_0(kt)$ $\qquad$ $\dfrac{1}{\sqrt{s^2 + k^2}}$

46. $I_0(kt)$ $\qquad$ $\dfrac{1}{\sqrt{s^2 - k^2}}$

47. $\dfrac{e^{bt} - e^{at}}{t}$ $\qquad$ $\ln\left(\dfrac{s - a}{s - b}\right)$

48. $\dfrac{2(1 - \cos kt)}{t}$ $\qquad$ $\ln\left(\dfrac{s^2 + k^2}{s^2}\right)$

49. $\dfrac{2(1 - \cosh kt)}{t}$ $\qquad$ $\ln\left(\dfrac{s^2 - k^2}{s^2}\right)$

50. $\dfrac{\sin kt}{t}$ $\qquad$ $\arctan\left(\dfrac{k}{s}\right)$

51. $\dfrac{\sin at \cos bt}{t}$ $\qquad$ $\dfrac{1}{2}\arctan\left(\dfrac{a + b}{s}\right) + \dfrac{1}{2}\arctan\left(\dfrac{a - b}{s}\right)$

52. $\dfrac{\sin(2\sqrt{kt})}{\sqrt{\pi k}}$ $\qquad$ $\dfrac{e^{-k/s}}{s\sqrt{s}}$

53. $\dfrac{\cos(2\sqrt{kt})}{\sqrt{\pi t}}$ $\qquad$ $\dfrac{e^{-k/s}}{\sqrt{s}}$

54. $\dfrac{\sinh(2\sqrt{kt})}{\sqrt{\pi k}}$ $\qquad$ $\dfrac{e^{k/s}}{s\sqrt{s}}$

55. $\dfrac{\cosh(2\sqrt{kt})}{\sqrt{\pi t}}$ $\qquad$ $\dfrac{e^{k/s}}{\sqrt{s}}$

56. $\dfrac{1}{\sqrt{\pi t}} e^{-a^2/4t}$ $\qquad$ $\dfrac{e^{-a\sqrt{s}}}{\sqrt{s}}$

57. $\dfrac{a}{2\sqrt{\pi t^3}}\,e^{-a^2/4t}$

$e^{-a\sqrt{s}}$

58. $\mathrm{erfc}\!\left(\dfrac{a}{2\sqrt{t}}\right)$

$\dfrac{e^{-a\sqrt{s}}}{s}$

59. $\mathrm{erf}(at)$

$\dfrac{e^{s^2/4a^2}}{s}\,\mathrm{erfc}\!\left(\dfrac{s}{2a}\right)$

60. $\mathrm{erf}(\sqrt{at})$

$\dfrac{\sqrt{a}}{s\sqrt{s+a}}$

61. $e^{a^2t}\mathrm{erf}(a\sqrt{t})$

$\dfrac{a}{\sqrt{s}(s-a^2)}$

62. $\mathrm{erfc}(\sqrt{at})$

$\dfrac{\sqrt{s+a}-\sqrt{a}}{s\sqrt{s+a}}$

63. $e^{a^2t}\mathrm{erfc}(a\sqrt{t})$

$\dfrac{1}{s+a\sqrt{s}}$

64. $\dfrac{1}{\sqrt{\pi t}}+ae^{a^2t}\mathrm{erf}(a\sqrt{t})$

$\dfrac{\sqrt{s}}{s-a^2}$

65. $\dfrac{1}{\sqrt{\pi t}}-ae^{a^2t}\mathrm{erfc}(a\sqrt{t})$

$\dfrac{1}{\sqrt{s}+a}$

66. $2\sqrt{\dfrac{t}{\pi}}\,e^{-a^2/4t}-a\,\mathrm{erfc}\!\left(\dfrac{a}{2\sqrt{t}}\right)$

$\dfrac{e^{-a\sqrt{s}}}{s\sqrt{s}}$

67. $-e^{ab}e^{b^2t}\mathrm{erfc}\!\left(b\sqrt{t}+\dfrac{a}{2\sqrt{t}}\right)+\mathrm{erfc}\!\left(\dfrac{a}{2\sqrt{t}}\right)$

$\dfrac{be^{-a\sqrt{s}}}{s(\sqrt{s}+b)}$

68. $e^{ab}e^{b^2t}\mathrm{erfc}\!\left(b\sqrt{t}+\dfrac{a}{2\sqrt{t}}\right)$

$\dfrac{e^{-a\sqrt{s}}}{s+b\sqrt{s}}$

69. $f(at)$

$\dfrac{1}{a}F\!\left(\dfrac{s}{a}\right)$

70. $e^{at}f(t)$

$F(s-a)$

71. $\mathcal{U}(t-a)$

$\dfrac{e^{-as}}{s}$

72. $f(t-a)\mathcal{U}(t-a)$

$e^{-as}F(s)$

73. $g(t)\mathcal{U}(t-a)$

$e^{-as}\mathcal{L}\{g(t+a)\}$